Hand Book of

Mathematics
for Biosciences & Paramedical Students

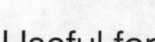

Useful for

The students of
Biosciences, Social Sciences, Industrial Chemistry, Business Management,
Biochemistry, Zoology, Botany, Microbiology, Nursing, Environmental Biology,
Biotechnology and Paramedical Courses

Dr. SUDHIR KUMAR PUNDIR

M.Sc., M.Phil, NET, Ph.D.
Associate Professor
Department of Mathematics
S.D. (P.G.) College,
Muzaffarnagar (U.P.)

CBS Publishers & Distributors Pvt. Ltd.

New Delhi • Bengaluru • Chennai • Kochi • Kolkata • Mumbai
Hyderabad • Nagpur • Patna • Pune • Vijayawada

Hand Book of
Mathematics
for Biosciences & Paramedical Students

ISBN: 978-93-85915-13-0

First Edition: 2016

Published by:
Satish Kumar Jain for CBS Publishers & Distributors Pvt. Ltd.,
4819/XI Prahlad Street, 24 Ansari Road, Daryaganj, New Delhi - 110002
delhi@cbspd.com, cbspubs@airtelmail.in • www.cbspd.com
Ph.: 23289259, 23266861, 23266867 • Fax: 011-23243014
Corporate Office: 204 FIE, Industrial Area, Patparganj, Delhi - 110 092
Ph: 49344934 • Fax: 011-49344935
E-mail: publishing@cbspd.com • publicity@cbspd.com
Branches:
• *Bengaluru:* 2975, 17th Cross, K.R. Road, Bansankari 2nd Stage,
  Bengaluru - 70 • Ph: +91-80-26771678/79 • Fax: +91-80-26771680
  E-mail: cbsbng@gmail.com, bangalore@cbspd.com
• *Chennai:* No. 7, Subbaraya Street, Shenoy Nagar, Chennai - 600030
  Ph: +91-44-26681266, 26680620 • Fax: +91-44-42032115
  E-mail: chennai@cbspd.com
• *Kochi:* Ashana House, 39/1904, A.M. Thomas Road, Valanjambalam,
  Ernakulum, Kochi • Ph: +91-484-4059061-65
  Fax: +91-484-4059065 • E-mail: cochin@cbspd.com
• *Kolkata:* 6-B, Ground Floor, Rameshwar Shaw Road, Kolkata - 700014
  Ph: +91-33-22891126/7/8 • E-mail: kolkata@cbspd.com
• *Mumbai:* 83-C, Dr. E. Moses Road, Worli, Mumbai - 400018
  Ph: +91-9833017933, 022-24902340/41 • E-mail: mumbai@cbspd.com
Representatives:
• Hyderabad: 0-9885175004   • Nagpur: 0-9021734563
• Patna: 0-9334159340   • Pune: 0-9623451994
• Vijayawada: 0-9000660880

Printed at:
India Binding House, Noida (U.P.)

Preface

HAND BOOK OF MATHEMATICS FOR BIOSCIENCES & PARAMEDICAL STUDENTS is meant for the students of biosciences, social sciences, chemistry, industrial chemistry, biochemistry, business management, zoology, botany, microbiology, environmental biology, biotechnology and paramedical courses. Various aspects of biomathematics and biostatistics prescribed in the syllabi of UGC, NET and Civil Services have also been included.

This book consists of twenty-five chapters and one appendix. In each chapter, an ample amount of theory is given which is supported by solved examples followed by exercises along with their answers.

The book intends to provide a sound basic knowledge of mathematics and biostatistics. Different concepts have been explained with the help of solved examples. There is plenty of scope in the form of exercises for the reader to try and solve the problems on his own.

I express my gratitude to the authors and publishers of various books I consulted during the preparation of the book.

I wish to sincerely thank **Sh. S.K. Jain**, MD, CBS Publishers & Distributors Pvt. Ltd., New Delhi, for his encouragement and help in bringing out this publication in a present nice form.

My special thanks to Sh. B.M. Singh, Sh. Sunil Dutt, Sh. Hariom and entire team of CBS Publishers & Distributors Pvt. Ltd., New Delhi, whose encouragement and unstinted support enabled me to complete this book. Mr. Peeyush Goel, M/S. Dreamshapers also deserves special mention for nice typesetting.

I must record my appreciation due to my wife Dr. Rimple, daughter Rijuta, and son Shrish for their understanding and love during the long period that I have taken to complete this book.

Above all, I am thankful to The Almighty God, without whose grace nothing is possible for any one.

Readers are welcome to point out errors, if any, and send their valuable suggestions for improving the quality of the book.

Dr. SUDHIR KUMAR PUNDIR

Preface

HAND BOOK OF MATHEMATICS FOR BIOSCIENCES & PARAMEDICAL STUDENTS is meant for the students of biosciences, social sciences, chemistry, industrial chemistry, biochemistry, business management, zoology, botany, microbiology, environmental biology, biotechnology and paramedical courses. Various aspects of biomathematics and biostatistics presented in the syllabi of UGC, NET and Civil Services have also been included.

This book consists of twenty five chapters and one appendix. In each chapter, an ample amount of theory is given which is supported by solved examples followed by exercises along with their answers.

The book intends to provide a sound basic knowledge of mathematics and biostatistics. Different concepts have been explained with the help of solved examples. There is plenty of scope in the form of exercises for the reader to try and solve the problems on his own.

I express my gratitude to the authors and publishers of various books I consulted during the preparation of the book.

I wish to sincerely thank Sh. S.K. Jain, MD, CBS Publishers & Distributors Pvt. Ltd., New Delhi, for his encouragement and help in bringing out this publication in a present time form.

My special thanks to Sh. B.M. Singh, Sh. Sunil Dutt, Sh. Hariom and entire team of CBS Publishers & Distributors Pvt. Ltd., New Delhi, whose encouragement and unstinted support enabled me to complete this book. Mr Peeyush Goel, M/S Dreamshapers also deserves special mention for nice typesetting.

I must record my appreciation due to my wife Dr Himple, daughter Rijuta and son Shloli for their understanding and love during the long period that I have taken to complete this book.

Above all, I am thankful to The Almighty God, without whose grace nothing is possible for any one.

Readers are welcome to point out errors, if any, and send their valuable suggestions for improving the quality of the book.

Dr. SUDHIR KUMAR PUNDIR

Contents

CHAPTER 1

Set Theory

1.1 INTRODUCTION

The concept of set is fundamental in all branches of mathematics. It was developed by German mathematician George Cantor. This chapter introduces the notations and terminology of set theory. Classical set theory, also termed as crisp set theory, is fundamental to the study of pure mathematics.

1.2 NUMBER SYSTEM

The number system plays a key role in mathematics. The real number system **R** is one of the most important and beautiful mathematical system. There are different ways of introducing the real number system, but the most common way is to start with Peano's Axioms for the natural numbers. The axioms for natural numbers, discovered by the Italian Mathematician Peano are:

   (i) 1 is a natural number

   (ii) Each natural number n has a successor $(n+1)$.

   (iii) Two natural numbers are equal if their successors are equal.

   (iv) Except 1, each natural number is a successor of natural number.

   (v) Any set of natural numbers which contains 1 and the successor of every natural number $(k+1)$ whenever it contains k in the set **N** of natural numbers.

⋙ REMARKS

⟹ Axiom (v) is commonly known as the axiom of induction or principle of finite induction.

⟹ The above axioms completely define the set of natural numbers.

Definition. *The numbers 1, 2, 3, ... are called natural numbers*. We represent the set of natural numbers by **N**.

i.e., $\qquad\qquad$ **N** = {1,2,3, ...}

The Peano's axioms can be used to extend the set **N** of natural numbers to another large system, known as the set of integers.

Definition. *The numbers ... , –3, –2, –1, 0, 1, 2, 3... are called integers. We represent the set of integers by* **Z**.

i.e., $\qquad\qquad$ **Z** = { ... –3, –2, –1, 0,1,2,3,.. }

Integers can be used to define the rational numbers.

Definition. *Any number of the form p/q, where $p,q \in$* **Z***, $q \neq 0$ and p,q have no common factor (except $\pm$ 1) is called a rational number.*

The set of rational numbers is denoted by **Q**.

$$\therefore \qquad \mathbf{Q} = \left\{ \frac{p}{q}; \ p,q \in \mathbf{Z}, q \neq 0 \right\}$$

⮑ REMARK

➠ The set of rational numbers consists of integers and fractions.

Definition: *Any number which is not rational, is called an irrational number.*

For example, $\sqrt{2}, \sqrt{3}$ etc. It should be noted that every rational number can be expressed as a terminating or recurring decimal whereas every irrational number can be expressed as a non-terminating infinite decimal.

1.2.1 REAL NUMBER

A number which is either rational or irrational is called a real number. The set of real numbers is denoted by **R.**

1.2.2 INTEGRAL POWERS OF A REAL NUMBER

Let $a \in \mathbf{R}$, and n be any positive integer then we can define $a^n = a.a.a...n$ times.

In particular $\qquad a = a$

$$a^2 = a.a$$

$$a^3 = a.a.a = a^2.a \quad \text{and so on.}$$

Also, if n is any negative integer, then we have $x^{-n} = (x^n)^{-1} = (x^{-1})^n$

GENERAL DEFINITIONS

(i) A real number a is called positive, if $a > 0$ and the set of all positive real numbers, denoted by $\mathbf{R}^+$, is given by $\mathbf{R}^+ = \{x : x \in \mathbf{R}, x > 0\}$

(ii) A real number a is called negative if $a < 0$ and the set of all negative real numbers, denoted by $\mathbf{R}^-$, is given by $\mathbf{R}^- = \{x : x \in \mathbf{R}, x < 0\}$

1.3 INTERVAL

A subset S of **R** is called an interval if a, b ∈ S, x ∈ **R** such that a < x < b implies x ∈ S. There are following four type of intervals.

(i) $a \circ\!\!-\!\!-\!\!-\!\!-\!\!-\!\!-\!\!\circ \; b \Rightarrow \;]a, b[= \{x : a < x < b\}$

(ii) $a \bullet\!\!-\!\!-\!\!-\!\!-\!\!-\!\!-\!\!\bullet \; b \Rightarrow \; [a, b] = \{x : a \le x \le b\}$

(iii) $a \circ\!\!-\!\!-\!\!-\!\!-\!\!-\!\!-\!\!\bullet \; b \Rightarrow \;]a, b] = \{x : a < x \le b\}$

(iv) $a \bullet\!\!-\!\!-\!\!-\!\!-\!\!-\!\!-\!\!\circ \; b \Rightarrow \; [a, b[= \{x : a \le x < b\}$

OBSERVATIONS

➠ The set $]a, b[$ in which the end points are not included, is called an open interval.

➠ The set $[a, b]$ also contains both its end points, is called a closed interval.

➠ The sets $[a, b[$ and $]a, b]$ are called half open (or half closed) intervals or semi-open (or semi-closed) as they contain only one end point.

Apart from the four types of intervals listed above; there are a few more types: These are

(i) $[a, \infty[\; = \; \{x : a < x\}$ (open right ray)

(ii) $[a, \infty[\; = \; \{x : a \le x\}$ (closed right ray)

(iii) $]-\infty, b[\; = \; \{x : x < b\}$ (open left ray)

(iv) $]\infty, b[\; = \; \{x : x \le b\}$ (closed left ray)

(v) $]-\infty, \infty[\; = \;$ (open interval)

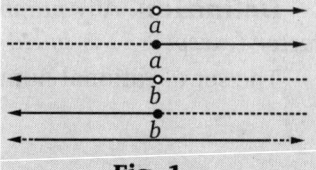

Fig. 1

REMARKS

➠ If S is any interval and if c and d are two elements of S, then all numbers lying between c and d are also elements of S.

➠ The proper use of a bracket, for example, parenthesis' for open and square brackets for closed and end points, itself specifies the interval. As such, to emphasize the nature of an interval, we shall drop the used 'description' and shall simply express the interval by using the appropriate brackets.

1.3.1 LENGTH OF AN INTERVAL

The number $b - a$ is called length of the intervals $]a, b[, [a, b[,]a, b]$ *and* $[a, b]$. If the length of the interval is finite, the interval is said to be finite and if the length is infinite, then it is known as infinite interval.

1.3.2 ABSOLUTE VALUE OF A REAL NUMBER

The absolute value of a real number a denoted by $|a|$ is the real number a, $-a$ or 0 according as a is positive, negative or zero, i.e.,

$$|a| = \begin{cases} a & if \quad a \geq 0 \\ -a & if \quad a < 0 \end{cases}$$

From the above definition, it is clear that

(i) $|a| = \max.\{a - a\}$ (ii) $-|a| = \min.\{a - a\}$ (iii) $|a| \geq a \geq -|a|$

1.3.3 SOME USEFUL RESULTS

(i) $|xy| = |x| \cdot |y|$ (ii) $|x+y| \leq |x| + |y|$

(iii) $|x - y| \geq | |x| - |y| |$ (iv) $|x - y| \leq |x| + |y|$

(v) $\left|\dfrac{x}{y}\right| = \dfrac{|x|}{|y|}$ (vi) If $\in > 0$, then $|x - y| < \varepsilon \Leftrightarrow y - \in < x < y + \in$

1.4 CONCEPT OF SETS

The theory of sets is one of the most important tools of pure mathematics. Pure mathematics is the study of sets equipped with assigned structures, known as mathematical systems. In this section, we shall study some fundamental concept of set theory.

Definition: *'A set is a well defined collection of objects'.*

The objects of a set are called the elements or members of that set and their membership is defined by certain conditions.

The basic concept used can be defined. Suppose, for example, one defines the set by "A set is well defined collection of objects" then what is meant by a collection. Perhaps, then one defines "A collection is an aggregate of things". What then is an aggregate? How our language is finite, so after sometime we will run out of new words to use and have to repeat some words already questioned. The definition is then circular and obviously worthless.

Mathematicians realize that there must be some undefined or primitive concept at the moment they have agreed that set shall be such a primitive concept.

The sets are usually denoted by the capital letters of English alphabets: Say $A, B, C, \ldots, X, Y, Z$.

For example :

(i) The collection of the letters $a, b, c, d,..$

(ii) The collection of all natural numbers denoted by **N**.

(iii) The students of M.Sc., Mathematics in C.C.S. University, Meerut.

(iv) The collection of vowels in English alphabet. This set containing only five elements, namely a, e, i, o, u.

(v) The collection of all states in Indian union.

If S is a set, an object a in the collection S is called an element of S. This fact is expressed in symbol as $a \in S$ (read as a is in S or a belongs to S). If a is not in S, we write $a \notin S$. For example, $4 \in \mathbf{R}$, the set of real numbers, but $\sqrt{-2} \notin \mathbf{R}$.

Here, Greek letter $\in$ denotes 'belongs to'. It is the abbreviation of the Greek word meaning 'is'.

🎗 REMARKS

➠ By the term 'well defined' we mean that we are given a collection of objects, with certain definite property, so that we are able to determine whether a given object belongs to our collection or not. Thus, every collection of objects is not a set.

➠ Set and aggregate both have the same meaning.

➠ The elements of a set must be distinguished from one another. The collection of sand particles does not form a set.

➠ The collection of rich persons of a city is not a set. However the collection of those persons of city whose wealth exceeds, a fixed amount, say rupees fifty thousands, is a set.

➠ The order is not preserved in case of a set, whereas order is necessarily preserved in case of sequence. That is to say, each of the sets {1,2,3}, {3,2,1}, {1,3,2} denotes the same sets.

➠ The repetition of an element does not change the nature of a set, *i.e.*, each of the sets {1,2,3}, {1,2,2,3}, {1,3,3,2} denotes the same sets.

1.4.1. REPRESENTATION OF A SET

There are two ways of representing a set:

(i) Roster or tabulation method

(ii) Set-builder or rule method

Roster Method: In this method, the elements of the set are listed within braces, and separated by comma.

For example:

(i) A= {1,2,3,4,5,6}

(ii) The set of vowels of English alphabet may be represent as {a, e, i, o, u}.

(iii) The set of a natural numbers from 1 to 100 may be written as **N**= {1,2,3, ..., 100}. We use three dots in the middle to include the missing elements.

(iv) The set of positive integers, which is a non-ending set may be written as $\mathbf{Z}^+$ = {1,2,3,4,5, ...}. The three dots in the end means that the elements continue in the same manner.

(v) The set of prime number is written as P = {2, 3, 5, 7, 11, 13, 17, 19, ...}

Set-Builder Method: In this method, we first try to find a property which characterizes the elements of a set, that is, a property P, which all the elements of the set possess and which no other objects possess. Then, we describe the set as {$x : x$ has property P}.

This is to be read as "the set of all x such that x has property P".

For example:

(i) The set of all integers can be written as **Z** = $\{x : x \text{ is an integer}\}$

(ii) The set $A = \{1, 2, 3, 4, 5\}$ can be written as **A** = $\{x \in \mathbf{N} : x \leq 5\}$.

(iii) The set of complex numbers can be written as **C** = $\{a + ib : a, b \in \mathbf{R}\}$

(iv) The set $A = \{1, 8, 27, ...\}$ can be written as $A = \{x^3 : x \in \mathbf{Z}^+\}$.

SOLVED EXAMPLES

EXAMPLE 1. *Use the Roster method to identify each set:*

(a) *The set of possible integers greater than 8 and less than 14.*

(b) *The set of numbers whose elements are the first five positive odd integers.*

(c) *The set of even positive integers.*

(d) *The set of even positive integers that are divisible by 10.*

(e) *The set of all vowels in English alphabets which precedes r.*

SOLUTION.
(a) $\{9, 10, 11, 12, 13\}$ (b) $\{1, 3, 5, 7, 9\}$

(c) $\{2, 4, 6, 8, 10 ...\}$ (d) $\{10, 20, 30, 40, 50 ...\}$

(e) $\{a, e, i, o\}$

EXAMPLE 2. *Use the set-builder method, identify the following sets :*

(a) $A = \{1, 3, 5, 7, 9, ...\}$ (b) $B = \left\{1, \dfrac{1}{4}, \dfrac{1}{9}, \dfrac{1}{16}, \dfrac{1}{25}, ...\right\}$

(c) $C = \{0, 1, 2, 3, ...\}$ (d) $D = \left\{\dfrac{1}{2}, \dfrac{2}{3}, \dfrac{3}{4}, \dfrac{4}{5}, ...\right\}$

SOLUTION.
(a) The set of odd positive integers.

(b) Here, elements of the set B are the reciprocals of the squares of the natural numbers.

So, the set $B = \left\{\dfrac{1}{n^2} : n \in \mathbf{N}\right\}$

(c) The set of whole numbers.

(d) Here, each element in the given set has the denominator one more than the numerator. Hence,

$$D = \left\{x : x = \dfrac{n}{n+1} : n \in \mathbf{N}\right\}$$

EXAMPLE 3. *Write the set* $\left\{\dfrac{1}{2}, \dfrac{2}{5}, \dfrac{3}{10}, \dfrac{5}{26}, ...\right\}$ *in the set-builder form.*

SOLUTION. We observe that each element in the given set has the denominator one more than the square of the numerator. Also, the numerator begins with 1. Hence, in the set builder form, the given set can be written as

$$\left\{x : x = \dfrac{n}{n^2 + 1} : n \in \mathbf{N}\right\}$$

Exercise 1.1

1. Which of the following collections are sets?
   (i) All mathematics students in your college.
   (ii) All poor hockey players in the college.
   (iii) All odd numbers less than 20.
   (iv) The collection of good teachers in your college.
   (v) All successful and rich people in your city.
   (vi) The people in your immediate family (father, mother, sister, brother).

2. Write the members of each of following sets by the Roster method.
   (i) $\{x : x$ is odd whole number less than 14$\}$
   (ii) $\{x : x^2 < 36$ and $x \in \mathbf{N}\}$
   (iii) $\{x :$ squares of all whole numbers less than 8$\}$
   (iv) $\{x : x$ is a prime number, $10 < x < 20\}$
   (v) $\{x: x$ is a composite number less than 20$\}$
   (vi) $\{x : x < x \}$

3. Rewrite the following sets using set-builder method.
   (i) $A = \{2, 4, 6, 8, ...\}$
   (ii) $B = \left\{1, \dfrac{1}{2}, \dfrac{1}{3}, \dfrac{1}{4}, ...\right\}$

   (iii) $C = \{0, 3, 6, 9, 12, ...\}$
   (iv) $D = \{0, 4, 6, 8, 10, ...\}$

4. List the elements of the following sets.
   (i) $A = \{x : x^2 \leq 16 : x \in \mathbf{Z}\}$
   (ii) $B = \{x : 1 \leq x \leq 5$ and $x \in \mathbf{N}\}$
   (iii) $C = \{x : x \in \mathbf{N}$ and x is a factor of 15$\}$
   (iv) $D = \{x : x$ is a month of year having 31 days$\}$
   (v) $E = \{x : x \in \mathbf{Z}$ and $3x - 2 = 3\}$
   (vi) $E = \{x : x$ is an integer lying between $-1/2$ and $1/2 \}$

5. Use the appropriate symbols $\in$ or $\notin$ to fill in the blanks below:
   (i) 12 ... the set of all numbers dividing 84.
   (ii) K ... the set of all vowels of the English alphabets.
   (iii) $\dfrac{1}{2}$ ... the set of natural number.
   (iv) India ... the set of members of UNO.
   (v) $\sqrt{2}$ ... The set of rational number
   (vi) 15 ... the set of multiples of 3.

ANSWERS

1. (i), (iii), (vi)
2. (i) $\{1, 3, 5, 7, 9, 11, 13\}$ (ii) $\{1, 2, 3, 4, 5\}$ (iii) $\{0, 1, 4, 9, 16, 25, 36, 49\}$
   (iv) $\{11, 13, 17, 19\}$ (v) $\{1, 4, 6, 8, 9, 10, 12, 14, 15, 16, 18\}$ (vi) ϕ
3. (i) $A = \{x : x = 2n : n \in \mathbf{N}\}$ (ii) $\{1/n : n \in \mathbf{N}\}$
   (iii) $\{x : x = 3n, n$ is the whole number   (iv)$\{x : x = 2n, n$ is the whole number$\}$
4. (i) $\{-4, -3, -2, -1, 0, 1, 2, 3, 4\}$ (ii) $\{1, 2, 3, 4, 5\}$ (iii) $\{3, 5\}$
   (iv) $\{$Jan, March, May, July, August, October, December$\}$
5. (i) $\in$ (ii) $\notin$ (iii) $\notin$ (iv) $\in$ (v) $\notin$ (vi) $\in$

1.5 TYPE OF SETS

(i) **Empty Set:** *A set containing no elements is called empty set and is denoted by the symbol* ϕ.

For example:
(i) $\phi = \{x : x$ is a negative integer whose square is $-1\}$
(ii) $\phi = \{x : x$ is a natural number lying between 2 and 3$\}$
(iii) $\phi = \{$the set of such persons, who never die$\}$
(iv) $\phi = \{x : x$ is a real number, $x^2 < 0\}$
(v) $\phi = \{x : x$ is an even prime number greater than five$\}$
(vi) $\phi = \{$the set of real numbers which are solution of equation $x^2 + 1 = 0 \}$
(vii) $\phi = \{x : x$ is a straight ling passing through three distinct points on a circle$\}$

REMARKS

➟ The empty set is also known as null set or void set.

➟ The Roster method, the empty set is denoted by {}.

➟ To describe the null set, we can use any property, which is not true for any element.

➟ It is wrong to use the expression 'an empty' or 'a null set' as there is one and only one empty set through, it may have many-many descriptions. We shall always call 'The empty or the null set.'

➟ A set consisting of at least one element is called a non-empty or non-void set.

➟ $\{\phi\}$ is not a null set.

(ii) Singleton Set: *Set containing only one element is a singleton set.* The set {a} is a singleton set.

REMARKS

➟ {0} is not a null set, since it contains 0 as its member. It is a singleton set.

➟ A room containing only one man is not same thing as a man. In a similar way, the singleton set {a} is not the same thing as the element a.

(iii) Finite Set: *A set is said to be finite if it consists of only finite number of elements.* Here, the process of counting the different elements comes to an end.

   For example:

   (i)   Set of natural numbers less than 50.

   (ii) Set of all persons in a city.

   (iii) Set of English alphabets.

   (iv) Set of all persons on the earth.

(iv) Infinite Set: *A set which is not finite, i.e., it contains infinite number of elements.* Here, process of counting the different elements never comes to an end.

   For example:

   (i)   Set of natural numbers $\mathbf{N} = \{1,2,3, \ldots\}$

   (ii) Set of all points of plane.

   (iii) Set of all even integers.

   (iv) Set of rational numbers lying between two integers.

(v) Equal Sets: *Two sets are said to be equal if they contain exactly the same elements.*

   For example:

   $A = \{x : x$ is a letter in the word 'Area'$\}$, i.e., $A = \{a, r, e\}$

And $B = \{y : y$ is a letter in the word 'ear'$\}$, i.e., $B = \{a, r, e\}$

Here A and B are equal sets.

1.5.1 CARDINAL NUMBER OF A SET

The number of distinct elements contained in a finite set A is called cardinal number of A and is denoted by $n(A)$.

1.5.2 EQUIVALENT SETS

Two finite sets are said to be equivalent if they have the same cardinal number.

REMARKS

➟ Equivalent sets are not always equal but equal sets are always equivalent.

➟ The number of distinct elements in a finite set is also called the order of the set. If the order of a set is zero, the set is empty.

➟ If the order of a set is one, the set is singleton.

➟ The order of an infinite set is never defined.

1.6 SUBSET

Let A and B be two sets. *The set A is said to be a subset of the set B if every element of A is also an element of B.* Symbolically, we write $A \subseteq B$.

When A is subset of B, it means that 'A is contained in B' or 'B contains A'. Here B is called superset of A and is written as $B \not\subset A$.

REMARKS

⇒ Every set is a subset of itself.

⇒ Empty set is a subset of every set.

⇒ If A is not a subset of B, we write $A \nsubseteq B$.

⇒ An element cannot be a subset of a set, only a set can be subset of a set.

1.6.1 PROPER SUBSET

We know that for A to be a subset of B all that is needed is that every element of A is in B. It is possible that every element of B may or may not be in A. If it so happens that every element of B is also in A, then we will have $B \subset A$. Obviously, then A and B are the same set, so that we have $A \subset B$ and $B \subset A \Leftrightarrow A = B$.

If every element of A is in B, but every element of B is not in A , i.e., if $A \subset B$ and $B \not\subset A$, then A is said to be a proper subset of B.

For example:

(i) $\{a, b\}$ is a proper subset of $\{a, b, c\}$.

(ii) Set of natural number $\mathbf{N}$ is a proper subset of set $\mathbf{Z}$ of integer.

REMARKS

⇒ Here, it follows that every element of A is an element of B and B contains at least one element which does not belong to A.

⇒ If the subset is not proper, it is called **improper subset.** $A \subseteq A$ and $\phi \subseteq A$ are improper subsets.

1.6.2 NUMBER OF SUBSETS OF A SET

If A is a set contains n distinct element. Let $0 < r \le n$. If we consider those subsets of A that have r elements each, then we know that the number of ways in which r elements can be choose out of n elements is $^{n}C_{r}$. Therefore, the number of subsets of A having r elements each is $^{n}C_{r}$.

Hence, the total number of subsets of A is equal to

$$^{n}C_{0} + {}^{n}C_{1} + {}^{n}C_{2} + ... + {}^{n}C_{n} = (1+1)^{n} = 2^{n}$$

For example:

(i) If a set A has one element, then it has $2^{1} = 2$ subsets.

(ii) If a set A has two elements, then it has $2^{2} = 4$ subsets.

REMARKS

⇒ The number of proper subsets of a set with n elements is 2^{n-1}.

⇒ The collection of all possible subsets of a given set A is called power set. It is denoted by $P(A)$. For example : If A = $\{1,2,3\}$ then the power set $P(A) = \{ \phi , \{1\}, \{2\}, \{3\}, \{1,2\}, \{1,3\}, \{2,3\}, \{1,2,3\}\}$.

⇒ $P(\phi) = \{\phi\}$

⇒ The power set of any given set is always non-empty.

1.7 UNIVERSAL SET

In any discussion , we are given particular set and we consider different subsets of the given set. This given set is called Universal Set. It is denoted by U.

For Example:

(i) The universal set is of real numbers **R**, while considering the set of natural numbers, whole numbers, integers and rational numbers.

(ii) The set of alphabets is the universal set from which the letters of any word may be chosen to form a set.

(iii) In geometry, we discuss set of lines, triangles and circles, then the universal set is the plane, in which the lines, triangles and circles lie.

☰ REMARKS

➠ Universal set is a super set of each of the given sets.

➠ The universal set is not unique.

1.7.1 COMPLEMENT OF A SET

Let U be the universal set and the set $A \subseteq U$. Complement of set A with respect to the universal set U is the set of all those elements of U which are not the elements of A and is denoted by A' or A^c,

$$A' = \{x : x \in U \text{ and } x \notin A\}$$

For example:

(i) If $U = \{1, 2, 3, 4, 5, 6, 7, 8, 9, 11\}$ and $A = \{1, 2, 3\}$
then $A' = \{4, 5, 6, 7, 8, 9, 11\}$.

☰ REMARKS

➠ Complement of the universal set is the null set and *vice-versa*.

➠ $(A')' = A$

➠ If $A \subseteq B$, then $B' \subseteq A'$.

➠ $x \in A' \Leftrightarrow x \notin A$

☛ RECAPITULATIONS

- A set containing no element is called empty set.
- Set containing finite number of elements is called finite otherwise infinite.
- The number of distinct elements contained in a finite set is called its cardinality.
- Two finite sets are said to be equivalent if they have same cardinality.
- A set A is said to be subset of a set B if every elements of A belongs to B.
- Total number of subsets of a set A of n elements is 2^n.
- $A' = \{x : x \in A' \text{ and } x \notin A\}$

SOLVED EXAMPLES

EXAMPLE 1. Let $A = \{1,2,3\}$, then find $P(A)$.

SOLUTION. Since $A = \{1, 2, 3\}$ then,

$$P(A) = \{\phi, \{1\}, \{2\}, \{3\}, \{1,2\}, \{1,3\}, \{2,3\}, \{1,2,3\}\}$$

EXAMPLE 2. Let $A = \{a,b,c,d\}$, $B = \{a,b,c\}$ and $C = \{b,d\}$, find all sets X such that
(i) $X \subset B$ and $X \subset C$ (ii) $X \subset A$ and $X \not\subset B$

SOLUTION.   (i) Here, we have

$$P(B) = \{\phi, \{a\}, \{b\}, \{c\}, \{a, b\}, \{a,c\}, \{b, c\}, \{a, b, c\}\}.$$

And $P(C) = \{\phi, \{b\}, \{d\}, \{b, d\}\}$, then $X \subset B$ and $X \subset C$ implies

$$X \in P(B) \text{ and } X \in P(C)$$

$$X = \{\phi, \{b\}\}$$

(ii) Here, we have, $X \subset A$ and $X \not\subset B$, which implies that

$$X \in P(A) \text{ and } X \in P(B)$$

Therefore $X = \{\{d\}, \{a,b,d\}, \{b,c,d\}, \{a,c,d\}, \{a,d\}, \{b,d\}, \{c,d\}, \{a,b,c,d\}\}$

EXAMPLE 3.   *Write down all the subsets of the following sets.*

   (i) $\{a\}$ *(ii)* $\{a,b\}$ *(iii)* $\{a,b,c\}$ *(iv)* ϕ

SOLUTION.   (i) Let $A = \{a\}$. Since A contains only one element, therefore, the total number of subsets is $2^1 = 2$, which are given by ϕ and $\{a\}$.

(ii) Here, total number of subsets, $= 2^2 = 4$, which are given by $\phi, \{a\}, \{b\}, \{a, b\}$

(iii) Here, total number of subsets $= 2^3 = 8$, given by

$$\phi, \{a\}, \{b\}, \{c\}, \{a,b\}, \{a,c\}, \{b,c\}, \{a,b,c\}$$

(iv) since ϕ contains no element therefore the number of subsets $= 2^0 = 1$. The only subset is ϕ.

EXAMPLE 4. *Which of the following sets are empty. Also, give the reason.*

   (i) $A = \{x : x \neq x, \text{ is a real number}\}$.

   (ii) $B = \{x : x + 4 = 4\}$

   (iii) $C = \{x : x^3 - 3 = 0 \text{ and } x \text{ is rational number}\}$

SOLUTION.   (i) Here, $A = \{x : x \neq x, x \text{ is a real number}\}$. Since $x \neq x$ is not true

$$\Rightarrow \quad A = \phi$$

(ii) $B = \{x : x + 4 = 4\} = \{x : x = 0\} = \{0\}$

$$\Rightarrow \quad B \text{ has one element } 0, \text{ therefore } B \neq \phi.$$

(iii) Since there is no rational number whose square is 3, so $x^3 - 3 = 0$ is not satisfied for any rational numbers. Therefore, C is an empty set.

EXAMPLE 5. *Which of the following sets are finite and which are infinite.*

   (i) The set of natural numbers divisible by 2.

   (ii) The set of natural numbers less then 8.

   (iii) The set of integers whose square is even.

   (iv) The set of integers greater than −18.

   (v) The set of lines passing through a point.

   (vi) The set of points of a plane at a fixed distance from a given point in the plane.

   (vii) The set of points common to two given parallel lines.

   (viii) The set of the roots of a polynomial of nth degree.

SOLUTION.   (i) The given set is $\{2, 4, 6, 8, \ldots\}$. It has an infinite number of finite number of elements, therefore it is an infinite set.

(ii) The given set is $\{1,2,3,4,5,6,7\}$. It has seven elements, *i.e.*, finite number of elements. Hence, it is a finite set.

(iii) The given set is $\{\ldots, -8, -6, -4, -2, 0, 2, 4, 6, 8, \ldots\}$. It has infinite number of elements, therefore it is an infinite set.

(iv) Here, the given set is {–17, –16, ... , 0, 1, 2 ...}. It has infinite number of elements therefore, it is an infinite set.

(v) Since infinite number of lines can pass through a fixed point, therefore the given set is an infinite set.

(vi) Since the points in a plane at a fixed distance from a given point in the plane lie on a circle with the given point as centre and the number of points on a circle is infinite. Therefore, the given set is an infinite set.

(vii) Since two parallel lines cannot meet anywhere, therefore, the set of points common to two given parallel lines is empty, therefore the given set cannot be infinite. Hence, it is a finite set.

(viii) Since, a polynomial of nth degree always have almost n roots.
Therefore, the given set is always a finite set.

EXAMPLE 6. *Which of the following sets are equivalent* ϕ ,{0} *and* { ϕ }.

SOLUTION. Since ϕ has no element. Also, {0} and { ϕ }, each contains one element namely 0 and ϕ respectively. Hence, {0} and { ϕ } are equivalent.

EXAMPLE 7. *Which of the following sets are equal ?*
$$A = \{1,2,3\}, B = \{2,3,4\}, C = \{3,2,1\}, D = \{2,3,5\}$$

SOLUTION. Since $1 \in A$ but $1 \notin B$, therefore $A \neq B$. A and C have exactly the same element, therefore $A = C$.

Also,

$$
\begin{array}{lll}
1 \in C & \text{but } 1 \notin D & \Rightarrow \quad C \neq D \\
4 \in B & \text{but } 4 \notin C & \Rightarrow \quad B \neq C \\
4 \in B & \text{but } 4 \notin C & \Rightarrow \quad B \neq C \\
1 \in A & \text{but } 1 \notin D & \Rightarrow \quad A \neq D
\end{array}
$$

Hence, only A and C are equal sets.

Exercise 1.2

1. Fill in the blanks:

(i) A set which contains no element is called ... set.

(ii) If $A = \{1,2,3\}$ and $B = \{3,2,1\}$ then they are said to be ...

(iii) If $A = \{a, b, c\}$ and $B = \{c, d, e\}$ then they are said to be ...

(iv) If every element of a set B is also an element of A, then B is said to be ... of A.

(v) The empty set is a ... of every set.

(vi) Every set is a .... of itself.

(vii) The set **Z** of integers is a ... of set of natural numbers **N**.

2. Which of the followings sets are equal?

(i) $A = \{1,2,3\}$

(ii) $B = \{1,2,2,3\}$

(iii) $C = (x \in \mathbf{R} : x^3 - 6x^2 + 11x - 6 = 0)$

3. Which of the following sets are equivalent to the set {4,7,11,17,20}?

(i) {5,1,2,3,4}

(ii) {all odd numbers less then 10}

(iii) {the months of a year of 30 days}

(iv) {all the prime numbers which lie between 10 and 25}.

4. Which of the following sets are finite and which are infinite ?

(i) $\{x \in \mathbf{N} : x > 10\}$

(ii) $\{x \in \mathbf{N} : x < 100\}$

(iii) $\{x \in \mathbf{R} : 1 \leq x \leq 2\}$

(iv) Set of vowels in English alphabets.

(v) The set of prime numbers less than 100.

(vi) The set of multiple of 8.

5. Which of the following statements are true? Give the reason.

(i) For any two sets A and B either $A \subseteq B$ or $B \subseteq A$

(ii) Every subset of a finite set is finite.

(iii) A subset of an infinite set may be finite.

(iv) Every set has a proper subset.

(v) A set containing n elements have 2^n subsets.

(vi) If $A = \{1,2,3,4,5,6\}$ and $B = \{$whole numbers less than 6$\}$, then $A = B$.

(vii) The empty set has no proper subset.

6. Examine which of the following sets are empty?

 (i) The set of tigers in your class.

 (ii) The set of triangles having three equal sides.

 (iii) The set of all numbers which, when added to zero, yield sum greater than the original.

 (iv) The set of odd numbers which are divisible by 2.

 (v) The set of men, who never die.

7. Which of the following statements are true?

 (i) If $x \in A$ and $A \subset B$, then $x \in B$

 (ii) If $A \subset B$ and $B \subset C$, then $A \subset C$

 (iii) If $A \not\subset B$ and $B \not\subset C$, then $A \not\subset C$

 (iv) If $x \in A$ and $A \not\subset B$, then $x \in B$

(v) If $A \subset B$ and $x \notin B$, then $x \notin A$

8. Are the following sets, i.e., (A and B) are equal.

 (i) $A = \{x : x$ is a letter of the word 'LITTLE'$\}$
    $B = \{x : x$ is a letter in the word 'TITLE'$\}$

 (ii) $A = \{x : x$ is a letter in the word 'FOLLOW'$\}$
    $B = \{x : x$ is a letter in the word 'WOLF'$\}$

 (iii) $A = \{x : x$ is a letter in the word 'LOYAL'$\}$
    $B = \{x : x$ is a letter In the word 'ALLOY'$\}$

9. Write down all possible subsets of each of the following sets.

 (i) $\{a\}$ (ii) $\{0,1\}$ (iii) $\{a, b, c\}$

 (iv) $\{1, \{1\}\}$ (v) ϕ

10. Which of the following statements are true?

 (i) $\{a, \phi\} \in \{a, \{a, \phi\}\}$

 (ii) If $A \subseteq B$ and $B \subseteq C$, then $A \subseteq C$

 (iii) If $A \in B$ and $B \subseteq C$, then $A \in C$

 (iv) If $A \subset B$ and $B \in C$, then $A \in C$

 (v) If $A \subseteq B$ and $B \in C$, then $A \subseteq C$

ANSWERS

1. (i) Empty (ii) equal (iii) equivalent (iv) subset (v) subset (vi) subset (vii) super set.

2. $A = B = C$ 3. (i), (ii), (iv)   4. (ii), (iv), (v) are finite sets and (i), (iii), (vi) are infinite.

5. (i) F   (ii) T   (iii) T   (iv) F (v) T   (vi) F (vii) T   6. (i), (iii), (iv), (v)

7. (i), (ii), (v)   8. (i) Equal, (ii) Equal, (iii) Equal

9. (i) ϕ, $\{a\}$; (ii) ϕ, $\{0\}$, $\{1\}$, $\{0,1\}$; (iii) ϕ, $\{a\}$, $\{b\}$, $\{c\}$, $\{a, b\}$, $\{b,c\}$, $\{a,c\}$, $\{a,b,c\}$

 (iv) $\{1\}$; $\{1\}$, $\{\{1\}\}$, $\{1,\{1\}\}$;

10. (i), (ii), (iii), (iv), (v)

1.8 VENN DIAGRAMS

A set can be represented by closed figures like circles, triangles, rectangles, etc. The point in the interior of the figure represents the elements of the set. Such a representations is called a Venn diagram. In Venn diagram, the universal set is usually represented by a rectangular region and its subset by closed bounded regions inside the rectangular region. For example, if A is a subset of B, i.e., $A \subset B$. This is shown in figure 2.

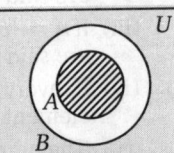

Fig. 2

REMARKS

➡ The diagrams drawn to represent sets are called Venn diagram or Venn-Euler diagrams, after the name of British mathematician **Venn.**

➡ If A and B are two sets, which are not equal, but have common elements, then to represent A and B, We draw two intersecting circles.

➡ Two disjoint sets are represented by two-intersecting circles.

➡ Venn diagrams are to be used for clarity and are no substitute for precise proof.

1.9 OPERATIONS ON SETS

1.9.1 UNION AND INTERSECTION OPERATIONS

(i) Union of Two sets

Let A and B be two sets. Then Union of A and B, denoted by $A \cup B$ is the set of all those elements, which either belongs to A or B or to both A and B.

It should be noted that the common elements are to be taken only once.

Symbolically: $A \cup B = \{x : x \in A \text{ and } x \in B\}$ It is shown in the adjoining figure 3.

For example:

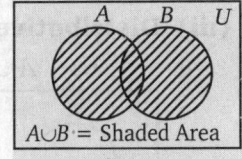

$A \cup B$ = Shaded Area

Fig. 3

(i) Let $A = \{3,4,5,6,7\}$ and $B = \{5,6,7,8,9\}$

Then $A \cup B = \{3, 4, 5, 6, 7, 8, 9\}$

(ii) Let $A = \{x : x = 2n, n = 1, 2, 3, ...\} = \{2, 4, 6, 8, ...\}$

$B = \{x : x = 3n, n = 1, 2, 3, ...\} = \{3, 6, 9, 12, ...\}$

Then $A \cup B = \{x : x \text{ is multiple of 2 or a multiple of 3}\}$

$= \{2, 3, 4, 6, 8, 10, 12, ...\}$

(iii) Let A = set of even natural numbers = $\{2, 4, 6, 8, ...\}$

and B = set of natural numbers = $\{1, 2, 3, 4, 5, ...\}$

Then $A \cup B = \{1, 2, 3, 4, ...\}$

⮞ REMARKS

➡ $x \in (A \cup B) \Leftrightarrow x \in A$ or $x \in B$.

➡ $x \notin (A \cup B) \Leftrightarrow x \notin A$ and $x \notin B$

➡ $A \cup B = B \cup A$, i.e., union of sets is commutative.

➡ $A \cup A' = U$ and $A \cup U = U$

➡ $A \cup \phi = A$

➡ If $A, B, C, D, ..., Z$ is a finite family of sets, then their union is denoted by $A \cup B \cup C \cup D ... \cup Z$.

➡ $(A \cup B) \cup C = A \cup (B \cup C)$, i.e., a union of sets is associative.

(ii) Intersection of Two sets

Let A and B be two sets. Then intersection of A and B, denoted by $A \cap B$ is the set of all those elements, which belongs to both A and B.

Symbolically: $A \cap B = \{x : x \in A \text{ and } x \in B\}$ It is shown in the adjoining figure 4.

For example:

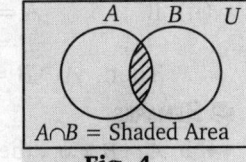

$A \cap B$ = Shaded Area

Fig. 4

(i) Let $A = \{2, 4, 6, 8, 10\}$ and $B = \{1, 2, 3, 4, 5\}$

Then $A \cap B = \{2, 4\}$

(ii) If $A = \{x : x = 3n, n \in \mathbf{Z}\}$

$B = \{x : x = 4n, n \in \mathbf{Z}\}$

Then $A \cap B = \{x : x \text{ is multiple of 3 and } x \text{ is a multiple of 4}\}$

$= \{x : x \text{ is multiple of 3 and 4 both}\}$

$= \{x : x = 12n, n \in \mathbf{Z}\}$

REMARKS

➡ $x \in (A \cap B) \Leftrightarrow x \in A$ and $x \in B$.

➡ $x \notin (A \cap B) \Leftrightarrow x \notin A$ and $x \notin B$

➡ $A = A \cap A$, *i.e.*, intersection of sets is idempotent.

➡ $A \cap \phi = \phi$

➡ $A \cap U = A$, where U is a universal set.

➡ $A \cap B = B \cap A$, *i.e.*, intersection of sets is commutative.

➡ $(A \cap B) \cap C = A \cap (B \cap C)$ intersection of sets is associative.

➡ If $A, B, C, D, ..., Z$ is a finite family of sets, then their intersection is denoted by $A \cap B \cap C ... \cap Z$.

(iii) Distributive Property of Union and Intersection

(i) $A \cup (B \cap C) = (A \cup B) \cap (A \cup C)$ (ii) $A \cap (B \cup C) = (A \cap B) \cup (A \cap C)$

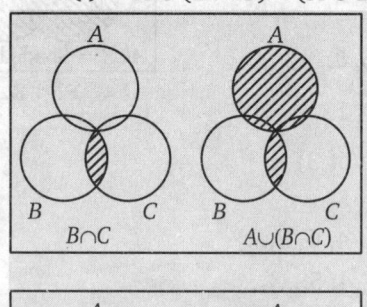

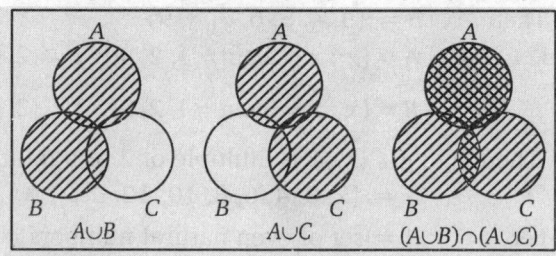

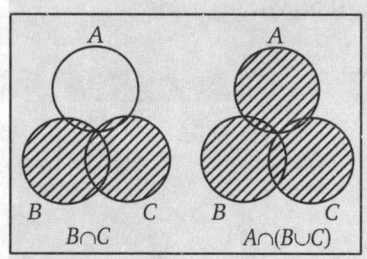

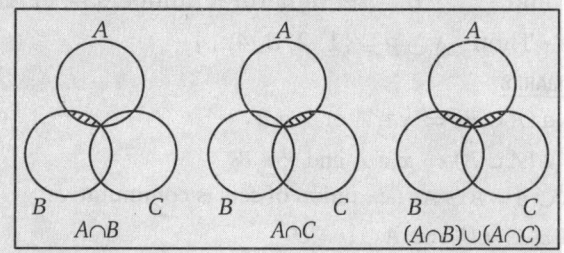

Fig. 5

1.9.2. DISJOINT SETS

When two sets have no common elements, they are called disjoint sets. Thus, if $A \cap B = \not\subset$, then A and B are disjoint. It is shown in the adjoining figure 6.

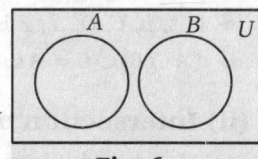

Fig. 6

For example:

(i) If $A = \{2, 4, 6, 8\}$ and $B = \{1, 3, 5, 7, 9\}$

Then, $A \cap B = \not\subset$

(ii) If A = Boys in school

B = Girls in school

Then, $A \cap B = \not\subset$

REMARKS

➡ If $A \cap B \neq \phi$, , then A and B are said to be intersecting or overlapping sets.

➡ A family of sets is said to be pair-wise disjoint family of sets if and only if any two sets of this family are disjoint. For example, classes of A_2, A_3, A_5 and A_7 defined as $A_2 = \{2, 2^2, 2^3, ...\}$; $A_3 = \{3, 3^2, 3^3, ...\}$; $A_5 = \{5, 5^2, 5^3, ...\}$ and $A_7 = \{7, 7^2, 7^3, ...\}$ are pair-wise disjoint.

➡ $\phi \cap A = \phi$, *i.e.*, null set is disjoint from every subset.

1.9.3 DIFFERENCE OF TWO SETS

If A and B are two sets, then the set of all elements which belong to A but do not belong to B is called the difference of sets A and B and is denoted by A ~ B. The set of all elements which belong to B but do not belong to A is called the difference of sets B and A and is denoted by B ~ A.

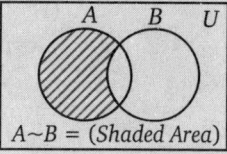

$A{\sim}B$ = (Shaded Area)

Fig. 7

Therefore,

$$A{\sim}B = \{x : x \in A \text{ and } x \notin B\} = A \cap B'$$

And $B{\sim}A = \{x : x \notin A \text{ and } x \in B\} = B \cap A'$

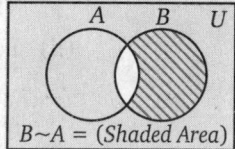

$B{\sim}A$ = (Shaded Area)

Fig. 8

For example:

(i) Let $\quad A = \{1, 2, 3, 4, 5\}$

And $\qquad B = \{-1, 0, 1, 2\}$

Then, $A{\sim}B = \{3, 4, 5\}$

And $\quad B{\sim}A = \{-1, 0\}$

▶ REMARKS

➡ $x \in (A - B) \Leftrightarrow x \in A$ and $x \notin B$.

➡ $x \notin (A - B) \Leftrightarrow x \notin A$ and $x \in B$

➡ $A - B \neq B \sim A$, *i.e.*, , difference of two sets is not commutative.

➡ $A \subset B$ then $A \sim B = \phi$

➡ The sets $A \sim B$, $A \cap B$ and $B \sim A$ are mutually disjoint.

➡ Difference of a set with the universal set is known as complementation.

➡ $A \sim B$ is a subset of A and $B \sim A$ is a subset of B.

1.9.4 SYMMETRIC DIFFERENCE OF TWO SETS

If A and B are two sets, then the symmetric difference of two sets A and B is denoted by $A\Delta B$ is given by $A \Delta B = (A \sim B) \cup (B \sim A)$

Symbolically: $\quad A \Delta B = \{x : (x \in A \text{ and } x \notin B) \text{ or } (x \in B \text{ and } x \notin A)\}$

For example:

(i) If $\qquad A = \{1,2,3,4,5,6,7,8\}$ and $B = \{1,3,5,6,7,8,9\}$

Then $A \sim B = \{2, 4\}$ and $B{\sim} A = \{9\}$

and $\quad A \Delta B = \{2, 4, 9\}$

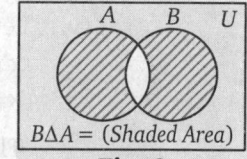

$B\Delta A$ = (Shaded Area)

Fig. 9

Equivalent Sets: Two finite sets A and B are equivalent if their cardinal numbers are same , *i.e.*, $n(A) = n(B)$.

1.9.5 LAW OF EXCLUDED MIDDLE AND LAW OF CONTRADICTION

Two special properties of set operations are known as the excluded middle axioms and law of contradiction. The excluded middle axioms are very important because they are the only set operations described here that are not valid for both classical sets and fuzzy sets. Let A be any subset of universal set X. Then , we define.

(i) Axiom of the excluded middle: $A \cup \overline{A} = X$

(ii) Axiom of the contradiction: $A \cap \overline{A} = \phi$

THEOREM 1.

$\qquad$ (i) $A \cup \phi = A \qquad$ (ii) $A \cap \phi = \phi \qquad$ (iii) $A \cup A = A$

$\qquad$ (iv) $A \cap A = A \qquad$ (v) $A \cup B = B \cup A \qquad$ (vi) $A \cap B = B \cap A$

PROOF.

(i) Let x be an arbitrary element of $A \cap \phi$.

i.e., $x \in A \cup \phi$

Then, by definition $x \in A \cup B \Leftrightarrow x \in A$ or $x \in B$

i.e., $x \in A \cup \phi \Rightarrow x \in A$ or $x \in \phi$

$\Leftrightarrow$ $x \in A$ ($\because \phi$ is a null set $\Rightarrow x \notin \phi$)

Therefore, $A \cup \phi = A$

(ii) Let x be an arbitrary element of $A \cap \phi$.

$x \in A \cap \phi \Leftrightarrow x \in A$ and $x \in \phi$ ($\because \phi$ is a null set)

Therefore, $A \cap \phi = \phi$

(iii) Let x be an arbitrary element of $A \cup A$,

$x \in A \cup A \Leftrightarrow x \in A$ or $x \in A$ (Repeated statement)

$\Leftrightarrow x \in A$

Therefore, $A \cup A = A$

(iv) Let x be an arbitrary element of $A \cap A$,

$x \in A \cap A \Leftrightarrow x \in A$ or $x \in A$ (Repeated statement)

$\Leftrightarrow x \in A$

Therefore, $A \cap A = A$

(v) Let x be an arbitrary element of $A \cup B$,

$x \in A \cup B \qquad \Leftrightarrow x \in A$ or $x \in A$ (Writing in reverse order)
$\Leftrightarrow x \in B$ or $x \in A \Leftrightarrow x \in B \cup A$

Therefore, $A \cup B = B \cup A$

(vi) Let x be an arbitrary element of $A \cap B$

$x \in A \cap B \qquad \Leftrightarrow x \in A$ and $x \in A$ (Writing in reverse order)
$\Leftrightarrow x \in B$ and $x \in A \Leftrightarrow x \in B \cap A$

Therefore, $A \cap B = B \cap A$

THEOREM 2.   *For any three sets A, B and C*

(*i*) $A \cup (B \cup C) = (A \cup B) \cup C$ (ii) $A \cap (B \cap C) = (A \cap B) \cap C$

(*iii*) $A \cup (B \cap C) = (A \cup B) \cap (A \cup C)$ (iv) $A \cap (B \cup C) = (A \cap B) \cup (A \cap C)$

PROOF.

(i) Let x be an arbitrary element of $A \cup (B \cup C)$, then

$x \in A \cup (B \cup C)$

$\Leftrightarrow \qquad x \in A$ or $x \in (B \cup C) \Leftrightarrow x \in A$ or $(x \in B$ or $x \in C)$

$\Leftrightarrow \qquad (x \in A$ or $x \in B)$ or $x \in C$ (By associativity)

$\Leftrightarrow \qquad x \in (A \cup B)$ or $x \in C \Leftrightarrow x \in A \cup (B \cup C)$

Therefore, $A \cup (B \cup C) = (A \cup B) \cup C$

(ii) Let x be an arbitrary element of $A \cap (B \cap C)$, then

$x \in A \cap (B \cap C)$

$$\Leftrightarrow \quad x \in A \quad \text{and} \Leftrightarrow x \in (B \cap C) \Leftrightarrow x \in A \text{ and } (x \in B \text{ and } x \in C)$$

$$\Leftrightarrow \quad (x \in A \quad \text{and} \quad x \in B) \text{ and } x \in C \qquad \text{(By associativity)}$$

$$\Leftrightarrow \quad x \in (A \cap B) \text{ and } x \in C \Leftrightarrow x \in (A \cap B) \cap C$$

Therefore, $A \cap (B \cap C) = (A \cap B) \cap C$

(iii) Let x be an arbitrary element of $A \cup (B \cup C)$, then

$$x \in A \cup B (\cap C)$$

$$\Leftrightarrow \quad x \in A \text{ or } x \in (B \cap C) \Leftrightarrow x \in A \text{ or } (x \in B \text{ and } x \in C)$$

$$\Leftrightarrow \quad (x \in A \text{ or } x \in B) \text{ and } (x \in A \text{ or } x \in C) \Leftrightarrow x \in (A \cup B) \text{ and } x \in (A \cup C)$$

$$\Leftrightarrow \quad x \in (A \cup B) \cap (A \cup C)$$

Therefore, $A \cup (B \cap C) = (A \cup C) \cap (A \cup C)$

(iv) Let x be an arbitrary element of $A \cap (B \cup C)$, then

$$x \in A \cap (B \cup C)$$

$$\Leftrightarrow \quad x \in A \text{ and } x \in (B \cup C) \Leftrightarrow x \in A \text{ and } (x \in B \text{ or } x \in C)$$

$$\Leftrightarrow \quad (x \in A \text{ and } x \in B) \text{ or } (x \in A \text{ and } x \in C) \Leftrightarrow x \in (A \cap B) \text{ or } x \in (A \cap C)$$

Therefore, $A \cap (B \cup C) = (A \cap B) \cup (A \cap C)$

THEOREM 3.

(i) $(A')' = A$ (ii) $A \cup A' = U$, where U is the universal set.

(iii) $A \cap A' = \phi$ (iv) $(A \cup B)' = A' \cap B'$ (De' Morgan's Law)

(v) $(A \cap B)' = A' \cup B'$ (De' Morgan's Law)

PROOF.

(i) Let x be an arbitrary element of $(A')'$,

$$x \in (A')' \Leftrightarrow x \notin A' \Leftrightarrow x \in A$$

Therefore, $(A')' = A$

(ii) Let x be an arbitrary element of $(A \cup A')$,

$$x \in (A \cup A') \qquad \Leftrightarrow x \in A \text{ or } x \in A' \Leftrightarrow x \in A \text{ or } x \in U - A$$

$$\Leftrightarrow \quad x \in A \text{ or } (x \in U, x \notin A) \Leftrightarrow x \in U$$

Therefore, $A \cup A' = U$

(iii) Let x be an arbitrary element of $(A \cap A')$,

$$x \in (A \cap A') \qquad \Leftrightarrow x \in A \text{ and } x \in A' \text{ but if } x \in A \text{ then } x \notin A'$$

Therefore, $A \cap A' = \phi$

(iv) Let x be an arbitrary element of $(A \cup B)'$,

$$x \in (A \cup B)' \qquad \Leftrightarrow x \notin (A \cup B) \qquad \Leftrightarrow x \notin A \text{ and } x \notin B$$

$$\Leftrightarrow \quad x \in A' \text{ and } x \in B' \Leftrightarrow x \in A' \cap B'$$

Therefore, $(A \cup B)' = A' \cap B'$

(v) Let x be an arbitrary element of $(A \cap B)'$,

$$x \in (A \cap B)' \qquad \Leftrightarrow x \notin (A \cap B) \qquad \Leftrightarrow x \notin A \text{ or } x \notin B$$

$$\Leftrightarrow \quad x \in A' \text{ or } x \in B' \Leftrightarrow x \in A' \cup B'$$

Therefore, $(A \cap B)' = A' \cup B'$

☛ RECAPITULATIONS

- $A \cup B = \{x : x \in A \text{ or } x \in B)$
- $x \notin A \cap B \Leftrightarrow x \notin A \text{ or } x \notin B$
- $x \in A - B \Leftrightarrow x \in A \text{ and } x \notin B$
- $A \cup \phi = A$
- $(A')' = A$
- $A \cap A' = \phi$
- $(A \cap B)' = A' \cup B'$

- $A \cap B = \{x : x \in A \text{ or } x \in B)$
- $x \notin A \cup B \Leftrightarrow x \notin A \text{ and } x \notin B$
- $A \Delta B = (A \sim B) \cup (B \sim A)$
- $A \cap \phi = \phi$
- $A \cup A' = \cup$
- $(A \cup B)' = (A' \cap B')$

SOLVED EXAMPLES

EXAMPLE 1. *Show that (i) $A \subset (A \cup B)$, (ii) $(A \cap B) \subset A$.*

SOLUTION. (i) Let $x \in A$ be arbitrary then $x \in A$ certainly but may or may not belong to B.

$\Rightarrow \qquad x \in A \cup B$

Therefore, $\qquad x \in A \qquad \Rightarrow \qquad x \in A \cup B$ gives $A \subset A \cup B$

(ii) Let $\qquad x \in A \cap B \qquad$ where x is arbitrary

$\qquad\qquad x \in A \cap B \qquad \Rightarrow \qquad x \in A \text{ and } x \in B$

In particular, $\quad x \in A \cap B \qquad \Rightarrow \qquad x \in A$

Therefore, $\qquad (A \cap B) \subset A$

⮑ REMARK

➟ Similarly we can show that (i) $B \subset (A \cup B)$ and (ii) $A \cap B \subset B$.

EXAMPLE 2. *Let A and B be two sets, if $A \cap X = B \cap X = \phi$ and $A \cup X = B \cup X$ for some set X, prove that $A = B$.*

SOLUTION. Given that $A \cup X = B \cup X$

$\Rightarrow \quad A \cap (A \cup X) = A \cap (B \cup X) \qquad$ (taking intersection by A on both sides)

$\Rightarrow \quad A = A \cap (B \cup X) \qquad\qquad\qquad (\because A \cap (A \cup X) = A)$

$\Rightarrow \quad A = (A \cap B) \cup (A \cap X) \qquad\qquad$ (By distributive law)

$\Rightarrow \quad A = (A \cap B) \cup \phi \quad \Rightarrow \quad A = A \cap B$

$\Rightarrow \quad A \subset (A \cap B) \qquad\qquad \Rightarrow \quad A \subset B \qquad\qquad\qquad$ ...(1)

Again consider, $A \cup X = B \cup X$

$\Rightarrow \quad B \cap (A \cup X) = B \cap (B \cup X) \qquad\qquad$ (taking intersection with B)

$\Rightarrow \quad B \cap (A \cup X) = B$

$\Rightarrow \quad (B \cap A) \cup (B \cap X) = B \qquad\qquad\qquad$ (By distributive law)

$\Rightarrow \quad (B \cap A) \cup \phi = B \qquad\qquad\qquad$ (Given $B \cap X = \phi$)

$\Rightarrow \quad (B \cap A) = B \qquad\qquad\qquad (\because A \cap B = B \cap A)$

$\Rightarrow \quad A \cap B = B \qquad\qquad \Rightarrow B \subset A \cap B \Rightarrow \quad B \subset A \qquad$ ...(2)

Hence, (1) and (2) gives $A \subset B$ and $B \subset A$.

$\Rightarrow \qquad A = B$

EXAMPLE 3. *For any two sets A and B, show that*

(i) $P(A \cap B) = P(A) \cap P(B)$, (ii) $P(A) \cup P(B) \subset P(A \cup B)$

SOLUTION. (i) Let $\quad X \in P(A \cap B) \qquad \Rightarrow X \subset A \cap B$

$\Rightarrow \qquad X \subset A \text{ and } x \subset B \Rightarrow X \in P(A) \text{ and } X \in P(B)$

$\Rightarrow \qquad\qquad X \in P(A) \cap P(B)$

Therefore, $\qquad P(A \cap B) \subset P(A) \cap P(B) \qquad\qquad\qquad$ ...(1)

Now, let $X \in P(A) \cap P(B) \Rightarrow X \in P(A)$ and $X \in P(B)$

$\Rightarrow \quad X \subset A$ and $X \subset B \Rightarrow X \subset A \cap B$

$\Rightarrow \quad X \in P(A \cap B)$

Therefore, $\quad P(A) \cap P(B) \subset P(A \cap B)$ ...(2)

From (1) and (2), we conclude that

$P(A \cap B) \subset P(A) \cap P(B)$ and $P(A) \cap P(B) \subset P(A \cap B)$ which gives

$P(A \cap B) = P(A) \cap P(B)$

(ii) Let $\quad X \in P(A) \cup P(B) \Rightarrow X \in P(A)$ and $X \in P(B)$

$\Rightarrow \quad X \subset A$ or $x \subset B \quad \Rightarrow X \subset A \cup B$

$\Rightarrow \quad X \in P(A \cup B)$

Therefore, $\quad P(A) \cup P(B) \subset P(A \cup B)$

REMARK

➠ Converse of the result (ii) is not necessarily true. For example, let $A = \{1,2\}$ and $B = \{4,5,6\}$, then we find that $x = \{1,2,3,5\}$ which is a subset of $A \cup B$. Therefore, $x \in P(A \cup B)$. But $x \notin P(A), x \notin P(B)$. So,

$$x \notin P(A) \cup P(B) \Rightarrow P(A \cup B) \not\subset P(A) \cup P(B)$$

1.9.6 SOME MORE RESULTS

1. If A and B are any two sets, then

   (i) $A - B = A \cap B'$

   (ii) $A - B = A \Leftrightarrow A \cap B = \phi$

   (iii) $(A - B) \cup B = A \cup B$

   (iv) $A \subset B \Leftrightarrow B' \subset A'$

   (v) $(A - B) \cup (B - A) = (A \cup B) - (A \cap B)$

2. If A and B are any two sets, then

   (i) $A - (B \cap C) = (A - B) \cup (A - C)$

   (ii) $A - (B \cup C) = (A - B) \cap (A - C)$

   (iii) $A \cap (B - C) = (A \cap B) = (A \cap C)$

Exercise 1.3

1. Let $A = \{a, b\}$, $B = \{a, b, c\}$. Is $A \subset B$. Find $A \cup B$ and $A \cap B$.

2. If $A = \{1,2,3,4\}$, $B = \{2,4,6,8\}$, $C = \{3,4,5,6\}$ and universal set $U = \{1,2,3,4,...9\}$. Verify that $A \cap (B \cup C) = (A \cap B) \cup (A \cap C)$.

3. If A, B, C are subsets of a set X, then show that $A \subseteq B$ and $B \subseteq C \Rightarrow A \subseteq C$.

4. Find the union of the following sets:

   (i) $A = \{x : x$ is an even integer$\}$,
   $B = \{x : x$ is an odd integer$\}$.

   (ii) $A = \{x : x$ is a multiple of 2$\}$,
   $B = \{x : x$ is a multiple of 3$\}$.

   (iii) $A = \{x : x$ is a rational number$\}$,
   $B = \{x : x$ is an irrational number$\}$.

   (iv) $A = \{x : x$ is a negative integer$\}$,
   $B = \{x : x$ is a non-negative integer$\}$

5. Find the intersection of the following sets.

   (i) $A = \{x : x$ is an even integer$\}$,
   $B = \{x : x$ is an odd integer$\}$

   (ii) $A = \{x : x$ is a rational number$\}$,
   $B = \{x : x$ is an irrational number$\}$.

   (iii) $A = \{x : x$ is a multiple of 5$\}$,
   $B = \{x : x$ is a multiple of 2$\}$

   (iv) $A = \{x : x$ is a rational number$\}$,
   $B = \{x : x$ is a real number$\}$

6. If $A = \{1,2,3,4\}$, $B = \{2,4,6,8\}$ and $C = \{3,4,5,6\}$, find

   (i) $(A \cup B) \cap C$

   (ii) $A \cup (B \cap C)$

7. Write T for true and F for false statement.

   (i) $A \in (A \cup B)$ (T/F)

   (ii) $(A \cup B) \in B$ (T/F)

   (iii) $(A \cap B) \in A$ (T/F)

   (iv) $A \cup A = A$ and $A \cap A = A$ (T/F)

   (v) If $A \cap B = \phi$, then $A \cap \phi = B$ (T/F)

   (vii) If A and B are disjoint sets, then intersection of their union and intersection is the null set. (T/F)

(viii) If A is the proper subset of U, then the union of A and A' is U.   (T/F)

(ix) $U' = \phi$ and $\phi' = U$ (T/F)

(x) $(A \cup B)' = A' \cap B'$ (T/F)

(xi) $A \cap A'$ is always empty   (T/F)

(xii) $(A \cap B)' = A' \cup B'$ (T/F)

8. If $A = \{1, 2, 3, 4, 5, 6, 7, 8\}$ and $B = \{1, 3, 5, 6, 7, 8, 9\}$, then show that
$$A \triangle B = \{2, 4, 9\}$$

9. Let $A = \{x : x \in \mathbf{N}\}$,
$B = \{x : x = 2n : n \in \mathbf{N}\}$,
$C = \{x : x = 2n-1 : n \in \mathbf{N}\}$
and $D = \{x : x \text{ is a prime natural number}\}$.
Find
(i) $A \cap B$ (ii) $A \cap C$
(iii) $A \cap D$ (iv) $B \cap C$
(v) $B \cap D$ (vi) $C \cap D$

10. For any two sets A and B, prove that $P(A) = P(B)$ implies that $A = B$

11. For any two sets A and B, show that
(i) $A \cup (A \cap B) = A$
(ii) $A \cap (A \cup B) = A$
(iii) $(A \cup B) \cap (A \cap B') = A$

(iv) $A' \cup B = U \Rightarrow A \subset B$

(v) $A \subset B \Leftrightarrow B' \subset A'$

(vi) $B \subset B \subset A \Leftrightarrow A \cap B = B$

12. Let $A = \{1, 2, 3, 4\}$, $B = \{2, 3, 4, 5\}$ and $C = \{4, 5, 6, 7\}$. Verify that
(i) $A \cup (B \cap C) = (A \cup B) \cap (A \cup C)$
(ii) $A \cap (B \cup C) = (A \cap B) \cup (A \cap C)$
(iii) $A \cap (B - C) = (A \cap B) - (A \cap C)$
(iv) $A - (B \cup C) = (A - B) \cap (A - C)$
(v) $A - (B \cap C) = (A - B) \cup (A - C)$

13. Show that
(i) If a sets has only even element, then it has 2 subsets.
(ii) If $B \subset A$ and B has one element less than that of A, show that A has twice as many subset as B has.
(iii) A set with 2 element has 2^2 subsets, a set with 3 elements has 2^3 subsets and so on.

14. If $X = \{4^n - 3n - 1 : n \in \mathbf{N}\}$ and $Y = \{9(n-1) : n \in \mathbf{N}\}$, show that $X \subset Y$.

15. Show that $A - B, A \cap B$ and $B - A$ are pairwise disjoint.

16. Show that $A \cup B \subseteq A \cap B$ implies that $A = B$.

ANSWERS

1. (i) Yes. $\{a, b, c\}, \{a, b\}$;

4. (i) $A \cup B = \{x : x \text{ is non-zero integer}\}$ (ii) $A \cup B = \{x : x \text{ is a multiple of 2 or 3}\}$
(iii) $A \cup B = \{x : x \text{ is a real number}\}$ (iv) $A \cup B = \{x : x \text{ is an integer}\}$

5. (i) ϕ (ii) ϕ (iii) 10   (iv) $\{x : x \text{ is a rational number}\}$

6. (i) $\{3, 4, 6\}$, (ii) $\{1, 2, 3, 4, 6\}$

7. (i) T   (ii) F   (iii) T   (iv) T   (v) F   (vi) T   (vii) T
(viii) T   (ix) T   (x) T   (xi) T   (xii) T

9. (i) B   (ii) C   (iii) D   (iv) ϕ (v) 2   (vi) $D - \{2\}$

1.10 SOME RESULTS ON VENN DIAGRAMS

If A is a finite set, and $n(A) = $ No. of element in the set A.

The following results may be remembered for direct application :

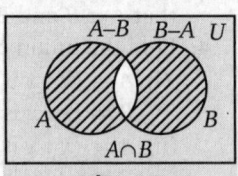

Fig. 10

(i) $n(A \cup B) = n(A) + n(B) - n(A \cap B)$

(ii) $n(A \cup B) = n(A) + n(B)$, provided A and B are disjoints, i.e., if $n(A \cap B) = 0$

(iii) $n(A \cap B') = n(A) - n(A \cap B)$

(iv) $n(B \cap A') = n(B) - n(A \cap B)$

(v) $n(A \cup B) = n(A \cap B') + n(B \cap A') + n(A \cap B)$

(vi) $n(A \triangle B) = n(A) + n(B) - 2n(A \cap B)$

(vii) $n(A' \cup B') = n[(A \cap B)'] = n(U) - n(A \cap B)$

(viii) $n(A' \cap B') = n[(A \cup B)'] = n(U) - n(A \cup B)$

(ix) $n(A - B) = n(A) - n(A \cap B) \Rightarrow n(A - B) + n(A \cap B) = n(A)$

(x) $n(A \cup B \cup C) = n(A) + n(B) + n(C) - n(A \cap B) - n(B \cap C)$
$$- n(A \cap C) + n(A \cap B \cap C)$$

SOLVED EXAMPLES

EXAMPLE 1. *In a group of athletic teams in a school, 21 are in the basket ball, 26 in the hockey team and 29 in the football team. If 14 play hockey and basket ball, 12 play football and basket ball, 15 play hockey and football and 8 play all the three games. Find (i) how many players are there in all (ii) how many play football only.*

SOLUTION. Let A, B and C denote the set of players, who play basket ball, hockey and football respectively. Then, according to question, we have

$$n(A) = 21, n(B) = 26, n(C) = 29$$

$$n(A \cap B) = 14, n(A \cap C) = 12, n(B \cap C) = 15 \text{ and } (A \cap B \cap C) = 8$$

Therefore, $\quad n(A \cup B \cup C) = [n(A) + n(B) + n(C) + n(A \cap B \cap C)]$
$$- [n(A \cap B) + n(A \cap C) + n(B \cap C)]$$
$$= [21 + 26 + 29 + 8] - [14 + 12 + 15] = 43$$

Hence, the total number of players is 43. Now, the number of players playing football only is $[29 - (7 + 8 + 4)] = 10$.

EXAMPLE 2. *In a canteen, out of 123 students, 42 students buy ice-cream, 36 buy burst and 10 buy cakes, 15 students buy ice-cream and 11 buy ice-cream and buns but no cakes. Draw Venn diagram to illustrate the above information and find (i) how many students buy nothing at all (ii) how many students buy at least two items. (iii) how many students buy all three items.*

SOLUTION. Define the sets A, B and C such that

A = Set of students who buy cakes

B = Set of students who buy ice-cream

C = Set of students who buy buns

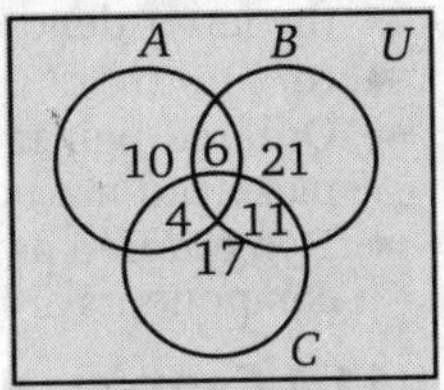

Fig. 11

According to question, we have,

$$n(A) = 10; \ n(B) = 42; \ n(C) = 36; \ n(B \cap C) = 15;$$

$$n(A \cap B) = 10; \ n[(A \cap C) - B] = 4;$$

$n[(B \cap C) - A] = 11$ and $n[A - B \cup C] = 10$

Now we have $n(B \cup C) = n(B) + n(C) - n(B \cap C)$
$$= 42 + 36 - 15 = 63$$

$$(B \cup C) - n(B) = 63 - 42 = 21$$

and $\quad n(B \cup C) - n(C) = 63 - 36 = 27$

The above distribution of the students can be illustrated by Venn diagram (Figure 11). Now, total number of students buying something.

$$= 10 + 6 + 21 + 4 + 4 + 11 + 17 = 73$$

(i) Number of students who did not buy anything = $123 - 73 = 50$

(ii) Number of students buying at least two items = $6 + 4 + 4 + 11 = 25$

And (iii) Number of students buying all three items = 4

Exercise 1.4

1. Out of 80 students who secured first class marks in Mathematics or in Physics, 50 obtained first class marks in Mathematics, 10 in both Physics and Mathematics. How many students secured first class marks in Physics only?

2. The Mathematics club in a school held an open house on three afternoons 115, 110 and 135 students attended both the first, second and third afternoons respectively. 25 attended just the first, 30 attended both the first and second days, 80 attended both the first and third days, and 60 attended both the second and third days. How many attended (i) all three days (ii) just the second day (iii) just the third day?

3. In a school of 250 pupils, 100 are girls, and 200 pupils stay at school for lunch. If 40 girls go home for lunch. Find the number of boys who go home for lunch.

4. In a class of 150 students, the following results were obtained in a certain examination. 45 students failed in Maths; 50 students failed in Physics, 48 students failed in Chemistry, 35 failed in both Maths and Chemistry, 25 failed in the three subject. Find the number of students who have failed in at least one subject.

ANSWERS

1. 30   **2.** 20, 30, 15   **3.** 10   **4.** 71

1.11 ORDERED PAIR

Sometimes, there are situations in which order is very important. Some results may be affected by order and other are not.

Definition: *An ordered pair is a pair of entries whose components occur in a specific order. It is written by listing the two components in the specific order, separating them by a comma and enclosing the pair in parentheses.*

Symbolically: If A and B are two sets, then by ordered pair of elements, we must mean a pair $(a,b): a \in A, b \in B$ in that order.

REMARKS
- It may be noted that (a, b) is not the same as $\{a, b\}$. The former denotes an ordered pair whereas the latter denotes a set.
- $(a, b) \neq (b, a)$ unless $a = b$.
- Ordered pair may have the same first and second components, *i.e.*, two elements of an ordered pair need not be district.
- Two ordered pairs are said to be equal when both the first components are equal and their second components are also equal.

1.11.1 CARTESIAN PRODUCT OF TWO SETS

The set of all ordered pairs of elements (a,b), $a \in A$, $b \in B$ is called the cartesian product of two sets A and B. It is denoted by $A \times B$.

Symbolically: $A \times B = \{(a, b) : a \in A, b \in B\}$
For example :
If $A = \{2, 3\}$ and $B = \{4,5,6\}$, then
$$A \times B = \{(2, 4), (2, 5), (2, 6), (3, 4), (3, 5), (3, 6)\}$$

REMARKS
- $A \times B = \phi \Leftrightarrow A = \phi$ or $B = \phi$
- If A and B are finite sets, then $n(A \times B) = n(A) \cdot n(B)$
- If either A or B are infinite sets, then $A \times B$ is an infinite set.

1.11.2 ORDERED TRIPLET

If A, B, C are three sets, then by ordered triple product of elements, we mean a triplet $(a, b, c) : a \in A, b \in B, c \in C$ in that order.

This is also called **ordered 3-tuple.**

The set of all ordered triplets $(a, b, c): a \in A, b \in B, c \in C$ is also called the cartesian triple product of three sets A, B and C and is denoted by $(A \times B \times C)$

Symbolically: $\qquad A \times B \times C = \{(a, b, c): a \in A, b \in B, c \in C\}$

REMARK

➡ In general, the cartesian product on n sets $A, A_2, ..., A_n$ is a ordered n tuples $(a_1, a_2,....,a_n)$, where $a_1 \in A_1, a_2 \in A_2, ..., a_n \in A_n$. It is denoted by $A_1 \times A_2 ... \times A_n$ or briefly by $\prod\limits_{i=1}^{n} A_i$ where $\prod$ stands for the product.

SOLVED EXAMPLES

EXAMPLE 1. *If A = {1, 2} and B = {a, b, c}, find the value of A×B, B×A, A×A, B×B.*

SOLUTION. We have $A = \{1, 2\}$ and $B = \{a, b, c\}$.
Therefore,
$$A \times B = \{(1, a), (1, b), (1, c), (2, a), (2, b), (2, c)\}$$
$$B \times A = \{(a, 1), (a, 2), (b, 1), (b, 2), (c, 1), (c, 2),\}$$
$$A \times A = \{(1, 1), (1, 2), (2, 1), (2, 2)\}$$
$$B \times B = \{(a, a), (a, b), (a, c), (b, a)\ (b, b), (b, c), (c, a), (c, b), (c, c)\}$$

EXAMPLE 2. *If $A = \{1, 2, 3\}$, $B = \{a, b, c, d\}$ and $C = \{-1, -2\}$, find $A \times B$, $B \times A$ and $C \times (B \cup C)$.*

SOLUTION. Given that $A = \{1, 2, 3\}$, $B = \{a, b, c, d\}$ and $C = \{-1, -2\}$.
Therefore,
$$A \times B = \{(1, a), (1, b), (1, c), (1, d), (2, a), (2, b), (2, c), (2, d),$$
$$(3, a), (3, b), (3, c), (3, d)\}$$
$$B \times A = \{(a, 1), (b, 1), (c, 1), (d, 1), (a, 2), (b, 2), (c, 2), (d, 2), (a, 3),$$
$$(b, 3), (c, 3), (d, 3)\}$$
Also, $\quad B \cup C = \{a, b, c, d, -1, -2\}$
Therefore,
$$C \times (B \cup C) = \{(-1, a), (-1, b), (-1, c), (-1, d), (-1, -1), (-1, -2), (-2, a),$$
$$(-2, b), (-2, c), (-2, d), (-2, -1), (-2, -2)\}$$

EXAMPLE 3. *Find the values of a and b if $(4a-2, b+4) = (2a, 4)$.*

SOLUTION. Since we know that two ordered pairs (a_1, b_1) and (a_2, b_2) are said to be equal if $a_1 = a_1$ and $b_1 = b_2$. Therefore, for the equality of two given ordered pairs, we have $\qquad 4a - 2 = 2a$ and $b + 4 = 4$
Therefore, $4a - 2a = 2 \quad \Rightarrow \quad a = 1$ and $b + 4 = 4 \Rightarrow b = 0$

EXAMPLE 4. *If $A = \{1, 2, 3, 4\}$ and $B = \{4, 5\}$, represent $A \times B$, $B \times A$ and $B \times B$ pictorially and find their values.*

SOLUTION. Given $\quad A = \{1, 2, 3, 4\}$ and $B = \{4, 5\}$
$$A \times B = \{(1, 4), (1, 5), (2, 4), (2, 5), (3, 4), (3, 5), (4, 4), (4, 5)\}$$
$$B \times A = \{(4, 1), (5, 1), (4, 2), (5, 2), (4, 3), (5, 3), (4, 4), (5, 4)\}$$
And $B \times B = \{(4, 4), (4, 5), (5, 4), (5, 5)\}$
Pictorially, $A \times B$, $B \times B$ and $B \times A$ can be represented as shown in figure 12.

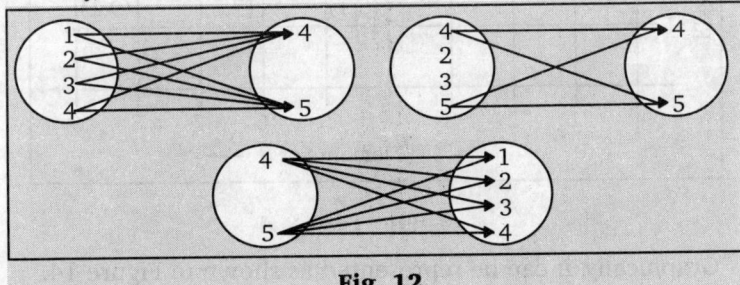

Fig. 12

EXAMPLE 5. *Let A = {1, 2, 3, 4} and B = {5, 7, 9}. Determine (i) A×B, (ii) B×A. Also represent A×B and B×A graphically.*

SOLUTION. (i) Given $A = \{1, 2, 3, 4\}$ and $B = \{5, 7, 9\}$. Then,

$$A \times B = \{(1, 5), (1, 7), (1, 9), (2, 5), (2, 7), (2, 9) (3, 5), (3, 7), (3, 9),$$
$$(4, 5), (4, 7), (4, 9)\}$$

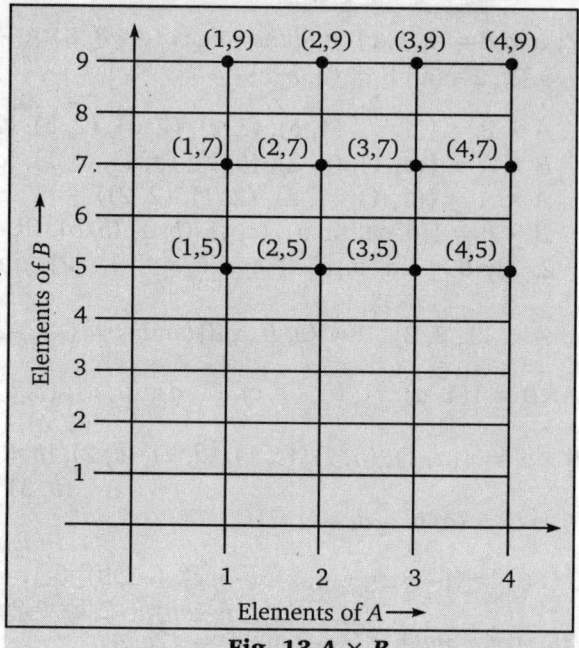

Fig. 13 A × B

Graphically, it can be represented as shown in Figure 13.

Now, $B \times A = \{(5, 1), (5, 2), (5, 3) (5, 4) (7, 1), (7, 2), (7, 3) (7, 4) (9, 1),$
$(9, 2), (9 3) (9, 4)\}$

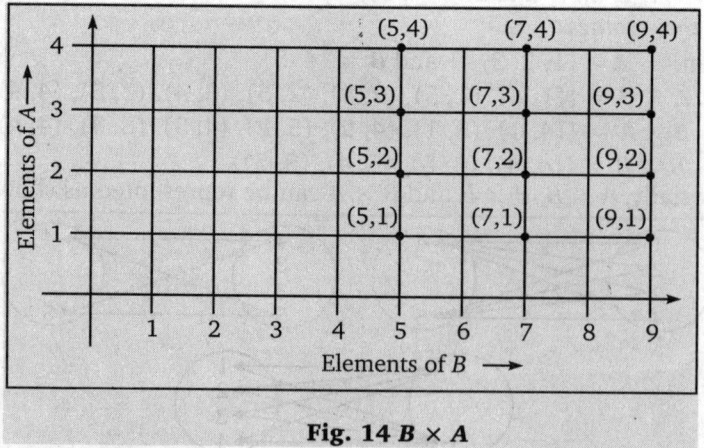

Fig. 14 B × A

Graphically, it can be represented as shown in Figure 14.

THEOREM 1.   *For any three subsets A, B and C , we have.*

(i) $A \times (B \cap C) = (A \times B) \cap (A \times C)$ (ii) $A \times (B \cup C) = (A \times B) \cup (A \times C)$

PROOF.   (i)   If $(x, y) \in A \times (B \cap C)$

$\Rightarrow$ Then, $x \in A$ and $y \in (B \cap C)$

$\Rightarrow$ $x \in A$ and $y \in B$ and $y \in C \Rightarrow x \in A, y \in B$ and $x \in A, y \in C$

$\Rightarrow$ $(x, y) \in A \times B$ and $(x, y) \in (A \times C) \Rightarrow (x, y) \in (A \times B) \cap (A \times C)$

But (x, y) is arbitrary, therefore

$$A \times (B \cap C) \subset (A \times B) (A \times C) \qquad \text{... (1)}$$

Conversely ,

If $(x, y) \in (A \times B) \cap (A \times C)$

Then, $(x, y) \in A \times B$ and $(x, y) \in A \times C$

$\Rightarrow$ $x \in A, y \in B$ and $x \in A, y \in C \Rightarrow x \in A, y \in B$ and $y \in C$

$\Rightarrow$ $x \in A$ and $y \in (B \cap C)$ $\Rightarrow$ $(x, y) \in A \times (B \cap C)$

But (x, y) is arbitrary, therefore

$$(A \times B) \cap (A \times C) \subseteq A \times (B \cap C) \qquad \text{... (2)}$$

From (1) and (2), we conclude that

$$A \times (B \cap C) = (A \times B) \cap (A \times C)$$

(ii) $(x, y) \in A \times (B \cup C)$

Then, $x \in A$ and $y \in (B \cup C)$

$\Rightarrow$ $x \in A$ and $y \in B$ or $y \in C$

$\Rightarrow$ $(x \in A$ and $y \in B)$ or $(x \in A$ and $y \in C)$

$\Rightarrow \{(x, y) \in (A \times B)\}$ or $\{(x, y) \in (A \times C)\}$

$\Rightarrow$ $(x, y) \in (A \times B) \cup (A \times C)$

Since (x, y) is arbitrary, therefore

$$A \times (B \cup C) \subseteq (A \times B) \cup (A \times C) \qquad \text{... (1)}$$

Conversely,

If $(x, y) \in (A \times B) \cup (A \times C)$

Then, $(x, y) \in (A \times B)$ or $(x, y) \in (A \times C)$

$\Rightarrow$ $(x \in A$ and $y \in B)$ or $(x \in A$ and $y \in C)$

$\Rightarrow$ $x \in A$ and $(y \in B$ or $y \in C) \Rightarrow (x, y) \in A \times (B \cup C)$

But (x, y) is arbitrary, therefore

$$(A \times B) \cup (A \times C) \subseteq A \times (B \cup C) \qquad \text{...(2)}$$

From (1) and (2), we conclude that

$$A \times (B \cup C) = (A \times B) \cup (A \times C).$$

THEOREM 2.   *For any sets A, B, C, D we have* $(A \times B) \cap (C \times D) = (A \cap C) \times (B \cap D)$

PROOF.   If $(a, b) \in (A \times B) \cap (C \times D)$, then

$\Rightarrow$ $(a, b) \in (A \times B)$ and $(a, b) \in (C \times D)$

$\Rightarrow$ $(a \in A$ and $b \in B)$ and $(a \in C$ and $(b \in D)$

$\Rightarrow$ $(a \in A$ and $a \in C)$ and $(b \in B$ and $b \in D)$

$\Rightarrow$ $a \in (A \cap C)$ and $b \in (B \cap D) \Rightarrow (a, b) \in (A \cap C) \times (B \cap D)$

Since (a, b) is arbitrary, therefore

$$(A \times B) \cap (C \times D) \subseteq (A \cap C) \times (B \cap D) \qquad \text{... (1)}$$

Now, let $(a, b) \in (A \cap C) \times (B \cap D)$

$\Rightarrow \quad a \in (A \cap C)$ and $b \in (B \cap D) \Rightarrow (a \in A$ and $a \in C)$ and $(b \in B$ and $b \in D)$

$\Rightarrow \quad (a \in A$ and $b \in B)$ and $(a \in C$ and $b \in D)$

$\Rightarrow \quad (a, b) \in (A \times B) \cap (C \times D)$

Since, (a, b) is arbitrary, therefore

$$(A \cap C) \times (B \cap D) \subseteq (A \times B) \cap (C \times D) \qquad \text{... (2)}$$

From (1) and (2), we conclude that

$$(A \times B) \cap (C \times D) = (A \cap C) \times (B \cap D)$$

REMARKS

➠ $(A \times B) \cap (B \times A) = (A \cap B) \times (B \cap A)$

➠ $A \times (B' \cup C')' = A \times (B \cap C) = (A \times B) \cap (A \times C)$

➠ $A \times (B' \cap C')' = A \times (B \cup C) = (A \times B) \cup (A \times C)$

THEOREM 3. *If A and B are two non-empty sets having n elements in common, then $A \times B$ and $B \times A$ have n^2 elements in common.*

PROOF. We know that $(A \times B) \cap (C \times D) = (A \cap C) \times (B \cap D)$

$$(A \times B) \cap (B \times A) = (A \cap B) \times (B \cap A)$$

$$(A \times B) \cap (B \times A) = (A \cap B) \times (A \cap B)$$

Since $(A \times B)$ has n elements, therefore $(A \cap B) \times (B \cap A)$ has n^2 elements.

$$(A \times B) \cap (B \times A) = (A \cap B) \times (B \cap A) \text{ has } n^2 \text{ elements.}$$

Hence, $(A \times B)$ and $(B \times A)$ have n^2 elements in common.

REMARKS

➠ For any three sets, A, B, C, we have $A \times (B - C) = (A \times B) - (A \times C)$

➠ If A and B are any two non-empty sets, then $A \times B = B \times A$ iff $A = B$.

➠ If $A \subseteq B$, then $A \times A \subseteq (A \times B) \cap (B \times A)$

➠ If $A \subseteq B$, then $A \times C \subseteq B \times C$ for any set C.

➠ If $A \subseteq B$ and $C \subseteq D$, then $A \times C \subseteq B \times D$.

➠ $A \times B = A \times C \Rightarrow B = C$

Exercise 1.5

1. If $A = \{a, b, c\}$, $B = \{d\}$, $C = \{2\}$, then verify
   (i) $A \times (B \cup C) = (A \times B) \cup (A \times C)$
   (ii) $A \times (B \cap C) = (A \times B) \cap (A \times C)$
   (iii) $A \times (B - C) = (A \times B) - (A \times C)$
   (iv) $(A \cap B) \times C = (A \times C) \cap (B \times C)$

2. If $A = \{2, 3\}$, $B = \{1, 2, 3\}$, $C = \{2, 3, 4\}$, show that $A \times A = (B \times B) \cap (C \times C)$.

3. If $A = \{1, 2, 3\}$, $B = \{4, 5\}$ and $C = \{1, 2, 3, 4, 5\}$, then show that $(C \times B) - (A \times B) = B \times B$.

4. The ordered pairs $(2, 7)$, $(4, 8)$ and $(5, 9)$ and among nine elements of the set $A \times B$. Determine the other six elements of $A \times B$.

5. Let $A = \{2, 3, 5, 7\}$, $B = \{1, 12, 13, 15\}$. How many elements are there in $A \times B$? In $B \times A$? Is $A \times B = B \times A$? Is $n(A \times B) = n(B \times A)$?

6. Let A and B be two sets. Show that the sets $A \times B$ and $B \times A$ have an element in common if and only if the sets A and B have an element in common.

7. Some elements of $A \times B$ are (a, x), (a, y), (d, z). If $A : \{a, b, c\ d\}$, find the remaining elements of $A \times B$ such that $n(A \times B)$ is least.

8. If A and B are two sets having 3 elements in common. If $n(A) = 5$, $n(B) = 4$, find $n(A \times B)$ and $n\{(A \times B) \cap (B \times A)\}$.

9. The ordered pairs $(1, 1)$, $(2, 2)$ and $(3, 3)$ are among the elements in the set $A \times B$. If A and B have elements each, how many elements in all does the set $A \times B$ have? Also find the remaining elements.

10. If A and B are two sets such that $n(A) = 3$ and $n(B) = 2$. If $(x, 1)$, $(y, 2)$, $(z, 1)$ are in $A \times B$, find A and B, where x, y, z are distinct.

11. Write 'T' for true and 'F' for false statement:

   (a) If $A = (a, b)$ and $B = (b, a)$, then $A \times B = \{(a, b)\ (b, a)\}$ **(T/F)**

   (b) $\{(a, x), (a, y), (b, x), (b, y)\}$ is product set. **(T/F)**

   (c) If $n(A) = x$ and $n(B) = y$ and $A \cap B = \phi$,

   then $n(A \times B) = xy$ **(T/F)**

   (d) If A and B are non-empty sets, then $A \times B$ is a non-empty set of ordered pairs (x, y) such that $x \in A$ and $y \in A$. **(T/F)**

12. (a) If $A = \{1, 2, 3\}$, $B = \{4, 5\}$ and $C = \{1, 2, 3, 4, 5\}$. Find

   (i) $A \times B$, (ii) $C \times B$, (iii) $B \times B$

   (b) If $A = \{1, 2, 3, 4\}$ and $B = \{5, 7, 9\}$, find $(A \times B) \cap (A \cap B)$.

ANSWERS

4. $(2, 8)$, $(2, 9)$, $(4, 7)$, $(4, 9)$, $(5, 7)$, $(5, 8)$ 5. $16, 16$, No, yes

7. (a, y), $(a, 2)$, (b, x), (b, y), (b, z), (c, x), (c, z), (d, x), (d, y) 8. $20, 9$

9. 9, $(1, 2)$, $(1, 3)$, $(2, 1)$, $(2, 3)$, $(3, 1)$, $(3, 2)$

10. $A = \{x, y, z\}$, $B = \{1, 2\}$,   11. (a) F   (b) T   (c) T   (d) F

12. (a) (i) $A \times B = (1, 4)$, $(1, 5)$, $(2, 4)$, $(2, 5)$, $(3, 4)$, $(3, 5)$

   (ii) $C \times B = \{(1, 4), (1, 5), (2, 4), (2, 5), (3, 4), (3, 5), (4, 4), (4, 5), (5, 4), (5, 5), \}$

   (iii) $B \times B = \{(4, 4), (4, 5), (5, 5)\}$ (b) ϕ

1.12 RELATION

Let us take two sets of natural numbers N_1 and N_2. We define R as a relation between them such that N_1 is a square of N_2. Then we can write $1R1$, $2R4$, $3R9$, ...

In terms of ordered pair, we can write

$$R = \{(1, 1), (2, 4), (3, 9), (4, 16), ...\} = \{(x, y : x, y \in \mathbf{N} \text{ and } y = x^2\}$$

The relation from set $\mathbf{N}$ to $\mathbf{N}$ is a subset of $\mathbf{N} \times \mathbf{N}$ such that $y = x^2$.

Definition: *Let A and B be two sets. Then a relation R from A to B is a subset of $A \times B$.*

Symbolically: R is a relation from A to $B \Leftrightarrow R \subseteq A \times B$.

⬆ REMARKS

⟼ If R is a relation from A to B, then A is called the domain and B the range of R.

⟼ If R is a relation from a non-empty set A to a non-empty set B and if $(a, b) \in R$, then we write aRb, read as "a is related to b by the relation R." On the other hand, if $(a, b) \notin R$, we write $a\bar{R}b$ and say that 'a is not related to b by the relation R'.

⟼ In particular, any subset $A \times A$ defined a relation in A, known as **Binary relation.**

☛ ILLUSTRATIONS

(i) If $a, b \in \mathbf{N}$ and R is defined as "a is divisor of b" then R is relation on $\mathbf{N}$.

The subset $\mathbf{N} \times \mathbf{N}$, which corresponds to the relation R is $S = \{(n, r): n \in \mathbf{N}, r \in \mathbf{N}\}$

Here, it is clear that $(1, 3)$, $(2, 4)$, $(3, 9)$ $(4, 8)$, $(4, 4)$, are in S, whereas $(2, 3)$, $(4, 5)$, $(5, 6)$ are not in S.

(ii) If R is a relation from set $A = \{1, 2, 3\}$ to the set $B = \{-1, -2\}$ defined by $x + y = 0$, then $R = \{(1, -1), (2, -2)\}$

Here, domain of R is $\{1, 2\}$ and Range = $\{-1, -2\}$.

(iii) If $A = \{a, b, c, d, e\}$ and $B = \{f, g, h, i\}$ and let $R = \{(a, g), (a, i), (d, h), (e, f)\}$ by a relation from A to B then

Domain of $R = \{a, d, e\}$ and Range of $R = \{g, i, h, f\}$

(iv) If $a, b \in \mathbf{R}$, the set of real numbers and $\mathbf{R}$ is "$|a - b|$ is a rational number" then R is a relation on $\mathbf{R}$. The subset S of $\mathbf{R} \times \mathbf{R}$ which corresponds to the relation is

$$S = \{(a, b + a): a \in \mathbf{R}, b \in \mathbf{Q}\}$$

It is observed that $\left(1, 2\frac{1}{2}\right), \left(\pi, 1 - \frac{1}{2}\right)$ belongs to S, while $(\sqrt{2}, \pi + \sqrt{2}) \notin S$.

(v) If $A = \{2, 3, 4\}$ and $B = \{a, b, c\}$, then $R = \{(2, b), (3, c), (2, a), (4, a)\}$ being a subset of $A \times B$, is a relation from $A \times B$. Here $(2, b), (3, c), (2, a), (4, a) \in R$, so we may write $2Rb, 3Rc, 2Ra, 4Ra$. But $(3, b) \notin R$ therefore, $3 \, R \, b$.

(vi) If $a, b \in \mathbf{N}$ and R is defined by "$a - b$ is divisible by a number $n \in \mathbf{N}$", then R is a relation on $\mathbf{N}$. The subset S of $\mathbf{N} \times \mathbf{N}$ corresponding to the relation by

$$S = \{n, n + rm : n \in \mathbf{N}, r \in \mathbf{N}\}$$

Here, $m = 3, (2, 8), (5, 11) \in S$ [$\because 2 - 8 = 6$, which is divisible by 3]

While $(3, 8) \in S$ [$\because 3 - 8 = 5$, which is not divisible by 3]

1.12.1 TOTAL NUMBER OF RELATIONS

Let A and B be two non-empty finite sets consisting p and q elements respectively, then $A \times B$ consists of $p \, q$ ordered pairs. Therefore, total number of subset of $A \times B$ is 2^{pq}.

⮊ REMARKS

⮕ For a non-empty set A, $\phi \in A \times A$, therefore it is a relation on A, called **void** or **empty** relation on A.

⮕ The void relation ϕ and the universal relation $A \times B$ are called trivial relation from A to B.

⮕ The void and universal relation on set A respectively the smallest and the largest relation on A.

1.12.2 IDENTITY RELATION

Let A be a set. The identity relation on A is the relation $I_A = \{(x, x) : x \in A\}$ on A.

For example : If $A = \{a, b, c\}$ then the relation $I_A = \{(a, a), (b, b), (c, c)\}$ is the identity relation. $R = \{(a, a), (b, b)\}$ is not an identity relation as $(c, c) \notin R$.

1.12.3 INVERSE OF A RELATION

Let A, B be two non-empty sets and R be a relation from a set A to B and let (x, y), number of the subset D of $A \times B$ corresponding to the relation R from A to B.

To the relation R from the set A to the set B, there corresponds a relation from the set B to the set A called the inverse of the relation, denoted by R^{-1} such that the subset $B \times A$ corresponding to the relation R^{-1} is $= \{(y, x) : (x, y) \in D\}$.

i.e., $yR^{-1}x \Leftrightarrow xRy$

For example:

(i) Let $A = \{a, b, c\}$ and $B = \{1, 2, 3\}$ be two sets and let $R = \{(a, 1), (a, 2), (b, 1), (b, 2)\}$ be a relation from A to B then $R^{-1} = \{(1, a), (2, a), (1, b), (2, b)\}$

(ii) If $A = \{1, 2, 3\}, B = \{5, 6, 7\}$ and let $R = \{(1, 5), (2, 5), (2, 7)\}$ be a relation from A to B.

Then $R^{-1} = \{(5, 1), (5, 2), (7, 2)\}$ which is a relation from B to A.

Also, Domain $(R) = \{1, 2\} =$ Range (R^{-1})

And, Range $(R) = \{5, 7\} =$ Domain (R^{-1})

(iii) The inverse of the relation "*is less than*" In $\mathbf{R}$ "*is greater than*".

⮊ REMARK

⮕ It may be noted than sometimes, the inverse of a relation coincides with the relation itself. For example, the inverse of the relation "perpendicular to" in the set of straight lines coincides with itself.

1.13 CLASSIFICATION OF RELATIONS

(a) Reflexive Relation: Let R be a relation on a set A.

"*A relation R is said to be reflexive if $(x, x) \in R \ \forall \ x \in A$*"

i.e., $$x \, R \, x \ \forall \ x \in A$$

☛ **ILLUSTRATIONS**

(i) In a set of integers, a relation R defined by $x \, R \, y$ iff $x - y$ is divisible by 4, then R is a reflexive relation because $x - x = 0$ which is a divisible by 4.

(ii) The universal relation on a non-empty set A is reflexive.

(iii) The relation "is less than," i.e., '<' in the set of rational number is not reflexive, because no member have the relation is less than to itself.

(iv) The relation "is a factor of" in the set of rational number is reflexive, since every rational number is a factor of itself.

(v) The relation "is less than or equal to." i.e., $\leq$ is in the set of natural number is reflexive.

$$n \leq n \ \forall \ n \in \mathbf{N}$$

(b) Symmetric Relation. *A relation R on a set A is said to be symmetric if*

$$(y, x) \in R \ whenever \ (x, y) \in R \ \forall \ x, y \in R$$

i.e., $$x \, R \, y \Leftrightarrow y \, R \, x \ \forall \ x, y \in R$$

☛ **ILLUSTRATIONS**

(i) Let l_1, l_2 be two lines such that l_1 is perpendicular to l_2,

i.e., $l_1 \perp l_2$. Then $l_1 \perp l_2 \Rightarrow l_2 \perp l_1$. Therefore the relation $\perp$ is symmetric.

(ii) The identity and the universal relation on a non-empty set are symmetric relations.

(iii) Consider the set $\mathbf{N}$ of natural numbers and the relation 'is less than'. This relation is not symmetric. Since if $2 < 3$ then $3 \nless 2$.

Let $A = \{1, 2, 3\}$ and relations R_1 and R_2 defined by

$R_1 = \{(1, 2), (1, 3), (3, 1), (2, 1)\}$ and $R_2 = \{(1, 2), (2, 3), (3, 1)\}$

Then R_1 is a symmetric relation, but R_2 is not symmetric.

(c) Transitive Relation: *A relation R on a set A is said to be transitive iff $(x, y) \in R$ and $(y, z) \in R \Rightarrow (x, z) \in R \ \forall \ x, y, z \in A$, i.e., $x \, R \, y, y \, R \, z \Rightarrow xRz$.*

☛ **ILLUSTRATIONS**

(i) Let a, b, c be three numbers such that a is a factor of b and b is a factor of c, then obviously a is a factor of c. Therefore, 'is a factor of' is a transitive relation.

(ii) If l_1, l_2, l_3 are three lines such that $l_1 \perp l_2$ and $l_2 \perp l_3$ then it is obvious that l_1 is parallel to l_3. Therefore the relation "$\perp$" is not transitive.

(iii) The identity and universal relation on a non-empty set are transitive.

(iv) Let l_1, l_2, l_3 be three straight lines, such that l_1 is parallel to l_2 and l_2 is parallel to l_3 then it is clear that l_1 is parallel to l_3. Therefore, 'is parallel to' is a transitive relation.

(d) Anti-symmetric Relation. *A relation R on a non-empty set A is said to be an anti-symmetric relation iff $(x, y) \in R$ and $(y, x) \in R \Rightarrow x = y \ \forall \ x, y \in R$*

☷ **REMARKS**

⟹ The identity relation R on a set A is an anti – symmetric relation.

⟹ If $(x, y) \in R$ and $(y, x) \notin R$, then it may be noted that $x = y$.

⟹ The universal relation on a set A containing at least two elements is not anti – symmetric.

1.13.1 EQUIVALENCE RELATIONS

A relation R on a set E is said to be equivalence if it is

(i) Reflexive, (ii) Symmetric and (iii) Transitive

For example :

(i) In a set of integers, a relation R is defined by $x R y$ if and only if $x - y$ is divisible by 4. Then R is an equivalence relation. Since

   (a) For $x R x, x - x = 0$ is divisible by 4. Therefore, it is reflexive.

   (b) For $x R y$. Let $x - y = 4m$ so $y - x = 4m$, which is also divisible by 4. Therefore, it is symmetric.

   (c) For $x R y$, let $x - y = 4m$; for $y R z$, let $y - z = 4n$. By adding these two equations, we get $x - z = -4(m + n)$,

        which is divisible by 4. Therefore it is transitive.

(ii) Let R be a relation on the set of all lines in a plane L defined by $(l_1, l_2) \in R$ if and only if line l_1 is parallel to l_2, then R is an equivalence relation because

   (a) For each line $l \in L$, we have l is parallel to l.

        $\Rightarrow lRl \Rightarrow R$ is reflexive.

   (b) Let $l_1, l_2 \in L$ such that $(l_1, l_2) \in R$, then

        $\Rightarrow (l_1, l_2) \in R \Rightarrow l_1$ is parallel to $l_2 \Rightarrow l$ is symmetric.

   (c) Let $l_1, l_2, l_3 \in L$ such that (l_1, l_2) and $(l_2, l_3) \in R$, then obviously $(l_1, l_3) \in R$ because if l_1 is parallel to l_2 and l_2 is parallel to l_3, then l_3 should be parallel to l_1.

1.13.2 CONGRUENCE MODULO 'm'

Let m be an arbitrary but fixed integer. If $x - y$ is divisible by m, then two integers x and y are said to be congruence modulo m of one another.

Symbolically: $x \equiv y \pmod{m}$ is $x - y$ divisible by m.

For example: $32 \equiv 2 \pmod{3}$, as $32 - 2 = 30$ which is divisible by 3.

1.13.3 COMPOSITION OF RELATIONS

Let R_1 and R_2 be two relations from sets A to B and B to C respectively, then we can define a relation $R_1 o R_2$ from A to C, such that $(x, z) \in R_1 o R_2$ if and only if there exist $y \in Y$ such that $(x, y) \in R_1$ and $(y, z) \in R_2$.

This relation is called composition of R_1 and R_2.

☞ REMARKS

➠ $R_1 o R_2 \neq R_2 o R_1$

➠ $(R_2 o R_1)^{-1} = R_1^{-1} o R_2^{-1}$

For example : Let A, B, C be three sets such that

$A = \{-1, -2\}, B = \{p, q, r\}$ and $C = \{\alpha, \beta, \gamma\}$

Also, $R_1 = \{(-1, p), (-1, r), (-2, q)\}$ is a relation from A and B and

$R_2 = \{(p, \alpha), (q, \beta), (r, \gamma)\}$ and is a relation from set to B to C.

Then $R2 o R1$ is a relation from A to C given by

$R_2 o R_1 = \{(-1, \alpha), (-1, \gamma), (-z, \beta)\}$

THEOREM 4. *The intersection of two equivalence relations on a set is an equivalence relation.*

PROOF. Let R_1, R_2 be two equivalence relation on a set A. To show $(R_1 \cap R_2)$ also an equivalence relation.

(i) Let $a \in A$ and a is arbitrary.

Since R_1 and R_2 both are reflexive on A.

$\therefore (a, a) \in R_1$ and $(a, a) \in R_2 \Rightarrow (a, a) \in R_1 \cap R_2$

Therefore, $(R_1 \cap R_2)$ is reflexive.

(ii) Let $a, b \in A$ such that $(a, b) \in R_1 \cap R_2$

$$(a, b) \in R_1 \cap R_2 \Rightarrow (a, b) \in R_1 \text{ and } (a, b) \in R_2$$

Also, R_1 and R_2 both are symmetric on A.

Therefore, $(b, a) \in R_1$ and $(b, a) \in R_2 \Rightarrow (b, a) \in R_1 \cap R_2 \Rightarrow (R_1 \cap R_2)$ is symmetric on A.

(iii) Let $a, b, c \in A$ such that $(a, b) \in R_1 \cap R_2$, $(b, c) \in R_1 \cap R_2$

Then, $(a, b) \in R_1 \cap R_2$ and $(b, c) \in R_1 \cap R_2$

$\Rightarrow \quad \{(a, b) \in R_1 \text{ and } (a, b) \in R_2 \text{ and } \{(b, c) \in R_1 \text{ and } (b, c) \in R_2\}$

$\Rightarrow \quad \{(a, b) \in R_1, (b, c) \in R_1\} \text{ and } \{(a, b) \in R_2, (b, c) \in R_2\}$

$\Rightarrow \quad (a, c) \in R_1 \text{ and } (a, c) \in R_2 \qquad [\because R_1 \text{ and } R_2 \text{ both are transitive.}]$

$\Rightarrow \quad (a, c) \in R_1 \cap R_2$

Therefore, $(R_1 \cap R_2)$ is transitive on A.

From (i), (ii) and (iii), we have that $R_1 \cap R_2$ is reflexive, symmetric and transitive, and hence $R_1 \cap R_2$ is an equivalence relation.

☙ REMARK

⇒ The union of two equivalence relations on a set is not necessarily an equivalence relation.

THEOREM 5. *If R is an equivalence relation, then R^{-1} is also an equivalence relation.*

PROOF. Let R be an equivalence relation on a set A. Then by definition of relation on a set, we have

$$R \subseteq A \times A \Rightarrow R^{-1} \subseteq A \times A$$

Therefore, R^{-1} is a relation on A.

Now, to show R^{-1} is an equivalence relation.

(i) Let $a \in A$, then $(a, a) \in R$

$(\because R \text{ is an equivalence relation and hence reflexive})$

$\Rightarrow \quad (a, a) \in R^{-1}$

Thus, $(a, a) \in R^{-1} \forall a \in R \Rightarrow R^{-1}$ is reflexive on A.

(ii) Let $(a, b) \in R^{-1}$, then $(a, b) \in R^{-1} \Rightarrow (b, a) \in R$

$\Rightarrow \quad (a, b) \in R \qquad (\because R \text{ is symmetric})$

$\Rightarrow \quad (b, a) \in R^{-1}$

Therefore R^{-1} is symmetric .

(iii) Let $(a, b) \in R^{-1}$ and $(b, c) \in R^{-1}$ then $(a, b) \in R^{-1} \Rightarrow (b, c) \in R$

and $(b, c) \in R^{-1} \Rightarrow (c, b) \in R$

Now, $(c, b) \in R$ and $(b, a) \in R$

$(c, a) \in R \qquad (\because R \text{ is transitive})$

$(a, c) \in R^{-1}$

Therefore R^{-1} is transitive .

From (i), (ii) and (iii), we conclude that R^{-1} is an equivalence relation.

☞ RECAPITULATIONS

- If $n(A) = p$, $n(B) = q$ then total number of subsets of $A \times B = 2^{pq}$.

- **Reflexive Relation:** $xRx, \ \forall \ x \in A$

- **Symmetric relation:** $xRy \Leftrightarrow yRx \ \forall \ x, y \in R$

- **Transitive relation:** $xRy, yRz \Rightarrow xRz$

- **Anti-symmetric relation:** $xRy \Rightarrow yRx \Leftrightarrow x = y$

- **Equivalence relation:** Reflexive, symmetric and transitive **(RST).**

- **Partial ordered relation:** Reflexive, anti-symmetric and transitive **(RAT)**

- R is equivalence $\Rightarrow R^{-1}$ is equivalence.

- Intersection of two equivalence relations on a set is again equivalence.

SOLVED EXAMPLES

EXAMPLE 1. *Let **Z** be the set of integers. Define a relation R on **Z** such that x R y holds if and only if x − y is divisible by 5, x ∈ **Z**, y ∈ **Z**. Show that it is an equivalence relation.*

SOLUTION.   (i)   For each $x \in \mathbf{Z}$, $x - x$ i.e., 0 is divisible by 5.

Therefore, for all $x \in \mathbf{Z}$, $x R x \Rightarrow x$ is reflexive.

(ii)   Let $x R y \Rightarrow x - y$ is divisible by 5.

$\Rightarrow y - x$ is divisible by 5.

Thus $xRy = yRx$

Therefore R is symmetric.

(iii)   Let us suppose xRy and yRz, then $(x - y)$ and $(y - z)$ are both divisible by 5. Hence, 5 is also a divisor of $(x - y) + (y - z)$.

5 is a divisor of $(x - z)$.

Therefore, $xRy, yRz \Rightarrow xRz \Rightarrow R$ is transitive.

From (i), (ii) and (iii), we conclude that R is an equivalence relation.

EXAMPLE 2. *Let **N×N** be the set of ordered pairs of natural numbers. Also, let R be the relation in **N×N**, defined by (a, b) R (c, d) if and only if a+d = b+c. Show that R is an equivalence relation.*

SOLUTION :   (i)   For all $(a, b) \in \mathbf{N \times N}$, we have $a + b = b + a$, i.e., $(a, b) R (b, a)$.

Therefore, R is reflexive.

(ii)   Let $(a, b) R (c, d)$, then, by definition of R

$(a+d) = (b+c)$ or $(c+b) = (d+a)$

$(c, d) R (a, b) \Rightarrow R$ is symmetric.

(iii)   Let us suppose $(a, b) R (c, d)$ and $(c, d) R (e, f)$, then

$a + d = b + c$ and $c + f = d + e$

$\Rightarrow \quad (a + d) + (c + f) = (b + c) + (d + e) \Rightarrow a + f = b + e$

$\Rightarrow \quad (a, b) R (e, f)$

Therefore, R is transitive.

Hence, from (i), (ii) and (iii), we conclude that R is an equivalence relation.

EXAMPLE 3. *If R is the relation for natural number defined by x+4y = 20. Find the domain and range of the relation R.*

SOLUTION.   Let $x + 4y = 20 \quad \Rightarrow \quad y = \dfrac{20 - x}{4}$

For $x = 4, y = 4$ and for $x = 8, y = 3$.

For $x = 16, y = 1$ and for $x = 12, y = 2$

Therefore, Domain = {4, 8, 12, 16} and range = {4, 3, 2, 1}

EXAMPLE 4. *A relation R defined on the set of integers **Z**, as follows*

$$(x, y) \in R \Rightarrow x^2 + y^2 = 25$$

Express R and R^{-1} as the sets of ordered pairs and hence find their respective domains.

SOLUTION. Since $(x, y) \in R \Leftrightarrow x^2 + y^2 = 25 \Rightarrow y = \pm\sqrt{25 - x^2}$

If $x = 0 \Rightarrow y = 5$.

Therefore, $(0, 5) \in R$ and $(0, -5) \in R$

Now, $x = 3 \Rightarrow y = \sqrt{25 - 9} = \pm 4$

$(3, 4) \in R, (-3, 4) \in R, (3, -4) \in R$ and $(-3, -4) \in R$

$x = \pm 4 \Rightarrow y = \pm 3$

Therefore, $(4, 3) \in R, (-4, 3) \in R, (4, -3) \in R$ and $(-4, -3) \in R$

$x = \pm 5 \Rightarrow y = \sqrt{25 - 25} = 0 \quad \therefore (5, 0) \in R$ and $(-5, 0) \in R$

Here, it is clear that for any other integral value of x, y is not an integer. Therefore,

$R = \{(0, 5), (0, -5), (3, 4), (-3, 4), (3, -4), (-3, -4), (4, 3), (-4, 3), (4, -3),$
$(-4, -3), (5, 0), (-5, 0)\}$

and $R^{-1} = \{(5, 0), (-5, 0), (4, 3), (4, -3), (-4, 3), (-4, -3), (3, 4), (3, -4),$
$(-3, 4), (-3, -4), (0, 5), (0, -5)\}$

Also, Domain $(R) = \{(0, 3, -3, 4, -4, 5, -5\} =$ domain of (R^{-1}).

EXAMPLE 5. *Consider the set $A = \{a, b, c\}$. Give an example of a relation R on A which is*

(i) *reflexive and symmetric but not transitive.*

(ii) *symmetric and transitive, but not reflexive.*

(iii) *reflexive and transitive, but not symmetric.*

SOLUTION. (i) Given $A = \{a, b, c\}$

Let $R = \{(a, a), (a, b), (b, a), (b, c), (c, b), (b, b), (c, c)\}$ on A.

Clearly, R is reflexive and symmetric but not transitive.

(ii) Let $R = \{(a, a), (a, b), (b, a), (b, b)\}$ on A.

Here, R is symmetric and transitive but not reflexive.

(iii) Let $R = \{(a, a), (b, b), (c, c), (a, b)\}$ on A.

Here, R is reflexive, transitive but not symmetric.

EXAMPLE 6. *If R is a relation in $N \times N$, show that the relation **R** defined by $(a, b) R (c, d)$ if and only if $ad = bc$ is an equivalence relation.*

SOLUTION. (i) Since $ab = ba \ \forall \ a, b \in N$.

Therefore, $(a, b) R (a, b) \forall a, b \in N \Rightarrow R$ is reflexive.

(ii) We have $(a, b) R (c, d)$ iff $ad = bc \ \forall \ a, b, c, d \in N$

Now, $(c, d) R (a, b)$ iff $cb = da \ \forall \ a, b, c, d \in N \Rightarrow R$ is symmetric.

(iii) We have $(a, b) R (c, d)$ iff $ad = bc \ \forall \ a, b, c, d \in N$

Therefore, $(a, b) R (c, d), (c, d) R (e, f) \Rightarrow (a, b) R (e, f) \ \forall a, b, c, d \in N$

Using $(a, d), (c, f) = (b, c)(d, e)$

$\Rightarrow \quad (a, f) = (b, e) \Rightarrow R$ is transitive

Hence, from (i), (ii) and (iii), we conclude that R is an equivalence relation.

EXAMPLE 7. *Let R_1 and R_2 be two relations on a set A, where $A = \{1, 2, 3, 5\}$ such that*
$$R_1 = \{(1, 1), (1, 2), (1, 5), (2, 1), (2, 5)\}$$
and $\quad R_2 = \{(3, 3), (3, 2), (2, 3), (1, 2), (2, 1)\}$
Then, which of the following statement is false :
(i) $R_1 \cup R_2$ *is symmetric*   (ii) $R_1 \cap R_2$ *is transitive*
(iii) $R_1 \cap R_2$ *is symmetric*   (iv) $R_1 \cup R_2$ *is transitive.*

SOLUTION.   (i) As $(1, 2) \in R_1$, also $(2, 1) \in R_1$, therefore, it is symmetric and as $(1, 2) \in R_2$, also $(2, 1) \in R_2 \Rightarrow R_2$ is symmetric.

       Now, $R_1 \cup R_2 = \{(1, 1), (1, 2), (1, 5), (2, 1), (2, 5), (3, 3), (3, 2), (2, 3)\}$

       In $R_1 \cup R_2$, as $(1, 2) \in R_1 \cup R_2$, also $(2, 1) \in R_1 \cup R_2 \Rightarrow R_1 \cup R_2$ is symmetric Therefore, (i) is true.

   (ii) We have $R_1 \cap R_2 = \{(1, 2), (2, 1)\}$
$\Rightarrow$ $(1, 1)$ should also belong to $R_1 \cap R_2$.
       But in this case $(1, 1) \notin R_1 \cap R_2$ is not transitive.
       Therefore, (ii) is false.

   (iii) We have, $R_1 \cap R_2 = \{(1, 2), (2,1)\}$
$$(1, 2) \in R_1 \cap R_2 \text{ and also } (2, 1) \in R_1 \cap R_2.$$
       Therefore, (iii) is true.

   (iv) In $R_1 \cup R_2$, $(1, 2) \in R_1 \cup R_2$
       and $(2, 5) \in R_1 \cup R_2$, also $(1, 5) \in R_1 \cup R_2$
$\Rightarrow$ $R_1 \cup R_2$ is transitive
       Therefore, (iv) is true.

EXAMPLE 8. *If A be the set of all triangles in a plane and $R = \{(a, b) : \Delta a = \Delta b\}$, i.e.,*

aRb $\Leftrightarrow$ Area of triangle a = Area of triangle b, then show that R is an equivalence relation.

SOLUTION.   (i) Since, for all $a \in A$ we have $\Delta a = \Delta a$
       Therefore, $aRa \Rightarrow R$ is reflexive.
   (ii) For any $a, b \in A$, we have $(a, b) \in R \Rightarrow \Delta a = \Delta b$
$\Rightarrow$ $\Delta b = \Delta a \Rightarrow (b, a) \in R$
       Therefore, $(b, a) \in R$, i.e., $bRa \Rightarrow R$ is symmetric.
   (iii) For all $a, b, c \in A$, we have $(a, b) \in R$, $(b, c) \in R$
          $\Delta a = \Delta b$ and $\Delta b = \Delta c \Rightarrow \Delta a = \Delta c \Rightarrow (a, c) \in R$
       Therefore, R is transitive.
       Hence, from (i), (ii) and (iii), we conclude that R is an equivalence relation.

EXAMPLE 9. *If Z be a set of non-zero integers and a relation R defined by $xRy \Rightarrow x^y = y^x \ \forall \ x, y \in Z$, then show that R is not an equivalence relation on Z.*

SOLUTION.   (i) Let $x \in Z$, then $x^x = x^x$, $\forall \ x \in Z$
$\Rightarrow$ xRx, $\forall \ x \in Z$
       Therefore, R is reflexive.
   (ii) Let $x, y \in Z$, such that xRy, i.e., $x^y = y^x$
$\Rightarrow$ $x^y = y^x \Rightarrow y^x = x^y$
       Therefore, $xRy \Rightarrow yRx$, $\forall \ x, y \in Z$
$\Rightarrow$ R is symmetric.

(iii) Let $x, y, z \in \mathbf{Z}$ such that xRy and yRz

i.e., $x^y = x^y$ and $y^z = z^y$ which does not give $x^z = z^x$

$\Rightarrow$ R is not transitive.

Hence, we conclude that R is not an equivalence relation.

EXAMPLE 10. *Let $A = \mathbf{R} \times \mathbf{R}$ ($\mathbf{R}$ is the set of real numbers) and define the following relation on $A : (a, b) R (c, d)$ iff $a^2 + b^2 = c^2 + d^2$*

 (*i*) *verify that (A, R) is an equivalence relation.*

 (*ii*) *describe geometrically what the equivalence classes are for this reason.*

SOLUTION. (i) we have $(a, b)R(c, d)$ $\Rightarrow$ $a^2 + b^2 = c^2 + d^2$

$\Rightarrow$ $c^2 + d^2 = a^2 + b^2$ $\Rightarrow$ $(c, d)R(a, b)$ ...(1)

$\Rightarrow$ R is symmetric.

Now, $(a, b)R(c, d)$ and $(c, d)R(x, y) \Rightarrow a^2 + b^2 = c^2 + d^2$

and $c^2 + d^2 = x^2 + y^2$

$\Rightarrow$ $a^2 + b^2 = x^2 + y^2$ $\Rightarrow$ $(a, b)R(x, y)$ ...(2)

$\Rightarrow$ R is transitive.

Again $(a, b)R(a, b) \Leftrightarrow a^2 + b^2 = a^2 + b^2$ ...(3)

$\Rightarrow$ R is reflexive.

Hence, from (1), (2) and (3), we conclude that R is an equivalence relation.

(ii) For any point (a, b), the sum $a^2 + b^2$ is the square of the distance from the origin. The equivalence classes are, therefore, the set of points in the place which have the same distance from the origin. Hence, the equivalence classes are concentric circles centered on the origin.

EXAMPLE 11. *Let R be the binary relation defined as $R = \{(a, b) \in R^2 : a - b \le 3\}$. Determine whether R is reflexive, symmetric, anti symmetric and transitive.*

SOLUTION. We have $(a, b) \in R^2 : a - b \le 3$.

$\Rightarrow$ $(a, a) \in R^2 : a - a \le 3$ *i.e.,* $0 \le 3$, which is true. So, R is reflexive.

In a similar way, we can easily show that R is neither symmetric, anti symmetric nor transitive.

1.13.4 RELATIONS OTHER THAN EQUIVALENCE

Let R be a given relation on the set X. Then R is

  (i) non-reflexive if $\exists x$, such that $(x, x) \notin R$.

 (ii) anti-reflexive or reflexive if $i_x \cap R = \phi$ (where i_x is the identity relation on X or $\forall n \in X :$ $(x, x) \notin R$

(iii) non-symmetrical if for some $(x, y) \in R$, we have $(y, x) \notin R$

 (iv) anti-symmetric if $R \cap R^{-1} = i$, i.e., $(x, y) \in R$ and $(y, x) \in R \Rightarrow x = y$

  (v) asymmetric if $R \cap R^{-1} = \phi$, i.e., $(x, y) \in R \Rightarrow (y, x) \notin R$

 (vi) non-transitive if $R \circ R \notin R$

(vii) anti-transitive if $(R \circ R) \cap R = \phi$

(viii) A reflexive and symmetric, but not transitive relation is called a tolerance relation.

 (ix) A non-symmetric transitive relation is called an ordered relation.

  (x) A reflexive, anti-symmetric and transitive relation is called partial-ordered relation.

Exercise 1.6

1. If R is the relation 'is less than' from $A = \{1, 2, 3, 4, 5\}$ to $B = \{1, 4, 5\}$, find the set of ordered pairs corresponding to R. Also find R^{-1}.

2. A relation R defined from a set $A = \{2, 3, 4, 5\}$ to a set $B = \{3, 6, 7, 10\}$ as follows : $(x, y) \in R \Rightarrow x$ divides y. Write R as a set of ordered pairs and determine the domain and range of R. Also find R^{-1}.

3. Find the domain and range of $A = \{1, 2, 3, 4, 5, 6\}$ when the relation are defined as
   (i) xR_1y if and only if $x - y > 0$
   (ii) xR_2y if and only if $x + y < 0$

4. Two sets A and B are given by $A = \{1, 2, 8, 9\}$ and $B = \{2, 3, 4, 6, 7\}$ and if R is the relation form A to B given by $\{(1,2), (1,3), (2,4), (2,6)\}$, then which of the following statement is true?
   (i) Domain (R) = Range (R^{-1}) and
       Range (R) = Domain (R^{-1})
   (ii) Domain (R) = Domain (R^{-1}) and
       Range (R) = Range (R^{-1})
   (iii) Domain (R) = Range (R^{-1}) and
       Range (R) = Domain (R^{-1})
   (iv) Domain (R) = Range (R)

5. If R is a relation on a set A, then which of the following statement is not true?
   (i) If R is reflexive then R^{-1} is reflexive.
   (ii) If R is symmetric then R^{-1} is symmetric.
   (iii) If R is transitive, then R^{-1} is transitive.
   (iv) None of these

6. Find the domain and range of the following relations:
   (i) $R = \{(x + 1, x + 5)\} : x \in \{0, 1, 2, 3, 4, 5\}$
   (ii) $R = \{(x, x^3) : x$ is a prime number, less than 10\}
   (iii) $R = \{(a, b) : a \in \mathbf{N}, a < 5, b = 4\}$
   (iv) $R = \{(a, b) : b = |a - l|, a \in \mathbf{Z},$ and $|a| \le 3\}$

7. Let R_1 be the relation defined on the set of reals $\mathbf{R}$ such as $(a, b) \in R_1$ if and only if $1 + ab > 0$ for all $a, b \in \mathbf{R}$. Show that R_1 is reflexive, symmetric but not transitive.

8. Let R be relation on $\mathbf{N} \times \mathbf{N}$, defined by $(a, b) R (c, d)$ if and only if $ad (b + c) = bc (a + d)$. Show that R is an equivalence relation.

9. Show that the relation 'congruence modulo m' on the set of integers is an equivalence relation.

10. Let R_1 be a relation on the set of reals defined by $R_1 = \{(a, b) \in R \times R : a^2 + b^2 = 1\}$ Show that R_1 is not an equivalence relation on R.

11. In a set L of all straight lines in a plane, discuss which of the following two relations are equivalence relations L.
    (i) $R_1 = \{(x, y) : x, y \in L$ and x is parallel to $y\}$
    (ii) $R_2 = \{(x, y) : x, y \in L$ and x is perpendicular to $y\}$.

12. Show that the relation $R = \{(a, b) : a - b = $ even integr $\forall a, b \in \mathbf{Z}\}$, i.e., $aRb \Leftrightarrow a - b = $ even integer, is an equivalence relation.

13. Show that the relation R in $\mathbf{N}$, the set of natural numbers, defined by xRy if $x^2 - 4xy + 3y^2 = 0$, $(x, y \in \mathbf{N})$ is reflexive, not symmetric and not transitive.

14. For the given relation R on a set S, determine which are equivalence relations:
    (i) S is the set of all rational numbers, aRb if and only if $a = b$
    (ii) S is the set of all real numbers iff
        (a) $|a| = |b|$ (b) $a \ge b$
    (iii) S is the set of all triangles in a plane, aRb iff a is congruent to b.
    (iv) S is the set of all triangles in a plane, aRb iff a and b have equal perimeters.

15. An integer m is said to be related to another integer n if m is a multiple of n. Show that this relation is reflexive and transitive but not symmetric.

16. Let R be a relation defined on the set of natural number $\mathbf{N}$ as $R = \{(x, y) : x, y \in \mathbf{N}, 2x + y = 41\}$. Find the domain and range of R.

17. Let O be the origin. Define a relation between two points P and Q in a plane if $PO = OQ$. Show that the relation is an equivalence relation.

18. Given the relation $R = \{(1, 2), (2, 3)\}$ on the set of natural number $\mathbf{N}$, add a minimum of ordered pairs so that the enlarged relation is symmetric, transitive and reflexive.

19. Let $\mathbf{N}$ denote the set of all natural numbers and R be the relation on $\mathbf{N} \times \mathbf{N}$ defined by $(a, b) R (c, d) \Leftrightarrow ad (b + c) = bc (a + d)$. Show that R is an equivalence relation.

20. Show that the relation, which is symmetric and transitive, is not necessarily reflexive.

ANSWERS

1. $aRb = \{(1, 4), (1, 5), (2, 4), (3, 4), (2, 5), (3, 5), (4, 5)\}$,

  $R^{-1} = \{(4, 1), (5, 1), (4, 2), (5, 2), (4, 3), (5, 3), (5, 4)\}$

2. Domain $(R) = \{2, 3, 5\}$, Range $(R) = \{3, 6, 10\}$, $R^{-1} = \{(6, 2), (10, 2), (3, 3), (6, 3), (10, 5)\}$

3. (i) $\{2, 3, 4, 5, 6\}$, $\{1, 2, 3, 4, 5\}$, (ii) ϕ, ϕ **4.** (iii) **5.** (iv) **6.** (i) Domain $(R) = \{1, 2, 3, 4, 5, 6\}$, Range $(R) = \{5, 6, 7, 8, 9, 10\}$ (ii) Domain $(R) = \{2, 3, 5, 7\}$, Range $(R) = \{8, 27, 125, 243\}$ (iii) Domain $(R) = \{1, 2, 3, 4\}$, Range $(R) = \{4\}$ (iv) Domain $(R) = \{0, -1, -2, -3, 1, 2, 3\}$, Range $(R) = \{1, 2, 3, 4, 0, 1, 2\}$ **11.** R_1 = Equivalence relation, R_2 = Not equivalence **14.** (i), (ii)

16. Domain $(R) = \{1, 2, ..., 19, 20\}$, Range $(R) = \{39, 37, 35, ..., 5, 3, 1\}$ **18.** $\{(1, 2), (2, 1), (2, 3), (3, 2), (1, 3), (3, 1), (1, 1), (2, 2), (3, 3), (4, 4), ...\}$

1.14 FUNCTIONS

Definition: *Let A and B be two sets, then the rule or corresponding, which associates each element of A to a unique element to B, is called a function from set A to set B.*

If a general element of set A is denoted by x, and of set B is denoted by y, then we say that y is a function of x if, for every $x \in A$, one and only one value of $y \in B$ can be determined.

Symbolically: If f is a function from a set A to a set B, then we write $f : A \to B$, read as f is a function from A to B or f maps A to B.

1.14.1 RANGE AND DOMAIN OF A FUNCTION

Let an element $y \in B$ be corresponded by an element $x \in A$, then y is called the image of x and is denoted by $f(x)$. Here, x is defined as the pre-image of y.

The set A is called the domain and the set B is called the co-domain of the function f.

The set of all f-images of the element of A, is called image set or the range of f and is denoted by

$$f(A) \quad \text{or} \quad \{f(x) : x \in A\}$$

Evidently, $f(A) \subseteq B$.

Thus, a mapping $f : A \to B$ is the set of ordered pairs $\{(a, b) : a \in A, b \in B\}$, so that no two ordered pairs have the same finite element.

$$f = \{(a, b) : a \in A, b \in B, b = f(x) \,\forall\, a \in A\}$$

For example: Let $A = \{-2, -1, 0, 1, 2\}$ and B is the set of natural numbers for every $x \in A$, $f(x) \in B$ and $f(x) = x^2$.

Here, A is the domain and B is the co-domain.

$f(a)$ is the value of the function $f(x)$, when x takes the value a, *i.e.*, when x is replaced by a.

The elements of the co-domain which is equal to $f(x)$ form the range.

When $x = -2, f(-2) = (-2)^2 = 4$

When $x = -1, f(-1) = 1$

When $x = 0, f(0) = 0$

When $x = 1, f(1) = 1$

When $x = 2, f(2) = 4$.

Which can be illustrated in the figure (15).

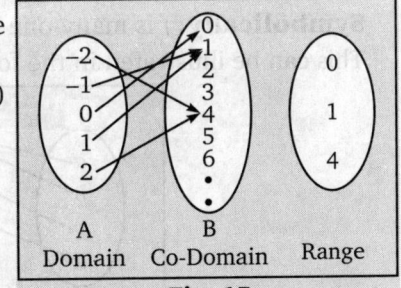

A B
Domain Co-Domain Range

Fig. 15

⬱ REMARKS

➠ If $f : A \to B$ then a single element in A cannot have more than one image in B. However, two or more elements in A may have the same images in B.

➠ Every element in A must have its image in B, but every element in B may not have it pre-image in A.

➠ To each element x in A, there exists a unique element y in B such that $y = f(x)$.

➠ The unique element y of B is called the value of f at x (the image of f under x), and written as $y = f(x)$.

➠ The range of f consist of those elements in B which appear as the image of at least one element in A.

➠ Range of a function is the image of its domain.

➠ Range is a subset of co-domain.

1·15 TYPE OF FUNCTIONS

(a) One-One Function: *A function f from A to B, i.e., f : A → B is said to be one-one (or injective) iff distinct elements of A have distinct images.*

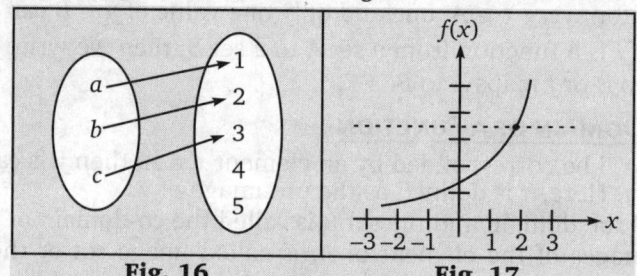

Fig. 16   **Fig. 17**

Symbolically: f is one-one if for $x_1, x_2 \in A$, we have

$$x_1 \neq x_2 \quad \Rightarrow \quad f(x_1) \neq f(x_2) \; \forall \; x_1, x_2 \in A$$

or $f(x_1) = f(x_2) \Rightarrow \quad x_1 = x_2 \; \forall \; x_1, x_2 \in A$

It is also called **Univalent function.**

Graphically, a function is one-one if and only if no line parallel to x-axis meets the graph of the function in more than one point.

(b) Many-One Function: *A function f : A → B is called many-one, if at least one element of co-domain B has two or more than two pre-images in domain A.*

Symbolically: f is many-one if for $x_1, x_2 \in A$, we have $x_1 \neq x_2 \Rightarrow f(x_1) = f(x_2)$

This can be illustrated in the following figures.

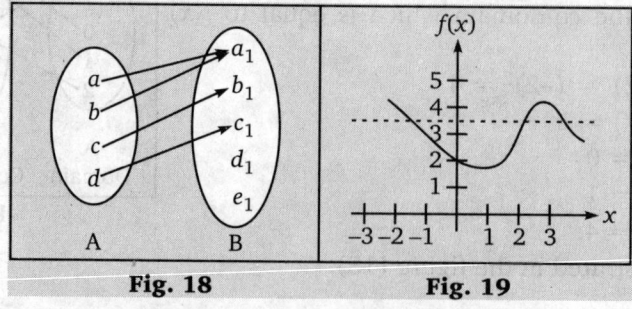

Fig. 18   **Fig. 19**

Graphically, a function is many-one if and only if a line parallel to x-axis meets the graph of the function in more than one point.

REMARK

⟶ One-many function does not exist.

(c) Onto Function: *A function $f : A \rightarrow B$ is called an onto function, if there is no element of B which is not an image of some element of A, i.e., every element of B appears as the image of at least one element of A.* This is illustrated in Figure 20.

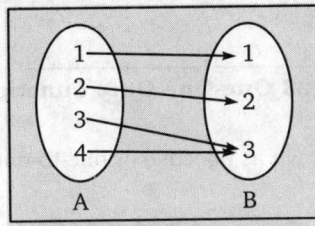

Fig. 20 Onto Function

REMARKS

⟶ In an onto function, Range = Co-domain

⟶ Onto function is also called subjective.

(d) Into Function: *A function $f : A \rightarrow B$ is called an into function, i.e., if there is at least one element of set B which has no pre-image in the set A.* This is illustrated in Figure 21.

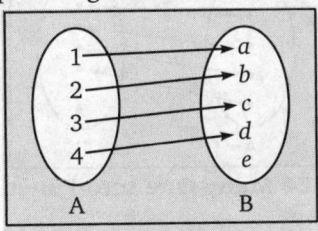

Fig. 21 Into Function

REMARK

⟶ In an into function, Range $\subset$ Co-domain.

(e) One-One Into Function: *A function $f : A \rightarrow B$ is called a one-one into function, if it is both one-one and into, i.e., the different points in A are joined to different points in B and there are some points in B which are not joined to any point in A.* This is illustrated in Figure 22.

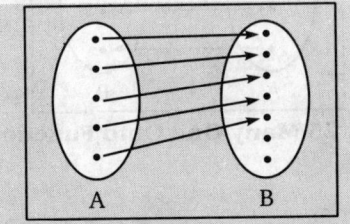

Fig. 22 One-One Into Function

Symbolically : One-one into function is defined as

(i) Range $\subset$ Co-domain.

(ii) $f(x_1) \neq f(x_2) \Rightarrow x_1 \neq x_2$.

(f) One-One Onto Function: *A function* $f : A \rightarrow B$ *is both one-one and onto, i.e., the different points in A are joined to different points in B and no point in B is left vacant. This is illustrated in Figure 23.*

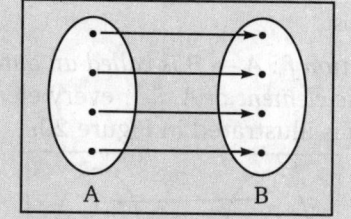

Fig. 23 One-One Onto Function

⮞ REMARKS

⮞ One-one onto mapping is also known as bijective or one-to-one.

⮞ For a one-one onto function

⮞ Range = Co-domain, and $x_1 \neq x_2 \Rightarrow f(x_1) \neq f(x_2)$ or $f(x_1) = f(x_2) \Rightarrow x_1 = x_2$

(g) Many-One Into Function: *A function* $f : A \rightarrow B$ *which is both many-one and into function is called a many-one into function, i.e., two or more points in A are joined to some points in B and there are some point in B which are not joined to any point in A. Therefore, for many-one into function.*

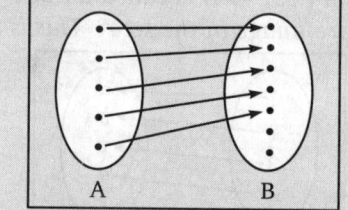

Fig. 24 Many-One Into Function

(i) Range $\subset$ Co-domain.

(ii) $x_1 \neq x_2 \Rightarrow f(x_1) = f(x_2)$

(h) Many-One Onto Function: *If function* $f : A \rightarrow B$ *is both many-one and onto function is called a many one onto function, i.e., in B one point is joined to at least one point in A and two or more points in A are joined to some points in B. Therefore, for many-one onto function.*

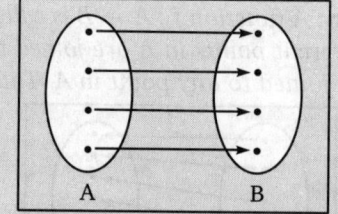

Fig. 25 Many-One Onto Function

(i) Range = Co-domain.

(ii) $x_1 \neq x_2$
$\Rightarrow f(x_1) = f(x_2)$

WORKING PROCEDURE

1. For checking the Injectivity (One-One) of the function

Let x and y be two arbitrary elements in the domain of f.

Step 1. Take $f(x) = f(y)$

Step 2. If we get $x = y$, after solving $f(x) = f(y)$. Then, $f : A \to B$ is one-one.

2. For checking the surjectivity (Onto) of a function

Step 1. Take an arbitrary element y in the co-domain.

Step 2. Put $f(x) = f(y)$

Step 3. Solve $f(x) = y$ for x and obtain x in terms of y.

Step 4. Get the equation of the form $x = g(y)$

Step 5. If $x = g(y)$ belongs to domain f, for all values of y, then f is onto.

☞ RECAPITULATIONS

- For a function $f : A \to B$, A = domain, B = co-domain.

- **For one-one function:** $x_1 \neq x_2 \Rightarrow f(x_1) \neq f(x_2) \; \forall \; x_1, x_2 \in A$ or $f(x) = f(x_2) \Rightarrow x_1 = x_2 \; \forall \; x_1, x_2 \in A$

- **For many-one function:** $x_1 \neq x_2 \Rightarrow f(x_1) = f(x_2), x_1, x_2 \in A$

- **For onto function:** Range = co-domain

- **For into function:** Range $\subseteq$ co-domain

- **For one-one into function:** (i) Range $\subseteq$ co-domain
  (ii) $f(x_1) \neq f(x_2)$

- **For one-one onto function:** (i) Range = codomain
  (ii) $x_1 \neq x_2 f(x_1) \neq f(x_2)$ or $f(x_1) = f(x_2) \Rightarrow x_1 = x_2$

- **For many-one into function:** (i) Range = co-domain   (ii) $x_1 = x_2 \Rightarrow f(x_1) = f(x_2)$

- **For many-one onto function:** (i) Range = co-domain (ii) $x_1 \neq x_2 \Rightarrow f(x_1) = f(x_2)$

SOLVED EXAMPLES

EXAMPLE 1. Let $f : \mathbf{R} \to \mathbf{R}$ be a function defined by

$$f(x) = \begin{cases} 3x - 1 & \text{when} \quad x > 3 \\ x^2 - 1 & \text{when} -2 \leq x \leq 3 \\ 2x + 3 & \text{when} \quad x < -2 \end{cases}$$

Find (i) $f(2)$,   (ii) $f(4)$,   (iii) $f(-1)$,   (iv) $f(-3)$

SOLUTION.   (i) $f(2) = (2)^2 - 2 = 4 - 2 = 2$

(ii) $f(4) = 3(4) - 1 = 12 - 1 = 11$

(iii) $f(-1) = (-1)^2 - 2 = 1 - 2 = -1$

(iv) $f(-3) = 2(-3) + 3 = -6 + 3 = -3$

EXAMPLE 2. For $y = +\sqrt{x}$, say whether it is a function or not. If it is a function, find its domain and range.

SOLUTION.   Here we have $y = +\sqrt{x}$ ...(1)

Since y is real if $x \geq 0$ and is unique and finite for each $x \geq 0$.

Therefore, (1) is a function with domain $[0, \infty[$.

Again from (1), $y \geq 0 \; \forall \; x \geq 0$

Hence, range = $[\, 0, \infty \,[$

EXAMPLE 3. Find the domain of $f(x) = \dfrac{x^3 - x^2 + 4x + 2}{3x + 11}$

SOLUTION. Since f is defined for all real values of x except when $3x + 11 = 0$

i.e., when, $x = -\dfrac{11}{3}$

Hence, domain of $f = \mathbf{R} - \left\{ -\dfrac{11}{3} \right\}$

EXAMPLE 4. Let $f : \mathbf{N} - \{1\} \to \mathbf{N}$ be defined by $f(n) =$ the highest prime factor of n. Show that f is neither one-one nor onto. Also, find the range f.

SOLUTION. Since we have

$f(6) =$ the highest prime factor of $6 = 3$

$f(9) =$ the highest prime factor of $9 = 3$

$f(12) =$ the highest prime factor of $12 = 3$

Therefore, f is a many-one function.

Clearly, image of any $n \in \mathbf{N} - \{1\}$ is the largest prime number that divides n. So the range of f consists of prime number only. Consequently, range of $f \neq \mathbf{N}$ (Co-domain)

$\Rightarrow$ f is not onto function.

Hence, f is neither one-one nor onto. The range of f is the set of all prime numbers.

EXAMPLE 5. Let $A = \{1, 2\}$. Find all one-to-one function from A to A.

SOLUTION. Let $f : A \to A$ be a one-one function.

Then, for $f(1)$, there are two choices, i.e., 1 or 2.

Let us first suppose $f(1) = 1$.

As $f : A \to A$ is one-one, $f(2) = 2$

Therefore, we have $f(1) = 1, f(2) = 2$

Now, let $f(1) = 2$

Since, $f : A \to A$ is one-one, therefore $f(2) = 1$.

Therefore, we have $f(1) = 2$ and $f(2) = 1$.

Hence, we have two one-one function say f and g form A and A given by $f(1) = 1, f(2) = 2$ and $f(2) = 1$ and $f(1) = 2$.

EXAMPLE 6. Let $\{x \in \mathbf{R} : -1 \leq x \leq 1\} = B$. Show that $f : A \to B$ given by $f(x) = x\,|x|$ is one-one and onto.

SOLUTION. Let x, y be any two elements in A, then

$$x \neq y \Rightarrow x|x| \neq y|y| \Rightarrow f(x) \neq f(y).$$

Therefore, f is one-one.

Since, range of $f = f(A) = B$ so $f : A \to B$ is onto mapping. Hence f is one-one and onto.

EXAMPLE 7. Find the domain and range of the function.

$$f(x) = -\sqrt{-5 - 6x - x^2}$$

SOLUTION. Given that, $f(x) = -\sqrt{-5 - 6x - x^2}$

For f to be real, $-5 - 6x - x^2 \geq 0$ $\Rightarrow$ $x^2 + 6x + 5 \leq 0$

$\Rightarrow$ $x^2 + 6x \leq -5$ $\Rightarrow$ $x^2 + 6x + 9 \leq -5 + 9$

$\Rightarrow \quad (x + 3)^2 \le 4$ $\qquad\qquad \Rightarrow \qquad |x + 3|^2 \le 4$

$\Rightarrow \quad |x + 3| \le 2$ $\qquad\qquad \Rightarrow \qquad -2 \le x + 3 \le 2$

$\Rightarrow \quad -2 -3 \le x \le 2 -3$ $\qquad \Rightarrow \qquad -5 \le x \le -1$

Therefore, domain of $f(x) = [-5, -1]$

To find the range of $f(x)$, put $y = f(x)$

Therefore, $f(x) = -\sqrt{-5 - 6x - x^2}$, $y \le 0$

$\Rightarrow \qquad y^2 = -5 - 6x - x^2$ $\qquad \Rightarrow \quad x^2 + 6x + (y^2 + 5) = 0$

For real x, discriminant ≥ 0, $\qquad$ i.e., $(6)^2 - 4 \times 1 \times (y^2 + 5) \ge 0$

$\Rightarrow \qquad 36 - 4y^2 - 20 \ge 0$ $\qquad \Rightarrow \qquad -4y^2 \ge -16$

$\Rightarrow \qquad\qquad y^2 \ge 4$ $\qquad\qquad \Rightarrow \qquad |y|^2 \le 4$

$\Rightarrow \qquad\qquad |y| \le 2$ $\qquad$ i.e., $\qquad\qquad -2 \le y \le 2$

But $y \le 0$ therefore, $-2 \le y \le 0$.

Hence, Range of $f = [-2, 0]$

EXAMPLE 8. *For a finite set A, if $f : A \to A$ is a one-one function, show that f is onto.*

SOLUTION. Let $A = \{a_1, a_2, ..., a_n\}$ be a finite set.

Since $f : A \to A$ is one-one function, therefore $f(a_1), f(a_2), ..., f(a_n)$ are distinct elements of the set A, but A has only n elements. Therefore,

$$A = \{f(a_1), f(a_2), ..., f(a_n)\}$$

$\Rightarrow \qquad$ Co-domain = Range

Hence, every element in A (co-domain) has its pre-image in the domain A.

$\Rightarrow \quad f : A \to A$ is onto.

🕮 REMARK

➡ For a finite set A, if $f : A \to A$ is onto function, then f is one-one.

EXAMPLE 9. *If $f : \mathbf{R} \to \mathbf{R}$ be a function defined by $f(x) = 4x^3 - 7$, show that the function f is bijective.*

SOLUTION. Given that $f(x) = 4x^3 - 7; x \in \mathbf{R}$

f **is one-one :** Let $x_1, x_2 \in \mathbf{R}$

Now, $\quad f(x_1) = f(x_2)$

$\Rightarrow \quad 4x_1^3 - 7 = 4x_2^3 - 7$ $\qquad\qquad\qquad \Rightarrow 4x_1^3 = 4x_2^3$

$\Rightarrow \qquad\qquad x_1^3 = x_2^3$ $\qquad\qquad\qquad \Rightarrow x_1^3 - x_2^3 = 0$

$\Rightarrow \quad (x_1 - x_2)(x_1^2 + x_1 x_2 + x_2^2) = 0$

$\Rightarrow \quad (x_1 - x_2)\left[\left(x_1 + \dfrac{x_2}{2}\right)^2 + \dfrac{3x_2^2}{4}\right]$ $\qquad \left\{\because \left[\left(x_1 + \dfrac{x_2}{2}\right)^2 + \dfrac{3x_2^2}{4} \ne 0\right]\right\}$

$\Rightarrow \quad (x_1 - x_2) = 0$ $\qquad\qquad\qquad\qquad \Rightarrow x_1 = x_2$

Therefore, f is one-one.

f **is one-one :** Let $c \in \mathbf{R}$

$$f(x) = c \quad \Rightarrow 4x^3 - 7 = c \qquad \Rightarrow x = \left(\dfrac{c + 7}{4}\right)^{1/3}$$

Now, $\left(\dfrac{c + 7}{4}\right)^{1/3} \in \mathbf{R}$ and $f\left\{\left(\dfrac{c + 7}{4}\right)^{1/3}\right\} = 4\left[\left(\dfrac{c + 7}{4}\right)^{1/3}\right]^3 - 7 = c + 7 - 7 = c$

Which implies that c is the image of $\left(\dfrac{c+7}{4}\right)^{1/3}$

Therefore, f is onto. Hence, f is objective function.

EXAMPLE 10. *Let A and B be two sets. Prove that $f : A \times B \to B \times A$ difined by $f(a, b) = (b, a)$ is one-one and onto.*

SOLUTION.   **f is one-one :** Let (a_1, b_1) and $(a_2, b_2) \in A \times B$ such that

$$f(a_1, b_1) = f(a_2, b_2)$$
$$\Rightarrow \qquad (b_1, a_1) = (b_2, a_2)$$
$$\Rightarrow \qquad b_1 = b_2 \text{ and } a_1 = a_2$$

Therefore, $(a_1, b_1) = (a_2, b_2)$

Thus, $f(a_1, b_1) = f(a_2, b_2)$
$$\Rightarrow \qquad (a_1, b_1) = (a_2, b_2) \; \forall \; (a_1, b_1), (a_2, b_2) \in A \times B$$
$$\Rightarrow \quad f \text{ is one-one.}$$

f is onto : Let $(b, a) \in B \times A$ such that $b \in B$ and $a \in A$.
$$\Rightarrow \qquad (a, b) \in A \times B$$

Therefore, for all $(b, a) \in B \times A$, there exist $(a, b) \in A \times B$ such that $f(a, b) = (b, a)$
$$\Rightarrow \quad f \text{ is onto. Hence } f \text{ is one-one and onto.}$$

Exercise 1.7

1. Let $A = \{-2, -1, 0, 1, 2\}$ and $f : A \to \mathbf{Z}$ given by $f(x) = x^2 - 2x - 3$. Find :
   (i) the range of f,
   (ii) pre-image of 6, –3 and 5.

2. Find the domain and range of the following function
$$f(x) = \sqrt{(x-1)(3-x)}$$

3. Find the range of the following function
$$f(x) = \dfrac{1}{(2x-3)(x+1)}$$

4. Find the domain and range of the following functions :
   (i) $f(x) = \dfrac{x^2 - 1}{x - 1}$ (ii) $y = -|x|$
   (iii) $f(x) = \dfrac{|x-1|}{x-1}$ (iv) $y = \sqrt{x-3}$

5. If $A = \{-1, 0, 2, 5, 6, 11\}$,
   $B = \{-2, -1, 0, 18, 25, 108\}$
   and $f(x) = x^2 - x - 2$, find $f(A)$.

6. Let A be the set of two positive integers. Let $f : A \to \mathbf{Z}^+$, set of positive integers be defined by $f(n) = p$, where p is the highest prime factor of n. If range of $f = \{3\}$, find A.

7. Find the domain for which the function $f(x) = 2x^2 - 1$ and $g(x) = 1 - 3x$ are equal.

8. Let $f_1 : \mathbf{R} \to \mathbf{R}$ and $f_2 : \mathbf{C} \to \mathbf{C}$ be two functions defined as $f_1(x) = x^3$ and $f_2(x) = x^3$. Show that they are not equal.

9. Let $A = \{p, q, r, s\}$ and $B = \{1, 2, 3\}$. Which of the following relations from A to B not a funcion?
   (i) $R_1 = \{(p, 1), (q, 2), (r, 1), (s, 2)\}$
   (ii) $R_2 = \{(p, 1), (q, 1), (r, 1), (s, 1)\}$
   (iii) $R_3 = \{(p, 1), (q, 2), (r, 2), (s, 3)\}$
   (iv) $R_4 = \{(p, 2), (q, 3), (r, 2), (s, 2)\}$

10. Write the following relations as sets of ordered pairs and find which of them are functions :
    (i) $\{(x, y) : y = 3x, x \in (1, 2, 3),$
    $y \in (3, 6, 9, 12)\}$
    (ii) $\{(x, y) : y > x + 1, x = 1, 2$ and $y = 2, 4, 6\}$
    (iii) $\{(x, y) : x + y = 3$
    $x, y \in (0, 1, 2, 3)\}$

11. Express the following functions as sets of ordered pairs, and find their range :
    (i) $f_1 : A \to \mathbf{R} : f_1(x) = x^2 + 1$
    where $A = \{-1, 0, 2, 4\}$
    (ii) $f_2 : A \to \mathbf{N} : f_2(x) = 2x$
    where $A = \{x : x \in \mathbf{N}, x \le 10\}$

12. Let $f : \mathbf{R} \to \mathbf{R}$ be a function such that $f(x) = 2^x$. Determine :
    (i) range of f
    (ii) $\{x : f(x) = 1\}$
    (iii) whether $f(x + y) = f(x) \cdot f(y)$ holds

13. Let $f : \mathbf{R}^+ \to \mathbf{R}$, be a function such that $f(x) =$ log x. Determine :

  (i) the image set of domain of f

  (ii) $\{x : f(x) = -2\}$

  (iii) whether $f(xy) = f(x) + f(y)$ holds

14. Give an example of a map which is :

  (i) one-to-one but not onto

  (ii) not one to one, but onto

  (iii) neither one-to-one nor onto

ANSWERS

1. (i) $f(A) = \{-4, -3, 0, 5\}$, (ii) ϕ, $\{1, 2\}$, -2 **2.** Domain $= [1, 3]$, Range $= [-1, 1]$ **3.** $\left]-\infty \dfrac{-8}{25}\right] \cup [0, \infty[$

4. (i) $\mathbf{R} - \{1\}$, $\mathbf{R} - \{2\}$, (ii) $\mathbf{R} : \mathbf{R} - \mathbf{R}^+$, (iii) $\mathbf{R} - \{1\}$, $\{-1, 1\}$, (iv) $[3, \infty[$, $[0, \infty]$ **5.** $f(A) = \{1, -2, 18, 28, 108\}$ **6.** $A = \{3, 6\}$ or $(3, 9)$ or $[3, 12]$ etc. **7.** $(-2, 1/2)$ **9.** (iii) **10.** (i) $\{(1, 3), (2, 6), (3, 9)\}$, function, (ii) $\{(1, 4), (1, 6), (3, 4), (3, 6)\}$, not function (iii) $\{(0, 3), (1, 2), (2, 1), (3, 0)\}$, function **11.** (i) $f_1 = \{x, f(x) : x \in A\} = \{(-1, 2), (0, 1), (2, 5), (4, 17)\}$ (ii) $f_2 = \{(x, g(x)) : x \in A\} = \{(1,2),(2,4), (3, 6), ..., (10, 20)\}$ **12.** (i) Range of $f = \mathbf{R}^+$, the set of positive real numbers, (ii) $(x : f(x) = 1) = \{0\}$, (iii) $f(x+y) = f(x) . f(y)$ holds for all $x, y \in \mathbf{R}$

14. (i) $n \to n^2 : \mathbf{N} \to \mathbf{N}$ (ii) $n \to |n| : \mathbf{Z} \to \mathbf{N} \cup \{0\}$ (iii) $n \to |n|^2 : \mathbf{Z} \to \mathbf{N} \cup \{0\}$

Objective Evaluation

⌘ MULTIPLE CHOICE QUESTIONS

Choose the most appropriate one.

1. Let R_1 and R_2 be two equivalence relation on a set. Consider the following assertion :
   (i) $R_1 \cup R_2$ is an equivalence relation.
   (ii) $R_1 \cap R_2$ is an equivalence relation.
   Which of the following is correct?
   (a) Both assertions are true.
   (b) Assertion (i) is true but assertion (ii) is not true.
   (c) Assertion (ii) is true but assertion (i) is not true.
   (d) Neither (i) or (ii) is true.

2. The 'subset' relation on a set of set is :
   (a) a partial ordering
   (b) an equivalence relation
   (c) transitive and symmetric only
   (d) transitive and anti-symmetric only

3. Let R be a symmetric and transitive relation on a set A, then :
   (a) R is reflexive and hence an equivalence relation.
   (b) R is reflexive and hence a partial order.
   (c) R is not reflexive and hence is not an equivalence relation .
   (d) none of the above

4. The number of equivalence relations of the set $\{1, 2, 3, 4\}$ is :
   (a) 4   (b) 15
   (c) 16   (d) 24

5. Suppose A is a finite set with n elements. The number of elements in the large equivalence relation of A is :
   (a) 1   (b) n
   (c) $n + 1$ (d) n^2

6. The binary relation $S = \phi$ on the set $A = \{1, 2, 3\}$ is :
   (a) neither reflexive nor symmetric
   (b) symmetric and reflexive
   (c) transitive and reflexive
   (d) transitive and symmetric

7. Let $f(x) = x^2 + x$ and $g(x) = x + 1$ then fog is :
   (a) $x^2 + 3x + 2$ (b) $x^2 + x + 1$
   (c) $(x+1)^2 + (x+1)$ (d) None of these

8. Let A and B be sets with cardinalities m and n respectively. The number of one-to-one mapping from A to B where $m < n$ is :
   (a) m^n (b) ^nP_m
   (c) ^mC_n (d) ^nC_m

9. The number of functions from m element set to n element set is :
   (a) $m + n$ (b) m^n
   (c) n^m (d) $m * n$

10. _____ is an unordered collection of elements where an element can occur as a member more than once :
   (a) Multiset   (b) Ordered set
   (c) Set   (d) None of these

11. The number of substrings of all lengths that can be formed from a character string of length $n =$ _____
   (a) n (b) n^2
   (c) $\dfrac{n(n-1)}{1}$ (d) $\dfrac{n(n+1)}{2}$

12. In a room containing 28 females, there are 18 females who speak English, 15 females speak French and 22 speak German. 9 females speak both English and French, 11 Females speak both French and German whereas 13 speak both German and English. How many females speak all the three languages?
   (a) 9   (b) 8
   (c) 7   (d) 6

13. Consider the following statements :
   S_1 : There exist infinite set A, B and C such that $A \cap (B \cap C)$ is finite.
   S_2 : There exist two irrational numbers x and y such that $(x + y)$ is rational.
   Which of the following is True about S_1 and S_2?
   (a) Only S_1 is correct.
   (b) only S_2 is correct.
   (c) Both S_1 and S_2 are correct.
   (d) None of the S_1 and S_2 is correct.

14. The power set 2^S of the set $S = \{3, (1, 4), 5\}$ is:

    (a) $\{5, 3, 1, 4, (1, 3, 5), (1, 4, 5), (3, 4), \phi\}$

    (b) $\{5, 3\ (1, 4), 5\}$

    (c) $\{5, (3), [3, (1,4)], (3, 5), \phi\}$

    (d) None of the above.

15. Let A be a finite set of size n, the number of elements in the power set of $A \times A$ is :

    (a) 2^n (b) 2^{n^2}

    (c) $(2^n)^2$ (d) None of these

16. Let S be an infinite set and $S_1, S_2, S_3, \ldots S_n$ be the sets such that $S_1 \cup S_2 \cup S_3 \cup \ldots \cup S_n = S$. Then :

    (a) at least one of the set S_i is a finite set.

    (b) not more than one of the set S_i can be finite.

    (c) at least one of the sets S_i is an infinite set.

    (d) None of the above

17. The Number of elements in the power set $P(S)$ of the set $S = \{\{\phi\}, 1, \{2, 3\}\}$ is :

    (a) 2   (b) 4

    (c) 8   (d) none of these

18. Let A and B be sets and A' and B' denote the complements of the sets A and B. The set $(A - B) \cup (B - A) \cup (A \cap B)$ is equal to :

    (a) $A \cup B$ (b) $A' \cup B'$

    (c) $A \cap B$ (d) $A' \cap B'$

19. Let $P(S)$ denote the power set of set S which of the following is always TRUE?

    (a) $P(P(S)) = P(S)$

    (b) $P(S) \cap S = P(S)$

    (c) $P(S) \cap P(P(S)) = (\phi)$

    (d) $S \in P(S)$

☞ TRUE/FALSE

Write 'T' for True and 'F' for False statement.

1. A binary relation is a set.   **(T/F)**

2. A void set defines a relation.   **(T/F)**

3. The total number of relations from a set containing m elements to a finite set containing n elements is 2^{mn}.   **(T/F)**

4. Every relation is a function.   **(T/F)**

5. Every function is a relation.   **(T/F)**

6. The total number of bijections from a set containing n elements to a set containing n elements is n^n.   **(T/F)**

7. Every equivalence relation is symmetric.   **(T/F)**

8. Every symmetric relation is equivalence.   **(T/F)**

9. Every anti-symmetric relation is symmetric.   **(T/F)**

10. The composition of functions is commutative. **(T/F)**

11. Reflexivity is redundant in the definition of an equivalence relation on a set A, because by symmetry $(a, b) \in R \Rightarrow (b, a) \in R$ and by transitivity $(a, b) \in R$ and $(b, a) \in R \Rightarrow (a, a) \in R$ **(T/F)**

12. The relation $R = \{(1, 2), (1, 3)\}$ is a transitive relation on a set $A = \{1, 2, 3\}$.   **(T/F)**

13. The identity relation on a finite set A is the smallest equivalence relation on A.   **(T/F)**

☞ FILL IN THE BLANKS

1. A relation R on a set A is symmetric iff $R = $ _____ .

2. Let R be an anti-symmetric relation on a set A such that $(a, b) \in R$ and $(b, a) \in R$. Then _____ .

3. Let R be a relation on a set A such that $R = R^{-1}$. Then R is _____ .

4. Let $A = \{1, 2, 3\}$, then the smallest equivalence relation on A is _____ .

5. Let A be a finite set. Then the smallest equivalence relation on A is the _____ relation on A.

6. The void relation on a set is _____ and _____ but not _____ .

7. Let R be a relation defined by $R = \{(4, 5), (1, 4), (4, 6), (7, 6), (3, 7)\}$ on N. Then $R \circ R^{-1} = $ _____ .

8. Let $R = \{(a, a), (b, c), (a, b)\}$ be a relation on a set $A = \{a, b, c\}$. Then the minimum number of ordered pairs which when added to R make it transitive is _____ .

ANSWERS

☜ MULTIPLE CHOICE QUESTIONS

1. (c) **2.** (a) **3.** (d) **4.** (c) **5.** (d) **6.** (d) **7.** (a) **8.** (b) **9.** (c)
10. (a) **11.** (d) **12.** (d) **13.** (c) **14.** (d) **15.** (b) **16.** (c) **17.** (c) **18.** (a)
19. (d)

☜ TRUE/ FALSE

1. T **2.** T **3.** F **4.** F **5.** T **6.** F **7.** T **8.** F **9.** F
10. F **11.** F **12.** T **13.** T

☜ FILL IN THE BLANKS

1. $R = R$ **2.** $a = b$ **3.** Symmetric **4.** $\{(1, 1), (2, 2), (3, 3)\}$ **5.** The identity relation on A.
6. Symmetric, transitive but not reflexive **7.** $\{(5, 5), (4, 4), (6, 5), (6, 6), (5, 6), (7, 7)\}$
8. Many-One into

□□□□□

CHAPTER 2

Quadratic Equations

2.1 INTRODUCTION

(i) **Polynomial:** A function f defined by $f(x) = a_0 + a_1 x + a_2 x^2 + ... + a_n x^n, x \in R$ where $a_0, a_1, a_2, ..., a_n \in R$ is called a polynomial of a real variable with real coefficients.

REMARKS

➡ If $a_n \neq 0$ then the degree of the polynomial is n.

➡ If $a_0, a_1, ..., a_n \in C$, the set of complex numbers and $x \in R$, then the polynomial is called complex polynomial.

(ii) **Polynomial Equation:** Let $f(x)$ be a polynomial, then $f(x) = 0$ is called the polynomial equation. Generally, a polynomial equation of degree two is called quadratic equation.

(iii) **Degree of an Equation:** The degree of an equation is the index of the highest power of variable quantity involved in the equation, when the equation has been expressed to the rational integral form (radical free form).

(iv) **Roots of an Equation:** Let $f(x) = 0$ be a quadratic equation. A real or complex numbers α is said to be a root or solution of a quadratic equation $f(x) = ax^2 + bx + c = 0$, if $f(\alpha) = a\alpha^2 + b\alpha + c = 0$, i.e., α satisfies the given quadratic equation.

(v) **Solution Set:** The set of all roots of an equation, is called the solution set of the given equation.

(vi) **Identity:** An expression involving equality and a variable is called an identity, if it is satisfied by every value of the variable.

REMARK

➡ A root of an equation is also called zero.

2.2 LINEAR EQUATION

An equation of the form $ax + b = 0$ is called a linear equation in x, where x is unknown variable (quantity) and a and b are any constants. Here, $a \neq 0$ because if $a = 0$ then the equation gives $b = 0$, which has no unknown.

For example: $x - 2 = 0, 3x - 9 = 0, x + \dfrac{7}{2} = 0, x = 2, x = 0$ are the linear equation in x.

REMARK

➡ To find the degree of an equation, the unknown variable x must be in the numerator only in the equation and power of x must be a positive integer.

2.2.1 SOLUTION OF A LINEAR EQUATION

All those values of the unknown variable, which are involved in the equation, for which equation is satisfied, are called the solutions (or roots) of the equation. To find the solution of a linear equation we write the equation in standard form ($ax + b = 0$) if it is not.

Method: Consider a linear equation

$$ax + b = 0, a \neq 0 \qquad \qquad \qquad \text{...(1)}$$

Here, we want to find the value of x, so, add $-b$ to both sides in (1), we get

$$ax + b - b = 0 - b$$

$$\Rightarrow \qquad \qquad ax = -b$$

Now, dividing both sides by a, we get

$$x = -\frac{b}{a}, a \neq 0$$

Hence, the solution of the given equation is $x = -\dfrac{b}{a}, a \neq 0$

For example:

Consider the equation $2x + 3 = 0$. First adding -3 in both sides of the given equation, we get

$$2x + 3 - 3 = 0 - 3$$

$$\Rightarrow \qquad \qquad 2x = -3$$

Now, dividing both sides by 2, we get $x = -\dfrac{3}{2}$, which is the required solution.

SOLVED EXAMPLES

EXAMPLE 1.   *Solve the following equation*

$$11 - \frac{y}{10} - \frac{y-10}{5} = y$$

SOLUTION.   Here, the given equation is

$$11 - \frac{y}{10} - \frac{y-10}{5} = y \qquad \qquad \text{...(1)}$$

Clearly, y is the unknown variable in this equation.

Now, multiplying the equation (1) by 10, we get

$$10\left(11 - \frac{y}{10} - \frac{y-10}{5}\right) = 10y$$

$$\Rightarrow \qquad 110 - y - \frac{10(y-10)}{5} = 10y$$

$$\Rightarrow \qquad 110 - y - 2y + 20 = 10y$$

$$\Rightarrow \qquad \qquad 130 - 3y = 10y \qquad \qquad \text{...(2)}$$

Now, adding $3y$ on both sides of (2), we get

$$130 - 3y + 3y = 10y + 3y$$

$$\Rightarrow \qquad \qquad 130 = 13y$$

$$\Rightarrow \qquad \qquad y = \frac{130}{13} = 10$$

Hence, $y = 10$

EXAMPLE 2.   *Solve the equation $3(x + 2) = 5$.*

SOLUTION.   Here, the given equation is $3(x + 2) = 5$ and x is unknown variable in the given equation.

So, $3(x + 2) = 5$

$\Rightarrow$ $3x + 3 \times 2 = 5$

$\Rightarrow$ $3x + 6 = 5$

Now, adding -6 on both sides of the above equation, we get

$$3x + 6 - 6 = 5 - 6$$

$\Rightarrow$ $3x = -1$

$\Rightarrow$ $x = -\dfrac{1}{3}$

EXAMPLE 3.   *Solve the equation $x + 2 = 2x - 8$.*

SOLUTION.   Here, the given equation is $x + 2 = 2x - 8$. Adding 8 on both sides of the above equation, we get

$$x + 2 + 8 = 2x - 8 + 8$$

$\Rightarrow$ $x + 10 = 2x$

$\Rightarrow$ $10 = 2x - x$

$\Rightarrow$ $10 = x$ or $x = 10$

EXAMPLE 4.   *Solve : $\dfrac{x}{2} + 3 = \dfrac{x - 6}{8}$.*

SOLUTION.   Here, the given equation is $\dfrac{x}{2} + 3 = \dfrac{x - 6}{8}$

$\Rightarrow$ $8\left(\dfrac{x}{2} + 3\right) = x - 6$

$\Rightarrow$ $\dfrac{8x}{2} + 8 \times 3 = x - 6$

$\Rightarrow$ $4x + 24 = x - 6$

Now, adding -24 on both sides of the above equation, we get

$$4x + 24 - 24 = x - 6 - 24$$

$\Rightarrow$ $4x = x - 30$

$\Rightarrow$ $4x - x = -30$

$\Rightarrow$ $3x = -30$

$\Rightarrow$ $x = -\dfrac{30}{3} = -10$

$\Rightarrow$ $x = -10$

THEOREM 1.   *A quadratic equation can not have more than two roots.*

PROOF.   Let, if possible, α, β, γ be the three distinct roots of the quadratic equation $ax^2 + bx + c = 0$, then we have

$$a\alpha^2 + b\alpha + c = 0 \qquad \qquad \qquad \qquad \text{...(1)}$$
$$a\beta^2 + b\beta + c = 0 \qquad \qquad \qquad \qquad \text{...(2)}$$

and $a\gamma^2 + b\gamma + c = 0$ ...(3)

Using (1) and (2), we get

$$a(\alpha^2 - \beta^2) + b(\alpha - \beta) = 0$$

$\Rightarrow$ $a(\alpha - \beta)(\alpha + \beta) + b(\alpha - \beta) = 0$

$\Rightarrow$ $(\alpha - \beta)[a(\alpha + \beta) + b] = 0$

$\Rightarrow$ $a(\alpha + \beta) + b = 0$ $(\because \alpha \neq \beta \Rightarrow \alpha - \beta \neq 0)$

$$\Rightarrow \qquad \alpha + \beta = -\frac{b}{a} \qquad \qquad \ldots(4)$$

Similarly, using (2) and (3), we get

$$\beta + \gamma = -\frac{b}{a} \qquad \qquad \ldots(5)$$

Equations (4) and (5) gives

$$\alpha + \beta = \beta + \gamma \qquad \Rightarrow \qquad \alpha = \gamma$$

which is a contradiction. $\qquad\qquad$ ($\because$ α and γ both are distinct)

Hence, the quadratic equation cannot have more than two roots.

2.3 SOLUTION OF QUADRATIC EQUATIONS

There are four methods to solve a complete quadratic equation :

(a) Method of factorization
(b) Method of completing the square
(c) Method of formula
(d) Method of Identity

SOLVED EXAMPLES

Type (a) Method of Factorization:

EXAMPLE 1. *Solve $x^2 - 4x + 3 = 0$.*

SOLUTION. Here, we have $x^2 - 4x + 3 = 0$

$$\Rightarrow \qquad x^2 - 3x - x + 3 = 0$$
$$\Rightarrow \qquad x(x-3) - 1(x-3) = 0$$
$$\Rightarrow \qquad (x-3)(x-1) = 0$$
$$\Rightarrow \qquad x = 1, 3$$

EXAMPLE 2. *Solve $x^2 - 2x - 3 = 0$.*

SOLUTION. Here, we have $x^2 - 2x - 3 = 0$

$$\Rightarrow \qquad x^2 - 3x + x - 3 = 0$$
$$\Rightarrow \qquad x(x-3) + 1(x-3) = 0$$
$$\Rightarrow \qquad (x-3)(x+1) = 0$$
$$\Rightarrow \qquad x = -1, 3$$

Type (b) Method of Completing the Square:

This method can be employed when the factorization method fails, *i.e.*, when the quadratic expression is not factorize or difficult to factorize.

EXAMPLE 3. *Solve $4x^2 - 3x - 1 = 0$.*

SOLUTION. Here, we have $\qquad 4x^2 - 3x - 1 = 0$

$$\Rightarrow \qquad 4\left(x^2 - \frac{3}{4}x\right) = 1$$

$$\Rightarrow \qquad x^2 - \frac{3}{4}x = \frac{1}{4}$$

$$\Rightarrow \qquad x^2 - \frac{3}{4}x + \left(-\frac{3}{8}\right)^2 = \frac{1}{4} + \left(-\frac{3}{8}\right)^2$$

[By adding both sides (1/2 the coeff. of x)]

$$\Rightarrow \qquad \left(x - \frac{3}{8}\right)^2 = \frac{1}{4} + \frac{9}{64} = \frac{25}{64} = \left(\frac{5}{8}\right)^2$$

$$\Rightarrow \qquad x - \frac{3}{8} = \pm \frac{5}{8} \qquad \text{[By taking square root of both sides]}$$

$$\Rightarrow \qquad x = \frac{3}{8} \pm \frac{5}{8}$$

$$\Rightarrow \qquad x = \frac{3}{8} + \frac{5}{8} \quad \text{or} \quad x = \frac{3}{8} - \frac{5}{8}$$

$$\Rightarrow \qquad x = 1 \quad \text{or} \quad -\frac{1}{4}$$

Hence, the solution set of the given equation is $\left\{ 1, -\frac{1}{4} \right\}$.

EXAMPLE 4. *Solve* $2x^2 - 7x + 6 = 0$.

SOLUTION. Here, the given equation is $2x^2 - 7x + 6 = 0$

$$\Rightarrow \qquad 2x^2 - 7x = -6$$

$$\Rightarrow \qquad x^2 - \frac{7}{2}x = -3$$

$$\Rightarrow \qquad x^2 - \frac{7}{2}x + \left(-\frac{7}{4}\right)^2 = -3 + \left(-\frac{7}{4}\right)^2$$

$$\left(\text{By adding both sides} \left\{ \frac{1}{2} \text{ the coeff. of } x \right\}^2 \right)$$

$$\Rightarrow \qquad x^2 - \frac{7}{2}x + \frac{49}{16} = -3 + \frac{49}{16}$$

$$\Rightarrow \qquad \left(x - \frac{7}{4} \right)^2 = \frac{1}{16}$$

$$\Rightarrow \qquad x - \frac{7}{4} = \pm \frac{1}{4} \qquad \text{(By taking square root of both sides)}$$

Therefore, $x = \frac{7}{4} \pm \frac{1}{4} = \frac{7}{4} + \frac{1}{4} = \frac{8}{4} = 2$ and $x = \frac{7}{4} - \frac{1}{4} = \frac{6}{4} = \frac{3}{2}$

Hence, the solution set of the given equation is $\left\{ \frac{3}{2}, 2 \right\}$.

Type (c) Method of formula

Consider a quadratic equation $ax^2 + bx + c = 0$, $a \neq 0$ ...(1)
Then solution of equation (1) is given by

$$x = \frac{-b \pm \sqrt{b^2 - 4ac}}{2a}$$

EXAMPLE 5. *Solve each of the following equation by quadratic formula.*

  (i) $5x^2 - 15x + 11 = 0$ (ii) $x^2 - 3x + 5 = 0$

SOLUTION. (i) Comparing the given equation with $ax^2 + bx + c = 0$, we get,

$$a = 5, b = -15, c = 11$$

Therefore, $$x = \frac{-b \pm \sqrt{b^2 - 4ac}}{2a}$$

$$= \frac{-(-15) \pm \sqrt{(-15)^2 - 4 \times 5 \times 11}}{2 \times 5} = \frac{15 \pm \sqrt{225 - 220}}{10}$$

$$= \frac{15 \pm \sqrt{5}}{10} = \frac{15 + \sqrt{5}}{10} \text{ or } \frac{15 - \sqrt{5}}{10}$$

Hence, the solution set is $\left[\dfrac{15 + \sqrt{5}}{10}, \dfrac{15 - \sqrt{5}}{10} \right]$.

(ii) Comparing the given equation $x^2 - 3x + 5 = 0$ with the standard quadratic equation $ax^2 + bx + c = 0$, we get

$$a = 1, b = -3, c = 5$$

Therefore,

$$x = \frac{-b \pm \sqrt{b^2 - 4ac}}{2a}$$

$$= \frac{-(-3) \pm \sqrt{(-3)^2 - 4 \times 1 \times 5}}{2 \times 1} = \frac{3 \pm \sqrt{9 - 20}}{2}$$

$$= \frac{3 \pm \sqrt{-11}}{2} = \frac{3 \pm i\sqrt{11}}{2} \qquad\qquad (\because i^2 = -1)$$

Hence, the required solution set is given by $\left\{ \dfrac{3 + i\sqrt{11}}{2}, \dfrac{3 - i\sqrt{11}}{2} \right\}$.

EXAMPLE 6. *Solve the following equations by quadratic method.*

　(i) $2x^2 - 4x + 3 = 0$ 　　　　　　(ii) $25x^2 - 30x + 11 = 0$

SOLUTION. (i) Here the given equation is $2x^2 - 4x + 3 = 0$

Comparing with the standard quadratic equation $ax^2 + bx + c = 0$, we get

$$a = 2, b = -4, c = 3$$

Now,

$$x = \frac{-b \pm \sqrt{b^2 - 4ac}}{2a} = \frac{4 \pm \sqrt{16 - 24}}{2 \times 2}$$

$$= \frac{4 \pm \sqrt{-8}}{4} = \frac{4 \pm \sqrt{8i^2}}{4} = \frac{4 \pm i\sqrt{8}}{4}$$

$$= \frac{4 \pm 2i\sqrt{2}}{4} = 1 \pm \frac{1}{\sqrt{2}} i$$

$$= 1 + \frac{1}{\sqrt{2}} i \text{ or } 1 - \frac{1}{\sqrt{2}} i$$

Hence, the solution set is $\left\{ 1 + \dfrac{1}{\sqrt{2}} i, 1 - \dfrac{1}{\sqrt{2}} i \right\}$.

(ii) Comparing the given equation $25x^2 - 30x + 11 = 0$ with the standard quadratic equation, we get

$$a = 25, b = -30, c = 11$$

Therefore,

$$x = \frac{-b \pm \sqrt{b^2 - 4ac}}{2a} = \frac{30 \pm \sqrt{900 - 1100}}{50}$$

$$= \frac{30 \pm \sqrt{-200}}{50} = \frac{30 \pm \sqrt{200i^2}}{50}$$

$$= \frac{30 \pm 10i\sqrt{2}}{50} = \frac{3}{5} \pm \frac{\sqrt{2}}{5} i$$

$$= \frac{3}{5} + \frac{\sqrt{2}}{5}i \text{ or } \frac{3}{5} - \frac{\sqrt{2}}{5}i$$

Hence, the required solution set is $\left\{ \frac{3}{5} + \frac{\sqrt{2}}{5}i, \frac{3}{5} - \frac{\sqrt{2}}{5}i \right\}$.

EXAMPLE 7. *Solve the following equations :*

(i) $\dfrac{1}{x+1} + \dfrac{2}{x+2} = \dfrac{4}{x+4}$ (ii) $2x^{1/3} + 2x^{-1/3} = 5$

SOLUTION. (i) Here, $\dfrac{1}{x+1} + \dfrac{2}{x+2} = \dfrac{4}{x+4}$

$$\Rightarrow \quad \frac{(x+2) + 2(x+1)}{(x+1)(x+2)} = \frac{4}{x+4}$$

$$\Rightarrow \quad (3x+4)(x+4) = 4(x+1)(x+2)$$

$$\Rightarrow \quad 3x^2 + 16x + 16 = 4(x^2 + 3x + 2)$$

$$\Rightarrow \quad x^2 - 4x - 8 = 0$$

Put all these values in $x = \dfrac{-b \pm \sqrt{b^2 - 4ac}}{2a}$, we get

$$x = \frac{-(-4) \pm \sqrt{(-4)^2 - 4.1(-8)}}{2.1}$$

$$= \frac{4 \pm \sqrt{16 + 32}}{2} = 2(1 \pm \sqrt{3})$$

(ii) Here, the given equation is $2x^{1/3} + 2x^{-1/3} = 5$

So put $x^{1/3} = y$, we get $2y + \dfrac{2}{y} = 5$

$$\Rightarrow \quad 2y^2 - 5y + 2 = 0$$

$$\Rightarrow \quad (y-2)(2y-1) = 0$$

$$\Rightarrow \quad y = 2, \frac{1}{2}$$

Now, $x = y^3 = (2)^3 = 8$

or, $\left(\dfrac{1}{2}\right)^3 = \dfrac{1}{8}$

Hence, $x = 8$ or $x = \dfrac{1}{8}$.

EXAMPLE 8. *Solve the equation $x + \sqrt{x} = \dfrac{6}{25}$.*

SOLUTION. Put $\sqrt{x} = y$ in the given equation, we get $y^2 + y = \dfrac{6}{25}$.

$$\Rightarrow \quad 25y^2 + 25y - 6 = 0$$

$$\Rightarrow \quad (5y+6)(5y-1) = 0$$

$$\Rightarrow \quad y = -\frac{6}{5} \text{ or } \frac{1}{5}$$

Now,
$$x = y^2 = \left(-\frac{6}{5}\right)^2 = \frac{36}{25}$$

or
$$\left(\frac{1}{5}\right)^2 = \frac{1}{25}$$

Hence,
$$x = \frac{36}{25} \text{ or } x = \frac{1}{25}.$$

Exercise 2.1

1. Solve the following equations by factorization method:
   (i) $x^2 + x + 1 = 0$ (ii) $x^2 - x - 12 = 0$
   (iii) $x^2 + 1 = 0$ (iv) $x^2 - 4x + 3 = 0$

2. Solve the following equations by the quadratic method:
   (i) $x^2 - 7x + 12 = 0$ (ii) $x^2 - 4x + 7 = 0$
   (iii) $2x^2 - 3x + 1 = 0$
   (iv) $27x^2 - 10x + 1 = 0$
   (v) $3x^2 - x - 10 = 0$ (vi) $9x^2 + 12x + 4 = 0$

3. Solve the equation $\dfrac{x}{6} - \dfrac{x}{5} = \dfrac{x}{15} - \dfrac{x}{3} + 7$

4. Solve the following equations :
   (i) $\dfrac{x^2 - 3x}{7} + 2x = 6$
   (ii) $2x^2 - 10x = 3x - 15$

5. Solve the following equations :
   (i) $x^2 + x - 6 = 0$

   (ii) $\dfrac{2x^2 - 3x - 5}{3} = \dfrac{4x + 9}{5}$

6. Solve the following equations :
   (i) $\dfrac{3x - 7}{5} = \dfrac{16}{x - 5}$
   (ii) $x^4 - 10x^2 + 9 = 0$

7. Solve the equation :
   $$pqx^2 - (p^2 + q^2)x + pq = 0$$

8. Solve : $\sqrt{\{2(x^2 - x + 1)\}} = 3x - 4$

9. Solve : $5^{2x} - 126.5^x + 3 = 0$

10. Solve the following equations :
    (i) $\sqrt{(3x + 10)} = 9 - \sqrt{(9x + 7)}$
    (ii) $\sqrt{3x^2 + 1} + \dfrac{4}{\sqrt{3x^2 + 1}}$
    (iii) $\sqrt{\left(\dfrac{x}{1-x}\right)} + \sqrt{\left(\dfrac{1-x}{x}\right)} = \dfrac{13}{6}$

HINT TO SELECTED PROBLEMS

2. (i) $x^2 - 7x + 12 = 0$
   Here, $a = 1, b = -7, c = 12$
   $$x = \frac{-b \pm \sqrt{b^2 - 4ac}}{2a}$$
   $$x = \frac{7 \pm \sqrt{(-7)^2 - 4 \times 1 \times 12}}{2 \times 1} = \frac{7 \pm \sqrt{49 - 48}}{2}$$
   $$= \frac{7 + 1}{2} \text{ or } \frac{7 - 1}{2}$$
   $x = 4$ or 3

   (ii) $x^2 - 4x + 7 = 0$
   Here, $a = 1, b = -4, c = 7$
   $$x = \frac{-b \pm \sqrt{b^2 - 4ac}}{2a}$$

   $$x = \frac{4 \pm \sqrt{(-4)^2 - 4 \times 1 \times 7}}{2 \times 1}$$
   $$= \frac{4 \pm \sqrt{16 - 28}}{2} = \frac{4 \pm \sqrt{-12}}{2}$$
   $$= \frac{4 \pm 2\sqrt{3}i}{2} = 2 \pm \sqrt{3}i$$

3. $\dfrac{x}{6} - \dfrac{x}{5} = \dfrac{x}{15} - \dfrac{x}{3} + 7$
   $$\frac{5x - 6x}{30} = \frac{x - 5x}{15} + 7$$
   $$\frac{-x}{30} = \frac{-4x}{15} + 7$$
   $$\frac{4x}{15} - \frac{x}{30} = 7$$

$$\frac{8x - x}{30} = 7$$

$$\Rightarrow \frac{7x}{30} = 7 \quad \Rightarrow \quad x = 30$$

4. (i) $\dfrac{x^2 - 3x}{7} + 2x = 6$

$$\Rightarrow \frac{x^2 - 3x + 14x}{7} = 6$$

$$\Rightarrow x^2 + 11x - 42 = 0$$

$$\Rightarrow x^2 + 14x - 3x - 42 = 0$$

$$\Rightarrow x(x + 14) - 3(x + 14) = 0$$

$$\Rightarrow (x + 14)(x - 3) = 0$$

$$\Rightarrow \qquad x = -14$$

or $\qquad\qquad x = +3$

5. (ii) $\dfrac{2x^2 - 3x - 5}{3} = \dfrac{4x + 9}{5}$

$$\Rightarrow 10x^2 - 15x - 25 = 12x + 27$$

$$\Rightarrow 10x^2 - 27x - 52 = 0$$

$$\Rightarrow a = 10, b = -27, c = -52$$

$$x = \frac{-b \pm \sqrt{b^2 - 4ac}}{2a}$$

$$= \frac{27 \pm \sqrt{(-27)^2 - 4 \times 10 \times (-52)}}{2 \times 10}$$

$$= \frac{27 \pm \sqrt{729 + 2080}}{20}$$

$$= \frac{27 \pm \sqrt{2809}}{20} = \frac{27 \pm 53}{20}$$

$$= \frac{27 + 53}{20} \text{ or } \frac{27 - 53}{20} = 4 \text{ or } \frac{-13}{10}$$

6. (i) $\dfrac{3x - 7}{5} = \dfrac{16}{x - 5}$

$$\Rightarrow (3x - 7)(x - 5) = 80$$

$$\Rightarrow 3x^2 - 15x - 7x + 35 = 80$$

$$\Rightarrow 3x^2 - 22x - 45 = 0$$

$$\Rightarrow 3x^2 - 27x + 5x - 45 = 0$$

$$\Rightarrow 3x(x - 9) + 5(x - 9) = 0$$

$$\Rightarrow (3x + 5)(x - 9) = 0$$

$$\Rightarrow 3x + 5 = 0 \text{ or } x - 9 = 0$$

$$\Rightarrow x = \frac{-5}{3} \text{ or } x = 9$$

7. $\quad pqx^2 - (p^2 + q^2)x + pq = 0$

$$\Rightarrow pqx^2 - p^2x - q^2x + pq = 0$$

$$\Rightarrow px(qx - p) - q(qx - p) = 0$$

$$\Rightarrow (qx - p)(px - q) = 0$$

$$\Rightarrow \text{either } qx - p = 0 \text{ or } px - q = 0$$

$$\Rightarrow x = \frac{p}{q} \text{ or } x = \frac{q}{p}$$

8. $\quad \sqrt{\{2(x^2 - x + 1)\}} = 3x - 4$

Squaring both sides,

$$2(x^2 - x + 1) = (3x - 4)^2$$

$$\Rightarrow 2x^2 - 2x + 2 = 9x^2 + 16 - 24x$$

$$\Rightarrow 7x^2 - 22x + 14 = 0$$

$$\Rightarrow x = \frac{+22 \pm \sqrt{(22)^2 - 4 \times 7 \times 14}}{2 \times 7}$$

$$= \frac{22 \pm \sqrt{484 - 392}}{14}$$

$$= \frac{22 \pm \sqrt{92}}{14} = \frac{22 \pm 2\sqrt{23}}{14} = \frac{11 \pm \sqrt{23}}{7}$$

9. $5^{2x} - 126.5^x + 3 = 0$

let $5^x = y$

$$\Rightarrow y^2 - 126y + 3 = 0$$

$$\Rightarrow y = \frac{126 \pm \sqrt{(126)^2 - 4 \times 1 \times 3}}{2 \times 1} = \pm 126$$

ANSWERS

1. (i) $-\dfrac{1}{2} + \dfrac{i\sqrt{3}}{2}, -\dfrac{1}{2} - \dfrac{i\sqrt{2}}{2}$ (ii) $4, -3$ (iii) $(i, -i)$ (iv) $1, 3$

2. (i) $3, 4$ (ii) $2 + \sqrt{3}i, 2 - \sqrt{3}i$ (iii) $\dfrac{1}{2}, 1$ (iv) $\dfrac{5}{27} \pm \dfrac{\sqrt{2}}{27}i$ (v) $2, -\dfrac{5}{3}$ (vi) $-\dfrac{2}{3}$

3. 30 **4.** (i) $3, -14$ (ii) $\dfrac{3}{2}, 5$ **5.** (i) $2, -3$ (ii) $4, -\dfrac{13}{10}$

6. (i) $9, -\dfrac{5}{3}$ (ii) $+3, -3$ **7.** $\dfrac{p}{q}, \dfrac{q}{p}$ **8.** $\dfrac{11 \pm \sqrt{23}}{7}$ **9.** 0.3

10. (i) 2 (ii) $0, \pm\sqrt{5}$ (iii) $\dfrac{9}{13}, \dfrac{4}{13}$

2.4 EQUATIONS REDUCIBLE TO QUADRATICS

2.4.1 TYPE I: EQUATION OF THE FORM $ax^n + bx^{n-2} + c = 0$, WHERE x IS AN EXPRESSION IN x

WORKING PROCEDURE

In order to solve such type of equation we use the following steps :

Step 1. Put $X^n = y$ and obtain the equation $ay^2 + by + c = 0$.

Step 2. Now solve the obtained equation for y.

Step 3. Finally get the value of x, by using the relation $X^n = y$.

SOLVED EXAMPLES

EXAMPLE 1. Solve $x^4 - 9x^2 - 10 = 0$.

SOLUTION. The given equation is $x^4 - 9x^2 - 10 = 0$ ...(1)

Put, $x^2 = y$ in (1), we get $y^2 - 9y - 10 = 0$, which is a quadratic equation in y.

$$y^2 - 9y - 10 = 0$$
$$\Rightarrow \quad y^2 - 10y + y - 10 = 0$$
$$\Rightarrow \quad y(y - 10) + 1(y - 10) = 0$$
$$\Rightarrow \quad (y - 10)(y + 1) = 0$$
$$\Rightarrow \quad y = -1, 10$$

Now $\quad y = 10$

$$\Rightarrow \quad x^2 = 10 \quad \Rightarrow \quad x = \pm\sqrt{10}$$

and $\quad y = -1$

$$\Rightarrow \quad x^2 = -1 \quad \Rightarrow \quad x = \pm i$$

Hence, the solution set is given by $(\sqrt{10}, -\sqrt{10}, i, -i)$.

EXAMPLE 2. Solve $\left(\dfrac{2x+1}{x-1}\right)^4 - 10\left(\dfrac{2x+1}{x-1}\right)^2 + 9 = 0$. **[UPTU(B.Pharma)–2005]**

SOLUTION. We have $\left(\dfrac{2x+1}{x-1}\right)^4 - 10\left(\dfrac{2x+1}{x-1}\right)^2 + 9 = 0$.

$$\left[\text{Putting} \left(\dfrac{2x+1}{x-1}\right)^2 = y\right]$$

$$\Rightarrow \quad y^2 - 10y + 9 = 0$$
$$\Rightarrow \quad y^2 - 9y - y + 9 = 0$$
$$\Rightarrow \quad y(y - 9) - 1(y - 9) = 0$$
$$\Rightarrow \quad (y - 1)(y - 9) = 0$$
$$\Rightarrow \quad y = 1 \text{ or } y = 9$$

Now $\quad y = 1 \Rightarrow \left(\dfrac{2x+1}{x-1}\right)^2 = 1 \Rightarrow \dfrac{2x+1}{x-1} = \pm 1$

$$\Rightarrow \quad \dfrac{2x+1}{x-1} = 1 \quad \text{or} \quad \dfrac{2x+1}{x-1} = -1$$
$$\Rightarrow \quad 2x + 1 = x - 1 \quad \text{or} \quad 2x + 1 = -x + 1$$
$$\Rightarrow \quad x = -2 \quad \text{or} \quad x = 0$$

Again $\quad y = 9 \Rightarrow \left(\dfrac{2x+1}{x-1}\right)^2 = 9 \Rightarrow \dfrac{2x+1}{x-1} = \pm 3$

$$\Rightarrow \qquad \frac{2x+1}{x-1} = 3 \quad \text{or} \quad \frac{2x+1}{x-1} = -3$$

$$\Rightarrow \qquad 2x + 1 = 3x - 3 \text{ or} \quad 2x + 1 = -3x + 3$$

$$x = 4 \qquad \text{or} \qquad x = \frac{2}{5}$$

Hence, the solution of the given equation is $\left\{-2, 0, 4, \dfrac{2}{5}\right\}$.

EXAMPLE 3.   *Solve the equation* $\left(x - \dfrac{x}{x+1}\right)^2 + 2x\dfrac{x}{x+1} = 3$.

SOLUTION.   Here, the given equation is $\left(x - \dfrac{x}{x+1}\right)^2 + 2x\dfrac{x}{x+1} = 3$

which can be written as $\left(\dfrac{x^2}{x+1}\right)^2 + 2\dfrac{x^2}{x+1} = 3$.   ...(1)

Now putting $\dfrac{x^2}{x+1} = y$, the equation (1) reduces to $y^2 + 2y = 3$.

$$\Rightarrow \qquad y^2 + 2y - 3 = 0 \qquad \Rightarrow \qquad y^2 + 3y - y - 3 = 0$$

$$\Rightarrow \quad y(y + 3) - 1(y + 3) = 0 \qquad \Rightarrow \qquad (y + 3)(y - 1) = 0$$

$$\Rightarrow \qquad \qquad y = 1, -3$$

Now, $y = 1$

$$\Rightarrow \qquad \frac{x^2}{x+1} = 1 \qquad \Rightarrow \qquad x^2 = x + 1$$

$$\Rightarrow \qquad x^2 - x - 1 = 0$$

$$\Rightarrow \qquad x = \frac{1 \pm \sqrt{1+4}}{2} = \frac{1 \pm \sqrt{5}}{2}$$

Also, $y = -3$

$$\Rightarrow \qquad \frac{x^2}{x+1} = -3 \qquad \Rightarrow \qquad x^2 = -3x - 3$$

$$\Rightarrow \qquad x^2 + 3x + 3 = 0$$

$$\Rightarrow \qquad x = \frac{-3 \pm \sqrt{9-12}}{2} = \frac{-3 \pm i\sqrt{3}}{2}$$

Hence, the solution set is $\left\{\dfrac{1 \pm \sqrt{5}}{2}, \dfrac{-3 \pm i\sqrt{3}}{2}\right\}$.

EXAMPLE 4.   *Solve the following equation* $4^x - 5.2^x + 4 = 0$.

SOLUTION.   The given equation is $4^x - 5.2^x + 4 = 0$

which can be written as $(2^x)^2 - 5(2^x) + 4 = 0$

Put $2^x = y$, we get

$$y^2 - 5y + 4 = 0 \qquad \qquad \qquad \text{...(1)}$$

which is quadratic in y.

Solving (1) for y, we get

$$(y - 4)(y - 1) = 0$$

$$\Rightarrow \qquad y = 1, 4$$

Now, $\qquad y = 1$

$\Rightarrow \qquad 2^x = 1 \qquad \Rightarrow \qquad 2^x = 2^0$

$\Rightarrow \qquad x = 0$

Also, $\qquad y = 4$

$\Rightarrow \qquad 2^x = 4 \qquad \Rightarrow \qquad 2^x = 2^2$

$\Rightarrow \qquad x = 2$

Hence, the solution set of the given equation is $(0, 2)$.

EXAMPLE 5. *Solve the equation $x^{2/3} + x^{1/3} - 2 = 0$.*

SOLUTION. Here, the given equation is $x^{2/3} + x^{1/3} - 2 = 0$

$\Rightarrow \qquad (x^{1/3})^2 + x^{1/3} - 2 = 0$

Put $x^{1/3} = y$, we get $y^2 + y - 2 = 0$, which is quadratic in y.

Solving for y, we get

$$(y + 2)(y - 1) = 0$$

$\Rightarrow \qquad y = 1, -2$

Now, $\qquad y = 1$

$\Rightarrow \qquad x^{1/3} = 1 \qquad \Rightarrow \qquad x = 1$

also, $\qquad y = -2 \qquad \Rightarrow \qquad x^{1/3} = -2$

$\Rightarrow \qquad x = (-2)^3 \qquad \Rightarrow \qquad x = -8$

Hence, the solution set of the given equation is $(1, -8)$.

2.4.2 Type II: Equation of the form $aX + \dfrac{b}{X} + c = 0$, where X is an expression in x

WORKING PROCEDURE

In order to solve such type of equations, we use the following steps :

Step 1. Put $X = y$ and obtain the equation in y.

Step 2. Solve the quadratic equation for y.

Step 3. Finally get the value of x by using the relation $X = y$.

SOLVED EXAMPLES

EXAMPLE 1. *Solve $\sqrt{\dfrac{x}{x-1}} + \sqrt{\dfrac{x-1}{x}} = \dfrac{13}{6}, (x \neq 1, x \neq 0)$.*

SOLUTION. Here, the given equation is $\sqrt{\dfrac{x}{x-1}} + \sqrt{\dfrac{x-1}{x}} = \dfrac{13}{6}$.

Put $\sqrt{\dfrac{x}{x-1}} = y$, then the above equation reduces to

$$y + \frac{1}{y} = \frac{13}{6} \qquad \Rightarrow \qquad 6y^2 - 13y + 6 = 0$$

Solving for y, we get

$$(2y - 3)(3y - 2) = 0$$

$\Rightarrow \qquad y = \dfrac{3}{2}, \dfrac{2}{3}$

Now, $\qquad y = \dfrac{3}{2} \Rightarrow \sqrt{\dfrac{x}{x-1}} = \dfrac{3}{2} \Rightarrow \dfrac{x}{x-1} = \left(\dfrac{3}{2}\right)^2 \Rightarrow x = \dfrac{9}{5}$

also,
$$y = \frac{2}{3} \Rightarrow \sqrt{\frac{x}{x-1}} = \frac{2}{3} \Rightarrow \frac{x}{x-1} = \left(\frac{2}{3}\right)^2 \Rightarrow x = -\frac{4}{5}$$

Hence, the solution set is given by $\left\{\frac{9}{5}, -\frac{4}{5}\right\}$.

EXAMPLE 2. *Solve* $8x^{3/2} - \dfrac{8}{x^{3/2}} = 63$.

SOLUTION. Here, the given equation is $8x^{3/2} - \dfrac{8}{x^{3/2}} = 63$.

Put $x^{3/2} = y$, we get $8y - \dfrac{8}{y} = 63$

$\Rightarrow \qquad 8y^2 - 63y - 8 = 0$
$\Rightarrow \qquad 8y^2 - 64y + y - 8 = 0$
$\Rightarrow \qquad 8y(y-8) + 1(y-8) = 0$
$\Rightarrow \qquad (y-8)(8y+1) = 0$

$\Rightarrow \qquad y = 8, -\dfrac{1}{8}$

Now, $\qquad y = 8, \; x^{3/2} = 8$
$\Rightarrow \qquad x = (8)^{2/3} = (2^{3 \times 1/3})^2 = 2^2 = 4$

Also, $\qquad y = -\dfrac{1}{8} \Rightarrow x^{3/2} = -\dfrac{1}{8}$

$\Rightarrow \qquad x = \left(-\dfrac{1}{8}\right)^{2/3} = \dfrac{1}{4}$

Hence, the solution set is given by $\left\{4, \dfrac{1}{4}\right\}$.

EXAMPLE 3. *Solve the equation* $7^{1+x} + 7^{1-x} = 50$ [UPTU(B.Pharma)–2008]

SOLUTION. Here, the given equation can be written as
$$7 \cdot 7^x + 7 \cdot 7^{-x} = 50$$

$\Rightarrow \qquad 7 \cdot 7^x + \dfrac{7}{7^x} = 50$

Put $7^x = y$, we get

$$7y + \frac{7}{y} = 50$$

$\Rightarrow \quad 7y^2 - 50y + 7 = 0$, which is quadratic in y.

Solving for y, we get
$$7y^2 - 49y - y + 7 = 0$$
$\Rightarrow \qquad 7y(y-7) - 1(y-7) = 0$
$\Rightarrow \qquad (y-7)(7y-1) = 0$

$\Rightarrow \qquad y = 7, \dfrac{1}{7}$

Now, $\qquad y = 7$
$\Rightarrow \qquad 7^x = 7$
$\Rightarrow \qquad x = 1$ (By Putting the value of y)

also,
$$y = \frac{1}{7} \Rightarrow 7^x = \frac{1}{7} \Rightarrow x = -1$$

Hence, the solution set is given by {1, –1}.

Example 4. If $\frac{4x+1}{x+1} + \frac{x+1}{4x+1} = \frac{5}{2} = 0$, find the value of x. [UPTU(B.Pharma)–2008, 12]

Solution. We have $\frac{4x+1}{x+1} + \frac{x+1}{4x+1} = \frac{5}{2}$

Let $\frac{4x+1}{x+1} = y$

Then $y + \frac{1}{y} = \frac{5}{2}$

$\Rightarrow \qquad y^2 + 1 = \frac{5}{2}y$

$\Rightarrow \qquad 2y^2 + 2 = 5y$

$\Rightarrow \qquad 2y^2 - 5y + 2 = 0$

$\Rightarrow \qquad 2y^2 - 4y - y + 2 = 0$

$\Rightarrow \qquad 2y(y-2) - 1(y-2) = 0$

$\Rightarrow \qquad (y-2)(2y-1) = 0$

$\Rightarrow \qquad y = 2 \text{ or } y = 1/2$

If $y = 2$, then

$\Rightarrow \qquad \frac{4x+1}{x+1} = 2$

$4x + 1 = 2(x + 1)$

$\Rightarrow \qquad 4x + 1 = 2x + 2$

$\Rightarrow \qquad 2x = 1 \qquad \Rightarrow \qquad x = 1/2.$

Exercise 2.2

Solve the following equations:

1. $x^4 - 8x^2 - 9 = 0$

2. $4x^4 - 5x^2 + 1 = 0$

3. $3x^{-2} + 7x^{-1} + 5 = 0$

4. $\left(\frac{x-a}{x+a}\right)^2 - 5\left(\frac{x-a}{x+a}\right) + 6 = 0$

5. $(x^2 - 3x + 3)^2 - (x-1)(x-2) = 7$

6. $(x^2 - 5x)^2 - 30(x^2 - 5x) - 216 = 0$

7. $3x^{-2} + 7x^{-1} + 5 = 0$

8. $(x^2 - 5x + 7)^2 - (x-2)(x-3) = 1$

9. $x^{-2} - 12 = -x^{-1}$

10. $\frac{3x+1}{x+1} + \frac{x+1}{3x+1} = \frac{5}{2} (x \in \mathbf{R})$

11. $\frac{x}{1+x} + \frac{1+x}{x} = \frac{13}{6}$

12. $5^{1+x} + 5^{1-x} = 26$

13. $5^{x+1} + 5^{2-x} = 5^3 + 1$

14. $3x + 3^{-x} - 2 = 0$

15. $2^{2x+8} - 8.2^{x+2} + 1 = 0$

16. $\frac{4x-1}{4x+1} + \frac{4x+1}{4x-1} = \frac{10}{3}$

17. $8\sqrt{\frac{x}{x+3}} - \sqrt{\frac{x+3}{x}} = 2$

18. $\sqrt{3x^2 + 1} + \frac{4}{\sqrt{3x^2 + 1}} = 5$

HINT TO SELECTED PROBLEMS

1. $x^4 - 8x^2 - 9 = 0$

Let $x^2 = y \Rightarrow y^2 - 8y - 9 = 0$

$\Rightarrow y^2 - 9y + y - 9 = 0$

$\Rightarrow y(y - 9) + 1(y - 9) = 0$

$\Rightarrow \qquad (y - 9)(y + 1) = 0$

$\Rightarrow \quad y - 9 = 0 \text{ or } y + 1 = 0$

$\Rightarrow \qquad y = 9 \quad \text{or} \quad y = -1$

$\Rightarrow \qquad x^2 = 9 \quad \text{or} \quad x^2 = -1$

$\Rightarrow \qquad x = \pm 3 \text{ or } \quad x = \pm i$

3. Substitute $x^{-1} = y$ to obtain $3y^2 + 7y + 5 = 0$

4. Substitute $\left(\dfrac{x-a}{x+a}\right) = y$ to obtain $y^2 - 5y + 6 = 0$

5. $(x^2 - 3x + 3)^2 - (x - 1)(x - 2) = 7$

$\Rightarrow (x^2 - 3x + 3)^2 - (x^2 - 3x + 2) - 7 = 0$

$\Rightarrow (x^2 - 3x + 3)^2 - (x^2 - 3x + 3) - 6 = 0$

$\Rightarrow$ substitute $x^2 - 3x + 3 = y$, to obtain

$\qquad\qquad y^2 - y - 6 = 0$

9. $2^x = 4^{2x - 1}$

$\Rightarrow 2^x = 2^{2(2x - 1)} \Rightarrow 2^x = (2^x)^4 \cdot 2^{-2}$

$\Rightarrow$ Let $2^x = y \quad \Rightarrow \quad y = \dfrac{y^4}{4}$

11. Substitute $x^{-1} = y$

Then $\qquad\qquad y^2 - 12 = -y$

$\Rightarrow \qquad\qquad y^2 + y - 12 = 0$

$\Rightarrow \qquad\qquad y^2 + 4y - 3y - 12 = 0$

$\Rightarrow \qquad\qquad (y + 4)(y - 3) = 0$

$\Rightarrow \qquad\qquad y = 3 \quad \text{or} \quad y = -4$

$\Rightarrow \qquad\qquad x^{-1} = 3 \quad \text{or } x^{-1} = -4$

$\qquad\qquad\qquad x = \dfrac{1}{3} \quad \text{or} \quad x = -1$

ANSWERS

1. $(\pm 3, \pm i)$

2. $\pm\dfrac{1}{2}, \pm 1$

3. $-\dfrac{7}{6} \pm \dfrac{\sqrt{11}}{10} i$

4. $-2a, -3a$

5. $0, 3, \dfrac{3 \pm i\sqrt{11}}{2}$

6. $2, 3, -4, 9$

7. $\pm 2, \pm 3$

8. $2, 3, \dfrac{5 \pm i\sqrt{3}}{2}$

9. $\dfrac{1}{3}, -\dfrac{1}{4}$

10. $-\dfrac{1}{5}, 1$

11. $-3, 2$

12. $-1, 1$

13. $-1, 2$

14. 0

15. -4

16. $\pm\dfrac{1}{2}$

17. 1

18. $0, 0, \pm\sqrt{5}$

2.4.3 TYPE III: EQUATION OF THE FORM $(x + a)(x + b)(x + c)(x + d) = e$, WHERE a, b, c, d ARE CONSTANT SUCH THAT $a + b = c + d$

WORKING PROCEDURE

In order to solve such type of equations, we use the following steps :

Step 1. Put $x^2 + (a + b)x = y$ and obtain the quadratic equation in y.

Step 2. Solve this quadratic equation for y.

Step 3. Finally get the value of x by putting $x^2 + (a + b)x = y$.

SOLVED EXAMPLES

EXAMPLE 1. *Solve* $(x + 1)(x + 2)(x + 3)(x + 4) = 120$.

SOLUTION. Here, the given equation is

$\qquad (x + 1)(x + 2)(x + 3)(x + 4) = 120$

which can be written as

$\qquad [(x + 1)(x + 4)][(x + 2)(x + 3)] = 120$

$\Rightarrow \qquad (x^2 + 5x + 4)(x^2 + 5x + 6) = 120 \qquad\qquad\qquad \dots(1)$

put $x^2 + 5x = y$, then equation (1) reduces to

$$(y + 4)(y + 6) = 120$$

$\Rightarrow \qquad y^2 + 10y + 24 - 120 = 0$

$\Rightarrow \quad y^2 + 10y - 96 = 0$, which is quadratic in y, solving for y, we get

$$y(y + 16) - 6(y + 16) = 0$$

$\Rightarrow \qquad\qquad (y - 6)(y + 16) = 0$

$\Rightarrow \qquad\qquad\qquad y = 6, -16$

Now, $\qquad\qquad\qquad\qquad y = 6$

$\Rightarrow \qquad\qquad\qquad x^2 + 5x = 6$

$\Rightarrow \qquad\qquad\qquad x^2 + 5x - 6 = 0$

$\Rightarrow \qquad\qquad\qquad\qquad x = 1, -6$

Also, $\qquad\qquad\qquad\qquad y = -16$

$\Rightarrow \qquad\qquad\qquad x^2 + 5x = -16$

$\Rightarrow \qquad\qquad\qquad x^2 + 5x + 16 = 0$

$\Rightarrow \qquad\qquad\qquad\qquad x = \dfrac{-5 \pm i\sqrt{39}}{2}$

Hence, the solution set of the given equation is $\left\{1, -6, \dfrac{-5 \pm i\sqrt{39}}{2}\right\}$.

EXAMPLE 2. *Solve* $(2x - 7)(x^2 - 9)(2x + 5) = 91$.

SOLUTION. Here, the given equation is

$$(2x - 7)(x^2 - 9)(2x + 5) = 91$$

which can be written as

$$(2x - 7)(x - 3)(x + 3)(2x + 5) = 91$$

$\Rightarrow [(2x - 7)(x + 3)][(x - 3)(2x + 5)] = 91$

$\Rightarrow \qquad [2x^2 - x - 21][2x^2 - x - 15] = 91$

Put $2x^2 - x = y$, we get

$$(y - 21)(y - 15) = 91$$

$\Rightarrow \qquad\qquad y^2 - 36y + 315 - 91 = 0$

$\Rightarrow \qquad\qquad y^2 - 36y + 224 = 0$

$\Rightarrow \qquad\qquad (y - 8)(y - 28) = 0$

$\Rightarrow \qquad\qquad\qquad y = 8, 28$

Now, $\qquad\qquad\qquad\qquad y = 8$

$\Rightarrow \qquad\qquad\qquad 2x^2 - x = 8$

$\Rightarrow \qquad\qquad\qquad 2x^2 - x - 8 = 0$

$\Rightarrow \qquad\qquad\qquad\qquad x = \dfrac{1 \pm \sqrt{65}}{4}$

Also, $\qquad\qquad\qquad\qquad y = 28$

$\Rightarrow \qquad\qquad\qquad 2x^2 - x = 28$

$\Rightarrow \qquad\qquad\qquad 2x^2 - x - 28 = 0$

$\Rightarrow \qquad\qquad\qquad\qquad x = \dfrac{1 \pm 15}{4} = 4, -\dfrac{7}{2}$

Hence, the solution set of the given equation is $\left\{4, -\dfrac{7}{2}, \dfrac{1 \pm \sqrt{65}}{4}\right\}$.

EXAMPLE 3. *Solve* $(x^2 - 5x + 7)^2 - (x - 2)(x - 3) = 1$.

SOLUTION. Here, the given equation can be written as

$$(x^2 - 5x + 7)^2 - (x^2 - 5x + 6) = 1$$

Put $x^2 - 5x = y$, we get
$$(y + 7)^2 - (y + 6) = 1$$
$$\Rightarrow \qquad y^2 + 14y + 49 - y - 6 - 1 = 0$$
$$\Rightarrow \qquad y^2 + 13y + 42 = 0$$
$$\Rightarrow \qquad y^2 + 6y + 7y + 42 = 0$$
$$\Rightarrow \qquad y(y + 6) + 7(y + 6) = 0$$
$$\Rightarrow \qquad (y + 6)(y + 7) = 0$$
$$\Rightarrow \qquad y = -6, -7$$
Now, $\qquad\qquad\qquad y = 6$
$$\Rightarrow \qquad x^2 - 5x + 6 = 0$$
$$\Rightarrow \qquad (x - 2)(x - 3) = 0$$
$$\Rightarrow \qquad x = 2, 3$$
also, $\qquad\qquad y = -7 \qquad \Rightarrow \qquad x^2 - 5x + 7 = 0$
$$\Rightarrow \qquad x = \frac{5 \pm i\sqrt{3}}{2}$$

Hence, the solution set of the given equation is $\left\{ 2, 3, \dfrac{5 \pm i\sqrt{3}}{2} \right\}$.

2.4.4 TYPE IV: EQUATION OF THE TYPE $\sqrt{ax + b} + \sqrt{cx + d} = \sqrt{ex + f}$ WHERE a, b, c, d, e, f ARE CONSTANT

WORKING PROCEDURE

In order to solve such type of equation we use the following steps:

Step 1. Square both the sides of the given equation.

Step 2. Put the rational terms on one side and irrational terms on other side.

Step 3. Again squaring and obtain the quadratic terms.

Step 4. Solve the obtained quadratic equation.

SOLVED EXAMPLES

EXAMPLE 1. Solve $\sqrt{x + 5} + \sqrt{x + 21} = \sqrt{6x + 40}$

SOLUTION. Here, the given equation is $\sqrt{x + 5} + \sqrt{x + 21} = \sqrt{6x + 40}$

On squaring both the sides, we get
$$(x + 5) + (x + 21) + 2\sqrt{x + 5}.\sqrt{x + 21} = 6x + 40$$
$$\Rightarrow \qquad 2\sqrt{(x + 5)(x + 21)} = 4x + 14$$
$$\Rightarrow \qquad \sqrt{(x + 5)(x + 21)} = 2x + 7$$

Again squaring, we get
$$(x + 5)(x + 21) = (2x + 7)^2$$
$$\Rightarrow \qquad x^2 + 26x + 105 = 4x^2 + 28x + 49$$
$$\Rightarrow \qquad 3x^2 + 2x - 56 = 0$$
$$\Rightarrow \qquad (3x + 14)(x - 4) = 0$$
$$\Rightarrow \qquad (3x + 14) \text{ or } (x - 4) = 0$$
$$\Rightarrow \qquad x = -\frac{14}{3} \text{ or } x = 4$$

Now, we check, whether the obtained values $x = 4$ and $x = -\dfrac{14}{3}$ satisfy the given equation or not.

When, $x = 4$, we have,

$$\sqrt{4+5} + \sqrt{4+21} = \sqrt{24+40}$$

$\Rightarrow$ $\qquad$ $3 + 5 = 8$, which is true.

Hence, $x = 4$ is solution of the given equation.

When $x = -\dfrac{14}{3}$, then

$$\sqrt{-\dfrac{14}{3}+5} + \sqrt{-\dfrac{14}{3}+21} = \sqrt{6\left(-\dfrac{14}{3}\right)+40}$$

$\Rightarrow$ $\qquad$ $\sqrt{\dfrac{1}{3}} + \sqrt{\dfrac{49}{3}} = \sqrt{\dfrac{36}{3}} \Rightarrow 1 + 7 \neq 6$, which is not true.

Therefore, $x = -\dfrac{14}{3}$ is not a solution. It is an extraneous root and so reject it.

Hence, the solution is 4.

⊜ REMARK

➠ A root which is obtained by solving an equation but does not satisfy it, is called an extraneous root. Such roots enter the equation in the process of squaring because this process is irreversible.

EXAMPLE 2. Solve $\sqrt{(x+5)} + \sqrt{(x+12)} = \sqrt{2x+41}$.

SOLUTION. Here, the given equation is

$$\sqrt{x+5} + \sqrt{x+12} = \sqrt{2x+41}$$

Squaring both the sides, we get

$$(x+5) + (x+12) + 2\sqrt{(x+5)(x+12)} = 2x+41$$

$\Rightarrow$ $\qquad$ $2\sqrt{(x+5)(x+12)} = 24$

$\Rightarrow$ $\qquad$ $\sqrt{(x+5)(x+12)} = 12$

Again squaring, we get

$$(x+5)(x+12) = 144$$

$\Rightarrow$ $\qquad$ $x^2 + 17x + 60 = 144$

$\Rightarrow$ $\qquad$ $x^2 + 17x - 84 = 0$

$\Rightarrow$ $\qquad$ $x = 4, -21$

Therefore, $x = 4$ is a root, because it satisfy the given equation. Also $x = -21$ is an extraneous root, because it does not satisfy the given equation.

EXAMPLE 3. Solve $\sqrt{5x^2 - 6x + 8} - \sqrt{5x^2 - 6x - 7} = 1$

SOLUTION. Here, the given equation reduce to $\sqrt{5x^2 - 6x + 8} - \sqrt{5x^2 - 6x - 7} = 1$

Let $5x^2 - 6x = y$, then given equation reduce to

$$\sqrt{y+8} - \sqrt{y-7} = 1$$

Squaring both the sides, we get

$$(y+8) + (y-7) - 2\sqrt{(y+8)(y-7)} = 1$$

$\Rightarrow$ $\qquad$ $y = \sqrt{y^2 + y - 56}$

Again squaring, we get

$$y^2 = y^2 + y - 56$$

$$\Rightarrow \qquad y = 56$$
$$\Rightarrow \qquad 5x^2 - 6x = 56$$
$$\Rightarrow \qquad 5x^2 - 6x - 56 = 0$$
$$\Rightarrow \qquad (5x + 14)(x - 4) = 0$$
$$\Rightarrow \qquad x = 4, -\frac{14}{5}$$

Since, both the obtained values $\left(x = 4 \text{ and } -\frac{14}{5} \right)$ satisfies the given equation.

Hence, the solution set of the given equation is $\left\{ 4, -\frac{14}{5} \right\}$.

EXAMPLE 4. Solve $\sqrt{x + 4} + \sqrt{x + 20} = 2\sqrt{x + 11}$.

SOLUTION. Here, the given equation is $\sqrt{x + 4} + \sqrt{x + 20} = 2\sqrt{x + 11}$

Squaring both sides, we get

$$(x + 4) + (x + 20) + 2\sqrt{(x + 4)(x + 20)} = 4(x + 11)$$
$$\Rightarrow \qquad 2\sqrt{(x + 4)(x + 20)} = 2x + 20$$
$$\Rightarrow \qquad \sqrt{(x + 4)(x + 20)} = x + 10$$

Again squaring, we get

$$(x + 4)(x + 20) = (x + 10)^2 = x^2 + 20x + 100$$
$$\Rightarrow \qquad x^2 + 24x + 80 = x^2 + 20x + 100$$
$$\Rightarrow \qquad 4x = 20$$
$$\Rightarrow \qquad x = 5$$

Clearly, $x = 5$, satisfy the given equation. Hence, $x = 5$ is the required root of the given equation.

Exercise 2.3

Solve the following equations:

1.   (i) $x(x + 1)(x + 3)(x + 4) = 180$
  (ii) $(2x + 3)(2x + 5)(x - 1)(x - 2) = 30$
  (iii) $(x - 5)(x - 7)(x + 4)(x + 6) = 504$
  (iv) $x(2x + 1)(x - 2)(2x - 3) = 63$
  (v) $(x^2 - 3x - 10)(x^2 - 5x - 6) = 144$
  (vi) $(x + 2)(3x + 4)(3x + 7)(x + 3) = 2600$

2.   (i) $\sqrt{3x - 1} - \sqrt{x - 1} = 2$
  (ii) $\sqrt{2x + 8} + \sqrt{x + 5} = 7$
  (iii) $\sqrt{x + 4} + \sqrt{x + 20} = 2\sqrt{x + 11}$
  (iv) $\sqrt{x + 1} - \sqrt{x - 1} = \sqrt{4x - 1}$
  (v) $\sqrt{5x + 7} - \sqrt{3x + 1} = \sqrt{x + 3}$

HINT TO SELECTED PROBLEMS

1. $(2x + 3)(2x + 5)(x - 1)(x - 2) = 30$
$\Rightarrow [(2x + 3)(x - 1)][(2x + 5)(x - 2)] = 30$
$\Rightarrow \qquad (2x^2 + x - 3)(2x^2 + x - 10) = 30$
Let $2x^2 + x = y$
$\Rightarrow \qquad (y - 3)(y - 10) = 30$
$\Rightarrow \qquad y^2 - 13y + 30 = 30$
$\Rightarrow \qquad y^2 - 13y = 0$
$\Rightarrow \qquad y(y - 13) = 0$

$\Rightarrow \qquad y = 0 \text{ or } y - 13 = 0$
$\Rightarrow \qquad 2x^2 + x = 0 \text{ or } 2x^2 + x - 13 = 0$
$\Rightarrow \qquad x(2x + 1) = 0$

or $\qquad x = \dfrac{-1 \pm \sqrt{(1)^2 - 4 \times 2 \times (-13)}}{2 \times 2}$

$\Rightarrow \qquad x = 0, -\dfrac{1}{2}, \dfrac{-1 \pm \sqrt{1 + 104}}{4}$

$\qquad = \dfrac{-1 \pm \sqrt{105}}{4}$

2. (ii) $\sqrt{2x+8} + \sqrt{x+5} = 7$

Squaring both the sides

$$(2x+8)+(x+5)+2\sqrt{2x+8}\sqrt{x+5} = 49$$

$\Rightarrow \quad 3x+13+2\sqrt{(2x+8)(x+5)} = 49$

$\Rightarrow \quad 3x-36 = -2\sqrt{(2x+8)(x+5)}$

Squaring again, we get

$$(3x-36)^2 = 4(2x+8)(x+5)$$

$\Rightarrow \quad 9(x-12)^2 = 4[2x^2+10x+8x+40]$

$\Rightarrow \quad 9(x^2+144-24x) = 4[2x^2+18x+40]$

$\Rightarrow \quad 9x^2+144\times9-24\times9x = 8x^2+160+72x$

$\Rightarrow \quad x^2-144x+1136 = 0$

On solving we get the required result.

2. (v) $\sqrt{5x+7} - \sqrt{3x+1} = \sqrt{x+3}$

Squaring both the sides

$$(5x+7)+(3x+1)-2\sqrt{5x+7}\sqrt{3x+1} = x+3$$

$\Rightarrow \quad 8x+8-2\sqrt{5x+7}\sqrt{3x+1} = x+3$

$\Rightarrow \quad 7x+5 = 2\sqrt{5x+7}\sqrt{3x+1}$

Squaring both the sides

$$(7x+5)^2 = 4(5x+7)(3x+1)$$

$\Rightarrow 49x^2+25+70x = 4(15x^2+5x+21x+7)$

$\Rightarrow 49x^2+25+70x = 60x^2+100x+28$

$\Rightarrow \quad 11x^2+30x+3 = 0$

$\Rightarrow \quad x = \dfrac{-30\pm\sqrt{900-4\times11\times3}}{2\times11}$

$\qquad = \dfrac{-30\pm\sqrt{900-132}}{22} = \dfrac{-30\pm\sqrt{768}}{22}$

$\qquad = \dfrac{-30\pm16\sqrt{3}}{22} = \dfrac{-15\pm8\sqrt{3}}{11}$

ANSWERS

1. (i) $-6, 2, -2i\sqrt{11}$　　(ii) $0, -\dfrac{1}{2}, \dfrac{-1\pm0\sqrt{105}}{4}$　　(iii) $-7, -2, 3, 8$

(iv) $-\dfrac{3}{2}, 3, \dfrac{3\pm i\sqrt{47}}{4}$　　(v) $-3, 2, 2, 7$　　(vi) $-\dfrac{19}{3}, 2, \dfrac{13\pm\sqrt{-599}}{6}$

2. (i) $1, 5$　　(ii) 4　　(iii) 5　　(iv) $\dfrac{5}{4}$　　(vi) $-\dfrac{1}{11}$

2.5 REMOVAL OF COMMON FACTOR THROUGHOUT IN AN IRRATIONAL EQUATION

WORKING PROCEDURE

In this method, we use the following steps :

Step 1. Factorize each of the given expression.

Step 2. Put common factor equal to zero and find one value of x.

Step 3. Solve the remaining equation by the method discussed in (a).

SOLVED EXAMPLES

EXAMPLE 1. *Solve* $\sqrt{x^2-16} - \sqrt{x^2-5x+4} = x-4 \cdot$

SOLUTION. Here, the given equation is $\sqrt{x^2-16} - \sqrt{x^2-5x+4} = x-4 \cdot$

$\Rightarrow \qquad \sqrt{(x-4)(x+4)} - \sqrt{(x-1)(x-4)} = (x-4)$

$\Rightarrow \qquad \sqrt{(x-4)}[\sqrt{(x+4)} - \sqrt{(x-1)} - \sqrt{(x-4)}] = 0$

Now, either　$\sqrt{(x-4)} = 0 \Rightarrow x-4 = 0 \Rightarrow x = 4$

or　　　　$\sqrt{(x+4)} - \sqrt{(x-1)} - \sqrt{(x-4)} = 0$

$\Rightarrow \qquad\qquad\qquad \sqrt{(x+4)} - \sqrt{(x-1)} = \sqrt{(x-4)}$

On squaring, we get

$$4(x^2 + 3x - 4) = (x + 7)^2 = x^2 + 14x + 49$$

$\therefore$ $\qquad$ $4x^2 + 12x - 16 = x^2 + 14x + 49$

$\Rightarrow$ $\qquad$ $3x^2 - 2x - 65 = 0$

$\Rightarrow$ $\qquad$ $x = \dfrac{2 \pm \sqrt{4 + 780}}{6} = \dfrac{2 \pm 28}{6} = 5, -\dfrac{13}{3}$

Here, it is clear that $x = 5$, satisfy the given equation. Although $x = -\dfrac{13}{3}$ does not

satisfy the given equation, therefore $x = -\dfrac{13}{3}$ is an extraneous root.

Hence, the solution set of the given equation is {4, 5}.

EXAMPLE 2. *Solve* $\sqrt{x^2 + 2x - 3} + \sqrt{x^2 - x} = \sqrt{5(x - 1)}$.

SOLUTION. Here, the given equation is $\sqrt{x^2 + 2x - 3} + \sqrt{x^2 - x} = \sqrt{5(x - 1)}$

$\Rightarrow$ $\qquad$ $\sqrt{(x - 1)(x + 3)} + \sqrt{x(x - 1)} = \sqrt{5(x - 1)}$

$\Rightarrow$ $\qquad$ $\sqrt{(x - 1)}[\sqrt{x + 3} + \sqrt{x} - \sqrt{5}] = 0$

Now, either $\sqrt{(x - 1)} = 0 \Rightarrow x - 1 = 0 \Rightarrow x = 1$

or $\sqrt{x + 3} + \sqrt{x} - \sqrt{5} = 0 \Rightarrow \sqrt{x + 3} + \sqrt{x} = \sqrt{5}$

Again squaring, we get

$$(x + 3) + x + 2\sqrt{x}\sqrt{x + 3} = 5$$

$\Rightarrow$ $\qquad$ $2\sqrt{x(x + 3)} = 2 - 2x$

Again squaring $\qquad$ $4x(x + 3) = (2 - 2x)^2$

$\Rightarrow$ $\qquad$ $4x^2 + 12x = 4 - 8x + 4x^2$

$\Rightarrow$ $\qquad$ $20x = 4$

Therefore, $\qquad$ $x = \dfrac{4}{20} = \dfrac{1}{5}$

Hence, the solution set of the given equation is $\left\{1, \dfrac{1}{5}\right\}$.

2.6 EQUATION OF THE FORM $ax^2 + bx + c + p\sqrt{ax^2 + bx + c} = q$

WORKING PROCEDURE

In order to solve such type of equation, we use the following steps:

Step 1. Assume $\sqrt{ax^2 + bx + c} = y$ and obtain the quadratic equation in y.

Step 2. Solve the obtained equation for y.

Step 3. Finally obtain the value of x by putting $\sqrt{ax^2 + bx + c} = y$.

SOLVED EXAMPLE

EXAMPLE. *Solve* $x^2 - 4x - 12\sqrt{x^2 - 4x + 19} = -51$.

SOLUTION. Here, the given equation can be written as

$$(x^2 - 4x + 19) - 12\sqrt{x^2 - 4x + 19} + (51 - 19) = 0$$

$$\Rightarrow \qquad (x^2 - 4x + 19) - 12\sqrt{x^2 - 4x + 19} + 32 = 0 \qquad \qquad \ldots(1)$$

Put $\sqrt{x^2 - 4x + 19} = y$ in (1), we get

$\quad$ $y^2 - 12y + 32 = 0$, which is a quadratic equation in y.

Solving for y, we get

Now, $\qquad \qquad \qquad \qquad \qquad y = 4 \Rightarrow \sqrt{x^2 - 4x + 19} = 4$

$\Rightarrow \qquad \qquad \qquad \qquad x^2 - 4x + 19 = 16$

$\Rightarrow \qquad \qquad \qquad \qquad x^2 - 4x + 3 = 0$

$\Rightarrow \qquad \qquad \qquad (x - 1)(x - 3) = 0 \Rightarrow x = 1, 3$

Also, $\qquad \qquad \qquad \qquad \qquad y = 8 \Rightarrow \sqrt{x^2 - 4x + 19} = 8$

$\Rightarrow \qquad \qquad \qquad \qquad x^2 - 4x + 19 = 64$

$\Rightarrow \qquad \qquad \qquad \qquad x^2 - 4x - 45 = 0$

$\Rightarrow \qquad \qquad \qquad \quad x^2 + 5x - 9x - 45 = 0$

$\Rightarrow \qquad \qquad \qquad \quad x(x + 5) - 9(x + 5) = 0$

$\Rightarrow \qquad \qquad \qquad (x + 5)(x - 9) = 0 \Rightarrow x = 9, -5$

Hence, the solution set of the given equation is $\{1, 3, -5, 9\}$

(d) Method of Identity : Equation of the form $\sqrt{ax^2 + bx + c} + \sqrt{dx^2 + ex + f} = k$

In order to solve such type of equation we proceed as the following example:

SOLVED EXAMPLES

EXAMPLE 1. *Solve* $\sqrt{5x^2 - 6x + 8} - \sqrt{5x^2 - 6x - 7} = 1$

SOLUTION. Here, the given equation is

$$\sqrt{5x^2 - 6x + 8} - \sqrt{5x^2 - 6x - 7} = 1$$

Let $\sqrt{5x^2 - 6x + 8} = A$ and $\sqrt{5x^2 - 6x - 7} = B$, then given equation reduces to

$$A - B = 1 \qquad \qquad \ldots(1)$$

Also, $\qquad \qquad A^2 - B^2 = (5x^2 - 6x + 8) - (5x^2 - 6x - 7) = 15$

$\Rightarrow \qquad \qquad (A - B)(A + B) = 15(1)$

$\Rightarrow \qquad \qquad \qquad 1.(A + B) = 15 \qquad \qquad \ldots(2) \qquad$ [Using (1)]

Solving (1) and (2), we get

$$2A = 16$$

$\Rightarrow \qquad \qquad \qquad \qquad A = 8$

$\Rightarrow \qquad \qquad \qquad \sqrt{5x^2 - 6x + 8} = 8$

$\Rightarrow \qquad \qquad \qquad 5x^2 - 6x + 8 = 64$

$\Rightarrow \qquad \qquad \qquad 5x^2 - 6x - 56 = 0$

$\Rightarrow \qquad \qquad \qquad \qquad x = \dfrac{6 \pm \sqrt{36 + 1120}}{10} = \dfrac{6 \pm 34}{10}$

$\Rightarrow$ $\qquad$ $x = 4, -\dfrac{14}{5}$

Hence, the solution set of the given equation is $\left\{ 4, -\dfrac{14}{5} \right\}$.

EXAMPLE 2. *Solve* $\sqrt{2x^2 - 3x - 5} - \sqrt{x^2 - 3x + 4} = x + 3$.

SOLUTION. Here, the given equation is $\sqrt{2x^2 - 3x - 5} - \sqrt{x^2 - 3x + 4} = (x + 3)$

Let $\sqrt{2x^2 - 3x - 5} = A$ and $\sqrt{x^2 - 3x + 4} = B$, then given equation reduces to

$$A - B = x + 3 \qquad \qquad \dots(1)$$

Also, $\qquad A^2 - B^2 = (2x^2 - 3x - 5) - (x^2 - 3x + 4) = x^2 - 9$

$\Rightarrow \qquad (A - B)(A + B) = x^2 - 9$

$\Rightarrow \qquad (x + 3)(A + B) = (x - 3)(x + 3) \qquad \dots(2) \qquad$ [Using (1)]

$\Rightarrow \qquad \qquad A + B = x + 3$

$\qquad \qquad \qquad 2A = 2x$

$\Rightarrow \qquad \qquad \quad A = x$

Therefore, $\qquad \sqrt{2x^2 - 3x - 5} = x$

On squaring, $\qquad 2x^2 - 3x - 5 = x^2$

$\Rightarrow \qquad \qquad x^2 - 3x - 5 = 0$

$\Rightarrow \qquad \qquad x = \dfrac{3 \pm \sqrt{9 + 20}}{2} = \dfrac{3 \pm \sqrt{29}}{2}$

Hence, the solution set of the given equation is $\left\{ \dfrac{3 \pm \sqrt{29}}{2} \right\}$.

Exercise 2.4

Solve the following equations:

1. $\sqrt{x^2 - 5x + 6} + \sqrt{x^2 - 9} = \sqrt{2x^2 - 11x + 15}$

2. $\sqrt{x^2 + 2x - 3} + \sqrt{x^2 - x} = \sqrt{5(x - 1)}$

3. $\sqrt{2x^2 - 5x - 2} - \sqrt{2x^2 - 5x - 9} = 1$

4. $\sqrt{4x^2 - 7x - 15} - \sqrt{x^2 - 3x} = \sqrt{x^2 - 9}$

5. $\dfrac{3x - 2}{2} + \sqrt{2x^2 - 5x + 3} = \dfrac{(x + 1)^2}{3}$

6. $\sqrt{3x^2 - 4x + 34} + \sqrt{3x^2 - 4x - 11} = 9$

7. $\sqrt{x^2 + ax + b} - \sqrt{x^2 + 9x + 6} = \sqrt{b} + \sqrt{c}$

8. $x(x + 3) + 3\sqrt{2x^2 + 6x + 5} = 25$

HINT TO SELECTED PROBLEMS

1. $\sqrt{x^2 - 5x + 6} + \sqrt{x^2 - 9} = \sqrt{2x^2 - 11x + 15}$

$\Rightarrow \sqrt{(x - 2)(x - 3)} + \sqrt{(x - 3)(x + 3)}$

$\qquad \qquad = \sqrt{2x^2 - 6x - 5x + 15}$

$\Rightarrow \sqrt{(x - 3)}[\sqrt{x - 2} + \sqrt{x + 3}]$

$\qquad \qquad = \sqrt{2x(x - 3) - 5(x - 3)}$

$\Rightarrow \sqrt{(x - 3)}[\sqrt{x - 2} + \sqrt{x + 3}]$

$\qquad \qquad = \sqrt{(x - 3)(2x - 5)}$

$\Rightarrow \sqrt{(x - 3)}[\sqrt{x - 2} + \sqrt{x + 3} - \sqrt{2x - 5}] = 0$

either $\sqrt{(x - 3)} = 0 \Rightarrow x = 3$

or $\sqrt{x - 2} + \sqrt{x + 3} - \sqrt{2x - 5}$

Squaring both the sides,

$x - 2 + x + 3 + 2\sqrt{x - 2}\sqrt{x + 3} = 2x - 5$

$\Rightarrow \qquad 1 + 5 = -2\sqrt{x - 2}\sqrt{x + 3}$

$\qquad \qquad 6 = -2\sqrt{x - 2}\sqrt{x + 3}$

$\Rightarrow \qquad 3 = -\sqrt{x - 2}\sqrt{x + 3}$

Squaring again, we get

$\qquad \qquad 9 = (x - 2)(x + 3)$

$\Rightarrow \qquad 9 = x^2 - x - 6$

$\Rightarrow \qquad x^2 - x - 15 = 0$

$\Rightarrow \qquad x = \dfrac{1 \pm \sqrt{(1)^2 + 4 \times 1 \times 15}}{2 \times 1}$

$\Rightarrow \qquad x = \dfrac{1 \pm \sqrt{61}}{2}$

4. $\quad \sqrt{4x^2 - 7x - 15} - \sqrt{x^2 - 3x} = \sqrt{x^2 - 9}$

$\Rightarrow \sqrt{4x^2 - 12x + 5x - 15} - \sqrt{x(x-3)}$
$\qquad\qquad = \sqrt{(x-3)(x+3)}$

$\Rightarrow \sqrt{4x(x-3) + 5(x-3)} - \sqrt{x(x-3)}$
$\qquad\qquad = \sqrt{(x-3)(x+3)}$

$\Rightarrow \sqrt{(x-3)(4x+5)} - \sqrt{x(x-3)}$
$\qquad\qquad = \sqrt{(x-3)(x+3)}$

$\Rightarrow \sqrt{x-3}\,[\sqrt{4x+5} - \sqrt{x} - \sqrt{(x+3)}] = 0$

either $\qquad \sqrt{x-3} = 0 \Rightarrow x = 3$

or $\qquad \sqrt{4x+5} - \sqrt{x} - \sqrt{x+3} = 0$

$\Rightarrow \qquad \sqrt{4x+5} - \sqrt{x} = \sqrt{x+3}$

Squaring both the sides

$\qquad 4x + 5 + x - 2\sqrt{x}\sqrt{4x+5} = x + 3$

$\qquad 4x + 5 - 3 = 2\sqrt{x}\sqrt{4x+5}$

$\Rightarrow \qquad 4x + 2 = 2\sqrt{x}\sqrt{4x+5}$

$\Rightarrow \qquad 2x + 1 = \sqrt{x}\sqrt{4x+5}$

$\Rightarrow \qquad 4x^2 + 1 + 4x = x(4x+5)$

$\Rightarrow \qquad 4x^2 + 1 + 4x = 4x^2 + 5x$

$\Rightarrow \qquad 5x - 4x = 1$

$\Rightarrow \qquad x = 1$

ANSWERS

1. $3, \dfrac{1 \pm \sqrt{61}}{2}$	**2.** $1, \dfrac{1}{5}$	**3.** $-2, \dfrac{9}{4}$	**4.** 3
5. $2, \dfrac{1}{2}$	**6.** $3, -\dfrac{5}{3}$	**7.** $0, a$	**8.** $2, -5$

2.7 RECIPROCAL EQUATIONS

An equation which remains unchanged when x is changed to $\dfrac{1}{x}$ is called reciprocal equation.

⇶ REMARKS

⇒ The roots of a reciprocal equations occur in pairs.

⇒ If a is a root of reciprocal equation, then $\dfrac{1}{a}$ also a root of the given equation.

WORKING PROCEDURE

To solve such type of reciprocal equation, we use the following steps:

(A) For Even Degree (say, degree = 4)

Step 1. Divide both sides by x^2.

Step 2. Put $x + \dfrac{1}{x}$ or $x - \dfrac{1}{x} = y$ and solve for y.

Step 3. Finally obtained the value of x by putting $x + \dfrac{1}{x} = y$ or $x - \dfrac{1}{x} = y$.

(B) For Odd Degree

Step 1. If the coefficient of the terms equidistant from the starting and end are equal in magnitude as well as in sign. Then by inspection, we have that -1 is a root and then taking $(x + 1)$ as a common factor and get even degree equation.

Step 2. If the coefficients of the terms equidistant from the starting and end are equal in magnitude but opposite in sign. Then, by inspection take $x = 1$ is a root and take $(x - 1)$ as a common factor and get even degree equation, which can be easily solved.

SOLVED EXAMPLES

EXAMPLE 1.   *Solve $x^4 - x^3 + 2x^2 - x + 1 = 0$.*

SOLUTION.   Here, the given equation is $x^4 - x^3 + 2x^2 - x + 1 = 0$

Divide throughout by x^2, we get

$$x^2 - x + 2 - \frac{1}{x} + \frac{1}{x^2} = 0$$

$$\Rightarrow \quad \left(x^2 + \frac{1}{x^2}\right) - \left(x + \frac{1}{x}\right) = 2 \qquad \qquad \ldots(1)$$

Put $\left(x + \frac{1}{x}\right) = y, i.e., x^2 + \frac{1}{x^2} = y^2 - 2$ in (1), we get

$$y^2 - 2 - y + 2 = 0$$

$$\Rightarrow \quad y^2 - y = 0$$

$$\Rightarrow \quad y(y - 1) = 0$$

$$\Rightarrow \quad y = 0, 1$$

Now, if $\qquad\qquad\qquad y = 0 \Rightarrow x + \frac{1}{x} = 0$

$$\Rightarrow \quad x^2 + 1 = 0$$

$$\Rightarrow \quad x = \pm i$$

or $\qquad\qquad\qquad y = 1 \Rightarrow x + \frac{1}{x} = 1$

$$\Rightarrow \quad x^2 + 1 = x$$

$$\Rightarrow \quad x^2 - x + 1 = 0$$

$$\Rightarrow \quad x = \frac{1 \pm i\sqrt{3}}{2}$$

Here, the solution set of the given equation is $\left\{\pm i, \dfrac{1 \pm i\sqrt{3}}{2}\right\}$.

EXAMPLE 2.   *Solve $\left(x + \frac{1}{x}\right)^2 - \frac{3}{2}\left(x - \frac{1}{x}\right) = 4, x \neq 0$.*   **[UPTU(B.Pharma)–2004]**

SOLUTION.   We have $\qquad \left(x + \frac{1}{x}\right)^2 - \frac{3}{2}\left(x - \frac{1}{x}\right) = 4$

$$\Rightarrow \quad \left[\left(x - \frac{1}{x}\right)^2 + 4\right] - \frac{3}{2}\left(x - \frac{1}{x}\right) = 4$$

$$\Rightarrow \quad y^2 + 4 - \frac{3}{2}y = 4 \qquad \qquad \left[\text{putting } \left(x - \frac{1}{x}\right) = y\right]$$

$$\Rightarrow \quad y^2 - \frac{3}{2}y = 0 \Rightarrow y\left(y - \frac{3}{2}\right) = 0$$

$$\Rightarrow \quad y = 0 \text{ or } y - \frac{3}{2} = 0$$

$$\Rightarrow \quad y = 0 \text{ or } y = 3/2$$

Now $\qquad\qquad\qquad\qquad y = 0$

$$\Rightarrow \qquad x - \frac{1}{x} = 0 \qquad \Rightarrow \qquad x^2 - 1 = 0$$

$$\Rightarrow \qquad x^2 = 1 \qquad \Rightarrow \qquad x = \pm 1$$

Again

$$y = \frac{3}{2} \Rightarrow x - \frac{1}{x} = \frac{3}{2} \Rightarrow \frac{x^2 - 1}{x} = \frac{3}{2}$$

$$\Rightarrow \qquad 2x^2 - 2 = 3x$$

$$\Rightarrow \qquad 2x^2 - 3x - 2 = 0$$

$$\Rightarrow \qquad 2x(x - 2) + 1(x - 2) = 0 \qquad \Rightarrow \quad (x - 2)(2x + 1) = 0$$

$$\Rightarrow \qquad x - 2 = 0 \text{ or } 2x + 1 = 0 \qquad \Rightarrow \quad x = 2 \text{ or } x = -\frac{1}{2}$$

Hence, the solution set of the given equation is $\left\{ 1, -1, 2, -\frac{1}{2} \right\}$.

EXAMPLE 3.   *Solve* $\left(x + \dfrac{1}{x} \right)^2 - 2\left(x - \dfrac{1}{x} + 4 \right) - 11 = 0$

SOLUTION.   Here, the given equation is

$$\left(x + \frac{1}{x} \right)^2 - 2\left(x - \frac{1}{x} + 4 \right) - 11 = 0 \qquad \qquad \text{...(1)}$$

Put $x - \dfrac{1}{x} = y$, *i.e.*, $\left(x + \dfrac{1}{x} \right)^2 = \left(x - \dfrac{1}{x} \right)^2 + 4 = y^2 + 4$

Then, equation (1) becomes $(y^2 + 4) - 2(y + 4) - 11 = 0$

or $\qquad\qquad\qquad\qquad y^2 - 2y - 15 = 0 \qquad \Rightarrow \qquad (y - 5)(y + 3) = 0$

$$\Rightarrow \qquad\qquad\qquad y = -3, 5$$

Now, if $\qquad\qquad\qquad\qquad y = 5 \Rightarrow x - \dfrac{1}{x} = 5$

$$\Rightarrow \qquad\qquad x^2 - 5x - 1 = 0 \qquad \Rightarrow \qquad\qquad x = \frac{5 \pm \sqrt{29}}{2}$$

or if $\qquad\qquad y = -3$, then $x - \dfrac{1}{x} = -3 \qquad \Rightarrow \quad x^2 + 3x - 1 = 0$

$$\Rightarrow \qquad\qquad\qquad\qquad x = \frac{-3 \pm \sqrt{9 + 4}}{2} = \frac{-3 \pm \sqrt{13}}{2}$$

Hence, the solution set of the given equation is $\left\{ \dfrac{5 \pm \sqrt{29}}{2}, \dfrac{-3 \pm \sqrt{13}}{2} \right\}$

Exercise 2.5

Solve the following equations:

1. $2x^4 - x^3 - 11x^2 - x + 2 = 0$

2. $2x^3 - 3x^2 - 3x + 2 = 0$

3. $4x^4 - 4x^3 - 7x^2 - 4x + 4 = 0$

4. $\left(x + \dfrac{1}{x} \right)^2 - 6\left(x - \dfrac{1}{x} + 1 \right) - 5 = 0$

5. $2x^4 - 2x^3 + 14x^2 - 9x + 2 = 0$

6. $x^4 + 1 - 3(x^3 + x) = 2x^2$

7. $\left(x + \dfrac{1}{x} \right)^2 - \dfrac{3}{2}\left(x - \dfrac{1}{x} \right) - 4 = 0$

8. $2x^5 - 3x^4 - 10x^2 + 3x - 2 = 0$

9. $x^6 - x^5 + x^4 - x^2 - x - 1 = 0$

HINT TO SELECTED PROBLEMS

1. $2x^4 - x^3 - 11x^2 - x + 2 = 0$

divide by x^2, we get

$$2x^2 - x - 11 - \frac{1}{x} + \frac{2}{x^2} = 0$$

$$2\left(x^2 + \frac{1}{x^2}\right) - \left(x + \frac{1}{x}\right) - 11 = 0$$

$$\Rightarrow 2\left(x^2 + \frac{1}{x^2} + 2\right) - 4 - \left(x + \frac{1}{x}\right) - 11 = 0$$

$$\Rightarrow 2\left(x + \frac{1}{x}\right)^2 - \left(x + \frac{1}{x}\right) - 15 = 0$$

Let $\left(x + \frac{1}{x}\right) = y$

$$\Rightarrow \quad 2y^2 - y - 15 = 0$$

$$\Rightarrow \quad 2y^2 - 6y + 5y - 15 = 0$$

$$\Rightarrow \quad 2y(y - 3) + 5(y - 3) = 0$$

$$\Rightarrow \quad (y - 3)(2y + 5) = 0$$

$$\Rightarrow \quad y - 3 = 0 \text{ or } 2y + 5 = 0$$

$$\Rightarrow \quad x + \frac{1}{x} = 3 \text{ or } y = -\frac{5}{2}$$

$$\Rightarrow \quad x^2 + 1 = 3x \text{ or } x + \frac{1}{x} = -\frac{5}{2}$$

$$\Rightarrow \quad x^2 - 3x + 1 = 0 \text{ or } \frac{x^2 + 1}{x} = -\frac{5}{2}$$

$$\Rightarrow \quad x = \frac{+3 \pm \sqrt{9 - 4 \times 1 \times 1}}{2 \times 1} \text{ or } 2x^2 + 2 = -5x$$

$$\Rightarrow \quad = \frac{3 \pm \sqrt{5}}{2} \text{ or } 2x^2 + 5x + 2 = 0$$

$$2x^2 + 4x + x + 2 = 0$$

$$2x(x + 2) + 1(x + 2) = 0$$

$$\Rightarrow \quad x = -2 \text{ or } -\frac{1}{2}$$

4. $\left(x + \frac{1}{x}\right)^2 - 6\left(x - \frac{1}{x} + 1\right) - 5 = 0$

$$\Rightarrow x^2 + \frac{1}{x^2} + 2 - 6\left(x - \frac{1}{x}\right) - 6 - 5 = 0$$

$$\Rightarrow \quad x^2 + \frac{1}{x^2} - 6\left(x - \frac{1}{x}\right) - 9 = 0$$

$$\Rightarrow \quad \left(x^2 + \frac{1}{x^2} - 2\right) - 6\left(x - \frac{1}{x}\right) - 7 = 0$$

$$\Rightarrow \quad \left(x - \frac{1}{x}\right)^2 - 6\left(x - \frac{1}{x}\right) - 7 = 0$$

Let $x - \frac{1}{x} = y$

$$\Rightarrow \quad y^2 - 6y - 7 = 0$$

6. $x^4 + 1 - 3(x^3 + x) = 2x^2$

$$x^4 + 1 - 2x^2 - 3x^3 - 3x = 0$$

Divide by x^2, we get

$$x^2 + \frac{1}{x^2} - 2 - 3x - \frac{3}{x} = 0$$

$$\Rightarrow \left(x^2 + \frac{1}{x^2} - 2\right) - 3\left(x + \frac{1}{x}\right) - 4 = 0$$

$$\Rightarrow \left(x + \frac{1}{x}\right)^2 - 3\left(x + \frac{1}{x}\right) - 4 = 0 \qquad \ldots(1)$$

$\Rightarrow$ Let $x + \frac{1}{x} = y$

Then eqn (1) reduced to

$$y^2 - 3y - 4 = 0$$

ANSWERS

1. $-\frac{1}{2}, -2, \frac{3 \pm \sqrt{5}}{2}$ **2.** $-1, 2, \frac{1}{2}$ **3.** $\frac{1}{2}, 2, \frac{-3 \pm \sqrt{-7}}{4}$ **4.** $\frac{7 \pm \sqrt{53}}{2}, \frac{1 \pm \sqrt{5}}{2}$ **5.** $1, 2, \frac{1}{2}$

6. $2 \pm \sqrt{3}, \frac{1}{2}(1 \pm \sqrt{-3})$ **7.** $-1, -\frac{1}{2}, 1, 2$ **8.** $1, -2, 1, \frac{3 \pm \sqrt{5}}{2}$ **9.** $\pm i, \frac{1 \pm \sqrt{-3}}{2}$

2.8 NATURE OF ROOTS OF A QUADRATIC EQUATION

The roots of the quadratic equation $ax^2 + bx + c = 0$ are

$$\frac{-b \pm \sqrt{b^2 - 4ac}}{2a} \qquad \ldots(1)$$

Here the expression $D = b^2 - 4ac$ is called discriminant.

The nature of the roots, depend upon the value of D as given below :

(a) If $b^2 - 4ac \geq 0$, then roots are real

    (i) If $b^2 - 4ac > 0$, then roots are real and distinct

    (ii) If $b^2 - 4ac = 0$, then roots of the equation are real and equal.

        In this case, each root $= \dfrac{-b \pm 0}{2a} = \dfrac{-b}{2a}$

    (iii) Also, if $b^2 - 4ac$ is a perfect square, then the roots are rational and in this case it

        can't be a perfect square, then the roots are irrational.

(b) If $b^2 - 4ac < 0$ then $\sqrt{b^2 - 4ac}$ is imaginary.

Therefore, the roots are imaginary and unequal.

2.9 SYMMETRIC FUNCTION OF THE ROOTS

If α, β are the roots of the quadratic equation $ax^2 + bx + c = 0$, then

$$\alpha + \beta = -\frac{b}{a} \text{ (sum of the roots)} \qquad \qquad \text{...(1)}$$

and

$$\alpha\beta = \frac{c}{a} \text{ (product of the roots)} \qquad \qquad \text{...(2)}$$

Therefore,

$$\alpha^2 + \beta^2 = (\alpha + \beta)^2 - 2\alpha\beta = \frac{b^2 - 2ac}{a^2}$$

$$\alpha - \beta = \sqrt{\{(\alpha + \beta)^2 - 4\alpha\beta\}} = \frac{\sqrt{b^2 - 4ac}}{a}$$

$$\alpha^2 - \beta^2 = (\alpha + \beta)(\alpha - \beta) = -\frac{b}{a}\frac{\sqrt{b^2 - 4ac}}{a}$$

$$\alpha^3 + \beta^3 = (\alpha + \beta)^3 - 3\alpha\beta(\alpha + \beta) = \frac{b^3}{a^3} + \frac{3bc}{a^2} = \frac{-b(b^2 - 3ac)}{a^3}$$

and

$$\alpha^4 + \beta^4 = (\alpha^2 + \beta^2)^2 - 2\alpha^2\beta^2 = \left(\frac{b^2 - 2ac}{a^2}\right) - \frac{2c^2}{a^2}$$

2.9.1 THE SIGN OF EXPRESSION $(x - a)(x - b), (a < b)$

Here we have the following cases:

Case I. $(x - a)(x - b) = +\text{ve}$

It is possible if either both factors are positive or both negative

i.e., if $x - a > 0, x - b > 0,$ *i.e.*, $x > a, x > b$

Therefore, $x > b \, (\because a > b)$ ...(1)

Then, $x < a$ ...(2)

From (1) and (2), we conclude that $(x - a)(x - b)$ is positive if either $x < a$ or $x > b$. In other words, we mean that x does not lie between a and b $(a < b)$.

Case II. $(x - a)(x - b) = -\text{ve}$

It is possible if one factor is positive and the other is negative.

Let $x - a = +\text{ve} > 0, x - b = -\text{ve} < 0$

Therefore, $x > a, x < b$ or $a < x < b,$ *i.e.*, x lies between a and b $(a < b)$

or $x - a = -ve < 0, x - b = + ve > 0$

i.e., $x < a$ and $x > b$ which is not possible.

Therefore, $(x - a)(x - b) = $ positive if x does not lie between a and b and is negative if x lies between a and b.

For example:

Consider the expression.

$$(x + 3)(x - 5) = [x - (-3)](x - 5)$$

Here, $a = -3, b = 5 \Rightarrow a < b$

It is positive if x does not lie between -3 and 5 and is negative if x lies between -3 and 5.

2.9.2 THE SIGN OF EXPRESSION $ax^2 + bx + c$

Here, we consider the following cases:

Case I.

Let the roots of the equation $ax^2 + bx + c = 0$. (1) be imaginary.

Then we can write

$$ax^2 + bx + c = a\left(x^2 + \frac{b}{a}x + \frac{c}{a}\right) = a\left[\left(x + \frac{b}{2a}\right)^2 + \frac{4ac - b^2}{4a^2}\right]$$

Now, since the roots of equation (1) are imaginary, therefore $b^2 - 4ac < 0$, i.e., $4ac - b^2 > 0$.

Hence, the expression $\left(x + \frac{b}{2a}\right)^2 + \frac{4ac - b^2}{4a^2}$ is positive for all real values of x.

Therefore, $ax^2 + bx - c$ has same sign for all real value of x.

Case II.

Let the roots of the equations (1) are real and distinct, denoted by α and β. Let $\alpha > \beta$, then we have the identity.

$$ax^2 + bx + c = 0 = a(x - \alpha)(x - \beta) \qquad \qquad ...(3)$$

If $\beta < x < \alpha$, then $x - \alpha > 0$ and $x - \beta < 0$ so that $(x - \alpha)(x - \beta) < 0$. It follows that the sign of $ax^2 - bx + c$ is opposite to that of a.

If $x > \alpha$ or $x < \beta$, then $(x - \alpha)(x - \beta) > 0$ since the factor $(x - \alpha)$ and $(x - \beta)$ are either both positive or both negative.

Hence, in this case the sign of $ax^2 + bx + c$ is the same as that of a.

Case III.

Let the roots a, b be equal. Then

$$ax^2 + bx + c = (x - \alpha)^2$$

and $(x - \alpha)^2$ is positive for all real values of x and therefore, $ax^2 + bx + c$ has the same sign as a.

⬡ REMARK

➠ From above three cases, we conclude that, for all real values of x, the expression $ax^2 + bx + c$ has the same sign as a except when the roots of the equation $ax^2 + bx + c = 0$ are real and unequal, and x has a value lying between them.

2.10 RELATION BETWEEN ROOTS AND COEFFICIENTS

Let us consider the quadratic equation

$$ax^2 + bx + c = 0, a \neq 0, a, b, c \in R$$

2.10.1 TO FIND THE SUM AND PRODUCT OF THE ROOTS IN TERMS OF THE COEFFICIENTS a, b AND c

Consider the quadratic equation

$$ax^2 + bx + c = 0, (a \neq 0) \qquad \qquad ...(1)$$

If α, β be the roots, then by the theory of equation, we have

$$\alpha = \frac{-b + \sqrt{b^2 - 4ac}}{2a} \text{ and } \beta = \frac{-b - \sqrt{b^2 - 4ac}}{2a}$$

(i) The sum of the roots

$$= \alpha + \beta = \frac{-b + \sqrt{b^2 - 4ac}}{2a} + \frac{-b - \sqrt{b^2 - 4ac}}{2a} = -\frac{b}{a}$$

(ii) The product of the roots

$$= \alpha\beta = \left(\frac{-b + \sqrt{b^2 - 4ac}}{2a} \right) \left(\frac{-b - \sqrt{b^2 - 4ac}}{2a} \right)$$

$$= \frac{(-b)^2 + (\sqrt{b^2 - 4ac})^2}{4a^2} = \frac{b^2 - b^2 + 4ac}{4a^2} = \frac{c}{a}$$

Hence, we have sum of roots $= -\dfrac{b}{a} = -\dfrac{\text{coefficient of } x}{\text{coefficient of } x^2}$

and product of the roots $= \dfrac{c}{a} = \dfrac{\text{Constant term}}{\text{Coefficient of } x^2}$

2.11 FORMATION OF EQUATIONS

To find the equation whose roots are α, β :

Let the equation be

$$ax^2 + bx + c = 0, a \neq 0 \qquad \qquad ...(1)$$

Then by theory of equation,

Sum of roots : $\qquad \alpha + \beta = -\dfrac{b}{a}$

Product of the roots : $\qquad \alpha\beta = \dfrac{c}{a}$

Now equation (1) can be written as

$$x^2 + \frac{b}{a}x + \frac{c}{a} = 0$$

$$\Rightarrow \qquad x^2 - (\alpha + \beta)x + \alpha\beta = 0 \qquad \text{or} \qquad x(x - \alpha) - \beta(x - \alpha) = 0$$

$$\Rightarrow \qquad (x - \alpha)(x - \beta) = 0$$

☞ REMARK

⇒ Let $S = \alpha + \beta$, $P = \alpha\beta$, then the required equation is $x^2 - Sx + P = 0$.

2.12 COMMON ROOTS

(i) Condition for One Common Root

Consider two quadratic equations, such that

$$ax^2 + bx + c = 0 \qquad \qquad ...(1)$$

and $\qquad a'x^2 + b'x + c' = 0$ ...(2)

Let a be the common root then equations (1) and (2) gives

$$a\alpha^2 + b\alpha + c = 0 \qquad ...(3)$$

$$a'\alpha^2 + b'\alpha + c' = 0 \qquad ...(4)$$

Solving (3) and (4), we get

$$\frac{\alpha^2}{bc' - b'c} = \frac{\alpha}{ca' - c'a} = \frac{1}{ab' - a'b} \qquad ...(5)$$

Now taking first two numbers, we get

$$\frac{\alpha^2}{bc' - b'c} = \frac{\alpha}{ca' - c'a} \Rightarrow \alpha = \frac{bc' - b'c}{ca' - c'a} \qquad ...(6)$$

Taking last two numbers, we get

$$\Rightarrow \qquad \alpha = \frac{ca' - c'a}{ab' - a'b}$$

Now (5) and (6) gives.

$$\frac{bc' - b'c}{ca' - c'a} = \frac{ca' - c'a}{ab' - a'b}$$

$$\Rightarrow \qquad (ab' - a'b)(bc' - b'c) = (ca' - c'a)^2$$

which is the required condition for one common root.

(ii) Condition for Both Roots Common

Here, the given equations are

$$ax^2 + bx + c = 0 \qquad ...(1)$$

and $\qquad a'x^2 + b'x + c' = 0$ ...(2)

Let α, β be the common roots, then from (1), we heve

$$\alpha + \beta = -\frac{b}{a} \qquad ...(3)$$

and

$$\alpha\beta = \frac{c}{a} \qquad ...(4)$$

From (2), we have $\qquad \alpha + \beta = -\frac{b'}{a'}$ ...(5)

and

$$\alpha\beta = \frac{c'}{a'} \qquad ...(6)$$

Now (3) and (5) gives

$$-\frac{b}{a} = -\frac{b'}{a'} \Rightarrow \frac{b}{a} = \frac{b'}{a'} \Rightarrow \frac{a}{a'} = \frac{b}{b'} \qquad ...(7)$$

Equation (4) and (6) gives

$$\frac{c}{a} = \frac{c'}{a'} \Rightarrow \frac{a}{a'} = \frac{c}{c'} \qquad ...(8)$$

Combining (7) and (8), we have

$$\frac{a}{a'} = \frac{b}{b'} = \frac{c}{c'}$$

which is the required condition for both common roots.

SOLVED EXAMPLES

EXAMPLE 1.   *If α, β are the roots of $ax^2 + bx + c = 0$, find the value of the following :*

    (i) $\dfrac{1}{a\alpha + b} + \dfrac{1}{a\beta + b}$ (ii) $\dfrac{\beta}{a\alpha + b} + \dfrac{\alpha}{a\beta + b}$

   (iii) $(a\alpha + b)^{-3} + (a\beta + b)^{-3}$ (iv) $(a\alpha + b)^{-2} + (a\beta + b)^{-2}$

SOLUTION.   Since α, β are the roots of $ax^2 + bx + c = 0$

      Then $a\alpha^2 + b\alpha + c = 0 \Rightarrow a\alpha + b = \dfrac{-c}{\alpha}$

             $a\beta^2 + b\beta + c = 0 \Rightarrow a\beta + b = \dfrac{-c}{\beta}$

      Also, $\alpha + \beta = -\dfrac{b}{a}, \alpha\beta = \dfrac{c}{a}$

    (i) $\dfrac{1}{a\alpha + b} + \dfrac{1}{a\beta + b} = -\dfrac{\alpha}{c} - \dfrac{\beta}{c} = -\dfrac{1}{c}(\alpha + \beta) = -\dfrac{1}{c}\left(-\dfrac{b}{a}\right) = \dfrac{b}{ac}$

   (ii) $\dfrac{\beta}{a\alpha + b} + \dfrac{\alpha}{a\beta + b} = -\dfrac{\alpha\beta}{c} - \dfrac{\alpha\beta}{c} = -\dfrac{2}{c} \cdot \dfrac{c}{a} = -\dfrac{2}{a}$

   (iii) $(a\alpha + b)^{-3} + (a\beta + b)^{-3} = -\dfrac{\alpha^3 + \beta^3}{c^3}$

                             $= -\dfrac{1}{c^3}[(\alpha + \beta)^3 - 3\alpha\beta(\alpha + \beta)] = \dfrac{b^3 - 3abc}{a^3 c^3}$

   (iv) $(a\alpha + b)^{-2} + (a\beta + b)^{-2} = \dfrac{\alpha^2 + \beta^2}{c^2}$

                             $= \dfrac{1}{c^2}[(\alpha + \beta)^2 - 2\alpha\beta] = \dfrac{b^2 - 2ac}{a^2 c^2}$

EXAMPLE 2.   *If α and β are the roots of $ax^2 + bx + c = 0$. Find the equation whose roots are as given below :*

    (i) $\dfrac{1}{\alpha + \beta}, \dfrac{1}{\alpha} + \dfrac{1}{\beta}$ (ii) $\alpha^2 + \beta^2, \dfrac{1}{\alpha^2} + \dfrac{1}{\beta^2}$

SOLUTION.   Let α, β be the roots of the given quadratic equation

         $ax^2 + bx + c = 0$

    Then, we have $\alpha + \beta = -\dfrac{b}{a}$ and $\alpha\beta = \dfrac{c}{a}$

    (i) Sum $= \dfrac{1}{\alpha + \beta} + \dfrac{\alpha + \beta}{\alpha\beta} = -\dfrac{a}{b} - \dfrac{b}{c} = -\dfrac{(ac + b^2)}{bc}$

       Product $= \dfrac{1}{\alpha + \beta} \cdot \dfrac{\alpha + \beta}{\alpha\beta} = \dfrac{1}{\alpha\beta} = \dfrac{a}{c}$

    Now, consider the equation

         $x^2 - Sx + P = 0$

$$\Rightarrow \quad x^2 - x\left(-\frac{(ac+b^2)}{bc}\right) + \frac{a}{c} = 0$$

$$\Rightarrow \quad bc.x^2 + (b^2 + ac)x + ab = 0$$

(ii) $S = \alpha^2 + \beta^2 + \frac{1}{\alpha^2} + \frac{1}{\beta^2} = (\alpha^2 + \beta^2)\left(1 + \frac{1}{\alpha^2\beta^2}\right) = \frac{b^2 - 2ac}{a^2} \cdot \frac{a^2 + c^2}{c^2}$

$$P = (\alpha^2 + \beta^2) \cdot \frac{(\alpha^2 + \beta^2)}{\alpha^2\beta^2} = \frac{(b^2 - 2ac)^2}{a^2c^2}$$

Therefore, $x^2 - Sx + P = 0$ gives

$$a^2c^2x^2 - (b^2 - 2ac)(a^2 + c^2)x + (b^2 - 2ac)^2 = 0$$

EXAMPLE 3. *If α, β are the roots of $x^2 - p(x + 1) - c = 0$, show that*
$$(\alpha + 1)(\beta + 1) = 1 - c$$

Hence show that $\dfrac{\alpha^2 + 2\alpha + 1}{\alpha^2 + 2\alpha + c} + \dfrac{\beta^2 + 2\beta + 1}{\beta^2 + 2\beta + c} = 1.$

SOLUTION. Here, the given equation is
$$x^2 - p(x + 1) - c = 0$$
Therefore, $\alpha + \beta = p$ and $\alpha\beta = -(p + c)$ ...(1)
Now $(\alpha + 1)(\beta + 1) = \alpha\beta + (\alpha + \beta) + 1$
$$= -\pi - c + p + 1 = 1 - c \qquad ...(2)$$

Also, $\dfrac{\alpha^2 + 2\alpha + 1}{\alpha^2 + 2\alpha + c} + \dfrac{\beta^2 + 2\beta + 1}{\beta^2 + 2\beta + c} = \dfrac{(\alpha + 1)^2}{(\alpha + 1)^2 - (1 - c)} + \dfrac{(\beta + 1)^2}{(\beta + 1)^2 - (1 - c)}$ [Using (1)]

$$= \frac{\alpha + 1}{\alpha - \beta} + \frac{\beta + 1}{\beta - \alpha} = \frac{(\alpha + 1) - (\beta + 1)}{\alpha - \beta} = 1$$

EXAMPLE 4. *If α be a root of the equation $4x^2 + 2x - 1 = 0$, prove that $4\alpha^3 - 3\alpha$ is the other root.*
[UPTU(B.Pharma)–2002]

SOLUTION. Given α be a root of the equation $4x^2 + 2x - 1 = 0$.
Then $4\alpha^2 + 2\alpha - 1 = 0$ ...(1)
Let β be the other root of the given equation

Then $\alpha + \beta = -\dfrac{1}{2} \Rightarrow \beta = -\dfrac{1}{2} - \alpha$

We have to show $\beta = 4\alpha^3 - 3\alpha$

Now $4\alpha^3 - 3\alpha = \alpha(4\alpha^2 + 2\alpha - 1) - \dfrac{1}{2}(4\alpha^2 + 2\alpha - 1) - \alpha - \dfrac{1}{2}$

$$= \alpha.0 - \frac{1}{2}0 - \alpha - \frac{1}{2} = -\alpha - \frac{1}{2} = \beta$$

Hence $4\alpha^3 - 3\alpha$ is the other root of the given equation.

EXAMPLE 5. *Two students solve an equation. In solving, one commits a mistake in constant term and find the roots 8 and 2. Other commits a mistake in the coefficient of x and find the roots −9 and −1. Find the correct roots.*

SOLUTION. Let the correct equation be

$$x^2 + ax + b = 0 \qquad \ldots(1)$$

Roots found by first students are 8, 2, i.e., $S = 10, P = 16$

$\therefore$ Equation is $x^2 - 10x + 16 = 0 \qquad \ldots(2)$

Since, the committed mistake only in constant term. $\therefore a = -10$

Roots found by second student are $-9, -1$, i.e., $S = -10, P = 9$

Therefore equation is $x^2 + 10x + 9 = 0$

Since the committed mistake is in the coefficient of x. Therefore, $b = 9$

Having found $a = -10, b = 9$, the required equation is

$$x^2 - 10x + 9 = 0 \text{ or } (x - 9)(x - 1) = 0$$

i.e., $\qquad\qquad x = 1, 9$

Hence, the correct roots are 1 and 9.

EXAMPLE 6. *If α, β be the roots of $ax^2 + bx + c = 0$ and γ, δ those of $lx^2 + mx + n = 0$, then find the equation whose roots are $\alpha\gamma + \beta\delta$ and $\alpha\delta + \beta\gamma$.*

SOLUTION. Here, the given equations are

$$ax^2 + bx + c = 0 \qquad \ldots(1)$$

and $\qquad lx^2 + mx + n = 0 \qquad \ldots(2)$

Now, we have $\alpha + \beta = -\dfrac{b}{a}, \alpha\beta = \dfrac{c}{a}, \gamma + \delta = -\dfrac{m}{l}, \gamma\delta = \dfrac{n}{l}$

Now, $\qquad \alpha^2 + \beta^2 = (\alpha + \beta)^2 - 2\alpha\beta = \dfrac{b^2 - 2ac}{a^2}$

and $\qquad \gamma^2 + \delta^2 = \dfrac{m^2 - 2nl}{l^2}$

Now, $\qquad S = (\alpha\gamma + \beta\delta) + (\alpha\delta + \beta\gamma) = \alpha(\gamma + \delta) + \beta(\gamma + \delta)$

$$= (\alpha + \beta)(\gamma + \delta) = \dfrac{bm}{al}$$

and $\qquad P = (\alpha\gamma + \beta\delta)(\alpha\delta + \beta\gamma) = \alpha^2\gamma\delta + \alpha\beta\delta^2 + \alpha\beta\gamma^2 + \beta^2\gamma\delta$

$$= (\alpha^2 + \beta^2)(\gamma\delta) + \alpha\beta(\gamma^2 + \delta^2) = \dfrac{n}{l}\left(\dfrac{b^2 - 2ac}{a^2}\right) + \dfrac{c}{a}\left(\dfrac{m^2 - 2nl}{l^2}\right)$$

$$= \dfrac{nl(b^2 - 2ac) + ac(m^2 - 2nl)}{a^2l^2} = \dfrac{b^2nl + m^2ac - 4acnl}{a^2l^2}$$

Therefore, the required equation is given by

$$x^2 - xS + P = 0$$

$$\Rightarrow \quad x^2 - x\left(\dfrac{bm}{al}\right) + \dfrac{b^2nl + m^2ac - 4acnl}{a^2l^2} = 0$$

$$\Rightarrow \quad a^2l^2x^2 - xalbm + b^2nl + m^2ac - 4acnl = 0$$

EXAMPLE 7. *If p and q be the roots of $2x^2 - 6x + 3 = 0$, find the value of*

$$(p^3 + q^3) - 3pq(p^2 + q^2) - 3pq(p + q)$$

SOLUTION. Since p and q are roots of $2x^2 - 6x + 3 = 0$

Therefore, sum of the roots, $p + q = -\dfrac{-6}{2} = 3$

and product of the roots, $pq = \dfrac{3}{2}$

Therefore, $(p^3 + q^3) - 3pq(p^2 + q^2) - 3pq(p + q)$

$$= (p + q)^3 - 3pq(p + q) - 3pq[(p + q)^2 - 2pq] - 3pq(p + q)$$

$$= 27 - 3.\dfrac{3}{2}(3) - 3.\dfrac{3}{2}\left[9 - 2.\dfrac{3}{2}\right] - 3.\dfrac{3}{2}.3$$

$$= 27 - \dfrac{27}{2} - 27 - \dfrac{27}{2} = -27$$

EXAMPLE 8. *Solve the equation $x^2 + px + 45 = 0$, given the squared differences of its roots is equal to 144.*

SOLUTION. Here, the equation $x^2 + px + 45 = 0$ ...(1)

Let α, β be its roots such that $\alpha > \beta$

∴ $\qquad \alpha + \beta = -p,\ \alpha\beta = 45$

From the given condition $(\alpha - \beta)^2 = 144$

∴ $\qquad (\alpha + \beta)^2 - 4\alpha\beta = 144$

$\Rightarrow \qquad p^2 - 4.45 = 144$

$\Rightarrow \qquad p^2 = 324$

$\Rightarrow \qquad p = \pm 18$

When $p = 18$, equation (1) becomes $x^2 + 18x + 45 = 0$

$\Rightarrow \qquad (x + 3)(x + 15) = 0$

$\Rightarrow \qquad x = -3, -15$

When $p = -18$, equation (1) becomes

$\qquad x^2 - 18x + 45 = 0$

$\Rightarrow \qquad (x - 3)(x - 15) = 0$

$\Rightarrow \qquad x = 3, 15$

Hence, the roots of the given equation is 3 and 15 or –3 and –15.

EXAMPLE 9. *If the sum of the roots of the equation $\dfrac{1}{x+a} + \dfrac{1}{x+b} = \dfrac{1}{c}$ is zero. Show that the product of roots is $-\dfrac{1}{2}(a^2 + b^2)$.*

SOLUTION. Here, the given equation is

$$\dfrac{1}{x+a} + \dfrac{1}{x+b} = \dfrac{1}{c}$$

$\Rightarrow \qquad c(x + b) + c(x + a) = (x + a)(x + b)$

$\Rightarrow \qquad cx + cb + cx + ca = x^2 + ax + bx + ab$

$\Rightarrow \quad x^2 + (a + b - 2c)x + (ab - bc - ca) = 0$

Given that, the sum of its roots is equal to zero.

∴ $\qquad -\dfrac{(a + b - 2c)}{1} = 0$

$\Rightarrow \qquad 2c - a - b = 0 \qquad\qquad$ or $\qquad\qquad c = \dfrac{a+b}{2}$ ...(1)

$$\text{Product of roots} = \frac{ab - bc - ca}{1} = ab - c(a + b)$$

$$= ab - \left(\frac{a + b}{2}\right)(a + b) = \frac{2ab - (a + b)^2}{2}$$

$$= -\frac{1}{2}[(a + b)^2 - 2ab] = -\frac{1}{2}(a^2 + b^2)$$

EXAMPLE 10. *If α, β and γ, δ be the roots of the equations $x^2 + px - r = 0$ and $x^2 + px + r = 0$ respectively, show that $(\alpha - \gamma)(\alpha - \delta) = (\beta - \gamma)(\beta - \delta)$.*

SOLUTION. Since α, β are the roots of $x^2 + px - r = 0$

$$\therefore \qquad\qquad \alpha + \beta = -p \text{ and } \alpha\beta = -r$$

Now since γ, δ are the roots of $x^2 + px + r = 0$

$$\therefore \qquad\qquad \alpha^2 + p\alpha - r = 0 \qquad\qquad\qquad\qquad \dots(1)$$

$$\text{and} \qquad\qquad \beta^2 + p\beta - r = 0 \qquad\qquad\qquad\qquad \dots(2)$$

Subtracting (2) from (1), we get

$$(\alpha^2 - \beta^2) + p(\alpha - \beta) = 0$$

$$\therefore \qquad\qquad \alpha^2 + p\alpha = \beta^2 + p\beta$$

$$\Rightarrow \qquad \alpha^2 - (\gamma + \delta)\alpha = \beta^2 - (\gamma + \delta)\beta \qquad\qquad (\because p = -(\gamma + \delta))$$

$$\Rightarrow \quad \alpha^2 - (\gamma + \delta)\alpha + \gamma\delta = \beta^2 - (\gamma + \delta)\beta + \gamma\delta \quad \text{(by adding } \gamma\delta \text{ on both sides)}$$

$$\Rightarrow \qquad \alpha^2 - \gamma\alpha - \delta\alpha + \gamma\delta = \beta^2 - \beta\gamma - \beta\delta + \gamma\delta$$

$$\therefore \qquad \alpha(\alpha - \gamma) - \delta(\alpha - \gamma) = \beta(\beta - \gamma) - \delta(\beta - \gamma)$$

$$\Rightarrow \qquad\qquad (\alpha - \gamma)(\alpha - \delta) = (\beta - \gamma)(\beta - \delta)$$

EXAMPLE 11. *If α, β are the roots of $x^2 - 2x + 3 = 0$. Form an equation whose roots are $\alpha + 2$, $\beta + 2$.*

SOLUTION. Here, the given equation is

$$x^2 - 2x + 3 = 0 \qquad\qquad\qquad\qquad\qquad\qquad \dots(1)$$

Now, since α, β are the roots of (1), therefore $\alpha + \beta = 2$, $\alpha\beta = 3$

We are to form an equation whose roots $\alpha + 2$, $\beta + 2$

$$S = (\alpha + 2) + (\beta + 2) = (\alpha + \beta) + 4 = 2 + 4 = 6$$

$$\text{and} \qquad P = (\alpha + 2)(\beta + 2) = \alpha\beta + 2(\alpha + \beta) + 4$$

$$= 3 + 2.2 + 4 = 3 + 4 + 4 = 11$$

Hence, the required equation is

$$x^2 - Sx + P = 0$$

$$\Rightarrow \qquad\qquad x^2 - 6x + 11 = 0$$

Exercise 2.6

1. Find the equation whose roots are :

   (i) $2, \dfrac{1}{2}, -5$ (ii) $\dfrac{4}{5}, \dfrac{5}{4}$

   (iii) $-5, -3$ (iv) $2 \pm \sqrt{3}$

   (v) $3 \pm \sqrt{5}$ (vi) $5 \pm \sqrt{7}$

   (vii) $\dfrac{3}{4}, \dfrac{4}{3}$

2. If α, β are the roots of the equation $ax^2 + bx + c = 0$ then find the value of :

   (i) $\dfrac{1}{\alpha^2} + \dfrac{1}{\beta^2}$ (ii) $\dfrac{\alpha^3}{\beta} + \dfrac{\beta^3}{\alpha}$

   (iii) $\alpha^3\beta + \beta^3\alpha$ (iv) $\alpha^3 + \beta^3$

   (v) $\alpha^2 + \beta^2$

3. Find the equation whose roots are α^2, β^2 where α, β are the roots of the equation $x^2 + x + 1 = 0$.

4. If α, β are the roots of the equation $x^2 - 7x + 12 = 0$. Find the equation whose roots are $\dfrac{1}{\alpha}, \dfrac{1}{\beta}$.

5. If α, β are the roots of the equation $ax^2 + bx + c = 0$, then find the equation whose roots are :

(i) α^3, β^3

(ii) $\alpha^3\beta, \beta^3\alpha$

(iii) α^2, β^2

(iv) $\dfrac{\alpha}{\beta}, \dfrac{\beta}{\alpha}$

(v) $\dfrac{1}{\alpha+\beta}, \dfrac{1}{\alpha}+\dfrac{1}{\beta}$

6. If one of the root of $x^2 + px + q = 0$ is the square of the other, then show that
$$p^3 - q(3p - 1) + q^2 = 0$$

7. (i) If α be a root of the equation $4x^2 + 2x - 1 = 0$, prove that $4\alpha^3 - 3\alpha$ is the other root.

(ii) Form a quadratic equation whose roots are $\dfrac{a}{\sqrt{a \pm \sqrt{a - b}}}$.

8. If α, β be the root of $x^2 - px + q = 0$ and α', β' be the roots of $x^2 - p'x + q = 0$, find the value of
$$(\alpha - \alpha')^2 + (\beta - \alpha')^2 + (\alpha - \beta')^2 + (\beta - \beta')^2$$

9. If α and β are the roots of $x^2 + px + 1 = 0$ and γ, δ are the roots of $x^2 + qx + 1 = 0$,

show that
$$q^2 - p^2 = (\alpha - \gamma)(\beta - \gamma)(\alpha + \delta)(\beta + \delta)$$

10. If the roots $px^2 + qx + 2 = 0$ are reciprocals of each other, then

(a) $p = 0$

(b) $p = -2$

(c) $q = 0$

(d) $p = 2$

11. If the ratio of the roots of the equation $x^2 + px + q = 0$ be equal to ratio of the roots of $x^2 + lx + m = 0$, then prove that $p^2m = l^2q$.

12. If the ratio of the roots of the equation $lx^2 + mx + n = 0$ be $p : q$, then prove that
$$\sqrt{\dfrac{p}{q}} + \sqrt{\dfrac{q}{p}} + \sqrt{\dfrac{n}{l}} = 0$$

13. (a) Find the value of p for which $x + 1$ is a factor of
$$x^4 + (p - 3)x^3 - (3p - 5)x^2 + (2p - q)x + 6$$
Find the remaining factors for this value of p.

(b) If $x^2 - 3x + 2$ is a factor of $x^4 - px^2 + q = 0$, prove $p = 5, q = 4$.

14. Knowing that 2 and 3 are the roots of the equation $2x^3 + mx^2 - 13x + n = 0$, determine m and n and find the third root of the equation.

15. Find all the roots of the equation $4x^4 - 24x^3 + 57x^2 + 18x - 45 = 0$ if one of them is $3 + i\sqrt{6}$.

HINT TO SELECTED PROBLEMS

3. $x^2 + x + 1 = 0$ Given that its roots are α and β.

So
$$\alpha + \beta = -1$$
$$\alpha\beta = 1$$

We have to form quadratic equation whose roots are α^2 and β^2.

$$\alpha^2 + \beta^2 = (\alpha + \beta)^2 - 2\alpha\beta$$
$$= (-1)^2 - 2 \times 1 = -1$$
$$\alpha^2\beta^2 = (\alpha\beta)^2 = (1)^2 = 1$$

Required equation
$$\Rightarrow \quad x^2 - (\alpha^2 + \beta^2)x + \alpha\beta = 0$$
$$\Rightarrow \quad x^2 + 1x + 1 = 0$$
$$\Rightarrow \quad x^2 + x + 1 = 0$$

6. Given equation is $x^2 + px + q = 0$

Let the roots be α and α^2

Then
$$\alpha + \alpha^2 = -p$$
$$\Rightarrow \quad \alpha(1 + \alpha) = -p \qquad \qquad ...(1)$$
$$\alpha\alpha^2 = q$$
$$\Rightarrow \quad \alpha^3 = q \qquad \qquad ...(2)$$

Cubing equation (1), we get
$$\alpha^3(1 + \alpha)^3 = -p^3$$
$$\alpha^3[1 + \alpha^3 + 3\alpha(1 + \alpha)] = -p^3$$
$$q[1 + q + 3(-p)] = -p^3$$
$$\Rightarrow \quad q[1 + q - 3p] = -p^3$$
$$\Rightarrow \quad p^3 - q(3p - 1) + q^2 = 0$$

ANSWERS

1. (i) $2x^2 + 5x - 25 = 0$

(ii) $20x^2 - 41x + 20 = 0$

(iii) $x^2 + 8x + 15 = 0$

(iv) $x^2 - 4x + 1 = 0$

(v) $x^2 - 6x + 4 = 0$

(vi) $x^2 - 10x + 18 = 0$

(vii) $12x^2 - 25x + 12 = 0$

2. (i) $\dfrac{b^2 - 2ac}{c^2}$ (ii) $\dfrac{(b^2 - 2ac)^2 - 2a^2c^2}{a^3c}$ (iii) $\dfrac{c(b^2 - 2ac)}{a^3}$

(iv) $\dfrac{-b(b^2 - 3ac)}{a^3}$ (v) $\dfrac{b^2 - 2ac}{a^2}$

3. $x^2 + x + 1 = 0$ **4.** $12x^2 - 7x + 1 = 0$

5. (i) $a^2x^2 + b(b^2 - 3ac)x + c^2 = 0$ (ii) $a^4x^2 - ca(b^2 - 2ac)x + c^4 = 0$

(iii) $a^2x^2 - (b^2 - 2ac)x + c^2 = 0$ (iv) $acx^2 - (b^2 - 2ac)x + ac = 0$

(v) $bcx^2 + (b^2 + ac)x + ab = 0$ **7.** (i) $4\alpha^3 - 3\alpha$ (ii) $bx - 2a\sqrt{ax} + a^2 = 0$

8. $2[p^2 - 2q + p'^2 - 2a' - pp']$ **10.** d **13.** (a) $(x - 1), (x - 2), (x + 3)$

14. $m = -5, n = 30, r = -\dfrac{5}{2}$ **15.** $3 \pm i\sqrt{6}, \pm \dfrac{\sqrt{3}}{2}$

2.13 APPLICATION OF QUADRATIC EQUATIONS

EXAMPLE 1. *Divide 57 into two parts such that their product is 782.*

SOLUTION. Let the required parts be x and $(57 - x)$.

Then their product $= x(57 - x)$

$\therefore \qquad x(57 - x) = 782$

$\Rightarrow \qquad x^2 - 57x + 782 = 0$

$\Rightarrow \qquad x = \dfrac{57 \pm \sqrt{(57)^2 - 3128}}{2} = \dfrac{57 \pm \sqrt{3249 - 3128}}{2}$

$x = \dfrac{57 \pm \sqrt{121}}{2} = \dfrac{57 \pm 11}{2}$

$x = \dfrac{57 + 11}{2} = \dfrac{68}{2} = 34$

or $x = \dfrac{57 - 11}{2} = \dfrac{46}{2} = 23$

Hence, the two parts are 34 and 23.

EXAMPLE 2. *The sum of the squares of two consecutive whole numbers is 61. Find the numbers.*

SOLUTION. Let the required consecutive whole numbers be x and $x + 1$.

$\therefore \qquad x^2 + (x + 1)^2 = 61$

$\Rightarrow \qquad x^2 + x^2 + 1 + 2x = 61$

$\Rightarrow \qquad 2x^2 + 2x - 60 = 0$

$\Rightarrow \qquad x^2 + x - 30 = 0$

$\Rightarrow \qquad (x + 6)(x - 5) = 0$

$\Rightarrow \qquad x = -6 \text{ or } x = 5$

$\therefore \quad x$ is a whole number

$\therefore \qquad x = 5$

Hence, required numbers are x and $x + 1$ as 5 and $5 + 1 = 6$.

EXAMPLE 3. *The sum of the squares of two consecutive odd integers is 394. Find the integers.*

SOLUTION. Let the required integers be $(2x + 1)$ and $(2x + 3)$.

$$(2x + 1)^2 + (2x + 3)^2 = 394$$

$$\Rightarrow \qquad 8x^2 + 16x - 384 = 0$$

$$\Rightarrow \qquad x^2 + 2x - 48 = 0$$

$$\Rightarrow \qquad x^2 + 8x - 6x - 48 = 0$$

$$\Rightarrow \qquad x(x + 8) - 6(x + 8) = 0$$

$$\Rightarrow \qquad (x + 8)(x - 6) = 0$$

$$\Rightarrow \qquad x + 8 = 0 \quad \text{or } x - 6 = 0$$

$$\Rightarrow \qquad x = -8 \text{ or } \quad x = 6$$

When $x = -8$, the required integers are -15 and -13.

When $x = 6$, the required integers are 13 and 15.

Hence, the required integers are $(-15, -13)$ or $(13, 15)$.

EXAMPLE 4. *The sum of two natural numbers is 8. If the sum of their reciprocals is $\dfrac{8}{15}$, find the two numbers.*

SOLUTION. Let the numbers be x and $(8 - x)$.

$$\therefore \qquad \frac{1}{x} + \frac{1}{8 - x} = \frac{8}{15} \qquad \Rightarrow \qquad \frac{8 - x + x}{x(8 - x)} = \frac{8}{15}$$

$$\Rightarrow \qquad \frac{8}{8x - x^2} = \frac{8}{15}$$

i.e., $\qquad\qquad 120 = 64x - 8x^2$

$$\Rightarrow \qquad 8x^2 - 64x + 120 = 0 \qquad \Rightarrow \qquad x^2 - 8x + 15 = 0$$

$$\Rightarrow \qquad (x - 5)(x - 3) = 0 \qquad \Rightarrow \qquad x = 5 \text{ or } x = 3$$

When $x = 5$, the numbers are x and $8 - x = 5$ and 3.

When $x = 3$, the numbers are x and $8 - x = 3$ and 5.

$\therefore \qquad$ Required numbers are 5 and 3.

EXAMPLE 5. *A number consists of two digits whose product is 18. When 27 is subtracted from the number, the digits interchange their places. Find the numbers.*

SOLUTION. Let the tens digit be x. Then the units digit $= \dfrac{18}{x}$

Number formed $= \left(10x + \dfrac{18}{x}\right)$

Number formed on reversing the digits $= \left(10 \times \dfrac{18}{x} + x\right)$

$$\therefore \qquad \left(10x + \frac{18}{x}\right) - 27 = \left(10 \times \frac{18}{x} + x\right)$$

$$\Rightarrow \qquad 9x - \frac{162}{x} - 27 = 0$$

$$\Rightarrow \qquad 9x^2 - 27x - 162 = 0$$

$$\Rightarrow \qquad x^2 - 3x - 18 = 0$$

$$\Rightarrow \qquad x^2 - 6x + 3x - 18 = 0$$

$$\Rightarrow \qquad x(x - 6) + 3(x - 6) = 0$$

$$\Rightarrow \qquad (x - 6)(x + 3) = 0$$
$$\Rightarrow \qquad x = 6 \text{ or } x = -3$$
$$\Rightarrow \qquad x = 6 \qquad \text{[Since a digit can never be negative]}$$

Thus, we have the tens digit = 6, the unit digit $= \dfrac{18}{6} = 3$.

Hence, the required number = 63.

EXAMPLE 6. *Two positive numbers are in ratio 2 : 5. If the difference between the squares of these numbers is 189. Find the numbers.*

SOLUTION. Let the numbers be $2x$ and $5x$.
$$\therefore \qquad (5x)^2 - (2x)^2 = 189$$
$$\Rightarrow \qquad 25x^2 - 4x^2 = 189$$
$$\Rightarrow \qquad 21x^2 = 189$$
$$\Rightarrow \qquad x^2 = \dfrac{189}{21} = 9$$
$$\Rightarrow \qquad x = \pm 3$$

Since the required numbers are positive. $\therefore x = 3$.

And required numbers = $2x$ and $5x = 2 \times 3$ and 5×3
$$= 6 \text{ and } 15$$

EXAMPLE 7. *One year ago, a man was 8 times as old as his son. Now his age is equal to the square of his son's age. Find their present age.*

SOLUTION. Let the son's age one year ago be x years.

Then, the man's age one year ago = $(8x)$ years

Present age of the son = $(x + 1)$ years

Present age of the man = $(8x + 1)$ years
$$\therefore \qquad (8x + 1) = (x + 1)^2$$
$$\Rightarrow \qquad x^2 - 6x = 0$$
$$\Rightarrow \qquad x = 0 \text{ or } x = 6$$
$$\therefore \qquad x = 6 \qquad \text{[Son's age can not be zero]}$$

Present age of the son = $(x + 1)$ years = 7 years

and the present age of the man = $(8x + 1)$ years = 49 years

EXAMPLE 8. *The sides (in cm) of a right triangle are $x - 1$, x and $x + 1$. Find the sides of triangle.*

SOLUTION. It is clear that the largest side $x + 1$ is hypotenuse of that triangle.

According to Pythagoras theorem, we have

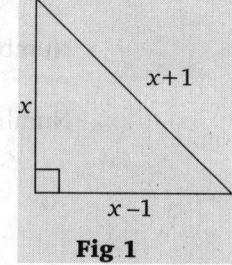

$$x^2 + (x - 1)^2 = (x + 1)^2$$
$$\Rightarrow \qquad x^2 + x^2 - 2x + 1 = x^2 + 2x + 1$$
$$\Rightarrow \qquad x^2 - 4x = 0$$
$$\Rightarrow \qquad x(x - 4) = 0$$
i.e., $\qquad x = 0 \text{ or } x = 4$

Since with $x = 0$ the triangle is not possible, hence $x = 4$.

Sides are $x - 1$, x and $x + 1 = 4 - 1$, 4 and $4 + 1$

Fig 1

i.e., 3 cm, 4 cm and 5 cm.

EXAMPLE 9. *Rs. 250 was divided equally among a certain number of children. If there were 25 more children, each would have received 50 paise less. Find the number of children.*

SOLUTION. Let the number of required children be x.

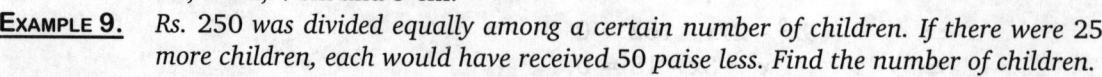

The share of each $= \dfrac{25000}{x}$ paise

If there were $(x + 25)$ children, share of each $= \left(\dfrac{25000}{x + 25}\right)$ paise

$\therefore \qquad \dfrac{25000}{x} - \dfrac{25000}{x + 25} = 50$

$\Rightarrow \qquad \dfrac{1}{x} - \dfrac{1}{x + 25} = \dfrac{1}{500}$

$\Rightarrow \qquad \dfrac{x + 25 - x}{x(x + 25)} = \dfrac{1}{500}$

$\Rightarrow \qquad x(x + 25) = 12500$

$\Rightarrow \qquad x^2 + 25x - 12500 = 0$

$\Rightarrow \qquad (x + 125)(x - 100) = 0$

$\Rightarrow \qquad x = -125 \text{ or } x = 100$

$\Rightarrow \qquad x = 100 \text{[since number of children can not be negative]}$

Hence, the number of children $= 100$.

EXAMPLE 10. *The hypotenuse of right triangle is 1 m less than twice the shortest side. If the third side is 1 m more than the shortest side, find the sides of the triangle.*

SOLUTION. Let the shortest side be x m.

$\therefore$ Hypotenuse $= (2x - 1)$m and third side $= (x + 1)$m

Applying Pythagoras theorem, we get

$$(2x - 1)^2 = x^2 + (x + 1)^2$$

$\Rightarrow \qquad 4x^2 - 4x + 1 = x^2 + x^2 + 2x + 1$

$\Rightarrow \qquad 2x^2 - 6x = 0$

$\Rightarrow \qquad x^2 - 3x = 0$

$\Rightarrow \qquad x(x - 3) = 0$

$\Rightarrow \qquad x = 0 \text{ or } x = 3$

Since, $x = 0$ makes the triangle impossible. Therefore, $x = 3$

and sides of the triangle are x, $2x - 1$ and $x + 1 = 3$, $2 \times 3 - 1$ and $3 + 1$

$$= 3m, 5m \text{ and } 1m$$

EXAMPLE 11. *A passenger train takes 3 hours less for journey of 360 km if its speed is increased by 10 km/hr. What is the usual speed?*

SOLUTION. Let the usual speed of the train be x km/hr.

Time taken to cover 360 km at x km/hr $= \dfrac{360}{x}$ hr .

New speed $= (x + 10)$km/hr

Time taken to hour 360 km at $(x + 10)$ km/hr $= \dfrac{360}{(x + 10)}$ hr

$\therefore \qquad \dfrac{360}{x} - \dfrac{360}{x + 10} = 3$

$\Rightarrow \qquad \dfrac{1}{x} - \dfrac{1}{x + 10} = \dfrac{1}{120} \qquad \Rightarrow \qquad \dfrac{x + 10 - x}{x(x + 10)} = \dfrac{1}{120}$

$\Rightarrow \qquad x(x + 10) = 1200 \qquad\qquad \Rightarrow \qquad x^2 + 10x - 1200 = 0$

$\Rightarrow \quad x^2 + 40x - 30x - 1200 = 0 \qquad \Rightarrow \quad x(x+40) - 30(x + 40) = 0$

$\Rightarrow \qquad\qquad (x + 40)(x - 30) = 0$

$\Rightarrow \qquad\qquad\qquad x = -40 \text{ or } x = 30$

$\Rightarrow \qquad\qquad\qquad x = 30 \qquad\qquad [\because \text{ speed cannot be negative}]$

Hence, the usual speed of the train is 30 km/hr.

EXAMPLE 12. *A plane left* 30 *minutes later than the scheduled time and in order to reach its destination* 1500 *km away in time, it has to increase its speed by* 250 *km/hr from its usual speed. Find its usual speed.*

SOLUTION. Let the usual speed to the plane = x km/hr

$\Rightarrow$ The increased speed of the plane = $(x + 250)$ km/hr

Usual time taken by the plane to cover 1500 km = $\dfrac{1500}{x}$ hrs

and the new time taken to cover 1500 km = $\dfrac{1500}{x + 250}$ hrs

From the given statement, it is clear that the new time taken is 30 minutes, *i.e.,*

$\dfrac{1}{2}$ hrs less than the usual time.

$$\dfrac{1500}{x} - \dfrac{1500}{x + 250} = \dfrac{1}{2} \qquad \Rightarrow \qquad \dfrac{1500(x + 250) - 1500x}{x(x + 250)} = \dfrac{1}{2}$$

i.e., $\quad \dfrac{1500x + 375000 - 1500x}{x^2 + 250x} = \dfrac{1}{2} \qquad \Rightarrow \qquad \dfrac{375000}{x^2 + 250x} = \dfrac{1}{2}$

On cross multiplying we get

$$x^2 + 250x = 750000$$

$\Rightarrow \qquad x^2 + 250x - 750000 = 0$

On factorising we get

$$(x + 1000)(x - 750) = 0$$

$$x = -1000 \text{ or } x = 750$$

Rejecting the negative values of x, we get $x = 750$

The usual speed of the plane = 750 km/hr

Exercise 2.7

1. Two numbers differ by 3 and their product is 504. Find the numbers.

2. The sum of two numbers is 18 and their product is 56. Find the numbers.

3. The sum of two numbers is 15 and the sum of their reciprocals is $\dfrac{3}{10}$. Find the numbers.

4. The sum of the squares of three consecutive positive integers is 50. Find the integers.

5. Find two consecutive even integers whose squares have the sum 340.

6. Find two consecutive positive odd integers whose squares have the sum 290.

7. A two digit number is 5 times the sum of its digits and is also equal to 5 more than twice the product of its digits. Find the number.

8. A two digit number is such that product of the digits is 35. When 18 is added to this number

the digits interchange their places. Find the number.

9. The sides (in cm) of a right triangle containing the right angle are $5x$ and $3x - 1$. If the area of the triangle is 60 cm2, find the sides of the triangle.

10. The hypotenuse of a right angle triangle is 6 meters more than thrice the shortest side. If the third side is 2 meters less than the hypotenuse, find the sides of the triangle.

11. The area of a right angle triangle is 600 sq. cm. If the base of the triangle exceeds the altitude by 10 cm, find the dimensions of the triangle.

12. If the perimeter of a rectangular plot is 68 m and its diagonal is 26 m, find its area.

13. A fast train takes 3 hours less than a slow train for a journey of 600 km. If the speed of the slow train is 10 km/hr less than that of the fast train, find the speeds of the two trains.

HINT TO SELECTED PROBLEMS

1. Let the required no. are x and $x - 3$

Their product $= x(x - 3)$

$$x(x - 3) = 504$$

On solving, we get $x = 24$ then $x - 3 = -21$

and $x = -21$ then $x - 3 = -24$

3. Let the required numbers be x and $(15 - x)$

Then $\dfrac{1}{x} + \dfrac{1}{15 - x} = \dfrac{3}{10}$

On solving we get $x = 10$ and 5

6. Let the required two consecutive positive odd integers be x and $x + 2$.

Then $x^2 + (x + 2)^2 = 290$

7. Let the numbers is $(10x + y)$

Then $(10x + y) = 5(x + y)$　...(1)

and $10x + y = 2xy + 5$　...(2)

From (1), we get $5x = 4y$

$\Rightarrow \qquad x = \dfrac{4}{5}y$

Substituting $x = \dfrac{4}{5}y$ in (2), we get

$$\left(10 \times \dfrac{4y}{5}\right) + y = \left(2 \times \dfrac{4y}{5} \times y\right) + 5$$

$\Rightarrow 8y^2 - 45y + 25 = 0$

$\Rightarrow \qquad y = 5$ or $\dfrac{5}{8}$

since $y \neq 5/8$

Hence, $\qquad y = 5$

Then $\qquad x = 4$

9. It is clear from the figure,

$\dfrac{1}{2} \times 5x \times (3x - 1) = 60$

$3x^2 - x - 24 = 0$

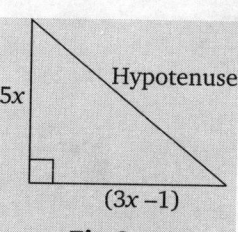

Fig 3

$x = -\dfrac{8}{3}$ or -3.

Since, sides of the triangle can not be negative $x = 3$

and sides $= 5x$ and $3x - 1$

$= 3 \times 5$ and $3 \times 3 - 1$

$= 15$ cm and 6 cm

Applying Pythagoras theorem, we get

$$H^2 = 15^2 + 8^2 = 289 = (17)^2$$

$$H = 17 \text{ cm}$$

10. Let the length of the shortest be x metres

Then, the hypotenuse $= (2x + 6)$ metres

and the third side $= (2x + 4)$

By Pythagoras Theorem, we have

$$(2x + 6)^2 = x^2 + (2x + 4)^2$$

$\Rightarrow x^2 - 8x - 20 = 0$

$\Rightarrow \qquad x = 10$ or $x = -2$

$\Rightarrow \qquad x = 10$

[since side of a triangle can never negative]

Shortest side $= 10$ m

Hypotenuse $= 2x + 6 = 26$ m

Third side $= 24$ m

13. Let the speed of the train $= x$ Km/hr

and speed of the slow train $= (x - 10)$ Km/hr

Time taken by fast train to cover 600 Km

$$= \dfrac{600}{x} \text{ hrs}$$

Time taken by slow train to cover 600 Km

$$= \dfrac{600}{x - 10} \text{ hrs}$$

$$\dfrac{600}{x - 10} - \dfrac{600}{x} = 3$$

$$\frac{600x - 600(x-10)}{x(x-10)} = 3$$

$\Rightarrow \quad x^2 - 10x - 2000 = 0$

$\Rightarrow \quad x = 50$ or $x = -40$

Neglecting $x = -40$, we get $x = 50$

and $x - 10 = 50 - 10 = 40$

Speed of fast train = 50 Km/hr

Speed of slow train = 40 Km/hr

ANSWERS

1. (24 and 21) or (–21 and –24)
2. 4 and 14
3. 10 and 5
5. 12 and 14
6. 11 and 13
7. 45   8. 57
9. 15 cm, 8 cm and 17 cm
10. 10m, 26 m and 24 m
11. Altitude = 30 cm, base = 40 cm
12. 240 m^2
13. 50 Km/hr and 40 Km/hr

Objective Evaluation

✑ MULTIPLE CHOICE QUESTIONS

Choose the most appropriate one.

1. Which of the following is linear equation?
   (a) $ax^2 + bx + c = 0$
   (b) $ax + \dfrac{b}{x} = 0$
   (c) $ax + 2 = 0$
   (d) none of these

2. Which of the following are quadratic equation?
   (a) $x - \dfrac{1}{x} = 0$
   (b) $x^3 + 3x^2 + x - 1 = 0$
   (c) $2x + 3 = 0$
   (d) none of these

3. The degree of the equation $2x - 3 = 0$ is :
   (a) 1
   (b) 2
   (c) 3
   (d) none of these

4. Solution of the equation $2x - 3 = 0$ is :
   (a) 2/5
   (b) 3/2
   (c) 1
   (d) 0

5. The root of the quadratic equation $2x^2 - 3x - 1 = 0$ is :
   (a) 2
   (b) 1
   (c) 3
   (4) none of these

6. The roots of equation $3x^2 - 4x - 7 = 0$ is :
   (a) $\dfrac{7}{3}$ or 1
   (b) $\dfrac{3}{7}$ or 2
   (c) $\dfrac{7}{3}$ or -1
   (d) none of these

7. If $ax^2 + bx + c = 0$ $(a \neq 0)$ is a quadratic equation and $b^2 - 4ac = 0$ then the roots are:
   (a) real and distinct
   (b) real and equal
   (c) imaginary
   (d) none of these

8. Solution of the equation $16x^2 = 25$ is :
   (a) $\pm \dfrac{4}{5}$
   (b) $\pm \dfrac{5}{4}$
   (c) $\dfrac{3}{4}$
   (d) $\pm \dfrac{4}{3}$

9. If the discriminant of $ax^2 + bx + c = 0$ $(a \neq 0)$ is greater than zero, then root are :
   (a) real and distinct
   (b) real and equal
   (c) imaginary
   (d) none of these

10. The root of the equation $x^2 + ax + b = 0$ are equal if :
    (a) $a^2 = 3b$
    (b) $a^2 = 4b$
    (c) $a^2 = b$
    (d) none of these

11. Solution of the equation $x^2 + 16x + 60 = 0$ is :
    (a) -10 or 6
    (b) 10 or -6
    (c) -10 or -6
    (d) none of these

12. The sum and product of the roots of the equation $ax^2 + bx + c = 0$ is :
    (a) $\dfrac{b}{a}, \dfrac{c}{a}$
    (b) $-\dfrac{b}{a}, -\dfrac{c}{a}$
    (c) $-\dfrac{b}{a}, \dfrac{c}{a}$
    (d) none of these

13. The sum of the roots of the equation $6x^2 + x - 2 = 0$ is :
    (a) $\dfrac{1}{6}$
    (b) $-\dfrac{1}{6}$
    (c) $\dfrac{5}{4}$
    (d) 6

14. If α, β are the roots of the equation $ax^2 + bx + c$ $(a \neq 0)$, then the value of $\alpha^3 + \beta^3$ is :
    (a) $\dfrac{3abc - c^3}{a^3}$
    (b) $\dfrac{3abc - b^3}{a^3}$
    (c) $\dfrac{3abc - a^3}{c^3}$
    (d) none of these

15. The equation whose roots are $\sqrt{3}$ and $3\sqrt{3}$ is:
    (a) $x^2 + 4\sqrt{3}x + 9 = 0$
    (b) $x^2 - 4\sqrt{3}x + 9 = 0$
    (c) $x^2 - 4\sqrt{3}x - 9 = 0$
    (d) none of these

16. The quadratic equation whose roots are 4 and 5 is :
    (a) $x^2 - 9x - 20 = 0$
    (b) $x^2 - 9x + 20 = 0$
    (c) $x^2 + 9x + 20 = 0$
    (d) none of these

17. If the sum and product of the roots of the equation $ax^2 - 5x + c = 0$ is 10 then the value of a and c is :
    (a) $a = \dfrac{1}{2}, c = \dfrac{1}{5}$
    (b) $a = 5, c = \dfrac{1}{2}$
    (c) $a = \dfrac{1}{2}, c = 5$
    (c) none of these

18. The roots of the equation $ax^2 + bx + c = 0$, $a \neq 0$ is :
    (a) $\dfrac{b \pm \sqrt{b^2 - 4ac}}{2a}$
    (b) $\dfrac{-b \pm \sqrt{b^2 + 4ac}}{2a}$
    (c) $\dfrac{-b \pm \sqrt{b^2 - 4ac}}{2a}$
    (d) none of these

19. The discriminant of the equation $5x^2 + 16x + 3 = 0$ is :
    (a) 195
    (b) 196
    (c) 197
    (d) none of these

20. The discriminant of the equation $3\sqrt{7}x^2 + 4x - \sqrt{7} = 0$ is :
    (a) 98
    (b) 99
    (c) 100
    (d) 101

∽ TRUE/FALSE

Write 'T' for True and 'F' for False statement.

1. The solution of the linear equation $3x + 6 = 0$ is –2. **(T/F)**

2. The solution of the equation $x - \dfrac{1}{x} = 0$ is ±1. **(T/F)**

3. The degree of the equation $x - \dfrac{1}{2x} - 3 = 0$ is 1. **(T/F)**

4. The sum of the roots of the equation is $6x^2 + x - 20 = 0$ is $\dfrac{1}{6}$. **(T/F)**

5. The roots of the equation $x^2 - 5x + 6 = 0$ is 2, 3. **(T/F)**

6. $ax + \dfrac{b}{3x} - 2 = 0$ is a linear equation. **(T/F)**

7. $x = -5$ is a root of $\dfrac{x^2}{-5} + 2x - 10 = 0$. **(T/F)**

8. The discriminant of $(x - 1)(x + 2) = 0$ is 9. **(T/F)**

9. If the roots of the equations $x^2 + 2x + ab = 0$ are real and unequal then the equation $x^2 - 2(z + b)x + a^2 + b^2 + 2c^2 = 0$ has no real roots. **(T/F)**

10. The roots of $\sqrt{2}x^2 - 5x - 3 = 0$ will be equal. **(T/F)**

11. For a quadratic equation $ax^2 + bx + c = 0$, $a \neq 0$ if $b^2 - 4ac < 0$ the two distinct real roots will exist. **(T/F)**

12. $\sqrt{2}x + \dfrac{1}{x} = \sqrt{3}$ is not a quadratic equation. **(T/F)**

13. The roots of the quadratic equation $ax^2 + bx + c = 0$, $a \neq 0$ can be found by using the quadratic formula $\dfrac{+b \pm \sqrt{b^2 - 4ac}}{2a}$, provided $b^2 - 4ac \geq 0$. **(T/F)**

14. The equation $(x - 1)^3 = x^3 - 2x + 1$ is not a quadratic equation. **(T/F)**

15. $\dfrac{1}{x-1} + \dfrac{1}{2x+1} = 3$ has two distinct real roots. **(T/F)**

∽ FILL IN THE BLANKS

1. If $x = 1$ is a common root of the equations $px^2 + px + 3 = 0$ and $x^2 + x + q = 0$ then $pq = $ _____ .

2. Discriminant for $x + \dfrac{2}{x} - 1 = 0$ is _____ .

3. If the roots of the equation $4x^2 - Kx + 9 = 0$ are not real this K should be _____ .

4. The roots of a quadratic equations can be found by using the method of _____ .

5. If α, β are the roots of the equation $x^2 - 2x + 1 = 0$, then $α - β$ is _____ .

ANSWERS

∽ MULTIPLE CHOICE QUESTIONS

1. (c)	**2.** (a)	**3.** (b)	**4.** (b)	**5.** (b)	**6.** (c)	**7.** (b)	**8.** (b)	**9.** (a)
10. (b)	**11.** (c)	**12.** (c)	**13.** (b)	**14.** (b)	**15.** (b)	**16.** (b)	**17.** (c)	**18.** (c)
19. (b)	**20.** (c)							

∽ TRUE / FALSE

1. T	**2.** T	**3.** F	**4.** F	**5.** T	**6.** F	**7.** F	**8.** T	**9.** T
10. F	**11.** F	**12.** F	**13.** F	**14.** F	**15.** T			

∽ FILL IN THE BLANKS

1. 3	**2.** –7	**3.** $-6\sqrt{2} < K < 6\sqrt{2}$	**4.** Factorization	**5.** 0

❑❑❑❑❑❑

CHAPTER 3

Simultaneous Linear Equations

3.1 INTRODUCTION

If a, b and c are three real numbers with $a \neq 0$ and x and y are two variables, then the equation of the type $ax + by + c = 0$ or $ax + by = c$ is called a linear equation in two variables.

For Example.

(i) $3x + 5y - 7 = 0$

(ii) $5x - 8y = 15$.

3.1.1 SOLUTION OF A LINEAR EQUATION

We say that $x = \alpha$ and $y = \beta$ is a solution of $ax + by + c = 0$, if $a\alpha + b\beta + c = 0$.

3.1.2 SIMULTANEOUS LINEAR EQUATIONS IN TWO VARIABLES

Two linear equations in two unknown x and y are said to form a system of simultaneous linear equations if each of them is satisfied by the same pair of values of x and y.

For Example.

The pair of linear equations $3x + 2y = 7$ and $4x - 8y - 2 = 0$ in two variables x and y form a system of simultaneous linear equations.

3.2 SOLUTION OF A GIVEN SYSTEM OF SIMULTANEOUS EQUATIONS

The values of x and y which satisfies each equation of the given system of linear equations is called its solution.

Example 1. *Show that $x = 5$, $y = 2$ is a solution of the system of linear equations.*

$$2x + 3y = 16, x - 2y = 1.$$

Solution. The given equations are

$$2x + 3y = 16 \qquad \qquad ...(1)$$

and $\qquad \qquad x - 2y = 1 \qquad \qquad ...(2)$

Putting $x = 5$ and $y = 2$ in (1), we get

$$\text{LHS} = 2 \times 5 + 3 \times 2 = 16 = \text{RHS}.$$

Putting $x = 5$ and $y = 2$ in (2), we get

$$\text{LHS} = 5 - 2 \times 2 = 1 = \text{RHS}.$$

Thus, $x = 5$ and $y = 2$ satisfy both (1) and (2)

Hence, $x = 5$, $y = 2$ is a solution of the given system of equations.

3.2.1 CONSISTENT / INCONSISTENT PAIR OF LINEAR EQUATIONS

A pair of linear equation in two variables which has a solution is called a consistent pair of linear equations. A pair of linear equations in two variables which has no solution is called an inconsistent pair of linear equations.

❧ REMARKS

➠ A system of simultaneous linear equations in two variables will have either
  (i) only one solution
  (ii) no solution
  (iii) an infinite number of solutions.

➠ It must be taken into memory that no system of linear equations in two variables will have only two solutions, only three solutions, only four solutions, etc. Infect, if any system of linear equations has two or more solution, it will always have an infinite number of solutions.

3.3 GRAPHICAL METHOD FOR SOLVING SIMULTANEOUS LINEAR EQUATIONS

To solve the system of linear equations in two variables use the following steps.

WORKING PROCEDURE

Step 1. On the same graph paper, draw graph (straight line) for each given equation.

Step 2. If the lines drawn intersect each other at a unique point, read the values of x and y for this point. The values of x and y so obtained, gives the required solution of the given system of equation.

3.3.1 DESCRIPTION OF METHOD

Let the given system of linear equations be
$$a_1 x + b_1 y + c_1 = 0 \qquad \qquad \dots(1)$$
and
$$a_2 x + b_2 y + c_2 = 0 \qquad \qquad \dots(2)$$
We draw the graph of each of the given linear equations on the same graph paper.

Let the lines L_1 and L_2 represent these graphs.

Case1: When the lines L_1 and L_2 intersect at a point. Let the graph L_1 and L_2 intersect at a point $P(\alpha, \beta)$.

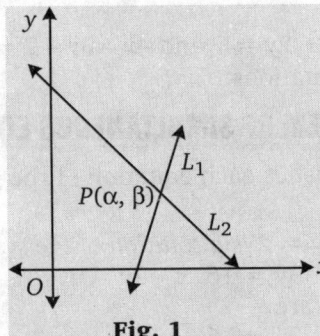

Fig. 1

Then $x = \alpha, y = \beta$ is the unique solution of the given system of equations.

Case 2: When the lines L_1 and L_2 are coincident.

In this case, the given system has infinitely many solutions.

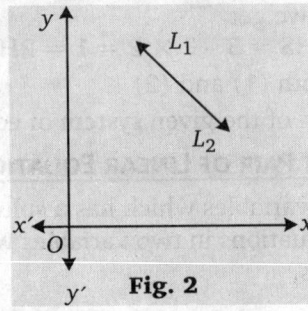

Fig. 2

Case 3: When the lines L_1 and L_2 are parallel.

In this case, there is no common solution of the given equation, as shown in figure, *i.e.*, the given system of equations has no solution. Thus, in this case, the system of given equations is inconsistent.

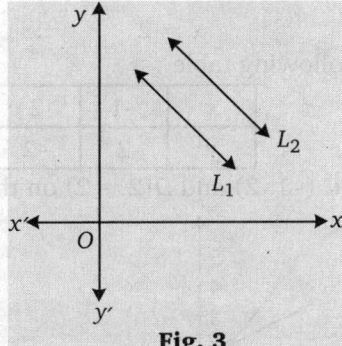

Fig. 3

⮞ REMARKS

➧ A system of two linear equations in x and y has
  (i)   a unique solution if the graph lines intersect at point.
  (ii) infinitely many solutions in the two graph lines coincide.
  (iii) no solution if the two graph lines are parallel.

➧ Straight lines as graphs of linear equations $a_1x + b_1y + c_1 = 0$ and $a_2x + b_2y + c_2 = 0$ intersect each other at a point if
$$\frac{a_1}{a_2} \neq \frac{b_1}{b_2}$$

➧ Straight lines as graphs of linear equations $a_1x + b_1y + c_1 = 0$ and $a_2x + b_2y + c_2 = 0$ are parallel each (*i.e.*, do not intersect) if
$$\frac{a_1}{a_2} = \frac{b_1}{b_2} \neq \frac{c_1}{c_2}$$

➧ Straight lines as graphs of linear equations $a_1x + b_1y + c_1 = 0$ and $a_2x + b_2y + c_2 = 0$ coincide (*i.e.*, becomes single line) if
$$\frac{a_1}{a_2} = \frac{b_1}{b_2} = \frac{c_1}{c_2}$$

Here one equation is a constant multiple of the other equation.

SOLVED EXAMPLES

EXAMPLE 1. *Solve graphically the system of equations :*
$$x + 2y = 3, \quad 4x + 3y = 2.$$

SOLUTION. We have $\qquad\qquad x + 2y = 3 \qquad \Rightarrow \qquad y = \frac{1}{2}(3 - x)$

Now $\qquad\qquad\qquad\qquad x = 1 \qquad \Rightarrow \qquad y = \frac{1}{2}(3 - 1) = 1.$

$\qquad\qquad\qquad\qquad\qquad x = 3 \qquad \Rightarrow \qquad y = \frac{1}{2}(3 - 3) = 0.$

Thus we have the following table:

x	1	3
y	1	0

Now, plot the point $A(1, 1)$ and $B(3, 0)$ on a graph paper. Join AB and produce it on both sides.

Now, we have $\qquad 4x + 3y = 2$

$$\Rightarrow \qquad y = \frac{1}{3}(2-4x).$$

Now, $\qquad\qquad\qquad x = -1 \qquad \Rightarrow \qquad y = \frac{1}{3}[2-4(-1)] = 2.$

$$x = 2 \qquad \Rightarrow \qquad y = \frac{1}{3}[2-4\times2] = -2.$$

Thus, we have the following table :

x	-1	2
y	2	-2

Now, plot the point C (-1, 2) and $D(2, -2)$ on the same graph paper.

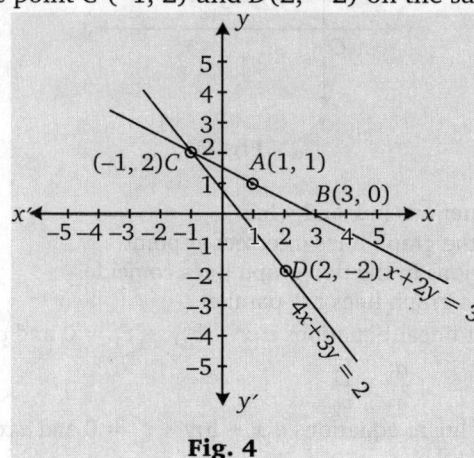

Fig. 4

Join CD and produce it on both sides.

The two graph lines AB and CD intersect at the point C (-1, 2).

$\therefore \qquad x = -1, y = 2$ is the solution of given system of equation.

EXAMPLE 2.   *Show graphically that the system of equations :*

$$3x - y = 2, \quad 9x - 3y = 6.$$

has an infinite number of solutions.

SOLUTION.   Graph $\qquad\qquad 3x - y = 2 \qquad\quad \Rightarrow \qquad y = 3x - 2$

$$x = 1 \qquad\quad \Rightarrow \qquad y = 3 \times 1 - 2 = 1$$
$$x = 2 \qquad\quad \Rightarrow \qquad y = 3 \times 2 - 2 = 4$$

Thus, we have the following table:

x	1	2
y	1	4

Now plot the points $A(1, 1)$ and $B(2, 4)$ on the graph paper. Join AB and produce it on both sides.

Graph of $9x - 3y = 6$:

$$9x - 3y = 6 \qquad \Rightarrow \qquad y = \frac{1}{3}(9x - 6)$$

Now $\qquad\qquad\qquad x = 0 \qquad \Rightarrow \qquad y = \frac{1}{3}(9 \times 0 - 6) = -2.$

$$x = 3 \qquad \Rightarrow \qquad y = \frac{1}{3}(9 \times 3 - 6) = 7.$$

Thus, we have the following table

x	0	3
y	–2	7

Now plot the points $C(0, –2)$ and $D(3, 7)$ on the same graph paper

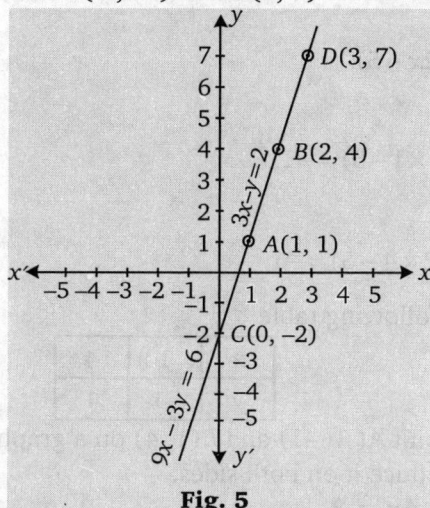

Fig. 5

We find that both the points C and D lie on the line AB.

Since the graph lines AB and CD coincide, the given system has an infinite number of solutions.

EXAMPLE 3. *Solve the following system of linear equations graphically:*

$$2x + y = 6, x – 2y + 2 = 0.$$

Find the vertices of the triangle formed by the above two lines and the x-axis. Also, find the area of the triangle.

SOLUTION.
$$2x + y = 6 \qquad \qquad ...(1)$$

$$y = 6 – 2x$$

x	1	4	2
y	4	–2	2

$$x – 2y + 2 = 0 ...(2)$$

$$y = \frac{x + 2}{2}$$

x	0	4	2
y	1	–3	2

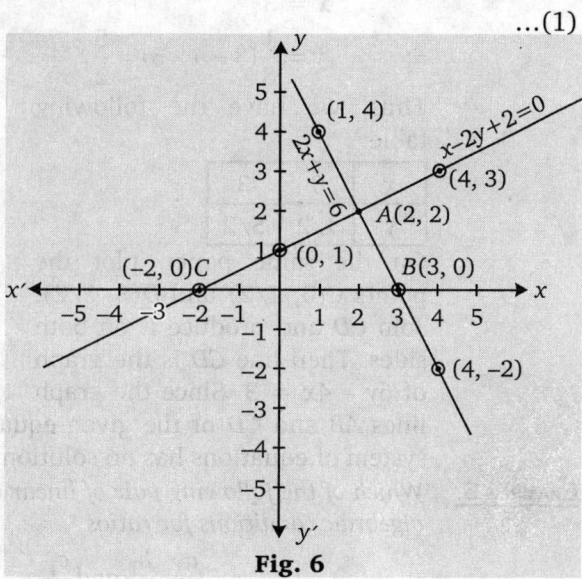

Fig. 6

The two straight lines intersect at $A(2, 2)$.

$\therefore x = 2, y = 2$ is the solution.

Vertices of the triangle are $A(2, 2)$, $B(3, 0)$ and $C(–2, 0)$
$AL \perp BC$. $BC = 5$ units, $AL = 2$ units.

Area of $\triangle ABC = \frac{1}{2} \times (5 \times 2) = 5$ square units.

EXAMPLE 4. *Show graphically that the system of linear equations*
$$2x - 3xy = 5, \quad 6y - 4x = 3$$
has no solution.

SOLUTION. **Graph of $2x - 3y = 5$.**
$$2x - 3y = 5$$
$$\Rightarrow \quad y = \frac{1}{3}(2x - 5)$$
Now $x = 1$
$$\Rightarrow \quad y = \frac{1}{3}(2 \times 1 - 5) = -1.$$
$$x = 4$$
$$\Rightarrow \quad y = \frac{1}{3}(2 \times 4 - 5) = -1.$$

Thus, we have following table

x	1	4
y	–1	1

Now plot the point $A(1, -1)$ and $B(1, 4)$ on a graph paper.
Join AB and produce it on both sides.

Graph of $6y - 4x = 3$
$$6y - 4x = 3$$
$$\Rightarrow \quad y = \frac{1}{6}(3 + 4x)$$
$$x = 0$$
$$\Rightarrow \quad y = \frac{1}{6}(3 + 4 \times 0) = \frac{1}{2}.$$
$$x = 3$$
$$\Rightarrow \quad y = \frac{1}{6}(3 + 4 \times 3) = \frac{5}{2}.$$

Thus, we have the following table:

x	0	3
y	1/2	5/2

On the same paper, plot the points $C(0, 1/2)$ and $D(3, 5/2)$. Join CD and produce it on both sides. Then line CD is the graph of $6y - 4x = 3$. Since the graph

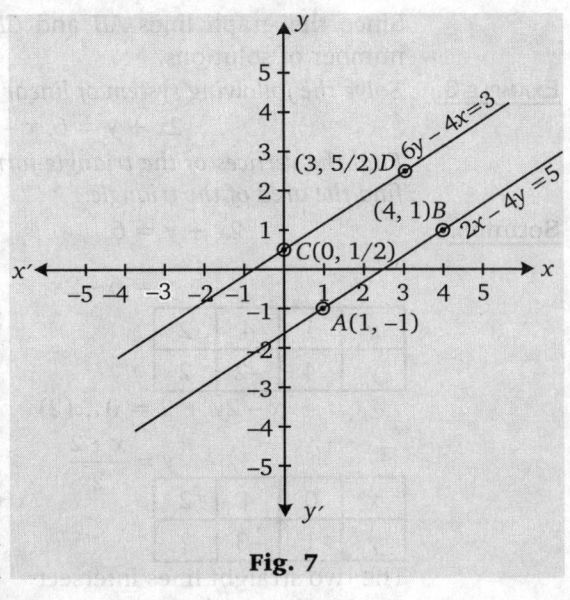

Fig. 7

lines AB and CD of the given equations are parallel, it follows that the given system of equations has no solution.

EXAMPLE 5. *Which of the following pair of linear equation are consistent / inconsistent? Use the algebraic conditions for ratios*
$$\frac{a_1}{a_2}, \frac{b_1}{b_2} \text{ and } \frac{c_1}{c_2}$$

    (i) $5x - y = 7, x - y + 1 = 0.$

    (ii) $3x + 2y - 4 = 0, 3x + 2y + 1 = 0.$

    (iii) $3x + 4y + 5 = 0, 6x + 8y + 10 = 0.$

SOLUTION.   (i) $5x - y - 7 = 0$ ...(1)

and $x - y + 1 = 0$ ...(2)

$$\Rightarrow \quad \frac{a_1}{a_2} \neq \frac{b_1}{b_2}$$

So, the pair of linear equations has unique solution and therefore, the equations are consistent.

    (ii) $3x + 2y - 4 = 0$ ...(1)

             $3x + 2y + 1 = 0$ ...(2)

$$\frac{a_1}{a_2} = \frac{3}{3} = 1; \frac{b_1}{b_2} = \frac{2}{2} = 1; \frac{c_1}{c_2} = \frac{-4}{1} = -4.$$

$$\frac{a_1}{a_2} = \frac{b_1}{b_2} \neq \frac{c_1}{c_2}$$

So, the linear equations are inconsistent because the pair of linear equations has no solutions.

    (iii) $3x + 4y + 5 = 0$ ...(1)

            $6x + 8y + 10 = 0$ ...(2)

$$\frac{a_1}{a_2} = \frac{3}{6} = \frac{1}{2}; \frac{b_1}{b_2} = \frac{4}{8} = \frac{1}{2}; \frac{c_1}{c_2} = \frac{5}{10} = \frac{1}{2}.$$

$$\Rightarrow \quad \frac{a_1}{a_2} = \frac{b_1}{b_2} = \frac{c_1}{c_2}.$$

There are infinitely many solution for the given pair of lines. Hence, the equations are consistent.

EXAMPLE 6.   *Find the values of k for which the system of equations are*
$$x - 2y = 3, 3x + ky = 1$$
has a unique solution.

SOLUTION.   The given system of equations are
$$x - 2y = 3, 3x + ky = 1$$
These equations are of the form:
$$a_1x + b_1y + c_1 = 0, a_2x + b_2y + c_2 = 0$$
where $a_1 = -1, b_1 = -2, c_1 = -3$, and $a_2 = 3, b_2 = k, c_2 = -1$
for a unique solution we must have
$$\frac{a_1}{a_2} \neq \frac{b_1}{b_2} \Rightarrow \frac{1}{3} \neq -\frac{2}{k}.$$
This happens when $k \neq -6$.
Thus, for all real values of k other than -6; the given system of equations will have a unique solution.

EXAMPLE 7.   *Find the values of k for which the system of equations*
$$3x + y = 1$$
$$(2k - 1)x + (k - 1)y = (2k + 1)$$
has no solution.

SOLUTION.   The given equations are
$$3x + y - 1 = 0$$
$$(2k - 1)x + (k - 1)y - (2k + 1) = 0$$

These equations are of the form

$$a_1x + b_1y + c_1 = 0, a_2x + b_2y + c_2 = 0$$

where

$$a_1 = 3, b_1 = 1, c_1 = 1$$
$$a_2 = (2k - 1), b_2 = (k - 1), c_2 = -(2k + 1)$$

for no solution, we must have $\dfrac{a_1}{a_2} = \dfrac{b_1}{b_2} \neq \dfrac{c_1}{c_2}$.

$$\Leftrightarrow \qquad \frac{3}{2k-1} = \frac{1}{k-1} \neq \frac{-1}{-(2k+1)}$$

$$\Leftrightarrow \qquad \frac{3}{2k-1} = \frac{1}{k-1} \text{ and } \frac{1}{k-1} \neq \frac{1}{2k+1}$$

$$\Leftrightarrow \qquad (3k - 3 = 2k - 1) \text{ and } (2k + 1 \neq k - 1)$$

$$\Leftrightarrow \qquad k = 2 \text{ and } k \neq -2$$

Thus, $\dfrac{a_1}{a_2} = \dfrac{b_1}{b_2} \neq \dfrac{c_1}{c_2}$ holds when $k = 2$.

Hence, the given system of equation has no solution when $k = 2$.

Exercise 3.1

Solve the following system of equations graphically.

1. $x + y = 3, 2x + 5y = 12$
2. $2x + y = 3, 2x - 3y = 7$.
3. $2x + 3y = 4, 3x - y = -5$.
4. $4x + 3y = 5, 2y - x = 7$
5. $3x + y + 1 = 0, 2x - 3y \neq 8 = 0$.
6. Solve the following system of linear equations graphically.

    $$2x - 3y - 17 = 0, 4x + y - 13 = 0$$

    Shade the region between the lines and the x-axis.
7. Solve the following system of linear equations graphically.

    $$2x + y - 5 = 0, x + y - 3 = 0.$$

    Find the points where the graph meet the x-axis.
8. Solve the following system of linear equations graphically.

    $$2x - y - 4 = 0; x + y + 1 = 0.$$

9. Solve the following system of linear equations graphically.

    $$x - y + 1 = 0, 3x + 2y - 12 = 0.$$

    Calculate the area bounded by these lines and the x-axis.
10. In each of the following systems of linear equations, find whether it has a unique solution, an infinite number of solution or no solution.

    (i) $3x + 5y = 13, 5x + 3y = 4$

    (ii) $2x - 3y = 5, 6x - 9y = 15$

    (iii) $6x - 10y = 3, 3x - 5y = 7$.

    (iv) $\dfrac{x}{3} + \dfrac{y}{2} = 3, x - 2y = 2$

11. For what value of k, the system of equations

    $$kx + 2y = 5, 3x - 4y = 10$$

    has (i) a unique solution (ii) no solution?
12. Find the value of k for which the following system of equation has no solution.

    $$3x + ky = 1, (2k - 1)x + (k - 1)y = (2k + 1)$$

ANSWERS

1. $x = 1, y = 2$ 2. $x = 2, y = -1$ 3. $x = -1, y = 2$ 4. $x = -1, y = 3$
5. $x = -1, y = 2$ 6. $x = 4, y = -3$ 7. $x = 2, y = 1, P(0, 5)$ and $Q(0, 3)$.
8. $x = 1, y = -2, P(2, 0)$ and $Q(-1, 0)$ 9. $x = 2, y = 3, 75$ sq. units
10. (i) Unique   (ii) Infinite   (iii) No solution   (iv) Unique.
11. (i) All real values except $-\dfrac{3}{2}$ (ii) $k = -\dfrac{3}{2}$ 12. $k = 2$.

3.4 ALGEBRAIC METHODS OF SOLVING A PAIR OF LINEAR EQUATIONS IN TWO VARIABLES

It is difficult to examine non-integer solutions of linear equations correctly by graphical method. In such cases only approximate values of x and y can be obtained from the graph. Algebraic methods provide exact and correct solution for each and every pair of linear equations

$$a_1x + b_1y + c_1 = 0 \text{ and } a_2x + b_2y + c_2 = 0$$

The most commonly used algebraic methods of solving a pair of linear equations in two variables are as under.

1. By substitution method
2. By elimination method
3. By cross-multiplication method

3.4.1 SUBSTITUTION METHOD

WORKING PROCEDURE

Suppose we are given two linear equations in x and y. To solve these we proceed using the following steps:

Step 1. Express y in terms of x, taking one of the given equations.
Step 2. Substitute this value of y in terms of x in the other equation to solve it for x.
Step 3. Substitute the value of x in the relation taken in step (1) solve it for y.

⮯ REMARK
➡ We may interchange the role of x and y in the above method.

SOLVED EXAMPLES

EXAMPLE 1. Solve $2x - y - 3 = 0$, $4x - y - 5 = 0$ by substitution method.

SOLUTION. Here, the given equations are

$$2x - y - 3 = 0 \quad \text{...(1)}$$
$$4x - y - 5 = 0 \quad \text{...(2)}$$

From (1): $y = 2x - 3 \quad \text{...(3)}$

Substitute y from (3) in (2), we get

$$4x - (2x - 3) - 5 = 0$$
$$4x - 2x - 3 - 5 = 0$$
$$2x = 2 \quad \Rightarrow \quad x = 1$$

Substituting $x = 1$ in (3), we get

$$y = 2 - 3 \quad \Rightarrow \quad y = -1$$

The required solution is $x = 1$, $y = -1$.

3.4.2 ELIMINATION METHOD

WORKING PROCEDURE

In this method, we eliminate one of the unknown variables and proceed using the following steps.

Step 1. Multiply the given equations by suitable numbers so as to make the coefficients of one of the unknown variable numerically equal.
Step 2. If the numerically equal coefficients are opposite in sign, then add the new equations. Otherwise subtract them.
Step 3. The resulting equation is linear in one unknown variable. Solve it to get the value of one of the unknown variable.
Step 4. Substitute the value in any of the given equations.
Step 5. Solve it to get the value of the other unknown variable.

EXAMPLE 2. *Solve*
$$10x + 3y = 75$$
$$6x - 5y = 11$$

SOLUTION. The given system of equations is
$$10x + 3y = 75 \qquad \ldots(1)$$
$$6x - 5y = 11 \qquad \ldots(2)$$

We multiply equation (1) by 3 and equation (2) by 5 so that coefficients of x in both the equations become equal and we get the following equation
$$30x + 9y = 225 \qquad \ldots(3)$$
$$30x - 25y = 55 \qquad \ldots(4)$$

On subtracting (4) from (3), we get

$$34y = 170 \quad \Rightarrow \quad y = \frac{170}{34} = 5.$$

Substituting $y = 5$ in (1) we get
$$10x + 3 \times 5 = 75 \quad \Rightarrow \quad 10x = 60$$
$$\Rightarrow \quad x = \frac{60}{10} = 6.$$

Hence, the solution is $x = 6$, $y = 5$.

EXAMPLE 3. *Solve $2x + 3y - 9 = 0$, $4x + 6y - 18 = 0$ by substitution method.*

SOLUTION. We have
$$2x + 3y - 9 = 0 \qquad \ldots(1)$$
$$4x + 6y - 18 = 0 \qquad \ldots(2)$$

From (1),
$$3y = 9 - 2x$$
$$y = \frac{9 - 2x}{3} \qquad \ldots(3)$$

Substituting y from (3) in (2), we get

$$\Rightarrow \quad 4x + 6\left(\frac{9 - 2x}{3}\right) - 18 = 0 \quad \Rightarrow \quad 4x + 2(9 - 2x) - 18 = 0$$

$$\Rightarrow \quad 4x - 4x + 18 - 18 = 0 \quad \Rightarrow \quad 0 = 0$$

Which is a true statement.

Hence, the given pair of linear equations has infinitely many solutions. Let us find solutions, put $x = k$ (any real constant) in (3), we get

$$y = \frac{9 - 2k}{3}$$

Hence, $x = k$,
$$y = \frac{9 - 2k}{3}$$

EXAMPLE 4. *Solve the following system of linear equations :*
$$\sqrt{2}x - \sqrt{3}y = 0, \sqrt{5}x + \sqrt{2}y = 0$$

SOLUTION. The given system of equations is
$$\sqrt{2}x - \sqrt{3}y = 0 \qquad \ldots(1)$$
$$\sqrt{5}x + \sqrt{2}y = 0 \qquad \ldots(2)$$

Multiplying (1) by $\sqrt{2}$ and (2) by $\sqrt{3}$, we get
$$2x - \sqrt{6}y = 0 \qquad \ldots(3)$$
$$\sqrt{15}x + \sqrt{6}y = 0 \qquad \ldots(4)$$

Adding (3) and (4), we get

$$(2x + \sqrt{15})x = 0 \qquad \Rightarrow \qquad x = 0$$

Substituting $x = 0$ in (2), we get

$$\sqrt{2}y = 0 \qquad \Rightarrow \qquad y = 0$$

Hence, the solution is $x = 0, y = 0$.

Example 5. *Solve $78x + 91y = 53, 65x + 117y = 60$ by elimination method.*

Solution.

$$78x + 91y = 53 \qquad \qquad \ldots (1)$$
$$65x + 117y = 60 \qquad \qquad \ldots (2)$$

Multiplying (1) by 5 and (2) by 6.

We get

$$390x + 455y = 265 \qquad \qquad \ldots (3)$$
$$390x + 702y = 360 \qquad \qquad \ldots (4)$$

Subtracting (3) from (4), we get

$$702y - 455y = 360 - 265$$

$$247y = 95 \Rightarrow y = \frac{95}{247} = \frac{5}{13} \qquad \Rightarrow \qquad y = \frac{5}{13}$$

Substituting $y = \frac{5}{13}$ in (1) we get

$$78x + 91 \times \frac{5}{13} = 53$$

$$\Rightarrow \qquad 78x + 35 = 53$$

$$\Rightarrow \qquad x = \frac{18}{78} = \frac{3}{13} \Rightarrow x = \frac{3}{13}.$$

Hence, $x = \dfrac{3}{13}, y\dfrac{5}{13}.$

Example 6. *Solve $\dfrac{1}{2x} - \dfrac{1}{y} = -1, \dfrac{1}{x} + \dfrac{1}{2y} = 8.$*

Solution. Taking $\dfrac{1}{x} = u$ and $\dfrac{1}{y} = v$, the given equation becomes

$$\frac{u}{2} - v = -1 \qquad \Rightarrow \qquad u - 2v = -2 \qquad \ldots (1)$$

$$u + \frac{v}{2} = 8 \qquad \Rightarrow \qquad 2u + v = 16 \qquad \ldots (2)$$

Multiplying (2) by 2 and adding the result with (1), we get

$$5u = 30 \qquad \Rightarrow \qquad u = \frac{30}{5} = 6.$$

Substituting $u = 6$ in (1) we get

$$6 - 2v = -2 \qquad \Rightarrow \qquad 2v = 8$$
$$\Rightarrow \qquad v = 4$$

Now

$$u = 6 \qquad \Rightarrow \frac{1}{x} = 6 \Leftrightarrow x = \frac{1}{6}.$$

$$v = 4 \qquad \Rightarrow \frac{1}{y} = 4 \Leftrightarrow y = \frac{1}{4}$$

Hence, the solution is $x = \dfrac{1}{6}, y = \dfrac{1}{4}.$

3.4.3 METHOD OF CROSS MULTIPLICATION

THEOREM.   *The system of two linear equations*

$$a_1x + b_1y + c_1 = 0, \quad a_2x + b_2y + c_2 = 0 \quad where \frac{a_1}{a_2} \neq \frac{b_1}{b_2}.$$

has a unique solution, given by

$$x = \frac{b_1c_2 - b_2c_1}{a_1b_2 - a_2b_1}, \quad y = \frac{c_1a_2 - c_2a_1}{a_1b_2 - a_2b_1}.$$

PROOF.   The given equations are

$$a_1x + b_1y + c_1 = 0 \qquad \qquad \dots (1)$$
$$a_2x + b_2y + c_2 = 0 \qquad \qquad \dots (2)$$

Multiplying equation (1) by b_2 and (2) by b_1 and subtracting, we get

$$(a_1b_2 - a_2b_1)x = (b_1c_2 - b_2c_1).$$

$$\Rightarrow \qquad \qquad x = \frac{b_1c_2 - b_2c_1}{a_1b_2 - a_2b_1} \qquad \qquad \left[\because \frac{a_1}{a_2} \neq \frac{b_1}{b_2} \Rightarrow a_1b_2 - a_2b_1 \neq 0\right]$$

Multiplying equation (2) by a_1, (1) by a_2 and subtracting, we get

$$(a_1b_2 - a_2b_1)y = (c_1a_2 - c_2a_1).$$

$$\Rightarrow \qquad \qquad y = \frac{c_1a_2 - c_2a_1}{a_1b_2 - a_2b_1} \qquad \qquad \left[\because (a_1b_2 - a_2b_1) \neq 0\right]$$

Hence, a unique solution exists, which is given by

$$x = \frac{b_1c_2 - b_2c_1}{a_1b_2 - a_2b_1}, \quad y = \frac{c_1a_2 - c_2a_1}{a_1b_2 - a_2b_1}$$

This can be written as

$$\frac{x}{b_1c_2 - b_2c_1} = \frac{y}{c_1a_2 - c_2a_1} = \frac{1}{a_1b_2 - a_2b_1}$$

REMARK

➡ The following diagram helps in remembering the above solution.

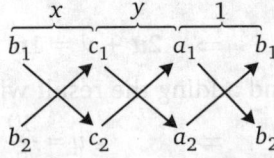

Numbers with downward arrows are multiplied first and from this product, the product of number with upward arrows is subtracted.

SOLVED EXAMPLES

EXAMPLE 1.   Solve the system of equations $2x + 3y = 17$, $3x - 2y = 6$ by the method of cross multiplication.

SOLUTION.   The given equations may be written as

$$2x + 3y - 17 = 0 \qquad \qquad \dots (1)$$
$$3x - 2y - 6 = 0 \qquad \qquad \dots (2)$$

By cross multiplication, we have:

$$\begin{array}{cccc} x & y & 1 \\ 3 \quad -17 & 2 \quad 3 \end{array}$$

$$\begin{array}{cccc} -2 & -6 & 3 & -2 \end{array}$$

$$\therefore \quad \frac{x}{\{3 \times (-6) - (-2) \times (-17)\}} = \frac{y}{\{(-17) \times 3 - (-6) \times 2\}} = \frac{1}{\{2 \times (-2) - 3 \times 3\}}$$

$$\Rightarrow \qquad \frac{x}{-18 - 34} = \frac{y}{-51 + 12} = \frac{1}{-4 - 9}$$

$$\Rightarrow \qquad \frac{x}{-52} = \frac{y}{-39} = \frac{1}{-13}$$

$$\Rightarrow \qquad x = \frac{-52}{-13} = 4, y = \frac{-39}{-13} = 3.$$

Hence, $x = 4, y = 3$ is the required solution.

EXAMPLE 2. *Solve* $3x - y - 2 = 0$; $2x + y - 8 = 0$ *by method of cross multiplication.*

SOLUTION.
$$3x - y - 2 = 0 \qquad \qquad \ldots(1)$$
$$2x + y - 8 = 0 \qquad \qquad \ldots(2)$$

are the given equations.

By cross multiplication method

$$\begin{array}{cccc} x & y & 1 \\ -1 \quad -2 & -2 \quad 3 & 3 \quad -1 \end{array}$$

$$\begin{array}{cccccc} 1 & -8 & 8 & 2 & 2 & 1 \end{array}$$

$$\Rightarrow \quad \frac{x}{\{(-1)(-8) - (1)(-2)\}} = \frac{y}{\{(-2)(2) - (-8)(3)\}} = \frac{1}{\{(3)(1) - (2)(-1)\}}$$

$$\Rightarrow \qquad \frac{x}{8 + 2} = \frac{y}{-4 + 24} = \frac{1}{3 + 2}$$

$$\Rightarrow \qquad \frac{x}{10} = \frac{y}{20} = \frac{1}{5}$$

$$\Rightarrow \qquad \frac{x}{10} = \frac{1}{5} \text{ and } \frac{y}{20} = \frac{1}{5}$$

$$\Rightarrow \qquad x = \frac{10}{5} \text{ and } y = \frac{20}{5}$$

$$\Rightarrow \qquad x = 2 \text{ and } y = 4$$

Hence, $x = 2, y = 4$ is the required solution.

EXAMPLE 3. *Solve* $\dfrac{x}{a} + \dfrac{y}{b} = a + b$; $\dfrac{x}{a^2} + \dfrac{y}{b^2} = 2$.

SOLUTION. The given equations may be written as

$$\frac{1}{a}x + \frac{1}{b}y - (a + b) = 0 \qquad \qquad \ldots(1)$$

$$\frac{1}{a^2}x + \frac{1}{b^2}y - 2 = 0 \qquad \qquad \ldots(2)$$

By cross multiplication

$$\cfrac{x}{\begin{matrix}\frac{1}{b} & & -(a+b)\\ & \diagdown\diagup & \\ \frac{1}{b^2} & & -2\end{matrix}} = \cfrac{y}{\begin{matrix}-(a+b) & & \frac{1}{a}\\ & \diagup\diagdown & \\ -2 & & \frac{1}{a^2}\end{matrix}} = \cfrac{1}{\begin{matrix}\frac{1}{a} & & \frac{1}{b}\\ & \diagdown\diagup & \\ \frac{1}{a^2} & & \frac{1}{b^2}\end{matrix}}$$

$$\cfrac{x}{\left[-\dfrac{2}{b}+\dfrac{a+b}{b^2}\right]} = \cfrac{y}{\left[\dfrac{-(a+b)}{a^2}+\dfrac{2}{a}\right]} = \cfrac{1}{\left[\dfrac{1}{ab^2}-\dfrac{1}{a^2 b}\right]}$$

$$\Rightarrow \qquad \cfrac{x}{\dfrac{a-b}{b^2}} = \cfrac{y}{\dfrac{a-b}{a^2}} = \cfrac{1}{\dfrac{a-b}{a^2 b^2}}$$

$$\Rightarrow \qquad x = \left[\dfrac{(a-b)}{b^2}\times\dfrac{a^2 b^2}{(a-b)}\right] = a^2$$

and

$$y = \dfrac{(a-b)}{a^2}\times\dfrac{a^2 b^2}{(a-b)} = b^2$$

Hence, $x = a^2$, and $y = b^2$ is the required solution.

EXAMPLE 4. *Solve* : $\dfrac{a}{x}-\dfrac{b}{y}=0, \dfrac{ab^2}{x}+\dfrac{a^2 b}{y}=(a^2+b^2); where\ x \neq 0, y \neq 0$.

SOLUTION. Taking $\dfrac{1}{x}=u$ and $\dfrac{1}{y}=v$ the given equations become.

$$au - bv \neq 0 = 0 \qquad\qquad\qquad \dots(1)$$

$$ab^2 u + a^2 bv - (a^2+b^2) = 0 \qquad\qquad \dots(2)$$

By cross multiplication, we have

$$\cfrac{u}{\begin{matrix}-b & & 0\\ a^2 b & & -(a^2+b^2)\end{matrix}} = \cfrac{v}{\begin{matrix}0 & & a\\ -(a^2+b^2) & & ab^2\end{matrix}} = \cfrac{1}{\begin{matrix}a & & -b\\ ab^2 & & a^2 b\end{matrix}}$$

$$\therefore \qquad \cfrac{u}{[b(a^2+b^2)-0]} = \cfrac{v}{[0+a(a^2+b^2)]} = \cfrac{1}{a^3 b+ab^3}$$

$$\therefore \qquad u = \dfrac{b(a^2+b^2)}{a^3 b+ab^3}, v = \dfrac{a(a^2+b^2)}{a^3 b+ab^3}$$

$$\Rightarrow \qquad u = \dfrac{1}{a}, v = \dfrac{1}{b} \qquad \Rightarrow \qquad \dfrac{1}{x}=\dfrac{1}{a}, \dfrac{1}{y}=\dfrac{1}{b}$$

$$\Rightarrow \qquad x = a, y = b$$

Hence, $x = a, y = b$ is the solution. Hence $x = 4, y = 3$ is the required solution.

Exercise 3.2

Solve each of the following systems of equations by using the method of cross multiplication.

1. $2x - y - 3 = 0$
$4x + y - 3 = 0$

2. $3x - 2y + 3 = 0$
$4x + 3y - 47 = 0.$

3. $2x + y = 35$
$3x + 4y = 65$

4. $3x - 5y + 25 = 0$
$2x + y + 10 = 0$

5. $4x - 7y + 28 = 0$
$5y - 7x + 9 = 0.$

6. Solve $\dfrac{2}{x} + \dfrac{3}{y} = 13$ and $\dfrac{5}{x} - \dfrac{4}{y} = -2$ where $x \neq 0$, $y \neq 0$.

7. $ax + by = (a - b)$ **8.** $\dfrac{x}{a} = \dfrac{y}{b}$
$bx - ay = (a + b)$ $ax + by = a^2 + b^2$

9. $\dfrac{x}{a} + \dfrac{y}{b} = (0 + b)$, $\dfrac{x}{a^2} + \dfrac{y}{b^2} = 2$

10. $\dfrac{1}{x} + \dfrac{1}{y} = 7$

$\dfrac{2}{x} + \dfrac{3}{y} = 17 \ (x \neq 0, y \neq 0)$

HINT TO SELECTED PROBLEMS

5. Given equations are:

$4x - 7y + 28 = 0$ ... (1)

$-7x + 5y + 9 = 0$ ... (2)

By cross multiplication, we have

$$\frac{x}{-7 \times 9 - 5 \times 28} = \frac{y}{28 \times (-7) - 9 \times 4}$$

$$= \frac{1}{4 \times 5 - (-7)(-7)}$$

$$\Rightarrow \quad \frac{x}{-63 - 140} = \frac{y}{-196 - 36} = \frac{1}{20 - 49}.$$

$$\Rightarrow \quad \frac{x}{-203} = \frac{y}{-232} = \frac{1}{-29}.$$

$$\Rightarrow \quad x = \frac{-203}{-29} = 7 \text{ and } y = \frac{-232}{-29} = 8$$

Hence, $x = 7$, $y = 8$ is the required solution.

6. Taking $\dfrac{1}{x} = u$ and $\dfrac{1}{y} = v$

We get $2u + 3v - 13 = 0$ ...(1)
                $5u - 4v + 2 = 0$ ...(2)

8. The given equations are
$$bx - ay = 0$$
$$ax + by - (a^2 + b^2) = 0$$

9. The given equations are
$$bx + ay - ab(a + b) = 0$$
$$b^2x + a^2y - 2a^2b^2 = 0$$

10. Put $\dfrac{1}{x} = u$ and $\dfrac{1}{y} = v$

ANSWERS

1. $x = 1, y = -1$ **2.** $x = 5, y = 9$ **3.** $x = 15, y = 5$ **4.** $x = \dfrac{125}{41}, y = \dfrac{-89}{41}$

5. $x = 7, y = 8$ **6.** $x = \dfrac{1}{2}, y = \dfrac{1}{3}$ **7.** $x = 1, y = -1$ **8.** $x = a, y = b$

9. $x = a^2, y = b^2$ **10.** $x = \dfrac{1}{4}, y = \dfrac{1}{3}$

3.5 WORD PROBLEMS ON SIMULTANEOUS LINEAR EQUATIONS

EXAMPLE 1. *7 audio cassettes and 3 video cassettes cost Rs1395, while 5 audio cassettes and 4 video cassettes cost Rs1665. Find the cost of an audio cassettes and that of a video cassette.*

SOLUTION. Let the cost of each audio cassette be Rs x and that of each video cassette be Rs y. Then

$$7x + 3y = 1395 \quad \ldots(1)$$
$$5x + 4y = 1665 \quad \ldots(2)$$

Multiplying (1) by 4 and (2) by 3 and subtracting, we get

$$13x = 585 \qquad \Rightarrow \qquad x = 45$$

Substituting $x = 45$ in (1) we get

$$(7 \times 45) + 3y = 1395$$
$$\Rightarrow \qquad 3y = (1395 - 315) = 1080$$
$$\Rightarrow \qquad y = 360$$

cost of 1 audio cassette = Rs 45

and cost of 1 video cassette = Rs 360.

EXAMPLE 2. *Five years ago, a man was seven times as old as his son, and five years hence, the man's age will be thr 2 times his son's age. Find their present ages.*

SOLUTION. Let the present ages of the man and his son be x years and y years respectively.

The man's age 5 years ago = $(x - 5)$ years

The son's age 5 years ago = $(y - 5)$ years

$$\therefore \qquad (x - 5) = 7(y - 5) \qquad \Rightarrow \qquad x - 7y = 30 \qquad \ldots(1)$$

The man's age 5 years hence = $(x + 5)$ years

The son's age 5 years hence = $(y + 5)$ years

$$\therefore \qquad (x + 5) = 3(y + 5)$$
$$\Rightarrow \qquad x - 3y = 10$$

On subtracting (1) from (2)

$$4y = 40 \qquad \Rightarrow \qquad y = 10$$

Putting $y = 10$ in (2), we get

$$x - 3 \times 10 = 10 \qquad \Rightarrow \qquad x = 40$$

$\therefore$ The man's present age = 40 years

$\therefore$ The son's present age = 10 years

EXAMPLE 3. *Four years ago mother was four times as old as her daughter. Six years later, the mother will be two and half times as old as her daughter, form the pair of linear equations for the situation and determine the present ages of mother and her daughter in years, solving the linear equations by substitution method.*

SOLUTION. Present age of mother = x years

Present age of daughter = y years

Four years ago:

The age of mother = $(x - 4)$ years

The age of daughter = $(y - 4)$ years

$$(x - 4) = 4(y - 4)$$
$$\Rightarrow \qquad x - 4y + 12 = 0 \qquad \ldots(1)$$

Six years later:

The age of mother $= (x + 6)$ years

The age of daughter $= (y + 6)$ years

$$(x + 6) = \frac{5}{2}(y + 6)$$

$$2x - 5y - 18 = 0 \qquad \qquad \qquad \qquad \text{...(2)}$$

from (1) $\qquad\qquad x = 4y - 12 \qquad\qquad\qquad\qquad \text{...(3)}$

Substituting x from (3) in (2), we get

$$2(4y - 12) - 15y - 18 = 0$$

or $\qquad 8y - 24 - 5y - 18 = 0$

or $\qquad\qquad\qquad\quad 3y = 42$

or $\qquad\qquad\qquad\quad y = 14$

Substituting $y = 14$ in (2), we get

$$x = 4 \times 14 - 12 \quad \Rightarrow \quad x = 44$$

Therefore age of mother $= 44$ years and the age of daughter $= 14$ years.

EXAMPLE 4. *A two digit number is 4 more than 6 times the sum of its digits. If 18 is subtracted from the number, the digits are reversed. Find the number.*

SOLUTION. Let the tens digit of the required number be x and let its units digit be y.

Then the number $= (10x + y)$

$\therefore \qquad\qquad\qquad 10x + y = 6(x + y) + 4$

$\Rightarrow \qquad\qquad\qquad 4x - 5y = 4 \qquad\qquad\qquad\qquad \text{...(1)}$

Number formed on reversing the digits $= (10y + x)$.

$\therefore \qquad\qquad 10x + y - 18 = 10y + x$

$\Rightarrow \qquad\qquad 9(x - y) = 18 \qquad\qquad \Rightarrow \qquad x - y = 2 \qquad \text{...(2)}$

Multiplying (2) by 5 and subtracting (1) from the result.

We get $\qquad\qquad\qquad x = 6$

Substituting $x = 6$ in (2), we get

$$y = (6 - 2) = 4$$

Thus $\qquad\qquad\qquad x = 6 \text{ and } y = 4$

tens digit $= 6$ and units digit $= 4$

Hence, the required number $= 64$.

EXAMPLE 5. *Seven times a two digit number is equal to four times the number obtained by reversing the order of its digits. If the difference the digits is 3, find the number.*

SOLUTION. Let the tens and units digits of the required number be x and y respectively.

Then, the number $= 10x + y$. The number obtained reversing the digits $= 10y + x$

$\therefore \qquad\qquad 7(10x + y) = 4(10y + x)$

$\Rightarrow \qquad\qquad 33(2x - y) = 0 \qquad\qquad \Rightarrow \qquad 2x - y = 0$

$\Rightarrow \qquad\qquad\qquad y = 2x \qquad\qquad\qquad\qquad\qquad\qquad \text{...(1)}$

Thus, unit digit $= 2 \times$ tens digit (unit digit) $>$ (tens digit) so $y > x$

$$y - x = 3 \qquad \qquad \text{...(2)}$$

Using (1) in (2), we get

$$(2x - x) = 3, x = 3$$

On substituting $x = 3$ in (1), we get $y = 2 \times 3 = 6$

Hence, the required number = 36.

EXAMPLE 6. *The sum of the present ages of Kamal and his mother is 89 years. After 11 years, mother's will be Kamal's age. Find their present ages.* **[uptu(b.pharma)–2001]**

SOLUTION. Let the present age of Kamal be x years and that of his mother be y years. Then according to the question, we have

$$x + y = 89 \qquad \qquad \text{...(1)}$$

and $\qquad (y + 11) = 2(x + 11)$ or $2x - y = -11 \qquad \qquad \text{...(2)}$

Adding (1) and (2), we get

$$3x = 78$$
$$x = 26$$

Putting $x = 26$ in (1), we get $y = 89 - 26 = 63$.

Hence, Kamal's present age is 26 years and his mother's present age is 63 years.

EXAMPLE 7. *In a given fraction, if the number is multiplied by 3 and the denominator is reduced by 1, we get $\dfrac{3}{2}$, but if the numerator is increased by 12 and the denominator is multiplied by 3, we get $\dfrac{1}{3}$, find the fraction.* **[Uptu(B.pharma)–2007]**

SOLUTION. Let the fraction be $\dfrac{x}{y}$. Then according to question,

$$\frac{3x}{y-1} = \frac{3}{2} \quad \Rightarrow \quad 6x = 3y - 3$$

$$\Rightarrow \qquad 6x - 3y = -3$$
$$\Rightarrow \qquad 2x - 3y = -1 \qquad \qquad \text{...(1)}$$

and $\qquad \dfrac{x+12}{3y} = \dfrac{1}{3} \quad \Rightarrow \quad 3x + 36 = 3y$

$$\Rightarrow \qquad 3x - 3y = -36$$
$$\Rightarrow \qquad x - y = -12 \qquad \qquad \text{...(2)}$$

Subtracting (2) from (1), we get $x = 11$, putting $x = 11$ in (1), we get

$$2 \times 11 - y = -1 \quad \Rightarrow \quad y = -1 - 22$$
$$\Rightarrow \qquad y = 23$$

Hence, the required fraction is $\dfrac{11}{23}$.

EXAMPLE 8. *If the numerator of a fraction is multiplied by 2 and denominator is increased by 1, it becomes 1. However, if the numerator is increased by 4 and denominator is multiplied by 2, then the ratio of the numerator and denominator is 1:2, from a pair of linear equations for the problem and solve by substitution method and hence find the fraction.*

SOLUTION. Let the given fraction be $\dfrac{x}{y}$. According to the given conditions

$$\frac{2x}{y+1} = 1 \text{ and } \frac{x+4}{2y} = \frac{1}{2}$$

$\Rightarrow \qquad\qquad 2x = y + 1 \text{ and } x + 4 = y$

Thus we get the required pair of linear equations:

$$2x - y - 1 = 0 \qquad\qquad\qquad ...(1)$$
$$x - y + 4 = 0 \qquad\qquad\qquad ...(2)$$

from (1) $\qquad\qquad\qquad y = 2x - 1 \qquad\qquad ...(3)$

Substituting y from (3) in (2), we get

$$x - (2x - 1) + 4 = 0 \qquad \Rightarrow \qquad x = 5$$

Substituting $x = 5$ in (3), we get

$$y = 10 - 1 = 9$$

Hence, the required fraction is $\dfrac{5}{9}$.

Exercise 3.3

1. Find two numbers such that the sum of twice the first and thrice the second is 92; and four times the first exceeds seven times the second by 2.

2. If 2 is added to each of two given numbers, their ratio becomes 1:2. However, if 4 is subtracted from each of the given numbers, the ratio becomes 5:11. Find the numbers.

3. The monthly incomes of A and B are in the ratio 8:7 and their expenditures are in the ratio 19:16. If each saves Rs. 2500 per month, find the monthly income of each.

4. A mother is three times as old as her daughter five years later, the mother will be two and a half times as old as her daughter, find the age of the daughter and that of her mother in years.

5. A fraction becomes 1/2 if 5 is subtracted from its numerator and 3 is subtracted from its denominator. However, if we divide the numerator by 2 and add 7 to the denominator, the fraction becomes 1/4. Determine the fraction.

6. Of the numbers, one number is greater than thrice the other number by 2 and 4 times the smaller number exceeds the larger number by 5. Find the numbers.

7. The sum of the digits of a two digit number is 93. The number obtained by interchanging the digits of the given number exceeds that number by 27, find the number.

8. Harsh purchased 4 chairs and 3 tables for Rs 1650. From the same place and at the same rates Sunit purchased 3 chairs and 2 tables for Rs 1150. Find the cost per chair and per table.

9. The sum of the digits of a two digit number is 12. If the digits are reversed, the new number is 12 less than twice the original number. Find the original number.

10. If 1 is added to both the numerator and the denominator of again fraction, it becomes 4/5. If however, 5 is subtracted from both the numerator and the denominator, the fraction becomes 1/2. Find the fraction.

HINT TO SELECTED PROBLEMS

3. Let the monthly income of A and B be Rs $8x$ and Rs $7x$ respectively and let their expenditures be Rs $19y$ and Rs. $16y$ respectively.

Then A's monthly saving = Rs$(8x - 19y)$
and B's monthly saving = Rs$(7x - 16y)$

$\therefore$ $8x - 19y = 2500$ ...(1)

       $7x - 16y = 2500$ ...(2)

6. Let the larger number $= x$ and the smaller number $= y$

Then according to the given conditions, we have

(large number) $- 3 \times$ (smaller number) $= 2$

and $4 \times$ (small number) $-$ (larger number) $= 5$

       $x - 3y = 2$ ...(1)

       $-x + 4y = 5$ ... (2)

Adding (1) and (2) we get $y = 7$

Substituting $y = 7$ in (1) we get

       $x - 3 \times 7 = 2 \implies x = 23$

the larger number $= 23$ and the smaller number $= 7$.

9. Let the tens digit and units digit be x and y respectively.

Then $x + y = 12$ ...(1)

Original number $= 10x + y$.

Number obtained on reversing the digits

       $= (10y + x)$

$10y + x = 2(10x + y) - 12$

$19x - 8y = 12$ ...(2)

10. Let the required fraction be $\dfrac{x}{y}$. Then according to the given condition, we have

$\dfrac{x+1}{y+1} = \dfrac{4}{5}$ and $\dfrac{x-5}{y-5} = \dfrac{1}{2}$.

$\implies 5(x+1) = 4(y+1)$ and $2(x-5) = (y-5)$.

$\implies -5x - 4y = -1$ ...(1)

       $2x - y = 5$ ...(2)

ANSWERS

1. 25, 14   **2.** 34, 70   **3.** Rs 1200 and Rs 10500

4. Age of daughter = 15 years, Age of mother = 45 years   **5.** 3/7   **6.** 23 and 7   **7.** 58

8. cost of chair = Rs. 150, cost of table = Rs. 350   **9.** 48   **10.** 7/9

Objective Evaluation

MULTIPLE CHOICE QUESTIONS

Choose the most appropriate one.

1. Solution of $3x - 4y = 1$, $4x - 3y - 6 = 0$ is :
   (a) $x = 3, y = 2$ (b) $x = 2, y = 3$
   (c) $x = 2, y = 2$ (d) $x = 3, y = 3$

2. The value of k for which $x - 2y = 3$, $ky + 3x = 1$ represent parallel lines is :
   (a) $k = 6$ (b) $k = -6$
   (c) $k \neq 6$ (d) $k \neq -6$

3. The system of linear equations $2x - 5 = 3y$ and $4x - 6y = 3$ represents a :
   (a) intersecting lines   (b) parallel lines.
   (c) coincident lines   (d) none of these

4. The system of linear equations $3x + 2y - 4 = 0$ and $ax - y - 3 = 0$, will represent intersecting lines if :
   (a) $a = -3/2$ (b) $a \neq -3/2$
   (c) $a = -2/3$ (d) $a \neq -2/3$

5. If the system of equations $2x + 3y = 7$ and $(a + b)x + (2a - b)y = 21$ has infinitely many solutions then :
   (a) $a = 1, b = 5$ (b) $a = 5, b = 1$
   (c) $a = -1, b = 5$ (d) $a = 5, b = -1$

6. The system of linear equations $ax + by + c = 0$ and $dx + ey + f = 0$ will represent coincident lines if :
   (a) $ae = bd$ and $bf = ec$
   (b) $ae = bd$ and $bf \neq ec$
   (c) $ab = de$ and $bc = ef$
   (d) $ab = de$ and $bc \neq ef$

7. Equation of line which is parallel to $\sqrt{2}x - \sqrt{3}y = 5$ is :
   (a) $\sqrt{8}x - 2\sqrt{3}y = 5$ (b) $\sqrt{8}x - 2\sqrt{3}y = 1$
   (c) $-\sqrt{8}x + 2\sqrt{3}y = 5$ (d) $-\sqrt{8}x + 2\sqrt{3}y = 1$

8. The system of linear equations $2x = 3(y - 3)$ and $6x - 9y = 5$ represent a :
   (a) parallel lines   (b) coincident lines
   (c) intersecting lines   (d) none of these

9. Equation of a line which is parallel to $\frac{2}{3}x + 3y + 1 = 0$ is :
   (a) $\frac{2}{3}x + 3y + 1 = 0$ (b) $\frac{2}{3}x - 3y + 5 = 0$
   (c) $\frac{3}{2}x + 3y + 1 = 0$ (d) $\frac{3}{2}x - 3y + 5 = 0$

10. Value of k for which $2(k - 1)x + y = 1$, $3x - y = 1$ represents parallel lines is :
    (a) $k \neq 5/2$ (b) $k = 5/2$
    (c) $k = 1/2$ (d) $k \neq -1/2$

TRUE / FALSE

Write 'T' for True and 'F' for False statement.

1. The lines $2x - 3y = 9$ and $8 - 2y = x$ are parallel. **(T/F)**

2. The lines $5x - 1 = 2y$ and $y = -\frac{1}{2} + \frac{5}{2}x$ are coincident. **(T/F)**

3. Equation of line which is coincident to $5(x - y) = 3$ is $10x - 10y - 3 = 0$. **(T/F)**

4. The lines $2(x - 3) = 5y$ and $4x - 1 = 10y$ represents intersecting lines. **(T/F)**

5. If $2x - 3y = 7$ and $(a + b)x - (a + b - 3)y = -4a + b$ represent coincident lines then $a + 5b = 0$. **(T/F)**

6. The solution of $2x - 3y = 5$ and $\frac{x}{4} - 3y = 1$ is $x = 4, y = 1$. **(T/F)**

FILL IN THE BLANKS

1. The lines $2x - 3y = 9$ and $kx - 9y = 18$ will be parallel if k is _____.

2. Value of k for which $x - 2y = 5$, $3x + ky + 15 = 0$ is inconsistent is _____.

3. The value of p for which the system of equations $3x + 5y = 0$ and $px + 10y = 0$ has a non-zero solution is _____.

4. The value of k for which the system of equations $kx - 2 = y$ and $-2y + 3 = 6x$ is consistent with unique solution is _____.

5. Solution of $\sqrt{2}x - \sqrt{5}y = 0, 2\sqrt{3}x - \sqrt{7}y = 0$ is _____.

6. $\overline{ax + by = c}$, $lx + my = n$ has a unique solution, if _____.

7. The equation $\dfrac{2}{3}x - 5y + 1 = 0$ and $\dfrac{3}{2}x + 3 = 5y$

are _____.

8. The system of equations $4x + py + 8 = 0$ and $2x + 2y + 2 = 0$ will have a unique solution if _____.

ANSWERS

⊷ MULTIPLE CHOICE QUESTIONS

1. (a) **2.** (b) **3.** (a) **4.** (b) **5.** (b) **6.** (a) **7.** (a) **8.** (a) **9.** (a)
10. (c)

⊷ TRUE/ FALSE

1. F **2.** T **3.** F **4.** F **5.** F **6.** F

⊷ FILL IN THE BLANKS

1. $k = 6$ **2.** $k = 6$ **3.** $6 \neq p$ **4.** $k \neq 6$ **5.** $x = 0, y = 0$
6. $am \neq bl$ **7.** intersecting **8.** $p \neq 4$

❑❑❑❑❑

CHAPTER 4

Binomial Theorem

4.1 INTRODUCTION

In previous classes, we have studied the following results

$$(x + y)^0 = 1$$
$$(x + y)^1 = x + y$$
$$(x + y)^2 = x^2 + 2xy + y^2$$
$$(x + y)^3 = x^3 + 3x^2y + 3xy^2 + y^3$$

These formulae can be obtained by direct multiplication of $(x + y)$ to $(x + y)$ as many times as we desired. Newton established a general formula for finding any positive integral power of $(x + y)$, *i.e.*, $(x + y)^n$.

4.2 BINOMIAL EXPRESSION

An algebraic expression which contains only two terms is called a binomial expression.

For example. $x^2 + y^2, x + y, 2x + 3y$ etc.

4.3 BINOMIAL THEOREM

For any positive integer n,

$$(x + a)^n = {}^nC_0 x^n + {}^nC_1 x^{n-1} \cdot a + {}^nC_2 x^{n-2} \cdot a^2 + ... + {}^n C_r x^{n-r} \cdot a^r + ...$$
$$+ {}^nC_{n-1} \cdot xa^{n-1} + {}^nC_n x^0 a^n$$

Proof. We shall prove this theorem by principle of mathematical induction.

Let $P(n) : (x + a)^n$

$$= {}^nC_0 x^n \cdot a^0 + {}^nC_1 x^{n-1} \cdot a + {}^nC_2 x^{n-2} \cdot a^2 + ... + {}^nC_r x^{n-r} \cdot a^r + ... + {}^nC_{n-1} \cdot x \cdot a^{n-1}$$
$$+ {}^nC_n x^0 a^n \qquad ...(1)$$

Step 1. For $n = 1$,

L.H.S. $= (x + y)^1 = x + a$

R.H.S. $= {}^1C_0 x^1 a^0 + {}^1C_1 x^0 a = 1 \cdot x \cdot 1 + 1 \cdot 1 \cdot a = (x + a) =$ L.H.S.

$\Rightarrow P(1)$ is true.

Step 2. Let $P(m)$ be true, *i.e.*,

$P(m) : (x + a)^m$

$$= {}^mC_0 x^m a^0 + {}^mC_1 \cdot x^{m-1} \cdot x^{m-1}a + {}^mC_2 x^{m-2} \cdot a^2 + ... + {}^mC_{m-1} x^1 \cdot a^{m-1}$$
$$+ {}^mC_m x^0 a^m \qquad ...(2)$$

Step 3. We have to prove that $P(n)$ is true for $n = m+1$.

$$P(m+1):(x+a)^{m+1} = (x+a)^m \cdot (x+a)$$

$$= (x+a)[{}^mC_0x^m \cdot a^0 + {}^mC_1x^{m-1} \cdot a + {}^mC_2x^{m-2} \cdot a^2 + ... + {}^mC_rx^{m-r} \cdot a^r$$

$$+ ... + {}^mC_{m-1}x \cdot a^{m-1} + {}^mC_mx^0a^m] \qquad (\because P(m) \text{ is true.})$$

$$= {}^mC_0x^{m+1} \cdot a^0 + ({}^mC_1 + {}^mC_0)x^ma^1 + ({}^mC_2 + {}^mC_1)x^{m-1} \cdot a^2 + ...$$

$$+ ({}^mC_r + {}^mC_{r-1})x^{m-r+1}a^r + ... + ({}^mC_{m-1} + {}^mC_m)x^1a^m + {}^mC_ma^{m+1}$$

$$= {}^{m+1}C_0x^{m+1} \cdot a^0 + {}^{m+1}C_1x^m \cdot a^1 + {}^{m+1}C_2x^{m-1}a^2 + ... + {}^{m+1}C_rx^{m+1-r}a^r$$

$$+ ... + {}^{m+1}C_mx^1a^m + {}^{m+1}C_{m+1}a^{m+1}] \quad (\because {}^mC_{r-1} + {}^mC_r = {}^{m+1}C_r)$$

$\Rightarrow P(m+1)$ is true.

Hence, by principle of mathematical induction $P(n)$ is true for all n.

4.3.1 SOME IMPORTANT POINTS TO BE REMEMBER

For the expansion of $(x+a)^n$

(1) The number of terms in the expansion of $(x+a)^n$ is $(n+1)$.

(2) In each term of the expansion, the sum of the exponents (powers of x and y) is n (i.e., each term is of degree n).

(3) The binomial coefficient of terms from the beginning and from the end are equal since
$${}^nC_r = {}^nC_{n-r}.$$

(4) Replacing a by $-a$, we get

$$(x-a)^n = {}^nC_0x^na^0 - {}^nC_1x^{n-1}a^1 + {}^nC_2x^{n-2}a^2 - {}^nC_3x^{n-3}a^3$$

$$+ ... + (-1)^r \, {}^nC_rx^{n-r}a^r + ... + (-1)^n \cdot {}^nC_nx^0a^n$$

i.e., $$(x-a)^n = \sum_{r=0}^{n} (-1)^r \cdot {}^nC_r \cdot x^{n-r} \cdot a^r$$

Thus the terms in the expansion of $(x-a)^n$ are alternatively positive and negative, the last term is positive or negative according as n is even or odd.

(5) Putting $x = 1$ and $a = x$ in the expansion of $(x+a)^n$, we get

$$(1+x)^n = {}^nC_0 + {}^nC_1x + {}^nC_2x^2 + ... + {}^nC_rx^r + ... + {}^nC_nx^n$$

i.e., $(1+x)^n = \sum_{r=0}^{n} {}^nC_rx^r$, this is the expansion of $(1+x)^n$ in ascending powers of x.

(6) Putting $a = 1$ in the expansion of $(x+a)^n$, we get

$$(x+1)^n = {}^nC_0x^n + {}^nC_1x^{n-1} + {}^nC_2x^{n-2} + ... + {}^nC_rx^{n-r} + ... + {}^nC_{n-1}x + {}^nC_nx^0$$

i.e., $$(x+1)^n = \sum_{r=0}^{n} {}^nC_rx^{n-r}$$

(7) Putting $x = 1$ and $a = -x$ in the expansion of $(x+a)^n$, we get

$$(1-x)^n = {}^nC_0 - {}^nC_1x + {}^nC_2x^2 - {}^nC_3x^3 + ... + (-1)^r \, {}^nC_rx^r + ... + (-1)^n \, {}^nC_nx^n$$

$\Rightarrow \qquad (1-x)^n = \sum_{r=0}^{n} (-1)^r \cdot {}^n C_r \cdot x^r$

(8) The coefficient of $(r+1)$th term in the expansion of $(1+x)^n$ is ${}^nC_r$.

(9) The coefficient of x^r in the expansion of $(1+x)^n$ is ${}^nC_r$.

(10) $(x+a)^n + (x-a)^n = 2[\,^nC_0 x^n a^0 +\,^n C_2 x^{n-2} a^2 +\,^n C_4 x^{n-4} a^4 + ...]$

(11) $(x+a)^n - (x-a)^n = 2[\,^nC_1 x^{n-1} a^1 +\,^n C_3 x^{n-3} a^3 + ...]$

⇒ REMARKS

⇒ In the above points (10) and (11) if n is odd, then $[(x+a)^n + (x-a)^n]$ and $[(x+a)^n - (x-a)^n]$ both have the same number of terms equal to $\dfrac{n+1}{2}$..

⇒ When n is even, then $[(x+a)^n + (x-a)^n]$ has $\left(\dfrac{n}{2}+1\right)$ terms and $[(x+a)^n - (x-a)^n]$ has $\dfrac{n}{2}$ terms.

⇒ $^nC_0,\,^n C_1,\,^nC_2,...,\,^nC_n$ are called the binomial coefficients.

⇒ ^nC_r is greatest for $r = \dfrac{n}{2}$ if n is even and for $r = \dfrac{n-1}{2}$ and $r = \dfrac{n+1}{2}$ if n is odd.

4.4 PASCAL'S TRIANGLE

We know that $(x+a)^0 = 1$

$$(x+a)^1 = x+a$$
$$(x+a)^2 = x^2 + 2xa + a^2$$
$$(x+a)^3 = x^3 + 3x^2 a + 3a^2 x + a^3$$
$$(x+a)^4 = x^4 + 4x^3 a + 6x^2 a^2 + 4xa^3 + a^4$$

The coefficient of above expansion can be written as follows :

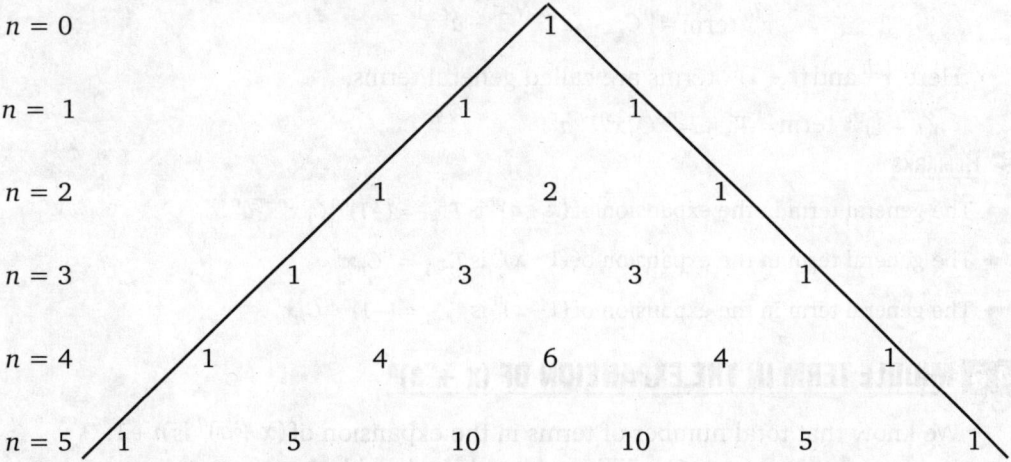

$n = 0$						1					
$n = 1$					1		1				
$n = 2$				1		2		1			
$n = 3$			1		3		3		1		
$n = 4$		1		4		6		4		1	
$n = 5$	1		5		10		10		5		1

This representation of numbers seen like a triangle. Therefore it is called Pascal's triangle.

Pascal's triangle in the form of combinations

$n = 0$						$C(0,0)$					
$n = 1$					$C(1,0)$		$C(1,1)$				
$n = 2$				$C(2,0)$		$C(2,1)$		$C(2,2)$			
$n = 3$			$C(3,0)$		$C(3,1)$		$C(3,2)$		$C(3,3)$		
$n = 4$		$C(4,0)$		$C(4,1)$		$C(4,2)$		$C(4,3)$		$C(4,4)$	
$n = 5$	$C(5,0)$		$C(5,1)$		$C(5,2)$		$C(5,3)$		$C(5,4)$		$C(5,5)$

From the Pascal's triangle, it is clear that

➠ Each row begins and ends with 1.

➠ Each coefficient in any row is the sum of the two coefficients on the two sides of it in the preceding row.

➠ The coefficient from the beginning and end are equal.

➠ The coefficient of an expansion can be obtained from the coefficients of the previous expansion.

➠ Coefficient of any term in the expansion of $(a + x)^n$

$$= \frac{\text{coefficient of preceding term} \times \text{power of } a \text{ in the preceding term}}{\text{Total number of preceding terms}}$$

4.5 GENERAL TERM IN THE EXPANSION OF $(x + a)^n$

We know that

$$(x+a)^n = {}^nC_0 x^n a^0 + {}^nC_1 x^{n-1} a^1 + {}^nC_2 x^{n-2} a^2 + \dots + {}^nC_{n-1} x^1 \cdot a^{n-1} + {}^nC_n x^0 \cdot a^n$$

Here,   first term $= {}^n C_0 x^n a^0$

second term $= {}^nC_1 x^{n-1} a^1$

third term $= {}^nC_2 x^{n-2} a^2$

fourth term $= {}^nC_3 x^{n-3} a^3$

$- - - - - - - - - - - - - - - - - - - -$ etc.

r^{th} term $= {}^nC_{r-1} \cdot x^{n-(r-1)} \cdot a^{r-1}$

Here, r^{th} and $(r+1)^{th}$ terms are called general terms.

$\therefore (r+1)^{th}$ term $= T_{r+1} = {}^n C_r x^{n-r} a^r$

⚛ REMARKS

➠ The general term in the expansion of $(x-a)^n$ is $T_{r+1} = (-1)^{rn} C_r x^{n-r} a^r$.

➠ The general term in the expansion of $(1+x)^n$ is $T_{r+1} = {}^nC_r x^r$.

➠ The general term in the expansion of $(1-x)^n$ is $T_{r+1} = (-1)^r \; {}^nC_r x^r$.

4.6 MIDDLE TERM IN THE EXPANSION OF $(x + a)^n$

We know that total number of terms in the expansion of $(x+a)^n$ is $n+1$.

If n is even, then $(n + 1)$ will be odd and if n is odd, then $(n + 1)$ is even.

Case I. If n is even

Let $n = 2m \implies m = \dfrac{n}{2}$

Total number of terms $= n + 1 = 2m + 1$ ($\because n = 2m$)

Clearly, middle term is $(m+1)^{th}$ term

$\therefore$ $T_{m+1} = {}^n C_m x^{n-m} a^m = {}^nC_{n/2} x^{n-n/2} a^{n/2} = {}^nC_{n/2} x^{n/2} a^{n/2}$

Case II. If n is odd

Then, let $n = 2m + 1 \implies m = \dfrac{n-1}{2}$

Total number of terms $(n + 1) = (2m + 1) + 1 = (2m + 2)$

So, there are two middle terms, *i.e.*, $(m + 1)^{th}$ and $(m + 2)^{th}$ terms.

$$\therefore \quad T_{m+1} = {}^nC_m x^{n-m} a^m = {}^nC_{\frac{(n-1)}{2}} x^{\frac{(n+1)}{2}} a^{\frac{(n-1)}{2}}$$

and
$$T_{m+2} = {}^nC_{m+1} x^{n-m-1} a^{m+1} = {}^nC_{\frac{n+1}{2}} \cdot x^{\frac{n-1}{2}} a^{\frac{n+1}{2}}$$

4.7 SOME PARTICULAR TERMS IN BINOMIAL COEFFICIENT

(i) p^{th} **term from the last in the expansion of** $(x + a)^n$:

$$= [(n + 1) - (p - 1)]^{th} \text{ term from beginning.}$$
$$= (n - p + 2)^{th} \text{ term from the beginning.}$$
$$= T_{n-p+2}$$

(ii) **The terms independent of x, in the expansion of** $(x + a)^n$:

To find it, use the following procedure.

WORKING PROCEDURE

Step 1. Find T_{r+1} and solve it.

Step 2. Equating power of x to 0 and find the value of r.

Step 3. Putting this value of r in T_{r+1}, we get the term independent of x.

(iii) **Total number of terms in the expansion of** $(x + y)^n = {}^{n+1}C_1$

Total number of terms in the expansion of $(x + y + z)^n = {}^{n+2}C_2$

Total number of terms in the expansion of $(x + y + z + w)^n = {}^{n+3}C_3$

4.8 GREATEST COEFFICIENT IN BINOMIAL EXPANSION

We know that in a binomial expansion, the coefficient of the middle term is greatest. If there are two middle terms, then their coefficients are equal and maximum.

Proof : Let ${}^nC_r$ be the greatest coefficient in the expansion of $(x + a)^n$.

Then
$$\frac{{}^nC_r}{{}^nC_{r+1}} = \frac{n!}{r!(n-r)!} \cdot \frac{(r+1)!(n-r-1)!}{n!} = \frac{r+1}{n-r} \qquad \text{...(1)}$$

$\therefore {}^nC_r$ is maximum, therefore

$$\frac{r+1}{n-r} \geq 1 \qquad \Rightarrow \qquad r + 1 \geq n - r$$

$$\Rightarrow \qquad 2r \geq n - 1$$

$$\Rightarrow \qquad r \geq \frac{n-1}{2} \qquad \text{...(2)}$$

Putting $(r - 1)$ for r in eqn. (1), we get

$$\frac{{}^nC_{r-1}}{{}^nC_r} = \frac{r}{n-r+1}$$

For ^nC_r to be maximum, we must have

$$\frac{r}{n-r+1} \le 1 \qquad \Rightarrow \qquad r \le \frac{n+1}{2} \qquad \qquad ...(3)$$

From eqn. (2) and (3), we get $\frac{n-1}{2} \le r \le \frac{n+1}{2}$ ...(4)

Case I. If n is an even, i.e., $n = 2m$

Then from eqn. (4), $m - \frac{1}{2} \le r \le m + \frac{1}{2} \qquad \Rightarrow \qquad r = m$

∴ Greatest coefficient $= {}^nC_r = {}^nC_m = {}^{2m}C_m$

= coefficient of $(m + 1)^{th}$ term.

= coefficient of middle term.

Case II. If n is odd, i.e., $n = 2m + 1$

Then from eqn. (4), $m \le r \le m + 1 \Rightarrow r = m, m + 1$

∴ Maximum coefficient $= {}^nC_r = {}^{2m+1}C_m$ or ${}^{2m+1}C_{m+1}$

= coefficient of $(m+1)^{th}$ or $(m+2)^{th}$ terms

= coefficient of two middle terms.

WORKING PROCEDURE : TO FIND THE GREATEST TERM IN THE EXPANSION OF $(x + a)^n$

Step 1. Find the value of $\frac{T_{r+1}}{T_r} = \frac{n-r+1}{r} \cdot \frac{a}{x}$.

Step 2. Solve $\frac{T_{r+1}}{T_r} \ge 1$ for r and find $r \le m$.

Step 3. If m is a positive integer then m^{th} and $(m + 1)^{th}$ term will be greatest. If m is an integer and is a sum of proper fraction, then $(m + 1)^{th}$ term will be greatest.

REMARKS

➡ In the expansion of $(x + a)^n$ if n is even, then $\left(\frac{n}{2}+1\right)^{th}$ will be the middle term.

➡ In the expansion of $(x + a)^n$ if n is odd, then $\left(\frac{n+1}{2}\right)^{th}$ and $\left(\frac{n+3}{2}\right)^{th}$ terms are the middle terms.

SOLVED EXAMPLES

EXAMPLE 1. *Find the number of terms in the expansion of the following.*

(i) $(5x - 4y)^9$ (ii) $(1 + 3\sqrt{2}x)^9 + (1 - 3\sqrt{2}x)^9$

(iii) $(\sqrt{x} + \sqrt{y})^{10} + (\sqrt{x} - \sqrt{y})^{10}$ (iv) $(2x + 3y - 4z)^n$

(v) $[(3x + y)^8 - (3x - y)^8]$ (vi) $(1 + 2x + x^2)^{20}$

SOLUTION. (i) We know that, the number of terms in the expansion of $(x + a)^n$ is $(n + 1)$.

∴ Number of terms in the expansion of $(5x - 4y)^9 = 9 + 1 = 10$

(ii) If n is odd, then number of terms in the expansion of $[(x + a)^n + (x - a)^n]$ is $\dfrac{n+1}{2}$.

$\therefore$ Total number of terms in the expansion of $(1 + 3\sqrt{2}x)^9 + (1 - 3\sqrt{2}x)^9$

$$= \frac{9+1}{2} = \frac{10}{2} = 5$$

(iii) If n is even, then in the expansion of $[(x + a)^n + (x - a)^n]$, total number of terms $= \left(\dfrac{n}{2} + 1\right)$

$\therefore$ Therefore, in the expansion of $(\sqrt{x} + \sqrt{y})^{10} + (\sqrt{x} - \sqrt{y})^{10}$,

$$\text{total number of terms} = \left(\frac{10}{2} + 1\right) = 6$$

(iv) Number of terms in the expansion of $(x + y + z)^n = \dfrac{(n+1)(n+2)}{1 \cdot 2}$

therefore, number of terms in the expansion of

$$(2x + 3y - 4z)^n = \frac{(n+1)(n+2)}{2}$$

(v) If n is even, then number of terms in the expansion of $[(x + a)^n - (x - a)^n] = \dfrac{n}{2}$

therefore, number of terms in the expansion of $[(3x + y)^8 - (3x - y)^8] = \dfrac{8}{2} = 4$

(vi) $(1 + 2x + x^2)^{20} = [(1 + x)^2]^{20} = (1 + x)^{40}$

Hence, as in (i) total number of terms $= 40 + 1 = 41$

EXAMPLE 2. *Explain $(x^2 + 2a)^5$ by binomial expansion.*

SOLUTION. Expanding, by binomial theorem

$$(x^2 + 2a)^5 = {}^5C_0(x^2)^5(2a)^0 + {}^5C_1(x^2)^4(2a)^1 + {}^5C_2(x^2)^3(2a)^2 + {}^5C_3(x^2)^2(2a)^3$$
$$+ {}^5C_4(x^2)(2a)^4 + {}^5C_5(x^2)^0(2a)^5$$
$$= x^{10} + 5(x^8)(2a) + 10(x^6)(4a^2) + 10(x^4)(8a^3) + 5(x^2)(16a^4) + 32a^5$$
$$= x^{10} + 10x^8a + 40x^6a^2 + 80x^4a^3 + 80x^2a^4 + 32a^5$$

EXAMPLE 3. *Expand the following by binomial theorem :*

(i) $(2x - 3y)^4$ (ii) $(1 + x + x^2)^3$ (iii) $(1 - x + x^2)^4$

(iv) $\left(x + \dfrac{1}{y}\right)^{11}$ (v) $\left(x^2 - \dfrac{2}{x}\right)^7$ (vi) $(1 + x - x^2)^4$

SOLUTION. (i) Using binomial theorem, we have

$$(2x - 3y)^4 = \{2x + (-3y)\}^4$$

$$= {}^4C_0(2x)^4(-3y)^0 + {}^4C_1(2x)^3(-3y) + {}^4C_2(2x)^2(-3y)^2$$
$$+ {}^4C_3(2x)^1(-3y)^3 + {}^4C_4(2x)^0(-3y)^4$$

$$= 16x^4 + 4(8x^3)(-3y) + 6(4x^2)(9y^2) + 4(2x)(-27y^3) + 81y^4$$
$$= 16x^4 - 96x^3y + 216x^2y^2 - 216xy^3 + 81y^4$$

(ii) Let $y = x + x^2$, then

$$(1 + x + x^2)^3 = (1 + y)^3 = {}^3C_0 + {}^3C_1 y + {}^3C_2 y^2 + {}^3C_3 y^3$$

$$= 1 + 3y + 3y^2 + y^3 = 1 + 3(x + x^2) + 3(x + x^2)^2 + (x + x^2)^3$$

$$= 1 + 3(x + x^2) + 3(x^2 + 2x^3 + x^4) + \{{}^3C_0 x^3 (x^2)^0 + {}^3C_1 x^{3-1} (x^2)^1$$

$$+ {}^3C_2 x^{3-2} (x^2)^2 + {}^3C_3 x^0 (x^2)^3\}$$

$$= 1 + 3(x + x^2) + 3(x^2 + 2x^3 + x^4) + (x^3 + 3x^4 + 3x^5 + x^6)$$

$$= x^6 + 3x^5 + 6x^4 + 7x^3 + 6x^2 + 3x + 1$$

(iii) Let $y = -x + x^2$, then

$$(1 - x + x^2)^4 = (1 + y)^4 = {}^4C_0 + {}^4C_1 y + {}^4C_2 y^2 + {}^4C_3 y^3 + {}^4C_4 y^4$$

$$= 1 + 4y + 6y^2 + 4y^3 + y^4$$

$$= 1 + 4(-x + x^2) + 6(-x + x^2)^2 + 4(-x + x^2)^3 + (-x + x^2)^4$$

$$= 1 - 4x + 4x^2 + 6x^2(1 - 2x + x^2) - 4x^3(1 - 3x + 3x^2 - x^3)$$

$$+ x^4(1 - 4x + 6x^2 - 4x^3 + x^4)$$

$$= 1 - 4x + 4x^2 + 6x^2 - 12x^3 + 6x^4 - 4x^3 + 12x^4 - 12x^5 + 4x^6$$

$$+ x^4 - 4x^5 + 6x^6 - 4x^7 + x^8$$

$$= 1 - 4x + 10x^2 - 16x^3 + 19x^4 - 16x^5 + 10x^6 - 4x^7 + x^8$$

(iv) Using binomial expansion, we get

$$\left(x + \frac{1}{y}\right)^{11} = {}^{11}C_0 x^{11} \left(\frac{1}{y}\right)^0 + {}^{11}C_1 x^{10} \left(\frac{1}{y}\right) + {}^{11}C_2 x^9 \left(\frac{1}{y}\right)^2 + {}^{11}C_3 x^8 \left(\frac{1}{y}\right)^3$$

$$+ {}^{11}C_4 x^7 \left(\frac{1}{y}\right)^4 + {}^{11}C_5 x^6 \left(\frac{1}{y}\right)^5 + {}^{11}C_6 x^5 \left(\frac{1}{y}\right)^6 + {}^{11}C_7 x^4 \left(\frac{1}{y}\right)^7$$

$$+ {}^{11}C_8 x^3 \left(\frac{1}{y}\right)^8 + {}^{11}C_9 x^2 \left(\frac{1}{y}\right)^9 + {}^{11}C_{10} x \left(\frac{1}{y}\right)^{10} + {}^{11}C_{11} \left(\frac{1}{y}\right)^{11}$$

$$= x^{11} + 11 \frac{x^{10}}{y} + 55 \frac{x^9}{y^2} + 165 \frac{x^8}{y^3} + 330 \frac{x^7}{y^4} + 462 \frac{x^6}{y^5} + 462 \frac{x^5}{y^6} + 330 \frac{x^4}{y^7}$$

$$+ 165 \frac{x^3}{y^8} + 55 \frac{x^2}{y^9} + 11 \frac{x}{y^{10}} + \frac{1}{y^{11}}$$

(v) $\left(x^2 - \dfrac{2}{x}\right)^7 = {}^7C_0 (x^2)^7 + {}^7C_1 (x^2)^6 \left(-\dfrac{2}{x}\right) + {}^7C_2 (x^2)^5 \left(-\dfrac{2}{x}\right)^2$

$$+ {}^7C_3 (x^2)^4 \left(-\frac{2}{x}\right)^3 + {}^7C_4 (x^2)^3 \left(-\frac{2}{x}\right)^4 + {}^7C_5 (x^2)^7 \left(-\frac{2}{x}\right)^5$$

$$+ {}^7C_6 (x^2)^1 \left(-\frac{2}{x}\right)^6 + {}^7C_7 \left(-\frac{2}{x}\right)^7$$

$$= x^{14} - 7x^{12}\frac{2}{x} + 21x^{10}\frac{4}{x^2} - 35x^8\frac{8}{x^3} + 35x^6\frac{16}{x^4}$$

$$- 21x^4\frac{32}{x^5} + 7x^2\frac{64}{x^6} - \frac{128}{x^7}$$

$$= x^{14} - 14x^{11} + 84x^8 - 280x^5 + 560x^2 - 672x^{-1} + 448x^{-4} - 128x^{-7}$$

(vi) Let $x - x^2 = y$

Now, $(1 + x - x^2)^4 = (1 + y)^4 = {}^4C_0 + {}^4C_1 y + {}^4C_2 y^2 + {}^4C_3 y^3 + {}^4C_4 y^4$

$$= 1 + 4y + 6y^2 + 4y^3 + y^4$$

$$= 1 + 4(x - x^2) + 6(x - x^2)^2 + 4(x - x^2)^3 + (x - x^2)^4$$

$$= 1 + 4(x - x^2) + 6(x^2 - 2x^3 + x^4) + 4(x^3 - 3x^4$$
$$+ 3x^5 - x^6) + x^4(1 - x)^4$$

$$= 1 + 4(x - x^2) + 6(x^2 - 2x^3 + x^4) + 4(x^3 - 3x^4$$
$$+ 3x^5 - x^6) + x^4(1 - 4x + 6x^2 - 4x^3 + x^4)$$

$$= 1 + 4x + 2x^2 - 8x^3 - 5x^4 + 8x^5 + 2x^6 - 4x^7 + x^8$$

EXAMPLE 4. *Using binomial theorem, expand* $\{(x + y)^5 + (x - y)^5\}$ *and find the value of* $\{(\sqrt{2} + 1)^5 + (\sqrt{2} - 1)^5\}$.

SOLUTION. We have $(x + y)^5 + (x - y)^5 = 2[{}^5C_0 x^5 + {}^5C_2 x^3 y^2 + {}^5C_4 x^1 y^4]$

$$= 2[x^5 + 10x^3 y^2 + 5xy^4]$$

Putting $x = \sqrt{2}$ and $y = 1$,

$$(\sqrt{2} + 1)^5 + (\sqrt{2} - 1)^5 = 2[(\sqrt{2})^5 + 10(\sqrt{2})^3 + 5\sqrt{2}]$$

$$= 2[4\sqrt{2} + 20\sqrt{2} + 5\sqrt{2}] = 58\sqrt{2}$$

EXAMPLE 5. *Let O be the sum of odd terms and E be the sum of even terms. Prove that*

(i) $O^2 - E^2 = (x^2 - a^2)^n$ (ii) $4OE = (x + a)^{2n} - (x - a)^{2n}$

(iii) $2(O^2 + E^2) = (x + a)^{2n} + (x - a)^{2n}$

SOLUTION. $(x + a)^n = {}^nC_0 x^n a^0 + {}^nC_1 x^{n-1} a^1 + {}^nC_2 x^{n-2} a^2 + \ldots + {}^nC_{n-1} x a^{n-1} + {}^nC_n a^n$

$\Rightarrow$ $(x + a)^n = ({}^nC_0 x^n a^0 + {}^nC_2 x^{n-2} a^2 + \ldots) + ({}^nC_1 x^{n-1} a^1 + {}^nC_3 x^{n-3} a^3 + \ldots)$

$\Rightarrow$ $(x + a)^n = O + E$ ...(1)

and $(x - a)^n = {}^nC_0 x^n - {}^nC_1 x^{n-1} a^1 + {}^nC_2 x^{n-2} a^2 - {}^nC_3 x^{n-3} a^3 + \ldots$

$$+ {}^nC_{n-1} x(-1)^{n-1} a^{n-1} + {}^nC_n(-1)^n a^n$$

$\Rightarrow$ $(x - a)^n = ({}^nC_0 x^n + {}^nC_2 x^{n-2} a^2 + \ldots) - ({}^nC_1 x^{n-1} a^1 + {}^nC_3 x^{n-3} a^3 + \ldots)$

$\Rightarrow$ $(x - a)^n = O - E$ ...(2)

Multiplying eqn. (1) and (2), we get

(i) $(x + a)^n (x - a)^n = (O + E)(O - E)$ $\Rightarrow$ $(x^2 - a^2)^n = O^2 - E^2$

(ii) $4(OE) = [(O + E)^2 - (O - E)^2]$

 $4(OE) = [(x + a)^n]^2 - [(x - a)^n]^2$

 $4(OE) = (x + a)^{2n} - (x - a)^{2n}$

(iii) Squaring and then adding eqn. (1) and (2), we get

$$(x+a)^{2n} - (x-a)^{2n} = (O+E)^2 + (O-E)^2 = 2(O^2 + E^2)$$

Example 6. *In the expansion of* $\left(\dfrac{4x}{5} - \dfrac{5}{2x}\right)^9$ *. Find the value of the following :*

(i) *6th term from the beginning*   (ii) *4th term from the end.*

Solution. We know that $T_{r+1} = (-1)^r \, {}^nC_r x^{n-r} a^r$

$\therefore$ In $\left(\dfrac{4x}{5} - \dfrac{5}{2x}\right)^9$, $T_{r+1} = (-1)^r \, {}^9C_r \left(\dfrac{4x}{5}\right)^{9-r}\left(\dfrac{5}{2x}\right)^r$

(i) 6th term from the beginning

$$T_6 = T_{5+1} = (-1)^5 \, {}^9C_5 \left(\dfrac{4x}{5}\right)^{9-5}\left(\dfrac{5}{2x}\right)^5 \qquad (\because r = 5)$$

$$= -\dfrac{9\cdot8\cdot7\cdot6}{1\cdot2\cdot3\cdot4} \times \dfrac{4^4 x^4}{5^4} \times \dfrac{5^5}{2^5 x^5} = -\dfrac{5040}{x} \qquad (\because {}^9C_5 = {}^9C_4)$$

(ii) pth term from the end in the expansion of $(x-a)^n$

$= (n-p+2)$th term from the beginning $= T_{n-p+2}$

$\therefore$ 4th term from the end in the expansion of $\left(\dfrac{4x}{5} - \dfrac{5}{2x}\right)^9$

$$= T_{9-4+2} = T_{6+1} = (-1)^6 \, {}^9C_6 \left(\dfrac{4x}{5}\right)^{9-6}\left(\dfrac{5}{2x}\right)^6 \qquad (\because r = 6)$$

$$= \dfrac{9\cdot8\cdot7}{1\cdot2\cdot3} \times \dfrac{4^3 x^3}{5^3} \times \dfrac{5^6}{2^6 x^6} = -\dfrac{10500}{x^3} \qquad (\because {}^9C_6 = {}^9C_3)$$

Example 7. (i) *Find the general term in the expansion of* $(x^2 - y)^6$.

(ii) *Find the 10th term in the expansion of* $\left(2x^2 + \dfrac{1}{x}\right)^{12}$.

(iii) *Find the 9th term in the expansion of* $\left(\dfrac{x}{a} - \dfrac{3a}{x^2}\right)^{12}$.

Solution. (i) We have $(x^2 - y)^6 = [x^2 + (-y)]^6$

General term $T_{r+1} = {}^6C_r (x^2)^{6-r}(-y)^r \qquad (\because T_{r+1} = {}^nC_r x^{n-r}a^r)$

$= (-1)^r \, {}^6C_r x^{12-2r} \cdot y^r$

(ii) $(r+1)$th term in the expansion of $(x+a)^n = T_{r+1} = {}^nC_r x^{n-r}a^r$

$\therefore$ In the expansion of $\left(2x^2 + \dfrac{1}{x}\right)^{12}$

$$T_{10} = T_{9+1} = {}^{12}C_9 (2x^2)^{12-9}\left(\dfrac{1}{x}\right)^9 = {}^{12}C_9 (2x^2)^3 \cdot \dfrac{1}{x^9} = {}^{12}C_9 \, 2^3 \left(\dfrac{1}{x^3}\right)$$

$$= {}^{12}C_3 \dfrac{8}{x^3} = \dfrac{12\times11\times10}{3\times2\times1} \times \dfrac{8}{x^3} = \dfrac{1760}{x^3} \qquad (\because {}^{12}C_9 = {}^{12}C_3)$$

(iii) Similarly, as above 9th term in the expansion of $\left(\dfrac{x}{a} - \dfrac{3a}{x^2}\right)^{12}$

$$T_9 = T_{8+1} = {}^{12}C_8\left(\dfrac{x}{a}\right)^{12-8}\left(\dfrac{-3a}{x^2}\right)^8 = {}^{12}C_8\left(\dfrac{x}{a}\right)^4\left(\dfrac{-3a}{x^2}\right)^8$$

$$= {}^{12}C_4 \cdot 3^8 \cdot \dfrac{a^4}{x^{12}} = ({}^{12}C_4 x^{-12}a^4)3^8 = 3247695 x^{-12}a^4$$

EXAMPLE 8. *Find the sum of the coefficients of x^{-17} and x^{32} in the expansion of $\left(x^4 - \dfrac{1}{x^3}\right)^{15}$.*

SOLUTION. The general term in the expansion of

$$\left(x^4 - \dfrac{1}{x^3}\right)^{15} = T_{r+1} = (-1)^r \; {}^{15}C_r (x^4)^{15-r}\left(\dfrac{1}{x^3}\right)^r$$

$$= (-1)^r \; {}^{15}C_r x^{60-7r} \qquad \qquad \dots(1)$$

(i) For x^{32}, $60 - 7r = 32 \implies r = 4$

    $\therefore$ Coefficient of $x^{32} = (-1)^4 \cdot {}^{15}C_4 = {}^{15}C_4$

(ii) For x^{-17}, $60 - 7r = -17 \implies r = 11$

    $\therefore$ Coefficient of $x^{-17} = (-1)^{11} \; {}^{15}C_{11} = -{}^{15}C_4$

    $\therefore$ Sum of the coefficient of x^{-17} and $x^{32} = {}^{15}C_4 + (-{}^{15}C_4)$

$$= {}^{15}C_4 - {}^{15}C_4 = 0$$

EXAMPLE 9. *Find the middle term in the expansion of the following :*

(i) $\left(3x - \dfrac{2}{x^2}\right)^{15}$ (ii) $\left(\dfrac{x}{3} + 9y\right)^8$ (iii) $\left(\dfrac{2}{3}x^2 - \dfrac{3}{2x}\right)^{20}$ (iv) $\left(3x - \dfrac{x^3}{6}\right)^7$

SOLUTION. (i) $\left(3x - \dfrac{2}{x^2}\right)^{15}$

Here $n = 15$, which is an odd number. Therefore, there are two middle terms given by $\left(\dfrac{n+1}{2}\right)^{th}$ and $\left(\dfrac{n+3}{2}\right)^{th}$, *i.e.*, T_8 and T_9.

Now, $T_8 = T_{7+1} = (-1)^7 \; {}^{15}C_7 (3x)^{15-7}\left(\dfrac{2}{x^2}\right)^7 = -{}^{15}C_7 \cdot \dfrac{3^8 \cdot 2^7}{x^6}$

and $T_9 = T_{8+1} = (-1)^8 \; {}^{15}C_8 (3x)^{15-8}\left(\dfrac{2}{x^2}\right)^8 = {}^{15}C_8 \cdot \dfrac{3^7 \cdot 2^8}{x^9}$

(ii) Here $n = 8$, which is an even number so there is only one middle term.

Middle term $= T_{\frac{n}{2}+1} = T_5$

r^{th} term of $\left(\dfrac{x}{3} + 9y\right)^8 = T_{r+1} = {}^8C_r\left(\dfrac{x}{3}\right)^{8-r}(9y)^r$

$$\Rightarrow \quad T_5 = T_{4+1} = {}^8C_4\left(\frac{x}{3}\right)^{8-4}(9y)^4 = \frac{8\cdot7\cdot6\cdot5}{1\cdot2\cdot3\cdot4}\cdot\frac{x^4}{3^4}\cdot9^4y^4 = 5670x^4y^4$$

(iii) Here $n = 20$, which is an even number, therefore there is only one middle term.

$$\Rightarrow \quad \left(\frac{20}{2}+1\right)^{th} \text{ term } = 11^{th} \text{ term will be the middle term}$$

$$\therefore \quad T_{11} = T_{10+1} = {}^{20}C_{10}\left(\frac{2}{3}x^2\right)^{20-10}\left(\frac{-3}{2x}\right)^{10} = {}^{20}C_{10}x^{10}$$

(iv) Here $n = 7$, which is an odd number, therefore $\left(\frac{7+1}{2}\right)^{th}$ and $\left(\frac{7+1}{2}+1\right)^{th}$ are the middle terms.

$$\therefore \quad T_4 = T_{3+1} = {}^7C_3(3x)^{7-3}\left(\frac{-x^3}{6}\right)^3 = (-1)^3\,{}^7C_3\cdot(3x)^4\cdot\left(\frac{x^3}{6}\right)^3$$

$$= -35\times81x^4\times\frac{x^9}{216} = \frac{-105x^{13}}{8}$$

and
$$T_5 = T_{4+1} = {}^7C_4(3x)^{7-4}\left(\frac{-x^3}{6}\right)^4 = {}^7C_4(3x)^3\left(\frac{-x^3}{6}\right)^4$$

$$= 35\times27x^3\times\frac{x^{12}}{1296} = \frac{35x^{15}}{48}$$

Hence, required term $= \dfrac{-105x^{13}}{8}$ and $\dfrac{35x^{15}}{48}$.

EXAMPLE 10. *Prove that the middle term in the expansion of $(1+x)^{2n}$ is $\dfrac{1\cdot3\cdot5\ldots(2n-1)}{n!}2^n\cdot x^n$.*

SOLUTION. We have to find the expansion of $(1+x)^{2n}$. Here $2n$ is an even number.

Middle term $= \left(\dfrac{2n}{2}+1\right)^{th}$ term $= (n+1)^{th}$ term

$$\Rightarrow \quad \text{Middle term} = T_{n+1} = {}^{2n}C_n\cdot1^{2n-n}\cdot x^n = {}^{2n}C_n\cdot x^n$$

$$= \frac{2n!}{(2n-n)!n!}\cdot x^n = \frac{1\cdot2\cdot3\ldots(2n-n)(2n-1)(2n)}{n!n!}x^n$$

$$= \frac{[1\cdot3\cdot5\ldots(2n-3)(2n-1)][2\cdot4\cdot6\cdot\ldots(2n-2)(2n)]}{n!n!}x^n$$

$$= \frac{[1\cdot3\cdot5\ldots(2n-3)(2n-1)][1\cdot2\cdot3\ldots(n-1)n]2^n}{n!n!}x^n$$

$$= \frac{[1\cdot3\cdot5\ldots(2n-3)(2n-1)]n!}{n!n!}2^n\cdot x^n$$

$$= \frac{1\cdot3\cdot5\ldots(2n-1)2^n\cdot x^n}{n!}$$

Example 11. *Prove that the coefficient of middle term in the expansion of $(1+x)^{2n}$ is equal to the sum of the coefficient of middle terms in the expansion of $(1+x)^{2n-1}$.*

Solution. Using above example, the middle term in the expansion of $(1+x)^{2n}$.

$$T_{n+1} = {}^{2n}C_n \cdot x^n$$

$\therefore$ Coefficient of middle term of the expansion of $(1+x)^{2n} = {}^{2n}C_n$ ...(1)

Now, we have to find the middle term in the expansion of $(1+x)^{2n-1}$.

Here, $2n-1$ is odd, so there are two middle terms.

i.e., $\left[\dfrac{(2n-1)+1}{2}\right]^{th}$ and $\left[\dfrac{(2n-1)+1}{2}+1\right]^{th}$ terms.

i.e., n^{th} and $(n+1)^{th}$ terms.

Therefore, $T_n = T_{(n-1)+1} = {}^{2n-1}C_{n-1} \cdot 1^{(2n-1)-(n-1)} \cdot x^{n-1} = {}^{2n-1}C_{n-1} \cdot x^{n-1}$

and $T_{n+1} = {}^{2n-1}C_{n-1} \cdot (1)^{(2n-1)-n} \cdot x^n = {}^{2n-1}C_n \cdot x^n$

Sum of the coefficients of middle terms $= {}^{2n-1}C_{n-1} + {}^{2n-1}C_n$

$$= {}^{(2n-1)+1}C_n = {}^{2n}C_n \qquad \text{...(2)}$$

Hence, result is obvious from eqn. (1) and (2).

Example 12. *Find the term independent of x, in the expansion of $\left(\sqrt{x} + \dfrac{1}{3x^2}\right)^{10}$.*

Solution. In the expansion of $\left(\sqrt{x} + \dfrac{1}{3x^2}\right)^{10}$

$$T_{r+1} = {}^{10}C_r(\sqrt{x})^{10-r} \cdot \left(\frac{1}{3x^2}\right)^r = {}^{10}C_r \cdot \frac{1}{3^r} x^{\frac{(10-r)}{2}-2r}$$

$$= {}^{10}C_r \cdot \frac{1}{3^r} \cdot x^{\frac{(10-5r)}{2}} \qquad \text{...(1)}$$

For the term, independent of x, we must have

$$\frac{1}{2}(10 - 5r) = 0$$

$\Rightarrow \qquad r = 2$

Putting the value of r in eqn. (1), we get

$$T_3 = T_{2+1} = {}^{10}C_2 \cdot \frac{1}{3^2} \cdot x^0 = \frac{10 \cdot 9}{1 \cdot 2} \cdot \frac{1}{9} = 5.$$

Example 13. (*i*) *Find the coefficient of x^{10} in the expansion of $\left(2x^3 - \dfrac{3}{x}\right)^{11}$ when $x \neq 0$. Also, prove that there is no term of x^6.*

(*ii*) *Find the coefficient of x^{40} in the expansion of $(1 + 2x + x^2)^{27}$.*

(*iii*) *Find the coefficient of x^5 in the expansion of $(1 + 2x)^6 (1 - x)^7$.*

Solution. (*i*) Let in the expansion of $\left(2x^2 - \dfrac{3}{x}\right)^{11}$, $(r+1)^{th}$ terms contain x^{10}.

$$T_{r+1} =^{11} C_r (2x^2)^{11-r} \left(\frac{-3}{x}\right)^r = (-1)^r \cdot {}^{11}C_r \cdot 2^{11-r} \cdot 3^r \cdot x^{22-3r} \qquad ...(1)$$

which contains x^{10} if $22 - 3r = 10 \qquad \Rightarrow \quad r = 4$

$\Rightarrow \qquad r + 1 = 4 + 1 = 5^{\text{th}}$ term contains x^{10}.

Putting $r = 4$ in eqn. (I), we get

$$T_5 = (-1)^4 \cdot {}^{11}C_4 \cdot (2)^{11-4} \cdot 3^4 \cdot x^{10} = {}^{11}C_4 2^7 \cdot 3^4 \cdot x^{10}$$

Coefficient of $= {}^{11}C_4 \cdot 2^7 \cdot 3^4$

We have to prove that, there is no term of x^6 in the expansion of $\left(2x^2 - \dfrac{3}{x}\right)^{11}$.

Let in the expansion of $\left(2x^2 - \dfrac{3}{x}\right)^{11}, (r+1)^{\text{th}}$ term has x^6.

Then $\qquad T_{r+1} = {}^{11}C_r (2x^2)^{11-r} \left(\dfrac{-3}{x}\right)^r = {}^{11}C_r (-1)^r \cdot 2^{11-r} \cdot 3^r \cdot x^{22-3r}$

If there is a term of x^6, then $22 - 3r = 6$

$\Rightarrow r = \dfrac{16}{3}$, which is fractional and hence not possible.

Hence, there is no term of x^6.

(ii) We have to find the coefficient of x^{40} in the expansion of $(1 + 2x + x^2)^{27}$.

$$(1 + 2x + x^2)^{27} = [(1+x)^2]^{27} = (1+x)^{54}$$

Let $(r+1)^{\text{th}}$ term be contains x^{40}.

$\therefore \qquad\qquad T_{r+1} = {}^{54}C_r \cdot x^r$

For the term x^{40}, we have $r = 40$

$\therefore \qquad$ Required coefficient $= {}^{54}C_{40}$

(iii) We have to find the coefficient of x^5 in the expansion of $(1 + 2x)^6 (1 - x)^7$.

$$(1+2x)^6(1-x)^7 = [1 + {}^6C_1(2x) + {}^6C_2(2x)^2 + {}^6C_3(2x)^3 + {}^6C_4(2x)^4 + {}^6C_5(2x)^5$$

$$+ {}^6C_6(2x)^6] \times [1 - {}^7C_1 x + {}^7C_2 x^2 - {}^7C_3 x^3 + {}^7C_4 x^4 - {}^7C_5 x^5 + ...]$$

$$= (1 + 12x + 60x^2 + 160x^3 + 240x^4 + 192x^5 + ...)$$
$$\times (1 - 7x + 21x^2 - 35x^3 + 35x^4 - 21x^5 + ...)$$

$\Rightarrow$ Coefficient of $x^5 = 1 \times (-21) + 12 \times 35 + 60 \times (-35) + 160 \times 21$
$$+ 240 \times (-7) + 192 \times 1$$

$$= -21 + 420 - 2100 + 3360 - 1680 + 192 = 171$$

EXAMPLE 14. *Find the term independent of x in the following expansions :*

(i) $\left(3x^2 - \dfrac{1}{2x^3}\right)^{10}$ $\qquad$ (ii) $\left(x - \dfrac{1}{x}\right)^{12}$ $\qquad$ (iii) $\left(2x - \dfrac{1}{x}\right)^{10}$

Solution. (i) Let $(r + 1)^{th}$ term is independent of x.

Now, $T_{r+1} = {}^{10}C_r (3x^2)^{10-r} \left(-\dfrac{1}{2x^3}\right)^r = {}^{10}C_r \cdot 3x^{10-r} \left(-\dfrac{1}{2}\right)^r x^{20-5x}$ ...(1)

This term is independent of x if $20 - 5r = 0 \Rightarrow r = 4$

$\therefore$ $(4 + 1) = 5^{th}$ term is independent of x.

Putting $r = 4$ in eqn. (1), we get

$$T_5 = {}^{10}C_4 \cdot 3^6 \cdot \left(-\frac{1}{2}\right)^4 = \frac{10 \cdot 9 \cdot 8 \cdot 7}{4 \cdot 3 \cdot 2 \cdot 1} \times \frac{729}{16} = \frac{76545}{8}$$

$\therefore$ Required term $= \dfrac{76545}{8}$, which is a fraction.

(ii) Let $(r + 1)^{th}$ term be independent of x.

Now, $T_{r+1} = {}^{12}C_r x^{12-r} \left(-\dfrac{1}{x}\right)^r = {}^{12}C_r (-1)^r x^{12-2r}$ ...(2)

This term is independent of x if

$$12 - 2r = 0$$
$$\Rightarrow r = 6$$

Therefore, $(6 + 1)^{th} = 7^{th}$ term is independent of x.

Putting $r = 6$ in equation (2), we get

$$T_7 = {}^{12}C_6 (-1)^6 = {}^{12}C_6$$

$\therefore$ Required term $= {}^{12}C_6$

(iii) Let $(r + 1)^{th}$ term be independent of x.

Now, $T_{r+1} = {}^{10}C_r (2x)^{10-r} \left(-\dfrac{1}{x}\right)^r$

$$= (-1)^r \, {}^{10}C_r \cdot 2^{10-r} \cdot x^{10-2r}$$ ...(3)

Which is independent of x if $10 - 2r = 0$

$$\Rightarrow r = 5$$
$$\Rightarrow (5 + 1)^{th} = 6^{th} \text{ term will be independent of } x.$$

Putting $r = 5$ in eqn. (3),

$$T_6 = (-1)^5 \cdot {}^{10}C_5 2^{10-5} = -{}^{10}C_5 \times 2^5$$

$$= -\frac{10 \cdot 9 \cdot 8 \cdot 7 \cdot 6}{5 \cdot 4 \cdot 3 \cdot 2 \cdot 1} \times 32 = -8064$$

Hence, required term $= -8064$

Example 15. (i) *In the expansion of $(1 + x)^n$ if 462, 330 and 165 be three consecutive coefficients, find n.*

(ii) *If $3^{rd}, 4^{th}$ and 5th terms in the expansion of $(x + a)^n$ are given by 84, 280 and 560. Find x, a and n.*

Solution. (i) The general term in the expansion of $(1 + x)^n$, $T_{r+1} = {}^nC_r x^r$

Its coefficient $= {}^nC_r$

Let three successive coefficient be $^nC_{r-1}, ^nC_r$ and $^nC_{r+1}$.

Then, $^nC_{r-1} = 462, ^nC_r = 330$ and $^nC_{r+1} = 165$

$\Rightarrow \quad \dfrac{^nC_{r-1}}{^nC_r} = \dfrac{462}{330}$ and $\dfrac{^nC_r}{^nC_{r+1}} = \dfrac{330}{165}$

$\Rightarrow \dfrac{n!}{(r-1)!(n-r+1)!} \cdot \dfrac{r!(n-r)!}{n!} = \dfrac{7}{5}$ and $\dfrac{n!}{r!(n-r)!} \cdot \dfrac{(r+1)!(n-r-1)!}{n!} = 2$

$\Rightarrow \quad \dfrac{r}{n-r+1} = \dfrac{7}{5}$ and $\dfrac{r+1}{n-r} = 2$

$\Rightarrow \quad 7n - 12r + 7 = 0$ and $2n - 3r - 1 = 0$

On solving both the equations, we get $n = 11$

(ii) As per given : $T_3 = 84, T_4 = 280$ and $T_5 = 560$

$$T_3 = 84 \Rightarrow {}^nC_2 x^{n-2} a^2 = 84 \qquad \ldots(1)$$

$$T_4 = 280 \Rightarrow {}^nC_3 x^{n-3} \cdot a^3 = 280 \qquad \ldots(2)$$

$$T_5 = 560 \Rightarrow {}^nC_4 x^{n-4} \cdot a^4 = 560 \qquad \ldots(3)$$

Therefore, $\dfrac{T_3 \times T_5}{(T_4)^2} = \dfrac{84 \times 560}{(280)^2} = \dfrac{3}{5}$

$\Rightarrow \quad \dfrac{T_3}{T_4} \cdot \dfrac{T_5}{T_4} = \dfrac{3}{5}$

$\Rightarrow \left(\dfrac{3}{n-2} \times \dfrac{x}{a} \right) \times \left(\dfrac{n-3}{4} \times \dfrac{a}{x} \right) = \dfrac{3}{5}$

$$\left(\because \dfrac{T_{r+1}}{T_r} = \dfrac{n-r+1}{r} \cdot \dfrac{a}{x} \text{ and } \dfrac{T_r}{T_{r+1}} = \dfrac{r}{n-r+1} \cdot \dfrac{x}{a} \right)$$

$\Rightarrow \quad 5n - 15 = 4n - 8 \Rightarrow n = 7$

Now $\dfrac{T_{r+1}}{T_r} = \dfrac{n-r+1}{r} \cdot \dfrac{a}{x} \Rightarrow \dfrac{T_4}{T_3} = \dfrac{n-2}{3} \cdot \dfrac{a}{x} \Rightarrow \dfrac{280}{84} = \dfrac{7-2}{3} \cdot \dfrac{a}{x}$

$\Rightarrow \quad \dfrac{10}{3} = \dfrac{5}{3}\left(\dfrac{a}{x} \right) \Rightarrow x = \dfrac{a}{2}$

$\therefore \quad T_3 = 84 \Rightarrow {}^nC_2 x^{n-2} a^2 = 84 \Rightarrow {}^7C_2 x^5 a^2 = 84$

$$21\left(\dfrac{a}{2} \right)^5 \cdot a^2 = 84$$

$\Rightarrow \quad a^7 = 2^7 \Rightarrow a = 2 \Rightarrow x = 1$

Hence, $x = 1, a = 2$ and $n = 7$

EXAMPLE 16. *If the coefficients of x^{r-1}, x^r and x^{r+1} in the binomial expansion of $(1+x)^n$ are in A.P, prove that $n^2 - n(4r + 1) + 4r^2 - 2 = 0$.*

SOLUTION. Here $T_{r+1} = {}^nC_r x^r$ and coefficient of $x^r = {}^nC_r$

By hypothesis $^nC_{r-1}, ^nC_r$ and $^nC_{r+1}$ are in A.P.

Therefore, $\quad {}^nC_{r-1} + {}^n C_{r+1} = 2 \cdot {}^nC_r$

$\Rightarrow \quad \dfrac{n!}{(r-1)!(n-r+1)!} + \dfrac{n!}{(r+1)!(n-r-1)!} = \dfrac{2 \times n!}{r!(n-r)!}$

$\Rightarrow \quad \dfrac{1}{(r-1)!(n-r-1)!} \left[\dfrac{n!}{(n-r)(n-r-1)} + \dfrac{1}{r(r+1)} \right] = \dfrac{2}{(r-1)!(n-r-1)![r(n-r)]}$

$\Rightarrow \quad \dfrac{1}{(n-r)(n-r+1)} + \dfrac{1}{r(r+1)} = \dfrac{2}{r(n-r)}$

$\Rightarrow \quad r(r+1) + (n-r)(n-r+1) = 2(n-r+1)(r+1)$

$\Rightarrow \quad r^2 + r + n^2 - 2nr + r^2 + n - r = 2(nr + n - r^2 + 1)$

$\Rightarrow \quad n^2 - 4nr + 4r^2 - n - 2 = 0$

$\Rightarrow \quad n^2 - n(4r+1) + 4r^2 - 2 = 0$

EXAMPLE 17. *In the binomial expansion of $(1+a)^{m+n}$, prove that the coefficient of a^m and a^n are equal.*

SOLUTION. Coefficient of a^m in the expansion of $(1+a)^{m+n} = {}^{m+n}C_m \dfrac{(m+n)!}{m!n!}$ $\qquad$ ...(1)

Similarly, coefficient of a^n in the expansion of $(1+a)^{m+n} = {}^{m+n}C_n = \dfrac{(m+n)!}{n!m!}$ $\qquad$ ...(2)

From eqn. (1) and (2), it is clear that coefficient of a^m and a^n in the expansion of $(1+a)^{m+n}$ are equal.

EXAMPLE 18. *If a_1, a_2, a_3, a_4 are the coefficients of four consecutive terms in expansion of $(1+x)^n$, then prove that*

$$\dfrac{a_1}{a_1 + a_2} + \dfrac{a_3}{a_3 + a_4} = \dfrac{2a_2}{a_2 + a_3}$$

SOLUTION. Let a_1, a_2, a_3 and a_4 be the coefficients of four consecutive terms r^{th}, $(r+1)^{\text{th}}$, $(r+2)^{\text{th}}$ and $(r+3)^{\text{th}}$.

Then $\qquad\qquad a_1 = {}^nC_{r-1}, a_2 = {}^nC_r, a_3 = {}^nC_{r+1}$ and $a_4 = {}^nC_{r+2}$

Now, $\qquad\qquad a_1 + a_2 = {}^nC_{r-1} + {}^nC_r = {}^{n+1}C_r$

$\qquad\qquad\qquad a_2 + a_3 = {}^nC_r + {}^nC_{r+1} = {}^{n+1}C_{r+1}$

and $\qquad\qquad a_3 + a_4 = {}^nC_{r+1} + {}^nC_{r+2} = {}^{n+1}C_{r+2}$

Therefore, $\qquad \dfrac{a_1}{a_1 + a_2} + \dfrac{a_3}{a_3 + a_4} = \dfrac{{}^nC_{r-1}}{{}^{n+1}C_{r+2}} + \dfrac{{}^nC_{r+1}}{{}^{n+1}C_{r+2}}$

$\qquad\qquad\qquad = \dfrac{n!}{(r-1)!(n-r+1)!} \cdot \dfrac{r!(n+1-r)!}{(n+1)!}$

$\qquad\qquad\qquad + \dfrac{n!}{(r+1)!(n-r-1)!(n+1)!} \cdot (r+2)!(n-r-2)!$

$\qquad\qquad\qquad = \dfrac{2\,{}^nC_r}{{}^{n+1}C_{r+1}} = \dfrac{2a_2}{a_2 + a_3}$

EXAMPLE 19. *If a and b are distinct integers, using binomial expansion prove that $(a^n - b^n)$ is divisible by $(a-b)$ whenever n is a positive integer.*

SOLUTION. $\qquad\qquad a^n = [(a-b) + b]^n$

$$= {}^nC_0(a-b)^n + {}^nC_1(a-b)^{n-1}b + \ldots + {}^nC_{n-1}(a-b)b^{n-1} + {}^nC_n b^n$$

$$= (a-b)^n + {}^nC_1(a-b)^{n-1} \cdot b + \ldots + {}^nC_{n-1}(a-b)b^{n-1} + b^n$$

$$[\because {}^nC_0 = {}^nC_n = 1]$$

$$\Rightarrow \quad a^n - b^n = (a-b)^n + {}^nC_1(a-b)^{n-1} \cdot b + \ldots + {}^nC_{n-1}(a-b)b^{n-1}$$

$$= (a-b)[(a-b)^{n-1} + {}^nC_1(a-b)^{n-2} \cdot b + \ldots + {}^nC_{n-1}b^{n-1}]$$

Clearly R.H.S. is divisible by $(a-b)$.

$\Rightarrow$ L.H.S. $= (a^n - b^n)$ is divisible by $a - b$.

EXAMPLE 20. (i) *Write the binomial expansion of* $(1+x)^{n+1}$, *when* $x = 8$. *Deduce that* $9^{n+1} - 8n - 9$ *is divisible by 64, where n is a positive integer.*

(ii) *Using binomial theorem, prove that* $6^n - 5n$ *always leaves the remainder 1, when divided by 25.*

SOLUTION : (i) We have $(1+x)^{n+1} = {}^{n+1}C_0 + {}^{n+1}C_1 x + {}^{n+1}C_2 x^2 + {}^{n+1}C_3 x^3 + \ldots + {}^{n+1}C_{n+1} x^{n+1}$

Putting $x = 8$, we get

$$(1+8)^{n+1} = {}^{n+1}C_0 + {}^{n+1}C_1(8) + {}^{n+1}C_2(8)^2 + {}^{n+1}C_3(8)^3 + \ldots + {}^{n+1}C_{n+1}(8)^{n+1}$$

$$\ldots(I)$$

Which is the required binomial expansion.

From eqn. (I)

$$9^{n+1} = 1 + (n+1) \times 8 + {}^{n+1}C_2(8)^2 + {}^{n+1}C_3(8)^3 + \ldots + {}^{n+1}C_{n+1} 8^{n+1}$$

$$\Rightarrow \quad 9^{n+1} - 8n - 9 = 8^2[{}^{n+1}C_2 + {}^{n+1}C_3(8) + {}^{n+1}C_4(8)^2 + \ldots + {}^{n+1}C_{n+1} 8^{n-1}$$

$$= 64 \times \text{an integer}$$

$\Rightarrow$ $9^{n+1} - 8n - 9$ is divisible by 64.

(ii) $$6^n - 5n = (1+5)^n - 5n$$

$$= [{}^nC_0 + {}^nC_1 \cdot 5 + {}^nC_2(5)^2 + {}^nC_3(5)^3 + \ldots + {}^nC_n(5)^n] - 5n$$

$$= 1 + 5n + {}^nC_2(5)^2 + {}^nC_3(5)^3 + \ldots + {}^nC_n(5)^n - 5n$$

$$\Rightarrow \quad 6^n - 5n - 1 = {}^nC_2(5)^2 + {}^nC_3(5)^3 + \ldots + {}^nC_n(5)^n$$

$$= 5^2[{}^nC_2 + {}^nC_3(5) + {}^nC_4(5)^2 + \ldots + {}^nC_n(5)^{n-2}]$$

$$= 25 \times \text{an integer.}$$

$$\Rightarrow \quad 6^n - 5n = (25 \times \text{an integer}) + 1$$

$$\Rightarrow \quad (6^n - 5n) \text{ leaves the remainder 1, when divisible by 25.}$$

EXAMPLE 21. *If* x^p *occurs in the expansion of* $\left(x^2 + \dfrac{1}{x}\right)^{2n}$. *Prove that its coefficient is*

$$\left[\frac{(2n)!}{\left(\dfrac{4n-p}{3}\right)!\left(\dfrac{2n+p}{3}\right)!}\right].$$

SOLUTION. Suppose x^p occurs in $(r+1)$th term in the expansion of $\left(x^2 + \dfrac{1}{x}\right)^{2n}$.

Then $$T_{r+1} = {}^{2n}C_r (x^2)^{2n-r}\left(\frac{1}{r}\right)^r = {}^{2n}C_r x^{4n-3r}$$

For this term to contain x^p, we have

$$4n - 3r = p$$

$$\Rightarrow \qquad r = \frac{4n - p}{3}$$

$$\therefore \qquad \text{Coefficient of } x^p = {}^{2n}C_r \text{ where } r = \frac{4n - p}{3}$$

$$= \frac{(2n)!}{\left\{2n - \left(\dfrac{4n - p}{3}\right)!\dfrac{4n - p}{3}\right\}!} = \frac{(2n)!}{\left(\dfrac{2n + p}{3}\right)!\left(\dfrac{4n - p}{3}\right)!}$$

EXAMPLE 22. *Using binomial theorem, compute the following :*

(i) $(99)^5$ (ii) $(102)^6$ (iii) $(10.1)^5$

SOLUTION. (i)

$$(99)^5 = (100 - 1)^5$$

$$= {}^5C_0 \times (100)^5 - {}^5C_1 \times (100)^4 + {}^5C_2 (100)^3 - {}^5C_3 (100)^2$$

$$+ {}^5C_4 \times (100)^1 - {}^5C_5 \times (100)^0$$

$$= (100)^5 - 5 \times (100)^4 + 10 \times (100)^3 - 10 \times (100)^2 + 5 \times 100 - 1$$

$$= 10^{10} - 5 \times 10^8 + 10^7 - 10^5 + 5 \times 10^2 - 1$$

$$= (10^{10} + 10^7 + 5 \times 10^2) - (5 \times 10^8 + 10^5 + 1)$$

$$= 10010000500 - 500100001 = 9509900499$$

(ii)

$$(102)^6 = (100 + 2)^6$$

$$= {}^6C_0 \times (100)^6 + {}^6C_1 (100)^5 \times 2 + {}^6C_2 \times (100)^4 \times 2^2$$

$$+ {}^6C_3 \times (100)^3 + 2^3 + {}^6C_4 \times (100)^2 \times 2^4 + {}^6C_5 \times (100)^1 \times 2^5$$

$$+ {}^6C_6 (100)^0 \times 2^6$$

$$= 10^{12} + 12 \times 10^{10} + 6 \times 10^9 + 16 \times 10^7 + 24 \times 10^5$$

$$+ 192 \times 10^2 + 64 = 1126162419264$$

(iii) $(10.1)^5 = (10 + 0.1)^5$

$$= {}^5C_0 (10)^5 (0.1)^0 + {}^5C_1 \times (10)^4 \times (0.1) + {}^5C_2 \times (10)^3 \times (0.1)^2$$

$$+ {}^5C_3 \times (10)^2 \times (0.1)^3 + {}^5C_4 \times (10)^1 \times (0.1)^4 + {}^5C_5 \times (10)^0 \times (0.1)^5$$

$$= (10)^5 + 5 \times 10^4 \times 0.1 + 10 \times 10^3 \times (0.1)^2 + 10 \times (10)^2 \times (0.1)^3 + 5 \times 10$$

$$\times (0.1)^4 + (0.1)^5$$

$$= 10^5 + 5 \times 10^3 + 10^2 + 1 + 5 \times 0.001 + 0.00001$$

$$= 100000 + 5000 + 100 + 1 + 0.005 + 0.00001 = 105101.00501$$

EXAMPLE 23. *Find the coefficient of x^7 in $\left(ax^2 + \dfrac{1}{bx}\right)^{11}$ and x^{-7} in $x^{-7}\left(ax - \dfrac{1}{bx^2}\right)^{11}$. Also find the relation between a and b so that these coefficients are equal.*

SOLUTION. The coefficient of x^7 in $\left(ax^2 + \dfrac{1}{bx}\right)^{11}$ is ${}^{11}C_5 \left(\dfrac{a^6}{b^5}\right)$.

The coefficient of x^{-7} in $\left(ax - \dfrac{1}{bx^2}\right)^{+11}$ is $^{11}C_6\left(\dfrac{a^5}{b^6}\right)$.

In case these coefficients are equal, we have $\dfrac{a^6}{b^5} = \dfrac{a^5}{b^6} \Rightarrow a = \dfrac{1}{b}$

$\Rightarrow \qquad ab = 1$

EXAMPLE 24. (i) *Show that the coefficient of x^4 in the expansion of $(1 + x + x^2 + x^3)^n$ is*

$$^nC_4 + {}^nC_2 + {}^nC_1 \cdot {}^nC_2$$

(ii) *Which term in the expansion of the binomial $\left[3\sqrt{\left(\dfrac{a}{\sqrt{b}}\right)} + \sqrt{\left(\dfrac{b}{3\sqrt{a}}\right)}\right]^{21}$ contains a and b to the same power.*

SOLUTION. (i) We have $(1 + x + x^2 + x^3)^n = (1 + x)^n (1 + x^2)^n$

$\qquad = (C_0 + C_1 x + C_2 x^2 + C_3 x^3 + C_4 x^4 + \ldots)(C_0 + C_1 x^2 + C_2 (x^2)^2 + \ldots)$

$\Rightarrow \qquad$ Coefficient of $x^4 = C_0 C_2 + C_2 C_1 + C_4 C_0$

$\qquad\qquad\qquad = {}^nC_2 + {}^nC_2 \cdot {}^nC_1 + {}^nC_4$

(ii) On simplification, we get

$$T_{r+1} = {}^{21}C_r a^{7 - \frac{r}{2}} b^{\frac{2r}{3} - \frac{7}{2}}$$

Since the powers of a and b are same, therefore we have

$$7 - \frac{r}{2} = \frac{2r}{3} - \frac{7}{2}$$

$\Rightarrow \qquad\qquad r = 9$

Hence, $(9 + 1) = 10^{th}$ term is the required term.

EXAMPLE 25. (i) *Let $(1 + x^2)^2 (1 + x)^n = \sum\limits_{k=0}^{n+4} a_k x^k$. If a_1, a_2 and a_3 are in A.P., then find n.*

(ii) *In the expansion of $(1 + x)^n$, the binomial coefficients of three consecutive terms are respectively 220, 495 and 792, find the value of n.*

(iii) *If the coefficients of rth, $(r + 1)^{th}$ and $(r + 2)^{th}$ terms in the expansion of $(1 + x)^{14}$ are in A.P., find r.*

(iv) *Show that three consecutive binomial coefficients can not be in (a) G.P. (b) H.P.*

(v) *If the coefficient of rth, $(r + 1)^{th}$ and $(r + 2)^{th}$ terms in the expansion of $(1 + x)^n$ be in H.P. prove that n is a root of the equation*

$$x^2 - (4r - 1)x + 4r^2 = 0$$

SOLUTION. (i) $\text{L.H.S} = (1 + 2x^2 + x^4)(1 + C_1 x + C_2 x^2 + C_3 x^3 + \ldots)$

$\qquad \text{R.H.S} = a_0 + a_1 x + a_2 x^2 + a_3 x^3 + \ldots$

$\qquad$ Equating, we get $a_1 = C_1, a_2 = C_2 + 2, a_3 = C_3 + 2C_1$ ...(1)

$\qquad$ Now, $\qquad\qquad 2a_2 = a_1 + a_3 \text{(A.P)}$

$\qquad\qquad\qquad 2({}^nC_2 + 2) = {}^nC_1 + ({}^nC_3 + 2 \cdot {}^nC_1)$

$\Rightarrow \qquad 2\dfrac{n(n-1)}{2} + 4 = 3n + \dfrac{n(n-1)(n-2)}{6}$

$$\Rightarrow \qquad n^3 - 9n^2 + 26n - 24 = 0$$

$$\Rightarrow \qquad (n-2)(n^2 - 7n + 12) = 0$$

$$\Rightarrow \qquad (n-2)(n-3)(n-4) = 0$$

$$\Rightarrow \qquad n = 2, 3, 4$$

(ii) Let the successive terms be T_r, T_{r+1} and T_{r+2}. whose coefficients are given by 220, 495 and 792.

Then $\qquad \dfrac{T_{r+1}}{T_r} = \dfrac{n-r+1}{r} x$

$$\Rightarrow \qquad \frac{n-r+1}{r} = \frac{495}{220} = \frac{9}{4}$$

Putting $r+1$ for r, we get

$$\frac{T_{r+2}}{T_{r+1}} = \frac{n-r}{r+1} \cdot x \Rightarrow \frac{n-r}{r+1} = \frac{792}{495} = \frac{8}{5}$$

$$\therefore \qquad 4n - 13r + 4 = 0$$
$$5n - 13r - 8 = 0$$

On solving, we get $n = 12, r = 4$

(iii) The coefficients of T_r, T_{r+1} and T_{r+2} are in A.P.

Then $\dfrac{T_r}{T_{r+1}}, 1, \dfrac{T_{r+2}}{T_{r+1}}$ are in A.P.

$$\Rightarrow \qquad \frac{T_{r+1}}{T_r} = \frac{n-r+1}{r} \cdot x \qquad \qquad \dots(1)$$

Coefficient of $\dfrac{T_{r+1}}{T_r} = \dfrac{14-r+1}{r} = \dfrac{15-r}{r} \qquad \dots(2)$

Putting $r+1$ for r, we get

$$\frac{T_{r+2}}{T_{r+1}} = \frac{15-(r+1)}{r+1} = \frac{14-r}{r+1} \qquad \dots(3)$$

From eqn. (2), $\dfrac{T_r}{T_{r+1}} = \dfrac{r}{15-r} \qquad \qquad \dots(4)$

From eqn. (1), (3) and (4), we can say that

$$\frac{r}{15-r}, 1, \frac{14-r}{r+1} \text{ are in A.P.}$$

$$\therefore \qquad 2 = \frac{r}{15-r} + \frac{14-r}{r+1}$$

$$\Rightarrow \quad 2(15 + 14r - r^2) = r^2 + r + 210 - 29r + r^2$$

$$\Rightarrow \quad 4r^2 - 56r + 180 = 0$$

$$\Rightarrow \qquad r^2 - 14r + 45 = 0$$

$$\Rightarrow \qquad (r-5)(r-9) = 0 \qquad \Rightarrow \qquad r = 5, 9$$

(iv) (a) T_r, T_{r+1}, T_{r+2}, are in G.P.

$$\therefore \quad \frac{T_r}{T_{r+1}}, 1, \frac{T_{r+2}}{T_{r+1}} \text{ are in G.P.}$$

$\Rightarrow \quad \dfrac{r}{n-r+1}, 1, \dfrac{n-r}{r+1},$ are in G.P.

$\therefore \qquad 1^2 = \dfrac{r(n-r)}{(n-r+1)(r+1)}$

$\Rightarrow \qquad n(r+1) - (r^2-1) = nr - r^2 \qquad \Rightarrow \qquad n+1 = 0$

$n = -1,$, which is not possible.

(b) $T_r, T_{r+1}, T_{r+2},$ are in H.P.

$\Rightarrow \qquad \dfrac{1}{T_r}, \dfrac{1}{T_{r+1}}, \dfrac{1}{T_{r+2}},$ are in A.P.

$\Rightarrow \qquad \dfrac{T_{r+1}}{T_r}, 1, \dfrac{T_{r+1}}{T_{r+2}},$ are in A.P.

$\Rightarrow \qquad 2 = \dfrac{n-r+1}{r} + \dfrac{r+1}{n-r}$

$\Rightarrow \quad 2r(n-r) = (n-r)^2 + (n-r) + r^2 + r$

$\Rightarrow \quad 2rn - 2r^2 = n^2 - 2nr + r^2 + n - r + r^2 + r$

$\Rightarrow \quad n^2 + 4r^2 - 4nr + n = 0$

$\Rightarrow \quad (n - 2r)^2 + n = 0$ this is not possible as $(n - 2r)^2$ and n both are positive.

(v) We have done in part (iv) that

$$n^2 + 4r^2 - 4nr + n = 0$$

Clearly, n is a root of the equation

$$x^2 - (4r-1)x + 4r^2 = 0$$

EXAMPLE 26. (i) Show that $(\sqrt{2}+1)^6 + (\sqrt{2}-1)^6 = 198$. Hence, prove that the integral part of $(\sqrt{2}+1)^6$ is 197.

(ii) Evaluate $[x + \sqrt{(x^2-1)}]^6 + [x - \sqrt{(x^2-1)}]^6$.

(iii) Prove that $\sqrt{10}[(\sqrt{10}+1)^{100} - (\sqrt{10}-1)^{100}$ is a whole number.

(iv) Which is larger in $99^{50} + 100^{50}$ and 101^{50}?

SOLUTION. (i) $(x+a)^n + (x-a)^n = 2(x^n + {}^nC_2 x^{n-2} \cdot a^2 + {}^nC_4 x^{n-4} \cdot a^4 + {}^nC_6 x^{n-6} \cdot a^6 + \ldots)$

Here $n = 6, {}^6C_2 = 15, {}^6C_4 = 15, {}^6C_6 = 1, x = \sqrt{2}, a = 1$

$\therefore \quad (\sqrt{2}+1)^6 + (\sqrt{2}-1)^6 = 2[[\sqrt{2}]^6 + 15[\sqrt{2}]^4 \cdot 1 + 15[\sqrt{2}]^2 \cdot 1 + 1 \cdot 1$

$\qquad\qquad = 2[8 + 15 \times 4 + 15 \times 2 + 1] = 2(99) = 198 \qquad \ldots(1)$

$\Rightarrow \qquad\qquad [\sqrt{2}+1]^6 = 198 - (\sqrt{2}-1)^6$

Now $\qquad\qquad (\sqrt{2}-1)^6 < 1$ and positive.

$\therefore \qquad\qquad 0 < (\sqrt{2}-1)^6 < 1 \qquad\qquad \ldots(2)$

From eqn. (1) and (2),

$\qquad\qquad (\sqrt{2}+1)^6 = 198 -$ (some thing positive but less than 1)

$\therefore$ Integral part of $(\sqrt{2}+1)^6 = 197$

(ii) As above, $\qquad\qquad x = a, a = \sqrt{x^2 - 1}$

Therefore, $[(x + \sqrt{x^2 - 1})]^6 + [(x - \sqrt{x^2 - 1})]^6$

$$= 2\left[x^6 + 15x^4\left(\sqrt{(x^2 - 1)}\right)^2 + 15x^2\left(\sqrt{(x^2 - 1)}\right)^4 + \left(\sqrt{(x^2 - 1)}\right)^6\right]$$

$$= 2[x^6 + 15x^4(x^2 - 1) + 15(x^2 - 1)^2 + (x^2 - 1)^3]$$

$$= 2[x^6 + 15(x^6 - x^4) + 15(x^6 - 2x^4 + x^2) + (x^6 - 3x^4 + 3x^2 - 1)]$$

$$= 64x^6 - 96x^4 + 36x^2 - 2$$

(iii) $x[(x + 1)^n - (x - 1)^n] = x.2[C_1 x^{n-1} + C_3 x^{n-3} + C_5 x^{n-5} + ...]$

$$= 2[C_1 x^n + C_3 x^{n-2} + C_5 x^{n-4} + ...]$$

Now, $n = 100, n - 2, n - 4, ...$ are all an even.

$$x = \sqrt{10}$$

$\therefore$ x^n, x^{n-2}, x^{n-4} are all integers and $C_1, C_3, C_5, ...$ are also integers. Hence, given expression is a whole number.

(iv) $(101)^{50} = (100 + 1)^{50}$

$$= 100^{50} + 50.100^{49} + \frac{50.49}{1.2}.100^{48} + ... \qquad ...(1)$$

and $(99)^{50} = (100 - 1)^{50}$

$$= 100^{50} + 50.100^{49} + \frac{50 \cdot 49}{1 \cdot 2} \cdot 100^{48} + ... \qquad ...(2)$$

From eq. (1) and (2), we have

$$(101)^{50} - (99)^{50} = 2[50.100^{49} + \frac{50 \cdot 49 \cdot 48}{1 \cdot 2 \cdot 3} \cdot 100^{47} + ...]$$

$$= 100^{50} + 2\frac{50 \cdot 49 \cdot 48}{1 \cdot 2 \cdot 3} \cdot 100^{47} + ... > 100^{50}$$

Therefore, $101^{50} > 99^{50} + 100^{50}$

EXAMPLE 27. *Find numerically the greatest term in the expansion of* $(3 - 2x)^9$, *when* $x = 1$.

SOLUTION. Let T_{r+1} be the greatest term in the expansion of $(3 - 2x)^9$.

Now $T_{r+1} = {}^9C_r \cdot 3^{9-r}(-2x)^r$

and $T_r = {}^9C_{r-1} 3^{9-(r-1)}(-2x)^{r-1}$

Their numerical values at $x = 1$ are respectively given by

$${}^9C_r 3^{9-r} 2^r \text{ and } {}^9C_{r-1} 3^{10-r} 2^{r-1}$$

$\because$ As the term T_{r+1} is greatest,

$$T_{r+1} \geq T_r$$

$\Rightarrow$ ${}^9C_r \cdot 3^{9-r} 2^r \geq {}^9C_{r-1} 3^{10-r} \cdot 2^{r-1}$

$\Rightarrow$ $\dfrac{9!}{r!(9-r)!} \cdot 3^{9-r} \cdot 2^r \geq \dfrac{9!}{(r-1)!(10-r)!} \cdot 3^{10-r} \cdot 2^{r-1}$

$\Rightarrow$ $\dfrac{10-r}{r} \cdot \dfrac{2}{3} \geq 1 \qquad \Rightarrow \qquad r \leq 4$

The numerical value of T_{r+1} is greatest if $r = 4$ or 3.

$\Rightarrow$ Numerically, the greatest terms are T_5 and T_4.

$\therefore$ Required greatest value of $T_5 = {}^9C_4 3^{9-4}(-2x)^4 = \dfrac{9 \cdot 8 \cdot 7 \cdot 6}{1 \cdot 2 \cdot 3 \cdot 4} \cdot 3^5 \cdot 2^4 = 489888$

EXAMPLE 28. *Find the value of the greatest term in the expansion of* $\sqrt{3}\left(1 + \dfrac{1}{\sqrt{3}}\right)^{20}$.

SOLUTION. We know that

$$\frac{T_{r+1}}{T_r} = \left(\frac{n - r + 1}{r}\right)x = \frac{20 - r + 1}{r} \cdot \frac{1}{\sqrt{3}} = \frac{21 - r}{r\sqrt{3}}$$

$$\Rightarrow \qquad \frac{T_{r+1}}{T_r} > 1 \Rightarrow 21 - r > r\sqrt{3}$$

$$\Rightarrow \qquad r < \frac{21}{(\sqrt{3} + 1)} \Rightarrow r < \left(\frac{21}{2}\right)(\sqrt{3} - 1)$$

$$\Rightarrow \qquad r < 7.686, \text{ i.e., } T_{r+1} > T_r \text{ if } r < 7.686$$

$\therefore$ But r is an integer, therefore maximum value of $r = 7$

$\Rightarrow$ $T_{7+1} = T_8$ is the greatest term.

If $r > 7.686$, then $T_{r+1} < T_r$

$\Rightarrow$ T_r is greatest when $r = 8$ (integer just greater than 7.686)

$\therefore$ T_8 is the greatest term.

$$\Rightarrow \qquad T_8 = \sqrt{3} \cdot {}^{20}C_7 \left(\frac{1}{\sqrt{3}}\right)^7 = \frac{25840}{9}$$

EXAMPLE 29.   (i) *Find numerically the greatest term in the expansion of* $(3 - 5x)^{15}$, *when* $x = \dfrac{1}{5}$.

           (ii) *Prove that the greatest term in the expansion of* $(1 + x)^{2n}$ *will have the greatest coefficients of* x *lies between* $\dfrac{n}{n+1}$ *and* $\dfrac{n+1}{n}$.

SOLUTION.   (i)

$$\frac{T_{r+1}}{T_r} = \frac{n - r + 1}{r}\left(\frac{5x}{3}\right) \qquad \left[\because (3 - 5x)^{15} = 3^{15}\left(1 - \frac{5}{3}x\right)^{15}\right]$$

$$\Rightarrow \frac{15 - r + 1}{r}\left(\frac{5}{3} \cdot \frac{1}{5}\right) \geq 1 \qquad\qquad\qquad \left(\because x = \frac{1}{5}\right)$$

$$\Rightarrow \qquad 16 - r \geq 3r \qquad \Rightarrow \qquad r \leq 4$$

$$\therefore \qquad r = 3, 4$$

$\Rightarrow$ T_4 and T_5 are numerically equal to each other and greater than any other terms.

Now, $T_4 = 3^{15} \cdot {}^{15}C_3\left(\dfrac{5}{3} \cdot \dfrac{1}{5}\right)^3 = 3^{15} \cdot \dfrac{15!}{(12)!3!} \cdot \dfrac{1}{3^3}$

$$= 3^{12} \cdot \frac{15 \cdot 14 \cdot 13}{1 \cdot 2 \cdot 3} = (455)3^{12}$$

and $T_5 = 3^{15} \cdot {}^{15}C_4\left(\dfrac{5}{3} \cdot \dfrac{1}{5}\right)^4 = 3^{15} \cdot \dfrac{15!}{11!4!} \cdot \dfrac{1}{3^4}$

$$= 3^{11} \times \frac{15 \cdot 14 \cdot 13 \cdot 12}{1 \cdot 2 \cdot 3 \cdot 4} = 3^{12} \cdot 5 \cdot 7 \cdot 13 = 455(3)^{12}$$

(ii) We know that $^N C_r$ is greatest term, when $r = \dfrac{N}{2}$ (N, even). Here $N = 2n$ (even)

$\therefore \quad ^{2n} C_n$ is greatest because $r = \dfrac{N}{2} = \dfrac{2n}{2} = n$

$\Rightarrow T_{n+1}$ have greatest coefficient.

It will be greater term if $\dfrac{T_{n+1}}{T_n} \geq 1$ and $\dfrac{T_{n+1}}{T_{n+2}} \geq 1$

Now $\quad \dfrac{T_{n+1}}{T_n} = \dfrac{n-r+1}{r} \cdot x$

Putting, $r = n, n+1$, and $n = 2n$, we get

$\dfrac{2n - n + 1}{n} x \geq 1 \quad$ and $\quad \dfrac{2n - (n+1) + 1}{n} x \leq 1$

$\dfrac{n+1}{n} x \geq 1 \quad$ and $\quad \dfrac{n}{n+1} x \leq 1$

$\therefore \qquad\qquad x \leq \dfrac{n+1}{n} \quad$ and $\qquad\qquad x \geq \dfrac{n}{n+1}$

$\Rightarrow \qquad\qquad \dfrac{n}{n+1} \leq x \leq \dfrac{n+1}{n}$

Exercise 4.1

1. Using binomial theorem, expand the following :

(i) $\left(\dfrac{2x}{3} - \dfrac{3}{2x} \right)^6$ (ii) $\left(2x - \dfrac{3}{y} \right)^5$

(iii) $(1 - x + x^2)^4$ (iv) $(1 + x + x^2)^3$

(v) $(\sqrt{x} + \sqrt{y})^{10}$ (vi) $(\sqrt[3]{x} - \sqrt[3]{y})^6$

2. Evaluate :

(i) $(x + y)^6 + (x - y)^6$

(ii) $(\sqrt{2} + 1)^5 + (\sqrt{2} - 1)^5$

(iii) $(\sqrt{3} + 1)^7 - (\sqrt{3} - 1)^7$

(iv) $(1001)^5$ (v) $(101)^5$

(vi) $(994)^4$

3. Find :

(i) general term in the expansion of $(a^2 - b^2)^6$.

(ii) sixth term in the expansion of $\left(\dfrac{4x}{5} + \dfrac{8}{3x} \right)^7$.

(iii) third term in the expansion of $(\sqrt{a} - \sqrt{b})^{17}$.

(iv) term independent of x in the expansion of $\left(x - \dfrac{1}{x} \right)^{12}$.

(v) term independent of x in the expansion of $\left(3x^2 - \dfrac{1}{3x} \right)^9$.

4. In the expansion of $(1 + 3x)^9$, find the third term from the last.

5. Find the coefficient of x^{12} in the expansion of $\left(x^3 - \dfrac{1}{x^3} \right)^{12}$.

6. Find the middle term in the expansion of the following :

(i) $\left(3x - \dfrac{x^3}{6} \right)^7$ (ii) $\left(x^4 - \dfrac{1}{x^3} \right)^{11}$

(iii) $\left(a - \dfrac{1}{2b} \right)^{10}$ (iv) $\left(1 + \dfrac{x^2}{2} \right)^{14}$

7. (i) Find the coefficient of x^6 in the expansion of $\left(3x^2 - \dfrac{1}{3x} \right)^9$.

(ii) Find the coefficient of x^{18} in the expansion of $\left(x^2 + \dfrac{2a}{x} \right)^{15}$.

(iii) Find the coefficient of x^5 in the expansion of $(1+x)^3(1-x)^6$.

(iv) Find the coefficient of x^{-n} in the expansion of $\left(x - \dfrac{1}{x}\right)^{3n}$.

(v) Find the coefficient of x^3 in the expansion of $\left(x + \dfrac{1}{x}\right)^7$.

8. (i) Prove that in the expansion of $\left(3x - \dfrac{1}{2x}\right)^6$, there is no term of x^3.

(ii) Prove that in the expansion of $\left(2x^2 - \dfrac{3}{x}\right)^{11}$, there is no term of x^6.

9. Find the ratio of the term independent of x in the expansion of $\left(x - \dfrac{2}{x}\right)^{10}$ to the coefficient of x^{10} in the expansion of $(1-x^2)^{10}$.

10. Prove that the coefficient of x^n in the expansion of $(1+x)^{2n}$ is twice the coefficient of x^n in the expansion of $(1+x)^{2n-1}$.

11. Prove that the coefficient of middle term in the expansion of $(1+x)^{2n}$ is same as the sum of coefficient of two middle terms in the expansion of $(1+x)^{2n-1}$.

12. Prove that in the expansion of $(1+x)^{m+n}$, the coefficient of x^m and x^n are equal.

13. If in the expansion of $(a+b)^n$ the coefficient of 4th and 13th term are equal. Find the value of n.

14. The ratio of coefficient of three consecutive terms in the expansion of $(1+x)^n$ are in the ratio $6:33:110$. Find the value of n.

15. Find the largest term in the expansion of
(i) $(2x+3y)^6$ if $x=2, y=3$
(ii) $(1+4x)^5$ if $x = \dfrac{3}{4}$

16. Prove that $2^{3n} - 7n - 1, n \in N$ has a factor 49.

17. Prove that when $6^n - 5n, n \in N$ divided by 25, 1 is always remainder.

18. Prove that in the expansion of $\left(x^2 - \dfrac{1}{x^2}\right)^8$, the term independent of x is 70.

19. If in the expansion of $(1+x)^{2n}$, the coefficient of $(p+1)$th term is same as the coefficient are $(p+3)$th term, then prove that $p = n - 1$.

20. If in the expansion of $(1+x)^{10}$, the coefficient of $(4r+5)$th term is equal to the coefficient of $(2r+1)$th term. Prove that $r = 1$.

21. Which term in the expansion of $\left(x^3 - \dfrac{3}{x^2}\right)^{15}$ is independent of x?

22. (i) Find the 7th term in the expansion of $\left(x^4 - \dfrac{4}{x^3}\right)^9$.

(ii) Find the coefficient of 25th term in the expansion of $\left(x^4 - \dfrac{1}{x^3}\right)^{15}$.

(iii) Find the middle term in the expansion of $(3+x)^6$.

(iv) Find the coefficient of x^{10} in the expansion of $(1-x^2)^{10}$.

(v) Find the coefficient of x in the expansion of $\left(x^2 + \dfrac{2}{x}\right)^5$.

(vi) Find the middle term in the expansion of $\left(\dfrac{a}{x} + \dfrac{x}{a}\right)^{10}$.

(vii) Find the 7th term in the expansion of $(1-2x)^7$.

(viii) In the expansion of $\left(2x^4 - \dfrac{1}{3x^7}\right)^{11}$, find the constant term.

(ix) Find the constant term in the expansion of $\left(\sqrt{x} + \dfrac{1}{3x^2}\right)^{10}$.

ANSWERS

1. (i) $\dfrac{64x^2}{729} - \dfrac{32x^4}{27} + \dfrac{20x^2}{3} - 20 + \dfrac{135}{4x^2} - \dfrac{243}{8x^4} + \dfrac{729}{64x^6}$,

(ii) $32x^5 - \dfrac{240x^4}{y} + \dfrac{720x^3}{y^2} - \dfrac{1080x^2}{y^3} + \dfrac{810x}{y^4} - \dfrac{243}{y^5}$

(iii) $1 - 4x + 10x^2 - 16x^3 + 19x^4 - 16x^5 + 10x^6 - 4x^7 + x^8$

(iv) $1 + 3x + 6x^2 + 7x^3 + 6x^4 + 3x^5 + x^6$

(v) $x^5 + 10x^{9/2}y^{1/2} + 45x^4y + 120x^{7/2}y^{3/2} + 210x^3y^2 + 252x^{5/2}y^{5/2}$
$$+ 210x^2y^3 + 120^{3/2}y^{7/2} + 45xy^4 + 10x^{1/2}y^{9/2} + y^5$$

(vi) $x^2 - 6x^{5/3}y^{1/3} + 15x^{4/3}y^{2/3} - 20xy + 15x^{2/3}y^{4/3} - 6x^{1/3}y^{5/3} + y^2$

2. (i) $2(x^6 + 15x^4y^2 + 15x^2y^4 + y^6)$ (ii) $234\sqrt{2}$ (iii) 1136
(iv) 1005010010005001 (v) 10510100501 (vi) 976215137296

3. (i) $(-1)^r {}^6C_r a^{12-2r}b^{2r}$ (ii) $\dfrac{36,70,016}{2,025x^3}$, (iii) $136a^{15/2}b$ (iv) 924, (v) $\dfrac{28}{9}$

4. $78732\,x^7$ **5.** 495 **6.** (i) $-\dfrac{105}{8}x^{13}, \dfrac{35}{48}x^{15}$, (ii) $-462x^9, 462x^2$ (iii) $\dfrac{63a^5}{8b^5}$,

(iv) $\dfrac{429}{16}x^{14}$ **7.** (i) 378, (ii) $21,840a^4$, (iii) -6, (iv) $\dfrac{(3n)!}{n!(2n)!}$ (v) 21

9. $32:1$ **13.** 15 **14.** 12 **15.** (i) 15,74,640, (ii) 405 **21.** 10th

22. (i) $\dfrac{84 \times 4^6}{x^6}$ (ii) -3003 (iii) $540x^3$ (iv) -252 (v) 80 (vi) 252

(vii) $448x^6$ (viii) $\dfrac{14080}{27}$, (ix) 5

4.9 BINOMIAL COEFFICIENTS

We know that $(1+x)^n = {}^nC_0 + {}^nC_1x + {}^nC_2x^2 + \ldots + {}^nC_nx^n$, when n is any positive integer.

Throughout this section, we shall write C_r for ${}^nC_r\,(r = 0,1,2,\ldots,n)$ unless stated otherwise. Clearly, $C_1, C_3, C_5\ldots$ are odd coefficients and $C_0, C_2, C_4,\ldots$ denote the even coefficients.

THEOREM 1. *In the expansion of $(1+x)^n$ the coefficients of terms equidistant from the beginning and the end are equal.*

PROOF. We have $(1+x)^n = C_0 + C_1x + C_2x^2 + \ldots + C_rx^r + \ldots + C_{n-1}x^{n-1} + C_nx^n$

In this expansion, the coefficient of $(r+1)$th term from the beginning is ${}^nC_r$. The $(r+1)$th term from the end is $(n-r+1)$th term from the beginning. Therefore, its coefficient is ${}^nC_{n-r}$. But ${}^nC_r = {}^nC_{n-r}$. Hence, the coefficients of the terms equidistant from the beginning and end are equal.

THEOREM 2. *The sum of the binomial coefficients in the expansion of $(1+x)^n$ is 2^n.*

PROOF. We know that $(1+x)^n = C_0 + C_1x + C_2x^2 + C_3x^3 + \ldots + C_{n-1}x^{n-1} + C_nx^n$ ...(1)
Putting $x = 1$ in eq. (I), we get
$$(1+1)^n = 2^n = C_0 + C_1 + C_2 + \ldots + C_{n-1} + C_n$$
Hence, sum of the binomial coefficients $= 2^n$.

THEOREM 3. *The sum of the coefficients of the odd terms in the expansion of $(1+x)^n$ is equal to the sum of the coefficients of the even terms and each is equal to 2^{n-1}.*

PROOF. We have $(1+x)^n = C_0 + C_1x + C_2x^2 + C_3x^3 + \ldots + C_{n-1}x^{n-1} + C_nx^n$
Putting $x = 1$ and $x = -1$ respectively in the above expansion, we get
$$(2)^n = C_0 + C_1 + C_2 + C_3 + \ldots + C_{n-1} + C_n$$
$$0 = C_0 - C_1 + C_2 - C_3 + \ldots + (-1)^{n-1}C_{n-1} + (-1)^n C_n$$
Adding and subtracting these two equations, we get
$$\Rightarrow \qquad 2^n = 2(C_0 + C_2 + C_4 + \ldots)$$

and
$$2^n = 2(C_1 + C_3 + C_5 + ...)$$
$$\Rightarrow \quad C_0 + C_2 + C_4 + ... = 2^{n-1} = C_1 + C_3 + C_5 + ...$$

THEOREM 4. *In the expansion of* $(1 + x)^n$ *the sum of the square of the coefficients is* $\left(\dfrac{2n!}{(n!)^2}\right)$.

PROOF. We have $\quad (1+x)^n = C_0 + C_1 x + C_2 x^2 + ... + C_n x^n$ ...(1)

and $\quad (x+1)^n = C_0 x^n + C_1 x^n + C_1 x^{n-1} + C_2 x^{n-2} + ... + C_n$ ...(2)

Multiplying eqn. (1) and (2), we get

$$(1+x)^{2n} = (C_0 + C_1 x + C_2 x^2 + ... + C_n x^n)$$
$$(C_0 x^n + C_1 x^{n-1} + C_2 x^{n-1} + C_2 x^{n-2} + ... + C_n) \quad ...(3)$$

Comparing the coefficients of x^n on both the sides of eqn. (3), we get

$$^{2n}C_n = C_0^2 + C_1^2 + C_2^2 + ... + C_n^2$$

Hence, sum of the squares of the coefficients $= \dfrac{2n!}{(n!)^2}$.

THEOREM 5. $\quad ^nC_r = \dfrac{n}{r} \cdot {}^{n-1}C_{r-1} = \dfrac{n}{r} \cdot \dfrac{n-1}{r-1} \cdot {}^{n-2}C_{r-2}$

PROOF. We have $\quad ^nC_r = \dfrac{n}{r!(n-r)!} = \dfrac{(n-1)!}{r!(r-1)![(n-1)-(r-1)]!} = \dfrac{n}{r} \cdot {}^{n-1}C_{r-1}$

Similarly, $\quad ^{n-1}C_{r-1} = \dfrac{n-1}{r-1} \cdot {}^{n-2}C_{r-2}$

$$\Rightarrow \quad ^nC_r = \dfrac{n}{r} \cdot {}^{n-1}C_{r-1} = \dfrac{n}{r} \cdot \dfrac{n-1}{r-1} \cdot {}^{n-2}C_{r-2}$$

REMARKS

➠ Use the following results also

➠ $\displaystyle\sum_{r=0}^{n} {}^nC_r = 2^n; \sum_{r=1}^{n} {}^nC_r = 2^n - 1$ or $\displaystyle\sum_{r=0}^{n} {}^{n-1}C_r = 2^{n-1}$

➠ $\displaystyle\sum_{r=0}^{n} {}^{n-1}C_{r-2} = 2^{n-2}$ or $\displaystyle\sum_{r=0}^{n-2} {}^{n-2}C_r = 2^{n-2}$

➠ $\displaystyle\sum_{r=0}^{n} (-1)^r C_n = 0$

☞ RECAPITULATIONS

- In the expansion of $(1 + x)^n$, the coefficients of terms equidistant from the beginning and the end are equal.

- The sum of the binomial coefficients in the expansion of $(1 + x)^n$ is 2^n.

- The sum of the coefficients of the odd terms in the expansions of $(1 + x)^n$ is equal to the sum of the coefficients of the even terms and each is equal to 2^{n-1}.

- In the expansions of $(1 + x)^n$, the sum of the square of the coefficients is $\dfrac{2n!}{(n!)^2}$.

SOLVED EXAMPLES

EXAMPLE 1. *If* $C_0, C_1, C_2, ..., C_n$ *are the coefficients in the expansion of* $(1 + x)^n$, *prove that—*

(i) $C_1 + 2C_2 + 3C_3 + ... + nC_n = n2^{n-1}$ (ii) $C_0 + 2C_1 + 3C_2 + ... + (n+1)C_n = (n+2)2^{n-1}$

(iii) $C_0 + 3C_1 + 5C_2 + ... + (2n+1)C_n = (n+1)2^n$

SOLUTION.   (i) $C_1 + 2C_2 + 3C_3 + ... + nC_n = \sum_{r=1}^{n} r \cdot C_r = \sum_{r=1}^{n} r \, {}^nC_r$ $(\because C_r = {}^nC_r)$

$$= \sum_{r=1}^{n} r \cdot \frac{n}{r} {}^{n-1}C_{r-1} \qquad \left(\because {}^nC_r = \frac{n}{r} {}^{n-1}C_{r-1} \right)$$

$$= n \sum_{r=1}^{n} {}^{n-1}C_{r-1}$$

$$= n({}^{n-1}C_0 + {}^{n-1}C_1 + {}^{n-1}C_2 + ... + {}^{n-1}C_{n-1}) = n \cdot 2^{n-1}$$

(ii) $C_0 + 2C_1 + 3C_2 + ... + (n+1)C_n = \sum_{r=0}^{n} (r+1) {}^nC_r$

$$= \sum_{r=0}^{n} (r \cdot {}^nC_r + {}^nC_r) = \sum_{r=0}^{n} r \cdot {}^nC_r + \sum_{r=0}^{n} {}^nC_r$$

$$= \sum_{r=0}^{n} r \cdot \frac{n}{r} {}^{n-1}C_{r-1} + \sum_{r=0}^{n} {}^nC_r \qquad \left(\because {}^nC_r = \frac{n}{r} {}^{n-1}C_{r-1} \right)$$

$$= n \left(\sum_{r=1}^{n} {}^{n-1}C_{r-1} \right) + \left(\sum_{r=0}^{n} {}^nC_r \right)$$

$$= n({}^{n-1}C_0 + {}^{n-1}C_1 + ... + {}^{n-1}C_{n-1}) + ({}^nC_0 + {}^nC_1 + ... + {}^nC_n)$$

$$= n2^{n-1} + 2^n = n \cdot 2^{n-1} + 2 \cdot 2^{n-1} = (n+2)2^{n-1}$$

(iii) $C_0 + 3C_1 + 5C_2 + ... + (2n+1)C_n$

$$= \sum_{r=0}^{n} (2r+1)C_r = \sum_{r=0}^{n} (2r+1) {}^nC_r \qquad (\because C_r = {}^nC_r)$$

$$= \sum_{r=0}^{n} (2r \, {}^nC_r + {}^nC_r)$$

$$= \sum_{r=0}^{n} 2r \, {}^nC_r + \sum_{r=0}^{n} {}^nC_r = 2 \sum_{r=0}^{n} r \cdot {}^n C_r + \sum_{r=0}^{n} {}^nC_r$$

$$= 2 \sum_{r=1}^{n} r \cdot \frac{n}{r} {}^{n-1}C_{r-1} + \sum_{r=0}^{n} {}^nC_r \qquad \left(\because {}^nC_r = \frac{n}{r} \cdot {}^{n-1}C_{r-1} \right)$$

$$= 2n \left(\sum_{r=1}^{n} {}^{n-1}C_{r-1} \right) + \sum_{r=0}^{n} {}^nC_r = 2n2^{n-1} + 2^n = n2^n + 2^n$$

$$= 2^n (n+1)$$

EXAMPLE 2. *Prove the following :*

(i) $1^2 \cdot C_1 + 2^2 \cdot C_2 + 3^2 \cdot C_3 + ... + n^2 \cdot C_n = n(n+1)2^{n-2}$

(ii) $aC_0 + (a+b)C_1 + (a+2b)C_2 + ... + (a+nb)C_n = (2a+nb)2^{n-1}$

(iii) $C_3 + 2C_4 + 3C_5 + ... + (n-2)C_n = (n-4)2^{n-1} + n + 2, where\, n > 3$

(iv) $C_0 - C_1 + C_2 - C_3 + ... + (-1)^n C_n = 0$

SOLUTION.   (i) $1^2 \cdot C_1 + 2^2 \cdot C_2 + 3^2 \cdot C_3 + \ldots + n^2 \cdot C_n$

$$= \sum_{r=1}^{n} r^2 C_r = \sum_{r=1}^{n} r^2 \, {}^nC_r = \sum_{r=1}^{n} [r(r-1) + r]^n C_r$$

$$= \sum_{r=1}^{n} r(r-1)^n C_r + \sum_{r=1}^{n} r \, {}^nC_r$$

$$= \sum_{r=2}^{n} r \cdot (r-1) \frac{n}{r} \frac{(n-1)}{(r-1)}{}^{n-2}C_{r-2} + \sum_{r=1}^{n} r \cdot \frac{n}{r}{}^{n-1}C_{r-1}$$

$$= n(n-1)\left(\sum_{r=2}^{n} {}^{n-2}C_{r-2} \right) + n\left(\sum_{r=1}^{n} {}^{n-1}C_{r-1} \right)$$

$$= n(n-1)[{}^{n-2}C_0 + {}^{n-2}C_1 + \ldots + {}^{n-2}C_{n-2}] + n[{}^{n-1}C_0 + {}^{n-1}C_1 + \ldots + {}^{n-1}C_{n-1}]$$

$$= n(n-1)2^{n-2} + n \cdot 2^{n-1} = n(n-1+2)2^{n-2}$$

$$= n(n+1)2^{n-2}$$

(ii) $aC_0 + (a+b)C_1 + (a+2b)C_2 + \ldots + (a+nb)C_n$

$$= \sum_{r=0}^{n} (a+rb)^n C_r = \sum_{r=0}^{n} a \cdot {}^n C_r + \sum_{r=0}^{n} rb^n C_r = a\left(\sum_{r=0}^{n} {}^nC_r \right) + b\left(\sum_{r=0}^{n} r \cdot {}^n C_r \right)$$

$$= \left(a \sum_{r=0}^{n} {}^nC_r \right) + b\left(\sum_{r=1}^{n} r \cdot \frac{n}{r}{}^{n-1}C_{r-1} \right) = a\left(\sum_{r=0}^{n} {}^nC_r \right) + bn\left(\sum_{r=1}^{n} {}^{n-1}C_{r-1} \right)$$

$$= a2^n + bn2^{n-1} \qquad \left[\because \sum_{r=0}^{n} {}^nC_r = 2^n, \sum_{r=1}^{n} {}^{n-1}C_{r-1} = 2^{n-1} \right]$$

$$= (2a + bn)2^{n-1}$$

(iii) $C_3 + 2C_4 + 3C_5 + \ldots + (n-2)C_n = \sum_{r=3}^{n} (r-2)C_r$

$$= \sum_{r=3}^{n} (r-2)^n C_r = \sum_{r=3}^{n} r \cdot {}^n C_r - \sum_{r=3}^{n} 2 \cdot {}^n C_r = \sum_{r=3}^{n} r \cdot \frac{n}{r}{}^{n-1}C_{r-1} - 2\sum_{r=3}^{n} {}^n C_r$$

$$= n\left(\sum_{r=3}^{n} {}^{n-1}C_{r-1} \right) - 2\left(\sum_{r=3}^{n} C_r \right)$$

$$= n({}^{n-1}C_2 + {}^{n-1}C_3 + \ldots + {}^{n-1}C_{n-1}) - 2({}^n C_3 + {}^n C_4 + \ldots + {}^n C_n)$$

$$= n[({}^{n-1}C_0 + {}^{n-1}C_1 + {}^{n-1}C_2 \ldots + {}^{n-1}C_{n-1}) - ({}^{n-1}C_0 + {}^{n-1}C_1)]$$

$$-2[({}^n C_0 + {}^n C_1 + {}^n C_2 + {}^n C_3 + \ldots + {}^n C_n) - ({}^n C_0 + {}^n C_1 + {}^n C_2)]$$

$$= n[2^{n-1} - \{1 + (n-1)\}] - 2\left[2^n - \left(1 + n + \frac{n(n-1)}{2} \right) \right]$$

$$= n[2^{n-1} - n] - 2\left[2^n - \left(\frac{n^2 + n + 2}{2} \right) \right]$$

$$= n \cdot 2^{n-1} - n^2 - 2 \cdot 2^n + n^2 + n + 2 = n \cdot 2^{n+1} - 2^{n+1} + n + 2$$

$$= (n-4)2^{n-1} + n + 2$$

(iv) $(1+x)^n = C_0 + C_1 x + C_2 x^2 + C_3 x^3 + ... + C_n x^n$

Putting $x = -1, C_0 - C_1 + C_2 - C_3 + ... + (-1)^n C_n = 0$

EXAMPLE 3. *Prove the following :*

(i) $C_1 - 2C_2 + 3C_3 - 4C_4 + ... + n(-1)^{n-1} C_n = 0$

(ii) $a - (a-1)C_1 + (a-2)C_2 - (a-3)C_3 + ... + (-1)^n(a-n)C_n = 0$

(iii) $aC_0 - (a+d)C_1 + (a+2d)C_2 - (a+3d)C_3 + ... + (-1)^n(a+nd)C_n = 0$

(iv) $C_0 - 2^2 C_1 + 3^2 C_2 - 4^2 C_3 + ... + (-1)^n (n+1)^2 C_n = 0, n > 0$

SOLUTION.

(i) $C_1 - 2C_2 + 3C_3 - 4C_4 + ... + n(-1)^{n-1} C_n$

$$= \sum_{r=1}^{n} (-1)^{r-1} r \cdot C_r = \sum_{r=1}^{n} (-1)^{r-1} r \cdot {}^n C_r$$

$$= \sum_{r=1}^{n} (-1)^{r-1} r \cdot \frac{n}{r} {}^{n-1}C_{r-1} = n \sum_{r=1}^{n} (-1)^{r-1} ({}^{n-1}C_{r-1})$$

$$= n[{}^{n-1}C_0 - {}^{n-1}C_1 + {}^{n-1}C_2 - {}^{n-1}C_3 + ... + (-1)^{n-1} {}^{n-1}C_{n-1}] = n \times 0 = 0$$

(ii) $a - (a-1)C_1 + (a-2)C_2 - (a-3)C_3 + ... + (-1)^n(a-n)C_n$

$$= \sum_{r=0}^{n} (-1)^r (a-r)C_r = \sum_{r=0}^{n} (-1)^r (a-r) {}^n C_r$$

$$= \sum_{r=0}^{n} (-1)^r a \cdot {}^n C_r = \sum_{r=0}^{n} (-1)^r \cdot r \cdot {}^n C_r$$

$$= a \left(\sum_{r=0}^{n} (-1)^r {}^n C_r \right) + \sum_{r=1}^{n} (-1)^r \cdot r \cdot \frac{n}{r} {}^{n-1}C_{r-1}$$

$$= a \left(\sum_{r=0}^{n} (-1)^r {}^n C_r \right) + n \sum_{r=1}^{n} (-1)^r {}^{n-1}C_{r-1}$$

$$= a[{}^n C_0 - {}^n C_1 + {}^n C_2 - {}^n C_3 + ... + (-1)^n {}^n C_n]$$

$$+ n[-{}^{n-1}C_0 + {}^{n-1}C_1 - {}^{n-1}C_2 + ... + (-1)^n {}^{n-1}C_{n-1}]$$

$$= a \times 0 + n \times 0 = 0$$

(iii) $aC_0 - (a+b)C_1 + (a+2d)C_2 - (a+3d)C_3 + ... + (-1)^n(a+nd)C_n$

$$= \sum_{r=0}^{n} (-1)^r (a+rd)C_r = \sum_{r=0}^{n} (-1)^r (a+rd) {}^n C_r \qquad (\because C_r = {}^n C_r)$$

$$= \sum_{r=0}^{n} (-1)^r \cdot a \cdot {}^n C_r + \sum_{r=0}^{n} (-1)^r \cdot rd \, {}^n C_r$$

$$= a \left(\sum_{r=0}^{n} (-1)^r {}^n C_r \right) + d \left(\sum_{r=0}^{n} (-1)^r r \cdot \frac{n}{r} \cdot {}^{n-1} C_{r-1} \right)$$

$$\left(\because {}^n C_r = \frac{n}{r} {}^{n-1} C_{r-1} \right)$$

$$= a \left(\sum_{r=0}^{n} (-1)^r {}^n C_r \right) + dn \left(\sum_{r=0}^{n} (-1)^r \cdot {}^{n-1} C_{r-1} \right)$$

$$= a[^nC_0 - {}^nC_1 + {}^nC_2 - {}^nC_3 + ... + (-1)^n \, {}^nC_n]$$

$$+ nd[-^{n-1}C_0 + ^{n-1}C_1 - ^{n-1}C_2 + ^{n-1}C_3 + ... + (-1)^n \, {}^{n-1}C_{r-1}]$$

$$= a \times 0 + nd \times 0 = 0$$

(iv) $C_0 - 2^2 C_1 + 3^2 C_2 - 4 \, {}^2 C_3 + ... + (-1)^n (n+1)^2 C_n$

$$= \sum_{r=0}^{n} (-1)^r (r+1)^2 C_r = \sum_{r=0}^{n} (-1)^r (r+1)^2 \, {}^nC_r = \sum_{r=0}^{n} (-1)^r (r^2 + 2r + 1) \, {}^nC_r$$

$$= \sum_{r=0}^{n} (-1)^r [r(r-1) + 3r + 1] \cdot {}^nC_r$$

$$= \sum_{r=0}^{n} (-1)^r r(r-1) \, {}^nC_r + \sum_{r=0}^{n} (-1)^r \cdot 3r \cdot {}^nC_r + \sum_{r=0}^{n} (-1)^r \cdot {}^n C_r$$

$$= \sum_{r=0}^{n} (-1)^r \cdot r \cdot (r-1) \frac{n\,(n-1)}{r\,(r-1)} \, {}^{n-2}C_{r-2}$$

$$+ 3 \sum_{r=1}^{n} (-1)^r r \cdot \frac{n}{r} \, {}^{n-1}C_{r-1} + \sum_{r=0}^{n} (-1)^r \, {}^nC_r$$

$$= n(n-1) \left(\sum_{r=2}^{n} (-1)^r \, {}^{n-2}C_{r-2} \right) + 3n \left(\sum_{r=1}^{n} (-1)^r \, {}^{n-1}C_{r-1} \right) + \left(\sum_{r=0}^{n} (-1)^r \, {}^nC_r \right)$$

$$= n(n+1) \times 0 + 3n \times 0 + 0 = 0$$

EXAMPLE 4.   *Prove the following :*

(i) $(C_0 + C_1)(C_1 + C_2)(C_2 + C_3)(C_3 + C_4)...(C_{n-1} + C_n) = \dfrac{C_0 C_1 C_2 ... C_{n-1}(n+1)^n}{n!}$

(ii) $\dfrac{C_1}{C_0} + 2\dfrac{C_2}{C_1} + 3\dfrac{C_3}{C_2} + ... + n\dfrac{C_n}{C_{n-1}} = \dfrac{n(n+1)}{2}$

(iii) $C_0^2 + C_1^2 + C_2^2 + ... + C_n^2 = \dfrac{(2n)!}{(n!)^2} = \dfrac{1 \cdot 3 \cdot 5 ... (2n-1)}{n!} \cdot 2^n$

(iv) $C_0 C_1 + C_1 C_2 + C_2 C_3 + ... + C_{n-1} C_n = \dfrac{2(n)!}{(n-1)!(n+1)!} = \dfrac{1 \cdot 3 \cdot 5 ... (2n-1)}{(n+1)!} n \cdot 2^n$

(v) $C_0 C_r + C_1 C_{r+1} + C_2 C_{r+2} + ... + C_{n-r} C_n = \dfrac{2(n)!}{(n-r)!(n+r)!}$

SOLUTION.   (i) We have to prove that

$$(C_0 + C_1)(C_1 + C_2)(C_2 + C_3)(C_3 + C_4)...(C_{n-1} + C_n) = \frac{C_0 C_1 C_2 ... C_{n-1}(n+1)^n}{n!}$$

or $\left(\dfrac{C_0 + C_1}{C_0} \right) \cdot \left(\dfrac{C_1 + C_2}{C_1} \right) \cdot \left(\dfrac{C_2 + C_3}{C_2} \right) \cdot \left(\dfrac{C_3 + C_4}{C_3} \right) ... \left(\dfrac{C_{n-1} + C_n}{C_{n-1}} \right) = \dfrac{(n+1)^n}{n!}$

Now $\left(\dfrac{C_0 + C_1}{C_0} \right) \cdot \left(\dfrac{C_1 + C_2}{C_1} \right) \cdot \left(\dfrac{C_2 + C_3}{C_2} \right) \cdot \left(\dfrac{C_3 + C_4}{C_3} \right) ... \left(\dfrac{C_{n-1} + C_n}{C_{n-1}} \right)$

$$= \left(1 + \frac{C_1}{C_0}\right) \cdot \left(1 + \frac{C_2}{C_1}\right) \cdot \left(1 + \frac{C_3}{C_2}\right) \cdot \left(1 + \frac{C_4}{C_3}\right) \cdots \left(1 + \frac{C_n}{C_{n-1}}\right)$$

$$= \left(1 + \frac{n}{1}\right) \cdot \left(1 + \frac{n-1}{2}\right) \cdot \left(1 + \frac{n-2}{3}\right) \cdot \left(1 + \frac{n-3}{4}\right) \cdots \left(1 + \frac{1}{n}\right)$$

$$= \frac{n+1}{1} \cdot \frac{n+1}{2} \cdot \frac{n+1}{3} \cdot \frac{n+1}{4} \cdots \frac{n+1}{n}$$

$$= \frac{(n+1)^n}{n!} \qquad \left[\because \frac{{}^nC_r}{{}^nC_{r-1}} = \frac{n-r+1}{r}, r = 1,2,3,\ldots n\right]$$

(ii) $\quad \dfrac{{}^nC_r}{{}^nC_{r-1}} = \dfrac{n!}{(n-r)!\,r!} \times \dfrac{(n-r+1)!\,(r-1)!}{n!}$

$$\Rightarrow \frac{{}^nC_r}{{}^nC_{r-1}} = \frac{n-r+1}{r}, r = 1,2,3\ldots,n \Rightarrow r\frac{{}^nC_r}{{}^nC_{r-1}} = n-r+1, r = 1,2,3,\ldots,n$$

$$\Rightarrow \qquad \sum_{r=1}^{n} r \cdot \frac{C_r}{C_{r-1}} = \sum_{r=1}^{n}(n-r+1)$$

$$\Rightarrow \qquad \sum_{r=1}^{n} r \cdot \frac{C_r}{C_{r-1}} = n + (n-1) + (n-2) + \ldots + 3 + 2 + 1$$

$$\Rightarrow \qquad \frac{C_1}{C_0} + 2 \cdot \frac{C_2}{C_1} + 3 \cdot \frac{C_3}{C_2} + \ldots + n \cdot \frac{C_n}{C_{n-1}} = \frac{n(n+1)}{2}$$

(iii) On expanding

$$(1+x)^n = C_0 + C_1 x + C_2 x^2 + \ldots + C_r x^r + \ldots + C_n x^n \qquad \ldots(1)$$

and $\quad (x+1)^n = C_0 x^n + C_1 x^{n-1} + C_2 x^{n-2} + \ldots + C_r x^{n-r} + \ldots + C_{n-1} \cdot x + C_n$

$$\ldots(2)$$

Multiplying eq. (1) and (2), we get

$$(1+x)^{2n} = (C_0 + C_1 x + C_2 x^2 + \ldots + C_r x^r + \ldots + C_n x^n)$$

$$(C_0 x^n + C_1 x^{n-1} + C_2 x^{n-2} + \ldots + C_r x^{n-r} + \ldots + C_{n-1} x + C_n)$$

$$\ldots(3)$$

Comparing the coefficients of x^n, we get

$$C_0^2 + C_1^2 + C_2^2 + \ldots + C_n^2 = {}^{2n}C_n \Rightarrow C_0^2 + C_1^2 + C_2^2 + \ldots + C_n^2 = \frac{(2n)!}{n!\,n!}$$

$$\frac{(2n)!}{n!\,n!} = \frac{1 \cdot 2 \cdot 3 \cdot 4 \cdot 5 \ldots (2n-2)(2n-1)(2n)}{n!\,n!}$$

$$= \frac{\{1 \cdot 3 \cdot 5 \ldots (2n-1)\}\{2 \cdot 4 \cdot 6 \ldots (2n-2)2n\}}{n!\,n!}$$

$$= \frac{[1 \cdot 3 \cdot 5 ... (2n-1)] \times 2^n [1 \cdot 2 \cdot 3 ... (n-1)n]}{n!n!}$$

$$= \frac{[1 \cdot 3 \cdot 5 ... (2n-1)] \times 2^n n!}{n!n!} = \frac{1 \cdot 3 \cdot 5 ... (2n-1)}{n!} 2^n$$

(iv) Comparing the coefficients of x^{n-1}, we get

$$C_0 C_1 + C_1 C_2 + ... + C_{n-1} C_n = {}^{2n}C_{n-1}$$

$$\Rightarrow \quad C_0 C_1 + C_1 C_2 + ... + C_{n-1} C_n = \frac{(2n)!}{(n+1)!(n-1)!}$$

Now, $\quad \dfrac{(2n)!}{(n+1)!(n-1)!} = \dfrac{1 \cdot 2 \cdot 3 \cdot 4 .... (2n-2)(2n-1)(2n)}{(n+1)!(n-1)!}$

$$= \frac{1 \cdot 3 \cdot 5 ... (2n-1)}{(n+1)!} 2^n \frac{n(n-1)!}{(n-1)!}$$

$$= \frac{1 \cdot 3 \cdot 5 ... (2n-1)}{(n+1)!} n 2^n$$

(v) Comparing the coefficients of x^{n-r}, we get

$$C_0 C_r + C_1 C_{r+1} + ... + C_n C_{n-r} = {}^{2n}C_{n-r} = \frac{2n!}{(n-r)!(n+r)!}$$

EXAMPLE 5. *Prove that—*

(i) $\quad C_0 + \dfrac{C_1}{2} + \dfrac{C_2}{3} + ... + \dfrac{C_n}{n+1} = \dfrac{2^{n+1}-1}{n+1}$

(ii) $\quad C_0 - \dfrac{C_1}{2} + \dfrac{C_2}{3} - \dfrac{C_3}{4} + ... + (-1)^n \dfrac{C_n}{n+1} = \dfrac{1}{n+1}$

SOLUTION. (i) $\quad C_0 + \dfrac{C_1}{2} + \dfrac{C_2}{3} + ... + (-1)^n \dfrac{C_n}{n+1}$

$$= \sum_{r=0}^{n} \frac{{}^n C_r}{r+1} = \sum_{r=0}^{n} \frac{1}{r+1} {}^n C_r$$

$$= \sum_{r=0}^{n} \frac{1}{(n+1)} \cdot \frac{(n+1)}{(r+1)} {}^n C_r = \frac{1}{(n+1)} \sum_{r=0}^{n} \frac{(n+1)}{(r+1)} {}^n C_r$$

$$= \frac{1}{(n+1)} \sum_{r=0}^{n} {}^{n+1} C_{r+1} \qquad \left(\because {}^{n+1} C_{r+1} = \frac{n+1}{r+1} {}^n C_r \right)$$

$$= \frac{1}{n+1} [{}^{n+1} C_1 + {}^{n+1} C_2 + {}^{n+1} C_3 + ... + {}^{n+1} C_{n+1}]$$

$$= \frac{1}{n+1} [({}^{n+1} C_0 + {}^{n+1} C_1 + ... + {}^{n+1} C_{n+1}) - ({}^{n+1} C_0)]$$

$$= \frac{1}{n+1} [2^{n+1} - 1]$$

(ii) $C_0 - \dfrac{C_1}{2} + \dfrac{C_2}{3} - \dfrac{C_3}{4} + ... + (-1)^n \dfrac{C_n}{n+1}$

$$= \sum_{r=0}^{n} (-1)^r \frac{{}^nC_r}{r+1} = = \sum_{r=0}^{n} \frac{(-1)^r}{(r+1)} {}^nC_r = \sum_{r=0}^{n} \frac{(-1)^r}{(n+1)} \times \frac{(n+1)}{(r+1)} {}^nC_r$$

$$= \frac{1}{(n+1)} \sum_{r=0}^{n} (-1)^r {}^{n+1}C_{r+1} \qquad \left(\because {}^{n+1}C_{r+1} = \frac{n+1}{r+1} {}^nC_r \right)$$

$$= \frac{1}{(n+1)} [{}^{n+1}C_1 - {}^{n+1}C_2 + {}^{n+1}C_3 - {}^{n+1}C_4 + ... + (-1)^n {}^{n+1}C_{n+1}]$$

$$= -\frac{1}{(n+1)} [-{}^{n+1}C_1 + {}^{n+1}C_2 - {}^{n+1}C_3 + {}^{n+1}C_4 - ... + (-1)^{n+1} {}^{n+1}C_{n+1}]$$

$$= -\frac{1}{(n+1)} [({}^{n+1}C_0 - {}^{n+1}C_1 + {}^{n+1}C_2 - {}^{n+1}C_3 + ... + (-1)^{n+1} {}^{n+1}C_{n+1}) - {}^{n+1}C_0]$$

$$= -\frac{1}{(n+1)} [0 - {}^{n+1}C_0] = \frac{1}{n+1}$$

EXAMPLE 6. *In the expansion of* $(1+x)^n$ *if* $C_0, C_1, ..., C_n$ *are the binomial coefficients, prove that*

$$C_0^2 - C_1^2 + C_2^2 - C_3^2 + ... + (-1)^n C_n^2 = \begin{cases} 0, & \text{if } n \text{ is odd} \\ (-1)^{n/2} \cdot {}^nC_{n/2}, & \text{if } n \text{ is even} \end{cases}$$

SOLUTION. We have $(1+x)^n = C_0 + C_1 x + C_2 x^2 + ... + C_n x^n$ ...(1)

and $(x+1)^n = C_0 x^n + C_1 x^{n-1} + C_2 x^{n-2} + ... + C_{n-1} x + C_n$ ...(2)

Putting $-x$ for x in eq. (1), we get

$(1-x)^n = C_0 - C_1 x^2 + C_2 x^2 - C_3 x^3 + ... (-1)^n C_n x^n$ ...(3)

Multiplying eqns. (2) and (3), we get

$$(C_0 - C_1 x + C_2 x^2 - C_3 x^3 + ... + (-1)^n \cdot C_n x^n)(C_0 x^n + C_1 x^{n-1} + C_2 x^{n-2} + ... + C_n)$$

$$= (1+x)^n (1-x)^n$$

$$\Rightarrow (C_0 - C_1 x + C_2 x^2 - C_3 x^3 + ... + (-1)^n \cdot C_n x^n)(C_0 x^n + C_1 x^{n-1} + C_2 x^{n-2} + ... + C_n)$$

$$= (1-x^2)^n \qquad ...(4)$$

Comparing the coefficients of x^n on both the sides,

$$C_0^2 - C_1^2 + C_2^2 - C_3^2 + ... + (-1)^n C_n^2 = \text{coefficients of } x^n \text{ in } (1-x^2)^n \qquad ...(5)$$

Now R.H.S. of eqn. (5) contains only even powers of x.

Therefore coefficients of x^n in $(1-x^2)^n = 0$ if n is odd natural number.

If n is even suppose $(r+1)^{\text{th}}$ term in the binomial expansion of $(1-x^2)^n$ contains x^n.

Then $T_{r+1} = {}^nC_r (-1)^r (x^2)^r = {}^nC_r (-1)^r x^{2r}$

For this term to contain x^n, we must have $2r = n \Rightarrow r = \dfrac{n}{2}$

Coefficient of $x^n = {}^nC_{n/2}(-1)^{n/2}$

Therefore, $C_0^2 - C_1^2 + C_2^2 - ... + (-1)^n C_n^2 = \begin{cases} 0, & \text{if } n \text{ is odd} \\ {}^nC_{n/2}(-1)^{n/2}, & \text{if } n \text{ is even} \end{cases}$

EXAMPLE 7. *Find the coefficient of* x^4 *in the expansion of* $(1+x)^n (1-x)^n$ *and prove that*

$$C_2 = C_0 C_4 - C_1 C_3 + C_2 C_2 - C_3 C_1 + C_4 C_0$$

SOLUTION. $(1+x)^n(1-x)^n =$

$\Rightarrow$ $(C_0 + C_1 x + C_2 x^2 + C_3 x^3 + ... + C_n x^n)$
$(C_0 - C_1 x + C_2 x^2 - C_3 x^3 + ... + (-1)^n C_n x^n)$

Coefficient of x^4 in R.H.S. $= C_0 C_4 - C_1 C_3 + C_2 C_2 - C_3 C_1 + C_4 C_0$

and $(1+x)^n(1-x)^n = (1-x^2)^n$

$= {}^n C_0 - {}^n C_1 x^2 + {}^n C_2 x^4 - {}^n C_3 x^6 + ... + (-1)^n C_n (-1)^n x^{2n}$

$\therefore$ Coefficient of $x^4 = {}^n C_2 = C_2$

Hence, $C_0 C_4 - C_1 C_3 + C_2 C_2 - C_3 C_1 + C_4 C_0 = C_2$

EXAMPLE 8.   *In the expansion of $(1+x)^n$, $P(n)$ stands for the product of all the binomial coefficients, then prove that* $\dfrac{P(n+1)}{P(n)} = \dfrac{(n+1)^n}{n!}$.

SOLUTION.   L.H.S. $= \dfrac{{}^{n+1}C_0 \cdot {}^{n+1}C_1 \cdot {}^{n+1}C_2 ... {}^{n+1}C_r ... {}^{n+1}C_{n+1}}{{}^n C_0 \cdot {}^n C_1 \cdot {}^n C_2 ... {}^n C_r ... {}^n C_n}$

Using ${}^{n+1}C_0 = {}^n C_0 = 1$ and ${}^{n+1}C_{n+1} = 1$,

$\therefore$ $\dfrac{P(n+1)}{P(n)} = \dfrac{{}^{n+1}C_1}{{}^n C_1} \cdot \dfrac{{}^{n+1}C_2}{{}^n C_2} ... \dfrac{{}^{n+1}C_r}{{}^n C_r} ... \dfrac{{}^{n+1}C_n}{{}^n C_n}$

Therefore, $\dfrac{{}^{n+1}C_r}{{}^n C_r} = \dfrac{(n+1)!}{r!(n-r+1)!} \cdot \dfrac{r!(n-r)!}{n!} = \dfrac{(n+1)}{(n-r+1)}$, $(1 \le r \le n)$

Putting $1, 2, 3, ..., n$ and then multiply each, we get

$$\frac{P(n+1)}{P(n)} = \frac{(n+1)(n+1) \cdots}{n(n-1) \cdots 2 \cdot 1} = \frac{(n+1)^n}{n!}$$

EXAMPLE 9.   *If $S_n = 1 + q + q^2 + ... + q^n$*

$S_n = 1 + \dfrac{q+1}{2} + \left(\dfrac{q+1}{2}\right)^2 + ... + \left(\dfrac{q+1}{2}\right)^n$, $q \ne 1$, *prove that*

${}^{n+1}C_1 + {}^{n+2}C_2 S_1 + {}^{n+1}C_3 S_2 + ... + {}^{n+1}C_{n+1}S_n = 2^n S_n$

SOLUTION.   Clearly, $S_n = \dfrac{1-q^{n+1}}{1-q}$ ...(1)

and $S_n = \dfrac{1 - \left(\dfrac{q+1}{2}\right)^{n+1}}{1 - \dfrac{q+1}{2}} = \dfrac{2^{n+1} - (q+1)^{n+1}}{2^n(1-q)}$ ...(2)

Therefore,

${}^{n+1}C_1 + {}^{n+1}C_2 \cdot S_1 + {}^{n+1}C_3 S_2 + ... + {}^{n+1}C_{n+1}S_n$

$= \dfrac{1}{(1-q)}[{}^{n+1}C_1(1-q) + {}^{n+1}C_2(1-q^2) + {}^{n+1}C_3(1-q^3)$

$+ ... + {}^{n+1}C_{n+1}(1-q^{n+1})]$

$$= \frac{1}{(1-q)}[^{n+1}C_1 + {}^{n+1}C_2 + \ldots + {}^{n+1}C_{n+1}]$$
$$- [^{n+1}C_1 \cdot q + {}^{n+1}C_2 \cdot q^2 + \ldots + {}^{n+1}C_{n+1} \cdot q^{n+1}]$$
$$= \frac{1}{(1-q)}[(2^{n+1} - 1) - \{(1+q)^{n+1} - 1\}] = \frac{2^{n+1} - (1+q)^{n+1}}{(1-q)}$$
$$= 2^n S_n \qquad\qquad \text{[From eqn. (2)]}$$

EXAMPLE 10. *Find the sum of the following series :*

$$\sum_{r=0}^{n} (-1)^r \cdot {}^nC_r \left[\frac{1}{2^r} + \frac{3^r}{2^{2r}} + \frac{7^r}{2^{3r}} + \frac{15^r}{2^{4r}} + \ldots upto \, m \, terms \right]$$

SOLUTION. $$\sum_{r=0}^{n} (-1)^r \cdot {}^nC_r \left[\frac{1}{2^r} + \frac{3^r}{2^{2r}} + \frac{7^r}{2^{3r}} + \frac{15^r}{2^{4r}} + \ldots \text{upto } m \text{ terms} \right]$$

$$= \sum_{r=0}^{n} (-1)^r \cdot {}^nC_r \left[\left(\frac{1}{2}\right)^r + \left(\frac{3}{4}\right)^r + \left(\frac{7}{8}\right)^r + \left(\frac{15}{16}\right)^r + \ldots m \text{ terms} \right]$$

Now, $\displaystyle\sum_{r=0}^{n} (-1)^r \cdot {}^nC_r \left(\frac{1}{2}\right)^r = 1 - {}^nC_1 \cdot \frac{1}{2} + {}^nC_2 \cdot \frac{1}{2^2} - {}^nC_3 \cdot \frac{1}{2^3} + \ldots = \left(1 - \frac{1}{2}\right)^n = \frac{1}{2^n}$

Therefore, $\displaystyle\sum_{r=0}^{n} (-1)^r \cdot {}^nC_r \left(\frac{3}{4}\right)^r = \left(1 - \frac{3}{4}\right)^n = \frac{1}{4^n}$

$\therefore$ Given series $= \dfrac{1}{2^n} + \dfrac{1}{4^n} + \dfrac{1}{8^n} + \dfrac{1}{16^n} + \ldots$ upto m terms $\qquad$ which is a G.P.

$$= \frac{\dfrac{1}{2^n}\left[\left(1 - \dfrac{1}{2^n}\right)^m\right]}{\left(1 - \dfrac{1}{2^n}\right)} = \frac{2^{mn} - 1}{2^{mn}(2^n - 1)}$$

EXAMPLE 11. *If $T_0, T_1, T_2, \ldots, T_n$, represents the terms in the expansion of $(x + a)^n$, prove that*
$$(T_0 - T_2 + T_4 - \ldots)^2 + (T_1 - T_3 + T_5 \ldots)^2 = (x^2 + a^2)^n$$

SOLUTION. Given $\qquad (x + a)^n = T_0 + T_1 + T_2 + \ldots + T_n \qquad\qquad \ldots(1)$
Replacing a by ai and $- ai$ respectively in eqn. (1),
$$(x + ai)^n = (T_0 - T_2 + T_4 + \ldots) + i(T_1 - T_3 + T_5 - \ldots) \qquad \ldots(2)$$
and $\qquad (x - ai)^n = (T_0 - T_2 + T_4 - \ldots) - i(T_1 - T_3 + T_5 - \ldots) \qquad \ldots(3)$
Multiplying eqn. (2) and (3), we get
$$(x^2 + a^2) = (T_0 - T_2 + T_4 - \ldots)^2 + (T_1 - T_3 + T_5 - \ldots)^2$$

EXAMPLE 12. *If $a_0, a_1, a_2, \ldots$ are the coefficients in the expansion of $(1 + x + x^2)^n$ in ascending powers, then prove that*

(i) $a_r = a_{2n-r}$ $\qquad$ (ii) $a_0 + a_1 + a_2 + \ldots + a_{n-1} = \dfrac{1}{2}(3^n - a_n)$

(iii) $(r+1)a_{r+1} = (n-r)a_r + (2n - r + 1)a_{r-1}$ $\qquad\qquad (0 < r < 2n)$

(iv) $a_0^2 - a_1^2 + a_2^2 - a_3^2 + ... - a_{2n-1}^2 + a_{2n}^2 = a_n$

(v) $a_0^2 - a_1^2 + a_2^2 - a_3^2 + ... + (-1)^{n-1} a_{2n-1}^2 = \dfrac{1}{2} a_n [1 - (-1)^n a_n]$

(vi) $(a_0 + a_3 + a_6 + ...) = (a_1 + a_4 + a_7 + ...) = (a_2 + a_5 + a_8 + ...) = 3^{n-1}$

SOLUTION.

(i) $(1 + x + x^2)^n = a_0 + a_1 x + a_2 x^2 + ... + a_{2n} x^{2n} = \displaystyle\sum_{r=0}^{2n} a_r x^r$ 　　　...(1)

Replacing x by $\dfrac{1}{x}$, we get

$$\left(1 + \frac{1}{x} + \frac{1}{x^2}\right)^n = a_0 + a_1 \left(\frac{1}{x}\right) + a_2 \left(\frac{1}{x}\right)^2 + ...$$

$\Rightarrow \qquad \dfrac{(1 + x + x^2)^n}{x^{2n}} = a_0 + a_1 \left(\dfrac{1}{x}\right) + a_2 \left(\dfrac{1}{x}\right)^2 + ...$

$\Rightarrow \qquad (1 + x + x^2)^n = a_0 x^{2n} + a_1 x^{2n-1} + ... + a_r x^{2n-r} + ...$ 　　　...(2)

Therefore, from eqn. (1) and (2)

$$\sum_{r=0}^{2n} a_r x^r = \sum_{r=0}^{2n} a_r x^{2n-r}$$

Comparing the coefficients of x^{2n-r} on both the sides

$$a_{2n-r} = a_r$$

or 　　　　　　　$a_r = a_{2n-r}$ where $0 < r < 2n$

(ii) Putting eqn. $x = 1$ in eqn. (1), we get

$$a_0 + a_1 + a_2 + ... + a_r + ... + a_{2n} = 3^n \qquad ...(3)$$

From (I), 　$a_r = a_{2n-r} \Rightarrow a_0 = a_{2n}, a_1 = a_{2n-1}...$etc.

$\therefore \qquad 2(a_0 + a_1 + a_2 + ... + a_{n-1}) + a_n = 3^n$

$\Rightarrow \qquad\qquad a_0 + a_1 + a_2 + ... + a_{n-1} = \dfrac{1}{2}(3^n - a_n)$

(iii) Differentiating both sides of eqn. (1), w.r.t. x, we get

$$n(1 + x + x^2)^{n-1}(1 + 2x) = \sum_{r=0}^{2n} r a_r x^{r-1}$$

Multiply both sides by $(1 + x + x^2)$ and using $(1 + x + x^2)^n = \displaystyle\sum_{r=0}^{2n} a_r x^r$, we get

$$n(2x + 1) \sum_{r=0}^{2n} a_r x^r = (1 + x + x^2) \sum_{r=0}^{2n} r a_r x^{r-1}$$

Comparing the coefficients of x^r on both sides, we get

$$n[2a_{r-1} + a_r] = \{(r+1)a_{r+1} + r a_r + (r-1)a_{r-1}\}$$

$\Rightarrow \qquad (r+1)a_{r+1} = (n-r)a_r + (2n - r + 1)a_{r-1} \qquad (0 < x < 2n)$

(iv) Replacing x by $-x$ in eqn. (1), we get

$$(1 - x + x^2)^n = a_0 - a_1 x + a_2 x^2 + ... + (-1)^n a_n x^n + ... + a_{2n} x^{2n}$$

$$= a_{2n} - a_{2n-1} \cdot x ... a_1 x^{2n-1} + a_0 x^{2n} \qquad ...(4)$$

Multiplying eqns. (1) and (4), we get

$$[(1 + x^2)^2 - x^2]^n = (1 + x^2 + x^4)^n$$

Coefficient of $x^{2n} = a_0^2 - a_1^2 + a_2^2 - ... + a_{2n}^2$

$\Rightarrow$ Coefficient of $(x^2)^n = a_0^2 - a_1^2 + a_2^2 - ... + a_{2n}^2$

Putting $x^2 = t$,

$$a_0^2 - a_1^2 + ... + a_{2n}^2 = \text{Coeff. of } x^{2n} \text{ in the expansion of } (1 + t + t^2)^n = a^n$$

$$\Rightarrow \quad a_0^2 - a_1^2 + a_2^2 - a_3^2 + ... - a_{2n-1}^2 + a_{2n}^2 = a_n$$

(v) $\qquad a_{2n} = a_0, a_{2n-1} = a_1$

$\therefore \qquad 2[a_0^2 - a_1^2 + ... + (-1)^{n-1} a_{n-1}^2] + (-1)^n a_n^2 = a_n$

$$a_0^2 - a_1^2 + ... + (-1)^{n-1} a_{n-1}^2 = \frac{1}{2} a_n [1 - (-1)^n a_n]$$

(vi) Putting $x = \omega$ and using $1 + \omega + \omega^2 = 0$, we get

$$\left(\text{where } \omega = -\frac{1}{2} + \frac{i\sqrt{3}}{2} \text{ and } \omega^2 = -\frac{1}{2} - \frac{i\sqrt{3}}{2} \right)$$

$$\Delta = (a_0 + a_3 + a_6 + ...) + \omega(a_1 + a_4 + ...) + \omega^2(a_2 + a_5 + ...)$$

$$= A + B\omega + C\omega^2$$

Equating real and imaginary parts of both the sides, we get

$$A - \frac{B}{2} - \frac{C}{2} = 0, \quad \frac{\sqrt{3}}{2}(B - C) = 0 \Rightarrow A = B = C$$

Now, putting $x = 1, 3^n = $ sum of all coefficients $= A + B + C$

$\Rightarrow \qquad 3^n = 3A$

$\Rightarrow \qquad A = 3^{n-1} = B = C$

EXAMPLE 13. *Find the sum of the infinite series* $a_1 + a_2 + a_3 + ...$

where $\qquad a_n = (\log 3)^n \sum\limits_{k=1}^{n} \dfrac{2k+1}{k!(n-k)!}$

SOLUTION. $\qquad a_n = (\log 3)^n \sum\limits_{k=1}^{n} \dfrac{2k+1}{k!(n-k)!}$

$$= \frac{(\log 3)^n}{n!} \sum_{k=1}^{n} (2k+1) \,^n C_k = \frac{(\log 3)^n}{n!} \left[\sum_{k=1}^{n} 2k \,^n C_k + \sum_{k=1}^{n} {}^n C_k \right]$$

$$= \frac{(\log 3)^n}{n!} [2(n2^{n-1}) + (2^n - 1)] = \frac{(2\log 3)^n n}{n!} + \frac{(2\log 3)^n}{n!} - \frac{(\log 3)^n}{n!}$$

$$\sum_{n=1}^{\infty} a_n = \sum_{n=1}^{\infty} \left[\frac{x^n}{(n-1)!} + \frac{y^n}{n!} - \frac{z^n}{n!} \right] = \sum_{n=1}^{\infty} \left[x \cdot \frac{x^{n-1}}{(n-1)!} + \frac{y^n}{n!} - \frac{z^n}{n!} \right].$$

We know that $e^x = 1 + x + \dfrac{x^2}{2!} + \dfrac{x^3}{3!} + ... + \dfrac{x^n}{n!} + ...$

$\Rightarrow \qquad e^x - 1 = \sum_{n=1}^{\infty} \dfrac{x^n}{n!}$

$\therefore \qquad \sum_{n=1}^{\infty} a_n = x e^x + (e^y - 1) + (e^z - 1) = 2\log 3 e^{2\log 3} + (e^{2\log 3} - 1) + (e^{\log 3} - 1)$

$$= (2\log 3)e^{\log 9} + (e^{\log 9} - 1) - (e^{\log 3} - 1)$$

$$= 9(2\log 3) + (9 - 1) - (3 - 1) = 18\log 3 + 6$$

EXAMPLE 14. *Prove that* $\sum_{r=0}^{n} r(n-r)C_r^2 = n^2(^{2n-2}C_n)$

SOLUTION. We have $(1 + x)^n = C_0 + C_1 x + C_2 x^2 + ... + C_r x^r + ... + C_n x^n$ ...(1)

Now $r(n-r)C_r^2 = r \cdot C_r \cdot (n-r)C_r$

and $\qquad C_r = C_{n-r}$

$$(x + 1)^n = C_0 x^n + C_1 x^{n-1} + ... + C_r x^{n-r} + ... + C_n \qquad\qquad ...(2)$$

Differentiating eqn. (1), we get

$$n(1 + x)^{n-1} = C_1 + 2C_2 x + ... + rC_r x^{r-1} + ... \qquad\qquad ...(3)$$

Now, differentiating eqn. (2), we get

$$n(1 + x)^{n-1} = {}^nC_0 x^{n-1} + ... + (n-r)C_r x^{n-r-1} + ... \qquad\qquad ...(4)$$

Multiplying eqn. (3) and (4), we get

$$n^2(1 + x)^{2n-2} = (r \cdot C_r)(n-r)C_r x^{r-1} x^{n-r-1} = \sum r(n-r)C_r^2 x^{n-2}$$

$\therefore$ Coefficient of $x^{n-2} = n^2 \cdot {}^{2n-2}C_{n-2} = n^2 \cdot (^{2n-2}C_n)$

EXAMPLE 15. *Prove that* $\dfrac{3!}{2(n+3)} = \sum_{r=0}^{n} (-1)^r \left(\dfrac{{}^nC_r}{{}^{r+3}C_r} \right)$

SOLUTION. We have to prove that $\dfrac{3!}{2(n+3)} = \sum_{r=0}^{n} (-1)^r \left(\dfrac{{}^nC_r}{{}^{r+3}C_r} \right)$

On putting $r = 0, 1, 2, ..., n$, the given Σ is equal to

$$\frac{{}^nC_0}{{}^3C_0} - \frac{{}^nC_1}{{}^4C_1} + \frac{{}^nC_2}{{}^5C_2} + ... + (-1)^n \cdot \frac{{}^nC_n}{{}^{n+3}C_n}$$

$$= 1 - \frac{n}{4} + \frac{n(n-1)}{5 \cdot 4} - \frac{n(n-1)(n-2)}{6 \cdot 5 \cdot 4} + \ldots + (-1)^n \frac{3!}{(n+3)(n+2)(n+1)}$$

$$= \frac{3!}{(n+3)(n+2)(n+1)} \left[\frac{(n+3)(n+2)(n+1)}{3!} - \frac{(n+3)(n+2)(n+1)(n)}{3!4} \right.$$

$$\left. + \frac{(n+3)(n+2)(n+1)n(n-1)}{3!5 \cdot 4} + \ldots + (-1)^n \right]$$

Now, putting $n + 3 = N$,

$$\text{R.H.S.} = \frac{3!}{N(N-1)(N-2)} \times \left[\frac{N(N-1)(N-2)}{3!} - \frac{N(N-1)(N-2)(N-3)}{4!} \right.$$

$$\left. + \frac{N(N-1)(N-2)(N-3)(N-4)}{5!} - \ldots + (-1)^{N-3} \right]$$

$$= \frac{3!}{N(N-1)(N-2)} [{}^N C_3 - {}^N C_4 + {}^N C_5 - \ldots + (-1)^{N-3}] \qquad \ldots(1)$$

Now, $C_0 - C_1 + C_2 - C_3 + \ldots = (1-1)^n = 0$

$\Rightarrow \quad C_0 - C_1 + C_2 = C_3 - C_4 + C_5$ where $C_3 = {}^N C_3$

$\Rightarrow \quad 1 - N + \frac{N(N-1)}{2} = C_3 - C_4 + C_5 - \cdots$

Putting this value in eqn. (1),

$$\text{R.H.S.} = \frac{3!}{N(N-1)(N-2)} \left[\frac{2 - 2N + N^2 - N}{2} \right]$$

$$= \frac{3!}{2N(N-1)(N-2)} [N^2 - 3N + 2]$$

$$= \frac{3!}{2N(N-1)(N-2)} (N-1)(N-2) = \frac{3!}{2N} = \frac{3!}{2(n+3)} \qquad (\because N = n+3)$$

$$= \text{L.H.S.}$$

Exercise 4.2

(A) Prove that—

1. ${}^8 C_0 + {}^8 C_1 + {}^8 C_2 + \ldots + {}^8 C_8 = 256$

2. ${}^9 C_0 + {}^9 C_2 + {}^9 C_4 + \ldots + {}^9 C_8 = 256$

3. ${}^{15} C_1 + {}^{15} C_3 + \ldots + {}^{15} C_{15} = 16384$

4. ${}^{12} C_1 + {}^{12} C_2 + \ldots + {}^{12} C_{12} = 4095$

(B) If $C_0, C_1, \ldots, C_n$, are the binomial coefficients in the expansion of $(1+x)^n$. Prove the following :

5. $C_0 + 3C_1 + 3^2 C_2 + \ldots + 3^n C_n = 4^n$

6. $C_1 + C_2 + C_3 + \ldots + C_n = 1 + 2 + 2^2 + \ldots + 2^{n-1}$

7. $C_0 + 2C_1 + 2^2 C_2 + \ldots + 2^n C_n = 3^n$

8. $\frac{C_1}{2} + \frac{C_3}{4} + \frac{C_5}{6} + \ldots = \frac{2^n - 1}{n+1}$

9. $C_1^2 + 2C_2^2 + 3C_3^2 + \ldots + nC_n^2 = \frac{(2n-1)!}{[(n-1)!]^2}$

10. $C_0^2 + 3C_1^2 + 5C_2^2 + \ldots + (2n+1)C_n^2 = \frac{(n+1)2n!}{(n!)^2}$

11. $p.C_0 + p^2 \frac{C_1}{2} + p^3 \frac{C_2}{3} + \ldots + p^{n+1} \cdot \frac{C_n}{n+1}$

$$= \frac{(p+1)^{n+1} - 1}{n+1}$$

12. $2C_0 + 5C_1 + 8C_2 + \ldots + (3n+2)C_n$

$$= (3n+4)2^{n-1}$$

13. If $(1+x+x^2)^n = C_0 + C_1 x + C_2 x^2 + ...,$ prove

that $C_0 C_1 - C_1 C_2 + C_2 C_3 - ... = 0$

14. Find the sum of the coefficients in the expansion of $(5a - 4b)^n, n \in N.$

15. If $C_0, C_1, ..., C_n$ are the binomial coefficients in the expansion of $(1+x)^n,$, prove that

(i) $2C_0 + \dfrac{2^2 C_1}{2} + \dfrac{2^3 C_2}{3} + ... + \dfrac{2^{n+1} \cdot C_n}{n+1}$

$\quad = \dfrac{3^{n+1} - 1}{n+1}$

(ii) $2C_0 + 2C_1 + 4C_2 + 6C_3 + ... + 2nC_n$

$\quad = 2(1 + n2^{n-1})$

(iii) $2C_0 + 2C_1 + 4C_2 + 6C_3 + ... + 2nC_n$

$\quad = 1 + n \cdot 2^n$

Objective Evaluation

∞ MULTIPLE CHOICE QUESTIONS

Choose the most appropriate one.

1. If A_r, B_r and C_r are the coefficients of x^r in the expansion of $(1+x)^{10}$, $(1+x)^{20}$ and $(1+x)^{30}$, $r = 0, 1, ..., 10$. Then value of
$$\sum_{r=1}^{10} A_r (B_{10} B_r - C_{10} A_r) =$$
   (a) $B_{10} - C_{40}$
   (b) $A_{10}(B_{10}^2 - C_{40} A_{10})$
   (c) 0
   (d) $C_{10} - B_{10}$

2. Which is greater $99^{50} + 100^{50}$ and 101^{50}?
   (a) 101^{50}
   (b) $99^{50} + 100^{50}$
   (c) both are equal
   (d) none of these

3. In the expansion of $(1 + x)^{20}$, if the coefficient of rth and $(r + 1)^{th}$ term is equal, then value of r is :
   (a) 7
   (b) 8
   (c) 9
   (d) 10

4. The remainder after dividing 5^{124} by 124 is:
   (a) 1
   (b) 2
   (c) 3
   (d) 5

5. $\dfrac{^{50}C_0}{1} + \dfrac{^{50}C_2}{3} + \dfrac{^{50}C_4}{5} + ... + \dfrac{^{50}C_{50}}{51} =$
   (a) $\dfrac{2^{50}}{51}$
   (b) $\dfrac{2^{50}-1}{51}$
   (c) $\dfrac{2^{51}}{51}$
   (d) $\dfrac{2^{51}-1}{51}$

6. In the expansion of $(1 + x + x^2 + x^3)^6$, the coefficient of x^{14} is :
   (a) 115
   (b) 120
   (c) 125
   (d) 130

7. For natural numbers m and n, if $(1-y)^m (1+y)^n = 1 + a_1 y + a_2 y^2 + ...$ and $a_1 = a_2 = 10$, then value of $(m, n) =$
   (a) (35, 20)
   (b) (45, 35)
   (c) (35, 45)
   (d) (20, 45)

8. In the expansion of $\left(x^4 - \dfrac{1}{x^3}\right)^{15}$, the coefficient of x^{32} is :
   (a) $^{15}C_2$
   (b) $^{15}C_3$
   (c) $^{15}C_5$
   (d) $^{15}C_4$

9. In the expansion of $\dfrac{3-2x}{(1+3x)^3}$, the coefficient of x^3 is :
   (a) −272
   (b) −540
   (c) −870
   (d) −918

10. $1 + \dfrac{2}{4} + \dfrac{2 \cdot 5}{4 \cdot 8} + \dfrac{2 \cdot 5 \cdot 8}{4 \cdot 8 \cdot 12} + \dfrac{2 \cdot 5 \cdot 8 \cdot 11}{4 \cdot 8 \cdot 12 \cdot 16} + ... =$
   (a) $3\sqrt{4}$
   (b) $3\sqrt{16}$
   (c) $4^{-2/3}$
   (d) $4^{3/2}$

11. If $(2x^2 - x - 1)^5 = a_0 + a_1 x + a_2 x^2 + ... + a_{10} x^{10}$, then $a_2 + a_4 + a_6 + a_8 + a_{10} =$
   (a) 32
   (b) 17
   (c) 16
   (d) 15

12. The last two digit (unit's and 10th) of the number 9^{200} is :
   (a) 19
   (b) 21
   (c) 01
   (d) 41

13. If $|x| < 1/2$, then
$$1 + n\left(\dfrac{x}{1-x}\right) + \dfrac{n(n+1)}{2!}\left(\dfrac{x}{1-x}\right)^2 + ... \infty =$$
   (a) $(1-x)^n$
   (b) $\left(\dfrac{1}{1-x}\right)^n$
   (c) $\left(\dfrac{1-2x}{1-x}\right)^n$
   (d) $\left(\dfrac{1-x}{1-2x}\right)^n$

14. $\begin{pmatrix}30\\0\end{pmatrix}\begin{pmatrix}30\\10\end{pmatrix} - \begin{pmatrix}30\\1\end{pmatrix}\begin{pmatrix}30\\11\end{pmatrix} + ... + \begin{pmatrix}30\\20\end{pmatrix}\begin{pmatrix}30\\30\end{pmatrix} =$
   (a) $^{30}C_{11}$
   (b) $^{60}C_{10}$
   (c) $^{30}C_{10}$
   (d) $^{65}C_{55}$

15. In the expansion of $(1 + 3x + 2x^2)^6$, the coefficient of $x^{11} =$
   (a) 216
   (b) 576
   (c) 288
   (d) $6 \cdot 2^{11}$

16. In the expansion of $\left(\dfrac{x}{2} - \dfrac{3}{x^2}\right)^{10}$, the coefficient of $x^4 =$
    - (a) 405/256
    - (b) 504/259
    - (c) 450/263
    - (d) none of these

17. The digit at unit place of the number 7^{289} is :
    - (a) 1
    - (b) 3
    - (c) 9
    - (d) 7

18. The middle term in the expansion of $\left(x - \dfrac{1}{x}\right)^{18}$ is :
    - (a) $^{18}C_{10}$
    - (b) $-^{18}C_{10}$
    - (c) $^{18}C_9$
    - (d) $-^{18}C_9$

19. If $^{n-1}C_r = (R^2 - 3)\,^n C_{r+1}$, then k lies in the interval :
    - (a) $(-\infty, \sqrt{2})$
    - (b) $[2, \infty)$
    - (c) $[-\sqrt{3}, \sqrt{3}]$
    - (d) $[\sqrt{3}, 2]$

20. In the expansion of $(1 + x)(1 - x)^n$, the coefficient of x^n is :
    - (a) $(-1)^{n-1}(n-1)^2$
    - (b) $(1-n)$
    - (c) $(-1)^n(1-n)$
    - (d) $(-1)^{n-1}n$

☞ TRUE/FALSE

Write 'T' for True and 'F' for False statement.

1. An expression which contains only two term is called binomial expression. **(T/F)**
2. In the expansion of $(a + x)^n$, the coefficients of rth term from the beginning and end are not equal. **(T/F)**
3. Middle term has greatest coefficient and if there are two middle terms their coefficient will be equal. **(T/F)**
4. In binomial expansion, 'greatest term' mean the numerically greatest term. **(T/F)**
5. The last two digit of the number 9^{200} is 21. **(T/F)**

☞ FILL IN THE BLANKS

1. The middle term in the expansion of $\left(x - \dfrac{1}{x}\right)^{18}$ is _____ .
2. The digit at unit place of the number 7^{289} is _____ .
3. $^nC_0 - {}^nC_1 + {}^nC_2 - {}^nC_3 + \ldots + (-1)^n\,{}^nC_n =$ _____ .
4. $\displaystyle\sum_{r=0}^{n} 3^r \cdot {}^nC_r =$ _____ .
5. The general term in the expansion of $(x - a)^n$ is $T_{r+1} =$ _____ .

ANSWERS

☞ MULTIPLE CHOICE QUESTIONS

1. (d)	**2.** (a)	**3.** (c)	**4.** (d)	**5.** (a)	**6.** (b)	**7.** (c)	**8.** (d)	**9.** (d)
10. (b)	**11.** (d)	**12.** (c)	**13.** (d)	**14.** (c)	**15.** (b)	**16.** (a)	**17.** (d)	**18.** (d)
19. (d)	**20.** (c)							

☞ TRUE/ FALSE

1. T	**2.** F	**3.** T	**4.** T	**5.** F

☞ FILL IN THE BLANKS

1. $-^{18}C_9$	**2.** 7	**3.** $\dfrac{1}{n+1}$	**4.** 4^n	**5.** $(-1)^{rn}\,C_r x^{n-r} a^r$

❑❑❑❑❑❑

CHAPTER 5

Logarithms

5.1 INTRODUCTION

Sometimes, to simplify the numerical expression involving multiplication, division or rational powers, we use the logarithms. It is very useful for such type of lengthy and typical calculations.

Definition. *Let there be a number $a > 0$ and $a \neq 1$. A number x is called the logarithm of another variable $y > 0$ to the base a if $a^x = y$.*　　　　[Meerut(B.Sc)Biotech–2005, 06]

$$\therefore \quad a^x = y \qquad \Leftrightarrow \qquad x = \log_a y. \qquad \qquad ...(1)$$

For example.

(1)   (i) $2^4 = 16$ $\Leftrightarrow$ $\log_2 16 = 4$

      (ii) $10^2 = 100$ $\Leftrightarrow$ $\log_{10} 100 = 2$

      (iii) $8^0 = 1$ $\Leftrightarrow$ $\log_8 1 = 0$

      (iv) $(64)^{1/6} = 2$ $\Leftrightarrow$ $\log_{64} 2 = \dfrac{1}{6}$

(2)   (i) $\log_2 128 = $ A real number x such that $2^x = 128$ $\Rightarrow$ $x = 7$

      (ii) $\log_4 2 = $ A real number x such that $4^x = 2$ $\Rightarrow$ $x = \dfrac{1}{2}$ $[\because 4^{1/2} = 2]$

⮞ REMARKS

⮞ Logarithm of a number satisfying the condition (1) is unique. For, if α, β are two distinct logarithms of the number y to a base a, then by definition, we have
$$a^\alpha = y \text{ and } a^\beta = y, \text{ when } a^\alpha = a^\beta. \qquad ...(2)$$
But by properties of powers with positive base different from 1, we conclude from (2) that $\alpha = \beta$. This, if the number y has a logarithm to base a, this logarithm is unique. We denote it by the definition
$$x = \log_a y \text{ if } a^x = y.$$
⮞ 'log' is the abbreviation of "logarithm".

⮞ The logarithm of a number to a given positive real number ($\neq 1$) as base is the index or the power to which the base must be raised in order to make it equal to the given number.

5.2 PROPERTIES OF LOGARITHMS

Let $a > 0$, $a \neq 1$, $m > 0$, $n > 0$

1. $a^x = y$, then $x = \log_a y$.

     Here, L.H.S. is called exponential form, whereas R.H.S. is corresponding logarithmic form.

2. $a^1 = a$, $b^1 = b$ etc., therefore, $\log_a a = \log_b b = 1$.

3. $a^0 = 1$, $b^0 = 1 \Rightarrow \log_a 1 = 0$, $\log_b 1 = 0$.

4. $\log_b a \cdot \log_a b = 1$ or $\log_b a = \dfrac{1}{\log_a b}$

5. Base change formula

$$\log_b a = \log_c a \cdot \log_b c \quad \text{or} \quad \log_b a = \frac{\log_c a}{\log_c b}$$

6. The log of the product of two numbers is equal to the sum of their logs.

🕮 **REMARK**

➠ If $x_1, x_2, ..., x_n$ are positive rational numbers then
$\log(x_1 \cdot x_2 \cdot ... \cdot x_n) = \log_a x_1 + \log_a x_2 + ... + \log_a x_n$.

7. The log of the ratio of two numbers is equal to the difference of their logs.

8. $\log_a m^n = n \log_a m$.

9. $\log_a n^{p/q} = \dfrac{p}{q} \log_a n$.

10. $a \log a^n = n$.

11. If $a > 1$, then $0 < \alpha < \beta \Rightarrow \log_a \alpha < \log_a \beta$.

12. If $0 < a < 1$, then $0 < \alpha < \beta \Rightarrow \log_a \alpha > \log_a \beta$.

13. If $a > 1$, $\alpha > 1$, then $\log_a \alpha > 0$.

14. If $0 < a < 1$, $0 < \alpha < 1$, then $\log_a \alpha > 0$.

15. If $0 < a < 1$, $\alpha > 1$, then $\log_a \alpha < 0$.

16. If $a > 1$, $0 < \alpha < 1$, then $\log a^\alpha < 0$.

17. If $a > 1$, $\alpha > 1$ and $\alpha < a$, then $0 < \log a^\alpha < 1$.

18. If $a > 1$, $\alpha > 1$ and $\alpha > a$, then $\log_a \alpha > 1$.

19. If $0 < a < 1$, $0 < \alpha < 1$ and $\alpha > a$ then $0 < \log_a \alpha < 1$.

20. If $0 < a < 1$, $0 < \alpha < 1$ and $\alpha < a$ then $\log_a \alpha > 1$.

5.3 SYSTEM OF LOGARITHMS

(a) Common Logarithm: In this system we take the base 10. This is also known as Bring's system.

For Example. $\log_{10} 10 = 1$, $\log_{10} 100 = \log_{10} 10^2 = 2$, $\log_{10} 1000 = 3$.

🕮 **REMARK**

➠ If no base is mentioned, the base is always taken as 10.

(b) Natural Logarithm: In this system, we take the base e, where e is an irrational number lying between 2 and 3 and is given by

$$e = 1 + \frac{1}{1!} + \frac{1}{2!} + \frac{1}{3!} + ...$$

5.4 STANDARD FORM OF DECIMAL

To calculate the logarithm of any positive number in decimal form, we always express the given positive number in decimal form a_0, the product of an integral power of 10 and a number between 1 and 10.

i.e., $\qquad c = m \times 10^k$

where k is an integer and $1 \le m \le 10$.

For Example.

(1) 1234.56 can be written as $1.23456 \times 1000 = 1.23456 \times 10^3$

(2) $0.0023 = (0.0023 \times 1000) \times 10^{-3} = 23 \times 10^{-3}$

5.5 CHARACTERISTIC AND MANTISSA

The integral part of a logarithm is called the characteristic and the decimal part is called the mantissa. Logarithm to the base 10 is called common logarithm. The characteristics of common logarithms can be written by inspection, using the following rule.

"The characteristic of the logarithm (base 10) of a number greater than 1 is less by one than the number of digits in the integral part, and is positive. The characteristic of the logarithm of a positive decimal fraction less than 1, is greater by unity than the number of consecutive zeroes immediately after the decimal point and is negative."

On the other hand, to find the mantissa, we used the table of logarithms of numbers. The position of the decimal point in a number is immaterial for finding the mantissa. To find the mantissa of a number, we consider first four digits from the left most side of the number. If the number in the decimal form is less than one and has four or more consecutive zeroes to the right of the decimal point, then mantissa is calculated with the help of number formed by digits, starting with the first non-zero digits.

Significant Digit: The digit which are used to find the mantissa of a given number are known as significant digits.

SOLVED EXAMPLES

EXAMPLE 1. *Express each of the following in exponential form*

(i) $\log_2 64 = 6$ (ii) $\log_{10} 0.01 = -2$

SOLUTION. (i) $\log_2 64 = 6$ $\Rightarrow$ $2^6 = 64$

(ii) $\log_{10} 0.01 = -2$ $\Rightarrow$ $10^{-2} = 0.01$

EXAMPLE 2. *Find the values of each of the following from:*

(i) $\log_9 81$ (ii) $\log_{\sqrt{2}} 4$

SOLUTION. (i) Let $\log_9 81 = x$

Then $x = \log_9 81$ $\Rightarrow 9^x = 81 \Rightarrow 9^x = 9^2$

$\Rightarrow$ $x = 2$

(ii) Let $\log_{\sqrt{2}} 4 = x$

Then, $\log_{\sqrt{2}} 4 = x$ $\Rightarrow$ $(\sqrt{2})^x = 4$

$\Rightarrow$ $[(2)^{1/2}]^x = 4 = 2^2$

$\Rightarrow$ $2^{x/2} = 2^2$

$\Rightarrow$ $\dfrac{x}{2} = 2$ $\Rightarrow$ $x = 4$.

EXAMPLE 3. *Rewrite the following equations in the logarithm form:*

(i) $4^{3/2} = 8$ (ii) $5^0 = 1$ (iii) $(2\sqrt{2})^{-2/3} = \dfrac{1}{2}$

SOLUTION. (i) $4^{3/2} = 8$ can be written as $\log_4 8 = \dfrac{3}{2}$

(ii) $5^0 = 1$, can be written as $\log_5 1 = 0$

(iii) $(2\sqrt{2})^{-2/3} = \dfrac{1}{2}$, can be written as $\log_{2\sqrt{2}} \dfrac{1}{2} = -\dfrac{2}{3}$

EXAMPLE 4. *Rewrite the following equations in the exponential form.*

(i) $\log_2 32 = 5$ (ii) $\log_3 \dfrac{1}{243} = -5$

(iii) $\log_{5\sqrt{5}} 5 = \dfrac{2}{3}$ (iv) $\log_{100}(0.1) = -\dfrac{1}{2}$

SOLUTION.

Logarithmic Form	Exponential Form
(i) $\log_2 32 = 5$	$32 = 2^5$
(ii) $\log_3\left(\dfrac{1}{243}\right) = -5$	$\dfrac{1}{243} = 3^{-5}$
(iii) $\log_{5\sqrt{5}}(5) = \dfrac{2}{3}$	$5 = (5\sqrt{5})^{2/3}$
(iv) $\log_{100}(0.1) = -\dfrac{1}{2}$	$0.1 = 100^{-1/2}$

EXAMPLE 5. If $\log_{10} x = a$, Find the value of 10^{a-1} in terms of x.

SOLUTION. Here, we have
$$\log_{10} x = a \implies x = 10^a.$$
Now, $10^{a-1} = 10^a \times 10^{-1} = \dfrac{10^a}{10} = \dfrac{x}{10}$

EXAMPLE 6. If $\log_5 x = a$ and $\log_2 y = a$. Find 100^{2a-1} in terms of x and y.

SOLUTION. Here, we have
$\log_5 x = a$ and $\log_2 y = a$.
Therefore, $x = 5^a$ and $y = 2^a$
Now, $100^{2a-1} = (5^2 \times 2^2)^{2a-1} = (5^2)^{2a-1} \times 2^{4a-2} = 5^{4a-2} \times 2^{4a-2}$
$$= \dfrac{5^{4a}}{5^2} \times \dfrac{2^{4a}}{2^2} = \dfrac{(5^a)^4}{5^2} \times \dfrac{(2^a)^4}{2^2} = \dfrac{x^4}{5^2} \times \dfrac{y^4}{2^2} = \dfrac{x^4 y^4}{100}$$

EXAMPLE 7. Evaluate each of the following:
(i) $\log 5 + \log 2$ (ii) $\log 500 - \log 5$
(iii) $4 \log 5 + 2 \log 4$
(iv) $\log 6 + 2 \log 5 + \log 4 - \log 3 - \log 2$
(v) $\dfrac{1}{2} \log 36 + \log 5 - \log 30$
(vi) $\log 5 + 2 \log 0.5 + 3 \log 2$

SOLUTION. (i) $\log 5 + \log 2 = \log(5 \times 2) = \log 10 = 1$.
[By using $\log(mn) = \log m + \log n$]
(ii) $\log 500 - \log 5 = \log\left(\dfrac{500}{5}\right) = \log 100 = 2$.
(iii) $4 \log 5 + 2 \log 4$
$= \log 5^4 + \log 4^2$
$= \log 625 + \log 16 = \log (625 \times 16)$
$= \log 10000 = 4$. [$\because \log 10000 = \log 10^4 = 4$]
(iv) $\log 6 + 2 \log 5 + \log 4 - \log 3 - \log 2$
$= \log 6 + \log 5^2 + \log 4 - \log 3 - \log 2$
$= \log 6 + \log 25 + \log 4 - (\log 3 + \log 2)$

$$= \log (6 \times 25 \times 4) - \log (3 \times 2) = \log \left(\frac{6 \times 25 \times 4}{3 \times 2} \right)$$

$$= \log 100 = 2.$$

(v) $\dfrac{1}{2} \log 36 + \log 5 - \log 30$

$$= \log (36)^{1/2} + \log 5 - \log 30 = \log 6 + \log 5 - \log 30$$

$$= \log (6 \times 5) - \log 30 = \log 30 - \log 30 = 0.$$

(vi) $\log 5 + 2 \log 0.5 + 3 \log 2$

$$= \log 5 + \log (0.5)^2 + \log 2^3 = \log 5 + \log (0.25) + \log 8$$

$$= \log (5 \times 0.25 \times 8) = \log 10 = 1.$$

EXAMPLE 8. *If* $\log (m + n) = \log m + \log n$. *Show that* $m = \dfrac{n}{n-1}$ [RGPV(B. Phrama)–2004]

SOLUTION. Given $\log (m + n) = \log m + \log n$

$\Rightarrow$ $\log (m + n) = \log mn$

$\Rightarrow$ $m + n = mn$

$\Rightarrow$ $n = mn - m$

$\Rightarrow$ $n = m(n - 1)$

$\Rightarrow$ $m = \dfrac{n}{n-1}$

EXAMPLE 9. *If* $\log (mn) = \log m - \log n$. *Show that* $n = 1$ (RGPV(B. Phrama)–2005)

SOLUTION. Gives $\log (mn) = \log m - \log n$

$\Rightarrow$ $\log m + \log n = \log m - \log n$

$\Rightarrow$ $2 \log n = 0$

$\Rightarrow$ $\log n = 0$

$\Rightarrow$ $\log n = \log 1$

$\Rightarrow$ $n = 1$

EXAMPLE 10. *Show that*

(i) $7 \log \dfrac{16}{15} + 5 \log \dfrac{25}{24} + 3 \log \dfrac{81}{80} = \log 2$

(ii) $\log \dfrac{70}{33} + \log \dfrac{22}{135} - \log \dfrac{7}{18} = 3 \log 2 - 2 \log 3$

SOLUTION. (i) We have $7 \log \dfrac{16}{15} + 5 \log \dfrac{25}{24} + 3 \log \dfrac{81}{80}$

$$= 7 (\log 16 - \log 15) + 5 (\log 25 - \log 24) + 3 (\log 81 - \log 80)$$

$$= 7 [\log 2^4 - \log (3 \times 5)] + 5 [\log 5^2 - \log (2^3 + 3)]$$
$$+ 3 [\log 3^4 - \log (2^4 \times 5)]$$

$$= 7 [\log 2^4 - (\log 3 + \log 5)] + 5 [\log 5^2 - (\log 2^3 + \log 3)]$$
$$+ 3 [\log 3^4 - (\log 2^4 + \log 5)]$$

$$= 7 [4 \log 2 - \log 3 - \log 5] + 5 (2 \log 5 - 3 \log 2 - \log 3)$$
$$+ 3 (4 \log 3 - 4 \log 2 - \log 5)$$

$$= 28 \log 2 - 7 \log 3 - 7 \log 5 + 10 \log 5 - 15 \log 2$$
$$- 5 \log 3 + 12 \log 3 - 12 \log 2 - 3 \log 5$$

$$= 28 \log 2 - 15 \log 2 - 12 \log 2 - 7 \log 3 - 5 \log 3 + 12 \log 3$$
$$- 7 \log 5 + 10 \log 5 - 3 \log 5$$

$$= \log 2.$$

(ii) $\log\dfrac{70}{33} + \log\dfrac{22}{135} - \log\dfrac{7}{18} = \log\left(\dfrac{70}{33} \times \dfrac{22}{135}\right) - \log\dfrac{7}{18}$

$$= \log\left(\dfrac{\dfrac{70}{33} \times \dfrac{22}{135}}{\dfrac{7}{18}}\right) = \log\left(\dfrac{70}{33} \times \dfrac{22}{135} \times \dfrac{7}{18}\right)$$

$$= \log\left(\dfrac{8}{9}\right) = \log 8 - \log 9$$

$$= \log 2^3 - \log 3^2 = 3\log 2 - 2\log 3.$$

EXAMPLE 11. *Find the values of x in each of the following:*

(i) $\dfrac{\log 144}{\log 12} = \log x$ (ii) $\dfrac{\log 125}{\log 25} = x$

(iii) $\log_x 4 + \log_x 16 + \log_x 64 = 12.$

SOLUTION.   (i) Here, we have

$$\dfrac{\log 144}{\log 12} = \log x \qquad\Rightarrow\qquad \dfrac{\log 12^2}{\log 12} = \log x$$

$$\Rightarrow\qquad \dfrac{2\log 12}{\log 12} = \log x \qquad\Rightarrow\qquad \log x = 2$$

$$i.e., \qquad \log_{10} x = 2 \qquad\Rightarrow\qquad x = 10^2 = 100.$$

(ii) Here, we have

$$\dfrac{\log 125}{\log 25} = x \qquad\Rightarrow\qquad \dfrac{\log 5^3}{\log 5^2} = x$$

$$\Rightarrow\qquad \dfrac{3\log 5}{2\log 5} = x \qquad\Rightarrow\qquad \dfrac{3}{2} = x$$

$$i.e., \qquad x = \dfrac{3}{2}$$

(iii) $\qquad\qquad \log_x 4 + \log_x 16 + \log_x 64 = 12$

$$\Rightarrow\qquad \log_x 2^2 + \log_x 2^4 + \log_x 2^6 = 12$$

$$\Rightarrow\qquad 2\log_x 2 + 4\log_x 2 + 6\log_x 2 = 12$$

$$\Rightarrow\qquad\qquad\qquad\qquad 12\log_x 2 = 12$$

$$\Rightarrow\qquad\qquad\qquad\qquad \log_x 2 = 1$$

$$i.e., \qquad\qquad\qquad\qquad x^1 = 2$$

$$\Rightarrow\qquad\qquad\qquad\qquad x = 2.$$

EXAMPLE 12. *If* $\dfrac{\log x}{b-c} = \dfrac{\log y}{c-a} = \dfrac{\log z}{a-b}$, *then show that* $x^{b+c-a} y^{c+a-b} z^{a+b-c} = 1.$

SOLUTION.   Let us suppose

$$\dfrac{\log x}{b-c} = \dfrac{\log y}{c-a} = \dfrac{\log z}{a-b} = k$$

which gives

$$\log x = k(b-c),\ \log y = k(c-a),\ \log z = k(a-b)$$

$$\Rightarrow\qquad (b+c-a)\log x + (c+a-b)\log y + (a+b-c)\log z$$

$$= (b+c-a)k(b-c) + (c+a-b)k(c-a)$$

$$+ (a+b-c)\,k(a-b)$$

$$= k\{(b^2 - c^2) - a(b - c)\} + k\{(c^2 - a^2) - b(c - a)$$
$$+ k\{(a^2 - b^2) - c(a - b)\}$$
$$= k\{b^2 - c^2 + c^2 - a^2 + a^2 - b^2\}$$
$$- k\{a(b - c) + b(c - a) + c(a - b)\}$$
$$= k \cdot 0 - k \cdot 0 = 0.$$

Therefore, $\log x^{b + c - a} + \log y^{c + a - b} \log z^{a + b - c} = 0$

$\Rightarrow \qquad \log (x^{b + c - a} \cdot y^{c + a - b} \cdot z^{a + b - c}) = 0 = \log 1.$

Hence $\qquad x^{b + c - a} \cdot y^{c + a - b} \cdot z^{a + b - c} = 1.$

EXAMPLE 13. *If* $\dfrac{\log a}{b - c} = \dfrac{\log b}{c - a} = \dfrac{\log c}{a - b}$ *prove that* $a^a \cdot b^b \cdot c^c = 1.$

SOLUTION. Let $\qquad \dfrac{\log a}{b - c} = \dfrac{\log b}{c - a} = \dfrac{\log c}{a - b} = k$

Then $\qquad \log a = k(b - c)$

$\qquad \log b = k(c - a)$

and $\qquad \log c = k(a - b)$

Adding all those, after multiplication by a, b, c respectively, we get

$a \log a + b \log b + c \log c = ak(b - c) + bk(c - a) + ck (a - b) = 0.$

Therefore, $\qquad \log a^a b^b c^c = 0 = \log 1$

$\qquad a^a b^b c^c = 1.$

EXAMPLE 14. *Solve the equation* $a^{2x} = b^{x - c} c^{x + 5}$, $a, b, c, > 0$ *but* $\neq 1.$

[Meerut(B.Sc Biotech)–2005]

SOLUTION. Take logarithm of both sides, we get

$$\log a^{2x} = \log b^{x - c} + \log c^{x + 5}$$

$\Rightarrow \qquad 2x \log a = (x - c) \log b + (x + 5) \log c$

$\Rightarrow \quad x[2 \log a - \log b - \log c] = 5 \log c - c \log b$

$\Rightarrow \qquad\qquad x \left[\log \dfrac{a^2}{bc} \right] = \log \dfrac{c^5}{b^c}$

$$x = \dfrac{\log \left(\dfrac{c^5}{bc} \right)}{\log \left(\dfrac{a^2}{bc} \right)}.$$

EXAMPLE 15. *Solve the system of equations*

(i) $5(\log_y x + \log_x y) = 26$

(ii) $xy = 64.$

[Meerut(B. Sc. Biotech)–2005]

SOLUTION. We know that $\log_x y = \dfrac{1}{\log_y x}$, therefore

(i) $\Rightarrow \qquad 5 \left(\log_y x + \dfrac{1}{\log_y x} \right) = 26$

Putting $\qquad \log_y x = p$, it gives $5 \left(p + \dfrac{1}{p} \right) = 26$

or $\qquad 5p^2 - 26p + 5 = 0$

$$5p^2 - 25p - p + 5 = 0$$
$$\Rightarrow \quad 5p(p-5) - 1(p-5) = 0$$
$$\Rightarrow \quad (p-5)(5p-1) = 0$$
$$\Rightarrow \quad p = 5, 1/5.$$

When $\qquad p = 5,$
$$\log_y x = 5 \qquad \Rightarrow x = y^5.$$
$$\therefore \text{(ii)} \quad \Rightarrow \quad x \cdot y = y^5 \cdot y = 64 \text{ or } y^6 = 64 \Rightarrow \quad y^3 = \pm 8$$
$$\Rightarrow \qquad y^3 = +8 = 2^3 \text{ or } y^3 = -8 = (-2)^3$$
$$\Rightarrow \qquad y = 2 \qquad \text{or } y = -2.$$

Then (ii) $\Rightarrow x = \dfrac{64}{y} = \dfrac{64}{2} = 32$ or $x = \dfrac{64}{-2} = -32$

But y and x both are used as base in equation (i), so x, y cannot be negative
$$\Rightarrow \qquad \text{one solution is} \qquad x = 32, y = 2.$$

When $\quad p = \dfrac{1}{5},$ then $\log_y x = \dfrac{1}{5} \Rightarrow x = y^{1/5}$ or $x^5 = y.$

$\therefore$ (ii) $\Rightarrow x, y = x^5 = 64$ or $x^6 = 64$ or $x^3 \neq 8$
$$\Rightarrow \quad x = 2 \text{ or } x = -2 \qquad \qquad \text{[reject negative value]}$$

Again (ii) $\Rightarrow \quad x = \dfrac{64}{y} = \dfrac{64}{2} = 32.$

This gives another solution.
Thus, we get two solutions of the given system of equations as follows:
$$x = 32, y = 2 \text{ or } x = 2, y = 32.$$

Exercise 5.1

1. Write the following into logarithmic form:
   (i) $2^8 = 256$ (ii) $10^3 = 1000$
   (iii) $7^3 = 243$ (iv) $4^{-4} = \dfrac{1}{256}$
   (v) $4^{3/2} = 8$

2. Write the following into exponential form:
   (i) $\log_5 25 = 2$ (ii) $\log_{10} 0.001 = -3.$

3. Find the value of b which satisfies :
   (i) $\log_{\sqrt{8}} b = 3\dfrac{1}{3}$
   (ii) $\log_e 2 \cdot \log_b 625 = \log_{10} 16 \log_e 10$
   (iii) $\log_{\sqrt{3}} x = 4$ (iv) $\log_4 x = 15$
   (v) $\log_{125} x = \dfrac{1}{6}$

4. Find:
   (i) $\log_6 16$, if $\log_{12} 27 = a$
   (ii) $\log_{25} 24$, if $\log_6 15 = \alpha$ and $\log_{12} 18 = \beta$
   (iii) $\log_{30} 8$, if $\log_{30} 3 = a$ and $\log_{30} 5 = b$

5. If $\log_{12} 18 = \alpha$ and $\log_{24} 54 = \beta$, show that $\alpha\beta + 5(\alpha - \beta) = 1.$

6. Without using the table, show that:
   (i) $\dfrac{1}{\log_2 \pi} + \dfrac{1}{\log_4 \pi} > 2$

   (ii) $\log_2 17 \cdot \log_{1/5} 2 \cdot \log(1/5) > 2$
   (iii) $|\log_b a + \log_a b| \geq 2$, where a and b are positive integer not unity.

7. Compute, without using tables
   (i) $\log_3 4 \cdot \log_4 5 \cdot \log_5 6 \cdot \log_6 7 \cdot \log_7 8 \cdot \log_8 9$
   (ii) $\log_3 2 \cdot \log_4 3 \cdot \log_5 4 \ldots \log_{15} 14 \cdot \log_{16} 15$

8. Show that $\log_2 3$ is an irrational number.

9. Prove the following :
   (i) $\dfrac{\log_a n}{\log_{ab} n} = 1 + \log_a b$
   (ii) $\log_{ab} x = \dfrac{\log_a x \log_b x}{\log_a x + \log_b x}$

10. Show that :
    (i) $\log 2 + 2 \log 5 - \log 3 - 2 \log 7 = \log \dfrac{50}{147}$
    (ii) $\log\left(\dfrac{9}{14}\right) + \log\left(\dfrac{35}{24}\right) - \log\left(\dfrac{15}{16}\right) = 0$
    (iii) $\dfrac{1}{2} \log 25 - 2 \log 3 + \log 18 = 1$
    (iv) $\dfrac{1}{3} \log_2 54 + \log_2 10 - \log_2 625 = 1$
    (v) $\log_{10} 10 + \log_{10} 100 + \log_{10} 1000 + \log_{10} 10000 = 10$

(vi) $\dfrac{1}{2}\log 9 + 2\log 6 + \dfrac{1}{4}\log 81 - \log 12$

$$= 3\log 3$$

11. Evaluate each of the following :

(i) $2\log_3 5 - 5\log_3 2$

(ii) $(81)^{1/\log_5 3} + 27^{\log_9 36} + 3^{(4\log_9 7)}$

(iii) $\log 15 + 2\log 0.5 + 3\log 2$
$$- \log 3 - \log 5$$

(iv) $\log 21 + \log 4 + 2\log 5 - \log 3 - \log 7$

12. Show that:

(i) $a\log_a 1 + 2\log_a 2 + 3\log_a 3 + \ldots + n\log_a n$
$$= 2^2 \cdot 3^3 \cdot 4^4 \ldots n^n$$

(ii) $a\log_a 1 + 2\log_a 2 + 2\log_a 3 + \ldots$
$$+ 2\log_a n = (n!)^2$$

(iii) $\log_{10}\tan 1° \log_{10}\tan 2° \log_{10}\tan 3° \ldots$
$$\log_{10}\tan 50° = 0$$

(iv) $\log_{10}\tan 1° + \log_{10}\tan 2° + \ldots$
$$+ \log_{10}\tan 89° = 0$$

ANSWERS

1. (i) $\log_2 256 = 8$ (ii) $\log_{10}1000 = 3$ (iii) $\log_7 243 = 3$ (iv) $\log_4(1/256) = -4$

(v) $\log_4 8 = 3/2$ 2. (i) $5^2 = 25$ (ii) $10^{-3} = 0.001$ 3. (i) 32 (ii) $(\sqrt{2})^{1/2}$

4. (i) $\dfrac{4(3-a)}{3+a}$ (ii) $\dfrac{5-\beta}{2\alpha\beta + 2\alpha - 4\beta + 2}$ (iii) $3(1-a-b)$ 7. (i) 2 (ii) $\dfrac{1}{4}$

11. (i) 0 (ii) 890 (iii) $\log 2$ (iv) 2

5.6 METHOD TO DETERMINE THE CHARACTERISTIC AND MANTISSA

5.6.1 METHOD TO DETERMINE THE CHARACTERISTIC

The characteristic is determined by using the following rule :

(i) The characteristic of the logarithm of any number greater than 1 is one less than the number of digits to the left of the decimal point in the given number.

(ii) The characteristic of the logarithm of any number less than 1 is negative and numerically one more than the number of zeroes of the right of the decimal point.

For Example. See the following table :

S. No.	Given number	Characteristic	Explanation
1.	63 389.6 3986 6.36	1 2 3 0	One less than the number of digits to the left of the decimal point.
2.	0.4 0.04 0.004×10^{-1}	-1 -2 -3	One more than the number of zeroes to the right immediately after the decimal point.

5.6.2 METHOD TO DETERMINE THE MANTISSA

The mantissa is determined by using the following rule :

(i) The mantissa is the same for the same significant figure in the same order and does not depend on the position of the decimal point.

(ii) The mantissa is always taken as positive.

For Example. See the following table :

Given number	Characteristic	Mantissa	Logarithm
5978	3	0.7766	3.7766
597.8	2	0.7766	2.7766
0.005978	− 3	0.7766	− 3 + 0.7766

The given number contain the same significant figures, namely 5, 9, 7, 8 in the same order and so the mantissa of their logarithm is same, they differ only in the characteristic.

In log 0.005978, which is equal to − 3 + 0.07766, the characteristic − 3 is negative and the mantissa 0.7666 is positive. To indicate that the negative sign applies to the characteristic only, the ' − ' sign is put above the characteristic. Thus,

$$\log 0.005975 = -3 + 0.7666 = \overline{3}.7766$$

It is read as "bar 3 point 7766".

☞ REMARK

➡ To find the mantissa of the logarithm of a number which contain less or more than four digits, make it a four digit number by having zeroes on its right or by condensing it by the rule of approximation which is given below :

(i) "$\frac{1}{2}$ or more than $\frac{1}{2}$ is takes as 1; and less than $\frac{1}{2}$ is neglected."

(ii) "5 or more than 5 is taken as 10; and less than 5 is neglected."

For Example.

(i) 6.76236 = 6.7624, upto four decimal places.

(ii) 6.7634 = 6.763, upto four decimal places.

5.6.3 MANTISSA OF THE LOGARITHM OF A GIVEN NUMBER

To find the mantissa of logarithm of a given number we use the standard table of logarithms. The table of logarithms consist of 90 rows and 20 columns. Every row begins with a two digit number 10, 11, ..., 98, 99 and every column is headed by a one digit number 0, 1, 2, ... 9. On the right of the table, a big column is divided into 9 sub columns headed by the digits 1, 2, 3, ... , 9, known as column of mean differences. To find the mantissa of a number, consider first four digits from the left most side of the number. If the number is the decimal point, then mantissa is calculated with the help of the number formed by digits beginning with the first non-zero digits.

☞ REMARK

➡ To find the log of a given number x, use the formula log x = characteristic + mantissa.

SOLVED EXAMPLES

EXAMPLE 1. *Find log 756.8.*

SOLUTION. By neglecting the decimal point, we obtain 7568, which is a four digit number. See the number 75 in the extreme left hand column in the logarithm table. In the horizontal line of 75 and under 6 (next digit in the number) we found the number 8785. In the same horizontal line and under 8 (4th digit of the given number), the number found is 8. Adding 8 to the already obtained number 8785, we get 8793. Therefore, the mantissa is 0.8793

⇒ log 756.8 = 2.8793   [∵ characteristic is 2]

Similarly log 75.68 = 1.8793   [characteristic is 1]

log 0.07568 = $\overline{2}$.8793   [characteristic is − 2]

EXAMPLE 2.   *Find the logarithm of the following number :*

  (*i*) 5395   (*ii*) 0.002359

  (*iii*) 25795   (*iv*) 0.005

SOLUTION.   (i) The given number 5395 is a four digit number. See the number 53 in the extreme left column is the logarithm table. In the horizontal line of 53 and under 9 (next digit in the given number) we found the number 73.16. In the same horizontal line and under 5 (4th digit of the number), the number found is 4. Add this number 4 to 7316 to get 7320, which is the required mantissa of log 5395. Also the characteristic of 5395 is 3.

Therefore, log 5395 = characteristic + mantissa = 3.7320

  (ii) Firstly, find the four digit number, by getting the first four digits beginning with the first non-zero digit on the right of the decimal point which is 2359. The mantissa of the given number can be determined by the procedure discussed in (i) and given by

mantissa of 2359 = 3711 + 17 = 3728.

Also, the characteristic of 0.00359 is –3.

Hence, log 0.002359 = $\bar{3}.3728$

  (iii) Clearly, the characteristic of the logarithm of 25795 is 4. To find the mantissa of the given number 25795, consider the four digit number 2579 and apply the same process, we get

mantissa of 2579 is 4114.

Therefore, the logarithm of the given number 2579 is 4.4114.

  (iv) The characteristic of 0.005 is – 3. To find the mantissa, consider the number 50. See in the row 50, under the column headed by 0 and get the number 6990.

Therefore, log 0.005 = – 3 + 0.6990 = $\bar{3}.6990$

EXAMPLE 3.   *Find* log 11.648.

SOLUTION.   The characteristic of 11.648 is 1. Now leaving the decimal point, the given number consists of five digits. Condensing it to a four digital number, by the rule of approximation, we get the number 1165. Now follow the same procedure, as above, the mantissa of 1165 is 0.0664.

Hence, log(11.647) = log(11.65) = 1.0664.

5.7 ANTILOGARITHM

If log x = n, then x is called the antilogarithm of n and is written as

$$x = \text{antilog } (n)$$

For Example:

(1) log 10 = 1   ⇔   antilog (1) = 10.

(2) log 0.0681 = $\bar{2}.8331$ ⇔   antilog ($\bar{2}.8331$) = 0.0681

5.7.1 METHOD OF FINDING THE ANTILOGARITHM

To find the antilog of a given number, we make use of the table of antilogarithms. The method is almost the same as that for finding the logarithm of a number. The table is divided into three similar sets of columns and the four digits are to be taken in the same manner starting with digits immediately to the right of the decimal point, not excluding zero.

The following points must be kept in mind for convenience :

  (i) If the given number is negative, first make it positive by adding one to the decimal part and by subtracting one from the integral part.

  (ii) Apply the method, same as that used for finding the logarithm of a number.

(iii) If the characteristic of the given number is positive and is equal to n, then insert the decimal point after $(n + 1)$ digits in the obtained number.
If $n > 4$, then write zeroes on the right side to get $(n + 1)$ digits.

(iv) If the characteristic of a given number is negative and is equal to $-n$ or n then on the right side to the decimal point, write $(n - 1)$ consecutive zeroes and then write the obtained number.

SOLVED EXAMPLES

EXAMPLE 1.   *Find* antilog 2.3456.

SOLUTION.   The mantissa is positive and is equal to 0.3456. Now look into the line starting with 0.34. In the horizontal line of 0.34 and under 5 (the next digit of the mantissa), the number obtained is 2213. In the same horizontal line and under 6 (the fourth digit of the given number) in the mean difference columns, the number found 3. Adding 3 to 2213, we get 2216. Now, since the characteristic is 2, the required number must have 3 digit in the integral part.

Therefore,   antilog 2.3456 = 221.6

Similarly, we can find

antilog 0.3456 = 2.216, antilog $\overline{1}$.3456 = 0.2216

antilog $\overline{2}$.3456 = 0.02216, antilog $\overline{5}$.3456 = 0.00002216

EXAMPLE 2.   *Find* antilog $\overline{3}$.0675.

SOLUTION.   In the horizontal line of 0.06 and under 7 we get 1167. In the same horizontal line and under 5 in the mean difference column, we get 1. Adding 1 to 1167 we get 1168. Since, the characteristic is $\overline{3}$, the required number must have 2 zeroes immediately to the right of the decimal part.

Therefore, antilog($\overline{3}$.0675) = 0.001168.

EXAMPLE 3.   *Find the antilog, each of the following:*

   (i) -1.2084 (ii) -0.62.

SOLUTION.   (i)   The given number is -1.2084.

Here, we observe that the mantissa of the given number is negative. First we make it positive by adding 1 in following manner

$$-1.2084 = -1 - 0.2084 = -1 - 1 + 1 - 0.2084$$
$$= -2 + 0.7916 = \overline{2}.7916$$

Now, using the antilogarithm table, we find that the number corresponding to the mantissa 0.7961 is 6189.

Since, the characteristic is $\overline{2}$, *i.e.*, -2, put one zero just after the decimal point to get the antilogarithm of the given number.

Therefore, antilog (-1.2084) = antilog $(\overline{2}.7916)$ = 0.06189

   (ii)   Consider the given number -0.62

$$= -1 + 1 - 0.62 = -1 + 0.28 = \overline{1}.28$$

Now, using the antilogarithm table, the number corresponding to 0.23 is 1905 also, the characteristic is $\overline{1}$, *i.e.*, -1.

Therefore, antilog (-0.62) = antilog $(\overline{1}.28)$ = 0.1905.

EXAMPLE 4.   *Find the values of the following*

   (i) $\overline{2}.76 \times 4$ (ii) 6.45×981.4

   (iii) $\overline{6}.42 \div 3$ (iv) 0.0064×1.507.

SOLUTION.   (i) $\overline{2}.76 \times 4 = (-2 + 0.76) \times 4 = -8 + 3.04 = -8 + 3 + 0.04 = -5 + 0.04 = \overline{5}.04$

(ii) Let $x = 6.45 \times 981.4$

$\Rightarrow \qquad\qquad \log x = \log 6.45 + \log 981.4$

$$= 0.8096 + 2.9919 = 3.8015$$

Therefore, $\qquad\qquad x = \text{antilog}\,(3.8015) = 6331.$

(iii) $\overline{6}.42 \div 3 = \dfrac{1}{3}(-6 + 0.42) = \dfrac{-6}{3} + \dfrac{0.42}{3} = -2 + 0.14 = \overline{2}.14$

(iv) Let $x = 0.0064 \times 1.507$, then

$$\log x = \log(0.0064 \times 1.507)$$

$$= \overline{3}.8062 + 0.1781 = \overline{3}.9843$$

$\Rightarrow \qquad\qquad x = \text{antilog}(\overline{3}.9843) = 0.09645$

EXAMPLE 5. *Find* $\log\{(27)^3 \times (0.81)^{4/5}\} \div (90)^{5/4}\}$, *where* $\log 3 = 0.4771213$.

SOLUTION. We have

$$\log\{(27)^3 \times (0.81)^{4/5} \div (90)^{5/4}\}$$

$$= 3\log\frac{27}{10} + \frac{4}{5}\log\frac{81}{100} - \frac{5}{4}\log 90$$

$$= 3(\log 3^3 - 1) + \frac{4}{5}(\log 3^4 - 2) - \frac{5}{4}(\log 3^2 + 1)$$

$$= \left(9 + \frac{16}{5} - \frac{5}{2}\right)\log 3 - \left(3 + \frac{8}{5} + \frac{5}{4}\right)$$

$$= \frac{97}{10}\log 3 - 5\frac{17}{20} = 4.6280766 - 5.85 = \overline{2}.7780766$$

EXAMPLE 6. *Using logarithmic table, evaluate the following*

(i) $\sqrt{\dfrac{41.32 \times 20.18}{12.69}}$ (ii) $\sqrt[3]{\dfrac{(45.4)^2}{(3.2)^2 \times (5.6)^3}}$

SOLUTION. (i) Let $\qquad x = \left(\dfrac{41.32 \times 20.18}{12.69}\right)^{1/2}$

Then, we have

$$\log x = \log\left(\frac{41.32 \times 20.18}{12.69}\right)^{1/2} = \frac{1}{2}\log\left(\frac{41.32 \times 20.18}{12.69}\right)$$

$$= \frac{1}{2}\,(\log 41.32 + \log 20.18 - \log 12.69)$$

$$= \frac{1}{2}\,(1.6162 + 1.3049 - 1.1035)$$

$$= \frac{1}{2}\,(2.9211 - 1.1035) = \frac{1}{2}\,(1.8176)$$

$$= 0.9088$$

Therefore, $\;x = \text{antilog}\,(0.9088) = 8.106$

(ii) Let $\qquad x = \left(\dfrac{(45.4)^2}{(3.2)^2 \times (5.6)^3}\right)^{1/3}$

Then, we have

$$\log x = \frac{1}{3}\log\left(\frac{(45.4)^2}{(3.2)^2 \times (5.6)^3}\right)$$

$$= \frac{1}{3}[2\log(45.4) - 2\log(3.2) - 3\log(5.6)]$$

$$= \frac{1}{3}(2 \times 1.6571 - 2 \times 0.5051 - 3 \times 0.7482)$$

$$= \frac{1}{3}(3.3142 - 1.0102 - 2.2446) = \frac{1}{3}(0.0594) = 0.0198.$$

Therefore, $x = $ antilog $(0.0198) = 1.047$

Exercise 5.2

1. Find the antilogarithm of each of the following :
   - (i) 25.795
   - (ii) 1270
   - (iii) 0.005
   - (iv) 431.5
   - (v) 0.0074
   - (vi) 0.002598
   - (viii) 0.3582

2. Find the antilogarithm of each of the following :
   - (i) 1.4114
   - (ii) 3.1038
   - (iii) $\bar{3}.6990$
   - (iv) 2.6350
   - (v) $\bar{3}.8692$
   - (vi) $\bar{3}.4146$
   - (vii) $\bar{1}.5541$

3. If $\log x = \bar{1}.4914$ and $\log y = 2.4669$, find the value of each of the following :
   - (i) xy
   - (ii) $\dfrac{x}{y}$
   - (iii) $\dfrac{x^2}{y^3}$
   - (iv) $\dfrac{x^3}{y^2}$
   - (v) $\dfrac{x^2}{y}$

4. Using logarithmic table, find the value of each of the following:
   - (i) $\dfrac{76.03 \times 9.08}{101.2 \times 63.17}$
   - (ii) $\dfrac{(73.56)^3 \times (0.0371)^2}{68.21}$
   - (iii) $\dfrac{(25.36)^2 \times 0.4569}{847.5}$
   - (iv) $\sqrt{\dfrac{8^{1/3} \times 7^{3/4}}{(7.29)^{1/3} \times 3.26}}$ (v) $(0.00001427)^{1/7}$
   - (vi) $\dfrac{(6.45)^3 \times (0.0034)^{1/3} \times 981.4}{(9.37)^2 \times (8.93)^{1/4}(0.0617)}$

5.8 APPLICATION OF LOGARITHM IN PHARMACEUTICAL PROBLEMS

(1) Half-life Period

The half-life period T of C^{14} gives the value of disintegration constant k from the equation $k = \dfrac{0.693}{T}$. The disintegration equation is $k = \dfrac{2.303}{t}\log\dfrac{a}{a-x}$.

EXAMPLE 1. *The amount of C^{14} isotope in a piece of wood found to be one sixth of its amount present in a fresh piece of wood. Calculate the age of wood.*

SOLUTION.   Half-life of $C^{14} = 5577$ years.

Half-life period $\Rightarrow T = \dfrac{0.693}{k}$

$$k = \frac{0.693}{T} = \frac{0.693}{5577}.$$

The disintegration equation, namely $k = \dfrac{2.303}{t} \log \dfrac{a}{a-x}$

$$\frac{0.693}{5577} = \frac{2.303}{t} \times \log \frac{1}{1/6}$$

$$t = \frac{2.303 \times 5577 \times \log 6}{0.693} = \frac{2.303 \times 5577 \times 0.7782}{0.693}$$

$$t = 20170 \text{ year}$$

(2) Calculation of Boiling point or Freezing point

If the freezing point or the boiling point of a liquid at one pressure is known, it is possible to calculate it at another pressure by the use of the Chaperone clauses equation,

$$\log_{10} \frac{P_2}{P_1} = \frac{\Delta Hv}{4.576}\left[\frac{1}{T_1} - \frac{1}{T_2}\right]$$

(3) Calculation of Equilibrium Constants and Concentrations

If the concentration of reactants and products are known at equilibrium is a reaction, the equilibrium constant can be calculated.

EXAMPLE 2. *An equilibrium system for the reaction between Hydrogen and Iodine to given Hydrogen Iodine at 670 K in a 5 litre flask contains 0.4 mole of Hydrogen, 0.4 mole of Iodine and 2.4 moles of Hydrogen Iodine calculate equilibrium constant.*

SOLUTION. For the reaction

$$H_{2(g)} + I_{2(g)} \rightleftharpoons 2HI_{(g)}$$

$$k = \frac{[HI]^2}{[H_2][I_2]}$$

Molar concentration of various species at equilibrium

$$[H_2] = \frac{0.4}{5}, [I_2] = \frac{0.4}{5} = 0.08 \text{ mol L}^{-1}$$

$$[HI] = \frac{2.4}{5} = 0.48 \text{ mol L}^{-1}$$

$$\therefore \quad k = \frac{(0.48)^2}{(0.08)(0.08)}$$

$$\log k = \log(0.48)^2 - \log(0.08)^2 = 2\log 0.48 - 2\log 0.08$$
$$= 2\log 4.8 \times 10^{-1} - 2\log 8.0 \times 10^{-2}$$
$$= 2(-1 + 0.6812) - 2(-2 + 0.9031) = 1.5562$$
$$k = 36$$

(4) pH Equation

$$pH = k_a + \log \frac{[Salt]}{[Acid]}$$

EXAMPLE 3. *Some of 0.2 m acetic acid are mixed with 50 ml of 0.2 m Sodium acetate solution. What will be the pH of the mixture if $k_a = 1.85 \times 10^{-5}$?*

SOLUTION.

$$[Acid] = \frac{0.2}{100}$$

$$[Salt] = \frac{0.2}{100}$$

We have
$$pH = k_a - \log \frac{[Salt]}{[Acid]} = -\log 1.85 \times 10^{-5} + \log \frac{0.2/100}{0.2/100}$$
$$pH = -\log 1.85 \times 10^{-5}$$
$$pH = 4.7325.$$

EXAMPLE 4. *Calculate the pH Value of :*

(i) 0.001 M HCl   (ii) 0.001 M NaOH

SOLUTION. (i) Since HCl is a strong Acid , H_3O^+ ion concentration is equal to that of the acid.

$$[H_3O^+] = [HCl] = 0.001\ m = 1 \times 10^{-3}\ m$$
$$pH = -\log[H_3O^+]$$

or $pH = -\log [1 \times 10^{-3}] = -(-3)\log 10 = 3.$

(ii) Since NaOH is a strong base, it completely ionizes.

∴ Hydroxyl ion concentration is equal to that of the base
$$[OH^{-1}] = [NaOH] = 0.01\ m = 1 \times 10^{-2}$$
$$k_w = [H_3O^+][OH^-]$$

$$[H_3O^+] = \frac{k}{[OH^-]} = \frac{1 \times 10^{-14}}{1 \times 10^{-2}} = 1 \times 10^{-12}$$
$$pH = -\log [H_3O^+] = -\log (1 \times 10^{-12}) = 12.$$

Exercise 5.3

1. Calculate the pH of 0.5 molar solution of sulphuric acid.
2. Given that the half life period of Radium is 1580 years. Calculate the disintegration constant and average life.
3. The equilibrium constant for the reaction

$$N_{2(g)} + 3H_{2(g)} = 2\ NH_{3(g)}$$

at 715 K is 6.0×10^{-2}. If in a particular reaction there are $0.25\ mol\ L^{-1}$ of H_2 and $0.06\ mol\ L^{-1}$ of NH_3 present, calculate the concentration of N_2 at equilibrium.

4. At what temperature will water boil under a pressure of 787 mm? The latent heat of vaporization is 536 cal per gram.

5. The pH of a soft drink is 4.4. Calculate $[H_3O^+]$ and $[OH^-]$.

HINT TO SELECTED PROBLEMS

1. $H_2SO_4 \rightleftharpoons 2H^+ + SO_4^{2-}$

In a 0.5 M H_2SO_4 solution
$$[H^+] = 2 \times 0.5\ mol\ L^{-1}$$
$$pH = -\log[H^+] = -1 \times \log 1 = 0.$$

2. Half-life period $T_{1/2} = \dfrac{0.693}{k}$

$$k = \frac{0.693}{T_{1/2}} = \frac{0.693}{1580}$$

Average life $\lambda = \dfrac{1}{k} = \dfrac{1}{4.3 \times 10^{-4}}$

4. $\log_{10} \dfrac{P_2}{P_1} = \dfrac{\Delta H_v}{4.576} \left[\dfrac{1}{T_1} - \dfrac{1}{T_2} \right]$

$$\log_{10} \frac{787}{762} = \frac{536 \times 18}{4.576} \left[\frac{T_2 - 373}{373 T_2} \right]$$

$$\log_{10} 787 - \log 762 = \frac{536 \times 18}{4.576} \left[\frac{T_2 - 373}{373 T_2} \right]$$

$$T_2 = 374\ K$$

5. $pH = -\log [H_3O^+]$
$$-\log [H_3O^+] = 4.4$$
$$\log [H_3O^+] = -4.4$$

$$[OH^-] = \frac{k_w}{[H_3O^+]} = \frac{1 \times 10^{-14}}{3.98 \times 10^{-5}}\ mol\ L^{-1}$$

ANSWERS

1. 0
2. 4.3×10^{-4} years, 2325 years
3. 3.84
4. 374 K or 101°C
5. $[H_3O^+] = 3.98 \times 10^{-5}\ mol\ L^{-1}$, $[OH^-] = 2.5 \times 10^{-10}\ mol\ L^{-1}$

Objective Evaluation

∞ MULTIPLE CHOICE QUESTIONS

Choose the most appropriate one.

1. If $a^x = b^y = c^z$ and $\log_b a = \log_c b$, then which one of the following will hold true?
   - (a) $y = x - z$
   - (b) $y = x + z$
   - (c) $x^2 = yz$
   - (d) $y^2 = xz$

2. The domain of the function $\sqrt{\log_{0.5} x}$ is :
   - (a) $(0.5, \infty)$
   - (b) $(1, \infty)$
   - (c) $(0, 1)$
   - (d) $(0.5, 1)$

3. If $\log_a 6 = m$ and $\log_a 3 = n$, then $\log_a\left(\dfrac{a}{2}\right)$ is equal to :
   - (a) $1 - \log_a 2$
   - (b) $1 + m + n$
   - (c) $1 - m - n$
   - (d) $1 - m + n$

4. If the logarithm of a number to the base $\sqrt{8}$ is 6, then, the number is :
   - (a) 512
   - (b) 343
   - (c) 216
   - (d) 36

5. The equation $\log_3(3x - 8) = 2 - x$ has the solution :
   - (a) $x = 4$
   - (b) $x = 3$
   - (c) $x = 2$
   - (d) $x = 1$

6. Evaluate $\log \tan 1° + \log \tan 2° + \dots + \log \tan 89°$:
   - (a) 0
   - (b) 1
   - (c) 1/2
   - (d) $\sqrt{2}$

7. Simplify $\log \dfrac{14}{15} + \log \dfrac{28}{27} = \log \dfrac{405}{196}$
   - (a) 1
   - (b) 1092
   - (c) 1/2
   - (d) $\sqrt{2}$

8. Solve for x if $\log_{16} x + (\log_{16} x)^2 + (\log_{16} x)^3 + \dots + \text{to } \infty = 1/3$.
   - (a) 1/2
   - (b) –1/2
   - (c) 2
   - (d) 3/5

9. If $\log_{10}(x^3 y^3) = 3a + 2b$ and $2 \log_{10}(x^2 y^3) = 2a + 3b$, then the value of x at $a = 3$ is :
   - (a) 100
   - (b) 1000
   - (c) $\log_{10} 2$
   - (d) $\log_{10} 3$

10. If $\log_{90} 2 = a$, then $\log_{10} 25$ is equal to :
   - (a) $(1 + a)$
   - (b) $1 - a$
   - (c) $2(1 - a)$
   - (d) $2(1 + a)$

∞ TRUE/FALSE

Write 'T' for True and 'F' for False statement.

1. The equation $\dfrac{1}{2}\log(x^2 + 2x) - \log\sqrt{x} + 2 = 0$ has the solution $x = 1$. **(T/F)**

2. If $\log\left(\dfrac{a+b}{2}\right) = \dfrac{1}{2}[\log a + \log b]$ $a = b$. **(T/F)**

3. If $\log_a x = p$ and $\log_b x = q$, then $\log_{(a+b)} x = \dfrac{bq}{q-p}$. **(T/F)**

4. $a^{\log_y 2} \cdot a^{\log_z 2} \cdot a^{\log_a 2} = z$ **(T/F)**

5. The number $\log_2 7$ is a rational number. **(T/F)**

6. The value of $3^{2\log 3}$ is equal to 49. **(T/F)**

7. The value of $\log_{10} 50000 - \log_{10} 5$ is equal to 10. **(T/F)**

∞ FILL IN THE BLANKS

1. If $\log_{\sqrt{8}} x = 3\dfrac{1}{3}$. The value of x in _____ .

2. If $\log 2 = 0.3010$ and $\log 3 = 0.4771$ the value of $\log 25$ _____ .

3. $\log_9 27 - \log_{27} 9 = $ _____ .

4. $\log_7 \log_5(\sqrt{x+5} + \sqrt{x}) = a$ then x _____ .

5. The value of $\log_{3\sqrt{2}}(5832)$ is equal to _____ .

6. The value of $64^{\log_8 5}$ is _____ .

7. If $\log_{10} 5 + \log_{10}(5x + 1) = \log_{10}(x + 5) + 1$ then x is equal to _____ .

8. If $\log 27 = 1.431$ then value of $\log 9$ is _____ .

9. $2 \log_{10} 5 + \log_{10} 8 - \dfrac{1}{2}\log_{10} 4 = $ _____

10. $\left[\dfrac{1}{(\log_a bc + 1)} + \dfrac{1}{(\log_b ca + a)} + \dfrac{1}{(\log_c ab + 1)}\right]$ is equal to _____ .

ANSWERS

☞ **MULTIPLE CHOICE QUESTIONS**

1. (d)   **2.** (c)   **3.** (d)   **4.** (a)   **5.** (c)   **6.** (c)   **7.** (b)   **8.** (c)   **9.** (b)
10. (c)

☞ **TRUE/ FALSE**

1. T   **2.** T   **3.** T   **4.** T   **5.** F   **6.** F   **7.** F

☞ **FILL IN THE BLANKS**

1. 32   **2.** 1.398   **3.** 5/6   **4.** 4   **5.** 6   **6.** 25   **7.** 3   **8.** 0.954   **9.** 2
10. 1

□□□□□

CHAPTER 6

Trigonometry

6.1 INTRODUCTION

The word trigonometry is derived from two Greek word "trigon" and "metron", means "triangle" and "to measure" respectively. Therefore, trigonometry means to measure a triangle, *i.e.*,

"Trigonometry is that branch of mathematics which deals with angles, whether of a triangle or any other figure".

6.2 ANGLES AND QUADRANTS

Consider the Fig. 1, the angle is obtained by rotating a given ray about its end points. The original ray is called the initial side and the ray about which the initial sides rotates is called the terminal side of the angle.

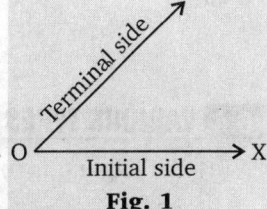

Fig. 1

☞ REMARKS

➠ The measure of an angle is the amount of rotation required to get the terminal side from initial side.

➠ If the revolving line revolves in anti-clockwise direction, then the angle is positive and if revolving line revolves in clockwise direction, then it is called negative angle. This may be clear in the following Fig. 2

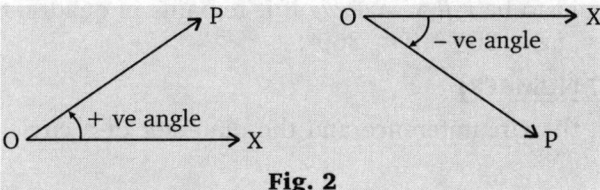

Fig. 2

6.3 MEASUREMENT OF ANGLES

Generally, we measure the angles in degrees or in radian, which are defined as follows :

6.3.1 SEXAGESIMAL SYSTEM OR ENGLISH SYSTEM (DEGREE MEASURE)

We can divide the right angle into 90 equal parts, and each small part is known as degree. Thus, a right angle is equal to 90 degrees. Similarly we can say that, the circumference of a circle can be divided into 360 equal parts. One degree is denoted by 1°.

Again we can divide a degree into sixty equal parts. Each small part is known as a minute and is denoted by $1'$.

i.e., $\qquad\qquad 1^\circ = 60'$ (sixty minutes).

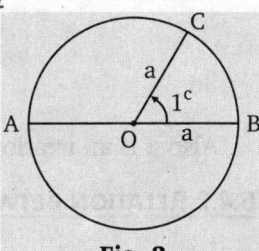

Fig. 3

A minute can also be divided into sixty equal parts and each small part is known as second and is denoted by 1˝.

i.e., $1' = 60˝$ (sixty seconds).

6.3.2 RADIANS (CIRCULAR SYSTEM)

Let us take a circle of radius a then 'a radian' is an angle subtended at the centre of a circle by an arc equal in length to the radius of the circle. One radian angle is denoted by as 1^c. In the fig 3 $\angle BOC = 1^c$.

➱ REMARK

➠ Relation between degree and radian : π radian = 180 degree

GRADE MEASURE

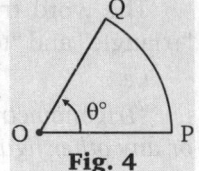

If we divide the right angle into 100 equal parts, then each small part is said to be grade. If we subdivide each grade into 100 equal part, then each part is known as minute and when we divide the minute into 100 parts then each part is known as seconds.

So,   1 right angle = 100^g

$1^g = 100'$ and $1' = 100˝$

π radian = $180° = 200^g$.

Fig. 4

6.4 VARIOUS TYPES OF ANGELES

(a) An angle θ is to be *acute angle* if it remains in quadrant number 1

i.e., $0° \leq \theta < 90°$.

(b) An angle θ is said to be *right angle* if $\theta = 90°$

(c) An angle is said to be *obtuse angle* if it remains in quadrant number 2

i.e., $90° < \theta < 180°$

(d) An angle θ is said to be *straight angle,* if $\theta = 180°$.

(e) An angle θ is said to be *reflexive angle* if it remains in quadrant number 3 and 4, i.e., $180° < \theta < 360°$.

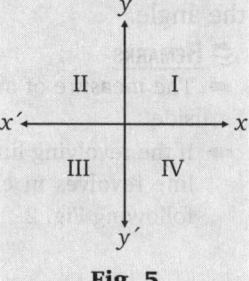

Fig. 5

6.4.1 π (A CONSTANT NUMBER)

The ratio between the circumference and the diameter of a circle is known as constant number π.

So, $\pi = \dfrac{circumference}{diameter}$

Let us take a circle of radius a, then

$\pi = \dfrac{circumference}{2a}$

So   circumference $= 2\pi a$

Also π is an irrational number. We use $\dfrac{22}{7}$ for π, which is its appropriate value.

6.4.2 RELATION BETWEEN AN ARC AND AN ANGLE

arc = radius × angle in radians

From figure 6, we have

$$l = \theta^c \cdot a.$$

6.4.3 AREA OF THE SECTOR

Let OPQ be the sector having an angle θ^c at the centre. This sector

form the circle of radius a, then area of sector $OPQ = \dfrac{1}{2}a^2\theta^c$.

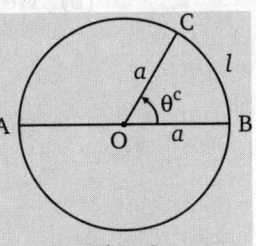

Fig. 6

☛ RECAPITULATIONS

- π Radian = 180 degree = 200^g
- $\pi = \dfrac{\text{Circumference}}{\text{diameter}}$

- Arc = radius × angles in radians
- π is an irrational number.

SOLVED EXAMPLES

EXAMPLE 1. *Find the degree measure of the given radian measure :*

    (a) $\left(\dfrac{5\pi}{12}\right)^c$ (b) $\left(\dfrac{1}{6}\right)^c$ (c) -3^c

SOLUTION. (a) We know that

$$1 \text{ radian} = \frac{180}{\pi} \text{ degree}$$

So, $\left(\dfrac{5\pi}{12}\right)^c = \left(\dfrac{5\pi}{12} \times \dfrac{180}{\pi}\right)^\circ = 75^\circ$

(b) We know that

$$1 \text{ radian} = \frac{180}{\pi} \text{ degree}$$

Therefore, $\left(\dfrac{1}{6}\right)^c = \left(\dfrac{1}{6} \times \dfrac{180}{\pi}\right)^\circ$ [taking the value of π as $\dfrac{22}{7}$]

$$= 9^\circ 32' 43.6'' = 9^\circ 32' 44''.$$

(c) We know that

$$1 \text{ radian} = \frac{180}{\pi} \text{ degree}$$

Therefore, $-3^c = \left[(-3) \times \dfrac{180}{\pi}\right]^\circ = \left[(-3) \times \dfrac{180}{22} \times 7\right]^\circ$

[taking the value of π as $\dfrac{22}{7}$]

$$= -171^\circ 49' 6''$$

EXAMPLE 2. *Find the radian measure of the given degree measure :*

    (a) 210° (b) $-22^\circ 30'$ (c) 135°

SOLUTION. (a) We know that $1 \text{ degree} = \dfrac{\pi}{180} \text{ radian}$

Therefore, $210^\circ = \left(210 \times \dfrac{\pi}{180}\right)^c = \left(\dfrac{7\pi}{6}\right)^c$

(b) We know that

$$1 \text{ degree} = \frac{\pi}{180} \text{ radian}$$

Therefore, $-22°30' = -\left(\frac{45}{2} \times \frac{\pi}{180}\right)^c$

$$\left[\because 30' = \left(\frac{1}{2}\right)^° , \text{ so } 22°30' = 22° + \left(\frac{1}{2}\right)^° = \left(\frac{45}{2}\right)^°\right]$$

$$= -\left(\frac{\pi}{8}\right)^c$$

(c) We know that

$$1 \text{ degree} = \frac{\pi}{180} \text{ radian}$$

Therefore, $135° = \left(135 \times \frac{\pi}{180}\right)^c = \left(\frac{3\pi}{4}\right)^c$

EXAMPLE 3. *Find the degree of the angle subtended at the centre of a circle of diameter 50 cm by an arc of length 11 cm.*

SOLUTION. Given that the diameter of the circle = 50 cm

i.e.,   radius of the circle = 25 cm

and   arc length s = 11 cm

We know that

$$\theta = \left(\frac{s}{r}\right)^c$$

Therefore, $\theta = \left(\frac{11}{25}\right)^c$

$$\Rightarrow \quad \theta = \left(\frac{11}{25} \times \frac{180}{\pi}\right)^° \qquad \left[\because 1 \text{ radian} = \left(\frac{180}{\pi}\right)^°\right]$$

$$\Rightarrow \quad \theta = \left(\frac{11}{25} \times \frac{180}{22} \times 7\right) \qquad \left[\text{Taking } \pi = \frac{22}{7}\right]$$

$$\Rightarrow \quad \theta = \left(\frac{126°}{5}\right) = \left(25\frac{1}{5}\right)^° = \left(25 + \frac{1}{5}\right)^°$$

$$= 25°12' \qquad \left[1 \text{ degree} = 60', \text{so} \left(\frac{1}{5}\right)^° = 12'\right]$$

EXAMPLE 4. *Find the area of a sector which has a central angle of 60° and a radius of 3 cm.*

SOLUTION. We know that the area of a circle which subtends an angle of 2π radians at the centre = πr^2.

Again $60° = \frac{\pi}{3} \text{ radian}$

Therefore, area of a sector which has a central angle of $60°\left(i.e., \dfrac{\pi}{3} \text{ radian}\right)$

$$= \dfrac{\pi r^2}{2\pi} \times \dfrac{\pi}{3} = \dfrac{\pi}{6} \times 3^2 \qquad \text{[since } r = 3 \text{ cm]}$$

$$= \dfrac{22}{7} \times \dfrac{1}{6} \times 9 = 4.714 \text{ square cm (app.)}$$

EXAMPLE 5. *Find the length of an arc of a circle of radius 5 cm subtending a central angle 15°.*

SOLUTION. Let s be the length of the arc subtending an angle θ at the center of a circle of radius r. Then $\theta = s/r$.

Here, $\qquad r = 5 \text{ cm and } \theta = 15° = 15 \times \dfrac{\pi}{160} = \left(\dfrac{\pi}{12}\right)^c$

$$\theta = \dfrac{s}{r} \quad \Rightarrow \quad s = \theta \times r = \left(\dfrac{\pi}{12} \times 5\right) \text{cm} = \dfrac{5\pi}{12} \text{cm}$$

EXAMPLE 6. *What is the ratio of two circles at the centres of which two arcs of same length subtend angles of 60° and 75°.*

SOLUTION. We know that, $\qquad \text{radius} = \dfrac{\text{arc}}{\text{angle}}$

Here, it is given that the length of arc of two circles is same.

Let the length of arc be s.

Now given $\qquad \theta_1 = 60° = 60 \times \dfrac{\pi}{180} = \dfrac{\pi}{3} \text{ radian}$

$$\theta_2 = 75° = 75 \times \dfrac{\pi}{180} = \dfrac{5\pi}{12} \text{ radian}$$

Also, $\qquad r_1 = \dfrac{s}{\theta_1} \text{ and } r_2 = \dfrac{s}{\theta_2}$

So, $\qquad \dfrac{r_1}{r_2} = \dfrac{\theta_2}{\theta_1} = \dfrac{5\pi/12}{\pi/3} = \dfrac{5}{4}$

Hence $\qquad r_1 : r_2 = 5 : 4$

EXAMPLE 7. *A horse is tied to a pole by a rope. If the horse moves along a circular path always keeping the rope tight and describes 88 metres when it has traced out 72° at the centre. Find the length of the rope.*

SOLUTION. In the fig. 7 the position of pole is indicated by O. Let P be the initial condition of the horse and OP is the rope by which the horse is tied by the rope and the pole is in tight condition. The horse moves along the path PQ.

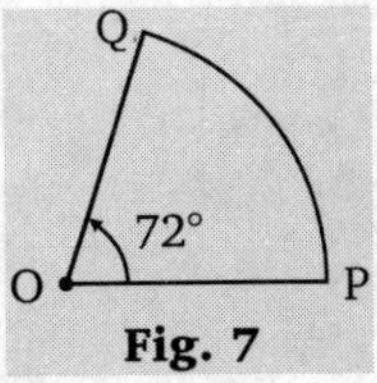

Now, $\qquad PQ = 88\text{m}$

Also $\qquad OP = OQ = r \text{ (say)}$

and $\qquad \angle QOP = 72° = \left(72 \times \dfrac{\pi}{180}\right) \text{ radian}$

Fig. 7

$$= \left(\dfrac{2\pi}{5}\right) \text{ radian}$$

Now we know that $\qquad r = \dfrac{s}{\theta}$

Therefore, $\qquad r = \left(\dfrac{88}{2\pi} \times 5\right) m \qquad \Rightarrow \quad r = \dfrac{88}{2 \times 22} \times 7 \times 5$

$\qquad\qquad\qquad\qquad\qquad\qquad\qquad \Rightarrow \quad r = 70 \text{ metres}$

EXAMPLE 8. *The minute hand of a clock is 10 cm long. How far does the tip of the hand move in 20 minutes?*

SOLUTION. We know that the minute hand of a clock describe a circle in 60 minute and given that the length of the circle is 10 cm.

So minute hand describe a circle of radius 10 cm.

Therefore, the distance covered by the minute hand in 60 minute

$$= 2\pi r = 2\pi \times 10 \text{ cm}$$

So, the distance covered by the minute hand in 20 minute

$$= \dfrac{22}{60} \times 2\pi \times 10 = \dfrac{1}{3} \times 20 \times \dfrac{22}{7} \qquad \text{(taking } \pi = 22/7)$$

$$= 20.95 \text{ cm} = 21 \text{ cm (appr.)}$$

EXAMPLE 9. *Find the number of degrees in the angle subtended at the centre of a circle of radius 10 ft. by an arc of length 20 ft.* **[UPTU(B.Pharma)–2005]**

SOLUTION. Using the relation $\quad s = r\theta$, we get

$$\theta^c = \text{angle subtended at the centre}$$

$$= \dfrac{20}{10} = 2 \text{ radians}$$

$$= 2 \times \left(\dfrac{180}{\pi}\right)^{\circ} = \left(2 \times \dfrac{180}{22} \times 7\right)^{\circ} = 114^{\circ}32'43''.$$

EXAMPLE 10. *An electric fan makes 100 revolution per minute. Find the angular speed in radian per second. Find also the linear speed in feet per second of the tips of the blades. The tips being 10 inches from the centre.*

SOLUTION. (i) Given that the number of revolution per minute = 1000.

So number of revolution per second $= \dfrac{1000}{60} = \dfrac{100}{6} = \dfrac{50}{3}$

Now, the angle described by the blade in one revolution = 2π radian

So, angular speed $(\omega) = 2\pi$ (number of revolution per second)

$$= 2\pi\left(\dfrac{50}{3}\right) = \dfrac{100\pi}{3} \text{ radian/sec.}$$

(ii) Also given that, radius of the blade = 10 inches

$$= \dfrac{10}{12} = \dfrac{5}{6} \text{ feet}$$

[Since, 1 feet = 12 inches]

So, linear speed of the tip of the blade = $r\omega$

[Since linear speed = radius × angular speed]

$$= \dfrac{5}{6} \times \dfrac{100\pi}{3} = \dfrac{5}{6} \times \dfrac{100}{3} \times \dfrac{22}{7}$$

Hence, $\qquad$ linear speed = 87.3 feet/sec.

EXAMPLE 11. *If G, D and C be the number of grade, degrees and radians in an angle. Prove that*

(a) $\dfrac{G}{100} = \dfrac{D}{90} = \dfrac{2C}{\pi}$ (b) $G - D = \dfrac{20C}{\pi}$

SOLUTION.   Let the measure of the given angle be x right angle.

Then, 1 right angle $= 100^g$ $\Rightarrow$ x right angle s $= (100x)^g$ ...(1)

1 right angle $= 90°$ $\Rightarrow$ x right angle s $= (90x)°$ ...(2)

2 right angle s $= \pi^c$ $\Rightarrow$ x right angle s $= \left(\dfrac{\pi}{2}x\right)^c$ ...(3)

(a) Now from equation (1), (2), (3), we have

$$100x = G,\ 90x = D,\ \dfrac{\pi x}{2} = C,$$

Hence, $\dfrac{G}{100} = \dfrac{D}{90} = \dfrac{2C}{\pi} = x$ $\Rightarrow$ $\dfrac{G}{100} = \dfrac{D}{90} = \dfrac{2C}{\pi}$

(b) Again $G = 100x,\ x = \dfrac{2c}{\pi}$ $\Rightarrow G - D = 100x - 90x$

$$= 10x = 10 \times \dfrac{2C}{\pi} = \dfrac{20C}{\pi}$$

$\Rightarrow$ $G - D = \dfrac{20C}{\pi}$

EXAMPLE 12. *The angles of a triangles are in A.P. The number of grades in the least is to the number of radians in the greatest as 40 : π. Find the angles of degrees.*

SOLUTION.   Let the angles be $(a - d)°,\ a°,\ (a + d)°$.

As we know that the sum of the angle of a triangles is 180°.

So, $a - d + a + a + d = 180$ $\Rightarrow$ $3a = 180°$

$\Rightarrow$ $a = 60°$

Therefore, the angles are $(60 - d)°,\ 60°,\ (60 + d)°$.

Clearly, the least angle is $(60 - d)°$ and the greater is $(60 + d)°$.

Now $(60 - d)° = \left[(60 - d) \times \dfrac{200}{180}\right]^g$

$$(60 - d)° = \left[\dfrac{600 - 10d}{9}\right]^g$$

and $(60 + d)° = \left[(60 + d)\dfrac{\pi}{180}\right]^c$

Now according to the given condition, we have

$\dfrac{(600 - 10d)}{9} \times \dfrac{180}{\pi(60 + d)} = \dfrac{40}{\pi}$ $\Rightarrow$ $120 + 2d = 600 - 10d$

$\Rightarrow$ $d = 40$

Hence, the angles of the given triangles are 20°, 60° and 100°.

EXAMPLE 13. *The circumference of a circle is divided into 5 parts in A.P. and if the greatest part is 6 times the least, find in radian the magnitudes of the angles that the parts subtend at the centre of the circle.*

Solution. Let r be the radius of the circle. If the five parts of the circumference of the circle are in A.P.

Then we can take five parts as $x - 2y, x - y, x, x + y, x + 2y$.

Also the sum of five part = circumference of the circles

i.e., $x - 2y + x - y + x + x + y + x + 2y = 2\pi r$

$$5x = 2\pi r$$

$$x = \frac{2\pi r}{5}$$

According to the given condition, we have

greatest part = $6 \times$ least part

$$(x + 2y) = 6(x - 2y)$$

$$x + 2y = 6x - 12y$$

$$5x = 14y$$

$$x = \frac{14y}{5}$$

or

$$y = \frac{5x}{14}$$

Putting the value of x in equation (2) from (1)

$$y = \frac{\pi r}{7}$$

∴ Five parts in A.P. of perimeter of the circle are $\dfrac{4\pi r}{35}, \dfrac{9\pi r}{35}, \dfrac{2\pi r}{5}, \dfrac{19\pi r}{35}, \dfrac{24\pi r}{35}$

Again we know that θ (angle at the centre) $= \dfrac{\text{arc}}{\text{radius}}$

So, angles in radians subtended at the centre of the circle by the five parts which are in A.P.

$$= \frac{4\pi}{35}, \frac{9\pi}{35}, \frac{2\pi}{35}, \frac{19\pi}{35}, \frac{24\pi}{35}$$

Exercise 6.1

1. Find the degree measure of the given radian measure :

   (a) $\left(\dfrac{7\pi}{12}\right)^c$ (b) $\left(\dfrac{1}{4}\right)^c$

   (c) -2^c

2. Find the radian measure of the given degree measure :

   (a) $240°$ (b) $15°$

   (c) $5°37'30''$.

3. Find the radius of a circle in which a central angle of $45°$ intercepts an arc of 187 cm.

4. The large hand of a big clock is 70 cm long. How many cms does its extremity move in 6 minute time.

5. Find the angle between the minute hand and the hour hand of a clock when the time is 7:20.

[**Hint.** Angle making by hour hand at 7:20

i.e., $\dfrac{22}{3} = \theta_1$

Angle making by minute hand in 20 minute $= \theta_2$

Angle between two hands $= \theta_1 - \theta_2$]

6. A point on a turbine wheel of radius 3 metres moves with a linear speed of 15 metres per second. Find the rate at which the wheel turns (i.e., the angular speed) in a radian per second.

7. The perimeter of a certain sector of a circle is equal to the length of the arc of a semicircle having the same radius; express the angle of the sector in degrees, minutes and seconds.

ANSWERS

1. (a) 105°   (b) 14°19′5″   (c) −114°32′44″

2. (a) $\left(\dfrac{4\pi}{3}\right)^c$ (b) $\left(\dfrac{\pi}{12}\right)^c$ (c) $\left(\dfrac{\pi}{32}\right)^c$ **3.** 238 cm   **4.** 44 cm

5. 100°   **6.** 5 rad/sec   **7.** $\left(\dfrac{720}{11}\right)^\circ$

6.5 TRIGONOMETRIC RATIO OR FUNCTIONS

In a right angled triangle ABC, if $\angle CAB = \theta$

$\Rightarrow$ BC = perpendicular (y)

and AC = Base (x)

                             AB = Hypotenuse (r)

We define the following trigonometric ratios which are also known as trigonometric functions :

(i) $\text{sine }\theta = \dfrac{\text{Perpendicular}}{\text{Hypotenuse}} = \dfrac{y}{r}$ and is written as sin θ.

(ii) $\text{cosine }\theta = \dfrac{\text{Base}}{\text{Hypotenuse}} = \dfrac{x}{r}$ and is written as cos θ.

(iii) $\text{tangent }\theta = \dfrac{\text{Perpendicular}}{\text{Base}} = \dfrac{y}{x}$ and is written as tan θ.

(iv) $\text{cosecant }\theta = \dfrac{\text{Hypotenuse}}{\text{Perpendicular}} = \dfrac{r}{y}$ and is written as cosec θ.

(v) $\text{secant }\theta = \dfrac{\text{Hypotenuse}}{\text{Base}} = \dfrac{r}{x}$ and is written as sec θ.

(vi) $\text{cotangent }\theta = \dfrac{\text{Base}}{\text{Perpendicular}} = \dfrac{x}{y}$ and is written as cot θ.

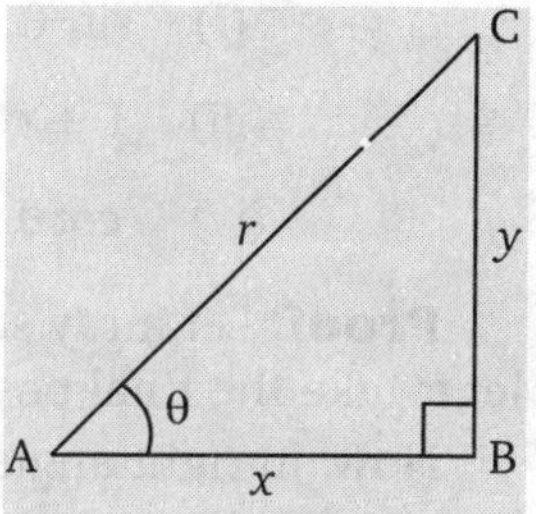

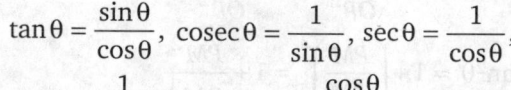

Fig. 8

From the above definitions, it follows some definitions :

$$\tan\theta = \dfrac{\sin\theta}{\cos\theta}, \quad \text{cosec}\,\theta = \dfrac{1}{\sin\theta}, \quad \sec\theta = \dfrac{1}{\cos\theta},$$

and $$\cot\theta = \dfrac{1}{\tan\theta} \quad \text{or} \quad \dfrac{\cos\theta}{\sin\theta}$$

REMARKS

⇒ Sine and cosine functions are called primary functions whereas tangent, cotangent, secant and cosecant functions are called secondary trigonometric functions.

⇒ When the terminal ray coincides with x-axis, cosec θ and cot θ are not defined.

⇒ When the terminal ray coincides with y-axis, sec θ and tan θ are not defined.

⇒ The domain of the sine function is the set of real number, whereas its range is the set of real numbers from − 1 to 1.

⇒ The domain of the cosine function is the set of all real numbers and range is the set of real number from − 1 to 1.

⇒ The domain of the tangent function is the set of all real numbers except odd multiples of π/2 and its range is the set of all real numbers.

⇒ The student should not commit the mistake of regarding sine as sin × θ; sin θ means the sine of angle θ, it is absolutely wrong to perform such operation as :

$$\sin (A + B) = \sin A + \sin B$$
$$\sin 2x + \sin x = \sin (2x + x)$$

⇒ Power notation for trigonometric function $(\sin\theta)^2$ is written as $\sin^2\theta$ and is read as sin square θ, $(\sin\theta)^3$ is written as $\sin^3\theta$ and is read as sin cube θ.

6.6 TRIGONOMETRIC IDENTITIES AND EQUATIONS

An expression involving trigonometric function which is true for all those values of θ for which the functions are defined is called a trigonometric identity. Otherwise, it is a trigonometric equation.

For example.

(1) $\sec\theta = \dfrac{1}{\cos\theta}$, $\operatorname{cosec}\theta = \dfrac{1}{\sin\theta}$ etc. are true for all values of θ except those for which $\cos\theta = 0$, $\sin\theta = 0$. So these are called trigonometric identities.

(2) $\sin\theta = \cos\theta$ is an expression which does not hold for all values of θ. So it is an equation and not identity.

6.7 FUNDAMENTAL TRIGONOMETRIC IDENTITIES

For any angle θ, we have

    (i) $\sin^2\theta + \cos^2\theta = 1$ (ii) $1 + \tan^2\theta = \sec^2\theta$

    (iii) $1 + \cot^2\theta = \operatorname{cosec}^2\theta$ (iv) $\tan\theta = \dfrac{\sin\theta}{\cos\theta}$

    (iv) $\cot\theta = \dfrac{\cos\theta}{\sin\theta}$

Proof. Let a ray starting from OX, trace out any angle θ in any of the four quadrants and let it take the final position OP. From P draw PM perpendicular to x-axis.

Now, in right angled triangle $\triangle OMP$, we have

$$OP^2 = OM^2 + PM^2$$

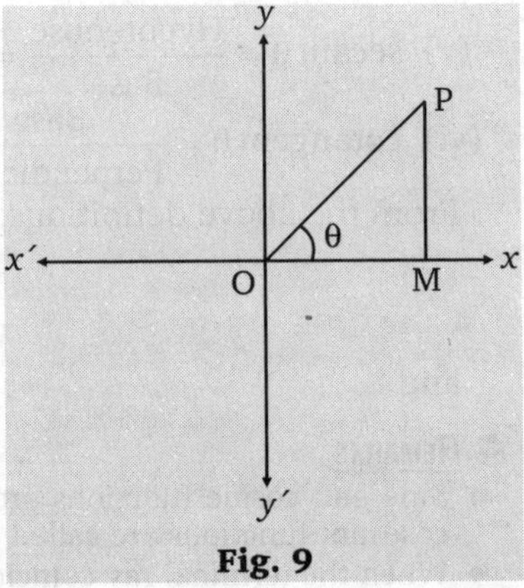

(i) $\sin^2\theta + \cos^2\theta = \left(\dfrac{PM}{OP}\right)^2 + \left(\dfrac{OM}{OP}\right)^2$

$$= \dfrac{PM^2 + OM^2}{OP^2} = \dfrac{OP^2}{OP^2} = 1$$

(ii) $1 + \tan^2\theta = 1 + \left(\dfrac{PM}{OM}\right)^2 = 1 + \dfrac{PM^2}{OM^2}$

$$= \dfrac{OM^2 + PM^2}{OM^2} = \dfrac{OP^2}{OM^2} = \left(\dfrac{OP}{OM}\right)^2$$

Fig. 9

(iii) $1 + \cot^2\theta = 1 + \left(\dfrac{OM}{PM}\right)^2$

$$= 1 + \dfrac{OM^2}{PM^2} = \dfrac{PM^2 + OM^2}{PM^2} = \dfrac{OP^2}{PM^2} = \operatorname{cosec}^2\theta$$

(iv) $\dfrac{\sin\theta}{\cos\theta} = \dfrac{PM}{OP} \times \dfrac{OP}{OM} = \dfrac{PM}{OM} = \tan\theta$

(v) $\dfrac{\cos\theta}{\sin\theta} = \dfrac{OM}{OP} \times \dfrac{OP}{PM} = \dfrac{OM}{PM} = \cot\theta$

(vi) $\cot\theta = \dfrac{1}{\tan\theta}$ $\Rightarrow$ $\tan\theta\cot\theta = 1$

6.7.1 TRIGONOMETRIC RATIOS OF STANDARD ANGLES

The values of trigonometric ratio of trigonometric angles of $\frac{\pi}{6}$ (30°), $\frac{\pi}{4}$ (45°), $\frac{\pi}{3}$ (60°), $\frac{\pi}{2}$ (90°) given in the following table 1.

Table 1

θ	0	π/6	π/4	π/3	π/2	π	3π/2	2π
sin	0	$\frac{1}{2}$	$\frac{1}{\sqrt{2}}$	$\frac{\sqrt{3}}{2}$	1	0	−1	0
cos	1	$\frac{\sqrt{3}}{2}$	$\frac{1}{\sqrt{2}}$	$\frac{1}{2}$	0	−1	0	1
tan	0	$\frac{1}{\sqrt{3}}$	1	$\sqrt{3}$	∞	0	∞	0
cot	∞	$\sqrt{3}$	1	$\frac{1}{\sqrt{3}}$	0	∞	0	∞
sec	1	$\frac{2}{\sqrt{3}}$	$\sqrt{2}$	2	∞	−1	∞	1
cosec	∞	2	$\sqrt{2}$	$\frac{2}{\sqrt{3}}$	1	∞	−1	∞

SOLVED EXAMPLES

EXAMPLE 1. *Show that* $\sqrt{\dfrac{1+\cos\theta}{1-\cos\theta}} = \operatorname{cosec}\theta + \cot\theta$.

SOLUTION. Here, we have

$$\text{L.H.S.} = \sqrt{\frac{1+\cos\theta}{1-\cos\theta}} = \sqrt{\frac{1+\cos\theta}{1-\cos\theta} \times \frac{1+\cos\theta}{1+\cos\theta}}$$

$$= \sqrt{\frac{(1+\cos\theta)^2}{1-\cos^2\theta}} = \sqrt{\frac{(1+\cos\theta)^2}{\sin^2\theta}}$$

$$= \frac{1+\cos\theta}{\sin\theta} = \frac{1}{\sin\theta} + \frac{\cos\theta}{\sin\theta} = \operatorname{cosec}\theta + \cot\theta = \text{R.H.S.}$$

EXAMPLE 2. *Show that* $\dfrac{\tan\theta - \cot\theta}{\sin\theta\cos\theta} = \sec^2\theta - \operatorname{cosec}^2\theta$.

SOLUTION. $\text{L.H.S.} = \dfrac{\tan\theta - \cot\theta}{\sin\theta\cos\theta} = \dfrac{\dfrac{\sin\theta}{\cos\theta} - \dfrac{\cos\theta}{\sin\theta}}{\sin\theta \cdot \cos\theta} = \dfrac{\dfrac{\sin^2\theta - \cos^2\theta}{\sin\theta\cos\theta}}{\sin\theta\cos\theta}$

$$= \frac{\sin^2\theta - \cos^2\theta}{\sin^2\theta\cos^2\theta}$$

$$= \frac{\sin^2\theta}{\sin^2\theta\cos^2\theta} - \frac{\cos^2\theta}{\sin^2\theta\cos^2\theta} = \frac{1}{\cos^2\theta} - \frac{1}{\sin^2\theta}$$

$$= \sec^2\theta - \csc^2\theta = \text{R.H.S.}$$

EXAMPLE 3.   *Show that* $(\sec A - \tan A)^2 = \dfrac{1-\sin A}{1+\sin A}$ [RGPV(B. Pharma)–2002, 14]

SOLUTION.   Consider   L.H.S. $= (\sec A - \tan A)^2 = \left(\dfrac{1}{\cos A} - \dfrac{\sin A}{\cos A}\right)^2 = \left(\dfrac{1-\sin A}{\cos A}\right)^2$

$$= \frac{(1-\sin A)^2}{\cos^2 A} = \frac{(1-\sin A)^2}{(1-\sin^2 A)} = \frac{(1-\sin A)^2}{(1-\sin A)(1+\sin A)}$$

$$= \frac{1-\sin A}{1+\sin A} = \text{R.H.S.}$$

EXAMPLE 4.   *Show that* $\dfrac{\tan\theta + \sec\theta - 1}{\tan\theta - \sec\theta + 1} = \dfrac{1+\sin\theta}{\cos\theta} = \dfrac{\cos\theta}{1-\sin\theta}$

[RGPV(B. Pharma)–2004, UPTU(B.Pharma)–2004]

SOLUTION.   Consider   L.H.S. $= \dfrac{\tan\theta + \sec\theta - 1}{\tan\theta - \sec\theta + 1}$

Using $1 + \tan^2\theta = \sec^2\theta$, *i.e.*, $\sec^2\theta - \tan^2\theta = 1$, we get

$$\frac{\tan\theta + \sec\theta - 1}{\tan\theta - \sec\theta + 1} = \frac{\tan\theta + \sec\theta - (\sec^2\theta - \tan^2\theta)}{\tan\theta - \sec\theta + 1}$$

$$= \frac{(\tan\theta + \sec\theta) - (\sec\theta - \tan\theta)(\sec\theta + \tan\theta)}{\tan\theta - \sec\theta + 1}$$

$$= \frac{(\sec\theta + \tan\theta)(1 - \sec\theta + \tan\theta)}{1 - \sec\theta + \tan\theta} = \sec\theta + \tan\theta$$

$$= \frac{1}{\cos\theta} + \frac{\sin\theta}{\cos\theta} = \frac{1+\sin\theta}{\cos\theta}$$

$$= \frac{(1+\sin\theta)}{\cos\theta} \times \frac{(1-\sin\theta)}{(1-\sin\theta)} = \frac{\cos^2\theta}{\cos\theta(1-\sin\theta)}$$

$$= \frac{\cos\theta}{1-\sin\theta} = \text{R.H.S.}$$

EXAMPLE 5.   *Show that* $2(\sin^6\theta + \cos^6\theta) - 3(\sin^4\theta + \cos^4\theta) + 1 = 0$.

[UPTU(B.Pharma)–2005, 11]

SOLUTION.   Consider   L.H.S. $= 2(\sin^6\theta + \cos^6\theta) - 3(\sin^4\theta + \cos^4\theta) + 1$

$$[\because a^3 + b^3 = (a+b)^3 - 3ab(a^2+b^2)]$$

$$= 2[(\sin^2\theta)^3 + (\cos^2\theta)^3] - 3[(\sin^2\theta)^2 + (\cos^2\theta)] + 1$$

$$= 2[(\sin^2\theta + \cos^2\theta)^3 - 3\sin^2\theta\cos^2\theta(\sin^2\theta + \cos^2\theta)$$

$$- 3[(\sin^2\theta + \cos^2\theta)^2 - 2\sin^2\theta\cos^2\theta] + 1$$

$$= 2[(1)^3 - 3\sin^2\theta\cos^2\theta \cdot 1] - 3[(1)^2 - 2\sin^2\theta\cos^2\theta] + 1$$

$$= 2[1 - 3\sin^2\theta\cos^2\theta] - 3[1 - 2\sin^2\theta\cos^2\theta] + 1$$

$$= 2 - 6\sin^2\theta\cos^2\theta - 3 + 6\sin^2\theta\cos^2\theta + 1$$

$$= 0 = \text{R.H.S.}$$

EXAMPLE 6. *Show that* $\dfrac{\sin\theta}{\cot\theta + \mathrm{cosec}\,\theta} = 2 + \dfrac{\sin\theta}{\cot\theta - \mathrm{cosec}\,\theta}$

SOLUTION. Consider   L.H.S. $= \dfrac{\sin\theta}{\cot\theta + \mathrm{cosec}\,\theta} = \dfrac{\sin\theta}{\dfrac{\cos\theta}{\sin\theta} + \dfrac{1}{\sin\theta}}$

$$= \dfrac{\sin\theta}{\dfrac{\cos\theta + 1}{\sin\theta}} = \dfrac{\sin^2\theta}{1 + \cos\theta} = \dfrac{1 - \cos^2\theta}{1 + \cos\theta}$$

$$= \dfrac{(1 - \cos\theta)(1 + \cos\theta)}{1 + \cos\theta} = (1 - \cos\theta) \qquad \ldots(1)$$

Now,   R.H.S. $= 2 + \dfrac{\sin\theta}{\cot\theta - \mathrm{cosec}\,\theta}$

$$= 2 + \dfrac{\sin\theta}{\dfrac{\cos\theta}{\sin\theta} - \dfrac{1}{\sin\theta}} = 2 + \dfrac{\sin\theta}{\dfrac{\cos\theta - 1}{\sin\theta}} = 2 + \dfrac{\sin^2\theta}{\cos\theta - 1}$$

$$= 2 + \dfrac{1 - \cos^2\theta}{\cos\theta - 1} = 2 - \dfrac{(1 - \cos\theta)(1 + \cos\theta)}{1 - \cos\theta}$$

$$= 2 - (1 + \cos\theta) = 2 - 1 - \cos\theta = 1 - \cos\theta \qquad \ldots(2)$$

From (1) and (2), we conclude that

$$\dfrac{\sin\theta}{\cot\theta + \mathrm{cosec}\,\theta} = 2 + \dfrac{\sin\theta}{\cot\theta - \mathrm{cosec}\,\theta}$$

EXAMPLE 7. *Show that* $\dfrac{1}{\sec\theta + \tan\theta} - \dfrac{1}{\cos\theta} = \dfrac{1}{\cos\theta} - \dfrac{1}{\sec\theta - \tan\theta}$.

SOLUTION. $\dfrac{1}{\sec\theta + \tan\theta} + \dfrac{1}{\sec\theta - \tan\theta} = \dfrac{1}{\cos\theta} + \dfrac{1}{\cos\theta}$

i.e., if $\dfrac{\sec\theta - \tan\theta + \sec\theta + \tan\theta}{(\sec + \tan\theta)(\sec\theta - \tan\theta)} = \dfrac{2}{\cos\theta}$

i.e., if $\dfrac{2\sec\theta}{\sec^2\theta - \tan^2\theta} = \dfrac{2}{\cos\theta}$

i.e., if $\dfrac{2\sec\theta}{1} = 2\sec\theta$, which is always true.

EXAMPLE 8. *Prove that the expression* $2(\sin^6\theta + \cos^6\theta) - 3(\sin^4\theta + \cos^4\theta)$ *is independent of* θ.

SOLUTION. We have

$$\sin^6\theta + \cos^6\theta = (\sin^2\theta + \cos^2\theta)(\cos^4\theta + \sin^4\theta - \sin^2\theta\cos^2\theta)$$
$$= 1\cdot(\cos^4\theta + \sin^4\theta + 2\sin^2\theta\cdot\cos^2\theta$$
$$- 2\sin^2\theta\cos^2\theta - \sin^2\theta\cos^2\theta)$$
$$= (\cos^2\theta + \sin^2\theta)^2 - 3\sin^2\theta\cos^2\theta.$$

Now using this result in L.H.S we get

$$2(1 - 3\sin^2\theta\cos^2\theta) - 3(\sin^4\theta + \cos^4\theta)$$
$$= 2(1 - 3\sin^2\theta\cos^2\theta) - 3[\sin^4\theta + \cos^4\theta$$
$$+ 2\sin^2\theta\cos^2\theta - 2\sin^2\theta\cos^2\theta]$$

$$= 2 - 6\sin^2\theta \, \cos^2\theta - 3[(\sin^2\theta + \cos^2\theta)^2$$
$$- 2\sin^2\theta\cos^2\theta]$$
$$= 2 - 6\sin^2\theta \, \cos^2\theta - 3 + 6\sin^2\theta\cos^2\theta = -1$$

which is independent of θ.

EXAMPLE 9. *Prove that* $\dfrac{(1 + \sin\theta - \cos\theta)^2}{(1 + \sin\theta + \cos\theta)^2} = \dfrac{1 - \cos\theta}{1 + \cos\theta}$
(Meerut(B.Sc. Biotech)–2006)

SOLUTION. Consider L.H.S. $= \dfrac{1 + \sin^2\theta + \cos^2\theta + 2\sin\theta - 2\cos\theta - 2\sin\theta\cos\theta}{1 + \sin^2\theta + \cos^2\theta + 2\sin\theta + 2\cos\theta + 2\sin\theta\cos\theta}$

$$= \dfrac{1 + 1 + 2\sin\theta - 2\cos\theta(1 + \sin\theta)}{1 + 1 + 2\sin\theta + 2\cos\theta(1 + \sin\theta)}$$

$$= \dfrac{2(1 + \sin\theta) - 2\cos\theta(1 + \sin\theta)}{2(1 + \sin\theta) + 2\cos\theta(1 + \sin\theta)}$$

$$= \dfrac{2(1 + \sin\theta)(1 - \cos\theta)}{2(1 + \sin\theta)(1 + \cos\theta)} = \dfrac{1 - \cos\theta}{1 + \cos\theta} = \text{R.H.S.}$$

EXAMPLE 10. *If* $\tan\theta + \sin\theta = m$ *and* $\tan\theta - \sin\theta = n$, *show that* $(m^2 - n^2) = 16mn$.

SOLUTION. Here the given equations are

$$\tan\theta + \sin\theta = m \qquad\qquad\qquad …(1)$$
and $$\tan\theta - \sin\theta = n \qquad\qquad\qquad …(2)$$

Adding (1) and (2), we get

$$2\tan\theta = m + n.$$

$\therefore \qquad\qquad \tan\theta = \dfrac{m + n}{2} \qquad \Rightarrow \qquad \cot\theta = \dfrac{2}{m + n}$

Subtracting (2) from (1), we get

$$2\sin\theta = m - n \qquad \Rightarrow \qquad \sin\theta = \dfrac{m - n}{2}$$

$\therefore \qquad\qquad \operatorname{cosec}\theta = \dfrac{2}{m - n}$

Putting the values of $\cot\theta$, $\operatorname{cosec}\theta$ in equation

$$\operatorname{cosec}^2\theta - \cot^2\theta = 1$$

We get $\dfrac{4}{(m - n)^2} - \dfrac{4}{(m + n)^2} = 1$

or $4(m + n)^2 - 4(m - n)^2 = [(m - n)(m + n)]^2$

Therefore, $4(m^2 + n^2 + 2mn) - 4(m^2 + n^2 - 2mn) = [(m - n)(m + n)]^2$

$\Rightarrow \qquad 4m^2 + 4n^2 + 8mn - 4m^2 - 4n^2 + 8mn) = (m^2 - n^2)^2$

$\Rightarrow \qquad\qquad 16\,mn = (m^2 - n^2)^2$

or $\qquad\qquad (m^2 - n^2)^2 = 16mn$

EXAMPLE 11. *Show that* $(1 + \cot\theta + \tan\theta)(\sin\theta - \cos\theta) = \dfrac{\sec\theta}{\operatorname{cosec}^2\theta} - \dfrac{\operatorname{cosec}\theta}{\sec^2\theta}$

SOLUTION. Consider L.H.S. $= (1 + \cot\theta + \tan\theta)(\sin\theta - \cos\theta)$

$$= \left(1 + \dfrac{\cos\theta}{\sin\theta} + \dfrac{\sin\theta}{\cos\theta}\right)(\sin\theta - \cos\theta)$$

$$= \left(\frac{\sin\theta\cos\theta + \cos^2\theta + \sin^2\theta}{\sin\theta\cos\theta} \right) (\sin\theta - \cos\theta)$$

$$[\because a^3 - b^3 = (a-b)(a^2 + b^2 + ab)]$$

$$= \frac{(\sin\theta - \cos\theta)(\sin^2\theta + \cos^2\theta + \sin\theta\cos\theta)}{\sin\theta\cos\theta}$$

$$= \frac{\sin^3\theta - \cos^3\theta}{\sin\theta\cos\theta} \qquad \qquad \qquad \ldots(1)$$

$$\text{R.H.S.} = \frac{\sec\theta}{\cosec^2\theta} - \frac{\cosec\theta}{\sec^2\theta}$$

$$= \frac{\dfrac{1}{\cos\theta}}{\dfrac{1}{\sin^2\theta}} - \frac{\dfrac{1}{\sin\theta}}{\dfrac{1}{\cos^2\theta}} = \frac{\sin^2\theta}{\cos\theta} - \frac{\cos^2\theta}{\sin\theta}$$

$$= \frac{\sin^3\theta - \cos^3\theta}{\sin\theta\cos\theta} \qquad \qquad \qquad \ldots(2)$$

From (1) and (2), we get

$$(\sin\theta - \cos\theta)(1 + \cot\theta + \tan\theta) = \frac{\sec\theta}{\cosec^2\theta} - \frac{\cosec\theta}{\sin^2\theta}$$

EXAMPLE 12. *If* $\sec\theta = x + \dfrac{1}{4x}$, *show that* $\sec\theta + \tan\theta = 2x$ *or* $\dfrac{1}{2x}$.

SOLUTION.   We have

$$\sec\theta = x + \frac{1}{4x} \qquad \Rightarrow \qquad \sec\theta = \frac{4x^2 + 1}{4x} \qquad \ldots(1)$$

Now, $\qquad \tan^2\theta = \sec^2\theta - 1$

$$= \left(\frac{4x^2 + 1}{4x} \right)^2 - 1 = \frac{(4x^2 + 1)^2 - 16x^2}{16x^2} = \frac{(4x^2 - 1)^2}{16x^2}$$

$$\therefore \qquad \tan\theta = \pm \frac{4x^2 - 1}{4x} \qquad \qquad \qquad \ldots(2)$$

Adding (1) and (2), we get

$$\sec\theta + \tan\theta = \frac{4x^2 + 1}{4x} \pm \frac{4x^2 - 1}{4x}$$

$$= \left(\frac{4x^2 + 1}{4x} + \frac{4x^2 - 1}{4x} \right) \text{or} \left(\frac{4x^2 + 1}{4x} - \frac{4x^2 - 1}{4x} \right)$$

$$= \frac{4x^2 + 1 + 4x^2 - 1}{4x} \text{or} \frac{4x^2 + 1 - 4x^2 + 1}{4x} = \frac{8x^2}{4x} \text{or} \frac{2}{4x}$$

Hence, $\qquad \sec\theta + \tan\theta = 2x$ or $\dfrac{1}{2x}$

Exercise 6.2

1. Show that :

(a) $\sqrt{\dfrac{1-\sin\theta}{1+\sin\theta}} = \sec\theta - \tan\theta$

(b) $\sqrt{\dfrac{1-\cos\theta}{1+\cos\theta}} = \operatorname{cosec}\theta - \cot\theta$

(c) $2\sec^2\theta - \sec^4\theta - 2\operatorname{cosec}^2\theta + \operatorname{cosec}^4\theta$
$\quad\quad = \cot^4\theta - \tan^4\theta$

(d) $\sec^6\theta = \tan^6\theta + 3\tan^2\theta\sec^2\theta + 1$

(e) $\sec^4\theta - \sec^2\theta = \tan^4\theta + \tan^2\theta$

(f) $\tan^2\theta - \sin^2\theta = \tan^2\theta\cdot\sin^2\theta$

(g) $\dfrac{1+\cos\theta}{1-\cos\theta} = \dfrac{\tan^2\theta}{(\sec\theta-1)^2}$

(h) $\dfrac{\sec\theta-\tan\theta}{\sec\theta+\tan\theta}$
$\quad\quad = 1 - 2\sec\theta\tan\theta + 2\tan^2\theta.$

(i) $\dfrac{1-\cos\theta}{\sin\theta} - \dfrac{\sin\theta}{1+\cos\theta}$

(j) $\tan^2\theta + \cot^2\theta + 2 = \sec^2\theta\cdot\operatorname{cosec}^2\theta$

[UPTU(B. Pharma)–2001]

2. Eliminate θ from the following equations :

(a) $x = a\cos^3\theta, y = b\sin^3\theta$

(b) $x = a\sec^3\theta, y = b\tan^3\theta$

(c) $\sec\theta + \tan\theta = m, \sec\theta - \tan\theta = \pi$

(d) $a\cot\theta + b\operatorname{cosec}\theta = x^2,$
$\quad b\cot\theta + d\operatorname{cosec}\theta = y^2$

3. Prove that : $\dfrac{\cos\theta}{1-\tan\theta} + \dfrac{\sin\theta}{1-\cot\theta} = \sin\theta + \cos\theta$

[RGPV(B.Pharma)–2001]

4. Prove the following :

(a) $\sin^8\theta - \cos^8\theta$
$\quad\quad = (\sin^2\theta - \cos^2\theta)(1 - 2\sin^2\theta\cos^2\theta)$

(b) $\dfrac{\sin\theta}{1-\cos\theta} + \dfrac{\tan\theta}{1+\cos\theta} = \sec\theta\operatorname{cosec}\theta + \cot\theta$

(c) $\dfrac{\sin\theta}{1+\cos\theta} + \dfrac{1+\cos\theta}{\sin\theta} = 2\operatorname{cosec}\theta$

(d) $\dfrac{1}{\operatorname{cosec}\theta+\cot\theta} - \dfrac{1}{\sin\theta}$

$\quad\quad = \dfrac{1}{\sin\theta} - \dfrac{1}{\operatorname{cosec}\theta-\cot\theta}$

5. (a) If $\cos\theta = \dfrac{21}{29}$ and θ lies in the fourth quadrant, find $\sin\theta$ and $\tan\theta$.

(b) If $\cos\theta\operatorname{cosec}\theta = -1$ and θ lies in the fourth quadrant, find $\cos\theta$ and $\operatorname{cosec}\theta$.

6. Prove that :

(a) $\sin\theta\cot\theta + \sin\theta\operatorname{cosec}\theta = 1 + \cos\theta$

(b) $\sec\theta(1-\sin\theta)(\sec\theta+\tan\theta) = 1$

(c) $\dfrac{\tan\theta}{1-\cot\theta} + \dfrac{\cot\theta}{1-\tan\theta} = 1 + \sec\theta\operatorname{cosec}\theta$

(d) $(1+\cot\theta-\operatorname{cosec}\theta)$
$\quad\quad (1+\tan\theta+\sec\theta) = 2$

(e) $(\operatorname{cosec}\theta-\sin\theta)(\sec\theta-\cos\theta)$
$\quad\quad (\tan\theta+\cot\theta) = 1$

(f) $(\sec\theta-\cos\theta)(\operatorname{cosec}\theta-\sin\theta)$

$\quad\quad = \dfrac{1}{\tan\theta+\cot\theta}$

(g) $\sqrt{\sec^2\theta+\operatorname{cosec}^2\theta} = \tan\theta + \cot\theta$
$\quad\quad = \sec\theta\operatorname{cosec}\theta$

(h) $(\sin\theta+\operatorname{cosec}\theta)^2 + (\cos\theta+\sec\theta)^2$
$\quad\quad = \tan^2\theta + \cot^2\theta = 7$

(i) $\dfrac{\operatorname{cosec}\theta}{\operatorname{cosec}\theta-1} + \dfrac{\operatorname{cosec}\theta}{\operatorname{cosec}\theta+1} = 2\sec^2\theta$

(j) $\dfrac{\sin A-\sin B}{\cos A+\cos B} + \dfrac{\cos A-\cos B}{\sin A+\sin B} = 0$

(k) $\dfrac{1}{\sec\theta-\tan\theta} = \dfrac{1+\sin\theta}{\cos\theta} = \sec\theta + \tan\theta.$

7. If $\dfrac{\sin A}{\sin B} = m$ and $\dfrac{\cos A}{\cos B} = n$, show that

$\tan A = \pm\dfrac{m}{n}\sqrt{\dfrac{1-n^2}{m^2-1}}$

8. (a) If $\sin\theta = \dfrac{21}{29}$, show that $\sec\theta + \tan\theta = \dfrac{5}{2}$,

if θ lies between 0 and $\dfrac{\pi}{2}$.

(b) What will be the value of the expression when θ lies between $\dfrac{\pi}{2}$ and π.

HINT TO SELECTED PROBLEMS

1. (e) $\sec^4\theta - \sec^2\theta = \sec^2\theta(\sec^2\theta-1)$

$\quad\quad = (1+\tan^2\theta)[\tan^2\theta]$

$\quad\quad = \tan^2\theta + \tan^4\theta$

$\quad\quad = \tan^4\theta + \tan^2\theta$

(h) $\dfrac{\sec\theta - \tan\theta}{\sec\theta + \tan\theta}$

$= \dfrac{(\sec\theta - \tan\theta)\times(\sec\theta - \tan\theta)}{(\sec\theta + \tan\theta)(\sec\theta - \tan\theta)}$

$= \dfrac{(\sec\theta - \tan\theta)^2}{\sec^2\theta - \tan^2\theta}$

$= \dfrac{\sec^2\theta + \tan^2\theta - 2\sec\theta\tan\theta}{1}$

$= 1 + \tan^2\theta + \tan^2\theta - 2\sec\theta\tan\theta$

$= 1 - 2\sec\theta\tan\theta + 2\tan^2\theta$

2. (a) $\quad x = a\cos^3\theta,\ y = \sin^3\theta$

$\dfrac{x}{a} = \cos^3\theta,\ \dfrac{y}{b} = \sin^3\theta$

$\cos\theta = \left(\dfrac{x}{a}\right)^{1/3},\ \sin\theta = \left(\dfrac{y}{b}\right)^{1/3}$

$\cos^2\theta + \sin^2\theta = 1$

$\left(\dfrac{x}{a}\right)^{2/3} + \left(\dfrac{y}{b}\right)^{2/3} = 1$

(c) $\sec\theta + \tan\theta = m$ ...(1)

$\sec\theta - \tan\theta = n$ ...(2)

Eq^n. (1) $\times$ (2)

$\Rightarrow (\sec\theta + \tan\theta)(\sec\theta - \tan\theta)$

$= m \times n$

$= \sec^2\theta - \tan^2\theta = mn$

$\Rightarrow \quad mn = 1$

4. (a) $\sin^8\theta - \cos^8\theta$

$= (\sin^4\theta + \cos^4\theta)(\sin^4\theta - \cos^4\theta)$

$= [(\sin^2\theta + \cos^2\theta) - 2\sin^2\theta\cos^2\theta)]$

$\quad [(\sin^2\theta + \cos^2\theta)(\sin^2\theta - \cos^2\theta)]$

$= (1 - 2\sin^2\theta\cos^2\theta)(\sin^2\theta - \cos^2\theta)$

$= (\sin^2\theta - \cos^2\theta)(1 - 2\sin^2\theta\cos^2\theta)$

(d) $\dfrac{1}{\operatorname{cosec}\theta + \cot\theta} - \dfrac{1}{\sin\theta}$

$= \dfrac{1}{\sin\theta} - \dfrac{1}{\operatorname{cosec}\theta - \cot\theta}$

or $\dfrac{1}{\operatorname{cosec}\theta + \cot\theta} + \dfrac{1}{\operatorname{cosec}\theta - \cot\theta} = \dfrac{2}{\sin\theta}$

5. (a) If $\cos\theta = \dfrac{21}{29}$ and

θ lies in fourth quadrant.

Then in fourth quadrant sin and tan both are negative.

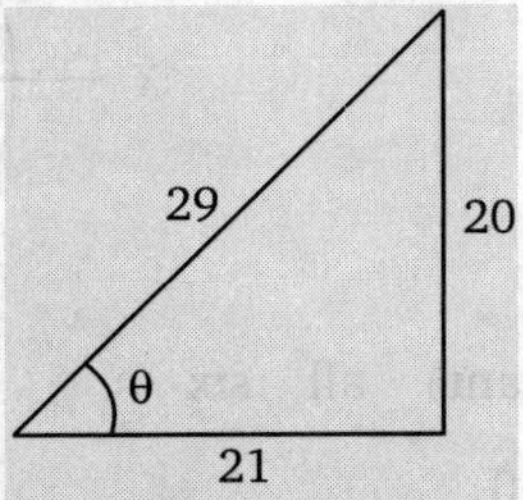

Fig. 10

$\sin\theta = -\dfrac{20}{29}$

$\tan\theta = -\dfrac{20}{21}$

6. (c) $\dfrac{\tan\theta}{1 - \cot\theta} + \dfrac{\cot\theta}{1 - \tan\theta} = 1 + \sec\theta\operatorname{cosec}\theta$

L.H.S $= \dfrac{\dfrac{\sin\theta}{\cos\theta}}{1 - \dfrac{\cos\theta}{\sin\theta}} + \dfrac{\dfrac{\cos\theta}{\sin\theta}}{1 - \dfrac{\sin\theta}{\cos\theta}}$

$= \dfrac{\sin^2\theta}{\cos\theta(\sin\theta - \cos\theta)} + \dfrac{\cos^2\theta}{\sin\theta(\cos\theta - \sin\theta)}$

$= \dfrac{\sin^3\theta - \cos^3\theta}{\sin\theta\cos\theta(\sin\theta - \cos\theta)}$

$= \dfrac{(\sin\theta - \cos\theta)(\sin^2\theta + \cos^2\theta + \sin\theta\cos\theta)}{\sin\theta\cos\theta(\sin\theta - \cos\theta)}$

$= \dfrac{1 + \sin\theta\cos\theta}{\sin\theta\cos\theta}$

(g) $\sqrt{\sec^2\theta + \operatorname{cosec}^2\theta} = \tan\theta + \cot\theta$

$= \sec\theta\operatorname{cosec}\theta$

L.H.S. $= \sqrt{\sec^2\theta + \operatorname{cosec}^2\theta}$

$= \sqrt{\dfrac{1}{\cos^2\theta} + \dfrac{1}{\sin^2\theta}}$

$= \sqrt{\dfrac{\sin^2\theta + \cos^2\theta}{\sin^2\theta\cos^2\theta}} = \dfrac{1}{\sin\theta\cos\theta}$

$= \sec\theta\operatorname{cosec}\theta = $ R.H.S

Middle Term $= \tan\theta + \cot\theta$

$= \dfrac{\sin\theta}{\cos\theta} + \dfrac{\cos\theta}{\sin\theta}$

$= \dfrac{\sin^2\theta + \cos^2\theta}{\sin\theta\cos\theta} = \dfrac{1}{\sin\theta\cos\theta}$

$= \sec\theta\operatorname{cosec}\theta = $ R.H.S.

(k) $\dfrac{1}{\sec\theta - \tan\theta} = \dfrac{1 + \sin\theta}{\cos\theta} = \sec\theta + \tan\theta$

$\Rightarrow$ LHS $= \dfrac{1}{\sec\theta - \dfrac{\sin\theta}{\cos\theta}}$

$= \dfrac{\cos\theta}{1 - \sin\theta} = \dfrac{\cos\theta(1 + \sin\theta)}{(1 - \sin\theta)(1 + \sin\theta)}$

$= \dfrac{\cos\theta(1 + \sin\theta)}{(1 - \sin^2\theta)}$

$= \dfrac{\cos\theta(1 + \sin\theta)}{\cos^2\theta}$

$$= \frac{1+\sin\theta}{\cos\theta} = \text{Middle Term}$$

$$= \frac{1}{\cos\theta} + \frac{\sin\theta}{\cos\theta} = \sec\theta + \tan\theta = \text{RHS}$$

ANSWERS

2. (a) $\left(\dfrac{x}{a}\right)^{2/3} + \left(\dfrac{y}{b}\right)^{2/3} = 1$ (b) $\left(\dfrac{x}{a}\right)^{2/3} - \left(\dfrac{y}{b}\right)^{2/3} = 1$ (c) $mn = 1$

(d) $x^4 - y^4 = b^2 - a^2$ **5.** (a) $-\dfrac{20}{29}, -\dfrac{20}{21}$ (b) $\dfrac{1}{\sqrt{2}}, -\dfrac{\sqrt{2}}{1}$

6.8 SIGN OF TRIGONOMETRIC FUNCTIONS

(a) When the angle is $-x$ radians ($x > 0$) :

Take a circle with centre O at the origin and radius equal to 1 unit.

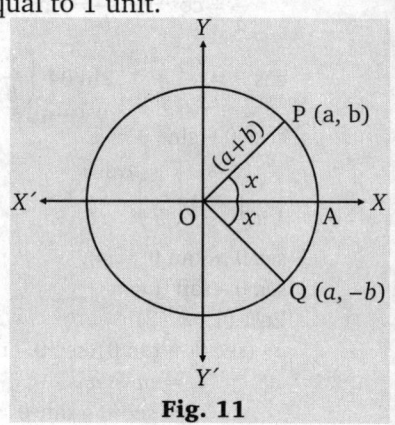

Take $\qquad \angle AOP = x$ radian

$\qquad\qquad a = \cos x, \ b = \sin x$

Let $\qquad \angle AOQ = -x$

$\qquad$ [clockwise rotation from OX]

Co-ordinates of point Q are $(a, -b)$.

$\therefore \qquad$ x-co-ordinates $= \cos(-x)$

$\qquad\qquad a = \cos(-x)$

$\qquad\qquad \cos(-x) = a = \cos x \qquad$ [By (1)]

$\therefore \qquad$ y-coordinate $= \sin(-x)$

$\Rightarrow \qquad\qquad -b = \sin(-x)$

$\Rightarrow \qquad\qquad \sin(-x) = -b = -\sin x$

Fig. 11

Now $\qquad \tan(-x) = \dfrac{\sin(-x)}{\cos(-x)} = \dfrac{-\sin x}{\cos x} = -\tan x$

Hence, $\qquad \sin(-x) = -\sin x, \ \cos(-x) = \cos x, \ \tan(-x) = -\tan x.$

(b) Trigonometric Ratio for $0 < x < \pi/2$ (First quadrant)

The terminal side OP lies in quadrant I and as such both a and b are positive, i.e., > 0

$\qquad \cos x = x\text{-co-ordinate of } P = a > 0$

$\qquad \sin x = y\text{-co-ordinate of } P = b > 0$

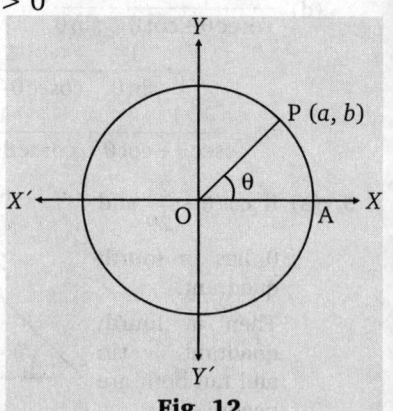

$$\tan x = \frac{\sin x}{\cos x} = \frac{1}{a} > 0$$

Taking the reciprocals, we get

$$\sec x = \frac{1}{a} > 0$$

$$\operatorname{cosec} x = \frac{1}{b} > 0$$

$$\cot x = \frac{a}{b} > 0$$

Hence, for $0 < x < \dfrac{\pi}{2}$ (first quadrant) all six

trigonometric ratios are positive, i.e., > 0

For, $\qquad\qquad\qquad 0 < x < \pi/2$

Fig. 12

$$\sin x > 0, \cos x > 0, \tan x > 0,$$
$$\operatorname{cosec} x > 0, \sec x > 0, \cot x > 0$$

Also for every $P(a, b)$ on the circle of radius 1 unit, we have
$$-1 \le a \le 1 \text{ and } -1 \le b \le 1$$
$$-1 \le \cos x \le 1 \text{ and } -1 \le \sin x \le 1.$$

✎ REMARK

⟾ For any angle x^c,
$$-1 \le \sin x \le 1 \qquad \text{and} \qquad -1 \le \cos x \le 1.$$

(c) Trigonometric Ratios for $\dfrac{\pi}{2} < x < \dfrac{3\pi}{2}$ (Second quadrant)

For $\dfrac{\pi}{2} < x < \pi$, i.e., $90° < x < 180°$, the terminal side CQ of $\angle AOQ$ lies in quadrant if, when co-ordinates of Q are $(-a, b)$.

$\Rightarrow$
$$\cos x = -a < 0$$
$$\sin x = b > 0$$
$$\tan x = \frac{b}{-a} < 0$$
$$\sec x = \frac{1}{a} < 0$$
$$\operatorname{cosec} x = \frac{1}{b} > 0$$
$$\cot x = \frac{-a}{b} < 0$$

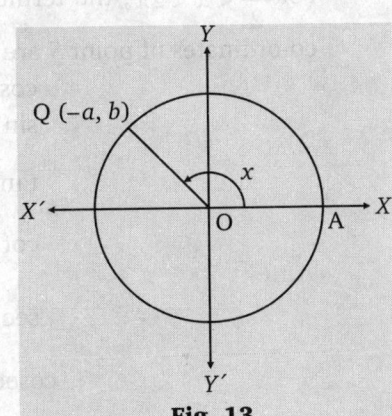

Fig. 13

Hence, for $\dfrac{\pi}{2} < x < \pi$ (quadrant II)

$\sin x$ and $\operatorname{cosec} x$ are positive and remaining four ratios are negative.

For, $\dfrac{\pi}{2} < x < \pi$
$$\sin x > 0, \cos x < 0, \tan x < 0.$$
$$\operatorname{cosec} x > 0, \sec x < 0, \cot x < 0.$$

(d) Trigonometric Ratios for $\pi < x < \dfrac{3\pi}{2}$ (Third quadrant)

For $\pi < x < \dfrac{3\pi}{2}$, the terminal side OR of the angle lies in quadrant III where the co-ordinates of point R are $(-a, -b)$.

$\Rightarrow$
$$\cos x = -a < 0$$
$$\sin x = -b < 0$$
$$\tan x = \frac{\sin x}{\cos x} = \frac{-b}{-a} = \frac{b}{a} > 0$$
$$\operatorname{cosec} x = \frac{1}{-b} < 0$$

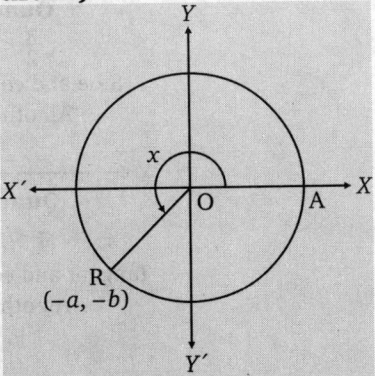

Fig. 14

$$\sec x = \frac{1}{-a} < 0$$

$$\cot x = \frac{a}{b} > 0$$

$\Rightarrow$ only $\tan x$ and $\cot x$ are positive and the remaining four ratios are negative in sign.

For $\pi < x < \dfrac{3\pi}{2}$

$$\sin x < 0; \cos x < 0; \tan x > 0$$
$$\operatorname{cosec} x < 0; \sec x < 0; \cot x > 0.$$

(e) Trigonometric Ratios for $\dfrac{3\pi}{2} < x < 2\pi$ (Fourth quadrant)

For $\dfrac{3\pi}{2} < x < 2\pi$, the terminal side OS of angle x lies in quadrant IV and as such the co-ordinates of point S are $(a, -b)$

$$\cos x = a > 0$$

$$\sin x = -b < 0$$

$$\tan x = \frac{-b}{a} < 0$$

$$\cot x = \frac{a}{-b} < 0$$

$$\sec x = \frac{1}{a} > 0$$

$$\operatorname{cosec} x = \frac{1}{-b} < 0$$

Only $\cos x$ and $\sec x$ are positive and the remaining four trigonometric ratios are negative.

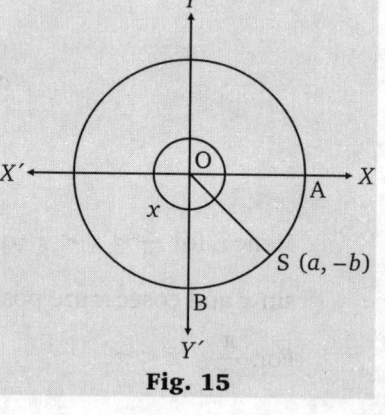

For $\dfrac{3\pi}{2} < x < 2\pi$

$$\sin x < 0; \cos x > 0; \tan x < 0$$
$$\operatorname{cosec} x < 0; \sec x > 0; \cot x < 0.$$

Fig. 15

The above results may by summarized as :

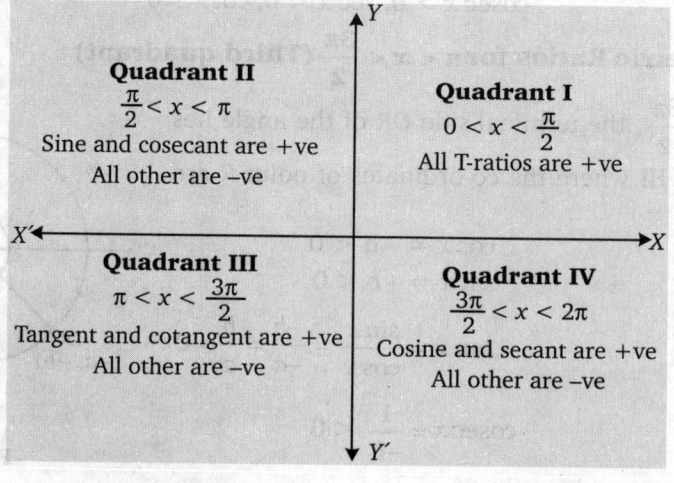

Quadrant II
$\dfrac{\pi}{2} < x < \pi$
Sine and cosecant are +ve
All other are –ve

Quadrant I
$0 < x < \dfrac{\pi}{2}$
All T-ratios are +ve

Quadrant III
$\pi < x < \dfrac{3\pi}{2}$
Tangent and cotangent are +ve
All other are –ve

Quadrant IV
$\dfrac{3\pi}{2} < x < 2\pi$
Cosine and secant are +ve
All other are –ve

We have already shown that one complete revolution followed by a turn of x radians brings the terminal line in the same position OP as in the case with only a turn of x radians.

$\Rightarrow$ For both angle x and angle $2\pi + x$, the co-ordinates of point P are (a, b).

$\therefore$
$$\cos x = a;\ \cos(2\pi + x) = a$$
$$\sin x = b;\ \sin(2\pi + x) = b$$

This may be followed up by any number of revolutions to give $\cos(4\pi + x) = a$. $\sin(4\pi + x) = b$ ...

Hence,
$$\cos(2n\pi + x) = \cos x = a, n \in Z$$
$$\sin(2n\pi + x) = \sin x = b, n \in Z$$
$$\sec(2n\pi + x) = \sec x,\ n \in Z$$
$$\operatorname{cosec}(2n\pi + x) = \operatorname{cosec} x, n \in Z$$

However in case of $\tan x$ and $\cot x$
$$\tan(n\pi + x) = \tan x, n \in Z$$
$$\cot(n\pi + x) = \cot x, n \in Z$$

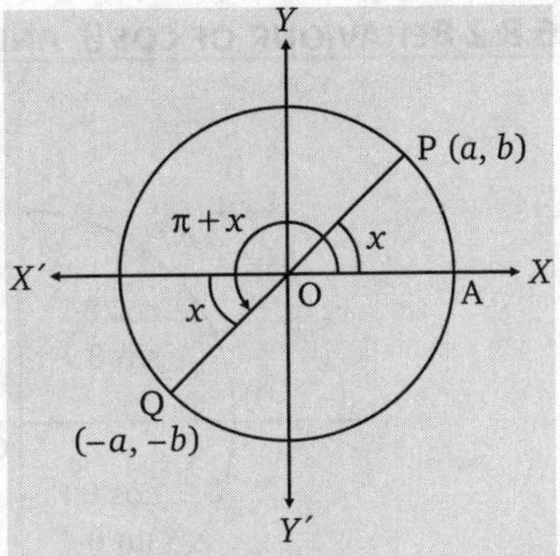

As shown below :
$$\cos x = a$$
$$\sin x = b$$
$$\tan x = \frac{b}{a}$$

Fig. 16

$\Rightarrow$
$$\cot x = \frac{a}{b}\ \text{so,}\quad \sec x = \frac{1}{a}\ \text{and}\quad \operatorname{cosec} x = \frac{1}{b}$$
$$\cos(\pi + x) = -a \neq \cos x \qquad \text{and}\ \sin(\pi + x) = -b \neq \sin x$$

$\Rightarrow$
$$\sec(\pi + x) = \frac{1}{-a} \neq \operatorname{cosec} x$$
$$\operatorname{cosec}(\pi + x) = \frac{1}{-b} \neq \operatorname{cosec} x$$
$$\tan x = \frac{-b}{-a} = \frac{b}{a} = \tan x$$
$$\cot x = \frac{-a}{-b} = \frac{a}{b} = \cot x$$

$\therefore$ The value of $\tan x$ and $\cot x$ repeated a rotation of an angle π.

Hence,
$$\tan(n\pi + x) = \tan x, n \in Z$$
$$\cot(n\pi + x) = \cot x, n \in Z$$

REMARKS

➡ In trigonometric ratios of sine, cosine, secant, cosecant we may add or subtracts a multiple of 2π to the angle without changing the value of the T-ratios.

➡ In trigonometric ratios of tangent and cotangent, we may add or subtract a multiple of π and this will not change the value of the T-ratios.

For Example.

(1) $\sin(765°) = \sin(765° - 2 \times 360°) = \sin 45° = \dfrac{1}{\sqrt{2}}$ [From tables]

(2) $\cos\left(-\dfrac{23\pi}{6}\right) = \cos\left(-\dfrac{23\pi}{6} + 2(2\pi)\right) = \cos\left(\dfrac{\pi}{6}\right) = \cos 30° = \dfrac{\sqrt{3}}{2}$

(3) $\tan (420°) = \tan [420° - 2(180°)] = \tan 60° = \sqrt{3}$

(4) $\cot \left(-\dfrac{11\pi}{4} \right) = \cot \left(-\dfrac{22\pi}{4} + 3\pi \right) = \cot \left(\dfrac{\pi}{4} \right) = 45° = 1.$

6.8.1 SOME IMPORTANT RESULTS

1. $\cos (2n\pi + \theta) = \cos \theta$ and $\sin(2n\pi + \theta) = \sin \theta$, where n is any integer.

2. $\cos (-\theta) = \cos \theta$ and $\sin (-\theta) = -\sin \theta$, for all values of θ.

6.8.2 BEHAVIOUR OF COS θ AND SIN θ AS θ VARIES FROM 0 TO 2π

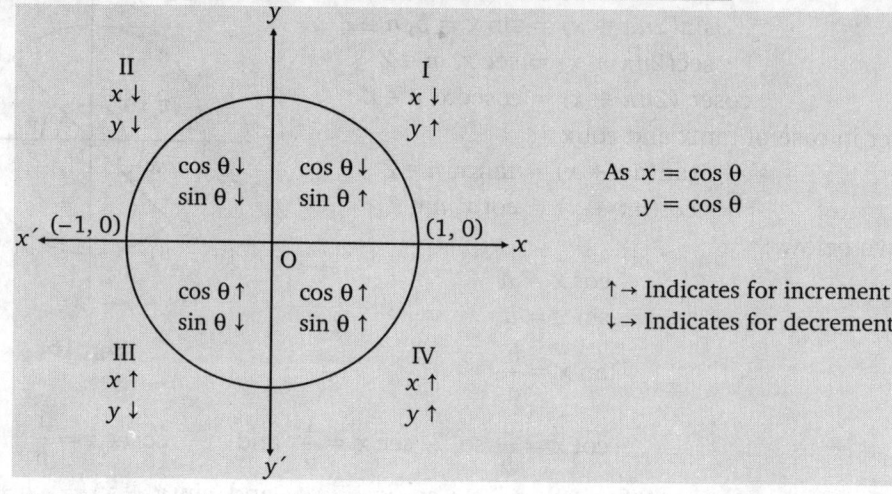

3. $\cos \theta = x$ and $\cos (-\theta) = x \cos (-\theta) = \cos \theta$ for all real numbers x :

(a) $\cos \left(\dfrac{\pi}{2} - x \right) = \sin x$ 　　　　　(b) $\cos \left(\dfrac{\pi}{2} + x \right) = -\sin x$

(c) $\sin \left(\dfrac{\pi}{2} - x \right) = \cos x$ 　　　　　(d) $\sin \left(\dfrac{\pi}{2} + x \right) = \cos x$

4. For all real number x :

(a) $\cos(\pi - x) = -\cos x$ 　　　　　(b) $\cos (\pi + x) = -\cos x$

(c) $\sin(\pi - x) = \sin x$ 　　　　　(d) $\sin (\pi + x) = -\sin x$

5. For all real number x :

(a) $\cos(x + 2\pi) = \cos x$ 　　　　　(b) $\sin(x + 2\pi) = \sin x$

6. For all real number x and y :

(a) $\sin (x + y) = \sin x \cos y + \cos x \sin y$ 　(b) $\sin (x - y) = \sin x \cos y - \cos x \sin y$

7. For all real number x and y :

(a) $\cos (x + y) = \cos x \cos y - \sin x \sin y$

(b) $\cos (x - y) = \cos x \cos y + \sin x \sin y$

8. For all real x :

(a) $\sin 2x = 2 \sin x \cos x$

(b) $\cos 2x = \cos^2 x - \sin^2 x = 2 \cos^2 x - 1 = 1 - 2 \sin^2 x.$

9. For all real x :

(a) $\sin 3x = 3 \sin x - 4 \sin^3 x$ 　　　　　(b) $\cos 3x = 4 \cos^3 x - 3 \cos x$

10. For all real number x and y :

(a) $2 \cos x \cos y = \cos(x + y) + \cos(x - y)$

(b) $2 \sin x \cos y = \sin(x + y) + \sin(x - y)$

(c) $2 \sin x \sin y = \cos(x - y) - \cos(x + y)$

(d) $2 \cos x \sin y = \sin(x + y) - \sin(x - y)$

Subtracting 6(a) and 6(b) we get

$2 \cos x \sin y = \sin(x + y) - \sin(x - y)$ this proves the result (d).

11. For all $x, y \in \mathbf{R}$

(a) $\sin x + \sin y = 2 \sin\left(\dfrac{x + y}{2}\right) \cos\left(\dfrac{x - y}{2}\right)$

(b) $\sin x - \sin y = 2 \cos\left(\dfrac{x + y}{2}\right) \sin\left(\dfrac{x - y}{2}\right)$

(c) $\cos x + \cos y = 2 \cos\left(\dfrac{x + y}{2}\right) \cos\left(\dfrac{x - y}{2}\right)$

(d) $\cos x - \cos y = 2 \sin\left(\dfrac{x + y}{2}\right) \sin\left(\dfrac{y - x}{2}\right) = -2 \sin\left(\dfrac{x + y}{2}\right) \sin\left(\dfrac{x - y}{2}\right)$

12. Prove for all real number x and y

(a) $\sin(x + y) \sin(x - y) = \sin^2 x - \sin^2 y$ **(UPTU(B. Pharma)–2001)**

(b) $\cos(x + y) \cos(x - y) = \cos^2 x - \sin^2 y$

Proof.

(a) We have

$\sin(x + y) \sin(x - y) = [\sin x \cos y + \cos x \sin y][\sin x \cos y - \cos x \sin y]$

$= \sin^2 x \cos^2 y - \cos^2 x \sin^2 y = \sin^2 x (1 - \sin^2 y) - (1 - \sin^2 x)\sin^2 y$

$= \sin^2 x - \sin^2 y$

(b) We have

$\cos(x + y) \cos(x - y) = [\cos x \cos y - \sin x \sin y][\cos x \cos y + \sin x \sin y]$

$= \cos^2 x \cos^2 y - \sin^2 x \sin^2 y = \cos^2 x (1 - \sin^2 y) - (1 - \cos^2 x)\sin^2 y$

$= \cos^2 x - \sin^2 y$

13. For all real number x prove that $\tan(-x) = -\tan x$.

Proof. We have $\tan(-x) = \dfrac{\sin(-x)}{\cos(-x)} = \dfrac{-\sin x}{\cos x} = -\tan x$.

14. For all $x \in R$, prove that :

(a) $\tan\left(\dfrac{\pi}{2} - x\right) = \cot x$ (b) $\tan\left(\dfrac{\pi}{2} + x\right) = -\cot x$

(c) $\tan(\pi - x) = -\tan x$ (d) $\tan(\pi + x) = \tan x$.

Proof.

(a) We have $\tan\left(\dfrac{\pi}{2} - x\right) = \dfrac{\sin\left(\dfrac{\pi}{2} - x\right)}{\cos\left(\dfrac{\pi}{2} - x\right)} = \dfrac{\cos x}{\sin x} = \cot x$

(b) We have $\tan\left(\dfrac{\pi}{2}+x\right) = \dfrac{\sin\left(\dfrac{\pi}{2}+x\right)}{\cos\left(\dfrac{\pi}{2}+x\right)} = \dfrac{\cos x}{-\sin x} = -\cot x.$

(c) We have $\tan(\pi - x) = \dfrac{\sin(\pi - x)}{\cos(\pi - x)} = \dfrac{\sin x}{-\cos x} = -\tan x$

(d) We have $\tan(\pi + x) = \dfrac{\sin(\pi + x)}{\cos(\pi + x)} = \dfrac{-\sin x}{-\cos x} = \tan x$

Table 2 gives the sine, cosine, tangent of some angles less than 90°

Table 2

θ	0	15°	18°	22.5°	36°	67.5°
sin	0	$\dfrac{\sqrt{6}-\sqrt{2}}{4}$	$\dfrac{\sqrt{5}-\sqrt{1}}{4}$	$\dfrac{\sqrt{2-\sqrt{2}}}{4}$	$\dfrac{\sqrt{10-2\sqrt{5}}}{4}$	$\dfrac{\sqrt{\sqrt{2}+1}}{\sqrt{2\sqrt{2}}}$
cos	1	$\dfrac{\sqrt{6}+\sqrt{2}}{4}$	$\dfrac{\sqrt{10+2\sqrt{5}}}{4}$	$\dfrac{\sqrt{\sqrt{2}+1}}{\sqrt{2\sqrt{2}}}$	$\dfrac{\sqrt{5}+1}{4}$	$\dfrac{\sqrt{2-\sqrt{2}}}{2}$
tan	0	$2-\sqrt{3}$	$\dfrac{\sqrt{25-10\sqrt{5}}}{5}$	$\sqrt{2}-1$	$\sqrt{5-2\sqrt{5}}$	$\sqrt{2}+1$

SOLVED EXAMPLES

EXAMPLE 1. *Given that* $\sin(A + B) = \dfrac{\sqrt{3}}{2}$ *and* $\cos(A - B) = \dfrac{\sqrt{3}}{2}$, *find A and B, where A and B are positive acute angles.*

SOLUTION. Here, the given equations are

$$\sin(A + B) = \frac{\sqrt{3}}{2} \qquad\qquad\qquad ...(1)$$

and $\qquad\cos(A - B) = \dfrac{\sqrt{3}}{2} \qquad\qquad\qquad ...(2)$

From (1), we have
$$A + B = 60° \qquad\qquad\qquad ...(3)$$
Also, from (2), we get
$$A - B = 30° \qquad\qquad\qquad ...(4)$$
Adding (3) and (4), we get
$$2A = 90° \quad \Rightarrow \quad A = 45°$$
Put this value in (4), finally, we get $B = 15°$.

EXAMPLE 2. *Show that* $\dfrac{\cos\theta}{\sin(90+\theta)} + \dfrac{\sin(-\theta)}{\sin(180+\theta)} - \dfrac{\tan(90+\theta)}{\cot\theta} = 3.$

SOLUTION. The L.H.S. of the given equation is

$$\frac{\cos\theta}{\sin(90+\theta)} + \frac{\sin(-\theta)}{\sin(180+\theta)} - \frac{\tan(90+\theta)}{\cot\theta} = \frac{\cos\theta}{\cos\theta} + \frac{-\sin}{-\sin\theta} - \frac{-\cot\theta}{\cot\theta}$$

$$= 1 + 1 + 1 = 3 = \text{R.H.S.}$$

EXAMPLE 3. *Show that* $\cos 24° + \cos 55° + \cos 125° + \cos 204° + \cos 300° = \dfrac{1}{2}$.

SOLUTION. L.H.S. $= \cos 24° + \cos 55° + \cos 125° + \cos 204° + \cos 300°$

$= \cos 24° + \cos 55° + \cos(180° - 55°) + \cos(180° + 24°)$
$\qquad + \cos(360° - 60°)$

$= \cos 24° + \cos 55° - \cos 55° - \cos 24° + \cos 60° = \cos 60° = \dfrac{1}{2} = $ R.H.S.

EXAMPLE 4. *Calculate :*

    (*i*) $\cos 15°$ **[UPTU (B. Pharma)–2004]**

    (*ii*) $\cos 75°$

SOLUTION.   (*i*) We know that

$\cos 15° = \cos(45° - 30°) = \cos 45° \cos 30° + \sin 45° \sin 30°$

$= \dfrac{1}{\sqrt{2}} \times \dfrac{\sqrt{3}}{2} + \dfrac{1}{\sqrt{2}} \times \dfrac{1}{2} = \dfrac{\sqrt{2}+1}{2\sqrt{2}}$

  (*ii*) We have

$\cos 75° = \cos(45° + 30°) = \cos 45° \cos 30° - \sin 45° \sin 30°$

$= \dfrac{1}{\sqrt{2}} \times \dfrac{\sqrt{3}}{2} - \dfrac{1}{\sqrt{2}} \times \dfrac{1}{2} = \dfrac{\sqrt{3}-1}{2\sqrt{2}}$

EXAMPLE 5. *Find the value of* $\cot 75°$.   **[UPTU(B. Pharma)–2002]**

SOLUTION. $\cot 75° = \cot(45° + 30°) = \dfrac{\cot 45° \cot 30° - 1}{\cot 45° + \cot 30°} = \dfrac{1 \cdot \sqrt{3} - 1}{1 + \sqrt{3}} = \dfrac{\sqrt{3}-1}{\sqrt{3}+1}$

EXAMPLE 6. *Show that* $\sin \dfrac{5\pi}{18} - \cos \dfrac{4\pi}{9} = \sqrt{3} \sin \dfrac{\pi}{9}$.   **[UPTU(B. Pharma)–2003, 10]**

SOLUTION. We have

$\sin \dfrac{5\pi}{18} - \cos \dfrac{4\pi}{9} = \sin \dfrac{5\pi}{18} - \sin\left(\dfrac{\pi}{2} - \dfrac{4\pi}{9}\right)$

$= \sin \dfrac{5\pi}{18} - \sin \dfrac{\pi}{18} = 2\cos\left(\dfrac{\dfrac{5\pi}{18} + \dfrac{\pi}{18}}{2}\right) \sin\left(\dfrac{\dfrac{5\pi}{18} - \dfrac{\pi}{18}}{2}\right)$

$= 2\cos \dfrac{\pi}{6} \sin \dfrac{\pi}{9} = \left(2 \times \dfrac{\sqrt{3}}{2}\right) \sin \dfrac{\pi}{9} = \sqrt{3} \sin \dfrac{\pi}{9}$

EXAMPLE 7. *Find the value of* $\sec(-1500°) \times \sin 390°$.

SOLUTION. $\sec(-1500°) \times \sin 390° = \sec(-4 \times 360° - 60°) \times \sin(360° + 30°)$

$= \sec(-60°) \times \sin 30° = \sec 60° \times \sin 30° = 2 \times \dfrac{1}{2} = 1$

EXAMPLE 8. *Prove that* $2\cos \dfrac{\pi}{13} \cos \dfrac{9\pi}{13} + \cos \dfrac{3\pi}{13} + \cos \dfrac{5\pi}{13} = 0$.   **[UPTU(B. Pharma)–2007, 13]**

SOLUTION. We have $2\cos \dfrac{\pi}{13} \cos \dfrac{9\pi}{13} + \cos \dfrac{3\pi}{13} + \cos \dfrac{5\pi}{13}$

$= \cos\left(\dfrac{\pi}{13} + \dfrac{9\pi}{13}\right) + \left(\dfrac{\pi}{13} - \dfrac{9\pi}{13}\right) + \cos \dfrac{3\pi}{13} + \cos \dfrac{5\pi}{13}$

$$= \cos\frac{10\pi}{13} + \cos\frac{8\pi}{13} + \cos\frac{3\pi}{13} + \cos\frac{5\pi}{13}$$

$$= \cos\left(\pi - \frac{3\pi}{13}\right) + \cos\frac{8\pi}{13} + \cos\frac{3\pi}{13} + \cos\left(\pi - \frac{8\pi}{13}\right)$$

$$= -\cos\frac{3\pi}{13} + \cos\frac{8\pi}{13} + \cos\frac{3\pi}{13} - \cos\frac{8\pi}{13} = 0$$

EXAMPLE 9. *If* $\tan A + \tan B = a$ *and* $\cot A + \cot B = b$, *prove that*

$$\cot(A + B) = \frac{1}{a} - \frac{1}{b}.$$

SOLUTION. We have $\cot(A + B) = \dfrac{1}{\tan(A + B)}$

$$= \frac{1 - \tan A \tan B}{\tan A + \tan B} = \frac{1}{\tan A + \tan B} - \frac{\tan A \tan B}{\tan A + \tan B}$$

$$= \frac{1}{\tan A + \tan B} - \frac{1}{\dfrac{\tan A + \tan B}{\tan A \tan B}}$$

$$= \frac{1}{\tan A + \tan B} - \frac{1}{\dfrac{\tan A}{\tan A \tan B} + \dfrac{\tan B}{\tan A \tan B}}$$

$$= \frac{1}{\tan A + \tan B} - \frac{1}{\dfrac{1}{\tan B} + \dfrac{1}{\tan A}}$$

$$= \frac{1}{\tan A + \tan B} - \frac{1}{\cot A + \cot B} = \frac{1}{a} - \frac{1}{b}$$

EXAMPLE 10. *Obtain the value of* :

(*i*) $\sin 15°$ [UPTU(B. Pharma)–2004]

(*ii*) $\tan 15°$ [UPTU(B. Pharma)–2006]

SOLUTION. (*i*) We have $\sin 15° = \sin(45° - 30°) = \sin 45° \cos 30° - \cos 45° \sin 30°$

$$= \frac{1}{\sqrt{2}}\frac{\sqrt{3}}{2} - \frac{1}{\sqrt{2}} \cdot \frac{1}{2} = \frac{\sqrt{3} - 1}{2\sqrt{2}}$$

(*ii*) We have $\tan 15° = \tan(45° - 30°) = \dfrac{\tan 45° - \tan 30°}{1 + \tan 45° \tan 30°}$

$$= \frac{1 - \dfrac{1}{\sqrt{3}}}{1 + \dfrac{1}{\sqrt{3}}} = \frac{\sqrt{3} - 1}{\sqrt{3} + 1}$$

EXAMPLE 11. *Prove that* $\dfrac{\cos 8° + \sin 8°}{\cos 8° - \sin 8°} = \cot 37°$. [UPTU(B. Pharma)–2002]

SOLUTION. We have

$$\frac{\cos 8° + \sin 8°}{\cos 8° - \sin 8°} = \frac{1 + \tan 8°}{1 - \tan 8°}$$ [Dividing Nr and Dr by $\cos 8°$]

$$= \frac{\tan 45° + \tan 8°}{1 - \tan 45° \tan 8°} = \tan (45° + 8°)$$

$$= \tan 53° = \tan (90° - 37°) = \cot 37°$$

EXAMPLE 12. *If* $\sin A = \frac{4}{5}$ *and* $\cos B = \frac{5}{13}$. *Find the value of* $\sin (A - B)$ *and* $\cos (A + B)$ *where*

$0 < A < \pi/2$ *and* $0 < B < \pi/2$. **[UPTU(B. Pharma)–2004]**

SOLUTION. We have $\qquad \sin A = \frac{4}{5}$ and $0 < A < \frac{\pi}{2}$

$$\cos A = \sqrt{1 - \sin^2 A} = \sqrt{1 - \left(\frac{4}{5}\right)^2} = \sqrt{1 - \frac{16}{25}} = \sqrt{\frac{9}{25}} = \frac{3}{5}$$

Again $\qquad \cos B = \frac{5}{13}$ and $0 < B < \frac{\pi}{2}$

$$\sin B = \sqrt{1 - \cos^2 B} = \sqrt{1 - \left(\frac{5}{13}\right)^2} = \sqrt{\frac{144}{169}} = \frac{12}{13}.$$

$$\sin (A - B) = \sin A \cos B - \cos A \sin B = \left(\frac{4}{5} \times \frac{5}{13}\right) - \left(\frac{3}{5} \times \frac{12}{13}\right)$$

$$= \frac{20}{65} - \frac{36}{65} = \frac{-16}{65}$$

Also $\qquad \cos (A + B) = \cos A \cos B - \sin A \sin B = \left(\frac{3}{5} \times \frac{5}{13}\right) - \left(\frac{4}{5} \times \frac{12}{13}\right)$

$$= \frac{15}{65} - \frac{48}{65} = \frac{-33}{65}$$

EXAMPLE 13. *Prove the following identity :*

$$\sin^6\theta + \cos^6 \theta = 1 - 3 \sin^2 \theta \cos^2 \theta \qquad \textbf{(UPTU B. Pharma)–2007}$$

SOLUTION. We have $\qquad$ LHS $= \sin^6 \theta + \cos^6\theta = (\sin^2\theta)^3 + (\cos^2\theta)^3$

$$= (\sin^2\theta + \cos^2\theta)[(\sin^4\theta + \cos^4\theta - \sin^2\theta \cos^2\theta]$$

$$= \sin^4\theta + \cos^4\theta - \sin^2\theta \cos^2\theta$$

$$= [(\sin^2\theta + \cos^2\theta)^2 - 2 \sin^2\theta \cos^2 \theta] - \sin^2\theta \cos^2\theta$$

$$= 1 - 3 \sin^2\theta \cos^2\theta = \text{RHS}$$

EXAMPLE 14. *If* $\cos \theta + \sin \theta = \sqrt{2} \cos \theta$, *show that* $\cos \theta - \sin \theta = \sqrt{2} \sin \theta$.

(UPTU(B. Pharma)–2005, RGPV(B. Pharma)–2006, 09)

SOLUTION. We have $\cos \theta + \sin \theta = \sqrt{2} \cos \theta$

$\Rightarrow \qquad \sin \theta = (\sqrt{2} - 1) \cos \theta$

$\Rightarrow \qquad (\sqrt{2} + 1) \sin \theta = (\sqrt{2} + 1)(\sqrt{2} - 1) \cos \theta$

$\Rightarrow \qquad (\sqrt{2} + 1) \sin \theta = (2 - 1) \cos \theta$

$\Rightarrow \qquad \sqrt{2} \sin\theta + \sin\theta = \cos\theta$

$\Rightarrow \qquad \cos\theta - \sin\theta = \sqrt{2} \sin\theta$

EXAMPLE 15. *Show that* $\cos 15° - \sin 15° = \frac{1}{\sqrt{2}}$.

SOLUTION. $\qquad$ L.H.S. $= \cos 15° - \sin 15° = \sqrt{2}\left(\frac{1}{\sqrt{2}}\cos 15° - \frac{1}{\sqrt{2}}\sin 15°\right)$

$$= \sqrt{2} \ (\cos 45° \cos 15° - \sin 45° \sin 15°)$$

$$= \sqrt{2} \ \cos(45° + 15°) = \sqrt{2} \cos 60° = \sqrt{2} \times \frac{1}{2} = \frac{1}{\sqrt{2}} = \text{R.H.S.}$$

EXAMPLE 16. *Show that* $\tan 13A - \tan 9A - \tan 4A = \tan 13A \cdot \tan 4A \cdot \tan 9A.$

[UPTU(B. Pharma)–2002, 11]

SOLUTION.   We can write
$$13A = 9A + 4A.$$
Therefore, $\tan 13A = \tan (9A + 4A)$

$$\Rightarrow \qquad \frac{\tan 13A}{1} = \frac{\tan 9A + \tan 4A}{1 - \tan 9A \cdot \tan 4A}$$

$$\Rightarrow \qquad \tan 13A - \tan 13A \cdot \tan 9A \cdot \tan 4A = \tan 9A + \tan 4A.$$

$$\Rightarrow \qquad \tan 13A - \tan 9A - \tan 4A = \tan 13A \cdot \tan 9A \cdot \tan 4A.$$

EXAMPLE 17. *Show that* $\sin^2 \dfrac{\pi}{8} + \sin^2 \dfrac{3\pi}{8} + \sin^2 \dfrac{5\pi}{8} + \sin^2 \dfrac{7\pi}{8} = 2.$

SOLUTION.   Consider,   L.H.S. $= \sin^2 \dfrac{\pi}{8} + \sin^2 \dfrac{3\pi}{8} + \sin^2 \dfrac{5\pi}{8} + \sin^2 \dfrac{7\pi}{8}$

$$= \sin^2 \frac{\pi}{8} + \sin^2 \frac{3\pi}{8} + \sin^2 \left(\pi - \frac{3\pi}{8} \right) + \sin^2 \left(\pi - \frac{\pi}{8} \right)$$

$$= \sin^2 \frac{\pi}{8} + \sin^2 \frac{3\pi}{8} + \sin^2 \frac{3\pi}{8} + \sin^2 \frac{\pi}{8}$$

$$= 2\left(\sin^2 \frac{\pi}{8} + \sin^2 \frac{3\pi}{8} \right) = 2\left[\sin^2 \frac{\pi}{8} + \sin^2 \left(\frac{\pi}{2} - \frac{\pi}{8} \right) \right]$$

$$= 2\left[\sin^2 \frac{\pi}{8} + \cos^2 \frac{\pi}{8} \right] = 2(1) = 2.$$

EXAMPLE 18. *Show that* $\cos^2 \dfrac{\pi}{8} + \cos^2 \dfrac{3\pi}{8} + \cos^2 \dfrac{5\pi}{8} + \cos^2 \dfrac{7\pi}{8} = 2.$

SOLUTION.   Consider   L.H.S $= \cos^2 \dfrac{\pi}{8} + \cos^2 \dfrac{3\pi}{8} + \cos^2 \dfrac{5\pi}{8} + \cos^2 \dfrac{7\pi}{8}$

$$= \cos^2 \frac{\pi}{8} + \cos^2 \frac{3\pi}{8} + \cos^2 \left(\pi - \frac{3\pi}{8} \right) + \cos^2 \left(\pi - \frac{\pi}{8} \right)$$

$$= \cos^2 \frac{\pi}{8} + \cos^2 \frac{3\pi}{8} + \cos^2 \frac{3\pi}{8} + \cos^2 \frac{7\pi}{8}$$

$$= 2\left(\cos^2 \frac{\pi}{8} + \cos^2 \frac{3\pi}{8} \right) = 2\left[\cos^2 \frac{\pi}{8} + \cos^2 \left(\frac{\pi}{2} - \frac{\pi}{8} \right) \right]$$

$$= 2\left[\cos^2 \frac{\pi}{8} + \sin^2 \frac{\pi}{8} \right] = 2.$$

EXAMPLE 19. *Prove that* $\sin (A + B) \sin (A - B) = \sin^2 A - \sin^2 B = \cos^2 B - \cos^2 A.$

(RGPV(B. Pharma)–2004)

SOLUTION. $\sin (A + B) \sin (A - B) = (\sin A \cos B + \cos A \sin B)(\sin A \cos B - \cos A \sin B)$

$$= \sin^2 A \cos^2 B - \cos^2 A \sin^2 B$$

$$= \sin^2 A (1 - \sin^2 B) - (1 - \sin^2 A) \sin^2 B$$

$$= \sin^2 A - \sin^2 A \sin^2 B - \sin^2 B + \sin^2 A \sin^2 B$$
$$= \sin^2 A - \sin^2 B$$
$$= (1 - \cos^2 A) - (1 - \cos^2 B) = \cos^2 B - \cos^2 A.$$

EXAMPLE 20. *Solve* $\sin 75° + \cos 75° + \sin 15° + \cos 15°$ **[RGPV(B. Pharma)–2001]**

SOLUTION. $\sin 75° + \cos 75° + \sin 15° + \cos 15°$

$$= \sin(30° + 45°) + \cos(30° + 45°) + \sin(45° - 30°) + \cos(45° - 30°)$$
$$= (\sin 30° \cos 45° + \cos 30° \sin 45°) + (\cos 30° \cos 45° - \sin 30° \sin 45°)$$
$$+ (\sin 45° \cos 30° - \sin 30° \cos 45°) + (\cos 30° \cos 45° - \sin 30° \sin 45°)$$

$$= \frac{1}{2} \times \frac{1}{\sqrt{2}} + \frac{\sqrt{3}}{2} \times \frac{1}{\sqrt{2}} + \frac{\sqrt{3}}{2} \times \frac{1}{2} - \frac{1}{2} \times \frac{1}{\sqrt{2}}$$

$$+ \frac{1}{2} \times \frac{\sqrt{3}}{2} - \frac{1}{2} \times \frac{1}{\sqrt{2}} + \frac{1}{\sqrt{2}} \times \frac{\sqrt{3}}{2} + \frac{1}{\sqrt{2}} \times \frac{1}{2}$$

$$= \frac{1}{2\sqrt{2}} + \frac{\sqrt{3}}{2\sqrt{2}} + \frac{\sqrt{3}}{2\sqrt{2}} - \frac{1}{2\sqrt{2}} + \frac{\sqrt{3}}{2\sqrt{2}} - \frac{1}{2\sqrt{2}} + \frac{\sqrt{3}}{2\sqrt{2}} + \frac{1}{2\sqrt{2}} = \frac{2\sqrt{3}}{\sqrt{2}}$$

EXAMPLE 21. *Prove that* $\sin 105° + \cos 105° = \cos 45°$ **[RGPV(B. Pharma)–2003]**

SOLUTION.   L.H.S $= \sin 105° + \cos 105° = \sin(60° + 45°) + \cos(60° + 45°)$

$$= \sin 60° \cos 45° + \cos 60° \sin 45° + \cos 60° \cos 45° - \sin 60° \sin 45°$$

$$= \frac{\sqrt{3}}{2} \times \frac{1}{\sqrt{2}} + \frac{1}{2} \times \frac{1}{\sqrt{2}} + \frac{1}{2} \times \frac{1}{\sqrt{2}} - \frac{\sqrt{3}}{2} \times \frac{1}{\sqrt{2}} = \frac{2}{2\sqrt{2}} = \frac{1}{\sqrt{2}} = \cos 45° = \text{L.H.S.}$$

EXAMPLE 22. *Show that* $\tan 50° = \tan 40° + 2 \tan 10°$.

SOLUTION.   Since, $50° = 40° + 10°$.

Therefore, $\tan 50° = \tan(40° + 10°)$

$$\Rightarrow \qquad \frac{\tan 50°}{1} = \frac{\tan 40° + \tan 10°}{1 - \tan 40° \tan 10°}$$

$$\Rightarrow \qquad \tan 50° - \tan 50° \tan 40° \tan 10° = \tan 40° + \tan 10°$$

$$\Rightarrow \qquad \tan 50° - \tan(90° - 40°) \tan 40° \tan 10° = \tan 40° + \tan 10°$$

$$\Rightarrow \qquad \tan 50° - \frac{1}{\tan 40°} \tan 40° \tan 10° = \tan 40° + \tan 10°$$

$$\Rightarrow \qquad \tan 50° - \tan 10° = \tan 40° + \tan 10°$$

$$\Rightarrow \qquad \tan 50° = \tan 40° + 2\tan 10°$$

EXAMPLE 23. *Using* $\tan(\theta - \phi) = \dfrac{\tan \theta - \tan \phi}{1 + \tan \theta \tan \phi}$, *evaluate* $\tan \dfrac{13\pi}{12}$.

SOLUTION.   Consider $\tan \dfrac{13\pi}{12} = \tan\left(\pi + \dfrac{\pi}{12}\right) = \tan \dfrac{\pi}{12} = \tan 15°$

$$= \tan(45° - 30°) = \frac{\tan 45° - \tan 30°}{1 + \tan 45° \tan 30°} = \frac{1 - \dfrac{1}{\sqrt{3}}}{1 + \dfrac{1}{\sqrt{3}}}$$

$$= \frac{\sqrt{3} - 1}{\sqrt{3} + 1} = \frac{\sqrt{3} - 1}{\sqrt{3} + 1} \times \frac{\sqrt{3} - 1}{\sqrt{3} - 1}$$

$$= \frac{3 + 1 - 2\sqrt{3}}{3 - 1} = \frac{4 - 2\sqrt{3}}{2} = 2 - \sqrt{3}.$$

EXAMPLE 24. *Show that* $\dfrac{\sin(A-B)}{\sin A \sin B} + \dfrac{\sin(B-C)}{\sin B \sin C} + \dfrac{\sin(C-A)}{\sin C \sin A} = 0.$

SOLUTION. Consider

$$\text{LHS} = \frac{\sin(A-B)}{\sin A \sin B} + \frac{\sin(B-C)}{\sin B \sin C} + \frac{\sin(C-A)}{\sin C \sin A}$$

$$= \frac{\sin A \cos B - \cos A \sin B}{\sin A \sin B} + \frac{\sin B \cos C - \cos B \sin C}{\sin B \sin C} + \frac{\sin C \cos A - \cos C \sin A)}{\sin C \sin A}$$

$$= \frac{\sin A \cos B}{\sin A \sin B} - \frac{\cos A \sin B}{\sin A \sin B} + \frac{\sin B \cos C}{\sin B \sin C} - \frac{\cos B \sin C}{\sin B \sin C} + \frac{\sin C \cos A}{\sin C \sin A} - \frac{\cos C \sin A}{\sin C \sin A}$$

$$= \cot B - \cot A + \cot C - \cot B + \cot A - \cot C = 0 = \text{R.H.S.}$$

EXAMPLE 25. *If* $A + B = \pi/4$, *prove that* $(1 + \tan A)\cdot(1 + \tan B) = 2.$ **[UPTU(B. Pharma)–2005]**

SOLUTION. Given $\qquad\qquad\qquad A + B = \pi/4$

$$\tan(A + B) = \tan(\pi/4) = 1$$

$$\Rightarrow \qquad\qquad \frac{\tan A + \tan B}{1 - \tan A \tan B} = 1$$

$$\Rightarrow \qquad\qquad \tan A + \tan B = 1 - \tan A \tan B$$

$$\Rightarrow \quad \tan A + \tan B + \tan A \tan B = 1$$

Adding (1) to both sides

$$1 + \tan A + \tan B + \tan A \tan B = 2$$

$$(1 + \tan A)(1 + \tan B) = 2$$

EXAMPLE 26. *Prove that* $\dfrac{\cos 8A \cos 5A - \cos 12A \cos 9A}{\sin 8A \cos 5A + \cos 12A . \sin 9A} = \tan 4A$ **[Meerut(B.Sc. Biotech)–2003]**

SOLUTION. Multiplying numerator and denominator by 2, we get

$$\text{L.H.S.} = \frac{2\cos 8A \cos 5A - 2\cos 12A \cos 9A}{2\sin 8A \cos 5A + 2\cos 12A \cdot \sin 9A}$$

$$= \frac{\cos(8A+5A) + \cos(8A-5A) - [\cos(12A+9A) + \cos(12A-9A)]}{[\sin(8A+5A) + \sin(8A-5A)] + [\sin 12A+9A) - \sin(12A-9A)]}$$

$$= \frac{(\cos 13A + \cos 3A) - (\cos 21A + \cos 3A)}{(\sin 13A + \sin 3A) - (\sin 21A - \sin 3A)} = \frac{\cos 13A - \cos 21A}{\sin 13A + \sin 21A}$$

$$= \frac{2\sin\left(\dfrac{13A+21A}{2}\right) \cdot \sin\left(\dfrac{21A-13A}{2}\right)}{2\sin\left(\dfrac{13A+21A}{2}\right) \cdot \cos\left(\dfrac{21A-13A}{2}\right)}$$

$$= \frac{\sin 17A \sin 4A}{\sin 17A \cos 4A} = \tan 4A = \text{R.H.S.}$$

Exercise 6.3

1. Given angle C of a triangle ABC to be obtuse, find all angles when :

$$\sin(A + B) = \frac{\sqrt{3}}{2} \text{ and } \cos(A - B) = \frac{1}{\sqrt{2}}$$

2. Show that

(i) $\dfrac{\cos(90° + \theta)\sec(-\theta)\tan(180° - \theta)}{\sec(360° - \theta)\sin(180° + \theta)\cot(90° - \theta)}$

$$= -1$$

(ii) $\dfrac{\sin(180° - \theta)}{\tan(90° + \theta)} \cdot \dfrac{\cos(360° - \theta)}{\tan(180° + \theta)}$

$\cdot \dfrac{\cot(90° - \theta)}{\sin(-\theta)} = \sin\theta.$

(iii) $\sin 75° - \sin 15° = \cos 105° + \cos 15°.$

(iv) $\sin 105° + \cos 105° = \cos 45°$

(v) $\sin(45° - A)\cos(45° - B) - \cos(45° - A)$

$\qquad = \sin(A + B).$

(vi) $\sin(n + 1)A \sin(n + 2)A + \cos(n + 1)A$

$\qquad \cos(n + 2)A = \cos A.$

(vii) $\cos(45° + \theta)\cos(10° + \theta)$

$\qquad + \sin(40° + \theta)\sin(10° + \theta) = \dfrac{\sqrt{3}}{2}$

(viii) $\cos\theta - \sin\theta = \sqrt{2}\cos\left(\theta + \dfrac{\pi}{4}\right)$

3. Show that $\dfrac{\cos 18° + \sin 18°}{\cos 18° - \sin 18°} = \cot 27°$

4. Show that

(i) $\tan\left(\dfrac{\pi}{4} + \theta\right)\tan\left(\dfrac{\pi}{4} - \theta\right) = 1$

(ii) $\tan 3\theta \tan 2\theta \tan\theta$

$\qquad = \tan 3\theta - \tan 2\theta - \tan\theta$

(iii) $\cos 2\theta \cos 2\phi + \sin^2(\theta - \phi) - \sin^2(\theta + \phi)$

$\qquad = \cos(2\theta + 2\phi).$

5. If A, B, C, D are the angles of a cyclic quadrilateral in order, show that

$$\cos A + \cos B + \cos C + \cos D = 0.$$

6. If $\sin\alpha = \dfrac{15}{17}$ and $\cos\beta = \dfrac{12}{13}$, find the values of $\sin(\alpha + \beta)$, $\cos(\alpha - \beta)$ and $\tan(\alpha + \beta)$.

7. (i) Find all positive values of x less than 2π which satisfy the equation $\cos^2 x = \dfrac{1}{4}$.

(ii) Find all positive values of x less than 2π which satisfy the equation $3\tan^2 x = 1$.

8. Evaluate

$\tan\dfrac{13\pi}{12}$ using $\tan(\theta + \phi) = \dfrac{\tan\theta + \tan\phi}{1 - \tan\theta\tan\phi}$.

9. Show that

(i) $\cos 70° \cos 10° + \sin 70° \sin 10° = \dfrac{1}{2}$

(ii) $\cos 130° \cos 40° + \sin 130° \sin 40° = 0.$

10. If $\cos A = \dfrac{1}{7}$, $\cos B = \dfrac{13}{14}$, (A and B, being positive acute angles), show that $A - B = 60°$.

11. If $\theta + \phi = 45°$, show that $(\cot\theta - 1)(\cot\phi - 1) = 2$.

12. If $\theta - \phi = \dfrac{\pi}{4}$, show that

(i) $(1 + \tan\theta)(1 - \tan\theta) = 2$

(ii) $(1 + \tan\theta)(1 + \tan\theta) = 2\tan\theta.$

13. Show that :

(i) $\dfrac{\cos(360° - \theta)\ \csc(180° + \theta)\cot(90° - \theta)}{\sec(90° + \theta)\cos(-\theta)} = \csc\theta.$

(ii) $\cos(270° - \theta)\sec(-\theta)\tan(180° - \theta)$

$\qquad + \sec(360° + \theta)\sin(180° + \theta)$

$\qquad \cot(90° - \theta) = 0.$

(iii) $\sin 420° \cos 390°$

$\qquad + \cos(-660°)\sin(-390°) = \dfrac{1}{2}$

(iv) $\dfrac{\sin 135° - \cos 120°}{\sin 135° + \cos 120°} = 3 + \sqrt{2}$

14. Show that :

(i) $\cot\left(\dfrac{\pi}{4} + \theta\right)\cot\left(\dfrac{\pi}{4} - \theta\right) = 1$

(ii) $\dfrac{\cos 15° - \sin 15°}{\cos 15° + \sin 15°} = \dfrac{1}{\sqrt{3}}$

(iii) $\cos\theta \cos\phi = \cos^2\dfrac{\theta - \phi}{2} - \sin^2\dfrac{\theta + \phi}{2}$

(iv) $\tan 2\theta - \tan\theta = \tan\theta \sec^2\theta.$

15. If $\tan B = \dfrac{n\sin A\cos A}{1 - n\sin^2 A}$, show that $(A - B) = (1 - n)\tan A$.

ANSWERS

1. $A = 52\dfrac{1}{2}°, B = 7\dfrac{1}{2}°, C = 120°$

6. $\dfrac{220}{221}, \dfrac{171}{221}, \dfrac{220}{221}$

7. (i) $\dfrac{\pi}{3}, \dfrac{2\pi}{3}, \dfrac{4\pi}{3}, \dfrac{5\pi}{3}$

(ii) $\dfrac{\pi}{6}, \dfrac{5\pi}{6}, \dfrac{7\pi}{6}, \dfrac{11\pi}{6}$

8. $2 - \sqrt{3}$

6.9 TRIGONOMETRICAL RATIO OF COMPOUND ANGLES

1. (a) $\sin (A + B) = \sin A \cos B + \cos A \sin B$

(b) $\sin (A - B) = \sin A \cos B - \cos A \sin B$

(c) $\cos (A + B) = \cos A \cos B - \sin A \sin B$

(d) $\cos (A - B) = \cos A \cos B + \sin A \sin B$

(e) $\tan (A + B) = \dfrac{\tan A + \tan B}{1 - \tan A \tan B}$ (f) $\tan (A - B) = \dfrac{\tan A - \tan B}{1 + \tan A \tan B}$

(g) $\cot (A + B) = \dfrac{\cot A \cot B - 1}{\cot A + \cot B}$ (h) $\cot (A - B) = \dfrac{\cot A \cot B + 1}{\cot A - \cot B}$

2. (a) $\sin (A + B) \sin (A - B) = \sin^2 A - \sin^2 B = \cos^2 B - \cos^2 A.$

(b) $\cos(A + B) \cos (A - B) = \cos^2 A - \sin^2 B = \cos^2 B - \sin^2 A.$

3. $\tan (A + B + C) = \dfrac{\tan A + \tan B + \tan C - \tan A \tan B \tan C}{1 - (\tan A \tan B + \tan B \tan C + \tan C \tan A)}$

4. (a) $\sin 2A = 2 \sin A \cos A = \dfrac{2 \tan A}{1 + \tan^2 A}$

(b) $\cos 2A = \cos^2 A - \sin^2 A = 2 \cos^2 A - 1 = 1 - 2 \sin^2 A = \dfrac{1 - \tan^2 A}{1 + \tan^2 A}$

(c) $\tan 2A = \dfrac{2 \tan A}{1 - \tan^2 A}$

5. (a) $\sin 3A = 3 \sin A - 4 \sin^3 A$ (b) $\cos 3A = 4 \cos^3 A - 3 \cos A$

(c) $\tan 3A = \dfrac{3 \tan A - \tan^3 A}{1 - 3 \tan^3 A}$

6. (a) $2 \sin A \cos B = \sin (A + B) + \sin(A - B)$

(b) $2 \cos A \sin B = \sin (A + B) - \sin (A - B)$

(c) $2 \cos A \cos B = \cos (A + B) + \cos (A - B)$

(d) $2 \sin A \sin B = \cos (A - B) - \cos (A + B)$

7. (a) $\sin C + \sin D = 2 \sin \dfrac{C + D}{2} \cos \dfrac{C - D}{2}$

(b) $\sin C - \cos D = 2 \cos \dfrac{C + D}{2} \sin \dfrac{C - D}{2}$

(c) $\cos C + \cos D = 2 \sin \dfrac{C + D}{2} \cos \dfrac{C - D}{2}$

(d) $\cos C - \cos D = 2 \sin \dfrac{C + D}{2} \sin \dfrac{D - C}{2}$

8. (a) $\sin (A + B + C) = \sin A \cos B \cos C + \cos A \sin B \cos C$
$$+ \cos A \cos B \sin C - \sin A \sin B \sin C$$

(b) $\cos (A + B + C) = \cos A \cos B \cos C - \cos A \sin B \sin C$
$$- \sin A \cos B \sin C - \sin A \sin B \cos C$$

(c) $\tan (A + B + C) = \dfrac{\tan A + \tan B + \tan C - \tan A \tan B \tan C}{1 - \tan A \tan B - \tan B \tan C - \tan C \tan A}$

6.10 TRIGONOMETRICAL RATIO FOR HALF ANGLES (OBTAINED BY REPLACING A BY $\frac{A}{2}$ IN THE ABOVE FORMULAE)

1. $\sin A = 2\sin\dfrac{A}{2}\cos\dfrac{A}{2}$

2. $\cos A = \cos^2\dfrac{A}{2} - \sin^2\dfrac{A}{2} = 2\cos^2\dfrac{A}{2} - 1 = 1 - 2\sin^2\dfrac{A}{2}$

3. $\sin\dfrac{A}{2} = \pm\sqrt{\dfrac{1-\cos A}{2}}$ 4. $\cos\dfrac{A}{2} = \pm\sqrt{\dfrac{1+\cos A}{2}}$

5. $\tan A = \dfrac{2\tan\dfrac{A}{2}}{1-\tan^2\dfrac{A}{2}}$ 6. $\sin A = \dfrac{2\tan\dfrac{A}{2}}{1+\tan^2\dfrac{A}{2}}$

7. $\cos A = \dfrac{1-\tan^2\dfrac{A}{2}}{1+\tan^2\dfrac{A}{2}}$

SOLVED EXAMPLES

EXAMPLE 1. *Show that* $\cos^2 A + \cos^2 B - 2\cos A \cos B \cos(A+B) = \sin^2(A+B)$.

SOLUTION. Consider,   L.H.S. $= \cos^2 A + \cos^2 B - 2\cos A \cos B \cos(A+B)$

$= \cos^2 A + \cos^2 B - [\cos(A+B) + \cos(A-B)]\cos(A+B)$

$= \cos^2 A + \cos^2 B - [\cos^2(A+B) + \cos(A+B)\cos(A-B)]$

$= \cos^2 A + \cos^2 B - \cos^2(A+B) - \cos^2 A + \sin^2 B$

$= (\cos^2 B + \sin^2 B) - \cos^2(A+B) = 1 - \cos^2(A+B)$

$= \sin^2(A+B) = $ R. H.S.

EXAMPLE 2. *Show that :*

(i) $\dfrac{\sin A + \sin B}{\cos A + \cos B} = \dfrac{\tan(A+B)}{2}$ **(UPTU B. Pharma)–2001)**

(ii) $\dfrac{\sin A + \sin 3A}{\cos A + \cos 3A} = \tan 2A$ **(Rgpv(B. Pharma)–2001)**

SOLUTION.   (i) Consider L.H.S. $= \dfrac{\sin A + \sin B}{\cos A + \cos B} = \dfrac{2\sin\dfrac{A+B}{2}\cos\dfrac{A-B}{2}}{2\cos\dfrac{A+B}{2}\cos\dfrac{A-B}{2}} = \dfrac{\sin\dfrac{A+B}{2}}{\cos\dfrac{A+B}{2}}$

$= \tan\dfrac{A+B}{2} = $ R.H.S

(ii) Consider L.H.S $= \dfrac{\sin A + \sin 3A}{\cos A + \cos 3A} = \dfrac{\sin 3A + \sin A}{\cos 3A + \cos A}$

$= \dfrac{2\sin\dfrac{3A+A}{2}\cos\dfrac{3A-A}{2}}{2\cos\dfrac{3A+A}{2}\cos\dfrac{3A-A}{2}} = \dfrac{\sin 2A}{\cos 2A} = \tan 2A = $ R.H.S

EXAMPLE 3. *Show that :*

(i) $(\cos A + \cos B)^2 + (\sin A + \sin B)^2 = 4\cos^2\dfrac{A-B}{2}$

(ii) $(\cos A - \cos B)^2 + (\sin A - \sin B)^2 = 4\sin^2\dfrac{A-B}{2}.$

SOLUTION. (i) Consider L.H.S $= (\cos A + \cos B)^2 + (\sin A + \sin B)^2$

$$= \left(2\cos\dfrac{A+B}{2}\cos\dfrac{A-B}{2}\right)^2 + \left(2\sin\dfrac{A+B}{2}\cos\dfrac{A-B}{2}\right)^2$$

$$= 4\cos^2\dfrac{A+B}{2}\cos^2\dfrac{A-B}{2} \times \sin^2\dfrac{A+B}{2}\cos^2\dfrac{A-B}{2}$$

$$= 4\cos^2\dfrac{A-B}{2}\left[\cos^2\dfrac{A+B}{2} + \sin^2\dfrac{A+B}{2}\right]$$

$$= 4\cos^2\dfrac{A-B}{2}[1] = 4\cos^2\dfrac{A-B}{2} = \text{R.H.S.}$$

(ii) Consider L.H.S $= (\cos A - \cos B)^2 + (\sin A - \sin B)^2$

$$= \left(-2\sin\dfrac{A+B}{2}\cos\dfrac{A-B}{2}\right)^2 + \left(2\cos\dfrac{A+B}{2}\sin\dfrac{A-B}{2}\right)^2$$

$$= 4\sin^2\dfrac{A+B}{2}\sin^2\dfrac{A-B}{2} + \cos^2\dfrac{A+B}{2}\sin^2\dfrac{A-B}{2}$$

$$= 4\sin^2\dfrac{A-B}{2}\left[\cos^2\dfrac{A+B}{2} + \sin^2\dfrac{A+B}{2}\right]$$

$$= 4\sin^2\dfrac{A-B}{2}\cdot 1 = 4\sin^2\dfrac{A-B}{2} = \text{R.H.S}$$

EXAMPLE 4. *Show that* $\sin A + \left(\sin A + \dfrac{2\pi}{3}\right) + \sin\left(A + \dfrac{4\pi}{3}\right) = 0.$

SOLUTION. Consider L.H.S $= \sin A + \sin\left(A + \dfrac{2\pi}{3}\right) + \sin\left(A + \dfrac{4\pi}{3}\right)$

$$= \sin A + \left[\sin\left(A + \dfrac{4\pi}{3}\right) + \sin\left(A + \dfrac{2\pi}{3}\right)\right]$$

$$= \sin A + \left[2\sin\dfrac{A+\dfrac{4\pi}{3}+A+\dfrac{2\pi}{3}}{2}\cos\dfrac{A+\dfrac{4\pi}{3}-A+\dfrac{2\pi}{3}}{2}\right]$$

$$= \sin A + \left[2\sin(A+B)\cos\dfrac{\pi}{3}\right] = \sin A + \left[2(-\sin A)\times\dfrac{1}{2}\right]$$

$$= \sin A - \sin A = 0 = \text{R.H.S.}$$

EXAMPLE 5. *Show that* $\tan\left(\dfrac{\pi}{4}+\theta\right) - \left(\tan\dfrac{\pi}{4}-\theta\right) = 2\tan 2\theta.$

SOLUTION. Consider L.H.S. $= \tan\left(\dfrac{\pi}{4}+\theta\right) - \left(\tan\dfrac{\pi}{4}-\theta\right) = \dfrac{\sin\left(\dfrac{\pi}{4}+\theta\right)}{\cos\left(\dfrac{\pi}{4}+\theta\right)} - \dfrac{\sin\left(\dfrac{\pi}{4}-\theta\right)}{\cos\left(\dfrac{\pi}{4}-\theta\right)}$

$$= \frac{\sin\left(\frac{\pi}{4}+\theta\right)\cos\left(\frac{\pi}{4}-\theta\right) - \cos\left(\frac{\pi}{4}+\theta\right)\sin\left(\frac{\pi}{4}-\theta\right)}{\cos\left(\frac{\pi}{4.}+\theta\right)\cos\left(\frac{\pi}{4}-\theta\right)}$$

$$= \frac{2\sin\left[\left(\frac{\pi}{4}+\theta\right)-\left(\frac{\pi}{4}-\theta\right)\right]}{2\cos\left(\frac{\pi}{4}+\theta\right)\cos\left(\frac{\pi}{4}-\theta\right)}$$

$$= \frac{2\sin 2\theta}{\cos\left(\frac{\pi}{4}+\theta+\frac{\pi}{4}-\theta\right) + \cos\left(\frac{\pi}{4}+\theta-\frac{\pi}{4}+\theta\right)}$$

$$= \frac{2\sin 2\theta}{\cos\frac{\pi}{2}+\cos 2\theta} = \frac{2\sin 2\theta}{0+\cos 2\theta} = 2.\frac{\sin 2\theta}{\cos 2\theta} = 2\tan 2\theta = \text{R.H.S.}$$

EXAMPLE 6. *Show that* $\sin 10^\circ \sin 30^\circ \sin 50^\circ \sin 70^\circ = \frac{1}{16}.$ **[UPTU(B. Pharma)–2003]**

SOLUTION. Consider   L.H.S $= \sin 10^\circ \sin 30^\circ \sin 50^\circ \sin 70^\circ$

$$= \frac{1}{2}\sin 10^\circ \sin 50^\circ \sin 70^\circ \qquad [\because \sin 30^\circ = 1/2]$$

$$= \frac{1}{4}\sin 10^\circ [2\sin 50^\circ \sin 70^\circ]$$

$$= \frac{1}{4}\sin 10^\circ [\cos(70^\circ - 50^\circ) - \cos(70^\circ + 50^\circ)]$$

$$= \frac{1}{4}\sin 10^\circ[\cos 20^\circ - \cos 120^\circ]$$

$$= \frac{1}{4}\sin 10^\circ\left[\cos 20^\circ + \frac{1}{2}\right] \qquad [\because \cos 120^\circ = -\frac{1}{2}]$$

$$= \frac{1}{8}\sin 10^\circ[2\cos 20^\circ + 1] = \frac{1}{8}[2\cos 20^\circ \sin 10^\circ + \sin 10^\circ]$$

$$= \frac{1}{8}[\sin(20^\circ + 10^\circ) - \sin(20^\circ - 10^\circ) + \sin 10^\circ]$$

$$= \frac{1}{8}[\sin 30^\circ - \sin 10^\circ + \sin 10^\circ]$$

$$= \frac{1}{8}\sin 30^\circ = \frac{1}{8}\times\frac{1}{2} = \frac{1}{16} = \text{R.H.S.}$$

EXAMPLE 7. *Show that* $\dfrac{\sin 8\theta\cos\theta - \cos 3\theta\sin 6\theta}{\cos 2\theta\cos\theta - \sin 3\theta\sin 4\theta} = \tan 2\theta$

SOLUTION. Consider

$$\text{L.H.S.} = \frac{\sin 8\theta\cos\theta - \cos 3\theta\sin 6\theta}{\cos 2\theta\cos\theta - \sin 3\theta\sin 4\theta} = \frac{2\sin 8\theta\cos\theta - 2\sin 6\theta\cos 3\theta}{2\cos 2\theta\cos\theta - 2\sin 4\theta\sin 3\theta}$$

$$= \frac{[\sin(8\theta+\theta)+\sin(8\theta-\theta)] - [\sin(6\theta+3\theta)+\sin(6\theta-3\theta)]}{[\cos(2\theta+\theta)+\cos(2\theta-\theta)] - [\cos(4\theta-3\theta)-\cos(4\theta+3\theta)]}$$

$$= \frac{\sin 9\theta + \sin 7\theta - \sin 9\theta - \sin 3\theta}{\cos 3\theta + \cos\theta - \cos\theta + \cos 7\theta} = \frac{\sin 7\theta - \sin 3\theta}{\cos 7\theta + \cos 3\theta}$$

$$= \frac{2\cos\dfrac{7\theta + 3\theta}{2}\sin\dfrac{7\theta - 3\theta}{2}}{2\cos\dfrac{7\theta + 3\theta}{2}\cos\dfrac{7\theta - 3\theta}{2}}$$

$$= \frac{2\cos 5\theta \cdot \sin 2\theta}{2\cos 5\theta \cos 2\theta} = \frac{\sin 2\theta}{\cos 2\theta} = \tan 2\theta = \text{R.H.S.}$$

EXAMPLE 8. *If* $\cos(\theta + 2\alpha) = m\cos\theta$, *show that* $\cot\theta = \dfrac{1+m}{1-m}\tan(\theta + \alpha)$.

SOLUTION. Given that $\qquad \cos(\theta + 2\alpha) = m\cos\theta$

$$\Rightarrow \qquad \frac{\cos(\theta + 2\alpha)}{\cos\theta} = \frac{m}{1}$$

Using componendo and dividendo, we get

$$\frac{\cos(\theta + 2\alpha) + \cos\theta}{\cos\theta - \cos(\theta + 2\alpha)} = \frac{1+m}{1-m}$$

$$\Rightarrow \quad \frac{2\cos\dfrac{\theta + 2\alpha + \theta}{2}\cos\dfrac{\theta + 2\alpha - \theta}{2}}{2\sin\dfrac{\theta + \theta + 2\alpha}{2}\sin\dfrac{\theta + 2\alpha - \theta}{2}} = \frac{1+m}{1-m}$$

$$\Rightarrow \qquad \frac{\cos(\theta + \alpha)\cos\alpha}{\sin(\theta + \alpha)\sin\alpha} = \frac{1+m}{1-m}$$

$$\Rightarrow \qquad \frac{\cot\alpha}{\tan(\theta + \alpha)} = \frac{1+m}{1-m}$$

Hence $\qquad\qquad \cos\alpha = \dfrac{1+m}{1-m}\tan(\theta + \alpha)$

EXAMPLE 9. *Show that* $\cos 7\theta + \cos 5\theta + \cos 3\theta + \cos\theta = 4\cos\theta \cdot \cos 2\theta \cdot \cos 4\theta$.

SOLUTION. Consider

$$\text{L.H.S.} = \cos 7\theta + \cos 5\theta + \cos 3\theta + \cos\theta$$

$$= (\cos 7\theta + \cos\theta) + (\cos 5\theta + \cos 3\theta)$$

$$= 2\cos\frac{7\theta + \theta}{2}\cos\frac{7\theta - \theta}{2} + 2\cos\frac{5\theta + 3\theta}{2}\cos\frac{5\theta - 3\theta}{2}$$

$$= 2\cos 4\theta \cos 3\theta + 2\cos 4\theta \cos\theta$$

$$= 2\cos 4\theta(\cos 3\theta + \cos\theta)$$

$$= 2\cos 4\theta\left(2\cos\frac{3\theta + \theta}{2}\cos\frac{3\theta - \theta}{2}\right)$$

$$= 2\cos 4\theta(2\cos 2\theta \cos\theta)$$

$$= 4\cos\theta \cos 2\theta \cos 4\theta = \text{R.H.S.}$$

EXAMPLE 10. *Prove that* $\cos 20° \cos 40° \cos 80° = \dfrac{1}{8}$ [UPTU(B. Pharma)–2005]

SOLUTION. We have $\qquad \text{L.H.S} = (\cos 40° \cos 20°)\cos 80°$

$$= \frac{1}{2}[2\cos 40° \cos 20°]\cos 80°$$

$$= \frac{1}{2}(\cos 60° + \cos 20°)\cos 80° = \frac{1}{2}\left[\frac{1}{2} + \cos 20°\right]\cos 80°$$

$$= \frac{1}{4}\cos 80° + \frac{1}{4}(2\cos 80°\cos 20°)$$

$$= \frac{1}{4}\cos 80° + \frac{1}{4}(\cos 100° + \cos 60°)$$

$$= \frac{1}{4}\cos 80° - \frac{1}{4}\cos 80° + \frac{1}{4}\cos 60°$$

$$= \frac{1}{4}\cdot\frac{1}{2} = \frac{1}{8} = \text{R.H.S}$$

EXAMPLE 11. *Calculate* $\sin\dfrac{\pi}{10°}$ *and* $\cos\dfrac{\pi}{5}$. [UPTU(B. Pharma)–2003]

SOLUTION. We have
$$\cos\frac{3\pi}{10} = \sin\left(\frac{\pi}{2} - \frac{3\pi}{10}\right) = \sin\frac{2\pi}{10}$$

Now, $$\cos\frac{3\pi}{10} = \sin\frac{2\pi}{10}$$

$$\Rightarrow \quad 4\cos^3\frac{\pi}{10} - 3\cos\frac{\pi}{10} = 2\sin\frac{\pi}{10}\cos\frac{\pi}{10}$$

$$[\because \cos 3A = 4\cos^3 A - 3\cos A \text{ and } \sin 2A = 2\sin A\cos A]$$

$$\Rightarrow \quad \left(\cos\frac{\pi}{10}\right)\left[4\cos^2\frac{\pi}{10} - 3 - 2\sin\frac{\pi}{10}\right] = 0$$

$$\Rightarrow \quad 4\cos^2\frac{\pi}{10} - 3 - 2\sin\frac{\pi}{10} = 0$$

$$\Rightarrow \quad 4\left(1 - \sin^2\frac{\pi}{10}\right) - 3 - 2\sin\frac{\pi}{10} = 0$$

$$\Rightarrow \quad 4\sin^2\frac{\pi}{10} + 2\sin\frac{\pi}{10} - 1 = 0$$

$$\sin\frac{\pi}{10} = \frac{-2 \pm \sqrt{4 + 16}}{8}$$

$$\left[\text{solving the quadratic equation in } \sin\frac{\pi}{10}\right]$$

$$= \frac{-1 \pm \sqrt{5}}{4}$$

Since, $0 < \dfrac{\pi}{10} < \dfrac{\pi}{2}$, therefore $\sin\dfrac{\pi}{10}$ is positive.

Hence, $$\sin\frac{\pi}{10} = \sin 18° = \frac{-1 + \sqrt{5}}{4} = \frac{\sqrt{5} - 1}{4}$$

Now, $$\cos\frac{\pi}{5} = 1 - 2\sin^2\frac{\pi}{10}$$

$$= 1 - 2\cdot\left(\frac{\sqrt{5} - 1}{4}\right)^2 = 1 - \frac{5 + 1 - 2\sqrt{5}}{8} = \frac{8 - 6 + 2\sqrt{5}}{8}$$

$$= \frac{2(\sqrt{5}+1)}{8} = \frac{\sqrt{5}+1}{4}$$

Hence $\qquad \cos\dfrac{\pi}{5} = \cos 36° = \dfrac{\sqrt{5}+1}{4}.$

EXAMPLE 12. *Prove that* $\dfrac{\cos\theta}{1+\sin\theta} = \tan\left(45° - \dfrac{\theta}{2}\right).$ [UPTU(B. Pharma)–2003]

SOLUTION. We have $\qquad \dfrac{\cos\theta}{1+\sin\theta} = \dfrac{\sin(90° - \theta)}{1+\cos(90° - \theta)}$

$$= \frac{2\sin\left(\dfrac{90° - \theta}{2}\right)\cos\left(\dfrac{90° - \theta}{2}\right)}{2\cos^2\left(\dfrac{90° - \theta}{2}\right)} = \frac{\sin\left(45° - \dfrac{\theta}{2}\right)}{\cos\left(45° - \dfrac{\theta}{2}\right)} = \tan\left(45° - \frac{\theta}{2}\right)$$

EXAMPLE 13. *If* $\tan\dfrac{\theta}{2} = \dfrac{p}{q}.$ *Find the value of* $(p\sin\theta + q\cos\theta).$ [UPTU(B.Pharma)–2007]

SOLUTION. We have $\qquad p\sin\theta + q\cos\theta = p\dfrac{2\tan\theta/2}{1+\tan^2\theta/2} + q\dfrac{1-\tan^2\theta/2}{1+\tan^2\theta/2}$

$$= p\frac{2\dfrac{p}{q}}{1+\dfrac{p^2}{q^2}} + q\frac{1-\dfrac{p^2}{q^2}}{1+\dfrac{p^2}{q^2}} = p.\frac{2pq}{q^2+p^2} + q.\frac{q^2-p^2}{q^2+p^2}$$

$$= \frac{2p^2 q + q^3 - p^2 q}{q^2+p^2} = \frac{q^3 + p^2 q}{q^2+p^2} = q.$$

EXAMPLE 14. *Prove that* $\sin 20° \sin 40° \sin 60° \sin 80° = \dfrac{3}{16}.$ [RGPV(B.Pharma)–2002]

SOLUTION. $\qquad$ L.H.S $= \sin 20° \sin 40° \sin 60° \sin 80°$

$$= \sin 20° \sin 40° \times \frac{\sqrt{3}}{2} \times \sin 80°$$

$$= \frac{\sqrt{3}}{2} \times \sin 20° \, (\sin 40° \sin 80°)$$

$$= \frac{\sqrt{3}}{4} \sin 20°[\cos(40° - 80°) - \cos(40° + 80°)]$$

$$= \frac{\sqrt{3}}{4} \sin 20° \cos 40° + \frac{\sqrt{3}}{8} \sin 20°$$

$$= \frac{\sqrt{3}}{8}(2\sin 20° \cos 40°) + \frac{\sqrt{3}}{8}\sin 20°$$

$$= \frac{\sqrt{3}}{8}[\sin 60° + \sin(-20°)] + \frac{\sqrt{3}}{8}\sin 20°$$

$$= \frac{\sqrt{3}}{8} \times \frac{\sqrt{3}}{2} - \frac{\sqrt{3}}{8}\sin 20° + \frac{\sqrt{3}}{8}\sin 20° = \frac{3}{16} = \text{R.H.S.}$$

EXAMPLE 15. *Prove that* $\dfrac{\sin A - \sin 5A + \sin 9A - \sin 13A}{\cos A - \cos 5A + \cos 9A - \cos 13A} = \cot 4A$ **[RGPV(B. Pharma)–2002]**

SOLUTION.

$$\text{L.H.S.} = \frac{\sin A - \sin 5A + \sin 9A - \sin 13A}{\cos A - \cos 5A + \cos 9A - \cos 13A}$$

$$= \frac{2\sin\dfrac{A-5A}{2}\cos\dfrac{A+5A}{2} + 2\sin\dfrac{9A-13A}{2}\cos\dfrac{9A+13A}{2}}{2\sin\dfrac{A+5A}{2}\sin\dfrac{5A-A}{2} + 2\sin\dfrac{13A+9A}{2}\sin\dfrac{9A-13A}{2}}$$

$$= \frac{2\sin(-2A)\cos 3A + 2\sin A(-2A)\cos 11A}{2\sin 3A\sin 2A + 2\sin 11A\sin(-2A)}$$

$$= \frac{-2\sin 2A\cos 3A - 2\sin 2A\cos 11A}{2\sin 3A\sin 2A - 2\sin 2A\sin 11A}$$

$$= \frac{-2\sin 2A(\cos 3A + \cos 11A)}{-2\sin 2A(-\sin 3A + \sin 11A)} = \frac{\cos 3A + \cos 11A}{\sin 11A - \sin 3A}$$

$$= \frac{2\cos 7A \times \cos(-4A)}{2\sin 4A \times \cos 7A} = \frac{\cos 4A}{\sin 4A} = \cot 4A = \text{R.H.S.}$$

EXAMPLE 16. *Prove that* $\dfrac{\cos 3\theta + 2\cos 5\theta + \cos 7\theta}{\cos\theta + 2\cos 3\theta + \cos 5\theta} = \dfrac{\cos 5\theta}{\sin 3\theta}$ **[RGPV(B. Pharma)–2002]**

SOLUTION.

$$\text{L.H.S.} = \frac{\cos 3\theta + 2\cos 5\theta + \cos 7\theta}{\cos\theta + 2\cos 3\theta + \cos 5\theta} = \frac{(\cos 3\theta + \cos 7\theta) + 2\cos 5\theta}{(\cos\theta + \cos 5\theta) + 2\cos 3\theta}$$

$$= \frac{2\cos 5\theta\cos(-2\theta) + 2\cos 5\theta}{2\cos 3\theta\cos(-2\theta) + 2\cos 3\theta} = \frac{2\cos 5\theta(\cos 2\theta + 1)}{2\cos 3\theta(\cos 2\theta + 1)}$$

$$= \frac{\cos 5\theta}{\cos 3\theta} = \text{R.H.S}$$

EXAMPLE 17. *Prove that* $4 \sin A \times \sin (60° + A) \times \sin (60° - A) = \sin 3A.$

[RGPV(B. Pharma)–2001]

SOLUTION.

$$\text{L.H.S.} = 4 \sin A \sin (60° + A) \sin(60° - A)$$

$$= 4 \sin A\,(\sin^2 60° - \sin^2 A) = 4\sin A\left(\frac{3}{4} - \sin^2 A\right)$$

$$= 3 \sin A - 4 \sin^3 A = \sin 3A = \text{R.H.S.}$$

EXAMPLE 18. *Prove that* $\cos 4x = 1 - 8 \sin^2 x \cdot \cos^2 x.$ **[RGPV(B. Pharma)–2003]**

SOLUTION.

$$\text{L.H.S.} = \cos 4x = \cos 2(2x) = 1 - \sin^2 2x$$

$$= 1 - 2[2 \sin x \cos x]^2 = 1 - 8 \sin^2 x \cos^2 x = \text{R.H.S.}$$

EXAMPLE 19. *If* $0 \le x \le 2\pi$, *find* $\sin\dfrac{x}{2}, \cos\dfrac{x}{2}$ *when* $\tan x = \dfrac{-4}{3}$, *x lies in second quadrant.*

[RGPV(B. Pharma)–2001]

SOLUTION. As $\tan x = \dfrac{-4}{3}$ the value of $\cos x$ is negative in second quadrant

$$\cos x = -\frac{3}{5}$$

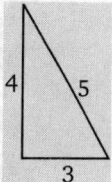

Now $\sin\dfrac{x}{2} = \sqrt{\dfrac{1-\cos x}{2}} = \sqrt{\dfrac{1+\dfrac{3}{5}}{2}} = \dfrac{2}{\sqrt{5}}$

and $\qquad \cos \dfrac{x}{2} = \sqrt{\dfrac{1 + \cos x}{2}} = \sqrt{\dfrac{1 - \dfrac{3}{5}}{2}} = \dfrac{1}{\sqrt{5}}.$

EXAMPLE 20. *Prove that* $\sec \alpha + \tan \alpha = \tan\left(\dfrac{\pi}{4} + \dfrac{\alpha}{2}\right)$ **[UPTU(B. Pharma)–2003, 2006]**

SOLUTION.

$$\text{L.H.S.} = \frac{1}{\cos\alpha} + \frac{\sin\alpha}{\cos\alpha} = \frac{1 + \sin\alpha}{\cos\alpha}$$

$$= \frac{\left(\cos^2\dfrac{\alpha}{2} + \sin^2\dfrac{\alpha}{2}\right) + 2\sin\dfrac{\alpha}{2}.\cos\dfrac{\alpha}{2}}{\cos^2\dfrac{\alpha}{2} - \sin^2\dfrac{\alpha}{2}}$$

$$= \frac{\left(\cos\dfrac{\alpha}{2} + \sin\dfrac{\alpha}{2}\right)^2}{\left(\cos\dfrac{\alpha}{2} + \sin\dfrac{\alpha}{2}\right)\left(\cos\dfrac{\alpha}{2} - \sin\dfrac{\alpha}{2}\right)}$$

$$= \frac{\cos\dfrac{\alpha}{2} + \sin\dfrac{\alpha}{2}}{\cos\dfrac{\alpha}{2} - \sin\dfrac{\alpha}{2}} \qquad \text{[Divide Nr and Dr by } \cos \alpha/2]$$

$$= \frac{\left(\dfrac{\cos\dfrac{\alpha}{2} + \sin\dfrac{\alpha}{2}}{\cos\dfrac{\alpha}{2}}\right)}{\left(\dfrac{\cos\dfrac{\alpha}{2} - \sin\dfrac{\alpha}{2}}{\cos\dfrac{\alpha}{2}}\right)} = \frac{1 + \tan\dfrac{\alpha}{2}}{1 - \tan\dfrac{\alpha}{2}} = \frac{\tan\dfrac{\pi}{4} + \tan\dfrac{\alpha}{2}}{1 - \tan\dfrac{\pi}{4}\tan\dfrac{\alpha}{2}}$$

$$[\because \tan \pi/4 = 1]$$

$$= \tan\left(\frac{\pi}{4} + \frac{\alpha}{2}\right) = \text{R.H.S.}$$

EXAMPLE 21. *Prove that* $\tan\left(\dfrac{\pi}{4} - \dfrac{A}{2}\right) = \sqrt{\dfrac{1 - \sin A}{1 + \sin A}}$ **[UPTU(B. Pharma)–2003, 2006]**

SOLUTION.

$$\text{R.H.S.} = \sqrt{\frac{(\cos^2 A/2 + \sin^2 A/2) - 2\sin A/2.\cos A/2}{(\cos^2 A/2 + \sin^2 A/2) + 2\sin A/2\cos A/2}}$$

$$= \sqrt{\frac{(\cos A/2 - \sin A/2)^2}{(\cos A/2 + \sin A/2)^2}} = \frac{\cos A/2 - \sin A/2}{\cos A/2 + \sin A/2}$$

$$= \frac{1 - \tan A/2}{1 + \tan A/2}$$

$$= \frac{\tan \pi/4 - \tan A/2}{\cos \pi/2 + \sin A/2} \qquad \text{[Divide Nr and Dr by } \cos A/2]$$

$$= \tan(\pi/4 - A/2) = \text{L.H.S.}$$

EXAMPLE 22. $\dfrac{\tan^3 A}{1 + \tan^2 A} + \dfrac{\cot^3 A}{1 + \cot^2 A} = \sec A \cdot \mathrm{cosec}\, A - \sin 2A.$ **[Meerut(B.Sc. Biotech)–2006]**

SOLUTION.

$$\mathrm{L.H.S} = \frac{\sin^3 A}{\cos^3 A} \cdot \frac{1}{\sec^2 A} + \frac{\cos^3 A}{\sin^3 A} \cdot \frac{1}{\mathrm{cosec}^2 A}$$

$$= \frac{\sin^3 A}{\cos A} + \frac{\cos^3 A}{\sin A} = \frac{\sin^4 A + \cos^4 A}{\sin A \cos A}$$

$$= \frac{\sin^4 A + \cos^4 A + 2\sin^2 A \cos^2 A - 2\sec^2 A \cos^2 A}{\sin A \cos A}$$

$$= \frac{(\cos^2 A + \sin^2 A)^2}{\sin A \cos A} - \frac{2\sin^2 A \cos^2 A}{\sin A \cos A}$$

$$= \frac{1}{\sin A \cos A} - 2\sin A \cos A$$

$$= \sec A \, \mathrm{cosec}\, A - \sin 2A = \mathrm{R.H.S.}$$

EXAMPLE 23. If $\sin 2A = \dfrac{4}{5}$, find the value of $\tan A$ $(0 \le A \le \pi/4)$ **[UPTU(B. Pharma)–2006]**

SOLUTION. We have $\sin 2A = \dfrac{4}{5}$

$$\Rightarrow \qquad \frac{2\tan A}{1 + \tan^2 A} = \frac{4}{5}$$

$$\Rightarrow \qquad 4 + 4\tan^2 A = 10\tan A$$

$$\Rightarrow \qquad 2\tan^2 A - 5\tan A + 2 = 0$$

$$\Rightarrow \qquad (\tan A - 2)(2\tan A - 1) = 0$$

$$\Rightarrow \qquad \tan A = 2 \text{ or } 1/2$$

But $0 \le A \le \pi/4$, so rejecting value 2. Hence, $\tan A = 1/2$.

Exercise 6.4

1. Show that :
   (i) $\tan(45^\circ + \theta) + \tan(45^\circ - \theta) = 2\sec 2\theta$
   (ii) $\sin^2 A + \sin^2(A - B) - 2\sin A \cos B$
   $$\sin(A - B) = \sin 2B$$
   (iii) $\dfrac{\sin 5A - \sin 3A}{\cos 3A + \cos 5A} = \tan A$
   (iv) $\dfrac{\cos 4A + \cos 3A + \cos 2A}{\sin 4A + \sin 3A + \sin 2A} = \cot 3A$

2. Show that :
   (i) $(\sin 3A + \sin A)\sin A + (\cos 3A - \cos A)$
   $$\cos A = 0$$
   (ii) $2\cos\dfrac{\pi}{13}\cos\dfrac{9\pi}{13} + \cos\dfrac{3\pi}{13} + \cos\dfrac{5\pi}{13} = 0$
   (iii) $(\cos A - \cos B)^2 + (\sin A - \sin B)^2$
   $$= 4\sin^2\dfrac{A - B}{2}$$

(iv) $\cos 2A \cdot \cos\dfrac{A}{2} - \cos 3A\cos\dfrac{9A}{2}$
   $$= \sin 5A \sin\dfrac{5A}{2}$$

3. Show that :
   (i) $\cos 20^\circ + \cos 100^\circ + \cos 140^\circ = 0$
   (ii) $\cos 20^\circ \cos 40^\circ \cos 60^\circ \cos 80^\circ = 1/16$
   (iii) $\cos 10^\circ \cos 50^\circ \cos 60^\circ \cos 70^\circ = \dfrac{\sqrt{3}}{16}$
   (iv) $\cos 52^\circ + \cos 68^\circ + \cos 172^\circ = 0.$

4. Show that :
   $$\dfrac{\sin A + \sin 3A + \sin 5A + \sin 7A}{\cos A + \cos 3A + \cos 5A + \cos 7A} = \tan 4A$$

5. Show that :
   (i) $\dfrac{2\cos 2A + 1}{2\cos 2A - 1}$
   $$= \tan(60^\circ + A)\tan(60^\circ - A)$$

(ii) $\cos A + \cos\left(A + \dfrac{2\pi}{2}\right) + \cos\left(A - \dfrac{2\pi}{3}\right) = 0$

$\qquad = \cos(A + B + C) + \cos(B + C - A)$
$\qquad\quad + \cos(C + A - B) + \cos(A + B - C).$

6. Show that :

$\cot 4\theta(\sin 5\theta + \sin 3\theta) + \cot\theta(\sin 5\theta - \sin 3\theta)$

9. If $\cos(A + B)\sin(C - D) = \cos(A - B)\sin(C + D),$
then show that $\tan A \tan B \tan C + \tan D = 0.$

7. Show that :

$\dfrac{\sin\theta + \sin 2\theta + \sin 4\theta + \sin 5\theta}{\cos\theta + \cos 2\theta + \cos 4\theta + \cos 5\theta} = \tan 3\theta$

8. Show that :

$4 \cos A \cos B \cos C$

10. If $\sin x + \sin y = a$ and $\cos x + \cos y = b.$
Show that :

(i) $\tan\dfrac{x + y}{2} = \dfrac{a}{b}$

(ii) $\tan\left(\dfrac{x - y}{2}\right) = \sqrt{\dfrac{4 - a^2 - b^2}{a^2 + b^2}}$

ADDITIONAL SOLVED EXAMPLES

(BASED ON '2A' AND '3A' FORMULAE)

EXAMPLE 1. *Show that* $\sin 18^\circ = \dfrac{\sqrt{5} - 1}{4}$

[UPTU(B. Pharma)–2001]

SOLUTION. Let $\qquad\qquad \theta = 18^\circ$

$\Rightarrow \qquad\qquad 5\theta = 90^\circ \qquad \Rightarrow \qquad 2\theta + 3\theta = 90^\circ$

$\Rightarrow \qquad\qquad 2\theta = 90^\circ - 3\theta \qquad \Rightarrow \qquad \sin 2\theta = \sin(90^\circ - 3\theta) = \cos 3\theta$

$\Rightarrow \qquad 2\sin\theta\cos\theta = 4\cos^3\theta - 3\cos\theta$

$\Rightarrow \qquad 2\sin\theta = 4\cos^2\theta - 3 \Rightarrow \qquad 2\sin\theta = 4(1 - \sin^2\theta) - 3$

$\Rightarrow 4\sin^2\theta + 2\sin\theta - 1 = 0$

$\therefore \qquad\qquad \sin\theta = \dfrac{-2 \pm \sqrt{4 + 16}}{8} = \dfrac{-2 \pm 2\sqrt{5}}{8} = \dfrac{-1 \pm \sqrt{5}}{4}$

Since, 18° lies in the first quadrant, therefore $\sin 18^\circ$ is positive.

Rejecting the value $\dfrac{-1 - \sqrt{5}}{4} < 0$, we get

$\qquad\qquad \sin 18^\circ = \dfrac{\sqrt{5} - 1}{4}$

⬛ REMARK

➠ Similarly, we can find $\cos 18^\circ = \dfrac{\sqrt{10 + 25}}{4}, \cos 36^\circ \dfrac{\sqrt{5} + 1}{4}, \sin 36^\circ = \dfrac{\sqrt{10 - 2\sqrt{5}}}{4}.$

EXAMPLE 2. *Show that* $\dfrac{1 + \sin 2\theta - \cos 2\theta}{1 + \sin 2\theta + \cos 2\theta} = \tan\theta.$

SOLUTION. Consider $\qquad$ L.H.S. $= \dfrac{(1 - \cos 2\theta) + \sin 2\theta}{(1 + \cos 2\theta) + \sin 2\theta} = \dfrac{2\sin^2\theta + 2\sin\theta\cos\theta}{2\cos^2\theta + 2\sin\theta\cos\theta}$

$\qquad\qquad = \dfrac{2\sin\theta(\sin\theta + \cos\theta)}{2\cos\theta(\cos\theta + \sin\theta)} = \dfrac{\sin\theta}{\cos\theta} = \tan\theta =$ R.H.S.

EXAMPLE 3. *Show that*

(i) $\dfrac{\sin 2A}{1 + \cos 2A} = \tan A$ $\qquad$ (ii) $\operatorname{cosec} 2A + \cot 2A = \cot A$

(iii) $\sin 3A + \sin 2A - \sin A = 4\sin A \cos\dfrac{A}{2}\cos\dfrac{3A}{2}$

Solution.   (i)   L.H.S. $= \dfrac{\sin 2A}{1+\cos 2A} = \dfrac{2\sin A\cos A}{2\cos^2 A} = \dfrac{\cos A}{\sin A} = \tan A = $ R.H.S.

(ii)   Consider   L.H.S $= \operatorname{cosec} 2A + \cot 2A = \dfrac{1}{\sin 2A} + \dfrac{\cos 2A}{\sin 2A}$

$$= \dfrac{1+\cos 2A}{\sin 2A} = \dfrac{2\cos^2 A}{2\sin A\cos A} = \dfrac{\cos A}{\sin A} = \cot A = \text{RHS}$$

(iii)   Consider   L.H.S. $= \sin 3A + \sin 2A - \sin A$

$$= (\sin 3A + \sin 2A) - \sin A$$

$$= 2\sin\dfrac{5A}{2}\cdot\cos\dfrac{A}{2} - 2\sin\dfrac{A}{2}\cdot\cos\dfrac{A}{2}$$

$$= 2\cos\dfrac{A}{2}\left[\sin\dfrac{5A}{2} - \sin\dfrac{A}{2}\right]$$

$$= 2\cos\dfrac{A}{2}\left(2\cos\dfrac{\dfrac{5A}{2}+\dfrac{A}{2}}{2}\sin\dfrac{\dfrac{5A}{2}-\dfrac{A}{2}}{2}\right)$$

$$= 2\cos\dfrac{A}{2}\left[2\sin\dfrac{3A}{2}\sin A\right] = 4\sin A\cos\dfrac{A}{2}\cos\dfrac{3A}{2}$$

$$= \text{R.H.S.}$$

Example 4.   *If $\tan\theta = \dfrac{b}{a}$, find the value of $a\cos 2\theta + b\sin 2\theta$.*

Solution.   Consider $a\cos 2\theta + b\sin 2\theta$

$$= a\left(\dfrac{1-\tan^2\theta}{1+\tan^2\theta}\right) + b\left(\dfrac{2\tan\theta}{1+\tan^2\theta}\right)$$

$$= a\left(\dfrac{1-\dfrac{b^2}{a^2}}{1+\dfrac{b^2}{a^2}}\right) + b\left(\dfrac{2\cdot\dfrac{b}{a}}{1+\dfrac{b^2}{a^2}}\right)$$

$$= a\left(\dfrac{a^2-b^2}{a^2+b^2}\right) + b\left(\dfrac{2ab}{a^2+b^2}\right) = \dfrac{a^3-ab^2}{a^2+b^2} + \dfrac{2ab^2}{a^2+b^2}$$

$$= \dfrac{a^3+b^3}{a^2+b^2} = \dfrac{a(a^2+b^2)}{a^2+b^2} = a$$

Example 5.   *Find the value of $\tan 22° 30′$.*

Solution.   We know that $\tan\dfrac{A}{2} = \sqrt{\dfrac{1-\cos A}{1+\cos A}}$

Put $A = 45°$, we get

$$\tan 22\dfrac{1}{2}° = \sqrt{\dfrac{1-\cos 45°}{1+\cos 45°}} = \dfrac{\sqrt{1-\dfrac{1}{\sqrt 2}}}{\sqrt{1+\dfrac{1}{\sqrt 2}}} = \dfrac{\sqrt{2}-1}{\sqrt{2}+1}$$

EXAMPLE 6.   *Show that* $\tan 4\theta = \dfrac{4\tan\theta(1-\tan^2\theta)}{1-6\tan^2\theta+\tan^4\theta}$.

SOLUTION. $\text{L.H.S} = \tan 4\theta = \tan(2\times 2\theta) = \dfrac{2\tan 2\theta}{1-\tan^2 2\theta} = \dfrac{2\times \dfrac{2\tan\theta}{1-\tan^2\theta}}{1-\left(\dfrac{2\tan\theta}{1-\tan^2\theta}\right)^2}$

$$= \dfrac{\dfrac{4\tan\theta}{1-\tan^2\theta}}{\dfrac{(1-\tan^2\theta)^2 - 4\tan^2\theta}{(1-\tan^2\theta)^2}} = \dfrac{4\tan\theta(1-\tan^2\theta)}{1-6\tan^2\theta+\tan^4\theta}$$

$$= \text{R.H.S.}$$

Exercise 6.5

1. Show that :

(i) $\dfrac{\tan 5\theta + \tan 3\theta}{\tan 5\theta - \tan 3\theta} = 4\cos 2\theta \cos 4\theta$

(ii) $\dfrac{\sin\theta + \sin 2\theta}{1 + \cos\theta + \cos 2\theta} = \tan\theta$

2. Show that :

(i) $\sin 4A = 4\sin A \cos^3 A - 4\cos A \sin^3 A$

(ii) $\cos 4A = 1 - 8\sin^2 A \cos^2 A$

(iii) $\cos 5A = 16\cos^5 A - 20\cos^3 A + 5\cos A$

(iv) $\cos 6A = 32\cos^6 A - 48\cos^4 A + 18\cos^2 A - 1$.

3. Show that :

(i) $\sin^2 72° - \sin^2 60° = \dfrac{\sqrt{5}-1}{8}$

(ii) $\sin\dfrac{\pi}{10} + \sin\dfrac{13\pi}{10} = -\dfrac{1}{2}$

(iii) $\sin\dfrac{\pi}{5}\sin\dfrac{2\pi}{5}\sin\dfrac{3\pi}{5}\sin\dfrac{4\pi}{5} = \dfrac{5}{16}$

(iv) $2\cos\theta = \sqrt{2+\sqrt{2+2\cos 4\theta}}$.

4. Find $\sin 7\tfrac{1}{2}°, \cos 7\tfrac{1}{2}°$ and $\tan 11\tfrac{1}{4}°$.

5. Show that : $\tan 15° + \cot 15° = 4$.

6. Prove that :

(i) $\sec 2\theta - \tan 2\theta = \dfrac{\cos\theta - \sin\theta}{\cos\theta + \sin\theta}$

(ii) $\operatorname{cosec}\theta - 2\sin\theta = 2\cot 2\theta \cos\theta$.

7. If $\tan^2\theta = 2\tan^2\phi + 1$, then show that $\cos 2\theta + \sin^2\phi = 0$.

8. If $x + \dfrac{1}{x} = 2\cos\theta$, show that $x^3 + \dfrac{1}{x^3} = 2\cos 3\theta$.

9. If $\cos\theta = \dfrac{\cos\phi - e}{1 - e\cos\phi}$, show that

$$\tan\dfrac{\theta}{2} = \pm\sqrt{\dfrac{1+e}{1-e}}\cdot\tan\dfrac{\phi}{2}.$$

10. If $2\tan\alpha = 3\tan\beta$, show that $\tan(\alpha - \beta) = \dfrac{\sin 2\beta}{5 - \cos 2\beta}$.

11. Show that :

(i) $\operatorname{cosec} 2A + \cos 2A = \cot A$

(ii) $\sec 2A + \tan 2A = \dfrac{\cos A + \sin A}{\cos A - \sin A}$

$$= \tan\left(\dfrac{\pi}{4} + A\right)$$

(iii) $\sqrt{\dfrac{1-\sin A}{1+\sin A}} = \tan\left(\dfrac{\pi}{4} - \dfrac{A}{2}\right)$

12. If $\cos\theta = \dfrac{1}{2}\left(a + \dfrac{1}{a}\right)$, show that

$$2\cos 4\theta = a^4 + \dfrac{1}{a^4}.$$

ANSWERS

4. $\dfrac{\sqrt{4-\sqrt{6+\sqrt{2}}}}{2\sqrt{2}}, \dfrac{\sqrt{4+\sqrt{6+\sqrt{2}}}}{2\sqrt{2}}, -(\sqrt{2}+1) + \sqrt{4+2\sqrt{2}}$

6.11 CONDITIONAL IDENTITIES

Type 1. Identities which involve sine and cosines.
Type 2. Identities which involve square of sines and cosines.
Type 3. Identities which involve tangents and cotangents.

BASED ON TYPE 1

EXAMPLE 1. *If $A + B + C = \pi$, prove that*

$$\sin 2A + \sin 2B + \sin 2C = 4 \sin A \sin B \sin C.$$ **[Meerut(B.Sc. Biotech)–2005]**

SOLUTION. We have,

$$\sin 2A + \sin 2B + \sin 2C = 2 \sin (A + B) \cos (A - B) + \sin 2C$$
$$= 2 \sin C \cos (A - B) + \sin 2C$$

$$[\text{Given } A + B + C = \pi$$
$$C = \pi - (A + B)$$
$$\sin C = \sin [\pi - (A + B)]]$$

$$= 2 \sin C \cos (A - B) + 2 \sin C \cos C$$
$$= 2 \sin C [\cos (A - B) + \cos C]$$
$$= 2 \sin C [\cos (A - B) - \cos(A + B)]$$

$$[\cos C = \cos [\pi - (A + B) = - \cos (A + B)]$$

$$= 2 \sin C[2 \sin A \sin B]$$
$$= 4 \sin A \sin B \sin C.$$

EXAMPLE 2. *Prove that $\sin (B + C - A) + \sin (C + A - B) + \sin (A + B - C) = 4 \sin A \sin B \sin C$ if $A + B + C = \pi$.*

SOLUTION. We have

$$\sin (B + C - A) + \sin (C + A - B) + \sin (A + B - C)$$
$$= \sin (\pi - 2A) + \sin(\pi - 2B) + \sin(\pi - 2C)$$

$$[\text{since } B + C = \pi - A, C + A = \pi - B, A + B = \pi - C]$$

$$= \sin 2A + \sin 2B + \sin 2C$$
$$= 4 \sin A \sin B \sin C \qquad \text{[From Ex. no. 1]}$$

EXAMPLE 3. *Prove that $\cos 2A + \cos 2B - \cos 2C = 1 - 4 \cos A \cos B \cos C$ if $A + B + C = \pi$.*

SOLUTION. We have,

$$\cos 2A + \cos 2B - \cos 2C = 2 \cos(A + B) \cos (A - B) - \cos 2C$$
$$= 2 \cos (\pi - C) \cos (A - B) - 2 \cos^2 C + 1$$
$$= - 2 \cos C \cos (A - B) - 2 \cos^2 C + 1$$
$$= 1 - 2 \cos C[\cos (A - B) + \cos C]$$
$$= 1 - 2 \cos C [\cos (A - B) + \cos \{\pi - (A + B)\}]$$
$$= 1 - 2 \cos C [\cos (A - B) - \cos (A + B)]$$
$$= 1 - 2 \cos C [2 \cos A \cos B]$$
$$= 1 - 4 \cos A \cos B \cos C.$$

EXAMPLE 4. *If $A + B + C = \pi$, prove that $\sin A + \sin B - \sin C = 4 \sin \dfrac{A}{2} \sin \dfrac{B}{2} \cos \dfrac{C}{2}$.*

SOLUTION 4. We have $\sin A + \sin B - \sin C = 2 \sin \dfrac{A + B}{2} \cos \dfrac{A - B}{2} - 2 \sin \dfrac{C}{2} \cos \dfrac{C}{2}$

$$= 2 \sin \left(\dfrac{\pi}{2} - \dfrac{C}{2} \right) \cos \dfrac{A - B}{2} - 2 \sin \dfrac{C}{2} \cos \dfrac{C}{2}$$

$$= 2 \cos \dfrac{C}{2} \cos \dfrac{A - B}{2} - 2 \sin \dfrac{C}{2} \cos \dfrac{C}{2}$$

$$= 2\cos\frac{C}{2}\left[\cos\frac{A-B}{2} - \sin\frac{C}{2}\right]$$

$$= 2\cos\frac{C}{2}\left[\cos\frac{A-B}{2} - \sin\left\{\frac{\pi}{2} - \frac{A+B}{2}\right\}\right]$$

$$= 2\cos\frac{C}{2}\left[\cos\frac{A-B}{2} - \cos\frac{A+B}{2}\right]$$

$$= 2\cos\frac{C}{2}\left[2\sin\frac{A}{2}\sin\frac{B}{2}\right]$$

$$= 4\sin\frac{A}{2}\sin\frac{B}{2}\cos\frac{C}{2}$$

EXAMPLE 5. If $A + B + C = \pi$, prove that $\cos A + \cos B - \cos C = 4\cos\frac{A}{2}\cos\frac{B}{2}\sin\frac{C}{2} - 1$.

SOLUTION. We have

$$\cos A + \cos B - \cos C = 2\cos\frac{A+B}{2}\cos\frac{A-B}{2} - \left(1 - 2\sin^2\frac{C}{2}\right)$$

$$= 2\cos\left\{\frac{\pi}{2} - \frac{C}{2}\right\}\cos\frac{A-B}{2} - 1 + 2\sin^2\frac{C}{2}$$

$$= -2\sin\frac{C}{2}\cos\frac{A-B}{2} + 2\sin^2\frac{C}{2} - 1$$

$$= 2\sin\frac{C}{2}\left[\left\{\sin\left\{\frac{\pi}{2} - \left(\frac{A+B}{2}\right)\right\}\right\} - \cos\frac{A-B}{2}\right] - 1$$

$$= 2\sin\frac{C}{2}\left[\cos\frac{A+B}{2} - \cos\frac{A-B}{2}\right] - 1$$

$$= 2\sin\frac{C}{2}\left[2\cos\frac{A}{2}\cos\frac{B}{2}\right] - 1$$

$$= 4\cos\frac{A}{2}\cos\frac{B}{2}\sin\frac{C}{2} - 1$$

EXAMPLE 6. If $A + B + C = \pi$, prove that $\sin(B + 2C) + \sin(C + 2A) + \sin(A + 2B) = 4\sin\frac{B-C}{2}\cos\frac{C-A}{2}\sin\frac{A-B}{2}$.

SOLUTION. We have, $\sin(B + 2C) + \sin(C + 2A) + \sin(A + 2B)$

$$\sin(B + 2C) = \sin[(A + B + C) - (A - C)]$$
$$= \sin[\pi - (A - C)] = \sin(A - C)$$

Similarly, $\sin(C + 2A) = \sin(B - C)$
$$\sin(A + 2B) = \sin(C - A)$$

Therefore, L.H.S. can be written as

$\sin(A - C) + \sin(B - A) + \sin(C - B)$

$$= 2\sin\frac{1}{2}(B - C)\cos\frac{1}{2}(2A - B - C) - 2\sin\frac{1}{2}(B - C)\cos\frac{1}{2}(B - C)$$

$$= \sin\frac{1}{2}(B - C)\left[\cos\frac{1}{2}(2A - B - C) - \cos\frac{1}{2}(B - C)\right]$$

$$= 2\sin\frac{1}{2}(B-C)\left[2\sin\frac{1}{2}(A-C)\sin\frac{1}{2}(B-A)\right]$$

$$= 4\sin\frac{B-C}{2}\sin\frac{C-A}{2}\sin\frac{A-B}{2}$$

EXAMPLE 7. *If* $A + B + C = \dfrac{\pi}{2}$, *Prove that* : $\cos^2 A + \cos^2 B + \cos^2 C = 2 + 2\sin A \sin B \sin C$.

[UPTU(B. Pharma)– 2004, 2006]

SOLUTION. We know that $\cos 2A = 2\cos^2 A - 1, \cos 2B = 2\cos^2 B - 1$

$$\Rightarrow \qquad \cos^2 A = \frac{(1+\cos 2A)}{2}, \cos^2 B = \frac{(1+\cos 2B)}{2}$$

$$\therefore \qquad \text{L.H.S.} = \frac{1}{2}(1+\cos 2A) + \frac{1}{2}(1+\cos 2B) + \cos^2 C$$

$$= 1 + \frac{1}{2}\left(\cos 2A + \cos 2B\right) + \cos^2 C$$

$$= 1 + \frac{1}{2}\left[2\cos\left(\frac{2A+2B}{2}\right)\cdot\cos\left(\frac{2A-2B}{2}\right)\right] + \cos^2 C$$

$$= 1 + \cos(A+B)\cdot\cos(A-B) + \cos^2 C$$

$$= 1 + \cos\left(\frac{\pi}{2}-C\right)\cos(A-B) + \cos^2 C$$

as $A + B + C = \pi/2$

$$= 1 + \sin C \cdot \cos(A-B) + (1-\sin^2 C)$$

as $\cos\left(\dfrac{\pi}{2}-C\right) = \sin C$

$$= 2 + \sin C[\cos(A-B) - \sin C]$$

$$= 2 + \sin C\left[\cos(A-B) - \sin\left(\frac{\pi}{2}-(A+B)\right)\right]$$

$$= 2 + \sin C[\cos(A-B) - \cos(A+B)],$$

as $\sin\left(\dfrac{\pi}{2}-\theta\right) = \cos\theta$

$$= 2 + \sin C\left[2\sin\left(\frac{A-B+A+B}{2}\right)\sin\left(\frac{A+B-A-B}{2}\right)\right]$$

$$= 2 + 2\sin A \sin B \sin C = \text{R.H.S.}$$

EXAMPLE 8. *Prove that* $\dfrac{\cos A}{\sin B \sin C} + \dfrac{\cos B}{\sin C \sin A} + \dfrac{\cos C}{\sin A \sin B} = 2.$

SOLUTION. Taking

$$\text{L.H.S.} = \frac{\sin A \cos A + \sin B \cos B + \sin C \cos C}{\sin A \sin B \sin C}$$

$$= \frac{2\sin A \cos A + 2\sin B \cos B + 2\sin C \cos C}{2\sin A \sin B \sin C}$$

$$= \frac{\sin 2A + \sin 2B + \sin 2C}{2\sin A \sin B \sin C}$$

$$= \frac{4\sin A \sin B \sin C}{2\sin A \sin B \sin C} = 2 = \text{R.H.S.}$$

EXAMPLE 9. *Prove that* $\cos 4A + \cos 4B + \cos 4C = -1 + 4 \cos 2A \cos 2B \cos 2C.$

SOLUTION.

$$\text{L.H.S.} = \cos 4A + \cos 4B + \cos 4C$$
$$= 2 \cos(2A + 2B) \cos(2A - 2B) + 2 \cos^2 2C - 1$$
$$= 2 \cos(2\pi - 2C) \cos(2A - 2B) + 2 \cos^2 2C - 1$$
$$= 2 \cos 2C \cos(2A - 2B) + 2 \cos^2 2C - 1$$
$$= 2\cos 2C[\cos(2A - 2B) + \cos\{2\pi - (2A + 2B)\}] - 1$$
$$= 2 \cos 2C[\cos(2A - 2B) + \cos(2A + 2B)] - 1$$
$$= 2 \cos 2C[2 \cos 2A \cos 2B] - 1$$
$$= -1 + 4 \cos 2A \cos 2B \cos 2C = \text{R.H.S.}$$

BASED ON TYPE 2

EXAMPLE 10. *Show that* $\sin^2 \dfrac{A}{2} + \sin^2 \dfrac{B}{2} + \sin^2 \dfrac{C}{2} = 1 - 2\sin\dfrac{A}{2}\sin\dfrac{B}{2}\sin\dfrac{C}{2}.$

Given that $\qquad A + B + C = \pi$

SOLUTION. We have

$$\sin^2 \frac{A}{2} + \sin^2 \frac{B}{2} + \sin^2 \frac{C}{2} = 1 - \cos^2 \frac{A}{2} + \sin^2 \frac{B}{2} + \sin^2 \frac{C}{2}$$

$$= 1 - \left[\cos\frac{A+B}{2}\cos\frac{A-B}{2}\right] + \sin^2\left[\frac{\pi}{2} - \frac{A+B}{2}\right]$$

$$(\because \cos^2 A - \sin^2 B = \cos(A+B)\cos(A-B))$$

$$= 1 - \cos\left(\frac{A+B}{2}\right)\cos\left(\frac{A-B}{2}\right) - \cos^2\left(\frac{A+B}{2}\right)$$

$$= 1 - \cos\left(\frac{A+B}{2}\right)\left[\cos\left(\frac{A-B}{2}\right) - \cos\left(\frac{A+B}{2}\right)\right]$$

$$= 1 - \cos\left(\frac{A+B}{2}\right)\left[2\sin\frac{A}{2}\sin\frac{B}{2}\right]$$

$$= 1 - \cos\left[\frac{\pi}{2} - \frac{C}{2}\right]\left[2\sin\frac{A}{2}\sin\frac{B}{2}\right]$$

$$= 1 - 2\sin\frac{A}{2}\sin\frac{B}{2}\sin\frac{C}{2}$$

EXAMPLE 11. *If* $A + B + C = \pi.$ *Prove that*

$$\cos^2 \frac{A}{2} + \cos^2 \frac{B}{2} + \cos^2 \frac{C}{2} = 2 + 2\sin\frac{A}{2}\sin\frac{B}{2}\sin\frac{C}{2}$$

SOLUTION. We have, $\cos^2 \dfrac{A}{2} + \cos^2 \dfrac{B}{2} + \cos^2 \dfrac{C}{2} = 1 - \sin^2 \dfrac{A}{2} + \cos^2 \dfrac{B}{2} + \cos^2 \dfrac{C}{2}$

$$= 1 + \cos^2 \frac{B}{2} - \sin^2 \frac{A}{2} + \cos^2 \frac{C}{2}$$

$$= 1 + \left[\cos^2 \frac{A+B}{2}\cos\frac{B-A}{2}\right] + \cos^2 \frac{C}{2}$$

$$= 1 + \cos\left(\frac{A+B}{2}\right)\cos\left(\frac{A-B}{2}\right) + 1 - \sin^2 \frac{C}{2}$$

$$= 1 + \cos\left[\frac{\pi}{2} - \frac{C}{2}\right]\cos\left(\frac{A-B}{2}\right) - \sin^2\frac{C}{2}$$

$$= 2 + \sin\frac{C}{2}\cos\left(\frac{A-B}{2}\right) - \sin^2\frac{C}{2}$$

$$= 2 + \sin\frac{C}{2}\left[\cos\left(\frac{A-B}{2}\right) - \sin\frac{C}{2}\right]$$

$$= 2 + \sin\frac{C}{2}\left[\cos\left(\frac{A-B}{2}\right) - \sin\left\{\frac{\pi}{2} - \left(\frac{A+B}{2}\right)\right\}\right]$$

$$= 2 + \sin\frac{C}{2}\left[\cos\left(\frac{A-B}{2}\right) - \cos\left(\frac{A+B}{2}\right)\right]$$

$$= 2 + \sin\frac{C}{2}\left[\cos\left(\frac{A-B}{2}\right) - \sin\left(\frac{A+B}{2}\right)\right]$$

$$= 2 + \sin\frac{C}{2}\left[2\sin\frac{A}{2}\sin\frac{B}{2}\right] = 2 + 2\sin\frac{A}{2}\sin\frac{B}{2}\sin\frac{C}{2}$$

BASED ON TYPE 3

EXAMPLE 12. *If* $A + B + C = \pi$, *prove that* $\tan A + \tan B + \tan C = \tan A \tan B \tan C$.

[UPTU(B. Pharma)–2004, 07]

SOLUTION. Given that

$$A + B + C = \pi$$

$$\Rightarrow \qquad A + B = \pi - C$$

$$\Rightarrow \qquad \tan(A + B) = \tan \pi - C$$

$$\Rightarrow \qquad \frac{\tan A + \tan B}{1 - \tan A \tan B} = -\tan C$$

$$\Rightarrow \qquad \tan A + \tan B = -\tan C + \tan A \tan B \tan C$$

$$\Rightarrow \qquad \tan A + \tan B + \tan C = \tan A \tan B \tan C.$$

EXAMPLE 13. *If* $A + B + C = \pi$, *show that* $\dfrac{\cot B + \cot C}{\tan B + \tan C} + \dfrac{\cot C + \cot A}{\tan C + \tan A} + \dfrac{\cot A + \cot B}{\tan A + \tan B} = 1$.

SOLUTION. We have $\dfrac{\cot B + \cot C}{\tan B + \tan C} + \dfrac{\cot C + \cot A}{\tan C + \tan A} + \dfrac{\cot A + \cot B}{\tan A + \tan B}$

$$= \frac{\dfrac{1}{\tan B} + \dfrac{1}{\tan C}}{\tan B + \tan C} + \frac{\dfrac{1}{\tan C} + \dfrac{1}{\tan A}}{\tan C + \tan A} + \frac{\dfrac{1}{\tan A} + \dfrac{1}{\tan B}}{\tan A + \tan B}$$

$$= \frac{1}{\tan B \tan C} + \frac{1}{\tan C \tan A} + \frac{1}{\tan A \tan B}$$

$$= \cot B \cot C + \cot A \cot C + \cot A \cot B \qquad \dots (1)$$

Again, $\qquad A + B + C = \pi$

$$\Rightarrow \qquad A + B = \pi - C$$

$$\Rightarrow \qquad \cot (A + B) = \cot(\pi - C)$$

$$\Rightarrow \qquad \frac{\cot A \cot B - 1}{\cot A + \cot B} = -\cot C .$$

$$\Rightarrow \qquad \cot A \cot B - 1 = -\cot C \cot A - \cot B \cot C.$$

$$\Rightarrow \qquad \cot A \cot B + \cot B \cot C + \cot C \cot A = 1.$$

Hence, $\dfrac{\cot A + \cot B}{\tan A + \tan B} + \dfrac{\cot B + \cot C}{\tan B + \tan C} + \dfrac{\cot C + \cot A}{\tan C + \tan A}$

$$= \cot A \cot B + \cot B \cot C + \cot C \cot A = 1$$

EXAMPLE 14. *If* $A + B + C = \pi$,

$$\dfrac{\tan A}{\tan B} + \dfrac{\tan B}{\tan C} + \dfrac{\tan C}{\tan A} + \dfrac{\tan B}{\tan A} + \dfrac{\tan C}{\tan B} + \dfrac{\tan A}{\tan C}$$

$$= \cos A \sec B \sec C + \cos B \sec C \sec A + \cos C \sec A \sec B.$$

SOLUTION.   We have, $\left(\dfrac{\tan A}{\tan B} + \dfrac{\tan C}{\tan B}\right) + \left(\dfrac{\tan B}{\tan C} + \dfrac{\tan A}{\tan C}\right) + \left(\dfrac{\tan C}{\tan A} + \dfrac{\tan B}{\tan A}\right)$

$$= \dfrac{\dfrac{\sin A}{\cos A} + \dfrac{\sin C}{\cos C}}{\dfrac{\sin B}{\cos B}} + \dfrac{\dfrac{\sin B}{\cos B} + \dfrac{\sin A}{\cos A}}{\dfrac{\sin C}{\cos C}} + \dfrac{\dfrac{\sin C}{\cos C} + \dfrac{\sin B}{\cos B}}{\dfrac{\sin A}{\cos A}}$$

$$= \dfrac{\sin(A+C)\cos B}{\cos A \cos C \sin B} + \dfrac{\sin(A+B)\cos C}{\cos A \cos B \sin C} + \dfrac{\sin(B+C)\cos A}{\cos B \cos C \sin A}$$

$$= \dfrac{\sin(\pi - B)\cos B}{\cos A \cos C \sin B} + \dfrac{\sin(\pi - C)\cos C}{\cos A \cos B \sin C} + \dfrac{\sin(\pi - A)\cos A}{\cos B \cos C \sin A}$$

$$= \dfrac{\sin B \cos B}{\cos A \cos C \sin B} + \dfrac{\sin C \cos C}{\cos A \cos B \sin C} + \dfrac{\sin A \cos A}{\cos B \cos C \sin A}$$

$$= \cos B \sec A \sec C + \cos C \sec A \sec B$$
$$+ \cos A \sec B \sec C.$$

Exercise 6.6

If $A + B + C = \pi$, prove that

1. $\sin A + \sin B + \sin C = 4 \cos \dfrac{A}{2} \cos \dfrac{B}{2} \cos \dfrac{C}{2}$

2. $\cos 2A + \cos 2B + \cos 2C$
   $$= -1 - 4 \cos A \cos B \cos C.$$

3. $\cos A + \cos B + \cos C$
   $$= 1 + 4 \sin \dfrac{A}{2} \sin \dfrac{B}{2} \sin \dfrac{C}{2}$$

4. $\dfrac{\sin 2A + \sin 2B + \sin 2C}{\sin A + \sin B + \sin C} = 8 \sin \dfrac{A}{2} \sin \dfrac{B}{2} \sin \dfrac{C}{2}$

5. $\sin 3A + \sin 3B + \sin 3C$
   $$= 4 \cos \dfrac{3A}{2} \cos \dfrac{3B}{2} \cos \dfrac{3C}{2}$$

6. $\cos \dfrac{A}{2} + \cos \dfrac{B}{2} + \cos \dfrac{C}{2}$
   $$= 4 \cos \dfrac{\pi - A}{2} \cos \dfrac{\pi - B}{2} \cos \dfrac{\pi - C}{2}$$

7. $\sin \dfrac{A}{2} + \sin \dfrac{B}{2} + \sin \dfrac{C}{2}$
   $$= 1 + 4 \sin\left(\dfrac{\pi - A}{2}\right) \sin\left(\dfrac{\pi - B}{2}\right) \sin\left(\dfrac{\pi - C}{2}\right)$$

8. $\sin^2 \dfrac{A}{2} + \sin^2 \dfrac{B}{2} - \sin^2 \dfrac{C}{2}$
   $$= 1 - 2 \cos \dfrac{A}{2} \cos \dfrac{B}{2} \sin \dfrac{C}{2}$$

9. $\cos^2 \dfrac{A}{2} + \cos^2 \dfrac{B}{2} - \cos^2 \dfrac{C}{2}$
   $$= 2 \cos \dfrac{A}{2} \cos \dfrac{B}{2} \sin \dfrac{C}{2}$$

10. $\cos^2 \dfrac{A}{2} - \sin^2 \dfrac{B}{2} - \sin^2 \dfrac{C}{2}$
    $$= 2 \sin \dfrac{A}{2} \sin \dfrac{B}{2} \sin \dfrac{C}{2}$$

11. $\sin^2 A + \sin^2 B - \sin^2 C = 2 \sin A \sin B \cos C.$

12. $\cot B \cot C + \cot C \cot A + \cot A \cot B = 1.$

13. $\tan 2A + \tan 2B + \tan 2C$
    $$= \tan 2A \tan 2B \tan 2C$$

14. $\tan \dfrac{A}{2} \tan \dfrac{B}{2} + \tan \dfrac{B}{2} \tan \dfrac{C}{2} + \tan \dfrac{C}{2} \tan \dfrac{A}{2} = 1$

Objective Evaluation

∞ MULTIPLE CHOICE QUESTIONS

Choose the most appropriate one.

1. In correct statement is :
   (a) $\sin \theta = -1/5$ (b) $\cos \theta = 1$
   (c) $\sec \theta = 1/2$ (d) $\tan \theta = 20$.

2. The value of $\sin^2 20° + \sin^2 70°$ is :
   (a) 1   (b) 0
   (c) –1   (d) 1/2

3. If $\sin \theta = -(1/2)$ and $\cos \theta = (\sqrt{3}/2)$, then θ lies in the quadrant :
   (a) I   (b) II
   (c) III   (d) IV

4. If $\tan \theta + \cot \theta = 2$, then $\sin \theta$ is :
   (a) 1   (b) $\pm\sqrt{2}$
   (c) $\sqrt{2}$ (d) $\pm 1/\sqrt{2}$

5. Value of θ which satisfy the equation $\sin \theta + \cos \theta = 2 \sin \theta \cos \theta$, are :
   (a) $0, \dfrac{2\pi}{3}, \dfrac{3\pi}{2}$ (b) $\dfrac{\pi}{3}, \dfrac{\pi}{2}$
   (c) $\dfrac{2\pi}{3}, \dfrac{4\pi}{2}, 2\pi$ (d) none of these

6. The value of $\dfrac{\cos 54°}{\tan 36°} + \dfrac{\tan 20°}{\cot 70°}$ is :
   (a) 3   (b) 1
   (c) 0   (d) 2

7. The value of $\sin 4620°$ is :
   (a) 1/2   (b) $\sqrt{3}/2$
   (c) $-\sqrt{3}/2$ (d) –1/2

8. If $\tan A = \dfrac{1}{2}$ and $\tan B = \dfrac{1}{3}$, then $(A + B)$ equal to :
   (a) $\pi/6$ (b) 0
   (c) $\pi/4$ (d) π

9. If $\sin A = \dfrac{4}{5}$, $\sin B = \dfrac{5}{13}$, then $\sin (A + B)$ is :
   (a) $\pm\dfrac{35}{65}$ or $\pm\dfrac{45}{65}$ (b) $\pm\dfrac{35}{65}$ or $\pm\dfrac{63}{65}$
   (c) $\pm\dfrac{63}{65}$ or $\pm\dfrac{33}{65}$ (d) $\pm\dfrac{45}{65}$ or $\pm\dfrac{33}{65}$

10. Which of the following in not true?
    (a) $\sin^2\theta = 1 - \cos^2\theta$
    (b) $1 + \tan^2\theta = \sec^2\theta$
    (c) $\cot^2\theta - \csc^2\theta = 1$
    (d) $\sin^2\theta + \cos^2\theta = 1$

11. If $\sin 3A = \cos (A - 26°)$ where $3A$ is an acute angle, then value of A is :
    (a) $A = 29°$ (b) $A = 39°$
    (c) $A = 64°$ (d) none of these

12. If $\sin (\theta + 36°) = \cos \theta$, $(\theta + 36°)$ is an acute angle. Then θ is equal to :
    (a) $25°$ (b) $54°$
    (c) $27°$ (d) $29°$

13. If $5 \tan \theta - 4 = 0$, then the value of
    $\dfrac{5 \sin \theta - 4 \cos \theta}{5 \sin \theta + 4 \cos \theta}$ is :
    (a) $\dfrac{5}{3}$ (b) $\dfrac{5}{6}$
    (c) 0   (d) $\dfrac{1}{6}$

14. Value of $\tan 5° \tan 25° \tan 30° \tan 65° \tan 85°$ is :
    (a) $\sqrt{3}$ (b) $\dfrac{1}{\sqrt{3}}$
    (c) 1   (d) 0

15. If $8 \tan x = 15$ then $\sin x - \cos x$ is equal to :
    (a) $\dfrac{8}{17}$ (b) $\dfrac{17}{7}$
    (c) $\dfrac{1}{17}$ (d) $\dfrac{7}{17}$

16. If $\tan \theta = \dfrac{3}{4}$, then the value of $\dfrac{1 - \cos \theta}{1 + \cos \theta}$ is :
    (a) $\dfrac{5}{9}$ (b) $\dfrac{9}{5}$
    (c) $\dfrac{1}{5}$ (d) $\dfrac{1}{9}$

17. $\dfrac{2 \tan 30°}{1 + \tan^2 30°}$ is equal to :
    (a) $\sin 60°$ (b) $\cos 60°$
    (c) $\tan 60°$ (d) $\sin 30°$

18. If $8 \tan x = 15$ then $\sin^2 x - \cos^2 x$ equal to :
    (a) $\dfrac{8}{17}$ (b) $\dfrac{7}{17}$
    (c) $\dfrac{30}{17}$ (d) none of these

19. If $\tan \theta = \dfrac{12}{5}$, then $\dfrac{1+\sin\theta}{1-\sin\theta}$ is :

(a) $\dfrac{5}{12}$ (b) $\dfrac{25}{13}$

(c) $\dfrac{13}{25}$ (d) 25

20. $x = \cot A + \cos A$ and $y = \cot A - \cos A$, then

$\left(\dfrac{x-y}{x+y}\right)^2 + \left(\dfrac{x-2}{2}\right)^2$ is :

(a) 0   (b) -1

(c) 1   (d) none of these

✑ TRUE/FALSE

Write 'T' for True and 'F' for False statement.

1. If $\theta = 30°$, then $\tan 2\theta = \dfrac{2\tan\theta}{1-\tan^2\theta}$. **(T/F)**

2. A, B, C are interior angles of ΔABC, then $\cos\dfrac{B+C}{2} = \sin\dfrac{A}{2}$. **(T/F)**

3. An equation involving trigonometric ratios of an angle is called a trigonometric identity if it is true for all values of the angle. **(T/F)**

4. $1 - \sec^2\theta = \tan^2\theta$. **(T/F)**

5. If A, B, C are interior angles of ΔABC, then $\tan\left(\dfrac{C+A}{2}\right) = \tan\dfrac{B}{2}$. **(T/F)**

6. The measure of an angle is the amount of rotation from the initial side to the terminal side. **(T/F)**

7. $\tan(A+B) = \tan A + \tan B$. **(T/F)**

8. $\sec 5A = \operatorname{cosec}(A - 36°)$, where $5A$ is an acute angle then A is $21°$. **(T/F)**

9. $\operatorname{cosec}\sqrt{1-\cos^2\theta} = 1$. **(T/F)**

✑ FILL IN THE BLANKS

1. If $2\sin\dfrac{x}{2} = 1$, then x is equal to _____.

2. If θ is a positive acute angle such that $\sec\theta = \operatorname{cosec}60°$, then $2\cos^2\theta - 1$ is _____.

3. $(1 + \tan\theta + \sec\theta)(1 + \cot\theta - \operatorname{cosec}\theta)$ is equal to _____.

4. In a ΔABC if $\angle B = 90°$ and $\sin A = \dfrac{3}{4}$, $\tan A$ is equal to _____.

5. If $x = a\cos\theta$ and $y = b\sin\theta$, $b^2x^2 + a^2y^2$ is _____.

6. If $6 + A = 3$, then $\operatorname{cosec} A$ equal to _____.

ANSWERS

✑ MULTIPLE CHOICE QUESTIONS

1. (c)	**2.** (a)	**3.** (d)	**4.** (d)	**5.** (c)	**6.** (d)	**7.** (c)	**8.** (c)	**9.** (c)
10. (c)	**11.** (d)	**12.** (c)	**13.** (c)	**14.** (b)	**15.** (d)	**16.** (d)	**17.** (a)	**18.** (d)
19. (d)	**20.** (a)							

✑ TRUE / FALSE

1. F	**2.** T	**3.** T	**4.** F	**5.** T	**6.** T	**7.** F	**8.** T	**9.** T

✑ FILL IN THE BLANKS

1. $60°$		**2.** $1/2$	**3.** 2	**4.** $3/\sqrt{7}$	**5.** a^2b^2	**6.** $\sqrt{10}$

⬚⬚⬚⬚⬚⬚

CHAPTER 7

Determinants

7.1 INTRODUCTION

Consider two homogeneous linear equations

$$a_1 x + b_1 y = 0$$
$$a_2 x + b_2 y = 0;$$

Multiplying the first equation by b_2 and second by b_1 and then subtracting and dividing by x, we obtained

$$a_1 b_2 - a_2 b_1 = 0$$

This result is sometimes written as

$$\begin{vmatrix} a_1 & b_1 \\ a_2 & b_2 \end{vmatrix} = 0$$

and the expression on the left is called determinant.

A determinant also is an arrangement of numbers in rows and columns but it always has a square form and can be reduced to a single value. Therefore, a determinant is distinct from matrix in the sense that the determinant is always in square shape and it has a numerical value. The arrangement of the numbers of a determinant is enclosed within two vertical parallel lines.

7.1.1 ORDER OF A DETERMINANT

The determinant of a square matrix of order n is known as determinant of order n.

7.2 DETERMINANT OF ORDER TWO

Let $a_{11}, a_{12}, a_{21}, a_{22}$ be any four number (real or complex). Then

$$|A| = \begin{vmatrix} a_{11} & b_{12} \\ a_{21} & b_{22} \end{vmatrix}$$

represent the number $a_{11} a_{22} - a_{21} a_{12}$ and is called a determinant of order two.

For Example $\quad |A| = \begin{vmatrix} 5 & 2 \\ 3 & -7 \end{vmatrix} = (5)(-7) - (3)(2) = -35 - 6 = -41$

7.3 DETERMINANT OF ORDER THREE

Let $\qquad |A| = \begin{vmatrix} a_{11} & a_{12} & a_{13} \\ a_{21} & a_{22} & a_{23} \\ a_{31} & a_{32} & a_{33} \end{vmatrix}$

is called a determinant of order 3 and its value can be obtained as follows:

$$|A| = a_{11} \begin{vmatrix} a_{22} & a_{23} \\ a_{32} & a_{33} \end{vmatrix} - a_{12} \begin{vmatrix} a_{21} & a_{23} \\ a_{31} & a_{33} \end{vmatrix} + a_{13} \begin{vmatrix} a_{21} & a_{22} \\ a_{31} & a_{32} \end{vmatrix}$$

$$= a_{11}(a_{22}a_{33} - a_{32}a_{23}) - a_{12}(a_{21}a_{33} - a_{31}a_{23}) + a_{13}(a_{21}a_{32} - a_{31}a_{22})$$

For Example, $|A| = \begin{vmatrix} 2 & 3 & 5 \\ -1 & 2 & 3 \\ 4 & -2 & 1 \end{vmatrix} = 2\begin{vmatrix} 2 & 3 \\ -2 & 1 \end{vmatrix} - 3\begin{vmatrix} -1 & 3 \\ 4 & 1 \end{vmatrix} + 4\begin{vmatrix} -1 & 2 \\ 4 & -2 \end{vmatrix}$

$$= 2(2 + 6) - 3(-1 - 12) + 4(2 - 8) = 16 + 39 - 24 = 31$$

⮚ REMARKS

➠ The value of a determinant is not changed if it is expanded along any row or column.

➠ When no reference of the corresponding matrix is needed, we may denote a determinant by D.

➠ The determinant of a square zero matrix is zero.

SOLVED EXAMPLES

EXAMPLE 1. *Find the value of* $|A|$ *if A is given by* $\begin{vmatrix} \cos\alpha & -\sin\alpha \\ \sin\alpha & \cos\alpha \end{vmatrix}$.

SOLUTION. We have $|A| = \begin{vmatrix} \cos\alpha & -\sin\alpha \\ \sin\alpha & \cos\alpha \end{vmatrix} = \cos^2\alpha - (-\sin^2\alpha) = \cos^2\alpha + \sin^2\alpha = 1$

EXAMPLE 2. *Find the value of* $\begin{vmatrix} 1 & \omega \\ \omega & -\omega \end{vmatrix}$.

SOLUTION. We have $|A| = \begin{vmatrix} 1 & \omega \\ \omega & -\omega \end{vmatrix} = -\omega - \omega^2 = -(\omega + \omega^2) = (-1) = 1$.

EXAMPLE 3. *Solve for x.*

$$\begin{vmatrix} x & 3 \\ 5 & 2x \end{vmatrix} = \begin{vmatrix} 5 & -4 \\ 5 & 3 \end{vmatrix}$$

SOLUTION. We have $\begin{vmatrix} x & 3 \\ 5 & 2x \end{vmatrix} = \begin{vmatrix} 5 & -4 \\ 5 & 3 \end{vmatrix}$

$\Rightarrow \qquad 2x^2 - 15 = 15 + 20 \qquad \Rightarrow \qquad 2x^2 = 50$

$\Rightarrow \qquad\qquad x^2 = 25 \qquad\qquad \Rightarrow \qquad\qquad x = \pm 5$

EXAMPLE 4. *Find the value of* $\begin{vmatrix} a & h & g \\ h & b & f \\ g & f & c \end{vmatrix}$.

SOLUTION. Let $\Delta = \begin{vmatrix} a & h & g \\ h & b & f \\ g & f & c \end{vmatrix}$

We expand Δ along first row, we get

$$\Delta = a(-1)^2 \begin{vmatrix} b & f \\ f & c \end{vmatrix} + h(-1)^3 \begin{vmatrix} h & f \\ g & c \end{vmatrix} + g(-1)^4 \begin{vmatrix} h & b \\ g & f \end{vmatrix}$$

$$= a(bc - f^2) - h(hc - fg) + g(hf - bg)$$

$$= abc - af^2 - ch^2 + fgh + fgh - bg^2$$

$$= abc + 2fgh - af^2 - bg^2 - ch^2$$

EXAMPLE 5. *Find the value of* $\begin{vmatrix} 0 & 1 & \sec\theta \\ \tan\theta & -\sec\theta & \tan\theta \\ 1 & 0 & 1 \end{vmatrix}$.

SOLUTION. Let
$$\Delta = \begin{vmatrix} 0 & 1 & \sec\theta \\ \tan\theta & -\sec\theta & \tan\theta \\ 1 & 0 & 1 \end{vmatrix}$$

Expand along R_1

$$\Delta = 0 + 1(-1)^3 \begin{vmatrix} \tan\theta & \tan\theta \\ 1 & 1 \end{vmatrix} + \sec\theta(-1)^4 \begin{vmatrix} \tan\theta & -\sec\theta \\ 1 & 0 \end{vmatrix}$$

$$= -(\tan\theta - \tan\theta) + \sec\theta(0 + \sec\theta)$$

$$= \sec\theta(0 + \sec\theta) = \sec^2\theta.$$

7.4 CO–FACTORS AND MINORS OF AN ELEMENT

If in the expansion of a determinant $|a_{ij}|$, all the containing a_{ij} as a factor, are collected and their sum a is denoted by $a_{ij}A_{ij}$ then the factor A_{ij} is called the co-factor of the element a_{ij}. Hence, in a determinant of order n

$$|a_{ij}| = a_{i1}A_{i1} + a_{i2}A_{i2} + \ldots + a_{in}A_{in} = \sum_{j-1}^{n} a_{ij}A_{ij}$$

Now, let M_{ij} be the $(n-1) \times (n-1)$ sub-matrix of $|a_{ij}|_{n \times n}$ obtained by deleting the ith row and j^{th} column. Then $|M_{ij}|$ is called the minor of the element a_{ij} the determinant $|a_{ij}|$ of order n. Thus, we can express the determinant as a linear combination of the minors of the elements of any row or any column.

🖘 REMARK

➠ $(-1)^{t+j}$ is 1 or –1 according as $t + j$ is even or odd.

∴ A_{ij} and M_{ij} coincides if $i + j$ is even and if $i + j$ is odd then we have $A_{ij} = -M_{ij}$.

SOLVED EXAMPLES

EXAMPLE 1. *Find the minors and cofactors of elements of the determinant* $\begin{vmatrix} 5 & -2 \\ 3 & 7 \end{vmatrix}$.

SOLUTION. Minor of the element a_{11} is $M_{11} = |7| = 7$

Minor of the element a_{12} is $M_{12} = 3$

Minor of the element a_{21} is $M_{21} = -2$

Minor of the element a_{22} is $M_{22} = 5$

Hence, $A_{11} = (-1)^{1+1}M_{11} = 7$

$A_{12} = (-1)^{1+2}M_{12} = -3$

$A_{21} = (-1)^{2+1}M_{21} = 2$

$A_{22} = (-1)^{2+2}M_{22} = 5$

EXAMPLE 2. *Find all the minors and cofactors of the elements in following determinants*
$$\begin{vmatrix} 4 & 3 & 1 \\ 1 & 3 & 2 \\ 2 & 1 & 5 \end{vmatrix}$$

SOLUTION. Here, $a_{11} = 4, a_{12} = 3, a_{13} = 1$

$a_{21} = 1, a_{22} = 3, a_{23} = 2$

$a_{31} = 2, a_{32} = 1, a_{33} = 5$

$$M_{11} = \begin{vmatrix} 3 & 2 \\ 1 & 5 \end{vmatrix} = 15 - 2 = 13 \qquad M_{12} = \begin{vmatrix} 1 & 2 \\ 2 & 5 \end{vmatrix} = 5 - 4 = 1$$

$$M_{13} = \begin{vmatrix} 1 & 3 \\ 2 & 1 \end{vmatrix} = 1 - 6 = -5 \qquad M_{21} = \begin{vmatrix} 3 & 1 \\ 1 & 5 \end{vmatrix} = 15 - 1 = 14$$

$$M_{22} = \begin{vmatrix} 4 & 1 \\ 2 & 5 \end{vmatrix} = 20 - 2 = 18 \qquad M_{23} = \begin{vmatrix} 4 & 3 \\ 2 & 1 \end{vmatrix} = 4 - 6 = -2$$

$$M_{31} = \begin{vmatrix} 3 & 1 \\ 3 & 2 \end{vmatrix} = 6 - 3 = 3 \qquad M_{32} = \begin{vmatrix} 4 & 1 \\ 1 & 2 \end{vmatrix} = 8 - 1 = 7$$

$$M_{33} = \begin{vmatrix} 4 & 3 \\ 1 & 3 \end{vmatrix} = 12 - 3 = 9.$$

The co-factors are

$$A_{11} = (-1)^{1+1} M_{11} = 1 \times 13 = 13$$
$$A_{12} = (-1)^{1+2} M_{12} = -1 \times 1 = -1$$
$$A_{13} = (-1)^{1+3} M_{13} = 1 \times (-5) = -5$$
$$A_{21} = (-1)^{2+1} M_{21} = -1 \times 14 = -14$$
$$A_{22} = (-1)^{2+2} M_{22} = 1 \times 18 = 18$$
$$A_{23} = (-1)^{2+3} M_{23} = -1 \times (-2) = 2$$
$$A_{31} = (-1)^{3+1} M_{31} = 1 \times 3 = 3$$
$$A_{32} = (-1)^{3+2} M_{32} = -1 \times 7 = -7$$
$$A_{33} = (-1)^{3+3} M_{33} = 1 \times 9 = 9.$$

EXAMPLE 3. *Find the minor and co-factors of elements of the following determinant*

$$\begin{vmatrix} 2 & -3 & 5 \\ 6 & 0 & 4 \\ 1 & 5 & -7 \end{vmatrix}.$$

SOLUTION. We have $M_{11} = \begin{vmatrix} 0 & 4 \\ 5 & -7 \end{vmatrix} = 0 - 20 = -20 \qquad A_{11} = -20$

$$M_{12} = \begin{vmatrix} 6 & 4 \\ 1 & -7 \end{vmatrix} = -42 - 4 = -46 \qquad A_{12} = 46$$

$$M_{13} = \begin{vmatrix} 6 & 0 \\ 1 & 5 \end{vmatrix} = 30 - 0 = 30 \qquad A_{13} = 30$$

$$M_{21} = \begin{vmatrix} -3 & 5 \\ 5 & -7 \end{vmatrix} = 21 - 25 = -4 \qquad A_{21} = 4$$

$$M_{22} = \begin{vmatrix} 2 & 5 \\ 1 & -7 \end{vmatrix} = -14 - 5 = -19 \qquad A_{22} = -19$$

$$M_{23} = \begin{vmatrix} 2 & -3 \\ 1 & 5 \end{vmatrix} = 10 + 3 = 13 \qquad A_{23} = -13$$

$$M_{31} = \begin{vmatrix} -3 & 5 \\ 0 & 4 \end{vmatrix} = -12 - 0 = -12 \qquad A_{31} = -12$$

$$M_{32} = \begin{vmatrix} 2 & 5 \\ 6 & 4 \end{vmatrix} = 8 - 30 = -22 \qquad A_{32} = 22$$

$$M_{33} = \begin{vmatrix} 2 & -3 \\ 6 & 0 \end{vmatrix} = 0 + 18 = 18 \qquad A_{33} = -18.$$

EXAMPLE 4. *Write the co-factors of elements of the second row of the following determinants and hence evaluate them* $\begin{vmatrix} 1 & a & bc \\ 1 & b & ca \\ 1 & c & ab \end{vmatrix}$

SOLUTION. Let $\Delta = \begin{vmatrix} 1 & a & bc \\ 1 & b & ca \\ 1 & c & ab \end{vmatrix}$

$$A_{21} = (-1)^{2+1} \begin{vmatrix} a & bc \\ c & ab \end{vmatrix} = -(a^2b - bc^2)$$

$$A_{22} = (-1)^{2+2} \begin{vmatrix} 1 & bc \\ 1 & ab \end{vmatrix} = ab - bc$$

$$A_{23} = (-1)^{2+3} \begin{vmatrix} 1 & a \\ 1 & c \end{vmatrix} = -(c - a) = a - c$$

$$\Delta = -(a^2b - bc^2) + b(ab - bc) + ca(a - c)$$

$$= bc^2 - a^2b + ab^2 - b^2c + a^2c - ac^2.$$

7.5 PROPERTIES OF DETERMINANTS

THEOREM 1. *The value of a determinant does not change when rows and columns are interchanged.*

PROOF. Let $\quad |A| = \begin{vmatrix} a_1 & b_1 & c_1 \\ a_2 & b_2 & c_2 \\ a_3 & b_3 & c_3 \end{vmatrix}$ be a determinant of order three.

Expanding $|A|$ along the first row, we get

$$|A| = a_1(b_2c_3 - b_3c_2) - b_1(a_2c_3 - a_3c_2) + c_1(a_2b_3 - a_3b_2)$$

$$= a_1(b_2c_3 - b_3c_2) - a_2(b_1c_3 - b_3c_1) + a_3(b_1c_2 - b_2c_1)$$

$$\text{(by rearrangement of terms)}$$

$$= \begin{vmatrix} a_1 & a_2 & a_3 \\ b_1 & b_2 & b_3 \\ c_1 & c_2 & c_3 \end{vmatrix}$$

Hence, the theorem is proved.

THEOREM 2. *If any two rows [or columns] of a determinant are interchanged, the sign of the determinant is changed.*

PROOF. Let $\quad |A| = \begin{vmatrix} a_1 & b_1 & c_1 \\ a_2 & b_2 & c_2 \\ a_3 & b_3 & c_3 \end{vmatrix}$ be a determinant of order three.

Expanding $|A|$ along the first row, we get

$$|A| = a_1(b_2c_3 - b_3c_2) - b_1(a_2c_3 - a_3c_2) + c_1(a_2b_3 - a_3b_2)$$

$$= -a_3(b_2c_1 - b_1c_2) - b_3(a_2c_1 - a_1c_2) + c_3(a_2b_1 - a_1b_2)$$

$$\text{(by rearrangement of terms)}$$

$$= -\begin{vmatrix} a_1 & a_2 & a_3 \\ b_1 & b_2 & b_3 \\ c_1 & c_2 & c_3 \end{vmatrix} = (-1)|A|$$

THEOREM 3. *If two rows or two columns of the determinant are identical, then the value of the determinant vanishes, i.e.,*

$$|A| = \begin{vmatrix} a_1 & b_1 & c_1 \\ a_2 & b_2 & c_2 \\ a_3 & b_3 & c_3 \end{vmatrix} = 0.$$

PROOF. We have $|A|$ is a determinant of order 3 whose first and third row are identical. If we interchange the two identical rows, then obviously there will be no change in the value of $|A|$. But by theorem 2, the value of A is multiplied by -1 if we interchange two rows. Therefore, we get

$$|A| = -|A|$$

$$2|A| = 0 \qquad \text{or} \qquad |A| = 0$$

THEOREM 4. *If all the elements of any row, or any column, of a determinant are multiplied by the same number then the determinant is multiplied by that number.*

PROOF. Let $|A| = \begin{vmatrix} a_{11} & a_{12} & \cdots & a_{1n} \\ a_{21} & a_{22} & \cdots & a_{2n} \\ \cdots & \cdots & \cdots & \cdots \\ a_{n1} & a_{n2} & \cdots & a_{nn} \end{vmatrix}$ be a determinant of order n

We have $\begin{vmatrix} ma_{11} & a_{12} & \cdots & a_{1n} \\ ma_{21} & a_{22} & \cdots & a_{2n} \\ \cdots & \cdots & \cdots & \cdots \\ ma_{n1} & a_{n2} & \cdots & a_{nn} \end{vmatrix} = ma_{i1}A_{i1} + ma_{i2}A_{i2} + \ldots ma_{in}A_{in} = m|A|$

(where $A_{i1}, A_{i2} \ldots, A_{in}$ be the cofactor of elements $a_{i1}, a_{i2}, \ldots, a_{in}$ of ith row of $|A|$)

THEOREM 5. *If in the determinant, the elements of a row are added in and m times the corresponding elements of the another rows (or column), the value of the determinant does not change in particular;*

$$\begin{vmatrix} a_1 + mb_1 + nc_1 & b_1 & c_1 \\ a_2 + mb_2 + nc_2 & b_2 & c_2 \\ a_3 + mb_3 + nc_3 & b_3 & c_3 \end{vmatrix} = \begin{vmatrix} a_1 & b_1 & c_1 \\ a_2 & b_2 & c_2 \\ a_3 & b_3 & c_3 \end{vmatrix}$$

PROOF. We have

$$\begin{vmatrix} a_1 + mb_1 + nc_1 & b_1 & c_1 \\ a_2 + mb_2 + nc_2 & b_2 & c_2 \\ a_3 + mb_3 + nc_3 & b_3 & c_3 \end{vmatrix} = \begin{vmatrix} a_1 & b_1 & c_1 \\ a_2 & b_2 & c_2 \\ a_3 & b_3 & c_3 \end{vmatrix} + \begin{vmatrix} mb_1 & b_1 & c_1 \\ mb_2 & b_2 & c_2 \\ mb_3 & b_3 & c_3 \end{vmatrix} + \begin{vmatrix} nc_1 & b_1 & c_1 \\ nc_2 & b_2 & c_2 \\ nc_3 & b_3 & c_3 \end{vmatrix}$$

$$= \begin{vmatrix} a_1 & b_1 & c_1 \\ a_2 & b_2 & c_2 \\ a_3 & b_3 & c_3 \end{vmatrix} + m\begin{vmatrix} b_1 & b_1 & c_1 \\ b_2 & b_2 & c_2 \\ b_3 & b_3 & c_3 \end{vmatrix} + n\begin{vmatrix} c_1 & b_1 & c_1 \\ c_2 & b_2 & c_2 \\ c_3 & b_3 & c_3 \end{vmatrix}$$

(By theorem 4)

$$= \begin{vmatrix} a_1 & b_1 & c_1 \\ a_2 & b_2 & c_2 \\ a_3 & b_3 & c_3 \end{vmatrix} + m(0) + n(0) \qquad \text{(By theorem 3)}$$

$$= \begin{vmatrix} a_1 & b_1 & c_1 \\ a_2 & b_2 & c_2 \\ a_3 & b_3 & c_3 \end{vmatrix}$$

☞ RECAPITULATIONS

- The value of a determinant does not change when rows and columns are interchanged.

- If any two rows [or columns] of a determinant are interchanged, the sign of the determinant is changed.

- If two rows or two columns of the determinant are identical, then the value of the determinant vanishes, i.e., $|A| = \begin{vmatrix} a_1 & b_1 & c_1 \\ a_2 & b_2 & c_2 \\ a_3 & b_3 & c_3 \end{vmatrix} = 0.$

- If all the elements of any row, or any column, of a determinant are multiplied by the same number then the determinant is multiplied by that number.

- If in the determinant, the elements of a row are added in and m times the corresponding elements of the another rows (or column), the value of the determinant does not change in particular, $\begin{vmatrix} a_1 + mb_1 + nc_1 & b_1 & c_1 \\ a_2 + mb_2 + nc_2 & b_2 & c_2 \\ a_3 + mb_3 + nc_3 & b_3 & c_3 \end{vmatrix} = \begin{vmatrix} a_1 & b_1 & c_1 \\ a_2 & b_2 & c_2 \\ a_3 & b_3 & c_3 \end{vmatrix}$

SOLVED EXAMPLES

EXAMPLE 1. *Evaluate the following determinant :* $\begin{vmatrix} 3 & -2 \\ 4 & 5 \end{vmatrix}$.

SOLUTION. We have $|A| = \begin{vmatrix} 3 & -2 \\ 4 & 5 \end{vmatrix} = 3 \times 5 - 4 \times (-2) = 15 + 8 = 23$.

EXAMPLE 2. *Find the value of the determinant of the matrix*

$$A = \begin{bmatrix} 1 & 2 & 3 \\ 2 & 3 & 1 \\ 3 & 1 & 2 \end{bmatrix}$$

SOLUTION. We have $\quad |A| = \begin{vmatrix} 1 & 2 & 3 \\ 2 & 3 & 1 \\ 3 & 1 & 2 \end{vmatrix}$

On expanding the determinant along the first row, we get

$$= 1 \begin{vmatrix} 3 & 1 \\ 1 & 2 \end{vmatrix} - 2 \begin{vmatrix} 2 & 1 \\ 3 & 2 \end{vmatrix} + 3 \begin{vmatrix} 2 & 3 \\ 3 & 1 \end{vmatrix}$$

$$= 1 \cdot (6 - 1) - 2 \cdot (4 - 3) + 3 \cdot (2 - 9) = -18.$$

EXAMPLE 3. *Evaluate the determinant of* $\begin{vmatrix} 4 & 1 & 4 \\ 0 & 1 & 0 \\ 1 & 2 & 1 \end{vmatrix}$.

SOLUTION. We have $\quad |A| = \begin{vmatrix} 4 & 1 & 4 \\ 0 & 1 & 0 \\ 1 & 2 & 1 \end{vmatrix}$

On expanding the determinant along the first column, we get

$$= 4 \begin{vmatrix} 1 & 0 \\ 2 & 1 \end{vmatrix} - 0 \begin{vmatrix} 1 & 4 \\ 2 & 1 \end{vmatrix} + 1 \begin{vmatrix} 1 & 4 \\ 1 & 0 \end{vmatrix}$$

$$= 4(1 - 0) - 0 + 1(0 - 4) = 4 - 4 = 0.$$

EXAMPLE 4. *Show that:*

$$\begin{vmatrix} 1 & x & y \\ 0 & \cos x & \sin y \\ 0 & \sin x & \cos y \end{vmatrix} = \cos(x + y).$$

SOLUTION. We have $\begin{vmatrix} 1 & x & y \\ 0 & \cos x & \sin y \\ 0 & \sin x & \cos y \end{vmatrix}$

On expanding the determinant along first column, we get

$$= 1 \begin{vmatrix} \cos x & \sin y \\ \sin x & \cos y \end{vmatrix} - 0 \begin{vmatrix} x & y \\ \sin x & \sin y \end{vmatrix} + 0 \begin{vmatrix} x & y \\ \cos x & \cos y \end{vmatrix}$$

$$= \cos x \cos y - \sin x \sin y = \cos(x + y)$$

EXAMPLE 5. *Show that* $\begin{vmatrix} 1 & 1 & 1 \\ 1 & 1+x & 1 \\ 1 & 1 & 1+y \end{vmatrix} = xy$

SOLUTION. We have $\quad$ L.H.S. $= \begin{vmatrix} 1 & 1 & 1 \\ 1 & 1+x & 1 \\ 1 & 1 & 1+y \end{vmatrix}$

Applying $C_2 - C_1$ and $C_3 - C_1$ in the given determinant, we get

$$= \begin{vmatrix} 1 & 0 & 0 \\ 1 & x & 0 \\ 1 & 0 & y \end{vmatrix}$$

On expanding the determinant along the first row, we get

$$= 1\begin{vmatrix} x & 0 \\ 0 & y \end{vmatrix} - 0\begin{vmatrix} 1 & 0 \\ 1 & y \end{vmatrix} - 0\begin{vmatrix} 1 & x \\ 1 & 0 \end{vmatrix} = xy = \text{R.H.S.}$$

EXAMPLE 6. Without expanding, show that $\begin{vmatrix} b-c & c-a & a-b \\ c-a & a-b & b-c \\ a-b & b-c & c-a \end{vmatrix} = 0$.

SOLUTION. We have $\begin{vmatrix} b-c & c-a & a-b \\ c-a & a-b & b-c \\ a-b & b-c & c-a \end{vmatrix} = \begin{vmatrix} 0 & c-a & a-b \\ 0 & a-b & b-c \\ 0 & b-c & c-a \end{vmatrix}$

$$\text{(Operating } C_1 \to C_1 + C_2 + C_3, \text{ we get)} = 0.$$

EXAMPLE 7. Without expanding, show that $\begin{vmatrix} b^2c^2 & bc & b+c \\ c^2a^2 & ca & c+a \\ a^2b^2 & ab & a+c \end{vmatrix} = 0$.

SOLUTION. Consider $\begin{vmatrix} b^2c^2 & bc & b+c \\ c^2a^2 & ca & c+a \\ a^2b^2 & ab & a+c \end{vmatrix} = \dfrac{abc}{abc}\begin{vmatrix} b^2c^2 & bc & b+c \\ c^2a^2 & ca & c+a \\ a^2b^2 & ab & a+b \end{vmatrix}$

$$\text{(Multiplying } R_1 \text{ by } a, R_2 \text{ by } b \text{ and } R_3 \text{ by } c)$$

$$= \frac{1}{abc}\begin{vmatrix} ab^2c^2 & abc & ab+ca \\ bc^2a^2 & abc & bc+ab \\ ca^2b^2 & abc & ca+bc \end{vmatrix}$$

$$\text{(Take } abc \text{ out from } C_1 \text{ and } C_2)$$

$$= \frac{abc.abc}{abc}\begin{vmatrix} bc & 1 & ab+ca \\ ca & 1 & bc+ab \\ ca & 1 & ca+bc \end{vmatrix}$$

$$= abc\begin{vmatrix} bc & 1 & ab+bc+ca \\ ca & 1 & ca+bc+ab \\ cb & 1 & ab+ca+bc \end{vmatrix}$$

$$\text{(Operate } C_3 \to C_3 + C_1)$$

$$= abc(ab+bc+ca)\begin{vmatrix} bc & 1 & 1 \\ ca & 1 & 1 \\ cb & 1 & 1 \end{vmatrix}$$

$$= abc(ab+bc+ca) \times 0 = 0$$

EXAMPLE 8. If a, b, c are in A.P., prove that $\begin{vmatrix} x+1 & x+2 & x+a \\ x+2 & x+3 & x+b \\ x+3 & x+4 & x+c \end{vmatrix} = 0$.

SOLUTION. Given a, b, c are in A.P. therefore $a + c = 2b$

$\Rightarrow \quad a + b - 2b = 0$

Operating $R_1 \to R_1 + R_3 - 2R_2$, we get

$$\begin{vmatrix} x+1 & x+2 & x+a \\ x+2 & x+3 & x+b \\ x+3 & x+4 & x+c \end{vmatrix} = \begin{vmatrix} 0 & 0 & a+c-2b \\ x+2 & x+3 & x+b \\ x+3 & x+4 & x+c \end{vmatrix} = \begin{vmatrix} 0 & 0 & 0 \\ x+2 & x+3 & x+b \\ x+3 & x+4 & x+c \end{vmatrix} = 0$$

EXAMPLE 9. *Prove that*

$$\begin{vmatrix} a & b & c \\ a^2 & b^2 & c^2 \\ a^3 & b^3 & c^3 \end{vmatrix} = abc \begin{vmatrix} 1 & 1 & 1 \\ a & b & c \\ a^2 & b^2 & c^2 \end{vmatrix} = abc(a-b)(b-c)(c-a).$$

SOLUTION. We have $\quad |A| = \begin{vmatrix} a & b & c \\ a^2 & b^2 & c^2 \\ a^3 & b^3 & c^3 \end{vmatrix} = abc \begin{vmatrix} 1 & 1 & 1 \\ a & b & c \\ a^2 & b^2 & c^2 \end{vmatrix}$

Now again $\quad |A| = abc \begin{vmatrix} 1 & 1 & 1 \\ a & b & c \\ a^2 & b^2 & c^2 \end{vmatrix}$

Applying $C_2 - C_1$ and $C_3 - C_1$, we get

$$= abc \begin{vmatrix} 1 & 0 & 0 \\ a & b-a & c-a \\ a^2 & b^2-a^2 & c^2-a^2 \end{vmatrix}$$

On expanding along the first row, we get

$$= abc \begin{vmatrix} b-a & c-a \\ b^2-a^2 & c^2-a^2 \end{vmatrix}$$

$$= abc[(b-a)(c^2-a^2) - (b^2-a^2)(c-a)]$$

$$= abc[(b-a)(c-a)\{(c+a)-(b+a)\}]$$

$$= abc(b-a)(c-a)(c+a-b-a) = abc(a-b)(b-c)(c-a)$$

EXAMPLE 10. *Prove that* $\begin{vmatrix} a+b+2c & a & b \\ c & b+c+2a & b \\ c & a & c+a+2b \end{vmatrix} = 2(a+b+c)^3.$

SOLUTION. Let $\quad |A| = \begin{vmatrix} a+b+2c & a & b \\ c & b+c+2a & b \\ c & a & c+a+2b \end{vmatrix}$

Adding C_2 and C_3 in C_1, we get

$$= \begin{vmatrix} 2(a+b+c) & a & b \\ 2(a+b+c) & b+c+2a & b \\ 2(a+b+c) & a & c+a+2b \end{vmatrix}$$

$$= 2(a+b+c) \begin{vmatrix} 1 & a & b \\ 1 & b+c+2a & b \\ 1 & a & c+a+2b \end{vmatrix}$$

Applying $(R_2 - R_1)$ and $(R_3 - R_1)$, we get

$$= 2(a+b+c)\begin{vmatrix} 1 & a & b \\ 0 & b+c+a & 0 \\ 0 & 0 & c+a+b \end{vmatrix}$$

On expanding the determinant along the first column, we get

$$= 2(a+b+c)\begin{vmatrix} b+c+a & 0 \\ 0 & a+b+c \end{vmatrix}$$

$$= 2(a + b + c)(a + b + c)^2$$

$$= 2(a + b + c)^3$$

EXAMPLE 11. *Prove that* $\begin{vmatrix} 1+a & 1 & 1 \\ 1 & 1+b & 1 \\ 1 & 1 & 1+c \end{vmatrix} = abc\left(1+\dfrac{1}{a}+\dfrac{1}{b}+\dfrac{1}{c}\right)$ [UPTU(B.Pharma)–2000, 06]

SOLUTION.   Operating $C_1 \to C_1 - C_3$ and $C_2 \to C_2 - C_3$, we get

$$\begin{vmatrix} 1+a & 1 & 1 \\ 1 & 1+b & 1 \\ 1 & 1 & 1+c \end{vmatrix} = \begin{vmatrix} a & 0 & 1 \\ 0 & b & 1 \\ -c & -c & 1+c \end{vmatrix}$$

$$= a[b \cdot (1 + c) - (-c) \cdot 1] + 1[0 \cdot (- c) - (- c)b]$$

$$= a(b + bc + c) + bc$$

$$= abc + bc + ca + ab = abc\left(1+\frac{1}{a}+\frac{1}{b}+\frac{1}{c}\right).$$

EXAMPLE 12. *Prove that* $\begin{vmatrix} a-b-c & 2a & 2a \\ 2b & b-c-a & 2b \\ 2c & 2c & c-a-b \end{vmatrix} = (a+b+c)^3.$

SOLUTION.   Operating $R_1 \to R_1 + R_2 + R_3$, we get

$$\begin{vmatrix} a-b-c & 2a & 2a \\ 2b & b-c-a & 2b \\ 2c & 2c & c-a-b \end{vmatrix} = \begin{vmatrix} a+b+c & a+b+c & a+b+c \\ 2b & b-c-a & 2b \\ 2c & 2c & c-a-b \end{vmatrix}$$

[Take $(a + b + c)$ out from R_1]

$$= (a+b+c)\begin{vmatrix} 1 & 1 & 1 \\ 2b & b-c-a & 2b \\ 2c & 2c & c-a-b \end{vmatrix}$$

(Operate $C_2 \to C_2 - C_1$ and $C_3 \to C_3 - C_1$)

$$= (a+b+c)\begin{vmatrix} 1 & 0 & 0 \\ 2b & -b-c-a & 0 \\ 2c & 0 & a-b-c \end{vmatrix}$$

(expand by R_1)

$$= (a + b + c)1(- a - b - c)(- a - b - c)$$

$$= (a + b + c)^3.$$

EXAMPLE 13. *Without expanding the determinant, show that*

$$\begin{vmatrix} 0 & b & -c \\ -b & 0 & a \\ c & -a & 0 \end{vmatrix} = 0$$

[UPTU(B.Pharma)–2001]

SOLUTION. Let
$$\Delta = \begin{vmatrix} 0 & -b & -c \\ b & 0 & a \\ c & -a & 0 \end{vmatrix}$$

By changing columns into rows :
$$\Delta = \begin{vmatrix} 0 & -b & c \\ b & 0 & -a \\ -c & a & 0 \end{vmatrix} = (-1)^3 \begin{vmatrix} 0 & b & -c \\ -b & 0 & a \\ c & -a & 0 \end{vmatrix}$$

(taking (–1) Common from each column)
$$= (-1)^3 \Delta = -\Delta.$$
$$2\Delta = 0 \quad \text{or} \quad \Delta = 0.$$

EXAMPLE 14. *Without expanding the determinant, show that*

$$\begin{vmatrix} 1 & a & bc \\ 1 & b & ca \\ 1 & c & ab \end{vmatrix} = \begin{vmatrix} 1 & a & a^2 \\ 1 & b & b^2 \\ 1 & c & c^2 \end{vmatrix} \text{ and evaluate it.} \qquad \text{[UPTU(B.Pharma)–2001, 08]}$$

SOLUTION. Let
$$\Delta = \begin{vmatrix} 1 & a & bc \\ 1 & b & ca \\ 1 & c & ab \end{vmatrix}$$

Multiplying the 1st, 2nd and 3rd rows by a, b, c respectively. we get

$$\Delta = \frac{1}{abc} \begin{vmatrix} a & a^2 & abc \\ b & b^2 & bca \\ c & c^2 & cab \end{vmatrix} = \frac{abc}{abc} \begin{vmatrix} a & a^2 & 1 \\ b & b^2 & 1 \\ c & c^2 & 1 \end{vmatrix}$$

Taking abc common from 3rd column

$$= \begin{vmatrix} a & 1 & a^2 \\ b & 1 & b^2 \\ c & 1 & c^2 \end{vmatrix} \qquad\qquad \text{applying } C_2 \leftrightarrow C_3$$

$$= \begin{vmatrix} 1 & a & a^2 \\ 1 & b & b^2 \\ 1 & c & c^2 \end{vmatrix} \qquad\qquad \text{. applying } C_1 \leftrightarrow C_2$$

Applying $R_2 \rightarrow R_2 - R_1$ and $R_3 \rightarrow R_3 - R_1$, we get

$$\Delta = \begin{vmatrix} 1 & a & a^2 \\ 0 & b-a & b^2-a^2 \\ 0 & c-a & c^2-a^2 \end{vmatrix} = \begin{vmatrix} b-a & b^2-a^2 \\ c-a & c^2-a^2 \end{vmatrix}$$

On expanding the determinant along C_1

$$= (b-a)(c-a) \begin{vmatrix} 1 & b+a \\ 1 & c+a \end{vmatrix}$$

taking $(b-a)$ common from R_1 and $(c-a)$ common from R_2
$$= (b-a)(c-a)[c+a-(b+a)] = (b-a)(c-a)(c-b)$$
$$= (a-b)(b-c)(c-a)$$

EXAMPLE 15. *Without expanding the determinant show that* $(a + b + c)$ *is a factor of following determinant.*

$$\Delta = \begin{vmatrix} a & b & c \\ b & c & a \\ c & a & b \end{vmatrix}$$

[UPTU(B.Pharma)–2003]

SOLUTION. Applying $C_1 \to C_1 + C_3$, we get

$$\Delta = \begin{vmatrix} a+b+c & b & c \\ a+b+c & c & a \\ a+b+c & a & b \end{vmatrix} = (a+b+c)\begin{vmatrix} 1 & b & c \\ 1 & c & a \\ 1 & a & b \end{vmatrix}$$

$$= (a+b+c)\begin{vmatrix} 1 & b & c \\ 0 & c-b & a-c \\ 0 & a-b & b-c \end{vmatrix}$$

Applying $R_2 \to R_2 - R_1, R_3 \to R_3 - R_1$

$$= (a+b+c)\begin{vmatrix} c-b & a-c \\ a-b & b-c \end{vmatrix}$$

$$= (a + b + c)\{-(b-c)^2 - (a-b)(a-c)\}$$

$$= (a + b + c)(-a^2 - b^2 - c^2 + ab + bc + ca)$$

Thus $(a + b + c)$ is 0 factor of Δ.

EXAMPLE 16. *Show that* $\begin{vmatrix} a & b & c \\ a-b & b-c & c-a \\ b+c & c+a & a+b \end{vmatrix} = a^3 + b^3 + c^3 - 3abc$.

SOLUTION. Operating $R_2 \to R_2 - R_1$ and $R_3 \to R_3 + R_1$, we get

$$\begin{vmatrix} a & b & c \\ a-b & b-c & c-a \\ b+c & c+a & a+b \end{vmatrix} = \begin{vmatrix} a & b & c \\ -b & -c & -a \\ a+b+c & a+b+c & a+b+c \end{vmatrix}$$

[Take $(a + b + c)$ out from R_3 and (-1) from R_2]

$$= -(a+b+c) \cdot \begin{vmatrix} a & b & c \\ b & c & a \\ 1 & 1 & 1 \end{vmatrix} \qquad \text{(Expand by } R_3)$$

$$= -(a + b + c) \cdot [1 \cdot (ab - c^2) - 1(a^2 - bc) + 1 \cdot (ca - b^2)]$$

$$= -(a + b + c) \cdot (ab + bc + ca - a^2 - b^2 - c^2)$$

$$= (a + b + c) \cdot (a^2 + b^2 + c^2 - ab - bc - ca)$$

$$= a^3 + b^3 + c^3 - 3abc.$$

EXAMPLE 17. *Find the value of x if* $\begin{vmatrix} 3+x & 5 & 2 \\ 1 & 7+x & 6 \\ 2 & 5 & 3+x \end{vmatrix} = 0$.

SOLUTION. We have $\begin{vmatrix} 3+x & 5 & 2 \\ 1 & 7+x & 6 \\ 2 & 5 & 3+x \end{vmatrix} = 0$

Applying $(R_1 - R_3)$, we get

$$\begin{vmatrix} 1+x & 0 & -1-x \\ 1 & 7+x & 6 \\ 2 & 5 & 3+x \end{vmatrix} = 0$$

Applying $C_3 \to C_3 + C_1$, we get

$$\begin{vmatrix} 1+x & 0 & 0 \\ 1 & 7+x & 7 \\ 2 & 5 & 5+x \end{vmatrix} = 0$$

On expanding the determinant along the first row, we get

$$(1+x)\begin{vmatrix} 7+x & 7 \\ 5 & 5+x \end{vmatrix} = 0$$

$$(1+x)[(7+x)(5+x) - 35] = 0$$

or $\qquad (1+x)(x^2 + 12x) = 0$

$$x(1+x)(x+12) = 0$$

$$x = 0, -1, -12.$$

EXAMPLE 18. *Evaluate:* $|A| = \begin{vmatrix} 3 & 2 & 1 & 4 \\ 15 & 29 & 2 & 14 \\ 16 & 19 & 3 & 17 \\ 23 & 39 & 8 & 38 \end{vmatrix}$

SOLUTION. Applying $C_1 \to C_1 - 3C_2, C_2 \to C_2 - 3C_3, C_4 \to C_4 - 4C_3$, we get

$$|A| = \begin{vmatrix} 0 & 0 & 1 & 0 \\ 9 & 25 & 2 & 6 \\ 7 & 13 & 3 & 5 \\ 9 & 23 & 8 & 6 \end{vmatrix}$$

On expanding the determinant along first row, we get

$$= 1\begin{vmatrix} 9 & 25 & 6 \\ 7 & 13 & 5 \\ 9 & 23 & 6 \end{vmatrix}$$

Applying $R_1 \to R_1 - R_3$, we get

$$= 1\begin{vmatrix} 0 & 2 & 0 \\ 7 & 13 & 5 \\ 9 & 23 & 6 \end{vmatrix}$$

On expanding the determinant along the first row, we get

$$= -2\begin{vmatrix} 7 & 5 \\ 9 & 6 \end{vmatrix} = -2(42 - 45) = 6$$

EXAMPLE 19. *Using properties of determinants, solve the following determinant for x.*

$$\begin{vmatrix} a+x & a-x & a-x \\ a-x & a+x & a-x \\ a-x & a-x & a+x \end{vmatrix}$$

SOLUTION. Given $\begin{vmatrix} a+x & a-x & a-x \\ a-x & a+x & a-x \\ a-x & a-x & a+x \end{vmatrix} = 0$ $\qquad$ (Operate $C_1 \to C_1 + C_2 + C_3$)

$$\Rightarrow \begin{vmatrix} 3a-x & a-x & a-x \\ 3a-x & a+x & a-x \\ 3a-x & a-x & a+x \end{vmatrix} = 0$$

$$\Rightarrow \quad (3a-x)\begin{vmatrix} 1 & a-x & a-x \\ 1 & a+x & a-x \\ 1 & a-x & a+x \end{vmatrix} = 0 \qquad (\text{Operate } R_2 \to R_2 - R_1, R_3 \to R_3 - R_1)$$

$$\Rightarrow \quad (3a-x)\begin{vmatrix} 1 & a-x & a-x \\ 0 & 2x & 0 \\ 0 & 0 & 2x \end{vmatrix} = 0 \qquad \text{(Expand by } C_1)$$

$$\Rightarrow \quad (3a-x)1.\begin{vmatrix} 2x & 0 \\ 0 & 2x \end{vmatrix} = 0$$

$$\Rightarrow \quad (3a-x)\cdot(4x^2 - 0) = 0$$

$$\Rightarrow \quad 4x^2(3a-x) = 0$$

$$\Rightarrow \quad x^2 = 0 \text{ or } 3a - x = 0$$

$$\Rightarrow \quad x = 0, 0, 3a$$

Hence, the values of x are $0, 0, 3a$.

EXAMPLE 20. *Using properties of determinant, prove that*

$$\begin{vmatrix} 1 & 1 & 1 \\ \alpha & \beta & \gamma \\ \beta\gamma & \gamma\alpha & \alpha\beta \end{vmatrix} = (\alpha - \beta)(\beta - \gamma)(\gamma - \alpha)$$

SOLUTION. Operate $C_2 \to C_2 - C_1$ and $C_3 \to C_3 - C_1$, we get

$$\begin{vmatrix} 1 & 1 & 1 \\ \alpha & \beta & \gamma \\ \beta\gamma & \gamma\alpha & \alpha\beta \end{vmatrix} = \begin{vmatrix} 1 & 0 & 0 \\ \alpha & \beta - \alpha & \gamma - \alpha \\ \beta\gamma & \gamma(\alpha - \beta) & \beta(\alpha - \gamma) \end{vmatrix}$$

[Take $(\alpha - \beta)$ out from C_2 and $(\gamma - \alpha)$ out from C_3]

$$= (\alpha - \beta)(\gamma - \alpha)\begin{vmatrix} 1 & 0 & 0 \\ \alpha & -1 & 1 \\ \beta\gamma & \gamma & -\beta \end{vmatrix} \qquad \text{(Expand by } C_1)$$

$$= (\alpha - \beta)(\gamma - \alpha)\cdot 1\cdot \begin{vmatrix} -1 & 1 \\ \gamma & -\beta \end{vmatrix}$$

$$= (\alpha - \beta)(\gamma - \alpha)(\beta - \gamma)$$

$$= (\alpha - \beta)(\beta - \gamma)(\gamma - \alpha)$$

EXAMPLE 21. *Prove that* $\begin{vmatrix} a^2+1 & ab & ac \\ ab & b^2+1 & bc \\ ac & bc & c^2+1 \end{vmatrix} = 1 + a^2 + b^2 + c^2.$

SOLUTION. We have $\quad |A| = \begin{vmatrix} a^2+1 & ab & ac \\ ab & b^2+1 & bc \\ ac & bc & c^2+1 \end{vmatrix}$

Now multiply the column 1st, 2nd and 3rd by a, b and c respectively, we get

$$|A| = \frac{1}{abc}\begin{vmatrix} a(a^2+1) & ab^2 & ac^2 \\ a^2 b & b(b^2+1) & bc^2 \\ a^2 c & b^2 c & c(c^2+1) \end{vmatrix}$$

To take a, b, c common from 1st, 2nd and 3rd rows respectively, we get

$$= \frac{abc}{abc}\begin{vmatrix} a^2+1 & b^2 & c^2 \\ a^2 & b^2+1 & c^2 \\ a^2 & b^2 & c^2+1 \end{vmatrix}$$

Now apply $C_1 \rightarrow C_1 + C_2 + C_3$, we get

$$= \begin{vmatrix} a^2 + b^2 + c^2 + 1 & b^2 & c^2 \\ a^2 + b^2 + c^2 + 1 & b^2 + 1 & c^2 \\ a^2 + b^2 + c^2 + 1 & b^2 & c^2 + 1 \end{vmatrix}$$

$$= (a^2 + b^2 + c^2 + 1) \begin{vmatrix} 1 & b^2 & c^2 \\ 1 & b^2 + 1 & c^2 \\ 1 & b^2 & c^2 + 1 \end{vmatrix}$$

Now applying $R_2 \rightarrow R_2 - R_1$ and $R_3 \rightarrow R_3 - R_1$, we get

$$= (a_2 + b_2 + c_2 + 1) \begin{vmatrix} 1 & 0 \\ 0 & 1 \end{vmatrix}$$

$$= a_2 + b_2 + c_2 + 1$$

EXAMPLE 22. *Prove that :* $\begin{vmatrix} b+c & c+a & a+b \\ q+r & r+p & p+q \\ y+z & z+x & x+y \end{vmatrix} = 2 \begin{vmatrix} a & b & c \\ p & q & r \\ x & y & z \end{vmatrix}$.

SOLUTION. We have L.H.S. $= \begin{vmatrix} b+c & c+a & a+b \\ q+r & r+p & p+q \\ y+z & z+x & x+y \end{vmatrix}$

Applying $C_1 \rightarrow C_1 + C_2 - 2C_3$, we get

$$= \begin{vmatrix} 2c & c+a & a+b \\ 2r & r+p & p+q \\ 2z & z+x & x+y \end{vmatrix} = 2 \begin{vmatrix} c & c+a & a+b \\ r & r+p & p+q \\ z & z+x & x+y \end{vmatrix}$$

Now applying $C_2 \rightarrow C_2 - C_1$, we get

$$= 2 \begin{vmatrix} c & a & a+b \\ r & p & p+q \\ z & x & x+y \end{vmatrix}$$

Applying $C_3 \rightarrow C_3 - C_2$, we get

$$= 2 \begin{vmatrix} c & a & b \\ r & p & q \\ z & x & y \end{vmatrix} = 2 \begin{vmatrix} a & b & c \\ p & q & r \\ x & y & z \end{vmatrix}$$

(by Interchanging the columns)

$$= \text{R.H.S.}$$

EXAMPLE 23. *If x, y, z are all different and* $\begin{vmatrix} x & x^2 & 1+x^3 \\ y & y^2 & 1+y^3 \\ z & z^2 & 1+z^3 \end{vmatrix} = 0$. *Show that* $xyz = -1$.

SOLUTION. Given $\begin{vmatrix} x & x^2 & 1+x^3 \\ y & y^2 & 1+y^3 \\ z & z^2 & 1+z^3 \end{vmatrix} = 0$.

$$\Rightarrow \quad \begin{vmatrix} x & x^2 & 1 \\ y & y^2 & 1 \\ z & z^2 & 1 \end{vmatrix} + \begin{vmatrix} x & x^2 & x^3 \\ y & y^2 & y^3 \\ z & z^2 & z^3 \end{vmatrix} = 0$$

[Take x, y, z out from R_1, R_2 and R_3 respectively from the second determinant]

$$\Rightarrow \quad \begin{vmatrix} 1 & x & x^2 \\ 1 & y & y^2 \\ 1 & z & z^2 \end{vmatrix} + xyz \begin{vmatrix} 1 & x & x^2 \\ 1 & y & y^2 \\ 1 & z & z^2 \end{vmatrix} = 0$$

$$\Rightarrow \quad \begin{vmatrix} 1 & x & x^2 \\ 1 & y & y^2 \\ 1 & z & z^2 \end{vmatrix}(1 + xyz) = 0$$

$$\Rightarrow \quad (x - y)\cdot(y - z)\cdot(z - x)\cdot(1 + xyz) = 0$$
$$\Rightarrow \quad (1 + xyz) = 0$$

(Because x, y, z are distinct, so $x - y \neq 0, y - z \neq 0, z - x \neq 0$).
$$\Rightarrow \quad xyz = -1.$$

EXAMPLE 24. *Evaluate the value of x for which* $\begin{vmatrix} 4x & 6x+2 & 8x+1 \\ 6x+2 & 9x+3 & 12x \\ 8x+1 & 12x & 16x+2 \end{vmatrix} = 0$.

SOLUTION. We have $\begin{vmatrix} 4x & 6x+2 & 8x+1 \\ 6x+2 & 9x+3 & 12x \\ 8x+1 & 12x & 16x+2 \end{vmatrix} = 0$

Applying $\left(C_2 \to C_2 - \dfrac{3}{2}C_1\right)$ and $C_3 \to C_3 - 2C_1$, we get

$$\begin{vmatrix} 4x & 2 & 1 \\ 6x+2 & 0 & -4 \\ 8x+1 & -3/2 & 0 \end{vmatrix} = 0$$

Now applying $R_2 \to R_2 + 4R_1$

$$\Rightarrow \quad \begin{vmatrix} 4x & 2 & 1 \\ 22x+2 & 8 & 0 \\ 8x+1 & -3/2 & 0 \end{vmatrix} = 0$$

On expanding the determinants along 3rd column, we get

$$1\begin{vmatrix} 22x+2 & 8 \\ 8x+1 & -3/2 \end{vmatrix} = 0$$

$$\Rightarrow \quad -33x - 3 - 64x - 8 = 0$$
or $\quad -97x = 11$ or $x = \dfrac{-11}{97}$

EXAMPLE 25. *Without expanding show that the value of the determinant given below is zero*

$$\begin{vmatrix} \sin\alpha & \cos\alpha & \sin(\alpha + \delta) \\ \sin\beta & \cos\beta & \sin(\beta + \delta) \\ \sin\gamma & \cos\gamma & \sin(\gamma + \delta) \end{vmatrix}$$

SOLUTION. Let
$$\Delta = \begin{vmatrix} \sin\alpha & \cos\alpha & \sin(\alpha+\delta) \\ \sin\beta & \cos\beta & \sin(\beta+\delta) \\ \sin\gamma & \cos\gamma & \sin(\gamma+\delta) \end{vmatrix}$$

Using $\sin(A + B) = \sin A \cos B + \cos A \sin B$, we get

$$\Delta = \begin{vmatrix} \sin\alpha & \cos\alpha & \sin\alpha\cos\delta + \cos\alpha\sin\delta \\ \sin\beta & \cos\beta & \sin\beta\cos\delta + \cos\beta\sin\delta \\ \sin\gamma & \cos\gamma & \sin\gamma\cos\delta + \cos\gamma\sin\delta \end{vmatrix}$$

$$= \begin{vmatrix} \sin\alpha & \cos\alpha & 0 \\ \sin\beta & \cos\beta & 0 \\ \sin\gamma & \cos\gamma & 0 \end{vmatrix}$$

Using $C_3 \to C_3 (\cos\delta)C_1 - (\sin\delta)C_2 = 0$

EXAMPLE 26. *Show that*

$$\begin{vmatrix} (b+c)^2 & a^2 & bc \\ (c+a)^2 & b^2 & ca \\ (a+b)^2 & c^2 & ab \end{vmatrix} = (a^2 + b^2 + c^2)(a + b + c)(b - c)(c - a)(a - b).$$

SOLUTION. Let
$$\Delta = \begin{vmatrix} (b+c)^2 & a^2 & bc \\ (c+a)^2 & b^2 & ca \\ (a+b)^2 & c^2 & ab \end{vmatrix}$$

Applying $C_1 \to C_1 - 2C_3$, we get

$$= \begin{vmatrix} b^2+c^2+a^2 & a^2 & bc \\ c^2+a^2+b^2 & b^2 & ca \\ a^2+b^2+c^2 & c^2 & ab \end{vmatrix}$$

Operating $C_1 \to C_1 + C_2$, we get

$$= (a^2+b^2+c^2)\begin{vmatrix} 1 & a^2 & bc \\ 1 & b^2 & ca \\ 1 & c^2 & ab \end{vmatrix}$$

Operating $R_2 \to R_2 - R_1$ and $R_3 \to R_3 - R_2$

$$= (a^2+b^2+c^2)\begin{vmatrix} 1 & a^2 & bc \\ 0 & b^2-a^2 & (ca-bc) \\ 0 & c^2-a^2 & (ab-bc) \end{vmatrix}$$

$$= (a^2 + b^2 + c^2)(b-c)(c-a)\begin{vmatrix} 1 & a^2 & bc \\ 0 & b+a & -c \\ 0 & c+a & -b \end{vmatrix}$$

$R_3 \to R_3 - R_2$, we get

$$= (a^2 + b^2 + c^2)(b-a)(c-a)\begin{vmatrix} 1 & a^2 & bc \\ 0 & b+a & -c \\ 0 & c-a & c-b \end{vmatrix}$$

$$= (a^2 + b^2 + c^2)(b-a)(c-a)(c-b)\begin{vmatrix} 1 & a^2 & bc \\ 0 & b+a & -c \\ 0 & 1 & 1 \end{vmatrix}$$

Expanding along first column, we get

$$\Delta = (a^2 + b^2 + c^2)(b-a)(c-a)(c-b)(a+b+c)$$

EXAMPLE 27. *Show that* $\begin{vmatrix} a+b & b+c & c+a \\ b+c & c+a & a+b \\ c+a & a+b & b+c \end{vmatrix} = 2\begin{vmatrix} a & b & c \\ b & c & a \\ c & a & b \end{vmatrix}.$

SOLUTION. Let

$$\Delta = \begin{vmatrix} a+b & b+c & c+a \\ b+c & c+a & a+b \\ c+a & a+b & b+c \end{vmatrix}$$

Applying $C_1 \to C_1 + C_2 + C_3$, we get

$$= \begin{vmatrix} 2(a+b+c) & b+c & c+a \\ 2(a+b+c) & c+a & a+b \\ 2(a+b+c) & a+b & b+c \end{vmatrix}$$

$$= 2\begin{vmatrix} a+b+c & -a & -b \\ a+b+c & -b & -c \\ a+b+c & -c & -a \end{vmatrix}$$

Applying $C_2 \to C_2 - C_1, \ C_3 \to C_3 - C_1$

We get

$$= 2(-1)(-1)\begin{vmatrix} a+b+c & a & b \\ a+b+c & b & c \\ a+b+c & c & a \end{vmatrix}$$

Applying $C_1 \to C_1 - C_2 - C_3$, we get

$$= 2\begin{vmatrix} c & a & b \\ a & b & c \\ b & c & a \end{vmatrix}$$

$$= 2\begin{vmatrix} a & c & b \\ b & a & c \\ c & b & a \end{vmatrix} (C_2 \to C_3) = 2\begin{vmatrix} a & b & c \\ b & c & a \\ c & a & b \end{vmatrix}$$

EXAMPLE 28. *If a, b, c (all positive) are the pth, q^{th} and r^{th} terms respectively of a geometric progression, show that*

$$\begin{vmatrix} \log a & p & 1 \\ \log b & q & 1 \\ \log c & r & 1 \end{vmatrix} = 0$$

SOLUTION. Consider the terms of G.P. which are $A, AR, AR^2, ...,$

$$a = T_p = AR^{p-1}, b = T_q = AR^{q-1}, c = T_r = AR^{r-1}$$

Consider

$$\begin{vmatrix} \log a & p & 1 \\ \log b & q & 1 \\ \log c & r & 1 \end{vmatrix} = \begin{vmatrix} \log AR^{p-1} & p & 1 \\ \log AR^{q-1} & q & 1 \\ \log AR^{r-1} & r & 1 \end{vmatrix}$$

$$= \begin{vmatrix} \log A + (p-1)\log R & p & 1 \\ \log A + (q-1)\log R & q & 1 \\ \log A + (r-1)\log R & r & 1 \end{vmatrix}$$

$$= \begin{vmatrix} \log A & q & 1 \\ \log A & p & 1 \\ \log A & r & 1 \end{vmatrix} + \begin{vmatrix} (p-1)\log R & p & 1 \\ (q-1)\log R & q & 1 \\ (r-1)\log R & r & 1 \end{vmatrix}$$

$$= \log A \begin{vmatrix} 1 & p & 1 \\ 1 & q & 1 \\ 1 & r & 1 \end{vmatrix} + \log R \begin{vmatrix} p-1 & p & 1 \\ q-1 & q & 1 \\ r-1 & r & 1 \end{vmatrix}$$

$$= \log A \times 0 + \log R \begin{vmatrix} p & p & 1 \\ q & q & 1 \\ r & r & 1 \end{vmatrix} = 0 + \log R \times 0 = 0$$

Exercise 7.1

Evaluate the following determinants (1 to 7):

1. $\begin{vmatrix} \frac{1}{2} & 8 \\ 4 & 2 \end{vmatrix}$

2. $\begin{vmatrix} -2 & 3 \\ 4 & -9 \end{vmatrix}$

3. $\begin{vmatrix} \cos\theta & -\sin\theta \\ \sin\theta & \cos\theta \end{vmatrix}$

4. $\begin{vmatrix} x^2-x+1 & x-1 \\ x+1 & x+1 \end{vmatrix}$

5. $\begin{vmatrix} 1 & 0 & 6 \\ 3 & 4 & 15 \\ 5 & 6 & 21 \end{vmatrix}$

6. $\begin{vmatrix} 23 & 12 & 11 \\ 36 & 10 & 26 \\ 63 & 26 & 37 \end{vmatrix}$

7. $\begin{vmatrix} 3 & 1 & -4 \\ 3 & 2 & 5 \\ 1 & 1 & 3 \end{vmatrix}$

Write the minor and co-factors of each element of the following determinants and also evaluate the determinants in each case (8 to 11) :

8. $\begin{vmatrix} 5 & -10 \\ 0 & 3 \end{vmatrix}$

9. $\begin{vmatrix} 1 & 3 & -2 \\ 4 & -5 & 6 \\ 3 & 5 & 2 \end{vmatrix}$

10. $\begin{vmatrix} 1 & 0 & 0 \\ 0 & 1 & 0 \\ 0 & 0 & 1 \end{vmatrix}$

11. $\begin{vmatrix} 1 & 0 & 4 \\ 3 & 5 & -1 \\ 0 & 1 & 2 \end{vmatrix}$

12. Evaluate $\begin{vmatrix} x+1 & x+2 & x+4 \\ x+5 & x+6 & x+8 \\ x+7 & x+10 & x+14 \end{vmatrix}$

13. Evaluate $\begin{vmatrix} 1 & a & bc \\ 1 & b & ca \\ 1 & c & ab \end{vmatrix}$

14. Evaluate $\begin{vmatrix} x+\lambda & x & x \\ x & x+\lambda & x \\ x & x & x+\lambda \end{vmatrix}$

15. Evaluate $\begin{vmatrix} b+c & a & a \\ b & c+a & b \\ c & c & a+b \end{vmatrix}$

16. Prove that $\begin{vmatrix} 1 & x & x^2 \\ 1 & y & y^2 \\ 1 & z & z^2 \end{vmatrix} = (x-y)(y-z)(z-x).$

17. Prove that $\begin{vmatrix} -a^2 & ab & ac \\ ba & -b^2 & bc \\ ac & bc & -c^2 \end{vmatrix} = 4\,a^2b^2c^2.$

18. Prove that
$$\begin{vmatrix} x & y^2 & yz \\ y & y^2 & zx \\ z & z^2 & xy \end{vmatrix} = (x-y)(y-z)(z-x)(xy+yz+zx)$$

19. Using properties of determinants, prove that
$$\begin{vmatrix} y+z & x & y \\ z+x & z & x \\ x+y & y & z \end{vmatrix} = (x+y+z)(x-z)^2.$$

20. Using properties of determinants, prove that
$$\begin{vmatrix} a-b-c & 2a & 2a \\ 2b & b-c-a & 2b \\ 2c & 2c & c-a-b \end{vmatrix} = (a+b+c)^3$$

21. Solve the following determinants
$$\begin{vmatrix} x-2 & 2x-3 & 3x-4 \\ x-4 & 2x-9 & 3x-16 \\ x-8 & 2x-27 & 3x-64 \end{vmatrix} = 0$$

[UPTU(B.Pharma)–2007]

22. Prove that using properties of determinants
$$\begin{vmatrix} 1+a^2-b^2 & 2ab & -2b \\ 2ab & 1-a^2+b^2 & 2a \\ 2b & -2a & 1-a^2-b^2 \end{vmatrix}$$
$$= (1+a^2+b^2)^3$$

23. Prove that $\begin{vmatrix} x & x^2 & 1+px^3 \\ y & y^2 & 1+py^3 \\ z & z^2 & 1+pz^3 \end{vmatrix}$

$$= (1 + pxyz)(x-y)(y-z)(z-x)$$

24. Prove that using properties of determinants

$\begin{vmatrix} 3a & -a+b & -a+c \\ -b+a & 3b & -b+c \\ -c+a & -c+b & 3c \end{vmatrix}$

$$=3(a + b + c)(ab + bc + ca)$$

25. Prove that

$\begin{vmatrix} \sin\alpha & \cos\alpha & \cos(\alpha+\delta) \\ \sin\beta & \cos\beta & \cos(\beta+\delta) \\ \sin\gamma & \cos\gamma & \cos(\gamma+\delta) \end{vmatrix} = 0$

HINT TO SELECTED PROBLEMS

1. (i) $\begin{vmatrix} 1/2 & 8 \\ 4 & 2 \end{vmatrix} = \frac{1}{2} \times 2 - 8 \times 4 = 1 - 32 = -31.$

3. We have $\begin{vmatrix} \cos\theta & -\sin\theta \\ \sin\theta & \cos\theta \end{vmatrix} = \cos^2\theta + \sin^2\theta = 1$

4. On expanding, we get

$\Rightarrow (x^2 - x + 1)(x - 1) - (x - 1)(x + 1)$
$= (x - 1)(x^2 - x + 1 - x - 1)$
$= (x - 1)(x^2 - 2x)$
$= x^3 - 2x^2 - x^2 + 2x$
$|A| = x^3 - 3x^2 + 2x$

5. $|A| = \begin{vmatrix} 1 & 0 & 6 \\ 3 & 4 & 15 \\ 5 & 6 & 21 \end{vmatrix} = \begin{vmatrix} 1 & 0 & 3\cdot2 \\ 3 & 4 & 3\cdot5 \\ 5 & 6 & 3\cdot7 \end{vmatrix}$

$= 3 \begin{vmatrix} 1 & 0 & 2 \\ 3 & 4 & 5 \\ 5 & 6 & 7 \end{vmatrix}$

$= 3 \begin{vmatrix} 1 & 0 & 2 \\ 3 & 2\cdot2 & 5 \\ 5 & 2\cdot3 & 7 \end{vmatrix} = 6 \begin{vmatrix} 1 & 0 & 2 \\ 3 & 2 & 5 \\ 5 & 3 & 7 \end{vmatrix}$

$= 6 [(14 - 15) + 2(9 - 10)] = -18.$

8. $\begin{vmatrix} 5 & -10 \\ 0 & 3 \end{vmatrix}$

Minor of the element a_{11} is $M_{11} = |3| = 3.$

Minor of the element a_{12} is $M_{12} = 0.$

Minor of the element a_{21} is $M_{21} = -10.$

Minor of the element a_{22} is $M_{22} = 5.$

Hence cofactors are as

$A_{11} = (-1)^{1+1}M_{11} = 3$
$A_{12} = (-1)^{1+2}M_{12} = 0$
$A_{21} = (-1)^{2+1}M_{21} = 0$
$A_{22} = (-1)^{2+2}M_{22} = 0$

$|A| = \begin{vmatrix} 5 & -10 \\ 0 & 3 \end{vmatrix} = 15 - 0 = 15$

9. Minor of $a_{11} = \begin{vmatrix} -5 & 6 \\ 5 & 2 \end{vmatrix} = -40$

Minor of $a_{12} = \begin{vmatrix} 4 & 6 \\ 3 & 2 \end{vmatrix} = -10$

Minor of $a_{13} = \begin{vmatrix} 4 & -5 \\ 3 & 5 \end{vmatrix} = 35$

Minor of $a_{21} = \begin{vmatrix} 3 & -2 \\ 5 & 2 \end{vmatrix} = 16$

Minor of $a_{22} = \begin{vmatrix} 1 & -2 \\ 3 & 2 \end{vmatrix} = 8$

Minor of $a_{23} = \begin{vmatrix} 1 & 3 \\ 3 & 5 \end{vmatrix} = -4$

Minor of $a_{31} = \begin{vmatrix} 3 & -2 \\ -5 & 6 \end{vmatrix} = 8$

Minor of $a_{32} = \begin{vmatrix} 1 & -2 \\ 4 & 6 \end{vmatrix} = 14$

Minor of $a_{33} = \begin{vmatrix} 1 & 3 \\ 4 & -5 \end{vmatrix} = -17$

Now Cofactors are :

$A_{11} = (-1)^{1+1}M_{11} = -40$
$A_{12} = (-1)^{1+2}M_{12} = 10$
$A_{13} = (-1)^{1+3}M_{13} = 35$
$A_{21} = (-1)^{2+1}M_{21} = 16$
$A_{22} = (-1)^{2+2}M_{22} = 8$
$A_{23} = (-1)^{2+3}M_{23} = 4$
$A_{31} = (-1)^{3+1}M_{31} = 8$
$A_{32} = (-1)^{3+2}M_{32} = -14$
$A_{33} = (-1)^{3+3}M_{33} = -17$

12. Applying the following operations and then expanding

$$R_3 \to R_3 - R_1, \; R_2 \to R_2 - R_1$$

and $C_3 \to C_3 - C_1, \; C_2 \to C_2 - C_1.$

13. Applying $R_2 \to R_2 - R_1$ and $R_3 \to R_3 - R_1$. And expanding along a_{11}, we get the required results.

14. Applying $R_1 \to R_1 + R_2 + R_3$.

Then $C_2 \to C_2 - C_1$ and $C_3 \to C_3 - C_1$ and expending we get the required result.

17. Taking a, b, c common from the first, second and third columns respectively, we get

$$\Delta = abc \begin{vmatrix} -a & a & a \\ b & -b & b \\ c & c & -c \end{vmatrix} = a^2 b^2 c^2 \begin{vmatrix} -1 & 1 & 1 \\ 1 & -1 & 1 \\ 1 & 1 & -1 \end{vmatrix}$$

Taking a, b, c common from 1st, 2nd and 3rd row respectively

$$= a^2 b^2 c^2 \begin{vmatrix} -1 & 1 & 1 \\ 0 & 0 & 2 \\ 0 & 2 & 0 \end{vmatrix}$$

applying $R_2 \to R_2 + R_1$, $R_3 \to R_3 + R_1$

$$= a^2 b^2 c^2 (-1)(-4) = 4a^2 b^2 c^2.$$

18. Multiplying the first, second and third rows of the determinant on the L.H.S. by x, y and z respectively. We get,

$$= \frac{1}{xyz} \begin{vmatrix} x^2 & x^3 & xyz \\ y^2 & y^3 & xyz \\ z^2 & z^3 & xyz \end{vmatrix}$$

$$= \frac{xyz}{xyz} \begin{vmatrix} x^2 & x^3 & 1 \\ y^2 & y^3 & 1 \\ z^2 & z^2 & 1 \end{vmatrix} = \begin{vmatrix} 1 & 1 & 1 \\ x^2 & y^2 & z^2 \\ x^3 & y^3 & z^3 \end{vmatrix}$$

Applying $C_2 \to C_2 - C_1$, $C_3 \to C_3 - C_1$ to the determinant. We get

$$= \begin{vmatrix} 1 & 0 & 0 \\ x^2 & y^2 - x^2 & z^2 - x^2 \\ x^3 & y^3 - x^3 & z^3 - x^3 \end{vmatrix}$$

$$= \begin{vmatrix} (y-x)(y+x) & (z-x)(z+x) \\ (y-x)(y^2 + xy + x^2) & (z-x)(z^2 + zx + x^2) \end{vmatrix}$$

$$= (y-x)(z-x) \begin{vmatrix} y+x & z+x \\ y^2 + xy + x^2 & z^2 + zx + x^2 \end{vmatrix}$$

[Taking $(y-x)$ common from the first column and $(z-x)$ from the second column]

Now Applying $C_2 \to C_2 - C_1$

We get

$$\Rightarrow (y-x)(z-x) \begin{vmatrix} y+x & z-y \\ y^2 + xy + x^2 & (z^2 - y^2) + zx - xy \end{vmatrix}$$

$$= (y-x)(z-x) \begin{vmatrix} y+x & z-y \\ y^2 + xy + x^2 & (z-y)(x+y+z) \end{vmatrix}$$

$$= (y-x)(z-x)(z-y) \begin{vmatrix} y+x & 1 \\ y^2 + xy + x^2 & x+y+z \end{vmatrix}$$

$$= (x-y)(y-z)(z-x)(xy + yz + zx)$$

20. Applying $R_1 \to R_1 + R_2 + R_3$

$C_2 \to C_2 - C_1$

$C_3 \to C_2 - C_1$ we get the required result.

21. Applying $R_2 \to R_2 - R_1$, $R_3 \to R_3 - R_1$, the given equation becomes

$$\begin{vmatrix} x-2 & 2x-3 & 3x-4 \\ -2 & -6 & -12 \\ -6 & -24 & -60 \end{vmatrix} = 0.$$

or $$\begin{vmatrix} x-2 & 2x-3 & 3x-4 \\ 1 & 3 & 6 \\ 1 & 4 & 10 \end{vmatrix} = 0.$$

Expanding the determinant along the first row, the above equation becomes :

$$(x-2)[30-24] - (2x-3)[10-6]$$
$$+ (3x-4)(4-3) = 0$$
$$6x - 12 - 8x + 12 + 3x - 4 = 0, x = 4$$

23. $$\begin{vmatrix} x & x^2 & 1+px^3 \\ y & y^2 & 1+py^3 \\ z & z^2 & 1+pz^3 \end{vmatrix} = \begin{vmatrix} x & x^2 & 1 \\ y & y^2 & 1 \\ z & z^2 & 1 \end{vmatrix} + \begin{vmatrix} x & x^2 & px^3 \\ y & y^2 & py^3 \\ z & z^2 & pz^3 \end{vmatrix}$$

$$= \begin{vmatrix} x & x^2 & 1 \\ y & y^2 & 1 \\ z & z^2 & 1 \end{vmatrix} + xyz \begin{vmatrix} 1 & x & px^2 \\ 1 & y & py^2 \\ 1 & z & pz^2 \end{vmatrix}$$

$$= \begin{vmatrix} x & x^2 & 1 \\ y & y^2 & 1 \\ z & z^2 & 1 \end{vmatrix} + pxyz \begin{vmatrix} 1 & x & x^2 \\ 1 & y & y^2 \\ 1 & z & z^2 \end{vmatrix}$$

$$= \begin{vmatrix} 1 & x & x^2 \\ 1 & y & y^2 \\ 1 & z & z^2 \end{vmatrix} + pxyz \begin{vmatrix} 1 & x & x^2 \\ 1 & y & y^2 \\ 1 & z & z^2 \end{vmatrix}$$

$$= (1 + pxyz) \begin{vmatrix} 1 & x & x^2 \\ 1 & y & y^2 \\ 1 & z & z^2 \end{vmatrix}$$

$$= (1 + pxyz)(x-y)(y-z)(z-x).$$

25. $$\begin{vmatrix} \sin\alpha & \cos\alpha & \cos(\alpha+\delta) \\ \sin\beta & \cos\beta & \cos(\beta+\delta) \\ \sin\gamma & \cos\gamma & \cos(\gamma+\delta) \end{vmatrix}$$

$$= \begin{vmatrix} \sin\alpha & \cos\alpha & \cos\alpha\cos\delta - \sin\alpha\sin\delta \\ \sin\beta & \cos\beta & \cos\beta\cos\delta - \sin\beta\sin\delta \\ \sin\gamma & \cos\gamma & \cos\gamma\cos\delta - \sin\gamma\sin\delta \end{vmatrix}$$

Applying $C_3 \rightarrow C_3 + (\sin \delta)C_1 - (\cos \delta)C_2$.

We get $\begin{vmatrix} \sin \alpha & \cos \alpha & 0 \\ \sin \beta & \cos \beta & 0 \\ \sin \gamma & \cos \gamma & 0 \end{vmatrix} = 0$

ANSWERS

1. -31 **2.** 6 **3.** 1 **4.** $x^3 - x^2 + 2$ **5.** -18 **6.** 0 **7.** 49

8. $M_{11} = 3, M_{12} = 0, M_{21} = -10, M_{22} = 5, A_{11} = 3, A_{12} = 0, A_{21} = 10, A_{22} = 5; 15$

9. $M_{11} = -40, M_{12} = -10, M_{13} = 35, M_{21} = 16, M_{22} = 8, M_{23} = -4, M_{31} = 8, M_{32} = 14, M_{33} = -17$

    $A_{11} = -40, A_{12} = 10, A_{13} = 35, A_{21} = -16, A_{22} = 8, A_{23} = 4, A_{31} = 8, A_{32} = -14, A_{33} = -17; -80$

10. $M_{11} = 1, M_{12} = 0, M_{13} = 0, M_{21} = 0, M_{22} = 1, M_{23} = 0, M_{31} = 0, M_{32} = 0, M_{33} = 1$

    $A_{11} = 1, A_{12} = 0, A_{13} = 0, A_{21} = 0, A_{22} = 1, A_{23} = 0, A_{31} = 0, A_{32} = 0, A_{33} = 1; 1$

11. $M_{11} = 11, M_{12} = 6, M_{13} = 3, M_{21} = -4, M_{22} = 2, M_{23} = 1, M_{31} = -20, M_{32} = -13, M_{33} = 5$

    $A_{11} = 11, A_{12} = -6, A_{13} = 3, A_{21} = 4, A_{22} = 2, A_{23} = -1, A_{31} = 20, A_{32} = 13, A_{33} = 5; 23$

12. -24 **13.** $(a-b)(b-c)(c-a)$ **14.** $\lambda^2(3x + \lambda)$ **15.** $4abc$ **21.** $x = 4$

7.6 CRAMER'S RULE

Consider the system of linear equations

$$a_1 x + b_1 y + c_1 z = d_1$$
$$a_2 x + b_2 y + c_2 z = d_2$$
$$a_3 x + b_3 y + c_3 z = d_3 \qquad \qquad \ldots(1)$$

We define $\Delta = \text{determinant coefficients} = \begin{vmatrix} a_1 & b_1 & c_1 \\ a_2 & b_2 & c_2 \\ a_3 & b_3 & c_3 \end{vmatrix}$.

Now we define Δ_x which is obtained by suppressing the column of coefficients of x and replacing it by the column of constant terms d_1, d_2, d_3 on right hand side

$$\therefore \qquad \Delta_x = \begin{vmatrix} d_1 & b_1 & c_1 \\ d_2 & b_2 & c_2 \\ d_3 & b_3 & c_3 \end{vmatrix}.$$

Similarly, we obtained

$$\Delta_y = \begin{vmatrix} a_1 & d_1 & c_1 \\ a_2 & d_2 & c_2 \\ a_3 & d_3 & c_3 \end{vmatrix} \text{ and } \Delta_z = \begin{vmatrix} a_1 & b_1 & d_1 \\ a_2 & b_2 & d_2 \\ a_3 & b_3 & d_3 \end{vmatrix}$$

Now

Case I. If $\Delta \neq 0$ solution of system (1) is given by

$$x = \frac{\Delta_x}{\Delta}, y = \frac{\Delta_y}{\Delta}, z = \frac{\Delta_z}{\Delta}$$

and system is called consistent.

Case II. $\Delta = 0$ but at least one of $\Delta_x, \Delta_y, \Delta_z \neq 0$, then, the system does not posses any common solution and system is called inconsistent.

Case III. $\Delta = 0$, also $\Delta_x = \Delta_y = \Delta_z = 0$ and at least one cofactor of $\Delta \neq 0$, then system has infinitely many solution and the system then be solved by elimination method.

Elimination of one unknown from three equations gives any one equations in two unknowns therefore two unknowns can be found in terms of the other, we give this unknown an arbitrary value.

If $\Delta = \Delta_x = \Delta_y = \Delta_z = 0$ and all cofactor of Δ, Δx, Δy and Δz are zero then system is equivalent to only one equation in three unknowns and then we give any two unknowns arbitrary values and find the remaining unknown in terms of three constants.

SOLVED EXAMPLES

EXAMPLE 1. *Using the Cramer's rule, solve the following system of equations*
$$x + y - 4 = 0, 2x - 3y - 8 = 0.$$

SOLUTION. The given equation is
$$x + y - 4 = 0 \qquad \qquad ...(1)$$
$$2x - 3y - 8 = 0 \qquad \qquad ...(2)$$

Here,
$$\Delta = \begin{vmatrix} 1 & 1 \\ 2 & -3 \end{vmatrix} = -5 \neq 0$$

$$\Delta_x = \begin{vmatrix} 4 & 1 \\ 3 & -3 \end{vmatrix} = -15$$

$$\Delta_y = \begin{vmatrix} 1 & 4 \\ 2 & 3 \end{vmatrix} = -5$$

$\therefore$ By Cramer's rule
$$x = \frac{\Delta_x}{\Delta} = 3, y = \frac{\Delta_y}{\Delta} = 1$$

EXAMPLE 2. *Show that the system of equations $x + y - 2 = 0$, $2x + 3y - 5 = 0$, $4x - y - 3 = 0$ is consistent. Find the solution using Cramer's rule.*

SOLUTION. The given system of equation is
$$x + y - 2 = 0$$
$$2x + 3y - 5 = 0$$
$$4x - y - 3 = 0$$
is consistent (*i.e.*, have common solution), if the determinant
$$\Delta^* = \begin{vmatrix} 1 & 1 & -2 \\ 2 & 3 & -5 \\ 4 & -1 & -3 \end{vmatrix} = 0, i.e., \begin{vmatrix} 1 & 0 & 0 \\ 2 & 1 & -1 \\ 4 & -5 & 5 \end{vmatrix} = 0$$

$\therefore$ $\Delta^* = 5 - 5 = 0$, hence, the system is consistent, so it is sufficient to solve any two equations by Cramer's rule.

Let us consider equation (1) and (2)
$$\Delta = \begin{vmatrix} 1 & 1 \\ 2 & 3 \end{vmatrix} = 1 \ (\neq 0)$$

$$\Delta_x = \begin{vmatrix} 2 & 1 \\ 5 & 3 \end{vmatrix} = 6 - 5 = 1$$

$$\Delta_y = \begin{vmatrix} 1 & 2 \\ 2 & 5 \end{vmatrix} = 5 - 4 = 1$$

$$x = \frac{\Delta_x}{\Delta} = 1, y = \frac{\Delta_y}{\Delta} = 1$$

Hence the required solution is given by $x = y = 1$.

EXAMPLE 3. *Solve the following by Cramer's rule*
$$x + y + z = 6, x - y + z = 2, 3x + 2y - 4z = -5.$$

SOLUTION.   We have $\Delta = \begin{vmatrix} 1 & 1 & 1 \\ 1 & -1 & 1 \\ 3 & 2 & -4 \end{vmatrix} = \begin{vmatrix} 1 & 0 & 0 \\ 1 & -2 & 0 \\ 3 & -1 & 7 \end{vmatrix} = 14 \neq 0$

$$\Delta_x = \begin{vmatrix} 6 & 1 & 1 \\ 2 & -1 & 1 \\ -5 & 2 & 4 \end{vmatrix} = \begin{vmatrix} 6 & 1 & 1 \\ -4 & -2 & 0 \\ 19 & 6 & 0 \end{vmatrix} = 14$$

$$\Delta_y = \begin{vmatrix} 1 & 6 & 1 \\ 1 & 2 & 1 \\ 3 & -5 & -4 \end{vmatrix} = = \begin{vmatrix} 1 & 6 & 1 \\ 0 & -4 & 0 \\ 0 & -23 & -7 \end{vmatrix} = 28$$

$$\Delta_z = \begin{vmatrix} 1 & 1 & 6 \\ 1 & -1 & 2 \\ 3 & 2 & -5 \end{vmatrix} = \begin{vmatrix} 1 & 1 & 6 \\ 0 & -2 & -4 \\ 0 & -1 & -23 \end{vmatrix} = 42$$

Hence, by Cramer's rule

$$x = \frac{\Delta_x}{\Delta} = 1, y = \frac{\Delta_y}{\Delta} = 2, z = \frac{\Delta_z}{\Delta} = 3.$$

Hence, solution is given by $x = 1, y = 2, z = 3$.

EXAMPLE 4.   *Solve the following system equations with the help of Cramer's rule.*

$$3x - 4y + 5z = -6, x + y - 2z = -1, 2x + 3y + z = 5.$$

[UPTU(B.Pharma)–2004]

SOLUTION.   Let $\Delta = \begin{vmatrix} 3 & -4 & 5 \\ 1 & 1 & -2 \\ 2 & 3 & 1 \end{vmatrix} = 3(1+6) + 4(1+4) + 5(3-2) = 46 \neq 0.$

Since, $\Delta \neq 0$, therefore the given system has a unique solution given by

$$\frac{x}{\Delta_x} = \frac{y}{\Delta_y} = \frac{z}{\Delta_z} = \frac{1}{\Delta}$$

Now, $\Delta_x = \begin{vmatrix} -6 & -4 & 5 \\ -1 & 1 & -2 \\ 5 & 3 & 1 \end{vmatrix}$ by $R_1 \to R_1 + 4R_2, R_3 \to R_3 - 3R_2$.

$$= \begin{vmatrix} -10 & 0 & -3 \\ -1 & 1 & -2 \\ 8 & 0 & 7 \end{vmatrix} = \begin{vmatrix} -10 & -3 \\ 8 & 7 \end{vmatrix} = -70 + 24 = -46.$$

$$\Delta_y = \begin{vmatrix} 3 & -6 & 5 \\ 1 & -1 & -2 \\ 2 & 5 & 1 \end{vmatrix} \text{ by } R_1 \to R_1 - 3R_2, R_3 \to R_3 - 2R_2$$

$$= \begin{vmatrix} 0 & -3 & 11 \\ 1 & -1 & -2 \\ 0 & 7 & 5 \end{vmatrix} = -\begin{vmatrix} -3 & 11 \\ 7 & 5 \end{vmatrix} = 92$$

$$\Delta_z = \begin{vmatrix} 3 & -4 & -6 \\ 1 & 1 & -1 \\ 2 & 3 & 5 \end{vmatrix} \text{ by } R_1 \to R_1 - 3R_2, R_3 \to R_3 - 2R_2$$

$$= \begin{vmatrix} 0 & -7 & -3 \\ 1 & 1 & -1 \\ 0 & 1 & 7 \end{vmatrix} = -\begin{vmatrix} -7 & -3 \\ 1 & 7 \end{vmatrix} = 46$$

The solution of the given system is

$$x = \frac{\Delta_x}{\Delta} = \frac{-46}{46} = -1, \ y = \frac{\Delta_y}{\Delta} = \frac{92}{46} = 2 \text{ and } z = \frac{\Delta_z}{\Delta} = \frac{46}{46} = 1$$

Hence, the required solution is $x = -1, y = 2, z = 1$.

EXAMPLE 5. *Solve using Cramer's rule*

$$x + y = 5, \ y + z = 3, \ z + x = 4. \qquad \text{(UPTU(B.Pharma)–2001, 07)}$$

SOLUTION. Let

$$\Delta = \begin{vmatrix} 1 & 1 & 0 \\ 0 & 1 & 1 \\ 1 & 0 & 1 \end{vmatrix} = 1(1 - 0) - 1(0 - 1) = 1 + 1 = 2$$

Since $\Delta \neq 0$, therefore the given systems has a unique solution given by

$$\frac{x}{\Delta_x} = \frac{y}{\Delta_y} = \frac{z}{\Delta_z} = \frac{1}{\Delta}$$

Now

$$\Delta_x = \begin{vmatrix} 5 & 1 & 0 \\ 3 & 1 & 1 \\ 4 & 0 & 1 \end{vmatrix} = 6$$

$$\Delta_y = \begin{vmatrix} 1 & 5 & 0 \\ 0 & 3 & 1 \\ 1 & 4 & 1 \end{vmatrix} = 4$$

$$\Delta_z = \begin{vmatrix} 1 & 1 & 5 \\ 0 & 1 & 3 \\ 1 & 0 & 4 \end{vmatrix} = 2.$$

The solution of the given system is

$$x = \frac{\Delta_x}{\Delta} = \frac{6}{2} = 3, \ y = \frac{\Delta_y}{\Delta} = \frac{4}{2} = 2, \ z = \frac{\Delta_z}{\Delta} = \frac{2}{2} = 1.$$

EXAMPLE 6. *Solve the following by using Cramer's rule.*

$$x - 2y + 3z = 2, \ 2x - 3z = 3, \ x + y + z = 6. \qquad \text{[UPTU(B.Pharma)–2002]}$$

SOLUTION. Let

$$\Delta = \begin{vmatrix} 1 & -2 & 3 \\ 2 & 0 & -3 \\ 1 & 1 & 1 \end{vmatrix} = \begin{vmatrix} 1 & 0 & 0 \\ 2 & 4 & -9 \\ 1 & 3 & -2 \end{vmatrix} \text{ by } R_2 + 2R_1, R_3 - 3R_1$$

$$= \begin{vmatrix} 4 & -9 \\ 3 & -2 \end{vmatrix} = -8 + 27 = 19 \neq 0$$

since $\Delta \neq 0$, therefore the given system has a unique solution given by

$$\frac{x}{\Delta_x} = \frac{y}{\Delta_y} = \frac{z}{\Delta_z} = \frac{1}{\Delta_x}.$$

Now

$$\Delta_x = \begin{vmatrix} 2 & -2 & 3 \\ 3 & 0 & -3 \\ 6 & 1 & 1 \end{vmatrix} = \begin{vmatrix} 2 & 0 & 5 \\ 3 & 3 & 0 \\ 6 & 7 & 7 \end{vmatrix} \text{ by } R_2 + R_1, R_3 + R_1$$

$$= 2 \begin{vmatrix} 3 & 0 \\ 7 & 7 \end{vmatrix} + 0 + 5 \begin{vmatrix} 3 & 3 \\ 6 & 7 \end{vmatrix} = 57.$$

$$\Delta_y = \begin{vmatrix} 1 & 2 & 3 \\ 2 & 3 & -3 \\ 1 & 6 & 1 \end{vmatrix} = \begin{vmatrix} 1 & 0 & 0 \\ 2 & -1 & -9 \\ 1 & 4 & -2 \end{vmatrix} \text{ by } R_2 - 2R_1, R_3 - 3R_1$$

$$= 1 \begin{vmatrix} -1 & -9 \\ 4 & -2 \end{vmatrix} = 38.$$

$$\Delta_z = \begin{vmatrix} 1 & -2 & 2 \\ 2 & 0 & 3 \\ 1 & 1 & 6 \end{vmatrix} = \begin{vmatrix} 1 & 0 & 0 \\ 2 & 4 & -1 \\ 1 & 3 & 4 \end{vmatrix} \text{ by } R_2 + 2R_1, R_3 - 2R_1$$

$$= 1 \begin{vmatrix} 4 & -1 \\ 3 & 4 \end{vmatrix} = 16 + 3 = 19.$$

The solution of the given system is

$$x = \frac{\Delta_x}{\Delta} = \frac{57}{19} = 3, \ y = \frac{\Delta_y}{\Delta} = \frac{38}{19} = 1 \text{ and } z = \frac{\Delta_z}{\Delta} = \frac{19}{19} = 1$$

Hence, the required solution is $x = 3, y = 2, z = 1$.

EXAMPLE 7. *Find the value of λ for which the system of equations $x + y - 2z = 0$, $2x - 3xy + z = 0$, $x - 5y + 4z = \lambda$ are consistent and find the solutions for all such value of λ.*

SOLUTION. The given system of equations is

$$x - 5y + 4z = \lambda \qquad \qquad \dots(1)$$
$$x + y - 2z = 0 \qquad \qquad \dots(2)$$
$$2x - 3y + z = 0 \qquad \qquad \dots(3)$$

$$\Delta = \begin{vmatrix} 1 & -5 & 4 \\ 1 & 1 & -2 \\ 2 & -3 & 1 \end{vmatrix} = \begin{vmatrix} 1 & -5 & 4 \\ 0 & 6 & -6 \\ 0 & 7 & -7 \end{vmatrix} = 0.$$

Hence, system is consistent only when

$$\Delta_x = \Delta_y = \Delta_z = 0$$

Now $\qquad \Delta_x = \begin{vmatrix} \lambda & -5 & 4 \\ 0 & 1 & -2 \\ 0 & -3 & 1 \end{vmatrix} = -5\lambda = 0 \qquad \Rightarrow \qquad \lambda = 0$

For $\lambda = 0$, clearly $\Delta_y = \Delta_z = 0$.

$\therefore$ System is consistent if $\lambda = 0$, then on eliminating x from (1), (2) and (3), we have

$$6y - 6z = 0, y - z = 0$$

and $\qquad 7y - 7z = 0$ or $y = z$.

Let $y = z = k \in R$, then from (1), we have $x = 5k - 4k = k$.

Hence, solution is given by $x = y = z = k \in R$.

EXAMPLE 8. *Solve the equations by Cramer's rule*

$$\frac{4}{x+5} + \frac{3}{y+7} = -1$$

$$\frac{6}{x+5} + \frac{6}{y+7} = -5$$

SOLUTION. The given system of equation is

$$\frac{4}{x+5} + \frac{3}{y+7} = -1$$

$$\frac{6}{x+5} + \frac{6}{y+7} = -5$$

Now putting $\dfrac{1}{x+5} = a, \dfrac{1}{y+7} = b$, the equations becomes

$$4a + 3b = -1$$
$$6a - 6b = -5$$

$$\Delta = \begin{vmatrix} 4 & 3 \\ 6 & -6 \end{vmatrix} = -42 \neq 0$$

$$\Delta_a = \begin{vmatrix} -1 & 3 \\ 5 & -6 \end{vmatrix} = 21, \Delta_b = \begin{vmatrix} 4 & -1 \\ 6 & -5 \end{vmatrix} = -14.$$

So by Cramer's rule

$$a = \dfrac{\Delta_a}{\Delta} = \dfrac{21}{-42} = \dfrac{1}{2}, \ b = \dfrac{\Delta_b}{\Delta} = \dfrac{-14}{-42} = \dfrac{1}{3}$$

$\therefore \qquad\qquad x + 5 = -2, y + 7 = 3$

or $\qquad\qquad\qquad x = -7, y = -4.$

Hence, the solution is $x = -7, y = -4$.

EXAMPLE 9. *Using Cramer's rule solve the following equation :*

$$x + 2y + 3z = 6$$
$$2x + 4y + z = 17$$
$$3x + 2y + 9z = 2$$

SOLUTION. We have

$$\Delta = \begin{vmatrix} 1 & 2 & 3 \\ 2 & 4 & 1 \\ 3 & 4 & 9 \end{vmatrix} = -20$$

$$\Delta_x = \begin{vmatrix} 6 & 2 & 3 \\ 17 & 4 & 1 \\ 2 & 2 & 9 \end{vmatrix} = -20$$

$$\Delta_y = \begin{vmatrix} 1 & 6 & 3 \\ 2 & 17 & 1 \\ 3 & 2 & 9 \end{vmatrix} = -80$$

$$\Delta_z = \begin{vmatrix} 1 & 2 & 6 \\ 2 & 4 & 17 \\ 3 & 2 & 2 \end{vmatrix} = -20$$

Then by Cramer's rule, we have

$$x = \dfrac{\Delta_x}{\Delta} = \dfrac{-20}{-20} = 1$$

$$y = \dfrac{\Delta_y}{\Delta} = \dfrac{-80}{-20} = 4$$

$$z = \dfrac{\Delta_z}{\Delta} = \dfrac{20}{-20} = -1$$

Exercise 7.2

1. (a) Using Cramer's rule, solve the following equations

$$x + y + z = 6$$
$$2x + y - z = 1$$
$$x + y - 2z = -3.$$

(b) $\qquad x + y + z = 6$
$\qquad x - y + z = 2$
$\qquad 3x + 2y - 9z = -5$

2. Find the value of k if the following equations are consistent :
$$x + y - 3 = 0$$
$$(1 + k)x + (2 + k)y - 8 = 0$$
$$x - (1 + k)y + (2 + k) = 0.$$

3. Find the value of k if the system of equations
$$(k + 1)^3 x + (k + 2)^3 y = (k + 3)^3$$
$$(k + 1)x + (k + 2)y = (k + 3)$$
$$x + y = 1 : \text{is consistent.}$$

4. If the system of equations
$x + 2y = 5, 2x - y = 5, x + 3y = 6$ is consistent, solve it.

5. Solve the following by Cramer's rule

$$x + y + z = 11$$
$$2x - 6y - z = 0$$
$$3x + 4y + 2z = 0.$$

6. Show that the system of equations
$$3x - y + 4z = 3$$
$$x + 2y - 3z = -2$$
$$6x + 5y + \lambda z = -3$$

has at least one solution for any real number λ. Find the set of solution if $\lambda = -5$.

7. Using Cramer's rule to solve the following system of linear equations.
$$2x - 3y + z = 7$$
$$2x + y + z = 1$$
$$4y + 3z = -11$$

ANSWERS

1. (a) $x = 1, y = 2, z = 3$ (b) $x = \dfrac{9}{4}, y = 2, z = \dfrac{21}{12}$

2. $k = 1$ or $-5/3$ 3. $k = -2$ 5. $x = -8, y = -7, z = 26.$

Objective Evaluation

⇒ MULTIPLE CHOICE QUESTIONS

Choose the most appropriate one.

1. The value of $\begin{vmatrix} a+ib & c+id \\ -c+id & a-ib \end{vmatrix}$ is :

   (a) $(a^2 + b^2 - c^2 - d^2)$

   (b) $(a^2 - b^2 + c^2 - d^2)$

   (c) $(a^2 + b^2 + c^2 + d^2)$

   (d) none of these

2. If ω is a cube root of unity then the value of

   $\begin{vmatrix} 1 & \omega & 1+\omega \\ 1+\omega & 1 & \omega \\ \omega & 1+\omega & 1 \end{vmatrix}$ is :

   (a) 4   (b) 2

   (c) 1   (d) 3

3. The value of $\begin{vmatrix} 1^2 & 2^2 & 3^2 \\ 2^2 & 3^2 & 4^2 \\ 3^2 & 4^2 & 5^2 \end{vmatrix}$ is :

   (a) 8   (b) 16

   (c) –8   (d) 142

4. The value of $\begin{vmatrix} a-b & b-c & c-a \\ b-c & c-a & a-b \\ c-a & a-b & b-c \end{vmatrix}$ is :

   (a) $(a + b + c)$ (b) $3(a + b + c)$

   (c) $3abc$ (d) 0

5. The value of $\begin{vmatrix} 1 & 1+p & 1+p+q \\ 2 & 3+2p & 1+3p+2q \\ 3 & 6+3p & 1+6p+3q \end{vmatrix}$ is :

   (a) 0   (b) –1

   (c) 1   (d) none of these

6. The value of $\begin{vmatrix} \sin\alpha & \cos\alpha & \sin(\alpha+\delta) \\ \sin\beta & \cos\beta & \sin(\beta+\delta) \\ \sin\gamma & \cos\gamma & \sin(\gamma+\delta) \end{vmatrix}$ is :

   (a) 1

   (b) 0

   (c) $\sin(\alpha + \delta) + \sin(\beta + \delta) + \sin(\gamma + \delta)$

   (d) none of these

7. If a, b, c be distinct positive real numbers then

   the value of $\begin{vmatrix} a & b & c \\ b & c & a \\ c & a & b \end{vmatrix}$ is :

   (a) positive   (b) negative

   (c) a perfect square   (d) 0

8. The value of $\begin{vmatrix} a & a+2b & a+2b+3c \\ 3a & 4a+6b & 5a+7b+9c \\ 6a & 9a+12b & 11a+15b+18c \end{vmatrix}$ is :

   (a) a^3 (b) $-a^3$

   (c) 0   (d) none of these

9. The value of $\begin{vmatrix} x+1 & x+2 & x+4 \\ x+3 & x+5 & x+8 \\ x+7 & x+10 & x+14 \end{vmatrix}$ is :

   (a) –2   (b) 2

   (c) $x^2 - 2$ (d) $x^2 + 2$

10. The value of $\begin{vmatrix} bc & b+c & 1 \\ ca & c+a & 1 \\ ab & a+b & 1 \end{vmatrix}$ is :

    (a) $(a - b)(b - c)(c - a)$

    (b) $-(a - b)(b - c)(c - a)$

    (c) $(a + b)(b + c)(c + a)$

    (d) None of these

11. The solution set of the equation $\begin{vmatrix} x & 3 & 7 \\ 2 & x & 2 \\ 7 & 6 & x \end{vmatrix} = 0$

    is :

    (a) $[2, 7, -9]$ (b) $[-2, 3, -7]$

    (c) $[2, -3, 7]$ (d) none of these

12. The solution set of the equation

    $\begin{vmatrix} a+x & a-x & a-x \\ a-x & a+x & a-x \\ a-x & a-x & a+x \end{vmatrix} = 0$ is :

    (a) $\{a, 0\}$ (b) $\{3a, 0\}$

    (c) $\{a, 3a\}$ (d) none of these

13. The vertices of ABC are $A(-2, 4)$, $B(2, -6)$ and $C(5, 4)$. The area of ABC is :

    (a) 17.5 sq. units   (b) 35 sq. units

    (c) 32 sq. units   (d) 28 sq. units

14. If the points $A(3, -2)$, $B(k, 2)$ and $C(8, 8)$ are collinear then the value of k is :

 (a) 2   (b) -3

 (c) 5   (d) 4

15. The value of $\begin{vmatrix} a^2 + 2a & 2a+1 & 1 \\ 2a+1 & a+2 & 1 \\ 3 & 3 & 1 \end{vmatrix}$ is :

 (a) $(a-1)$ (b) $(a-1)^2$

 (c) $(a-1)^3$ (d) none of these

16. The value of $\begin{vmatrix} \cos 70° & \sin 20° \\ \sin 70° & \cos 20° \end{vmatrix}$ is :

 (a) 0   (b) 1

 (c) $\cos 50°$ (d) $\sin 50°$

17. The value of $\begin{vmatrix} \sin 23° & -\sin 7° \\ \cos 23° & \cos 7° \end{vmatrix}$ is :

 (a) $\dfrac{\sqrt{3}}{2}$ (b) $\dfrac{1}{2}$

 (c) $\sin 16°$ (d) $\cos 16°$

☞ TRUE/FALSE

Write 'T' for True and 'F' for False statement.

1. The value of $\begin{vmatrix} \sin 10° & -\cos 10° \\ \sin 80° & \cos 80° \end{vmatrix}$ is 1.   **(T/F)**

2. The value of a determinant remains unchanged if its rows and columns are interchanged.   **(T/F)**

3. The value of $\begin{vmatrix} a-b & b-c & c-a \\ b-c & c-a & a-b \\ c-a & a-b & b-c \end{vmatrix}$ is 1.   **(T/F)**

4. The value of $\begin{vmatrix} 2 & 3 & 4 \\ 5 & 6 & 8 \\ 6x & 9x & 12x \end{vmatrix}$ is 0.   **(T/F)**

5. The value of $\begin{vmatrix} a & b & c \\ a-b & b-c & c-a \\ b+c & c+a & a+b \end{vmatrix}$ i $a^3 + b^3 + c^3 + 3abc$.   **(T/F]**

☞ FILL IN THE BLANKS

1. The value of $\begin{vmatrix} a & a+b & a+b+c \\ 2a & 3a+2b & 4a+3b+2c \\ 3a & 6a+3b & 10a+6b+3c \end{vmatrix}$ is _____.

2. The value of $\begin{vmatrix} 1 & a & a^2 \\ 1 & b & b^2 \\ 1 & c & c^2 \end{vmatrix}$ is _____.

3. If $A + B + C = \pi$, then the value of $\begin{vmatrix} \sin^2 A & \sin A \cos A & \cos^2 A \\ \sin^2 B & \sin B \cos B & \cos^2 B \\ \sin^2 C & \sin C \cos C & \cos^2 C \end{vmatrix}$ is _____.

4. If A is a 2×2 matrix such that $|A| \neq 0$ and $|A| = 5$ then the value of $|4A|$ is _____.

5. If A_{ij} is the cofactor of the element a_{ij} o $\begin{vmatrix} 2 & -3 & 5 \\ 6 & 0 & 4 \\ 1 & 5 & -7 \end{vmatrix}$, then the value of $(a_{32} A_{32})$ i _____.

6. The value of $\begin{vmatrix} x^2-x+1 & x-1 \\ x+1 & x+1 \end{vmatrix}$ is _____

ANSWERS

☞ MULTIPLE CHOICE QUESTIONS

 1. (c)   **2.** (a)   **3.** (c)   **4.** (d)   **5.** (c)   **6.** (b)   **7.** (b)   **8.** (b)   **9.** (a)

10. (a)   **11.** (a)   **12.** (b)   **13.** (b)   **14.** (c)   **15.** (c)   **16.** (a)   **17.** (b)

☞ TRUE/ FALSE

 1. T   **2.** T   **3.** F   **4.** T   **5.** F

☞ FILL IN THE BLANKS

 1. a^3 **2.** $(a-b)(b-c)(c-a)$ **3.** $-\sin(A-B)\sin(B-C)\sin(C-A)$ **4.** 80

 5. 110   **6.** $x^3 - x^2 + 2$

☐☐☐☐☐☐

CHAPTER 8

Matrices

8.1 INTRODUCTION

Matrix is an ordered rectangular array of numbers (real or complex) in horizontal and vertical lines called rows and columns respectively. Matrix plays an important role in various branches of mathematics, electrical engineering, genetic and sociology etc. The word 'matrix' was first used by British mathematician J.J. Silvestor in 1850. Another British mathematician Arthur Cayley formulated the general theory of matrix in 1857. Matrix is useful in every branch of science and engineering.

8.2 DEFINITION OF MATRIX

Definition. *A set of mn numbers* $a_{11}, a_{12}, ..., a_{mn}$ *arranged in a rectangular array of m rows and n columns is called a matrix of order* $m \times n$. *Generally it is denoted by* [ ] *or* ( ) *or* $\| \ \|$

A matrix of order $m \times n$ can be illustrated as follows :

$$A_{m \times n} = \begin{bmatrix} a_{11} & a_{12} & \cdots & a_{1j} & \cdots & a_{1n} \\ a_{21} & a_{22} & \cdots & a_{2j} & \cdots & a_{2n} \\ \vdots & & & & & \\ a_{i1} & a_{i2} & \cdots & a_{ij} & \cdots & a_{in} \\ \vdots & & & & & \\ a_{m1} & a_{m2} & \cdots & a_{mj} & \cdots & a_{mn} \end{bmatrix}$$

The quantities a_{ij} ($i = 1, 2, ..., m, j = 1, 2, ..., n$) are called the elements of the matrix A. An element occurring in the ith row (horizontal lines) and j^{th} column (vertical lines) of a matrix A is called $(i, j)^{th}$ element of A and is denoted by a_{ij}.

For example: Let $A = \begin{bmatrix} 2 & 3 & -5 \\ 4 & 3 & 8 \end{bmatrix}$

Then
$$a_{11} = 2, \qquad a_{12} = 3, \qquad a_{13} = -5$$
$$a_{21} = 4, \qquad a_{22} = 3, \qquad a_{23} = 8$$

8.3 KIND OF MATRICES

(1) **Horizontal Matrix :** A matrix of order $m \times n$ is called a horizontal matrix if $m < n$, *i.e.*, if number of rows is less than the number of columns.

For example: The matrix $\begin{bmatrix} a & b & c \\ d & e & f \end{bmatrix}$ is a horizontal matrix because it has two rows and three columns.

(2) **Vertical Matrix:** A matrix of order $m \times n$ is called a vertical matrix if $m > n$, *i.e.*, if number of rows is greater than the number of columns.

For example: $\begin{bmatrix} a & b \\ c & d \\ e & f \end{bmatrix}$ is a vertical matrix because it has three rows and two columns.

(3) Square Matrix : A matrix in which number of rows is equal to number of columns is called square matrix. In such type of matrix $m = n$.

For example: $\begin{bmatrix} a & b \\ c & d \end{bmatrix}$ and $\begin{bmatrix} a_{11} & a_{12} & a_{13} \\ a_{21} & a_{22} & a_{23} \\ a_{31} & a_{32} & a_{33} \end{bmatrix}$ are square matrix of order 2×2 and 3×3 respectively.

REMARK

➠ Because the matrix of order $m \times m$ is a square matrix. Therefore, we say it is of order m instead of order $m \times m$.

(4) Row Matrix or Row Vector : A matrix having only one row is called a row matrix or row vector.

For example: $\begin{bmatrix} 1 & 2 & 3 & 4 \end{bmatrix}$ and $\begin{bmatrix} a & b & c \end{bmatrix}$ are the row matrices of order 1×4 and 1×3 respectively.

(5) Column Matrix or Column Vector : A matrix having only one column is called a column matrix or column vector.

For example: $\begin{bmatrix} a \\ b \end{bmatrix}$ and $\begin{bmatrix} 1 \\ 3 \\ 5 \end{bmatrix}$ are the column matrices of order 2×1 and 3×1 respectively.

(6) Zero or Null Matrix : A matrix in which every element is zero is called the null or zero matrix. It is denoted by O.

For example: $\begin{bmatrix} 0 & 0 & 0 \end{bmatrix}$ and $\begin{bmatrix} 0 & 0 & 0 \\ 0 & 0 & 0 \\ 0 & 0 & 0 \\ 0 & 0 & 0 \end{bmatrix}$ are the null matrices of order 1×3 and 4×3 respectively.

(7) Unit or Identity Matrix : A square matrix in which all the elements along the principal diagonal are one (1) and all elements not occurring along the principal diagonal are zero. It is denoted by I.

For example: $\begin{bmatrix} 1 & 0 \\ 0 & 1 \end{bmatrix}, \begin{bmatrix} 1 & 0 & 0 \\ 0 & 1 & 0 \\ 0 & 0 & 1 \end{bmatrix}$ and $\begin{bmatrix} 1 & 0 & 0 & 0 \\ 0 & 1 & 0 & 0 \\ 0 & 0 & 1 & 0 \\ 0 & 0 & 0 & 1 \end{bmatrix}$ are the unit matrices of order 2×2, 3×3 and 4×4 respectively.

In other words A square matrix $A = [a_{ij}]$ is a unit matrix if

$$a_{ij} = \begin{cases} 1 & \text{if } i = j \\ 0 & \text{if } i \neq j \end{cases}$$

Principal Diagonal: Every square matrix has two diagonals in which the diagonal starting from the first element down to last element is said to be main diagonal.

For example :
$$\begin{bmatrix} a & b & c \\ d & e & f \\ g & h & i \end{bmatrix} \longrightarrow \text{main diagonal}$$

☞ REMARK

➠ Only square matrix has diagonals. The elements lying on main diagonal are called diagonal elements.

(8) Diagonal Matrix : A square matrix which has all its elements are zero except the diagonal elements, is said to be diagonal matrix.

For example: $\begin{bmatrix} 3 & 0 \\ 0 & 2 \end{bmatrix}$ and $\begin{bmatrix} 1 & 0 & 0 \\ 0 & 5 & 0 \\ 0 & 0 & 7 \end{bmatrix}$ are diagonal matrices.

(9) Scalar Matrix: A square matrix is said to be scalar if all elements along the principal diagonal are equal and non-diagonal elements are zero.

For example: $\begin{bmatrix} 2 & 0 & 0 \\ 0 & 2 & 0 \\ 0 & 0 & 2 \end{bmatrix}$ and $\begin{bmatrix} a & 0 & 0 \\ 0 & a & 0 \\ 0 & 0 & a \end{bmatrix}$ are scalar matrices.

(10) Sub Matrix : The matrix obtained by leave some row and column of the given matrix is called sub matrix.

For example: $\begin{bmatrix} 1 & 2 \\ 3 & 9 \end{bmatrix}$ is the sub matrix of the matrix $\begin{bmatrix} 5 & 4 & 8 \\ 6 & 1 & 2 \\ 7 & 3 & 9 \end{bmatrix}$.

(11) Triangular Matrices

   (i) Upper Triangular Matrix : A square matrix $A = [a_{ij}]_{m \times n}$ is called an upper triangular matrix if $a_{ij} = 0 \ \forall \ i > j$. Therefore, a square matrix is said to be upper triangular if all the elements below the main diagonal are equal to zero.

   For example: $\begin{bmatrix} 5 & 4 & 0 \\ 0 & 3 & 2 \\ 0 & 0 & 1 \end{bmatrix}$ and $\begin{bmatrix} 1 & 2 & 4 & 8 \\ 0 & 4 & 3 & 2 \\ 0 & 0 & 7 & 8 \\ 0 & 0 & 0 & 9 \end{bmatrix}$ are upper triangular matrices.

   (ii) Lower Triangular Matrix : A square matrix $A = [a_{ij}]$, is called a lower triangular matrix if $a_{ij} = 0$ for all $i < j$. Therefore, a square matrix is said to be lower triangular matrix if all the elements above the main diagonal are equal to zero.

   For example: $\begin{bmatrix} a & 0 \\ b & c \end{bmatrix}$ and $\begin{bmatrix} 5 & 0 & 0 & 0 \\ 8 & 1 & 0 & 0 \\ 4 & 2 & 9 & 0 \\ 3 & 5 & 1 & 7 \end{bmatrix}$ are lower triangular matrices.

(12) Comparable Matrices : Two matrices A and B are said to be comparable if they are of same order.

For example: $A = \begin{bmatrix} 1 & 4 & 7 \\ 6 & 0 & 8 \end{bmatrix}$ and $B = \begin{bmatrix} 5 & 2 & 7 \\ 4 & 8 & 9 \end{bmatrix}$ are two matrices of same order 2×3.

Hence, they are comparable.

8.3.1 EQUALITY OF MATRICES

Two matrices A and B are said to be equal (*i.e.*, $A = B$)

if   (i) they are of same order.

and (ii) their corresponding elements are equal.

Symbolically : Two matrices $A = [a_{ij}]_{m \times n}$ and $B = [b_{ij}]_{r \times s}$ are said to be equal if

(i) No. of rows in A = no. of rows in B.

(ii) No. of column in A = no. of column in B.

(iii) $(i, j)^{\text{th}}$ element of $A = (i, j)^{\text{th}}$ element of B, i.e., $a_{ij} = b_{ij} \ \forall \ i, j$

For example: $A = \begin{bmatrix} 2 & 4 \\ 6 & 8 \end{bmatrix}$ and $B = \begin{bmatrix} 2 & 4 \\ 6 & 8 \end{bmatrix}$. Then $A = B$.

8.4 OPERATIONS ON MATRICES

(i) **Addition of Matrices:** Let A and B two matrices of same order $m \times n$. Then addition of A and B denoted by $A + B$ is obtained by adding each element of A to the corresponding element of B.

If $A = [a_{ij}], B = [b_{ij}],$ $i = 1, 2, ..., m$
                                                                              $j = 1, 2, ..., n$

Then $A + B = [a_{ij} + b_{ij}],$ $i = 1, 2, ..., m$
                                                                              $j = 1, 2, ..., n$

In other words $[a_{ij}]_{m \times n} + [b_{ij}]_{m \times n} = [a_{ij} + b_{ij}]_{m \times n}$

For example:

Let $A = \begin{bmatrix} 2 & 3 & 4 \\ 0 & 2 & -1 \end{bmatrix}, B = \begin{bmatrix} -3 & 4 & 5 \\ 6 & 2 & 9 \end{bmatrix}$

Then $A + B = \begin{bmatrix} 2-3 & 3+4 & 4+5 \\ 0+6 & 2+2 & -1+9 \end{bmatrix} = \begin{bmatrix} -1 & 7 & 9 \\ 6 & 4 & 8 \end{bmatrix}$

✎ REMARK
➧ Operation of adding is a binary operation in set of matrices of same order.

(ii) **Negative of a Matrix:** Let $A = [a_{ij}]m \times n$ be a matrix. Negative of A is represented by $-A$ and is given by $(-a_{ij})$.

For example:

Let $A = \begin{bmatrix} 1 & 3 & 5 \\ -8 & 9 & 7 \\ 6 & -4 & 0 \end{bmatrix}$ and $-A = \begin{bmatrix} -1 & -3 & -5 \\ 8 & -9 & -7 \\ -6 & 4 & 0 \end{bmatrix}$

✎ REMARK
➧ If order of A is of $m \times n$ then order $-A$ also be of $m \times n$.

(iii) **Difference of Two Matrices:** Let A and B be two matrices of order $m \times n$, then difference between A and B represented by $A - B$ and it is defined as $A - B = A + (-B)$. Order of $A - B$ will be same as the order of A and B. Subtract each elements of B from corresponding elements of A to get $A - B$.

For example: Let $A = \begin{bmatrix} 6 & 7 & 8 \\ -4 & 3 & 7 \\ 6 & 5 & -2 \end{bmatrix}$ and $B = \begin{bmatrix} 4 & 2 & 5 \\ 4 & 6 & 4 \\ 9 & 8 & 3 \end{bmatrix}$

Then $A - B = \begin{bmatrix} 6-4 & 7-2 & 8-5 \\ -4-4 & 3-6 & 7-4 \\ 6-9 & 5-8 & -2-3 \end{bmatrix} = \begin{bmatrix} 2 & 5 & 3 \\ -8 & -3 & 3 \\ -3 & -3 & -5 \end{bmatrix}$

8.5 PROPERTIES OF MATRICES ADDITION

Property 1. Commutative law of Addition:

If A and B are two matrices of order m × n, then A + B = B + A.

Proof. Let $A = [a_{ij}]_{m \times n}$ and $B = [b_{ij}]_{m \times n}$, then

$$A + B = [a_{ij}]_{m \times n} + [b_{ij}]_{m \times n}$$
$$= [a_{ij} + b_{ij}]_{m \times n} \qquad \text{(Addition rule for matrix)}$$
$$= [b_{ij} + a_{ij}]_{m \times n} \qquad \text{(Commutative law of numbers)}$$
$$= [b_{ij}]_{m \times n} + [a_{ij}]_{m \times n}$$

Hence, $A + B = B + A$

Property 2. Associative law of Addition:

If A, B and C are three matrices of order m × n, then

$$A + (B + C) = (A + B) + C$$

Proof. Let $A = [a_{ij}]_{m \times n}$, $B = [b_{ij}]_{m \times n}$ and $C = [c_{ij}]_{m \times n}$, then

$$(A + B) + C = \{[a_{ij}]_{m \times n} + [b_{ij}]_{m \times n}\} + [c_{ij}]_{m \times n}$$
$$= [a_{ij} + b_{ij}]_{m \times n} + [c_{ij}]_{m \times n}$$
$$= [(a_{ij} + b_{ij}) + c_{ij}]_{m \times n}$$
$$= [a_{ij} + (b_{ij} + c_{ij})]_{m \times n}$$

$$\text{(Associative rule of addition of numbers)}$$

$$= [a_{ij}]_{m \times n} + \{[b_{ij}]_{m \times n} + [c_{ij}]_{m \times n}\}$$
$$= A + (B + C)$$

Hence, $(A + B) + C = A + (B + C)$

Property 3. Existence of Additive Identity:

Let A be a matrix of order m × n and there exists a null matrix O such that A + O = O + A = A.

Proof. Let $A = [a_{ij}]_{m \times n}$ and $O = [b_{ij}]_{m \times n}$

where $b_{ij} = 0$, $i = 1, 2, ..., m$ and $j = 1, 2, ..., n$

then
$$A + O = [a_{ij}]_{m \times n} + [b_{ij}]_{m \times n}$$
$$= [a_{ij} + b_{ij}]_{m \times n}$$
$$= [a_{ij} + 0]_{m \times n} \qquad [\because b_{ij} = 0]$$
$$= [a_{ij}]_{m \times n} = A$$

$\therefore$ $\qquad A + O = A$

Similarly $\qquad O + A = A$

Therefore, $\qquad A + O = O + A = A$

⮞ REMARK

⮕ Null matrix of $m \times n$ order is called additive identity for all $m \times n$ order matrices.

Property 4. Existence of Additive Inverse: *For a matrix A*

$$A + (-A) = (-A) + A = O, \text{ where } O \text{ is null matrix.}$$

Proof. Let $A = [a_{ij}]_{m \times n}$. Then $(-A) = [-a_{ij}]_{m \times n}$

$$A + (-A) = [a_{ij}]_{m \times n} + [-a_{ij}]_{m \times n}$$
$$= [a_{ij} + (-a_{ij})]_{m \times n}$$
$$= O, \text{ where } O \text{ is null matrix.}$$

So $\qquad A + (-A) = O$

Similarly, $\qquad (-A) + A = O$

Hence, $\qquad A + (-A) = (-A) + A = O$

REMARK

➤ Matrix (–A) is called additive inverse of A.

Property 5. Cancellation law for matrix addition:

If A, B and C are three matrices of order m × n. Then

(i) $A + B = A + C \Rightarrow B = C$ (Left cancellation law)

(ii) $B + A = C + A \Rightarrow B = C$ (Right cancellation law)

Proof: We have $A + B = A + C$

$\Rightarrow$ $(-A) + A + B = (-A) + A + C$

$\Rightarrow$ $(-A + A) + B = (-A + A) + C$ (Associativity)

$\Rightarrow$ $O + B = O + C$ $(-A + A = O)$

$\Rightarrow$ $B = C$

So, $A + B = A + C$ $\Rightarrow$ $B = C$

Similarly, we can show that

$B + A = C + A$ $\Rightarrow$ $B = C$

8.6 MULTIPLICATION OF A MATRIX BY A SCALAR

Scalar multiplication of a matrix A is obtained by multiplying of each element of matrix by the given scalar.

i.e., $kA = [ka_{ij}]$

For example: If $A = \begin{bmatrix} 1 & -3 \\ 0 & 3 \end{bmatrix}$

Then $3A = 3\begin{bmatrix} 1 & -3 \\ 0 & 3 \end{bmatrix} = \begin{bmatrix} 3 & -9 \\ 0 & 9 \end{bmatrix}$

Similarly, $-6A = -6\begin{bmatrix} 1 & -3 \\ 0 & 3 \end{bmatrix} = \begin{bmatrix} -6 & +18 \\ 0 & -18 \end{bmatrix}$

8.6.1 PROPERTIES OF SCALAR MULTIPLICATION

THEOREM. *In matrix addition, the scalar multiplication is distributive over addition, i.e., If A and B, are two matrices of same order and k is a scalar, then k(A + B) = kA + kB.*

PROOF. Let $A = [a_{ij}]_{m \times n}$ and $B = [b_{ij}]_{m \times n}$

$A + B = [a_{ij} + b_{ij}]_{m \times n}$

$\Rightarrow$ $k(A + B) = [ka_{ij} + kb_{ij}]_{m \times n}$

$= [ka_{ij}]_{m \times n} + [kb_{ij}]_{m \times n}$

$= k[a_{ij}]_{m \times n} + k[b_{ij}]_{m \times n}$

$= kA + kB$

Hence, $k(A + B) = kA + kB$

For example: If $A = \begin{bmatrix} a & b \\ c & d \end{bmatrix}$ and $B = \begin{bmatrix} x & y \\ z & u \end{bmatrix}$

Then $k(A + B) = k\begin{bmatrix} a + x & b + y \\ c + z & d + u \end{bmatrix} = \begin{bmatrix} ka + kx & kb + ky \\ kc + kz & kd + ku \end{bmatrix}$

$= \begin{bmatrix} ka & kb \\ kc & kd \end{bmatrix} + \begin{bmatrix} kx & ky \\ kz & ku \end{bmatrix} = k\begin{bmatrix} a & b \\ c & d \end{bmatrix} + k\begin{bmatrix} x & y \\ z & u \end{bmatrix} = kA + kB$

$\Rightarrow$ $k(A + B) = kA + kB$

EXAMPLE 1. *Write $A = [a_{ij}]_{2 \times 3}$ where $a_{ij} = 2i - 3j$.*

SOLUTION. Given

$$a_{ij} = 2i - 3j$$
$$a_{11} = 2 \times 1 - 3 \times 1 = -1$$
$$a_{12} = 2 \times 1 - 3 \times 2 = -4$$
$$a_{13} = 2 \times 1 - 3 \times 3 = -7$$
$$a_{21} = 2 \times 2 - 3 \times 1 = 1$$
$$a_{22} = 2 \times 2 - 3 \times 2 = -2$$
$$a_{23} = 2 \times 2 - 3 \times 3 = -5$$

So

$$A = \begin{bmatrix} -1 & -4 & -7 \\ 1 & -2 & -5 \end{bmatrix}$$

EXAMPLE 2. *If $\begin{bmatrix} 3x-2 & 4y-8 \\ z-2 & a+11 \end{bmatrix} = \begin{bmatrix} 10 & 8 \\ 11 & 9 \end{bmatrix}$ find x, y, z and a.*

SOLUTION. Comparing the corresponding elements

$$3x - 2 = 10 \qquad \Rightarrow \qquad x = 4$$
$$4y - 8 = 8 \qquad \Rightarrow \qquad y = 4$$
$$z - 2 = 11 \qquad \Rightarrow \qquad z = 13$$
$$a + 11 = 9 \qquad \Rightarrow \qquad a = -2$$

EXAMPLE 3. *If $A = \begin{bmatrix} 2 & 3 & -1 \\ 0 & -1 & 5 \end{bmatrix}$ and $B = \begin{bmatrix} 1 & 2 & -6 \\ 0 & -1 & +3 \end{bmatrix}$ then find*

(i) $5A + 2B$ \qquad (ii) $3A - 4B$

SOLUTION. (i)

$$5A = 5 \begin{bmatrix} 2 & 3 & -1 \\ 0 & -1 & 5 \end{bmatrix} = \begin{bmatrix} 10 & 15 & -5 \\ 0 & -5 & 25 \end{bmatrix}$$

and

$$2B = 2 \begin{bmatrix} 1 & 2 & -6 \\ 0 & -1 & 3 \end{bmatrix} = \begin{bmatrix} 2 & 4 & -12 \\ 0 & -2 & 6 \end{bmatrix}$$

$$5A + 2B = \begin{bmatrix} 10 & 15 & -5 \\ 0 & -5 & 25 \end{bmatrix} + \begin{bmatrix} 2 & 4 & -12 \\ 0 & -2 & 6 \end{bmatrix} = \begin{bmatrix} 12 & 19 & -17 \\ 0 & -7 & 31 \end{bmatrix}$$

(ii)

$$3A = 3 \begin{bmatrix} 2 & 3 & -1 \\ 0 & -1 & 5 \end{bmatrix} = \begin{bmatrix} 6 & 9 & -3 \\ 0 & -3 & 15 \end{bmatrix}$$

and

$$4B = 4 \begin{bmatrix} 1 & 2 & -6 \\ 0 & -1 & 3 \end{bmatrix} = \begin{bmatrix} 4 & 8 & -24 \\ 0 & -4 & 12 \end{bmatrix}$$

$$3A - 4B = \begin{bmatrix} 6 & 9 & -3 \\ 0 & -3 & 15 \end{bmatrix} - \begin{bmatrix} 4 & 8 & -24 \\ 0 & -4 & 12 \end{bmatrix} = \begin{bmatrix} 2 & 1 & 21 \\ 0 & 1 & 3 \end{bmatrix}$$

EXAMPLE 4. *Find the value of x, y, z and a, for which*

$$\begin{bmatrix} x+3 & 2y+x \\ z-1 & 4a-6 \end{bmatrix} = \begin{bmatrix} 0 & -7 \\ 3 & 2a \end{bmatrix}$$

SOLUTION.

$$\begin{bmatrix} x+3 & 2y+x \\ z-1 & 4a-6 \end{bmatrix} = \begin{bmatrix} 0 & -7 \\ 3 & 2a \end{bmatrix}$$

$$x + 3 = 0 \qquad \Rightarrow \qquad x = -3$$
$$2y + x = -7 \qquad \Rightarrow \qquad y = -2$$

$$z - 1 = 3 \qquad \Rightarrow \qquad z = 4$$
$$4a - 6 = 2a \qquad \Rightarrow \qquad a = 3$$

EXAMPLE 5. If $A = \begin{bmatrix} 1 & 0 & 2 \\ 0 & 2 & 3 \\ 1 & 2 & 3 \end{bmatrix}$ and $B = \begin{bmatrix} 3 & 1 & 1 \\ 1 & 2 & 3 \\ 0 & 1 & 2 \end{bmatrix}$ find $3A + 6B$.

SOLUTION. We have $\qquad 3A = 3 \begin{bmatrix} 1 & 0 & 2 \\ 0 & 2 & 3 \\ 1 & 2 & 3 \end{bmatrix} = \begin{bmatrix} 3 & 0 & 6 \\ 0 & 6 & 9 \\ 3 & 6 & 9 \end{bmatrix}$

and $\qquad 6B = 6 \begin{bmatrix} 3 & 1 & 1 \\ 1 & 2 & 3 \\ 0 & 1 & 2 \end{bmatrix} = \begin{bmatrix} 18 & 6 & 6 \\ 6 & 12 & 18 \\ 0 & 6 & 12 \end{bmatrix}$

Therefore, $\qquad 3A + 6B = \begin{bmatrix} 3 & 0 & 6 \\ 0 & 6 & 9 \\ 3 & 6 & 9 \end{bmatrix} + \begin{bmatrix} 18 & 6 & 6 \\ 6 & 12 & 18 \\ 0 & 6 & 12 \end{bmatrix} = \begin{bmatrix} 21 & 6 & 12 \\ 6 & 18 & 27 \\ 3 & 12 & 21 \end{bmatrix}$

EXAMPLE 6. If $A = \begin{bmatrix} a & b \\ -b & a \end{bmatrix}, B = \begin{bmatrix} -a & b \\ -b & -a \end{bmatrix}$ find $A + B$.

SOLUTION. We have $\qquad A + B = \begin{bmatrix} a & b \\ -b & a \end{bmatrix} + \begin{bmatrix} -a & b \\ -b & -a \end{bmatrix} = \begin{bmatrix} a-a & b+b \\ -b-b & a-a \end{bmatrix} = \begin{bmatrix} 0 & 2b \\ -2b & 0 \end{bmatrix}$

EXAMPLE 7. Find x and y if $x + y = \begin{bmatrix} 7 & 0 \\ 2 & 5 \end{bmatrix}$ and $x - y = \begin{bmatrix} 3 & 0 \\ 0 & 3 \end{bmatrix}$.

SOLUTION. We have $\quad x + y + x - y = 2x$

$$2x = (x + y) + (x - y) = \begin{bmatrix} 7 & 0 \\ 2 & 5 \end{bmatrix} + \begin{bmatrix} 3 & 0 \\ 0 & 3 \end{bmatrix} = \begin{bmatrix} 10 & 0 \\ 2 & 8 \end{bmatrix}$$

$$\Rightarrow \qquad x = \frac{1}{2} \begin{bmatrix} 10 & 0 \\ 2 & 8 \end{bmatrix} = \begin{bmatrix} 5 & 0 \\ 1 & 4 \end{bmatrix}$$

Again $\qquad 2y = (x + y) - (x - y) = \begin{bmatrix} 7 & 0 \\ 2 & 5 \end{bmatrix} - \begin{bmatrix} 3 & 0 \\ 0 & 3 \end{bmatrix} = \begin{bmatrix} 4 & 0 \\ 2 & 2 \end{bmatrix}$

$$y = \frac{1}{2} \begin{bmatrix} 4 & 0 \\ 2 & 2 \end{bmatrix} = \begin{bmatrix} 2 & 0 \\ 1 & 1 \end{bmatrix}$$

EXAMPLE 8. If $A = \begin{bmatrix} 1 & 4 & 3 & 6 \\ -3 & 7 & 0 & 2 \end{bmatrix}$ and $B = \begin{bmatrix} 2 & -3 & 4 & -1 \\ 0 & 6 & 5 & -7 \end{bmatrix}$ find matrix C, if $A - C = 3B$.

SOLUTION. Given, $\qquad A - C = 3B \qquad \Rightarrow \qquad C = A - 3B$

$$3B = 3 \begin{bmatrix} 2 & -3 & 4 & -1 \\ 0 & 6 & 5 & -7 \end{bmatrix} = \begin{bmatrix} 6 & -9 & 12 & -3 \\ 0 & 18 & 15 & -21 \end{bmatrix}$$

$$C = \begin{bmatrix} 1 & 4 & 3 & 6 \\ -3 & 7 & 0 & 2 \end{bmatrix} - \begin{bmatrix} 6 & -9 & 12 & -3 \\ 0 & 18 & 15 & -21 \end{bmatrix}$$

$$= \begin{bmatrix} -5 & 13 & -9 & 9 \\ -3 & -11 & -15 & 23 \end{bmatrix}$$

EXAMPLE 9. *Two farmers Radheyshyam and Hari Prasad cultivates three varieties of rice namely, Basmati, Permal and Naura. The sale (in Rs) of these varieties of rice by both the farmers in the month of September and October are given by the following matrices A and B.*

$$September\ sales \qquad A = \begin{bmatrix} Basmati & Permal & Naura \\ 10,000 & 20,000 & 30,000 \\ 50,000 & 30,000 & 10,000 \end{bmatrix} \begin{matrix} Radheyshyam \\ Hari\,Prasad \end{matrix}$$

$$October\ sales \qquad B = \begin{bmatrix} Basmati & Permal & Naura \\ 5,000 & 10,000 & 6,000 \\ 20,000 & 10,000 & 10,000 \end{bmatrix} \begin{matrix} Radheyshyam \\ Hari\,Prasad \end{matrix}$$

Find :

(i) *What are combined sales in September and October for each farmer in each variety?*

(ii) *What was the change in sales from September to October?*

(iii) *If the farmer receive 2% profit on gross rupees sales, compute the profit for each farmer and for each variety sold in October.*

SOLUTION.

(i) The combined sales is given by

$$A + B = \begin{bmatrix} Basmati & Permal & Naura \\ 15,000 & 30,000 & 36,000 \\ 70,000 & 40,000 & 20,000 \end{bmatrix} \begin{matrix} Radheyshyam \\ Hari\,Prasad \end{matrix}$$

(ii) Change in the sales from September to October is given by

Now

$$A - B = \begin{bmatrix} Basmati & Permal & Naura \\ 5,000 & 10,000 & 24,000 \\ 30,000 & 20,000 & 0 \end{bmatrix} \begin{matrix} Radheyshyam \\ Hari\,Prasad \end{matrix}$$

(iii) Since both the farmers receive 2% profit. Hence, profit for each farmer for each variety sold in October will be the entries of the matrix $\dfrac{2}{100}B$.

Now 2% of $B = \dfrac{2}{100} \times B = 0.02B$

$$= 0.02 \begin{bmatrix} Basmati & Permal & Naura \\ 5,000 & 10,000 & 6,000 \\ 20,000 & 10,000 & 10,000 \end{bmatrix} \begin{matrix} Radheyshyam \\ Hari\,Prasad \end{matrix}$$

$$= \begin{bmatrix} Basmati & Permal & Naura \\ 100 & 200 & 120 \\ 400 & 200 & 200 \end{bmatrix} \begin{matrix} Radheyshyam \\ Hari\,Prasad \end{matrix}$$

Hence, in October, Radheyshyam receives Rs. 100, Rs. 200, and Rs. 120 as profit in the sales of each variety of rice respectively and Hariprasad receives profit of Rs. 400, 200 and 200 in each variety of rice respectively.

Exercise 8.1

1. (a) If a matrix has five rows and each rows contains 3 elements. Find the order of matrix.

   (b) If a matrix has 12 elements, find all possible order of matrix.

   (c) If a matrix has 5 elements, find all possible order of matrix.

2. Construct a matrix $[a_{ij}]_{2 \times 2}$ where $a_{ij} = i + 2j$.

3. If $\begin{bmatrix} x & 3x - y \\ 2x + z & 3y - \omega \end{bmatrix} = \begin{bmatrix} 3 & 2 \\ 4 & 7 \end{bmatrix}$ then find x, y, z, ω.

4. Find $(A + B)$,

   if $A = \begin{bmatrix} 1 & 4 & 3 \\ 2 & 1 & 8 \\ 1 & 1 & 2 \end{bmatrix}$ and $B = \begin{bmatrix} 2 & 1 & 2 \\ 0 & 4 & 8 \\ 6 & 1 & 4 \end{bmatrix}$.

5. If $A = \begin{bmatrix} 2 & 3 & 4 \\ -3 & 0 & 2 \end{bmatrix}, B = \begin{bmatrix} 3 & -4 & -5 \\ 1 & 2 & 1 \end{bmatrix},$

$C = \begin{bmatrix} 5 & -1 & 0 \\ 7 & 0 & 3 \end{bmatrix}$, then find $A + B + C$.

6. Find $3A - 2B$,

$A = \begin{bmatrix} 1 & 6 & 2 \\ 4 & 3 & -5 \end{bmatrix}, B = \begin{bmatrix} 2 & 9 & -6 \\ 4 & -5 & 3 \end{bmatrix}$.

7. Find the matrix x and y of order 2×2 where

$2x - 3y = \begin{bmatrix} 2 & 5 \\ 3 & 1 \end{bmatrix}, 3x + 2y = \begin{bmatrix} 7 & 1 \\ 4 & 5 \end{bmatrix}$.

8. If $A = \begin{bmatrix} \cos^2 \alpha & \sin^2 \alpha \\ \cos \alpha & \sin \alpha \end{bmatrix}$

and $B = \begin{bmatrix} \sin^2 \alpha & \cos^2 \alpha \\ \sin \alpha & \cos \alpha \end{bmatrix}$. Find $(A + B)$.

9. If $A = \begin{bmatrix} 2 & 1 & 1 \\ 3 & -1 & 0 \\ 0 & 2 & 4 \end{bmatrix}, B = \begin{bmatrix} 9 & 7 & -1 \\ 3 & 5 & 4 \\ 2 & 1 & 6 \end{bmatrix}$

and $C = \begin{bmatrix} 2 & -4 & 3 \\ 1 & -1 & 0 \\ 9 & 4 & 5 \end{bmatrix}$. Verify associative law for

matrix addition.

10. If $A = \begin{bmatrix} -1 & 0 & 2 \\ 3 & 1 & 4 \end{bmatrix}, B = \begin{bmatrix} 0 & -2 & 5 \\ 1 & -3 & 1 \end{bmatrix}$

and $C = \begin{bmatrix} 1 & -5 & 2 \\ 6 & 0 & -4 \end{bmatrix}$,

then find $(2A - 3B + 4C)$.

11. Simplify :

$\cos\theta \begin{bmatrix} \cos\theta & \sin\theta \\ -\sin\theta & \cos\theta \end{bmatrix} + \sin\theta \begin{bmatrix} \sin\theta & -\cos\theta \\ \cos\theta & \sin\theta \end{bmatrix}$

12. Find the value of x, y and z if :

(i) $\begin{bmatrix} 3 & x \\ 4 & y \end{bmatrix} = 2 \begin{bmatrix} 1.5 & 1 \\ z & 1 \end{bmatrix}$

(ii) $\begin{bmatrix} 4 & 3 \\ x & 5 \end{bmatrix} = \begin{bmatrix} y & z \\ 1 & 5 \end{bmatrix}$

13. If $A = \begin{bmatrix} 2+i & -i \\ 3 & 4i \end{bmatrix}$ and $B = \begin{bmatrix} 1-i & 2i \\ 2i & 3 \end{bmatrix}$, prove

that $A + B = \begin{bmatrix} 3 & i \\ 3+2i & 4i+3 \end{bmatrix}$.

14. If $A = \begin{bmatrix} 2 & 3 \\ 4 & -5 \end{bmatrix}, B = \begin{bmatrix} 8 & 9 \\ 6 & 7 \end{bmatrix}$, find

(i) $4A$ (ii) $5B$

(iii) $2A + 3B$ (iv) $5A - 3B$.

15. If $x + y = \begin{bmatrix} 2 & 1 \\ 1 & 2 \end{bmatrix}$ and $2x - y = \begin{bmatrix} 1 & 2 \\ 2 & 1 \end{bmatrix}$, prove

that $x = \begin{bmatrix} 1 & 1 \\ 1 & 1 \end{bmatrix}$.

Answers

1. (a) 5×3 (b) $1 \times 12, 2 \times 6, 3 \times 4, 4 \times 3, 6 \times 2, 12 \times 1$ (c) $1 \times 5, 5 \times 1$

2. $\begin{bmatrix} 3 & 5 \\ 4 & 6 \end{bmatrix}$ **3.** $x = 3, y = 7, z = -2, \omega = 14$ **4.** $\begin{bmatrix} 3 & 5 & 5 \\ 2 & 5 & 16 \\ 7 & 2 & 6 \end{bmatrix}$ **5.** $\begin{bmatrix} 10 & -2 & -1 \\ 5 & 2 & 6 \end{bmatrix}$

6. $\begin{bmatrix} -1 & 0 & 18 \\ 4 & 19 & -21 \end{bmatrix}$ **7.** $x = \dfrac{1}{13}\begin{bmatrix} 25 & 13 \\ 18 & 17 \end{bmatrix}, y = \dfrac{1}{13}\begin{bmatrix} 8 & -13 \\ -1 & 7 \end{bmatrix}$ **8.** $\begin{bmatrix} 1 & 1 \\ \cos\alpha + \sin\alpha & \cos\alpha + \sin\alpha \end{bmatrix}$

10. $\begin{bmatrix} 2 & -14 & -3 \\ 27 & 11 & -11 \end{bmatrix}$ **11.** $\begin{bmatrix} 1 & 0 \\ 0 & 1 \end{bmatrix}$ **12.** (i) 2, 2, 2   (ii) 1, 4, 3

14. (i) $\begin{bmatrix} 8 & 12 \\ 16 & -20 \end{bmatrix}$ (ii) $\begin{bmatrix} 40 & 45 \\ 30 & 35 \end{bmatrix}$ (iii) $\begin{bmatrix} 28 & 33 \\ 26 & 11 \end{bmatrix}$ (iv) $\begin{bmatrix} -14 & -12 \\ 2 & -46 \end{bmatrix}$

8.7 MULTIPLICATION OF MATRICES

Definition. *Two matrices A and B can be multiplied only when*

No. of columns in A = No. of rows in B

In this way matrices A and B are said to be conformal to AB.

So, If $A = [a_{ik}]_{m \times p}$ and $B = [b_{kj}]_{p \times n}$ are two matrices, then order of AB is of $m \times n$ in which $(i, j)^{\text{th}}$ element is the sum of product of element in i^{th} row of A with corresponding element of jth column of B.

i.e., $AB = [a_{ik}]_{m \times p}[b_{kj}]_{p \times n}$

$$= \begin{bmatrix} a_{11} & a_{12} & \cdots & a_{1j} & \cdots & a_{1p} \\ a_{21} & a_{22} & \cdots & a_{2j} & \cdots & a_{2p} \\ \cdots & \cdots & \cdots & \cdots & \cdots & \cdots \\ a_{i1} & a_{i2} & \cdots & a_{ij} & \cdots & a_{ip} \\ \cdots & \cdots & \cdots & \cdots & \cdots & \cdots \\ a_{m1} & a_{m2} & \cdots & a_{mj} & \cdots & a_{mp} \end{bmatrix} \times \begin{bmatrix} b_{11} & b_{12} & \cdots & b_{1j} & \cdots & b_{1n} \\ b_{21} & b_{22} & \cdots & b_{2j} & \cdots & b_{2n} \\ \cdots & \cdots & \cdots & \cdots & \cdots & \cdots \\ b_{i1} & b_{i2} & \cdots & b_{ij} & \cdots & b_{ip} \\ \cdots & \cdots & \cdots & \cdots & \cdots & \cdots \\ b_{p1} & b_{p2} & \cdots & b_{pj} & \cdots & b_{pn} \end{bmatrix}$$

$$= [c_{ij}]_{m \times n}$$

where $c_{ij} = (i, j)^{th}$ element of product AB

$$= a_{i1}b_{1j} + a_{i2}b_{2j} + a_{i3}b_{3j} + \ldots + a_{ip}b_{pj} = \sum_{k=1}^{p} a_{ik}b_{kj}$$

= Row matrix of ith row of A × Column matrix of jth column of B

$$= \begin{bmatrix} a_{i1} & a_{i2} & a_{i3} & \cdots & a_{ip} \end{bmatrix} \begin{bmatrix} b_{1j} \\ b_{2j} \\ b_{3j} \\ \vdots \\ b_{pj} \end{bmatrix}$$

WORKING PROCEDURE

If matrix A and B are conformal to AB, then AB is obtained as follows :

Step 1. The first element of AB is obtained by sum of product of element in first row of A with corresponding element of first column of B.

Step 2. Again the second element of first row of AB is obtained by sum of product of elements in first row of A with corresponding elements of second column of B.

Step 3. The third element of first row of AB is obtained by sum of product of elements of first row of A with corresponding elements of third column of B.

Step 4. Continue the same procedure to find remaining elements of AB in first row.

Step 5. As above repeat the procedure to get the elements in second, third, ... rows of AB.

For example: If $A = \begin{bmatrix} 1 & 0 & 5 \\ -1 & 2 & 4 \\ 3 & -2 & 6 \end{bmatrix}_{3\times3}, B = \begin{bmatrix} 4 & -1 \\ 2 & -2 \\ 5 & 3 \end{bmatrix}_{3\times2}$

Now $AB = \begin{bmatrix} 1\times4+0\times2+5\times5 & 1(-1)+0(-2)+5\times3 \\ (-1)\times4+2\times2+4\times5 & (-1)(-1)+2(-2)+4\times3 \\ 3\times4+(-2)\times2+6\times5 & 3(-1)+(-2)(-2)+6\times3 \end{bmatrix}_{3\times2}$

$$= \begin{bmatrix} 29 & 14 \\ 20 & 9 \\ 38 & 19 \end{bmatrix}_{3\times2}$$

8.8 PROPERTIES OF MATRIX MULTIPLICATION

(1) Commutative Law

In general matrix multiplication does not obey commutative law, *i.e.*, it is not always possible that $AB = BA$.

This is clearified by following facts :

(a) If AB is well defined, then it is not necessary that BA is also defined for product *e.g.*, AB is possible for A of order 4×4 and B of order 4×3. But BA is not possible for this.

(b) If AB and BA both are possible and are of same order then it is not always possible that $AB = BA$.

⇌ REMARK

➠ If A and B are two matrices such that AB and BA both exist. Then order of AB and BA are not necessarily equal.

(2) Associative Law

Multiplication of matrix follow associative law provided they are conformal.

i.e., $\qquad\qquad A(BC) = (AB)C$

are possible only when A, B and C are of order $m \times n$, $n \times p$ and $p \times q$ respectively.

Proof. Let $A = [a_{ij}]_{m \times n}$, $B = [b_{jk}]_{n \times p}$ and $C = [c_{ki}]_{p \times q}$ be of order $m \times n$, $n \times p$ and $p \times q$ respectively.

Then first find BC which is a matrix of order $n \times q$.

$$BC = \left[\sum_{k=1}^{p} b_{jk} c_{ki} \right]$$

Similarly, by definition $A(BC)$ will be a matrix of the order $m \times q$, such that

$$A(BC) = \left[\sum_{j=1}^{n} a_{ij} \left(\sum_{k=1}^{p} b_{jk} c_{ki} \right) \right]$$

In this there is addition of elements follow the associative law, *i.e.*,

$$A(BC) = \left[\sum_{k=1}^{p} \left(\sum_{j=1}^{n} a_{ij} b_{jk} \right) c_{ki} \right] = (AB)C$$

So $\qquad\qquad A(BC) = (AB)C$

(3) Distributive Law

Multiplication of matrix is distributive over addition.

i.e., $\qquad\qquad A(B + C) = AB + AC$

If A, B, C are of the order $m \times n$, $n \times p$, $n \times p$ respectively.

Proof. Let $A = [a_{ij}]_{m \times n}$, $B = [b_{jk}]_{n \times p}$ and $C = [c_{jk}]_{n \times p}$.

$$A(B + C) = \left[\sum_{j=1}^{n} a_{ij} (b_{jk} + c_{jk}) \right] = \left[\sum_{j=1}^{n} a_{ij} b_{jk} + \sum_{j=1}^{n} a_{ij} c_{jk} \right]$$

$$= \left[\sum_{j=1}^{n} a_{ij} b_{jk} \right]_{m \times p} + \left[\sum_{j=1}^{n} a_{ij} c_{jk} \right]_{m \times p} = AB + AC$$

So $\qquad\qquad A(B + C) = AB + AC$

(4) If product of two matrices is a null matrix then it is not necessary that one of the two matrix is null.

If $\qquad\qquad A = \begin{bmatrix} 0 & 1 \\ 0 & 0 \end{bmatrix}, B = \begin{bmatrix} 1 & 0 \\ 0 & 0 \end{bmatrix}$

Then $\qquad\qquad AB = \begin{bmatrix} 0 & 1 \\ 0 & 0 \end{bmatrix} \begin{bmatrix} 1 & 0 \\ 0 & 0 \end{bmatrix} = \begin{bmatrix} 0 & 0 \\ 0 & 0 \end{bmatrix}$

So AB is null matrix even when both A and B are not null.

(5) Cancellation law

Cancellation law does not hold for matrix multiplication.

Proof. Let $A = \begin{bmatrix} 0 & 1 \\ 0 & 0 \end{bmatrix}$, $B = \begin{bmatrix} 2 & 0 \\ 0 & 0 \end{bmatrix}$ and $C = \begin{bmatrix} 1 & 0 \\ 0 & 0 \end{bmatrix}$

$$AB = \begin{bmatrix} 0 & 1 \\ 0 & 0 \end{bmatrix}\begin{bmatrix} 2 & 0 \\ 0 & 0 \end{bmatrix} = \begin{bmatrix} 0 & 0 \\ 0 & 0 \end{bmatrix}$$

$$AC = \begin{bmatrix} 0 & 1 \\ 0 & 0 \end{bmatrix}\begin{bmatrix} 1 & 0 \\ 0 & 0 \end{bmatrix} = \begin{bmatrix} 0 & 0 \\ 0 & 0 \end{bmatrix}$$

So $\quad\quad\quad\quad\quad AB = AC$. But $B \neq C$

So cancellation law does not hold.

SOLVED EXAMPLES

EXAMPLE 1. If $A = \begin{bmatrix} 2 & 5 \\ 1 & 3 \end{bmatrix}$, $B = \begin{bmatrix} 1 & -1 \\ -3 & 2 \end{bmatrix}$. *Find AB and BA. Is AB = BA?*

SOLUTION. We have $\quad AB = \begin{bmatrix} 2 & 5 \\ 1 & 3 \end{bmatrix}\begin{bmatrix} 1 & -1 \\ -3 & 2 \end{bmatrix} = \begin{bmatrix} 2\times1+5\times(-3) & 2\times(-1)+5\times2 \\ 1\times1+3\times(-3) & 1\times(-1)+3\times2 \end{bmatrix}$

$\Rightarrow \quad\quad\quad AB = \begin{bmatrix} -13 & 8 \\ -8 & 5 \end{bmatrix}$ ...(1)

and $\quad\quad\quad BA = \begin{bmatrix} 1 & -1 \\ -3 & 2 \end{bmatrix}\begin{bmatrix} 2 & 5 \\ 1 & 3 \end{bmatrix} = \begin{bmatrix} 1\times2+(-1)\times1 & 1\times5+(-1)\times3 \\ (-3)\times2+2\times1 & (-3)\times5+2\times3 \end{bmatrix}$

$\Rightarrow \quad\quad\quad BA = \begin{bmatrix} 1 & 2 \\ -4 & -9 \end{bmatrix}$ ...(2)

By eqn. (1) and (2), $\quad AB \neq BA$.

EXAMPLE 2. If $A = \begin{bmatrix} ab & b^2 \\ -a^2 & -ab \end{bmatrix}$, *show that $A^2 = 0$.*

SOLUTION. We have $\quad\quad A^2 = A \cdot A = \begin{bmatrix} ab & b^2 \\ -a^2 & -ab \end{bmatrix}\begin{bmatrix} ab & b^2 \\ -a^2 & -ab \end{bmatrix}$

$$= \begin{bmatrix} a^2b^2 - a^2b^2 & ab^3 - ab^3 \\ -a^3b + a^3b & -a^2b^2 + a^2b^2 \end{bmatrix} = \begin{bmatrix} 0 & 0 \\ 0 & 0 \end{bmatrix}$$

EXAMPLE 3. *Solve for the value of x and y.*

$$\begin{bmatrix} 3 & -4 \\ 1 & 2 \end{bmatrix}\begin{bmatrix} x \\ y \end{bmatrix} = \begin{bmatrix} 3 \\ 11 \end{bmatrix}$$

SOLUTION. Given $\quad \begin{bmatrix} 3 & -4 \\ 1 & 2 \end{bmatrix}\begin{bmatrix} x \\ y \end{bmatrix} = \begin{bmatrix} 3 \\ 11 \end{bmatrix}$

$\Rightarrow \quad\quad\quad \begin{bmatrix} 3x - 4y \\ x + 2y \end{bmatrix} = \begin{bmatrix} 3 \\ 11 \end{bmatrix}$

$\Rightarrow \quad\quad\quad 3x - 4y = 3$

$\quad\quad\quad\quad\quad x + 2y = 11$

On solving, we get $\quad x = 5, y = 3$.

EXAMPLE 4. If $A = \begin{bmatrix} 1 & 2 & 3 \\ 3 & 4 & 5 \end{bmatrix}$ and $B = \begin{bmatrix} 2 & 3 & 1 \\ 5 & 4 & 3 \\ 2 & 1 & 1 \end{bmatrix}$, find AB. Is AB = BA?

SOLUTION. We have

$$AB = \begin{bmatrix} 1 & 2 & 3 \\ 3 & 4 & 5 \end{bmatrix}_{2 \times 3} \begin{bmatrix} 2 & 3 & 1 \\ 5 & 4 & 3 \\ 2 & 1 & 1 \end{bmatrix}_{3 \times 3} = \begin{bmatrix} 18 & 14 & 10 \\ 36 & 30 & 20 \end{bmatrix}$$

$$BA = \begin{bmatrix} 2 & 3 & 1 \\ 5 & 4 & 3 \\ 2 & 1 & 1 \end{bmatrix}_{3 \times 3} \begin{bmatrix} 1 & 2 & 3 \\ 3 & 4 & 5 \end{bmatrix}_{2 \times 3} \text{ is not defined.}$$

So, $\qquad AB \ne BA.$

EXAMPLE 5. If $A = \begin{bmatrix} 1 & 1 & -1 \\ 2 & -3 & 4 \\ 3 & -2 & 3 \end{bmatrix}, B = \begin{bmatrix} -1 & -2 & -1 \\ 6 & 12 & 6 \\ 5 & 10 & 5 \end{bmatrix}, C = \begin{bmatrix} -1 & -1 & 1 \\ 2 & 2 & -2 \\ -3 & -3 & 3 \end{bmatrix}$. Prove that AB and CA are null matrices.

SOLUTION. We have

$$AB = \begin{bmatrix} 1 & 1 & -1 \\ 2 & -3 & 4 \\ 3 & -2 & 3 \end{bmatrix} \begin{bmatrix} -1 & -2 & -1 \\ 6 & 12 & 6 \\ 5 & 10 & 5 \end{bmatrix}$$

$$= \begin{bmatrix} -1+6-5 & -2+12-10 & -1+6-5 \\ -2-18+20 & -4-36+40 & -2-18+20 \\ -3-12+15 & -6-24+30 & -3-12+15 \end{bmatrix}$$

$$= \begin{bmatrix} 0 & 0 & 0 \\ 0 & 0 & 0 \\ 0 & 0 & 0 \end{bmatrix}$$

and

$$CA = \begin{bmatrix} -1 & -1 & 1 \\ 2 & 2 & -2 \\ -3 & -3 & 3 \end{bmatrix} \begin{bmatrix} 1 & 1 & -1 \\ 2 & -3 & 4 \\ 3 & -2 & 3 \end{bmatrix}$$

$$= \begin{bmatrix} -1-2+3 & -1+3-2 & 1-4+3 \\ 2+4-6 & 2-6+4 & -2+8-6 \\ -3-6+9 & -3+9-6 & 3-12+9 \end{bmatrix}$$

$$= \begin{bmatrix} 0 & 0 & 0 \\ 0 & 0 & 0 \\ 0 & 0 & 0 \end{bmatrix}$$

EXAMPLE 6. If $A = \begin{bmatrix} 0 & -\tan\dfrac{\alpha}{2} \\ \tan\dfrac{\alpha}{2} & 0 \end{bmatrix}$, and I be identity matrix of order 2 then show that

$$(I + A) = (I - A) \begin{bmatrix} \cos\alpha & -\sin\alpha \\ \sin\alpha & \cos\alpha \end{bmatrix}.$$

SOLUTION. We have $\cos\alpha = \dfrac{1 - \tan^2(\alpha/2)}{1 + \tan^2(\alpha/2)} = \dfrac{1-t^2}{1+t^2}, \sin\alpha = \dfrac{2\tan(\alpha/2)}{1 + \tan^2(\alpha/2)} = \dfrac{2t}{1+t^2}$

where $\qquad \tan\dfrac{\alpha}{2} = t$

Again $$(I + A) = \begin{bmatrix} 1 & 0 \\ 0 & 1 \end{bmatrix} + \begin{bmatrix} 0 & -t \\ t & 0 \end{bmatrix} = \begin{bmatrix} 1 & -t \\ t & 1 \end{bmatrix}$$

$$(I - A) = \begin{bmatrix} 1 & 0 \\ 0 & 1 \end{bmatrix} - \begin{bmatrix} 0 & -t \\ t & 0 \end{bmatrix} = \begin{bmatrix} 1 & t \\ -t & 1 \end{bmatrix}$$

Now $(I - A) \begin{bmatrix} \cos\alpha & -\sin\alpha \\ \sin\alpha & \cos\alpha \end{bmatrix} = \begin{bmatrix} 1 & t \\ -t & 1 \end{bmatrix} \begin{bmatrix} \dfrac{1-t^2}{1+t^2} & \dfrac{-2t}{1+t^2} \\ \dfrac{2t}{1+t^2} & \dfrac{1-t^2}{1+t^2} \end{bmatrix}$

$$= \begin{bmatrix} \dfrac{1-t^2}{1+t^2} + \dfrac{2t^2}{1+t^2} & \dfrac{-2t}{1+t^2} + \dfrac{t(1-t^2)}{1+t^2} \\ \dfrac{-t(1-t^2)}{1+t^2} + \dfrac{2t}{1+t^2} & \dfrac{2t^2}{1+t^2} + \dfrac{1-t^2}{1+t^2} \end{bmatrix}$$

$$= \begin{bmatrix} 1 & -t \\ t & 1 \end{bmatrix} = (I + A)$$

Hence, $$(I + A) = (I - A) \begin{bmatrix} \cos\alpha & -\sin\alpha \\ \sin\alpha & \cos\alpha \end{bmatrix}$$

EXAMPLE 7. If $\begin{bmatrix} 4 \\ 1 \\ 3 \end{bmatrix} X = \begin{bmatrix} -4 & 8 & 4 \\ -1 & 2 & 1 \\ -3 & 6 & 3 \end{bmatrix}$. *Find X.*

SOLUTION. Let $$A = \begin{bmatrix} 4 \\ 1 \\ 3 \end{bmatrix} \text{ and } B = \begin{bmatrix} -4 & 8 & 4 \\ -1 & 2 & 1 \\ -3 & 6 & 3 \end{bmatrix}$$

Given $AX = B$ ...(1)

We have to find X.

Since A is a matrix of order 3×1 and B is a matrix of order 3×3 so order of X is 1×3.

Let $X = \begin{bmatrix} a & b & c \end{bmatrix}$

Now $AX = B$

$$\begin{bmatrix} 4 \\ 1 \\ 3 \end{bmatrix} \begin{bmatrix} a & b & c \end{bmatrix} = \begin{bmatrix} -4 & 8 & 4 \\ -1 & 2 & 1 \\ -3 & 6 & 3 \end{bmatrix}$$

$$\begin{bmatrix} 4a & 4b & 4c \\ a & b & c \\ 3a & 3b & 3c \end{bmatrix} = \begin{bmatrix} -4 & 8 & 4 \\ -1 & 2 & 1 \\ -3 & 6 & 3 \end{bmatrix}$$

Since both matrices are equal so corresponding elements must be equal. Therefore

$$a = -1, b = 2, c = 1$$

So, $$X = \begin{bmatrix} -1 & 2 & 1 \end{bmatrix}$$

EXAMPLE 8. If $A = \begin{bmatrix} 1 & 0 \\ 1 & 1 \end{bmatrix}, B = \begin{bmatrix} 2 & 0 \\ 1 & 1 \end{bmatrix}$ and $C = \begin{bmatrix} -1 & 2 \\ 3 & 1 \end{bmatrix}$. *Then prove that* $A(B + C) = AB + AC$.

SOLUTION.   Here
$$B + C = \begin{bmatrix} 2 & 0 \\ 1 & 1 \end{bmatrix} + \begin{bmatrix} -1 & 2 \\ 3 & 1 \end{bmatrix} = \begin{bmatrix} 1 & 2 \\ 4 & 2 \end{bmatrix}$$

$$A \cdot (B + C) = \begin{bmatrix} 1 & 0 \\ 1 & 1 \end{bmatrix} \begin{bmatrix} 1 & 2 \\ 4 & 2 \end{bmatrix} = \begin{bmatrix} 1 & 2 \\ 5 & 4 \end{bmatrix}_{2 \times 2} \qquad \ldots(1)$$

and
$$AB = \begin{bmatrix} 1 & 0 \\ 1 & 1 \end{bmatrix}_{2 \times 2} \times \begin{bmatrix} 2 & 0 \\ 1 & 1 \end{bmatrix}_{2 \times 2} = \begin{bmatrix} 2 & 0 \\ 3 & 1 \end{bmatrix}_{2 \times 2}$$

Now
$$AC = \begin{bmatrix} 1 & 0 \\ 1 & 1 \end{bmatrix}_{2 \times 2} \times \begin{bmatrix} -1 & 2 \\ 3 & 1 \end{bmatrix}_{2 \times 2} = \begin{bmatrix} -1 & 2 \\ 2 & 3 \end{bmatrix}_{2 \times 2}$$

$$AB + AC = \begin{bmatrix} 2 & 0 \\ 3 & 1 \end{bmatrix} + \begin{bmatrix} -1 & 2 \\ 2 & 3 \end{bmatrix} = \begin{bmatrix} 1 & 2 \\ 5 & 4 \end{bmatrix}_{2 \times 2} \cdot \qquad \ldots(2)$$

It is clear from eqⁿ. (1) and (2) that
$$A \cdot (B + C) = AB + AC$$

REMARK

➠ Similarly we can prove that
$$(AB)C = A(BC)$$

EXAMPLE 9.   If $A = \begin{bmatrix} 1 & 2 & 2 \\ 2 & 1 & 2 \\ 2 & 2 & 1 \end{bmatrix}$ verify that $A^2 - 4A - 5I = 0$.

SOLUTION.   We have
$$A = \begin{bmatrix} 1 & 2 & 2 \\ 2 & 1 & 2 \\ 2 & 2 & 1 \end{bmatrix}$$

$$A^2 = A \cdot A = \begin{bmatrix} 1 & 2 & 2 \\ 2 & 1 & 2 \\ 2 & 2 & 1 \end{bmatrix} \begin{bmatrix} 1 & 2 & 2 \\ 2 & 1 & 2 \\ 2 & 2 & 1 \end{bmatrix} = \begin{bmatrix} 9 & 8 & 8 \\ 8 & 9 & 8 \\ 8 & 8 & 9 \end{bmatrix}$$

and
$$4A = 4 \begin{bmatrix} 1 & 2 & 2 \\ 2 & 1 & 2 \\ 2 & 2 & 1 \end{bmatrix} = \begin{bmatrix} 4 & 8 & 8 \\ 8 & 4 & 8 \\ 8 & 8 & 4 \end{bmatrix}$$

Also
$$5I = 5 \begin{bmatrix} 1 & 0 & 0 \\ 0 & 1 & 0 \\ 0 & 0 & 1 \end{bmatrix} = \begin{bmatrix} 5 & 0 & 0 \\ 0 & 5 & 0 \\ 0 & 0 & 5 \end{bmatrix}$$

Now
$$A^2 - 4A - 5I = \begin{bmatrix} 9 & 8 & 8 \\ 8 & 9 & 8 \\ 8 & 8 & 9 \end{bmatrix} - \begin{bmatrix} 4 & 8 & 8 \\ 8 & 4 & 8 \\ 8 & 8 & 4 \end{bmatrix} - \begin{bmatrix} 5 & 0 & 0 \\ 0 & 5 & 0 \\ 0 & 0 & 5 \end{bmatrix}$$

$$= \begin{bmatrix} 0 & 0 & 0 \\ 0 & 0 & 0 \\ 0 & 0 & 0 \end{bmatrix} = 0 \text{ (Null matrix)}$$

EXAMPLE 10.   If $A = \begin{bmatrix} -4 & 1 \\ 3 & 2 \end{bmatrix}$, find $f(A)$, if $f(x) = x^2 - 2x + 3$.

SOLUTION.   Since, $f(x) = x^2 - 2x + 3$
Therefore, $f(A) = A^2 - 2A + 3I$

$$= \begin{bmatrix} -4 & 1 \\ 3 & 2 \end{bmatrix} \begin{bmatrix} -4 & 1 \\ 3 & 2 \end{bmatrix} - 2 \begin{bmatrix} -4 & 1 \\ 3 & 2 \end{bmatrix} + 3 \begin{bmatrix} 1 & 0 \\ 0 & 1 \end{bmatrix}$$

$$= \begin{bmatrix} 19 & -2 \\ -6 & 7 \end{bmatrix} - \begin{bmatrix} -8 & 2 \\ 6 & 4 \end{bmatrix} + \begin{bmatrix} 3 & 0 \\ 0 & 3 \end{bmatrix} = \begin{bmatrix} 30 & -4 \\ -12 & 6 \end{bmatrix}$$

EXAMPLE 11. If $A = \begin{bmatrix} 1 & 1 & -1 \\ 2 & 0 & 3 \\ 3 & -1 & 2 \end{bmatrix}, B = \begin{bmatrix} 1 & 3 \\ 0 & 2 \\ -1 & 4 \end{bmatrix}$ and $C = \begin{bmatrix} 1 & 2 & 3 \\ 2 & 0 & -2 \end{bmatrix}$. Prove that $(AB)C = A(BC)$.

SOLUTION.   We have $AB = \begin{bmatrix} 1 & 1 & -1 \\ 2 & 0 & 3 \\ 3 & -1 & 2 \end{bmatrix}_{3\times 3} \times \begin{bmatrix} 1 & 3 \\ 0 & 2 \\ -1 & 4 \end{bmatrix}_{3\times 2}$

$$= \begin{bmatrix} 1\times 1 + 1\times 0 + (-1)(-1) & 1\times 3 + 1\times 2 - 1\times 4 \\ 2\times 1 + 0\times 0 + 3\times(-1) & 2\times 3 + 0\times 2 + 3\times 4 \\ 3\times 1 + (-1)\times 0 + 2\times(-1) & 3\times 3 + (-1)\times 2 + 2\times 4 \end{bmatrix}_{3\times 2}$$

$$= \begin{bmatrix} 2 & 1 \\ -1 & 18 \\ 1 & 15 \end{bmatrix}_{3\times 2}$$

and $(AB)C = \begin{bmatrix} 2 & 1 \\ -1 & 18 \\ 1 & 15 \end{bmatrix}_{3\times 2} \times \begin{bmatrix} 1 & 2 & 3 \\ 2 & 0 & -2 \end{bmatrix}_{2\times 3}$

$$= \begin{bmatrix} 2\times 1 + 1\times 2 & 2\times 2 + 1\times 0 & 2\times 3 + 1\times(-2) \\ -1\times 1 + 18\times 2 & -1\times 2 + 18\times 0 & -1\times 3 + 18\times(-2) \\ 1\times 1 + 15\times 2 & 1\times 2 + 15\times 0 & 1\times 3 + 15\times(-2) \end{bmatrix}_{3\times 3}$$

$$= \begin{bmatrix} 4 & 4 & 4 \\ 35 & -2 & -39 \\ 31 & 2 & -27 \end{bmatrix}_{3\times 3}$$

Again $BC = \begin{bmatrix} 7 & 2 & -3 \\ 4 & 0 & -4 \\ 7 & -2 & -11 \end{bmatrix}_{3\times 3}$

$$A(BC) = \begin{bmatrix} 1 & 1 & -1 \\ 2 & 0 & 3 \\ 3 & -1 & 2 \end{bmatrix} \begin{bmatrix} 7 & 2 & -3 \\ 4 & 0 & -4 \\ 7 & -2 & -11 \end{bmatrix}_{3\times 3}$$

$$= \begin{bmatrix} 4 & 4 & 4 \\ 35 & -2 & 39 \\ 31 & 2 & -27 \end{bmatrix}_{3\times 3}$$

Hence, $A(BC) = (AB)C$.

EXAMPLE 12. If $A = \begin{bmatrix} \cos\theta & -\sin\theta \\ \sin\theta & \cos\theta \end{bmatrix}$ then show that $A^n = \begin{bmatrix} \cos n\theta & -\sin n\theta \\ \sin n\theta & \cos n\theta \end{bmatrix}$ where n is positive integer.

SOLUTION.   For this we shall apply mathematical induction

Let $P(n) : A^n = \begin{bmatrix} \cos n\theta & -\sin n\theta \\ \sin n\theta & \cos n\theta \end{bmatrix}$

For $n = 1$,   L.H.S. $= A = \begin{bmatrix} \cos\theta & -\sin\theta \\ \sin\theta & \cos\theta \end{bmatrix}$

Now   R.H.S. $= \begin{bmatrix} \cos\theta & -\sin\theta \\ \sin\theta & \cos\theta \end{bmatrix}$

So, $P(1)$ is true.   ...(1)

Let $P(m)$ be true $A^m = \begin{bmatrix} \cos m\theta & -\sin m\theta \\ \sin m\theta & \cos m\theta \end{bmatrix}$ ...(2)

We have to prove that $P(m+1)$ is true, i.e.,

$$A^{m+1} = \begin{bmatrix} \cos(m+1)\theta & -\sin(m+1)\theta \\ \sin(m+1)\theta & \cos(m+1)\theta \end{bmatrix}$$ ...(3)

Multiply both side of eqn. (2) by matrix A

$$A^{m+1} = A\begin{bmatrix} \cos m\theta & -\sin m\theta \\ \sin m\theta & \cos m\theta \end{bmatrix}$$

$$= \begin{bmatrix} \cos\theta & -\sin\theta \\ \sin\theta & \cos\theta \end{bmatrix}\begin{bmatrix} \cos m\theta & -\sin m\theta \\ \sin m\theta & \cos m\theta \end{bmatrix}$$

$$= \begin{bmatrix} \cos\theta\cos m\theta - \sin\theta\sin m\theta & -\cos\theta\sin m\theta - \sin\theta\cos m\theta \\ \sin\theta\cos m\theta + \cos\theta\sin m\theta & -\sin\theta\sin m\theta + \cos\theta\cos m\theta \end{bmatrix}$$

$$= \begin{bmatrix} \cos(m+1)\theta & -\sin(m+1)\theta \\ \sin(m+1)\theta & \cos(m+1)\theta \end{bmatrix}$$

Therefore, $P(m+1)$ is true. ...(4)

From eqn. (1) and (4), it is clear that $P(n)$ is true for all natural no. n.

≋ REMARK

➠ In the above question, if $n = 2$, then, we get

$$A^2 = \begin{bmatrix} \cos 2\alpha & -\sin 2\alpha \\ \sin 2\alpha & \cos 2\alpha \end{bmatrix}$$

EXAMPLE 13. *In a co-operative stock there are* 10 *dozen physics books,* 8 *dozen chemistry books and* 5 *dozen mathematics books. Selling price of each book is Rs.* 8.30, 3.45 *and* 4.50 *respectively. Using matrix, find the total sale of all books.*

SOLUTION. Let column matrix X represents the no. of books

$$X = 12\begin{bmatrix} 10 \\ 8 \\ 5 \end{bmatrix}\begin{matrix} \text{Physics} \\ \text{Chemistry} \\ \text{Mathematics} \end{matrix} \Rightarrow X = \begin{bmatrix} 120 \\ 96 \\ 60 \end{bmatrix}$$

Now let row matrix Y represent selling price of books.

$$Y = \begin{bmatrix} \text{Physics} & \text{Chemistry} & \text{Mathematics} \\ 8.30 & 3.45 & 4.50 \end{bmatrix}$$

Therefore, total sale $= YX = \begin{bmatrix} 8.30 & 3.45 & 4.50 \end{bmatrix}\begin{bmatrix} 120 \\ 96 \\ 60 \end{bmatrix}$

$$= 8.30 \times 120 + 3.45 \times 96 + 4.50 \times 60$$

$$= 996 + 331.20 + 270 = \text{Rs. } 1597.20$$

EXAMPLE 14. *If* $A = \begin{bmatrix} 1 & 2 \\ 3 & 0 \\ 4 & 1 \end{bmatrix}$ *and* $B = \begin{bmatrix} 0 & 1 & 0 \\ 0 & 2 & 1 \\ 2 & 3 & 0 \end{bmatrix}$, *find the value of BA.*

SOLUTION. We have $BA = \begin{bmatrix} 0 & 1 & 0 \\ 0 & 2 & 1 \\ 2 & 3 & 0 \end{bmatrix}\begin{bmatrix} 1 & 2 \\ 3 & 0 \\ 4 & 1 \end{bmatrix}$

$$= \begin{bmatrix} 0\times1+1\times3+0\times4 & 0\times2+1\times0+0\times1 \\ 0\times1+2\times3+1\times4 & 0\times2+2\times0+1\times1 \\ 2\times1+3\times3+0\times4 & 2\times2+3\times0+0\times1 \end{bmatrix}$$

$$= \begin{bmatrix} 0+3+0 & 0+0+0 \\ 0+6+4 & 0+0+1 \\ 2+9+0 & 4+0+0 \end{bmatrix} = \begin{bmatrix} 3 & 0 \\ 10 & 1 \\ 11 & 4 \end{bmatrix}$$

Exercise 8.2

1. Test whether the following matrices is conformal for multiplication or not:

(i) $A = \begin{bmatrix} -1 & 2 \\ 3 & 4 \end{bmatrix}, B = \begin{bmatrix} 5 \\ 6 \end{bmatrix}$

(ii) $A = \begin{bmatrix} -5 & -7 \\ 6 & -8 \end{bmatrix}, B = \begin{bmatrix} 3 & 4 \end{bmatrix}$

(iii) $A = \begin{bmatrix} 1 & 2 & 3 \\ 4 & 0 & 5 \\ 6 & 7 & 0 \end{bmatrix}, B = \begin{bmatrix} 1 & 0 & 2 & 4 \\ 3 & 7 & 0 & 0 \end{bmatrix}$

2. Find matrix B if $A = \begin{bmatrix} 1 & 4 \\ 3 & 2 \end{bmatrix}$ and $A + 2B = A^2$.

3. If $A = \begin{bmatrix} 2 & 3 & 1 \\ -1 & 2 & 3 \\ 2 & 0 & 10 \end{bmatrix}$, find A^2.

4. If $A = \begin{bmatrix} 0 & 1 \\ 1 & 0 \end{bmatrix}$ and $B = \begin{bmatrix} 1 & 0 \\ 0 & -1 \end{bmatrix}$ prove that :

(i) $A^2 = B^2 = I$ (ii) $AB = -BA$

5. If $A = \begin{bmatrix} 0 & 0 & 1 \\ 2 & -3 & 0 \\ 1 & 1 & -1 \end{bmatrix}$, find $A^3 + 4A^2 - A$.

6. If $A = \begin{bmatrix} 2 & 0 & -1 \\ 2 & 3 & -2 \\ 0 & -2 & 5 \end{bmatrix}$, find $A^2 + 2A - 7I$.

7. Multiply $\begin{bmatrix} 1 & -2 & 3 \\ -4 & 2 & 5 \end{bmatrix} \times \begin{bmatrix} 2 & 3 \\ 4 & 5 \\ 2 & 1 \end{bmatrix}$.

8. If $\theta - \phi = \pi/2$, prove that

$$\begin{bmatrix} \cos^2\theta & \cos\theta\sin\theta \\ \cos\theta\sin\theta & \sin^2\theta \end{bmatrix} \begin{bmatrix} \cos^2\phi & \cos\phi\sin\phi \\ \cos\phi\sin\phi & \sin^2\phi \end{bmatrix} = 0$$

9. If $A = \begin{bmatrix} 1 & 3 & -1 \\ 2 & 2 & -1 \\ 3 & 0 & -1 \end{bmatrix}$ and $B = \begin{bmatrix} -2 & 3 & -1 \\ -1 & 2 & -1 \\ -6 & 9 & -4 \end{bmatrix}$, show that $AB = BA$.

10. If $A = \begin{bmatrix} 1 & 0 \\ 1 & 1 \end{bmatrix}, B = \begin{bmatrix} 2 & 0 \\ 1 & 1 \end{bmatrix}$ and $C = \begin{bmatrix} -1 & 2 \\ 3 & 1 \end{bmatrix}$, show that

(i) $(AB)C = A(BC)$

(ii) $A(B + C) = AB + AC$

(iii) $(B + C)A = BA + CA$

11. If $F(x) = \begin{bmatrix} \cos x & -\sin x & 0 \\ \sin x & \cos x & 0 \\ 0 & 0 & 1 \end{bmatrix}$. Show that

$$F(x)\cdot F(y) = F(x + y)$$

12. If $\begin{bmatrix} x & 0 \\ 2 & x+y \end{bmatrix}\begin{bmatrix} 1 & 0 \\ 0 & 1 \end{bmatrix} = \begin{bmatrix} 2 & 1 \\ 1 & 3 \end{bmatrix} - 2\begin{bmatrix} 0 & \frac{1}{2} \\ -\frac{1}{2} & \frac{3}{2} \end{bmatrix}$ find

the value of x and y.

13. If ω is the cube root of unity, prove that

$$\left\{ \begin{bmatrix} 1 & \omega & \omega^2 \\ \omega & \omega^2 & 1 \\ \omega^2 & 1 & \omega \end{bmatrix} + \begin{bmatrix} \omega & \omega^2 & 1 \\ \omega^2 & 1 & \omega \\ \omega & \omega^2 & 1 \end{bmatrix} \right\} \begin{bmatrix} 1 \\ \omega \\ \omega^2 \end{bmatrix} = \begin{bmatrix} 0 \\ 0 \\ 0 \end{bmatrix}$$

14. A manufacturer makes three items A, B, C which are sold in Delhi and Mumbai. Annual sale of these items are given as follow. If selling price of A, B and C are Rs. 2, 3 and 4 respectively then using matrix find total sale at each place.

Items			
	A	**B**	**C**
Delhi	5,000	75,000	15,000
Mumbai	9,000	12,000	87,000

15. If $A = \begin{bmatrix} 1 & -1 \\ -1 & 1 \end{bmatrix}$ and $B = \begin{bmatrix} 1 & 1 \\ 1 & 1 \end{bmatrix}$. Prove that AB is null matrix.

16. If $A_\alpha = \begin{bmatrix} \cos\alpha & \sin\alpha \\ -\sin\alpha & \cos\alpha \end{bmatrix}$. Prove that

(i) $A_\alpha \cdot A_\beta = A_{\alpha+\beta}$

(ii) $A_\alpha \cdot A_{(-\alpha)} = I$

17. If $A = \begin{bmatrix} 1 & -1 \\ -1 & 1 \end{bmatrix}$. Prove that $A^3 = 4A$.

18. If $A = \begin{bmatrix} 3 & 3 & 5 \\ 2 & 3 & 4 \\ 5 & 2 & 3 \end{bmatrix}$. Prove that $AI_3 = I_3A$.

19. If $A = \begin{bmatrix} 1 & 2 & 5 \\ 3 & 4 & 6 \end{bmatrix}$, $B = \begin{bmatrix} 4 & 0 \\ 2 & 1 \\ 1 & 5 \end{bmatrix}$.

Prove that $AB = \begin{bmatrix} 13 & 27 \\ 26 & 34 \end{bmatrix}$.

20. If $A = \begin{bmatrix} 2 & 0 & 0 \\ 0 & 2 & 0 \\ 0 & 0 & 2 \end{bmatrix}$ and $B = \begin{bmatrix} x_1 & y_1 & z_1 \\ x_2 & y_2 & z_2 \\ x_3 & y_3 & z_3 \end{bmatrix}$.

Prove that $AB = 2B$.

21. (i) If $A = \begin{bmatrix} 0 & i \\ i & 0 \end{bmatrix}$, $i^2 = -1$, prove that

$$A^2 = \begin{bmatrix} -1 & 0 \\ 0 & -1 \end{bmatrix}.$$

(ii) If $A = \begin{bmatrix} 0 & 1 \\ 1 & 0 \end{bmatrix}$, prove that $A^4 = \begin{bmatrix} 1 & 0 \\ 0 & 1 \end{bmatrix}$.

ANSWERS

1. (i) Yes   **(ii)** No   **(iii)** No   **2.** $B = \begin{bmatrix} 6 & 4 \\ 3 & 7 \end{bmatrix}$ **3.** $\begin{bmatrix} 3 & 12 & 21 \\ 2 & 1 & 35 \\ 24 & 6 & 102 \end{bmatrix}$

5. $\begin{bmatrix} 5 & 0 & -3 \\ -6 & 14 & 0 \\ -3 & -3 & 8 \end{bmatrix}$ **6.** $\begin{bmatrix} 1 & 2 & -9 \\ 14 & 12 & -22 \\ -4 & -20 & 32 \end{bmatrix}$ **7.** $\begin{bmatrix} 0 & -4 \\ 10 & 3 \end{bmatrix}$

12. $x = 2, y = -2$ **14.** Delhi Rs. 2,95,000, Mumbai Rs. 402000

8.9 TRANSPOSE OF A MATRIX

Let A be the given matrix then the transpose of A denoted by A^T or A' is a matrix obtained by interchanging row and columns of A.

For example. Let $A = \begin{bmatrix} 1 & 2 & 3 \\ 4 & 5 & -1 \end{bmatrix}$, then $A' = \begin{bmatrix} 1 & 4 \\ 2 & 5 \\ 3 & -1 \end{bmatrix}$.

REMARKS

⟹ The $(i, j)^{\text{th}}$ element of $A = (j, i)^{\text{th}}$ elements of A'.

⟹ If the order of A is $m \times n$ then order of its transpose A' is $n \times m$.

⟹ The transpose of the column matrix is a row matrix and transpose of the row matrix is a column matrix.

8.10 PROPERTIES OF THE TRANSPOSE OF MATRICES

(I) $(A + B)' = A' + B'$

(II) $(A')' = A$

(III) If $A = [a_{ij}]_{m \times n}$ and $B = [b_{jk}]_{n \times p}$, then $(AB)' = B'A'$.

Proof. Let $A = [a_{ij}]_{m \times n}$ and $B = [b_{jk}]_{n \times p}$, then

$$A' = [c_{ji}]_{n \times m} \text{ where } c_{ji} = a_{ij}$$

and $$B' = [d_{kj}]_{p \times n} \text{ where } d_{kj} = b_{jk}$$

Clearly, AB is a matrix of order $m \times p$, therefore $(AB)'$ is a matrix of order $p \times m$ and since the order of B' is $p \times n$ and order of A' is $n \times m$, therefore the order of $B'A'$ is $p \times m$.

$\therefore$ $(AB)'$ and $B'A'$ are of the same order.

$\therefore$ $(k, i)^{\text{th}}$ element of $(AB)' = (i, k)^{\text{th}}$ element of $AB = \sum\limits_{j=1}^{n} a_{ij}b_{jk}$

$$= \sum_{j=1}^{n} c_{ji}d_{kj} = \sum_{j=1}^{n} d_{kj}c_{ji} = (k,i)^{\text{th}} \text{ element of } B'A'.$$

$\because$ $(AB)'$ and $B'A'$ are of the same order and their $(k, i)^{\text{th}}$ element are equal.

Hence $(AB)' = B'A'.$

8.11 SYMMETRIC, SKEW-SYMMETRIC AND ORTHOGONAL MATRICES

1. **Symmetric Matrix:** A square matrix $A = [a_{ij}]$ is said to be a symmetric matrix if $a_{ij} = a_{ji}$, for all possible values of i and j.

For example: The matrices $A = \begin{bmatrix} 1 & 3 & 5 \\ 3 & 2 & 0 \\ 5 & 0 & -1 \end{bmatrix}, B = \begin{bmatrix} 1 & 3 \\ 3 & 0 \end{bmatrix}$ are symmetric.

⮞ REMARK
⟱ A square matrix A is symmetric If $A' = A$.

2. **Skew-Symmetric Matrix:** A square matrix $A = [a_{ij}]$ is said to be skew-symmetric if $a_{ij} = -a_{ji}$ for all possible values of i and j.

For example: $A = \begin{bmatrix} 0 & 2 \\ -2 & 0 \end{bmatrix}, B = \begin{bmatrix} 0 & 2 & -1 \\ -2 & 0 & -3 \\ 1 & 3 & 0 \end{bmatrix}$

⮞ REMARKS
⟱ A square matrix is skew symmetric if and only if $A' = -A$.
⟱ In skew symmetric matrix the elements of main diagonal are zero.

3. **Orthogonal Matrix:** A matrix A is said to be orthogonal if $A'A = I$ where A' is the transpose of A.

SOLVED EXAMPLES

EXAMPLE 1. *Write the transpose of matrix A.*

$$A = \begin{bmatrix} 5 & 1 & 2 \\ 6 & 4 & 5 \end{bmatrix}_{2 \times 3}$$

SOLUTION. Given $A = \begin{bmatrix} 5 & 1 & 2 \\ 6 & 4 & 5 \end{bmatrix}_{2 \times 3}$

$\therefore$ Transpose $A' = \begin{bmatrix} 5 & 6 \\ 1 & 4 \\ 2 & 5 \end{bmatrix}_{3 \times 2}$

EXAMPLE 2. *If $A = \begin{bmatrix} 0 & 1 \\ 2 & 3 \end{bmatrix}$ and $B = \begin{bmatrix} 1 & 2 \\ 3 & 4 \end{bmatrix}$, then verify $(A + B)' = A' + B'$ and find $(AB)'$.*

SOLUTION. We have $A' = \begin{bmatrix} 0 & 2 \\ 1 & 3 \end{bmatrix}, B' = \begin{bmatrix} 1 & 3 \\ 2 & 4 \end{bmatrix}$ ...(1)

and $A + B = \begin{bmatrix} 0 & 1 \\ 2 & 3 \end{bmatrix} + \begin{bmatrix} 1 & 2 \\ 3 & 4 \end{bmatrix} = \begin{bmatrix} 1 & 3 \\ 5 & 7 \end{bmatrix}$

$\Rightarrow$ $(A + B)' = \begin{bmatrix} 1 & 5 \\ 3 & 7 \end{bmatrix}$ ...(2)

From eqn. (1), $A' + B' = \begin{bmatrix} 0 & 2 \\ 1 & 3 \end{bmatrix} + \begin{bmatrix} 1 & 3 \\ 2 & 4 \end{bmatrix} = \begin{bmatrix} 1 & 5 \\ 3 & 7 \end{bmatrix}$ ...(3)

From eqn. (2) and (3),

$$(A + B)' = A' + B'$$

$$(AB)' = B'A' = \begin{bmatrix} 1 & 3 \\ 2 & 4 \end{bmatrix}\begin{bmatrix} 0 & 2 \\ 1 & 3 \end{bmatrix}$$

$$= \begin{bmatrix} 0+3 & 2+9 \\ 0+4 & 4+12 \end{bmatrix} = \begin{bmatrix} 3 & 11 \\ 4 & 16 \end{bmatrix}$$ ...(4)

EXAMPLE 3. If $A = \begin{bmatrix} 2 & 4 & 1 \\ -1 & 0 & 2 \end{bmatrix}, B = \begin{bmatrix} 3 & 4 \\ -1 & 2 \\ 2 & 1 \end{bmatrix}$. Show that $(AB)' = B'A'$.

SOLUTION. We have $AB = \begin{bmatrix} 2 & 4 & 1 \\ -1 & 0 & 2 \end{bmatrix}_{2\times3} \begin{bmatrix} 3 & 4 \\ -1 & 2 \\ 2 & 1 \end{bmatrix}_{3\times2} = \begin{bmatrix} 0 & 15 \\ 1 & -2 \end{bmatrix}_{2\times2}$

$$(AB)' = \begin{bmatrix} 0 & 1 \\ 15 & -2 \end{bmatrix}_{2\times2}$$ ...(1)

Again $B' = \begin{bmatrix} 3 & -1 & 2 \\ 4 & 2 & 1 \end{bmatrix}_{2\times3}, A' = \begin{bmatrix} 2 & -1 \\ 4 & 0 \\ -1 & 2 \end{bmatrix}_{3\times2}$

$$B'A' = \begin{bmatrix} 3 & -1 & 2 \\ 4 & 2 & 1 \end{bmatrix}\begin{bmatrix} 2 & -1 \\ 4 & 0 \\ -1 & 2 \end{bmatrix} = \begin{bmatrix} 0 & 1 \\ 15 & -2 \end{bmatrix}_{2\times2}$$ ...(2)

From eqn. (1) and (2),
$$(AB)' = B'A'$$

EXAMPLE 4. If $A = \begin{bmatrix} 1 & 2 & 3 \\ -1 & 2 & 4 \\ 0 & 0 & 1 \end{bmatrix}$ and $B = \begin{bmatrix} -1 & -2 & 0 \\ 0 & 1 & 3 \\ -1 & 2 & 0 \end{bmatrix}$. Prove that $(AB)' = B'A'$.

SOLUTION. By multiplication of matrices, we have

$$AB = \begin{bmatrix} -4 & 6 & 6 \\ -3 & 12 & 6 \\ -1 & 2 & 0 \end{bmatrix}_{3\times3}$$

$$\Rightarrow \qquad (AB)' = \begin{bmatrix} -4 & -3 & -1 \\ 6 & 12 & 2 \\ 6 & 6 & 0 \end{bmatrix}_{3\times3}$$ ...(1)

Again $A' = \begin{bmatrix} 1 & -1 & 0 \\ 2 & 2 & 0 \\ 3 & 4 & 1 \end{bmatrix}$ and $B' = \begin{bmatrix} -1 & 0 & -1 \\ -2 & 1 & 2 \\ 0 & 3 & 0 \end{bmatrix}$

Again by multiplication of matrices, we have

$$B'A' = \begin{bmatrix} -4 & -3 & -1 \\ 6 & 12 & 2 \\ 6 & 6 & 0 \end{bmatrix}$$ ...(2)

From eqn. (1) and (2), it is clear

$$(AB)' = B'A'$$

EXAMPLE 5. *Express the following matrix as a sum of symmetric and skew-symmetric matrices.*

$$\begin{bmatrix} 1 & 3 & 5 \\ -6 & 8 & 3 \\ -4 & 6 & 5 \end{bmatrix}$$

SOLUTION. For any square matrix, we can always write

$$A = \frac{1}{2}(A + A') + \frac{1}{2}(A - A')$$

where $\frac{1}{2}(A + A')$ is symmetric matrix and $\frac{1}{2}(A - A')$ is skew-symmetric matrix.

Given that

$$A = \begin{bmatrix} 1 & 3 & 5 \\ -6 & 8 & 3 \\ -4 & 6 & 5 \end{bmatrix} \qquad \therefore \qquad A' = \begin{bmatrix} 1 & -6 & -4 \\ 3 & 8 & 6 \\ 5 & 3 & 5 \end{bmatrix}$$

$$\Rightarrow \qquad \frac{1}{2}(A + A') = \frac{1}{2}\begin{bmatrix} 1+1 & 3-6 & 5-4 \\ -6+3 & 8+8 & 3+6 \\ -4+5 & 6+3 & 5+5 \end{bmatrix}$$

$$= \frac{1}{2}\begin{bmatrix} 2 & -3 & 1 \\ -3 & 16 & 9 \\ 1 & 9 & 10 \end{bmatrix} = \begin{bmatrix} 1 & \frac{-3}{2} & \frac{1}{2} \\ \frac{-3}{2} & 8 & \frac{9}{2} \\ \frac{1}{2} & \frac{9}{2} & 5 \end{bmatrix}$$

Which is a symmetric matrix.

Again

$$\frac{1}{2}(A - A') = \frac{1}{2}\begin{bmatrix} 1-1 & 3+6 & 5+4 \\ -6-3 & 8-8 & 3-6 \\ -4-5 & 6-3 & 5-5 \end{bmatrix}$$

$$= \frac{1}{2}\begin{bmatrix} 0 & 9 & 9 \\ -9 & 0 & -3 \\ -9 & 3 & 0 \end{bmatrix} = \begin{bmatrix} 0 & \frac{9}{2} & \frac{9}{2} \\ \frac{-9}{2} & 0 & \frac{-3}{2} \\ \frac{-9}{2} & \frac{3}{2} & 0 \end{bmatrix}$$

Which is a skew symmetric matrix.

Exercise 8.3

1. If $A = \begin{bmatrix} 1 & 0 & 5 \\ 2 & -1 & 3 \\ 4 & 1 & 0 \end{bmatrix}$ and $B = \begin{bmatrix} 3 & 2 & 1 \\ 0 & 2 & 1 \\ 3 & 2 & 5 \end{bmatrix}$. Show that

    (a) $(A + B)' = A' + B'$

    (b) $(2A)' = 2A'$

2. If $A = \begin{bmatrix} 1 & 0 \\ 3 & 2 \end{bmatrix}$ and $B = \begin{bmatrix} 4 & 0 \\ 6 & 5 \end{bmatrix}$. Show that

    $(AB)' = B'A'$.

3. (i) If $A = \begin{bmatrix} 2 & 5 \\ 3 & 1 \end{bmatrix}$ and $B = \begin{bmatrix} 5 & 7 \\ 2 & 0 \end{bmatrix}$. Show that

    $(AB)' = B'A'$.

    (ii) If $A = \begin{bmatrix} 2 & 3 \\ 0 & 1 \end{bmatrix}$ and $B = \begin{bmatrix} 3 & 4 \\ 2 & 1 \end{bmatrix}$. Prove that

    $(AB)' = B'A'$.

(iii) If $A = \begin{bmatrix} 2 & 4 \\ 3 & 1 \end{bmatrix}$ and $B = \begin{bmatrix} 5 & 6 \\ 2 & 0 \end{bmatrix}$. Prove that $(AB)' = B'A'$.

4. If $A = \begin{bmatrix} 3 \\ 5 \\ 4 \\ 2 \end{bmatrix}$. Then find $A'A$ and AA'.

5. If $A = \begin{bmatrix} 4 & 1 \\ 2 & 3 \end{bmatrix}$ and $B = \begin{bmatrix} 3 & 2 \\ 5 & 1 \\ 3 & 5 \end{bmatrix}$. Find AB' and BA.

6. For the following matrices, show that
   (i) $(A + B)^T = B^T + A^T = A^T + B^T$
   (ii) $(AB)^T = B^T A^T$

$$A = \begin{bmatrix} 1 & 0 & 5 \\ 6 & 2 & 1 \\ 7 & 1 & 5 \end{bmatrix}, B = \begin{bmatrix} 5 & 14 & 10 \\ 4 & 2 & 3 \\ 0 & 9 & 5 \end{bmatrix}.$$

7. If $A = \begin{bmatrix} 1 & -1 & 0 \\ 2 & 1 & 3 \\ 4 & 1 & 8 \end{bmatrix}$ and $B = \begin{bmatrix} 4 & 1 & 0 \\ 2 & -3 & 1 \\ 1 & 1 & -1 \end{bmatrix}$.
   Show that $(AB)' = B'A'$.

8. Given that $A = \begin{bmatrix} 1 & 0 & 3 \\ 2 & -1 & 1 \end{bmatrix}, B = \begin{bmatrix} 3 & 4 & 1 \\ 0 & -1 & 5 \\ 1 & 2 & -2 \end{bmatrix}$
   and $C = \begin{bmatrix} 2 \\ -1 \\ 4 \end{bmatrix}$, find
   (a) $(AB)'$ (b) $B'A'$
   (c) $(BC)'$ (d) $C'A'$

ANSWERS

1. $A'A = [3^2 + 5^2 + 4^2 + 2^2] = 54, AA' = \begin{bmatrix} 9 & 15 & 12 & 6 \\ 15 & 25 & 20 & 10 \\ 12 & 20 & 16 & 8 \\ 6 & 10 & 8 & 4 \end{bmatrix}$

5. $AB' = \begin{bmatrix} 14 & 21 & 17 \\ 12 & 13 & 21 \end{bmatrix}, BA = \begin{bmatrix} 16 & 9 \\ 22 & 8 \\ 22 & 18 \end{bmatrix}$

8. (a) $\begin{bmatrix} 6 & 7 \\ 10 & 11 \\ -5 & -5 \end{bmatrix}$ (b) $\begin{bmatrix} 6 & 7 \\ 10 & 11 \\ -5 & -5 \end{bmatrix}$ (c) $[6 \quad 21 \quad -8]$ (d) $[14 \quad 9]$

8.12 ELEMENTARY OPERATIONS (OR TRANSFORMATION) ON MATRICES

The elementary transformation on matrices are as follows:

(i) **The interchange of any two rows or columns:** The process to interchange i and j^{th} rows is written as R_{ij} or $R_i \leftrightarrow R_j$.
   i.e., R_{ij} = Interchange i^{th} & j^{th} row $\equiv R_i \leftrightarrow R_j$.
   Similarly C_{ij} = Interchange i^{th} & j^{th} column $\equiv C_i \leftrightarrow C_j$.

(ii) **The multiplication of the elements of any row (or column) by a non-zero number:** In symbol, if we multiply i^{th} row by a scalar k, then we write $R_i(k)$, $k \neq 0$.
   i.e., $R_i(k) \equiv$ multiply i^{th} row by a scalar $k \equiv R_i \to kR_i$
   Similarly, $C_i(k) \equiv$ multiply i^{th} column by a scalar $k \equiv C_i \to kC_i$

(iii) **The addition to the elements of any row or column, the corresponding elements of any other row or column multiplied by a non-zero number.**
   i.e., $R_{ij}(k) \equiv k^{th}$ times the jth row plus i^{th} row $\equiv R_i \to R_i + kR_j$
   $R_{ij}(-k) \equiv i^{th}$ row minus k^{th} times j^{th} column $\equiv R_i \to R_i - kR_j$
   $R_{ij}(1) \equiv$ addition of ith and j^{th} row $\equiv R_i \to R_i + R_j$

$R_{ij}(-1) \equiv$ Subtract j^{th} row from i^{th} row $\equiv R_i \rightarrow R_i - R_j$

Similarly, $C_{ij}(k) \equiv$ Add k items the jth column to i^{th} column $\equiv C_i \rightarrow C_i + kC_j$

$C_{ij}(-k) \equiv$ Subtract k times the jth column from i^{th} column $\equiv C_i \rightarrow C_i - kC_j$

REMARK

➡ Elementary transformations are applied either on columns or row but not on both.

8.13 EQUIVALENT MATRICES AND ELEMENTARY MATRIX

Equivalent Matrices: Two matrices A and B are said to be equivalent if one is obtained from other by applying finite no. of elementary operations. It is written as $A \sim B$.

For example: Let
$$A = \begin{bmatrix} 1 & 2 & 3 \\ 2 & 3 & 4 \\ 2 & 0 & 5 \end{bmatrix} \text{ and } B = \begin{bmatrix} 2 & 3 & 8 \\ 1 & 2 & 6 \\ 2 & 0 & 10 \end{bmatrix}$$

Now
$$A = \begin{bmatrix} 1 & 2 & 3 \\ 2 & 3 & 4 \\ 2 & 0 & 5 \end{bmatrix} \sim \begin{bmatrix} 2 & 3 & 4 \\ 1 & 2 & 3 \\ 2 & 0 & 5 \end{bmatrix} \qquad (R_1 \leftrightarrow R_2)$$

$$\sim \begin{bmatrix} 2 & 3 & 8 \\ 1 & 2 & 6 \\ 2 & 0 & 10 \end{bmatrix} = B \qquad (C_3 \rightarrow 2C_3)$$

Here $A \sim B$, since B is obtained by applying two elementary operations on A.

Elementary Matrix: Elementary matrix is obtained on applying only one elementary operation on identity matrix.

For example:
$$I = \begin{bmatrix} 1 & 0 \\ 0 & 1 \end{bmatrix}$$

Then
$$A = \begin{bmatrix} 2 & 0 \\ 0 & 1 \end{bmatrix} \qquad R_1 \rightarrow 2R_1$$

Clearly, A is elementary matrix.

Singular square Matrix

A square matrix is said to be singular if value of determinant is zero.

i.e., if $|A| = 0$, then A is singular.

Non-Singular Square Matrix

A square matrix is said to be non-singular if value of its determinant is non-zero.

i.e., if $|A| \neq 0$ then A is non-singular.

8.14 INVERSE MATRIX OR INVERSE OF A MATRIX

Inverse of a square matrix A (non-singular) is a square matrix which gives identity matrix on multiplying with A. In other words if for two square matrices $AB = I$, then B is said to be the inverse of A. Inverse of A is written as A^{-1}.

Similarly if $AB = I$. Then $\quad B = A^{-1}$

$$AA^{-1} = I.$$

8.14.1 IMPORTANT PROPERTIES OF INVERSE MATRIX

(i) If A^{-1} is inverse of A, then $AA^{-1} = A^{-1}A = I$.

(ii) Inverse of a matrix is always unique.

(iii) If A and B are two invertible matrices with same order, then AB is also invertible and $(AB)^{-1} = B^{-1}A^{-1}$.

(iv) For three matrices A, B, C $(ABC)^{-1} = C^{-1}B^{-1}A^{-1}$.

(v) $(A')^{-1} = (A^{-1})'$

8.15 INVERSE OF A MATRIX BY ELEMENTARY OPERATIONS

8.15.1 ELEMENTARY OPERATION ON MATRIX

Let A, B and X be three matrices such that

$$X = AB \qquad \qquad ...(1)$$

Matrix equation is valid when a row operation is applied on X (Left hand side) and pre matrix A of (Right hand side) on same row. Similarly column operation is valid on matrix equation when it is applied on X and post, matrix B of right hand side on same column. Row and column operations are applied on alternate order.

By using above method we can find the inverse of a matrix A if A is invertible.

8.15.2 TO FIND A^{-1}

Let A be a n order square matrix.

Then $\qquad\qquad A = I_n A = IA \qquad\qquad ...(1)$

Now apply only elementary row operation on $A = IA$ until left hand side A became I. A in right hand side remains same. Same row operation are applied on I on right hand side. After elementary operations it became a new matrtix B.

i.e., $\qquad\qquad I = BA$

By definition : $\qquad\qquad B = A^{-1}$

SOLVED EXAMPLES

EXAMPLE 1. *Using elementary operation find A^{-1} where* $A = \begin{bmatrix} 6 & -3 \\ -2 & 1 \end{bmatrix}$

SOLUTION. Let $\qquad\qquad A = \begin{bmatrix} 6 & -3 \\ -2 & 1 \end{bmatrix}$

To find A^{-1} we use elementary row operations.

Take the matrix equation

$$A = IA$$

Then $\qquad \begin{bmatrix} 6 & -3 \\ -2 & 1 \end{bmatrix} = \begin{bmatrix} 1 & 0 \\ 0 & 1 \end{bmatrix} A$

$\Rightarrow \qquad \begin{bmatrix} 1 & -1/2 \\ -2 & 1 \end{bmatrix} = \begin{bmatrix} 1/6 & 0 \\ 0 & 1 \end{bmatrix} A \qquad\qquad \left[R_1 \to \frac{1}{6} R_1 \right]$

$\qquad\qquad \begin{bmatrix} 1 & -1/2 \\ 0 & 0 \end{bmatrix} = \begin{bmatrix} 1/6 & 0 \\ 1/3 & 1 \end{bmatrix} A \qquad\qquad [R_2 \to R_2 + 2R_1]$

A^{-1} is not possible because all elements of second row of matrix in left are zero.

EXAMPLE 2. *Find the inverse of A by elementary transformation, where* $A = \begin{bmatrix} 1 & 2 & 3 \\ 1 & 3 & 5 \\ 1 & 5 & 12 \end{bmatrix}$.

SOLUTION. Considering the identity

$$A = IA$$

$$\begin{bmatrix} 1 & 2 & 3 \\ 1 & 3 & 5 \\ 1 & 5 & 12 \end{bmatrix} = \begin{bmatrix} 1 & 0 & 0 \\ 0 & 1 & 0 \\ 0 & 0 & 1 \end{bmatrix} A$$

By elementary row transformation, we get

$$\begin{bmatrix} 1 & 2 & 3 \\ 0 & 1 & 2 \\ 0 & 3 & 9 \end{bmatrix} \sim \begin{bmatrix} 1 & 0 & 0 \\ -1 & 1 & 0 \\ -1 & 0 & 1 \end{bmatrix} A \qquad (R_2 \rightarrow R_2 - R_1, R_3 \rightarrow R_3 - R_1)$$

$$\begin{bmatrix} 1 & 0 & -1 \\ 0 & 1 & 2 \\ 0 & 0 & 3 \end{bmatrix} \sim \begin{bmatrix} 3 & -2 & 0 \\ -1 & 1 & 0 \\ 2 & -3 & 1 \end{bmatrix} A \qquad \begin{matrix} (R_1 \rightarrow R_1 - 2R_2) \\ (R_3 \rightarrow R_3 - 3R_2) \end{matrix}$$

$$\Rightarrow \quad \begin{bmatrix} 1 & 0 & -1 \\ 0 & 1 & 2 \\ 0 & 0 & 1 \end{bmatrix} \sim \begin{bmatrix} 3 & -2 & 0 \\ -1 & 1 & 0 \\ 2/3 & -1 & 1/3 \end{bmatrix} A \qquad (R_3 \rightarrow 1/3 \, R_3)$$

$$\Rightarrow \quad \begin{bmatrix} 1 & 0 & 0 \\ 0 & 1 & 0 \\ 0 & 0 & 1 \end{bmatrix} \sim \begin{bmatrix} 11/3 & -3 & 1/3 \\ -7/3 & 3 & -2/3 \\ 2/3 & -1 & 1/3 \end{bmatrix} A \qquad \begin{matrix} (R_1 \rightarrow R_1 + R_3) \\ (R_2 \rightarrow R_2 - 2R_3) \end{matrix}$$

$$A^{-1} = \begin{bmatrix} 11/3 & -3 & 1/3 \\ -7/3 & 3 & -2/3 \\ 2/3 & -1 & 1/3 \end{bmatrix} \text{ or } \frac{1}{3} \begin{bmatrix} 11 & -9 & 1 \\ -7 & 9 & -2 \\ 2 & -3 & 1 \end{bmatrix}$$

EXAMPLE 3. *Using elementary transformation find the inverse of the following matrix.*

$$\begin{bmatrix} 1 & 3 & -2 \\ -3 & 0 & -5 \\ 2 & 5 & 0 \end{bmatrix}$$

SOLUTION. Let $\qquad A = \begin{bmatrix} 1 & 3 & -2 \\ -3 & 0 & -5 \\ 2 & 5 & 0 \end{bmatrix}$

Now $\qquad A = I_3 A$

$$\begin{bmatrix} 1 & 3 & -2 \\ -3 & 0 & -5 \\ 2 & 5 & 0 \end{bmatrix} = \begin{bmatrix} 1 & 0 & 0 \\ 0 & 1 & 0 \\ 0 & 0 & 1 \end{bmatrix} A$$

$$\begin{bmatrix} 1 & 3 & -2 \\ 0 & 9 & -11 \\ 0 & -1 & 4 \end{bmatrix} = \begin{bmatrix} 1 & 0 & 0 \\ 3 & 1 & 0 \\ -2 & 0 & 1 \end{bmatrix} A \qquad \begin{matrix} (R_2 \rightarrow R_2 + 3R_1) \\ (R_3 \rightarrow R_3 - 2R_1) \end{matrix}$$

$$\begin{bmatrix} 1 & 3 & -2 \\ 0 & 1 & 21 \\ 0 & -1 & 4 \end{bmatrix} = \begin{bmatrix} 1 & 0 & 0 \\ -13 & 1 & 8 \\ -2 & 0 & 1 \end{bmatrix} A \qquad (R_2 \rightarrow R_2 + 8R_3)$$

$$\begin{bmatrix} 1 & 0 & -65 \\ 0 & 1 & 21 \\ 0 & 0 & 25 \end{bmatrix} = \begin{bmatrix} 40 & -3 & -24 \\ -13 & 1 & 8 \\ -15 & 1 & 9 \end{bmatrix} A \qquad \begin{matrix} (R_1 \rightarrow R_1 - 3R_2) \\ (R_3 \rightarrow R_3 + R_2) \end{matrix}$$

$$\begin{bmatrix} 1 & 0 & -65 \\ 0 & 1 & 21 \\ 0 & 0 & 1 \end{bmatrix} = \begin{bmatrix} 40 & -3 & -24 \\ -13 & 1 & 8 \\ \dfrac{-3}{5} & \dfrac{1}{25} & \dfrac{9}{25} \end{bmatrix} A \qquad R_3 \rightarrow \frac{1}{25} R_3$$

$$\begin{bmatrix} 1 & 0 & 0 \\ 0 & 1 & 0 \\ 0 & 0 & 1 \end{bmatrix} = \begin{bmatrix} 1 & \dfrac{-2}{5} & \dfrac{-3}{5} \\ \dfrac{-2}{5} & \dfrac{4}{25} & \dfrac{11}{25} \\ \dfrac{-3}{5} & \dfrac{1}{25} & \dfrac{9}{25} \end{bmatrix} A \qquad \begin{array}{l}(R_1 \to R_1 + 65R_3) \\ (R_2 \to R_2 - 21R_3)\end{array}$$

$$I_3 = BA \text{ where } B = \begin{bmatrix} +1 & \dfrac{-2}{5} & \dfrac{-3}{5} \\ \dfrac{-2}{5} & \dfrac{4}{25} & \dfrac{11}{25} \\ \dfrac{-3}{5} & \dfrac{1}{25} & \dfrac{9}{25} \end{bmatrix} = A^{-1}$$

EXAMPLE 4. *Find the inverse of matrix* $A = \begin{bmatrix} 1 & 1 & 1 \\ 1 & -1 & 1 \\ 2 & 1 & -1 \end{bmatrix}$.

SOLUTION. Consider, $A = IA$

$$\Rightarrow \quad \begin{bmatrix} 1 & 1 & 1 \\ 1 & -1 & 1 \\ 2 & 1 & -1 \end{bmatrix} = \begin{bmatrix} 1 & 0 & 0 \\ 0 & 1 & 0 \\ 0 & 0 & 1 \end{bmatrix} A$$

$$\Rightarrow \quad \begin{bmatrix} 1 & 0 & -2 \\ 1 & -1 & 1 \\ 2 & 1 & -1 \end{bmatrix} = \begin{bmatrix} -1 & 0 & 1 \\ 0 & 1 & 0 \\ 0 & 0 & 1 \end{bmatrix} A \qquad R_1 \to R_3 - R_1$$

$$\Rightarrow \quad \begin{bmatrix} 1 & 0 & -2 \\ 0 & -1 & 3 \\ 0 & 1 & 3 \end{bmatrix} = \begin{bmatrix} -1 & 0 & 1 \\ 1 & 1 & -1 \\ 2 & 0 & -1 \end{bmatrix} A, \qquad R_2 \to R_2 - R_1 \text{ and } R_3 \to R_3 - 2R_1$$

$$\Rightarrow \quad \begin{bmatrix} 1 & 0 & -2 \\ 0 & -1 & 3 \\ 0 & 0 & 6 \end{bmatrix} = \begin{bmatrix} -1 & 0 & 1 \\ 1 & 1 & -1 \\ 3 & 1 & -2 \end{bmatrix} A \qquad R_3 \to R_2 + R_3$$

$$\Rightarrow \quad \begin{bmatrix} 1 & 0 & -2 \\ 0 & -1 & 3 \\ 0 & 0 & 1 \end{bmatrix} = \begin{bmatrix} -1 & 0 & 1 \\ 1 & 1 & -1 \\ 1/2 & 1/6 & -1/3 \end{bmatrix} A \qquad R_3 \to \frac{1}{6} R_3$$

$$\Rightarrow \quad \begin{bmatrix} 1 & 0 & 0 \\ 0 & -1 & 0 \\ 0 & 0 & 1 \end{bmatrix} = \begin{bmatrix} 0 & 1/3 & 1/3 \\ -1/2 & 1/2 & 0 \\ 1/2 & 1/6 & -1/3 \end{bmatrix} A \qquad \begin{array}{l}(R_1 \to R_1 + 2R_3) \\ (R_2 \to R_2 - 3R_3)\end{array}$$

$$\Rightarrow \quad \begin{bmatrix} 1 & 0 & 0 \\ 0 & 1 & 0 \\ 0 & 0 & 1 \end{bmatrix} = \begin{bmatrix} 0 & 1/3 & 1/3 \\ 1/2 & -1/2 & 0 \\ 1/2 & 1/6 & -1/3 \end{bmatrix} A \qquad (R_2 \to -R_2)$$

$$A^{-1} = \begin{bmatrix} 0 & 1/3 & 1/3 \\ 1/2 & -1/2 & 0 \\ 1/2 & 1/6 & -1/3 \end{bmatrix}$$

Exercise 8.4

Using elementary transformation, find the inverse of matrix A.

1. $A = \begin{bmatrix} 2 & 1 \\ 7 & 4 \end{bmatrix}$ **2.** $A = \begin{bmatrix} 2 & 1 \\ 1 & 1 \end{bmatrix}$

3. $A = \begin{bmatrix} 2 & 3 \\ 1 & 4 \end{bmatrix}$ **4.** $A = \begin{bmatrix} 3 & -1 \\ -4 & 2 \end{bmatrix}$

5. $A = \begin{bmatrix} 2 & 1 \\ 4 & 2 \end{bmatrix}$ **6.** $A = \begin{bmatrix} 10 & -2 \\ -5 & 1 \end{bmatrix}$

7. $A = \begin{bmatrix} -1 & 1 & 2 \\ 2 & 4 & 3 \\ 1 & 3 & 2 \end{bmatrix}$ **8.** $A = \begin{bmatrix} -1 & 2 & 1 \\ -1 & 0 & 2 \\ 2 & 1 & -3 \end{bmatrix}$

9. $A = \begin{bmatrix} 2 & -3 & 3 \\ 2 & 2 & 3 \\ 3 & -2 & 2 \end{bmatrix}$ **10.** $A = \begin{bmatrix} 2 & -1 & 3 \\ -5 & 3 & 1 \\ -3 & 2 & 3 \end{bmatrix}$

11. $A = \begin{bmatrix} 1 & 3 & 3 \\ 1 & 4 & 3 \\ 1 & 3 & 4 \end{bmatrix}$

12. If $A = \begin{bmatrix} \cos\alpha & -\sin\alpha \\ \sin\alpha & \cos\alpha \end{bmatrix}$, prove that

$$A^{-1} = \begin{bmatrix} \cos\alpha & \sin\alpha \\ -\sin\alpha & \cos\alpha \end{bmatrix}.$$

13. If $A = \begin{bmatrix} 3 & -3 & 4 \\ 2 & -3 & 4 \\ 0 & -1 & 1 \end{bmatrix}$, prove that $A^3 = A^{-1}$.

14. If $A = \begin{bmatrix} 1 & 1 & 1 \\ 1 & 2 & 1 \\ 1 & 1 & 2 \end{bmatrix}$, then prove that

$$A^{-1} = \begin{bmatrix} 3 & -1 & -1 \\ -1 & 1 & 0 \\ -1 & 0 & 1 \end{bmatrix}.$$

ANSWERS

1. $A^{-1} = \begin{bmatrix} 4 & -1 \\ -7 & 2 \end{bmatrix}$ **2.** $A^{-1} = \begin{bmatrix} 1 & -1 \\ -1 & 2 \end{bmatrix}$ **3.** $A^{-1} = \begin{bmatrix} 4/5 & -3/5 \\ -1/5 & 2/5 \end{bmatrix}$ **4.** $A^{-1} = \begin{bmatrix} 1 & \frac{1}{2} \\ 2 & \frac{3}{2} \end{bmatrix}$

5. Inverse does not exist **6.** Inverse does not exist **7.** $A^{-1} = \frac{1}{4}\begin{bmatrix} -1 & 4 & -5 \\ -1 & -4 & 7 \\ 2 & 4 & -6 \end{bmatrix}$

8. $A^{-1} = \frac{1}{3}\begin{bmatrix} -2 & 7 & 4 \\ 1 & 1 & 1 \\ -1 & 5 & 2 \end{bmatrix}$ **9.** $A^{-1} = \begin{bmatrix} -\frac{2}{5} & 0 & \frac{3}{5} \\ -\frac{1}{5} & \frac{1}{5} & 0 \\ \frac{2}{5} & \frac{1}{5} & -\frac{2}{5} \end{bmatrix}$ **10.** $A^{-1} = \begin{bmatrix} -7 & -9 & 10 \\ -12 & -15 & 17 \\ 1 & 1 & -1 \end{bmatrix}$

11. $A^{-1} = \begin{bmatrix} 7 & -3 & -3 \\ -1 & 1 & 0 \\ -1 & 0 & 1 \end{bmatrix}$

Objective Evaluation

∽ MULTIPLE CHOICE QUESTIONS

Choose the most appropriate one.

1. The number of 3×3 non-singular matrices with entries 0 and 1 for which the system

$$A = \begin{bmatrix} x \\ y \\ z \end{bmatrix} = \begin{bmatrix} 1 \\ 0 \\ 0 \end{bmatrix} \text{ has two different solutions is :}$$

   (a) 0   (b) $2^9 - 1$
   (c) 168   (d) 2

2. How many solution for the following system of linear equations is possible?
$$x_1 + 2x_2 + x_3 = 3$$
$$2x_1 + 3x_2 + x_3 = 3$$
$$3x_1 + 5x_2 + 2x_3 = 1$$

   (a) only 3 solutions   (b) unique solution
   (c) no solution   (d) infinite solutions

3. Which one of the following is an orthogonal matrix?

   (a) $\begin{bmatrix} \cos\alpha & 2\sin\alpha \\ -2\sin\alpha & \cos\alpha \end{bmatrix}$

   (b) $\begin{bmatrix} \cos\alpha & \sin\alpha \\ -\sin\alpha & \cos\alpha \end{bmatrix}$

   (c) $\begin{bmatrix} \cos\alpha & \sin\alpha \\ \sin\alpha & \cos\alpha \end{bmatrix}$

   (d) $\begin{bmatrix} 1 & 1 \\ 1 & 1 \end{bmatrix}$

4. If matrix $\begin{bmatrix} 0 & +1 & -2 \\ -1 & 0 & 3 \\ \lambda & -3 & 0 \end{bmatrix}$ is not invertible, value

   of λ is :
   (a) -2 (b) -1
   (c) 1   (d) 2

5. If $A = \begin{bmatrix} 2x & 0 \\ x & x \end{bmatrix}$ and $A^{-1} = \begin{bmatrix} 1 & 0 \\ -1 & 2 \end{bmatrix}$, then value

   of x is :
   (a) 2   (b) $-1/2$
   (c) 1   (d) $1/2$

6. For how many values of x in closed interval

   $[-4, -1]$ the matrix $\begin{bmatrix} 3 & -1+x & 2 \\ 3 & -1 & x+2 \\ x+3 & -1 & 2 \end{bmatrix}$ is

not invertible?
   (a) 2   (b) 0
   (c) 3   (d) 1

7. Which one of the following is not true ?
   (a) addition is commutative for matrix
   (b) addition is associative for matrix
   (c) multiplication is commutative for matrix
   (d) multiplication is associative for matrix

8. Let $A = \begin{bmatrix} 5 & 5\alpha & \alpha \\ 0 & \alpha & 5\alpha \\ 0 & 0 & 5 \end{bmatrix}$ if $|A|^2 = 25$, then value

   of $|\alpha|$ is equal to :
   (a) 5^2 (b) 1
   (c) $1/5$ (d) 5

9. If system of linear equations
$$\begin{bmatrix} 1 & 2 & 4 \\ 2 & 1 & 2 \\ 1 & 2 & \alpha-4 \end{bmatrix}\begin{bmatrix} x \\ y \\ z \end{bmatrix} = \begin{bmatrix} 6 \\ 4 \\ \alpha \end{bmatrix}$$

   have unique solution, then :
   (a) $\alpha \in R$ (b) $\alpha \in Z$
   (c) $\alpha = 8$ (d) $\alpha \neq 8$

10. If $A = \begin{bmatrix} \cos\alpha & \sin\alpha \\ -\sin\alpha & \cos\alpha \end{bmatrix}$, then A^n is equal to :

   (a) $\begin{bmatrix} \cos n\alpha & -\sin n\alpha \\ \sin n\alpha & \cos n\alpha \end{bmatrix}$

   (b) $\begin{bmatrix} \cos n\alpha & \sin n\alpha \\ -\sin n\alpha & \cos n\alpha \end{bmatrix}$

   (c) $\begin{bmatrix} \sin n\alpha & \cos n\alpha \\ -\cos n\alpha & \sin n\alpha \end{bmatrix}$

   (d) $\begin{bmatrix} 1 & 0 \\ 0 & 1 \end{bmatrix}$

11. Value of $\begin{bmatrix} 1 \\ -1 \\ 2 \end{bmatrix}\begin{bmatrix} 2 & 1 & -1 \end{bmatrix}$ is :

   (a) $\begin{bmatrix} 2 \\ -1 \\ -2 \end{bmatrix}$ (b) $\begin{bmatrix} 2 & 1 & -1 \\ -2 & -1 & 1 \\ 4 & 2 & -2 \end{bmatrix}$

   (c) $[-1]$ (d) not define

12. If $A = \begin{bmatrix} 1 & 2 \\ 3 & -5 \end{bmatrix}$, then value of A^{-1} is :

(a) $\begin{bmatrix} -5 & -2 \\ -3 & 1 \end{bmatrix}$

(b) $\begin{bmatrix} 5/11 & 2/11 \\ 3/11 & -1/11 \end{bmatrix}$

(c) $\begin{bmatrix} -5/11 & -2/11 \\ -3/11 & -1/11 \end{bmatrix}$

(d) $\begin{bmatrix} 5 & 2 \\ 3 & -1 \end{bmatrix}$

13. If U is a 3×3 matrix where $U = \begin{bmatrix} 1 & 2 & 2 \\ -2 & -1 & -1 \\ 1 & -4 & -3 \end{bmatrix}$ then $\begin{bmatrix} 3 & 2 & 0 \end{bmatrix} U \begin{bmatrix} 3 \\ 2 \\ 0 \end{bmatrix}$ is equal to :

(a) 4
(b) 5
(c) 3/2
(d) 5/2

14. If A and B are such n order square matrix that $A^2 - B^2 = (A - B)(A + B)$, then which is true of the following?

(a) $A = B$
(b) $AB = BA$
(c) any one of A or B is a null matrix
(d) either A or B is identity matrix

15. Let $A = \begin{bmatrix} 1 & 2 \\ 3 & 4 \end{bmatrix}$ and $B = \begin{bmatrix} a & 0 \\ 0 & b \end{bmatrix}, a, b \in N$ then:

(a) any B is not such that $AB = BA$
(b) more then one but finite no. of B are such that $AB = BA$
(c) only one B will be such that $AB = BA$
(d) there are infinite no. of B such that $AB = BA$

16. If $A = \begin{bmatrix} 6 & 8 & 5 \\ 4 & 2 & 3 \\ 9 & 7 & 1 \end{bmatrix}$ is the sum of a symmetric matrix B and skew symmetric matrix C, then B is :

(a) $\begin{bmatrix} 0 & 2 & -2 \\ -2 & 5 & -2 \\ 2 & 2 & 0 \end{bmatrix}$

(b) $\begin{bmatrix} 6 & 6 & 7 \\ 6 & 2 & 5 \\ 7 & 5 & 1 \end{bmatrix}$

(c) $\begin{bmatrix} 6 & 6 & 7 \\ -6 & 2 & -5 \\ -7 & 5 & 1 \end{bmatrix}$

(d) $\begin{bmatrix} 0 & 6 & -2 \\ 2 & 0 & -2 \\ -2 & -2 & 0 \end{bmatrix}$

17. If B is a matrix obtained by interchanging any two rows of a square matrix, then $|A + B|$ is equal to :

(a) $2|A|$
(b) $2|B|$
(c) 0
(d) $|A| - |B|$

18. Let $A = \begin{bmatrix} 1 & 0 \\ 0 & -1 \end{bmatrix}$ and $B = \begin{bmatrix} 1 & x \\ 0 & 1 \end{bmatrix}$. If $AB = BA$, then x is equal to :

(a) -1
(b) 0
(c) 1
(d) Any real no.

19. If A and B are square matrices of order 2 then which is true of the following?

(a) $(A + B)^2 = A^2 + B^2 + 2AB$
(b) $(A - B)^2 = A^2 + B^2 - 2AB$
(c) $(A - B)(A + B) = A^2 + AB - BA - B^2$
(d) $(A + B)(A - B) = A^2 - B^2$

20. Consider $A = \begin{bmatrix} 1 & 1 & 1 \\ 1 & 1 & 1 \\ 1 & 1 & 1 \end{bmatrix}$, find a positive integer n for which A^n is equal to :

(a) A
(b) $3^n A$
(c) $3^{n-1} A$
(d) $3A$

21. If A is invertible square matrix of order 2, then $adj(adj\, A)$ is equal to :

(a) A^2
(b) A
(c) A^{-1}
(d) none of these

22. A is matrix of order 3×3 such that $|A| = 5$ if $B = 4A^2$, then value of $|B|$ is equal to :

(a) 20
(b) 100
(c) 320
(d) 1600

23. If A is square matrix such that $A - A^T = 0$, then which is true of following ?

(a) A is null matrix
(b) A is unit matrix
(c) A is scalar matrix
(d) none of these

24. If $A = \begin{bmatrix} a & b \\ b & -a \end{bmatrix}$ and for any matrix M, $MA = A^{2n}$, $m \in N$, then M is equal to :

(a) $\begin{bmatrix} a^{2na} & b^{2n} \\ b^{2m} & -a^{2m} \end{bmatrix}$

(b) $(a^2 + b^2)^m \begin{bmatrix} 1 & 0 \\ 0 & 1 \end{bmatrix}$

(c) $(a^m + b^m) \begin{bmatrix} 1 & 0 \\ 0 & 1 \end{bmatrix}$

(d) $(a^2 + b^2)^{m-1} \begin{bmatrix} a & b \\ b & -a \end{bmatrix}$

25. If $A = \begin{bmatrix} 1 & 0 & 0 \\ 0 & 1 & 1 \\ 0 & -2 & 4 \end{bmatrix}, 6A^{-1} = A^2 + cA + dI$, then (c, d) is :

(a) $(-6, 11)$
(b) $(-11, 6)$
(c) $(11, 6)$
(d) $(6, 11)$

☞ TRUE/FALSE

Write 'T' for True and 'F' for False statement.

1. m rows and n columns of a matrix represent $m \times n$ order of matrix. **(T/F)**
2. If no. of rows (m) is equal to the no. of columns (n), then such matrix is called rectangular matrix. **(T/F)**
3. If no. of rows is equal to the no. of columns, then such matrix is called square matrix. **(T/F)**

4. If each element in leading diagonal of a square matrix is unity and all other elements are zero such matrix is called diagonal matrix. **(T/F)**
5. If each element except in diagonal of a square matrix is zero, then such matrix is called scalar matrix. **(T/F)**
6. If some rows of a matrix are left then remaining matrix is called sub-matrix of the matrix. **(T/F)**

☞ FILL IN THE BLANKS

1. Multiplication of matrix does not obey _____ law.
2. Multiplication of matrix does not obey _____ law provided they are conformal.
3. Inverse of the inverse of a matrix is _____ also.

4. The inverse of a invertible matrix is _____ .
5. If each element in leading diagonal of a square matrix is unity and all other element are zero such matrix is called _____ matrix.

ANSWERS

☞ MULTIPLE CHOICE QUESTIONS

1. (a)	**2.** (c)	**3.** (b)	**4.** (d)	**5.** (d)	**6.** (d)	**7.** (a)	**8.** (c)	**9.** (c)
10. (b)	**11.** (b)	**12.** (b)	**13.** (b)	**14.** (b)	**15.** (d)	**16.** (b)	**17.** (c)	**18.** (b)
19. (c)	**20.** (c)	**21.** (b)	**22.** (d)	**23.** (d)	**24.** (d)	**25.** (a)		

☞ TRUE/ FALSE

1. T	**2.** F	**3.** T	**4.** F	**5.** F	**6.** T

☞ FILL IN THE BLANKS

1. commutative	**2.** associative	**3.** matrix	**4.** unique	**5.** identity

❑❑❑❑❑

CHAPTER 9

Differentiation

9.1 INTRODUCTION

Let $y = f(x)$ be a function of x where x is an independent variable and y is dependent variable. Let δx be any increment in the value of independent variable x and δy be the corresponding increment in the value of dependent variable y, then $\dfrac{\delta y}{\delta x}$ is known as the rate of change of y with respect to x.

Definition. *Derivative of a function $f(x)$ is the limiting value of $\dfrac{\delta y}{\delta x}$ as $\delta x \to 0$ provided the limit exist finitely and it is denoted by $\dfrac{dy}{dx}$.*

Thus, $\dfrac{dy}{dx} = \lim\limits_{\delta x \to 0} \dfrac{\delta y}{\delta x} = \lim\limits_{\delta x \to 0} \dfrac{f(x + \delta x) - f(x)}{\delta x}$, provided the limit exist.

Here, the increments δx and δy can be positive or negative.

⮞ REMARKS

➡ The differential coefficient and instantaneous rate of change are also used for derivative.

➡ $\dfrac{dy}{dx}, \dfrac{d}{dx}(y), y', y_1, \dfrac{d}{dx}(f(x)), f'(x)$ or $Df(x)$ have the same meaning.

➡ The derivative of a function $f(x)$ at a point $x = a$ is denoted by $f'(a)$ or $\left[\dfrac{d}{dx} f(x)\right]_{x=a}$.

9.1.1 GEOMETRICAL MEANING OF A DERIVATIVE

Let us consider a function $y = f(x)$ in a rectangular coordinates system. We also consider a point $P(x, y)$ on the curve.

If a point corresponding to an increased value of the argument $x + \Delta x$ is considered its ordinates value is given by

$$y + \Delta y = f(x + \Delta x)$$

The point $(x + \Delta x, y + \Delta y)$ is represent by A. Hence PA is the secant to the curve.

Now, on $\Delta x, \Delta y \to 0$

$\Rightarrow PA \to 0$ (*i.e.*, the distance PA tends to zero and to a single point P)

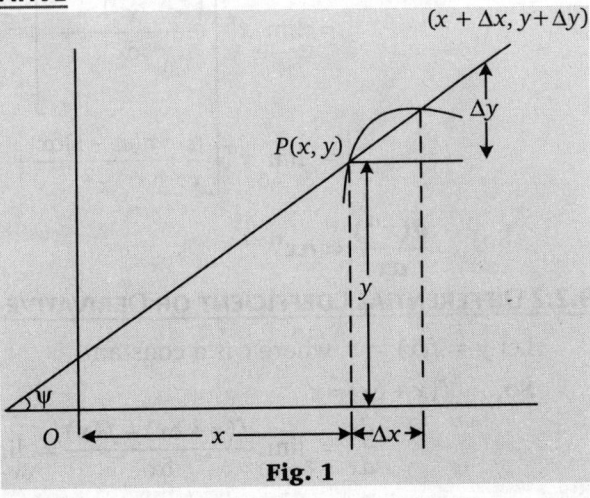

Fig. 1

$\Rightarrow \lim_{\Delta x \to 0}$ (slope of chord PA) $\to$ (slope of the tangent at P)

or $\qquad \tan \psi = \lim_{\Delta x \to 0} \tan \theta$

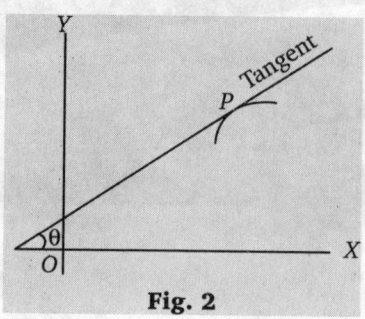

Fig. 2

$\Rightarrow \qquad \lim_{\Delta x \to 0} \frac{\Delta y}{\Delta x} = \frac{dy}{dx}$ or $f'(x)$

which means that the value of the derivative $f'(x)$ for a given value of x is equal to the tangent of the angle formed by the line tangent to the graph of function $y = f(x)$ at a point $P(x, y)$ with the positive direction.

9.2 METHOD FOR FINDING THE DERIVATIVE USING FIRST PRINCIPLE

WORKING PROCEDURE

To find the derivative of a function from the first principle, we use the following steps.

Step 1. First we take the function in the form of $y = f(x)$.

Step 2. Take the small increment δx in x and let corresponding increment in y is δy such that $y + \delta y = f(x + \delta x)$.

Step 3. Find out the increment $\delta y = f(x + \delta x) - f(x)$.

Step 4. Now dividing both sides of $\delta y = f(x + \delta x) - f(x)$ by δx and takes the limit $\delta x \to 0$, we get

$$\frac{dy}{dx} = \lim_{\delta x \to 0} \frac{\delta y}{\delta x} = \lim_{\delta x \to 0} \frac{f(x + \delta x) - f(x)}{\delta x}$$

9.2.1 DIFFERENTIAL COEFFICIENT OR DERIVATIVE OF $y = x^n$

Let $\qquad y = f(x) = x^n$ so $f(x + \delta x) = (x + \delta x)^n$

$\therefore \qquad \dfrac{dy}{dx} = \lim_{\delta x \to 0} \dfrac{f(x + \delta x) - f(x)}{\delta x} = \lim_{\delta x \to 0} \dfrac{(x + \delta x)^n - x^n}{\delta x}$

$$= \lim_{\delta x \to 0} x^n \left[\frac{\left(1 + \frac{\delta x}{x}\right)^n - 1}{\delta x} \right] = \lim_{\delta x \to 0} x^n \frac{\left[1 + n\frac{\delta x}{x} + \frac{n(n-1)}{2!}\left(\frac{\delta x}{x}\right)^2 + \ldots - 1\right]}{\delta x}$$

$$= \lim_{\delta x \to 0} x^n \left[\frac{n}{x} + \frac{n(n-1)\delta x}{x^2} + \ldots \right] = x^n \frac{n}{x} = nx^{n-1}$$

So, $\qquad \dfrac{d(x^n)}{dx} = nx^{n-1}$

9.2.2 DIFFERENTIAL COEFFICIENT OR DERIVATIVE OF A CONSTANT FUNCTION

Let $y = f(x) = c$, where c is a constant.

So, $\qquad f(x + \delta x) = c$

$\therefore \qquad \dfrac{dy}{dx} = \lim_{\delta x \to 0} \dfrac{f(x + \delta x) - f(x)}{\delta x} = \lim_{\delta x \to 0} \dfrac{c - c}{\delta x} = 0.$

So, $\dfrac{d(c)}{dx} = 0$.

Hence, derivative of a constant function is always zero.

9.2.3 DERIVATIVE OF THE PRODUCT OF A CONSTANT WITH A FUNCTION

The derivate of the product of a constant with a function is equal to the product of constant and the derivate of the function.

Let $y = Cf(x)$, where C is a constant.

So, $\dfrac{dy}{dx} = \lim_{\delta x \to 0}\left(\dfrac{Cf(x + \delta x) - Cf(x)}{\delta x}\right) = C\lim_{\delta x \to 0}\dfrac{f(x + \delta x) - f(x)}{\delta x} = C\dfrac{d[f(x)]}{dx}$

Hence, $\dfrac{d}{dx}[Cfx] = C\dfrac{d}{dx}[f(x)]$.

9.2.4 DERIVATIVE OF SIN X

Let $y = f(x) = \sin x$, then $f(x + \delta x) = \sin(x + \delta x)$

So, $\dfrac{dy}{dx} = \lim_{\delta x \to 0}\dfrac{f(x + \delta x) - f(x)}{\delta x} = \lim_{\delta x \to 0}\dfrac{\sin(x + \delta x) - \sin x}{\delta x}$

$= \lim_{\delta x \to 0}\dfrac{2\cos\left(x + \dfrac{\delta x}{2}\right)\sin\dfrac{\delta x}{2}}{\delta x}\qquad\left[\because \sin C - \sin D = 2\cos\left(\dfrac{C + D}{2}\right)\sin\left(\dfrac{C - D}{2}\right)\right]$

$= \lim_{\delta x \to 0}\dfrac{\cos\left(x + \dfrac{\delta x}{2}\right)\sin\dfrac{\delta x}{2}}{\dfrac{\delta x}{2}} = \lim_{\delta x \to 0}\cos\left(x + \dfrac{\delta x}{2}\right)\lim_{\delta x \to 0}\dfrac{\sin\dfrac{\delta x}{2}}{\dfrac{\delta x}{2}}$

$= \cos x \cdot 1 = \cos x\qquad\qquad\left[\text{Since }\lim_{x \to 0}\dfrac{\sin x}{x} = 1\right]$

$\therefore\quad \dfrac{d}{dx}(\sin x) = \cos x$

9.2.5 DERIVATIVE OF COS X

Let $y = f(x) = \cos x$, then $f(x + \delta x) = \cos(x + \delta x)$

So $\dfrac{dy}{dx} = \lim_{\delta x \to 0}\dfrac{f(x + \delta x) - f(x)}{\delta x} = \lim_{\delta x \to 0}\dfrac{\cos(x + \delta x) - \cos x}{\delta x}$

$= \lim_{\delta x \to 0}\dfrac{2\sin\left(x + \dfrac{\delta x}{2}\right)\sin\left(-\dfrac{\delta x}{2}\right)}{\delta x}$

$\left[\text{Since }\cos C - \cos D = \sin\left(\dfrac{C + D}{2}\right)\sin\left(\dfrac{D - C}{2}\right)\right]$

$= \lim_{\delta x \to 0}\left[\dfrac{-\sin\left(x + \dfrac{\delta x}{2}\right)\sin\left(\dfrac{\delta x}{2}\right)}{\dfrac{\delta x}{2}}\right]$

$$= \lim_{\delta x \to 0}\left[-\sin\left(x + \frac{\delta x}{2}\right)\right]\lim_{\delta x \to 0}\frac{\sin\dfrac{\delta x}{2}}{\dfrac{\delta x}{2}} = -\sin x \cdot 1 = -\sin x$$

Hence, $\dfrac{d}{dx}(\cos x) = -\sin x$

9.2.6 DERIVATIVE OF TAN X

Let $y = f(x) = \tan x$, then $f(x + \delta x) = \tan(x + \delta x)$

So, $\dfrac{dy}{dx} = \lim_{\delta x \to 0}\dfrac{f(x + \delta x) - f(x)}{\delta x}$

$$= \lim_{\delta x \to 0}\frac{\tan(x + \delta x) - \tan x}{\delta x} = \lim_{\delta x \to 0}\left[\frac{\dfrac{\sin(x + \delta x)}{\cos(x + \delta x)} - \dfrac{\sin x}{\cos x}}{\delta x}\right]$$

$$= \lim_{\delta x \to 0}\left[\frac{\sin(x + \delta x)\cos x - \cos(x + \delta x)\sin x}{\delta x \cos(x + \delta x)\cos x}\right]$$

$$= \lim_{\delta x \to 0}\frac{\sin(x + \delta x - x)}{\cos(x + \delta x)\cos x.\delta x}$$

$$[\text{Since} \sin A \cos B - \cos A \sin B = \sin(A - B)]$$

$$= \lim_{\delta x \to 0}\left(\frac{\sin \delta x}{\delta x}\right)\frac{1}{\cos(x + \delta x)\cos x}$$

$$= \lim_{\delta x \to 0}\left(\frac{\sin \delta x}{\delta x}\right)\lim_{\delta x \to 0}\frac{1}{\cos(x + \delta x)\cos x} = 1\frac{1}{\cos^2 x} = \sec^2 x$$

Hence, $\dfrac{d}{dx}(\tan x) = \sec^2 x$.

9.2.7 DERIVATIVE OF COT X

Let $y = f(x) = \cot x$, then $f(x + \delta x) = \cot(x + \delta x)$

So, $\dfrac{dy}{dx} = \lim_{\delta x \to 0}\dfrac{f(x + \delta x) - f(x)}{\delta x} = \lim_{\delta x \to 0}\left(\dfrac{\cot(x + \delta x) - \cot x}{\delta x}\right)$.

Proceeding in the same way, we may get

$$\frac{d}{dx}(\cot x) = -\mathrm{cosec}^2 x.$$

9.2.8 DERIVATIVE OF SEC X

Let $y = f(x) = \sec x$ then $f(x + \delta x) = \sec(x + \delta x)$

So, $\dfrac{dy}{dx} = \lim_{\delta x \to 0}\dfrac{f(x + \delta x) - f(x)}{\delta x}$

$$= \lim_{\delta x \to 0}\frac{\sec(x + \delta x) - \sec x}{\delta x} = \lim_{\delta x \to 0}\frac{\left(\dfrac{1}{\cos(x + \delta x)} - \dfrac{1}{\cos x}\right)}{\delta x}$$

$$= \lim_{\delta x \to 0}\frac{\cos x - \cos(x + \delta x)}{\delta x \cos x \cos(x + \delta x)}$$

$$= \lim_{\delta x \to 0} \frac{2\sin\left(x + \dfrac{\delta x}{2}\right)\sin\dfrac{\delta x}{2}}{\delta x \cos x \cos(x + \delta x)} = \lim_{\delta x \to 0} \frac{\sin\left(x + \dfrac{\delta x}{2}\right)}{\cos x \cos(x + \delta x)} \lim_{\delta x \to 0} \frac{\sin\dfrac{\delta x}{2}}{\dfrac{\delta x}{2}}$$

$$= \frac{\sin x}{\cos^2 x} = \sec x \tan x$$

Hence, $\dfrac{d}{dx}(\sec x) = \sec x \tan x$.

9.2.9 DERIVATIVE OF e^x

Let $y = f(x) = e^x$ then $f(x + \delta x) = e^{x + \delta x}$

So, $\qquad \dfrac{dy}{dx} = \lim_{\delta x \to 0} \dfrac{f(x + \delta x) - f(x)}{\delta x} = \lim_{\delta x \to 0} \dfrac{e^{x+\delta x} - e^x}{\delta x}$

$$= \lim_{\delta x \to 0} \frac{e^x e^{\delta x} - e^x}{\delta x} = \lim_{\delta x \to 0} \frac{e^x[e^{\delta x} - 1]}{\delta x} = \lim_{\delta x \to 0} e^x \left[1 + \frac{(\delta x)}{2!} + \frac{(\delta x)^2}{3!} + \dots\right]$$

$$= e^x$$

Hence, $\quad \dfrac{d}{dx}(e^x) = e^x$.

9.2.10 DERIVATIVE OF $\log_e x$

Let $y = f(x) = \log_e x$, then $f(x + \delta x) = \log_e(x + \delta x)$

So, $\qquad \dfrac{dy}{dx} = \lim_{\delta x \to 0} \dfrac{f(x + \delta x) - f(x)}{\delta x} = \lim_{\delta x \to 0} \dfrac{\log_e(x + \delta x) - \log x}{\delta x}$

$$= \lim_{\delta x \to 0} \frac{\log_e\left(\dfrac{x + \delta x}{x}\right)x - \log_e x}{\delta x} = \lim_{\delta x \to 0} \frac{\log_e x\left(1 + \dfrac{\delta x}{x}\right) - \log_e x}{\delta x}$$

$$= \lim_{\delta x \to 0} \frac{\log_e x + \log\left(1 + \dfrac{\delta x}{x}\right) - \log_e x}{\delta x} = \lim_{\delta x \to 0} \frac{\log_e\left(1 + \dfrac{\delta x}{x}\right)}{\delta x}$$

$$= \lim_{\delta x \to 0}\left[\frac{1}{x} - \frac{1}{2}\frac{\delta x}{x^2} + \dots\right] \qquad \left[\text{Since, } \log(1 + x) = x - \frac{x^2}{2} + \frac{x^3}{3}\dots\right]$$

$$= \frac{1}{x}$$

Hence, $\dfrac{d}{dx}(\log_e x) = \dfrac{1}{x}$.

9.2.11 DERIVATIVE OF $\log_a x$

Let $y = f(x) = \log_a x$

We know $\quad \log_a x = \log_e x . \log_a e$

So, $\quad \dfrac{d}{dx}(\log_a x) = \dfrac{d}{dx}[\log_e x . \log_a e] = \log_a e \dfrac{d}{dx}[\log_e x] = \log_a e \dfrac{1}{x}$

Hence, $\dfrac{d}{dx}(\log_a x) = \dfrac{1}{x}\log_a e$.

9.2.12 DERIVATIVE OF a^x

Let $y = f(x) = a^x$ or $f(x) = e^{x\log a}$

then $\quad f(x + \delta x) = e^{(x+\delta x)\log a}$

So, $\quad \dfrac{dy}{dx} = \lim\limits_{\delta x \to 0} \dfrac{e^{(x+\delta x)\log a} - e^{x\log a}}{\delta x}$

$$= \lim_{\delta x \to 0} \frac{e^{x\log a} \cdot e^{\delta x \log a} - e^{x\log a}}{\delta x} = \lim_{\delta x \to 0} \frac{e^{x\log a}[e^{\delta x \log a} - 1]}{\delta x}$$

$$= \lim_{\delta x \to 0} \frac{e^{x\log a}\left[\dfrac{\delta x \log a}{1!} + \dfrac{\delta x (\log a)^2}{2!} + \dots\right]}{\delta x}$$

$$= \lim_{\delta x \to 0} e^{x\log a}\left[\log a + \frac{\delta x (\log a)^2}{2!} + \dots\right]$$

$$= e^{x\log a} \cdot \log a = a^x \log a$$

Hence, $\quad \dfrac{d}{dx}(a^x) = a^x \cdot \log_e a.$

9.3 DERIVATIVE OF THE SUM OF TWO FUNCTIONS

"The derivative of the sum of two functions is equal to the sum of their derivatives."

Let $\quad y = f(x) = \phi_1(x) + \phi_2(x)$, then

$$f(x + \delta x) = \phi_1(x + \delta x) + \phi_2(x + \delta x)$$

So, $\quad \dfrac{dy}{dx} = \dfrac{df(x)}{dx} = \lim\limits_{\delta x \to 0} \dfrac{f(x + \delta x) - f(x)}{\delta x}$

$$= \lim_{\delta x \to 0} \frac{\{\phi_1(x + \delta x) + \phi_2(x + \delta x)\} - \{\phi_1(x) + \phi_2(x)\}}{\delta x}$$

$$= \lim_{\delta x \to 0} \frac{\{\phi_1(x + \delta x) - \phi_1(x)\} + \{\phi_2(x + \delta x) - \phi_2(x)\}}{\delta x}$$

$$= \lim_{\delta x \to 0} \frac{\phi_1(x + \delta x) - \phi_1(x)}{\delta x} + \lim_{\delta x \to 0} \frac{\phi_2(x + \delta x) - \phi_2(x)}{\delta x}$$

$$= \frac{d}{dx}\phi_1(x) + \frac{d}{dx}\phi_2(x).$$

Hence, $\dfrac{d}{dx}\{\phi_1(x) + \phi_2(x)\} = \dfrac{d}{dx}\phi_1(x) + \dfrac{d}{dx}\phi_2(x).$

❧ REMARK
➡ The above result can be generalized as follows

$$\frac{d}{dx}[\phi_1(x) + \phi_2(x) + \dots + \phi_n(x)] = \frac{d}{dx}\phi_1(x) + \frac{d}{dx}\phi_2(x) + \dots + \frac{d}{dx}\phi_n(x)$$

9.4 DERIVATIVE OF THE DIFFERENCE OF TWO FUNCTIONS

"The derivative of the difference of two functions is the difference of their derivatives."

Let $\quad f(x) = \phi_1(x) - \phi_2(x)$, then

$$f(x + \delta x) = \phi_1(x + \delta x) - \phi_2(x + \delta x)$$

So, $\dfrac{df(x)}{dx} = \lim\limits_{\delta x \to 0} \dfrac{f(x + \delta x) - f(x)}{\delta x}$

$$\lim\limits_{\delta x \to 0} \dfrac{\{\phi_1(x + \delta x) - \phi_2(x + \delta x)\} - \{\phi_1(x) - \phi_2(x)\}}{\delta x}$$

$$= \lim\limits_{\delta x \to 0} \dfrac{\{\phi_1(x + \delta x) - \phi_1(x)\} - \{(\phi_2(x + \delta x) - \phi_2(x)\}}{\delta x}$$

$$= \lim\limits_{\delta x \to 0} \dfrac{\phi_1(x + \delta x) - \phi_1(x)}{\delta x} - \lim\limits_{\delta x \to 0} \dfrac{\phi_2(x + \delta x) - \phi_2(x)}{\delta x}$$

$$= \dfrac{d}{dx}\phi_1(x) - \dfrac{d}{dx}\phi_2(x)$$

Hence, $\dfrac{d}{dx}\{\phi_1(x) - \phi_2(x)\} = \dfrac{d}{dx}\phi_1(x) - \dfrac{d}{dx}\phi_2(x)$.

⚜ REMARK

➠ The above result can be generalized as follows

$$\dfrac{d}{dx}\{f_1(x) - f_2(x) - \ldots - f_n(x)\} = \dfrac{d}{dx}f_1(x) - \dfrac{d}{dx}f_2(x) - \ldots - \dfrac{d}{dx}f_n(x).$$

➠ From the above discussion, we have, if $y = f(x) \pm g(x) \pm h(x)\ldots$ then

$$\dfrac{dy}{dx} = \dfrac{d[f(x)]}{dx} \pm \dfrac{d[g(x)]}{dx} \pm \dfrac{d[h(x)]}{dx} \pm \ldots$$

SUMMARY OF STANDARD DERIVATIVES

	Function	Derivative		Function	Derivative
1.	x^n	nx^{n-1}	14.	$\log\lvert \operatorname{cosec} x - \cot x\rvert$	$\operatorname{cosec} x$
2.	e^x	e^x	15.	$\sin^{-1} x$	$\dfrac{1}{\sqrt{1-x^2}}$
3.	$\log\lvert x\rvert$	$1/x$	16.	$\cos^{-1} x$	$-1/\sqrt{1-x^2}$
4.	a^x	$a^x \log_e a,$ $a > 0, a \neq 1$	17.	$\tan^{-1} x$	$1/(1+x^2)$
5.	$\cos x$	$-\sin x$	18.	$\cot^{-1} x$	$-1/(1+x^2)$
6.	$\sin x$	$\cos x$	19.	$\sec^{-1} x$	$1/(x\sqrt{x^2-1})$
7.	$\tan x$	$\sec^2 x$	20.	$\operatorname{cosec}^{-1} x$	$-1/(x\sqrt{x^2-1})$
8.	$\cot x$	$-\operatorname{cosec}^2 x$	21.	$\sin^{-1}(x/a)$	$1/\sqrt{a^2-x^2}$
9.	$\sec x$	$\sec x \tan x$	22.	$\cos^{-1}(x/a)$	$-1/\sqrt{a^2-x^2}$
10.	$\operatorname{cosec} x$	$-\operatorname{cosec} x \cot x$	23.	$(1/a)\tan^{-1}(x/a)$	$1/(a^2+x^2)$
11.	$\log\lvert\sin x\rvert$	$\cot x$	24.	$(1/a)\cot^{-1}(x/a)$	$-1/(a^2+x^2)$
12.	$-\log\lvert\cos x\rvert$	$\tan x$	25.	$(1/a)\sec^{-1}(x/a)$	$1/(x\sqrt{x^2-a^2})$
13.	$\log\lvert\sec x + \tan x\rvert$	$\sec x$	26.	$(1/a)\operatorname{cosec}^{-1}(x/a)$	$-1/(x\sqrt{x^2-a^2})$

SOLVED EXAMPLES

EXAMPLE 1. *Find the differential coefficient of $\dfrac{1}{3x+4}$ from the first principles.*

SOLUTION. Let $\qquad y = f(x) = \dfrac{1}{3x+4}$

$\Rightarrow \qquad y + \delta y = \dfrac{1}{3(x+\delta x)+4}$

$\therefore \qquad \delta y = (y+\delta y) - y = \dfrac{1}{3(x+\delta x)+4} - \dfrac{1}{3x+4}$

$\qquad\qquad = \dfrac{3x+4-3x-3\delta x-4}{[3(x+\delta x)+4](3x+4)} = \dfrac{-3\delta x}{[3(x+\delta x)+4](3x+4)}$

$\therefore \qquad \dfrac{\delta y}{\delta x} = \dfrac{-3}{[3(x+\delta x)+4](3x+4)}$

Now, $\displaystyle\lim_{\delta x\to 0}\dfrac{\delta y}{\delta x} = \lim_{\delta x\to 0}\dfrac{-3}{[3(x+\delta x)+4](3x+4)} = \dfrac{-3}{[3(x+0)+4](3x+4)}$

$\therefore \qquad \dfrac{dy}{dx} = \dfrac{-3}{(3x+4)^2}$

EXAMPLE 2. *From the first principle find the derivative of $\dfrac{x^2+1}{x}$ w.r. to x.*

SOLUTION. We have

$$f(x) = \dfrac{x^2+1}{x} = x + \dfrac{1}{x}$$

$\therefore \qquad f(x+\delta x) = x + \delta x + \dfrac{1}{x+\delta x}$

or $\quad f(x+\delta x) - f(x) = \left[x+\delta x+\dfrac{1}{x+\delta x}\right] - \left[x+\dfrac{1}{x}\right] = (x+\delta x-x) + \left[\dfrac{1}{x+\delta x} - \dfrac{1}{x}\right]$

$\therefore \qquad \dfrac{f(x+\delta x)-f(x)}{\delta x} = \dfrac{\delta x}{\delta x} + \dfrac{x-x-\delta x}{x(x+\delta x)\delta x}$

$\displaystyle\lim_{\delta x\to 0}\dfrac{f(x+\delta x)-f(x)}{\delta x} = 1 + \lim_{\delta x\to 0}\dfrac{-\delta x}{x(x+\delta x)\delta x} = 1 - \lim_{\delta x\to 0}\dfrac{1}{x(x+\delta x)}$

$\therefore \qquad \dfrac{df}{dx} = 1 - \dfrac{1}{x^2} = \dfrac{x^2-1}{x^2}$

EXAMPLE 3. *From the first principle find the differential coefficient of $\sqrt{x}$.*

SOLUTION. Let $\qquad y = \sqrt{x}$

then $\qquad y + \delta y = \sqrt{x+\delta x}$

$\therefore \qquad \delta y = \sqrt{x+\delta x} - \sqrt{x} = \dfrac{(\sqrt{x+\delta x}-\sqrt{x})(\sqrt{x+\delta x}+\sqrt{x})}{\sqrt{x+\delta x}+\sqrt{x}}$

or $\qquad \delta y = \dfrac{(x+\delta x)-x}{\sqrt{x+\delta x}+\sqrt{x}}$

or $\qquad \delta y = \dfrac{\delta x}{\sqrt{x+\delta x}+\sqrt{x}}$

$$\Rightarrow \qquad \frac{\delta y}{\delta x} = \frac{1}{\sqrt{x + \delta x} + \sqrt{x}}$$

$$\lim_{\delta x \to 0} \frac{\delta y}{\delta x} = \lim_{\delta x \to 0} \frac{1}{\sqrt{x + \delta x} + \sqrt{x}}$$

$$\therefore \qquad \frac{dy}{dx} = \frac{1}{\sqrt{x} + \sqrt{x}} = \frac{1}{2\sqrt{x}}$$

EXAMPLE 4. *From the first principle find the differential coefficient of* $\dfrac{1}{\sqrt{x}}$.

SOLUTION. Let $\qquad y = \dfrac{1}{\sqrt{x}}$

$$\therefore \qquad y + \delta y = \frac{1}{\sqrt{x + \delta x}}$$

$$\therefore \qquad \delta y = \frac{1}{\sqrt{x + \delta x}} - \frac{1}{\sqrt{x}} = \frac{\sqrt{x} - \sqrt{x + \delta x}}{\sqrt{x}\sqrt{x + \delta x}} \times \frac{\sqrt{x} + \sqrt{x + \delta x}}{\sqrt{x} + \sqrt{x + \delta x}}$$

(Rationalising the numerator)

$$\text{or} \qquad \frac{\delta y}{\delta x} = \frac{x - (x + \delta x)}{\delta x (\sqrt{x}\sqrt{x + \delta x})(\sqrt{x} + \sqrt{x + \delta x})}$$

$$= \frac{\delta x}{\delta x (\sqrt{x}\sqrt{x + \delta x})(\sqrt{x} + \sqrt{x + \delta x})}$$

Taking the limit as $\delta x \to 0$. We get

$$\lim_{\delta x \to 0} \frac{\delta y}{\delta x} = \lim_{\delta x \to 0} \frac{-1}{(\sqrt{x}\sqrt{x + \delta x})(\sqrt{x} + \sqrt{x + \delta x})}$$

$$\therefore \qquad \frac{dy}{dx} = \frac{-1}{(\sqrt{x}\sqrt{x + 0})(\sqrt{x} + \sqrt{x + 0})} = \frac{-1}{x \times 2\sqrt{x}} = \frac{-1}{2x^{3/2}}$$

EXAMPLE 5. *Find the differential coefficient of* $\sqrt{2x + 3}$ *w.r. to x.*

SOLUTION. Let $\qquad y = \sqrt{2x + 3} = (2x + 3)^{1/2}$

$$\therefore \qquad y + \delta y = [2(x + \delta x) + 3]^{1/2}$$

$$\therefore \qquad \delta y = [2(x + \delta x) + 3]^{1/2} - (2x + 3)^{1/2}$$

$$= \frac{[2(x + \delta x) + 3]^{1/2} - (2x + 3)^{1/2}}{1} \times \frac{[2(x + \delta x) + 3]^{1/2} + (2x + 3)^{1/2}}{[2(x + \delta x) + 3]^{1/2} + (2x + 3)^{1/2}}$$

$$= \frac{[2(x + \delta x) + 3] - (2x + 3)}{[2(x + \delta x) + 3]^{1/2} + (2x + 3)^{1/2}} \times \frac{2x + 2\delta x + 3 - 2x - 3}{(2x + 2\delta x + 3)^{1/2} + (2x + 3)^{1/2}}$$

$$\text{or} \qquad \frac{\delta y}{\delta x} = \frac{2\delta x}{[2x + 3 + 2\delta x]^{1/2} + (2x + 3)^{1/2}]\delta x}$$

Taking the limit as $\delta x \to 0$, we get

$$\lim_{\delta x \to 0} \frac{\delta y}{\delta x} = \lim_{\delta x \to 0} \frac{2}{(2x + 3 + 2\delta x)^{1/2} + (2x + 3)^{1/2}}$$

$$\Rightarrow \qquad \frac{dy}{dx} = \frac{2}{(2x + 3)^{1/2} + (2x + 3)^{1/2}} = \frac{2}{2(2x + 3)^{1/2}} = \frac{1}{\sqrt{2x + 3}}$$

EXAMPLE 6. *Differentiate* $\tan^{-1} x^2$ *from first principle.*

SOLUTION. Let $\qquad y = \tan^{-1} x^2 \Rightarrow x^2 = \tan y$

Let $\qquad\qquad t = \tan y$

$\Rightarrow \qquad\qquad t + \delta t = \tan(y + \delta y)$

$\Rightarrow \qquad\qquad \delta t = \tan(y + \delta y) - \tan y$

$\Rightarrow \qquad \dfrac{\delta t}{\delta y} = \dfrac{\sin(y + \delta y)\cos y - \sin y \cos(y + \delta y)}{\delta y \cos y . \cos(y + \delta y)} = \dfrac{\sin(y + \delta y - y)}{\delta y \cos y \cos(y + \delta y)}$

$\Rightarrow \qquad \dfrac{\delta y}{\delta t} = \dfrac{\delta y \cos y \cdot \cos(y + \delta y)}{\sin \delta y}$

Taking limit as $\delta t \to 0$, we get

$$\frac{dy}{dt} = \lim_{\delta t \to 0} \frac{\delta y}{\delta t} = \lim_{\delta t \to 0} \left[\frac{\delta y}{\sin \delta y} \cdot \cos(y + \delta y) \cdot \cos y \right]$$

$$= \left[\lim_{\delta x \to 0} \frac{\delta y}{\sin \delta y} \right] \left[\lim_{\delta y \to 0} \cos(y + \delta y) \right] \cos y$$

$$\frac{dy}{dt} = 1 \cdot \cos y \cdot \cos y = \cos^2 y$$

$\Rightarrow \qquad \dfrac{dy}{dx} = 2x \cos^2 y \qquad\qquad\qquad [\because x^2 = t \Rightarrow 2x dx = dt]$

$$= \frac{2x}{\sec^2 y} = \frac{2x}{1 + \tan^2 y} = \frac{2x}{1 + (x^2)^2}$$

Hence $\dfrac{d}{dx}(\tan^{-1} x^2) = \dfrac{2x}{1 + x^4}$.

9.5 DERIVATIVE OF THE PRODUCT TO TWO FUNCTIONS

Let $\quad y = f(x) = \phi_1(x)\phi_2(x)$, then

$\qquad f(x + \delta x) = \phi_1(x + \delta x)\phi_2(x + \delta x)$.

So, $\qquad \dfrac{dy}{dx} = \dfrac{df(x)}{dx} = \lim_{\delta x \to 0} \dfrac{f(x + \delta x) - f(x)}{\delta x}$

$$= \lim_{\delta x \to 0} \frac{\phi_1(x + \delta x)\phi_2(x + \delta x) - \phi_1(x)\phi_2(x)}{\delta x}$$

$$\qquad\qquad\qquad [\text{Adding and subtracting } \phi_1(x + \delta x)\phi_2(x)]$$

$$= \lim_{\delta x \to 0} \frac{\{\phi_1(x + \delta x)\phi_2(x) - \phi_1(x)\phi_2(x)\} + \{\phi_1(x + \delta x)\phi_2(x + \delta x) - \phi_1(x + \delta x) - \phi_2(x)\}}{\delta x}$$

$$= \lim_{\delta x \to 0} \frac{\{\phi_1(x + \delta x) - \phi_1(x)\}\phi_2(x) + \phi_1(x + \delta x)\{\phi_2(x + \delta x) - \phi_2(x)\}}{\delta x}$$

$$= \lim_{\delta x \to 0} \frac{\phi_1(x + \delta x) - \phi_1(x)}{\delta x}\phi_2(x) + \lim_{\delta x \to 0} \phi_1(x + \delta x)\left\{\frac{\phi_2(x + \delta x) - \phi_2(x)}{\delta x}\right\}$$

$$= \frac{d\phi_1(x)}{dx}\phi_2(x) + \phi_1(x)\frac{d}{dx}\phi_2(x) = \phi_1(x)\frac{d\phi_2(x)}{dx} + \frac{d\phi_1 x}{dx}\phi_2(x).$$

Thus $\quad \dfrac{d}{dx}\{\phi_1(x) \cdot \phi_2(x)\} = \phi_1(x)\dfrac{d[\phi_2(x)]}{dx} + \dfrac{d}{dx}\phi_1(x) \cdot \phi_2(x).$

Hence, The derivative of the product of two functions

$\qquad\qquad$ = First function × derivative of the second function

$\qquad\qquad\qquad$ + derivative of the first function × second function.

9.6 DERIVATIVE OF THE QUOTIENT OF TWO FUNCTIONS

Let $\qquad\qquad y = f(x) = \dfrac{\phi_1(x)}{\phi_2(x)} \quad$ then $\quad y = f(x + \delta x) = \dfrac{\phi_1(x + \delta x)}{\phi_2(x + \delta x)}$

So, $\qquad \dfrac{dy}{dx} = \dfrac{df(x)}{dx} = \lim_{\delta x \to 0} \dfrac{f(x + \delta x) - f(x)}{\delta x} = \lim_{\delta x \to 0} \dfrac{\dfrac{\phi_1(x + \delta x)}{\phi_2(x + \delta x)} - \dfrac{\phi_1(x)}{\phi_2(x)}}{\delta x}$

$\qquad\qquad = \lim_{\delta x \to 0} \dfrac{\dfrac{\phi_1(x + \delta x)}{\phi_2(x + \delta x)} - \dfrac{\phi_1(x)}{\phi_2(x + \delta x)} + \dfrac{\phi_1(x)}{\phi_2(x + \delta x)} + \dfrac{\phi_1(x)}{\phi_2(x)}}{\delta x}$

$$\left[\text{Adding and subtracting } \dfrac{\phi_1(x)}{\phi_2(x + \delta x)}\right]$$

$\qquad\qquad = \lim_{\delta x \to 0} \dfrac{\dfrac{1}{\phi_2(x + \delta x)}\{\phi_1(x + \delta x) - \phi_1(x)\} - \dfrac{\phi_1(x)}{\phi_2(x + \delta x)\phi_2(x)}\{\phi_2(x + \delta x) - \phi_2(x)\}}{\delta x}$

$\qquad\qquad = \lim_{\delta x \to 0} \dfrac{1}{\phi_2(x + \delta x)}\left\{\dfrac{\phi_1(x + \delta x) - \phi_1(x)}{\delta x}\right\}$

$\qquad\qquad\qquad - \lim_{\delta x \to 0} \dfrac{\phi_1(x)}{\phi_2(x + \delta x)\phi_2(x)}\left\{\dfrac{\phi_2(x + \delta x) - \phi_2(x)}{\delta x}\right\}$

$\qquad\qquad = \dfrac{1}{\phi_2(x)}\dfrac{d}{dx}\phi_1(x) - \dfrac{\phi_1(x)}{\phi_2(x)\phi_2(x)}\dfrac{d}{dx}\phi_2(x)$

$\qquad\qquad = \dfrac{\phi_2(x)\dfrac{d}{dx}\phi_1(x) - \phi_1(x)\dfrac{d}{dx}\phi_2(x)}{[\phi_2(x)]^2}$

Thus, $\qquad \dfrac{d}{dx}\left[\dfrac{\phi_1(x)}{\phi_2(x)}\right] = \dfrac{\phi_2(x)\dfrac{d}{dx}\phi_1(x) - \phi_1(x)\dfrac{d}{dx}\phi_2(x)}{[\phi_2(x)]^2}.$

SOLVED EXAMPLES

EXAMPLE 1. *Find the derivative of the following functions.*

$\quad$ (i) $\ y = x^6$ $\qquad\qquad$ (ii) $\ y = x^3 + 2x$ $\qquad$ (iii) $\ y = -2x^2 + 4x + 6$

SOLUTION. $\quad$ (i) Here, we have $y = x^6$

$\qquad\qquad$ So, $\quad \dfrac{dy}{dx} = \dfrac{d}{dx}(x^6) = 6x^5$ $\qquad\qquad\qquad \left[\because \dfrac{d}{dx}(x^n) = nx^{n-1}\right]$

(ii) Here, we have $y = x^3 + 2x$

$$\frac{dy}{dx} = \frac{d}{dx}(x)^3 + 2\frac{d}{dx}x$$

$$= 3x^2 + 2 \qquad \qquad \left[\because \frac{d}{dx}(x) = 1\right]$$

(iii) Here, $\qquad y = -2x^2 + 4x + 6$

$$\frac{dy}{dx} = -2\frac{d}{dx}(x^2) + 4\frac{d}{dx}(x) + \frac{d}{dx}(6)$$

$$= -4x + 4 \qquad \qquad \left[\because \frac{d}{dx}(6) = 0\right]$$

EXAMPLE 2.   *Find the differential coefficients or derivatives of the following functions :*

     (*i*) ax^7 (*ii*) $8e^x$ (*iii*) $4\log x$

     (*iv*) $5\log_{10} x$ (*v*) 10^x

SOLUTION.   (*i*) Here, $\qquad y = ax^7$.

        So, $\qquad \dfrac{dy}{dx} = a\dfrac{d}{dx}(x^7) = 7ax^6$

    (*ii*) Here, $\qquad y = 8e^x$

        So, $\qquad \dfrac{dy}{dx} = 8\dfrac{d}{dx}e^x = 8e^x$.

    (*iii*) Here, $\qquad y = 4\log x$.

        So, $\qquad \dfrac{dy}{dx} = 4\dfrac{d}{dx}(\log x) = \dfrac{4}{x}$

    (*iv*) Here, $\qquad y = 5\log_{10} x$

        So, $\qquad \dfrac{dy}{dx} = 5\dfrac{d}{dx}\log_{10} x = 5\dfrac{1}{x}\log_{10}(e)$

    (*v*) Here, $\qquad y = 10^x$

        So, $\qquad \dfrac{dy}{dx} = \dfrac{d}{dx}(10^x) = 10^x \log_e 10$

EXAMPLE 3.   *Find the differential coefficient or derivative of the following functions.*

     (*i*) $4\cos x$ (*ii*) $5\sin^{-1} x$ (*iii*) 7^x

     (*iv*) ab^x (*v*) $7\operatorname{cosec} x$

SOLUTION.   (*i*) Here, $\qquad y = 4\cos x$.

        So, $\qquad \dfrac{dy}{dx} = 4\dfrac{d}{dx}(\cos x) = -4\sin x$

    (*ii*) Here, $\qquad y = 5\sin^{-1} x$

        So, $\qquad \dfrac{dy}{dx} = 5\dfrac{d}{dx}(\sin^{-1} x) = \dfrac{5}{\sqrt{1-x^2}}$.

    (*iii*) Here, $\qquad y = 7^x$

        So, $\qquad \dfrac{dy}{dx} = \dfrac{d}{dx}(7^x) = 7^x \log_e 7$

(iv) Here, $y = ab^x$

So, $\dfrac{dy}{dx} = a\dfrac{d}{dx}b^x = ab^x \log_e b$

(v) Here, $y = 7\operatorname{cosec} x$

So, $\dfrac{dy}{dx} = 7\dfrac{d}{dx}(\operatorname{cosec} x) = -7\operatorname{cosec} x \cot x$

EXAMPLE 4. *Find the derivative of the following functions.*

(i) $y = e^x + \cos x + 5\log x$ (ii) $y = 8^x + 8\log_a x + \cos^{-1} x$

(iii) $y = 3x^2 + 6x + 10$ (iv) $y = \tan x + \sin x + \cos x.$

SOLUTION. (i) Here, we have $y = e^x + \cos x + 5\log x.$

So, $\dfrac{dy}{dx} = \dfrac{d}{dx}e^x + \dfrac{d}{dx}\cos x + 5\dfrac{d}{dx}\log x$

$= e^x - \sin x + \dfrac{5}{x}.$

(ii) We have $y = 8^x + 8\log_a x + \cos^{-1} x$

So, $\dfrac{dy}{dx} = \dfrac{d}{dx}8^x + 8\dfrac{d}{dx}\log_a x + \dfrac{d}{dx}(\cos^{-1} x)$

$= 8^x \log_e 8 + 8\dfrac{1}{x}\log_a e - \dfrac{1}{\sqrt{1-x^2}}$

(iii) We have $y = 3x^2 + 6x + 10$

So, $\dfrac{dy}{dx} = 3\dfrac{d}{dx}x^2 + 6\dfrac{d}{dx}x + \dfrac{d}{dx}10 = 6x + 6$ $\left[\because \dfrac{d}{dx}(10) = 0\right]$

(iv) Here, We have $y = \tan x + \sin x + \cos x.$

So, $\dfrac{dy}{dx} = \dfrac{d}{dx}\tan x + \dfrac{d}{dx}\sin x + \dfrac{d}{dx}\cos x$

$= \sec^2 x + \cos x - \sin x$

EXAMPLE 5. *Find the differential coefficient of the following functions.*

(i) $a^x + \log x + a^2 e^x$ (ii) $x^3 \sin x$ (iii) $x^3 \log x$

(iv) $x^{-5} \cot x$ (v) $e^x \cos x$

SOLUTION. (i) $\dfrac{d}{dx}(a^x + \log x + a^2 e^x) = \dfrac{d}{dx}a^x + \dfrac{d}{dx}\log x + a^2\dfrac{d}{dx}e^x$

$= a^x \log_e a + \dfrac{1}{x} + a^2 e^x$

(ii) $\dfrac{d}{dx}(x^3 \sin x) = x^3\dfrac{d}{dx}\sin x + \dfrac{d}{dx}(x^3)\sin x = x^3 \cos x + 3x^2 \sin x.$

(iii) $\dfrac{d}{dx}(x^3 \log x) = x^3\dfrac{d}{dx}(\log x) + \dfrac{d}{dx}(x^3)\log x = x^3\dfrac{1}{x} + 3x^2 \log x$

$= x^2 + 3x^2 \log x = x^2.(1 + 3\log x)$

(iv) $\dfrac{dy}{dx}(x^{-5} \cot x) = x^{-5}\dfrac{d}{dx}\cot x + \dfrac{d}{dx}(x^{-5})\cot x - 5x^{-6}\cot x$

$= x^{-5}(-\operatorname{cosec}^2 x) = x^{-5}\operatorname{cosec}^2 x - 5x^{-6}\cot x$

$= -x^6(x\operatorname{cosec}^2 x + 5\cot x).$

(v) $\quad \dfrac{d}{dx}(e^x \cos x) = e^x \dfrac{d}{dx}\cos x + \cos x \dfrac{d}{dx}(e^x)$

$$= e^x(-\sin x) + e^x \cos x = e^x(\cos x - \sin x)$$

EXAMPLE 6. *Find the differential coefficient or derivative of the following functions.*

(i) $\dfrac{\log x}{x}$ $\qquad$ (ii) $\dfrac{\sin x}{x^2}$ $\qquad$ (iii) $\tan x$ $\qquad$ (iv) $\dfrac{x}{(x+1)^2}$.

SOLUTION. (i) $\quad \dfrac{d}{dx}\left(\dfrac{\log x}{x}\right) = \dfrac{x\dfrac{d}{dx}(\log x) - \log x \dfrac{d}{dx}(x)}{x^2}$

$$= \dfrac{x\dfrac{1}{x} - \log x \cdot 1}{x^2} = \dfrac{1 - \log x}{x^2}.$$

(ii) $\quad \dfrac{d}{dx}\left(\dfrac{\sin x}{x^2}\right) = \dfrac{x^2 \dfrac{d}{dx}\sin x - \sin x \dfrac{d}{dx}x^2}{x^4} = \dfrac{x^2 \cos x - \sin x\, 2x}{x^4}$

$$= \dfrac{x(x\cos x - 2\sin x)}{x^4} = \dfrac{x\cos x - 2\sin x}{x^3}$$

(iii) $\quad \dfrac{d}{dx}(\tan x) = \dfrac{d}{dx}\left(\dfrac{\sin x}{\cos x}\right) = \dfrac{\cos x \dfrac{d}{dx}\sin x - \sin x \dfrac{d}{dx}\cos x}{\cos^2 x}$

$$= \dfrac{\cos^2 x + \sin^2 x}{\cos^2 x} = \dfrac{1}{\cos^2 x} = \sec^2 x \quad (\because\ \sin^2 x + \cos^2 x = 1)$$

(iv) $\quad \dfrac{d}{dx}\left(\dfrac{x}{(x+1)^2}\right) = \dfrac{(x+1)^2 \dfrac{d}{dx}(x) - x\dfrac{d}{dx}(x+1)^2}{(x+1)^4}$

$$= \dfrac{(x+1)^2 1 - x.2(x+1)}{(x+1)^4} = \dfrac{(x+1) - 2x}{(x+1)^3} = \dfrac{1-x}{(x+1)^3}.$$

9.7 DERIVATIVE OF FUNCTIONS OF A FUNCTION (CHAIN RULE)

If y is a function of u, say $y = f(u)$, where u is a function of x, say $u = g(x)$, then y is called a function of a function or a composite function of x.

WORKING PROCEDURE

To find the derivative of a composite function of x, say $y = f[g(x)]$, we use the following steps

Step 1. Put inner function $g(x) = u$, *i.e.,* $y = f(u)$.

Step 2. Differentiate with respect to u, find $\dfrac{dy}{du}$.

Step 3. Differentiate u with respect to x and find $\dfrac{du}{dx}$.

Step 4. Multiply $\dfrac{dy}{du}$ and $\dfrac{du}{dx}$ and find $\dfrac{dy}{dx} = \dfrac{dy}{du}\cdot\dfrac{du}{dx}$.

Step 5. Put the value of $u = g(x)$, we get the required derivative.

THEOREM 1. *If y is a function of u and u is a function of x, then*

$$\frac{dy}{dx} = \frac{dy}{du} \cdot \frac{du}{dx}.$$

PROOF. Here, y is a function of u and u is a function of x.

i.e., $\qquad y = f(u)$ where $u = g(x)$.

To find the derivative of $y = f[g(x)]$. Let δx be the increment in the value of x and δy be the corresponding increment in the value of y. Now, since $y = f(u)$ then $y + \delta y = f(u + \delta u)$

$$\Rightarrow \qquad \delta y = f(u + \delta u) - f(u)$$

So, $\qquad \dfrac{\delta y}{\delta x} = \dfrac{f(u + \delta u) - f(u)}{\delta u} \dfrac{\delta u}{\delta x}.$

Taking $\delta u \neq 0$ when $\delta x \neq 0$ if $\delta x \to 0$ then δu is also tends to 0.

Then, we have

$$\lim_{\delta x \to 0} \frac{\delta y}{\delta x} = \lim_{\delta x \to 0} \frac{f(u + \delta u) - f(u)}{\delta u} \lim_{\delta x \to 0} \frac{\delta u}{\delta x}$$

$$\Rightarrow \qquad \frac{dy}{dx} = \frac{d}{du}[f(u)]\frac{du}{dx} = \frac{dy}{du} \cdot \frac{du}{dx}$$

Hence, $\qquad \dfrac{dy}{dx} = \dfrac{dy}{du} \cdot \dfrac{du}{dx}.$

 REMARK

➠ The above result can be generalised as follows:

If y is a function of u, u is a function of v, v is a function of w and so on, then we have

$$\frac{dy}{dx} = \frac{dy}{du} \cdot \frac{du}{dv} \cdot \frac{dv}{dw} \cdots \frac{d}{dx}.$$

9.7.1 PARAMETRIC EQUATIONS

If both x and y are the function of the third variable u, such that $x = f(u)$ and $y = g(u)$ where u is a parameter and the above two relations is known as parametric equations.

In parametric equations the value of $\dfrac{dy}{dx}$ is $\dfrac{dy / du}{dx / du}$

i.e., $\qquad \dfrac{dy}{dx} = \dfrac{dy / du}{dx / du},$

where u is a parameter.

THEOREM 2. *If x is a function of t, i.e., x = f(t) and y is also a function of t, i.e., y = g(t) then*

$$\frac{dy}{dx} = \frac{dy / dt}{dx / dt}, \text{where t is a parameter.}$$

PROOF. Let δt be the increment in the value of t and let δx and δy be the corresponding increment in x and y. Now, if $\delta t \to 0$ then $\delta x \to 0$ and $\delta y \to 0$. Now, we have

$$\frac{\delta y}{\delta x} = \frac{\delta y}{\delta t} \cdot \frac{\delta t}{\delta x}.$$

Take the limit as $\delta t \to 0$ then $\delta x \to 0$ and $\delta y \to 0$, we have

$$\lim_{\delta x \to 0} \frac{\delta y}{\delta x} = \lim_{\delta t \to 0} \frac{\delta y / \delta t}{\delta x / \delta t} = \frac{\displaystyle\lim_{\delta t \to \infty} \delta y / \delta t}{\displaystyle\lim_{\delta t \to 0} \delta x / \delta t}$$

So, $\qquad \dfrac{dy}{dx} = \dfrac{dy / dt}{dx / dt}.$

SOLVED EXAMPLES

EXAMPLE 1. *Find the derivative of the following functions.*

        (i) $\log \sin x$ (ii) $\cos x^3$ (iii) $\sin 5x$.

SOLUTION.   (i) Here, $y = \log \sin x$.

           Let $\sin x = t$

           then $y = \log t$.

           Now, $\dfrac{dy}{dt} = \dfrac{1}{t}$ and $\dfrac{dt}{dx} = \cos x$

           So, $\dfrac{dy}{dx} = \dfrac{dy}{dt}\dfrac{dt}{dx} = \dfrac{\cos x}{t} = \dfrac{\cos x}{\sin x} = \cot x$

     (ii) Here, $y = \cos x^3$.

           Put $x^3 = t$ then $y = \cos t$

           Now, $\dfrac{dy}{dt} = -\sin t$ and $\dfrac{dt}{dx} = 3x^2$

           So, $\dfrac{dy}{dx} = \dfrac{dy}{dt} \cdot \dfrac{dt}{dx} = -\sin t\, 3x^2$

                        $= -3x^2 \sin t = -3x^2 \sin x^3$

    (iii) Here, $y = \sin 5x$

           Put $5x = t$ then $y = \sin t$

           Now, $\dfrac{dy}{dt} = \cos t$ and $\dfrac{dt}{dx} = 5$

           So, $\dfrac{dy}{dx} = \dfrac{dy}{dt}\dfrac{dt}{dx} = \cos t \cdot 5 = 5 \cos 5x$.

EXAMPLE 2. *Find the differential coefficient (or derivative) of the following with respect to x :*

        (i) $\tan^3 x$ (ii) $y = (3x^3 - 5x + 8)^3$.

SOLUTION.   (i) Here, $y = \tan^3 x$

           Put $\tan x = t$ then; $y = t^3$

           Now, $\dfrac{dy}{dt} = 3t^2$ and $\dfrac{dt}{dx} = \sec^2 x$

           Therefore, $\dfrac{dy}{dx} = \dfrac{dy}{dt}\dfrac{dt}{dx} = 3t^2 \sec^2 x = 3\tan^2 x \sec^2 x$

     (ii) Here, $y = (3x^3 - 5x^2 + 8)^3$

           Put $t = 3x^3 - 5x^2 + 8$;

           then $y = t^3$.

           Now, $\dfrac{dy}{dt} = 3t^2$ and $\dfrac{dt}{dx} = 9x^2 - 10x$.

           So, $\dfrac{dy}{dx} = \dfrac{dy}{dt}\dfrac{dt}{dx} = 3t^2(9x^2 - 10x)$

           Hence, $\dfrac{dy}{dx} = 3(3x^3 - 5x^2 + 8)^2(9x^2 - 10x)$

EXAMPLE 3.   Find $\dfrac{dy}{dx}$ of the following functions

(i) $x = at^2, y = 2at$ (ii) $x = a\cos t, y = a\sin t$

(iii) $x = a(\cos t + \log\tan t/2), y = a\sin t$

(iv) $x = a\cos^2 t, y = a\sin^2 t.$

SOLUTION.   (i) Here $x = at^2$ and $y = 2at$

then, $\dfrac{dx}{dt} = 2at$ and $\dfrac{dy}{dt} = 2a.$

Therefore, $\dfrac{dy}{dx} = \dfrac{dy/dt}{dx/dt} = \dfrac{2a}{2at} = \dfrac{1}{t}.$

(ii) Here, $x = a\cos t$ and $y = a\sin t.$

Now, $\dfrac{dx}{dt} = -a\sin t$ and $\dfrac{dy}{dt} = a\cos t.$

So, $\dfrac{dy}{dx} = \dfrac{dy/dt}{dx/dt} = \dfrac{a\cos t}{-a\sin t} = -\cot t$

(iii) Here, $x = a(\cos t + \log\tan t/2), y = a\sin t.$

Now, $\dfrac{dx}{dt} = a\left(-\sin t + \dfrac{1}{\tan t/2}\dfrac{d}{dx}(\tan t/2)\right)$ and $\dfrac{dy}{dt} = a\cos t$

$= a\left(-\sin t + \dfrac{1}{\tan t/2}\sec^2 t/2 \cdot \dfrac{1}{2}\right)$

$= a\left(-\sin t + \dfrac{1}{2\tan t/2\cos^2 t/2}\right)$

$= a\left(-\sin t + \dfrac{1}{2\sin t/2\cos t/2}\right) = a\left(-\sin t + \dfrac{1}{\sin t}\right)$

$= a\left(\dfrac{1-\sin^2 t}{\sin t}\right) = \dfrac{a\cos^2 t}{\sin t}$

So, $\dfrac{dy}{dx} = \dfrac{dy/dt}{dx/dt} = \dfrac{a\cos t}{\dfrac{a\cos^2 t}{\sin t}} = \dfrac{\sin t}{\cos t} = \tan t.$

(iv) Here, $x = a\cos^2 t, y = a\sin^2 t$

Now $\dfrac{dx}{dt} = 2a\cos t\dfrac{d}{dt}\cos t = 2a\cos t(-\sin t)$

$= -a2\sin t\cos t = -a\sin 2t$

and $\dfrac{dy}{dt} = 2a\sin t\left(\dfrac{d}{dt}\sin t\right) = 2a\sin t\cos t = a\sin 2t$

So, $\dfrac{dy}{dx} = \dfrac{dy/dt}{dx/dt} = \dfrac{a\sin 2t}{-a\sin 2t} = -1$

So, $\dfrac{dy}{dx} = -1$

EXAMPLE 4. *Find the derivative of the following functions w.r. to x*

(i) $\sin e^x \log x$　　(ii) $\sqrt{(x^2 + 5x + 6)}$　　(iii) $\dfrac{1 + e^x}{1 - e^x}$

(iv) $\sin^2 x \cos^2 x$　　(v) $(1 + \sin x)^3$.

SOLUTION.　(i) Here,　$y = \sin e^x \log x$

So,　$\dfrac{dy}{dx} = \dfrac{d}{dx}(\sin e^x \log x)$

$$= \sin e^x \dfrac{d}{dx} \log x + \dfrac{d}{dx}\{\sin(e^x)\}\log x$$

$$= \sin e^x \dfrac{1}{x} + \cos(e^x)\dfrac{d}{dx}e^x \cdot \log x$$

$$= \dfrac{1}{x}\sin e^x + \cos(e^x)\log x \cdot e^x.$$

(ii) Here,　$y = \sqrt{(x^2 + 5x + 6)}$

Put $x^2 + 5x + 6 = t$ then $y = \sqrt{t}$

$\dfrac{dy}{dt} = \dfrac{1}{2}t^{-1/2}$ and $\dfrac{dt}{dx} = 2x + 5$.

So,　$\dfrac{dy}{dx} = \dfrac{dy}{dt}\cdot\dfrac{dt}{dx} = \dfrac{1}{2}t^{-1/2}(2x+5) = \dfrac{1}{2}(x^2 + 5x + 6)^{-1/2}(2x + 5).$

(iii) Here,　$y = \dfrac{1 + e^x}{1 - e^x}$

So,　$\dfrac{dy}{dx} = \dfrac{d}{dx}\left(\dfrac{1 + e^x}{1 - e^x}\right) = \dfrac{(1 - e^x)\dfrac{d}{dx}(1 + e^x) - (1 + e^x)\dfrac{d}{dx}(1 - e^x)}{(1 - e^x)^2}$

$$= \dfrac{(1 - e^x)e^x - (1 + e^x)(-e^x)}{(1 - e^x)^2} = \dfrac{e^x - e^{2x} + e^x + e^{2x}}{(1 - e^x)^2} = \dfrac{2e^x}{(1 - e^x)^2}$$

(iv) Here,　$y = \sin^2 x \cos^2 x$

So,　$\dfrac{dy}{dx} = \dfrac{d}{dx}(\sin^2 x \cos^2 x) = \sin^2 x \dfrac{d}{dx}(\cos^2 x) + \dfrac{d}{dx}(\sin^2 x)\cos^2 x$

$$= \sin^2 x\, 2\cos x \dfrac{d}{dx}(\cos x) + 2\sin x \dfrac{d}{dx}(\sin x)\cos^2 x$$

$$= 2\sin^2 x \cos x(-\sin x) + 2\sin x \cos x \cos^2 x$$

$$= -2\sin^3 x \cos x + 2\sin x \cos^3 x = 2\sin x \cos x(\cos^2 x - \sin^2 x)$$

$$= \sin 2x \cos 2x$$

$\qquad\qquad (\because \cos^2 x - \sin^2 x = \cos 2x$ and $2\sin x \cos x = \sin 2x)$

(v) Here,　$y = (1 + \sin x)^3$

So,　$\dfrac{dy}{dx} = \dfrac{d}{dx}(1 + \sin x)^3 = 3(1 + \sin x)^2 \dfrac{d}{dx}(\sin x)$

$$= 3(1 + \sin x)^2 \cos x$$

IInd method: Put $1 + \sin x = t$ then $y = t^3$.

Now,　$\dfrac{dy}{dx} = 3t^2$ and $\dfrac{dt}{dx} = \cos x$.

So, $\dfrac{dy}{dx} = \dfrac{dy}{dt}\dfrac{dt}{dx} = 3t^2\cos x = 3(1+\sin x)^2 \cos x$

So, $\dfrac{d}{dx}(1+\sin x)^3 = 3(1+\sin x)^2 \cos x$

EXAMPLE 5. If $y = \sin\sqrt{\sin x}$ find $\dfrac{dy}{dx}$.

SOLUTION. We have $y = \sin\sqrt{\sin x}$

Differentiating w.r.t x we get

$$\dfrac{dy}{dx} = (\cos\sqrt{\sin x})\cdot\dfrac{1}{2\sqrt{\sin x}}\cdot\cos x = \dfrac{\cos x(\cos\sqrt{\sin x})}{2\sqrt{\sin x}}$$

9.8 DIFFERENTIATION OF IMPLICIT FUNCTIONS

A function which can be expressed in terms of independent variable x is known as explicit function. On the other hand, a function which is not explicit, is known as implicit function, or we can say a function which cannot be expressed directly in terms of independent variable x is implicit function. For example, $x^y + y^x = a$ is an implicit function, because this function cannot be expressed in terms of x. To find the $\dfrac{dy}{dx}$ of implicit function, we differentiate each term with respect to x treating y as a function of x and then separating $\dfrac{dy}{dx}$. A method by which we find the $\dfrac{dy}{dx}$ of implicit function is known as implicit differentiation.

SOLVED EXAMPLES

EXAMPLE 1. Find $\dfrac{dy}{dx}$ of the implicit function

$$ax^2 + 2hxy + by^2 + 2gx + 2fy + c = 0.$$

SOLUTION. Differentiating the given equation with respect to x, we get

$$2ax + 2h\left(x\dfrac{dy}{dx} + y\right) + 2by\dfrac{dy}{dx} + 2g + 2f\dfrac{dy}{dx} = 0$$

$$\Rightarrow \dfrac{dy}{dx}(hx + by + f) = -(ax + hy + g)$$

$$\Rightarrow \dfrac{dy}{dx} = -\dfrac{ax + hy + g}{hx + by + f}.$$

EXAMPLE 2. Find $\dfrac{dy}{dx}$ of the function $x\sqrt{1+y} + y\sqrt{1+x} = 0$.

SOLUTION. Here, $x\sqrt{1+y} + y\sqrt{1+x} = 0$

$$\Rightarrow x\sqrt{1+y} = -y\sqrt{1+x}.$$

On squaring, we get

$$x^2(1+y) = y^2(1+x) \quad \text{or} \quad (x^2-y^2) + (x^2y - xy^2) = 0$$

$$\Rightarrow (x^2-y^2) + xy(x-y) = 0$$

$$\Rightarrow (x+y)(x-y) + xy(x-y) = 0$$

$$\Rightarrow x + y + xy = 0$$

or $\quad y(1+x) + x = 0$

$$\Rightarrow \qquad y = -\frac{x}{1+x}.$$

Now, differentiating both sides, with respect to x, we get

$$\frac{dy}{dx} = \frac{(1+x)\dfrac{d}{dx}(-x)-(-x)\dfrac{d}{dx}(1+x)}{(1+x)^2} = \frac{-(1+x)+x}{(1+x)^2} = -\frac{1}{(1+x)^2}.$$

EXAMPLE 3. Find $\dfrac{dy}{dx}$ of the function $x^3 - xy^2 + 3y^2 + 2 = 0$.

SOLUTION. Differentiating each term of the given equation with respect to x, we get

$$3x^2 - \left(x \cdot 2y\frac{dy}{dx} + y^2\right) + 6y\frac{dy}{dx} = 0$$

$$\Rightarrow \qquad 3x^2 - 2xy\frac{dy}{dx} - y^2 + 6y\frac{dy}{dx} = 0$$

$$\Rightarrow \qquad 3x^2 - y^2 + \frac{dy}{dx}(6y - 2xy) = 0$$

$$\Rightarrow \qquad \frac{dy}{dx} = \frac{y^2 - 3x^2}{6y - 2xy}.$$

Hence, $\qquad \dfrac{dy}{dx} = \dfrac{y^2 - 3x^2}{2y(3-x)}.$

EXAMPLE 4. Find $\dfrac{dy}{dx}$ of the function $y = x^{x^{x \dots \infty}}$.

SOLUTION. Here, $\qquad y = x^{x^{x \dots \infty}}$

$$\Rightarrow \qquad y = x^y \Rightarrow \log y = y \log x$$

Now, differentiating with respect to x, we get

$$\frac{1}{y}\frac{dy}{dx} = y\frac{d}{dx}(\log x) + \frac{dy}{dx}\log x$$

or $\qquad \dfrac{1}{y}\dfrac{dy}{dx} = y\dfrac{1}{x} + \dfrac{dy}{dx}\log x$

$$\Rightarrow \qquad \left(\frac{1}{y} - \log x\right)\frac{dy}{dx} + \frac{y}{x} = 0$$

$$\Rightarrow \qquad \frac{dy}{dx} = \frac{y^2}{(1 - y\log x)x}.$$

EXAMPLE 5. Find $\dfrac{dy}{dx}$ of the function $\log xy = x^2 + y^2$.

SOLUTION. Here, $\log(xy) = x^2 + y^2$ or $\log x + \log y = x^2 + y^2$.
Now, differentiating both sides with respect to x, we get

$$\frac{1}{x} + \frac{1}{y}\frac{dy}{dx} = 2x + 2y\frac{dy}{dx}$$

or $\qquad \left(\dfrac{1}{y} - 2y\right)\dfrac{dy}{dx} = 2x - \dfrac{1}{x}$

or $\quad \dfrac{(1-2y^2)}{y}\dfrac{dy}{dx}=\dfrac{(2x^2-1)}{x}.$

Therefore, $\quad \dfrac{dy}{dx}=\dfrac{(2x^2-1)\,y}{(1-2y^2)\,x}.$

EXAMPLE 6. If $x^3+y^3=3axy$; find $\dfrac{dy}{dx}$.

SOLUTION. We have $\quad x^3+y^3=3axy$

Differentiating both sides with respect to x, we get

$$\dfrac{d}{dx}x^3+\dfrac{d}{dx}y^3=\dfrac{d}{dx}(3axy)$$

$$3x^2+3y^2\dfrac{dy}{dx}=3a\left\{x\dfrac{dy}{dx}+y\cdot1\right\}$$

$\Rightarrow \quad (3y^2-3ax)\dfrac{dy}{dx}=3ay-3x^2$

$\Rightarrow \quad 3(y^2-ax)\dfrac{dy}{dx}=3(ay-x^2)$

Hence, $\quad \dfrac{dy}{dx}=\dfrac{ay-x^2}{y^2-ax}$

EXAMPLE 7. Find $\dfrac{dy}{dx}$ when x and y are connected by following relations :

(i) $\tan(x+y)+\tan(x-y)=1$ (ii) $xy\log(x+y)=1$

SOLUTION. (i) We have $\quad \tan(x+y)+\tan(x-y)=1$

Differentiating both sides with respect to x, we get

$$\sec^2(x+y)\left[1+\dfrac{dy}{dx}\right]+\sec^2(x-y)\left[1-\dfrac{dy}{dx}\right]=0$$

$\Rightarrow \quad \dfrac{dy}{dx}[\sec^2(x+y)-\sec^2(x-y)=-[\sec^2(x+y)+\sec^2(x-y)]$

$\Rightarrow \quad \dfrac{dy}{dx}=-\dfrac{[\sec^2(x+y)+\sec^2(x-y)]}{\sec^2(x+y)-\sec^2(x-y)}$

(ii) $xy\log(x+y)=1$

Differentiating both sides w.r.t x, we get

$$xy\cdot\dfrac{1}{x+y}\left[1+\dfrac{dy}{dx}\right]+x\log(x+y)\dfrac{dy}{dx}+y\cdot\log(x+y)=0$$

$\Rightarrow \quad \dfrac{dy}{dx}\left[\dfrac{xy}{x+y}+x\log(x+y)\right]=\dfrac{-xy}{x+y}-y\log(x+y)$

$\Rightarrow \quad \dfrac{dy}{dx}[xy+x(x+y)\log(x+y)]=-[xy+y(x+y)\log(x+y)]$

$\Rightarrow \quad \dfrac{dy}{dx}=\dfrac{-[xy+y(x+y)\log(x+y)]}{[xy+(x+y)x\log(x+y)]}$

EXAMPLE 8. If $\sqrt{1-x^2} + \sqrt{1-y^2} = a(x-y)$, then prove that

$$\frac{dy}{dx} = \frac{\sqrt{1-y^2}}{\sqrt{1-x^2}}$$

SOLUTION.   We have $\sqrt{1-x^2} + \sqrt{1-y^2} = a(x-y)$

Putting $x = \sin\theta$ and $y = \sin\phi$, we get

$$\sqrt{1-\sin^2\theta} + \sqrt{1-\sin^2\phi} = a(\sin\theta - \sin\phi)$$

$$\Rightarrow \qquad\qquad\qquad \cos\theta + \cos\phi = a(\sin\theta - \sin\phi)$$

$$\Rightarrow \quad 2\cos\left(\frac{\theta+\phi}{2}\right)\cos\left(\frac{\theta-\phi}{2}\right) = a\left\{2\cos\left(\frac{\theta+\phi}{2}\right)\sin\left(\frac{\theta-\phi}{2}\right)\right\}$$

$$\Rightarrow \qquad\qquad\qquad \cot\left(\frac{\theta-\phi}{2}\right) = a$$

$$\Rightarrow \qquad\qquad\qquad \frac{\theta-\phi}{2} = \cot^{-1}a$$

$$\Rightarrow \qquad\qquad \sin^{-1}x - \sin^{-1}y = 2\cot^{-1}a$$

Differentiating w.r.t. x, we get

$$\frac{1}{\sqrt{1-x^2}} - \frac{1}{\sqrt{1-y^2}}\frac{dy}{dx} = 0$$

$$\frac{dy}{dx} = \sqrt{\frac{1-y^2}{1-x^2}}$$

EXAMPLE 9.   Find $\dfrac{dy}{dx}$; if $x^m y^n = (x+y)^{m+n}$.

SOLUTION.   We have $x^m y^n = (x+y)^{m+n}$

Taking log on both sides, we get

$$m\log x + n\log y = (m+n)\log(x+y)$$

Differentiating w.r.t. x both sides, we get

$$\frac{m}{x} + \frac{n}{y}\frac{dy}{dx} = (m+n)\cdot\frac{1}{x+y}\left(1+\frac{dy}{dx}\right)$$

$$\Rightarrow \qquad \left(\frac{n}{y} - \frac{m+n}{x+y}\right)\frac{dy}{dx} = \frac{m+n}{x+y} - \frac{m}{x}$$

$$\frac{n(x+y)-(m+n)y}{y(x+y)}\frac{dy}{dx} = \frac{(m+n)x - m(x+y)}{x(x+y)}$$

$$\Rightarrow \qquad\qquad \frac{nx - my}{y}\frac{dy}{dx} = \frac{nx - my}{x}$$

$$\frac{dy}{dx} = \frac{y}{x}$$

EXAMPLE 10. If $\sin y = x\sin(a+y)$, then prove that $\dfrac{dy}{dx} = \dfrac{\sin^2(a+y)}{\sin a}$.

SOLUTION. We have $\sin y = x \sin(a + y)$ ...(1)
Differentiating w.r.t x both sides, we get

$$\Rightarrow \qquad \cos y \frac{dy}{dx} = 1 \cdot \sin(a + y) + x \cos(a + y) \cdot \frac{d}{dx}(a + y)$$

$$\Rightarrow \qquad \cos y \frac{dy}{dx} = \sin(a + y) + x \cos(a + y) \frac{dy}{dx}$$

$$\Rightarrow \qquad \{\cos y - x \cos(a + y)\}\frac{dy}{dx} = \sin(a + y)$$

$$\Rightarrow \qquad \left\{\cos y - \frac{\sin y}{\sin(a + y)}\cos(a + y)\right\}\frac{dy}{dx} = \sin(a + y) \qquad \text{[From (1)]}$$

$$\Rightarrow \qquad \left\{\frac{\cos y \sin(a + y) - \sin y \cos(a + y)}{\sin(a + y)}\right\}\frac{dy}{dx} = \sin(a + y)$$

$$\Rightarrow \qquad \frac{\sin(a + y - y)}{\sin(a + y)}\frac{dy}{dx} = \sin(a + y)$$

$$\Rightarrow \qquad \frac{dy}{dx} = \frac{\sin^2(a + y)}{\sin a}$$

EXAMPLE 11. Find $\frac{dy}{dx}$, if $y^x = x^{\sin y}$

SOLUTION. We have $y^x = x^{\sin y}$
Taking log on both sides, we get

$$x \log y = \sin y \cdot \log x \qquad \qquad \text{...(1)}$$

Differentiating w.r.t x both sides, we get

$$1 \cdot \log y + \frac{x}{y}\frac{dy}{dx} = \frac{\sin y}{x} + \log x \cdot \cos y \cdot \frac{dy}{dx}$$

$$\Rightarrow \quad \frac{dy}{dx}\left(\frac{x}{y} - \log x . \cos y\right) = \frac{\sin y}{x} - \log y$$

$$\Rightarrow \quad \frac{dy}{dx}\left(\frac{x - y \log x . \cos y}{y}\right) = \frac{\sin y - x \log y}{x}$$

Hence, $$\frac{dy}{dx} = \frac{y(\sin y - x \cdot \log y)}{x(x - y \log x \cdot \cos y)}$$

EXAMPLE 12. If $x\sqrt{1 + y} + y\sqrt{1 + x} = 0$, prove that $\frac{dy}{dx} = -\frac{1}{(x + 1)^2}$.

SOLUTION. If $x\sqrt{1 + y} + y\sqrt{1 + x} = 0$

$$x\sqrt{1 + y} = -y\sqrt{1 + x}$$

Squaring both sides

$$x^2(1 + y) = y^2(1 + x)$$

$$x^2 - y^2 = y^2 x - xy^2$$

$$(x + y)(x - y) = -xy(x - y)$$

$$\Rightarrow \qquad x + y = -xy$$

$$\Rightarrow \qquad y + xy = -x$$

$$y = -\frac{x}{1 + x}$$

Differentiating w.r.t. x, we get

$$\frac{dy}{dx} = -\left\{\frac{(1+x)\cdot 1 - x(0+1)}{(1+y)^2}\right\} = -\frac{1}{(1+x)^2}$$

EXAMPLE 13. If $x^2 + y^2 = t - \dfrac{1}{t}$ and $x^4 + y^4 = t^2 + \dfrac{1}{t^2}$ then prove that

$$\frac{dy}{dx} = \frac{1}{x^3 y}.$$

SOLUTION. We have $x^2 + y^2 = t - \dfrac{1}{t}$

On Squaring

$$(x^2 + y^2) = \left(t - \frac{1}{t}\right)^2 = t^2 + \frac{1}{t^2} - 2$$

$$x^4 + y^4 + 2x^2 y^2 = x^4 + y^4 - 2$$

$$\Rightarrow \qquad 2x^2 y^2 = -2$$

$$\Rightarrow \qquad x^2 y^2 = -1$$

$$\Rightarrow \qquad y^2 = -\frac{1}{x^2}$$

Differentiating w.r.t. x both sides, we get

$$2y\frac{dy}{dx} = +\frac{2}{x^3}$$

$$\frac{dy}{dx} = \frac{1}{x^3 y}$$

EXAMPLE 14. If $\sqrt{y+x} + \sqrt{y-x} = c$ then prove that

$$\frac{dy}{dx} = \frac{y}{x} - \sqrt{\frac{y^2}{x^2} - 1}.$$

SOLUTION. We have $\sqrt{y+x} + \sqrt{y-x} = c$

Differentiating both sides w.r.t. x, we get

$$\frac{1}{2\sqrt{y+x}}\frac{d}{dx}(y+x) + \frac{1}{2\sqrt{y-x}}\frac{d}{dx}(y-x) = 0$$

$$\Rightarrow \qquad \frac{1}{2\sqrt{y+x}}\left(\frac{dy}{dx}+1\right) + \frac{1}{2\sqrt{y-x}}\left(\frac{dy}{dx}-1\right) = 0$$

$$\frac{dy}{dx}\left\{\frac{1}{\sqrt{y+x}} + \frac{1}{\sqrt{y-x}}\right\} = \frac{1}{\sqrt{y-x}} - \frac{1}{\sqrt{y+x}}$$

$$\Rightarrow \qquad \frac{dy}{dx}\left\{\frac{\sqrt{y-x}+\sqrt{y+x}}{\sqrt{y^2-x^2}}\right\} = \frac{\sqrt{y+x}-\sqrt{y-x}}{\sqrt{y^2-x^2}}$$

$$\Rightarrow \qquad \frac{dy}{dx} = \frac{\sqrt{y+x}-\sqrt{y-x}}{\sqrt{y+x}+\sqrt{y-x}} \times \frac{\sqrt{y+x}-\sqrt{y-x}}{\sqrt{y+x}-\sqrt{y-x}}$$

$$= \frac{(y+x)+(y-x)-2\sqrt{y^2-x^2}}{(y+x)-(y-x)} = \frac{2y-2\sqrt{y^2-x^2}}{2x}$$

$$= \frac{y}{x} - \sqrt{\frac{y^2}{x^2}-1}$$

EXAMPLE 15. If $x^{2/3} + y^{2/3} = 2$ find $\dfrac{dy}{dx}$ at $(1, 1)$.

SOLUTION.   We have $x^{2/3} + y^{2/3} = 2$

Differentiating both sides w.r.t. x, we get

$$\frac{2}{3}x^{-1/3} + \frac{2}{3}y^{-1/3}\frac{dy}{dx} = 0 \quad \Rightarrow \quad x^{-1/3} + y^{-1/3}\frac{dy}{dx} = 0$$

$$\frac{dy}{dx} = \frac{-x^{-1/3}}{y^{-1/3}} = -\left(\frac{y}{x}\right)^{1/3}$$

$$\Rightarrow \qquad \left(\frac{dy}{dx}\right)_{(1,1)} = -\left(\frac{1}{1}\right)^{1/3} = -1$$

EXAMPLE 16. If $y = \cos^{-1}\left(\dfrac{2\cos x + 3\sin x}{\sqrt{13}}\right)$, find $\dfrac{dy}{dx}$.

SOLUTION.   Differentiating the given function w.r.t. x, we get

$$\frac{dy}{dx} = \frac{-1}{\sqrt{1 - \dfrac{1}{13}(2\cos x + 3\sin x)^2}} \cdot \frac{1}{\sqrt{13}}(-2\sin x + 3\cos x)$$

$$= -\frac{(-2\sin x + 3\cos x)}{\sqrt{13}\sqrt{\dfrac{13 - (4\cos^2 x + 9\sin^2 x + 12\sin x \cos x)}{13}}}$$

$$= -\frac{(-2\sin x + 3\cos x)}{\sqrt{13 - 4 - 5\sin^2 x - 12\sin x \cos x}} = -\frac{(-2\sin x + 3\cos x)}{\sqrt{9 - 5\sin^2 x - 12\sin x \cos x}}$$

$$= -\frac{(-2\sin x + 3\cos x)}{\sqrt{(9\cos^2 x + 4\sin^2 x - 12\sin x \cos x)}}$$

$$= \frac{(2\sin x - 3\cos x)}{\sqrt{(2\sin x - 3\cos x)^2}} = \frac{2\sin x - 3\cos x}{2\sin x - 3\cos x} = 1$$

9.9 LOGARITHMIC DIFFERENTIATION

To find the derivative of a function, which is of the form of the product of functions or quotient of function or a function of the form $(f(x))^{g(x)}$. In this case, we take the logarithms on both sides and then differentiate. This process is known as Logarithmic differentiation.

SOLVED EXAMPLES

EXAMPLE 1. Find $\dfrac{dy}{dx}$ of the function $y = x^{x^x}$.

SOLUTION. Here, $y = x^{x^x}$.

Taking logarithms on both sides, we get

$$\log y = \log(x^{x^x}) \quad \text{or} \quad \log y = x^x \log x.$$

Now, differentiating w.r. to x, we get

$$\frac{1}{y}\frac{dy}{dx} = x^x \frac{d}{dx}(\log x) + \frac{d}{dx}(x^x)\log x$$

$$\frac{1}{y}\frac{dy}{dx} = x^x \frac{1}{x} + x^x(1 + \log x)\log x \qquad \left[\because \frac{d}{dx}(x^x) = x^x(1 + \log x)\right]$$

$$\frac{dy}{dx} = y[x^{x-1} + x^x \log x(1 + \log x)].$$

Hence, $\qquad \dfrac{dy}{dx} = x^{x^x}[x^{x-1} + x^x \log x(1 + \log x)]$

EXAMPLE 2. Find $\dfrac{dy}{dx}$ of the following functions :

(i) $y = ax^{-3} + bx^3 + cx^{9/2}\sin x$ 　　　　　　(ii) $y = x^x$

(iii) $y = (1 + x)^x$ 　　(iv) $y = (ax + b)^x$ 　　(v) $y = (\cos x)^{\log x}$

SOLUTION. (i) We have, $y = ax^{-3} + bx^3 + cx^{9/2}\sin x$

So, $\qquad \dfrac{dy}{dx} = \dfrac{d}{dx}(ax^{-3} + bx^3 + cx^{9/2}\sin x)$

$$= a\frac{d}{dx}(x^{-3}) + b\frac{d}{dx}(x^3) + c\frac{d}{dx}(x^{9/2}\sin x)$$

$$= -3ax^{-4} + 3bx^2 + c\left(x^{9/2}\frac{d}{dx}\sin x + \frac{d}{dx}(x^{9/2})\sin x\right)$$

$$= -3ax^{-4} + 3bx^2 + c(x^{9/2}\cos x + \frac{9}{2}x^{7/2}\sin x)$$

(ii) Here, $\qquad y = x^x$.

Taking logarithms on both sides, we get

$$\log y = x \log x.$$

Now, differentiating w.r. to x, we get

$$\frac{1}{y}\frac{dy}{dx} = x\frac{d}{dx}(\log x) + \frac{d}{dx}(x)\log x$$

or $\qquad \dfrac{1}{y}\dfrac{dy}{dx} = x\dfrac{1}{x} + 1\log x$

or $\qquad \dfrac{dy}{dx} = y(1 + \log x)$

Hence, $\qquad \dfrac{dy}{dx} = x^x(1 + \log x)$

(iii) Here, $\qquad y = (1 + x)^x$.

Taking logarithms on both sides, we get

$$\log y = x \log(1 + x).$$

Now, differentiating both sides with respect to x, we get

$$\frac{1}{y}\frac{dy}{dx} = x\frac{d}{dx}[\log(1+x)] + \frac{d}{dx}(x)\log(1+x)$$

$$\Rightarrow \quad \frac{1}{y}\frac{dy}{dx} = x\frac{1}{1+x} + 1.\log(1+x)$$

$$\Rightarrow \quad \frac{dy}{dx} = y\left[\frac{x}{1+x} + \log(1+x)\right].$$

Hence, $\quad \dfrac{dy}{dx} = (1+x)^x\left(\dfrac{x}{x+1} + \log(1+x)\right).$

(iv) Here, $\qquad y = (ax+b)^x$

Taking logarithms on both sides, we get

$$\log y = x\log(ax+b)$$

Now, differentiating both sides w.r. to x, we get

$$\frac{1}{y}\frac{dy}{dx} = x\frac{d}{dx}\log(ax+b) + \frac{d}{dx}(x)\cdot\log(ax+b)$$

or $\quad \dfrac{1}{y}\dfrac{dy}{dx} = x\dfrac{1}{(ax+b)}\dfrac{d}{dx}(ax+b) + 1\cdot\log(ax+b)$

or $\quad \dfrac{1}{y}\dfrac{dy}{dx} = x\dfrac{1}{(ax+b)}a + \log(ax+b)$

or $\quad \dfrac{dy}{dx} = y\left[\dfrac{ax}{ax+b} + \log(ax+b)\right].$

Hence, $\quad \dfrac{dy}{dx} = (ax+b)^x\left[\dfrac{ax}{ax+b} + \log(ax+b)\right].$

(v) Here, $\qquad y = (\cos x)^{\log x}.$

Taking logarithms on both sides, we get

$$\log y = \log x\log(\cos x).$$

Now, differentiating both sides w.r. to x, we get

$$\frac{1}{y}\frac{dy}{dx} = \log x\frac{d}{dx}\log(\cos x) + \frac{d}{dx}\log x\cdot\log(\cos x)$$

or $\quad \dfrac{1}{y}\dfrac{dy}{dx} = \log x\dfrac{1}{\cos x}\dfrac{d}{dx}(\cos x) + \dfrac{1}{x}\log(\cos x)$

or $\quad \dfrac{1}{y}\dfrac{dy}{dx} = -\log x\dfrac{\sin x}{\cos x} + \dfrac{1}{x}\log(\cos x)$

$$\Rightarrow \quad \frac{1}{y}\frac{dy}{dx} = \frac{1}{x}\log(\cos x) - \tan x\log x$$

or $\quad \dfrac{dy}{dx} = (\cos x)^{\log x}\left[\dfrac{1}{x}\log(\cos x) - \tan x\log x\right]$

EXAMPLE 3. *Find* $\dfrac{dy}{dx}$ *of the following functions :*

(i) $xy + \tan y = \log x$ $\qquad$ (ii) $y = \dfrac{x^2\sqrt{(4x+3)}}{3x+1}.$

SOLUTION.　(i)　We have,

$$xy + \tan y = \log x$$

Differentiating both sides w.r. to x, we get

$$x\frac{dy}{dx} + y + \sec^2 y \frac{dy}{dx} = \frac{1}{x} \quad\quad \text{or} \quad\quad (x + \sec^2 y)\frac{dy}{dx} = \frac{1}{x} - y$$

or

$$\frac{dy}{dx} = \frac{1 - xy}{x(x + \sec^2 y)}.$$

(ii)　Here,

$$y = \frac{x^2 \sqrt{4x + 3}}{3x + 1}.$$

Taking logarithms on both sides, we get

$$\log y = \log x^2 + \log\{\sqrt{(4x + 3)}\} - \log(3x + 1)$$

or

$$\log y = 2\log x + \frac{1}{2}\log(4x + 3) - \log(3x + 1).$$

Now, differentiating both sides with respect to x, we get

$$\frac{1}{y}\frac{dy}{dx} = 2\frac{d}{dx}\log x + \frac{1}{2}\frac{d}{dx}\log(4x + 3) - \frac{d}{dx}[\log(3x + 1)]$$

or

$$\frac{1}{y}\frac{dy}{dx} = 2\frac{1}{x} + \frac{1}{2}\frac{1}{(4x + 3)}\frac{d}{dx}(4x + 3) - \frac{1}{3x + 1}\frac{d}{dx}(3x + 1)$$

or

$$\frac{1}{y}\frac{dy}{dx} = \frac{2}{x} + \frac{2}{4x + 3} - \frac{3}{3x + 1}$$

or

$$\frac{dy}{dx} = \frac{x^2 \sqrt{(4x + 3)}}{3x + 1}\left[\frac{2}{x} + \frac{2}{4x + 3} - \frac{3}{3x + 1}\right]$$

EXAMPLE 4.　*Find the derivative (or differential coefficient) of the following functions.*

　　(i)　$(ax)^n + b$　　　　　　　(ii)　$\dfrac{2}{x^2} - \dfrac{a}{x} + be^x + (ax)^m$

　(iii)　$x\log x + e^x \sin x + 2x$

SOLUTION.　(i)　We have,　　　　$y = (ax)^n + b.$

　　　　　−So,　　　　　　$\dfrac{dy}{dx} = \dfrac{d}{dx}(ax)^n + \dfrac{d}{dx}b$

$$= n(ax)^{n-1}\frac{d}{dx}(ax) + 0 = n(ax)^{n-1}a = na^n x^{n-1}$$

(ii)　We have,　　　　$y = \dfrac{2}{x^2} - \dfrac{a}{x} + be^x + (ax)^m.$

　　　So,　　　　$\dfrac{dy}{dx} = 2\dfrac{d}{dx}x^{-2} - a\dfrac{d}{dx}x^{-1} + b\dfrac{d}{dx}e^x + \dfrac{d}{dx}(ax)^m$

$$= 2(-2x^{-3}) - a(-x^{-2}) + be^x + m(ax)^{m-1}a$$

$$= -4x^{-3} + ax^{-2} + be^x + ma^m x^{m-1}$$

(iii)　We have,　　　　$y = x\log x + e^x \sin x + 2x$

　　　So,　　　　$\dfrac{dy}{dx} = \dfrac{d}{dx}(x\log x) + \dfrac{d}{dx}(e^x \sin x) + 2\dfrac{d}{dx}(x)$

$$= x\frac{d}{dx}(\log x)+\frac{d}{dx}(x)\log x+e^x\frac{d}{dx}\sin x+\frac{d}{dx}(e^x)\sin x+2$$

$$= x\frac{1}{x}+\log x+e^x\cos x+e^x\sin x+2$$

$$= 3+\log x+e^x(\cos x+\sin x)$$

EXAMPLE 5. *Find the derivative of the following functions.*

   (i) $(x^n+a)(x^m+b)$ (ii) $\log_e x\log_a x$.

SOLUTION.   (i) We have, $y=(x^n+a)\cdot(x^m+b)$

   So, $\dfrac{dy}{dx}=\dfrac{d}{dx}\{(x^n+a)(x^m+b)\}.$

$$=(x^n+a)\frac{d}{dx}(x^m+b)+\frac{d}{dx}(x^n+a)\cdot(x^m+b)$$

$$=(x^n+a)mx^{m-1}+nx^{n-1}(x^m+b)$$

   (ii) Here, $y=\log_e x\log_a x.$

   So, $\dfrac{dy}{dx}=\dfrac{d}{dx}(\log_e x\log_a x)$

$$=\log_e x\frac{d}{dx}(\log_a x)+\frac{d}{dx}(\log_e x)\log_a x$$

$$=\log_e x\frac{1}{x}\log_a(e)+\frac{1}{x}\log_a x$$

$$=\frac{1}{x}[\log_e x\log_a e+\log_a x].$$

EXAMPLE 6. *Find the $\dfrac{dy}{dx}$ of the following functions.*

   (i) $e^x\log\sin 2x$ (ii) $\dfrac{1+\cos x}{\sin x}$

   (iii) $\sin x(x^2+3)$ (iv) 3^{x+8}.

SOLUTION.   (i) Here, $y=e^x\log\sin 2x$

$$\frac{dy}{dx}=\frac{d}{dx}(e^x\log\sin 2x)=e^x\frac{d}{dx}(\log\sin 2x)+\frac{d}{dx}(e^x)\log\sin 2x$$

$$=e^x\frac{1}{\sin 2x}\frac{d}{dx}\sin 2x+e^x\log\sin 2x$$

$$=e^x\frac{1}{\sin 2x}\cos 2x\frac{d}{dx}(2x)+e^x\log\sin 2x$$

$$=2e^x\cot 2x+e^x\log\sin 2x.$$

   (ii) Here, $y=\dfrac{1+\cos x}{\sin x}$

   So, $\dfrac{dy}{dx}=\dfrac{d}{dx}\left(\dfrac{1+\cos x}{\sin x}\right)=\dfrac{\sin x\dfrac{d}{dx}(1+\cos x)-(1+\cos x)\dfrac{d}{dx}\sin x}{(\sin x)^2}$

$$=\frac{\sin x(-\sin x)-(1+\cos x)(\cos x)}{(\sin x)^2}$$

$$= \frac{-\sin^2 x - \cos x - \cos^2 x}{(\sin x)^2} = \frac{-(1 + \cos x)}{1 - \cos^2 x}$$

$$= \frac{-(1 + \cos x)}{(1 + \cos x)(1 - \cos x)} = \frac{-1}{1 - \cos x}$$

(iii) Here, $y = \sin x (x^2 + 3)$.

So, $\dfrac{dy}{dx} = \dfrac{d}{dx}[\sin x (x^2 + 3)] = \sin x \dfrac{d}{dx}(x^2 + 3) + (x^2 + 3)\dfrac{d}{dx}(\sin x)$

$$= \sin x \cdot 2x + \cos x (x^2 + 3) = 2x \sin x + (x^2 + 3)\cos x$$

(iv) Here, $y = 3^{x+8}$

So, $\dfrac{dy}{dx} = \dfrac{d}{dx}(3^{x+8}) = 3^{x+8} \cdot \log_e 3$

EXAMPLE 7. Find $\dfrac{dy}{dx}$ of the following functions.

(i) $y = \sin x^3 - 4\cos x^2$ (ii) $y = x^4 \sec x + e^x \operatorname{cosec} x$

(iii) $y = \log(\log x)$.

SOLUTION. (i) Here, $y = \sin x^3 - 4\cos x^2$

$$\frac{dy}{dx} = \frac{d}{dx}(\sin x^3 - 4\cos x^2) = \frac{d}{dx}(\sin x^3) - 4\frac{d}{dx}(\cos^2)$$

$$= \cos x^3 \frac{d}{dx}x^3 - 4(-\sin x^2)\frac{d}{dx}(x^2)$$

$$= 3x^2 \cos x^3 + 4(2x)\sin x^2 = 3x^2 \cos x^3 + 8x \sin x^2$$

(ii) Here, $y = x^4 \sec x + e^x \operatorname{cosec} x$

$$\frac{dy}{dx} = \frac{d}{dx}(x^4 \sec x) + \frac{d}{dx}(e^x \operatorname{cosec} x)$$

$$= x^4 \frac{d}{dx}(\sec x) + \frac{d}{dx}(x^4)\sec x + e^x \frac{d}{dx}(\operatorname{cosec} x) + \frac{d}{dx}e^x \operatorname{cosec} x$$

$$= x^4 \sec x \tan x + 4x^3 \sec x + e^x(-\operatorname{cosec} x \cot x) + e^x \operatorname{cosec} x$$

$$= x^3 \sec x(x \tan x + 4) + e^x \operatorname{cosec} x(1 - \cot x)$$

(iii) Here $y = \log(\log x)$.

So, $\dfrac{dy}{dx} = \dfrac{d}{dx}[\log(\log x)] = \dfrac{1}{(\log x)}\dfrac{d}{dx}(\log x) = \dfrac{1}{(\log x)}\dfrac{1}{x} = \dfrac{1}{(\log x)}$.

EXAMPLE 8. Find the derivative of the following functions.

(i) $ax + by + c = 0$ (ii) $y^2 = 4ax$

(iii) $x^2 + y^2 = a^2$ (iv) $x^2 + y^2 = 4ax^2$.

SOLUTION. (i) The given equation is $ax + by + c = 0$.

Differentiating w.r. to x, taking y as a function of x,

$$a\frac{d}{dx}(x) + b\frac{dy}{dx} = 0 \qquad \Rightarrow \qquad a + b\frac{dy}{dx} = 0$$

$$\Rightarrow \qquad \frac{dy}{dx} = -\frac{a}{b}.$$

(ii) The given equation is $y^2 = 4ax$

Differentiating both sides w.r. to x, we get

$$2y\frac{dy}{dx} = 4a \qquad \Rightarrow \qquad \frac{dy}{dx} = \frac{4a}{2y}$$

$$\Rightarrow \qquad \frac{dy}{dx} = \frac{2a}{y}$$

(iii) The given equation is $x^2 + y^2 = a^2$

Differentiating both sides w.r.t. to x, we get

$$2x + 2y\frac{dy}{dx} = 0 \qquad \Rightarrow \qquad 2x = -2y\frac{dy}{dx}$$

$$\Rightarrow \qquad \frac{dy}{dx} = -\frac{x}{y}.$$

(iv) The given equation is $x^2 + y^2 = 4ax^2$.

Differentiating both sides with respect to x, we get

$$2x + 2y\frac{dy}{dx} = 8ax$$

$$2y\frac{dy}{dx} = 8ax - 2x = 2x(4a - 1)$$

$$\frac{dy}{dx} = \frac{x(4a - 1)}{y}.$$

EXAMPLE 9. *Find $\dfrac{dy}{dx}$ of the following functions.*

(i) $x^3 + y^3 = 3ax^2$ (ii) $\dfrac{x^m}{a^m} + \dfrac{y^m}{b^m} = 1$

(iii) $x = a(t - \sin t), y = a(1 - \cos t)$

SOLUTION. (i) Here, the given equation is $x^3 + y^3 = 3ax^2$.

Differentiating both sides, w.r. to x, we get

$$3x^2 + 3y^2\frac{dy}{dx} = 6ax$$

or $\qquad 3y^2\dfrac{dy}{dx} = 6ax - 3x^2 = 3x(2a - x)$

or $\qquad \dfrac{dy}{dx} = \dfrac{x(2a - x)}{y^2}.$

(ii) Here, the given equation is $\dfrac{x^m}{a^m} + \dfrac{y^m}{b^m} = 1.$

Differentiating both sides w.r. to x, we get

$$\frac{1}{a^m}\frac{d}{dx}(x^m) + \frac{1}{b^m}\frac{d}{dx}(y^m) = \frac{d}{dx}(1)$$

$$\frac{mx^{m-1}}{a^m} + \frac{1}{b^m}my^{m-1}\frac{dy}{dx} = 0.$$

or $\qquad b^m mx^{m-1} + a^m my^{m-1} \dfrac{dy}{dx} = 0$

or $\qquad \dfrac{dy}{dx} = -\dfrac{b^m mx^{m-1}}{a^m my^{m-1}} = -\left(\dfrac{b}{a}\right)^m \left(\dfrac{x}{y}\right)^{m-1}$

or $\qquad \dfrac{dy}{dx} = -\left(\dfrac{b}{a}\right)^m \left(\dfrac{x}{y}\right)^{m-1}.$

(iii) Here, $\qquad x = a(t - \sin t)$ and $\qquad y = a(1 - \cos t)$

So, $\qquad \dfrac{dx}{dt} = a(1 - \cos t)$ and $\dfrac{dy}{dt} = a[-(-\sin t)]$

or $\qquad \dfrac{dy}{dt} = a \sin t.$

So, $\qquad \dfrac{dy}{dx} = \dfrac{dy/dt}{dx/dt} = \dfrac{a \sin t}{a(1 - \cos t)} = \dfrac{\sin t}{1 - \cos t}$

$\qquad\qquad = \dfrac{2 \sin t/2 \cos t/2}{2 \sin^2 t/2} = \dfrac{\cos t/2}{\sin t/2} = \cot \dfrac{t}{2}.$

So, $\qquad \dfrac{dy}{dx} = \cot \dfrac{t}{2}.$

EXAMPLE 10. Find $\dfrac{dy}{dx}$ of the following functions :

$\quad$ (i) $\; y = (x-1)(x+1)(x-\sqrt{5})(x+\sqrt{5})$ $\quad$ (ii) $\; y = \tan^3 x$

$\quad$ (iii) $\; y = (x + e^x + \cos x) \log x$

SOLUTION. $\quad$ (i) Here, $\qquad y = (x-1)(x+1)(x-\sqrt{5})(x+\sqrt{5})$

$\qquad$ or $\qquad y = (x^2 - 1)(x^2 - 5).$

$\qquad$ So, $\qquad \dfrac{dy}{dx} = \dfrac{d}{dx}\{(x^2 - 1)(x^2 - 5)\}$

$\qquad\qquad\qquad = (x^2 - 1)\dfrac{d}{dx}(x^2 - 5) + \dfrac{d}{dx}(x^2 - 1)(x^2 - 5)$

$\qquad\qquad\qquad = (x^2 - 1)2x + 2x(x^2 - 5) = 4x^3 - 12x$

$\quad$ (ii) Here, $\qquad y = \tan^3 x$

$\qquad$ So, $\qquad \dfrac{dy}{dx} = \dfrac{d}{dx}(\tan^3 x) = 3 \tan^2 x \dfrac{d}{dx} \tan x$

$\qquad\qquad\qquad = 3 \tan^2 x \sec^2 x$

$\quad$ (iii) Here, $\qquad y = (x + e^x + \cos x) \log x.$

$\qquad$ So, $\qquad \dfrac{dy}{dx} = \dfrac{d}{dx}\{(x + e^x + \cos x) \log x\}$

$\qquad\qquad\qquad = (x + e^x + \cos x)\dfrac{d}{dx}(\log x) + \dfrac{d}{dx}(x + e^x + \cos x)\log x$

$\qquad\qquad\qquad = (x + e^x + \cos x)\dfrac{1}{x} + \dfrac{d}{dx}(1 + e^x - \sin x)\log x$

EXAMPLE 11. If $y = Ax^2 e^x - x + 8$ and for $x = 1, \dfrac{dy}{dx} = 2$, then obtain the value of A.

SOLUTION. Here $y = Ax^2e^x - x + 8$ and for $x = 1, \dfrac{dy}{dx} = 2$

So, $\qquad \dfrac{dy}{dx} = A\dfrac{d}{dx}(x^2e^x) - \dfrac{d}{dx}(x) + \dfrac{d}{dx}(8)$

$$= A\left\{x^2\dfrac{d}{dx}e^x + \dfrac{d}{dx}x^2e^x - 1\right\}$$

or $\qquad \dfrac{dy}{dx} = A\{x^2e^x + 2xe^x\} - 1$

Now, put $x = 1, \dfrac{dy}{dx} = 2$

$$2 = A\{e + 2e\} - 1$$

$\Rightarrow \qquad A = \dfrac{3}{3e} = \dfrac{1}{e} \qquad \Rightarrow \qquad A = \dfrac{1}{e}.$

EXAMPLE 12. Find $\dfrac{dy}{dx}$ of the following functions.

(i) $y = \log\tan\left(\dfrac{\pi}{4} + \dfrac{x}{2}\right)$ $\qquad$ (ii) $y = \tan^{-1}(\sec x + \tan x)$

(iii) $y = e^{\sin x}$ $\qquad\qquad\qquad$ (iv) $y = \log(x^2 + x)$

SOLUTION. (i) Here, $\qquad y = \log\tan\left(\dfrac{\pi}{4} + \dfrac{x}{2}\right).$

So, $\qquad \dfrac{dy}{dx} = \dfrac{d}{dx}\left[\log\tan\left(\dfrac{\pi}{4} + \dfrac{x}{2}\right)\right] = \dfrac{1}{\tan\left(\dfrac{\pi}{4} + \dfrac{x}{2}\right)}\dfrac{d}{dx}\left[\tan\left(\dfrac{\pi}{4} + \dfrac{x}{2}\right)\right]$

$$= \dfrac{1}{\tan\left(\dfrac{\pi}{4} + \dfrac{x}{2}\right)}\sec^2\left(\dfrac{\pi}{4} + \dfrac{x}{2}\right)\dfrac{d}{dx}\left(\dfrac{\pi}{4} + \dfrac{x}{2}\right) = \dfrac{\sec^2\left(\dfrac{\pi}{4} + \dfrac{x}{2}\right)}{\tan\left(\dfrac{\pi}{4} + \dfrac{x}{2}\right)}\dfrac{1}{2}$$

$$= \dfrac{1}{2\cos\left(\dfrac{\pi}{4} + \dfrac{x}{2}\right)\sin\left(\dfrac{\pi}{4} + \dfrac{x}{2}\right)} = \dfrac{1}{\sin 2\left(\dfrac{\pi}{4} + \dfrac{x}{2}\right)} = \dfrac{1}{\sin\left(\dfrac{\pi}{2} + x\right)}.$$

(ii) $\qquad y = \tan^{-1}(\sec x + \tan x)$

So, $\qquad \dfrac{dy}{dx} = \dfrac{d}{dx}[\tan^{-1}(\sec x + \tan x)]$

$$= \dfrac{1}{1 + (\sec x + \tan x)^2}\dfrac{d}{dx}(\sec x + \tan x)$$

$$= \dfrac{1}{1 + (\sec x + \tan x)^2}(\sec x\tan x + \sec^2 x)$$

(iii) Here, $\qquad y = e^{\sin x}.$

So, $\qquad \dfrac{dy}{dx} = \dfrac{d}{dx}(e^{\sin x}) = e^{\sin x}\dfrac{d}{dx}\sin x = e^{\sin x}\cos x.$

(iv) Here, $y = \log(x^2 + x)$.

So, $\dfrac{dy}{dx} = \dfrac{d}{dx}[\log(x^2 + x)] = \dfrac{1}{(x^2 + x)} \dfrac{d}{dx}(x^2 + x)$

$$= \dfrac{1}{(x^2 + x)}(2x + 1) = \dfrac{2x + 1}{(x^2 + x)}$$

EXAMPLE 13. Find $\dfrac{dy}{dx}$ of the following functions

(i) $y \cos x = 1$
(ii) $xy = e^{x+y}$

(iii) $2xy + y^2 = x^3 + y^3$
(iv) $x^3 + y^3 + x^3 y^3 + 3x^2 + 6y^2 + 7 = 0$

SOLUTION.   (i) Here, $y \cos x = 1$

Differentiating both sides w.r. to x, we get

$$\dfrac{d}{dx}(y \cos x) = \dfrac{d}{dx}(1)$$

or $y \dfrac{d}{dx} \cos x + \dfrac{dy}{dx} \cos x = 0$

or $-y \sin x + \dfrac{dy}{dx} \cos x = 0$

or $\dfrac{dy}{dx} = \dfrac{y \sin x}{\cos x} = y \tan x$

So, $\dfrac{dy}{dx} = \sec x \tan x$ $\left(\because y = \dfrac{1}{\cos x} = \sec x \right)$

(iii) Here, $xy = e^{x+y}$.

Taking logarithms on both sides, we have

$$\log(xy) = \log e^{x+y}$$

or $\log x + \log y = x + y$.

Now, differentiating both sides, w.r. to x, we get

$$\dfrac{1}{x} + \dfrac{1}{y} \dfrac{dy}{dx} = 1 + \dfrac{dy}{dx}$$

$\Rightarrow$ $\left(\dfrac{1}{y} - 1 \right) \dfrac{dy}{dx} = 1 - \dfrac{1}{x}$

$\Rightarrow$ $\dfrac{dy}{dx} = \dfrac{(x-1)y}{x(1-y)}$

(iii) Here, $2xy + y^2 = x^3 + y^3$ is given.

So, differentiating both sides w.r.to x, taking y as a function of x, we get

$$2 \left(x \dfrac{dy}{dx} + \dfrac{d}{dx}(x)y \right) + 2y \dfrac{dy}{dx} = 3x^2 + 3y^2 \dfrac{dy}{dx}$$

or $2x \dfrac{dy}{dx} + 2y + 2y \dfrac{dy}{dx} = 3x^2 + 3y^2 \dfrac{dy}{dx}$

or $\qquad (2x + 2y - 3y^2)\dfrac{dy}{dx} = 3x^2 - 2y$

or $\qquad \dfrac{dy}{dx} = \dfrac{3x^2 - 2y}{(2x + 2y - 3y^2)}.$

(iv) Here, $x^3 + y^3 + x^3y^3 + 3x^2 + 6y^2 + 7 = 0$ is given.
Differentiating w.r. to x, we get

$$3x^2 + 3y^2\dfrac{dy}{dx} + x^3\dfrac{d}{dx}y^3 + \left(\dfrac{d}{dx}x^3\right)\cdot y^3 + 6x + 12y\dfrac{dy}{dx} = 0$$

or $\qquad 3x^2 + 3y^2\dfrac{dy}{dx} + x^3 3y^2\dfrac{dy}{dx} + 3x^2y^3 + 6x + 12y\dfrac{dy}{dx} = 0$

or $\qquad (3y^2 + 3x^3y^2 + 12y)\dfrac{dy}{dx} = -(3x^2y^3 + 6x - 3x^2)$

or $\qquad \dfrac{dy}{dx} = -\dfrac{(3x^2y^3 + 6x - 3x^2)}{(3y^2 + 3x^3y^2 + 12y)} = -\dfrac{(x^2y^3 + 2x - x^2)}{(y^2 + x^3y^2 + 4y)}.$

EXAMPLE 14. *Find the derivative of the following functions.*

(i) $\quad y = \dfrac{(1 - x^2)}{\sqrt{(1 + x^2)}}$

(ii) $\quad y = \tan^{-1}\left(\dfrac{2x}{1 - x^2}\right)$

(iii) $\quad y = \dfrac{\sin x + \cos x}{\sin x - \cos x}$

(iv) $\quad y = \sin^{-1}\left(\dfrac{2x}{1 + x^2}\right)$

SOLUTION. (i) Here, $\quad y = \dfrac{(1 - x^2)}{\sqrt{(1 + x^2)}}$

So, $\quad \dfrac{dy}{dx} = \dfrac{d}{dx}\left(\dfrac{1 - x^2}{\sqrt{1 + x^2}}\right) = \dfrac{\sqrt{1 + x^2}\,\dfrac{d}{dx}(1 - x^2) - (1 - x^2)\dfrac{d}{dx}\sqrt{1 + x^2}}{(1 + x^2)}$

$$= \dfrac{\sqrt{1 + x^2}(-2x) - (1 - x^2)\dfrac{1}{2}(1 + x^2)^{-1/2}}{(1 + x^2)} = \dfrac{-2x\sqrt{1 + x^2} - \dfrac{1}{2}\dfrac{(1 - x^2)}{\sqrt{1 + x^2}}}{(1 + x^2)}$$

$$= \dfrac{-2x(1 + x^2) - \dfrac{1}{2}(1 - x^2)}{(1 + x^2)^{3/2}} = \left(-2x - 2x^3 - \dfrac{1}{2} + \dfrac{1}{2}x^2\right)(1 + x^2)^{3/2}$$

(ii) Here, $\quad y = \tan^{-1}\left(\dfrac{2x}{1 - x^2}\right).$

Differentiating with respect to x, we get

$$\dfrac{dy}{dx} = \dfrac{1}{1 + \left(\dfrac{2x}{1 - x^2}\right)^2}\dfrac{d}{dx}\dfrac{2x}{(1 - x^2)}$$

$$= \frac{(1-x^2)^2}{\{(1-x^2)^2 + 4x^2\}} \cdot \frac{(1-x^2)\dfrac{d}{dx}(2x) - 2x\dfrac{d}{dx}(1-x^2)}{(1-x^2)^2}$$

$$= \frac{(1-x^2)^2\{2(1-x^2) + 4x^2\}}{\{(1-x^2)^2 + 4x^2\}(1-x^2)^2}$$

$$= \frac{2 + 2x^2}{\{(1-x^2)^2 + 4x^2\}} = \frac{2(1+x^2)}{(1+x^4 - 2x^2 + 4x^2)}$$

$$= \frac{2(1+x^2)}{1 + x^4 + 2x^2} = \frac{2(1+x^2)}{(1+x^2)^2} = \frac{2}{1+x^2}.$$

(iii) Here, $y = \dfrac{\sin x + \cos x}{\sin x - \cos x}$

So, $\dfrac{dy}{dx} = \dfrac{d}{dx}\left(\dfrac{\sin x + \cos x}{\sin x - \cos x}\right)$

$$= \frac{(\sin x - \cos x)\dfrac{d}{dx}(\sin x + \cos x) - (\sin x + \cos x)\dfrac{d}{dx}(\sin x - \cos x)}{(\sin x - \cos x)^2}$$

$$= \frac{(\sin x - \cos x)(\cos x - \sin x) - (\sin x + \cos x)(\cos x + \cos x)}{(\sin x - \cos x)^2}$$

$$= \frac{-(\sin x - \cos x)^2 - (\sin x + \cos x)^2}{(\sin x - \cos x)^2}$$

$$= \frac{-(\sin^2 x + \cos^2 x - 2\sin x \cos x) - (\sin^2 x + \cos^2 x + 2\sin x \cos x)}{(\sin x - \cos x)^2}$$

$$= \frac{-1-1}{(\sin x - \cos x)^2} = \frac{-2}{(\sin x - \cos x)^2}.$$

(iv) Here, $y = \sin^{-1}\left(\dfrac{2x}{1+x^2}\right).$

Differentiating w.r.t. x, we get

$$\frac{dy}{dx} = \frac{d}{dx}\left\{\sin^{-1}\left(\frac{2x}{1+x^2}\right)\right\} = \frac{1}{\sqrt{1 - \left(\dfrac{2x}{1+x^2}\right)^2}} \cdot \frac{d}{dx}\left(\frac{2x}{1+x^2}\right)$$

$$= \frac{(1+x^2)}{\sqrt{(1+x^2)^2 - 4x^2}} \cdot \frac{(1+x^2)\dfrac{d}{dx}(2x) - 2x\dfrac{d}{dx}(1+x^2)}{(1+x^2)^2}$$

$$= \frac{\{2(1+x^2) - 4x^2\}}{\sqrt{(1-x^2)^2(1+x^2)}}$$

$$= \frac{2 - 2x^2}{(1+x^2)} = \frac{2(1-x^2)}{1+x^2\sqrt{(1-3x^2)}} = \frac{2(1-x^2)}{(1-x^2)(1+x^2)} = \frac{2}{1+x^2}.$$

EXAMPLE 15. *Find $\dfrac{dy}{dx}$ of the following functions*

(i) $y = \dfrac{1 - \tan x}{1 + \tan x}$ (ii) $y = \dfrac{x \sin x}{1 + \cos x}$

(iii) $y = x \log x + e^x \sin x - a^x x^3$

SOLUTION. (i) Here, $y = \dfrac{1 - \tan x}{1 + \tan x}$

Differentiating w.r.t. x, we get

$$\dfrac{dy}{dx} = \dfrac{(1 + \tan x) \dfrac{d}{dx}(1 - \tan x) - (1 - \tan x) \dfrac{d}{dx}(1 + \tan x)}{(1 + \tan x)^2}$$

$$= \dfrac{-(1 + \tan x) \sec^2 x - (1 - \tan x) \sec^2 x}{(1 + \tan x)^2} = \dfrac{-2 \sec^2 x}{(1 + \tan x)^2}.$$

(ii) Here, $y = \dfrac{x \sin x}{1 + \cos x}$.

Differentiating w.r.t. to x, we get

$$\dfrac{dy}{dx} = \dfrac{(11 + \cos x) \dfrac{d}{dx}(x \sin x) - x \sin x \dfrac{d}{dx}(1 + \cos x)}{(1 + \cos x)^2}$$

$$= \dfrac{(1 + \cos x) \left\{ x \dfrac{d}{dx} \sin x + \dfrac{d}{dx}(x) \sin x \right\} + x \sin^2 x}{(1 + \cos x)^2}$$

$$= \dfrac{(1 + \cos x) \{ x \cos x + \sin x \} + x \sin^2 x}{(1 + \cos x)^2}$$

$$= \dfrac{x \cos x + \sin x + x \cos^2 x + \sin x \cos x + x \sin^2 x}{(1 + \cos x)^2}$$

$$= \dfrac{x \cos x + \sin x + \sin x \cos x + x}{(1 + \cos x)^2} = \dfrac{\cos x (x + \sin x) + 1 (x + \sin x)}{(1 + \cos x)^2}$$

$$= \dfrac{(\cos x + 1)(x + \sin x)}{(1 + \cos x)^2} = \dfrac{x + \sin x}{(1 + \cos x)}.$$

(iii) Here, $y = x \log x + e^x \sin x - a^x x^3$.

Differentiating w.r.t. x, we get

$$\dfrac{dy}{dx} = \dfrac{d}{dx}(x \log x) + \dfrac{d}{dx}(e^x \sin x) - \dfrac{d}{dx}(a^x x^3)$$

$$= x \dfrac{d}{dx}(\log x) + \dfrac{d}{dx}(x) \log x + e^x \dfrac{d}{dx}(\sin x) + \dfrac{d}{dx}(e^x) \sin x$$

$$\qquad - \left(a^x \dfrac{d}{dx}(x^3) + \dfrac{d}{dx}(a^x) x^3 \right)$$

$$= x \dfrac{1}{x} + \log x + e^x \cos x + e^x \sin x - (3 a^x x^2 + x^3 a^x \log_e a)$$

$$= 1 + \log x + e^x \cos x + e^x \sin x - 3a^x x^2 + x^3 \log_e a \cdot a^x$$

$$= 1 + \log x + e^x (\sin x + \cos x) - a^x (3x^2 + x^3 \log_e a)$$

EXAMPLE 16. *Find the differential coefficient at $x = 0$, if $y = a^x + \sqrt{\dfrac{1+x}{1-x}}$.*

SOLUTION.   We have $y = a^x + \sqrt{\dfrac{1+x}{1-x}}$

Differentiating w.r.t. x, we get

$$\frac{dy}{dx} = a^x \log a + \frac{1}{2}\left(\frac{1+x}{1-x}\right)^{1/2} \left\{ \frac{1(1-x) - (1+x)(-1)}{(1-x)^2} \right\}$$

$$= a^x \log a + \frac{1}{2}\sqrt{\frac{1-x}{1+x}} \left\{ \frac{1 - x + 1 + x}{(1-x)^2} \right\}$$

$$= a^x \log a + \frac{1}{2}\sqrt{\frac{1-x}{1+x}} \left\{ \frac{2}{(1-x)^2} \right\}$$

putting $x = 0$

$$\left(\frac{dy}{dx}\right)_{x=0} = a^0 \log a + \frac{1}{2}\sqrt{\frac{1-0}{1+0}} \left\{ \frac{2}{(1-0)^2} \right\} = \log a + \frac{1}{2} \cdot \frac{2}{1^2} = \log a + 1$$

EXAMPLE 17. *If $\sqrt{x} + \sqrt{y} = 5$, prove that*

$$\left[\frac{dy}{dx}\right]_{(4,\,9)} = -\frac{3}{2}$$

SOLUTION.   We have $\sqrt{x} + \sqrt{y} = 5$

Differentiating w.r.t. x, we get

$$\frac{1}{2\sqrt{x}} + \frac{1}{2\sqrt{y}}\frac{dy}{dx} = 0$$

$$\Rightarrow \qquad \frac{dy}{dx} = -\sqrt{\frac{y}{x}}$$

Thus $\left[\dfrac{dy}{dx}\right]_{(4,\,9)} = -\sqrt{\dfrac{9}{4}} = -\dfrac{3}{2}$

MISCELLANEOUS SOLVED EXAMPLES

EXAMPLE 1.   *Differentiate $\tan^{-1}\left(\dfrac{2x}{1-x^2}\right)$ w.r.t. $\sin^{-1}\left(\dfrac{2x}{1+x^2}\right)$.*

SOLUTION.   Let $y = \tan^{-1}\left(\dfrac{2x}{1-x^2}\right)$

Put $x = \tan\theta$

$$\Rightarrow \qquad \frac{2x}{1-x^2} = \frac{2\tan\theta}{1-\tan^2\theta} = \tan 2\theta$$

Also $y = \tan^{-1}(\tan 2\theta) = 2\theta = 2\tan^{-1} x$

$$\therefore \quad \frac{dy}{dx} = \frac{2}{1+x^2} \quad \ldots(1)$$

Further let $\quad z = \sin^{-1} \dfrac{2x}{1+x^2}$

Putting $\quad x = \tan\theta$

$$\therefore \quad \frac{2x}{1+x^2} = \frac{2\tan\theta}{1+\tan^2\theta} = \sin 2\theta$$

$$\Rightarrow \quad \sin^{-1}(\sin 2\theta) = 2\tan^{-1} x$$

Thus $\quad \dfrac{dz}{dx} = 2 \cdot \dfrac{1}{1+x^2}$

From (1) and (2) we conclude that

$$\frac{dy}{dz} = \frac{dy/dx}{dz/dx} = 1$$

Hence, $\quad \dfrac{dy}{dz} = 1$

EXAMPLE 2. *Differentiate* $\sin^{-1}\left(\dfrac{1-x}{1+x}\right)$ *w.r.t.* $\sqrt{x}$.

SOLUTION. Let $\quad u = \sin^{-1}\left(\dfrac{1-x}{1+x}\right), v = \sqrt{x}$

Differentiating w.r.t. x

$$\frac{du}{dx} = \frac{1}{\sqrt{1-\left(\dfrac{1-x}{1+x}\right)^2}} \frac{-(1+x)-(1-x)}{(1+x)^2}, \frac{dv}{dx} = \frac{1}{2\sqrt{x}}$$

$$= \frac{1+x}{\sqrt{(1+x)^2-(1-x)^2}}\left(\frac{-2}{(1+x)^2}\right), \qquad \frac{dv}{dx} = \frac{1}{2\sqrt{x}}$$

$$= -\frac{1}{\sqrt{x}(1+x)}, \frac{dv}{dx} = \frac{1}{2\sqrt{x}}$$

Therefore, $\quad \dfrac{du}{dx} = -\dfrac{1}{\sqrt{x}(1+x)} \bigg/ \dfrac{1}{2\sqrt{x}} = \dfrac{-2}{1+x}$

EXAMPLE 3. *Find the derivative of* $\tan^{-1}\dfrac{\sqrt{1+x^2}-1}{x}$ *w.r.t.* $\tan^{-1}x$.

SOLUTION. Let $\quad u = \tan^{-1}\left(\dfrac{\sqrt{1+x^2}-1}{x}\right)$ and $v = \tan^{-1} x$

Putting $x = \tan\theta$, we have

$$u = \tan^{-1}\left[\frac{\sqrt{1+\tan^2\theta}-1}{\tan\theta}\right] = \tan^{-1}\left[\frac{\sec\theta-1}{\tan\theta}\right]$$

$$= \tan^{-1}\left[\left(\frac{1}{\cos\theta}-1\right)\cdot\frac{\cos\theta}{\sin\theta}\right] = \tan^{-1}\left[\frac{1-\cos\theta}{\sin\theta}\right]$$

$$= \tan^{-1}\left[\frac{1-\left(1-2\sin^2\frac{\theta}{2}\right)}{2\sin\frac{\theta}{2}\cdot\cos\frac{\theta}{2}}\right] = \tan^{-1}\left(\tan\frac{\theta}{2}\right) = \frac{\theta}{2}$$

Also $\qquad v = \tan^{-1}(\tan\theta) = \theta$

$\Rightarrow \qquad u = \dfrac{v}{2}$

Hence $\qquad \dfrac{du}{dv} = \dfrac{1}{2}$

EXAMPLE 4. *Find the derivative of $(\log x)^{\tan x}$ w.r.t. $\sin(m\cos^{-1} x)$.*

SOLUTION. Let $\qquad u = (\log x)^{\tan x}$

$\Rightarrow \qquad \log u = \tan x \log(\log x)$

Diff. w.r.t. x, we get

$$\frac{1}{u}\cdot\frac{du}{dx} = \tan x\cdot\frac{1}{\log x}\cdot\frac{1}{x} + [\log(\log x)]\sec^2 x$$

$$\Rightarrow \qquad \frac{du}{dx} = u\left[\frac{\tan x}{x\cdot\log x} + \sec^2 x\log(\log x)\right]$$

$$= (\log x)^{\tan x}\left[\frac{\tan x}{x\cdot\log x} + \sec^2 x\log(\log x)\right]$$

Further, let $\qquad v = \sin(m\cos^{-1} x)$

$$\Rightarrow \qquad \frac{dv}{dx} = \cos(m\cos^{-1} x)\cdot m\left(\frac{-1}{\sqrt{1-x^2}}\right) = -\frac{m}{\sqrt{1-x^2}}\cos(m\cos^{-1} x)$$

Hence $\qquad \dfrac{du}{dv} = \dfrac{du}{dx}\cdot\dfrac{dx}{dv} = \dfrac{(\log x)^{\tan x}\left[\dfrac{\tan x}{x\cdot\log x} + \sec^2 x\cdot\log(\log x)\right]}{-\dfrac{m}{\sqrt{1-x^2}}\cos(m\cos^{-1} x)}$

EXAMPLE 5. *If $y = \sqrt{\dfrac{x}{a}} + \sqrt{\dfrac{a}{x}}$. Then show that*

$$\frac{dy}{dx} = \frac{x-a}{2x\sqrt{ax}}$$

SOLUTION. We have $\qquad y = \sqrt{\dfrac{x}{a}} + \sqrt{\dfrac{a}{x}}$

Differentiating w.r.t. x, we get

$$\frac{dy}{dx} = \frac{1}{\sqrt{a}}\frac{1}{2\sqrt{x}} + \sqrt{a}\left(-\frac{1}{2x^{3/2}}\right)$$

$$= \frac{1}{2x^{3/2}a^{1/2}}(x-a) = \frac{x-a}{2x\sqrt{ax}}.$$

EXAMPLE 6. $y = e^{x+e^{x+e^{x+\cdots^{\infty}}}}$, *then show that*

$$\frac{dy}{dx} = \frac{y}{1-y}.$$

SOLUTION. We have $y = e^{x+e^{x+e^{x+\cdots^{\infty}}}}$

$$y = e^{x+y} \qquad \qquad \dots(1)$$

Diff. w.r.t. x, we get

$$\frac{dy}{dx} = e^{x+y}\left(1+\frac{dy}{dx}\right) = y\left(1+\frac{dy}{dx}\right) \qquad \text{[From (1)]}$$

$$\frac{dy}{dx}(1-y) = y$$

$$\frac{dy}{dx} = \frac{y}{1-y}$$

EXAMPLE 7. (a) *Find* $\dfrac{dy}{dx}$ *if* $y = \tan^{-1}\sqrt{\dfrac{1-\cos x}{1+\cos x}}$

(b) *Find* $\dfrac{dy}{dx}$ *if* $y = x^{\sin x} + (\sin x)^x$

(c) *If* $y = (x+\sqrt{(x^2-1)})^m$. *Prove that* $(x^2-1)\left(\dfrac{dy}{dx}\right)^2 = m^2 y^2$

SOLUTION. (a) We have $y = \tan^{-1}\sqrt{\dfrac{1-\cos x}{1+\cos x}}$

$$\left(\because \cos x = 1-2\sin^2\frac{x}{2}, \cos x = 2\cos^2\frac{x}{2}-1\right)$$

$$= \tan^{-1}\sqrt{\frac{1-\left(1-2\sin^2\frac{x}{2}\right)}{1+\left(2\cos^2\frac{x}{2}-1\right)}} = \tan^{-1}\sqrt{\frac{\sin^2\frac{x}{2}}{\cos^2\frac{x}{2}}} = \tan^{-1}\left(\tan\frac{x}{2}\right)$$

$$\Rightarrow \qquad y = \frac{x}{2}$$

Diff. w.r.t. x, we get

$$\frac{dy}{dx} = \frac{1}{2}$$

(b) We have

$$y = x^{\sin x} + (\sin x)^x$$

Diff. w.r.t. x, we get

$$\frac{dy}{dx} = \frac{d}{dx}(x^{\sin x}) + \frac{d}{dx}(\sin x)^x \qquad \dots(1)$$

Let $u = x^{\sin x}$

Taking logarithm of both sides, we get

$$\log u = \sin x \log x$$

Differentiating w.r.t. x, we get

$$\frac{1}{u}\cdot\frac{du}{dx} = \cos x \log x + \frac{\sin x}{x}$$

or

$$\frac{du}{dx} = x^{\sin x}\left(\cos x \log x + \frac{\sin x}{x}\right) \qquad \ldots(2)$$

and $\qquad v = \sin x^x$

$\Rightarrow \qquad \log v = x \log \sin x$

Differentiating w.r.t. x, we get

$$\frac{1}{v}\frac{dv}{dx} = \log \sin x + \frac{x}{\sin x}\cos x$$

or

$$\frac{dv}{dx} = \sin x^x (\log \sin x + x \cot x) \qquad \ldots(3)$$

Substituting (2) and (3) in (1), we get

$$\frac{dy}{dx} = x^{\sin x}\left(\cos x \log x + \frac{\sin x}{x}\right) + \sin x^x (\log \sin x + x \cot x)$$

(c) We have $\qquad y = \left(x + \sqrt{x^2 - 1}\right)^m$

Differentiating w.r. to x, we get

$$\frac{dy}{dx} = m\left(x + \sqrt{x^2 - 1}\right)^{m-1}\left(1 + \frac{x}{(x^2-1)^{1/2}}\right)$$

$$= m\left(x + \sqrt{x^2 - 1}\right)^{m-1}\left(\frac{x + \sqrt{x^2-1}}{\sqrt{x^2-1}}\right)$$

or $(\sqrt{x^2-1})\dfrac{dy}{dx} = m(x + \sqrt{x^2-1})^m$

or $(\sqrt{x^2-1})\dfrac{dy}{dx} = my$ $\qquad\qquad\qquad$ $[\because y = (x + \sqrt{x^2-1})^m]$

Squaring, we get

$$(x^2 - 1)\left(\frac{dy}{dx}\right)^2 = m^2 y^2$$

Exercise 9.1

1. Find the derivatives (or differential coefficients) of the following functions :

  (i) $y = \dfrac{(x+1)(x+2)}{(x+3)(x+4)}$

  (ii) $y = \log \sin x^2$

  (iii) $y = \dfrac{x^4 - 5x^2}{5x^6 + 7x}$ (iv) $y = \dfrac{1 + \tan x}{1 - \tan x}$

2. Find dy/dx of the following functions :

  (i) $y = \sqrt{\left(\dfrac{1-x}{1+x}\right)}$ (ii) $y = \sqrt{\left(\dfrac{1+x}{x}\right)}$

  (iii) $y = \log[\sqrt{(x+1)} - \sqrt{(x-1)}]$

  (iv) $y = (2x^2 + 5x + 7)^{-2}$.

3. Find the derivatives of the following functions :

  (i) $y = (x-2)(x+2)(x-3)(x+3)$

  (ii) $y = (x+1)(2x^3 - 21)$

  (iii) $y = \sin^{-1}(\tan x)$ (ii) $y = (\tan x)^x$.

4. Differentiate the following functions:

  (i) $y = (\sin x)^{\cos x}$

(ii) $y = \sin(e^x) + \dfrac{1}{x} + \log(x^2 + x)$

(iii) $y = \sin x \log x$ (iv) $y = e^x \tan^4 x$

5. Find $\dfrac{dy}{dx}$ of the following functions :

   (i) $y = \log \sin x + \cos^{-1}(e^x) + x^4 \sec x$

(ii) $y = (\log x)^x$ (iii) $\log\left(\dfrac{x}{y}\right) = x + y$

(iv) $y = \sec(x^2 - 2x + 1)$

6. Find the derivatives of the following functions:

   (i) $e^{2x} \cos 3x$ (ii) $\log(\sin^{-1} x^4)$

   (iii) $x^3 - y^3 - 3axy = 0$.

HINT TO SELECTED PROBLEMS

1. (i) $y = \dfrac{(x+1)(x+2)}{(x+3)(x+4)}$

Taking log on both sides, we get

$\log y = \log(x+1) + \log(x+2)$

Diff. w.r.t. x, $- \log(x+3) - \log(x+4),$

$\dfrac{1}{y}\dfrac{dy}{dx} = \dfrac{1}{(x+1)} + \dfrac{1}{(x+2)} - \dfrac{1}{(x+3)} - \dfrac{1}{(x+4)}$

$\Rightarrow \dfrac{dy}{dx} = y\left\{\dfrac{2x+3}{(x+1)(x+2)} - \dfrac{2x+7}{(x+3)(x+4)}\right\}$

$= y\left\{\dfrac{\begin{array}{c}2x(x^2+7x+12) \\ -x^2-3x-2) \\ +3(x^2+7x+12) \\ -7(x^2+3x+2)\end{array}}{(x^2+3x+2)(x^2+7x+12)}\right\}$

$= \dfrac{(x+1)(x+2)}{(x+3)(x+4)}\left\{\dfrac{4x^2+20x+22}{(x^2+3x+2)(x^2+7x+12)}\right\}$

$= \dfrac{4x^2+20x+22}{(x^2+7x+12)^2}$

(ii) $y = \log \sin x^2$

Diff.w.r.t. x we get

$\dfrac{dy}{dx} = \dfrac{1}{\sin x^2} \cdot (\cos x^2)2x = 2x \cot x^2$

(iv) $y = \dfrac{1 + \tan x}{1 - \tan x}$

Diff. both sides by Quotient rule

$\dfrac{dy}{dx} = \dfrac{\sec^2(1 - \tan x) + \sec^2(1 + \tan x)}{(1 - \tan x)^2}$

$= \dfrac{2\sec^2 x}{(1 - \tan x)^2}$

2. (i) $y = \sqrt{\left(\dfrac{1-x}{1+x}\right)} = \dfrac{(1-x)^{1/2}}{(1+x)^{1/2}}$

Diff. both sides by Quotient Rule

$\dfrac{dy}{dx} = \dfrac{-\dfrac{1}{2}\dfrac{1}{\sqrt{1-x}}\sqrt{1+x} - \dfrac{1}{2}\sqrt{1-x}\dfrac{1}{\sqrt{1+x}}}{(1+x)}$

$= -\dfrac{1}{2}\dfrac{\{(1+x)+(1-x)\}}{(1+x)\sqrt{1-x^2}} = \dfrac{-1}{(1+x)\sqrt{1-x^2}}$

(ii)   Do as Question No.2 (i)

(iii) $y = \log[\sqrt{x+1} - \sqrt{x-1}]$, diff. w.r.t. x both sides we have

$\dfrac{dy}{dx} = \dfrac{1}{\sqrt{x+1} - \sqrt{x-1}}\left(\dfrac{1}{2\sqrt{x+1}} - \dfrac{1}{2\sqrt{x-1}}\right)$

$= \dfrac{1}{2}\dfrac{-(\sqrt{x+1} - \sqrt{x-1})}{(\sqrt{x+1} - \sqrt{x-1})\sqrt{x^2-1}} = \dfrac{1}{2\sqrt{x^2-1}}$

(iv) $y = (2x^2 + 5x + 7)^{-2}$, diff, both sides w.r.t. x

$\dfrac{dy}{dx} = \dfrac{-2}{(2x^2 + 5x + 7)^3}(4x + 5)$

$= \dfrac{-2(4x + 5)}{(2x^2 + 5x + 7)^3}$

3. (i) $y = (x-2)(x+2)(x-3)(x+3)$

$= (x^2 - 4)(x^2 - 9)$

Diff. w.r.t. x by product rule

$\dfrac{dy}{dx} = 2x(x^2 - 9) + 2x(x^2 - 4)$

$= 4x^3 - 26x$

(ii)   Do as in above Question

(iii) $y = \sin^{-1}(\tan x)$, diff w.r.t. x

$\dfrac{dy}{dx} = \dfrac{1}{\sqrt{1 - \tan^2 x}}\sec^2 x$

$= \dfrac{\cos x}{\sqrt{\cos^2 x - \sin^2 x}}\sec^2 x$

$= \dfrac{\sec x}{\sqrt{\cos 2x}} = \sec x\sqrt{\sec 2x}$

(iv) $y = (\tan x)^x$, take log on both the sides

$\log y = x \log(\tan x)$,

Diff w.r.t. x

$\dfrac{1}{y}\dfrac{dy}{dx} = \log(\tan x) + \dfrac{x}{\tan x} \sec^2 x$

$\dfrac{dy}{dx} = y\{\log \tan x + x \sec x \operatorname{cosec} x\}$

$= (\tan x)^x \{\log \tan x + x \sec x \operatorname{cosec} x\}$

4. (i) $y = (\sin x)^{\cos x}$ taking log on both the sides

$\log y = \cos x \log \sin x$,

diff. both the sides w.r.t. x, we get

$\dfrac{1}{y}\dfrac{dy}{dx} = -\sin x \log \sin x + \cos x \dfrac{\cos x}{\sin x}$

$\dfrac{dy}{dx} = y(\cos x \cot x - \sin x \log \sin x)$

$= (\sin x)^{\cos x}(\cos x \cot x - \sin x \log \sin x)$

(ii) $y = \sin(e^x) + \dfrac{1}{x} + \log(x^2 + x)$, diff w.r.t. x

$\dfrac{dy}{dx} = \cos(e^x) e^x - \dfrac{1}{x^2} + \dfrac{1+2x}{x^2 + x}$

(iii) $y = \sin x(\log x)$, diff w.r.t. x by product rule

$\dfrac{dy}{dx} = \dfrac{\sin x}{x} + \cos x \,(\log x)$

(iv) $y = e^x \tan^4 x$, diff w.r.t. x by product rule

$\dfrac{dy}{dx} = e^x \tan^4 x + 4\tan^3 x \sec^2 x\, e^x$

$= e^x \tan^x(\tan^3 x + 4\sec^2 x)$

5. (i) $y = \log \sin x + \cos^{-1}(e^x) + x^4 \sec x$,

diff. w.r.t. x

$\dfrac{dy}{dx} = \dfrac{\cos x}{\sin x} - \dfrac{e^x}{\sqrt{1 - e^{2x}}} + 4x^3 \sec x$

$\quad + x^4 \sec x \tan x$

$= \cot x - \dfrac{e^x}{\sqrt{1 - e^{2x}}} + x^3 \sec x(x\tan x + 4)$

(ii) $y = (\log x)^x$

Taking log on both sides

$\log y = x \log(\log x)$

Diff. w.r.t. x, we get

$\dfrac{1}{y}\dfrac{dy}{dx} = \dfrac{x}{\log x}\left(\dfrac{1}{x}\right) + \log(\log x)$

$\dfrac{dy}{dx} = y\left\{\dfrac{x}{\log x} + \log(\log x)\right\}$

$= (\log x)^x \left\{\dfrac{1}{\log x} + \log(\log x)\right\}$

(iii) $\log\left(\dfrac{x}{y}\right) = x + y$

$\log x - \log y = x + y$,

Diff. both the sides w.r.t. x, we get

$\dfrac{1}{x} - \dfrac{1}{y}\dfrac{dy}{dx} = 1 + \dfrac{dy}{dx}$

$\dfrac{dy}{dx}\left(1 + \dfrac{1}{y}\right) = -1 + \dfrac{1}{x}$

$\dfrac{dy}{dx} = \dfrac{y}{x}\dfrac{(1-x)}{(1+y)}$

(iv) $y = \sec(x^2 - 2x + 1)$, diff. w.r.t. x both sides

$\dfrac{dy}{dx} = 2(x-1)\sec(x^2 - 2x + 1)$

$\tan(x^2 - 2x + 1)$

$= 2(x - 1)\sec(x-1)^2 \tan(x-1)^2$

6. (i) $y = e^{2x} \cos 3x$, diff. w.r.t. x

$\dfrac{dy}{dx} = 2e^{2x}\cos 3x + (-3\sin 3x)e^{2x}$

$= e^{2x}(2\cos 3x - 3\sin 3x)$

(ii) Let $y = \log(\sin^{-1} x^4)$ diff. both sides w.r.t. x

$\dfrac{dy}{dx} = \dfrac{1}{\sin^{-1} x^4} \times \dfrac{4x^3}{\sqrt{1 - x^8}}$

$= \dfrac{4x^3}{\sin^{-1} x^4 \sqrt{1 - x^8}}$

(iii) $x^3 - y^3 - 3axy = 0$, diff. w.r.t. x

$3x^2 - 3y^2 \dfrac{dy}{dx} - 3a\left(y + x\dfrac{dy}{dx}\right) = 0$

$-\dfrac{dy}{dx}(y^2 + ax) + x^2 - ay = 0$

$\dfrac{dy}{dx} = \dfrac{x^2 - ay}{y^2 + ax}.$

ANSWERS

1. (i) $\dfrac{4x^2 + 20x + 22}{(x^2 + 7x + 12)^2}$ (ii) $2x \cot x^2$ (iii) $\dfrac{x^2(-10x^7 + 100x^5 + 21x^2 - 35)}{(5x^6 + 7x)^2}$

(iv) $\dfrac{2\sec^2 x}{(1 - \tan x)^2}$ **2.** (i) $-\dfrac{1}{\sqrt{(1 - x^2)}}$ (ii) $-\dfrac{1}{2x\sqrt{x(x+1)}}$ (iii) $\dfrac{-1}{2\sqrt{x^2 - 1}}$

2. (iv) $\dfrac{-2(4x + 5)}{(2x^2 + 5x + 7)^3}$ **3.** (i) $4x^3 - 26x$ (ii) $8x^3 + 6x^2 - 21$ (iii) $\sec x \sqrt{(\sec 2x)}$

3. (iv) $(\tan x)^x (x \sec x \, \text{cosce}\, x + \log \tan x)$ **4.** (i) $(\sin x)^{\cos x}(\cos x \cot x - \sin x \log \sin x)$

4. (ii) $e^x \cos(e^x) - \dfrac{1}{x^2} + \dfrac{2x + 1}{x^2 + x}$ (iii) $\dfrac{\sin x}{x} + \cos x \log x$ (iv) $e^x \tan^3 x(4\sec^2 x + \tan x)$

5. (i) $\cot x - \dfrac{e^x}{\sqrt{1 - e^{2x}}} + x^3 \sec x(x \tan x + 4)$ (ii) $(\log x)^x \left[\dfrac{1}{\log x} + \log(\log x)\right]$ (iii) $\dfrac{(1 - x)y}{(1 + y)x}$

5. (iv) $2(x - 1)\sec(x^2 - 2x + 1)\tan(x^2 - 2x + 1)$ **6.** (i) $e^{2x}(2\cos 3x - 3\sin 3x)$

6. (ii) $\dfrac{4x^3}{\sin^{-1} x^4 \sqrt{(1 - x^8)}}$ (iii) $\dfrac{x^2 - ay}{ax + y^2}$

Exercise 9.2

Find $\dfrac{dy}{dx}$ for the following implicit functions.

1. $xy = x + y$ **2.** $(x^2 + y^2)^2 = xy$

3. $\sin(xy) + \dfrac{x}{y^2} = x^2 - y$

4. $ye^x + 2^x \sin x = \cos y$

5. $\sin(x + y) = \log(x + y)$

6. $e^{x - y} = \log\left(\dfrac{x}{y}\right)$ **7.** $x \cdot 2^y + 2^x = y$

8. $\tan^{-1}(x^2 + y^2) = 9$

9. $3\sin(xy) + 4\cos(xy) = 5$

10. $y \cos x = x - y$

11. If $\sqrt{\dfrac{y}{x}} + \sqrt{\dfrac{x}{y}} = 6$, show that $\dfrac{dy}{dx} = \dfrac{x - 17y}{17x - y}$.

12. If $\sin y = x \cos(a + y)$, prove that

$$\dfrac{dy}{dx} = \dfrac{\cos^2(a + y)}{\cos a}$$

13. If $\sqrt{1 - x^6} + \sqrt{1 - y^6} = a(x^3 - y^3)$ prove that

$$\dfrac{dy}{dx} = \dfrac{x^2 \sqrt{1 - y^6}}{y^2 \sqrt{1 - x^6}}$$

14. If $\log \sqrt{x^2 + y^2} = \tan^{-1} \dfrac{y}{x}$, prove that

$$\dfrac{dy}{dx} = \dfrac{x + y}{x - y}$$

15. If $e^x + e^y = e^{x + y}$, prove that $\dfrac{dy}{dx} = -\dfrac{e^x(e^y - 1)}{e^y(e^x - 1)}$

16. If $\cos y = x \cos(b + y)$, show that

$$\dfrac{dy}{dx} = \dfrac{\cos^2(b + y)}{\sin b}$$

17. If $y = \sqrt{\sin x \sqrt{\sin x \sqrt{\sin x \ldots \infty}}}$, show that

$$\dfrac{dy}{dx} = \dfrac{\cos x}{2y - 1}$$

HINT TO SELECTED PROBLEMS

1. $xy = x + y$

$\Rightarrow y = \dfrac{x}{x - 1} = 1 + \dfrac{1}{x - 1}$

Diff. w.r.t. x $\dfrac{dy}{dx} = -\dfrac{1}{(x - 1)^2}$

2. $(x^2 + y^2)^2 = xy$

Diff. w.r.t. x $2(x^2 + y^2)\left\{2x + 2y\dfrac{dy}{dx}\right\} = x\dfrac{dy}{dx} + y$

$\Rightarrow \{4y(x^2 + y^2) - x\}\dfrac{dy}{dx} = y - 4x(x^2 + y^2)$

$$\dfrac{dy}{dx} = \dfrac{y - 4x(x^2 + y^2)}{4y(x^2 + y^2) - x}$$

3. $\sin(xy) + \dfrac{x}{y^2} = x^2 - y$ ...(1)

Diff. w.r.t x

$$\cos(xy)\left(y + x\dfrac{dy}{dx}\right) + \dfrac{1}{y^2} - \dfrac{2x}{y^3}\dfrac{dy}{dx} = 2x - \dfrac{dy}{dx}$$

$$\Rightarrow \left(x\cos(xy) - \dfrac{2x}{y^3} + 1\right)\dfrac{dy}{dx} = 2x - \dfrac{1}{y^2} - y\cos(xy)$$

$$\Rightarrow \left(xy^2\cos(xy) - 2\dfrac{x}{y} + y^2\right)\dfrac{dy}{dx}$$

$$= 2xy^2 - 1 - y^3\cos(xy)$$

$$\text{(Multiplying by } y^2)$$

$$\Rightarrow \{xy^2\cos(xy) - 2(xy^2 - y^2$$

$$- y\sin(xy) + y^2)\dfrac{dy}{dx}$$

$$= 2xy^2 - 1 - y^3\cos(xy)$$

Putting the value of $\dfrac{x}{y}$ from (1)

$$\Rightarrow \dfrac{dy}{dx} = \dfrac{2xy^2 - y^3\cos(xy) - 1}{2y\sin(xy) + xy^2\cos(xy) - 2xy^2 + 3y^2}$$

4. $ye^x + 2^x \sin x = \cos y$

Diff. w.r.t. x

$$ye^x + e^x\dfrac{dy}{dx} + 2^x\log 2\sin x + 2^x\cos x$$

$$= -\sin y\dfrac{dy}{dx}$$

$$\Rightarrow \dfrac{dy}{dx} = \dfrac{-[ye^x + 2^x\cos x + 2^x\sin x\log 2]}{(e^x + \sin y)}$$

5. $\sin(x + y) = \log(x + y)$

Diff w.r.t. x

$$\cos(x + y)\left\{1 + \dfrac{dy}{dx}\right\} = \dfrac{1}{x + y}\left\{1 + \dfrac{dy}{dx}\right\}$$

$$\Rightarrow \left\{\cos(x + y) - \dfrac{1}{x + y}\right\}\dfrac{dy}{dx}$$

$$= -\left\{\cos(x + y) - \dfrac{1}{x + y}\right\}$$

$$\Rightarrow \dfrac{dy}{dx} = -1$$

6. $e^{x-y} = \log\dfrac{x}{y} = \log x - \log y$

Diff w.r.t x, we have

$$e^{x-y}\left(1 - \dfrac{dy}{dx}\right) = \dfrac{1}{x} - \dfrac{1}{y}\dfrac{dy}{dx}$$

$$\dfrac{dy}{dx}\left(\dfrac{1}{y} - e^{x-y}\right) = \dfrac{1}{x} - e^{x-y}$$

$$\dfrac{dy}{dx} = \dfrac{y(1 - xe^{x-y})}{x(1 - ye^{x-y})}$$

7. $x \cdot 2^y + 2^x = y$

Diff w.r.t. x, we have

$$2^y + x2^y\log 2\dfrac{dy}{dx} + 2^x\log 2 = \dfrac{dy}{dx}$$

$$2^y + 2^x\log 2 = \dfrac{dy}{dx}(1 - x \cdot 2^y\log 2)$$

$$\dfrac{dy}{dx} = \left[\dfrac{2^y + 2^x\log 2}{1 - x2^y\log 2}\right]$$

8. $\tan^{-1}(x^2 + y^2) = 9 \Rightarrow x^2 + y^2 = \tan 9$

Diff. w.r.t. x, we get

$$2x + 2y\dfrac{dy}{dx} = 0 \Rightarrow \dfrac{dy}{dx} = -\dfrac{x}{y}$$

9. $3\sin(xy) + 4\cos(xy) = 5$

Diff. w.r.t. x

$$3\cos(xy)\left\{y + x\dfrac{dy}{dx}\right\}$$

$$- 4\sin(xy)\left\{y + x\dfrac{dy}{dx}\right\} = 0$$

$$\dfrac{dy}{dx}\{3x\cos(xy) - 4x\sin(xy)$$

$$= -y\{3\cos(xy) - 4\sin(xy)\}$$

$$\dfrac{dy}{dx} = -\left(\dfrac{y}{x}\right)$$

10. $y\cos x = x - y$ ...(1)

Diff. w.r.t. x

$$-y\sin x + (\cos x)\dfrac{dy}{dx} = 1 - \dfrac{dy}{dx}$$

$$\dfrac{dy}{dx}(1 + \cos x) = 1 + y\sin x$$

$$\dfrac{dy}{dx} = \dfrac{1 + y\sin x}{1 + \cos x}$$

$$= \left(1 + \dfrac{x\sin x}{1 + \cos x}\right)\dfrac{1}{(1 + \cos x)}$$

$$= \dfrac{1 + x\sin x + \cos x}{(1 + \cos x)^2}$$

11. $\sqrt{\dfrac{x}{y}} + \sqrt{\dfrac{y}{x}} = 6$

$$\Rightarrow y + x = 6\sqrt{xy} \Rightarrow x^2 + y^2 - 34xy = 0$$

Diff. w.r.t. x

$$2x + 2y\frac{dy}{dx} - 34\left(x\frac{dy}{dx} + y\right) = 0$$

$$x + \frac{dy}{dx}(y - 17x) - 17y = 0$$

$$\frac{dy}{dx} = \frac{x - 17y}{17x - y}$$

12. $\sin y = x\cos(a + y)$ ...(1)

Diff. w.r.t. x

$$\frac{dy}{dx}(\cos y) = \cos(a + y) - x\sin(a + y)\left(\frac{dy}{dx}\right)$$

$$\frac{dy}{dx}\{\cos y + x\sin(a + y)\} = \cos(a + y)$$

$$\frac{dy}{dx}\left\{\cos y + \frac{\sin y}{\cos(a + y)}\sin(a + y)\right\} = \cos(a + y)$$

From (1)

$$\frac{dy}{dx}\{\cos(a + y)\cos y + \sin y \sin(a + y)\}$$

$$= \cos^2(a + y)$$

$$\frac{dy}{dx} = \frac{\cos^2(a + y)}{\cos a}$$

14. $\frac{1}{2}\log(x^2 + y^2) = \tan^{-1}\frac{y}{x}$

$$\frac{1}{2(x^2 + y^2)}\left(2x + 2y\frac{dy}{dx}\right) = \frac{1}{1 + \frac{y^2}{x^2}}\left(\frac{x\frac{dy}{dx} - y}{x^2}\right)$$

$$x + y\frac{dy}{dx} = x\frac{dy}{dx} - y$$

$$x + y = (x - y)\frac{dy}{dx}$$

$$\frac{dy}{dx} = \frac{x + y}{x - y}$$

15. $e^x + e^y = e^{x+y}$,

Diff w.r.t. x

$$e^x + e^y\frac{dy}{dx} = e^{x+y}\left\{1 + \frac{dy}{dx}\right\}$$

$$e^x - e^{x+y} = \frac{dy}{dx}(-e^y + e^{x+y})$$

$$\frac{dy}{dx} = \frac{e^x(1 - e^y)}{e^y(e^x - 1)}$$

ANSWERS

1. $\dfrac{1 - y}{x - 1}$ or $\dfrac{-1}{(x - 1)^2}$

2. $\dfrac{y - 4x(x^2 + y^2)}{4y(x^2 + y^2) - (x)}$

3. $\dfrac{2xy^2 - y^3\cos(xy) - 1}{2y\sin xy + xy^2\cos xy - 2xy^2 + 3y^2}$

4. $\dfrac{-[ye^x + 2^x\cos x + 2^x\sin x \log 2]}{e^x + \sin y}$

5. -1

6. $\dfrac{y}{x}\left[\dfrac{xe^{x-y} - 1}{ye^{x-y} - 1}\right]$

7. $\left[\dfrac{2^y + 2^x\log 2}{1 - x2^y\log 2}\right]$

8. $(-x/y)$

9. $-(y/x)$

10. $\dfrac{1 + y\sin x}{1 + \cos x}$

9.10 SECOND ORDER DERIVATIVES

It is known that derivative of y w.r.t. x (if exists) is denoted by $\dfrac{dy}{dx}$ and is called the first derivative of y,

Further, derivative of $\dfrac{dy}{dx}$ w.r.t. x (if it exists) is denoted by $\dfrac{d^2y}{dx^2}$ and is called the second derivative of y,

Thus, $\dfrac{d^2y}{dx^2} = \dfrac{d}{dx}\left(\dfrac{dy}{dx}\right) =$ second derivative of y w.r.t. x.

Similarly, $\dfrac{d^3y}{dx^3} = \dfrac{d}{dx}\left(\dfrac{d^2y}{dx^2}\right) =$ derivative of $\dfrac{d^2y}{dx^2}$ w.r.t. x.

In general $\dfrac{d^ny}{dx^n}$ denoted the nth derivative of y w.r.t. x.

OTHER SYMBOLS

(1) $\dfrac{dy}{dx}$ is also denoted by y_1 or y'.

(2) $\dfrac{d^2 y}{dx^2}$ is also denoted by y_2 or y''.

(3) $\dfrac{dy}{dx}$ is also denoted by Dy, where D is the operator $\dfrac{d}{dx}$.

$\dfrac{d^2 y}{dx^2}$ is also denoted by $D^2 y$, where D^2 is the operator $\dfrac{d^2}{dx^2}$,

(4) Let $f(x)$ be a differentiable function, then $f'(x)$ denotes the first derivative of $f(x)$ w.r.t.

x. Thus $f'(x) = \dfrac{d}{dx}\{f(x)\}$

Similarly $f''(x) = \dfrac{d}{dx}\{f'(x)\} = \dfrac{d^2\{f(x)\}}{dx^2}$ = second derivative of $f(x)$ w.r.t. x.

WORKING PROCEDURE

To Find the Higher Ordered Derivatives

Step 1. Let the given function be y.

Step 2. (i) Differentiate the given functions w.r.t x to get $\dfrac{dy}{dx}$.

(ii) If both base and power in the given function are variables, then first take logarithm and then differentiate to get $\dfrac{dy}{dx}$.

Step3. Now differentiate $\dfrac{dy}{dx}$ w.r.t. x to get $\dfrac{d^2 y}{dx^2}$.

Step 4. If a particular expression is to be obtained, simplify the expression involved after obtaining first derivative making use of the given relation between x and y and if required also use the expression for first derivative obtained.

After simplification find the higher derivative.

SOLVED EXAMPLES

EXAMPLE 1. *Find the second derivatives of the following functions*

      (i) $y = x^3 \log x$ (ii) $e^{6x} \cos 3x$

SOLUTION.   (i)   Let $y = x^3 \log x$

Differentiating w.r.t. x we get

$$\therefore \qquad \frac{dy}{dx} = x^3 \cdot \frac{1}{x} + 3x^2 \cdot \log x = x^2 + 3x^2 \log x$$

Again differentiating w.r.t. x, we get

$$\frac{d^2 y}{dx^2} = 2x + 3x^2 \cdot \frac{1}{x} + 6x \cdot \log x = 5x + 6x \log x$$

(ii) Let $\qquad y = e^{6x}\cos 3x$ $\qquad\qquad$ ...(1)

Differentiating (1) w.r.t x, we get

$\therefore \qquad \dfrac{dy}{dx} = e^{6x}\cdot 6\cos 3x + e^{6x}(-\sin 3x)\cdot 3$

$\qquad\qquad = 6e^{6x}\cos 3x - 3e^{6x}\sin 3x$ $\qquad\qquad$ ...(2)

Again differentiating (2) w.r.t. x, we get

$$\dfrac{d^2 y}{dx^2} = 6[e^{6x}\cdot 6\cos 3x + e^{6x}(-3\sin 3x)] - 3[e^{6x}\cdot 6\sin 3x + e^{6x}\cdot 3\cos 3x]$$

$$= e^{6x}[36\cos 3x - 18\sin 3x - 18\sin 3x - 9\cos 3x]$$

$$= e^{6x}(27\cos 3x - 54\sin 3x) = 27e^{6x}(\cos 3x - 2\sin 3x)$$

EXAMPLE 2. *If* $y = e^{\tan x}$, *prove that*

$$\cos^2 x\,\dfrac{d^2 y}{dx^2} - (1 + \sin 2x)\dfrac{dy}{dx} = 0$$

SOLUTION. $\qquad$ Given, $\qquad y = e^{\tan x}$

$\therefore \qquad\qquad \log y = \tan x$ $\qquad\qquad\qquad\qquad\qquad$ ...(1)

$\therefore \qquad\qquad \dfrac{1}{y}\dfrac{dy}{dx} = \sec^2 x \quad$ or $\quad \dfrac{dy}{dx} = y\sec^2 x$ $\qquad$ ...(2)

or $\qquad\qquad \cos^2 x\,\dfrac{dy}{dx} = y$ $\qquad\qquad\qquad\qquad$ ...(3)

Differentiating again w.r.t. x, we get

$$\cos^2 x\,\dfrac{d^2 y}{dx^2} - 2\cos x\sin x\,\dfrac{dy}{dx} = \dfrac{dy}{dx}$$

or $\quad \cos^2 x\,\dfrac{d^2 y}{dx^2} - (1 + \sin 2x)\dfrac{dy}{dx} = 0.$

EXAMPLE 3. *If* $y = a\cos(\log x) + b\sin(\log x)$, *show that* $x^2\dfrac{d^2 y}{dx^2} + x\dfrac{dy}{dx} + y = 0$.

SOLUTION. $\quad$ Given $\qquad\qquad y = a\cos(\log x) + b\sin(\log x)$ $\qquad$ ...(1)

Differentiating (1) w.r.t x, we get

$\therefore \qquad\qquad \dfrac{dy}{dx} = -a\sin(\log x)\cdot\dfrac{1}{x} + b\cos(\log x)\cdot\dfrac{1}{x}$

or $\qquad\qquad x\cdot\dfrac{dy}{dx} = -a\sin(\log x) + b\cos(\log x)$

Again differentiating w.r.t. x, we get

$$x\dfrac{d^2 y}{dx^2} + 1\cdot\dfrac{dy}{dx} = -a\cos(\log x)\cdot\dfrac{1}{x} - b\sin(\log x)\cdot\dfrac{1}{x}$$

or $\qquad x^2\dfrac{d^2 y}{dx^2} + x\dfrac{dy}{dx} = -[a\cos(\log x) + b\sin(\log x)] = -y$ $\qquad$ [From (1)]

or $\quad x^2\dfrac{d^2 y}{dx^2} + x\dfrac{dy}{dx} + y = 0.$

EXAMPLE 4. If $x = (\sin^{-1} x)^2$, prove that $(1 - x^2)\dfrac{d^2 y}{dx^2} = x\dfrac{dy}{dx} + 2$.

SOLUTION. Given $\qquad\qquad y = (\sin^{-1} x)^2 \qquad\qquad\qquad$ ...(1)

Differentiating (1) w.r.t. x, we get

$\therefore \qquad\qquad \dfrac{dy}{dx} = 2\sin^{-1} x \cdot \dfrac{1}{\sqrt{1 - x^2}} \quad$ or $\quad \sqrt{1 - x^2}\dfrac{dy}{dx} = 2\sin^{-1} x$

Squaring, we get

$$(1 - x^2)\left(\dfrac{dy}{dx}\right)^2 = 4(\sin^{-1} x)^2 = 4y \qquad\qquad\qquad \text{[From (1)]}$$

Again differentiating both sides w.r.t. x, we get

$$(1 - x^2)2\dfrac{dy}{dx} \cdot \dfrac{d^2 y}{dx^2} + (-2x)\left(\dfrac{dy}{dx}\right)^2 = 4\dfrac{dy}{dx}$$

Dividing both sides by $2\dfrac{dy}{dx}$, we have

$$(1 - x^2)\dfrac{d^2 y}{dx^2} = x \cdot \dfrac{dy}{dx} + 2.$$

EXAMPLE 5. If $y = e^{ax}\sin bx$, prove that $\dfrac{d^2 y}{dx^2} - 2a\dfrac{dy}{dx} + (a^2 + b^2)y = 0$.

SOLUTION. Given, $\qquad\qquad y = e^{ax}\sin bx \qquad\qquad\qquad$ ...(1)

$\therefore \qquad\qquad \dfrac{dy}{dx} = e^{ax} \cdot b\cos bx + a \cdot e^{ax} \cdot \sin bx$

$$= be^{ax}\cos bx + ay \qquad\qquad \text{[From (1)]} \qquad \text{...(2)}$$

Again differentiating both sides w.r.t. x, we have

$$\dfrac{d^2 y}{dx^2} = bae^{ax}(\cos bx) + be^{ax} \cdot b(-\sin bx) + a\dfrac{dy}{dx}$$

$$= a(be^{ax}\cos bx) - b^2(e^{ax}\sin bx) + a\dfrac{dy}{dx}$$

$$= a\left(\dfrac{dy}{dx} - ay\right) - b^2 y + a\dfrac{dy}{dx} \qquad\qquad \text{[From (1) and (2)]}$$

$$= 2a\dfrac{dy}{dx} - (a^2 + b^2)y$$

$\therefore \qquad\qquad \dfrac{d^2 y}{dx^2} - 2a\dfrac{dy}{dx} + (a^2 + b^2)y = 0.$

EXAMPLE 6. If $y = x\sin x$, prove that $x^2\dfrac{d^2 y}{dx^2} - 2x\dfrac{dy}{dx} + (x^2 + 2)y = 0$.

SOLUTION. Given $\qquad\qquad y = x\sin x \qquad\qquad\qquad$ ...(1)

$\therefore \qquad\qquad \dfrac{dy}{dx} = x\cos x + \sin x \qquad\qquad\qquad$ ... (2)

Again differentiating both sides w.r.t. x, we get

$$\dfrac{d^2 y}{dx^2} = x(-\sin x) + 1 \cdot \cos x + \cos x = 2\cos x - x\sin x$$

Multiplying by x^2 we get

$$x^2 \frac{d^2y}{dx^2} = 2x^2 \cos x - x^2 \cdot x \sin x = 2x^2 \cos x - x^2 y \quad \text{[From (1)]}$$

$$= 2x\left(\frac{dy}{dx} - \sin x\right) - x^2 y \quad \text{[From (2)]}$$

$$= 2x\frac{dy}{dx} - 2x\sin x - x^2 y = 2x\frac{dy}{dx} - 2y - x^2 y \quad \text{[From (1)]}$$

$$\therefore \qquad x^2 \frac{d^2y}{dx^2} - 2x\frac{dy}{dx} + (x^2 + 2)y = 0.$$

EXMAPLE 7. *If* $x = a(\cos t + t\sin t),\ y = a(\sin t - t\cos t)$

Find $\dfrac{d^2y}{dx^2}.$

SOLUTION. We have

$$x = a(\cos t + t\sin t)$$
$$y = a(\sin t - t\cos t)$$

$$\frac{dx}{dt} = a(-\sin t + t\cos t + \sin t) = at\cos t$$

$$\frac{dy}{dt} = a(\cos t + t\sin t - \cos t) = at\sin t$$

$$\frac{dx}{dt} = \frac{dy/dt}{dx/dt} = \frac{at\sin t}{at\cos t} = \tan t$$

$$\frac{d^2y}{dx^2} = \frac{d}{dx}\left(\frac{dy}{dx}\right) = \frac{d}{dt}\left(\frac{dy}{dx}\right) \cdot \frac{dt}{dx}$$

$$= \frac{d}{dt}(\tan t) \cdot \frac{1}{at\cos t} = \sec^2 t \cdot \frac{1}{at\cos t}$$

$$\frac{d^2y}{dx^2} = \frac{\sec^3 t}{at}$$

EXAMPLE 8. *If* $y = x + \tan x,$ *show that*

$$\cos^2 x \cdot \frac{d^2y}{dx^2} - 2y + 2x = 0$$

SOLUTION. We have $\qquad y = x + \tan x$

$$\frac{dy}{dx} = 1 + \sec^2 x$$

$$\Rightarrow \qquad \frac{d^2y}{dx^2} = 2\sec x \cdot \sec x \tan x = 2\sec^2 x \tan x$$

Now, consider LHS

$$\cos^2 x \cdot \frac{d^2y}{dx^2} - 2y + 2x$$

$$= \cos^2 x(2\sec^2 x \tan x) - 2(x + \tan x) + 2x$$

$$= 2\tan x - 2x - 2\tan x + 2x = 0 = \text{RHS}$$

Exercise 9.3

1. Find the second order derivative of the following functions :
   (i) $\log x$
   (ii) $x^2 + 3x + 2$
   (iii) $x \cos x$
   (iv) $e^x \sin 5x$
   (v) $\sin(\log x)$

2. If $x = a(\theta - \sin \theta), y = a(1 - \cos \theta)$, find $\dfrac{dy}{dx}$.

   Also find $\dfrac{d^2 y}{dx^2}$.

3. If $\cos x = \dfrac{1-t^2}{1+t^2}$ and $\sin y = \dfrac{2t}{1+t^2}, 0 \le t \le 1$.

   Show that $\dfrac{d^2 y}{dx^2}$ is independent of t.

4. If $x = 3\sin t - \sin 3t, y = 3\cos t - \cos 3t$,

   find $\dfrac{d^2 y}{dx^2}$ at $t = \dfrac{\pi}{3}$.

5. If $y = x^3 + \tan x$, show that

   $$\dfrac{d^2 y}{dx^2} = 6x + 2\sec^2 x \tan x$$

ANSWERS

1. (i) $-\dfrac{1}{x^2}$ (ii) 2 (iii) $-x \cos x - 2\sin x$ (iv) $2e^x(5\cos 5x - 12\sin 5x)$

1. (v) $\dfrac{-\sin(\log x) + \cos(\log x)}{x^2}$ 2. $\dfrac{dy}{dx} = \cot \dfrac{\theta}{2}, \dfrac{d^2 y}{dx^2} = -\dfrac{1}{4a} \operatorname{cosec} \dfrac{4\theta}{2}$ 4. $-\dfrac{8}{9}$.

Objective Evaluation

⊙ MULTIPLE CHOICE QUESTIONS

Choose the most appropriate one.

1. If $y = \dfrac{1}{3x+4}$, then $\dfrac{dy}{dx}$ is :

(a) $\dfrac{3}{(3x+4)^2}$ (b) $\dfrac{-3}{(3x+4)^2}$

(c) $\dfrac{3}{(3x-4)^2}$ (d) $\dfrac{-3}{(3x-4)^2}$

2. If $y = -2x^2 + 4x + 6$, then $\dfrac{dy}{dx}$ is :

(a) $-4x + 4$ (b) $4x + 4$

(c) $-4x - 4$ (d) $-2x + 6$

3. If $y = 5\sin^{-1}x$, then $\dfrac{dy}{dx}$ is :

(a) $\dfrac{2}{\sqrt{1-x^2}}$ (b) $\dfrac{2}{\sqrt{1+x^2}}$

(c) $\dfrac{5}{\sqrt{1-x^2}}$ (d) $\dfrac{5}{\sqrt{1+x^2}}$

4. If $y = a^x + \log x + a^2 e^x$, then $\dfrac{dy}{dx}$ is :

(a) $a^x \log_e a + \dfrac{1}{x} + a^2 e^x$

(b) $a^x \log_e a + \dfrac{1}{x} - a^2 e^x$

(c) $a^x \log_e a - \dfrac{1}{x} - a^2 e^x$

(d) none of these

5. If $y = \cos x^3$, then $\dfrac{dy}{dx}$ is :

(a) $-3x^2 \sin x^3$ (b) $3x^2 \sin x^3$

(c) $2x^2 \sin x^3$ (d) none of these

6. If $y = x^{x^{x^{\cdots\infty}}}$, then $\dfrac{dy}{dx}$ is :

(a) $\dfrac{y^2}{(1 - y \log x)x}$ (b) $\dfrac{-y^2}{(1 - y \log x)x}$

(c) $\dfrac{y^2}{(1 + y \log x)x}$ (d) none of these

7. If $y = (ax)^n + b$, then $\dfrac{dy}{dx}$ is :

(a) $na^n x^{n+1}$ (b) $na^n x^n$

(c) $na^n x^{n-1}$ (d) none of these

8. If $x = a(t - \sin t)$, $y = a(1 - \cos t)$, then $\dfrac{dy}{dx}$ is :

(a) $\sin t/2$ (b) $\cos t/2$

(c) $\tan t/2$ (d) $\cot t/2$

9. If $\sqrt{x} + \sqrt{y} = 5$, then $\left[\dfrac{dy}{dx}\right]_{(4,9)}$ is :

(a) $\dfrac{2}{3}$ (b) $-\dfrac{2}{3}$

(c) $-\dfrac{3}{2}$ (d) $\dfrac{3}{2}$

10. If $y = \tan^{-1}\sqrt{\dfrac{1 - \cos x}{1 + \cos x}}$, then $\dfrac{dy}{dx}$ is :

(a) $\dfrac{1}{2}$ (b) $\dfrac{3}{2}$

(c) $\dfrac{5}{2}$ (d) $\dfrac{7}{2}$

⊙ TRUE/FALSE

Write 'T' for True and 'F' for False statement.

1. $\dfrac{d}{dx}(\cos x) = \sin x$ **(T/F)**

2. $\dfrac{d}{dx}(\tan x) = \sec^2 x$ **(T/F)**

3. $\dfrac{d}{dx}(e^x) = e^x$ **(T/F)**

4. $\dfrac{d}{dx}(4\cos x) = 4\sin x$ **(T/F)**

5. $\dfrac{d}{dx}(x^3 \log x) = x^2 \cdot (1 + 3\log x)$ **(T/F)**

⊙ FILL IN THE BLANKS

1. If $y = 3x^2 + 6x + 10$, then $\dfrac{dy}{dx}$ is _____ .

2. If $y = \sin 5x$, then $\dfrac{dy}{dx}$ is _____ .

3. If $y = x^x$, then $\dfrac{dy}{dx}$ is _____ .

4. $\dfrac{d}{dx}(\cot x) =$ _____ .

5. $\dfrac{d}{dx}\log_a x =$ _____ .

ANSWERS

⊛ MULTIPLE CHOICE QUESTIONS

1. (b)   **2.** (a)   **3.** (c)   **4.** (a)   **5.** (a)   **6.** (a)   **7.** (c)   **8.** (d)   **9.** (c)
10. (a)

⊛ TRUE/ FALSE

1. F   **2.** T   **3.** T   **4.** F   **5.** T

⊛ FILL IN THE BLANKS

1. $6x + 6$ **2.** $5\cos 5x$ **3.** $x^x(1 + \log x)$ **4.** $-\csc^2 x$

5. $\dfrac{1}{x}\log_a e$

□□□□□

CHAPTER 10

Integration

10.1 INTRODUCTION

We have already discuss the methods of finding the derivatives of a function $f(x)$. We have notice that the derivative of a function is also a function (may be the function of independent variable or constant). In this chapter we dealt with the converse of the derivative.

Consider the following questions :

(i) What is the function, whose derivative is $8x^7$.

(ii) What is the function, whose derivative is 8.

Obviously, x^8 and $8x$ be the functions, whose derivative is $8x^7$ and 8 respectively, which are called the integrals of the given functions. Further, integral is the inverse operation of differentiation.

For example: If the derivative of $\tan x$ is $\sec^2 x$ then integration of $\sec^2 x$ is $\tan x$. Now if $f(x)$ is the function of x which is differentiable, such that $\dfrac{d}{dx}[F(x)] = f(x)$. Then $F(x)$ is known as the integral of $f(x)$ and denoted by $F(x) = \int f(x)\,dx$ and the function $f(x)$ is called the integrand of $\int f(x)\,dx$ and the function $\int f(x)\,dx$ is read as the integral of $f(x)$ w.r.t. x".

Thus, integration is the process of finding the integral of a given function.

⮞ REMARKS

⟼ Integral is also known as primitive or antiderivative or an indefinite integral.

⟼ The symbol $\int ... dx$, is stands for integral and taking separately the symbols is meaningless.

⟼ In general, integral is defined as the sum of the certain infinite terms of a function at very small interval.

10.2 INDEFINITE INTEGRAL

To define the indefinite, integral, let $F(x)$ be a function and C be a constant.

So, $\qquad \dfrac{d}{dx}[F(x) + C] = \dfrac{d}{dx}[F(x)] + \dfrac{dc}{dx} = f(x)$

$[\because$ derivative of a constant function is always zero$]$

Hence, $\int f(x)dx = F(x) + C$, where, the symbol '$\int$' is an integral sign, and C is the constant of integration. The constant C may have any value, but in general C is omitted and the function $F(x)$ is called the indefinite integral of integrand $f(x)$.

SOME STANDARD RESULTS

(i) $\int x^n dx = \dfrac{x^{n+1}}{n+1} + C, (\text{if } n \neq -1)$	(ii) $\int e^x dx = e^x + C,$
(iii) $\int \dfrac{1}{x} dx = \dfrac{\log_a x}{\log_a e} + C$	(iv) $\int \dfrac{1}{x} dx = \log_e x + C$
(v) $\int a^x dx = \dfrac{a^x}{\log_e a} + C$	(vi) $\int \cos x \, dx = \sin x + C$
(vii) $\int \sin x \, dx = -\cos x + C$	(viii) $\int \text{cosec}^2 x \, dx = -\cot x + C$
(ix) $\int \sec^2 x \, dx = \tan x + C$	(x) $\int \sec x \tan x \, dx = \sec x + C$
(xi) $\int \text{cosec}\, x \cdot \cot x \, dx = -\text{cosec}\, x + C$	(xii) $\int \dfrac{dx}{\sqrt{(1-x^2)}} = \sin^{-1} x + C$
(xiii) $\int \dfrac{(-1)dx}{\sqrt{(1-x^2)}} = \cos^{-1} x + C$	(xiv) $\int \dfrac{dx}{x\sqrt{(x^2-1)}} = \sec^{-1} x + C$
(xv) $\int -\dfrac{dx}{x\sqrt{(x^2-1)}} = \text{cosec}^{-1} x + C$	(xvi) $\int \dfrac{dx}{1+x^2} = \tan^{-1} x + C$
(xvii) $\int \dfrac{-dx}{1+x^2} = \cot^{-1} x + C$	(xviii) $\int c \, dx = cx + d,$ where c is a constant

THEOREM 1. *The integral of the product of a function with a constant is equal to the product of the constant and integral of that function.*

PROOF. Let $\int f(x) \, dx = F(x)$

$\therefore \qquad \dfrac{d}{dx} F(x) = f(x)$

Then, by differential calculus, we have

$\qquad \dfrac{d}{dx}\{cF(x)\} = c\dfrac{d}{dx} F(x) = cf(x)$

So, $\qquad \int c \cdot f(x) = c \cdot F(x) = c\int f(x) \, dx$

Hence, $\qquad \int c \cdot f(x) = c\int f(x) \, dx$

THEOREM 2. *The integral of the sum or difference of two functions is equal to the sum or difference of their integral.*

i.e., $\qquad \int \{f_1(x) \pm f_2(x)\}dx = \int f_1(x)dx \pm \int f_2(x)dx$

PROOF. Consider the function

$\qquad \int \{f_1(x)dx \pm f_2(x)\}dx$

then $\dfrac{d}{dx}\left[\int f_1(x)dx \pm \int f_2(x)dx\right] = \dfrac{d}{dx}\int f_1(x)dx \pm \dfrac{d}{dx}\int f_2(x)dx = f_1(x) \pm f_2(x)$

Hence, $\qquad \int [f_1(x) \pm f_2(x)]dx = \int f_1(x)dx \pm \int f_2(x)dx.$

⮷ **REMARK**
⟹ From the above discussion, we can write

$$\int [f_1(x) \pm f_2(x) \pm ...] dx = \int f_1(x)\,dx \pm \int f_2(x)\,dx \pm .... \,.$$

SOLVED EXAMPLES

EXAMPLE 1. *Evaluate the following integrals.*

    (i) $\int 5x^4\,dx$ (ii) $\int x^7\,dx$ (iii) $\int x^{-n}\,dx;\ n \neq 1$

    (iv) $\int (ax^5 + bx^3 + cx + d)\,dx$ (v) $\int \dfrac{1}{\sqrt{1-x^2}}\,dx$

    (vi) $\int \left(\sqrt{x} + \dfrac{1}{\sqrt{x}} \right)^2 dx$ (vii) $\int \dfrac{1}{t\sqrt{t^2-1}}\,dt.$

SOLUTION.   (i) Here, $\int 5x^4\,dx = 5\int x^4\,dx = 5\dfrac{x^{4+1}}{4+1} = 5 \cdot \dfrac{x^5}{5} = x^5 + C.$

    (ii) $\int x^7\,dx = \dfrac{x^{7+1}}{7+1} = \dfrac{x^8}{8} + C$

    (iii) $\int x^{-n}\,dx = \dfrac{x^{-n+1}}{-n+1} + C;\ n \neq 1$

    (iv) $\int (ax^5 + bx^3 + cx + d)\,dx = a\int x^5\,dx + b\int x^3\,dx + c\int x\,dx + \int d\,dx$

$$= a \cdot \dfrac{x^{5+1}}{5+1} + b \cdot \dfrac{x^{3+1}}{3+1} + c\dfrac{x^{1+1}}{1+1} + d \cdot x + C$$

$$= a\dfrac{x^6}{6} + b\dfrac{x^4}{4} + c\dfrac{x^2}{2} + d \cdot x + C$$

    (v) $\int \dfrac{1}{\sqrt{(1-x^2)}}\,dx = \sin^{-1} x + C$

    (vi) $\int \left(\sqrt{x} + \dfrac{1}{\sqrt{x}} \right)^2 dx = \int \left(x + \dfrac{1}{x} + 2 \right) dx = \int x\,dx + \int \dfrac{1}{x}\,dx + \int 2\,dx$

$$= \dfrac{x^{1+1}}{1+1} + \log x + 2x + C = \dfrac{x^2}{2} + \log x + 2x + C$$

    (vi) $\int \dfrac{1}{t\sqrt{t^2-1}}\,dt = \sec^{-1} t + C.$

EXAMPLE 2. *Evaluate the following integrals.*

    (i) $\int (\operatorname{cosec} x \cot x - 3\sec^2 x)\,dx$ (ii) $\int \left(\sqrt{bx} + \dfrac{1}{\sqrt{bx}} \right) dx$

    (iii) $\int \tan^2 x\,dx$ (iv) $\int \dfrac{(x+a)^3}{2\sqrt{x}}\,dx$

    (v) $\int \sqrt{(1 + \cos 2x)}\,dx$ (vi) $\int \left(\dfrac{4\sin x}{5\cos^2 x} + \dfrac{3}{5\sin^2 x} \right) dx.$

SOLUTION.   (i) Here, we have

$$\int (\operatorname{cosec} x \cot x - 3\sec^2 x)\,dx = \int \operatorname{cosec} x. \cot x\,dx - 3\int \sec^2 x\,dx$$

$$= -\operatorname{cosec} x - 3\tan x + C.$$

(ii) We have $I = \int\left(\sqrt{bx} + \dfrac{1}{\sqrt{bx}}\right)dx = \sqrt{b}\int x^{1/2}dx + \dfrac{1}{\sqrt{b}}\int x^{-1/2}dx$

$\qquad = \sqrt{b}\,\dfrac{x^{1/2+1}}{\dfrac{1}{2}+1} + \dfrac{1}{\sqrt{b}}\dfrac{x^{-1/2+1}}{-\dfrac{1}{2}+1} + C$

$\qquad = \dfrac{2\sqrt{b}x^{3/2}}{3} + \dfrac{2x^{1/2}}{\sqrt{b}} + C.$

(iii) We have $I = \int \tan^2 x\,dx = \int(\sec^2 x - 1)dx$

$\qquad = \int\sec^2 x\,dx - \int dx = \tan x - x + C.$

(iv) We have $I = \int\dfrac{(x+a)^3}{2\sqrt{x}}dx = \dfrac{1}{2}\left(\dfrac{x^3 + 3ax^2 + 3a^2x + a^3}{\sqrt{x}}\right)dx$

$\qquad = \dfrac{1}{2}\int\dfrac{x^3}{\sqrt{x}}dx + \dfrac{3}{2}a\int\dfrac{x^2}{\sqrt{x}}dx + \dfrac{3}{2}a^2\int\dfrac{x}{\sqrt{x}}dx + \dfrac{1}{2}a^3\int\dfrac{dx}{\sqrt{x}}$

$\qquad = \dfrac{1}{2}\int x^{5/2}dx + \dfrac{3}{2}a\int x^{3/2}dx + \dfrac{3}{2}a^2\int x^{1/2}dx + \dfrac{1}{2}a^3\int x^{-1/2}dx$

$\qquad = \dfrac{1}{2}\dfrac{x^{5/2+1}}{\dfrac{5}{2}+1} + \dfrac{3}{2}a\cdot\dfrac{x^{3/2+1}}{\dfrac{3}{2}+1} + \dfrac{3}{2}a^2\dfrac{x^{1/2+1}}{\dfrac{1}{2}+1} + \dfrac{1}{2}a^3\dfrac{x^{-1/2+1}}{-\dfrac{1}{2}+1} + C$

$\qquad = \dfrac{1}{2}\dfrac{x^{7/2}}{\dfrac{7}{2}} + \dfrac{3}{2}a\cdot\dfrac{x^{5/2}}{\dfrac{5}{2}} + \dfrac{3}{2}a^2\dfrac{x^{3/2}}{\dfrac{3}{2}} + \dfrac{1}{2}a^3\dfrac{x^{1/2}}{\dfrac{1}{2}} + C$

$\qquad = \dfrac{1}{7}x^{7/2} + \dfrac{3a}{5}x^{5/2} + a^2x^{3/2}. + a^3x^{1/2} + C$

(v) We have, $\int\sqrt{1 + \cos 2x}\,dx = \int\sqrt{(1 + 2\cos^2 x - 1)}dx$

$\qquad\qquad (\because \cos 2x = 2\cos^2 x - 1)$

$\qquad = \int\sqrt{(2\cos^2 x)}dx = \sqrt{2}\int\cos x\,dx$

$\qquad = \sqrt{2}\sin x + C$

(vi) We have

$I = \int\left(\dfrac{4\sin x}{5\cos^2 x} + \dfrac{3}{5\sin^2 x}\right)dx = \dfrac{4}{5}\int\dfrac{\tan x}{\cos x}dx + \dfrac{3}{5}\int \text{cosce}^2\, x\,dx$

$\qquad = \dfrac{4}{5}\int\tan x.\sec x\,dx + \dfrac{3}{5}\int\text{cosec}^2\, x\,dx$

$\qquad = \dfrac{4}{5}\sec x + \dfrac{3}{5}(-\cot x) + c = \dfrac{4}{5}\sec x - \dfrac{3}{5}\cot x + C.$

EXAMPLE 3. If $\dfrac{d}{dx}[f(x)] = 3x^2 + 2x + 1$, then find $f(x)$.

SOLUTION. Since $\dfrac{d}{dx}f(x) = 3x^2 + 2x + 1$

then $\qquad f(x) = \int(3x^2 + 2x + 1)dx = 3\int x^2 dx + 2\int x\,dx + \int 1.dx$

$\qquad = 3\dfrac{x^{2+1}}{2+1} + \dfrac{2\cdot x^{1+1}}{1+1} + x + C = x^3 + x^2 + x + C$

EXAMPLE 4. *Evaluate the following integrals.*

(i) $\int (ae^x + x^3)\, dx$ (ii) $\int (1+x)(2-5x)\, dx$

(iii) $\int \left(\dfrac{x^3 + 5x - 6}{x^2} \right) dx$ (iv) $\int (x-a)^2\, dx$

(v) $\int \left(\dfrac{ax^3 + bx^2 + cx + d}{x} \right) dx$ (vi) $\int \left(\dfrac{4 + 3\sin x}{\cos^2 x} \right) dx.$

SOLUTION.

(i) We have $I = \int (ae^x + x^3)\, dx = a\int e^x dx + \int x^3 dx$

$$= ae^x + \frac{x^{3+1}}{3+1} = ae^x + \frac{x^4}{4} + C.$$

(ii) We have $I = \int (1+x)(2-5x)\, dx = \int (2 - 3x - 5x^2)\, dx$

$$= 2\int dx - 3\int x\, dx - 5\int x^2 dx$$

$$= 2x - \frac{3x^{1+1}}{1+1} - \frac{5x^{2+1}}{2+1} = 2x - \frac{3}{2}x^2 - \frac{5}{3}x^3 + C$$

(iii) We have $I = \int \left(\dfrac{x^3 + 5x - 6}{x^2} \right) dx = \int \left(x + \dfrac{5}{x} - \dfrac{6}{x^2} \right) dx$

$$= \int x\, dx + 5\int \frac{1}{x} - 6\int x^{-2} dx$$

$$= \frac{x^{1+1}}{1+1} + 5 \cdot \log_e x - 6\frac{x^{-2+1}}{-2+1} = \frac{x^2}{2} + 5\log_e x + \frac{6}{x} + C.$$

(iv) We have $I = \int (x-a)^2\, dx = \int (x^2 + a^2 - 2ax)\, dx$

$$= \int x^2 dx + a^2 \int dx - 2a\int x\, dx = \frac{x^{2+1}}{2+1} + a^2 x - 2a \cdot \frac{x^{1+1}}{1+1} + C$$

$$= \frac{x^3}{3} + a^2 x - ax^2 + C.$$

(v) We have $I = \int \left(\dfrac{ax^3 + bx^2 + cx + d}{x} \right) dx = \int \left(ax^2 + bx + c + \dfrac{d}{x} \right) dx$

$$= a\int x^2 dx + b\int x\, dx + c\int 1.dx + d\int \frac{1}{x} dx$$

$$= a\frac{x^{2+1}}{2+1} + b\frac{x^{1+1}}{1+1} + cx + d\log_e x + C$$

$$= \frac{ax^3}{3} + \frac{bx^2}{2} + cx + d\log_e x + C$$

(vi) We have $I = \int \left(\dfrac{4 + 3\sin x}{\cos^2 x} \right) dx = \int \left(\dfrac{4}{\cos^2 x} + \dfrac{3\sin x}{\cos^2 x} \right) dx$

$$= \int (4\sec^2 x + 3\tan x \cdot \sec x)\, dx$$

$$= 4\int \sec^2 x\, dx + 3\int \tan x. \sec x\, dx = 4\tan x + 3\sec x + C$$

Example 5.   *Evaluate* $\int \dfrac{x^4}{x^2+1} dx$.

Solution.   Consider $I = \int \dfrac{x^4}{x^2+1} dx = \int \dfrac{(x^4-1)+1}{x^2+1} dx = \int \dfrac{x^4-1}{x^2+1} dx + \int \dfrac{dx}{x^2+1}$

$$= \int \dfrac{(x^2-1)(x^2+1)}{x^2+1} dx + \int \dfrac{dx}{x^2+1} = \int (x^2-1) dx + \int \dfrac{dx}{x^2+1}$$

$$= \dfrac{x^3}{3} - x + \tan^{-1} x + C$$

10.3 METHOD OF INTEGRATION

10.3.1 INTEGRATION BY SUBSTITUTION

In this method, we transform the given integral into the another standard integral of some other independent variable. The method of changing the variable is called substitution method. To explain the method of substitution,

Consider the integral $\int f\{\phi(x)\}\phi'(x) dx$.

Put $\phi(x) = t$ so $\phi'(x) dx = dt$.

So, $\int f\{\phi(x)\} \cdot \phi'(x) dx = \int f(t) dt$

Now, $\int f(t) dt$ can be easily evaluated, then substitute $t = \phi(x)$ and get the required result.

SOME IMPORTANT SUBSTITUTION

Expression	Substitution
(i) $a^2 + x^2$	$x = a\tan\theta$ or $a\cot\theta$
(ii) $a^2 - x^2$	$x = a\sin\theta$ or $a\cos\theta$
(iii) $x^2 - a^2$	$x = a\sec\theta$ or $a\,\mathrm{cosec}\,\theta$
(iv) $\sqrt{\dfrac{a-x}{a+x}}$ or $\sqrt{\dfrac{a+x}{a-x}}$	$x = a\cos 2\theta$
(v) $\sqrt{\dfrac{x-a}{x-b}}$ or $\sqrt{(x-a)(x-b)}$	$x = a\cos^2\theta + b\sin^2\theta$

SOLVED EXAMPLES

Example 1.   *Obtain the following integrals*

(i) $\int x^2 \sin x^3 dx$ (ii) $\int (4x+5)^6 dx$

(iii) $\int x(x^2+4)^5 dx$ (iv) $\int 2x^3 \sqrt{(x^2+4)}\ dx$.

Solution.   (i)   The given integral is $\int x^2 \sin x^3 dx$.

Put $x^3 = t$ so $3x^2 dx = dt$.

So, $\int x^2 \sin x^3 dx = \dfrac{1}{3} \int \sin t \cdot dt = \dfrac{1}{3}(-\cos t) = -\dfrac{\cos t}{3} + C$

$$= -\frac{1}{3}\cos x^3 + C \qquad\qquad (\because\ t = x^3)$$

(ii) The given integral is $\int (4x+5)^6 dx$. Put $4x+5 = t \Rightarrow 4\,dx = dt$

So, $\qquad \int t^6\,\frac{dt}{4} = \frac{1}{4}\int t^6 dt = \frac{1}{4}\frac{t^7}{7} + C$

$$= \frac{1}{4}\frac{(4x+5)^7}{7} + C \qquad\qquad (\text{On putting } t = 4x+5)$$

(iii) The given integral is $\int x(x^2+4)^5 dx$.

Put $x^2 + 4 = t$ so, $2x\,dx = dt$

So, $\qquad \int t^5\,\frac{dt}{2} = \frac{1}{2}\int t^5 dt = \frac{1}{2}\frac{t^{5+1}}{5+1} = \frac{1}{12}t^6 + C$

$$= \frac{1}{2}(x^2+4)^6 + C \qquad\qquad (\text{On putting } t = x^2+4)$$

(iv) The given integral is $\int 2x^3\sqrt{x^2+4}\;dx$.

Put $(x^2+4) = t$ so $2x\,dx = dt$ and $x^2 = t-4$.

So, $\int 2x^2 \cdot x\sqrt{x^2+4}\;dx = \int(t-4)\sqrt{t}\;dt = \int t^{3/2}dt - 4\int t^{1/2}dt$

$$= \frac{t^{3/2+1}}{\frac{3}{2}+1} - 4\frac{t^{1/2+1}}{\frac{1}{2}+1} + C = \frac{2}{5}t^{5/2} - \frac{8}{3}t^{3/2} + C$$

$$= \frac{2}{5}(x^2+4)^{5/2} - \frac{8}{3}(x^2+4)^{3/2} + C$$

$$(\text{On putting } t = x^2+4)$$

EXAMPLE 2. *Evaluate* $\int \dfrac{dx}{5+4\cos x}$.

SOLUTION. Let

$$I = \int \frac{dx}{5+4\cos x} = \int \frac{dx}{5+4\left(2\cos^2\dfrac{x}{2}-1\right)} = \int \frac{fx}{1+8\cos^2\dfrac{x}{2}}$$

Divide numerator and denominator by $\cos^2\dfrac{x}{2}$, we get

$$I = \int \frac{\sec^2\dfrac{x}{2}dx}{\sec^2\dfrac{x}{2}+8} = \int \frac{\sec^2\dfrac{x}{2}dx}{1+\tan^2\dfrac{x}{2}+8} = \int \frac{\sec^2 x/2\,dx}{9+\tan^2 x/2}$$

Putting $\tan\dfrac{x}{2} = t$, and differenting w.r.t. x, we get

$$\frac{1}{2}\sec^2\frac{x}{2}dx = dt$$

Then $\qquad I = \int \dfrac{2dt}{3^2 + t^2} = \dfrac{2}{3} \tan^{-1} \dfrac{t}{3} + C = \dfrac{2}{3} \tan^{-1} \left(\dfrac{\tan \dfrac{x}{2}}{3} \right) + C$

10.3.2 SOME SPECIAL CASES OF METHOD OF SUBSTITUTION

Case 1. Functions of linear function of x, i.e., $(ax + b)$

If the integrand is a function of the form $(ax + b)$, then the integral of the same form of the function is obtained by dividing the integral of function by the coefficient of x in the $ax + b$, i.e., if,

$$\int f(x)\,dx = \phi(x),$$

then we have $\int f(ax + b)\,dx = \dfrac{\phi(ax + b)}{a}.$

EXAMPLE 3.   Obtain the integral $\int -\dfrac{dx}{1 - (cx + d)^2}.$

SOLUTION.   Here, the given integral is $\int \dfrac{dx}{\sqrt{1 - (cx + d)^2}}$

Put $cx + d = t$, so $c\,dx = dt$ or $dx = \dfrac{dt}{c}$

So, $\int \dfrac{dx}{\sqrt{1 - (cx + d)^2}} = \dfrac{1}{c} \int \dfrac{dt}{\sqrt{1 - t^2}} = \dfrac{1}{c} \sin^{-1} t$

$\qquad\qquad\qquad = \dfrac{1}{c} \sin^{-1}(cx + d) + K \qquad$ (On putting $t = cx + d$)

EXAMPLE 4.   Obtain $\int 10^{6x}\,dx$.

SOLUTION.   Put $6x = t$ so $\quad 6\,dx = dt$

or $\qquad\qquad\qquad dx = \dfrac{dt}{6}$

So, $\qquad \int 10^{6x}\,dx = \int 10^t \dfrac{dt}{6} = \dfrac{1}{6} \int 10^t\,dt$

$\qquad\qquad\qquad = \dfrac{1}{6} \dfrac{10^t}{\log_e 10} = \dfrac{1}{6} \dfrac{10^{6x}}{\log_e 10} + C \qquad$ (On putting $t = 6x$)

10.3.3 SOME IMPORTANT RESULTS

(i) $\int (ax + b)^n\,dx = \dfrac{1}{a} \dfrac{(ax + b)^{n+1}}{n + 1}, n \ne -1$

(ii) $\int e^{ax}\,dx = \dfrac{1}{a} e^{ax}, a \ne 0$ $\qquad$ (iii) $\int -\dfrac{dx}{ax + b} = \dfrac{1}{a} \log_e(ax + b), a \ne 0$

(iv) $\int e^{ax+b}\,dx = \dfrac{e^{ax+b}}{a}, a \ne 0$

(v) $\int \dfrac{1}{(ax + b)^n}\,dx = -\dfrac{1}{a(n - 1)(ax + b)^{n-1}}, a \ne 0, n \ne 1$

(vi) $\int \cos(ax + b)\,dx = \dfrac{\sin(ax + b)}{a}, a \ne 0$

(vii) $\int \sin(ax+b)\,dx = -\dfrac{1}{a}\cos(ax+b), a \neq 0$

(viii) $\int \sec(ax+b)\tan(ax+b)\,dx = \dfrac{1}{a}\sec(ax+b), a \neq 0$

(ix) $\int \csc^2(ax+b)\,dx = -\dfrac{1}{a}\cot(ax+b), a \neq 0$

(x) $\int a^{cx+d}dx = \dfrac{1}{c}\dfrac{a^{cx+d}}{\log_e a}, c \neq 0$ (xi) $\int -\csc(ax+b)\cot(ax+b)\,dx = \dfrac{1}{a}\csc(ax+b)$

Case II. Integral of the type $\int [\phi(x)]^n \phi'(x)\,dx$:

In this case, we put $\phi(x) = t$. So, $\phi'(x)\,dx = dt$, then we get,

$$f(x) = \int t^n dt = \frac{t^{n+1}}{n+1} + C, (n \neq 1)$$

Now, put $t = \phi(x)$ in above result, we get

$$\int \phi(x)^n \phi'(x)\,dx = \frac{[\phi(x)]^{n+1}}{n+1} + C.$$

EXAMPLE 5. *Evaluate the integral* $\int \tan^3 x \sec^2 x\,dx$.

SOLUTION. Here, the given integral is $\int \tan^3 x \cdot \sec^2 x\,dx$.

Put $\tan x = t$, so $\sec^2 x\,dx = dt$

So, we get $\int \tan^3 x \sec^2 x\,dx = \int t^3 dt = \dfrac{t^{3+1}}{3+1} + C = \dfrac{t^4}{4} + C$

$$= \frac{\tan^4 x}{4} + C \qquad \text{(Putting } t = \tan x\text{)}$$

Case III. Function of the type $(a^2 \pm x^2)$

In this case, put $x = at$ so, $dx = a\,dt$ and then substitute these value in given integral and then integrate. After integration, put, $t = \dfrac{x}{a}$, we get the required result.

EXAMPLE 6. *Evaluate* $\int \dfrac{1}{a^2 + x^2}\,dx$.

SOLUTION. The given integral is $\int \dfrac{1}{a^2 + x^2}\,dx$.

Put $x = at \Rightarrow dx \Rightarrow a\,dt$

So, $\int \dfrac{1}{a^2 + x^2}\,dx = \int \dfrac{a\,dt}{a^2 + a^2 t^2} = \int \dfrac{a\,dt}{a^2(1+t^2)}$

$$= \frac{1}{a}\int \frac{dt}{1+t^2} = \frac{1}{a}\tan^{-1} t$$

Now, put $t = \dfrac{x}{a}$, we get

$$\int \frac{1}{a^2 + x^2}\,dx = \frac{1}{a}\tan^{-1}\left(\frac{x}{a}\right).$$

Similarly, we can obtain the following results:

(i) $\int \dfrac{1}{\sqrt{a^2 - x^2}} dx = \sin^{-1} \dfrac{x}{a}$
          (ii) $\int \dfrac{-1}{\sqrt{a^2 - x^2}} dx = \cos^{-1}\left(\dfrac{x}{a}\right)$

(iii) $\int \dfrac{1}{x\sqrt{x^2 - a^2}} dx = \dfrac{1}{a} \sec^{-1}\left(\dfrac{x}{a}\right)$
   (iv) $\int \dfrac{-dx}{x\sqrt{x^2 - a^2}} = \dfrac{1}{a} \text{cosec}^{-1}\left(\dfrac{x}{a}\right)$

(v) $\int \dfrac{-dx}{a^2 + x^2} = \dfrac{1}{a} \cot^{-1}\left(\dfrac{x}{a}\right)$.

Case IV. Integral of the type $\int x^{n-1} f(x^n)\, dx$.

In this case, put $x^n = t \Rightarrow nx^{n-1} dx = dt$, substitute these values in given integral and then integrate and after integration, put $t = x^n$, to get the required result.

EXAMPLE 7.   *Evaluate the integral $\int x^4 \tan x^5 dx$.*

SOLUTION.   The given integral is $\int x^4 \tan x^5 dx$.

Put $x^5 = t$ so, $5x^4 dx = dt \Rightarrow x^4 dx = \dfrac{dt}{5}$.

So, we get $\int x^4 \tan x^5 dx = \dfrac{1}{5} \int \tan t\, dt = \dfrac{1}{5} \log \cos t + C$

$$= \dfrac{1}{5} \log \cos x^5 + C \qquad\qquad \text{(On putting } t = x^5)$$

EXAMPLE 8.   *Evaluate the integral $\int x^3 \sin x^4 dx$.*

SOLUTION.   The given integral is $\int x^3 \sin x^4 = dx$.

Put $x^4 = t \Rightarrow 4x^3 dx = dt$

or $x^3 dx = \dfrac{dt}{4}$

So, we get $\int x^3 \sin x^4 dx = \dfrac{1}{4} \int \sin t\, dt = \dfrac{1}{4}(-\cos t) + C$

$$= -\dfrac{1}{4} \cos x^4 + C \qquad\qquad \text{(On putting } t = x^4)$$

Case V. Integral of the type $\int \dfrac{f'(x)}{f(x)}\, dx$.

In this case, put $f(x) = t$. So, $f'(x)dx = dt$

So, we get $\int \dfrac{f'(x)}{f(x)} dx = \int \dfrac{dx}{t} = \log t$

$$= \log f(x) \qquad\qquad [\text{On putting } t = f(x)]$$

For example, Evaluate the integral $\int \dfrac{nx^{n-1}}{x^n} dx$.

Solution. Put $x^n = t$ so $nx^{n-1}dx = dt$

So,
$$\int \frac{nx^{n-1}}{x^n}dx = \int \frac{dt}{t} = \log t + C$$
$$= \log x^n + C \qquad \text{(On putting } t = x^n\text{)}$$

MORE IMPORTANT RESULTS

(i) $\int \cot x \, dx = \log \sin x$ (ii) $\int \sec x \, dx = \log(\sec x + \tan x) = \log \tan\left(\frac{\pi}{4} + \frac{x}{2}\right)$

(iii) $\int \tan x \, dx = -\log \cos x = \log \sec x$ (iv) $\int \operatorname{cosec} x \, dx = -\log(\operatorname{cosec} x + \cot x) = \log \tan\left(\frac{x}{2}\right)$

EXAMPLE 9. Evaluate the integral $\int \dfrac{e^x + e^{-x}}{e^x - e^{-x}}dx$.

SOLUTION. Put $e^x - e^{-x} = t$ so, $(e^x + e^{-x})dx = dt$

Therefore,
$$\int \frac{e^x + e^{-x}}{e^x - e^{-x}}dx = \int \frac{dt}{t} = \log t + C$$
$$= \log(e^x - e^{-x}) + C \qquad \text{(On putting } t = e^x - e^{-x}\text{)}$$

EXAMPLE 10. Evaluate $\int \dfrac{3x^2 + 2x}{x^3 + x^2 + 1}dx$.

SOLUTION. Let
$$I = \int \frac{3x^2 + 2x}{x^3 + x^2 + 1}dx$$

Put $x^3 + x^2 + 1 = t$, differentiating w.r.t. x

We get $(3x^2 + 2x)dx = dt$

Then
$$I = \int \frac{dt}{t} = \log t + C = \log(x^3 + x^2 + 1) + C$$

EXAMPLE 11. Evaluate the following integrals.

(i) $\int \dfrac{2x \sin x^2}{\cos x^2}dx$ (ii) $\int \left(\dfrac{e^{2\log x} - 1}{e^{2\log x} + 1}\right)\dfrac{1}{x}dx$

SOLUTION. (i) Here, the given integral is $\int \dfrac{2x \sin x^2}{\cos x^2}dx$.

Put $x^2 = t$ so $2x\, dx = dt$

Therefore, $\int \dfrac{2x \sin x^2}{\cos x^2}dx = \int \dfrac{\sin t}{\cos t}dt$.

Again put $\cos t = v$, so $-\sin t \, dt = dv$

Therefore,
$$\int \frac{\sin t}{\cos t}dt = \int \frac{-dv}{v} = -\log v$$
$$= -\log \cos t + C \qquad \text{(On putting } v = \cos t\text{)}$$
$$= -\log \cos x^2 + C \qquad \text{(On putting } t = x^2\text{)}$$
$$= \log \sec x^2 + C$$

(ii) The given integral is $\int \left(\dfrac{e^{2\log x} - 1}{e^{2\log x} + 1} \right) \dfrac{1}{x} dx$

Put $\log x = t$ so, $1/x \, dx = dt$

Therefore,

$$\int \left(\dfrac{e^{2\log x} - 1}{e^{2\log x} + 1} \right) \dfrac{1}{x} dx = \int \dfrac{e^{2t} - 1}{e^{2t} + 1} dt = \int \dfrac{e^t - e^{-t}}{e^t + e^{-t}} dt$$

(Dividing numerator and denominator by e^t)

$$= \log(e^t + e^{-t}) + C$$

$$= \log(e^{\log x} + e^{-\log x}) + C \qquad \text{(On putting } t = \log x)$$

$$= \log(x + x^{-1}) + C$$

EXAMPLE 12. *Evaluate the following integrals :*

(i) $\displaystyle\int \dfrac{dx}{\tan^{-1} x(1+x^2)}$　　(ii) $\displaystyle\int \dfrac{\cos^2 x}{\sin^4 x} dx$　　(iii) $\displaystyle\int \dfrac{n x^{n-1}}{\sqrt{(1-x^{2n})}} dx.$

SOLUTION.　　(i) The given integral is $\displaystyle\int \dfrac{dx}{\tan^{-1} x(1+x^2)}$

Put $\tan^{-1} x = t$, so $\dfrac{1}{1+x^2} dx = dt$

So, $\displaystyle\int \dfrac{dx}{\tan^{-1} x(1+x^2)} = \int \dfrac{dt}{t} = \log t + C$

$$= \log(\tan^{-1}x) + C \qquad \text{(On putting } t = \tan^{-1} x)$$

(ii) The given integral can be written as $\int \csc^2 x \cot^2 x \, dx$

Put $\cot x = t$, so $-\csc^2 x \, dx = dt$

Therefore,

$$\int \csc^2 x . \cot^2 x \, dx = \int -t^2 \, dt$$

$$= -\dfrac{t^3}{3} + C = -\dfrac{\cot^3 x}{3} + C \qquad \text{(On putting } t = \cot x)$$

(iii) The given integral is $\displaystyle\int \dfrac{n x^{n-1}}{\sqrt{(1-x^{2n})}} dx.$

Put $x^n = t$ so, $n x^{n-1} dx = dt$

Therefore, $\displaystyle\int \dfrac{n x^{n-1}}{\sqrt{(1-x^{2n})}} dx = \int \dfrac{dt}{\sqrt{1-t^2}} = \sin^{-1} t + C$

$$= \sin^{-1}(x^n) + C \qquad \text{(On putting } t = x^n)$$

EXAMPLE 13. *Find*

(i) $\displaystyle\int \dfrac{1+x^2}{1+x^4} dx$　　**(Meerut(BCA)–2008)**　　(ii) $\displaystyle\int \dfrac{x + \sin x}{1 + \cos x} dx$

SOLUTION. (i) Consider $I = \int \dfrac{1+x^2}{1+x^4}\,dx = \int \dfrac{1+\dfrac{1}{x^2}}{x^2 + \dfrac{1}{x^2}}\,dx = \int \dfrac{\left(1+\dfrac{1}{x^2}\right)}{\left(x-\dfrac{1}{x}\right)^2 + (\sqrt{2})^2}\,dx \cdot$

Let $x - \dfrac{1}{x} = t$. Then $\left(1 + \dfrac{1}{x^2}\right)dx = dt$

$$I = \int \frac{dt}{t^2 + (\sqrt{2})^2} = \frac{1}{\sqrt{2}}\tan^{-1}\frac{t}{\sqrt{2}} + C$$

$$I = \frac{1}{\sqrt{2}}\tan^{-1}\left(\frac{x^2+1}{\sqrt{2}x}\right) + C$$

(ii) We have $I = \int \dfrac{x + \sin x}{1 + \cos x}\,dx$

$$I = \int \frac{x}{1+\cos x}\,dx + \int \frac{\sin x}{1+\cos x}\,dx$$

$$= \int \frac{x}{2\cos^2 x/2}\,dx + \int \frac{\sin x}{1+\cos x}\,dx$$

$$= \frac{1}{2}\int x\sec^2 x/2\,dx + \int\left(-\frac{1}{t}\right)dt$$

where $1 + \cos x = t \Rightarrow -\sin x\,dx = dt$

$$= \frac{1}{2}\left[x \cdot \frac{\tan x}{2} \times 2 - \int 1 \cdot \frac{\tan x}{2} \cdot 2\right] - \log t\,dt$$

$$= x \cdot \frac{\tan x}{2} - \log\left(\frac{\sec x}{2}\right) \times 2 - \log(1+\cos x) + C$$

$$= x\frac{\tan x}{2} - 2\log\frac{\sec x}{2} - \log(1+\cos x) + C$$

EXAMPLE 14. *Evaluate the following integral.*

(i) $\displaystyle\int \frac{\log x}{x}\,dx$ (ii) $\displaystyle\int \frac{\sin x \cos x}{1+\sin^2 x}\,dx$

SOLUTION. (i) Here, the given integral is $\int \log x\,dx$.

Put $\log x = t$, so $\dfrac{1}{x}\,dx = dt$

So, $\displaystyle\int \frac{\log x}{x}\,dx = \int t\,dt = \frac{t^2}{2} + C = \frac{(\log x)^2}{2} + C$ (On putting $t = \log x$)

(ii) Here, put $1 + \sin^2 x = t$

Put $2\sin x.\cos x = dt$

Therefore, $\displaystyle\int \frac{\sin x \cos x}{1+\sin^2 x}\,dx = \frac{1}{2}\int\frac{dt}{t} = \frac{1}{2}\log t = \frac{1}{2}\log(1+\sin^2 x)$

EXAMPLE 15. *Evaluate $\displaystyle\int \frac{f(x)}{(x+1)^2}\,dx$ where $f(x)$ is a polynomial of degree 2 in x such that $f(0) = 1$, $f(1) = 3$ and $f(-1) = 5$.*

SOLUTION.   Consider a second degree polynomial

$$f(x) = ax^2 + bx + c$$

Now,

$$f(0) = 1 \quad \Rightarrow \quad c = 1$$
$$f(1) = 3 \quad \Rightarrow \quad a + b + c = 3$$
$$f(-1) = 5 \quad \Rightarrow \quad a - b + c = 5$$

On solving we get $a = 3, b = -1, c = 1$

$$\therefore \qquad\qquad f(x) = 3x^2 - x + 1$$

$\therefore$ Consider $\displaystyle\int \frac{3x^2 - x + 1}{(x+1)^2} dx = \int \frac{3(x+1)^2 - 7x - 2}{(x+1)^2} dx = \int 3\, dx - \int \frac{7x+2}{(x+1)^2} dx$

$$= \int 3\, dx - \int \left\{ \frac{7(x+1) - 5}{(x+1)^2} \right\} dx$$

$$= \int 3\, dx - \int \frac{7}{x+1} dx + 5 \int \frac{1}{(x+1)^2} dx$$

$$= 3x - 7 \log(x+1) + \frac{5(-1)}{(x+1)} + C$$

$$= 3x - 7 \log(x+1) - \frac{5}{(x+1)} + C$$

10.4 INTEGRATION BY PARTS

The method of 'integration by parts' is very powerful method of finding the integral. If the integrand is the product of two different functions, then the method of "integration by parts" can be used.

If $f(x)$ and $g(x)$ are two functions of x, then

$$\int f(x)g(x)dx = f(x)\int g(x)dx - \int \frac{d}{dx} f(x) \left\{ \int g(x)dx \right\} dx$$

Proof. Let $f(x)$ and $g(x)$ be two functions of x, then

$$\frac{d}{dx}[f(x)g(x)] = f(x)\frac{d}{dx}g(x) + \frac{d}{dx}f(x) \cdot g(x).$$

Now integrating both the sides w.r.t x, we get

$$f(x)g(x) = \int \left\{ f(x)\frac{d}{dx}g(x) \right\} dx + \int \left\{ \frac{d}{dx}f(x).g(x) \right\} dx$$

So, $\displaystyle\int \left\{ f(x) \cdot \frac{d}{dx}g(x) \right\} dx = f(x) \cdot g(x) - \int \left\{ \frac{d}{dx}f(x) \cdot g(x) \right\} dx$

or $\displaystyle\int f(x)g(x)dx = f(x)\int g(x)dx - \int \left\{ \frac{d}{dx}f(x) \left(\int g(x)dx \right) \right\} dx$

The method of 'integration by parts, can be written into words as follows:
The integral of the product of two functions
   = First functions × integral of the second functions
            – integral of (the derivative of first function
                 × integral of the second function)

WORKING PROCEDURE

There is no general rule to choose the first and second function, but remember the following points:

Step 1. If the second function is not given then unity may be taken as the second function.

Step 2. The integral of second function must be known.

Step 3. If necessary, then the above formula can be applied more than once.

Step 4. If the integral is of the form $\int x^n f(x)dx$ where n is positive integer, then x^n must be taken as the first function.

Step 5. Trick using here is ILATE which stands for

$I \to$ Inverse Trigonometric function $L \to$ Logarithmetic function

$A \to$ Algebraic function $T \to$ Trigonometric function

$E \to$ Exponential function

SOLVED EXAMPLES

EXAMPLE 1. *Evaluate the following integrals.*

(*i*) $\int x^2 \sin x\, dx$ (*ii*) $\int \log x\, dx$ (*iii*) $\int x^2 e^{3x} dx.$

SOLUTION. (*i*) Here, the given integral is $\int x^2 \sin x\, dx.$

Now, integrating by parts, taking x^2 as first function, we get

$$\int x^2 \sin xdx = x^2 \int \sin xdx - \int \left\{\frac{d}{dx}x^2 \int \sin xdx\right\}dx$$

$$= -x^2 \cos x - \int \{2x.(-\cos x)\}dx$$

$$= -x^2 \cos x + 2\int x.\cos x\, dx$$

$$= -x^2 \cos x + 2\left[x.\int \cos x\, dx - \int\left\{\frac{d}{dx}(x).\int \cos x\, dx\right\}dx\right]$$

$$= -x^2 \cos x + 2[x\sin x - \int \sin xdx]$$

$$= -x^2 \cos x + 2x\sin x + 2\cos x + C,$$

where C the constant of integration.

(ii) Here, the given integral is $\int \log x\, dx.$

Since the second function is not given, therefore we take the second function as unity (*i.e.*, 1).

$$\int \log x.1dx = \log x\int 1.dx - \int\left\{\frac{d}{dx}\log x\int 1.dx\right\}dx$$

$$= x\log x - \int \frac{1}{x}x\, dx = x(\log x - 1) + C = x\log x - x + C$$

(iii) Here, the given integral is $\int x^2 e^{3x} dx.$

Now, integrating by parts, taking x^2 as the first function, we get

$$\int x^2 e^{3x} dx = x^2 \int e^{3x} dx - \int\left\{\frac{d}{dx}x^2 \int e^{3x} dx\right\}dx$$

$$= \frac{x^2 e^{3x}}{3} - \int 2x\frac{e^{3x}}{3}dx = \frac{x^2 e^{3x}}{3} - \frac{2}{3}\int x.e^{3x} dx.$$

Again integrating by parts, taking x as the first function, we get

$$\int x^2 e^{3x} dx = \frac{x^2 e^{3x}}{3} - \frac{2}{3}\left[x.\int e^{3x} dx - \int\left\{1.\int e^{3x} dx\right\} dx\right]$$

$$= \frac{x^2 e^{3x}}{3} - \frac{2}{3}\left[x.\frac{e^{3x}}{3} - \frac{1}{3}\int e^{3x} dx\right]$$

$$= \frac{x^2 e^{3x}}{3} - \frac{2}{3}\left[x.\frac{e^{3x}}{3} - \frac{e^{3x}}{9}\right] + C = \frac{x^2 e^{3x}}{9} - \frac{2xe^{3x}}{9} + \frac{2e^{3x}}{27} + C$$

EXAMPLE 2. *Evaluate the following integrals.*

(i) $\int \dfrac{1}{(a^2 - x^2)^{3/2}} dx$ \qquad\qquad (ii) $\int (\log_e x)^2 dx$

(iii) $\int x^2 e^x dx$ \qquad\qquad\qquad (iv) $\int x^n \log x\, dx$.

SOLUTION. (i) The given integral is $\int \dfrac{1}{(a^2 - x^2)^{3/2}} dx$.

Put $\quad x = a\sin\theta \Rightarrow dx = a\cos\theta\, d\theta$

So, $\int \dfrac{1}{(a^2 - x^2)^{3/2}} dx = \int \dfrac{a\cos\theta\, d\theta}{(a^2 - a^2\sin^2\theta)^{3/2}}$

$$= \frac{1}{a^2}\int \frac{\cos\theta\, d\theta}{(1 - \sin^2\theta)^{3/2}} = \frac{1}{a^2}\int \sec^2\theta\, d\theta$$

$$= \frac{1}{a^2}\tan\theta + C = \frac{1}{a^2}\left\{\frac{x}{\sqrt{a^2 - x^2}}\right\} + C$$

(ii) The given integral is $\int (\log_e x)^2 dx$.

Here, the second function is not given so, we take second function as unity. Therefore,

$$\int (\log_e x)^2 \cdot 1\, dx = (\log_e x)^2 \int 1\, dx - \int\left\{\frac{d}{dx}(\log_e x)^2 \int 1\cdot dx\right\} dx$$

$$= x(\log_e x)^2 - \int 2\log_e x \frac{1}{x}.\, x\ dx$$

$$= x(\log_e x)^2 - 2\int \log_e x \cdot 1\, dx$$

$$= x(\log_e x)^2 - 2\left[\log_e x \int 1\, dx - \int\left\{\frac{d}{dx}\log_e x \int 1\, dx\right\} dx\right]$$

$$= x(\log_e x)^2 - 2x\log_e x + 2\int \frac{1}{x} x\, dx$$

$$= x(\log_e x)^2 - 2x\log_e x + 2x + C$$

(iii) The given integral is $\int x^2 e^x dx$.

Integrating by parts, taking x^2 as the first function, we get

$$\int x^2 e^x dx = x^2 \int e^x dx - \int\left\{\frac{d}{dx}x^2 \int e^x dx\right\} dx$$

$$= x^2 e^x - 2 \int x \cdot e^x \, dx$$

$$= x^2 e^x - 2 \left[x \int x \cdot e^x \, dx - \int \left\{ \frac{d}{dx} x \int e^x \, dx \right\} dx \right].$$

$$= x^2 e^x - 2x e^x + 2 \int e^x \cdot dx = x^2 e^x - 2x e^x + 2 e^x + C$$

$$= e^x (x^2 - 2x + 2) + C$$

(iv) Here, given integral is $\int x^n \log x \, dx$. Now, integrating by parts, taking $\log x$ as the first function, we get

$$\int x^n \log x \, dx = \log x \int x^n \, dx - \int \left\{ \frac{d}{dx} \log x \int x^n \, dx \right\} dx$$

$$= \log x \frac{x^{n+1}}{n+1} - \int \frac{1}{x} \frac{x^{n+1}}{n+1} \, dx = \log x \frac{x^{n+1}}{n+1} - \int \frac{x^n}{n+1} \, dx$$

$$= \log x \frac{x^{n+1}}{n+1} - \frac{1}{n+1} \frac{x^{n+1}}{(n+1)} + C$$

$$= \frac{1}{n+1} \left(x^{n+1} \log x - \frac{x^{n+1}}{n+1} \right) + C$$

EXAMPLE 3. *Evaluate the following integrals :*

(i) $\int \cos^{-1} \left(\frac{1-x^2}{1+x^2} \right) dx$
(ii) $\int x \sin x \, dx$

(iii) $\int x \sec^2 x \, dx$
(iv) $\int \sec^3 x \, dx$

SOLUTION. (i) Here, given integral is $\int \cos^{-1} \left(\frac{1-x^2}{1+x^2} \right) dx$

Put $x = \tan \theta \Rightarrow dx = \sec^2 \theta \, d\theta$

So, $\int \cos^{-1} \left(\frac{1-x^2}{1+x^2} \right) dx = \int \cos^{-1} \left(\frac{1-\tan^2 \theta}{1+\tan^2 \theta} \right) \sec^2 \theta \, d\theta$

$$= \int \cos^{-1} (\cos 2\theta) \sec^2 \theta \, d\theta = 2 \int \theta \sec^2 \theta \, d\theta$$

$$= 2 \left[\theta \int \sec^2 \theta \, d\theta - \int \left\{ \frac{d}{d\theta} \theta \int \sec^2 \theta \, d\theta \right\} d\theta \right]$$

$$= 2\theta \tan \theta - 2 \int 1 \cdot \tan \theta \cdot d\theta + C$$

$$= 2\theta \tan \theta - 2 \log \sec \theta + C$$

$$= 2x \tan^{-1} x - 2 \log \sqrt{1+x^2} + C \text{ (Putting the value of } \theta)$$

(ii) Here, given integral is $\int x \sin x \, dx$.

Integrating by parts, talking x as the first function.

We get $\int x \sin x \, dx = x \int \sin x \, dx - \int \left(\frac{d}{dx} x \int \sin x \, dx \right) dx$

$$= -x \cos x + \int \cos x \, dx + C = -x \cos x + \sin x + C$$

TWO IMPORTANT FORMULAE

(i) $\int e^{ax} \cos bx \, dx = \dfrac{e^{ax}}{a^2 + b^2}(a \cos bx + b \sin bx) + C$

(ii) $\int e^{ax} \sin bx \, dx = \dfrac{e^{ax}}{a^2 + b^2}(a \sin bx - b \cos bx) + C$

EXAMPLE 4.   *Evaluate the following integrals :*

     (i) $\int x^2 \cos x \, dx$ (ii) $\int x^3 \cos x \, dx$

     (iii) $\int x^2 \sin x \, dx$ (iv) $\int e^x \cos x \, dx$

SOLUTION.   (i) Here, the given integral is $\int x^2 \cos x \, dx$. Integrating by parts, taking x^2 as the first function, we get

$$\int x^2 \cos x \, dx = x^2 \int \cos x \, dx - \int \left[\frac{d}{dx}(x^2) . \int \cos x \, dx\right] dx$$

$$= x^2 \sin x - 2\int x \sin x \, dx$$

$$= x^2 \sin x - 2\left[x(-\cos x) - \int \left\{\frac{d}{dx}(x) \int \sin x \, dx\right\} dx\right]$$

$$= x^2 \sin x - 2[-x \cos x + \int \cos x \, dx]$$

$$= x^2 \sin x + 2x \cos x - 2 \sin x + C$$

(ii) Here, the given integral is $\int x^3 \cos x \, dx$.

Integrating by parts, taking x^3 as the first function, we get

$$\int x^3 \cos x \, dx = x^3 \int \cos x \, dx - \int \left[\frac{d}{dx}(x)^3 \int \cos x \, dx\right] dx$$

$$= x^3 \sin x - 3\int x^2 \sin x \, dx$$

$$= x^3 \sin x - 3[x^2 \int \sin x \, dx - \int \{2x . \int \sin x \, dx\} \, dx]$$

$$= x^3 \sin x + 3x^2 \cos x - 6[x . \sin x - \int \sin x \, dx]$$

$$= x^3 \sin x + 3x^2 \cos x - 6x \sin x - 6 \cos x + C,$$

where C is the constant of integration.

$$= (x^3 - 6x) \sin x + (3x^2 - 6) \cos x + C$$

(iii) Here, the given integral is $\int x^2 \sin x \, dx$.

Integrating by parts, taking x^2 as the first function, we get

$$\int x^2 \sin x \, dx = x^2 \int \sin x \, dx - \int (2x \int \sin x \, dx) \, dx$$

$$= -x^2 \cos x + 2\int x \cos x \, dx$$

$$= -x^2 \cos x + 2[x \sin x - \int \sin x \, dx]$$

$$= -x^2 \cos x + 2x \sin x + 2 \cos x + C$$

(iv) Here, the given integral is $\int e^x . \cos x \, dx$

Let $I = \int e^x . \cos x \, dx.$

Integrating by parts, taking $\cos x$ as the first function, we get

$$\int e^x \cos x \, dx = \cos x . e^x + \int \sin x . e^x \, dx.$$

Again integrating by parts, we get

$$\int e^x \cos x \, dx = e^x \cos x + \sin x \, e^x - \int \cos x \, e^x dx$$

$$I = e^x \cos x + e^x \sin x - I$$

or $\qquad 2I = e^x \cos x + e^x \sin x$

$$I = \frac{1}{2}(e^x \cos x + e^x \sin x) = \frac{e^x}{2}(\cos x + \sin x) + C$$

EXAMPLE 5. *Evaluate* $\int \dfrac{x}{1 + \sin x} \, dx$

SOLUTION. Consider

$$\int \frac{x}{1 + \sin x} \, dx = \int \frac{x(1 - \sin x)}{\cos^2 x} \, dx$$

$$= \int x(\sec^2 x - \tan x \sec x) dx$$

$$= \int \underset{\text{I \quad II}}{x \sec^2 x} - \int \underset{\text{I \quad II}}{x \tan x \sec x} \, dx$$

$$= x \tan x - \int \tan x \, dx - x \sec x + \int \sec x \, dx$$

$$= x \tan x + \log \cos x - x \sec x + \log(\sec x + \tan x) + c$$

$$= x(\tan x - \sec x) + \log \cos x (\sec x + \tan x) + c$$

$$= x(\tan x - \sec x) + \log (1 + \sin x) + c$$

10.5 INTEGRATION BY PARTIAL FRACTIONS

If the given function is of the type $\dfrac{f(x)}{g(x)}$, where $f(x)$ and $g(x)$ both are polynomials, then we may assume that the numerator $f(x)$ and denominator $g(x)$ have no common polynomial factor and that the degree of $f(x)$ is less than the degree of $g(x)$. If the degree of numerator $f(x)$ is greater than or equal to the degree of denominator $g(x)$, then we have to divide $f(x)$ by $g(x)$ so that $\dfrac{f(x)}{g(x)} = h(x) + \dfrac{R(x)}{g(x)}$, where $h(x)$ is the quotient (a polynomial) and $R(x)$ is the remainder whose degree is less than the degree of $g(x)$. The method in which we change the $\dfrac{f(x)}{g(x)}$ into partial fractions, depends on factors of the denominator. We know that every polynomial can be expressed as a product of linear and irreducible quadratic factors with real coefficient. So, we have the following cases :

(i) If all the factor of denominator are linear and non repeated, then for each linear non-repeated factor $(ax + b)$, there corresponds a fraction of the form $\dfrac{A}{ax + b}$, where A is a constant, is to be obtained.

(ii) If all the factors in the denominator are linear but some of them are repeated, then for each linear factor $(ax + b)^n$ repeating n times, there correspond the sum of r partial fractions as:

$$\frac{A_1}{ax + b} + \frac{A_2}{(ax + b)^2} + \dots + \frac{A_n}{(ax + b)^n}, \text{ where } A_1, A_2, \dots, A_n \text{ are constants.}$$

(iii) If the factor in the denominator are linear and irreduciable but quadratic factor and are non repeated, then for each irreduciable quadratic $ax^2 + bx + c$, which occurs only one in the denominator, there correspond a partial fraction of the form $\dfrac{Ax + B}{ax^2 + bx + c}$.

SOLVED EXAMPLES

EXAMPLE 1. *Evaluate the integral* $\int \dfrac{x+4}{3+2x-x^2} dx$.

SOLUTION. Here, the given integral is $\int \dfrac{x+4}{3+2x-x^2} dx$.

Consider $\dfrac{x+4}{3+2x-x^2} = \dfrac{x+4}{(3-x)(1+x)} = \dfrac{A}{3-x} + \dfrac{B}{1+x}$.

So, $x+4 = (1+x)A + (3-x)B$

Putting $x = 3$, we get $A = \dfrac{7}{4}$ and putting, $x = -1$, we get $B = \dfrac{3}{4}$

So, we have $\dfrac{x+4}{3+2x-x^2} = \dfrac{7}{4(3-x)} + \dfrac{3}{4(1+x)}$

Hence, $\int \dfrac{x+4}{3+2x-x^2} dx = \dfrac{7}{4} \int \dfrac{1}{3-x} dx + \dfrac{3}{4} \int \dfrac{1}{1+x} dx$; Integrating

$$= -\dfrac{7}{4} \log(x-3) + \dfrac{3}{4} \log(1+x) + C$$

EXAMPLE 2. *Evaluate the integral* $\int \dfrac{3x+2}{(x-2)(x+1)^2} dx$.

SOLUTION. Here, the given integral is $\int \dfrac{3x+2}{(x-2)(x+1)^2} dx$.

Consider $\dfrac{3x+2}{(x-2)(x+1)^2} = \dfrac{A}{(x-2)} + \dfrac{B}{(x+1)} + \dfrac{C}{(x+1)^2}$

or $(3x+2) = A(x+1)^2 + B(x-2)(x+1) + C(x-2)$

On putting, $x = 2$, we get $A = \dfrac{8}{9}$, and putting $x = -1$, we get $C = \dfrac{1}{3}$.

Now, on comparing the coefficient of x^2, , we get $A + B = 0$

i.e., $B = -A = -\dfrac{8}{9}$

So, we have $\dfrac{3x+2}{(x-2)(x+1)^2} = \dfrac{8}{9(x-2)} - \dfrac{8}{9(x+1)} + \dfrac{1}{3(x+1)^2}$

Hence, $\int \dfrac{3x+2}{(x-2)(x+1)^2} dx = \dfrac{8}{9} \int \dfrac{1}{x-2} dx - \dfrac{8}{9} \int \dfrac{1}{x+1} dx + \dfrac{1}{3} \int \dfrac{1}{(x+1)^2} dx$

$$= \dfrac{8}{9} \log(x-2) - \dfrac{8}{9} \log(x+1) - \dfrac{1}{3} \dfrac{1}{(x+1)} + C$$

$$= \dfrac{8}{9} \log\left(\dfrac{x-2}{x+1}\right) - \dfrac{1}{3(x+1)} + C$$

EXAMPLE 3. *Evaluate the integral $\int \dfrac{1}{x(1-x^2)} dx$.*

SOLUTION. Here, the given integral is $\int \dfrac{1}{x(1-x^2)} dx$

Consider $\quad \dfrac{1}{x(1-x^2)} = \dfrac{1}{x(x+1)(1-x)} = \dfrac{A}{x} + \dfrac{B}{(1-x)} + \dfrac{C}{(1+x)}$

$\Rightarrow \qquad\qquad 1 = (1-x)(1+x)A + x(1+x)B + x(1-x)C$

Now, putting $x = 0 \Rightarrow A = 1$ and putting $x = 1 \Rightarrow B = \dfrac{1}{2}$ and

putting $\qquad\qquad x = -1 \Rightarrow C = -\dfrac{1}{2}.$

Thus, $\qquad \dfrac{1}{x(1-x^2)} = \dfrac{1}{x} + \dfrac{1}{2(1-x)} - \dfrac{1}{2}(1+x)$

Hence, $\quad \int \dfrac{1}{x(1-x)^2} dx = \int \dfrac{1}{x} dx + \dfrac{1}{2}\int \dfrac{1}{(1-x)} dx - \dfrac{1}{2}\int \dfrac{1}{1+x} dx$

$\qquad\qquad = \log x + \dfrac{1}{2}[-\log(1-x) - \log(1+x)] + C$

$\qquad\qquad = \dfrac{1}{2}[2\log x - \log\{(1-x)(1+x)\}] + C$

$\qquad\qquad = \dfrac{1}{2}\log\left\{\dfrac{x^2}{1-x^2}\right\} + C$

EXAMPLE 4. *Evaluate the integral $\int \dfrac{7x^2 + 3x + 1}{x(x+1)} dx$.*

SOLUTION. Here, the given integral is $\int \dfrac{7x^2 + 3x + 1}{x(x+1)}$.

To change the integrand into partial fraction, first we divide $(7x^2 + 3x + 1)$ by $(x^2 + x)$, because the degree of numerator must be less than the degree of denominator. So, we get

$\qquad \dfrac{7x^2 + 3x + 1}{x^2 + x} = 7 + \dfrac{1 - 4x}{x^2 + x}$

So, $\quad \int \dfrac{7x^2 + 3x + 1}{x^2 + x} = 7\int dx + \int \dfrac{1-4x}{x^2+1} dx = 7x + \int \dfrac{1-4x}{x^2+x} dx \qquad \ldots(1)$

Now, to find $\int \dfrac{1-4x}{x^2+x} dx$.

Let $\qquad\qquad \dfrac{1-4x}{x^2+x} = \dfrac{A}{x} + \dfrac{B}{x+1}$

$\Rightarrow \qquad\qquad 1 - 4x = A(x-1) + Bx \qquad\qquad \ldots(2)$

Now, putting $x = -1 \Rightarrow B = -5$ and putting $x = 0 \Rightarrow A = 1$

Thus, $$\frac{1-4x}{x^2+x}=\frac{1}{x}-\frac{5}{x+1}$$

So, $$\int\frac{1-4x}{x^2+x}dx=\int\frac{1}{x}-5\int\frac{1}{x+1}dx=\log x-5\log(x+1)+C$$

Hence, by equation (1), we get

$$\int\frac{7x^2+3x+1}{x^2+x}dx=7x+\log x-5\log(x+1)+C.$$

EXAMPLE 5. *Evaluate the integral* $\int\frac{x}{(x-1)(2x+1)}dx.$

SOLUTION. Here, the given integral is $\int\frac{x}{(x-1)(2x+1)}dx.$

Consider, $$\frac{x}{(x-1)(2x+1)}=\frac{A}{(x-1)}+\frac{B}{(2x+1)}$$

$$\Rightarrow\quad x=(2x+1)A+(x-1)B.$$

Now, putting $x=1$, we get $A=\dfrac{1}{3}$, and putting $x=-\dfrac{1}{2}$, we get $B=\dfrac{1}{3}$

Thus, $$\frac{x}{(x-1)(2x+1)}=\frac{1}{3(x-1)}+\frac{1}{3(2x+1)}$$

Hence, $$\int\frac{x}{(x-1)(2x+1)}dx=\frac{1}{3}\int\frac{1}{x-1}dx+\frac{1}{3}\int\frac{1}{2x+1}dx$$

$$=\frac{1}{3}\log(x-1)+\frac{1}{3}\cdot\frac{1}{2}\log(2x+1)+C$$

$$=\frac{1}{3}\log(x-1)+\frac{1}{6}\log(2x+1)+C$$

EXAMPLE 6. *Evaluate the integral* $\int\dfrac{1}{x[(\log x)^2-5\log x+6]}dx.$

SOLUTION. Here, the given integral is $\int\dfrac{1}{x[(\log x)^2-5\log x+6]}dx.$

Let $$\log x=u\Rightarrow\frac{1}{x}dx=du$$

So, $\int\dfrac{1}{x[(\log x)^2-5\log x+6]}dx=\int\dfrac{1}{u^2-5u+6}du=\int\dfrac{1}{(u-2)(u-3)}du$

Now let $$\frac{1}{(u-2)(u-3)}=\frac{A}{(u-2)}+\frac{B}{(u-3)}$$

$$\Rightarrow\quad 1=(u-3)A+(u-2)B\Rightarrow 1=(u-3)A+(u-2)B.$$

Now, putting $u=3$, we get $B=1$ and putting $u=2$, we get $A=-1$

Thus, $$\frac{1}{(u-2)(u-3)}=\frac{1}{u-3}-\frac{1}{u-2}$$

So, $\int\dfrac{1}{x[(\log x)^2-5\log x+6]}dx=\int\dfrac{1}{(u-2)(u-3)}du$

$$= \int \frac{1}{u-3} du - \int \frac{1}{u-2} = \log(u-3) - \log(u-2) + C$$

$$= \log\left(\frac{u-3}{u-2}\right) + C$$

Now, put $u = \log x$, we get

$$\int \frac{1}{x[(\log x)^2 - 5\log x + 6]} dx = \log\left(\frac{\log x - 3}{\log x - 2}\right) + C$$

EXAMPLE 7. *Evaluate the integral* $\int \frac{x^3 - 5x}{(x^2 - 9)(x^2 + 1)} dx$.

SOLUTION. Here, the given integral is $\int \frac{x^3 - 5x}{(x^2 - 9)(x^2 + 1)} dx$.

Consider $\dfrac{x^3 - 5x}{(x^2 - 9)(x^2 + 1)} = \dfrac{x^3 - 5x}{(x-3)(x+3)(x^2+1)} = \dfrac{A}{x-3} + \dfrac{B}{x+3} + \dfrac{Cx+D}{x^2+1}$

$\Rightarrow \quad x^3 - 5x = (x+3)(x^2+1)A + (x-3)$

$$(x^2+1)B + (x^2-9)(Cx+D) \qquad \ldots(1)$$

Now, putting $x = 3$, we get $A = 1/5$ and putting $x = -3$, we get $B = \dfrac{1}{5}$.

Now, comparing the coefficient of x^3 and constants in (1), we get

$$A + B + C = 1 \quad \text{or} \quad C = 1 - A - B = \frac{3}{5}$$

and $\quad 3A - 3B - 9D = 0 \quad \text{or} \quad 9D = 3A - 3B = 0$

$\Rightarrow \qquad\qquad D = 0$

Thus, we have $\dfrac{x^3 - 5x}{(x^2 - 9)(x^2 + 1)} = \dfrac{1}{5(x-3)} + \dfrac{1}{5(x+3)} + \dfrac{3x}{5(x^2+1)}$.

Hence, $\int \dfrac{x^3 - 5x}{(x^2-9)(x^2+1)} dx = \dfrac{1}{5}\int \dfrac{1}{x-3}dx + \dfrac{1}{5}\int \dfrac{1}{x+3}dx + \dfrac{3}{5}\int \dfrac{x}{x^2+1}dx$

$$= \frac{1}{5}\log(x-3) + \frac{1}{5}(x+3) + \frac{3}{5}\cdot\frac{1}{2}\log(x^2+1) + C$$

$$= \frac{1}{5}[\log\{(x-3)(x+3)\}] + \frac{3}{10}\log(x^2+1) + C$$

$$= \frac{1}{5}\log(x^2-9) + \frac{3}{10}\log(x^2+1) + C.$$

EXAMPLE 8. *Evaluate the integral* $\int \dfrac{1 - \cos x}{\cos x(1 + \cos x)} dx$

SOLUTION. Here, the given integral is $\int \dfrac{1 - \cos x}{\cos x (1 + \cos x)} dx$.

Let $\cos x = t$, then we have

$$\frac{1 - \cos x}{\cos x(1 + \cos x)} = \frac{1 - t}{t(1 + t)} = \frac{A}{t} + \frac{B}{(1+t)}$$

$$\Rightarrow \qquad (1 - t) = (1 + t)A + tB.$$

Putting $t = 0$, we get $A = 1$ and putting $t = -1$, we get $B = -2$ thus, we have

$$\frac{1-t}{t(1+t)} = \frac{1}{t} - \frac{2}{(1+t)}$$

$\therefore$

$$\frac{1-\cos x}{\cos x(1+\cos x)} = \frac{1}{\cos x} - \frac{2}{(1+\cos x)}$$

$$= \sec x - \frac{2}{2\cos^2 \frac{x}{2}} = \sec x - \sec^2 \frac{x}{2}$$

Hence, $\int \dfrac{1-\cos x}{\cos x(1+\cos x)} dx = \int \sec x \, dx - \int \sec^2 \frac{x}{2} dx$

$$= \log(\sec x + \tan x) - 2\tan\frac{x}{2} + C$$

EXAMPLE 9. *Evaluate* $I = \int \dfrac{dx}{\sin(x-a)\sin(x-b)} dx$

SOLUTION. We have

$$I = \int \frac{dx}{\sin(x-a)\sin(x-b)} = \int \frac{\sin(a-b)}{\sin(a-b)\sin(x-a)\sin(x-b)} dx$$

$$= \int \frac{\sin\{(a-x)-(b-x)\}}{\sin(a-b)\sin(x-a)\sin(x-b)} dx$$

$$= \int \frac{\sin(a-x)\cos(b-x) - \cos(a-x)\sin(b-x)}{\sin(a-b)\sin(x-a)\sin(x-b)} dx$$

$$= \frac{1}{\sin(a-b)}\int \frac{-\sin(x-a)\cos(x-b) + \sin(x-b)\cos(x-a)}{\sin(x-a)\sin(x-b)} dx$$

$$= \frac{1}{\sin(a-b)}\int \{-\cot(x-b) + \cot(x-a)\} dx$$

$$= \frac{1}{\sin(a-b)}\{-\log\sin(x-b) + \log\sin(x-a)\} + C$$

$$= \frac{1}{\sin(a-b)}\log\left\{\frac{\sin(x-a)}{\sin(x-b)}\right\} + C$$

EXAMPLE 10. *Evaluate the integral* $\int \dfrac{1+x}{(1+x^2)(1-x+x^2)} dx$.

SOLUTION. Here, the given integral is $\int \dfrac{1+x}{(1+x^2)(1-x+x^2)} dx$.

Let $\dfrac{1+x}{(1+x^2)(1-x+x^2)} = \dfrac{Ax+B}{1+x^2} + \dfrac{Cx+D}{1-x+x^2}$

$\Rightarrow \dfrac{1+x}{(1+x^2)(1-x+x^2)} = \dfrac{Ax+B}{1+x^2} + \dfrac{Cx+D}{1-x+x^2}$

$\Rightarrow \qquad 1+x = (1-x+x^2)(Ax+B) + (1+x^2)(Cx+D) \qquad \dots(1)$

Here, to find the four constants A, B, C and D, the method of comparing the coefficient is lengthy. So, we equate each quadratic factor equal to zero and find the value of x^2 and put this value of x^2 in both sides of (1).

So, $1 + x^2 = 0 \Rightarrow x^2 = -1.$

Then, from (1), we get

$$1 + x = (Ax + B)(1 - x - 1) = -(Ax^2 + Bx).$$

Again put $x^2 = -1$. Now, comparing x and constant, we get

$$B = -1, A = 1$$

Again, $1 - x + x^2 = 0 \Rightarrow x^2 = x - 1$

Then from (1), we get

$$1 + x = C(x - 1) + D(x) = (C + D)x - C$$

Comparing x and constants, we get

$$C = -1 \text{ and } D + C = 1 \Rightarrow D = 2.$$

So, $\dfrac{1 + x}{(1 + x^2)(1 - x + x^2)} = \left\{ \dfrac{x - 1}{(1 + x^2)} + \dfrac{(-x + 2)}{1 - x + x^2} \right\}$

Hence,

$$\int \frac{1 + x}{(1 + x^2)(1 - x + x^2)} dx = \int \left(\frac{x - 1}{1 + x^2} - \frac{(x - 2)}{1 - x + x^2} \right) dx$$

$$= \int \left(\frac{1}{2} \cdot \frac{2(x - 1)}{1 + x^2} - \frac{1}{2} \frac{(2x - 1) - 3}{1 - x + x^2} \right) dx$$

$$= \frac{1}{2} \int \frac{2x}{1 + x^2} - \int \frac{2}{1 + x^2} - \frac{1}{2} \int \frac{(2x - 1) - 3}{1 - x + x^2} dx$$

$$= \frac{1}{2} \log(1 + x^2) - \tan^{-1} x - \frac{1}{2} \log(1 - x + x^2)$$

$$+ \frac{3}{2} \int \frac{dx}{\left(x - \frac{1}{2} \right)^2 + \left(\frac{\sqrt{3}}{2} \right)^2}$$

$$= \frac{1}{2} \log(1 + x^2) - \tan^{-1} x - \frac{1}{2} \log(1 - x + x^2)$$

$$+ \frac{3}{2} \cdot \frac{2}{\sqrt{3}} \tan^{-1} \frac{x - \frac{1}{2}}{\sqrt{3}/2} + C$$

$$= \frac{1}{2} \log\left(\frac{1 + x^2}{1 - x + x^2} \right) - \tan^{-1} x + \sqrt{3} \tan^{-1} \left(\frac{2x - 1}{\sqrt{3}} \right) + C$$

EXAMPLE 11. *Evaluate the integral* $\int \dfrac{x^2}{(x - 1)^3 (x - 2)} dx.$

SOLUTION. Here, the given integral is $\int \dfrac{x^2}{(x - 1)^3 (x - 2)} dx.$

Let $\dfrac{x^2}{(x - 1)^3 (x - 2)} = \dfrac{A}{x - 1} + \dfrac{B}{(x - 1)^2} + \dfrac{C}{(x - 1)^3} + \dfrac{D}{x - 2}$

$\Rightarrow \quad x^2 = (x - 1)^2 (x - 2)A + (x - 1)(x - 2)B + (x - 2)C + (x - 1)^3 D$ ...(1)

Put $x = 1 \quad \Rightarrow \quad C = -1$

Put $x = 2$ $\Rightarrow$ $D = 4$

Now, we want to find the coefficient A and B.

On comparing the coefficient of x^3 in (1), we get

$$A + D = 0$$

Now comparing the constant terms in (1), we get

$$-2A + 2B - 2C - D = 0$$

Putting the value of C and D, we get

$$A = -4, B = -3$$

Thus, $$\frac{x^2}{(x-1)^3(x-2)} = -\frac{4}{x-1} - \frac{3}{(x-1)^2} - \frac{1}{(x-1)^3} + \frac{4}{x-2}$$

Hence, $$\int \frac{x^2}{(x-1)^3(x-2)}\,dx = -4\int \frac{1}{x-1}\,dx - 3\int \frac{1}{(x-1)^2}\,dx$$

$$-\int \frac{1}{(x-1)^3}\,dx + 4\int \frac{1}{x-2}\,dx$$

$$= -4\log(x-1) + \frac{3}{x-1} + \frac{1}{2(x-1)^2} + 4\log(x-2) + C$$

$$= 4\log\left(\frac{x-2}{x-1}\right) + \frac{3}{x-1} + \frac{1}{2(x-1)^3} + C.$$

MORE SOLVED EXAMPLES BASED ON INTEGRAL BY SUBSTITUTION

EXAMPLE 1.   *Evaluate* $\int \dfrac{1}{1+e^{-x}}\,dx$.

SOLUTION.

$$I = \int \frac{1}{1+e^{-x}}\,dx = \int \frac{1}{1+\dfrac{1}{e^x}}\,dx = \int \frac{e^x}{1+e^x}\,dx$$

Putting $1 + e^x = t$ $\Rightarrow$ $e^x dx = dt$

$$I = \int \frac{dt}{t} = \log|t| + C = \log|1 + e^x| + C$$

EXAMPLE 2.   *Evaluate* $\int \dfrac{\sin 2x}{a^2 \sin^2 x + b^2 \cos^2 x}\,dx$

SOLUTION.   We have $$I = \int \frac{\sin 2x\, dx}{a^2 \sin^2 x + b^2 \cos^2 x}$$ ...(1)

Let $a^2 \sin^2 x + b^2 \cos^2 x = t$

$\Rightarrow$ $(2a^2 \sin x \cos x - 2b^2 \sin x \cos x)\,dx = dt$

$\Rightarrow$ $(a^2 \sin 2x - b^2 \sin 2x)\,dx = dt$

$\Rightarrow$ $\sin 2x\, dx = \dfrac{dt}{a^2 - b^2}$

Putting in (1), we get $I = \int \dfrac{dt}{t(a^2 - b^2)} = \dfrac{1}{a^2 - b^2}\log|t| + C$

$$= \frac{1}{a^2 - b^2}\log\left|a^2 \sin^2 x + b^2 \cos^2 x\right| + C$$

EXAMPLE 3. *Find* $\int \dfrac{dx}{x(1+x^n)}$

SOLUTION. Let $\qquad I = \int \dfrac{dx}{x(1+x^n)} = \int \dfrac{x^{n-1}dx}{x^n(1+x^n)}$ (Multiply by x^{n-1} into Nr. and Dr.)

Put $\qquad (1+x^n) = t \qquad \Rightarrow \qquad x^n = t-1$

$\Rightarrow \qquad nx^{n-1}dx = dt \qquad \Rightarrow \quad x^{n-1}dx = \dfrac{dt}{n}$

Therefore, $\qquad I = \dfrac{1}{n}\int \dfrac{dt}{(t-1)t}$

$\qquad\qquad = \dfrac{1}{n}\int \left(\dfrac{1}{t-1} - \dfrac{1}{t}\right)dt \qquad\qquad$ (By partial fraction)

Integrating $\qquad I = \dfrac{1}{n}\{\log|t-1| - \log|t|\} + C = \dfrac{1}{n}\log\left|\dfrac{t-1}{t}\right| + C$

$\qquad\qquad = \dfrac{1}{n}\log\left|\dfrac{x^n}{x^n+1}\right| + C \qquad\qquad$ (By putting the value of t)

EXAMPLE 4. *Evaluate* $\int \dfrac{(2x+5)}{\sqrt{x^2+3x+1}}dx$

SOLUTION. Let $\qquad \int \dfrac{(2x+5)}{\sqrt{x^2+3x+1}}dx = \int \dfrac{(2x+3)+2}{\sqrt{x^2+3x+1}}dx$

$\qquad\qquad = \underbrace{\int \dfrac{(2x+3)}{\sqrt{x^2+3x+1}}dx}_{I_1} + \underbrace{\int \dfrac{2}{\sqrt{x^2+3x+1}}dx}_{I_2} \qquad$ ...(1)

Suppose $\qquad I_1 = \int \dfrac{(2x+3)}{\sqrt{x^2+3x+1}}dx$

Put $\qquad x^2+3x+1 = t \qquad \Rightarrow \qquad (2x+3)dx = dt$

Now $\qquad I_1 = \int \dfrac{dt}{\sqrt{t}} = \int t^{-1/2}dt = \dfrac{t^{-1/2+1}}{-\dfrac{1}{2}+1} + C_1$

$\qquad\qquad = 2t^{1/2} + C_1 = 2\sqrt{x^2+3x+1} + C_1$

$\qquad I_2 = \int \dfrac{2dx}{\sqrt{x^2+3x+1}} = 2\int \dfrac{dx}{\left(x^2+3x+\dfrac{9}{4}-\dfrac{9}{4}+1\right)^{1/2}}$

$\qquad\qquad = 2\int \dfrac{dx}{\left\{\left(x+\dfrac{3}{2}\right)^2 - \left(\dfrac{\sqrt{5}}{2}\right)^2\right\}^{1/2}}$

$\qquad\qquad = 2\log\left|\left(x+\dfrac{3}{2}\right) + \sqrt{x^2+3x+1}\right| + C_2$

Putting the value of I_1 and I_2 in (1), we get

$$I = 2\sqrt{x^2 + 3x + 1} + C_1 + 2\log\left|\left(x + \frac{3}{2}\right) + \sqrt{x^2 + 3x + 1}\right| + C_2$$

$$= 2\sqrt{x^2 + 3x + 1} + 2\log\left|\left(x + \frac{3}{2}\right) + \sqrt{x^2 + 3x + 1}\right| + C$$

where, $\qquad C = C_1 + C_2$

EXAMPLE 5. Evaluate $\int \dfrac{\cot x}{\log \sin x} dx$.

SOLUTION. Let $\qquad I = \int \dfrac{\cot x}{\log \sin x} dx$

Putting $\quad \log \sin x = t \quad \Rightarrow \quad \dfrac{1}{\sin x} \cos x\, dx = dt$

$\Rightarrow \qquad \cot x\, dx = dt$

Then $\qquad I = \int \dfrac{dt}{t} = \log|t| + C = \log|\log \sin x| + C$

EXAMPLE 6. Evaluate $\int \dfrac{1}{\sqrt{x}\sqrt{x} + 1} dx$

SOLUTION. Let $\qquad I = \int \dfrac{1}{\sqrt{x}\sqrt{x} + 1} dx$

Putting $\sqrt{x} + 1 = t$, differentiating w.r.t. x

$$\dfrac{1}{2\sqrt{x}} dx = dt$$

$$\dfrac{dx}{\sqrt{x}} = 2dt$$

Then $\qquad I = \int \dfrac{2dt}{t} = 2\log|t| + C = 2\log|\sqrt{x} + 1| + C$

EXAMPLE 7. Evaluate $\int \cot x\, e^{m\log(\sin x)} dx$

SOLUTION. Let $\qquad I = \int \cot x\, e^{m\log(\sin x)} dx$

Putting $\log \sin x = t$ and diff. w.r.t. x, we get

$$\dfrac{\cos x}{\sin x} dx = dt \qquad \Rightarrow \qquad \cot x\, dx = dt$$

Then $\qquad I = \int e^{mt} dt$.

Integrating $\qquad I = \dfrac{e^{mt}}{m} + C = \dfrac{e^{m\log(\sin x)}}{m} + C \qquad$ (Putting value of t)

EXAMPLE 8. Evaluate $\int \dfrac{3x^2 + 1}{x(x^2 + 1)} dx$

SOLUTION. Let $\qquad I = \int \dfrac{3x^2 + 1}{x(x^2 + 1)} dx = \int \dfrac{3x^2 + 1}{x^3 + x} dx$

Putting $\qquad x^3 + x = t \qquad \Rightarrow \qquad (3x^2 + 1)\,dx = dt$

Then $\qquad I = \int \dfrac{dt}{t} = \log|t| + C = \log\left|x^3 + x\right| + C$

EXAMPLE 9. *Evaluate* $\int \dfrac{\sin x}{\sin(x - a)}\,dx.$

SOLUTION. Let $\qquad I = \int \dfrac{\sin x}{\sin(x - a)}\,dx$

Putting $x - a = t$, *i.e.*, $dx = dt$ we get

$$I = \int \frac{\sin(a + t)}{\sin t}\,dt = \int \frac{\sin a \cos t + \cos a \sin t}{\sin t}\,dt$$

$$= \sin a \int \cot t\,dt + \cos a \int 1\,dt$$

$$= \sin a \log \sin t + t \cos a + C.$$

Putting the value of t, we have

$$I = \sin a \log \sin(x - a) + (x - a)\cos a + C.$$

EXAMPLE 10. *Evaluate* $\int x^5 \sqrt{a^3 + x^3}\,dx$

SOLUTION. Let $\qquad I = \int x^5 \sqrt{a^3 + x^3}\,dx = \int x^3 \, x^2 \sqrt{a^3 + x^3}\,dx$

Putting $a^3 + x^3 = t$, differentiating w.r.t. x,

$$3x^2 dx = dt$$

$$x^2 dx = \frac{dt}{3}$$

Then, $\qquad I = \dfrac{1}{3}\int (t - a^3)\sqrt{t}\,dt$

$$= \frac{1}{3}\int (t^{3/2} - a^3 t^{1/2})\,dt, \qquad \text{(On integrating)}$$

$$= \frac{1}{3} \cdot \frac{t^{5/2}}{5/2} - \frac{a^3}{3}\frac{t^{3/2}}{3/2} + C$$

Putting the value of t, we have

$$I = \frac{2}{15}(a^3 + x^3)^{5/2} - \frac{2}{9}a^3(a^3 + x^3)^{3/2} + C$$

EXAMPLE 11. *Evaluate* $\int \dfrac{\log \tan x/2}{\sin x}\,dx.$

SOLUTION. Let $\qquad I = \int \dfrac{\log \tan \dfrac{x}{2}}{\sin x}\,dx$

Putting, $\log \tan \dfrac{x}{2} = t$, and differentiating, we get

$$\frac{1}{\tan x/2}\sec^2 x/2 \cdot \frac{1}{2}dx = dt$$

$$\Rightarrow \qquad \frac{1}{2\dfrac{\sin x/2}{\cos x/2} \cdot \cos^2 x/2}\,dx = dt$$

$$\Rightarrow \quad \frac{1}{2\sin x/2\cos x/2}dx = dt$$

$$\Rightarrow \quad \frac{1}{\sin x}dx = dt$$

Then $\quad I = \int t\,dt = \dfrac{t^2}{2} + C = \dfrac{(\log \tan x/2)^2}{2} + C$

EXAMPLE 12. Evaluate $\int \sec^3 x \tan x\,dx$.

SOLUTION. Let

$$I = \int \sec^3 x \tan x\,dx$$

$$I = \int \sec^2 x (\sec x \tan x)\,dx$$

Putting, $\sec x = t$, differentiating w.r.t. x, we get

$$\sec x \tan x\,dx = dt$$

Then $\quad I = \int t^2 dt = \dfrac{t^3}{3} + C,$

Now, putting the value of t, we have

$$I = \frac{\sec^3 x}{3} + C$$

EXAMPLE 13. Evaluate $\int \dfrac{1}{\sin x \cos^3 x}dx$

SOLUTION. Let

$$I = \int \frac{dx}{\dfrac{\sin x}{\cos x}\cos^4 x} = \int \frac{dx}{\tan x \cos^4 x} = \int \frac{\sec^4 x\,dx}{\tan x}$$

$$= \int \frac{\sec^2 x \sec^2 x}{\tan x}dx = \int \frac{(1 + \tan^2 x)\sec^2 x}{\tan x}dx$$

Putting, $\tan x = t$

$$\Rightarrow \quad \sec^2 x\,dx = dt$$

Then $\quad I = \int \dfrac{1 + t^2}{t}dt = \int\left(\dfrac{1}{t} + t\right)dt$, Integrating

and putting the value of t, we get

$$I = \log|t| + \frac{t^2}{2} + C = \log|\tan x| + \frac{\tan^2 x}{2} + C$$

EXAMPLE 14. Evaluate $\int \dfrac{\log x^2}{x}dx$.

SOLUTION. Let

$$I = \int \frac{\log x^2}{x}dx$$

$$I = \int \frac{\log x^2}{x}dx = \int \frac{2\log x}{x}dx$$

Putting $\log x = t$, differentiating

$$\frac{1}{x}dx = dt$$

Then $$I = 2\int t\, dt = \frac{2t^2}{2} + C = t^2 + C$$

After putting the value of t, we have $I = (\log x)^2 + C$

EXAMPLE 15. *Evaluate* $\int \dfrac{x^2 \tan^{-1} x^3}{1 + x^6}\, dx$

SOLUTION. Let $$I = \int \frac{x^2 \tan^{-1} x^3}{1 + x^6}\, dx$$

Let us put $\tan^{-1} x^3 = t$, differentiating

$$\frac{1}{1 + (x^3)^2} \cdot 3x^2\, dx = dt$$

$$\Rightarrow \qquad \frac{x^2}{1 + x^6}\, dx = \frac{dt}{3}$$

Then $$I = \frac{1}{3}\int t\, dt = \frac{t^2}{6} + C$$

On integrating and putting the value of t, we get

$$I = \frac{(\tan^{-1} x^3)^2}{6} + C$$

EXAMPLE 16. *Evaluate* $\int \tan x \sec^2 x \sqrt{1 - \tan^2 x}\, dx$

SOLUTION. Let $$I = \int \tan x \sec^2 x \sqrt{1 - \tan^2 x}\, dx$$

Putting $\quad 1 - \tan^2 x = t$

$\Rightarrow -2 \tan x \sec^2 x\, dx = dt$

$$\tan x \sec^2 x\, dx = -\frac{dt}{2}$$

Then $$I = -\frac{1}{2}\int \sqrt{t}\, dt = -\frac{1}{2}\frac{t^{3/2}}{3/2} + C = -\frac{1}{3}(1 - \tan^2 x)^{3/2} + C$$

(After putting the value of t)

EXAMPLE 17. *Evaluate* $\int \dfrac{e^{m\sin^{-1} x}}{\sqrt{1 - x^2}}\, dx.$

SOLUTION. Let $$I = \int \frac{e^{m\sin^{-1} x}}{\sqrt{1 - x^2}}\, dx$$

Putting, $\sin^{-1} x = t$, differentiating w.r.t. x, we get

$$\frac{1}{\sqrt{1 - x^2}}\, dx = dt$$

Then $$I = \int e^{mt}\, dt = \frac{e^{mt}}{m} + C = \frac{e^{m\sin^{-1} x}}{m} + C$$

(After putting the value of t)

EXAMPLE 18. *Evaluate* $\int \dfrac{\cos^9 x}{\sin x} dx$

SOLUTION. Let $I = \int \dfrac{\cos^9 x}{\sin x} dx = \int \dfrac{\cos^8 x \cos x}{\sin x} dx = \int \dfrac{(\cos^2 x)^4 \cos x\, dx}{\sin x}$

$$= \int \dfrac{(1 - \sin^2 x)^4 \cos x}{\sin x} dx$$

Put $\sin x = t \quad \Rightarrow \quad \cos x\, dx = dt$

Then $I = \int \dfrac{(1 - t^2)^4 dt}{t} = \int \dfrac{1 - 4t^2 + 6t^4 - 4t^6 + t^8}{t} dt$

$$= \int \left(\dfrac{1}{t} - 4t + 6t^3 - 4t^5 + t^7 \right) dt$$

$$= \log|t| - 2t^2 + \dfrac{3}{2} t^4 - \dfrac{2}{3} t^6 + \dfrac{1}{8} t^8 + C, \qquad \text{(On integrating)}$$

$$= \log|\sin x| - 2\sin^2 x + \dfrac{3}{2} \sin^4 x - \dfrac{2}{3} \sin^6 x + \dfrac{1}{8} \sin^8 x + C$$

$$\text{(After putting the value of } t)$$

EXAMPLE 19. *Evaluate* $\int \sec^4 x\, dx$

SOLUTION. Let $I = \int \sec^4 x\, dx = \int \sec^2 x \sec^2 x\, dx$

$$= \int \sec^2 x(1 + \tan^2 x) dx = \int (\sec^2 x + \sec^2 x \tan^2 x) dx$$

$$= \int \sec^2 x\, dx + \int \sec^2 x \tan^2 x\, dx$$

$$= \tan x + \int \sec^2 x \tan^2 x\, dx$$

Put $\tan x = t$, in 2nd integral, we have

$$\sec^2 x\, dx = dt$$

Then $I = \tan x + \int t^2 dt + C = \tan x + \dfrac{t^3}{3} + C = \tan x + \dfrac{\tan^3 x}{3} + C$

EXAMPLE 20. *Evaluate* $\int \dfrac{x^2}{\sqrt{1+x}} dx$

SOLUTION. Let $I = \int \dfrac{x^2}{\sqrt{1+x}} dx$

put $1 + x = t^2 \quad \Rightarrow \quad x = t^2 - 1$

$$dx = 2t\, dt$$

Then $I = \int \dfrac{(t^2 - 1)^2 . 2t\, dt}{\sqrt{t^2}} = 2\int \dfrac{(t^2 - 1)^2 t\, dt}{t}$

$$= 2\int (t^2 - 1)^2 dt = 2\int (t^4 - 2t^2 + 1) dt$$

$$= 2\left[\dfrac{t^5}{5} - \dfrac{2t^3}{3} + t \right] + C;$$

$$= \dfrac{2}{5}(1+x)^{5/2} - \dfrac{4}{3}(1+x)^{3/2} + 2\sqrt{1+x} + C$$

EXAMPLE 21. *Evaluate* $\int \cos^3 \sqrt{x}\, dx$

SOLUTION. Consider $\int \cos^3 \sqrt{x}\, dx$

Putting $\sqrt{x} = t \implies x = t^2$

$\implies dx = 2t\, dt$

$\therefore \int \cos^3 \sqrt{x}\, dx = 2\int \cos^3 t \cdot t\, dt$

Now using $3x = 4\cos^3 x - 3\cos x$, we have

$$= 2\int t \left[\frac{\cos 3t + 3\cos t}{4}\right] dt = \frac{1}{2}\int t\cos 3t\, dt + \frac{3}{2}\int t \cos t\, dt$$

$$= \frac{1}{2}\left[t\frac{\sin 3t}{3} + \frac{1}{9}\cos 3t\right] + \frac{3}{2}[t\sin t + \cos t] + c$$

$$= \frac{1}{6}\left[t\sin 3t + \frac{1}{3}\cos 3t\right] + \frac{3}{2}[t\sin t + \cos t] + c$$

$$= \frac{1}{6}\left[\sqrt{x}\sin 3\sqrt{x} + \frac{1}{3}\cos 3\sqrt{x}\right] + \frac{3}{2}(\sqrt{x}\sin\sqrt{x} + \cos\sqrt{x}) + C$$

$$(\because \ \sqrt{x} = t)$$

Exercise 10.1

Evaluate the following Integrals

1. $\int \dfrac{\sin\sqrt{x}}{\sqrt{x}}\, dx$

2. $\int \dfrac{1}{x^2}\cos^2\left(\dfrac{1}{x}\right) dx$

3. $\int (4x+2)\sqrt{x^2+x+1}\, dx$

4. $\int \cosec x \cdot \log(\cosec x - \cot x)\, dx$

5. $\int \dfrac{\sec^2(2\tan^{-1}x)}{1+x^2}\, dx$

6. $\int \dfrac{1-\sin 2x}{x+\cos^2 x}\, dx$

7. $\int \dfrac{\cosec x}{\log\tan\dfrac{x}{2}}\, dx$

8. $\int \dfrac{1-\cot x}{1+\cot x}\, dx$

9. $\int 4x^3\sqrt{5-x^2}\, dx$

10. $\int \dfrac{(x+1)e^x}{\cos^2(xe^x)}\, dx$

11. $\int \dfrac{dx}{\sin\dfrac{x}{2} + \tan\dfrac{x}{2}}$

HINT TO SELECTED PROBLEMS

1. $I = \int \dfrac{\sin\sqrt{x}}{\sqrt{x}}\, dx$

Put $\sqrt{x} = t \implies \dfrac{1}{2\sqrt{x}}dx = dt \implies \dfrac{dx}{\sqrt{x}} = 2dt$

$\therefore \ I = 2\int \sin t\, dt = -2\cos t + C = -2\cos\sqrt{x} + C$

2. $I = \int \dfrac{1}{x^2}\cos^2\left(\dfrac{1}{x}\right) dx$

Put $\dfrac{1}{x} = t \implies \dfrac{dt}{dx} = -\dfrac{1}{x^2} \implies \dfrac{dx}{x^2} = -dt$

$I = -\int \cos^2 t\, dt = -\int\left(\dfrac{1+\cos 2t}{2}\right) dt$

$\left(\because \ \cos^2 x = \dfrac{1+\cos 2x}{2}\right)$

$= -\dfrac{1}{2}\int (1+\cos 2t)\, dt,$

Integrating

$= -\dfrac{1}{2}\left(t + \dfrac{\sin 2t}{2}\right) + C = -\dfrac{1}{2x} - \dfrac{1}{4}\sin\dfrac{2}{x} + C$

3. $I = \int (4x+2)\sqrt{x^2+x+1}\, dx$

Put $x^2 + x + 1 = 0$

$\implies (2x+1)\, dx = dt$

$I = 2\int \sqrt{t}\, dt$

$= 2\dfrac{t^{3/2}}{3/2} + C = \dfrac{4}{3}t^{3/2} = \dfrac{4}{3}(x^2+x+1)^{3/2} + C$

4. $I = \int \operatorname{cosec} x \cdot \log|\operatorname{cosec} x - \cot x| \, dx$

Put $\log(\operatorname{cosec} x - \cot x) = t$

$$\frac{1}{\operatorname{cosec} x - \cot x}(-\operatorname{cosec} x \cot^2 x$$
$$+ \operatorname{cosec}^2 x) dx = dt$$

$\Rightarrow \qquad \operatorname{cosec} x \, dx = dt$

$$I = \int t \, dt = \frac{t^2}{2} + C$$

$$= \frac{1}{2}\{\log|\operatorname{cosec} x - \cot x|\}^2 + C$$

5. $I = \int \dfrac{\sec^2(2\tan^{-1} x)}{1+x^2} dx$

Put $\tan^{-1} x = t$

$$\frac{1}{1+x^2} dx = dt$$

$I = \int \sec^2(2t) \, dt.$ Integrating

$$= \frac{\tan 2t}{2} + C$$

putting the value of t, we have

$$= \frac{1}{2}\tan(2\tan^{-1} x) + C$$

6. $I = \int \dfrac{1 - \sin 2x}{x + \cos^2 x} dx$

Put $x + \cos^2 x = t$

$(1 - 2\sin x \cos x) dx = dt$

$(1 - \sin 2x) dx = dt$

$I = \int \dfrac{dt}{t},$

Integrating $= \log t + C = \log|x + \cos^2 x| + C$

7. $I = \int \dfrac{\operatorname{cosec} x}{\log \tan \dfrac{x}{2}} dx$

put $\log \tan \dfrac{x}{2} = t$, differentiating

We get $\dfrac{1}{\tan \dfrac{x}{2}}\left(\sec^2 \dfrac{x}{2}\right)\dfrac{1}{2} dx = dt$

$\Rightarrow \qquad \dfrac{1}{2\sin x / 2 \cos x / 2} dx = dt$

$\Rightarrow \qquad \operatorname{cosec} x \, dx = dt$

$I = \int \dfrac{dt}{t},$ Integrating

$= \log t + C$ putting the value of t

$$= \log\left|\log \tan \frac{x}{2}\right| + C$$

8. $I = \int \dfrac{1 - \cot x}{1 + \cot x} dx = \int \dfrac{\sin x - \cos x}{\sin x + \cos x} dx$

Put $\cos x + \sin x = t \Rightarrow -(\sin x - \cos x) dx = dt$

$$I = -\int \frac{dt}{t} = -\log t + C,$$

Putting the value of t

$$I = -\log|\cos x + \sin x| + C$$

9. $I = \int 4x^3\sqrt{5 - x^2}\,dx = \int 4x^2 \cdot x\sqrt{5 - x^2}\,dx;$

Put $5 - x^2 = t \Rightarrow -x \, dx = \dfrac{dt}{2}$

$$I = 4\int (5 - t)\sqrt{t}\left(-\frac{dt}{2}\right)$$

$$= -2\int (5\sqrt{t} - t^{3/2})dt, \text{ Integrating}$$

$$= 2\left\{5\frac{t^{3/2}}{3/2} - \frac{2}{5}5/2\right\} + C$$

$$= \frac{-20}{3}t^{3/2} + \frac{4}{5}t^{5/2} + C, \text{ putting the value of } t$$

$$= -\frac{20}{3}(5 - x^2)^{3/2} + \frac{4}{5}(5 - x^2)^{5/2} + C$$

10. $I = \int \dfrac{(x+1)e^x}{\cos^2(xe^x)} dx$

Put $x e^x = t \Rightarrow (e^x + xe^x) dx = dt$

$\Rightarrow e^x(x+1)dx = dt$

$I = \int \dfrac{dt}{\cos^2 t} = \int \sec^2 t \, dt;$ Integrating

$= \tan t + C$

$= \tan(xe^x) + C,$ putting value of t

ANSWERS

1. $-2\cos\sqrt{x}$

2. $-\dfrac{1}{2x}-\dfrac{1}{4}\sin\dfrac{2}{x}+C$

3. $\dfrac{4}{3}(x^2+x+1)^{3/2}+C$

4. $\dfrac{1}{2}\{\log|\mathrm{cosec}\,x-\cot x|\}^2+C$ 5. $\dfrac{1}{2}\tan(2\tan^{-1}x)+C$

6. $\log|x+\cos^2 x|+C$

7. $\log\left|\log\tan\dfrac{x}{2}\right|+C$ 8. $-\log|\cos x+\sin x|+C$

9. $\dfrac{4}{5}(5-x^2)^{5/2}-\dfrac{20}{3}(5-x^2)^{3/2}+C$

10. $\tan(xe^x)+C$

11. $\log\tan\dfrac{x}{4}-\dfrac{\left(\sec\dfrac{x}{4}\right)^2}{2}+C$

Exercise 10.2

1. Evaluate the following integrals :

   (i) $\displaystyle\int\left(\dfrac{2+3\log x}{x}\right)dx$ (ii) $\displaystyle\int 3\sec^2(3x+9)dx$

   (iii) $\displaystyle\int\dfrac{1}{(8-6x)^2}dx$ (iv) $\displaystyle\int\dfrac{dx}{2x+3}$

2. Evaluate the following integrals :

   (i) $\displaystyle\int\dfrac{\sin 2x}{\cos^2 2x}dx$ (ii) $\displaystyle\int\sec^7 x\tan x\,dx$

   (iii) $\displaystyle\int\dfrac{1}{x\log x}dx$ (iv) $\displaystyle\int x\sin x^2 dx$.

3. Evaluate the following integrals :

   (i) $\displaystyle\int\dfrac{a}{b+ce^x}dx$ (ii) $\displaystyle\int\dfrac{e^x}{1+e^{2x}}dx$

   (iii) $\displaystyle\int\dfrac{\sec^2(\log x)}{x}dx$ (iv) $\displaystyle\int\dfrac{\sec^2 x}{1+\tan x}dx$.

4. Evaluate the following integrals :

   (i) $\displaystyle\int\mathrm{cosec}\,x\,dx$ (ii) $\displaystyle\int\dfrac{\cot x}{\log\sin x}dx$

   (iii) $\displaystyle\int\dfrac{4x^3}{1+x^8}dx$.

5. Evaluate the following integrals :

   (i) $\displaystyle\int\dfrac{1}{x}\log\log x\,dx$ (ii) $\displaystyle\int\dfrac{1}{x\sqrt{1+x}}dx$

   (iii) $\displaystyle\int\sqrt{(a^2-x^2)}\,dx$ (iv) $\displaystyle\int x\sin x\cos x\,dx$.

6. Evaluate the following integrals :

   (i) $\displaystyle\int x\cos 2x\,dx$ (ii) $\displaystyle\int x\cos 2x\sin x\,dx$

   (iii) $\displaystyle\int\dfrac{1}{x^2\sqrt{(1+x^2)}}dx$

7. Evaluate the following integrals :

   (i) $\displaystyle\int\dfrac{\log\sin^{-1}x}{\sqrt{(1-x^2)}}dx$ (ii) $\displaystyle\int\dfrac{\sqrt{2}dx}{\sqrt{(1+\cos 2x)}}$

8. Evaluate the following integrals :

   (i) $\displaystyle\int\dfrac{x\,dx}{\sqrt{1+x}}$ (ii) $\displaystyle\int\sqrt{1+\sin x}\,dx$

   (iii) $\displaystyle\int\cos^4 x\,dx$ (iv) $\displaystyle\int\dfrac{\log(1+x^2)}{x^2}dx$.

9. Evaluate the following integrals :

   (i) $\displaystyle\int xe^{3x}dx$ (ii) $\displaystyle\int\sqrt{x}\log x\,dx$

   (iii) $\displaystyle\int\dfrac{1+x}{(2+x)^2}dx$

   (iv) $\displaystyle\int\dfrac{e^x}{x}[x(\log x)^2+2\log x]dx$.

10. Evaluate the following integrals :

    (i) $\displaystyle\int\dfrac{\tan x}{\sec x+\cos x}dx$ (ii) $\displaystyle\int\dfrac{e^x}{x}(x\log x+1)dx$

    (iii) $\displaystyle\int\dfrac{\sin 2x+1}{\sqrt{(x+\sin^2 x)}}$

    (iv) $\displaystyle\int\cos x\cdot\cos 2x\cdot\cos 3x\,dx$

11. Evaluate the following integrals :

    (i) $\displaystyle\int\sqrt{1+\cos x}\,dx$ (ii) $\displaystyle\int\dfrac{\cos 2x}{\sin x}dx$.

12. Evaluate the following integrals :

    (i) $\displaystyle\int\dfrac{\cot x\,dx}{\log\sin x}$ (ii) $\displaystyle\int\dfrac{x+3}{\sqrt{(1-x^2)}}dx$

    (iii) $\displaystyle\int\dfrac{\cos x-\sin x}{(\cos x+\sin x)^2}dx$

    (iv) $\displaystyle\int(\sin x-\cos x)^2 dx$.

13. Evaluate the following integrals :

    (i) $\displaystyle\int\dfrac{1}{x\{1+(\log x)^2\}}dx$

    (ii) $\displaystyle\int\dfrac{6x-8}{3x^2-8x+5}$ (iii) $\displaystyle\int\dfrac{\cos\sqrt{x}}{\sqrt{x}}dx$.

14. Evaluate the following integrals :

(i) $\int \dfrac{xe^x dx}{(x+1)^2}$
(ii) $\int \dfrac{4x^3}{5x^4+7} dx$

(iii) $\int \dfrac{\cos 2x}{(\cos x+\sin x)} dx$

(iv) $\int \dfrac{\sin(x-a)}{\sin x} dx.$

15. Evaluate the following integrals by partial fraction:

(i) $\int \dfrac{x^3}{(x-a)(x-b)(x-c)} dx$

(ii) $\int \dfrac{x-1}{(x+1)(x^2+1)} dx$

(iii) $\int \dfrac{3x\,dx}{(x-1)^2(x+3)}$
(iv) $\int \dfrac{x^2}{(x-a)(x-b)} dx.$

16. Evaluate the following integrals :

(i) $\int \dfrac{3x^2}{(x+1)(x+2)(x+3)} dx.$

(ii) $\int \dfrac{1}{e^x+e^{2x}} dx$
(iii) $\int \dfrac{2x+1}{(x+2)(x-1)^2} dx$

(iv) $\int \dfrac{x^2+8}{x^3+4x} dx.$

17. Evaluate the following integrals :

(i) $\int \dfrac{x^2+4x+1}{x^3+2x^2-x-2} dx$

(ii) $\int \dfrac{x}{(x-1)(2x+1)} dx$

(iii) $\int \dfrac{2x}{(x^2+1)(x^2+3)}.dx$

18. Evaluate the following integrals :

(i) $\int x.\sin^{-1}x\,dx$
(ii) $\int x\sin^3 x\,dx$

(iii) $\int \sin x \log(\cos x)\,dx$

(iv) $\int \sin x \log(\sec x+\tan x)\,dx.$

19. Evaluate the following integrals :

(i) $\int \dfrac{4x+7}{2x+3} dx$
(ii) $\int \dfrac{x^2}{1+x^3} dx$

20. Evaluate the following integrals :

(i) $\int \sec^4 x \tan x\,dx.$
(ii) $\int \dfrac{(\sin^{-1}x)^2}{\sqrt{1-x^2}} dx$

HINT TO SELECTED PROBLEMS

2. (ii) $\int \sec^7 x \tan x\,dx = \int \sec^6 x \sec x \tan x\,dx,$
Put $\sec x = t$ and then integrate

(iii) $\int \dfrac{1}{x\log x} dx$ Put $\log x = t \Rightarrow \dfrac{1}{x} dx = dt$

(iv) $I = \int x \sin x^2 dx$ Put $x^2 = t \Rightarrow x\,dx = \dfrac{dt}{2}$

$I = \int \dfrac{1}{2} \sin t\,dt = -\dfrac{1}{2} \cos t + C$

$= -\dfrac{1}{2} \cos x^2 + C$

3. (i) $I = \int \dfrac{a\,dx}{b+ce^x} = \int \dfrac{ae^{-x}}{be^{-x}+c} dx$

[Divide Nr and Dr by e^x]

Put $be^{-x}+c = t \Rightarrow -be^{-x} dx = dt$

$I = -\dfrac{a}{b} \int \dfrac{dt}{t} = -\dfrac{a}{b} \log t + D$

$= -\dfrac{a}{b} \log(c+be^{-x}) + D$

(ii) $I = \int \dfrac{e^x}{1+e^{2x}} dx$

Put $e^x = t \Rightarrow e^x dx = dt,$

$I = \int \dfrac{dt}{1+t^2},$ Integrating

$= \tan^{-1} t + C = \tan^{-1} e^x + C$

(iii) $I = \int \dfrac{\sec^2(\log x)}{x} dx$

Put $\log x = t \Rightarrow \dfrac{1}{x} dx = dt$

$I = \int \sec^2 t\,dt,$ Integrating

$= \tan t + C = \tan(\log x) + C$

(iv) $I = \int \dfrac{\sec^2 x\,dx}{1+\tan x}$ Putting $1+\tan x = t$

4. (i) $I = \int \text{cosec } x\,dx = \int \dfrac{1}{\sin x} dx$

$= \int \dfrac{dx}{2\sin\dfrac{x}{2}\cos\dfrac{x}{2}}$

$= \dfrac{1}{2} \int \dfrac{\sec^2 x/2}{\tan x/2} dx$

[Dividing Nr and Dr by $\cos^2 x/2$]

Put $\tan\dfrac{x}{2} = t$

$\dfrac{1}{2} \sec^2 \dfrac{x}{2} dx = dt$

$I = \int \dfrac{dt}{t},$ Integrating

$= \log t + C = \log\left(\tan\dfrac{x}{2}\right) + C$

5. (ii) $I = \int \dfrac{dx}{x\sqrt{1+x}}$

Put $x = \tan^2 \theta$

$dx = 2\tan\theta \sec^2 \theta \, d\theta$

$I = \int \dfrac{2\tan\theta \sec^2 \theta \, d\theta}{\tan^2 \theta \sqrt{1+\tan^2 \theta}} = 2 \int \dfrac{\sec\theta}{\tan\theta} d\theta$

$= 2\int \operatorname{cosec}\theta \, d\theta$, Integrating, we get

$= 2\log(\operatorname{cosec}\theta - \cot\theta) + C$

$= -2\log(\operatorname{cosec}\theta + \cot\theta) + C$

(iii) $\int \sqrt{a^2 - x^2} \, dx = \int \underset{\text{I}}{\sqrt{a^2 - x^2}} \cdot \underset{\text{II}}{1} \, dx,$

Integrating by parts

$= \sqrt{a^2 - x^2} \cdot x - \int \dfrac{1}{2}(a^2 - x^2)^{-\frac{1}{2}}(-2x)x \, dx$

$= x\sqrt{a^2 - x^2} - \int \dfrac{-x^2 dx}{\sqrt{a^2 - x^2}}$

$I = x\sqrt{a^2 - x^2} - \int \dfrac{a^2 - x^2 - a^2}{\sqrt{a^2 - x^2}} dx$

$= x\sqrt{a^2 - x^2} - \int \sqrt{a^2 - x^2} + a^2 \int \dfrac{dx}{\sqrt{a^2 - x^2}}$

$= x\sqrt{a^2 - x^2} - I + a^2 \sin^{-1}\dfrac{x}{a} + C$

$2I = x\sqrt{a^2 - x^2} + a^2 \sin^{-1}\dfrac{x}{a} + C$

$I = \dfrac{x}{2}\sqrt{a^2 - x^2} + \dfrac{a^2}{2}\sin^{-1}\dfrac{x}{a} + C$

(iv) $I = \int x \sin x \cos x \, dx$

$= \dfrac{1}{2}\int x(2\sin x \cos x) dx = \dfrac{1}{2}\int \underset{\text{I}}{x} \underset{\text{II}}{\sin 2x} \, dx$

(Integrating by parts)

$= \dfrac{1}{2}\int \left[x\left(\dfrac{-\cos 2x}{2}\right) - \int 1\left(\dfrac{-\cos 2x}{2}\right) dx \right]$

$= -\dfrac{x\cos 2x}{4} + \dfrac{\sin 2x}{8} + C$

$= \dfrac{1}{8}(\sin 2x - 2x \cos 2x) + C$

6. (ii) $I = \int x \cos 2x \sin x \, dx$

$= \int x(2\cos^2 x - 1)\sin x \, dx$

Put $\cos x = t \Rightarrow -\sin x \, dx = dt$

$I = \dfrac{1}{2}\int x(2\cos 2x \sin x) dx$

$= \dfrac{1}{2}\int x(\sin 3x - \sin x) dx$

$= \dfrac{1}{2}\left[\int x \sin 3x \, dx - \int x \sin x \, dx \right]$

$= \dfrac{1}{2}\left[x\dfrac{(-\cos 3x)}{3} - \int 1.\left(\dfrac{-\cos 3x}{3}\right) dx \right.$

$\left. -x(-\cos x) - \int 1.(-\cos x) dx \right]$

$= -\dfrac{x\cos x}{6} + \dfrac{\cos 3x}{18} + \dfrac{x\cos x}{2} - \dfrac{\sin x}{2} + C$

(iii) $\int \dfrac{dx}{x^2\sqrt{1+x^2}}$, put $x = \tan\theta$ and integrate

7. (i) $I = \int \dfrac{\log(\sin^{-1} x)}{\sqrt{1-x^2}} dx$, put $\sin^{-1} x = t$ and

integrate by parts

Taking 1 as II function

(ii) $I = \int \dfrac{\sqrt{2} \, dx}{\sqrt{1 + \cos 2x}} = \int \dfrac{\sqrt{2} \, dx}{\sqrt{2\cos^2 x}}$

$= \int \sec x \, dx = \log(\sec x + \tan x) + C$

8. (i) $I = \int \dfrac{x \, dx}{\sqrt{1+x}} = \int \dfrac{\{(x+1) - 1\}}{\sqrt{1+x}} dx$

$= \int \left\{ \sqrt{1+x} - \dfrac{1}{\sqrt{1+x}} \right\}$

$= \left[\dfrac{2}{3}(1+x)^{3/2} - 2(1+x)^{1/2} \right]$

(ii) $\int \sqrt{1 + \sin x} \, dx$,

Use : $1 + \sin x = \left(\cos\dfrac{x}{2} + \sin\dfrac{x}{2} \right)^2$

(iii) $\int (\cos^4 x) dx$,

Use : $\cos^4 x = (\cos^2 x)^2 = \left(\dfrac{1 + \cos 2x}{2} \right)^2$

(iv) $I = \int \dfrac{\log(1+x^2)}{x^2} dx$

$= \int \underset{\text{I}}{\log(1+x)^2} . \underset{\text{II}}{\dfrac{1}{x^2}} dx$, integrating by parts

$= -\dfrac{1}{x}\log(1+x^2) - \int -\dfrac{1}{x}\dfrac{2x}{(1+x^2)} dx$

$= -\dfrac{1}{x}\log(1+x^2) + \int \dfrac{2 \, dx}{1+x^2}$

$= -\dfrac{1}{x}\log(1+x^2) + 2\tan^{-1}x + C$

9. (i) $I = \int x \, e^{3x} dx$, Integrating by parts taking x as first function.

(ii) $I = \int x^{1/2} \log x \, dx$. Integrate by parts taking $\log x$ as first function

(iii) $I = \int \dfrac{1+x}{(2+x)^2} dx = \int \dfrac{2+x-1}{(2+x)^2} dx$

$= \int \left\{ \dfrac{1}{2+x} - \dfrac{1}{(2+x)^2} \right\} dx$, Integrating

$= \log(2+x) + \dfrac{1}{(2+x)} + C$

(iv) $I = \int e^x [x(\log x)^2 + 2\log x] \, dx$

$= \int e^x \left[(\log x)^2 + \dfrac{2}{x} \log x \right] dx$

$= \int e^x (\log x)^2 dx + \int \dfrac{2e^x}{x} \log x \, dx$.

Integrating by parts, I Integral

$= \int e^x (\log x)^2 - \int \dfrac{2}{x} (\log x) e^x dx$

$+ \int \dfrac{2e^x}{x} (\log x) dx = e^x (\log x)^2 + C \cdot$

10. (i) $I = \int \dfrac{\tan x}{\sec x + \cos x} dx$

$= \int \dfrac{\sin x}{\cos x (\sec x + \cos x)} dx = \int \dfrac{\sin x \, dx}{1 + \cos^2 x}$

put $\cos x = t \Rightarrow \sin x \, dx = -dt$

$= \int \dfrac{-dt}{1+t^2}$,

Integrating $= \cot^{-1} t + C = \cot^{-1}(\cos x) + C$

(iii) $I = \int \dfrac{\sin 2x + 1}{\sqrt{x + \sin^2 x}} dx$

Put $x + \sin^2 x = t$

$\Rightarrow (1 + 2\sin x \cos x) \, dx = dt$

$\Rightarrow (1 + \sin x) \, dx = dt$

$I = \int \dfrac{dt}{\sqrt{t}}$; Integrating $= 2\sqrt{x + \sin^2 x} + C$

(iv) $I = \int \cos x \cos 2x \cos 3x \, dx$

$= \dfrac{1}{2} \int (2\cos x \cos 2x) \cos 3x \, dx$

$= \dfrac{1}{2} \int (\cos 3x + \cos x) \cos 3x \, dx$

$= \dfrac{1}{4} \int (2\cos^2 3x + 2\cos x \cos 3x) \, dx$

$= \dfrac{1}{4} \int \{(1 + \cos 6x) + \cos 4x + \cos 2x\} \, dx$

Integrating

$= \dfrac{1}{4} \left(x + \dfrac{\sin 6x}{6} + \dfrac{\sin 4x}{4} + \dfrac{\sin 2x}{2} \right) + C$

$= \dfrac{1}{48} (12x + 2\sin 6x + 3\sin 4x$

$+ 6\sin 2x) + C$

11. (i) $I = \int \sqrt{1 + \cos x} \, dx$

Use $1 + \cos x = 2\cos^2 \dfrac{x}{2}$

(ii) $I = \int \dfrac{\cos 2x}{\sin x} dx$

Use : $\cos 2x = \cos^2 x - \sin^2 x$

12. (i) $I = \int \dfrac{\cot x}{\log \sin x} dx$;

put $\log \sin x = t \Rightarrow \cot x \, dx = dt$

$I = \int \dfrac{dt}{t}$, Integrating

$= \log t + C = \log \log x + C$

(ii) $I = \int \dfrac{x+3}{\sqrt{1-x^2}} dx = \int \dfrac{x}{\sqrt{1-x^2}} dx + \int \dfrac{3}{\sqrt{1-x^2}} dx$

$= \int \dfrac{x \, dx}{\sqrt{1-x^2}} + 3\sin^{-1} x$

Put $1 - x^2 = t$ in I integrating $x \, dx = -\dfrac{dt}{2}$

$= -\dfrac{1}{2} \int \dfrac{dt}{\sqrt{t}} + 3\sin^{-1} x + C$

$= -\dfrac{1}{2} 2\sqrt{t} + 3\sin^{-1} x + C$

$= -\sqrt{1-x^2} + 3\sin^{-1} x + C$

(iii) $I = \int \dfrac{\cos x - \sin x}{(\cos x + \sin x)^2} dx$,

put $\cos x + \sin x = t$ and then integrating

13. (i) $I = \int \dfrac{dx}{x\{1 + (\log x)^2\}}$,

put $\log x = t \Rightarrow \dfrac{1}{x} dx = dt$

$= \int \dfrac{dt}{1+t^2}$, integrating

$= \tan^{-1} t + C = \tan^{-1}(\log x) + C$

(ii) $I = \int \dfrac{6x-8}{3x^2 - 8x + 5} dx$

Put $3x^2 - 8x + 5 = t$ and integrating

(iii) $\int \frac{\cos \sqrt{x}}{\sqrt{x}} dx$

Put $\sqrt{x} = t$ and integrate

14. (i) $I = \int \frac{x e^x}{(x+1)^2} dx = \int \frac{(x+1-1)e^x}{(x+1)^2}$

$= \int e^x \left(\frac{1}{x+1} - \frac{1}{(x+1)^2} \right) dx$

Now, do same as in Question 9(iv).

(ii) $I = \int \frac{\cos 2x}{\cos x + \sin x} dx$

Use : $\cos 2x = \cos^2 x - \sin^2 x$

(iv) $I = \int \frac{\sin(x-a)}{\sin x} dx$

$= \int \frac{\sin x \cos a - \sin a \cos x}{\sin x} dx$

$= \int (\cos a - \sin a \cot x) dx$

$= x \cos a - \sin a \log \sin x + C$

15. (i) $\int \frac{x^3}{(x-a)(x-b)(x-c)} dx$

$= \int \left\{ 1 + \frac{a^3}{(a-b)(a-c)(x-a)} \right.$

$+ \frac{b^3}{(b-a)(b-c)(x-b)}$

$\left. + \frac{c^3}{(c-a)(c-b)(x-c)} \right\} dx$

After breaking into partial fraction

$= x + \frac{a^3}{(a-b)(a-c)} \log(x-a)$

$+ \frac{b^3}{(b-a)(b-c)} \log(x-b)$

$+ \frac{c^3}{(c-a)(c-b)} \log(x-c) + C$

(ii) $I = \int \frac{x-1}{(x+1)(x^2+1)} dx$

$= \int \left(\frac{-1}{x+1} + \frac{x}{x^2+1} \right) dx$

By partial fraction

Now integrating, we have

$I = -\log(x+1) + \frac{1}{2} \log(x^2+1) + C$

(iii) $I = \int \frac{x^2 dx}{(x-a)(x-b)}$

$= \int \left\{ 1 + \frac{a^2}{(a-b)(x-a)} - \frac{b^2}{(a-b)(x-b)} \right\} dx$

$= x + \frac{a^2}{a-b} \log(x-a) - \frac{b^2}{a-b} \log(x-b) + C$

16. (i) $I = \int \frac{3x^2}{(x+1)(x+2)(x+3)} dx$

$= \int \left\{ \frac{3}{2(x+1)} - \frac{12}{(x+2)} + \frac{27}{2(x+3)} \right\} dx$,

Integrating

$= \frac{3}{2} \log(x+1) - 12 \log(x+2)$

$+ \frac{27}{2} \log(x+3) + C$

(ii) $\int \frac{dx}{e^x + e^{2x}} = \int \frac{dx}{e^x(1+e^x)}$

$= \int \left(\frac{1}{e^x} - \frac{1}{1+e^x} \right) dx = \int \left(e^{-x} - \frac{e^{-x}}{e^{-x}-1} \right) dx$

Integrating

$= -\frac{1}{e^x} + \log(e^{-x} + 1) + C$

$= -\frac{1}{e^x} + \log \left(\frac{e^x + 1}{e^x} \right) + C$

(iii) $I = \int \frac{x^2+8}{x^3+4x} dx = \int \frac{x^2+8}{x(x^2+4)} dx$

$= \int \left(\frac{2}{x} - \frac{x}{x^2+4} \right) dx$

(By breaking into partial fraction)

$= 2 \log x - \frac{1}{2} \log(x^2+4) + C$

17. (iii) $I = \int \frac{2x}{(x^2+1)(x^2+3)} dx$;

Put $x^2 = t \Rightarrow 2x \, dx = dt$

$I = \int \frac{dt}{(t+1)(t+3)} = \int \left(\frac{1}{2(t+1)} - \frac{1}{2(t+3)} \right) dt$

$= \frac{1}{2} \log(t+1) - \frac{1}{2} \log(t+3) + C$

$= \frac{1}{2} \log(x^2+1) - \frac{1}{2} \log(x^3+3) + C$

$= \frac{1}{2} \log \left(\frac{x^2+1}{x^2+3} \right) + C$

18. (i) $I = \int x \sin^{-1} x \, dx$, Integrating by parts, we get

$$= \frac{x^2}{2} \sin^{-1} x - \int \frac{x^2}{2\sqrt{1-x^2}} dx$$

$$= \frac{x^2}{2} \sin^{-1} x + \frac{1}{2} \int \frac{1-x^2-1}{\sqrt{1-x^2}} dx$$

$$= \frac{x^2}{2} \sin^{-1} x + \frac{1}{2} \left\{ \int \frac{1-x^2}{\sqrt{1-x^2}} dx - \int \frac{dx}{\sqrt{1-x^2}} \right\}$$

$$= \frac{x^2}{2} \sin^{-1} x + \frac{1}{2} \left\{ \int \sqrt{1-x^2} dx - \int \frac{dx}{\sqrt{1-x^2}} \right\}$$

$$= \frac{x^2}{2} \sin^{-1} x + \frac{1}{2} \left[\left\{ \frac{x}{2} \sqrt{1-x^2} + \frac{1}{2} \sin^{-1} x \right\} - \sin^{-1} x \right] + C$$

$$= \frac{x^2}{2} \sin^{-1} x + \frac{1}{4} x \sqrt{1-x^2} - \frac{1}{4} \sin^{-1} x + C$$

(ii) $\int x \sin^3 x \, dx$

Use : $\sin 3x = 3 \sin x - 4 \sin^3 x$ and then integrating by parts.

(iv) $I = \int \underset{\text{II}}{\sin x} \underset{\text{I}}{\log(\sec x + \tan x)} dx$,

Integrating by parts

$$= -\cos x \log(\sec x + \tan x) - \int -\cos x \sec x \, dx$$

$$= -\cos x \log(\sec x + \tan x) + x + C.$$

20. (i) $\int \sec^4 x \tan x \, dx = \int \sec^3 x (\sec x \tan x) dx$

Put $\sec x = t \Rightarrow \sec x \tan x \, dx = dt$

$$I = \int t^3 dt = \frac{t^4}{4} + C = \frac{\sec^4 x}{4} + C$$

ANSWERS

1. (i) $\dfrac{(2 + 3 \log x)^2}{6}$,   (ii) $\tan(3x + 9)$,   (iii) $\dfrac{1}{6(8 - 6x)}$,   (iv) $\dfrac{1}{2} \log(2x + 3)$.

2. (i) $\dfrac{1}{2} \sec 2x$,   (ii) $\dfrac{1}{7} \sec^7 x$,   (iii) $\log \log x$,   (iv) $-\dfrac{1}{2} \cos x^2$.

3. (i) $-\dfrac{a}{b} \log(c + be^{-x})$,   (ii) $\tan^{-1} e^x$,   (iii) $\tan \log x$,   (iv) $\log(1 + \tan x)$.

4. (i) $\log \tan \dfrac{x}{2}$,   (ii) $\log(\log \sin x)$,   (iii) $\tan^{-1} x^4$.   **5.** (i) $\log x (\log \log x - 1)$

5. (ii) $-2 \log(\text{cosec}\, \theta + \cot \theta)$ where $x = \tan^2 \theta$,   (iii) $\dfrac{x}{2} \sqrt{a^2 - x^2} + \dfrac{a^2}{2} \sin^{-1} \left(\dfrac{x}{a} \right)$,

5. (iv) $\dfrac{1}{8} (\sin 2x - 2x \cos 2x)$.   **6.** (i) $\dfrac{1}{4} (2x \sin 2x + \cos 2x)$,

6. (ii) $\dfrac{1}{18} (\sin 3x - 3x \cos 3x) + \dfrac{1}{2} (x \cos x - \sin x)$ (iii) $-\dfrac{\sqrt{1 + x^2}}{x}$,

7. (i) $\sin^{-1} x [\log \sin^{-1} x - 1]$ (ii) $\log(\sec x + \tan x)$,   **8.** (i) $\left[\dfrac{2}{3} (1 + x)^{3/2} - 2(1 + x)^{1/2} \right]$

8. (ii) $2 \left(\sin \dfrac{x}{2} - \cos \dfrac{x}{2} \right)$ (iii) $\dfrac{\pi}{4}$,   (iv) $-\dfrac{1}{x} \log(1 + x^2) + 2 \tan^{-1} x$.   **9.** (i) $\dfrac{e^{3x}}{3} \left(x - \dfrac{1}{3} \right)$,

9. (ii) $\dfrac{2}{3} x^{3/2} \log x - \dfrac{4}{9} x^{3/2}$,   (iii) $\dfrac{1}{2 + x} + \log(2 + x)$ (iv) $(\log x)^2 e^x$.   **10.** (i) $\cot^{-1} \cos x$,

10. (ii) $e^x \log x$,   (iii) $2\sqrt{(x + \sin^2 x)}$,   (iv) $\dfrac{1}{48} (2 \sin 6x + 3 \sin 4x + 6 \sin 2x + 12x)$.

11. (i) $2\sqrt{2} \sin \dfrac{x}{2}$,   (ii) $\log \tan \dfrac{x}{2} + 2 \cos x$.   **12.** (i) $\log \log \sin x$,

12. (ii) $3 \sin^{-1} x - \sqrt{(1 - x^2)}$.   (iii) $-\dfrac{1}{\sin x + \cos x}$,   (iv) $x + \dfrac{1}{2} \cos 2x$.   **13.** (i) $\tan^{-1} \log x$,

13. (ii) $\log(3x^2 - 8x + 5)$,   (iii) $2 \sin \sqrt{x}$.   **14.** (i) $\dfrac{e^x}{1 + x}$,   (ii) $\dfrac{1}{5} \log(5x^4 + 7)$,

14. (iii) $\sin x + \cos x$,   (iv) $x \cos a - \sin a . \log \sin x$.

15. (i) $x + \dfrac{a^3}{(a-b)(a-c)} \log(x-a) + \dfrac{b^3}{(b-a)(b-c)} \log(x-b) + \dfrac{c^3}{(c-a)(c-b)} \log(x-c),$

15. (ii) $\dfrac{1}{2} \log(x^2+1) - \log(x+1),$ (iii) $\dfrac{9}{16} \log\left(\dfrac{x-1}{x+3}\right) - \dfrac{3}{4} \dfrac{1}{x-1}.$

15. (iv) $\dfrac{a^2}{a-b} \log(x-a) - \dfrac{b^2}{a-b} \log(x-b) + c$ **16.** (i) $\dfrac{3}{2} \log(x+1) - 12\log(x+2) + \dfrac{27}{2} \log(x+3),$

16. (ii) $\log\left(\dfrac{1+e^x}{e^x}\right) - \dfrac{1}{e^x},$ (iii) $\dfrac{1}{3} \log\left(\dfrac{x-1}{x+2}\right) - \dfrac{1}{x-1},$ (iv) $2\log x - \dfrac{1}{2} \log(x^2+4).$

17. (i) $\log\left(\dfrac{x^2-1}{x+2}\right),$ (ii) $\dfrac{1}{3} \log(x-1) + \dfrac{1}{2} \log(2x+1),$ (iii) $\dfrac{1}{2} \log \dfrac{x^2+1}{x^2+3} + c.$

18. (i) $\dfrac{x^2}{2} \sin^{-1} x + \dfrac{x}{4} \sqrt{1-x^2} - \dfrac{1}{4} \sin^{-1} x,$ (ii) $\dfrac{1}{36}[3x(\cos 3x - 9\cos x) - \sin 3x + 27\sin x],$

18. (iii) $\cos x (1 - \log \cos x),$ (iv) $x - \cos x \log(\sec x + \tan x).$ **19.** (i) $2x + \dfrac{1}{2} \log(2x+3),$

19. (ii) $\dfrac{1}{3} \log(1+x^3)$ **20.** (i) $\dfrac{1}{4} \sec^4 x,$ (ii) $\dfrac{1}{3}(\sin^{-1} x)^3,$

10.6 DEFINITE INTEGRAL

If $f(x)$ is a continuous and non-negative function over a closed interval $[a,b]$, then $\int_a^b f(x)\,dx$ is called the definite integral of $f(x)$ between the limits a and $b\,(b > a)$..

If $\int f(x)\,dx = F(x) + c$, then $\int_a^b f(x)\,dx = \left[F(x) + c\right]_a^b = F(b) - F(a)$ is a definite value.

Here, a is called the lower limit and b is called the upper limit and the interval $[a,b]$ is called the range of integration.

REMARKS

➟ $\int_a^b f(x)\,dx$ represents the area bounded by the lines $x = a$ and $x = b$.

➟ If $F(b) - F(a)$ is not a definite value, then the integral $\int_a^b f(x)\,dx$ is indefinite.

10.7 PROPERTIES OF DEFINITE INTEGRALS

1. $\int_a^a f(x)\,dx = 0.$

Proof. Let $\qquad \int f(x)\,dx = F(x),$

then $\qquad \int_a^a f(x)\,dx = [F(x)]_a^a = F(a) - F(a) = 0$

2. The value of definite integral is independent of the variable of integration.

i.e., $\qquad \int_a^b f(x)\,dx = \int_a^b f(u)\,du.$

Proof. Let $\qquad \int f(x)\,dx = F(x),$

then $\qquad \int_a^b f(x)\,dx = \left[F(x)\right]_a^b = F(b) - F(a) = \int_a^b f(u)\,du$

3. $\int_a^b f(x)\,dx = -\int_b^a f(x)\,dx.$

Proof. Let $\qquad \int f(x)\,dx = F(x),$

then $\qquad \int_a^b f(x)\,dx = \left[F(x)\right]_a^b = F(b) - F(a) = -\left[F(x)\right]_b^a = \int_b^a f(x)\,dx$

4. $\int_a^c f(x)dx + \int_c^b f(x)dx = \int_a^b f(x)dx$ where $a < c < b$.

Proof. Let $\int f(x)dx = F(x),$

then $\int_a^c f(x)dx + \int_c^b f(x)dx = [F(x)]_a^c + [F(x)]_c^b = [F(c) - F(a)] + [F(b) - F(c)]$

$$= F(b) - F(a) = [F(x)]_a^b = \int_a^b f(x)dx.$$

5. $\int_0^a f(a - x)dx = \int_0^a f(x)dx$

Proof. We have $\int_0^a f(a-x)dx$

Let $a - x = t,$ then $-dx = dt.$ If $x = 0 \Rightarrow t = a$ and $x = a, t = 0.$

So, $\int_0^a f(a - x)dx = -\int_a^0 f(t)dt = \int_0^a f(x)dx$ (By property 2)

6. If $f(x)$ is an even function of x, then $\int_{-a}^a f(x)\,dx = 2\int_0^a f(x)\,dx$ and if $f(x)$ is an odd function then $\int_{-a}^a f(x)\,dx = 0$.

Proof. Consider $\int_{-a}^a f(x)\,dx.$

Then, $\int_{-a}^a f(x)\,dx = \int_{-a}^0 f(x)x + \int_0^a f(x)\,dx.$

Now let $I = \int_{-a}^0 f(x)\,dx$

Put $x = -t$, so $x = -a \Rightarrow t = a$ and $x = 0 \Rightarrow t = 0.$

So, $\int_{-a}^0 f(x)\,dx = -\int_a^0 f(-t)dt = \int_0^a f(-t)dt = \int_0^a f(-x)dx$ (By 2)

Thus, we have $\int_{-a}^a f(x)\,dx = \int_0^a f(-x)dx + \int_0^a f(x)\,dx = \int_0^a [f(-x)] + f(x)\,dx.$

Now, if $f(x)$ is an even function, i.e., $f(-x) = f(x),$ then we get

$$\int_{-a}^a f(x)\,dx = \int_0^a [f(x) + f(x)]dx = 2\int_0^a f(x)\,dx$$

and, if $f(x)$ an odd function, i.e., $f(-x) = -f(x),$ then we get

$$\int_{-a}^a f(x)\,dx = \int_0^a [-f(x) + f(x)]dx = 0.$$

7. $\int_0^{2a} f(x)\,dx = 2\int_0^a f(x)\,dx$ if $f(2a - x) = f(x)$ and $\int_0^{2a} f(x) = 0$ if $f(2a - x) = -f(x)$.

Proof. This integral can be written as

$$\int_0^{2a} f(x)dx = \int_0^a f(x)dx + \int_a^{2a} f(x)dx \qquad \ldots(1)$$

Now, consider the integral $\int_a^{2a} f(x)\,dx$

Put $x = 2a - t,$ then $dx = -dt$

And if $x = a$ then $t = a$ and if $x = 2a$ then $t = 0.$

So, $\int_a^{2a} f(x)\,dx = -\int_a^0 f(2a - t)dt = \int_0^a f(2a - t)dt = \int_0^a f(2a - x)dx$ (By 2)

Therefore, from (1), we have

$$\int_0^{2a} f(x)dx = \int_0^a f(x)dx + \int_0^a f(2a - x)dx = \int_0^a [f(x) + f(2a - x)]dx. \qquad \ldots(2)$$

Now, if $f(2a - x) = f(x)$ then from (2), we get

$$\int_0^{2a} f(x)\,dx = 2\int_0^a f(x)\,dx$$

and if $f(2a - x) = -f(x),$ then from (2), we get

$$\int_0^{2a} f(x)\,dx = \int_0^a [f(x) - f(x)]\,dx = 0.$$

8. $\int_0^{na} f(x)\,dx = n\int_0^a f(x)\,dx,$ if $f(x + ma) = f(x)$ for all integral values of m and n is a positive integer.

Proof. The given integral can be written as

$$\int_0^{na} f(x)\,dx = \int_0^a f(x)\,dx + \int_0^{2a} f(x)\,dx + \dots + \int_{(m-1)a}^{ma} f(x)\,dx + \dots$$

$$+ \int_{(n-1)a}^{na} f(x)\,dx \qquad \dots (1)$$

Now, consider the integral $\int_{(m-1)a}^{ma} f(x)\,dx,$

Put $$x = y + ma \implies dx = dy$$

So, $$\int_{(m-1)a}^{ma} f(x)\,dx = \int_0^a f(y + ma)\,dy.$$

But $f(y + ma) = f(y)$ is given, so we have

$$\int_{(m-1)a}^{ma} f(x)\,dx = \int_0^a f(y)\,dy = \int_0^a f(x)\,dx.$$

Now from (1) we have

$$\int_0^{na} f(x)\,dx = \int_0^a f(x)\,dx + \int_0^a f(x)\,dx + \dots + \int_0^a f(x)\,dx = n\int_0^a f(x)\,dx.$$

Hence, $\int_0^{na} f(x)\,dx = n\int_0^a f(x)\,dx$ if $f(x + ma) = f(x)$ for all integral value of m.

☞ RECAPITULATIONS

- $\int_a^a f(x)\,dx = 0.$

- The value of definite integral is independent of the variable of integration.

  i.e., $\int_a^b f(x)\,dx = \int_a^b f(u)\,du.$

- $\int_a^b f(x)\,dx = -\int_b^a f(x)\,dx$

- $\int_a^c f(x)\,dx + \int_c^b f(x)\,dx = \int_a^b f(x)\,dx$ where $a < c < b.$

- $\int_0^a f(a - x)\,dx = \int_0^a f(x)\,dx$

- If $f(x)$ is an even function of $x,$ then $\int_{-a}^a f(x)\,dx = 2\int_0^a f(x)\,dx$ and if $f(x)$ is an odd function then

  $\int_{-a}^a f(x)\,dx = 0.$

- $\int_0^{2a} f(x)\,dx = 2\int_0^a f(x)\,dx$ if $f(2a - x) = f(x)$ and $\int_0^{2a} f(x) = 0$ if $f(2a - x) = -f(x).$

- $\int_0^{na} f(x)\,dx = n\int_0^a f(x)\,dx,$ if $f(x + ma) = f(x)$ for all integral values of m and n is a positive integer.

SOLVED EXAMPLES

EXAMPLE 1. *Evaluate the following integrals*

(i) $\int_0^{\pi/2} \log \tan x\,dx$ (ii) $\int_0^{\pi/4} \log(1 + \tan\theta)\,d\theta$

(iii) $\int_0^\pi \dfrac{x \sin x}{1 + \sin x}\,dx$ (iv) $\int_0^{\pi/2} \log \sin x\,dx$

SOLUTION. (i) Consider, $I = \int_0^{\pi/2} \log \tan x\,dx.$

Now, $\qquad I = \int_0^{\pi/2} \log \tan\left(\dfrac{\pi}{2} - x\right) dx$

$I = \int_0^{\pi/2} \log \cot x\, dx \qquad \left(\because \int_0^a f(x)\,dx = \int_0^a f(a-x)\,dx\right)$

On adding, we get

$$2I = \int_0^{\pi/2} \log \tan x\, dx + \int_0^{\pi/2} \log \cot x\, dx$$

$$= \int_0^{\pi/2} \log(\tan x \cot x)\, dx = \int_0^{\pi/2} \log 1 = 0$$

Hence, $\qquad 2I = 0 \quad \Rightarrow \quad I = 0.$

(ii) Consider $\qquad I = \int_0^{\pi/4} \log(1 + \tan\theta)\, d\theta$

$$I = \int_0^{\pi/4} \log\left[1 + \tan\left(\dfrac{\pi}{4} - \theta\right)\right] d\theta$$

$$\left[\because \int_0^a f(x)\,dx = \int_0^a f(a-x)\,dx\right]$$

$$= \int_0^{\pi/4} \log\left[1 + \dfrac{1 - \tan\theta}{1 + \tan\theta}\right] d\theta = \int_\theta^{\pi/4} \log\left(\dfrac{2}{1 + \tan\theta}\right) d\theta$$

$$= \int_0^{\pi/4} [\log 2 - \log(1 + \tan\theta)]\, d\theta$$

$$= \int_0^{\pi/4} \log 2\, d\theta - \int_0^{\pi/4} \log(1 + \tan\theta)\, d\theta = \dfrac{\pi}{4} \log 2 - I$$

So, $\qquad 2I = \dfrac{\pi}{4} \log 2$

Hence $\qquad I = \dfrac{\pi}{8} \log 2$

(iii) Here $\qquad I = \int_0^\pi \dfrac{x \sin x}{1 + \sin x}\, dx$

Now $\qquad I = \int_0^\pi \dfrac{(\pi - x) \sin(\pi - x)}{1 + \sin(\pi - x)}\, dx \quad \left(\because \int_0^a f(x)\,dx = \int_0^a f(a-x)\,dx\right)$

$$= \int_0^\pi \dfrac{(\pi - x) \sin x}{1 + \sin x}\, dx = \int_0^\pi \dfrac{\pi \sin x}{1 + \sin x}\, dx - \int_0^\pi \dfrac{x \sin x}{1 + \sin x}\, dx$$

$$= \int_0^\pi \dfrac{\pi \sin x}{1 + \sin x}\, dx - I$$

So, $\qquad 2I = \int_0^\pi \dfrac{\pi \sin x}{1 + \sin x} = \pi \int_0^\pi \dfrac{\sin x}{1 + \sin x}$

$$= \pi \cdot \int_0^\pi \left(1 - \dfrac{1}{1 + \sin x}\right) dx = \pi \int_0^\pi \left(1 - \dfrac{1 - \sin x}{\cos^2 x}\right) dx$$

$$= \pi \int_0^\pi [1 - \sec^2 x + \sec x \tan x]\, dx$$

$$= \pi [x - \tan x + \sec x]_0^\pi = \pi(\pi - 2)$$

$\therefore \qquad 2I = \pi(\pi - 2)$

Hence, $\qquad I = \pi\left(\dfrac{\pi}{2} - 1\right)$

(iv) Here $\qquad I = \int_0^{\pi/2} \log \sin x \, dx.$

Also, $\qquad I = \int_0^{\pi/2} \log \sin\left(\dfrac{\pi}{2} - x\right) dx \quad \left(\because \int_0^a f(x)\,dx = \int_0^a f(a-x)\,dx\right)$

$\therefore \qquad I = \int_0^{\pi/2} \log \cos x \, dx.$

On adding, we get

$$2I = \int_0^{\pi/2} (\log \sin x)\,dx + \int_0^{\pi/2} \log \cos x \, dx$$

$$= \int_0^{\pi/2} \log\left(\dfrac{\sin 2x}{2}\right) dx = \int_0^{\pi/2} \log \sin 2x \, dx - \int_0^{\pi/2} \log 2\,dx$$

Let $2x = t$ for first integral, then on differentiating, we get $2dx = dt$

Now $\qquad 2I = \dfrac{1}{2}\int_0^{\pi} \log \sin t \, dt - \big[(x \log 2)\big]_0^{\pi/2}$

$$= \dfrac{1}{2}\int_0^{\pi} \log \sin t \, dt - \dfrac{\pi}{2}\log 2 = \int_0^{\pi/2} \log \sin t \, dt - \dfrac{\pi}{2}\log 2$$

$\therefore \qquad 2I = I - \dfrac{\pi}{2}\log 2$

Hence, $\qquad I = -\dfrac{\pi}{2}\log 2.$

EXAMPLE 2. *Evaluate the following integral :*

(i) $\int_0^3 |3x - 1|\,dx$ $\qquad$ (ii) $\int_0^{\pi} |\cos x|\,dx$ $\qquad$ (iii) $\int_0^6 |x + 2|\,dx$

SOLUTION. (i) Given integral is $\int_0^3 |3x - 1|\,dx.$

Now, $\qquad |3x - 1| = \begin{cases} 3x - 1; & \text{when } x \geq \dfrac{1}{3} \\ -(3x - 1) & \text{when } x < \dfrac{1}{3} \end{cases}$

So, $\qquad \int_0^3 |3x - 1|\,dx = \int_0^{1/3} -(3x - 1)\,dx + \int_{1/3}^3 (3x - 1)\,dx$

$$= \left[-\dfrac{3x^2}{2} + x\right]_0^{1/3} + \left[\dfrac{3x^2}{2} - x\right]_{1/3}^3 = \dfrac{65}{6}.$$

(ii) Here, the given integral is $\int_0^{\pi} |\cos x|\,dx.$

Now $\qquad |\cos x| = \begin{cases} \cos x, & 0 \leq x \leq \dfrac{\pi}{2} \\ -\cos x, & \dfrac{\pi}{2} \leq x \leq \pi \end{cases}$

So, $\qquad \int_0^{\pi} |\cos x|\,dx = \int_0^{\pi/2} \cos x \, dx + \int_{\pi/2}^{\pi} (-\cos x)\,dx$

$$= \int_0^{\pi/2} \cos x \, dx - \int_{\pi/2}^{\pi} \cos x \, dx$$

$$= \big[\sin x\big]_0^{\pi/2} - \big[\sin x\big]_{\pi/2}^{\pi} = 2$$

(iii) Let $\int_0^6 |x+2| dx = \int_0^6 (x+2) dx = \int_0^6 x \, dx + \int_0^6 2 \, dx$

$$= \left(\frac{x^2}{2}\right)_0^6 + (2x)_0^6 = \frac{36}{2} - 0 + 2 \times 6 = 18 + 12 = 30$$

EXAMPLE 3. *Evaluate the integral $\int_0^\pi \log(1 + \cos x) dx$.*

SOLUTION. We have, $\qquad I = \int_0^\pi \log(1 + \cos x) dx$

Now, $\qquad I = \int_0^\pi \log\{1 + \cos(\pi - x)\} dx = \int_0^\pi \log(1 - \cos x) dx$

On adding, $\qquad 2I = \int_0^\pi \log(1 + \cos x) dx + \int_0^\pi (1 - \cos x) dx$

$$= \int_0^\pi \log(1 + \cos x)(1 - \cos x) dx$$

$$= \int_0^\pi \log \sin^2 x \, dx = 2 \int_0^\pi \log \sin x \, dx$$

$$= 4 \int_0^{\pi/2} \log \sin x \, dx \qquad \text{(Using property 7)}$$

$\therefore \qquad 2I = 4\left(-\frac{\pi}{2} \log 2\right)$

So, $\qquad I = -\pi \log 2.$

Hence, $\qquad I = \pi \log\left(\frac{1}{2}\right)$

EXAMPLE 4. *Evaluate the integral $\int_0^1 \frac{\sin^{-1} x}{x} dx$.*

SOLUTION. We have, $\qquad I = \int_0^1 \frac{\sin^{-1} x}{x} dx.$

Putting $\qquad x = \sin\theta$

$\Rightarrow \qquad dx = \cos\theta \, d\theta.$

So, $\qquad I = \int_0^{\pi/2} \theta \cot\theta \, d\theta.$

Now, integrating by parts w.r.t. θ, we get

$$I = \left[\theta \cdot \log \sin\theta\right]_0^{\pi/2} - \int_0^{\pi/2} \log \sin\theta \, d\theta$$

$$= 0 - \left[-\frac{\pi}{2} \log 2\right] = \frac{\pi}{2} \log 2.$$

EXAMPLE 5. *Evaluate the integral $\int_0^\infty \log\left(\frac{1+x^2}{x}\right) \frac{dx}{1+x^2}.$*

SOLUTION. We have, $\qquad I = \int_0^\infty \log\left(\frac{1+x^2}{x}\right) \frac{dx}{1+x^2}.$

Now, putting $x = \tan\theta \quad \Rightarrow \quad \theta = \tan^{-1} x, \, d\theta = \frac{dx}{1+x^2}.$

So, $I = \int_0^{\pi/2} \log\left(\dfrac{\sec^2\theta}{\tan\theta}\right).d\theta$

$\qquad = \int_0^{\pi/2} \log\left(\dfrac{1}{\sin\theta.\cos\theta}\right) d\theta = \int_0^{\pi/2} \log\left(\dfrac{2}{2\sin\theta\cos\theta}\right) d\theta$

$\qquad = \int_0^{\pi/2} \log 2\, d\theta - \int_0^{\pi/2} \log\sin 2\theta \; d\theta$

$\qquad = \dfrac{\pi}{2}.\log 2 - \int_0^{\pi/2} \log\sin 2\theta \; d\theta$

Let $\quad 2\theta = t \qquad \Rightarrow \qquad 2d\theta = dt.$

Then $\qquad I = \dfrac{\pi}{2}\log 2 - \dfrac{1}{2}\int_0^{\pi} \log\sin t \; dt = \dfrac{\pi}{2}.\log 2 - \int_0^{\pi/2} \log\sin x \; dx$

$\qquad\qquad\qquad\qquad\qquad\qquad\qquad$ (Using property 7)

$\qquad = \dfrac{\pi}{2}.\log 2 - \left(-\dfrac{\pi}{2}\log 2\right) = \pi\log 2.$

$\qquad\qquad\qquad\qquad\qquad \left(\because \int_0^{\pi/2}\log\sin x\, dx = -\dfrac{\pi}{2}\log 2\right)$

EXAMPLE 6. *Evaluate the integral $\int_{-1}^{1} f(x)\,dx$, where*

$$f(x) = \begin{cases} e^x, & -1 \le x \le 0 \\ 1, & 0 \le x \le \dfrac{1}{2} \\ 3^x, & \dfrac{1}{2} \le x \le 1 \end{cases}$$

SOLUTION. We can write $\int_{-1}^{1} f(x)\,dx = \int_{-1}^{0} f(x)\,dx + \int_{0}^{1/2} f(x)\,dx + \int_{1/2}^{1} f(x)\,dx$

$\qquad = \int_{-1}^{0} e^x dx + \int_0^{1/2} 1\, dx + \int_{1/2}^{1} 3^x \, dx$

$\qquad = \left[e^x\right]_{-1}^{0} + \left[x\right]_0^{1/2} + \left[\dfrac{3^x}{\log 3}\right]_{1/2}^{1}$

$\qquad = (e^0 - e^{-1}) + \dfrac{1}{2} + \dfrac{1}{\log 3}(3 - \sqrt{3})$

$\qquad = 1 - \dfrac{1}{e} + \dfrac{1}{2} + \dfrac{1}{\log 3}(3 - \sqrt{3})$

EXAMPLE 7. *Evaluate the integral $\int_0^{\pi} \dfrac{x\tan x}{\sec x + \tan x}\,dx.$*

SOLUTION. Here, $\qquad I = \int_0^{\pi} \dfrac{x\tan x}{\sec x + \tan x}\,dx \qquad\qquad ...(1)$

Now, $\qquad I = \int_0^{\pi} \dfrac{(\pi - x)\tan(\pi - x)}{\sec(\pi - x) + \tan(\pi - x)}\,dx$

$\qquad\qquad\qquad\qquad\qquad \left(\because \int_0^{a} f(x)\,dx = \int_0^{a} f(a - x)\,dx\right)$

$$= \int_0^\pi \frac{(\pi - x)\tan x}{\sec x + \tan x}. \qquad \qquad \dots(2)$$

On adding, (1) and (2) we get

$$2I = \int_0^\pi \frac{x\tan x}{\sec x + \tan x}dx + \int_0^\pi \frac{(\pi - x)\tan x}{\sec x \tan x}dx$$

$$= \int_0^\pi \frac{x\tan x}{\sec x + \tan x}dx$$

$$= \int_0^\pi \frac{\pi \tan x(\sec x - \tan x)}{\sec^2 x - \tan^2 x}dx \qquad (\because \sec^2 \theta - \tan^2 \theta = 1)$$

$$= \pi \int_0^\pi [\sec x \tan x - \sec^2 x + 1]dx = \pi[\sec x - \tan x + x]_0^\pi$$

$$= \pi[\{\sec \pi - \tan \pi + \pi\} - \{\sec 0 - \tan 0 + 0\}]$$

$$= \pi(-1 + \pi - 1) = \pi(\pi - 2)$$

So; $\qquad \qquad I = \pi\left(\dfrac{\pi}{2} - 1\right).$

EXAMPLE 8. *Evaluate* $\int_0^{\pi/2} \dfrac{dx}{5 + 4\sin x}.$

SOLUTION. Let $\qquad \qquad I = \int_0^{\pi/2} \dfrac{dx}{5 + 8\sin x/2 \cos x/2}$

Divide Nr and Dr by $\cos^2 x/2$ we have

$$I = \int_0^{\pi/2} \frac{\sec^2 x/2\, dx}{5\sec^2 \dfrac{x}{2} + 8\tan \dfrac{x}{2}} = \int_0^{\pi/2} \frac{\sec^2 x/2\, dx}{5\left(1 + \tan^2 \dfrac{x}{2}\right) + 8\tan x/2}$$

Put $\tan \dfrac{x}{2} = t \quad \Rightarrow \quad \dfrac{1}{2}\sec^2 \dfrac{x}{2}dx = dt \quad \Rightarrow \sec^2 \dfrac{x}{2}dx = 2\,dt$

When $x = 0 \quad \Rightarrow \qquad \qquad t = 0$

$\qquad x = \dfrac{\pi}{2} \quad \Rightarrow \qquad \qquad t = 1$

Then $\qquad I = \int_0^1 \dfrac{2dt}{5t^2 + 8t + 5} = \dfrac{2}{5}\int_0^1 \dfrac{dt}{t^2 + \dfrac{8}{5}t + 1} = \dfrac{2}{5}\int_0^1 \dfrac{dt}{\left(t + \dfrac{4}{5}\right)^2 + \left(\dfrac{3}{4}\right)^2}$

$$= \frac{2}{5} \cdot \frac{4}{3}\left[\tan^{-1}\left(\frac{t + \dfrac{4}{5}}{3/4}\right)\right]_0^1 = \frac{8}{15}\left[\tan^{-1}\frac{36}{15} - \tan^{-1}\frac{16}{15}\right]$$

EXAMPLE 9. *Evaluate* $\int_0^1 \sin^{-1} x\, dx.$

SOLUTION. Let $\qquad \qquad I = \int_0^1 \sin^{-1} x\, dx$

Putting $\qquad x = \sin\theta \quad \Rightarrow \quad dx = \cos\theta\, d\theta$

When $\qquad x = 0 \quad \Rightarrow \quad \theta = 0 \quad$ and $\quad x = 1 \quad \Rightarrow \qquad \theta = \dfrac{\pi}{2}$

Then $\quad I = \int_0^{\pi/2} \theta \cos\theta \, d\theta;$ Integrating by parts, we get

$$I = (\theta \sin\theta)_0^{\pi/2} - \int_0^{\pi/2} \sin\theta \, d\theta$$

$$= \frac{\pi}{2} + [\cos\theta]_0^{\pi/2} = \frac{\pi}{2} - 1 = \frac{\pi - 2}{2}$$

EXAMPLE 10. *Evaluate* $\int_0^a \dfrac{x^4 . dx}{\sqrt{a^2 - x^2}}$

SOLUTION.   Let $\quad I = \int_0^a \dfrac{x^4 . dx}{\sqrt{a^2 - x^2}}$

Put $\quad\quad\quad x = a\sin\theta \quad\Rightarrow\quad\quad\quad dx = a\cos\theta \, d\theta$

When $x = 0 \quad\Rightarrow\quad \theta = 0 \quad$ and $\quad x = a \quad\Rightarrow\quad\quad\quad \theta = \pi/2$

Then $\quad\quad I = \int_0^{\pi/2} \dfrac{a^4 \sin^4\theta \cdot a\cos\theta \, d\theta}{\sqrt{a^2 - a^2\sin^2\theta}} = a^4 \int_0^{\pi/2} \sin^4 d\theta$

$$= a^4 \cdot \frac{3.1}{4.2} \frac{\pi}{2} \quad\quad\quad\quad\quad\quad \text{By Walli's formula}$$

$$= \frac{3\pi}{16} a^4.$$

EXAMPLE 11. *Evaluate* $\int_2^3 \dfrac{dx}{\sqrt{5x - 6 - x^2}}.$

SOLUTION.   Let

$$I = \int_2^3 \frac{dx}{\sqrt{5x - 6 - x^2}} = \int_2^3 \frac{dx}{2\sqrt{-(x^2 - 5x + 6)}}$$

$$= \int_2^3 \frac{dx}{\sqrt{-\left\{\left(x - \frac{5}{2}\right)^2 - \frac{1}{4}\right\}}} = \int_2^3 \frac{dx}{\sqrt{\left(\frac{1}{2}\right)^2 - \left(x - \frac{5}{2}\right)^2}}$$

$$= \left[\sin^{-1}\left(\frac{x - 5/2}{1/2}\right)\right]_2^3$$

$$= [\sin^{-1}(2x - 5)]_2^3 = [\sin^{-1} 1 - \sin^{-1}(-1)]$$

$$= [\sin^{-1} 1 + \sin^{-1} 1] = 2\sin^{-1} 1 = 2 \cdot \frac{\pi}{2} = \pi$$

EXAMPLE 12. *Evaluate* $\int_0^\infty \dfrac{dx}{(x^2 + a^2)(x^2 + b^2)}$

SOLUTION.   Let $\quad \dfrac{1}{(x^2 + a^2)(x^2 + b^2)} = \dfrac{Ax + B}{(x^2 + a^2)} + \dfrac{Cx + D}{x^2 + b^2}$

$\Rightarrow \quad\quad\quad\quad 1 = (Ax + B)(x^2 + b^2) + (Cx + D)(x^2 + a^2)$ ...(1)

Equating like powers of x on both the sides of (1) we get

$$0 = A + C, \, 0 = B + D$$

$$0 = b^2 A + a^2 C$$

$$1 = b^2 B + a^2 D$$

Solving these equations we get

$$A = C = 0 \quad \text{and} \quad B = -D = \frac{1}{b^2 - a^2}$$

Thus, we can write

$$\int_0^\infty \frac{dx}{(x^2 + a^2)(x^2 + b^2)}$$

$$= \frac{1}{b^2 - a^2} \int_0^\infty \frac{dx}{x^2 + a^2} + \frac{1}{a^2 - b^2} \int_0^\infty \frac{dx}{x^2 + b^2}$$

$$= \frac{1}{a(b^2 - a^2)} \left[\tan^{-1} \frac{x}{a} \right]_0^\infty + \frac{1}{b(a^2 - b^2)} \left[\tan^{-1} \frac{x}{b} \right]_0^\infty$$

$$= \frac{1}{a(b^2 - a^2)} \left(\tan^{-1} \frac{\infty}{a} - \tan^{-1} \frac{0}{a} \right) + \frac{1}{b(a^2 - b^2)} \left(\tan^{-1} \frac{\infty}{a} + \tan^{-1} \frac{0}{b} \right)$$

$$= \frac{1}{a(b^2 - a^2)} \left[\frac{\pi}{2} - 0 \right] + \frac{1}{b(a^2 - b^2)} \left(\frac{\pi}{2} - 0 \right) = \frac{\pi}{2ab(a + b)}$$

EXAMPLE 13. *Evaluate* $\int_{-1}^1 \log \left(\frac{2 - x}{2 + x} \right) dx$.

SOLUTION. Let

$$I = \int_{-1}^1 \log \left(\frac{2 - x}{2 + x} \right) dx$$

Clearly $f(x) = \log \left(\frac{2 - x}{2 + x} \right)$ is an odd function because

$$f(-x) = -f(x)$$

Therefore, $\int_{-1}^1 \left(\frac{2 - x}{2 + x} \right) dx = 0$

Exercise 10.3

1. Evaluate the integral $\int_0^{\pi/2} \frac{\sin^4 x \, dx}{\sin^4 x + \cos^4 x}$.

2. Evaluate the integral $\int_0^{\pi/2} \frac{\sin x - \cos x}{1 + \sin x \cos x} dx$.

3. Evaluate the integral $\int_0^4 f(x) \, dx$, where

$$f(x) = \begin{cases} 2x + 3, & 0 \le x \le 3 \\ 3x, & 3 \le x \le 4 \end{cases}$$

4. Evaluate $\int_0^{\pi/2} \frac{x \sin x \cos x}{\cos^4 x + \sin^4 x} dx$

5. Evaluate the integral $\int_0^\pi \frac{x \sin x}{1 + \cos^2 x} dx$.

6. Show that $\int_0^\pi \frac{x \, dx}{a^2 \cos^2 x + b^2 \sin^2 x} = \frac{\pi^2}{2ab}$.

7. Evaluate the integral $\int_0^{\pi/2} \log(\tan x + \cot x) dx$.

8. Evaluate $\int_0^1 e^{\sin^{-1} x} dx$.

HINT TO SELECTED PROBLEMS

1. $I = \int_0^{\pi/2} \frac{\sin^4 x \, dx}{\sin^4 x + \cos^4 x}$ ...(1)

$$= \int_0^{\pi/2} \frac{\sin^4 \left(\frac{\pi}{2} - x \right)}{\sin^4 \left(\frac{\pi}{2} - x \right) + \cos^4 \left(\frac{\pi}{2} - x \right)} dx$$

$$= \int_0^{\pi/2} \frac{\cos^4 x}{\sin^4 x + \cos^4 x} dx \qquad ...(2)$$

Adding (1) and (2), we get

$$2I = \int_0^{\pi/2} dx = \frac{\pi}{2}$$

$$I = \frac{\pi}{4}$$

2. $I = \int_0^{\pi/2} \frac{\sin x - \cos x}{1 + \cos x \sin x}$ ...(1)

$= \int_0^{\pi/2} \frac{\sin\left(\frac{\pi}{2} - x\right) - \cos\left(\frac{\pi}{2} - x\right)}{1 + \cos\left(\frac{\pi}{2} - x\right)\sin\left(\frac{\pi}{2} - x\right)} dx$

$= \int_0^{\pi/2} \frac{\cos x - \sin x}{1 + \sin x \cos x} dx$ ...(2)

Adding (1) and (2)

$2I = 0 \quad \Rightarrow \quad I = 0$

3. $I = \int_0^4 f(x)\,dx = \int_0^3 (2x+3)dx + \int_3^4 3x\,dx$

$= [x^2 + 3x]_0^3 + \left[\frac{3x^2}{2}\right]_3^4$

$= 18 + 24 - \frac{27}{2} = 42 - \frac{27}{2} = \frac{57}{2}$

4. $I = \int_0^{\pi/2} \frac{x \sin x \cos x\,dx}{\cos^4 x + \sin^4 x}$

$= \int_0^{\pi/2} \frac{\left(\frac{\pi}{2} - x\right)\sin\left(\frac{\pi}{2} - x\right)\cos\left(\frac{\pi}{2} - x\right)}{\cos^4\left(\frac{\pi}{2} - x\right) + \sin^4\left(\frac{\pi}{2} - x\right)} dx$

$= \int_0^{\pi/2} \frac{\pi}{2} \frac{\sin x \cos x}{\sin^4 x + \cos^4 x} dx$

$\qquad - \int_0^{\pi/2} \frac{x \sin x \cos x\,dx}{\sin^4 x + \cos^4 x}$

$\Rightarrow 2I = \frac{\pi}{2} \int_0^{\pi/2} \frac{\sin x \cos x\,dx}{\sin^4 x + \cos^4 x}$

divide Nr and Dr by $\cos^4 x$

$2I = \frac{\pi}{2} \int_0^{\pi/2} \frac{\tan x \sec^2 x}{1 + \tan^4 x} dx$

Put $\tan^2 x = t \Rightarrow 2\tan x \sec^2 x\,dx = dt$

$\qquad \Rightarrow \tan x \sec^2 x\,dx = \frac{dt}{2}$

$= \frac{\pi}{2} \int_0^\infty \frac{1}{2} \frac{dt}{1 + t^2}$ When $x = 0 \Rightarrow t = 0$

$\qquad\qquad$ When $x = \pi/2 \Rightarrow t = \infty$

$= \frac{\pi}{4}[\tan^{-1} t]_0^\infty = \frac{\pi}{4}\left(\frac{\pi}{2} - 0\right) = \frac{\pi^2}{8}$

So, $\quad I = \frac{\pi^2}{16}$

5. $\quad I = \int_0^\pi \frac{x \sin x}{1 + \cos^2 x} dx$ ...(1)

$= \int_0^\pi \frac{(\pi - x)\sin(\pi - x)}{1 + \cos^2(\pi - x)} dx$ ...(2)

Adding (1) and (2)

$2I = \int_0^\pi \frac{(x + \pi - x)\sin x}{1 + \cos^2 x} dx = \pi\int_0^\pi \frac{\sin x}{1 + \cos^2 x} dx$

Let $\qquad \cos x = t$

$\Rightarrow \qquad \sin x\,dx = -dt$

$\qquad x = 0 \Rightarrow t = 1$

$\qquad x = \pi \Rightarrow t = -1$

Now $2I = \pi\int_1^{-1} \frac{dt}{1+t^2} = -\pi[\tan^{-1} t]_1^{-1}$

$= \pi\left[-\frac{\pi}{4} - \frac{\pi}{4}\right]$

$2I = \frac{\pi^2}{2} \quad \Rightarrow \quad I = \frac{\pi^2}{4}$

6. $I = \int_0^\pi \frac{x\,dx}{a^2 \cos^2 x + b^2 \sin^2 x}$ ...(1)

$= \int_0^\pi \frac{(\pi - x)\,dx}{a^2 \cos^2(\pi - x) + b^2 \sin^2(\pi - x)}$

$= \int_0^\pi \frac{(\pi - x)}{a^2 \cos^2 x + b^2 \sin^2 x} dx$ ...(2)

Adding (1) and (2) we get

$2I = \int_0^\pi \frac{(x + \pi - x)\,dx}{a^2 \cos^2 x + b^2 \sin^2 x}$

$= \pi\int_0^\pi \frac{dx}{a^2 \cos^2 x + b^2 \sin^2 x}$

$= 2\pi\int_0^{\pi/2} \frac{dx}{a^2 \cos^2 x + b^2 \sin^2 x}$

$\left[\int_0^{2a} f(x)dx = 2\int_0^a f(x)\,dx; f(2a - x) = f(x)\right]$

Divide Nr and Dr by $\cos^2 x$

$I = \pi\int_0^{\pi/2} \frac{\sec^2 x\,dx}{a^2 + b^2 \tan^2 x}$

Put $\tan x = t$

$\sec^2 x\,dx = dt$

When $x = 0 \Rightarrow t = 0$

$\qquad x = \frac{\pi}{2} \Rightarrow t = \infty$

$I = \pi\int_0^\infty \frac{dt}{a^2 + b^2 t^2} = \frac{\pi}{b^2}\int_0^\infty \frac{dt}{\left(\frac{a}{b}\right)^2 + t^2}$

$= \frac{\pi}{b^2} \frac{1}{a/b}\left[\tan^{-1}\frac{t}{a/b}\right]_0^\infty$

$I = \frac{\pi}{ab}\left[\tan^{-1}\frac{bt}{a}\right]_0^\infty = \frac{\pi}{ab}\left(\frac{\pi}{2} - 0\right) = \frac{\pi^2}{2ab}$

7. $I = \int_0^{\pi/2} \log(\tan x + \cot x) dx$

$= \int_0^{\pi/2} \log\left(\dfrac{\sin^2 x + \cos^2 x}{\sin x \cos x}\right) dx$

$= \int_0^{\pi/2} \log\left(\dfrac{1}{\sin x \cos x}\right) dx$

$= \int_0^{\pi/2} - (\log\cos x + \log\sin x) dx$

$= -\left[\int_0^{\pi/2} \log\sin x \, dx + \int_0^{\pi/2} \log\cos x \, dx\right]$

$= -2\int_0^{\pi/2} \log\sin x \, dx$

$\left[\because \int_0^{\pi/2} \log\sin x \, dx = \int_0^{\pi/2} \log\cos x \, dx = -\dfrac{\pi}{2}\log 2\right]$

$= -2\left(-\dfrac{\pi}{2}\log 2\right) = \pi\log 2$

8. $I = \int_0^1 e^{\sin^{-1} x} dx$

Put $\sin^{-1} x = t \Rightarrow \sin t = x \Rightarrow \cos t \, dt = dx$

When $x = 0 \Rightarrow t = 0$

$\quad\quad\quad x = 1 \Rightarrow t = \pi/2$

$I = \int_0^{\pi/2} e^t \cos t \, dt = \left[\dfrac{1}{1^2 + 1^2} \int [e^t \cos t + e^t \sin t]\right]_0^{\pi/2}$

$= \dfrac{1}{2}[e^{\pi/2}(0+1) - 1]$

$\left[\text{Since } \int_0^{\pi/2} e^{ax} \cos bx \, dx\right.$

$= \dfrac{1}{a^2 + b^2}[a\, e^{ax}\cos bx + b\, e^{ax}\sin bx]$

$= \dfrac{1}{2}(e^{\pi/2} - 1)$

ANSWERS

1. $\dfrac{\pi}{4}$ **2.** 0 **3.** $\dfrac{57}{2}$ **4.** $\dfrac{\pi^2}{16}$ **5.** $\dfrac{\pi^2}{4}$ **7.** $\pi\log 2$ **8.** $\dfrac{e^{\pi/2} - 1}{2}$

SOME MORE SOLVED PROBLEMS RELATED TO DEFINITE INTEGRALS

EXAMPLE 1. *Evaluate* $\int_0^\pi \dfrac{x\sin x}{(1 + \cos^2 x)} dx$.

SOLUTION. Let $I = \int_0^\pi \dfrac{x\sin x}{1 + \cos^2 x} dx$ ...(1)

$= \int_0^\pi \dfrac{(\pi - x)\sin(\pi - x)}{1 + \cos^2(\pi - x)} dx$ $\left[\because \int_0^a f(x)\,dx = \int_0^a f(a - x)\,dx\right]$

$= \int_0^\pi \dfrac{(\pi - x)\sin x}{1 + \cos^2 x} dx$ ...(2)

Adding (1) and (2), we get

$$2I = \int_0^\pi \dfrac{\pi\sin x}{1 + \cos^2 x} dx = 2\pi\int_0^{\pi/2} \dfrac{\sin x}{1 + \cos^2 x} dx.$$

Now put $\cos x = t \Rightarrow -\sin x \, dx = dt.$

Also $\quad t = 1$ at $x = 0$ and $t = 0$ at $x = \pi/2.$

Therefore, we have

$$I = \pi\int_1^0 -\dfrac{dt}{1 + t^2} = \pi\int_0^1 \dfrac{dt}{1 + t^2} = \pi[\tan^{-1} t]_0^1 = \dfrac{\pi^2}{4}.$$

EXAMPLE 2. *Evaluate* $\int_0^\pi x\sin^6 x\cos^4 x\,dx$.

SOLUTION. Here $I = \int_0^\pi x\sin^6 x\cos^4 x\,dx = \int_0^\pi (\pi - x)\sin^6(\pi - x)\cos^4(\pi - x)\,dx$

$= \int_0^\pi (\pi - x)\sin^6 x\cos^4 x\,dx$

$$= \int_0^\pi \pi \sin^6 x \cos^4 x \, dx - \int_0^\pi x \sin^6 x \cos^4 x \, dx$$

$$= \pi \int_0^\pi \sin^6 x \cos^4 x \, dx - I$$

Hence, $\quad 2I = \pi \int_0^\pi \sin^6 x \cos^4 x \, dx = 2\pi \int_0^{\pi/2} \sin^6 x \cos^4 x \, dx$

$$I = \pi \int_0^{\pi/2} \sin^6 x \cos^4 x \, dx$$

$$= \pi \cdot \frac{5 \cdot 3 \cdot 1 \cdot 3 \cdot 1}{10 \cdot 8 \cdot 6 \cdot 4 \cdot 2} \cdot \frac{\pi}{2} = \frac{3\pi^2}{512}. \qquad \text{(By Walli's formula)}$$

EXAMPLE 3. *Evaluate $\int_0^\pi \sin^3 \theta (1 + 2\cos\theta)(1 + \cos\theta)^2 \, d\theta$*

SOLUTION. Let $\qquad I = \int_0^\pi \sin^3 \theta (1 + 2\cos\theta)(1 + \cos\theta)^2 \, d\theta$

$$= \int_0^\pi \sin^3 \theta (1 + 2\cos\theta)(1 + 2\cos\theta + \cos^2\theta) \, d\theta$$

$$= \int_0^\pi (\sin^3 \theta + 4\sin^3 \theta \cos\theta + 5\sin^3 \theta \cos^2\theta + 2\sin^3 \theta \cos^3\theta) \, d\theta.$$

Now $\int_0^\pi \sin^m \theta \cos^n \theta \, d\theta = 2\int_0^{\pi/2} \sin^m \theta \cos^n \theta \, d\theta$, if n is even

$$= 0, \text{ if } n \text{ is odd.}$$

Hence, $\qquad I = 2 \cdot \dfrac{2}{3 \cdot 1} + 10 \cdot \dfrac{2 \cdot 1}{5 \cdot 3 \cdot 1} = \dfrac{4}{3} + \dfrac{4}{3} = \dfrac{8}{3}$

EXAMPLE 4. *Evaluate $\int_0^{\pi/2} \log \sin 2x \, dx$.*

SOLUTION. Let $\qquad I = \int_0^{\pi/2} \log \sin 2x \, dx.$ $\qquad\qquad\qquad$ ...(1)

Put $2x = t \implies 2dx = dt$, we get

$$I = \frac{1}{2} \int_0^\pi \log \sin t \, dt = \frac{1}{2} \cdot 2 \int_0^{\pi/2} \log \sin t \, dt$$

$$= \int_0^{\pi/2} \log \sin t \, dt = \int_0^{\pi/2} \log \sin\left(\frac{\pi}{2} - t\right) dt = \int_0^{\pi/2} \log \cos t \, dt. \quad \text{...(2)}$$

Adding (1) and (2), we get

$$\implies \quad 2I = \int_0^{\pi/2} (\log \sin t) \, dt + \int_0^{\pi/2} \log \cos t \, dt = \int_0^{\pi/2} \log\left(\frac{\sin 2t}{2}\right) dt$$

$$= \int_0^{\pi/2} \log \sin 2t \, dt - \int_0^{\pi/2} \log 2 \, dt = I - \frac{\pi}{2} \log 2$$

Hence, $\qquad I = -\dfrac{\pi}{2} \log 2$

EXAMPLE 5. *Show that $\int_0^{\pi/2} \dfrac{\sqrt{\sin x}}{\sqrt{\sin x} + \sqrt{\cos x}} \, dx = \dfrac{\pi}{4}$.*

SOLUTION. Let $\qquad I = \int_0^{\pi/2} \dfrac{\sqrt{\sin x}}{\sqrt{\sin x} + \sqrt{\cos x}} \, dx$ $\qquad\qquad$ ...(1)

$$= \int_0^{\pi/2} \frac{\sqrt{\sin\left(\frac{\pi}{2} - x\right)}}{\sqrt{\sin\left(\frac{\pi}{2} - x\right)} + \sqrt{\cos\left(\frac{\pi}{2} - x\right)}} \, dx$$

$$= \int_0^{\pi/2} \frac{\sqrt{\cos x}}{\sqrt{\cos x} + \sqrt{\sin x}} \, dx. \qquad \dots (2)$$

Adding (1) and (2), we get

$$2I = \int_0^{\pi/2} \left[\frac{\sqrt{\sin x} + \sqrt{\cos x}}{\sqrt{\sin x} + \sqrt{\cos x}} \right] dx = \int_0^{\pi/2} 1 . dx = \frac{\pi}{2}$$

Hence, $I = \dfrac{\pi}{4}.$

EXAMPLE 6. *Show that* $\int_0^{\pi/2} \dfrac{\sin^2 x}{(\sin x + \cos x)} \, dx = \dfrac{1}{\sqrt{2}} \log(\sqrt{2} + 1).$

SOLUTION.   Let $I = \int_0^{\pi/2} \dfrac{\sin^2 x}{(\sin x + \cos x)} \, dx$ $\dots (1)$

$$= \int_0^{\pi/2} \frac{\sin^2\left(\frac{\pi}{2} - x\right)}{\sin\left(\frac{\pi}{2} - x\right) + \cos\left(\frac{\pi}{2} - x\right)} \, dx = \int_0^{\pi/2} \frac{\cos^2 x}{\cos x + \sin x} \, dx \qquad \dots (2)$$

Adding (1) and (2), we get

$$2I = \int_0^{\pi/2} \frac{\sin^2 x}{\sin x + \cos x} \, dx + \int_0^{\pi/2} \frac{\cos^2 x}{\cos x + \sin x} \, dx$$

$$= \int_0^{\pi/2} \frac{dx}{\sin x + \cos x} = \int_0^{\pi/2} \frac{(1/\sqrt{2}) \, dx}{\left(\frac{1}{\sqrt{2}} \sin x + \frac{1}{\sqrt{2}} \cos x \right)}$$

$$= \frac{1}{\sqrt{2}} \int_0^{\pi/2} \frac{dx}{\cos(x - \pi/4)} = \frac{1}{\sqrt{2}} \int_0^{\pi/2} \sec\left(x - \frac{\pi}{4} \right) dx$$

$$= \frac{1}{\sqrt{2}} \log \left[\sec\left(x - \frac{\pi}{4} \right) + \tan\left(x - \frac{\pi}{4} \right) \right]_0^{\pi/2}$$

$$= \frac{1}{\sqrt{2}} \left[\log\left(\sec\frac{\pi}{4} + \tan\frac{\pi}{4} \right) - \log\left\{ \sec\left(-\frac{\pi}{4} \right) + \tan\left(-\frac{\pi}{4} \right) \right\} \right]$$

$$= \frac{1}{\sqrt{2}} \log \left[\frac{(\sqrt{2} + 1)(\sqrt{2} + 1)}{\sqrt{2} - 1)(\sqrt{2} + 1)} \right] = \frac{1}{\sqrt{2}} \log(\sqrt{2} + 1)^2$$

$$= \frac{1}{\sqrt{2}} \cdot 2\log(\sqrt{2} + 1)$$

Hence, $I = \dfrac{1}{\sqrt{2}} \log(\sqrt{2} + 1)$

Exercise 10.4

Show that

1. $\int_0^\pi x \log \sin x \, dx = \frac{1}{2}\pi^2 \log \frac{1}{2}$.

2. $\int_0^{\pi/2} x \cot x \, dx = \frac{\pi}{2}\log 2$.

3. $\int_0^{\pi/2}\left[\frac{\theta}{\sin\theta}\right]^2 d\theta = \pi \log 2$.

4. $\int_0^1 \frac{\sin^{-1}x}{x} dx = \frac{\pi}{2}\log 2$.

5. $\int_0^{\pi/4}\log(1+\tan\theta)d\theta = \frac{\pi}{8}\log 2$.

6. $\int_0^\infty \frac{x\,dx}{(1+x)(1+x^2)} = \frac{\pi}{4}$.

7. $\int_0^{\pi/2}\frac{\sqrt{\tan x}}{\sqrt{\tan x}+\sqrt{\cot x}}dx = \frac{\pi}{4}$.

8. $\int_0^{\pi/2}\frac{\cos^2 x\,dx}{(\sin x+\cos x)} = \frac{1}{\sqrt{2}}\log(\sqrt{2}+1)$.

9. $\int_0^\pi \sin^m x \cos^{2m+1} x \, dx = 0$.

10. $\int_0^\pi \frac{x^2 \sin 2x \sin\left(\frac{\pi}{2}\cos x\right)}{2x-\pi} = \frac{8}{\pi}$.

HINT TO SELECTED PROBLEMS

1. $I = \int_0^\pi x \log \sin x \, dx = \int_0^\pi (\pi-x)\log\sin(\pi-x)dx$

$= \int_0^\pi \pi \log \sin x \, dx - \int_0^\pi x \log \sin x \, dx$

$\Rightarrow 2I = \pi\int_0^\pi \log \sin x \, dx$

$= 2\pi \int_0^{\pi/2}\log \sin x \, dx$

$$\left(\because \int_a^{2a} f(x)dx = 2\int_0^a f(x)dx\right.$$
$$\left. \text{if } f(2a-x)=f(x)\right)$$

$= 2\pi\left(\frac{\pi}{2}\log 2\right) = \pi^2 \log\frac{1}{2}$

$I = \frac{\pi^2}{2}\log\frac{1}{2}$

3. $I = \int_0^{\pi/2}\left(\frac{\theta}{\sin\theta}\right)^2 d\theta = \int_0^{\pi/2}\theta^2 \csc^2\theta\,d\theta$

$\cdot -(-\theta^2\cot\theta)_0^{\pi/2} - \int_0^{\pi/2}2\theta\cot\theta\,d\theta$

$= (-0+0) - 2\int_0^{\pi/2}-\theta\cot\theta\,d\theta$

$= +2\left[(\theta\log\sin\theta)_0^{\pi/2} - \int_0^{\pi/2}\log\sin\theta\,d\theta\right]$

$= +2\left[(0-0) - \int_0^{\pi/2}\log\sin\theta\,d\theta\right]$

$= +2\left(\frac{\pi}{2}\log 2\right)$

$\qquad$ [Since $\int_0^{\pi/2}\log\sin\theta = \pi\log 2$]

$= \pi\log 2$

5. $I = \int_0^{\pi/4}\log(1+\tan\theta)d\theta$

$= \int_0^\pi \log\left\{1+\log\left(\frac{\pi}{4}-\theta\right)\right\}d\theta$

$= \int_0^{\pi/4}\log\left\{1+\frac{\tan\pi/4-\tan\theta}{1+\tan\pi/4\tan\theta}\right\}d\theta$

$= \int_0^{\pi/4}\log\left\{1+\frac{1-\tan\theta}{1+\tan\theta}\right\}d\theta$

$= \int_0^{\pi/4}\log\left\{\frac{2}{1+\tan\theta}\right\}d\theta$

$= \int_0^{\pi/4}\log 2\,dx - \int_0^{\pi/4}\log(1+\tan\theta)\,d\theta$

$2I = \log 2[x]_0^{\pi/4} = \frac{\pi}{4}\log 2$

$I = \frac{\pi}{8}\log 2$

7. $I = \int_0^{\pi/2}\frac{\sqrt{\tan x}}{\sqrt{\tan x}+\sqrt{\cot x}}dx$ ...(1)

$= \int_0^{\pi/2}\frac{\sqrt{\tan\left(\frac{\pi}{2}-x\right)}}{\sqrt{\tan\left(\frac{\pi}{2}-x\right)}+\sqrt{\cot\left(\frac{\pi}{2}-x\right)}}dx$

$= \int_0^{\pi/2}\frac{\sqrt{\cot x}}{\sqrt{\cot x}+\sqrt{\tan x}}$ ...(2)

Adding (1) and (2)

$2I = \int_0^{\pi/2}dx = \frac{\pi}{2}$

$I = \frac{\pi}{4}$

8. $I = \int_0^{\pi/2}\frac{\cos^2 x\,dx}{\sin x+\cos x}$ ...(1)

$= \int_0^{\pi/2}\frac{\cos^2\left(\frac{\pi}{2}-x\right)}{\sin\left(\frac{\pi}{2}-x\right)+\cos\left(\frac{\pi}{2}-x\right)}dx$

$$= \int_0^{\pi/2} \frac{\sin^2 x}{\cos x + \sin x} dx \qquad \ldots(2)$$

Adding (1) and (2)

$$2I = \int_0^{\pi/2} \frac{\cos^2 x + \sin^2 x}{\cos x + \sin x} dx$$

$$= \int_0^{\pi/2} \frac{dx}{\cos x + \sin x}$$

$$= \int_0^{\pi/2} \frac{dx}{1 - 2\sin^2 \frac{x}{2} + 2\sin \frac{x}{2}\cos \frac{x}{2}}$$

Divide Nr and Dr by $\cos^2 \frac{x}{2}$

$$2I = \int_0^{\pi/2} \frac{\sec^2 x / 2 \, dx}{1 + 2\tan \frac{x}{2} - \tan^2 \frac{x}{2}}$$

Let $\tan \frac{x}{2} = t \implies \sec^2 \frac{x}{2} dx = 2dt$

Also when $x = 0 \implies t = 0$

$$x = \frac{\pi}{2} \implies t = 1$$

$$2I = \int_0^1 \frac{2dt}{2t + 1 - t^2} = 2 \int_0^1 \frac{dt}{(\sqrt{2})^2 - (t-1)^2}$$

$$= 2 \times \frac{1}{2\sqrt{2}} \left[\log \left| \frac{\sqrt{2} + t - 1}{\sqrt{2} - t + 1} \right| \right]_0^1$$

$$= \frac{1}{\sqrt{2}} \left[\log\left(\frac{\sqrt{2}}{\sqrt{2}}\right) - \log\left(\frac{\sqrt{2}-1}{\sqrt{2}+1}\right) \right]$$

$$= \frac{1}{\sqrt{2}} \log\left\{ \frac{\sqrt{2}-1}{\sqrt{2}+1} \right\} = \frac{1}{\sqrt{2}} \log \frac{\sqrt{2}+1}{\sqrt{2}-1}$$

$$= \frac{1}{\sqrt{2}} \log\left\{ \frac{(\sqrt{2}+1)(\sqrt{2}+1)}{(\sqrt{2}-1)(\sqrt{2}-1)} \right\}$$

$$= \frac{1}{\sqrt{2}} \log(\sqrt{2}+1)^2 = \frac{2}{\sqrt{2}} \log(\sqrt{2}+1)$$

So, $I = \frac{1}{\sqrt{2}} \log(\sqrt{2}+1)$

9. $I = \int_0^{\pi} \sin^m x \cos^{2m+1} x \, dx = f(x)$, say

Here $f(x) = \sin^m x \cos^{2m+1}$

$$f(\pi - x) = \sin^m(\pi - x)\cos^{(2m+1)}(\pi - x)$$

$$= -\sin^m x \cos^{2m+1} x = f(-x)$$

So, $I = 0$

Since, $\int_0^{2a} f(x) = 0$ if $f(2a - x) = -f(x)$

10.8 DEFINITE INTEGRAL AS THE LIMIT OF THE SUM

It is always possible to regard a definite integral as the limit of the sum of certain number of terms, when the number of terms tends to infinity and each term tends to zero.

Here, we define the definite integral as follows:

$$\int_a^b f(x)\,dx = \lim h[f(a) + f(a + h) + f(a + 2h) + \ldots + f\{a + (n-1)h\}]$$

when $n \to \infty$, $h \to 0$ and $nh \to b - a$.

SOLVED EXAMPLES

EXAMPLE 1. Evaluate $\int_a^b x^2 dx$, directly from the definition of integral as the limit of a sum.

SOLUTION. We know that

$$\int_a^b f(x)\,dx = \lim_{n \to \infty} h[f(a) + f(a + h) + f(a + 2h) + \ldots + f\{a + (n-1)h\}] \qquad \ldots(1)$$

Here $f(x) = x^2 \qquad f(a) = a^2$

$$f(a + h) = (a + h)^2$$

$$\ldots \ldots \ldots \text{and so on.}$$

Put all these values in (1), we get

$$\int_a^b x^2 dx = \lim_{n \to \infty} h[a^2 + (a + h)^2 + (a + 2h)^2 + \ldots + \{a + (n-1)h\}^2]$$

where $h \to 0$ as $n \to \infty$ and $nh \to b - a$

$$= \lim_{h \to \infty} h[na^2 + 2ah\{1 + 2 + 3 + ... + (n-1)\}] + h^2[1^2 + 2^2 + ... + (n-1)^2]$$

Using $\sum n = \dfrac{n(n+1)}{2}$ and $\sum n^2 = \dfrac{n(n+1)(2n+1)}{6}$

$\therefore \quad \displaystyle\int_a^b x^2 dx = \lim_{n \to \infty} h\left[na^2 + 2ah\dfrac{(n-1)n}{2} + \dfrac{h^2}{6}(n-1)n(2n-1) \right]$

$$= \lim_{n \to \infty} \left[(nh)a^2 + a(nh)(n-1)h + \dfrac{1}{6}(nh)(n-1)h(2n-1)h \right]$$

$$= \lim_{n \to \infty} \left[(nh)a^2 + a(nh)^2\left(1 - \dfrac{1}{n}\right) + \dfrac{1}{6}2(nh)^3\left(1 - \dfrac{1}{n}\right)\left(1 - \dfrac{1}{2n}\right) \right]$$

$$= (b-a)a^2 + a(b-a)^2 + \dfrac{1}{3}(b-a)^3 \quad (\because \text{ as } n \to \infty, h \to 0, nh \to b - a)$$

$$= \dfrac{1}{3}(b-a)[3a^2 + 3(b-a)a + b^2 - 2ab + a^2]$$

$$= \dfrac{1}{3}(b-a)(a^2 + ab + b^2) = \dfrac{1}{3}(b^3 - a^3)$$

EXAMPLE 2. *From the definition of a definite integral as the limit of a sum, evaluate $\int_a^b e^x dx$.*

SOLUTION. Here, we have $f(x) = e^x$.

Therefore $f(a) = e^a$

$\qquad f(a+h) = e^{a+h}$

$\qquad$ ... ... etc.

Now $\displaystyle\int_a^b e^x dx = \lim_{h \to 0} h[e^a + e^{a+h} + e^{a+2h} + ... + e^{a+(n-1)h}]$

where, $nh = b - a$ and $n \to \infty$ as $h \to 0$

$$= \lim_{h \to 0} he^a[1 + e^h + e^{2h} + ... + e^{(n-1).h}]$$

$$= \lim_{h \to 0} he^a\left\{ \dfrac{(e^h)^n - 1}{e^h - 1} \right\} = \lim_{h \to 0} he^a\left\{ \dfrac{e^{nh} - 1}{e^h - 1} \right\}$$

$$= \lim_{h \to 0} he^a\left[\dfrac{e^{b-a} - 1}{e^h - 1} \right] \qquad\qquad [\because nh = b - a]$$

$$= \lim_{n \to 0} e^a\left[\dfrac{e^{b-a} - 1}{\dfrac{e^h - 1}{h}} \right] = e^b - e^a \qquad\qquad \left(\because \lim_{h \to 0} \dfrac{e^h - 1}{h} = 1 \right)$$

10.9 SUMMATION OF SERIES WITH THE HELP OF DEFINITE INTEGRAL

We know that

$$\int_a^b f(x)\, dx = \lim_{n \to \infty} h[f(a) + f(a+h) + ... + f\{a+(n-1)h\}]$$

$$= \lim_{n \to \infty} \sum_{r=0}^{n-1} f(a+rh), \qquad \text{where } na = b - a$$

Now putting $a = 0$ and $b = 1$ so that $h = 1/n$, we get

$$\int_0^1 f(x)dx = \lim_{n \to \infty} \frac{1}{n} \sum_{r=0}^{n-1} f\left(\frac{r}{n}\right).$$

WORKING PROCEDURE

Step 1. Write the rth term of the series.

Step 2. Write the rth term in the form of $\dfrac{1}{n} f\left(\dfrac{r}{n}\right)$

Step 3. Replace $\dfrac{r}{n}$ by x, $\dfrac{1}{n}$ by dx and $\lim\limits_{x \to \infty}$ by f.

Then, lower limit of the definite integral will be value of $\dfrac{r}{n}$ for the first term as $n \to \infty$ and the upper limit will be the value of $\dfrac{r}{n}$ for the last term as $n \to \infty$.

SOLVED EXAMPLES

EXAMPLE 1. *Evaluate the following :* $\lim\limits_{n \to \infty}\left[\dfrac{1}{n+1} + \dfrac{1}{n+2} + \ldots + \dfrac{1}{2n}\right].$

SOLUTION. The general term is given by $(r^{th}$ term$) = \dfrac{1}{n+r}$

We have to find $\lim\limits_{n \to \infty} \sum\limits_{r=1}^{n} \dfrac{1}{n+r} = \lim\limits_{n \to \infty} \dfrac{1}{n[1+r/n]} = \lim\limits_{n \to \infty} \dfrac{1}{n} \sum\limits_{r=1}^{n} \dfrac{1}{[1+r/n]}$

Since the limit of r in the summation are 1 to n, therefore the lower limit of integration $= \lim\limits_{n \to \infty} \dfrac{1}{n} = 0.$

Also, the upper limit of integration $= \lim\limits_{n \to \infty} \dfrac{n}{n} = 1.$

Hence, the required limit

$$\int_0^1 \frac{1}{1+x} dx = \left[\log(1+x)\right]_0^1 = \log 2$$

EXAMPLE 2. *Evaluate :* $\lim\limits_{n \to \infty} n\left[\dfrac{1}{(n+1)(n+2)} + \dfrac{1}{(n+2)(n+4)} + \ldots + \dfrac{1}{6n^2}\right].$

SOLUTION. The given limit $= \lim\limits_{n \to \infty} n \sum\limits_{r=1}^{n} \dfrac{1}{(n+r)(n+2r)} = \lim\limits_{n \to \infty} \dfrac{n}{n^2} \sum\limits_{r=1}^{n} \dfrac{1}{(1+r/n)(1+2r/n)}$

$$= \lim_{n \to \infty} \frac{1}{n} \sum_{r=1}^{n} \frac{1}{\left(1+\dfrac{r}{n}\right)\left(1+\dfrac{2r}{n}\right)} = \int_0^1 \frac{1}{(1+x)(1+2x)} dx$$

$$= \int_0^1 \left[\frac{-1}{1+x} + \frac{2}{1+2x}\right] dx \qquad \text{(Resolving into partial fraction)}$$

$$= \left[-\log(1+x) + \log(1+2x)\right]_0^1 = \log\left[\frac{(1+2x)}{(1+x)}\right]_0^1 = \log\frac{3}{2} - \log 1 = \log\frac{3}{2}.$$

EXAMPLE 3. *Evaluate* $\displaystyle\lim_{n\to\infty}\left[\left(1+\frac{1}{n^2}\right)\left(1+\frac{2^2}{n^2}\right)\left(1+\frac{3^2}{n^2}\right)\cdots\left(1+\frac{n^2}{n^2}\right)\right]^{1/n}$

SOLUTION. Let $\displaystyle A=\lim_{n\to\infty}\left[\left(1+\frac{1}{n^2}\right)\left(1+\frac{2^2}{n^2}\right)\left(1+\frac{3^2}{n^2}\right)\cdots\left(1+\frac{n^2}{n^2}\right)\right]^{1/n}$

$\Rightarrow\quad \log A=\displaystyle\lim_{n\to\infty}\frac{1}{n}\left[\log\left(1+\frac{1}{n^2}\right)+\log\left(1+\frac{2^2}{n^2}\right)+\log\left(1+\frac{3^2}{n^2}\right)+\ldots+\log\left(1+\frac{n^2}{n^2}\right)\right]$

$\displaystyle=\lim_{n\to\infty}\frac{1}{n}\sum_{r=1}^{\infty}\log\left(1+\frac{r^2}{n^2}\right)=\int_0^1(1+x^2)\,dx$

$\displaystyle=\int_0^1\log(1+x^2).1\,dx=[x\log(1+x^2)]_0^1-\int_0^1\frac{2x.x\,dx}{1+x^2}$

$\displaystyle=\log 2-2\int_0^1\frac{(1+x^2)-1}{1+x^2}\,dx=\log 2-2\int_0^1\left[1-\frac{1}{(1+x^2)}\right]dx$

$\displaystyle=\log 2-2\left[x-\tan^{-1}x\right]_0^1=\log 2-2\left(1-\frac{\pi}{4}\right)$

Therefore,

$\log A=\log 2+\dfrac{1}{2}(\pi-4)$

$\Rightarrow\quad \log\dfrac{A}{2}=\dfrac{1}{2}(\pi-4)$

$\Rightarrow\quad A=2e^{(\pi-4)/2}$

EXAMPLE 4. *Find the limit of* $\left[\dfrac{n!}{n^n}\right]^{1/n}$ *when* $n\to\infty$.

SOLUTION. Let $\displaystyle A=\lim_{n\to\infty}\left[\frac{n!}{n^n}\right]^{1/n}=\lim_{n\to\infty}\left[\frac{1\cdot 2\cdot 3\cdot 4\ldots n}{n\cdot n\cdot n\ldots n}\right]^{1/n}$

$\Rightarrow\quad \log A=\displaystyle\lim_{n\to\infty}\frac{1}{n}\left[\log\left(\frac{1}{n}\right)+\log\left(\frac{2}{n}\right)+\log\left(\frac{3}{n}\right)+\ldots+\log\left(\frac{n}{n}\right)\right]$

$\displaystyle=\lim_{n\to\infty}\sum_{r=1}^{n}\frac{1}{n}\log\left(\frac{r}{n}\right)=\int_0^1\log x\,dx=\int_0^1\log x.1\,dx$

$\displaystyle=[(\log x)\cdot x]_0^1-\int_0^1\frac{1}{x}\cdot x\,dx=0-[x]_0^1=-1$

Hence, $A=e^{-1}=\dfrac{1}{e}$.

Exercise 10.5

1. Show that the limit of the sum

$$\frac{1}{n}+\frac{1}{n+1}+\frac{1}{n+2}+\ldots+\frac{1}{6n}$$

when n is indefinitely increased is $\log 6$.

2. Evaluate $\int_a^b x^2\,dx$ directly from the definition of the integral as the limit of the sum.

3. Evaluate by summation $\int_1^2 x\,dx$.

4. Evaluate by summation $\int_a^b \sin x \, dx$.

5. Evaluate by summation $\int_0^{\pi/2} \sin x \, dx$.

6. Show that the limit (when $n \to \infty$) of the series

$$\frac{n}{(n+1)^2} + \frac{n}{(n+2)^2} + \dots + \frac{n}{(n+n)^2} \text{ is } \frac{1}{2}.$$

7. Show that

$$\lim_{n \to \infty}\left[\frac{n}{n^2} + \frac{n}{n^2+1^2} + \frac{n}{n^2+n^2} + \dots \right.$$
$$\left. + \frac{n}{n^2+(n+1)^2} \right] = \frac{\pi}{4}.$$

8. Show that

$$\lim_{n \to \infty}\left[\frac{n}{n^2+1^2} + \frac{n}{n^2+2^2} + \dots + \frac{n}{2n} \right] = \frac{\pi}{4}.$$

9. Show that

$$\lim_{n \to \infty}\left[\frac{1}{n^3}(1+4+9+\dots+n^2) \right] = \frac{1}{3}.$$

10. Show that

$$\lim_{n \to \infty}\left[\frac{1}{n} + \frac{n^2}{(n+1)^3} + \frac{n^2}{(n+2)^2} + \dots + \frac{1}{8n} \right] = \frac{3}{8}.$$

11. Show that

$$\lim_{n \to \infty}\left[\frac{1}{n} + \frac{1}{\sqrt{n^2-1^2}} + \frac{1}{\sqrt{n^2-2^2}} + \dots \right.$$
$$\left. + \frac{1}{\sqrt{n^2-(n-1)^2}} \right] = \frac{\pi}{2}.$$

12. Show that

$$\lim_{n \to \infty}\left[\frac{1}{n^2}\sec^2\frac{1}{n^2} + \frac{2}{n^2}\sec^4\frac{4}{n^2} + \frac{3}{n^2}\sec^2\frac{9}{n^2} \right.$$
$$\left. + \dots + \frac{1}{n}\sec^2 1 \right] = \frac{1}{2}\tan 1.$$

13. Show that

$$\lim_{n \to \infty}\left[\frac{1}{\sqrt{n^2-1^2}} + \frac{1}{\sqrt{n^2-2^2}} + \dots \right.$$
$$\left. + \frac{1}{\sqrt{n^2-(n-1)^2}} \right] = \frac{\pi}{2}.$$

14. Show that

$$\lim_{n \to \infty}\left[\frac{n^{1/2}}{n^{3/2}} + \frac{n^{1/2}}{(n+3)^{3/2}} + \frac{n^{1/2}}{(n+6)^{3/2}} + \dots \right.$$
$$\left. + \frac{n^{1/2}}{\{n+3(n+1)^{3/2}\}} \right] = \frac{1}{3}.$$

HINT TO SELECTED PROBLEMS

1. $\lim_{n \to \infty}\left[\frac{1}{n} + \frac{1}{n+1} + \frac{1}{n+2} + \dots + \frac{1}{6n} \right]$

$$= \lim_{n \to \infty}\sum_{r=0}^{5n}\left[\frac{1}{n+r} \right]$$

$$= \lim_{n \to \infty}\frac{1}{n}\sum_{r=0}^{5n}\left[\frac{1}{1+\frac{r}{n}} \right] = \int_0^5 \frac{1}{1+x}\,dx$$

$$= [\log(1+x)]_0^5 = \log 6 - \log 1 = \log 6$$

4. Let $I = \int_a^b \sin x \, dx$ Here $f(x) = \sin x$

Let $h = \dfrac{b-a}{n}, n \in N$

$I = \lim_{n \to \infty} h[f(a) + f(a+h) + \dots + f(a+(n-1)h]$

$= \lim_{n \to \infty} h[\sin a + \sin(a+h) + \dots$
$\qquad + \sin(a+(n-1)h)]$

Now $\sin a + \sin(a+h) + \dots + \sin(a+(n-1)h)$

$= \dfrac{1}{2\sin\dfrac{h}{2}}\left[2\sin a \sin\dfrac{h}{2} + 2\sin(a+h).\sin\dfrac{h}{2} \right.$

$\left. + \dots + 2\sin(a+(n-1)\sin\dfrac{h}{2} \right]$

$= \dfrac{1}{2\sin h/2}\left[\left\{ \cos\left(a-\dfrac{h}{2} \right) - \cos\left(a+\dfrac{h}{2} \right) \right\} \right.$

$\qquad + \left\{ \cos\left(a+\dfrac{h}{2} \right) - \cos\left(a+\dfrac{3h}{2} \right) \right\}$

$\qquad + \dots + \left\{ \cos\left(a+\left(n-\dfrac{3}{2} \right)h \right) - \cos\left(a+\left(n-\dfrac{1}{2} \right)h \right) \right\} \Big]$

$= \dfrac{1}{2\sin h/2}\left[\cos\left(a-\dfrac{h}{2} \right) - \cos\left(a+\left(n-\dfrac{1}{2} \right)h \right) \right]$

$= \dfrac{1}{2\sin h/2}\left[\cos\left(a-\dfrac{h}{2} \right) - \cos\left(b-\dfrac{h}{2} \right) \right]$

$b = a + nh$

$I = \lim_{h \to \infty} h.\dfrac{1}{2\sin h/2}\left[\cos\left(a-\dfrac{h}{2} \right) - \cos\left(b-\dfrac{h}{2} \right) \right]$

$= \lim_{h \to \infty}\dfrac{h/2}{\sin h/2}\lim_{h \to \infty}\left[\cos\left(a-\dfrac{h}{2} \right) - \cos\left(b-\dfrac{h}{2} \right) \right]$

$= 1.[\cos(a-0-\cos(b-0)] = \cos a - \cos b$

6. $\lim_{n \to \infty}\left[\dfrac{n}{(n+1)^2} + \dfrac{n}{(n+2)^2} + \dfrac{n}{(n+3)^2} + \dots \right.$
$\left. + \dfrac{n}{(n+n)^2} \right]$

$$= \lim_{n\to\infty} \sum_{r=0}^{n+1} \frac{n}{(n+r)^2}$$

$$= \lim_{n\to\infty} \frac{1}{n} \sum_{r=0}^{n+1} \frac{1}{\left(1+\frac{r}{n}\right)^2} = \int_0^1 \frac{1}{(1+x)^2} dx$$

$$= \left[-\frac{1}{(1+x)}\right]_0^1 = \frac{1}{2}$$

7. $\lim_{n\to\infty} \left[\frac{n}{n^2} + \frac{n}{n^2+1^2} + \frac{n}{n^2+2^2} + \ldots \right.$

$$\left. + \frac{n}{n^2+(n+1)^2}\right]$$

$$= \lim_{n\to\infty} \sum_{r=0}^{n} \frac{n}{n^2+r^2} = \lim_{n\to\infty} \frac{1}{n} \sum_{r=0}^{n+1} \frac{1}{1+\left(\frac{r}{h}\right)^2}$$

$$= \int_0^1 \frac{1}{1+x^2} dx = (\tan^{-1} x)_0^1 = \tan^{-1} 1 - \tan^{-1} 0$$

$$= \pi/4$$

8. $\lim_{n\to\infty} \left[\frac{1}{n^3}(1+4+9+\ldots+n^2)\right]$

$$= \lim_{n\to\infty} [1^2 + 2^2 + 3^2 + \ldots + n^2]$$

$$= \lim_{n\to\infty} \sum_{r=0}^{n} \frac{1}{n^3} r^2 = \lim_{n\to\infty} \frac{1}{n} \sum_{r=0}^{n} \left(\frac{r}{n}\right)^2$$

$$= \int_0^1 x^2 dx = \left(\frac{x^3}{3}\right)_0^1 = \frac{1}{3}$$

11. $\lim_{n\to\infty} \left[\frac{1}{n} + \frac{1}{\sqrt{n^2-1^2}} + \frac{1}{\sqrt{n^2-2^2}} + \ldots \right.$

$$\left. + \frac{1}{\sqrt{n^2-(n-1)^2}}\right]$$

$$= \lim_{n\to\infty} \sum_{r=0}^{n-1} \frac{1}{\sqrt{n^2-r^2}}$$

$$= \lim_{n\to\infty} \frac{1}{n} \sum_{r=0}^{n-1} \frac{1}{\sqrt{1-\left(\frac{r}{h}\right)^2}}$$

$$= \int_0^1 \frac{1}{\sqrt{1-x^2}} = (\sin^{-1} x)_0^1$$

$$= \sin^{-1}(1) - \sin^{-1}(0) = \pi/2$$

12. $\lim_{n\to\infty} \left[\frac{1}{n^2}\sec^2 \frac{1}{n^2} + \frac{2}{n^2}\sec^2 \frac{4}{n^2} + \frac{3}{n^2}\sec^2 \frac{9}{n^2}\right.$

$$\left. + \ldots + \frac{1}{n}\sec^2 1\right]$$

$$= \lim_{n\to\infty} \sum_{r=0}^{n} \frac{r}{n^2}\sec^2 \frac{r^2}{n^2}$$

$$= \lim_{n\to\infty} \frac{1}{n} \sum_{r=0}^{n} \left(\frac{r}{n}\right)\sec^2 \left(\frac{r}{n}\right)^2$$

$$= \int_0^1 x \cdot \sec^2 x^2 dx$$

Let $x^2 = t \Rightarrow 2x\, dt = dt$

$$= \frac{1}{2} \int_0^1 \sec^2 t\, dt = \frac{1}{2}.(\tan t)_0^1 = \frac{1}{2}\tan 1$$

14. $\lim_{n\to\infty} \frac{n^{1/2}}{n^{3/2}} + \frac{n^{1/2}}{(n+3)^{3/2}} + \frac{n^{1/2}}{(n+6)^{3/2}}$

$$+ \ldots + \frac{n^{1/2}}{\{n+3(n+1)\}^{3/2}}$$

$$= \lim_{n\to\infty} \sum_{r=0}^{n} \frac{n^{1/2}}{(n+3r)^{3/2}}$$

$$= \lim_{n\to\infty} \frac{1}{n} \sum_{r=0}^{n} \frac{1}{\left(1+\frac{3r}{n}\right)^{\frac{3}{2}}} = \int_0^1 \frac{1}{(1+3x)^{3/2}} dx$$

$$= \int_0^1 (1+3x)^{-3/2} dx = \left[\frac{(1+3x)^{-1/2}}{-1/2 \times 3} \times 3\right]_0^1$$

$$= -\frac{2}{3}[(1+3x)^{-1/2}]_0^1 = -\frac{2}{3}[4^{-1/2} - 1^{-1/2}]$$

$$= -\frac{2}{3}\left[\frac{1}{2-1}\right] = -\frac{2}{3}\left[-\frac{1}{2}\right] = \frac{1}{3}.$$

ANSWERS

2. $\frac{1}{3}(b^3 - a^3)$ **3.** $\frac{3}{2}$ **4.** $\cos b - \cos a$ **5.** 1

Objective Evaluation

☞ MULTIPLE CHOICE QUESTIONS

Choose the most appropriate one.

1. The value of $\int x^{5/3}\,dx$ is :

   (a) $\dfrac{3}{5}x^{2/3}+c$ (b) $\dfrac{3}{8}x^{8/3}+c$

   (c) $\dfrac{5}{3}x^{8/3}+c$ (d) $\dfrac{8}{3}x^{8/3}+c$

2. The value of $\int 2^{\log x}dx$ is :

   (a) $\dfrac{x^{(\log 2+1)}}{(\log 2+1)}+c$ (b) $\dfrac{2^{\log x}}{\log 2}+c$

   (c) $\dfrac{2^{\log x}}{2}+c$ (d) $\dfrac{2^{\log x+1}}{(\log x+1)}+c$

3. The value of $\int\dfrac{\cos 2x}{\cos^2 x\sin^2 x}\,dx$ is :

   (a) $-\cot x-\tan x+c$
   (b) $-\cot x+\tan x+c$
   (c) $\cot x-\tan x+c$
   (d) $\cot x+\tan x+c$

4. The value of $\int\sec x\,(\sec x+\tan x)dx$ is :
   (a) $\tan x-\sec x+c$ (b) $-\tan x+\sec x+c$
   (c) $\tan x+\sec x+c$ (d) $-\tan x-\sec x+c$

5. The value of $\int\dfrac{\sec^2 x}{\csc^2 x}\,dx$ is :

   (a) $\tan x+x+c$ (b) $\tan x-x+c$
   (c) $-\tan x+x+c$ (d) $-\tan x-x+c$

6. The value of $\int\cos^{-1}\left(\dfrac{1-\tan^2 x}{1+\tan^2 x}\right)dx$ is :

   (a) $-x^2+c$ (b) x^2+c

   (c) $\dfrac{1}{\sqrt{1-x^2}}+c$ (d) $\dfrac{1}{\sqrt{1+x^2}}+c$

7. The value of $\int\dfrac{(\sin^3 x+\cos^2 x)}{\sin^2 x\cos^2 x}\,dx$ is :

   (a) $\sin x-\cos x+c$
   (b) $\tan x-\cos x+c$
   (c) $\sec x-\csc x+c$
   (d) none of these

8. The value of $\int\dfrac{\sin x}{\sin(x-\alpha)}\,dx$ is :

   (a) $x\cos\alpha+(\sin\alpha)\log|\sin(x-\alpha)|+c$
   (b) $x\sin\alpha+(\sin\alpha)\log|\sin(x-\alpha)|+c$
   (c) $x\cos\alpha-(\sin\alpha)\log|\sin(x-\alpha)|+c$
   (d) $x\sin\alpha-(\sin\alpha)\log|\sin(x-\alpha)|+c$

9. The value of $\int\sqrt{ax+b}\,dx$ is :

   (a) $\dfrac{2(ax+b)^{3/2}}{3a}+c$ (b) $\dfrac{3(ax+b)^{3/2}}{2a}+c$

   (c) $\dfrac{1}{2\sqrt{ax+b}}+c$ (d) none of these

10. The value of $\int\tan^2\dfrac{x}{2}\,dx$ is :

    (a) $\tan\dfrac{x}{2}+x+c$ (b) $\tan\dfrac{x}{2}-x+c$

    (c) $2\tan\dfrac{x}{2}+x+c$ (d) $2\tan\dfrac{x}{2}-x+c$

11. The value of $\int\sqrt{1+\sin x}\,dx$ is :

    (a) $-\sqrt{2}\sin\left(\dfrac{\pi}{4}-\dfrac{x}{2}\right)+c$

    (b) $\sqrt{2}\sin\left(\dfrac{\pi}{4}-\dfrac{x}{2}\right)+c$

    (c) $-2\sqrt{2}\sin\left(\dfrac{\pi}{4}-\dfrac{x}{2}\right)+c$

    (d) none of these

12. The value of $\int\dfrac{e^{\sqrt{x}}}{\sqrt{x}}\,dx$ is :

    (a) $e^{\sqrt{x}}+c$ (b) $\dfrac{1}{2}e^{\sqrt{x}}+c$

    (c) $2e^{\sqrt{x}}+c$ (d) none of these

13. The value of $\int(\sqrt{\sin x})\cos x\,dx$ is :

    (a) $\dfrac{3}{2}(\cos x)^{3/2}+c$ (b) $\dfrac{2}{3}(\sin x)^{3/2}+c$

    (c) $\dfrac{3}{2}(\sin x)^{3/2}+c$ (d) $\dfrac{2}{3}(\cos x)^{3/2}+c$

14. The value of $\int\dfrac{1}{x\cos^2(1+\log x)}\,dx$ is :

    (a) $\tan(1+\log x)+c$
    (b) $\cot(1+\log x)+c$
    (c) $\sec(1+\log x)+c$
    (d) none of these

15. The value of $\int\csc^3(2x+1)\cot(2x+1)dx$ is :

    (a) $\dfrac{1}{4}\csc^4(2x+1)+c$

    (b) $-\dfrac{1}{3}\csc^3(2x+1)+c$

(c) $-\dfrac{1}{6}\text{cosec}^3(2x+1)+c$

(d) $\dfrac{1}{2}\text{cosec}(2x+1)\cot(2x+1)+c$

16. The value of $\int\dfrac{\cos x}{(1+\cos x)}dx$ is :

(a) $x+\tan\dfrac{x}{2}+c$ (b) $-x+\tan\dfrac{x}{2}+c$

(c) $x-\tan\dfrac{x}{2}+c$ (d) none of these

17. The value of $\int\dfrac{e^x}{(e^{2x}+1)}dx$ is :

(a) $\cot^{-1}(e^x)+c$ (b) $\tan^{-1}(e^x+c)$

(c) $2\tan^{-1}(e^x)+c$ (d) none of these

18. The value of $\int\dfrac{dx}{(e^x+e^{-x})}$ is :

(a) $\tan^{-1}(e^{-x})+c$ (b) $\tan^{-1}(e^x)+c$

(c) $-\tan^{-1}(e^{-x})+c$ (d) none of these

19. The value of $\int\dfrac{\sin 2x}{(\sin^4 x+\cos^4 x)}dx$ is :

(a) x^2+c (b) $\tan^{-1}(\tan^2 x)+c$

(c) $-\tan^{-1}(\tan^2 x)+c$ (d) none of these

20. The value of $\int\dfrac{dx}{(\cos^2 x-3\sin^2 x)}$ is :

(a) $\dfrac{1}{\sqrt{3}}\log\left|\dfrac{\sqrt{3}+\tan x}{\sqrt{3}-\tan x}\right|+c$

(b) $\dfrac{1}{\sqrt{3}}\log\left|\dfrac{1-\sqrt{3}\tan x}{1+\sqrt{3}\tan x}\right|+c$

(c) $\dfrac{1}{2\sqrt{3}}\log\left|\dfrac{1+\sqrt{3}\tan x}{1-\sqrt{3}\tan x}\right|+c$

(d) none of these

21. The value of $\int_0^2\dfrac{dx}{\sqrt{4-x^2}}$ is :

(a) $\sin^{-1}(1/2)$ (b) $\pi/4$

(c) 1 (d) none of these

22. The value of $\int_1^e\dfrac{(\log x)^2}{x}dx$ is :

(a) $1/3$ (b) $1/3(e^3)$

(c) $1/3(e^3-1)$ (d) none of these

23. The value of $\int_0^{\pi/2}\dfrac{\cos x}{(1+\sin^2 x)}dx$ is :

(a) π (b) $\pi/2$

(c) $\pi/4$ (d) none of these

24. The value of $\int_{-a}^{a}\sqrt{\dfrac{a-x}{a+x}}dx$ is :

(a) $a\pi$ (b) $a\pi/2$

(c) $2a\pi$ (d) none of these

25. The value of $\int_0^{\pi/2}\dfrac{\sqrt[3]{\tan x}}{(\sqrt[3]{\tan x}+\sqrt[3]{\cot x})}dx$ is :

(a) 0 (b) π

(c) $\pi/2$ (d) $\pi/4$

☞ TRUE/FALSE

Write 'T' for True and 'F' for False statement.

1. $\int k\cdot f(x)dx = k\cdot\int f(x)dx$, where k is a constant.

(T/F)

2. The value of $\int_0^{\pi/2}\dfrac{\sin x}{\sqrt{1+\cos x}}dx$ is $2(\sqrt{2}+1)$.

(T/F)

3. The value of $\int_0^{\pi/2}\dfrac{dx}{(4+9\cos^2 x)}dx$ is $\dfrac{\pi}{4\sqrt{13}}$.

(T/F)

4. The value of $\int_0^1 x\cdot\sqrt{\dfrac{1-x^2}{1+x^2}}dx$ is $\left(\dfrac{\pi}{6}-\dfrac{1}{2}\right)$.

(T/F)

5. The value of $\int_0^{\pi/2}\{\sqrt{\tan x}+\sqrt{\cot x}\}dx$ is $\sqrt{2}\pi$.

(T/F)

6. The value of $\int\dfrac{\cos x}{\cos 3x}dx$ is

$\dfrac{1}{2\sqrt{3}}\log\left|\dfrac{1+\sqrt{3}\tan x}{1-\sqrt{3}\tan x}\right|+c$. **(T/F)**

☞ FILL IN THE BLANKS

1. The value of $\int 2^{(x+3)}dx$ is _____.

2. The value of $\int\cos x\cos 2x\cos 3x\,dx$ is _____.

3. $\int\dfrac{x^2}{(1-x^6)}dx=$ _____.

4. $\int\dfrac{\cos x}{(1-\sin x)(2-\sin x)}dx=$ _____.

5. The value of $\int_0^{\pi/2}\dfrac{dx}{(5+4\sin x)}$ is _____.

6. The value of $\int_1^2\dfrac{dx}{x(1+\log x)^2}$ is _____.

Answers

∞ Multiple Choice Questions

1. (b)	**2.** (a)	**3.** (a)	**4.** (c)	**5.** (b)	**6.** (b)	**7.** (c)	**8.** (a)	**9.** (a)
10. (d)	**11.** (c)	**12.** (c)	**13.** (d)	**14.** (a)	**15.** (c)	**16.** (c)	**17.** (b)	**18.** (b)
19. (b)	**20.** (c)	**21.** (d)	**22.** (a)	**23.** (c)	**24.** (a)	**25.** (d)		

∞ True/ False

1. T   **2.** F   **3.** T   **4.** F   **5.** T   **6.** T

∞ Fill in the Blanks

1. $\dfrac{2^{(x+3)}}{\log 2}+c$

2. $\dfrac{x}{4}+\dfrac{\sin 6x}{24}+\dfrac{\sin 4x}{16}+\dfrac{\sin 2x}{8}+c$

3. $\dfrac{1}{6}\log\left|\dfrac{1+x^3}{1-x^3}\right|+c$

4. $\log\left|\dfrac{2-\sin x}{1-\sin x}\right|+c$

5. $\dfrac{2}{3}\tan^{-1}\left(\dfrac{1}{3}\right)$

6. $\dfrac{\log 2}{(1+\log 2)}$

CHAPTER 11

Geometry : System of Coordinates

11.1 INTRODUCTION

Coordinate Geometry is the branch of mathematics in which two numbers are used to represent the position of a point with respect to two mutually perpendicular number lines called coordinate axes.

The French mathematician and philosopher Rene Descaotes first published his book La Geometric in 1637 in which he used algebra in the study of geometry. This he did by representing points in the plane by ordered pairs of real number called cartesian coordinates and representing lines and curves by algebraic equations.

11.1.1 COORDINATE AXES

The adjoining figure 1 shows two number lines XOX' and YOY' intersecting each other at their zeros.

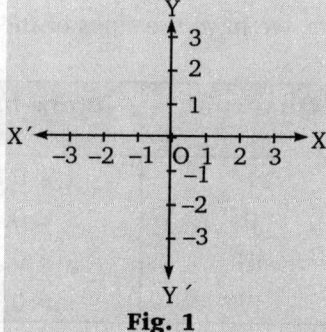

Fig. 1

XOX' and YOY' are called coordinate axes out of which XOX' is called the x-axis, YOY' is called y-axis and their point of intersection is called the origin.

☝ REMARK

⟹ Number lines XOX' and YOY' are sometimes also called rectangular axes as they are perpendicular to each other.

11.1.2 CONVENTION OF SIGNS

The distance measured along OX and OY are taken as positive and those along OX' and OY' are taken as negative as shows in figure 1.

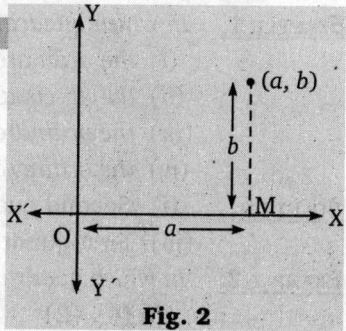

Fig. 2

11.1.3 COORDINATE OF A POINT IN A PLANE

Let P be a point in a plane. Let the distance of P from the y-axis $= a$ units.

And, the distance of P from the x-axis $= b$ units. Then we say that the coordinate of P are (a, b). a is called the x-coordinate or abscissa of P and b is called the y-coordinate or ordinate of P.

☙ REMARKS

➥ (x, y) and (y, x) do not represent the same point unless $x = y$.

    e.g. (5, 4) and (4, 5) represent two different points.

➥ In stating the co-ordinates of a point the abscissa proceeds the ordinate. The two are separated by a comma and enclosed in a bracket. Thus a point, whose abscissa is x and whose ordinate is y designated by the notation (x, y), i.e., (abscissa, ordinate)

➥ Since at origin the value of x-coordinate is 0 and the value of y-coordinate is also 0, therefore, the coordinate of origin = (0, 0).

➥ Since for every point on x-axis, its distance from x-axis is 0, i.e., ordinate is 0.

Therefore, the coordinate of a point on x-axis are taken as $(x, 0)$.

➥ In the same way, for every point on y-axis its distance from y-axis is 0, i.e., abscissa is 0, therefore, the coordinate of a point on y-axis are taken as $(0, y)$.

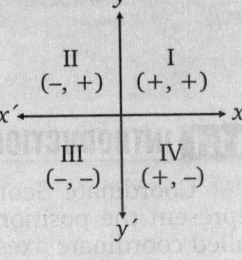

Fig. 3

11.1.4 QUADRANTS

Let $X'OX$ and YOY' be the coordinate axes.

These axes divide the plane of the paper into four regions, called Quadrants. The Regions XOY', YOX', $X'OY'$ and $Y'OY'$ are respectively known as first, second, third and fourth quadrants.

Using the convention of sings, we have the sings of the coordinates in various quadrant given below.

Region	Quadrant	Nature of x and y	Signs of coordinate
XOY	I	$x > 0, y > 0$	$(+, +)$
YOX'	II	$x < 0, y > 0$	$(-, +)$
$X'OY'$	III	$x < 0, y < 0$	$(-, -)$
$Y'OX$	IV	$x > 0, y < 0$	$(+, -)$

SOLVED EXAMPLES

EXAMPLE 1.   *In which quadrant will the point lie if*

     (i) *the ordinate is 3 and the abscissa is – 4 ?*

     (ii) *the abscissa is – 5 and the ordinate is – 3 ?*

     (iii) *the ordinate is 4 and the abscissa is 5 ?*

     (iv) *the ordinate is 4 and the abscissa is – 8?*

SOLUTION.   (i)   Second quadrant   (ii) Third quadrant

            (iii)   First quadrant   (v) Fourth quadrant.

EXAMPLE 2.   *In which quadrant do the given point lie?*

     (i) (4 , –2)   (ii) (– 3 , 7)   (iii) (– 1, –2)   (iv) (3, 6).

SOLUTION.   (i)   Fourth quadrant   (ii) Second quadrant

            (iii)   Third quadrant   (iv) First quadrant

EXAMPLE 3. *On which axes do the given points lie?*
      (i) (7, 0)   (ii) (0, –3)
      (iii) (0, 6)   (iv) (–5, 0)

SOLUTION.
    (i) In (7, 0), we have the ordinate = 0.   ∴ (7, 0) lies on the x-axis.
    (ii) In (0, – 3) we have the abscissa = 0.   ∴ (0, – 3) lies on the y-axis
    (iii) In (0, 6) we have the abscissa = 0.   ∴ (0, 6) lies on the y-axis
    (iv) In (–5, 0) we have the ordinate = 0.   ∴ (– 5, 0) lies on the x-axis.

EXAMPLE 4. *Plot the points (– 3, 0), (2 ,3), (– 4, 3) and (3,– 5) in a rectangular coordinate system.*

SOLUTION.

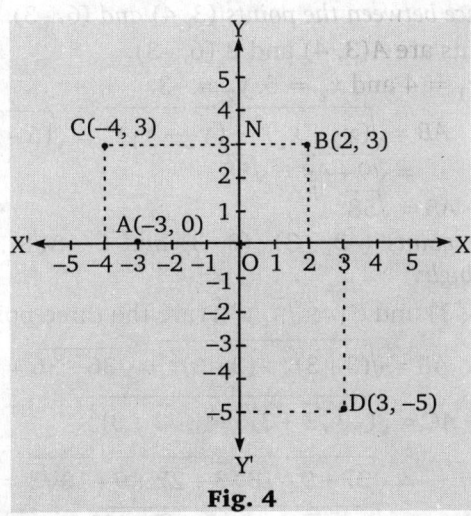

Fig. 4

11.2 DISTANCE BETWEEN TWO POINTS

THEOREM. *The distance between two points $A(x_1, y_1)$ and $B(x_2, y_2)$ is given by the formula*

$$AB = \sqrt{(x_2 - x_1)^2 + (y_2 - y_1)^2}$$

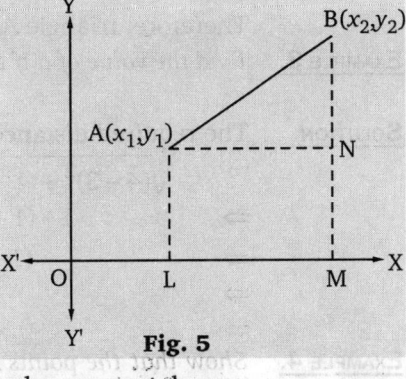

PROOF.   Let $A(x_1, y_1)$ and $B(x_2, y_2)$ be the given points.
Let $X'OX$ and YOY' be the coordinate axes.
Draw $AL \perp OX$, $BM \perp OX$ and $AN \perp BM$. Then
    $OL = x_1$, $OM = x_2$, $AN = y_1$ and $BM = y_2$
    $AN = LM = OM - OL = (x_2 - x_1)$
    $BN = BM - NM = BM - AC = (y_2 - y_1)$
                              $[\because NM = AC]$

Fig. 5

Now, $\triangle ANB$ is a right-angled, so by Pythagoras theorem, we have
$$AB^2 = AN^2 + BN^2 = (x_2 - x_1)^2 + (y_2 - y_1)^2$$
$$AB = \sqrt{(x_2 - x_1)^2 + (y_2 - y_1)^2}$$

REMARK
➡ The distance of a point $P(x, y)$ from the point (0, 0) is given by
$$OP = \sqrt{(x - 0)^2 + (y - 0)^2} = \sqrt{x^2 + y^2}$$

WORKING PROCEDURE

In order to prove that a figure is a :

Step 1. Square, prove that four sides are equal and the diagonals are also equal.

Step 2. Rhombus, prove that the four sides are equal.

Step 3. Rectangle, prove that opposite sides are equal and the diagonals are also equal.

Step 4. A parallelogram, prove that the opposite sides are equal.

SOLVED EXAMPLES

EXAMPLE 1. *Find the distance between the points* (3, 4) *and* (6, –3).

SOLUTION. The given points are A(3, 4) and B (6, –3).

Here $x_1 = 3, y_1 = 4$ and $x_2 = 6, y_2 = -3$.

$$AB = \sqrt{(x_2 - x_1)^2 + (y_2 - y_1)^2} = \sqrt{(6-3)^2 + (-3-4)^2}$$
$$= \sqrt{9 + 49} = \sqrt{58}$$

Therefore $AB = \sqrt{58}$

EXAMPLE 2. *Show that the points* (–3, –3), (3, 3) *and* $C(-3\sqrt{3}, 3\sqrt{3})$ *are the vertices of an equilateral triangle.*

SOLUTION. $A(-3, -3), B(3, 3)$ and $C(-3\sqrt{3}, 3\sqrt{3})$ are the three points.

$$AB = \sqrt{(3+3)^2 + (3+3)^2} = \sqrt{36 + 36} = \sqrt{72} = 6\sqrt{2}$$

$$AC = \sqrt{(-3\sqrt{3}+3)^2 + (3\sqrt{3}+3)^2}$$

$$= \sqrt{27 + 9 - 18\sqrt{3} + 27 + 9 + 18\sqrt{3}} = \sqrt{72} = 6\sqrt{2}$$

$$BC = \sqrt{(-3\sqrt{3}-3)^2 + (3\sqrt{3}-3)^2} = \sqrt{72} = 6\sqrt{2}$$

Then $AB = BC = AC$.

Therefore, triangle ABC is an equilateral triangle.

EXAMPLE 3. *Find the value of a if the distance between the points* (3, 0) *and* (4, 1) *is* $\sqrt{10}$.

[UPTU(B. Pharma)–2003]

SOLUTION. The required distance is given by

$$\sqrt{(4-3)^2 + (1-a)^2} = \sqrt{10}$$

$$\Rightarrow \qquad 1 + (1-a)^2 = 10$$

$$\Rightarrow \qquad (1-a)^2 = 9$$

$$\Rightarrow \qquad 1 - a = \pm 3$$

$$\Rightarrow \qquad a = 1 \pm 3 = -2, 4$$

EXAMPLE 4. *Show that the points* $A(2, -2), B(8, 4), C(5, 7)$ *and* $D(-1, 1)$ *are the vertices of a rectangle.*

[UPTU(B. Pharma)–2007]

SOLUTION. Here,

$$AB = \sqrt{(8-2)^2 + (4+2)^2} = \sqrt{6^2 + 6^2} = \sqrt{72}$$

$$BC = \sqrt{(5-8)^2 + (7-4)^2} = \sqrt{(-3)^2 + 3^2} = \sqrt{18}$$

$$CD = \sqrt{(-1-5)^2 + (1-7)^2} = \sqrt{(-6)^2 + (-6)^2} = \sqrt{72}$$

$$DA = \sqrt{(2+1)^2 + (-2-1)^2} = \sqrt{3^2 + (-3)^2} = \sqrt{18}$$

$$AC = \sqrt{(5-2)^2 + (7+2)^2} = \sqrt{3^2 + 9^2} = \sqrt{90}$$
$$BD = \sqrt{(-1-8)^2 + (1-4)^2} = \sqrt{(-9)^2 + (-3)^2} = \sqrt{90}$$

Thus, the opposite sides of the quadrilateral $ABCD$ are equal and its diagonals are also equal. Hence, it is a rectangle.

EXAMPLE 5. *Find the point on the y-axis which is equidistant from the points (3, 4) and (6, 7).*

SOLUTION. Given points are $A(3, 4)$ and $B(6, 7)$. Required point P is on the y-axis. Its abscissa = 0. Suppose its ordinate = y.

Then coordinate of P are $(0, y)$.

Now $PA = PB$

$\Rightarrow \sqrt{(0-3)^2 + (y-4)^2} = \sqrt{(0-6)^2 + (y-7)^2}$

$\Rightarrow \quad 9 + y^2 - 8y + 16 = 36 + y^2 - 14y + 49$

$\Rightarrow \quad\quad -8y + 14y = 36 + 49 - 9 - 16$

$\Rightarrow \quad\quad\quad 6y = 60$

$\Rightarrow \quad\quad\quad\quad y = 10$

Therefore, the required point is $(0, 10)$.

EXAMPLE 6. *If the point (x, y) is equidistant from the points (a + b, b − a) and (a − b, a + b) prove that bx = ay.*

SOLUTION. Let $P(x, y)$, $Q(a + b, a − b)$ and $R(a − b, a + b)$ be the given points.

Then $PQ = PR$ (given)

$\Rightarrow \sqrt{[x-(a+b)]^2 + [y-(b-a)]^2} = \sqrt{[x-(a-b)]^2 + [y-(b+a)]^2}$

$\Rightarrow \quad [x-(a+b)]^2 + [y-(b-a)]^2 = [x-(a-b)]^2 + [y-(b+a)]^2$

$\Rightarrow \quad x^2 - 2x(a+b) + (a+b)^2 + y^2 - 2y(b-a) + (b-a)^2$

$$= x^2 + (a-b)^2 - 2x(a-b) + y^2 - 2y(a+b) + (a+b)^2$$

$\Rightarrow \quad\quad -2x(a+b) - 2y(b-a) = -2x(a-b) - 2y(a+b)$

$\Rightarrow \quad\quad ax + bx + by - ay = ax - bx + ay + by$

$\Rightarrow \quad\quad\quad\quad 2\,bx = 2\,ay$

$\Rightarrow \quad\quad\quad\quad bx = ay$

EXAMPLE 7. *Find the value of x, if the distance between the points (x, −1) and (3, 2) is 5.*

SOLUTION. Let $P(x, −1)$ and $Q(3, 2)$ be the given points, then :

$$PQ = 5 \quad\quad\quad \text{[Given]}$$

$\Rightarrow \sqrt{(x-3)^2 + (-1-2)^2} = 5$

$\Rightarrow \quad\quad (x-3)^2 + 9 = 5^2$

$\Rightarrow \quad\quad (x^2 - 6x + 18) = 25$

$\Rightarrow \quad\quad x^2 - 6x - 7 = 0$

$\Rightarrow \quad\quad (x-7)(x+1) = 0$

$\Rightarrow \quad\quad\quad x = 7 \text{ or } x = -1$

11.3 COLLINEAR POINTS

Three points A, B, C are said to be collinear if they lie on the same straight line.

TEST FOR COLLINEARITY OF THREE POINTS

WORKING PROCEDURE

In order to show that three given points A, B, C are collinear. We find distances AB, BC and AC. If the sum of any two of these distance is equal to the third distance then the given points are collinear.

SOLVED EXAMPLES

EXAMPLE 1. *Using distance formula, show that the points* (–3, 2), (1, –2) *and* (9, –10) *are collinear.*

SOLUTION. $A(-3, 2)$, $B(1, -2)$ and $C(9, -10)$ are the given points.

Now,
$$AB = \sqrt{(1+3)^2 + (-2-2)^2} = \sqrt{16+16} = 4\sqrt{2} \text{ units}$$

$$BC = \sqrt{(9-1)^2 + (-10+2)^2} = \sqrt{64+64} = 8\sqrt{2} \text{ units}$$

$$AC = \sqrt{(9+3)^2 + (-10-2)^2} = \sqrt{144+144} = 12\sqrt{2} \text{ units}$$

$$\Rightarrow \qquad AB + BC = 4\sqrt{2} + 8\sqrt{2} = 12\sqrt{2} = AC.$$

Therefore, the three points are collinear.

EXAMPLE 2. *Show that the points* (1, 1), (–2, 7) *and* (3, –3) *are collinear.*

SOLUTION. Let $A(1, 1)$, $B(-2, 7)$ and $C(3, -3)$ be the given points.

Then we have
$$AB = \sqrt{(-2-1)^2 + (7-1)^2} = \sqrt{9+36} = 3\sqrt{5}$$

$$BC = \sqrt{(3+2)^2 + (-3-7)^2} = \sqrt{25+100} = 5\sqrt{5}$$

and
$$AC = \sqrt{(3-1)^2 + (-3-1)^2} = \sqrt{4+16} = 2\sqrt{5}$$

Clearly
$$BC = AB + AC.$$

Hence, A, B and C are collinear.

Exercise 11.1

1. Find the distance between the points :
   (i) $A(7, 13)$, $B(10, 9)$
   (ii) $P(-4, 7)$ and $Q(1, -5)$

2. Find the distance of the point $P(6, -6)$ from the origin.

3. Find the value or values of k for which the distance between the point $A(k, -5)$ and $B(2, 7)$ is 13 units.

4. Prove that the points $A(-3, 0)$, $B(1, -3)$ and $C(4, 1)$ are the vertices of an isosceles right angled triangle. Find the area of this triangle.

5. Prove that the points $A(a, a)$, $B(-a, -a)$, $C(-\sqrt{3a}, \sqrt{3a})$ are the vertices of an equilateral triangle. Calculate the area of this rectangle.

6. Prove that the points $A(1, -3)$, $B(13, 9)$, $C(10, 12)$, $D(-2, 0)$ taken in order are the angular points of a rectangle. Find the area of the rectangle.

7. Prove that the points $A(1, 1)$, $B(-2, 7)$ and $C(3, -3)$ are collinear.

8. If $P(2, -1)$, $Q(3, 4)$, $R(-2, 3)$ and $S(-3, -1)$ be four points in a plane. Show that $PQRS$ is a rhombus but not a square. Find the area of the rhombus.

HINT TO SELECTED PROBLEMS

5. Let $A(a, a)$, $B(-a, a)$ and $C(-\sqrt{3}a, \sqrt{3}a)$

$AB = 2\sqrt{2}\,a$ units, $BC = 2\sqrt{2}\,a$ units

$AC = 2\sqrt{2}\,a$ units, $AB = BC = AC = 2\sqrt{2}a$

Area of $\triangle ABC = \dfrac{\sqrt{3}}{4}(\text{side})^2 = \dfrac{\sqrt{3}}{4} \times (2\sqrt{2}a)^2$

$= (2\sqrt{3}a^2)$ sq. units.

6. $AB = 12\sqrt{2}$ units, $BC = 3\sqrt{2}$ units,

$DC = 12\sqrt{2}$ units, $AD = 3\sqrt{2}$ units

$AB = DC$ & $BC = AD$, $AC = 3\sqrt{34}$ units

$BD = 3\sqrt{34}$ units, $AC = BD$

Hence, $ABCD$ is a rectangle.

ANSWERS

1. (i) 5 units   (ii) 13 units   **2.** $6\sqrt{2}$ units   **3.** $k = -3$ or $k = 7$

4. 12.5 sq units   **5.** $(2\sqrt{3}a^2)$ sq units   **8.** 24 sq units

11.4 SECTION FORMULA

THEOREM.   *The coordinates of the point $P(x, y)$ which divides the line segment joining $A(x_1, y_1)$ and $B(x_2, y_2)$ internally in the ratio $m : n$ are given by*

$$x = \frac{mx_2 + nx_1}{m+n}, y = \frac{my_2 + ny_1}{m+n}$$

PROOF.   Let $X'OX$ and YOY' be the coordinate axes.

Let $A(x_1, y_1)$ and $B(x_2, y_2)$ be the end points of the given line segments AB.
If $P(x, y)$ is the point which divides AB in the ratio $m : n$.

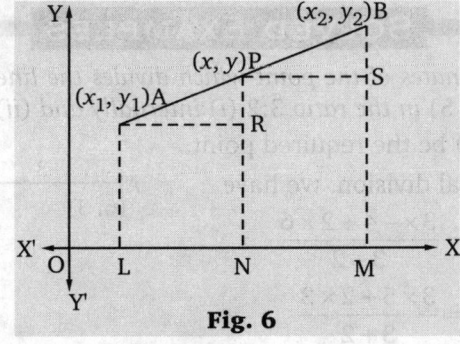

Fig. 6

Then $\dfrac{AP}{PB} = \dfrac{m}{n}$

Draw $AL \perp OX; BM \perp OX; PN \perp OX$

$AR \perp PN; PS \perp BM$

Now $AR = LN = ON - OL = (x - x_1)$

$PS = NM = OM - ON = (x_2 - x)$

$PR = PN - RN = PN - AL = (y - y_1)$

$BS = BM - SM = BM - PN = y_2 - y$

Clearly, $\triangle ARP$ and $\triangle PSB$ are similar and therefore, their sides are proportional.

$$\frac{AP}{PB} = \frac{AR}{PS} = \frac{PR}{BS}$$

$\Rightarrow$ $\dfrac{m}{n} = \dfrac{x - x_1}{x_2 - x} = \dfrac{y - y_1}{y_2 - y}$

$$\Rightarrow \quad \frac{m}{n} = \frac{x - x_1}{x_2 - x} \quad \text{and} \quad \frac{m}{n} = \frac{y - y_1}{y_2 - y}$$

$$\Rightarrow \quad mx_2 - mx = nx - nx_1 \quad \text{and} \quad my_2 - my = ny - ny_1$$

$$\Rightarrow \quad (m+n)x = mx_2 + nx_1 \quad \text{and} \quad (m+n)y = my_2 + ny_1$$

$$x = \frac{mx_2 + nx_1}{m+n}, y = \frac{my_2 + ny_1}{m+n}$$

Hence, the coordinate of P are $\left(\dfrac{mx_2 + nx_1}{m+n}, \dfrac{my_2 + ny_1}{m+n} \right)$

11.5 MID POINT FORMULA

The coordinates of the midpoint M on a line segment AB with end points $A(x_1, y_1)$ and $B(x_2, y_2)$ are $\left(\dfrac{x_1 + x_2}{2}, \dfrac{y_1 + y_2}{2} \right)$.

📖 REMARKS

➠ The coordinate of the point P, which divides the line-segment joining $A(x_1, y_1)$ and $B(x_2, y_2)$ internally, in the ratio $k : 1$ are given by $\left(\dfrac{kx_2 + x_1}{k+1}, \dfrac{ky_2 + y_1}{k+1} \right)$.

➠ The coordinate of the point which divides the line segment joining the points (x_1, y_1) and (x_2, y_2) externally in the ratio $m : n$ are given by $x = \dfrac{mx_2 - nx_1}{m-n}, y = \dfrac{my_2 - ny_1}{m-n}$.

SOLVED EXAMPLES

EXAMPLE 1. *Find the coordinates of the point which divides the line segment joining the points (6, 3) and (–4, 5) in the ratio 3:2 (i) internally and (ii) externally.*

SOLUTION. Let $P(x, y)$ be the required point.

(i) For internal division, we have

$$x = \frac{3 \times -4 + 2 \times 6}{3+2}$$

and $\quad y = \dfrac{3 \times 5 + 2 \times 3}{3+2}$

$x = 0$ and $y = \dfrac{21}{5}$

So the coordinates of P are $\left(0, \dfrac{21}{5} \right)$.

(ii) For external division, we have

$$x = \frac{3 \times -4 - 2 \times 6}{3-2}$$

and $\quad y = \dfrac{3 \times 4 - 2 \times 6}{3-2}$

$x = -24$ and $y = 9$.

So, the coordinates of point are $(-24, 9)$.

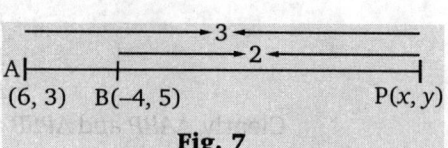

Fig. 7

EXAMPLE 2. *In what ratio does the point (–2, 3) divide the line segment joining the points (–3, 5) and (4, –9).*

SOLUTION. Let the required ratio be $k : 1$. 
Compairing x-coordinate

$$\frac{k \times 4 + 1 \times (-3)}{k+1} = -2$$

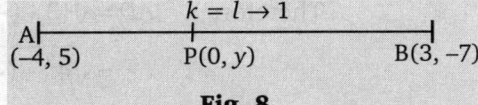

$\Rightarrow \qquad \dfrac{4k-3}{k+1} = -2 \qquad \Rightarrow \qquad 4k - 3 = -2k - 2$

$\Rightarrow \qquad\qquad 6k = 1 \qquad \Rightarrow \qquad k = 1/6$

Compairing y-coordinate

$$\frac{k \times (-9) + (1) \times 5}{k+1} = 3$$

$\Rightarrow \qquad \dfrac{-9k+5}{k+1} = 3 \qquad \Rightarrow \qquad -9k + 5 = 3k + 3$

$\Rightarrow \qquad\qquad 12k = 2 \qquad \Rightarrow \qquad k = 1/6$

Hence, the required ratio is 1 : 6.

EXAMPLE 3. *In what ratio does the y-axis divide the line segment joining the point (–4, 5) and (3, –7)?*

SOLUTION. The line segment joining the points $A(-4, 5)$ and $B(3, -7)$ is divided by the y-axis at the point $P(0, y)$ in the ratio $k : 1$.

A $\vdash$ $k = l \rightarrow 1$ $\dashv$
(-4, 5) $P(0, y)$ B(3, -7)

Fig. 8

Compairing x-coordinate, we have

$$\frac{k \times 3 + 1 \times (-4)}{k+1} = 0 \qquad \Rightarrow \qquad 3k - 4 = 0$$

$\Rightarrow \qquad\qquad 3k = 4 \qquad \Rightarrow \qquad k = 4/3.$

Therefore, the required ratio is 4 : 3.

EXAMPLE 4. *Find the ratio in which the line 3x + y – 9 = 0 divides the line segment joining A(1, 3) and B(2, 7).*

SOLUTION. The equation of the given line is
$$3x + y - 9 = 0 \qquad ...(1)$$
meets the line segment joining $A(1, 3)$ and $B(2, 7)$ at the point $P(x, y)$ and divides the segment internally in the ratio $k : 1$.

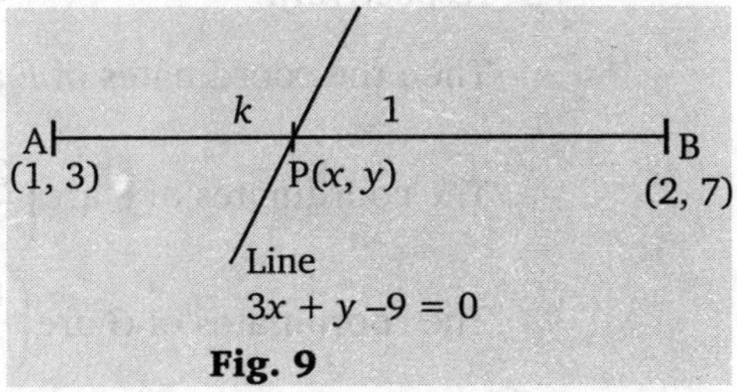

By section formula :

$$x = \frac{2k+1}{k+1}, \, y = \frac{7k+3}{k+1}$$

A $\vdash$ k / 1 $\dashv$ B
(1, 3) $P(x, y)$ (2, 7)

Line
$3x + y - 9 = 0$

Fig. 9

i.e., Coordinate of P are $\left(\dfrac{2k+1}{k+1}, \dfrac{7k+3}{k+1} \right)$

The point P lies on the line whose equation is given by (1)

Therefore, $\left(\dfrac{2k+1}{k+1} \right) + \left(\dfrac{7k+3}{k+1} \right) = 0.$

$\Rightarrow \qquad 6k + 3 + 7k + 3 - 9k - 9 = 0$

$\Rightarrow \qquad\qquad\qquad 4k - 3 = 0$

$\Rightarrow \qquad\qquad\qquad\qquad k = 3/4$

Hence, the required ratio is 3 : 4.

EXAMPLE 5. *Find the coordinates of point which divides the line joining the point* (1, 2) *and* (−3, 4) *in the ratio* 2 : 3 *internally.* **[RGPV(B. Pharma)–2005]**

SOLUTION. Let $A(1, 2)$ and $B(−3, 4)$ be the given points.

Point P divides A and B in the ratio 2 : 3

$\Rightarrow \qquad\qquad PA : PB = 2:3$

$\therefore \quad$ Coordinates of P are $\left(\dfrac{2 \times -3 + 3 \times 1}{2+3}, \dfrac{2 \times 4 + 3 \times 2}{2+3} \right)$

i.e., $\left(\dfrac{-6+3}{5}, \dfrac{8+6}{5} \right)$, *i.e.,* $\left(\dfrac{-3}{5}, \dfrac{14}{5} \right)$.

EXAMPLE 6. *If* $A(−1, −3)$, $B(1, −1)$ *and* $C(5, 1)$ *are the vertices of a triangle ABC, find the length of median through A.* **[UPTU(B. Pharma)–2005]**

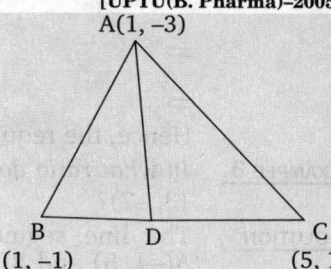

A(1, −3)

B (1, −1)   D   C (5, 1)

Fig. 10

SOLUTION. Let D be the middle point of BC.

Then AD is the median through A and D is the

point $\left(\dfrac{1+5}{2}, \dfrac{-1+1}{2} \right)$, *i.e.,* (3, 0)

Therefore, $\quad AD = \sqrt{\{3-(-1)\}^2 + \{0-3\}^2}$

$= \sqrt{4^2 + (-3)^2} = 5.$

EXAMPLE 7. *A quadrilateral has the vertices at the points* (−4, 2), (2, 6), (8, 5) *and* (9, −7). *Show that the mid-points of the sides of this quadrilateral are the vertices of a parallelogram.* **[UPTU(B. Pharma)–2004]**

SOLUTION. Let $ABCD$ be the given quadrilateral with vertices $A(−4, 2)$, $B(2, 6)$, $C(8, 5)$ and $D(9, −7)$. Let E, F, G and H be the mid-points of the sides AB, BC, CD and DA respectively.

Then the coordinates of E are $\left(\dfrac{-4+2}{2}, \dfrac{2+6}{2} \right)$, *i.e.,* (−1, 4).

The coordinates of F are $\left(\dfrac{2+8}{2}, \dfrac{6+5}{2} \right)$, *i.e.,* $\left(5, \dfrac{11}{2} \right)$.

The coordinates of G are $\left(\dfrac{8+9}{2}, \dfrac{5-7}{2} \right)$, *i.e.,* $\left(\dfrac{17}{2}, -1 \right)$.

and the coordinates of H are $\left(\dfrac{-4+9}{2}, \dfrac{2-7}{2} \right)$, *i.e.,* $\left(\dfrac{5}{2}, \dfrac{-5}{2} \right)$.

Now, the coordinates of the mid-point of EG are $\left(\dfrac{-1 + \dfrac{17}{2}}{2}, \dfrac{4-1}{2} \right)$, *i.e.,* $\left(\dfrac{15}{4}, \dfrac{3}{2} \right)$.

and the coordinate of the mid-point of FH are $\left(\dfrac{5 + \dfrac{5}{2}}{2}, \dfrac{\dfrac{11}{2} - \dfrac{5}{2}}{2} \right)$, *i.e.,* $\left(\dfrac{15}{4}, \dfrac{3}{2} \right)$

Thus, we see that the diagonals *EG* and *FH* of the quadrilateral *EFGH* bisect each other. Hence, *EFGH* is a parallelogram.

EXAMPLE 8. *Three consecutive vertices of a parallelogram are A(1, 2), B(1, 0) and C(4, 0). Find the fourth vertex D.*

SOLUTION. Let the coordinates of the vertex *D* be (x, y). Diagonals *AC* and *BD* of the parallelogram *ABCD* bisect each other at *M*, *i.e.*, *M* is mid-point of *AC* as well as of *BD*.

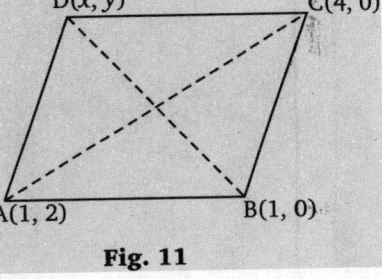

Mid-point of $AC = \left(\dfrac{1+4}{2}, \dfrac{2+0}{2}\right) = \left(\dfrac{5}{2}, 1\right)$ ...(1)

Mid-point of $BD = \left(\dfrac{x+1}{2}, \dfrac{y+0}{2}\right)$ ...(2)

Fig. 11

Eq. (1) and (2) have same coordinates of the point *M*.

Then $\dfrac{x+1}{2} = \dfrac{5}{2}$ and $\dfrac{y}{2} = 1$

$\Rightarrow$ $x = 4$ and $y = 2$. Therefore, the coordinates of *D* are (4, 2).

Exercise 11.2

1. Find the coordinates of the point which divides the line segment joining the points $A(4, -3)$ and $B(9, 7)$ in the ratio 3 : 2.

2. Find the coordinates of the mid-point of the line segment joining the points $A(-5, 4)$ and $B(7, -8)$.

3. Find the ratio in which the point $P(m, 6)$ divides the joining $A(-4, 3)$ and $B(2, 8)$. Also, find the value of m.

4. In what ratio does point $P(2, -5)$ divide the line segment joining $A(-3, 5)$ and $(4, -9)$.

5. In what ratio is the line segment joining the point $A(6, 3)$ and $B(-2, -5)$ divide by the *x*-axis.
   Also, find the coordinates of the point of intersection of *AB* and the *x*-axis.

6. Find the ratio in which the *y*-axis divides the line segment joining the points $A(-4, 10)$ and $B(7, -1)$.
   Also, find the coordinates of their point of intersection.

7. The coordinates of one end point of diameter *AB* of a circle are $A(4, -1)$ and the coordinates of the centre of the circle are $C(1, -3)$. Find the coordinates of *B*.

8. The three vertices of a parallelogram *ABCD*, taken in order are $A(1, -2)$, $B(3, 6)$ and $C(5, 10)$. Find the coordinates of the fourth vertex *D*.

9. Find the lengths of the medians of a $\triangle ABC$ whose vertices are $A(7, -3)$, $B(5, 3)$ and $C(3, -1)$.

10. Let $D(3, -2)$, $E(-3, 1)$ and $F(4, -3)$ be the mid-points of the sides *BC*, *CA* and *AB* respectively of $\triangle ABC$. Then, find the coordinates of the vertices *A*, *B* and *C*.

ANSWERS

1. (7, 3), (3, 5)   2. (1, –2)   3. $m = \dfrac{-2}{5}$ 4. 5 : 2

5. (3, 0)   6. 4 : 7, (0, 6)   7. (2, –5)   8. (3, 2)

9. $5, 5, \sqrt{10}$ 10. $A(-2, 0), B(10, -6), C(-4, 2)$

11.6 AREA OF A TRIANGLE

THEOREM. *The Area of a $\triangle ABC$ with vertices $A(x_1, y_1)$, $B(x_2, y_2)$ and $C(x_3, y_3)$ is given by*

$$\text{area }(\triangle ABC) = \left| \frac{1}{2}[x_1(y_2 - y_3) + x_2(y_3 - y_1) + x_3(y_1 - y_2)] \right|$$

PROOF. Let $A(x_1, y_1)$, $B(x_2, y_2)$ and $C(x_3, y_3)$ be the vertices of the given $\triangle ABC$. Draw AL, BM and CN perpendicular to the x-axis.

Then $ML = (x_1 - x_2)$, $LN = (x_3 - x_1)$ and $MN = (x_3 - x_2)$

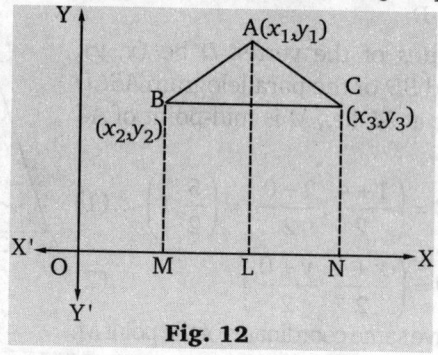

Fig. 12

Area of $\triangle ABC$

$$= \text{area (trap. } BMLA) + \text{area (trap. } ALNC) - \text{area (trap. } BMNC)$$

$$= \left[\frac{1}{2}(AL + BM) \times ML \right] + \left[\frac{1}{2}(AL + CN) \times LN \right] - \left[\frac{1}{2}(BM + CN) \times MN \right]$$

$$= \frac{1}{2}(y_1 + y_2)(x_1 - x_2) + \frac{1}{2}(y_1 + y_3)(x_3 - x_1) - \frac{1}{2}(y_2 + y_3)(x_3 - x_2)$$

$$= \frac{1}{2}[x_1(y_1 + y_2 - y_1 - y_3) + x_2(y_2 + y_3 - y_1 - y_2) + x_3(y_1 + y_3 - y_2 - y_3)]$$

$$= \frac{1}{2}[x_1(y_2 - y_3) + x_2(y_3 - y_1) + x_3(y_1 - y_2)]$$

Since, the area is never negative. Therefore,

$$\text{area}(\triangle ABC) = \frac{1}{2}[x_1(y_2 - y_3) + x_2(y_3 - y_1) + x_3(y_1 - y_2)]$$

REMARKS

➡ The area of a triangle is always taken as positive real quantity. Some times the result from the area formula given negative value in that case we reject the negative sign.

➡ The three points (x_1, y_1), (x_2, y_2) and (x_3, y_3) are collinear, i.e., in a line if $x_1(y_2 - y_3) + x_2(y_3 - y_1) + x_3(y_1 - y_2) = 0$ because in this situation the area of the triangle reduces to zero.

SOLVED EXAMPLES

EXAMPLE 1. Find the area of the triangle whose vertices are (4, 3), (5, 4) and (11, 2).

SOLUTION. Let $A(4, 3)$, $B(5, 4)$ and $C(11, 2)$ be the three vertices of the given triangle.

$$x_1 = 4, x_2 = 5, x_3 = 11$$
$$y_1 = 3, y_2 = 4, y_3 = 2$$

$$\text{Area of } \triangle ABC = \frac{1}{2}[x_1(y_2 - y_3) + x_2(y_3 - y_2) + x_3(y_1 - y_2)]$$

$$= \frac{1}{2}[4(4 - 2) + 5(2 - 3) + 11(3 - 4)] = \frac{1}{2}[8 - 5 - 11] = -4$$

Rejecting negative sign, we have the area of the given triangle equal to 4 square units.

EXAMPLE 2. *Find the value of k so that the point A(−2, 3), B(3, −1) and C(5, k) be collinear.*

SOLUTION. We have $x_1 = -2, x_2 = 3, x_3 = 5$
$y_1 = 3, y_2 = -1, y_3 = k.$

Area of $\triangle ABC = \dfrac{1}{2}[x_1(y_2 - y_3) + x_2(y_3 - y_1) + x_3(y_1 - y_2)]$

$= \dfrac{1}{2}[-2(-1 - k) + 3(k - 3) + 5(3 + 1)]$

$= \dfrac{1}{2}[-2 + 2k + 3k - 9 + 20 = \dfrac{1}{2}[5k + 9]$

Now, the three points are collinear if the area of $\triangle ABC = 0$

i.e., $\dfrac{1}{2}(5k + 9) = 0$

$5k + 9 = 0$

$k = -9/5.$

EXAMPLE 3. *Find the area of the quadrilateral ABCD whose vertices are respectively A(1, 1), B(7, −3), C(12, 2) and D(7, 21).*

SOLUTION. Area of quadrilateral $ABCD = |\text{Area of } (\triangle ABC)| + |\text{Area of } \triangle ACD|$

Now Area of $\triangle ABC = \dfrac{1}{2}|1 \times (-3 - 2) + 7(2 - 1) + 12 \times (1 + 3)| = \dfrac{1}{2}|-5 + 7 + 48|$

$= 25 \text{ sq units}$

Area of $\triangle ACD = \dfrac{1}{2}|1 \times (2 - 21) + 12(21 - 1) + 7(1 - 2)|$

$= \dfrac{1}{2}|-19 + 240 - 7| = 107 \text{ sq. units.}$

∴ Area of quadrilateral $ABCD = 25 + 107 = 132 \text{ sq. units.}$

EXAMPLE 4. *For what value of k the points (k, 2 − 2k), (−k + 1, 2k) and (− 4 − k, 6 − k) are collinear.*

SOLUTION. Let the three points be $A(x_1, y_1) = (k, 2 - 2k), B(x_2, y_2) = (-k + 1, 2k)$ and $C(x_3, y_3) = (-4 - k, 6 - 2k).$
If the given points are collinear, then

$$x_1(y_2 - y_3) + x_2(y_3 - y_1) + x_3(y_1 - y_2) = 0$$

$\Rightarrow k(2k - 6 + 2k) + (-k + 1)(6 - 2k - 2 + 2k) + (-4 - k)(2 - 2k - 2k) = 0$

$\Rightarrow \qquad\qquad k(4k - 6) - 4(k - 1) + (4 + k)(4k - 2) = 0$

$\Rightarrow \qquad\qquad 4k^2 - 6k - 4k + 4 + 4k^2 + 14k - 8 = 0$

$\Rightarrow \qquad\qquad\qquad\qquad 8k^2 + 4k - 4 = 0$

$\Rightarrow \qquad\qquad\qquad\qquad 2k^2 + k - 1 = 0$

$\Rightarrow \qquad\qquad\qquad\qquad (2k - 1)(k + 1) = 0$

$\Rightarrow \qquad\qquad k = 1/2 \text{ or } k = -1.$

Hence, the given points are collinear for $k = 1/2$ or $k = -1.$

EXAMPLE 5. *If the vertices of a triangle have integral coordinates prove that the triangle cannot be equilateral.*

SOLUTION. Let $A(x_1, y_1), B(x_2, y_2)$ and $C(x_3, y_3)$ be the vertices of triangle ABC, then the area of $\triangle ABC$ is given by

$$\Delta = \dfrac{1}{2}[x_1(y_2 - y_3) + x_2(y_3 - y_1) + x_3(y_1 - y_2)]$$

$= \text{A rational number}$

If possible let the triangle ABC be an equilateral triangle, then its area is given by

$$\Delta = \frac{\sqrt{3}}{4}(\text{side})^2 = \frac{\sqrt{3}}{4}(AB)^2$$

$$= \frac{\sqrt{3}}{4} \times \text{a positive number}$$

$$= \text{an irrational number}$$

This is a contradiction to the fact that the area is a rational number. Hence, the triangle cannot be equilateral.

EXAMPLE 6. *Prove that the following points are collinear :* $(-3, 0)$, $(0, -9)$ and $(-2, -3)$.

[RGPV(B. Pharma)–2004]

SOLUTION. The given points will be collinear if the area of the triangle formed by these points is zero. Now area of the triangle

$$= \frac{1}{2}[x_1(y_2 - y_3) + x_2(y_3 - y_1) + x_3(y_1 - y_2)]$$

$$= \frac{1}{2}[-3(-9 + 3) + 0(3 - 0) + (-2)(0 - (-9))]$$

$$= \frac{1}{2}[18 - 18] = 0$$

EXAMPLE 7. *The co-ordinates of vertices B and C of triangle are* $(1, -2)$, $(2, 4)$ *lies on the line* $2x + y - 2 = 0$. *The area of the triangle is 8 units. Then find the vertices coordinates of A.*

[RGPV(B. Pharma)–2001]

SOLUTION. Given points are $B(1, 2)$, $C(2, 3)$ and $A(x, y)$ lie on the line $2x + y - 2 = 0$
The coordinate of A are $(x, 2 - 2x)$

$$\text{Area of } \Delta ABC = \frac{1}{2}[x_1(y_2 - y_3) + x_2(y_3 - y_1) + x_3(y_1 - y_2)]$$

$$\pm 8 = \frac{1}{2}[x(-2 - 3) + 1(2 - (2 - 2x)) + 2(2 - 2x + 2)]$$

$$\pm 16 = [-5x + 1 + 2x + 8 - 4x]$$

$$\pm 16 = -7x + 9$$

$$16 - 9 = -7x \text{ (taking positive sign)}$$

$$7 = -7x$$

$$x = -1 \text{ put } x = -1 \text{ in } y = 2 - 2x \text{ and get } y = 2 + 2 = 4.$$

∴   Coordinate of A are $(-1, 4)$.

Taking negative sign $-16 = -7x + 9$

$$-16 - 9 = -7x \quad \Rightarrow \quad x = \frac{25}{7}$$

Put $x = 25/7$ in $y = 2 - 2x$

We get $y = 2 - 2 \times \frac{25}{7} = \frac{14 - 50}{7} = \frac{36}{9}$

∴   Coordinate of A are $\left(\frac{25}{7}, \frac{36}{7}\right)$ or $(-1, 4)$.

EXAMPLE 8. *Find the area of the triangle whose vertices are* $(0, 5)$, $(2, 3)$ *and* $(4, 5)$.

[UPTU(B. Pharma)–2002]

SOLUTION. Here $x_1 = 0$, $y_1 = 5$, $x_2 = 2$, $y_2 = 3$, $x_3 = 4$, $y_3 = 5$.

$\therefore$ Area of the triangle

$$= \frac{1}{2}[(x_1 y_2 + x_2 y_3 + x_3 y_1) - (y_1 x_2 + y_2 x_3 + y_3 x_1)]$$

$$= \frac{1}{2}[(0 \times 3 + 2 \times 5 + 4 \times 5) - (-5 \times 2 + 3 \times 4 + 5 \times 0)]$$

$$= \frac{1}{2}[(0 + 10 + 20) - (10 + 12 + 0)]$$

$$= \frac{1}{2}[30 - 22] = \frac{1}{2} \times 8 = 4 \text{ sq. units}$$

EXAMPLE 9.   *Prove that the points $(a, b + c)$, $(b, c + a)$, $(c, a + b)$ are collinear.*

                                                          [UPTU(B. Pharma)–2001, 07]

SOLUTION.   Here $x_1 = a, y_1 = b + c, y_2 = c + a, x_3 = c, y_3 = a + b, x_2 = b$

Now area of the triangle formed by the given points

$$= \frac{1}{2}[(x_1 y_2 + x_2 y_3 + x_3 y_1) - (y_1 x_2 + y_2 x_3 + y_3 x_1)]$$

$$= \frac{1}{2}[\{a(c + a) + b(a + b) + c(b + c)\}$$

$$\qquad - \{(b + c)b + (c + a)c + (a + b)a\}]$$

$$= \frac{1}{2}[ac + a^2 + ab + b^2 + bc + c^2 - b^2$$

$$\qquad - bc + c^2 - ac - a^2 - ab]$$

$$= 0$$

Hence, the given points are collinear.

EXAMPLE 10.   *Find the area of a triangle formed by the lines:*

$$y = 2x, y = x \text{ and } y = 3x + 4.$$ **(UPTU(B. Pharma)–2001)**

SOLUTION.   Let the equations of the sides AB, BC and CA of $\triangle ABC$ be $y - x = 0$, $y - 2x = 0$ and $y - 3x - 4 = 0$ respectively.

Solving these equations in pairs, the coordinates of A, B and C are $(-2, -2)$, $(0, 0)$ and $(-4, -8)$ respectively.

$$\text{Area of } \triangle ABC = \frac{1}{2}[x_1(y_2 - y_3) + x_2(y_3 - y_1) + x_3(y_1 - y_2)]$$

$$= \frac{1}{2}[(-2)(0 + 8) + 0.(-8 + 2) + (-4)(-2 - 0)]$$

$$= \frac{1}{2}[-16 + 8]$$

$$= \frac{1}{2}[-8] = 4 \text{ sq. units, neglecting the negative sign.}$$

EXAMPLE 11.   *Four points $A(6, 3)$, $B(-3, 5)$, $C(4, -2)$ and $D(x, 3x)$ are given in such a way that*

$$\frac{\triangle DBC}{\triangle ABC} = \frac{1}{2}, \text{ find } x.$$ **[UPTU(B.Pharma)–2006]**

SOLUTION. $$\frac{\text{Area of } \triangle DBC}{\text{Area of } \triangle ABC} = \frac{\frac{1}{2}[x(5 + 2) - 3(-2 - 3x) + 4(3x - 5)]}{\frac{1}{2}[6(5 + 2) - 3(-2 - 3) + 4(3 - 5)]}$$

$$\frac{1}{2} = \frac{7x + 6 + 9x + 12x - 20}{42 + 15 - 8}$$

$$\frac{1}{2} = \frac{28x - 14}{49}$$

$$\Rightarrow \qquad \frac{1}{2} = \frac{4x - 2}{7}$$

$$\Rightarrow \qquad \frac{7}{2} = 4x - 2$$

$$\Rightarrow \qquad x = \frac{11}{8}$$

Exercise 11.3

1. Find the area of the triangle whose vertices are $A(2, 7)$, $B(3, -1)$ and $C(-5, 6)$.
2. Find the value of k for which the area formed by the triangle with vertices $A(k, 2k)$, $B(-2, 6)$, $C(3, 1)$ is 5 square units.
3. Show that the points $A(-1, 1)$, $B(5, 7)$ and $C(8, 10)$ are collinear.
4. For what value of k the points $A(1, 5)$, $B(k, 1)$ and $C(4, 11)$ are collinear ?
5. If the vertices of a triangle are $A(1, k)$, $B(4, -3)$ and $C(-9, 7)$ and its area is 15 sq. units, then find the value of k.

ANSWERS

1. 28.5 sq. units   **2.** $k = 2, k = 2/3$ **4.** $k = -1$ **5.** $k = -3$ or $k = \dfrac{21}{13}$

11.7 LOCUS AND EQUATION TO A LOCUS

11.7.1 LOCUS

The curve described by a point which moves under given condition or conditions is called its locus.

For example :

(i) Suppose C is a point in the plane of the paper and P is a variable point in the plane of the paper such that its distance from C is always equal to r(say). Obviously all the positions of the moving point P lie on the circumference of a circle whose radius is r. The circumference of this circle is therefore the Locus of the point P when it moves under the condition that its distance from point C is always equal to constant r.

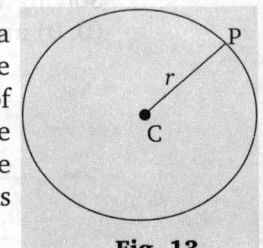

Fig. 13

11.7.2 EQUATION OF THE LOCUS OF A POINT

The equation of the locus of a point is the relation which is satisfied by the coordinates of every point on the locus of the point.

WORKING PROCEDURE

Step 1. Assume the coordinates of the point say (h, k) whose locus is to be found.

Step 2. Write the given condition in mathematical form involving h, k.

Step 3. Eliminate the variables, if any.

Step 4. Replace h by x and k by y in the result obtained in step-3. The equation so obtained is the locus of the point which moves under some stated conditions.

SOLVED EXAMPLES

EXAMPLE 1. *Find the locus of a point P such that the sum of the squares of abscissa and ordinate is equal to the product of abscissa and ordinate.*

SOLUTION. Let $P(h, k)$ be any point on the locus.

∴ h is abscissa and k is ordinate of P.

By the given geometrical condition, we get

$$h^2 + k^2 = hk$$

Hence, locus of (h, k) is $x^2 + y^2 = xy$.

EXAMPLE 2. *Find the equation to the locus of a point equidistant from the points $A(1, 3)$ and $B(-2, 1)$.*

SOLUTION. Let $P(h, k)$ be any point on the locus. Then

$$PA = PB \text{ (given)}$$

$$\Rightarrow \qquad PA^2 = PB^2$$

$$\Rightarrow \qquad (h - 1)^2 + (k - 3)^2 = (h + 2)^2 + (k - 1)^2$$

$$\Rightarrow \qquad 6h + 4k = 5$$

Hence, locus (h, k) is $6x + 4y - 5$.

EXAMPLE 3. *A point moves so that the sum of its distances from $(ae, 0)$ and $(-ae, 0)$ is $2a$, prove that the equation to its locus is*

$$\frac{x^2}{a^2} + \frac{y^2}{b^2} = 1 \text{ where } b^2 = a^2(1 - e^2).$$

SOLUTION. Let $P(h, k)$ be the moving point such that the sum of its distance from $A(ae, 0)$ and $B(-ae, 0)$ is $2a$.

Then $\qquad PA + PB = 2a$

$$\Rightarrow \quad \sqrt{(h - ae)^2 + (k - 0)^2} + \sqrt{(h + ae)^2 + (k - 0)^2} = 2a$$

$$\Rightarrow \quad \sqrt{(h - ae)^2 + k^2} = 2a - \sqrt{(h + ae)^2 + k^2}$$

$$\Rightarrow \quad (h - ae) + k^2 = 4a^2 + (h + ae)^2 + k^2 - 4a\sqrt{(h + ae)^2 + k^2}$$

$$\text{(squaring both sides)}$$

$$\Rightarrow \qquad -4aeh - 4a^2 = -4a\sqrt{(h + ae)^2 + k^2}$$

$$\Rightarrow \qquad (eh + a) = \sqrt{(h + ae)^2 + k^2}$$

$$\Rightarrow \qquad (eh + a)^2 = (h + ae)^2 + k^2$$

$$\Rightarrow \quad e^2h^2 + a^2 + 2aeh = h^2 + a^2e^2 + 2aeh + k^2$$

$$\Rightarrow \qquad h^2(1 - e^2) + k^2 = a^2(1 - e^2)$$

$$\Rightarrow \qquad \frac{h^2}{a^2} + \frac{k^2}{a^2(1 - e^2)} = 1$$

Hence, locus of (h, k) is given by

$$\frac{h^2}{a^2} + \frac{k^2}{a^2(1 - e^2)} = 1$$

or $\qquad \dfrac{x^2}{a^2} + \dfrac{y^2}{b^2} = 1$, where $b^2 = a^2(1 - e^2)$

EXAMPLE 4. *A rod of length l slides with its ends on two perpendicular lines, find the locus of its mid-point.*

SOLUTION. Let the two perpendicular lines be the coordinate axes. Let AB be a rod of length l. Let the coordinate of A and B be $(a, 0)$ and $(0, b)$ respectively. As the rod slides, the values of a and b change. So a and b are two variables.

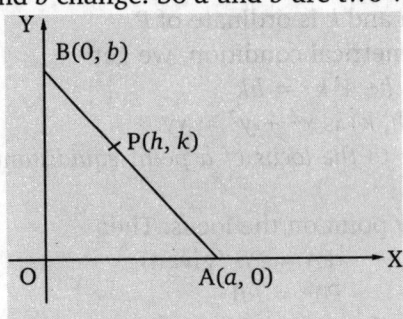

Fig. 14

Let $P(h, k)$ be the mid point of the rod AB in one of the infinite position it attains. Then

$$h = \frac{a+0}{2} \text{ and } k = \frac{0+b}{2} \Rightarrow h = \frac{a}{2} \text{ and } k = \frac{b}{2} \qquad \ldots (1)$$

From $\triangle OAB$, we have $AB^2 = OA^2 + OB^2$
$$= a^2 + b^2 = l^2$$
$$\Rightarrow \qquad (2h)^2 + (2k)^2 = l^2 \qquad \text{(From (1))}$$
$$\Rightarrow \qquad 4h^2 + 4k^2 = l^2$$

Hence, the locus of (h, k) is $4x^2 + 4y^2 = l^2$.

EXAMPLE 5. *If O is the origin and Q is a variable point on $x^2 = 4y$. Find the locus of the mid-point of OQ.*

SOLUTION. Let the coordinates of Q be (a, b) and let $P(h, k)$ be the mid-point of OQ. Then

$$h = \frac{a+0}{2} = \frac{a}{2} \text{ and } k = \frac{0+b}{2} = \frac{b}{2} \Rightarrow a = 2h \text{ and } b = 2k. \qquad \ldots (1)$$

Here a and b are two variables which are to be eliminated. Since (a, b) lies on
$$x^2 = 4y$$
Therefore, $\qquad a^2 = 4b \quad \Rightarrow \quad (2h)^2 = 4(2k)$
$$\Rightarrow \qquad h^2 = 2k \qquad \text{[using (1)]}$$

Hence, the locus of (h, k) is $x^2 = 2y$.

EXAMPLE 6. *A point moves so that its distance from $(3, 0)$ is twice the distance from $(-3, 0)$. Find the equation of the locus.* **[UPTU(B.Pharma)–2005]**

SOLUTION. Let A represent the point $(3, 0)$, B the point $(-3, 0)$. Further, point $(-3, 0)$ and $P(h, k)$ be the moving point.

According to the question:
$$PA = 2PB$$
or $\qquad (PA)^2 = 4(PB)^2$
$$[(h-3)^2 + (k-0)^2] = 4[(h+3)^2 + (k-0)^2]$$
$$\Rightarrow \qquad h^2 + 9 - 6h + k^2 = 4h^2 + 36 + 24h + 4k^2$$
$$\Rightarrow \qquad 3h^2 + 3k^2 + 30h + 27 = 0$$

Hence, the required locus is $3x^2 + 3y^2 + 30x + 27 = 0$.

EXAMPLE 7. *Find the locus of a point such that the line segments having end points (2, 0) and (–2, 0) subtend a right angle at that point.* **[UPTU(B.Pharma)–2006]**

SOLUTION. Let $A(2, 0)$ and $B(–2, 0)$ be the given points and $P(h, k)$ be the variable point.

According to the question

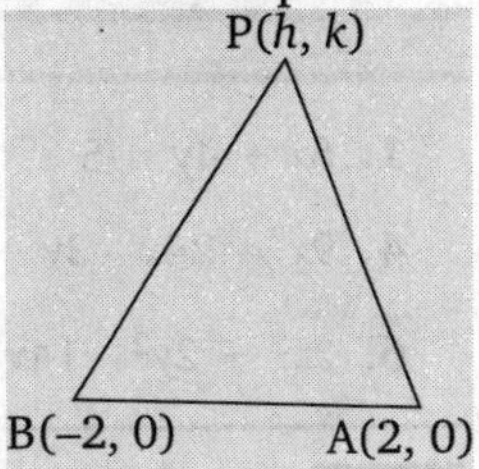

Fig. 15

$$\angle APB = 90°$$

$\therefore$ *i.e.,* $\triangle APB$ is a right angle.

$$AB^2 = PA^2 + PB^2$$

$$[2 – (–2)]^2 + [0 – 0]$$

$$= [(2 – h)^2 + (0 – k)^2]$$
$$+ [(–2 – h)^2 + (0 – k)^2]$$

$$16 = (2 – h)^2 + k^2 + (–2 – h)^2 + k^2$$

$$16 = 4 + h^2 – 4h + 2k^2 + 4 + h^2 + 4h$$

$$16 = 2h^2 + 2k^2 + 8$$

$$h^2 + k^2 = 4.$$

Hence, the required locus is $x^2 + y^2 = 4$.

EXAMPLE 8. *Find the equation to the locus of a point which moves so that the sum of its distance from (3, 0) and (–3, 0) is less then 9.* **[UPTU(B.Pharma)–2003, 04]**

SOLUTION. Let $A(3, 0)$ and $B(–3, 0)$ be the two given points and (h, k) be the coordinates of the moving point P whose locus is to be found. According to the question

$$PA + PB < 9$$

or $\qquad \sqrt{(h-3)^2 + k^2} + \sqrt{(h+3)^2 + k^2} < 9$

$$\sqrt{(h-3)^2 + k^2} < \left\{ 9 - \sqrt{(h+3)^2 + k^2} \right\}$$

Squaring both sides, we get

$$(h-3)^2 + k^2 < 81 - 18\sqrt{(h+3)^2 + k^2} + (h+3)^2 + k$$

$$-12h - 81 < -18\sqrt{(h+3)^2 + k^2}$$

or $\qquad 4h + 27 > 6\sqrt{(h+3)^2 + k^2}$

Again squaring both the sides, we get

$$16h^2 + 729 + 216h > 36[h^2 + 9 + 6h + k^2]$$

$$20h^2 + 36k^2 < 405$$

The required locus of the point (h, k) is $20x^2 + 36y^2 < 405$.

Exercise 11.4

1. Find the equation to the locus of a point equidistant from the points $A(1, 3)$ and $B(–2, 1)$.

2. $A = (2, –2)$, B is a point on the locus $y^2 = 4x$. Find the equation of the locus of a point which divides AB internally in the ratio 1:2.

3. $A = (–3, 0)$ and $B(3, 0)$ are two given points. Find the equation of the locus of point P such that $PA + PB = 10$.

4. If B is $(4, –3)$, $C(0, 2)$ and point A lies on the locus $y = 1 + x$. Then find the equation of locus of centroid of triangle ABC.

5. Find the equation of the circle having segment AB as a diameter where $A(–1, 4)$ and $B(3, –2)$.

6. Find the equation of the locus of a point such that the sum of the squares of its distance from $(5, –3)$ and $(2, –2)$ is 20.

7. Let $A = (0, 5)$ and $B(0, -5)$. Find the equation of the locus of point P such that $PA - PB = 4$.

8. $A(1, 2)$ and $B(4, -5)$ are two vertices of $\triangle ABC$. Find the locus of the third vertex C if the centroid lie on the locus $2x + 3y = 11$.

ANSWERS

1. $6x + 4y = 5$

2. $9y^2 - 12x + 24y + 32 = 0$

3. $\dfrac{x^2}{25} + \dfrac{y^2}{16} = 1$

4. $9x^2 - 24x - 3y + 16 = 0$

5. $x^2 + y^2 - 2x - 2y - 11 = 0$

6. $2x^2 + 2y^2 - 14x + 8y + 19 = 0$

7. $\dfrac{-x^2}{21} + \dfrac{y^2}{4} = 1$

8. $2x + 3y = 32$

11.8 THE STRAIGHT LINES

A straight line is the locus of all those points which are collinear with two given points. Since, we know that one and only one line can be drawn from any two given points. So straight line is a curve such that every point on the line segment joining any two points on it lies on it.

REMARKS

➡ Every first degree equation in x, y represents a straight line.

➡ The x-axis and all lines parallel to it are called horizontal lines.

➡ The y-axis and all lines parallel to it are called vertical lines.

11.9 SLOPE OR GRADIENT OF A LINE

Geometrically, tangent of the angle that a line makes with the positive direction of the x-axis is anticlockwise sense is called the slope or gradient of the line.

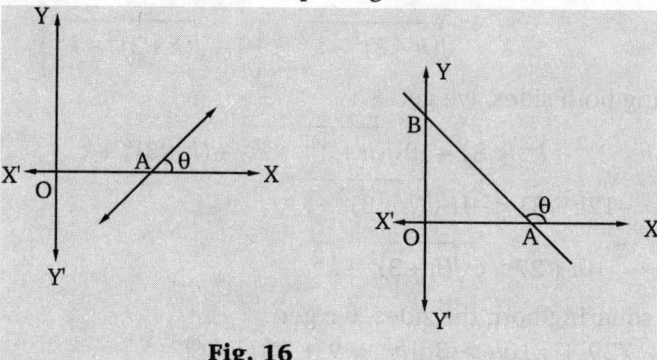

Fig. 16

The slope of a line is generally denoted by m. Thus $m = \tan \theta$.

REMARKS

➡ Slope of any line parallel to x-axis is zero.

➡ Slope of any line parallel to y-axis, i.e., perpendicular to x-axis is infinite. Thus, not defined.

➡ Slope of the general equation of line $ax + by + c = 0$ is

$$m = -\frac{a}{b}, = -\frac{\text{coefficient of } x}{\text{coefficient of } y}.$$

➡ Slope of a line equally inclined with axes is 1 or -1 as it makes $45°$ or $135°$ angle with x-axis.

➡ The angle of inclination of a line with the positive direction of x-axis is anticlockwise sense always lies between $0°$ and $180°$.

➡ When $m < 0$, $\tan \theta < 0$, i.e., $\theta > 90°$. Therefore, the angle of inclination is obtuse.

➡ When $m = 0$, $\tan \theta = 0$, i.e., $\theta = 0°$. Therefore, the line is parallel to x-axis.

➡ When $m > 0$, $\tan \theta > 0$, i.e., $\theta < 90°$. Therefore, the angle of inclination is acute.

11.10 SLOPE OF A LINE THROUGH TWO POINTS

Let L be the line through two fixed points $A(x_1, y_1)$ and $B(x_2, y_2)$ respectively. Let us draw perpendicular from A on x and y axis respectively so as to meet at $M(x_1, 0)$ and $P(0, y_1)$. Also through B perpendiculars on x and y axes meet at $N(x_2, 0)$ and $Q(0, y_2)$.

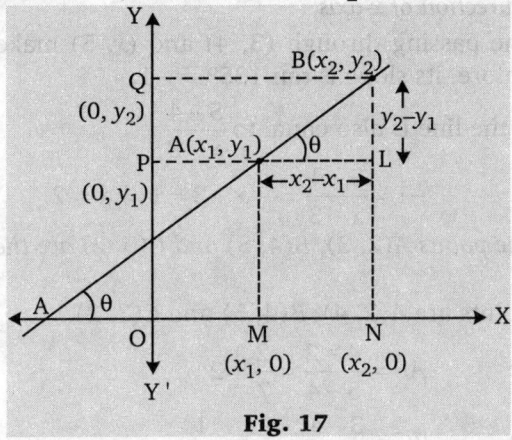

Fig. 17

Now $\qquad MN = (x_2 - x_1) = AC \qquad$ and $\qquad PQ = (y_2 - y_1) = BC$.

Let the line L makes an angle Q with the positive direction of x-axis. Then

$$\tan\theta = \frac{BC}{AC} \Rightarrow \tan\theta = \frac{y_2 - y_1}{x_2 - x_1}$$

$$m = \tan\theta = \frac{y_2 - y_1}{x_2 - x_1}$$

Which is the required slope of the line joining the points $A(x_1, y_1)$ and $B(x_2, y_2)$.

REMARK

⟹ Let $A(x, y)$ and $O = (0, 0) = B$. Then slope of line $AB = \dfrac{y - 0}{x - 0} = \dfrac{y}{x} = $ slope of OA.

11.10.1 CONDITION OF PARALLELISM OF LINES

It two lines of slopes m_1 and m_2 are parallel, then angle θ made by them with positive direction of x-axis are equal.

So $\qquad m_1 = m_2$

11.10.2 CONDITION FOR PERPENDICULARITY OF TWO LINES

If m_1 and m_2 are the slopes of two mutually perpendicular lines

then $\qquad m_1 = -\dfrac{1}{m_2}$ or $m_1 m_2 = -1$

SOLVED EXAMPLES

EXAMPLE 1. *Find the slope of a line which makes an angle of 30° with the positive direction of x-axis.*

SOLUTION. Let m be the slope of the line.

Then $\qquad m = \tan\theta = \tan 30° = \dfrac{1}{\sqrt{3}}$.

EXAMPLE 2. *Find the slope of the line joining A(3, 4) and B(6, 8).*

SOLUTION. Slope of line $AB = \dfrac{y_2 - y_1}{x_2 - x_1} = \dfrac{8-4}{6-3} = \dfrac{4}{3}$.

EXAMPLE 3. *Determine x so that the line passing through (3, 4) and (x, 5) makes 135° angle with the positive direction of x-axis.*

SOLUTION. Since the line passing through (3, 4) and (x, 5) makes an angle of 135° with x-axis. Therefore, its slope is tan 135° = –1.

But slope of the line is also equal to $\dfrac{5-4}{x-3}$.

$\therefore$ $\qquad -1 = \dfrac{5-4}{x-3} \Rightarrow -x + 3 = 1 \Rightarrow x = 2.$

EXAMPLE 4. *Show that the points A(2, 3), B(4, 5) and C(3, 2) are the vertices of a right angled triangle.*

SOLUTION. The given points are A(2, 4), B(4, 5) and C(3, 2).

Slope of line $\qquad AB = \dfrac{4-2}{5-4} = \dfrac{2}{7} = 2$

Slope of line $\qquad BC = \dfrac{3-4}{2-5} = \dfrac{-1}{-3} = \dfrac{1}{3}$

and slope of line $\qquad AC = \dfrac{3-2}{2-4} = \dfrac{-1}{2}$

$\therefore$ (Slope of line AB) × (Slope of line AC) $= 2 \times \left(\dfrac{-1}{2}\right) = -1$

$\Rightarrow$ line AB is perpendicular to line AC.

$\therefore$ ABC is a triangle right angled at A.

EXAMPLE 5. *Find k, if the points (–1, 3), (8, k) and (2, 1) are collinear.*

SOLUTION. The points (–1, 3), (8, k) and (2, 1) are collinear then slope of AB = slope of AC.

Slope of $AB = \dfrac{k-3}{8-(-1)} = \dfrac{k-3}{9}$

and Slope of $AC = \dfrac{1-3}{2-(-1)} = \dfrac{-2}{3}$

$\therefore$ $\qquad \dfrac{k-3}{9} = \dfrac{-2}{3} \Rightarrow 3(k-3) = -18$

$\Rightarrow$ $\qquad 3k - 9 = -18$

$\qquad k = -3$

EXAMPLE 6. *Find angle made by the lines x cos 30° + y sin 30° + sin 120° = 0 with the positive direction of x-axis.* **[UPTU (B.Pharma)–2007]**

SOLUTION. The equation of the given line is

$x \cos 30° + y \sin 30° + \sin 120° = 0$ $\qquad$ ...(1)

$y = \dfrac{-\cos 30° x}{\sin 30°} - \dfrac{\sin 120° x}{\sin 30°}$

$y = -\cot 30° x - \dfrac{\sqrt{3}/2}{1/2}$

$\Rightarrow y = \tan 120° x - \sqrt{3}$, which is the slope intercept form. Hence, the angle made by the given line with the positive direction of x-axis is 120°.

EXAMPLE 7. *Reduce $4x + 3y - 9 = 0$ to the normal form and find the distance (perpendicular distance p) from origin.* **[UPTU(B.Pharma)–2004]**

SOLUTION. We have $4x + 3y - 9 = 0$ or $4x + 3y = 9$

Dividing both sides by $\sqrt{(4)^2 + (3)^2} = 5$, we get

$$\frac{4}{5}x + \frac{3}{5}y = \frac{9}{5}, \text{ which is the normal form.}$$

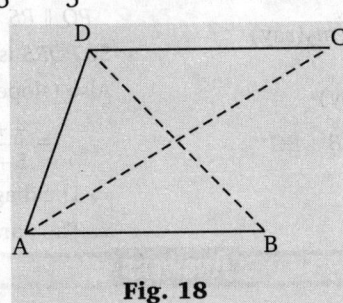

Fig. 18

Hence, the length of perpendicular from the origin to the line is $p = \frac{9}{5}$.

EXAMPLE 8. *Prove that the points $(-1, 0)$, $(3, 1)$, $(2, 2)$ and $(-2, 1)$ are the vertices of a parallelogram.*

SOLUTION. Let $A(-1, 0)$, $B(3, 1)$, $C(2, 2)$ and $D(-2, 1)$ be the vertices of the parallelogram, ABCD taken in order. The mid-points of diagonals AC and BD are

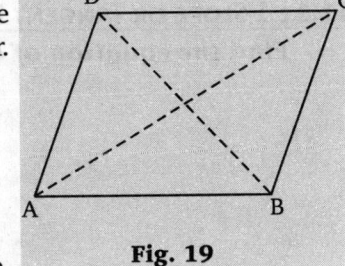

$$AC = \left(\frac{-1+2}{2}, \frac{0+2}{2}\right) = \left(\frac{1}{2}, 1\right)$$

and $$BD = \left(\frac{3-2}{2}, \frac{1+1}{2}\right) = \left(\frac{1}{2}, 1\right).$$

Fig. 19

Since, the mid-points of AC and BD are same. Also, slope of AB × slope of AD ≠ – 1.
and slope of AC × slope of BD ≠ – 1.
Hence, ABCD is a parallelogram.

Exercise 11.5

1. Find the slope of a line whose inclination to the positive direction of x-axis in anticlockwise sense is :
   (i) 60°   (ii) 0°
   (iii) 150°   (iv) 120°
2. Find the slope of the line passing through (2, 3) and (1, 4).
3. Show that the points (1, 3), (2, 5) and (4, 9) are collinear.
4. Let $A(6, 4)$ and $B(2, 12)$ be two given points. Find the slope of a line perpendicular to AB.

5. Determine x so that 2 is the slope of the line through (2, 5) and (x, 3).
6. Without using Pythagoras theorem, show that the points (1, 2), (4, 5) and (6, 3) represent the vertices of a right angle triangle.
7. Show that the points $P(-4, -5)$, $Q(-2, 2)$, $R(5, 4)$ and $S(3, -3)$ are the vertices of a rhombus.
8. Show that the following points represent a rectangle (0, 0), (0, 5), (6, 5), (6, 0).

9. Show that the following points represent a square (3, 2), (0, 5), (–3, 2), (0 – 1).

10. Prove that the lines.

(i) $x + 3y + 4 = 0$ and $2x + 6y – 7 = 0$ are parallel.

(ii) $2x + 3y + 3 = 0$ and $3x – 2y + 5 = 0$ are perpendicular.

HINT TO SELECTED PROBLEMS

6. Let $A(1, 2)$, $B(4, 5)$ and $C(6, 3)$ be the vertices of the given triangle.

Slope of $AB = \dfrac{5-2}{4-1} = \dfrac{3}{3} = 1 = m_1$ (say)

Slope of $BC = \dfrac{3-5}{6-4} = \dfrac{-2}{2} = -1 = m_2$ (say)

Slope of $AC = \dfrac{3-2}{6-1} = \dfrac{1}{5} = m_2$ (say)

$m_1 \times m_2 = 1 \times -1 = -1 \Rightarrow AB \perp BC$

Hence, $\triangle ABC$ is right angled.

7. Slope of $PQ = \dfrac{7}{2}$, Slope of $RS = \dfrac{7}{2}$

Slope of $QR = \dfrac{2}{7}$, Slope of $PS = \dfrac{2}{7}$

$PQ \parallel RS$ and $QR \parallel PS$

So $PQRS$ is a parallelogram.

Also (slope of PR) × (slope of SQ)

$= \dfrac{4+5}{5+4} \times \dfrac{-3-2}{3-(-2)} = \dfrac{9}{9} \times \dfrac{-5}{5} = -1$

⇒ The diagonals PR and QS are perpendicular.

⇒ The parallelogram $PQRS$ is rhombus.

ANSWERS

1. (i) $\sqrt{3}$ (ii) 0   (iii) $-\dfrac{1}{\sqrt{3}}$ (iv) $-\sqrt{3}$

2. 7   **4.** $\dfrac{1}{2}$ **5.** $x = 1$

11.11 EQUATION OF LINES IN STANDARD FORM

11.11.1 SLOPE OR TANGENT FORM

Find the equation of a line whose y-intercept 'c' and slope 'm' are given.

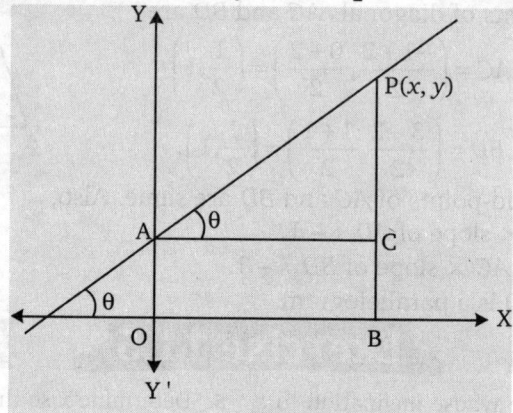

Fig. 20

Let the given line meet y-axis in A and let $P(x, 4)$ be any point on it. As the y-intercept of the line is C.

∴ Coordinates of A are $(0, C)$.

Draw PB ⊥ to x-axis and $AC \perp PB$.

Then $\tan\theta = \dfrac{PC}{AC} = \dfrac{PB - BC}{DB}$

⇒ $\tan\theta = \dfrac{BP - OA}{OB}$ ⇒ $m = \dfrac{y - c}{x}$

$\Rightarrow$ $$y = mx + c$$

which is called the slope intercept form of the equation of a straight line.

REMARKS

➡ If c becomes zero, the equation $y = mx + c$ reduces to $y = mx$ which is the equation of a line through the origin.

➡ If $m = 0$, $c \neq 0$, then equation $y = mx + c$ reduces to $y = c$ which is the equation of a line parallel to x-axis at a distance c from it.

➡ If $m = 0$, $c = 0$, then the equation becomes $y = 0$ which represents the x-axis.

11.11.2 POINT SLOPE FORM

To find the equation of a line passing through the given point (x_1, y_1) and having slope m:

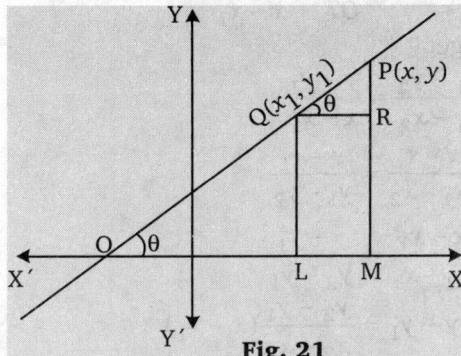

Fig. 21

Let the given point (x_1, y_1) be represented by Q. Let $P(x, y)$ be any point on the line.

Draw PM and QL perpendicular to x-axis from points P and Q and $QR \perp MP$. Then

$$PR = MP - MR = MP - QL = y - y_1$$

and $$QR = LM = OM - OL = x - x_1$$

Then $$\tan\theta = \frac{PR}{QR} = \frac{y - y_1}{x - x_1}$$

$$m = \frac{y - y_1}{x - x_1}$$

$$y - y_1 = m(x - x_1)$$

which is the equation of the line in the point slope form.

11.11.3 TWO POINTS FORM

To find the equation of the straight line passing through two given points:

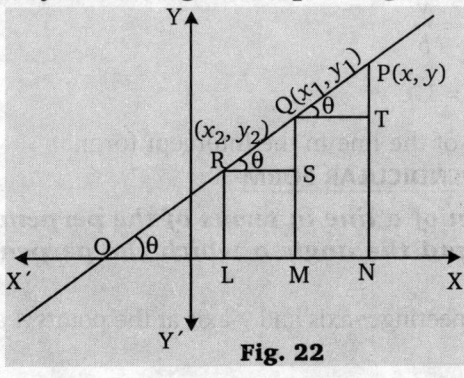

Fig. 22

Let the two given points be $Q(x_1, y_1)$ and $R(x_2, y_2)$. Let $P(x, y)$ be any point on the line. Draw RL, QM and PN perpendiculars to x-axis from points R, Q and P respectively. Let $RS \perp QM$ and $QT \perp PN$.

Then
$$RS = LM = ON - OL = x_1 - x_2$$
$$QS = MQ - MS = MQ - RL = y_1 - y_2$$
$$QT = MN = ON - OM = x - x_1$$
$$PT = NP - NT = NP - MQ = y - y_1$$

In $\Delta RQS, \tan\theta = \dfrac{QS}{RS} = \dfrac{y_1 - y_2}{x_1 - x_2}$ ...(1)

In $\Delta QTP, \tan\theta = \dfrac{PT}{QT} = \dfrac{y - y_1}{x - x_1}$ ...(2)

From (1) and (2), we get
$$\frac{y_1 - y_2}{x_1 - x_2} = \frac{y - y_1}{x - x_1}$$

$$\frac{x - x_1}{x_1 - x_2} = \frac{y - y_1}{y_1 - y_2}$$

$$\frac{x - x_1}{x_2 - x_1} = \frac{y - y_1}{y_2 - y_1}$$

$$y - y_1 = \frac{y_2 - y_1}{x_2 - x_1}(x - x_1)$$

which is the required equation of line in two point form.

11.11.4 INTERCEPT FORM

To find the equation of the line which cuts off intercepts a and b on x-axis and y-axis respectively.

Let the line meet x-axis at point A and y-axis is at point B. As the respective intercepts are a and b. So $OA = a$ and $OB = b$.

Coordinates of A and B are $(a, 0)$ and $(0, b)$ respectively.

Using two point form, the equation of line is
$$\frac{x - a}{0 - a} = \frac{y - 0}{b - 0}$$

$$\frac{-x}{a} + 1 = \frac{y}{b}$$

$$\frac{x}{a} + \frac{y}{b} = 1.$$

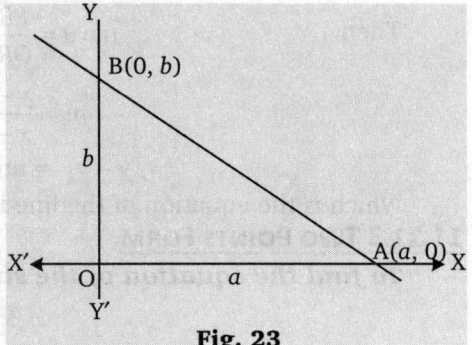

Fig. 23

Which is the equation of the line in the intercept form.

11.11.5 NORMAL OR PERPENDICULAR FORM

To find the equation of a line in terms of the perpendicular segment p, from the origin to the lines and the angle a which the perpendicular segment makes with the x-axis.

Let l be the given line meeting x-axis and y-axis at the points A and B respectively. Let $OC \perp l$ and $\angle AOC = \alpha$, $OC = p$.

Now
$$\frac{OA}{OC} = \sec\alpha$$

$$\frac{OA}{P} = \sec\alpha$$

$$OA = p\sec\alpha$$

Again
$$\frac{OB}{OC} = \operatorname{cosec}\alpha$$

$$\frac{OB}{P} = \operatorname{cosec}\alpha$$

$$OB = p\operatorname{cosec}\alpha$$

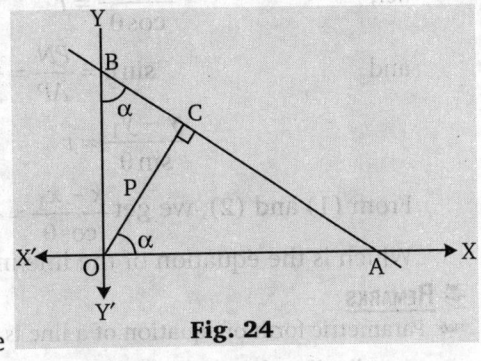

Fig. 24

Using the intercept form of the equation of the line, the equation of the given line is

$$\frac{x}{OA} + \frac{y}{OB} = 1$$

$$\frac{x}{p\sec\alpha} + \frac{y}{p\operatorname{cosec}\alpha} = 1$$

or $x\cos\alpha + y\sin\alpha = p$

which is the required equation of the line.

11.11.6 PARAMETRIC FORM

To find the equation of a straight line in the parametric form:

$$\frac{x - x_1}{\cos\theta} = \frac{y - y_1}{\sin\theta} = r, \textit{where } r \textit{ is the parameter.}$$

Let the given line passes through the point $A(x_1, y_1)$ and be inclined at an angle θ with the positive direction of x-axis.

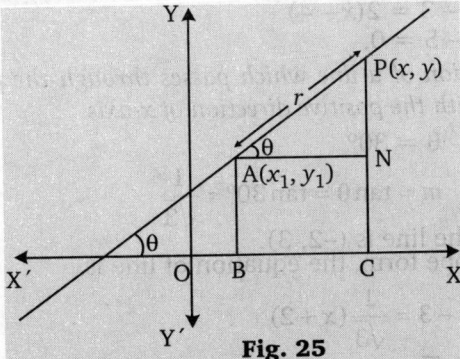

Fig. 25

Let $P(x, y)$ be any point and $AP = r$.

Draw AB and PC perpendiculars to x-axis from A and P respectively and $AN \perp PC$.

Now
$$AN = BC = OC - OB = x - x_1$$

$$PN = PC - CN = PC - AB = y - y_1$$

Also
$$AP = r.$$

In right angle triangle $\triangle ANP$.

$$\cos\theta = \frac{AN}{AP} = \frac{x - x_1}{r}$$

i.e.,
$$\frac{x - x_1}{\cos \theta} = r \qquad \ldots(1)$$

and
$$\sin \theta = \frac{PN}{AP} = \frac{y - y_1}{r}$$

$$\frac{y - y_1}{\sin \theta} = r \qquad \ldots(2)$$

From (1) and (2), we get $\dfrac{x - x_1}{\cos \theta} = \dfrac{y - y_1}{\sin \theta} = r$

Which is the equation of the line in the parametric form.

☞ REMARKS

⇒ Parametric form of equation of a line is also known as symmetrical form of equation.

⇒ From $\dfrac{x - x_1}{\cos \theta} = \dfrac{y - y_1}{\sin \theta} = r$, we have

$x = x_1 + r \cos \theta$, $y = y_1 + r \sin \theta$, thus the coordinates of any point at a distance r from (x_1, y_1) are $(x_1 + r \cos \theta, y_1 + r \sin \theta)$.

SOLVED EXAMPLES

EXAMPLE 1. *Find the equation of a line which cuts off an intercept –2 on the axis of y and makes an angle of 45° with the positive direction of x-axis.*

SOLUTION. Here, $c = -2$, and $m = \tan 45° = 1$.
Substituting these values in $y = mx + c$, we get $y = x - 2$ which is required equation of the line.

EXAMPLE 2. *Find the equation of a line through (4, 3) with slope 2.*

SOLUTION. Equation of line passing through (x_1, y_1) and with slope m is
$$y - y_1 = m(x - x_1)$$
∴ The required equation of the line is
$$y - 3 = 2(x - 4)$$
$$\Rightarrow \qquad 2x - y - 5 = 0.$$

EXAMPLE 3. *Find the equation of a line which passes through the point (–2, 3) and makes an angle of 30° with the positive direction of x-axis.*

SOLUTION. Here $\theta = 30°$

$$\Rightarrow \qquad m = \tan \theta = \tan 30° = \frac{1}{\sqrt{3}}$$

The point on the line is $(-2, 3)$.
Using point slope form, the equation of line is
$$y - 3 = \frac{1}{\sqrt{3}}(x + 2)$$
$$\Rightarrow \qquad \sqrt{3}y - 3\sqrt{3} = x + 2$$
$$\Rightarrow x - \sqrt{3}y + (3\sqrt{3} + 2) = 0.$$

EXAMPLE 4. *Find the ratio in which the line segment joining the points (2, 3) and (4, 5) is divided by the line joining the points (6, 8) and (–3, 2).*

SOLUTION. The equation of the line joining the points (6, 8) and (–3, –2) is
$$\frac{y - 8}{-2 - 8} = \frac{x - 6}{-3 - 6} \quad \text{(Two points form)}.$$
$$\Rightarrow \qquad \frac{y - 8}{-10} = \frac{x - 6}{-9}$$

or $\quad 9y - 72 = 10x - 60$

$\qquad 10x - 9y + 12 = 0$ ...(1)

Let this line divide the join of (2, 3) and (4, 5) at the point P in the ratio of $k : 1$.

Then the coordinates of P are $\left(\dfrac{4k+2}{k+1}, \dfrac{5k+3}{k+1}\right)$

Now, the point P on the line (1)

Therefore, $10 \times \dfrac{4k+2}{k+1} - 9 \times \dfrac{5k+3}{k+1} + 12 = 0$

$\qquad 40k + 20 - 45k - 27 + 12k + 12 = 0$

$\qquad\qquad\qquad\qquad\qquad 7k = -5$

$\qquad\qquad\qquad\qquad\qquad k = -5/7.$

Since, the value of k is negative, the line is divided externally.

Hence, the required ratio is 5 : 7 externally.

EXAMPLE 5. *Find the equation of the line which passes through the point* (3, 4) *and the sum of its intercept on the axes is* 14. **[UPTU(B.Pharma)–2008]**

SOLUTION. Let the intercept made by the line on x-axis be a.

Then intercept on y-axis = $14 - a$.

∴ Equation of the line is given by

$$\frac{x}{a} + \frac{y}{14-a} = 1 \qquad\qquad ...(1)$$

As the point (3, 4) lies on it, we have

$$\frac{3}{a} + \frac{4}{14-a} = 1$$

$\qquad 3(14 - a) + 4a = 14a - a^2$

$\Rightarrow\quad 42 - 3a + 4a = 14a - a^2$

$\Rightarrow\quad a^2 - 13a + 42 = 0$

$\Rightarrow\quad (a - 7)(a - 6) = 0$

$\qquad\qquad\qquad a = 7, 6.$

Putting these values of a in (1), we get equation of the lines

$$\frac{x}{7} + \frac{y}{7} = 1 \quad \text{or} \quad x + y = 7$$

and $\qquad \dfrac{x}{6} + \dfrac{y}{8} = 1 \quad \text{or} \quad 4x + 3y = 24$

EXAMPLE 6. *A line is such that its segment between the axes is bisected at the point* (x_1, y_1). *Prove that the equation of line is*

$$\frac{x}{2x_1} + \frac{y}{2y_1} = 1$$

SOLUTION. Let l be the given line which meets x-axis at A and y-axis at B. Then segment AB is bisected at the point $P(x_1, y_1)$.

Let the equation of the line be

$$\frac{x}{a} + \frac{y}{b} = 1 \qquad\qquad ...(1)$$

∴ The coordinates of A and B are $(a, 0)$ and $(0, b)$ respectively. As P is the mid-point of AB.

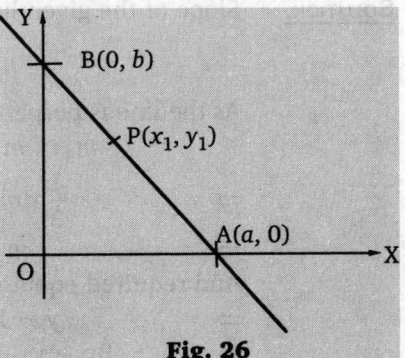

Fig. 26

Its coordinate, therefore are given by $\left(\dfrac{a}{2}, \dfrac{b}{2}\right)$.

We are given that coordinates of P are (x_1, y_1).

$\therefore \qquad \dfrac{a}{2} = x_1$ and $\dfrac{b}{2} = y_1$

or $\qquad a = 2x_1$ and $b = 2y_1$

$\therefore$ Substituting the value of a and b in (1), equation of the line l becomes

$$\frac{x}{2x_1} + \frac{y}{2y_1} = 1.$$

EXAMPLE 7. *Find the equation of the straight line passing through the points $(-3, 4)$ and $(1, -3)$.*

[RGPV(B.Pharma)–2004]

SOLUTION. Equation of the straight line passing through (x_1, y_1) and (x_2, y_2) is given by

$$y - y_1 = \frac{y_2 - y_1}{x_2 - x_1}(x - x_1) \qquad \qquad ...(1)$$

Now, we have $x_1 = -3, y_1 = 4$

$\qquad \qquad x_2 = 1, y_2 = -3$

Putting these values in equation (1) we have

$$y - 4 = \frac{-3 - 4}{1 + 3}(x + 3)$$

$\Rightarrow \qquad 4(y - 4) = -7x - 21$

$\Rightarrow \qquad 4y - 16 = -7x - 21$

$\Rightarrow \qquad 7x + 4y + 5 = 0.$

EXAMPLE 8. *Find the equation of a line which passes through $(2, 3)$ and whose m is 3.*

[UPTU(B.Pharma)–2005]

SOLUTION. Equation of line passing through one point is

$$(y - y_1) = m(x - x_1) \qquad \qquad ...(1)$$

$$m = 3, x_1 = 2, y_1 = 3$$

Putting the values in (1), the equation of required line is given by

$$y - 3 = 3(x - 2)$$

$\Rightarrow \qquad y - 3 = 3x - 6$

$\Rightarrow \qquad 3x - y - 3 = 0.$

EXAMPLE 9. *Find the equation of a line passing through the point $(3, -2)$ and perpendicular to the line $x - 3y + 5 = 0$.*   **[UPTU(B.Pharma)–2006]**

SOLUTION. Slope of the given line $x - 3y + 5 = 0$ is

$$m_1 = \frac{1}{3}$$

As the line is perpendicular to line passing through $(3, -2)$.

$$m_1 \times m_2 = -1.$$

$\Rightarrow \qquad \dfrac{1}{3} \times m_2 = -1$

$\Rightarrow \qquad m_2 = -3.$

And required equation is $(y + 2) = -3(x - 3)$

$\Rightarrow \qquad y + 2 = -3x + 9$

$\Rightarrow \qquad 3x + y - 7 = 0$

EXAMPLE 10. *Find the equation of perpendicular bisector of the line segment joining the points A(2, 3) and B(6, –5).*

SOLUTION. $$\text{Slope} = \frac{-5-3}{6-2} = \frac{-8}{4} = -2.$$

∴ Slope of a line perpendicular to the line $AB = \dfrac{1}{2}.$ $[\because m_1 m_2 = -1]$

The coordinates of the middle point M of AB are

$$\left(\frac{2+6}{2}, \frac{3+(-5)}{2} \right), i.e., (4, -1).$$

Hence, the equation of the perpendicular bisector of AB, *i.e.*, the equation of the line passing through m and perpendicular to AB is

$$y + 1 = \frac{1}{2}(x - 4)$$

$$\Rightarrow \qquad x - 2y = 6.$$

EXAMPLE 11. *Find the equation of the straight line which passes through (1, 2) and is perpendicular to the line 4x – 3y = 8.*   **[UPTU(B.Pharma)–2001]**

SOLUTION. The equation of any straight line perpendicular to the line $4x - 3y - 8 = 0$ is

$$3x + 4y + \lambda = 0 \qquad\qquad \text{...(1)}$$

If the line (1) passes through the point (1, 2) then

$$3 + 8 + \lambda = 0$$
$$\lambda = -11.$$

Putting $\lambda = -11$ in (1) the required equation of the line is

$$3x + 4y - 11 = 0$$

EXAMPLE 12. *Find the equation of the straight line passing through the point $(a \cos^3\theta, a \sin^3\theta)$ and perpendicular to the line $x \sec\theta + y \csc\theta = a \cos 2\theta$.*   **[UPTU(B.Pharma)–2007]**

SOLUTION. The slope of the given line $x \sec\theta + y \csc\theta = a$ is

$$\frac{-\sec\theta}{\csc\theta}, i.e., \frac{-\sin\theta}{\cos\theta}$$

∴ The slope of a line perpendicular to the given line $= \dfrac{\cos\theta}{\sin\theta}$

Now, the equation of the straight line which passes through the point $(a \cos^3\theta, a \sin^3\theta)$ and whose slope is $\dfrac{\cos\theta}{\sin\theta}$, is

$$y - a\sin^3\theta = \frac{\cos\theta}{\sin\theta}(x - a\cos^3\theta)$$

$$\Rightarrow \quad x \cos\theta - y \sin\theta = a(\cos^4\theta - \sin^4\theta)$$

or $\quad x \cos\theta - y \sin\theta = a(\cos^2\theta + \sin^2\theta)(\cos^2\theta - \sin^2\theta)$

Hence, $x \cos\theta - y \sin\theta = a \cos 2\theta.$

EXAMPLE 13. *Find the equation of the straight line which makes equal intercepts on the axes and passes through the point (3, –5).*   **[UPTU(B.Pharma)–2002]**

SOLUTION. Let the equation of the straight line be

$$\frac{x}{a} + \frac{y}{b} = 1 \qquad\qquad \text{...(1)}$$

The line (1) makes equal intercepts on the axes, *i.e.*, $a = b$.

∴ $$\frac{x}{a} + \frac{y}{a} = 1 \text{ or } x + y = a$$

If this line passes through the point $(3, -5)$, then
$$3 - 5 = a \text{ or } a = -2$$
Hence, the required equation is $x + y = -2$
or $x + y + 2 = 0$.

EXAMPLE 14. *Find the equation of the straight line, the portion of which intercepted between the axes is divided by the point $(-2, 6)$ in the ratio $3 : 2$.*   [UPTU(B.Pharma)–2007]

SOLUTION.   Let the equation of the straight line be
$$\frac{x}{a} + \frac{y}{b} = 1 \qquad \qquad \dots(1)$$
The line (1) meet x-axis at the point $A(a, 0)$ and y-axis at the point $B(0, b)$. Then the point $(-2, 6)$ divides the line AB in the ratio $3 : 2$.
By section formula, we have
$$(-2, 6) \equiv \left(\frac{2a + 3 \times 0}{2 + 3}, \frac{2 \times 0 + 3 \times b}{2 + 3} \right)$$
$$\Rightarrow \qquad -2 = \frac{2a}{5} \text{ and } 6 = \frac{3b}{5}$$
or $a = -5, b = 10$.
Putting the value of a and b in (1), the required equation of the line is
$$\frac{x}{-5} + \frac{y}{10} = 1$$
or $y - 2x = 10$

EXAMPLE 15. *A straight line, drawn through the point $A(2, 1)$ makes an angle $\frac{\pi}{4}$ with positive x-axis and intersects another line $x + 2y + 1 = 0$ at point B. Find the length AB.*
                                                                        [UPTU(B.Pharma)–2003]

SOLUTION.   The equation of any line passing through the given point $A(2, 1)$ and making an angle $\frac{\pi}{4}$ with x-axis is
$$\frac{x - 2}{\cos 45°} = \frac{y - 1}{\sin 45°} = r(\text{say}) \qquad \qquad \dots(1)$$
Where r represents the distance of any point B on this line from the given point $A(2, 1)$. The coordinates (x, y) of any point B on the line (1) are
$$(2 + r\cos 45°, 1 + r\sin 45°), i.e., \left(2 + r.\frac{1}{\sqrt{2}}, 1 + r.\frac{1}{\sqrt{2}} \right)$$
If the point B lies on the line $x + 2y + 1 = 0$, then
$$\left(2 + r.\frac{1}{\sqrt{2}} \right) + 2\left(1 + r.\frac{1}{\sqrt{2}} \right) + 1 = 0$$
$$\left(5 + r.\frac{3}{\sqrt{2}} \right) = 0 \text{ or } r = -\frac{5}{3}\sqrt{2}$$
Hence, the length $AB = -\frac{5\sqrt{2}}{3}$

EXAMPLE 16. *Find the equation of the line passing through the points $(4, 3)$ and $(7, 8)$.*
                                                                        [UPTU(B.Pharma)–2001]

SOLUTION. The two points are $(x_1, y_1) = (4, 3)$ and $(x_2, y_2) = (7, 8)$

Using
$$y - y_1 = \frac{y_2 - y_1}{x_2 - x_1}(x - x_1)$$

We get
$$y - 3 = \frac{8-3}{7-4}(x-4)$$

$$y - 3 = \frac{5}{3}(x-4)$$

$$\Rightarrow \qquad 5x - 3y - 11 = 0.$$

EXAMPLE 17. *Find the slope and the equation of the straight line joining the points* $(2, -5)$ *and* $(4, 1)$. **[UPTU(B.Pharma)–2002]**

SOLUTION. The slope of the line joining the points $(2, -5)$ and $(4, 1)$ is

$$= \frac{1-(-5)}{4-2} = \frac{6}{2} = 3.$$

Now, the equation of the straight line joining the points $(2, -5)$ and $(4, 1)$ and whose slope is 3, is

$$y - (-5) = 3(x - 2)$$
$$\Rightarrow \qquad y + 5 = 3x - 6$$
$$\Rightarrow \qquad 3x - y = 11.$$

EXAMPLE 18. *Find the equation of the straight line which divides the line joining the point* $(5, -2)$ *and* $(-5, 8)$ *in the ratio 3 : 4 and is also perpendicular to it.* **[UPTU(B.Pharma)–2006]**

SOLUTION. The equation of the line joining the points $(2, 3)$ and $(-5, 8)$ is

$$y - 3 = \frac{8-3}{-5-2}(x-2) \quad \text{or} \quad y - 3 = \frac{5}{-7}(x-2)$$

$$\Rightarrow \qquad -7y + 21 = 5x - 10$$
$$\Rightarrow \qquad 5x + 7y = 31 \qquad \qquad \ldots(1)$$

The slope of line (1) is $\frac{-5}{7}$ and so the slope of the line perpendicular to it will

be $\frac{7}{5}$. The coordinates (h, k) of the point dividing line (1) in the ratio 3 : 4 are given by

$$h = \frac{3 \times (-5) + 4 \times 2}{3+4} \quad \text{and} \quad k = \frac{3 \times 8 + 4 \times 3}{3+4}$$

i.e., $\qquad h = 1 \text{ and } k = \frac{36}{7}$

Hence, the equation of the line passing through (h, k) and having slope $\frac{7}{5}$ is

$$y - k = \frac{7}{5}(x - h)$$

$$\Rightarrow \qquad y - \frac{36}{7} = \frac{7}{5}(x-(-1)) \quad \text{or} \quad 49x - 35y + 229 = 0.$$

EXAMPLE 19. *Find the equation of a line at a distance of 3 units from the origin such that the perpendicular from the origin to the line makes an angle* $\tan^{-1}\left(\frac{3}{4}\right)$ *with the positive direction of x-axis.* **[UPTU(B.Pharma)–2006]**

SOLUTION. We have $p = 3$ and $\alpha = \tan^{-1}\dfrac{3}{4}$

$\Rightarrow$ $\tan\alpha = \dfrac{3}{4}$

$\Rightarrow$ $\cos\alpha = \dfrac{4}{5}$ and $\sin\alpha = \dfrac{3}{5}$

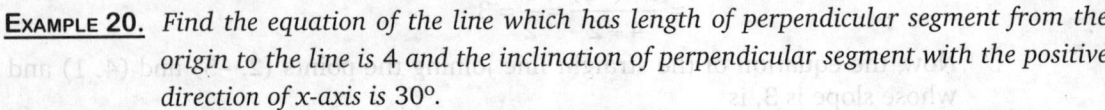

Fig. 27

Hence, the equation of the line in normal form is

$$x\cos\alpha + y\sin\alpha = p$$

or $\qquad x \times \dfrac{4}{5} + y \times \dfrac{3}{5} = 3$

$\Rightarrow \qquad 4x + 3y = 15$

EXAMPLE 20. *Find the equation of the line which has length of perpendicular segment from the origin to the line is 4 and the inclination of perpendicular segment with the positive direction of x-axis is 30°.*

SOLUTION. The normal form of the equation of a line is

$$x\cos\alpha + y\sin\alpha = p$$

Now here, $\qquad p = 4$ and $\alpha = 30°$

∴ Equation of line is

$$x\cos 30° + y\sin 30° = 4$$

$\Rightarrow \qquad x \times \dfrac{\sqrt{3}}{2} + y \times \dfrac{1}{2} = 4$

$\Rightarrow \qquad \sqrt{3}x + y = 8$

Hence, $\sqrt{3}x + y - 8 = 0$ is the required equation of the line.

EXAMPLE 21. *Through the point P(3, –5), a line is drawn inclined at 45° with the positive direction of x-axis. It meets the line x + y – 6 = 0 at the point Q. Find the length of PQ.*

SOLUTION. The equation of the line through (3, –5) inclined at an angle of 45° (by using parametric form) is

$$\frac{x-3}{\cos 45°} = \frac{y+5}{\sin 45°} = r$$

$\Rightarrow \qquad \dfrac{x-3}{\dfrac{1}{\sqrt{2}}} = \dfrac{y+5}{\dfrac{1}{\sqrt{2}}} = r$

$\Rightarrow \qquad x = \dfrac{r}{\sqrt{2}} + 3$ and $y = \dfrac{r}{\sqrt{2}} - 5$

Now, the point $Q\left(\dfrac{r}{\sqrt{2}} + 3, \dfrac{r}{\sqrt{2}} - 5\right)$ lies on the line $x + y - 6 = 0$.

∴ $\qquad \dfrac{r}{\sqrt{2}} + 3, \dfrac{r}{\sqrt{2}} - 5 - 6 = 0$

$\Rightarrow \qquad \dfrac{2r}{\sqrt{2}} = 8$

$$r = \frac{8\sqrt{2}}{2} = 4\sqrt{2}$$

Hence, required length of $PQ = 4\sqrt{2}$.

Exercise 11.6

1. The x-intercept of a line is double to its y-intercept. If it passes through $(2, 3)$, find its equation.

2. A line makes equal intercept on the coordinate axes and passes through $(1, 3)$. Find its equation.

3. Find the equation of the line passing through the points $(2, 3)$ and $(-1, -4)$.

4. If length and inclination of the perpendicular from the origin on the line is 4 and 135° respectively. Find the equation of the line.

5. If $A(0, 2)$, $B(4, 1)$, $C(1, 3)$ are the vertices of a $\triangle ABC$, find the equation of (i) side AB (ii) median CF and (iii) attitude on side BC.

6. Find the equation of the line which passes through the point $(-3, 8)$ and the sum of its

intercept on the axes is 7.

7. Find the equation of the line through $(2, 3)$ so that the segment of the line intercepted between the axes is bisected at this point.

8. The length of the perpendicular from the origin to a line is 6 and the line makes an angle of 30° with the positive direction of y-axis. Find the equation of the line.

9. Find the equation of the line through the point $(2, 3)$ and making an angle of 45° with the x-axis. Also determine the length of intercept on it between A and the line $x + y + 1 = 0$.

10. If p be the length of the perpendicular drawn from the origin to the line $bx + ay = ab$ show that $\frac{1}{a^2} + \frac{1}{b^2} = \frac{1}{p^2}$.

HINT TO SELECTED PROBLEMS

5. Given $A = (0, 2)$, $B(4, 1)$ and $C(1, 3)$

   (i) Equation of AB is

   $$\frac{y-2}{2-1} = \frac{x-0}{0-4}$$

   $$\Rightarrow \quad y - 2 = \frac{x}{-4}$$

   $$\Rightarrow -4(y - 2) = x$$

   $$\Rightarrow \quad -4y + 8 = x$$

   $$\Rightarrow \quad x + 4y = 8.$$

   $\therefore$ $x + 4y = 8$ is the eqn. of side AB.

   (ii) Median CF :

   $$F = \text{the mid of } AB = \left(\frac{0+4}{2}, \frac{2+1}{2}\right)$$

   $$F = \left(2, \frac{3}{2}\right) \text{ and } C = (1, 3)$$

   Using two point form $\dfrac{y - y_1}{y_1 - y_2} = \dfrac{x - x_1}{x_1 - x_2}$

   $$\Rightarrow \quad \frac{y - 3/2}{3/2 - 3} = \frac{x - 2}{2 - 1}$$

   $$\Rightarrow \quad \frac{2y - 3}{3 - 6} = \frac{x - 2}{1}$$

   $$\Rightarrow \quad 2y - 3 = -3(x - 2)$$

   $$\Rightarrow \quad 2y - 3 = -3x + 6$$

   $\Rightarrow 3x + 2y - 9 = 0$ is the required equation of median CF.

   (iii) Altitude $AD \perp BC$

   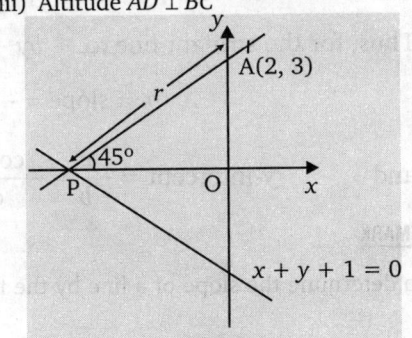

   Fig. 28

   Slope of $BC = \dfrac{3-1}{1-4} = -\dfrac{2}{3}$

   Slope of $AD = \dfrac{3}{2} (\because AD \perp BC)$

Equation of AD is $y - y_1 = m(x - x_1)$

$$y - 2 = \frac{3}{2}(x - 0)$$

$$y - 2 = \frac{3}{2}x$$

$$2y - 4 = 3x$$

$$\Rightarrow 3x - 2y + 4 = 0.$$

The equation of a line through A and making an angle of $45°$ with the x-axis is

$$\frac{x - 2}{\cos 45°} = \frac{y - 3}{\sin 45°}$$

$$\Rightarrow \frac{x - 2}{\frac{1}{\sqrt{2}}} = \frac{y - 3}{\frac{1}{\sqrt{2}}}$$

$$\Rightarrow x - 2 = y - 3$$

$$\Rightarrow x - y + 1 = 0.$$

Suppose the line meets the line $x + y + 1 = 0$

at P such that $AP = r$.

Then the coordinates of P are given by

$$\frac{x - 2}{\cos 45°} = \frac{y - 3}{\sin 45°} = r$$

$$\Rightarrow x = 2 + \frac{r}{\sqrt{2}}, \ y = 3 + \frac{r}{\sqrt{2}}$$

Thus, the coordinate of P are

$$\left(2 + \frac{r}{\sqrt{2}}, 3 + \frac{r}{\sqrt{2}}\right)$$

Since P lies on $x + y + 1 = 0$

$$\therefore \ 2 + \frac{r}{\sqrt{2}} + 3 + \frac{r}{\sqrt{2}} + 1 = 0$$

$$\Rightarrow \sqrt{2}r = -6$$

$$r = -3\sqrt{2}$$

$$\Rightarrow \text{length of } AP = |r| = 3\sqrt{2}$$

Thus, the length of intercept $= 3\sqrt{2}$

ANSWERS

1. $x + 2y + 4 = 0$ **2.** $x + y + 4 = 0$ **3.** $7x - 3y + 2 = 0$ **4.** $x - y - 4\sqrt{2} = 0$

5. (i) $x + 4y = 8$ (ii) $3x + 2y - 9 = 0$ (iii) $3x - 2y + 4 = 0$ **6.** $4x + 3y = 12$

7. $3x + 2y - 12 = 0$ **8.** $\sqrt{3}x + y - 12 = 0$ **9.** $x - y + 1 = 0, 3\sqrt{2}$

11.12 TRANSFORMATION OF GENERAL EQUATION IN DIFFERENT STANDARD FORMS

The general equation of a straight line is $ax + by + c = 0$ which can be transformed to various standard forms as discussed below :

(1) Transformation of $ax + by + c = 0$ in the slope intercept form ($y = mx + c$)

We have $ax + by + c = 0 \Rightarrow by = -ax - c \Rightarrow y = \left(-\frac{a}{b}\right)x + \left(-\frac{c}{b}\right)$

This 0 is of the form $y = mx + c$, where $m = -\frac{a}{b}$ and $c = -\frac{c}{b}$

Thus, for the straight line $ax + by + c = 0$

$$m = \text{slope} = -\frac{a}{b} = \frac{\text{coefficient of } x}{\text{coefficient of } y}$$

and $y\text{-intercept} = -\frac{c}{b} = -\frac{\text{constant term}}{\text{coeff. of } y}$

❧ REMARK

➠ To determine the slope of a line by the formula $m = \dfrac{\text{coefficient of } x}{\text{coefficient of } y}$ we must transfer all terms in

the equation on one side. Transformation of $Ax + By + C = 0$ intercept form $\left(\dfrac{x}{a} + \dfrac{y}{b} = 1\right)$:

We have $Ax + By + C = 0 \Rightarrow Ax + By = -C$

$$\Rightarrow \frac{Ax}{-C} + \frac{By}{-C} = 1$$

$$\Rightarrow \quad \frac{x}{\left(-\dfrac{C}{A}\right)} + \frac{y}{\left(-\dfrac{C}{B}\right)} = 1.$$

This is of the form $\dfrac{x}{a} + \dfrac{y}{b} = 1$. Thus, for the straight line $Ax + By + C = 0$.

$$\text{Intercept on } x\text{-axis} = \frac{-C}{A} = \frac{-\text{constant term}}{\text{coefficient of } x}$$

$$\text{Intercept on } y\text{-axis} = \frac{-C}{B} = \frac{-\text{constant term}}{\text{coefficient of } y}$$

(2) Transformation of $Ax + By + C = 0$ in normal form $(x \cos \alpha + y \sin \alpha = p)$

We have $\quad\quad Ax + By + C = 0$ \hfill ...(1)

Let $\quad\quad x \cos \alpha + y \sin \alpha - p = 0$ \hfill ...(2)

be the normal form of $Ax + By + C = 0$. Then (1) and (2) represent the same straight line.

$$\frac{A}{\cos \alpha} = \frac{B}{\sin \alpha} = \frac{C}{-p}$$

$$\Rightarrow \quad\quad \cos \alpha = \frac{-Ap}{C} \text{ and } \sin \alpha = \frac{-Bp}{C} \hspace{2cm} ...(3)$$

$$\Rightarrow \quad\quad \cos^2 \alpha + \sin^2 \alpha = \frac{A^2 p^2}{C^2} + \frac{B^2 p^2}{C^2}$$

$$\Rightarrow \quad\quad 1 = \frac{p^2}{C^2}(A^2 + B^2)$$

$$\Rightarrow \quad\quad p = \pm \frac{C}{\sqrt{A^2 + B^2}}$$

But, p denotes the length of the perpendicular from the origin to the line and is always positive.

$$\therefore \quad\quad p = \frac{C}{\sqrt{A^2 + B^2}}$$

Putting the value of p in (3) we get

$$\cos \alpha = \frac{-A}{\sqrt{A^2 + B^2}}, \quad \sin \alpha = \frac{-B}{\sqrt{A^2 + B^2}}$$

So, the equation (2) takes the form

$$\frac{-A}{\sqrt{A^2 + B^2}} x - \frac{B}{\sqrt{A^2 + B^2}} y - \frac{C}{\sqrt{A^2 + B^2}} x = 0.$$

$$\frac{-A}{\sqrt{A^2 + B^2}} x - \frac{B}{\sqrt{A^2 + B^2}} y - \frac{C}{\sqrt{A^2 + B^2}} x$$

This is the required normal form of the line $Ax + By + C = 0$.

SOLVED EXAMPLES

EXAMPLE 1. *Reduce $3x - 4y + 5 = 0$ to slope form and find its intercept on y-axis.*

SOLUTION. The given equation $3x - 4y + 5 = 0$ can be written as

$$4y = 3x + 5$$

$$\Rightarrow \qquad y = \frac{3}{4}x + \frac{5}{4}$$

Intercept on y-axis $= \dfrac{5}{4}$

EXAMPLE 2. *Reduce the lines $3x - 4y + 4 = 0$ and $4x - 3y + 12 = 0$ to the normal form and hence determine which line is nearer to the origin.*

SOLUTION. We have $3x - 4y + 4 = 0$

$$\Rightarrow \qquad -3x + 4y = 4$$

$$\Rightarrow \qquad -\frac{3x}{\sqrt{(-3) + 4^2}} + \frac{4y}{\sqrt{(-3)^2 + 4^2}} = \frac{4}{\sqrt{(-3)^2 + 4^2}}$$

$$\Rightarrow \qquad -\frac{3}{5}x + \frac{4}{5}y = \frac{4}{5}$$

This is the normal form of $3x - 4y + 4 = 0$ and the length of the perpendicular from the origin to it is given by

$$p_1 = \frac{4}{5}$$

Now $\quad 4x - 3y + 12 = 0$

$$-4x + 3y = 12$$

$$\Rightarrow \qquad \frac{-4x}{\sqrt{(-4) + 3^2}} + \frac{3y}{\sqrt{(-4)^2 + (3)^2}} = \frac{12}{\sqrt{(-4) + 3^2}}$$

$$\Rightarrow \qquad -\frac{4}{5}x + \frac{3}{5}y = \frac{12}{5}$$

This is the normal form of $4x - 3y + 12 = 0$ and the length of the perpendicular from origin to it is given by $p_2 = \dfrac{12}{5}$.

Clearly $p_2 > p_1$ therefore, line $3x - 4y + 4 = 0$ is nearer to the origin.

EXAMPLE 3. *Reduce $3x + 5y + 4 = 0$ to the intercept form and find the y-intercept.*

SOLUTION. $\qquad 3x + 5y + 4 = 0$

$$\Rightarrow \qquad 3x + 5y = -4$$

or $\qquad \dfrac{3x}{-4} + \dfrac{5y}{-4} = \dfrac{-4}{-4} \qquad\qquad \Rightarrow \qquad \dfrac{x}{(-4/3)} + \dfrac{y}{(-4/5)} = 1$

which is the required intercept form.

Hence, y-intercept is $-\dfrac{4}{5}$.

11.13 POINT OF INTERSECTION OF TWO LINES

Let the two lines be

$$A_1 x + B_1 y + C_1 = 0 \qquad\qquad\qquad \dots(1)$$
$$A_2 x + B_2 y + C_2 = 0 \qquad\qquad\qquad \dots(2)$$

Let (x_1, y_1) be the point of intersection of these two lines.
Then $\qquad A_1x_1 + B_1y_1 + C_1 = 0 \qquad \qquad \qquad \qquad$ ...(3)
and $\qquad A_2x_1 + B_2y_1 + C_2 = 0 \qquad \qquad \qquad \qquad$ ...(4)
From (3) and (4), we have

$$\frac{x_1}{B_1C_2 - B_2C_1} = \frac{y_1}{C_1A_2 - C_2A_1} = \frac{1}{A_1B_2 - A_2B_1}$$

$$\Rightarrow \qquad x_1 = \frac{B_1C_2 - B_2C_1}{A_1B_2 - A_2B_1} \text{ and } y_1 = \frac{C_1A_2 - C_2A_1}{A_1B_2 - A_2B_1}$$

Hence, the coordinates of the point of intersection of the two lines (1) and (2) are

$$\left(\frac{B_1C_2 - B_2C_1}{A_1B_2 - A_2B_1}, \frac{C_1A_2 - C_2A_1}{A_1B_2 - A_2B_1} \right)$$

⟳ REMARKS

➠ To find the coordinates of the point of intersection of two non-parallel lines, we solve the given equations simultaneously and the values of x and y so obtained determine the coordinates of the point of intersection.

➠ The coordinates of the point of intersection determined above do not exist if

$$A_1B_2 - A_2B_1 = 0, \textit{ i.e., if } \qquad \frac{A_1}{A_2} = \frac{B_1}{B_2} \ne \frac{C_1}{C_2}$$

➠ If $\dfrac{A_1}{A_2} = \dfrac{B_1}{B_2} = \dfrac{C_1}{C_2}$, then the lines are coincident.

➠ If there is only one point which satisfied both equation the system of equations is called consistent.

In that case $\dfrac{A_1}{A_2} \ne \dfrac{B_1}{B_2} \ne \dfrac{C_1}{C_2}$.

11.14 CONDITION OF CONCURRENCY OF THREE GIVEN LINES

Let the equation of the three lines be

$$a_1x + b_1y + c_1 = 0 \qquad \qquad \qquad \text{...(1)}$$
$$a_2x + b_2y + c_2 = 0 \qquad \qquad \qquad \text{...(2)}$$
$$a_3x + b_3y + c_3 = 0 \qquad \qquad \qquad \text{...(3)}$$

For given lines to be concurrent, no two of these lines can be parallel or coincident, *i.e.,*

$$\frac{a_1}{b_1} \ne \frac{a_2}{b_2} \ne \frac{a_3}{b_3} \qquad \qquad \qquad \text{...(4)}$$

and the point of intersection of any two lines must lie on the third line.

Now, the point of intersection of (1) and (2) can be obtained as below:

$$\frac{x}{b_1c_2 - b_2c_1} = \frac{y}{c_1a_2 - a_1c_2} = \frac{1}{a_1b_2 - a_2b_1}$$

$$x = \frac{b_1c_2 - b_2c_1}{a_1b_2 - a_2b_1}$$

$$y = \frac{c_1a_2 - a_1c_2}{a_1b_2 - a_2b_1}$$

Now, the point $\left(\dfrac{b_1c_2 - b_2c_1}{a_1b_2 - a_2b_1}, \dfrac{c_1a_2 - a_1c_2}{a_1b_2 - a_2b_1}\right)$ lies on (3) because the lines are concurrent

$$a_3\left(\dfrac{b_1c_2 - b_2c_1}{a_1b_2 - a_2b_1}\right) + b_3\left(\dfrac{c_1a_2 - a_1c_2}{a_1b_2 - a_2b_1}\right) + c_3 = 0.$$

$\Rightarrow\ a_3(b_1c_2 - b_2c_1) + b_3(c_1a_2 - a_1c_2) + c_3(a_1b_2 - a_2b_1) = 0.$

$\Rightarrow\ a_1(b_2c_3 - b_3c_2) + b_1(c_2a_3 - a_2c_3) + c_1(a_2b_3 - a_3b_2) = 0.$ \hfill ...(5)

Thus, for the given three lines to be concurrent, the condition (4) and (5) must hold.

SOLVED EXAMPLES

EXAMPLE 1. *Find the coordinates of the point of intersection of the lines* $2x - y + 3 = 0$ *and* $x + 2y - 4 = 0$.

SOLUTION. Solving simultaneously the equation $2x - y + 3 = 0$ and $x + 2y - 4 = 0$, we obtain

$$\frac{x}{4-6} = \frac{y}{3+8} = \frac{1}{4+1}$$

$\Rightarrow \qquad \dfrac{x}{-2} = \dfrac{y}{11} = \dfrac{1}{5}\ \Rightarrow\ x = \dfrac{-2}{5}, y = \dfrac{11}{5}.$

Hence, $(-2/5, 11/5)$ is the required point of intersection.

EXAMPLE 2. *Show that lines* $x - y - 6 = 0$, $4x - 3y - 20 = 0$ *and* $6x + 5y + 8 = 0$ *are concurrent. Also, find their common point of intersection.*

SOLUTION. The given lines are

$$x - y - 6 = 0 \qquad\qquad ...(1)$$
$$4x - 3y - 20 = 0 \qquad\qquad ...(2)$$
$$6x + 5y + 8 = 0 \qquad\qquad ...(3)$$

Solving (1) and (2) by cross multiplication, we get

$$\frac{x}{20-18} = \frac{y}{-24+20} = \frac{1}{-3+4}$$
$$x = 2, y = -4$$

Thus, the two lines intersect at the point $(2, -4)$. Putting $x = 2, y = -4$ in (3), we get

$\qquad 6 \times 2 + 5x \times (-4) + 8 = 0$

So $(2, -4)$ lies on (3).

Hence, the given lines are concurrent and their common point of intersection is $(2, -4)$.

EXAMPLE 3. *Prove that the lines* $2x + 3y - 13 = 0$, $x + 2y - 8 = 0$ *and* $3x - y - 3 = 0$ *are concurrent.*

SOLUTION. Solving the equations

$$2x + 3y - 13 = 0 \text{ and } x + 2y - 8 = 0$$

We have $\qquad \dfrac{x}{-24+26} = \dfrac{y}{-13+16} = \dfrac{1}{4-3}.$

or $\qquad\qquad x = 2, y = 3$

The lines will be concurrent if the point $(2, 3)$ satisfies the equation of third line. Putting the coordinates $(2, 3)$ in $3x - y - 3 = 0$, we have

$\qquad\qquad 3(2) - 3 - 3 = 0$

$\qquad\qquad\qquad 0 = 0$, which is true.

Hence, the lines are concurrent.

EXAMPLE 4. *Find the value of k, so that the lines*
$$x - 2y + 1 = 0; 2x - 5y + 3 = 0 \text{ and } 5x - 4y + k = 0 \text{ are concurrent.}$$

SOLUTION. The equation of the lines are :
$$x - 2y + 1 = 0 \qquad \qquad \text{...(1)}$$
$$2x - 5y + 3 = 0 \qquad \qquad \text{...(2)}$$
$$5x - 4y + k = 0 \qquad \qquad \text{...(3)}$$

Solving (1) and (2) by cross multiplication method, we get
$$\frac{x}{-6+5} = \frac{y}{2-3} = \frac{1}{-5+4}$$
$$\Rightarrow \qquad \frac{x}{-1} = \frac{y}{-1} = \frac{1}{-1}$$
$$\Rightarrow \qquad x = 1, y = 1$$

∴ The point of intersection of (1) and (2) is (1, 1).

This point will lie on (3) if $5 - 4 + k = 0$ or $k = -1$.

Thus, for concurrency of (1), (2) and (3), $k = -1$.

EXAMPLE 5. *Find the equation of the line which is perpendicular to the line $3x - 2y + 4 = 0$ and passes through the point of intersection of the lines $x + 2y + 1 = 0$ and $y = x + 7$.*

[UPTU (B.Pharma)–2008, 12]

SOLUTION. The equation of a line perpendicular to $3x - 2y + 4 = 0$ is
$$2x + 3y + \lambda = 0 \qquad \qquad \text{...(1)}$$

Point of intersection of $x + 2y + 1 = 0$ and $y = x + 7$ is
$$x = -5, y = 2 \text{ is } (-5, 2).$$

Line (1) passes through this point so
$$2(-5) + 3(2) + \lambda = 0$$
$$-10 + 6 + \lambda = 0$$
$$\lambda = 4$$

Putting in (1), we get
$$2x + 3y + 4 = 0.$$

11.15 ANGLE BETWEEN TWO INTERSECTING LINES

THEOREM 1. *The angle θ between the line $y = m_1 x + c_1$ and $y = m_2 x + c_2$ is given by*
$$\tan \theta = \frac{m_1 - m_2}{1 + m_1 m_2}.$$

PROOF. Let l_1 and l_2 be two lines $y = m_1 x + c_1$ and $y = m_2 x + c_2$ respectively. Let l_1 intersect l_2 at P making an angle θ between them. Let l_1 and l_2 meet x-axis at R and Q respectively and l_1 and l_2 making an angle α and β respectively, with the positive direction of x-axis.

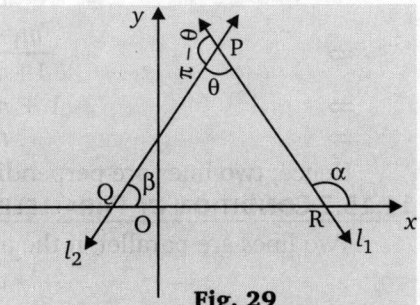

Fig. 29

The exterior angle $\alpha = \theta + \beta$
$$\theta = \alpha - \beta$$
so, $$\tan \theta = \tan (\alpha - \beta)$$
$$\Rightarrow \qquad \tan \theta = \frac{\tan \alpha - \tan \beta}{1 - \tan \alpha \tan \beta}$$

or $\tan \theta = \dfrac{m_1 - m_2}{1 + m_1 m_2}$

$\Rightarrow$ $\theta = \tan^{-1} \dfrac{m_1 - m_2}{1 + m_1 m_2}$

🔖 REMARK

➡ The value of $\tan \theta$ can be both positive and negative because between two lines there are two distinct angles. If this value is +ve, then the angle between the lines is acute and if it is –ve the angle is obtuse.

THEOREM 2. *The angle θ between the lines $a_1 x + b_1 y + c_1 = 0$ and $a_2 x + b_2 y + c_2 = 0$ is given by*

$$\tan \theta = \left| \frac{a_2 b_1 - a_1 b_2}{a_1 a_2 + b_1 b_2} \right|.$$

PROOF. Let m_1 and m_2 be the slopes of the lines $a_1 x + b_1 y + c_1 = 0$ and $a_2 x + b_2 y + c_2 = 0$. Then

$$m_1 = \frac{-a_1}{b_1} \text{ and } m_2 = \frac{-a_2}{b_2}$$

Now $\tan \theta = \left| \dfrac{m_1 - m_2}{1 + m_1 m_2} \right|$

$\Rightarrow$ $\tan \theta = \left| \dfrac{-\dfrac{a_1}{b_1} + \dfrac{a_2}{b_2}}{1 + \left(-\dfrac{a_1}{b_1}\right)\left(-\dfrac{a_2}{b_2}\right)} \right|$

$\Rightarrow$ $\tan \theta = \left| \dfrac{a_2 b_1 - a_1 b_2}{a_1 a_2 + b_1 b_2} \right|$

Hence $\theta = \tan^{-1} \left| \dfrac{a_2 b_1 - a_1 b_2}{a_1 a_2 + b_1 b_2} \right|.$

11.15.1 CONDITION OF PERPENDICULARITY

Two lines are perpendicular, if the angle between them is a right angle, *i.e.*, $\alpha = 90°$

$$\tan \alpha = \tan 90° = \infty$$

$\Rightarrow$ $\dfrac{m_1 - m_2}{1 + m_1 m_2} = \infty$

$\Rightarrow$ $1 + m_1 m_2 = -1$

$\Rightarrow$ $m_1 m_2 = -1$

Hence, two lines are perpendicular if the product of their slopes is –1.

11.15.2 CONDITION OF PARALLELISM

Two lines are parallel, if the angle between them is either 0 or π, *i.e.*, $\alpha = 0$ or π

$\therefore$ $\tan \alpha = \tan 0 \text{ or } \tan \pi = 0$

$\Rightarrow$ $\dfrac{m_1 - m_2}{1 + m_1 m_2} = 0$

$\Rightarrow$ $m_1 - m_2 = 0$

$\Rightarrow$ $m_1 = m_2$

SOLVED EXAMPLES

EXAMPLE 1. *Find the acute angle between the line $9x + 3y - 5 = 0$ and $2x + 4y + 3 = 0$.*

SOLUTION. We have $9x + 3y - 5 = 0$ ...(1)

$2x + 4y + 3 = 0$ ...(2)

Slope of (1) $\Rightarrow$ $m_1 = -\dfrac{9}{3} = -3$

Slope of (2) $\Rightarrow$ $m_2 = -\dfrac{2}{4} = -\dfrac{1}{2}$

The acute angle between the lines is given by

$$\tan\theta = \left|\frac{m_1 - m_2}{1 + m_1 m_2}\right|$$

$$\tan\theta = \left|\frac{-3 + 1/2}{1 + 3/2}\right|$$

$\Rightarrow$ $\tan\theta = \left|\dfrac{-5}{5}\right|$ $\Rightarrow$ $\tan\theta = 1$ $\Rightarrow$ $\theta = 45°$.

EXAMPLE 2. *If $A(-2, 1)$, $B(2, 3)$ and $C(-2, -4)$ are three points, find the angle between BA and BC.*

SOLUTION. Let m_1 and m_2 be the slope of *BA* and *BC* respectively. Then

$$m_1 = \frac{3-1}{2-(-2)} = \frac{2}{4} = \frac{1}{2}$$

and $$m_2 = \frac{-4-3}{-2-2} = \frac{7}{4}$$

Let θ be the angle between *BA* and *BC*. Then

$$\tan\theta = \frac{m_1 - m_2}{1 + m_1 m_2}$$

$\Rightarrow$ $\left|\dfrac{7/4 - (1/2)}{1 + 7/4(1/2)}\right| = \left|\dfrac{10/8}{15/8}\right| = \pm\dfrac{2}{3}$

$\Rightarrow$ $\theta = \tan^{-1}\left(\dfrac{2}{3}\right)$.

EXAMPLE 3. *Find the angle between the following lines:*

(i) $x \cos\alpha_1 + y \sin\alpha_1 = p_1$ and $x \cos\alpha_2 + y \sin\alpha_2 = p_2$

(ii) $\dfrac{x}{a} + \dfrac{y}{b}$ and $\dfrac{x}{b} - \dfrac{y}{a} = 1$

SOLUTION. (i) The slope of $x \cos\alpha_1 + y \sin\alpha_1 = p_1$ is $m_1 = -\dfrac{\cos\alpha_1}{\sin\alpha_1} = -\cot\alpha_1$

The slope of $x \cos\alpha_2 + y \sin\alpha_2 = p_2$ is $m_2 = -\dfrac{\cos\alpha_2}{\sin\alpha_2} = -\cot\alpha_2$

Now, $\tan\theta = \left|\dfrac{m_1 - m_2}{1 + m_1 m_2}\right| = \dfrac{-\cot\alpha_1 + \cot\alpha_2}{1 + \cot\alpha_1 \cot\alpha_2}$

$$= \frac{-\dfrac{1}{\tan\alpha_1} + \dfrac{1}{\tan\alpha_2}}{1 + \dfrac{1}{\tan\alpha_1}\dfrac{1}{\tan\alpha_2}} = \frac{\tan\alpha_1 - \tan\alpha_2}{1 + \tan\alpha_1\tan\alpha_2} = \tan(\alpha_1 - \alpha_2)$$

$$\Rightarrow \qquad \theta = \alpha_1 - \alpha_2$$

(ii) Slope of $\dfrac{x}{a} + \dfrac{y}{b} = 1$ is $m_1 = -\dfrac{(1/a)}{1/b} = -\dfrac{b}{a}$.

Slope of $\dfrac{x}{b} - \dfrac{y}{a} = 1$ is $m_2 = \dfrac{-(1/b)}{(-1/a)} = -\dfrac{a}{b}$.

Hence $\qquad m_1 m_2 = \left(-\dfrac{b}{a}\right)\cdot\left(\dfrac{a}{b}\right) = -1$

$\Rightarrow$ The lines are at right angles.

$\Rightarrow \qquad \theta = 90°$

EXAMPLE 4. *The angle between two lines is 45°. If the slope of one of them is 1/4. Find the slope of other.*

SOLUTION. Here, $\theta = 45°$, $m_1 = 1/4$. Let the slope of the required line be m_2.

Now $\qquad \tan\theta = \left|\dfrac{m_1 - m_2}{1 + m_1 m_2}\right|$

$\Rightarrow \qquad \tan 45° = \dfrac{|m_1 - m_2|}{1 + m_1 m_2}$

$\Rightarrow \qquad 1 = \dfrac{|1/4 - m|}{1 + m/4}$

$\Rightarrow \qquad 1 + \dfrac{m}{4} = \pm\left(\dfrac{1}{4} - m\right)$

For +ve sign $\quad 1 + \dfrac{m}{4} = \dfrac{1}{4} - m \Rightarrow \dfrac{5}{4}m = \dfrac{-3}{4} \Rightarrow m = \dfrac{-3}{5}$

For −ve sign $\quad 1 + \dfrac{m}{4} = -\dfrac{1}{4} + m \Rightarrow \dfrac{5}{4} = \dfrac{3}{4}m \Rightarrow m = \dfrac{5}{3}$

The possible slope of the lines are $\dfrac{5}{3}, \dfrac{-3}{5}$.

EXAMPLE 5. *Find the angle between the lines:*

$$x - y\sqrt{3} - 5 = 0 \text{ and } \sqrt{3}x + y - 7 = 0 \qquad \text{[UPTU(B.Pharma)–2001]}$$

SOLUTION. The given two lines are:

$$x - y\sqrt{3} - 5 = 0 \qquad \qquad \qquad \dots(1)$$

and $\qquad \sqrt{3}x + y - 7 = 0 \qquad \qquad \qquad \dots(2)$

Here $\qquad m_1 = \text{Slope of the line (1)} = -\dfrac{1}{-\sqrt{3}} = \dfrac{1}{\sqrt{3}}$

$\qquad m_2 = \text{Slope of the line (2)} = \dfrac{-\sqrt{3}}{1} = -\sqrt{3}$

Clearly, $\quad m_1 \times m_2 = -1$.

Hence, the two lines are at right angles.

EXAMPLE 6. *The line joining (–5, 7) and (0, –2) is perpendicular to the line joining (1, 3) and (4, x). Then find x.* **[UPTU(B.Pharma)–2003]**

SOLUTION. Here m_1 = Slope of the line joining the points (–5, 7) and (0, –2).

$$= \frac{-2-7}{0-(-5)} = \frac{-9}{5}.$$

and m_2 = Slope of the line joining the points (1, 3) and (4, x).

$$= \frac{x-3}{4-1} = \frac{x-3}{3}.$$

If the given two lines are perpendicular, then

$$m_1 m_2 = -1$$

$$\Rightarrow \qquad \left(-\frac{9}{5}\right)\left(\frac{x-3}{3}\right) = -1 \quad \Rightarrow \quad -9(x-3) = -15$$

$$\Rightarrow \qquad x = 14/3.$$

11.15.3 DISTANCE OF A POINT FROM A LINE

Let $ax + by + c = 0$ be any equation of the line and $P(x, y)$ be any point in space, then the perpendicular distance d of the point p from the line is

$$d = \left| \frac{ax_1 + by_1 + c}{\sqrt{a^2 + b^2}} \right|$$

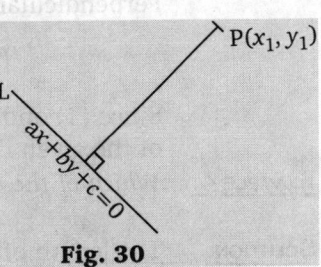

Fig. 30

11.15.4 DISTANCE BETWEEN TWO PARALLEL LINES

Let $ax + by + c_1 = 0$ and $ax + by + c_2 = 0$ be two equation of parallel lines, then the distance between the two lines is given by

$$d = \left| \frac{c_2 - c_1}{\sqrt{a^2 + b^2}} \right|$$

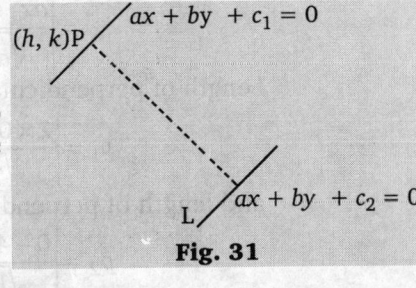

Fig. 31

SOLVED EXAMPLES

EXAMPLE 1. *Find the length of the perpendicular drawn from the point (–2, 3) on the line $12x – 5y + 1 = 0$.*

SOLUTION. We know that the length of the perpendicular segment from point (x_1, y_1) on

$ax + by + c = 0$ is $\left| \dfrac{ax_1 + by_1 + c}{\sqrt{a^2 + b^2}} \right|$

∴ Here $a = 12$, $b = -5$, $c = 1$, $x_1 = 2$, $y_1 = 3$

∴ Length of the required perpendicular segment

$$= \left| \frac{12 \times (-2) + (-5) \times 3 + 1}{\sqrt{(12)^2 + (-5)^2}} \right| = \left| \frac{-24 - 15 + 1}{13} \right| = \left| \frac{-38}{13} \right| = \frac{38}{13}.$$

EXAMPLE 2. *Find the distance between the parallel lines $2x – 3y + 9 = 0$ and $4x – 6y + 1 = 0$.*

SOLUTION. As the given lines are parallel, they have same distance between them through out. So, we shall find the distance of any point on the first line from the second line (0, 3) is a point on the line $2x - 3y + 9 = 0$.

Perpendicular distance of the point (0, 3) from $4x - 6y + 1 = 0$ is

$$\left| \frac{4 \times -6 \times 3 + 1}{\sqrt{4^2 + (-6)^2}} \right| = \left(\frac{-17}{\sqrt{52}} \right) = \frac{17}{2\sqrt{13}}$$

Hence, the distance between the given lines is $\dfrac{17}{2\sqrt{13}}$.

EXAMPLE 3. *Are the points (2, –4) and (0, 5) on the same or opposite sides of the line* $2x - 5y + 6 = 0$?

SOLUTION. Perpendicular distance of $(2, -4)$ from the given line is

$$P_1 = \frac{2 \times 2 - 5(-4) + 6}{\sqrt{4 + 25}} = \frac{4 + 20 + 6}{\sqrt{29}} = \frac{30}{\sqrt{29}}. \qquad \ldots(1)$$

Perpendicular distance of $(0, 5)$ from the given line is

$$P_2 = \frac{2 \times 0 - 5 \times 5 + 6}{\sqrt{4 + 25}} = \frac{-25 + 6}{\sqrt{29}} = \frac{-19}{\sqrt{29}}. \qquad \ldots(2)$$

Since (1) and (2) are of opposite signs, therefore, the point are on opposite sides of the given line.

EXAMPLE 4. *Which of the lines* $2x - y + 3 = 0$ *and* $x - 4y - 7 = 0$ *is farther from the origin?*

[RGPV(B.Pharma)–2001]

SOLUTION. The length of perpendicular from (x_1, y_1) on $ax + by + c = 0$ is

$$= \left| \frac{ax_1 + by_1 + c}{\sqrt{a^2 + b^2}} \right|$$

Length of perpendicular of $2x - y + 3 = 0$ from origin

$$P_1 = \left| \frac{2 \times 0 - 0 + 3}{\sqrt{4 + 1}} \right| = \frac{3}{\sqrt{5}}.$$

and length of perpendicular of $x - 4y - 7 = 0$ from origin

$$P_2 = \left| \frac{0 - 4 \times 0 - 7}{\sqrt{1 + 16}} \right| = \frac{7}{\sqrt{17}}.$$

as $\qquad P_1 > P_2$

$\therefore \quad 2x - y + 3 = 0$ is farther from origin.

EXAMPLE 5. *Find the distance between the two parallel straight lines* $y = mx + c$ *and* $y = mx + d$.

[RGPV(B.Pharma)–2002]

SOLUTION. Putting $y = 0$ in $y = mx + c$, we get $x = -c/m$. Thus $\left(-\dfrac{c}{m}, 0 \right)$ is a point on the

line $y = mx + c$. Length of perpendicular from $\left(-\dfrac{c}{m}, 0 \right)$ to $y = mx + d$ is given by

$$p = \left| \frac{m \times -\dfrac{c}{m} - 0 + d}{\sqrt{m^2 + 1}} \right| = \left| \frac{d - c}{\sqrt{m^2 + 1}} \right|$$

EXAMPLE 6. *Find the distance between the parallel lines* $3x + 4y = 12$ *and* $3x + 4y = 3$.

[UPTU(B.Pharma)–2004]

SOLUTION. The given lines are $3x + 4y = 12$ ...(1)

$$3x + 4y = 3 \qquad \text{...(2)}$$

Putting $x = 0$ we get $y = 3$. Thus $(0, 3)$ is a point on the line (1). The perpendicular distance between the lines (1) and (2) is

= the length of perpendicular from the point $(0, 3)$ to the line (2)

$$= \frac{3 \times 0 + 4 \times 3 - 3}{\sqrt{9 + 16}} = \frac{9}{5}.$$

Exercise 11.7

1. Find the length of the perpendicular from the origin on the line $4x - 3y = 7$.
2. Find the distance of the point $(3, -2)$ from the line $7x - 5y - 29 = 0$. Determine whether the point lies on the origin side of the line.
3. For what value of k will the point $(3, k)$ lie on the origin side of the line $2x + 3y + 6 = 0$.
4. Find the foot of the perpendicular drawn from the point $(-2, -1)$ on to the line

$3x + 2y - 5 = 0$.
5. Show that the point $(1, 2)$ is equidistant from the lines $5x - 2y - 9 = 0$ and $5x - 2y + 7 = 0$.
6. Find the distance between the pair of parallel lines $2x - 3y + 4 = 0$ and $4x - 6y - 5 = 0$.
7. If a and b are the intercepts of a line on the x and y axis respectively and P be its perpendicular distance from the origin then show that $\dfrac{1}{p^2} = \dfrac{1}{a^2} + \dfrac{1}{b^2}$.

HINT TO SELECTED PROBLEMS

4. Let $P(-2, -1) = (x_1, y_1)$ and $M = (h, k)$ be the foot of the perpendicular on to $3x + 2y - 5 = 0$.

Now (h, k) are given by

$$\frac{h - x_1}{a} = \frac{k - y_1}{b} = -\frac{(ax_1 + by_1 + c)}{a^2 + b^2}$$

$$\Rightarrow \frac{h + 2}{3} = \frac{k + 1}{2} = \frac{-(-6 - 2 - 5)}{9 + 4}$$

$$\Rightarrow \frac{h + 2}{3} = \frac{k + 1}{2} = 1$$

$$\Rightarrow h + 2 = 3, k + 1 = 2$$

$\Rightarrow h = 1, k = 1$

$\therefore$ The foot of perpendicular $(1, 1)$.

7. $\dfrac{x}{a} + \dfrac{y}{b} = 1 = bx + ay = ab$

$\Rightarrow bx + ay - ab = 0$ ...(1)

$$p = \left| \frac{-ab}{\sqrt{a^2 + b^2}} \right| = p^2 = \frac{a^2 \times b^2}{a^2 + b^2}$$

$$\Rightarrow \frac{1}{p^2} = \frac{a^2 + b^2}{a^2 \times b^2} = \frac{1}{a^2} + \frac{1}{b^2}.$$

ANSWERS

1. $7/5$

2. $\dfrac{2}{\sqrt{74}}$, origin lie on the opposite side of the line

3. $k < 4$

4. $(1, 1)$

6. $\dfrac{\sqrt{13}}{2}$

Objective Evaluation

☞ MULTIPLE CHOICE QUESTIONS

Choose the most appropriate one.

1. Value of k for which $(8, 1)$, $(k, -4)$ and $(2, -5)$ are collinear is :
   - (a) $k = 2$
   - (b) $k = -3$
   - (c) $k = 3$
   - (d) $k = -2$

2. The ratio in which y-axis divides the join of $(5, -6)$ and $(-1, -4)$ is :
   - (a) $1 : 3$
   - (b) $1 : 5$
   - (c) $5 : 1$
   - (d) $3 : 1$

3. AB is diameter of a circle whose center is $(2, -3)$. If coordinates of B are $(1, 4)$ then coordinates of A are :
   - (a) $(3, -10)$
   - (b) $(-10, 3)$
   - (c) $(5, 2)$
   - (d) $(-3, 10)$

4. A point on y-axis equidistant from $(6, 5)$ and $(-4, 3)$ is :
   - (a) $(9, 0)$
   - (b) $(0, 3)$
   - (c) $(0, 9)$
   - (d) $(0, -9)$

5. The area of triangle PQR where coordinates of P, Q and R are respectively $(4, 5)$, $(1, -6)$ and $(-4, -5)$ is :
   - (a) 38 sq. units
   - (b) 19 sq. units
   - (c) $\frac{19}{2}$ sq. units
   - (d) none of these

6. The value of p for which $(-5, 1)$, $(1, p)$ and $(4, -2)$ are collinear is
   - (a) $p = -1$
   - (b) $p = -2$
   - (c) $p = 1$
   - (d) none of these

7. The points $(1, 7)$, $(4, 2)$, $(-1, 1)$ and $(-4, 4)$ are vertices of a :
   - (a) square
   - (b) P. rhombus
   - (c) rectangle
   - (d) parallelogram

8. Area of ΔABC, where $A(2, 3)$, $B(-2, 1)$ and $C(3, -2)$ is :
   - (a) 10 sq. units
   - (b) 22 sq. units
   - (c) 11 sq. units
   - (d) 24 sq. units

9. If $A(5, 3)$, $B(11, -5)$ and $P(12, y)$ are vertices of a right angled triangle, right angled at P then y is :
   - (a) $-2, 4$
   - (b) $-3, 4$
   - (c) $2, -4$
   - (d) $2, 4$

10. In the figure 32 if ΔOBA is equilateral then coordinates of vertex B are :

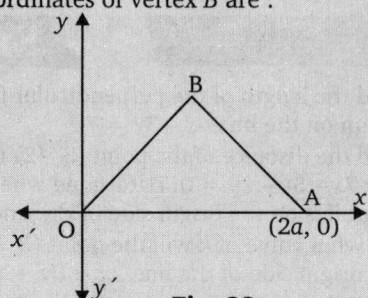

Fig. 32

   - (a) (a, a)
   - (b) $(a, 2a)$
   - (c) $(a, \sqrt{3}a)$
   - (d) $(-a, -\sqrt{3}a)$

11. Value of p for which $(-5, 1)$, $(1, p)$ and $(4, -2)$ are collinear :
   - (a) $p = 1$
   - (b) $p = -1$
   - (c) $p = 2$
   - (d) none of these

12. If the point $(2, 3)$ lies on the locus whose equation is $y = 2x + c$, find c.
   - (a) 1
   - (b) -1
   - (c) 0
   - (d) 2

13. If $A = (a, 1)$ is a point on the locus whose equation is $4x + 3y = 5$, find a.
   - (a) $1/2$
   - (b) $3/2$
   - (c) $2/3$
   - (d) $4/5$

14. Find the equation of the locus of points P, such that distance of P from the $(-1, 2)$ is equal to 3.
   - (a) $(x - 1)^2 + (y + 2)^2 = 7$
   - (b) $(x + 1)^2 - (y + 2)^2 = 5$
   - (c) $(x + 1)^2 + (y - 2)^2 = 9$
   - (d) $(x - 1)^2 + (y + 2)^2 = 9$

15. Find the equation of the locus of points P, such that twice the ordinate of P exceeds three times its abscissa by 5.
   - (a) $3x - 2y + 5$
   - (b) $2x - 3y + 5 = 0$
   - (c) $3x + 2y - 5 = 0$
   - (d) none of these

16. The equation of the straight line which passes through the point $(3, 4)$ and whose intercept on y-axis is twice, *i.e.*, on x-axis is :

(a) $x + 2y = 10$ (b) $x - 2y = 10$

(c) $2x - y = 10$ (d) $2x + y = 10$

17. The equation of the straight line which makes an angle of 15° with the positive direction of x-axis and cuts an intercept of length 4 on the negative direction of y-axis is :

(a) $(2 + \sqrt{3})x - y - 4 = 0$

(b) $(2 - \sqrt{3})x + y - 4 = 0$

(c) $(2 - \sqrt{3})x - y - 4 = 0$

(d) $(2 + \sqrt{3})x + y + 4 = 0$

18. The ratio in which the line $y - x + 2 = 0$ divides the line joining $(3, -1)$ and $(8, 9)$ is :

(a) $3 : 2$ (b) $2 : 3$

(c) $-3 : 2$ (d) $1 : 3$

19. Find the acute angle between the pairs of lines $3x + 2y = 5$ and $2x + y + 7 = 0$.

(a) $\tan\theta = \dfrac{5}{3}$ (b) $\tan\theta = \dfrac{3}{2}$

(c) $\tan\theta = \dfrac{7}{4}$ (d) $90°$

20. If the lines $13x - 6y + 7 = 0$, $x - y + 9 = 0$ and $2x + 5y = k$ are concurrent, then value of k is given by :

(a) $k = 1$ (b) $k = 2$

(c) $k = 3$ (d) $k = 4$

21. Find the equation of the line passing through the point $(-3, 7)$ and having equal intercept on the coordinate axis :

(a) $x - y = 2$ (b) $x + y = 3$

(c) $x + y = 4$ (d) $x - y = 4$

22. Find the co-ordinates of the point of intersection of the lines $x + 3 = 0$ and $3y - 4 = 0$.

(a) $\left(2, \dfrac{1}{3}\right)$ (b) $\left(3, -\dfrac{2}{3}\right)$

(c) $\left(1, \dfrac{1}{2}\right)$ (d) $\left(-3, \dfrac{4}{3}\right)$

23. Find the equation of the line passing through the point $(-3, -7)$ and perpendicular to the y-axis.

(a) $y = -3$ (b) $y = -5$

(c) $y = -7$ (d) $y = -9$

24. Find the equation of the line passing through the point $P(-5, 7)$ and parallel to the coordinate axis.

(a) $x = 5$ (b) $x = -5$

(c) $y = 5$ (d) $y = -5$

25. Find the slope of a line which makes equal intercepts of opposite sign on the coordinate axis.

(a) 3 (b) 2

(c) 1 (d) $1/2$

26. The equation of the line having slope m and y-intercept c.

(a) $y = mx + c$ (b) $x = my + c$

(c) $y = -mx + c$ (d) none of these

27. The slope of line which makes an angle of 60° with the positive x-axis.

(a) $\sqrt{3}$ (b) $\sqrt{2}$

(c) 1 (d) 2

28. The value of k, if the slope of the line containing the points $(k, 3)$ and $(-2, 5)$ is $4/5$.

(a) $5/2$ (b) $-7/3$

(c) $-9/2$ (d) $11/4$

29. Find the slope of the line passing through the points $(-2, 3)$ and $(5, -7)$.

(a) $-5/6$ (b) $-6/7$

(c) $-9/10$ (d) $-10/7$

30. The value of k so that the lines $2x - 3y + k = 0$, $3x - 4y - 13 = 0$ and $8x - 11y - 33 = 0$ are concurrent is :

(a) 6 (b) -6

(c) 7 (d) -7

☞ TRUE/FALSE

Write 'T' for True and 'F' for False statement.

1. The distance of point $(5, 3)$ from origin is 5 units.   **(T/F)**

2. If the vertices of a triangle have integral co-ordinates then the triangle cannot be equilateral.   **(T/F)**

3. The ratio in which $(4, 5)$ divides the join of $(2, 3)$ and $(7, 8)$ is $2 : 3$.   **(T/F)**

4. The abscissa and ordinate of a given point are the distances of the point from y-axis respectively.   **(T/F)**

5. The distance between the points $(\cos\theta, \sin\theta)$ and $(-\cos\theta, -\sin\theta)$ is $\sqrt{2}$.   **(T/F)**

6. If the points $(k, 2k)$, $(3k, 3k)$ and $(3, 1)$ are collinear, then $k = 1/3$.   **(T/F)**

⤷ FILL IN THE BLANKS

1. Coordinates of any point lying on x-axis are _____.

2. The distance between $A(2, 4)$ and $B(-2, 3)$ is _____.

3. If $(a, 0)$, $(0, b)$ and $(1, 1)$ are collinear, then $\dfrac{1}{a} + \dfrac{1}{b}$ is equal to _____.

4. If $A(-1, p)$ lies on BC, where $B = (-5, 3)$ and $C = (0, 3)$. Then p is equal to _____.

5. If the points $(t, 2t)$, $(-2, 6)$ and $(3, 1)$ are collinear, then t is equal to _____.

6. If centroid of the triangle formed by $(7, x)$, $(y, -6)$ and $(9, 10)$ is at $(6, 3)$ then (x, y) is equal to _____.

7. Distance of $(1, 2)$ from the mid point of the line segment whose end points are $(6, 8)$ and $(2, 4)$ is _____.

8. If area of triangle formed by $(x, 2x)$, $(-2, 6)$ and $(3, 1)$ is 5 sq. units, then x is equal to _____.

9. $A(3, 2)$, $B(-2, 1)$ are two vertices of ΔABC, whose centered G has the coordinate $\left(\dfrac{5}{3}, \dfrac{1}{3}\right)$. Then the co-ordinates of the third vertex C are _____.

10. Area of triangle whose vertices are $(1, 3)$, $(-2, 4)$ and $(0, 6)$ is _____.

ANSWERS

⤷ MULTIPLE CHOICE QUESTIONS

1. (c)	2. (c)	3. (a)	4. (c)	5. (b)	6. (a)	7. (b)	8. (c)	9. (c)
10. (c)	11. (b)	12. (a)	13. (c)	14. (a)	15. (a)	16. (d)	17. (c)	18. (b)
19. (c)	20. (d)	21. (c)	22. (d)	23. (c)	24. (b)	25. (c)	26. (a)	27. (a)
28. (c)	29. (d)	30. (d)						

⤷ TRUE/ FALSE

1. F	2. T	3. T	4. T	5. T	6. F

⤷ FILL IN THE BLANKS

1. $(x, 0)$	2. $\sqrt{17}$	3. 0	4. 2	5. $\dfrac{4}{3}$	6. $(5, 2)$	7. 5 units	8. 2/3
9. $(4, -4)$		10. 4 sq. units.					

⬜⬜⬜⬜⬜

CHAPTER 12

Limit and Continuity

12.1 INTRODUCTION

Consider the following example: $\dfrac{1}{3} = 0.\overline{3} = 0.33333...$

We know that exact value of $0.3333...$ is not $\dfrac{1}{3}$. But as we increased the digit 3 in $0.3333...$ the difference between $\dfrac{1}{3}$ and $0.3333...$ will decrease. Then we say that $\dfrac{1}{3}$ is the limiting value of 0.3333.

Consider another example of an infinite G.P. given by

$$1 + \frac{1}{2} + \frac{1}{2^2} + \frac{1}{2^3} + \frac{1}{2^4} + ...\infty$$

First term of the series $= 1$

Sum of first two terms of this series $= 1 + \dfrac{1}{2} = \dfrac{3}{2} = 1.5$

Sum of first three terms of this series $= 1 + \dfrac{1}{2} + \dfrac{1}{2^2} = 1.75$

Sum of first four terms of this series $= 1 + \dfrac{1}{2} + \dfrac{1}{2^2} + \dfrac{1}{2^3} = 1.875$

Sum of first five terms of this series $= 1 + \dfrac{1}{2} + \dfrac{1}{2^2} + \dfrac{1}{2^3} + \dfrac{1}{2^4} = 1.9375$

It is very clear from above that as we increased the number of terms of this series, the sum will approaches to 2. We may decrease the difference between the sum of the series and 2 as small as we please, but the sum of the series may not be equal to 2, whatever be the number of terms we sum. Hence, we conclude that the value of a function may not be obtained but value of the function for the approximate value of the variable may be known, then such value of the function is known as limit of the function.

12.2 LIMIT OF A SEQUENCE

Consider the sequence $1, \dfrac{1}{2}, \dfrac{1}{3}, \dfrac{1}{4}, ...$

Clearly the nth term of this sequence is $S_n = \dfrac{1}{n}$. We observe that as the value of n increases, the value of S_n will decrease.

For example. If we take $n = 10^6$ then $S_n = 10^{-6}$ and if we take $n = 10^{10}$ then $S_n = 10^{-10}$. It is clear that as n tends to ∞, the value of S_n tends to zero.

Therefore, the limit of $|S_n|$ is zero. Symbolically, it can be written as $\lim\limits_{n \to \infty} |S_n| = 0$.

Consider another sequence whose n^{th} term $S_n = \dfrac{n+3}{n}$. We observe that as the value of n increases, the value of S_n will approaches to 1. Therefore, we shall make the difference $|S_n - 1|$ as small as we please. For this sequence $\lim\limits_{n \to \infty} |S_n| = 1$.

Definition. *Let S_1, S_2, ... be the given sequence. Then a number A is said to be the limit of this sequence if the difference between S_n and A can be made as small as we please by making n large. Symbolically, it can be written as $\lim\limits_{n \to \infty} |S_n| = A$.*

12.3 SOME IMPORTANT EXPLANATIONS

(i) Meaning of $x \to 0$, i.e., x tends to 0

Let x be a real variable which takes the values as required. The meaning of $x \to 0$ is that x is very close to 0 but it is never equal to zero.

(ii) Meaning of $x \to \infty$, i.e., x tends to ∞

The meaning of $x \to \infty$ is that
   (i) x is larger than any number however large
   (ii) x is not a fixed number

(iii) Meaning of $x \to a$, i.e., x tends to a

By $x \to a$ we mean that x assumes successively values (either less than or greater than a) whose numerical difference from a can be made as small as we please. Here x is very close to a but not equal to a. i.e., $x \to a$ implies that
   (i) $x \neq a$
   (ii) x assumes values nearer and nearer to a

REMARK
➭ In the above definition x tends to a either from left or from right.

12.4 MODULUS

In the above discussion we have seen that the difference between variable x and a constant a can be made as small as we please, but we do not know that, which quantity (among $x - a$ or $a - x$) is positive. In such situation we express the difference either in the form of $x \sim a$ or $|x - a|$.

For example. The value of both $|6 - 2|$ and $|2 - 6|$ is 4. Therefore when $x \to a$, we expressed it as $|x - a| \to 0$.

12.5 DEFINED AND UNDEFINED FUNCTIONS

(i) Defined function. Let $y = f(x)$ be a function. Putting $x = a$ in $f(x)$ to get $f(a)$. If $f(a)$ is a finite quantity, then we say that $f(x)$ is a defined function at $x = a$.

For example. Let $f(x) = 2x^3 + 5x^2 - 9x + 7$

Then value of $f(x)$ at $x = 2$, *i.e.*,

$$f(2) = 2(2)^3 + 5(2)^2 - 9 \times 2 + 7$$
$$= 16 + 20 - 18 + 7 = 25$$

$\Rightarrow f(2) = 25$, which is a finite quantity. Hence, the given function $f(x)$ is a defined function at $x = 2$.

(ii) Undefined function: A function $y = f(x)$ is said to be undefined function at a point $x = a$ if we get any one of the following indeterminate form $\dfrac{0}{0}, \infty \times \infty, \infty - \infty \ldots$ when we substitute $x = a$ in $f(x)$.

12.6 CONCEPT OF THE LIMIT OF A FUNCTION

Let $y = f(x)$ be a given function. Suppose that for any value of $x = a$ (say) if the value of $f(x)$, *i.e.*, $f(a)$ becomes infinite or undefined then we compute the limiting value of f(x) which is very nearer to the actual value of $f(x)$. This limiting value of the function is called limit of the function.

This is somewhat strengthened by considering the following example?

$$f(x) = \frac{x^2 - 4}{x - 2}$$

Clearly at $x = 2$, function is of the form $\dfrac{0}{0}$.

If $x = 1.9$, then $f(1.9) = \dfrac{(1.9)^2 - 4}{1.9 - 2} = 3.9$

If $x = 1.99$, then $f(1.99) = \dfrac{(1.99)^2 - 4}{1.99 - 2} = 3.99$

If $x = 1.999$, then $f(1.999) = \dfrac{(1.999)^2 - 4}{1.999 - 2} = 3.999$

If $x = 1.9999$, then $f(1.9999) = \dfrac{(1.9999)^2 - 4}{1.9999 - 2} = 3.9999$

x	1.9	1.99	1.999	1.9999	...
y = f(x)	3.9	3.99	3.999	3.9999	...

Clearly when $x \to 2$, $f(x) = \dfrac{x^2 - 4}{x - 2}$ tends to 4

Again if $x = 2.1$, then $f(2.1) = \dfrac{(2.1)^2 - 4}{(2.1) - 2} = 4.1$

If $x = 2.01$, then $f(2.01) = \dfrac{(2.01)^2 - 4}{(2.01) - 2} = 4.01$

If $x = 2.001$, then $f(2.001) = \dfrac{(2.001)^2 - 4}{(2.001) - 2} = 4.001$

and $x = 2.0001$, then $f(2.0001) = \dfrac{(2.0001)^2 - 4}{(2.0001 - 2)} = 4.0001 \ldots$ etc.

x	2.1	2.01	2.001	2.0001	...
$y = f(x)$	4.1	4.01	4.001	4.0001	...

So, it is clear that as $x \to 2$ the value of $f(x) \to 4$.

Hence, we conclude that if x is slightly greater than 2, *i.e.*, $2 + \varepsilon$, (where ε is very small positive number) will reach near to 2 after decreasing or if the value of x is slightly less than 2, *i.e.*, $2 - \varepsilon$ will reach near to 2 after increasing. In both the case, value of $f(x)$ approaches to 4. This definite value 4 is called the limit of $\dfrac{x^2 - 4}{x - 2}$ at $x = 2$. Symbolically it can be written as $\lim\limits_{x \to 2} \dfrac{x^2 - 4}{x - 2} = 4$.

12.7 LIMIT OF THE FUNCTION

Let $y = f(x)$ be a function. The value of x approaches to a quantity a after increasing from left or decreasing from right. As the value of the function will tends to a finite quantity A, then this finite quantity A is called the limit of the function.

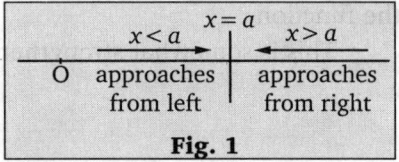

Fig. 1

Symbolically it can be written as $\lim\limits_{x \to a} f(x) = A$.

12.8 THEOREMS RELATED TO LIMITS

Some theorems related to 'limits' are given below:

If $\lim\limits_{x \to a} f(x) = A$ and $\lim\limits_{x \to a} \phi(x) = B$, then

(i) $\lim\limits_{x \to a} [f(x) \pm \phi(x)] = \lim\limits_{x \to a} f(x) \pm \lim\limits_{x \to a} \phi(x) = A \pm B$

(ii) $\lim\limits_{x \to a} [k f(x)] = k \lim\limits_{x \to a} f(x) = k \cdot A$

(iii) $\lim\limits_{x \to a} [f(x) \cdot \phi(x)] = \lim\limits_{x \to a} f(x) \cdot \lim\limits_{x \to a} \phi(x) = A \cdot B$

(iv) $\lim\limits_{x \to a} \dfrac{f(x)}{\phi(x)} = \dfrac{\lim\limits_{x \to a} f(x)}{\lim\limits_{x \to a} \phi(x)} = \dfrac{A}{B}$; if $B \neq 0$ (v) $\lim\limits_{x \to a} f(x) = \lim\limits_{h \to 0} f(a + h)$

12.9 METHOD OF FINDING THE LIMIT OF A FUNCTION

To find the limit of the function we have to calculate left hand limit (LHL) and right hand limit (RHL) of the function.

Case-I. To find the limit of rational function $f(x) = \dfrac{\phi(x)}{\psi(x)}$

(i) If at $x = a$, $\phi(a) = 0$ and $\psi(a) = 0$, then function is of the form $\dfrac{0}{0}$. In such cases use the following working steps to calculate the limit of function.

REMARK

➠ If limit is of the form $\lim\limits_{x \to a} \dfrac{f(x)}{g(x)} = \dfrac{0}{0}$, then

We have,　　　　$\lim\limits_{x \to a} \dfrac{f(x)}{g(x)} = \lim\limits_{x \to a} \dfrac{f'(x)}{g'(x)}$ (L-Hospital Rule)

WORKING PROCEDURE

Step 1. Factorize numerator and denominator by assuming $x \neq a$.

Step 2. Cancel the common factors from numerator and denominator.

Step 3. Substitute $x = a$ in the function obtained in step-2.

SOLVED EXAMPLES

EXAMPLE 1. *Find the value of* $\lim\limits_{x \to a} \dfrac{x^3 - a^3}{x - a}$

SOLUTION.　We have, $\lim\limits_{x \to a} \dfrac{x^3 - a^3}{x - a}$

$$= \lim_{x \to a} \frac{(x-a)(x^2 + ax + a^2)}{(x-a)} \qquad \text{(If } x \neq a\text{)}$$

$$= \lim_{x \to a}(x^2 + ax + a^2) = a^2 + a \cdot a + a^2 \qquad \text{(Putting } x = a\text{)}$$

$$= 3a^2$$

Aliter: We have to calculate

$$\lim_{x \to a} \frac{(x^3 - a^3)}{x - a}$$

Let $x = a + h$ when $x = a$, then $h = 0$

∴　　　　　$x \to a \;\Rightarrow\; h \to 0$

Therefore,

$$\lim_{x \to a} \frac{x^3 - a^3}{x - a} = \lim_{h \to 0} \frac{(a+h)^3 - a^3}{a + h - a} = \lim_{h \to 0} \frac{a^3 + 3a^2 h + 3ah^2 + h^3 - a^3}{h}$$

$$= \lim_{h \to 0} \frac{h(3a^2 + 3ah + h^2)}{h} = \lim_{h \to 0}(3a^2 + 3ah + h^2)$$

$$= 3a^2 + 0 + 0 \qquad \text{(Putting } h = 0\text{)}$$

$$= 3a^2$$

Case II. To find the limit of irrational functions

Let $f(x)$ be an irrational function, *i.e.*, $f(x) = \dfrac{\phi(x)}{\psi(x)}$, such that it becomes indefinite when

$x = a$. To find the limit of such function, first rationalize the given function and then use the same method as in case I.

EXAMPLE 2. *Evaluate* $\lim\limits_{x\to 0} \dfrac{x}{\sqrt{(1+x)}-1}$

SOLUTION. We have, $\lim\limits_{x\to 0} \dfrac{x}{\sqrt{(1+x)}-1}$ $\left(\text{Form } \dfrac{0}{0}\right)$

Multiply numerator and denominator by the conjugate of denominator $\sqrt{1+x}+1$, we get

$$= \lim_{x\to 0} \frac{x[\sqrt{(1+x)}+1]}{[\sqrt{(1+x)}-1][\sqrt{(1+x)}+1]} = \lim_{x\to 0} \frac{x[\sqrt{(1+x)}+1]}{(1+x)-1}$$

$$= \lim_{x\to 0} \frac{x[\sqrt{(1+x)}+1]}{x} = \lim_{x\to 0} \sqrt{(1+x)}+1] = \sqrt{(1+0)}+1 = 2$$

EXAMPLE 3. *Prove* $\lim\limits_{x\to 0} \dfrac{(1+x)^{1/2}-(1-x)^{1/2}}{x} = 1$

SOLUTION. We have, $\lim\limits_{x\to 0} \dfrac{(1+x)^{1/2}-(1-x)^{1/2}}{x}$ $\left(\text{Form } \dfrac{0}{0}\right)$

Multiply numerator and denominator by the conjugate, $i.e., [(1+x)^{1/2}+(1-x)^{1/2}]$

$$= \lim_{x\to 0} \frac{[(1+x)^{1/2}-(1-x)^{1/2}][(1+x)^{1/2}+(1-x)^{1/2}]}{x[(1+x)^{1/2}+(1-x)^{1/2}]}$$

$$= \lim_{x\to 0} \frac{(1+x)-(1-x)}{x[(1+x)^{1/2}+(1-x)^{1/2}]}$$

$$= \lim_{x\to 0} \frac{2x}{x[(1+x)^{1/2}+(1-x)^{1/2}]}$$

$$= \lim_{x\to 0} \frac{2}{(1+x)^{1/2}+(1-x)^{1/2}}$$

$$= \frac{2}{(1+0)^{1/2}+(1-0)^{1/2}} \qquad \text{(Putting } x = 0)$$

$$= \frac{2}{2} = 1$$

Case III. If limit of the function $x \to \infty$

To find the limit of such function, put $\dfrac{1}{z}$ for x and then use $z \to 0$

$$\left[\because \text{When } x \to \infty, \frac{1}{x} \to 0, i.e., z \to 0\right]$$

EXAMPLE 4. *Evaluate* $\lim\limits_{x\to\infty} \left[\dfrac{x^2+x+1}{3x^2+2x}\right]$

SOLUTION. $\lim\limits_{x\to\infty} \dfrac{x^2+x+1}{3x^2+2x}$

Let $x = \dfrac{1}{z}$ or $z = \dfrac{1}{x}$

$$x \to \infty \Rightarrow z \to \frac{1}{\infty} \Rightarrow z \to 0$$

Therefore,

$$\lim_{x \to \infty} \frac{x^2 + x + 1}{3x^2 + 2x} = \lim_{z \to 0} \frac{\dfrac{1}{z^2} + \dfrac{1}{z} + 1}{\dfrac{3}{z^2} + \dfrac{2}{z}}$$

$$= \lim_{z \to 0} \frac{\dfrac{1 + z + z^2}{z^2}}{\dfrac{3 + 2z}{z^2}} = \lim_{z \to 0} \frac{1 + z + z^2}{3 + 2z}$$

$$= \frac{1 + 0 + 0}{3 + 0} \qquad \text{(Putting } z = 0\text{)}$$

$$= \frac{1}{3}$$

Aliter:

We have,

$$\lim_{x \to \infty} \frac{x^2 + x + 1}{3x^2 + 2x} = \lim_{x \to \infty} \frac{x^2 \left(1 + \dfrac{1}{x} + \dfrac{1}{x^2}\right)}{x^2 \left(3 + \dfrac{2}{x}\right)}$$

$$= \lim_{x \to \infty} \frac{1 + \dfrac{1}{x} + \dfrac{1}{x^2}}{3 + \dfrac{2}{x}} = \frac{1 + \dfrac{1}{\infty} + \dfrac{1}{\infty}}{3 + \dfrac{2}{\infty}} = \frac{1 + 0 + 0}{3 + 0} = \frac{1}{3}$$

12.9.1 LIMIT BASED ON THE EXPANSION OF THE FUNCTIONS

EXAMPLE 5. *Evaluate* $\lim\limits_{x \to 0} \left(\dfrac{e^x - e^{-x}}{x} \right)$

SOLUTION. We know that,

$$e^x = 1 + x + \frac{x^2}{2!} + \frac{x^3}{3!} + \frac{x^4}{4!} + \dots \infty$$

and

$$e^{-x} = 1 - x + \frac{x^2}{2!} + \frac{x^3}{3!} + \frac{x^4}{4!} + \dots \infty$$

$$\therefore \quad e^x - e^{-x} = 2 \left(x + \frac{x^3}{3!} + \frac{x^5}{5!} + \dots \infty \right) = 2x \left(1 + \frac{x^2}{3!} + \frac{x^4}{5!} + \dots \infty \right)$$

$$\therefore \quad \frac{e^x - e^{-x}}{x} = 2 \left[1 + \frac{x^2}{3!} + \frac{x^4}{5!} + \dots \infty \right]$$

Therefore,

$$\lim_{x \to 0} \frac{e^x - e^{-x}}{x} = \lim_{x \to 0} 2 \left[1 + \frac{x^2}{3!} + \frac{x^4}{5!} + \dots \infty \right]$$

$$= 2[1 + 0 + 0 + \dots] = 2$$

EXAMPLE 6. *Evaluate* $\lim\limits_{x\to 0}\dfrac{(1+x)^{1/n}-1}{x}$

SOLUTION. Using Binomial expansion, we have

$$(1+x)^n = 1 + nx + \frac{n(n-1)}{2!}x^2 + \dots$$

Putting $\dfrac{1}{n}$ for n, we get

$$(1+x)^{1/n} = 1 + \frac{1}{n}x + \frac{\frac{1}{n}\left(\frac{1}{n}-1\right)}{2!}x^2 + \frac{\frac{1}{n}\left(\frac{1}{n}-1\right)\left(\frac{1}{n}-2\right)}{3!}x^3 + \dots$$

$\Rightarrow \qquad (1+x)^{1/n} - 1 = \dfrac{1}{n}x + \dfrac{\frac{1}{n}\left(\frac{1}{n}-1\right)}{2!}x^2 + \dfrac{\frac{1}{n}\left(\frac{1}{n}-1\right)\left(\frac{1}{n}-2\right)}{3!}x^3 + \dots$

$\Rightarrow \qquad \dfrac{(1+x)^{1/n}-1}{x} = \dfrac{1}{n}\left[1 + \dfrac{\frac{1}{n}-1}{2!}x + \dfrac{\left(\frac{1}{n}-1\right)\left(\frac{1}{n}-2\right)}{3!}x^2 + \dots\right]$

Therefore,

$$\lim_{x\to 0}\left[\frac{(1+x)^{1/n}-1}{x}\right] = \lim_{x\to 0}\left[\frac{1}{n}\left\{1 + \frac{\left(\frac{1}{n}-1\right)}{2!}x + \frac{\left(\frac{1}{n}-1\right)\left(\frac{1}{n}-2\right)}{3!}x^2 + \dots\right\}\right]$$

$$= \frac{1}{n}[1 + 0 + 0 + \dots]$$

$\therefore \quad \lim\limits_{x\to 0}\dfrac{(1+x)^{1/n}-1}{x} = \dfrac{1}{n}$

EXAMPLE 7. *Evaluate* $\lim\limits_{x\to 1}\dfrac{\log x}{x-1}$

SOLUTION. Let $x = 1 + h$. If $x = 1$, then $h = 0$

Therefore, $x \to 1 \Rightarrow h \to 0$

$\Rightarrow \qquad \lim\limits_{x\to 1}\dfrac{\log x}{x-1} \Rightarrow \lim\limits_{h\to 0}\dfrac{\log(1+h)}{1+h-1} = \lim\limits_{h\to 0}\dfrac{\log(1+h)}{h}$

We know that,

$$\log(1+x) = x - \frac{1}{2}x^2 + \frac{1}{3}x^3 - \frac{1}{4}x^4 + \dots\infty$$

Therefore,

$$\lim_{h\to 0}\frac{\log(1+h)}{h} = \lim_{h\to 0}\frac{h - \dfrac{1}{2}h^2 + \dfrac{1}{3}h^3 - \dfrac{1}{4}h^4 + \dots\infty}{h}$$

$$= \lim_{h\to 0}\frac{h\left[1 - \dfrac{1}{2}h + \dfrac{1}{3}h^2 - \dfrac{1}{4}h^3 + \dots\infty\right]}{h}$$

$$= \lim_{h \to 0}\left[1 - \frac{1}{2}h + \frac{1}{3}h^2 - \frac{1}{4}h^3 + ...\infty\right] = 1 - 0 + 0 - 0 + ...$$

$$\therefore \quad \lim_{x \to 1}\frac{\log x}{x - 1} = 1$$

EXAMPLE 8. *Evaluate* $\lim\limits_{x \to a}\dfrac{x^m - a^m}{x - a}$

SOLUTION. Consider $\lim\limits_{x \to a}\dfrac{x^m - a^m}{x - a}$

Let $x = a + h$. If $x = a$, then $h = 0$

$\therefore \quad x \to a \implies h \to 0$

$$\therefore \quad \lim_{x \to a}\frac{x^m - a^m}{x - a} = \lim_{h \to 0}\frac{(a + h)^m - a^m}{a + h - a}$$

$$= \lim_{h \to 0}\frac{a^m\left(1 + \dfrac{h}{a}\right)^m - a^m}{h}$$

$$= \lim_{h \to 0}\frac{a^m}{h}\left[\left(1 + \frac{h}{a}\right)^m - 1\right] \qquad \text{(By Binomial theorem)}$$

$$= \lim_{h \to 0}\frac{a^m}{h}\left[1 + m\left(\frac{h}{a}\right) + \frac{m(m - 1)}{2!}\left(\frac{h}{a}\right)^2 + ... - 1\right]$$

$$= \lim_{h \to 0}\frac{a^m}{h}\left[m\left(\frac{h}{a}\right) + \frac{m(m - 1)}{2!}\left(\frac{h}{a}\right)^2 + ...\right]$$

$$= \lim_{h \to 0}\frac{a^m}{h}m\left(\frac{h}{a}\right)\left[1 + \frac{(m - 1)}{2!}\left(\frac{h}{a}\right) + ...\right]$$

$$= \lim_{h \to 0}ma^{m-1}\left[1 + \frac{(m - 1)}{2!}\left(\frac{h}{a}\right) + ...\right]$$

$$= m \cdot a^{m-1}[1 + 0 + 0 + ...] = ma^{m-1}$$

EXAMPLE 9. *Prove that* $\lim\limits_{\theta \to 0}\dfrac{\sin\theta}{\theta} = 1$

SOLUTION. **Method-1** Let *OA* be the initial position of the imaginary axis. After rotating anticlockwise direction, it comes in the position *OB* when $\angle AOB = \theta$.

Draw perpendicular *AC* from *A* on *OB*.

$\therefore$ in $\triangle OAC$

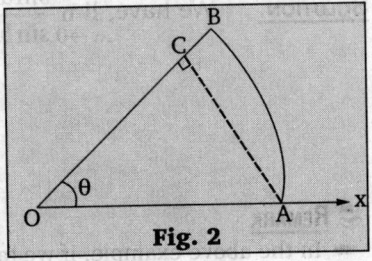

Fig. 2

$$\sin\theta = \frac{AC}{OA} \text{ and } \theta = \frac{\text{Arc}}{\text{radius}} = \frac{\text{Arc } AB}{OA}$$

$$\therefore \quad \frac{\sin\theta}{\theta} = \frac{AC}{OA} \div \frac{\text{Arc } AB}{OA} = \frac{AC}{OA} \times \frac{OA}{\text{Arc } AB}$$

$$\therefore \quad \frac{\sin\theta}{\theta} = \frac{AC}{\text{Arc } AB}$$

As θ decrease, the difference between AC and arc AB will decrease and at the end, they becomes equal

$$\therefore \quad \lim_{\theta \to 0} \frac{\sin\theta}{\theta} = \lim_{\theta \to 0} \frac{AC}{AB} = \frac{AC}{AC} = 1$$

$$\therefore \quad \lim_{\theta \to 0} \frac{\sin\theta}{\theta} = 1$$

🕮 REMARK

⟾ When imaginary line rotate in clockwise direction then θ will be negative acute angle. Let $\theta = -\alpha$ therefore,

$$\frac{\sin\theta}{\theta} = \frac{\sin(-\alpha)}{-\alpha} = \frac{-\sin\alpha}{-\alpha} = \frac{\sin\alpha}{\alpha}$$

$$\therefore \quad \lim_{\theta \to 0} \frac{\sin\theta}{\theta} = \lim_{\alpha \to 0} \frac{\sin\alpha}{\alpha} = 1$$

Method-2 We know that (from trigonometry)

$$\sin\theta = \theta - \frac{\theta^3}{3!} + \frac{\theta^5}{5!} + \frac{\theta^7}{7!} + \ldots \infty$$

$$= \theta\left[1 - \frac{\theta^2}{3!} + \frac{\theta^4}{5!} + \frac{\theta^6}{7!} + \ldots \infty\right]$$

$$\therefore \quad \frac{\sin\theta}{\theta} = 1 - \frac{\theta^2}{3!} + \frac{\theta^4}{5!} - \frac{\theta^6}{7!} + \ldots \infty$$

$$\therefore \quad \lim_{\theta \to 0} \frac{\sin\theta}{\theta} = \lim_{\theta \to 0}\left[1 - \frac{\theta^2}{3!} + \frac{\theta^4}{5!} - \frac{\theta^6}{7!} + \ldots \infty\right] = 1$$

🕮 REMARK

⟾ Use the following as a formula

$$\lim_{\theta \to 0} \frac{\sin\theta}{\theta} = 1; \quad \lim_{\theta \to 0} \frac{\theta}{\sin\theta} = 1; \quad \lim_{\theta \to 0} \cos\theta = 1$$

EXAMPLE 10. *Evaluate* $\lim\limits_{x \to 0} \dfrac{\sin ax}{\sin bx}$.

SOLUTION. We have, $\lim\limits_{x \to 0} \dfrac{\sin ax}{\sin bx} = \lim\limits_{x \to 0} \sin ax \lim\limits_{x \to 0} \dfrac{1}{\sin bx} = \lim\limits_{x \to 0} \dfrac{\sin ax}{ax} \cdot xa \cdot \lim\limits_{x \to 0} \dfrac{bx}{bx \sin bx}$

$$= \frac{a}{b} \lim_{x \to 0} \frac{\sin ax}{ax} \cdot \lim_{x \to 0} \frac{bx}{\sin bx}$$

$$= \frac{a}{b} \cdot 1 \cdot 1 = \frac{a}{b} \qquad \left(\because \lim_{\theta \to 0} \frac{\sin\theta}{\theta} = 1 \text{ and } \lim_{\theta \to 0} \frac{\theta}{\sin\theta} = 1\right)$$

🕮 REMARK

⟾ In the above example, if we take $a = 4$ and $b = 6$, then

$$\lim_{x \to 0} \frac{\sin 4x}{\sin 6x} = \frac{4}{6} = \frac{2}{3}$$

EXAMPLE 11. *Prove that* $\lim\limits_{x\to 0} \dfrac{\tan x - \sin x}{x^3} = \dfrac{1}{2}$.

SOLUTION. We have, $\lim\limits_{x\to 0} \dfrac{\tan x - \sin x}{x^3}$

$$= \lim_{x\to 0} \frac{\sin x - \sin x \cos x}{x^3 \cos x} = \lim_{x\to 0} \frac{\sin x(1 - \cos x)}{x^3 \cos x}$$

$$= \lim_{x\to 0} \frac{\sin x}{x} \cdot \lim_{x\to 0} \frac{1 - \cos x}{x^2 \cos x} = \lim_{x\to 0} \frac{\sin x}{x} \lim_{x\to 0} \frac{1 - 1 + 2\sin^2 x/2}{x^2 \cos x}$$

$$= \lim_{x\to 0} \frac{\sin x}{x} \lim_{x\to 0} \frac{2}{4} \left(\frac{\sin x/2}{x/2}\right)^2 \cdot \lim_{x\to 0} \frac{1}{\cos x}$$

$$= \frac{1}{2} \lim_{x\to 0} \frac{\sin x}{x} \lim_{x\to 0} \left(\frac{\sin x/2}{x/2}\right)^2 \lim_{x\to 0} \frac{1}{\cos x}$$

$$= \frac{1}{2} \times 1 \times 1 \times 1 = \frac{1}{2} \qquad \left(\because \lim_{x\to 0} \frac{\sin \theta}{\theta} = 1\right)$$

EXAMPLE 12. *Evaluate* $\lim\limits_{x\to \pi/4} \dfrac{\sin x - \cos x}{x - \pi/4}$.

SOLUTION. Let $x = \pi/4 + h$. If $x = \pi/4 \Rightarrow h = 0$

Therefore $x \to \dfrac{\pi}{4} \Rightarrow h \to 0$

$$\therefore \lim_{x\to \pi/4} \frac{\sin x - \cos x}{x - \pi/4} = \lim_{h\to 0} \frac{\sin(\pi/4 + h) - \cos((\pi/4 + h)}{\dfrac{\pi}{4} + h - \dfrac{\pi}{4}}$$

$$= \lim_{h\to 0} \frac{\sin(\pi/4\cos h + \cos \pi/4 \sin h) - (\cos \pi/4 \cos h - \sin \pi/4 \sin h)}{h}$$

$$= \lim_{h\to 0} \frac{1}{\sqrt{2}} \frac{(\cos h + \sin h - \cos h + \sin h)}{h} \qquad \left(\cos \pi/4 = \sin \pi/4 = \frac{1}{\sqrt{2}}\right)$$

$$= \frac{1}{\sqrt{2}} \lim_{h\to 0} \frac{2\sin h}{h} = \sqrt{2} \lim_{h\to 0} \frac{\sin h}{h}$$

$$= \sqrt{2} \times 1 = \sqrt{2} \qquad \left(\because \lim_{h\to 0} \frac{\sin h}{h} = 1\right)$$

EXAMPLE 13. *Prove that* $\lim\limits_{n\to \infty} \dfrac{n}{2} r^2 \sin \dfrac{2\pi}{n} = \pi r^2$.

SOLUTION. If $n = \infty$, then function is of the form $\infty \times 0$.

$$\therefore \lim_{n\to \infty} \frac{n}{2} r^2 \sin \frac{2\pi}{n} = \lim_{n\to \infty} \frac{r^2 \sin \dfrac{2\pi}{n}}{2/n} = \lim_{n\to \infty} \frac{\pi r^2 \sin \dfrac{2\pi}{n}}{\dfrac{2\pi}{n}} = \pi r^2 \lim_{n\to \infty} \frac{\sin \dfrac{2\pi}{n}}{\dfrac{2\pi}{n}}$$

Let $\dfrac{2\pi}{n} = \theta$. If $n = \infty$, then $\theta = 0$

$$\therefore \quad n \to \infty \quad \Rightarrow \quad \theta \to 0$$

$$\therefore \pi r^2 \lim_{n\to \infty} \frac{\sin(2\pi/n)}{2\pi/n} = \pi r^2 \lim_{\theta\to 0} \frac{\sin \theta}{\theta} = \pi r^2 \times 1 \qquad \left(\because \lim_{\theta\to 0} \frac{\sin \theta}{\theta} = 1\right)$$

$$= \pi r^2$$

EXAMPLE 14. *Evaluate* $\lim\limits_{x\to\infty} x\sin\dfrac{1}{x}$

SOLUTION.
$$\lim_{x\to\infty} x\sin\frac{1}{x} = \lim_{x\to\infty}\frac{\sin(1/x)}{1/x}.$$

Let $\dfrac{1}{x} = \theta$. Therefore, if $x = \infty$ then $\theta = 0$

$\therefore \quad x\to\infty \implies \theta\to 0$

$\therefore \quad \lim\limits_{x\to\infty}\dfrac{\sin 1/x}{(1/x)} = \lim\limits_{\theta\to 0}\dfrac{\sin\theta}{\theta} = 1$

Exercise 12.1

Evaluate the following limits:

1. $\lim\limits_{x\to 0}(7x^2 - 5x + 1)$

2. $\lim\limits_{x\to 0}(6x^3 - 5x^2 - 7x + 8)$

3. $\lim\limits_{x\to 0}\dfrac{\cos x}{x+1}$

4. $\lim\limits_{x\to 1}\dfrac{x^2 - 1}{x - 1}$

5. $\lim\limits_{x\to -2}\dfrac{x^3 + 8}{x + 2}$

6. $\lim\limits_{x\to 0}\dfrac{e^x - 1}{e^x}$

7. $\lim\limits_{x\to 0}\dfrac{3e^x - 3}{x}$

8. $\lim\limits_{x\to 0}\dfrac{a^x - 1}{x}, a > 0$

9. $\lim\limits_{x\to 0}\dfrac{\log\left(1 - \dfrac{x}{2}\right)}{x}$

10. $\lim\limits_{x\to 0}\dfrac{1 - \cos x}{x^2}$

11. $\lim\limits_{x\to 0}\dfrac{x}{\sqrt{2+x} - \sqrt{2}}$

12. $\lim\limits_{\theta\to 0}\dfrac{\sin(\theta/4)}{\theta}$

13. $\lim\limits_{x\to 0}\dfrac{\log(1+x)}{x}$

14. $\lim\limits_{x\to\infty}\cos\log\left(\dfrac{x-1}{x}\right)$

15. $\lim\limits_{x\to 0}\dfrac{\cos x}{x+2}$

16. $\lim\limits_{x\to 0}\dfrac{\sin x}{x+5}$

17. $\lim\limits_{x\to\infty}\dfrac{1^2 + 2^2 + 3^2 + \dots + n^2}{n^3}$

18. $\lim\limits_{x\to\pi/2}(\sec x - \tan x)$

Prove the following:

19. $\lim\limits_{x\to\infty}\dfrac{9x^2 + 3x + 7}{5x^2 + 2x + 1} = \dfrac{9}{5}$

20. $\lim\limits_{x\to 0}\dfrac{\log_e(1+x) - x}{x^2} = \dfrac{-1}{2}$

21. $\lim\limits_{x\to 1}\dfrac{(2x-3)(\sqrt{x}-1)}{(2x^2 + x - 3)} = \dfrac{-1}{10}$

22. $\lim\limits_{x\to 1}\dfrac{x-1}{2x^2 - 7x + 5} = -\dfrac{1}{3}$

23. $\lim\limits_{x\to 0}\dfrac{\sqrt[3]{(1+x)} - 1}{x} = \dfrac{1}{3}$

24. $\lim\limits_{x\to a}\dfrac{x^m - a^m}{x^n - a^n} = \dfrac{m}{n}a^{m-n}$ if $(m > n)$

25. $\lim\limits_{x\to\infty}\left(1 + \dfrac{a}{x}\right)^x = e^a$

26. $\lim\limits_{x\to\infty}\left[\dfrac{\sin x}{x}\right] = 0$

[Hint: Let $x = \dfrac{1}{y}$ then if $x\to\infty \implies y\to 0$**]**

27. $\lim\limits_{n\to\infty}\dfrac{1+2+3+4+\dots+n}{n^2} = \dfrac{1}{2}$

28. $\lim\limits_{x\to 1}\dfrac{\sqrt{(4+x)} - \sqrt{5}}{x - 1} = \dfrac{\sqrt{5}}{10}$

29. $\lim\limits_{x\to 0}\left[\dfrac{1 - \cos mx}{1 - \cos nx}\right] = \dfrac{m^2}{n^2}$

30. $\lim\limits_{x\to 0}\left[\dfrac{a^x + a^{-x} - 2}{a^x - a^{-x}}\right] = 0$

31. $\lim\limits_{x\to 0}\dfrac{a^x - b^x}{x} = \log_e\left(\dfrac{a}{b}\right)$

32. $\lim\limits_{x\to 2}\dfrac{\sqrt{(3-x)} - 1}{2 - x} = \dfrac{1}{2}$

33. $\lim\limits_{\theta\to 0}\left(\dfrac{\sin\theta}{\sin\theta/2}\right) = 2$

34. (i) $\lim\limits_{x\to\infty}\left[\left(1 + \dfrac{1}{x}\right)^x\right] = e$

   (ii) $\lim\limits_{x\to 0}(1+x)^{1/x} = \dfrac{5}{2}$

35. If $f(x) = \dfrac{x^2}{1+x^2}$, prove that $\lim\limits_{x\to\infty} f(x) = 1$.

36. Prove that $\lim\limits_{\theta\to\pi/2}\dfrac{2\cos\theta}{\pi - 2\theta} = 1$

   [Hint: Put $\theta = \dfrac{\pi}{2} + \phi$ **]**

37. Prove that $\lim\limits_{x \to \infty} \left(1 + \dfrac{2}{x}\right)^x = e^2$

38. Prove that $\lim\limits_{x \to 0} \dfrac{\tan x°}{x} = \dfrac{\pi}{180}$

[**Hint:** function $\dfrac{\tan x \cdot \dfrac{\pi}{180}}{x}$]

39. Prove that $\lim\limits_{n \to \infty} \left(\dfrac{1^2}{n^3} + \dfrac{2^2}{n^3} + \dfrac{3^2}{n^3} + ... + \dfrac{n^2}{n^3}\right) = \dfrac{1}{3}$

40. Prove that $\lim\limits_{x \to 0} \dfrac{\sin^{-1} x}{x} = 1$.

ANSWERS

1. 1 **2.** 8 **3.** 1 **4.** 2 **5.** 12 **6.** 0 **7.** 3 **8.** $\log_e a$ **9.** $\dfrac{-1}{2}$

10. $\dfrac{1}{2}$ **11.** $2\sqrt{2}$ **12.** $\dfrac{1}{4}$ **13.** 1 **14.** 1 **15.** $\dfrac{1}{2}$ **16.** 0 **17.** $\dfrac{1}{3}$ **18.** 0

12.10 RIGHT AND LEFT HAND LIMIT OF A FUNCTION

Let $f(x)$ be a function defined in some interval I containing a point a but may or may not be defined at a itself. Consider the behaviour of $f(x)$ as $x \to a$. It may happen that the values of f become closer and closer to a number l as $x \to a$. Therefore, there are two methods by which x tends to a. As a result, there are two types of limits.

1. **Left hand limit:** A function $f(x)$ is said to approaches the limit l as x approaches from the left, if corresponding to an arbitrary positive number ε there exists a positive number $\delta > 0$ such that
$$|f(x) - l| < \varepsilon \text{ whenever } |x - a| < \delta$$
It is written as $\lim\limits_{x \to a^-} f(x)$ or $f(a - 0)$

Therefore $\qquad f(a - 0) = \lim\limits_{x \to a^-} f(x)$

WORKING PROCEDURE

To find the limit from left, put $a - h$ for x in $f(x)$ and then take limit as $h \to 0$

i.e., $\lim\limits_{x \to a-0} f(x) = \lim\limits_{x \to a^-} f(x) = \lim\limits_{h \to 0} f(a - h)$

2. **Right hand limit:** A function f is said to approaches the limit l as x approaches a from the right, if corresponding to an arbitrary positive number ε then there exists a positive number $\delta > 0$ such that
$$|f(x) - l| < \varepsilon \text{ whenever } |x - a| < \delta$$
We say that right hand limit of $f(x)$ as x tends to a exists and is equal to l if as x approaches a, always remaining greater than a, the value of $f(x)$ approaches a definite unique real number l.
It is written as $\lim\limits_{x \to a^+} f(x)$ or $f(a + 0)$

Therefore $\qquad f(a + 0) = \lim\limits_{x \to a^+} f(x)$.

WORKING PROCEDURE

To find the limit from right, put $a + h$ for x in $f(x)$ and then take limit as $h \to 0$

i.e., $\lim\limits_{x \to a+0} f(x) = \lim\limits_{h \to 0} f(a + h) = f(a + 0)$

12.11 EXISTENCE OF THE LIMIT OF A FUNCTION

Let $y = f(x)$ be a function defined in an interval I. Let $a \in I$. The limit of $f(x)$ at $x = a$ exists if

right hand limit of $f(x)$ = left hand limit of $f(x)$

i.e.,
$$\lim_{x \to a^-} f(x) = \lim_{x \to a^+} f(x)$$

or
$$f(a - 0) = f(a + 0)$$

REMARK

➡ It is clear from above that limit of a function is unique.

SOLVED EXAMPLES

EXAMPLE 1. *Find right hand and left hand limits of the function $f(x) = \dfrac{2}{5+x}$ at $x = 2$.*

SOLUTION.   **(Left hand limit)**   **(Right hand limit)**

$f(x) = \dfrac{2}{5+x}$ $f(x) = \dfrac{2}{5+x}$

Let $x = 2 - h$ Let $x = 2 + h$

∴ when $x = 2, h = 0$ when $x = 2, h = 0$

∴ $x \to 2 \Rightarrow h \to 0$ ∴ $x \to 2 \Rightarrow h \to 0$

∴ $\displaystyle\lim_{x \to 2^-} f(x) = \lim_{h \to 0} f(2 - h)$ ∴ $\displaystyle\lim_{x \to 2^+} f(x) = \lim_{h \to 0} f(2 + h)$

$= \displaystyle\lim_{h \to 0} \dfrac{2}{5 + (2 - h)} = \lim_{h \to 0} \dfrac{2}{7 - h}$ $= \displaystyle\lim_{h \to 0} \dfrac{2}{5 + (2 + h)} = \lim_{h \to 0} \dfrac{2}{7 + h}$

$= \dfrac{2}{7 - 0} = \dfrac{2}{7}$ $= \dfrac{2}{7 + 0} = \dfrac{2}{7}$

∴ $f(2 - 0) = \dfrac{2}{7}$ ∴ $f(2 + 0) = \dfrac{2}{7}$

EXAMPLE 2. *Show that right hand and left hand limits of the function $\dfrac{\log x}{x - 1}$, when $x \to 1$ are equal to 1.*

SOLUTION.   **Left hand limit of the function**

$$f(x) = \frac{\log x}{x - 1}$$

Let $x = 1 - h$ therefore, if $x = 1$ then $h = 0$ ∴ $x \to 1 \Rightarrow h \to 0$

∴ $\displaystyle\lim_{x \to 1} f(x) = \lim_{h \to 0} f(1 - h) = \lim_{h \to 0} \frac{\log(1 - h)}{1 - h - 1}$

$= \displaystyle\lim_{h \to 0} \frac{\log(1 - h)}{-h} = \lim_{h \to 0} \frac{-[h + \dfrac{1}{2}h^2 + \dfrac{1}{3}h^3 + ...\infty]}{-h}$

$$[\because \log(1 - x) = -[x - \frac{x^2}{2} + \frac{x^3}{3} + ...\infty]]$$

$= \displaystyle\lim_{h \to 0} \frac{-h\left[1 + \dfrac{h}{2} + \dfrac{h^2}{3} + \dfrac{h^3}{4} + ...\infty\right]}{-h}$

$$= \lim_{h \to 0} \left[1 + \frac{h}{2} + \frac{h^2}{3} + \frac{h^3}{4} + \dots \infty \right] = [1 + 0 + 0 + \dots] = 1$$

$\therefore$ Therefore, left hand limit of the function = 1

Right hand limit of the function

Let $\qquad x = 1 + h$

If $x = 1$, then $h = 0$

$\therefore \qquad \lim_{x \to 1^+} f(x) = \lim_{h \to 0} f(1+h) = \lim_{h \to 0} \frac{\log(1+h)}{(1+h-1)}$

$$= \lim_{h \to 0} \frac{\log(1+h)}{h} = \lim_{h \to 0} \frac{h - \dfrac{h^2}{2} + \dfrac{h^3}{3} - \dfrac{h^4}{4} + \dots}{h}$$

$$= \lim_{h \to 0} \frac{h \left[1 - \dfrac{h}{2} + \dfrac{h^2}{3} - \dfrac{h^3}{4} + \dots \infty \right]}{h}$$

$$= \lim_{h \to 0} \left[1 - \frac{h}{2} + \frac{h^2}{3} - \frac{h^3}{4} + \dots \infty \right] = 1 - 0 + 0 - 0 + \dots = 1$$

$\therefore$ Right hand limit of the function = 1

Hence, $\qquad$ RHL = LHL = 1

EXAMPLE 3. *Find RHL and LHL of the function $f(x) = x \cos\left(\dfrac{1}{x}\right)$ at $x = 0$.*

SOLUTION. $\quad$ Given $\qquad f(x) = x \cos\left(\dfrac{1}{x}\right)$

Left hand limit of the function

Let $x = 0 - h$, if $x = 0 \Rightarrow h = 0$

$$\lim_{x \to 0^-} f(x) = \lim_{h \to 0} f(0-h) = \lim_{h \to 0} (0-h) \cos\left(\frac{1}{0-h}\right)$$

$$= \lim_{h \to 0} (-h) \cos\left(\frac{1}{-h}\right) = \lim_{h \to 0} (-h) \lim_{h \to 0} \cos\left(-\frac{1}{h}\right)$$

$$= 0 \times \{\text{a finite quantity lying between 1 and } -1\}$$

$$= 0$$

Right hand limit of the function

Let $x = 0 + h$, if $x = 0 \Rightarrow h = 0$

Therefore, $x \to 0 \Rightarrow h \to 0$

$$\lim_{x \to 0^+} f(x) = \lim_{h \to 0} f(0+h) = \lim_{h \to 0} (0+h) \cos\left(\frac{1}{0+h}\right)$$

$$= \lim_{h \to 0} h \cos\left(\frac{1}{h}\right) = \lim_{h \to 0} h \times \lim_{h \to 0} \cos\left(\frac{1}{h}\right)$$

$$= 0 \times \{\text{a finite quantity lying between 1 and } -1\}$$

$$= 0$$

Hence, $\qquad$ RHL = LHL = 0

EXAMPLE 4. *Prove that RHL and LHL of the function* $\dfrac{1+\cos x}{\tan^2 x}$ *when* $x \to \pi$ *are equal. Hence, evaluate* $\lim\limits_{x \to \pi} \dfrac{1+\cos x}{\tan^2 x}$.

SOLUTION. Given that, $f(x) = \dfrac{1+\cos x}{\tan^2 x}$

Right hand limit of the function:

Clearly x tends to π from right

Let $x = \pi + h$, when $x = \pi$ then $h = 0$

$$\therefore \quad \lim_{x \to \pi^+} f(x) = \lim_{h \to 0} f(\pi + h) = \lim_{h \to 0} \frac{1 + \cos(\pi + h)}{\tan^2(\pi + h)}$$

$$= \lim_{h \to 0} \frac{1 - \cos h}{\tan^2 h}$$

$$[\because \cos(\pi + h) = -\cos h \text{ and } \tan^2(\pi + h) = (\tan h)^2 = \tan^2 h]$$

$$= \lim_{h \to 0} \frac{1 - \left\{1 - \dfrac{h^2}{2!} + \dfrac{h^4}{4!} - \dfrac{h^6}{6!} + \ldots \infty\right\}}{\left(h + \dfrac{1}{3}h^3 + \dfrac{2}{15}h^5 + \ldots \infty\right)}$$

$$\left(\because \cos\theta = 1 - \frac{\theta^2}{2!} + \frac{\theta^4}{4!} - \frac{\theta^6}{6!} + \ldots \infty \text{ and } \tan\theta = \theta + \frac{\theta^3}{3} + \frac{2}{15}\theta^5 + \ldots \infty\right)$$

$$= \lim_{h \to 0} \frac{h^2\left[\dfrac{1}{2!} - \dfrac{h^2}{4!} + \ldots \infty\right]}{h^2\left[1 + \dfrac{h^2}{3} + \dfrac{2}{15}h^4 + \ldots \infty\right]^2}$$

$$= \lim_{h \to 0} \frac{\dfrac{1}{2!} - \dfrac{h^2}{4!} + \ldots \infty}{\left(1 + \dfrac{h^2}{3} + \dfrac{2}{15}h^4 + \ldots \infty\right)^2} = \frac{\dfrac{1}{2!} - 0 + 0 - 0 + \ldots}{(1 + 0 + 0 + \ldots)^2}$$

$$= \frac{1}{2!} = \frac{1}{2}$$

Left hand limit of the function:

In this case, x tends to π from left

Let $x = \pi - h$, when $x = \pi$ then $h = 0$

$$\therefore \quad \lim_{x \to \pi^-} f(x) = \lim_{h \to 0} f(\pi - h) = \lim_{h \to 0} \frac{1 + \cos(\pi - h)}{\tan^2(\pi - h)} = \lim_{h \to 0} \frac{1 - \cos h}{\tan^2 h}$$

$$= \lim_{h \to 0} \frac{1 - \left\{1 - \dfrac{h^2}{2!} + \dfrac{h^4}{4!} - \dfrac{h^6}{6!} + \ldots \infty\right\}}{\left(h + \dfrac{h^3}{3} + \dfrac{2}{15}h^5 + \ldots \infty\right)^2}$$

$$= \lim_{h\to 0} \frac{h^2 \left[\frac{1}{2!} - \frac{h^2}{4!} + \ldots\infty\right]}{h^2 \left(1 + \frac{h^2}{3} + \frac{2}{15}h^4 + \ldots\infty\right)^2} = \lim_{h\to 0} \frac{\frac{1}{2!} - \frac{h^2}{4!} + \ldots\infty}{\left(1 + \frac{h^2}{3} + \frac{2}{15}h^4 + \ldots\infty\right)}$$

$$= \frac{\frac{1}{2!} - 0 + \ldots}{(1 + 0 + \ldots)^2} = \frac{1}{2!} = \frac{1}{2}$$

Therefore when $x \to \pi$, RHL of $f(x)$ = LHL of $f(x)$

But we know that

RHL of $f(x)$ = LHL of $f(x)$

Hence, $\lim\limits_{x\to\pi} \dfrac{1 + \cos x}{\tan^2 x} = \dfrac{1}{2}$

MISCELLANEOUS EXAMPLES

EXAMPLE 1. *Evaluate* $\lim\limits_{x\to 1} \dfrac{(2x - 3)(\sqrt{x} - 1)}{3x^2 + 3x - 6}$.

SOLUTION. We have, $\lim\limits_{x\to 1} \dfrac{(2x - 3)(\sqrt{x} - 1)}{3x^2 + 3x - 6} = \lim\limits_{x\to 1} \dfrac{(2x - 3)(\sqrt{x} - 1)(\sqrt{x} + 1)}{3(x^2 + x - 2)(\sqrt{x} + 1)}$

$$= \lim_{x\to 1} \frac{2x - 3}{3(x + 2)(x - 1)} \left(\frac{x - 1}{\sqrt{x} + 1}\right) = \lim_{x\to 1} \frac{2x - 3}{3(x + 2)(\sqrt{x} + 1)} = \frac{-1}{18}$$

EXAMPLE 2. *Evaluate* $\lim\limits_{x\to 4} \dfrac{x^3 - 2x^2 - 9x + 4}{x^2 - 2x - 8}$

SOLUTION. We have, $\lim\limits_{x\to 4} \dfrac{(x^3 - 2x^2 - 9x + 4)}{x^2 - 2x - 8} = \lim\limits_{x\to 4} \dfrac{(x - 4)(x^2 + 2x - 1)}{(x - 4)(x + 2)}$

$$= \lim_{x\to 4} \frac{x^2 + 2x - 1}{x + 2} = \frac{(4)^2 + 2(4) - 1}{4 + 2} = \frac{23}{6}$$

EXAMPLE 3. *Evaluate the following limits :*

(i) $\lim\limits_{x\to 1} \left(\dfrac{1}{x^2 - 1} - \dfrac{2}{x^4 - 1}\right)$ (ii) $\lim\limits_{x\to a} \dfrac{\sqrt{a + 2x} - \sqrt{3x}}{\sqrt{3a + x} - 2\sqrt{x}}, a \neq 0$

SOLUTION. (i) We have, $\lim\limits_{x\to 1} \left(\dfrac{1}{x^2 - 1} - \dfrac{2}{x^4 - 1}\right) = \lim\limits_{x\to 1} \left(\dfrac{1}{(x^2 - 1)} - \dfrac{2}{(x^2 - 1)(x^2 + 1)}\right)$

$$= \lim_{x\to 1} \left[\frac{x^2 + 1 - 2}{(x^2 - 1)(x^2 + 1)}\right]$$

$$= \lim_{x\to 1} \frac{(x^2 - 1)}{(x^2 - 1)(x^2 + 1)} = \lim_{x\to 1} \frac{1}{x^2 + 1} = \frac{1}{1^2 + 1} = \frac{1}{2}$$

(ii) We have, $\lim\limits_{x \to a} \dfrac{\sqrt{a+2x}-\sqrt{3x}}{\sqrt{3a+x}-2\sqrt{x}}, a \neq 0$

$$= \lim_{x \to a} \dfrac{\dfrac{(\sqrt{a+2x}-\sqrt{3x})(\sqrt{a+2x}+\sqrt{3x})}{(\sqrt{a+2x}+\sqrt{3x})}}{\dfrac{(\sqrt{3a+x}-2\sqrt{x})(\sqrt{3a+x}+2\sqrt{x})}{(\sqrt{3a+x}+2\sqrt{x})}}$$

$$= \lim_{x \to a} \left\{ \dfrac{\dfrac{a+2x-3x}{\sqrt{a+2x}+\sqrt{3x}}}{\dfrac{3a+x-4x}{\sqrt{3a+x}+2\sqrt{x}}} \right\}$$

$$= \lim_{x \to a} \left[\dfrac{(a-x)}{\sqrt{a+2x}+\sqrt{3x}} \cdot \dfrac{\sqrt{3a+x}+2\sqrt{x}}{3(a-x)} \right]$$

$$= \lim_{x \to a} \dfrac{\sqrt{3a+x}+2\sqrt{x}}{3(\sqrt{a+2x}+\sqrt{3x})} = \dfrac{\sqrt{4a}+2\sqrt{a}}{3(\sqrt{3a}+\sqrt{3a})}$$

$$= \dfrac{4\sqrt{a}}{3 \cdot 2\sqrt{3a}} = \dfrac{2}{3\sqrt{3}}$$

EXAMPLE 4. *Evaluate the following limits:*

    (i) $\lim\limits_{x \to 0} \dfrac{\sin ax}{\tan bx}$ (ii) $\lim\limits_{x \to 0} \dfrac{\operatorname{cosec} x - \cot x}{x}$

SOLUTION.   (i) We have,

$$\lim_{x \to 0} \dfrac{\sin ax}{\tan bx} = \lim_{x \to 0} \left\{ \dfrac{\left(\dfrac{\sin ax}{ax}\right).ax}{\left(\dfrac{\tan bx}{bx}\right)bx} \right\} = \lim_{x \to 0} \dfrac{\left(\dfrac{\sin ax}{ax}\right)}{\left(\dfrac{\tan bx}{bx}\right)} \cdot \dfrac{a}{b}$$

$$= \dfrac{1 \cdot a}{1 \cdot b} = \dfrac{a}{b} \qquad \left(\because \lim_{x \to 0} \dfrac{\sin ax}{ax} = 1, \lim_{x \to 0} \dfrac{\tan bx}{bx} = 1 \right)$$

(ii) We have,

$$\lim_{x \to 0} \dfrac{\operatorname{cosec} x - \cot x}{x} = \lim_{x \to 0} \left\{ \dfrac{\dfrac{1}{\sin x} - \dfrac{\cos x}{\sin x}}{x} \right\} = \lim_{x \to 0} \dfrac{1 - \cos x}{x \sin x}$$

$$= \lim_{x \to 0} \dfrac{2\sin^2 \dfrac{x}{2}}{x \cdot 2\sin\dfrac{x}{2}.\cos\dfrac{x}{2}} = \lim_{x \to 0} \dfrac{\sin\dfrac{x}{2}}{x \sin\dfrac{x}{2}} \cdot \dfrac{\sin\dfrac{x}{2}}{\cos\dfrac{x}{2}}$$

$$= \lim_{x \to 0} \dfrac{\tan\dfrac{x}{2}}{x} = \lim_{x \to 0} \dfrac{\tan\dfrac{x}{2}}{2 \cdot \dfrac{x}{2}} = \lim_{x \to 0} \dfrac{1}{2} \cdot \left(\dfrac{\tan\dfrac{x}{2}}{\dfrac{x}{2}} \right) = \dfrac{1}{2}$$

EXAMPLE 5. *Evaluate* $\lim\limits_{x \to 0}\left(\dfrac{\tan 2x - x}{3x - \sin x}\right)$.

SOLUTION. We have, $\lim\limits_{x \to 0}\left(\dfrac{\tan 2x - x}{3x - \sin x}\right) = \lim\limits_{x \to 0}\dfrac{\left(\dfrac{\tan 2x}{2x}\right)2x - x}{3x - \left(\dfrac{\sin x}{x}\right).x}$

$$= \lim\limits_{x \to 0}\dfrac{\left(\dfrac{\tan 2x}{2x}\right)2 - 1}{3 - \dfrac{\sin x}{x}} = \dfrac{2 \cdot 1 - 1}{3 - 1} = \dfrac{1}{2}$$

EXAMPLE 6. *Evaluate* $\lim\limits_{x \to 0}\dfrac{8}{x^8}\left[1 - \cos\dfrac{x^2}{2} - \cos\dfrac{x^2}{4} + \cos\dfrac{x^2}{2}\cos\dfrac{x^2}{4}\right]$

SOLUTION. We have, $\lim\limits_{x \to 0}\dfrac{8}{x^8}\left[1 - \cos\dfrac{x^2}{2} - \cos\dfrac{x^2}{4} + \cos\dfrac{x^2}{2}\cos\dfrac{x^2}{4}\right]$

$$= \lim\limits_{x \to 0}\dfrac{8}{x^8}\left[\left(1 - \cos\dfrac{x^2}{2}\right) - \cos\dfrac{x^2}{4}\left(1 - \cos\dfrac{x^2}{2}\right)\right]$$

$$= \lim\limits_{x \to 0}\dfrac{8}{x^8}\left(1 - \cos\dfrac{x^2}{2}\right)\left(1 - \cos\dfrac{x^2}{4}\right)$$

$$= \lim\limits_{x \to 0}\dfrac{8}{x^8} \cdot 2\sin^2\dfrac{x^2}{4} \cdot 2\sin^2\dfrac{x^2}{8}$$

$$= \lim\limits_{x \to 0}\dfrac{32}{x^8}\left[\dfrac{\sin\dfrac{x^2}{4}}{\dfrac{x^2}{4}}\right]^2\left(\dfrac{x^2}{4}\right)^2 \cdot \left[\dfrac{\sin\dfrac{x^2}{8}}{\dfrac{x^2}{8}}\right]^2\left(\dfrac{x^2}{8}\right)^2 = \dfrac{1}{32}$$

EXAMPLE 7. *Evaluate* $\lim\limits_{x \to 0}\dfrac{\sin(a + b)x + \sin(a - b)x + \sin 2ax}{\cos 2bx - \cos 2ax}.x$

SOLUTION. We have, $\lim\limits_{x \to 0}\dfrac{\sin(a + b)x + \sin(a - b)x + \sin 2ax}{\cos 2bx - \cos 2ax}.x$

$$= \lim\limits_{h \to 0}\left\{\dfrac{\dfrac{\sin(a + b)x}{(a + b)x}.(a + b)x + \dfrac{\sin(a - b)x}{(a - b)x}.(a - b)x + \dfrac{\sin 2ax}{2ax}.2ax}{\dfrac{2\sin(a + b)x}{(a + b)x}.(a + b)x.\sin\dfrac{(a - b)x}{(a - b)x}(a - b)x}\right\}.x$$

$$= \lim\limits_{x \to 0}\left\{\dfrac{\dfrac{\sin(a + b)x}{(a + b)x}.(a + b) + \dfrac{\sin(a - b)x}{(a - b)x}.(a + b) + \dfrac{\sin 2ax}{2ax}.2a}{\dfrac{2\sin(a + b)x}{(a + b)x}.(a + b).\dfrac{\sin(a - b)x}{(a - b)x}(a - b)}\right\}$$

$$= \dfrac{1 \cdot (a + b) + 1(a - b) + 1.2a}{2 \cdot 1 \cdot (a + b) \cdot 1 \cdot (a - b)} = \dfrac{4a}{2(a^2 - b^2)} = \dfrac{2a}{(a^2 - b^2)}$$

Example 8. *Evaluate* $\lim\limits_{x \to 0} \dfrac{1 - \cos x \sqrt{\cos 2x}}{x^2}$.

Solution. We have, $\lim\limits_{x \to 0} \left(\dfrac{1 - \cos x \sqrt{\cos 2x}}{x^2} \right)$

$$= \lim_{x \to 0} \left\{ \frac{(1 - \cos x \sqrt{\cos 2x})(1 + \cos x \sqrt{\cos 2x})}{x^2 (1 + \cos x \sqrt{\cos 2x})} \right\}$$

$$= \lim_{x \to 0} \left(\frac{1 - \cos^2 x \times \cos 2x}{x^2 (1 + \cos x \sqrt{\cos 2x})} \right)$$

$$= \lim_{x \to 0} \left(\frac{1 - (1 - \sin^2 x)(1 - 2\sin^2 x)}{x^2 (1 + \cos x \sqrt{\cos 2x})} \right)$$

$$= \lim_{x \to 0} \left(\frac{1 - (1 - 3\sin^2 x + 2\sin^4 x)}{x^2 (1 + \cos x \sqrt{\cos 2x})} \right)$$

$$= \lim_{x \to 0} \left(\frac{\sin^2 x (3 - 2\sin^2 x)}{x^2 (1 + \cos x \sqrt{\cos 2x})} \right)$$

$$= \lim_{x \to 0} \left(\frac{\sin x}{x} \right)^2 \left(\frac{3 - 2\sin^2 x}{1 + \cos x \sqrt{\cos 2x}} \right) = 1^2 \left(\frac{3}{1 + 1} \right) = \frac{3}{2}$$

Example 9. *Evaluate* $\lim\limits_{\theta \to \frac{\pi}{2}} (\sec \theta - \tan \theta)$

Solution. We have,

$$\lim_{\theta \to \frac{\pi}{2}} (\sec \theta - \tan \theta) = \lim_{\theta \to \frac{\pi}{2}} \left(\frac{1}{\cos \theta} - \frac{\sin \theta}{\cos \theta} \right) = \lim_{\theta \to \frac{\pi}{2}} \left(\frac{1 - \sin \theta}{\cos \theta} \right)$$

$$= \lim_{\theta \to \frac{\pi}{2}} \left(\frac{(1 - \sin \theta) \cos \theta}{\cos^2 \theta} \right) = \lim_{\theta \to \frac{\pi}{2}} \frac{(1 - \sin \theta) \cos \theta}{(1 - \sin^2 \theta)}$$

$$= \lim_{\theta \to \frac{\pi}{2}} \frac{\cos \theta}{1 + \sin \theta} = \frac{0}{1 + 1} = 0$$

Example 10. *Evaluate* $\lim\limits_{x \to 0} \left(\dfrac{1 - \cos x \cos 2x \cos 3x}{\sin^2 2x} \right)$.

Solution. We have,

$$\cos x \cos 2x \cos 3x = \frac{1}{2}(2 \cos x \cos 3x \cos 2x) = \frac{1}{2}[(\cos 2x + \cos 4x) \cos 2x]$$

$$= \frac{1}{4}[(2 \cos^2 2x + 2 \cos 4x \cos 2x)]$$

$$= \frac{1}{4}[1 + \cos 4x + \cos 2x + \cos 6x]$$

Therefore,

$$\lim_{x \to 0}\left[\frac{1 - \cos x \cos 2x \cos 3x}{\sin^2 2x}\right] = \lim_{x \to 0}\left[\frac{1 - \frac{1}{4}(1 + \cos 4x + \cos 2x + \cos 6x)}{\sin^2 2x}\right]$$

$$= \lim_{x \to 0}\left[\frac{1 - \cos 2x + 1 - \cos 4x + 1 - \cos 6x}{4 \sin^2 2x}\right]$$

$$= \lim_{x \to 0}\frac{2\sin^2 x + 2\sin^2 2x + 2\sin^2 3x}{4\sin^2 2x}$$

$$= \lim_{x \to 0}\left\{\frac{2\left(\dfrac{\sin x}{x}\right)^2 . x^2 + 2\left(\dfrac{\sin 2x}{2x}\right)^2 . 4x^2 + 2\left(\dfrac{\sin 3x}{3x}\right)^2 . 9x^2}{4\left(\dfrac{\sin 2x}{2x}\right)^2 . 4x^2}\right\}$$

$$= \frac{28}{16} = \frac{7}{4}$$

EXAMPLE 11. *Evaluate* $\lim\limits_{x \to y} \dfrac{\tan x - \tan y}{x - y}$.

SOLUTION. We have,

$$\lim_{x \to y}\frac{\tan x - \tan y}{x - y} = \lim_{h \to 0}\frac{\tan(y + h) - \tan y}{y + h - y}$$

$$\lim_{h \to 0}\frac{1}{h}\left[\frac{\sin(y + h)}{\cos(y + h)} - \frac{\sin y}{\cos y}\right] = \lim_{h \to 0}\frac{\sin(y + h)\cos y - \cos(y + h)\sin y}{h\cos(y + h)\cos y}$$

$$= \lim_{h \to 0}\frac{\sin(y + h - y)}{h\cos(y + h)\cos y}$$

$$= \lim_{h \to 0}\frac{\sin h}{h} \cdot \frac{1}{\cos(y + h)\cos y} = 1 \cdot \frac{1}{\cos^2 y} = \sec^2 y$$

EXAMPLE 12. *Evaluate* $\lim\limits_{x \to 0}\left(\dfrac{\sqrt{1 + 2x} - \sqrt{1 - 2x}}{\sin x}\right)$.

SOLUTION. We have, $\lim\limits_{x \to 0}\left(\dfrac{\sqrt{1 + 2x} - \sqrt{1 - 2x}}{\sin x}\right)$

$$= \lim_{x \to 0}\frac{(\sqrt{1 + 2x} - \sqrt{1 - 2x})(\sqrt{1 + 2x} + \sqrt{1 - 2x})}{\sin x(\sqrt{1 + 2x} + \sqrt{1 - 2x}}$$

$$= \lim_{x \to 0}\frac{(1 + 2x) - (1 - 2x)}{\sin x(\sqrt{1 + 2x} + \sqrt{1 - 2x})} = \lim_{x \to 0}\frac{4x}{\sin x(\sqrt{1 + 2x} + \sqrt{1 - 2x})}$$

$$= 4\lim_{x \to 0}\frac{x}{\sin x} \cdot \lim_{x \to 0}\frac{1}{(\sqrt{1 + 2x} + \sqrt{1 - 2x})} = 4 \times 1\frac{1}{\sqrt{1 + 0} + \sqrt{1 - 0}}$$

$$= 4 \times \frac{1}{2} = 2$$

EXAMPLE 13. *Evaluate* $\displaystyle\lim_{\theta \to \frac{\pi}{6}} \left(\frac{\sin\left(\theta - \frac{\pi}{6}\right)}{\frac{\sqrt{3}}{2} - \cos\theta} \right).$

SOLUTION. Let $\theta = h + \dfrac{\pi}{6}$

$$\therefore \lim_{\theta \to \frac{\pi}{6}} \frac{\sin\left(\theta - \frac{\pi}{6}\right)}{\frac{\sqrt{3}}{2} - \cos\theta} = \lim_{h \to 0} \frac{\sin h}{\frac{\sqrt{3}}{2} - \cos\left(\frac{\pi}{6} + h\right)}$$

$$= \lim_{h \to 0} \frac{\sin h}{\cos\frac{\pi}{6} - \cos\left(\frac{\pi}{6} + h\right)} = \lim_{h \to 0} \frac{\sin h}{2\sin\left(\frac{\pi}{6} + \frac{h}{2}\right)\sin\frac{h}{2}}$$

$$= \lim_{h \to 0} \left\{ \frac{\left(\frac{\sin h}{h} \cdot h\right)}{2\sin\left(\frac{\pi}{6} + \frac{h}{2}\right)\left(\frac{\sin(h/2)}{h/2}\right)h/2} \right\} = \frac{1}{\sin\frac{\pi}{6}} = 2$$

EXAMPLE 14. *If* α, β *are the roots of the equation* $ax^2 + bx + c = 0$, *find the value of*

$$\lim_{x \to \frac{1}{\alpha}} \sqrt{\frac{1 - \cos(cx^2 + bx + a)}{2(1 - \alpha x)^2}}.$$

SOLUTION. Since α, β are the roots of the equation $ax^2 + bx + c = 0$ then roots of the equation $cx^2 + bx + a = 0$ will be $\dfrac{1}{\alpha}$ and $\dfrac{1}{\beta}$.

$$\Rightarrow \quad cx^2 + bx + a = c\left(x - \frac{1}{\alpha}\right)\left(x - \frac{1}{\beta}\right)$$

$$\therefore \quad \lim_{x \to \frac{1}{\alpha}} \sqrt{\frac{1 - \cos(cx^2 + bx + a)}{2(1 - \alpha x)^2}}$$

$$= \lim_{x \to \frac{1}{\alpha}} \sqrt{\frac{1 - \cos\left\{c\left(x - \frac{1}{\alpha}\right)\left(x - \frac{1}{\beta}\right)\right\}}{2(1 - \alpha x)^2}} = \lim_{x \to \frac{1}{\alpha}} \left| \frac{\sin\left\{\frac{c}{2}\left(x - \frac{1}{\alpha}\right)\left(x - \frac{1}{\beta}\right)\right\}}{1 - \alpha x} \right|$$

$$= \lim_{x \to \frac{1}{\alpha}} \left| \frac{\sin\left\{\frac{c}{2}\left(x - \frac{1}{\alpha}\right)\left(x - \frac{1}{\beta}\right)\right\}}{\frac{c}{2}\left(x - \frac{1}{\alpha}\right)\left(x - \frac{1}{\beta}\right)} \cdot \frac{c(\alpha x - 1)(\beta x - 1)}{2\alpha\beta(1 - \alpha x)} \right|$$

$$= \left| \frac{c}{2\alpha\beta}\left(\frac{\beta}{\alpha} - 1\right) \right| = \left| \frac{c}{2\alpha}\left(\frac{1}{\alpha} - \frac{1}{\beta}\right) \right|$$

EXAMPLE 15. *Evaluate* $\lim\limits_{x \to \theta} \dfrac{x \sin\theta - \theta \sin x}{x - \theta}$.

SOLUTION. Let $x = \theta + h \Rightarrow x \to \theta \Rightarrow h \to 0$

Now $\lim\limits_{x \to \theta} \dfrac{x \sin\theta - \theta \sin x}{x - \theta} = \lim\limits_{h \to 0} \dfrac{(\theta + h)\sin\theta - \theta \sin(\theta + h)}{\theta + h - \theta}$

$$= \lim\limits_{h \to 0} \dfrac{\theta \sin\theta + h \sin\theta - \theta \sin(\theta + h)}{h}$$

$$= \lim\limits_{h \to 0} \dfrac{\theta[\sin\theta - \sin(\theta + h)] + h \sin\theta}{h}$$

$$= \lim\limits_{h \to 0} \left[\dfrac{\theta \cdot 2 \cdot \cos\dfrac{2\theta + h}{2} \cdot \sin\left(-\dfrac{h}{2}\right)}{h} + \dfrac{h \sin\theta}{h} \right]$$

$$= \lim\limits_{h \to 0} \left[\dfrac{\theta \cdot 2\cos\dfrac{2\theta + h}{2} \cdot \dfrac{\sin\left(-\dfrac{h}{2}\right)}{\left(-\dfrac{h}{2}\right)}\left(-\dfrac{h}{2}\right)}{h} + \sin\theta \right]$$

$$= -\theta\cos\theta + \sin\theta = \sin\theta - \theta\cos\theta$$

EXAMPLE 16. *If* $f(x) = \dfrac{ax^2 + b}{x^2 + 1}$, $\lim\limits_{x \to 0} f(x) = 1$ *and* $\lim\limits_{x \to \infty} f(x) = 1$.

prove that $f(-2) = f(2) = 1$.

SOLUTION. We have, $f(x) = \dfrac{ax^2 + b}{x^2 + 1}$ ...(1)

and $\lim\limits_{x \to 0} f(x) = 1 \Rightarrow \lim\limits_{x \to 0} \dfrac{ax^2 + b}{x^2 + 1} = 1 \Rightarrow b = 1$

Also $\lim\limits_{x \to \infty} f(x) = 1 \Rightarrow \lim\limits_{x \to 0} \dfrac{ax^2 + b}{x^2 + 1} = 1 \Rightarrow \lim\limits_{x \to 0} \dfrac{a + \dfrac{b}{x^2}}{1 + \dfrac{1}{x^2}} = 1 \Rightarrow a = 1$

Now, putting the values of a and b in eqn. (1) we get

$$f(x) = \dfrac{x^2 + 1}{x^2 + 1} = 1$$

$\Rightarrow$ $\quad f(2) = f(-2) = 1$

EXAMPLE 17. *Evaluate* $\lim\limits_{n\to\infty} \dfrac{n!}{(n+1)!-n!}$.

SOLUTION. We have, $\lim\limits_{n\to\infty} \dfrac{n!}{(n+1)!-n!} = \lim\limits_{n\to\infty} \dfrac{\dfrac{n!}{(n+1)!}}{1-\dfrac{n!}{(n+1)!}} = \lim\limits_{n\to\infty} \dfrac{\dfrac{1}{n+1}}{1-\dfrac{1}{n+1}} = \dfrac{0}{1-0} = 0$

EXAMPLE 18. *Evaluate* $\lim\limits_{n\to\infty}\left(\dfrac{1^2}{n^3}+\dfrac{2^2}{n^3}+\dfrac{3^2}{n^3}+...+\dfrac{n^2}{n^3}\right)$.

SOLUTION. We have, $\lim\limits_{n\to\infty}\left[\dfrac{1^2}{n^3}+\dfrac{2^2}{n^3}+\dfrac{3^2}{n^3}+...+\dfrac{n^2}{n^3}\right] = \lim\limits_{n\to\infty}\dfrac{1^2+2^2+3^2+...+n^2}{n^3}$

$$= \lim\limits_{n\to\infty}\dfrac{n(n+1)(2n+1)}{6n^3} = \lim\limits_{n\to\infty}\dfrac{(n+1)(2n+1)}{6n^2}$$

$$= \lim\limits_{n\to\infty}\left(\dfrac{2n^2+3n+1}{6n^2}\right) = \lim\limits_{n\to\infty}\left(\dfrac{2+\dfrac{3}{n}+\dfrac{1}{n^2}}{6}\right) = \dfrac{2}{6} = \dfrac{1}{3}$$

EXAMPLE 19. *Evaluate* $\lim\limits_{x\to\infty}(\sqrt{x^2+x+1}-x)$.

SOLUTION. We have, $\lim\limits_{x\to\infty}(\sqrt{x^2+x+1}-x)$

$$= \lim\limits_{x\to\infty}\dfrac{(\sqrt{x^2+x+1}-x)(\sqrt{x^2+1}+x+x)}{(\sqrt{x^2+x+1}+x)}$$

$$= \lim\limits_{x\to\infty}\dfrac{x^2+x+1-x^2}{\sqrt{x^2+x+1}+x} = \lim\limits_{x\to\infty}\dfrac{x+1}{\sqrt{x^2+x+1}+x}$$

$$= \lim\limits_{x\to\infty}\dfrac{x+1}{x\sqrt{1+\dfrac{1}{x}+\dfrac{1}{x^2}}+x} = \lim\limits_{x\to\infty}\dfrac{x\left(1+\dfrac{1}{x}\right)}{x\left(\sqrt{1+\dfrac{1}{x}+\dfrac{1}{x^2}}+1\right)}$$

$$= \lim\limits_{x\to\infty}\dfrac{1+\dfrac{1}{x}}{x\sqrt{1+\dfrac{1}{x}+\dfrac{1}{x^2}}+1} = \dfrac{1}{2}$$

EXAMPLE 20. *Evaluate* $\lim\limits_{x\to 0^-}\left(\dfrac{x^2-3x+2}{x^3-2x^2}\right)$.

SOLUTION. We have, $\lim\limits_{x\to 0^-}\left(\dfrac{x^2-3x+2}{x^3-2x^2}\right) = \lim\limits_{x\to 0^-}\dfrac{(x-1)(x-2)}{x^2(x-2)} = \lim\limits_{x\to 0^-}\dfrac{x-1}{x^2} = -\infty$

EXAMPLE 21. *Evaluate* $\lim\limits_{x\to 0}\left(\dfrac{1}{|x|}\right)$.

SOLUTION. By definition, when $x\to 0^-, |x|\to 0^+ \Rightarrow \lim\limits_{x\to 0^-}\left(\dfrac{1}{|x|}\right) = \infty$ $\qquad$ ...(1)

and when $\qquad x \to 0^+, |x| \to 0^+ \Rightarrow \lim_{x \to 0^+}\left(\dfrac{1}{|x|}\right) = \infty \qquad \qquad \qquad ...(2)$

From eqn. (1) and (2), we can say that

$$\lim_{x \to 0}\left(\dfrac{1}{|x|}\right) = \infty$$

EXAMPLE 22. *If* $f(x) = \begin{cases} \dfrac{e^{1/x}}{1+e^{1/x}} & , \quad x \ne 0 \\ 0 & , \quad x = 0 \end{cases}$; *Then find* $\lim\limits_{x \to 0} f(x).$

SOLUTION. If $x < 0$, $\qquad$ LHL $= \lim\limits_{x \to 0-0}\left(\dfrac{1}{x}\right) = \infty$

$\Rightarrow \qquad \lim\limits_{x \to 0-0} f(x) = \lim\limits_{x \to 0} \dfrac{e^{1/x}}{1+e^{1/x}} = \dfrac{0}{1+0} = 0 \qquad [\because e^{-\infty} = 0] \qquad ...(1)$

Similarly RHL $(x > 0)$

$\Rightarrow \qquad \lim\limits_{x \to 0+0}\left(\dfrac{1}{x}\right) = \infty$

$\Rightarrow \qquad \lim\limits_{x \to 0+0} f(x) = \lim\limits_{x \to 0} \dfrac{e^{1/x}}{1+e^{1/x}}$

$\qquad \qquad = \lim\limits_{x \to 0} \dfrac{1}{\dfrac{1}{e^{1/x}}+1} = \dfrac{1}{0+1} = 1 \qquad [\because \lim\limits_{x \to 0} e^{-1/x} = 0] \qquad ...(2)$

From eqn. (1) and (2), we have

$$\lim\limits_{x \to 0-0} f(x) \ne \lim\limits_{x \to 0+0} f(x)$$

Hence, $\lim\limits_{x \to 0} f(x)$ does not exist.

EXAMPLE 23. *If* $f(x) = \begin{cases} mx^2 + n & ; \quad x < 0 \\ nx + m & ; \quad 0 \le x \le 1 \\ nx^3 + m & ; \quad x > 1 \end{cases}$

Then for what integral values of m and n $\lim\limits_{x \to 0} f(x)$ *and* $\lim\limits_{x \to 1} f(x)$ *exist?* $\qquad$ **(NCERT)**

SOLUTION. We have

$$f(x) = \begin{cases} mx^2 + n & ; \quad x < 0 \\ nx + m & ; \quad 0 \le x \le 1 \\ nx^3 + m & ; \quad x > 1 \end{cases}$$

Therefore $\lim\limits_{x \to 0^-} f(x) = \lim\limits_{x \to 0^-} (mx^2 + n) = n$

and $\qquad \lim\limits_{x \to 0^+} f(x) = \lim\limits_{x \to 0^+} (nx + m) = m$

Clearly, for the existence of $\lim\limits_{x \to 0} f(x)$

We must have $\lim\limits_{x \to 0^-} f(x) = \lim\limits_{x \to 0^+} f(x)$

i.e., m must be equal to n.

Similarly,
$$\lim_{x \to 1^-} f(x) = \lim_{x \to 1^-} (nx + m) = n + m$$

$$\lim_{x \to 1^+} f(x) = \lim_{x \to 1^+} (nx^3 + m) = n + m$$

For the existence of $\lim_{x \to 1} f(x)$, we have

$$\lim_{x \to 1^-} f(x) = \lim_{x \to 1^+} f(x)$$

Hence, for the existence of both $\lim_{x \to 0} f(x)$ and $\lim_{x \to 1} f(x)$ we must have

$$n + m = m + n \Rightarrow m = n.$$

EXAMPLE 24. *If the function $f(x)$ satisfies $\lim_{x \to 1} \dfrac{f(x) - 2}{x^2 - 1} = \pi$, find the value of $\lim_{x \to 1} f(x)$.* **(NCERT)**

SOLUTION. We have, $\lim_{x \to 1} \dfrac{f(x) - 2}{x^2 - 1} = \pi$ ...(1)

Clearly, $\lim_{x \to 1} (x^2 - 1) = 0$

Therefore, if $\lim_{x \to 1} f(x) - 2 \neq 0$, then $\lim_{x \to 1} \dfrac{f(x) - 2}{x^2 - 1} = \infty$ or $-\infty$ which is not possible by eqn. (1).

Therefore $\lim_{x \to 1} \{f(x) - 2\} = 0$

$\Rightarrow \qquad \lim_{x \to 1} f(x) = 2$

EXAMPLE 25. *If $f(x) = \begin{cases} a + bx & ; \quad x < 1 \\ 4 & ; \quad x = 1 \\ b - ax & ; \quad x > 1 \end{cases}$ and $\lim_{x \to 1} f(x) = f(1)$, find the value of a and b.* **(NCERT)**

SOLUTION. We have, $f(x) = \begin{cases} a + bx & ; \quad x < 1 \\ 4 & ; \quad x = 1 \\ b - ax & ; \quad x > 1 \end{cases}$

$\Rightarrow \qquad f(1) = 4$ ...(1)

$$\lim_{x \to 1^-} f(x) = \lim_{x \to 1^-} (a + bx) = a + b \qquad ...(2)$$

and $\lim_{x \to 1^+} f(x) = \lim_{x \to 1^+} (b - ax) = b - a \qquad ...(3)$

It is given that $\lim_{x \to 1} f(x)$ exists.

Therefore, $\lim_{x \to 1^-} f(x) = \lim_{x \to 1^+} f(x)$

$\Rightarrow \qquad a + b = b - a$

$\Rightarrow \qquad 2a = 0 \qquad \Rightarrow \qquad a = 0$

Similarly, $\lim_{x \to 1} f(x) = f(1)$

$\Rightarrow \qquad a + b = 4 \qquad \Rightarrow \qquad b = 4 \qquad (\because a = 0)$

Hence, $\qquad a = 0 \qquad$ and $\qquad b = 4.$

EXAMPLE 26. Let $f(x) = \begin{cases} x & 0 \le x < 1 \\ 3 & x = 1 \\ 3 - x & x > 1 \end{cases}$

Does $\lim\limits_{x \to 1} f(x)$ exists.

SOLUTION.
$$\text{RHS} = \lim_{h \to 0} f(1+h) = \lim_{h \to 0} 3 - (1+h)$$

$$= \lim_{h \to 0}(2 - h) = 2 - 0 = 2$$

$$\text{LHS} = \lim_{h \to 0} f(1-h) = \lim_{h \to 0}(1-h) = 1 - 0 = 1$$

Clearly   LHS $\ne$ RHS at $x = 1$.

Hence, $\lim\limits_{x \to 1} f(x)$ does not exists.

Exercise 12.2

1. Evaluate the following limits:

(i) $\lim\limits_{x \to 1} \dfrac{x^2 + 3x + 2}{x^2 + 1}$

(ii) $\lim\limits_{x \to 64} \dfrac{x^{1/6} - 2}{x^{1/3} - 4}$

(iii) $\lim\limits_{x \to 0} \dfrac{\sqrt{1 + x^2} - \sqrt{1 + x}}{x}$

(iv) $\lim\limits_{x \to 1} \dfrac{x^2 - \sqrt{x}}{\sqrt{x} - 1}$

(v) $\lim\limits_{x \to 1} \dfrac{x^4 - 3x^2 + 2}{x^3 - 5x^2 + 3x + 1}$

(vi) $\lim\limits_{x \to -1} \dfrac{x^3 - 4x + 1}{x - 1}$

2. Evaluate the following limits:

(i) $\lim\limits_{x \to -1}(1 + x + x^2 + \dots + x^{10})$

(ii) $\lim\limits_{x \to \pi}\left(x - \dfrac{22}{7}\right)$

(iii) $\lim\limits_{x \to 2} \dfrac{x^4 - 16}{x - 2}$

(iv) $\lim\limits_{x \to -3} \dfrac{x^3 + 27}{x^5 + 243}$

(v) $\lim\limits_{x \to 1} \dfrac{x^{1/3} - 1}{x^{1/6} - 1}$

(vi) $\lim\limits_{x \to a} \dfrac{x - a}{x^{3/2} - a^{3/2}}$

(vii) $\lim\limits_{x \to \infty}\left(1 + \dfrac{p}{x}\right)^x$

(viii) $\lim\limits_{x \to 1}\left(\dfrac{1 - x^{1/3}}{1 - x^{-2/3}}\right)$

(ix) $\lim\limits_{x \to 3} \dfrac{x^n - 3^n}{x - 3}$

(x) $\lim\limits_{x \to 0}\left(\dfrac{\sqrt{a + x} - \sqrt{a}}{x\sqrt{a(a + x)}}\right)$

(xi) $\lim\limits_{x \to 2} \dfrac{x^3 - 2x^2}{x^2 - 5x + 6}$

(xii) $\lim\limits_{x \to 1} \dfrac{\sqrt{x^2 + 8} - \sqrt{10 - x^2}}{\sqrt{x^2 + 3} - \sqrt{5 - x^2}}$

(xiii) $\lim\limits_{x \to 2}\left[\dfrac{1}{x - 2} - \dfrac{2(2x - 3)}{x^3 - 3x^2 + 2x}\right]$

(xiv) $\lim\limits_{x \to 1}\left[\dfrac{x - 2}{x^2 - x} - \dfrac{1}{x^3 - 3x^2 + 2x}\right]$

3. Evaluate the following limits:

(i) $\lim\limits_{x \to 0} \dfrac{\sin 2x + 3x}{2x + \sin 3x}$

(ii) $\lim\limits_{x \to 0} \dfrac{x(\cos x + \cos 2x)}{\sin x}$

(iii) $\lim\limits_{x \to 0} \dfrac{\sin x - 2\sin 3x + \sin 5x}{x}$

(iv) $\lim\limits_{x \to 0} \dfrac{\tan x - \sin x}{x^3}$

(v) $\lim\limits_{x \to 0} \dfrac{1 - \sqrt{\cos x}}{x^2}$

(vi) $\lim\limits_{y \to 0} \dfrac{(x + y)\sec(x + y) - x\sec x}{y}$

(vii) $\lim\limits_{x \to \frac{\pi}{2}} \dfrac{\cot x}{\frac{\pi}{2} - x}$

(viii) $\lim\limits_{x \to \frac{\pi}{4}} \dfrac{\sec^2 x - 2}{\tan x - 1}$

(ix) $\lim\limits_{x \to \frac{\pi}{6}} \dfrac{\sqrt{3}\sin x - \cos x}{x - \dfrac{\pi}{6}}$

(x) $\lim\limits_{x \to 0} \dfrac{3\sin x - \sin 3x}{x^3}$

(xi) $\lim\limits_{x \to 0} x(3\,\text{cosec}\,2x - 2\cot 3x)$

(xii) $\lim\limits_{x \to 0} \dfrac{\cos 7x - \cos 9x}{\cos 3x - \cos 5x}$

(xiii) $\lim\limits_{x \to 0} \dfrac{\tan x - \sin x}{x^3}$

(xiv) $\lim\limits_{x \to 0} \dfrac{\cos ax - \cos bx}{\cos x - 1}$

(xv) $\lim\limits_{\theta \to 0} \dfrac{\sin \theta^n}{(\sin \theta)^m}, n > m > 0$

(xvi) $\lim\limits_{x \to \pi} \dfrac{\sin(\pi - x)}{\pi(\pi - x)}$

(xvii) $\lim\limits_{x \to 1} (1 - x) \tan \dfrac{\pi x}{2}$

(xviii) $\lim\limits_{\theta \to \frac{\pi}{2}} \dfrac{\sec \theta - \tan \theta}{\pi - 2\theta}$

4. Evaluate the following limits:

(i) $\lim\limits_{x \to y} \dfrac{\sin x - \sin y}{x - y}$

(ii) $\lim\limits_{x \to \pi} \dfrac{1 - \sin \dfrac{x}{2}}{\cos \dfrac{x}{2}\left(\cos \dfrac{x}{4} - \sin \dfrac{x}{4}\right)}$

(iii) $\lim\limits_{\theta \to \pi/4} \left(\dfrac{\sin \theta - \cos \theta}{\theta - \dfrac{\pi}{4}}\right)$

(iv) $\lim\limits_{x \to a} \dfrac{\sin x - \sin a}{\sqrt{x} - \sqrt{a}}$

(v) $\lim\limits_{y \to x} \dfrac{y \cos x - x \cos y}{y - x}$

(vi) $\lim\limits_{\theta \to \frac{\pi}{4}} \dfrac{1 - \tan \theta}{1 - \sqrt{\tan \theta}}$

5. Evaluate the following limits:

(i) $\lim\limits_{x \to \infty} \dfrac{x^3 + x^2 - 6x + 8}{4x^3 + 5x - 8}$

(ii) $\lim\limits_{x \to -\infty} (\sqrt{x^2 + 4x} - \sqrt{x^2 - 4x})$

(iii) $\lim\limits_{x \to \infty} \left(\dfrac{1}{n^2} + \dfrac{2}{n^2} + \dots + \dfrac{n}{n^2}\right)$

(iv) $\lim\limits_{n \to \infty} \left(\dfrac{1^3}{n^4} + \dfrac{2^3}{n^4} + \dfrac{3^3}{n^4} + \dots + \dfrac{n^3}{n^4}\right)$

(v) $\lim\limits_{n \to \infty} \left[\dfrac{1 \cdot 2 + 2 \cdot 3 + \dots + n \cdot (n+1)}{n^3}\right]$

(vi) $\lim\limits_{x \to -\infty} (\sqrt{x^2 + ax} - \sqrt{x^2 - ax})$

(vii) $\lim\limits_{n \to \infty} \dfrac{2^n - 1}{2^n + 1}$

(viii) $\lim\limits_{n \to \infty} \dfrac{(n+2)! + (n+1)!}{(n+3)!}$

(ix) $\lim\limits_{n \to \infty} \dfrac{\sqrt{n^2 + 1} + n)^2}{(n^6 + 1)^{1/3}}$

(x) $\lim\limits_{x \to \infty} \dfrac{(x+1)^{10} + (x+2)^{10} + (x+3)^{10}}{x^{10} + (x+3)^{10}}$

6. Evaluate the following limits: where [ ] denotes greatest integer function

(i) $\lim\limits_{x \to \frac{\pi}{2}^-} (\tan x)$

(ii) $\lim\limits_{x \to a^+} [x]$

(iii) $\lim\limits_{x \to 2-0} [x]$

(iv) $\lim\limits_{x \to a^-} \dfrac{|x^3|}{|x|}$

(v) $\lim\limits_{x \to \frac{5^-}{2}} [x]$

(vi) $\lim\limits_{x \to \frac{7}{3}} [-x]$

(vii) $\lim\limits_{x \to 0^-} (2 - \cot x)$

(viii) $\lim\limits_{x \to 0^-} (1 + \csc x)$

(ix) $\lim\limits_{x \to -3^+} \dfrac{[x - 7]}{[x + 4]}$

(x) $\lim\limits_{x \to -2^-} \dfrac{x - 3}{x^2 - 4}$

7. Let $f(x)$ be a function such that $f(-x) = -f(x)$ and $\lim\limits_{x \to 0} f(x)$ exists, prove that

$$\lim\limits_{x \to 0} f(x) = 0$$

8. Prove that $\lim\limits_{x \to 3^+} \left(\dfrac{x}{[x]}\right) \neq \lim\limits_{x \to 3^-} \dfrac{x}{[x]}$.

9. Prove that $\lim\limits_{x \to 1^+} \left(\dfrac{1}{x - 1}\right) \neq \lim\limits_{x \to 1^-} \left(\dfrac{1}{x - 1}\right)$

10. If $f(x) = \begin{cases} 3 - x^2 & ; & x \le -2 \\ ax + b & ; & -2 < x < 2 \\ \dfrac{x^2}{2} & ; & x \ge 2 \end{cases}$

then for the existence of the $\lim\limits_{x \to 2} f(x)$ and $\lim\limits_{x \to -2} f(x)$, prove that $a = \dfrac{3}{4}, b = \dfrac{1}{2}$.

11. A function $f(x)$ is defined as follows

$$f(x) = x^2 \quad \text{when } x < 1$$
$$= \dfrac{5}{2} \quad \text{when } x = 1$$
$$= x^2 + 2 \quad \text{when } x > 1$$

Does $\lim\limits_{x \to 1} f(x)$ exists.

ANSWERS

1. (i) 3 (ii) $\dfrac{1}{4}$ (iii) $-\dfrac{1}{2}$ (iv) 3 (v) $\dfrac{1}{2}$ (vi) –2 **2.** (i) 1 (ii) $\left(\pi - \dfrac{22}{7}\right)$ (iii) 32

(iv) $\dfrac{1}{15}$ (v) 2 (vi) $\dfrac{2}{3\sqrt{a}}$ (vii) e^p (viii) $-\dfrac{1}{2}$ (ix) $n.3^{n-1}$ (x) $\dfrac{1}{2}a^{-3/2}$ (xi) – 4

(xii) $\dfrac{2}{3}$ (xiii) $-\dfrac{1}{2}$ (xiv) 2 **3.** (i) 1 (ii) 2 (iii) 0 (iv) $\dfrac{1}{2}$ (v) $\dfrac{1}{4}$

(vi) $\sec x(1+x\tan x)$ (vii) 1 (viii) 2 (ix) 2 (x) 4 (xi) $\dfrac{5}{6}$ (xii) 2

(xiii) $\dfrac{1}{2}$ (xiv) $a^2 - b^2$ (xv) 0 (xvi) $\dfrac{1}{\pi}$ (xvii) $\dfrac{2}{\pi}$ (xviii) $\dfrac{1}{4}$

4. (i) $\cos y$ (ii) $\dfrac{1}{\sqrt{2}}$ (iii) $\sqrt{2}$ (iv) $2\sqrt{a}\cos a$ (v) $\cos x + x\sin x$ (vi) 2

5. (i) $\dfrac{1}{4}$ (ii) 4 (iii) $\dfrac{1}{2}$ (iv) $\dfrac{1}{4}$ (v) $\dfrac{1}{3}$ (vi) a (vii) 1 (viii) 0 (ix) 4 (x) $\dfrac{3}{2}$

6. (i) ∞ (ii) $[a]$ (iii) 1 (iv) a^2 (v) 2 (vi) –3 (vii) ∞ (viii) $-\infty$ (ix) –10 (x) ∞

11. limit does not exist

12.12 INTRODUCTION

Geometrically, a continuous process is one that goes on smoothly without any sudden change. Continuity of a function can also be interpreted in a similar way when we say that a function $f(x)$ is continuous at $x = a$ we mean that there is no interruption (cut) in the graph of $f(x)$ at $x = a$, i.e., the graph of $f(x)$ is unbroken at $x = a$ and there is no hole, gap or jump in the graph. On the other hand, if there is a sudden jump in the value of the function at $x = a$ and the value of the function changes gradually for change in the value of the independent variable. Consider the following graph.

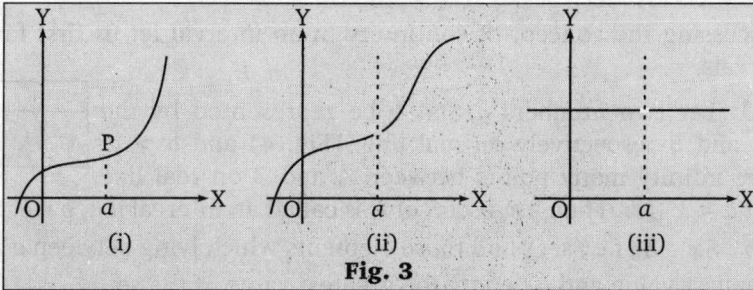

Fig. 3

If we draw a perpendicular from a on x-axis, then it will cut the curve at any point say P, i.e., at $x = a$, there is no cut in the graph, but if graph of the function has a sudden cut at $x = a$, then we say that graph is broken.

The graph of the function in figure (i) proceed smoothly while in figure (ii) graph has a cut at the point $x = a$. Hence, we say that the function of fig. (i) is continuous while function of fig. (ii) is discontinuous at $x = a$.

12.13 CONTINUITY OF A FUNCTION AT A POINT

A function $f(x)$ is said to be continuous at $x = a$ if $f(x)$ is defined at $x = a$, $\lim_{x\to a} f(x)$ exists and $\lim_{x\to a} f(x) = f(a)$, i.e., $f(x)$ is said to be continuous at $x = a$ if

$$\lim_{x\to a^+} f(x) = \lim_{x\to a^-} f(x) = f(a)$$

$\Rightarrow$ Right hand limit of $f(x)$ at $x = a$ = Left hand limit of $f(x)$ at $x = a$ = Value of the function at $x = a$

i.e., $$f(a + 0) = f(a - 0) = f(a)$$

WORKING PROCEDURE

To check the continuity of the function $f(x)$ at $x = a$, use the following steps :

Step 1. Find the value of $f(x)$ at $x = a$, i.e., $f(a)$

Step 2. Evaluate LHL = $\lim\limits_{x \to a^-} f(x) = f(a - 0)$

Step 3. Evaluate RHL = $\lim\limits_{x \to a^+} f(x) = f(a + 0)$

Step 4. If $f(a) = f(a + 0) = f(a - 0)$, then $f(x)$ is said to be continuous at $x = a$.

☞ REMARK

➠ If any of the value $f(a + 0), f(a - 0)$ or $f(a)$ is different, the function is not continuous at $x = a$.

Hence, to check the continuity of a function $f(x)$ at a point $x = a$ we have to calculate all three values, i.e., $f(a + 0), f(a - 0)$ and $f(a)$.

12.14 CONTINUITY OF THE FUNCTION FROM THE LEFT AND FROM THE RIGHT

Let $f(x)$ be a function. Then $f(x)$ is said to be continuous at $x = a$ from left if

$$\lim\limits_{x \to a^-} f(x) = f(a)$$

Again, $f(x)$ is said to be continuous at $x = a$ from right if

$$\lim\limits_{x \to a^+} f(x) = f(a)$$

12.15 CONTINUITY OF A FUNCTION IN ANY INTERVAL

Before discussing the concept of continuity in an interval let us first familiar with the concept of intervals.

(i) Interval: Let two numbers a and b be represented by the points A and B respectively on real line. (Fig. 4) and $b > a$. There are infinity many points between A and B on real line such that $a < x < b$. Then a subset S of $\mathbf{R}$ is called an interval if $a, b \in S, x \in \mathbf{R}$ such that $a < x < b \Rightarrow x \in S$, i.e., set of all those elements which lying between a and b, where a is the smallest value and b denote the greatest value of the set.

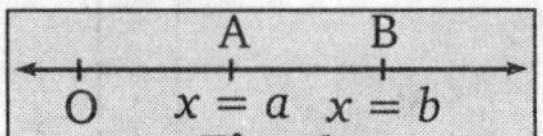

Fig. 4

(a) Open Interval: The set $]a, b[$ in which the end points a and b are not included is called an open interval. It is represented by $a < x < b$ or (a, b) or $]a, b[$ on real line, to represented the open interval its end points are represented by empty small circle (O).

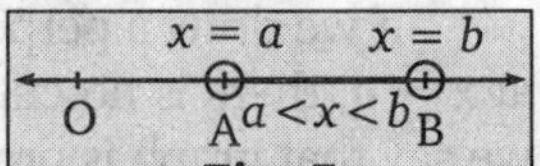

Fig. 5

(b) Closed Interval: On the real line, the set of all elements between $A(x = a)$ and $B(x = b)$ including a and b is called closed interval. It is represented by $a \le x \le b$. Symbolically, it can be written as $[a, b]$.

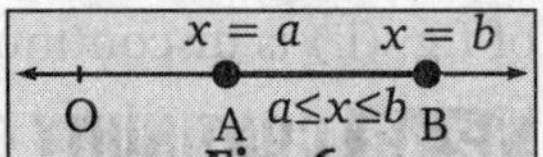

In the closed interval, we represent the end points by black circle and a dark line between them.

Fig. 6

12.16 CONTINUITY OF A FUNCTION IN ANY INTERVAL

A function $f(x)$ is said to be continuous in an interval $[a, b]$ if

(i) $f(x)$ is continuous at $x = a$ from right, i.e., $\lim\limits_{x \to a^+} f(x) = f(a)$

(ii) $f(x)$ is continuous at $x = b$ from left, i.e., $\lim\limits_{x \to b^-} f(x) = f(b)$

(iii) $f(x)$ is continuous from both sides for all x lying between a and b.

12.17 DISCONTINUITY OF A FUNCTION IN AN INTERVAL

A function $f(x)$ is said to be discontinuous in an interval I if it is not continuous at one or more points of the interval I.

12.18 PROPERTIES OF CONTINUOUS FUNCTIONS

We shall present here some theorems related to continuous functions (without proof).

(i) If $f(x)$ and $g(x)$ be two continuous functions at $x = a$, then sum and difference of $f(x)$ and $g(x)$ are also continuous at $x = a$.

i.e., $[f(x) \pm g(x)]$ is continuous at $x = a$.

(ii) If $f(x)$ and $g(x)$ be two continuous functions at $x = a$, then their product and quotient are also continuous at $x = a$, i.e., $[f(x) \cdot g(x)]$ and $\left[\dfrac{f(x)}{g(x)}\right]$, $g(a) \neq 0$, are also continuous.

(iii) If $f(x)$ is continuous at $x = a$, then $kf(x)$ is also continuous at $x = a$, $k \in \mathbf{R}$.

(iv) If $f(x)$ is continuous at $x = a$, then $\dfrac{1}{f(x)}$ is also continuous at $x = a$ provided $f(a) \neq 0$.

(v) If $f(x)$ is continuous at $x = a$ then $|f(x)|$ is also continuous at $x = a$.

(vi) Every polynomial function is continuous.

(vii) Every constant function is continuous.

(viii) If fog is defined for two continuous functions f and g then fog is also continuous.

☞ RECAPITULATIONS

- A function $f(x)$ is said to be continuous at $x = a$ if $f(x)$ is defined at $x = a$, $\lim\limits_{x \to a} f(x)$ exists and $\lim\limits_{x \to a} f(x) = f(a)$.

- Every polynomial function is continuous.

- Every constant function is continuous.

- Composite of two continuous functions is again continuous.

- Sum, difference, product and quotient of two continuous functions is again continuous.

SOLVED EXAMPLES

EXAMPLE 1. Prove that $f(x) = \dfrac{1}{x - a}$ is not continuous at $x = a$.

SOLUTION. **Left Hand Limit (LHL)**

Let $x = a - h$

If $x = a$ then $h = 0$

$\therefore \quad x \to a \qquad \Rightarrow \qquad h \to 0$

$$\therefore \quad \lim_{x \to a^-} f(x) = \lim_{x \to 0}(a - h)$$

$$= \lim_{x \to 0} \frac{1}{a - h - a} = \lim_{h \to 0} \frac{1}{-h} = -\infty$$

Right Hand Limit (RHL)

Let $x = a + h$ If $x = a$ then $h = 0$

$$\therefore \qquad\qquad \lim_{x \to a} \Rightarrow \lim_{h \to 0}$$

$$\therefore \quad \lim_{x \to a^+} f(x) = \lim_{h \to 0} f(a + h) = \lim_{h \to 0} \frac{1}{a + h - a} = \lim_{h \to 0} \frac{1}{h} = \infty$$

$$\therefore \qquad\qquad \text{LHL} \ne \text{RHL}$$

Hence $f(x)$ is not continuous at $x = a$.

EXAMPLE 2. *Prove that*

$$f(x) = \begin{cases} x \sin \dfrac{1}{x} & , \quad x \ne 0 \\ 0 & , \quad x = 0 \end{cases}$$

is continuous at $x = 0$.

SOLUTION: Given, $\qquad f(0) = 0$ \qquad\qquad ...(1)

LHL

Let $x = 0 - h$ if $x = 0 \Rightarrow h = 0$

$$\therefore \qquad\qquad \lim_{x \to 0} \Rightarrow \lim_{h \to 0}$$

$$f(0 - 0) = \lim_{x \to 0^-} f(x) = \lim_{x \to 0} f(0 - h) = \lim_{h \to 0}(0 - h)\sin\left(\frac{1}{0 - h}\right)$$

$$= \lim_{h \to 0}(-h)\sin\left(\frac{1}{-h}\right) = \lim_{h \to 0} h \sin\left(\frac{1}{h}\right)$$

$$= \lim_{h \to 0} h \lim_{h \to 0} \sin\left(\frac{1}{h}\right)$$

$$= 0 \times (\text{a finite quantity lying between } -1 \text{ and } 1)$$

$$= 0 \qquad\qquad ...(2)$$

RHL

Let $x = 0 + h$ If $x = 0 \Rightarrow h = 0$

$$\therefore \qquad\qquad \lim_{x \to 0} \Rightarrow \lim_{h \to 0}$$

$$f(0 + 0) = \lim_{x \to 0^+} f(x) = \lim_{h \to 0} f(0 + h)$$

$$= \lim_{h \to 0}(0 + h)\sin\left(\frac{1}{0 + h}\right)$$

$$= \lim_{h \to 0} h \sin\left(\frac{1}{h}\right) = \lim_{h \to 0} h \cdot \lim_{h \to 0} \sin\left(\frac{1}{h}\right)$$

$$= 0 \times [\text{a finite quantity lying between } -1 \text{ and } 1]$$

$$= 0 \qquad\qquad ...(3)$$

$$\therefore \qquad\qquad f(0 + 0) = 0$$

It is clear from (1), (2) and (3) that
$$f(0) = f(0-0) = f(0+0)$$
$\Rightarrow$ $f(x)$ is continuous at $x = 0$.

EXAMPLE 3. *For what value of k, the function*

$$f(x) = \begin{cases} \dfrac{x^2 - 9}{x - 3} & , \quad x \neq 3 \\ k & , \quad x = 3 \end{cases}$$

is continuous at $x = 3$?

SOLUTION. Given, $f(3) = k$

LHL at $x = 3$

Let $x = 3 - h$. If $x = 3 \Rightarrow h = 0$

$\therefore$ $\displaystyle\lim_{x \to 3^-} \Rightarrow \lim_{h \to 0}$

$\therefore$ $\displaystyle\lim_{x \to 3^-} f(x) = \lim_{h \to 0} f(3 - h) = \lim_{h \to 0} \dfrac{(3-h)^2 - 9}{3 - h - 3}$

$\qquad = \displaystyle\lim_{h \to 0} \dfrac{(3 - h + 3)(3 - h - 3)}{-h}$

$\qquad = \displaystyle\lim_{h \to 0} \dfrac{(6 - h)(-h)}{(-h)}$

$\qquad = \displaystyle\lim_{h \to 0} (6 - h) = 6$ $\qquad (\because h = 0)$

RHL at $x = 3$

Let $x = 3 + h$, if $x = 3 \Rightarrow h = 0$

$\therefore$ $\displaystyle\lim_{x \to 3} \Rightarrow \lim_{h \to 0}$

$\therefore$ $\displaystyle\lim_{x \to 3^+} f(x) = \lim_{h \to 0}(3 + h) = \lim_{h \to 0} \dfrac{(3 + h)^2 - 9}{3 + h - 3}$

$\qquad = \displaystyle\lim_{h \to 0} \dfrac{(3 + h + 3)(3 + h - 3)}{h}$

$\qquad = \displaystyle\lim_{h \to 0} \dfrac{(6 + h)h}{h} = \lim_{h \to 0}(6 + h)$

$\qquad = 6 + 0 = 6$ $\qquad (\because h = 0)$

$\because$ $f(x)$ is continuous at $x = 3$.

$\therefore$ $f(3) = f(3 - h) = f(3 + h)$

$\Rightarrow$ $k = 6 = 6$

$\therefore$ $k = 6$

EXAMPLE 4. *A function $f(x)$ is defined as follows :*

$$f(x) = \begin{cases} x, & \text{if} \quad 0 \leq x < \dfrac{1}{2} \\ 0, & \text{if} \quad x = \dfrac{1}{2} \\ 1 - x, & \text{if} \quad \dfrac{1}{2} < x \leq 1 \end{cases}$$

Find the value of $\displaystyle\lim_{x \to 1/2} f(x)$.

SOLUTION. **LHL of the function**

Let $x = \dfrac{1}{2} - h$ if $x = \dfrac{1}{2} \Rightarrow h = 0$

$\therefore \qquad \lim\limits_{x \to \frac{1}{2}} \Rightarrow \lim\limits_{h \to 0}$

$\therefore \qquad \lim\limits_{x \to \left(\frac{1}{2}\right)^-} f(x) = \lim\limits_{h \to 0} f\left(\dfrac{1}{2} - h\right) = \lim\limits_{h \to 0}\left[\dfrac{1}{2} - h\right] = \left(\dfrac{1}{2} - 0\right) = \dfrac{1}{2}$ $\qquad (\because h = 0)$

RHL of the function

Let $x = \dfrac{1}{2} + h$ if $x = \dfrac{1}{2} \Rightarrow h = 0$

$\therefore \qquad \lim\limits_{x \to \left(\frac{1}{2}\right)^+} \Rightarrow \lim\limits_{h \to 0}$

$\therefore \qquad \lim\limits_{x \to \left(\frac{1}{2}\right)^+} f(x) = \lim\limits_{h \to 0} f\left(\dfrac{1}{2} + h\right) = \lim\limits_{h \to 0}\left[1 - \left(\dfrac{1}{2} + h\right)\right] = \lim\limits_{h \to 0}\left[\dfrac{1}{2} - h\right]$

$$= \left(\dfrac{1}{2} - 0\right) = \dfrac{1}{2} \qquad\qquad (\because h = 0)$$

Clearly $\qquad$ LHL $=$ RHL $= \dfrac{1}{2}$

$\therefore \qquad \lim\limits_{x \to \frac{1}{2}} f(x) = \dfrac{1}{2}$

EXAMPLE 5. *Check the continuity of the function*

$$f(x) = \begin{cases} \dfrac{|x|}{x} & ,\text{if} \quad x \neq 0 \\ 1 & ,\text{if} \quad x = 0 \end{cases} \ at\ x = 0.$$

SOLUTION. **LHL of the function**

Let $x = 0 - h$, if $x = 0 \Rightarrow h = 0$

$\therefore \qquad \lim\limits_{x \to 0} \Rightarrow \lim\limits_{h \to 0}$

$\therefore \qquad \lim\limits_{x \to 0^-} f(x) = \lim\limits_{h \to 0} f(0 - h) = \lim\limits_{h \to 0} \dfrac{|0 - h|}{(0 - h)} = \lim\limits_{h \to 0} \dfrac{|-h|}{-h} = \lim\limits_{h \to 0} \dfrac{h}{-h} = -1$

RHL of the function

Let $x = 0 + h$, if $x = 0 \Rightarrow h = 0$

$\therefore \qquad \lim\limits_{x \to 0} \Rightarrow \lim\limits_{h \to 0}$

$\therefore \qquad \lim\limits_{x \to 0^+} f(x) = \lim\limits_{h \to 0} f(0 + h) = \lim\limits_{h \to 0} \dfrac{|0 + h|}{(0 + h)} = \lim\limits_{h \to 0} \dfrac{|h|}{h} = \lim\limits_{h \to 0} \dfrac{h}{h} = 1$

Also, the value of $f(x)$ at $x = 0, f(0) = 1$

$\Rightarrow \qquad$ LHL $\neq$ RHL $= f(0)$

Hence, $f(x)$ is not continuous at $x = 0$.

EXAMPLE 6. *Check the continuity of the function*

$$f(x) = \begin{cases} -x^2 & \text{,if } x \le 0 \\ 5x - 4 & \text{,if } 0 < x \le 1 \\ 4x^2 - 3x & \text{,if } 1 < x < 2 \\ 3x + 4 & \text{,if } x \ge 2 \end{cases}$$

at x = 0, 1, 2.

SOLUTION. We have to check the continuity of $f(x)$ at $x = 0$

(i) Continuity at $x = 0$

LHL of the function

Let $x = 0 - h$ if $x = 0 \Rightarrow h = 0$

$\therefore$ $\lim_{x \to 0} \Rightarrow \lim_{h \to 0}$

$\therefore$ $\lim_{x \to 0^-} f(x) = \lim_{h \to 0} f(0 - h)$

$\qquad\qquad = \lim_{h \to 0} [-(0 - h)^2]$ $\qquad [\because \text{Here } f(x) = -x^2]$

$\qquad\qquad = \lim_{h \to 0} [-h^2] = 0$ $\qquad\qquad\qquad$ ...(1)

RHL of the function

Let $x = 0 + h$ if $x = 0 \Rightarrow h = 0$

$\therefore$ $\lim_{x \to 0} \Rightarrow \lim_{h \to 0}$

$\therefore$ $\lim_{x \to 0^+} f(x) = \lim_{h \to 0} f(0 + h) = \lim_{h \to 0} [5(0 + h) - 4]$

$\qquad\qquad\qquad\qquad\qquad\qquad [\because \text{Here } f(x) = 5x - 4]$

$\qquad\qquad = \lim_{h \to 0} [5h - 4] = 5 \times 0 - 4 = -4$ $\qquad$ ...(2)

From eqn. (1) and (2) it is clear that RHL ≠ LHL

$\Rightarrow$ The function $f(x)$ is not continuous at $x = 0$.

(ii) Continuity at $x = 1$

LHL of the function

Let $x = 1 - h$ if $x = 1 \Rightarrow h = 0$

$\therefore$ $\lim_{x \to 1} \Rightarrow \lim_{h \to 0}$

$\therefore$ $\lim_{x \to 1^-} f(x) = \lim_{h \to 0} f(1 - h)$

$\qquad\qquad = \lim_{h \to 0} [5(1 - h) - 4]$ $\qquad [\because \text{Here } f(x) = 5x - 4]$

$\qquad\qquad = \lim_{h \to 0} (5 - 5h - 4) = \lim_{h \to 0} (1 - 5h) = 1 - 5 \times 0$

$\qquad\qquad = 1 - 0 = 1$

RHL of the function

Let $x = 1 + h$ if $x = 1$, then $h = 0$

$\therefore$ $\lim_{x \to 1} \Rightarrow \lim_{h \to 0}$

$\therefore$ $\lim_{x \to 1^+} f(x) = \lim_{h \to 0} f(1 + h)$

$$= \lim_{h \to 0}[4(1+h)^2 - 3(1+h)]$$

$$[\because \text{ Here } f(x) = (4x^2 - 3x)]$$

$$= \lim_{h \to 0}(4h^2 + 8h + 4 - 3 - 3h)$$

$$= \lim_{h \to 0}[4h^2 + 5h + 1] = 0 + 0 + 1 = 1$$

At $x = 1; f(x) = 5x - 4 \quad \therefore \quad f(1) = 5 - 4 = 1$

$\therefore \qquad\qquad\qquad \text{LHL} = \text{RHL} = f(1)$

$\Rightarrow \; f(x)$ is continuous at $x = 1$.

(iii) Continuity of $f(x)$ at $x = 2$

LHL of the function

Let $x = 2 - h$ if $x = 2 \Rightarrow h = 0$

$$\therefore \qquad\qquad \lim_{x \to 2} \Rightarrow \lim_{h \to 0}$$

$$\therefore \qquad \lim_{x \to 2^-} f(x) = \lim_{h \to 0} f(2 - h)$$

$$= \lim_{h \to 0}[4(2 - h)^2 - 3(2 - h)]$$

$$[\because \text{ Here } f(x) = (4x^2 - 3x)]$$

$$= \lim_{h \to 0}(4h^2 - 16h + 16 - 6 + 3h)$$

$$= \lim_{h \to 0}[4h^2 - 13h + 10]$$

$$= 0 - 0 + 10 = 10$$

RHL of the function

Let $x = 2 + h$ if $x = 2 \Rightarrow h = 0$

$$\therefore \qquad\qquad \lim_{x \to 2} \Rightarrow \lim_{h \to 0}$$

$$\therefore \qquad \lim_{x \to 2^+} f(x) = \lim_{h \to 0} f(2 + h)$$

$$= \lim_{h \to 0}[3(2 + h) + 4] \qquad [\because \text{ Here } f(x) = 3x + 4]$$

$$= \lim_{h \to 0}[10 + 3h] = 10 + 0 = 10$$

$\because$ At $x = 2; f(x) = 3x + 4 \quad \therefore \quad f(2) = 3 \times 2 + 4 = 10$

$\therefore$ At $x = 2$

$$\text{LHL} = \text{RHL} = f(2)$$

$\Rightarrow f(x)$ is continuous at $x = 2$.

EXAMPLE 7. *Show that* $f(x) = \begin{cases} \dfrac{|x - 3|}{x - 3} & , \quad x \neq 3 \\ 0 & , \quad x = 3 \end{cases}$

is continuous at all points except at $x = 3$.

SOLUTION. **Continuity at $x = 3$**

LHL of the function

Let $x = 3 - h$, if $x = 3 \Rightarrow h = 0$

$$\therefore \qquad\qquad \lim_{x \to 3} \Rightarrow \lim_{h \to 0}$$

$$\therefore \quad \lim_{x \to 3^-} f(x) = \lim_{h \to 0} f(3-h)$$

$$= \lim_{h \to 0} \frac{|(3-h)-3|}{3-h-3} \quad \left[\text{Here } f(x) = \frac{|x-3|}{x-3} \because x \neq 3 \right]$$

$$= \lim_{h \to 0} \frac{|-h|}{-h} = \lim_{h \to 0} \frac{h}{-h} = -1$$

RHL of the function

Let $x = 3 + h$, if $x = 3 \Rightarrow h = 0$

$$\therefore \quad \lim_{x \to 3} \Rightarrow \lim_{h \to 0} \qquad \left[\because x \neq 3 \therefore f(x) = \frac{|x-3|}{x-3} \right]$$

$$\therefore \quad \lim_{x \to 3^+} = \lim_{h \to 0} f(3+h)$$

$$= \lim_{h \to 0} \frac{|(3+h)-3|}{3+h-3}$$

$$= \lim_{h \to 0} \frac{|h|}{h} = \lim_{x \to 0} \frac{h}{h} = 1$$

At $x = 3, f(x) = 0$

i.e., $\qquad f(3) = 0$

$\Rightarrow$ At $x = 3$, $\qquad\qquad$ LHL $\neq$ RHL $\neq f(3)$

$\Rightarrow f(x)$ is not continuous at $x = 3$.

Now we check the continuity of $f(x)$ at $x = a$ when $a \neq 3$.

$$\because \quad a \neq 3, \qquad\qquad f(x) = \frac{|x-3|}{x-3}$$

$$\therefore \qquad\qquad \lim_{x \to a} f(x) = \lim_{x \to a} \frac{|x-3|}{x-3} = \frac{|a-3|}{a-3}$$

Again for $a \neq 3$; $\qquad\qquad f(x) = \frac{|x-3|}{x-3}$

$$\therefore \qquad\qquad f(a) = \frac{|a-3|}{a-3}$$

$$\therefore \qquad\qquad \lim_{x \to a} f(x) = f(a)$$

i.e., $f(x)$ is continuous at $x = a$ while $a \neq 3$, and at $x = 3$ the function $f(x)$ is not continuous.

✎ REMARK

➡ If function $f(x) = \dfrac{|x-a|}{x-a}$ is continuous for all x except at $x = a$. Therefore, we have to check the continuity at $x = a$ only.

EXAMPLE 8. *Let $f(x)$ be a function in $[0, 1]$ defined by*

$$f(0) = 0$$

$$f(x) = \frac{1}{2} - x \text{ if } 0 < x < \frac{1}{2}$$

$$f\left(\frac{1}{2}\right) = \frac{1}{2}$$

$$f(x) = \frac{2}{3} - x \text{ if } \frac{1}{2} < x < 1$$

$$f(1) = 1$$

Find the point of discontinuity.

SOLUTION. (i) At $x = 0$

$$f(0) = 0$$

RHL of $f(x)$

Let $x = 0 + h$ if $x = 0 \Rightarrow h = 0$

$$\therefore \quad \lim_{x \to 0} \Rightarrow \lim_{h \to 0}$$

$$\therefore \lim_{x \to 0^+} f(x) = \lim_{h \to 0} f(0 + h) = \lim_{h \to 0} \left[\frac{1}{2} - (0 + h) \right] \qquad \left(\because f(x) = \frac{1}{2} - x \right)$$

$$= \frac{1}{2} \neq f(0)$$

$$\therefore \qquad f(x) \text{ is not continuous at } x = 0.$$

(ii) At $x = \frac{1}{2}$

$$f\left(\frac{1}{2}\right) = \frac{1}{2}$$

RHL of the function

Let $x = \frac{1}{2} + h$ if $x = \frac{1}{2} \Rightarrow h = 0$

$$\therefore \quad \lim_{x \to \frac{1}{2}} \Rightarrow \lim_{h \to 0}$$

$$\therefore \lim_{x \to \left(\frac{1}{2}\right)^+} f(x) = \lim_{h \to 0} f\left(\frac{1}{2} + h\right) = \lim_{h \to 0} \left[\frac{2}{3} - \left(\frac{1}{2} + h\right) \right] \qquad \left(\because x > \frac{1}{2} \therefore f(x) = \frac{2}{3} - x \right)$$

$$= \lim_{h \to 0} \left[\frac{2}{3} - \frac{1}{2} - h \right] = \frac{2}{3} - \frac{1}{2} - 0 = \frac{1}{6} \neq f\left(\frac{1}{2}\right)$$

$$\therefore \text{ The function } f(x) \text{ is discontinuous at } x = \frac{1}{2}.$$

(iii) We have $\qquad\qquad f(1) = 1$

LHL of the function

Let $x = 1 - h$ if $x = 1$, then $h = 0$

$$\therefore \quad \lim_{x \to 1} \Rightarrow \lim_{h \to 0}$$

$$\therefore \lim_{x \to 1^-} f(x) = \lim_{h \to 0} f(1 - h) = \lim_{h \to 0} \left[\frac{2}{3} - (1 - h) \right] \qquad \left(\because x < 1 \therefore f(x) = \frac{2}{3} - x \right)$$

$$= \lim_{h \to 0} \left[\frac{2}{3} - 1 + h \right] = \lim_{h \to 0} \left[-\frac{1}{3} + h \right] = -\frac{1}{3} + 0 = -\frac{1}{3}$$

$$\Rightarrow f(x) \text{ is not continuous at } x = 1.$$

EXAMPLE 9. *Show that $f(x) = \sin x$ is continuous for all values of x.*

SOLUTION. Let a be an arbitrary value of x.

Therefore $f(a) = \sin a$

$$\therefore |f(x) - f(a)| = |\sin x - \sin a| = \left| 2 \cos\left(\frac{x + a}{2}\right) \sin\left(\frac{x - a}{2}\right) \right|$$

$$= \left| 2 \cos\left(\frac{x + a}{2}\right) \right| \left| \sin\left(\frac{x - a}{2}\right) \right| < 2 \left| \sin\left(\frac{x - a}{2}\right) \right| \qquad \left[\because \left| \cos\left(\frac{x + a}{2}\right) \right| < 1 \right]$$

$$< 2\left(\frac{x-a}{2}\right)$$

$$< (x-a)$$

$$\left[\because \sin\theta < \theta \text{ and } 0 < \theta < \frac{\pi}{2}\right]$$

Therefore, $|f(x) - f(a)| < (x-a)$

$\therefore \quad |f(x) - f(a)| < \epsilon$ if $|x-a| < \epsilon$

$\therefore \quad$ If for given $\epsilon > 0 \exists \lambda (=\epsilon)$ such that $|f(x) - f(a)| < \epsilon$ when $0 < |x-a| < \lambda$

$\Rightarrow \quad f(x)$ is continuous at $x = a$. Now, since, a is arbitrary. Hence $f(x)$ is continuous for all x.

Exercise 12.3

1. A function $f(x)$ is defined as follows :

$$f(x) = \begin{cases} 1 & \text{if } x > 0 \\ -1 & \text{if } x < 0 \\ 0 & \text{if } x = 0 \end{cases}$$

Prove that $\lim_{x \to 0} f(x)$ does not exist.

2. Show that $\lim_{x \to 2} \dfrac{|x-2|}{(x-2)}$ does not exist.

3. A function $f(x)$ is defined as follows :

$$f(x) = \begin{cases} x^2, & \text{if } x \neq 1 \\ 1, & \text{if } x = 1 \end{cases}$$

Prove that $\lim_{x \to 1} f(x) = 1$.

4. If $f(x) = |x|$, prove that $\lim_{x \to 0} f(x) = 0 \cdot$

5. If $f(x) = \dfrac{|x-1|}{x-1}$, prove that $\lim_{x \to 1} f(x)$ does not exist.

6. If $f(x) = \dfrac{1}{|x|}$, prove that $\lim_{x \to 0} f(x)$ does not exist.

7. Prove that RHL and LHL of the function $f(x) = \dfrac{1 + \cos x}{\tan^2 x}$ at $x \to \pi$ are equal. Also find the $\lim_{x \to \pi} f(x)$.

8. Find RHL and LHL of the function $f(x) = \dfrac{x^2 - 1}{|x-1|}$ at $x \to 1$.

9. Find RHL and LHL of the function $f(x) = \dfrac{|\sin x|}{x}$ at $x \to 0$.

10. Check the continuity of the function

$$f(x) = \begin{cases} 1 + x^2, & \text{if } 0 \leq x \leq 1 \\ 1 - x, & \text{if } x > 1 \end{cases} \text{ at } x = 1.$$

11. Check the continuity of the function

$$f(x) = \begin{cases} x, & \text{if } x \geq 2 \\ x^2, & \text{if } x < 2 \end{cases}.$$

12. Prove that $f(x) = \sin x$ is continuous at $x = \dfrac{\pi}{2}$.

13. If $f(x) = \begin{cases} x - 4, & \text{if } x \geq 5 \\ 5x - 24, & \text{if } x < 5 \end{cases}$,

prove that $f(x)$ is continuous at $x = 5$.

14. For what value of k, the function

$$f(x) = \begin{cases} kx^2, & \text{if } x \leq 2 \\ 3, & \text{if } x > 2 \end{cases}$$

is continuous?

15. For what value of k, the function

$$f(x) = \begin{cases} \dfrac{x^2 - 16}{x - 4}, & \text{if } x \neq 4 \\ k, & \text{if } x = 4 \end{cases}$$

is continuous at $x = 4$?

16. Check the continuity of $f(x)$ at $x = 0$ if

$$f(x) = \begin{cases} x^2 - 1, & \text{if } x \leq 0 \\ 0, & \text{if } x > 0 \end{cases}$$

17. Check the continuity of $f(x)$ at $x = 1$ if

$$f(x) = \begin{cases} 2|x|, & \text{if } |x| \leq 1 \\ 0, & \text{if } |x| > 1 \end{cases}$$

18. Check the continuity of $f(x)$ at $x = 1$ and $x = 2$ if $f(x) = \begin{cases} 0, & \text{if } 0 < x < 1 \\ x, & \text{if } 1 \leq x < 2 \\ x^3/4, & \text{if } 2 \leq x < 3 \end{cases}$.

19. Check the continuity of $f(x)$ at $x = a$ if

$$f(x) = \begin{cases} \dfrac{x^2 - a^2}{x - a}, & \text{if } x \neq a \\ 2a, & \text{if } x = a \end{cases}$$

20. Prove that the function $f(x) = x^2 - 7x + 3$ is continuous at $x = 1, x = 2, x = 3$.

21. Check the continuity of the function $f(x)$ at

$$x = 0 \text{ if } f(x) = \begin{cases} x^5, & \text{if } x \neq 0 \\ 1, & \text{if } x = 0 \end{cases}.$$

22. A function $f(x)$ is defined as follows

$$f(x) = \begin{cases} 1+x, & \text{if } x > 0 \\ 0, & \text{if } x = 0 \\ x^2+1, & \text{if } x < 0 \end{cases}$$

Is $f(x)$ continuous at $x = 0$? If not, what minimum correction should be made to make $f(x)$ continuous at $x = 0$?

ANSWERS

7. $\dfrac{1}{2}$ **8.** $2, -2$ **9.** $1, -1$ **10.** discontinuous

11. discontinuous at $x = 2$ and continuous for all other points   **14.** $k = \dfrac{3}{4}$ **15.** $k = 8$

16. discontinuous   **17.** discontinuous   **18.** discontinuous at $x = 1$, continuous at $x = 2$

19. Continuous at $x = a$ **21.** discontinuous   **22.** Take $f(0) = 1$

MISCELLANEOUS EXAMPLES

EXAMPLE 1.   *Check the continuity of the following function :*

$$f(x) = \begin{cases} x+2, & \text{if } x \leq 1 \\ x-2, & \text{if } x > 1 \end{cases}$$

SOLUTION.   The given function is defined for all real numbers.

Case-I: If $a < 1$, then $f(a) = a + 2$

$\therefore \quad \lim\limits_{x \to a} f(x) = \lim\limits_{x \to a} (x + 2) = a + 2 = f(a)$

$\Rightarrow f(x)$ is continuous for all real numbers less than 1.

Case-II: If $a > 1$, then $f(a) = a - 2$

$\therefore \quad \lim\limits_{x \to a} f(x) = \lim\limits_{x \to a} (x - 2) = a - 2 = f(a)$

$\Rightarrow f(x)$ is continuous for all real numbers greater than 1.

Case-III: If $a = 1$, then

$\qquad$ LHL $= f(1 - 0)$

$\qquad\qquad = \lim\limits_{x \to 1^-} f(x) = \lim\limits_{x \to 1^-} (x + 2) = 1 + 2 = 3$

and RHL

$\qquad f(1 + 0) = \lim\limits_{x \to 1^+} f(x) = \lim\limits_{x \to 1^+} (x - 2) = 1 - 2 = -1$

Clearly $f(1 + 0) \neq f(1 - 0)$

$\Rightarrow f(x)$ is not continuous at $x = 1$.

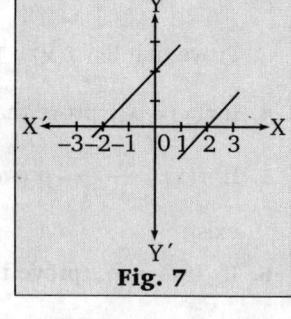

Fig. 7

EXAMPLE 2.   *Find the point of discontinuity of the following function :*

$$f(x) = \begin{cases} x+2, & \text{if } x < 1 \\ 0, & \text{if } x = 1 \\ x-2, & \text{if } x > 1 \end{cases}$$

SOLUTION.   From above example (1) we can say that $f(x)$ is continuous for all values of x except $x = 1$.

Continuity at $x = 1$

$\qquad$ LHL $= f(1 - 0) = \lim\limits_{x \to 1^-} (x + 2) = 1 + 2 = 3$

$\qquad$ RHL $= f(1 + 0) = \lim\limits_{x \to 1^+} (x - 2) = 1 - 2 = -1$

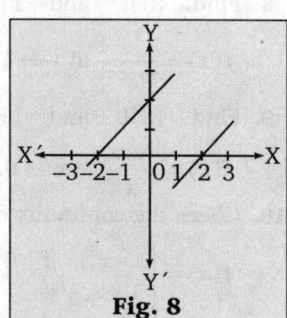

Fig. 8

$\Rightarrow \quad f(1+0) \neq f(1-0)$

$\Rightarrow \quad f(x)$ is not continuous at $x = 1$.

$\Rightarrow$ The only point of discontinuity of $f(x)$ is $x = 1$.

EXAMPLE 3. *Check the continuity of the function*

$$f(x) = \begin{cases} 2x - 1, \text{ if } x < 0 \\ 2x + 1, \text{ if } x \geq 0 \end{cases}$$

SOLUTION. **Case-I :** When $x < 0$, then $f(x) = 2x - 1$, which is a polynomial function and hence continuous at all point $x < 0$.

Case-II : When $x > 0$ then $f(x) = 2x + 1$, which is again a polynomial function and hence continuous for all point $x > 0$.

Case-III : When $x = 0$, then $f(0) = 2 \times 0 + 1 = 1$

and $\quad$ RHL $= f(0+0) = \lim\limits_{x \to 0^+} f(x)$

$\qquad = \lim\limits_{h \to 0} f(0+h) = \lim\limits_{h \to 0} f(h) = \lim\limits_{h \to 0}(2h + 1) = 1$

Similarly, $\quad$ LHL $= f(0-0) = \lim\limits_{x \to 0^-} f(x) = \lim\limits_{h \to 0} f(0-h) = \lim\limits_{h \to 0} f(-h)$

$\qquad = \lim\limits_{h \to 0}[(2(-h) - 1)] = -1$

Clearly $\quad f(0+0) \neq f(0-0)$

$\Rightarrow \quad f(x)$ is not continuous at $x = 0$.

EXAMPLE 4. *Check the continuity of the function* $f(x) = |x| + |x - 1|$ *in the interval* $[-1, 2]$.

SOLUTION. The given function can be defined as follows:

$$f(x) = \begin{cases} 1 - 2x; & \text{if } -1 \leq x \leq 0 \\ 1; & \text{if } 0 < x \leq 1 \\ 2x - 1 & \text{if } 1 < x \leq 2 \end{cases}$$

We know that a polynominal function and constant function are always continuous. So $f(x)$ is continuous at $-1 \leq x \leq 0, 0 < x \leq 1$ and $1 < x \leq 2$. So, we have to check the continuity at $x = -1, 0, 1$ and 2.

Case-I: At $x = -1$
$\qquad$ RHL $= f(-1+0) = \lim\limits_{x \to -1^+} f(x) = \lim\limits_{x \to -1^+}(1 - 2x) = 1 + 2 = 3$

and $f(x) = 1 - 2x$, when $x = -1$

$\therefore \qquad f(-1) = 1 - 2(-1) = 1 + 2 = 3$

$\Rightarrow \qquad \lim\limits_{x \to -1^+} f(x) = f(-1)$

$\Rightarrow \quad f(x)$ is continuous at $x = -1$.

Case-II: At $x = 0$

$\qquad \lim\limits_{x \to 0^-} f(x) = \lim\limits_{x \to 0^-}(1 - 2x) = 1 - 2 \times 0 = 1$

and $\qquad \lim\limits_{x \to 0^+} f(x) = \lim\limits_{x \to 0^+}(1) = 1$

Also $\qquad f(0) = 1$

Clearly, $\qquad \lim\limits_{x \to 0^-} f(x) = \lim\limits_{x \to 0^+} f(x) = f(0) = 1$

$\Rightarrow \quad f(x)$ is continuous at $x = 0$.

Case-III: At $x = 1$

We have, $\quad \lim\limits_{x \to 1^-} f(x) = \lim\limits_{x \to 1^-} (1) = 1$

and $\quad\quad \lim\limits_{x \to 1^+} f(x) = \lim\limits_{x \to 1^+} (2x - 1) = 1$

Also, $\quad\quad\quad f(1) = 1$

Clearly, $\quad \lim\limits_{x \to 1^-} f(x) = \lim\limits_{x \to 1^+} f(x) = f(1) = 1$

Hence, $f(x)$ is continuous at $x = 1$.

Case-IV: At $x = 2$

$$\lim\limits_{x \to 2^-} f(x) = \lim\limits_{x \to 2^-} (2x - 1) = 4 - 1 = 3$$

and $\quad\quad\quad f(2) = 2 \times 2 - 1 = 3$

$\Rightarrow \quad f(x)$ is continuous at $x = 2$.

$\Rightarrow \quad f(x)$ is continuous at each point of the interval $[-1, 2]$.

EXAMPLE 5. *Check the continuity of the function*

$$f(x) = \begin{cases} x^3 - 2, & \text{if } x \le 0 \\ 1, & \text{if } x > 0 \end{cases}$$

SOLUTION. If $x = 0$, then $\quad f(0) = 0^3 - 2 = -2$

$$\text{RHL} = \lim\limits_{x \to 0^+} f(x) = \lim\limits_{h \to 0} f(0 + h) = \lim\limits_{h \to 0} 1 = 1$$

$$\text{LHL} = \lim\limits_{x \to 0^-} f(x) = \lim\limits_{h \to 0} f(0 - h) = \lim\limits_{h \to 0} [(-h)^3 - 2] = -2$$

$\Rightarrow \quad\quad \text{RHL} \ne \text{LHL}$

$\Rightarrow \quad f(x)$ is not continuous at $x = 0$.

EXAMPLE 6. *Check the point of discontinuity of the function*

$$f(x) = \begin{cases} \dfrac{x^4 - 16}{x - 2}, & \text{if } x \ne 2 \\ 16, & \text{if } x = 2 \end{cases}$$

SOLUTION. At $x \ne 2$, $f(x) = \dfrac{x^4 - 16}{x - 2}$ is a rational function.

$\Rightarrow \quad f(x)$ is continuous $\forall x \in \mathbf{R} (x \ne 2)$

When $x = 2$, $\quad\quad f(x) = f(2) = 16$

Now $\quad \lim\limits_{x \to 2^+} f(x) = \lim\limits_{h \to 0} f(2 + h) = \lim\limits_{h \to 0} \dfrac{(2 + h)^4 - 16}{(2 + h) - 2}$

$$= \lim\limits_{h \to 0} \dfrac{2^4 + 4 \times 2^3 h + 6 \times 2^2 h^2 + 8h^3 + h^4 - 16}{h}$$

$$= \lim\limits_{h \to 0} \dfrac{32h + 24h^2 + 8h^3 + h^4}{h}$$

$$= \lim\limits_{h \to 0} (32 + 24h + 8h^2 + h^3) = 32$$

$\Rightarrow \quad\quad\quad\quad f(2) \ne \lim\limits_{x \to 2^+} f(x)$

$\Rightarrow$ $f(x)$ is not continuous at $x = 2$.

$\Rightarrow$ Point $x = 2$ is the point of discontinuity of the function $f(x)$.

EXAMPLE 7. *For what value of k, the following function is continuous at $x = 2$?*

$$f(x) = \begin{cases} 2x+1, & \text{if } x < 2 \\ k, & \text{if } x = 2 \\ 3x-1, & \text{if } x > 2 \end{cases}$$

SOLUTION.

$$\text{LHL} = \lim_{x \to 2^-} f(x) = \lim_{h \to 0} f(2-h)$$

$$= \lim_{h \to 0} 2(2-h) + 1 = 4 + 1 = 5$$

and

$$\text{RHL} = \lim_{x \to 2^+} f(x) = \lim_{h \to 0} f(2+h)$$

$$= \lim_{h \to 0} 3(2+h) - 1 = 6 - 1 = 5$$

Given, $f(2) = k$.

For the continuity of $f(x)$ at $x = 2$.

$$\text{LHL} = \text{RHL} = f(2) = k$$

$\Rightarrow$ $f(2) = 5 = k$

$\Rightarrow$ $k = 5$

EXAMPLE 8. *Check the continuity of the following function at $x = 1$:*

$$f(x) = \begin{cases} x^2 + 1, & \text{if } x \neq 1 \\ 3, & \text{if } x = 1 \end{cases}$$

SOLUTION. Given, $f(1) = 3$

Now, $\text{LHL} = \lim_{x \to 1^-} f(x) = \lim_{h \to 0} f(1-h) = \lim_{h \to 0} [(1-h)^2 + 1]$

$$= \lim_{h \to 0} (1-h)^2 + 1 = 1 + 1 = 2$$

and $\text{RHL} = \lim_{x \to 1^+} f(x) = \lim_{h \to 0} (1+h) = \lim_{h \to 0} [(1+h)^2 + 1] = 1 + 1 = 2$

Clearly, $f(1) \neq \text{LHL} = \text{RHL}$

$\Rightarrow$ $f(x)$ is not continuous at $x = 1$.

EXAMPLE 9. *Prove that* $f(x) = \begin{cases} \dfrac{e^{1/x}}{1 + e^{1/x}}, & \text{if } x \neq 0 \\ 0, & \text{if } x = 0 \end{cases}$

is not continuous at $x = 0$.

SOLUTION.

$$\text{LHL} = \lim_{x \to 0^-} f(x) = \lim_{h \to 0} f(0-h)$$

$$= \lim_{h \to 0} \frac{e^{\frac{1}{(0-h)}}}{1 + e^{\frac{1}{(0-h)}}} = \lim_{h \to 0} \frac{e^{-1/h}}{1 + e^{-1/h}} = \lim_{h \to 0} \frac{1}{e^{1/h} + 1} = 0$$

Similarly, $\text{RHL} = \lim_{x \to 0^+} f(x) = \lim_{h \to 0} f(0+h)$

$$= \lim_{h \to 0} \frac{e^{\frac{1}{(0+h)}}}{1 + e^{\left(\frac{1}{0+h}\right)}} = \lim_{h \to 0} \frac{e^{1/h}}{1 + e^{1/h}} = \lim_{h \to 0} \frac{1}{1 + e^{-1/h}} = \frac{1}{1 + 0} = 1$$

Clearly, $\text{RHL} \neq \text{LHL}$

$\Rightarrow$ $f(x)$ is not continuous at $x = 0$.

EXAMPLE 10. *Prove that* $\cos x$ *is continuous.*

SOLUTION. Let $f(x) = \cos x$ and $a \in \mathbf{R}$

Then $f(a) = \cos a$

LHL $f(a-0) = \lim_{x \to a^-} f(x) = \lim_{h \to 0} f(a-h) = \lim_{h \to 0} \cos(a-h) = \cos a$

Similarly,

RHL $f(a+0) = \lim_{x \to a^+} f(x) = \lim_{h \to 0} f(a+h) = \lim_{h \to 0} \cos(a+h) = \cos a$

Clearly $f(a) = f(a+0) = f(a-0)$

$\Rightarrow$ $f(x)$ is continuous at $x = a$ for every value of a.

$\Rightarrow$ $f(x)$ is continuous.

EXAMPLE 11. *Prove that* $f(x) = x - [x]$ *is discontinuous for all integral value of* x, *here* $[x]$ *denotes the greatest integer function.*

SOLUTION. Given, $f(x) = x - [x]$

Let r be an integer, then

$$f(r) = r - [r]$$

LHL : $x < r \Rightarrow [x] = r - 1$

$\therefore$ $\text{LHL} = \lim_{x \to r^-} f(x) = \lim_{x \to r^-} (x - [x]) = r - (r-1) = 1$

RHL : $x > r \Rightarrow [x] = r$

$\therefore$ $\lim_{x \to r^+} f(x) = \lim_{x \to r^+} (x - [x]) = r - r = 0$

Clearly $\text{LHL} \neq \text{RHL}$

$\Rightarrow$ $f(x)$ is not continuous at $x = r$ for each value of r.

EXAMPLE 12. *Prove that* $|\sin x|$ *is continuous.*

SOLUTION. Let $h(x) = |\sin x|$

and $f(x) = \sin x$ and $g(x) = |x|$

$\therefore$ $(gof)(x) = g[f(x)]$

$$= g(\sin x) = |\sin x|$$

$\Rightarrow$ $(gof)(x) = h(x)$

Now, since $f(x) = \sin x$ and $g(x) = |x|$ are continuous therefore gof is also continuous.

$\therefore$ $h(x) = |\sin x|$ is a continuous function.

EXAMPLE 13. *Find the point of discontinuity of the greatest integer function $f(x) = [x]$.*

SOLUTION. Let $f(x) = [x]$, then we have

$$f(x) = \begin{cases} -2 & ; & -2 \le x < -1 \\ -1 & ; & -1 \le x < 0 \\ 0 & ; & 0 \le x < 1 \\ 1 & ; & 1 \le x < 2 \\ 2 & ; & 2 \le x < 3 \\ \vdots & & \vdots \end{cases}$$

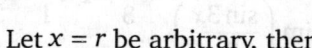

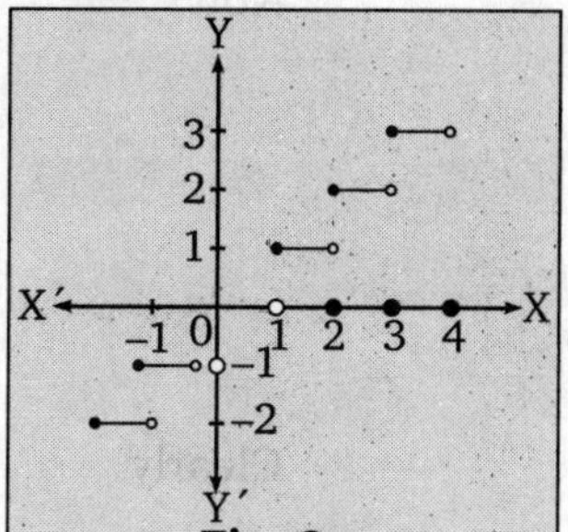

Fig. 9

Let $x = r$ be arbitrary, then

$$f(r) = [r] = r$$

LHL : Here $\qquad x < r \Rightarrow f[x] = r - 1$

$\therefore \qquad \lim_{x \to r^-} f(x) = \lim_{x \to r^-} [x] = r - 1$

RHL : Here $\qquad x > r \Rightarrow [x] = r$

Then $\qquad \lim_{x \to r^+} f(x) = \lim_{x \to r^+} [x] = r$

$\Rightarrow \qquad \lim_{x \to r^-} f(x) \ne \lim_{x \to r^+} f(x)$

$\Rightarrow \quad f(x)$ is discontinuous at $x = r$ for all integral value of r.

$\Rightarrow \quad$ All integers are the point of discontinuity of $f(x)$.

EXAMPLE 14. *Find the point of discontinuity of the function*

$$f(x) = \begin{cases} \dfrac{x}{|x|}, & \text{if } \ x < 0 \\ -1, & \text{if } \ x > 0 \end{cases}$$

SOLUTION. We know that, $\quad |x| = \begin{cases} -x, & \text{when } \ x < 0 \\ x, & \text{when } \ x > 0 \end{cases}$

Therefore, $\qquad f(x) = \begin{cases} \dfrac{-x}{x} = -1, & \text{if } \ x < 0 \\ \dfrac{x}{x} = 1, & \text{if } \ x > 0 \\ 0, & \text{if } \ x = 0 \end{cases}$

$\Rightarrow$ We have to check the continuity at $x = 0$.

Now, $\qquad f(0) = 0$

$$f(0 - 0) = \text{LHL} = \lim_{x \to 0^-} f(x) = \lim_{x \to 0^-} (-1) = -1$$

Clearly $\qquad f(0) \ne f(0 - 0)$

$\Rightarrow \quad f(x)$ is not continuous at $x = 0$.

$\Rightarrow \quad x = 0$ is the point of discontinuity of $f(x)$.

EXAMPLE 15. *If* $f(x) = \begin{cases} \dfrac{\sin 2x}{\sin 3x} & , \ x \ne 0 \\ 2 & , \ x = 0 \end{cases}$

Prove that $f(x)$ is not continuous at $x = 0$.

SOLUTION.   Given, $f(0) = 2$

Now $\displaystyle \lim_{x \to 0} f(x) = \lim_{x \to 0} \frac{\sin 2x}{\sin 3x} = \lim_{x \to \infty} \left(\frac{\sin 2x}{2x} \cdot \frac{3x}{\sin 3x} \right) \cdot \frac{2}{3}$

$$= \frac{2}{3} \lim_{x \to 0} \left(\frac{\sin 2x}{2x} \right) \cdot \lim_{x \to 0} \left(\frac{3x}{\sin 3x} \right)$$

$$= \frac{2}{3} \lim_{x \to 0} \left(\frac{\sin 2x}{2x} \right) \cdot \frac{1}{\displaystyle \lim_{x \to 0} \left(\frac{\sin 3x}{3x} \right)} = \frac{2}{3} \times 1 \times \frac{1}{1} = \frac{2}{3}$$

Clearly $\displaystyle f(0) \neq \lim_{x \to 0} f(x)$

$\Rightarrow$ $f(x)$ is not continuous at $x = 0$.

EXAMPLE 16. *Find the point of discontinuity of the function f defined by*

$$f(x) = \begin{cases} |x| + 3, & \text{if} & x \leq -3 \\ -2x, & \text{if} & -3 < x < 3 \\ 6x + 2, & \text{if} & x \geq 3 \end{cases}$$

SOLUTION.   If $x \leq -3$, then $|x| = -x$ $(\because |x| = -x; \text{ if } x < 0)$

$\therefore$ We can defined $f(x)$ as follows :

$$f(x) = \begin{cases} -x + 3, & \text{if} & x \leq -3 \\ -2x, & \text{if} & -3 < x < 3 \\ 6x + 2, & \text{if} & x \geq 3 \end{cases}$$

$\Rightarrow$ We have to check the continuity at $x = -3$ and $x = 3$.

(i) Continuity at x = -3

$f(x) = -x + 3 \text{ if } x \leq -3$

$\Rightarrow$ $f(-3) = -(-3) + 3 = 6$

LHL: If $x < -3$ then $f(x) = -x + 3$

$\therefore$ $\displaystyle \lim_{x \to -3^-} f(x) = \lim_{x \to -3^-} (-x + 3) = -(-3) + 3 = 6$

RHL: when $x > -3$ then $f(x) = -2x$

$\therefore$ $\displaystyle \lim_{x \to -3^+} f(x) = \lim_{x \to -3^+} (-2x) = -2(-3) = 6$

It is clear that

$$\lim_{x \to -3^-} f(x) = \lim_{x \to -3^+} f(x) = f(-3)$$

$\Rightarrow$ $f(x)$ is continuous at $x = -3$.

(ii) Continuity at x = 3

Given $f(x) = 6x + 2 \text{ if } x \geq 3$

$\Rightarrow$ $f(3) = 6 \times 3 + 2 = 20$

Now $\displaystyle \lim_{x \to 3^-} f(x) = \lim_{x \to 3^-} (-2x) = -6$

and $\displaystyle \lim_{x \to 3^+} f(x) = \lim_{x \to 3^+} (6x + 2) = 20$

$\Rightarrow$ $\displaystyle \lim_{x \to 3^-} f(x) \neq f(3)$

$\Rightarrow$ $\qquad$ $f(x)$ is not continuous at $x = 3$.

$\Rightarrow$ $\qquad$ $x = 3$ is the point of discontinuity of $f(x)$.

EXAMPLE 17. Let $f(x) = \begin{cases} \dfrac{\sin x}{x}, & \text{if } x < 0 \\ (x+1), & \text{if } x \geq 0 \end{cases}$

Check the continuity of $f(x)$.

SOLUTION. We know that $\sin x$ and one-one function x are always continuous.

$\Rightarrow \dfrac{\sin x}{x}, x < 0$ is continuous.

Similarly, $x + 1$ is continuous being the polynomial function.

Continuity at $x = 0$

$$f(0) = 0 + 1 = 1$$

$$\text{RHL} = f(0+0) = \lim_{h \to 0} f(0+h) = \lim_{h \to 0}(h+1) = 1$$

$$\text{LHL} = f(0-0) = \lim_{h \to 0} f(0-h) = \lim_{h \to 0} f(-h) = \lim_{h \to 0}\left(\frac{\sin(-h)}{-h}\right) = \lim_{h \to 0}\frac{\sin h}{h} = 1$$

$\Rightarrow$ $\qquad$ $f(0+0) = f(0-0) = f(0) = 1$

$\Rightarrow$ $\qquad$ $f(x)$ is continuous at $x = 0$.

$\Rightarrow$ $\qquad$ $f(x)$ is continuous at each point.

EXAMPLE 18. Let $f(x) = \begin{cases} x^2 + ax + b, & \text{if } 0 \leq x < 2 \\ 3x + 2, & \text{if } 2 \leq x \leq 4 \\ 2ax + 5b, & \text{if } 4 < x \leq 8 \end{cases}$

Find the value of a and b if $f(x)$ is continuous in the interval $[0,8]$.

SOLUTION. Since the given function $f(x)$ is continuous in the interval $[0,8]$, so it is also continuous in the interval $]0,2[,]2,4[$ and $]4,8[$.

$\Rightarrow$ $\qquad$ $f(x)$ is continuous at $x = 0$.

$\Rightarrow$ $\qquad$ $\displaystyle\lim_{x \to 0^+} f(x) = f(0)$

$\Rightarrow$ $\qquad$ $\displaystyle\lim_{x \to 0^+}(x^2 + ax + b) = 0^2 + a \times 0 + b = b$

$\because$ $\qquad$ $f(x)$ is continuous at $x = 2$. Therefore

$\qquad\qquad\qquad = \displaystyle\lim_{x \to 2^-} f(x) = \lim_{x \to 2^+} f(x) = f(2)$

$\Rightarrow$ $\qquad$ $\displaystyle\lim_{x \to 2^-}(x^2 + ax + b) = \lim_{x \to 2^+}(3x + 2) = 3 \times 2 + 2$

$\Rightarrow$ $\qquad$ $4 + 2a + b = 3 \times 2 + 2$

$\Rightarrow$ $\qquad$ $2a + b = 4$ $\qquad\qquad\qquad\qquad\qquad\qquad$ ...(1)

Similarly, $f(x)$ is continuous at $x = 4$.

$\Rightarrow$ $\qquad$ $\displaystyle\lim_{x \to 4^+} f(x) = \lim_{x \to 4^-} f(x) = f(4)$

$\Rightarrow$ $\qquad$ $\displaystyle\lim_{x \to 4^+}(2ax + 5b) = \lim_{x \to 4^-}(3x + 2) = 3 \times 4 + 2$

$\Rightarrow$ $\qquad$ $8a + 5b = 12 + 2$

$\Rightarrow \qquad\qquad 8a + 5b = 14$ $\qquad\qquad\qquad$ ...(2)

On solving eqn. (1) and (2) we get
$$a = 3, b = -2$$

EXAMPLE 19. If the function $f(x) = \begin{cases} 3ax + b, & \text{if } x > 1 \\ 11, & \text{if } x = 1 \\ 5ax - 2b, & \text{if } x < 1 \end{cases}$

is continuous at $x = 1$. Find the value of a and b.

SOLUTION. Given, $f(x)$ is continuous at $x = 1$.

$\therefore \qquad\qquad f(1 + 0) = f(1 - 0) = f(1)$

Now, $\qquad f(1 - 0) = \lim_{x \to 1^-} f(x) = \lim_{x \to 1^-} (5ax - 2b) = 5a - 2b$ $\qquad$ ...(1)

$\qquad\qquad f(1 + 0) = \lim_{x \to 1^+} f(x) = \lim_{x \to 1^+} (3ax + b) = 3a + b$ $\qquad$ ...(2)

and $\qquad\qquad f(1) = 11$ $\qquad\qquad\qquad\qquad$ ...(3)

On solving eqns. (1), (2) and (3) we get
$$a = 3, b = 2$$

EXAMPLE 20. Let $f(x) = |x - 1| + |x + 2|, x \in \textbf{R}$. Show that the function $f(x)$ is continuous at $x = 1$ and $x = -2$.

SOLUTION. **(i) Right hand limit at $x = 1$**

Given $\qquad\qquad f(x) = |x - 1| + |x + 2|$

$\Rightarrow \qquad\qquad f(x + h) = |x + h - 1| + |x + h + 2|$

$\Rightarrow \qquad\qquad f(1 + h) = |1 + h - 1| + |1 + h + 2|$

$\Rightarrow \qquad\qquad \lim_{h \to 0} f(1 + h) = \lim_{h \to 0} [|h| + |3 + h|] = |0| + |3 + 0| = 3$

(ii) Left hand limit at $x = 1$

Here, $\qquad\qquad f(1 - h) = |1 - h - 1| + |1 - h + 2| = |-h| + |3 - h|$

$\therefore \qquad\qquad \lim_{h \to 0} f(1 - h) = \lim_{h \to 0} [|h| + |3 - h|] = |0| + |3 - 0| = 3$

$\Rightarrow \qquad\qquad \text{RHL} = \text{LHL}$

and $\qquad\qquad f(1) = |1 - 1| + |1 + 2| = |0| + |3| = 3$

$\Rightarrow \qquad\qquad f(1) = \lim_{h \to 0} f(1 - h) = \lim_{h \to 0} f(1 + h)$

Hence, $f(x)$ is continuous at $x = 1$.

Now, **LHL at $x = -2$**

$\qquad f(-2 - h) = |-2 - h - 1| + |-2 - h + 2| = |-3 - h| + |-h|$
$\qquad\qquad\qquad = |3 + h| + |h| = |3 + 0| + |0| = 3$

RHL at $x = -2$

$\qquad f(-2 + h) = |-2 + h - 1| + |-2 + h + 2|$
$\qquad\qquad\qquad = |h - 3| + |h|$

$\Rightarrow \qquad \lim_{h \to 0} f(-2 + h) = |-3| + |0| = 3$

and $\qquad\qquad f(-2) = |-2 - 1| + |-2 + 2| = |-3| + |0| = 3$

$$\Rightarrow \qquad f(-2) = \lim_{h \to 0} f(-2+h) = \lim_{h \to 0} f(-2-h)$$

Hence, we conclude that $f(x)$ is continuous at $x = -2$ also.

Exercise 12.4

1. Prove that

$f(x) = \begin{cases} x, & x \geq 0 \\ 1, & x < 0 \end{cases}$, is not continuous at $x = 0$.

2. Prove that

$f(x) = \begin{cases} \dfrac{\sin x}{x}; & x \neq 0 \\ 1; & x = 0 \end{cases}$, is continuous at $x = 0$.

3. Prove that

$f(x) = \begin{cases} x+2, & x \leq 3 \\ 8-x, & x > 3 \end{cases}$, is continuous at $x = 3$.

4. Prove that

$f(x) = \begin{cases} x-1, & \text{if } x \leq 2 \\ 2x-3, & \text{if } x > 2 \end{cases}$, is continuous at every point.

5. Prove that

$f(x) = \begin{cases} \dfrac{\sin^{-1} x}{x} + e^x, & x \neq 0 \\ 2, & x = 0 \end{cases}$, is continuous at $x = 0$.

6. Let the function $f(x) = \begin{cases} 1, & \text{if } x \leq 3 \\ ax+b, & \text{if } 3 < x < 5 \\ 7, & \text{if } 5 \leq x \end{cases}$

be continuous, find the value of a and b.

7. Prove that $f(x) = 2x - |x|$ is continuous at $x = 0$.

8. Prove that $f(x) = \cos x^2$ is a continuous function.

9. Prove that $f(x) = |1 - x + |x||$ is a continuous function at $x = 0$.

10. Prove that $f(x) = x^n$, is continuous at $x = 0$.

11. If $f(x) = \begin{cases} x-4, & x \leq 5 \\ 5x-24, & x > 5 \end{cases}$ then show that $f(x)$ is continuous at $x = 5$.

12. Prove that

$f(x) = \begin{cases} \dfrac{x^2}{a} - a, & 0 < x < a \\ 0, & x = a \\ a - \dfrac{a^3}{x^2}, & x > a \end{cases}$ is continuous at $x = a$.

13. Check the continuity of the following functions:

(i) $f(x) = \begin{cases} \dfrac{1}{1-e^{1/x}}, & x \neq 0 \\ 0, & x = 0 \end{cases}$

(ii) $f(x) = \begin{cases} e^{1/x}, & x \neq 0 \\ 0, & x = 0 \end{cases}$

(iii) $f(x) = \begin{cases} \dfrac{1}{1+e^{1/x}}, & x \neq 0 \\ 0, & x = 0 \end{cases}$

14. Check the continuity of the following functions:

(i) $f(x) = \begin{cases} x \cos \dfrac{1}{x}, & x \neq 0 \\ 0, & x = 0 \end{cases}$

(ii) $f(x) = \begin{cases} \cos \dfrac{1}{x}, & x \neq 0 \\ 1, & x = 0 \end{cases}$

(iii) $f(x) = \begin{cases} \sin \dfrac{1}{x}, & x \neq 0 \\ 0, & x = 0 \end{cases}$

(iv) $f(x) = \begin{cases} \dfrac{\cos x}{\dfrac{\pi}{2} - x}, & x \neq \pi/2 \\ 1, & x = \pi/2 \end{cases}$

15. Prove that

$f(x) = \begin{cases} \dfrac{\sin^2 ax}{1}, & x \neq 0 \\ 1, & x = 0 \end{cases}$ is not continuous at $x = 0$.

16. If $f(x) = x^2 + 1$, when $x \neq 1$ and $f(x) = 3$ when $x = 1$. Prove that $f(x)$ is not continuous at $x = 1$.

17. If $f(x) = \begin{cases} 4x+a, & x < 1 \\ 6, & x = 1 \\ 3x-b, & x > 1 \end{cases}$ is continuous at $x = 1$.

Find the value of a and b.

ANSWERS

6. $a = 3, b = -8$

17. $a = 2, b = -3$

Objective Evaluation

☞ **MULTIPLE CHOICE QUESTIONS**

Choose the most appropriate one.

1. Let $f : \mathbf{R} \to \mathbf{R}$ be a positive increasing function such that $\lim\limits_{x \to \infty} \dfrac{f(3x)}{f(x)} = 1$ then $\lim\limits_{x \to \infty} \dfrac{f(2x)}{f(x)} =$

  (a) 1   (b) 2/3

  (c) 3/2   (d) 3

2. Let $l = \lim\limits_{x \to 0} \dfrac{a - \sqrt{a^2 - x^2} - \dfrac{x^2}{4}}{x^4}, a > 0$, if l is finite, then:

  (a) $a = 2$ (b) $a = 1$

  (c) $l = \dfrac{1}{64}$ (d) $l = \dfrac{1}{32}$

3. $\lim\limits_{x \to \pi/2} \dfrac{\cos x - \operatorname{cosec} x}{(\pi - 2x)^3} =$

  (a) $\dfrac{1}{16}$ (b) $\dfrac{1}{8}$

  (c) $\dfrac{1}{4}$ (d) $\dfrac{\pi}{2}$

4. $\lim\limits_{x \to 0} (\cos x)^{\cot^2 x} =$

  (a) e^{-1} (b) $e^{-1/2}$

  (c) 1   (d) does not exist

5. $\lim\limits_{x \to 0} \log_e (\sin x) =$

  (a) -1 (b) $\log_e 1$

  (c) 1   (d) none of these

6. $\lim\limits_{x \to 0} \left\{ \dfrac{1 + \tan x}{1 + \sin x} \right\}^{\operatorname{cosec} x} =$

  (a) $1/e$ (b) 1

  (c) e (d) e^2

7. $\lim\limits_{x \to \pi/4} \dfrac{\int_2^{\sec^2 x} f(t)\,dt}{x^2 - \dfrac{\pi^2}{16}} =$

  (a) $\dfrac{8}{\pi} f(2)$ (b) $\dfrac{2}{\pi} f(2)$

  (c) $\dfrac{2}{\pi} f\left(\dfrac{1}{2}\right)$ (d) $4 f(2)$

8. $\lim\limits_{x \to \infty} \dfrac{(2x - 3)(3x - 4)}{(4x - 5)(5x - 6)} =$

  (a) $\dfrac{1}{10}$ (b) 0

  (c) $\dfrac{1}{5}$ (d) $\dfrac{3}{10}$

9. $\lim\limits_{x \to \infty} \dfrac{x^2 + bx + 4}{x^2 + ax + 5} =$

  (a) $\dfrac{b}{a}$ (b) 0

  (c) 1   (d) $\dfrac{4}{5}$

10. $\lim\limits_{x \to 0} \left[(\sin x)^{1/x} + \left(\dfrac{1}{x}\right)^{\sin x} \right], \forall\, x > 0 =$

  (a) 0   (b) -1

  (c) 1   (d) 2

11. $\lim\limits_{x \to \infty} \dfrac{\sin x}{x} =$

  (a) ∞ (b) 1

  (c) 0   (d) does not exist

12. $\lim\limits_{x \to \infty} \left[\dfrac{1}{n^2} \sec^2 \dfrac{1}{n^2} + \dfrac{2}{n^2} \sec^2 \dfrac{4}{n^2} + \ldots \right.$
$\left. + \dfrac{1}{n} \sec^2 1 \right] =$

  (a) $\dfrac{1}{2} \sec 1$ (b) $\dfrac{1}{2} \operatorname{cosec} 1$

  (c) $\tan 1$ (d) $\dfrac{1}{2} \tan 1$

13. If α, β are two unequal roots of the equation $ax^2 + bx + c$, then $\lim\limits_{x \to \alpha} \dfrac{1 - \cos(ax^2 + bx + c)}{(x - a)^2} =$

  (a) $\dfrac{\alpha^2}{2} (\alpha - \beta)^2$ (b) 0

  (c) $\dfrac{\alpha^2}{2} (\alpha - \beta)$ (d) $\dfrac{1}{2} (\alpha - \beta)^2$

14. $\lim\limits_{x \to 0} x \log \sin x =$

  (a) 0   (b) ∞

  (c) 1   (d) none of these

15. $\lim\limits_{x \to 0} \dfrac{e^x + \log(1 + x) - (1 - x)^{-2}}{x^2} =$

(a) 0   (b) −3

(c) −1   (d) ∞

16. $\lim\limits_{x\to 0}(\operatorname{cosec} x)^{1/\log x} =$

(a) 0   (b) 1

(c) $1/e$ (d) none of these

17. $\lim\limits_{x\to 2}\dfrac{2^x - x^2}{x^x - 2^2} =$

(a) $\dfrac{\log 2 - 1}{\log 2 + 1}$ (b) $\dfrac{\log 2 + 1}{\log 2 - 1}$

(c) 1   (d) −1

18. $\lim\limits_{x\to 0}\dfrac{2\sin^2 3x}{x^2} =$

(a) 9   (b) 2

(c) 18   (d) 3

19. $\lim\limits_{x\to 0}\dfrac{(2a)^x - (3b)^x}{x} =$

(a) $\log ab$ (b) $\log(2a/3b)$

(c) $\log(b/a)$ (d) none of these

20. If $\lim\limits_{x\to 1}\dfrac{x^4 - 1}{x - 1} = \lim\limits_{x\to k}\dfrac{x^3 - k^3}{x^2 - k^2}$, then value of k is:

(a) $\dfrac{2}{3}$ (b) $\dfrac{4}{3}$

(c) $\dfrac{8}{3}$ (d) none of these

21. $\lim\limits_{x\to 1}\dfrac{x^3 + x^2 - 2}{\sin(x-1)} =$

(a) 2   (b) 3

(c) 4   (d) 5

22. $\lim\limits_{x\to\infty}\dfrac{x^3 - 3x + 2}{2x^3 + x - 3} =$

(a) 2   (b) 1/2

(c) 0   (d) does not exist

23. $\lim\limits_{x\to 1}(\log(2x))^{\log_x 5} =$

(a) $\log_2 5$ (b) $e^{\log_2 5}$

(c) e (d) 0

24. $\lim\limits_{x\to \pi/2}\dfrac{-2x + \pi}{\cos x} =$

(a) 1   (b) −2

(c) 2   (d) 0

25. $\lim\limits_{x\to 0}\dfrac{\cos(\sin x) - \cos x}{x^4} =$

(a) $\dfrac{1}{5}$ (b) $\dfrac{1}{6}$

(c) $\dfrac{1}{4}$ (d) $\dfrac{1}{2}$

26. The value of x for which

$$f(x) = \frac{\sqrt{a^2 - ax + x^2} - \sqrt{a^2 + ax + x^2}}{\sqrt{a + x} - \sqrt{a - x}}$$

is continuous is:

(a) $a^{3/2}$ (b) $a^{1/2}$

(c) $-a^{1/2}$ (d) $-a^{3/2}$

27. If $f(x) = \begin{cases} \cos\dfrac{1}{x}, & x \neq 0 \\ k, & x = 0 \end{cases}$, is continuous at $x = 0$

then value of k is :

(a) 1   (b) −1

(c) 0   (d) none of these

28. If $f(x) = \begin{cases} (x^2 + e^{2x})^{-1}, & x \neq 2 \\ k, & x = 2 \end{cases}$, is continuous

at $x = 2$ from right then value of k:

(a) 0   (b) $\dfrac{1}{4}$

(c) $-\dfrac{1}{2}$ (d) None of these

29. Function $f : \mathbf{R} - \{0\} \to \mathbf{R}$ is given such that

$f(x) = \dfrac{1}{x} - \dfrac{2}{e^{2x} - 1}$ can be made continuous

at $x = 0$, then value of $f(0)$:

(a) −1   (b) 0

(c) 1   (d) 2

30. Function $f(x) = \begin{cases} x - 1, & x < 2 \\ 2x - 3, & x \geq 2 \end{cases}$, is continuous:

(a) for $x = 2$ only

(b) for all real values of x when $x \neq 2$

(c) for all real values of x

(d) for all integral values of x

∞ True / False

Write 'T' for True and 'F' for False statement.

1. $\lim\limits_{x\to a}[f(x) \pm \phi(x)] = \lim\limits_{x\to a} f(x) \pm \lim\limits_{x\to a} \phi(x)$. **(T/F)**

2. The value of $\lim\limits_{x\to a}\dfrac{x^3 - a^3}{x - a}$ is $3a^2$. **(T/F)**

3. The value of $\lim\limits_{x\to 1}\dfrac{\log}{x - 1}$ is 2. **(T/F)**

4. $\lim\limits_{\theta\to 0}\dfrac{\sin\theta}{\theta} = 1$. **(T/F)**

5. $\lim\limits_{x \to 4} \dfrac{x^3 - 2x^2 - 9x + 4}{x^2 - 2x - 8} = \dfrac{21}{5}$. **(T/F)** **6.** $f(x) = \dfrac{1}{x-a}$ is continuous at $x = a$. **(T/F)**

FILL IN THE BLANKS

1. $\lim\limits_{x \to a} \dfrac{f(x)}{\phi(x)} = $ _____.

2. $\lim\limits_{x \to a} [kf(x)] = $ _____.

3. $\lim\limits_{n \to \infty} \dfrac{n}{2} r^2 \sin \dfrac{2\pi}{n} = $ _____.

4. $\lim\limits_{x \to \infty} \dfrac{9x^2 + 3x + 7}{5x^2 + 2x + 1} = $ _____.

5. $\lim\limits_{x \to \infty} \left[\dfrac{\sin x}{x} \right] = $ _____.

ANSWERS

MULTIPLE CHOICE QUESTIONS

1. (a)	**2.** (a, c)	**3.** (a)	**4.** (b)	**5.** (b)	**6.** (b)	**7.** (a)	**8.** (d)	**9.** (c)
10. (c)	**11.** (c)	**12.** (d)	**13.** (a)	**14.** (a)	**15.** (b)	**16.** (c)	**17.** (a)	**18.** (c)
19. (b)	**20.** (c)	**21.** (d)	**22.** (b)	**23.** (b)	**24.** (c)	**25.** (b)	**26.** (c)	**27.** (d)
28. (b)	**29.** (c)	**30.** (c)						

TRUE / FALSE

1. T	**2.** T	**3.** F	**4.** T	**5.** F	**6.** F

FILL IN THE BLANKS

1. $\dfrac{\lim\limits_{x \to a} f(x)}{\lim\limits_{x \to a} \phi(x)}$ **2.** $k \lim\limits_{x \to a} f(x)$ **3.** πr^2 **4.** $\dfrac{9}{5}$ **5.** 0

❑❑❑❑❑

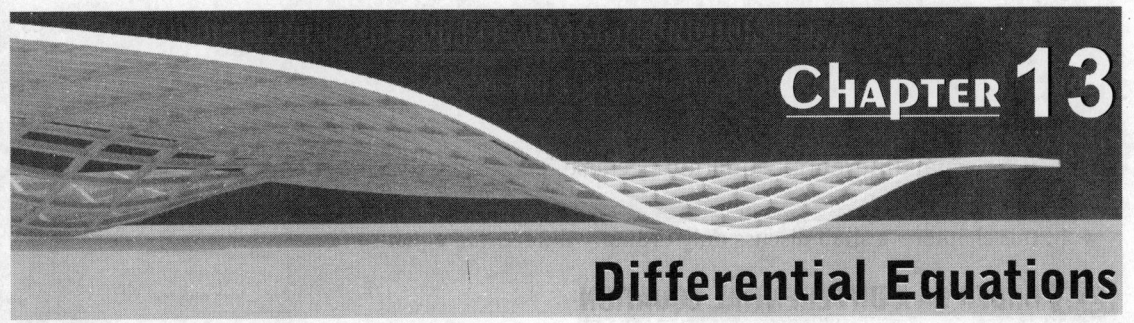

CHAPTER 13

Differential Equations

13.1 INTRODUCTION

Sometimes in the field of Physics, Chemistry, Engineering and Biology, it is necessary to prepare a Mathematical model which represents some specific problems. In these Mathematical Models generally we tries to find such functions who satisfies the given equation which contain the unknown function and some of its differential coefficients. These equations are called Differential equation.

13.2 DIFFERENTIAL EQUATION

An equation is said to be differential equation if it contains not only the dependent variable and independent variables but their differential coefficients also.

For example.

$$\frac{dy}{dx} = x + 9 \qquad \ldots(1)$$

$$\frac{dy}{dx} + x \cos x = 0 \qquad \ldots(2)$$

$$\frac{d^2y}{dx^2} = 0 \qquad \ldots(3)$$

$$\frac{dy}{dx} = \sin x \qquad \ldots(4)$$

$$\frac{d^2y}{dx^2} = e^x \qquad \ldots(5)$$

$$\frac{d^2y}{dx^2} + 4y^2 = x \qquad \ldots(6)$$

$$\left(\frac{d^2y}{dx^2}\right)^3 + \left(\frac{dy}{dx}\right)^3 = x^3 + 7y + 9 \qquad \ldots(7)$$

are differential equations.

13.3 KINDS OF DIFFERENTIAL EQUATIONS

(i) **Ordinary Differential Equation:** A differential equation is said to be ordinary differential equation if it contains only one independent variable. In article 13.2 the equations from (1) to (7) are all ordinary differential equation because in these equations only one variable x is used.

(ii) **Partial Differential Equation:** A differential equation is said to be partial differential equation if it contains more than one independent variable and derivatives. **For example.**

$$y^2 \frac{\partial z}{\partial x} + y \frac{\partial z}{\partial y} = x$$

is a partial differential equation because it contains two independent variables x and y.

☒ REMARK
⟱ In this chapter we shall discuss only ordinary differential equations.

13.4 ORDER OF A DIFFERENTIAL EQUATION

The order of a differential equation is the order of the derivative of the highest order appearing in it.

For example. The equation $\frac{dy}{dx} = x + 7$ is a differential equation of order 1 while the

differential equation $\left(\frac{d^2y}{dx^2}\right)^2 + 5\frac{dy}{dx} = x^2 + 6y$ is a differential equation of second order.

13.4.1 DEGREE OF A DIFFERENTIAL EQUATION

The degree of a differential equation is the degree (power) of the highest order differential coefficient appearing in it. **For example.** In article 13.2,

Equation (1) is a differential equation of first order and first degree.

Equation (2) is a differential equation of first order and first degree.

Equation (3) is a differential equation of second order and first degree.

Equation (4) is a differential equation of first order and first degree.

Equation (5) is a differential equation of second order and first degree.

Equation (6) is a differential equation of second order and second degree.

Equation (7) is a differential equation of second order and third degree.

☒ REMARK
⟱ To find the order and degree of a differential equation, it should be free from radicals and fractions.
For Example.

$$\sqrt{\left[\left(\frac{d^2y}{dx^2}\right)^2 + y\right]} = \sin x$$

$$\Rightarrow \qquad\qquad \left(\frac{d^2y}{dx^2}\right)^2 + y = \sin^2 x$$

$\Rightarrow$ order = 2, degree = 2

13.5 SOLUTION OF A DIFFERENTIAL EQUATION

Any relation between the dependent and independent variable which satisfies the given differential equation is called the solution of the given differential equation.

13.5.1 GENERAL SOLUTION OF A DIFFERENTIAL EQUATION

If the solution of nth order differential equation contains n arbitrary constant, then it is called general solution or complete primitive.

13.5.2 Particular Solution of a Differential Equation

A solution obtained from the general solution by giving particular values to the arbitrary constants in the general solution is called particular solution or particular integral.

13.6 FORMATION OF A DIFFERENTIAL EQUATION

If an equation contains n arbitrary constants, a differential equation of nth order can be obtained by eliminating these arbitrary constants from the given equations and n equations obtained by differentially the given equation n times. Let us suppose there is an equation representing a family of curves, containing n arbitrary constants. Then in order to find its differential equation we proceed as follows.

Working Procedure

Step 1. Differentiate the given equation of family of curves n times to get n more equations containing n arbitrary constants and derivatives.

Step 2. Eliminating all the n constants from the above $(n + 1)$ equation to get an equation containing a nth order derivative, which is the required differential equation of the family of curves.

13.7 METHOD OF FORMING A DIFFERENTIAL EQUATION

Making of a differential equation corresponding to a dependent variable is depend upon the given situation. Let us consider the following situations.

(a) When a geometrical fact is represented by differential calculus: Let the slope of a straight line be m. Then this fact can be represented by the differential equation as follows:

$$\frac{dy}{dx} = m$$

(b) If a scientific fact is represented by a differential equation: Let the acceleration of a moving particle be f. Then we represent it in the form of differential equation as follows:

$$\frac{d^2x}{dt^2} = f$$

(c) By eliminating arbitrary constant from the given equation:

$$y = 4\,ax \qquad \qquad ...(1)$$

$$\therefore \qquad \frac{dy}{dx} = 4a \qquad \qquad ...(2)$$

The differential equation obtained from (1) and (2) is

$$x\frac{dy}{dx} = 4ax = y$$

$$\Rightarrow \qquad x\frac{dy}{dx} = y$$

Working Procedure

To make a differential equation from the equation of x and y use the following steps:

Step 1. Write the given equation.

Step 2. Differentiate the given equation w.r.t. x as many times as the number of constants.

Step 3. Solve the equations of step (1) and step (2).

13.8 SOME SPECIAL EXAMPLES

(a) Differential equation of Simple Harmonic Motion

We know that S.H.M. is a motion in which a particle moved in straight line such that the direction of its acceleration is always towards a fixed point. Let O be the centre, known as origin. Let at A, the particle be at rest. It start to move from A to O such that $OA = a$.

Let at time t, the particle be at P such that $OP = x$.

By definition of S.H.M.

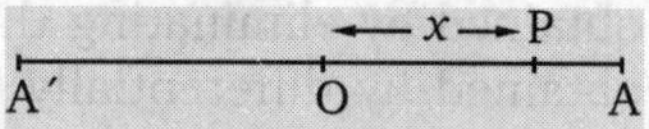

$$\frac{dv}{dt} \propto x \Rightarrow \frac{dv}{dt} = \mu x$$

where μ is the constant of proportionality called intensity of force.

$$\frac{dv}{dt} = -\mu x$$

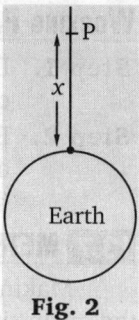

Fig. 1

Fig. 2

(b) Differential equation for the motion under gravity

Let a particle be falling from the height h to the earth. At time t, its height from the earth be x. Then by Newton's second law of motion, we have

$$m\frac{dv}{dt} = -\frac{\mu}{x^2}$$

where g is the gravitational force.

13.9 GENERAL AND PARTICULAR SOLUTIONS OF A DIFFERENTIAL EQUATION

We know that if the solution of nth order differential equation contains n arbitrary constants, then it is called general solution but if we give some particular values to the arbitrary constants, then solution thus obtained is called particular solution.

For Example. Consider the differential equation

$$\frac{d^3y}{dx^3} = 0 \qquad \qquad ...(i)$$

Its primitives is $y = Ax^2 + Bx + C$, which is the general solution of the given equation. If $A = 0$, $C = 0$ and $B = 5$, then particular solution of (1) y given by $y = 5x$ and if $A = 3$, $B = 0 = C$, then particular solution of (1) y given by $y = 3x^2$

SOLVED EXAMPLES

EXAMPLE 1. *Find the order and degree of the following differential equation.*

    (i) $\dfrac{dy}{dx} = 3x + 7$ (ii) $\dfrac{d^2y}{dx^2} + \dfrac{dy}{dx} + P_x = y$

    (iii) $\left(\dfrac{dy}{dx}\right)^3 + \dfrac{dy}{dx} + \sin x = 0$ (iv) $\dfrac{dy}{dx} = xy - \sin x$

SOLUTION.   (i) First order and first degree
              (ii) Second order and first degree
              (iii) First order and third degree
              (iv) First order and first degree.

EXAMPLE 2. *The slope of a curve at every point (x, y) is two times the sum of its coordinates. Represent it by differential equation.*

SOLUTION. The coordinates of $P = (x, y)$

Sum of coordinates $= (x + y)$

Slope of the curve at $P = 2(x + y)$

Slope of the curve at $P = \dfrac{dy}{dx}$

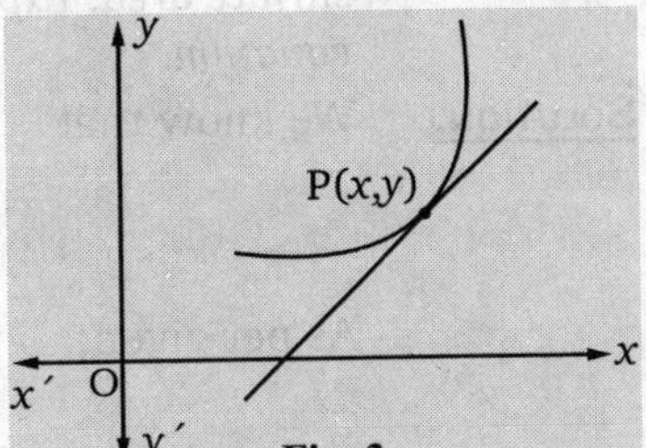

As per given $\dfrac{dy}{dx} = 2(x + y)$

which is the required differential equation.

Fig. 3

EXAMPLE 3. *In a city, the rate of increase of population, equal to the product of $(P - 75000)$ and P where P is the population. Express it by differential equation.*

SOLUTION. Rate of change of increase in population $= \dfrac{dP}{dt}$

As per given $\dfrac{dP}{dt} = P(P - 75000)$

Which is the required differential equation.

EXAMPLE 4. *A particle of m unit is falling under gravity. Represent it by differential equation.*

SOLUTION. See Fig. 2, Let at any time t, the particle be at P, whose distance from earth is x.

By gravitational law $F \propto \dfrac{1}{x^2} = \dfrac{k}{x^2}$ where k is the constant of proportionality.

$\therefore$ Required equation $m\dfrac{d^2x}{dt^2} = -\dfrac{k}{x^2}$

EXAMPLE 5. *The rate of diffusion of radius is proportional to the amount present Q. Express this statement in the form of differential equation.*

SOLUTION. The amount present $= Q$

Rate of diffusion $= \dfrac{dQ}{dt}$
As per given,

$\dfrac{dQ}{dt} \propto Q$

$\Rightarrow \quad \dfrac{dQ}{dt} = kQ$, where k is the constant of proportionality.

EXAMPLE 6. *The rate of change of 100 gm sugar to dextrose is directly proportional to the original quantity. Represent the rate of change at time t in the form of differential equation.*

SOLUTION. Let in time t, m gm sugar converted into dextrose.

$\therefore$ Remains sugar $= (100 - m)$ gm

Rate of change $= \dfrac{dm}{dt}$
Now as per given

$\dfrac{dm}{dt} \propto (100 - m) \ \ldots(1)$

$\Rightarrow \quad \dfrac{dm}{dt} = k(100 - m)$, where k is a constant of proportionality.

Which is the required differential equation.

EXAMPLE 7. *The rate of evaporation of a spherical drop of rain water is directly proportional to its surface area. Express the rate of change of spherical drop in the form of differential equation.*

SOLUTION. We know that

$$V = \frac{4}{3}\pi r^3 \text{ and } S = 4\pi r^2$$

As per given, $\quad\quad\quad\quad \dfrac{dV}{dt} \propto S$

$\Rightarrow\quad\quad\quad\quad\quad\quad \dfrac{dV}{dt} = k4\pi r^2$

Where k is a constant of proportionality.

$\Rightarrow\quad \dfrac{d}{dt}\left(\dfrac{4}{3}\pi r^3\right) = k4\pi r^2 \quad\quad\quad \Rightarrow \quad \dfrac{4}{3}\pi 3r^2 \dfrac{dr}{dt} = k4\pi r^2$

$\Rightarrow\quad\quad\quad\quad\quad\quad \dfrac{dr}{dt} = k$

Which is the required differential equation.

EXAMPLE 8. *Find the differential equation of the following equations:*

(i) $xy = c^2$ \quad\quad\quad\quad\quad\quad (ii) $y = (c_1 + c_2 x)e^x$

SOLUTION. (i) \quad\quad\quad\quad\quad $xy = c^2$

Differentiating w.r.t. x, we get

$$x\frac{dy}{dx} + y \cdot 1 = 0$$

$\Rightarrow\quad\quad \dfrac{dy}{dx} = -\dfrac{y}{x}$, which is the required differential equation.

(ii) \quad\quad\quad\quad\quad $y = (c_1 + c_2 x)e^x$ \quad\quad\quad\quad\quad\quad ...(1)

Differentiating w.r.t. x, we get

$$\frac{dy}{dx} = (c_1 + c_2 x)e^x + c_2 e^x$$

$\Rightarrow\quad\quad\quad \dfrac{dy}{dx} = y + c_2 e^x$ \quad\quad\quad\quad\quad\quad ...(2)

Again differentiating

$\Rightarrow\quad\quad\quad \dfrac{d^2 y}{dx^2} = \dfrac{dy}{dx} + c_2 e^x$ \quad\quad\quad\quad\quad\quad ...(3)

On subtracting (2) from (3),

$$\frac{d^2 y}{dx^2} - \frac{dy}{dx} = -y + \frac{dy}{dx}$$

$\Rightarrow\quad \dfrac{d^2 y}{dx^2} - 2\dfrac{dy}{dx} + y = 0$, which is the required differential equation.

EXAMPLE 9. *Represent the circles by differential equation whose centre is on x-axis and radius is r.*

SOLUTION. The equation of the circles whose centre is on x-axis and radius is r, is given by

$$(x-a)^2 + (y-0)^2 = r^2$$

$$\Rightarrow \qquad (x-a)^2 + y^2 = r^2 \qquad \qquad \ldots(1)$$

Differentiating (1) w.r.t. x, we get

$$2(x-a) + 2y\frac{dy}{dx} = 0$$

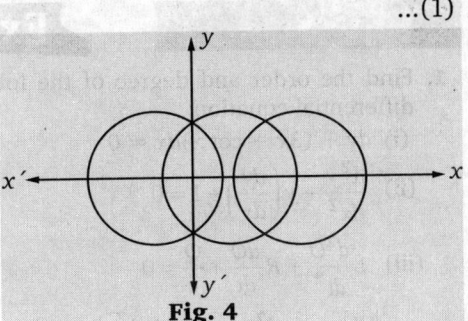

Fig. 4

$$\Rightarrow \qquad (x-a) + y\frac{dy}{dx} = 0$$

Again differentiating,

$$1 - 0 + \left(\frac{dy}{dx}\right)^2 + y\left(\frac{d^2y}{dx^2}\right) = 0$$

$$\Rightarrow \qquad y\frac{d^2y}{dx^2} + \left(\frac{dy}{dx}\right)^2 + 1 = 0$$

which is the required differential equation.

EXAMPLE 10. *Find the differential equation of all circles in first quadrant of xy-plane which touches with the axes.*

SOLUTION. Let centre of the circle be (a, a) and radius is a.
Equation of the circle

$$(x-a)^2 + (y-a)^2 = a^2 \qquad \ldots(1)$$

where a is arbitrary constant.

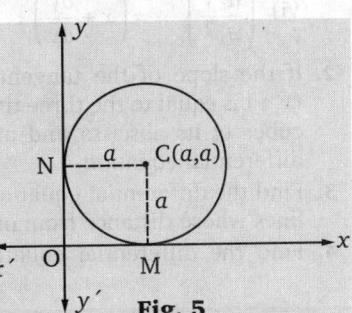

Fig. 5

$$2(x-a) + 2(y-a)\frac{dy}{dx} = 0$$

$$\Rightarrow \qquad (x-a) = -(y-a)\frac{dy}{dx} \qquad \ldots(2)$$

$$\Rightarrow \qquad x - a + y\frac{dy}{dx} - a\frac{dy}{dx} = 0$$

$$\Rightarrow \qquad x + y\frac{dy}{dx} = a\left(1 + \frac{dy}{dx}\right)$$

$$\Rightarrow \qquad a = \frac{x + y\dfrac{dy}{dx}}{1 + \dfrac{dy}{dx}} \qquad \qquad \ldots(3)$$

From eq. (1) and (2), we get

$$(y-a)^2\left(\frac{dy}{dx}\right)^2 + (y-a)^2 = a^2$$

$$\Rightarrow \qquad (y-a)^2\left[1 + \left(\frac{dy}{dx}\right)^2\right] = a^2 \qquad \ldots(4)$$

Putting the value of a from eq. (3) in (4), we get

$$\left[y - \frac{x + y\dfrac{dy}{dx}}{1 + \left(\dfrac{dy}{dx}\right)}\right]^2 \left[1 + \left(\frac{dy}{dx}\right)^2\right] = \left[\frac{x + y\dfrac{dy}{dx}}{1 + \dfrac{dy}{dx}}\right]^2$$

which is the required differential equation.

Exercise 13.1

1. Find the order and degree of the following differential equation.

   (i) $dy + (3x + \cot x)dx = 0$

   (ii) $\dfrac{d^2y}{dx^2} + y\left(\dfrac{dy}{dx}\right) + 1 = 0$

   (iii) $L\dfrac{d^2Q}{dt^2} + R\dfrac{dQ}{dt} + \dfrac{Q}{c} = 0$

   (iv) $\dfrac{d^3y}{dx^3} + x\dfrac{d^2y}{dx^2} + 2y\left(\dfrac{dy}{dx}\right)^2 + xy = 0$

   (v) $\dfrac{d^2r}{d\theta^2} = \sqrt[4]{1 + \left(\dfrac{dr}{d\theta}\right)^2}$

   (vi) $\left(\dfrac{d^2y}{dx^2}\right)^{3/2} = \left(x + \dfrac{dy}{dx}\right)^{1/2}$

2. If the slope of the tangent to the curve at (x, y) is equal to the three times to sum of the cubes of its abscissa and ordinate. Find the differential equation.

3. Find the differential equation for the straight lines whose distance from origin is 1 unit.

4. Find the differential equation of all circles who touches y-axis at origin.

5. Find the differential equation of all parabolas whose vertex is at origin and foci are on y-axis.

6. Find the differential equation of all ellipse $\dfrac{x^2}{a^2} + \dfrac{y^2}{b^2} = 1$, where a and b are constant.

7. Prove that $y = \dfrac{A}{x} + B$ is the solution of the differential equation $\dfrac{d^2y}{dx^2} + \dfrac{2}{x}\dfrac{dy}{dx} = 0$.

8. Prove that $y = A \cos mx + B \sin mx$ is the solution of the differential equation $\dfrac{d^2y}{dx^2} + m^2y = 0$.

9. Find the differential equation of $y = a + bx + cx^2$.

10. Find the differential equation of $xy = ae^x + be^{-x}$.

11. Find the differential equation of $x^2 + y^2 = a^2$.

12. Find the differential equation of $y = e^x (A \cos x + B \sin x)$.

13. Find the differential equation of $y = A \cos x^2 + B \sin x^2$.

ANSWERS

1. (i) order = 1, degree = 1 (ii) order = 2, degree = 1 (iii) order = 2 degree = 1
   (iv) order = 3, degree = 1 (v) order = 2, degree = 4 (vi) order = 2 degree = 3

2. $\dfrac{dy}{dx} = 3(x^3 + y^3)$ 3. $\left(x\dfrac{dy}{dx} - y\right)^2 = 1 + \left(\dfrac{dy}{dx}\right)^2$ 4. $y^2 - x^2 = 2xy\dfrac{dy}{dx}$ 5. $x\dfrac{dy}{dx} = 2y$

6. $y\dfrac{dy}{dx} = x\left[y\dfrac{d^2y}{dx^2} + \left(\dfrac{dy}{dx}\right)^2\right]$ 9. $\dfrac{d^3y}{dx^3} = 0$ 10. $x\dfrac{d^2y}{dx^2} + 2\dfrac{dy}{dx} = xy$

11. $\dfrac{dy}{dx} = -\dfrac{x}{y}$ 12. $\dfrac{d^2y}{dx^2} - 2\left(\dfrac{dy}{dx}\right) + 2y = 0$ 13. $x\dfrac{d^2y}{dx^2} - \dfrac{dy}{dx} + 4x^3y = 0$

13.10 SOLUTION OF DIFFERENTIAL EQUATIONS

(I) Type I. $\dfrac{dy}{dx} = f(x)$ 　　　　　　　　　　　　　　　　　　...(1)

We know that integration is the reverse process of differentiation. To solve the equation (1), integrate both sides.

WORKING PROCEDURE

Step 1. Write the given equation of the form $dy = f(x)dx$.

Step 2. Integrate and use the constant of integration.

SOLVED EXAMPLES

EXAMPLE 1. *Solve the differential equation* $\dfrac{dy}{dx} = \cos x$.

SOLUTION. We have, $\dfrac{dy}{dx} = \cos x$

$\Rightarrow \qquad dy = \cos x \, dx$

Integrating both sides, we get

$$\int dy = \int \cos x \, dx + c$$

$\Rightarrow \qquad y = \sin x + c$, where c is any constant.

EXAMPLE 2. *Solve the differential equation* $\dfrac{dy}{dx} = \sec^2 x + 3x^2$.

SOLUTION. We have $\dfrac{dy}{dx} = \sec^2 x + 3x^2$

$\Rightarrow \qquad dy = \sec^2 x \, dx + 3x^2 \, dx$

Integrating both sides, we get

$$\int dy = \int \sec^2 x \, dx + 3 \int x^2 dx + c$$

$\Rightarrow \qquad y = \tan x + x^3 + c$, where c is any constant.

EXAMPLE 3. *Solve the differential equation* $\dfrac{dy}{dx} = \sec x (2 \sec x + \tan x)$.

SOLUTION. We have $\dfrac{dy}{dx} = \sec x (2 \sec x + \tan x)$

$\Rightarrow \qquad dy = 2\sec^2 x \, dx + \sec x \tan x \, dx$

Integrating both sides, we get

$$\int dy = 2 \int \sec^2 x \, dx + \int \sec x \tan x dx + c$$

$\Rightarrow \qquad y = 2 \tan x + \sec x + c$, where c is any constant.

EXAMPLE 4. *Solve the differential equation* $\dfrac{dy}{dx} = \sin(5x + 9)$.

SOLUTION. We have $\dfrac{dy}{dx} = \sin(5x + 9)$

$\Rightarrow \qquad dy = \sin(5x + 9)dx$

On integrating both sides, we get

$$\int dy = \int \sin(5x + 9) \, dx + c$$

$\Rightarrow \qquad y = -\dfrac{1}{5} \cos(5x + 9) + c$, where c is any constant.

EXAMPLE 5. *Solve the differential equation* $\dfrac{dy}{dx} = x^3 + \sin 4x$

SOLUTION. We have $\dfrac{dy}{dx} = x^3 + \sin 4x$

$\Rightarrow \qquad dy = x^3 \, dx + \sin 4x \, dx$

Integrating both sides, we get

$$\int dy = \int x^3 dx + \int \sin 4x \, dx + c$$

$\Rightarrow \qquad y = \dfrac{1}{4} x^4 - \dfrac{1}{4} \cos 4x + c$, where c is any constant.

EXAMPLE 6. *Solve the differential equation* $\dfrac{dy}{dx} = \dfrac{\cos x}{2 - \cos^2 x}$.

SOLUTION. We have
$$\frac{dy}{dx} = \frac{\cos x}{2-\cos^2 x} = \frac{\cos x}{2-(1-\sin^2 x)}$$

$$\Rightarrow \quad \frac{dy}{dx} = \frac{\cos x}{1+\sin^2 x}$$

$$\Rightarrow \quad dy = \frac{\cos x\, dx}{1+\sin^2 x}$$

Integrating both sides, we get

$$\int dy = \int \frac{\cos x\, dx}{1+\sin^2 x} + c$$

Let $\sin x = t$ $\therefore$ $\cos x\, dx = dt$

$$\Rightarrow \quad \int dy = \int \frac{dt}{1+t^2} + c$$

$$\Rightarrow \quad y = \tan^{-1} t + c$$

$$\Rightarrow \quad y = \tan^{-1}(\sin x) + c$$

Where c is an arbitrary constant.

EXAMPLE 7. *Solve the differential equation $\cos y\, dy + \cos x \sin y\, dx = 0$.*

Given : $x = \frac{\pi}{2}$, if $y = \frac{\pi}{2}$

SOLUTION. We have $\cos y\, dy + \cos x \sin y\, dx = 0$

$$\Rightarrow \quad \frac{\cos y}{\sin y} dy + \cos x\, dx = 0$$

On integrating, we get

$$\int \frac{\cos y}{\sin y} dy + \int \cos x\, dx = c$$

$$\Rightarrow \quad \log \sin y + \sin x = c$$

$x = \frac{\pi}{2}$ if $y = \frac{\pi}{2}$ (given)

$$\Rightarrow \quad \log \sin \frac{\pi}{2} + \sin \frac{\pi}{2} = c$$

$$\therefore \quad c = 1$$

$$\therefore \quad \log \sin y + \sin x = 1$$

Exercise 13.2

Solve the following differential equations.

1. $\frac{dy}{dx} = e^x$

2. $\frac{dy}{dx} = \frac{1}{x}$

3. $\frac{dy}{dx} = \tan x$

4. $\frac{dy}{dx} = \cot x$

5. $\frac{dy}{dx} = \text{cosec}^2 x + \sec^2 x + \cos x$

6. $\frac{dy}{dx} = \sec^2 x + 4x$

7. $\frac{dy}{dx} = x^2 + \sin 4x$

8. $\frac{dy}{dx} = x^3 + x^2 + 6x + 9$

9. $\frac{dy}{dx} = \sin^8 x \cos x$

10. $(1+x^2)\frac{dy}{dx} = x$

11. $\frac{dy}{dx} + \frac{1+x^2}{x} = 0$

12. $\frac{dy}{dx} = \text{cosec}\, x(3\text{cosec}\, x + 4\cot x)$

13. $\frac{dy}{dx} = \cos(ax+b)$

ANSWERS

1. $y = e^x + c$ 2. $y = \log x + c$ 3. $y = \log \sec x + c$ 4. $y = \log \sin x + c$

5. $y = -\cot x + \tan x - \sin x$ **6.** $y = \tan x + 2x^2 + c$ **7.** $\dfrac{x^3}{3} - \dfrac{\cos 4x}{4} + c$

8. $\dfrac{x^4}{4} + \dfrac{x^3}{3} + 3x^2 + 9x + c$ **9.** $y = \dfrac{\sin^9 x}{9} + c$ **10.** $y = \dfrac{1}{2}\log(1 + x^2) + c$

11. $y + \log x + \dfrac{1}{2}x^2 = c$ **12.** $y = -3\cot x - 4\operatorname{cosec} x + c$ **13.** $y = -\dfrac{1}{a}\sin(ax + b) + c$

(II) Type II. Equation in which variables are separable

Here, the equation is of the form

$$\frac{dy}{dx} = \frac{f(x)}{\phi(y)}$$

Using transposition of terms, we get

$$f(x)\,dx = \phi(y)\,dy$$

We separate all the terms of x with dx and all the terms of y with dy.

Then adding the constant of integration after integrate.

WORKING PROCEDURE

Step 1. Write the given equation in the form of $\dfrac{dy}{dx} = f(x) \cdot g(y)$

Step 2. Write all the terms of x with dx, and all terms of y with dy, *i.e.*,

$$f(x)dx = \frac{dy}{g(y)}$$

Step 3. Integrating both the sides.

SOLVED EXAMPLES

EXAMPLE 1. *Solve the differential equation* $\dfrac{dy}{dx} = \dfrac{x}{y}$.

SOLUTION. We have $\dfrac{dy}{dx} = \dfrac{x}{y}$

$\Rightarrow$ $y\,dy = x\,dx$

On integrating, we get

$$\int y\,dy = \int x\,dx + c \qquad \Rightarrow \qquad \frac{y^2}{2} = \frac{x^2}{2} + c$$

$\Rightarrow$ $y^2 = x^2 + k$, where k is an arbitrary constant.

EXAMPLE 2. *Solve the differential equation* $\dfrac{dy}{dx} = xy + x + y + 1$.

SOLUTION. We have $\dfrac{dy}{dx} = xy + x + y + 1 = x(y + 1) + 1(y + 1)$

$\Rightarrow$ $\dfrac{dy}{dx} = (x + 1)(y + 1)$ $\Rightarrow$ $\dfrac{dy}{y + 1} = (x + 1)dx$

On integrating, we get

$$\int \frac{dy}{y + 1} = \int (x + 1)dx + c$$

$\Rightarrow$ $\log(y + 1) = \dfrac{x^2}{2} + x + c$, where c is an arbitrary constant.

EXAMPLE 3. *Solve the differential equation* $\dfrac{dy}{dx} + \sqrt{\left(\dfrac{1-y^2}{1-x^2}\right)} = 0$.

SOLUTION. We have $\dfrac{dy}{dx} + \sqrt{\left(\dfrac{1-y^2}{1-x^2}\right)} = 0$

$\Rightarrow \qquad \dfrac{dy}{\sqrt{1-y^2}} + \dfrac{dx}{\sqrt{1-x^2}} = 0$

On integrating, we get

$\displaystyle \int \dfrac{dy}{\sqrt{1-y^2}} + \int \dfrac{dx}{\sqrt{1-x^2}} = c$

$\Rightarrow \quad \sin^{-1} y + \sin^{-1} x = c$, where c is an arbitrary constant.

EXAMPLE 4. *Solve the differential equation* $\dfrac{dy}{dx} = e^{x-y} + x^2 e^{-y}$.

SOLUTION. We have $\qquad \dfrac{dy}{dx} = e^{x-y} + x^2 e^{-y} = \dfrac{e^x}{e^y} + \dfrac{x^2}{e^y}$

$\Rightarrow \qquad\qquad e^y dy = e^x dx + x^2 dx$

On integrating, we get

$\displaystyle \int e^y dy = \int e^x dx + \int x^2 dx + c$

$\Rightarrow \qquad\qquad e^y = e^x + \dfrac{1}{3}x^3 + c$, where c is an arbitrary constant.

EXAMPLE 5. *Solve the differential equation* $(1+x^2)\sec^2 ydy + 2x \tan ydx = 0$.

Given that $y = \dfrac{\pi}{4}$ *when* $x = 1$.

SOLUTION. We have $(1+x^2)\sec^2 ydy + 2x\tan ydx = 0$

$\Rightarrow \qquad \dfrac{\sec^2 y}{\tan y}dy + \dfrac{2x}{1+x^2}dx = 0$

On integrating,

$\displaystyle \int \dfrac{\sec^2 y}{\tan y}dy + \int \dfrac{2x}{1+x^2}dx = 0$

$\Rightarrow \log \tan y + \log(1+x^2) = c$

If $y = \dfrac{\pi}{4}$ and $x = 1$ then $c = \log \tan \dfrac{\pi}{4} + \log(1+1)$

$\Rightarrow \qquad\qquad\qquad c = \log 2$

$\therefore \ \log \tan y + \log(1+x^2) = \log 2$

$\Rightarrow \qquad \log[\tan y (1+x^2)] = \log 2$

$\therefore \qquad\qquad\qquad \tan y = \dfrac{2}{1+x^2}$

EXAMPLE 6. *Solve the differential equation* $(x - y^2 x)dx - y(1-x^2)dy = 0$.

SOLUTION. We have $(x - y^2 x)dx - y(1-x^2)dy = 0$

$\Rightarrow \qquad\qquad x(1-y^2)dx = y(1-x^2)dy$

$$\Rightarrow \qquad \frac{x}{1-x^2}dx = \left(\frac{y}{1-y^2}\right)dy$$

$$\Rightarrow \qquad \frac{xdx}{(1-x)(1+x)} = \frac{ydy}{(1-y)(1+y)}$$

$$\Rightarrow \qquad \frac{1}{2}\left[\frac{1}{1-x} - \frac{1}{1+x}\right]dx = \frac{1}{2}\left[\frac{1}{1-y} - \frac{1}{1+y}\right]dy$$

$$\Rightarrow \qquad \int\left(\frac{1}{1-x} - \frac{1}{1+x}\right)dx = \int\left(\frac{1}{1-y} - \frac{1}{1+y}\right)dy$$

On integrating,

$$\log(1+x) + \log(1-x) = \log(1+y) + \log(1-y) + \log c$$
$$\log(1-x^2) = \log(1-y^2) + \log c$$

$$\Rightarrow \qquad \log\left(\frac{1-x^2}{1-y^2}\right) = \log c$$

$$\Rightarrow \qquad (1-x^2) = c(1-y^2), \text{ where } c \text{ is an arbitrary constant.}$$

EXAMPLE 7. *Solve the differential equation* $\left(y - x\dfrac{dy}{dx}\right) = a\left(y^2 + \dfrac{dy}{dx}\right)$.

SOLUTION. We have $\qquad \left(y - x\dfrac{dy}{dx}\right) = a\left(y^2 + \dfrac{dy}{dx}\right)$

$$\Rightarrow \qquad y - ay^2 = (x+a)\frac{dy}{dx}$$

$$\Rightarrow \qquad y(1-ay) = (x+a)\frac{dy}{dx}$$

$$\Rightarrow \qquad \frac{dx}{x+a} = \frac{dy}{y(1-ay)}$$

$$\Rightarrow \qquad \frac{dx}{x+a} = \left(\frac{a}{1-ay} + \frac{1}{y}\right)dy \text{ ,(On resolving into partial fraction)}$$

On integrating, we get

$$\log y - \log(1-ay) = \log(a+x) + \log c$$

$$\Rightarrow \log\left(\frac{y}{1-ay}\right) - \log(a+x) = \log c$$

$$\Rightarrow \qquad \log\frac{y}{(1-ay)(a+x)} = \log c$$

$$\Rightarrow \qquad y = c(1-ay)(a+x), \text{ where } c \text{ is an arbitrary constant.}$$

EXAMPLE 8. *Solve the differential equation* $(3x^2y - xy)dx + (2x^3y^2 + x^3y^4)dy = 0$.

SOLUTION. We have $(3x^2y - xy)dx + (2x^3y^2 + x^3y^4)dy = 0$

$$\Rightarrow \qquad y(3x^2 - x)dx + x^3(2y^2 + y^4)dy = 0$$

$$\Rightarrow \qquad \left(\frac{3x^2 - x}{x^3}\right)dx + \left(\frac{2y^2 + y^4}{y}\right)dy = 0$$

$$\Rightarrow \qquad \left(\frac{3}{x} - \frac{1}{x^2}\right)dx + (2y + y^3)dy = 0$$

On integrating, we get

$$\int \left(\frac{3}{x} - x^{-2} \right) dx + \int (2y + y^3) dy = 0$$

$\Rightarrow \quad 3 \log x + \frac{1}{x} + \frac{2y^2}{2} + \frac{y^4}{4} = c$, where c is an arbitrary constant.

EXAMPLE 9. *Solve the differential equation* $x^2(y + 1)dx + y^2(x - 1)dy = 0$.

SOLUTION. We have $\quad x^2(y + 1)dx + y^2(x - 1)dy = 0$

$$\Rightarrow \quad \left(\frac{x^2}{x-1} \right) dx + \left(\frac{y^2}{y+1} \right) dy = 0$$

$$\Rightarrow \quad \left(\frac{x^2 - 1 + 1}{x - 1} \right) dx + \left(\frac{y^2 - 1 + 1}{y + 1} \right) dy = 0$$

$$\Rightarrow \quad \int \left(x + 1 + \frac{1}{x-1} \right) dx + \int \left(y - 1 + \frac{1}{y+1} \right) dy = 0$$

On integrating, we get

$$\int \left(x + 1 + \frac{1}{x-1} \right) dx + \int \left(y - 1 + \frac{1}{y+1} \right) dy = 0$$

$$\Rightarrow \quad \frac{x^2}{2} + x + \log(x-1) + \frac{y^2}{2} - y + \log(y+1) = c$$

$$\Rightarrow \quad \frac{1}{2}(x^2 + y^2) + (x - y) + \log(x-1)(y+1) = c$$

$$\Rightarrow \quad (x^2 + y^2) + 2(x - y) + 2\log(x-1)(y+1) = k,$$

when k is an arbitrary constant.

EXAMPLE 10. *Solve the differential equation* $\frac{dy}{dx} = e^{ax} \cos y$, *if* $y(0) = 0$.

SOLUTION. We have $\quad \frac{dy}{dx} = e^{ax} \cos y$

$$\Rightarrow \quad \frac{dy}{\cos y} = e^{ax} dx$$

$$\Rightarrow \quad \sec y \, dy = e^{ax} \, dx$$

On integrating, we get

$$\int \sec y \, dy = \int e^{ax} dx + c$$

$$\Rightarrow \quad \log(\sec y + \tan y) = \frac{1}{a} e^{ax} + c$$

$$\Rightarrow a \log(\sec y + \tan y) = e^{ax} + k \qquad \qquad \qquad \text{...(1)}$$

Now if $x = 0$, then $y = 0$

$\Rightarrow a \log (\sec 0 + \tan 0) = e^0 + k$

$\Rightarrow \qquad a \log (1 + 0) = 1 + k$

$\Rightarrow \qquad \qquad k = -1$

Hence, from eq. (1), we get

$\quad a \log (\sec y + \tan y) = e^{ax} - 1$

EXAMPLE 11. *Solve the differential equation* $(1 + e^{2x})\,dy + (1 + y^2)\,e^x\,dx = 0$, *where* $y = 1$ *if* $x = 0$.

SOLUTION. We have $(1 + e^{2x})dy + (1 + y^2)e^x dx = 0$

$$\Rightarrow \qquad \frac{dy}{1+y^2} + \frac{e^x}{1+e^{2x}}dx = 0$$

On integrating, we get

$$\int \frac{dy}{1+y^2} + \int \frac{e^x dx}{1+e^{2x}} = 0 \qquad\qquad \text{...(1)}$$

$$\Rightarrow \qquad \tan^{-1}y + \tan^{-1}(e^x) = c$$

Now, when $x = 0$ then $y = 1$

$$\Rightarrow \qquad \tan^{-1}1 + \tan^{-1}(e^x) = c$$

$$\Rightarrow \qquad \frac{\pi}{4} + \frac{\pi}{4} = c \quad \Rightarrow \quad c = \frac{\pi}{2}$$

Hence, from eq. (1), we get

$$\tan^{-1}y + \tan^{-1}(e^x) = \frac{\pi}{2}$$

EXAMPLE 12. *Solve the differential equation* $\dfrac{dy}{dx} + \dfrac{1+y^3}{xy^2(1+x^2)} = 0$.

SOLUTION. We have $\dfrac{dy}{dx} + \dfrac{1+y^3}{xy^2(1+x^2)} = 0$

$$\Rightarrow \qquad \frac{y^2 dy}{1+y^3} + \frac{dx}{x(1+x^2)} = 0$$

$$\Rightarrow \qquad \frac{y^2 dy}{1+y^3} + \frac{x\,dx}{x^2(1+x^2)} = 0$$

On integrating, we get

$$\int \frac{y^2 dy}{1+y^3} + \int \frac{x\,dx}{x^2(1+x^2)} = c \qquad\qquad \text{...(1)}$$

Let $1 + y^3 = t_1 \qquad \Rightarrow \quad 3y^2 dy = dt_1$,

$$\Rightarrow \qquad y^2 dy = \frac{1}{3}dt_1$$

and $x^2 = t_2 \qquad \Rightarrow \quad x\,dx = \frac{1}{2}dt_2$

Therefore from eq. (1),

$$\frac{1}{3}\int \frac{dt_1}{t_1} + \frac{1}{2}\int \frac{dt_2}{t_2(1+t_2)} = c$$

$$= \frac{1}{3}\int \frac{dt_1}{t_1} + \frac{1}{2}\int \left(\frac{1}{t_2} - \frac{1}{1+t_2}\right)dt_2 = c$$

(Second function is resolved by partial fraction.)

$$\Rightarrow \frac{1}{3}\log t_1 + \frac{1}{2}[\log t_2 - \log(1+t_2)] = c$$

$$\Rightarrow \frac{1}{3}\log t_1 + \frac{1}{2}\left[\log\left(\frac{t_2}{1+t_2}\right)\right] = c$$

$$\Rightarrow 2\log t_1 + 3\log\left(\frac{t_2}{1+t_2}\right) = 6c$$

$$\Rightarrow \log t_1^2 + \log\left(\frac{t_2}{1+t_2}\right)^3 = \log k \qquad \text{(Say)}$$

$$\Rightarrow \log\left(\frac{t_1^2 t_2^3}{(1+t_2)^3}\right) = \log k$$

$$\Rightarrow t_1^2 t_2^3 = k(1+t_2)^3$$

$$\Rightarrow (1 + y^3)^2 x^6 = k(1 + x^2)^3$$

EXAMPLE 13. *Solve the differential equation* $\sec^2 x \tan y \, dx + \sec^2 y \tan x \, dy = 0$.

SOLUTION. We have $\sec^2 x \tan y \, dx + \sec^2 y \tan x \, dy = 0$

$$\Rightarrow \frac{\sec^2 x}{\tan x}dx + \frac{\sec^2 y}{\tan y}dy = 0$$

On integrating, we get

$$\Rightarrow \int\frac{\sec^2 x}{\tan x}dx + \int\frac{\sec^2 y}{\tan y}dy = \log k$$

$$\Rightarrow \log \tan x + \log \tan y = \log k$$

$$\Rightarrow \log \tan x \tan y = \log k$$

$$\Rightarrow \tan x \tan y = k$$

Where k is an arbitrary constant.

EXAMPLE 14. *Solve* $\dfrac{dy}{dx} = \tan^{-1} x$.

SOLUTION. Given

$$\frac{dy}{dx} = \tan^{-1} x$$

$$\Rightarrow \int\frac{dy}{dx}dx = \int \tan^{-1}x \, dx + c$$

$$\Rightarrow \int dy = \int \tan^{-1}x \cdot 1 \, dx + c$$

$$\Rightarrow y = \tan^{-1}x \cdot x + \int\left[\frac{1}{1+x^2}\cdot x\right]dx + c$$

$$\Rightarrow y = x\tan^{-1}x - \frac{1}{2}\log(1+x^2) + c$$

Exercise 13.3

Solve the following differential equations:

1. $\dfrac{dy}{dx} = -\dfrac{y}{x}$

2. $\dfrac{dy}{dx} = \dfrac{1}{y + \sin y}$

3. $3 e^x \tan y \, dx + (1 - e^x)\sec^2 y \, dy = 0$

4. $(e^x + 1)y \, dy + (y + 1) \, dx = 0$

5. $(1+x^2)\dfrac{dy}{dx}+1+y^2=0$

6. $\dfrac{dy}{dx}=e^{x+y}+x^2e^y$

7. $y\sec^2 x+(y+7)\tan x\dfrac{dy}{dx}=0$

8. $\dfrac{dy}{dx}=\dfrac{1+y^2}{1+x^2}$

9. $\dfrac{dy}{dx}=\sin x\sin y$

10. $\sqrt{(a+x)}\dfrac{dy}{dx}+x=0$ **11.** $\dfrac{dy}{dx}+\dfrac{1+y^2}{y}=0$

12. $(1+x)\,ydx+(1+y)\,xdy=0$

13. $(1+x^2)\dfrac{dy}{dx}=2x$ **14.** $x\dfrac{dy}{dx}+y=y^2$

15. $\dfrac{dy}{dx}=\dfrac{x(2\log x+1)}{\sin y+y\cos y}$

16. $e^{2x-3y}dx+e^{2y-3x}dy=0$

ANSWERS

1. $xy=c$ 　　**2.** $\dfrac{y^2}{2}-\cos y=x+c$ 　　**3.** $\log\tan y=3\log(1-e^x)+\log c$

4. $y=\log(1+y)+\log(1+e^{-x})+c$ 　　**5.** $\tan^{-1}x+\tan^{-1}y=c$

6. $e^x+e^{-y}+\dfrac{1}{3}x^3=c_1$ 　　**7.** $y^7\tan x=e^{c-y}$ 　　**8.** $\dfrac{y-x}{1+xy}=c$

9. $\log(\operatorname{cosec}y-\cot y)+\cos x=c$ 　　**10.** $y=\dfrac{2}{3}(a-x)\sqrt{a+x}+c$

11. $\dfrac{1}{2}\log(1+y^2)+x=c$ 　　**12.** $xy=c_1e^{x+y}$ 　　**13.** $y=\log(1+x^2)+c$

14. $y-1=cxy$ 　　**15.** $y\sin y=x^2\log x+c$ 　　**16.** $e^{5x}+e^{5y}=c$

(III) Type III: Solution of the differential equation of the form $\dfrac{d^2y}{dx^2}=f(x)$

To solve such type of differential equation, integrate the given equation two times.

SOLVED EXAMPLES

EXAMPLE 1. *Solve the differential equation* $\dfrac{d^2y}{dx^2}=\cos x-\sin x$.

SOLUTION. The given differential equation is

$$\frac{d}{dx}\left(\frac{dy}{dx}\right)=\cos x-\sin x$$

Integrating both the sides w.r.t. x, we get

$$\frac{dy}{dx}=\int(\cos x-\sin x)dx+c_1$$

where c_1 is an arbitrary constant.

$$\Rightarrow\qquad \frac{dy}{dx}=\sin x+\cos x+c_1$$

$$\Rightarrow\qquad dy=(\sin x+\cos x+c_1)dx$$

Again, integrating both the sides w.r.t. x, we get

$$y=\int\sin xdx+\int\cos xdx+c_1\int dx+c_2$$

where c_2 is an arbitrary constant.

$$\Rightarrow\qquad y=-\cos x+\sin x+c_1x+c_2$$

EXAMPLE 2. *Solve the differential equation* $\dfrac{d^2y}{dx^2}=x^2+e^x$.

SOLUTION. The given differential equation

$$\frac{d}{dx}\left(\frac{dy}{dx}\right) = x^2 + e^x$$

Integrating both the sides w.r.t. x, we get

$$\frac{dy}{dx} = \int(x^2 + e^x)dx + c_1,$$

where c_1 is an arbitrary constant.

$$\Rightarrow \quad \frac{dy}{dx} = \frac{x^3}{3} + e^x + c_1$$

$$\Rightarrow \quad dy = \left(\frac{x^3}{3} + e^x + c_1\right)dx$$

$$\Rightarrow \quad dy = \frac{x^3}{3}dx + e^x dx + c_1 dx$$

Again, integrating both the sides w.r.t. x, we get

$$y = \frac{1}{3}\int x^3 dx + \int e^x dx + c_1\int dx + c_2;$$

where c_2 is an arbitrary constant.

$$\Rightarrow \quad y = \frac{x^4}{12} + e^x + c_1 x + c_2$$

EXAMPLE 3. *Solve the differential equation* $\dfrac{d^2y}{dx^2} = xe^x$.

SOLUTION. The given differential equation

$$\frac{d}{dx}\left(\frac{dy}{dx}\right) = xe^x$$

Integrating both the sides w.r.t. x, we get

$$\frac{dy}{dx} = \int xe^x dx = x\int e^x dx - \int\left[\frac{d}{dx}x\int e^x dx\right]dx + c_1$$

(On integrating by parts)

$$= xe^x - \int e^x dx + c_1$$

where c_1 is an arbitrary constant.

$$\Rightarrow \quad \frac{dy}{dx} = xe^x - e^x + c_1$$

$$\Rightarrow \quad dy = xe^x dx - e^x dx + c_1 dx$$

Again, integrating both the sides w.r.t. x, we get

$$\int dy = \int xe^x dx - \int e^x dx + c_1\int dx + c_2,$$

where c_2 is an arbitrary constant.

$$y = x\int e^x dx - \int\left[\frac{d}{dx}x\int e^x dx\right]dx - e^x + c_1 x + c_2$$

(On integrating by parts)

$$\Rightarrow \quad y = xe^x - e^x - e^x + c_1 x + c_2$$
$$= (x-2)e^x + c_1 x + c_2$$

EXAMPLE 4. *Solve the differential equation* $\dfrac{d^2y}{dx^2} = \log x$.

$\quad$ *Given* : $y = 1, \dfrac{dy}{dx} = -1$ *when* $x = 1$

SOLUTION. The given differential equation is

$$\frac{d}{dx}\left(\frac{dy}{dx}\right) = \log x$$

Integrating both the sides w.r.t. x, we get

$$\frac{dy}{dx} = \log x \int 1 dx - \int\left[\frac{d}{dx}\log x \int 1 dx\right] dx + c_1$$

$$\text{(On integrating by parts)}$$

$$\Rightarrow \qquad \frac{dy}{dx} = x \log x - x + c_1$$

when $x = 1, \dfrac{dy}{dx} = -1$

$\therefore \qquad\qquad -1 = 0 - 1 + c_1 \quad \Rightarrow \quad c_1 = 0$

So, $\qquad\qquad \dfrac{dy}{dx} = x \log x - x$

Again integrating, we get

$$y = \log x \int x dx - \int\left[\frac{d}{dx}\log x \cdot \int x dx\right] dx - \int x dx + c_2$$

$$\Rightarrow \qquad y = \frac{x^2}{2}\log x - \frac{1}{2}\int x dx - \frac{x^2}{2} + c_2$$

$$y = \frac{x^2}{2}\log x - \frac{x^2}{4} - \frac{x^2}{2} + c_2$$

$$\Rightarrow \qquad y = \frac{x^2}{2}\log x - \frac{3}{4}x^2 + c_2$$

Given $x = 1 \Rightarrow y = 1$

$$\therefore \qquad 1 = \frac{1}{2}\times 1^2 \log 1 - \frac{(3\times 1^2)}{4} + c_2$$

$$\Rightarrow \qquad c_2 = \frac{7}{4}$$

Hence, $\qquad y = \dfrac{1}{2}x^2 \log x - \dfrac{3x^2}{4} + \dfrac{7}{4}$

Exercise 13.4

Solve the following differential equations:

1. $\dfrac{d^2y}{dx^2} = 1$

2. $\dfrac{d^2y}{dx^2} = \sin x$

3. $\dfrac{d^2y}{dx^2} = x^7$

4. $\dfrac{d^2y}{dx^2} = e^{4x}$

5. $\dfrac{d^2y}{dx^2} = xe^x + \cos x$

6. $\dfrac{d^2y}{dx^2} = \cos x + x$

7. $\dfrac{d^2y}{dx^2} = x^2 e^x$

8. $\dfrac{d^2y}{dx^2} = x \sin x + e^x$

9. $\dfrac{d^2y}{dx^2} = x \sin x$

10. $\dfrac{d^2y}{dx^2} = x \cos x$

11. $\dfrac{d^2y}{dx^2} = x + \sin x$, given at $y = 0$, $\dfrac{dy}{dx} = -1$ if $x = 0$.

12. $\dfrac{d^2y}{dx^2} = x^2 \sin x$, given at $x = 0$ if $y = 0, \dfrac{dy}{dx} = 1$

13. $\dfrac{d^2y}{dx^2} + \sin 2x = (x^2 + 1)e^{2x}$

14. $\dfrac{d^2y}{dx^2} = \cos x + e^{3x} + x^3$

15. $\dfrac{d^2y}{dx^2} = \dfrac{1}{1+x^2}$ when $x = 0, y = 0$ and $\dfrac{dy}{dx} = 0$

16. Solve the following differential equation

$$x\dfrac{d^2y}{dx^2} + \dfrac{dy}{dx} + x = 0$$

[**Hint:** Let $p = \dfrac{dy}{dx} \Rightarrow \dfrac{dp}{dx} = \dfrac{d^2y}{dx^2}$

Which can be reduce as

$$x\dfrac{dp}{dx} + p + x = 0$$
$$\Rightarrow \qquad xdp + pdx = -xdx$$
$$\Rightarrow \qquad d(px) = -xdx$$

On integrating,

$$px = -\dfrac{x^2}{2} + c$$
$$\Rightarrow \qquad p = -\dfrac{x}{2} + cx^{-1}$$
$$\Rightarrow \qquad \dfrac{dy}{dx} = -\dfrac{x}{2} + cx^{-1}$$
$$\Rightarrow \qquad dy = -\dfrac{x}{2}dx + cx^{-1}dx$$

Again integrating, we get

$$y = -\dfrac{x^2}{4} + c\log x + d\ ]$$

ANSWERS

1. $y = \dfrac{x^2}{2} + c_1 x + c_2$ 2. $y = c_1 x - \sin x + c_2$ 3. $y = \dfrac{x^9}{72} + c_1 x + c_2$ 4. $y = \dfrac{e^{4x}}{16} + c_1 x + c_2$

5. $y = xe^x - 2e^x - \cos x + c_1 x + c_2$ 6. $y = -\cos x + \dfrac{x^3}{6} + c_1 x + c_2$

7. $(x^2 - 4x + 6)e^x + c_1 x + c_2$ 8. $y = -x \sin x - 2 \cos x + e^x + c_1 x + c_2$

9. $y = -x \sin x - 2 \cos x + c_1 x + c_2$ 10. $y = -x \cos x + 2 \sin x + c_1 x + c_2$

11. $y = \dfrac{x^3}{6} - \sin x$ 12. $y = -x^2 \sin x - 4x \cos x + 6 \sin x - x$

13. $y = \dfrac{1}{8}(2x^2 - 4x + 5)e^{2x} + \dfrac{1}{4}\sin 2x + c_1 x + c_2$ 14. $y = -\cos x + \dfrac{1}{9}e^{3x} + \dfrac{1}{20}x^5 + c_1 x + c_2$

15. $y = x \tan^{-1} x - \dfrac{1}{2}\log(1+x^2)$ 16. $y = -\dfrac{x^2}{4} + c_1 \log x + c_2$

13.11 HOMOGENEOUS DIFFERENTIAL EQUATIONS

We know that a function $f(x, y)$ is said to be homogenous function of degree n if each term in $f(x, y)$ is of degree n.

Definition. *A differential equation of the form* $\dfrac{dy}{dx} = \dfrac{f(x,y)}{g(x,y)}$ *is said to be homogeneous if* $f(x, y)$ *and* $g(x, y)$ *are homogenous function of same degree.*

For Example. $\qquad \dfrac{dy}{dx} = \dfrac{x^2 + xy}{y^2}$

13.11.1 SOLUTION OF THE HOMOGENEOUS DIFFERENTIAL EQUATION

Let $\qquad\qquad \dfrac{dy}{dx} = f(x, y)$ ...(1)

be a homogeneous equation.

To solve the above equation, put $y = vx$, *i.e.*, $\dfrac{dy}{dx} = v + x\dfrac{dv}{dx}$. Then applying the method of separation of variables, find the required solution.

Putting $y = vx \Rightarrow \dfrac{dy}{dx} = v + x\dfrac{dv}{dx}$ in eq. (1), we get

$$v + x\frac{dv}{dx} = f(v)$$

$$\Rightarrow \qquad x\frac{dv}{dx} = f(v) - v$$

$$\Rightarrow \qquad \int \frac{dv}{f(v) - v} = \int \frac{dx}{x} + c \text{, where } c \text{ is a constant of integration.}$$

After integrating put $v = \dfrac{y}{x}$.

WORKING PROCEDURE

Step 1. Write the given equation in the form of $\dfrac{dy}{dx} = f(x, y)$.

Step 2. Check the homogeneity of the given equation, *i.e.*, $f(kx, ky) = f(x, y)$.

Step 3. Putting $y = vx$, *i.e.*, $\dfrac{dy}{dx} = v + x\dfrac{dv}{dx}$.

Step 4. Separate the variables and then integrates.

Step 5. Finally put $v = \dfrac{y}{x}$.

SOLVED EXAMPLES

EXAMPLE 1. Solve $x\dfrac{dy}{dx} - y = \sqrt{x^2 + y^2}$. $\qquad$ **(NCERT)**

SOLUTION. We have $\qquad x\dfrac{dy}{dx} - y = \sqrt{x^2 + y^2}$

$$\Rightarrow \qquad x\frac{dy}{dx} = y + \sqrt{x^2 + y^2}$$

$$\Rightarrow \qquad \frac{dy}{dx} = \frac{y + \sqrt{x^2 + y^2}}{x}, \text{ which is a homogeneous equation.}$$

Putting $y = vx \Rightarrow \dfrac{dy}{dx} = v + x\dfrac{dv}{dx}$, we get

$$v + x\frac{dv}{dx} = \frac{vx + \sqrt{x^2 + v^2 x^2}}{x} = \frac{vx + x\sqrt{1 + v^2}}{x} = v + \sqrt{1 + v^2}$$

$$\Rightarrow \qquad x\frac{dv}{dx} = \sqrt{1 + v^2} \qquad \Rightarrow \qquad \frac{dv}{\sqrt{1 + v^2}} = \frac{dx}{x}$$

Integrating both the sides, we get

$$\int \frac{dv}{\sqrt{1 + v^2}} = \int \frac{dx}{x}$$

$$\Rightarrow \qquad \log|v + \sqrt{1 + v^2}| = \log|x| + \log c_1 \qquad \Rightarrow \qquad |v + (\sqrt{1 + v^2})| = c_1|x|$$

$$\Rightarrow \qquad v + \sqrt{1+v^2} = \pm c_1 x = cx \qquad\qquad\qquad (c = \pm c_1)$$

Again, putting $v = \dfrac{y}{x}$

$$\frac{y}{x} + \sqrt{1 + \frac{y^2}{x^2}} = cx$$

which is the required solution of the given differential equation.

EXAMPLE 2. *Solve* $(x^2 + xy)dy = (x^2 + y^2)dx$.   **(NCERT)**

SOLUTION. Given

$$(x^2 + xy)dy = (x^2 + y^2)dx$$

$$\Rightarrow \qquad \frac{dy}{dx} = \frac{x^2 + y^2}{x^2 + xy} \qquad\qquad\qquad \dots(1)$$

which is a homogeneous equation.

Putting $y = vx$, i.e., $\dfrac{dy}{dx} = v + x\dfrac{dv}{dx}$ in the above equation, we get

$$v + x\frac{dv}{dx} = \frac{x^2 + v^2 x^2}{x^2 + x \cdot vx} = \frac{1 + v^2}{1 + v}$$

$$\Rightarrow \qquad x\frac{dv}{dx} = \frac{1 + v^2}{1 + v} - v = \frac{1 - v}{1 + v}$$

$$\Rightarrow \qquad \frac{1 + v}{1 - v}dv = \frac{dx}{x}$$

Integrating both the sides, we get

$$\int\left[-1 + \frac{2}{1-v}\right]dv = \int\frac{dx}{x} + c$$

$$\Rightarrow \qquad -v - 2\log|(1 - v)| = \log|x| + c$$

$$\Rightarrow \qquad -\frac{y}{x} - 2\log\left|\left(1 - \frac{y}{x}\right)\right| = \log|x| + c \qquad \left(\because v = \frac{y}{x}\right)$$

which is the required solution of the given differential equation.

EXAMPLE 3. *Solve* $(y^2 - x^2)dy = 3xydx$.

SOLUTION. Given

$$(y^2 - x^2)dy = 3xydx$$

$$\Rightarrow \qquad \frac{dy}{dx} = \frac{3xy}{y^2 - x^2}, \text{ which is a homogeneous equation.}$$

Putting $y = vx$, i.e., $\dfrac{dy}{dx} = v + x\dfrac{dv}{dx}$ in the above equation, we get

$$v + x\frac{dv}{dx} = \frac{3x \cdot vx}{v^2 x^2 - x^2} = \frac{3v}{v^2 - 1}$$

$$\Rightarrow \qquad x\frac{dv}{dx} = \frac{3v}{v^2 - 1} - v = \frac{4v - v^3}{v^2 - 1}$$

On separating the variables, we get

$$\frac{v^2 - 1}{v(4 - v^2)}dv = \frac{dx}{x}$$

Integrating both the sides, we get

$$\int \frac{v^2 - 1}{v(2-v)(2+v)} dv = \int \frac{dx}{x} + c$$

$$\Rightarrow \int \left[-\frac{1}{4v} + \frac{3}{8} \cdot \frac{1}{2-v} - \frac{3}{8} \cdot \frac{1}{2+v} \right] dv = \int \frac{dx}{x} + c$$

$$\Rightarrow -\frac{1}{4} \log|v| - \frac{3}{8} \log|2-v| - \frac{3}{8} \log|2+v| = \log|x| + c$$

$$\Rightarrow 2\log|v| + 3(\log|2-v| + 3\log|2+v|) = -8\log|x| - 8c$$

$$\Rightarrow \log|v^2 \cdot (2-v)^3 (2+v)^3| = -8\log|x| + \log c_1 \qquad (-8c = \log c_1)$$

$$\Rightarrow \log|v^2(4-v^2)^3| = \log\left|\frac{c_1}{x^8}\right|$$

$$\Rightarrow v^2(4-v^2)^3 = \pm \frac{c_1}{x^8}$$

Now, putting $v = \dfrac{y}{x}$, we get

$$\frac{y^2}{x^2}\left(4 - \frac{y^2}{x^2}\right)^3 = \pm \frac{c_1}{x^8}$$

$$\Rightarrow y^2(4x^2 - y^2)^3 = c_1, \text{ which is a homogeneous equation.}$$

EXAMPLE 4.   *Solve* $(x^3 + y^3)dy - x^2y\,dx = 0$.

SOLUTION.   Given $(x^3 + y^3)dy - x^2y\,dx = 0$

$$\Rightarrow \qquad \frac{dy}{dx} = \frac{x^2 y}{x^3 + y^3}, \qquad\qquad ...(1)$$

which is a homogeneous equation.

Putting $y = vx$, i.e., $\dfrac{dy}{dx} = v + x\dfrac{dv}{dx}$ in eq. (1), we get

$$v + x\frac{dv}{dx} = \frac{x^2 \cdot vx}{x^3 + v^3 x^3} = \frac{v}{1+v^3}$$

$$\Rightarrow \qquad x\frac{dv}{dx} = \frac{v}{1+v^3} - v = -\frac{v^4}{1+v^3}$$

On separating the variables, we get

$$\frac{1+v^3}{v^4} dv + \frac{dx}{x} = 0$$

On integrating,

$$\int \frac{1+v^3}{v^4} dv + \int \frac{dx}{x} = 0 \quad \Rightarrow \quad \int \left[\frac{1}{v^4} + \frac{1}{v}\right] dv + \int \frac{dx}{x} = 0$$

$$\Rightarrow \quad -\frac{1}{3v^3} + \log|v| + \log|x| = c \quad \Rightarrow \quad -\frac{1}{3v^3} + \log|vx| = c$$

Again, putting $v = \dfrac{y}{x}$, we get

$$-\frac{x^3}{3y^3} + \log|y| = c, \text{ which is the required solution.}$$

EXAMPLE 5. *Solve* $2xy\,dx + (x^2 + 2y^2)dy = 0$.

SOLUTION. Given

$$2xy\,dx + (x^2 + 2y^2)dy = 0$$

$$\Rightarrow \qquad \frac{dy}{dx} = -\frac{2xy}{x^2 + 2y^2}, \qquad\qquad\qquad ...(1)$$

which is a homogeneous equation.

Putting $y = vx$, i.e., $\dfrac{dy}{dx} = v + x\dfrac{dv}{dx}$ in eq. (1), we get

$$v + x\frac{dv}{dx} = -\frac{2x(vx)}{x^2 + 2v^2x^2} = -\frac{2v}{1 + 2v^2}$$

$$\Rightarrow \qquad x\frac{dv}{dx} = -\frac{2v}{1 + 2v^2} - v = -\frac{(3v + 2v^3)}{1 + 2v^2}$$

$$\Rightarrow \qquad \frac{1 + 2v^2}{3v + 2v^3}dv + \frac{dx}{x} = 0$$

On integrating, we get

$$\int \frac{1 + 2v^2}{3v + 2v^3}dv + \int \frac{dx}{x} = 0$$

$$\Rightarrow \qquad \frac{1}{3}\int \frac{dt}{t} + \log|x| = \log c_1 \qquad \left[\text{Let } t = 3v + 2v^3 \Rightarrow \frac{dt}{3} = (1 + 2v^2)dv\right]$$

$$\Rightarrow \qquad \frac{1}{3}\log|t| + \log|x| = \log c_1$$

$$\Rightarrow \qquad |t|^{1/3} \cdot x = c_1 \qquad\qquad \Rightarrow \qquad tx^3 = c_1^3 = c$$

$$\Rightarrow \qquad (3v + 2v^3)x^3 = c \qquad\qquad \Rightarrow \qquad 3vx^3 + 2v^3x^3 = c$$

Again putting $v = \dfrac{y}{x}$, we get

$$3x^2y + 2y^3 = c, \text{ which is a homogeneous equation.}$$

EXAMPLE 6. *Solve* $x^2 \cdot \dfrac{dy}{dx} = 2xy + y^2$. **(ncert)**

SOLUTION. Given

$$x^2 \cdot \frac{dy}{dx} = 2xy + y^2$$

$$\Rightarrow \qquad \frac{dy}{dx} = \frac{2xy + y^2}{x^2} \qquad\qquad\qquad ...(1)$$

which is a homogeneous equation.

Putting $y = vx$, i.e., $\dfrac{dy}{dx} = v + x\dfrac{dv}{dx}$ in eq. (1), we get

$$v + x\frac{dv}{dx} = \frac{2x \cdot vx + (vx)^2}{x^2} = 2v + v^2$$

$$\Rightarrow \qquad x\frac{dv}{dx} = v + v^2 = v(1 + v)$$

On separating the variables, we get

$$\frac{dv}{v(1 + v)} = \frac{dx}{x}$$

Integrating both the sides, we get

$$\Rightarrow \quad \int \left[\frac{1}{v} - \frac{1}{1+v} \right] dv = \int \frac{dx}{x}$$

$$\log |v| - \log |1 + v| = \log |x| + \log c_1$$

$$\Rightarrow \quad \left| \frac{v}{1+v} \right| = c_1 |x|$$

Again putting $v = \dfrac{y}{x}$, we get

$$\frac{y}{x+y} = \pm c_1 x = cx \qquad\qquad (c = \pm c_1)$$

$$\Rightarrow \quad y = cx(x + y)$$

which is the required solution.

EXAMPLE 7. Solve $y^2 dx + (x^2 - xy + y^2) dy = 0$.

SOLUTION. Given

$$y^2 dx + (x^2 - xy + y^2) dy = 0$$

$$\Rightarrow \quad \frac{dy}{dx} = -\frac{y^2}{(x^2 - xy + y^2)} \qquad\qquad ...(1)$$

Putting $y = vx$, i.e., $\dfrac{dy}{dx} = v + x\dfrac{dv}{dx}$ in eq. (1), we get

$$v + x\frac{dv}{dx} = -\frac{v^2}{(1 + v^2 - v)}$$

$$\Rightarrow \quad x\frac{dv}{dx} = -\left[\frac{v^2}{(1 - v + v^2)} + v \right] = \frac{-(v + v^3)}{(1 - v + v^2)}$$

Separating the variables, we get

$$\frac{(1 - v + v^2)}{v(1 + v^2)} dv = -\frac{1}{x} dx$$

$$\Rightarrow \quad \left[\frac{1 + v^2}{v(1 + v^2)} - \frac{v}{v(1 + v^2)} \right] dv = -\frac{1}{x} dx$$

Integrating both the sides, we get

$$\Rightarrow \quad \int \left(\frac{1}{v} - \frac{1}{(1 + v^2)} \right) dv = -\int \frac{1}{x} dx$$

$$\log |v| - \tan^{-1} v + \log |x| = \log c$$

$$\Rightarrow \quad \tan^{-1} v = \log \frac{|vx|}{c}$$

Again putting $v = \dfrac{y}{x}$, we get

$$\tan^{-1} \frac{y}{x} = \log \frac{(|y|)}{c}$$

$$\Rightarrow \quad \frac{|y|}{c} = e^{\tan^{-1} \frac{y}{x}}$$

$$\Rightarrow \qquad y = c_1 e^{\tan^{-1}\frac{y}{x}} \qquad\qquad (c_1 = \pm c)$$

which is the required solution of the given differential equation.

EXAMPLE 8.   Solve $(3xy + y^2)dx + (x^2 + xy)dy = 0$.

SOLUTION.   Given

$$(3xy + y^2)dx + (x^2 + xy)dy = 0$$

$$\Rightarrow \qquad \frac{dy}{dx} = -\frac{(3xy + y^2)}{(x^2 + xy)} \qquad\qquad ...(1)$$

Putting $y = vx$, i.e., $\frac{dy}{dx} = v + x\frac{dv}{dx}$ in eq. (1), we get

$$v + x\frac{dv}{dx} = -\frac{(3vx^2 + v^2 x^2)}{x^2 + vx^2} \quad\Rightarrow\quad v + x\frac{dv}{dx} = -\frac{(3v + v^2)}{(1+v)}$$

$$\Rightarrow \quad x\frac{dv}{dx} = \left[-\frac{(3v + v^2)}{(1+v)} - v\right] \quad\Rightarrow\quad x\frac{dv}{dx} = \frac{-2(2v + v^2)}{(1+v)}$$

$$\Rightarrow \quad \frac{(1+v)}{(2v + v^2)}dv = -\frac{2}{x}dx$$

On integrating both the sides, we get

$$\int \frac{1+v}{(2v + v^2)}dv + \int\frac{2}{x}dx = \log c \qquad \Rightarrow \frac{1}{2}\log|2v + v^2| + 2\log|x| = \log c$$

$$\Rightarrow \qquad \log|x^2\sqrt{2v + v^2}| = \log c \qquad \Rightarrow |x^2\sqrt{2v + v^2}| = c$$

Again putting $v = \frac{y}{x}$, we get

$$x^2\sqrt{\frac{2y}{x} + \frac{y^2}{x^2}} = c \qquad\qquad \Rightarrow \qquad x\sqrt{(2xy + y^2)} = c$$

$$\Rightarrow \qquad x^2(2xy + y^2) = c^2$$

which is the required solution.

EXAMPLE 9.   Solve $(y + x)\frac{dy}{dx} = (y - x)$.

SOLUTION.   Given

$$(y + x)\frac{dy}{dx} = (y - x)$$

$$\Rightarrow \qquad \frac{dy}{dx} = \frac{(y - x)}{(y + x)} \qquad\qquad ...(1)$$

Putting $y = vx$, i.e., $\frac{dy}{dx} = v + x\frac{dv}{dx}$ in eq. (1), we get

$$v + x\frac{dv}{dx} = \frac{(vx - x)}{(vx + x)} = \frac{(v - 1)}{(v + 1)}$$

$$\Rightarrow \qquad x\frac{dv}{dx} = \frac{(v - 1)}{(v + 1)} - v = \frac{-(1 + v^2)}{(1 + v)}$$

On separating the variables, we get

$$\frac{(v+1)}{(v^2+1)} \cdot dv + \frac{dx}{x} = 0$$

Integrating both the sides, we get

$$\int \frac{(v+1)}{(v^2+1)} dv + \int \frac{dx}{x} = c$$

$$\Rightarrow \quad \int \frac{v}{(v^2+1)} dv + \int \frac{1}{(v^2+1)} dv + \int \frac{dx}{x} = c$$

$$\Rightarrow \quad \frac{1}{2} \log(v^2+1) + \tan^{-1} v + \log|x| = c$$

$$\Rightarrow \quad \frac{1}{2}[\log(v^2+1) + 2\log(x)] + \tan^{-1} v = c$$

Again putting $v = \dfrac{y}{x}$, we get

$$\frac{1}{2} \log\left(\left(\frac{y}{x}\right)^2 + 1\right) x^2 + \tan^{-1}\left(\frac{y}{x}\right) = c$$

$$\Rightarrow \quad \frac{1}{2} \log(x^2+y^2) + \tan^{-1}\left(\frac{y}{x}\right) = c$$

which is the required solution.

EXAMPLE 10. *Solve* $(x^3 - 3xy^2)dx = (y^3 - 3x^2y)dy$.

SOLUTION. Given $\qquad (x^3 - 3xy^2)dx = (y^3 - 3x^2y)dy$

$$\Rightarrow \qquad\qquad \frac{dy}{dx} = \frac{x^3 - 3xy^2}{y^3 - 3x^2y} \qquad\qquad\qquad ...(1)$$

Putting $y = vx$, *i.e.,* $\dfrac{dy}{dx} = v + x\dfrac{dv}{dx}$ in eq. (1), we get

$$v + x\frac{dv}{dx} = \frac{x^3 - 3v^2 x^3}{v^3 x^3 - 3vx^3}$$

$$\Rightarrow \qquad\qquad v + x\frac{dv}{dx} = \frac{1 - 3v^2}{v^3 - 3v}$$

$$\Rightarrow \qquad\qquad x\frac{dv}{dx} = \left(\frac{1 - 3v^2}{v^3 - 3v} - v\right) = \frac{1 - v^4}{v^3 - 3v}$$

On separating the variables, we get

$$\frac{3v - v^3}{v^4 - 1} dv = \frac{dx}{x}$$

$$\Rightarrow \quad \left[\frac{1}{2(v+1)} + \frac{1}{2(v-1)} - \frac{2v}{(v^2+1)}\right] dv = \frac{dx}{x}$$

On integrating both the sides, we get

$$\int \left[\frac{1}{2(v+1)} + \frac{1}{2(v-1)} - \frac{2v}{(v^2+1)}\right] dv = \int \frac{dx}{x}$$

$$\Rightarrow \frac{1}{2}\log|v+1| + \frac{1}{2}\log|v-1| - \log|v^2+1| = \log|x| + \log c$$

$$\Rightarrow \log\left|\frac{(\sqrt{v}+1)(\sqrt{v}-1)}{x(v^2+1)}\right| = \log c$$

$$\Rightarrow (v^2-1) = c^2 \cdot x^2(v^2+1)^2$$

Again putting $v = \dfrac{y}{x}$, we get

$$(y^2 - x^2) = c^2(y^2 + x^2)^2$$

which is the required solution.

EXAMPLE 11. *Solve* $x\dfrac{dy}{dx} = y - x\tan\dfrac{y}{x}$.

SOLUTION.   Given

$$x\frac{dy}{dx} = y - x\tan\frac{y}{x}$$

$$\Rightarrow \frac{dy}{dx} = \frac{y}{x} - \tan\frac{y}{x} \qquad \qquad \ldots(1)$$

Putting $y = vx$, *i.e.,* $\dfrac{dy}{dx} = v + x\dfrac{dv}{dx}$ in eq. (1), we get

$$v + x\frac{dv}{dx} = v - \tan v$$

$$\Rightarrow x\frac{dv}{dx} = -\tan v$$

On separating the variables, we get

$$\frac{dv}{\tan v} = -\frac{dx}{x}$$

$$\Rightarrow \cot v \cdot dv = -\frac{dx}{x}$$

$$\Rightarrow \frac{\cos v}{\sin v} \cdot dv = -\frac{dx}{x}$$

On integrating both the sides, we get

$$\int \frac{\cos v}{\sin v} \cdot dv = -\int \frac{dx}{x}$$

$$\Rightarrow \log|\sin v| + \log|x| = \log c$$

$$\Rightarrow \log|x\sin v| = \log c$$

$$\Rightarrow |x\sin v| = c$$

Again putting $v = \dfrac{y}{x}$, we get

$$x\sin\frac{y}{x} = c_1 \qquad \qquad (c_1 = \pm c)$$

which is the required solution.

EXAMPLE 12. *Solve* $\left(x\cos\dfrac{y}{x} + y\sin\dfrac{y}{x}\right)y\,dx = \left(y\sin\dfrac{y}{x} - x\cos\dfrac{y}{x}\right)x\,dy$.

SOLUTION. Given

$$\left(x\cos\frac{y}{x}+y\sin\frac{y}{x}\right)ydx=\left(y\sin\frac{y}{x}-x\cos\frac{y}{x}\right)xdy$$

$$\Rightarrow \qquad \frac{dy}{dx}=\frac{\left[x\cos\left(\dfrac{y}{x}\right)+y\sin\left(\dfrac{y}{x}\right)\right]y}{\left[y\sin\left(\dfrac{y}{x}\right)-x\cos\left(\dfrac{y}{x}\right)\right]x}$$

Divide numerator and denominator by x^2

$$\frac{dy}{dx}=\frac{\left[\cos\left(\dfrac{y}{x}\right)+\left(\dfrac{y}{x}\right)\sin\left(\dfrac{y}{x}\right)\right]\left(\dfrac{y}{x}\right)}{\left[\left(\dfrac{y}{x}\right)\sin\left(\dfrac{y}{x}\right)-\cos\left(\dfrac{y}{x}\right)\right]} \qquad ...(1)$$

Putting $y=vx$, i.e., $\dfrac{dy}{dx}=v+x\dfrac{dv}{dx}$ in eq. (1), we get

$$v+x\frac{dv}{dx}=\frac{v(\cos v+v\sin v)}{(v\sin v-\cos v)}$$

$$\Rightarrow \qquad x\frac{dv}{dx}=\left[\frac{v(\cos v+v\sin v)}{(v\sin v-\cos v)}-v\right]=\frac{2v\cos v}{v\sin v-\cos v}$$

On separating the variables, we get

$$\frac{v\sin v-\cos v}{v\cos v}dv=\frac{2}{x}dx$$

Integrating both the sides, we get

$$\int\frac{v\sin v-\cos v}{v\cos v}dv=\int\frac{2}{x}dx+\log c$$

$$\Rightarrow \qquad \int\tan vdv-\int\frac{dv}{v}=\int\frac{2}{x}dx+\log c$$

$$\Rightarrow \qquad \log|\cos v|+\log|v|+2\log|x|=\log c$$

$$\Rightarrow \qquad \log|x^2v\cos v|=\log c$$

$$\Rightarrow \qquad x^2v\cos v=c$$

Again putting $v=\dfrac{y}{x}$, we get

$$x^2\cdot\frac{y}{x}\cos\frac{y}{x}=c$$

$$\Rightarrow \qquad xy\cos\left(\frac{y}{x}\right)=c \text{, which is the required solution.}$$

EXAMPLE 13. *Solve* $2ye^{x/y}dx+(y-2xe^{x/y})dy=0$.
Given, $y=1$ then $x=0$.

SOLUTION. The given equation can be written as

$$\frac{dx}{dy}=\frac{2xe^{x/y}-y}{2ye^{x/y}}=\frac{2\cdot\dfrac{x}{y}\cdot e^{x/y}-1}{2e^{x/y}} \qquad ...(1)$$

Putting $x = vy$, i.e., $\dfrac{dx}{dy} = v + y\dfrac{dv}{dy}$ in eq. (1), we get

$$v + y\frac{dv}{dy} = \frac{2ve^v - 1}{2e^v}$$

$$\Rightarrow \qquad y\frac{dv}{dy} = \frac{2ve^v - 1}{2e^v} - v = \frac{-1}{2e^v}$$

On separating the variables, we get

$$2e^v \cdot dv = -\frac{dy}{y}$$

On integrating both the sides, we get

$$\int 2e^v \cdot dv = -\int \frac{dy}{y}$$

$$\Rightarrow \qquad 2e^v = -\log|y| + c$$

Again putting $v = \dfrac{x}{y}$, we get

$$2e^{x/y} + \log|y| = c \qquad\qquad\qquad\qquad …(2)$$

Given $x = 0 \Rightarrow y = 1$, therefore

$$2e^0 + \log 1 = c$$

$$\Rightarrow \qquad 2 \cdot 1 + 0 = c$$

$$\Rightarrow \qquad c = 2$$

Putting the value of c in eq. (2), we get

$$2e^{x/y} + \log|y| = 2$$

which is the required solution of the given differential equation.

EXAMPLE 14. *Solve* $(x\sqrt{x^2 + y^2} - y^2)dx + xy\,dy = 0$.

SOLUTION. Given

$$(x\sqrt{x^2 + y^2} - y^2)dx + xy\,dy = 0$$

$$\Rightarrow \qquad \frac{dy}{dx} = \frac{y^2 - x\sqrt{x^2 + y^2}}{xy} \qquad\qquad …(1)$$

Putting $y = vx$, i.e., $\dfrac{dy}{dx} = v + x\dfrac{dv}{dx}$ in eq. (1), we get

$$v + x\frac{dv}{dx} = \frac{v^2x^2 - x\sqrt{x^2 + v^2x^2}}{vx^2}$$

$$\Rightarrow \qquad x\frac{dv}{dx} = \left(\frac{v^2 - \sqrt{1 + v^2}}{v} - v\right)$$

$$\Rightarrow \qquad x\frac{dv}{dx} = \frac{-\sqrt{1 + v^2}}{v}$$

On separating the variables, we get

$$\int \frac{v}{\sqrt{1 + v^2}}dv = -\int \frac{dx}{x}$$

$$\Rightarrow \qquad \sqrt{1+v^2} = -\log|x| + c$$

Again putting $v = \dfrac{y}{x}$, we get

$$\sqrt{x^2 + y^2} + x\log|x| = cx \text{, which is the required solution.}$$

EXAMPLE 15. *Solve* $x^2\left(\dfrac{dy}{dx}\right) = x^2 - 2y^2 + xy$.

SOLUTION. Given

$$x^2\frac{dy}{dx} = x^2 - 2y^2 + xy$$

$$\Rightarrow \qquad \frac{dy}{dx} = \frac{x^2 - 2y^2 + xy}{x^2} \qquad\qquad \text{...(1)}$$

Putting $y = vx$, *i.e.*, $\dfrac{dy}{dx} = v + x\dfrac{dv}{dx}$ in eq. (1), we get

$$v + x\frac{dv}{dx} = \frac{x^2 - 2v^2x^2 + vx^2}{x^2} = 1 - 2v^2 + v$$

$$\Rightarrow \qquad x\frac{dv}{dx} = 1 - 2v^2$$

On separating the variables, we get

$$\frac{dv}{1 - 2v^2} = \frac{dx}{x}$$

Integrating both the sides, we get

$$\int\frac{dv}{1 - 2v^2} = \int\frac{dx}{x} \qquad\qquad \Rightarrow \qquad \frac{1}{2}\int\frac{dv}{\frac{1}{2} - v^2} = \int\frac{dx}{x}$$

$$\Rightarrow \qquad \frac{1}{2}\int\frac{dv}{\left(\frac{1}{\sqrt{2}}\right)^2 - v^2} = \int\frac{dx}{x}$$

$$\Rightarrow \qquad \frac{1}{2}\cdot\frac{1}{2\times\frac{1}{\sqrt{2}}}\log\left|\frac{\frac{1}{\sqrt{2}} + v}{\frac{1}{\sqrt{2}} - v}\right| = \log|x| + c$$

Again putting $v = \dfrac{y}{x}$, we get

$$\frac{1}{2\sqrt{2}}\log\left|\frac{x + \sqrt{2}y}{x - \sqrt{2}y}\right| = \log|x| + c \text{, which is the required solution.}$$

EXAMPLE 16. *Solve* $2x^2\dfrac{dy}{dx} - 2xy + y^2 = 0$ *when* $y(e) = e$.

SOLUTION. Given

$$2x^2\frac{dy}{dx} - 2xy + y^2 = 0$$

$$\Rightarrow \qquad \frac{dy}{dx} = \frac{2xy - y^2}{2x^2} \qquad\qquad \dots(1)$$

Which is a homogeneous equation.

Putting $y = vx$, i.e., $\dfrac{dy}{dx} = v + x\dfrac{dv}{dx}$ in eq. (1), we get

$$v + x\frac{dv}{dx} = \frac{2x(vx) - (vx)^2}{2x^2} = \frac{2v - v^2}{2} = v - \frac{1}{2}v^2$$

$$\Rightarrow \qquad x\frac{dv}{dx} = v - \frac{1}{2}v^2 - v = -\frac{1}{2}v^2$$

On separating the variables, we get

$$\Rightarrow \qquad \frac{dv}{v^2} + \frac{1}{2}\frac{dx}{x} = 0 \qquad\qquad \Rightarrow \qquad \int\frac{dv}{v^2} + \frac{1}{2}\int\frac{dx}{x} = 0$$

$$\Rightarrow \qquad -\frac{1}{v} + \frac{1}{2}\log|x| = c$$

Again putting $v = \dfrac{y}{x}$,

$$-\frac{x}{y} + \frac{1}{2}\log|x| = c \qquad\qquad \dots(2)$$

Given, $\qquad\qquad y(e) = e \qquad\qquad \dots(3)$

Putting $x = e$ and $y = e$ in eq. (2), we get

$$-\frac{e}{e} + \frac{1}{2}\log e = c \qquad \Rightarrow \qquad -1 + \frac{1}{2}\cdot 1 = c$$

$$\Rightarrow \qquad\qquad c = -\frac{1}{2}$$

Putting the value of c in eq. (2), we get

$$-\frac{x}{y} + \frac{1}{2}\log|x| = -\frac{1}{2}$$

$$\Rightarrow \qquad \frac{1}{2}(\log x + 1) = \frac{x}{y}$$

$$\Rightarrow \qquad \frac{1}{2}(\log x + \log e) = \frac{x}{y}$$

$$\Rightarrow \qquad y\log ex = 2x, \text{ which is the required solution.}$$

EXAMPLE 17. *Solve* $(1 + e^{x/y})dx + e^{x/y}\left(1 - \dfrac{x}{y}\right)dy = 0$

SOLUTION. Given

$$(1 + e^{x/y})dx + e^{x/y}\left(1 - \frac{x}{y}\right)dy = 0$$

$$\Rightarrow \qquad \frac{dx}{dy} = \frac{-e^{x/y}\left(1 - \dfrac{x}{y}\right)}{1 + e^{x/y}} \qquad\qquad \dots(1)$$

Putting $x = vy$, i.e., $\dfrac{dx}{dy} = v + y\dfrac{dv}{dy}$ in eq. (1), we get

$$v + y\frac{dv}{dy} = -\frac{e^v(1-v)}{1+e^v}$$

$$\Rightarrow \qquad y\frac{dv}{dy} = -\frac{v+e^v}{1+e^v}$$

Integrating after separating the variables, we get

$$\Rightarrow \qquad \int\frac{1+e^v}{v+e^v}\cdot dv = -\int\frac{dy}{y}$$

$$\Rightarrow \qquad \log|v + e^v| = -\log|y| + c$$

$$\Rightarrow \qquad \log|y(v + e^v)| = c$$

$$\Rightarrow \qquad |y(v+ e^v)| = e^c = c_1 \qquad \text{(say)}$$

Again putting $v = \dfrac{x}{y}$, we get

$$y\left(\frac{x}{y} + e^{x/y}\right) = c_1$$

$$\Rightarrow \qquad x + ye^{x/y} = c_1, \text{ which is the required solution.}$$

EXAMPLE 18. *Solve* $2xy + y^2 - 2x^2\cdot\dfrac{dy}{dx} = 0,$ *if* $y(1) = 2$

SOLUTION. Given

$$2xy + y^2 - 2x^2\cdot\frac{dy}{dx} = 0$$

$$\Rightarrow \qquad \frac{dy}{dx} = \frac{2xy + y^2}{2x^2} \qquad\qquad ...(1)$$

Putting $y = vx$, i.e., $\dfrac{dy}{dx} = v + x\dfrac{dv}{dx}$ in eq. (1), we get

$$v + x\frac{dv}{dx} = \frac{2x\cdot vx + v^2x^2}{2x^2} = \frac{2v + v^2}{2} = v + \frac{1}{2}v^2$$

$$\Rightarrow \qquad x\frac{dv}{dx} = \frac{1}{2}v^2$$

On integrating after separating the variables, we get

$$\int -\frac{2}{v^2}dv + \int\frac{dx}{x} = 0$$

$$\Rightarrow \qquad \frac{2}{v} + \log|x| = c$$

Again putting $v = \dfrac{y}{x}$, we get

$$2\cdot\frac{x}{y} + \log|x| = c \qquad\qquad ...(2)$$

Given $x = 1 \Rightarrow y = 2$, putting this value in eq. (2), we get

$$\frac{2 \times 1}{2} + \log 1 = c$$

$$\Rightarrow \qquad c = 1$$

Putting the value of c in eq. (2), we get

$$\frac{2x}{y} + \log|x| = 1$$

$$2x - y + y \log|x| = 0, \text{ which is the required solution.}$$

EXAMPLE 19. *Solve* $x\dfrac{dy}{dx} - y = \sqrt{x^2 + y^2}$.

SOLUTION. Given

$$x\frac{dy}{dx} - y = \sqrt{x^2 + y^2}$$

$$\Rightarrow \qquad \frac{dy}{dx} = \frac{y + \sqrt{x^2 + y^2}}{x} \qquad \qquad ...(1)$$

Putting $y = vx$, *i.e.*, $\dfrac{dy}{dx} = v + x\dfrac{dv}{dx}$ in eq. (1), we get

$$v + x\frac{dv}{dx} = \frac{vx + \sqrt{x^2 + v^2 x^2}}{x} = v + \sqrt{1 + v^2}$$

$$\Rightarrow \qquad x\frac{dv}{dx} = \sqrt{1 + v^2}$$

On integrating after separating the variables, we get

$$\int \frac{dv}{\sqrt{1 + v^2}} = \int \frac{dx}{x}$$

$$\Rightarrow \qquad \log|(v + \sqrt{1 + v^2})| = \log|x| + \log c_1$$

$$\Rightarrow \qquad \log\left|\frac{v + \sqrt{1 + v^2}}{|x|}\right| = \log c_1$$

$$\Rightarrow \qquad \left|\frac{v + \sqrt{1 + v^2}}{|x|}\right| = c_1$$

$$\Rightarrow \qquad v + \sqrt{1 + v^2} = cx \qquad \qquad (c = \pm c_1)$$

Again putting $v = \dfrac{y}{x}$, we get

$$y + \sqrt{x^2 + y^2} = cx^2, \text{ which is the required solution.}$$

Exercise 13.5

Solve the following differential equations:

1. $x\dfrac{dy}{dx} = x + y$

2. $\dfrac{dy}{dx} = \dfrac{2xy}{x^2 - y^2}$

3. $\dfrac{dy}{dx} = \dfrac{x + y}{x - y}$

4. $x(x - y)dy + y^2\, dx = 0$

5. $(x - y)\dfrac{dy}{dx} = x + 2y$

6. $2xy\dfrac{dy}{dx} = x^2 + y^2$

7. $x^2 \cdot \dfrac{dy}{dx} = x^2 + xy + y^2$

8. $ye^{x/y}dx = (xe^{x/y} + y)dy$

9. $x\dfrac{dy}{dx} = y - x\cos^2\left(\dfrac{y}{x}\right)$

10. $x\dfrac{dy}{dx} - y = 2\sqrt{y^2 - x^2}$

11. $(x^3 + 3xy^2)dx + (y^3 + 3x^2y)dy = 0$

12. $x\dfrac{dy}{dx} = y(\log y - \log x + 1)$

13. $(1 + 2e^{x/y})dx + 2e^{x/y}\left(1 - \dfrac{x}{y}\right)dy = 0$

14. $y^2dx + (x^2 + xy + y^2)dy = 0$

15. $ydx + x\log\left(\dfrac{y}{x}\right)dy - 2xdy = 0$

16. $x\dfrac{dy}{dx} - y + x\sin\dfrac{y}{x} = 0$

17. $x^2dy + y(x + y)dx = 0$

18. $(x - \sqrt{xy})dy = ydx$

19. $x\dfrac{dy}{dx} - y = x\tan\dfrac{y}{x}$

20. $\left(x\sin^2\dfrac{y}{x} - y\right)dx + xdy = 0,\ y(1) = \dfrac{\pi}{4}$

21. $(x^2 - y^2)dx + 2xydy = 0,\ y(1) = 1$

22. Prove that the family of curve for which the slope of the tangent at the point (x, y) is $\dfrac{x^2 + y^2}{2xy}$ is given by $x^2 - y^2 = cx$.

ANSWERS

1. $y = x\log|x| + cx$ **2.** $y = c(x^2 + y^2)$ **3.** $\tan^{-1}\dfrac{y}{x} = \dfrac{1}{2}\log\left(\dfrac{x^2 + y^2}{x}\right) + c$ **4.** $y = ce^{y/x}$

5. $\log|x^2 + xy + y^2| = 2\sqrt{3}\tan^{-1}\left(\dfrac{x + 2y}{\sqrt{3}\cdot x}\right) + c$ **6.** $x = c(x^2 - y^2)$

7. $\tan^{-1}\dfrac{y}{x} = \log|x| + c$ **8.** $e^{x/y} = \log cy$ **9.** $\tan\dfrac{y}{x} = c - \log|x|$

10. $y + \sqrt{y^2 - x^2} = cx^3$ **11.** $y^4 + 6x^2y^2 + x^4 = c$ **12.** $y = xe^{cx}$

13. $x + 2ye^{x/y} = c\cdot x\cdot y$ **14.** $\log|y| + \dfrac{x}{x + y} = c$ **15.** $1 + \log\dfrac{x}{y} = cy$

16. $x\tan\left(\dfrac{y}{x}\right) = c$ **17.** $x^2y = c(y + 2x)$ **18.** $2\sqrt{\dfrac{x}{y}} - \log|y| = c$

19. $\left|\sin\dfrac{y}{x}\right| = c|x|$ **20.** $\cot\left(\dfrac{y}{x}\right) = \log|x| + 1$ **21.** $x^2 + y^2 = 2x$

13.12 LINEAR DIFFERENTIAL EQUATIONS

Definition. *A differential equation of the form*

$$\frac{dy}{dx} + Py = Q$$

where P and Q are function of x only or constant is called linear differential equation.

13.12.1 SOLUTION OF THE LINEAR DIFFERENTIAL EQUATIONS

Let the given equation be

$$\frac{dy}{dx} + Py = Q \qquad \qquad \qquad \dots(1)$$

First of all, find the value of integrating factor (I.F.) $= e^{\int Pdx}$.

Multiply both sides of eq. (1) by $e^{\int Pdx}$

$$e^{\int Pdx}\cdot\frac{dy}{dx} + Pye^{\int Pdx} = Qe^{\int Pdx}$$

$$\Rightarrow \qquad e^{\int Pdx}dy + Pye^{\int Pdx}dx = Qe^{\int Pdx} \cdot dx$$

$$\Rightarrow \qquad d(y \cdot e^{\int Pdx}) = Qe^{\int Pdx} \cdot dx$$

On integrating both the sides, we get

$$ye^{\int Pdx} = \int ye^{\int Pdx}dx + c,$$

which the required solution of the given equation.

WORKING PROCEDURE

Step 1. Compare the given equation with $\dfrac{dy}{dx} + Py = Q$ and find the value of P and Q.

Step 2. Find the value of $e^{\int Pdx}$.

Step 3. Putting the value of Q and $e^{\int Pdx}$ in the following formula

$$y(e^{\int Pdx}) = \int Qe^{\int Pdx}dx + c$$

SOLVED EXAMPLES

EXAMPLE 1. Solve $\dfrac{dy}{dx} - \dfrac{y}{x} = 2x^2$.

SOLUTION. Given

$$\frac{dy}{dx} - \frac{y}{x} = 2x^2 \qquad \qquad ...(1)$$

Clearly, it is a linear differential equation.

Comparing eq. (1) with $\dfrac{dy}{dx} + Py = Q$, we get

$$P = -\frac{1}{x}, Q = 2x^2$$

$\therefore$ Integrating factor (I.F.) $= e^{\int Pdx} = e^{\int -\frac{1}{x}dx} = e^{-\log x} = e^{\log \frac{1}{x}} = \dfrac{1}{x}$

Hence, solution of the given equation is

$$y \cdot e^{\int Pdx} = \int Qe^{\int Pdx} \cdot dx + c$$

$$\Rightarrow \qquad y \cdot \frac{1}{x} = \int 2x^2 \cdot \frac{1}{x}dx + c = 2\int xdx + c = x^2 + c$$

$$\Rightarrow \qquad y = x^3 + cx, \text{ which is the required solution.}$$

EXAMPLE 2. Solve $x\dfrac{dy}{dx} - y = x^2$.

SOLUTION. Given

$$x\frac{dy}{dx} - y = x^2$$

$$\Rightarrow \qquad \frac{dy}{dx} - \frac{y}{x} = x \qquad \qquad ...(1)$$

which is a linear differential equation.

Comparing eq. (1) with $\dfrac{dy}{dx} + Py = Q$, we get

$$P = -\frac{1}{x}, Q = x$$

$$\therefore \qquad \text{I.F.} = e^{\int Pdx} = e^{\int -\frac{1}{x}dx} = e^{-\log x} = e^{\log\frac{1}{x}} = \frac{1}{x}$$

Hence, solution of the given equation is

$$y \cdot e^{\int Pdx} = \int Qe^{\int Pdx} \cdot dx + c$$

$$\Rightarrow \qquad y \cdot \frac{1}{x} = \int x \cdot \frac{1}{x}dx + c = \int dx + c = x + c$$

$$\Rightarrow \qquad y = x^2 + cx, \text{ which is the required solution.}$$

EXAMPLE 3. Solve $\dfrac{dy}{dx} + \sec x \cdot y = \tan x \, ; \, 0 \le x < \dfrac{\pi}{2}$.

SOLUTION. Given

$$\frac{dy}{dx} + \sec x \cdot y = \tan x \qquad\qquad ...(1)$$

Comparing eq. (1) with $\dfrac{dy}{dx} + Py = Q$, we get

$$P = \sec x, \, Q = \tan x$$

Now, $\qquad \text{I.F.} = e^{\int Pdx} = e^{\int \sec x dx} = e^{\log(\sec x + \tan x)} = \sec x + \tan x$

Hence, solution of the given equation is

$$y \cdot e^{\int Pdx} = \int Qe^{\int Pdx} \cdot dx + c$$

$$\Rightarrow \qquad y \cdot (\sec x + \tan x) = \int \tan x (\sec x + \tan x) dx + c$$

$$= \int \sec x \tan x dx + \int \tan^2 x dx + c$$

$$= \int \sec x \tan x dx + \int \sec^2 x dx - \int 1 dx + c$$

$$\Rightarrow \qquad y(\sec x + \tan x) = \sec x + \tan x - x + c, \text{ which is the required solution.}$$

EXAMPLE 4. Solve $\dfrac{dy}{dx} - y = x \cdot e^x$.

SOLUTION. Given

$$\frac{dy}{dx} - y = x \cdot e^x \qquad\qquad ...(1)$$

Comparing eq. (1) with $\dfrac{dy}{dx} + Py = Q$, we get

$$P = -1, \, Q = xe^x$$

Now, $\qquad \text{I.F.} = e^{\int Pdx} = e^{\int -1 dx} = e^{-x}$

Hence, required solution is given by

$$y \cdot e^{\int Pdx} = \int Qe^{\int Pdx} \cdot dx + c$$

$$\Rightarrow \qquad y \cdot e^{-x} = \int xe^x \cdot e^{-x} \cdot dx + c = \int x dx + c = \frac{x^2}{2} + c$$

$$\Rightarrow \qquad y = \left(\frac{x^2}{2} + c\right) e^x, \text{ which is the required solution.}$$

EXAMPLE 5. Solve $\dfrac{dy}{dx} + y \cot x = 2\cos x$.

SOLUTION. Given

$$\frac{dy}{dx} + y \cot x = 2\cos x \qquad\qquad ...(1)$$

Comparing eq. (1) with $\dfrac{dy}{dx} + Py = Q$, we get

$$P = \cot x, \ Q = 2 \cos x$$

Now, $\quad \text{I.F.} = e^{\int Pdx} = e^{\int \cot x dx} = e^{\int \frac{\cos x}{\sin x}dx} = e^{\log \sin x} = \sin x$

∴ Required solution is given by

$$y \cdot e^{\int Pdx} = \int Q e^{\int Pdx} \cdot dx + c$$

$$\Rightarrow \quad y \cdot \sin x = \int 2\cos x \sin x dx + c = \int \sin 2x dx + c$$

$$= -\frac{1}{2}\cos 2x + c$$

$$\Rightarrow \quad y = -\frac{1}{2}\cos 2x \operatorname{cosec} x + c\operatorname{cosec} x$$

Which is the required solution.

EXAMPLE 6. Solve $\dfrac{dy}{dx} - y \tan x = 2 \sin x$.

SOLUTION. Given

$$\frac{dy}{dx} - y \tan x = 2\sin x \qquad \qquad ...(1)$$

Comparing eq. (1) with $\dfrac{dy}{dx} + Py = Q$, we get

$$P = -\tan x, \ Q = 2\sin x$$

Now, $\quad \text{I.F.} = e^{\int Pdx} = e^{-\int \tan x dx} = e^{\log \cos x} = \cos x$

∴ Required solution is given by

$$y \cdot e^{\int Pdx} = \int Q e^{\int Pdx} \cdot dx + c$$

$$\Rightarrow \quad y \cdot \cos x = \int 2\sin x \cos x dx + c = \int \sin 2x dx + c = -\frac{1}{2}\cos 2x + c$$

$$\Rightarrow \quad y = -\frac{1}{2}\cos 2x \sec x + c \sec x$$

Which is the required solution.

EXAMPLE 7. Solve $(1 + x^2)\dfrac{dy}{dx} + y = e^{\tan^{-1} x}$.

SOLUTION. Given

$$\frac{dy}{dx} + \frac{y}{1 + x^2} = \frac{e^{\tan^{-1} x}}{1 + x^2} \qquad \qquad ...(1)$$

Comparing eq. (1) with $\dfrac{dy}{dx} + Py = Q$, we get

$$P = \frac{1}{1 + x^2}, Q = \frac{e^{\tan^{-1} x}}{1 + x^2}$$

Now, $\quad \text{I.F.} = e^{\int Pdx} = e^{\int \frac{1}{1+x^2}dx} = e^{\tan^{-1} x}$

Hence, required solution is given by

$$y \cdot (\text{I.F.}) = \int Q \cdot (\text{I.F.})dx + c$$

$$\Rightarrow \quad y \cdot e^{\tan^{-1} x} = \int \frac{e^{\tan^{-1} x}}{1 + x^2} \cdot e^{\tan^{-1} x} \cdot dx + c$$

$$= \int \frac{e^{2\tan^{-1}x}}{1+x^2}dx + c \qquad \qquad ...(2)$$

Let $\tan^{-1}x = t \Rightarrow \frac{1}{1+x^2}dx = dt$

Then from eq. (2),

$$y \cdot e^t = \int e^{2t}dt + c = \frac{1}{2}e^{2t} + c$$

$$\therefore \qquad y = \frac{1}{2}\frac{e^{2t}}{e^t} + \frac{c}{e^t} = \frac{1}{2}e^t + ce^{-t}$$

$$\Rightarrow \qquad y = \frac{1}{2}e^{\tan^{-1}x} + ce^{-\tan^{-1}x}$$

Which is the required solution.

EXAMPLE 8. *Solve* $\sin x \dfrac{dy}{dx} + \cos x \cdot y = \cos x \sin^2 x$.

SOLUTION. Given

$$\sin x \frac{dy}{dx} + \cos x \cdot y = \cos x \sin^2 x$$

$$\Rightarrow \qquad \frac{dy}{dx} + \frac{\cos x}{\sin x}y = \frac{\cos x \sin^2 x}{\sin x}$$

$$\Rightarrow \qquad \frac{dy}{dx} + y \cot x = \cos x \sin x \qquad \qquad ...(1)$$

Comparing eq. (1) with $\dfrac{dy}{dx} + Py = Q$, we get

$$P = \cot x, \ Q = \cos x \sin x$$

Now, $\qquad \text{I.F.} = e^{\int Pdx} = e^{\int \cot x dx} = e^{\log \sin x} = \sin x$

$\therefore$ Solution of the given equation is

$$\Rightarrow \qquad y \cdot \sin x = \int \cos x \cdot \sin x \cdot \sin x dx + c = \int \sin^2 x \cos x dx + c$$

$$= \frac{\sin^3 x}{3} + c$$

$$\Rightarrow \qquad y = \frac{1}{3}\sin^2 x + c \operatorname{cosec} x, \text{ which is the required solution.}$$

EXAMPLE 9. *Solve* $(1+x^2)\dfrac{dy}{dx} - 2xy = (x^2+2)(x^2+1)$.

SOLUTION. Given

$$(1+x^2)\frac{dy}{dx} - 2xy = (x^2+2)(x^2+1)$$

$$\Rightarrow \qquad \frac{dy}{dx} - \frac{2x}{(1+x^2)}y = \frac{(x^2+1)(x^2+2)}{(x^2+1)} = x^2+2 \qquad \qquad ...(1)$$

Comparing eq. (1) with $\dfrac{dy}{dx} + Py = Q$, we get

$$P = \frac{-2x}{1+x^2}, \ Q = x^2+2$$

Now, $\qquad \text{I.F.} = e^{\int Pdx} = e^{-\int \frac{2x}{1+x^2}dx} = e^{-\log(1+x^2)}$

$$= e^{\log\left(\frac{1}{1+x^2}\right)} = \frac{1}{1+x^2}$$

∴ Solution of the equation is given by

$$y \cdot (\text{I.F.}) = \int Q(\text{I.F.})dx + c$$

$$\Rightarrow \qquad y \cdot \frac{1}{1+x^2} = \int (x^2+2)\frac{1}{x^2+1}dx + c = \int \left(1 + \frac{1}{1+x^2}\right)dx + c$$

$$= \int dx + \int \frac{1}{1+x^2}dx + c = x + \tan^{-1}x + c$$

$$\Rightarrow \qquad y = x(1+x^2) + (1+x^2)\tan^{-1}x + c(1+x^2)$$

Which is the required solution.

EXAMPLE 10. *Solve* $\dfrac{dy}{dx} + 2\tan x \cdot y = \sin x$.

SOLUTION. Given

$$\frac{dy}{dx} + 2\tan x \cdot y = \sin x \qquad \qquad \ldots(1)$$

Comparing eq. (1) with $\dfrac{dy}{dx} + Py = Q$, we get

$$P = 2\tan x, \ Q = \sin x$$

Now, $\qquad \text{I.F.} = e^{\int P dx} = e^{\int 2\tan x dx} = e^{2\log\sec x} = e^{\log\sec^2 x} = \sec^2 x$

∴ Solution of the equation is given by

$$y \cdot (\text{I.F.}) = \int Q(\text{I.F.})dx + c$$

$$\Rightarrow \qquad y \cdot \sec^2 x = \int \sin x \cdot \sec^2 x dx + c = \int \sec x \tan x\, dx + c$$

$$= \sec x + c$$

$$y = \frac{\sec x}{\sec^2 x} + \frac{c}{\sec^2 x} = \cos x + c\cos^2 x$$

Which is the required solution.

EXAMPLE 11. *Solve* $x\dfrac{dy}{dx} - ay = x + 1$.

SOLUTION. Given

$$x\frac{dy}{dx} - ay = x + 1$$

$$\Rightarrow \qquad \frac{dy}{dx} - \frac{a}{x} \cdot y = \frac{x+1}{x} \qquad \qquad \ldots(1)$$

Comparing eq (1) with $\dfrac{dy}{dx} + Py = Q$, we get

$$P = -\frac{a}{x}, \ Q = \frac{x+1}{x}$$

Now, $\qquad \text{I.F.} = e^{\int P dx} = e^{-\int \frac{a}{x}dx} = e^{-a\log x} = e^{\log x^{-a}} = x^{-a}$

∴ Solution of the given equation is

$$\Rightarrow \qquad y \cdot e^{\int P dx} = \int (Q e^{\int P dx})dx + c$$

$$\Rightarrow \qquad y \cdot x^{-a} = \int \frac{x+1}{x} \cdot x^{-a} \cdot dx + c = \int \frac{x+1}{x^{a+1}}dx + c$$

$$= \int [x^{-a} + x^{-(a+1)}] dx + c = \frac{x^{-a+1}}{-a+1} + \frac{x^{-a}}{-a} + c$$

$$\Rightarrow \qquad y = \frac{x}{1-a} - \frac{1}{a} + cx^a, \text{ which is the required solution.}$$

EXAMPLE 12. *Solve* $(1 + x^2)\dfrac{dy}{dx} + y = \tan^{-1} x$.

SOLUTION. Given

$$(1 + x^2)\frac{dy}{dx} + y = \tan^{-1} x$$

$$\Rightarrow \qquad \frac{dy}{dx} + \frac{y}{(1+x^2)} = \frac{\tan^{-1} x}{1+x^2} \qquad \qquad \dots(1)$$

Comparing equation (1) with $\dfrac{dy}{dx} + Py = Q$, we get

$$P = \frac{1}{1+x^2}, Q = \frac{\tan^{-1} x}{1+x^2}$$

Now, $\qquad \qquad$ I.F. $= e^{\int P dx} = e^{\int \frac{1}{1+x^2} dx} = e^{\tan^{-1} x}$

$\therefore$ Solution of the given equation is

$$\Rightarrow \qquad y \cdot e^{\int P dx} = \int Q e^{\int P dx} \cdot dx + c$$

$$\Rightarrow \qquad y \cdot e^{\tan^{-1} x} = \int \frac{e^{\tan^{-1} x} \cdot \tan^{-1} x}{1+x^2} dx + c$$

Let $\tan^{-1} x = t \Rightarrow \dfrac{1}{1+x^2} dx = dt$

$$\therefore \qquad y \cdot e^{\tan^{-1} x} = \int t e^t dt + c = e^t (t-1) + c$$

$$= e^{\tan^{-1} x}(\tan^{-1} x - 1) + c$$

$$\Rightarrow \qquad y = -1 + \tan^{-1} x + ce^{-\tan^{-1} x},$$

which is the required solution.

EXAMPLE 13. *Solve* $x\dfrac{dy}{dx} - y = \log x$.

SOLUTION. Given

$$x\frac{dy}{dx} - y = \log x$$

$$\Rightarrow \qquad \frac{dy}{dx} - \frac{y}{x} = \frac{\log x}{x} \qquad \qquad \dots(1)$$

Comparing equation (1) with $\dfrac{dy}{dx} + Py = Q$, we get

$$P = -\frac{1}{x}, Q = \frac{\log x}{x}$$

Now, $\qquad \qquad$ I.F. $= e^{\int P dx} = e^{\int -\frac{1}{x} dx} = e^{-\log x} = e^{\log\left(\frac{1}{x}\right)} = \frac{1}{x}$

$\therefore$ Solution of the given equation is

$$y \cdot e^{\int Pdx} = \int Q e^{\int Pdx} \cdot dx + c$$

$$\Rightarrow \qquad y \cdot \frac{1}{x} = \int \frac{\log x}{x} \cdot \frac{1}{x} dx + c = \log x \left(-\frac{1}{x}\right) - \int \frac{1}{x}\left(-\frac{1}{x}\right) dx + c$$

$$= -\frac{\log x}{x} + \int \frac{1}{x^2} dx + c = -\frac{\log x}{x} - \frac{1}{x} + c$$

$$y = cx - (\log x + 1), \text{ which is the required solution.}$$

EXAMPLE 14. Solve $\dfrac{dy}{dx} + \dfrac{y}{x} = e^x$, $x > 0$.

SOLUTION.   Given

$$\frac{dy}{dx} + \frac{y}{x} = e^x \qquad\qquad\qquad ...(1)$$

Comparing equation (1) with $\dfrac{dy}{dx} + Py = Q$, we get

$$P = \frac{1}{x}, Q = e^x$$

Now,   I.F. $= e^{\int Pdx} = e^{\int \frac{1}{x} dx} = e^{\log x} = x$

$\therefore$ Solution of the given equation is

$$y \cdot (\text{I.F.}) = \int Q(\text{I.F.}) dx + c$$

$$\Rightarrow \qquad\qquad y \cdot x = \int e^x \cdot x dx + c$$

$$= e^x (x - 1) + c, \text{ which is the required solution.}$$

EXAMPLE 15. Solve $\dfrac{dy}{dx} + y \cot x = x^2 \cot x + 2x$, when $y\left(\dfrac{\pi}{2}\right) = 0$

SOLUTION.   Given

$$\frac{dy}{dx} + y \cot x = x^2 \cot x + 2x \qquad\qquad ...(1)$$

Comparing equation (1) with $\dfrac{dy}{dx} + Py = Q$, we get

$$P = \cot x, Q = x^2 \cot x + 2x$$

Now,   I.F. $= e^{\int Pdx} = e^{\int \cot x dx} = e^{\log \sin x} = \sin x$

$\therefore$ Solution of the given equation is

$$y \cdot (\text{I.F.}) = \int Q(\text{I.F.}) dx + c$$

$$\Rightarrow \qquad\qquad y \cdot \sin x = \int (x^2 \cot x + 2x) \sin x dx + c$$

$$= \int (x^2 \cot x \sin x + 2x \sin x) dx + c$$

$$= \int (x^2 \cos x + 2x \sin x) dx + c = \int d(x^2 \sin x) dx + c$$

$$= x^2 \sin x + c$$

$$\Rightarrow \qquad\qquad y = x^2 + c \, \text{cosec} \, x \qquad\qquad\qquad ...(2)$$

Given: $y = 0$ when $x = \dfrac{\pi}{2}$

Hence from equation (2), $c = -\dfrac{\pi^2}{4}$

Putting the value of c in eq. (2), we get

$$y = x^2 - \frac{\pi^2}{4} \text{cosec} \, x \text{ , which is the required solution.}$$

EXAMPLE 16. *Solve* $\dfrac{dy}{dx} + 2y = xe^{4x}$.

SOLUTION. Given

$$\frac{dy}{dx} + 2y = xe^{4x} \qquad \qquad ...(1)$$

Comparing equation (1) with $\dfrac{dy}{dx} + Py = Q$, we get

$$P = 2,\ Q = xe^{4x}$$

Now,

$$\text{I.F.} = e^{\int Pdx} = e^{\int 2dx} = e^{2x}$$

∴ Solution of the given equation is

$$y \cdot (\text{I.F.}) = \int Q(\text{I.F.})dx + c$$

$$\Rightarrow \qquad y \cdot e^{2x} = \int xe^{4x} \cdot e^{2x} dx + c = \int xe^{6x} dx + c = x\frac{e^{6x}}{6} - \frac{e^{6x}}{36} + c$$

$$\Rightarrow \qquad y = \frac{1}{6}xe^{4x} - \frac{1}{36}e^{4x} + ce^{-2x},$$

which is the required solution.

EXAMPLE 17. *Solve* $\cos^2 x \dfrac{dy}{dx} + y = \tan x$, $0 \le x \le \dfrac{\pi}{2}$

SOLUTION. Given

$$\cos^2 x \frac{dy}{dx} + y = \tan x$$

$$\Rightarrow \qquad \frac{dy}{dx} + \frac{y}{\cos^2 x} = \frac{\tan x}{\cos^2 x}$$

$$\Rightarrow \qquad \frac{dy}{dx} + \sec^2 x \cdot y = \tan x \sec^2 x \qquad \qquad ...(1)$$

Comparing equation (1) with $\dfrac{dy}{dx} + Py = Q$, we get

$$P = \sec^2 x,\ Q = \tan x \sec^2 x$$

Now,

$$\text{I.F.} = e^{\int Pdx} = e^{\int \sec^2 xdx} = e^{\tan x}$$

∴ Solution of the given equation is

$$y \cdot (\text{I.F.}) = \int Q(\text{I.F.})dx + c$$

$$\Rightarrow \qquad y \cdot e^{\tan x} = \int \tan x \sec^2 x \cdot e^{\tan x} dx + c$$

$$= \int e^t \cdot t dt \qquad\qquad \text{(Let } t = \tan x$$

$$= e^t \cdot t - e^t + c \qquad\qquad \Rightarrow dt = \sec^2 xdx)$$

$$= e^t(t-1) + c = e^{\tan x}(\tan x - 1) + c$$

$$\Rightarrow \qquad y = \tan x - 1 + ce^{-\tan x}, \text{ which is the required solution.}$$

EXAMPLE 18. *Solve* $x\dfrac{dy}{dx} + y = x^3$, *when at* $x = 2, y = 1$.

SOLUTION. Given

$$x\frac{dy}{dx} + y = x^3$$

$$\Rightarrow \qquad \frac{dy}{dx} + \frac{y}{x} = x^2 \qquad\qquad ...(1)$$

Comparing eq. (1) with $\dfrac{dy}{dx} + Py = Q$, we get

$$P = \dfrac{1}{x}, Q = x^2$$

Now, $\text{I.F.} = e^{\int P dx} = e^{\int \frac{1}{x} dx} = e^{\log x} = x$

∴   Solution of the given equation is

$$y \cdot (\text{I.F.}) = \int Q(\text{I.F.}) dx + c$$

⇒ $y \cdot x = \int x^2 \cdot x dx + c = \int x^3 dx + c$

⇒ $yx = \dfrac{x^4}{4} + c$ ...(2)

Given: when $x = 2, y = 1$
Putting this value in eq. (2), we get

$$c = -2$$

Putting the value of c in eq. (2), we get

$$xy = \dfrac{x^4}{4} - 2$$

⇒ $y = \dfrac{x^3}{4} - 2x^{-1}$, which is the required solution.

EXAMPLE 19.   Solve $\dfrac{dy}{dx} + \dfrac{1}{x} \cdot y = y^3$.

SOLUTION.   Given

$$\dfrac{dy}{dx} + \dfrac{1}{x} \cdot y = y^3$$

⇒ $\dfrac{1}{y^3} \dfrac{dy}{dx} + \dfrac{1}{x \cdot y^2} = 1$ ...(1)

Let $\dfrac{1}{y^2} = t \Rightarrow -\dfrac{2}{y^3} \dfrac{dy}{dx} = \dfrac{dt}{dx}$

$$\Rightarrow \dfrac{1}{y^3} \dfrac{dy}{dx} = -\dfrac{1}{2} \cdot \dfrac{dt}{dx}$$

∴   From eq. (1),

$$-\dfrac{1}{2} \dfrac{dt}{dx} + t \dfrac{1}{x} = 1$$

⇒ $\dfrac{dt}{dx} - \dfrac{2}{x} t = -2$ ...(2)

which is linear equation of the form $\dfrac{dt}{dx} + Pt = Q$

where $P = \dfrac{-2}{x}, Q = -2$

Now, $\text{I.F.} = e^{\int P dx} = e^{\int -\frac{2}{x} dx} = e^{-2\log x} = e^{\log\left(\frac{1}{x^2}\right)} = \dfrac{1}{x^2}$

∴   Solution of the given equation

⇒ $t \cdot \dfrac{1}{x^2} = \int (-2) \cdot \dfrac{1}{x^2} dx + c = \dfrac{2}{x} + c$

$$\Rightarrow \qquad \frac{1}{y^2} \cdot \frac{1}{x^2} = \frac{2}{x} + c \qquad \left(\because t = \frac{1}{y^2} \right)$$

$$\Rightarrow \qquad 2xy^2 + cx^2y^2 = 1, \text{ which is the required solution.}$$

EXAMPLE 20. *Solve* $\dfrac{dy}{dx} + 2y \tan x = \cos 2x$.

SOLUTION. The given equation is

$$\frac{dy}{dx} + 2y \tan x = \cos 2x$$

Which is a linear differential equation of the type $\dfrac{dy}{dx} + py = Q$

Whose $\qquad P = 2 \tan x, \ Q = \cos 2x$

$$\text{I.F.} = e^{\int P dx} = e^{\int 2 \tan x dx} = e^{-2 \log \cos x} = e^{\log (\cos x)^{-2}} = \frac{1}{\cos^2 x}$$

Hence, solution is given by

$$y \cdot (\text{I.F.}) = \int Q \cdot (\text{I.F.}) dx + c$$

$$\Rightarrow \qquad y \cdot \frac{1}{\cos^2 x} = \int \left(\cos 2x \cdot \frac{1}{\cos^2 x} \right) dx + c$$

$$= \int \left\{ (2 \cos^2 x - 1) \cdot \frac{1}{\cos^2 x} \right\} dx + c$$

$$= \int 2 dx - \int \sec^2 x dx + c = 2x - \tan x + c$$

$$\Rightarrow \qquad y = 2x \cos^2 x - \cos^2 x + \tan x + c \cos^2 x$$

$$= (2x + c) \cos^2 x - \cos x \sin x$$

⮂ REMARKS

⮕ The equation given in the above example is not a linear differential equation originally but it can be made linear by using some suitable transformation.

⮕ The equation of the form $\dfrac{dy}{dx} + Py = Qy^n$ is called **Bernaulli's equation**, which can be reduce to

a linear equation by dividing both sides by y^n and then by putting $t = \dfrac{1}{y^{n-1}}$.

Exercise 13.6

Solve the following differential equations:

1. $2x \dfrac{dy}{dx} + y = 6x^3$

2. $(1 - x^2) \dfrac{dy}{dx} - xy = x$

3. $\dfrac{dy}{dx} + y \cot x = \cos x$

4. $x \cos x \dfrac{dy}{dx} + y(x \sin x + \cos x) = 1$

5. $4 \dfrac{dy}{dx} + 8y = 5e^{-3x}$

6. $x \log x \dfrac{dy}{dx} + y = 2 \log x$

7. $x \dfrac{dy}{dx} + 2y = x^2 \log x$

8. $\dfrac{dy}{dx} + y = \sin x$ \qquad **9.** $\dfrac{dy}{dx} + y = e^{-2x}$

10. $(1 + x^2) \dfrac{dy}{dx} + 2xy = 4x^2$

11. $(x^2 - 1) \dfrac{dy}{dx} + 2xy = \dfrac{2}{x^2 - 1}$

12. $x \dfrac{dy}{dx} + 2y = x^2, x \neq 0$

13. $\sec x \cdot \dfrac{dy}{dx} = y + \sin x$

14. $\dfrac{dy}{dx} + y = e^x$ **15.** $\dfrac{dy}{dx} + \dfrac{y}{x} = e^x$

16. $(1 - x^2)\dfrac{dy}{dx} + xy = ax$

17. $\dfrac{dy}{dx} - y \tan x = e^x \sec x$

18. $\dfrac{dy}{dx} + y \cot x = x$

19. $(1 + x^2)\dfrac{dy}{dx} + 2xy = \cos x$

20. $\dfrac{dy}{dx} + y \tan x = 2x + x^2 \tan x$

21. $(x^2 - 1)\dfrac{dy}{dx} + 2(x + 2)y = 2(x + 1)$

22. $x\dfrac{dy}{dx} + 2y = \sin x$

23. $(1 + x^2)dy + 2xy\,dx = \cot x\,dx,\ x \neq 0$

24. $\left(\dfrac{e^{-2\sqrt{x}}}{\sqrt{x}} - \dfrac{y}{\sqrt{x}}\right)\dfrac{dx}{dy} = 1, x \neq 0$

25. $\dfrac{dy}{dx} + y \cos x = \sin x \cos x$

26. $x\dfrac{dy}{dx} + y - x + xy \cot x = 0, x \neq 0$

27. $x\dfrac{dy}{dx} - y = x + 1$

28. $x\dfrac{dy}{dx} + 2y = x \cos x$

29. $\dfrac{dy}{dx} + y \cot x = 2x + x^2 \cot x$; when $y(0) = 0$

30. $\dfrac{dy}{dx} - 2y = \cos 3x$

31. $x\dfrac{dy}{dx} = y(\log y - \log x - 1)$

32. $(x \log x)\dfrac{dy}{dx} + y = \dfrac{2}{x} \log x$

33. $\dfrac{dy}{dx} + 2y \cot x = 3x^2 \text{cosec}^2 x$

34. $\dfrac{dy}{dx} + 2y = \sin x$

35. $x\dfrac{dy}{dx} - y = 2x^2 \sec x$

36. $\dfrac{dy}{dx} - 3y \cot x = \sin 2x$; where at $x = \dfrac{\pi}{2}, y = 2$

37. $y\dfrac{dy}{dx} \sin x = \cos x\left(\sin x - \dfrac{y^2}{2}\right)$; $y = 1$ when at $x = \dfrac{\pi}{2}$

38. $\dfrac{dy}{dx} + y \cot x = 4x\,\text{cosec}\,x$, $x = 0$ and at $x = \dfrac{\pi}{2}$, $y = 0$

39. $(1 + x^2)\dfrac{dy}{dx} + 2xy = \dfrac{1}{1 + x^2}$

(Given: $y = 0$ when $x = 1$)

40. $(x + y)\dfrac{dy}{dx} = 1$ **41.** $(x + 2y^2)\dfrac{dy}{dx} = y$

42. $\dfrac{dy}{dx} + xy = xy^3$

43. $y\,dx + (x - y^2)dy = 0, y > 0$

44. $x\dfrac{dy}{dx} + y = y^2 \log x$

45. $\tan y \cdot \dfrac{dy}{dx} + \tan x = \cos y \cos^2 x$

ANSWERS

1. $y = \dfrac{6}{7}x^3 + cx^{-1/2}$ **2.** $y = -1 + \dfrac{c}{\sqrt{1 - x^2}}$ **3.** $y \sin x = \dfrac{\sin^2 x}{2} + c$ **4.** $yx \sec x = \tan x + c$

5. $ye^{2x} = -\dfrac{5}{4}e^{-x} + c$ **6.** $y \log x = (\log x)^2 + c$ **7.** $y = \dfrac{x^2}{16}(4 \log x - 1) + \dfrac{c}{x^2}$

8. $y = ce^{-x} + \dfrac{1}{2}(\sin x - \cos x)$ **9.** $ye^x = -e^{-x} + c$ **10.** $y(1 + x^2) = \dfrac{4x^3}{3} + c$

11. $y(x^2 - 1) = 2 \log\left|\dfrac{x - 1}{x + 1}\right| + c$ **12.** $y = \dfrac{x^2}{4} + \dfrac{c}{x^2}$ **13.** $y = ce^{\sin x} - (1 + \sin x)$

14. $y = \dfrac{e^x}{2} + ce^{-x}$ **15.** $y = e^x - \dfrac{e^x}{x} + \dfrac{c}{x}$ **16.** $y = a + c\sqrt{1 - x^2}$ **17.** $y \cos x = e^x + c$

18. $(y - 1)\sin x + x \cos x = c$ **19.** $y(1 + x^2) = \sin x + c$ **20.** $y \sec x = x^2 \sec x + c$

21. $y = \dfrac{(x + 1)}{(x - 1)^3}[x^2 - 6x + 8 \log(x + 1)]$ **22.** $x^2 y = -x \cos x + \sin x + c$

23. $y(1 + x^2) = \log \sin x + c$ **24.** $y = (2\sqrt{x} + c)e^{-2\sqrt{x}}$ **25.** $y = \sin x - 1 + ce^{-\sin x}$

26. $xy \sin x = -x \cos x + \sin x + c$ **27.** $y = x \log x - 1 + cx$

28. $y = \sin x + \dfrac{2}{x} \cos x - \dfrac{2}{x^2} \sin x + \dfrac{c}{x^2}$ **29.** $y = x^2$ **30.** $y = \dfrac{(3\sin 3x - 2\cos 3x)}{13} + ce^{2x}$

31. $y = xe^x$ **32.** $y(\log x) = \dfrac{-2}{x}(\log x + 1) + c$ **33.** $y(\sin^2 x) = x^3 + c$

34. $y = \dfrac{1}{5}(2\sin x - \cos x) + ce^{-2x}$ **35.** $y = cx + 2x \log(\sec x + \tan x)$

36. $y = -2\sin^2 x + 4\sin^3 x$ **37.** $y^2 = \sin x$ **38.** $y \sin x = 2x^2 - \dfrac{\pi^2}{2}$

39. $y(x^2 + 1) = \tan^{-1} x - \dfrac{\pi}{4}$ **40.** $x + y + 1 = ce^y$ **41.** $x = 2y^2 + cy$ **42.** $\dfrac{1}{y^2} = 1 + ce^{x^2}$

43. $3xy = y^3 + c$ **44.** $\dfrac{1}{xy} = \dfrac{\log x + 1}{x} + c$ **45.** $\sec x \sec y = \sin x + c$

13.13 LINEAR DIFFERENTIAL EQUATIONS WITH CONSTANT COEFFICIENTS

The equation $\dfrac{d^n y}{dx^n} + A_1 \dfrac{d^{n-1} y}{dx^{n-1}} + A_2 \dfrac{d^{n-2} y}{dx^{n-2}} + \dots + A_n y = B$ ...(1)

having $A_1, \dots, A_n$ and B either constant or function of x, is called the linear differential

equation of nth order.

If $A_1, A_2, \dots, A_n$ are all constants and B may not be constant, then equation (1) is said to be linear differential equation of nth degree with constant coefficients.

If we take $B = 0$, then the corresponding equation is called homogeneous equation.

Using the symbols $D, D^2, \dots, D^n$ for $\dfrac{d}{dx}, \dfrac{d^2}{dx^2}, \dots, \dfrac{d^n}{dx^n}$ respectively in (1), then we get

$$D^n y + A_1 D^{n-1} y + A_2 D^{n-2} y + \dots + A_n y = B$$

$$\Rightarrow \quad (D^n + A_1 D^{n-1} + A_2 D^{n-2} + \dots + A_n)y = B \quad \Rightarrow \quad f(D)y = B \quad \dots(2)$$

where, $f(D) = D^n + A_1 D^{n-1} + A_2 D^{n-2} + \dots + A_n$.

Now, consider the homogeneous differential equation

$$f(D)y = 0 \quad \dots(3)$$

(Obtained by putting right hand side, i.e., B equal to zero).

Now, we shall show that if $y_1, y_2, \dots, y_n$ are n linearly independent solutions of (3), then $(C_1 y_1 + C_2 y_2 + \dots + C_n y_n)$ is also a solution of (3), where $C_1, C_2, \dots, C_n$ are arbitrary constants.

Since, we assumed that $y_1, y_2, \dots, y_n$ are solution of (3) $\Rightarrow y_1, y_2, \dots, y_n$ must satisfy (3).

which gives $\qquad \left.\begin{array}{l} f(D)y_1 = 0 \\ f(D)y_2 = 0 \\ \dots\dots\dots\dots \\ f(D)y_n = 0 \end{array}\right\}$ ...(4)

Now consider

$$f(D)(C_1 y_1 + C_2 y_2 + \dots + C_n y_n) = f(D)(C_1 y_1) + f(D)(C_2 y_2) + \dots + f(D)(C_n y_n)$$

$$= C_1 f(D)y_1 + C_2 f(D)y_2 + \dots + C_n f(D)y_n$$

$$= C_1 . 0 + C_2 . 0 + \dots + C_n . 0 \qquad \text{(By using (4))}$$

Therefore, we have $f(D)[C_1y_1 + C_2y_2 + ... C_ny_n] = 0$ ...(5)

$\Rightarrow \quad (C_1y_1 + C_2y_2 + ... + C_ny_n)$ satisfies (3).

$\Rightarrow \quad (C_1y_1 + C_2y_2 + ... + C_ny_n)$ is also a solution of (3).

Hence, we can say that if $y_1, y_2, ..., y_n$ are n linearly independent solution of (3), then $(C_1y_1 + C_2y_2 + ... + C_ny_n)$ is also a solution of (3) known as complete or general solution of (3), containing n arbitrary constants $C_1, C_2, ..., C_n$.

Now, let us suppose $(C_1y_1 + C_2y_2 + ... + C_ny_n) = u$ (say).

Then, from (5), we have

$$f(D)u = 0 \qquad ...(6)$$

Again, let v be any particular solution of (2). Therefore, we have

$$f(D)v = B \qquad ...(7)$$

Now, $\qquad f(D)(u + v) = f(D)u + f(D)v \qquad$ (Using (6) and (7))

which shows that $(u + v)$, *i.e.*, $\{(C_1y_1 + C_2y_2 + ... + C_ny_n) + v\}$ is the general solution of (2).

WORKING PROCEDURE

Step 1. Firstly, we find the general solution of (2), which is called the complimentary function (C.F), contains as many arbitrary constants as is the order of the given differential equation.

Step 2. Next, find the solution of (1), with no arbitrary constant which is called the particular integral (P.I.).

Step 3. To find the general solution of (1), add C.F. and P.I. obtained in (1) and (2), *i.e.*, $y = u + v = $ C.F. + P.I.

REMARKS

➠ Here, the operator D stands for d / dx, D^2 for d^2 / dx^2 and so on.

➠ The operator D^{-1} stands for integration.

➠ Since, the symbol D satisfies the fundamental laws of algebra, therefore it can be regarded as an algebraic quantity.

➠ The general solution of (1) is $y = C.F. + P.I.$, where C.F. involves n arbitrary constants and P.I. does not involve any arbitrary constant.

➠ Since P.I. appears due to B in (1), therefore, if a linear differential equation with constant coefficients is given with $B = 0$, then its general solution will not involve P.I. and hence the general solution of the differential equation is given by $y = $ C.F.

➠ The method (discussed above) of solving these type of equations, is given by Euler and D'Alembert.

13.13.1 AUXILIARY EQUATION

Consider the differential equation (1) with $B = 0$, *i.e.*,

$$(D^n + A_1D^{n-1} + A_2D^{n-2} + ... + A_n)y = 0 \qquad \text{or} \qquad f(D)y = 0. \qquad ...(1)$$

Substitute $y = e^{mx}$ on the trial basis, then we get $e^{mx}(m^n + A_1m^{n-1} + A_2m^{n-2} + ... + A_n) = 0$ which holds if

$$m^n + A_1m^{n-1} + A_2m^{n-2} + ... + A_n = 0 \qquad \text{or} \qquad f(m) = 0. \qquad ...(2)$$

Equation (2) is called the auxiliary equation.

From (1) and (2), we observe that the auxiliary equation $f(m) = 0$ will give the same value of m as the equation $f(D) = 0$ gives the value of D.

13.14 METHOD OF FINDING THE COMPLEMENTARY FUNCTION (C.F.)

To find the C.F., the roots of the auxiliary equation (2) are to be considered. Three different cases arise :

(i) The roots of auxiliary equation (2) are real.

(ii) The roots of auxiliary equation (2) are complex, $i.e.$, $\alpha \pm i\beta$ type.

(iii) The roots of auxiliary equation (2) are surds, $i.e.$, $\alpha \pm \sqrt{\beta}$ type.

Case (i) : (a) Suppose that the auxiliary equation (2) has n distinct roots $m_1, m_2, ..., m_n$, then C.F. is given by

$$C_1 e^{m_1 x} + C_2 e^{m_2 x} + ... + C_n e^{m_n x}$$

where $C_1, C_2, ..., C_n$ are arbitrary constants.

(b) If the auxiliary equation having r roots are equal to m_1(say) and remaining roots are distinct, then the C.F. is given by

$$[C_1 + C_2 x + C_3 x^2 + ... + C_r x^{r-1}] e^{m_1 x} + C_{r+1} e^{m_{r+1} x} + ... + C_n e^{m_n x}.$$

Case (ii): If some of the roots of the auxiliary equation are complex, then we shall use the following procedure.

Let $\alpha \pm i\beta$ be the roots of the auxiliary equation, then the corresponding part becomes

$$= C_1 e^{(\alpha + i\beta)x} + C_2 e^{(\alpha - i\beta)x} = C_1 e^{\alpha x} . e^{i\beta x} + C_2 e^{\alpha x} . e^{-i\beta x}$$

$$= e^{\alpha x} [C_1 \cos \beta x + i C_1 \sin \beta x] + e^{\alpha x} [C_2 \cos \beta x - i C_2 \sin \beta x]$$

$$= e^{\alpha x} [(C_1 + C_2) \cos \beta x + (i C_1 - i C_2) \sin \beta x]$$

$$\text{C.F.} = e^{\alpha x} [B_1 \cos \beta x + B_2 \sin \beta x] \qquad \qquad ...(1)$$

where, B_1, B_2 are arbitrary constants.

The expression (1) can also be written as

(a) $B_1 e^{\alpha x} \cos(\beta x + B_2)$ (b) $B_1 e^{\alpha x} \sin(\beta x + B_2)$.

If, the equation has two equal pair of complex roots $\alpha + i\beta$ and $\alpha - i\beta$, say, occur twice, then the corresponding part of C.F. is written as

$$e^{\alpha x} [(B_1 + B_2 x) \cos \beta x + (B_3 + B_4 x) \sin \beta x] .$$

In general, if $\alpha \pm i\beta$ occur k times, then the corresponding part of the C.F. can be written as

$$e^{\alpha x} \{(B_1 + B_2 x + ... + B_k x^{k-1}) \cos \beta x + (B_{k+1} + B_{k+2} x + ... + B_{2k} x^{k-1})\} \sin \beta x$$

where $B_1, B_2, ..., B_k, B_{k+1}, ..., B_{2k}$ are arbitrary constants.

Case (iii) : If a pair of the roots of the auxiliary equation involves surds, say $\alpha \pm \sqrt{\beta}$, where $\beta > 0$, then the corresponding part of C.F. is one of the following three forms

(a) $e^{\alpha x} [B_1 \cosh(x\sqrt{\beta}) + B_2 \sinh(x\sqrt{\beta})]$ (b) $B_1 e^{\alpha x} \cosh(x\sqrt{\beta} + B_2)$

(c) $B_1 e^{\alpha x} \sinh(x\sqrt{\beta} + B_2)$

⮞ REMARKS

➠ The results obtained in case (iii), are exactly similar to those of case (ii) except that sin and cos replaced by sinh and cosh respectively.

➠ The method of finding the complimentary function (C.F.) of the following differential equation of the form

$$(D^n + A_1 D^{n-1} + A_2 D^{n-2} + ... + A_n)y = 0$$

can be concluded as follows :

S. No.	Nature of the Roots	Solution
1.	Real and distinct say $m_1, m_2, ..., m_n$	$y = B_1 e^{m_1 x} + B_2 e^{m_2 x} ... + B_n e^{m_n x}$
2.	Real and equal, say n	$y = (B_1 + B_2 x + B_3 x^2 + ... + B_n x^{n-1}) e^{nx}$
3.	Non-repeated roots : $\alpha \pm i\beta$	(a) $y = (B_1 \cos \beta x + B_2 \sin \beta x) e^{\alpha x}$ (b) $y = B_1 e^{\alpha x} \cos(\beta x + B_2)$
4.	Repeated roots : $\alpha \pm i\beta$, r times	$y = \{(B_1 + B_2 x + ... + B_r x^{r-1}) \cos \beta x$ $+ (B_1' + B_2' x + ... + B_r' x^{r-1}) \sin \beta x\} e^{\alpha x}$
5.	Irrational roots : $\alpha \pm \sqrt{\beta}$	(a) $y = B_1 e^{\alpha x} \cosh(x\sqrt{\beta} + B_2)$ (b) $y = B_1 e^{\alpha x} \sinh(x\sqrt{\beta} + B_2)$

SOLVED EXAMPLES

EXAMPLE 1. *Solve* $[D^3 + 6D^2 + 11D + 6] y = 0$. 　　　　　　　　**[UPTU(B.Pharma)–2001]**

SOLUTION. Here, the given differential equation is

$$[D^3 + 6D^2 + 11D + 6] y = 0$$

To find the auxiliary equation, replace D by m, then (1) becomes,

$$m^3 + 6m^2 + 11m + 6 = 0 \qquad \Rightarrow \qquad (m+1)(m^2 + 5m + 6) = 0$$

$$\Rightarrow \qquad (m+1)(m+2)(m+3) = 0$$

$$\Rightarrow \qquad m = -1, -2, -3$$

i.e., Roots are real and unequal. Hence, the general solution is

$$y = C_1 e^{-x} + C_2 e^{-2x} + C_3 e^{-3x}.$$

EXAMPLE 2. *Solve* $[D^4 + 2D^3 - 3D^2 - 4D + 4] y = 0$.

SOLUTION. Here, the auxiliary equation is

$$m^4 + 2m^3 - 3m^2 - 4m + 4 = 0 \qquad \text{or} \qquad (m-1)(m^3 + 3m^2 - 4) = 0$$

$$\Rightarrow (m-1)(m-1)(m^2 + 4m + 4) = 0 \qquad \Rightarrow \qquad (m-1)(m-1)(m+2)^2 = 0$$

$$\Rightarrow \qquad m = +1, +1, -2, -2$$

$$\Rightarrow \qquad \text{Repeated real roots exist.}$$

Hence, general solution is

$$y_1 = (C_1 + C_2 x) e^x + (C_3 + C_4 x) e^{-2x}.$$

EXAMPLE 3. *Solve* $(D^4 + k^4) y = 0$.

SOLUTION. Here, the auxiliary equation is

$$m^4 + k^4 = 0 \qquad \text{or} \qquad (m^2 + k^2)^2 - 2k^2 m^2 = 0$$

$$\Rightarrow \qquad (m^2 + k^2)^2 - (\sqrt{2} \cdot km)^2 = 0$$

$$\Rightarrow \qquad (m^2 + k^2 - \sqrt{2} \cdot km)(m^2 + k^2 + \sqrt{2} \cdot km) = 0$$

$$\Rightarrow \qquad m^2 - \sqrt{2} \cdot km + k^2 = 0$$

and $\quad m^2 + k^2 + \sqrt{2}\,.\,km = 0$

$\Rightarrow \qquad m = \dfrac{\sqrt{2}\,k \pm \sqrt{(2k^2 - 4k^2)}}{2}$

and $\qquad m = \dfrac{-\sqrt{2}\,k \pm \sqrt{(2k^2 - 4k^2)}}{2}$

$\Rightarrow \quad m = \dfrac{k}{\sqrt{2}} \pm i\,\dfrac{k}{\sqrt{2}} \quad$ and $\quad m = -\dfrac{k}{\sqrt{2}} \pm i\,\dfrac{k}{\sqrt{2}}$

Hence, the solution is

$$y = e^{kx/\sqrt{2}} \{C_1 \cos(kx/\sqrt{2}) + C_2 \sin(kx/\sqrt{2})$$
$$+ e^{-kx/\sqrt{2}} \{C_3 \cos(kx/\sqrt{2}) + C_4 \sin(kx/\sqrt{2})\}$$

EXAMPLE 4. Solve $[D^4 - 4D^3 + 8D^2 - 8D + 4]\,y = 0$.

SOLUTION. Here, the auxiliary equation is

$$m^4 - 4m^3 + 8m^2 - 8m + 4 = 0$$

$\Rightarrow \quad (m^2 - 2m + 2)^2 = 0 \Rightarrow m = 1 \pm i,\ 1 \pm i$

$\Rightarrow \quad$ Repeated complex roots exist.

Hence, the solution of the given equation is

$$y = e^x \{(C_1 + C_2 x)\cos x + (C_3 + C_4 x)\sin x\}.$$

EXAMPLE 5. Solve $(D^2 + 6D + 4)y = 0$.

SOLUTION. Here, the auxiliary equation is

$$m^2 + 6m + 4 = 0 \quad \Rightarrow \quad m = -3 \pm \sqrt{5}.$$

$\Rightarrow \quad$ Irrational roots exist.

Hence, the solution of the given equation is

$$y = e^{-3x}\,(C_1 \cosh x\sqrt{5} + C_2 \sinh x\sqrt{5}).$$

Exercise 13.7

Solve the following equations :

1. $\dfrac{d^2 y}{dx^2} + 3\dfrac{dy}{dx} + 2y = 0$

2. $(D^3 - 9D^2 + 23D - 15)y = 0$

3. $(D^4 - D^3 - 9D^2 - 11D - 4)y = 0$

4. $(D^2 + 1)^2 (D - 1)^2 y = 0$ **[UPTU(B.Pharma)–2002]**

5. $(D^3 - D^2 - 12D)y = 0$

6. $(D^4 + 2n^2 D^2 + n^4)y = 0$

7. $[D^2 - 2\lambda D + (\lambda^2 + \mu^2)]y = 0$

8. $(D^4 + D^3 + 2D^2 - D + 3)\,y = 0$

9. $(D^5 - 13D^3 + 26D^2 + 82D + 104)y = 0$

10. $\dfrac{d^4 y}{dx^4} + y = 0$

11. $(D^3 - 3D^2 + 4)y = 0$

12. $(D^2 - 2D + 4)^2 y = 0$

13. $\dfrac{d^2 x}{dt^2} + 5\dfrac{dx}{dt} + 6x = 0$, given $\quad x(0) \quad = \quad 0,$

$\dfrac{dx(0)}{dt} = 15$

14. $(D^4 - 4D + 4)y = 0$

15. $(D^2 + 1)^3 y = 0$, where $D \equiv d/dx$

16. $\dfrac{d^2 x}{dt^2} - 4\dfrac{dx}{dt} + 13x = 0, x(0), \dfrac{dx(0)}{dt} = 2$

17. $\dfrac{d^3 y}{dx^3} + y = 0$

18. $\dfrac{d^4 y}{dx^4} + 8\dfrac{d^2 y}{dx^2} + 16y = 0$

19. $(4D^4 - 8D^3 - 7D^2 + 11D + 6)y = 0$

HINT TO SELECTED PROBLEMS

1. $m = -1, -2$ **2.** $m = 1, 5, 3$

9. $m = -1 \pm i, \ -3 \pm 2i, \ -4$

3. $m = -1, -1, -1, 4$ **4.** $m = \pm i, \pm i, 1, 1$

10. $m = \dfrac{-1 \pm i}{\sqrt{2}}, \dfrac{1 \pm i}{\sqrt{2}}$ **11.** $m = -1, 2, 2$

5. $m = 0, 4, 3$ **6.** $m = \pm ni, \ \pm ni$

7. $m = \lambda \pm \mu i$ **8.** $m = -1 \pm i\sqrt{2}, \dfrac{1}{2} \pm i\dfrac{\sqrt{3}}{2}$

12. $m = 1 \pm \sqrt{3}\,i, \ 1 \pm \sqrt{3}\,i$

ANSWERS

1. $y = C_1 e^{-x} + C_2 e^{-2x}$ **2.** $y = C_1 e^x + C_2 e^{3x} + C_3 e^{5x}$

3. $y = e^{-x}(C_1 + C_2 x + C_3 x^2) + C_4 e^{4x}$ **4.** $y = (C_1 + C_2 x)\sin x + (C_3 + C_4 x)\cos x + (C_5 + C_6 x)e^x$

5. $y = C_1 + C_2 e^{4x} + C_3 e^{-3x}$ **6.** $y = (C_1 + C_2 x)\cos nx + (C_3 + C_4 x)\sin nx$

7. $y = e^{\lambda x}(C_1 \cos \mu x + C_2 \sin \mu x)$

8. $y = e^{-x}\left[C_1 \cos(\sqrt{2}x) + C_2 \sin(\sqrt{2}x) + e^{x/2}\left(C_3 \cos\dfrac{\sqrt{3}}{2}x + C_4 \sin\dfrac{\sqrt{3}}{2}x \right) \right]$

9. $y = C_1 e^{-x}\cos(x + \alpha) + C_2 e^{-3x}\cos(2x + \beta) + C_3 e^{-4x}$

10. $y = C_1 e^{x/\sqrt{2}}\cos\left(\dfrac{x}{\sqrt{2}} + C_2 \right) + C_3 e^{-x/\sqrt{2}}\cos\left(\dfrac{x}{\sqrt{2}} + C_4 \right)$

11. $y = C_1 e^{-x} + (C_2 + C_3 x)e^{2x}$ **12.** $y = e^x[(C_1 + C_2 x)\cos\sqrt{3}x + (C_3 + C_4 x)\sin\sqrt{3}\,x]$

13. $x = 15(e^{-2t} - e^{-3t})$ **14.** $y = ((C_1 + C_2 x)e^{\sqrt{2}x} + (C_3 + C_4 x)e^{-\sqrt{2}x})$

15. $y = (C_1 + C_2 x + C_3 x^2)\cos x + (C_4 + C_5 x + C_6 x^2)\sin x$ **16.** $\dfrac{2}{3}e^{2t}\sin 3t$

17. $y = C_1 e^{-x} + e^{x/2}\left(C_2 \cos\dfrac{\sqrt{3}x}{2} + C_3 \sin\dfrac{\sqrt{3}x}{2} \right)$ **18.** $y = (C_1 + C_2 x)\cos 2x + (C_3 + C_4 x)\sin 2x$

19. $y = C_1 e^{-x} + C_2 e^{2x} + e^{x/2}\left(C_3 \cos\dfrac{x}{\sqrt{2}} + C_4 \sin\dfrac{x}{\sqrt{2}} \right)$

13.15 PARTICULAR INTEGRAL

Consider the differential equation $f(D)y = B \ \Rightarrow \ y = \dfrac{1}{f(D)} \cdot B$

Let $\dfrac{1}{f(D)} \cdot B$ denote some function of x, which when operated upon by $f(D)$ produces B.

Hence, P.I. $= \dfrac{1}{f(D)} \cdot B$.

13.15.1 GENERAL METHOD OF FINDING P.I.

THEOREM 1.   *If B is a function of x, then* $\dfrac{1}{D - a}B = e^{ax}\int B e^{-ax}\, dx$.

PROOF.   Let $y = \dfrac{1}{D - a}B \ \Rightarrow \ (D - a)y = B \ \Rightarrow \ \left(\dfrac{d}{dx} - a \right)y = B \ \Rightarrow \ \dfrac{dy}{dx} - ay = B$

which is the linear differential equation. I.F. $= e^{-\int a\,dx} = e^{-ax}$.

Hence, solution is given by $ye^{-ax} = \int B\,e^{-ax}dx$.

(Since we find the P.I., therefore we omit the constant of integration.)

$$\therefore \quad y = e^{ax} \int Be^{-ax}\,dx \qquad \Rightarrow \qquad \frac{1}{D-a}\cdot B = e^{ax}\int Be^{-ax}dx$$

REMARKS

➠ P.I. never contains any arbitrary constant.

➠ The method discussed above can be used to evaluate P.I. in any problem. It does not depend upon the form of B.

➠ The method discussed above must be used when B is of the form $\sec ax$, $\operatorname{cosec} ax$, $\tan ax$, etc.

➠ Here, the operator $\dfrac{1}{f(D)}$ (known as increase operator) having the following properties :

(a) If $B = u_1 + u_2 + ... + u_n$, then $\dfrac{1}{f(D)}\cdot B = \dfrac{1}{f(D)}\cdot u_1 + \dfrac{1}{f(D)}\cdot u_2 + ... \dfrac{1}{f(D)}\cdot u_n$

(b) $\dfrac{1}{f(D)}(KB) = \dfrac{K}{f(D)}\cdot B$ (c) $\dfrac{1}{f(D)}$ can be resolved into factors.

(d) $\dfrac{1}{f(D)}$ can be broken into partial fractions.   (e) $\dfrac{1}{f(D)}\cdot B$ is a particular integration.

SOLVED EXAMPLES

EXAMPLE 1. *Solve* $D^2 - 5D + 6 = e^{3x}$.   [UPTU(B.Pharma)–2002, 04]

SOLUTION. The given equation can be written as

$$(D-3)(D-2)y = e^{3x}$$

$$\text{C.F.} = C_1 e^{3x} + C_2 e^{2x}$$

and

$$\text{P.I.} = \frac{1}{D-3}\cdot\frac{1}{D-2}e^{3x} = \frac{1}{D-3}e^{2x}\int e^{3x}e^{-2x}dx$$

$$= \frac{1}{D-3}e^{2x}.e^x = e^{3x}\int e^{3x}.e^{-3x}dx = xe^{3x} .$$

Now, general solution = C.F. + P.I.

$$\Rightarrow \quad y = C_1 e^{3x} + C_2 e^{2x} + xe^{3x} .$$

EXAMPLE 2. *Solve* $(D^2 + 1)y = \sec^2 x$.

SOLUTION. Here, the given equation is

$$(D^2 + 1)y = \sec^2 x \qquad\qquad\qquad ...(1)$$

The auxiliary equation of (1) is given by

$$m^2 + 1 = 0 \Rightarrow m = \pm i$$

$$\Rightarrow \text{C.F.} = C_1 \cos x + C_2 \sin x$$

$$\text{P.I.} = \frac{1}{D^2+1}\sec^2 x = \frac{1}{(D+i)(D-i)}\sec^2 x = \frac{1}{2i}\left[\frac{1}{D-i} - \frac{1}{D+i}\right]\sec^2 x$$

$$= \frac{1}{2i}\left[e^{xi}\int e^{-ix}\sec^2 x\,dx - e^{-ix}\int e^{ix}\sec^2 x\,dx\right]$$

$$= \frac{1}{2i}\left\{e^{ix}\int \frac{\cos x - i\sin x}{\cos^2 x}dx - e^{-ix}\int \frac{\cos x + i\sin x}{\cos^2 x}dx\right\}$$

$$= \frac{1}{2i}\left\{e^{ix}\int (\sec x - i\sec x \tan x)dx - e^{-ix}\int (\sec x + i\sec x \tan x)dx\right\}$$

$$= \frac{1}{2i}\left\{(e^{ix} - e^{-ix})\int \sec x dx - i(e^{ix} + e^{-ix})\int \tan x \sec x dx\right\}$$

$$= \frac{1}{2i}\left\{2i\sin x \log(\sec x + \tan x) - 2i\cos x \sec x\right\}$$

$$= \sin x \log(\sec x + \tan x) - 1.$$

Hence, the general solution is $y = $ C.F. + P.I.

$$\Rightarrow \quad y = C_1 \cos x + C_2 \sin x + \sin x \log(\sec x + \tan x) - 1.$$

EXAMPLE 3. Solve $(D^2 + 9)y = \sec 3x$.

[UPTU(B.Pharma)–2002]

SOLUTION. Auxiliary equation is $m^2 + 9 = 0 \Rightarrow m = \pm 3i$

$\therefore \quad$ C.F. $= c_1 \cos 3x + c_2 \sin 3x$

$$\text{P.I.} = \frac{\sec 3x}{D^2 + 9} = \frac{\sec 3x}{(D + 3i)(D - 3i)} = \frac{1}{6i}\left[\frac{1}{D - 3i} - \frac{1}{D + 3i}\right]\sec 3x$$

$$= \frac{1}{6i}\left[e^{3ix}\int e^{-3ix}\sec 3x dx - e^{-3ix}\int e^{3ix}\sec 3x dx\right]$$

$$= \frac{1}{6i}\left[e^{3ix}\left\{\int \left(1 - i\frac{\sin 3x}{\cos 3x}\right)dx\right\} - e^{-3ix}\left\{\int \left(1 + i\frac{\sin 3x}{\cos 3x}\right)dx\right\}\right]$$

$$= \frac{1}{6i}\left[e^{3ix}\left\{x + \frac{i}{3}\log \cos 3x\right\} - e^{-3ix}\left\{x - \frac{i}{3}\log \cos 3x\right\}\right]$$

$$= \frac{1}{6i}\left[(\cos 3x + i\sin 3x)\left(x + \frac{i}{3}\log \cos 3x\right)\right.$$

$$\left. - (\cos 3x - i\sin 3x)\left(x - \frac{i}{3}\log \cos 3x\right)\right]$$

$$= \frac{1}{6i}\left[\frac{2i}{3}\cos 3x \log \cos 3x + 2ix \sin 3x\right].$$

$$= \frac{1}{9}\cos 3x \log \cos 3x + \frac{x}{3}\sin 3x$$

Hence, $y = $ C.F. + P.I.

$$= c_1 \cos 3x + c_2 \sin 3x + \frac{x}{3}\sin 3x + \frac{1}{9}\cos 3x \log \cos 3x.$$

Exercise 13.8

Solve the following differential equations :

1. $(D^2 + a^2)y = \sec ax$

2. $(D^2 + a^2)y = \tan ax$

3. $(D^2 + 1)y = \csc x$

4. $(D^2 + n^2)y = \cot nx$

5. $(D^2 + n^2)y = \tan nx$

HINT TO SELECTED PROBLEMS

1. $m = \pm ai \Rightarrow$ C.F. $= C_1 \cos ax + C_2 \sin ax$

P.I. $= \dfrac{1}{D^2 + a^2} \sec ax = \dfrac{1}{(D + ai)(D - ai)} \sec ax$

$= \dfrac{1}{2ai} \left[\dfrac{1}{(D - ai)} - \dfrac{1}{(D + ai)} \right] \sec ax$

$= \dfrac{1}{2ai} \left\{ e^{iax} \int e^{-iax} \sec ax \, dx \right.$

$\left. - e^{-iax} \int e^{iax} \sec ax \, dx \right\}$

2. P.I. $= \dfrac{1}{D^2 + a^2} \tan ax$

$= \dfrac{1}{2ai} \left[\dfrac{1}{D - ia} - \dfrac{1}{D + ia} \right] \tan ax.$

Then proceed as above.

3. P.I. $= \dfrac{1}{D^2 + 1} \operatorname{cosec} x$

$= \dfrac{1}{(D + i)(D - i)} \operatorname{cosec} x$

$= \dfrac{1}{2i} \left[\dfrac{1}{D - i} + \dfrac{1}{D + i} \right] \operatorname{cosec} x \cdot$

Now proceed as above.

4. C.F. $= C_1 \cos nx + C_2 \sin nx$

P.I. $= \dfrac{1}{D^2 + n^2} \cot nx = \dfrac{1}{(D - in)(D + in)} \cot nx$

$= \dfrac{1}{2in} \left[\dfrac{1}{D - in} \cot nx - \dfrac{1}{D + in} \cot nx \right]$

ANSWERS

1. $y = C_1 \cos ax + C_2 \sin ax + \dfrac{x}{a} \sin ax + \dfrac{1}{a^2} \cos ax \log \cos ax$

2. $y = C_1 \cos ax + C_2 \sin ax - \dfrac{1}{a^2} \cos ax \log \tan \left(\dfrac{\pi}{4} + \dfrac{ax}{2} \right)$

3. $y = C_1 \cos x + C_2 \sin x + \sin x \log \sin x - x \cos x$

4. $y = C_1 \cos nx + C_2 \sin nx + \dfrac{1}{n^2} \sin nx \log(\operatorname{cosec} nx - \cot nx)$

5. $y = C_1 \cos nx + C_2 \sin nx - \dfrac{1}{n^2} \cos nx \log(\sec nx + \tan nx)$

13.15.2 SHORT METHODS OF GETTING P.I.

The general method for getting P.I. discussed above requires lot of calculations. In certain cases, the P.I. can be obtained by methods which are shorter than the general method.

(1) To evaluate P.I., when B is of the form e^{ax} :

Here, we want to evaluate $\dfrac{1}{f(D)} e^{ax}$

where, $f(D) = A_0 D^n + A_1 D^{n-1} + ... + A_n$ with $f(a) \neq 0$.

Here, $B = e^{ax}$, we have

$D(e^{ax}) = a e^{ax}$

$D^2(e^{ax}) = a^2 e^{ax}$

$\cdots\cdots\cdots\cdots\cdots$

$\cdots\cdots\cdots\cdots\cdots$

$D^n(e^{ax}) = a^n e^{ax}$

$\Rightarrow \qquad f(D) e^{ax} = (A_0 D^n + A_1 D^{n-1} + ... + A_n) e^{ax} = A_0 D^n e^{ax} + A_1 D^{n-1} e^{ax} + ... + A_n e^{ax}$

$$= A_0 a^n e^{ax} + A_1 a^{n-1} e^{ax} + \ldots + A_n e^{ax} = (A_0 a^n + A_1 a^{n-1} + \ldots + A_n) e^{ax}$$

$\Rightarrow \qquad f(D) e^{ax} = f(a) e^{ax}.$

Operating upon both sides with $\dfrac{1}{f(D)}$, we get $\dfrac{1}{f(D)} \cdot f(D) . e^{ax} = \dfrac{1}{f(D)} \cdot f(a) e^{ax}$

$\Rightarrow \qquad e^{ax} = f(a) \dfrac{1}{f(D)} e^{ax}$

$\Rightarrow \qquad \dfrac{1}{f(D)} e^{ax} = \dfrac{e^{ax}}{f(a)}$, provided $f(a) \neq 0$.

SOLVED EXAMPLES

EXAMPLE 1. *Solve* $(D^2 - 3D + 2)y = e^{5x}$.

SOLUTION. The given equation is

$\qquad (D^2 - 3D + 2)y = e^{5x}$

Auxiliary equation is $m^2 - 3m + 2 = 0$

$\Rightarrow \quad (m-1)(m-2) = 0 \quad \Rightarrow \quad m = 1, 2.$

$\therefore \qquad \text{C.F.} = C_1 e^x + C_2 e^{2x}$

Now, $\qquad \text{P.I.} = \dfrac{1}{D^2 - 3D + 2} . e^{5x} = \dfrac{1}{25 - 3 \times 5 + 2} e^{5x} = \dfrac{1}{12} e^{5x}$

Hence, the general solution is

$\qquad y = \text{C.F.} + \text{P.I.}$

$\Rightarrow \qquad y = C_1 e^x + C_2 e^{2x} + \dfrac{1}{12} . e^{5x}.$

EXAMPLE 2. *Solve* $(D^3 + 1)y = (e^x + 1)^2$.

SOLUTION. The given equation is $(D^3 + 1)y = (e^x + 1)^2$

The auxiliary equation is $m^3 + 1 = 0$

$\Rightarrow (m+1)(m^2 - m + 1) = 0 \Rightarrow m = -1, \dfrac{1}{2} \pm \dfrac{i\sqrt{3}}{2}$

Therefore,

$\qquad \text{C.F.} = C_1 e^{-x} + e^{x/2} \left[C_2 \cos\left(\dfrac{x\sqrt{3}}{2}\right) + C_3 \sin\left(\dfrac{x\sqrt{3}}{2}\right) \right].$

Now,

$\qquad \text{P.I.} = \dfrac{1}{(D^3 + 1)} [e^x + 1]^2 = \dfrac{1}{(D^3 + 1)} (e^{2x} + 2e^x + 1)$

$\qquad = \dfrac{1}{D^3 + 1} (e^{2x} + 2e^x + e^{0x})$

$\qquad = \dfrac{1}{D^3 + 1} e^{2x} + 2\dfrac{1}{D^3 + 1} e^x + \dfrac{1}{D^3 + 1} e^{0x}.$

$\qquad = \dfrac{1}{2^3 + 1} e^{2x} + 2\dfrac{1}{1^3 + 1} e^x + \dfrac{1}{0 + 1} e^{0x} = \dfrac{1}{9} e^{2x} + e^x + 1$

Here, the general solution is
$$y = \text{C.F.} + \text{P.I.}$$

$$\Rightarrow \quad y = C_1 e^{-x} + e^{x/2}\left[C_2 \cos\left(\frac{x\sqrt{3}}{2}\right) + C_3 \sin\left(\frac{x\sqrt{3}}{2}\right)\right] + \frac{1}{9}e^{2x} + e^x + 1$$

Exercise 13.9

Solve the following differential equations :

1. $(D^2 - 4D + 1)y = e^{2x} - e^{-x}$

2. $(D^2 + 5D + 6)y = e^{2x}$

3. $(4D^2 + 4D - 3)y = e^{2x}$

4. $(D^2 - 2D + 1)y = 2e^{5x/2}$

5. $(D^2 + D + 1)y = e^{-x}$

6. $D^2(D+1)^2(D^2 + D + 1)^2 y = e^x$

7. $[D^2 + 2pD + (p^2 + q^2)]y = e^{ax}$

8. $(4D^2 + 12D + 9)y = 144e^{-3x}$

9. $(D^2 - 4D + 3)y = e^{3x}$

10. $(D^2 - a^2)y = e^{ax} - e^{-ax}$

11. $(D^2 + D + 1)y = (1 + e^x)^2$

12. $\dfrac{d^3y}{dx^3} - 3\dfrac{d^2y}{dx^2} + 3\dfrac{dy}{dx} - y = e^x + 2$

HINT TO SELECTED PROBLEMS

1. $\quad m = 2 \pm \sqrt{3}$

$\Rightarrow$ C.F. $= e^{2x}[C_1 \cosh x\sqrt{3} + C_2 \sinh x\sqrt{3}]$

$$\text{P.I.} = \frac{1}{D^2 - 4D + 1}[e^{2x} - e^{-x}]$$

$$= \frac{1}{D^2 - 4D + 1}e^{2x} - \frac{1}{(D^2 - 4D + 1)}e^{-x}$$

$$= \frac{e^{2x}}{2^2 - 4\times 2 + 1} - \frac{e^{-x}}{(-1)^2 - 4(-1) + 1}$$

$$= -\frac{e^{2x}}{3} - \frac{e^{-x}}{6}$$

4. $m = 1, 1 \Rightarrow$ C.F. $= (C_1 + C_2 x)e^x$

$$\text{P.I.} = \frac{1}{D^2 - 2D + 1}(2e^{5x/2}) = 2.\frac{e^{5x/2}}{\frac{25}{4} - 4}.$$

6. $m = 0, 0, -1, -1, -\dfrac{1}{2} \pm \dfrac{i\sqrt{3}}{2}, \ -\dfrac{1}{2} \pm \dfrac{i\sqrt{3}}{2}.$

7. $m = -p \pm iq$.

8. $m = -\dfrac{3}{2}, -\dfrac{3}{2} \Rightarrow$ C.F. $= (C_1 + C_2 x)\,e^{-3x/2}$

$$\text{P.I.} = 144\left(\frac{1}{4D^2 + 12D + 9}\right)e^{-3x} = \frac{144\,e^{-3x}}{9}$$

9. P.I. $= \dfrac{1}{D^2 - 4D + 3}e^{3x} = \dfrac{1}{2D - 4}e^{3x} = \dfrac{x}{2}.e^{3x}$

10. $m = \pm a$

11. $m = -\dfrac{1}{2} \pm \dfrac{\sqrt{3}}{2}i$

12. $m = 1, 1, 1$

ANSWERS

1. $y = e^{2x}(C_1 \cosh x\sqrt{3} + C_2 \sinh x\sqrt{3}) - \dfrac{1}{3}e^{2x} - \dfrac{1}{6}e^{-x}$

2. $y = C_1 e^{-2x} + C_2 e^{-3x} + \dfrac{1}{20}e^{2x}$

3. $y = C_1 e^{x/2} + C_2 e^{-3x/2} + \dfrac{1}{21}e^{2x}$

4. $y = (C_1 + C_2 x)\,e^x + \dfrac{8}{9}e^{5x/2}$

5. $y = e^{-x/2}\left[C_1 \cos\left(\dfrac{1}{2}x\sqrt{3}\right) + C_2 \sin\left(\dfrac{1}{2}x\sqrt{3}\right)\right] + e^{-x}$

6. $y = (C_1 + C_2 x)e^{0x} + (C_3 + C_4 x)e^{-x} + e^{-x/2}\left[(C_5 + C_6 x)\cos\left(\dfrac{1}{2}\sqrt{3}x\right) + (C_7 + C_8 x)\sin\left(\dfrac{1}{2}\sqrt{3}x\right)\right] + \dfrac{1}{36}e^x$

7. $y = e^{-px}(C_1 \cos qx + C_2 \sin qx) + \dfrac{e^{ax}}{[(p+a)^2 + q^2]}$

8. $y = (C_1 + C_2 x)\,e^{-3x/2} + 16e^{-3x}$

9. $y = C_1 e^x + C_2 e^{3x} + \dfrac{x}{2} e^{3x}$

10. $y = C_1 e^{ax} + C_2 e^{-ax} + \dfrac{x}{9} \cosh ax$

11. $y = e^{-x/2} \left[C_1 \cos \dfrac{\sqrt{3}}{2} x + C_2 \sin \dfrac{\sqrt{3}}{2} x \right] + 1 + \dfrac{1}{7} e^{2x} + \dfrac{2}{3} e^x$

12. $y = (C_1 + C_2 x + C_2 x^2) e^x + \dfrac{x^3}{6} e^x - 2$

(2) To evaluate P.I., when B is of the form $\sin ax$ or $\cos ax$:

Case (I) : If $f(D)$ contains even power of D :

Let us suppose

$$f(D^2) = A_0 (D^2)^n + A_1 (D^2)^{n-1} + \ldots + A_n.$$

Here, we observe that

$$D^2 \sin ax = -a^2 \sin ax$$

$$D^4 \sin ax = (-a^2)^2 \sin ax$$

$$D^6 \sin ax = (-a^2)^3 \sin ax$$

$$\cdots\cdots\cdots\cdots\cdots\cdots\cdots$$

$$(D^2)^n \sin ax = (-a^2)^n \sin ax$$

Consider

$$f(D^2) \sin ax = [A_0 (D^{2n}) + A_1 (D^{2n-2}) + \ldots + A_n] \sin ax$$

$$= A_0 D^{2n} \sin ax + A_1 D^{2n-2} \sin ax + \ldots + A_n \sin ax$$

$$= A_0 (-a^2)^n \sin ax + A_1 (-a^2)^{n-1} \sin ax + \ldots + A_n \sin ax$$

$$= f(-a^2) \sin ax$$

Now, operating on both sides with $\dfrac{1}{f(D^2)}$, we get

$$\frac{1}{f(D^2)} \cdot f(D^2) \sin ax = f(-a^2) \frac{1}{f(D^2)} \sin ax$$

$$\Rightarrow \quad \sin ax = f(-a^2) \left[\frac{1}{f(D^2)} \sin ax \right] \Rightarrow \frac{1}{f(D^2)} \sin ax = \frac{1}{f(-a^2)} \sin ax.$$

Case (II) : If $f(D)$ contains odd power of D :

Let us suppose, it be put in the form $f_1(D^2) + D f_2(D^2)$, then

$$\frac{1}{f(D)} \sin ax = \frac{1}{f_1(D^2) + f_2(D^2)D} \sin ax = \frac{1}{f_1(-a^2) + f_2(-a^2)D} \sin ax$$

$$= \frac{1}{p + qD} \sin ax \text{ (say)} \qquad \text{(Where } p = f_1(-a^2), q = f_2(-a^2)]$$

$$= (p - qD) \left[\frac{1}{(p - qD)(p + qD)} \sin ax \right] = (p - qD) \left[\frac{1}{p^2 - q^2 D^2} \sin ax \right]$$

$$= (p - qD) \left[\frac{1}{p^2 + q^2 a^2} \sin ax \right] \qquad \text{(By putting } D^2 = -a^2)$$

$$= \frac{(p - qD)\sin ax}{(p^2 + a^2 q^2)} = \frac{p\sin ax - qa\cos ax}{p^2 + a^2 q^2}$$

$$\Rightarrow \quad \frac{1}{f(D)}\sin ax = \frac{f_1(-a^2)\sin ax - f_2(-a^2)a\cos ax}{\{f_1(-a^2)\}^2 + a^2\{f_2(-a^2)\}^2}$$

REMARKS

➜ To find P.I. $= \dfrac{1}{f(D)}\sin ax$, replace D^2 by $-a^2$ provided $f(-a^2) \neq 0$.

➜ If the linear factors of D contains the odd powers of D, then first multiplying the numerator and denominator by the conjugate factors $(P \pm qD)$ and then putting D^2 by $(-a^2)$.

➜ Similar results are true for $\dfrac{1}{f(D)}\cos ax$.

SOLVED EXAMPLES

EXAMPLE 1. *Solve* $\dfrac{d^2 y}{dx^2} - 3\dfrac{dy}{dx} + 2y = \cos 3x$.

SOLUTION. The given differential equation can be written as

$$(D^2 - 3D + 2)y = \cos 3x \qquad \qquad ...(1)$$

To find C.F., the auxiliary equation is

$$m^2 - 3m + 2 = 0$$
$$\Rightarrow \quad (m-1)(m-2) = 0$$

which gives $m = 1$ and $m = 2$.

Therefore, C.F. $= C_1 e^x + C_2 e^{2x}$.

Now, $\quad$ P.I. $= \dfrac{1}{D^2 - 3D + 2}\cos 3x = \dfrac{1}{-9 - 3D + 2}\cos 3x \qquad [\because D^2 = -a^2 = -9]$

$$= \frac{1}{-7 - 3D}\cos 3x = -\frac{(7 - 3D)}{(7^2 - 9D^2)}\cos 3x$$

$$= -\frac{(7 - 3D)}{7^2 - 9(-9)}\cos 3x = -\frac{1}{130}[7\cos 3x - 3D\cos 3x]$$

$$= -\frac{7}{130}\cos 3x - \frac{9}{130}\sin 3x = -\frac{1}{130}(7\cos 3x + 9\sin 3x).$$

Hence, the general solution of (1) is given by

$$y = \text{C.F.} + \text{P.I.}$$
$$\Rightarrow \quad y = C_1 e^x + C_2 e^{2x} - \frac{1}{130}[7\cos 3x + 9\sin 3x].$$

EXAMPLE 2. *Solve* $\dfrac{d^2 y}{dx^2} - \dfrac{dy}{dx} - 2y = \sin 2x$.

SOLUTION. Here, the given equation can be written as

$$(D^2 - D - 2)y = \sin 2x \qquad \qquad ...(1)$$

To find the C.F. of (1), the auxiliary equation is

$$m^2 - m - 2 = 0$$

which gives, $(m+1)(m-2) = 0 \Rightarrow m = -1, 2$

$$\text{C.F.} = C_1 e^{-x} + C_2 e^{2x}.$$

Now,   P.I. $= \dfrac{1}{D^2 - D - 2} \sin 2x = \dfrac{1}{-4 - D - 2} \sin 2x$ $[\because D^2 = -a^2 = -4]$

$$= -\dfrac{1}{D + 6} \sin 2x = -\dfrac{(D-6)}{(D+6)(D-6)} \sin 2x$$

$$= -\dfrac{(D-6)}{D^2 - 36} \sin 2x = -\dfrac{(D-6)}{-4 - 36} \sin 2x = \dfrac{1}{40}[(D-6)\sin 2x]$$

$$= \dfrac{1}{40}[D \sin 2x - 6 \sin 2x]$$

$$= \dfrac{1}{40}[2\cos 2x - 6\sin 2x] = \dfrac{1}{20}\cos 2x - \dfrac{3}{20}\sin 2x$$

Hence, the complete solution is given by

$$y = \text{C.F.} + \text{P.I.}$$

$\Rightarrow$ $y = C_1 e^{-x} + C_2 e^{2x} + \dfrac{1}{20}\cos 2x - \dfrac{3}{20}\sin 2x.$

EXAMPLE 3.   *Solve* $(D^2 + 4)y = \cos^2 x$.   [MTU(B.Pharma)–2011]

SOLUTION.   Here, the given differential equation is

$$(D^2 + 4)y = \cos^2 x \qquad \qquad \qquad \qquad \text{...(1)}$$

To find the C.F. of (1), the auxiliary equation is

$$m^2 + 4 = 0 \quad \Rightarrow \quad m = \pm 2i$$

$\therefore$ $\text{C.F.} = C_1 \cos 2x + C_2 \sin 2x.$

Now,   P.I. $= \dfrac{1}{D^2 + 4}\cos^2 x = \dfrac{1}{2}\left[\dfrac{1}{D^2 + 4}(1 + \cos 2x)\right]$

$$= \dfrac{1}{2}\left[\dfrac{1}{D^2 + 4}(e^{0x}) + \dfrac{1}{D^2 + 4}(\cos 2x)\right] = \dfrac{1}{2}\left[\dfrac{1}{4} + x \cdot \dfrac{1}{2D}(\cos 2x)\right]$$

$$= \dfrac{1}{2}\left[\dfrac{1}{4} + \dfrac{x}{4}\sin 2x\right] = \dfrac{1}{8}(1 + x + \sin 2x)$$

Hence, the general solution of (1) is given by

$$y = \text{C.F.} + \text{P.I.} = C_1 \cos 2x + C_2 \sin 2x + \dfrac{1}{8}(1 + x + \sin 2x)$$

Exercise 13.10

Solve the following differential equations :

1. $(D^2 + 9)y = \cos 4x$ [UPTU(B.pharma)–2002]

2. $(D^2 - 2D + 5)y = \sin 3x$

3. $(D^2 - 3D + 2)y = \sin 3x$

4. $(D^4 + 2D^3 - 3D^2)y = 3e^{2x} + 4\sin x$

5. $(D^3 - 2D^2 + 3)y = \cos x$

6. $(D^2 + 16)y = \sin 2x$, given that $y = 0$ and $\dfrac{dy}{dx} = \dfrac{5}{6}$ when $x = 0$.

7. $(D^4 - 2D^2 + 1)y = \cos x$

8. $(D^2 + 2D + 2)y = \cos 2x$

9. $(D^2 - 9)y = \sin x + \cos x$

10. Solve $\dfrac{d^2y}{dx^2} + 2\dfrac{dy}{dx} + 10y + 37\sin 3x = 0$ and

find the value of y when $x = \dfrac{\pi}{2}$ being given

that $y = 3, \dfrac{dy}{dx} = 0$ when $x = 0$.

11. $\dfrac{d^2y}{dx^2} + 4y = e^x + \sin 2x$

[UPTU(B.Pharma)–2009, 2010]

12. $(D^2 + 5D - 6)y = \sin 3x + \cos 2x$

13. $(D^2 + 5D - 6)y = \sin 4x \sin x$

HINT TO SELECTED PROBLEMS

1. $m = \pm 3i \Rightarrow \text{C.F.} = C_1\cos 3x + C_2\sin 3x$

$\text{P.I.} = \dfrac{1}{D^2 + 9}\cos 4x$

$= \dfrac{1}{-4^2 + 9}\cos 4x = -\dfrac{1}{7}\cos 4x.$

5. $m = -1, \dfrac{3}{2} \pm \dfrac{i\sqrt{3}}{2}$

$\Rightarrow \text{C.F.} = C_1 e^{-x} + e^{3/2.x}\left[C_2\cos\left(\dfrac{\sqrt{3}}{2}x\right)\right.$

$\left. + C_3\sin\left(\dfrac{\sqrt{3}}{2}x\right)\right]$

$\text{P.I.} = \dfrac{1}{(D+1)(D^2 - 3D + 3)}\cos x.$

9. $\text{C.F.} = C_1 e^{3x} + C_2 e^{-3x}$

$\text{P.I.} = \dfrac{1}{D^2 - 9}(\sin x + \cos x)$

$= \dfrac{1}{D^2 - 9}\sin x + \dfrac{1}{D^2 - 9}\cos x$

$= \dfrac{1}{-1 - 9}\sin x + \dfrac{1}{-1 - 9}\cos x$

$= -\dfrac{1}{10}\sin x - \dfrac{1}{10}\cos x$

ANSWERS

1. $y = C_1\cos 3x + C_2\sin 3x - \dfrac{1}{7}\cos 4x$

2. $y = e^x[C_1\cos 2x + C_2\sin 2x] + \dfrac{1}{26}(3\cos 3x - 2\sin 3x)$

3. $y = C_1 e^x + C_2 e^{2x} + \dfrac{1}{130}(9\cos 3x - 7\sin 3x)$

4. $y = (C_1 + C_2 x) + C_3 e^x + C_4 e^{-3x} + \dfrac{3}{20}e^{2x} + \dfrac{4}{5}\sin x + \dfrac{2}{5}\cos x$

5. $y = C_1 e^{-x} + \left\{C_2\cos\left(\dfrac{x\sqrt{3}}{2}\right) + C_3\sin\left(\dfrac{x\sqrt{3}}{2}\right)\right\}e^{3x/2} + \dfrac{1}{26}[5\cos x - \sin x]$

6. $y = \dfrac{1}{6}\sin 4x + \dfrac{1}{12}\sin 2x$

7. $y = (C_1 + C_2 x)e^x + (C_3 + C_4 x)e^{-x} + \dfrac{1}{4}\cos x$

8. $y = e^{-x}[C_1\cos x + C_2\sin x] - \dfrac{1}{10}(\cos 2x - 2\sin 2x)$

9. $y = C_1 e^{3x} + C_2 e^{-3x} - \dfrac{1}{10}[\sin x + \cos x]$

10. $y = e^{-x}(C_1\cos 3x + C_2\sin 3x) + 6\cos 3x - \sin 3x$ and $y = 1$ at $x = \pi/2$

11. $y = C_1\cos 2x + C_2\sin 2x + \dfrac{1}{5}e^x - \dfrac{x}{4}\cos 2x$

12. $y = C_1 e^x + C_2 e^{-6x} - \dfrac{1}{30}(\cos 3x + \sin 3x) + \dfrac{1}{20}(\sin 2x - \cos 2x)$

13. $y = C_1 e^x + C_2 e^{-6x} + \dfrac{1}{2}\left[\dfrac{\sin 3x - \cos 3x}{30} + \dfrac{31\cos 5x - 25\sin 5x}{1586}\right]$

(3) To evaluate P.I., when B is of the form x^m, when m is positive integer :

i.e., to evaluate $\dfrac{1}{f(D)}x^m$, $m \in Z^+$ and $f(D) = A_0 D^n + A_1 D^{n-1} + \ldots + A_n$

Let us consider $\dfrac{1}{D - a}x^m$

i.e., $\dfrac{1}{(D-a)} x^m = e^{ax} \int e^{-ax} x^m \, dx$

$$= e^{ax} \left\{ \dfrac{e^{-ax} x^m}{a} - \dfrac{mx^{m-1} e^{-ax}}{a^2} - \dfrac{m(m-1)x^{m-2} e^{-ax}}{a^3} - \dots - \dfrac{m(m-1)\dots 2.1 \, e^{-ax}}{a^{m+1}} \right\}$$

...(1)

If we expand $\dfrac{1}{D-a}$ in powers of D, we get

$$\dfrac{1}{(D-a)} x^m = -\dfrac{1}{a(1-D/a)} x^m = -\dfrac{1}{a}\left[1 + \dfrac{D}{a} + \dfrac{D^2}{a^2} + \dots \right] x^m$$

$$\dfrac{1}{D-a} x^m = -\dfrac{1}{a}\left[x^m + \dfrac{mx^{m-1}}{a} + \dfrac{m(m-1)x^{m-2}}{a^2} + \dots + \dfrac{m(m-1)\dots 2.1}{a^m} \right] \quad ...(2)$$

Here, we observe that (1) and (2) are the same.

Working Procedure

Take the lowest degree term from $f(D)$ and remaining factor will be of the form $[1 + f(D)]$ or $[1 - f(D)]$. Now, this factor can be taken in the numerator with a negative index, which can be expanded by Binomial theorem. Here, it should be noted that the expansion is to be carried upto the term D^m, since we always have $D^{m+1} x^m = 0$, $D^{m+2} x^m = 0$ and all other higher differential coefficients of x^m are zero.

Some Important Expansions (to be used directly)

1. $[1+x]^n = 1 + nx + \dfrac{n(n-1)}{2!} x^2 + \dfrac{n(n-1)(n-2)}{3!} x^3 + \dots$
2. $(1+x)^{-1} = 1 - x + x^2 - x^3 + x^4 - x^5 + \dots$
3. $(1-x)^{-1} = 1 + x + x^2 + x^3 + x^4 + \dots$　　4. $(1-x)^{-2} = 1 + 2x + 3x^2 + 4x^3 + \dots$
5. $(1+x)^{-2} = 1 - 2x + 3x^2 - 4x^3 + \dots$

SOLVED EXAMPLES

Example 1. *Solve* $(D^2 + D - 2)y = x + \sin x$.

Solution. The given equation is $(D^2 + D - 2)y = x + \sin x$　　　...(1)

To find C.F., the auxiliary equation is

$$m^2 + m - 2 = 0$$

$\Rightarrow \quad (m-1)(m+2) = 0 \Rightarrow m = 1, -2$

$\therefore \quad$ C.F. $= C_1 e^x + C_2 e^{-2x}$

Now, P.I. $= \dfrac{1}{(D^2 + D - 2)}(x + \sin x) = \dfrac{1}{(D^2 + D - 2)} x + \dfrac{1}{(D^2 + D - 2)} \sin x$

$$= \dfrac{1}{-2\left(1 - \dfrac{1}{2}D - \dfrac{1}{2}D^2\right)} x + \dfrac{1}{-1 + D - 2} \sin x$$

$$= -\frac{1}{2}\left[1 - \left(\frac{1}{2}D + \frac{1}{2}D^2\right)\right]^{-1} x + \frac{(D+3)}{(D-3)(D+3)}\sin x$$

$$= -\frac{1}{2}\left(1 + \frac{1}{2}D + \ldots\right)x + \frac{(D+3)}{D^2 - 9}\sin x = -\frac{1}{2}\left(x + \frac{1}{2}\right) + \frac{D+3}{-1-9}\sin x$$

$$= -\frac{1}{2}\left(x + \frac{1}{2}\right) - \left(\frac{1}{10}\right)[D(\sin x) + 3\sin x]$$

$$= -\frac{1}{2}x - \frac{1}{4} - \frac{1}{10}(\cos x + 3\sin x).$$

Hence, the complete solution is given by
$$y = C.F. + P.I.$$

$$\therefore \quad y = C_1 e^x + C_2 e^{-2x} - \frac{1}{2}x - \frac{1}{4} - \frac{1}{10}(\cos x + 3\sin x)$$

EXAMPLE 2. Solve $(D^2 - 4)y = x^2$.

SOLUTION. The differential equation is
$$(D^2 - 4)y = x^2 \qquad \ldots(1)$$
To find the C.F. of (1), the auxiliary equation is
$$m^2 - 4 = 0 \implies m = \pm 2$$
$$\therefore \quad C.F. = C_1 e^{2x} + C_2 e^{-2x}$$

Now, P.I. $= \dfrac{1}{D^2 - 4}x^2 = \dfrac{1}{-4\left[1 - \frac{1}{4}D^2\right]}x^2 = -\frac{1}{4}\left[1 - \frac{1}{4}D^2\right]^{-1}x^2$

$$= -\frac{1}{4}\left[1 + \frac{1}{4}D^2 + \ldots\right]x^2 = -\frac{1}{4}\left[x^2 + \frac{1}{4}D^2(x^2)\right] = -\frac{1}{4}\left[x^2 + \frac{1}{2}\right]$$

Hence, the complete solution is given by
$$y = C.F. + P.I.$$

$$\implies \quad y = C_1 e^{2x} + C_2 e^{-2x} - \frac{1}{4}\left[x^2 + \frac{1}{2}\right].$$

EXAMPLE 3. Solve $(D^2 - 4D + 4)y = x^2 + e^x + \cos 2x$.   **(UPTU(B.Pharma)SUM-2009)**

SOLUTION. The given differential equation is
$$(D^2 - 4D + 4)y = x^2 + e^x + \cos 2x \qquad \ldots(1)$$
To find C.F., the auxiliary equation is given by
$$m^2 - 4m + 4 = 0 \implies (m-2)^2 = 0 \implies m = 2, 2$$
$$\therefore \quad C.F. = (C_1 + C_2 x)\,e^{2x}$$

Now,   P.I. $= \dfrac{1}{(D^2 - 4D + 4)}(x^2 + e^x + \cos 2x)$

$$= \frac{1}{(D-2)^2}x^2 + \frac{1}{(D-2)^2}e^x + \frac{1}{(D^2 - 4D + 4)}\cos 2x$$

$$= \frac{1}{4\left(1 - \frac{D}{2}\right)^2}x^2 + \frac{1}{(1-2)^2}e^x + \frac{1}{(-2^2 - 4D + 4)}\cos 2x$$

$$= \frac{1}{4}\left(1 - \frac{D}{2}\right)^{-2} x^2 + \frac{e^x}{1} - \frac{1}{4D} \cos 2x$$

$$= \frac{1}{4}\left[1 + D + \frac{3}{4}D^2 + ...\right] x^2 + e^x - \frac{1}{4}\int \cos 2x \, dx$$

$$= \frac{1}{4}\left(x^2 + D(x^2) + \frac{3}{4}D^2(x^2)\right) + e^x - \frac{1}{4} \cdot \frac{1}{2} \sin 2x$$

$$= \frac{1}{4}\left[x^2 + 2x + \frac{3}{2}\right] + e^x - \frac{1}{8} \sin 2x \, .$$

Hence, the complete solution is given by
$$y = \text{C.F.} + \text{P.I.}$$

$$\Rightarrow \qquad y = (C_1 + C_2 x)e^{2x} + \frac{1}{4}\left(x^2 + 2x + \frac{3}{2}\right) + e^x - \frac{1}{8}\sin 2x \, .$$

Exercise 13.11

Solve the following differential equations :

1. $(D^3 - D^2 - 6D)y = x^2 + 1$

2. $(D^4 - a^4)y = x^4$

3. $(D^3 + 2D^2 + D)y = e^{2x} + x^2 + x$

4. $(D^3 - 3D - 2)\,y = x^3$

5. $(D^3 - 3D^2 + 2D)\,y = 4 + 60e^{5x}$

6. $(D^3 + 1)\,y = \sin 3x - \cos^2 \dfrac{x}{2}$

7. $(D^2 - 2D + 3)y = \cos x + x^2$

8. $(D^2 - 5D + 6)y = x + e^{mx}$

9. $(D^2 + 16)y = \cos 3x + e^{3x} + x^4$.

10. $(D^2 + 4)y = \sin 3x + x^2$

11. $\dfrac{d^2y}{dx^2} - \dfrac{dy}{dx} + 4y = x^2 + e^x$

(UPTU(B.Pharma)SUM–2010)

12. If $\dfrac{d^2x}{dt^2} + \dfrac{g}{b}(x - a) = 0; a, b$ and g are positive numbers and $x = a', \dfrac{dx}{dt} = 0$ when t = 0, show that $x = a + (a' - a)\cos\sqrt{\dfrac{g}{b}}t$

HINT TO SELECTED PROBLEMS

1. $m = 0, -2, 3 \Rightarrow$ C.F $= C_1 e^{0x} + C_2 e^{-2x} + C_3 e^{3x}$

$$\text{P.I.} = \frac{1}{D^3 - D^2 - 6D}(1 + x^2)$$

$$= -\frac{1}{6D}\left[1 + \frac{D}{6} - \frac{D^2}{6}\right]^{-1}(1 + x^2)$$

$$= -\frac{1}{6D}\left[1 - \frac{1}{6}(-D + D^2)\right]^{-1}(1 + x^2)$$

Now expand by binomial theorem and D for differentiation and $1/D$ for integration.

7. $m = 1 \pm i\sqrt{2}$

$$\Rightarrow \text{C.F.} = e^x[C_1 \cos\sqrt{2}x + C_2 \sin\sqrt{2}x]$$

$$\text{P.I.} = \frac{1}{D^2 - 2D + 3}(\cos x + x^2)$$

$$= \frac{1}{D^2 - 2D + 3}\cos x + \frac{1}{D^2 - 2D + 3}x^2$$

$$= \frac{1}{-1 - 2D + 3}\cos x + \frac{1}{3}\left[1 - \left(\frac{2D}{3} - \frac{D^2}{3}\right)\right]^{-1}x^2.$$

10. $m = \pm 2i \Rightarrow$ C.F $= C_1 \cos 2x + C_2 \sin 2x$

$$\text{P.I.} = \frac{1}{D^2 + 4}(\sin 3x + x^2)$$

$$= \frac{1}{D^2 + 4}\sin 3x + \frac{1}{D^2 + 4} \cdot x^2$$

$$= \frac{1}{-9 + 4}\sin 3x + \frac{1}{4\left(1 + \dfrac{D^2}{4}\right)} \cdot x^2$$

$$= -\frac{1}{5}\sin 3x + \frac{1}{4}\left(1 + \frac{D^2}{4}\right)^{-1} \cdot x^2$$

Now expand by Binomial expansion.

ANSWERS

1. $y = C_1 + C_2 e^{3x} + C_3 e^{-2x} - \dfrac{25}{108}x - \dfrac{1}{18}x^3 + \dfrac{1}{36}x^2$

2. $y = C_1 e^{ax} + C_2 e^{-ax} + C_3 \cos ax + C_4 \sin ax - \dfrac{x^4}{a^4} - \dfrac{24}{a^8}$

3. $y = C_1 + (C_2 + C_3 x)e^{-x} + \dfrac{1}{18}e^{2x} + \dfrac{1}{3}x^3 - \dfrac{3}{2}x^2 + 4x$

4. $y = (C_1 + C_2 x)e^{-x} + C_3 e^{2x} - \dfrac{1}{2}x^3 + \dfrac{9}{4}x^2 \dfrac{27}{4}x + 15$ 5. $y = C_1 + C_2 e^x + C_3 e^{2x} + 2x + e^{5x}$

6. $y = C_1 e^{-x} + e^{x/2}\left\{ C_2 \cos\dfrac{x\sqrt{3}}{2} + C_3 \sin\dfrac{x\sqrt{3}}{2} \right\} + \dfrac{1}{730}[\sin 3x + 27\cos 3x] - \dfrac{1}{2} - \dfrac{1}{4}(\cos x - \sin x)$

7. $y = e^x[C_1 \cos(x\sqrt{2}) + C_2 \sin(x\sqrt{2})] + \dfrac{1}{4}(\cos x - \sin x) + \dfrac{x^2}{3} + \dfrac{4}{9}x + \dfrac{2}{27}$

8. $y = C_1 e^{2x} + C_2 e^{3x} + \dfrac{1}{6}\left[x + \dfrac{5}{6}\right] + [e^{mx}/(m^2 - 5m + 6)]$

9. $y = C_1 \cos 4x + C_2 \sin 4x + \dfrac{1}{7}\cos 3x + \dfrac{1}{25}e^{3x} + \dfrac{1}{16}x^4 - \dfrac{3}{64}x^2 + \dfrac{3}{512}$

10. $y = C_1 \cos 2x + C_2 \sin 2x - \dfrac{1}{5}\sin 3x + \dfrac{1}{4}x^2 - \dfrac{1}{8}$

11. $y = e^{x/2}\left(C_1 \cos\dfrac{\sqrt{15}}{2}x + C_2 \sin\dfrac{\sqrt{15}}{2}x \right) + \dfrac{1}{4}\left(e^x + x^2 + \dfrac{x}{2} - \dfrac{3}{8} \right)$ 12. $x = (a' - a)\cos\sqrt{\dfrac{g}{b}}t + a$

(4) To evaluate $\dfrac{1}{f(D)}e^{ax}.X$, where X is any function of x :

Let us consider any function X_1 of x. Then, by simple differentiation, we get

$$D(e^{ax}.X_1) = e^{ax}D(X_1) + X_1 a e^{ax} = e^{ax}(D+a)X_1. \qquad \ldots(1)$$

Now, let us assume

$$D^n[e^{ax}.X_1] = e^{ax}(D+a)^n.X_1 \qquad \ldots(2)$$

Then, consider $D^{n+1}[e^{ax}.X_1] = D[D^n(e^{ax}.X_1)] = D[e^{ax}(D+a)^n.X_1]$

$$= ae^{ax}(D+a)^n.X_1 + e^{ax}.D(D+a)^n.X_1$$

$$= e^{ax}(D+a)^{n+1}.X_1$$

Therefore, by the method of induction, we have $D^n[e^{ax}.X_1] = e^{ax}(D+a)^n X_1$, for all positive integer n

$\therefore \qquad f(D)e^{ax}.X_1 = e^{ax}f(D+a)X_1. \qquad \ldots(3)$

Now, operating on equation (3) with $\dfrac{1}{f(D)}$, we get

$$\dfrac{1}{f(D)}.f(D)e^{ax}.X_1 = \dfrac{1}{f(D)}e^{ax}f(D+a).X_1$$

$$\Rightarrow \qquad e^{ax}.X_1 = \frac{1}{f(D)} e^{ax} f(D+a).X_1. \qquad \qquad ...(4)$$

Let
$$X = f(D+a).X_1 \Rightarrow X_1 = \frac{X}{f(D+a)}.$$

Now, (4) becomes

$$e^{ax}.\frac{X}{f(D+a)} = \frac{1}{f(D)} e^{ax} . \frac{X}{f(D+a)}.f(D+a)$$

$$\Rightarrow \qquad \frac{1}{f(D)}[e^{ax}.X] = e^{ax}\left[\frac{1}{f(D+a)}.X\right]$$

✎ REMARKS

⟱ Here, we observe that if e^{ax} is brought to the left from the right of $\frac{1}{f(D)}$, then D should be replaced by $(D+a)$.

⟱ This method will be used if X is $\cos ax$, $\sin ax$ or x^m or a polynomial of degree m.

⟱ This method is also capable to find $\left\{\frac{1}{f(D)} e^{ax}\right\}$, when $f(a) = 0$.

WORKING PROCEDURE

Replace D by $(D+a)$ and brought e^{ax} before the operator $\frac{1}{f(D)}$. After that, determine $\frac{1}{f(D+a)}.X$ as usual.

SOLVED EXAMPLES

EXAMPLE 1. *Solve* $(D^2 + 4D - 12)y = (x-1)e^{2x}$.

SOLUTION. The given differential equation is

$$(D^2 + 4D - 12)y = (x-1)e^{2x} \qquad \qquad ...(1)$$

To find C.F. of (1), the auxiliary equation is

$$m^2 + 4m - 12 = 0 \Rightarrow (m-2)(m+6) = 0$$

which gives $m = 2$ and $m = -6$.

$$\therefore \quad \text{C.F.} = C_1 e^{2x} + C_2 e^{-6x}$$

Now, P.I. $= \dfrac{1}{(D^2 + 4D - 12)} e^{2x}(x-1) = e^{2x}\dfrac{1}{[(D+2)^2 + 4(D+2) - 12]}(x-1)$

$$= e^{2x}\frac{1}{(D^2 + 8D)}(x-1) = e^{2x}\frac{1}{8D\left(1+\dfrac{D}{8}\right)}(x-1)$$

$$= \frac{1}{8}e^{2x}\frac{1}{D}\left(1+\frac{1}{8}D\right)^{-1}(x-1) = \frac{1}{8}e^{2x}\frac{1}{D}\left(1-\frac{1}{8}D+...\right)(x-1)$$

$$= \frac{1}{8}e^{2x}\frac{1}{D}\left(x-1-\frac{1}{8}\right) = \frac{1}{8}e^{2x}\frac{1}{D}\left(x-\frac{9}{8}\right) = \frac{1}{8}e^{2x}\int\left(x-\frac{9}{8}\right)dx$$

$$= \frac{1}{8}e^{2x}\left(\frac{x^2}{2} - \frac{9}{8}x\right).$$

Hence, the general solution of (1), is given by

$$y = \text{C.F.} + \text{P.I.}$$

$$\Rightarrow \qquad y = C_1 e^{2x} + C_2 e^{-6x} + \frac{1}{8}e^{2x}\left[\frac{x^2}{2} - \frac{9}{8}x\right].$$

EXAMPLE 2.   Solve $(D^2 - 2D + 4)y = e^x \cos x$.

SOLUTION.   The differential equation is

$$(D^2 - 2D + 4)y = e^x \cos x \qquad\qquad\qquad\qquad\qquad\qquad ...(1)$$

To find the C.F. of (1), the auxiliary equation is

$$m^2 - 2m + 4 = 0 \quad \Rightarrow \quad m = 1 \pm i\sqrt{3}.$$

Therefore, C.F. $= e^x(C_1 \cos\sqrt{3}.x + C_2 \sin\sqrt{3}.x)$

Now, $\qquad$ P.I. $= \dfrac{1}{(D^2 - 2D + 4)}e^x \cos x = e^x \dfrac{1}{[(D+1)^2 - 2(D+1) + 4]}\cos x$

$$= e^x \frac{1}{(D^2 + 3)}\cos x = e^x \frac{1}{-1^2 + 3}\cos x = \frac{1}{2}e^x \cos x.$$

Hence, the complete solution of (1) is given by

$$y = \text{C.F.} + \text{P.I.}$$

$$\Rightarrow \qquad y = e^x [C_1 \cos\sqrt{3}.x + C_2 \sin\sqrt{3}.x] + \frac{1}{2}e^x \cos x.$$

EXAMPLE 3.   Solve $(D^2 - 5D + 6)y = e^{2x} \sin 2x$.

SOLUTION.   The given differential equation is

$$(D^2 - 5D + 6)y = e^{2x} \sin 2x \qquad\qquad\qquad\qquad\qquad\qquad ...(1)$$

To find the C.F. of (1), the auxiliary equation is given by

$$m^2 - 5m + 6 = 0 \quad \Rightarrow \quad (m-2)(m-3) = 0$$
which gives, $m = 2$ and $m = 3$.

$$\therefore \qquad\qquad \text{C.F.} = C_1 e^{2x} + C_2 e^{3x}$$

Now, $\qquad$ P.I. $= \dfrac{1}{D^2 - 5D + 6}e^{2x} \sin 2x = e^{2x} \dfrac{1}{[(D+2)^2 - 5(D+2) + 6]}\sin x$

$$= e^{2x} \frac{1}{D^2 - D}\sin 2x = e^{2x} \frac{1}{-2^2 - D}\sin 2x$$

$$= e^{2x} \frac{1}{-4 - D}\sin 2x = -e^{2x} \frac{1}{(4 + D)}\sin 2x$$

$$= -e^{2x} \frac{(D-4)}{(D+4)(D-4)}\sin 2x$$

$$= -e^{-2x}\left[\frac{D-4}{D^2 - 16}\right]\sin 2x = -e^{2x}\left[\frac{D-4}{-4-16}\right]\sin 2x$$

$$= \frac{e^{2x}}{20}(D-4)\sin 2x = \frac{e^{2x}}{20}[D\sin 2x - 4\sin 2x]$$

$$= \frac{e^{2x}}{20}[2\cos 2x - 4\sin 2x]$$

Hence, the complete solution of (1) is given by

$$y = \text{C.F.} + \text{P.I.}$$

$$\Rightarrow \qquad y = C_1 e^{2x} + C_2 e^{3x} + \frac{e^{2x}}{20}[2\cos 2x - 4\sin 2x]$$

Exercise 13.12

Solve the following differential equations :

1. $(D^2 - 2D + 1)y = e^x \cdot x^2$ [UPTU(B.Pharma)–2004]

2. $(D^2 - 5D + 6)y = x^3 \cdot e^{2x}$

3. $(D^2 - 1)y = e^x(1 + x^2)$ [UPTU(B.Pharma)–2001]

4. $(D^2 - 4D + 1)y = e^{2x}\sin x$

5. $(D^2 - 2D + 1)y = x^2 e^{3x}$

6. $(D^2 - 1)y = e^x \cos x$

7. $(D^2 - 2D + 5)y = e^{2x}\sin x$

8. $(D^2 - 2D + 6)y = e^x \cos x$

9. $(D^2 - 1)y = \cosh x \cos x + a^x$

10. $(D^2 - 4D - 5)y = xe^{-x}$ given that $y = 0$ and $\frac{dy}{dx} = 0$ at $x = 0$.

11. $(D^2 - 4D + 4)y = e^x \cos x$ **(GBTU(CO)–2010)**

12. $\frac{d^2 y}{dx^2} - 2\frac{dy}{dx} + 4y = e^{2x}\cos x$

13. $(D^2 - 3D + 2)y = xe^x + \sin 2x$

14. $(D^2 - 1)y = xe^x + \cos^2 x$

15. $(D^2 - 1)y = x \sin x + x^2 e^x$

16. $(D^2 - 2D + 1)y = x \sin x$

17. $\frac{d^2 y}{dx^2} + 2\frac{dy}{dx} + y = x^2 e^{-x}\cos x$

HINT TO SELECTED PROBLEMS

1. $m = 1, 1, \therefore \text{C.F.} = (C_1 + C_2 x)e^x$

$$\text{P.I.} = \frac{1}{D^2 - 2D + 1}e^x \cdot x^2 = \left[\frac{1}{(D-1)^2}e^x \cdot x^2\right]$$

$$= e^x \left[\frac{1}{[(D+1)-1]^2} \cdot x^2\right]$$

$$= e^x \cdot \frac{1}{D^2} \cdot x^2 = \frac{e^x \cdot x^4}{12}$$

3. $m = \pm 1, \therefore \text{C.F.} = C_1 e^x + C_2 e^{-x}$

$$\text{P.I.} = \frac{1}{D^2 - 1}e^x(1 + x^2)$$

$$= e^x \frac{1}{(D+1)^2 - 1}(1 + x^2)$$

$$= e^x \left[\frac{1}{D^2 + 2D + 1 - 1}\right] \cdot (1 + x^2)$$

$$= e^x \cdot \frac{1}{D^2 + 2D}(1 + x^2) = \frac{e^x}{2D}\left[1 + \frac{D}{2}\right]^{-1}[1 + x^2]$$

Expand by binomial expansion.

7. $m = 1 \pm 2i, \therefore \text{C.F.} = e^x(C_1 \cos 2x + C_2 \sin 2x)$

$$\text{P.I.} = \frac{1}{D^2 - 2D + 5}e^{2x}\sin x$$

$$= e^{2x}\frac{1}{(D+2)^2 - 2(D+2) + 5} \cdot \sin x$$

$$= e^{2x} \cdot \frac{1}{D^2 + 2D + 5}\sin x.$$

Now,

9. $\text{C.F.} = C_1 e^x + C_2 e^{-x}$

$$\text{P.I.} = \frac{1}{D^2 - 1}\cosh x \cos x + \frac{1}{D^2 - 1}a^x$$

$$= \frac{1}{D^2 - 1}\left(\frac{e^x + e^{-x}}{2}\right)\cos x + \frac{1}{(D^2 - 1)}e^{\log a^x}$$

$$= \frac{1}{2}e^x \left\{\frac{1}{(D+1)^2 - 1}\cos x + \frac{1}{2}e^{-x}\right.$$

$$\left.\frac{1}{(D-1)^2 - 1}\cos x + \frac{1}{(\log a)^2 - 1}e^{x \log a}\right\}.$$

ANSWERS

1. $y = (C_1 + C_2 x)e^x + \dfrac{1}{12}e^x . x^4$

2. $y = C_1 e^{2x} + C_2 e^{3x} - e^{2x}\left[\dfrac{x^4}{4} + x^3 + 3x^2 + 6x\right]$

3. $y = C_1 e^x + C_2 e^{-x} + \dfrac{1}{12}e^x[9x + 2x^3 - 3x^2]$

4. $y = C_1 e^{(2+\sqrt{3})x} + C_2 e^{(2-\sqrt{3})x} - \dfrac{1}{4}e^{2x}\sin x$

5. $y = (C_1 + C_2 x)e^x + \dfrac{1}{8}e^{3x}(2x^2 - 4x + 3)$

6. $y = C_1 e^x + C_2 e^{-x} - \dfrac{1}{5}e^x(\cos x - 2\sin x)$

7. $y = e^x[C_1 \cos 2x + C_2 \sin 2x] - \dfrac{1}{10}e^{2x}(\cos x - 2\sin x)$

8. $y = e^x[C_1 \cos \sqrt{5}.x + C_2 \sin \sqrt{5}.x] + \dfrac{1}{4}e^x \cos x$

9. $y = C_1 e^x + C_2 e^{-x} + \dfrac{1}{10}e^x[2\sin x - \cos x] - \dfrac{1}{10}e^{-x}(2\sin x + \cos x) + \dfrac{a^x}{(\log a)^2 - 1}$

10. $y = -\dfrac{1}{216}e^{-x} + \dfrac{1}{216}e^{5x} - \dfrac{1}{36}xe^{-x} - \dfrac{1}{12}x^2 e^{-x}$

11. $y = (C_1 + C_2 x)e^{2x} - \dfrac{e^x}{2}\sin x$

12. $y = e^x(C_1 \cos \sqrt{3}x + C_2 \sin \sqrt{3}x) + \dfrac{1}{13}e^{2x}(2\sin x + 3\cos x)$

13. $y = C_1 e^x + C_2 e^{2x} - e^x\left(\dfrac{x^2}{2} + x\right) + \dfrac{1}{20}(3\cos 2x - \sin 2x)$

14. $y = C_1 e^x + C_2 e^{-x} + \dfrac{1}{4}e^x(x^2 - x) - \dfrac{1}{2} - \dfrac{1}{10}\cos 2x$

15. $y = C_1 e^x + C_2 e^{-x} - \dfrac{1}{2}(x\sin x + \cos x) + \dfrac{xe^x}{12}(2x^2 - 3x + 3)$

16. $y = (C_1 + C_2 x)e^x + \dfrac{1}{2}[(x+1)\cos x - \sin x]$

17. $y = (C_1 + C_2 x)e^{-x} + e^{-x}(-x^2 \cos x + 4x \sin x + 6\cos x)$

(5) To evaluate $\dfrac{1}{f(D)}e^{ax}$, **when** $f(a) = 0$:

Let us suppose $f(a) = 0$. In this case $(D - a)$ is at least one factor of $f(D)$.

Let $f(D) = (D - a)^r \, g(D)$, where $g(a) \neq 0$.

Then, $\dfrac{1}{f(D)}e^{ax} = \dfrac{1}{(D-a)^r} \cdot \dfrac{1}{g(a)}e^{ax} = \dfrac{1}{g(a)} \cdot \dfrac{1}{(D-a)^r}e^{ax}$

$= \dfrac{1}{g(a)} \cdot \dfrac{1}{(D-a)^{r-1}}e^{ax}\int e^{ax}.e^{-ax}dx$

$= \dfrac{1}{g(a)} \cdot \dfrac{1}{(D-a)^{r-1}}xe^{ax} = \dfrac{1}{g(a)} \cdot \dfrac{1}{(D-a)^{r-2}}e^{ax}\int xe^{ax} . e^{-ax}dx$

$= \dfrac{1}{g(a)} \cdot \dfrac{1}{(D-a)^{r-2}} \cdot \dfrac{x^2}{2!}e^{ax}.$

Proceeding in the same way, finally, we get $\dfrac{1}{f(D)}e^{ax} = \dfrac{1}{g(a)} \cdot \dfrac{x^r}{r!}e^{ax}.$

📖 REMARKS

➡ Substitute $D = a$ in those factors of $f(D)$ which do not vanish for $D = a$ and then make the question as P.I. of a product of e^{ax} and 1, which is calculated by previous section and reduce to the calculation of $\frac{1}{D} \cdot 1$ or $\frac{1}{D^2} \cdot 1$ or $\frac{1}{D^3} \cdot 1$ and so on.

➡ Here, $\frac{1}{D^n}$ implies n times integral of 1, with respect to x.

SOLVED EXAMPLES

EXAMPLE 1. Solve $(D^2 + D - 6)y = e^{2x}$.

[UPTU(B.Pharma)–2002]

SOLUTION. The given equation is

$$(D^2 + D - 6)y = e^{2x} \qquad \ldots(1)$$

To find C.F. of (1), the auxiliary equation is

$$m^2 + m - 6 = 0$$

$$\Rightarrow \quad (m + 3)(m - 2) = 0 \Rightarrow m = 2, -3$$

$$\therefore \qquad \text{C.F.} = C_1 e^{2x} + C_2 e^{-3x}$$

Now, $\qquad \text{P.I.} = \dfrac{1}{D^2 + D - 6} e^{2x} = \dfrac{1}{(D + 3)(D - 2)} e^{2x}$

$$= \dfrac{1}{(2 + 3)(D - 2)} e^{2x} = \dfrac{1}{5(D - 2)} e^{2x} \cdot 1$$

$$= \dfrac{1}{5} e^{2x} \dfrac{1}{(D + 2) - 2} \cdot 1 = \dfrac{1}{5} e^{2x} \dfrac{1}{D} \cdot 1 = \dfrac{1}{5} x e^{2x}.$$

Hence, the complete solution of (1) is given by

$$y = \text{C.F.} + \text{P.I.}$$

$$\Rightarrow \qquad y = C_1 e^{2x} + C_2 e^{-3x} + \dfrac{1}{5} x e^{2x}.$$

EXAMPLE 2. Solve $\dfrac{d^2 y}{dx^2} - 3\dfrac{dy}{dx} + 2y = e^x$.

SOLUTION. The given differential equation can be written as

$$(D^2 - 3D + 2)y = e^x \qquad \ldots(1)$$

To find the C.F. of (1), the auxiliary equation is

$$m^2 - 3m + 2 = 0$$

$$\Rightarrow \quad (m - 1)(m - 2) = 0 \Rightarrow m = 1, 2$$

$$\therefore \qquad \text{C.F.} = C_1 e^x + C_2 e^{2x}$$

Now, $\qquad \text{P.I.} = \dfrac{1}{(D^2 - 3D + 2)} e^x = \dfrac{1}{(D - 2)(D - 1)} e^x = \dfrac{1}{(1 - 2)(D - 1)} e^x$

(By putting 1 for D in $(D - 2)$, because at $D = 1$ $(D - 2) \neq 0$)

$$= -\dfrac{1}{D - 1} e^x = -\dfrac{1}{D - 1} e^x \cdot 1 = -e^x \dfrac{1}{(D + 1) - 1} \cdot 1 = -e^x \cdot \dfrac{1}{D} \cdot 1 = -e^x \cdot x$$

Hence, the complete solution of (1) is given by

$$y = \text{C.F.} + \text{P.I.} \implies y = C_1 e^x + C_2 e^{2x} - x e^x.$$

EXAMPLE 3. Solve $(D^3 + 3D^2 + 3D + 1)\, y = e^{-x}$.

SOLUTION. The given differential equation is

$$(D^3 + 3D^2 + 3D + 1)\, y = e^{-x} \qquad \ldots(1)$$

To find the C.F. of (1), the auxiliary equation is given by $(m+1)^3 = 0$

$$\implies m = -1, -1, -1$$

$$\therefore \qquad \text{C.F.} = (C_1 + C_2 x + C_3 x^2)\, e^{-x}$$

Now, $\quad \text{P.I.} = \dfrac{1}{(D+1)^3} e^{-x} = e^{-x} \dfrac{1}{(D-1+1)^3}.1 = e^{-x} \cdot \dfrac{1}{D^3}.1 = e^{-x} \cdot \dfrac{x^3}{3!}.$

Hence, the complete solution of (1) is given by

$$y = \text{C.F.} + \text{P.I.}$$

$$\implies \qquad y = (C_1 + C_2 x + C_3 x^2)e^{-x} + e^{-x} \cdot \dfrac{x^3}{3!}.$$

Exercise 13.13

Solve the following differential equations :

1. $(D^2 + 4D + 3)\, y = e^{-3x}$

2. $(D^2 + 6D + 9)y = 2e^{-3x}$

3. $(D^4 + D^3 + D^2 - D - 2)y = e^x$

4. $(D^2 - 9D + 18)y = \cosh 3x$

5. $(D-1)^2 (D^2 + 1)^2 y = e^x$

6. $(D^2 - 3D + 2)y = e^x$ when $y = 3$, $\dfrac{dy}{dx} = 3$ at $x = 0$

7. $(D-1)^3 (D+1)y = e^x + e^{-x}$

8. $(D^2 - 6D + 9)y = 4e^{3x}$

9. $(D^2 - 1)y = \cosh x$

10. $(D^2 - 4D + 4)y = 8(x^2 + e^{2x} + \sin 2x)$

HINT TO SELECTED PROBLEMS

1. $\text{C.F.} = C_1 e^{-x} + C_2 e^{-3x}$

$$\text{P.I.} = \dfrac{1}{D^2 + 4D + 3} e^{-3x} = \dfrac{1}{(D+1)(D+3)} e^{-3x}$$

$$= \dfrac{1}{(-3+1)(D+3)} e^{-3x} = -\dfrac{1}{2(D+3)} e^{-3x}.1$$

$$= -\dfrac{1}{2} e^{-3x} \dfrac{1}{[(D-3)+3]}.1$$

$$= -\dfrac{1}{2} e^{-3x} \cdot \dfrac{1}{D}.1 = -\dfrac{1}{2} e^{-3x} .x$$

4. $\text{C.F.} = C_1 e^{3x} + C_2 e^{6x}$

$$\text{P.I.} = \dfrac{1}{D^2 - 9D + 18} \cosh 3x$$

$$= \dfrac{1}{D^2 - 9D + 18} \left(\dfrac{e^{3x} + e^{-3x}}{2} \right)$$

$$= \dfrac{1}{2(D-3)(D-6)} (e^{3x} + e^{-3x})$$

7. $\text{C.F.} = (C_1 + C_2 x + C_3 x^2)\, e^x + C_1 e^{-x}$

$$\text{P.I.} = \dfrac{1}{(D-1)^3 (D+1)} (e^x + e^{-x})$$

$$= \dfrac{1}{2} \dfrac{1}{(D-1)^3} e^x.1 - \dfrac{1}{8} \dfrac{1}{(D+1)} e^{-x}.1.$$

10. $\text{P.I.} = \dfrac{1}{(D^2 - 4D + 4)} (8x^2 + 8e^{2x} + 8\sin 2x)$

$$= \dfrac{1}{(D-2)^2} 8x^2 + \dfrac{1}{(D-2)^2} 8e^{2x}$$

$$+ \dfrac{1}{(D-2)^2}.8 \sin 2x.$$

ANSWERS

1. $y = C_1 e^{-x} + C_2 e^{-3x} - \dfrac{x}{2} e^{-3x}$ **2.** $y = (C_1 + C_2 x)\, e^{-3x} + x^2 e^{-3x}$

3. $y = C_1 e^x + C_2 e^{-x} + e^{-x/2}\left[C_3 \cos\left(\dfrac{\sqrt{7}}{2} x\right) + C_4 \sin\left(\dfrac{\sqrt{7}}{2} x\right)\right] + \dfrac{1}{8} x e^x$

4. $y = C_1 e^{3x} + C_2 e^{6x} - \dfrac{1}{6} x e^{3x} + \dfrac{1}{108} e^{-3x}$

5. $y = (C_1 + C_2 x) e^x + (C_3 + C_4 x)\cos x + (C_5 + C_6 x)\sin x + \dfrac{1}{8} x^2 e^x$

6. $y = 2e^x + e^{2x} - x e^x$ **7.** $y = (C_1 + C_2 x + C_3 x^2) e^x + C_4 e^{-x} + \dfrac{1}{12} x^3 e^x - \dfrac{x}{8} e^{-x}$

8. $y = (C_1 + C_2 x)\, e^{3x} + 2x^2 e^{3x}$ **9.** $y = C_1 e^x + C_2 e^{-x} + \dfrac{1}{2} x \sinh x$

10. $(C_1 + C_2 x)\, e^{2x} + 2x^2 + 3 + 4x + 4x^2 e^{2x} + \cos 2x$

(6) To evaluate $\dfrac{1}{f(D^2)}\sin ax$ or $\cos ax$, when $f(-a^2)=0$:

To find the particular integral of such cases, we shall calculate P.I. for e^{iax} instead of $\sin ax$ or $\cos ax$.

Here, we have $e^{iax} = \cos ax + i\sin ax$.

Thus, P.I. for e^{iax} = P.I. for $(\cos ax + i\sin ax)$

$\Rightarrow$ P.I. for $\cos ax$ = Real part of P.I. for e^{iax} and P.I. for $\sin ax$ = imaginary part of P.I. for e^{iax}.

Therefore, $\dfrac{\cos ax}{D^2 + a^2}$ and $\dfrac{\sin ax}{D^2 + a^2}$ are respectively, real and imaginary part of $\dfrac{e^{iax}}{D^2 + a^2}$

$$= \dfrac{e^{iax}}{(D + ai)(D - ai)} = \dfrac{e^{iax}}{(ai + ai)(D - ai)}$$

(By putting ai in $(D + ai)$ because at $D = ai$ it does not vanish.)

$$= \dfrac{e^{iax}}{2ai}\left[\dfrac{1}{D + ai - ai} \cdot 1\right] = \dfrac{e^{iax}}{2ai} \cdot \dfrac{1}{D} \cdot 1 = \dfrac{x}{2ai}(e^{aix})$$

$$= -\dfrac{ix(\cos ax + i\sin ax)}{2a} = -\dfrac{ix}{2a}\cos ax + \dfrac{x}{2a}\sin ax$$

$\Rightarrow$ $\dfrac{1}{D^2 + a^2}\sin ax = -\dfrac{x}{2a}\cos ax = \dfrac{x}{2}\int \sin ax\, dx$

and $\dfrac{1}{D^2 + a^2}\cos ax = \dfrac{x}{2a}\sin ax = \dfrac{x}{2}\int \cos ax\, dx$.

SOLVED EXAMPLES

EXAMPLE 1.   Solve $(D^2 + a^2)y = \sin ax$.

SOLUTION.   The given equation is

$$(D^2 + a^2)y = \sin ax$$

...(1)

To find the C.F. of (1), the auxiliary equation is
$$m^2 + a^2 = 0 \implies m = 0 \pm ai$$

$\therefore$ $\quad$ C.F. $= e^{0x}[C_1 \cos ax + C_2 \sin ax] = [C_1 \cos ax + C_2 \sin ax]$

Now, $\quad$ P.I. $= \dfrac{1}{D^2 + a^2} \sin ax$

$= $ Imaginary part of $\left[\dfrac{1}{D^2 + a^2}(\cos ax + i \sin ax) \right]$

$= $ Imaginary part of $\left[\dfrac{1}{D^2 + a^2} e^{iax} \right]$

$= $ Imaginary part of $\left[\dfrac{1}{(D+ai)(D-ai)} e^{iax} \right]$

$= $ Imaginary part of $\left[\dfrac{1}{(ai+ai)(D-ai)} e^{iax} \right]$

$= $ Imaginary part of $\left[\dfrac{1}{2ai} \cdot \dfrac{1}{(D-ai)} e^{iax} \right]$

$= $ Imaginary part of $\dfrac{1}{2ai} \dfrac{1}{(D-ai)} e^{iax} . 1$

$= $ Imaginary part of $\dfrac{1}{2ai} e^{iax} \dfrac{1}{[(D+ia) - ia]} . 1$

$= $ Imaginary part of $\dfrac{1}{2ai} e^{iax} \dfrac{1}{D} . 1$

$= $ Imaginary part of $\dfrac{1}{2ai} e^{iax} . x$

$= $ Imaginary part of $\dfrac{1}{2ai} . x (\cos ax + i \sin ax)$

$= $ Imaginary part of $\dfrac{1}{2a} x \left[\dfrac{1}{i} \cos ax + \sin ax \right]$

$= $ Imaginary part of $\dfrac{1}{2a} x \left[\dfrac{i}{i^2} \cos ax + \sin ax \right]$

$= $ Imaginary part of $\dfrac{1}{2a} x[-i \cos ax + \sin ax] = -\dfrac{x}{2a} \cos ax$.

Hence, the complete solution of (1) is given by
$$y = \text{C.F.} + \text{P.I.}$$

$\implies \qquad y = C_1 \cos ax + C_2 \sin ax - \dfrac{x}{2a} \cos ax$.

EXAMPLE 2. $\quad$ *Solve* $(D^2 + 1)y = \sin x \sin 2x$.

SOLUTION. $\quad$ The given equation is
$$(D^2 + 1)y = \sin x \sin 2x \qquad \qquad \qquad ...(1)$$

To find the C.F. of (1), the auxiliary equation is
$$(m^2 + 1) = 0 \implies m = \pm i$$

$$\therefore \qquad \text{C.F.} = C_1 \cos x + C_2 \sin x$$

Now, $\qquad$ P.I. $= \dfrac{1}{(D^2+1)}(\sin x \sin 2x)$

$$= \dfrac{1}{(D^2+1)} \cdot \dfrac{1}{2}[2\sin x \sin 2x] = \dfrac{1}{2} \cdot \dfrac{1}{D^2+1}[\cos x - \cos 3x]$$

$$= \dfrac{1}{2}\left[\dfrac{1}{D^2+1}\cos x - \dfrac{1}{D^2+1}\cos 3x\right].$$

Now, $\dfrac{1}{D^2+1}\cos 3x = \dfrac{1}{-3^2+1}\cos 3x = -\dfrac{1}{8}\cos 3x$

Again, $\dfrac{1}{D^2+1}\cos x = \text{Real part of}\left[\dfrac{1}{D^2+1}e^{ix}\right]$

$$= \text{Real part of}\left[\dfrac{1}{(D+i)(D-i)}e^{ix}\right]$$

$$= \text{Real part of}\left[\dfrac{1}{(i+i)(D-i)}e^{ix}\right]$$

$$= \text{Real part of}\left[\dfrac{1}{2i}\cdot\dfrac{1}{(D-i)}e^{ix}\right] = \text{Real part of}\left[\dfrac{1}{2i}\dfrac{1}{D-i}e^{ix}\cdot 1\right]$$

$$= \text{Real part of}\left[\dfrac{1}{2i}e^{ix}\cdot\dfrac{1}{(D+i-i)}\cdot 1\right]$$

$$= \text{Real part of}\left[\dfrac{1}{2i}e^{ix}\dfrac{1}{D}\cdot 1\right] = \text{Real part of}\left[\dfrac{1}{2i}\cdot e^{ix}\cdot x\right]$$

$$= \text{Real part of}\left[\dfrac{x}{2i}(\cos x + i\sin x)\right]$$

$$= \text{Real part of}\left[-i.\dfrac{1}{2}x\cos x + \dfrac{1}{2}x\sin x\right] = \dfrac{1}{2}x\sin x$$

Hence, the complete solution of (1) is given by

$$y = \text{C.F.} + \text{P.I.}$$

$$\Rightarrow \qquad y = C_1\cos x + C_2\sin x + \dfrac{x}{4}\sin x - \dfrac{1}{16}\cos 3x.$$

EXAMPLE 3. *Find the solution of the equation*

$$\dfrac{d^2y}{dx^2} + 4y = 8\cos 2x \text{ given that } y = 0 \text{ and } \dfrac{dy}{dx} = 2 \text{ when } x = 0.$$

SOLUTION. The given differential equation can be written as $(D^2+4)y = 8\cos 2x$ ...(1)

To find the C.F. of (1), the auxiliary equation is given by $m^2+4 = 0 \Rightarrow m = \pm 2i$

$$\therefore \qquad \text{C.F.} = C_1\cos 2x + C_2\sin 2x$$

Now, P .I. $= 8.\dfrac{1}{D^2+4}\cos 2x = 8.\text{Real part of}\left[\dfrac{1}{D^2+4}(\cos 2x + i\sin 2x)\right]$

$$= 8.\text{Real part of}\left[\dfrac{1}{D^2+4}e^{2ix}\right] = 8.\text{Real part of}\left[\dfrac{1}{(D+2i)(D-2i)}e^{2ix}\right]$$

$$= 8 . \text{Real part of} \left[\frac{1}{(2i + 2i)(D - 2i)} e^{2ix} \right]$$

$$= 8 . \text{Real part of} \left[\frac{1}{4i} \frac{1}{(D - 2i)} e^{2ix} . 1 \right]$$

$$= 8 . \text{Real part of} \left[\frac{1}{4i} \frac{1}{D - 2i} e^{2ix} . 1 \right]$$

$$= 8 . \text{Real part of} \left[\frac{1}{4i} e^{2ix} \frac{1}{D} . 1 \right] = 8 . \text{Real part of} \left[\frac{1}{4i} e^{2ix} . x \right]$$

$$= 8 . \text{Real part of} \left[\frac{1}{4i} (\cos 2x + i \sin 2x) . x \right]$$

$$= 8 . \text{Real part of} \frac{1}{-4} [i \cos 2x - \sin 2x] . x = 8 . \frac{x}{4} \sin 2x = 2x \sin 2x$$

Hence, the complete solution of (1) is given by
$$y = \text{C.F.} + \text{P.I.}$$

$$\Rightarrow \qquad y = C_1 \cos 2x + C_2 \sin 2x + 2x \sin 2x$$

$$\Rightarrow \qquad \frac{dy}{dx} = -2C_1 \sin 2x + 2C_2 \cos 2x + 2 \sin 2x + 4x \cos 2x.$$

Now, using the given conditions

$$y = 0 = \frac{dy}{dx} \text{ at } x = 0 \text{ gives } C_1 = 0 \text{ and } C_2 = 1$$

∴ General solution is $y = \sin 2x + 2x \sin 2x$

EXAMPLE 4. *Solve the given differential equation*

$$\frac{d^2 y}{dx^2} + 9y = 2 \sin 3x + \cos 3x .$$

SOLUTION. The given differential equation can be written as

$$(D^2 + 9)y = 2 \sin 3x + \cos 3x \qquad \qquad \qquad ...(1)$$

To find the C.F. of (1), the auxiliary equation is given by $m^2 + 9 = 0 \Rightarrow m = \pm 3i$

∴ C.F. $= C_1 \cos 3x + C_2 \sin 3x$

Now, P.I. $= \dfrac{1}{D^2 + 9}(2 \sin 3x + \cos 3x) = 2 \dfrac{1}{D^2 + 9} \sin 3x + \dfrac{1}{D^2 + 9} \cos 3x.$

Now, $\dfrac{1}{D^2 + 9} \cos 3x + i . \dfrac{1}{D^2 + 9} \sin 3x$

$$= \frac{1}{D^2 + 9}(\cos 3x + i \sin 3x) = \frac{1}{D^2 + 9} e^{i3x}$$

$$= \frac{1}{(D + 3i)(D - 3i)} e^{i3x} = \frac{1}{(3i + 3i)(D - 3i)} e^{3ix} . 1$$

$$= \frac{e^{i3x}}{6i} . \frac{1}{[(D + 3i) - 3i]} . 1 = \frac{1}{6i} . e^{i3x} \frac{1}{D} . 1 = \frac{1}{6i} e^{i3x} . x$$

$$= \frac{x}{6i}[\cos 3x + i \sin 3x] = \frac{x}{6} \sin 3x - i \frac{x}{6} \cos 3x \,.$$

Now, equating real and imaginary part of both sides, we get

$$\frac{1}{D^2+9}\cos 3x = \frac{x}{6}\sin 3x \quad \text{and} \quad \frac{1}{D^2+9}\sin 3x = -\frac{x}{6}\cos 3x \,.$$

$$\therefore \text{Required} \quad \text{P.I.} = 2\left[-\frac{x}{6}\cos 3x\right] + \frac{x}{6}\sin 3x = -\frac{x}{3}\cos 3x + \frac{x}{6}\sin 3x.$$

Hence, the complete solution of (1) is given by

$$y = \text{C.F.} + \text{P.I.}$$

$$\Rightarrow \qquad y = C_1 \cos 3x + C_2 \sin 3x - \frac{1}{3}x\cos x + \frac{1}{6}x \sin 3x$$

Exercise 13.14

Solve the following differential equations :

1. $(D^2 + a^2)y = \cos ax$

2. $(D^2 + 4)y = \cos 2x$

3. $(D^2 + 4)y = e^x + \sin 2x$

4. $(D^3 + a^2 D)y = \sin ax$ **[UPTU(B.Pharma)–2005]**

5. $(D^3 + 1)y = \cos 2x$

6. $(D^4 + D^2 + 1)y = e^{-x/2}\cos\dfrac{x\sqrt{3}}{2}$

7. $(D^2 + 2D + 2)y = 2e^{-x}\sin x$

8. $(D^4 + 2D^2 + 1)y = \cos x$

9. $(D^2 + 4)y = 4 + \sin 2x$

HINT TO SELECTED PROBLEMS

1. $m = \pm ai \Rightarrow \text{C.F.} = C_1 \cos ax + C_2 \sin ax$

$$\text{P.I.} = \frac{1}{D^2 + a^2}\cos ax$$

$$= \text{Real part of}\left[\frac{1}{D^2 + a^2}e^{iax}\right].$$

3. $m = \pm 2i, \text{C.F.} = C_1 \cos 2x + C_2 \sin 2x$

$$\text{P.I.} = \frac{1}{(D^2 + 4)}(e^x + \sin 2x)$$

$$= \frac{e^x}{D^2 + 4} + \frac{1}{D^2 + 4}\cdot \sin 2x$$

$$= \frac{e^x}{1+4} + \text{Imag. part of}\left[\frac{1}{D^2+4}\cdot e^{2ix}\right]$$

4. $\text{P.I.} = \frac{1}{D^2 + a^2}\sin ax$

$$= \text{Imaginary part of}\left[\frac{1}{(D+ai)(D-ai)}e^{iax}\right]$$

ANSWERS

1. $y = C_1 \cos ax + C_2 \sin ax + \dfrac{x}{2a}\sin ax$ **2.** $y = C_1 \cos 2x + C_2 \sin 2x + \dfrac{x}{4}\sin 2x$

3. $y = C_1 \cos 2x + C_2 \sin 2x + \dfrac{1}{5}e^x - \dfrac{1}{4}x\cos 2x$

4. $y = C_1 + C_2 \cos ax + C_3 \sin ax - \dfrac{1}{2a^2}x\sin ax$

5. $y = C_1 e^{-x} + e^{x/2}\left[C_2 \cos \dfrac{\sqrt{3}}{2}x + C_3 \sin \dfrac{\sqrt{3}}{2}x\right] + \dfrac{1}{65}(\cos 2x - 8\sin 2x)$

6. $y = e^{-x/2}\left[C_1\cos\frac{1}{2}x\sqrt{3} + C_2\sin\frac{1}{2}x\sqrt{3}\right] + e^{x/2}\left[C_3\cos\frac{1}{2}x\sqrt{3} + C_4\sin\frac{1}{2}x\sqrt{3}\right]$

$\qquad - \frac{1}{12}\sqrt{3}\,x\,e^{-x/2}\sin\frac{\sqrt{3}x}{2} + \frac{x}{4}e^{-x/2}\cos\frac{\sqrt{3}}{2}x$

7. $y = e^{-x}(C_1\cos x + C_2\sin x) - xe^{-x}\cos x$

8. $y = (C_1 + C_2 x)\cos x + (C_3 + C_4 x)\sin x - \frac{1}{8}x^2\cos x$

9. $y = C_1\cos 2x + C_2\sin 2x + 1 - \frac{x}{\cdot 4}\cos 2x$

(7) To evaluate $\frac{1}{f(D)}x.X$, where X is any function of x (except e^{ax}) :

Consider $D^n(x.X) = xD^n.X + {}^nC_1\,D^{n-1}.X$ \hfill (By Leibnitz's theorem)

We have $f(D)\,(xX) = x\,f(D)X + f'(D).X$

Now, taking the inverse operator, we have $\dfrac{1}{f(D)}(xX) = x.\dfrac{1}{f(D)}X + \left[\dfrac{d}{dD}\dfrac{1}{f(D)}\right]X$

But we have $\dfrac{d}{dD}\left[\dfrac{1}{f(D)}\right] = -\dfrac{f'(D)}{\{f(D)\}^2}$.

Therefore, $\dfrac{1}{f(D)}(x.X) = x.\dfrac{1}{f(D)}X - \dfrac{f'(D)}{\{f(D)\}^2}X$.

♻ Remark

▪▪▶ If we want to find P.I. when B is of the form $x^m.X$, where X is any function of x, then there are two cases

(a) If $X = x^n$, then $x^m.X = x^{m+n}$.

Then B is of the form x^{m+n} (Polynomial). Here, we should apply the method of finding P.I. for polynomial discussed earlier.

(b) If $X = e^{ax}$, then $x^m.X = x^m.e^{ax}$ and we should apply the method, discussed earlier.

(c) If $X = \cos ax$, then $x^m.X = x^m\cos ax$

Then P.I. $= \dfrac{1}{f(D)}x^m\cos ax = \dfrac{1}{f(D)}$ (Real part of $x^m.e^{iax}$) $=$ Real part of $\dfrac{1}{f(D)}x^m.e^{iax}$, which can

be easily calculated.

Similar results hold if $X = \sin ax$, then taking imaginary part.

SOLVED EXAMPLES

Example 1. Solve $(D^2 + 2D + 1)y = x\cos x$.

Solution. The given equation is

$\qquad (D^2 + 2D + 1)y = x\cos x$ \hfill ...(1)

To find the C.F. of (1), the auxiliary equation is

$\qquad m^2 + 2m + 1 = 0 \ \Rightarrow\ m = -1, -1$

$\therefore \quad$ C.F. $= (C_1 + C_2 x)\,e^{-x}$

Now, P.I. $= \dfrac{1}{(D^2 + 2D + 1)} \cdot x \cos x$

$= x \cdot \dfrac{1}{D^2 + 2D + 1} \cos x - \dfrac{2D + 2}{(D^2 + 2D + 1)^2} \cos \dot{x}$

$= x \cdot \dfrac{1}{2D} \cos x - \dfrac{2D + 2}{4D^2} \cos x = \dfrac{x}{2} \sin x + \dfrac{(D + 1)}{2} \cos x$

$= \dfrac{x}{2} \sin x + \dfrac{\cos x}{2} - \dfrac{\sin x}{2}$

Hence, the complete solution of (1) is given by
$$y = \text{C.F.} + \text{P.I.}$$

$\Rightarrow \qquad y = (C_1 + C_2 x)e^{-x} + \dfrac{x}{2} \sin x + \dfrac{\cos x}{2} - \dfrac{\sin x}{2}.$

EXAMPLE 2. *Solve* $(D^2 - 4D + 4) = 8x^2 e^{2x} \sin 2x$.

SOLUTION. The given equation is
$$(D^2 - 4D + 4) = 8x^2 e^{2x} \sin 2x \qquad \qquad \dots(1)$$

To find the C.F. of (1), the auxiliary equation is
$$m^2 - 4m + 4 = 0 \quad \Rightarrow \quad m = 2, 2$$

$\therefore \qquad \text{C.F.} = (C_1 + C_2 x)\, e^{2x}$

Now, P.I. $= 8 \cdot \dfrac{1}{(D - 2)^2} e^{2x} (x^2 \sin 2x) = 8e^{2x} \cdot \dfrac{1}{(D + 2 - 2)^2} \cdot (x^2 \sin 2x)$

$= 8e^{2x} \cdot \dfrac{1}{D^2}(x^2 \sin 2x) = 8\, e^{2x} \cdot I_1$

where, $I_1 = \dfrac{1}{D^2}(x^2 \sin 2x) = $ Imaginary part of $\dfrac{1}{D^2} x^2 e^{2ix}$

$= $ Imaginary part of $e^{2ix} \dfrac{1}{(D + 2i)^2} x^2$

$= $ Imaginary part of $\dfrac{e^{2ix}}{4i^2} \left(1 + \dfrac{D}{2i}\right)^{-2} x^2$

$= $ Imaginary part of $\dfrac{e^{2ix}}{-4} \left(1 - \dfrac{iD}{2}\right)^{-2} x^2$

$= $ Imaginary part of $\dfrac{e^{2ix}}{-4} \left[1 + 2\left(\dfrac{iD}{2}\right) + 3\left(\dfrac{iD}{2}\right)^2 + \dots\right] x^2$

$= $ Imaginary part of $\dfrac{e^{2ix}}{-4} \left[1 + Di - \dfrac{3}{4}D^2 + \dots\right] x^2$

$= $ Imaginary part of $\dfrac{e^{2ix}}{-4} \left[x^2 + 2ix - \dfrac{3}{2}\right]$

$= $ Imaginary part of $\left\{-\dfrac{1}{4}(\cos 2x + i \sin 2x)\left(x^2 + 2ix - \dfrac{3}{2}\right)\right\}$

$$= -\frac{1}{4}\left[\left(x^2 - \frac{3}{2}\right)\sin 2x + 2x\cos 2x\right] = -\frac{1}{8}[(2x^2 - 3)\sin 2x + 4x\cos 2x]$$

$$\therefore \quad \text{P.I.} = 8e^{2x}.I_1 = 8e^{2x}\left[-\frac{1}{8}\{(2x^2 - 3)\sin 2x + 4x\cos 2x\}\right]$$

$$= -e^{2x}[(2x^2 - 3)\sin 2x + 4x\cos 2x].$$

Hence, the complete solution of (1) is given by

$$y = \text{C.F.} + \text{P.I.}$$

$$\Rightarrow \quad y = e^{2x}[C_1 + C_2 x + 3\sin 2x - 2x^2\sin 2x - 4x\cos 2x].$$

EXAMPLE 3. *Solve* $(D^2 - 2D + 1)\, y = xe^x\,\sin x$.

SOLUTION. The given differential equation can be written as

$$(D^2 - 2D + 1)\, y = xe^x\,\sin x \qquad\qquad\qquad ...(1)$$

To find the C.F. of (1), the auxiliary equation is given by $m^2 - 2m + 1 = 0 \Rightarrow m = 1, 1$

$$\therefore \quad \text{C.F.} = (C_1 + C_2 x)\, e^x$$

Now, $\text{P.I.} = \dfrac{1}{(D^2 - 2D + 1)} xe^x \sin x = \dfrac{1}{(D-1)^2} e^x (x\sin x)$

$$= e^x\,\frac{1}{(D+1)^2 - 2(D+1) + 1}\, x\sin x = e^x\,\frac{1}{D^2}(x\sin x) = e^x.\frac{1}{D}\int x\,\sin x\,dx$$

$$= e^x\left(\frac{1}{D}\right)(-x\cos x + \sin x) = e^x\left[\int -x\cos x\,dx + \int \sin x\,dx\right]$$

$$= e^x\,[-x\sin x - 2\cos x].$$

Hence, the complete solution of (1) is given by

$$y = \text{C.F.} + \text{P.I.}$$

$$\Rightarrow \quad y = (C_1 + C_2 x)\, e^x - e^x(x\sin x + 2\cos x).$$

EXAMPLE 4. *Solve* $\dfrac{d^2 y}{dx^2} + 4y = x\,\sin x$.

SOLUTION. The given differential equation can be written as $(D^2 + 4)y = x\sin x \qquad ...(1)$

To find the C.F. of (1), the auxiliary equation is given by $m^2 + 4 = 0 \Rightarrow m = \pm 2i$

$$\therefore \quad \text{C.F.} = C_1\cos 2x + C_2\sin 2x$$

Now, $\text{P.I.} = \dfrac{1}{D^2 + 4} x\sin x = x\dfrac{1}{D^2 + 4}\sin x - \dfrac{2D}{(D^2 + 4)^2}\sin x$

$$= \frac{x\sin x}{-1^2 + 4} - \frac{2D}{(-1^2 + 4)^2}\sin x = \frac{1}{3}x\sin x - \frac{2}{9}D(\sin x)$$

$$= \frac{1}{3}x\sin x - \frac{2}{9}\cos x$$

Hence, the complete solution of (1) is given by

$$y = \text{C.F.} + \text{P.I.}$$

$$\Rightarrow \quad y = C_1\cos 2x + C_2\sin 2x + \frac{x}{3}\sin x - \frac{2}{9}\cos x.$$

Exercise 13.15

Solve the following differential equations :

1. $(D^2 - 2D + 1)y = x \sin x$

2. $(D^2 + m^2)y = x \cos mx$

3. $(D^2 - 1) = x^2 \sin x$

4. $(D^2 + a^2)^2 y = \sin ax$

5. $(D^4 - 1)y = x \sin x$

6. $(D^4 - 1)y = e^x \cos x$

7. $(D^4 + 2D^2 + 1)y = x^2 \cos x$

8. $(D^2 + 1)y = x^2 \sin 2x$

9. $(D^2 + 1)^2 y = 24x \cos x$, given that

$x = 0, y = 0, Dy = 0, \ D^2y = 0, D^3y = 0$.

10. $(D^2 + 4)y = \cos^2 x$ **[UPTU(B.Pharma)–2003]**

HINT TO SELECTED PROBLEMS

1. C.F. $= (C_1 + C_2 x) e^x$

P.I. $= \dfrac{1}{D^2 - 2D + 1} x \sin x$

$= x \dfrac{1}{D^2 - 2D + 1} \sin x - \dfrac{(2D-2)}{(D^2 - 2D + 1)^2} \sin x$

$= x. \dfrac{1}{-1 - 2D + 1} \sin x - \dfrac{(2D-2)}{(-1 - 2D + 1)^2} \sin x$

$= x. \dfrac{1}{-2D} \sin x - \dfrac{(2D-2)}{4D^2} \sin x$.

2. P.I. $= \dfrac{1}{D^2 + m^2} x \cos mx$

$= $ Real part of $\dfrac{1}{(D^2 + m^2)} x \, e^{imx}$

$= $ Real part of $e^{imx} \left[\dfrac{1}{(D + im)^2 + m^2} . x \right]$

8. P.I. $= \dfrac{1}{D^2 + 1} x^2 \sin 2x$

$= $ Imaginary part of $\dfrac{1}{D^2 + 1} . xe^{2ix}$

$= $ Imag. part of $e^{2ix} \left[\dfrac{1}{(D + 2i)^2 + 1} \right]. x$.

10. P.I. $= \dfrac{1}{(D^2 + 1)^2} 24x \cos x = 24 \dfrac{1}{(D^2 + 1)^2} x \cos x$

$= $ Real part of $24e^{ix} \dfrac{1}{[(D + i)^2 + 1]^2} . x$

ANSWERS

1. $y = (C_1 + C_2 x) e^x + \dfrac{1}{2}(x \cos x + \cos x - \sin x)$

2. $y = C_1 \cos mx + C_2 \sin mx + \dfrac{x^2}{4m} \sin mx + \dfrac{x}{4m^2} \cos mx$

3. $y = C_1 e^x + C_2 e^{-x} - x \cos x - \dfrac{1}{2}(x^2 - 1) \sin x$

4. $y = (C_1 + C_2 x) \cos ax + (C_3 + C_4 x) \sin ax - \dfrac{1}{8a^2}(x^2 \sin ax)$

5. $y = C_1 e^x + C_2 e^{-x} + C_3 \cos x + C_4 \sin x + \dfrac{1}{8}(x^2 \cos x - x \sin x)$

6. $y = C_1 e^x + C_2 e^{-x} + C_3 \cos x + C_4 \sin x - \dfrac{1}{5} e^x \cos x$

7. $y = (C_1 + C_2 x) \cos x + (C_3 + C_4 x) \sin x - \dfrac{1}{48}(x^4 - 9x^2) \cos x + \dfrac{1}{12} x^3 \sin x$

8. $y = C_1 \cos x + C_2 \sin x - \dfrac{1}{27}[24x \cos 2x + (9x^2 - 26) \sin 2x]$ **9.** $y = 3x^2 \sin x - x^3 \cos x$

10. $y = C_1 e^{2x} + C_2 e^{-2x} - \dfrac{1}{8} - \dfrac{1}{16} \cos 2x$

13.16 ORDINARY SIMULTANEOUS LINEAR DIFFERENTIAL EQUATIONS

In this section, we shall discuss the ordinary differential equations involving two or more dependent variables. Here, we shall discuss the case when there are as many simultaneous equations as there are dependent variables with one independent variable, by the process of elimination. After solving the derived equation, we substitute back to get the other dependent variable.

Consider the simultaneous equation as

$$\left.\begin{array}{l} f_1(D)x + f_2(D)y = f(t) \\ \text{and} \quad g_1(D)x + g_2(D)y = g(t) \end{array}\right] \quad \text{with} \quad D \equiv \frac{d}{dt} \qquad \ldots(1)$$

where, x and y are functions of t and $f_1(D)$, $f_2(D)$, $g_1(D)$ and $g_2(D)$ are rational integral functions with constant coefficients, $f(t)$ and $g(t)$ are the functions of the independent variable t. Now define the determinant Δ such as

$$\Delta = \begin{vmatrix} f_1(D) & f_2(D) \\ g_1(D) & g_2(D) \end{vmatrix} \qquad \ldots (2)$$

then we can say Δ involves the operator coefficients of x and y in (1).

The equation (2) can be solved by the usual methods.

REMARKS

➡ To solve the simultaneous equations completely, we always require as many simultaneous equations as are the number of dependent variables.

➡ The method of solving the simultaneous differential equations with constant coefficients is similar to that of solving a set of simultaneous equations in Algebra.

➡ The number of arbitrary constants appearing in the general solution of the system (1) is equal to the degree in D of the determinant Δ given by (2), provided determinant is non-zero.

➡ The determinant Δ, defined by (2), involves the operator coefficients of x and y in (1).

13.17 METHOD OF SOLVING SIMULTANEOUS LINEAR DIFFERENTIAL EQUATION WITH CONSTANT COEFFICIENTS

Let x and y be the dependent variables and t be the independent variable. Generally, there are two methods for the solution of simultaneous linear differential equations with constant coefficients.

METHOD-1: SYMBOLIC METHOD WITH USE OF D

Consider the simultaneous equation such as

$$f_1(D)x + g_1(D)y = T_1 \qquad \ldots (1)$$

and

$$f_2(D)x + g_2(D)y = T_2 \qquad \ldots (2)$$

where T_1 and T_2 are the functions of independent variable t and f_1, f_2, g_1, g_2 are polynomial functions with constant coefficients.

Operate on both sides of equation (1) by $g_2(D)$ and equation (2) by $g_1(D)$, we get

$$g_2(D)f_1(D)x + g_2(D)g_1(D)y = g_2(D)T_1 \qquad \ldots (3)$$

and

$$g_1(D) f_2(D)x + g_1(D)g_2(D)y = g_1(D) T_2 \qquad \ldots (4)$$

Now, since $g_1(D)$ and $g_2(D)$ both have the constant coefficients then

$$g_1(D) g_2(D)y = g_2(D) g_1(D)y$$

therefore, from (3) and (4)

$$[g_2(D)f_1(D) - g_1(D)f_2(D)]x = g_2(D)T_1 - g_1(D)T_2 \qquad \ldots (5)$$

Equation (5) is an ordinary differential equation with one dependent variable and can be solved by the usual methods. Thus, x can be obtained as a function of t. The value of y is then obtained by substituting the value of x in any of the given equations and integrating the resulting equation, if necessary. If however, y is obtained by an independent elimination as in the case of x, the values of x and y are to be substituted in given equation (1) and (2) and the arbitrary constants in x and y are to be so adjusted that the given equations are satisfied. Here, the number of independent arbitrary constants entering in the general solution is the index of the highest power of D in

$$[g_2(D) f_1(D) - g_1(D) f_2(D)]$$

METHOD-2: USE OF DIFFERENTIATION

If two equations containing $x, y, \dfrac{dx}{dt}$ and $\dfrac{dy}{dt}$ are given. Then we can obtain more equations containing $x, y, \dfrac{dx}{dt}, \dfrac{dy}{dt}, \dfrac{d^2x}{dt^2}$ and $\dfrac{d^2y}{dt^2}$ by differentiating the given equations with respect to t.

From these equations, we can obtain an equation containing x (or y) and its derivative, by eliminating x (or y) and its derivatives. Now, solve this new equation for x (or y) and substituting the value of x (or y) in any of the given equation and if necessary, solve the resulting equation.

REMARKS

➡ The method of differentiation will be used when found very necessary.

➡ Generally t will be the independent variable and x and y will be dependent variables. In some problems any other variable, say x, will be given as the independent variable and y and z as the dependent variables.

SOLVED EXAMPLES

EXAMPLE 1. *Solve* $\qquad \dfrac{dx}{dt} - 7x + y = 0$

$$\dfrac{dy}{dt} - 2x - 5y = 0. \qquad \text{[UPTU (B.Pharma)–2002]}$$

SOLUTION. The given equation can be written as

$$(D - 7)x + y = 0 \qquad \ldots (1)$$

and $\qquad (D - 5)y - 2x = 0 \qquad \ldots (2)$

Now, eliminating y, we get

$$(D - 7)(D - 5)x + 2x = 0$$

$$\Rightarrow \qquad \dfrac{d^2x}{dt^2} - 12\dfrac{dx}{dt} + 37x = 0 \qquad \ldots (3)$$

The auxiliary equation is

$$m^2 - 12m + 37 = 0 \quad \Rightarrow \quad m = 6 \pm i$$

$$\therefore \qquad x = e^{6t}[C_1 \cos t + C_2 \sin t]$$

Putting this value of x in (1), we get

$$e^{6t}[-C_1 \sin t + C_2 \cos t] + 6e^{6t}[C_1 \cos t + C_2 \sin t] - 7e^{6t}(C_1 \cos t + C_2 \sin t) + y = 0$$

$$\Rightarrow \quad y = e^{6t}(C_1 - C_2)\cos t + (C_1 + C_2)\sin t$$

Hence, the solution is

$$x = e^{6t}[C_1 \cos t + C_2 \sin t]$$

$$y = e^{6t}[(C_1 - C_2)\cos t + (C_1 + C_2)\sin t]$$

EXAMPLE 2. Solve $\dfrac{dx}{dt} = -\omega y$ ... (1)

$$\frac{dy}{dt} = \omega x \qquad \qquad \text{... (2)}$$

SOLUTION. Differentiating (1) with respect to t, we get

$$\frac{d^2x}{dt^2} + \omega\frac{dy}{dt} = 0 \quad \Rightarrow \quad \frac{d^2x}{dt^2} + \omega^2 x = 0 \qquad \text{[By using (2)]}$$

$$\therefore \quad x = C_1 \cos \omega t + C_2 \sin \omega t$$

Putting this value of x in (1), we get

$$y = -(1/\omega)[-C_1\omega \sin \omega t + C_2\omega \cos \omega t] = -C_2 \cos \omega t + C_1 \sin \omega t$$

EXAMPLE 3. Solve $\dfrac{dx}{dt} + 2\dfrac{dy}{dt} - 2x + 2y = 3e^t$, $3\dfrac{dx}{dt} + \dfrac{dy}{dt} + 2x + y = 4e^{2t}$.

[UPTU(B.Pharma)–2003, 06, 07]

SOLUTION. The given equation can be written as

$$(D-2)x + 2(D+1)y = 3e^t \qquad \text{... (1)}$$

and $\qquad (3D+2)x + (D+1)y = 4e^{2t} \qquad \text{... (2)}$

Eliminating y between (1) and (2), we obtain

$$[2(3D+2)-(D-2)]x = 8e^{2t} - 3e^t$$

or $\qquad\qquad\qquad (5D+6)x = 8e^{2t} - 3e^t$

or $\qquad\qquad\qquad \dfrac{dx}{dt} + \dfrac{6}{5}x = \dfrac{8}{5}e^{2t} - \dfrac{3}{5}e^t$

which is a linear differential equation with

$$\text{I.F.} = e^{\int 6/5\,dt} = e^{6t/5}$$

Now, solution becomes

$$x.e^{6t/5} = \int e^{6t/5}\left[\frac{8}{5}e^{2t} - \frac{3}{5}e^t\right]dt + C_1 = \frac{8}{5}\int e^{16t/5} - \frac{3}{5}\int e^{11t/5}dt + C_1$$

$$= \frac{1}{2}e^{16t/5} - \frac{3}{11}e^{11t/5} + C_1$$

$$\therefore \qquad x = \frac{1}{2}e^{2t} - \frac{3}{11}e^t + C_1e^{-6t/5}$$

Now, $\qquad \dfrac{dx}{dt} = e^{2t} - \dfrac{3}{11}e^t - \dfrac{6}{5}C_1e^{-6t/5}$

Putting this value in (1), we get

$$2Dy + 2y + e^{2t} - \frac{3}{11}e^t - \frac{6}{5}C_1e^{-6t/5} - e^{2t} + \frac{6}{11}e^t - 2C_1e^{-6t/5} = 3e^t$$

or $\quad 2\dfrac{dy}{dt} + 2y = \dfrac{30}{11}e^t + \dfrac{16}{5}C_1 e^{-6t/5}$

or $\quad \dfrac{dy}{dt} + y = \dfrac{15}{11}e^t + \dfrac{8}{5}C_1 e^{-6t/5}$

which is a linear differential equation with

$$\text{I.F.} = e^{\int dt} = e^t$$

The solution is

$\therefore \qquad y \cdot e^t = \dfrac{15}{11}\displaystyle\int e^{2t}dt + \dfrac{8}{5}C_1 \int e^{-t/5}dt + C_2 = \dfrac{15}{22}e^{2t} - 8C_1 e^{-t/5} + C_2$

$\therefore \qquad y = C_2 e^{-t} + \dfrac{15}{22}e^t - 8C_1 e^{-6t/5}$

Hence, the solution is given by

$$x = \dfrac{1}{2}e^{2t} - \dfrac{3}{11}e^t + C_1 e^{-6t/5}$$

$$y = \dfrac{15}{22}e^t - 8C_1 e^{-6t/5} + C_1 e^{-t}$$

EXAMPLE 4. Solve $\dfrac{dx}{dt} + \dfrac{dy}{dt} - 2y = 2\cos t - 7\sin t$

$$\dfrac{dx}{dt} - \dfrac{dy}{dt} + 2x = 4\cos t - 3\sin t$$

[UPTU(B.Pharma)–2005]

SOLUTION. The given equation can be written as

$\qquad Dx + (D-2)y = 2\cos t - 7\sin t$ ... (1)

and $\quad (D+2)x - Dy = 4\cos t - 3\sin t$ ... (2)

Eliminating y between (1) and (2), we obtain

$[D^2 + (D-2)(D+2)]x = D(2\cos t - 7\sin t) + (D-2)(4\cos t - 3\sin t)$

or $\qquad (D^2 - 2)x = -9\cos t$

Auxiliary equation is $m^2 - 2 = 0 \Rightarrow m = \pm\sqrt{2}$

$$\text{C.F.} = C_1 e^{\sqrt{2}t} + C_2 e^{-\sqrt{2}t}$$

and $\qquad \text{P.I.} = -9\dfrac{1}{D^2 - 2}\cos t = \dfrac{-9\cos t}{D^2 - 2} = 3\cos t$

$\therefore \qquad x = C_1 e^{\sqrt{2}t} + C_2 e^{-\sqrt{2}t} + 3\cos t$

$\qquad \dfrac{dx}{dt} = \sqrt{2}\,C_1 e^{\sqrt{2}t} - C_2\sqrt{2}\,e^{-\sqrt{2}t} - 3\sin t$

Now adding (1) and (2), we get

$\qquad 2Dx + 2x - 2y = 6\cos t - 10\sin t$

or $\; y = \dfrac{dx}{dt} + x - 3\cos t + 5\sin t$

$\qquad = \sqrt{2}C_1 e^{\sqrt{2}t} - C_2\sqrt{2}e^{-\sqrt{2}t} - 3\sin t + C_1 e^{\sqrt{2}t} + C_2 e^{-\sqrt{2}t} + 3\cos t - 3\cos t + 5\sin t$

$\qquad = (\sqrt{2}+1)C_1 e^{\sqrt{2}t} + (1-\sqrt{2})C_2 e^{-\sqrt{2}t} + 2\sin t$

Hence, the solution is given as

$$x = C_1 e^{\sqrt{2}t} + C_2 e^{-\sqrt{2}t} + 3\cos t, \; y = (\sqrt{2}+1)C_1 e^{\sqrt{2}t} + (1-\sqrt{2})C_2 e^{-\sqrt{2}t} + 2\sin t$$

EXAMPLE 5. *Solve* $\dfrac{dx}{dt} = ax + by, \; \dfrac{dy}{dt} = a'x + b'y$.

SOLUTION. The given equation can be written as

$$(D-a)x - by = 0 \qquad \qquad \text{... (1)}$$

and $-a'x + (D-b')y = 0$ $\qquad \qquad \text{... (2)}$

Eliminating y between (1) and (2), we get

$$[(D-a)(D-b') - a'b]x = 0$$

or $\quad [D^2 - (a+b')D + (ab'-a'b)]x = 0$

$\therefore$ Auxiliary equation is

$$m^2 - (a+b')m + (ab'-a'b) = 0$$

$$m = \frac{(a+b') \pm \sqrt{[(a+b')^2 - 4(ab'-a'b)]}}{2}$$

$$= \frac{(a+b') \pm \sqrt{(a-b')^2 + 4a'b}}{2}$$

where roots m_1 and m_2 is

$$\left.\begin{array}{l} m_1 = \dfrac{(a+b') + \sqrt{(a-b')^2 + 4a'b}}{2} \\[4mm] m_2 = \dfrac{(a+b') - \sqrt{(a-b')^2 + 4a'b}}{2} \end{array}\right] \qquad \text{...(3)}$$

$\therefore \qquad x = C_1 e^{m_1 t} + C_2 e^{m_2 t}$

$$\frac{dx}{dt} = C_1 m_1 e^{m_1 t} + C_2 m_2 e^{m_2 t}$$

$\therefore$ From (1), we get

$$y = \frac{1}{b}\left[\frac{dx}{dt} - ax\right] = \frac{1}{b}[(m_1 - a)C_1 e^{m_1 t} + (m_2 - a)C_2 e^{m_2 t}]$$

Hence, the solution is

$$x = C_1 e^{m_1 t} + C_2 e^{m_2 t}$$

$$y = \frac{1}{b}[(m_1 - a)C_1 e^{m_1 t} + (m_2 - a)C_2 e^{m_2 t}]$$

EXAMPLE 6. *Solve* $\dfrac{dx}{dt} + 4x + 3y = t, \quad \dfrac{dy}{dt} + 2x + 5y = e^t$

SOLUTION. The given equation can be written as

$$(D+4)x + 3y = t \qquad \qquad \text{... (1)}$$

and $\qquad 2x + (D+5)y = e^t$ $\qquad \qquad \text{... (2)}$

Eliminating y between (1) and(2), we get

$$[(D+4)(D+5) - 6]x = (D+5)t - 3e^t$$

or $\qquad (D^2 + 9D + 14)x = 1 + 5t - 3e^t$

$\therefore$ Auxiliary equation is $m^2 + 9m + 14 = 0$

$\therefore$ $m = -2, -7$

Complementary function $= C_1 e^{-2t} + C_2 e^{-7t}$

$$\text{Particular integral} = \frac{1}{14 + 9D + D^2}(1 + 5t) - \frac{1}{14 + 9D + D^2}\, 3e^t$$

$$= \frac{1}{14}\left[1 + \frac{9}{14}D + \frac{1}{14}D^2\right]^{-1}(1 + 5t) - \frac{3e^t}{14 + 9 + 1}$$

[By replacing D by 1]

$$= \frac{1}{14}\left[1 - \frac{9}{14}D - \frac{1}{14}D^2 + \ldots\right](1 + 5t) - \frac{1}{8}e^t$$

$$= \frac{1}{14}\left[1 + 5t - \frac{9}{14}.5\right] - \frac{1}{8}e^t = \frac{1}{14}\left[5t - \frac{31}{14}\right] - \frac{1}{8}e^t$$

$\therefore$

$$x = C_1 e^{-2t} + C_2 e^{-7t} + \frac{5}{14}t - \frac{1}{8}e^t - \frac{31}{196}$$

$$\frac{dx}{dt} = -2C_1 e^{-2t} - 7C_2 e^{-7t} + \frac{5}{14} - \frac{1}{8}e^t$$

Putting above value in (1), we get

$$3y = -\frac{dx}{dt} - 4x + t = -2C_1 e^{-2t} + 3C_2 e^{-7t} - \frac{10}{7}t + t - \frac{5}{14} + \frac{31}{49} + \frac{1}{8}e^t + \frac{1}{2}e^t$$

or

$$y = \frac{1}{3}\left[-2C_1 e^{-2t} + 3C_2 e^{-7t} - \frac{3}{7}t + \frac{27}{98} + \frac{5}{8}e^t\right]$$

or

$$x = C_1 e^{-2t} + C_2 e^{-7t} + \frac{5}{14}t - \frac{31}{196} - \frac{1}{8}e^t$$

EXAMPLE 7.   *Solve* $t\,dx = (t - 2x)dt = 0$ *and* $t\,dy = (tx + ty + 2x - t)dt$.

SOLUTION.   The given equation can be written as

$$t\,dx - (t - 2x)dt = 0 \qquad \ldots (1)$$

and $t\,dy - (tx + ty + 2x - t)dt = 0$ $\ldots (2)$

From (1), we get $\dfrac{dx}{dt} + \dfrac{2}{t}x = 1$

which is a linear equation

I.F. $= e^{\int 2/t\, dt} = e^{2\log t} = t^2$ $\ldots(3)$

$\therefore$ $xt^2 = \int t^2.1\,dt + C_1 = \dfrac{1}{3}t^3 + C_1$

$\therefore$

$$x = \frac{1}{3}t + C_1 t^{-2}$$

Now adding (1) and (2), we get

$$t(dx + dy) = t(x + y)dt \quad \text{or} \quad = \frac{dx + dy}{x + y} = dt$$

Integrating, $\log(x + y) = t + \log C_2$

$$y = C_2 e^t - C_1 t^{-2} - \frac{1}{3}t$$

Hence, solution is . $x = \dfrac{1}{3}t + C_1 t^{-2}$; $y = C_2 e^t - C_1 t^{-2} - \dfrac{1}{3}t$

EXAMPLE 8. *Solve* $x\dfrac{dy}{dx} + z = 0$, $x\dfrac{dz}{dx} + y = 0$, *both equations be simultaneous differential equations.*

SOLUTION. Differentiating first equation w.r.t. x , we get

$$x\frac{d^2y}{dx^2} + \frac{dy}{dx} + \frac{dz}{dx} = 0$$

or $\quad x^2\dfrac{d^2y}{dx^2} + x\dfrac{dy}{dx} + x\dfrac{dz}{dx} = 0$ $\qquad\qquad$ ... (A)

and from second equation,

$$x\frac{dz}{dx} = -y$$

Put this value of $x\dfrac{dz}{dx}$ in equation (A), we get

$$x^2\frac{d^2y}{dx^2} + x\frac{dy}{dx} - y = 0 \qquad\qquad ...\text{(B)}$$

This equation is a homogeneous equation.

Put $x = e^t$ so that $\dfrac{dy}{dx} = \dfrac{dy}{dt}\cdot\dfrac{dt}{dx} = \dfrac{1}{x}\dfrac{dy}{dt}$

$\therefore \quad$ $x\dfrac{dy}{dx} = \dfrac{dy}{dt}$ or $x\dfrac{d}{dx} \equiv \dfrac{d}{dt}$

$$x\frac{d}{dx}\left[x\frac{dy}{dx}\right] = x^2\frac{d^2y}{dx^2} + x\frac{dy}{dx}$$

or $\quad x^2\dfrac{d^2y}{dx^2} = \left(x\dfrac{d}{dx} - 1\right)x\dfrac{dy}{dx} = (D-1)Dy$ $\qquad\left[\because \dfrac{d}{dt} \equiv D\right]$

$\therefore \quad$ From (B), we get

$$\{(D-1)D + (D-1)\}y = 0 \text{ or } (D^2-1)y = 0$$

$\therefore \qquad\qquad y = C_1e^t + C_2e^{-t}$ or $y = C_1x + C_2x^{-1}$ $\qquad\qquad$ ... (1)

so that $\qquad \dfrac{dy}{dx} = C_1 - \dfrac{C_2}{x^2}$

$\therefore$ From first equation, we get $z = -x\dfrac{dy}{dx}$

$$z = -C_1x + C_2x^{-1}$$

or Desired solution is

$$y = C_1x + C_2x^{-1}, \quad z = -C_1x + C_2x^{-1}$$

EXAMPLE 9. *Solve* $\dfrac{d^2x}{dt^2} + m^2y = 0$, $\dfrac{d^2y}{dt^2} - m^2x = 0$.

SOLUTION. The given equation can be written as

$$D^2x + m^2y = 0 \qquad\qquad ...\text{(1)}$$

$$-m^2 x + D^2 y = 0 \qquad \cdots (2)$$

Eliminating y between (1) and (2),

$$(D^4 + m^4)x = 0.$$

Auxiliary equation is $M^4 + m^4 = 0$

or $\qquad (M^2 + m^2)^2 - 2M^2 m^2 = 0$

or $\quad (M^2 - \sqrt{2}Mm + m^2)(M^2 + \sqrt{2}Mm + m^2) = 0$

$\therefore \qquad\qquad M^2 - \sqrt{2}mM + m^2 = 0$

or $\qquad\qquad M^2 + \sqrt{2}mM + m^2 = 0$

$$\therefore \quad M = \frac{\sqrt{2}m \pm \sqrt{(2m^2 - 4m^2)}}{2},$$

$$M = \frac{-\sqrt{2}m \pm \sqrt{(2m^2 - 4m^2)}}{2} = \frac{m}{\sqrt{2}} \pm \frac{m}{\sqrt{2}}i = -\frac{m}{\sqrt{2}} \pm \frac{m}{\sqrt{2}}i$$

$$\therefore \quad x = e^{mt/\sqrt{2}}\left[C_1 \cos\frac{mt}{\sqrt{2}} + C_2 \sin\frac{mt}{\sqrt{2}}\right] + e^{-mt/\sqrt{2}}\left[C_3 \cos\frac{mt}{\sqrt{2}} + C_4 \sin\frac{mt}{\sqrt{2}}\right] \quad \cdots (3)$$

so that

$$\frac{dx}{dt} = \frac{m}{\sqrt{2}}e^{mt/\sqrt{2}}\left[C_1 \cos\frac{mt}{\sqrt{2}} + C_2 \sin\frac{mt}{\sqrt{2}}\right] + e^{-mt/\sqrt{2}}\frac{m}{\sqrt{2}}\left[-C_1 \sin\frac{mt}{\sqrt{2}} + C_2 \cos\frac{mt}{\sqrt{2}}\right]$$

$$+ \left(-\frac{m}{\sqrt{2}}\right)e^{-mt/\sqrt{2}}\left[C_3 \cos\frac{mt}{\sqrt{2}} + C_4 \sin\frac{mt}{\sqrt{2}}\right]$$

$$+ e^{-mt/\sqrt{2}}\frac{m}{\sqrt{2}}\left[-C_3 \sin\frac{mt}{\sqrt{2}} + C_4 \cos\frac{mt}{\sqrt{2}}\right]$$

and $\frac{d^2 x}{dt^2} = \frac{m^2}{2}e^{mt/\sqrt{2}}\left[C_1 \cos\frac{mt}{\sqrt{2}} + C_2 \sin\frac{mt}{\sqrt{2}}\right]$

$$+ \frac{m^2}{2}e^{mt/\sqrt{2}}\left[-C_1 \sin\frac{mt}{\sqrt{2}} + C_2 \cos\frac{mt}{\sqrt{2}}\right]$$

$$- \frac{m^2}{2}e^{mt/\sqrt{2}}\left[C_1 \cos\frac{mt}{\sqrt{2}} + C_2 \sin\frac{mt}{\sqrt{2}}\right]$$

$$+ \frac{m^2}{2}e^{-mt/\sqrt{2}}\left[C_3 \cos\frac{mt}{\sqrt{2}} + C_4 \sin\frac{mt}{\sqrt{2}}\right]$$

$$- \frac{m^2}{2}e^{-mt/\sqrt{2}}\left[-C_3 \sin\frac{mt}{\sqrt{2}} + C_4 \cos\frac{mt}{\sqrt{2}}\right]$$

$$- \frac{m^2}{2}e^{-mt/\sqrt{2}}\left[-C_3 \sin\frac{mt}{\sqrt{2}} + C_4 \cos\frac{mt}{\sqrt{2}}\right]$$

$$- \frac{m^2}{2}e^{-mt/\sqrt{2}}\left[C_3 \cos\frac{mt}{\sqrt{2}} + C_4 \sin\frac{mt}{\sqrt{2}}\right]$$

$$= m^2 e^{mt/\sqrt{2}}\left[-C_1 \sin\frac{mt}{\sqrt{2}} + C_2 \cos\frac{mt}{\sqrt{2}}\right]$$

$$-m^2 e^{-mt/\sqrt{2}}\left[-C_3 \sin\frac{mt}{\sqrt{2}} + C_4 \cos\frac{mt}{\sqrt{2}}\right]$$

∴   From (1)

$$y = -\frac{1}{m^2}\frac{d^2x}{dt^2}$$

$$= e^{mt/\sqrt{2}}\left[C_1 \sin\frac{mt}{\sqrt{2}} - C_2 \cos\frac{mt}{\sqrt{2}}\right] + e^{-mt/\sqrt{2}}\left[-C_3 \sin\frac{mt}{\sqrt{2}} + C_4 \cos\frac{mt}{\sqrt{2}}\right]\ ...(4)$$

Hence, (3) and (4) be complete solution.

Exercise 13.16

Solve the following simultaneous differential equations :

1. $\dfrac{d^2x}{dt^2} - 3x - 4y = 0,\quad \dfrac{d^2y}{dt^2} + x + y = 0$

2. $\dfrac{dx}{dt} = 3x + 2y,\quad \dfrac{dy}{dt} = 5x + 3y$

3. $\dfrac{d^2x}{dt^2} + 4x + y = te^{3t},\quad \dfrac{d^2y}{dt^2} + y - 2x = \cos^2 t$

4. $\dfrac{dx}{dt} + 5x + y = e^t,\quad \dfrac{dy}{dt} - x + 3y = e^{2t}$

5. $\dfrac{dx}{dt} = 3x + 2y,\quad \dfrac{dy}{dt} + 5x + 3y = 0$

6. $\dfrac{dx}{dt} + 2x - 3y = t,\quad \dfrac{dy}{dt} - 3x + 2y = e^{2t}$

7. $(D-17)y + (2D-8)z = 0, (13D-53)y - 2z = 0$

8. $2\dfrac{d^2y}{dx^2} - \dfrac{dz}{dx} - 4y = 2x,\quad 2\dfrac{dy}{dx} + 4\dfrac{dz}{dx} - 3z = 0$

9. $\dfrac{dx}{dt} + \dfrac{2}{t}(x-y) = 1,\quad \dfrac{dy}{dt} + \dfrac{1}{t}(x+5y) = t$

10. $\dfrac{dx}{dt} + 5x - 2y = t,\quad \dfrac{dy}{dt} + 2x + y = 0$

11. $\dfrac{d^2x}{dt^2} + y = \sin t, \dfrac{d^2y}{dt^2} + x = \cos t$

12. $(D^2 - 1)x + 8Dy = 16e^t,\ Dx + 3(D^2 + 1)y = 0$

13. $\dfrac{dx}{dt} + 7x - y = 0, \dfrac{dy}{dt} + 2x + 5y = 0)$

14. $\dfrac{dx}{dt} + x - 2y = 0, \dfrac{dy}{dt} + x + 4y = 0;\ x(0) = y(0) = 1$

15. $\dfrac{dx}{dt} = 3x + 8y, \dfrac{dy}{dt} = -x - 3y;\ x(0) = 6, y(0) = -2$

16. $\dfrac{dx}{dt} = y + 1, \dfrac{dy}{dt} = x + 1$

17. $\dfrac{dx}{dt} - y = e^t, \dfrac{dy}{dt} + x = \sin t;\ x(0) = 1, y(0) = 0$

18. $\dfrac{dx}{dt} + 5x + y = e^t, \dfrac{dy}{dt} + x + 5y = e^{5t}$

19. $\dfrac{dx}{dt} + \dfrac{dy}{dt} + 2x + y = 0, \dfrac{dy}{dt} + 5x + 3y = 0$

20. $\dfrac{dx}{dt} = -4(x+y), \dfrac{dx}{dt} + 4\dfrac{dy}{dt} = -4y$ with conditions $x(0) = 1, y(0) = 0$

21. $\dfrac{dx}{dt} + y = \sin t, \dfrac{dy}{dt} + x = \cos t$ given that $x = 2$ and $y = 0$ when $t = 0$

22. $\dfrac{dx}{dt} + 2x + 3y = 0,\ \ 3x + \dfrac{dy}{dt} + 2y = 2e^{2t}$

23. $\dfrac{dx}{dt} + 2y = e^t, \dfrac{dy}{dt} - 2x = e^{-t}$

24. $\dfrac{d^2x}{dt^2} - 3x - 4y = 0, \dfrac{d^2y}{dt^2} + x + y = 0$

25. $\dfrac{d^2x}{dt^2} + y = \sin t, \dfrac{d^2y}{dt^2} + x = \cos t$

HINT TO SELECTED PROBLEMS

1. Eliminating y, we get

$$[(D^2 + 1)(D^2 - 3) + 4]x = 0$$

$$\Rightarrow \qquad (D^2 - 1)^2 x = 0$$

The auxiliary equation is $(m^2 - 1)^2 = 0$

$$\Rightarrow m = \pm 1, \pm 1$$

$$x = (C_1 + C_2 t)e^{-t} + (C_3 + C_4 t)\,e^t$$

$$\frac{dx}{dt} = -(C_1 + C_2 t)e^{-t} + C_2 e^{-t} + (C_3 + C_4 t)e^t + C_4 e^t$$

$$\frac{d^2x}{dt^2} = (C_1 + C_2 t)e^{-t} - 2C_2 e^{-t} + (C_3 + C_4 t)^t + 2C_4 e^t$$

Now, for given equation

$$4y = D^2x - 3x$$

$$= (C_1 + C_2t)\,e^{-t} - 2C_2e^{-t} + (C_3 + C_4t)\,e^t$$

$$\quad + 2C_4e^t - 3(C_1 + C_2t)e^{-t} - 3(C_2t + C_4t)\,e^t$$

$$= -(2C_1 + 2C_2 + 2C_2t)e^{-t}$$

$$\quad + (-2C_3 + 2C_4 - 2C_4t)e^t$$

$$= -\frac{1}{2}(C_1 + C_2 + C_2t)\,e^{-t} + \frac{1}{2}(C_4 + C_3 - C_5t)\,e^t$$

3. The given equation can be written as

$$(D^4 + 4)x + y = t\,e^{3t} \qquad \text{...(1)}$$

$$-2x + (D^2 + 1)y = \cos^2 t \qquad \text{...(2)}$$

Eliminating y, we get

$$\left[(D^2 + 1)(D^2 + 4) + 2\right]x = (D^2 + 1)t\,e^{3t} - \cos^2 t$$

$$(D^4 + 5D^2 + 6)x = 10t\,e^{3t} - \cos^2 t + 6e^{3t}$$

$$m^4 + 5m^2 + 6 = 0$$

$$(m^3 + 3)(m^2 + 2) = 0$$

$$m = \pm\sqrt{3}\,i,\ \pm\sqrt{2}\,i$$

C.F. $= (C_1 \cos\sqrt{3}t + C_2 \sin\sqrt{3}t)$

$$\quad + (C_3 \cos\sqrt{2}t + C_4 \sin\sqrt{2}t)$$

P.I. $= \dfrac{10}{D^4 + 5D^2 + 6}t\,e^{3t} + \dfrac{6}{D^4 + 5D^2 + 6}e^{3t}$

$$\quad - \frac{1}{D^4 - 5D^2 + 6}\cos^2 t$$

$$= 10\,e^{3t} \cdot \frac{1}{(D+3)^4 + 5(D+3)^2 + 6}t$$

$$\quad + \frac{6e^{3t}}{3^4 + 5.3^2 + 6}$$

$$\quad - \frac{1}{(D^4 + 5D^2 + 6)} \cdot \frac{1}{2}(1 + \cos 2t)$$

$$= 10e^{3t} \cdot \frac{1}{131 + 138D + 59D^2 + \ldots}t + \frac{1}{22}e^{3t}$$

$$\quad - \frac{1}{6 + 5D^2 + D^4} \cdot \frac{1}{2} - \frac{1}{D^4 + 5D^2 + 6}\left(\frac{1}{2}\cos 2t\right)$$

$$= 10^{3t}\frac{1}{132}\left(1 + \frac{23}{22}D + \frac{59}{132}D^2 + \ldots\right)^{-1}t$$

$$\quad + \frac{e^{3t}}{22} - \frac{1}{6}\left(1 + \frac{5D^2}{6} + \frac{D^4}{6}\right)^{-1}$$

$$\quad \frac{1}{2} - \frac{\dfrac{1}{2}\cos 2t}{(-2)^2 + 5(-2)^2 + 6}$$

$$= \frac{5}{66}te^{3t} - \frac{49}{1452}e^{3t} - \frac{1}{12} - \frac{1}{4}\cos 2t$$

$$x = (C_1 \cos\sqrt{3}t + C_2 \sin\sqrt{3}t)$$

$$\quad + (C_3 \cos\sqrt{2}t + C_4 \sin\sqrt{2}t)$$

$$\quad + \frac{5}{66}t\,e^{3t} - \frac{49}{1452}e^{3t} - \frac{1}{12} - \frac{1}{4}\cos 2t \qquad \text{...(3)}$$

$$\frac{dx}{dt} = \left(-C_1\sqrt{3}\sin\sqrt{3}t + C_2\sqrt{3}\cos\sqrt{3}t\right)$$

$$\quad + \left(-C_3\sqrt{3}\sin\sqrt{2}t + C_4\sqrt{2}\cos\sqrt{2}t\right)$$

$$\quad + \frac{5}{66}\left(3t\,e^{3t} + e^{3t}\right) - \frac{49}{1452}3e^{3t} + \frac{1}{2}\sin 2t$$

$$\frac{d^2x}{dt^2} = -3(C_1\cos\sqrt{3}t + C_2\sin\sqrt{3}t)$$

$$\quad - 2(C_3\cos\sqrt{2}t + C_4\sin\sqrt{2}t)$$

$$\quad + \frac{5}{66}(9te^{3t} + 6e^{3t}) - \frac{49}{1452}9e^{3t} + \cos 2t$$

Substituting in (1), we get

$$y = -\frac{d^2x}{dt^2} - 4x + te^{3t}$$

$$y = \left(C_1\cos\sqrt{3}t + C_2\sin\sqrt{3}t\right)$$

$$\quad - 2(C_3\cos\sqrt{2}t + C_4\sin\sqrt{2}t)$$

$$\quad + \frac{1}{66}te^{2t} - \frac{23}{1452}e^{3t} + \frac{1}{3} \qquad \text{...(4)}$$

6. The given equation can be written as

$$(D + 2)x - 3y = t \qquad \text{...(1)}$$

$$-3x + (D + 2)y = e^{2t} \qquad \text{...(2)}$$

Eliminating y, we get

$$[(D + 2)^2 - 9]x = (D + 2)t + 3e^{2t}$$

$$(D^2 + 4 + 4D - 9)x = (D + 2)t + 3e^{2t}$$

$$= (1 + 2t) + 3e^{2t}$$

A.E. $\quad m^2 + 4m - 5 = 0 \qquad m = 1, -5$

C.F. $= C_1e^{-5t} + C_2e^t$

P.I. $= \dfrac{1 + 2t}{(D - 1)(D + 5)} + 3\dfrac{e^{2t}}{(D - 1)(D + 5)}$

$$= -\frac{1}{5}\left[1 - \frac{4}{5}D - \frac{1}{5}D^2\right]^{-1}(1 + 2t) + \frac{3e^{2t}}{(2 - 1)(2 + 5)}$$

$$= -\frac{1}{5}\left[1 + \frac{4}{5}D + \frac{1}{5}D^2 + \ldots\right](1 + 2t) + \frac{3}{7}e^{2t}$$

$$= -\frac{13}{25} - \frac{2}{5}t + \frac{3}{7}e^{2t}$$

$$x = C_1 e^{-5t} + C_2 e^t + \frac{3}{7} e^{2t} - \frac{2}{5} t - \frac{13}{25}$$

$$\frac{dx}{dt} = -5C_1 e^{-5t} + C_2 e^t + \frac{6}{7} e^{2t} - \frac{2}{5}$$

From (1) :

$$3y = -5C_1 e^{-5t} + C_2 e^t + \frac{6}{7} e^{2t} - \frac{2}{5} + 2C_1 e^{-5t}$$

$$+ 2C_2 e^t + \frac{6}{7} e^{2t} - \frac{4}{5} t - \frac{26}{25} t$$

$$y = -C_1 e^{-5t} + C_2 e^t + \frac{4}{7} e^{2t} - \frac{12}{25} - \frac{3}{5} t$$

9. The given equation can be written as

$$t \frac{dx}{dt} + 2(x - y) = t \qquad \ldots (1)$$

$$t \frac{dy}{dt} + x + 5y = t^2 \qquad \ldots (2)$$

Differentiating (1) w.r.t. t, we have

$$t \frac{d^2 x}{dt^2} + \frac{dx}{dt} + 2 \frac{dx}{dt} - 2 \frac{dy}{dt} = 1$$

$$t^2 \frac{d^2 x}{dt^2} + t \frac{dx}{dt} + 2t \frac{dx}{dt} - 2t \frac{dy}{dt} = t \qquad \ldots (3)$$

Substituting the value of $t \dfrac{dy}{dt}$ from (2) in (1), we get

$$t^2 \frac{d^2 x}{dt^2} + 3t \frac{dx}{dt} + 2x + 5\left(t \frac{dx}{dt} + 2x - t\right) - 2t^2 = t$$

$$\Rightarrow t \frac{d^2 x}{dt^2} + 8t \frac{dx}{dt} + 12x = 2t^2 + 6t \qquad \ldots (4)$$

which is a homogeneous linear equation.

Put $t = e^z$

$$\frac{dx}{dt} = \frac{dx}{dz} \cdot \frac{dz}{dt} = \frac{dx}{dz} \cdot \frac{1}{t}$$

$$t \frac{d}{dt}\left(t \frac{dx}{dt}\right) = t^2 \frac{d^2 x}{dt^2} + t \frac{dx}{dt}$$

$$t^2 D^2 x = (D - 1)Dx$$

Equation (4) gives

$$[(D - 1)D + 8D + 12] x = 2e^{2z} + 6e^z$$

$$(D^2 + 7D + 12)x = 2e^{2z} + 6e^z$$

A.E. is $m^2 + 7m + 12 = 0 \Rightarrow m = -3, -4$

C.F. $= C_1 e^{-3z} + C_2 e^{-4z}$

P.I. $= \dfrac{2}{(D^2 + 7D + 12)} e^{2z} + \dfrac{6}{(D^2 + 7D + 12)} e^z$

$$= \frac{2}{30} e^{2z} + \frac{6}{20} e^z = \frac{1}{15} e^z + \frac{3}{10} e^z$$

$$x = C_1 e^{-3z} + C_2 e^{-4z} + \frac{1}{15} e^{2z} + \frac{3}{10} e^z$$

$$x = \frac{C_1}{t^3} + \frac{C_2}{t^4} + \frac{t^2}{15} + \frac{3}{10} t$$

$$\frac{dx}{dt} = -\frac{3C_1}{t^4} - \frac{4C_2}{t^5} + \frac{2t}{15} + \frac{3}{10}$$

From (1) :

$$2y = -\frac{3C_1}{t^3} - \frac{4C_2}{t^4} + \frac{2t^2}{15} + \frac{3t}{10}$$

$$+ \frac{2C_1}{t^3} + \frac{2C_2}{t^3} + \frac{2t^2}{15} + \frac{6}{10} t - t$$

$$y = -\frac{C_1}{2t^3} - \frac{C_2}{t^4} + \frac{2t^2}{15} + \frac{3}{10}$$

ANSWERS

1. $x = (C_1 + C_2 t)e^{-t} + (C_3 + C_4 t) e^t$, $y = -\dfrac{1}{2}(C_1 + C_2 + C_2 t)e^{-t} + \dfrac{1}{2}(C_4 - C_3 - C_4 t)e^t$

2. $x = C_1 e^{(3 + \sqrt{10})t} + C_2 e^{(3 - \sqrt{10})t}$, $y = \dfrac{1}{2}\sqrt{10} [C_1 e^{(3 + \sqrt{10})t} - C_2 e^{(3 - \sqrt{10})t}]$

3. $x = (C_1 \cos\sqrt{3}t + C_2 \sin\sqrt{3}t) + (C_3 \cos\sqrt{2}t + C_4 \sin\sqrt{2}t) + \dfrac{5}{66} t e^{3t} - \dfrac{49}{1452} e^{3t} - \dfrac{1}{12} - \dfrac{1}{4} \cos 2t$

   $y = (C_1 \cos\sqrt{3}t + C_2 \sin\sqrt{3}t) - (C_3 \cos\sqrt{2}t + C_4 \sin\sqrt{2}t) + \dfrac{1}{60} t e^{3t} - \dfrac{23}{1452} e^{3t} + \dfrac{1}{3}$

4. $x = (C_1 + C_2 t)e^{-4t} + \dfrac{4}{25} e^t - \dfrac{1}{36} e^{2t}$; $y = -(C_1 + C_2 + C_3 t) e^{-4t} + \dfrac{7}{36} e^{2t} + \dfrac{1}{25} e^t$

5. $x = C_1 \cos t + C_2 \sin t$, $y = \dfrac{1}{2}(C_2 - 3C_1)\cos t - \dfrac{1}{2}(C_1 + 3C_2)\sin t$

6. $x = C_1 e^{-5t} + C_2 e^t + \dfrac{3}{7} e^{2t} - \dfrac{2}{5} t - \dfrac{13}{25}$, $y = -C_1 e^{-5t} + C_2 e^t + \dfrac{4}{7} e^{2t} - \dfrac{3}{5} t - \dfrac{12}{25}$

7. $y = C_1 e^{3x} + C_2 e^{5x}$, $z = -7C_1 e^{3x} + 6C_2 e^{5x}$

8. $y = (C_1 + C_2 x) e^x + C_3 e^{-3x/2} - \dfrac{1}{2}x$, $z = -2(C_1 + C_2 x - 3C_2)e^x - \dfrac{1}{3}C_3 e^{-3x/2} - \dfrac{1}{3}$

9. $x = \dfrac{C_1}{t^3} + \dfrac{C_2}{t^4} + \dfrac{t^2}{15} + \dfrac{3}{10}t$, $y = -\dfrac{C_1}{2t^3} - \dfrac{C_2}{t^4} + \dfrac{2t^2}{15} - \dfrac{t}{20}$

10. $x = -\dfrac{1}{27}(1 + 6t)e^{-3t} + \dfrac{1}{27}(1 + 3t)$, $y = -\dfrac{2}{27}(2 + 3t)e^{-3t} + \dfrac{2}{27}(2 - 3t)$

11. $x = C_1 e^t + C_2 e^{-t} + C_3 \cos t + C_4 \sin t + \dfrac{t}{4}(\sin t - \cos t)$,

$y = -C_1 e^t - C_2 e^{-t} + C_3 \cos t + C_4 \sin t + \dfrac{1}{4}(2 + t)(\sin t - \cos t)$

12. $y = C_1 \cos \dfrac{t}{\sqrt{3}} + C_2 \sin \dfrac{t}{\sqrt{3}} + C_3 \cosh \sqrt{3}.t + C_4 \sinh \sqrt{3}t + 2e^t$; $x = \sqrt{3}C_1 \sin \dfrac{t}{\sqrt{3}} - \sqrt{3}C_2 \cos \dfrac{t}{\sqrt{3}}$

$-3\sqrt{3}C_3 \sinh \sqrt{3}t - 3\sqrt{3}C_4 \cosh \sqrt{3}t - 6e^t - 3t$

13. $x = e^{-6t}(A\cos t + B\sin t)$, $y = e^{-6t}[(A + B)\cos t - (A - B)\sin t]$

14. $x = 4e^{-2t} - 3e^{-3t}$, $y = -2e^{-2t} + 3e^{-3t}$ **15.** $x = 4e^t + 2e^{-t}$, $y = -e^t - e^{-t}$

16. $x = C_1 e^t + C_2 e^{-t} - 1$, $y = C_1 e^t - C_2 e^{-t} - 1$

17. $x = 2\sin t + \dfrac{3}{2}\cos t + \dfrac{t}{2}\cos t - \dfrac{1}{2}e^t$, $y = \dfrac{1}{2}\cos t - \dfrac{3}{2}\sin t + \dfrac{t}{2}\sin t - \dfrac{1}{2}e^t$

18. $x = C_1 e^{-6t} + C_2 e^{-4t} + \dfrac{6e^t}{35} - \dfrac{e^{5t}}{99}$, $y = C_1 e^{-6t} - C_2 e^{-4t} - \dfrac{1}{35}e^t + \dfrac{10}{99}e^{5t}$

19. $x = \left(\dfrac{C_1 - 3C_2}{5}\right)\sin t - \left(\dfrac{C_2 + 3C_1}{5}\right)\cos t$, $y = C_1 \cos t + C_2 \sin t$ **20.** $x = (1 - 2t)e^{-2t}$, $y = te^{-2t}$

21. $x = e^t + e^{-t}$, $y = e^{-t} - e^t + \sin t$ **22.** $x = C_1 e^t + C_2 e^{-st} + \dfrac{6}{7}e^{2t}$; $y = C_2 e^{-st} - C_1 e^t + \dfrac{8}{7}e^{2t}$

23. $x = \dfrac{1}{5}e^t + \dfrac{2}{5}e^{-t} - C_1 \sin 2t + C_2 \cos 2t$, $y = \dfrac{2}{5}e^t + \dfrac{1}{5}e^{-t} + C_1 \cos 2t + C_2 \sin 2t$

24. $x = (C_1 + C_2 t)e^{-t} + (C_3 + C_4 t)e^t$, $y = -\dfrac{1}{2}[C_1 + C_2(1 + t)]e^{-t} + \dfrac{1}{2}[C_4(1 - t) - C_3]e^t$

25. $x = C_1 e^t + C_2 e^{-t} + C_3 \cos t + C_4 \sin t - \dfrac{t}{4}\cos t + \dfrac{t}{4}\sin t$,

$y = -C_1 e^t - C_2 e^{-t} + C_3 \cos t + C_4 \sin t + \dfrac{1}{4}(2 + t)(\sin t - \cos t)$

13.18 SIMULTANEOUS EQUATIONS IN DIFFERENT FORM

Consider the equations of the type $P_1 dx + Q_1 dy + R_1 dz = 0$

$P_2 dx + Q_2 dy + R_2 dz = 0$ ... (1)

where P_1, P_2, Q_1, Q_2, R_1 and R_2 are functions of x, y, z.

Equation (1) can be written as

$$P_1 \frac{dx}{dz} + Q_1 \frac{dy}{dz} + R_1 = 0 \,, P_2 \frac{dx}{dz} + Q_2 \frac{dy}{dz} + R_2 = 0 \,.$$

Solving the above equations for $\frac{dx}{dz}, \frac{dy}{dz}$, we get $\quad \frac{dx}{dz} = \frac{Q_1 R_2 - Q_2 R_1}{P_1 Q_2 - Q_1 P_2} \,, \frac{dy}{dz} = \frac{R_1 P_2 - P_1 R_2}{P_1 Q_2 - Q_1 P_2}$

Hence, $$\frac{dx}{Q_1 R_2 - Q_2 R_1} = \frac{dy}{R_1 P_2 - R_2 P_1} = \frac{dz}{P_1 Q_2 - P_2 Q_1}$$

i.e., the equation can be put in the form $\dfrac{dx}{P} = \dfrac{dy}{Q} = \dfrac{dz}{R}$

where P, Q and R the functions of x, y and z.

WORKING PROCEDURE

Method-I

Step 1. Take any two member of an equation (1) say $\dfrac{dx}{P} = \dfrac{dy}{Q}$ (say)

After integrating it, we may get an equation.

Step 2. Again take two member of equation (1) say $\dfrac{dy}{Q} = \dfrac{dz}{R}$ (say)

After integrating it, we also get an equation.

Step 3. The solution obtained from (i) and (ii) give the required general solution.

Method-II

Step 1. The given equation is $\dfrac{dx}{P} = \dfrac{dy}{Q} = \dfrac{dz}{R}$

If we choose l, m and n such that $\dfrac{dx}{P} = \dfrac{dy}{Q} = \dfrac{dz}{R} = \dfrac{l \, dx + m \, dy + n \, dz}{lP + mQ + nR}$

If $lP + mQ + nR = 0$, then, $l \, dx + m \, dy + n \, dz = 0$

If it is an exact differential, say du, then $u = a$ is one equation of the complete solution.

REMARKS

- To find a solution of the given equation, we choose l, m, n such that $l \, dx + m \, dy + n \, dz$ is differential of $lP + mQ + nR$.
- If we have obtained one solution, then this solution can be used to simplify the other differential equations in the integrable form.
- Sometimes, we use only one set of multiples, but in some cases, we have a need of more than one set of multipliers.
- We can obtain one relation, say $u = a$ by the first method and the second relation by the second method.

13.18.1 Geometrical meaning of $\dfrac{dx}{P} = \dfrac{dy}{Q} = \dfrac{dz}{R}$

Since, we know that the direction cosines of the tangent to a curve at any point (x, y, z) are $\dfrac{dx}{ds}, \dfrac{dy}{ds}, \dfrac{dz}{ds}$ or proportional to dx, dy, dz. Therefore, geometrically the above situations represents a system of curves in such a way that the direction-ratios of the tangent from it at any point $A(x, y, z)$ are proportional to P, Q and R. If $u = a$ and $v = b$ are the complete solutions $\dfrac{dx}{P} = \dfrac{dy}{Q} = \dfrac{dz}{R}$, then system of curves is intersection of the surfaces $u = a$, $v = b$. It is also clear that since a, b are arbitrary constants, the system of curves represented by the equations is doubly infinite.

SOLVED EXAMPLES

EXAMPLE 1. *Solve the simultaneous equations.*

$$\frac{a\,dx}{(b - c)\,yz} = \frac{b\,dy}{(c - a)zx} = \frac{c\,dz}{(a - b)\,xy}.$$

[UPTU(B.Pharma)–2001]

SOLUTION. Let us take the x, y, z as multipliers.

Each fraction $= \dfrac{ax\,dx + by\,dy + cz\,dz}{0}$

$\therefore \qquad ax\,dx + by\,dy + cz\,dz = 0$

Integrating $\quad ax^2 + by^2 + cz^2 = C_1$ ... (1)

Now taking ax, by, cz as multipliers.

Each fraction $= \dfrac{a^2 x\,dx + b^2 y\,dy + c^2 z\,dz}{0}$

$\therefore \quad a^2 x\,dx + b^2 y\,dy + c^2 z\,dz = 0$

On integrating,

$$a^2 x^2 + b^2 y^2 + c^2 z^2 = 0 \qquad \text{... (2)}$$

Hence, complete solution is

$$\phi(ax^2 + by^2 + cz^2, \; a^2 x^2 + b^2 y^2 + c^2 z^2) = 0$$

EXAMPLE 2. *Solve* $\dfrac{xdx}{z^2 - 2yz - y^2} = \dfrac{dy}{y + z} = \dfrac{dz}{y - z}$.

SOLUTION. Let us take $1, y, z$ as multiplier, we get

Each fraction $= \dfrac{xdx + ydy + zdz}{0}$

$xdx + ydy + zdz = 0$

Integrating, $\quad x^2 + y^2 + z^2 = C_1$ ... (1)

Again, last two members, we get

$$\frac{dy}{y + z} = \frac{dz}{y - z}$$

$$ydy - zdy = ydz + zdz$$

or $ydy - (ydz + zdy) - zdz = 0$

Integrating $y^2 - 2yz - z^2 = C_2$ ... (2)

Complete solution is

$$x^2 + y^2 + z^2 = C_1$$

$$y^2 - 2yz - z^2 = C_2$$

EXAMPLE 3. *Solve the simultaneous equation*

$$\frac{dx}{y^2 + z^2 - x^2} = \frac{dy}{-2xy} = \frac{dz}{-2xz}.$$

SOLUTION. From last two members, we get

$$\frac{dy}{y} = \frac{dz}{z}$$

$\therefore$ $y = C_1 z$ ... (1)

Now, taking x, y, z as multiplier, we get

Each fraction $= \dfrac{dz}{-2xz} = \dfrac{xdx + ydy + zdz}{-x(x^2 + y^2 + z^2)}$

$$\frac{dz}{z} = \frac{2xdx + 2ydy + 2zdz}{x^2 + y^2 + z^2}$$

Integrating $\log z + \log C_2 = \log(x^2 + y^2 + z^2)$

$$x^2 + y^2 + z^2 = C_2 z$$ ... (2)

Complete solution is $y = C_1 z$

$$x^2 + y^2 + z^2 = C_2 z$$

EXAMPLE 4. *Solve* $\dfrac{dx}{x^2 + y^2 + yz} = \dfrac{dy}{x^2 + y^2 - xz} = \dfrac{dz}{z(x + y)}.$

SOLUTION. Given equation can change to be new form as

$$\frac{dx - dy}{z(x + y)} = \frac{dz}{z(x + y)} \quad \text{or} \quad dx - dy = dz$$

Integrating $x - y - z = C_1$ ... (1)

Again from the given equation, we get

$$\frac{xdx + ydy}{(x + y)(x^2 + y^2)} = \frac{dz}{z(x + y)}$$

or $\dfrac{xdx + ydy}{x^2 + y^2} = \dfrac{dz}{z}.$

Integrating, $\log(x^2 + y^2) = 2\log z + \log C_2$

$$\therefore \qquad x^2 + y^2 = z^2 C_2 \qquad \qquad \dots (2)$$

From (1) and (2), we get complete solution

$$x - y - z = C_1$$

$$x^2 + y^2 = z^2 C_2$$

EXAMPLE 5.   *Solve* $\dfrac{dx}{x(y^2 - z^2)} = \dfrac{dy}{-y(z^2 + x^2)} = \dfrac{dz}{z(x^2 + y^2)}.$

SOLUTION.   Taking $\dfrac{1}{x}, -\dfrac{1}{y}, -\dfrac{1}{z}$ as multipliers, we get

$$\text{Each fraction} = \frac{\dfrac{dx}{x} - \dfrac{dy}{y} - \dfrac{dz}{z}}{0}$$

$$\therefore \quad \frac{dx}{x} - \frac{dy}{y} - \frac{dz}{z} = 0 \quad \text{or} \quad \frac{dy}{y} + \frac{dz}{z} = \frac{dx}{x}$$

Integrating, $\log y + \log z = \log x + \log C_1.$

$$\therefore \qquad yz = C_1 x \qquad \qquad \dots (A)$$

Again using x, y, z as multipliers, we get

$$\text{Each fraction} = \frac{xdx + ydy + zdz}{0}$$

$$\therefore \qquad xdx + ydy + zdz = 0$$

Integrating $x^2 + y^2 + z^2 = C_2$. $\dots (B)$

From (A) and (B), we obtain complete solution.

EXAMPLE 6.   *Solve* $\dfrac{dx}{y^3 x - 2x^4} = \dfrac{dy}{2y^4 - x^3 y} = \dfrac{dz}{9z(x^3 - y^3)}$

SOLUTION.   Taking first two members, we get

$$(2y^4 - x^3 y)dx = (y^3 x - 2x^4)dy \qquad \dots (1)$$

Dividing (1) by $x^3 y^3$,

$$\left[\frac{2y}{x^3} - \frac{1}{y^2} \right] dx = \left[\frac{1}{x^2} - \frac{2x}{y^3} \right] dy$$

or $\qquad \left[\dfrac{1}{x^2} dy - \dfrac{2y}{x^3} dx \right] + \left[\dfrac{1}{y^2} dx - \dfrac{2x}{y^3} dy \right] = 0$

Integrating $\qquad \dfrac{y}{x^2} + \dfrac{x}{y^2} = C_1 \qquad \qquad \dots (1)$

Again using $\dfrac{1}{x} \cdot \dfrac{1}{y} \cdot \dfrac{1}{3z}$ as multipliers, we get

$$= \frac{\frac{1}{x}dx + \frac{1}{y}dy + \frac{1}{3z}dz}{(y^3 - 2x^3) + (2y^3 - x^3) + 3(x^3 - y^3)}$$

$$= \frac{\frac{1}{x}dx + \frac{1}{y}dy + \frac{1}{3z}dz}{0}$$

$$\therefore \qquad \frac{1}{x}dx + \frac{1}{y}dy + \frac{1}{3z}dz = 0$$

On integrating, we get

$$\log x + \log y + \frac{1}{3}\log z = \log C_2 \qquad \qquad \ldots (2)$$

or $\qquad \qquad xyz^{1/3} = C_2$

Hence, complete solution is given by

$$\frac{y}{x^2} + \frac{x}{y^2} = C_1, xyz^{1/3} = C_2$$

EXAMPLE 7. *Solve* $\dfrac{dx}{\cos(x+y)} = \dfrac{dy}{\sin(x+y)} = \dfrac{dz}{z}$.

SOLUTION.

$$\frac{dx}{\cos(x+y)} = \frac{dy}{\sin(x+y)} = \frac{dz}{z}.$$

$$\Rightarrow \qquad \frac{dx}{\cos(x+y)} = \frac{dy}{\sin(x+y)} = \frac{dz}{z} = \frac{dx + dy}{\cos(x+y) + \sin(x+y)}$$

Taking $\qquad \dfrac{dz}{z} = \dfrac{dx + dy}{\cos(x+y) + \sin(x+y)}$

$$\Rightarrow \qquad \frac{dz}{z} = \frac{du}{\cos u + \sin u}, u = x + y$$

$$\Rightarrow \qquad \frac{dz}{z} = \frac{du}{\sqrt{2}\sin(u + \pi/4)}$$

$$\Rightarrow \qquad \frac{dz}{z} = \frac{1}{\sqrt{2}}\mathrm{cosec}(u + \pi/4)du.$$

Integrating both sides, we get

$$\log z = \frac{1}{\sqrt{2}}\log\tan\left(\frac{u + \pi/4}{2}\right) + \frac{1}{\sqrt{2}}\log C_1$$

$$z^{\sqrt{2}} = c_1 \tan\left(\frac{u}{2} + \frac{\pi}{8}\right) = c_1 \tan\left(\frac{x+y}{2} + \frac{\pi}{8}\right)$$

Also each fraction is equal to

$$\frac{dx + dy}{\cos(x+y) + \sin(x+y)} = \frac{dx - dy}{\cos(x+y) - \sin(x+y)}$$

or $\dfrac{\cos(x+y) - \sin(x+y)}{\cos(x+y) + \sin(x+y)}(dx + dy) = dx - dy$

Integrating, we get

$$\log[\cos(x+y) + \sin(x+y)] = x - y + \log c_2$$

or $\quad \cos(x+y) + \sin(x+y)]e^{y-x} = c_2$

Hence the solution is $f\left\{\{\cos(x+y) + \sin(x+y)\}e^{y-x} z^{\sqrt{2}} \cot\left(\dfrac{x+y}{2} + \dfrac{\pi}{8}\right)\right\} = 0$

Exercise 13.17

Solve the following simultaneous differential equations :

1. $\dfrac{dx}{xy} = \dfrac{dy}{y^2} = \dfrac{dz}{zyx - 2x^2}$

2. $\dfrac{dx}{x^2 + y^2} = \dfrac{dy}{2xy} = \dfrac{dz}{(x+y).z}$

3. $\dfrac{dx}{(x^2 - yz)} = \dfrac{dy}{y^2 - zx} = \dfrac{dz}{z^2 - xy}$

4. $\dfrac{dx}{yz} = \dfrac{dy}{zx} = \dfrac{dz}{xy}$

5. $\dfrac{dx}{y+z} = \dfrac{dy}{z+x} = \dfrac{dz}{x+y}$

6. $\dfrac{dx}{mz - ny} = \dfrac{dy}{nx - lz} = \dfrac{dz}{ly - mx}$

7. $\dfrac{dx}{z(x+y)} = \dfrac{dy}{z(x-y)} = \dfrac{dz}{x^2 + y^2}$

8. $\dfrac{dx}{z} = \dfrac{dy}{-z} = \dfrac{dz}{z^2 + (x+y)^2}$

9. $\dfrac{dx}{x(y-z)} = \dfrac{dy}{y(z-x)} = \dfrac{dz}{z(x-y)}$

10. $\dfrac{dx}{x^2 + y^2} = \dfrac{dy}{2xy} = \dfrac{dz}{(x+y)^2}$

11. $\dfrac{dx}{y^2 + yz + z^2} = \dfrac{dy}{z^2 + zx + x^2} = \dfrac{dz}{x^2 + xy + y^2}$

12. $\dfrac{dx}{\cos(x+y)} = \dfrac{dy}{\sin(x+y)} = \dfrac{dz}{z}$

HINT TO SELECTED PROBLEMS

1. Taking the first two members, we get

$$\dfrac{dx}{xy} = \dfrac{dy}{y^2} \quad \text{or} \quad \dfrac{dx}{x} = \dfrac{dy}{y}$$

Integrating $\log x = \log y + \log C_1$

$$x = C_1 y \qquad \qquad \dots(1)$$

Again taking the last two members, we have

$$\dfrac{dy}{y^2} = \dfrac{dz}{zxy - 2x^2}$$

$$\dfrac{dy}{y^2} = \dfrac{dz}{zC_1 y^2 - 2C_1^2 y^2} \qquad \text{from (1)}$$

$$dy = \dfrac{dz}{zC_1 - 2C_1^2} \quad \text{or} \quad C_1 dy = \dfrac{dz}{z - 2C_1}$$

Integrating, we get

$$C_1 y = \log(z - 2C_1) + C_2$$

$$x = \log\left(z - \dfrac{2x}{y}\right) + C_2 \qquad [\because C_1 y = x]$$

$$x = \log(zy^{-2}) - \log y + C_3 \qquad \dots (2)$$

3. Obviously, each of the given ratios

$$= \dfrac{dx - dy}{x^2 yz - y^2 + zx} \qquad \dots (1)$$

$$= \dfrac{dy - dz}{y^2 - zx - z^2 + xy} \qquad \dots (2)$$

$$= \dfrac{dz - dx}{x^2 - xy - x^2 + yz} \qquad \dots (3)$$

From (1) and (2)

$$\dfrac{dx - dy}{(x-y)(x+y+z)} = \dfrac{dy - dz}{(y-z)(x+y+z)}$$

$$\dfrac{dx - dy}{x - y} = \dfrac{dy - dz}{y - z}$$

Integrating, we get

$$\log(x - y) = \log(y - z) + \log a$$

$$\frac{x - y}{y - z} = a \qquad \ldots (4)$$

From (2) and (3), similarly

$$\frac{y - z}{z - x} = b \qquad \ldots (5)$$

From (4) and (5), the complete solution of the equation is

$$\phi\left(\frac{x - y}{y - z}, \frac{y - z}{z - x}\right) = 0$$

7. Using $x, -y, -z$ as multipliers, we have

Each fraction

$$= \frac{x\,dx - y\,dy - z\,dz}{xz(x + y) - yz(x - y) - z(x^2 + y^2)}$$

$$= \frac{x\,dx + y\,dy - z\,dz}{0}$$

$$x\,dx - y\,dy - z\,dz = 0$$

Integrating, $x^2 - y^2 - z^2 = C_1$ $\qquad \ldots (1)$

Similarly, using $y, x, -z$ as multipliers, we get

Each fraction

$$= \frac{y\,dx + x\,dy - z\,dz}{yz(x + y) + xz(x - y) - z(x^2 + y^2)}$$

$$= \frac{y\,dx + x\,dy - z\,dz}{0}$$

$\therefore \quad y\,dx + x\,dy - z\,dz = 0$

Integrating, $2xy - z^2 = C_2$

9. Obviously, $\dfrac{dx}{x(y - z)} = \dfrac{dy}{y(z - x)} = \dfrac{dz}{z(x - y)}$

$$= \frac{dx + dy + dz}{xy - xz + yz - yx + zx - zy}$$

$$dx + dy + dz = 0$$

Integrating, $x + y + z = C_1$ $\qquad \ldots (1)$

Now using $\dfrac{1}{x}, \dfrac{1}{y}, \dfrac{1}{z}$ as multipliers, we have

$$\frac{\frac{1}{x}dx}{y - z} = \frac{\frac{1}{y}dy}{z - x} = \frac{\frac{1}{z}dz}{x - y}$$
$$\quad I \qquad\quad II \qquad\quad III$$

$$I = II = III \;=\; \frac{\frac{1}{x}dx + \frac{1}{y}dy + \frac{1}{z}dz}{y - z + z - x + x - y}$$

$$\frac{1}{x}dx + \frac{1}{y}dy + \frac{1}{z}dz = 0$$

On integrating, we get

$$\log x + \log y + \log z = \log C_2$$

$$xyz = C_2 \qquad \ldots (2)$$

10. Obviously,

$$\frac{dx + dy}{x^2 + y^2 + 2xy} = \frac{dx - dy}{x^2 + y^2 - 2xy} = \frac{dz}{(x + y)^2}$$

$$\frac{dx + dy}{(x + y)^2} = \frac{dx - dy}{(x - y)^2} = \frac{dz}{(x + y)^2}$$
$$\quad I \qquad\qquad II \qquad\qquad III$$

Taking first two members

$$\frac{dx + dy}{(x + y)^2} = \frac{dx - dy}{(x - y)^2}$$

Integrating, we get

$$-(x + y)^{-1} = -(x - y)^{-1} + C_1$$

$$\frac{1}{x - y} - \frac{1}{x + y} = C_1$$

$$\frac{2y}{x^2 - y^2} = C_1 \qquad \ldots (1)$$

Now, taking first and last members

$$\frac{dx + dy}{(x + y)^2} = \frac{dz}{(x + y)^2}$$

$$dx + dy - dz = 0$$

Integrating, we get $x + y - z = C_2$ $\qquad \ldots (2)$

From equation (1) and (2), the complete solution is given by

$$\phi\left(\frac{2y}{x^2 - y^2}, \; x + y - z\right) = 0$$

ANSWERS

1. $\phi\left[\left(\dfrac{x}{y}\right), x - \log(zy - 2) + \log y\right] = 0$

2. $\phi\left(\dfrac{x + y}{z}, \dfrac{2y}{y^2 - x^2}\right) = 0$

3. $\phi\left(\dfrac{x-y}{y-z},\ \dfrac{y-z}{z-x}\right) = 0$

4. $\phi(x^2 - y^2,\ x^2 - z^2) = 0$

5. $\phi\left[\left(\dfrac{y-x}{z-y}\right),\ (x-y)^2(x+y+z)\right] = 0$

6. $\phi(lx + my + nz,\ x^2 + y^2 + z^2) = 0$

7. $\phi(x^2 - y^2 - z^2,\ 2xy - z^2) = 0$

8. $\phi[x + y,\ \log\{z^2 + (x+y)^2\} - 2x] = 0$

9. $\phi(x + y + z,\ xyz) = 0$

10. $\phi\left(\dfrac{-2y}{x^2 - y^2},\ x + y - z\right) = 0$

11. $\phi\left(\dfrac{y-x}{z-x},\ \dfrac{y-x}{z-y}\right) = 0$

12. $f\left\{(\cos(x+y) + \sin(x+y))\right\}\, e^{y-x}\, z^{\sqrt{2}}\, \cot\left(\dfrac{x+y}{2} + \dfrac{\pi}{8}\right) = 0$

Objective Evaluation

MULTIPLE CHOICE QUESTIONS
Choose the most appropriate one.

1. The solution of $(D^2 + 1)y = 0$ is $y =$:
   - (a) $A\cos x - B\sin x$
   - (b) $-A\cos x - B\sin x$
   - (c) $A\cos x + B\sin x$
   - (d) $-A\cos x + B\sin x$

2. The general solution of the differential equation $(D^2 + 1)y = 0$ is :
   - (a) $y = \cos x$
   - (b) $y = C\cos x$
   - (c) $y = C_1\cos(x + C_2)$
   - (d) $C_1\cos(C_2 + C_3 x)$

3. The complete solution of $(D^2 - 3D + 4)y = 0$ is :
   - (a) $y = C_1 e^{-x} + C_2 e^{4x}$
   - (b) $y = C_1 x + C_2 \cdot x$
   - (c) $y = (C_1 + C_2 x)e^x$
   - (d) none of these

4. The C.F. of $(D^2 + 2D + 1)\,y = (x - 1)$ is :
   - (a) $y = (C_1 + C_2 x)e^x$
   - (b) $y = (C_1 + C_2 x)e^{-x}$
   - (c) $y = C_1 e^x + C_2 e^{-x}$
   - (d) none of these

5. The particular integral of nth order differential equation contains :
   - (a) $(n + 1)$ arbitrary constants
   - (b) n arbitrary constants
   - (c) one arbitrary constant
   - (d) none of these

6. The complete primitive can be obtained by :
   - (a) C.F
   - (b) P.I.
   - (c) C.F. + P.I.
   - (d) none of these

7. The P.I. of $(D^2 - 5D + 6)y = e^{mx}$ is :
   - (a) e^{mx}
   - (b) $\dfrac{e^{mx}}{m^2 - 5m + 6}$
   - (c) $m^2 - 5m + 6$
   - (d) none of these

8. To find $\dfrac{1}{f(D)}\,e^{ax}.X$, we bring e^{ax} to the left from right of $\dfrac{1}{f(D)}$, then D must be replaced by :
   - (a) $D - a$
   - (b) a
   - (c) m
   - (d) $D + a$

9. The general solution of $(2D + 1)^2 y = 4e^{-x/2}$ is $y =$:
   - (a) $(C_1 + C_2 x)\,e^{-x/2}$
   - (b) $e^{-x/2}\left(\dfrac{x^2}{2}\right)$
   - (c) (a) + (b)
   - (d) none of these

10. The general solution of $(D^2 + a^2)y = \cos ax$ is $y =$:
    - (a) $C_1 \cos ax + C_2 \sin ax$
    - (b) $\cos ax$
    - (c) $\sin ax$
    - (d) (a) + (c)

TRUE/FALSE
Write 'T' for True and 'F' for False statement.

1. The particular integral of an nth order differential equation contains n independent arbitrary constants. **(T/F)**

2. The complementary functions of a differential equation of order n contains n independent arbitrary constant. **(T/F)**

3. If $f(D) = 0$, then we cannot find the complementary function. **(T/F)**

4. If P.I. $= \dfrac{1}{f(D)}e^{ax}$, then we put a for D in $f(D)$ and we get the required P.I. provided $F(a) \neq 0$. **(T/F)**

5. The C.F. of the differential equation $(D^2 - 3D + 2)y = e^{\sqrt{x}}$ is $C_1 e^x + C_2 e^{2x}$. **(T/F)**

6. To find the P.I. when $Q = e^{ax}.X$, where X is a function of x, then we replace D by $(D + a)$ and bring e^{ax} before the operator $1/f(D)$. **(T/F)**

7. The method discussed in (6) cannot be used to find $\dfrac{1}{f(D)}e^{ax}$ when $f(a) = 0$. **(T/F)**

8. The general solution of a differential equation cannot be found with the particular integral and complimentary function. **(T/F)**

⊕ FILL IN THE BLANKS

1. If the order of the given differential equation is n, then C.F. of this equation contains _____ arbitrary constant.
2. If the order of the given differential equation is n, then _____ of this equation does not contain any arbitrary constants.
3. The _____ can be obtained by adding complementary function and particular integral.
4. The auxiliary equation can be obtained by replacing D by _____.
5. If the auxiliary equation having two roots namely m_1 and m_2 such that $m_1 \neq m_2$, then C.F. is _____.
6. If the auxiliary equation having two equal roots namely $m_1 = m_2 = m$, then C.F. is _____.
7. The imaginary roots of an equation always occurs in _____.
8. If the roots of auxiliary equation are $\alpha \pm i\beta$, then C.F. is _____.
9. The value of $\frac{1}{f(D)} e^{ax} = \frac{1}{f(a)} e^{ax}$ provided _____.
10. The particular integral of the equation is $(D^2 - 2D + 5)y = e^{-x}$ is _____.

ANSWERS

⊕ MULTIPLE CHOICE QUESTIONS

1. (c) **2.** (c) **3.** (a) **4.** (b) **5.** (d) **6.** (c) **7.** (b) **8.** (d) **9.** (c)
10. (d)

⊕ TRUE/FALSE

1. T **2.** F **3.** F **4.** T **5.** T **6.** T **7.** F **8.** F

⊕ FILL IN THE BLANKS

1. n **2.** particular integral **3.** general solution **4.** m **5.** $C_1 e^{m_1 x} + C_2 e^{m_2 x}$
6. $(C_1 + C_2 x) e^{mx}$ **7.** pairs **8.** $e^{\alpha x}[C_1 \cos\beta x + C_2 \sin\beta x]$ **9.** $f(a) \neq 0$
10. $\frac{1}{8} e^{-x}$

❑❑❑❑❑

CHAPTER 14

The Laplace Transform

14.1 INTRODUCTION

An integral of the type $\int_{-\infty}^{\infty} k(p,t)F(t)dt$ is defined as the integral transform of $F(t)$, provided it is convergent. It is denoted by $f(p)$ or $T[F(t)]$.

Therefore, $$f(p) = TF(t) = \int_{-\infty}^{\infty} k(p,t)F(t)dt$$

☞ REMARK

➠ The function $k(p,t)$ appearing in the integral is called kernel of the transform. Here p is a parameter and is independent of $t.p$ may be real or complex number.

Definition 1. *If $F(t)$ be a function of t defined for all values of t, then Laplace transform of $f(t)$, denoted by $L\{F(t)\}$ or $f(p)$ is defined by*

$$L\{F(t)\} = f(p) = \int_0^\infty e^{-pt}\, F(t)\, dt \qquad \qquad ...(1)$$

☞ REMARKS

➠ If the integral (1) converges for some value of p, then only the Laplace transform of $f(t)$ exists otherwise not.

➠ L is called Laplace transform operator.

Definition 2. *A function $f(x)$ is said to be of exponential order a as $x \to \infty$ if* $\lim_{x \to \infty} e^{-ax} f(x) = a$ *finite quantity, i.e., for a given positive integer n, if a real number M such that* $|e^{-ax}\, f(x)| < M,\ \forall\ x \geq n$ *which can be written as $f(x) = O(e^{-ax}),\ x \to \infty$.*

Definition 3. *A function $f(x)$ is called sectionally continuous over the closed interval $x_1 \leq x \leq x_2$ if the closed interval can be divided into a finite number of subintervals $a \leq x \leq b$ such that*

(i) *$f(x)$ is continuous in the closed interval (a, b).*

(ii) *$\lim_{x \to a+0} f(x)$ and $\lim_{x \to b-0} f(x)$ both exist.*

Definition 4. *A function which is sectionally (or piecewise) continuous over every finite interval in the range $t \geq 0$ and ω of exponential order as $t \to \infty$ is called a function of class A.*

14.2 LINEARITY PROPERTY

THEOREM. *The Laplace transformation is a linear transformation*
$$L\{a_1\, F_1(t) + a_2\, F_2(t)\} = a_1\, L\{F_1(t)\} + a_2\, L\{F_2(t)\}$$

if a_1 and a_2 be constants.

PROOF. We know that $L = \{F(t)\} = \int_0^\infty e^{-pt} F(t)\, dt$.

Therefore,
$$L\{a_1 F_1(t) + a_2 F_2(t)\} = \int_0^\infty e^{-pt} \{a_1 F_1(t) + a_2 F_2(t)\}\, dt$$

$$= a_1 \int_0^\infty e^{-pt} F_1(t)\, dt + a_2 \int_0^\infty e^{-pt} F_2(t) dt$$
$$= a_1\, L\, \{F_1(t)\} + a_2 L\{F_2(t)\}\,.$$

14.3 EXISTENCE OF LAPLACE TRANSFORM

THEOREM.   *If $F(t)$ is a function which is piecewise continuous on every finite interval in the range $t \geq 0$ and satisfies $|F(t)| \leq M\, e^{at}$ for all $t \geq 0$ and for constant a and M then the Laplace transform of $f(t)$ exists for all $p > a$.*

PROOF.   We know that $L\{F(t)\} = \int_0^\infty e^{-pt} F(t)\, dt = \int_0^{t_0} F(t) e^{-pt} dt + \int_{t_0}^\infty F(t) e^{-pt} dt$ ...(1)

Now, $\int_0^{t_0} F(t)\, e^{-pt} dt$ exists since $F(t)$ is sectionally continuous on every finite interval $0 \leq t \leq t_0$

and $\left| \int_{t_0}^\infty F(t)\, e^{-pt}\, dt \right| \leq \int_{t_0}^\infty |F(t)\, e^{-pt}|\, dt \leq \int_{t_0}^\infty e^{-pt}\, M\, e^{at}\, dt\ [\because |F(t)| \leq Me^{at}]$

$$= \int_{t_0}^\infty e^{(a-p)t}\, M\, dt$$

$$= M \left[\frac{e^{-(p-a)t}}{-(p-a)} \right]_{t_0}^\infty = \frac{M}{p-a}\, e^{-(p-a)t_0}\,, \text{ if } p > a$$

$$\Rightarrow \quad \left| \int_{t_0}^\infty F(t)\, e^{-pt}\, dt \right| \leq \frac{M}{p-a}\, e^{-(p-a)t_0}\,, \text{if } p > a\,.$$

Now, $\dfrac{Me^{-(p-a)t_0}}{p-a}$ can be made small as we please by taking t_0 sufficiently large.

Hence, from (1), we conclude that $L\{f(t)\}$ exists for all $p < a$.

📖 REMARK

➠ The above conditions are sufficient but not necessary for the existence of the Laplace transform. If these conditions are satisfied, the Laplsce transform must exist. If these conditions are not satisfied, the Laplace transform may or may not exist.

14.4 LAPLACE TRANSFORM OF SOME ELEMENTARY FUNCTIONS

(i) $F(t) = 1$

SOLUTION.   We have $L\,\{F(t)\} = \int_0^\infty e^{-pt} F(t)\, dt$ ...(1)
Here $F(t) = 1$.
Therefore, from (1)

$$L\{1\} = \int_0^\infty e^{-pt} \cdot 1\, dt\ = \left[-\frac{e^{-pt}}{p} \right]_0^\infty = \frac{1}{p}\,,\ p > 0$$

Hence, $L\{1\} = \dfrac{1}{p}\,.$

(ii) $F(t) = t^n$

SOLUTION.   We have $L\{F(t)\} = \int_0^\infty e^{-pt} F(t)\, dt$

$$\Rightarrow \qquad L\{t^n\} = \int_0^\infty e^{-pt}\, t^n\, dt = \int_0^\infty e^{-pt} \cdot t^{(n+1)-1}\, dt$$

$$= \frac{\Gamma(n+1)}{p^{n+1}} = \frac{n!}{p^{n+1}}, \ p > 0 \qquad \left[\because \int_0^\infty e^{-u} u^n du = \Gamma(n+1) \right]$$

Hence, $\qquad L\{t^n\} = \dfrac{n!}{p^{n+1}}$.

(iii) $F(t) = t$

SOLUTION. We have $\qquad L\{F(t)\} = \int_0^\infty e^{-pt} \cdot t \ dt$

$$= \left[-\frac{1}{p} t \ e^{-pt} \right]_0^\infty + \frac{1}{p} \int_0^\infty e^{-pt} dt = \frac{1}{p^2}, \ p > 0$$

(iv) $F(t) = e^{at}$

SOLUTION. We have $\qquad L\{F(t)\} = \int_0^\infty e^{-pt} e^{at} \ dt = \int_0^\infty e^{-(p-a)t} \ dt$.

If $p \le a$, integral diverges. For $p > a$, the integral converges. Hence, for $p > a$,

$$L\{e^{at}\} = \int_0^\infty e^{-(p-a)t} \ dt = \left[-\frac{e^{-(p-a)t}}{p-a} \right]_0^\infty = 0 + \frac{1}{p-a}$$

$$= \frac{1}{p-a}, \ p > a.$$

(v) $F(t) = \sin at$

SOLUTION.

$$L\{\sin at\} = \int_0^\infty e^{-pt} \sin at \ dt = \left[\frac{e^{-pt}(-p \sin at - a \cos at)}{p^2 + a^2} \right]_0^\infty$$

$$\because \int e^{ax} \sin bx dx = e^{ax} \frac{\left[a \sin bx - b \cos bx \right]}{a^2 + b^2}$$

$$= \frac{a}{p^2 + a^2}, \ p > 0$$

Hence, $\qquad L\{\sin at\} = \dfrac{a}{p^2 + a^2}$.

(vi) $F(t) = \cos at$

SOLUTION. We know that

$$\int e^{ax} \cos bx \ dx = \frac{e^{ax} (a \cos bx + b \sin bx)}{a^2 + b^2}$$

Therefore, we have

$$L\{\cos at\} = \int_0^\infty e^{-pt} \cos at \ dt = \left[\frac{e^{-pt}(-p \cos at + a \sin at)}{a^2 + p^2} \right]_0^\infty$$

$$= \frac{p}{p^2 + a^2}, \ p > 0.$$

(vii) $F(t) = \sinh at$

SOLUTION. Consider $L\{\sinh at\} = L\left\{ \dfrac{e^{at} - e^{-at}}{2} \right\} = \dfrac{1}{2} L\{e^{at}\} - \dfrac{1}{2} L\{e^{-at}\}$ (Using (iv))

$$= \frac{1}{2} \cdot \frac{1}{p-a} - \frac{1}{2} \cdot \frac{1}{p+a} = \frac{a}{p^2 - a^2}$$

Hence, $\qquad L\{\sinh at\} = \dfrac{a}{p^2 - a^2}$.

(viii) $F(t) = \cosh at$

SOLUTION. Consider
$$L\{\cosh at\} = L\left[\frac{1}{2}(e^{at} + e^{-at})\right] = \frac{1}{2}L\{e^{at}\} + \frac{1}{2}L\{e^{-at}\}$$

$$= \frac{1}{2}\cdot\frac{1}{p-a} + \frac{1}{2}\cdot\frac{1}{p+a}, \quad p > a \text{ and } p > -a$$

$$= \frac{p}{p^2 - a^2}, \quad p > |a|$$

Hence, $\quad L\{\cosh at\} = \dfrac{p}{p^2 - a^2}$.

☛ **RECAPITULATIONS**

$F(t)$	$L(F(t))$	$F(t)$	$L(F(t))$
1	$1/p$	$\cos at$	$\dfrac{p}{p^2 + a^2}, p > 0$
t^n	$\dfrac{n!}{p^{n+1}}$	$\sinh at$	$\dfrac{a}{p^2 - a^2}$
e^{at}	$\dfrac{1}{p-a}, p > a$	$\cosh at$	$\dfrac{p}{p^2 - a^2}$
$\sin at$	$\dfrac{a}{p^2 + a^2}, p > 0$		

SOLVED EXAMPLES

EXAMPLE 1. Find the Laplace transform of the function $F(t) = \dfrac{e^{at} - 1}{a}$.

SOLUTION. We have $\quad L\{F(t)\} = L\left\{\dfrac{e^{at} - 1}{a}\right\} = L\left\{\dfrac{1}{a}e^{at} - \dfrac{1}{a}\right\} = \dfrac{1}{a}L\{e^{at}\} - \dfrac{1}{a}L\{1\}$

$$= \frac{1}{a}\left(\frac{1}{p-a}\right) - \frac{1}{a}\left(\frac{1}{p}\right) = \frac{1}{p(p-a)}.$$

EXAMPLE 2. Find $L\{(t^2 + 1)^2\}$.

SOLUTION. $\quad\quad L\{(t^2 + 1)^2\} = L\{t^4 + 2t^2 + 1\}$

$$= L\{t^4\} + 2L\{t^2\} + L(1) \quad\quad \text{(By linearity property)}$$

$$= \frac{4!}{p^5} + 2.\frac{2!}{p^3} + \frac{1}{p} = \frac{24 + 4p^2 + p^4}{p^5}, \quad p > 0.$$

EXAMPLE 3. Find $L\{F(t)\}$ where $F(t) = (\sin t - \cos t)^2$.

SOLUTION. Consider
$$L\{(\sin t - \cos t)^2\} = L\{\sin^2 t + \cos^2 t - 2\sin t \cos t\}$$

$$= L\{1 - \sin 2t\} = L\{1\} - L\{\sin 2t\}$$

$$= \frac{1}{p} - \frac{2}{p^2 + 2^2}, \quad p > 0 = \frac{p^2 - 2p + 4}{p(p^2 + 4)}, \quad p > 0$$

EXAMPLE 4. *Find* $L \{6 \sin 2t - 5 \cos 2t\}$.

SOLUTION. $L\{6 \sin 2t - 5 \cos 2t\} = 6L\{\sin 2t\} - 5L\{\cos 2t\}$

$$= 6.\frac{2}{p^2 + 2^2} - 5.\frac{p}{p^2 + 2^2}, \quad p > 0 = \frac{12 - 5p}{p^2 + 4}, \quad p > 0.$$

EXAMPLE 5. *Find* $L \{7e^{2t} + 9e^{-2t} + 5 \cos t + 7t^3 + 5 \sin 3t + 2\}$.

SOLUTION. $L \{7e^{2t} + 9e^{-2t} + 5 \cos t + 7t^3 + 5 \sin 3t + 2\}$

$$= 7L \{e^{2t}\} + 9L\{e^{-2t}\} + 5L\{\cos t\} + 7.L\{t^3\} + 5.L\{\sin 3t\} + 2.L\{1\}$$

$$= \frac{7}{p-2} + \frac{9}{p+2} + \frac{5p}{p^2+1} + \frac{4^2}{p^4} + \frac{15}{p^2+9} + \frac{q}{p}$$

$$= \frac{4(4p-1)}{p^2-4} + \frac{5p}{p^2+1} + \frac{42}{p^4} + \frac{15}{p^2+4} + \frac{q}{p}.$$

EXAMPLE 6. *Find* $L \{2e^{3t} - e^{-3t}\}$.

SOLUTION. $L\{2e^{3t} - e^{-3t}\} = 2L\{e^{3t}\} - L\{e^{-3t}\}$

$$= 2.\frac{1}{p-3} - \frac{1}{p+3}, \quad p > 3 \text{ and } p > -3$$

$$= \frac{p+9}{p^2-9}, \quad p > |3|.$$

EXAMPLE 7. *Find* $L\{F(t)\}$, *if* $F(t) = \begin{cases} e^t, & 0 < t \le 1 \\ 0, & t > 1 \end{cases}$.

SOLUTION. $L\{f(t)\} = \int_0^\infty e^{-pt} F(t) \, dt = \int_0^1 e^{-pt}.e^t \, dt + \int_1^\infty e^{-pt}.0 \, dt$

$$= \int_0^1 e^{-(p-1)t}.dt = \left[-\frac{e^{-(p-1)t}}{p-1} \right]_0^1$$

$$= \frac{1}{(p-1)} [1 - e^{-(p-1)}], \quad p \ne 1.$$

EXAMPLE 8. *Find* $L\{F(t)\}$, *where* $F(t) = \begin{cases} 0, & 0 < t < 1 \\ t, & 1 < t < 2 \\ 0, & t > 2 \end{cases}$.

SOLUTION. We have that $F(t)$ is not defined at $t = 0, 1$ and 2.

$$\therefore \quad L\{F(t)\} = \int_0^\infty e^{-pt} F(t) \, dt = \int_0^1 e^{-pt}.0 \, dt + \int_1^2 e^{-pt}.t \, dt + \int_2^\infty e^{-pt}.0 \, dt$$

$$= \int_1^2 e^{-pt}.tdt = \left[-t\frac{e^{-pt}}{p} - \frac{e^{-pt}}{p^2} \right]_1^2, p \ne 0$$

$$= -\left(\frac{2}{p} + \frac{1}{p^2} \right) e^{-2p} + \left(\frac{1}{p} + \frac{1}{p^2} \right) e^{-p}, p \ne 0.$$

EXAMPLE 9. *Find the Laplace transform of* $F(t) = \begin{cases} t^2, & 0 < t < 2 \\ t-1, & 2 < t < 3 \\ 7, & t > 3 \end{cases}$

SOLUTION. $L[F(t)] = \int_0^\infty e^{-pt} F(t)dt$

$$= \int_0^2 t^2 e^{-pt} dt + \int_2^3 (t-1)e^{-pt} dt + \int_3^\infty 7e^{-pt} dt$$

$$= \left[t^2 \frac{e^{-pt}}{(-p)} - 2t \frac{e^{-pt}}{(-p)^2} + 2\frac{e^{-pt}}{(-p)^3} \right]_0^2 + \left[(t-1)\left(\frac{e^{-pt}}{(-p)} \right) - \frac{e^{-pt}}{(-p)^2} \right]_2^3 + 7\left[\frac{e^{-pt}}{-p} \right]_3^\infty$$

$$= \left[-4\left(\frac{e^{-2p}}{p} \right) - 4\left(\frac{e^{-2p}}{p^2} \right) + \frac{2}{p^3} \right] + \left[2\left(\frac{e^{-3p}}{-p} \right) - \left(\frac{e^{-3p}}{p^2} \right) + \left(\frac{e^{-2p}}{p} \right) + \frac{e^{-2p}}{p^2} \right]$$

$$+ 7\left[0 + \frac{e^{-3p}}{p} \right]$$

$$= \frac{2}{p^3} + e^{-2p}\left[-\frac{4}{p} - \frac{4}{p^2} - \frac{2}{p^3} \right] + e^{-3p}\left[-\frac{2}{p} - \frac{1}{p^2} \right] + e^{-2p}\left[\frac{1}{p} + \frac{1}{p^2} \right] + e^{-3p}\left[\frac{7}{p} \right]$$

$$= \frac{2}{p^3} + e^{-2p}\left[-\frac{4}{p} - \frac{4}{p^2} - \frac{2}{p^3} + \frac{1}{p} + \frac{1}{p^2} \right] + e^{-3p}\left[-\frac{2}{p} - \frac{1}{p^2} + \frac{7}{p} \right]$$

$$= \frac{2}{p^3} - \frac{e^{-2p}}{p^3}(2 + 3p + 3p^2) + \frac{e^{-3p}}{p^2}(5p - 1)$$

EXAMPLE 10. Find $L\{F(t)\}$, if $F(t) = \begin{cases} 1, & 0 < t < 2 \\ t, & t > 2 \end{cases}$.

SOLUTION. $L\{F(t)\} = \int_0^\infty F(t) e^{-pt} dt = \int_0^2 1.e^{-pt} dt + \int_2^\infty t e^{-pt} dt$

$$= \left[\frac{e^{-pt}}{-p} \right]_0^2 + \left[\left(\frac{e^{-pt}}{-p} \right)t - \left(\frac{e^{-pt}}{p^2} \right).1 \right]_2^\infty$$

$$= -\frac{e^{-2p}}{p} + \frac{1}{p} - \frac{1}{p}\lim_{t\to\infty}\frac{t}{e^{pt}} + \frac{2}{p}e^{-2p} - \frac{1}{p^2}\lim_{t\to\infty}e^{-pt} + \frac{e^{-2p}}{p^2}$$

$$= \frac{1}{p}(1 + e^{-2p}) + \frac{1}{p^2}e^{-2p} - \frac{1}{p}\lim_{t\to\infty}\frac{t}{e^{pt}} - \frac{1}{p^2}\lim_{t\to\infty}e^{-pt}.$$

Now, if $p > 0$, we have

$$\lim_{t\to\infty} e^{-pt} = 0 \text{ and } \lim_{t\to\infty}\frac{t}{e^{pt}}$$ (Form ∞ / ∞)

$$= \lim_{t\to\infty}\frac{1}{pe^{pt}} = 0$$

Hence, $L\{F(t)\} = \frac{1}{p}[1 + e^{-2p}] + \frac{1}{p^2}e^{-2p}, p > 0$.

EXAMPLE 11. Show that $L\left\{ \frac{1}{\sqrt{\pi t}} \right\} = \frac{1}{\sqrt{p}}$.

SOLUTION. We have $L\left\{\dfrac{1}{\sqrt{\pi t}}\right\} = \int_0^\infty e^{-pt} \cdot \dfrac{1}{\sqrt{\pi t}} \cdot dt = \dfrac{1}{\sqrt{\pi}} \int_0^\infty e^{-pt} \cdot \dfrac{1}{\sqrt{t}} dt$

$$= \dfrac{1}{\sqrt{\pi}} \int_0^\infty e^{-pt} \cdot t^{-1/2} \, dt = \dfrac{1}{\sqrt{\pi}} \int_0^\infty e^{-pt} \cdot t^{-1/2-1} \, dt$$

$$= \dfrac{1}{\sqrt{\pi}} \dfrac{\boxed{1/2}}{p^{1/2}}, \quad p > 0 \qquad \text{(Using gamma function)}$$

$$= \dfrac{1}{\sqrt{\pi}} \cdot \dfrac{\sqrt{\pi}}{p^{1/2}} \qquad\qquad \left[\because \Gamma\left(\tfrac{1}{2}\right) = \sqrt{\pi}\right]$$

$$= \dfrac{1}{\sqrt{p}}.$$

EXAMPLE 12. *Show that* $L\left\{\dfrac{\cos\sqrt{t}}{\sqrt{t}}\right\} = \sqrt{\left(\dfrac{\pi}{p}\right)} e^{-1/4p}.$ **(Mumbai–2009)**

SOLUTION. Here, we have

$$\dfrac{\cos\sqrt{t}}{\sqrt{t}} = \dfrac{1}{\sqrt{t}}\left\{1 - \dfrac{1}{2!}(\sqrt{t})^2 + \dfrac{1}{4!}(\sqrt{t})^4 - \dfrac{1}{6!}(\sqrt{t})^6 + \ldots\right\}$$

$$= t^{-1/2} - \dfrac{1}{2!}t^{1/2} + \dfrac{1}{4!}t^{3/2} - \dfrac{1}{6!}t^{5/2} + \ldots$$

Therefore,

$$L\left\{\dfrac{\cos\sqrt{t}}{\sqrt{t}}\right\} = L\{t^{-1/2}\} - \dfrac{1}{2!}L\{t^{1/2}\} + \dfrac{1}{4!}L\{t^{3/2}\} - \dfrac{1}{6!}L\{t^{5/2}\} + \ldots$$

$$= \dfrac{\Gamma(\tfrac{1}{2})}{p^{1/2}} - \dfrac{1}{2!}\dfrac{\Gamma(\tfrac{3}{2})}{p^{3/2}} + \dfrac{1}{4!}\dfrac{\Gamma(\tfrac{5}{2})}{p^{5/2}} - \dfrac{1}{6!}\dfrac{\Gamma(\tfrac{7}{2})}{p^{7/2}} + \ldots, p > 0$$

$$= \dfrac{\sqrt{\pi}}{p^{1/2}} - \dfrac{1}{1.2}\cdot\dfrac{\tfrac{1}{2}\cdot\sqrt{\pi}}{p^{3/2}} + \dfrac{\tfrac{3}{2}\cdot\tfrac{1}{2}\cdot\sqrt{\pi}}{1.2.3.4}\cdot\dfrac{1}{p^{5/2}} - \dfrac{\tfrac{5}{2}\cdot\tfrac{3}{2}\cdot\tfrac{1}{2}\cdot\sqrt{\pi}}{1.2.3.4.5.6}\cdot\dfrac{1}{p^{7/2}} + \ldots$$

$$= \sqrt{\left(\dfrac{\pi}{p}\right)}\left[1 - \dfrac{1}{1!}\left(\dfrac{1}{4p}\right) + \dfrac{1}{2!}\left(\dfrac{1}{4p}\right)^2 - \dfrac{1}{3!}\left(\dfrac{1}{4p}\right)^3 + \ldots\right]$$

$$= \sqrt{\left(\dfrac{\pi}{p}\right)} \cdot e^{-1/4p}.$$

Exercise 14.1

Find the Laplace transform of the following functions : (Ques. 1 to 11)

1. $\sin t \cos t$

2. $4\cos^2 2t$

3. $\sin^2 at$

4. $3\cosh 5t - 4\sinh 5t$ **(Nagarjuna–2006)**

5. $3t^4 - 2t^3 + 4e^{-3t} - 2\sin 5t + 3\cos 2t$

6. $e^{-2t} - e^{-3t}$

7. $F(t) = \begin{cases} \sin t, & 0 < t < \pi \\ 0, & t > \pi \end{cases}$ **(Madras–2000S)**

8. $F(t) = \begin{cases} (t-1)^2, & t > 1 \\ 0, & 0 < t < 1 \end{cases}$

9. $F(t) = \begin{cases} e^t, & 0 < t < 5 \\ 3, & t > 5 \end{cases}$

10. $F(t) = \begin{cases} t, & 0 < t < 4 \\ 5, & t > 4 \end{cases}$

11. $F(t) = \sin \sqrt{t}$

12. Show that t^2 is of exponential order 3.

13. Show that the function e^{t^2} is not of exponential order as $t \to \infty$.

14. Show that the Laplace transforms of the function $F(t) = t^n, -1 < n < 0$ exists, although it is not a function of class A.

15. Find the Laplace transform of
$$F(t) = \begin{cases} e^t, & 0 < t < 1 \\ 0, & t > 1 \end{cases}.$$

HINT TO SELECTED PROBLEMS

1. The given function can be written as
$$F(t) = \sin t . \cos t = \frac{1}{2}\sin 2t .$$

2. $L\{4\cos^2 2t\} = L\{2(1 + \cos 4t)\}$
$$= 2[L\{1\} + L\{\cos 4t\}].$$

7. $L\{F(t)\} = \int_0^\infty e^{-pt} F(t) dt$

$$= \int_0^\pi e^{-pt} . \sin t \, dt + \int_\pi^\infty e^{-pt} . 0 \, dt$$

8. $L\{F(t)\} = \int_0^\infty F(t) e^{-pt} dt$
$$= \int_0^1 0 . e^{-pt} dt + \int_0^\infty (t-1)^2 e^{-pt} dt$$

11. $L\{\sin\sqrt{t}\} = L\left[\sqrt{t} - \frac{(\sqrt{t})^3}{3!} + \frac{(\sqrt{t})^5}{5!} - \frac{(\sqrt{t})^7}{7!} + ...\right]$

ANSWERS

1. $\dfrac{1}{p^2+4}, \ p > 0$ **2.** $\dfrac{4(p^2+8)}{p(p^2+16)}, \ p > 0$ **3.** $\dfrac{2a^2}{p(p^2+4a^2)}, \ p > 0$ **4.** $\dfrac{3p-20}{p^2-25}, \ p > 5$

5. $\dfrac{72}{p^5} - \dfrac{12}{p^4} + \dfrac{4}{p+3} - \dfrac{10}{p^2+25} + \dfrac{3p}{p^2+4}, \ p > 0$ **6.** $\dfrac{1}{p^2+5p+6}, \ p > -2$ **7.** $\dfrac{e^{-p\pi}+1}{p^2+1}$

8. $\dfrac{2e^{-p}}{p^3}, \ p > 0$ **9.** $\dfrac{1-e^{-5(p-1)}}{p-1} + \dfrac{3}{p}e^{-5p}, \ p > 0$ **10.** $\dfrac{1+(p-1)e^{-4p}}{p^2}, \ p > 0$

11. $\dfrac{\sqrt{\pi}}{2p^{3/2}} e^{-1/4p}$ **15.** $\dfrac{e^{1-p}-1}{1-p}$

14.5 TRANSLATION OR SHIFTING THEOREMS

THEOREM 1. **(First Translation or Shifting Theorem).** *If $f(p)$ is the Laplace transform of $F(t)$, then $f(p-a)$ is the Laplace transforms of $e^{at} F(t)$, i.e., if $L\{F(t)\} = f(P)$, when $p > a$, then $L\{e^{at}F(t)\} = f(p-a), \ p > a$.*

PROOF. We have, by definition of Laplace transform

$$L\{F(t)\} = f(p) = \int_0^\infty e^{-pt} F(t) \, dt$$

Therefore, $L\{e^{at} F(t)\} = \int_0^\infty e^{-pt} . e^{at} F(t) \, dt = \int_0^\infty e^{-(p-a)t} . F(t) \, dt$

$$= \int_0^\infty e^{-ut} F(t) \, dt ,$$

where $\qquad u = p - a > 0$

$$= f(u) \qquad\qquad \text{(By definition)}$$
$$= f(p-a) .$$

THEOREM 2. **(Second Translation or Heaviside's Shifting Theorem).**

If $L\{F(t)\} = f(p)$ and $G(t) = \begin{cases} F(t-a), & t > a \\ 0, & t < a \end{cases}$ then, $L\{G(t)\} = e^{-ap} f(p).$

(UPTU–2006, 08)

PROOF. Let $\qquad L\{F(t)\} = f(p) \qquad$ and $\qquad G(t) = \begin{cases} F(t-a), & \text{if } t > a \\ 0, & \text{if } t < a \end{cases}$

Then, $\qquad L\{G(t)\} = \int_0^\infty e^{-pt} G(t) \, dt$

$$= \int_0^a e^{-pt} \, G(t) \, dt + \int_a^\infty e^{-pt} \, G(t) \, dt$$

$$= \int_0^a e^{-pt} . 0 \, dt \ + \ \int_a^\infty e^{-pt} \, F(t-a) \, dt$$

$$= 0 + \int_a^\infty e^{-pt} \, F(t-a) \, dt \, .$$

Let $t - a = u$, therefore $dt = du$.

If $t = a$, then $u = t - a \ = \ a - a \ = \ 0$ and if $t = \infty$, then $u = \infty - a = \infty$.

Hence, $L\{G(t)\} = \int_0^\infty e^{-p(u+a)} \, F(u) \, du = e^{-pa} \int_0^\infty e^{-pu} \, F(u) \, du = e^{-pa} \, f(p)$.

THEOREM 3. **(Change of Scale Property).** *If* $L\{F(t)\} \ = \ f(p)$, *then* $L\{F(at)\} = \dfrac{1}{a} f\left(\dfrac{p}{a}\right)$.

PROOF. By definition

$$L\{F(at)\} \ = \ \int_0^\infty e^{-pt} \, F(at) \, dt = \int_0^\infty e^{-pu/a} \, F(u) \, \frac{du}{a} \qquad \text{(where } at = u \text{)}$$

$$= \frac{1}{a} \int_0^\infty e^{-pu/a} F(u) \, du \ = \frac{1}{a} \int_0^\infty e^{-su} \, F(u) \, du, \quad \text{where } s = \frac{p}{a}$$

$$= \frac{1}{a} \, f(s) \ = \ \frac{1}{a} f\left(\frac{p}{a}\right).$$

☛ RECAPITULATIONS

- If $f(p)$ is the Laplace transform of $F(t)$, then $f(p - a)$ is the Laplace transforms of $e^{at}F(t)$, *i.e.*, if $L\{F(t)\} = f(P)$, when $p > a$, then $L\{e^{at}F(t)\} = f(p - a)$, $p > a$.

- If $L\{F(t)\} = f(P)$ and $G(t) = \begin{cases} F(t-a), & t > a \\ 0, & t < a \end{cases}$ then, $L\{G(t)\} = e^{-ap}f(p)$.

- If $L\{F(t)\} = f(P)$, then $L\{F(at)\} = \dfrac{1}{a} f\left(\dfrac{p}{a}\right)$.

SOLVED EXAMPLES

EXAMPLE 1. *Find* $L\left\{\dfrac{e^{-at} \, t^{n-1}}{(n-1)!}\right\}$.

SOLUTION. We have $L\left\{\dfrac{t^{n-1}}{(n-1)!}\right\} = \dfrac{1}{(n-1)!} . \dfrac{(n-1)!}{p^n} = \dfrac{1}{p^n}$.

Therefore, using first shifting theorem, we have

$$L\left\{e^{-at} \, \frac{t^{n-1}}{(n-1)!}\right\} = f(p+a) = \frac{1}{(p+a)^n} \, .$$

EXAMPLE 2. *Find* $L \, \{e^t \, \cos^2 t\}$.

SOLUTION. We have

$$L\{\cos^2 t\} = L\left\{\frac{1}{2}(1 + \cos 2t)\right\} = \frac{1}{2}[L\{1\} + L\{\cos 2t\}]$$

$$= \frac{1}{2}\left\{\frac{1}{p} + \frac{p}{p^2 + 2^2}\right\} = \frac{p^2 + 2}{p(p^2 + 4)} = f(p) \qquad\qquad \text{(say)}$$

Using first shifting theorem, we have

$$L\{e^t \, \cos^2 t\} = f(p-1) = \frac{(p-1)^2 + 2}{(p-1)(p-1)^2 + 4} = \frac{p^2 - 2p + 3}{(p-1)(p^2 - 2p + 5)}$$

EXAMPLE 3. *Find* $L\{e^{-t}(3\sin 2t - 5\cosh 2t)\}$.

SOLUTION. We have
$$L\{3\sin 2t - 5\cosh 2t\} = 3 \cdot \frac{2}{p^2 + 2^2} - \frac{5p}{p^2 - 2^2} = f(p) \qquad \text{(say)}.$$

Using first shifting theorem, we have
$$L\{e^{-t}(3\sin 2t - 5\cosh 2t)\} = f(p+1)$$
$$= \frac{6}{(p+1)^2 + 4} - \frac{5(p+1)}{(p+1)^2 - 4} = \frac{6}{p^2 + 2p + 5} - \frac{5(p+1)}{p^2 + 2p - 3}.$$

EXAMPLE 4. *Find* $L\{e^{-t}(3\sinh 2t - 5\cosh 2t)\}$.

SOLUTION. We have
$$L\{3\sinh 2t - 5\cosh 2t\} = 3 \cdot \frac{2}{p^2 - 2^2} - 5\frac{p}{p^2 - 2^2} = \frac{6 - 5p}{p^2 - 4} = f(p) \qquad \text{(say)}$$

Using first shifting theorem, we have
$$L\{e^{-t}(3\sinh 2t - 5\cosh 2t)\} = f(p+1)$$
$$= \frac{6 - 5(p+1)}{(p+1)^2 - 4} = \frac{1 - 5p}{p^2 + 2p - 3}.$$

EXAMPLE 5. *Find* $L\{e^t(t+3)^2\}$.

SOLUTION. We have
$$L\{(t+3)^2\} = L\{t^2 + 6t + 9\}$$
$$= \frac{2!}{p^3} + 6 \cdot \frac{1!}{p^2} + \frac{9}{p} = \frac{2 + 6p + 9p^2}{p^3} = f(p) \qquad \text{(say)}$$

Using first shifting theorem, we have
$$L\{(t+3)^2 e^t\} = f(p-1) = \frac{2 + 6(p-1) + 9(p-1)^2}{(p-1)^3} = \frac{9p^2 - 12p + 5}{(p-1)^3}.$$

EXAMPLE 6. *If* $L\{\cos^2 t\} = \dfrac{p^2 + 2}{p(p^2 + 4)}$, *find* $L[\cos^2 at]$. **(UPTU-2006)**

SOLUTION. We have $L\{\cos^2 t\} = \dfrac{p^2 + 2}{p(p^2 + 4)}$

By change of scale property, we have
$$L\{\cos^2 at\} = \frac{1}{a} \cdot \frac{\left(\dfrac{p}{a}\right)^2}{\left(\dfrac{p}{a}\right)\left[\left(\dfrac{p}{a}\right)^2 + 4\right]} = \frac{1}{p}\left[\frac{p^2 + 2a^2}{\dfrac{p}{a}(p^2 + 4a^2)}\right] = \frac{p^2 + 2a^2}{p(p^2 + 4a^2)}.$$

EXAMPLE 7. *Given* $L\{F(t)\} = \dfrac{p^2 - p + 1}{(2p+1)^2(p-1)}$.

Applying the change of scale property, show that $L\{F(2t)\} = \dfrac{p^2 - 2p + 4}{4(p+1)(p-2)}$.

SOLUTION. Given that $\quad L\{F(t)\} = \dfrac{p^2 - p + 1}{(2p+1)^2(p-1)} = f(p)$ (say)

By using change of scale property, we have

$$L\{F(2t)\} = \frac{1}{2} f\left(\frac{p}{2}\right) = \frac{1}{2} \cdot \frac{\left(\dfrac{p}{2}\right)^2 - \left(\dfrac{p}{2}\right) + 1}{\left[2 \cdot \left(\dfrac{p}{2}\right) + 1\right]^2 \left(\dfrac{p}{2} - 1\right)} = \frac{p^2 - 2p + 4}{4(p+1)^2 (p-2)}.$$

EXAMPLE 8. Find $L\{F(t)\}$, where $F(t) = \begin{cases} \cos\left(t - \dfrac{2}{3}\pi\right), & t > \dfrac{2\pi}{3} \\ 0, & t < \dfrac{2\pi}{3} \end{cases}$

SOLUTION. Let $F(t) = \cos t$

Then, $G(t) = \begin{cases} F\left(t - \dfrac{2\pi}{3}\right), & t > 2\pi/3 \\ 0, & t < 2\pi/3 \end{cases}$

We have $L\{F(t)\} = L\{\cos t\} = \dfrac{p}{p^2 + 1} = f(p)$ (say)

Using second shifting theorem, we have

$$L\{G(t)\} = e^{\left(-\frac{2\pi}{3}\right)p} \cdot f(p) = e^{-2\pi p/3} \cdot \frac{p}{p^2 + 1}.$$

EXAMPLE 9. Find $L\{G(t)\}$, where $G(t) = \begin{cases} e^{t-a}, & t > a \\ 0, & t < a \end{cases}$. **(UPTU–2008)**

SOLUTION. By second shifting theorem, we have

$$L\{F(t)\} = f(p) \text{ and } G(t) = \begin{cases} F(t-a), & t > a \\ 0, & t < a \end{cases}$$

Then, $L\{G(t)\} = e^{-ap} f(p)$

Let $F(t) = e^t$

Then, $L\{F(t)\} = L\{e^t\} = \int_0^\infty e^{-pt} \cdot e^t dt = \dfrac{1}{p-1}, p > 1 = f(p)$ (say)

Now, let $G(t) = \begin{cases} F(t-a) = e^{t-a}, & t > a \\ 0, & t < a \end{cases}$

Then, $L\{G(t)\} = e^{-ap} f(p) = \dfrac{e^{-ap}}{p-1}, p > 1$.

Exercise 14.2

1. Find $L\{t^3 e^{-3t}\}$.
2. Find $L\{e^{3t} \cos 5t\}$.
3. Find $L\{e^{-t} \sin^2 t\}$. **(Mumbai–2009)**
4. Find $L\{e^t \sin^2 t\}$
5. Find $L\{e^{-4t} \cosh 2t\}$
6. Find $L\{e^{-2t}(3\cos 6t - 5\sin 6t)\}$.
7. Using first shifting theorem, find the value of

$L\{e^{6t}(t+2)^2\}$
8. If $L\{F(t)\} = f(p)$, find $L\{F(t) \cos \omega t\}$
9. Applying change of scale property, find
   (i) $L\{\sinh 3t\}$, (ii) $L\{\cos 5t\}$
10. Find $L\{F(t)\}$, where

$$F(t) = \begin{cases} \sin\left(t - \dfrac{\pi}{3}\right), & t > \pi/3 \\ 0, & t < \pi/3 \end{cases}$$

11. Find $L\{F(t)\}$, where

$$F(t) = \begin{cases} \sin\left(t - \dfrac{2}{3}\pi\right), & t > 2\pi/3 \\ 0, & t < 2\pi/3 \end{cases}$$

12. If $\{F(t)\} = \dfrac{1}{p}e^{-1/p}$, show that

$$L\{e^{-t}F(3t)\} = \dfrac{e^{-3/(p+1)}}{p+1}.$$

13. Find the Laplace transform of $e^t\, t^{-1/2}$.

[UPTU(Q. Bank)–2001]

14. Find the Laplace transform of :

(i) $t^2 e^t \sin 4t$ **[UPTU(Sp.)–2001]**

(ii) $t\, e^{-t} \sin 2t$ **[UPTU(Sp.)–2002]**

15. Find the Laplace transform of

$$F(t) = \begin{cases} 1, & 0 \le t < 1 \\ t, & 1 \le t < 2 \\ t^2, & 2 \le t < \infty \end{cases}$$

[UPTU(Q. Bank)–2001]

HINT TO SELECTED PROBLEMS

1. $L\{t^3\} = \dfrac{3!}{p^4}$, then

$$L\{t^3 e^{-3t}\} = f(p+3) = \dfrac{6}{(p+3)^4}$$

5. $L(\cosh 2t) = \dfrac{p}{p^2 - 2^2} = \dfrac{p}{p^2 - 4} = f(p)$, then

apply first shifting theorem.

7. $L\{(t+3)^2\} = L\{t^2 + 6t + 9\}$

$$= \dfrac{2!}{p^3} + 6.\dfrac{1}{p^2} + \dfrac{9}{p} = f(p) \text{ (say)},$$

then applying first shifting theorem.

9. (i) $L\{\sinh t\} = \dfrac{1}{(p^2 - 1)} = f(p)$, then by

change of scale property

$$L\{\sinh 3t\} = \dfrac{1}{3} f\left(\dfrac{p}{3}\right)$$

$$= \dfrac{1}{3}.\dfrac{1}{(p/3)^2 - 1} = \dfrac{3}{p^2 - 9}$$

10. Let $G(t) = \sin t$, then

$$F(t) = \begin{cases} G\left(t - \dfrac{\pi}{3}\right), & t > \pi/3 \\ 0, & t < \pi/3 \end{cases}$$

Then, $L\{G(t)\} = L\{\sin t\} = \dfrac{1}{p^2 + 1} = f(p)$ (say)

then apply second shifting theorem.

ANSWERS

1. $\dfrac{6}{(p+3)^4}$

2. $\dfrac{p-3}{p^2 - 6p + 25}$

3. $\dfrac{2}{(p+1)(p^2 + 2p + 5)}$

4. $\dfrac{2}{(p-1)(p^2 - 2p + 5)}$

5. $\dfrac{p+4}{p^2 + 8p + 12}$

6. $\dfrac{3p - 24}{p^2 + 4p + 40}$

7. $\dfrac{4p^2 - 44p + 122}{(p-6)^3}$

8. $\dfrac{1}{2}[f(p - i\omega) + f(p + i\omega)]$

9. (i) $\dfrac{3}{p^2 - 9}$ (ii) $\dfrac{p}{p^2 + 25}$, $p > 0$ 10. $e^{-\pi p/3}.\dfrac{1}{p^2 + 1}$, $p > 0$ 11. $\dfrac{e^{-2\pi p/3}}{p^2 + 1}$, $p > 0$

13. $\dfrac{\sqrt{\pi}}{\sqrt{p-1}}$ 14. (i) $\dfrac{8(3p^2 - 6p - 13)}{(p^2 - 2p + 17)^3}$ (ii) $\dfrac{4p + 4}{(p^2 + 2p + 5)^2}$

15. $\dfrac{1}{p} + \dfrac{2}{p}e^{-2p} + \dfrac{e^{-p}}{p^2} + \dfrac{3}{p^2}e^{-2p} + \dfrac{2}{p^3}e^{-2p}$

14.6 LAPLACE TRANSFORM OF DERIVATIVES

THEOREM 1. *Let $F(t)$ be continuous for all $t \ge 0$ and be of exponential order as $t \to \infty$ and if $F'(t)$ is of class A, the Laplace transforms of derivatives $F'(t)$ exists when $p > a$ and*

$$L\{F'(t)\} = p\, L\{F(t)\} - F(0).$$

PROOF. By definition, we have

$$L\{F'(t)\} = \int_0^\infty e^{-pt} F'(t)\, dt = \left[e^{-pt} F(t)\right]_0^\infty + p \int_0^\infty e^{-pt} F(t)\, dt$$

(On integrating by parts)

$$= -F(0) + pL\{F(t)\} \qquad \left[\because \lim_{t\to\infty} e^{-pt} F(t) = 0\right]$$

$$= pL\{F(t)\} - F(0).$$

REMARK

⟹ Proceeding same as above, we get

$$L\{F''(t)\} = pL\{F'(t)\} - F'(0) = p[p\,L\{F(t)\} - F(0)] - F'(0)$$

$$= p^2\,L\{F(t)\} - p\,F(0) - F'(0) = p^2 f(p) - p\,F(0) - F'(0).$$

THEOREM 2. *If $F(t), F'(t), ..., F^{n-1}(t)$ are continuous for $t \geq 0$ and be of exponential order as $t \to \infty$ and if $F^n(t)$ is of class A and if $L\{F(t)\} = f(p)$, then*

$$L\{F^n(t)\} = p^n f(p) - p^{n-1}F(0) - p^{n-2}\,F(0)...pF^{(n-2)}(0) - F^{(n-1)}(0)$$

$$= p^n\,f(p) - \sum_{r=0}^{n-1} p^{n-1-r}\,F^r(0)$$

PROOF.   Using above theorem, we have

$$L\{F'(t)\} = pL\{F(t)\} - F(0) \quad \text{and} \quad L\{F''(t)\} = p^2 L\{F(t)\} - pF(0) - F'(0)$$

Similarly, we can find

$$L\{F'''(t)\} = pL\{F''(t)\} - F''(0) = p\,[p^2 L\{F(t)\} - p\,F(0) - F'(0)] - F''(0)$$

$$= p^3 L\{F(t)\} - p^2 F(0) - pF'(0) - F''(0).$$

Proceeding, similarly, we get

$$L\{F^n(t)\} = p^n L\{F(t)\} - p^{n-1}F(0) - p^{n-2}F'(0) - ... - F^{n-1}(0)$$

$$= p^n\,L\{F(t)\} - \sum_{r=0}^{n-1} p^{n-1-r}\,F^r(0).$$

THEOREM 3.   *If $F(t)$ is a function of class A and if $L\{F(t)\} = f(p)$, then $L\{t\,.\,F(t)\} = -f'(p)$*

PROOF.   We know that $f(p) = L\{F(t)\} = \int_0^\infty e^{-pt}\,F(t)\,dt$.

Therefore,

$$f'(p) = \frac{d}{dp}\int_0^\infty e^{-pt} F(t)dt = \int_0^\infty \frac{\partial}{\partial p}\{e^{-pt}F(t)\}\,dt$$

(By Leibnitz's rule of differentiation under the sign of integral)

$$= -\int_0^\infty t\,e^{-pt}\,F(t)\,dt = -\int_0^\infty e^{-pt}\,\{t\,F(t)\}\,dt$$

$$= -L\{t\,F(t)\}$$

$$\Rightarrow \qquad L\{t\,F(t)\} = -f'(p).$$

REMARKS

⟹ If $F(t)$ is a function of class A and if $L\{F(t)\} = f(p)$. Then, $L\{t^n F(t)\} = (-1)^n \dfrac{d^n}{dp^n} f(p)$.

⟹ Let a function $F(t)$ be periodic with period w, so that $F(t + nw) = F(t)$ for $n = 1, 2, 3, ...,$ then

$$L\{F(t)\} = \frac{1}{1 - e^{-pw}} \int_0^w e^{-pt} F(t)dt.$$

THEOREM 4.   **(Initial Value Theorem).** *Let $F(t)$ be continuous for all $t \geq 0$ and be of exponential order as $t \to \infty$ and if $F'(t)$ is of class A, then $\lim_{t\to 0} F(t) = \lim_{p\to\infty} pL\{F(t)\}$.*

PROOF.   We know that $L\{F'(t)\} = \int_0^\infty e^{-pt} F'(t)dt = pL\{F(t)\} - F(0)$.   ...(1)

Since $F'(t)$ is sectionally continuous and of exponential order.

Therefore, $\lim_{p \to \infty} \int_0^\infty e^{-pt} F'(t)\, dt = 0$

Now, taking limit as $p \to \infty$ in (1), we have

$$0 = \lim_{p \to \infty} pL\{F(t)\} - F(0)$$

$\Rightarrow$ $F(0) = \lim_{p \to \infty} pL\{F(t)\}$

$\Rightarrow$ $\lim_{t \to 0} F(t) = \lim_{p \to \infty} pL\{F(t)\}$.

THEOREM 5.   **(Final Value Theorem).** *Let $F(t)$ be continuous for all $t \geq 0$ and be of exponential order as $t \to \infty$ and if $F'(t)$ is of class A, then* $\lim_{t \to \infty} F(t) = \lim_{p \to 0} pL\{F(t)\}$.

PROOF.   We know that $L\{F'(t)\} = \int_0^\infty e^{-pt} F'(t)\, dt = pL\{F(t)\} - F(0)$.   ...(1)

Taking limit as $p \to 0$ in (1), we get

$$\lim_{p \to 0} \int_0^\infty e^{-pt} F'(t)dt = \lim_{p \to 0} [pL\{F(t)\} - F(0)]$$

$\Rightarrow$ $\int_0^\infty F'(t)\, dt = \lim_{p \to 0} pL\{F(t)\} - F(0)$

$\Rightarrow$ $\left[F(t)\right]_0^\infty = \lim_{p \to 0} pL\{F(t)\} - F(0)$

$\Rightarrow$ $\lim_{t \to \infty} F(t) - F(0) = \lim_{p \to 0} pL\{F(t)\} - F(0)$

$\Rightarrow$ $\lim_{t \to \infty} F(t) = \lim_{p \to 0} pL\{F(t)\}$.

THEOREM 6.   **(Laplace Transform of the Laplace Transform).** *We have*

$$L[L\{F(t)\}] = L\left\{\int_0^\infty e^{-pt} F(t)\, dt\right\} = \int_0^\infty e^{up}\left\{\int_0^\infty e^{-pt} F(t)\, dt\right\} dp.$$

PROOF.   The area of integration being the whole positive quadrant. Now, changing the order of integration, we get

$$L[L\{F(t)\}] = \int_0^\infty F(t)\left\{\int_0^\infty e^{-p(t+u)}dp\right\} dt$$

$$= \int_0^\infty F(t)\left\{\left[\frac{e^{-p(t+u)}}{-(t+u)}\right]_{p=0}^\infty\right\} dt = \int_0^\infty \frac{F(t)}{t+u}\, dt.$$

THEOREM 7.   **(Laplace Transforms of Integrals).** *If $F(t)$ is piecewise continuous and satisfies $|F(t)| \leq Me^{at}, \forall t \geq 0$ for some constant a and M, then*

$$L\left\{\int_0^t F(x)\, dx\right\} = \frac{1}{p} L\{F(t)\}   (p > 0, \ p > a)$$

PROOF.   Let $F(t)$ be piecewise continuous such that

$$|F(t)| \leq M e^{at}   ...(1)$$

for some constants a and M.

If (1) holds for some negative value of a, then it also holds for positive value of a. Therefore, suppose that a is positive.

Let $$G(t) = \int_0^t F(x)\,dx\,.$$

Then $G(t)$ is continuous. ($\because$ Integral of an integrable function is continuous)

Now, $|G(t)| \le \int_0^t |F(x)|\,dx \le \int_0^t Me^{ax}\,dx$

$\Rightarrow \qquad |G(t)| \le \dfrac{M}{a}(e^{at} - 1),\ a > 0$ \hfill ...(2)

Further, $G'(t) = F(t)$, except for points at which $F(t)$ is discontinuous. Therefore, $G'(t)$ is piecewise continuous on each finite interval.

$\therefore \qquad L\{G'(t)\} = pL\{G(t)\} - G(0) = pL\{G(t)\}$ \hfill $[\because G(0) = 0]$

$\Rightarrow \qquad L\{G(t)\} = \dfrac{1}{p}\,L\{G'(t)\}$

$\Rightarrow \qquad L\left\{\int_0^t F(x)\,dx\right\} = \dfrac{1}{p}\,L\{F(t)\}\,.$

THEOREM 8. **(Division by t).** If $L\{F(t)\} = f(p)$, then $L\left\{\dfrac{1}{t}F(t)\right\} = \int_p^\infty f(x)\,dx$ provided $\lim\limits_{t\to 0}\left\{\dfrac{1}{t}F(t)\right\}$ exists. \hfill **(UPTU–2005, 07)**

PROOF. Let $$G(t) = \dfrac{1}{t}F(t),\ i.e.,\ F(t) = t\,G(t)$$

Therefore, $L\{F(t)\} = L\{t\,G(t)\} = -\dfrac{d}{dp}L\{G(t)\}$

$\Rightarrow \qquad f(p) = -\dfrac{d}{dp}L\{G(t)\}\,.$

On integrating both sides with respect to p to ∞, we get

$$-\left[L\{G(t)\}\right]_p^\infty = \int_p^\infty f(p)\,dp$$

$\Rightarrow \quad -\lim\limits_{p\to\infty} L\{G(t)\} + L\{G(t)\} = \int_p^\infty f(p)\,dp$

$\Rightarrow \qquad 0 + L\{G(t)\} = \int_p^\infty f(p)\,dp\,,$

$$\left(\text{By using } \lim_{p\to\infty} L\{G(t)\} = \lim_{p\to\infty}\int_0^\infty e^{-pt}G(t)dt = 0\right)$$

$\Rightarrow \qquad L\left\{\dfrac{1}{t}F(t)\right\} = \int_p^\infty f(x)\,dx\,.$

SOLVED EXAMPLES

EXAMPLE 1. Find $L\{t\cos at\}$. \hfill **(Raipur–2005)**

SOLUTION. We know that $L\{\cos at\} = \dfrac{p}{p^2 + a^2},\ p > 0\,.$

Therefore, $L\{t\cos at\} = -\dfrac{d}{dp}L\{\cos at\} = -\dfrac{d}{dp}\left(\dfrac{p}{p^2 + a^2}\right). = \dfrac{p^2 - a^2}{(p^2 + a^2)^2}\,.$

EXAMPLE 2. Find $L\{t^2\sin at\}$. \hfill **[UPTU(Q. Bank)–2001]**

SOLUTION. We know that $L\{\sin at\} = \dfrac{a}{p^2 + a^2}$

Therefore, $L\{t^2 \sin at\} = (-1)^2 \dfrac{d^2}{dp^2} L\{\sin at\} = \dfrac{d^2}{dp^2}\left\{\dfrac{a}{p^2+a^2}\right\}$

$$= \dfrac{d}{dp}\left\{\dfrac{-2ap}{(p^2+a^2)^2}\right\} = \dfrac{2a(3p^2-a^2)}{(p^2+a^2)^3},\quad p>0.$$

EXAMPLE 3.   *Find $L\{(\sin at - at\cos at)\}$.*

SOLUTION.   Consider

$$L\{\sin at - at\cos at\} = L\{\sin at\} - aL\{t\cos at\} = \dfrac{a}{p^2+a^2} - a.(-1)\dfrac{d}{dp}[L\{\cos at\}]$$

$$= \dfrac{a}{p^2+a^2} + a\dfrac{d}{dp}\left(\dfrac{p}{p^2+a^2}\right)$$

$$= \dfrac{a}{p^2+a^2} + \dfrac{a(a^2-p^2)}{(p^2+a^2)^2} = \dfrac{2a^3}{(p^2+a^2)^2}.$$

EXAMPLE 4.   *Find the Laplace transform of the function $F(t) = t\,e^{-t}\sin 2t$*

(UPTU–2002, Kurukshetra–2005, 13)

SOLUTION.

$$L\{\sin 2t\} = \dfrac{2}{p^2+4},$$

$$L\{e^{-t}\sin 2t\} = \dfrac{2}{(p+1)^2+4} = f(p)\quad\text{(say)}$$

$$L\{te^{-t}\sin 2t\} = f'(p) = -\dfrac{d}{dp}\left[\dfrac{2}{(p+1)^2+4}\right].$$

$$= \dfrac{-2.2(p+1)}{[(p+1)^2+4]^2} = \dfrac{4(p+1)}{[(p+1)^2+4]^2}.$$

EXAMPLE 5.   *Obtain the Laplace transform of $t^2 e^t \sinh t$.*

(UPTU–2002)

SOLUTION.

$$L\{\sin 4t\} = \dfrac{4}{p^2+16},\quad L\{e^t\sin 4t\} = \dfrac{4}{(p-1)^2+16}$$

$$L\{te^t\sin 4t\} = -\dfrac{d}{dp}\left(\dfrac{4}{p^2-2p+17}\right) = \dfrac{4(2p-2)}{(p^2-2p+17)^2}$$

$$L\{t^2e^t\sin 4t\} = -\dfrac{d}{dp}\left(\dfrac{4(2p-2)}{(p^2-2p+17)^2}\right)$$

$$= \dfrac{-4(2p^2-4p+34-8p^2+16p-8)}{(p^2-2p+17)^3}$$

$$= \dfrac{-4(-6p^2+12p+26)}{(p^2-2p+17)^3} = \dfrac{8(3p^2-6p-13)}{(p^2-2p+17)^3}.$$

EXAMPLE 6. *Given $L\{\sin\sqrt{t}\} = \dfrac{\sqrt{\pi}}{2p^{3/2}} e^{-1/4p}$, show that $L\left\{\dfrac{\cos\sqrt{t}}{\sqrt{t}}\right\} = \sqrt{\left(\dfrac{\pi}{p}\right)}.e^{-1/4p}.$*

(Mumbai–2009)

SOLUTION. Let $\qquad F(t) = \sin\sqrt{t}$

Then we have $F'(t) = \cos\dfrac{\sqrt{t}}{2\sqrt{t}}$ and $F(0) = 0$

Put all these values in

$$L\{F'(t)\} = pL\{F(t)\} - F(0)$$

We get $L\left\{\dfrac{\cos\sqrt{t}}{2\sqrt{t}}\right\} = pL\{\sin\sqrt{t}\} = p\left[\dfrac{\sqrt{\pi}}{2p^{3/2}} e^{-1/4p}\right] = \dfrac{1}{2}\sqrt{\left(\dfrac{\pi}{p}\right)} e^{-1/4p}.$

Hence, $L\left\{\dfrac{\cos\sqrt{t}}{\sqrt{t}}\right\} = \sqrt{\left(\dfrac{\pi}{p}\right)}.e^{-1/4p}.$

EXAMPLE 7. *Show that $L\left\{\dfrac{\sin t}{t}\right\} = \tan^{-1}\dfrac{1}{p}$ and hence find $L\left\{\dfrac{\sin at}{t}\right\}$. Does the Laplace transform*

of $\dfrac{\cos at}{t}$ exists? (UPTU–2005, PTU–2010)

SOLUTION. Let $\qquad F(t) = \sin t$

Then, $\qquad \lim_{t\to 0}\dfrac{F(t)}{t} = \lim_{t\to 0}\dfrac{\sin t}{t} = 1.$

We know that $L\{\sin t\} = \dfrac{1}{p^2+1} = f(p)$ (say)

Then we have

$$L\left\{\dfrac{\sin t}{t}\right\} = \int_p^\infty f(x)\,dx = \int_p^\infty \dfrac{dx}{x^2+1} = \left(\tan^{-1} x\right)_p^\infty$$

$$= \dfrac{\pi}{2} - \tan^{-1} p = \cot^{-1} p = \tan^{-1}\left(\dfrac{1}{p}\right).$$

Now, $\qquad L\left\{\dfrac{\sin at}{t}\right\} = aL\left\{\dfrac{\sin at}{at}\right\} = a.\dfrac{1}{a}\tan^{-1}\left(\dfrac{1}{p/a}\right)$

$$\left[\because L\{f(at)\} = \dfrac{1}{a} f\left(\dfrac{p}{a}\right)\right]$$

$$= \tan^{-1}\left(\dfrac{a}{p}\right).$$

Also, since $\quad L\{\cos at\} = \dfrac{p}{p^2+a^2} = f(p)$ (say)

Then, $\qquad L\left\{\dfrac{\cos at}{t}\right\} = \int_p^\infty \dfrac{x}{x^2+a^2}\,dx = \left[\dfrac{1}{2}\log(x^2+a^2)\right]_p^\infty$

$$= \dfrac{1}{2}\lim_{x\to\infty}\log(x^2+a^2) - \dfrac{1}{2}\log(p^2+a^2)$$

which does not exist since $\lim_{x\to\infty}\log(x^2+a^2)$ is infinite.

Therefore, $L\left\{\dfrac{\cos at}{t}\right\}$ does not exist.

XAMPLE 8. *Find* $L\left\{\dfrac{\sinh t}{t}\right\}$.

Solution. Let

$$F(t) = \sinh t.$$

Now,

$$\lim_{t\to 0}\frac{F(t)}{t} = \lim_{t\to 0}\frac{\sinh t}{t} = 1.$$

Since

$$L\{\sinh t\} = \frac{1}{p^2-1} = f(p)\ (\text{say})$$

$\therefore$

$$L\left\{\frac{\sinh t}{t}\right\} = \int_p^\infty f(x)\,dx = \int_p^\infty \frac{dx}{x^2-1}$$

$$= \left[\frac{1}{2}\log\frac{x-1}{x+1}\right]_p^\infty = \left[\frac{1}{2}\log\frac{1-1/x}{1+1/x}\right]_p^\infty$$

$$= 0 - \frac{1}{2}\log\frac{p-1}{p+1} = \frac{1}{2}\log\frac{p+1}{p-1}$$

Example 9. *If* $F(t) = \dfrac{e^{at}-\cos bt}{t}$, *find the Laplace transform of* $F(t)$. **(UPTU–2003)**

Solution.

$$F(t) = \frac{e^{at}-\cos bt}{t} = \frac{e^{at}}{t} - \frac{\cos bt}{t}$$

We know that

$$L\left(e^{at}-\cos bt\right) = \left(\frac{1}{p-a} - \frac{p}{p^2+b^2}\right)$$

$\therefore$

$$L\left(\frac{e^{at}-\cos bt}{t}\right) = \int_p^\infty\left(\frac{1}{p-a} - \frac{p}{p^2+b^2}\right)ds$$

$$= \left[\log(p-a) - \frac{1}{2}\log(p^2+b^2)\right]_p^\infty$$

$$= \left[\frac{2\log(p-a)-\log(p^2+b^2)}{2}\right]_p^\infty$$

$$= \frac{1}{2}\left[\log(p-a)^2 - \log(p^2+b^2)\right]_p^\infty$$

$$= \frac{1}{2}\left[\log\frac{(p-a)^2}{p^2+b^2}\right]_p^\infty = \frac{1}{2}\left[\log\left[\frac{(1-(a/p))}{(1+(b^2/p^2))}\right]\right]_p^\infty$$

$$= \frac{1}{2}\left[0 - \log\frac{(1-(1/p))^2}{(1+(b^2/p^2))}\right] = \frac{1}{2}\left[\log\frac{p^2+b^2}{(p-a)^2}\right]$$

Exercise 14.3

1. Show that $L\{-a\sin at\} = -\dfrac{a^2}{p^2+a^2}$.

2. Evaluate

(i) $L\{t\cosh 3t\}$,   (ii) $L\{t\sinh at\}$

3. Show that $L\{t^2\cos at\} = \dfrac{2p(p^2-3a^2)}{(p^2+a^2)^3}$, $p>0$.

4. Show that $L(t^n\,e^{at}) = \dfrac{n!}{(p-a)^{n+1}}$, $p>a$.

5. Show that

$$L\{t(3\sin 2t - 2\cos 2t)\} = \dfrac{8+12p-2p^2}{(p^2+4)^2}.$$

6. Show that

$$L\{\sin\alpha t + t\cos\alpha t\} = \dfrac{(\alpha+1)p^2 + (\alpha-1)\alpha^2}{(p^2+\alpha^2)^2}.$$

7. Show that

$$L\{t^2 - 3t + 2\}\sin 3t$$

$$= \dfrac{6p^4 - 18p^3 + 126p^2 - 162p + 432}{(p^2+9)^3}.$$

8. If $L\{F(t), t\to p\} = f(p)$, show that

$$L\left\{\int_0^t \frac{F(u)}{u}\,du,\ t\to p\right\} = \frac{1}{p}\int_p^\infty f(y)\,dy.$$

Hence, show that

$$L\left\{\int_0^t \frac{\sin u}{u}\,du,\ t\to p\right\} = \frac{\cot^{-1} p}{p}.$$

9. Show that if $L\{F(t)\} = f(p)$, then

$$\int_0^\infty \frac{F(t)}{t}\,dt = \int_0^\infty f(x)\,dx,\ \text{provided that the}$$

integral converges.

10. If $L\{t\sin wt\} = \dfrac{2wp}{(p^2+w^2)^2}$, evaluate

$$L\{wt\cos wt + \sin wt\}.$$

11. Find the Laplace transform of

(i) $\int_0^t e^{-t}\cos t\,dt$,   (ii) $\int_0^t \dfrac{\sin t}{t}\,dt$

12. Find the Laplace transform of

(i) $te^{-t}\sin 2t$,   **(UPTU–2002)**

(ii) $t^2 e^t \sin 4t$ **[UPTU(Sp)–2001]**

HINT TO SELECTED PROBLEMS

1. $F(t) = -a\sin at$, $F'(t) = -a^2\cos at$,

$F''(t) = a^3\sin at$

$F'(0) = -a^2$ and $F(0) = 0$, then using

$$L\{F''(t)\} = p^2\,L\,F\{t\} - pF(0) - F'(0)$$

9. Use Theorem 10.

ANSWERS

2. (a) $\dfrac{p^2+9}{(p^2-9)^2}$ (b) $\dfrac{2ap}{(p^2-a^2)^2}$ **10.** $\dfrac{2wp^2}{(p^2+w^2)^2}$

11. (i) $\dfrac{p+1}{p(p^2+2p+2)}$, (ii) $\dfrac{1}{p}\cot^{-1}p$ **12.** (i) $\dfrac{4p+4}{(p^2+2p+5)^2}$, (ii) $\dfrac{8(3p^2-6p-13)}{(p^2-2p+17)^3}$

14.7 EVALUATION OF INTEGRALS

If $L\{F(t)\} = f(p)$, i.e., $\int_0^\infty e^{-pt}\,F(t)\,dt = f(p)$

By taking limit as $p\to 0$, we have $\int_0^\infty F(t)\,dt = f(0)$, provided the integral is convergent.

14.8 SOME IMPORTANT SPECIAL FUNCTIONS

(i) The sine and cosine integrals. The sine and cosine integrals, which are denoted by $S_i(t)$ and $C_i(t)$ respectively are defined by $S_i(t) = \int_0^t \dfrac{\sin u}{u}\,du$ and $C_i(t) = \int_t^\infty \dfrac{\cos u}{u}\,du$.

(ii) Error Function. The error function denoted by $erf(t)$, is defined by

$$erf(t) = \frac{2}{\sqrt{\pi}}\int_0^t e^{-u^2}\,du.$$

(iii) The Gamma function. If $n > 0$, the gamma function is defined by

$$\Gamma(n) = \int_0^\infty u^{n-1} e^{-u} \, du \, .$$

(iv) Heaviside's unit function. The unit step function or heaviside's unit function denoted by $H(t - a)$ is defined by

$$H(t - a) = \begin{cases} 0, & t < a \\ 1, & t \ge a \end{cases} .$$

(v) Bessel's functions. $J_n(t) = \dfrac{t^n}{2^n \, \Gamma(n+1)} \left[1 - \dfrac{t^2}{2(2n+2)} + \dfrac{t^4}{2.4(2n+2)(2n+4)} \cdots \right].$

SOLVED EXAMPLES

EXAMPLE 1.   Find $\int_0^\infty \dfrac{(e^{-at} - e^{-bt})}{t} \, dt \, .$ (SVTU–2009, Mumbai–2007, JNTU–2006)

SOLUTION.   Let $F(t) = e^{-at} - e^{-bt}$.

Thus, we have

$$L\{F(t)\} = L\{e^{-at}\} - L\{e^{-bt}\}$$

$$= \frac{1}{p+a} - \frac{1}{p+b} = f(p) \text{ (say)}.$$

Therefore, $L\left\{ \dfrac{F(t)}{t} \right\} = \int_p^\infty f(x) \, dx = \int_p^\infty \left(\dfrac{1}{x+a} - \dfrac{1}{x+b} \right) dx$

$$= \left[\log\left(\frac{x+a}{x+b} \right) \right]_p^\infty = \lim_{x \to \infty} \log \frac{x+a}{x+b} - \log \frac{p+a}{p+b}$$

$$= \lim_{x \to \infty} \log \frac{1 + a/x}{1 + b/x} - \log \frac{p+a}{p+b}$$

$$= 0 - \log \frac{p+a}{p+b} = \log \frac{p+b}{p+a}$$

Therefore, $L\left\{ \dfrac{F(t)}{t} \right\} = \int_0^\infty e^{-pt} . \dfrac{e^{-at} - e^{-bt}}{t} \, dt = \log \dfrac{p+b}{p+a}$

Hence, taking limit as $p \to 0$, we have

$$\int_0^\infty \frac{e^{-at} - e^{-bt}}{t} \, dt = \log \frac{b}{a} \, .$$

EXAMPLE 2.   Show that $\int_0^\infty t \, e^{-2t} \cos t \, dt = \dfrac{3}{25}$

SOLUTION.   We have $L\{t \cos t\} = -\dfrac{d}{dp} L\{\cos t\}$

or $\int_0^\infty e^{-pt} . t \cos t \, dt = -\dfrac{d}{dp} \left(\dfrac{p}{p^2 + 1} \right) = \dfrac{p^2 - 1}{(p^2 + 1)^2}$

Taking $p = 2$, we get

$$\int_0^\infty t\, e^{-2t} \cos t \, dt = \frac{3}{25}.$$

EXAMPLE 3. *Show that*

(i) $L\{\sinh at \cos at\} = \dfrac{a(p^2 - 2a^2)}{p^4 + 4a^4}$ (ii) $L\{\sinh at \sin at\} = \dfrac{2a^2 p}{p^4 + 4a^4}.$

SOLUTION. (i)   We know that

$$L\{\sinh at\} = \frac{a}{p^2 - a^2} = f(p) \quad \text{(say)}$$

Therefore, $L\{e^{iat} \sin at\} = f(p - ia)$

$$= \frac{a}{(p - ia)^2 - a^2} = \frac{a}{(p^2 - 2a^2) - 2iap}$$

$$= \frac{a\{(p^2 - 2a^2) + 2iap\}}{(p^2 - 2a^2)^2 - (2ipa)^2}.$$

$$\Rightarrow \quad L\{\sinh at(\cos at + i \sin at)\} = \frac{a(p^2 - 2a^2) + 2ia^2 p}{p^4 + 4a^4}$$

$$\Rightarrow \quad L\{\sinh at \cos at\} + iL\{\sinh at \sin at\} = \frac{a(p^2 - 2a^2)}{p^4 + 4a^4} + i\frac{2a^2 p}{p^4 + 4a^4}$$

Equating real and imaginary parts of both the sides, we get

$$L\{\sinh at \cos at\} = \frac{a(p^2 - 2a^2)}{p^4 + 4a^4}$$

and $$L\{\sinh at \sin at\} = \frac{2a^2 p}{p^4 + 4a^4}.$$

Exercise 14.4

1. Show that
$$L\{(1 + te^{-t})^3\} = \frac{1}{p} + \frac{3}{(p+1)^2} + \frac{6}{(p+2)^3} + \frac{6}{(p+3)^4}.$$

2. If $L\left\{2\sqrt{\dfrac{t}{\pi}}\right\} = \dfrac{1}{p^{3/2}}$, then show that

$$\frac{1}{p^{1/2}} = L\left\{\frac{1}{\sqrt{\pi t}}\right\}.$$ **(UPTU–2005, Madras–2003)**

3. Find (i) $L\{F(t)\}$ and (ii) $L\{F'(t)\}$ for the function defined by $F(t) = \begin{cases} 2t, & 0 \le t \le 1 \\ t, & t > 1 \end{cases}.$

4. Show that $\int_0^\infty \dfrac{\sin^2 t}{t^2}\, dt = \dfrac{\pi}{2}.$

5. Show that $\int_0^\infty \dfrac{\cos 6t - \cos 4t}{t}\, dt = \log\left(\dfrac{2}{3}\right).$
(Mumbai–2008, PTU–2006)

6. Show that

(i) $L\{J_0(t)\} = \dfrac{1}{\sqrt{1 + p^2}}$

(ii) $L\{t\, J_0(at)\} = \dfrac{p}{(p^2 + a^2)^{3/2}}.$

7. Show that $L\{J_1(t)\} = 1 - \dfrac{p}{\sqrt{p^2 + 1}}$, where $J_1(t)$ is the Bessel function of order one and hence deduce that $L\{tJ_1(t)\} = \dfrac{1}{(p^2 + 1)^{3/2}}.$

8. Prove that $L\{J_0(a\sqrt{t})\} = \dfrac{1}{p} e^{-a^2/4p}$.

9. Show that $L\{t.erf(2\sqrt{t})\} = \dfrac{3p+8}{p^2(p+4)^{3/2}}$.

10. Show that $L\{e^{3t} . erf\sqrt{t}\} = \dfrac{1}{(p-3)\sqrt{p-2}}$.

11. Show that $L\{c_i(t)\} = \dfrac{1}{2p} . \log(p^2+1)$, where

$c_i(t) = \int_0^\infty \dfrac{\cos u}{u} du$.

12. If $F(t) = t^2$, $0 < t < 2$ and $F(t+2) = F(t)$. Then

show that $L\{F(t)\} = \dfrac{-(4p^2 + 4p + 2)e^{-2p+2}}{p^3(1 - e^{-2p})}$

13. If $F(t) = \begin{cases} 3t, & 0 < t < 2 \\ 6, & 2 < t < 4 \end{cases}$ and $F(t)$ is a periodic

function of period 4, then show that

$L\{F(t)\} = \dfrac{3 - 3e^{-2p} - 6pe^{-4p}}{p^2(1 - e^{-4p})}$

14. Find the Laplace transform of the Heaviside's unit step function $H(t-a)$.

15. Show that $\int_0^\infty \dfrac{\sin t}{t} dt = \dfrac{\pi}{2}$.

16. Show that $\int_0^\infty \dfrac{e^{-t} - e^{-3t}}{t} dt = \log 3$.

17. Show that $\int_0^\infty t^3 e^{-t} \sin t \, dt = 0$.

18. Show that $\int_0^\infty e^{-t} \dfrac{\sin t}{t} dt = \dfrac{1}{4} \log \dfrac{p^2+4}{p^2}$.

(UPTU-2008, VTU-2009S)

HINT TO SELECTED PROBLEMS

1. Let $F(t) = 2\sqrt{\left(\dfrac{t}{\pi}\right)} \Rightarrow F'(t) = \dfrac{1}{\sqrt{\pi t}}$.

Then use $L\{F'(t)\} = pLF(t) - F(0)$.

5. (i) Use $J_0(t) = 1 - \dfrac{t^2}{2^2} + \dfrac{t^4}{2^2.4^2} - \dfrac{t^6}{2^2.4^2.6^2}$,

(ii) $L\{tJ_0(t)\} = -\dfrac{d}{dp} L\{J_0(t)\}$.

6. Since $J_0'(t) = -J_1(t)$.

Now using $L\{f'(t)\} = pL\{f(t)\} - f(0)$

$L\{J_1(t)\} = L\{-J_0'(t)\} = -L\{J_0'(t)\}$.

7. Use $J_0(t) = 1 - \dfrac{t^2}{2^2} + \dfrac{t^4}{2^2.4^2} - \dfrac{t^6}{2^2.4^2.6^2} + ...$

$\Rightarrow J_0(a\sqrt{t}) = 1 - \dfrac{a^2 t}{2^2} + \dfrac{a^4 t^2}{2^2.4^2} - \dfrac{a^6.t^3}{2^2.4^2.6^2} + ...$

8. Use $erf(\sqrt{t}) = \dfrac{2}{\sqrt{\pi}} \int_0^{\sqrt{t}} e^{-u^2} du$

$= \dfrac{2}{\sqrt{\pi}} \int_0^{\sqrt{t}} \left(1 - u^2 + \dfrac{u^4}{2!} - \dfrac{u^6}{3!} + ...\right) du$.

10. Using $L\{c_i(t)\} = L\left\{\int_t^\infty \dfrac{\cos u}{u} du\right\}$.

11. Using

$L\{F(t)\} = \dfrac{\int_0^T e^{-pt} F(t) dt}{1 - e^{-pT}} = \dfrac{\int_0^2 t^2 e^{-pt} dt}{1 - e^{-2p}}$.

ANSWERS

2. (i) $\dfrac{2}{p^2} - \left(\dfrac{1}{p} + \dfrac{1}{p^2}\right)e^{-p}$, $p > 0$ (ii) $\dfrac{1}{p}(2 - e^{-p})$

14.9 THE INVERSE LAPLACE TRANSFORM

If the Laplace transform of a function $F(t)$ is $f(p)$, i.e., if $L\{F(t)\} = f(p)$. Then, $F(t)$ is known as inverse Laplace transform of $F(p)$.

Symbolically, $F(t) = L^{-1}\{f(p)\}$. where, L^{-1} is called the inverse Laplace transformation operator.

For example, If $L\{e^{-2t}\} = \dfrac{1}{p+2}$. Then we can write $L^{-1}\left(\dfrac{1}{p+2}\right) = e^{-2t}$

14.9.1 NULL FUNCTION

A function $N(t)$ of t such that $\int_0^t N(t)\, dt = 0, \quad \forall t > 0$ is called null function.

14.9.2 UNIQUENESS OF INVERSE LAPLACE TRANSFORMS

Since, we know that the Laplace transform of a null function $N(t)$ is zero. Also, it is clearly that if $L\{F(t)\} = f(p)$, then also

$$L\{F(t) + N(t)\} = f(p)$$

It follows that we can have two different functions with same Laplace transform.

If we allow null functions, we see that the inverse Laplace transform is not unique. It is unique, however, if we disallow null functions.

14.9.3 LEARCH THEOREM

If we restrict ourselves to functions $F(t)$ which are sectionally continuous in every finite interval $0 \le t \le N$ and of exponential order for $t > N$, then the inverse Laplace transform of $f(p)$.

i.e., $L^{-1}\{f(p)\} = F(t)$, is unique.

14.10 SOME INVERSE LAPLACE TRANSFORMS

	$f(p)$	$L^{-1}[f(p)] = F(t)$		$f(p)$	$L^{-1}[f(p)] = F(t)$
(i)	$\dfrac{1}{p}$	1	**(vi)**	$\dfrac{1}{p^2 + a^2}$	$\cos at$
(ii)	$\dfrac{1}{p^2}$	t	**(vii)**	$\dfrac{1}{p^2 - a^2}$	$\dfrac{\sinh at}{a}$
(iii)	$\dfrac{1}{p^{n+1}}, n = 0, 1, 2 ...$	$\dfrac{t^n}{n!}$	**(viii)**	$\dfrac{p}{p^2 - a^2}$	$\cosh at$
(iv)	$\dfrac{1}{p - a}$	e^{at}	**(ix)**	$\dfrac{\Gamma(a+1)}{p^n}$	$t^a, \; a > -1$
(v)	$\dfrac{1}{p^2 + a^2}$	$\dfrac{\sin at}{a}$			

14.11 IMPORTANT PROPERTIES OF INVERSE LAPLACE TRANSFORM

(i) Linearity Property.

If C_1 and C_2 are any constants while $f_1(p)$ and $f_2(p)$ are the Laplace transform $F_1(t)$ and $F_2(t)$ respectively, then

$$L^{-1}\{c_1 f_1(p) + c_2 f_2(p)\} = c_1 L^{-1}\{f_1(p)\} + c_2 L^{-1}\{f_2(p)\}$$

Proof. We have

$$L\{c_1 F_1(t) + c_2 F_2(t)\} = c_1 L\{F_1(t)\} + c_2 L\{F_2(t)\} = c_1 f_1(p) + c_2 f_2(p)$$

$$\Rightarrow \quad L^{-1}\{c_1 f_1(p) + c_2 f_2(p)\} = c_1 F_1(t) + c_2 F_2(t) = c_1 L^{-1}\{f_1(p)\} + c_2 L^{-1}\{f_2(p)\}.$$

(ii) First Translation or Shifting Theorem.

If $\qquad L^{-1}\{f(p)\} = F(t)$, then

$$L^{-1}\{f(p - a)\} = e^{at}\, F(t) = e^{at} L^{-1}\{f(p)\}$$

Proof. We have $\qquad f(p) = \int_0^\infty e^{-pt} F(t)\, dt$

$\Rightarrow \qquad f(p-a) = \int_0^\infty e^{-(p-a)t} F(t)\, dt = \int_0^\infty e^{-pt} \{e^{at} F(t)\}\, dt$

$\qquad\qquad\qquad = L\{e^{at} F(t)\}$

Hence, $\qquad L^{-1}\{f(p-a)\} = e^{at} F(t) = e^{at} L^{-1}\{f(p)\}$.

(iii) Second Translation or Shifting Theorem. If $L^{-1}\{f(p)\} = F(t)$, then $L^{-1}\{e^{-ap} f(p)\} = G(t)$, where

$$G(t) = \begin{cases} F(t-a), & t > a \\ 0, & t < a \end{cases}$$

Proof. We know that $\qquad f(p) = \int_0^\infty e^{-pt} F(t)\, dt$

Therefore, $\qquad e^{-ap} f(p) = \int_0^\infty e^{-p(t+a)} F(t)\, dt$

$\qquad\qquad\qquad = \int_0^\infty e^{-px} F(x-a)\, dx$, putting $t + a = x \Rightarrow dt = dx$

$\qquad\qquad\qquad = \int_0^a e^{-pt} \cdot 0\, dt + \int_a^\infty e^{-pt} F(t-a)\, dt$

$\qquad\qquad\qquad = \int_0^\infty e^{-pt} G(t)\, dt = L\{G(t)\}$

where, $G(t) = \begin{cases} F(t-a), & t > a \\ 0, & t < a \end{cases}$ shows, $L^{-1}\{e^{ap} f(p)\} = G(t)$

(iv) Change of Scale Property.

If $L^{-1}\{f(p)\} = F(t)$, then $L^{-1}\{f(ap)\} = \dfrac{1}{a} F\left(\dfrac{t}{a}\right)$

Proof. We know that $\quad f(ap) = \dfrac{1}{a} \int_0^\infty e^{-px} F\left(\dfrac{x}{a}\right) dx \quad \Rightarrow \quad f(ap) = \int_0^\infty e^{-apt} F(t)\, dt$

Putting $\qquad at = x \quad \Rightarrow \quad dt = \dfrac{1}{a} dx$, we get

$$f(ap) = \dfrac{1}{a} \int_0^\infty e^{-px} F\left(\dfrac{x}{a}\right) dx = \dfrac{1}{a} \int_0^\infty e^{-pt} F\left(\dfrac{t}{a}\right) dt$$

[By the property of definite integral]

$$= \dfrac{1}{a} L\left\{F\left(\dfrac{t}{a}\right)\right\} = L\left\{\dfrac{1}{a} F\left(\dfrac{t}{a}\right)\right\}$$

Hence, $\qquad L^{-1}\{f(ap)\} = \dfrac{1}{a} F\left(\dfrac{t}{a}\right)$

SOLVED EXAMPLES

<u>EXAMPLE 1.</u>   *Find the inverse Laplace transforms of the following functions :*

(i) $\dfrac{2p+1}{p(p+1)}$ $\qquad$ (ii) $\dfrac{3p-8}{4p^2+25}$

SOLUTION.   (i) We have

$$L^{-1}\left\{\frac{2p+1}{p(p+1)}\right\} = L^{-1}\left\{\frac{p+(p+1)}{p(p+1)}\right\} = L^{-1}\left\{\frac{1}{p+1}\right\} + L^{-1}\left\{\frac{1}{p}\right\} = e^{-t}+1$$

(ii) Here, we have

$$L^{-1}\left\{\frac{3p-8}{4p^2+25}\right\} = \frac{3}{4}L^{-1}\left\{\frac{p}{p^2+\left(\frac{5}{2}\right)^2}\right\} - 2L^{-1}\left\{\frac{1}{p^2+\left(\frac{5}{2}\right)^2}\right\}$$

$$= \frac{3}{4}\cos\left(\frac{5}{2}t\right) - 2.\frac{2}{5}\sin\left(\frac{5}{2}t\right) = \frac{3}{4}\cos\left(\frac{5}{2}t\right) - \frac{4}{5}\sin\left(\frac{5}{2}t\right)$$

EXAMPLE. 2   *Show that* $\dfrac{1}{p^{1/2}} = L\left[\dfrac{1}{\sqrt{\pi t}}\right]$ (UPTU–2005)

SOLUTION.   We have to show that $\dfrac{1}{p^{1/2}} = L\left[\dfrac{1}{\sqrt{\pi t}}\right]$

Now, $$L^{-1}\left[\frac{1}{p^n}\right] = \frac{t^{n-1}}{(n-1)!} = \frac{t^{n-1}}{\Gamma(n)}$$

So, $$L^{-1}\left[\frac{1}{p^{1/2}}\right] = \frac{t^{\frac{1}{2}-1}}{\Gamma(1/2)} = \frac{t^{-1/2}}{\Gamma(1/2)} = \frac{t^{-1/2}}{\sqrt{\pi}}$$

$$L^{-1}\left[\frac{1}{p^{1/2}}\right] = \frac{1}{\sqrt{\pi t}} \Rightarrow \frac{1}{p^{1/2}} = L\left[\frac{1}{\sqrt{\pi t}}\right].$$

EXAMPLE 3.   *Find* $L^{-1}\left\{\dfrac{3p-2}{p^{5/2}} - \dfrac{7}{3p+2}\right\}$

SOLUTION.   Here, we have

$$L^{-1}\left\{\frac{3p-2}{p^{5/2}} - \frac{7}{3p+2}\right\} = 3L^{-1}\left\{\frac{1}{p^{3/2}}\right\} - 2L^{-1}\left\{\frac{1}{p^{5/2}}\right\} - \frac{7}{3}L^{-1}\left\{\frac{1}{p+(2/3)}\right\}$$

$$= 3\frac{t^{1/2}}{\Gamma\left(\frac{3}{2}\right)} - 2\frac{t^{3/2}}{\Gamma\left(\frac{5}{2}\right)} - \frac{7}{3}e^{\left(-\frac{2}{3}\right)t}$$

$$= 6\sqrt{\left(\frac{t}{\pi}\right)} - \frac{8}{3}t\sqrt{\left(\frac{t}{\pi}\right)} - \frac{7}{3}e^{-2t/3}.$$

EXAMPLE 4.   *Find* $L^{-1}\left\{\dfrac{3}{p^2-3} + \dfrac{3p+2}{p^3} - \dfrac{3p-27}{p^2+9} + \dfrac{6-30\sqrt{p}}{p^4}\right\}.$

SOLUTION.   We have $L^{-1}\left\{\dfrac{3}{p^2-3} + \dfrac{3p+2}{p^3} - \dfrac{3p-27}{p^2+9} + \dfrac{6-30\sqrt{p}}{p^4}\right\}$

$$= L^{-1}\left\{\frac{3}{p^2-3}+\frac{3}{p^2}+\frac{2}{p^3}-\frac{3p}{p^2+9}\right.$$

$$\left.+\frac{27}{p^2+9}+\frac{6}{p^4}-\frac{30}{p^{7/2}}\right\}$$

$$= 3L^{-1}\left\{\frac{1}{p^2-(\sqrt{3})^2}\right\}+3L^{-1}\left\{\frac{1}{p^2}\right\}$$

$$+ 2L^{-1}\left\{\frac{1}{p^3}\right\}-3L^{-1}\left\{\frac{p}{p^2+3^2}\right\}+27L^{-1}\left\{\frac{1}{p^2+3^2}\right\}$$

$$+ 6L^{-1}\left\{\frac{1}{p^4}\right\}-30L^{-1}\left\{\frac{1}{p^{7/2}}\right\}$$

$$= 3.\frac{1}{\sqrt{3}}\sinh\sqrt{3}.t+3.\frac{t^{2-1}}{(2-1)!}+2.\frac{t^{3-1}}{(3-1)!}-3\cos 3t+\frac{27}{3}\sin 3t$$

$$+ 6\frac{t^4-1}{(4-1)!}-30\frac{t^{7/2-1}}{\Gamma\left(\frac{7}{2}\right)}$$

$$= \sqrt{3}\sinh\sqrt{3}t+3t+t^2-3\cos 3t+9\sin 3t+t^3-16t^2\sqrt{\left(\frac{t}{\pi}\right)}.$$

EXAMPLE 5. *A function $f(t)$ obey the equation $f(t)+2\int_0^t f(t)\,dt=\cosh 2t$. Find the Laplace transformation of $f(t)$.* [UPTU 2006]

SOLUTION. We have $f(t)+2\int_0^t f(t)\,dt=\cosh 2t$

Taking Laplace transformation of both the sides, we get

$$L\{f(t)\}+2L\int_0^t f(t)\,dt=L(\cosh 2t)$$

$$\Rightarrow\qquad F(p)+2.\frac{1}{p}F(p)=\frac{p}{p^2-4}\qquad\Rightarrow\qquad F(p)\left[1+\frac{2}{p}\right]=\frac{p}{p^2-4}$$

$$\Rightarrow\qquad F(p).\left[\frac{p+2}{p}\right]=\frac{p}{p^2-4}\qquad\Rightarrow\qquad F(p)=\left(\frac{p}{p^2-4}\right).\left(\frac{p}{p+2}\right)$$

$$\Rightarrow\qquad F(p)=\frac{p^2}{(p^2-4)(p+2)}$$

EXAMPLE 6. *Show that $L^{-1}\left\{\frac{1}{p}\cos\frac{1}{p}\right\}=1-\frac{t^2}{(2!)^2}+\frac{t^4}{(4!)^2}-\frac{t^6}{(6!)^2}+\ldots$*

SOLUTION. $$L^{-1}\left\{\frac{1}{p}\cos\frac{1}{p}\right\}=L^{-1}\left\{\frac{1}{p}\left(1-\frac{(1/p)^2}{2!}+\frac{(1/p)^4}{4!}-\frac{(1/p)^6}{6!}+\ldots\right)\right\}$$

$$= L^{-1}\left\{\frac{1}{p}\right\} - \frac{1}{2!}L^{-1}\left\{\frac{1}{p^3}\right\}$$

$$+ \frac{1}{4!}L^{-1}\left\{\frac{1}{p^5}\right\} - \frac{1}{6!}L^{-1}\left\{\frac{1}{p^7}\right\} + \ldots$$

$$= 1 - \frac{t^2}{(2!)^2} + \frac{t^4}{(4!)^2} - \frac{t^6}{(6!)^2} + \ldots$$

EXAMPLE. 7   *Evaluate* $L^{-1}\left\{\dfrac{3p-2}{p^2-4p+20}\right\}$.

SOLUTION.   We have

$$L^{-1}\left\{\frac{3p-2}{p^2-4p+20}\right\} = L^{-1}\left\{\frac{3(p-2)+4}{(p-2)^2+16}\right\} = L^{-1}\left\{\frac{3(p-2)}{(p-2)^2+16} + \frac{4}{(p-2)^2+16}\right\}$$

$$= 3L^{-1}\left\{\frac{p-2}{(p-2)^2+4^2}\right\} + 4L^{-1}\left\{\frac{1}{(p-2)^2+4^2}\right\}$$

$$= 3e^{2t}L^{-1}\left\{\frac{p}{p^2+4^2}\right\} + 4e^{2t}L^{-1}\left\{\frac{1}{p^2+4^2}\right\}$$

$$= 3e^{2t}\cos 4t + e^{2t}\sin 4t$$

EXAMPLE 8.   *Evaluate the inverse Laplace transform* $L^{-1}\left\{\dfrac{p}{(p+1)^{5/2}}\right\}$.

SOLUTION. $L^{-1}\left\{\dfrac{p}{(p+1)^{5/2}}\right\} = L^{-1}\left\{\dfrac{(p+1)-1}{(p+1)^{5/2}}\right\} = e^{-t}L^{-1}\left\{\dfrac{p-1}{p^{5/2}}\right\}$

$$= e^{-t}L^{-1}\left[\left\{\frac{1}{p^{3/2}}\right\} - \frac{1}{p^{5/2}}\right]$$

$$= e^{-t}L^{-1}\left\{\frac{1}{p^{3/2}}\right\} - e^{-t}L^{-1}\left\{\frac{1}{p^{5/2}}\right\}$$

$$= e^{-t}\frac{t^{(3/2)-1}}{\Gamma\left(\dfrac{3}{2}\right)} - e^{-t}\frac{t^{(5/2)-1}}{\Gamma(5/2)}$$

$$= 2e^{-t}\sqrt{\left(\frac{t}{\pi}\right)} - \frac{4}{3}e^{-t}.t\sqrt{\left(\frac{t}{\pi}\right)} = \frac{2}{3}e^{-t}\sqrt{\left(\frac{t}{\pi}\right)}(3-2t)$$

EXAMPLE 9.   *Evaluate* $L^{-1}\left\{\dfrac{1}{(p+2)(p-1)^2}\right\}$

SOLUTION. $L^{-1}\left\{\dfrac{1}{(p+2)(p-1)^2}\right\} = L^{-1}\left\{\dfrac{1}{(p-1+3)(p-1)^2}\right\}$

$$= e^t L^{-1} \left\{ \frac{1}{p+3} \cdot \frac{1}{p^2} \right\} = e^t L^{-1} \left\{ \frac{1}{p^2} \left(\frac{1}{3} - \frac{1}{9} p + \frac{1}{9} \frac{p^2}{p+3} \right) \right\}$$

(Dividing 1 by $3 + p$ till p^2 is a common factor in the remainder)

$$= e^t L^{-1} \left\{ \frac{1}{3} \cdot \frac{1}{p^2} - \frac{1}{9} \cdot \frac{1}{p} + \frac{1}{9} \cdot \frac{1}{(p+3)} \right\} = e^t \left(\frac{1}{3} t - \frac{1}{9} + \frac{1}{9} e^{-3t} \right)$$

$$= \frac{1}{9} [(3t - 1)e^t + e^{-2t}]$$

EXAMPLE 10. *If* $L^{-1} \left\{ \dfrac{p}{(p^2+1)^2} \right\} = \dfrac{1}{2} t . \sin t$ *find* $L^{-1} \left\{ \dfrac{32p}{(16p^2+1)^2} \right\}$

SOLUTION. We have

$$L^{-1} \left\{ \frac{p}{(p^2+1)^2} \right\} = \frac{1}{2} t \sin t$$

$$\therefore \qquad L^{-1} \left\{ \frac{ap}{(a^2 p^2 + 1)^2} \right\} = \frac{1}{2} \cdot \frac{1}{a} \cdot \frac{t}{a} . \sin \frac{t}{a}$$

$$\Rightarrow \qquad L^{-1} \left\{ \frac{2a^2 p}{(a^2 p^2 + 1)^2} \right\} = \frac{t}{a} \sin \frac{t}{a}$$

Now putting $a = 4$, we get

$$L^{-1} \left\{ \frac{32p}{(16p^2 + 1)^2} \right\} = \frac{t}{4} \sin \frac{t}{4}$$

EXAMPLE 11. Evaluate $L^{-1} \left\{ \dfrac{e^{4-3p}}{(p+4)^{5/2}} \right\}$

SOLUTION. We have $L^{-1} \left\{ \dfrac{1}{(p+4)^{5/2}} \right\} = e^{-4t} L^{-1} \left\{ \dfrac{1}{p^{5/2}} \right\} = e^{-4t} \dfrac{t^{(5/2)-1}}{\Gamma \left(\dfrac{5}{2} \right)} = \dfrac{4t^{3/2} e^{-4t}}{3\sqrt{\pi}}$

Therefore, $L^{-1} \left\{ \dfrac{e^{4-3p}}{(p+4)^{5/2}} \right\} = e^4 L^{-1} \left\{ \dfrac{e^{-3p}}{(p+4)^{5/2}} \right\}$

$$= \begin{cases} e^4 \cdot \dfrac{4}{3\sqrt{\pi}} (t-3)^{3/2} e^{-4(t-3)}, & t > 3 \\ 0, & t < 3 \end{cases}$$

$$= \begin{cases} \dfrac{4}{3\sqrt{\pi}} (t-3)^{3/2} e^{-4(t-4)}, & t > 3 \\ 0, & t < 3 \end{cases}$$

$$= \dfrac{4}{3\sqrt{\pi}} (t-3)^{3/2} e^{-4(t-4)} . H(t-3).$$

EXAMPLE 12. *Find the inverse Laplace transform of* $\dfrac{e^{-cp}}{p^2(p+a)}$, $c > 0$

SOLUTION. We have $L^{-1}\left[\dfrac{e^{-cp}}{p^2(p+a)}\right] = L^{-1}\left[-\dfrac{e^{-cp}}{a^2 p} + \dfrac{e^{-cp}}{ap^2} + \dfrac{e^{-cp}}{a^2(p+a)}\right]$ [By Partial Fractions]

$$= L^{-1}\left[-\dfrac{1}{a^2}\dfrac{e^{-cp}}{p} + \left(\dfrac{1}{a}\right)\dfrac{e^{-cp}}{p^2} + \dfrac{1}{a^2}\cdot\dfrac{e^{-c(p+a)}}{e^{-ca}(p+a)}\right]$$

$$= -\dfrac{1}{a^2}u(t-c) + \dfrac{1}{a}(t-c)u(t-c) + \dfrac{1}{a^2 e^{-ca}}\cdot e^{-at}u(t-c)$$

$$= u(t-c)\left[-\dfrac{1}{a^2} + \dfrac{1}{a}(t-c) + \dfrac{1}{a^2}e^{-a(c+t)}\right]$$

Where $u(t-c)$ = unit step function.

EXAMPLE 13. *Find a function for which* $F(t) = L^{-1}\left\{\dfrac{3}{p} - \dfrac{4e^{-p}}{p^2} + \dfrac{4e^{-3p}}{p^2}\right\}$

SOLUTION. We have $F(t) = L^{-1}\left\{\dfrac{3}{p} - \dfrac{4e^{-p}}{p^2} + \dfrac{4e^{-3p}}{p^2}\right\}$

$$= 3L^{-1}\left\{\dfrac{1}{p}\right\} - 4L^{-1}\left[\dfrac{e^{-p}}{p^2}\right] + 4L^{-1}\left[\dfrac{e^{-3p}}{p^2}\right] \qquad ...(1)$$

Now, $L^{-1}\left[\dfrac{1}{p}\right] = 1, \quad L^{-1}\left\{\dfrac{1}{p^2}\right\} = t$

Therefore, $L^{-1}\left\{\dfrac{e^{-p}}{p^2}\right\} = (t-1)H(t-1)$

and $L^{-1}\left\{\dfrac{e^{-3p}}{p^2}\right\} = (t-3)H(t-3).$

Putting all these values in (1), we get
$$F(t) = 3 - 4(t-1)H(t-1) + 4(t-3)H(t-3)$$

EXAMPLE 14. *Evalute* $L^{-1}\left\{\dfrac{p+1}{p^2+6p+25}\right\}$

SOLUTION. We have
$$L^{-1}\left\{\dfrac{p+1}{p^2+6p+25}\right\} = L^{-1}\left\{\dfrac{(p+3)-2}{(p+3)^2+16}\right\} = e^{-3t}L^{-1}\left\{\dfrac{p-2}{p^2+16}\right\}$$

$$= e^{-3t}\left[L^{-1}\left\{\dfrac{p}{p^2+4^2}\right\} - 2L^{-1}\left\{\dfrac{1}{p^2+4^2}\right\}\right]$$

$$= e^{-3t}\left[\cos 4t - \dfrac{1}{2}\sin 4t\right].$$

EXAMPLE 15. *Evalute* $L^{-1}\left[\dfrac{e^{-p}-3e^{-3p}}{p^2}\right]$

SOLUTION.   We have

$$L^{-1}\left[\frac{e^{-p}-3e^{-3p}}{p^2}\right] = L^{-1}\left[\frac{e^{-p}}{p^2}-\frac{3e^{-3p}}{p^2}\right] \qquad \ldots(1)$$

We know that $Lu(t-a) = \dfrac{e^{-ap}}{p}$ and $L\big[(t-a)\,u(t-a)\big] = \dfrac{e^{-ap}}{p^2}$

Using these results in (1), we get

$$L^{-1}\left[\frac{e^{-p}-3e^{-3p}}{p^2}\right] = (t-1)u(t-1)-3(t-3)u(t-3).$$

EXAMPLE 16. *Evalute* $L^{-1}\left\{\dfrac{p+5}{(p+2)(p^2+4)}\right\}$

SOLUTION.   We have $L^{-1}\left\{\dfrac{p+5}{(p+2)(p^2+4)}\right\}$

$$= L^{-1}\left\{\frac{1}{8}\left(\frac{3}{p+2}-\frac{3p-14}{p^2+4}\right)\right\}$$

$$= \frac{1}{8}\left[3L^{-1}\left\{\frac{1}{p+2}\right\}-3L^{-1}\left\{\frac{p}{p^2+4}\right\}+14L^{-1}\left\{\frac{1}{p^2+4}\right\}\right]$$

$$= \frac{1}{8}\,(3e^{-2t}-3\cos 2t+7\sin 2t)$$

EXAMPLE 17. *Show that* $L^{-1}\left\{\dfrac{p}{p^4+p^2+1}\right\} = \dfrac{2}{\sqrt{3}}\sinh\dfrac{t}{2}.\sin\dfrac{1}{2}\sqrt{3}\,t$ (Raipur-2005)

SOLUTION.   We have

$$L^{-1}\left\{\frac{p}{p^4+p^2+1}\right\} = L^{-1}\left\{\frac{p}{(p^2+1)^2-p^2}\right\}$$

$$= L^{-1}\left\{\frac{p}{(p^2+p+1)(p^2-p+1)}\right\}$$

$$= L^{-1}\left\{\frac{1}{2}\frac{(p^2+p+1)-(p^2-p+1)}{(p^2-p+1)(p^2+p+1)}\right\}$$

$$= L^{-1}\left\{\frac{1}{2(p^2-p+1)}-\frac{1}{2(p^2+p+1)}\right\}$$

$$= \frac{1}{2}L^{-1}\left\{\frac{1}{\left(p-\frac{1}{2}\right)^2+\frac{3}{4}}\right\}-\frac{1}{2}L^{-1}\left\{\frac{1}{\left(p+\frac{1}{2}\right)^2+\frac{3}{4}}\right\}$$

$$= \frac{1}{2}e^{t/2}L^{-1}\left\{\frac{1}{p^2+\left(\frac{1}{2}\sqrt{3}\right)^2}\right\}-\frac{1}{2}e^{-t/2}L^{-1}.\left\{\frac{1}{p^2+\left(\frac{1}{2}\sqrt{3}\right)^2}\right\}$$

$$= \frac{1}{2}e^{t/2}\frac{2}{\sqrt{3}}\sin\left(\sqrt{3}\cdot\frac{t}{2}\right) - \frac{1}{2}e^{-t/2}\frac{2}{\sqrt{3}}\sin\left(\sqrt{3}\cdot\frac{t}{2}\right)$$

$$= \frac{1}{\sqrt{3}}(e^{t/2} - e^{-t/2})\sin\left(\sqrt{3}\cdot\frac{t}{2}\right) = \frac{2}{\sqrt{3}}\sinh\frac{t}{2}\sin\left(\sqrt{3}\cdot\frac{t}{2}\right).$$

EXAMPLE 18. *Find* $L^{-1}\left\{\dfrac{2p^3 + 2p^2 + 4p + 1}{(p^2 + 1)(p^2 + p + 1)}\right\}$

SOLUTION. We have $L^{-1}\left\{\dfrac{2p^3 + 2p^2 + 4p + 1}{(p^2 + 1)(p^2 + p + 1)}\right\}$

$$= L^{-1}\left\{\frac{p+2}{p^2 + 1} + \frac{p-1}{p^2 + p + 1}\right\}$$

$$= L^{-1}\left\{\frac{p}{p^2 + 1}\right\} + 2L^{-1}\left\{\frac{1}{p^2 + 1}\right\} + L^{-1}\left\{\frac{\left(p + \dfrac{1}{2}\right) - \dfrac{3}{2}}{\left(p + \dfrac{1}{2}\right)^2 + \dfrac{3}{4}}\right\}$$

$$= \cos t + 2\sin t + e^{-t/2}L^{-1}\left[\frac{p - 3/2}{p^2 + 3/4}\right]$$

$$= \cos t + 2\sin t + e^{-t/2}\left[L^{-1}\frac{p}{p^2 + \left(\dfrac{\sqrt{3}}{2}\right)^2} - \frac{3}{2}L^{-1}\left\{\frac{1}{p^2 + \left(\dfrac{\sqrt{3}}{2}\right)^2}\right\}\right]$$

$$= \cos t + 2\sin t + e^{-t/2}\left\{\cos\left(\frac{1}{2}\sqrt{3}t\right) - \frac{3}{2}\cdot\frac{2}{\sqrt{3}}\sin\left(\frac{1}{2}\sqrt{3}t\right)\right\}$$

$$= \cos t + 2\sin t + e^{-t/2}\left[\cos\left(\frac{1}{2}\cdot\sqrt{3}t\right) - \sqrt{3}\sin\left(\frac{1}{2}\sqrt{3}t\right)\right]$$

EXAMPLE 19. *Evaluate* $L^{-1}\left\{(p+1)\dfrac{e^{-\pi p}}{p^2 + p + 1}\right\}.$

SOLUTION. We get

$$L^{-1}\left\{\frac{p+1}{p^2 + p + 1}\right\} = L^{-1}\left\{\frac{\left(p + \dfrac{1}{2}\right) + \dfrac{1}{2}}{\left(p + \dfrac{1}{2}\right)^2 + \dfrac{3}{4}}\right\} = e^{-t/2}L^{-1}\left\{\frac{p + \dfrac{1}{2}}{p^2 + \dfrac{3}{4}}\right\}$$

$$= e^{-t/2}L^{-1}\left\{\frac{p}{p^2 + \left(\dfrac{\sqrt{3}}{2}\right)^2}\right\} + \frac{1}{2}e^{-t/2}L^{-1}\left\{\frac{1}{p^2 + \left(\dfrac{\sqrt{3}}{2}\right)^2}\right\}$$

$$= e^{-t/2}\cos\left(\frac{\sqrt{3}t}{2}\right) + \frac{1}{2}e^{-t/2}\left(\frac{2}{\sqrt{3}}\right)\sin\left(\frac{\sqrt{3}t}{2}\right)$$

$$\therefore \quad L^{-1}\left\{\frac{(p+1)e^{-\pi p}}{p^2+p+1}\right\}$$

$$= \begin{cases} \dfrac{e^{-(t-\pi)/2}}{\sqrt{3}}\left[\sqrt{3}\cos\dfrac{\sqrt{3}}{2}(t-\pi) \\ \qquad +\sin\dfrac{\sqrt{3}}{2}(t-\pi)\right], & t>\pi \\ 0, & t<\pi \end{cases}$$

$$= \frac{e^{-(t-\pi)/2}}{\sqrt{3}}\left[\sqrt{3}\cos\frac{\sqrt{3}}{2}(t-\pi)+\sin\frac{\sqrt{3}}{2}(t-\pi)\,H(t-\pi)\right].$$

EXAMPLE 20. *Find* $L^{-1}\left\{\dfrac{5p+3}{(p-1)(p^2+2p+5)}\right\}$

[UPTU-2005, Rohtak-2009]

SOLUTION. We have

$$\frac{5p+3}{(p-1)(p^2+2p+5)}=\frac{1}{p-1}+\frac{2-p}{p^2+2p+5} \qquad \text{[By partial fractions]}$$

$$\therefore \quad L^{-1}\left\{\frac{5p+3}{(p-1)(p^2+2p+5)}\right\}$$

$$= L^{-1}\left\{\frac{1}{p-1}+\frac{2-p}{p^2+2p+5}\right\}$$

$$= L^{-1}\left\{\frac{1}{p-1}-\frac{p+1}{(p+1)^2+4}+\frac{3}{(p+1)^2+4}\right\}$$

$$= L^{-1}\left\{\frac{1}{p-1}\right\}-L^{-1}\left\{\frac{p+1}{(p+1)^2+4}\right\}+3L^{-1}\left\{\frac{1}{(p+1)^2+4}\right\}$$

$$= e^t-e^{-t}L^{-1}\left\{\frac{p}{p^2+4}\right\}+3e^{-t}L^{-1}\left\{\frac{1}{p^2+4}\right\}$$

$$= e^t-e^{-t}\cos 2t+\frac{3}{2}e^{-t}\sin 2t = e^t-e^{-t}\left(\cos 2t-\frac{3}{2}\sin 2t\right).$$

Exercise 14.5

1. Find the inverse Laplace transform of the following functions :

(a) $\dfrac{1}{p^4}$

(b) $\dfrac{1}{p^2+4}$

(c) $\dfrac{4}{p-2}$

(d) $\dfrac{1}{\sqrt{p}}$

(e) $\dfrac{p}{p^2+2}+\dfrac{6p}{p^2-16}+\dfrac{3}{p-3}$

(f) $\dfrac{2p-5}{p^2-9}$

(g) $\dfrac{6}{2p-3}-\dfrac{3+4p}{9p^2-16}+\dfrac{8-6p}{16p^2+9}$ [UPTU-2001]

(h) $\dfrac{3(p^2-1)^2}{2p^5}+\dfrac{4p-18}{9-p^2}+\dfrac{(p+1)(2-\sqrt{p})}{p^{5/2}}$

(i) $\dfrac{1}{p}\sin\dfrac{1}{p}$

2. Find the inverse Laplace transform of the following functions :

(a) $\dfrac{1}{p^2-6p+10}$

(b) $\dfrac{p+b}{(p+b)^2+a^2}$

(c) $\dfrac{3p+7}{p^2-2p-3}$ (d) $\dfrac{1}{(p+a)^n}$

(e) $\dfrac{p}{(p+1)^5}$ (f) $\dfrac{p^2-2p+3}{(p-1)^2(p+1)}$

(g) $\dfrac{2p^2-1}{(p^2+1)(p^2+4)}$ **[UPTU-2004]**

(h) $\dfrac{2p^2-6p+5}{p^3-6p^2+11p-6}$ **[UPTU-2004; VTU-2007]**

3. If $L^{-1}\left\{\dfrac{e^{-1/p}}{p^{1/2}}\right\}=\dfrac{\cos 2\sqrt{t}}{\sqrt{\pi t}}$, evaluate

$$L^{-1}\left\{\dfrac{e^{-a/p}}{p^{1/2}}\right\}, a>0$$

4. Find the inverse Laplace transforms of the following functions

(a) $\dfrac{e^{-5p}}{(p-2)^4}$ (b) $\dfrac{e^{-4p}}{(p-3)^4}$

(c) $\dfrac{e^{-3p}}{p^3}$ (d) $\dfrac{p+8}{p^2+8p+5}$

5. Show that

(a) $L^{-1}\left\{\dfrac{pe^{-ap}}{p^2-w^2}\right\}$

$=\cosh w(t-a)H(t-a), a>0$

(b) $L^{-1}\left\{\dfrac{p+2}{p^2-2p+5}\right\}=e^t\left[\cos 2t+\dfrac{3}{2}\sin 2t\right]$

(c) $L^{-1}\left\{\dfrac{6p^2+22p+18}{p^3+6p^2+11p+6}\right\}$

$=e^{-t}+2e^{-2t}+3e^{-3t}$

(d) $L^{-1}\left\{\dfrac{4p+5}{(p-1)^2(p+2)}\right\}=3te^t+\dfrac{1}{3}e^t-\dfrac{1}{3}e^{-2t}$

(e) $L^{-1}\left\{\dfrac{4p+5}{(p-4)^2(p+3)}\right\}$

$=-\dfrac{1}{7}e^{-3t}+\dfrac{1}{7}e^{4t}+3te^{4t}$

(f) $L^{-1}\left\{\dfrac{5p^2-15p-11}{(p+1)(p-2)^3}\right\}$

$=-\dfrac{1}{3}e^{-t}+\dfrac{1}{3}e^{2t}+4te^{2t}+4te^{2t}-\dfrac{7}{2}t^2e^{2t}$

(g) $L^{-1}\left\{\dfrac{2p+1}{(p+2)^2(p-1)^2}\right\}=\dfrac{1}{3}t(e^t-e^{-2t})$

(h) $L^{-1}\left\{\dfrac{p}{(p^2-2p+2)(p^2+2p+2)}\right\}$

$=\dfrac{1}{2}\sin t\sinh t$

6. Show that

(a) $L^{-1}\left\{\dfrac{p^2}{p^4+4a^4}\right\}$

$=\dfrac{1}{2a}(\cosh at\sin at+\sinh at\cos at)$

(b) $L^{-1}\left\{\dfrac{p^2}{p^4+4a^4}\right\}=\dfrac{1}{2a^2}\sin at\sinh at$

(c) $L^{-1}\left\{\dfrac{1}{(p^2+4)(p+1)^2}\right\}$

$=\dfrac{1}{25}\left\{e^{-t}(2+5t)-2\cos 2t-\dfrac{3}{2}\sin 2t\right\}$

(d) $L^{-1}\left\{\dfrac{3p^3-3p^2-40p+36}{(p^2-4)^2}\right\}$

$=(5t+3)e^{-2t}-2te^{2t}$

HINT TO SELECTED PROBLEMS

1. (g) $L^{-1}\left[\dfrac{6}{2p-3}-\dfrac{3+4p}{9p^2-16}+\dfrac{8-6p}{16p^2+9}\right]$

$=3L^{-1}\left(\dfrac{1}{p-\left(\dfrac{3}{2}\right)}\right)-\dfrac{1}{3}L^{-1}\left(\dfrac{1}{p^2-\left(\dfrac{4}{3}\right)^2}\right)$

$-\dfrac{4}{9}L^{-1}$

$\left(\dfrac{p}{p^2-\left(\dfrac{4}{3}\right)^2}\right)+\dfrac{1}{2}L^{-1}\left(\dfrac{1}{p^2+\left(\dfrac{3}{4}\right)^2}\right)$

$-\dfrac{3}{8}L^{-1}\left(\dfrac{p}{p^2+\left(\dfrac{3}{4}\right)^2}\right)$

$$= 3e^{3t/2} - \frac{1}{4}\sinh\frac{4t}{3} - \frac{4}{9}\cosh\frac{4t}{3}$$

$$+ \frac{2}{3}\sin\frac{3t}{4} - \frac{3}{8}\cos\frac{3}{4}t$$

2. (a) $L^{-1}\left\{\dfrac{1}{p^2 - 6p + 10}\right\} = L^{-1}\left[\dfrac{1}{(p-3)^2 + 1}\right]$

$$= e^{3t}L^{-1}\left[\frac{1}{p^2 + 1}\right] = e^{3t}.\sin t$$

3. $L^{-1}\left\{\dfrac{e^{-1/p}}{p^{1/2}}\right\} = \dfrac{\cos 2\sqrt{t}}{\sqrt{\pi t}} \Rightarrow L^{-1}\left(\dfrac{e^{-1/pk}}{(pk)^{1/2}}\right)$

$$= \frac{1}{k}\frac{\cos 2\sqrt{t/k}}{\sqrt{\pi t/k}}.$$

Now taking $k = 1/a$

5. (a) Since $L^{-1}\left(\dfrac{p}{p^2 - w^2}\right) = \cosh wt$

$$\Rightarrow L^{-1}\left[\frac{pe^{-ap}}{p^2 - w^2}\right] = \cosh w(t-a) H(t-a)$$

6. (a) $L^{-1}\left[\dfrac{p^2}{p^4 + 4a^4}\right] = L^{-1}\left[\dfrac{p^2}{(p^2 + 2a^2)^2 - 4a^2 p^2}\right]$

$$= L^{-1}\left[\frac{p^2}{(p^2 - 2ap + 2a^2)(p^2 + 2ap + 2a^2)}\right]$$

$$= L^{-1}\left[\frac{1}{4a}\frac{\begin{array}{c}p(p^2 + 2ap + 2a^2)\\ -p(p^2 - 2ap + 2a^2)\end{array}}{(p^2 + 2ap + 2a^2)}\right]$$

$$= L^{-1}\left[\frac{p}{4a(p^2 - 2ap + 2a^2)}\right.$$

$$\left. - \frac{p}{4a(p^2 + 2ap + 2a^2)}\right]$$

$$= \frac{1}{4a}L^{-1}\left[\frac{(p-a)+a}{(p-a)^2 + a^2}\right]$$

$$- \frac{1}{4a}L^{-1}\left[\frac{(p+a)-a}{(p+a)^2 + a^2}\right]$$

$$= \frac{e^{at}}{4a}L^{-1}\left[\frac{p+a}{p^2 + a^2}\right] - \frac{1}{4a}e^{-at}L^{-1}\left[\frac{p-a}{p^2 + a^2}\right]$$

ANSWERS

1. (a) $\dfrac{t^3}{6}$; (b) $\dfrac{1}{2}\sin 2t$; (c) $4e^{2t}$; (d) $\dfrac{1}{\sqrt{\pi t}}$; (e) $\cos\sqrt{2}\,t + 6\cosh 4t + 3e^{3t}$;

(f) $2\cosh 3t - \dfrac{5}{3}\sinh 3t$; (g) $3e^{3t/2} - \dfrac{1}{4}\sinh\dfrac{4t}{3} - \dfrac{4}{9}\cosh\dfrac{4t}{3} + \dfrac{2}{3}\sin\dfrac{3t}{4} - \dfrac{3}{8}\cos\dfrac{3}{4}t$;

(h) $\dfrac{1}{2} - \dfrac{3}{2}t^2 + \dfrac{1}{16}t^4 - 4\cosh 3t + 6\sinh 3t + 4\sqrt{\dfrac{t}{\pi}} + \dfrac{8}{3}t\sqrt{\dfrac{t}{\pi}} - t$; (i) $t - \dfrac{t^3}{(3!)^2} + \dfrac{t^5}{(5!)^2} - \dfrac{t^7}{(7!)^2} + ...$

2. (a) $e^{3t}\sin t$; (b) $e^{-bt}\cos at$; (c) $4e^{3t} - e^{-t}$; (d) $e^{-at}\dfrac{t^{n-1}}{(n-1)!}, n \in \mathbf{Z}^+$;

(e) $e^{-t}(4t^3 - t^4)/24$; (f) $\left(t - \dfrac{1}{2}\right)e^t + \dfrac{3}{2}e^{-t}$ (g) $-\sin t + \dfrac{3}{2}\sin 2t$ (h) $\dfrac{1}{2}e^t - e^{2t} + \dfrac{5}{2}e^{3t}$

3. $\cos\dfrac{2\sqrt{at}}{\sqrt{\pi t}}$ **4.** (a) $\dfrac{1}{6}(t-5)^3 e^{2(t-5)}H(t-5)$ (b) $\dfrac{1}{6}(t-4)^3 e^{3(t-4)}.H(t-4)$

(c) $\dfrac{1}{2}(t-3)^2 H(t-3)$ (d) $e^{-4t}\left[\cosh(\sqrt{11}t) + \left(\dfrac{4}{\sqrt{11}}\right)\sinh(\sqrt{11}.t)\right]$

14.12 INVERSE LAPLACE TRANSFORMS OF DERIVATIVES

Theorem 1. If $L^{-1}\{f(p)\} = F(t)$, then $L^{-1}\{f(n)(p)\} = (-1)^n.t^n.F(t)$

Proof. Since we know that $L\{t^n F(t)\} = (-1)^n f^{(n)}(p)$

Therefore, $t^n F(t) = L^{-1}\{(-1)^n f^{(n)}(p)\} = (-1)^n L^{-1}\{f^{(n)}(p)\}$

Hence, $L^{-1}\{f^{(n)}(p)\} = (-1)^n t^n F(t)$

14.13 DIVISION BY p

Theorem 1. *If* $L^{-1}\{f(p)\} = F(t)$, *then* $L^{-1}\left\{\dfrac{f(p)}{p}\right\} = \displaystyle\int_0^t F(u)\, du$

Proof. Since we know that $\dfrac{f(p)}{p} = L\left\{\displaystyle\int_0^t F(u)\, du\right\}$

$$\Rightarrow \qquad L^{-1}\left\{\dfrac{f(p)}{p}\right\} = \int_0^t F(u)\, du$$

14.14 MULTIPLICATION BY POWERS OF p

Theorem 1. *If* $L^{-1}\{f(p)\} = F(t)$ *and* $F(0) = 0$, *then* $L^{-1}\{pf(p)\} = F'(t)$

Proof. We know that

$$L\{F'(t)\} = pL\{F(t)\} - F(0) = pL\,[F(t)] = p\,f(p) \qquad\qquad [\because\ f(0)=0].$$

Hence, $\qquad L^{-1}\{p\,f(p)\} = F'(t)$

14.15 INVERSE LAPLACE TRANSFORMS OF INTEGRALS

Theorem 1. *If* $L^{-1}\{f(p)\} = F(t)$, *then* $L^{-1}\left[\displaystyle\int_p^\infty f(x)\, dx\right] = \dfrac{F(t)}{t}$.

Proof. We know that

$$L\left\{\dfrac{1}{t}F(t)\right\} = \int_p^\infty f(x)\, dx \text{ provided } \lim_{t\to 0}\left\{\dfrac{F(t)}{t}\right\} \text{ exists.}$$

Hence, $\qquad L^{-1}\left\{\displaystyle\int_p^\infty f(x)\, dx\right\} = \dfrac{F(t)}{t}$.

☛ RECAPITULATIONS

- If $L^{-1}\{f(p)\} = F(t)$, then $L^{-1}\{f(n)(p)\} = (-1)^n . t^n . F(t)$

- If $L^{-1}\{f(p)\} = F(t)$, then $L^{-1}\left\{\dfrac{f(p)}{p}\right\} = \int_0^t F(u)du$

- If $L^{-1}\{f(p)\} = F(t)$ and $F(0) = 0$, then $L^{-1}\{pf(p)\} = F'(t)$

- If $L^{-1}\{f(p)\} = F(t)$, then $L^{-1}\left[\int_p^\infty f(x)dx\right] = \dfrac{F(t)}{t}$.

SOLVED EXAMPLES

EXAMPLE 1. Find $L^{-1}\left\{\dfrac{p}{(p^2 + a^2)^2}\right\}$.

(SVTU-2009, VTU-2010)

SOLUTION. We have

$$L^{-1}\left\{\dfrac{p}{(p^2 + a^2)^2}\right\} = L^{-1}\left\{-\dfrac{1}{2}\dfrac{d}{dp}\left(\dfrac{1}{p^2 + a^2}\right)\right\} = -\dfrac{1}{2}L^{-1}\left\{\dfrac{d}{dp}\left(\dfrac{1}{p^2 + a^2}\right)\right\}$$

$$= -\dfrac{1}{2}t(-1)L^{-1}\left\{\dfrac{1}{p^2 + a^2}\right\} = \dfrac{t}{2a}\sin at$$

EXAMPLE 2.   *Evaluate* $L^{-1}\left\{\log\left(1 - \dfrac{1}{p^2}\right)\right\}$.

SOLUTION.   Let us suppose $f(p) = \log\left(1 - \dfrac{1}{p^2}\right)$.

$$= \log\left(\frac{p^2 - 1}{p^2}\right) = -2\log p + \log(p^2 - 1)$$

$$\Rightarrow \qquad f'(p) = -2\left(\frac{1}{p} - \frac{p}{p^2 - 1}\right)$$

$$\Rightarrow \qquad L^{-1}\{f'(p)\} = -2(1 - \cosh t)$$

$$\Rightarrow \qquad -tL^{-1}\{f(p)\} = -2(1 - \cosh t)$$

$$\Rightarrow L^{-1}\left\{\log\left(1 - \frac{1}{p^2}\right)\right\} = \frac{2}{t}(1 - \cosh t).$$

EXAMPLE 3.   *Find the function whose Laplace transform is* $\log\left(1 + \dfrac{1}{p}\right)$ **[UPTU-2007]**

SOLUTION. $L^{-1}\left[\log\left(1 + \dfrac{1}{p}\right)\right] = -\dfrac{1}{t} L^{-1}\left[\dfrac{d}{dp}\log\left(\dfrac{p+1}{p}\right)\right] = -\dfrac{1}{t}L^{-1}\left[\left(\dfrac{p}{p+1}\right)\left(-\dfrac{1}{p^2}\right)\right]$

$$= -\frac{1}{t}L^{-1}\left[-\frac{1}{p(p+1)}\right] = -\frac{1}{t}L^{-1}\left[\frac{1}{p+1} - \frac{1}{p}\right]$$

$$= -\frac{1}{t}[e^{-t} - 1] = \frac{1}{t}[1 - e^{-t}].$$

EXAMPLE 4.   *Evaluate* $L^{-1}\left\{\dfrac{p+2}{p^2(p+3)}\right\}$

SOLUTION.   Consider $L^{-1}\left\{\dfrac{p+2}{p^2(p+3)}\right\} = L^{-1}\left\{\dfrac{(p+3)-1}{p^2(p+3)}\right\} = L^{-1}\left\{\dfrac{1}{p^2} - \dfrac{1}{p^2(p+3)}\right\}$

$$= L^{-1}\left\{\frac{1}{p^2}\right\} - L^{-1}\left\{\frac{1}{p^2(p+3)}\right\} \qquad \qquad \dots(1)$$

Since, $L^{-1}\left\{\dfrac{1}{p^2}\right\} = t$ and $L^{-1}\left\{\dfrac{1}{p+3}\right\} = e^{-3t} = F(t)$ (say)

Then, we have

$$L^{-1}\left\{\frac{1}{p(p+3)}\right\} = \int_0^t F(x)\,dx \qquad \qquad \dots(1)$$

$$= \int_0^t e^{-3x}dx = \frac{1}{3}(1 - e^{-3t}) = F_1(t) \qquad \qquad \text{(say)}$$

Therefore, $L^{-1}\left\{\dfrac{1}{p^2(p+3)}\right\} = \int_0^1 F_1(x)\,dx$

$$= \frac{1}{3}\int_0^t (1 - e^{-3x})\, dx = \frac{1}{3}t + \frac{1}{9}(e^{-3t} - 1)$$

Hence, from (1), we have

$$L^{-1}\left\{\frac{p+2}{p^2(p+3)}\right\} = t - \frac{t}{3} - \frac{1}{9}(e^{-3t} - 1) = \frac{2}{3}t - \frac{1}{9}e^{-3t} + \frac{1}{9}$$

EXAMPLE 5. *Find the inverse Laplace transform of* $f(p) = \log\dfrac{p+a}{p+b}$ **[UPTU-2003, Anna-2003]**

SOLUTION. We have $L^{-1}\log\left[\dfrac{p+a}{p+b}\right] = -\dfrac{1}{t}L^{-1}\left[\dfrac{d}{dp}\log\dfrac{p+a}{p+b}\right]$

$$= -\frac{1}{t}L^{-1}\left[\frac{d}{dp}\log(p+a) - \frac{d}{dp}\log(p+b)\right]$$

$$= -\frac{1}{t}L^{-1}\left[\frac{1}{p+a} - \frac{1}{p+b}\right]$$

$$= -\frac{1}{t}\left[e^{-at} - e^{-bt}\right] = \frac{1}{t}\left[e^{-bt} - e^{-at}\right].$$

EXAMPLE 6. *Evaluate* $L^{-1}\left\{\dfrac{1}{p^4(p^2+1)}\right\}$

SOLUTION. Since we know that $L^{-1}\left\{\dfrac{1}{p^2+1}\right\} = \sin t$

Therefore

$$L^{-1}\left\{\frac{1}{p(p^2+1)}\right\} = \int_0^t \sin x\, dx = 1 - \cos t$$

Also, $L^{-1}\left\{\dfrac{1}{p^2(p^2+1)}\right\} = \int_0^t (1 - \cos x)dx = t - \sin t$

and $L^{-1}\left\{\dfrac{1}{p^3(p^2+1)}\right\} = \int_0^t (x - \sin x)dx = \dfrac{t^2}{2} + \cos t - 1$

Hence, $L^{-1}\left\{\dfrac{1}{p^4(p^2+1)}\right\} = \int_0^t \left\{\dfrac{1}{2}x^2 + \cos x - 1\right\}dx = \dfrac{1}{6}t^3 + \sin t - t$

EXAMPLE 7. *Evaluate* $L^{-1}\left\{\dfrac{1}{p(p+1)^3}\right\}$

SOLUTION. We know that

$$L^{-1}\left\{\frac{1}{p(p+1)^3}\right\} = L^{-1}\left\{\frac{1}{(p+1-1)(p+1)^3}\right\}$$

$$= e^{-t}L^{-1}\left\{\frac{1}{(p-1)\,p^3}\right\} \qquad\qquad ...(1)$$

Also, since $\qquad L^{-1}\left\{\dfrac{1}{p-1}\right\} = e^t$

Therefore, $\qquad L^{-1}\left\{\dfrac{1}{p(p-1)}\right\} = \int_0^t e^x dx = (e^t - 1)$

and $\qquad L^{-1}\left\{\dfrac{1}{p^2(p-1)}\right\} = \int_0^t (e^x - 1)dx = e^t - t - 1$

and $\qquad L^{-1}\left\{\dfrac{1}{p^3(p-1)}\right\} = \int_0^t (e^x - x - 1)dx = e^t - \dfrac{1}{2}t^2 - t - 1$.

Therefore, from (1), we get

$$L^{-1}\left\{\dfrac{1}{p(p+1)^3}\right\} = e^{-t}\left\{e^t - \dfrac{1}{2}t^2 - 1 - t\right\} = 1 - e^{-t}\left(1 + t + \dfrac{t^2}{2}\right)$$

EXAMPLE 8. *Obtain the inverse Laplace transformation* $\cot^{-1}\left(\dfrac{p+3}{2}\right)$

SOLUTION. We know that $L^{-1}[f(p)] = -\dfrac{1}{t}L^{-1}\left[\dfrac{d}{dp}f(p)\right]$

$$\therefore \quad \left[\cot^{-1}\left(\dfrac{p+3}{2}\right)\right] = -\dfrac{1}{t}L^{-1}\left[\dfrac{d}{dp}\cot^{-1}\left(\dfrac{p+3}{2}\right)\right]$$

$$= -\dfrac{1}{t}L^{-1}\left[\dfrac{-\dfrac{1}{2}}{1+\left(\dfrac{p+3}{2}\right)^2}\right] = \dfrac{1}{2t}L^{-1}\left[\dfrac{4}{4+(p+3)^2}\right]$$

$$= \dfrac{1}{t}L^{-1}\left[\dfrac{2}{2^2+(p+3)^2}\right] = \dfrac{1}{t}e^{-3t}L^{-1}\left[\dfrac{2}{2^2+p^2}\right] = \dfrac{e^{-3t}}{t}\sin 2t.$$

EXAMPLE 9. *Evaluate*

(i) $L^{-1}\left\{\log\left(1+\dfrac{1}{p^2}\right)\right\}$ $\qquad$ (ii) $L^{-1}\left\{\dfrac{1}{p}\log\left(1+\dfrac{1}{p^2}\right)\right\}$

SOLUTION. (i) Let $\qquad f(p) = \log\left(1+\dfrac{1}{p^2}\right) = -\log\left(\dfrac{p^2}{p^2+1}\right)$

$$= -2\log p + \log(p^2+1)$$

Therefore, $\qquad f'(p) = -\dfrac{2}{p} + \dfrac{2p}{p^2+1}$

$\Rightarrow \qquad L^{-1}\{f'(p)\} = -2 + 2\cos t \quad \Rightarrow \quad -tL^{-1}\{f(p)\} = -2(1-\cos t)$

Hence, $L^{-1}\left\{\log\left(1+\dfrac{1}{p^2}\right)\right\} = \dfrac{2(1-\cos t)}{t}$

(ii) Since $L^{-1}\left\{\log\left(1+\dfrac{1}{p^2}\right)\right\}=\dfrac{2(1-\cos t)}{t}$

Therefore,

$$L^{-1}\left\{\dfrac{1}{p}\log\left(1+\dfrac{1}{p^2}\right)\right\}=L^{-1}\left\{\dfrac{1}{p}\,f(p)\right\}$$

$$=\int_0^t F(x)dx=\int_0^t\dfrac{2}{x}(1-\cos x)\,dx\,.$$

Exercise 14.6

1. Evaluate the following inverse Laplace transforms :

(a) $L^{-1}\left\{\dfrac{p}{(p^2-a^2)^2}\right\}$

(b) $L^{-1}\left\{\dfrac{p}{(p^2-16)^2}\right\}$

(c) $L^{-1}\left\{\dfrac{1}{(p-a)^3}\right\}$

(d) $L^{-1}\left\{\dfrac{p+1}{(p^2+2p+2)^2}\right\}$

(e) $L^{-1}\left\{\dfrac{p^2}{(p^2+4)^2}\right\}$ **[UPTU Q. Bank -2001]**

2. Show that :

(a) $L^{-1}\left\{\dfrac{1}{p^3(p+1)}\right\}=1-t+\dfrac{t^2}{2}-e^{-t}$

(b) $L^{-1}\left\{\dfrac{1}{p^3(p^2+1)}\right\}=\dfrac{t^2}{2}+\cos t-1$

(GBTU-2012)

(c) $L^{-1}\left\{\log\dfrac{p+2}{p+1}\right\}=\dfrac{1}{t}(e^{-t}-e^{-2t})$

(d) $L^{-1}\left\{\dfrac{1}{p}\log\dfrac{p+2}{p+1}\right\}=\int_0^t\dfrac{1}{x}(e^{-x}-e^{-2x})dx$

(e) $L^{-1}\left\{\dfrac{1}{p}\log\dfrac{p+3}{p+2}\right\}=\int_0^t\dfrac{1}{x}(e^{-2x}-e^{-3x})\,dx$

(f) $L^{-1}\left\{\tan^{-1}\left(\dfrac{2}{p^2}\right)\right\}=\dfrac{2}{t}\sin t\sinh t$

[UPTU Q. Bank–2001, Mumbai–2005, VTU–2011]

3. If $L^{-1}\left\{\dfrac{p}{(p^2+1)^2}\right\}=\dfrac{1}{2}t\sin t$, then show that

$$L^{-1}\left\{\dfrac{1}{(p^2+1)^2}\right\}=\dfrac{1}{2}(\sin t-t\cos t)\,.$$

HINT TO SELECTED PROBLEMS

1. (a) Since $\dfrac{d}{dp}(p^2-a^2)^{-1}=-\dfrac{2p}{(p^2-a^2)^2}$,

therefore $\dfrac{p}{(p^2-a^2)^2}=-\dfrac{1}{2}\dfrac{d}{dp}\left(\dfrac{1}{p^2-a^2}\right)$

$$\Rightarrow L^{-1}\left(\dfrac{p}{(p^2-a^2)^2}\right)=-\dfrac{1}{2}L^{-1}\left[\dfrac{d}{dp}\left(\dfrac{1}{p^2-a^2}\right)\right]$$

$$=-\dfrac{1}{2}(-1)'t\,L^{-1}\left(\dfrac{1}{p^2-a^2}\right)$$

$$=\dfrac{1}{2}t\dfrac{1}{a}\sinh at$$

2. (a) $L^{-1}\left[\dfrac{1}{p+1}\right]=e^{-t}=F(t)$ (say)

$$\therefore\ L^{-1}\left[\dfrac{1}{p(p+1)}\right]=\int_0^t F(x)\,dx$$

$$=\int_0^t e^{-x}dx=1-e^{-t}$$

$$\Rightarrow L^{-1}\left[\dfrac{1}{p^2(p+1)}\right]=\int_0^t(1-e^{-x})dx$$

$$=t+e^{-t}-1$$

and $L^{-1}\left[\dfrac{1}{p^3(p+1)}\right]=\int_0^t(x+e^{-x}-1)dx$

$$=1-t+\dfrac{t^2}{2}-e^{-t}$$

3. We have $L^{-1}\left[\dfrac{p}{(p^2+1)^2}\right]=\dfrac{1}{2}t\sin t=F(t)$ (say)

$$\Rightarrow L^{-1}\left[\dfrac{1}{(p^2+1)^2}\right]=L^{-1}\left[\dfrac{1}{p}\cdot\dfrac{p}{(p^2+1)^2}\right]$$

$$=\int_0^t F(x)dx$$

$$=\dfrac{1}{2}\int_0^t x.\sin x\,dx=\dfrac{1}{2}(\sin t-t\cos t).$$

Answers

1. (a) $\dfrac{t}{2a}\sinh at$;   (b) $\dfrac{t}{8}\sinh 4t$;   (c) $\dfrac{1}{2}t^2 e^{at}$;

(d) $\dfrac{t}{2}e^{-t}\sin t$;   (e) $\dfrac{1}{4}(\sin 2t + 2t\cos 2t)$

14.16 CONVOLUTION

If $L^{-1}\{f(p)\} = F(t)$ *and* $L^{-1}\{g(p)\} = G(t)$, *where* $F(t)$ *and* $G(t)$ *are two functions of class A,*
then   **[UPTU-2002]**

$$L^{-1}\{f(p).g(p)\} = \int_0^t F(u)\,G(t-u)\,du = F * G$$

We call $F * G$ *the convolution or falting of F and G.*

REMARKS

➠ $F * G$ is commutative, i.e., $F * G = G * F.$

➠ $F * G$ is associative.

➠ $F * G$ is distributive over addition.

14.17 THE HEAVISIDE EXPANSION FORMULA

Theorem.   *If* $F(P)$ *and* $G(P)$ *are polynomials in P, the degree of* $F(P)$ *being less than*
that of $G(P)$ *and if* $G(p) = (p - \alpha_1)(p - \alpha_2)...(p - \alpha_n)$

where, $\alpha_1, \alpha_2, ..., \alpha_n$ *are distinct constants, real or complex, then*

$$L^{-1}\left\{\frac{F(p)}{G(p)}\right\} = \sum_{r=1}^{n} \frac{F(\alpha_r)}{G'(\alpha_r)} e^{\alpha_r.t}$$

SOLVED EXAMPLES

EXAMPLE 1.   *Using convolution theorem, evaluate*

$$L^{-1}\left\{\frac{1}{(p-1)(p+2)}\right\}$$

SOLUTION.   We have $L^{-1}\left\{\dfrac{1}{p-1}\right\} = e^t = F_1(t)$ (say)

and $L^{-1}\left\{\dfrac{1}{p+2}\right\} = e^{-2t} = F_2(t)$ (say)

Using convolution theorem, we have

$$L^{-1}\left\{\frac{1}{p-1}\cdot\frac{1}{p+2}\right\} = F_1 * F_2 = \int_0^t F_1(x)\,F_2(t-x)\,dx$$

$$= \int_0^t e^x e^{-2(t-x)}dx \;=\; e^{-2t}\int_0^t e^{3x}dx = \frac{1}{3}(e^t - e^{-2t})$$

EXAMPLE 2.   *Use the convolution theorem to find* $L^{-1}\left\{\dfrac{p^2}{(p^2+a^2)^2}\right\}$ **(Hazaribag–2009)**

SOLUTION.   We know that $L^{-1}\left\{\dfrac{p}{(p^2+a^2)}\right\} = \cos at$

Therefore, by convolution theorem, we have

$$L^{-1}\left\{\frac{p^2}{(p^2+a^2)^2}\right\} = L^{-1}\left\{\frac{p}{p^2+a^2}\cdot\frac{p}{p^2+a^2}\right\} = \int_0^t \cos ax \, \cos a(t-x) \, dx$$

$$= \int_0^t \cos ax \, (\cos at \cos ax + \sin at \sin ax) \, dx$$

$$= \cos at \int_0^t \cos^2 ax \, dx + \sin at \int_0^t \cos ax \, \sin ax \, dx$$

$$= \frac{1}{2}\cos at \int_0^t (1+\cos 2ax) \, dx + \frac{1}{2}\sin at \int_0^t \sin 2ax \, dx$$

$$= \frac{1}{2}\cos at \left[x+\frac{1}{2a}\sin 2ax\right]_0^t + \frac{1}{2}\sin at \left[-\frac{1}{2a}\cos 2ax\right]_0^t$$

$$= \frac{1}{2}\cos at \left[t+\frac{1}{2a}\sin 2a\right] + \frac{1}{4a}\sin at(1-\cos 2at)$$

$$= \frac{1}{2}t\cos at + \frac{1}{4a}\sin at + \frac{1}{4a}(\sin 2at \cos at - \sin at \cos 2at)$$

$$= \frac{1}{2}t\cos at + \frac{1}{4a}[(\sin at + \sin(2at-at)]$$

$$= \frac{1}{2a}[at \cos at + \sin at]$$

EXAMPLE 3. *Evaluate* $L^{-1}\left[\dfrac{p}{(p^2+1)(p^2+4)}\right]$ [UPTU–2002]

SOLUTION. We know that

$$L^{-1}\frac{p}{p^2+1} = \cos x \text{ and } L^{-1}\frac{2}{p^2+2^2} = \sin 2x$$

$$L^{-1}\left[\frac{p}{(p^2+1)(p^2+4)}\right] = \frac{1}{2}L^{-1}\left[\left(\frac{p}{p^2+1}\right)\cdot\left(\frac{2}{p^2+4}\right)\right]$$

$$= \frac{1}{2}\int_0^t \sin 2x \, \cos(t-x)dx$$

$$= \int_0^t \sin x \cos x \, \{\cos t \cos x + \sin t \sin x\} \, dx$$

$$= \int_0^t (\sin x \cos^2 x \cos t + \sin^2 x \cos x \sin t) \, dx$$

$$= \left[-\frac{\cos^3 x}{3}\cos t + \frac{\sin^3 x}{3}\sin x\right]_0^t$$

$$= -\frac{\cos^4 t}{3} + \frac{\sin^4 t}{3} + \frac{\cos t}{3} = \frac{1}{3}\left[\sin^4 t - \cos^4 t\right] + \frac{\cos t}{3}$$

$$= \frac{1}{3}(\sin^2 t + \cos^2 t)(\sin^2 t - \cos^2 t) + \frac{\cos t}{3}$$

$$= \frac{1}{3}(\sin^2 t - \cos^2 t) + \frac{\cos t}{3} = -\frac{1}{3}\cos 2t + \frac{\cos t}{3}$$

$$= \frac{1}{3}(\cos t - \cos 2t)$$

EXAMPLE 4. *Using convolution theorem, prove that*

$$L^{-1}\left[\frac{1}{p^3(p^2+1)}\right] = \frac{t^2}{2} + \cos t - 1$$

[UPTU–2005, VTU–2007, GBTU–2012]

SOLUTION. We know that

$$L^{-1}\left\{\frac{1}{p^3}\right\} = \frac{t^2}{2!}$$

$$L^{-1}\left\{\frac{1}{p^2+1}\right\} = \sin t$$

Using convolution theorem

$$L^{-1}\left[\frac{1}{p^3(p^2+1)}\right] = \int_0^t \frac{(t-x)^2}{2!}\sin x\, dx = \frac{1}{2}\int_0^t (t^2+x^2-2tx)\sin x\, dx$$

$$= \frac{1}{2}\left[(t^2+x^2-2tx)(-\cos x) - \int (2x-2t)(-\cos x)dx\right]$$

$$= \frac{1}{2}\left[-\cos x(t^2+x^2-2tx) + 2\int (x-t)\cos x\, dx\right]_0^t$$

$$= \frac{1}{2}\left[-\cos x(t^2+x^2-2tx) + 2(x-t)\sin x + 2\cos x\right]_0^t$$

$$= \frac{1}{2}\left[-\cos x(t^2+x^2-2tx) + 2(x-t)\sin x + 2\cos x\right]_0^t$$

$$= \frac{1}{2}\left[-\cos t(t^2+t^2-2t^2) + 0 + 2\cos t + t^2\cos 0 - 2\cos 0\right]_0^t$$

$$= \cos t + \frac{t^2}{2} - 1 = \frac{t^2}{2} + \cos t - 1.$$

EXAMPLE 5. *Using the convolution theorem, find* $L^{-1}\left[\dfrac{p^2}{(p^2+a^2)(p^2+b^2)}\right]$, $a \neq b$.

[UPTU–2004, Mumbai–2007, Bhopal–2008, UKTU–2011, VTU–2011S]

SOLUTION. We have

$$L(\cos at) = \frac{p}{p^2+a^2}$$

$$L(\cos bt) = \frac{p}{p^2+b^2}$$

Hence, by convolution theorem,

$$L\left[\int_0^t \cos ax \cos(bt-x)dx\right] = \frac{p^2}{(p^2+a^2)(p^2+b^2)}$$

Therefore $L^{-1}\left[\dfrac{p^2}{(p^2+a^2)(p^2+b^2)}\right] = \int_0^t \cos ax \cos b(t-x)\, dx$

$$= \frac{1}{2}\int_0^t \{\cos(ax+bt-bx) + \cos(ax-bt+bx)\}dx$$

$$= \frac{1}{2}\int_0^t \cos\{(a-b)x+bt\}\,dx + \frac{1}{2}\int_0^t \cos[(a+b)x-bt]\,dx$$

$$= \frac{\sin at - \sin bt}{2(a-b)} + \frac{\sin at + \sin bt}{2(a+b)} = \frac{a\sin at - b\sin bt}{a^2 - b^2}.$$

Exercise 14.7

1. Using the convolution theorem, show that

(a) $L^{-1}\left\{\dfrac{1}{(p+1)(p-2)}\right\} = \dfrac{1}{3}[e^{2t}-e^{-t}]$

(b) $L^{-1}\left\{\dfrac{p}{(p^2+a^2)^2}\right\} = \dfrac{1}{2a}t\sin at$ **[UPTU–2008]**

(c) $L^{-1}\left\{\dfrac{1}{p(p^2+4)^2}\right\} = \dfrac{1}{16}(1-t\sin 2t - \cos 2t)$

(d) $L^{-1}\left\{\dfrac{1}{(p-2)(p^2+1)}\right\}$

$= \dfrac{1}{5}[e^{2t}-2\sin t - \cos t]$

(e) $\displaystyle\int_0^t \sin u\,\cos(t-u)\,du = \dfrac{t}{2}\sin t$

(f) $\displaystyle\int_0^t J_0(u)\,J_0(t-u)\,du = \sin t$

(g) $L^{-1}\left(\dfrac{p}{(p^2+4)^2}\right) = \dfrac{t}{4}\sin 2t$

[UPTU–2004, GBTU–2010]

(h) $L^{-1}\left\{\dfrac{1}{(p^2+1)(p^2+q)}\right\} = \dfrac{1}{8}\left(\sin t - \dfrac{1}{3}\sin 3t\right)$

(Mumbai-2005S)

(i) $L^{-1}\left\{\dfrac{p}{(p^2+1)(p^2+4)(p^2+q)}\right\}$

$= \dfrac{1}{12}\cos t - \dfrac{1}{10}\cos 2t + \dfrac{1}{60}\cos 3t$

(Mumbai-2006)

(j) $L^{-1}\left\{\dfrac{1}{(p-2)(p+2)^2}\right\}$

$= \dfrac{1}{16}(e^{2t}-e^{-2t}-4te^{-2t})$ **(Mumbai-2009)**

(k) $L^{-1}\left\{\dfrac{P}{(P+2)(P^2+q)}\right\}$

$= \dfrac{1}{13}(3\sin 3t + 2\cos 2t - 2e^{-2t})$

(VTU-2008 S)

(l) $L^{-1}\left\{\dfrac{1}{(P^2+4P+13)^2}\right\}$

$= \dfrac{e^{-2t}}{54}(\sin 3t - 3t\cos 3t)$

(Mumbai-2008)

2. Using the Heaviside formula, show that

(a) $L^{-1}\left\{\dfrac{p^2-6}{p^3+4p^2+3p}\right\} = -2 + \dfrac{5}{2}e^{-t} + \dfrac{1}{2}e^{-3t}$

(b) $L^{-1}\left\{\dfrac{19p+37}{(p+1)(p-2)(p+3)}\right\}$

$= -3e^{-t} + 5e^{2t} - 2e^{-3t}$

(c) $L^{-1}\left(\dfrac{1}{p^3-1}\right) = \dfrac{1}{3}\left[e^t - e^{-t/2}\right.$

$\left.\left\{\cos\left(\dfrac{1}{2}\sqrt{3}\,t\right) + \sqrt{3}\sin\left(\dfrac{1}{2}\sqrt{3}\,t\right)\right\}\right]$

(d) $L^{-1}\left(\dfrac{1}{p^3+1}\right) = \dfrac{1}{3}\left[e^{-t} - e^{t/2}\right.$

$\left.\cos\left\{\dfrac{1}{2}\sqrt{3}t - \sqrt{3}\sin\left(\dfrac{1}{2}\sqrt{3}t\right)\right\}\right]$

HINT TO SELECTED PROBLEMS

1. (a) $L^{-1}\left(\dfrac{1}{p+1}\right) = e^{-t} = F_1(t)$ (say)

and $L^{-1}\left(\dfrac{1}{p+2}\right) = e^{2t} = F_2(t)$ (say)

Then using convolution theorem.

2. (a) $f(p) = 2p^2 - 6p + 5$

and $G(p) = p^3 - 6p^2 + 11p - 6$

$= (p-1)(p-2)(p-3)$

$\Rightarrow G'(p) = 3p^2 - 12p + 11$

$\Rightarrow G(p)$ has 3 distinct roots $\alpha_1 = 1$, $\alpha_2 = 2$

and $\alpha_3 = 3$.

Then using Heavyside expansion formula.

2. (c) $f(p) = 19p + 37$

$G(p) = (p+1)(p-2)(p+3)$

$\Rightarrow \quad G(p)$ has three distinct roots $\alpha_1 = -1, \ \alpha_2 = 2, \ \alpha_3 = -3$.

Then by Heavyside's formula, we have

$$L^{-1}\left[\frac{19p+37}{(p+1)(p-2)(p+3)}\right] = \frac{f(-1)}{G'(-1)}e^{-t}$$
$$+ \frac{f(2)}{G'(2)}e^{2t} + \frac{f(-3)}{G'(-3)}e^{-3t}$$
$$= -3e^{-t} + 5e^{2t} - 2e^{-3t}$$

14.18 SOLUTION OF ORDINARY DIFFERENTIAL EQUATION WITH CONSTANT COEFFICIENTS

Consider a linear differential equation with constant coefficients

$$\frac{d^n y}{dt^n} + A_1 \frac{d^{n-1}y}{dt^{n-1}} + \dots + A_{n-1}\frac{dy}{dt} + A_n y = F(t) \qquad \dots(1)$$

where t is the independent variable and $F(t)$ is a function of t.

Let $\qquad y(0) = C_1, \ y'(0) = C_2, \dots, y^{n-1}(0) = C_{n-1}$ $\qquad\qquad\dots(2)$

be the given initial or boundary conditions, where $C_1, C_2, \dots, C_{n-1}$ are constants. Now, taking the Laplace transform of both sides of (1) and using the conditions given by (2), we get an algebraic equation from which $\bar{y}(p) = L\{y(t)\}$ is determined. The required solution is then obtained by finding the inverse Laplace transform of $\bar{y}(p)$.

⮞ REMARKS

⮕ The algebraic equation, obtained above is known as subsidiary equation.

⮕ The above method is easily extended to higher order differential equation.

WORKING PROCEDURE

Step 1. Taking Laplace transform of both the sides of the given differential equation and use given initial conditions.

Step 2. Solve the equation obtained in step (1) for $L\{y\}$.

Step 3. Taking inverse Laplace transform to find y.

SOLVED EXAMPLES

EXAMPLE 1. Solve $\dfrac{d^2 y}{dt^2} + y = 0$ under the condition that $y = 1, \dfrac{dy}{dt} = 0$ when $t = 0$.

SOLUTION. Here, the given equation is

$$\frac{d^2 y}{dt^2} + y = 0. \qquad \dots(1)$$

Taking the Laplace transform of both sides of the given differential equation, we get

$$L\{y''\} + L\{y\} = 0$$

$$\Rightarrow \quad p^2 L\{y\} - py(0) - y'(0) + L\{y\} = 0$$

$$\Rightarrow \quad (p^2 + 1)\, L\{y\} - p.1 - 0 = 0 \qquad\qquad \text{[Using the given conditions]}$$

$$\Rightarrow \qquad\qquad L\{y\} = \frac{p}{p^2 + 1}$$

Therefore, $\quad y = L^{-1}\left\{\dfrac{p}{p^2 + 1}\right\} = \cos t$.

EXAMPLE 2. Solve $(D^2 + 1)y = 6\cos 2t$ if $y = 3$, $Dy = 1$ when $t = 0$.

SOLUTION. The given equation can be written as

$$y'' + y = 6\cos 2t$$

Taking the Laplace transform of both the sides of the given differential equation, we get

$$L\{y''\} + L\{y\} = 6L\{\cos(2t)\}$$

$$\Rightarrow \quad p^2 L\{y\} - py(0) - y'(0) + L\{y\} = 6\frac{p}{p^2 + 2^2}$$

$$\Rightarrow \quad (p^2 + 1)\,L\{y\} - 3p - 1 = \frac{6p}{p^2 + 4} \qquad \text{[Using the given conditions]}$$

$$\Rightarrow \quad L\{y\} = \frac{3p}{p^2 + 1} + \frac{1}{p^2 + 1} + \frac{6p}{(p^2 + 1)\,(p^2 + 4)}$$

$$= \frac{3p}{p^2 + 1} + \frac{1}{p^2 + 1} + \frac{2p[(p^2 + 4) - (p^2 + 1)]}{(p^2 + 1)(p^2 + 4)}$$

$$= \frac{3p}{p^2 + 1} + \frac{1}{p^2 + 1} + 2p\left\{\frac{1}{p^2 + 1} - \frac{1}{p^2 + 4}\right\}$$

$$= \frac{5p}{p^2 + 1} + \frac{1}{p^2 + 1} - \frac{2p}{p^2 + 4}$$

Therefore, $\quad y = 5L^{-1}\left\{\dfrac{p}{p^2 + 1}\right\} + L^{-1}\left\{\dfrac{1}{p^2 + 1}\right\} - 2L^{-1}\left\{\dfrac{p}{p^2 + 4}\right\}$

$$\Rightarrow \quad y = 5\cos t + \sin t - 2\cos 2t$$

EXAMPLE 3. Using Laplace transforms, find the solution of the initial value problem :

$$y'' + 9y = 6\cos 3t \,,\ y(0) = 2,\ y'(0) = 0 \qquad \textbf{[UPTU–2006]}$$

SOLUTION. The given equation can be written as

$$y'' + 9y = 6\cos 3t \,,\ y(0) = 2,\ y'(0) = 0 \qquad \text{... (1)}$$

Taking Laplace transform of (1), we get

$$[p^2 L\{y\} - py(0) - y'(0)] + 9L\{y\} = 6\frac{p}{p^2 + 9}$$

Putting the value of $y(0)$ and $y'(0)$ in (2), we have

$$p^2 L\{y\} - 2p + 9L\{y\} = \frac{6p}{p^2 + 9}$$

$$(p^2 + 9)L\{y\} = 2p + \frac{6p}{p^2 + 9}$$

$$L\{y\} = \frac{2p}{p^2 + 9} + \frac{6p}{(p^2 + 9)^2}$$

$$\Rightarrow \quad y = L^{-1}\left\{\frac{2p}{p^2 + 9}\right\} + L^{-1}\left\{\frac{6p}{(p^2 + 9)^2}\right\} = 2\cos 3t + 3L^{-1}\frac{d}{dp}\left[-\frac{3}{p^2 + 9}\right]$$

$$= 2\cos 3t + t\sin 3t$$

EXAMPLE 4. *Solve* $(D^2 + 9)y = \cos 2t$ *if* $y(0) = 1$, $y\left(\dfrac{\pi}{2}\right) = -1$.　　　　**[UPTU–2002, 06, Bhopal–2008]**

SOLUTION. The given equation can be written as

$$y'' + 9y = \cos 2t \qquad \qquad \dots (1)$$

Taking the Laplace transform of both the sides of (1), we get

$$L\{y''\} + 9L\{y\} = L\{\cos 2t\}$$

$$\Rightarrow \quad p^2 L\{y\} - py(0) - y'(0) + 9L\{y\} = \dfrac{p}{p^2 + 4}$$

$$\Rightarrow \quad (p^2 + 9)\, L\{y\} - p - C = \dfrac{p}{p^2 + 4}, \text{ where } C = y'(0)$$

$$\therefore \quad L\{y\} = \dfrac{p + C}{p^2 + 9} + \dfrac{p}{(p^2 + 9)(p^2 + 4)} = \dfrac{p}{p^2 + 9} + \dfrac{C}{p^2 + 9} + \dfrac{p}{5(p^2 + 4)} - \dfrac{p}{5(p^2 + 9)}$$

Therefore, $y = L^{-1}\left\{\dfrac{p}{p^2 + 9}\right\} + CL^{-1}\left\{\dfrac{1}{p^2 + 9}\right\} + \dfrac{1}{5}L^{-1}\left\{\dfrac{p}{p^2 + 4}\right\} - \dfrac{1}{5}L^{-1}\left\{\dfrac{p}{p^2 + 9}\right\}$

$$= \cos 3t + \dfrac{1}{3}C \sin 3t + \dfrac{1}{5}\cos 2t - \dfrac{1}{5}\cos 3t$$

$$= \dfrac{4}{5}\cos 3t + \dfrac{1}{3}C \sin 3t + \dfrac{1}{5}\cos 2t \qquad \qquad \dots (2)$$

Now, since $y\left(\dfrac{\pi}{2}\right) = -1$, therefore, from (1), we have

$$-1 = \dfrac{4}{5}\cos\dfrac{3\pi}{2} + \dfrac{1}{3}C \sin\dfrac{3\pi}{2} + \dfrac{1}{5}\cos\pi$$

On solving, we get $C = \dfrac{12}{5}$

Put this value in (2), we get $y = \dfrac{4}{5}\cos 3t + \dfrac{4}{5}\sin 3t + \dfrac{1}{5}\cos 2t$.

EXAMPLE 5. *Solve using Laplace transform method*

$$y''(t) + 4y'(t) + 4y(t) = 6e^{-t}, \text{with } y(0) = -2, \ y'(0) = 8. \qquad \textbf{[UPTU–2007]}$$

SOLUTION. The given equation can be written as

$$y''(t) + 4y'(t) + 4y(t) = 6e^{-t}$$

Taking Laplace transform on both sides of the given equation, we get

$$[p^2 L\{y\} - py(0) - y'(0)] + 4[pL\{y\} - y(0)] + 4L\{y\} = \dfrac{6}{p + 1} \qquad \dots(1)$$

Putting $y(0) = -2$ and $y'(0) = 8$ in (1), we get

$$[p^2 L\{y\} - p(-2) - 8] + 4[pL\{y\} + 2] + 4L\{y\} = \dfrac{6}{p + 1}$$

$$\Rightarrow \quad (p^2 + 4p + 4)L\{y\} + 2p = \dfrac{6}{p + 1} \qquad \Rightarrow \quad (p^2 + 4p + 4)L\{y\} = -2p + \dfrac{6}{p + 1}$$

$$\Rightarrow \quad (p + 2)^2 L\{y\} = \dfrac{-2p^2 - 2p + 6}{(p + 1)} \qquad \Rightarrow \quad L\{y\} = \dfrac{-2p^2 - 2p + 6}{(p + 1)(p + 2)^2}$$

Let $\dfrac{-2p^2 - 2p + 6}{(p+1)(p+2)^2} = \dfrac{A}{p+1} + \dfrac{B}{p+2} + \dfrac{C}{(p+2)^2}$

$-2p^2 - 2p + 6 = A(p+2)^2 + B(p+1)(p+2) + C(p+1)$

$-2 + 2 + 6 = A(-1+2)^2 \Rightarrow A = 6$ [Putting $p = -1$]

$-8 + 4 + 6 = C(-2+1) \Rightarrow C = -2$ [Putting $p = -2$]

Comparing the coefficients of p^2 on both sides, we get

$-2 = A + B \Rightarrow -2 = 6 + B \Rightarrow B = -8$

$$L\{y\} = \frac{6}{p+1} + \frac{-8}{p+2} + \frac{2}{(p+2)^2}$$

$$y = L^{-1}\left[\frac{6}{p+1} - \frac{8}{p+2} - \frac{2}{(p+2)^2}\right]$$

Hence, $y = 6e^{-t} - 8e^{-2t} - 2e^{-2t}t$

EXAMPLE 6. *Solve* $(D^3 - 2D^2 + 5D)y = 0$ *given that* $y(0) = 0$, $y'(0) = 1$, $y\left(\dfrac{\pi}{8}\right) = 1$.

[UPTU Q. Bank–2001]

SOLUTION. The given equation can be written as

$$y''' - 2y'' + 5y' = 0 \qquad ...(1)$$

Taking the Laplace transforms of both sides of (1), we get

$L\{y'''\} - 2L\{y''\} + 5L\{y'\} = 0$

$\Rightarrow \quad p^3 L\{y\} - p^2 y(0) - p y'(0) - y''(0)$

$\qquad -2[p^2 L\{y\} - py(0) - y'(0)] + 5[pL\{y\} - y(0)] = 0$

$\Rightarrow \quad [p^3 - 2p^2 + 5p] L\{y\} - p - C - 2(-1) + 5.0 = 0,$

where $y''(0) = C$

$$L\{y\} = \frac{C - 2 + p}{p^3 - 2p^2 + 5p} = \frac{C-2}{p(p^2 - 2p + 5)} + \frac{1}{p^2 - 2p + 5}$$

$$= \frac{C-2}{5p} - \frac{C-2}{5} \cdot \frac{p-2}{p^2 - 2p + 5} + \frac{1}{p^2 - 2p + 5}$$

$$= \frac{C-2}{5p} - \frac{C-2}{5} \cdot \frac{(p-1)-1}{(p-1)^2 + 4} + \frac{1}{(p-1)^2 + 4}$$

$$= \frac{C-2}{5p} - \frac{C-2}{5} \cdot \frac{(p-1)}{(p-1)^2 + 4} + \frac{C+3}{10} \cdot \frac{2}{(p-1)^2 + 4}$$

Therefore, $y = \dfrac{C-2}{5} . L^{-1}\left\{\dfrac{1}{p}\right\} - \dfrac{C-2}{5} L^{-1}\left\{\dfrac{p-1}{(p-1)^2 + 4}\right\} + \dfrac{C+3}{10} L^{-1}\left\{\dfrac{2}{(p-1)^2 + 4}\right\}$

$$= \frac{C-2}{5} - \frac{C-2}{5} e^t \cos 2t + \frac{C+3}{10} e^t \sin 2t \qquad ...(2)$$

Now, since $y\left(\dfrac{\pi}{8}\right)=1$, therefore $1=\dfrac{C-2}{5}-\dfrac{C-2}{5}e^{\pi/8}\dfrac{1}{\sqrt{2}}+\dfrac{C+3}{10}e^{\pi/8}\cdot\dfrac{1}{\sqrt{2}}$

$\Rightarrow \qquad \dfrac{7-C}{5}=\dfrac{e^{\pi/8}}{10\sqrt{2}}(-2C+4+C+3)$

$\Rightarrow \qquad \left(\dfrac{7-C}{5}\right)\cdot\left(1-\dfrac{e^{\pi/8}}{2\sqrt{2}}\right)=0 \qquad \Rightarrow \qquad C=7$

Put this value of C in (2), we get $y=1+e^t(\sin 2t-\cos 2t)$.

EXAMPLE 7. *Solve* $(D^2-3D+2)y=1-e^{2t}$, $y=1$, $Dy=0$ *when* $t=0$.

SOLUTION. The given equation can be written as

$$y''-3y'+2y=1-e^{2t} \qquad \qquad \text{... (1)}$$

Taking Laplace transform of both the sides of (1), we get

$$L\{y''\}-3L\{y'\}+2L\{y\}=L\{1\}-L\{e^{2t}\}$$

$$p^2 L\{y\}-py(0)-y'(0)-3[pL\{y\}$$
$$-y(0)]+2L\{y\}=\dfrac{1}{p}-\dfrac{1}{p-2}$$

$$\Rightarrow \qquad (p^2-3p+2)L\{y\}-p+3=-\dfrac{2}{p(p-2)}$$

$$\Rightarrow \qquad (p-1)(p-2)L\{y\}=-\dfrac{2}{p(p-2)}+(p-3)$$

$$\Rightarrow \qquad L\{y\}=\dfrac{p^3-5p^2+6p-2}{p(p-1)(p-2)^2}=\dfrac{p^2-4p+2}{p(p-2)^2}$$

$$=\dfrac{1}{2p}+\dfrac{1}{2(p-2)}-\dfrac{1}{(p-2)^2}$$

Therefore, $y=\dfrac{1}{2}L^{-1}\left\{\dfrac{1}{p}\right\}+\dfrac{1}{2}L^{-1}\left\{\dfrac{1}{p-2}\right\}-L^{-1}\left\{\dfrac{1}{(p-2)^2}\right\}=\dfrac{1}{2}+\dfrac{1}{2}e^{2t}-te^{2t}$

Exercise 14.8

1. Solve $\dfrac{dy}{dt}+y=1$ if $y=2$ when $t=0$.

2. Show that the general solution of the equation $(D^2+k^2)y=0$ is $y=C_1\cos kt+C_2\sin kt$.

3. Solve $y''(t)+y(t)=t$ if $y'(0)=1$, $y(\pi)=0$.

4. Solve $(D^2-1)y=a\cosh nt$ if $y=Dy=0$, when $t=0$.

5. Solve $(D^2+m^2)x=a\cos nt$, $t\neq 0$, where x, Dx equal to x_0 and x_1, when $t=0, n\neq m$.

6. Solve $(D^2+m^2)y=a\cos nt$, $t>0$ if $y=0=Dy$, when $t=0$.

7. Solve $(D^2+m^2)x=a\sin nt$, $t>0$, where x, Dx equal to x_0 and x_1, when $t=0, n\neq m$.

8. Solve $(D+2)^2 y=4e^{-2t}$, $y(0)=-1$ and $y'(0)=4$.

9. Solve $(D^2+6D+9)y=\sin x$, where $y(0)=1$, $y'(0)=0$.

10. Solve $(D^2+4D+4)x=\sin\omega t$, $t>0$, where x_0 and x_1 are the values of x and Dx, when $t=0$.

11. Solve $(D^2+3D+2)y=0$, where $y=y_0$ and $Dy=y_1$ at $t=0$.

12. Solve $(D^2 + 9)y = 18t$, if $y(0) = 0$, $y\left(\dfrac{\pi}{2}\right) = 0$.

13. Solve $(D^2 + 2D + 1)y = 3te^{-t}$, $t > 0$ subject to the conditions $y = 4, Dy = 2$, when $t = 0$.

14. Solve $(D^2 + 1)y = \sin t \sin 2t, t > 0$ if $y = 1, Dy = 0$, when $t = 0$.
[UPTU–2001(Sp.), 02]

15. Solve $(D^2 + n^2)y = a \sin(nt + \alpha)$ if $y = Dy = 0$, when $t = 0$. **(GBTU(CO)–2010)**

16. Solve $(D^3 + 1)y = 1, t > 0$ if $y = Dy = D^2y = 0$, when $t = 0$.

17. Solve
$(D^3 - D)y = 2\cos t, y = 3, Dy = 2, D^2y = 1$,
when $t = 0$.

18. Solve
$(D^3 + D)y = e^{2t}$, $y(0) = y'(0) - y''(0) = 0$.

19. Solve $(D^4 - 1)y = 1$ if $y = Dy = D^2y = D^3y = 0$ at $t = 0$.

20. Solve $(D^4 + 2D^2 + 1)y = 0$ if $y(0) = 0$, $y'(0) = 1$, $y''(0) = 2$ and $y'''(0) = -3$.

21. Solve $(D^2 + D)y = t^2 + 2t$ if $y(0) = 4$, $y'(0) = -2$.

22. Solve $\dfrac{d^3y}{dt^3} - 3\dfrac{d^2y}{dt^2} + 3\dfrac{dy}{dt} - y = t^2e^t$ where

$y(0) = 1, y(0) = 1, \left(\dfrac{dy}{dt}\right)_{t=0} = 0, \left(\dfrac{d^2y}{dt^2}\right)_{t=0} = -2$
(UPTU (SUM)–2008, SVTU–2009)

23. Voltage Ee^{-at} is applied at $t = 0$ to a circuit of inductance L and resistance R. Show that the current at time t is $\dfrac{E}{R - aL}(e^{-at} - e^{-Rt/L})$
(UPTU (SUM)–2007, VTU–2000)

24. Solve $y'' + 4y' + 3y = e^{-t}, y(0) = y'(0) = 1$
(VTU–2008 S, Kurukshetra–2005)

25. Solve $y'' + y = t, y(0) = 1, y'(0) = 0$
(Mumbai–2009)

26. Solve $y'' - 3y' + 2y = e^{3t}$ when $y(0) = 1$ and $y'(0) = 0$
(VTU–2010)

27. Solve $(D^2 - 3D + 2)y = 4e^{2t}$ with $y(0) = -3, y(0) = 5$
(Mumbai–2008)

28. Solve $y'' + 25y = 10\cos 5t$ given that $y(0) = 2, y''(0) = 0$
(SVTU–2008)

29. Solve $\dfrac{d^2y}{dt^2} + 2\dfrac{dy}{dt} - 3y = \sin t, y = \dfrac{dy}{dt} = 0$ when $t = 0$
(Kurukshetra–2005, Madras–2003)

30. Solve $y'' + 2y' + 5y = 5(t - 2), y(0) = 0$, $y'(0) = 0$ **(PTU–2005 S)**

31. Solve $\dfrac{d^2x}{dt^2} + 9x = \sin 2t, x(0) = 1, x'(0) = 0$
(GBTU (CO)–2011)

32. Solve $\dfrac{d^2y}{dt^2} + 9x = \sin 3t$, given $y = 0, \dfrac{dy}{dt} = 0$ at $t = 0$ **(MTU–2012)**

33. Solve $\dfrac{d^2x}{dt^2} + 6\dfrac{dx}{dt} + 8x = e^{-3t} - e^{-5t}$;

$x(0) = 0, x'(0) = 0$ **(UPTU(CO)–2009)**

34. Solve $y'' + 3y' + 2y = te^{-t}; y(0) = 1, y'(0) = 0$
(GBTU–2012)

35. Solve $y'' + 2y' + y = te^{-t}; y(0) = 1, y'(0) = 2$
(MTU (SUM)–2011)

36. Solve $\dfrac{d^2x}{dt^2} + 3\dfrac{dx}{dt} + 2x = r(t)$

where $r(t) = \begin{cases} e^t & , \ 0 < t < 2 \\ 0 & , \ t > 2 \end{cases}$

and $x(0) = 1, x'(0) = -2$ **(GBTU(CO)–2010)**

37. Solve $\dfrac{d^2y}{dt^2} + 9y = r(t)$ with intial

conditions $y(0) = 0$ and $y'(0) = 4$ where

$r(t) = \begin{cases} 8\sin t & 0 < t < \pi \\ 0 & t > \pi \end{cases}$ **(GBTU–2011)**

38. A particle moves in a line so that its displacement x from a fixed point O at any time t, is given by

$$\dfrac{d^2x}{dt^2} + 4\dfrac{dx}{dt} + 5x = 80\sin 5t$$

Using Laplace transform, find its displacement at any time t if initially particle is at rest at $x=0$. **(UPTU (CO)–2009)**

39. An alternating e.m.f $E \sin \omega t$ is applied to circuit with an inductance L and a capacitance C in series. Show that the current in the circuit is $\dfrac{E\omega}{(n^2 - \omega^2)L}(\cos \omega t - \cos nt)$ where $n^2 = \dfrac{1}{LC}$. **(GBTU–2010)**

HINT TO SELECTED PROBLEMS

1. Taking the Laplace transform of the given equation, we get $L(y') + L(y) = L(1)$

$$\Rightarrow \quad pL\{y\} - y\{0\} + L\{y\} = \frac{1}{p}$$

$$\Rightarrow \quad L\{y\} = \frac{2p+1}{p(p+1)} = \frac{1}{p+1} + \frac{1}{p}.$$

Now taking inverse Laplace transform.

3. $L\{y''\} + L\{y\} = L\{t\}$

$$\Rightarrow p^2 L\{y\} - py(0) - y'(0) + L\{y\} = \frac{1}{p^2}$$

$$\Rightarrow (p^2 + 1)L\{y\} - pA - 1 = \frac{1}{p^2}.$$

$$[\because A = y(0), \ y'(0) = 1]$$

$$\Rightarrow \quad L\{y\} = \frac{pA}{p^2+1} + \frac{1}{p^2}$$

Now taking inverse Laplace transform.

4. Taking Laplace transform of the given equation and after simplification, we get

$$L\{y\} = \frac{ap}{(p^2-1)(p^2-n^2)}$$

$$= \frac{ap}{(n^2-1)}\left[\frac{1}{p^2-n^2} - \frac{1}{p^2-1}\right]$$

$$= \frac{1}{(n^2-1)}\left[\frac{p}{p^2-n^2} - \frac{p}{p^2-1}\right]$$

Now taking inverse Laplace transform of both the sides.

7. Proceeding as usual, we get

$$L\{x\} = \frac{px_0}{(p^2+m^2)} + \frac{x_1}{(p^2+m^2)}$$

$$+ \frac{a_n}{(p^2+m^2)(p^2+n^2)} = \frac{px_0}{(p^2+m^2)}$$

$$+ \frac{x_1}{(p^2+m^2)} + \frac{a}{(m^2-n^2)}\left[\frac{n}{p^2+n^2} - \frac{n}{p^2+m^2}\right]$$

Now taking inverse Laplace transform of both the sides.

9. We have $L\{y\} = \frac{(p+6)}{(p+3)^2} + \frac{1}{(p^2+1)(p+3)^2}$

$$= \frac{1}{(p+3)} + \frac{3}{(p+3)^2} +$$

$$\left[\frac{3}{50(p+3)} + \frac{1}{10(p+3)^2} - \frac{3p-4}{50(p^2+1)}\right]$$

$$= \frac{1}{50}\left[\frac{53}{p+3} + \frac{155}{(p+3)^2} - \frac{3p}{(p^2+1)} + \frac{4}{(p^2+1)}\right]$$

Now taking inverse Laplace transform of both the sides.

11. Proceeding as usual, we get

$$L\{y\} = \frac{p+3}{(p^2+3p+2)}y_0 + \frac{y_1}{(p^2+3p+2)}$$

$$= \frac{(p+3)}{(p+1)(p+2)}y_0 + \frac{y_1}{(p+1)(p+2)}$$

$$= \left[\frac{2}{p+1} - \frac{1}{p+2}\right]y_0 + \left[\frac{1}{p+1} - \frac{1}{p+2}\right]y_1$$

$$= \frac{(2y_0 + y_1)}{(p+1)} - \frac{(y_0 + y_1)}{(p+2)}$$

Now taking inverse Laplace transform of both the sides.

14. $(D^2 + 1)y = \sin t \sin 2t = \frac{1}{2}[\cos t - \cos 3t]$

$$\Rightarrow \quad L\{y''\} + L\{y\} = \frac{1}{2}[L\{\cos t\} - L\{\cos 3t\}]$$

After simplification, we get

$$L\{y\} = \frac{p}{p^2+1} + \frac{p}{2(p^2+1)^2}$$

$$- \frac{p}{16}\left[\frac{(p^2+9) - (p^2+1)}{(p^2+9)(p^2+1)}\right]$$

$$= \frac{p}{p^2+1} - \frac{1}{4}\left[\frac{d}{dp}\left(\frac{1}{p^2+1}\right)\right]$$

$$- \frac{p}{16(p^2+1)} + \frac{p}{16(p^2+9)}$$

Now taking inverse Laplace transform of both the sides.

15. The given equation can be written as

$$y'' + n^2 y = a[\sin nt \cos\alpha + \cos nt \sin\alpha]$$

Taking Laplace transform and simplifying, we get

$$L\{y\} = a\cos\alpha . \frac{n}{(p^2+n^2)^2} + a\sin\alpha \frac{p}{(p^2+n^2)^2}$$

$$\Rightarrow \quad y = a.n \cos\alpha L^{-1}\left\{\frac{1}{(p^2+n^2)^2}\right\}$$

$$+ a\sin\alpha L^{-1}\left\{\frac{p}{(p^2+n^2)^2}\right\}$$

$$= a.n\cos\alpha \int_0^t \left(\frac{1}{n}\sin nx\right)\frac{1}{n}\sin n(t-x)dx$$

$$-\frac{a\sin\alpha}{2}L^{-1}\left\{\frac{d}{dp}\frac{1}{(p^2+n^2)^2}\right\}$$

(Using convolution theorem)

$$=a\frac{\cos\alpha}{2n}\int_0^t[\cos n(t-2x)-\cos nt]dx$$

$$+\frac{a\sin\alpha}{2}t.L^{-1}\left\{\frac{1}{p^2+n^2}\right\}$$

16. Proceeding as usual, we get

$$L\{y\}=\frac{1}{p}-\frac{1}{3(p+1)}-\frac{2\left(p-\frac{1}{2}\right)}{3\left[\left(p-\frac{1}{2}\right)^2+\frac{3}{4}\right]}$$

$$\Rightarrow\ y=1-\frac{e^{-t}}{3}-\frac{2}{3}e^{t/2}L^{-1}\left\{\frac{p}{p^2+\left(\sqrt{3}/2\right)^2}\right\}$$

ANSWERS

1. $y=e^{-t}+1$ **3.** $y=\pi\cos t+t$ **4.** $y=\dfrac{a}{n^2-1}(\cosh nt-\cosh t)$

5. $x=x_0\cos mt+\dfrac{x_1}{m}\sin mt+\dfrac{a}{m^2-n^2}(\cos nt-\cos mt)$ **6.** $y=\dfrac{a}{m^2-n^2}(\cos nt-\cos mt)$

7. $x=x_0\cos mt+\dfrac{x_1}{m}\sin mt+\dfrac{a}{m^2-n^2}\left(\sin nt-\dfrac{n}{m}\sin mt\right)$ **8.** $y=e^{-2t}(2t^2+2t-1)$

9. $\dfrac{1}{50}[(53+155x)e^{-3x}-(3\cos x-4\sin x)]$

10. $x=e^{-2t}\left[x_0(1-2t)+(x_1+4x_0)+\dfrac{w}{(4+w^2)}t+\dfrac{4w}{(4+w^2)^2}\right]-\dfrac{4w}{(4+w^2)^2}\cos wt+\dfrac{(4-w^2)}{(4+w^2)^2}\sin wt$

11. $y=(2y_0+y_1)e^{-t}-(y_0+y_1)e^{-2t}$ **12.** $y=\pi\sin 3t+2t$

13. $y=\dfrac{1}{2}e^{-t}.t^3+4e^{-t}+6te^{-t}$ **14.** $y=\dfrac{15}{16}\cos t+\dfrac{1}{4}t\sin t+\dfrac{1}{16}\cos 3t$

15. $y=\dfrac{a}{2n^2}[\cos\alpha\sin nt-nt\cos(\alpha+nt)]$ **16.** $y=1-\dfrac{1}{3}e^{-t}-\dfrac{2}{3}e^{t/2}\cos\left(\dfrac{\sqrt{3}t}{2}\right)$

17. $y=3\sinh t-\sin t+\cosh t+2$ **18.** $y=\dfrac{1}{3}e^{-t}(\sin 2t+\sin t)$ **19.** $y=-1+\dfrac{1}{2}\cosh t+\dfrac{1}{2}\cos t$

20. $y=t(\sin t+\cos t)$ **21.** $y=\dfrac{1}{3}t^3+2+2e^{-t}$ **22.** $y=\left(1-t-\dfrac{t^2}{2}+\dfrac{t^5}{60}\right)e^t$

24. $y=\dfrac{7}{4}e^{-t}-\dfrac{3}{4}e^{-3t}-\dfrac{1}{2}te^{-t}$ **25.** $y=t-3\sin t+\cos t$ **26.** $y=2t+3+\dfrac{1}{2}(e^{3t}-e^t)-2e^{2t}$

27. $y=4e^{2t}(1+t)-7e^t$ **28.** $y=2\cos 5t+t\sin 5t$

29. $y=\dfrac{1}{8}e^t-\dfrac{1}{40}e^{-3t}-\dfrac{1}{10}(2\sin t+\cos t)$ **30.** $y=\dfrac{-12}{5}+\dfrac{12}{5}e^{-t}\cos 2t+\dfrac{7}{10}e^{-t}\sin 2t$

31. $x=\cos 3t+\dfrac{1}{5}\sin 2t-\dfrac{2}{15}\sin 3t$ **32.** $y=\dfrac{1}{18}(\sin 3t-3t\cos 3t)$

33. $x=\dfrac{1}{3}(e^{-2t}-e^{-5t})-e^{-3t}+e^{-4t}$ **34.** $y=3e^{-t}-2e^{-2t}+e^{-t}\left(\dfrac{t^2}{2}-t\right)$

35. $y=e^{-t}\left(1+3t+\dfrac{t^3}{6}\right)$

36. $x=\dfrac{4}{3}e^{-2t}+\dfrac{1}{6}e^t[1-u(t-2)]-\dfrac{1}{2}e^{-t}+\dfrac{1}{2}e^{4-t}u(t-2)-\dfrac{1}{3}e^{6-2t}u(t-2)$

37. $y=\sin 3t+\sin t+[\sin(t-\pi)-\dfrac{1}{3}\sin 3(t-\pi)]u(t-\pi)$

38. $x=e^{-2t}(2\cos t+14\sin t)-2\cos 5t-2\sin 5t$

14.19 SOLUTION OF ORDINARY DIFFERENTIAL EQUATION WITH VARIABLE COEFFICIENTS

The Laplace transform can also be used in solving some ordinary differential equations in which the coefficients are variable. A particular differential equation when the method proves useful is one in which the terms have the form $t^m y^n$ (t) whose Laplace transform is

$$(-1)^m \frac{d^m}{dp^m} [L\{y^n(t)\}]$$

SOLVED EXAMPLES

EXAMPLE 1. Solve $(tD^2 + D + 4t)y = 0$ if $y(0) = 3$, $y'(0) = 0$.

SOLUTION. The given equation can be written as

$$ty'' + y' + 4ty = 0 \qquad \qquad \text{... (1)}$$

Taking the Laplace transform of both sides of (1), we get

$$L\{ty''\} + L\{y'\} + 4L\{ty\} = 0$$

$$\Rightarrow \quad -\frac{d}{dp}L\{y''\} + L\{y'\} + 4(-1)\frac{d}{dp}L\{y\} = 0$$

$$\Rightarrow \quad -\frac{d}{dp}[p^2 L\{y\} - py(0) - y'(0)] + [pL\{y\} - y(0)] - 4\frac{d}{dp}L\{y\} = 0$$

$$\Rightarrow \quad -\frac{d}{dp}[p^2 L\{y\} - 3p] + (pL\{y\} - 3) - \frac{4d[L\{y\}]}{dp} = 0$$

$$\Rightarrow \quad -(p^2 + 4)\frac{d[L\{y\}]}{dp} - pL\{y\} = 0$$

$$\Rightarrow \quad \frac{d[L\{y\}]}{L\{y\}} + \frac{p}{p^2 + 4}dp = 0$$

On integrating, we get

$$\log[L\{y\}] + \frac{1}{2}\log(p^2 + 4) = \log C_1$$

$$\Rightarrow \quad L\{y\} = \frac{C_1}{\sqrt{p^2 + 4}}$$

Therefore, $\quad y = L^{-1}\left\{\dfrac{C_1}{\sqrt{p^2 + 4}}\right\} = C_1 J_0(2t)$.

EXAMPLE 2. Solve $[tD^2 + (t-1)D - 1]y = 0$ if $y(0) = 5, \ y(\infty) = 0$.

SOLUTION. The given equation can be written as

$$ty'' + ty' - y' - y = 0 \qquad \qquad \text{...(1)}$$

Taking the Laplace transforms of both sides of (1), we get

$$L\{ty''\} + L\{ty'\} - L\{y'\} - L\{y\} = 0$$

$$\Rightarrow \quad -\frac{d}{dp}[L\{y''\}] - \frac{d}{dp}[L\{y'\} - [pL\{y\} - y(0) - L\{y\}] = 0$$

$$\Rightarrow \quad -\frac{d}{dp}[p^2 L\{y\} - py(0) - y'(0)] - \frac{d}{dp}[pL\{y\} - y(0)] - pL\{y\} + 5 - L\{y\} = 0$$

$$\Rightarrow \quad -\frac{d}{dp}[p^2L\{y\} - 5p - A] - \frac{d}{dp}[pL\{y\} - 5] - (p+1)L\{y\} + 5 = 0,$$

where $A = y'\{0\}$

$$\Rightarrow \quad \frac{d[L\{y\}]}{dp} + \frac{3p+2}{p^2+p}L\{y\} = \frac{10}{p^2+p} \qquad \qquad ...(2)$$

which is a linear differential equation in $L\{y\}$.

Therefore,

$$\text{I.F.} = e^{\int\left\{\frac{3p+2}{(p^2+p)}\right\}dp} = e^{\int\left(\frac{2}{p}+\frac{1}{p+1}\right)dp}$$

$$= e^{[2\log p + \log(p+1)]} = p^2(p+1).$$

Hence, the solution of equation (2) is given by

$$L\{y\}.p^2(p+1) = C_1 + \int \frac{10}{p^2+p}.p^2(p+1)dp$$

$$= C_1 + 10\int p\, dp = C_1 + 5p^2$$

$$\Rightarrow \quad L\{y\} = \frac{C_1}{p^2(p+1)} + \frac{5}{p+1} = C_1\left\{\frac{1}{p^2} - \frac{1}{p} + \frac{1}{p+1}\right\} + \frac{5}{p+1}$$

Therefore,

$$y = C_1 L^{-1}\left\{\frac{1}{p^2} - \frac{1}{p} + \frac{1}{p+1}\right\}.$$

$$+ 5L^{-1}\left\{\frac{1}{p+1}\right\} = C_1(t-1+e^{-t}) + 5e^{-t}.$$

Now, using the given conditions $y(\infty) = 0$.

We must have $C_1 = 0$. Hence, $y = 5e^{-t}$ is the required solution.

EXAMPLE 3. Solve $\dfrac{d^2y}{dx^2} + 2\dfrac{dy}{dx} + 5y = e^{-x}\sin x$, where $y(0) = 0$, $y'(0) = 1$.

[UPTU–2004, 08, (SUM)–2009, PTU–2010, MTU–2011]

SOLUTION. We have $\dfrac{d^2y}{dx^2} + 2\dfrac{dy}{dx} + 5y = e^{-x}\sin x$

Taking the Laplace transform on both the sides,

we get $[p^2L\{y\} - py(0) - y'(0)] + 2[pL\{y\} - y(0)] + 5L\{y\} = L\{e^{-x}\sin x\}$

$$[p^2L\{y\} - py(0) - y'(0)] + 2[pL\{y\} - y(0)] + 5L\{y\} = \frac{1}{(p+1)^2+1} \qquad ...(1)$$

On substituting the values of $y(0)$ and $y'(0)$ in (1), we get

$$(p^2L\{y\} - 1) + 2pL\{y\} + 5L\{y\} = \frac{1}{p^2+2p+2}$$

$$(p^2+2p+5)L\{y\} = 1 + \frac{1}{p^2+2p+2} = \frac{p^2+2p+3}{p^2+2p+2}$$

$$L\{y\} = \frac{p^2+2p+3}{(p^2+2p+5)(p^2+2p+2)}$$

On resolving R.H.S. into partial fractions, we get

$$L\{y\} = \frac{2}{3} \cdot \frac{1}{p^2 + 2p + 5} + \frac{1}{3} \cdot \frac{1}{p^2 + 2p + 2}$$

On inversion, we obtain

$$y = \frac{2}{3}L^{-1}\frac{1}{p^2 + 2p + 5} + \frac{1}{3}L^{-1}\frac{1}{p^2 + 2p + 2}$$

$$y = \frac{1}{3}L^{-1}\frac{2}{(p+1)^2 + (2)^2} + \frac{1}{3}L^{-1}\frac{1}{(p+1)^2 + (1)^2}$$

$$\Rightarrow \quad y = \frac{1}{3}e^{-x}\sin 2x + \frac{1}{3}e^{-x}\sin x$$

$$y = \frac{1}{3} \cdot e^{-x}(\sin x + \sin 2x)$$

Example 4. Solve $(D^2 + 1)y = t\cos 2t$ subject to the condition $y = 0$, $\frac{dy}{dt} = 0$ when $t = 0$.

[UPTU–2005, UKTU–2012, Raipur–2005]

Solution. The given equation can be written as

$$y'' + y = t\cos 2t \qquad \ldots (1)$$

Taking the Laplace transform of both sides of (1), we get

$$L\{y''\} + L\{y\} = L\{t\cos 2t\}$$

$$\Rightarrow \quad p^2 L\{y\} - py(0) - y'(0) + L\{y\} = -\frac{d}{dp}[L\{\cos 2t\}]$$

$$\Rightarrow \quad (p^2 + 1)L\{y\} = -\frac{d}{dp}\left(\frac{p}{p^2 + 4}\right) = -\frac{1}{p^2 + 4} + \frac{2p^2}{(p^2 + 4)^2}$$

$$\therefore \quad L\{y\} = \frac{p^2 - 4}{(p^2 + 1)(p^2 + 4)^2} = -\frac{5}{9(p^2 + 1)} + \frac{5}{9(p^2 + 4)} + \frac{8}{3(p^2 + 4)^2}$$

[Resolving into partial fractions]

$$\Rightarrow \quad y = -\frac{5}{9}L^{-1}\left\{\frac{1}{p^2 + 1}\right\} + \frac{5}{9}L^{-1}\left\{\frac{1}{p^2 + 4}\right\} + \frac{8}{3}L^{-1}\left\{\frac{1}{(p^2 + 4)^2}\right\}$$

$$= -\frac{5}{9}\sin t + \frac{5}{18}\sin 2t + \frac{8}{3}\int_0^t \frac{1}{2}\sin 2x \cdot \frac{1}{2}\sin 2(t - x)dx$$

[By convolution theorem and using $L^{-1}\left\{\frac{1}{p^2 + 4}\right\} = \frac{1}{2}\sin 2t$]

$$= -\frac{5}{9}\sin t + \frac{5}{18}\sin 2t + \frac{1}{3}\int_0^t \{\cos(2t - 4x) - \cos 2t\}dx$$

$$= -\frac{5}{9}\sin t + \frac{5}{18}\sin 2t + \frac{1}{3}\left[-\frac{1}{4}\sin(2t - 4x) - x\cos 2t\right]_0^t$$

$$= -\frac{5}{9}\sin t + \frac{5}{18}\sin 2t + \frac{1}{12}\sin 2t - \frac{1}{3}t\cos 2t + \frac{1}{12}\sin 2t$$

$$= -\frac{5}{9}\sin t + \frac{4}{9}\sin 2t - \frac{1}{3}t\cos 2t$$

EXAMPLE 5. *Solve* $(D^3 - D^2 - D + 1)y = 8te^{-t}$ *if* $y = D^2y = 0$, $Dy = 0$ *when* $t = 0$.

SOLUTION. The given equation can be written as

$$y''' - y'' - y' + y = 8te^{-t} \qquad \qquad \ldots (1)$$

Taking the Laplace transforms of both sides of (1), we get

$$L\{y'''\} - L\{y''\} - L\{y'\} + L\{y\} = 8L\{te^{-t}\}$$

$$\Rightarrow \quad p^3 L\{y\} - p^2 y(0) - py'(0) - y''(0) - [p^2 L\{y\} - py(0) - y'(0)]$$

$$-[pL\{y\} - y(0)] + L\{y\} = -8\frac{d}{dp}L\{e^{-t}\}$$

or $\qquad (p^3 - p^2 - p + 1)L\{y\} - p + 1 = -8\dfrac{d}{dp}\left[\dfrac{1}{p+1}\right]$

$$\Rightarrow \qquad (p-1)^2(p+1)L\{y\} = p - 1 + \frac{8}{(p+1)^2}$$

$$\Rightarrow \qquad L\{y\} = \frac{1}{(p-1)(p+1)} + \frac{8}{(p-1)^2(p+1)^3}$$

$$= \frac{1}{2}\left(\frac{1}{p-1} - \frac{1}{p+1}\right) - \frac{3}{2(p-1)} + \frac{1}{(p-1)^2}$$

$$+ \frac{3}{2(p+1)} + \frac{2}{(p+1)^2} + \frac{2}{(p+1)^3}$$

$$= -\frac{1}{p-1} + \frac{1}{p+1} + \frac{1}{(p-1)^2} + \frac{2}{(p+1)^2} + \frac{2}{(p+1)^3}$$

Therefore,

$$y = -L^{-1}\left\{\frac{1}{p-1}\right\} + L^{-1}\left\{\frac{1}{p+1}\right\} + L^{-1}\left\{\frac{1}{(p-1)^2}\right\}$$

$$+ 2L^{-1}\left\{\frac{1}{(p+1)^2}\right\} + 2L^{-1}\left\{\frac{1}{(p+1)^3}\right\}$$

$$= -e^t + e^{-t} + e^t L^{-1}\left\{\frac{1}{p^2}\right\} + 2e^{-t}L^{-1}\left\{\frac{1}{p^2}\right\} + 2e^{-t}L^{-1}\left\{\frac{1}{p^3}\right\}$$

$$= -e^t + e^{-t} + e^t.t + 2e^{-t}.t + 2e^{-t}\left(\frac{t^2}{2!}\right)$$

$$= (1 + 2t + t^2)e^{-t} - (1 - t)e^t.$$

EXAMPLE 6. *Solve* $[tD^2 + (1 - 2t)D - 2]y = 0$, *where* $y(0) = 1$, $y'(0) = 2$. **[UPTU–2002, PTU–2002]**

SOLUTION. Here, $tD^2y + (1 - 2t)Dy - 2y = 0 \quad \Rightarrow \quad ty'' + y' - 2ty' - 2y = 0$

Taking Laplace transform of given differential equation, we get

$$L\{ty''\} + L\{y'\} - 2L\{ty'\} - 2L\{y\} = 0$$

$$\Rightarrow \quad -\frac{d}{dp}L\{y''\} + L\{y'\} + 2\frac{d}{dp}L\{y'\} - 2L\{y\} = 0$$

$$-\frac{d}{dp}\Big[p^2 L\{y\} - py(0) - y'(0)\Big] + [pL\{y\} - y(0)] + 2[pL\{y\} - y(0)] - 2L\{y\} = 0$$

Putting the values of $y(0)$ and $y'(0)$, we get

$$-\frac{d}{dp}(p^2 L\{y\} - p - 2) + (pL\{y\} - 1) + 2\frac{d}{dp}(pL\{y\} - 1) - 2L\{y\} = 0$$

$$[\because y(0) = 1, y'(0) = 2]$$

$$\Rightarrow \quad -p^2\frac{dL\{y\}}{dp} - 2pL\{y\} + 1 + pL\{y\} - 1 + 2\left(p\frac{dL\{y\}}{dp} + L\{y\}\right) - 2L\{y\} = 0$$

$$\Rightarrow \quad -(p^2 - 2p)\frac{dL\{y\}}{dp} - pL\{y\} = 0$$

$$\Rightarrow \quad -\frac{dL\{y\}}{\overline{y}} - \frac{1}{p-2}dp = 0 \qquad\qquad \text{[Separating the variables]}$$

$$\Rightarrow \quad \int \frac{dL\{y\}}{\overline{y}} + \int \frac{dp}{p-2} = 0$$

$$\Rightarrow \quad \log L\{y\} + \log(p-2) = \log C$$

$$\Rightarrow \quad \log L\{y\}(p-2) = \log C$$

$$\Rightarrow \quad L\{y\}(p-2) = C$$

$$\Rightarrow \quad L\{y\} = \frac{C}{p-2}$$

$$\Rightarrow \quad y = CL^{-1}\left\{\frac{1}{p-2}\right\} \Rightarrow y = Ce^{2t} \qquad\qquad\qquad ...(1)$$

At $\qquad\qquad\qquad x = 0, \ y(0) = Ce^0 \qquad\qquad\qquad\qquad\qquad ...(2)$

Putting $y(0) = 1$, in (2), we get

$$1 = Ce^0 \Rightarrow C = 1$$

Putting $C = 1$ in (1), we get $y = e^{2t}$. This is the required solution.

EXAMPLE 7. Solve $y'' - ty' + y = 1$ if $y(0) = 1, \ y'(0) = 2$.

SOLUTION. Taking the Laplace transforms of both sides of the given equation, we get

$$L\{y''\} - L\{ty'\} + L\{y\} = L\{1\}$$

$$\Rightarrow \quad p^2 L\{y\} - py(0) - y'(0) + \frac{d}{dp}[L\{y'\}] + L\{y\}] = \frac{1}{p}$$

$$\Rightarrow \quad p^2 L\{y\} - p - 2 + \frac{d}{dp}[pL\{y\} - y(0)] + L\{y\} = \frac{1}{p}$$

$$\Rightarrow \quad p^2 L\{y\} - p - 2 + \frac{d}{dp}[pL\{y\} - 1] + L\{y\} = \frac{1}{p}$$

$$\Rightarrow \quad p\frac{d[L\{y\}]}{dp} + (p^2 + 2)L\{y\} = p + 2 + \frac{1}{p}$$

$$\Rightarrow \quad d\frac{[L\{y\}]}{dp} + \left(p + \frac{2}{p}\right)L\{y\} = 1 + \frac{2}{p} + \frac{1}{p^2} \qquad\qquad ...(1)$$

which is a linear differential equation in $L(y)$.

$\therefore$ 　　　　I.F. $= e^{\int \left(p+\frac{2}{p}\right)dp} = e^{\frac{p^2}{2}+2\log p} = p^2 e^{p^2/2}$.

Therefore, solution of (1) is given by

$$p^2 e^{p^2/2} L\{y\} = C_1 + \int \left(1+\frac{2}{p}+\frac{1}{p^2}\right) p^2 e^{p^2/2} dp$$

Hence, the solution of (1) is given by

$$p^2 e^{p^2/2} L\{y\} = C_1 + \int \left(1+\frac{2}{p}+\frac{1}{p^2}\right) p^2 e^{p^2/2}.dp$$

$$= C_1 + \int (p^2 + 2p + 1) e^{p^2/2} dp$$

$$= C_1 + \int (p^2 + 1) e^{p^2/2} dp + 2\int p e^{p^2/2} dp$$

$$= C_1 + \int (2v + 1) e^v . \frac{dv}{\sqrt{2v}} + 2\int \sqrt{2v} . e^v . \frac{dv}{\sqrt{2v}}$$

$$\left[\text{where } \frac{p^2}{2} = v \Rightarrow pdp = dv, \text{ i.e., } dp = \frac{dv}{\sqrt{2v}}\right]$$

$$= C_1 + \sqrt{2v}.e^v - \int \frac{e^v}{\sqrt{2v}} dv + \int \frac{e^v}{\sqrt{2v}} dv + 2\int e^v dv$$

$$= C_1 + \sqrt{2v}\, e^v + 2e^v = C_1 + p e^{p^2/2} + 2e^{p^2/2}.$$

Therefore,

$$L\{y\} = \frac{C_1}{p^2} e^{-p^2/2} + \frac{1}{p} + \frac{2}{p^2} = \frac{C_1}{p^2}\left(1 - \frac{p^2}{2} + \frac{p^4}{4.2!} - ...\right) + \frac{1}{p} + \frac{2}{p^2}$$

$$= \frac{(2+C_1)}{p^2} - \frac{C_1}{2} + \frac{C_1}{8} p^2 ... + \frac{1}{p}$$

[On expanding the exponential function]

Hence, 　　　$y = (2+C_1)L^{-1}\left\{\frac{1}{p^2}\right\} - \frac{1}{2}C_1 L^{-1}\{1\} + \frac{1}{8}C_1 L^{-1}\{p^2\} + ... + L^{-1}\left\{\frac{1}{p}\right\}$

$$= (2+C_1)t + 1 \qquad [\because L^{-1}\{p^n\} = 0, \text{ for } n = 0,1,2,...]$$

Also, given that $y'(0) = 2$

$\therefore$ 　　　　$2 = 2 + C_1 \Rightarrow C_1 = 0$ which gives $y = 2t + 1$ is the required solution.

EXAMPLE 8. *Using Laplace transform, solve the following differential equation*

$$y'' + 2ty' - y = t \text{ where, } y(0) = 0 \text{ and } y'(0) = 1. \qquad \text{[UPTU–2003]}$$

SOLUTION. 　We have 　　　$y'' + 2ty' - y = t$ 　　　　　　　　...(1)

Taking Laplace transform of (1), we get

$$[p^2 L\{y\} - py(0) - y'(0)] - 2\frac{d}{dp}[pL\{y\} - y(0)] - L\{y\} = \frac{1}{p^2} \qquad ...(2)$$

On putting $y(0) = 0$ and $y'(0) = 1$ in (2), we get

$$(p^2 L\{y\} - 1) - 2\frac{d}{dp}(pL\{y\} - 0) - L\{y\} = \frac{1}{p^2}$$

$$\Rightarrow \quad (p^2 L\{y\} - 1) - 2L\{y\} - 2p\frac{dL\{y\}}{dp} - L\{y\} = \frac{1}{p^2}$$

$$\Rightarrow \quad -2p\frac{dL\{y\}}{dp} + (p^2 - 3)L\{y\} = \frac{1}{p^2} + 1 = \frac{1+p^2}{p^2}$$

$$\Rightarrow \quad \frac{dL\{y\}}{dp} - \frac{p^2 - 3}{2p}L\{y\} = \frac{1+p^2}{-2p^3}$$

$$\Rightarrow \quad \frac{dL\{y\}}{dp} - \left(\frac{p}{2} - \frac{3}{2p}\right)L\{y\} = -\frac{1}{2p^3} - \frac{1}{2p} \qquad \ldots(3)$$

Thus, (3) is a linear differential equation

$$\text{I.F.} = e^{\frac{1}{2}\int \left(\frac{3}{p} - p\right)dp} = e^{\frac{1}{2}\left(3\log p - \frac{p^2}{2}\right)} = e^{\frac{p^2}{4}} \cdot p^{3/2}$$

Solution of differential equation (3) is

$$L\{y\}\, e^{-p^2/4} \cdot p^{3/2} = \frac{1}{2}\int \left(\frac{1}{p^3} + \frac{1}{p}\right)p^{3/2} \cdot e^{-p^2/4}\, dp$$

$$= -\frac{1}{2}\int \left(\sqrt{p} + \frac{1}{p^{3/2}}\right)e^{-p^2/4}\, dp$$

Put $p^2 = ut \Rightarrow p = 2\sqrt{t}$ so that $dp = \dfrac{dt}{\sqrt{t}}$. Then we have

$$L\{y\}p^{3/2}e^{-p^2/4} = -\frac{1}{2}\int \left(\sqrt{2}\, t^{1/4} + \frac{1}{2\sqrt{2}}t^{-3/4}\right)e^{-t}\frac{dt}{\sqrt{t}}$$

$$= -\frac{1}{\sqrt{2}}\int \left(t^{-1/4} + \frac{1}{4}t^{-5/4}\right)e^{-t}\, dt$$

$$= -\frac{1}{\sqrt{2}}\int t^{-1/4}e^{-t}\, dt - \frac{1}{4\sqrt{2}}\int t^{-5/4}e^{-t}\, dt$$

$$= -\frac{1}{\sqrt{2}}\left[t^{-1/4}\frac{e^{-t}}{-1} + \int \left(-\frac{1}{4}\right)t^{-5/4}e^{-t}\, dt\right]$$

$$\qquad\qquad + \frac{1}{4\sqrt{2}}\int t^{-5/4}e^{-t}\, dt$$

$$= \frac{1}{\sqrt{2}}e^{-t} \cdot t^{-1/4} = \frac{1}{\sqrt{2}}e^{-p^2/4}\left(\frac{p^2}{4}\right)^{-1/4}$$

$$= \frac{1}{\sqrt{p}}e^{-p^2/4}$$

$$\Rightarrow \quad L\{y\} = \frac{1}{p^2} \quad \Rightarrow \quad L\{y\} = \frac{1}{p^2} + C \Rightarrow y = L^{-1}\left\{\frac{1}{p^2} + C\right\} = t + C.$$

14.20 SOLUTION OF SIMULTANEOUS ORDINARY DIFFERENTIAL EQUATIONS

The Laplace transform can be used to solve two or more simultaneous ordinary differential equations. The procedure is essentially the same as that described in previous sections.

SOLVED EXAMPLES

EXAMPLE 1. *Solve* $(D^2 + 2)x - Dy = 1$, $Dx + (D^2 + 2)y = 0$, *if* $x = 0 = Dx = y = Dy$, *when* $t = 0$.

SOLUTION. Taking Laplace transforms of both sides of the given equations, we have
$$L\{x''\} + 2L\{x\} - L\{y'\} = L\{1\}$$
and $L\{x'\} + 2L\{y''\} + 2L\{y\} = 0$

$$\Rightarrow \quad p^2L\{x\} - px(0) - x'(0) + 2L\{x\} - [pL\{y\} - y(0)] = \frac{1}{p}$$

and $pL\{x\} - x(0) + p^2L\{y\} - py(0) - y'(0) + 2L\{y\} = 0$

which gives $(p^2 + 2)L(x) - pL\{y\} = \dfrac{1}{p}$

and $pL\{x\} + (p^2 + 2)L\{y\} = 0$

Solving for $L\{x\}$ and $L\{y\}$, we have

$$L\{x\} = \frac{p^2 + 2}{p(p^4 + 5p^2 + 4)} = \frac{1}{2p} - \frac{1}{6}\left[\frac{2p}{p^2 + 1} + \frac{p}{p^2 + 4}\right]$$

and $\quad L\{y\} = \dfrac{-1}{p^4 + 5p^2 + 4} = \dfrac{1}{3}\left[\dfrac{1}{p^2 + 4} - \dfrac{1}{p^2 + 1}\right].$

Therefore, $x = \dfrac{1}{2}L^{-1}\left\{\dfrac{1}{p}\right\} - \dfrac{1}{6}\left[2L^{-1}\left\{\dfrac{p}{p^2 + 1}\right\} + L^{-1}\left\{\dfrac{p}{p^2 + 4}\right\}\right]$

$$= \frac{1}{2} - \frac{1}{6}[2\cos t + \cos 2t]$$

and $\quad y = \dfrac{1}{3}\left[\dfrac{1}{2}\sin 2t - \sin t\right] = \dfrac{1}{6}[\sin 2t - 2\sin t]$

EXAMPLE 2. *Solve* *the* *simultaneous* *equation* $\dfrac{dx}{dt} - y = e^t$, $\dfrac{dy}{dt} + x = \sin t$, *given*

$x(0) = 1$, $y(0) = 0$. **[UPTU–2006, Q.Bank–2001, GBTU (SUM)–2010, UKTU–2011, Delhi–2002]**

SOLUTION. Taking Laplace transforms of the given equations, we get

$$[p\bar{x} - x(0)] - \bar{y} = \frac{1}{p - 1}, \text{ where } \quad \bar{x} = L(x), \ \bar{y} = L(y)$$

i.e., $\quad p\bar{x} - 1 - \bar{y} = \dfrac{1}{p - 1}$ $\qquad\qquad$ [∵ $x(0) = 1$]

$$p\bar{x} - \bar{y} = \frac{p}{p - 1} \text{ and } [p\bar{y} - y(0)] + \bar{x} = \frac{1}{p^2 + 1}$$

i.e., $\quad \bar{x} + p\bar{y} = \dfrac{1}{p^2 + 1}$ $\qquad\qquad$ [∵ $y(0) = 0$] $\qquad\qquad$ ...(2)

Solving (1) and (2) for $\bar{x}$ and $\bar{y}$, we have

$$\bar{x} = \frac{p^2}{(p - 1)(p^2 + 1)} + \frac{1}{(p^2 + 1)^2}$$

$$= \frac{1}{2}\left[\frac{1}{p-1} + \frac{p}{p^2+1} + \frac{1}{p^2+1}\right] + \frac{1}{(p^2+1)^2}$$

$$\overline{y} = \frac{p}{(p^2+1)^2} - \frac{p}{(p-1)(p^2+1)}$$

$$= \frac{p}{(p^2+1)^2} - \frac{1}{2}\left[\frac{1}{p-1} - \frac{p}{p^2+1} + \frac{1}{p^2+1}\right]$$

Taking inverse Laplace transform of both sides, we get

$$x = \frac{1}{2}L^{-1}\left\{\frac{1}{p-1} + \frac{p}{p^2+1} + \frac{1}{p^2+1}\right\} + L^{-1}\left\{\frac{1}{(p^2+1)^2}\right\}$$

$$= \frac{1}{2}\left[e^t + \cos t + \sin t\right] + \frac{1}{2}(\sin t - t\cos t)$$

$$= \frac{1}{2}\left[e^t + \cos t + 2\sin t - t\cos t\right]$$

$$y = L^{-1}\left\{\frac{p}{(p^2+1)^2}\right\} - \frac{1}{2}L^{-1}\left\{\frac{1}{p-1} - \frac{p}{p^2+1} + \frac{1}{p^2+1}\right\}$$

$$= \frac{1}{2}t\sin t - \frac{1}{2}\left[e^t - \cos t + \sin t\right] = \frac{1}{2}\left[t\sin t - e^t + \cos t - \sin t\right]$$

Hence, $\qquad x = \frac{1}{2}(e^t + \cos t + 2\sin t - t\cos t)$

$$y = \frac{1}{2}(t\sin t - e^t + \cos t - \sin t)$$

EXAMPLE 3. *Using Laplace transformation, solve*
$$(D-2)x - (D+1)y = 6e^{3t}$$
$$(2D-3)x + (D-3)y = 6e^{3t}$$
Given $x = 3$, $y = 0$ *when* $t = 0$. $\qquad$ **[UPTU-2001]**

SOLUTION. Taking Laplace transformation of the given equations, we get

$$\left.\begin{array}{l} LDx - 2Lx - LDy - Ly = 6Le^{3t} \\ 2LDx - 3Lx + LDy - 3Ly = 6Le^{3t} \end{array}\right\}$$

$$\Rightarrow \left.\begin{array}{l} p\overline{x} - x(0) - 2\overline{x} - p\overline{y} + y(0) - \overline{y} = 6\dfrac{1}{p-3} \\ 2p\overline{x} - 2x(0) - 3\overline{x} + p\overline{y} - y(0) - 3\overline{y} = \dfrac{6}{p-3} \end{array}\right\}, \text{ where } \overline{x} = L(x)$$

$$\Rightarrow \left.\begin{array}{l} (p-2)\overline{x} - (p+1)\overline{y} - 3 = \dfrac{6}{p-3} \\ (2p-3)\overline{x} + (p-3)\overline{y} - 6 = \dfrac{6}{p-3} \end{array}\right\} \Rightarrow \left.\begin{array}{l} (p-2)\overline{x} - (p+1)\overline{y} = \dfrac{3p-3}{p-3} \\ (2p-3)\overline{x} + (p-3)\overline{y} = \dfrac{6p-12}{p-3} \end{array}\right\}$$

$$\left.\begin{array}{c}(p-3)(p-2)\bar{x}-(p-3)(p+1)\bar{y}\\=3p-3\\(p+1)(2p-3)\bar{x}+(p+1)(p-3)\bar{y}\\=\dfrac{(p+1)(6p-12)}{p-3}\end{array}\right\}$$

On adding, we get

$$(3p^2-6p+3)\bar{x}=3(p-1)+\frac{6(p^2-p-2)}{p-3}$$

$$\Rightarrow \quad \bar{x}=\frac{3(p-1)}{3(p-1)^2}+\frac{6(p^2-p-2)}{3(p-1)^2(p-3)}$$

So, $\quad x=L^{-1}\left\{\dfrac{1}{p-1}+\dfrac{2}{(p-1)^2}+\dfrac{2}{p-3}\right\}=e^t+2te^t+2e^{3t}$

Putting the value of x in (1), we get

$$(D-2)(e^t+2te^t+2e^{3t})-(D+1)y=6e^{3t}$$

$$\Rightarrow \quad e^t+2te^t+2e^t+6e^{3t}-2e^t-4te^t-4e^{3t}-(D+1)y=6e^{3t}$$

$$\Rightarrow \qquad\qquad (D+1)y=e^t-2te^t-4e^{3t} \qquad\qquad\qquad \text{... (2)}$$

Taking Laplace transform of (2), we get

$$p\bar{y}-y(0)+\bar{y}=\frac{1}{p-1}-\frac{2}{(p-1)^2}-\frac{4}{p-3}$$

$$\Rightarrow \qquad (p+1)\bar{y}=\frac{1}{p-1}-\frac{2}{(p-1)^2}-\frac{y}{p-3}$$

$$\bar{y}=\frac{1}{p^2-1}-\frac{2}{(p+1)(p-1)^2}-\frac{4}{(p+1)(p-3)}$$

$$\bar{y}=\frac{1}{p^2-1}-\frac{1/2}{p+1}+\frac{1/2}{p-1}-\frac{1}{(p-1)^2}+\frac{1}{p+1}-\frac{1}{p-3}$$

$$\bar{y}=\frac{1}{p^2-1}+\frac{1/2}{p+1}+\frac{1/2}{p-1}-\frac{1}{(p-1)^2}-\frac{1}{p-3}$$

$$\Rightarrow \qquad y=L^{-1}\left\{\frac{1}{p^2-1}+\frac{1}{2}\frac{1}{p+1}+\frac{1}{2}\frac{1}{p-1}-\frac{1}{p-3}-\frac{1}{(p-1)^2}\right\}$$

$$\Rightarrow \qquad y=\sinh t+\frac{1}{2}e^{-t}+\frac{1}{2}e^t-e^{3t}-te^t$$

$$\Rightarrow \qquad y=\sinh t+\cosh t-e^{-3t}-te^t$$

EXAMPLE 4. $Solve(D^2-1)x+5Dy=t,-Dx+(D^2-4)y=-2, when\, x=0=Dx=y=Dy,\, dt=0$.

SOLUTION. Taking the Laplace transforms of both sides of the given equations, we have

$$L\{x''\}-L\{x\}+5L\{y'\}=L\{t\}$$

and $\quad -2L\{x'\}+L\{y''\}-4L\{y\}=-2L\{1\}$

or $\quad p^2L\{x\}-px(0)-x'(0)-L\{x\}+5[pL\{y\}-y(0)]=1/p^2$

and $\quad -2[pL\{x\} - x(0)] + p^2 L\{y\} - py(0) - y'(0) - 4L\{y\} = -2/p$

which gives

$$(p^2 - 1)L\{x\} + 5pL\{y\} = 1/p^2 \qquad \qquad ...(1)$$

and $\quad -2pL\{x\} + (p^2 - 4)L\{y\} = -2/p \qquad \qquad ...(2)$

On solving (1) and (2) for $L(x)$ and $L(y)$, we get

$$L(x) = \frac{11p^2 - 4}{p^2(p^2 + 1)(p^2 + 4)} = -\frac{1}{p^2} + \frac{5}{p^2 + 1} - \frac{4}{p^2 + 4}$$

and $\quad L\{y\} = \dfrac{-2p^2 + 4}{p(p^2 + 1)(p^2 + 4)} = \dfrac{1}{p} - \dfrac{2p}{p^2 + 1} + \dfrac{p}{p^2 + 4}$

Therefore, we get

$$x = -L^{-1}\left\{\frac{1}{p^2}\right\} + 5L^{-1}\left\{\frac{1}{p^2 + 1}\right\} - 4L^{-1}\left\{\frac{1}{p^2 + 4}\right\} = -t + 5\sin t - 2\sin 2t$$

and $\quad y = L^{-1}\left\{\dfrac{1}{p}\right\} - 2L^{-1}\left\{\dfrac{p}{p^2 + 1}\right\} + L^{-1}\left\{\dfrac{p}{p^2 + 4}\right\} = 1 - 2\cos t + \cos 2t$.

EXAMPLE 5. *Solve* $Dx + Dy = t$; $D^2 x - y = e^{-t}$, *when* $x(0) = 3$, $x'(0) = -2$, $y(0) = 0$.

SOLUTION. Taking the Laplace transforms of both the sides of the given equations, we get

$$L\{x'\} + L\{y'\} = L\{t\}$$

and $\quad L\{x''\} - L\{y\} = L\{e^{-t}\}$

which gives $pL\{x\} - x(0) + pL\{y\} - y(0) = 1/p^2$

and $\quad p^2 L\{x\} - px(0) - x'(0) - L\{y\} = \dfrac{1}{p+1}$

or $\quad pL\{x\} + pL\{y\} = 3 + \dfrac{1}{p^2} \qquad \qquad ...(1)$

and $\quad p^2 L\{x\} - L\{y\} = 3p - 2 + \dfrac{1}{p+1} \qquad \qquad ...(2)$

Solving (1) and (2) for $L\{x\}$ and $L\{y\}$, we get

$$L\{x\} = \frac{2}{p} + \frac{1}{p^3} + \frac{1}{2(p+1)} + \frac{p}{2(1+p^2)} - \frac{3}{2(p^2+1)}$$

and $\quad L\{y\} = \dfrac{1}{p(p+1)(p^2+1)} + \dfrac{2}{p^2+1}$

$$= \frac{1}{p} - \frac{p}{2(p+1)} - \frac{p}{2(p^2+1)} - \frac{1}{2(p^2+1)} + \frac{2}{p^2+1} .$$

$$= \frac{1}{p} - \frac{1}{2(p+1)} - \frac{p}{2(p^2+1)} + \frac{3}{2(p^2+1)}$$

Therefore,

$$x = 2L^{-1}\left\{\frac{1}{p}\right\} + L^{-1}\left\{\frac{1}{p^3}\right\} + \frac{1}{2}L^{-1}\left\{\frac{1}{p+1}\right\} + \frac{1}{2}L^{-1}\left\{\frac{p}{p^2+1}\right\} - \frac{3}{2}L^{-1}\left\{\frac{1}{p^2+1}\right\}$$

$$= 2 + \frac{1}{2}t^2 + \frac{1}{2}e^{-t} + \frac{1}{2}\cos t - \frac{3}{2}\sin t$$

and $$y = L^{-1}\left\{\frac{1}{p}\right\} - \frac{1}{2}L^{-1}\left\{\frac{1}{p+1}\right\} - \frac{1}{2}L^{-1}\left\{\frac{p}{p^2+1}\right\} + \frac{3}{2}L^{-1}\left\{\frac{1}{p^2+1}\right\}$$

$$= 1 - \frac{1}{2}e^{-t} - \frac{1}{2}\cos t + \frac{3}{2}\sin t.$$

Exercise 14.9

Solve the following simultaneous equations :

1. $3\dfrac{dx}{dt} - y = 2t, \dfrac{dx}{dt} + \dfrac{dy}{dt} - y = 0$ with the conditions $x(0) = y(0) = 0$

2. $\dfrac{dx}{dt} + \dfrac{dy}{dt} + x + y = 1, \dfrac{dy}{dt} = 2x + y;$

$x(0) = 0, y(0) = 1$ **[MTU (SUM)–2011]**

3. $\dfrac{d^2x}{dt^2} - x = y, \dfrac{d^2y}{dt^2} + y = -x,$

given that $t = 0; x = 2, y = -1, \dfrac{dx}{dt} = 0$

and $\dfrac{dy}{dt} = 0$ **(PTU–2009S)**

4. $3\dfrac{dx}{dt} + \dfrac{dy}{dt} + 2x = 1, \dfrac{dx}{dt} + 4\dfrac{dy}{dt} + 3y = 0;$

given $x = 0, y = 0$ when $t = 0.$

(Madras–2003S)

ANSWERS

1. $x = \dfrac{t^2}{2} + \dfrac{t}{2} - \dfrac{3}{4}e^{2t/3} + \dfrac{3}{4}$

2. $x = e^{-t} - 1, y = 2 - e^{-t}$

3. $x = 2 + t^2/2, y = -1 - t^2/2$

4. $x = \dfrac{1}{10}(5 - 2e^{-t} - 3e^{-6t/11}), \; y = \dfrac{1}{5}(e^{-t} - e^{-6t/11})$

Objective Evaluation

∞ **MULTIPLE CHOICE QUESTIONS**

Choose the most appropriate one.

1. The Laplace transform of 1 is :
  (a) $1/p$
  (b) $1/p^2$
  (c) $1/\sqrt{p}$
  (d) none of these

2. The Laplace transform of t is :
  (a) $1/p$
  (b) $1/p^2$
  (c) $1/\sqrt{p}$
  (d) none of these

3. The Laplace transform of $t^{n-1}/(n-1)!$ is :
  (a) $\dfrac{1}{p^{n-1}}$
  (b) $1/p^n$
  (c) $\dfrac{1}{p^{n+1}}$
  (d) none of these

4. The Laplace transform of $\dfrac{t^{n-1}}{\Gamma(a)}$ is :
  (a) $\dfrac{1}{p^{n-1}}$
  (b) $1/p^n$
  (c) $\dfrac{1}{p^{n+2}}$
  (d) none of these

5. The Laplace transform of e^{at} is :
  (a) $\dfrac{1}{p-a}$
  (b) $\dfrac{1}{(p-a)^2}$
  (c) $\dfrac{1}{(p-a)^n}$
  (d) none of these

6. The Laplace transform of te^{at} is :
  (a) $\dfrac{1}{p-a}$
  (b) $\dfrac{1}{(p-a)^2}$
  (c) $\dfrac{1}{(p-a)^n}$
  (d) none of these

7. The Laplace transform of $\dfrac{1}{(n-1)!}t^{n-1}e^{at}$ is :
  (a) $\dfrac{1}{p-a}$
  (b) $\dfrac{1}{(p-a)^2}$
  (c) $\dfrac{1}{(p-a)^n}, n=1,2,3,...$
  (d) none of these

8. The Laplace transform of $\dfrac{1}{a-b}(e^{at}-e^{bt})$ is :
  (a) $\dfrac{1}{(p-a)(p-b)}, (a \neq b)$
  (b) $\dfrac{p}{(p-a)(p-b)}, (a \neq b)$
  (c) $\dfrac{p}{(p+a)(p+b)}, (a \neq b)$
  (d) none of these

9. If $L\{F(t)\} = f(p)$, then $L\{F'(t)\}$ is :
  (a) $L\{f'(t)\} = f(p)$
  (b) $L\{f'(t)\} = f(b) + f(0)$
  (c) $L\{f'(t)\} = pf(p) + f(0)$
  (d) $L\{f'(t)\} = pf(p) - f(0)$

10. If $L\{F(t)\} = f(p)$, then $L\{F''(t)\}$ is :
  (a) $p^2 f(p) - pF(0) - F'(0)$
  (b) $pf(p) - pF(0)$
  (c) $f''(p)$
  (d) none of these

11. The Laplace transform of $f(t)$ is $f(p)$, then :
  (a) $L\{t\,F(t)\} = f(p)$
  (b) $L\{tF(t)\} = -f(p)$
  (c) $L\{tF(t)\} = f'(p)$
  (d) none of these

12. If $u(x,t)$ is a function of two variables x and t and $L\{u(x,t)\} = U(x,p)$, then $L\left(\dfrac{\partial u}{\partial t}\right) = :$
  (a) $pU(x,p) - u(x,0)$
  (b) $pU(x,p)$
  (c) $u(x,0)$
  (d) none of these

13. The function whose Laplace transform is $\dfrac{1}{p^2+w^2}$ is :
  (a) $\dfrac{1}{w}\sin wt$
  (b) $\cos wt$
  (c) $\sin wt$
  (d) none of these

14. The funciton whose Laplace transform is $\dfrac{p}{p^2+a^2}$ is:
  (a) $\cos wt$
  (b) $\sin wt$
  (c) $\dfrac{1}{w}\cos wt$
  (d) none of these

15. The function whose Laplace transform is $\dfrac{1}{p^2 - a^2}$ is:

(a) $\dfrac{1}{a}\sinh at$ (b) $\dfrac{1}{a}\cosh at$

(c) $\sinh at$ (d) none of these

16. The function whose Laplace transform is $\dfrac{p-a}{(p-a)^2 + w^2}$ is:

(a) $e^{at}\cos wt$ (b) $\cos wt$

(c) $e^{at}\sin wt$ (d) none of these

17. The function whose Laplace transform is $\dfrac{1}{p(p^2 + w^2)}$ is:

(a) $\dfrac{1}{w^3}(wt - \sin wt)$

(b) $\dfrac{1}{2w^2}(\sin wt - wt\cos wt)$

(c) $\dfrac{1}{2w}\sin wt$

(d) none of these

18. The function whose Laplace transform is $\dfrac{p}{(p^2 + w^2)^2}$ is:

(a) $\dfrac{1}{w}(w - \sin wt)$ (b) $\dfrac{1}{2w}\sin wt$

(c) $\dfrac{1}{w}\sin wt$ (d) none of these

19. The function whose Laplace transform is $\dfrac{p^2}{(p^2 + w^2)^2}$ is:

(a) $\dfrac{1}{2w}(\sin wt + wt + \cos wt)$

(b) $\sin wt + wt\cos wt$

(c) $\sin wt$

(d) none of these

20. The function whose Laplace transform is $\dfrac{1}{p^4 - a^4}$ is:

(a) $\sinh at$ (b) $\cosh at$

(c) $\sinh at - \cosh at$ (d) none of these

❧ TRUE/FALSE

Write 'T' for True and 'F' for False statement.

1. $L\{\cos at\} = \dfrac{p}{p^2 + a^2}$.   **(T/F)**

2. $L\{\sin at\} = \dfrac{a}{p^2 + a^2}$.   **(T/F)**

3. $L\{e^{at}\} = \dfrac{1}{p-a}$.   **(T/F)**

4. $L\{\sinh at\} = \dfrac{p}{p^2 - a^2}$.   **(T/F)**

5. $L\{\cosh at\} = \dfrac{p}{p^2 - a^2}$.   **(T/F)**

6. $L^{-1}\left(\dfrac{1}{p^2}\right) = t$ **(T/F)**

7. $L^{-1}\left(\dfrac{1}{p^{n+1}}\right) = \dfrac{t^n}{n!}$ **(T/F)**

8. $L^{-1}\left(\dfrac{1}{p^2 - a^2}\right) = \dfrac{\sinh at}{a}$ **(T/F)**

9. $L^{-1}\left(\dfrac{p}{p^2 - a^2}\right) = \cos at$ **(T/F)**

10. $L^{-1}\left(\dfrac{1}{p-a}\right) = e^{at}$ **(T/F)**

❧ FILL IN THE BLANKS

1. An integral of the type $\int_{-\infty}^{\infty} k(p,t)\, F(t)\, dt$ is called _____ of $F(t)$.

2. The function $k(p,t)$ is known as _____ of the transform.

3. If the integral $\int_0^\infty e^{-pt} F(t)\, dt$ _____ for some value of p, then only the Laplace transform of $f(t)$ exists.

4. $L\{1\} = $ _____

5. $L\{t^n\} = \dfrac{\quad}{p^{n+1}}$

6. A function $N(t)$ of t such that $\int_0^t N(t) = 0 \ \forall t$ is called _____ function.

7. If $L^{-1}[f(p)] = F(t)$, then $L^{-1}\{f(p-a)\} = $ ____.

8. If $L^{-1}\{f(p)\} = F(t)$, then $L^{-1}\{f(ap)\} = $ _____ $F\left(\dfrac{t}{a}\right)$.

9. $L^{-1}\left[\dfrac{1}{p}\right] = $ _____.

10. $L^{-1}\left[\dfrac{p}{p^2 + a^2}\right] = $ _____.

ANSWERS

∽ **MULTIPLE CHOICE QUESTIONS**

1. (a)	**2.** (b)	**3.** (b)	**4.** (b)	**5.** (a)	**6.** (b)	**7.** (c)	**8.** (a)	**9.** (d)
10. (a)	**11.** (b)	**12.** (a)	**13.** (a)	**14.** (a)	**15.** (a)	**16.** (a)	**17.** (b)	**18.** (b)
19. (a)	**20.** (d)							

∽ **TRUE/FALSE**

1. T	**2.** T	**3.** T	**4.** F	**5.** T	**6.** T	**7.** T	**8.** T	**9.** F
10. T								

∽ **FILL IN THE BLANKS**

1. integral transform	**2.** kernel	**3.** exist	**4.** $1/p$	**5.** $n!$	**6.** null
7. $e^{at}F(t)$	**8.** $1/a$	**9.** 1		**10.** cos at	

□□□□□□

CHAPTER 15

An Introduction to Biostatistics and Biometrics

15.1 INTRODUCTION

The word "Statistics" have been derived from the Latin word status or the Italian word statista, both meaning a political state. The naturalists, the biologist, the astronomers, the administrators, the businessmen, and the economists all make use of statistical methods and facts. Its scope has become so wide today that few statisticians, if any, are expert in all branches.

Definitions:

A.L. Bowley defines "*Statistics may be called the science of counting*".

At another place he defines "*Statistics any be called the science of averages*".

According to King "*The science of statistics is the method of judging collective, natural or social phenomenon from the result obtained from the analysis of enumeration or estimates*".

Bodidington has defined "*Statistics as the science of estimate and probabilities*".

15.2 APPLICATIONS OF STATISTICS

(1) **To the Governance of Public bodies :** These are the days of planning and any plan, to be successful, must be based on statistics. An estimate of the revenue and expenditure for the ensuing year is necessary for the successful running of the Government machinery.

(2) **In Business and Commerce :** A manufacturer in order to be successful should make a study of the seasonal changes in the demand of his goods and the rates of interest for borrowing. Insurance companies in deciding upon the premium to be charged or the annuities to be granted have to consider the mortality of sickness etc.

(3) **In Medical Sciences :** Statistical methods are necessary in finding the effectiveness' of medicines and drugs for the prevention and cure of disease.

(4) **In Agricultural Research :** Much ingenuity and statistical knowledge is required in the design and analysis to test the effect of different types of manures, levels of irrigation and varieties of Crops.

(5) **In Meteorology :** Whether forecasting depends on statistical methods.

(6) In advantageous in education, anthropometry and higher sciences.

15.3 CHARACTERISTICS OF STATISTICS

1. Statistics is the aggregate of fails.
2. Statistics is numerically expressed.
3. Statistics is usually affected by multiplicity of causes and not by single cause.
4. Statistics must be related to some field of inquiry.
5. Statistics should be capable of being related to each other. So the some cause and effect relationship can be established.
6. The reasonable standard of accuracy should be maintained in statistics.

15.4 LIMITATION OF STATISTICS

1. Statistical laws are held to be true on the average and in the long run.
2. Statistics can be used to analyse only collective matters not individual events.

3. It is applicable only to quantitative data.
4. Statistical results are ascertained by samples. If the selection of samples is biased, errors will accumulate and results will not be reliable.
5. The greatest limitation of statistics is that only one who has an expert knowledge of statistical methods can efficiently handle statistical data.

15.5 INTRODUCTION TO BIOSTATISTICS

Statistics has wide application in almost all sciences—social as well as physical such as biology, medicines, agriculture, veterinary, economics, psychology, ethnology, business management etc. It plays a major role in bioscience because data of bioscience are of a variable nature. It is very difficult to draw a concrete distribution from biological experiments because of inherent differences between two individuals. Homozygous twins are even not exactly same in physiology and behaviour.

Most of the happening in life science depends upon counting or measurements. Plants and animals obtained by any hybridization experiment agree with Mendel's Law or not, can only be concluded by statistical test, i.e., χ^2 test or low blood pressure has no meaning values it is expressed in numbers.

Blood pressure, pulse rate, Hb%, rate of reproduction, rate of transpiration action of a drug on individuals or a group etc. Varies not only from person to person but also from group to group. The extent of this variability in a character in by way of chance, i.e., biological or normal is revealed by statistical methods.

15.6 BIO-STATISTICS

It is obvious that bio-statistics deals with the data collected in the field of biology and life sciences. Bio-statistics is simply the application of statistical methods to the solution of biological problems. It is sometimes also called biometrics. It involves collection, classification, analysis and interpretation of the numerical facts so as to draw scientific conclusions or to make effective decisions.

15.7 APPLICATIONS AND USES OF BIO-STATISTICS

1. **In physiology and anatomy :**
   (i) To define what is normal or healthy in a population and to find limits of normality in variables.
   (ii) To find the difference between the means and proportions of normal at two places or in different periods.
   (iii) To find out correlation between two variables X and Y as height and weight.
2. **In Pharmacology:**
   (i) **To find out the action of drug–** a drug is given to animals and humans to observe the change produced are due to the drug or by chance.
   (ii) To compare the action of two different drags or two successive dosages of the same drug.
   (iii) To find out the relative potency of a new drug with respect to a standard drug.
3. **In Medicine:**
   (i) To compare the efficiency of a particular drug. For this, the percentage of cured and died in the experiment and control groups.
   (ii) To find out an association between two attributes such as cancer and smoking.
   (iii) To identify signs and symptoms of a disease or syndrome.
       Cough and typhoid is found by chance and fever is found in almost every case.
4. **In Community Medicine and Public health :**
   (i) **To test usefulness of sera and vaccines in the field–** The percentage of attacks or deaths among the vaccinated subjects is compared with that among the unvaccinated ones to find whether the difference observed is statistically significant.

(ii) **In epidemiological studies–** The role of causative factors is statistically tested.

(iii) In public health, the measures adopted are evaluated.

15.8 SCOPE OF BIO-STATISTICS

Use of statistical methods are constantly increasing in biological sciences. The development of biological theories are closely associated with statistical methods. Heredity, one of the recent branches of biology is mainly based on biostatistics. Therefore for the students of biology the knowledge of biostatistics is a must.

15.9 ERROR AND APPROXIMATIONS

Approximations and errors are in integral part of our life. These are exist everywhere, and sometime are unavoidable.

A number of different types of errors arise during the process of numerical computing. These errors contribute to the total error in the final result. Also the numerical data used in solving the problems are usually not exact, and the numbers expressing such data are therefore not exact. They are merely approximations, true to three, four or more figures. Not only are the data of practical problems usually approximate, but sometimes the methods and process by which the desired result is to be obtained are also approximate. Therefore, an approximate calculation is one which involves approximate data, or approximate methods or both. Therefore, it is evident that the error in a computed result may be due to one or both sources, *i.e.*

(i) error in data   and   **(ii)** error in calculation.

The first type of error can not be decrease, but the second type can be made as small as we please, by taking the number to as many figure as we desired. Therefore, we can assume that the calculations are always carried out in such a manner as to make the errors of calculation negligible.

In this section, we examine the sources of various types of computational errors and their subsequent propagation.

15.10 ACCURACY OF NUMBERS

(i) Exact numbers: The numbers in which, there is no uncertainty and no approximation, is said to be exact numbers. For example : $5, 6, 7, \dfrac{8}{2}, \dfrac{1}{5}, \dots$ are exact numbers.

(ii) Approximate numbers: These are numbers which are not exact.

For example: 1.41421 ... 3.141592 ... are not exact numbers, since they contains infinitely many digits, are called approximate numbers.

⬗ REMARKS

⟹ The approximate number is a number which can not be expressed by a finite number of digits.

⟹ Although, the numbers $\pi, \sqrt{2}$, etc. are exact numbers, they can not be expressed exactly by a finite number of digits. But when we expressed these numbers in digital form 3.141592, 1.41421, etc. such numbers are therefore only approximation to the true values and in such cases are called approximate numbers.

⟹ Some authors always insist that one must say "approximate value" of a number in place of approximate number.

⟹ Here, we used the symbol $\simeq$ for approximately equal to.

⟹ Such numbers which represents the given numbers to a certain degree of accuracy are called approximate numbers.

(iii) Rounding-off a Number : If we divide 22 by 7 we get $\dfrac{22}{7} = 3.142857143\dots$ a quotient which is a non-terminating decimal fraction. For use this type of number in practical computation, it is to be cut-off to a manageable size such as 3.14, 3.143 ... . The process of cutting-off superfluous digits and retaining as many digits as desired is known as rounding off a number.

✎ REMARK
➠ To round off a number is to retain a certain number of digits, counted from the left and dropped the others. Thus, to round off π to three, four or five and six figures respectively, we have 3.14, 3.142, 3.1416, 3.14159.

WORKING PROCEDURE

To rounding off a number or digit to n significant figures, discard all digits to the right of the nth place using the following concepts.

(a) If this number is less than half a unit in the nth place, leave the nth digits as it is.

(b) If the discarded number is greater than half a unit in the nth place, add 1 to the nth digit.

(c) If the discarded number is exactly half a unit in the nth place, leave the nth digit unchanged.

For Example : The following numbers are rounded off correctly to four significant figures

(i) 38.63243 becomes 38.63

(ii) 91.8773 becomes 91.88

(iii) 21.64489 becomes 21.64

(iv) 87.495 becomes 87.50.

WORKING PROCEDURE

The old rule of rounding off the number says that when a 5 is dropped the preceding digit should always be increased by 1. It is not a good exercise and give inaccuracy in computations. Since, it is obvious that when a 5 is cut off, the preceding digit should be increased by one in only half the cases and should be left unchanged in the other half.

✎ REMARK
➠ The numbers rounded off to n significant figures are said to be correct to n significant figures.

(iv) Significant Figures : Here, we have that all the digits 1, 2 ... upto 9 are significant figures and 0 is a significant figure except when it is used to fix the decimal point or to fill the places of unknown digits, i.e., 0 may or may not be a significant figure. It depends upon the position in which zero has been used. As discussed earlier when zero is used to fixup the decimal point or to fill up the places of discarded digits, it is not a significant figure.

For example: Consider the numbers 0.00086 and 5800, correct to two significant figures. Then all zeros, which are used are insignificant. On the other hand, zero used in 430, correct to three significant figures, is a significant figure.

✎ REMARKS
➠ The zeroes used between two non-zero digits are always significant figure e.g. 408.
➠ To round off a number or figure to r significant digits, discard all the digits or replace by zeros to the right of rth digit according as the number to be rounded off is a decimal fraction or whole number. The rth digit to be increased by 1 or to be left unaltered, according as the portion to be discarded or replaced by zeroes as greater than or less than half of the unit at the rth places (counted from the left). In case the discarded portion is exactly half of the rth unit, then the rth unit is to be increased by 1, if it is odd, otherwise it is left unchanged.

WORKING PROCEDURE

(1) Significant digits are counted from left to right starting with the left most non-digits.

(2) The significant figure in a number in positional notations consists of

(a) all non-zero digits

(b) zero digits which
- lie between significant digits
- lie to the right of decimal points and at the same time, to the right of a non-zero digit.
- are specifically indicated to be significant

(3) The significant figure in a number written in scientific notation e.g. $M \times 10^k$ consists of all the digits explicitly in M.

For Example

(i) The number 8.3678235, when rounded to three places of decimal, we get it as 8.368. Because, we leave the portion 0.0008235 which is more than half of 0.001.

(ii) The number 83988235, when rounded to five significant digits, we get as 83988. Because the portion left out is 235, which is less than half of 1000.

(iii) The number 8.6325 when rounded to three decimal places, we get 8.632 as the rounded number.

(iv) 83675, rounded to four significant figures as obtained as 83680. Here the fourth place, when we counted from the left is 7 which is odd and the portion left out is exactly half of the unit at this place. Therefore we increase 7 by one.

SOLVED EXAMPLES

EXAMPLE 1. *Round-off the following numbers correct to four significant figures*
$$68.3643, 878.367, 8.7265, 56.395$$

SOLUTION. Here, we have to retain first four significant figures. Therefore
(i) 68.3643 becomes 68.36
(ii) 878.367 becomes 878.4
(iii) 8.7265 becomes 8.726 (Because the digit in the fourth place is even).
(iv) 56.395 becomes 56.40 (Because the digit in fourth place is odd).

EXAMPLE 2. *Find the sum of the following approximate numbers, each being correct to its last figures*
$$396.56, 657.2, 758.9826, 3.052$$

SOLUTION. Since the number 657.2 is correct to one decimal place. Therefore, it is not worth while to retain digits beyond two decimal places. Hence, we rounded off the given numbers to two decimal places, and then found the sum. Therefore, the required sum
$$= 396.56 + 657.20 + 758.98 + 3.05 = 1815.79 \simeq 1815.8$$

15.11 ERRORS AND THEIR ANALYSIS

Definition : *The quantity, True value – Approximate value is called the error.*

15.11.1 SOURCES OF ERRORS

Following are some sources of error in numerical computations.

(i) **Input Errors:** The input information is rarely exact. It comes from the experiments and any experiment can give results of any limited accuracy.

(ii) Algorithmic Errors: Sometimes, the direct algorithms based on a finite sequence of operations are used. Errors due to limited steps don't amplify the existing errors. Since the application of some formula is not possible for a infinite number of times, algorithm has to be stopped after a finite number of steps. Hence, the obtained results are not exact.

(iii) Computational Errors: Sometimes, when we performing elementary operations, the number of digits increases greatly. Therefore, the result can not be held fully in a register available in the given system.

15.11.2 TYPES OF ERROR

(i) Absolute error: If x^A is the approximate value of exact number x^T, then the absolute error denoted by E_a is defined by

$$E_a = \Delta x = |x^T - x^A|$$
$$\Rightarrow \qquad E_a = |x^T - x^A|$$

🍃 REMARK

⟼ In error analysis, the magnitude of the error is important, not the, sign of error. Therefore, we consider the absolute error generally.

(ii) Relative Error: In many cases, absolute error may not reflect its influence correctly as it does not take into account the order of magnitude of the value under consideration. For example an error of 1 gram is much more significant in the weight of 10 grams Gold, that in the weight of a bag of sugar. Due to this reason the concept of relative error is introduced.

The relative error is the absolute error divided by the true value of the given quantity. It is denoted by E_r and defined as

$$E_r = \left| \frac{x^T - x^A}{x^T} \right| = \frac{\text{Absolute error}}{\text{True value}}$$

(iii) Percentage Error: The percentage error in x^A, which is the approximate value of x^T is

$$E_p = 100 \times E_r = 100 \times \left| \frac{x^T - x^A}{x^T} \right|$$

🍃 REMARKS

⟼ The relative error is also known as normalized absolute error.

⟼ If $\bar{x}$ be a number such that $|x^T - x^A| \le \bar{x}$, then $\bar{x}$ is said to be an upper limit on the magnitude of absolute error and measures the absolute accuracy.

⟼ The relative and percentage errors are independent of the units of measurement, while absolute errors are expressed in terms of unit used.

⟼ If a number is correct to n significant figures then its absolute error can not be greater than half a unit in a nth places.

⟼ If a number is correct to n decimal places then the error $= \frac{1}{2} \cdot 10^{-n}$.

For example: If the number 8.869 correct to three decimal points its absolute error is not greater than $0.001 \times \frac{1}{2} = \frac{1}{2} \times 10^{-3} = 0.0005$.

SOLVED EXAMPLES

EXAMPLE 1. *Find the sum of* 392, 780.56, 64320, 72300, 23657 *assuming that the number* 72300 *is known to only three significant figures.*

SOLUTION. Since We have, that the number 72300 is known to hundred places.

Therefore, we round off other numbers correct to tens places and then find the sum, *i.e.,*

$$\text{Sum } S = 390 + 780 + 64320 + 72300 + 23660$$
$$= 161450 \simeq 161400$$

Here, we observe that, the last significant digit (counting from left) is 4 which is uncertain by one unit of this place.

GENERAL THEOREMS

THEOREM 1. *If the first significant figure of a number is r and the number is correct to n significant figures, then the relative error is less than* $\dfrac{1}{r \times 10^{n-1}}$.

PROOF. Let us suppose that N be any given exact number which contains n significant figures and m denotes the number of correct decimal places.

Then, there are following three cases :

Case (i): If $m < n$

In this case the number of digits in the integral part of N is given by $(n - m)$. Let us denote the first significant figure of N by r. Then, we have

Absolute error $\qquad E_a \leq \dfrac{1}{10^m} \times \dfrac{1}{2}$

and $\qquad N \geq r \times 10^{n-m-1} - \dfrac{1}{10^m} \times \dfrac{1}{2}$

which gives $\qquad E_r \leq \dfrac{\dfrac{1}{10^m} \times \dfrac{1}{2}}{r \times 10^{n-m-1} - \dfrac{1}{10^m} \times \dfrac{1}{2}}$

$$E_r = \dfrac{10^{-m}}{2r \times 10^{n-1} \times 10^{-m} - 10^{-m}}$$

$$= \dfrac{1}{2r \times 10^{n-1} - 1} = \dfrac{1}{2\left(r \times 10^{n-1} - \dfrac{1}{2}\right)}$$

Now, since n is any positive integer and r stands for any digits 0, 1, ..., 9. Then we have $2r \times 10^{n-1} > r \times 10^{n-1}$ in all cases except $r = 1$ and $n = 1$. (We can ignore this case, because it is a trivial case when $N = 1, 0.001, 0.0001$ etc., *i.e.,* the case in which N contains only one digit differerent from zero, which would not occur in common practice). Therefore, we may assume that

$$2r \times 10^{n-1} - 1 > r \times 10^{n-1} \text{ for all cases}$$

Then, the relative error $E_r < \dfrac{1}{r \times 10^{n-1}}$

Case (II): If $m = n$

Here we have N is a decimal and r is the first decimal figure, then we have

The absolute error $\quad E_a \leq \dfrac{1}{10^m} \times \dfrac{1}{2}$

and $\qquad\qquad N \geq r \times 10^{-1} - \dfrac{1}{10^m} \times \dfrac{1}{2}$

$$\Rightarrow \qquad E_r \leq \frac{10^{-m} \times \dfrac{1}{2}}{r \times 10^{-1} - 10^{-m} \times \dfrac{1}{2}}$$

$$= \frac{10^{-m}}{2r \times 10^{-1} - 10^{-m}} = \frac{1}{2r \times 10^{m-1} - 1}$$

$$= \frac{1}{2r \times 10^{m-1} - 1} < \frac{1}{r \times 10^{m-1}}$$

Case (III): If $m > n$

Here we have $m > n$, therefore, r occupies the $(m - n + 1)^{\text{th}}$ decimal place.

$$\Rightarrow \qquad N \geq r \times 10^{-(m-n+1)} - \frac{1}{10^m} \times \frac{1}{2} \text{ and } E_a \leq \frac{1}{10^m} \times \frac{1}{2}$$

Therefore, $\qquad E_r \leq \dfrac{10^{-m} \times \dfrac{1}{2}}{r \times 10^{-m} \times 10^{n-1} - 10^{-m} \times \dfrac{1}{2}}$

$$= \frac{10^{-m}}{2r \times 10^{-m} \times 10^{n-1} - 10^{-m}}$$

$$= \frac{1}{2r \times 10^{n-1} - 1} < \frac{1}{r \times 10^{n-1}}$$

Here, we can say that the theorem is true in all the three possible cases.

REMARKS

➠ Except in the case of approximate numbers of the form $r(1.000...) \times 10^k$, in which r is the only digit from zero, the relative error is less than $\dfrac{1}{2r \times 10^{n-1}}$.

➠ If $r \geq 5$ then the given approximate number is not of the form $r(1.000...) \times 10^k$, then $E_r < \dfrac{1}{10^n}$; for in the case $2r \geq 10$ and therefore $2r \times 10^{n-1} \geq 10^n$.

➠ If the relative error in an approximate number is less than $\left[\dfrac{1}{(r+1) \times 10^{n-1}} \right]$, the number is correct to n significant figures or at least is in error by less than a unit in the nth significant figures.

➠ If $E_r < \dfrac{1}{[2(r+1) \times 10^{n-1}]}$, then E_a is less than half a unit in the nth significant figures and the given number is correct to n^{th} significant figures.

➠ If the relative error of any number is not greater than $\dfrac{1}{(2 \times 10^n)}$, the number is certainly correct to n significant figures.

➠ The absolute error is always connected with the number of decimal places, whereas the relative error is connected with the number of significant figures.

SOLVED EXAMPLES

EXAMPLE 1. *Verify the theorem (1) for the number 875.32 correct to five significant figures.*

SOLUTION. The given number $N = 875.32$

We observe that $r = 8$ and $n = 5$

Since, we have the absolute error $E_a \not> 0.01 \times \dfrac{1}{2} = 0.005$

Therefore, the relative error $\leq \dfrac{0.005}{875.32} = \dfrac{5}{875320}$

$$= \dfrac{1}{2 \times 87532} < \dfrac{1}{2 \times 80000} = \dfrac{1}{2 \times 8 \times 10^4}$$

$$< \dfrac{1}{8 \times 10^4} \left(= \dfrac{1}{r \times 10^{n-1}} \right)$$

Hence, the theorem is verified.

EXAMPLE 2. *Round off the numbers 865250 and 37.46235 to four significant figures and compute E_a, E_r and E_p.*

SOLUTION. Here, the given numbers are (i) 865250 and (ii) 37.46235

(i) 865250

If we rounded off the given number to four significant figures, then we get 865200.

Therefore, the absolute error

$$E_a = \left| x^T - x^A \right| = \left| 865250 - 865200 \right| = 50$$

Now, the relative error

$$E_r = \dfrac{E_a}{x^T} = \dfrac{50}{865250} = 6.71 \times 10^{-5}$$

Also, the percentage error

$$E_p = E_r \times 100 = 6.71 \times 10^{-5} \times 100 = 6.71 \times 10^{-3}.$$

(ii) 37.46235

If we rounded off the given number to four significant figures, then we get 37.46.

Then $E_a = \left| 37.46235 - 37.46 \right| = 0.00235$

$$E_r = \dfrac{E_a}{x^T} = \dfrac{0.00235}{37.46235} = 6.27 \times 10^{-5}$$

and $E_p = E_r \times 100 = 6.27 \times 10^{-3}$

EXAMPLE 3. *If 0.333 is the approximate value of $\dfrac{1}{3}$, find the absolute, relative and percentage errors.*

SOLUTION. Here, we have

$$x^T = \dfrac{1}{3}, x^A = 0.333$$

Therefore,

(i) Absolute error

$$E_a = \left| x^T - x^A \right| = \left| \dfrac{1}{3} - 0.333 \right| = \left| \dfrac{1}{3} - \dfrac{333}{1000} \right| = \dfrac{1}{3000} = 0.00033$$

(ii) Relative Error

$$E_r = \frac{E_a}{x^T} = \frac{0.00033}{1/3} = 0.00099$$

(iii) Percentage error

$$E_p = 100 \times E_r = 100 \times 0.00099 = 0.099$$

EXAMPLE 4. Let $x = 0.005998$. Find the relative error if x is truncated to three decimal digits.

SOLUTION. Given that $x = 0.005998 = 0.5998 \times 10^{-2}$.

Now, $x_a = 0.599 \times 10^{-2}$ (after truncating to three decimal places)

$$\text{Relative error} = \left| \frac{x - x_a}{x} \right| = \left| \frac{0.5998 \times 10^{-2} - 0.599 \times 10^{-2}}{0.5998 \times 10^{-2}} \right|$$

$$= 0.00333 = 0.333 \times 10^{-2}.$$

EXAMPLE 5. If 1.414 is used as an approximation to $\sqrt{2}$. Find the absolute and relative errors.

SOLUTION. We have

True value $= \sqrt{2} = 1.41421356$

and approximate value $= 1.414$

Therefore, Error $=$ True value $-$ Approximate value

$$= \sqrt{2} - 1.414 = 1.41421356 - 1.414 = 0.00021356$$

The absolute error $= |0.00021356| = 0.21356 \times 10^{-3}$

Finally, the relative error $= \dfrac{\text{Absolute error}}{\text{True value}} = \dfrac{0.21356 \times 10^{-3}}{\sqrt{2}} = 0.151 \times 10^{-3}$.

EXAMPLE 6. Find the sum $S = \sqrt{3} + \sqrt{5} + \sqrt{7}$ to 4 significant digits and find its absolute and relative errors.

SOLUTION. It is known that

$$\sqrt{3} = 1.732, \sqrt{5} = 2.236, \sqrt{7} = 2.646$$

$$\therefore \qquad S = 1.732 + 2.236 + 2.646 = 6.614$$

Now, absolute error $E_a = 0.0005 + 0.0005 + 0.0005 = 0.0015$

The total absolute error shows that the sum is correct to 3 significant figures only.

Thus, we take $S = 6.61$

Then, we have relative error $= \dfrac{0.0015}{6.61} = 0.0002$

EXAMPLE 7. It is required to obtain the roots of $X^2 - 2X + \log_{10} 2$ to four decimal places. To what accuracy should $\log_{10} 2$ be given?

SOLUTION. The roots of the given equation are

$$X = \frac{2 \pm \sqrt{4 - 4\log_{10} 2}}{2} = 1 \pm \sqrt{1 - \log_{10} 2}$$

Then $$\left| \Delta X \right| = \frac{1}{2} \frac{\Delta(\log 2)}{\sqrt{1 - \log_{10} 2}} < 0.5 \times 10^{-4}$$

$$= \Delta(\log 2) < 2 \times 0.5 \times 10^{-4} (1 - \log 2)^{1/2} < 0.83604 \times 10^{-4}$$

$$= 8.3604 \times 10^{-5}$$

EXAMPLE 8. If $a = 10.00 \pm 0.05$, $b = 0.0356 \pm 0.0002$, $c = 15300 \pm 100$, $d = 62000 \pm 500$. Find the maximum value of absolute error in $a + b + c + d$.

Solution. We have

$$\text{Absolute error in } a = |\pm 0.05| = 0.05$$
$$\text{Absolute error in } b = |\pm 0.0002| = 0.0002$$
$$\text{Absolute error in } c = |\pm 100| = 100$$
$$\text{Absolute error in } d = |\pm 500| = 500$$

Hence, the maximum absolute error in $a + b + c + d$

$$= 0.05 + 0.0002 + 100 + 500$$
$$= 600.0502$$

Example 9. *Three approximated values of number $\dfrac{1}{3}$ are given as 0.30, 0.33 and 0.34. Which of these three is the best approximation?*

Solution. We know that the best approximation will be the one which has the least absolute error.

Here, true value $= \dfrac{1}{3} = 0.33333$

Case I. Approximate value = 0.30

$\therefore$ Absolute error $= |\text{True value} - \text{Approximate value}| = |0.33333 - 0.30|$
$$= 0.03333$$

Case II. Approximate value = 0.33

$\therefore$ Absolute error $= |\text{True value} - \text{Approximate value}| = |0.33333 - 0.33|$
$$= 0.00333$$

Case I. Approximate value = 0.34

$\therefore$ Absolute error $= |\text{True value} - \text{Approximate value}| = |0.33333 - 0.34|$
$$= |-0.00667| = 0.00667$$

We observe that, absolute error is least in case II. Hence, 0.33 is the best approximation.

Example 10. *Given the solution of a problem as $x_A = 35.25$ with the relative error in the solution atmost 2%. Find, to four decimal digits, the range of values within which the exact value of the solution must lie.*

Solution. It is given that

(i)   Maximum relative error in the solution = 2% = 0.02

(ii)   Approximate value of the solution is $x_A = 35.25$.

Let x be the exact value of the solution, then as per given, we have

$$\left| \frac{x - x_A}{x} \right| < 0.02, \ i.e., \left| 1 - \frac{x_A}{x} \right| < 0.02$$

$$\Rightarrow \qquad -0.02 < \left(1 - \frac{x_A}{x} \right) < 0.02$$

If $\left(1 - \dfrac{x_A}{x} \right) > -0.02$ then

$$-\frac{x_A}{x} > -1 - 0.02 \qquad \Rightarrow \qquad -\frac{x_A}{x} > -1.02$$

$$\Rightarrow \qquad \frac{x_A}{x} < 1.02 \qquad \Rightarrow \qquad x_A < 1.02x.$$

$$\Rightarrow \qquad x > \frac{x_A}{1.02} = \frac{35.25}{1.02} = 34.558823594$$

Also, if $\left(1 - \dfrac{x_A}{x}\right) < 0.02$ then

$$-\frac{x_A}{x} < -1 + 0.02 \qquad \Rightarrow \qquad -\frac{x_A}{x} > -0.98$$

$$\Rightarrow \qquad \frac{x_A}{x} > 0.98 \qquad \Rightarrow \qquad x_A < 0.98x$$

$$\Rightarrow \qquad x < \frac{x_A}{0.98} = \frac{35.25}{0.98} = 35.9693877551$$

Thus, we have

$$34.558823594 < x < 35.9693877551$$

Hence, the range of values within which the exact value of the solution lies, correct to four decimal places is given by

$$34.5588 < x < 35.9694.$$

Exercise 15.1

1. Round off the following numbers correct to four significant figures :
   (i) 58.3643　　　　(ii) 979.267
   (iii) 7.7265　　　　(iv) 0.065738
   (v) 3.26425　　　　(vi) 35.46735
   (vii) 7326583000　(viii) 18.265101

2. Find the relative error if 2/3 is approximated to 0.667.

3. If the number r is correct to 3 significant digits, what will be the maximum relative error.

4. A carpenter measures a 10-foot beam to the nearest eighth of an inch and a mechanist measures a $\dfrac{1}{2}$ inch bolt to the nearest

thousandth of an inch. Which measurement is more correct ?

5. The following numbers are all approximate and are correct as far as their last digit only. Find their sum 136.421, 28.3, 321, 68.243, 17.482.

6. If the number p is correct to three significant digits, what will be the maximum relative error ?

7. The height of an observation tower was estimated to be 47 m whereas it actual height was 45 m. Find the percentage relative error in the measurement.

8. If true value $= \dfrac{10}{3}$, approximate value $= 3.33$. Then, find absolute and relative errors.

ANSWERS

1. (i) 58.36　　(ii) 979.3　　(iii) 7.726　　　　(iv) 0.06574　　　　(v) 3.264
1. (vi) 35.45　(vii) 7327×10^6　(viii) 18.26　2. 0.0005　　3. 0.0005
4. Beam measurement　　5. 571　　　　6. 0.0005　　7. 4.44%
8. 0.003333, 0.000999

15.12 INHERENT ERRORS

The errors which are already present in the statement of a problem before its solution are called Inherent errors. These types of errors arise either due to the given data being approximated or due to limitations of the mathematical measurements.

The inherent error contains two components :

(i) Data errors: The data error arises when data are obtained by some experimental methods with limited accuracy and precision. This may be due to some special limitations in instrument or in reading.

(ii) Conversion errors: The conversion error arise due to the limitations of the computer to store the data exactly. Generally, it occurs in the floating- point representation which retains only a specified number of digits. The digits which are not retain gives the round off error.

REMARKS

➧ The inherent errors is also known as input errors.

➧ Data errors is also known as empirical errors.

➧ Conversion errors are also known as representation errors.

15.13 BOUNDING OFF ERROR

It occurs from the process of rounding off the numbers during the computations, *i.e.*, it occur when a fixed number of digits are used to represent exact numbers. Such types of errors are unavoidable in most of the calculations due to the limitations of the computing aids.

If a number x has the floating point representation of the form

$$x = d_1 d_2 \ldots d_t d_{t+1} \ldots \times B^e \qquad \ldots(1)$$

where $d_1, d_2, \ldots, d_t \ldots$ are integers and satisfies $0 \leq d_i \leq B$ and the e is the exponent. Then Rounding a number can be done by the following two ways :

(i) Chopping: Here, we neglect $d_{t+1}, d_{t+2} \ldots$ in (1) and obtain the number

$$= d_1 d_2 \ldots d_t \times B^e$$

(ii) Symmetric rounding: Here the fractional part in (1) is written as

$$d_1 d_2 \ldots d_t d_{t+1} + \frac{1}{2} B$$

and the first t digits are taken to write the floating point number.

For Example, find the sum of 0.223×10^3 and 0.556×10^2 and write the result in three digit mantissa.

Solution. Here, the number of the smaller magnitude is adjusted so that its exponent is same as that of the number of larger magnitude. We have

$$0.2230 \times 10^3$$
$$0.0556 \times 10^3$$
$$\overline{0.2786 \times 10^3}$$

$$\Rightarrow \quad \begin{cases} 0.278 \times 10^3, & \text{for chopping} \\ 0.279 \times 10^3, & \text{for rounding} \end{cases}$$

REMARKS

➧ In chopping, the extra digits are dropped, which is called truncating the number.

➧ In symmetric round off method, the last retained significant digit is rounded up by 1 if the first discarded digit is larger or equal to 5, otherwise the last retained digits is unchanged.

For example: The numbers 83.8893 becomes 83.89 and the number 86.6431 would become 86.64.

➧ The rounded off error can be reduced by retaining at least one more significant figure at each step than that given in the data and rounded off at the last step.

15.14 TRUNCATION ERROR

The truncation errors arises by using some approximations in place of an exact mathematical procedure.

For example, when we calculate the sine of an angle using the following series

$$\sin x = x - \frac{x^3}{3!} + \frac{x^5}{5!} - \frac{x^7}{7!} + \ldots$$

Then, we can not use the infinite terms of above series. After a certain number of terms, we terminate the process. Then, an error which is introduced here, is called truncation error.

REMARKS

⟹ Truncation error is a type of algorithm error.

⟹ In numerical computing, we used many iterative procedures, which are infinite. Therefore, a knowledge of the truncation error is very much important.

⟹ This error can be reduced by using a better numerical model which increases the number of arithmetic operations.

⟹ When we use a number of discrete steps in the solution of a differential equation, then the error which is introduced here, is called discretisation error.

SOLVED EXAMPLES

EXAMPLE 1. Obtain a second degree polynomial approximation to
$$f(x) = (1 + x)^{1/2}, x \in [0, 0.1]$$
Using the Taylor series expansion about $x = 0$. Use the expansion to approximate $f(0.05)$ and found the truncation error.

SOLUTION. Here, the given function is
$$f(x) = (1 + x)^{1/2}$$

Then, we get

$$f(x) = (1 + x)^{1/2} \implies f(0) = 1$$

$$f'(x) = \frac{1}{2}(1+x)^{-1/2} \implies f'(0) = \frac{1}{2}$$

$$f''(x) = -\frac{1}{4}(1+x)^{-3/2} \implies f''(0) = -\frac{1}{4}$$

$$f'''(x) = \frac{3}{8}(1+x)^{-5/2} \implies f'''(0) = \frac{3}{8}$$

Now, using the Taylor series expansion, we get
$$(1+x)^{1/2} = 1 + \frac{x}{2} - \frac{x^2}{8} + R_n$$

where R_n is the remainder term and given by
$$R_n = \frac{1}{16} \cdot \frac{x^3}{[(1+\theta)^{1/2}]^5}, 0 < \theta < 0.01$$

Then the truncation error is given by
$$T = (1+x)^{1/2} - \left(1 + \frac{x}{2} - \frac{x^2}{8}\right) = \frac{1}{16} \cdot \frac{x^3}{[(1+\theta)^{1/2}]^5}$$

Now, $$f(0.05) = 1 + \frac{0.05}{2} - \frac{(0.05)^2}{8} = 0.10246875 \times 10^1$$

Then, the bound of the truncation error for $x \in [0, 1]$ is given by

$$|T| \le \frac{(0.1)^3}{16[(1+8)^{1/2}]^5} \le \frac{(0.1)^3}{16} = 0.625 \times 10^{-4}$$

EXAMPLE 2. *Find the truncation error in the result of the following functions for $x = \dfrac{1}{5}$ when we use*

(a) *First three terms*   (b) *First four terms*

$$e^x = 1 + x + \frac{x^2}{2!} + \frac{x^3}{3!} + \frac{x^4}{4!} + \frac{x^5}{5!} + \frac{x^6}{6!}$$

SOLUTION. (a) Let T denote the truncation error. If we add first three terms then

$$T = \left(1 + x + \frac{x^2}{2!} + \dots + \frac{x^6}{6!}\right) - \left(1 + x + \frac{x^2}{2!}\right)$$

$$= \frac{x^3}{3!} + \frac{x^4}{4!} + \frac{x^5}{5!} + \frac{x^6}{6!}$$

Now, T at $x = \dfrac{1}{5} = \dfrac{(0.2)^3}{6} + \dfrac{(0.2)^4}{24} + \dfrac{(0.2)^5}{120} + \dfrac{(0.2)^6}{720} = 0.1402755 \times 10^{-2}$

(b) Now, we find the truncation error, when first four terms are added

$$T = \left(1 + x + \frac{x^2}{2!} + \dots + \frac{x^6}{6!}\right) - \left(1 + x + \frac{x^2}{2!} + \frac{x^3}{3!}\right)$$

$$= \frac{x^4}{4!} + \frac{x^5}{5!} + \frac{x^6}{6!}$$

Now, T at $x = \dfrac{1}{5} = \dfrac{(0.2)^4}{24} + \dfrac{(0.2)^5}{120} + \dfrac{(0.2)^6}{720} = 0.694222 \times 10^{-4}$

15.15 THE GENERAL FORMULA FOR ERRORS

Let $Y = f(x_1, x_2, \dots, x_n)$ be a function of n variables $x_1, x_2, \dots, x_n$. Suppose, ΔY is the error in Y due to the errors $\Delta x_1, \Delta x_2, \dots, \Delta x_n$ in $x_1, x_2, \dots, x_n$ respectively.

Then we have

$$Y + \Delta Y = f(x_1 + \Delta x_1, x_2 + \Delta x_2, \dots, x_n + \Delta x_n) \qquad \dots(1)$$

Expanding by Taylor series, we get

$$Y + \Delta Y = f(x_1, x_2, \dots, x_n) + \left(\Delta x_1 \frac{\partial Y}{\partial x_1} + \Delta x_2 \frac{\partial Y}{\partial x_2} + \dots + \Delta x_n \frac{\partial Y}{\partial x_n}\right)$$

$$+ \frac{1}{2}\left[(\Delta x_1)^2 \frac{\partial^2 Y}{\partial x_1^2} + (\Delta x_2)^2 \frac{\partial^2 Y}{\partial x_2^2} + \dots + (\Delta x_n)^2 \frac{\partial^2 Y}{\partial x_n^2} + 2\Delta x_1 \Delta x_2 \frac{\partial^2 Y}{\partial x_1 \partial x_2} + \dots\right] + \dots$$

$$\dots(2)$$

Now, since the errors $\Delta x_1, \Delta x_2, \dots, \Delta x_n$ all are very small. So, that we can neglect $(\Delta x_i)^2$ and higher order terms of Δx_i.

Then, we have

$$Y + \Delta Y = f(x_1, x_2, ..., x_n) + \left(\Delta x_1 \frac{\partial Y}{\partial x_1} + \Delta x_2 \frac{\partial Y}{\partial x_2} + ... + \Delta x_n \frac{\partial Y}{\partial x_n} \right) \qquad ...(3)$$

$$\Rightarrow \qquad \Delta Y = \Delta x_1 \frac{\partial Y}{\partial x_1} + \Delta x_2 \frac{\partial Y}{\partial x_2} + ... + \Delta x_n \frac{\partial Y}{\partial x_n} \qquad ...(4)$$

$$[\because Y = f(x_1, x_2, ..., x_n)]$$

Now, divide the equation (4) by Y, we get the relative error is

$$\frac{\Delta Y}{Y} = \frac{\partial x_1}{Y} \cdot \frac{\partial Y}{\partial x_1} + \frac{\partial x_2}{Y} \cdot \frac{\partial Y}{\partial x_2} + ... + \frac{\partial x_n}{Y} \cdot \frac{\partial Y}{\partial x_n} \qquad ...(5)$$

Now, taking the modulus of (4) and (5), the maximum absolute error and relative error are given by

$$\left| \Delta Y \right| \le \left| \Delta x_1 \frac{\partial Y}{\partial x_1} \right| + \left| \Delta x_2 \frac{\partial Y}{\partial x_2} \right| + ... + \left| \Delta x_n \frac{\partial Y}{\partial x_n} \right|$$

and

$$\left| \frac{\Delta Y}{Y} \right| \le \left| \frac{\partial x_1}{Y} \cdot \frac{\partial Y}{\partial x_1} \right| + \left| \frac{\partial x_2}{Y} \cdot \frac{\partial Y}{\partial x_2} \right| + ... + \left| \frac{\partial x_n}{Y} \cdot \frac{\partial Y}{\partial x_n} \right|$$

SOLVED EXAMPLES

EXAMPLE 1. *In a $\triangle ABC$, $a = 6$ cm, $c = 15$ cm and $\angle B = 90°$. Find the possible error in the computed value of A, if the errors in the measurement of a and c are 1 mm and 2 mm respectively.*

SOLUTION. Here, we have $a = 6$ cm

$$c = 15 \text{ cm}$$
$$\angle B = 90°$$

Then, we have the triangle given by fig. 1.

From figure 1, we have $A = \tan^{-1} \dfrac{a}{c}$

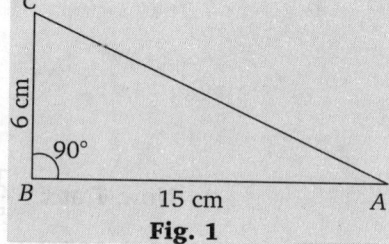

Fig. 1

$$\Rightarrow \qquad \Delta A = \Delta a \frac{\partial A}{\partial a} + \Delta c \frac{\partial A}{\partial c}$$

$$= (\Delta a) \cdot \frac{c}{(a^2 + c^2)} - \frac{a}{(a^2 + c^2)} \cdot \Delta c \qquad ...(1)$$

or $\qquad \left| \Delta A \right| \le \left| \Delta a \cdot \dfrac{c}{a^2 + c^2} \right| + \left| \Delta c \cdot \dfrac{a}{a^2 + c^2} \right|$

Given that $\Delta a = 1$ mm $= 0.1$ cm, $\Delta c = 2$ mm $= 0.2$ cm, $a = 6$ cm and $c = 15$ cm. Putting all these values in equation (1), we get

$$\left| \Delta A \right| \le \left| \frac{0.1 \times 15}{(6)^2 + (15)^2} \right| + \left| \frac{0.2 \times 6}{(6)^2 + (15)^2} \right| = \frac{1.5 + 1.2}{261} = \frac{2.7}{261} = 0.0103 \text{ Radians}$$

$$\Rightarrow \quad \left| \Delta A \right| \le 0.0103 \text{ radians}$$
or $\quad \left| \Delta A \right| \le 35'25''$

EXAMPLE 2. *If $u = \dfrac{4x^2 y^3}{z^4}$ and $\Delta x = \Delta y = \Delta z = 0.001$, compute the relative maximum error in u when $x = y = z = 1$.*

SOLUTION. Here, we have

$$u = \frac{4x^2y^3}{z^4} \qquad \qquad \dots(1)$$

From eq. (1), we have

$$\frac{\partial u}{\partial x} = \frac{8xy^3}{z^4}, \frac{\partial u}{\partial y} = \frac{12x^2y^2}{z^4} \text{ and } \frac{\partial u}{\partial z} = -\frac{16x^2y^3}{z^5}$$

Now, we have

$$\Delta u = \frac{\partial u}{\partial x}\Delta x + \frac{\partial u}{\partial y}\Delta y + \frac{\partial u}{\partial z}\Delta z \qquad \qquad \dots(2)$$

Now, putting the values of $\frac{\partial u}{\partial x}, \frac{\partial u}{\partial y}$ and $\frac{\partial u}{\partial z}$ in eq. (2), we get

$$\Delta u = \frac{8xy^3}{z^4}\Delta x + \frac{12x^2y^2}{z^4}\Delta y - \frac{16x^2y^3}{z^5}\Delta z$$

Now, $(\Delta u)_{\max} = \left|\frac{8xy^3}{z^4}\Delta x\right| + \left|\frac{12x^2y^2}{z^4}\Delta y\right| + \left|\frac{16x^2y^3}{z^5}\Delta z\right|$

$$= 8(0.001) + 12(0.001) + 16(0.001) = 0.036$$

Therefore, the maximum relative error is

$$= \frac{(\Delta u)_{\max}}{(u)_{\text{at } x=y=z=1}} = \frac{0.036}{4} = 0.009$$

EXAMPLE 3. *In a $\triangle ABC$, a = 30 cm, b = 80 cm, $\angle B = 90°$. Find the maximum error in the computed value of A, if possible errors in a and b are $\frac{1}{3}$% and $\frac{1}{4}$% respectively.*

SOLUTION. Here, we have

In $\triangle ABC$, $a = 30$ cm, $b = 80$ cm, $\angle B = 90°$
From figure 2, we have

$$\sin A = \frac{a}{b}$$

$\Rightarrow \qquad A = \sin^{-1}\frac{a}{b} \qquad \dots(1)$

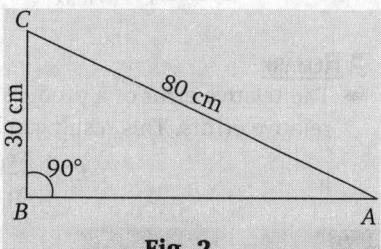

Fig. 2

Therefore, we have

$$|\Delta A| < \left|\Delta a \cdot \frac{\partial A}{\partial a}\right| + \left|\Delta b \cdot \frac{\partial A}{\partial b}\right| \qquad \dots(2)$$

Now, we have the possible errors in a and b are 1/3% and 1/4% respectively, then

$$\frac{\Delta a}{a} \times 100 = \frac{1}{3} \qquad \Rightarrow \qquad \Delta a = 0.1$$

and $\frac{\Delta b}{b} \times 100 = \frac{1}{4} \qquad \Rightarrow \qquad \Delta b = 0.2$

Also, from equation (1)

$$\frac{\partial A}{\partial a} = \frac{1}{\sqrt{b^2 - a^2}} \quad \text{and} \quad \frac{\partial A}{\partial b} = \frac{a}{b\sqrt{b^2 - a^2}}.$$

Putting all these values in equation (2), we get

$$|\Delta A| < |0.00135 + 0.00100| = 0.00235 \text{ radians}$$

$$\Rightarrow \qquad \Delta A < 8'5''$$

EXAMPLE 4. *Find the relative error in the function*

$$y = ax_1^{m_1} x_2^{m_2} \ldots x_n^{m_n}$$

SOLUTION. Here, we have

$$y = ax_1^{m_1} x_2^{m_2} \ldots x_n^{m_n} \qquad \qquad \ldots(1)$$

Taking log of both sides, we get

$$\log y = \log a + m_1 \log x_1 + m_2 \log x_2 + \ldots + m_n \log x_n \qquad \ldots(2)$$

Now, differentiating eq.(2), we get

$$\frac{1}{y} \cdot \frac{\partial y}{\partial x_1} = \frac{m_1}{x_1}$$

$$\frac{1}{y} \cdot \frac{\partial y}{\partial x_2} = \frac{m_2}{x_2}, \ldots \frac{1}{y} \cdot \frac{\partial y}{\partial x_n} = \frac{m_n}{x_n}$$

Therefore, the error

$$E_r = \frac{\partial y}{\partial x_1} \cdot \frac{\Delta x_1}{y} + \frac{\partial y}{\partial x_2} \cdot \frac{\Delta x_2}{y} + \ldots + \frac{\partial y}{\partial x_n} \cdot \frac{\Delta x_n}{y}$$

$$= m_1 \frac{\Delta x_1}{x_1} + m_2 \frac{\Delta x_2}{x_2} + \ldots + m_n \frac{\Delta x_n}{x_n}$$

Hence, $(E_r)_{\max} \leq m_1 \left| \frac{\Delta x_1}{x_1} \right| + m_2 \left| \frac{\Delta x_2}{x_2} \right| + \ldots + m_n \left| \frac{\Delta x_n}{x_n} \right|$

⟐ REMARK

➠ The relative error of a product of n numbers is approximately equal to the algebraic sum of their relative errors. This result can be verified easily by taking $a = 1$, $m_1 = m_2 = \ldots = m_n = 1$, then

$$E_r = \frac{\Delta x_1}{x_1} + \frac{\Delta x_2}{x_2} + \ldots \frac{\Delta x_n}{x_n}.$$

Exercise 15.2

1. What do you understand by the term 'Statistics'?

2. Discuss the application of "Bio-Statistics".

3. Mention the scope of Bio-Statistics.

4. Explain the limitation of 'Statistics'.

5. What do you understand by 'Error'?

6. Mention the scope of Biometrics.

□□□□□□

CHAPTER 16

Sampling and Inferences

16.1 INTRODUCTION

The collection of selected number of individuals, objects or results from the parent universe according to the given rule is known as sample. We know that it is not possible to discuss all the members of the universe separately, because it is very costly and time consuming. So, it becomes necessary to find a rule or process under which by examining only a selected part of universe, we get all the information about it. Thus, these selected members are known as sample.

For example: If we buy the sweets from a shop, we examine only a single piece of it. Then this single piece will be treated as a sample.

Size of Sample : Number of objects, results or members in the sample is the size of the sample.

16.2 SAMPLING

A manner in which we can form a sample from the parent universe (population) is known as sampling.
For example :
1. Population of male children born in a particular year.
2. Number of smokers in a particular locality.

16.3 TYPES OF SAMPLING

The sample of population can be chosen in four manners or it can be stated as the sampling is of four types :

1. **Random Sampling :** Random sampling is a well known method of sampling. In this method, while choosing the sample, if each member have the same probability of chosen out then this type of sampling is known as random sampling. In short, we can say that when the selection is taken at random, then the sampling is known as random sampling.
   For example : If we throw a dice then each number have the same probability of coming out.

2. **Simple Sampling**
   Simple sampling is a special case of random sampling. In this, each event has the equal probability of chosen out, in which the probability of choosing an event is free from the previous probability of successes or failure of an event.
   For example :
   1. In tossing of a coin, the probability of coming head does not depend upon the previous trial in any manner.
   2. If we want to select a boy from a group of 8 children containing 4 boys and 4 girls,

then the probability of choosing a boy is 4/8 and if the child selected is a girl and we does not replace it in the group, then the probability of choosing a boy will be 4/7. Then this sampling is known as simple sampling without replacement. And if we replace the girl to the group and then select the boy again, then this sampling is known as simple sampling with replacement.

3. **Purposive Sampling :** Purposive sampling is a sampling in which samples are taken under a particular consideration. In this sampling, the investigator select that part of universe by which it conclude its desired result. In this sampling personal individuals have a great chance of chosen out.

For example : If we want to choose the student of 55-60 Kg from a group of students, then the random selection will give the students of all weights. Then, if we choose the students of particular weight, then this is known as purposive sampling.

4. **Stratified Sampling**

The mixture of random sampling and purposive sampling is known as stratified sampling. Sometimes it is impossible to discuss a population by taking random sampling or purposive sampling as a representative sampling. In such cases, we divide the whole universe into distinct parts and then take the sample randomly according to the size of the part. Then this type of sampling is known as stratified sampling.

Tippett's Number : A table constructed by L.H.C. Tippett's which solve the problem of making random sampling such that there is no relation among the numbers used is known as Tipptet's number table.

This table consists of 41600 digit and gives 10400 four figure numbers which are useful in construction of random sampling. The numbers are chosen randomly from this table.

A small part of this table is shown below :

2952	6641	3992	9792	7979	5911
3170	5624	4167	9525	1545	1396
7203	5356	1300	2693	2370	7483
3408	2762	3563	6107	6913	7691
0560	5246	1112	9025	6008	8126

For example : If we want to select a sample of 8 people from a group of 5000 numbered from 1 to 5000, then we choose the first 8 numbers which not exceed 5000 as

2952, 3992, 3170, 4167, 1545, 1396, 1300, 2693

This is the random sampling according to Tippett.

16.4 CHARACTERISTICS OF A GOOD SAMPLE DESIGN

We can list down the characteristics of a good sample design as under :

1. Sample design must result in a truly representative sample.
2. Sample design must be such which results in a small sampling error.
3. Sample design must be viable in context of funds available for the research study.
4. Sample design must be such so that systematic bias can be controlled in a better way.
5. Sample should be such that the results of the sample study can be applied, in general, for the universe with a reasonable level of confidence.

16.5 TECHNIQUES FOR RANDOM SAMPLING

The techniques for random sampling can be of three types :

1. Random sampling by lottery system.
2. By arranging the whole numbers according to a rule and then selecting the individuals in a sequence
3. By random number method

1. **Random sampling by lottery system :** The lottery system is used by three ways. In first case, we make the pieces of a paper of same size such that we cannot differentiate them. Now, we numbered them according to the individuals and then we mix them. A chit is drawn out and the process will go on until we get the required number of chits equal to the sample size.

The individuals corresponding to the chits form a sample. This method of random sampling is known as chit method.

Now, the second method of lottery system is card method. In this, cards are used in place of chits. All the cards are numbered according to the sample and then shuffled. One card is drawn out and then the cards are reshuffled and the process will go on until we get the cards equal to the sample size. Now the individuals corresponding to card form a sample.

The third method is lottery system. This method is almost same as chit method. In this method, chits are placed in similar containers and these containers are rotated in a rotating drum. Then these containers are picked up one by one until the sample size is obtained. These individuals corresponding to container form a sample. This method is known as lottery system for random sampling.

2. **By arranging the whole numbers according to a rule and then selecting the individuals in a sequence :** Here, at first we arrange the whole universe in a particular manner and then select the individuals in a sequence.

For example : If we want to select 10 boys from a group of 250, then we arrange the whole group according to a rule (according to height, weight or names). Now, if we select every 25th boy, we get a random sampling of 10 boys.

3. **By random number method :** This method completely depends upon the random number table. Fisher, Kendall, Mahalnobis and Tippett have published such type of tables.

According to Tippett, if we want to find out a sample of 8 people from a group of 5000, then we number them from 1 to 5000 and then select first 8 numbers which are less than 5000. The people corresponding to these numbers form a random sample.

16.6 STRATIFIED RANDOM SAMPLING

Let us suppose the given population is heterogeneous (non-homogeneous) in nature. Then, entire heterogeneous population is divided into a number of homogeneous group, which is called strata or subpopulation.

Let population of N units be divided into l sub populations of $N_1, N_2, ..., N_l$ units respectively such that each subpopulations (strata) are non-overlapping and together they form the whole population. Thus, we can write $\sum_{i=1}^{l} N_i = N$

When the subpopulation (strata) have been determined a sample is drawn from each stratum. The drawing are made independently, in different strata, the sample sizes within the strata are denoted by $n_1, n_2, ..., n_k$ respectively, i.e., $\sum_{i=1}^{l} n_i = n$

REMARKS

➠ If simple random sampling is taken in each stratum, the whole procedure is known as "Stratified Random Sampling".

➠ The sample which is the set of all the sampling units drawn from each stratum is known as stratified sampling.

16.6.1 PRINCIPLES OF STRATIFICATION

In stratifying a population, following are the principles :

1. The stratification of population should be done such that strata are homogeneous.
2. The strata should be non-overlapping and together they must form the whole population.
3. Sometimes, when it is difficult to stratify with respect to characteristic under study, administrative convenience may be considered as the basis for stratification.
4. It will be better to treat each subpopulation as a stratum.

16.7 SYSTEMATIC RANDOM SAMPLING

In this sampling method, we used partly the arbitrariness and partly randomness. Let us suppose N units of populations are numbered $1, 2, ..., N$ in some order. Let $N = nk$, where n is the sample size and k is an integer (called the sampling interval). Then, a random number less than or equal to k be selected and every k^{th} unit thereafter. The resulting sample is called k^{th} systematic sample and such a procedure is known as linear systematic sampling.

Now, if $N \neq nk$ and every k^{th} unit be included in a circular manner till the whole list is exhausted, it is known as circular systematic sampling.

The systematic random sampling is used if a complete and up-to-date sampling is available. Also, under many situations, systematic sampling provides estimates which are more efficient than those obtained with simple random sampling without replacement.

16.8 LIMITATIONS OF SAMPLING

Some limitations of sampling theory are as follows :

1. Proper care should be taken in the planning, otherwise the results obtained might be inaccurate and misleading.
2. If the information is required about each and every unit of the universe, there is no way but to resort to complete enumeration. Also, if time and money are not important, a complete census may be better than any sampling method.
3. In the absence of the services of trained and qualified personnel and sophisticated equipments for its planning, execution and analysis, the result of sample survey are not trustworthy.

Exercise 16.1

1. Write short notes on the following :
   (i) Census method   (ii) Sampling method
   (iii) Systematic random sample
2. Write the advantages and disadvantages of the following :
   (i) Census method   (ii) Sampling method

3. Describe the followings :
   (i) Simple random sampling
   (ii) Stratified random sampling
4. Discuss various methods of selecting a random sample.
5. What are the essentials of sampling.

□□□□□□

CHAPTER 17

Collection of Data

17.1 INTRODUCTION

Collection of data plays a very important role in the study of statistics. If the data are properly collected, then we can find the required result and if it is not proper, then there will be a defect or fault in the result. Thus, we can say that all the results depend upon data.

Statistical data can be divided into two parts :

(a) Primary data

(b) Secondary data

Primary Data: The primary data are the data which are wholly collected by the investigator. Primary data contain all the new facts.

Secondary Data: Secondary data are the data which are investigated by a investigator and after some time when a new investigator, investigate a new thing, he assumes all these data true.

17.2 SOURCES OF DATA

In statistics, data sources may be of following two types :

1. External sources

2. Internal sources

External data sources may be further divided into primary sources and secondary sources. Primary sources may be taken from the field studies. Secondary data sources collected through primary sources for certain specific purpose. Statistical material obtained from secondary sources is not always as reliable as that from the primary sources.

17.3 METHODS OF COLLECTING PRIMARY DATA

The primary data can be collected by the following five methods :

1. Direct personal investigation

2. Indirect oral investigation

3. Investigation through local agencies

4. Questionnaires sent by hand or post

5. Informations with the help of an enumerator.

1. **Direct Personal Investigation:** Direct personal investigations are the investigations which are made by the investigator by examining the field himself. This method consists of, necessary informations only, required for the investigation, which are collected by the investigator.

   This method is applied when the field of investigation is small and personal contact is must.

Demerits: This method is very expensive, time consuming and require too much labour and this is not scientific.

2. **Indirect Oral Investigation:** This method is used when the investigator can not collect all the inquiries about the data himself. In this method, all the informations are collected from the persons who have knowledge about the data. Such a procedure of collecting the data is known as indirect oral investigation.

This method is very cheap and time saving and in this it is not necessary to make direct contact with the field.

Demerits: In this method, all the inquiries are collected by the persons related to it, so it can be inaccurate and may contain improper informations.

3. **Investigation through Local Agencies:** In this method, the investigator does not make contact with the persons. It appoints local agencies and agents in the area under survey and instruct them to give informations to the head office. But, in general, these agencies collect the informations to a limited extent and send all the informations by their own experiences.

This method is cheap and time saving. It is applied when the field of investigation is large.

Demerits: In this method, informations are send by the agencies on their own experience. So, it is not reliable. So, this method can be used only when the accuracy of data is not important.

4. **Questionnaires sent by Hand or Post:** In this method, a list of questions is made by the investigator, which is known as questionnaires and send by hand or by post to the related person with request to send answers and explaining them. The questionnaires contain all the doubtful and unknown questions about the investigation. Sometimes all the informations are kept secret on the request of investigator.

In this method the informer should be intelligent and have the complete knowledge about the investigation.

This method is cheap and used when the personal contact is not possible and the field is widespread.

Demerits:

1. The informer may be careless and irresponsible.

2. Intelligent informer is required so the field of inquiry is very limited.

5. **Informations with the help of an Enumerator:** In this method, some enumerators are appointed to different places such as they contact with the different informers and collect all the informations according to questionnaires.

This method is applied when the area of investigation is large. In this, informations can be obtained by the persons of all levels and informations will be homogeneous.

Demerits: If the enumerator is careless and irresponsible, the informations will be unreliable. This method is very costly and time consuming and need much labour.

17.4 SCHEDULE AND QUESTIONNAIRE

Generally, schedules and questionnaires are used in the collection of primary data. In practice, these two words are used in the same sense, but there is little difference between the two. Schedule is a list of questions, which is filled in by the enumerators, while questionnaire is filled in by the persons themselves.

17.4.1 Drafting of a Schedule or Questionnaire

To draft a questionnaire or schedule is a specific art. The success of a statistical investigation and the quality of its results depend upon tactful drafting of questionnaire. Because the task of eliciting information from human populations in desired form and with sufficient exactness to be useful in scientific analysis is the most difficult problem. People have their own whims and feelings of pride, desire and prejudice. The main problems in drafting a questionnaire into the following heads :

(i) Decisions regarding question-content.

(ii) Decisions regarding form of response to the question.

All the above points should help in determining the size, content, sequence, etc. of the questionnaire.

17.4.2 Qualities of Questionnaire

Since the value of the results from the investigation depends largely on the adequacy and appropriate drafting of the questionnaire, the following points should be borne in mind :

(i) The question should be brief.

(ii) Simple, clear and unambiguous questions.

(iii) Nature of questions.

(iv) Use of proper words in the question.

(v) The question should be such as the answers of which are known to the informants.

(vi) Questions capable of objective answers.

(vii) Should not affect pride or sentiment.

(viii) Some kind of questions should be avoided.

(ix) Sequence of questions.

(x) Setting of the questionnaire

(xi) To test the accuracy

17.5 METHODS OF COLLECTING SECONDARY DATA

Secondary data can be obtained by the following methods :

(a) **Government Publications:** Secondary data can be obtained by the Committees of government publications such as Central Statistical Organisation, Wanchoo Commission's Report on Taxation, etc.

(b) **Semi Government Publications:** Secondary data can be obtained by the Committee such as Municipalities, Khadi-Gramodyog Board, District Boards, Corporations and U.P. Educational Board, etc.

(c) **Business Publications:** Various banks, FICCI, Trade Unions, market reports of stock exchanges, cooperative societies, etc. are the business publications which give us the knowledge of secondary data.

(d) **Periodicals and News Agencies:** Magazines such as Times of India, Year Book and newspapers are the main sources of secondary data, e.g., Indian Finance, Financial Express, Indian Journal of Economics, etc.

(e) **Publication of International Bodies or Foreign Governments:** Secondary data can be collected by the publication of International bodies such as UNO publication and statistical abstract of United States.

All above are the sources of collecting published secondary data. Unpublished secondary data can be collected by research institutions and by researcher individually.

17.5.1 PRECAUTIONS IN COLLECTING THE SECONDARY DATA

1. There should be the same object of study between primary and secondary data.
2. The data collected by Government agencies will be more reliable and dependable than a private agency.
3. The timing of data collected should be accurate so that these are not too old to apply.
4. The data collected by the present investigator must tally with the past investigator in the sense of accuracy.

17.6 CLASSIFICATION OF DATA

When the enquiry of statistics started, there was a great number of details and data which are available there. But all of these are disarranged. It was too difficult to find the details about any data and come to any conclusion. So it became necessary to arrange all the data and details in a particular manner so as to find it easily. Thus, a particular arrangement of all the data and details is known as classification of data.

For example. If we consider that from a school 50 students have passed the examination of class X with at least 60 percent marks. Then we group all the students which have got marks in a particular range as per ten percent. Thus the data are arranged in a table.

17.7 OBJECTIVE OF CLASSIFICATION

Mainly, we have the following main objectives of classification :
1. To arrange the raw data in a scientific manner.
2. To check the similarities (or dissimilarities) of the individual of their characteristics.
3. Comparison of data.
4. To identify the most significant features of the data.
5. To make use of relevant information.
6. To give statistical treatment of the data collected for investigation.

17.8 TYPES OF CLASSIFICATION

Classification of data can be done by following two ways :
1. According to class interval
2. According to attributes.

1. **According to Class Interval:** When the data given are in numerical form, then we classify it according to class interval. For this, we divide all the data in the particular range intervals.

   For example. If the marks obtained by the students of a class of 15 is given by 8, 9, 6, 12, 15, 21, 25, 27, 29, 32, 37, 35, 39, 38, 31, then these can be divided as

Range (Marks obtained)	Number of students
0-10	3
10-20	2
20-30	4
30-40	6

2. **According to Attributes:** When the data are not given in numerical form, then these are classified according to their quality, then this type of classification is known

as classification according to attributes.

This classification is of two types :

    (i) Simple classification   (ii) Manifold classification

(i) Simple classification: Simple classification is the classification in which whole data are divided according to a single quality.

For example. If we want to classify all the flowers according to colours, then

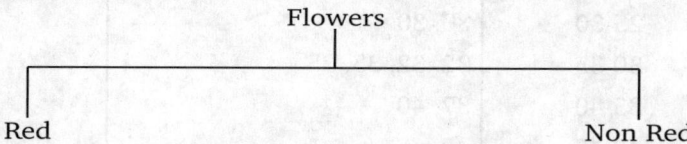

(ii) Manifold classification: Manifold classification is the classification in which simple classification is again subdivided according to any other property.

For example:

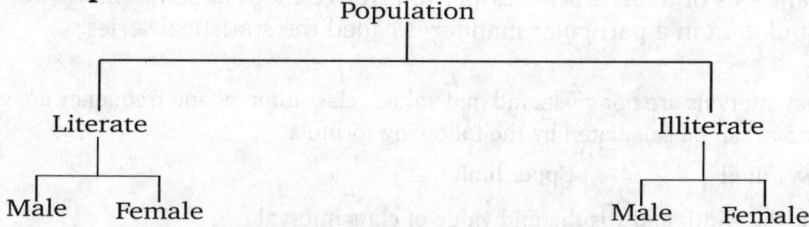

███ SOLVED EXAMPLES ███

EXAMPLE 1. *Marks obtained by 25 students in Statistics paper carrying 80 marks. Classify them by class-interval of ten marks :*

32, 31, 48, 53, 52, 65, 32, 46, 71, 73, 61, 79, 64, 35, 33, 43, 42, 57, 51, 62, 34, 64, 77, 65, 72

SOLUTION. We take the class interval of ten marks and intervals 30-40 have the data 32, 31, 32, 35, 33, 34 and so on.

Class Interval	Data	Frequency
30-40	32, 31, 32, 35, 33, 34	6
40-50	48, 46, 43, 42	4
50-60	53, 52, 57, 51	4
60-70	65, 62, 64, 65, 61, 64	6
70-80	71, 73, 79, 77, 72	5

EXAMPLE 2. *Consider the rent of forty houses given below :*

5, 6, 7, 8, 11, 15, 20, 8, 11, 25, 30, 15, 17, 11, 6, 22, 25, 20, 22, 15, 30, 32, 32, 8, 20, 25, 22, 22, 35, 37, 40, 20, 11, 25, 20, 10, 10, 15, 35, 42

Classify these by taking class interval of length 5.

SOLUTION.

Class Interval	Houses	Frequency
5-10	5, 6, 7, 8, 8, 6, 8, 10, 10	9
10-15	11, 15, 11, 15, 11, 15, 11, 15	8
15-20	20, 20, 17, 20, 20, 20	6
20-25	25, 22, 25, 22, 25, 22, 22, 25	8
25-30	30, 30	2
30-35	32, 32, 35, 35	4
35-40	37, 40	2
40-45	42	1

17.9 STATISTICAL SERIES

In the analysis of data, it is necessary to arrange them in a systematic manner. Thus the arrangement of data in a particular manner is called the statistical series.

REMARK

➡ If the class intervals are not given and mid values, class interval and frequency are given, then the class interval can be calculated by the following formula :

$$\text{Lower limit } l_1 = x - \frac{1}{2} y \text{ , Upper limit } l_2 = x + \frac{1}{2} y$$

where y is the width and x is the mid value of class interval.

17.9.1 CLASSIFICATION OF STATISTICAL SERIES

The whole statistical series can be studied into two groups :

1. Series according to nature
2. Series according to construction.

1. **Series according to nature:** According to nature, statistical series can be divided into three parts :

   (i) **Time Series:** When all the data are arranged, at a difference of particular time then this series will be time series. The difference in time taken can be year, month, week or day.

   For example. Let the marks obtained by a student in monthly test out of 20 be given by

Month	July	Sep	Nov	Jan	March	May
Marks obtained	18	15	19	12	17	16

   (ii) **Spatial Series:** Spatial series is formed when all the data are arranged in a manner based on location or geographical consideration.

   (iii) **Condition Series:** A series in which a condition is applied on the classification of data is known as condition series. For example, The members of a locality can be arranged according to their age.

2. **Series according to construction:** According to construction, series can also be divided into following three parts :

   (a) Individual series, 　　　　　(b) Continuous series,

   (c) Discrete series

   (a) **Individual Series:** The series which can not be classified according to a rule. In this series, each of the data is independent and separately taken :

Name of locality	U	V	W	X	Y	Z
Population	350	276	198	400	316	358

(b) Continuous Series: The series in which all the data are classified into the classes or groups with definite intervals. These classes are continuous and the point, at which the class of first interval ends, the second started. Thus the continuity of class interval is regular.

For example, the marks obtained by the students in a class that is given by

Marks	0-10	10-20	20-30	30-40
Frequency	5	8	3	4

(c) Discrete Series: The series in which data can not be classified into the groups and there is not a definite difference between two values of them. All the variables of this series are discrete.

For example, the rent of 20 houses in a city are given by
250, 400, 200, 500, 300, 375, 225, 475, 500, 435, 475, 400, 375, 300, 500, 200, 250, 250, 435, 375

Rent	No. of houses
200	2
225	1
250	3
300	2
375	3
400	2
435	2
475	2
500	3

17.10 TABULATION OF DATA

In the analysis of data, all the data are arranged in the tables so that accurate results can be obtained. Thus this arrangement of data into the tables is known as tabulation of data.

According to Goxton and Cowden, *"Either for one's own use or for the use of others, the data must be presented in some suitable form."*

17.10.1 OBJECTIVES OF TABULATION

1. The data which remain unchanged by tabulation is in compact form.
2. The presentation of classified data into arranged and ordered form.

17.10.2 MERITS OF TABULATION

1. The simple and compact form of any problem become easy for calculation and analysis by the tabulation.
2. Tabulation is a process by which a lot of informations can be expressed in a small space. Thus the space and the time can be saved by tabulation.
3. This process comparable study is easy.

17.10.3 DIFFERENCE BETWEEN CLASSIFICATION AND TABULATION

1. Classification is a process by which all the data are classified into classes or groups according to data, while in tabulation all the data are tabulated into rows and columns.
2. In general classification is the base of tabulation. First of all, the data are classified and then these classified data are presented in tabular form.
3. Tabulation is a process by which we can arrange the data in tables while classification is a method of statistical analysis.

17.10.4 TYPES OF TABULATION

There are mainly three types of tabulation stated as follows :

(a) Tabulation based on originality
(b) Tabulation based on purpose
(c) Tabulation based on construction

(a) Tabulation based on Originality: This tabulation can further be divided into two parts :

(i) **Primary table:** Primary table is the table in which all the data are put in the form in which they are originally collected. This type of table is also known as original table.

(ii) **Derivative table:** Derivative table is the table in which not only the original data are presented, but the derivative data from these data such as sum, ratio, percentage and mean, etc. are also presented.

(b) Tabulation based on Purpose: This tabulation can further be divided into two parts :

(i) **General purposive table:** General purposive table is the table which constitute almost all the material (from which a few steps are removed) of the source, from which it is collected.

(ii) **Special purposive table:** This table is also known as summary purposive table. When we construct a table from the general purposive table for a definite purpose or to explain any result, then this type of table is known as special purposive table.

(c) Tabulation based on Construction: This tabulation can further be divided into two parts :

(i) **Simple tabulation:** Simple tabulation is the tabulation in which whole the data presented in the table are based on a single quality. For example, all the population of a country can be divided into male and female population.

(ii) **Complex tabulation:** The tabulation in which the whole data presented in the table are based on more than one quality is known as complex tabulation. This tabulation can be divided into three parts :

(a) **Double tabulation:** The tabulation in which the whole data presented in the table are based on two quality is known as double tabulation. By this tabulation, two types of problems can be solved by a single table. For example, if we classify the students of a school according to their height and weight, then this tabulation will be a double tabulation.

(b) **Three fold tabulation:** The tabulation in which the whole data presented in the table are based on three quality is known as three fold tabulation or triple tabulation. For example :

S.N.	Crop	Area under cultivation						Grand total
		Irrigated area		Non-irrigated area		Total area		
		With fertilizer	Without fertilizer	With fertilizer	Without fertilizer	With fertilizer	Without fertilizer	
1	Wheat	—	—	—	—	—	—	—
2	Maize	—	—	—	—	—	—	—
3	Gram	—	—	—	—	—	—	—
4	Sugarcane	—	—	—	—	—	—	—
5	Miscellaneous	—	—	—	—	—	—	—
	Total							

(c) **Manifold tabulation:** The tabulation in which the whole data presented in the table are based on four or more than four quality is known as manifold table. For example, the study of a district on the basis of crop, use of insecticides, use of fertilizers and irrigation can be represented by the following manifold table :

S.N.	Crop	Area under cultivation												Grand total
		Irrigated area						Non-irrigated area						
		With fertilizer			Without fertilizer			With fertilizer			Without fertilizer			
		With insecticide	Without insecticide	Total	With insecticide	Without insecticide	Total	With insecticide	Without insecticide	Total	With insecticide	Without insecticide	Total	
1	Wheat	–	–	–	–	–	–	–	–	–	–	–	–	–
2	Maize	–	–	–	–	–	–	–	–	–	–	–	–	–
3	Gram	–	–	–	–	–	–	–	–	–	–	–	–	–
4	Sugarcane	–	–	–	–	–	–	–	–	–	–	–	–	–
5	Miscellaneous	–	–	–	–	–	–	–	–	–	–	–	–	–
	Total													

17.10.5 POINTS TO REMEMBER FOR THE CONSTRUCTION OF STATISTICAL TABLE

There are some points which should be remembered in the construction of statistical table :

1. **Heading or Title:** The first and most important thing for the construction of a table is its heading or title. The title should be simple, clear and brief that can give the knowledge about the table.
2. **Subheading:** Subheading is a necessary part of a table. It further divides the heading into necessary parts. With the help of subheading, we can briefly know about each point of a table.

3. **Sources:** The source is that thing from which the data are collected. It must be shown below the table so that everybody can know about the source.

4. **Columns and Rows:** The number of columns and rows must be such that these are suitable for the problem under consideration.

5. **Total:** For the grand total of columns and rows, space must always be provided in each table.

6. **Arrangement of items:** The data in the table must be taken in the sequence of their importance. The most important thing must be mentioned at first and other will follow it according to their sequence.

7. **Attractive form:** According to Dr. Bowley, *"In collection and tabulation, common sense is the chief requisite and experience is the teacher."* So, we can say that the table must be attractive, artistic and its size must suit the paper.

Thus, all these things make a table attractive and more suitable for the purpose.

17.11 FREQUENCY DISTRIBUTION

The tabular arrangement of data showing the frequency of each item is called a frequency distribution. Frequency distribution are of two types :

1. **Discrete frequency distribution:** The process of preparing this type of distribution is very simple. It is done by the use of the method of tally marks.

   A bar (|) called tally mark is put against the number when it occurs. When occurred 4 times, the firth occurrence is represented by putting diagonally a cross tally (\) on the first four tallies. This process will be clear from the following example of the number of children in 20 families: 1, 1, 2, 3, 4, 3, 2, 1, 1, 4, 5, 2, 4, 2, 2, 1, 3, 3, 2, 5

No. of Children	Tally Bars	Frequency
1	ﾄﾄﾄ	5
2	ﾄﾄﾄ	6
3	||||	4
4	|||	3
5	||	2

2. **Continuous frequency distribution:** A continuous frequency distribution is such a distribution in which data are arranged in classes or groups which are not exactly measurable. This process will be clear from the following example:

   If the number of observations in data is large and the difference between the greatest and smallest observations is large, then we condense the data into classes or groups:

   For example: 39, 25, 5, 33, 19, 21, 12, 48, 13, 21, 9, 1, 10, 8, 12, 17, 41, 19, 17, 40, 12, 46, 37, 17, 27, 30, 6, 2, 23, 19

Marks	Tally Bars	No. of students (Frequency)
0-10	ﾄﾄﾄ|	6
10-20	ﾄﾄﾄﾄﾄﾄ|	11
20-30	ﾄﾄﾄ	5
30-40	||||	4
40-50	||||	4

In the above table, the class 0-10 means the number of students between 0-10 including and excluding 10.

There are two types of series according to the class interval.

(1) Inclusive form: A frequency distribution in which each upper limit of each class is also included. For example:

Class Intervals	Frequency
0-4	5
5-9	6
10-14	5
15-19	3
20-24	2
25-29	4
	Total = 25

(2) Exclusive form: A frequency distribution in which the upper limit of one class is the lower limit of the next class interval. For example :

Class Intervals	Frequency
0-10	4
10-20	7
20-30	9
30-40	5
40-50	2
50-60	3
	Total = 30

17.12 GROUPED FREQUENCY DISTRIBUTION

Sometimes the class-marks of a distribution are given and we need to find out the class-size and various classes :

1. Class size: The difference between any two consecutive class marks is known as class size.

WORKING PROCEDURE

For arranging the given data into class intervals, we use the following steps:

Step 1. Find out the Range = maximum marks – minimum marks

Step 2. Decide the class size.

Step 3. Divide the range into groups, then number of classes is equal to $\dfrac{\text{Range}}{\text{Class size}}$.

Step 4. The minimum value of variate should be included in the first class and the maximum value of the variate should be included in the last class interval.

Step 5. Tally the mark in the proper class by the tally method given before the article

$$\text{Mid value of the class} = \frac{\text{Upper class limit} + \text{lower class limit}}{2}$$

SOLVED EXAMPLES

EXAMPLE 1. *The class marks of a distribution are 47, 52, 57, 62, 67, 72, 77, 82*
Determine the (i) class size, (ii) class limit, (iii) true class limits

SOLUTION. (i) Class size = The difference between any two consecutive class marks
$$= 52 - 47 = 5$$
(ii) Half the class size = $5/2 = 2.5$.

To find lower class limit, we subtract 2.5 from each class mark and to get upper limit, we add 2.5 to each class mark. Therefore, we get the following class limits

Class marks	Class limit
47	44.5-49.5
52	49.5-54.5
57	54.5-59.5
62	59.5-64.5
67	64.5-69.5
72	69.5-74.5
77	74.5-79.5
82	79.5-84.5

Since the classes are exclusive, hence true class limits are the same as the class limits.

EXAMPLE 2. *Given below are the ages of 25 students of class X in a school. Prepare a discrete frequency distribution*

15, 16, 16, 17, 17, 16, 15, 15, 16, 16, 17, 15, 16, 16, 14, 16, 15, 14, 15, 16, 16, 15, 14, 14, 15

SOLUTION. **Frequency distribution of ages**

Age	Tally marks	Frequency
14	||||	4
15	⊞ |||	8
16	⊞ ⊞	10
17	|||	3
	Total	25

EXAMPLE 3. *From the data given below of the ages of 30 students of class IX in a school, prepare a discrete frequency distribution.*

15, 16, 16, 14, 17, 17, 16, 15, 15, 16, 16, 17, 15, 16, 16, 14, 16, 15, 14, 15, 16, 16, 15, 14, 15, 15, 16, 17, 16, 15.

SOLUTION. **Discrete frequency distribution**

Age	Tally marks	Frequency
14	||||	4
15	⊞ ⊞	10
16	⊞ ⊞ ||	12
17	||||	4
	Total	30

EXAMPLE 4. *Form a discrete frequency distribution from the following scores :*
15, 18, 16, 20, 25, 24, 25, 20, 16, 15, 18, 18, 16, 24, 15, 20, 28, 30, 27, 16, 24, 25, 20, 18, 28, 27, 25, 24, 24, 18, 18, 25, 20, 16, 15, 20, 27, 28, 29, 16

SOLUTION.

Frequency distribution of scores

Variate	Tally marks	Frequency
15	||||	4
16	⑷||	6
18	⑷||	6
20	⑷||	6
24	⑷|	5
25	⑷|	5
27	|||	3
28	|||	3
29	|	1
30	|	1
	Total	40

EXAMPLE 5. *The following data gives marks out of 60 obtained by 30 students of a class in a test:*
50, 22, 50, 47, 27, 37, 40, 16, 12, 33, 29, 35, 49, 15, 43, 29, 31, 22, 51, 27, 29, 27, 22, 18, 20, 11, 19, 31, 23, 58
Arrange them in ascending order and present it as a grouped data (i) in inclusive form and (ii) in exclusive form.

SOLUTION. Arranging the marks in ascending order, we get
11, 12, 15, 16, 18, 19, 20, 22, 22, 22, 23, 27, 27, 27, 29, 29, 29, 31, 31, 33, 35, 37, 40, 43, 47, 49, 50, 51, 56, 58

(i) In inclusive form

Frequency distribution of marks

Marks (class interval)	Tally marks	Frequency
11-20	⑷ ||	7
21-30	⑷ ⑷	10
31-40	⑷ |	6
41-50	||||	4
51-60	|||	3
	Total	30

(ii) In exclusive form

Frequency distribution of marks

Marks (class interval)	Tally marks	Frequency
11-21	⑷ ||	7
21-31	⑷ ⑷	10
31-41	⑷ |	6
41-51	||||	4
51-61	|||	3
	Total	30

Here, class interval 11-21 means marks obtained are 11 and more but less than 21. Clearly, 21 does not belong to this class 11-21.

Example 6. *Form a grouped frequency distribution from a following data by inclusive method taking 4 as the magnitude of class intervals*

31, 23, 19, 29, 22, 20, 16, 10, 13, 34, 38, 33, 28, 21, 15, 18, 36, 24, 18, 15, 12, 30, 27, 23, 20, 17, 14, 32, 26, 25, 18, 29, 24, 19, 16, 11, 22, 15, 17, 10

Solution. Here the maximum and minimum values of the variable are 38 and 10 respectively

Hence, Range $= 38 - 10 = 28$

Since, here given magnitude of class intervals $= 4$

Therefore, number of class intervals $= 28/4 = 7$.

Thus, the frequency distribution is to be formed by inclusive method. So, the class intervals are

10-13, 14-17, 18-21, 22-25, 26-29, 30-33, 34-37, 38-41

Class interval	Tally marks	Frequency
10-13	ꟷꞪꞪꞪꞪꞪ	5
14-17	ꟷꞪꞪꞪꞪꞪ |||	8
18-21	ꟷꞪꞪꞪꞪꞪ |||	8
22-25	ꟷꞪꞪꞪꞪꞪ ||	7
26-29	ꟷꞪꞪꞪꞪꞪ	5
30-33	||||	4
34-37	||	2
38-41			1
	Total	40

Example 7. *The electricity bills (in Rs.) of 15 houses in a locality are given below. Construct a frequency distribution table with the class size 5.*

40, 53, 47, 62, 49, 51, 61, 45, 54, 67, 42, 55, 57, 43

Solution. Here,   Minimum observation $= 40$

Maximum observation $= 67$

Hence,   Range $= 67 - 40 = 27$.

Class-size $= 5$

Therefore, there will be six classes of size 5 each, first class being 40-45.

Frequency distribution table

Bills (in Rs.)	Tally marks	Frequency
40-45	|||	3
45-50	|||	3
50-55	|||	3
55-60	||	2
60-65	||	2
65-70	||	2
	Total	15

EXAMPLE 8. *The monthly wages of 30 workers in a factory are given below :*

830, 835, 890, 810, 835, 836, 869, 845, 898, 890, 820, 860, 832, 833, 855, 845, 804, 808, 812, 840, 885, 835, 836, 878, 840, 868, 890, 806, 840, 890

Represent the data in the form of a frequency distribution with class size 10.

SOLUTION. Here, Minimum monthly wage = Rs. 804.

Maximum monthly wage = Rs. 898.

Range = 898 – 804 = Rs. 94

Size of the class interval = 10

$$\frac{\text{Range}}{\text{Class size}} = \frac{94}{10} = 9.4$$

The first class interval is 804 – 814.

Frequency distribution of monthly wages

Monthly wages	Tally marks	Frequency
804-814	~~||||~~	5
814-824	|	1
824-834	|||	3
834-844	~~||||~~ |||	8
844-854	||	2
854-864	||	2
864-874	||	2
874-884	|	1
884-894	~~||||~~	5
894-904	|	1
	Total	30

EXAMPLE 9. *Form a grouped frequency distribution from the following data by taking 10 as the magnitude of class interval :*

Weights (in gms) of 30 apples

96, 92, 86, 70, 105, 108, 106, 76, 82, 116, 110, 108, 94, 188, 124, 86, 87, 90, 126, 112, 104, 118, 82, 104, 84, 75, 131, 81, 107

SOLUTION. Here, Minimum observation = 70, Maximum observation = 188

Range = 188 – 70 = 118, Class size = 10

Therefore, there will be 12 classes of size 10 each, first class being 70-80.

Frequency distribution table

Weights (in gms)	Tally marks	Frequency
70-80	|||	3
80-90	~~||||~~ ||	7
90-100	||||	4
100-110	~~||||~~ ||	7
110-120	~~||||~~	5
120-130	||	2
130-140	|	1
140-150	0	0
150-160	0	0
160-170	0	0
170-180	0	0
180-190	|	1
	Total	30

EXAMPLE 10. *Make a continuous frequency distribution series from the following data. The class interval is to be 5:*

10, 21, 20, 24, 20, 20, 34, 15, 13, 16, 21, 25, 29, 11, 17, 20, 11, 15, 30, 32, 16, 20, 18, 23, 10, 29, 26, 30

SOLUTION. Here, Minimum observation = 10

Maximum observation = 34

Range = 34 – 10 = 24

Class size = 5

Hence, there will be 5 classes, first class being 10-15.

Continuous frequency distribution table

Class intervals	Tally marks	Frequency
10-15	卌	5
15-20	卌 I	6
20-25	卌 III	8
25-30	卌	5
30-35	IIII	4
	Total	28

EXAMPLE 11. *Convert the given inclusive series into an exclusive series*

Inclusive series

Class intervals	Frequency
100-149	20
150-199	25
200-249	60
250-299	15
300-349	5
Total	125

SOLUTION. Difference between upper limit of first class and lower limit of second class = 150 – 149 = 1. On subtracting and adding this value from and to the respective lower and upper class limits, we get the following series.

Exclusive series

Class intervals	Frequency
99.5-149.5	20
149.5-199.5	25
199.5-249.5	60
249.5-299.5	15
299.5-349.5	5
Total	125

Exercise 17.1

1. 30 students in an examination got marks as follows :
20, 23, 28, 30, 32, 34, 35, 36, 36, 40, 41, 42, 43, 44, 45, 48, 49, 50, 50, 52, 53, 54, 56, 56, 58, 61, 62, 65, 65
Form a frequency distribution taking class intervals as 20-29, 30-39, 40-49, 50-59 and 60-69.

2. The water tax bills (in Rupees) of 30 houses in a locality are given below. Construct a grouped frequency distribution with class size of 10.
30, 32, 45, 54, 74, 78, 108, 112, 66, 76, 88, 40, 14, 20, 15, 35, 44, 66, 75, 84, 95, 96, 102, 110, 88, 74, 112, 14, 34, 44
Form a frequency distribution taking class intervals as 20-29, 30-39, 40-49, 50-59 and 60-69.

3. Construct a frequency table as the following data : 315, 320, 324, 317, 324, 319, 320, 317, 323, 325, 316, 318, 321, 322, 325, 319, 324, 321, 316, 319, 320, 320, 322, 323

4. Height of 16 students is given in inches. Using class interval of 5, construct a frequency distribution : 62, 70, 72, 64, 65, 71, 70, 60, 64, 72, 60, 61, 69, 70, 64, 71

5. The class marks of a distribution are : 104, 114, 124, 134, 144, 154 and 164.
Determine the class size and the class limits.

6. The electricity bills (in Rs.) as 40 houses in a locality are given below :
116, 127, 107, 100, 80, 82, 91, 101, 65, 95, 87, 81, 105, 129, 92, 75, 89, 78, 87, 81, 59, 52, 65, 101, 115, 108, 95, 65, 98, 62, 84, 76, 63, 128, 121, 61, 118, 108, 116, 130
Construct grouped frequency table.

7. The final marks in mathematics of 30 students are as follows : 53, 61, 48, 60, 78, 68, 55, 100, 67, 90, 75, 88, 77, 37, 84, 58, 60, 48, 62, 56, 44, 58, 52, 64, 98, 59, 70, 39, 50, 60

(i) Arrange these marks in the ascending order, 30 to 39 one group, 40 to 49 second group, etc.

(ii) What is the highest score?

(iii) What is the lowest score?

(iv) What is the range?

(v) If 40 is the pass mark, how many have failed?

(vi) How many have secured 75 or more?

(vii) Which observations between 50 and 60 have not actually appeared?

(viii) How many have secured less than 50?

8. Convert the given inclusive series into exclusive series :

Inclusive series

Class intervals	Frequency
10-19	3
20-29	5
30-39	2
40-49	9
50-59	5
Total	24

ANSWERS

1.

Marks	Tally marks	Frequency
20-29	|||	3
30-39	﷼|	6
40-49	﷼ ||||	9
50-59	﷼ |||	8
60-69	||||	4
	Total	30

3.

x	Tally marks	Frequency
315			1
316				2
317				2
318			1
319					3
320						4
321				2
322				2
323				2
324					3
325				2
	Total	24

2.

Bills (in Rs.)	Tally marks	Frequency
14-24	||||	4
24-34	||	2
34-44	|||	3
44-54	|||	3
54-64			1
64-74	||	2
74-84	﷼	5
84-94	|||	3
94-104	|||	3
104-114	||||	4
	Total	30

4.

Height (in inches)	Tally marks	Frequency
60-65	﷼ ||	7
65-70	||	2
70-75	﷼ ||	7
	Total	16

5. Class size = 10

Class limits are 100-110, 110-120, 120-130, 130-140, 140-150, 150-160, 160-170, 170-180

6.

Frequency distribution of electricity bills

Electricity Bills (in Rs.)	Tally marks	Frequency
52-62	|||	3
62-72	~~||||~~	5
72-82	~~||||~~		6
82-92	~~||||~~		6
92-102	~~||||~~ ||	7
102-112	||||	4
112-122	~~||||~~	5
122-132	||||	4
	Total	40

7. (ii) 100,   (iii) 37,   (iv) 63,   (v) 2,   (vi) 8,   (vii) 51, 54, 57,   (viii) 5

8.

Inclusive series

Class intervals	Frequency
9.5-19.5	3
19.5-29.5	5
29.5-39.5	2
39.5-49.5	9
49.5-59.5	5
Total	24

17.13 CUMULATIVE FREQUENCY

The cumulative frequency of a class interval is the sum of the frequencies upto and including that class.

There are two types of cumulative frequency :

(1) Less than series

(2) Greater than series

For less than cumulative frequencies, we add up the frequencies from above and for greater than cumulative frequencies, we add up the frequencies from below.

SOLVED EXAMPLES

EXAMPLE 1. *Following are the ages in years of 360 patients getting medical treatment in a hospital*

Age (in years)	10-20	20-30	30-40	40-50	50-60	60-70
No. of patients	90	50	60	80	50	30

Construct the cumulative table (less than type and more than type) for the above data.

SOLUTION.

Cumulative frequency distribution table (Less than type)

Age (in years)	No. of patients
Less than 20	90
Less than 30	140 = (90 + 50)
Less than 40	200 = (90 + 50 + 60)
Less than 50	280 = (90 + 50 + 60 + 80)
Less than 60	330 = (90 + 50 + 60 + 80 + 100)
Less than 70	360 = (90 + 50 + 60 + 80 + 100 + 30)

Cumulative frequency distribution table (More than type)

Age (in years)	No. of patients
More than 9	360 = (30 + 50 + 80 + 60 + 50 + 90)
More than 19	270 = (30 + 50 + 80 + 60 + 50)
More than 29	220 = (30 + 50 + 80 + 60)
More than 39	160 = (30 + 50 + 80)
More than 49	80 = (30 + 50)
More than 59	30

EXAMPLE 2. *Given below is a cumulative frequency distribution table showing the marks secured by 50 students of a class*

Marks	No. of students
Below 20	17
Below 40	22
Below 60	29
Below 80	37
Below 100	50

Form a frequency table from the above table.

SOLUTION.

Marks	C.F	No. of students
0-20	17	17
20-40	22	(22 – 17) = 5
40-60	29	(29 – 22) = 7
60-80	37	(37 – 29) = 8
80-100	50	(50 – 37) = 13
		$\Sigma f = 50$

EXAMPLE 3. *Convert the given simple frequency series into cumulative frequency series*

Marks	No. of students
0-10	4
10-20	8
20-30	15
30-40	20
40-50	13
Total	60

SOLUTION.

Cumulative frequency series

Marks	No. of students
Less than 10	4
Less than 20	12 = (4 + 8)
Less than 30	27 = (12 + 15)
Less than 40	47 = (27 + 20)
Less than 50	60 = (47 + 13)

Cumulative frequency series

Marks	No. of students
More than 0	60
More than 10	56 = (60 – 4)
More than 20	48 = (56 – 8)
More than 30	33 = (48 – 15)
More than 40	13 = (33 – 20)

EXAMPLE 4.   *Make a frequency table from the following*

Age	No. of persons
More than 100	0
More than 90	7
More than 80	18
More than 70	26
More than 60	41
More than 50	63
More than 40	75
More than 30	98
More than 20	117
More than 10	139
More than 0	150

SOLUTION.

Age	No. of persons
0-10	11 = (150 – 139)
10-20	22 = (139 – 117)
20-30	19 = (117 – 98)
30-40	23 = (98 – 75)
40-50	12 = (75 – 63)
50-60	22 = (63 – 41)
60-70	15 = (41 – 26)
70-80	8 = (26 – 18)
80-90	11 = (18 – 7)
90-100	7 = (7 – 0)

EXAMPLE 5. *The following data gives the marks scored by 378 students in an entrance examination*

Marks	0-10	10-20	20-30	30-40	40-50	50-60	60-70	70-80	80-90	90-100
No. of students	3	12	36	76	97	85	39	12	12	6

SOLUTION. (i) Less than cumulative frequency table :

Marks obtained	No. of students
Less than 10	3
Less than 20	15
Less than 30	51
Less than 40	127
Less than 50	224
Less than 60	309
Less than 70	348
Less than 80	360
Less than 90	372
Less than 100	378

(ii) More than cumulative frequency table :

Marks obtained	No. of students
More than 0	378
More than 9	375
More than 19	363
More than 29	327
More than 39	257
More than 49	154
More than 59	69
More than 69	30
More than 79	18
More than 89	6

EXAMPLE 6. *Find the unknown entries from the following frequency distribution table of marks of 24 students in a class*

Marks	Frequency	Cumulative Frequency
60-75	2	a
75-90	b	6
90-105	6	c
105-120	d	14
120-135	e	20
135-150	4	f
	Total = g	

SOLUTION. Since the given frequency distribution is of the marks of 24 students in a class

$\therefore$ $g = 24$

Now, $a = 2 ; b = 6 - a = 6 - 2 = 4 ; c = 2 + b + 6 = 2 + 4 + 6 = 12$
$$d = 14 - (2 + b + 6) = 14 - (2 + 4 + 6) = 14 - 12 = 2$$
$$e = 20 - (2 + b + 6 + d) = 20 - (2 + 4 + 6 + 2) = 20 - 14 = 6$$
$$f = 2 + b + 6 + d + e + 4 = 2 + 4 + 6 + 2 + 6 + 4 = 24$$
Hence, $a = 2, b = 4, c = 12, d = 2, e = 6, f = 24 \text{ and } g = 24$

Exercise 17.2

1. Convert the following "Less than" cumulative frequency series into simple frequency series:

Marks	Less than 5	Less than 10	Less than 15
Frequency	6	15	20
Marks	Less than 20	Less than 25	Less than 30
Frequency	27	35	48

2. Draw a cumulative frequency distribution :

Score	20-30	30-40	40-50	50-60
No. of students	20	35	40	32
Score	60-70	70-80	80-90	90-100
No. of students	24	27	18	34

3. The following is the distribution weights (in Kg) of 40 persons :

Weight (in Kg)	No. of persons
40-45	4
45-50	4
50-55	13
55-60	5
60-65	6
65-70	5
70-75	2
75-80	1
Total	40

(i) Determine the class mark of the class 40-45, 45-50.

(ii) Construct the cumulative frequency distribution table.

4. Find the unknown entries (a, b, c, d, e, f) from the following frequency distribution of heights of 50 students in a class :

Class intervals (Height in cm)	Frequency	Cumulative Frequency
150-155	12	a
155-160	b	25
160-165	10	c
165-170	d	43
170-175	e	48
175-180	2	f
Total	g	

5. Prepare a frequency distribution from the following data :

Mid-points	Frequency
5	4
15	8
25	15
35	20
45	13
Total	60

Also find out :

 (i) What is the size of class intervals?

 (ii) What is the class mark of the interval 20-30?

 (iii) What is the lower limit of the third class interval?

6. Following are the ages of 360 patients getting medical treatment in a hospital on a day :

Age in years	10-20	20-30	30-40
No. of patients	90	50	60
Age in years	40-50	50-60	60-70
No. of patients	80	50	30

Construct a cumulative frequency distribution.

7. The following cumulative frequency distribution table shows the marks secured by students of X class :

Marks obtained	No. of students
More than 100	0
More than 90	4
More than 80	17
More than 70	32
More than 60	50
More than 50	61
More than 40	73
More than 30	80
More than 20	84
More than 10	90
More than 0	92

Construct a frequency table.

8. Construct a cumulative frequency table from the following data :

Marks	0-20	20-40	40-60	60-80
Frequency	3	8	12	5

ANSWERS

1. Simple frequency series

Marks	Frequency
0-5	6
5-10	9 (15 – 6)
10-15	5 (20 – 15)
15-20	7 (27 – 20)
20-25	8 (35 – 27)
25-30	13 (48 – 35)
Total	48

2.

Score	Score more than	No. of candidates	c.f.
20-30	20	20	230
30-40	35	35	210
40-50	40	40	175
50-60	32	32	135
60-70	24	24	103
70-80	27	27	79
80-90	18	18	52

3. Cumulative frequency distribution table

Age (in years)	No. of patients
Less than 20	6
Less than 30	18 = (6 + 12)
Less than 40	46 = (6 + 12 + 28)
Less than 50	64 = (6 + 12 + 28 + 18)
Less than 60	74 = (6 + 12 + 28 + 18 + 10)
Less than 70	78 = (6 + 12 + 28 + 18 + 10 + 4)

4. $a = 12, b = 13, c = 35, d = 8, e = 5, f = 50$ and $g = 50$.

5. Frequency distribution

Class intervals	Frequency
0-10	4
10-20	8
20-30	15
30-40	20
40-50	13
Total	60

(i) Size of class interval = 10

(ii) Class mark of the interval 20-30 is 25

(iii) Lower limits of the third class interval is 20.

7. Frequency distribution table

Marks	No. of students
0-10	2
10-20	6
20-30	4
30-40	7
40-50	12
50-60	11
60-70	18
70-80	15
80-90	13
90-100	4

6.

Age (in years)	No. of students (f)	c.f.
10-20	90	90
20-30	50	140
30-40	60	200
40-50	80	280
50-60	50	330
60-70	30	360
Total	$\Sigma f = 360$	

8. Cumulative frequency table

Marks	Frequency	Cumulative frequency
0-20	3	3
20-40	8	11 (3 + 8)
40-60	12	23 (3 + 8 + 12)
60-80	5	28 (3 + 8 + 11 + 5)

Objective Evaluation

⚙ MULTIPLE CHOICE QUESTIONS

Choose the most appropriate one.

1. For the mid-values given below:
   25, 34, 43, 53, 61, 70
   The first class of the distribution is
   (a) 24.5-34.5   (b) 25-34
   (c) 20-30   (d) 20.5-29.5

2. In an exclusive type distribution, the limits excluded are :
   (a) lower limits
   (b) upper limits
   (c) either of the lower or upper limit
   (d) lower limit and upper limit both

3. Class interval is measured as :
   (a) the sum of the upper and lower limit
   (b) half of the sum of lower and upper limit
   (c) half of the difference between upper and lower limit
   (d) the difference between upper and lower limit

4. A simple table represents :
   (a) only one factor or variable
   (b) always two factors or variables
   (c) two or more numbers of factors or varaibles
   (d) all the above

5. A frequnecy distribution can be :
   (a) discrete
   (b) continuous
   (c) both (a) and (b)

6. In an individual series, each variate value :
   (a) has same frequency
   (b) has frequency one
   (c) has varied frequency
   (d) has frequency two

7. Frequency of a variable is always :
   (a) in percentage   (b) a fraction
   (c) an integer   (d) none of the above

8. Classification is applicable in case of :
   (a) quantitative characters
   (b) qualitative characters
   (c) both (a) and (b)
   (d) none of the above

9. The data given as 5, 7, 12, 17, 79, 84, 91 will be called as :
   (a) a continuous series
   (b) a discrete series
   (c) an individual series
   (d) time series

10. The following frequency distribution

x :	12	17	24	36	45	48	52
f :	2	5	3	8	9	6	1

   is classified as :
   (a) continuous distribution
   (b) discrete distribution
   (c) cumulative frequency distribution
   (d) none of the above

(d) none of (a) and (b)

⚙ TRUE/FALSE

Write 'T' for True and 'F' for False statement.

1. Tabulation follows classification. **(T/F)**
2. A general table in a repository of large amount of data. **(T/F)**
3. Percentage frequency is the relative frequency multiplied by 10. **(T/F)**
4. Frequencies added successively in an ordered series giving the number of items upto that values are called cumulative frequency.**(T/F)**
5. A frequency distribution with upper limits of classes and corresponding cumulative frequencies is known as more than type distribution. **(T/F)**
6. A frequency distribution with lower limits of classes and corresponding cumulative frequencies is known as more than type distribution. **(T/F)**
7. The graphs of less than type and more than type distribution intersect at median. **(T/F)**
8. A grouped series, in which either the lower limit of the first group or the upper limit of the last group is missing or both is called an open end series. **(T/F)**

9. A series arranged in accordance with each and every observation is known as individual series. **(T/F)**

10. The distribution of frequencies according to individual variable values is called discrete frequency distribution. **(T/F)**

⇨ FILL IN THE BLANKS

1. Classification in the _____ of facts that are distinguished by some significant_____.

2. For a good classification, the class should be _____ and _____.

3. Classification can be done according to _____.

4. Quantitative classification leads to _____.

5. The Census data published for citywise population in India will be known as _____ classification.

6. Class boundaries are also sometimes called _____ limits.

7. Mid-values of the classes are also called _____.

8. An arrangement of data in rows and columns is known as _____.

9. Tables helps in _____ of data.

10. Tabulation makes the data easily _____.

ANSWERS

⇨ MULTIPLE CHOICE QUESTIONS

1. (d) **2.** (c) **3.** (d) **4.** (a) **5.** (c) **6.** (b) **7.** (c) **8.** (c) **9.** (c)
10. (b)

⇨ TRUE/ FALSE

1. T **2.** T **3.** F **4.** T **5.** F **6.** T **7.** T **8.** T **9.** T
10. T

⇨ FILL IN THE BLANKS

1. grouping; characteristic **2.** exhaustive, mutually exclusive **3.** attributes
4. frequency distribution **5.** geographical **6.** mathematical **7.** class marks
8. tabulation **9.** analysis **10.** understandable

❑❑❑❑❑

CHAPTER 18

Graphical Representation of Data

18.1 INTRODUCTION

In general, all of us are familiar with the word 'graph'. Firstly, when we come in contact with the tabular form of frequency distribution, we are introduced with graph. "To draw the given data, with the help of lines, curves and bars, etc. are known as the graphical representation of data." A graph represents the nature of the data at each of its point. This is the easiest way to understand any data properly.

The diagrams are mainly distributed in the following categories.

18.2 TYPES OF DIAGRAMS

1. **One dimensional diagram:** The diagram or graphs which are represented in one dimension only such as the line graph or bar graph is known as one-dimensional diagram.
2. **Two dimensional diagram:** The diagram which are represented in two dimensions such as circle or rectangular form, etc., are known as two-dimensional diagram.
3. **Three dimensional diagram:** The diagrams which are represented in three dimensions such as cube, sphere in the proportional size is known as three-dimensional diagram.
4. **Cartogram:** These are known as the map-diagram. It is the geographical representation of the given distribution.
5. **Pictogram:** In this diagram, the data is represented by the picture.
   The other well-known diagrams are as follows :
6. **Histogram:** These graphs represent the time series. In these types of graphs, the data is plotted against the corresponding period.
7. **Line graph:** Line graph is the graph in which we mark a point corresponding to the given data and then draw a line by joining them.
8. **Pie graph:** In the pie graph, we use a circle divided into the required number of parts to represent the data.

18.3 ONE-DIMENSIONAL DIAGRAM

One-dimensional diagram or bar diagram is further divided into the following four categories :

(a) **Simple bar diagram:** These are the one-dimensional simple diagrams. These are used when the data corresponding to different classes do not coincide. In this, we draw the bars of equal width corresponding to the required length. Here, the scale must be shown on the top of the bar and the vertical scale must start with zero.

For example. Draw a simple bar diagram for the number of passed students from a commerce college :

Year	1994	1996	1998	2000	2002	2004
No. of students	200	400	600	850	1200	1650

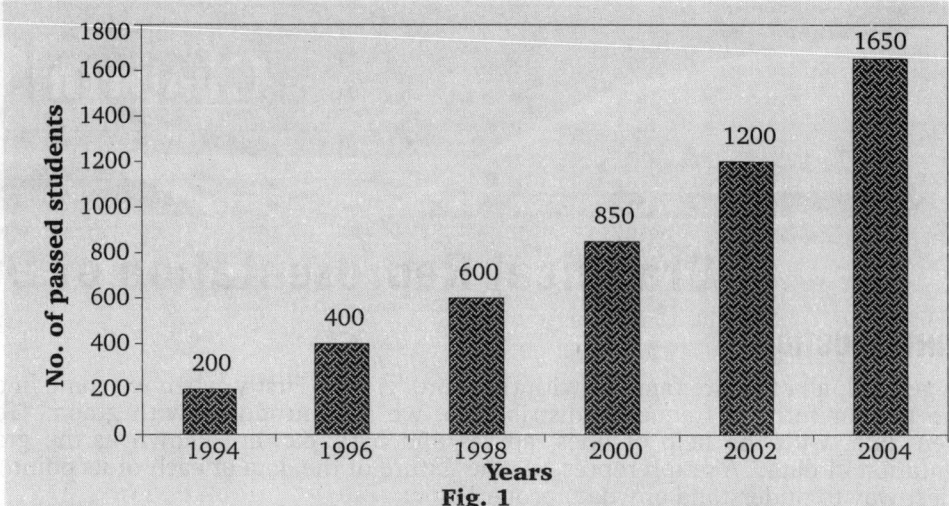

Fig. 1

(b) **Multiple bar diagram:** Multiple bar diagram is the extended form of simple bar diagram. In these diagrams, we can represent more than one aspects of data. We can separate these simultaneous bars by colouring them with different shades. This is a easy way to compare the different phenomena represented simultaneously. A table consisting of used colour or shades must be shown in the diagram.

For example. Draw a multiple bar diagram for the number of students passed competitive examinations as follows :

Year	1975	1976	1977	1978
Maths	250	280	310	310
Science	300	290	350	320
Commerce	350	320	400	400

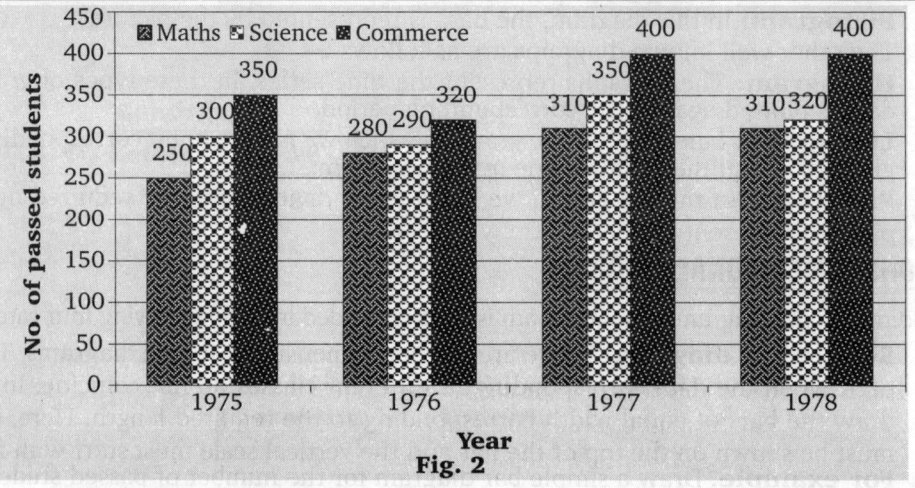

Fig. 2

(c) **Sub-divided bar diagram:** Sub-divided bar diagrams are used when the given phenomenon can be divided into different and distinct components and the summation of these different components must be equal to the total phenomenon. The length of the bars should be proportional to the different components. These components must be shown by using different colours and shades. An index consisting of used colours and shades should be shown with the diagram.

For example. Draw a multiple bar diagram for the number of students passed competitive examinations as follows :

Year	1975	1976	1977	1978	1979
Maths	250	250	275	280	150
Science	350	320	350	390	250
Commerce	470	400	425	450	400
Total	1070	970	1050	1120	800

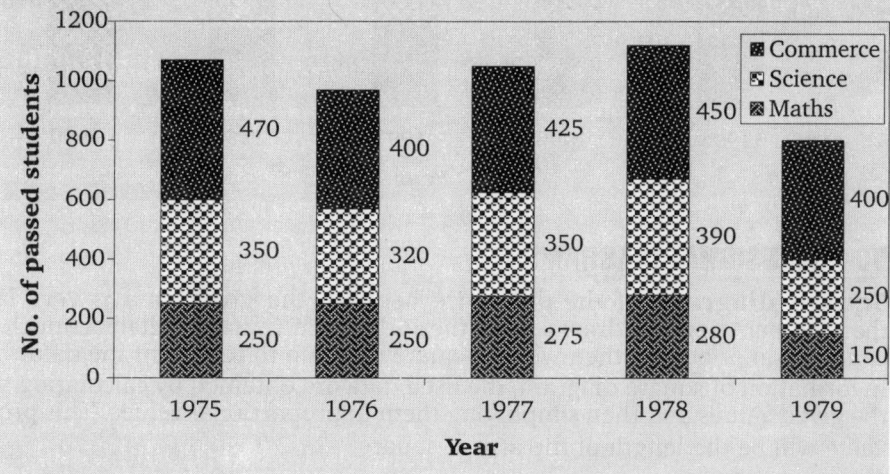

Fig. 3

(d) **Percentage sub-divided bar diagram:** If we want to compare the variation of the percentages of the component parts from the different figures, then the percentage sub-divided bar diagrams are used. In these diagrams, we assume that each total will be equal to 100 and the components shown in the figure will be the percentage value of the components. The length of each bar will be the same. An index consisting the used shades and colours must be shown in the diagram. The bars will be drawn same as in subdivided bar diagrams.

For example. Draw a multiple bar diagram for the number of students passed competitive examinations as follows :

Year	1995	1996	1997	1998	1999
Maths	250	250	275	280	150
Science	350	300	350	390	250
Commerce	470	390	425	450	400
Total	1070	940	1050	1120	800

Solution. First, convert this absolute table into the percentage table which is as follows :

Year	1995	1996	1997	1998	1999
Maths	23.36	26.60	26.19	25.00	18.75
Science	32.71	31.91	33.33	34.82	31.25
Commerce	43.93	41.49	40.48	40.18	50.00
Total	100	100	100	100	100

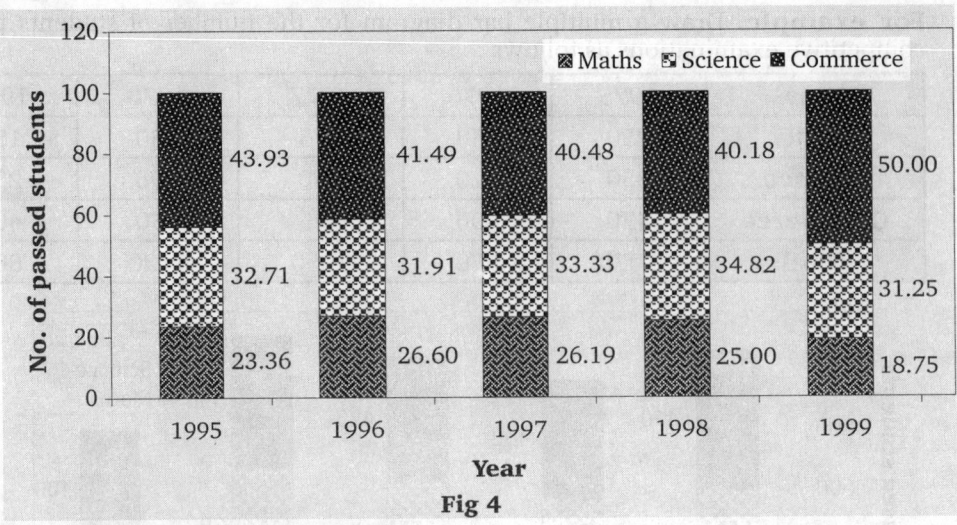

Fig 4

18.4 TWO-DIMENSIONAL DIAGRAMS

(a) **Square diagram:** If the difference between the given data is very large such that the bar diagram drawn from these data are much smaller or much larger in comparison of others, then we use square diagram to represent the data.

In formation of square diagram, the used data are obtained by calculating the root of the given values and then simplifying them in proportional values. This proportional value will be the length of the arm of square.

For example. Draw a square diagram for the data given below :

Year	1970	1980	1990	2000
No. of electrified villages	156729	256302	372603	498701

Solution.

Year	1970	1980	1990	2000
n	156729	256302	372603	498701
$\sqrt{n}$	395.89	506.26	610.412	706.18
Side of square (in cm)	1	1.30	1.5	1.8

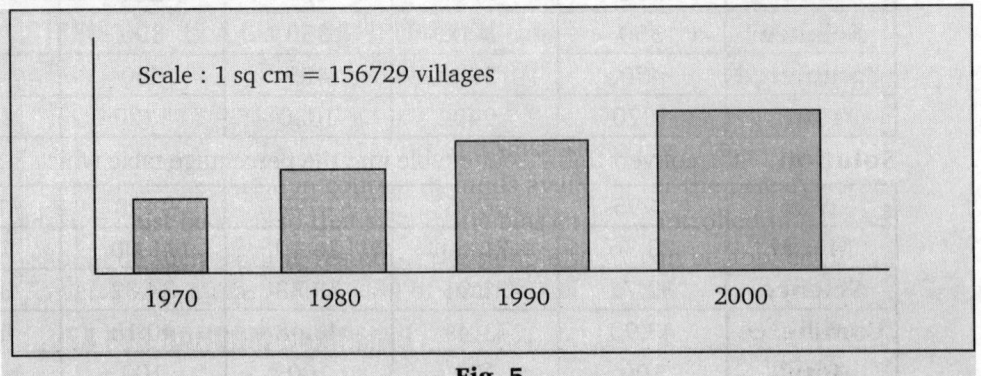

Fig. 5

(b) Rectangle diagram: Rectangle diagrams are used when the three quantities in which two are related to third one have to be drawn in the figure. In such type of graph we consider both the length and width of the bar.

For example. Draw a rectangular bar diagram for the data given below :

	Shelter	Food	Clothing	Fuel & lighting	Others
X (Earning : 200)	20	40	50	5	55
Y (Earning : 100)	15	50	25	2	8

Solution.

	Shelter	Food	Clothing	Fuel & lighting	Others
X (Earning : 200)	20	40	50	5	55
Percentage	10	35	25	2.5	27.5
Y (Earning : 100)	15	50	25	2	8
Percentage	15	50	25	2	8

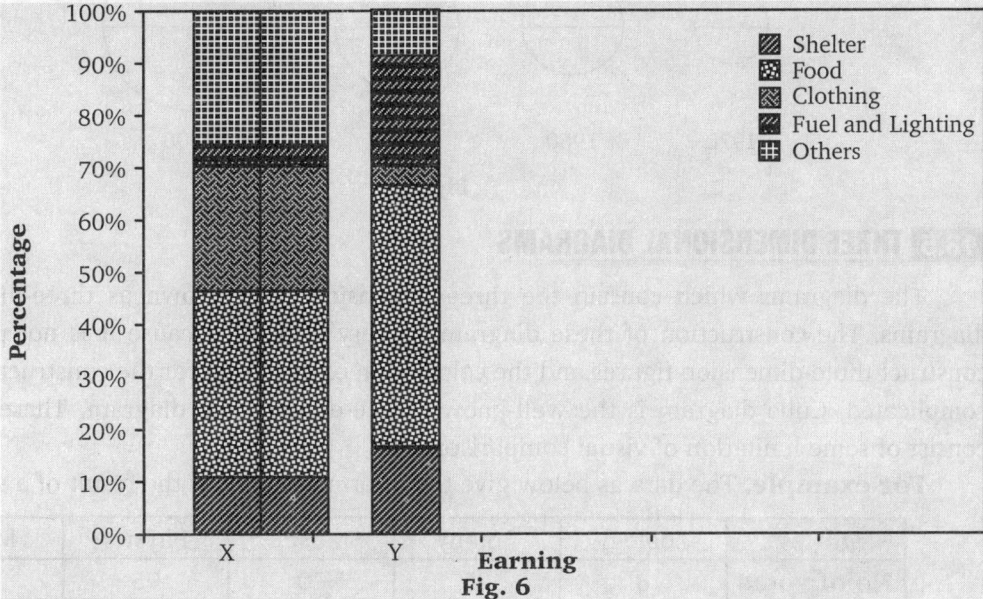

Fig. 6

(c) Circle diagram: When the difference between the given data is very large such that the bar diagrams drawn from these data are much smaller or much larger in comparison of other, then we use the circular diagram. In circle diagrams, the radii of the circles will be proportional to the square root of the given data. The centre of all the circles will lie in the straight line. If the summation of different components gives the value of the given total, then this circle diagram can be divided into the different sectors.

The value of the angle for different sectors are given by the following formula :

$$\text{Angle of sector} = \frac{\text{Value of the component}}{\text{Total value}} \times 360$$

For example. Draw a circle diagram for the data given below :

Year	1970	1980	1990	2000
No. of electrified villages	156729	256302	372603	498701

Solution.

Year	1970	1980	1990	2000
n	156729	256302	372603	498701
$\sqrt{n}$	395.89	506.26	610.412	706.18
Proportional value of radii of circle	0.7	0.9	1.1	1.3

where scale is given by $\dfrac{156729}{\pi(0.7)^2} = 101864.68$ villages.

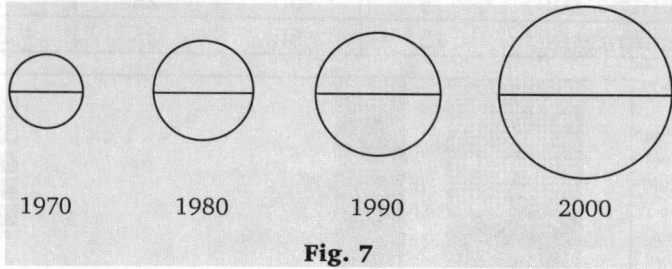

1970 1980 1990 2000

Fig. 7

18.5 THREE DIMENSIONAL DIAGRAMS

The diagrams which contain the three dimensions, are known as three-dimensional diagrams. The construction of these diagrams is very difficult because it is not possible to construct three-dimension figures and the calculation of third root for the construction is very complicated. Cube diagram is the well-known three-dimensional diagram. These diagrams consist of some limitation of visual comparison also.

For example. The data as below give the information about the result of a survey :

Subject	Zoology	Botany	Statistics	Physics	Maths
No of votes	8	25	70	95	125

Solution.

Subject	Zoology	Botany	Statistics	Physics	Maths
n	8	25	70	95	125
$\sqrt[3]{n}$	2	2.923	4.121	4.564	5
Proportional values of the side of cube	1	1.5	2	2.3	2.5

Scale : 1 cube cm = 8 votes

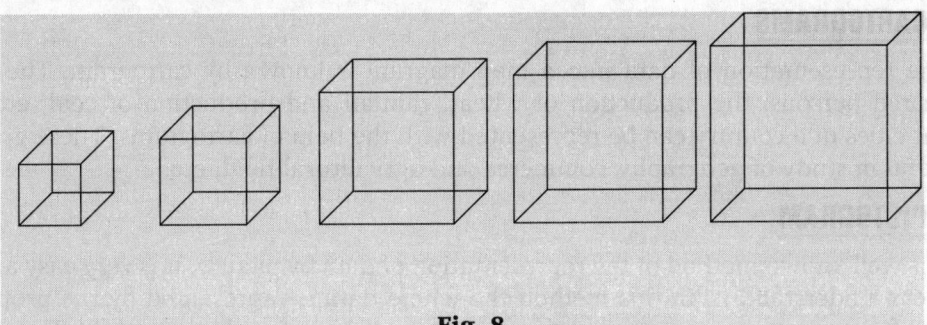

Fig. 8

18.6 PIE DIAGRAMS

When a circle diagram is divided into the required number of sectors in which the area of each sector will be proportional to the corresponding value of component. Then each sector looks like a pie and such diagrams are known by Pie diagrams. The angle of each sector can be given by the following formula

$$\text{Angle of sector} = \frac{\text{Value of component}}{\text{Total value}} \times 360$$

For example. The monthly average expenditure of a lower class family is given as follows:

Item	Milk	Ghee	Vegetables	Sugar	Fruits	Others
Expenditure	42	30	52	20	28	78

Draw a pie diagram.

Solution.

Item	Expenditure (n)	Angle
Milk	42	60
Ghee	30	43
Vegetable	52	75
Sugar	20	29
Fruits	28	40
Others	78	113
	250	360

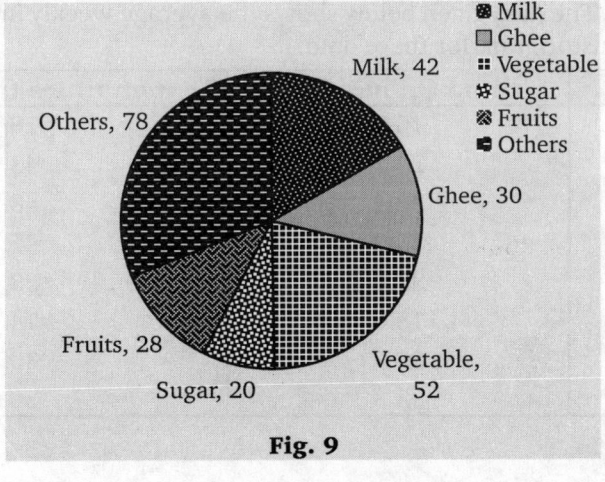

Fig. 9

18.7 CARTOGRAMS

The representation of data into a map diagram is known by cartogram. The number of educated persons, the production of wheat, rainfall and production of coal, etc. in the different cities of a country can be represented with the help of cartograms. These graphs are very useful in study of geography, commerce and agricultural field, etc.

18.8 PICTOGRAM

It is well known method of the representation of data by picture. It is very easy and every person can understand it. In this method the whole data is represented by the proportional number of pictures or we can say that the figures can be compared by means of the number of pictures. In many institutions this method of comparison is frequently used.

For example. Draw a pictogram for the data given as follows:

Age group	15-18	18-21	21-24
Married women	4000	13000	20000

Solution. Each picture represents 1000 married women.

Age group	Married women
15-18	
18-21	
21-24	

18.9 HISTORIGRAMS

Historigrams are the time series graphs. In these graphs, the time series is plotted at the x-axis and the corresponding data are plotted on y-axis. Then we join the plotted points by the straight line or a curve. By these graphs, the comparison between two different data can easily be observed.

For example: The data given below shows the average weekly income and expenditure of a labour. Draw a historigram for these data.

Year	Income (in hundreds)	Expenditure (in hundreds)
1942-43	16.9	28.1
1943-44	24.1	43.1
1944-45	32.4	48.5
1945-46	35.0	47.3
1946-47	33.0	33.1
1947-48	18.6	14.1
1948-49	35.9	30.8
1949-50	33.3	30.0

Solution.

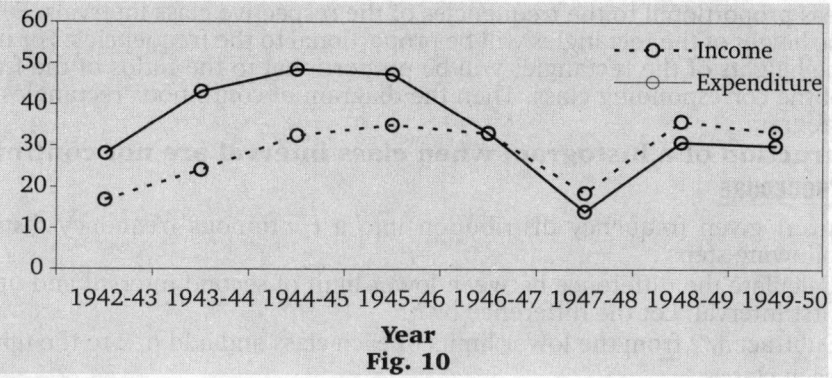

Year
Fig. 10

18.10 LINE GRAPHS

When the given data are continuous, then we plot the points corresponding to the given data and join them by straight lines. This graph is very useful to find the intervening value of the data.

If we want to find any intervening value, then we draw a perpendicular from this point on y-axis and this length of the y-axis will be the required value. This procedure of finding the values is known as interpolation.

If we want to find any value which is beyond the given series, then we increase the graph continuously and then draw a perpendicular. This procedure is known by extrapolation.

For example: Average gold prices in India in the year 1989 is given as follows:

Month	Jan	Feb	Mar	April	May
Price	2700	2900	2800	3200	3600

Draw a line graph.

Solution.

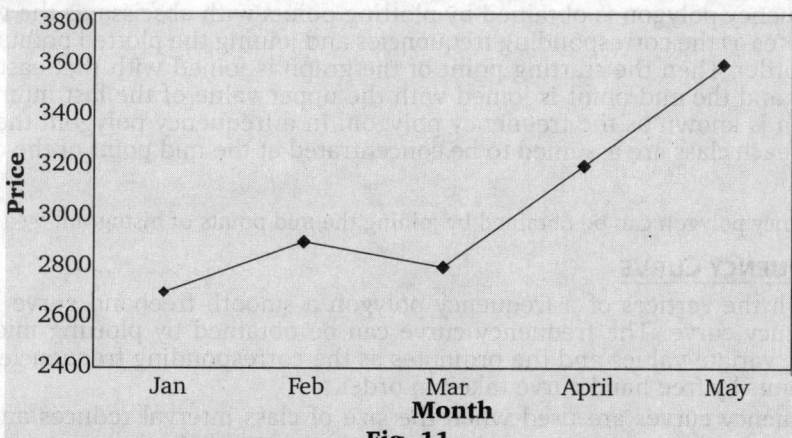

Month
Fig. 11

18.11 GRAPHS OF FREQUENCY DISTRIBUTIONS

It is often useful to represent a frequency distribution by means of a diagram which makes the unwisely data intelligible and conveys to the eye the general run of the observations. When data of two items is compared with one another, it is always easier to compare through graphs and diagrams. Here, we consider some important types of graphic representations.

18.11.1 HISTOGRAM

In drawing the histograms of a given grouped frequency distribution, first we mark off along a horizontal base line all the class intervals using a suitable scale, then draw rectangles with the areas proportional to the frequencies of the respective class intervals. For equal class intervals, the height of the rectangles will be proportional to the frequencies. For unequal class intervals, the heights of the rectangles will be proportional to the ratios of the frequencies to the width of the corresponding class. Then the diagram of continuous rectangles so obtained is called histogram.

Construction of a histogram when class interval are not continuous

WORKING PROCEDURE

To convert given frequency distribution into a continuous frequency distribution, we adopt the following steps :

Step 1. Calculate the difference between lower limit of second interval and upper limit of first interval. Let the difference by h.

Step 2. Subtract $h/2$ from the lower limits of each class and add $h/2$ to the upper limits of each class.

⮂ REMARKS

⮕ Histograms are useful when the class intervals are not of the same width. They are appropriate to cases in which the frequency changes rapidly.

⮕ To draw the histogram for an ungrouped frequency distribution of a variable, assume that the frequency corresponding to the variate value x is distributed over the interval $\left(x - \dfrac{h}{2} \right)$ to $\left(x + \dfrac{h}{2} \right)$, where h is the jump from one value to the next.

⮕ If the grouped frequency distribution is not continuous, convert it into continuous distribution and then draw the histogram.

⮕ The height of each rectangle is proportional to the frequency of the corresponding class, the height of a fraction of the rectangle is not proportional to the frequency of the corresponding fraction of the class, therefore, the histogram can not be directly used to read frequency over a fraction of a class interval.

18.11.2 FREQUENCY POLYGON

The frequency polygon is obtained by plotting points with abscissa as the variate values and the ordinates as the corresponding frequencies and joining the plotted points by a straight line taken in order. Then the starting point of the graph is joined with the least value of the lower interval and the end point is joined with the upper value of the last interval. Now the obtained graph is known as the frequency polygon. In a frequency polygon, the variables or individuals of each class are assumed to be concentrated at the mid point of the class interval.

⮂ REMARK

⮕ The frequency polygon can be obtained by joining the mid points of histogram by a straight line.

18.11.3 FREQUENCY CURVE

If through the vertices of a frequency polygon a smooth freehand curve is drawn, we get the frequency curve. The frequency curve can be obtained by plotting mid points with abscissa as the variate values and the ordinates as the corresponding frequencies and joining the plotted points by free hand curve taken in order.

The frequency curves are used when the size of class interval reduces and due to this number of observations increases and the frequency polygon becomes more close. Then it is easy to draw a smooth free hand curve.

18.11.4 CUMULATIVE FREQUENCY CURVE OR OGIVE

If the limits of the class taken as x-coordinate and the cumulative frequencies as the y-coordinate and the points are plotted, then these points, when joined by a freehand smooth curve gives the cumulative frequency curve or the ogive.

These are of two types :

(1) Less than ogive: If the less than cumulative frequency is plotted against the upper limit of the intervals, then the joined freehand smooth curve is known as less than ogive.

WORKING PROCEDURE

Step 1. Convert the frequency from inclusive to exclusive form.
Step 2. Obtain the cumulative frequency distribution.
Step 3. Mark upper class limits along *x*-axis.
Step 4. Mark cumulative frequencies along *y*-axis.
Step 5. Plot the point and join them by a free hand curve.

(2) More than ogive: If the more than cumulative frequency is plotted against the lower limit of the intervals, then the joined freehand smooth curve is known as more than ogive.

WORKING PROCEDURE

Step 1. Convert the frequency distribution into more than type cumulative frequency distribution, subtracting the frequency of each class.
Step 2. Mark the lower class on *x*-axis.
Step 3. Mark the cumulative frequency on *y*-axis.
Step 4. Plot the points and join them by a free hand curve.

REMARK

➠ We can also find the values of the quartiles from the ogive by drawing perpendicular to the corresponding value such as $N/4$, $N/2$, $3N/4$, ..., etc.

18.12 TYPES OF FREQUENCY CURVE

(1) U shaped frequency curve: When the frequency curve looks like *U* and exhibit maximum frequency at the end point and minimum frequency at the mid point of the curve is known by the *U*-shaped frequency curve.

(2) J shaped frequency curve: The frequency curves which looks like *J* and with the increase in variate, the frequency of the variables continuously decrease or increase is known by the *J*-shaped frequency curve.

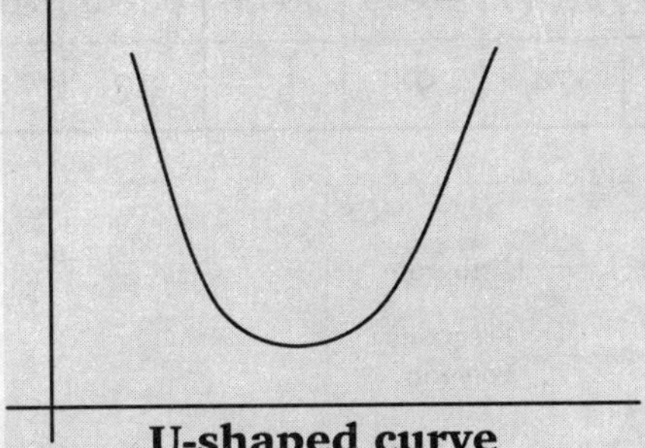

U-shaped curve
Fig. 12

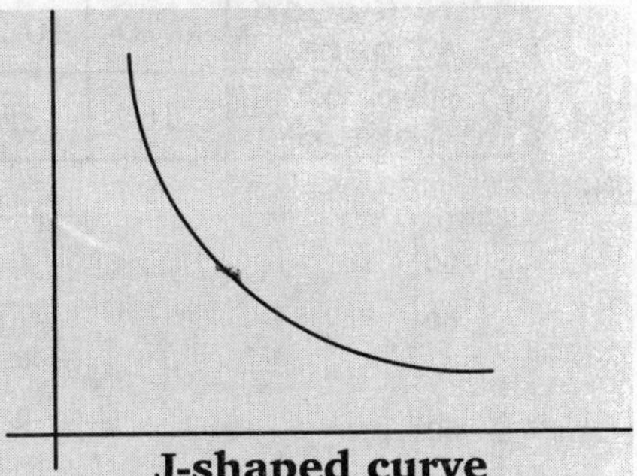

J-shaped curve
Fig. 13

(3) Unimodal frequency curve: A curve having only one maximum point is known by unimodal frequency curve.

(4) Bimodal frequency curve: A curve having two maximum point is known by bimodal frequency curve.

(5) Multimodal frequency curve: A curve having more than two maximum point is known by multimodal frequency curve.

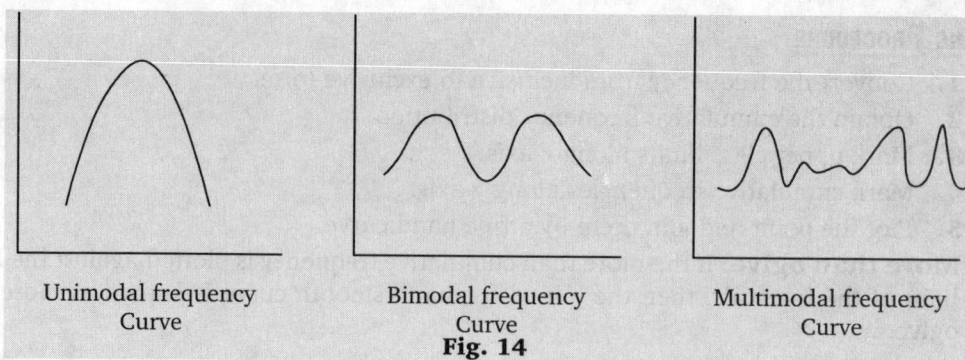

Unimodal frequency Curve   Bimodal frequency Curve   Multimodal frequency Curve

Fig. 14

(6) Symmetrical frequency curve: The frequency curves are said to be symmetrical in which the frequency increases with the increase in variate upto the maximum value and after obtaining the maximum value it starts to decrease with increase in variable. The curve will be symmetrical about the line of maximum frequency.

(7) Asymmetrical frequency curve: A curve which is not symmetrical about the line of maximum frequency is known as the asymmetrical frequency curve. It is also known as skewed curve.

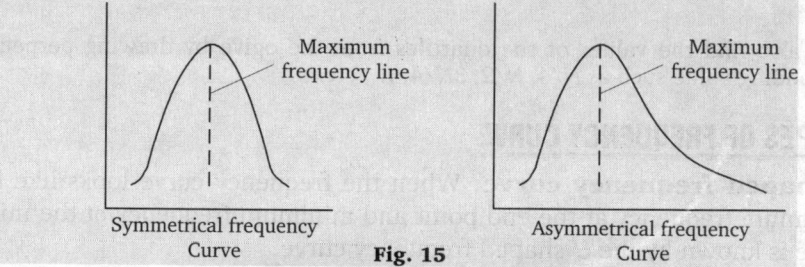

Symmetrical frequency Curve   **Fig. 15**   Asymmetrical frequency Curve

SOLVED EXAMPLES

EXAMPLE 1. *Draw a histogram and frequency polygon for the data given as follows :*

Production (tonnes)	0-10	10-20	20-30	30-40	40-50	50-60	60-70	70-80
No. of workers	16	36	46	74	94	52	32	10

SOLUTION.

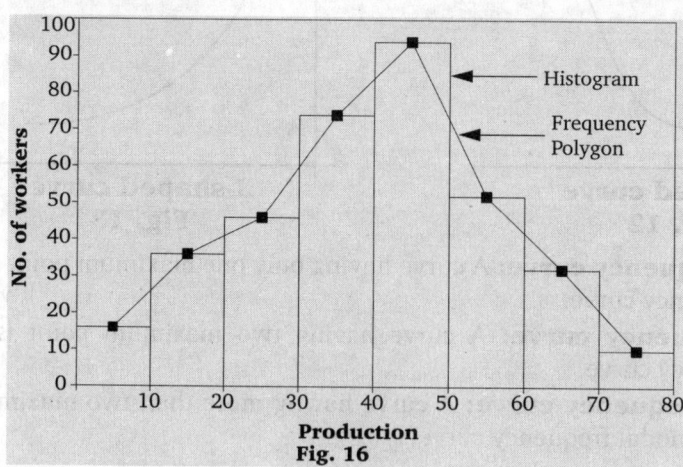

Fig. 16

EXAMPLE 2. *Draw a line diagram for the data given as follows :*

Oil seeds	Groundnut	Pine	Olive	Mustard	Custard
Production	75	115	90	277	103

SOLUTION.

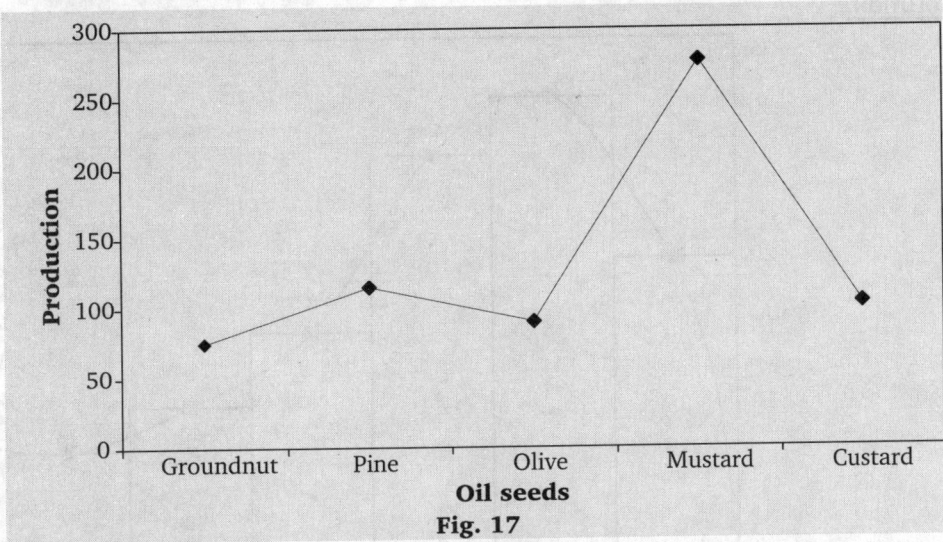

Fig. 17

EXAMPLE 3. *Draw a frequency curve for the data given below :*

Class interval	0-10	10-20	20-30	30-40	40-50	50-60
Frequency	25	35	70	55	30	20

SOLUTION.

Mid Value	5	15	25	35	45	55
Frequency	25	35	70	55	30	20

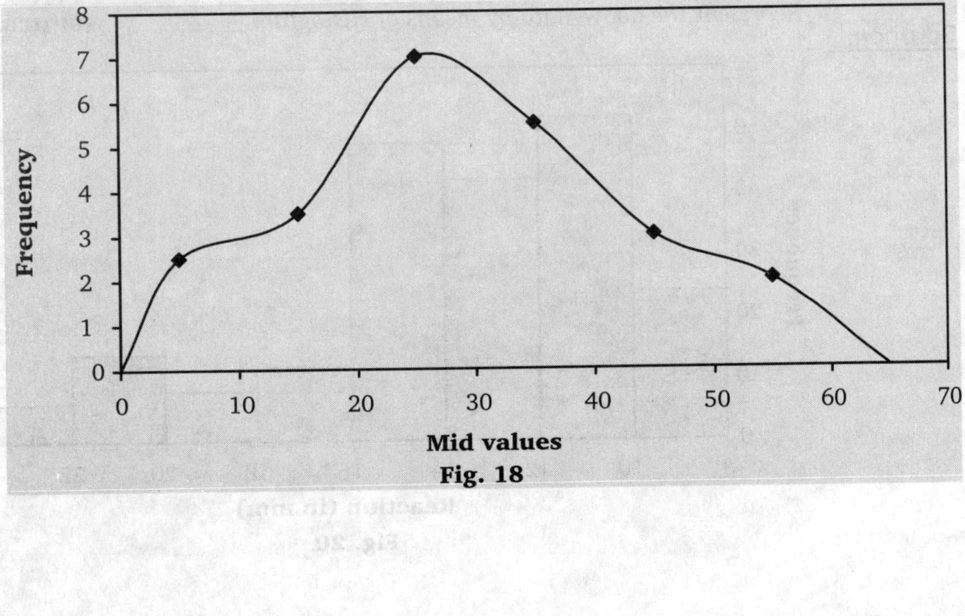

Fig. 18

EXAMPLE 4. *Draw a histogram and smooth frequency curve for the given frequency distribution:*

x	1-9	9-17	17-25	25-33	33-41	41-49	49-65
f	20	31	27	15	10	7	8

SOLUTION.

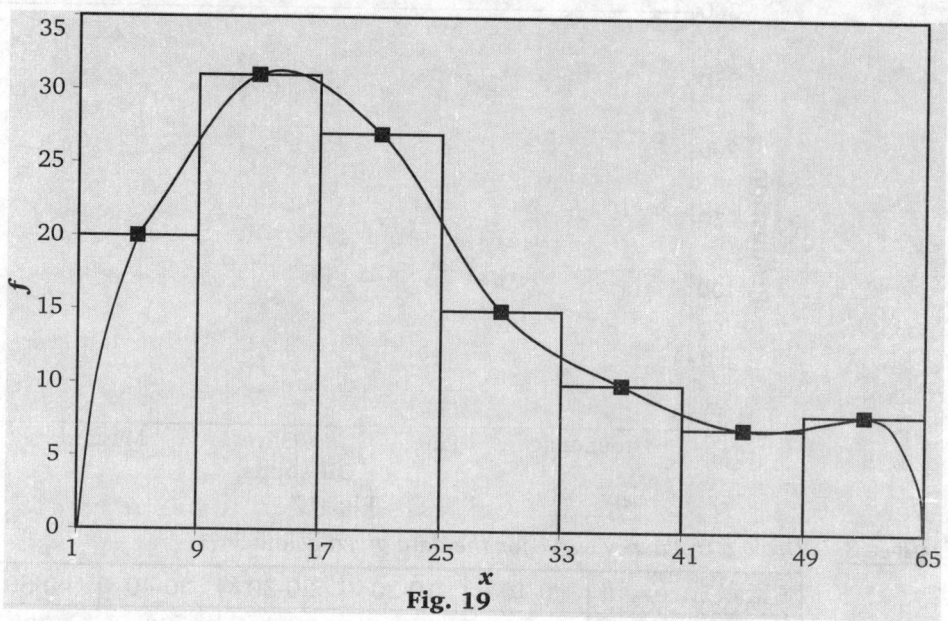

Fig. 19

EXAMPLE 5. *Tuberculin reaction measures in 206 persons is as follows :*

Reaction (in mm)	8-10	10-12	12-14	14-16	16-18	18-20	20-22	22-24
Frequency	24	52	42	48	12	08	14	06

Represent the above data by means of Histogram. **[UPTU(B.Pharma)–2001]**

SOLUTION.

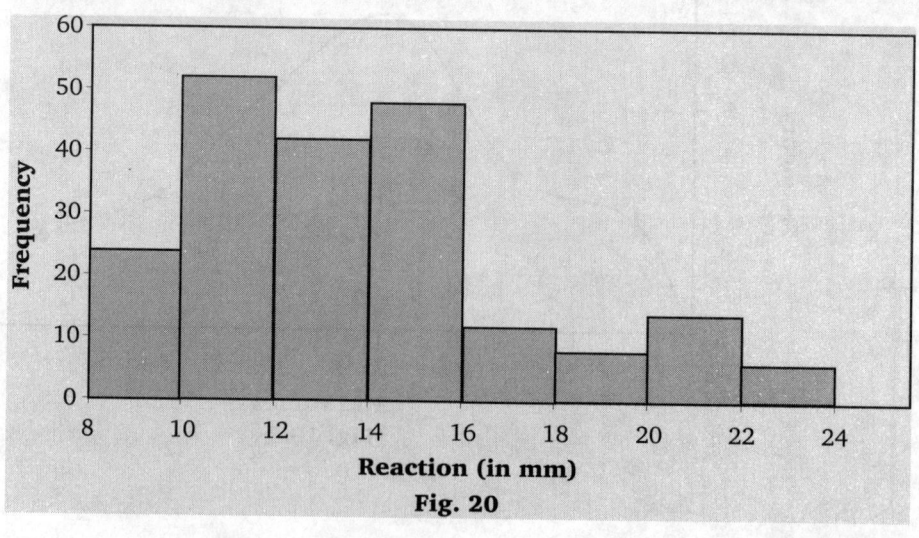

Fig. 20

Exercise 18.1

1. Construct a histogram for the following data :

Monthly school fee (in Rs.)	30-60	60-90	90-120	120-150
No. of Schools	5	12	14	18
Monthly school fee (in Rs.)	150-180	180-210	210-240	
No. of Schools	10	9	4	

2. The following table shows the number of illiterate persons in the age group (10-58 years) in a town :

Age group (in years)	10-16	17-23	24-30	31-37
No. of illiterate persons	175	325	100	150
Age group (in years)	38-44	45-51	52-58	
No. of illiterate persons	250	400	525	

Represent the data in the form of a histogram.

3. The following table gives the marks scored by 100 students in an entrance examination :

Marks	0-10	10-20	20-30	30-40
No. of students	4	10	10	22
Marks	40-50	50-60	60-70	70-80
No. of students	20	18	8	2

Represent the data in the form of a histogram.

4. Construct a frequency polygon from the following table :

Score	32-34	35-37	38-40
Frequency	13	10	20
Score	41-43	44-46	47-49
Frequency	10	12	8

5. Construct a frequency polygon from the following data :

Monthly pocket expenses of a student in Rs.	No. of students
0-5	10
5-10	16
10-15	30
15-20	40
20-25	50
25-30	30
30-35	16
35-40	12

6. Draw a histogram for the following data :

Class intervals	Frequency
50-54	4
55-59	9
60-64	17
65-69	22
70-74	18
75-79	8
80-84	2

7. Draw an ogive to represent following data, showing monthly cost of living index of a city :

Cost of living index	No. of months in a period
340-350	10
350-360	19
360-370	24
370-380	18
380-390	16
390-400	13
Total	100

8. The following is the distribution of total household expenditure (in Rs.) of workers in a city :

Expenditure (in Rs.)	Number of workers
100-150	25
150-200	40
200-250	33
250-300	28
300-350	30
350-400	22
400-450	16
450-500	8

9. Draw a frequency polygon of the following data :

Marks	20-25	25-30	30-35
No. of students	5	8	15
Marks	35-40	40-45	45-50
No. of students	6	4	2

10. Draw a bar diagram for the data given as follows :

Year	1971-72	1972-73	1973-74
General index	108	122	158
Year	1974-75	1975-76	1976-77
General index	163	191	229

11. Draw a bar diagrams of the data given as follows :

Year	1891	1901	1911	1921
Population (in crore)	23.59	23.83	25.20	25.13
Year	1931	1941	1951	1971
Population (in crore)	27.89	31.86	36.10	54.82

12. Draw the multiple bar diagram of the gross income and payment given as follows :

Year	1978	1981	1984
Gross income (in crore)	380	430	700
Payments (in crore)	230	300	500

Year	1987	1990	1993
Gross income (in crore)	1100	1180	1210
Payments (in crore)	800	920	970

13. The number of students passed the competitive examination in different subjects is given as follows. Draw a multiple bar diagram of this data.

Year	Mathematics		Biology	
	Total	Qualified	Total	Qualified
1975-76	1298	299	1020	126
1976-77	1602	430	1100	89
1977-78	1998	374	1273	162
1978-79	1782	520	1345	179

14. Draw a subdivided bar diagram of the data given below :

Year	1974-75	1975-76	1976-77
Current expenditure	580	729	409
Capital expenditure	473	669	349
Year	1977-78	1978-79	1979-80
Current expenditure	1108	1398	648
Capital expenditure	927	1068	514

15. The average monthly expenditure of food of a middle class family is given as follows. Draw a pie chart of this data.

Item	Milk	Sugar	Veg.	Fruits
Expenditure	50	26	30	20
Item	Ghee	Eggs	Tea	Others
Expenditure	26	20	22	6

Objective Evaluation

✦ MULTIPLE CHOICE QUESTIONS

Choose the most appropriate one.

1. Ogive curve occur for
   (a) more than type distribution
   (b) less than type distribution
   (c) both (a) and (b)
   (d) none of (a) and (b)

2. In an ogive curve, the points are plotted for
   (a) the values and frequencies
   (b) the valves and cumulative frequencies
   (c) frequencies and cumulative frequencies
   (d) none of the above

3. In a bar diagram, the bar line is
   (a) horizontal   (b) vertical
   (c) false base line   (d) any of the above

4. In a column chart, the base line is
   (a) horizontal   (b) vertical
   (c) at an angle of 45° (d) false base line.

5. In a column charts, bars are
   (a) horizontal   (b) vertical
   (c) slanting   (d) none of the above.

6. In a bar diagram, the bars are
   (a) horizontal   (b) vertical
   (c) slanting   (d) none of the above

7. In a case of frequency distribution with classes of unequal widths, the heights of bars of a histogram are proportional to :
   (a) class frequency
   (b) class intervals
   (c) frequencies in percentage
   (d) frequency densities

8. Year wise production of rice, wheat and maize for last ten years can be displayed by
   (a) simple column chart
   (b) sub divided column chart
   (c) broken bar diagram
   (d) multiple column chart

9. Profit and loss of a firm during various years can be displayed through.
   (a) simple bar diagram
   (b) duo-directional bar diagram
   (c) deviation bar chart
   (d) multiple bar diagram

10. Histogram is suitable for
    (a) time series data
    (b) chronological distribution
    (c) none of (a) or (b)
    (d) both (a) and (b)

✦ TRUE/FALSE

Write 'T' for True and 'F' for False statement.

1. Pictograms are non-dimensional diagrams. **(T/F)**

2. A zig-zag graph shows the fluctuations of a series. **(T/F)**

3. In rectangular diagrams, comparison is based on length of the rectangle. **(T/F)**

4. Line diagram is suitable when there too many variate values in the frequency distributions. **(T/F)**

5. Multiple bar diagrams has a group of bars for each year or place. **(T/F)**

6. When more than one factor is to be displayed for comparison during various years, multiple bar diagram is suitable. **(T/F)**

7. Two related factors having different units of measurements can be displayed suitably by line diagram. **(T/F)**

8. A graph of variate values and corresponding frequencies is known as frequency curve. **(T/F)**

9. A smoothened frequency polygon is known as frequency curve. **(T/F)**

10. The heights of bars with unequal class intervals are proportional to frequency densities. **(T/F)**

✦ FILL IN THE BLANKS

1. Diagrams are another form of _____.

2. It is comparatively easier to understand _____ than numerical figures.

3. The ogives of less than type and more than type distribution intersect at _____.

4. Histogram and histogram are _____.

5. In a histogram bars _____ each other.

6. A histogram is _____ for geographical classification.

7. Frequency polygon can be drawn with the help of _____.

8. Sub-divided bar diagram depicts the distribution of _____ of a factor.

9. _____ bar diagram is suitable for showing the differences between budget provisions and actual expenditure of PWD in the last ten years.

10. Pie-chart is always _____.

ANSWERS

∽ MULTIPLE CHOICE QUESTIONS

1. (c)   **2.** (b)   **3.** (b)   **4.** (a)   **5.** (b)   **6.** (a)   **7.** (d)   **8.** (d)   **9.** (c)
10. (d)

∽ TRUE / FALSE

1. T   **2.** T   **3.** F   **4.** T   **5.** T   **6.** T   **7.** F   **8.** F   **9.** T
10. T

∽ FILL IN THE BLANKS

1. tabulation   **2.** diagrams   **3.** median   **4.** not same   **5.** touch
6. not suitable   **7.** histogram   **8.** components   **9.** deviation
10. circular

❑❑❑❑❑❑

CHAPTER 19

Measure of Central Tendency

19.1 INTRODUCTION

In previous classes, we have studied about the classification and tabulation of data, but our study about classification and tabulation is not enough to get all the desirable results since when two or more series of same type are under observations, we cannot classify and tabulate them so we need an arithmetic idea or characteristic about the distribution. These characteristic are measure of central tendency, measure of dispersion and skewness and the peackedness.

In this chapter, we study the measure of central tendency or average according to Dr. Bowley statistics may rigidly be called the sequence of averages and averages are statistical constant which enable us to comprehend in a single effort the significance of the whole.

"Average is a point about which all the values of variate cluster."

19.2 KINDS OF STATISTICAL AVERAGES

The statistical averages can be divided into five parts.

1. Arithmetic mean (A.M.)　　　　　　2. Median (Md)
3. Mode (Mo)　　　　　　　　　　　　4. Geometric mean (G.M.)
5. Harmonic mean (H.M.)

1. Arithmetic mean

"*Arithmetic mean is the amount obtained by dividing the sum of values of the items in a series by their number.*

The arithmetic mean of some observations is the value which we can obtain by dividing the sum of all the numbers by the total number of items, *i.e.*,

$$\text{A.M.} = \bar{x} = \frac{\text{Sum of all the observations}}{\text{Total number of terms}}$$

Let $x_1, x_2, x_3, \ldots x_n$ are the observations. Then arithmetic mean is given by

$$\bar{x} = \frac{x_1 + x_2 + x_3 + \ldots\ldots + x_n}{n} = \frac{\Sigma x_i}{n}$$

$$\Rightarrow \qquad \bar{x} = \frac{\Sigma x}{n}$$

If the variate $x_1, x_2, \ldots x_n$ occurs $f_1, f_2, f_3, \ldots f_n$ times, then the arithmetic mean is known as weighted arithmetic mean and given by

$$\bar{x} = \frac{f_1 x_1 + f_2 x_2 + \ldots. + f_n x_n}{f_1 + f_2 + \ldots + f_n}, \quad \bar{x} = \frac{\Sigma f_i x_i}{\Sigma f_i}$$

19.3 METHODS OF CALCULATING ARITHMETIC MEAN IN INDIVIDUAL SERIES

We can calculate the arithmetic mean by following three methods :

1. Direct method
2. Short-cut method
3. Step-deviation method

1. Direct Method : In this method, the mean is calculated using the following formula

$$\bar{x} = \frac{x_1 + x_2 + \dots\dots + x_n}{n} = \frac{\sum\limits_{i=1}^{n} x_i}{n}$$

WORKING PROCEDURE

Step 1. Adding all the observations to find Σx_i .

Step 2. Divide this sum Σx_i by total number of observations, *i.e.*, n.

2. Short-cut Method : In this method, we assumed a middle number as an assumed mean. Here, we use the following formula

$$\bar{x} = A + \frac{\Sigma d}{n}$$

where, $\bar{x}$ = Arithmetic mean; A = Assumed mean; $d = (x - A)$; Σd = Sum of deviations

WORKING PROCEDURE

Step 1. Select the assumed mean, *i.e.*, A.

Step 2. Calculate the deviation from A, *i.e.*, $d = x - A$.

Step 3. Find the sum of deviation as Σd .

Step 4. Using the formula $\bar{x} = A + \dfrac{\Sigma d}{n}$, we get the required mean.

3. Step Deviation Method. Let us assume a number h called scale then $d' = \dfrac{x}{h}$ (x denotes the monthly income)

$$\Sigma d' = \frac{\Sigma x}{h}, \quad \Sigma x = h\Sigma d'$$

$$\frac{\Sigma x}{n} = \frac{h\Sigma d'}{n}, \quad \bar{x} = hd'$$

$$\text{A.M.} = \frac{h\,\Sigma d'}{n}$$

SOLVED EXAMPLES

EXAMPLE 1. *Calculate the arithmetic mean of*
129, 117, 112, 200, 172, 138, 183

SOLUTION. Arithmetic mean of the above data can be given by :

$$\bar{x}\,(\text{A.M.}) = \frac{129 + 117 + 112 + 200 + 172 + 138 + 183}{7} = \frac{1051}{7} = 150.14$$

So, A.M. = 150.14.

EXAMPLE 2. *Find the arithmetic mean of first n natural numbers.*

SOLUTION. Arithmetic mean can be given by :

$$\bar{x}\,(A.M.) = \frac{\text{Sum of all the observations}}{\text{Total number of terms}} = \frac{1+2+3+4+\ldots\ldots+n}{n} = \frac{\Sigma n}{n}$$

$$= \frac{1}{n}\frac{n(n+1)}{2} \qquad \left(\because \Sigma n = \frac{n(n+1)}{2} \right)$$

$$= \frac{1}{2}(n+1).$$

So the arithmetic mean of first n natural numbers is given by $\frac{1}{2}(n+1)$.

EXAMPLE 3. *Show that the arithmetic mean of the series* $1, 2, 2^2, 2^3, 2^4, \ldots 2^n$, *is given by* $(2^{n+1}-1)/n+1$.

SOLUTION. Arithmetic Mean $= \dfrac{\text{Sum of all the observations}}{\text{Total number of terms}} = \dfrac{1+2+2^2+2^3+\ldots\ldots+2^n}{n+1}$

$$= \frac{\text{Sum of the G.P.}}{n+1} = \frac{(2^{n+1}-1)}{(2-1)(n+1)} = \frac{(2^{n+1}-1)}{n+1}$$

Arithmetic mean of the given G.P. $= \dfrac{(2^{n+1}-1)}{n+1}$.

EXAMPLE 4. *A candidate obtain the following marks in an examination in a paper of 100 marks each*

Subject	English	Maths	Physics	Chemistry	Biology
Marks (out of 100)	48	82	70	64	60

It is agreed to give double weight to physics and mathematics as compared to other subjects. What is the arithmetic mean? **(UPTU B.Pharma-2005)**

SOLUTION. Since weights are given, we shall calculate weighted arithmetic mean in place of simple arithmetic mean.

Weights are given accordance with the statement given in the example.

Subject	Marks (X)	Weight (W)	WX
English	48	1	48
Maths	82	2	164
Physics	70	2	140
Chemistry	64	1	64
Biology	60	1	60
		$\Sigma W = 7$	$\Sigma WX = 476$

$$\bar{X}_W = \frac{\Sigma WX}{\Sigma W} = \frac{476}{7} = 68\%.$$

EXAMPLE 5. *A person makes a trip which entails travelling 900 kms by train at an average speed of 60 kph, 300 kms by bus at an average speed of 25 kph, 400 kms by plane at 350 kph and finally 15 kms by a taxi at 25 kph. What is the average speed of the entire distance?*

SOLUTION.

Journey by	Distance (Km)	Aveage speed (kph)	Time taken
Train	900	60	15
Bus	300	25	12
Plane	400	350	8/7
Taxi	15	25	3/5
Total	1615		1006/35

$$\text{Average speed} = \frac{\text{Distance}}{\text{Time}} = \frac{1615 \times 35}{1006}$$
$$= 56.18 \text{ kph.}$$

EXAMPLE 6. *A person travels n equal distances with vertices* $V_1, V_2, V_3, ..., V_n$. *Show that his average velocity cannot exceed* $(V_1 + V_2 + ... + V_n)/n$. *When it will be equal to this value?*

SOLUTION. Total distance = n equal distances.

Velocity per distance $V_1, V_2, V_3, ..., V_n$

$$\text{Average speed} = \frac{\text{Total distance}}{\text{Total time}}$$

Let equal distance = x

$$\text{Time} = \frac{\text{Distance}}{\text{Speed}}$$

$$\text{Per unit distance time} = \frac{\text{Speed}}{\text{Distance}} = \frac{V_1 + V_2 + V_3 + ... + V_n}{n}$$

This speed is maximum $\dfrac{V_1 + V_2 + V_3 + ... + V_n}{n} \leq$ total time

Thus maximum average speed can be $\dfrac{V_1 + V_2 + ... + V_n}{n}$.

When time taken by each distance is same with same speed then it will be equal.

EXAMPLE 7. *The mean of n numbers of a series is* $\bar{x}$ *and the sum of first* $(n-1)$ *numbers is* λ. *Find the value of the last number.*

SOLUTION. Mean of n numbers = $\bar{x}$

Sum of first $(n-1)$ number = λ

$$\text{Arithmetic Mean} = \frac{\text{Sum of all the observations}}{\text{Total number of terms}}$$

Sum of all the observations = Mean $\times$ Total no. of terms
$$= \bar{x} \times n$$

Sum of all the observations = $n\bar{x}$

Now, sum of first $(n-1)$ numbers = λ

Then the last number = (Sum of n terms) – (Sum of $n-1$ terms)
$$= n\bar{x} - \lambda$$

The value of the last number $= (n\bar{x} - \lambda)$

EXAMPLE 8. *Hemoglobin percentage (Hb%) of a patients of a ward of hospital was obtained as 6 mg, 7 mg, 5 mg, 4 mg, 8 mg, 7 mg, 9 mg, 6 mg and 8 mg. Find out the arithmetic mean of the data.*

SOLUTION. We have $\bar{x} = \dfrac{\Sigma x}{N}$

$$\bar{x} = \frac{6+7+5+4+8+7+9+6+8}{9} = \frac{60}{9} = 6.66 \ mg$$

EXAMPLE 9. *WBC's in number of 10 male frogs (Rana tigrina) are 8.19, 9.21, 10.40, 10.95, 12.14, 12.52, 13.41, 13.92, 14.78 and 15.74 lac/mm³. Find mean WBC's number.*

SOLUTION. Here, $\bar{x} = \dfrac{\Sigma x}{N}$

$$= \frac{\begin{array}{c}8.19+9.21+10.40+10.95+12.14+12.52\\+13.41+13.92+14.78+15.74\end{array}}{10}$$

$$= \frac{121.26}{10} = 12.13$$

19.4 CALCULATION OF ARITHMETIC MEAN IN A DISCRETE FREQUENCY DISTRIBUTION

1. **Direct Method:** In case of discrete frequency distribution, we multiply the values of a variable (x) by their respective frequencies (f). Then, we use the following formula

$$\bar{x} = \frac{\Sigma f.x}{\Sigma f} = \frac{\Sigma f.x}{n}$$

where, Σfx = The sum of products of observations with their respective frequencies
$\Sigma f = n$ = Total number of frequencies.

WORKING PROCEDURE

Step 1. Multiply the value of variable x by corresponding frequency f to find fx.
Step 2. Calculate sum Σfx of the product obtained in step (1).
Step 3. Putting the values in the formula

$$\bar{x} = \frac{\Sigma f.x}{\Sigma f}$$

2. **Short-cut Method :** Firstly, we shall assume a mean and then take deviation of the variable from this assumed mean. In this method, we use the following formula :

$$\bar{x} = A + \frac{\Sigma fd}{n}$$

where, A = Assumed mean; $d = x - A$; Deviation; f = Frequency

WORKING PROCEDURE

Step 1. Select the assumed mean, *i.e.,* A.
Step 2. Calculate the deviation from A, *i.e.,* $d = x - A$.
Step 3. Calculate $f.d$.
Step 4. Sum all the deviation to obtain $\Sigma f d$.
Step 5. Using the formula $\bar{x} = A + \dfrac{\Sigma fd}{n}$.

3. Step-Deviation Method : In this method, we divide our deviation by the common factor h. Therefore

$$\bar{x} = A + \frac{\Sigma fd'}{n} \times h,$$

where, h = Common factor of the deviation; $d' = \dfrac{d}{h} = \dfrac{x-A}{h}$

WORKING PROCEDURE

Step 1. Choose the assumed mean A.

Step 2. Calculate $d = x - A$.

Step 3. Calculate $d' = d / h$.

Step 4. Calculate $f.d'$

Step 5. Calculate $\Sigma f.d'$.

Step 6. Using the formula $\bar{x} = A + \dfrac{\Sigma fd'}{n} \times h$

19.5 CALCULATION OF ARITHMETIC MEAN IN A CONTINUOUS FREQUENCY DISTRIBUTION

In this case, we have to calculate the mid point of the various class intervals and denote it by x. Then, proceed same as above.

SOLVED EXAMPLES

EXAMPLE 1. *Compute the mean of the following data by direct and short-cut method:*

Height (cm)	195	198	201	204	207	210	213	216	219
Children	1	4	5	7	11	10	6	4	2

SOLUTION.

Heigh x (cm)	f	fx	$d = (x - A)$	fd
195	1	195	−12	− 12
198	4	792	−9	− 36
201	5	1005	−6	− 30
204	7	1428	−3	− 21
207	11	2277	0	0
210	10	2100	3	30
213	6	1278	6	36
216	4	864	9	36
219	2	438	12	24
Total	50	10377		27

Here, $A = 207$

By direct method, we can calculate the A.M. by the formula

$$\text{A.M.} \ \frac{\Sigma fx}{N} = \frac{10377}{50} = 207.54 \text{ cm}$$

By short-cut method, the formula is given by

$$\text{A.M.} = A + \frac{\Sigma fd}{N} = 207 + \frac{27}{50} = 207 + 0.54 = 207.54.$$

EXAMPLE 2. *Compute the mean of the following distribution by step deviation method:*

Class	0 - 11	11-22	22-33	33-44	44-55	55-66
Frequency	9	17	28	26	15	8

SOLUTION.

Class	Mid Value x	f	$(x - 38.5)$	$d' = \dfrac{x - 38.5}{11}$	fd'
0-11	5.5	9	−33	−3	−27
11-22	16.5	17	−22	−2	−34
22-33	27.5	28	−11	−1	−28
33-44	38.5	26	0	0	0
44-55	49.5	15	11	1	15
55-66	60.5	8	22	2	16
Total		$\Sigma f = 103$			$\Sigma fd' = -58$

Here, the assumed mean, $A = 38.5$ and $h = 11$.

A.M. by step deviation method is given by

$$M = A + h\,\frac{\Sigma fd'}{N} = 38.5 + \frac{11 \times (-58)}{103} = 38.5 - 6.194 = 32.306$$

Hence, A.M. = 32.306.

EXAMPLE 3. *Value of fecundity (Rate of reproduction) of 50 fishes of a species of fish (Macrognathus aculatus) were obtained and on the basis of that, a frequency table is given below. Calculate the mean value of fecundity by long method (Direct Method).*

Class Interval	1-10	11-20	21-30	31-40	41-50	51-60	61-70	71-80
Frequency	3	11	7	4	15	0	7	3

SOLUTION.

Class Interval	Mid Value x	Frequency f	fx
1-10	5.5	3	16.5
11-20	15.5	11	170.5
21-30	25.5	7	178.5
31-40	35.5	4	142.5
41-50	45.5	15	682.5
51-60	55.5	0	0
61-70	65.5	7	458.5
71-80	75.5	3	226.5
		$\Sigma f = 50$	$\Sigma fx = 1875$

$$\bar{x} = \frac{\Sigma fx}{\Sigma f} = \frac{1875}{50} = 37.5$$

EXAMPLE 4. *Calculate the arithmetic mean of the marks from the following table:*

[Meerut(B.Sc. Biotech)–2001]

Marks	0-10	10-20	20-30	30-40	40-50	50-60
Number of students	12	18	27	20	17	6

SOLUTION.

Marks	Mid Value x	Frequency f	fx
0-10	5	12	60
10-20	15	18	270
20-30	25	27	675
30-40	35	20	700
40-50	45	17	765
50-60	55	6	330
		$n = 100$	$\Sigma fx = 2800$

We know that $\qquad \bar{x} = \dfrac{\Sigma fx}{n} = \dfrac{2800}{100} = 28$

Hence, the arithmetic mean of the marks = 28.

EXAMPLE 5. *Find the arithmetic mean of the following frequency distribution :*

[Meerut(B.Sc. Biotech)–2000]

x	1	2	3	4	5	6	7
f	5	9	12	17	14	10	6

SOLUTION.

x	f	fx
1	5	5
2	9	18
3	12	36
4	17	68
5	14	70
6	10	60
7	6	42
	$\Sigma f = 73$	$\Sigma fx = 299$

We know that $\qquad \Sigma f = n$

and $\qquad \bar{x} = \dfrac{\Sigma fx}{n} = \dfrac{299}{73} = 4.095$

Hence, the arithmetic mean = 4.095.

EXAMPLE 6. *Find the mean for the following distribution :*   (UPTU–2001)

Class	0-9	10-19	20-29	30-39	40-49	50-59	60-69	70-79
Frequency	21	74	100	120	110	84	30	11

SOLUTION.

Class	Mid value x	f	(x – 44.5)	$d' = \dfrac{x-A}{10}$	fd'
0-9	4.5	21	– 40	– 4	– 84
10-19	14.5	74	– 30	– 3	– 222
20-29	24.5	100	– 20	– 2	– 200
30-39	34.5	120	– 10	– 1	– 120
40-49	44.5	110	0	0	0
50-59	54.5	84	10	1	84
60-69	64.5	30	20	2	60
70-79	74.5	11	30	3	33
		$\Sigma f = 550$			$\Sigma fd' = -449$

Hence, $A = 44.5$ and $h = 10$

A.M. by step deviation method is

$$M = A + h\frac{\Sigma fd'}{h} = 44.5 - \frac{449}{550} \times 10 = 44.5 - 8.16 = 36.34$$

19.6 CHARLIER'S ACCURACY CHECK

To check the accuracy of the results by short-cut method and step deviation method Charlier suggested the following two formulae

(a) $\Sigma f(x - A) = \Sigma[f\{(x - A) + 1\}] - \Sigma f$ (For short-cut method)

(b) $\Sigma f \dfrac{(x - A)}{h} = \Sigma\left[f\left\{ \dfrac{(x - A)}{h} + 1 \right\} \right] - \Sigma f$ (For step deviation method)

$\Sigma f(d') = \Sigma[f(d' + 1)] - \Sigma f$

SOLVED EXAMPLES

EXAMPLE 1.   *Compute arithmetic mean of the following by both short-cut and step deviation method and apply Charlier's accuracy check for both the result.*

Class	20-30	30-40	40-50	50-60	60-70
Frequency	8	26	30	20	16

SOLUTION.

Class	Mid Value x	f	(x – A)	(x – A)+1	f(x – A)	f(x – A)+1
20-30	25	8	–20	–19	– 160	–152
30-40	35	26	–10	–9	– 260	–234
40-50	45	30	0	1	0	30
50-60	55	20	10	11	200	220
60-70	65	16	20	21	320	336
		100			100	200

$$\text{A.M.} = A + \frac{\Sigma f(x - A)}{\Sigma f} = 45 + \frac{100}{100} = 45 + 1 = 46$$

Applying Charlier's check

$$\Sigma f(x - A) = \Sigma[f(x - A) + 1] - \Sigma f = [200] - 100 = 100$$

Thus, we get $\Sigma f(x - A) = 100$, which is exactly correct by our table.

Now, applying step deviation method for the given data

Class	Mid Value x	f	$(x - A)$	$\mu = \dfrac{x - A}{i}$	fd'	$d' + 1$	$f(d' + 1)$
20-30	25	8	−20	−2	−16	−1	−8
30-40	35	26	−10	−1	−26	0	0
40-50	45	30	0	0	0	1	30
50-60	55	20	10	1	20	2	40
60-70	65	16	20	2	32	3	48
		100			10		110

Applying Charlier's check for step deviation method

$$\Sigma fd' = \Sigma[f(d' + 1)] - \Sigma f = 110 - 100 = 10$$

which is correct since by the table, we see that $\Sigma fd' = 10$

How to Calculate Unknown Value :

In the given data or frequency of any term is missing and the arithmetic mean of the series is given, then we can calculate the term by using any of three methods (direct method, short-cut method, step deviation method).

EXAMPLE 2. *If the arithmetic mean of the following frequency distribution is 39.25, find the missing term.*

Daily wages	25	30	35	50	60	75
No. of labour	10	?	13	8	5	4

SOLUTION.

Daily wages x	No. of labours f	fx
25	10	250
30	z	$30z$
35	13	455
50	8	400
60	5	300
75	4	300
	$\Sigma f = z + 40$	$\Sigma fx = 30z + 1705$

We know that the arithmetic mean is given by

$$A.M. = \frac{\Sigma fx}{\Sigma f} \Rightarrow 39.25 = \frac{30z + 1705}{z + 40}$$

$$39.25z + 1570 = 30z + 1705$$

$$9.25z = 135, \quad z = \frac{135}{9.25}$$

$$z = 14.59 = 14.6.$$

EXAMPLE 3. *The mean marks of 100 students were found to be 40, later on it was discovered that a score of 53 was misread as 83. Find the corrected mean corresponding to the correct score.*

SOLUTION. Total number of students, $N = 100$.

The arithmetic mean of the marks of these students is given by $a = 40$.

Then the total marks obtained by the students

$$= N \times a = 100 \times 40 = 4000$$

Now a score of 53 was misread as 83 so the score calculated will be wrong. Then the correct score can be calculated

Correct score = Total previous marks – wrong score + correct score

$$= 4000 - 83 + 53 = 3970$$

Now, we have to find out the correct mean corresponding to the correct score.

$$\text{Mean} = \frac{\text{Correct score}}{N} = \frac{3970}{100}$$

Mean = 39.7 marks

19.7 PROPERTIES OF ARITHMETIC MEAN

1. If every variable is increased by a particular value a, then the arithmetic mean is also increased by a.

2. The algebraic sum of the deviations of all the variate values from their arithmetic mean is zero.

3. The sum of the squares of the deviations of all the values taken about their arithmetic mean is minimum.

4. If $M_1, M_2, ..., M_k$ be the arithmetic mean of k distributions with respective frequencies $N = n_1, n_2, ..., n_k$, then the mean M of the whole distribution with frequency $N = (n_1, n_2, ..., n_k)$ is given by $M = \dfrac{1}{N} \sum\limits_{r=1}^{k} n_r M_r$.

5. Arithmetic mean is not independent of the change of origin and scale.

SOLVED EXAMPLES

EXAMPLE 1. *The weight of 150 students in a certain class is 60 kilograms. The mean weight of boys in the class is 70 kilograms and that of the girls is 55 kilograms. Find the number of boys and girls in the class.*

SOLUTION. $N = 150$

Let mean weight of boys = $\bar{X}_1 = 70$

Let mean weight of girls = $\bar{X}_2 = 55$

Let number of boys = n_1

no. of girls = n_2

$$n_1 + n_2 = N = 150 \qquad \qquad ... (1)$$

$$\bar{X}_{12} = 60$$

We know that

$$\bar{X}_{12} = \frac{n_1 \bar{X}_1 + n_2 \bar{X}_2}{n_1 + n_2}$$

$$60 = \frac{70 n_1 + 55 n_2}{150}$$

$$70n_1 + 55n_2 = 9000$$

$$14n_1 + 11n_2 = 1800 \qquad \ldots(2)$$

$$14n_1 + 14n_2 = 2100 \qquad \ldots(3)$$

Subtracting eq. (2) and (3)

$$3n_2 = 300$$

$$n_2 = 100$$

$$n_1 = 150 - 100 = 50$$

Thus, number of boys = 50

Number of girls = 100.

EXAMPLE 2. *The average monthly wages of all the workers in a factory is Rs. 444. If the average wages paid to male and female workers are Rs. 480 and Rs. 360 respectively. Find the percentage of male and female workers employed by factory.*

SOLUTION. Let the total workers be n.

$$\therefore \qquad 444 = \frac{\Sigma x}{n}$$

$$444x = \Sigma x$$

$$480 = \frac{\Sigma x_1}{n_1} \qquad \qquad [n_1 = \text{Number of males}]$$

$$\Sigma x_1 = 480n_1$$

and $\qquad \Sigma x_2 = 360n_2 \qquad \qquad [n_2 = \text{Number of females}]$

$$n = n_1 + n_2$$

$$\Sigma x = \Sigma x_1 + \Sigma x_2$$

$$444n = 480n_1 + 360n - 360n_1$$

$$84n = 120n_1$$

$$7n = 10n_1$$

$$\text{Male percentage} = \frac{n_1}{n} \times 100 = \frac{n_1 \times 7}{10 \times n_1} \times 100 = 70\%$$

Female percentage = 30 %.

EXAMPLE 3. *The mean of 200 items is found to be 60. If at the time of calculation two items are wrongly taken as 3 and 67 instead of 13 and 17, find the correct mean.* **(UPTU–2003)**

SOLUTION. It is given that number of items, $n = 200$

and $\qquad \qquad \bar{x} = 60$

From the formula of mean, we have

$$\Sigma x = n\bar{x} = 200 \times 60 = 12000$$

Here, Σx is incorrect as it includes two values 3 and 67 which are actually 13 and 17. Therefore,

$$\Sigma x = 12000 - (3 + 67) + (13 + 17)$$

$$= 12000 - 70 + 30 = 11960$$

Hence, correct mean $\quad (\bar{x}) = \dfrac{11960}{200} = 59.8$

Exercise 19.1

1. Compute the arithmetic mean of first n natural numbers whose weights are equal to the corresponding number $\frac{1}{3}(2n+1)$.

2. Compute the mean marks of a student from the following table :

Marks	No. of students
Above 0	80
Above 10	77
Above 20	72
Above 30	65
Above 40	55
Above 50	43
Above 60	28
Above 70	16
Above 80	10
Above 90	8
Above 100	0

3. The rainfall of a certain town in centimeters for the first six months of the year are 102, 103, 95, 98, 100, 105. Compute the average rainfall of the town.

4. Compute the arithmetic average in rupees from the data given below :

Salary	100	150	200	250	300	500
No. of labours	30	20	15	10	4	1

5. Find the missing frequency from the following data, it is being given that 19.92 is the average number of the given data :

Tables	4-8	8-12	12-16	16-20	20-24
No. of persons cured	11	13	16	14	?

Tables	24-28	28-32	32-36	36-40
No. of persons cured	9	17	6	4

6. Calculate the mean marks of a student from the given data :

Marks	No. of students
Below 10	15
Below 20	35
Below 30	60
Below 40	84
Below 50	96
Below 60	127
Below 70	128
Below 80	250

7. Find the combined average daily wages for the workers of two factories :

No. of workers	250	200
Average	2.00	2.50

8. If the arithmetic average of data given below be 165 rupees, compute the missing term :

Monthly salary	100	150	200	-	300	500
No. of labourers	30	20	15	10	4	1

9. Compute the weighted arithmetic average wage rate of 31 building trade workers from the following table :

Kind of worker	Daily wages (Rs)	Frequency
Masons	15	4
Labourers	8	20
Carpenters	12	5
Painters	10	2

10. Compute the arithmetic average of the marks obtained by 9 students in a test :
75, 43, 52, 65, 48, 35, 40, 70, 40.

11. Compute the missing frequency term from the following data whose arithmetic average is given by 35.64 :

Class	20-25	25-30	30-35	35-40	40-45	45-50
Frequency	18	44	102	—	57	19

12. Compute the arithmetic average of the following data :

0-5	5-8	8-10	10-12	12-15
2	5	7	5	6

15-17	17-20	20-25	25-30	
4	4	9	6	

13. The arithmetic average of a group of 40 items is 100 and that of another group of 50 items is 70. Find the mean of the combined group of size 90.

14. Compute the arithmetic mean for the following data :

Class	0-5	5-10	10-15	15-20	20-25
Frequency	4	16	2	15	2

15. If the arithmetic average of the following frequency distribution is 7.85. Calculate missing frequency term.

Salary	5	6	7	10	12	15
Labourers	10	—	13	8	5	4

16. The average salary of 500 workers in a factory running in two shifts of 360 and 140 workers respectively is Rs. 70. The average salary of 360 workers working in day shift is Rs. 75. Find the average salary of 140 workers working in the night shift.

17. Find the mean of the following

Height (cm)	65	66	67	68	
Plants	1	4	5	7	
Height (cm)	69	70	71	72	73
Plants	11	10	6	4	2

18. Find the mean of the following distribution :

Class	Frequency
0-7	19
7-14	25
14-21	36
21-28	72
28-35	51
35-42	43
42-49	28

19. If $p + q = 1$, compute the mean of the following :

x	0	1	2
f	q^n	$^nC_1 q^{n-1}p$	$^nC_2 q^{n-2}p^2$
x	3	...	n
f	$^nC_3 q^{n-3}p^3$	...	p^n

ANSWERS

2. 51.75   **5.** 250   **6.** 50.4   **7.** 2.22   **8.** 250

9. 9.68   **11.** 160   **12.** 15.417   **13.** 83.33   **15.** 15.05   **19.** np

---

19.8 COMBINED MEAN

If $\bar{x}_1$ and $\bar{x}_2$ are the mean of two groups of sizes n_1 and n_2, then the mean $\bar{x}$ is the mean of two groups, given by

$$\bar{x} = \frac{n_1\bar{x}_1 + n_2\bar{x}_2}{n_1 + n_2}$$

Proof. Let $x_1, x_2, ..., x_n$ be the variates of a group of size n_1 and $y_1, y_2, ... y_n$ be the variates of a group of size n_2. Then

$$\bar{x}_1 = \frac{x_1 + x_2 + x_3 + ... + x_n}{n_1} \Rightarrow n_1\bar{x}_1 = x_1 + x_2 + x_3 + ... + x_n \qquad ...(1)$$

and $$\bar{x}_2 = \frac{y_1 + y_2 + y_3 + ... + y_n}{n_2} \Rightarrow n_2\bar{x}_2 = y_1 + y_2 + y_3 + ... + y_n \qquad ...(2)$$

Let $\bar{x}$ be the mean of these two groups, then

$$\bar{x} = \frac{(x_1 + x_2 + x_3 + ... + x_n) + (y_1 + y_2 + y_3 + ... + y_n)}{n_1 + n_2}$$

$$\Rightarrow \quad \bar{x} = \frac{n_1\bar{x}_1 + n_2\bar{x}_2}{n_1 + n_2} \qquad \qquad \text{[Using (1) and (2)]}$$

SOLVED EXAMPLES

EXAMPLE 1. *The mean of the marks secured by 25 students of section A of class B.Tech is 47, that of 35 students of section B is 51, and that of 30 students of section C is 53. Find the mean of marks secured by 90 students of class B.Tech.*

SOLUTION. Let n_1, n_2, n_3 be the numbers of students respectively in section A, B and C and $\bar{x}_1, \bar{x}_2$ and $\bar{x}_3$ be the mean of marks secured by them.

$\therefore \quad n_1 = 25, \ n_2 = 35, \ n_3 = 30$

and $\bar{x}_1 = 47, \ \bar{x}_2 = 51$ and $\bar{x}_3 = 53$

$\therefore$ Combined mean $\bar{x} = \dfrac{n_1\bar{x}_1 + n_2\bar{x}_2 + n_3\bar{x}_3}{n_1 + n_2 + n_3}$

$= \dfrac{24 \times 47 + 35 \times 51 + 30 \times 53}{25 + 35 + 30}$

$= \dfrac{4550}{90} = 50.56$

EXAMPLE 2. *The school has two sections. The mean marks of one section of size 40 is 60 and mean marks of other section of size 60 is 80. Find the combined mean of the students of the school.*

SOLUTION. Here given, $n_1 = 40, \ n_2 = 60$

$\bar{x}_1 = 60, \ \bar{x}_2 = 80$

$\therefore$ Combined mean $\bar{x} = \dfrac{n_1\bar{x}_1 + n_2\bar{x}_2}{n_1 + n_2} = \dfrac{40 \times 60 + 60 \times 80}{40 + 60}$

$\Rightarrow \quad \bar{x} = \dfrac{2400 + 4800}{100} = \dfrac{7200}{100} = 72.$

EXAMPLE 3. *The mean of 10 observations is 20 and that of another 15 observations is 16, find the mean of all the 25 observations.*

SOLUTION. Here,

$n_1 = 10, \ n_2 = 15, \ \bar{x}_1 = 20, \ \bar{x}_2 = 16$

$\therefore$ Combined mean $\bar{x} = \dfrac{n_1\bar{x}_1 + n_2\bar{x}_2}{n_1 + n_2} = \dfrac{10 \times 20 + 15 \times 16}{10 + 15}$

$\Rightarrow \quad \bar{x} = \dfrac{200 + 240}{25} = \dfrac{440}{25} = 17.6$

EXAMPLE 4. *The average score of boys in an examination of a school is 71 and that of girls is 73. The average score of school in that examination is 71.8. Find the ratio of number of boys to the number of girls appeared in the examination.*

SOLUTION. Let there be n_1 boys and n_2 girls in the school.

Here, $\bar{x}_1 = 71, \ \bar{x}_2 = 73$ and $\bar{x} = 71.8$

$\therefore$ Combined mean $\bar{x} = \dfrac{n_1\bar{x}_1 + n_2\bar{x}_2}{n_1 + n_2}$

$\Rightarrow \quad 71.8 = \dfrac{n_1 \times 71 + n_2 \times 73}{n_1 + n_2}$

$\Rightarrow \quad 71.8(n_1 + n_2) = 71n_1 + 73n_2$

$\Rightarrow \quad 71.8n_1 + 71.8n_2 = 71n_1 + 73n_2$

$$\Rightarrow \qquad 0.8n_1 = 1.2n_2$$

$$\Rightarrow \qquad 8n_1 = 12n_2 \quad \Rightarrow \quad \frac{n_1}{n_2} = \frac{12}{8} = \frac{3}{2}$$

Hence, $\qquad n_1 : n_2 = 3 : 2$.

Exercise 19.2

1. The mean wage of 150 workers of the first shift in a factory is Rs. 400. The mean wage of 75 workers of the second shift is Rs. 600. Find the combined mean wage of the workers of the factory.

2. There are 50 students in a class out of which 20 are girls. The average weight of 20 girls is 45 Kg and that of 30 boys is 52 Kg. Find the mean weight in Kg of the entire class.

3. The average marks obtained by 30 students of group I is 60 and average marks of 40 students of group II is 55 and that of 30 students of group III is 70. Find the combined average of students of all the three groups.

4. There are 100 students in a class. The mean height of the class is 150 cm. If the mean height of 60 boys is 170 cm. Find the mean height of the girls.

5. The mean weight of 150 students in a class is 60. The mean weight of boys is 70 Kg and that of girls is 55 Kg. Find the number of boys and girls in the class.

ANSWERS

1. 466.67	**2.** 49.2 Kg	**3.** 61	**4.** 120 cm	**5.** 50, 100

19.9 GEOMETRIC MEAN

Let $x_1, x_2, x_3, \ldots\ldots, x_n$ be the n variates of a variable x, then the geometric mean G of n variables is defined by

$$G = (x_1 . x_2 . x_3 \ldots\ldots x_n)^{1/n}$$

If $f_1, f_2, f_3, \ldots, f_n$ be the frequency of these variables and $N = f_1 + f_2 + f_3 + \ldots\ldots + f_n$

Then,

$$G = (x_1^{f_1} . x_2^{f_2} . x_3^{f_3} \ldots\ldots x_n^{f_n})^{1/N}$$

$$\log G = \frac{1}{N} \left[f_1 \log x_1 + f_2 \log x_2 + f_3 \log x_3 + \ldots + f_n \log x_n \right]$$

$$\log G = \frac{1}{N} \left[\sum_{i=1}^{n} f_i \log x_i \right]$$

Thus, we can say that the logarithm of the geometric mean can be calculated by taking weighted mean of the logarithm of the variables x_i.

WORKING PROCEDURE

Step 1. Find the logarithm of the variable x.

Step 2. Obtain $\Sigma f \log x$

Step 3. Obtain $\dfrac{\Sigma f \log x}{\Sigma f}$

Step 4. Calculate the antilog of $\dfrac{\Sigma f \log x}{\Sigma f}$

SOLVED EXAMPLES

EXAMPLE 1. *Calculate the geometric mean of the following data :*

2574, 0.005, 0.8, 0.0009, 5, 75, 475, 0.08

SOLUTION. We know that the geometric mean of the data can be calculated by the formula G.M. $= \dfrac{1}{N} \Sigma \log x_i$. Then, we will solve it by this formula.

x	log x
2574	3.4106
0.005	$\bar{3}.6990$
0.8	$\bar{1}.9031$
0.0009	$\bar{4}.9542$
5	0.6990
75	1.8751
475	2.6767
0.08	$\bar{2}.9031$
	2.1208

$$\log G.M. = \frac{1}{N} \Sigma \log x_i = \frac{1}{8} \times 2.1208$$

$$\log G.M. = 0.2651$$

Then, G.M. = Antilog 0.2651 = 1.841

So, geometric mean = 1.841.

EXAMPLE 2. *Calculate the geometric mean of the given data :*
8, 15, 36, 40, 45, 70, 75, 85, 250, 500

Solution.

x	log x
8	0.9031
15	1.1761
36	1.5563
40	1.6021
45	1.6532
70	1.8451
75	1.8751
85	1.9294
250	2.3979
500	2.6990
	17.6373

Now, the geometric mean is given by

$$\log G.M. = \frac{1}{N} \Sigma \log x_i = \frac{1}{10} \times 17.6373$$

$$\log G.M. = 1.7637$$

$$G.M. = \text{Antilog } 1.7637 = 58.03$$

EXAMPLE 3. *Compute the geometric mean from the following data :*

x	11	12	13	14	15
f	3	7	8	5	2

SOLUTION.

f	Frequency (f)	$\log x$	$f \log x$
11	3	$\log 11 = 1.0414$	3.1242
12	7	$\log 12 = 1.0792$	7.5544
13	8	$\log 13 = 1.1139$	8.9112
14	5	$\log 14 = 1.1461$	5.7305
15	2	$\log 15 = 1.1761$	2.3522
	25		27.6725

The geometric mean is given by

$$\log \text{G.M.} = \frac{1}{N} \Sigma f \log x_i = \frac{1}{25} \times 27.6725 = 1.1069$$

Then,   G.M. = Antilog 1.1069 = 12.79

Hence,   G.M. = 12.79

EXAMPLE 4. *Find the geometric mean of the given data :*

Marks	1-10	10-20	20-30	30-40	40-50
f	8	12	20	6	4

SOLUTION.

Marks	**Mid value** x	**Frequency** (f)	$\log x$	$f \log x$
0-10	5	8	$\log 5 = 0.6990$	5.592
10-20	15	12	$\log 15 = 1.1761$	14.1132
20-30	25	20	$\log 25 = 1.3979$	27.958
30-40	35	6	$\log 35 = 1.5441$	9.2646
40-50	45	4	$\log 45 = 1.6532$	6.6128
		50	17.6373	63.5406

Geometric mean can be calculated by

$$\log \text{G.M.} = \frac{1}{N} \Sigma f \log x_i = \frac{1}{50} \times 63.5406 = 1.2708$$

Then,   G.M. = Antilog 1.2708 = 18.65

So,   G.M = 18.65

EXAMPLE 5. Calculate the weighted geometric mean of the data given below :

Articles	Price	Weight
A	125	40
B	150	25
C	100	5
D	122	20
E	75	10

SOLUTION.

Articles	Price x	Weight (w)	$\log x$	$w \log x$
A	125	40	$\log 125 = 2.0969$	83.876
B	150	25	$\log 150 = 2.1761$	54.4025
C	100	5	$\log 100 = 2$	10
D	122	20	$\log 122 = 2.0864$	41.728
E	75	10	$\log 75 = 1.8751$	18.751
		100		208.7575

Weighted geometric mean can be calculated as

$$\log \text{(weighted G.M.)} = \frac{\Sigma w \log x}{\Sigma w} = \frac{1}{100} \times 208.7575 = 2.0875$$

Weighted G.M. = Antilog 2.0875 = 122.3

So, Weighted G.M. = 122.3

EXAMPLE 6. *Calculate the geometric mean of 3, 7, 8, 5, 2.*

SOLUTION. The geometric mean of the terms 3, 7, 8, 5, 2 is given by
$$\text{G.M.} = (3 \times 7 \times 8 \times 5 \times 2)^{1/5} = (1680)^{1/5} = 4.416$$

EXAMPLE 7. *On 1st March a baby weighted 14 lbs. On 1st May it weighted 20 lbs. What was the approximate weight of the said baby on 1st April.*

SOLUTION.
$$\text{G.M.} = (x_1 \cdot x_2 \cdot x_3 ... x_n)^{1/n}$$
$$\text{G.M.} = (x_1 \cdot x_2)^{1/2} = (14 \times 20)^{1/2}$$
$$= (2 \times 7 \times 2 \times 2 \times 5)^{1/2} = (2\sqrt{70}) = 2 \times 8.36 = 16.72$$

The weight of the baby was 16.72 lbs on 1st April.

EXAMPLE 8. *The number of Bascophiles (a kind of WBC) in blood of 30 patients of a hospital and their frequency were recorded as [sources 11, 14, 17, 19, 22 and frequencies 5, 6, 8, 7, 4]. Find out the G.M.*

SOLUTION.

Scores (x)	Frequencies (f)	$\log x$	$f \log x$
11	5	1.0414	5.2070
14	6	1.1461	6.8766
17	8	1.2304	9.8432
19	7	1.2788	8.9516
22	4	1.3424	5.3696
			36.2480

$$\text{G.M.} = \text{Antilog}\left(\frac{\Sigma f \log x}{\Sigma f}\right) = \text{antilog}\left(\frac{36.2480}{30}\right)$$

$$= \text{antilog}(1.20826) = 16.15$$

19.10 PROPERTIES OF GEOMETRIC MEAN

1. If we put the value of geometric mean in place of the each value of a series, then the product of the value of the series will be unchanged.

2. If G_1 is the geometric mean of the series $x_1, x_2, x_3..., x_n$, G_2 is the geometric mean of the series $y_1, y_2, y_3..., y_n$ and G is the geometric mean of the series obtained by the ratios of corresponding observations. Then G will be equal to G_1 / G_2, i.e., $G = \dfrac{G_1}{G_2}$.

3. Let us consider n series with frequencies $N_1, N_2, N_3, ..., N_n$ respectively and geometric means $G_1, G_2, G_3, ..., G_n$ respectively. Then the combined geometric mean of n series with frequency $N_1 + N_2 + N_3 + ... + N_n$ is given by

$$G = (G_1^{N_1} G_2^{N_2} G_3^{N_3} ...... G_n^{N_n})^{1/N}$$

4. Let us consider n sets of observations whose geometric means are respectively $G_1, G_2, ..., G_n$. Now, if G is the geometric mean of the product of these n sets, then the product of the geometric means of these series will be equal to the value of G.

$$G = G_1 . G_2 . G_3 ... G_n$$

$$\log G = \sum_{i=1}^{n} \log G_i$$

5. Let us consider a series $x_1, x_2, ..., x_p, x_{p+1}...x_n$ whose geometric mean is given by G.

In which G is greater than from the each value $x_1, x_2, ..., x_p$ and less than from each of the values $x_{p+1}, x_{p+2}...x_n$, then

$$G^n = (x_1 x_2 ... x_p . x_{p+1} ... x_n)$$

$\Rightarrow \qquad G^p . G^{n-p} = (x_1 . x_2 ... x_p)(x_{p+1} ... x_n)$

$\Rightarrow \qquad \dfrac{G}{x_1} . \dfrac{G}{x_2} ... \dfrac{G}{x_p} = \dfrac{x_{p+1}}{G} . \dfrac{x_{p+2}}{G} ... \dfrac{x_n}{G}$

Exercise 19.3

1. Find the geometric mean of the following data :
50, 100, 1920, 143740, 204980, 1206740, 154910

2. Find the geometric mean of the series :
$1, 2, 2^2, 2^3, 2^4, ...2^n$

3. The price of certain article rises 5% in first year, 8% in second year and 77% in third year. What is the average change per year?

4. The geometric mean of 10 data are calculated as 16.2. It was later found that one of the data was wrongly read as 12.9, in fact it was 21.9. Calculate the correct geometric mean.

5. Find the geometric mean of 2, 6, 18, 54, 162.

6. Find the geometric mean of the following series:

Class	0-10	10-20	20-30	30-40	40-50
Frequency	10	15	12	8	5

7. Find the geometric mean from the following table :

Marks obtained	5	7	9	11	13	15
No. of students	1	2	3	5	11	9

8. Calculate the average rate of increment in population which is increased by 20% in first year, 25% in second year and 44% in third year.

9. Find out the geometric mean of the following distribution :

Marks obtained	0-10	10-20	20-30	30-40
No. of students	5	8	3	4

1. 12700	**2.** $2^{n/2}$	**3.** 26%	**4.** 17.08	**5.** 18
6. 19.10	**7.** 11.86	**8.** 28.02%	**9.** 14.64	

19.11 HARMONIC MEAN

The harmonic mean of any series is given by the reciprocal of the arithmetic mean of the reciprocals of the variables.

For different type of series, it can be calculated by different method

(i) For individual series: Let $x_1, x_2, x_3 ... x_n$ be the n variables, then the harmonic mean of these variables is given by

$$H = \frac{n}{\dfrac{1}{x_1} + \dfrac{1}{x_2} + ... + \dfrac{1}{x_n}}.$$

(ii) For discrete series: Let $x_1, x_2, x_3 ... x_n$ be the n variables and $f_1, f_2 ..., f_n$ be the frequency of them. Then the harmonic mean is given by

$$\frac{1}{H} = \frac{1}{N} \sum_{i}^{n} \left(\frac{f}{x} \right), \qquad \text{where} \quad N = \Sigma f$$

$$H = N \sum_{i}^{n} \left(\frac{x}{f} \right)$$

(iii) For grouped series : When the grouped series are given, we take the mid value of each group and named them as $x_1, x_2, x_3 ... x_n$ and if the frequencies of these groups are $f_1, f_2, f_3, ..., f_n$, then the harmonic mean can be calculated by

$$H = \frac{N}{\Sigma(f / x)}.$$

SOLVED EXAMPLES

EXAMPLE 1. *Find the harmonic mean of the following data :* 12, 8, 6, 24

SOLUTION. Harmonic mean for individual series is given by

$$H = \frac{n}{\dfrac{1}{x_1} + \dfrac{1}{x_2} + ... + \dfrac{1}{x_n}} = \frac{4}{\dfrac{1}{12} + \dfrac{1}{8} + \dfrac{1}{6} + \dfrac{1}{24}}$$

$$= \frac{4}{0.0833 + 0.1250 + 0.1666 + 0.0416}$$

$$H = \frac{4}{0.4165} = 9.6038.$$

EXAMPLE 2. *Find the harmonic mean of the following data :* 4, 8, 16

SOLUTION.
$$H = \frac{n}{\dfrac{1}{x_1} + \dfrac{1}{x_2} + ... + \dfrac{1}{x_n}} = \frac{3}{\dfrac{1}{4} + \dfrac{1}{8} + \dfrac{1}{16}} = \frac{3}{0.25 + 0.125 + 0.0625}$$

$$= \frac{3}{0.4375} = 6.8571$$

So the harmonic mean is given by 6.8571.

EXAMPLE 3. Find the harmonic mean of the following frequency distribution :

Class	0-10	10-20	20-30	30-40	40-50
Frequency	4	5	11	6	4

SOLUTION.

Class	Mid Value x	Frequency	$1/x$	f/x
0-10	5	4	0.2	0.800
10-20	15	5	0.0666	0.333
20-30	25	11	0.04	0.440
30-40	35	6	0.0285	0.171
40-50	45	4	0.0222	0.088
		30		1.832

Now, harmonic mean for grouped series is given by

$$\frac{N}{\Sigma(f/x)} = \frac{30}{1.832} = 1.63755.$$

EXAMPLE 4. *Find the harmonic mean of the following frequency distribution :*

Class	11	12	13	14	15
Frequency	3	7	8	5	2

SOLUTION. We have

Marks (x)	Frequency (f)	$1/x$	f/x
11	3	0.0909	0.2727
12	7	0.0833	0.5831
13	8	0.0769	0.6152
14	5	0.0714	0.357
15	2	0.0666	0.1332
	25		1.9612

Now, harmonic mean for grouped series is given by

$$\text{H.M.} = \frac{N}{\Sigma(f/x)} = \frac{25}{1.9612} = 12.7472.$$

EXAMPLE 5. *A man drives a car for three days by covering a distance of 360 km per day. First day he drives for a time of 10 hours and drive with the speed of 36 km/h. On the second day, he drives 15 hours at a speed of 24 km/h and on the third day, he drives for 12 hours at a speed of 30 km/h. Calculate the average speed of the car.*

SOLUTION. It is given that he covers a constant distance of 360 km per day.
His speed on the first day is given by = 36 km/h
His speed on the second day is given by = 24 km/h
His speed on the third day is given by = 30 km/h
Since the distance is given to be constant so the average speed can be calculated by taking harmonic mean of the speeds.

So, average speed $= \dfrac{n}{\dfrac{1}{v_1}+\dfrac{1}{v_2}+\dfrac{1}{v_3}} = \dfrac{3}{\dfrac{1}{36}+\dfrac{1}{24}+\dfrac{1}{30}}$

$$= \dfrac{3}{0.0277+0.0416+0.0333} = \dfrac{3}{0.1026} = 29.2397 \text{ km/h}$$

⚙ REMARK

➡ Where the distance in each part of the journey is given to be constant, then average speed will be calculated by harmonic mean. In the case when time being constant, the average is given by arithmetic mean.

EXAMPLE 6. *In a certain factory a unit of work is completed by A in 4 minutes, by B in 6 minutes, by C in 8 minutes and by D in 12 minutes. What is their average rate of working? At this rate, how many units will they complete in a 8 hour day?*

SOLUTION. The average rate of working can be calculated by harmonic mean. So

$$H = \dfrac{n}{\dfrac{1}{x_1}+\dfrac{1}{x_2}+\dfrac{1}{x_3}+\dfrac{1}{x_4}} = \dfrac{4}{\dfrac{1}{4}+\dfrac{1}{6}+\dfrac{1}{8}+\dfrac{1}{12}}$$

$$= \dfrac{4}{0.25+0.166+0.125+0.083} = \dfrac{4}{0.624} = 6.41 \text{ minute}$$

So, 6.41 minute per unit is the average rate of working

$$8 \text{ hour} = 8 \times 60 = 480 \text{ minutes}$$

They will complete the units $= \dfrac{1}{6.41} \times 480$.

EXAMPLE 7. *A train travels first part of its journey of 100 kms at a speed of 20 km/h, second part of 100 kms at a speed of 25 km/h and the third part of the same distance at 30 km/h. Find its average speed.*

SOLUTION. Since the train travels a distance of 100 kms in each part. So, its average speed can be calculated by the harmonic mean.

$$H = \dfrac{n}{\dfrac{1}{x_1}+\dfrac{1}{x_2}+\dfrac{1}{x_3}} = \dfrac{3}{\dfrac{1}{20}+\dfrac{1}{25}+\dfrac{1}{30}} = \dfrac{3}{0.05+0.04+0.033}$$

$$= \dfrac{3}{0.123} = 24.390 .$$

EXAMPLE 8. *Hemoglobin percentage of five perosns were measured as 1, 5, 10, 15 and 25. Find out the Harmonic mean.*

SOLUTION.
$$\dfrac{1}{HM} = \dfrac{1}{5}\left(\dfrac{1}{1}+\dfrac{1}{5}+\dfrac{1}{10}+\dfrac{1}{15}+\dfrac{1}{25}\right)$$

$$\dfrac{1}{HM} = \dfrac{1}{5}\left[\dfrac{150+30+15+10+6}{150}\right] = \dfrac{1}{5}\times\dfrac{211}{150} = \dfrac{211}{750}$$

$$HM = \dfrac{750}{211} = 2.55$$

EXAMPLE 9. *Hb% and its frequencies in 10 members of a family was studied and following results were obtained. Find the HM of the given series :*

Hb% (mg / 100 ml)	12 mg	13 mg	14 mg	15 mg	16 mg
Frequencies	3	3	1	2	1

SOLUTION.

Hb% (mg / 100 ml)	Frequencie (f)	1/x	f/x
12	3	0.083	0.25
13	3	0.076	0.23
14	1	0.071	0.071
15	2	0.066	0.133
16	1	0.0625	0.0625
	$\Sigma f = 10$		$\Sigma(f/x) = 0.7465$

EXAMPLE 10. *A man motors from A to B a large part of the distance is uphill and gets a mileage of only 10 miles per gallon of gasoline. On the return trip he makes 15 miles per gallon. Find the harmonic mean of this mileage. Verify the fact that this is the proper average to use by assuming that the distance from A to B is 60 miles.*

SOLUTION. Let H be the harmonic mean of his mileage per gallon, then

$$\frac{1}{H} = \frac{\frac{1}{10} + \frac{1}{5}}{2} = \frac{0.1 + 0.066}{2} = \frac{0.1666}{2} = 0.0833$$

$\Rightarrow$ H.M. = 12 miles per gallon

Now,   A.M. $= \dfrac{10 + 15}{2} = 12.5$ miles per gallon.

In this case we have A.M. = 12.5. Then, H.M. is the proper average.

19.12 PROPERTIES OF HARMONIC MEAN

1. If x_1 and x_2 are any two observations, then A.H. $= G^2$
   where, A = Arithmetic mean; H = Harmonic mean; G = Geometric mean
2. If $x_1, x_2, x_3, ..., x_n$ be the n positive observations then $A \geq G \geq H$.
   The sign of equality will hold if the values of all observations under consideration are same.
3. A variate takes values $a, ar, ar^2, ..., ar^{n-1}$ each with the frequencies one. Then

$$AM = a(1 - r^n) / n(1 - r)$$

$$GM = ar^{(n-1)/2}$$

$$HM = a.n(1 - r)r^{n-1} / (1 - r)^n$$

Exercise 19.4

1. Find the harmonic mean of the following data : 5, 10, 15, 20, 25, 30, 35

2. Find the harmonic mean of the following data:
   0.00002853, 0.0003425, 0.004656, 0.07834, 0.676, 9.45, 78.3, 800

3. Calculate the A.M., G.M and H.M. of the observations and show that : A.M. > G.M. > H.M.

   37, 32, 36, 35, 43, 39, 41

4. Calculate the geometric mean of the following data:

Marks obtained	5	6	7	8	9
Frequency	3	4	8	7	2

5. Calculate the harmonic mean of the following frequency distribution:

Class	40-50	50-60	60-70
Frequency	19	25	36
Class	70-80	80-90	90-100
Frequency	72	51	43

6. Calculate the harmonic mean of the following frequency distribution:

Class	0-4	4-8	8-12	12-16	16-20
Frequency	4	12	20	9	5

7. Compute the harmonic mean of the following frequency distribution:

Class	40-50	50-60	60-70
Frequency	12	10	15
Class	70-80	80-90	90-100
Frequency	17	8	3

8. A car runs at the rate of 15 km/h during the first 30 km, at 20 km/h during the second 30 km and at the rate of 25 km/h during the third 30 km. Find out the average speed of the car.

9. A train starts from rest and travel a distance of 1 km in four parts each of 0.25 km with average speed 12, 16, 24 and 48 km/h. Explain the statement that the average speed over the whole journey of 1 km is 19.2 km/h and not 25 km/h.

10. A variate takes values $1, r, r^2, ..., r^{n-1}$ each with frequency unity. Show that

$$A = \frac{1-r^n}{n(1-r)}; \quad G = r^{(n-1)/2};$$

$$H = \frac{n(1-r) r^{n-1}}{1-r^n}$$

From the above observations, also show that $AH = G^2$ and $A > G > H$.

11. Find out the average speed of a car running at the rate of 20 km/h during the first 30 km; at 25 km/h during the second 30 km and at 30 km/h during the third 30 km.

12. Calculate the average speed of a train running at the rate of 20 km/h during the first 100 km, at 25 km/h during the second 100 km and at 30 km/h during the third 100 km.

ANSWERS

1. 13.5030 **2.** 0.0002095 **4.** 6.84 **5.** 82.5669 **6.** 7.246 **7.** 32.049
8. 1 **9.** 15 **11.** 24.32 km/h **12.** 24.39 km/h

19.13 MEDIAN

If we arrange the whole data in ascending or descending order, then the value of the middle variable is known as median.

In case when the number of variables are odd, then the middle value is known as median.

If the number of variables are even, *i.e.*, $(2n)$, the value of the mean of nth, $(n+1)^{th}$ variables will be median.

According to **Connor**, "The median is that value of the variable which divides the group into two equal parts, one part comprising all values greater and the other all values less than the median."

19.13.1 COMPUTATION OF MEDIAN

1. **Formula for individual series :** When the data given are ungrouped, then firstly, we arrange them in ascending or descending order. Then, if number of data are odd number, then the value of the middle variable will be median.

If number of data are even number $(2n)$, then the value of the mean of the nth and $(n+1)^{th}$ variable will give the median.

2. Formula for discrete series : Let us assume that $x_1, x_2..., x_n$ are the n observations

whose frequencies are given by $f_1, f_2...f_n$. To calculate the median of such series first of all we calculate the cumulative frequency and then calculate the sum of the frequency. Now, we calculate the median of series according to the N(sum of the frequency) is odd or even.

3. **Formula for Continuous Series:** In these type of questions all the data are divided into particular classes and their respective frequencies are given. Firstly, we calculate the cumulative frequencies. Then, we calculate the sum of the frequencies (N). According to N is even or odd, we find out the median. The class which contain this median is known as median class.

Now, the median for this series can be calculated by the formula

$$\text{Median} = l + \frac{\frac{1}{2}N - F}{f} \times i$$

where, l = Lower limit of the median class
           N = Sum of all the frequencies
           F = Sum of all the frequencies preceding the median class
           f = Frequency of median class
and i = Width of the median class

SOLVED EXAMPLES

EXAMPLE 1. *Find the median of the following data :* 20, 18, 22, 27, 25, 12, 15.

SOLUTION. Firstly, we will arrange these data in ascending orders 12, 15, 18, 20, 22, 25, 27
Here, $n = 7$, which is odd.

So,   Median = Value of the $\frac{(n+1)^{\text{th}}}{2}$ term = Value of $\frac{(7+1)^{\text{th}}}{2}$ th term
        = Value of 4th term.
Hence, Median = 20.

EXAMPLE 2. *Find the median in the following frequency distribution :*

Size	3	5	7	9	11	13	15
Frequency	7	3	12	28	10	9	6

SOLUTION.

(x)	(f)	Cumulative frequency
3	7	7
5	3	10
7	12	22
9	28	50
11	10	60
13	9	69
15	6	75
	$N = 75$	

Here, $N = 75$, which is odd.

So, the median = Value of the $\frac{(75+1)^{\text{th}}}{2}$ term
        = Value of the 38th term.

In the table, we see that the cumulative frequency 50 contain the 38th term. So the value of x for this column will be the median. Median = 9.

EXAMPLE 3. *Compute the median of the following frequency distribution :*

Size	8	10	12	14	16	18	20
Frequency	3	7	12	28	10	9	6

SOLUTION.

(x)	(f)	Cumulative frequency
8	3	3
10	7	10
12	12	22
14	28	50
16	10	60
18	9	69
20	6	75
	N = 75	

Here, $N = 75$, which is odd.

So, median is given by the value of $\dfrac{(N+1)}{2}^{th}$ term.

$$\text{Median} = \text{Value of the } \dfrac{(75+1)^{th}}{2} \text{ term.}$$

$$= \text{Value of the } 38^{th} \text{ term.}$$

The 38th term will fall in the cumulative frequency 50. So the median for this distribution is given by the value of x for this frequency.

Median = 14.

EXAMPLE 4. *Compute the median of the following frequency distribution :*

Wages	0-10	10-20	20-30	30-40	40-50
No. of workers	22	38	46	35	20

SOLUTION.

Wages (x)	Workers(f)	Cumulative frequency
0-10	22	22
10-20	38	60
20-30	46	106
30-40	35	141
40-50	20	161
	N = 161	

Here, $N = 161$, which is odd.

So, median is given by the value of $\dfrac{(N+1)^{th}}{2}$ term.

$$\Rightarrow \quad \dfrac{161+1}{2} = \dfrac{162}{2} = 81^{th} \text{ term.}$$

The 81th term is contained in the interval 20-30. So the class 20-30 will be the median class.

Lower limit of median class, $l = 20$

Sum of all the frequencies $N = 161$

Sum of all the frequencies preceding the median class, *i.e.*,

$$F = 60$$

Frequency of median class, $f = 46$

width of the median class $i = 10$

Then, median is given by

$$\text{Median} = l + \frac{\frac{1}{2}N - F}{f} \times i = 20 + \frac{\left(\frac{1}{2} \times 161 - 60\right)}{46} \times 10$$

$$= 20 + \frac{(80.5 - 60)}{46} \times 10 = 20 + \frac{205}{46} = 20 + 4.4565$$

$$= 24.4565$$

EXAMPLE 5.   *The height of the 11 students in inches of a hockey team is given. Find the median of these data*

$$65, 67, 69, 61, 60, 65, 66, 70, 71, 62, 72$$

SOLUTION.   First of all, we arrange all the terms in ascending or descending order

$$60, 61, 62, 65, 65, 66, 67, 69, 70, 71, 72$$

Number of terms are 11 which is odd.

So, median is given by the term $= \dfrac{(n+1)}{2} = \dfrac{(11+1)}{2} = 6$

Hence, the value of 6th term will be the median of these data.

$$\text{Median} = 66 \text{ inches}$$

EXAMPLE 6.   *RBC's number of 8 patients is 35, 44, 38, 36, 39, 40, 42 and 41 lac/mm³. Find out the median of this series.*

SOLUTION.   First of all data is ascending order, i.e., 35, 36, 38, 40, 41, 42 and 44 (lac/mm³)

$$\text{Median} = \frac{\left(\dfrac{n}{2}\right)^{\text{th}} \text{item} + \left(\dfrac{n}{2} + 1\right)^{\text{th}} \text{item}}{2} \qquad \text{(since } n = 8 \text{ is even)}$$

$$= \frac{\left(\dfrac{8}{2}\right)^{\text{th}} \text{item} + \left(\dfrac{8}{2} + 1\right)^{\text{th}} \text{item}}{2}$$

$$= \frac{4^{\text{th}} \text{item} + 5^{\text{th}} \text{item}}{2} = \frac{39 + 40}{2} = 39.5$$

EXAMPLE 7.   *The daily wages in Rupees of ten labourers of a factory are 4, 6, 9, 12, 11, 8, 5, 10, 11, 8. Calculate the median of these wages.*

SOLUTION.   Firstly we arrange the data in ascending order

$$4, 5, 6, 8, 8, 9, 10, 11, 11, 12$$

Here, $n = 10$, *i.e.*, number of data are even so the mean of $n/2$th and $\left(\dfrac{n}{2} + 1\right)^{\text{th}}$ value will be actual median.

$$\frac{n}{2} = 5, \frac{n}{2} + 1 = 5 + 1 = 6$$

5th term is given by $= 8$

6th term is given by = 9

$$\text{Median} = \frac{\text{value of } 5^{th} \text{ term} + \text{value of } 6^{th} \text{ term}}{2} = \frac{8+9}{2} = \frac{17}{2} = \text{Rs } 8.5$$

EXAMPLE 8. *Hb% of an animal was recorded as 6, 7, 4, 5, 5, 3 and 4 gm/100 ml. Calculate the median.*

SOLUTION. First of all above data is arranged in an ascending order, i.e., 3, 4, 5, 5, 6 and 7. The total number of scores is 7 (an odd number)

$$\text{Median} = \left(\frac{n+1}{2}\right)^{th} \text{item} = \left(\frac{7+1}{2}\right)^{th} \text{item}$$

$$= \left(\frac{8}{2}\right)^{th} \text{item} = 4^{th} \text{item} = 5$$

Median = 5 gm/100 ml

EXAMPLE 9. *Percentage of body water of 15 fishes and their frequencies given as follows. Find median of the given data.*

Water %	60	62	64	70	72	74	76	78	82	84	86
Frequency	1	1	1	2	1	1	2	1	1	3	1

SOLUTION.

Cumulative Frequency Table

Water %	Frequency (f)	Cumulative Frequency
60	1	1
62	1	2
64	1	3
70	2	5
72	1	6
74	1	7
76	2	9
78	1	10
82	1	11
84	3	14
86	1	15

Median will fall in $\left(\frac{N}{2}\right)^{th}$ item $= \left(\frac{15}{2}\right)^{th}$ item $= 7.5^{th}$ item

So median = 76.

EXAMPLE 10. *From the following table, calculate the median of the cost of living index.*

Cost of living index (Rs)	140-150	150-160	160-170	170-180	180-190	190-200
No. of weeks	5	10	20	9	6	2

(UPTU–2005)

SOLUTION. Calculation of median

Cost of living index	No. of weeks	Cf
140-150	5	5
150-160	10	15
160-170	20	35
170-180	9	44
180-190	6	50
190-200	2	52

Median number $\left(\dfrac{N}{2}\right)^{th}$ item $= \left(\dfrac{52}{2}\right)^{th}$ item $= 26^{th}$ item

which lies in the class 160-170. Thus, 160-170 is the median class in which $l_1 = 160, f = 20, F = 15, i = 10$.

Using the formula.

$$\text{Median} = l_1 + \frac{i}{f}\left(\frac{N}{2} - F\right) = 160 + \frac{10}{20}(26 - 15)$$

$$= 160 + \frac{11}{2} = 160 + 5.5 = 165.5$$

EXAMPLE 11. *Find the median for the following distribution :*

Size	10-15	15-20	20-25	25-30	30-35	35-40
Frequency	8	12	12	18	14	10

(UPTU–2002)

SOLUTION. Construct the table for cumulative frequency:

Size	Frequency (f)	Cumulative Frequency
10-15	8	8
15-20	12	20
20-25	12	32
25-30	18	50
30-35	14	64
35-40	10	74
	$\Sigma f = 74$	

Here, $N = 74$, which is even

So, median is given by $\left(\dfrac{N}{2}\right)^{th}$ term $= \left(\dfrac{74}{2}\right)^{th}$ term $= 37^{th}$ term

Hence, 37th term is contained in the interval 25-30.

So, 25-30 is the median class.

Then, the median is given by

$$\text{Median} = l + \frac{\frac{1}{2}N - F}{f} \times i = 25 + \frac{5}{18}(37 - 32)$$

$$= 25 + \frac{5}{18} \times 5 = 25 + 1.38 = 26.38$$

EXAMPLE 12. *Calculate the median for the following data :*

Marks	0-10	10-20	20-30	30-40	40-50	50-60	60-70	70-80
Frequency	6	4	5	4	5	7	17	12

(CCSU(B.Sc. Biotech)–1995)

SOLUTION.

Marks	Frequency	Cumulative Frequency
0-10	6	6
10-20	4	10
20-30	5	15
30-40	4	19
40-50	5	24
50-60	7	31
60-70	17	48
70-80	12	60

Here $\quad N = 60$

So, median is given by $\left(\dfrac{N}{2}\right)^{th}$ term $= \left(\dfrac{60}{2}\right)^{th}$ term $= 30^{th}$ term

Hence, 30th term is contained in the interval 50-60.

So, 50-60 is the median class.

Then, the median is given by

$$\text{Median} = l + \frac{\dfrac{N}{2} - F}{f} \times i = 50 + \frac{10}{7}(30 - 24)$$

$$= 50 + \frac{10}{7} \times 6 = 50 + 8.57 = 58.57$$

19.14 QUARTILES

If we arrange all the variates into ascending or descending order and this series is divided into four equal parts, then each part of this series is known as a quartile. The first part or first quartile contain first quarter of the series. The second quartile is said to be median and the third quartile is the $\left(\dfrac{3}{4}\right)^{th}$ term, which is known as upper quartile.

If Q_1, Q_2 and Q_3 are known as first, second and third quartile, then it can be calculated as

$$Q_k = l + \frac{\dfrac{1}{4}(kN) - F}{f} \times i, \quad k = 1, 2, 3$$

where, l = Lower limit of the quartile class.

N = Total frequency.

F = Cumulative frequency preceding the quartile class.

f = Frequency of quartile class.

and i = Class interval of quartile class.

19.14.1 Interquartile and Semi-Quartile

If Q_1, Q_2 and Q_3 are the first, second and third quartile, then $Q_3 - Q_1$ is called the interquartile range and $\frac{1}{2}(Q_3 - Q_1)$ is called the semi-interquartile range.

19.15 QUANTILES

If we arrange all the variates into ascending or descending order and divide this series into five equal parts, then each part of this series is known as quantiles.

If $Q_{n_1}, Q_{n_2}, Q_{n_3}$ and Q_{n_4} are the first, second, third and fourth quantile, then it can be calculated as

$$Q_{n_k} = l + \frac{\frac{1}{5}(kN) - F}{f} \times i, \quad k = 1, 2, 3, 4$$

where l = lower limit of the quantile class.

                     N = Total frequency.

                     F = Cumulative frequency preceding the quantile class.

and i = Class interval of quantile class.

19.16 DECILE AND PERCENTILE

In an ascending or descending series, if we divide the complete series into ten equal parts, then each part of the series is known as decile and if we divide the whole series into 100 equal parts, then each part of the series is known as percentile or centile. If $D_1, D_2, D_3, ..., D_9$ are the 9 deciles and $P_1, P_2, ..., P_{99}$ are 99 percentiles, then it can be calculated as

$$D_k = l + \frac{\frac{1}{10}kN - F}{f} \times i, \quad \text{where} \quad k = 1, 2, ..., 9$$

and

$$P_k = l + \frac{\frac{1}{100}kN - F}{f} \times i, \quad \text{where} = 1, 2, ..., 99$$

where l = Lower limit of decile or percentile class.

                     N = Total frequency.

                     F = Frequency preceding the decile or percentile class.

                     f = Frequency of decile or percentile class.

and i = Class width.

SOLVED EXAMPLES

EXAMPLE 1.   *Calculate the median, lower quartile and upper quartile of the data given below :*

          15, 37, 53, 18, 40, 54, 55, 40, 20, 36, 52, 53, 33, 75, 49, 49, 33, 35, 27, 44, 64, 61, 41, 26, 48, 70, 26, 29, 40, 59

SOLUTION.   Firstly, we arrange all the data in increasing order :

          15, 18, 20, 26, 27, 29, 33, 33, 35, 36, 37, 40, 40, 40, 41, 44, 48, 49, 49, 52, 53, 53, 54, 55, 59, 61, 64, 70, 75

          Here, $n = 30$, which is even.

So, the mean of the values of $\left(\dfrac{n}{2}\right)^{th}$ term and $\left(\dfrac{n}{2}+1\right)^{th}$ term

$$\dfrac{n}{2} = 15, \quad \dfrac{n}{2}+1 = 16$$

$$\text{Median} = \dfrac{\text{Value of }15^{th}\text{ term} + \text{Value of }16^{th}\text{ term}}{2} = \dfrac{40+41}{2} = 40.5$$

Now, we have to calculate the lower quartile and upper quartile.

Here, $N = 30$

$$\dfrac{1}{4}N = \dfrac{30}{4} = 7.5, \dfrac{3}{4}N = \dfrac{3}{4} \times 30 = 22.5$$

Thus, we see that the lower quartile lies between 7th and 8th term.

7th term $= 29$, 8th term $= 33$

So, the lower quartile,

$$Q_1 = 29 + \dfrac{1}{2}(33-29) = 29+2$$
$$Q_1 = 31$$

Now, $\dfrac{3}{4}N = 22.5$. So, the upper quartile will lie between 22nd and 23rd term.

22nd term $= 53$, 23rd term $= 53$

Hence, the upper quartile will lie between 53 and 53. So, it will be Q_3 term $= 53$.

EXAMPLE 2. *Calculate the median, lower quartile and upper quartile of the data given below :*

Classes	0-4	4-6	6-8	8-12	12-18	18-20
Frequency	4	6	8	12	7	2

SOLUTION. In this question, the classes are not equally divided. So, firstly, we tabulate them equally.

Classes	Frequency (f)	Cumulative frequency
0-4	4	4
4-8	14	18
8-12	12	30
12-16	5	35
16-20	4	39
	$N = 39$	

Here, $N = 39$, which is odd.

The median number is given by $= \dfrac{(N+1)}{2} = 20$

Median number 20 lies in the class 8-12. So, 8-12 will be the median class

Now, lower limit of median class, $l = 8$

Frequency of median class, $f = 12$

Frequency preceding the median class,
$\qquad F = 18$

Width of class interval, $i = 4$.

Then, median $= l + \dfrac{\left(\dfrac{1}{2}N - F\right) \times i}{f} = 8 + \dfrac{\left(\dfrac{1}{2} \times 39 - 18\right)}{12} \times 4$

$$= 8 + \dfrac{(19.5 - 18)}{12} \times 4$$

Median $= 8.5$

Now, we will calculate the value of lower quartile and upper quartile.

Firstly, we will calculate the lower quartile number.

Here, $N = 39$.

Lower quartile number $= \dfrac{1}{4}(39 + 1) = 10$

The lower quartile number 10 lies in the class 4-8 so it will be lower quartile class.

So, $l = 4, f = 14, F = 4, i = 4$

So,

$$Q_1 = l + \dfrac{\dfrac{1}{4}N - F}{f} \times i = 4 + \dfrac{\left(\dfrac{1}{4} \times 39 - 4\right) \times 4}{14} = 4 + 1.6428$$

So, $Q_1 = 5.6428$

Now, upper quartile number will be $= \dfrac{3}{4}(N + 1) = \dfrac{3}{4} \times 40 = 30$

The quartile number 30 lies in the class 8-12. So, it will be the upper quartile class.

For this, $l = 8, \ f = 12, \ F = 18, \ i = 4$

So, $Q_3 = l + \dfrac{\left(\dfrac{3}{4}N - F\right) \times i}{f} = 8 + \dfrac{\left(\dfrac{3}{4} \times 39 - 18\right) \times 4}{12} = 8 + 3.75$

Hence, $Q_3 = 11.75$

EXAMPLE 3.   *Calculate the value of median, quartile, 7th decile and eighty second percentile.*

Salary (in thousand)	0-10	10-20	20-30	30-40	40-50
Frequency	22	38	46	35	20

SOLUTION.

Class	Frequency (f)	Cumulative Frequency
0-10	22	22
10-20	38	60
20-30	46	106
30-40	35	141
40-50	20	161
	$N = 161$	

Now, for lower quartile

$$\dfrac{1}{4}N = \dfrac{1}{4} \times 161 = 40.25$$

The lower quartile number 40.25 lies in the class 10-20. So, it will be the lower quartile class.

$$l = 10, f = 38, F = 22, i = 10$$

$$Q = l + \frac{\left(\frac{1}{4}N - F\right)}{f} \times i = 10 + \frac{(40.25 - 22)}{38} \times 10$$

$$= 10 + 4.8026 = 14.8026$$

For the second quartile, the quartile number is given by $\frac{2N}{4} = \frac{2}{4} \times 161 = 80.5$

The quartile number 80.5 lies in the class 30-40, so it will be second quartile class.

$$l = 20, f = 46, F = 60, i = 10$$

Then
$$Q_2 = l + \frac{\left(\frac{2}{4}N - F\right)}{f} \times i = 20 + \frac{(80.5 - 60) \times 10}{46}$$

$$= 20 + 4.4565 = 24.4565 \text{ thousand}$$

For third quartile, the third quartile number is given by $\frac{3}{4}N = \frac{3}{4} \times 161 = 120.75$

The upper quartile number 120.75 lies in the class 30-40. So, it will be upper quartile class.

$$l = 30, f = 35, F = 106, i = 10$$

Then
$$Q_3 = l + \frac{\frac{3}{4}N - F}{f} \times i = 30 + \frac{(120.75 - 106)}{35} \times 10$$

$$= 30 + 4.2142 = 34.2142$$

$$Q_3 = 34.2142 \text{ thousand}$$

Now we have to calculate the 7th decile.

For this we calculate $\frac{7}{10}N = \frac{7}{10} \times 161 = 112.7$ thousand.

The decile number 112.7 lies in the class interval 30-40. So, it will be decile class.

For this
$$l = 30, f = 35, F = 106, i = 10$$

So, 7th decile D_7 is given by

$$D_7 = l + \frac{\frac{7}{10}N - F}{f} \times i = 30 + \frac{\frac{7}{10} \times 161 - 106}{35} \times 10$$

$$= 30 + \frac{112.7 - 106}{35} \times 10$$

$$D_7 = 31.9142 \text{ thousand}$$

For 82th percentile, calculate the percentile number.

Percentile number $= \frac{82}{100} \times 161 = 132.02$ thousand.

The percentile number 132.02 thousand lies in the class interval 30-40. So

$$l = 30, f = 35, F = 106, i = 10$$

$$P_{82} = l + \frac{\frac{82}{100}N - F}{f} \times i = 30 + \frac{\frac{82}{100} \times 161 - 106}{35} \times 10$$

$$= 30 + 7.4342 = 37.4342 \text{ thousand}$$

EXAMPLE 4. *Find the missing frequency in the following distribution :*

Class	10-20	20-30	30-40	40-50	50-60	60-70	70-80
Frequency	12	30	?	65	?	25	18

It is given that total frequency is 229 and the median is 46.

SOLUTION.

Class	Frequency (f)	Cumulative Frequency
10-20	12	12
20-30	30	42
30-40	x	$42 + x$
40-50	65	$107 + x$
50-60	y	$107 + x + y$
60-70	25	$132 + x + y$
70-80	18	$150 + x + y$
	$N = 150 + x + y$	

Now, it is given that the total frequency is given by 229.

So, $150 + x + y = 229$

$$x + y = 79 \qquad \qquad \qquad ...(1)$$

Now, it is also given that the median = 46.

So, median class will be 40-50. Then,

$$l = 40, F = 42 + x, f = 65, i = 10$$

Then,

Median, $M_d = l + \dfrac{\dfrac{1}{2}N - F}{f} \times i$

$$46 = 40 + \dfrac{\dfrac{1}{2} \times 229 - (42 + x)}{65} \times 10$$

$$46 = 40 + \dfrac{114.5 - 42 - x}{65} \times 10$$

$$\dfrac{72.5 - x}{65} \times 10 = 6$$

$$x = 72.5 - 39 = 33.5$$

or we can say that $x = 34$

Now, put this value in equation (1), we get

$$34 + y = 79$$

$$\Rightarrow \qquad \qquad y = 45$$

Exercise 19.5

1. Compute the median of the data : 9, 10, 15, 7, 11, 9, 8, 11, 7, 9, 10

2. The marks obtained by the ten students of class 8th is as follows:

75, 80, 96, 92, 89, 94, 100, 82, 63, 105

Find the median.

3. In a factory the daily wages of labourers are given by the following frequency distribution. Find the median:

Wages (Rs.)	6	8	10	12	14
No. of labourers	6	3	4	5	2

4. Compute the median for the following frequency distribution:

Age	5-7	8-10	11-13	14-16	17-19
No. of students	7	12	19	10	2

5. Compute the median for the following frequency distribution:

Variable	45-50	50-55	55-60	60-65	65-70
Frequency	2	3	5	7	9
Variable	70-75	75-80	80-85	85-90	90-95
Frequency	11	7	2	3	1

6. Calculate the median for the following frequency distribution:

Variable	0-5	5-10	10-15	15-20	20-25
Frequency	4	16	2	15	2

7. Calculate the median for the following frequency distribution:

Variable	0-10	10-20	20-30	30-40
Frequency	2	18	30	45
Variable	40-50	50-60	60-70	70-80
Frequency	35	20	6	3

8. Compute the median and quartiles of the following frequency distribution :

Age	15-19	20-24	25-29
Frequency	4	20	38
Age	30-34	35-39	40-44
Frequency	24	10	4

9. Compute the median and quartiles of the following frequency distribution :

Class	0-4	4-6	6-8
Frequency	4	6	8
Class	8-12	12-18	18-20
Frequency	12	7	2

10. Calculate quartiles, 3rd quantile, 3rd decile and 60th percentile from the following data :

Classes	0-10	10-20	20-30	30-40
Frequency	5	8	7	12
Classes	40-50	50-60	60-70	70-80
Frequency	28	20	10	10

11. Find the median from the following distribution: **(UPTU–2001)**

Class	0-9	10-19	20-29	30-39
Frequency	21	74	100	120
Class	40-49	50-59	60-69	70-79
Frequency	110	84	30	11

12. Calculate the median from the following data: **[CCSU(B.Sc Biotech)–2000]**

Monthly Rent (Rs.)	No. of families paying the rent
20-40	6
40-60	9
60-80	11
80-100	14
100-120	20
120-140	15
140-160	10
160-180	7

ANSWERS

1. 9 **2.** 90.5 **3.** Rs. 10 **4.** 11.447 **5.** 7.583 **6.** 36.559

7. 34.45 **8.** Median = 28.49; $Q_1 = 21.64$, $Q_2 = 32.89$ **9.** Median = 8.5; $Q_1 = 5.93$, , $Q_3 = 11.75$

10. $Q_1 = 37.5$, , $Q_2 = 46.43$, $Q_3 = 55.4$, $D_3 = 38.3$, $P_{60} = 50$ Quantile = 50

11. 36.17 **12.** 106

19.17 MODE

The variable whose frequency is maximum, is known as the mode of the distribution. In other words, we can say that the value which occurs most frequently in a distribution is known as the mode of the distribution.

19.17.1 COMPUTATION OF MODE

1. **For individual series :** For individual series, we can find the mode by inspection only. If number of variables are very large, then we arrange the data into discrete series and then we check the frequency for each variable to know the mode of the series.

If frequency for different variables are same in the frequency table, then we use the method of grouping to calculate the mode of the distribution.

2. **For discrete series:** Firstly, we arrange all the data in the frequency table. If the maximum frequency has the unique value, then it will be the mode of the series and if the maximum frequency occurs more than once, then mode can be calculated by grouping of data.

3. **For continuous series :** The class with maximum frequency is

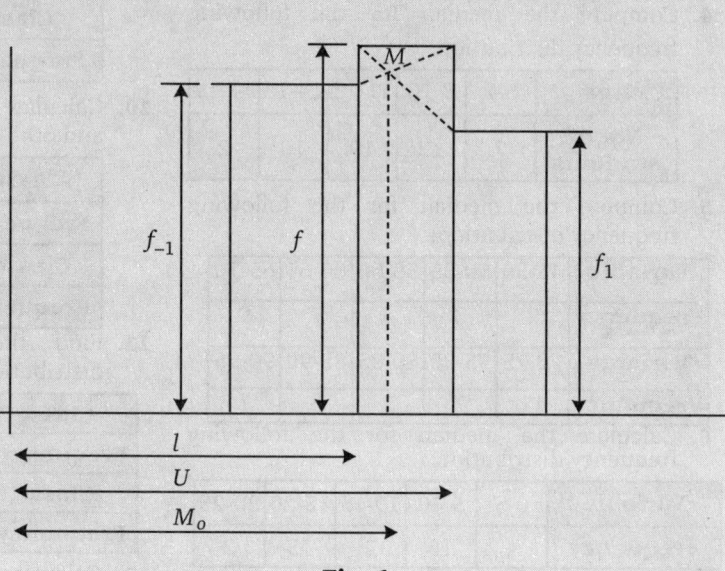

Representation of Mode of the Graph

Fig. 1

known as the modal class and we can obtain the mode of this series by calculating the formula :

$$\text{Mode} = l + \frac{f - f_{-1}}{2f - f_{-1} - f_1} \times i \qquad \text{[When } f > f^{-1}\text{]}$$

where,
l = Lower limit of the modal class.
f = Frequency of the modal class.
f_{-1} = Frequency preceding the modal class.
f_1 = Frequency succeeding the modal class.
i = Class width.

If f_{-1} and f_1 are both (or one) is greater than f, then we use the following formula

$$M_0 = l + \frac{f_1 \times i}{f_1 + f_{-1}}.$$

SOLVED EXAMPLES

EXAMPLE 1. *Find the mode of the data given below:*
0, 1, 6, 7, 2, 3, 7, 6, 6, 2, 6, 0, 5, 6, 0

SOLUTION. In the given data, we see that the frequency of 6 is 5, which is maximum and no other frequency is equal to this frequency. So, 6 is the mode of the given data.

EXAMPLE 2. *Find the mode of the data given below:*
25, 15, 23, 40, 27, 25, 23, 25, 20

SOLUTION. In the given data, the frequency for 25 is given by 3 which is maximum, since no other data has the same frequency or more than 3. So the mode of the given data is given by 25.

EXAMPLE 3. *Find the mode of the following frequency distribution :*

Mid Value	15	20	25	30	35	40	45	50	55
Frequency	2	22	19	14	3	4	6	1	1

SOLUTION. Here, the mid value for each classes is given. So, firstly, we have to convert the given data into grouped frequency. So,

Mid Value	Class	Frequency
15	12.5 - 17.5	2
20	17.5 - 22.5	22
25	22.5 - 27.5	19
30	27.5 - 32.5	14
35	32.5 - 37.5	3
40	37.5 - 42.5	4
45	42.5 - 47.5	6
50	47.5 - 52.5	1
55	52.5 - 57.5	1

Here, by inspection, we see that the maximum frequency is given by = 22. So the class 17.5 – 22.5 will be the modal class.
For the modal class
$f = 22, l = 17.5, f = 22, f_{-1} = 2, f = 19, i = 5$
Then,
mode $\quad M_0 = l + \dfrac{f - f_{-1}}{2f - f_{-1} - f_1} \times i = 17.5 + \dfrac{22 - 2}{2 \times 22 - 2 - 19} \times 5$

$$= 17.5 + \dfrac{20}{44 - 21} \times 5 = 17.5 + \dfrac{20}{23} \times 5 = 17.5 + 4.3478$$

Hence, Mode $M_0 = 21.8478$.

EXAMPLE 4. *Calculate the mode of the following frequency distribution:*

Size of item	4	5	6	7	8	9	10	11	12	13
Frequency	2	5	8	9	12	14	14	15	11	13

SOLUTION. In the given frequency distribution, we see that the frequency of item 11 is maximum, and the frequency for the item 10 and 12 is also not very different, but the effect of other frequencies may change the position of mode. So, we use the grouping method to know the mode of given distribution.

Size	Frequency					
	I	II	III	IV	V	VI
4	2		13			
5	5	7		15	22	
6	8		21			
7	9	17				
8	12		(28)	35	(40)	29
9	14	26				
10	14		26			(43)
11	(15)	(29)		(40)	39	
12	11					
13	13	24				

First of all, we write the corresponding frequencies for each item. Now, we grouped them in II column by taking two frequencies at a time. In III column, we add the frequencies two at a time after leaving the first frequency. In IV column, we add the frequencies taking three at a time. In column V, we add the frequencies taking three at a time after leaving the first frequency. In column VI, we add three frequencies taking at a time after leaving the first two frequencies. Now, we check the maximum frequency in each column and note the item having maximum frequency in each column. The item which occurs maximum number of times will be the mode of the series.

Column	Item with more frequency
I	11
II	10, 11
III	9, 10
IV	10, 11, 12
V	8, 9, 10
VI	9, 10, 11

Now frequency of items is given by

8	9	10	11	12
1	3	5	4	1

Here, we see that the number 10 occurs more times than any other items. The frequency of 10 is 5 which is maximum. So, 10 will be the mode of given series.

EXAMPLE 5. *The expenses of 100 families are given below. If the mode of this distribution is 48, then find the missing frequency.*

Expenses (Rs.)	0-20	20-40	40-60	60-80	80-100
No. of families	14	–	27	–	15

SOLUTION.

Expenses	Frequency
0-20	14
20-40	x
40-60	27
60-80	y
80-100	15
	$N = 56+x+y$

It is given that $N = 100$

So, $56 + x + y = 100$

$x + y = 44$ ... (1)

Now, it is given that mode of this distribution = 48 which falls in the class 40-60. So for this

$$l = 40, f = 27, f_{-1} = x, f_1, y, i = 20$$

Then mode,

$$M_0 = l + \frac{f - f_{-1}}{2f - f_{-1} - f_1} \times i$$

$$48 = 40 + \frac{27-x}{54-x-y} \times 20$$

$$8 = \frac{27-x}{54-x-y} \times 20$$

$$8(54 - x - y) = 20(27 - x)$$

$$2(54 - x - y) = 5(27 - x)$$

$$108 - 2x - 2y = 135 - 5x$$

$$3x - 2y = 27 \qquad \dots (2)$$

Solving equation (1) and (2), we get

$$5x = 115$$

$$x = \frac{115}{5} = 23$$

Put this value in equation (1)

$$23 + y = 44, \ y = 21$$

Hence, the missing frequency is given by = 23, 21.

19.17.2 EMPIRICAL FORMULA

The empirical relationship between mean, mode and median is given by:

Mode = 3 Median – 2 Mean

EXAMPLE 6. *A distribution* $x_1, x_2, ..., x_n$ *with frequencies* $f_1, f_2, ..., f_n$ *is transformed into the distribution with the same corresponding frequencies of the relation* $X_i = ax_i + b, \ i = 1, 2, ..., n$ *, where a and b are constant. Show that the mean, mode and median of the new distribution are given in terms of those of the first distribution by the same transformation.*

SOLUTION.

$$X_i = ax_i + b, \ i = 1, 2, ..., n$$

$$\frac{\Sigma f_i X_i}{\Sigma f_i} = \frac{a}{\Sigma f_i} \Sigma X_i + \frac{1}{\Sigma f_i} \Sigma b$$

$$\overline{X} = a\overline{x} + b \qquad [\because \Sigma b = bN \text{ and } \Sigma f_i = N]$$

Hence, A.M. is obtained by the same transformation.

In case of median, we find the value of the variable corresponding to the middle term and we know that it is not affected by the change of origin and change of scale. The median item in the frequency distribution will remain the median item in transformed distribution. This argument for mode holds good because mode is the most fashionable item of the series.

$$X_{mode} = ax_{mode} + b$$

EXAMPLE 7. *Calculate the mode of the following data:*

Wages (Rs.)	Below 100	100-200	200-300	300-400	400-500	500 and above
Frequency	8	12	25	15	10	5

SOLUTION. As the frequencies are regular and highest frequency 25 belongs to the class 200-300. Thus 200-300 is the modal class in which $l=200, f=25, f_{-1} = 12, f = 15$, $i = 100$

Using the formula

$$\text{Mode} = l + \frac{f - f_{-1}}{2f - f_{-1} - f_1} \times i = 200 + \frac{25 - 12}{50 - 12 - 15} \times 100$$

$$= 200 + \frac{1300}{23} = 200 + 5.652 = 205.652.$$

EXAMPLE 8. *Compute the mode of the following distribution:*

Class	10-20	20-30	30-40	40-50	50-60	60-70	70-80
Frequency	22	35	40	62	50	45	40

SOLUTION. Here, the frequencies are regular and highest frequency 62 belongs to the class 40-50. Thus 40-50 is the modal class in which $l = 40$, $f = 62$, $f_{-1} = 40$, $f_1 = 50$, $i = 10$.

Using the formula

$$\text{Mode} = l + \frac{f - f_{-1}}{2f - f_{-1} - f_1} \times i = 40 + \frac{62 - 40}{2 \times 62 - 40 - 50} \times 10$$

$$= 40 + \frac{220}{34} = 40 + 6.47 = 46.47.$$

EXAMPLE 9. *Obtain mean, median and mode of the following classes:*

Class	0-10	10-20	20-30	30-40	40-50
Frequency	13	22	30	20	15

SOLUTION. **Calculation of Mean**

Class	Mid Value x	f	(x – 25) = d	f.d
0-10	5	13	– 20	–260
10-20	15	22	– 10	–220
20-30	25	30	0	0
30-40	35	20	10	200
40-50	45	15	20	300
		100		20

$a = 25$

Then, Mean $= a + \frac{\Sigma fd}{\Sigma f} = 25 + \frac{20}{100} = 25.2$

Calculation for Median Value

Class	Frequency f	Cumulative frequency
0-10	13	13
10-20	22	35
20-30	30←Mode class	65←Mode class
30-40	20	85
40-50	15	100

Here, the total frequencies = 100

Size of median $= \frac{100}{2} = 50^{\text{th}}$

Observation in the data set. This observation lies in the class (20-30).

Thus,

$$\text{Median} = 20 + \frac{\left(\dfrac{100}{2}\right) - 35}{30} \times 10 = 20 + 5 = 25$$

Calculation of Modal Value

In the table, the largest frequency corresponds to the class interval 20-30, therefore it is the mode class. Then, we have

$$l = 20, \ f = 30, \ f_{-1} = 22, \ f_1 = 20, \ h = 10$$

$$\text{Mode} = l + \frac{f - f_{-1}}{2f - f_{-1} - f_1} \times h = 20 + \frac{30 - 22}{60 - 22 - 20} = 20 + \frac{80}{18}$$

$$= 20 + 4.44 = 24.44$$

EXAMPLE 10. *Water percentage of fifteen fishes of a fish wave recorded as 60, 64, 62, 76, 70, 84, 82, 72, 76, 84, 78, 84 and 86. Find the mode of this series.*

SOLUTION. First of all, data is arranged in ascending order. Not even single observation is spared. It comes as 60, 62, 64, 70, 72, 74, 76, 76, 78, 82, 84, 84, 84 and 86. By single observation one can say that 84 is the mode, because 84 is repeated maximum times (three times) in the above series.

EXAMPLE 11. *Find the mode of the following frequency distribution :*

Marks less than	2	4	6	8	10	12	14
No. of students	6	20	36	56	58	62	68

SOLUTION. Firstly, we convert the given cumulative frequency into the ordinary frequency.

Marks	No. of students					
	I	II	III	IV	V	VI
0-2	6					
2-4	14	20		36		
4-6	16		30		50	
6-8	(20)	36				38
8-10	2		22	26		
10-12	4	6			12	
12-14	6		10			

Column	Marks with maximum frequency
I	6-8
II	4-6, 6-8
III	2-4, 4-6
IV	0-2, 2-4, 4-6
V	2-4, 4-6, 6-8
VI	4-6, 6-8, 8-10

Now, the frequencies for each class is given by

0-2	2-4	4-6	6-8	8-10
1	3	5	4	1

So, the class 4-6 with frequency 5 will be the modal class.

For this $l = 4, f = 16, f_{-1} = 14, f_1 = 2, i = 2$

Then, mode is given by

$$M_0 = l + \frac{f - f_{-1}}{2f - f_{-1} - f_1} \times i = 4 + \frac{16 - 14}{2 \times 16 - 14 - 20} \times 2$$

$$= 4 + \frac{2}{32 - 14 - 20} \times 2 = 4 + \frac{2}{-2} \times 2 = 4 - 2 = 2$$

So, $M_0 = 2$

EXAMPLE 12. *Calculate the mode of the following data :*

Class Interval	0-10	10-20	20-30	30-40	40-50	50-60
Frequency	4	2	10	12	6	8

SOLUTION. Here, the frequencies are regular and highest frequency 12 belongs to the class 30-40. Thus, 30-40 is the modal class in which

$$l = 30, f = 12, f_{-1} = 10, f_1 = 6, i = 10$$

Using the formula

$$\text{Mode} = l + \frac{f - f_{-1}}{2f - f_{-1} - f_1} \times i = 30 + \frac{12 - 10}{2 \times 12 - 10 - 6} \times 10$$

$$= 30 + \frac{20}{8} = 30 + 2.5 = 32.5$$

EXAMPLE 13. *Calculate the mean, median and mode of the following data of 120 aricles.*

Weight in gms	0-10	10-20	20-30	30-40	40-50	50-60
No. of articles	14	17	22	26	23	18

SOLUTION.

Table for Calculation of Mean

Weight in gms	No. of articles f	Mid Value X	$d = X - 35$	$d' = \dfrac{d}{10}$	$f \cdot d'$
0-10	14	5	− 30	− 3	− 42
10-20	17	15	− 20	− 2	− 34
20-30	22	25	− 10	− 1	− 22
30-40	26	35	0	0	0
40-50	23	45	10	1	23
50-60	18	55	20	2	36
	$N = 120$		100		$\Sigma fd' = -39$

$$\bar{X} = A + \frac{\Sigma fd'}{N} \times i = 35 + \frac{(-39)}{120} \times 10 = 35 - 3.25 = 31.75$$

Table for Calculation of Median

Wt in grams	Number of articles	Cumulative Frequency
0-10	14	14
10-20	17	31
20-30	22	53
30-40	26	79
40-50	23	102
50-60	18	120
	120	

$$\text{Median No.} = \left(\frac{120}{2}\right)^{th} \text{item} = 60^{th} \text{item}$$

It belongs to 30-40 class. Then using the formula

$$\text{Median} = l + \frac{i}{f}\left(\frac{N}{2} - f\right) = 30 + \frac{10}{26}(60 - 53)$$

$$= 30 + \frac{10}{26} \times 7 = 30 + \frac{70}{26} = 32.69$$

Mode

Since highest frequency is 26 which belongs to 30-40 class. Thus, using the formula

$$\text{Mode} = l + \frac{f - f_{-1}}{2f - f_{-1} - f_1} \times h = 30 + \frac{26 - 22}{2 \times 26 - 22 - 23} \times 10$$

$$= 30 + \frac{40}{52 - 45} = 35.71$$

EXAMPLE 14. *Find the mode from the following distribution :*

Class	0-9	10-19	20-29	30-39	40-49	50-59	60-69	70-79
Frequency	21	74	100	120	110	84	30	11

(UPTU–2011)

SOLUTION.

Class	0-9.5	9.5-19.5	19.5-29.5	29.5-39.5
Frequency	21	74	100	120
Class	39.5-49.5	49.5-59.5	59.5-69.5	69.5-79.5
Frequency	110	84	30	11

Here, the highest frequency is 12, which lies in the interval 29.5-39.5

Therefore, $l = 29.5, f = 120, f_{-1} = 100, f_1 = 110$ and $i = 10$

$$\text{Mode} = l + \frac{f - f_{-1}}{2f - f_{-1} - f_1} \times i = 29.5 + \frac{120 - 100}{2 \times 120 - 100 - 110} \times 10$$

$$= 29.5 + \frac{20}{240 - 210} \times 10 = 29.5 + \frac{200}{30}$$

$$= 29.5 + 6.67 = 36.17$$

Exercise 19.6

1. Find the mode of the following data : 4, 5, 8, 6, 9, 8, 8, 6, 5, 11, 9, 8

2. Find the mode of the following data : 0, 1, 6, 7, 2, 3, 7, 6, 6, 2, 6, 0, 5, 6, 0

3. Find the mode of the following data :

Size	2	3	4	5	6	7
Frequency	3	8	10	12	16	14
Size	8	9	10	11	12	13
Frequency	10	8	17	5	4	1

4. Find the mode of the following data :

[CCSU(B.Sc. Biotech)–2000]

Monthly rent	20-40	40-60	60-80	80-100	100-120
No. of students	6	9	11	14	20
Monthly rent	120-140	140-160	160-180	180-200	
No. of students	15	10	8	7	

5. Find the mode of the following frequency distribution :

Height	120-124	125-129	130-134	135-139
Frequency	2	5	8	15
Height	140-144	145-149	150-154	
Frequency	20	10	5	

6. Find the mode of the following frequency distribution : [CCSU(B.Sc. Biotech)–1997]

Marks	10-25	25-40	40-55
Frequency	6	20	44
Marks	55-70	70-85	85-100
Frequency	26	7	1

7. Find the mode for the following frequency distribution :

Age	20-25	25-30	30-35	35-40
No. of persons	50	70	80	180
Age	40-45	45-50	50-55	55-60
No. of persons	150	120	70	50

8. Find the mode for the following distribution : (UPTU–2001)

x	0-10	10-20	20-30	30-40
f	4	6	10	20
x	40-50	50-60	60-70	
f	10	6	4	

9. Find the mode for the following distribution: (UPTU–2002)

Size	10-15	15-20	20-25
Frequency	8	12	12
Size	25-30	30-35	35-40
Frequency	18	14	10

ANSWERS

8. 0.35   **9.** 28

Objective Evaluation

MULTIPLE CHOICE QUESTIONS

Choose the most appropriate one.

1. Mean is a measure of :
   - (a) location
   - (b) dispersion
   - (c) correlation
   - (d) none of these

2. Which of the following is a measure of central value ?
   - (a) median
   - (b) standard deviation
   - (c) mean deviation
   - (d) quartile deviation

3. Which of the followign represents median ?
   - (a) first quartile
   - (b) fifth percentile
   - (c) sixth decile
   - (d) none of the above

4. If a constant value 50 is subtracted from each observation of a set, the mean of the set is :
   - (a) increased by 50
   - (b) decreased by 50
   - (c) is not affeected
   - (d) zero

5. If a constant 5 is added to each observation of a set, the mean of the set is :
   - (a) increased by 5
   - (b) decreased by 5
   - (c) 5 timese the original mean
   - (d) not affected

6. If the sum of N observations is 630 and their mean is 42, then the value of N is :
   - (a) 21
   - (b) 30
   - (c) 15
   - (d) 20

7. If the two observations are 10 and -10 then their Harmonic mean is :
   - (a) 10
   - (b) 0
   - (c) 5
   - (d) ∞

8. If the two observations are 5 and -5, their geometric mean is :
   - (a) 5
   - (b) -5
   - (c) 0
   - (d) none of the above

9. If the two observations are 20 and -20, their arithmetic mean is :
   - (a) 10
   - (b) 20
   - (c) 0
   - (d) none of the above

10. The average of $2n$ natural numbers from 1 to $2n$ is :
    - (a) $\dfrac{(n+1)}{2}$
    - (b) $\dfrac{(2n+1)}{2}$
    - (c) $\dfrac{n(n+1)}{2}$
    - (d) $\dfrac{n(2n+1)}{2}$

TRUE/FALSE

Write 'T' for True and 'F' for False statement.

1. Second quartile is same as median. **(T/F)**

2. Third quartile and 75th percentile are same. **(T/F)**

3. Second quartile and 5th decile are same. **(T/F)**

4. The decile which preceeds 72th percentile is D_7. **(T/F)**

5. The percentage of value which lie between P_{30} and P_{56} is 26 percent. **(T/F)**

6. 25th percentile is same as first quartile. **(T/F)**

7. Sixth decile is same as 60th percentile. **(T/F)**

8. Percentage of values lying between first and sixth decile is 50 percentage. **(T/F)**

9. Percentage of values which are greater than 66th percentile is 34 percent. **(T/F)**

10. In a frequency distribution fourth decile and 40th percentile are same. **(T/F)**

FILL IN THE BLANKS

1. The median of the series 3, 18, 7, 20, 11, 12, 9, 17, 22 is _____ .

2. Median is _____ for each and every distribution.

3. Two ogives, the one less than type and the other less than type intersect at _____ .

4. Median is more suited average for grouped data with _____ class.

5. The variate value having maximum frequency distribution is called _____ .

6. The mode of the distribution of values 5, 7, 9, 9, 8, 5, 6, 8, 7, 7, 5, 7, 9, 2, 7 is_____ .

7. _____ and _____ are least affected by the extreme values as a measure of central tendency.

8. Out of all the measures of central tendency _____ is the only measure which is not unique.

9. _____ is not approximate for further algebraic treatment as a measure of central value.

10. The distribution which has only one mode is called _____ .

ANSWERS

⌘ MULTIPLE CHOICE QUESTIONS

1. (a)   **2.** (a)   **3.** (b)   **4.** (b)   **5.** (a)   **6.** (c)   **7.** (d)   **8.** (d)   **9.** (c)
10. (b)

⌘ TRUE/ FALSE

1. T   **2.** T   **3.** T   **4.** T   **5.** T   **6.** T   **7.** T   **8.** T   **9.** T
10. T

⌘ FILL IN THE BLANKS

1. 12   **2.** unique **3.** median **4.** open end   **5.** mode   **6.** 7   **7.** median; mode
8. mode **9.** mode **10.** unimodal

❑❑❑❑❑

CHAPTER 20

Measure of Dispersion and Skewness

20.1 INTRODUCTION

Sometimes we see that the mean, mode and median of different series give the same value, so it becomes necessary to differentiate these series. For this, we check the deviation of variables from the mean value in both the series. Then the maximum deviation from the mean value in both the series gives the different values. Thus we can differentiate these series and such type of variation is known as dispersion.

For Example :

The marks of six students in I and II tests are given as follows :

I test	43	46	49	50	52	54
II test	38	48	49	51	55	53

The mean in both the cases is given by 49. Now we check that the deviation from mean in first series is given by 49-43 = 6, while in II series, it is given by 49 – 38 = 11. Thus we can differentiate these series with the same value of mean. Such variations in the series is known as dispersions.

According to Griffin, *"A measure of variation or dispersion describes the degree of scatter shown by the observations and is usually measured as an average deviation about some central value or by an order statistics."*

According to Dr. Bowley, *"Dispersion is the measure of the variation of the items."*

20.1.1 MERITS OF A GOOD MEASURE OF DISPERSION

1. It should be easy to calculate.
2. It should be well defined.
3. All the data must be used in calculation.
4. It should be capable of further mathematical and statistical treatment.
5. It should not be affected by the fluctuation of sampling.

20.1.2 MEASURE OF DISPERSION

The dispersion can be calculated by the following two methods :

1. Method of limit. 2. Average deviation method.

In the method of limit, the following measure can be done

1. Range 2. Interquartile range
3. Percentile range

In average deviation method, the following measure can be done :

1. Mean deviation 2. Standard deviation
3. Quartile deviation

20.2 RANGE

It is very simple measure of dispersion. It can be defined by the difference with maximum and minimum value of variables, in the distribution.

If $x_1, x_2, ..., x_n$ be any distribution, x_1 is the minimum and x_n is the maximum value of distribution, then range is defined by

$$\text{Range} = x_n - x_1$$

20.2.1 COEFFICIENT OF RANGE

The coefficient of the range is defined by the ratio of the difference between maximum and minimum value with the summation of these values. If $x_1, x_2, ..., x_n$ by any distribution and x_1, x_n be the minimum and maximum values respectively, then coefficient of range is given by

$$\text{Coefficient of Range} = \frac{\text{The difference of max and min value}}{\text{The sum of max and min value}} = \frac{x_n - x_1}{x_n + x_1}$$

20.2.2 MERITS OF RANGE

1. We can easily find the value of range.
2. It is very simple method to know the measure of dispersion.
3. This measure depends only on two values of the given series. So, it is very useful in quick computation.

20.2.3 DEMERITS OF RANGE

1. It suffers much fluctuation due to its dependence on only two (max or min) values of the series.
2. It does not give any idea about the series.
3. It depends only on two extreme values of the series.

SOLVED EXAMPLES

EXAMPLE 1. *Find the range and its coefficient in the given series :*
        110, 117, 129, 300, 357, 100, 500, 630, 750

SOLUTION. The largest value of the series, $x_n = 750$
        The smallest value of the series, $x_1 = 100$.
        Then, Range $x_n - x_1 = 750 - 100$
                Range $= 650$

$$\text{Coefficient of range} = \frac{x_n - x_1}{x_n + x_1} = \frac{750 - 100}{750 + 100} = \frac{650}{850}$$

        Hence, coefficient of range $= 0.7647$.

EXAMPLE 2. *Find the range and its coefficients in the following frequency distribution :*

x	6	12	18	24	30	36	42
f	4	7	9	18	15	10	5

SOLUTION. The maximum value of the series, $x_n = 42$
        The minimum value of the series, $x_1 = 6$
        Then,   Range $= x_n - x_1 = 42 - 6$
                Range $= 36$

$$\text{Coefficient of range} = \frac{x_n - x_1}{x_n + x_1} = \frac{42 - 6}{42 + 6} = \frac{36}{48}$$

Coefficient of range = 0.75.

EXAMPLE 3. *Find the range and its coefficients in the following frequency distribution :*

Marks Group	No. of Students
0-10	5
10-20	7
20-30	10
30-40	16
40-50	11
50-60	7
60-70	3
70-80	2
80-90	2
90-100	0

SOLUTION. The maximum value of the series, $x_n = 100$

The minimum value of the series, $x_1 = 0$

Then, Range = $x_n - x_1 = 100 - 0 = 100$

Range = 100

$$\text{Coefficient of range} = \frac{x_n - x_1}{x_n + x_1} = \frac{100 - 0}{100 + 0}$$

Coefficient of range = 1.

20.3 INTERQUARTILE RANGE AND SEMI-QUARTILE DEVIATION

If we arrange a series into ascending order and this series is divided into four equal parts, then each part of this series is known as a quartile. If Q_1, Q_2 and Q_3 are the first, second and third quartile, then $Q_3 - Q_1$ is called the interquartile range and $\frac{1}{2}(Q_3 - Q_1)$ is called the semi-interquartile range.

i.e., Interquartile range = $Q_3 - Q_1$

Semiquartile deviation = $\frac{1}{2}(Q_3 - Q_1)$.

☙ REMARK
➠ Semiquartile deviation is also known as quartile deviation.

20.3.1 COEFFICIENT OF QUARTILE DEVIATION

Coefficient of quartile deviation is given by $= \dfrac{\frac{1}{2}(Q_3 - Q_1)}{\frac{1}{2}(Q_3 + Q_1)}$

$$\text{Coefficient of quartile deviation} = \frac{Q_3 - Q_1}{Q_3 + Q_1}$$

Merits

1. It can be easily calculated.
2. It is easy to understand.
3. It does not depend only on the extreme values.

Demerits

1. Fluctuation of sampling affect this deviation very much.
2. We can get only a rough measure from it.
3. All the data are not used in the measure.

☛ **RECAPITULATIONS**

- Dispersion is the measure of variations of the items
- Range is the difference of maximum and minimum values of the given data.
- Coeff. of Range = $\dfrac{\text{difference of max. and min value}}{\text{Sum of max and min value}}$
- Interguartile range = $Q_3 - Q_1$
- Semiquartile devitation = $\dfrac{1}{2}(Q_3 - Q_1)$

SOLVED EXAMPLES

EXAMPLE 1. *Compute the quartile deviation and its coefficient from the following series:*

Item	1	2	3	4	5
Frequency	3	2	5	7	9
Item	6	7	8	9	10
Frequency	5	8	10	2	1

SOLUTION.

Item	1	2	3	4	5	6
Frequency	3	2	5	7	9	5
Cumulative Frequency	3	5	10	17	26	31
Item	7	8	9	10		
Frequency	8	10	2	1	52 Total	
Cumulative Frequency	39	49	51	52		

For first quartile, $\dfrac{N}{4} = \dfrac{52}{4} = 13$, which lies in the frequency of 4th item. So Q_1 is given by 4.

$$Q_1 = 4$$

Now, for third quartile, $\dfrac{3N}{4} = \dfrac{3}{4} \times 52 = 39$ *i.e.*, $Q_3 = 7$

EXAMPLE 2. *Calculate the quartile deviation for the marks of 63 students in mathematics given as follows :*

Marks	0-10	10-20	20-30	30-40	40-50
Frequency	5	7	10	16	11

Marks	50-60	60-70	70-80	80-90	90-100
Frequency	7	3	2	2	0

SOLUTION.

Marks	Frequency	Cumulative frequency
0-10	5	5
10-20	7	12
20-30	10	22
30-40	16	38
40-50	11	49
50-60	7	56
60-70	3	59
70-80	2	61
80-90	2	63
90-100	0	63
	63	

For first quartile, $\quad\quad N = 63$,

which is odd, so $\quad\quad \dfrac{N+1}{4} = \dfrac{63+1}{4} = 16^{\text{th}}$ student.

which is contained in the group 20-30.

So, 20-30 will be the first quartile class.

$$l = 20, \ N = 63, \ F = 12, \ f = 10, \ i = 10$$

Then, $\quad Q_1 = l + \dfrac{\dfrac{1}{4}N - F}{f} \times i = 20 + \dfrac{\dfrac{1}{4} \times 63 - 12}{10} \times 10 = 20 + 3.75$

$$Q_1 = 23.75 \text{ marks}$$

Now, for

$$Q_3 = \dfrac{3}{4}(N+1) = \dfrac{3}{4}(63+1) = \dfrac{3}{4} \times 64 = 48$$

48th student will lie in the class 40-50 so it will be the third or upper quartile class.

For Q_3,

$$l = 40, \ N = 63, \ F = 38, \ f = 11, \ i = 10$$

$$Q_3 = l + \dfrac{\dfrac{3}{4}N - F}{f} \times i = 40 + \dfrac{\dfrac{3}{4} \times 63 - 38}{11} \times 10$$

$$= 40 + 8.4090 = 48.4090$$

Now, we have to calculate the quartile deviation.

$$\text{Quartile deviation} = \dfrac{1}{2}(Q_3 - Q_1) = \dfrac{1}{2}(48.4090 - 23.75)$$

$$= 12.3295 \text{ marks}$$

Coefficients of quartile deviation is given by

$$\text{Coefficient of quartile deviation} = \dfrac{Q_3 - Q_1}{Q_3 + Q_1} = \dfrac{48.4090 - 23.75}{48.4090 + 23.75} = \dfrac{24.659}{72.159}$$

Coefficient of quartile deviation = 0.3417 marks.

EXAMPLE 3. *Water percentage in the body of a species of fish and their frequency is given in the following table. Compute the quartile deviation.*

Class Interval	Frequency
16-20	4
21-25	3
26-30	8
31-35	9
36-40	14
41-45	3
46-50	3
51-55	2
56-60	2
61-65	2
	$\Sigma f = 50$

SOLUTION.

Cumulative Frequency Table

Class Interval	Frequency	Cumulative Frequency
16-20	4	4
21-25	3	7
26-30	8	15
31-35	9	24
36-40	14	38
41-45	3	41
46-50	3	44
51-55	2	46
56-60	2	48
61-65	2	50

where $\qquad Q_1 = l + \dfrac{\left(\dfrac{N}{4} - F\right) \times i}{f}$

$$Q_1 = \text{size} \dfrac{N^{\text{th}}}{4} \text{ observation} = \dfrac{50}{4} = 12.5$$

$$Q_1 = 25.5 + \dfrac{(12.5 - 7)}{8} \times 5 = 25.5 + \dfrac{5.5}{8} \times 5$$

$$= 25.5 + 0.687 \times 5 = 25.5 + 3.437 = 28.93$$

$$Q_3 = \text{size} \dfrac{3N^{\text{th}}}{4} \text{ observation} = \dfrac{3 \times 50}{4} = 37.5$$

$$Q_3 = l + \frac{\left(\frac{3N}{4} - F\right) \times i}{f} = 35.5 + \frac{35.5 - 24}{14} \times 5$$

$$= 35.5 + \frac{13.5}{14} \times 5 = 35.5 + 0.964 \times 5$$

$$= 35.5 + 4.821 = 40.32$$

Quartile deviation $= (Q) = \dfrac{Q_3 - Q_1}{2} = \dfrac{40.32 - 28.93}{2} = \dfrac{11.39}{2} = 5.69$

Coefficient of quartile deviation

$$= \frac{Q_3 - Q_1}{Q_3 + Q_1} = \frac{40.32 - 28.93}{40.32 + 28.93} = \frac{11.39}{69.25} = 0.164$$

20.4 AVERAGE DEVIATION OR MEAN DEVIATION

In the given series, if we calculate the deviation of each point from a particular value (mean, mode or median), then the arithmetic average of these deviations is known as average deviation or mean deviation.

1. For ungrouped data, the mean deviation is given by the formula

   Mean deviation $= \dfrac{\Sigma \, |x - M|}{N}$

2. For discrete series, the deviation is given by

   Mean deviation $= \dfrac{\Sigma f \, |x - M|}{N}$

3. For continuous series, firstly we calculate the value of x as a mid value of the class and then apply the same formula for discrete series.

⬤ REMARK
⟹ Each value of the deviation will be positive.

20.4.1 MERITS OF AVERAGE DEVIATION

1. All the terms of the series are used to calculate this deviation.
2. It can be calculated from any of the value, mean, mode and median.
3. It is a simple method to calculate the measure of dispersion.

20.4.2 DEMERITS OF AVERAGE DEVIATION

1. Mathematically, this measure cannot give the absolute value since the signs (+*ve* and –*ve*) are ignored.
2. About mode, this measure does not gives any satisfaction.

SOLVED EXAMPLES

EXAMPLE 1. *The weight of five items in Kg is given by 30, 40, 45, 50, 55 Find the deviation from the median.*

SOLUTION. Here $N = 5$ which is odd. So the median is given by $\dfrac{1}{2}(5 + 1)$ th term.

$$= \frac{1}{2} \times 6 \text{ th term} = 3^{\text{rd}} \text{ term}$$

The 3rd term of the series is $= 45$

So, the median of the series $= 45$

x	$x - M_e$	$|x - M_e|$
30	−15	15
40	−5	5
45	0	0
50	5	5
55	10	10
		$\Sigma|x - M_e| = 35$

Then,   deviation is given by $= \dfrac{\Sigma|x - M_e|}{N} = 35/5$

So, the mean deviation about mean = 7.

EXAMPLE 2.   *Find the mean deviation of the numbers 3, 10, 9, 9, 4, 7, 14 from the mean.*

SOLUTION.   Mean $= \dfrac{\text{Sum of all the items}}{\text{Total no. of observations}} = \dfrac{3 + 10 + 9 + 9 + 4 + 7 + 14}{7} = \dfrac{56}{7}$

Mean = 8

x	$x - M$	$|x - M|$
3	−5	5
10	2	2
9	1	1
9	1	1
4	−4	4
7	−1	1
14	6	6
		$\Sigma|x - M| = 20$

Now the mean deviation from mean is given by $= \dfrac{\Sigma|x - M|}{N} = \dfrac{20}{7}$

So, the mean deviation about mean = 2.8571.

EXAMPLE 3.   *Find the mean deviation about mean for the given frequency distribution :*

x	30	31.5	33	34.5	36	37.5	39	40.5
f	4	19	30	63	66	29	18	1

SOLUTION.   Let us assume mean A is given by $A = 36$
Firstly we will calculate the arithmetic mean of the given series.

x	f	$x - A$	$\Sigma f(x - A)$	$x - M$	$f|x - M|$
30	4	− 6	−24	−5	20
31.5	19	− 4.5	−85.5	−3.5	66.5
33	30	− 3	−90	−2	60
34.5	63	−1.5	−94.5	−0.5	31.5
36	66	0	0	1	66
37.5	29	1.5	43.5	2.5	72.5
39	18	3	54	4	72.5
40.5	1	4.5	4.5	5.5	5.5
	$\Sigma f = 230$		$\Sigma = -192$		$\Sigma = 394$

Now arithmetic mean,

$$M = A + \frac{\Sigma f(x - A)}{\Sigma f} = 36 + \frac{(-192)}{230} = 36 - 0.8347$$

$$M = 35.1653$$

Let $M = 35$

Then, mean deviation about mean

$$= \frac{\Sigma f \, | \, x - M \, |}{N} = \frac{394}{230}$$

Mean deviation about mean = 1.7130

EXAMPLE 4. *Calculate the average deviation about median for the following frequency distribution:*

x	6	12	18	24	30	36	42
f	4	7	9	18	15	10	5

SOLUTION. Firstly, we will calculate the median for the given frequency distribution and then find the average deviation.

x	f	Cumulative frequency	$x - M_e$	$f \, | \, x - M_e \, |$
6	4	4	− 18	72
12	7	11	− 12	84
18	9	20	− 6	54
24	18	38	0	0
30	15	53	6	90
36	10	63	12	120
42	5	68	18	90
	N = 68			Σ = 510

Here, $N = 68$ which is even so the median is given by the value of $N \, / \, 2 = 34^{th}$ item.

Value of 34th item = 24

So the median, $M_e = 24$.

Now, the average deviation about mean

$$= \frac{\Sigma f \, | \, x - M_e \, |}{N} = \frac{510}{68}$$

Average deviation about median = 7.5

EXAMPLE 5. *Calculate the mean deviation about mean for the following frequency distribution*

Class	20-30	30-40	40-50	50-60	60-70
Frequency	120	201	150	75	25

SOLUTION. Firstly, we calculate the arithmetic mean of the given frequency distribution and then we can calculate the mean deviation about mean.

Class	Mid Value x	Frequency f	$d' = \dfrac{x-A}{h}$	fd'	$x-M$ $M=39$	$f\,|x-M|$
20-30	25	120	-2	-240	-14	1680
30-40	35	201	-1	-201	-4	804
40-50	45	150	0	0	6	900
50-60	55	75	1	75	16	1200
60-70	65	25	2	50	26	650
		571		-316		5234

Let the assumed mean = 45

and h (class width) = 10

The arithmetic mean for continuous series is given by the formula

$$M = A + \frac{\Sigma fd'}{N} \times h = 45 + \frac{(-316)}{571} \times 10$$

$$= 45 - 5.5341$$

$$M = 39.4659$$

Let $M = 39$

Now, mean deviation about mean

$$= \frac{\Sigma f\,|x-M|}{N} = \frac{5234}{571} = 9.1663$$

Hence, mean deviation about mean = 9.1663.

EXAMPLE 6.   *Calculate the mean deviation from median of the following frequency distribution:*

Age under	10	20	30	40	50	60	70	80
Frequency	15	30	53	75	100	110	115	125

SOLUTION.   Firstly, we will calculate the median of the given frequency distribution and then find the mean deviation about median.

Here $N = 125$ which is odd. So the median is given by the value of

$$\frac{N+1}{2} = \frac{125+1}{2} = \text{63}^{\text{rd}} \text{ term.}$$

63rd term falls in the class $30-40$. So, it will be the median class.

$$l = 30, \quad F = 53, \quad f = 22, \quad i = 10$$

Median is given by the formula,

$$M_e = l + \frac{\frac{1}{2}N - F}{f} \times i$$

$$M_e = 30 + \frac{\frac{125}{2} - 53}{22} \times 10 = 30 + 4.3181$$

$$M_e = 34.3181 \text{ years.}$$

Let $M_e = 34$ years (approximately)

Class	Mid Value x	Frequency f	Cumulative Frequency F	$x - M_e$ $M_e = 34$	$f \mid x - M_e \mid$
0-10	5	15	15	−29	435
10-20	15	15	30	−19	285
20-30	25	23	53	−9	207
30-40	35	22	75	1	22
40-50	45	25	100	11	275
50-60	55	10	110	21	210
60-70	65	5	115	31	155
70-80	75	10	125	41	410
		125			1999

Then mean deviation about median

$$= \frac{\Sigma f \mid x - M_e \mid}{N} = \frac{1999}{125} = 15.992 \text{ years.}$$

EXAMPLE 7. *Hb% of 10 patients of a ward of a hospital were obtained as 5, 7, 8, 10, 14, 12, 13, 5, 8, 8. Compute the mean deviation.*

SOLUTION.

$$\text{Mean} \frac{\Sigma X}{N} = \frac{5+7+8+10+14+12+13+5+8+8}{10} = \frac{90}{10} = 9$$

Now following table is prepared to obtain deviations between each score and mean.

Score X	Score mean $\mid X - \bar{X} \mid$	Deviation X
5	$\mid 5 - 9 \mid$	4
7	$\mid 7 - 9 \mid$	2
8	$\mid 8 - 9 \mid$	1
10	$\mid 10 - 9 \mid$	1
14	$\mid 14 - 9 \mid$	5
12	$\mid 12 - 9 \mid$	3
13	$\mid 13 - 9 \mid$	4
5	$\mid 5 - 9 \mid$	4
8	$\mid 8 - 9 \mid$	1
9	$\mid 9 - 9 \mid$	

$$\Sigma \mid X - \bar{X} \mid = 4 + 2 + 1 + 1 + 5 + 3 + 4 + 4 + 1 = 25$$

$$\text{MD} = \frac{\Sigma \mid X - \bar{X} \mid}{N} = \frac{25}{10} = 2.5$$

EXAMPLE 8. *Testis weight (g) of 50 fishes of a species and their frequency were obtained as*

$$\left[\frac{2}{2}, \frac{2.5}{1}, \frac{2.7}{1}, \frac{2.9}{2}, \frac{3}{3}, \frac{3.1}{1}, \frac{3.3}{3}, \frac{3.7}{2}, \frac{3.9}{4}, \frac{4}{3}, \frac{4.6}{2}, \frac{4.8}{3}, \frac{4.9}{3}, \frac{5}{2}, \frac{5.5}{3}, \frac{5.9}{3}, \frac{6}{3}, \frac{6.1}{3}, \frac{6.9}{3} \right]$$

Compute Mean Deviation.

SOLUTION.

C.I.	Mid point x	Frequency f	$f \cdot x$	d	$f \cdot d$
2-2.9	2.45	6	14.7	−2.14	−12.84
3-3.9	3.45	13	44.85	−1.14	−14.85
4-4.9	4.45	11	48.95	−0.14	−1.54
5-5.9	5.45	9	43.6	0.86	6.88
6-6.9	6.45	12	77.4	1.86	22.32
	$\Sigma x = 22.25$	$\Sigma f = 50$	$\Sigma fx = 229.5$		$\Sigma|fd| = 54.4$

$$\text{Mean } \bar{x} = \frac{\Sigma fx}{\Sigma f} = \frac{229.5}{50} = 4.59$$

$$MD = \frac{\Sigma f|d|}{\Sigma f} = \frac{58.4}{50} = 1.168$$

$$\text{Coefficient of mean deviation} = \frac{MD}{\text{Mean}} = \frac{1.168}{4.59} = 0.254$$

EXAMPLE 9. *Calculate mean deviation from mean of the following cumulative frequency (more than type) distribution :*

Age in year (more than)	0	10	20	30	40	50	60	70
Number of persons died	125	110	95	72	50	25	15	0

SOLUTION. Firstly, we calculate the arithmetic mean of the given frequency distribution and then we can calculate the mean deviation about mean.

Class	Mid Value x	Frequency f	$d' = \dfrac{x-A}{h}$	fd'	$\begin{array}{c}|x-M| \\ M=\end{array}$	$f|x-M|$
0-10	5	15	−3	−45	29	435
10-20	15	15	−2	−30	19	285
20-30	25	23	−1	−23	09	207
30-40	35	22	0	0	1	22
40-50	45	25	1	25	11	275
50-60	55	10	2	20	21	210
60-70	65	15	3	45	31	465
		125		−08		1899

Let us assumed mean = 45
and h (class width) = 10
The arithmetic mean for the continuous series is given by the formula

$$M = A + \frac{\Sigma fd'}{N} \times h = 35 + \frac{-8}{125} \times 10 = 35 - 0.64 = 34.36$$

Let $\qquad M = 34$

Now, mean deviation about mean $= \dfrac{\Sigma f \, |x - M|}{N} = \dfrac{1899}{125} = 15.192$

Hence, mean deviation about mean = 15.192.

Exercise 20.1

1. Find the average deviation from mean of the following frequency distribution :

(UPTU–2001)

Class	0-6	6-12	12-18	18-24	24-30
Frequency	8	10	12	9	5

2. Find the range and quartile deviation of the data given below :

139, 150, 151, 151, 157, 158, 160, 161, 162, 162, 173, 175

3. Find the range and its coefficients in the following series :

Size	3	4	5	6	7	8	9	10
Frequency	35	30	20	10	6	3	2	1

4. Find out the range and its coefficient in the following series :

Monthly average (Rs.)	100	150	200	250	300	500
Frequency	30	20	15	10	4	1

5. Find the range and its coefficient in the following frequency distribution:

Size	10-60	60-120	120-180
Frequency	3	5	6
Size	180-240	240-300	
Frequency	3	2	

6. Find the quartile deviation and its coefficient in the following frequency distribution :

Weight (Kg)	70-80	80-90	90-100	100-110
Frequency	12	18	35	42
Weight (Kg)	110-120	120-130	130-140	140-150
Frequency	50	45	25	8

7. Find the semi-interquartile range and its coefficient from the following data:

Size	6	7	8	9	10	11	12
Frequency	3	6	9	13	8	5	3

8. Find the mean deviation from the mean of the following frequency distribution :

Size	56	63	70	77	84	91	98
Frequency	3	6	14	16	13	6	2

9. Find the mean deviation from median of the following frequency distribution :

Size	4	6	8	10	12	14	16
Frequency	2	4	5	3	2	1	4

10. Find the mean deviation about mode for the following frequency distribution :

Class	14-15	15-16	16-17
Frequency	4	6	10
Class	17-18	18-19	19-20
Frequency	18	9	3

11. Find the mean deviation from mean of the following frequency distribution :

Age-under	10	20	30	40
Frequency	15	30	53	75
Age-under	50	60	70	80
Frequency	100	110	115	125

12. Find the mean deviation from mean of the following frequency distribution :

[CCSU(B.Sc Biotech)–1995]

Marks	0-10	10-20	20-30	30-40	40-50
No. of students	5	8	15	16	6

13. Find the mean deviation about median for the following data :   [CCSU(B.Sc Biotech)–1997]

Height (in inches)	58	60	62	64	66
Number of students	1	32	33	20	8
Height (in inches)	59	61	63	65	
Number of students	20	35	22	10	

ANSWERS				
1. 6.3	**2.** 36	**3.** 7, 0.53	**4.** 400, 0.666	
5. 290, 0.93	**6.** 13.03, 0.117	**7.** 1, 0.11	**8.** 76.53	**9.** 3.24
10. 0.986	**11.** 15.8	**12.** 9.44		

20.5 ABSOLUTE AND RELATIVE MEASURE OF DISPERSION

The absolute measure are expressed in terms of the units of the observation such as rupees, kilograms, degree, celsius, etc. These values may be used to compare the variations in the two distributions when the variables are expressed in the same units and are of same average size.

The relative measures of dispersion are useful for comparing two series expressed in different units. They are also useful when we are comparing the variations of two series which have quite different magnitudes even when the units of original measurements are same.

20.5.1 SOME RELATIVE MEASURES OF DISPERSION

(1) Range Coefficient of Dispersion

$$= \frac{\text{Difference between extreme values}}{\text{Sum of extreme values}}$$

(2) Quartile Coefficient of Dispersion

$$= \frac{Q_3 - Q_1}{Q_3 + Q_1}$$

(3) Coefficient of Mean Dispersion

$$= \frac{\text{Mean deviation about any point } a}{a}$$

20.6 STANDARD DEVIATION AND ROOT MEAN SQUARE DEVIATION

If in a given series we calculate the deviation of each point from the arithmetic mean then the arithmetic mean of the squares of these deviation is known as variance.

Variance $\qquad \sigma^2 = \dfrac{1}{N} \Sigma (x - M)^2$

Now the square root of this variance is known by the standard deviation and it is denoted by σ

$$\sigma = + \sqrt{\frac{1}{N} \Sigma (x - M)^2}$$

Root mean square deviation is the deviation in which we calculate the deviation of each point from an arbitrary point A.

In other words, if in a given series, we calculate the deviation of each point from an arbitrary point, then the arithmetic mean of the squares of these deviation is known as mean square deviation and the root of this mean square deviation is known by root mean square deviation. Hence,

mean square deviation, $\qquad S^2 = \dfrac{1}{N} \Sigma f(x - M)^2$

root mean square deviation, $\quad S = + \sqrt{\dfrac{1}{N} \Sigma f(x - M)^2}$

⮞ REMARK

⮕ The standard deviation is a particular case of mean square deviation when the arithmetic mean is taken in place of arbitrary point.

Merits

1. The standard deviation depends on all the values of observations.
2. In standard deviation, the signs ($+ve$ and $-ve$) are not ignored.
3. It is very useful to calculate the coefficient of correlation and other statistics calculations.
4. In calculating the standard deviation, we use the mean which is more suitable.

Demerits

1. A person with non-mathematical mind cannot understand it.
2. It wants a long calculation so sometimes it becomes very complicated to solve these problems.
3. It gives more weight to extreme values.

20.6.1 COEFFICIENT OF STANDARD DEVIATION AND VARIATION

1. Coefficient of standard deviation $= \dfrac{\sigma}{M}$

where, $\qquad \sigma =$ Standard deviation

$\qquad\qquad M =$ Arithmetic mean

2. Coefficient of variation $= \dfrac{\sigma}{M} \times 100$

where, $\qquad \sigma =$ Standard deviation,

$\qquad\qquad M =$ arithmetic mean

20.6.2 RELATION BETWEEN STANDARD DEVIATION AND ROOT MEAN SQUARE DEVIATION

We know that the variance is given by

$$\sigma^2 = \frac{1}{N} \Sigma f(x - M)^2 \text{, where } M \text{ is the arithmetic mean.}$$

$$\sigma^2 = \frac{1}{N} \Sigma f[x - A - (M - A)]^2 = \frac{1}{N} \Sigma f[x - A - d]^2 \quad [\because (M - A) = d]$$

$$= \frac{1}{N} \Sigma f\left[(x - A)^2 + d^2 - 2(x - A)d \right]$$

$$= \frac{1}{N} \Sigma f(x - A)^2 + \frac{1}{N} \Sigma f d^2 - \frac{2}{N} \Sigma f(x - A).d$$

$$= \frac{1}{N} \Sigma f(x - A)^2 + d^2 - 2.d \frac{1}{N} \Sigma f(x - A)$$

$$= \frac{1}{N} \Sigma f(x - A)^2 + d^2 - 2d \frac{1}{N} \Sigma f x + 2d \frac{1}{N} \Sigma f.A \qquad (\because \Sigma f = N)$$

$$= \frac{1}{N} \Sigma f(x - A)^2 + d^2 - 2d.M + 2d.A$$

$$= \frac{1}{N} \Sigma f(x - A)^2 + d^2 - 2d(M - A)$$

We know that $\quad (M - A) = d$.

So,
$$\sigma^2 = \frac{1}{N} \Sigma f(x - A)^2 + d^2 - 2d.d$$

$$= \frac{1}{N} \Sigma f(x - A)^2 + d^2 - 2d^2 = \frac{1}{N} \Sigma f(x - A)^2 - d^2$$

Hence,
$$\sigma^2 = S^2 - d^2 \qquad \left[\because S^2 = \frac{1}{N} \Sigma f(x - A)^2 \right]$$

THEOREM 1. *The sum of the squares of the deviation about arithmetic mean is least.*

PROOF. Consider a discrete distribution

x	x_1	x_2	...	x_n
f	f_1	f_2	...	f_n

Then arithmetic mean, $\bar{x} = \dfrac{\Sigma fx}{\Sigma f} = \dfrac{\Sigma fx}{N} \quad \Rightarrow \quad \Sigma fx = N\bar{x}$, where $\quad \Sigma f = N$

Let A be any arbitrary point and S be the sum of squares of deviation from A, then $S = \Sigma f(x - A)^2$

It is known that S will be minimum if
$$\frac{\partial S}{\partial A} = 0 \quad \text{and} \quad \frac{\partial^2 S}{\partial A^2} > 0$$

Now,
$$\frac{\partial S}{\partial A} = -2\Sigma f(x - A) = 0$$

$$\Rightarrow \qquad -2\Sigma fx + 2A\Sigma f = 0$$

$$\Rightarrow \qquad -\Sigma fx + AN = 0$$

$$\Rightarrow \qquad N\bar{x} = AN \quad \Rightarrow \quad \bar{x} = A$$

Also, $\dfrac{\partial^2 S}{\partial A^2} = 2\Sigma f = 2N > 0$. Hence, S is minimum where $A = \bar{x}$ (Mean).

SOLVED EXAMPLES

EXAMPLE 1. *Calculate the standard deviation of the data given as follows:*
3, 4, 9, 11, 13, 6, 8, 10

SOLUTION. We have

x	$x - M$	$(x - M)^2$
3	-5	25
4	-4	16
9	1	1
11	3	9
13	5	25
6	-2	4
8	0	0
10	2	4
		84

Firstly, we will calculate the mean of the given data

So, $$M = \frac{\text{Sum of all observations}}{\text{Total number of items}}$$

$$= \frac{3+4+9+11+13+6+8+10}{8} = 8$$

Now, the standard deviation is given by

$$\sigma = \sqrt{\frac{1}{N} \Sigma(x - M)^2}$$

$$= \sqrt{\frac{84}{8}} = \sqrt{10.5} = 3.25$$

EXAMPLE 2. *Calculate the standard deviation for the following frequency distribution :*

x	0	10	20	30	40	50	60	70	80
f	150	140	100	80	80	70	30	14	0

SOLUTION. Firstly we calculate the mean of the given frequency distribution and then find the standard deviation

x	f	fx	x − M	(x − M)²	f(x − M)²
0	150	0	− 23	529	79350
10	140	1400	−13	169	23660
20	100	2000	−3	9	900
30	80	2400	7	49	3920
40	80	3200	17	289	23120
50	70	3500	27	729	51030
60	30	1800	37	1369	41070
70	14	980	47	2209	30926
80	0	0	57	3249	0
	$\Sigma f = 664$	$\Sigma fx = 15280$			$\Sigma f(x - M)^2 = 253976$

Mean of the given distribution will be

$$M = \frac{\Sigma fx}{\Sigma f}$$

$$\Rightarrow \qquad M = \frac{15280}{664} = 23.0120 \approx 23$$

Now the standard deviation is given by

$$\sigma = \sqrt{\frac{\Sigma f(x - M)^2}{N}} = \sqrt{\frac{253976}{664}} = \sqrt{382.4939} \; ;$$

$$\sigma = 19.557$$

EXAMPLE 3. *Hb% of 10 patients of a ward was recorded as 7, 8, 9, 10, 11, 12, 13, 14.5, 15 and 15.5 g/100 ml. Find out the variance of data.*

SOLUTION.

Hb%	Deviation $d = x - \bar{x}$	d^2
7	$7 - 11.5 = -4.5$	20.25
8	$8 - 11.5 = -3.5$	12.25
9	$9 - 11.5 = -2.5$	6.25
10	$10 - 11.5 = -1.5$	2.25
11	$11 - 11.5 = -0.5$	0.25
12	$12 - 11.5 = 0.5$	0.25
13	$13 - 11.5 = 1.5$	2.25
14.5	$14.5 - 11.5 = 3.0$	9.0
15	$15 - 11.5 = 3.5$	12.0
15.5	$15.5 - 11.5 = 4.0$	16.0
$\Sigma x = 115$		$\Sigma d^2 = 80.75$

Now, $\bar{x} = \dfrac{\Sigma x}{N} = \dfrac{115}{10} = 11.5$

and variance $= \dfrac{\Sigma d^2}{N} = \dfrac{80.75}{10} = 8.075$

EXAMPLE 4. *Hemoglobin percentage g/100 ml of Heteropneustes fossils was recorded as 23, 22, 20, 24, 16, 17, 18, 19 and 21. Compute the standard deviation by Indirect method.*

SOLUTION. Here, $\Sigma x = 180, N = 9$, Mean $= \dfrac{180}{9} = 20$

Observation x	Observation − Mean $= x - \bar{x}$	Deviation (d)	(Deviation)^2 $= d^2$
16	$16 - 20$	-4	16
17	$17 - 20$	-3	9
18	$18 - 20$	-2	4
19	$19 - 20$	-1	1
20	$20 - 20$	0	0
21	$21 - 20$	1	1
22	$22 - 20$	2	4
23	$23 - 20$	3	9
24	$24 - 20$	4	16
			$\Sigma d^2 = 60$

Here, size of sample is less than 30. Therefore following formula is applicable.

$$\sigma = \sqrt{\frac{\Sigma d^2}{N-1}} = \sqrt{\frac{60}{9-1}} = \sqrt{\frac{60}{8}} = \sqrt{7.5} = 2.75$$

20.7 SHORT CUT METHOD TO CALCULATE THE STANDARD DEVIATION OF DISCRETE SERIES

We know that standard deviation, $\sigma = \sqrt{\dfrac{\Sigma f(x-M)^2}{N}}$

Now, $\sigma_x = \sqrt{\dfrac{1}{N}\Sigma f(x-\bar{x})^2}$ (Let us write $M = \bar{x}$) ... (1)

Now, let us assume that d is the deviation of each point x from an arbitrary point A.

Then, $d = x - A,\quad x = d + A\ ,\quad \bar{d} = \bar{x} - A,\ \bar{x} = \bar{d} + A$

Put these values in equation (1)

$$\sigma_x = \sqrt{\frac{1}{N}\Sigma f(d + A - \bar{d} - A)^2} = \sqrt{\frac{1}{N}\Sigma f(d + A - \bar{d} - A)^2}$$

$$\sigma_x = \sqrt{\frac{1}{N}\Sigma f(d - \bar{d})^2} \Rightarrow \sigma_x = \sqrt{\frac{1}{N}\Sigma f d^2 - \left(\frac{1}{N}\Sigma f d\right)^2}$$

$$\sigma_x = \sigma_d$$

Hence

Thus, we see that the standard deviation is independent from the change of origin.

20.7.1 STEP DEVIATION METHOD

The standard deviation for a given series is

$$\sigma_x = \sqrt{\frac{1}{N}\Sigma f(x-\bar{x})^2}\ , \text{ where } \bar{x} = M \qquad \text{... (1)}$$

Now, let us define a new variable μ such that $\mu = \dfrac{x-a}{h}, x = a + h u, \bar{\mu} = \dfrac{\bar{x}-a}{h}, \bar{x} = a + h\bar{u}$.

Put this value of x in equation (1)

$$\sigma_x = \sqrt{\frac{1}{N}\Sigma f\left[a + h\mu - (a + h\bar{\mu})\right]^2} = \sqrt{\frac{1}{N}\Sigma f(a + h\mu - a - h\bar{\mu})^2}$$

$$= \sqrt{\frac{1}{N}\Sigma f(\mu - \bar{\mu})^2 . h^2}\ ;\qquad \sigma_x = \sigma_\mu . h.$$

$$\sigma_x = h\,\sigma_\mu \qquad \text{...(2)}$$

$$\sigma_x = \sqrt{\frac{1}{N}\Sigma f(\mu - \bar{\mu})^2\, h^2} = \sqrt{\frac{1}{N}\Sigma f(\mu^2 - 2\mu\bar{\mu} + \mu^2).h^2}$$

$$= \sqrt{\left(\frac{1}{N}\Sigma f\mu^2 - \frac{2}{N}\Sigma f\mu\bar{\mu} + \frac{1}{N}\Sigma f\bar{\mu}^2\right)h^2} = \sqrt{\left(\frac{1}{N}\Sigma f\mu^2 - 2\bar{\mu}\bar{\mu} + \bar{\mu}^2\right)h^2}$$

$$= \sqrt{\left(\frac{1}{N}\Sigma f\mu^2 - \bar{\mu}^2\right)h^2}$$

$$\sigma_n = f_1\sqrt{\frac{1}{N}\Sigma f\mu^2 - \left(\frac{1}{N}\Sigma f\mu\right)^2}$$

Now, by equation (2), we see that the standard deviation is changed by the change of scale.

⇌ REMARK

⇒ Using the above two results, we conclude that

(i) Standard deviation is independent from the change of origin.

(ii) Standard deviation is changed by the change of scale.

Example 5. *Calculate the standard deviation for the following frequency distribution using short-cut method :*

Class	35-36	36-37	37-38	38-39	39-40	40-41	41-42
Frequency	14	20	42	54	45	18	7

Solution.

Class	Mid value x	f	Cumulative frequency	$d = x - A$	d^2	fd	fd^2
35-36	35.5	14	14	-3	9	-42	126
36-37	36.5	20	34	-2	4	-40	80
37-38	37.5	42	76	-1	1	-42	42
38-39	38.5	54	130	0	0	0	0
39-40	39.5	45	175	1	1	45	45
40-41	40.5	18	193	2	4	18	72
41-42	41.5	7	200	3	9	7	63
		$\Sigma f = 200$				$\Sigma fx = -54$	$\Sigma fd^2 = 428$

Let us assume mean $A = 38.5$

Then the standard deviation

$$= \left[\frac{1}{N} \Sigma fd^2 - \left(\frac{1}{N} \Sigma fd \right)^2 \right] = \frac{1}{200} \times 428 - \left[\frac{1}{200} (-54) \right]^2$$

$$= 2.14 - 0.0729$$

$$\sigma^2 = 2.0671$$

$$\sigma = \sqrt{2.0671} = 1.4.$$

Example 6. *Calculate the standard deviation for the following frequency distribution using short-cut method :*

Class	5-15	15-25	25-35	35-45	45-55	55-65
Frequency	15	32	51	78	97	109

Solution.

Class	Mid Value x	f	Cumulative Frequency	$d = x - A$	d^2	fd	fd^2
5-15	10	15	15	-25	625	-375	9375
15-25	20	32	47	-15	225	-480	7200
25-35	30	51	98	-5	25	-255	1275
35-45	40	78	176	5	25	390	1950
45-55	50	97	273	15	225	1455	21825
55-65	60	109	382	25	625	2725	68125
		382				3460	109750

Let the assumed mean $A = 35$

Then the standard deviation,

$$\sigma = \sqrt{\frac{1}{N}\Sigma fd^2 - \left(\frac{1}{N}\Sigma fd^2\right)^2}$$

$$\sigma = \sqrt{\frac{1}{382} \times 109750 - \left(\frac{1}{382} \times 3460\right)^2}$$

$$= \sqrt{287.3036 - 82.0383} = \sqrt{205.2653}$$

$$\sigma = 14.3269.$$

EXAMPLE 7. *Weight of testis of 50 frogs is given below with their frequency. Find the standard deviation.*

Wt. of Testis	2	2.5	2.7	2.9	3	3.1	3.3	3.7	3.9	4
Frequency	2	1	1	2	3	1	3	2	4	3
Wt. of Testis	4.6	4.8	4.9	5	5.5	5.9	6	6.1	6.7	6.9
Frequency	2	3	3	3	2	3	3	3	3	

SOLUTION.

Class Interval	Mid point x	Frequency f	$f \cdot x$	Deviation	Deviation squared d^2	fd^2
2-2.9	2.45	6	14.7	– 2.14	4.5796	27.47
3-3.9	3.45	13	44.85	– 1.14	1.2996	16.88
4-4.9	4.45	11	48.95	– 0.14	0.0196	0.21
5-5.9	5.45	8	43.6	0.86	0.7396	5.91
6-6.9	6.45	12	77.4	1.86	3.4596	41.4
		$\Sigma f = 50$	$\Sigma fx = 229.5$		$\Sigma d^2 = 10.000$	$\Sigma fd^2 = 91.87$

$$\sigma = \sqrt{\frac{\Sigma fd^2}{\Sigma f}} = \sqrt{\frac{91.87}{50}} = \sqrt{1.05} = 1.02$$

EXAMPLE 8. *Calculate the arithmetic mean and standard deviation from the following:*

(RGPV(B.PHARMA)–2004)

Marks (more than)	0	10	20	30	40	50	60	70
Number of students	100	90	75	50	25	15	5	0

SOLUTION. Firstly we calculate the mean of the given frequency distribution and then find the standard deviation.

Class	Mid Value x	frequency (f)	$d' = \dfrac{x-A}{h}$	$f \cdot d'$	d'^2	$f.d'^2$
0-10	5	10	-3	-30	9	90
10-20	15	15	-2	-30	4	60
20-30	25	25	-1	-25	1	25
30-40	35	25	0	0	0	0
40-50	45	10	1	10	1	10
50-60	55	10	2	20	4	40
60-70	65	5	3	15	9	45
		$N = 100$		-40		270

Let us assume mean = 35

and h (class width) = 10

The arithmetic mean for continuous series is given by the formula

$$M = A + \frac{\Sigma f d'}{N} \times h = 36 + \frac{(-40)}{100} \times 10 = 35 - 4 = 31$$

Now, standard deviation

$$\sigma = h \sqrt{\frac{\Sigma f d'^2}{N} - \left(\frac{\Sigma f d'}{N}\right)^2} = 10 \sqrt{\frac{270}{100} - \left(\frac{-40}{100}\right)^2}$$

$$= 10\sqrt{2.7 - 0.16} = 10 \times 1.5937 = 15.937$$

EXAMPLE 9. *Calculate the mean, median and standard deviation from the following table :*

[RGPV(B.PHARMA)–2004]

Defect in size (mm)	1-10	11-20	21-30	31-40	41-50	51-60
Number of item	3	6	26	31	16	8

SOLUTION.

Class	Mid Value x	Frequency	Cumulative frequency	$d' = \dfrac{x-A}{h}$	d'^2	$f \cdot d'$	$f.d'^2$
0.5-10.5	5.5	3	3	-2	4	-6	12
10.5-20.5	15.5	6	9	-1	1	-6	6
20.5-30.5	25.5	26	35	0	0	0	0
30.5-40.5	35.5	31	66	1	1	31	31
40.5-50.5	45.5	16	82	2	4	32	64
50.5-60.5	55.5	8	90	3	9	24	72
		$N = 90$				75	185

Let us assume mean = 25

and h (class width) = 10

The arithmetic mean is given by

$$M = A + \frac{\Sigma fd'}{N} \times h = 25.5 + \frac{75}{90} \times 10 = 25.5 + 8.33$$

$$M = 33.83$$

Now, median $= \left(\frac{N}{2}\right)^{th}$ term $= \frac{90}{2} = 45^{th}$ term

Then, 45th term lies in 30.5-40.5 class

Then, $h = 10, l = 30.5, F = 35$ and $f = 31$

$$M_d = l + \frac{\frac{N}{2} - F}{f} \times h = 30.5 + \frac{45 - 35}{31} \times 10 = 30.5 + 3.225$$

$\Rightarrow$ $M_d = 33.725$

Now, standard deviation

$$\sigma = h\sqrt{\frac{\Sigma fd'^2}{N} - \left(\frac{\Sigma fd'}{N}\right)^2} = 10\sqrt{\frac{180}{90} - \left(\frac{75}{90}\right)^2}$$

$$= 10\sqrt{2.0555 - 0.6944} = 10 \times 1.166$$

$$\sigma = 11.66$$

EXAMPLE 10. *Find the arithmetic mean and standard deviation of the height of 10 men given below :*

Height (in cms) : 160, 160, 161, 162, 163, 163, 163, 164, 164, 170

[CCSU(B.Sc Biotech)–1996]

SOLUTION.

Height (in cms) x	$d' = x - 162$	d'^2
160	– 2	4
160	– 2	4
161	– 1	1
162	0	0
163	1	1
163	1	1
163	1	1
164	2	4
164	2	4
170	8	64
	$\Sigma d' = 10$	$\Sigma d'^2 = 84$

Now, arithmetic mean $= A + \dfrac{\Sigma d'}{n} = 162 + \dfrac{10}{10} = 163$

and standard deviation $= \sqrt{\dfrac{\Sigma d'^2}{n} - \left(\dfrac{\Sigma d'}{n}\right)^2} = \sqrt{\dfrac{84}{10} - \dfrac{10}{10}} = \sqrt{8.4 - 1} = 2.72$

EXAMPLE 11. *Find the standard deviation for the following distribution :*

x	0-10	10-20	20-30	30-40	40-50	50-60	60-70
f	4	6	10	20	10	6	4

(UPTU–2001)

SOLUTION.

Size	Mid Value	Frequency	$d' = x - A$	$f \cdot d'$	$f \cdot d'^2$
0-10	5	4	– 30	– 120	3600
10-20	15	6	– 20	– 120	2400
20-30	25	10	– 10	– 100	1000
30-40	35	20	0	0	0
40-50	45	10	10	100	1000
50-60	55	6	20	120	2400
60-70	65	4	30	120	3600
		$N = 60$		$\Sigma fd' = 0$	$\Sigma fd'^2 = 14{,}000$

Hence, standard deviation

$$(\sigma) = \sqrt{\frac{\Sigma fd'^2}{n} - \left(\frac{\Sigma fd'}{n}\right)^2} = \sqrt{\frac{14{,}000}{60} - 0} = \sqrt{\frac{1400}{6}} = \sqrt{233.3} = 15.27$$

20.7.1 TERMS RELATED TO STANDARD DEVIATION

1. Probable error = $0.6745 \times \sigma$ **2.** Variance = $\sigma^2 = \dfrac{\Sigma f(x - \bar{x})^2}{N}$

3. Modulus (C) = $\sqrt{\dfrac{2\Sigma f(x - \bar{x})^2}{N}}$ or $\sigma\sqrt{2}$

i.e., $C = \sigma\sqrt{2}$

4. Fluctuation = $C^2 = 2\sigma^2$, where C is the modulus.

or fluctuation = (modulus)^2

5. Precision : $P = \dfrac{1}{C} = \dfrac{1}{\text{modulus}}$ (where σ is standard deviation) *i.e.,* $P = \dfrac{1}{\sigma\sqrt{2}}$

Exercise 20.2

1. Calculate the standard deviation and coefficient of variation for the given frequency distribution

x	1	2	3	4	5	6
f	31	37	33	30	35	34

2. Calculate the standard deviation of the data :
5, 7, 9, 11

3. Calculate the standard deviation and its coefficient from the given frequency distribution :

Class	20-25	25-30	30-35
Frequency	18	44	102
Class	35-40	40-45	45-50
Frequency	160	57	19

4. Calculate the standard deviation for the given frequency distribution using step deviation method :

Class	0-10	10-20	20-30	30-40	40-50
Frequency	5	10	15	20	4

5. Calculate the standard deviation for the given frequency distribution using short-cut method :

x	2	3	4	5	9	10	12	13	15
f	25	37	44	59	68	43	31	32	12

6. Calculate the standard deviation and its coefficient for the given frequency distribution using step deviation method :

Class	0-10	10-20	20-30
Frequency	15	17	19
Class	30-40	40-50	50-60
Frequency	27	19	12

7. Calculate the mean and standard deviation for the following:

Size of item	6	7	8	9	10	11	12
Frequency	3	6	9	13	8	5	4

8. Find the standard deviation from the following distribution: (UPTU–2001)

Class	0-9	10-19	20-29	30-39
Frequency	21	74	100	120
Class	40-49	50-59	60-69	70-79
Frequency	110	84	30	11

9. Calculate the standard deviation for the following

Marks more than	70	60	50
Number of students	7	18	40
Marks more than	40	30	20
Number of students	40	63	65

[CCSU(B.Sc Biotech)–1995]

ANSWERS

1. 1.71, 48.61	**2.** 2.23	**3.** 5.687, 0.159	**4.** 10.95	**5.** 3.76
6. 156, 0.52	**7.** 9, 1.607	**8.** 16.28	**9.** 14.56	

20.8 SUMMARY OF THE MATHEMATICAL PROPERTIES OF STANDARD DEVIATION

1. The sum of the squares of the deviation about mean is least.
2. The standard deviation of a series remain unchanged if each observations of the series is increased or decreased by same constant value.
3. If each observation of a series is multiplied or divided by the same constant value. The standard deviation can also be obtained by dividing or multiplying by the same constant value.
4. Standard deviation is not affected by the change of origin but is not independent of scale.
5. In a symmetrical distribution, the mean, median and mode are coincident. On the basis of mean and S.D., the number of observations lying within specific ranges can be measured in case of normal or symmetrical distribution, which is given as below:

   Mean ± 1 S.D. covers 68.27% parts of the distribution.

   Mean ± 2 S.D. covers 95.45% parts of the distribution.

   Mean ± 3 S.D. covers 99.73% parts of the distribution.

SOLVED EXAMPLES

EXAMPLE 1. Find the mean, variance and standard deviation for the first n natural numbers.

SOLUTION. We have to find the mean, variance and standard deviation for the series 1, 2, 3, ..., n.

The mean for this series is given by

$$M = \frac{\text{Sum of all the observations}}{\text{Total number of items}}$$

$$= \frac{1+2+3+...n}{n} = \frac{\Sigma x}{n} = \frac{n(n+1)}{2.n} = \frac{n+1}{2}$$

So, the mean M is given by $= \frac{n+1}{2}$

Now, we have to find the variance

Variance, $\sigma^2 = \frac{1}{n}\Sigma(x-\bar{x})^2 = \frac{1}{n}\Sigma x^2 - \left(\frac{1}{n}\Sigma x\right)^2$

$$= \frac{1}{n} \cdot \frac{n(n+1)(2n+1)}{6} - \left[\frac{1}{n} \cdot \frac{n.(n+1)}{2}\right]^2$$

$$= \frac{(n+1)(2n+1)}{6} - \frac{(n+1)^2}{4} = \frac{(n+1)}{2}\left[\frac{(2n+1)}{3} - \frac{n+1}{2}\right]$$

$$= \frac{(n+1)}{2}\left[\frac{2(2n+1)-3(n+1)}{6}\right] = \frac{(n+1)}{2}\left[\frac{4n+2-3n-3}{6}\right]$$

$$= \frac{(n+1)}{2}\left[\frac{(n-1)}{6}\right]$$

Variance, $\sigma^2 = \frac{(n^2-1)}{12}$

Hence, standard deviation, $\sigma = \left[\frac{(n^2-1)}{12}\right]^{1/2}$

EXAMPLE 2. *In a given distribution, show that the mean deviation is less than the standard deviation about mean.*

SOLUTION. Mean deviation $= \frac{1}{N}\Sigma|x-M|$ and standard deviation $= \sqrt{\frac{1}{N}\Sigma f(x-M)^2}$

To show that the mean deviation is less than the standard deviation, we will show

Standard deviation – Mean deviation ≥ 0

$$\text{S.D.} - \text{M.D.} = \sqrt{\frac{1}{N}\Sigma f(x-M)^2} - \frac{1}{N}\Sigma f|x-M|$$

Let $x-M = d$, then

$$\text{S.D.} - \text{M.D.} = \sqrt{\frac{1}{N}\Sigma fd^2} \geq \frac{1}{N}\Sigma f|d|$$

Since we have to prove that $\text{S.D} \geq \text{M.D}$

then $\sqrt{\frac{1}{N}\Sigma fd^2} \geq \frac{1}{N}\Sigma f|d|$

Squaring both sides, we get

$$\frac{1}{N}\Sigma fd^2 \geq \left[\frac{1}{N}\Sigma f|d|\right]^2$$

$$N\Sigma fd^2 \geq \left[\Sigma f|d|\right]^2$$

$$(f_1 + f_2 + f_3 + \ldots + f_n)(f_1 d_1^2 + f_2 d_2^2 + f_3 d_3^2 + \ldots)$$

$$\geq [f_1 \,|\, d_1 \,|\, + f_2 \,|\, d_2 \,|\, + f_3 \,|\, d_3 \,|\, + \ldots]^2$$

$$(f_1 d_1^2 + f_1 f_2 d_1^2 + f_1 f_2 d_2^2 + \ldots) + (f_1 f_2 d_1^2 + f_2^2 d_2^2 + f_2 f_3 d_3^2 + \ldots)$$

$$+ (f_1 f_3 d_1^2 + f_2 f_3 d_2^2 + f_3^2 d_3^2 + \ldots)$$

$$\geq (f_1^2 d_1^2 + f_2^2 d_2^2 + \ldots) + 2 f_1 f_2 d_1 d_2 + 2 f_2 f_3 d_2 d_3 + \ldots$$

$$f_1 f_2 (d_1^2 + d_2^2 - 2 d_1 d_2) + f_1 f_3 (d_1^2 + d_2^2 - 2 d_1 d_3) + \ldots \geq 0$$

$$\sum_{i,\, j=1}^{n} f_i f_j (d_i - d_j)^2 \geq 0$$

Since the square of a number will never be negative, so it proves that the mean deviation is less than the standard deviation about mean.

EXAMPLE 3.   *Compute the mean deviation from mean and standard deviation of the series a, $a + d$, $a + 2d$, ..., $a + 2nd$.*
and verify standard deviation > mean deviation.

SOLUTION.   The arithmetic series is

$a, a + d, a + 2d, \ldots, a + 2nd$

The summation of this series will be

$$= 2n + 1 \big[2a + \{2n + 1 - 1\}.d \big] = \frac{2n + 1}{2} [2a + 2nd]$$

$$= (2n + 1)(a + nd)$$

Now, the mean of this series

$$M = \frac{\text{Sum of all the observations}}{\text{Total number of items}} = \frac{(2n + 1)(a + nd)}{(2n + 1)}$$

$$M = a + nd$$

Now, we will calculate the mean deviation about mean.

Mean deviation about mean $= \dfrac{1}{N} \Sigma \,|\, x - M \,|$

$$= \frac{1}{N} \sum_{k=0}^{2n} |\, x_k - M \,| \qquad \qquad \ldots(1)$$

The total number of terms, $N = (2n + 1)$
We know that x_k takes value from $x = a \ldots a + 2nd$ and mean of this arithmetic series is given by $M = a + nd$.

$$\sum_{k=0}^{2n} |\, x_k - M \,| = |\, a - a - nd \,| + |\, a + d - a - nd \,| + \ldots |\, a + nd - a - nd \,|$$

$$+ |\, a + (n-1)d - a - nd \,| + \ldots |\, a + 2nd - a - nd \,| \qquad \ldots (2)$$

$$= nd + (n-1)d + \ldots + d + 0 + d + \ldots nd = 2[nd + (n-1)d + \ldots d]$$

$$= 2d\,[n + (n-1) + (n-2) + \ldots 1]$$

$$= 2d\,[1 + 2 + 3 + \ldots (n-2) + (n-1) + n] = 2d.\Sigma n$$

$$\sum_{k=0}^{2n} |x_k - M| = 2d \cdot \frac{n(n+1)}{2} = n(n+1).d$$

Put this value in equation (1), we get

$$\text{Mean deviation about mean} = \frac{1}{2n+1} \cdot n(n+1).d$$

$$= \frac{n(n+1)d}{(2n+1)} \qquad \qquad \dots (3)$$

Now, we have to calculate the standard deviation of the given arithmetic series

$$\sigma^2 = \frac{1}{N} \sum_{k=0}^{2n} (x_k - M)^2 \qquad \qquad \dots (4)$$

With the help of equation (2), we can write

$$\sum_{k=0}^{2n} (x_k - M)^2 = (a - a - nd)^2 + (a + d - a - nd)^2 + \dots + (a + nd - a - nd)^2$$

$$+ (a + (n+1)d - a - nd)^2 + \dots + (a + 2nd - a - nd)^2$$

$$= n^2 d^2 + (n-1)^2 d^2 + \dots d^2 + 0 + d^2 + \dots n^2 d^2$$

$$= 2[n^2 d^2 + (n-1)^2 d^2 + \dots + d^2]$$

$$= 2d^2 [n^2 + (n-1)^2 + (n-2)^2 + \dots + 3^2 + 2^2 + 1^2]$$

$$= 2d^2 [1^2 + 2^2 + \dots + (n-2)^2 + (n-1)^2 + n^2] = 2d^2 \cdot \Sigma n^2$$

$$\sum_{k=0}^{2n} (x_k - M)^2 = 2d^2 \cdot \frac{n(n+1)(2n+1)}{6} \qquad \left[\because \Sigma n^2 = \frac{n(n+1)(2n+1)}{6} \right]$$

But this value in equation (4)

$$\sigma^2 = \frac{1}{(2n+1)} \frac{2d^2 . n(n+1)(2n+1)}{6}$$

$$\sigma^2 = \frac{n(n+1) \cdot d^2}{3}$$

$$\text{S.D. } (\sigma) = d\sqrt{\frac{n(n+1)}{3}} \qquad \qquad \dots (5)$$

Now, we have to show that S.D. > M.D.

Put the value of S.D. and M.D. from equation (3) and (5), we have

$$d\sqrt{\frac{n(n+1)}{3}} > \frac{n(n+1)d}{2n+1}$$

Squaring both sides

$$\frac{n(n+1)}{3} > \frac{n^2(n+1)^2}{(2n+1)^2} ;$$

$$(2n+1)^2 > 3n(n+1)$$

$$4n^2 + 4n + 1 > 3n^2 + 3n ; \qquad n^2 + n + 1 > 0$$

which is true $\forall \, n > 0$. So the result is proved.

EXAMPLE 4. *Show that in a discrete distribution, the mean deviation is least when measured from the median.*

SOLUTION. Let us suppose that when number of terms are odd, then

$$(x_0 - nh), ..., (x_0 - 2h), (x_0 - h), (x_0 + h), (x_0 + 2h), ... + (x_0 + nh)$$

be the given discrete distribution. By inspection, we observe that x_0 will be the median of this distribution.

Let us define S_k such that

$$S_k = \Sigma |x_k - x_r|$$

$$S_0 = \Sigma |x_0 - x_r|$$

$$= |x_0 - x_0| + |x_0 - x_0 + h| + |x_0 - x_0 + 2h| + ... |x_0 - x_0 + nh|$$

$$|x_0 - x_0 - h| + |x_0 - x_0 - 2h| + |x_0 - x_0 - nh|$$

$$= h + 2h + ...(n-1)h + nh + h + 2h + ...(n-1)h + nh$$

$$= 2[h + 2h + ...(n-1)h + nh]$$

$$= 2h[1 + 2 + ... + n] = 2h \frac{n(n+1)}{2}$$

$$S_0 = n(n+1).h$$

$$S_1 = \Sigma |x_1 - x_r|$$

$$= |x_0 + h - x_0| + |x_0 + h - x_0 + h| + |x_0 + h - x_0 + 2h| + ...$$

$$+ |x_0 + h - x_0 + nh| + |x_0 + h - x_0 - h|$$

$$+ |x_0 + h - x_0 - 2h|$$

$$+ |x_0 + h - x_0 - 3h| + ... + |x_0 + h - x_0 - nh|$$

$$= h + 2h + 3h + ... + (n+1)h + h + 2h + ...(n-1)h$$

$$= h + 2h + 3h + ... + (n-1)h + nh + (n+1)h + h + 2h + ...(n-1)h$$

$$= 2[h + 2h + ...(n-1)h + nh] + (n+1)h - nh$$

$$S_1 - S_0 = nh(n+1) - nh + (n+1)h - n(n+1)h = h$$

$$S_1 - S_0 = h$$

Thus, by calculating $S_2, S_3, ..., S_n$, we get

$$S_2 - S_1 = 3h, \qquad S_3 - S_2 = 5h$$

So, $\qquad S_0 < S_1 < S_2$

In the same way, we can find $\quad S_0 < S_{-1} < S_{-2}$

Similarly, we can prove it for even number of terms. Thus, we can say that mean deviation is least when measured from the median.

EXAMPLE 5. *Show that if deviations x are small as compared with the mean M so that $\left(\dfrac{x}{M}\right)^3$ and higher power of $\dfrac{x}{M}$ may be neglected.*

(i) $G = M\left[1 - \dfrac{\sigma^2}{2M^2}\right]$ 	(ii) $M^2 - G^2 = \sigma^2$

(iii) $H = M\left[1 - \dfrac{\sigma^2}{M^2}\right]$ 	(iv) $H + M = 2G$

(v) Mean $\sqrt{X} = \sqrt{M}\left[1 - \dfrac{\sigma^2}{2M^2}\right]$ 	(vii) $MH = G^2$

where the symbols have their usual meanings.

SOLUTION.   Let us suppose that $X - M = x \implies X = M + x$

We know that geometric mean

$$G = \left(X_1^{f_1} \cdot X_2^{f_2} \cdot X_3^{f_3} \ldots X_n^{f_n} \right)^{1/N}$$

Taking log of both sides, we get

$$\log G = \log \left(X_1^{f_1} \cdot X_2^{f_2} \cdot X_3^{f_3} \ldots X_n^{f_n} \right)^{1/N}$$

$$= \frac{1}{N} \left[f_1 \log X_1 + f_2 \log X_2 + \ldots + f_n \log X_n \right]$$

$$= \frac{1}{N} \sum_{i=1}^{n} f \log X = \frac{1}{N} \Sigma f \log (M + x) = \frac{1}{N} \Sigma f \log \left[M \left(1 + \frac{x}{M} \right) \right]$$

$$= \frac{1}{N} \Sigma f \left[\log M + \log \left(1 + \frac{x}{M} \right) \right] = \frac{1}{N} \Sigma f \log M + \frac{1}{N} \Sigma f \log \left(1 + \frac{x}{M} \right)$$

$$= \log M + \frac{1}{N} \Sigma f \left(\frac{x}{M} - \frac{x^2}{2M^2} + \frac{x^3}{3M^3} - \ldots \right)$$

Now, it is given that $\left(\dfrac{x}{M} \right)^3$ and higher power of $\dfrac{x}{M}$ may be neglected then

$$\log G = \log M + \frac{1}{N} \Sigma f \left(\frac{x}{M} - \frac{x^2}{2M^2} \right)$$

$\because$ For $\dfrac{1}{N} \Sigma fx = 0, \quad \dfrac{1}{N} \Sigma fx^2 = \sigma^2$

Then, $\log G = \log M - \dfrac{1}{2M^2} \cdot \sigma^2$

$$\log G - \log M = - \frac{1}{2M^2} \cdot \sigma^2$$

$$\log \frac{G}{M} = - \frac{\sigma^2}{2M^2}$$

$$\frac{G}{M} = e^{-\sigma^2/2M^2} \qquad \qquad \ldots (1)$$

$$G = M e^{-\sigma^2/2M^2} = M \left(1 - \frac{\sigma^2}{2M^2} + \ldots \right)$$

Neglecting the higher power of σ/M, we get

$$G = M \left(1 - \frac{\sigma^2}{2M^2} \right)$$

(ii) $\boldsymbol{M^2 - G^2 = \sigma^2}$

From equation (1) $\dfrac{G}{M} = e^{-\sigma^2/2M^2}$

Squaring both sides

$$\frac{G^2}{M^2} = e^{-\sigma^2/M^2}$$

$$\frac{G^2}{M^2} = \left(1 - \frac{\sigma^2}{M^2} - \ldots\right)$$

Neglecting the terms $\dfrac{\sigma^3}{M^2}$ and higher power of $\dfrac{\sigma}{M}$

$$\frac{G^2}{M^2} = \left(1 - \frac{\sigma^2}{M^2}\right) \qquad\qquad \Rightarrow \qquad G^2 = M^2 - \sigma^2$$

$$\Rightarrow \qquad M^2 - G^2 = \sigma^2$$

(iii) $\quad \boldsymbol{H = M \left[1 - \dfrac{\sigma^2}{M^2}\right]}$

We know that harmonic mean H is given by

$$\frac{1}{H} = \frac{1}{N}\Sigma f\left(\frac{1}{X}\right)$$

$$\frac{1}{H} = \frac{1}{N}\Sigma f\left(\frac{1}{M+x}\right) = \frac{1}{N}\Sigma f(M+x)^{-1}$$

$$= \frac{1}{N}\Sigma f.\frac{1}{M}\left(1+\frac{x}{M}\right)^{-1} = \frac{1}{N}\Sigma f.\frac{1}{M}\left(1 - \frac{x}{M} + \frac{x^2}{M^2} - \ldots\right)$$

Neglecting higher power of $\dfrac{x}{M}$, we get

$$= \frac{1}{M}\frac{1}{N}\Sigma f\left(1 - \frac{x}{M} + \frac{x^2}{M^2}\right)$$

$$= \frac{1}{M}\left[\frac{1}{N}\Sigma f - \frac{1}{M}\frac{\Sigma fx}{N} + \frac{1}{M^2}.\frac{\Sigma fx^2}{N}\right]$$

$$= \frac{1}{M}\left[1 - \frac{1}{M}.0 + \frac{1}{M^2}.\sigma^2\right] \qquad \left(\text{For } \Sigma fx = 0, \ \frac{1}{N}\Sigma fx^2 = \sigma^2\right)$$

$$\frac{1}{H} = \frac{1}{M}\left[1 + \frac{\sigma^2}{M^2}\right]$$

Then, $\qquad H = \dfrac{M}{\left[1 + \dfrac{\sigma^2}{M^2}\right]} = M\left[1 + \dfrac{\sigma^2}{M^2}\right]^{-1} = M.\left[1 - \dfrac{\sigma^2}{M^2} + \dfrac{\sigma^3}{M^3} - \ldots\right]$

Neglecting the higher power of $\dfrac{x}{M}$ greater than 2

$$H = M\left[1 - \frac{\sigma^2}{M^2}\right]$$

(iv)   **H + M = 2G**

Consider LHS $= H + M$

Put the value of H from part (iii)

$$\Rightarrow \qquad M + M\left[1 - \frac{\sigma^2}{M^2}\right]$$

$$\Rightarrow \qquad M + M - \frac{\sigma^2}{M}$$

$$\Rightarrow \qquad 2M - \frac{\sigma^2}{M}$$

$$\Rightarrow \qquad 2M - \frac{2\sigma^2}{2M}$$

$$\Rightarrow \qquad 2\left(M - \frac{\sigma^2}{2M}\right) \Rightarrow 2.M\left[1 - \frac{\sigma^2}{2M^2}\right]..(2)$$

By part (i), we observe that

$$M\left[1 - \frac{\sigma^2}{2M^2}\right] = G$$

So, by equation (2)

$$H + M = 2G$$

(v) **Mean** $\quad \sqrt{\boldsymbol{X}} = \sqrt{\boldsymbol{M}}\left[\boldsymbol{1} - \dfrac{\boldsymbol{\sigma^2}}{\boldsymbol{2M^2}}\right]$

We know

Mean $\quad \sqrt{X} = \dfrac{1}{N}\Sigma f\sqrt{X} = \dfrac{1}{N}\Sigma f\sqrt{M + x}$ $\qquad\qquad (\because X = M + x)$

$$= \frac{1}{N}\sqrt{M}\ \Sigma f\sqrt{1 + \frac{x}{M}} = \sqrt{M}\cdot\frac{1}{N}\Sigma f\left(1 + \frac{x}{M}\right)^{1/2}$$

$$= \sqrt{M}\cdot\frac{1}{N}\ \Sigma f\left(1 + \frac{1}{2}\frac{x}{M} - \frac{1}{8}\cdot\frac{x^2}{M^2} + ...\right)$$

Neglecting the higher power of $\dfrac{x}{M}$

Mean $\quad \sqrt{X} = \sqrt{M}\cdot\dfrac{1}{N}\Sigma f\left(1 + \dfrac{1}{2}\cdot\dfrac{x}{M} - \dfrac{x^2}{8M^2}\right)$

$$= \sqrt{M}\left[\frac{1}{N}\Sigma f + \frac{1}{2}\cdot\frac{1}{M}\frac{1}{N}\Sigma fx - \frac{1}{8}\frac{1}{M^2}\frac{1}{N}\ \Sigma fx^2\right]$$

$$\left[\text{For } \frac{1}{N}\Sigma fx = 0,\ \frac{1}{N}\Sigma fx^2 = \sigma^2\right]$$

Mean $\sqrt{X} = \sqrt{M}\left[1 + 0 - \dfrac{1}{8M^2}.\sigma^2\right]$ $\qquad\qquad (\because N = \Sigma f)$

$$\text{Mean } \sqrt{X} = \sqrt{M}\left[1 - \frac{\sigma^2}{2M^2}\right]$$

(vi) $MH = G^2$

From part (iii), put the value of H

$$M.M\left[1 - \frac{\sigma^2}{M^2}\right]$$

$$\Rightarrow \quad M^2\left[1 - \frac{\sigma^2}{M^2}\right] \qquad \qquad \dots (3)$$

From relation (ii)

$$M^2 - G^2 = \sigma^2$$

Put in equation (3)

$$MH = M^2\left[1 - \frac{M^2 - G^2}{M^2}\right] = M^2\frac{\left[M^2 - M^2 + G^2\right]}{M^2}$$

$$MH = G^2$$

20.9 RELATION BETWEEN DIFFERENT MEASURES OF DISPERSION

1. Semi-interquartile range $= \dfrac{2}{3}$ standard deviation

2. Mean deviation $= \dfrac{4}{5}$ standard deviation

SOLVED EXAMPLES

EXAMPLE 1. *Compute the mean, variance and standard deviation for the series* $1^2, 2^2, 3^2 \dots n^2$.

SOLUTION. We know that

$$\text{Mean} = \frac{\Sigma x}{N} = \frac{1^2 + 2^2 + 3^2 + \dots + n^2}{n} = \frac{\Sigma n^2}{n} = \frac{n(n+1)(2n+1)}{6n}$$

Thus,   Mean $\dfrac{(n+1)(2n+1)}{6}$

Now, Variance, $\sigma^2 = \dfrac{1}{N}\Sigma(x - \overline{x})^2$;

$$\sigma^2 = \frac{1}{N}\Sigma x^2 - \left(\frac{1}{N}\Sigma x\right)^2 \qquad \qquad \dots (1)$$

We know that

$$\Sigma x = 1^2 + 2^2 + 3^2 + \dots + n^2 = \frac{n(n+1)(2n+1)}{6}$$

$$\Sigma x^2 = 1^4 + 2^4 + 3^4 + \dots + n^4$$

$$= \frac{n(n+1)(2n+1)(3n^2 + 3n - 1)}{30}$$

Putting these values in equation (1)

$$\sigma^2 = \frac{1}{n} n \frac{(n+1)(2n+1)(3n^2+3n-1)}{30} - \left(\frac{n(n+1)(2n+1)}{6n}\right)^2$$

$$\sigma^2 = \frac{(n+1)(2n+1)(3n^2+3n-1)}{30} - \left[\frac{(n+1)(2n+1)}{6}\right]^2$$

$$= \frac{(n+1)(2n+1)}{6}\left[\frac{3n^2+3n-1}{5} - \frac{(n+1)(2n+1)}{6}\right]$$

$$= \frac{(n+1)(2n+1)}{6}\left[\frac{18n^2+18n-6-5(2n^2+3n+1)}{30}\right]$$

$$= \frac{(n+1)(2n+1)}{6}\left[\frac{8n^2+3n-11}{30}\right]$$

$$\sigma^2 = \frac{(n+1)(2n+1)(8n^2+3n-11)}{180};$$

$$\sigma = \left[\frac{(n+1)(2n+1)(8n^2+3n-11)}{180}\right]^{1/2}$$

EXAMPLE 2. *The details of runs gained by two batsman A and B are as follows :*

A	24	79	31	114	14	02	68	01	110	07
B	05	18	42	53	09	47	52	17	81	56

(i) *Which of the two batsman is better run scorer?*

(ii) *Which of the two batsman has more consistency in the number of runs?*

SOLUTION. Let us assume the quantity x with respect to the runs of the batsman A and y with respect to the runs of batsman B.

X	$X - M_A$	$(X - M_A)^2$	Y	$Y - M_B$	$(Y - M_B)^2$
24	−21	441	05	−33	1089
79	34	1156	18	−20	400
31	−14	196	42	4	16
114	69	4761	53	15	225
14	−31	961	09	29	841
02	−43	1849	47	9	81
68	23	529	52	14	196
01	−44	1936	17	−21	441
110	65	4225	81	43	1849
07	−38	1444	56	18	324
450		17498	380		5462

For A

The total number of terms $N = 10$

Then, $M_A = \dfrac{\Sigma X}{N} = \dfrac{450}{10} = 45$

For B

Total number of terms $N = 10$

$$M_B = \frac{\Sigma Y}{N} = \frac{380}{10} = 38$$

Now, by the above calculation, it proves that

$$M_A > M_B$$

So, we can say that A is better scorer.

Now, we will calculate the standard deviation

$$\sigma_A = \sqrt{\frac{1}{N} \Sigma(X - M_A)^2}$$

$$= \sqrt{\frac{1}{10} \times 17498} = \sqrt{1749.8} = 41.8306 \text{ runs}$$

$$\sigma_B = \sqrt{\frac{1}{N} \Sigma(Y - M_B)^2} = \sqrt{\frac{1}{10} \times 5462} = \sqrt{546.2}$$

$$\sigma_B = 28.3709 \text{ runs}$$

To check the consistency in number of runs, we have to calculate the coefficient of variation.

$$V_A = \frac{\sigma_A}{M_A} \times 100 \qquad V_B = \frac{\sigma_B}{M_B} \times 100$$

$$V_A = \frac{41.8306}{45} \times 100 \qquad V_B = \frac{23.3709}{38} \times 100$$

$$V_A = 92.9568 \qquad V_B = 61.5023$$

We observe that $V_A > V_B$

So, B has the consistency in number of runs.

EXAMPLE 3. *An analysis of the monthly wages paid to workers in two firms A and B, belonging to the same industry gives the following results*

	Firm A	Firm B
Number of wage earners	586	648
Average monthly wages	Rs 525	Rs 475
Variance of the distribution	10,000	12,100

(i) *Which firm A or B pays out the larger amount as monthly wages.*

(ii) *In which firm A or B is there greater variability in individual wages.*

(iii) *What are measures of average monthly wages and the variability in individual wages of all the workers in the two firms A and B taken together?*

SOLUTION. Given $N_A = 586, \ N_B = 648$

$$\overline{X}_A = \text{Rs. } 525, \quad \overline{X}_B = \text{Rs. } 475$$

Standard deviation $= \sigma_A = \sqrt{10000} = 100$

$$\sigma_B = \sqrt{12100} = 110$$

(i) Total amount paid by firm $A = 525\ (586) =$ Rs. 307,650

Total amount paid by firm $B = 475\ (648) =$ Rs. 307,800

$\therefore$ Company B pays larger amount as monthly wages.

(ii) Coefficient of variation for firm A

$$VC_A = \frac{\sigma_A}{\overline{X}_A} \times 100 = \frac{100}{525} \times 100 = 19.05\%$$

$$VC_B = \frac{\sigma_B}{\overline{X}_B} \times 100 = \frac{110}{475} \times 100 = 23.16\%$$

Company B has greater variability in individual wages.

(iii) Measure of average taking A and B together

$$\overline{X}_{AB} = \frac{N_A \overline{X}_A + N_B \overline{X}_B}{N_A + N_B} = \frac{586(525) + 648(475)}{586 + 648}$$

$$= \frac{307650 + 307800}{1234} = 498.743$$

$$\overline{X}_{AB} = 498.743$$

$$\sigma_{AB} = \sqrt{\frac{N_A \sigma_A^2 + N_A(\overline{X}_A - \overline{X}_{AB})^2 + N_B \sigma_B^2 + N_B(\overline{X}_B - \overline{X}_{AB})}{N_A + N_B}}$$

$$= \sqrt{\frac{586(100)^2 + 586(689.43) + 648(110)^2 + 648(563.73)}{1234}}$$

$$= \sqrt{\frac{5860000 + 404005.98 + 7840800 + 365297.04}{1234}}$$

$$= \sqrt{11726.17749} = 108.287$$

EXAMPLE 4. *An analysis of monthly wages paid to workers in two firms A and B, belonging to the same industry gives the following results:*

	Firm A	Firm B
Number of workers	500	600
Average monthly wages (Rs.)	186	175
Variance of the distribution of wages (Rs.)	81	100

(i) *Which firm A or B has a larger wage bill?*

(ii) *In which firm A or B is there greater variability in individual wages.*

(iii) *Calculate the mean and variance of all the workers in the firms A and B taken together.*

SOLUTION. For firm A, total monthly wage $= 500 \times 186 =$ Rs. 93,000

For firm B, total monthly wage $= 600 \times 175 =$ Rs. 90,500

(i) Firm B pays larger wage as monthly wages.

(ii) Coefficient of variance for firm $A = \dfrac{\sigma}{\overline{X}} \times 100$

$$= \frac{\sqrt{81}}{186} \times 100 = \frac{9}{186} \times 100 = 4.83$$

Coefficient of variance for firm

$$B = \frac{\sqrt{100}}{175} \times 100 = \frac{10}{175} \times 100 = 5.71$$

There is greater variability in individual wages in firm B.

(iii) Combined mean $\bar{X}_{12} = \dfrac{N_1 \bar{X}_1 + N_2 \bar{X}_2}{N_1 + N_2} = \dfrac{500(186) + 600(175)}{500 + 600}$

$$= \frac{93000 + 105000}{1100} = 180$$

Combined mean $= 180$

and combined variance

$$\sigma_{12}^2 = \sqrt{\frac{N_1 \sigma_1^2 + N_1(\bar{X}_1 - \bar{X}_{12})^2 + N_2 \sigma_2^2 + N_2(\bar{X}_2 - \bar{X}_{12})}{N_1 + N_2}}$$

$$= \sqrt{\frac{500 \times 81 + 500(186-180)^2 + 600 \times 100 + 600(175-180)^2}{500+600}}$$

$$= \sqrt{\frac{40500 + 18000 + 60000 + 15000}{1100}} = \sqrt{\frac{133500}{1100}}$$

$$\Rightarrow \qquad \sigma_{12}^2 = 121.36$$

EXAMPLE 5. *Find the mean, median, mode and coefficient of variation for the following distribution:* **(UPTU–2002)**

Size	10-15	15-20	20-25	25-30	30-35	35-40
Frequency	8	12	12	18	14	10

SOLUTION.

Size	Mid value (x)	Frequency (f)	Cumulative Frequency	$f \cdot x$
10-15	12.5	8	8	100
15-20	17.5	12	20	210
20-25	22.5	12	32	270
25-30	27.5	18	50	495
30-35	32.5	14	64	455
35-40	37.5	10	74	375
		$N = 74$		$\Sigma fx = 1905$

$$\text{Median} = \bar{X} = \frac{\Sigma fx}{\Sigma f} = \frac{1905}{74} = 25.74$$

Now, Median number $= \left(\dfrac{N}{2}\right)^{\text{th}}$ term $= \dfrac{74}{2} = 37^{\text{th}}$ term

Hence, 37th term lies in the interval 25-30. Hence, 25-30 is the modal class in which $l_1 = 25, f = 18, i = 10$.

$$\text{Median} = l_1 + \frac{i}{f}\left(\frac{N}{4} - C\right)$$

$$= 25 + \frac{5}{18}(37 - 32) = 25 + \frac{5}{18} \times 5$$

$$= 25 + \frac{25}{18} = 26.389$$

Now, for mode see the highest frequency is 18 which lies in the interval 25-30. Hence, 25-30 is the modal class in which

$$l_1 = 25, f = 18, f_1 = 14, f_{-1} = 12 \text{ and } i = 5$$

$$\text{Mode} = l_1 + \frac{f - f_{-1}}{2f - f_{-1} - f_1} \times i$$

$$= 25 + \frac{18 - 12}{2 \times 18 - 12 - 14} \times 5 = 25 + \frac{6}{36 - 26} \times 5$$

$$= 25 + \frac{30}{10} = \frac{250 + 30}{10} = \frac{280}{10} = 28$$

Now, for coefficient of variation we construct the second table:

Class	Frequency (f)	Mid value (x)	$d = x - 27.5$	fd	fd^2
10-15	8	12.5	− 15.0	−120	1800
15-20	12	17.5	− 10.0	−120	1200
20-25	12	22.5	− 5.0	−60	300
25-30	18	27.5	0	0	0
30-35	14	32.5	5.0	70	350
35-40	10	37.5	10.0	100	1000
	$N = 74$			$\Sigma fd = -130$	$\Sigma fd^2 = 4650$

Standard deviation

$$= \sqrt{\frac{\Sigma fd^2}{N} - \left(\frac{\Sigma fd}{N}\right)^2} = \sqrt{\frac{4650}{74} - \left(\frac{-130}{74}\right)^2}$$

$$= \sqrt{62.83 - 3.10} = \sqrt{59.73}$$

$$\Rightarrow \qquad \sigma = 7.73$$

Now, coefficient of variation

$$= \frac{\sigma}{\overline{X}} \times 100 = \frac{7.73}{25.74} \times 100 = 30.03\%$$

EXAMPLE 6. *From the following data, find the coefficient of variation :* [CCSU(B.Sc Biotech)–1997]

Marks	0-10	10-20	20-30	30-40	40-50	50-60	60-70	70-80
Number of students	5	10	20	40	30	20	10	4

SOLUTION.

Marks	Number of Students (f)	Mid value x	$d = \dfrac{x - A}{h}$	fd	fd^2
0-10	5	5	-4	-20	80
10-20	10	15	-3	-30	90
20-30	20	25	-2	-40	80
30-40	40	35	-1	-40	40
40-50	30	45	0	0	0
50-60	20	55	1	20	20
60-70	10	65	2	20	40
70-80	4	75	3	12	30
	$N = 139$			$\Sigma fd = -78$	$\Sigma fd^2 = 386$

$$\text{Mean}\,(\overline{X}) = A + \frac{\Sigma fd}{N} \times i = 45 + \frac{(-78)}{139} \times 10$$

$$= 45 - \frac{780}{139} = 45 - 5.61 = 39.39$$

Standard deviation

$$(\sigma) = i\sqrt{\frac{\Sigma fd^2}{N} - \left(\frac{\Sigma fd}{N}\right)^2} = 10\sqrt{\frac{386}{139} - \left(\frac{-78}{139}\right)^2}$$

$$= 10\sqrt{2.77 - 0.31} = 10\sqrt{2.46} = 10 \times 1.568 = 15.68$$

Now, coefficient of variation

$$= \frac{\sigma}{\overline{X}} \times 100 = \frac{15.68}{39.39} \times 100 = 39.80\%$$

20.10 STANDARD DEVIATION OF TWO COMBINED SETS

Let us consider two sets with n_1 and n_2 number of terms and whose standard deviations are given by σ_1 and σ_2, respectively, measured from respective means M_1 and M_2. If we grouped these sets, then standard deviation σ of the combined sets is given by

$$\sigma = \left[\frac{n_1\sigma_1^2 + n_2\sigma_2^2}{n_1 + n_2} + \frac{n_1 n_2}{(n_1 - n_2)}(M_1 - M_2)^2\right]^{1/2}$$

PROOF.   Let us assume that $x_1, x_2, ..., x_{n_1}$ and $y_1, y_2, ..., y_{n_2}$ be the given two series with total number of terms n_1 and n_2. Then the mean of these series will be

$$M_1 = \frac{1}{n_1}\Sigma x_{n_1} \qquad\qquad M_2 = \frac{1}{n_2}\Sigma x_{n_2}$$

Let S_1^2 and S_2^2 be the mean square deviations of these series and A be the assumed mean.

Then, $$S_1^2 = \frac{1}{n_1}\Sigma(x_{n_1} - A)^2 \qquad\qquad\qquad\qquad ...(1)$$

$$S_2^2 = \frac{1}{n_2}\Sigma(x_{n_2} - A)^2 \qquad\qquad\qquad\qquad ...(2)$$

The mean square deviation for the combined series will be

$$S^2 = \frac{1}{n_1 + n_2}\left[\Sigma(x_{n_1} - A)^2 + \Sigma(x_{n_2} - A)^2\right] = \frac{1}{n_1 + n_2}\left[n_1 S_1^2 + n_2 S_2^2\right]$$

(with the help of equation (1) and (2))

Since we know that $\quad S^2 = \sigma^2 + d^2$

Then,

$$S^2 = \frac{1}{n_1 + n_2}\left[n_1(\sigma_1^2 + d_1^2) + n_2(\sigma_2^2 + d_2^2)\right]$$

$$= \frac{1}{n_1 + n_2}\left[n_1\sigma_1^2 + n_1 d_1^2 + n_2\sigma_2^2 + n_2 d_2^2\right]$$

$$S^2 = \frac{n_1\sigma_1^2 + n_2\sigma_2^2}{n_1 + n_2} + \frac{n_1 d_1^2 + n_2 d_2^2}{n_1 + n_2} \qquad \ldots (3)$$

For combined series, if $A = M$, then $S^2 = \sigma^2$.

Now, we have to calculate the value of d for grouped series.

The mean for the grouped series $\quad M = \dfrac{n_1 M_1 + n_2 M_2}{n_1 + n_2}$

Also $\qquad\qquad\qquad d_1 = M_1 - A \qquad\qquad d_2 = M_2 - A$

Put these values of d_1 and d_2 in equation (3)

$$S^2 = \frac{n_1\sigma_1^2 + n_2\sigma_2^2}{n_1 + n_2} + \frac{n_1(M_1 - A)^2 + n_2(M_2 - A)^2}{n_1 + n_2} \qquad \ldots (4)$$

For grouped series $A = M$, then

$$M_1 - A = M_1 - \frac{n_1 M_1 + n_2 M_2}{n_1 + n_2} = \frac{n_1 M_1 + n_2 M_1 - n_1 M_1 - n_2 M_2}{n_1 + n_2}$$

$$M_1 - A = \frac{n_2(M_1 - M_2)}{n_1 + n_2}$$

$$M_2 - A = M_2 - \frac{n_1 M_1 + n_2 M_2}{n_1 + n_2}$$

$$M_2 - A = \frac{n_1(M_2 - M_1)}{n_1 + n_2}$$

Put the value of $(M_1 - A)$ and $(M_2 - A)$ in equation (4).

$$\sigma^2 = \frac{n_1\sigma_1^2 + n_2\sigma_2^2}{n_1 + n_2} + \frac{n_1 n_2^2(M_1 - M_2)^2}{(n_1 + n_2)^3} + \frac{n_1^2 n_2(M_2 - M_1)^2}{(n_1 + n_2)^3}$$

$$\sigma^2 = \frac{n_1\sigma_1^2 + n_2\sigma_2^2}{n_1 + n_2} + \frac{n_1 n_2(M_1 - M_2)^2\,[n_2 + n_1]}{(n_1 + n_2)^3}$$

$$\sigma^2 = \frac{n_1\sigma_1^2 + n_2\sigma_2^2}{n_1 + n_2} + \frac{n_1 n_2(M_1 - M_2)^2}{(n_1 + n_2)^2}$$

So the standard deviation for the combined sample is given by

$$\sigma = \left[\frac{n_1\sigma_1^2 + n_2\sigma_2^2}{n_1 + n_2} + \frac{n_1 n_2(M_1 - M_2)^2}{(n_1 + n_2)^2}\right]^{1/2}$$

SOLVED EXAMPLES

EXAMPLE 1. *The mean and standard deviation of two samples are given as follows :*

$$n_1 = 50 \qquad M_1 = 54.1 \qquad \sigma_1 = 8$$
$$n_2 = 100 \qquad M_2 = 50.3 \qquad \sigma_2 = 7$$

Find the mean and standard deviation of the combined sample.

SOLUTION. Let M and σ be the mean and standard deviation of the combined sample

Then, $$M = \frac{M_1 n_1 + M_2 n_2}{n_1 + n_2} = \frac{54.1 \times 50 + 50.3 \times 100}{50 + 100} = \frac{2705 + 5030}{150}$$

$$M = 51.566$$

Now, we have to calculate the standard deviation of combined sample.

We know that S.D. for combined sample

$$\sigma = \left[\frac{n_1 \sigma_1^2 + n_2 \sigma_2^2}{n_1 + n_2} + \frac{n_1 n_2 (M_1 - M_2)^2}{(n_1 + n_2)^2} \right]^{1/2}$$

$$= \left[\frac{50 \times 64 + 100 \times 49}{50 + 100} + \frac{50 \times 100 (54.1 - 50.3)^2}{(50 + 100)^2} \right]^{1/2}$$

$$= \left[\frac{3200 + 4900}{150} + \frac{5000 \times 14.44}{(150)^2} \right]^{1/2}$$

$$= \left[\frac{8100}{150} + \frac{19000}{22500} \right]^{1/2} = [54 + 3.2088]^{1/2}$$

$$\sigma = \sqrt{57.2088} = 7.555$$

EXAMPLE 2. *Out of the two groups the number of observations are given by 100 and 150. The mean and the standard deviations of these groups and of combined sample is as follows.*

First group

$$n_1 = 100 \qquad M_1 = ? \qquad \sigma_1 = 7$$

Second group

$$n_2 = 150 \qquad M = 55 \qquad \sigma = ?$$

Combined group

$$n = ? \qquad M = 51 \qquad \sigma = \sqrt{130}$$

Find the missing values.

SOLUTION. We know that the total number of terms in the combined group $= n_1 + n_2$

$$n = 100 + 150 = 250$$

Now, the mean of the combined sample

$$M = \frac{M_1 n_1 + M_2 n_2}{n_1 + n_2}$$

$$51 = \frac{M_1 \times 100 + 55 \times 150}{100 + 150}$$

$$51 = \frac{100M_1 + 8250}{250}$$

$$100\,M_1 = 12750 - 8250$$

$$M_1 = 45$$

So the mean of the first group $M_1 = 45$.

Now, we have to calculate the standard deviation of the second group.

S.D. of combined sample,

$$\sigma = \left[\frac{n_1\sigma_1^2 + n_2\sigma_2^2}{n_1 + n_2} + \frac{n_1 n_2 (M_1 - M_2)^2}{(n_1 + n_2)^2} \right]^{1/2}$$

$$\sqrt{130} = \left[\frac{100 \times 49 + 150 \times \sigma_2^2}{250} + \frac{100 \times 150\,(45 - 55)^2}{(250)^2} \right]^{1/2}$$

$$130 = \frac{4900 + 150\sigma_2^2}{250} + \frac{1500000}{(250)^2}$$

$$106 = \frac{4900 + 150\sigma_2^2}{250}$$

$$150\,\sigma_2^2 = 21600$$

$$\sigma_2^2 = 144$$

$$\sigma_2 = 12.$$

EXAMPLE 3. *The means of the two sample of size 50 and 100 respectively 54.1 and 50.3 and standard deviations are 8 and 7. Obtain the mean and the standard deviation of combined sample of size 150.*

SOLUTION.

$$n_1 = 50, \quad n_2 = 100$$

$$\overline{X}_1 = 54.1, \quad \overline{X}_2 = 50.3$$

$$\sigma_1 = 8, \quad \sigma_2 = 7$$

Combined mean $= \overline{X}_{12} = \dfrac{n_1 \overline{X}_1 + n_2 \overline{X}_2}{n_1 + n_2} = \dfrac{50(54.1) + 100(50.3)}{150}$

$$\overline{X}_{12} = 51.57.$$

Combined standard deviation

$$\sigma_{12} = \sqrt{\frac{n_1\sigma_1^2 + n_1(\overline{X}_1 - \overline{X}_{12})^2 + n_2\sigma_2^2 + n_2(\overline{X}_2 - \overline{X}_{12})}{n_1 + n_2}}$$

$$= \sqrt{\frac{50(64) + 100(49) + 50(6.4) + 100(1.6129)}{150}}$$

$$= \sqrt{\frac{3200 + 4900 + 320 + 161.20}{150}}$$

$$\sigma_{12} = 7.564$$

EXAMPLE 4. *For a group of 200 candidates, the mean and standard deviation were found to be 40 and 15, respectively. Find the corrected mean and standard deviation to the correct figures.*

SOLUTION. Incorrect mean = 40.

Incorrect standard deviation = 15.

Correct value = 43

Incorrect value = 53

$$\bar{X} = \frac{\Sigma X}{N}$$

Incorrect : $\Sigma X = \bar{X}N = 400(200) = 8000$

Correct : ΣX = Incorrect ΣX – Incorrect value + Correct value

$$= 8000 - 53 + 43 = 7990$$

Correct $\bar{X} = \dfrac{7990}{200} = 39.95$

Standard deviation = $\sqrt{\dfrac{\Sigma X^2}{N} - \left(\dfrac{\Sigma X}{N}\right)^2}$

$$(15)^2 = \frac{\Sigma X^2}{200} - (40)^2$$

$$225 + 1600 = \frac{\Sigma X^2}{200}$$

Incorrect $\Sigma X^2 = 200(1825)$

Incorrect $\Sigma X^2 = 365000$

Correct ΣX^2 = Incorrect ΣX^2 – (Incorrect value)^2 + (Correct value)^2

$$= 365000 - (53)^2 + (43)^2 = 36440$$

Correct S.D. = $\sqrt{\dfrac{364040}{200} - (39.95)^2} = 15.185$

EXAMPLE 5. *The daily average wages of a group of 70 workers is Rs. 13.5 and the standard deviation is Rs 14. The daily average wages of another group of 80 workers is Rs 15. and variance is Rs 4. Find (i) Daily average wages, (ii) variance and (iii) coefficient of variation for all the 150 workers.*

SOLUTION. It is given that

$$n_1 = 70, \quad M_1 = 13.5, \quad \sigma_1 = 14$$
$$n_2 = 80, \quad M_2 = 15, \quad \sigma_2 = 2$$

The total number of workers in the combined group

$$= n_1 + n_2 = 70 + 80 = 150$$

Now, average wages of combined group

$$M = \frac{n_1 M_1 + n_2 M_2}{n_1 + n_2} = \frac{70 \times 13.5 + 80 \times 15}{150}$$

$$= \frac{945 + 1200}{150} = \frac{2145}{150} = 14.3$$

$$M = 14.3$$

Now, $\quad d_1 = M_1 - A = 13.5 - 14.3 = -0.8$

then $\quad d_1^2 = 0.64$

Now, $\quad d_2 = M_2 - A = 15 - 14.3 = 0.7$

then $\quad d_2^2 = 0.49$ $\qquad$ (where $M = A$)

Now, coefficient of variance

$$\sigma^2 = \frac{n_1(\sigma_1^2 + d_1^2) + n_2(\sigma_2^2 + d_2^2)}{n_1 + n_2}$$

$$\Rightarrow \qquad = \frac{70[(1.4)^2 + (0.64)] + 80[(2)^2 + (0.49)]}{150}$$

$$= \frac{182 + 359.2}{150} = \frac{541.2}{150}$$

$$\sigma^2 = 3.608$$

$$\Rightarrow \qquad \sigma = \sqrt{3.608} = 1.899$$

$$\Rightarrow \qquad \sigma = 1.9$$

Now, combined coefficient of variance

$$= \frac{\sigma}{M} = \frac{1.9}{14.3} = 0.1328 = 13.28\%$$

EXAMPLE 6. *The mean height of 27 boys in section A of a class is 150 cms and the mean height of 33 boys in section B of the same class is 140 cms. Find the overall mean height of 60 boys.*

SOLUTION. It is given that $n_1 = 27$, $M_1 = 150$ cm, $n_2 = 33$ cm, $M_2 = 140$ cm.

Now, mean height of 60 students

$$M = \frac{M_1 n_1 + M_2 n_2}{n_1 + n_2} = \frac{150 \times 27 + 140 \times 33}{27 + 33}$$

$$= \frac{4050 + 4620}{60} = \frac{8670}{60} = 144.5$$

$$M = 144.5$$

EXAMPLE 7. *A distribution consist of three components with frequency 200, 250 and 300 having means 26, 11 and 16 and standard deviation 3, 4 and 5 respectively. Find the mean variance of the combined distribution.*

$\qquad$ **(RGPV–2003)**

SOLUTION. It is given that

$$n_1 = 200, \qquad M_1 = 26, \qquad \sigma_1 = 3$$
$$n_2 = 250, \qquad M_2 = 11, \qquad \sigma_2 = 4$$
$$n_3 = 300, \qquad M_3 = 16, \qquad \sigma_3 = 5$$

The total number of workers in the combined group

$$= n_1 + n_2 + n_3 = 200 + 250 + 300 = 750$$

Now, combined mean

$$= \frac{n_1 M_1 + n_2 M_2 + n_3 M_3}{n_1 + n_2 + n_3}$$

$$= \frac{200 \times 26 + 250 \times 11 + 300 \times 16}{200 + 250 + 300}$$

$$= \frac{5200 + 2750 + 4800}{750} = \frac{12750}{750} = 17$$

Now, $\qquad d_1 = 26 - 17 = 9 \qquad \Rightarrow \qquad d_1^2 = 81$

$$d_2 = 11 - 17 = -6 \quad \Rightarrow \quad d_2^2 = 36$$
$$d_3 = 16 - 17 = -1 \quad \Rightarrow \quad d_3^2 = 1$$

Therefore, combined variance

$$\sigma^2 = \frac{200(9+81) + 250(16+36) + 300(25+1)}{750}$$

$$= \frac{18000 + 13000 + 7800}{750} = \frac{38800}{750} = 51.73$$

EXAMPLE 8. *Five samples of sizes 20, 25, 30, 35 and 40 were from a population and their means and S.D. calculated which are :* **(UPTU–2002)**

Mean (M)	17.8	18.2	16.9	17.3	17.7
S.D. (σ)	2.6	1.9	2.8	2.3	2.1

SOLUTION. It is given that

$$n_1 = 20, \, n_2 = 25, \, n_3 = 30, \, n_4 = 35 \text{ and } n_5 = 40$$

For combined group total number

$$= n_1 + n_2 + n_3 + n_4 + n_5 = 20 + 25 + 30 + 35 + 40 = 150$$

Now, combined mean

$$= \frac{n_1 M_1 + n_2 M_2 + n_3 M_3 + n_4 M_4 + n_5 M_5}{n_1 + n_2 + n_3 + n_4 + n_5}$$

$$= \frac{20 \times 17.8 + 25 \times 18.2 + 30 \times 16.9 + 35 \times 17.3 + 40 \times 17.7}{150}$$

$$= \frac{356 + 455 + 507 + 605.5 + 708}{150} = \frac{2631.5}{150} = 17.54$$

Now,
$$d_1 = 17.8 - 17.54 = 0.26 \quad \Rightarrow \quad d_1^2 = 0.676$$
$$d_2 = 18.2 - 17.54 = 0.66 \quad \Rightarrow \quad d_2^2 = 0.4356$$
$$d_3 = 16.9 - 17.54 = -0.64 \quad \Rightarrow \quad d_3^2 = 0.4096$$
$$d_4 = 17.3 - 17.54 = -0.24 \quad \Rightarrow \quad d_4^2 = 0.0576$$
$$d_5 = 17.7 - 17.54 = 0.16 \quad \Rightarrow \quad d_5^2 = 0.0256$$

Therefore combined mean

$$\sigma = \sqrt{\frac{\begin{array}{c} n_1(\sigma_1^2 + d_1^2) + n_2(\sigma_2^2 + d_2^2) + n_3(\sigma_3^2 + d_3^2) \\ + n_4(\sigma_4^2 + d_4^2) + n_5(\sigma_5^2 + d_5^2) \end{array}}{n_1 + n_2 + n_3 + n_4 + n_5}}$$

$$= \sqrt{\frac{\begin{array}{c} 20\{(2.6)^2 + 0.676\} + 25\{(1.9)^2 + 0.4356\} \\ + 30\{(2.8)^2 + 0.4096\} + 35\{(2.3)^2 + 0.576\} \\ + 40\{(2.1)^2 + 0.256\} \end{array}}{150}}$$

$$= \sqrt{\frac{\begin{array}{c} 20(6.76 + 0.676) + 25(3.61 + 0.4356) \\ + 30(7.84 + 0.4096) + 35(5.29 + 0.576) \\ + 40(4.41 + 0.256) \end{array}}{150}}$$

$$= \sqrt{\frac{\begin{array}{l}20 \times 6.8276 + 25 \times 4.0456 + 30 \times 8.2496 \\ + 35 \times 5.3476 + 40 \times 4.4356\end{array}}{150}}$$

$$= \sqrt{\frac{\begin{array}{l}136.552 + 101.14 + 247.488 \\ + 187.166 + 177.424\end{array}}{150}} = \sqrt{\frac{849.77}{150}} = \sqrt{5.665} = 2.38$$

Hence, combined mean and combined standard deviation are 17.54 and 2.38 respectively.

20.11 STANDARD VARIABLE

A variable with mean 0, variance unity and defined by $Z = \dfrac{X - M}{\sigma}$ is known as standard variable, in which M and σ are the mean and standard deviation of the variable X.

SOLVED EXAMPLE

EXAMPLE. *The marks of three students in an examination are given as follows :*

Student	Eng.	Math.	Science	Total
A	95	61	70	226
B	69	74	83	226
C	70	82	74	226
Mean	55	50	53	

Find the order of equitable while the standard deviation is given by 16, 11 and 12.

SOLUTION. To calculate the equitable order, we will find the standard variable for the marks obtained by each student in each subject

Student	Eng.	Math.	Science
A	$\dfrac{95-55}{16}$	$\dfrac{61-50}{11}$	$\dfrac{70-53}{12}$
B	$\dfrac{69-55}{16}$	$\dfrac{74-50}{11}$	$\dfrac{83-53}{12}$
C	$\dfrac{70-55}{16}$	$\dfrac{82-50}{11}$	$\dfrac{74-53}{12}$

Student	Eng.	Math.	Science	Total
A	2.5	1	1.416	4.916
B	0.875	2.182	2.5	5.557
C	0.9375	2.909	1.75	5.5965

So, the equitable order will be $A < B < C$.

Exercise 20.3

1. Calculate the standard deviation of the given two series and find which show greater deviation.

   Series A : 192, 288, 236, 229, 184, 260, 348, 291, 330, 243

   Series B : 83, 87, 93, 109, 124, 126, 126, 101, 102, 108

2. In the given series, if d_1 and d_2 represent the deviation from assumed mean 100, then calculate the coefficient of variation for two series for which

   $$n_1 = 150, \ \Sigma d_1 = 100, \ \Sigma d_1^2 = 245320$$

   $$n_2 = 200, \ \Sigma d_2 = 250, \ \Sigma d_2^2 = 245320$$

3. The fluctuation in the rate of two items A and B are given as follows. Calculate for which the rate is more consistent.

A	55	54	52	53	56
B	108	107	105	105	107
A	58	52	50	51	49
B	104	103	104	106	101

   and find the values of standard deviation for each item.

4. The table shows the number of workers in two factories whose weekly earnings are given. Determine the mean value and standard deviations in both the factory.

Range	Factory A	Factory B
4-6	74	71
6-8	376	379
8-10	304	303
10-12	110	112
12-14	18	18
14-16	0	1
16-18	9	3
18-20	9	9
20-22	0	4

5. Show that the variance can be defined in the form of half of mean of the squares of deviations of natural terms

   $$\sigma^2 = \frac{1}{2}\left[\frac{1}{n^2} \sum_{i=1}^{n} \sum_{j=1}^{n} (x_i - x_j)^2\right],$$

   where $x_1, x_2, ..., x_n$ are n variate value.

6. Discuss the effect of change of origin and scale on standard deviation.

7. Discuss the various measure of dispersion.

8. Establish a relationship between root-mean square deviation and standard deviation.

9. The number examined, the mean weight and the standard deviation in each group of examination by three medical examines are given below. Find the mean weight and standard deviation of the entire data when grouped together: **(UPTU–2003)**

Medical Examiner	Numbers Examined	Mean Weight	Standard Deviation
A	50	113	6
B	60	120	7
C	90	115	8

10. The first of the two samples has 100 items with mean 15 and standard deviation 3. If the whole group has 250 items with mean 15.6 and standard deviation $\sqrt{13.44}$, find the standard deviation of second group.

    (UPTU–2001)

11. The mean and standard deviation of 200 items are found to be 60 and 20. If at the time of calculation, two items are wrongly taken as 3 and 67 instead of 13 and 17. Find the correct mean and the correct standard deviation. **(UPTU–2003)**

12. Find out coefficient of variation of following table : **(CCSU(B.Sc BIOTECH)–1996)**

Income (in Rs)	Number of persons
900-1000	50
800-900	60
700-800	90
600-700	200
500-600	200
400-500	100
300-400	150
200-300	50
100-200	100

1. $\sigma_A = 51.6$, $\sigma_B = 14.96$. Series A shows greater deviation   **2.** 39.9, 14.6

3. $\sigma_A = 2.65$, $V_A = 4.99\%$, $\sigma_B = 2$, $V_B = 1.90\%$ **4.** $M_A = 8.34$, $\sigma_A = 2.34$, $M_B = 8.36$, $\sigma_B = 2.29$
10. 4   **11.** 59.8, 20.09   **12.** 0.47%

20.12 SKEWNESS

Skewness is the measure of asymmetry. A distribution which does not occur equidistant on both sides of the mean value is known as asymmetric or skewed distribution.

Positive and Negative Skewness

A frequency curve is said to be positive skewed if it has a longer tail towards the direction of higher values and the curve which has a longer tail towards the direction of smaller values is known as negative skewed.

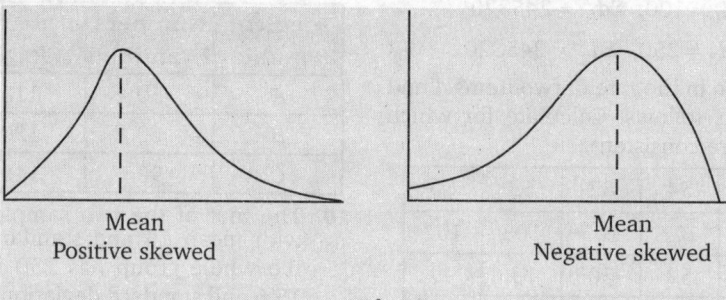

Mean
Positive skewed

Mean
Negative skewed

Fig. 1

For a skewed curve, the value of mean, mode and median will never be equal.

For a positive skewed : Mode < Median < Mean

For a negative skewed : Mean < Median < Mode

20.13 MEASURE OF SKEWNESS

The coefficient of skewness can be calculated by the following two methods

1. According to Karl Pearson,

$$\text{Coeff. of skewness} = \frac{\text{Mean} - \text{Mode}}{\text{Standard deviation}}$$

$$= \frac{3(\text{Mean} - \text{Median})}{\text{Standard deviation}}$$

The value of this coefficient lies between -3 to $+3$.

2. The second formula for coefficient of skewness is given by Bowley, which is as follows

$$\text{Coefficient of skewness} = \frac{Q_3 + Q_1 - 2Q_2}{Q_3 - Q_1}$$

where Q_1, Q_2, Q_3 are the first, second and third quartiles. It can also be written as

$$\text{Coefficient of skewness} = \frac{(Q_3 - Q_2) - (Q_2 - Q_1)}{(Q_3 - Q_2) + (Q_2 - Q_1)}$$

The value of this coefficient lies between -1 to $+1$.

20.13.1 MEASURE OF SKEWNESS BASED ON MOMENTS

1. If $v_1 = 0$ or if $\mu_3 = 0$, then there will be no skewness.

2. If $v_1 > 0$: positive skewed

3. If $v_1 < 0$: Negative skewed

SOLVED EXAMPLES

EXAMPLE 1. *Find the mean, mode, standard deviation and coefficient of skewness for the following frequency distribution.*

[RGPV(B. Pharma)–2001]

Years	0-10	10-20	20-30	30-40	40-50	50-60
Frequency	15	17	19	27	19	12

SOLUTION.

Class	Mid Value x	Frequency	fx	$(x-M)$	$(x-M)^2$	$f(x-M)^2$
0-10	5	15	75	−25	625	9375
10-20	15	17	255	−15	225	3825
20-30	25	19	475	−5	25	475
30-40	35	27	945	5	25	675
40-50	45	19	855	15	225	4275
50-60	55	12	660	25	625	7500
		109	3265			26125

$$M = \frac{\Sigma fx}{N} = \frac{3265}{109} = 29.95$$

$$M = 29.95 \quad \text{year}$$

Let $\quad M = 30$

Now, $\quad \sigma = \sqrt{\frac{1}{N}\Sigma f(x-M)^2} = \sqrt{\frac{1}{109}(26125)}$

$$= \sqrt{239.678} \; ; \; \sigma = 15.48 \; \text{year}$$

Now, we have to calculate the mode to obtain the coefficient of skewness.

Here, we observe that the frequency of class 30-40 is maximum. So, it will be the modal class.

$$l = 30, f_{-1} = 19, f = 27, f_1 = 19, i = 10$$

$$\text{Mode} = l + \frac{f - f_{-1}}{2f - f_{-1} - f_1} \times i$$

$$= 30 + \frac{27 - 19}{54 - 19 - 19} \times 10 = 30 + \frac{8}{16} \times 10$$

$$\text{Mode} = 35$$

The coefficient of skewness

$$= \frac{\text{Mean} - \text{Mode}}{\text{Standard deviation}}$$

$$= \frac{29.95 - 35}{15.48} = -0.3262$$

EXAMPLE 2. *Find the Karl Pearson's coefficient of skewness for the following frequency distribution:*

Class	0-4	4-8	8-12	12-16	16-20	20-24
Frequency	5	7	10	15	8	4

SOLUTION.

Class	Mid value x	f	$d = \dfrac{x-A}{h}$	d^2	fd	fd^2
0-4	2	5	-2	4	-10	20
4-8	6	7	-1	1	-7	7
8-12	10	10	0	0	0	0
12-16	14	15	1	1	15	15
16-20	18	8	2	4	16	32
20-24	22	4	3	9	22	36
		49			26	110

Let assumed mean $A = 10$

The mean $\quad M = A + \dfrac{\Sigma fd}{N} \times i = 10 + \dfrac{26 \times 4}{49} = 12.1224$

Standard deviation,

$$\alpha = i \sqrt{\frac{1}{N}\Sigma fd^2 - \left(\frac{1}{N}\Sigma fd\right)^2} = 4\sqrt{\frac{1}{49} \times 110 - \left(\frac{1}{49} \times 26\right)^2}$$

$$= 4\sqrt{2.2448 - 0.2815}$$

$$\sigma = 4\sqrt{1.9633} = 5.6047$$

Now, we have to calculate the mode of given distribution. We observe that the frequency of class 12-16 is maximum. So it will be the modal class.

$$l = 12, \ f_{-1} = 10, \ f = 15, \ f_1 = 8, \ i = 4$$

$$\text{Mode} = l + \frac{f - f_{-1}}{2f - f_{-1} - f_1} \times i = 12 + \frac{15 - 10}{30 - 10 - 8} \times 4 = 12 + \frac{5}{12} \times 4$$

$$\text{Mode} = 13.66$$

The coefficient of skewness

$$= \frac{\text{Mean} - \text{Mode}}{\text{Standard deviation}} = \frac{12.1224 - 13.66}{5.6047} = -0.27$$

EXAMPLE 3. *Calculate the Bowley's coefficient of skewness for the following frequency distribution:*

Class	0-10	10-20	20-30	30-40	40-50
Frequency	2	7	10	5	3

SOLUTION. To calculate the Bowley's coefficient of skewness, firstly we have to calculate the value of Q_1, Q_2 and Q_3.

Class	Frequency f	Cumulative Frequency
0-10	2	2
10-20	7	9
20-30	10	19
30-40	5	24
40-50	3	27
	27	

For Q_1

$$\frac{N}{4} = \frac{27}{4} = 6.75$$

Now, 6.75th item will lie in class 10-20. So, it will be the first quartile class.

$$l = 10, \ f = 7, \ f = 2, \ i = 10$$

$$Q_1 = l + \frac{\frac{1}{4}N - F}{f} \times i = 10 + \frac{\frac{1}{4} \times 27 - 2}{7} \times 10$$

$$= 10 + \frac{4.75 \times 10}{7}$$

$$Q_1 = 16.7857 \ \text{mark}$$

For Q_2

$$\frac{2N}{4} = \frac{N}{2} = \frac{27}{2} = 13.5$$

13.5th item will lie in class 20-30. So, it will be the second quartile class.

$$l = 20, \ f = 10, \ F = 9, \ i = 10$$

$$Q_2 = l + \frac{\frac{2}{4}N - F}{f} \times i = 20 + \frac{13.5 - 9}{10} \times 10 = 20 + 4.5$$

$$Q_2 = 24.5 \ \text{mark}$$

For Q_3

$$\frac{3N}{4} = \frac{3 \times 27}{4} = 20.25$$

Now, 20.25th item will lie in class 30-40. So, it will be the third quartile class.

$$l = 30, \ f = 5, \ F = 19, \ i = 10$$

$$Q_3 = l + \frac{\frac{3}{4}N - F}{f} \times i = 30 + \frac{20.25 - 19}{5} \times 10 = 30 + 2.5$$

$$Q_3 = 32.5 \ \text{mark}$$

EXAMPLE 4. *If Karl Pearson's coefficient of skewness of a distribution is 0.32, its S.D. is 6.5 mean 29.5. Find the mode of the distribution.*

SOLUTION. S.D. $(\sigma) = 6.5$, Coefficient of skewness $= 0.32$

Mean $= 29.5$

$$\text{Coefficient of skewness} = \frac{\text{Mean} - \text{Mode}}{\text{S.D.}}$$

$$\Rightarrow \quad 0.32 = \frac{29.5 - \text{Mode}}{6.5}$$

$$\Rightarrow \quad 2.080 = 29.5 - \text{Mode}$$

$$\Rightarrow \quad \text{Mode} = 29.5 - 2.080$$

$$\therefore \quad \text{Mode} = 27.420$$

EXAMPLE 5. *From a mederately skewed distribution of retail prices for mean's shoes, it is found that the mean price is Rs. 20 and the median price is Rs 17. If the coefficient of variation is 20%, find the pearson's coefficient of skewness of the distribution.*

SOLUTION.   It is given that

Mean $M = 20$ and median $M_d = 17$

and coefficient of variance = 20%

Therefore, coefficient of variance

$$= \frac{\sigma}{M} \times 100$$

$$\Rightarrow \qquad 20 = \frac{\sigma}{20} \times 100$$

$$\sigma = \frac{20 \times 20}{100} = \frac{400}{100} = 4$$

and coefficient of skewness

$$= \frac{3(\text{Mean} - \text{Median})}{\sigma}$$

$$= \frac{3(20 - 17)}{4} = \frac{9}{4} = 2.25$$

EXAMPLE 6. *Find out the coefficient of distribution and measures of skewness from the following table giving the wages of 230 persons :*   **(UPTU–2001)**

Wages (in Rs.)	70-80	80-90	90-100	100-110	110-120	120-130	130-140	140-150
Number of persons	12	18	35	42	50	45	20	8

SOLUTION.

Wages (in Rs)	Mid value x	Number of persons (f)	$\dfrac{x-A}{h} = d$	fd	fd^2
70-80	75	12	-4	-48	192
80-90	85	18	-3	-54	162
90-100	95	35	-2	-70	140
100-110	105	42	-1	-42	42
110-120	115	50	0	0	0
120-130	125	45	1	45	45
130-140	135	20	2	40	80
140-150	145	8	3	24	72
		$n = 230$		$\Sigma fd = -105$	$\Sigma fd^2 = 733$

Let assumed mean $A = 10$

The   mean $M = A + \dfrac{\Sigma fd}{N} \times i = 115 + \dfrac{(-105)}{230} \times 10$

$$= 115 - 4.56 = 110.44$$

Now, we have to calculate the mode of given distribution. We observe that the frequency of class 110-120 is maximum. So it will be the modal class.

$$l = 110, f = 50, f_{+1} = 45, f_{-1} = 42, i = 10$$

$$\text{Mode} = l + \frac{f - f_{-1}}{2f - f_{-1} - f_1} \times i = 110 + \frac{50 - 42}{2 \times 50 - 42 - 45} \times 10$$

$$= 110 + \frac{80}{87} = 110.91$$

Standard deviation

$$(\sigma) = i \sqrt{\frac{1}{N} \Sigma f d^2 - \left(\frac{1}{N} \Sigma f d\right)^2} = 10 \sqrt{\frac{733}{230} - \left(\frac{-105}{230}\right)^2}$$

$$= 10\sqrt{3.186 - 0.208} = 10\sqrt{2.978} = 10 \times 1.725 = 17.25$$

The coefficient of skewness

$$= \frac{\text{Mean} - \text{Mode}}{\text{Standard deviation}}$$

$$= \frac{110.44 - 110.91}{17.25} = \frac{-0.47}{17.25} = -0.0278$$

20.14 MOMENTS

If we calculate the deviation of each point of the given distribution from a particular point A, then the arithmetic mean of the nth power of deviation is known as moment about the point A.

It is defined by

$$\mu_n' = \frac{1}{N} \Sigma f(x - A)^n$$

If we consider the mean M in place of the point A, then the calculated moment will be the moment about mean and defined by

$$\mu_n = \frac{1}{N} \Sigma f(x - M)^n$$

20.14.1 FIRST FOUR MOMENTS ABOUT MEAN

Let $x_1, x_2, \ldots, x_n$ be the n observations with mean $\bar{x}$. Then, r^{th} moment about the actual mean of a variable for both grouped and ungrouped data are given as follows :

(i) For ungrouped data

$$\mu_r = \frac{1}{n} \Sigma(x - \bar{x})^r, \quad r = 1, 2, 3, 4$$

$$\Rightarrow \qquad \mu_1 = \frac{1}{n} \Sigma(x - \bar{x}) = 0 \; ; \qquad \mu_2 = \frac{1}{n} \Sigma(x - \bar{x})^2 = \sigma^2$$

$$\mu_3 = \frac{1}{n} \Sigma(x - \bar{x})^3 \; ; \qquad \mu_4 = \frac{1}{n} \Sigma(x - \bar{x})^4$$

(ii) For grouped data

$$\mu_r = \frac{1}{n} \Sigma f(x - \bar{x})^r, \quad r = 1, 2, 3, 4$$

$$\Rightarrow \qquad \mu_1 = \frac{1}{n} \Sigma f(x - \bar{x}) \; ; \qquad \mu_2 = \frac{1}{n} \Sigma f(x - \bar{x})^2$$

$$\mu_3 = \frac{1}{n} \Sigma f(x - \bar{x})^3 \; ; \qquad \mu_4 = \frac{1}{n} \Sigma(x - \bar{x})^4$$

20.14.2 RELATION BETWEEN THE MOMENTS ABOUT MEAN AND MOMENTS ABOUT A POINT

We know that

$$\mu'_n = \frac{1}{N}\Sigma f(x-A)^n \quad \text{and} \quad \mu_n = \frac{1}{N}\Sigma f(x-M)^n$$

$$\mu'_0 = \frac{1}{N}\Sigma f \qquad ; \qquad \mu_0 = \frac{1}{N}\Sigma f$$

$$\mu'_0 = 1 \qquad ; \qquad \mu_0 = 1$$

$$\mu'_1 = \frac{1}{N}\Sigma f(x-A) \quad ; \qquad \mu_1 = \frac{1}{N}\Sigma f(x-M)$$

$$\qquad = \frac{1}{N}\Sigma fx - \frac{1}{N}\Sigma fA \qquad = \frac{1}{N}\Sigma fx - \frac{1}{N}\Sigma f.M = M-M$$

$$\mu'_1 = M-A \qquad ; \qquad \mu_1 = 0$$

$$\mu'_2 = \frac{1}{N}\Sigma f(x-A)^2 \quad ; \qquad \mu_2 = \frac{1}{N}\Sigma f(x-M)^2$$

$$\mu'_2 = \frac{1}{N}\Sigma f(x-A)^2 \quad ; \qquad \mu_2 = \sigma^2$$

$$\mu'_3 = \frac{1}{N}\Sigma f(x-A)^3 \quad ; \qquad \mu_3 = \frac{1}{N}\Sigma f(x-M)^3$$

We know

$$\mu_n = \frac{1}{N}\Sigma f(x-M)^n = \frac{1}{N}\Sigma f(x-A-M+A)^n$$

$$= \frac{1}{N}\Sigma f[(x-A)-(M-A)]^n = \frac{1}{N}\Sigma f\left[x-A-\mu'_1\right]^n$$

Now expanding by binomial expansion

$$\mu_n = \frac{1}{N}\Sigma f\left[(x-A)^n - {}^nC_1(x-A)^{n-1}.\mu'_1 + {}^nC_2(x-A)^{n-2}.\mu'^2_1 + \ldots\right.$$

$$\left. \ldots + (-1)^n(\mu'_1)^n\right]$$

$$= \mu'_n - {}^nC_1\mu'_{n-1}\mu'_1 + {}^nC_2.\mu'_{n-2}\mu'^2_1 \ldots + (-1)^n(\mu'_1)^n \quad \left[\because \frac{1}{N}\Sigma f = 1\right]$$

$$\mu_n = \mu'_n - {}^nC_1\mu'_{n-1}\mu'_1 + {}^nC_2\mu'_{n-2}.\mu'^2_1 \ldots + (-1)^n(\mu'_1)^n$$

Now in particular cases

$$\mu_1 = \mu'_1 - {}^1C_1\mu'_0\mu'_1 \quad \Rightarrow \quad \mu_1 = \mu'_1 - \mu'_1 = 0 \quad \Rightarrow \quad \mu_1 = 0 \qquad (\because \mu'_0 = 1)$$

$$\mu_2 = \mu'_2 - 2(\mu'_1)^2 + \mu'_0(\mu'_1)^2$$

$$\mu_2 = \mu'_2 - 2\mu'^2_1 + \mu'^2_1 \qquad\qquad (\because \mu'_0 = 1)$$

$$\mu_2 = \mu'_2 - \mu'^2_1$$

$$\mu_3 = \mu'_3 - {}^3C_1\mu'_2\mu'_1 + {}^3C_2\mu'_1(\mu'_1)^2 - {}^3C_3\mu'_0(\mu'_1)^3$$

$$= \mu'_3 - 3\mu'_1\mu'_2 + 3\mu'^3_1 - \mu'^3_1$$

$$\mu_3 = \mu'_3 - 3\mu'_1\mu'_2 + 2\mu'^3_1$$

$$\mu_4 = \mu_4' - {}^4C_1\mu_3'\mu_1' + {}^4C_2\mu_2'(\mu_1')^2 - {}^4C_3\mu_1'(\mu_1')^3 + {}^4C_4\mu_0'(\mu_1')^4$$

$$= \mu_4' - 4\mu_1'\mu_3' + 6\mu_1'^2\mu_2' - 4(\mu_1')^4 + (\mu_1')^4$$

$$\mu_4 = \mu_4' - 4\mu_1'\mu_3' + 6\mu_1'^2\mu_2' - 3(\mu_1')^4$$

20.14.3 Moments About Any Point A in Terms of Moments About Mean

We know that the moment about any point A is

$$\mu_n' = \frac{1}{N}\Sigma f(x - A)^n = \frac{1}{N}\Sigma f(x - M + M - A)^n = \frac{1}{N}\Sigma f(x - M - d)^n$$

Now expanding by binomial expansion

$$\mu_n' = \frac{1}{N}\Sigma f\left[(x - M)^n + {}^nC_1(x - M)^{n-1}d + {}^nC_2(x - M)^{n-2}d^2 + \ldots + d^n\right]$$

$$= \frac{1}{N}\Sigma f(x - M)^n + {}^nC_1\frac{1}{N}\Sigma f(x - M)^{n-1}.d + {}^nC_2\frac{1}{N}\Sigma f(x - M)^{n-2}.d^2 + \ldots$$

$$\ldots + \frac{1}{N}\Sigma f.d^n$$

$$\mu_n' = \mu_n + {}^nC_1\mu_{n-1}.d + {}^nC_2\mu_{n-2}d^2 + \ldots + d^n \qquad (\because N = \Sigma f)$$

For particular cases

$$\mu_1' = \mu_1 + {}^1C_1\mu_0.d$$

$$\mu_1' = \mu_1 + d \qquad (\because \mu_0 = 1)$$

$$\mu_1' = 0 + (M - A)$$

$$\mu_1 = M - A$$

$$\mu_2' = \mu_2 + {}^2C_1\mu_1.d + {}^2C_2\mu_0.d^2 = \mu_2 + 2\mu_1.d + d^2$$

$$\mu_2' = \mu_2 + d^2 \qquad (\because \mu_1 = 0)$$

$$\mu_3' = \mu_3 + {}^3C_1\mu_2.d + {}^3C_2\mu_1d^2 + {}^3C_3\mu_0.d^3$$

$$\mu_3' = \mu_3 + 3\mu_2.d + d^3 \qquad [\because \mu_0 = 1, \ \mu_1 = 0]$$

$$\mu_4' = \mu_4 + {}^4C_1\mu_3.d + {}^4C_2\mu_2.d^2 + {}^4C_3\mu_1d^3 + {}^4C_4\mu_0.d^4$$

$$= \mu_4 + 4\mu_3.d + 6\mu_2.d^2 + d^4 \qquad (\because \mu_0 = 1, \ \mu_1 = 0)$$

$$\mu_4' = \mu_4 + 4\mu_3.d + 6\mu_2.d^2 + d^4$$

20.14.4 Effect of Changing the Origin and Scale on Moments

Let x be the previous variable and a new variable is related to x by

$$\mu = \frac{x - A}{h}, \quad x - A = h\mu, \quad x = A + h\mu$$

$$\bar{x} - A = h\bar{\mu}, \quad \bar{x} = A + h\bar{\mu}$$

We know that

$$\mu_n = \frac{1}{N}\Sigma f(x - \bar{x})^n$$

Put the value of x and $\bar{x}$ in the above equation, we get

$$\mu_n = \frac{1}{N} \Sigma f (A + h\mu - A - h\bar{\mu})^n$$

$$\mu_0' = 1 \qquad\qquad \mu_0 = 1$$

$$\mu_1' = d \qquad\qquad \mu_1 = 0$$

$$\mu_2 = \sigma^2$$

$$\mu_n = \frac{1}{N} \Sigma f (h\mu - h\bar{\mu})^n = \frac{1}{N}. h^n \Sigma f (\mu - \bar{\mu})^n$$

$$\mu_n = \frac{h^n}{N} \Sigma f (\mu - \bar{\mu})^n$$

Now, we consider the change on the moment about a point

$$\mu_n' = \frac{1}{N} \Sigma f (x - A)^n$$

Put the value $(x - A) = h\mu$ in the above equation

$$\mu_n' = \frac{1}{N} \Sigma f . h^n \mu^n \; ; \; \mu_n' = \frac{h^n}{N} \Sigma f . \mu^n$$

Thus, we can say that the central moment remains constant with respect to the change of origin.

20.14.5 PEARSON'S β AND γ COEFFICIENTS

Karl Pearson defined four coefficients β_1, β_2, γ_1 and γ_2 with the help of the central moments. These four coefficients are known by Karl Pearson's coefficient and are defined as follows :

$$\beta_1 = \frac{\mu_3^2}{\mu_2^3}, \beta_2 = \frac{\mu_4}{\mu_2^2}, \; \gamma_1 = \pm \sqrt{\beta_1} \text{ and } \gamma_2 = \beta_2 - 3$$

Indirectly the sign of γ_1 depends on γ_2. These coefficients give some information about the shape of the distribution curve. Here the coefficient β_1 and γ_1 tell us about the skewness of distribution and β_2 and γ_2 are the measure of peakedness or flatness of the top of the curve.

The formula for skewness can be given in the form of the Pearson's coefficient.

$$\text{Skewness} = \frac{\sqrt{\beta_1} \, (\beta_2 + 3)}{2(5\beta_2 - 6\beta_1 - 9)}$$

20.14.6 ABSOLUTE MOMENT

Absolute moment about the mean M is $\quad y_n = \frac{1}{N} \Sigma f \, | x - M |^n$

Absolute moment about the origin is $y_n' = \frac{1}{N} \Sigma f \, | x |^n$

20.14.7 FACTORIAL MOMENT

Factorial moment about the mean M is given by $\quad \mu_{(n)} = \frac{1}{N} \Sigma f (x - M)^{(n)}$

$$\mu_{(n)} = \frac{1}{N} \Sigma f \left[(x - M)(x - M - 1)(x - M - 2)...(x - M - n + 1) \right]$$

Factorial moment about the origin is given by

$$\mu'_{(n)} = \frac{1}{N} \Sigma f(x)^n$$

$$\mu'_{(n)} = \frac{1}{N} \Sigma f\left[x(x-1)(x-2)...(x-n+1)\right]$$

For particular values of n

$$\mu'_{(1)} = \frac{1}{N} \Sigma fx$$

$$\mu'_{(1)} = M$$

$$\mu'_{(2)} = \frac{1}{N} \Sigma f[x(x-1)] = \frac{1}{N} \Sigma f(x^2 - x) = \frac{1}{N} \Sigma fx^2 - \frac{1}{N} \Sigma fx$$

$$\mu'_{(2)} = \mu'_2 - \mu'_1$$

$$\mu'_{(3)} = \frac{1}{N} \Sigma f[x(x-1)(x-2)] = \frac{1}{N} \Sigma f[x(x^2 - 3x + 2)]$$

$$= \frac{1}{N} \Sigma f[x^3 - 3x^2 + 2x]$$

$$\mu'_{(3)} = \frac{1}{N} \Sigma fx^3 - 3.\frac{1}{N} \Sigma fx^2 + 2.\frac{1}{N} \Sigma f$$

$$\mu'_{(3)} = \mu'_3 - 3\mu'_2 + 2\mu'_1$$

$$\mu'_{(4)} = \frac{1}{N} \Sigma f[x(x-1)(x-2)(x-3)] = \frac{1}{N} \Sigma f[(x^2 - x)(x^2 - 5x + 6)]$$

$$= \frac{1}{N} \Sigma f[x^4 - 5x^3 + 6x^2 - x^3 + 5x^2 - 6x]$$

$$= \frac{1}{N} \Sigma f[x^4 - 6x^3 + 11x^2 - 6x]$$

$$= \frac{1}{N} \Sigma fx^4 - 6.\frac{1}{N} \Sigma fx^3 + 11.\frac{1}{N} \Sigma fx^2 - 6.\frac{1}{N} \Sigma fx$$

$$\mu'_{(4)} = \mu'_4 - 6\mu'_3 + 11\mu'_2 - 6\mu'_1$$

20.14.8 CHARLIER'S CHECK

Charlier suggested the formula known as Charlier check to check the mistakes while calculating mean, variance and other moments.

These are as follows

$$\Sigma f(\xi + 1) = \Sigma f\xi + \Sigma f \qquad \qquad ... (1)$$

$$\Sigma f(\xi + 1)^2 = \Sigma f\xi^2 + 2\Sigma f\xi + \Sigma f \qquad \qquad ... (2)$$

$$\Sigma f(\xi + 1)^3 = \Sigma f\xi^3 + 3\Sigma f\xi^2 + 3\Sigma f\xi + \Sigma f \qquad \qquad ... (3)$$

$$\Sigma f(\xi + 1)^4 = \Sigma f\xi^4 + 4\Sigma f\xi^3 + 6\Sigma f\xi^2 + 4\Sigma f\xi + \Sigma f \qquad \qquad ... (4)$$

If we calculate the values of $\Sigma f(\xi + 1)$ and $\Sigma f\xi$, then it will satisfy equation (1). The values of $\Sigma f(\xi + 1)^2$, $\Sigma f\xi^2$ and $\Sigma f\xi$ will satisfy equation (2). Thus it is a ready check against the mistakes in calculation.

20.14.9 SHEPPARD'S CORRECTION

While solving the values of mean, mode and variance, etc., in case of class interval the mid point of this interval is assumed to be representative of this class and the whole frequency is assumed to be concentrated on it. Thus the values obtained by these processes are not accurately correct. So W.F. Sheppard suggested the formula to obtain the accurate results. These are as follows :

$$\mu_{2(corrected)} = \mu_{2(calculated)} - \frac{h^2}{10}$$

$$\mu_{3(corrected)} = \mu_{3(calculated)}$$

$$\mu_{4(corrected)} = \mu_{4(calculated)} - \frac{1}{2}h^2\mu_{2(calculated)} + \frac{7}{240}h^4$$

where, h = class interval

SOLVED EXAMPLES

EXAMPLE 1. *The first three moments of a distribution about the point 2 of a variable are 1, 16 and −40. Show that the mean is 3, $\sigma^2 = 15$ and $\mu_3 = -86$. Also find the first three moments about zero.*

(VTU-2003 S)

SOLUTION. It is given that

$A = 2, \ \mu_1' = 1, \ \mu_2' = 16, \ \mu_3' = -40$

The moment about any point A is given by

$$\mu_n' = \frac{1}{N}\Sigma f(x - A)$$

$$\mu_1' = \frac{1}{N}\Sigma f(x - 2)$$

$$1 = \frac{1}{N}\Sigma fx - \frac{1}{N}\Sigma f.2$$

$$1 = M - 2 ; \quad M = 3 \qquad\qquad (\because N = \Sigma f)$$

We know that

$$\mu_2 = \mu_2'(2) - [\mu_1'(2)]^2$$

$$\mu_2 = 16 - 1$$

$\Rightarrow$ $\mu_2 = 15$

$$\mu_3 = \mu_3' - 3\mu_1'\mu_2' + 2\mu_1'^3 = -40 - 3 \times 1 \times 16 + 2 \times 1 = -40 - 48 + 2$$

$$\mu_3 = -86$$

Now, we have to find out the moments about zero.

$$\mu_1' = \frac{1}{N}\Sigma f(x - 0) = \frac{1}{N}\Sigma fx$$

$$\mu_1'(0) = \text{Mean}$$

$$\mu_1'(0) = 3$$

$$\mu_2' = \frac{1}{N}\Sigma f(x - 0)^2 \qquad\qquad\qquad ... (1)$$

To calculate the value of μ_2', we will convert the R.H.S. of (1) into the terms of central moments.

$$\mu_2' = \frac{1}{N}\Sigma f[x-3+3]^2 = \frac{1}{N}\Sigma f\left[(x-3)^2 + 6(x-3)+9\right]$$

$$= \frac{1}{N}\Sigma f(x-3)^2 + 6\frac{1}{N}\Sigma f(x-3) + \frac{9\times 1}{N}\Sigma f = \mu_2 + 6\mu_1 + 9$$

$$\mu_2'(0) = 24$$

$$\mu_3'(0) = \frac{1}{N}\Sigma f(x-0)^3 = \frac{1}{N}\Sigma\left[(x-3)+3\right]^3$$

$$= \frac{1}{N}\Sigma f\left[(x-3)^3 + 27 + 3(x-3)\times 3\,(x-3+3)\right]$$

$$= \frac{1}{N}\Sigma f\left[(x-3)^3 + 9x(x-3)+27\right]$$

$$= \frac{1}{N}\Sigma f\left[(x-3)^3 + 9(x-3+3)(x-3)+27\right]$$

$$= \frac{1}{N}\Sigma f\left[(x-3)^3 + 9(x-3)^2 + 27(x-3)+27\right]$$

$$= \frac{1}{N}\Sigma f\,(x-3)^3 + 9.\frac{1}{N}\Sigma f(x-3)^2 + 27.\frac{1}{N}\Sigma f(x-3) + 27.\frac{1}{N}\Sigma f$$

$$= \mu_3 + 9.\mu_2 + 27.\mu_1 + 27$$

$$= -86 + 9\times 15 + 27\times 0 + 27$$

$$= -86 + 135 + 27$$

$$\mu_3'(0) = 76$$

So, the moments about zero is

$$\mu_1'(0) = 3, \quad \mu_2'(0) = 24, \quad \mu_3'(0) = 76$$

EXAMPLE 2. *The first four moments about the value 5 of a variable are given by 2, 20, 40 and 50. Find the moments about mean.*

SOLUTION. It is given that

$$\mu_1' = 2, \ \mu_2' = 20, \ \mu_3' = 40, \ \mu_4' = 50$$

We know that

$$\mu_1' = \frac{1}{N}\Sigma f(x-5)$$

$$2 = \frac{1}{N}\Sigma fx - \frac{1}{N}\Sigma f.5 \qquad (\because \Sigma f = N)$$

$$= M - 5$$

$$M = 7$$

We know, $\quad \mu_1 = 0$

Second moment about mean

$$\mu_2 = \mu_2' - \mu_1'^2 = 20 - (2)^2$$

$$\mu_2 = 16$$

Third moment about mean

$$\mu_3 = \mu_3' - 3\mu_1'\mu_2' + 2\mu_1'^3 = 40 - 3 \times 2 \times 20 + 2 \times (2)^3$$

$$= 40 - 120 + 16 = -64$$

$$\mu_3 = -64$$

Fourth moment about mean

$$\mu_4 = \mu_4' - 4\mu_3'\mu_1' + 6\mu_2'\mu_1'^2 - 3(\mu_1')^4$$

$$= 50 - 4 \times 40 \times 2 + 6 \times 20 \times 4 - 3 \times 16$$

$$= 50 - 320 + 480 - 48$$

$$\mu_4 = 162$$

EXAMPLE 3. *The four moments of a distribution about the value −1.5, 17, −30 and 108. Find the moments about β_1 and β_2.*

SOLUTION. Moment about an arbitrary point are

$$\mu_1' = -1.5, \quad \mu_2' = 17, \quad \mu_3' = -30, \quad \mu_4' = 108$$

First moment about mean

$$\mu_1 = \mu_1' - \mu_1' = 0$$

Second moment about mean

$$\mu_2 = \mu_2' - \mu_1'^2 = 17 - (-1.5)^2$$

$$\mu_2 = 14.75$$

Third moment about mean

$$\mu_3 = \mu_3' - 3\mu_1'\mu_2' + 2\mu_1'^3 = -30 - 3(-1.5)(17) + 2(-1.5)^3$$

$$\mu_2 = 39.75$$

Fourth moment about mean

$$\mu_4 = \mu_4' - 4\mu_1'\mu_3' + 6\mu_1'\mu_2' - 3\mu_1'^4 = 142.3125$$

Now, $$\beta_1 = \frac{\mu_3^2}{\mu_2^3} = \frac{(39.75)^2}{(14.75)^3} = \frac{1580.0625}{3209.0469} = 0.4923$$

$$\beta_2 = \frac{\mu_4}{\mu_2^2} = \frac{142.3125}{(14.75)^2} = \frac{142.3125}{217.5625} = 0.6541224$$

EXAMPLE 4. *The first four moments of a distribution about the value 4 of a variable are −1.5, 17, −30 and 108. Find the moments about mean and the mean of the distribution.*

SOLUTION. It is given that

$$\mu_1' = -1.5, \quad \mu_2' = 17, \quad \mu_3' = -30, \quad \mu_4' = 108$$

We know that

$$\mu_1' = \frac{1}{N}\Sigma f(x - A)$$

$$-1.5 = \frac{1}{N}\Sigma f(x - 4)$$

$$-1.5 = \frac{1}{N}\Sigma fx - \frac{1}{N}\Sigma f.4 \qquad (\because \Sigma f = N)$$

$$\therefore \qquad -1.5 = \frac{1}{N}\Sigma fx - 4$$

$$2.5 = \frac{1}{N}\Sigma fx \qquad \left(\because \frac{1}{N}\Sigma fx = M\right)$$

$$\therefore \qquad M = 2.5$$

We know, $\mu_1 = 0$

Second moment about mean

$$\mu_2 = \mu_2' - {\mu_1'}^2 = 17 - 2.25$$

$$\mu_2 = 14.75$$

Third moment about mean

$$\mu_3 = \mu_3' - 3\mu_1'\mu_2' + 2{\mu_1'}^3 = -30 - 3\times17\times(-1.5) + 2\times(-1.5)^3$$

$$= -30 + 76.5 - 6.75$$

$$\mu_3 = 39.75$$

Fourth moment about mean

$$\mu_4 = \mu_4' - 4\mu_3'\mu_1' + 6\mu_2'{\mu_1'}^2 - 3(\mu_1')^4$$

$$= 108 - 4\times(-30)(-1.5) + 6\times17\times(-1.5)^2 - 3(-1.5)^4$$

$$= 108 - 180 + 229.5 - 15.1875$$

$$\mu_4 = 142.3125$$

Exercise 20.4

1. Calculate the coefficient of skewness for the values given below :

25, 40, 23, 23, 15, 27, 25, 20, 25

2. Calculate the coefficient of skewness for the given frequency distribution :

Class	0-5	5-10	10-15	15-20
Frequency	5	20	10	0
Class	20-25	25-30	30-35	35-40
Frequency	5	20	8	7

3. Calculate the coefficient of skewness for the following data :

Marks Above	0	10	20	30	40
Frequency	150	140	100	80	80
Marks Above	50	60	70	80	
Frequency	70	30	14	0	

4. Calculate the coefficient of skewness for the following data :

Salary	104.5	105.5	106.5	107.5
Frequency	35	40	48	100
Salary	108.5	109.5	110.5	111.5
Frequency	125	87	43	22

5. Calculate the coefficient of skewness for the given frequency distribution :

Class	0-5	5-10	10-15	15-20
Frequency	2	5	7	13
Class	20-25	25-30	30-35	35-40
Frequency	21	16	8	3

6. Check the skewness of the distribution for which $\mu_1 = 0$, $\mu_2 = 2.5$, $\mu_3 = 0.7$, $\mu_4 = 18.75$.

7. Find the value of mean, median, standard deviation and then the coefficient of skewness for the given frequency distribution :

Class	5-7	8-10	11-13	14-16	17-19
Frequency	7	12	19	10	2

8. For a discrete distribution prove that :

(i) $\beta_2 > 1$,   (ii) $\beta_2 > \beta_1$

9. Calculate the missing terms and coefficient of skewness for the given frequency distribution for which Mode = 54, A.M. = 53.4 :

Class	0-20	20-40	40-60	Total
Frequency	10	–	30	
Class	60-80	80-100		
Frequency	–	14		94

10. Calculate the first and second central moments about mean for the given frequency distribution :

Variable	0	1	2	3	4
Frequency	1	9	26	59	72
Variable	5	6	7	8	
Frequency	52	29	7	1	

11. Calculate the first four moments about 5 for the given datas : 2, 4, 6, 7, 9

12. Using short-cut method, calculate the mean, variance and third central moment of the following frequency distribution :

Variable	0	1	2	3	4
Frequency	1	9	26	59	72
Variable	5	6	7	8	
Frequency	52	29	7	1	

13. Calculate the first three moments about the assumed mean :

Variable	2.0	2.5	3.0	3.5
Frequency	5	38	65	92
Variable	4.0	4.5	5.0	
Frequency	70	40	10	

14. Calculate the first three central moments :

(VTU-2004, MADRAS-2003)

Variable	0	1	2	3	4
Frequency	1	8	28	56	70
Variable	5	6	7	8	
Frequency	56	28	8	1	

15. Calculate the first four moments about mean of the following frequency distribution :

Class	2-4	4-6	6-8	8-10	10-12
Frequency	4	2	8	6	1

16. Calculate the first, second and third central moment about zero and then find these moments about mean :

Variable	0	1	2	3	4	5	6
Frequency	15	38	55	82	60	40	10

17. The first four moments of a distribution about the value 4 is given by –1.5, 17, –30 and 108. Find the central moments.

18. In a frequency distribution, the class width of the interval is given by 3. Then find the corrected value of μ_2, μ_3 and μ_4 where $\mu_2 = 43.353, \mu_3 = -9.774, \mu_4 = 5508.567$.

19. Calculate the first four central moments and the value of β_1 and β_2 for the given frequency distribution :

Variable	2.0	2.5	3.0	3.5	4.0	4.5	5.0
Frequency	5	38	65	92	70	40	10

20. In a given frequency distribution, the mean of the distribution is given by 10, variance = 16, $\gamma_1 = 1, \beta_2 = 4$. Find the first four moments about origin.

21. The first four moments of a distribution about 4 is given by –1.5, 17, –30 and 108. Find the value of β_1 and β_2.

22. For a given distribution, the first four moments about the point 5 is given by 2, 20, 40 and 50. Find the value of mean and all other characteristics.

23. For any distribution, the value of mean, variance and γ_1 are given by 1, 1 and 1. Find first three moments about zero, when $\mu_3 > 0$.

ANSWERS

1. – 0.03 2. – 0.76 3. 0.8 4. – 0.245 5. – 0.148 6. $\beta_1 = 0.177$
7. Mean = 11.28, Median = 11.947, S.D. = 3.15, Coeff. of skewness = – 0.635
9. $f_2 = 16, f_4 = 24$, coefficient of skewness = –0.025 10. $\mu_2 = 1.979, \mu_3 = 0.015$
11. $\mu_1' = 0.6, \mu_2' = 6.2, \mu_3' = 9.0, \mu_4' = 71$ 12. 3.973, 1.979, 0.0115
13. For A = 3.5, 0.038, 0.455, 0.061 14. 0, 2, 0 15. 0, 5.3, -4.44, 62.45
16. 2.98, 11.05, 45.64, 0, 2.17, -0.22 17. 0, 14.75, 39.75, 142.3125
18. $\mu_2 = 42.603, \mu_3 = -9.774, \mu_4 = 5315.838$
19. 0, 18.13125, 0.0791, 8.033, $\beta_1 = 0.001, \beta_2 = 2.44$ 20. 10, 116, 1544, 23184
21. 0.4924, 0.6541
22. Mean = – 7, $\sigma = 4, \mu_1 = 0, \mu_2 = 16, \mu_3 = -64, \mu_4 = 162, \beta_1 = 1, \beta_2 = 0.63, \gamma_1 = 1, \gamma_2 = -2.37$
23. 1, 2, 5

Objective Evaluation

∞ MULTIPLE CHOICE QUESTIONS

Choose the most appropriate one.

1. Which of the following is not a measure of dispersion ?
   (a) mean deviation
   (b) quartile deviation
   (c) standard deviation
   (d) average deviation from mean

2. Which of the following is a unitless measure of dispersion ?
   (a) standard deviation
   (b) mean deviation
   (c) coefficient of variation
   (d) range

3. Which one of the given measures of dispersion is considered best ?
   (a) standard deviation
   (b) range
   (c) variance
   (d) coefficient of variation

4. For comparison of two different series, the best measure of dispersion is :
   (a) range
   (b) mean deviation
   (c) standard deviation
   (d) none of the above

5. Mean deviation is minimum when deviations are taken from :
   (a) mean
   (b) median
   (c) mode
   (d) zero

6. Sum of squares of the deviation is minimum when deviations are taken from :
   (a) mean
   (b) median
   (c) mode
   (d) zero

7. If a constant value 5 is subtracted from each observation of a set, the variance is :
   (a) reduced by 5
   (b) reduced by 25
   (c) unaltered
   (d) increased by 25

8. If the standard deviation of a distribution is 15, the quartile deviation of the distribution is :
   (a) 15
   (b) 12.5
   (c) 10
   (d) none of the above

9. If the quartile deviation of a series is 60, the mean deviation of the series is :
   (a) 72
   (b) 48
   (c) 50
   (d) 75

10. In a discrete set of values, the correct relation between mean deviation and standard deviation is :
    (a) M.D. > S.D.
    (b) M.D. < S.D.
    (c) M.D. ≤ S.D.
    (d) M.D. ≥ S.D.

∞ TRUE/FALSE

Write 'T' for True and 'F' for False statement.

1. Best measure of dispersion is standard deviation. **(T/F)**

2. Measure of dispersion suitable for comparing any two series is coefficient of variation. **(T/F)**

3. The relation between variance and standard deviation is S.D. = $\sqrt[3]{\text{variance}}$. **(T/F)**

4. Lesser the value of coefficient of range, better it is. **(T/F)**

5. Quartile deviation can be obtained in case of open end intervals. **(T/F)**

6. Quartile deviation is not affected by 50 percent extreme values. **(T/F)**

7. Quartile deviation is a positional measure of dispersion. **(T/F)**

8. The relation $2Q_2 = Q_3 + Q_1$ holds in case of symmetric distribution. **(T/F)**

9. Mean deviation is calculating by considering absolute deviations. **(T/F)**

10. Range is highly susceptible to sampling fluctuations. **(T/F)**

∞ FILL IN THE BLANKS

1. Inter-quartile range is equal to _____ .

2. Percentile range is given by _____ .

3. Percentile range in terms of decile is given as _____ .

4. Mean deviation is based on _____ .

5. Mean deviation is minimum about _____ .

6. Mean deviation suffers with the lacuna that it considers all differences _____.

7. Coefficient of mean deviation about a central value 'C' is given by _____.

8. The formula for calculating mean deviation about median for a discrete frequency distribution is _____.

9. Standard deviation is the _____ of volume.

10. Other names of standard deviation are _____.

ANSWERS

☞ MULTIPLE CHOICE QUESTIONS

1. (d) **2.** (c) **3.** (a) **4.** (d) **5.** (b) **6.** (a) **7.** (c) **8.** (c) **9.** (a)
10. (b)

☞ TRUE/ FALSE

1. T **2.** T **3.** F **4.** T **5.** T **6.** T **7.** T **8.** T **9.** T
10. T

☞ FILL IN THE BLANKS

1. $Q_3 - Q_1$ **2.** $P_{90} - P_{10}$ **3.** $D_9 - D_1$ **4.** all observations **5.** median
6. positive **7.** SC/C **8.** $\Sigma f |x - Md| / \Sigma f$ **9.** the square root **10.** mean error

❑❑❑❑❑❑

CHAPTER 21

Method of Least Squares

21.1 INTRODUCTION

If some values of a variate are collected in the arbitrary order, in which they occur, we cannot properly grasp the significance of the data.

For example, consider the marks of 50 students in Mathematics, arranged according to their roll numbers, the maximum marks being 100.

9, 70, 75, 15, 0, 33, 69, 66, 37, 99, 81, 12, 31, 22, 60, 79, 46, 73, 46, 79, 75, 65, 85, 22, 8, 12, 41, 87, 82, 72, 50, 22, 87, 50, 89, 28, 29, 50, 40, 36, 40, 30, 28, 87, 81, 90, 22, 15, 30, 35.

The data given in the above form is called ungrouped data. If the data arranged in ascending or descending order of magnitude, it is said to be arranged in an array. If arranged the given data into class intervals 0-10, 10-20, ..., 90-100. Then, this method is known as tally method.

If the identity of the individuals about whom, a particular information is taken is not relevant, nor the order in which the observations arise, then we divide the observed range of variables into a suitable number of class intervals and record the number of observation in each class.

For example in the above case, the data may be expressed as in the following table :

Such a table showing the distribution of the frequencies in the different classes is called a frequency table and the way in which the class frequencies are distributed over the class intervals is called the grouped frequency distribution of the variable.

The following points may be considerable for classifications :

(i) The class should be clearly defined and should not lead to any ambiguity.

(ii) The number of class should never be less than 6 and not more than 30. Because with less number of classes, the accuracy may be lost, and with more number of classes, the computations become lengthy.

Marks	No. of Students
0–10	3
10–20	4
20–30	7
30–40	7
40–50	5
50–60	3
60–70	4
70–80	7
80–90	8
90–100	2

(iii) The observation corresponding to common point of two classes should always be put in the higher class. For example, a number corresponding to the values 20 is to be put up in the class 20-30 and not in 20-30.

(iv) The classes should be of equal width.

21.1.1 MAGNITUDE OF THE CLASS INTERVAL

Having fixed the number of classes, divide the range by it and nearest integer to this value gives the magnitude of the class interval.

21.1.2 CLASS LIMITS

The class limit should be chosen in such a way that the mid value of the class interval and actual average of the observations in that class intervals are as near to each other as possible.

21.2 CONTINUOUS FREQUENCY DISTRIBUTION

If we deal with a continuous variable, it is not possible to arrange the data in the class interval.

Let us consider the distribution of age in years. If we take the intervals 15-19, 20-24, then the persons with ages between 19 and 20 years are not taken into consideration. In such a case, we form the class interval as follows.

Age in years
Below 5 :   5 or more but less than 10
                      10 or more but less than 15
                      15 or more but less than 20
                      20 or more but less than 25
....

where all the persons with any fraction of age are included in one group or the other. Practically, we re-write it as
                      0–5; 5–10; 10–15; 15–20; 20–25

This form of frequency distribution is known as continuous frequency distribution.

21.3 CURVE FITTING

In applied mathematics, several equations of different types can be obtained to express the given data approximately. But we want to find the equation of the curve of best fit which may be most suitable for predicting the unknown values. The process of finding such an equation of best fit is known as curve fitting.

By curve fitting we means an expression of the relationship between two variables by an equation. If there are n pair of observed values, then it is possible to fit the given data to an equation that contains n arbitrary constants for we can solve n simultaneous equation for n unknowns.

Let us consider m independent linear equations

$$\left.\begin{aligned}
a_{11}x_1 + a_{12}x_2 + \ldots + a_{1n}x_n &= b_1 \\
a_{21}x_1 + a_{22}x_2 + \ldots + a_{2n}x_n &= b_2 \\
\ldots\ldots\ldots \quad\quad \ldots\ldots\ldots\ldots\ldots \\
\ldots\ldots\ldots \quad\quad \ldots\ldots\ldots\ldots\ldots \\
a_{m1}x_1 + a_{m2}x_2 + \ldots + a_{mn}x_n &= b_m
\end{aligned}\right\} \quad\quad \ldots(1)$$

where $a's$ and $b's$ are constants and $x_1, x_2, \ldots, x_n$ are n variables. Now, there are two cases :

(i) If $m = n$, then there exists a unique set of values satisfying the given system of equations.

(ii) If $m > n$, which implies that the number of equations is greater than the number of variables, then there exists no solution. In this case we try to find these values of variables $x_1, x_2, \ldots, x_n$ which satisfy as closed as possible to the given equation. These obtained values are called the most plausible values or the best fit values.

➠ The general problem of finding equation of approximate curves which fit given set of data is called curve fitting.
➠ 'Curve fitting' is considered very important both from the point of view of theoretical and practical use.
➠ In theoretical statistics, the line of regression can be regarded as fitting of linear curves.
➠ The difference between the observed values and expected values is known as residual.

21.4 METHOD OF LEAST SQUARES

Consider m independent linear equations in n unknowns $x_1, x_2, ..., x_n$ where $m > n$ as

$$\left.\begin{array}{l} a_{11}x_1 + a_{12}x_2 + ... + a_{1n}x_n = b_1 \\ a_{21}x_1 + a_{22}x_2 + ... + a_{2n}x_n = b_2 \\ .... \qquad\qquad .... \\ .... \qquad\qquad .... \\ a_{m1}x_1 + a_{m2}x_2 + ... + a_{mn}x_n = b_m \end{array}\right\} \qquad ...(1)$$

with constants $a's$ and $b's$.
Let $x_1, x_2, ..., x_n$ be the most plausible values then, we have

$$E_i = (a_{i1}x_1 + a_{i2}x_2 + ... + a_{in}x_n - b_i)$$

(known as residual or deviation, or the error E_i)

$$\Rightarrow \qquad E_i = (a_{i1}x_1 + a_{i2}x_2 + ... + a_{in}x_n - b_i) \quad i = 1, 2, ... m. \qquad ...(2)$$

Let us suppose S is the sum of the squares of E_i.

Then, we get $\qquad S = \sum_{i=1}^{n} (a_{i1}x_1 + a_{i2}x_2 + ... + a_{in}x_n - b_i)^2 = \sum_{i=1}^{n} E_i^2 \qquad ...(3)$

To find the maximum or minimum values of S, we must have

$$\frac{\partial S}{\partial x_1} = 0, \frac{\partial S}{\partial x_2} = 0, ... \frac{\partial S}{\partial x_n} = 0$$

Then (3) $\Rightarrow \quad \sum_{i=1}^{m} a_{i1}E_i = 0, \sum_{i=1}^{m} a_{i2}E_i = 0, ... \sum_{i=1}^{m} a_{in}E_i = 0 \qquad ...(4)$

These n equations given by (4) are known as the normal equations and can be easily solved for the n variables $x_1, x_2, ..., x_n$. The values of $x_1, x_2, ..., x_n$ so obtained are the most plausible or best values.

➠ The principle of least squares, is first given by Gauss in 1795 but it was named and published for the first time in 1805 by Legendre.
➠ The method of least square does not help us to choose the degree of the curve to be fitted but helps us in finding the values of the constants when the form of the curve has already been given.

21.5.1 NORMAL EQUATIONS

Consider the curve of nth degree

$$y = a + bx + cx^2 + ... + kx^n \text{ with } (k \neq 0)$$

Then, the normal equation obtained from (4), are

$$\Sigma y = ma + b\Sigma x + ... + k\Sigma x^n$$

$$\Sigma xy = a\Sigma x + b\Sigma x^2 + ... + k\Sigma x^{n+1}$$

$$\Sigma x^2 y = a\Sigma x^2 + b\Sigma x^3 + ... + k\Sigma x^{n+2}$$

$$\Sigma x^n y = a\Sigma x^n + b\Sigma x^{n+1} + ... + k\Sigma x^{2n}$$

When second order partial derivatives are calculated and substituted these values, they gives a positive value of the function, which implies that S is minimum.

In particular, if we take $n = 1$ then we get the equation of straight line

$$y = a + bx$$

and the normal equations are

$$\Sigma y = ma + b\Sigma x + c\Sigma x^2$$

$$\Sigma xy = a\Sigma x + b\Sigma x^2 + c\Sigma x^3$$

$$\Sigma x^2 y = a\Sigma x^2 + b\Sigma x^3 + c\Sigma x^4$$

SOLVED EXAMPLES

EXAMPLE 1. *Find the normal equations and hence find the best fit values of x, y, z in the least square sense from the following equations*

$$x + 2y + z = 1$$
$$2x + y + z = 4$$
$$-x + y + 2z = 4$$
$$4x + 2y - 5z = -7$$

SOLUTION. The given equation can be written as

$$\left. \begin{array}{r} x + 2y + z - 1 = 0 \\ 2x + y + z - 4 = 0 \\ -x + y + 2z - 4 = 0 \\ 4x + 2y - 5z + 7 = 0 \end{array} \right] \qquad \text{...(1)}$$

Now, to obtain the normal equations of x, we multiply these equations by the coefficient of x in that equation and then add.

Then, the normal equations are

$$1(x + 2y + z - 1) + 2(2x + y + z - 4) + (-1)(-x + y + 2z - 4)$$
$$+ 4(4x + 2y - 5z + 7) = 0$$

$$\Rightarrow \qquad 22x + 11y - 19z + 23 = 0 \qquad \text{...(2)}$$

Similarly, for the normal equation of y, multiply these equations by the coefficient of y in that equation and then add, we get

$$2(x + 2y + z - 1) + 1(2x + y + z - 4) + 1(-x + y + 2z - 4)$$
$$+ 2(4x + 2y - 5z + 7) = 0$$

$$\Rightarrow \qquad 11x + 10y - 5z + 4 = 0 \qquad \text{...(3)}$$

Similarly, the normal equation for z is

$$1(x + 2y + z - 1) + 1(2x + y + z - 4) + 2(-x + y + 2z - 4)$$
$$- 5(4x + 2y - 5z + 7) = 0$$

$$\Rightarrow \qquad 19x + 5y - 31z + 48 = 0 \qquad \text{...(4)}$$

Solving (2), (3) and (4) for x, y and z we get

$$x = 0.910, y = -0.378 \text{ and } z = 2.045$$

EXAMPLE 2. *Find the normal equations for fitting the curve of type $y = ax + \dfrac{b}{x}$ by least square method.*

SOLUTION. The given equation can be written as

$$y = ax + bx^{-1}$$

Consider the n points

$$(x_1, y_1), (x_2, y_2), ..., (x_i, y_i), ..., (x_n, y_n).$$

Then the residual or deviation of the ith point (x_i, y_i) is

$$y_i = \left(ax_i + \frac{b}{x_i}\right)$$

The sum of the squares of the error is given by

$$S = \Sigma \left(y_i - ax_i - \frac{b}{x_i}\right)^2 \qquad \qquad ...(1)$$

For the maxima and minima of S, we have

$$\frac{\partial S}{\partial a} = 0, \frac{\partial S}{\partial b} = 0,$$

$$\frac{\partial S}{\partial a} = 0 \Rightarrow \Sigma x_i \left(y_i - ax_i - \frac{b}{x_i}\right) = 0$$

$$\Rightarrow \qquad \Sigma (x_i y_i - ax_i^2 - b) = 0$$

$$\Rightarrow \qquad \Sigma x_i y_i = a\Sigma x_i^2 + nb \qquad \qquad ...(2)$$

$$\frac{\partial S}{\partial b} = 0 \quad \Rightarrow \quad \Sigma \left[\frac{1}{x_i}\left(y_i - ax_i - \frac{b}{x_i}\right)\right] = 0$$

$$\Rightarrow \qquad \Sigma \left[\frac{y_i}{x_i} - a - \frac{b}{x_i^2}\right] = 0$$

$$\Rightarrow \qquad \Sigma \frac{y_i}{x_i} = na + b\Sigma \frac{1}{x_i^2} \qquad \qquad ...(3)$$

Equations (2) and (3) gives the required normal equations.

EXAMPLE 3. *Find the most plausible values of x and y from the following equations :*
x + y = 3.00, 2x − y = 0.5, x + 3y = 7.25,
3x + y = 4.95.

SOLUTION. The given equations can be written as

$$\left.\begin{array}{l} x + y - 3.00 = 0 \\ 2x - y - 0.5 = 0 \\ x + 3y - 7.25 = 0 \\ 3x + y - 4.95 = 0 \end{array}\right] \qquad \qquad ...(1)$$

Now sum of the square 'S' of the errors is given by

$$S = (x + y - 3.00)^2 + (2x - y - 0.5)^2$$
$$+ (x + 3y - 7.25)^2 + (3x + y - 4.95)^2 = 0$$

For the maxima and minima of S, we must have

$$\frac{\partial S}{\partial a} = 0, \frac{\partial S}{\partial b} = 0, \frac{\partial S}{\partial x} = 0$$

$$\Rightarrow \quad (x + y - 3.00) + 2(2x - y - 0.50)$$
$$+ 1(x + 3y - 7.25) + 3(3x + y - 4.95) = 0$$

$$\Rightarrow \qquad \qquad \qquad 15x + 5y - 26.10 = 0 \qquad \qquad ...(2)$$

$$\frac{\partial S}{\partial y} = 0$$

$\Rightarrow \quad (x+y-3.00)-(2x-y-0.50)$
$+3(x+3y-7.25)+3(3x+y-4.95)=0$
$\Rightarrow \qquad\qquad 13x+14y-39.20=0 \qquad\qquad ...(3)$

Solving (2) and (3), we get
$\qquad x=1.234, \ y=1.919$

EXAMPLE 4. *Find the most plausible values of x and y from the following equations*
$\qquad 3x+y=4.95$
$\qquad x+y=3.00$
$\qquad 2x-y=0.5$
$\qquad x+3y=7.25$

SOLUTION. Let $S=(3x+y-4.95)^2+(x+y-3.00)^2$
$\qquad +(2x-y-0.5)^2+(x+3y-7.25)^2=0 \qquad ...(1)$

$\Rightarrow \quad \dfrac{\partial S}{\partial x}=6(3x+y-4.95)+2(x+y-3.00)+4(2x-y-0.5)+2(x+3y-7.25)$
$\qquad = 30x+10y-52.2$

and $\dfrac{\partial S}{\partial y}=10x+24y-58.4$

We obtained the normal equation by putting
$$\frac{\partial S}{\partial x}=0 \ \text{and} \ \frac{\partial S}{\partial y}=0$$

Therefore, $\quad 3x+y=5.22$
$\qquad x+2.4y=5.84$
On solving, we get
$\qquad x=1.07871, y=1.98387$
which is the required plausible solution.

Exercise 21.1

1. Find the values of x and y which satisfy the following equations most satisfactorily with the help of normal equation of x, y
$\qquad x+2.5y=21, 3.2x-y=28,$
$\qquad 4x+1.2y=42.04, 1.5x+6.3y=40.$

2. Find the most plausible values of x and y from the four equations
$\qquad x-y+2z=3, 3x+2y-5z=5,$
$\qquad 4x+y+4z=21$ and $-x+3y+3z=14.$

3. Use the method of least squares to find the most plausible values of x and y from the following equations
$\qquad x+y=3.01, 2x-y=0.03, x+3y=7.03,$
$\qquad 3x+y=4.97$

4. Find the most plausible values of x and y from the following equations
$\qquad x+y=301, 2x-y=3, x+3y=703,$
$\qquad 3x+y=497.$

ANSWERS

1. $x=9.620, y=4.064$ 2. $x=2.47, y=3.55, z=1.92$ 3. $x=0.997, y=2.001$
4. $x=100, y=200$

21.5 METHOD OF CURVE-FITTING

Let us consider the rth degree curve

$$y = a + bx + cx^2 + ... + kx^r, \text{ with } k \neq 0 \qquad ...(i)$$

with the given values $(x_1, y_1), (x_2, y_2), ..., (x_m, y_m)$.

The curve given by (i) has $(r + 1)$ unknown $a, b, c, ..., k$ and so if $m = r + 1$, we get $(r + 1)$ equations when the values $(x_1, y_1), (x_2, y_2), ..., (x_m, y_m)$ are substituted for (x, y) in (i) and thus a unique solution of the values of the unknown $a, b, c, ..., k$ is possible.

Now let

$$y_i' = a + bx_i + cx_i^2 + ... + kx_i^r \qquad ...(ii)$$

and let y_i be the observed values y for x_i.

Then if u_i be the residual for this point, we have

$$u_i = y_i - y_i' = y_i - (a + bx_i + cx_i^2 + ... + kx_i^r), \qquad \text{from (ii)}$$

$$u_i = y_i - a - bx_i - cx_i^2 - ... - kx_i^r \qquad ...(iii)$$

Now in order to make the sum of the squares of the errors minimum, we define

$$S = \sum_{i=1}^{m} u_i^2 = \sum_{i=1}^{m} (y_i - a - bx_i - cx_i^2 - ... - kx_i^r) \qquad ...(iv)$$

$\therefore$ By the principle of maxima and minima, we must have

$$\frac{\partial S}{\partial a} = 0, \frac{\partial S}{\partial b} = 0, ..., \frac{\partial S}{\partial k} = 0 \qquad ...(v)$$

which gives the following $(r + 1)$ equations :

$$\frac{\partial S}{\partial a} = -2\sum_{i=1}^{m} (y_i - a - bx_i - cx_i^2 - ... - kx_i^r) = 0$$

$$\Rightarrow \qquad \sum_{i=1}^{m} (y_i - a - bx_i - cx_i^2 - ... - kx_i^r) = 0$$

$$\sum_{i=1}^{m} y_i - \sum_{i=1}^{m} a - \sum_{i=1}^{m} bx_i - \sum_{i=1}^{m} cx_i^2 - ... = 0$$

$$\Rightarrow \qquad \Sigma y_i = ma + b\Sigma x_i + c\Sigma x_i^2 + ... \qquad ...(1)$$

Now

$$\frac{\partial S}{\partial b} = 0 = -2\Sigma(y_i - a - bx_i - cx_i^2 - ... - kx_i^r)x_i$$

or

$$\Sigma(y_i - a - bx_i - cx_i^2 - ... - kx_i^r)x_i = 0$$

$\therefore$

$$\Sigma x_i y_i = a\Sigma x_i + b\Sigma x_i^2 + c\Sigma x_i^3 + ... \qquad ...(2)$$

Similarly,

$$\Sigma x_i^2 y_i = a\Sigma x_i^2 + b\Sigma x_i^3 + c\Sigma x_i^4 + ... \qquad ...(3)$$

The equations (1), (2) and (3) are known as normal equations and can be solved as simultaneous equations to evaluate $a, b, c, ... k$.

21.6 FITTING OF SOME SPECIAL CURVES

21.6.1 FITTING OF A STRAIGHT LINE

Let

$$y = a + bx \qquad ...(1)$$

be the equation of the straight line, to be fitted.

Then the normal equations can be obtained as follows.

Let

$$u = (y_i - a - bx_i) \quad \text{and} \quad S = u^2 = \Sigma(y_i - a - bx_i)^2$$

for the maxima and minima of S, we must have $\dfrac{\partial S}{\partial a} = 0$

$$\Rightarrow \qquad -2\Sigma(y_i - a - bx_i).1 = 0 \qquad \Rightarrow \qquad \Sigma(y_i - a - bx_i).1 = 0$$

$$\therefore \qquad \Sigma y_i = ma + b\Sigma x_i \qquad \qquad ...(2)$$

Now $$\frac{\partial S}{\partial a} = 0$$

$$-2\Sigma(y_i - a - bx_i)(-x) = 0 \qquad \Rightarrow \qquad \Sigma(y_i - a - bx_i)x = 0$$

$$\Sigma x_i y_i = a\Sigma x_i + b\Sigma x_i^2 \qquad \qquad ...(3)$$

Solving these two equations we can find the values of a and b.

Special Case : If the straight line to be fitted passes through the origin, then the equation of the straight line is given by

$$y = bx$$

$$\therefore \qquad \text{Its normal equation is } \Sigma xy = b\Sigma x^2$$

we can find the value of constant b easily.

21.6.2 FITTING OF A PARABOLIC CURVE

Let $$y = a + bx + cx^2 \qquad \qquad ...(1)$$

be the equations of the parabolic curve to be fitted.

Then the normal equations are

$$\left.\begin{array}{l} \Sigma y = ma + b\Sigma x + c\Sigma x^2 \\ \Sigma xy = a\Sigma x + b\Sigma x^2 + c\Sigma x^2 \\ \Sigma x^2 y = a\Sigma x^2 + b\Sigma x^3 + c\Sigma x^4 \end{array}\right\} \qquad ...(2)$$

Solving above three equations for a, b and c simultaneously we can find the values of a, b and c.

21.6.3 FITTING OF AN EXPONENTIAL CURVE

Let $$y = ae^{bx} \qquad \qquad ...(1)$$

be the equation of the exponential curve to be fitted.

Taking logarithms of both sides of (1) to the base, 10.

We get $$\log_{10} y = \log_{10} a + bx \log_{10} e \qquad \qquad ...(2)$$

This is of the form $$Y = A + Bx \qquad \qquad ...(3)$$

where $Y = \log_{10} y, A = \log_{10} a$ and $B = b \log_{10} e$.

Now applying the method of fitting a straight lines we can find the values A and B and hence the values of a and b.

21.6.4 FITTING OF THE CURVE OF THE TYPE $y = ab^x$

Here the method of fitting is the same as given in part 21.7.3 above.

21.6.5 FITTING OF THE LOGARITHMIC CURVE OF THE FORM $y = ax^b$

We have $$y = ax^b$$

Taking log of both the sides, we get

$$\log y = \log a + b \log x$$

Let us put $\log y = Y, \log a = A$ and $\log x = X$

Then, we have $$Y = A + bX \qquad \qquad ...(1)$$

Normal equation for (1) are

$$\Sigma Y = Ax + b\Sigma X$$

$$\Sigma XY = A\Sigma X + b\Sigma X^2$$

21.6.6 FITTING OF THE CURVE $y = a + b/x$

Let put $\dfrac{1}{x} = X$. Then given equations becomes

$$y = a + bx \qquad \qquad \text{...(1)}$$

The normal equation for (1) are

$$\Sigma y = an + b\Sigma x \qquad \qquad \text{...(2)}$$

$$\Sigma Xy = a\Sigma X + b\Sigma X^2 \qquad \qquad \text{...(3)}$$

Solving these equations we get the required value of a and b.

21.6.7 FITTING OF THE CURVE $y = ax + b/x$

Let the curve $y = ax + \dfrac{b}{x}$ passes through the point $(x_i, y_i), i = 1, 2, ..., n$. The error of estimate for i^{th} point (x_i, y_i) is

$$d_i = y_i - ax_i - \frac{b}{x_i}$$

By the principle of least squares, the sum of squares of the error should be minimum.

i.e.,
$$S = \sum_{i=1}^{n} d_i^2 \text{ should be minimum.}$$

$$\Rightarrow \qquad S = \sum_{i=1}^{n} \left(y_i - ax_i - \frac{b}{x_i} \right)^2 \text{ should be minimum.}$$

We obtained the normal equations by $\dfrac{\partial S}{\partial a} = 0$ and $\dfrac{\partial S}{\partial b} = 0$.

$$\therefore \qquad \frac{\partial \Sigma}{\partial a} = 0 \Rightarrow -2\sum_{i=1}^{n} x_i \left(y_i - ax_i - \frac{b}{x_i} \right) = 0 \Rightarrow \sum_{i=1}^{n} x_i y_i = a \sum_{i=1}^{n} x_i^2 + bx$$

$$\text{...(1)}$$

and
$$\frac{\partial \Sigma}{\partial b} = 0 \Rightarrow -2\sum_{i=1}^{n} \frac{1}{x_i} \left(y_i - ax_i - \frac{b}{x_i} \right) = 0 \Rightarrow \sum_{i=1}^{n} \frac{y_i}{x_i} = an + b \sum_{i=1}^{n} \frac{1}{x_i^2}$$

$$\text{...(2)}$$

Solving (1) and (2) we get the required values of a and b.

21.6.8 FITTING OF THE CURVE $y = a + b/x + c/x^2$

We have $S = \sum_{i=1}^{n} \left(y_i - a - \dfrac{b}{x_i} - \dfrac{b}{x_i^2} \right)$ should be minimum.

Normal equations can be obtained by $\dfrac{\partial S}{\partial a} = 0, \dfrac{\partial S}{\partial b} = 0, \dfrac{\partial S}{\partial c} = 0$

Now $\dfrac{\partial S}{\partial a} = 0 \Rightarrow -2\sum\limits_{i=1}^{n}\left(y_i - a - \dfrac{b}{x_i} - \dfrac{c}{x_i^2}\right) = 0 \Rightarrow \Sigma y_i = an + b\sum\limits_{i=1}^{n}\dfrac{1}{x_i} + c\sum\limits_{i=1}^{n}\dfrac{1}{x_i^2}$

$\dfrac{\partial S}{\partial b} = 0 \Rightarrow -2\sum\limits_{i=1}^{n}\dfrac{1}{x_i}\left(y_i - a - \dfrac{b}{x_i} - \dfrac{c}{x_i^2}\right) = 0 \Rightarrow \sum\limits_{i=1}^{n}\dfrac{y_i}{x_i} = a\sum\limits_{i=1}^{n}\dfrac{1}{x_i} + b\sum\limits_{i=1}^{n}\dfrac{1}{x_i^2} + c\sum\limits_{i=1}^{n}\dfrac{1}{x_i^3}$

and $\dfrac{\partial S}{\partial c} = 0 \Rightarrow -2\sum\limits_{i=1}^{n}\dfrac{1}{x_i^2}\left(y_i - a - \dfrac{b}{x_i} - \dfrac{c}{x_i^2}\right) = 0 \Rightarrow \sum\limits_{i=1}^{n}\dfrac{y_i}{x_i^2} = a\sum\limits_{i=1}^{n}\dfrac{1}{x_i^2} + b\sum\limits_{i=1}^{n}\dfrac{1}{x_i^3} + c\sum\limits_{i=1}^{n}\dfrac{1}{x_i^4}$

...(3)

Solving (1), (2) and (3), we get the required values of a and b.

21.6.9 Fitting of the Curve $y = ax^2 + b/x$

We have $\quad S = \sum\limits_{i=1}^{n}\left(y_i - ax_i^2 - \dfrac{b}{x_i}\right)^2$

Now $\quad \dfrac{\partial S}{\partial a} = 0 \Rightarrow -2\sum\limits_{i=1}^{n}x_i^2\left(y_i - ax_i^2 - \dfrac{b}{x_i}\right) = 0 \Rightarrow \sum\limits_{i=1}^{n}x_i^2 y_i = a\sum\limits_{i=1}^{n}x_i^4 + b\sum\limits_{i=1}^{n}x_i$ ...(1)

$\dfrac{\partial S}{\partial b} = 0 \Rightarrow -2\sum\limits_{i=1}^{n}\dfrac{1}{x_i}\left(y_i - ax_i^2 - \dfrac{b}{x_i}\right) = 0 \Rightarrow \sum\limits_{i=1}^{n}\dfrac{y_i}{x_i} = a\sum\limits_{i=1}^{n}x_i + b\sum\limits_{i=1}^{n}\dfrac{1}{x_i^2}$ ...(2)

Solving (1) and (2) we get the values of a and b.

21.6.10 Fitting of the Curve $y = a/x + b\sqrt{x}$

Here we have $\quad S = \sum\limits_{i=1}^{n}\left(y_i - ax_i^2 - b\sqrt{x_i}\right)^2$

Now $\quad \dfrac{\partial S}{\partial a} = 0 \Rightarrow -2\sum\limits_{i=1}^{n}\dfrac{1}{x_i}\left(y_i - \dfrac{a}{x_i} - b\sqrt{x_i}\right) = 0 \Rightarrow \sum\limits_{i=1}^{n}\dfrac{y_i}{x_i} = a\sum\limits_{i=1}^{n}\dfrac{1}{x_i^2} + b\sum\limits_{i=1}^{n}\dfrac{1}{\sqrt{x_i}}$

...(1)

$\dfrac{\partial S}{\partial b} = 0 \Rightarrow -2\sum\limits_{i=1}^{n}\sqrt{x_i}\left(y_i - \dfrac{a}{x_i} - b\sqrt{x_i}\right) = 0 \Rightarrow \sum\limits_{i=1}^{n}(\sqrt{x_i})y_i = a\sum\limits_{i=1}^{n}\dfrac{1}{\sqrt{x_i}} + b\sum\limits_{i=1}^{n}x_i$

...(2)

On solving (1) and (2) we get the required values of a and b.

SOLVED EXAMPLES

EXAMPLE 1. *Fit a straight line to the following data :*   (Rgpv–2001)

x	1	2	3	4	5
y	5	7	9	10	11

SOLUTION. Here, we have

Total

x	1	2	3	4	5	$\Sigma x = 15$
y	5	7	9	10	11	$\Sigma y = 42$
xy	5	14	27	40	55	$\Sigma xy = 141$
x^2	1	4	9	16	25	$\Sigma x^2 = 55$

the equation of the line be $y = a + bx$ ...(1)

Normal equation are

$$\Sigma y = ma + b\Sigma x$$

and $$\Sigma xy = a\Sigma x + b\Sigma x^2$$

Using the given values,

$$42 = 5a + 15b \qquad ...(2)$$
$$141 = 15a + 55b \qquad ...(3)$$

Solving (2) and (3) we get $b = \dfrac{3}{2}, a = \dfrac{39}{10}$

or $$b = 1.5, a = 3.9$$

Hence, required straight line is given by

$$y = (3.9) + (1.5)x$$

🐟 REMARK

➡ For the sake of convenience, it is sometimes easy to change the origin and scale with the substitution $x = \dfrac{x - A}{h}$ and $y = \dfrac{y - B}{h}$, where A and B are the middle values of x and y series respectively.

EXAMPLE 2. *Fit a straight line to the following data :*

x	0	5	10	15	20	25
y	12	15	17	22	24	30

SOLUTION. We observe that the number of values given is 6, *i.e.*, even and the difference in the values of x is 5. In such a case the calculations can be simplified if we take half the common distance which is 2.5 here as unit of measurement. Also the two mid-values are 10 and 15, so we take their mean, *i.e.*, $\dfrac{1}{2}(10+15)$, *i.e.*, 12.5 as the origin.

Therefore, introduce two new variables

$$u = \frac{x - 12.5}{2.5} \quad \text{and} \quad v = y - 20 \qquad ...(1)$$

Then we have

$$u : \frac{0 - 12.5}{2.5}, \frac{5 - 12.5}{2.5}, \frac{10 - 12.5}{2.5}, \frac{15 - 12.5}{2.5}, \frac{20 - 12.5}{2.5}, \frac{25 - 12.5}{2.5}$$

Total

u	-5	-3	-1	1	3	5	0
v	$12 - 20$	$15 - 20$	$17 - 20$	$22 - 20$	$24 - 20$	$30 - 20$	
v	-8	-5	-3	2	4	10	0
uv	40	15	3	2	12	50	122
u^2	25	9	1	1	9	25	70

Now let the equation of line be

$$v = a + bu \qquad ...(2)$$

Then its normal equations are

$$\Sigma v = ma + b\Sigma u$$

$$\Sigma vu = a\Sigma u + b\Sigma u^2$$

$$\Rightarrow \qquad 0 = 6.a + 0.b \qquad ...(3)$$

$$122 = 0.a + 70.b \qquad ...(4)$$

Solving (3) and (4) we get $a = 0$ and $b = 1.743$.

Substituting these values in (2), the equation of the line is

$$v = 0 + (1.743)u$$

or $$y - 20 = (1.743)\left[\frac{x - 12.5}{2.5}\right] \qquad \text{[Using (1)]}$$

or $$y - 20 = \left(\frac{1.743}{2.5}\right)x - \frac{1.743 \times 12.5}{2.5}$$

$$y - 20 = 0.7x - 8.715$$

$$\Rightarrow \qquad y = 0.7x + (20 - 8.715)$$

$$\Rightarrow \qquad y = 0.7x + 11.285$$

$$\Rightarrow \qquad y = 11.285 + 0.7x$$

☙ REMARK

⟫ If in such a problem, the number of values given is odd then the calculations are simplified provided we take the common difference as the unit of measurement and the mid-values as the assumed origin.

EXAMPLE 3. *Fit a parabola of the second degree of the following data :*

[UPTU(MCA)–2006, VTU–2009, Bhopal–2008]

x	1.0	1.5	2.0	2.5	3.0	3.5	4.0
y	1.1	1.3	1.6	2.6	2.7	3.4	4.1

SOLUTION. Here we observe that the number of values given here is 7 which is odd, so we take the middle value 2.5, *i.e.*, 4th value as our assumed mean and let

$$u = \frac{x - 2.5}{0.5} \quad \text{and} \quad v = y - 2.7$$

Let the second degree curve to be fitted is

$$v = a + bu + cu^2 \qquad ...(1)$$

Its normal equations are

$$\Sigma vu = a\Sigma u + b\Sigma u^2 + c\Sigma u^3 \qquad ...(2)$$

$$\Sigma vu^2 = a\Sigma u^2 + b\Sigma u^3 + c\Sigma u^4 \qquad ...(3)$$

Now from the given data, we have   Total

u	-3	-2	-1	0	1	2	3	0
v	-1.6	-1.4	-1.1	-0.1	0	0.7	1.4	-14.1
vu	4.8	2.8	1.1	0	0	1.4	4.2	14.3
u^2	9	4	1	0	1	4	9	2.8
vu^2	-14.4	-5.6	-1.1	0	0	2.8	12.6	-5.7
u^3	-27	-8	-1	0	1	8	27	0
u^4	81	16	1	0	1	16	81	196

Substituting the values of $\Sigma u, \Sigma v, \Sigma uv, \Sigma vu^2, \Sigma u^2, \Sigma u^3, \Sigma u^4$ in (2), (3) and (4), we get

$$-2.1 = 7a + 0.b + 27.c \qquad \qquad ...(5)$$
$$14.3 = 0.a + 27b + 0.c \qquad \qquad ...(6)$$
$$-5.7 = 27a + 0.b + 196c \qquad \qquad ...(7)$$

Solving (5), (6) and (7), we get $a = -0.04, b = 0.53, c = 0.03$.
Substituting these values of a, b and c in (1), we get

$$v = -0.04 + 0.53u + 0.03u^2$$

$$\Rightarrow \quad y - 2.7 = -0.04 + 0.53\left(\frac{x-2.5}{0.5}\right) + 0.03\left(\frac{x-2.5}{0.5}\right)^2$$

$$\Rightarrow \quad y - 2.7 = -0.04 + \frac{0.53x}{0.5} - \frac{2.5 \times 0.53}{0.5} + 0.03\left(\frac{x-2.5}{0.5}\right)^2$$

$$\Rightarrow \quad y - 2.7 = -0.04 + 1.06x - 2.65 + 0.03\left[\frac{x^2 + 6.25 - 5.0x}{0.5}\right]$$

$$\Rightarrow \quad y = -0.04 + 2.7 - 2.65 + 1.06x + 0.06(x^2 - 5x + 6.25)$$
$$\Rightarrow \quad y = 0.01 + 1.06x + 0.06x^2 - 0.30x + 0.375$$
$$\Rightarrow \quad y = 0.385 + 0.76x + 0.66x^2$$

EXAMPLE 4. *Fit a parabolic curve of regression of y on x to the seven pairs of values*
[MDU(B.E.)–2007]

x	1.0	1.5	2.0	2.5	3.0	3.5	4.0
y	1.1	1.3	1.6	2	2.7	3.4	4.1

SOLUTION. Let $X = \dfrac{x-2.5}{0.5} = 2x - 5$ and $Y = y$

Then

x	y	X	Y	X^2	XY	X^2Y	X^3	X^4
1.0	1.1	-3	1.1	9	-3.3	9.9	-27	81
1.5	1.3	-2	1.3	4	-2.6	5.2	-8	16
2.0	1.6	-1	1.6	1	-1.6	1.6	-1	1
2.5	2.0	0	2.0	0	0	0	0	0
3.0	2.7	1	2.7	1	2.7	2.7	1	1
3.5	3.4	2	3.4	4	6.8	6.9	8	16
4.0	4.1	3	4.1	9	12.3	12.3	27	81
Total		**0**	**16.2**	**28**	**14.3**	**69.9**	**0**	**196**

Let the equation of the parabolic curve is

$$Y = a + bX + cX^2 \qquad \ldots(1)$$

The normal equation of (1) are given by

$$\left.\begin{array}{l} \Sigma Y = ma + b\Sigma X + c\Sigma X^2 \\ \Sigma YX = a\Sigma X + b\Sigma X^2 + c\Sigma X^3 \\ \Sigma YX^2 = a\Sigma X^2 + b\Sigma X^3 + c\Sigma X^4 \end{array}\right\} \qquad \ldots(2)$$

Putting the values in (2), from the table, we get

$$16.2 = 7a + 27c$$
$$14.3 = 27b$$
$$69.9 = 28a + 196c$$
$$\Rightarrow \quad a = 2.07, b = 0.511, c = 0.061$$

Putting the values of a, b and c in (1), we get

$$Y = 2.07 + 0.511X + 0.061X^2$$
$$\Rightarrow \qquad y = 2.07 + 0.511(2x - 5) + 0.061(2x - 5)^2$$
$$\Rightarrow \qquad y = 1.04 - 0.193x + 0.243x^2$$

EXAMPLE 5. *Fit a second degree parabola to the following data*

x	1	2	3	4	5
y	1090	1220	1390	1625	1915

SOLUTION. Let us define u and v such that $u = x - 3, v = \dfrac{y - 1450}{5}$

and equation of the parabola is $v = a + bu + cu^2$ $\qquad \ldots(1)$

Therefore, we have the following table :

x	y	u	v	u²	v⁴	uv	u²v
1	1090	−2	−72	4	16	144	−288
2	1220	−1	−46	1	1	46	−46
3	1390	0	−12	0	0	0	0
4	1625	1	35	1	1	35	35
5	1915	2	93	4	16	186	372
Total		**0**	**−2**	**10**	**34**	**411**	**73**

Then putting values in the normal equation

$$\Sigma v = ma + b\Sigma u + c\Sigma u^2$$
$$\Sigma uv = a\Sigma u + b\Sigma u^2 + c\Sigma u^3$$
$$\Sigma u^2 v = a\Sigma u^2 + b\Sigma u^3 + c\Sigma u^4$$

we get

$$-2 = 59 + 0 + 10c$$
$$411 = 0 + 106 + 0$$
$$73 = 10a + 3 + 34c$$

Solving these equations for a, b and c we get

$$a = -11.4, b = 41.1, c = 5.5$$

Put in (1), we get $v = -11.4 + 41.1u + 5.5u^2$

Now changing the origin, we get the fit curve is

$$\Rightarrow \qquad y = 1024 + 40.5x + 27.5x^2$$

EXAMPLE 6. *If P is a pull required to lift a load W by means of a pulley block, find a linear law of the form P = mW + c connecting P and W, using the following data :*

P	12	15	21	25
W	50	70	100	120

Compute P where W = 150 Kg. **(UPTU–2007, VTU–2002)**

SOLUTION. The given equation is $P = mW + c$ ...(1)

The normal equations are

$$\left. \begin{array}{l} \Sigma P = 4c + m\Sigma W \\ \Sigma WP = c\Sigma W + m\Sigma W^2 \end{array} \right\} \quad ...(2)$$

Then, we have

W	P	W^2	WP
50	12	2500	600
70	15	4900	1050
100	21	10000	2100
120	25	14400	3000
340	73	31800	6750

Putting these values in (2), we get

$$73 = 4c + 340m$$
$$6750 = 340c + 31800m$$

which gives $m = 0.1879, c = 2.2785$

Hence, the line of best fit is

$$P = 2.2759 + 0.1879W$$

Now, for $W = 1500$ Kg $P = 30.4635$ Kg

EXAMPLE 7. *The pressure and the volume of a gas are related by the equation $pV^r = K, r$ and K being constants. Fit this equation to the following set of observations.*

P(Kg/cm²)	0.5	1.0	1.5	2.0	2.5	3.0
V(litres)	1.62	1.00	0.75	0.62	0.52	0.46

(VTU–2011)

SOLUTION. We have

$$\log_{10} p + r \log_{10} V = \log_{10} K$$

$$\Rightarrow \qquad \log_{10} V = \frac{1}{r}\log_{10} K - \frac{1}{r}\log_{10} P \qquad \Rightarrow \qquad y = A + BX$$

where

$$X = \log_{10} p, Y = \log_{10} V, A = \frac{1}{r}\log_{10} K,$$

Now

$$A = \frac{1}{r}\log_{10} K, B = -\frac{1}{r}$$

P	V	X	Y	XY	X²
0.5	1.62	−0.3010	0.2095	−0.0630	0.0906
1.0	1.00	0	0	0	0
1.5	0.75	0.1762	−0.1249	−0.0220	0.0310
2.0	0.62	0.3010	−0.2076	−0.0625	0.0906
2.5	0.52	0.3979	−0.2840	−0.1130	0.1583
3.0	0.46	0.4771	−0.3372	−0.1609	0.2276
Total		1.0511	−0.7442	−0.4214	0.5981

Now putting all these values in the normal equations and get the required fitted curve.

EXAMPLE 8. *Fit a second degree parabola to the following data taking as dependent variable.*

x	1	2	3	4	5	6	7	8	9
y	2	6	7	8	10	11	11	10	9

SOLUTION. Let us introduce two new variables X and Y such that $X = x - 5, Y = y - 7$

Also, let the curve to be fit be $Y = a + bX + cX^2$

Then, we have the following table.

x	y	X	Y	XY	X²	X²Y	X³	X⁴
1	2	−4	−5	20	16	−80	−64	256
2	6	−3	−1	3	9	−9	−27	81
3	7	−2	0	0	4	0	−8	16
4	8	−1	1	−1	1	1	−1	1
5	10	0	2	0	0	0	0	0
6	11	1	4	4	1	4	1	1
7	11	2	4	8	4	16	8	16
8	10	3	3	9	9	27	27	81
9	9	4	2	8	16	32	64	256
Total		**0**	**11**	**51**	**60**	**−9**	**0**	**708**

Therefore, the normal equations are

$$11 = 9a + 0 + 60c$$
$$51 = 0 + 60b$$
$$-9 = 60a + 0 + 708c$$

Solving these equations for a, b and c, we get $a = 3, b = 0.85$ and $c = -0.27$
Hence, the curve of fit is

$$Y = a + bX + cX^2 = 3 + 0.85X - 0.27X^2$$

$\Rightarrow \qquad y - 7 = 3 + 0.85(x - 5) - 0.27(x - 5)^2$

$$= 3 + 0.85x - 4.25 - 0.27x^2 + 2.7x - 6.75$$

$\Rightarrow \qquad y = -1 + 3.55x - 0.27x^2$

EXAMPLE 9. *Fit a parabola $y = ax^2 + bx + c$ by the method of least square using following data*

x	1	2	3	4	5
y	5	12	26	60	97

SOLUTION. The given equation of the parabola is $y = ax^2 + bx + c$
Therefore, the normal equations are

$$a\Sigma x^2 + b\Sigma x + n.c = \Sigma y$$
$$a\Sigma x^3 + b\Sigma x^2 + c\Sigma x = \Sigma xy \qquad \qquad ...(1)$$
$$a\Sigma x^4 + b\Sigma x^3 + c\Sigma x^2 = \Sigma x^2 y$$

Here, we have $\qquad n = 5$

Now, we construct the following table.

x	y	x^2	x^3	x^4	xy	x^2y
1	5	1	1	1	5	5
2	12	4	8	16	24	48
3	26	9	27	81	78	234
4	60	16	64	256	240	960
5	97	25	125	625	485	2425
$\Sigma x = 15$	$\Sigma y = 200$	$\Sigma x^2 = 55$	$\Sigma x^3 = 225$	$\Sigma x^4 = 979$	$\Sigma xy = 832$	$\Sigma x^2y = 3672$

Putting all the above values in (1), we obtained

$$55a + 15b + 5c = 200$$
$$225a + 55b + 15c = 832 \qquad \qquad \text{...(2)}$$
$$979a + 225b + 55c = 3672$$

Solving (2), we get

$$a = 5.7143, b = -11.0858, c = 10.4001$$

Hence, the required parabola for best fit is $y = 5.7143x^2 - 11.0858x + 10.4001$

EXAMPLE 10. *By the method of least squares, find the curve $y = ax + bx^2$ that best fits the following data :*

x	1	2	3	4	5
y	1.8	5.1	8.9	14.1	19.8

SOLUTION. Let error of estimate for i^{th} point (x_i, y_i) be $E_i = (y_i - ax_i - bx_i^2)$. By principle of least squares the values of a and b are such that

$$U = \sum_{i=1}^{5} E_i^2 = \sum_{i=1}^{5} (y_i - ax_i - bx_i^2) \text{ is minimum}$$

$\therefore$ Normal equation are given by $\dfrac{\partial u}{\partial a} = 0$.

$$\Rightarrow \qquad \sum_{i=1}^{5} x_i y_i = a \sum_{i=1}^{5} x_i^2 + b \sum_{i=1}^{5} x_i^3$$

and $\qquad \dfrac{\partial u}{\partial b} = 0$

Then, $\qquad \sum_{i=1}^{5} x_i^2 y_i = a \sum_{i=1}^{5} x_i^3 + b \sum_{i=1}^{5} x_i^4$

Delete the suffix i, then normal equations are

$$\Sigma xy = a\Sigma x^2 + b\Sigma x^3 \qquad \qquad \text{...(A)}$$

and $\qquad \Sigma x^2 y = a\Sigma x^3 + b\Sigma x^4 \qquad \qquad \text{...(B)}$

x	y	x^2	x^3	x^4	xy	x^2y
1	1.8	1	1	1	1.8	1.8
2	5.1	4	8	16	10.2	20.4
3	8.9	9	27	81	26.7	80.1
4	14.1	16	64	256	56.4	225.6
5	19.8	25	125	625	99	49.5
		$\Sigma x^2 = 55$	$\Sigma x^3 = 255$	$\Sigma x^4 = 979$	$\Sigma xy = 194.1$	$\Sigma x^2y = 822.9$

Putting all these value in equation (1) and (2), we get
$$1941 = 55a + 225b$$
and
$$822.9 = 255a + 979b$$
On solving, we get

$$a = \frac{83.85}{55} \simeq 1.52, b = \frac{317.4}{664} \simeq 0.49$$

Hence, required parabolic curve is
$$y = 1.52x + 0.49x^2.$$

EXAMPLE 11. *Find a relation of the form $y = AB^x$ for the following data by the method of least squares*

x	2	3	4	5	6
y	8.3	15.4	33.1	65.2	126.4

SOLUTION. The curve to be fitted is $y = A(B)^x$.

or, $y = a + bX$ where

$a = \log_{10} A, b = \log_{10} B$ and $h = \log_{10} y$

Therefore, the normal equations are $\Sigma Y = 5a + b\Sigma X$ and $\Sigma XY = a\Sigma X + b\Sigma X^2$

X	Y	$Y = \log_{10} y$	X^2	XY
2	8.3	0.1191	4	1.8382
3	15.4	1.1872	9	3.5616
4	33.1	1.5198	16	6.0792
5	65.2	1.8142	25	9.0710
6	127.4	2.1052	36	12.6312
$\Sigma X = 20$		$\Sigma Y = 7.5455$	$\Sigma X^2 = 90$	$\Sigma XY = 33.1812$

Therefore, the normal equations, becomes
$$7.5455 = 5a + 20b \text{ and } 33.1812 = 20a + 90b$$
$\Rightarrow$ $a = 0.31$ and $b = 0.3$
$\therefore$ $A = $ Antilog $a = 2.04$
and $B = $ Antilog $b = 1.995$

Hence, the required curve is $y = 2.04(1.995)^x$.

EXAMPLE 12. *Using the method of least square, fit the non-linear curve of the form $y = ae^{bx}$ to the following data :*

x	0	2	4
y	5.012	10	31.62

SOLUTION. The curve of the form $y = ae^{bx}$ is to fitted in the form
$$Y = A + Bx \qquad \qquad \text{...(1)}$$
where $Y = \log_{10} y, A = \log_{10} a$
and $B = b\log_{10} e.$
The normal equations for (1) are

$$\Sigma Y = An + b\Sigma x \qquad \qquad \text{...(2)}$$

and $\qquad \Sigma xY = A\Sigma x + b\Sigma x^2$ $\qquad$ ...(3)

The table is given below :

x	y	$Y = \log_{10}y$	xY	x^2
0	5.012	0.70001	0	0
2	10	1.00000	2.0000	4
4	31.62	1.49996	5.99984	16
6	46.632	3.19997	7.99984	20

Here $n = 3, \Sigma Y = 3.19997, \Sigma xY = 7.99984, \Sigma x^2 = 20$.

Substituting these values in (2) and (3), we get

$\qquad 3.19997 = 3A + 6B$ $\qquad$ ...(4)

and $\qquad 7.199984 = 6A + 20B$ $\qquad$ ...(5)

Solving (4) and (5), we get

$\qquad A = 0.66667, B = 0.19999,$

since $A = \log_{10} a, B = b\log_{10} e = (0.43429)b$

Therefore $\qquad a = 10^A, b = \dfrac{B}{0.43429}$

$\Rightarrow \qquad a = (10)^{0.66667}, b = \dfrac{0.19999}{0.43429}$

$\Rightarrow \qquad a = 4.64162, b = 0.6050$

Hence, the required curve of the best fit is given by $y = 4.64162e^{0.4605x}$

EXAMPLE 13. *Find the curve of the best fit of the type $y = ae^{bx}$ to the following data by the method of least square*

x	1	5	7	9	12
y	10	15	12	15	21

SOLUTION. The given equation of the curve can be written as

$\qquad Y = A + Bx$ $\qquad$ ...(1)

where $\qquad Y = \log_{10} y, A = \log_{10} a$

and $\qquad B = b\log_{10} e.$

The normal equations for (1) are

$\qquad \Sigma Y = An + b\Sigma x$ $\qquad$ ...(2)

and $\qquad \Sigma xY = A\Sigma x + b\Sigma x^2$ $\qquad$ ...(3)

x	y	$Y = \log_{10}y$	xY	x^2
1	10	1	1	1
5	15	1.17609	5.88045	25
7	12	1.07918	7.55426	49
9	15	1.17609	10.58481	81
12	21	1.32222	15.86664	144
34	73	5.75358	40.88616	300

Putting the values from table in (2) and (3), we get

$$5.75358 = 5A + 34B \qquad \qquad ...(4)$$

$$40.88616 = 34A + 300B \qquad \qquad ...(5)$$

On solving (4) and (5), we get

$$A = 0.97658, B = 0.02561$$

$$\Rightarrow \qquad a = 10^A = 10^{0.97658} = 9.47502, b = \frac{0.02561}{0.43429} = 0.05897$$

Hence, the required curve of the best fit is

$$y = 9.47502 e^{0.05897x}.$$

EXAMPLE 14. *Determine the constant a and b by the method of least squares such that $y = ae^{bx}$ fits the following data :*

x	2	4	6	8	10
y	4.077	11.084	30.128	81.897	222.62

SOLUTION. We have $\qquad y = ae^{bx}$

Taking log, we get

$$\log y = \log a + bx$$

i.e., $\qquad Y = A + bx \qquad \qquad ...(1)$

where $\qquad Y = \log y$

and $\qquad A = \log a$

Normal equations of equation (1) are

$$\Sigma Y = An + b\Sigma x \qquad \qquad ...(2)$$

and $\qquad \Sigma xY = A\Sigma x + b\Sigma x^2 \qquad \qquad ...(3)$

x	y	$Y = \log y$	xY	x^2
2	4.077	1.4054	2.8107	4
4	11.084	2.4055	9.6220	16
6	30.128	3.4054	20.4327	36
8	81.897	4.4055	35.2437	64
10	222.62	5.4055	54.0547	100
30		17.0272	122.1638	220

Putting all these above values in equations (2) and (3), we get

$$17.0272 = 5A + 30b \qquad \qquad ...(4)$$

$$122.1638 = 30A + 220b \qquad \qquad ...(5)$$

From (4) and (5), we get $b = 0.5, A = 0.4054$, i.e., $\qquad A = \log a = 0.4054$

Taking antilog, we get $a = 1.4999 \Rightarrow a = 1.5$ (app.)

EXAMPLE 15. *Fit a second degree parabola to the following data.*   (RGPV–2001, 2002)

x	0	1	2	3	4
y	1	5	10	22	38

SOLUTION. The equation of second degree parabola is

$$y = ax^2 + bx + c$$

Therefore, the normal equations are

$$\left.\begin{array}{c} a\Sigma x^2 + b\Sigma x + nc = \Sigma y \\ a\Sigma x^3 + b\Sigma x^2 + c\Sigma x = \Sigma xy \\ a\Sigma x^4 + b\Sigma x^3 + c\Sigma x^2 = \Sigma x^2 y \end{array}\right\} \quad ...(1)$$

Here, we have $n = 5$

Now, we construct the following table :

x	y	x^2	x^3	x^4	xy	$x^2 y$
0	1	0	0	0	0	0
1	5	1	1	1	5	5
2	10	4	8	16	20	40
3	22	9	27	81	66	198
4	38	16	64	256	152	608
10	76	30	100	354	243	851

Putting all these values in (1), we obtained

$$30a + 10b + 5c = 76$$
$$100a + 30b + 10c = 243$$
$$354a + 100b + 30c = 851$$

Solving these value for a, b and c, we get

$$a = 1.42, \ b = 0.26 \text{ and } c = 2.21$$

Hence, the required parabola for best fit is

$$y = 1.42x^2 + 0.26x + 2.21$$

21.7 CURVE FITTING BY SUM OF EXPONENTIALS

Consider the equation

$$y = A_1 e^{\lambda_1 x} + A_2 e^{\lambda_2 x} \quad ...(1)$$

It can be seen that the function given by (1) satisfy a differential equation of the type

$$\frac{d^2 y}{dx^2} = a_1 \frac{dy}{dx} + a_2 y \quad ...(2)$$

where a_1, a_2 are constants.

Assume that a is the initial value of x we get, by integrating (2) w.r.t. from a to x.

$$y'(x) - y'(a) = a_1 y(x) - a_1 y(a) + a_2 \int_a^x y(x)dx$$

$$\Rightarrow \qquad y'(x) = \frac{dy}{dx}$$

Again integrating (3) w.r.t. x from a to x, we get

$$y(x) - y(a) - (x-a)y'(a) = a_1 \int_a^x y(x)dx - a_1(x-a)y(a) + a_2 \int_a^x \int_a^x y dx dx \quad ...(4)$$

Now using the result

$$\int_a^x \cdots \int_a^x f(x)dx = \frac{1}{(n-1)!}\int_a^x (x-t)^{n-1} f(t)dt \qquad \ldots(5)$$
$$\underset{n \text{ times}}{}$$

(By Convolution theorem of integration)

The equation (5) can be written as

$$y(x_1) - y(a) - (x-a)y'(a) = a_1\int_a^x ydx - a_1(x-a)y(a) + a_2\int_a^x (x-t)y(t)dt \quad \ldots(6)$$

Now, choosing two points x_1 and x_2 such that $a - x_1 = x_2 - a$, then (6) gives

$$y(x_1) - y(a) - (x_1 - a)y'(a) = a_1\int_a^{x_1} y(x)dx - a_1(x_1 - a)y(a) + a_2\int_a^{x_1} (x_1 - t)y(t)dt$$

and $\quad y(x_2) - y(a) - (x_2 - a)y'(a) = a_1\int_a^{x_2} y(x)dx - a_1(x_2 - a)y(a) + a_2\int_a^{x_2} (x_2 - t)y(t)dt$

Adding above two equations and simplifying by using $a - x_1 = x_2 - a$, we get

$$y(x_1) + y(x_2) - 2y(a) = a_1\left[\int_a^{x_1} y(x)dx + \int_a^{x_2} y(x)dx\right]$$
$$+ a_2\left[\int_a^{x_1} (x_1 - t)y(t)dt + \int_a^{x_2} (x_2 - t)y(t)dt\right] \qquad \ldots(7)$$

Now the equation (7) can be used to setup a linear system of equations for a_1 and a_2 and then obtain λ_1 and λ_2 from the following characteristic equation $\lambda^2 = a_1\lambda + a_2$.

Finally, A_1 and A_2 can be obtained by the method of least square.

SOLVED EXAMPLE

EXAMPLE. Fit a function of the form $y = A_1 e^{\lambda_1 x} + A_2 e^{\lambda_2 x}$ to the following data :

x	1.0	1.1	1.2	1.3	1.4	1.5	1.6	1.7	1.8
y	1.54	1.67	1.81	1.97	2.15	2.35	2.58	2.83	3.11

SOLUTION. To fit the given curve we use the following steps :

Step I : Choose x_1 and x_2 such that

$$a - x_1 = x_2 - a \qquad \ldots(1)$$

where a is the initial value of x taken from $x_i : i = 1, 2, \ldots, n$.

Step II : Use $y(x_1) + y(x_2) - 2y(a)$

$$= a_1\left[\int_a^{x_1} y(x)dx + \int_a^{x_2} y(x)dx\right]$$
$$+ a_2\left[\int_a^{x_1} (x_1 - t)y(t)dt + \int_a^{x_2} (x_2 - t)y(t)dt\right] \qquad \ldots(2)$$

Step III : Repeat the step I and II by choosing another set of x_1, x_2 and a such that (2) is true. Therefore, we obtain a linear system of equations for a_1 and a_2. Solve these equation to get a_1 and a_2.

Step IV : Substitute the values of a_1 and a_2, obtained in step III in the characteristic equation $\lambda^2 = a_1\lambda + a_2$

Step V : Finally, use the method of least square to obtain A_1 and A_2 choosing $x_1 = 1.0, a = 1.2$ and $x_2 = 1.4$, we have

Now equation (2) gives

$$1.54 + 2.15 - 3.62$$
$$= a_1\left[\int_{1.2}^{1.0} ydx + \int_{1.2}^{1.4} ydx\right] + a_2\left[\int_{1.2}^{1.0}(1.0 - t)y(t)dt + \int_{1.2}^{1.4}(1.4 - t)ydt\right]$$

$$\Rightarrow 0.07 = a_1\left[-\int_{1.0}^{1.2} ydx + \int_{1.2}^{1.4} ydx\right] + a_2\begin{bmatrix} -\int_{1.0}^{1.2}(1.0 - t)ydt \\ +\int_{1.2}^{1.4}(1.4 - t)ydt \end{bmatrix}.$$

On simplification, we get

$$1.81a_1 + 2.180a_2 = 2.10 \qquad \dots(4)$$

Again choosing $x_1 = 1.4$, $a = 1.6$ and $x_2 = 1.8$, so that we have $a - x_1 = 0.2 = x_2 - a$ and using equation (2), we get

$$2.88a_1 + 3.104a_2 = 3.00 \qquad \dots(5)$$

Solving (4) and (5), we get $a_1 = 0.03204$, $a_2 = 0.9364$

Put these values in (3), we get $\lambda^2 - 0.03204\lambda - 0.9364 = 0$

Now, using the method of least squares, we get $A_1 = 0.499$, $A_2 = 0.491$

Hence, the required curve for best fit is $y = 0.499e^{0.99x} + 0.491e^{-0.96x}$.

Exercise 21.2

1. Show that the line of fit to the following data :

x	6	7	7	8	8	8	9	9	10
y	5	5	4	5	4	3	4	3	3

is given by $y = -0.5x + 8$.

2. Show that the best-fitting linear function for the points $(x_1, y_1), (x_2, y_2) \dots (x_n, y_n)$ may be expressed in the form

$$\begin{vmatrix} x & y & 1 \\ \Sigma x & \Sigma y & n \\ \Sigma x^2 & \Sigma y^2 & \Sigma x \end{vmatrix} = 0$$

Show also that the line passes through the mean points $(\bar{x}, \bar{y})$.

3. Find the parabola of the form $a + bx + cx^2$ which fit most closely with the observations : **(VTU–2006, JNTU–2000S)**

x	–3	–2	–1	0	1	2	3
y	4.63	2.11	0.67	0.09	0.63	2.15	4.58

4. Fit a second degree parabola to the following data

x	1.0	1.5	2.0	2.5	3.0	3.5	4.0
y	1.1	1.1	1.6	2.0	2.7	3.4	4.1

5. The following table gives the results of the measurement of train resistance; V is the velocity in miles per hour, R is the resistance in pounds per ton.

V	20	40	60	80	100	120
R	5.5	9.1	14.9	22.8	33.3	46.0

If R is related to V by the relation $R = a + bV + cV^2$, fit the curve. **(UPTU–2002)**

6. Fit the relation $R = a + bV^2$ from the following data : **(Indore–2008)**

V	10	20	30	40	50
R	8	10	15	21	30

7. Fit a least square geometric curve $y = ax^b$ to the following data :

x	1	2	3	4	5
y	0.5	2	4.5	8	12.5

8. The voltage V across a capacitor at time t seconds is given by the following table :

t	0	2	4	6	8
V	150	63	28	12	5.6

Use the method of least square to fit a curve of the form $V = ae^{kt}$ to the above data.

9. Applying the method of least squares to fit the curves $y = ax^2 + \dfrac{b}{x}$ to the following data : **(Madras–2003)**

x	1	2	3	4
y	–1.51	0.99	3.88	7.66

10. Use the method of least square to fit the curve $y = kx^m$ for the following data :

x	1	2	3	4	5
y	7.1	27.8	62.1	110	161

11. Use the method of least square to fit the straight line for the following data :

x	71	68	73	69	67	65	66	67
y	69	72	70	70	68	67	68	64

12. Fit an exponential curve of the form $y = ab^x$ to the following data : **(Tiruchirapalli–2001)**

x	1	2	3	4	5	6	7	8
y	1	12	1.8	2.5	3.6	4.7	6.6	9.1

13. Fit the curve $y = ae^{bx}$ to the following data : **(Coimbatore–1997)**

x	0	2	4
y	5.1	10	31.1

14. Fit a second degree parabola to the following data : **(UPTU–2009)**

x	1989	1990	1991	1992	1993
y	352	356	357	358	360
x	1994	1995	1996	1997	
y	361	361	360	359	

15. Find the best possible curve of the form $y = a + bx$, using method of least squares for the data : **(VTU–2011)**

x	1	3	4	6	8	9	11	14
y	1	2	4	4	5	7	8	9

16. Fit a second degree parabola to the following data : **(UPTU–2009)**

x	1	2	3	4	5
y	124	129	140	159	228
x	6	7	8	9	10
y	289	315	302	263	210

17. The velocity V of a liquid is known to vary with temperature according to a quadratic law $V = a + bT + cT^2$. Find the best values of a, b and c for the following table : **(UPTU(MCA)–2010)**

T	1	2	3	4	5	6	7
V	2.31	2.01	3.80	1.66	1.55	1.47	1.41

18. Find the least squares fit of the form $y = a_0 + a_1 x^2$ to the following data : **(UPTU–2008)**

x	–1	0	1	2
y	2	5	3	0

19. Predict the mean radiation dose at an altitude of 3000 feet by fitting an exponential curve to the given data : **(SVTU–2007, JNTU–2003)**

Amplitude (x)	50	450	780	1200
Dose of radiation (y)	28	30	32	36
Amplitude (x)	4400	4800	5300	
Dose of radiation (y)	51	58	69	

20. Fit the curve $y = ax + b/x$ to the following data : **(UPTU–2010)**

x	1	2	3	4
y	5.4	6.3	8.2	10.3
x	5	6	7	8
y	12.6	14.9	17.3	19.5

21. Predict y at $x = 3.75$, by fitting a power curve $y = ax^b$ to the given data : **(JNTU–2003)**

x	1	2	3	4	5	6
y	2.98	4.26	5.21	6.10	6.80	7.50

22. Fit the curve of the form $y = ae^{bx}$ to the following data : **(VTU–2011S, JNTU–2006)**

x	77	100	185	239	285
y	2.4	3.4	7.0	11.1	19.6

ANSWERS

3. $y = 1.243 - 0.004x + 0.22x^2$

4. $y = 10.4 - 0.198x + 0.244x^2$

5. $R = 3.48 - 0.002V + 0.003V^2$

6. $R = 6.32 + 0.0095V^2$

7. $a = 0.5012, b = 1.9977$

8. $a = 146.3, k = -0.412$

9. $a = 0.51, b = -2.04$

10. $k = 7.17, m = 1.95$

12. $y = 0.6823(1.384)^x$

13. $a = 4.1, b = 0.43$

14. $y = -1000106.41 + 1034.29x - 0.267x^2$

15. $a = 0.545, b = 0.636$

16. $y = 18.866 + 66.158x - 4.333$

17. $V = 2.593 - 0.326T + 0.023T^2$

18. $y = 4.167 - 1.111x^2$

19. $y = 44.9$ approx.

20. $a = 3, b = 2$

21. $y = 2.978x^{0.5143}; 5.8769$

22. $a = 0.1839, b = 0.0221$

□□□□□□

CHAPTER 22

Correlation and Regression

22.1 INTRODUCTION

Correlation is one of the most widely used statistical techniques. It is very useful in the field of biology, economics, agriculture, psychology, etc.

Here, there exists certain relationship between pairs of variables. These relationships enable us to predict certain thing. For example, an increase in rainfall results in increase in agriculture yield or increase in production of rice results in fall in price.

22.2 MULTIVARIATE AND BIVARIATE DATA

1. **Multivariate Data:** If for each unit of observation, we record the values of more than one variable, then this observation forms a multivariate data.
2. **Bivariate Data :** If for each unit of observation, we record the values of two variables, then the observation forms a bivariate data.

For example:
1. If we observe the age, height, weight and sex of each student of some college, then the observation so recorded forms a multivariate data.
2. If we measure the height and weight of a certain group of persons, we shall get a bivariate data.

22.3 CORRELATION

Whenever two variables are so related that a change in one variable results in direct or inverse change (positive or negative change) in the other and also greater the magnitude of change in the other, then the variables are said to be correlated and the relationship between the variables is known as correlation.

Definition. *When the relationship is of quantitative nature, the appropriate statistical tool for discovering and measuring the relationship and expressing it in brief formula is known as correlation.*

22.4 TYPES OF CORRELATION

22.4.1 POSITIVE AND NEGATIVE CORRELATION

The correlation is said to be positive when the values of the two variables increase or decrease together, *i.e.*, an increase in one variable correspond to an increase in the other and a decrease in one variable correspond to a decrease in the other.

On the other hand, it is said to be negative when an increase in one variable corresponds to a decrease in the other and a decrease in one variable correspond to an increase in the other.

⮧ REMARK
⟿ The positive correlation is also known as direct correlation.

22.4.2 LINEAR AND NON-LINEAR CORRELATION

If the change in one variable is always in a fixed ratio to the change in the other variable, the correlation is said to be linear. On the other hand, if the change in one variable is in a variable ratio to the change in the other variable, then the correlation is said to be non-linear or curvilinear.

For example:

1. If due to 10% increase in the currency results in a permanent increase of 30% in the general price level, then the correlation is linear.
2. If due to 10% increase in the currency results in increase sometimes of 15%, sometimes of 20% and sometimes of 30% in the general price level, then such a correlation is said to be non-linear.

✎ REMARK

➠ Graphically, the linear correlation is represented by a straight line.

22.5 PERFECT CORRELATION

When two variables change in the same direction and in the same ratio, then there is perfect positive correlation. On the other hand, if two variable changes in the same ratio but in the opposite direction, then there is a perfect negative correlation.

✎ REMARKS

➠ In case of perfect positive correlation, the coefficient of correlation is +1.
➠ In case of perfect negative correlation, the coefficient of correlation is –1.
➠ If the change in one variable has no effect on the other variable, then the correlation is completely absent and we say there is no correlation.
➠ In case of no correlation, the coefficient of correlation is always zero.

22.6 METHODS FOR FINDING THE CORRELATION

22.6.1 SCATTER DIAGRAM

It is the simplest way of diagrammatic representation of bivariate data. Thus for the bivariate distribution (x_i, y_i), $i = 1, 2, ..., n$. If the values of the variables X and Y be plotted along the x-axis and y-axis respectively in the XY-plane, then the diagram of dots so obtained is called 'scatter diagram'.

Following are the figures of the scattered data $r > 0, < 0, = 0, r = \pm 1$.

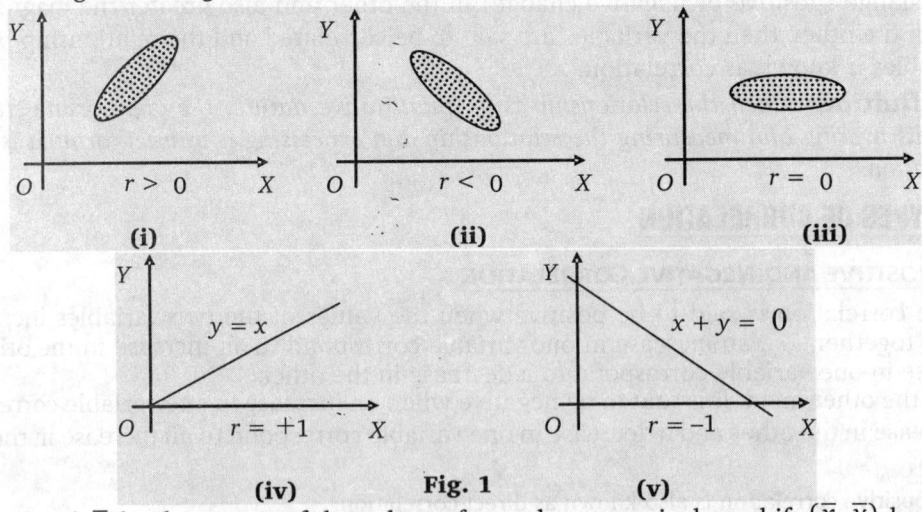

Fig. 1

If $\bar{x}$ and $\bar{y}$ be the means of the values of x and y respectively and if $(\bar{x}, \bar{y})$ be taken as origin of coordinate axis, then the points may be scattered around the origin. Then we observe that for points in first and third quadrants the product $(x - \bar{x})(y - \bar{y})$ is positive and for points in second and fourth quadrants, this is negative.

A measure of correlation between the values of x and y is given by $\sum\limits_{i=1}^{n} (x_i - \bar{x})(y_i - \bar{y})$.

REMARKS
⟹ If there is a positive correlation, then there is a cluster of points in first and third quadrants and if there is negative correlation then there is a cluster of points in second and fourth quadrants.
⟹ Large positive or negative values of x correspond to large positive or negative values of y in first and third quadrants and to large negative or positive values of y in second and fourth quadrants.
⟹ This method is not much scientific and does not give an exact amount of correlation found between the two variables.

22.6.2 KARL PEARSON'S COEFFICIENT OF CORRELATION
The coefficient of correlation r between two variables x and y is given by the relation.

$$r = \frac{\Sigma xy}{\sqrt{\Sigma x^2 . \Sigma y^2}} = \frac{Cov(x, y)}{\sqrt{Var(x). Var(y)}}$$

where x, y are the deviation about their respective means.

REMARK
⟹ This is also called the product moment correlation coefficient.

SOLVED EXAMPLES

EXAMPLE 1.   *Ten students got the following percentage of marks in Mathematics and Statistics:*

Roll No.	1	2	3	4	5	6	7	8	9	10
Marks in maths	78	36	98	25	75	82	90	62	65	39
Marks in Statistics	84	51	91	60	68	62	86	58	53	47

Calculate the coefficient of correlation.

SOLUTION.   Let marks of two subjects be denoted by X and Y respectively.

Then, Mean for X marks $= \dfrac{650}{10} = 65$ and Mean for Y marks $= \dfrac{660}{10} = 66$

If x and y are the deviations of X's and Y's from their respective means, then we have the following table:

X	Y	x	y	x^2	y^2	xy
78	84	13	18	169	324	234
36	51	−29	−15	841	225	435
98	91	33	25	1089	625	825
25	60	−40	−6	1600	36	240
75	68	10	2	100	4	20
82	62	17	−4	289	16	−68
90	86	25	20	625	400	500
62	58	−3	−8	9	64	24
65	53	0	−13	0	169	0
39	47	−26	−19	676	361	454
650	660	0	0	5398	2224	2704

Now, $r = \dfrac{\Sigma xy}{\sqrt{\Sigma x^2 . \Sigma y^2}} = \dfrac{2704}{\sqrt{5398 \times 2224}} = \dfrac{2704}{73.4 \times 47.1} = 0.78$

EXAMPLE 2. Find the Karl Pearson coefficient of correlation for the following data :

X	11	10	9	8	7	6	5
Y	20	18	12	8	10	5	4

SOLUTION. Here, we have

$$\bar{X} = \dfrac{\Sigma X}{n} = \dfrac{56}{7} = 8 \text{ and } \bar{Y} = \dfrac{\Sigma Y}{n} = \dfrac{77}{7} = 11$$

Then, we have the following table

X	$X - \bar{X} = x$	x^2	Y	$Y - \bar{Y} = y$	y^2	xy
11	3	9	20	9	81	27
10	2	4	18	7	49	14
9	1	1	12	1	1	1
8	0	0	8	−3	9	0
7	−1	1	10	−1	1	1
6	−2	4	5	−6	36	12
5	−3	9	4	−7	49	21
56	0	28	77	0	226	76

Then, Karl Pearson's coefficient of correlation is given by

Now, $r = \dfrac{\Sigma xy}{\sqrt{\Sigma x^2 . \Sigma y^2}} = \dfrac{76}{\sqrt{28 \times 226}} = \dfrac{76}{\sqrt{6328}} = \dfrac{76}{79.55} = 0.96$

EXAMPLE 3. Calculate Karl Pearson's coefficient of correlation for the following data relating to overhead expenses and cost of production:

Overhead expenses	8	9	10	11	12	13	14	15	16
Cost of production	15	15	16	19	17	18	16	18	19

SOLUTION. We have $\bar{X} = \dfrac{\Sigma X}{n} = \dfrac{108}{9} = 12$ and $\bar{Y} = \dfrac{\Sigma Y}{n} = \dfrac{153}{9} = 17$

Then, we have the following table :

X	$X - \bar{X} = x$	x^2	Y	$Y - \bar{Y} = y$	y^2	$(X - \bar{X})(Y - \bar{Y}) = xy$
8	− 4	16	15	− 2	4	8
9	−3	9	15	− 2	4	6
10	−2	4	16	− 1	1	2
11	−1	1	19	2	4	−2
12	0	0	17	0	0	0
13	1	1	18	1	1	1
14	2	4	16	−1	1	−2
15	3	9	18	1	1	3
16	4	16	19	2	4	8
108	0	60	153	0	20	24

Putting all these values in $r = \dfrac{\Sigma xy}{\sqrt{\Sigma x^2 \cdot \Sigma y^2}}$, we get

$$r = \dfrac{24}{\sqrt{60 \times 20}} = \dfrac{24}{24.641} = 0.693 .$$

22.6.3 STEP-DEVIATION METHOD

If we take A and B as assumed means for the values of x and y respectively, then set $u_i = x_i - A$ and $v_i = y_i - B$. In this case the coefficient of correlation r is given by

$$r = \dfrac{\Sigma u_i v_i - \dfrac{1}{n}(\Sigma u_i\ \Sigma v_i)}{\sqrt{\left[\left\{\Sigma u_i^2 - \dfrac{1}{n}(\Sigma u_i)^2\right\}\left\{\Sigma v_i^2 - \dfrac{1}{n}(\Sigma v_i)^2\right\}\right]}} .$$

SOLVED EXAMPLES

EXAMPLE 1. *Find the coefficient of correlation for the following data:*

x	10	14	18	22	26	30
y	18	12	24	6	30	36

SOLUTION. Let us define $v = \dfrac{y_i - 22}{6}$ and $v = \dfrac{y_i - 24}{6}$

where 22 and 24 are the assumed means for x and y respectively.

Now, we have the following table

x	y	u	v	uv	u^2	v^2
10	18	−3	−1	3	9	1
14	12	−2	−2	4	4	4
18	24	−1	0	0	1	0
22	6	0	−3	0	0	9
26	30	1	1	1	1	1
30	36	2	2	4	4	4
Total		−3	−3	12	19	19

Now,

$$r = \dfrac{\Sigma u_i v_i - \dfrac{1}{n}(\Sigma u_i\ \Sigma v_i)}{\sqrt{\left[\left\{\Sigma u_i^2 - \dfrac{1}{n}(\Sigma u_i)^2\right\}\left\{\Sigma v_i^2 - \dfrac{1}{n}(\Sigma v_i)^2\right\}\right]}}$$

$$= \dfrac{12 - \dfrac{(-3)(-3)}{6}}{\sqrt{\left[\left\{19 - \dfrac{(-3)^2}{6}\right\}\left\{19 - \dfrac{(-3)^2}{6}\right\}\right]}}$$

$$= \dfrac{12 - 1.5}{19 - 1.5} = \dfrac{10.5}{17.5} = 0.6 .$$

EXAMPLE 2. *Calculate the correlation coefficient from the following data:*

x	9	8	7	6	5	4	3	2	1
y	15	16	14	13	11	12	10	8	9

SOLUTION. Let us define $u = x - 5$ and $v = y - 12$. Then, we have the following table

x	y	u	v	uv	u^2	v^2
9	15	4	3	12	16	9
8	16	3	4	12	9	16
7	14	2	2	4	4	4
6	13	1	1	1	1	1
5	11	0	−1	0	0	1
4	12	−1	0	0	1	0
3	10	−2	−2	4	4	4
2	8	−3	−4	12	9	16
1	9	−4	−3	12	16	9
$n = 9$		0	0	57	60	60

Putting all these values in

$$r = \frac{\Sigma uv - \dfrac{1}{n}(\Sigma u\, \Sigma v)}{\sqrt{\left[\left\{\Sigma u^2 - \dfrac{1}{n}(\Sigma u)^2\right\}\left\{\Sigma v^2 - \dfrac{1}{n}(\Sigma v)^2\right\}\right]}}$$

We get $\qquad r = \dfrac{57 - 0}{\sqrt{[\{60 - 0\}\{60 - 0\}]}} = \dfrac{57}{60} = 0.95 .$

EXAMPLE 3. *Find the correlation coefficient between x and y for the following data:*

x	60	34	40	50	45	41	22	43
y	75	32	34	40	45	33	12	30

SOLUTION. Let us define $u = x - 40$ and $v = y - 40$. Then, we have the following table:

x	y	u	v	u^2	v^2	uv
60	75	20	35	400	1225	700
34	32	−6	−8	36	64	48
40	34	0	−6	0	36	0
50	40	10	0	100	0	0
45	45	5	5	25	25	25
41	33	1	−7	1	49	−7
22	12	−18	−28	324	784	504
43	30	3	−10	9	100	−30
$n = 8$		15	−19	895	2283	1240

Putting all these values in

$$r = \frac{\Sigma uv - \frac{1}{n}(\Sigma u \; \Sigma v)}{\sqrt{\left[\left\{\Sigma u^2 - \frac{1}{n}(\Sigma u)^2\right\}\left\{\Sigma v^2 - \frac{1}{n}(\Sigma v)^2\right\}\right]}}$$

$$= \frac{1240 - \dfrac{15 \times (-19)}{8}}{\sqrt{\left[\left\{895 - \dfrac{(15)^2}{8}\right\}\left\{2283 - \dfrac{(-19)^2}{8}\right\}\right]}}$$

$$= \frac{1240 + 35.625}{\sqrt{(895 - 28.125)(2283 - 45.125)}} = 0.9185.$$

EXAMPLE 4. *It is given that* $r = 0.8$, $\Sigma XY = 60$, $\Sigma X^2 = 90$ *and* $\sigma_y = 2.5$. *Find the number of items. Here, X and Y are deviations from the arithmetic mean.*

SOLUTION. It is given that

$$X = x - \overline{x} \qquad \text{and} \qquad Y = y - \overline{y}$$

$$\Rightarrow \qquad \Sigma XY = \Sigma(x - \overline{x})(y - \overline{y}) = 60 \qquad \text{and} \qquad \Sigma X^2 = \Sigma(x - \overline{x})^2 = 90$$

$$\sigma_x = \sqrt{\frac{\Sigma X^2}{n}} = \sqrt{\frac{90}{n}}$$

and $Cov(X, Y) = \dfrac{1}{n}\Sigma(x - \overline{x})(y - \overline{y}) = \dfrac{1}{n}\Sigma XY = \dfrac{60}{n}$

Now, using the formula

$$r = \frac{Cov\,(X, Y)}{\sigma_x \cdot \sigma_y} = \frac{60}{n\sqrt{\dfrac{90}{n}} \times 2.5}$$

$$\Rightarrow \qquad 0.8 = \frac{60}{\sqrt{90n} \times 2.5}$$

$$\Rightarrow \qquad 90n \times (2.5)^2 \times (0.8)^2 = (60)^2$$

$$\Rightarrow \qquad 360n = 3600$$

$$\Rightarrow \qquad n = 10$$

EXAMPLE 5. *A computer while calculating the correlation coefficient between two variables x and y from 25 pairs of observations obtained the following constants*

$n = 25$, $\Sigma x = 125$, $\Sigma x^2 = 650$, $\Sigma y = 100$, $\Sigma y^2 = 460$, $\Sigma xy = 508$

It was, however, later discovered at the time of checking that it had copied down two pairs as

x	6	8
y	14	6

while the correct values are

x	8	6
y	12	8

Obtain the correct value of correlation coefficient.

SOLUTION.   Let us find corrected values by subtracting incorrect values and adding correct values

Correct $\Sigma x = 125 - 6 - 8 + 8 + 6 = 125$

Correct $\Sigma y = 100 - 14 - 6 + 12 + 8 = 100$

Correct $\Sigma x^2 = 650 - 6^2 - 8^2 + 8^2 + 6^2 = 650$

Correct $\Sigma y^2 = 460 - 14^2 - 6^2 + 12^2 + 8^2 = 436$

Correct $\Sigma xy = 508 - 6 \times 14 - 8 \times 6 + 8 \times 12 + 6 \times 8 = 520$

Now,

$$r = \frac{\Sigma xy - \dfrac{1}{n}\Sigma x . \Sigma y}{\sqrt{\left[\Sigma x^2 - \dfrac{(\Sigma x)^2}{n}\right]\left\{\Sigma y^2 - \dfrac{(\Sigma y)^2}{n}\right\}}}$$

$$= \frac{520 - \dfrac{125 \times 100}{25}}{\sqrt{\left\{650 - \dfrac{(125)^2}{25}\right\}\left\{436 - \dfrac{(100)^2}{25}\right\}}}$$

$$= \frac{20}{\sqrt{(650 - 625)(436 - 400)}} = \frac{20}{\sqrt{25 \times 36}} = \frac{20}{30} = 0.67$$

EXAMPLE 6.   *Show that the Pearson's coefficient of correlation r lies between –1 and 1.*

SOLUTION.   Let $u = \Sigma(y - ax - b)^2$

Using principle of maxima and minima, u will be minimum if

$$\frac{\partial u}{\partial a} = -2\Sigma x(y - ax - b) = 0$$

$$\frac{\partial u}{\partial b} = -2\Sigma(y - ax - b) = 0$$

Since, $\Sigma y = 0 = \Sigma x$, we get $b = 0$

Therefore, $a = \dfrac{\Sigma xy}{\Sigma x^2}$

$$u = \Sigma(y - ax)^2 = \Sigma y^2 - 2a\Sigma xy + a^2 \Sigma x^2$$

$$= \Sigma y^2 - \frac{2\Sigma xy}{\Sigma xy} . \Sigma xy + \left(\frac{\Sigma xy}{\Sigma x^2}\right)^2 \Sigma x^2$$

$$= \Sigma y^2 - \frac{2(\Sigma xy)^2}{\Sigma x^2} + \left(\frac{\Sigma xy}{\Sigma x^2}\right)^2 = \Sigma y^2 - \frac{(\Sigma xy)^2}{\Sigma x^2}$$

$$= \Sigma y^2 \left\{1 - \frac{(\Sigma xy)^2}{\Sigma x^2 . \Sigma y^2}\right\} = \Sigma y^2 (1 - r^2) \qquad \left(\because r = \frac{\Sigma xy}{\sqrt{\Sigma x^2 . \Sigma y^2}}\right)$$

Since u is the sum of squares, so it will not be negative. Similarly, Σy^2 will not be negative. Thus

$$1 - r^2 \geq 0 \qquad \Rightarrow \qquad r^2 \leq 1 \qquad \Rightarrow \qquad -1 \leq r \leq 1$$

EXAMPLE 7. *Find the coefficient of correlation between industrial production and export using the following data :* **[RGPV (B. Pharma) 2003]**

Production (in crore quintal)	55	56	58	59	60	61	62
Exports (in crore quintal)	35	38	38	39	44	43	44

SOLUTION. We have $\bar{X} = \dfrac{\Sigma X}{n} = \dfrac{411}{7} = 58.71 = 59$ app.

$$\bar{Y} = \frac{\Sigma Y}{n} = \frac{281}{7} = 40.14 = 40 \text{ app.}$$

X	$X - \bar{X} = x$	x^2	Y	$Y - \bar{Y} = y$	y^2	$(X - \bar{X})(Y - \bar{Y})$ $= xy$
55	-4	16	35	-5	25	20
56	-3	9	38	-2	4	6
58	-1	1	38	-2	4	2
59	0	0	39	-1	1	0
60	1	1	44	4	16	4
61	2	4	43	3	9	6
62	3	9	44	4	16	12
$\Sigma X = 411$	$\Sigma x = -2$	$\Sigma x^2 = 40$	$\Sigma Y = 281$	$\Sigma y = +1$	$\Sigma y^2 = 75$	$\Sigma xy = 50$

Putting all these value in

$$r = \frac{\Sigma xy}{\sqrt{\Sigma x^2 . \Sigma y^2}}, \text{ we get}$$

$$r = \frac{50}{\sqrt{40 \times 75}} = \frac{50}{\sqrt{3000}} = 0.01839$$

EXAMPLE 8. *Find the coefficient of correlation from the data given below :*

x	65	66	67	67	68	69	70	72
y	67	68	65	68	72	72	69	71

[CCSU (B.Sc. Biotech)-2000]

SOLUTION. We have $\bar{X} = \dfrac{\Sigma x}{n} = \dfrac{544}{8} = 68$ and $\bar{Y} = \dfrac{\Sigma y}{n} = \dfrac{552}{8} = 69$

X	$X - \bar{X} = x$	x^2	Y	$Y - \bar{Y} = y$	y^2	$(X - \bar{X})(Y - \bar{Y})$ $= xy$
65	-3	9	67	-2	4	6
66	-2	4	68	-1	1	2
67	-1	1	65	-4	16	4
67	-1	1	68	-1	1	1
68	0	0	72	3	9	0
69	1	1	72	3	9	3
70	2	4	69	0	0	0
72	4	16	71	2	4	8
$\Sigma X = 544$	$\Sigma x = 0$	$\Sigma x^2 = 36$	$\Sigma Y = 552$	$\Sigma y = 0$	$\Sigma y^2 = 44$	$\Sigma xy = 24$

Putting all these value in

$$r = \frac{\Sigma xy}{\sqrt{\Sigma x^2 \cdot \Sigma y^2}}, \text{ we get}$$

$$r = \frac{24}{\sqrt{36 \times 44}} = \frac{24}{\sqrt{1584}} = \frac{24}{39.799} = 0.60$$

22.7 COEFFICIENT OF CORRELATION FOR GROUPED DISTRIBUTION

For a bivariate frequency distribution, we define

$$r = \frac{\Sigma fuv - \dfrac{\Sigma fu \, \Sigma fv}{\Sigma f}}{\sqrt{\left[\left\{ \Sigma fu^2 - \dfrac{(\Sigma fu)^2}{\Sigma f} \right\} \left\{ \Sigma fv^2 - \dfrac{(\Sigma fv)^2}{\Sigma f} \right\} \right]}}$$

where, f = Frequency, u = Deviation of x from any assumed value A.

v = Deviation of y from any assumed value A.

SOLVED EXAMPLE

EXAMPLE. *Calculate the coefficient of correlation for the following table*

x \ y	0-4	4-8	8-12	12-16
0-5	7			
5-10	6	8		
10-15		5		
15-20		7	3	
20-25			2	9

SOLUTION. Let us choose some convenient origin, say interval 8-12 for x's and 13-15 for y's. Let us find out the deviations u and v (dividing all by common factor). Find out fu, fv, fu^2 and fv^2. Now, we can obtain the second column from the right as

-14 in the row $u = -2$ is obtained $7 \times (-2)$

-20 in the row $u = -1$ is obtained $6 \times (-2) + 8(-1)$

-5 in the row $u = 0$ is obtained $7\,(-1) + 2 \times 0$

9 in the row $u = 2$ is obtained 9×1

In a similar manner, we can find all others. Then we get the following table :

x / y	0-4	4-8	8-12	12-16	f	u	fu	fu^2	fv	$u\Sigma fv$
0-5	④ 7				7	−2	−14	28	−14	28
5-10	6	① 8			14	−1	−14	14	−20	20
10-15		⓪ 5	⓪ 3		8	0	0	0	−5	0
15-20		⓵ 7	⓪ 2		9	1	9	9	−7	−7
20-25				② 9	9	2	18	36	9	18
f	13	20	5	9	47	total	−1	87	−37	59
v	−2	−1	0	1	total					
fv	−26	−20	0	9	−37					
fv^2	52	20	0	9	81					
fu	−20	−1	2	18	−1					
$v\Sigma fu$	40	1	0	18	59					

Here, $u = \dfrac{y - (10-15)}{5}$; $v = \dfrac{x - (8-12)}{4}$

Now put all these values in

$$r = \frac{\Sigma fuv - \dfrac{\Sigma fu \,\Sigma fv}{n}}{\sqrt{\left[\left\{\Sigma fu^2 - \dfrac{(\Sigma fu)^2}{n}\right\}\left\{\Sigma fv^2 - \dfrac{(\Sigma fv)^2}{n}\right\}\right]}}$$

$$= \frac{59 - \dfrac{(-37)(-1)}{47}}{\sqrt{\left\{87 - \dfrac{1}{47}\right\}\left\{81 - \dfrac{37 \times 37}{47}\right\}}} = \frac{2736}{3156.8} = 0.8 \cdot$$

Exercise 22.1

1. Calculate the Karl Pearson's coefficient of correlation from the data given below :

x	4	6	8	10	12
y	2	3	4	6	10

2. Calculate the coefficient of correlation for the following data: **(JNTU-2005)**

x	78	89	97	69	59	79	68	57
y	125	137	156	112	107	138	123	108

3. Find the correlation between x and y from the following data :

y \ x	10-40	40-70	70-100	Total
0-30	5	20	–	25
30-60	–	28	2	30
60-90	–	32	13	45
Total	5	80	15	100

4. Calculate the coefficient of correlation between x and y from the following data :
$n = 10, \Sigma x = 140, \Sigma y = 150, \Sigma(x-10)^2 = 180$
$\Sigma(y-15)^2 = 215$ and $\Sigma(x-10)(y-15) = 60$

5. Find the coefficient of correlation of the following data :

x	3	5	7	12	20	22	24
y	30	25	24	16	11	9	5

6. Find the coefficient of correlation between x and y when $Cov(x, y) = -16.5$, $Var(x) = 2.89$ and $Var(y) = 100$.

7. The following data regarding the heights (y) and weight (x) of 100 college students are given
$\Sigma x = 15000, \Sigma x^2 = 2272500, \Sigma y = 6800$
$\Sigma y^2 = 463025$ $\Sigma xy = 1022250$
Find the correlation coefficient between height and weight.

8. Calculate the coefficient of correlation between the height of father and height of son from the following data :

(CCSU(B.Sc. Biotech)–1997)

Height of father (in inches)	64	65	66	67
Height of sons (in inches)	66	67	65	68
Height of father (in inches)	68	69	70	
Height of sons (in inches)	70	68	72	

9. Find the coefficient of correlation from the data given below : (UPTU–2002)

x	67	69	71	75
y	95	90	87	80
x	85	93	87	73
y	79	75	80	85

10. In an attempt to plot a Dose Response Curve (DRC), a bioassay was performed with the following results :

Dose (mg)	Log Dose	Response
2	0.30	32
4	0.60	58
8	0.90	94
16	1.20	120
32	1.50	150
64	1.80	274
128	2.10	213

Calculate using the above data, the value of correlation coefficient.

11. Find the correlation coefficient between the height of father and son from the following data : [CCSU(B.Sc. Biotech)–1993]

Height of father (in cms.)	165	166	167	167
Height of son (in cms.)	167	168	165	168
Height of father (in cms.)	168	169	170	172
Height of son (in cms.)	172	172	169	171

12. Find the coefficient of correlation between the values of X and Y for the following :

(CCSU (B.Sc. Biotech)–1995)

x	1	3	5	7	8	10
y	6	12	15	17	18	20

13. Twelve students take tests in two subjects A and B with scores : [CCSU (B.Sc. Biotech)–2001]

A	47	51	58	60	63	65
B	44	63	45	71	58	62
A	71	78	79	80	86	92
B	71	80	63	75	57	74

Determine the coefficient of correlation.

1. 0.95	**2.** 0.96		**3.** 0.4571	**4.** 0.915	**5.** – 0.987	**6.** – 0.97 **7.** 0.6
8. 0.81	**9.** – 0.896		**10.** 0.9986		**11.** 0.60	**12.** 0.9689 **13.** 0.58

22.8 RANK CORRELATION

There are many cases, where the distribution of the variable x and y are unknown but a dependency between these can be observed. Suppose n students are examined in mathematics and the top scorer of marks is assigned a number 1, the second a number 2 and so on. Similarly, the same n students are graded according to the marks obtained by them in another subject. Then we find the correlation coefficient between the graded in the two subjects. This correlation coefficient is known as rank correlation.

Let us assume, no individuals are equal in either classification each of the individual takes the values 1, 2, ..., n. Therefore their arithmetic means are equal.

$$\frac{1+2+...+n}{n} = \frac{n+1}{2}$$

Let $x = X - Y = \left(X - \frac{n+1}{2}\right) - \left(Y - \frac{n+1}{2}\right) = x - y$, where x and y are the deviation from

the mean.

$$\Sigma x^2 = \Sigma\left(X - \frac{n+1}{2}\right)^2 = \Sigma x^2 - (n+1)\Sigma X + \Sigma\left(\frac{n+1}{2}\right)^2$$

$$= \frac{n(n+1)(2n+1)}{6} - \frac{(n+1)\,n\,(n+1)}{2} + \frac{n(n+1)^2}{2} = \frac{n(n^2-1)}{12}$$

Clearly, $\quad\quad\quad \Sigma X = \Sigma Y$ and $\Sigma X^2 = \Sigma Y^2$

Now, $\quad\quad\quad \Sigma Y^2 = \frac{n(n^2-1)}{12}$

$\therefore \quad\quad\quad \Sigma d^2 = \Sigma(x-y)^2 = \Sigma x^2 + \Sigma y^2 - 2\Sigma xy$

$$\Sigma xy = \frac{1}{2}(\Sigma x^2 + \Sigma y^2 - \Sigma d^2) = \frac{1}{2}\left(\frac{n(n^2-1)}{6} - \Sigma d^2\right)$$

$$= \frac{1}{12}n(n^2-1) - \frac{1}{2}\Sigma d^2$$

Putting all these values in $r = \dfrac{\Sigma xy}{\sqrt{\Sigma x^2 . \Sigma y^2}}$, we get $\quad r = 1 - \dfrac{6\,\Sigma d^2}{n(n^2-1)}$

REMARKS

➠ This formula is known as formula for Spearmen rank correlation coefficients.

➠ It implies substituting for the given quantities their rank. In each series the item with the largest size is ranked 1, next largest 2 and so on.

➠ If ties occur, then assigned the average of the ranks, they would have received. For example, if two items are tied for 4th rank, each may be ranked $\frac{4+5}{2} = 4.5$.

➠ If however, some values of x_i are equal, then the coefficient of rank correlation is given by

$$r = 1 - \frac{6\left[\Sigma d^2 + \dfrac{1}{12}\Sigma(m^3 - m)\right]}{n\,(n^2-1)}$$, where m is the number of times a particular x_i is reported.

22.8.1 LIMITS OF COEFFICIENT OF RANK

To show that rank correlation coefficient lies between 1 and –1.

Since we know that $r = 1 - \dfrac{6 \, \Sigma d^2}{n(n^2 - 1)}$

Now, we observe that r is maximum if Σd^2 is minimum.

Since Σd^2 is always positive, so it is minimum if each d is zero.

i.e., $\Sigma d^2 = 0$. Hence, the maximum value of r is $+1$.

Also, r is minimum if d is maximum. d will be maximum if the ranks of the n individuals are given in the following manner.

x	1	2	3	...	$n-1$	n
y	n	$n-1$	$n-2$	...	2	1

Case I : *If n is odd.*

If n is odd, then we can take $n = 2r+1$, then different d's are

$$(2r + 1) - 1, \left[(2r + 1 - 1) - 2\right], ... 4, 2, 0, -2, -4, ... -(2r - 2), -2r$$

i.e., $2r, 2r - 2, 2r - 4, ..., 2, 0, -2, -4, ... -(2r - 2), -2r$

$$\therefore \qquad \Sigma d^2 = 2\left\{(2r)^2 + (2r - 2)^2 + ..., 4^2 + 2^2\right\} = 8\left[r^2 + (r - 1)^2 + ... + 1^2\right]$$

$$= \frac{8r(r + 1)(2r + 1)}{6}$$

$$\therefore \qquad \frac{6 \, \Sigma d^2}{n \,(n^2 - 1)} = \frac{6 . 8r(r + 1)(2r + 1)}{6 . (2r + 1) (4r^2 + 4r + 1 - 1)} = 2$$

$$\Rightarrow \qquad r = 1 - \frac{6 \, \Sigma d^2}{n(n^2 - 1)} = 1 - 2 = -1$$

Case II : *If n is even*

In this case the values of d are

$$(2r - 1), (2r - 3), ..., 1, -1, -3, ..., -(2r - 3), -(2r - 1)$$

$$\therefore \qquad \Sigma d^2 = 2\left\{(2r - 1)^2 + (2r - 3)^2 + ... 1^2\right\}$$

$$= \left[(2r)^2 + (2r - 1)^2 + (2r - 2)^2 + ... + 3^2 + 2^2 + 1\right]$$

$$\qquad\qquad - \left[(2r)^2 + (2r - 2)^2 + ... + 4^2 + 2^2\right]$$

$$= 2\left[1^2 + 2^2 + 3^2 + ... + (2r)^2 + 2^2 r^2 + 2^r (r - 1)^2 + ... + 2^2\right]$$

$$= 2\left[\frac{1}{6} 2r(2r + 1)(4r + 1) - \frac{4}{6} r(r + 1)(2r + 1)\right] = \frac{2}{3} r(4r^2 - 1)$$

$$\Rightarrow \qquad 1 - \frac{6 \, \Sigma d^2}{n(n^2 - 1)} = 1 - \frac{4r \,(4r^2 - 1)}{2r \,(4r^2 - 1)} = -1 . \text{ Hence, we get } -1 \le r \le 1 .$$

SOLVED EXAMPLES

EXAMPLE 1. *Find the rank correlation coefficient from the following data*

x	10	12	15	14	19
y	40	41	48	60	50

SOLUTION. Here, the table is given as follows :

x	y	Rank in $x = R_1$	Rank in $y = R_2$	$d = R_1 \sim R_2$	d^2
10	40	5	5	0	0
12	41	4	4	0	0
15	48	2	3	–1	1
14	60	3	1	2	4
19	50	1	2	–1	1
				$\Sigma d = 0$	$\Sigma d^2 = 6$

Here, $n = 5$. Now, $r = 1 - \dfrac{6 \Sigma d^2}{n(n^2 - 1)} = 1 - \dfrac{6.6}{5(25 - 1)} = 1 - \dfrac{36}{120} = 0.7$

EXAMPLE 2. *Two judges ranked 10 beauty contestants as follows*

$(x_i, y_i) : (6, 4), (4, 1), (3, 6), (1, 7), (2, 5), (7, 8), (9, 10), (8, 9), (10, 3), (5, 2)$

Calculate the coefficient of rank correlation.

SOLUTION.

Rank by 1st judge, x_i	Rank by 2nd judge, x_i	$d = x_i - y_i$	d^2
6	4	–2	4
4	1	–3	9
3	6	3	9
1	7	6	36
2	5	3	9
7	8	1	1
9	10	1	1
8	9	1	1
10	3	–7	49
5	2	–3	9
			$\Sigma d^2 = 128$

Here, $n = 10$.

Now, $r = 1 - \dfrac{6 \Sigma d^2}{n(n^2 - 1)} = 1 - \dfrac{128}{10(100 - 1)}$

$= 1 - \dfrac{128}{9900} = 1 - 0.013 = 0.987$.

EXAMPLE 3. *Compute rank correlation from the following data*

Marks in Physics	15	20	27	13	45	60	20	75
Marks in Maths	50	30	55	30	25	10	30	70

SOLUTION. Here, we have the following table :

Marks in Physics	Marks in Maths	Rank in Physics x_i	Rank in Maths, y_i	$d = x_i - y_i$	d_i^2
15	50	7	3	4	16
20	30	5.5	5	0.5	0.25
27	55	4	2	2	4
13	30	8	5	3	9
45	25	3	7	−4	16
60	10	2	8	−6	36
20	30	5.5	5	0.5	0.25
75	70	1	1	0	0

Here, $\Sigma d^2 = 81.50$ and $n = 8$.

Also, the marks in Physics $m = 2$ and in Maths $M = 3$

where m is the number of times a particular mark has been repeated. Then

$$r = 1 - \frac{6\left[\Sigma d^2 + \frac{1}{12}(m^3 - m) + \frac{1}{12}(M^3 - M)\right]}{n(n^2 - 1)}$$

$$= 1 - \frac{6\left[81.5 + \frac{1}{12}(2^3 - 2) + \frac{1}{12}(3^3 - 3)\right]}{8(8^2 - 1)} = 0$$

EXAMPLE 4. *Seven methods of imparting education were ranked by B.Tech students of two universities as follows*

Method of teaching	I	II	III	IV	V	VI	VII
Rank by students of unit A	2	1	5	3	4	7	6
Rank by students of unit B	1	3	2	4	7	5	6

Calculate rank correlation coefficient

SOLUTION. We denote the rank by students of unit A as r_1 and those by unit B by r_2. Then, we have the following table :

Method of teaching	r_1	r_2	$d = r_1 - r_2$	d^2
I	2	1	1	1
II	1	3	−2	4
III	5	2	3	9
IV	3	4	−1	1
V	4	7	−3	9
VI	7	5	2	4
VII	6	6	0	0
$n = 7$				$\Sigma d_i^2 = 28$

Putting all these values in

$$r = 1 - \frac{6 \Sigma d_i^2}{n(n^2 - 1)} = 1 - \frac{6 \times 28}{7(7^2 - 1)} = 1 - \frac{6 \times 28}{48 \times 7}$$

$$= 1 - 0.5 = 0.5.$$

Exercise 22.2

1. Find the coefficient of correlation between the ranks obtained by ten students in mathematics and physics in an examination as given below :

Rank in Maths	1	3	2	5	4	7	6	9	8	10
Rank in Physics	3	5	10	2	1	4	9	7	8	9

2. The coefficient of rank correlation of marks obtained by 10 students in English and Mathematics was found to be 0.5. It was later discovered that the differences in ranks in the two subjects obtained by one of the students was wrongly taken as 3 instead of 7. Find the correct coefficient of rank correlation.

3. Find the rank correlation coefficient of the following data :

x	75	30	60	80	53
y	85	45	54	91	58
x	35	15	40	38	48
y	63	35	43	46	44

4. Find rank correlation coefficient of the following data:

x	85	60	73	40	90
y	93	75	65	50	80

5. Ten competitors in a beauty contest are ranked by three judges in the following order. Use the rank correlation coefficient to determine which pair of judges has the nearest approach to common tests in beauty.

Judge X	1	6	5	10	3	2	4	9	7	8
Judge Y	3	5	8	4	7	10	2	1	6	9
Judge Z	6	4	9	8	1	2	3	10	5	7

6. For two independent variable prove that Karl Pearson's correlation coefficient is zero. Show by an example that the converse is not true.

7. If x and y are two variables, σ is the coefficient of correlation between them, then show that

$$\rho = \frac{\sigma_x^2 + \sigma_y^2 - \sigma_{x-y}^2}{2\sigma_x \sigma_y}$$

where σ_x^2, σ_y^2 and σ_{x-y}^2 are the variance of x, y and $x - y$ respectively.

ANSWERS

1. 0.224   **2.** 0.2676   **3.** 0.685   **4.** 0.8   **5.** X and Z

22.9 REGRESSION

If we measure the heights and weights of a certain number of students, denote the quantity by x and y and plot them on a graph paper referring to two perpendicular axes. For each student, there shall be one point and thus we get scatter diagram.

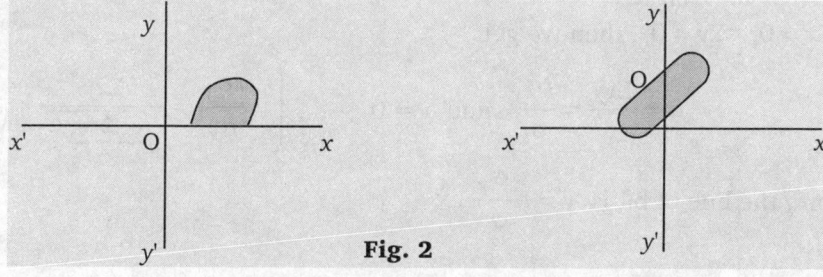

Fig. 2

If the origin of axes is taken as $(\bar{x}, \bar{y})$ where $\bar{x}, \bar{y}$ are the means of the values of x and y respectively, the points may be scattered all around the region. For points lying in I and III quadrants, the product $(x - \bar{x})(y - \bar{y})$ is positive and for those points which are lying in II and IV quadrants, it is negative.

Definition : *Let us suppose that the scatter diagram indicates some relationship between the two variables x and y, the dots of the scatter diagram will be more or less concentrated round a curve. This curve is called the curve regression.*

The straight line about which the various points may be considered as scattered is called the regression line.

⇌ Remark

⇒ It should be noted that one can predict exactly only if the two variables are perfectly related. In that case, there is no scatter in the data and the various points lie exactly on the regression line, but when the correlation is less than perfect, *i.e.*, there is a scatter of points on the scatter diagram, then the regression line is only a representation of the general trend.

22.9.1 Equation of the Line of Regression

Let $y = ax + b$ is the given equation of straight line. The method of least square can be used to fit a straight line to the set of points given on the scatter diagram. Now, transfer the origin to the points $(\bar{x}, \bar{y})$, where $\bar{x}$ and $\bar{y}$ are the means of x-series and y-series, respectively.

Suppose that x, y be the deviation from the respective means $\bar{x}$ and $\bar{y}$.

Therefore, $x = X - \bar{x}$ and $y = Y - \bar{y}$

Let $Y = aX + b$ be the equation of the line of best fit of x. Changing the origin to $(\bar{x}, \bar{y})$, we get the form

$$y = ax + b; \text{ where } y = Y - \bar{y} \text{ and } x = X - \bar{x}.$$

Let $P(x_r, y_r)$ be any dot, then the difference between P and the line is

$$y_r - ax_r - b$$

Let I denote the sum of the squares of such distances given by

$$I = \Sigma(y - ax - b)^2 \text{ for all values of } r.$$

Now, using the principle of least squares, choose a and b such that I is minimum. For minima of I, we must have

$$\frac{\partial I}{\partial a} = -2\Sigma x \,(y - ax - b) = 0 \Rightarrow \Sigma xy - a\Sigma x^2 - b\Sigma x = 0$$

and

$$\frac{\partial I}{\partial b} = -2\Sigma(y - ax - b) = 0 \Rightarrow \Sigma y - a\Sigma x - nb = 0$$

Since $\Sigma x = 0$, $\Sigma y = 0$, then we get

$$a = \frac{\Sigma xy}{\Sigma x^2} = \frac{r\sigma_y}{\sigma_x} \text{ and } b = 0 \qquad \left(\because \frac{r\sigma_y}{\sigma_x} = \frac{\Sigma xy}{\sqrt{\Sigma x^2 \, \Sigma y^2}} \times \sqrt{\frac{\Sigma y^2}{n} \, \frac{n}{\Sigma x^2}} \right)$$

Therefore, the line of bit is $y = r \dfrac{\sigma_y}{\sigma_x} . x$

Now, rechanging the origin, we get $Y - \bar{y} = r\dfrac{\sigma_y}{\sigma_x}(X - \bar{x})$

This is known as regression line of Y on X.
If X is taken to be dependent variable, then the regression line is

$$X - \bar{x} = r\frac{\sigma_x}{\sigma_y}(Y - \bar{y})$$

This is called the regression line of X on Y.

REMARKS

➠ If the straight line is chosen such that the sum of squares of deviation parallel to the axis of y is minimum, it is called the line of regression of y on x.

➠ The coefficients $r\dfrac{\sigma_y}{\sigma_x}$ and $r\dfrac{\sigma_x}{\sigma_y}$ are called the regression coefficients of y on x and of x on y, respectively.

➠ If $r = \pm 1$, the two regression lines will coincide.

22.9.2 LEAST SQUARE REGRESSION

The approach discussed above, is known as least square regression.

Here if $y = a + bx = f(x)$ is the given equation, then using the principle of least square and principle of maxima and minima, we may find the normal equations, given by

$$\Sigma x_i = na + b\Sigma y_i \quad \text{and} \quad \Sigma x_i y_i = a\Sigma x_i + b\Sigma x_i^2$$

Solving the above equations for a and b, we get

$$a = \frac{\Sigma y_i}{n} - b\frac{\Sigma x_i}{n} = \bar{y} - b\bar{x} \quad \text{and} \quad b = \frac{n\Sigma x_i y_i - \Sigma x_i \Sigma y_i}{\Sigma x_i^2 - (\Sigma x_i)^2}$$

where $\bar{x}, \bar{y}$ are the means of x-series and y-series respectively.

22.10 PROPERTIES OF REGRESSION COEFFICIENTS

1. Correlation coefficient is the geometric mean of the regression coefficients.
2. If one of the regression coefficient is greater than unity, the other must be less than unity.
3. Arithmetic mean of regression coefficients is greater than the correlation coefficient.
4. The regression coefficients are independent of the origin but not of scale.
5. The correlation coefficient and the two regression coefficients have same sign.

22.11 ANGLE BETWEEN TWO LINES OF REGRESSION

If θ is the acute angle between the two regression lines in the case of two variables x and y, then

$$\tan\theta = \frac{1 - r^2}{r} \cdot \frac{\sigma_x \sigma_y}{\sigma_x^2 + \sigma_y^2}$$

where, r, σ_x, σ_y have their usual meaning. **[UPTU-2007, VTU-2007]**

Proof. Equations of regression lines are given by

$$y - \bar{y} = r\frac{\sigma_y}{\sigma_x}(x - \bar{x}) \qquad \qquad ... (1)$$

and
$$x - \bar{x} = r\frac{\sigma_x}{\sigma_y}(y - \bar{y})$$
... (2)

Slopes of (1) and (2) are given by $m_1 = r\dfrac{\sigma_y}{\sigma_x}$ and $m_2 = \dfrac{\sigma_y}{r\sigma_x}$

Now
$$\tan\theta = \pm\frac{m_2 - m_1}{1 + m_2 m_1} = \frac{\dfrac{\sigma_y}{r\sigma_x} - \dfrac{r\sigma_y}{\sigma_x}}{1 + \dfrac{\sigma_y^2}{\sigma_x^2}}$$

$$= \pm\frac{1-r^2}{r} \cdot \frac{\sigma_y}{\sigma_x} \cdot \frac{\sigma_x^2}{\sigma_x^2 + \sigma_y^2} = \pm\frac{1-r^2}{r} \cdot \frac{\sigma_x \sigma_y}{\sigma_x^2 + \sigma_y^2}$$

Now, since $r^2 \le 1$ and σ_x, σ_y are positive, therefore, positive sign gives the acute angle between the lines.

Hence,
$$\tan\theta = \frac{1-r^2}{r} \cdot \frac{\sigma_x \sigma_y}{\sigma_x^2 + \sigma_y^2}$$

Also, when $r = 0$, $\theta = \dfrac{\pi}{2}$. Therefore, two lines of regression are perpendicular to each other. Thus, the estimated value of y is the same for all values of x and vice-versa.
When, $r = \pm 1$, $\tan\theta = 0$ or π.

Hence, the lines of regression coincide and there is a perfect correlation between two variates x and y.

SOLVED EXAMPLES

EXAMPLE 1. *Fit a straight line to the following set of data*

x	1	2	3	4	5
y	3	4	5	6	8

SOLUTION. We have

x_i	y_i	x_i^2	$x_i y_i$
1	3	1	3
2	4	4	8
3	5	9	15
4	6	16	24
5	8	25	40
15	26	55	90

Now,
$$b = \frac{n\Sigma x_i y_i - \Sigma x_i \Sigma y_i}{n\Sigma x_i^2 - (\Sigma x_i)^2}$$

$$= \frac{5 \times 90 - 15 \times 26}{5 \times 55 - 15 \times 15} = 1.20$$

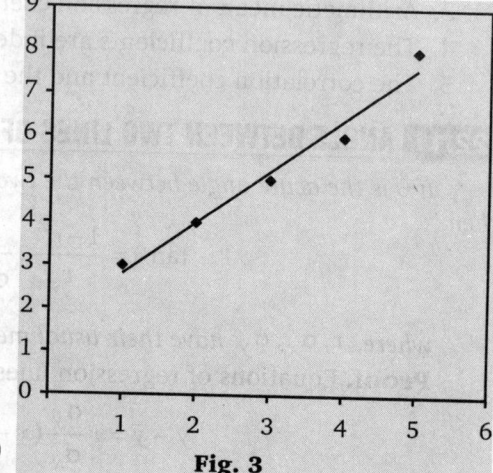

Fig. 3

and $\qquad a = \dfrac{\Sigma y_i}{n} - b\dfrac{\Sigma x_i}{n}$

$$= \dfrac{26}{5} - 1.20 \times \dfrac{5}{5} = 1.60$$

Therefore, the linear equation is

$$y = 1.6 + 1.2x$$

which can be shown in the adjoining figure.

EXAMPLE 2. *If the regression coefficients are 0.8 and 0.2, what would be the value of coefficient of correlation.*

SOLUTION. It is known that

$$r^2 = b_{yx} \cdot b_{xy} = 0.8 \times 0.2 = 0.16$$

Since r, b_{xy} and b_{yx} having same sign as both the regression coefficients b_{yx} and b_{xy}.

Hence, $\qquad r = \sqrt{0.16} = 0.4$

EXAMPLE 3. *Find linear regression from the following data:*

x	1	2	3	4	5	6	7	8
y	3	7	10	12	14	17	20	24

SOLUTION. The regression coefficients are given by

$$b_{yx} = \dfrac{n\Sigma xy - \Sigma x \Sigma y}{n\Sigma x^2 - (\Sigma x)^2} \qquad \text{... (1)}$$

$$b_{xy} = \dfrac{n\Sigma xy - \Sigma x \Sigma y}{n\Sigma y^2 - (\Sigma y)^2} \qquad \text{... (2)}$$

Here, we have the following table.

x	y	x^2	y^2	xy
1	3	1	9	3
2	7	4	49	14
3	10	9	100	30
4	12	16	144	48
5	14	25	196	70
6	17	36	289	102
7	20	49	400	140
8	24	64	576	192
36	107	204	1763	599

Also, $n = 8$

Putting these values from above table in (1) and (2), we get

$$b_{yx} = \dfrac{(8 \times 599) - (86 \times 107)}{(8 \times 204) - (36)^2} = 2.7976$$

$$b_{xy} = \dfrac{(8 \times 599) - (36 \times 107)}{(8 \times 1763) - (107)^2} = 0.3540 \, .$$

EXAMPLE 4. *The regression lines of y on x and x on y are respectively y = ax +b, x = cy+d. Show that*

$$\frac{\sigma_y}{\sigma_x} = \sqrt{\frac{a}{c}}, \quad \bar{x} = \frac{bc+d}{1-ac} \quad \text{and} \quad \bar{y} = \frac{ad+b}{1-ac}.$$

SOLUTION. The regression line of y on x is given by

$$y = ax + b \qquad \qquad \dots (1)$$

$$\therefore \qquad b_{yx} = a$$

Similarly, $b_{xy} = c \qquad \qquad \dots (2)$

Now, $\quad b_{yx} = r \dfrac{\sigma_y}{\sigma_x} \qquad \qquad \dots (3)$

and $\quad b_{xy} = r \dfrac{\sigma_x}{\sigma_y} \qquad \qquad \dots (4)$

Using (3) and (4), we get

$$\frac{b_{yx}}{b_{xy}} = \frac{\sigma_y^2}{\sigma_x^2} \Rightarrow \frac{a}{c} = \frac{\sigma_y^2}{\sigma_x^2}, \text{ i.e., } \frac{\sigma_y}{\sigma_x} = \sqrt{\frac{a}{c}}$$

Since, both the regression lines pass through the point $(\bar{x}, \bar{y})$, therefore

$$\bar{y} = a\bar{x} + b, \quad \bar{x} = c\bar{y} + d$$

$$\Rightarrow \qquad a\bar{x} - \bar{y} = -b$$

$$\qquad \qquad \dots (5)$$

$$\bar{x} - c\bar{y} = d$$

$$\qquad \qquad \dots (6)$$

Multiplying (6) by a and then subtracting from (5), we get

$$(ac - 1)\bar{y} = -ad - b$$

$$\Rightarrow \qquad \bar{y} = \frac{ad+b}{1-ac}$$

Similarly, we get

$$\bar{x} = \frac{bc+d}{1-ac}.$$

EXAMPLE 5. *For 10 observations on price x and supply y, the following data were obtained (in appropriate units) :*

$$\Sigma x = 130, \Sigma y = 220, \Sigma x^2 = 2288, \Sigma y^2 = 5506 \text{ and } \Sigma xy = 3467$$

Find the two lines of regression and estimate the supply when the price is 16 units.

SOLUTION. We have $n = 10$, $\bar{x} = \dfrac{\Sigma x}{n} = 13$, $\bar{y} = \dfrac{\Sigma y}{n} = 22$

Regression coefficient of y on x is

$$b_{yx} = \frac{n\Sigma xy - \Sigma x \Sigma y}{n\,\Sigma x^2 - (\Sigma x)^2} = \frac{(10 \times 3467) - (130 \times 220)}{(10 \times 2288) - (130)^2}$$

$$= \frac{34670 - 28600}{22880 - 16900} = \frac{6070}{5980} = 1.015$$

Therefore, regression line of y on x is

$$y - \bar{y} = b_{yx}(x - \bar{x})$$
$$\Rightarrow \qquad y - 22 = 1.015(x - 13)$$
$$\Rightarrow \qquad y = 1.015x + 8.805$$

Similarly, regression coefficient of x on y is

$$b_{xy} = \frac{n\Sigma xy - \Sigma x\,\Sigma y}{n\Sigma y^2 - (\Sigma y)^2} = \frac{(10 \times 3467) - (130 \times 220)}{(10 \times 5506) - (220)^2}$$

$$= \frac{6070}{6600} = 0.9114$$

$\therefore$ Regression line of x on y is given by

$$x - \bar{x} = b_{xy}(y - \bar{y})$$
$$\Rightarrow \qquad x - 13 = 0.9114\,(y - 22)$$
$$\Rightarrow \qquad x = 0.9114y - 7.0508$$

Now, since we have to estimate supply (y) when price (x) is given. Therefore, we have to use regression line of y on x.

$\therefore$ When $x = 16$

$$y = 1.015(16) + 8.805 = 24.045$$

EXAMPLE 6. *Find the line of regression of y on x for the following data :*

x	1.53	1.78	2.60	2.95	3.42
y	33.50	36.30	40.00	45.80	53.50

SOLUTION. We know that regression line of y on x is given by

$$y - \bar{y} = b_{yx}(x - \bar{x}) \qquad \qquad ...(1)$$

where $\qquad b_{yx} = \dfrac{n\Sigma xy - \Sigma x\,\Sigma y}{n\,\Sigma x^2 - (\Sigma x)^2} \qquad \qquad ...(2)$

Here, we have the following table

x	y	x^2	xy
1.53	33.50	2.3409	51.255
1.78	36.30	2.1684	64.614
2.60	40.00	6.76	104.00
2.95	45.80	8.7025	135.11
3.42	53.50	11.6664	182.97
12.28	209.1	32.6682	537.949

Also, $\qquad n = 5$.

Then, $\qquad b_{yx} = \dfrac{(5 \times 537.949) - (12.28)(209.1)}{(5 \times 32.6682) - (12.28)^2} = 9.726$

Also, Mean, $\qquad \bar{x} = \dfrac{\Sigma x}{n} = \dfrac{12.28}{5} = 2.456$

and $\qquad \bar{y} = \dfrac{\Sigma y}{n} = \dfrac{20.91}{5} = 41.82$

Putting all these values in (1), we get

$$y - 41.82 = 9.726(x - 2.456) = 9.726x - 23.887$$

$$\Rightarrow \qquad y = 17.932 + 9.726x$$

which is the required equation of regression.

EXAMPLE 7. *The equations of two regression lines, obtained in a correlation analysis of 60 observations are :*

$$5x = 6y + 24 \quad and \quad 1000y = 768x - 3608$$

What is the correlation coefficient? Show that the ratio of the coefficients of variability of x to that of y is 5/24. What is the ratio of variances of x and y?

SOLUTION. As per given, the regression line of x on y is

$$5x = 6y + 24$$

$$\Rightarrow \qquad x = \frac{6}{5}y + \frac{24}{5}$$

$$\Rightarrow \qquad b_{xy} = \frac{6}{5} \qquad\qquad\qquad \dots (1)$$

Also, regression line of y on x is

$$1000y = 768x - 3608 \qquad \Rightarrow \qquad y = 0.768x - 3.608$$

$$\Rightarrow \qquad b_{yx} = 0.768 \qquad\qquad\qquad \dots (2)$$

Using (1), we get

$$r\frac{\sigma_x}{\sigma_y} = \frac{6}{5} \qquad\qquad\qquad \dots (3)$$

From (2)

$$r\frac{\sigma_y}{\sigma_x} = 0.678 \qquad\qquad\qquad \dots (4)$$

Multiplying (3) and (4), we get

$$r^2 = 0.9216 \Rightarrow r = 0.96 \qquad\qquad\qquad \dots (5)$$

Now, dividing (4) by (3), we get

$$\frac{\sigma_x^2}{\sigma_y^2} = \frac{6}{5 \times 0.768} = 1.5625 \qquad\qquad\qquad \dots (6)$$

$$\Rightarrow \qquad \frac{\sigma_x}{\sigma_y} = 1.25 = \frac{5}{4}$$

Now, since the regression line passes through the point $(\overline{x}, \overline{y})$, then

$$5\overline{x} = 6\overline{y} + 24$$

and $\qquad 1000\overline{y} = 768\overline{x} - 3608$

On solving, we get $\overline{x} = 6$, $\overline{y} = 1$

Also, coefficient of variability of $x = \dfrac{\sigma_x}{\overline{x}}$

and coefficient of variability of $y = \dfrac{\sigma_y}{\overline{y}}$

Hence, required ratio $= \dfrac{\sigma_x}{\overline{x}} \times \dfrac{\overline{y}}{\sigma_y} = \dfrac{\overline{y}}{\overline{x}}\left(\dfrac{\sigma_x}{\sigma_y}\right) = \dfrac{1}{6} \times \dfrac{5}{4} = \dfrac{5}{24}$.

EXAMPLE 8. *The two regression equations of the variables x and y are x = 19.13 – 0.87 y and y = 11.64 – 0.50 x. find (i) mean of x's , (ii) mean of y's and (iii) the correlation coefficient between x and y.* **(VTU–2004, Anna–2003, Burdwan–2003)**

SOLUTION. Since the mean of x's and the mean of y's line on the two regression lines, we have

$$\bar{x} = 19.13 - 0.87\bar{y} \qquad \text{... (1)}$$
$$\bar{y} = 11.64 - 0.50\bar{x} \qquad \text{... (2)}$$

Multiplying (2) by 0.87 and subtracting from (1), we have,

$$[1 - (0.87)(0.50)]\bar{x} = 19.13 - (11.64)(0.87)$$

or $$0.57\bar{x} = 9.00$$

or $$\bar{x} = 15.79$$

∴ $$\bar{y} = 11.64 - (0.50)(15.79) = 3.74$$

∴ regression coefficient of y on x is -0.50 and that of x on y is -0.87.

Now since the coefficient of correlation is the geometric mean between the two regression coefficients.

∴ $$r = \sqrt{[(-0.50)(-0.87)]} = \sqrt{(0.43)} = -0.66$$

[–ve sign is taken since both the regression coefficients are negative].

EXAMPLE 9. *In a partially destroyed laboratory record, only the lines of regression of y on x and x on y are available as $4x - 5y + 33 = 0$ and $20x - 9y = 107$ respectively. Calculate $\bar{x}, \bar{y}$ and the coefficient of correlation between x and y.*

(SVTU–2009, UPTU–2009, VTU–2005, SRM–2014)

SOLUTION. Since the regression lines pass through $(\bar{x}, \bar{y})$.

∴ $$4\bar{x} - 5\bar{y} + 33 = 0, 20\bar{x} - 9\bar{y} = 107$$

Solving these equation, we get $\bar{x} = 13, \bar{y} = 17$

Rewriting the lines of regression of y on x as

$$y = \frac{4}{5}x + \frac{33}{5}, \text{ we get}$$

$$b_{yx} = r\frac{\sigma_y}{\sigma_x} = \frac{4}{5} \qquad \text{... (1)}$$

Rewriting the lines of regression of x on y as

$$x = \frac{9}{20}y + \frac{107}{9}, \text{ we get}$$

$$b_{xy} = r\frac{\sigma_y}{\sigma_x} = \frac{9}{20} \qquad \text{... (2)}$$

Multiplying (1) and (2), we get

$$r^2 = \frac{4}{5} \times \frac{9}{20} = 0.36$$

∴ $$r = 0.6$$

Hence, $r = 0.6$, the positive sign being taken as b_{yx} and b_{xy} both are positive.

EXAMPLE 10. *While Calculating correlation coefficient between two variables x and y from 25 pairs of observations, the following results were obtained.*

$$n = 25, \Sigma x = 125, \Sigma x^2 = 650, \Sigma y = 100, \Sigma y^2 = 460, \Sigma xy = 508.$$

Later it was discovered at the time of checking that the pairs of values

x	y
8	12
6	8

were copied down as $\dfrac{x\ |\ y}{\begin{array}{c|c} 6 & 14 \\ 8 & 6 \end{array}}$.

Obtain the correct value of correlation coefficient.   **(VTU-2011S, SVTU-2009)**

SOLUTION.   To get the correct results, we subtract the incorrect values and add the corresponding correct values.

The correct results would be

$$\Sigma n = 25, \Sigma x = 125 - 6 - 8 + 8 + 6 = 125,$$

$$\Sigma x^2 = 650 - 6^2 - 8^2 + 8^2 + 6^2 = 650$$

$$\Sigma y = 100 - 14 - 6 + 12 + 8 = 100,$$

$$\Sigma y^2 = 460 - 14^2 - 6^2 + 12^2 + 8^2 = 436$$

$$\Sigma xy = 508 - 6 \times 14 - 8 \times 6 + 8 \times 12 + 6 \times 8 = 520$$

$$r = \frac{n\Sigma xy - (\Sigma x)(\Sigma y)}{\sqrt{[\{n\Sigma x^2 - (\Sigma x)^2\}\{n\Sigma y^2 - (\Sigma y)^2\}]}}$$

$$= \frac{25 \times 520 - 125 \times 100}{\sqrt{[\{25 \times 650 - (125)^2\}\{25 \times 436 - (100)^2\}]}}$$

$$= \frac{20}{\sqrt{(25 \times 36)}} = \frac{2}{3}.$$

EXAMPLE 11.   *The following marks have been obtained by a class of students in statistics (out of 100).*

Paper I	80	45	55	56	58	60	65	68	70	75	85
Paper II	82	56	50	48	60	62	64	65	70	74	90

Compute the coefficient of correlation for the above data.

Find the equations of the lines of regression.   **[RGPV (B. Pharma)–2001]**

SOLUTION.   We have $\bar{X} = 65.2$ and $\bar{Y} = 65.5$ and $n = 11$.

X	X – 65 = x	x^2	Y	Y – 70 = y	y^2	$(X - \bar{X})(Y - \bar{Y}) = xy$
80	15	225	82	12	144	180
45	– 20	400	56	– 14	196	280
55	– 10	100	50	– 20	400	200
56	– 9	81	48	– 22	484	198
58	– 7	49	60	– 10	100	70
60	– 5	25	62	– 8	64	40
65	0	0	64	– 6	36	0
68	3	9	65	– 5	25	– 15
70	5	25	70	0	0	0
75	10	100	74	4	16	40
85	20	400	90	20	400	400
	$\Sigma x = 2$	$\Sigma x^2 = 1414$		$\Sigma y = -49$	$\Sigma y^2 = 1865$	$\Sigma xy = 1393$

Putting all these value in

$$r = \frac{\Sigma xy}{\sqrt{\Sigma x^2 \cdot \Sigma y^2}}, \text{ we get}$$

$$r = \frac{1393}{\sqrt{1414 \times 1865}} = 0.919$$

Now coefficients of regression are

$$b_{YX} = \frac{n\Sigma xy - \Sigma x \Sigma y}{n\Sigma x^2 - (\Sigma x)^2} \qquad \ldots (1)$$

$$b_{XY} = \frac{n\Sigma xy - \Sigma x \Sigma y}{n\Sigma y^2 - (\Sigma y)^2} \qquad \ldots (2)$$

$$b_{YX} = \frac{11 \times 1393 - (2 \times -49)}{11 \times 1414 - (2)^2} = \frac{15323 + 98}{15554 - 4}$$

$$= \frac{15421}{15550} = 0.992$$

Now, $$b_{XY} = \frac{11 \times 1393 - (2 \times -49)}{11 \times 1865 - (-49)^2} = \frac{15323 + 98}{20515 + 2401}$$

$$= \frac{15421}{22916} = 0.67294 = 0.673$$

Equation of regression line for Y is

$$Y - \overline{Y} = b_{YX}(X - \overline{X}) = Y - 65.55 - 0.992 \, (X - 65.2)$$

$\Rightarrow \qquad Y = 0.992X + 1.002$

Equation of regression line for X is

$\Rightarrow \qquad X - \overline{X} = b_{XY}(Y - \overline{Y})$

$\Rightarrow \qquad X - 65.2 = 0.67(Y - 65.5)$

$\Rightarrow \qquad X = 0.67 \, Y + 21.3$

22.12 SIMPLIFIED DETERMINATION OF REGRESSION ANALYSIS

We know that the regression line can be written as :

$$y - \overline{y} = b_{yx}(x - \overline{x}); \text{ for regression of } y \text{ on } x \text{ and} \qquad \ldots (1)$$

$$x - \overline{x} = b_{xy}(y - \overline{y}); \text{ for regression of } x \text{ on } y. \qquad \ldots (2)$$

This form leads to the following simplified equations :

$$b_{yx} = \frac{\dfrac{\Sigma(x - \overline{x})(y - \overline{y})}{n}}{\dfrac{\Sigma(x - \overline{x})^2}{n}}$$

SOLVED EXAMPLES

EXAMPLE 1. *A panel of two judges P and Q graded seven dramatic performance by independently awarding marks as follows :*

Performance	1	2	3	4	5	6	7
Marks by P	46	42	44	40	43	41	45
Marks by Q	40	38	36	35	39	37	41

The eight performance, which judge Q would not attend, was awarded 37 marks by judge P. If judge Q had been present, how many marks would be expected to have been awarded by him to the eighth performance.

SOLUTION. Here, we first calculate the means :

Mean marks by $P(x$ variable$) = \bar{x} = \dfrac{301}{7} = 43$

Mean marks by $Q(y$ variable$) = \bar{y} = \dfrac{266}{7} = 38$

To calculate by x, we have the following table

No.	x	$(x - \bar{x})$	$(x - \bar{x})^2$	y	$(y - \bar{y})$	$(y - \bar{y})(x - \bar{x})$
1	46	+3	9	40	+2	+6
2	42	−1	1	38	0	0
3	44	+1	1	36	−2	−2
4	40	−3	9	35	−3	+9
5	43	0	0	39	+1	0
6	41	−2	4	37	−1	+2
7	45	+2	4	41	+3	+6
	301		28	266		21

Now, we have
$$b_{yx} = \frac{\Sigma(x - \bar{x})(y - \bar{y})}{\Sigma(x - \bar{x})^2} = \frac{21}{28} = 0.75$$

Therefore, the regression line is
$$y - 38 = 0.75(x - 43) \quad \Rightarrow \quad y = 0.75x + 5.75$$
If $x = 37$, then we have
$$y = (0.75) \times (37) + 5.75 = 33.5$$
So, Judge Q was likely to award 33.5 marks.

EXAMPLE 2. *For 10 observations on price x and supply y, the following data was obtained*
$$\Sigma x = 130, \Sigma y = 220, \Sigma x^2 = 228, \Sigma y^2 = 5506, \Sigma xy = 3467$$
Obtain the line of regression of y on x and estimate the supply when the price is 16 units.

SOLUTION. The regression line of y on x is
$$y - \bar{y} = b_{yx}(x - \bar{x}) \qquad \qquad ... (1)$$

Here, $\bar{x} = \dfrac{\Sigma x}{n} = \dfrac{130}{10} = 13$

and $\bar{y} = \dfrac{\Sigma y}{n} = \dfrac{220}{10} = 22$

Now,
$$b_{yx} = \frac{\Sigma xy - \dfrac{(\Sigma x)(\Sigma y)}{n}}{\Sigma x^2 - \dfrac{(\Sigma x)^2}{n}} = \frac{3467 - \dfrac{130 \times 220}{10}}{2288 - \dfrac{(130)^2}{10}} = 1.02$$

Therefore, the regression line is
$$y - \bar{y} = b_{yx}(x - \bar{x})$$
$$\Rightarrow \qquad y - 22 = 1.20(x - 13)$$
$$\Rightarrow \qquad y = 1.20x + 8.74$$
where, $\qquad x = 16, \ y = 1.20 \times 16 + 8.74 = 25.06$

Hence, the supply is expected to be 25 units corresponding to the price 16 units.

22.13 REGRESSION ANALYSIS OF GROUPED DATA

The regression analysis of the grouped data will follow the same procedure as above, except the fact that all items falling with in a specified group are approximated as having a value equal to the mid-point value of the group. Since the grouping may be on both variables x and y, the data is usually organized in a two way matrix. The following formula may be used for the value of b_{yx}.

$$b_{yx} = \left[\frac{\left(\dfrac{\Sigma fxy}{n} - \dfrac{\Sigma fx}{n} \cdot \dfrac{\Sigma fy}{n} \right)}{\left(\dfrac{\Sigma fx^2}{n} - \left(\dfrac{\Sigma fx}{n} \right)^2 \right)} \right]$$

22.14 MULTIPLE LINEAR REGRESSION

If the dependent variable is a function of two or more variables, then we cannot fit the regression line by using our usual methods.

For example, the salary of a salesman may be expressed as : $y = 500 + 5x_1 + 8x_2$

where x_1 and x_2 are the number of units sold of product 1 and 2 respectively.

Now, we shall discuss an approach to fit the experimental data where the variable under consideration is a linear function of two independent variables.

Consider a two-variable linear function
$$y = a_1 + a_2 x + a_3 x \qquad \qquad \text{... (1)}$$

Let I denote the sum of the squares of errors, then I is given by
$$I = \sum_{i=1}^{n} (y_i - a_1 - a_2 x_i - a_3 z_i)^2 \qquad \qquad \text{... (2)}$$

Differentiating (2) with respect to a_1, a_2 and a_3, we get
$$\frac{\partial I}{\partial a_1} = -2\Sigma(y_i - a_1 - a_2 x_i - a_3 z_i) \qquad \qquad \text{... (3)}$$

$$\frac{\partial I}{\partial a_2} = -2\Sigma(y_i - a_1 - a_2 x_i - a_3 z_i) \, x_i \qquad \qquad \text{... (4)}$$

and $\qquad \dfrac{\partial I}{\partial a_3} = -2\Sigma(y_i - a_1 - a_2 x_i - a_3 z_i) \, y_i \qquad \qquad \text{... (5)}$

By the principle of maxima and minima of I, we must have
$$\frac{\partial I}{\partial a_1} = 0, \ \frac{\partial I}{\partial a_2} = 0 \ \text{and} \ \frac{\partial I}{\partial a_3} = 0.$$

Thus, (3), (4) and (5) gives

$$\left.\begin{array}{c} \Sigma y_i = na_1 + a_2\Sigma x_i + a_3\Sigma z_i \\ \Sigma y_i x_i = a_1\Sigma x_i + a_2\Sigma x_i^2 + a_3\Sigma x_i z_i \\ \Sigma y_i z_i = a_i\Sigma z_i + a_2\Sigma x_i z_i + a_3\Sigma z_i^2 \end{array}\right] \qquad \dots (6)$$

The system (6) can also be written in the matrix form as follows :

$$\begin{bmatrix} n & \Sigma x_i & \Sigma z_i \\ \Sigma x_i & \Sigma x_i^2 & \Sigma x_i z_i \\ \Sigma z_i & \Sigma x_i z_i & \Sigma z_i^2 \end{bmatrix} \begin{bmatrix} a_1 \\ a_2 \\ a_3 \end{bmatrix} = \begin{bmatrix} \Sigma y_i \\ \Sigma y_i x_i \\ \Sigma y_i z_i \end{bmatrix}$$

The above equation can be solved using any standard method.

✎ REMARKS

➟ Since, this is a two-dimensional case, therefore, we obtain a regression plan rather than regression line.

➟ This case can be easily extended to the more general case

$$y = a_1 + a_2 x_1 + a_3 x_2 + \dots + a_{n+1} x_m$$

SOLVED EXAMPLES

EXAMPLE 1. *Obtain a regression plane to fit the data, using the following table*

x	1	2	3	4
y	0	1	2	3
z	12	18	24	30

SOLUTION. Using the above table, we may get

x	z	y	x^2	z^2	xz	yx	yz
1	0	12	1	0	0	12	0
2	1	18	4	1	2	36	18
3	2	24	9	4	6	72	48
4	3	30	16	9	12	120	90
10	6	84	30	14	20	240	156

Now putting all these values in the following normal equations :

$$\begin{bmatrix} n & \Sigma x_i & \Sigma z_i \\ \Sigma x_i & \Sigma x_i^2 & \Sigma x_i z_i \\ \Sigma z_i & \Sigma x_i z_i & \Sigma z_i^2 \end{bmatrix} \begin{bmatrix} a_1 \\ a_2 \\ a_3 \end{bmatrix} = \begin{bmatrix} \Sigma y_i \\ \Sigma y_i x_i \\ \Sigma y_i z_i \end{bmatrix}$$

we get

$$4a_1 + 10a_2 + 5a_3 = 84$$

$$10a_1 + 30a_2 + 20a_3 = 240$$

$$6a_1 + 20a_2 + 14a_3 = 156$$

Solving these equations for a_1, a_2 and a_3, we get

$$a_1 = 10, a_2 = 2, \ a_3 = 4$$

Hence, the required regression plane is
$$y = 10 + 2x + 4z .$$

EXAMPLE 2. *In a partially destroyed laboratory records of an analysis of correlation data, the following results only are legible*

Variance of x is 9.

Regression equation: $8x - 10y + 66 = 0$ and $40x - 18y = 214$.

Calculate :

(a) *The mean value of x and y*   (b) *The standard deviation of y*

(c) *The coefficient of correlation between x and y.*

SOLUTION. We know that two regression lines pass through the point $(\bar{x}, \bar{y})$ where $\bar{x}$ and $\bar{y}$ are mean of x and y respectively

(a) To find $\bar{x}$ and $\bar{y}$, solve the equation

$$8\bar{x} - 10\bar{y} + 66 = 0 \qquad \qquad \dots (1)$$

and $40\bar{x} - 18\bar{y} - 214 = 0 \qquad \qquad \dots (2)$

Solving (1) and (2), we get

Hence, the values of x and y are respectively 13 and 17.

(b) and (c) The equations of the two line regression lines can also be put in the following form

$$y = 0.80x + 6.60, \ x = 0.45y + 5.35$$

The regression coefficient of y on x is b_{yx} and is given by

$$b_{yx} = r\frac{\sigma_y}{\sigma_x} = 0.80 \qquad \qquad \dots (3)$$

and the regression coefficient of x on y is b_{xy} and is given by

$$b_{xy} = r\frac{\sigma_x}{\sigma_y} = 0.45 \qquad \qquad \dots (4)$$

From (3) and (4), we get

$$r^2 = 0.80 \times 0.45 = 0.36$$

$\Rightarrow \qquad \qquad r = 0.60$

$\therefore$ The coefficient of correlation between x and y is 0.60.

Also, since variance (σ_x^2) of x is given as 9. Therefore, S.D. = 3.

Putting these values in (4), we get $\sigma_y = 4$.

Exercise 22.3

1. The following table gives the various values of two variables :

x	42	44	58	55	89	98	66
y	56	49	53	58	64	76	58

Determine the regression equations which may be associated with these values.

2. Obtain the line of regression of the following data:

x	1	2	3	4	5	6	7
y	9	8	10	12	11	13	14

3. Find the regression lines using the following data : Mean height = 50.07, Mean age = 9.98, Standard deviation for height = 5.26, Standard deviation of age = 2.59 and $r = 0.898$.

4. Find the mean of the variables x and y and the coefficients of correlation, given the following regression equations $2y - x = 50$, $3y - 2x = 10$.

5. Two lines of regression are given by $x + 2y - 5 = 0$ and $2x + 3y - 8 = 0$ and $\sigma_x^2 = 12$. Calculate the mean value of x and y, variance of y and the coefficient of correlation between x and y.

6. Use multiple linear regression to fit :

x_1	1	2	3	4	5
x_2	4	3	2	1	0
y	18	16	16	12	10

Compute coefficients and the error of estimate.

7. Obtain a regression plane to fit the following data:

x_1	5	4	3	2	1
x_2	3	− 2	− 1	4	0
y	15	− 8	− 1	26	8

8. The mean of bivariate frequency distribution are at (3, 4) and $r = 0.4$. The line of regression of y on x is parallel to the line $y = x$. Find the two lines of regression and estimate value of x when $y = 1$.

9. Find the multiple linear regression of x_1 on x_2 and x_3 using the following data:

x_1	4	6	7	9	23	15
x_2	15	12	8	6	4	3
x_3	30	24	20	14	10	4

10. Find the coefficient of correlation and line of regression of the data. **(RGPV–2001, 2002)**

x	2	4	4	7	5
y	8	8	5	6	2

11. Find the coefficient and the equations of regression lines for the following value of x and y.

x	1	2	3	4	5
y	2	5	3	8	7

12. Both plasma and urine data are required to estimate renal clearance. The renal clearance is the slope of the line. Find the renal clearance. **(UPTU–2001)**

Time Interval	AUC (x)	Amount excreted (y)
0 − 2	4.0	16.0
2 − 4	2.26	9.0
4 − 6	1.40	5.6
6 − 8	0.86	3.4
8 − 12	0.80	3.2
12 − 24	0.43	1.9

ANSWERS

1. $y = 0.372x + 35.27$, $x = 2.2y - 65.9$
2. $Y = 0.929x + 7.284$, $X = 0.929Y - 6.219$
3. $y = 0.422x - 12.15$, $x = 1.825y + 31.86$ 4. Mean of $X = 130$, Mean of $Y = 90$, $r = 0.866$
5. 1, 2, $\sigma_y^2 = 6$, $r = 0.86$ 8. $y = x + 1$, $x = 0.16y + 2.36$, $x = 2.52$
10. Coefficient of correlation $r = 0.420$, $y = - 0.575x + 8.3$ and $x = - 0.3064y + 6.177$
11. $r = 0.8$, $y = 1.290x + 1.13$, $x = 0.496y + 0.52$

Objective Evaluation

MULTIPLE CHOICE QUESTIONS

Choose the most appropriate one.

1. The term regression was introduced by :
   (a) R.A. Fisher
   (b) Sir Francis Gatton
   (c) Karl Pearson
   (d) None of the above

2. If X and Y are two variates, there can be at most :
   (a) one regression line
   (b) two regression line
   (c) three regression line
   (d) an infinite no. of regression line

3. In a regression line Y on X, the variable X is known as :
   (a) independent variable
   (b) estimating equation
   (c) explanatory variable
   (d) all the above

4. Regression equation is also named as :
   (a) prediction equation
   (b) estimating equation
   (c) line of average relationship
   (d) all the above

5. Scatter diagram of the variate values (X, Y) gives the idea about :
   (a) functional relationship
   (b) regression model
   (c) distribution of errors
   (d) none of the above

6. The estimate of β in the regression equation $Y = \alpha + \beta X + c$ by the method of least squares is :
   (a) biased
   (b) unbiased
   (c) consistent
   (d) efficient

7. The formula for the estimate of β in the regression eqn $Y = \alpha + \beta X + \varepsilon$ is :
   (a) $\mathrm{Cor}\,(X, Y)/v(x)$
   (b) $r\sigma_y/\sigma_x$
   (c) $\dfrac{\Sigma(X_i - \bar{X})(Y_i - \bar{Y})}{\Sigma(X_i - \bar{X})^2}$
   (d) all the above

8. In the regression line $Y = \alpha + \beta X$, β is called the :
   (a) slope of the line
   (b) intercept of the line
   (c) neither (a) nor (b)
   (d) both (a) and (b)

9. In the regression line $Y = \beta_0 + \beta_1 X$, β_0 is the :
   (a) slope of the line
   (b) intercept of the line
   (c) both (a) and (b)
   (d) neither (a) nor (b)

10. If β_{YX} and β_{XY} are two regression coefficient, they have :
    (a) same sign
    (b) opposite sign
    (c) either same or opposite sign
    (d) nothing can be said

FILL IN THE BLANKS

1. The functional relationship of a dependent variable with independent variable(s) is called _____.

2. The linear regression equation of Y on X is known as _____.

3. The independent variable in a multiple regression are also termed as _____ .

4. If there are two variables X and Y, there can be at most _____ regression lines.

5. Pearson's formula for correlation coefficient P_{xy} is _____.

6. Correlation ratio is defined as the ratio of _____ sum of squares to the _____ sum of squares.

7. The positive square root of the quantity R^2 is known as _____ correlation.

8. The formula for the coefficient of correlation which does not involve the covariance term is _____ .

9. Correlation ratio is suitable measure of correlation if the relation between X and Y is _____.

10. In correlation ratio, for each value of X, there can be _____ values of Y.

ANSWERS

∽ MULTIPLE CHOICE QUESTIONS

1. (b)   **2.** (b)   **3.** (d)   **4.** (d)   **5.** (a)   **6.** (b)   **7.** (d)   **8.** (a)   **9.** (b)
10. (a)

∽ FILL IN THE BLANKS

1. regression equation   **2.** regression line   **3.** regression   **4.** two

5. $Cor(X,Y)/\sqrt{V(X)V(Y)}$ **6.** among group; total   **7.** multiple

8. $(\sigma_X^2 + \sigma_Y^2 = \sigma_{X-Y}^2)/2\sigma_X\sigma_Y$ **9.** non-linear   **10.** many

□□□□□□

CHAPTER 23

Probability and Distributions

23.1 INTRODUCTION

The word 'probability' is very commonly used in our daily life in different forms. For example, most probably, he will pass with second division in this year or probably there may be very hot today. These sentences convey the sense of probability of happening of an event. Thus, we can define the probability as follows : *"A mathematical measure of uncertainty is known as probability"*.

(1) Sample space : Consider an experiment whose outcome is not predicted with certainty in advance. However, although the outcome of the experiment will not be known in advance, let us suppose the set of all possible outcomes is known. Then, the set of all possible outcomes of an experiment is known as the sample space.

For example :

1. If the experiment consists of two dice, then the sample space consists of 36 points, *i.e.*,
$$S = \{(i, j) : i, j = 1, 2, ..., 6\}$$
where outcome (i, j) is said to occur if i appears on the left most dice and j on the other dice.

2. If the outcome of an experiment consist in the determination of the sex of a newborn baby, then the sample space $S = \{g, b\}$, where the outcome g means the child is girl and b is that it is a boy.

(2) Trial and Events : Consider an experiment which though repeated under same conditions not give unique results, but may result in many one of the several possible outcomes. Then, the experiment is known as trial and these outcomes are known as events or cases.

For example :

1. Tossing of a coin is a trial and getting head or tail is an event.
2. Drawing two cards from a pack of cards is a trial and getting of a specific card (king, joker or queen) are events.

(3) Exhaustive Events : A set of events is said to be exhaustive, if it includes all possible events.

Definition. *The total number of possible outcomes in any trial is known as exhaustive events.*

For example :

1. In tossing a coin, there are two exhaustive cases, either a head or a tail and there is no any other possibility.
2. In throwing of a dice, there are 6 exhaustive cases, since any one of the 6 faces 1, 2, 3, ..., 6 may come uppermost.

(4) Mutually Exclusive Events : The events are said to be mutually exclusive if the happening of any one of them precludes the happening of all others, *i.e.*, if no two or more of them can happen simultaneously in the same trial.

For example :

1. In tossing of a coin, either head comes up or the tail and both cannot happen together. Thus, these are mutually exclusive events.
2. In throwing of a coin, all the six faces are mutually exclusive, since if any one of these faces comes, the possibility of other in the same trial is ruled out.

(5) Equally Likely Events : When there is no reason to except the happening of one event in preference to the other, then events are known as equally likely events.

For example :

1. In tossing of a coin, the coming of the head or the tail is equally likely.
2. In throwing a unbiased dice, all the six faces are equally likely.

(6) Independent Events : Some events are said to be independent if the happening (or non-happening) of an event is not affected by the supplementary knowledge concerning the occurrence of any number of the remaining events. **For example :** In tossing of a coin, the event of getting a head in the first toss is independent of getting a head in the second, third and so on..

(7) Complimentary Events : If A denotes an event, then $\bar{A}$ denote an event which includes all the sample points not included in A.

For example : The complimentary event of an odd number falling up in the throw of a dice is the coming up of an even number.

(8) Compound Event : When two or more events occur in relation with each other, they are known as compound events.

For example :

1. If a dice is rolled two times, the events of coming up of 1 in the first throw and six in the second throw is a compound event.
2. If four cards are drawn in succession from a pack of cards, the coming of the card of the same suit all the four times is a compound event.

✎ REMARKS

➥ Compound event may be independent or dependent.

➥ Every compound event can be expressed as the union of simple events corresponding to the same experiments.

(9) Favourable Events : The outcomes which make necessary the happening of an event in a trial are called favourable events.

For example : If two dice are thrown, the number of favourable events of getting a sum of 5 is four, *i.e.*, (1, 4),   (2, 3), (3, 2) and (4, 1).

23.2 CLASSICAL (OR PRIOR) PROBABILITY

If there are n exhaustive mutually exclusive and equally likely events of which m are favourable to an event A , the probability P of the happening of A is $P(A) = m/n$, *i.e.*,

$$P(A) = \frac{\text{Number of favourable cases}}{\text{Number of exhaustive cases}} = \frac{m}{n}$$

REMARK

➠ If m is the number of favourable cases out of n exhaustive mutually exclusive and equally likely cases, then the number of unfavourable cases is $n - m$.

23.2.1 ODDS IN FAVOUR OF AN EVENT A

As there are $(n-m)$ cases in which A will not happen, therefore the chance of A not happening is q or $P(A')$ so that

$$q = \frac{n-m}{n} = 1 - \frac{m}{n} = 1 - p$$

$$\Rightarrow \qquad P(A') = 1 - P(A)$$

$$\Rightarrow \qquad P(A) + P(A') = 1$$

Clearly, $P(A)$ and $P(A')$ are non-negative and can not exceed unity.

Hence, $\qquad 0 \le P(A) \le 1$ and $0 \le P(A') \le 1$.

The probability $\frac{m}{n}$ of happening of an event is sometimes expressed as 'the odds' are m to $n - m$ in favour of the event or $(n - m)$ to m against the events.

REMARKS

➠ Odds in favour of an event A is defined as the ratio of number of favourable cases to the number of unfavourable cases.

➠ If an event is certain to happen, then its probability is unity, while if it is certain not to happen, its probability is zero. In first case, the events are called certain or sure events while the later case, A is called impossible event.

SOLVED EXAMPLES

EXAMPLE 1. *In a single throw with two dice, what is the chance of throwing a sum of 5?*

SOLUTION. Since there are 6 numbers on each dice, therefore, the total number of possible outcome is $6 \times 6 = 36$. The sum of the numbers appearing on the upper faces of two dice can be in four ways, *i.e.*, (1, 4), (2, 3), (3, 2), (4, 1). Thus, the required probability of throwing 5

$$= \frac{\text{No. of favourable cases}}{\text{No. of exhaustive cases}} = \frac{4}{36} = \frac{1}{9}.$$

EXAMPLE 2. *A dice is thrown, find*

    (i) *P(Prime number)*   (ii) *P(a number ≥ 3)*

   (iii) *P(a number ≤ 4)*   (iv) *P(a number more than 6)*

    (v) *P(a number less than 6).*

SOLUTION. The number of exhaustive cases = 6.

    (i) The no. of favourable cases, $m = 3$ ($\because$ 2, 3, 5 are primes.)

$$\therefore \ P \text{ (prime number)} = \frac{m}{n} = \frac{3}{6} = \frac{1}{2}.$$

    (ii) m, the number of favourable cases = 4($\because$ possible numbers are 3, 4, 5, 6)

$$\therefore \ P \text{ (a number} \ge 3 \text{)} = \frac{m}{n} = \frac{4}{6} = \frac{2}{3}.$$

(iii) m, the number of favourable cases = 4

(∵ possible numbers are 1, 2, 3, 4.)

$$\therefore P \text{ (a number } \leq 4 \text{)} = \frac{m}{n} = \frac{4}{6} = \frac{2}{3}.$$

(iv) m, the number of favourable cases = 0   (∵ There is no number > 6.)

$$\therefore P \text{ (a number } > 6 \text{)} = \frac{m}{n} = \frac{0}{6} = 0.$$

(v) m, the number of favourable cases = 5

(∵ possible numbers are 1, 2, 3, 4, 5.)

$$\therefore P(\text{a number less than 6}) = \frac{m}{n} = \frac{5}{6}.$$

EXAMPLE 3.   *A bag contains 5 black and 3 white balls. Two balls are drawn at random. Find the probability of drawing*

(i) *2 black balls*   (ii) *Two white balls*

SOLUTION .   (i) Here, two black balls can be drawn out of 5 in $^5C_2 = \dfrac{5 \times 4}{1 \times 2} = 10$ ways.

Thus, m, the number of favourable cases = 10.

Total number of balls = 5 + 3 = 8.

Two balls can be drawn out of 8 in

$$^8C_2 = \frac{8 \times 7}{1 \times 2} = 28 \text{ ways.}$$

∴ n, the number of exhaustive cases = 28.

Thus, probability of drawing two black balls $= \dfrac{m}{n} = \dfrac{10}{28} = \dfrac{5}{14}$

(ii) Now, two white balls can be drawn out of 3 in $^3C_2$ ways.

$$i.e., \quad ^3C_2 = \frac{3 \times 2}{1 \times 2} = 2 \text{ ways.}$$

∴ The number of favourable cases $m = 3$

and the number of exhaustive cases = 28.   (same as part (i))

Hence, probability of drawing two white balls $= \dfrac{m}{n} = \dfrac{3}{28}.$

EXAMPLE 4.   *What is the chance that leap year, selected at random will contain 53 Sundays?*

(Madras–2003, RGPV–2001)

SOLUTION .   It is known that the total number of days in a leap year is 366. Thus, it contains 52 complete weeks and two extra days. The probability of combination of these two days is given as

(i) Monday and Tuesday   (ii) Tuesday and Wednesday

(iii) Wednesday and Thursday   (iv) Thursday and Friday

(v) Friday and Saturday   (vi) Saturday and Sunday

(vii) Sunday and Monday

Of these 7 equally likely cases, the last two contain Sunday. Therefore, last two are favourable. Hence, the required probability is given by 2/7 .

EXAMPLE 5. *Find the chance of throwing exactly* 10 *in one throw with 3 dice.*

SOLUTION . Possible chance of throwing exactly 10 with three dice only once are

$$(1, 3, 6), (1, 6, 3), (3, 1, 6), (6, 1, 3), (3, 6, 1), (6, 3, 1), (1, 4, 5), (1, 5, 4), (4, 1, 5), (5, 1, 4),$$
$$(5, 4, 1), (4, 5, 1), (2, 2, 6), (2, 6, 2), (6, 2, 2), (2, 3, 5), (2, 5, 3), (3, 2, 5),$$
$$(5, 3, 2), (3, 5, 2), (2, 4, 4), (4, 2, 4), (4, 4, 2)$$

$$\text{Total favourable} = 27$$

$$\text{Total possible ways} = 6 \times 6 \times 6 = 216$$

Thus, required chances $= \dfrac{27}{216} = \dfrac{9}{72} = \dfrac{1}{8}$

EXAMPLE 6. *What is the chance that a non leap year selected at random will contain* 53 *Sundays?*

SOLUTION . We know that a non-leap year consists of 365 days. Therefore in a non-leap year, there are 52 complete weeks and one extra which can be any one of the seven days of a week. Hence, the probability that single day is Sunday is $\dfrac{1}{7}$.

EXAMPLE 7. *Two dice are thrown simultaneously. Find the probability of getting eight as the sum.*

SOLUTION . The total number of exhaustive cases

$$n = 6 \times 6 = 36 .$$

Out of these, the following are the favourable cases

$$(2, 6), (6, 2), (3, 5), (5, 3), (4, 4)$$

So, the number of favourable cases, $m = 5$.

Hence, the required probability $= \dfrac{m}{n} = \dfrac{5}{36}$.

EXAMPLE 8. *In a single throw of three dice, find the probability of not showing the same number on all the dice.*

SOLUTION . The number of exhaustive cases

$$n = 6 \times 6 \times 6 = 216 .$$

Number of ways of getting same number on all dice $= 6$

and number of ways of not getting same number on all dice $= 216 - 6 = 210$.

Hence, required probability $= \dfrac{m}{n} = \dfrac{210}{216} = \dfrac{35}{36}$.

EXAMPLE 9. *In a throw of two coins, find the probability of getting both heads or both tails.*

SOLUTION. We have the sample space $= \{HH, HT, TH, TT\}$, where H denotes Head and T denotes Tail.

Therefore, the total number of exhaustive cases, $n = 4$.

Out of these, the number of favourable cases $m = 2$, *i.e.*, $[HH, TT]$

Hence, the required probability $= \dfrac{m}{n} = \dfrac{2}{4} = \dfrac{1}{2}$.

EXAMPLE 10. *Four coins are tossed simultaneously. Write the sample space and then complete the following table :*

No. of heads	0	1	2	3	4
Probability	—	—	—	—	—

SOLUTION . The sample space is given by

{HHHH, HHHT, HHTH, HTHH, THHH, HHTT, HTHT, HTTH, THHT, THTH, TTHH, HTTT, THTT, TTHT, TTTH, TTTT}

Now, $P(\text{no head}) = \dfrac{1}{16}$; $P(1 \text{ head}) = \dfrac{4}{16}$;

$P(2 \text{ head}) = \dfrac{6}{16}$; $P(3 \text{ head}) = \dfrac{4}{16}$;

$P(4 \text{ head}) = \dfrac{1}{16}$

Hence, the complete table is as follows

No. of heads	0	1	2	3	4
Probability	1/16	4/16	6/16	4/16	1/16

EXAMPLE 11. *A class consist of 10 boys and 8 girls. Three students are selected at random. Find the probability that the selected group has : (i) all boys, (ii) all girls, (iii) 2 boys and 1 girl.*

SOLUTION .   (i) Required probability P (all boys)

$$= \frac{^{10}C_3}{^{18}C_3} = \frac{\dfrac{10 \times 9 \times 8}{3 \times 2 \times 1}}{\dfrac{18 \times 17 \times 16}{3 \times 2 \times 1}} = \frac{5}{34}.$$

(ii) Required probability P (all girls)

$$= \frac{^{8}C_3}{^{18}C_3} = \frac{\dfrac{8 \times 7 \times 6}{3 \times 2 \times 1}}{\dfrac{18 \times 17 \times 16}{3 \times 2 \times 1}} = \frac{7}{102}$$

(iii) Required probability (2 boys and 1 girl)

$$= \frac{^{10}C_2 \times {}^{8}C_1}{^{18}C_3} = \frac{\dfrac{10 \times 9}{2 \times 1} \times 8}{\dfrac{18 \times 17 \times 16}{3 \times 2 \times 1}} = \frac{15}{34}.$$

EXAMPLE 12. *A five figure number is formed by the digit 0, 1, 2, 3, 4 without repetition. Find the probability that the number formed is divisible by 4.*

SOLUTION . The five digits can be arranged in 5! ways, out of which 4! begins with zero. Thus total number of 5 figure number formed = $5! - 4! = 96$.

Now, those numbers formed will be divisible by 4, which will have two extreme right digits divisible by 4, *i.e.*, number ending in 04, 12, 20, 24, 32, 40.

Now, Numbers ending in 04 = 3! = 6

Numbers ending in 12 = 3! − 2! = 4

Numbers ending in 20 = 3! = 6

Numbers ending in 24 = 3! – 2! = 4

Numbers ending in 32 = 3! – 2! = 4

and Numbers ending in 40 = 3! = 6.

Therefore, total number of favourable cases

$$= 6 + 4 + 6 + 4 + 4 + 6 = 30.$$

Hence, required probability is given by $\dfrac{30}{96} = \dfrac{5}{16}$.

EXAMPLE 13. *Two dice are thrown simultaneously. Find the probability of getting six as a product.*

SOLUTION . Here, total number of exhaustive cases $n = 6 \times 6 = 36$.

Out of these, the following are favourable cases

$(1, 6), (6, 1), (2, 3), (3, 2)$, *i.e.*, 4.

Hence, required probability $= \dfrac{4}{36} = \dfrac{1}{9}$.

EXAMPLE 14. *A bag contains 8 red, 3 white and 9 blue balls. Three balls are drawn at random from the bag. Find the probability that none of the balls is white.*

SOLUTION . Total number of exhaustive cases, $n = {}^{20}C_3$

and number of favourable cases, $m = {}^{17}C_3$ [∵ 17 balls are non-white]

Hence, required probability

$$= \dfrac{m}{n} = \dfrac{{}^{17}C_3}{{}^{20}C_3} = \dfrac{17 \times 16 \times 15}{20 \times 19 \times 18} = \dfrac{34}{57}$$

EXAMPLE 15. *Find the probability of 4 turning for at least once in two tosses of a fair dice.*

SOLUTION . Here, total number of possible outcomes, $n = 36$ and favourable outcomes to the occurrence of the event are

$(4, 1), (4, 2), (4, 3), (4, 4), (4, 5), (4, 6), (1, 4), (2, 4), (3, 4), (5, 4), (6, 4)$

Therefore, $m = 11$.

Hence, required probability is given by $\dfrac{m}{n} = \dfrac{11}{36}$.

EXAMPLE 16. *A certain team wins with probability 0.7, loses with probability 0.2 and tie with probability 0.1. The team plays three games. Find the probability that the team wins at least two of the games, but not loses.*

SOLUTION . Let us denote

W ≡ Wins

L ≡ Loses

T ≡ Ties

Then, $P(W) = 0.7$, $P(L) = 0.2$

and $P(T) = 0.1$.

Hence, the required probability $= P(WWT) + P(WTW) + P(TWW) + P(WWW)$

$= P(W)P(W)P(T) + P(W)P(T)P(W) + P(T)P(W)P(W) + P(W)P(W)P(W)$

$= \dfrac{7}{10} \times \dfrac{7}{10} \times \dfrac{1}{10} + \dfrac{7}{10} \times \dfrac{1}{10} \times \dfrac{7}{10} + \dfrac{1}{10} \times \dfrac{7}{10} \times \dfrac{7}{10} + \dfrac{7}{10} \times \dfrac{7}{10} \times \dfrac{7}{10}$

$= \dfrac{49 + 49 + 49 + 343}{1000} = \dfrac{490}{1000} = \dfrac{49}{100}$

EXAMPLE 17. *A bag contains 50 tickets numbered 1, 2, 3, ..., 50 of which five are drawn at random and arranged in ascending order of magnitude $(x_1 < x_2 < x_3 < x_4 < x_5)$. Find the probability that $x_3 = 30$.*

SOLUTION . We know that five tickets out of 50 can be drawn in $^{50}C_5$ ways. Thus, the number of exhaustive cases, $n = {}^{50}C_5$.

Now, since $x_1 < x_2 < x_3 < x_4 < x_5$

and $\qquad x_3 = 30$

$\Rightarrow \qquad x_1, x_2 < 30$

Therefore, x_1 and x_2 are to be drawn from 1 to 29, which can happen in $^{29}C_2$ ways.

Further, we have

$\qquad x_4, x_5 > 30$

Therefore, x_4 and x_5 are to be drawn from 31 to 50, which can happen in $^{20}C_2$ ways.

Thus, the number of favourable cases,

$$m = {}^{29}C_2 \times {}^{20}C_2$$

Hence, the required probability

$$= \frac{m}{n} = \frac{{}^{29}C_2 \times {}^{20}C_2}{{}^{50}C_5}$$

$$= \frac{\dfrac{29 \times 18}{1 \times 2} \times \dfrac{20 \times 19}{1 \times 2}}{\dfrac{50 \times 49 \times 48 \times 47 \times 46}{1 \times 2 \times 4 \times 5}} = \frac{551}{15134}$$

EXAMPLE 18. *A bag contain 20 tickets numbered 1 to 20. Two tickets are drawn at random. Find the probability that both the numbers on the tickets are prime.*

SOLUTION . We have

The number of exhaustive cases, $n = {}^{20}C_2$

Also, prime numbers from 1 to 20 are 2, 3, 5, 7, 11, 13, 17, 19, Total = 8

Thus, the number of favourable cases, $m = {}^{8}C_2$

Hence, required probability

$$= \frac{m}{n} = \frac{{}^{8}C_2}{{}^{20}C_2} = \frac{8 \times 7}{20 \times 19} = \frac{14}{95}.$$

EXAMPLE 19. *There are three events E_1, E_2 and E_3 one of which must and only one can happen. The odds are 7 to 4 against E_1 and 5 to 3 against E_2. Find odds against E_3.*

SOLUTION . It is given that one and only one of three events E_1, E_2 and E_3 can happen.

Therefore

$$P(E_1) + P(E_2) + P(E_3) = 1$$

Also, odds against E_1 are 7 : 4.

$\Rightarrow$ Odds in favour of E_1 are 4 : 7.

Therefore, $P(E_1) = \dfrac{4}{4+7} = \dfrac{4}{11}$.

Further, odds against E_2 are 5 : 3. Therefore, odds in favour of E_2 are 3 : 5. Thus

$$P(E_2) = \dfrac{3}{3+5} = \dfrac{3}{8}$$

Putting the values of $P(E_1)$ and $P(E_2)$ in

$P(E_1) + P(E_2) + P(E_3) = 1$, we get

$$\dfrac{4}{11} + \dfrac{3}{8} + P(E_3) = 1$$

$$\Rightarrow \qquad P(E_3) = 1 - \dfrac{4}{11} - \dfrac{3}{8} = \dfrac{23}{88}$$

Hence, odds against E_3 are

$$\left(1 - \dfrac{23}{88}\right) : \dfrac{23}{88} \ = \ 65 : 23 \cdot$$

EXAMPLE 20. *Four digit numbers are formed by using the digits 1, 2, 3, 4 and 5 without repeating any digit. Find the probability that a number chosen at random is an odd number.*

SOLUTION . We have
total number of exhaustive cases,

$$n = {}^5P_4 = 5 \times 4 \times 3 \times 2 \times 1 = 120$$

Let m be the favourable number of cases.

$$\begin{array}{cccc} \text{Th} & \text{H} & \text{T} & \text{U} \\ \times & \times & \times & \times \end{array}$$

Unit's place can be filled by any of 1, 3, 5 and 3 ways. Now, we are left with 4 digits. Therefore, remaining three places can be filled up in ${}^4P_3$ ways, *i.e.,* 24 ways.

Thus, total number of favourable cases

$$= 3 \times 24 = 72$$

Hence, the required probability $= \dfrac{m}{n} = \dfrac{72}{120} = \dfrac{3}{5}$.

EXAMPLE 21. *A card is drawn at random from a well-shuffled pack of 52 cards. Find the probability that it is neither an ace nor a king.*

SOLUTION . We know that, from a pack of 52 cards, 1 card can be drawn in ${}^{52}C_1$, *i.e.,* 52 ways.

Therefore, the number of exhaustive cases, $n = 52$.

Further, there are 4 aces and 4 kings, so the number (m) of the ways in which the card is neither an ace nor a king $= 52 - 8 = 44$.

Hence, required probability $= \dfrac{m}{n} = \dfrac{44}{52} = \dfrac{11}{13}$.

EXAMPLE 22. *From a pack of 52 cards 6 cards are drawn at random. Find the probability that 3 are red and 3 are black cards.* **(RGPV–2002)**

SOLUTION. It is given that the total number of cards = 52

Total number of cards are drawn out of 52, $n = {}^{52}C_6$

Number of event of getting 3 red and 3 black cards, $m = {}^{26}C_3 \times {}^{26}C_3$

Hence, required probability

$$= \frac{m}{n} = \frac{{}^{26}C_3 \times {}^{26}C_3}{{}^{52}C_6} = \frac{\dfrac{26 \times 25 \times 24}{3 \times 2 \times 1} \times \dfrac{26 \times 25 \times 24}{3 \times 2 \times 1}}{\dfrac{52 \times 51 \times 50 \times 49 \times 48 \times 47}{6 \times 5 \times 4 \times 3 \times 2 \times 1}}$$

$$= \frac{5200}{4071704} = 0.00127$$

Example 23. *Tickets are numbered from 1 to 100. They are well shuffled and a ticket drawn at random. What is the probability that the drawn ticket has*

  (a) *an even number.*

  (b) *a number 5 or multiple of 5.*

  (c) *a number which is a square.*

  (RGPV–2003)

Solution.   (a) The even number tickets are, $m_1 = 50$, $n_1 = 100$, total number of tickets,

  then probability $= \dfrac{m_1}{n_1} = \dfrac{50}{100} = \dfrac{1}{2}$

  (b) The number of tickets has number 5 and multiple of 5, $m_2 = 20$

  and n = Total number of tickets

  then, probability $= \dfrac{m_2}{n} = \dfrac{20}{100} = \dfrac{1}{5}$

  (c) The event of a number which is a square, $m_3 = 10$

  Then, probability $= \dfrac{m_3}{n} = \dfrac{10}{100} = \dfrac{1}{10}$

Exercise 23.1

1. The letters of the word 'SOCIETY' are placed at random in a row. What is the probability that the three vowels come together.

2. A box contains 5 white and 4 red balls. What is the probability of getting 3 white balls from the box.

3. Two cards are drawn from a well shuffled pack of 52 cards one after the other without replacement. Find the probability that one of these is a queen and other is a king of opposite colour.

4. Two dice are thrown together. What is the probability that the sum of the numbers on the two faces is neither 9 or 11?

5. The letter of the word 'CLIFTON' are placed at random in a row. What is the chance that two vowels come together?

6. Find the probability of getting the sum as a prime number when two dice are thrown together.

7. A box contains 9 balls, two of which are red, three blue and four black. Three balls are drawn from the box at random. What is the probability that

  (i) The three balls are of different colours.

  (ii) Two balls are of the same colour and third is different.

  (iii) The balls are of the same colour.

8. What is the probability that a number selected from 1, 2, 3, ..., 25 is a prime number if each of the 25 numbers is equally likely to be selected?

9. Three unbiased coins are tossed once. Find the probability of getting

  (i) two heads

  (ii) one head or two heads

  (iii) at least two heads

  (iv) at most two heads

10. Find the probability of event given in the following experiments :

  (i) An even number appears in the toss of a fair dice.

  (ii) One or more heads appear in the toss of three fair coins.

(iii) A red ball drawn in a random drawing of one ball from a box containing four white, three red and five blue balls.

1. 1/7 **2.** 5/42 **3.** 2/663 **4.** 5/6 **5.** 2/7 **6.** 5/12 **7.** (i) 2/7 (ii) 55/84 (iii) 5/84
8. 9/25 **9.** (i) 3/8 (ii) 3/4 (iii) 1/2 (iv) 7/8 **10.** (i) 1/2 (ii) 7/8 (iii) 1/4

23.3 THEOREMS ON PROBABILITY

THEOREM 1. *Let S be a sample space and A be an event in a random experiment. Then*
(i) $P(A) \geq 0$ (ii) $P(\phi) = 0$ (iii) $P(S) = 1$

PROOF. Since S be a sample space and A be an event, therefore $A \subseteq S$

Now, (i) $P(A) = \dfrac{n(A)}{n(S)} \geq 0$ (ii) $P(\phi) = \dfrac{n(\phi)}{n(S)} = \dfrac{0}{n(S)} = 0$ (iii) $P(S) = \dfrac{n(S)}{n(S)} = 1$

REMARKS

➠ From above theorem, we conclude that

(i) Probability of occurrence of an event is always non-negative.

(ii) Probability of an impossible event is 0.

(iii) Probability of sure event is 1.

THEOREM 2. *If A and B are mutually exclusive events, then* $P(A \cap B) = 0$.

PROOF. Let A and B be two mutually exclusive events in a sample space S. Now, since A and B are mutually exclusive. Therefore

$$A \cap B = \phi$$

$\Rightarrow$ $P(A \cap B) = P(\phi) \Rightarrow P(A \cap B) = 0$

THEOREM 3. *If A and B are two mutually exclusive events, then* $P(A) + P(B) = 1$.

PROOF. Let A and B be two mutually exclusive events of a sample space S. Since A and B are mutually exclusive, therefore $S = A \cup B$ and $A \cap B = \phi$

Therefore, $P(A) + P(B) = P(A \cup B) = P(S) = 1$

THEOREM 4. **(Addition Law of Probability).** *If A and B are two events. Then*
$$P(A \cup B) = P(A) + P(B) - P(A \cap B) \quad or \quad P(A + B) = P(A) + P(B) - P(AB)$$

PROOF. Let the number of elements in the sample space S be r and the number of elements in the events A and B be m_1 and m_2 respectively, *i.e.*,
$$n(S) = r, \quad n(A) = m_1 \text{ and } n(B) = m_2 \qquad \ldots(1)$$
Now, as A and B are two events (not necessarily mutually exclusive) so they will have some common elements and let us suppose there are k common elements between A and B . Therefore, the number of elements in $A \cap B$ is k.

$\Rightarrow$ Number of elements in $A \cup B = (m_1 + m_2) - k$

i.e., $n(A \cup B) = k$ and $n(A \cup B) = (m_1 + m_2) - k$

Therefore, $P(A \cup B) = \dfrac{n(A \cup B)}{n(S)} = \dfrac{m_1 + m_2 - k}{r}$

$$= \dfrac{m_1}{r} + \dfrac{m_2}{r} - \dfrac{k}{r} = \dfrac{n(A)}{n(S)} + \dfrac{n(B)}{n(S)} - \dfrac{n(A \cap B)}{n(S)}$$

$$= P(A) + P(B) - P(A \cap B)$$

Hence, $P(A \cup B) = P(A) + P(B) - P(A \cap B)$

☜ **REMARKS**

➠ If A and B are mutually exclusive elements, then there is no common element in A and B, i.e., $A \cap B = \phi \Rightarrow P(A \cap B) = 0$. Hence, $P(A + B) = P(A \cup B) = P(A) + P(B)$.

➠ The above result can be generalized as follows : "If $A_1, A_2, ..., A_n$ are n mutually exclusive events, then $P(A_1 + A_2 + ... + A_n) = P(A_1 \cup A_2 \cup ... \cup A_n) = P(A_1) + P(A_2) + ... + P(A_n)$.

THEOREM 5. *For any two events A and B, $P(A - B) = P(A) - P(A \cap B)$.*

PROOF. Let A and B be two compatible events in a sample space S. By set theory, we have

$$(A - B) \cap (A \cap B) = \phi \text{ and } (A - B) \cup (A \cap B) = A$$

Therefore, $\qquad P(A) = P[(A - B) \cup (A \cap B)] = P(A - B) + P(A \cap B)$

Hence, $\qquad P(A - B) = P(A) - P(A \cap B)$

THEOREM 6. *For each event A, $P(\overline{A}) = 1 - P(A)$, where $\overline{A}$ is the complementary event.*

PROOF. Let A be any event in a sample space S. We know that

$$(A - B) \cap (A \cap B) = \phi \text{ and } (A - B) \cup (A \cap B) = A$$

Then, $\qquad A \subseteq S$

Also, $\qquad A \cap \overline{A} = \phi$ and $A \cup \overline{A} = S$

Now, since A and $\overline{A}$ are mutually exclusive, therefore

$$P(A) + P(\overline{A}) = P(A \cup \overline{A}) = P(S) = 1$$

$\Rightarrow \qquad P(A) = 1 - P(\overline{A})$

THEOREM 7. *If A and B be two events such that $A \le B$. Then $P(A) \le P(B)$.*

PROOF. Since $A \subset B$, then

$$B = A \cup (B - A)$$

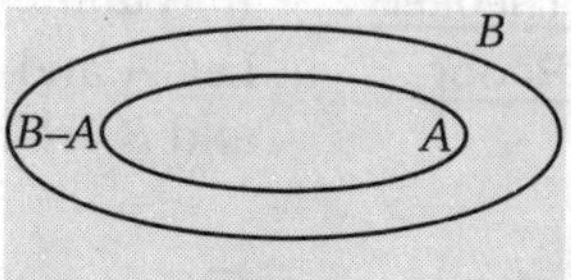

and $\qquad A \cap (B - A) = \phi$

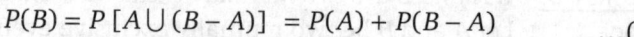

Fig. 1

$\therefore \qquad P(B) = P[A \cup (B - A)] = P(A) + P(B - A)$ ...(1)

Now, since $P(B - A) \ge 0$, so from (1) $P(A) \le P(B)$.

THEOREM 8. *If A is an event associated with a random experiment, then $0 \le P(A) \le 1$.*

PROOF. Let A be any event in a sample space S. Then we have $\phi \le A$ and $A \le S$

$\Rightarrow \qquad P(\phi) \le P(A)$ and $P(A) \le P(S)$ ...(1)

But we know that

$$P(S) = 1 \text{ and } P(\phi) = 0$$ ...(2)

From (1) and (2), we conclude that $0 \le P(A) \le 1$.

☞ **RECAPITULATIONS**

• Let S be a sample space and A be an event in a random experiment. Then
  (i) $P(A) \ge 0$ $\qquad\qquad$ (ii) $P(\phi) = 0$ $\qquad\qquad$ (iii) $P(S) = 1$

• If A and B are mutually exclusive events, then $P(A \cap B) = 0$.

• If A and B are two mutually exclusive events, then $P(A) + P(B) = 1$.

• **(Addition Law of Probability).** If A and B are two events. Then
  $P(A \cup B) = P(A) + P(B) - P(A \cap B)$ or $P(A + B) = P(A) + P(B) - P(AB)$

• For any two events A and B, $P(A - B) = P(A) - P(A \cap B)$.

• For each event A, $P(\overline{A}) = 1 - P(A)$, where $\overline{A}$ is the complementary event.

• If A and B be two events such that $A \le B$. Then $P(A) \le P(B)$.

• If A is an event associated with a random experiment, then $0 \le P(A) \le 1$.

23.4 CONDITIONAL PROBABILITY

If there be two events E_1 and E_2, then the happening of E_1 with the condition that E_2 has already happened is denoted by $E_1|E_2$ and the probability of the event E_1 when the event E_2 has already happened is known as conditional probability. It is denoted by $P(E_1/E_2)$ and read as 'probability of E_1 given that E_2 has occurred.'

In order to calculate $P(E_1/E_2)$, we take elementary events favourable to the occurrence of E_2 as the new sample space and then we find how many of these are favourable to the occurrence of E_1.

Therefore

$$P(E_1 \mid E_2) = \frac{\text{No. of elementary events favourable to } E_2 \text{ which are also favourable to } E_1}{\text{Number of elementary events favourable to } E_2}$$

23.5 DEPENDENT AND INDEPENDENT EVENTS

Let E_1 and E_2 be two events of any sample space S. If there is no effect of the occurrence of any of these events on the probability of the other, then two events are independent of others.

For example : If a bag contains 6 black and 7 white balls and if one ball is drawn at random from the bag and is not put back in the bag and a second ball is drawn, then drawing of the second ball is dependent on that of the first. But if the first ball is put back in the bag before second is drawn then the drawing of the second ball is independent of the drawing of the first.

Definition (1). *Two events are said to be independent if the occurrence of one does not depend upon the occurrence of the other.*

Definition (2). *Two random experiments are said to be independent if for each pair of events E_1 and E_2 associated with the first and second experiment respectively, the probability of simultaneous occurrence of E_1 and E_2 is the product of $P(E_1)$ and $P(E_2)$.*

THEOREM 1.   **(Multiplication Law of Probability).** *The probability of combined happening of two events A and B is the product of the probability of the event B and the conditional probability of A on the assumption that B had already happened.*

i.e., $P(A \cap B) = P(B) . P(A \mid B)$, *where* $B \neq \phi$

or $P(AB) = P(B) . P(A \mid B)$

PROOF.   Let S be the sample space and A, B be two events. Now, $B \subset S$, $B \neq \phi$ and B has happened.

All the elements of S have not happened and only those elements of S have happened which are included in B, thus, the reduced sample set is B.

Here, we consider the combined happening of the elements A and B so all the elements of A cannot happen and only those element of A can happen which are included in B. The set of such element is $A \cap B$. Therefore, the probability of A when B has already happened, means the probability of $A \cap B$ when the sample space is B.

$$\therefore \quad P(A \mid B) = \frac{n(A \cap B)}{n(B)} = \frac{\dfrac{n(A \cap B)}{n(S)}}{\dfrac{n(B)}{n(S)}} \quad \Rightarrow \quad P(A \mid B) = \frac{P(A \cap B)}{P(B)}$$

Hence, $P(A \cap B) = P(A \mid B) . P(B)$.

⤳ REMARKS

▸ If the events A and B are independent, *i.e.*, if the happening of B does not depend upon A. Then $P(B \mid A) = P(B)$ and $P(A \mid B) = P(A)$. Then $P(AB) = P(A) \cdot P(B)$.

▸ If P_1 and P_2 be the probabilities of happening of two independent events then

(i) The probability that the first event happen and the second fails is $p_1(1 - p_2)$.

(ii) The probability that both events fail to happen is $(1 - p_1)(1 - p_2)$.

(iii) The probability that at least one of the events happen is $1 - (1 - p_1)(1 - p_2)$ (known as cumulative probability).

THEOREM 2. If A and B are independent, then $P(A \cup B) = 1 - P(\overline{A}) P(\overline{B})$.

PROOF. Since A and B are independent, then

$$P(A \cap B) = P(A) \cdot P(B)$$

Now, $$P(A \cup B) = P(A) + P(B) - P(A \cap B)$$

$$= P(A) + P(B) - P(A) \cdot P(B)$$

$$= (1 - P(\overline{A})) + (1 - P(\overline{B})) - (1 - P(\overline{A}))(1 - P(\overline{B}))$$

$$= 1 - P(\overline{A}) + 1 - P(\overline{B}) - 1 + P(\overline{A}) + P(\overline{B}) - P(\overline{A}) P(\overline{B})$$

$$= 1 - P(\overline{A}) P(\overline{B})$$

THEOREM 3. If A and B are two independent events associated with a random experiments, the following pairs of events are also independent

(i) $\overline{A}, B$ (ii) $A, \overline{B}$ (iii) $\overline{A}, \overline{B}$

PROOF. (i) We have $P(B) P(\overline{A}) = P(B)(1 - P(A)) = P(B) - P(B) P(A)$

$$= P(B) - P(B \cap A)$$

$(\because A$ and B are independent events$)$

$$= P(B - (B \cap A)) = P(B - A) = P(B \cap \overline{A})$$

Here, $\overline{A}$ and B are independent.

(ii) By interchanging A and B in part (i), we get

$$P(A) P(\overline{B}) = P(A)(1 - P(B)) = P(A) - P(A) P(B)$$

$$= P(A) - P(A \cap B) \qquad (\because A \text{ and } B \text{ are independent})$$

$$= P(A - (A \cap B)) = P(A - B) = P(A \cap \overline{B})$$

Hence, A and $\overline{B}$ are independent events.

(iii) Since A and B are independent, therefore $\overline{A}$ and B are independent (using part (i)).

Now, on using result (ii) for the events $\overline{A}$ and B, we get $\overline{A}$ and $\overline{B}$ are independent events.

THEOREM 4. If the events A and B defined on the sample space S of a random experiment are independent, then

$$P(A \mid B) = P(A) \quad \text{and} \quad P(B \mid A) = P(B)$$

PROOF. We know that if A and B are independent events, then

$$P(A \text{ and } B) = P(A) \cdot P(B)$$

Therefore, $$P(A \mid B) = \frac{P(A \cap B)}{P(B)} = \frac{P(A) \cdot P(B)}{P(B)} = P(A)$$

Similarly, $$P(B \mid A) = \frac{P(A \cap B)}{P(A)} = \frac{P(A) \cdot P(B)}{P(A)} = P(B)$$

THEOREM 5. Let A and B are events in a sample space S, then

(i) $P(A) = P(A\overline{B}) + P(AB)$ (ii) $P(B) = P(\overline{A} B) + P(AB)$

(iii) $P(A + B) = P(AB) + P(A\overline{B}) + P(\overline{A}B)$

(iv) $P(AB) + P(\overline{A}B) + P(A\overline{B}) + P(\overline{A}\overline{B}) = 1$

PROOF.

(i) If there be two events A and B, then A can happen in two ways, *i.e.*, (i) B happens while A happens and (ii) B does not happen while A happens. The probability of happening of B while A happens $= P(A \cap B) = P(AB)$

Also, the probability of not happening of B while A happens

$$= P(A \cap \overline{B}) = P(A\overline{B})$$

Combining these two events, we get

$$P(A) = P(AB) + P(A\overline{B})$$

(ii) Do same as result (i).

(iii) It is known that

$$P(A \cup B) = P(A) + P(B) - P(A \cap B)$$

$$\Rightarrow \qquad P(A + B) = P(A) + P(B) - P(AB)$$

$$= \left[(P(AB) + P(A\overline{B})\right] + [P(\overline{A}B) + P(AB)] - P(AB)$$

[Using (i) and (ii)]

$$= P(AB) + P(A\overline{B}) + P(\overline{A}B)$$

(iv) Consider

$$P(AB) + P(\overline{A}B) + P(A\overline{B}) + P(\overline{A}\overline{B}) = P(A + B) + P(\overline{A}\overline{B}) \qquad \text{[Using (iii)]}$$

$$= P(A + B) + \text{ Probability of non-occurrence of both events } A \text{ and } B$$

$$= P(A + B) + \{1 - P(A + B)\} = 1$$

THEOREM 6. *If A and B are two events in a sample space S , then* $P(A \mid B) + P(\overline{A} \mid B) = 1$

PROOF. It is known that $P(AB) = P(A) \cdot P(B \mid A)$

$$\Rightarrow \qquad P(B \mid A) = \frac{P(AB)}{P(A)}$$

Therefore, $P(A \mid B) + P(\overline{A} \mid B) = \dfrac{P(AB)}{P(B)} + \dfrac{P(\overline{A}B)}{P(B)} = \dfrac{P(AB) + P(\overline{A}B)}{P(B)} = \dfrac{P(B)}{P(B)} = 1$

23.6 PROBABILITY OF HAPPENING OF AT LEAST ONE OF INDEPENDENT EVENTS

Let $A_1, A_2, ...A_r$ be r independent events. Also, let $P_1, P_1, ..., P_r$ be their respective probabilities of happening. Therefore, we have

$$P(A_1) = p_1, \qquad P(A_2) = p_2, \qquad ... P(A_r) = p_r$$

$$\Rightarrow \qquad P(\overline{A}_1) = 1 - p_1, \quad P(\overline{A}_2) = 1 - p_2, \quad ..., \quad P(\overline{A}_r) = 1 - p_r$$

Therefore, the probability when no event happens $= (1 - p_1)(1 - p_2)...(1 - p_r)$.

Hence, the probability when at least one event happens.

$$= 1 - (1 - p_1)(1 - p_2)...(1 - p_r)$$

$$= 1 - P(\overline{A}_1) P(\overline{A}_2) ... P(\overline{A}_r)$$

23.7 LAW OF TOTAL PROBABILITY AND BAYES' THEOREM

Let S be the sample space B_1, B_2 be two mutually exclusive events and A be the event that occur with B_1 or B_2.

Now, $S = B_1 \cup B_2$ and $S \cap A = (B_1 \cup B_2) \cap A = (B_1 \cap A) \cup (B_2 \cap A)$

$B_1 \cap A$ and $B_2 \cap A$ are mutually exclusive.

Hence,
$$P(A) = P(B \cap A) = P[(B_1 \cap E) \cup (B_2 \cap E)]$$
$$= P(B_1 \cap E) + P(B_2 \cap E)$$
$$= P(B_1) \cdot P(A \mid B_1) + P(B_2) \, P(A \mid B_2)$$

which is the law of total probability.

23.7.1 BAYES' THEOREM

If $B_1, B_2, ..., B_n$ are mutually disjoint events with $P(B_i) \neq 0$, then for an arbitrary event A, which is a subset of $\bigcup\limits_{i=1}^{n} B_i$ such that $P(B) > 0$. We have

$$P(B_i \mid A) = \frac{P(B_i) \, P(A \mid B_i)}{\sum\limits_{i=1}^{n} P(B_i) \, P(A \mid B_i)}$$

Proof. As per given, A is a subset of $\bigcup\limits_{i=1}^{n} B_i$, i.e., $A \subset \bigcup\limits_{i=1}^{n} B_i$.

Now,
$$A = A \cap \left(\bigcup_{i=1}^{n} B_i \right) = \bigcup_{i=1}^{n} (A \cap B) \qquad \text{(By distributively)}$$

Since $(A \cap B_i) \subset B_i$ $(i = 1, 2, ..., n)$ are mutually exclusive events, then by addition theorem on probability, we have

$$P(A) = P\left(\bigcup_{i=1}^{n} (A \cap B_i) \right) = \sum_{i=1}^{n} P(A \cap B_i)$$

$$= \sum_{i=1}^{n} P(B_i) \, P(A \mid B_i) \qquad \text{(By multiplication law of probability)}$$

Also, we have $\quad P(A \cap B_i) = P(A) \, P(B_i \mid A)$

$$\Rightarrow \qquad P(B_i \mid A) = \frac{P(A \cap B_i)}{P(A)} = \frac{P(B_i) \, P(A \mid B_i)}{\sum\limits_{i=1}^{n} P(B_i) \, P(A \mid B_i)}$$

☝ REMARKS

➟ Bayes' theorem is also known as 'Theorem of inverse probability'.

➟ Bayes' theorem enables us to connect the conditional probability of B_i given A, with the conditional probability of A, given B_i and the probability of B_i themselves.

➟ The probabilities $P(B_i)$ are called 'Priori probabilities' because they exist before getting any information from the experiment itself.

➟ The probabilities $P(A \mid B_i)$, $i = 1, 2, ..., n$ are called 'likelihoods' because they indicate how likely the events A under consideration is to occur given each and every priori probability.

➟ The probabilities $P(B_i \mid A)$, $i = 1, 2, ..., n$ are called 'posteriori probabilities' because they are determined after the results of the experiment are known.

☞ RECAPITULATIONS

- **(Multiplication Law of Probability).** The probability of combined happening of two events A and B is the product of the probability of the event B and the conditional probability of A on the assumption that B had already happened.

  i.e., $P(A \cap B) = P(B).P(A|B)$, where $B \neq \phi$ or $P(AB) = P(B).P(A|B)$

- If A and B are independent, then $P(A \cup B) = 1 - P(\overline{A}) P(\overline{B})$.

- If A and B are two independent events associated with a random experiments, the following pairs of events are also independent

  (i) $\overline{A}, B$ (ii) $A, \overline{B}$ (iii) $\overline{A}, \overline{B}$

- If the events A and B defined on the sample space S of a random experiment are independent, then $P(A|B) = P(A)$ and $P(B|A) = P(B)$.

- Let A and B are events in a sample space S, then

  (i) $P(A) = P(A\overline{B}) + P(AB)$ (ii) $P(B) = P(\overline{A}B) + P(AB)$

  (iii) $P(A + B) = P(AB) + P(A\overline{B}) + P(\overline{A}B)$ (iv) $P(AB) + P(\overline{A}B) + P(A\overline{B}) + P(\overline{A}\overline{B}) = 1$

- If A and B are two events in a sample space S, then $P(A|B) + P(\overline{A}|B) = 1$

- If $B_1, B_2, ..., B_n$ are mutually disjoint events with $P(B_i) \neq 0$, then for an arbitrary event A, which is a subset of $\bigcup\limits_{i=1}^{n} B_i$ such that $P(B) > 0$. We have $P(B_i | A) = \dfrac{P(B_i)P(A|B_i)}{\sum\limits_{i=1}^{n} P(B_i)P(A|B_i)}$

SOLVED EXAMPLES

EXAMPLE 1. *In a bag there are 6 balls of which 3 are white and 3 are black. They are drawn successively : (i) without replacement, (ii) with replacement. What is the chance that the colours alternate? It has been supposed that the number of the balls drawn remains the same?*

SOLUTION. (i) If balls are chosen without replacement, there will be two situations, first ball is white or first ball is black.

If first ball is white, 2nd black, 3rd white and so on, so the required probability

$$= \frac{3}{6} \times \frac{3}{5} \times \frac{2}{4} \times \frac{2}{3} \times \frac{1}{2} \times \frac{1}{1} = \frac{3}{60} = \frac{1}{20}$$

If first ball is black, 2nd white and so on

$$= \frac{3}{6} \times \frac{3}{5} \times \frac{2}{4} \times \frac{2}{3} \times \frac{1}{2} \times \frac{1}{1} = \frac{1}{20}$$

Both events are mutually exclusive. Only one can occur, so required probability

$$= \frac{1}{20} + \frac{1}{20} = \frac{2}{20} = \frac{1}{10}.$$

(ii) With replacement first white, 2nd black and so on

$$\text{Probability} = \frac{3}{6} \times \frac{3}{6} \times \frac{3}{6} \times \frac{3}{6} \times \frac{3}{6} \times \frac{3}{6} = \frac{1}{64}$$

If first is black, 2nd white and so on

$$\text{Probability} = \frac{1}{64}$$

$$\text{Required probability} = \frac{1}{64} + \frac{1}{64} = \frac{1}{32}.$$

EXAMPLE 2. *Four persons are chosen at random from a group containing 3 men, 2 women and 4 children. Obtain the chance that at the most 2 of them are children.*

SOLUTION. Four persons can be chosen in the following ways :

(i) 1 child, 2 men, 1 woman

$$P(E) = \frac{^4C_1 \times {}^3C_2 \times {}^2C_1}{^9C_4}$$

$$= \frac{4 \times 3 \times 2}{9 \times 8 \times 7 \times 6} \times 4 \times 3 \times 2 \times 1 = \frac{8}{42} = \frac{4}{21}$$

(ii) 1 child, 1 man, 2 women

$$P(E) = \frac{^4C_1 \times {}^3C_1 \times {}^2C_2}{^9C_4}$$

$$= \frac{4 \times 3 \times 1}{9 \times 8 \times 7 \times 6} \times 4 \times 3 \times 2 \times 1 = \frac{4}{42} = \frac{2}{21}$$

(iii) 2 children, 1 man, 1 woman

$$P(E) = \frac{^4C_2 \times {}^3C_1 \times {}^2C_1}{^9C_4}$$

$$= \frac{4 \times 3}{2 \times 1} \times \frac{3 \times 2}{9 \times 8 \times 7 \times 6} \times 4 \times 3 \times 2 \times 1 = \frac{2}{7}$$

(iv) 0 child, 3 men, 1 woman

$$P(E) = \frac{^4C_0 \times {}^3C_3 \times {}^2C_1}{^9C_4}$$

$$= \frac{1 \times 1 \times 2}{9 \times 8 \times 7 \times 6} \times 4 \times 3 \times 2 \times 1 = \frac{1}{63}$$

(v) 0 child, 2 men, 2 women

$$P(E) = \frac{^4C_0 \times {}^3C_2 \times {}^2C_2}{^9C_4} = \frac{1 \times 3 \times 1 \times 4 \times 3 \times 2 \times 1}{9 \times 8 \times 7 \times 6} = \frac{1}{42}$$

EXAMPLE 3. *Three groups of children contain 3 girls and 1 boy, 2 girls and 2 boys, 1 girl and 3 boys. One child is selected at random from each group. Show that chance that the three selected will consist of 1 girl and 2 boys is $\frac{13}{32}$.*

SOLUTION. One girl and 2 boys may be selected in the following cases :

(i) Girl from first group, boy from second, boy from third group

$$\text{Probability} = \frac{3}{4} \times \frac{2}{4} \times \frac{3}{4} = \frac{9}{32}$$

(ii) Boy from first group, girl from second and boy from third group

$$\text{Probability} = \frac{1}{4} \times \frac{2}{4} \times \frac{3}{4} = \frac{3}{32}$$

(iii) Boy from first, boy from second, girl from third

$$\text{Probability} = \frac{1}{4} \times \frac{2}{4} \times \frac{1}{4} = \frac{1}{32}$$

All these events are mutually exclusive, so required probability

$$= \frac{9}{32} + \frac{3}{32} + \frac{1}{32} = \frac{13}{32}.$$

EXAMPLE 4. *A bag contains 8 white and 6 red balls. Find the probability of drawing two balls of the same colour.*

SOLUTION. Since there are total 14 balls, therefore two balls out of 14 balls can be drawn in $^{14}C_2$ ways.

∴ Total number of outcomes = $^{14}C_2$.

Also, two white balls out of 8 can be drawn in $^8C_2$ ways. Therefore probability of drawing 2 white balls

$$= \frac{^8C_2}{^{14}C_2} = \frac{28}{91}.$$

Similarly, two red balls out of 6 can be drawn in $^6C_2$ ways. Therefore, probability of drawing 2 red balls

$$= \frac{^6C_2}{^{14}C_2} = \frac{15}{91}.$$

Hence, the probability of drawing 2 balls of the same colour (either both white of both red)

$$= \frac{28}{91} + \frac{15}{91} = \frac{43}{91}.$$

EXAMPLE 5. *Ram appears for an interview for two posts A and B for which selection is independent. The probability of his selection for post A is 1/6 and for post B is 1/7. Find the probability that Ram is selected for at least one post.*

SOLUTION. We have

$$P(A) = \frac{1}{6}, \qquad P(B) = \frac{1}{7}$$

Therefore, $\quad P(\bar{A}) = 1 - \frac{1}{6} = \frac{5}{6}$

and $\quad P(\bar{B}) = 1 - P(B) = 1 - \frac{1}{7} = \frac{6}{7}$

Hence, the probability that Ram is selected for at least one of the posts

$$= 1 - P(\bar{A})\, P(\bar{B}) = 1 - \frac{5}{6} \times \frac{6}{7} = 1 - \frac{5}{7} = \frac{2}{7}.$$

EXAMPLE 6. *Two dice are thrown. What is the probability that sum of the numbers appearing on the dice is 7 or 8 ?*

(Kurukshetra–2002, SVTU–2004)

SOLUTION. It is known that, the numbers which can appear on each dice are 1, 2, 3, ..., 6 which are 6 in numbers, which lie in equally likely manner.

Total number of outcomes = $6 \times 6 = 36 = n(S)$, where S is the sample space.

$$S = \{(1, 1), (1, 2), ..., (5, 6), (6, 1), ..., (6, 6)\}$$

where (6, 1) means 6 appearing on the first dice and 1 appearing on second dice.

Let B_1 be the event that sum of the numbers on the dice is 7, then
$$B_1 = \{(1, 6), (2, 5), (3, 4), (4, 3), (5, 2), (6, 1)\}$$
$$\Rightarrow \quad n(B_1) = 6$$
$$\therefore \quad P(B_1) = \frac{n(B_1)}{n(S)} = \frac{6}{36} = \frac{1}{6}$$

Further, if B_2 be the event that the sum of numbers on the dice is 8, then
$$B_2 = \{(2, 6), (3, 5), (4, 4), (5, 3), (6, 2)\}$$
$$\Rightarrow \quad n(B_2) = 5$$
$$\therefore \quad P(B_2) = \frac{n(B_2)}{n(S)} = \frac{5}{36}$$

Also, since the events B_1 and B_2 are mutually exclusive, so the probability that the events B_1 and B_2 may happen
$$= P(B_1 \cup B_2) = P(B_1) + P(B_2) = \frac{6}{36} + \frac{5}{36} = \frac{11}{36}$$

EXAMPLE 7. *Two dice are rolled once. Find the probability that the numbers on the two dice are different. What is the probability that the total is at least 4 ?*

SOLUTION. (i) Let A be the event that numbers on the two dice are different. Then ' not A' is the event where two dice have same numbers. Therefore
$$P(A) = \frac{6}{36} = \frac{1}{6} \quad \{(1, 1), (2, 2), ..., (6, 6)\}$$
$$\therefore \quad P(B) = 1 - P(A') = 1 - \frac{1}{6} = \frac{5}{6}.$$

(ii) Let B be the event when the total is at least 4.
$$B = \{(1,1),(1,2),(2,1),(3,1),(1,3),(2,2)\}$$
$$P(B) = \frac{6}{36} = \frac{1}{6}.$$

EXAMPLE 8. *Two dice are thrown. What is the probability that sum of the numbers appearing on the dice is 11, if 5 appears on the first dice ?*

SOLUTION. Let B_1 be the event that the sum of the numbers on the dice is 11 and let B_2 be the event of 5 appearing on the first dice. Then,
$B_2 = \{(5, 1), (5, 2), (5, 3), (5, 4), (5, 5), (5, 6)\}$ Now, the event B_1 happens after B_2 has happened, so that the sample space for the event B_1 is B_2 and there is only one element in B_2 , i.e., (5, 6) the sum whose number is 11.
$$\therefore \quad B_1 \cap B_2 = \{(5,6)\}$$
$$\therefore \quad n(B_2) = 6 \text{ and } n(B_1 \cap B_2) = 1.$$

Then, by multiplication law of probability, we have
$$P(B_1 \mid B_2) = \frac{n(B_1 \cap B_2)}{n(B_2)} = \frac{1}{6}.$$

EXAMPLE 9. *4 coins are tossed. Find the probability of getting at least one head.*

SOLUTION. The probability of getting a head with one coin $= \dfrac{1}{2}$

Also, probability of not getting a head with one coin $= 1 - \dfrac{1}{2} = \dfrac{1}{2}$.

Probability of not getting a head with 4 coins

$$= \dfrac{1}{2} \times \dfrac{1}{2} \times \dfrac{1}{2} \times \dfrac{1}{2} = \dfrac{1}{16}$$

Hence, probability of getting at least one head with 4 coins $= 1 - \dfrac{1}{16} = \dfrac{15}{16}$.

EXAMPLE 10. *A box contains 2 white and 4 black balls. Another box B contains 5 white and 7 black balls. A ball is transferred from the box A to the box B. Then a ball is drawn from the box B. Find the probability that it is white.* **(VTU–2004)**

SOLUTION. The probability of drawing a white ball from box B will depend on whether the transferred ball is black or white.

If the transferred ball is black, its probability is 4/6. There are now 5 white and 8 black balls in box B. The probability of drawing white ball from box B is 5/13. Therefore, the probability of drawing a white ball from B, if transferred ball is

black $= \dfrac{4}{6} \times \dfrac{5}{13} = \dfrac{10}{39}$.

Similarly, the probability of drawing a white ball from B, if transferred ball is

white $= \dfrac{2}{6} \times \dfrac{6}{13} = \dfrac{2}{13}$.

Hence, the required probability $= \dfrac{10}{39} + \dfrac{2}{13} = \dfrac{16}{39}$.

EXAMPLE 11. *A problem in Mathematics is given to three students whose chance of solving it are 1/2, 1/3, 1/4. What is the probability in the following cases :*

(i) that the problem is solved. **(VTU–2004, 10)**

(ii) only one of them solves it correctly.

(iii) at least one of them may solve it.

SOLUTION. Let A, B, C be three events, when a problem in mathematics is solved by three students.

As per given, we have

$$P(A) = \dfrac{1}{2}, \quad P(B) = \dfrac{1}{3} \quad \text{and} \quad P(C) = \dfrac{1}{4}$$

$\therefore \qquad P(\bar{A}) = 1 - \dfrac{1}{2} = \dfrac{1}{2}, \quad P(\bar{B}) = \dfrac{2}{3}$

and $\qquad P(\bar{C}) = \dfrac{3}{4}$

(i) The probability that the problem is solved.

$\qquad$ = Probability that the problem is solved by at least one student

$$= 1 - P(\bar{A}) \, P(\bar{B}) \, P(\bar{C}) = 1 - \dfrac{1}{2} \times \dfrac{2}{3} \times \dfrac{3}{4}$$

$$= 1 - \dfrac{1}{4} = \dfrac{3}{4}$$

(ii) The probability that only one solves it correctly

$$= P(A)\,P(\bar{B})\,P(\bar{C}) + P(\bar{A})P(B)\,P(\bar{C}) + P(\bar{A})\,P(\bar{B})\,PC)$$

$$= \frac{1}{2} \times \frac{2}{3} \times \frac{3}{4} + \frac{1}{2} \times \frac{1}{3} \times \frac{3}{4} + \frac{1}{2} \times \frac{2}{3} \times \frac{1}{4}$$

$$= \frac{1}{4} + \frac{1}{8} + \frac{1}{12} = \frac{11}{24}$$

(iii) The probability that at least one of them may solve the problem

$$= 1 - P(\bar{A})\,P(\bar{B})\,P(\bar{C}) = 1 - \frac{1}{2} \times \frac{2}{3} \times \frac{3}{4} = 1 - \frac{1}{4} = \frac{3}{4}$$

EXAMPLE 12. **(Huyghen's Problem).** *A and B throw alternately with a pair of dice. A wins if he throws 6 before B throws 7 and B wins if he throws 7 before A throws 6. If A begins, find his chance of winning.* (JNTU–2003, Madras–2006)

SOLUTION . We observe that the sum 6 can be obtained as follows

(1, 5), (2, 4), (3, 3), (4, 2), (5, 1) = Total 5 ways

Now, the probability of A's throwing 6 with 2 dice is 5/36.

The probability of B's not throwing 6 is $\dfrac{31}{36}$.

Similarly, the probability of B's throwing 7 is $\dfrac{6}{36}$, *i.e.*, $\dfrac{1}{6}$.

The probability of B's not throwing 7 is $\dfrac{5}{6}$.

Now, A can win if he throws 6 in the first, third, fifth, seventh, etc. throws.

Hence, the chance of A's winning

$$= \frac{5}{36} + \frac{31}{36} \times \frac{5}{6} \times \frac{5}{36} + \left(\frac{31}{36} \times \frac{5}{6}\right)^2 \times \frac{5}{36} + ...$$

$$= \frac{5}{36}\left[1 + \left(\frac{31}{36} \times \frac{5}{6}\right) + \left(\frac{31}{36} \times \frac{5}{6}\right)^2 + \left(\frac{31}{36} \times \frac{5}{6}\right)^3 + ...\right]$$

$$= \frac{5}{36}\left[\frac{1}{1 - \left(\frac{31}{36} \times \frac{5}{6}\right)}\right] \quad \left[\because \text{ Sum of infinite G.P.} = \frac{a}{1-r} \right]$$

$$= \frac{5}{36} \times \frac{36 \times 6}{61} = \frac{30}{61}.$$

EXAMPLE 13. *The probability that Ram gets scholarship is 0.5 and that Shyam will get is 0.8. What is the probability that at least one of them gets the scholarship?*

SOLUTION . We have $\quad P(A)$ = Probability that Ram gets scholarship
$$= 0.5 = 1/2$$

$\therefore \qquad P(\bar{A})$ = Probability that Ram does not get the scholarship

$$= 1 - \frac{1}{2} = 1/2$$

Now, $\qquad P(B)$ = Probability that Shyam gets scholarship

$$= 0.8 = \frac{8}{10} = \frac{4}{5}$$

$\therefore \qquad P(\bar{B})$ = Probability that Shyam does not get the scholarship

$$= 1 - \frac{4}{5} = \frac{1}{5}$$

Hence, required probability

$$= 1 - P(\bar{A})\, P(\bar{B}) = 1 - \frac{1}{2} \times \frac{1}{5} = 1 - \frac{1}{10}$$

$$= \frac{9}{10} = 0.9 \,.$$

EXAMPLE 14. *A man is known to speak the truth 3 out of 4 times. He throws a dice and reports that it is a six. Find the probability that it is actually a six.*

SOLUTION. Let B_1 and B_2 be the events when the man speaks the truth or not respectively.

Then, $\qquad P(B_1) = \dfrac{3}{4}$ and $P(B_2) = \dfrac{1}{4}$

Further, let A be the event when there is a six.

$\Rightarrow \qquad P(A \mid B_1) = \dfrac{1}{6}$ and $P(A \mid B_2) = \dfrac{5}{6}$

Then, by Bayes' theorem, we have

$$P(B_1 \mid E) = \frac{P(B_1)\, P(A \mid B_1)}{P(B_1)\, P(A \mid B_1) + P(B_2)\, P(A \mid B_2)}$$

$$= \frac{\dfrac{3}{4} \times \dfrac{1}{6}}{\dfrac{3}{4} \times \dfrac{1}{6} + \dfrac{1}{4} \times \dfrac{5}{6}} = \frac{3}{3+5} = \frac{3}{8}$$

EXAMPLE 15. *Three clerks process incoming copies of a certain form. The first clerk A processes 40% of the forms, the second clerk B 35% and the third clerk C processes 25%. The first clerk has an error rate of 0.04, the second clerk has an error rate of 0.06 and the third clerk has an error rate of 0.03. A form selected at random from a day's output is found to have an error. The supervisor wishes to know the probability that it was processed by the second clerk B.*

SOLUTION. Let B_1, B_2, B_3 be the events when the error is made by clerks A, B, C respectively and E be the event when the error is made.

$\therefore \qquad P(B_1) = \dfrac{40}{100}, \quad P(B_2) = \dfrac{35}{100}, \quad P(B_3) = \dfrac{25}{100}$

and $P(E \mid B_1) = 0.04$, $P(E \mid B_2) = 0.06$, $P(E \mid B_3) = 0.03$.

Then by Bayes' theorem, we have

$$P(B_2 \mid E) = \frac{P(B_2)P(E \mid B_2)}{P(B_1)P(E \mid B_1) + P(B_2)P(E \mid B_2) + P(B_3)P(E \mid B_3)}$$

$$= \frac{\dfrac{35}{100} \times 0.06}{\dfrac{40}{100} \times 0.04 + \dfrac{35}{100} \times 0.06 + \dfrac{25}{100} \times 0.03}$$

$$= \frac{35 \times 0.06}{40 \times 0.04 + 35 \times 0.06 + 25 \times 0.03}$$

$$= \frac{2.1}{1.6 + 2.1 + 0.75} = \frac{2.1}{4.45} = 0.47.$$

EXAMPLE 16. *Three urns are given, each containing red and black balls as indicated below :*
Urn 1 : 6 red and 4 black balls
Urn 2 : 2 red and 6 black balls
Urn 3 : 1 red and 8 black balls
An urn is chosen at random and a ball is drawn from the urn. The ball drawn is red.
Find the probability that the ball is drawn from urn 2 or urn 3.

SOLUTION . Let B_1, B_2, B_3 be the events of choosing urn 1, 2 and 3, respectively. Also, let A be the event that 'ball drawn is red'.

Then $P(B_1) = P(B_2) = P(B_3) = \dfrac{1}{3}$

and $P(A \mid B_1) = \dfrac{6}{10} = \dfrac{3}{5}, \quad P(A \mid B_2) = \dfrac{2}{8} = \dfrac{1}{4}, P(A \mid B_3) = \dfrac{1}{9}$

Hence, P (ball drawn from urn 2 or urn 3)

$$= P(B_2 \mid E) + P(B_3 \mid E) = 1 - P(B_1 \mid E)$$

$$= 1 - \frac{P(B_1)\, P(A \mid B_1)}{P(B_1)\, P(A \mid B_1) + P(B_2)\, P(A \mid B_2) + P(B_3)\, P(A \mid B_3)}$$

$$= 1 - \frac{\dfrac{1}{3} \times \dfrac{3}{5}}{\dfrac{1}{3} \times \dfrac{3}{5} + \dfrac{1}{3} \times \dfrac{1}{4} + \dfrac{1}{3} \times \dfrac{1}{9}} = 1 - \frac{108}{173} = \frac{65}{173}.$$

EXAMPLE 17. *From a bag containing 10 black and 5 white balls. A ball is drawn at random. What is the probability that it is white?*

SOLUTION. Here, total balls in the bag are 15. So total ways of drawing one ball are 15. Now the white ball can be drawn in 5 ways.

$\Rightarrow$ The required probability of drawing white ball $= \dfrac{5}{15} = \dfrac{1}{3}.$

EXAMPLE 18. *What is the probability of obtaining 9, 10 and 11 points with 3 dice?*

SOLUTION. Total ways of experiment of throwing 3 dice are $6 \times 6 \times 6 = 216$.

Out of these 216 ways, the 9 points can be obtained as follows :
(6, 1, 2), (6, 2, 1), (5, 1, 3), (5, 2, 2), (5, 3, 1), (4, 1, 4), (4, 2, 3), (4, 3, 2), (4, 4, 1), (3, 1, 5), (3, 2, 4), (3, 3, 3), (3, 4, 2), (3, 5, 1), (2, 1, 6), (2, 2, 5), (2, 3, 4), (2, 4, 3), (2, 5, 2), (2, 6, 1), (1, 2, 6), (1, 3, 5), (1, 4, 4), (1, 5, 3), (1, 6, 2).

$\Rightarrow$ Number of ways of getting 9 point $= 25$

$\Rightarrow$ Probability of 9 points $= \dfrac{25}{216}$

Similarly, 10 points can be obtained as follows :
(6, 1, 3), (6, 2, 2), (6, 3, 1), (5, 1, 4), (5, 2, 3), (5, 3, 2), (5, 4, 1), (4, 1, 5), (4, 2, 4), (4, 3, 3), (4, 4, 2), (4, 5, 1), (3, 1, 6), (3, 2, 5), (3, 3, 4), (3, 4, 3), (3, 5, 2), (3, 6, 1), (2, 2, 6), (2, 3, 5), (2, 4, 4), (2, 5, 3), (2, 6, 2), (1, 3, 6), (1, 4, 5), (1, 5, 4), (1, 6, 3)

$\Rightarrow$ Number of ways of getting 10 point $= 27$

$\Rightarrow$ Probability of 10 points $= \dfrac{27}{216} = \dfrac{1}{8}$

Similarly 11 points can be obtained as follows :

(6, 1, 4), (6, 2, 3), (6, 3, 2), (6, 4, 1), (5, 1, 5), (5, 2, 4), (5, 3, 3), (5, 4, 2),

(5, 5, 1), (4, 1, 6), (4, 2, 5), (4, 3, 4), (4, 4, 3), (4, 5, 4) (4, 5, 2), (4, 6, 1),

(3, 2, 6), (3, 3, 5), (3, 4, 4), (3, 5, 3), (3, 6, 2), (2, 3, 6), (2, 4, 5), (2, 5, 4),

(2, 6, 3), (1, 4, 6), (1, 5, 5), (1, 6, 4)

Number of ways getting 11 point = 27

$\Rightarrow$ Probability of 11 points $= \dfrac{27}{216} = \dfrac{1}{8}$

But events of getting 9 or 10 or 11 are mutually exclusive.

Probability of getting 9 or 10 or 11 points.

$= P(9 \text{ points}) + P(10 \text{ points}) + P(11 \text{ points})$

$= \dfrac{25}{216} + \dfrac{1}{8} + \dfrac{1}{8} = \dfrac{79}{216}$

EXAMPLE 19. *A book contains* 1000 *pages. A page is choosen at random. What is the chance that the sum of digits on the page is equal to* 9?

SOLUTION. Total ways of choosing one page at random are 1000.

Now sum of digits on the page is equal to 9 if page number is any of the following :

9, 18, 81, 27, 72, 36, 45, 54, 90,

108, 117, 126, 135, 144, 153, 162, 171, 180,

207, 216, 225, 234, 243, 252, 261, 270,

306, 315, 324, 333, 342, 351, 360,

405, 414, 423, 434, 441, 450,

504, 513, 522, 531, 540,

603, 614, 621, 630,

801, 810,

900

$\Rightarrow$ Total number of ways = 55

Required probability $= \dfrac{55}{100} = 0.55$.

EXAMPLE 20. *The odd against student 'A' solving a problem are 4 to 3 and odds in favour of 'B' solving the same problem are 7 to 8. What is the probability that the problem will be solved if they both try?*

(CCSU(B.Sc. Biotech)–1995)

SOLUTION. The odds against $A = 4 : 3$, then the probability by A, *i.e.*, $P(\bar{A}) = \dfrac{4}{7}$

$\Rightarrow \qquad P(A) = 1 - \dfrac{4}{7} = \dfrac{3}{7}$

and the odds against $B = 7 : 8$, thus the probability by B, *i.e.*, $P(B) = \dfrac{7}{15}$

$\Rightarrow \qquad P(\bar{B}) = 1 - P(B) = 1 - \dfrac{7}{15} = \dfrac{8}{15}$

Hence, required probability

$$= 1 - P(\bar{A})P(\bar{B}) = 1 - \frac{4}{7} \times \frac{8}{15}$$

$$= 1 - \frac{32}{105} = \frac{73}{15}$$

Exercise 23.2

1. In a class of 25 students with roll number 1 to 25, a student is picked up at random to answer a question. Find the probability that the roll number of the selected students is either a multiple of 5 or 7 ?

2. In a single throw of two dice, find the probability that neither a doublet nor a total of 10 will appear.

3. Tickets are numbered from 1 to 100. One ticket is picked up at random. Find the probability that the ticket picked up has a number which is divisible by 5 or 8.

4. A box contains 100 bolts and 50 nuts. It is given that 50% bolts and 50% nuts are rusted. Two objects are selected from the box at random. Find the probability that both are bolts or nuts or rusted.

5. Two dice are tossed once. Find the probability of getting an even number on the first dice or a total of 8.

6. In a race, the odds in favour of horses A, B, C, D are $1 : 3, 1 : 4, 1 : 5$ and $1 : 6$ respectively. Find the probability that one of them wins the race.

7. A natural number is chosen at random from amongst the first 500. What is the probability that the number so chosen is divisible by 3 or 5?

8. In a single throw of 2 dice, find the probability of not getting the same number on the two dice.

9. Let A, B, C be three mutually exclusive events associated with a random experiment. Find $P(A)$ given that $P(B) = \frac{3}{2} P(A)$ and $P(C) = \frac{1}{2} P(B)$.

10. One card is drawn from a pack of 52 cards, each of the 52 cards being equally likely to be drawn. Find the probability of (i) the card drawn is red, (ii) the card drawn is a king.

11. From a set of 17 cards numbered 1, 2, 3, 4, ..., 16, 17, one is drawn at random. Show that the chance that its number is divisible by 3 or 7 is 7/17.

12. The probability of student A passing an examination is 2/9 and of student B passing is 5/9. Assuming the two events : 'A passes', 'B passes' as independent, find the probability of (i) only A passing the examination, (ii) only one of them passing the examinations.

13. In a group, there are 3 men and 2 women, 3 persons are selected at random from this group. Find the probability that 1 man and 2 women or 2 men and 1 woman are selected.

14. A bag contains 3 red, 4 black and 2 green balls. Two balls are drawn at random from the bag. Find the probability that both balls are of different colours.

15. Ram is taking up subjects — Mathematics, Physics and Chemistry in the examination. His probability of getting Grade A in these subjects are 0.2, 0.3 and 0.5, respectively. Find the probability that he gets (i) grade A in all subjects, (ii) grade A in no subject, (iii) grade A in two subjects.

16. A bag contains 30 tickets, numbered from 1 to 30. Five tickets are drawn at random and arranged in ascending order. Find the probability that 'the third' number is 20.

17. There are 3 red and 3 black balls in a bag, 3 balls are taken at random from the bag. Find the probability of getting 2 red and 1 black balls or 1 red and 2 black balls.

18. A bag contains 5 red, 6 white and 7 black balls. Two balls are drawn at random. What is the probability that both balls are red or black?

19. A machine manufactured by a firm consists of two parts A and B. Out of 100 A's manufactured, 9 are likely to be defective and of 100 B's 5 are likely to be defective. Find the probability that a machine manufactured by the firm is free of any defect.

20. Three bags contain 7 white, 8 red, 9 white, 6 red and 5 white, 7 red balls respectively. One ball at random is drawn from each bag. Find the probability that all of them are of the same colour.

21. A bag contains 1 red and 4 blue balls. Two balls are drawn at random with replacement. Find the probability of getting one red and one blue ball.

22. A bag contains 4 red and 5 black balls. Another bag contains 3 red and 6 black balls. One ball is drawn from first bag and two balls are drawn from the second bag. Find the probability that out of the three, two are black and one is red.

23. Two cards are drawn from a well shuffled pack of 52 cards, one after another without replacement. Find the probability that one of them is an ace and other is a queen of opposite shade.

24. There are three urns A, B and C. Urn A contains 4 red balls and 3 black balls. Urn B contains 5 red balls and 4 black balls. Urn C contains 4 red balls and 4 black balls. One ball is drawn from each of these urns. What is the probability that the 3 balls drawn consists of 2 red balls and a black ball. **(JNTU–2003)**

25. A husband and wife appear in an interview for two vacancies in the same post. The probability of husband's selection is 1/3 and that of wife's selection is 1/5. Show that the probability that only one of them will be selected is 2/5.

26. The probability of A solving a problem is 3/7 and that of B solving it is 1/3. What is the probability that (i) at least one of them will solve the problem, (ii) only one of them will solve the problem.

27. A problem in Mathematics is given to three students whose chances of solving it are (i) 1/2, 1/4 and 1/3, (ii) 1/3, 1/4 and 1/5, (iii) 1/3, 1/5 and 1/6, (iv) 1/2 , 1/3 and 1/5. What is the probability that at least one of them may solve it.

28. A bag contains 1 white and 6 red balls and bag B contains 4 white and 3 red balls. One ball is drawn at random from one of the chosen bags and is found to be white. Find the probability that it was drawn from bag A.

29. In a bolt factory, machine A, B and C are producing 25%, 35% and 40% of total bolts. Of these, 4%, 3% and 2% respectively are defective. A bolt is drawn at random and found to be defective. What is the probability that it is produced by machine A?

30. Let us suppose 5 men out of 100 and 25 women out of 1000 are good orators. An orator is chosen at random. Find the probability that a male person is selected. Assume that there are equal number of men and women.

31. Ram speaks the truth 8 times out of 10 times. A dice is tossed. He reports that it was 5. What is the probability that it was actually 5?

32. A can hit a target 3 times in 5 shots, B 2 times in 5 shots and C 3 times in 4 shots. They fire a volley, what is the probability that

 (i) two shots hits?

(ii) atleast two shots hits?

(AMITE–2003, Madras–2003)

33. If on an average one ship in every 10 wrecked, find the probability that out of 5 ship expected to arrive, 4 at least will arive safely. **(UPTU–2001)**

34. Assume that a factory has two machines. Past records show that machine no. 1 produced 30% of the items of output and machine no. 2 produced 70% of the items. Further, 5% of the items produced by machine no. 1 were defective and only 1% produced by machine no. 2 were defective. If an item is drawn at random and found to be defective, what is the probability that the defective item was produced by machine no. 1 or by machine no. 2? **(UPTU–2002)**

ANSWERS

1. 8/25	**2.** 7/9	**3.** 0.3	**4.** 0.58	**5.** 5/9	**6.** 319/420	
7. 233/500	**8.** 5/6	**9.** 9/13	**10.** 1/2	**11.** 1/13	**12.** 8/81, 43/81	**13.** 9/10
14. 13/18	**15.** 0.03, 0.28, 0.22			**17.** 9/10	**18.** 31/153	**19.** 0.86
20. 217/900	**21.** 56/121		**22.** 25/54		**23.** 4/663	**24.** 17/42
26. 13/21, 10/21	**27.** 7/10, 3/5, 7/9, 11/15			**28.** 1/5	**29.** 0.26	**30.** 2/3
31. 4/7 **32.** (i) – 0.45			(ii) 0.63	**33.** 0.92	**34.** 1.0	

23.8 RANDOM VARIABLES AND PROBABILITY DISTRIBUTION

If we throw an ordinary dice, we shall get one of the numbers 1, 2, 3, 4, 5, 6 in each throw. Therefore, we get quantitative outcomes. In some cases, the outcome may be qualitative, for example, throw of a coin which may be head or tail. In such type of cases, we may denote the outcome 'head' of the tossing of a coin by 1 and the 'tail' by 0. In this way, each outcome of a random experiment, whether it is qualitative or quantitative, can be expressed by a real number. This numerical value associated with the outcome of a random experiment is known as a random variable.

Definition (1). *The variate which can take certain values depending on chance is called a random variable. It is also known as 'chance variate' or 'stochastic variate'.*

Definition (2). *A real valued function X, defined on a sample space S, there is a real number denoted by X(s), X is called a function defined on S. X(s) is called the value of the function at s.*

Definition (3). *A real valued function X, defined on a sample space S of a random experiment E, is called a random variable which assign to each sample, one and only one real numbers X(s) = x (say), where s $\in$ S .*

or

If E is a random experiment and S is a sample space associated with it, a function X(s), s $\in$ S is called a random variable.

Definition (4). *Let X be a random variable which takes the values $x_1, x_2, \ldots$, then the probability that x = x_i is denoted by P(X = x_i) or p(x_i) or p_i . The probability of the event X takes values from x_i to x_j is denoted by*

$$P(x_i \le X < x_j)$$

Also,

$$p(x_i) \ge 0 \quad \text{and} \quad \Sigma p(x_i) = 1 .$$

23.8.1 TYPES OF RANDOM VARIABLE

There are two types of random variables

(i) Discrete Random Variables

If a random variable assumes only a finite number or countably infinite number of values of X, then X is called discrete random variable.

☙ REMARK

➡ The possible values of X may be taken $x_1, x_2, \ldots, x_n$ in infinite case and $x_1, x_2, \ldots$ in countably infinite case. The countably infinite case will have an infinite sequence of distinct values and that sequence will be countable.

(ii) Continuous Random Variable

If a random variable assumes any value in some interval, it is called a continuous random variable, i.e., if a variate can take an infinite set of values in the given interval, say $a \le x \le b$, it is a continuous random variable and their distribution are accordingly known as continuous distribution.

23.9 PROBABILITY DISTRIBUTIONS

The distribution obtained by taking the possible values of a random variable together with their respective probabilities is called probability distribution.

23.9.1 PROBABILITY DISTRIBUTION OF A DISCRETE RANDOM VARIABLE

The probability distribution of a discrete random variable is the set of ordered pair $(x_i, p(x_i))$ provided

(i) $p(x_i) \ge 0$ (ii) $\Sigma p_i(x_i) = 1$

REMARKS

➡ $P(x \le x_i) = P(X = x_1) + P(X = x_2) + \ldots + P(X = x_i) = p_1 + p_2 + \ldots + p_i$

➡ $P(x \ge x_i) = P(X = x_i) + P(X = x_{i+1}) + \ldots + P(X = x_n) = p_{i+1} + p_{i+2} + \ldots + p_n$

SOLVED EXAMPLES

EXAMPLE 1. *Two cards are drawn one by one without replacement from a well shuffled pack of 52 cards. Find the probability distribution of the number of aces.*

SOLUTION . Let X denote the random variable, *i.e.*, X denote the number of aces. Therefore, X takes values 0, 1, 2.

Then, $P(X = 0) = \dfrac{48}{52} \times \dfrac{47}{51} = \dfrac{188}{221}$

$P(X = 1) = 2\left(\dfrac{4}{52} \times \dfrac{48}{51}\right) = 2\left(\dfrac{1}{13} \times \dfrac{16}{17}\right) = 2\left(\dfrac{16}{221}\right) = \dfrac{32}{221}$

Also, $P(X = 2) = \dfrac{4}{52} \times \dfrac{3}{51} = \dfrac{1}{13} \times \dfrac{1}{17} = \dfrac{1}{221}$.

Hence, the required probability distribution is given by

X	0	1	2
$P(X)$	$\dfrac{188}{221}$	$\dfrac{32}{221}$	$\dfrac{1}{221}$

EXAMPLE 2. *Find the probability distribution of the number of success in two tosses of a dice, when success is defined of getting a 5 and a 6.*

SOLUTION . We have

Probability of success $= \dfrac{2}{6} = \dfrac{1}{3}$

and Probability of failure $= \dfrac{4}{6} = \dfrac{2}{3}$

Let X be the random variable, therefore

$$P(X = 0) = \dfrac{2}{3} \times \dfrac{2}{3} = \dfrac{4}{9}$$

$$P(X = 1) = \dfrac{1}{3} \times \dfrac{2}{3} + \dfrac{2}{3} \times \dfrac{1}{3} = \dfrac{2}{9} + \dfrac{2}{9} = \dfrac{4}{9}$$

and $P(X = 2) = \dfrac{1}{3} \times \dfrac{1}{3} = \dfrac{1}{9}$

Hence, the required probability distribution is given by

X	0	1	2
$P(X)$	$\dfrac{4}{9}$	$\dfrac{4}{9}$	$\dfrac{1}{9}$

EXAMPLE 3. *A fair dice is tossed twice. If the number appearing on the top is less than 3, it is a success. Find the probability distribution of successes.*

SOLUTION . We have

$$\text{Probability of success} = \frac{2}{6} = \frac{1}{3}$$

and Probability of failure $= \dfrac{4}{6} = \dfrac{2}{3}$

If X is the random variable, then we have

$$P(0) = \frac{2}{3} \times \frac{2}{3} = \frac{4}{9}$$

$$P(1) = \frac{1}{3} \times \frac{2}{3} + \frac{2}{3} \times \frac{1}{3} = \frac{2}{9} + \frac{2}{9} = \frac{4}{9}$$

and

$$P(2) = \frac{1}{3} \times \frac{1}{3} = \frac{1}{9}$$

Hence, the required probability distribution is given by

X	0	1	2
$P(X)$	$\dfrac{4}{9}$	$\dfrac{4}{9}$	$\dfrac{1}{9}$

EXAMPLE 4. *Find the probability distribution of the number of heads when three coins are tossed.*

SOLUTION . Let X be the random variable, which denote the number of heads. Then X takes the values 0, 1, 2, 3. Let p and q be the probabilities of success and failure respectively then we have

$$p = \frac{1}{2}, \quad q = 1 - \frac{1}{2} = \frac{1}{2}$$

therefore

$$P(X = 0) = P(TTT) = \frac{1}{2} \times \frac{1}{2} \times \frac{1}{2} = \frac{1}{8}$$

$$P(X = 1) = P(HTT) + P(THT) + P(TTH) = 3 \left(\frac{1}{2} \times \frac{1}{2} \times \frac{1}{2} \right) = \frac{3}{8}$$

$$P(X = 2) = P(HHT) + P(HTH) + P(THH) = 3 \left(\frac{1}{2} \times \frac{1}{2} \times \frac{1}{2} \right) = \frac{3}{8}$$

and

$$P(X = 3) = \frac{1}{2} \times \frac{1}{2} \times \frac{1}{2} = \frac{1}{8}$$

Hence, the required probability of distribution is given by

X	0	1	2	3
$P(X)$	$\dfrac{1}{8}$	$\dfrac{3}{8}$	$\dfrac{3}{8}$	$\dfrac{1}{8}$

EXAMPLE 5. *A coin is tossed 4 times, if X is the number of heads observed, then find the probability distribution of X.*

SOLUTION . We have the sample space X given by

{*HHHH, HHHT, HTHH, THHH, HHTT, HTHT, HTTH, THHT,*

THTH, TTHH, HTTT, THTT, TTHT, TTTH, TTTT}

Also, X takes values 0, 1, 2, 3, 4.

Hence, the required probability distribution is given by

X	0	1	2	3	4
P(X)	$\dfrac{1}{16}$	$\dfrac{4}{16}$	$\dfrac{6}{16}$	$\dfrac{4}{16}$	$\dfrac{1}{16}$

EXAMPLE 6. *A random variable X has the following probability distribution*

X	0	1	2	3	4	5	6	7
P(X)	0	k	2k	2k	3k	k^2	$2k^2$	$7k^2+k$

(i) *Find k.*

(ii) *Evaluate* $P(x < 3)$, $P(X \leq 6)$ *and* $P(0 < X < 3)$ **(JNTU–2003, WBTU–2005)**

SOLUTION. (i) We know that

$$\Sigma P(X) = 1$$

Therefore,

$$0 + k + 2k + 2k + 3k + k^2 + 2k^2 + 7k^2 + k = 1$$

$$\Rightarrow \quad 10k^2 + 9k - 1 = 0$$

i.e.,

$$k = \frac{-9 \pm \sqrt{81 + 40}}{20} = \frac{-9 \pm 11}{20} = \frac{1}{10}, -1$$

We reject the case $k = -1$, because probability is always non-negative.

Hence, $$k = \frac{1}{10}$$

(ii) $$P(x < 3) = P(x = 0) + P(X = 1) + P(X = 2)$$

$$= 0 + k + 2k = 3k = 3 \times \frac{1}{10} = \frac{3}{10}$$

$$P(X \geq 6) = P(6) + P(7)$$

$$= 2k^2 + 7k^2 + k = 9k^2 + k$$

$$= \frac{9}{100} + \frac{1}{10} = \frac{19}{100}$$

and $$P(0 < X < 3) = P(X = 1) + P(X = 2)$$

$$= k + 2k = 3k = \frac{3}{10}$$

Exercise 23.3

1. A coin is tossed 5 times, X is the number of heads observed. Find the probability distribution of X.

2. Two cards are drawn successively with replacement from a well shuffled pack of 52 cards. Find the probability distribution of the number of (i) aces, (ii) kings.

3. Find the probability distribution of the number of heads when three coins are tossed.

4. Find the probability distribution of the sum of numbers obtained when two dice are thrown.

5. Four bad eggs are mixed with 10 good ones. If 3 eggs are drawn one by one without replacement, find the probability distribution of the number of bad eggs.

6. A coin is biased so that the head is 3 times likely to occur as a tail. If the coin is tossed twice, find the probability distribution for the number of tails.

7. A random variable X has the following probability distribution :

X	-2	-1	0	1	2	3
$P(X)$	0.1	k	0.2	$2k$	0.3	k^2

Find (i) The value of k, (ii) $P(X \le 1)$,
(iii) $P(X \ge 0)$.

8. A bag contains 2 white, 3 red and 4 blue balls. Two balls are drawn at random from the bag. If the random variable X denote the number of white balls among the two balls drawn, find the probability distribution of X.

9. Four rotten mangoes are mixed accidentally with 20 good mangoes. Obtain the probability distribution of the number of rotten mangoes in a random draw of 2 mangoes without replacement.

10. Two cards are drawn successively from a well-shuffled pack of 52 cards. Find the probability distribution of the number of queens one by one.

ANSWERS

1.

X	0	1	2	3	4	5
$P(X)$	1/32	5/32	10/32	10/32	5/32	1/32

2.

X	0	1	2
$P(X)$	144/169	24/169	1/169

3.

X	0	1	2	3
$P(X)$	1/8	3/8	3/8	1/8

4.

X	2	3	4	5	6	7	8	9	10	11	12
$P(X)$	1/36	2/36	3/36	4/36	5/36	6/36	5/36	4/36	3/36	2/36	1/36

5.

X	0	1	2	3
$P(X)$	1000/2197	900/2197	270/2197	27/2197

6.

X	0	1	2
$P(X)$	9/16	3/8	1/16

7. (i) $P(0) = \dfrac{188}{221}$, $P(1) = \dfrac{32}{221}$, $P(2) = \dfrac{1}{221}$

(ii) $P(0) = \dfrac{105}{221}$, $P(1) = \dfrac{96}{221}$, $P(2) = \dfrac{20}{221}$

(iii) $P(0) = \dfrac{19}{34}$, $P(1) = \dfrac{13}{34}$, $P(2) = \dfrac{2}{34}$

8.

X	0	1	2
$P(X)$	49/81	28/81	41/81

9.

X	0	1	2
$P(X)$	95/138	40/138	3/138

10.

X	0	1	2
$P(X)$	144/169	24/169	1/169

23.10 BINOMIAL DISTRIBUTION

Let p and q be the probability of the success and failure respectively of an event in one trial, so that $p + q = 1$. If the event can be tried n times and assume that these n trials are independent and the probability p of success is the same in each trial. The number of successor may be 0, 1, 2, ..., n in these n trials.

Definition. *The probability of r successes in a series of n trials is given by $^n C_r \, p^r \, q^{n-r}$, where r takes any integral value for 0 to n. The probabilities of $0, 1, 2, ..., r, ..., n$ successes are given by*

$$q^n, \, ^n C_1 p q^{n-1}, \, ^n C_2 p^2 q^{n-2}, ..., \, ^n C_r p^r q^{n-r}, ..., p^n.$$

Then the probability of the number of successes so obtained is called the binomial distribution.

REMARKS

➡ Binomial distribution contains two independent contents p (or $q = 1 - p$) and n which are known as parameter of Binomial distribution.

➡ Binomial distribution was discovered by Jacob Bernoulli. Due to this reason, it is also known as Bernoulli distribution.

➡ If $p = q = 1/2$, then this distribution is said to be symmetric.

➡ The distribution $f(r) = {}^nC_r\, p^r\, q^{n-r}$, $r = 0, 1, 2, ..., n$ determines a probability distribution, called the binomial distribution, since the total frequency $p + q = 1$.

➡ Binomial distribution is a discrete probability distribution.

➡ Binomial distribution is important not only because of its wide applicability, but because it gives rise to many other probability distribution.

23.11 APPLICABILITY OF BINOMIAL DISTRIBUTION

For a binomial distribution, the following conditions must be satisfied :

  (i) There should be a finite number of trials

  (ii) The trials are mutually exclusive and does not depend on each other.

  (iii) Each trial should have only two possibility, either a success or failure.

  (iv) The probability of success or failure is the same for each trial.

23.12 MOMENTS OF THE BINOMIAL DISTRIBUTION

23.12.1 MOMENT ABOUT ORIGIN

The probability distribution function for binomial distribution is

$$f(x) = {}^nC_x\, p^x\, q^{n-x}$$

(1) First Moment about Origin (Mean of Binomial Distribution)

$$\mu_1' = \sum_{x=0}^{n} x \cdot {}^nC_x\, p^x\, q^{n-x} = 0.q^n + 1.\, {}^nC_1\, q^{n-1} \cdot p + 2.\, {}^nC_2\, q^{n-2} \cdot p^2 + ... + n\, {}^nC_n\, p^n$$

$$= np\, (q^{n-1} + {}^{n-1}C_1\, q^{n-2} \cdot p + {}^{n-1}C_2\, q^{n-2}\, p^2 + ... + p^{n-1})$$

$$= np\, (q + p)^{n-1} \qquad\qquad (\because\ p + q = 1)$$

Therefore,   mean $= \mu_1' = np$

   ⇒   Mean of the binomial distribution is np.

(2) Second Moment about Origin

$$\mu_2' = \sum_{x=0}^{n} x^2 \cdot {}^nC_x\, p^x\, q^{n-x} = \sum_{x=0}^{n} [x(x-1) + x]\, {}^nC_x\, p^x\, q^{n-x}$$

$$= \sum_{x=0}^{n} x(x-1) \frac{n!}{x!(n-x)!}\, p^x\, q^{n-x} + \sum_{x=0}^{n} x \cdot {}^nC_x\, p^x\, q^{n-x}$$

$$= \sum_{x=1}^{n} x(x-1) \frac{\{n(n-1)(n-2)\}}{[x(x-1)(x-2)! \cdot \{(n-2) - (x-2)\}!]} \cdot p^2\, p^{x-2}\, q^{((n-2)-(x-2))}$$

$$+ \sum_{x=0}^{n} x \cdot {}^nC_x\, p^x\, q^{n-x}$$

$$= \sum_{x=1}^{n} n(n-1) \frac{(n-2)!}{(x-2)![(n-2)-(x-2)]!} \cdot p^{x-2} q^{((n-2)-(x-2))} \cdot p^2$$

$$+ \sum_{x=0}^{n} x \cdot {}^nC_x p^x q^{n-x}$$

$$= n(n-1) p^2 \sum_{x=1}^{n} {}^{n-2}C_{x-2} p^{x-2} q^{((n-2)-(x-2))} + np$$

$$= n(n-1) p^2 (q+p)^{n-2} + np$$

$$= n(n-1) p^2 + np = np[(n-1)p+1] = np[np+(1-p)]$$

$$= np(np+q)$$

$$\therefore \qquad \mu'_2 = n^2 p^2 + npq$$

(3) Third Moment about Origin

$$\mu'_3 = \sum_{x=0}^{n} x^3 \cdot {}^nC_x p^x q^{n-x} = \sum_{x=0}^{n} \{x(x-1)(x-2)+3x(x-1)+x\} {}^nC_x p^x q^{n-x}$$

$$[\because x^3 = x(x-1)(x-2)+3x(x-1)+x]$$

$$= n(n-1)(n-2)p^3(q+p)^{n-3} + 3n(n-1) p^2(q+p)^{n-2} + np$$

$$\mu'_3 = n(n-1)(n-2)p^3 + 3n(n-1)p^2 + np$$

(4) Fourth Moment about Origin

$$\mu'_4 = \sum_{x=0}^{n} x^4 \cdot {}^nC_x p^x q^{n-x}$$

$$= \sum_{x=0}^{n} \{x(x-1)(x-2)(x-3)+6x(x-1)(x-2)+7x(x-1)+x\} {}^nC_x p^x q^{n-x}$$

$$= n(n-1)(n-2)(n-3)p^4(q+p)^{n-4} + 6n(n-1)(n-2)p^3(q+p)^{n-3}$$

$$+ 7n(n-1) p^2(q+p)^{n-2} + np$$

$$\mu'_4 = n(n-1)(n-2)(n-3)p^4 + 6n(n-1)(n-2)p^3 + 7n(n-1)p^2 + np$$

23.12.2 Moment About Mean

(1) First Moment about Mean

$$\mu_1 = 0$$

(2) Second Moment about Mean (Variance of Binomial Distribution)

$$\mu_2 = \mu'_2 - (\mu'_1)^2 = (n^2 p^2 + npq) - (np)^2$$

or   Variance $= \mu_2 = npq$

or   S.D. $= \sqrt{\mu_2} = \sqrt{npq}$

(3) Third Moment about Mean

$$\mu_3 = \mu'_3 - 3\mu'_1 + \mu'^3_1$$

$$= np[(n-1)(n-2)p^2 + 3(n-1)p+1] - 3(n^2 p^2 + npq)np + 2(np)^3$$

$$= np[\{(n-1)(n-2) - 3n^2 + 2n^2\} p^2 + \{3(n-1) + 3nq\}p + 1]$$

$$= np[(-3n + 2)p^2 + 3(n-1)p - 3np(1-p) + 1] \qquad (\because p = 1 - q)$$

$$= np (2p^2 - 3p + 1) = np (1 - 2p) (1 - p)$$

$$= np(1-p) \{1 - 2(1-q)\} = npq(2q - 1) = npq [q - (1-q)]$$

$$\Rightarrow \qquad \mu_3 = npq (q - p)$$

(4) Fourth Moment about Mean

We know $\qquad \mu_4 = \mu_4' - 4\mu_3'\mu_1' + 6\mu_2'\mu_1'^2 - 3\mu_1'^4$

$$= [n(n-1)(n-2)(n-3)p^4 + 6n(n-1)(n-2)p^3 + 7n(n-1)p^2 + np]$$

$$- 4[n(n-1)(n-2)p^3 + 3n(n-1)p^2 + np]$$

$$+ 6[n(n-1)p^2 + np]n^2 p^2 - 3(np)^4$$

On simplifying, we get

$$\mu_4 = 3n^2(p^4 + 2p^3 + p^2) - n(6p^4 - 12p^3 + 6p^2) - np^2 + np$$

$$= 3n^2 p^2 (p^2 - 2p + 1) - 6np^2(1 - 2p + p^2) + np(1-p)$$

$$= 3n^2 p^2 (1-p)^2 - 6np^2(1-p)^2 + npq = 3n^2 p^2 q^2 - 6np^2 q^2 + npq$$

$$= 3n^2 p^2 q^2 + npq(1 - 6pq)$$

(C) Pearson's Coefficients

(1) $\beta_1 = \dfrac{\mu_3^2}{\mu_2^3} = \dfrac{[nq(q-p)]^2}{(npq)^3} \Rightarrow \beta_1 = \dfrac{(1-2p)^2}{npq}$

(2) $\beta_2 = \dfrac{\mu_4}{\mu_2^2} = \dfrac{3n^2 p^2 q^2 + npq(1-6pq)}{(npq)^2} = 3 + \dfrac{(1-6pq)}{npq}$

(3) $\gamma_1 = \sqrt{\beta_1} = \dfrac{(q-p)}{\sqrt{npq}}$ \qquad (4) $\gamma_2 = \beta_2 - 3 = \dfrac{1-6pq}{npq}$

☙ REMARK

⟼ In case of binomial distribution, mean > variance.

☛ RECAPITULATIONS

- **For Binomial Distribution**
  (1) Moments about origin

$$\mu_1' = np, \ \mu_2' = n^2 p^2 + npq$$

$$\mu_3' = n(n-1)(n-2)p^3 + 3n(n-1)p^2 + np$$

$$\mu_4' = n(n-1)(n-2)(n-3)p^4 + 6n(n-1)(n-2)p^3 + 7n(n-1)p^2 + np$$

  (2) Moments about mean

$$\mu_1 = 0, \mu_2 = npq$$

$$\mu_3 = npq(q-p), \mu_4 = 3n^2 p^2 q^2 + npq(1-6pq)$$

(3) Pearson's Coefficients $\beta_1 = \dfrac{\mu_3^2}{\mu_2^3}, \beta_2 = \dfrac{\mu_4}{\mu_2^2}, \gamma_1 = \sqrt{\beta_1}, \gamma_2 = \beta_2 - 3$

23.13 MOMENT GENERATING FUNCTION OF BINOMIAL DISTRIBUTION

For binomial distribution, we have $f(x) = {}^nC_x \, p^x \, q^{n-x}$. Therefore, the m.g.f. about origin will be given by

$$M_0(t) = \sum_{x=0}^{n} e^{tx} \cdot {}^nC_x \, p^x \, q^{n-x} = \sum_{x=0}^{n} {}^nC_x \, (pe^t)^x \, q^{n-x} = (q + pe^t)^n$$

$$= \left[q + p\left(1 + t + \frac{t^2}{2!} + \frac{t^3}{3!} + \ldots\right) \right]^n = \left[1 + pt + \frac{pt^2}{2!} + \frac{pt^3}{3!} + \ldots \right]^n$$

$$(\because p + q = 1)$$

∴   The m.g.f. about mean is given by

$$M_m(t) = e^{-mt} M_0(t) \text{ where } m = np$$

$$= e^{-npt}(q + pe^t)^n = [e^{-pt}(q + pe^t)]^n = [qe^{-pt} + pe^{(1-p)t}]^n$$

$$\text{where } p + q = 1$$

$$\Rightarrow \quad M_m(t) = [qe^{-pt} + pe^{qt}]^n$$

$$= \left(1 + pq\frac{t^2}{2!} + pq(q^2 - p^2)\frac{t^3}{3!} + pq(q^3 + p^3)\frac{t^4}{4!} + \ldots \right)^n$$

or $\quad 1 + \mu_1 t + \mu_2\frac{t^2}{2!} + \mu_3\frac{t^3}{3!} + \mu_4\frac{t^4}{4!} + \ldots$

$$= 1 + npq\frac{t^2}{2!} + npq(q - p)\frac{t^3}{3!} + npq[1 + 3(n - 2)pq]\frac{t^4}{4!} + \ldots \quad \ldots(1)$$

🗢 REMARKS

⇛ We can generate the moments of binomial distribution by equating the coefficients of like powers of t.

⇛ As the number of trials increase indefinitely, $\beta_1 \to 0$ and $\beta_2 \to 3$.

⇛ We can apply the binomial distribution in following types of problems :

(i) Number of defective items in a sample from production line

(ii) Estimation of reliability of systems

(iii) Number of rounds fired from a gun hitting a target

(iv) Radar detection

23.14 CUMULANT GENERATING FUNCTION OF BINOMIAL DISTRIBUTION

(A) About Mean $m = np$

We know that

$$M_m(t) = \sum_{x=0}^{n} e^{t(x-m)} \cdot {}^nC_x p^x q^{n-x} = \sum_{x=0}^{n} e^{tx-npt} \cdot {}^nC_x q^{n-x}$$

$$= e^{-npt} \sum_{x=0}^{n} {}^nC_x (pe^t)^x \, q^{n-x} = e^{-npt}(q + pe^t)^n$$

$$= (qe^{-pt} + pe^{(1-p)t}) = (qe^{-pt} + pe^{qt})^n$$

$$= \left[q\left(1 - pt + \frac{p^2t^2}{2!} - \frac{p^3t^3}{3!} + ...\right) + p\left(1 + qt + \frac{q^2t^2}{2!} + ...\right) \right]^n$$

$$= \left[(q+p) + \frac{1}{2!}pq(p+q)t^2 + \frac{1}{3!}pq(q^2 - p^2)t^3 + \frac{1}{4!}pq(q^3 + p^3)t^4 + ... \right]^n$$

$$\Rightarrow \quad M_m(t) = \left[1 + \frac{1}{2!}pqt^2 + \frac{1}{3!}pq(q^2 - p^2)t^3 + \frac{1}{4!}pq(q^3 + p^3)t^4 + ... \right]^n$$

Now,

$$K_m(t) = \log M_m(t) = n\log\left(1 + \frac{1}{2!}pqt^2 + \frac{1}{3!}pq(q^2 - p^2)t^3 + \frac{1}{4!}pq(q^3 + p^3)t^4 + ...\right)$$

which is the cumulative function of the binomial distribution about the mean $m = np$. Now, from (1), we have

$$K_m(t) = n\left[\left\{\frac{1}{2}pqt^2 + \frac{1}{6}pq(q^2 - p^2)t^3 + \frac{1}{24}pq(q^3 + p^3)t^4 + ...\right\}\right.$$

$$\left. - \frac{1}{2}\left\{\frac{1}{4}p^2q^2t^4 + ...\right\} + \frac{1}{3}\left\{\frac{1}{8}p^3q^3t^6 + ...\right\} + ...\right]$$

$$= n\left[pq\frac{t^2}{2!} + pq(q^2 - p^2)\frac{t^3}{3!} + \left\{pq(q^3 - p^3) - 3p^2q^2\right\}\frac{t^4}{4!} + ...\right]$$

The coefficients of $t, \frac{t^2}{2!}, \frac{t^3}{3!}, \frac{t^4}{4!}, ...$ gives the cumulants as

$$K_1 = 0, \quad K_2 = npq, \quad K_3 = npq(q - p)$$

and

$$K_4 = npq(q^3 + p^3) - 3np^2q^2 = npq\left[(q+p)(q^2 - qp + p^2) - 3pq\right]$$

$$= npq\left[(q+p)^2 - 4pq\right] = npq\left[(q^2 + p^2)^2 - 6pq\right] = npq\left[(1 - 6pq)\right]$$

SOLVED EXAMPLES

EXAMPLE 1. *For the binomial distribution, prove that*

$$\mu_{r+1} = pq\left(nr\mu_{r-1} + \frac{d\mu_r}{dp}\right).$$

SOLUTION. Since we know that

$$\mu_r = \sum_{x=0}^{n} \{f(x)\}(x - \mu_1')^r = \sum_{x=0}^{n} (^nC_x\, p^x q^{n-x})(x - np)^r$$

$$\therefore \quad \frac{d\mu_r}{dp} = \sum_{x=0}^{n} {}^nC_x\, .xp^{x-1}(1-p)^{n-x}(x - np)^r$$

$$+ \sum_{x=0}^{n} {}^nC_x\, p^x(n-x)(1-p)^{n-x-1}(-1)(x - np)^r$$

$$+ \sum_{x=0}^{n} {}^nC_x\, p^x(1-p)^{n-x}.r(x - np)^{r-1}(-n)$$

$$= \sum_{x=0}^{n} {}^{n}C_{x}\, x p^{x-1} q^{n-x}(x-np)^{r} - \sum_{x=0}^{n} {}^{n}C_{x}(n-x)\, p^{x} q^{n-x-1}(x-np)^{r}$$

$$- \sum_{x=0}^{n} {}^{n}C_{x}\, p^{x} q^{n-x}(x-np)^{r-1}.nr$$

$$= \sum_{x=0}^{n} {}^{n}C_{x}\, p^{x-1} q^{n-x-1}(x-np)^{r}[xq-(n-x)p] - nr \sum_{x=0}^{n} {}^{n}C_{x}\, p^{x} q^{n-x}(x-np)^{r-1}$$

$$= \sum_{x=0}^{n} {}^{n}C_{x}\, p^{x-1} q^{n-x-1}(x-np)^{r}[x(q+p)-np] - nr\mu_{r-1}$$

$$\Rightarrow \quad pq \frac{d\mu_{r}}{dp} = \sum_{x=0}^{n} {}^{n}C_{x}\, p^{x} q^{n-x}(x-np)^{r}(x-np) - npqr\mu_{r-1}$$

$$= \sum_{x=0}^{n} {}^{n}C_{x}\, p^{x} q^{n-x}(x-np)^{r+1} - npqr\mu_{r-1} = \mu_{r+1} - npqr\,\mu_{r-1}$$

$$\Rightarrow \quad \mu_{r+1} = pq\left(nr\mu_{r-1} + \frac{d\mu_{r}}{dp}\right).$$

EXAMPLE 2. *Obtain following recurrence relation for the binomial distribution*

$$f(x) = \frac{n-x+1}{x} \cdot \frac{p}{q} f(x-1)$$

SOLUTION . For binomial distribution, we have

$$f(x) = {}^{n}C_{x}\, p^{x}\, q^{n-x} = \frac{n!}{x!\,(n-x)!}\, p^{x}\, q^{n-x} \qquad \ldots(1)$$

$$\Rightarrow \quad f(x-1) = \frac{n!}{(x-1)!\,(n-x+1)!}.p^{x-1}\, q^{n-x+1} \qquad \ldots(2)$$

From (1) and (2),

$$\frac{f(x)}{f(x-1)} = \frac{n-x+1}{x} \cdot \frac{p}{q}$$

$$\Rightarrow \quad f(x) = \frac{n-x+1}{x} \cdot \frac{p}{q} f(x-1).$$

EXAMPLE 3. *The probability that evening college student will graduate is 0.4. Determine the probability that out of 5 students, (a) none, (b) one and (c) at least one will graduate.*

SOLUTION . Given that the probability that evening college student is graduate = 0.4.

$$p = \frac{4}{10} = \frac{2}{5},\ q = 1-\frac{2}{5} = \frac{3}{5},\ n = 5$$

If a random variable X denotes success, then probability of r success in 5 trials

$$= p(n) = {}^{5}C_{r}\, p^{r}\, q^{n-r}$$

(i) The probability of zero success

$$= {}^{5}C_{0}\left(\frac{2}{5}\right)^{0}\left(\frac{3}{5}\right)^{5-0} = 1\times1\times\left(\frac{3}{5}\right)^{5} = 0.046$$

(ii) The probability of one success

$$= {}^5C_0\left(\frac{2}{5}\right)^1\left(\frac{3}{5}\right)^{5-1} = 5\times\frac{2}{5}\times\frac{81}{625} = 0.2592$$

(iii) The probability of at least one success

$$= 1 - \text{probability of no success} = 1 - 0.046 = 0.954 \,.$$

EXAMPLE 4. *If the chance that a bus arrives safely at a bus stand is 9/10. Find the chance that out of 5 buses expected at least 4 will arrive safely.*

SOLUTION . We have $\quad P=\dfrac{9}{10}, \; n=5, \;$ Not arrive, $\quad q=1-\dfrac{9}{10}=\dfrac{1}{10}.$

Expected at least 4 will arrive

$$= P(x=4) + P(x=5)$$

$$= {}^5C_4\left(\frac{9}{10}\right)^4\left(\frac{1}{10}\right)^1 + {}^5C_5\left(\frac{9}{10}\right)^5\left(\frac{1}{10}\right)^0$$

$$= 5\times\frac{9\times9\times9\times9}{10\times10\times10\times10}\times\frac{1}{10}+1\times\left(\frac{9}{5}\right)^5$$

$$= \frac{9^4}{10^5}(5+9) = \frac{14\times9^4}{10^5}$$

EXAMPLE 5. *The mean and variance of binomial distribution are 4 and 4/3 respectively. Find $P(X >= 1)$.*

SOLUTION . Mean of binomial distribution $\quad m = np = 4 \,.$

$$\text{Variance} = npq = \frac{4}{3}$$

Putting, $\quad np = 4$

$$4q = \frac{4}{3} \quad\Rightarrow\quad q = \frac{1}{3},$$

$$p = 1-q = 1-\frac{1}{3} = \frac{2}{3}.$$

$$np = 4 \quad\Rightarrow\quad n = 4/p$$

$$n = 6$$

Thus, $\quad P(X \geq 1) = P(X > 0) = 1 - P(X = 0)$

$$= 1 - {}^6C_0\left(\frac{2}{3}\right)^0\left(\frac{1}{3}\right)^{6-0} = 1-\left(\frac{1}{3}\right)^6$$

$$P(X \geq 1) = \frac{728}{729}\,.$$

EXAMPLE 6. *A student is given a true false examination with 8 questions. If he gets 6 or more correct answers, he passes the examination. Given that he guesses the answer to each question, find the probability that he passes the examination.*

SOLUTION . Let probability of success be p.

$\Rightarrow p$ is the probability of answering a question correctly.

Then, $p = \dfrac{1}{2}$ and $q = \dfrac{1}{2}$.

Let X denote the number of successes in attempting the questions. Then X is a binomial variate with parameter $n = 8$ and $p = \dfrac{1}{2}$.

Therefore,

$$P(X - r) = {}^8C_r \left(\dfrac{1}{2}\right)^r \left(\dfrac{1}{2}\right)^{8-r} = {}^8C_r \left(\dfrac{1}{2}\right)^8, \quad r = 0, 1, 2, \ldots, 8$$

Probability of passing the examination

$$= P(X \geq 6) = P(X = 6 \text{ or } X = 7 \text{ or } X = 8)$$

$$= P(X = 6) + P(X = 7) + P(X = 8)$$

$$= {}^8C_6 \left(\dfrac{1}{2}\right)^8 + {}^8C_7 \left(\dfrac{1}{2}\right)^8 + {}^8C_8 \left(\dfrac{1}{2}\right)^8$$

$$= (28 + 8 + 1)\left(\dfrac{1}{2}\right)^8 = \dfrac{37}{256}.$$

EXAMPLE 7. *In sampling a large number of parts are manufactured by a machine, the mean number of defectives in a sample of 20 is 2. Out of 1000 such samples, how many would be expected to contain at least 3 defective parts?* **(VTU-2004)**

SOLUTION . The mean of the defectives $= 2 = np = 20p$

$$\Rightarrow \qquad p = \dfrac{2}{20} = 0.1$$

i.e., the probability of defective part is $p = 0.1$ and the probability of non-defective part is $q = 0.9$. Therefore, the probability of at least three defectives in a sample of 20

$$= 1 - (\text{prob. that either none or one or two are non-defective parts})$$

$$= 1 - ({}^{20}C_0 (0.9)^{20} + {}^{20}C_1 (0.1)(0.9)^{19} + {}^{20}C_2 (0.1)^2 (0.9)^{18})$$

$$= 1 - (0.9)^{18} \times 4.51 = 0.323$$

Thus the number of sampling having at least three defective parts out of 1000 samples

$$= 1000 \times 0.323 = 323$$

EXAMPLE 8. *A group of 50 aeroplanes are sent on an operation flight. The chance that an aeroplane fails to return from the flight is 3 percent. Find the probability that at most five planes do not return.*

SOLUTION . Let the event 'failure to return' be called a success. Then, $p = 0.03$ and $n = 50$. Therefore, probability p of x successes is

$$= {}^{50}C_x (0.03)^x (0.97)^{50-x}$$

Hence, the probability $P(x \leq 5)$ of at most 5 successors is given by

$$P = \sum_{x=0}^{5} [{}^{50}C_x (0.03)^x (0.97)^{50-x}]$$

$$= 0.218 + 0.337 + 0.256 + 0.126 + 0.046 + 0.013 = 0.995$$

EXAMPLE 9. *If 10% of the affecting aircrafts are expected to be shot down before reaching the target, what is the probability that out of 5 aircrafts at least 4 will be shot before they reach the target?*

SOLUTION . Let p and q be the probability of being shot before reaching and safely reaching the target. Then as per given, we have

$$p = \frac{1}{10}, \quad q = 1 - \frac{1}{10} = \frac{9}{10}$$

Thus, required probability of at least 4 aircrafts out of 5 being shot before reaching the target

$$= {}^5C_5 \left(\frac{1}{10}\right)^5 \left(\frac{9}{10}\right)^{5-5} + {}^5C_4 \left(\frac{1}{10}\right)^4 \left(\frac{9}{10}\right)^{5-4}$$

$$= \left(\frac{1}{10}\right)^5 + 5\left(\frac{1}{10}\right)^4 \left(\frac{9}{10}\right) = \frac{83}{50000}.$$

EXAMPLE 10. *An experiment succeeds twice as often as it fails. Find the chance that in the next six trials, there will be at least four successes.*

SOLUTION . Let p and q denote the chance of success and failure respectively. Then, we have

$$p = 2q \quad \text{or} \quad p = 2(1-p) \qquad\qquad (\because p + q = 1)$$

Therefore, $3p = 2 \Rightarrow p = 2/3$ and $q = 1/3$

Hence, the required probability of at least four successes

$$= \sum_{x=4}^{6} {}^6C_x \left(\frac{2}{3}\right)^x \left(\frac{1}{3}\right)^{6-x}$$

$$= {}^6C_4 \left(\frac{2}{3}\right)^4 \left(\frac{1}{3}\right)^2 + {}^6C_5 \left(\frac{2}{3}\right)^5 \left(\frac{1}{3}\right)^1 + {}^6C_6 \left(\frac{2}{3}\right)^6$$

EXAMPLE 11. *From a large lot of articles, with fraction defectives 1/10, a sample of 10 articles are taken. What is the probability that the same will not contain any defective articles.*

SOLUTION . Let us denote the event of obtaining a defective articles as a success. As the lot is large, the probability p of success for any trial is constant and equal to $p = \frac{1}{10}$.

$$\Rightarrow \qquad q = 1 - p = 1 - \frac{1}{10} = \frac{9}{10}$$

Hence, required probability is

$$= {}^nC_0 \left(\frac{1}{10}\right)^0 \left(\frac{9}{10}\right)^{10} = \left(\frac{9}{10}\right)^{10} = \left(1 - \frac{1}{10}\right)^{10}$$

$$= 1 - 10 \times \frac{1}{10} + \frac{10 \times 9}{2} \left(\frac{1}{10}\right)^2 = 0.45 .$$

EXAMPLE 12. *Two persons A and B throw with one dice for a stake of Rs. 11 which is to be won by the player who first throws 6. If A has the first throw, what are their respective expectations?*

SOLUTION . We know that in the first throw, the probability of A's success is $\frac{1}{6}$.

Further, each player must have failed once before A can have a second throw, therefore in second throw, the probability of A's success is

$$\left(\frac{5}{6} \times \frac{5}{6}\right) \times \frac{1}{6} = \left(\frac{5}{6}\right)^2 \times \frac{1}{6}$$

Similarly in his first throw, the probability of his success is $\left(\frac{5}{6}\right)^4 \times \frac{1}{6}$ because each player must failed twice and so on.

Proceeding in the same way, the probability of A's success is the sum of infinite series $\frac{1}{6}\left[1 + \left(\frac{5}{6}\right)^2 + \left(\frac{5}{6}\right)^4 + ...\right]$.

Similarly, the probability of B's success is the sum of infinite series

$$\frac{5}{6} \times \frac{1}{6}\left[1 + \left(\frac{5}{6}\right)^2 + \left(\frac{5}{6}\right)^4 + ...\right]$$

Hence, A's success to B's as 6 is to 5. Thus, their respective probabilities of success are $\frac{6}{11}$ and $\frac{5}{11}$. Therefore, respective expectations of A and B are Rs. 6 and Rs. 5 respectively.

EXAMPLE 13. *Six dice are thrown 729 times. How many times do you expect at least 3 dice to show 5 or 6?*

SOLUTION . The probability of throwing 5 or 6

$$= \frac{1}{6} + \frac{1}{6} = \frac{1}{3}$$

Now, probability of showing 5 or 6 at least 3 dice
= Sum of the probabilities of showing 5 or 6 by 3, 4, 5, 6 dice

$$= \sum_{x=3}^{6} {}^6C_x \left(\frac{1}{3}\right)^x \left(1 - \frac{1}{3}\right)^{6-x} = \sum_{x=3}^{6} {}^6C_x \left(\frac{1}{3}\right)^x \left(\frac{2}{3}\right)^{6-x}$$

$$= {}^6C_3 \left(\frac{1}{3}\right)^3 \left(\frac{2}{3}\right)^3 + {}^6C_4 \left(\frac{1}{3}\right)^4 \left(\frac{2}{3}\right)^2 + {}^6C_5 \left(\frac{1}{3}\right)^5 \left(\frac{2}{3}\right)$$

Hence, expected number out of 729 throws

$$= 729 \times \frac{233}{729} = 233.$$

EXAMPLE 14. *Three persons A, B and C in order to cut a pack of cards replacing them after each cut or the condition that the first who cuts a card of spade shall win a prize. Find their respective chances.*

SOLUTION . We know that the probability of getting a card of spade $= \frac{13}{52} = \frac{1}{4}$

Probability of getting a card other than spade $= \frac{39}{52} = \frac{3}{4}$.

Thus, A's chance of winning

      = Probability that he gets a spade in first chance

        + probability that he draws a spade in second chance

            when B and C fail in the first chance

+ probability that he draws a spade in 3rd chance

when B and C fail in 1st and 2nd chances + ...

$$= \frac{1}{4} + \left(\frac{3}{4}\right)\left(\frac{3}{4}\right)\left(\frac{3}{4}\right)\left(\frac{1}{4}\right) + \left(\frac{3}{4}\right)\left(\frac{3}{4}\right)\left(\frac{3}{4}\right)\left(\frac{3}{4}\right)\left(\frac{3}{4}\right)\left(\frac{3}{4}\right)\left(\frac{1}{4}\right) + ...$$

$$= \frac{1}{4} + \left(\frac{3}{4}\right)^3\left(\frac{1}{4}\right) + \left(\frac{3}{4}\right)^6\left(\frac{1}{4}\right) + ...$$

$$= \frac{1}{4}\left[1 + \left(\frac{3}{4}\right)^3 + \left(\frac{3}{4}\right)^6 + \left(\frac{3}{4}\right)^9 + ...\right], \text{ which is an infinite G.P.}$$

$$= \frac{1}{4}\left[\frac{1}{\left(1-(3/4)^3\right)}\right] = \frac{1}{4}\left[\frac{64}{64-27}\right] = \frac{16}{37}$$

$$\left[\because \text{Sum of infinite G.P.} = \frac{a}{1-r}\right]$$

Similarly, we have the B's chance of winning

$$= \frac{3}{4}\left(\frac{1}{4}\right) + \left(\frac{3}{4}\right)^4 \cdot \frac{1}{4} + \left(\frac{3}{4}\right)^7\left(\frac{1}{4}\right) + ...$$

$$= \frac{1}{4}\left[\frac{3}{4} + \left(\frac{3}{4}\right)^4 + \left(\frac{3}{4}\right)^7 + ...\right]$$

$$= \frac{1}{4}\left[\frac{3/4}{1-(3/4)^3}\right] = \frac{1}{4}\left[\frac{3/4}{(64-27)/64}\right]$$

$$= \frac{1}{4}\left[\frac{3/4}{37/64}\right] = \frac{1}{4}\left[\frac{3}{4} \times \frac{64}{37}\right] = \frac{12}{37}$$

Finally, C's chance of winning

$$= \left(\frac{3}{4}\right)^2\left(\frac{1}{4}\right) + \left(\frac{3}{4}\right)^5\left(\frac{1}{4}\right) + \left(\frac{3}{4}\right)^8\left(\frac{1}{4}\right) + ...$$

$$= \frac{1}{4}\left[\left(\frac{3}{4}\right)^2 + \left(\frac{3}{4}\right)^5 + \left(\frac{3}{4}\right)^8 + ...\right] = \frac{1}{4}\left[\frac{(3/4)^2}{1-(3/4)^3}\right] = \frac{9}{37}$$

EXAMPLE 15. *The probability of A wins a game against B is 2/3. Find the probability that A wins at least 3 game out of 5 games.*

SOLUTION . Let us suppose, p be the probability of A winning a game against B. Then as per given, we have

$$p = \frac{2}{3}, \quad q = 1 - \frac{2}{3} = \frac{1}{3}$$

Hence, required probability

= Probability of A winning 3 games

+ probability of A winning 4 games

+ probability of A winning 5 games

$$= \sum_{x=3}^{5} {}^{5}C_x \left(\frac{2}{3}\right)^x \left(\frac{1}{3}\right)^{5-x}$$

$$= {}^{5}C_3 \left(\frac{2}{3}\right)^3 \left(\frac{1}{3}\right)^2 + {}^{5}C_4 \left(\frac{2}{3}\right)^4 \left(\frac{1}{3}\right) + {}^{5}C_5 \left(\frac{2}{3}\right)^5 = \frac{64}{81}.$$

EXAMPLE 16. *A factory A produces 10% defective valves and another factory B produces 20% defectives. A bag contains 4 values of factory A and 5 valves of factory B. If two valves are drawn at random from the bag, find the probability that at least one valve is defective.*

SOLUTION . Let A_1, A_2, A_3 be the events of selecting 2 from the factory A, 1 from factory A, and 1 from B and 2 from factory B respectively.

Then, we have

$$P(A_1) = \frac{{}^{4}C_2}{{}^{9}C_2} = \frac{1}{6}, \; P(A_2) = \frac{{}^{4}C_1 - {}^{5}C_1}{{}^{9}C_2} = \frac{5}{9}, \; P(A_3) = \frac{{}^{5}C_2}{{}^{9}C_2} = \frac{5}{18}$$

Further, suppose that E be the event that both selections contain at least one defective valve. Then E' is the event that both selected valves are roots of affections.

Since A_1, A_2, A_3 are mutually exclusive and exhaustive events. Therefore

$$E' = E'A_1 \cup E'A_2 \cup E'A_3$$

$$= P(A_1) P(E' | A_1) + P(A_2) P(E' | A_2) + P(A_3) P(E' | A_3)$$

$$= \frac{1}{6} (0.9)^2 + \frac{5}{9}(0.9)(0.8) + \frac{5}{18}(0.8)^2 = \frac{12.83}{18}$$

Hence, required probability

$$P(E) = 1 - \frac{12.83}{18} = \frac{5.18}{18} = 0.2872$$

EXAMPLE 17. *In how many ways 3 girls and 9 boys can be seated in two vans, each having numbered seats 3 in front and 4 at back? How many seating arrangements are possible if 3 girls should sit together in a back row on adjacent seats. If all the seating arrangements are equally likely, what is the probability of 3 girls sitting together in back row on adjacent seats?*

SOLUTION . We have the number of ways that 14 seats can be filled by 12 people, *i.e.*, 3 girls and 9 boys

$$= {}^{14}C_{12} = 91$$

For each such way, the 12 seats are occupied. The arrangements are $\dfrac{12!}{3! \, 9!}$.

Thus, total number of arrangements

$$= 91 \times \frac{12!}{3! \, 9!}.$$

With 3 girls in the back row, there are 4 ways. For each such way, the remaining 11 seats can be occupied by 9 boys in $^{11}C_9$ ways. Therefore, total number of arrangement with girls in the adjacent seats

$$= 4 \times {}^{11}C_9 = 4 \times {}^{11}C_2$$

Hence, required probability

$$= \frac{4 \times {}^{11}C_2}{91 \times 12!} \times 3! \times 9! = \frac{4 \times 11!}{9!\,.\,2!}\,3!\,.\,9!\,.\,\frac{1}{91}\,.\,\frac{1}{12!}$$

$$= 12 \times 11! \times \frac{1}{91} \times \frac{1}{12!} = \frac{1}{91}.$$

EXAMPLE 18. *A cartoon contains* 100 *capsules,* 20 *of which are defectives,* 10 *are selected, indicate what is the probability that all are good?* **(RGPV–2004)**

SOLUTION.　The number of defective capsules, $m = 20$

The total number of capsules, $N = 100$

Now, the probability of 20 defective capsules

$$\frac{m}{N} = \frac{20}{100} = \frac{1}{5}$$

$$q = \frac{1}{5} \text{ and } p = 1 - q = 1 - \frac{1}{5} = \frac{4}{5}$$

Given that　　$n = 10$

$$p(n) = {}^nC_r \cdot q^{n-r} p^r$$

Probability for all capsules are good.

$$p(10) = {}^{10}C_{10}\left(\frac{1}{5}\right)^{10-10}\left(\frac{4}{5}\right)^{10} = {}^{10}C_{10}\left(\frac{1}{5}\right)^{0}\left(\frac{4}{5}\right)^{10} = 0.1073$$

Exercise 23.4

1. If on an average 1 vessels in every 10 is wrecked, find the probability that out of 5 vessels expected to arrive, at least 4 will arrive safely. **(PTU–2005)**

2. Out of 800 families with 5 children each, how many would you expect to have (a) 3 boys, (b) 5 girls, (c) either 2 or 3 boys? Assume equal probability for boys and girls.
   (VTU–2004)

3. Fit a binomial distribution for the following data and compare the theoretical frequencies with the actual ones. **(Bhopal–2006)**

x	0	1	2	3	4	5
f	2	14	20	34	22	8

4. The probability that a bomb dropped from a plane will strike the target is 1/5. If six bombs are dropped, find the probability that

   (i) exactly two will strike the target
   (ii) at least two will strike the target

5. If 10% of rivets produced by a machine are defective, find the probability that out of 5 rivets chosen at random
   (i) none will be defective
   (ii) one will be defective, and
   (iii) at least two will be defective

6. Find the binomial distribution whose mean is 5 and variance is 10/3.

7. If 10% of bolts produced by a machine are defective, calculate the probability that out of a sample selected at random of 7 bolts, not more than one bolt will be defective.

8. The sum and product of the mean and variance of a binomial distribution are 24 and 128 respectively. Find the distribution.

9. A dice is thrown 6 times. If getting an odd number is a success, find the probability of
   (i) 5 successes
   (ii) at least 5 successes
   (iii) at most 5 successes

10. If the probability that a new born child is a male is 0.6, find the probability that in a family of 5 children there are exactly 3 boys.
    (Kurukshetra–2005)

11. The probability that a pen manufactured by a company will be defective is 1/10. If 12 such pens are manufactured, find the probability that:
    (a) exactly two will be defective.
    (b) at least two will be defective
    (c) none will be defective.
    (VTU–2004, Burdwan–2003)

12. In 256 sets of 12 tosses of a coin, in how many cases one can except 8 heads and 4 tails. **(JNTU–2003)**

13. In a bombing action there is 50% chance that any bomb will strike the target. Two direct hits are needed to destroy the target completely. How many bombs are required to be dropped to give a 99% chance or better of completely destroying the target. **(VTU–2003S)**

14. 500 articles were selected at random out of a batch containing 10,000 articles and 30 were found to be defective. How many defectives articles would you reasonably expect to have in the whole batch? **(JNTU–2003)**

15. Fit a binomial distribution to the following frequency distribution :

x	0	1	2	3	4	5	6
f	13	25	52	58	32	16	4

(Kurukshetra–2009, SVTU–2007)

ANSWERS

1. 45927/50000   **2.** (a) 250,   (b) 25,   (c) 500   **3.** $100(0.432 + 0.568)^5$

4. (i) 0.246,   (ii) 0.345   **5.** (i) 0.59049,   (ii) 0.32805,   (iii) 0.08146

6. $P(r) \ ^{15}C_r \left(\dfrac{1}{3}\right)^r \left(\dfrac{2}{3}\right)^{15-r}$ **7.** $(1.6)(0.9)^6$

8. $P(X = r) = \ ^{32}C_r \left(\dfrac{1}{2}\right)^r \left(\dfrac{1}{2}\right)^{32-r}, r = 0, 1, 2, ..., 32$ **9.** (i) 3/22   (ii) 7/64   (iii) 63/64

10. 0.3456   **11.** (a) 0.2301,   (b) 0.3412   (c) 0.2833   **12.** 31 (say),

13. 11   **14.** 600   **15.** $200 (0.554 + 0.446)^6$

23.15 POISSON DISTRIBUTION

A variable x is said to have the Poisson distribution if it takes the values 0, 1, 2, ... to ∞ with probabilities $e^{-m}, \dfrac{m}{1!}e^{-m}, \dfrac{m^2}{2!}e^{-m}, \dfrac{m^3}{3!}e^{-m}, ...$ to ∞ respectively. The probability distribution function for Poisson distribution is given by

$$P(x, m) = P(X = x) = \begin{cases} \dfrac{e^{-m} \cdot m^x}{x!}, & x = 0, 1, 2 \\ 0, & \text{otherwise} \end{cases}$$

🕭 REMARKS
➠ If Poisson distribution was discovered by French mathematician S. D. Poisson.
➠ Poisson distribution has only one parameter m.
➠ Poisson distribution occurs when there are events which do not occur as outcomes of a definite number of trials of an experiment but which occur at random points of time and space where our interest lies only in the number of occurrence of the event, not its non-occurrence.
➠ It should be noted that $\displaystyle\sum_{x=0}^{\infty} P(X = x) = e^{-m} \sum_{x=0}^{\infty} \dfrac{m^x}{x!} = e^{-m} \cdot e^m = 1$

23.16 LIMITING FORM OF BINOMIAL DISTRIBUTION

THEOREM 1. *Poisson distribution is the limiting form of the binomial distribution* $(q + p)^n$ *when* $p(\text{or } q) \to 0$ *as* $n \to \infty$ *such that* np *(or* nq*) is a finite quantity say* m, *i.e.,* $np = m$.

PROOF. We know that, the probability of r success in a binomial distribution is

$$P(r) = {}^nC_r \, p^r \, q^{n-r}$$

$$= \frac{n(n-1)(n-2)\dots(n-r+1)}{r!} p^r \, q^{n-r}$$

$$= \frac{np(np-p)(np-2p)\dots(np-(r-1)p)}{r!}(1-p)^{n-r}$$

As $n \to \infty$, $p \to 0$ $(np = m)$, we get

$$P(r) = \frac{m^r}{r!} \lim_{n\to\infty} \frac{\left(1-\dfrac{m}{n}\right)^n}{\left(1-\dfrac{m}{n}\right)^r} = \frac{m^r}{r!} e^{-m}$$

Therefore, the probabilities of $0, 1, 2, \dots, r$ successes in a Poisson distribution are given by

$$e^{-m}, me^{-m}, \frac{m^2 e^{-m}}{2!}, \dots, \frac{m^r e^{-m}}{r!}, \dots$$

Hence, the limiting form of the binomial distribution $(q + p)^n$ when $p \to 0$ and $n \to \infty$ such that $np = m$ is the Poisson distribution.

23.17 MOMENTS OF POISSON DISTRIBUTION

23.17.1 MOMENT ABOUT ORIGIN

(1) First Moment about Origin (Mean of Poisson Distribution)

$$\mu'_1 = \sum_{x=0}^{\infty} \frac{e^{-m} m^x}{x!} \cdot x$$

$$= e^{-m}.0 + \left(e^{-m}.\frac{m}{1!}.1\right) + \left(e^{-m}.\frac{m^2}{2!}.2\right) + \dots + \left(e^{-m}.\frac{m^r}{r!}\right) + \dots$$

$$= m\,e^{-m}\left[1 + \frac{m}{1!} + \frac{m^2}{2!} + \dots + \frac{m^{r-1}}{(r-1)!} + \dots\right] = me^{-m}\,[e^m] = m$$

$\Rightarrow$ Mean of the Poisson distribution $= \mu'_1 = m$.

(2) Second Moment about Origin

$$\mu'_2 = \sum_{x=0}^{\infty} \frac{e^{-m}.m^x}{x!}.x^2 = \sum_{x=0}^{\infty} e^{-m} \frac{m^x}{x!}\{x(x-1)+x\}$$

$$= \sum_{x=0}^{\infty} \frac{e^{-m}.m^x}{(x-2)!} + \sum_{x=0}^{\infty} \frac{e^{-m}.m^x}{(x-1)!} = \sum_{x=0}^{\infty} \frac{m^2 e^{-m}.m^{x-2}}{(x-2)!} + \sum_{x=0}^{\infty} \frac{me^{-m}.m^{x-1}}{(x-1)!}$$

$$= m^2 e^{-m} \sum_{x=0}^{\infty} \frac{m^{x-2}}{(x-2)!} + me^{-m} \sum_{x=0}^{\infty} \frac{m^{x-1}}{(x-1)!} = m^2 e^{-m}e^m + me^{-m}e^m = m^2 + m$$

(3) Third Moment about Origin

$$\mu_3' = \sum_{x=0}^{\infty} \frac{e^{-m} \cdot m^x}{x!} \cdot x^3 = \sum_{x=0}^{\infty} e^{-m} \frac{m^x}{x!} \{x(x-1)(x-2) + 3x(x-1) + x\}$$

$$\Rightarrow \quad \mu_3' = m^3 + 3m^2 + m$$

(4) Fourth Moment about Origin

$$\mu_4' = \sum_{x=0}^{\infty} \frac{e^{-m} \cdot m^x}{x!} \cdot x^4$$

$$= \sum_{x=0}^{\infty} e^{-m} \frac{m^x}{x!} \{x(x-1)(x-2)(x-3) + 6x(x-1)(x-2) + 7x(x-1) + x\}$$

$$\Rightarrow \quad \mu_4' = m^4 + 6m^3 + 7m^2 + m$$

23.17.2 MOMENT ABOUT MEAN

(1) First Moment about Mean

$$\mu_1 = 0$$

(2) Second Moment about Mean (Variance of Poisson Distribution)

$$\mu_2 = \mu_2' - (\mu_1')^2 = m^2 + m - m^2 = m$$

$$\Rightarrow \text{Variance} = \mu_2 = m$$

and S.D. $= \sqrt{\mu_2} = \sqrt{m}$.

(3) Third Moment about Mean

$$\mu_3 = \mu_3' - 3\mu_2'\mu_1' + 2\mu_1'^3 = (m^3 + 3m^2 + m) - 3(m^2 + m)m + 2(m)^3$$

$$= m^3 + 3m^2 + m - 3m^3 - 3m^2 + 2m^3 = m$$

REMARK

⟹ For poisson distribution, mean = variance = m.

(4) Fourth Moment about Mean

We know that

$$\mu_4 = \mu_4' - 4\mu_3'\mu_1' + 6\mu_2'\mu_1'^2 - 3\mu_1'^4$$

$$= (m^4 + 6m^3 + 7m^2 + m) - 4(m^3 + 3m^2 + m)m + 6(m^2 + m)(m)^2 - 3(m)^4$$

$$= m^4 + 6m^3 + 7m^2 + m - 4m^4 - 12m^3 - 4m^3 + 6m^4 + 6m^3 - 3m^4$$

$$\Rightarrow \quad \mu_4 = 3m^2 + m$$

23.17.3 PEARSON'S COEFFICIENTS

(1) $\beta_1 = \dfrac{\mu_3^2}{\mu_2^3} = \dfrac{m^2}{m^3} = \dfrac{1}{m}$

(2) $\beta_2 = \dfrac{\mu_4}{\mu_2^2} = \dfrac{3m^2 + m}{m^2} = 3 + \dfrac{1}{m}$

(2) $\gamma_1 = \sqrt{\beta_1} = \dfrac{1}{\sqrt{m}}$

(4) $\gamma_2 = \beta_2 - 3 = \dfrac{1}{m}$

23.18 MOMENT GENERATING FUNCTION (M.G.F.) OF POISSON DISTRIBUTION

The probability distribution function of Poisson distribution is given by $P(x) = \dfrac{e^{-m} m^x}{x!}$

Therefore, m.g.f. about origin will be given by

$$M_0(t) = \sum_{x=0}^{\infty} e^{tx} \left(\frac{e^{-m} m^x}{x!} \right)$$

$$= e^{-m} \sum_{x=0}^{\infty} \frac{(m e^t)^x}{x!} = e^{-m} e^{m e^t} \qquad \left(\because e^x = \sum_{r=0}^{\infty} \left(\frac{x^r}{r!} \right) \right)$$

$$\Rightarrow \qquad M_0(t) = e^{m(e^t - 1)}$$

Also, we have the mean of Poisson distribution is m and $M_a(t) = e^{-at} M_0(t)$, where $M_0(t)$ is the m.g.f. about the point a.

Therefore, m.g.f. about mean will be given by $M_m(t) = e^{-mt} M_0(t) = e^{-mt} e^{m(e^t - 1)}$

$\Rightarrow M_m(t) = e^{m(e^t - 1 - t)}$.

23.19 CUMULANT GENERATING FUNCTION OF POISSON DISTRIBUTION

We know that the moment generating function of Poisson distribution is given by $M_m(t) = e^{m(e^t - 1 - t)}$.

Now, cumulant function $K_m(t) = \log M_m(t)$

$$= m(e^t - t - 1) = m \left[1 + t + \frac{t^2}{2!} + \frac{t^3}{3!} + \frac{t^4}{4!} + \dots - t - 1 \right] = m \left[\frac{t^2}{2!} + \frac{t^3}{3!} + \frac{t^4}{4!} + \dots \right]$$

Now, the coefficients of $\dfrac{t}{1!}, \dfrac{t^2}{2!}, \dfrac{t^3}{3!}, \dfrac{t^4}{4!}, \dots$ give the cumulant such that

$$K_1 = 0, \quad K_2 = m, \quad K_3 = m, \quad K_4 = m, \dots$$

23.19.1 SOME EXAMPLES OF POISSON VARIATE

(1) Number of suicides reported in a particular city.
(2) Number of air accidents in some unit of time.
(3) Number of printing mistakes at each page of the book.
(4) Number of telephone calls received at a particular telephone exchange in some unit of time.
(5) Number of car passing a crossing per hour during the busy hour of a day.
(6) Number of deaths from a disease.

☛ RECAPITULATIONS

For Poisson Distribution

(1) **Moments about origin** $\mu'_1 = m, \mu'_2 = m^2 + m, \mu'_3 = m^3 + 3m^2 + m, \mu'_4 = m^4 + 6m^3 + 7m^2 + m$

(2) **Moments about mean** $\mu_1 = 0, \mu_2 = m, \mu_3 = m, \mu_4 = 3m^2 + m$

(3) **Pearson's Coefficients** $\beta_1 = \dfrac{1}{m}, \beta_2 = 3 + \dfrac{1}{m}, \gamma_1 = \dfrac{1}{\sqrt{m}}, \gamma_2 = \dfrac{1}{m}$

SOLVED EXAMPLES

EXAMPLE 1. *Find the probability that at most 5 defective fuses will be found in a box of 200 fuses, if experience shows that 2 percent of such fuses are defective.*

SOLUTION. The probability of a fuse being defective is $\dfrac{2}{100}$

Here, $\qquad p = \dfrac{2}{100} = \dfrac{1}{50},\ \ n = 200\ \Rightarrow\ m = np = 200 \times \dfrac{1}{50} = 4$

$\therefore \qquad e^{-m} = e^{-4} = \dfrac{1}{e^4} = \dfrac{1}{(2.7)^4} = (0.3703)^4 = 0.019$

Now, the required probability of five or less than five defective fuses

$$= \sum_{x=0}^{5} \frac{e^{-m}.m^x}{x!} \qquad\qquad (\because\ x \le 5)$$

$$= \sum_{x=0}^{5} \frac{e^{-4}.4^x}{x!} \qquad\qquad (\because\ m = 4)$$

$$= \frac{e^{-4}.4^0}{0!} + \frac{e^{-4}.4^1}{1!} + \frac{e^{-4}.4^2}{2!} + \frac{e^{-4}.4^3}{3!} + \frac{e^{-4}.4^4}{4!} + \frac{e^{-4}.4^5}{5!}$$

$$= e^{-4}\left[1 + 4 + 8 + \frac{32}{3} + \frac{32}{3} + \frac{128}{15}\right]$$

$$= 0.019\left(\frac{643}{15}\right) = 0.814.$$

EXAMPLE 2. *Fit a Poisson distribution to the following and calculate theoretical frequencies.*

Death (x)	0	1	2	3	4
Frequency (f)	122	60	15	2	1

(Bhopal–2007S, VTU–2004, UPTU–2003)

SOLUTION. Since we know that

$$\frac{\Sigma fx}{\Sigma f} = \frac{(122 \times 0) + (60 \times 1) + (15 \times 2) + (2 \times 3) + (1 \times 4)}{122 + 60 + 15 + 2 + 1}$$

$$= \frac{60 + 30 + 6 + 4}{200} = \frac{100}{200} = \frac{1}{2} = 0.5$$

and $\qquad e^{-m} = e^{-0.5} = 0.61$.

The probability of x deaths is given by $\dfrac{Ne^{-m}m^x}{x!}$
where $N = 200$ and $m = 0.5$.

Therefore, the required theoretical frequencies of 0, 1, 2, 3 and 4 deaths are respectively given by

$$\frac{200 \times e^{-0.5}(0.5)^0}{0!},\ \frac{200 \times e^{-0.5}(0.5)^1}{1!},\ \frac{200 \times e^{-0.5}(0.5)^2}{2!},\ \frac{200 \times e^{-0.5}(0.5)^3}{3!}$$

and $\dfrac{200 \times e^{-0.5}(0.5)^4}{4!}$

i.e.,

$$200 \times 0.61,\ 200 \times 0.61 \times 0.5,\ \frac{200 \times 0.61 \times (0.5)^2}{2!}.$$

$$\frac{200 \times 0.61 \times (0.5)^3}{3!} \quad \text{and} \quad \frac{200 \times 0.61 \times (0.5)^4}{4!}$$

$$\Rightarrow \quad 122, 61, 15.25, 2.54 \quad \text{and} \quad 0.31,$$

i.e., $\quad$ 122, 61, 15, 3, 0

EXAMPLE 3. *A book of 585 pages contain 43 pages with misprints. If these pages are randomly distributed throughout the book, what is the probability that 10 pages, selected at random will be free from pages with misprint* $(e^{-0.735} = 0.4795)$.

SOLUTION . Let p denote the probability that a page selected at random is a page with misprints.

$$\Rightarrow \qquad p = \frac{43}{585} = 0.0735$$

Here, $\qquad n = 10$

$$\Rightarrow \qquad m = np = 10 \times 0.0735 = 0.735$$

Let X denote the number of pages with misprints in a sample of 10 pages. Then X is a Poisson variate with parameter $m = 0.735$.

$$\therefore \qquad P(X = r) = \frac{e^{-0.735} . (0.735)^r}{r!}, \quad r = 1, 2, 3, ..., 10$$

The probability that the sample of 10 pages will be free from pages with misprint

$$= P(X = 0) = e^{-0.735} = 0.4795 .$$

EXAMPLE 4. *Using Poisson distribution, find the probability that the ace of spades will be drawn from a pack of well shuffled cards at least once in 104 consecutive trials.*

SOLUTION . The probability of drawing an ace of spades from a pack of well-shuffled cards

$$= \frac{1}{52} p = \text{ the number of trials} = n = 104$$

$$\text{Mean} = m = np = 104 \times \frac{1}{52} = 2 .$$

The probability of drawing an ace at most once in 104 trials

= the sum of probabilities of drawing no ace

and exactly one ace in 104 trials

$$= \frac{e^{-m} m^o}{0!} + \frac{e^{-m} . m!}{1!} = e^{-m} [1 + m]$$

$$= e^{-2}(1 + 2) = 3e^{-2}$$

$$= 3(0.1353) = 0.4059 \qquad \text{(say)}$$

Hence, required probability of drawing an ace at least once in 4 trials

$$= 1 - P = 1 - 0.4059 = 0.5941 .$$

EXAMPLE 5. *Five coins are tossed 5400 times. Using the Poisson distribution, what is the approximate probability of getting 5 heads x times.*

SOLUTION . The probability of getting all the heads in a throw of 5 coins

$$= \left(\frac{1}{2}\right)^5 = \frac{1}{32} = (q)^m$$

$$\text{Mean} = np = 5400 \times \frac{1}{32} = \frac{2700}{16} = \lambda$$

Hence, the probability of getting 5 heads x times according to Poisson distribution is given by

$$p(x) = \frac{e^{-\lambda}\lambda^x}{x!} = \frac{e^{-(2700/16)}(2700/16)^x}{x!}$$

EXAMPLE 6. If a random variable x follows Poisson distribution such that $p(x = 1) = p(x = 2)$, then find

(i) the mean of the distribution

(ii) $p(x = 0)$

SOLUTION . We have

$$p(x) = \frac{e^{-\lambda}\lambda^x}{x!}; \quad x = 0, 1, 2, \ldots$$

We have

$$p(x = 1) = p(x = 2)$$

$$\Rightarrow \quad \frac{e^{-\lambda}\lambda}{1!} = \frac{e^{-\lambda}\lambda^2}{2!}$$

Hence, mean of the distribution is $\lambda = 2$.

Also, $\quad p(x) = \frac{e^{-\lambda}\lambda^x}{x!} = \frac{e^{-2}2^x}{x!}$

$$\Rightarrow \quad p(x = 0) = e^{-2} = 0.1353.$$

EXAMPLE 7. If m and μ_r denote the mean and r^{th} central moment of a Poisson distribution, then prove that

$$\mu_{r+1} = rm\mu_{r-1} + \frac{md\mu_r}{dm}$$

Hence, obtain β_1 and β_2 of Poisson distribution.

SOLUTION . We have $\dfrac{d\mu_r}{dm} = \dfrac{d}{dm}\left[\displaystyle\sum_{x=0}^{\infty}(x-m)^r \dfrac{e^{-m}m^x}{x!}\right]$

$$= -r\sum_{x=0}^{\infty}(x-m)^{r-1}\frac{e^{-m}m^x}{x!} + \sum_{x=0}^{\infty}\frac{(x-m)^r}{x!}[xm^{x-1}e^{-m} - m^x e^{-m}]$$

$$= -r\mu_{r-1} + \sum_{r=0}^{\infty}\frac{(x-m)^r e^{-m}x^{m-1}(x-m)}{x!}$$

Multiplying both sides by m

$$m\frac{d\mu_r}{dm} = -mr\mu_{r-1} + \sum_{x=0}^{\infty}\frac{(x-m)^{r+1}e^{-m}m^x}{x!}$$

$$m\frac{d\mu_r}{dm} = -mr\mu_{r-1} + \mu_{r+1}$$

Hence, $\quad \mu_{r+1} = mr\mu_{r-1} + m\dfrac{d\mu_r}{dm}$

Now, $\mu_2 = m\mu_0 + m\dfrac{d}{dm}\mu_1$ $\qquad\qquad [\mu_1 = 0]$

$\qquad\qquad = m.1 + 0 = m$

and $\qquad \mu_3 = m.2\mu_1 + m.\dfrac{d}{dm}\mu_2 = 0 + m.\dfrac{dm}{dm} = m$

Also, $\qquad \beta_1 = \dfrac{\mu_3^2}{\mu_2^3} = \dfrac{m^2}{m^3} = \dfrac{1}{m}$

$\qquad\qquad \beta_2 = \dfrac{\mu_4}{\mu_2^2} = \dfrac{3m^2 + m}{m^2} = \left(3 + \dfrac{1}{m}\right)$

EXAMPLE 8. *Derive Poisson distribution as an approximation of binomial distribution.*

SOLUTION. In a binomial distribution

$$P(r) = P(x = r) = {}^nC_r\, p^r\, q^{n-r} = {}^nC_r\, p^r\, (1-p)^{n-r}$$

$$= \dfrac{n(n-1)(n-2)\ldots(n-r+1)}{r!} p^r \left(1 - \dfrac{np}{n}\right)^{n-r}$$

$$= \dfrac{\left(1 - \dfrac{1}{n}\right)\left(1 - \dfrac{2}{n}\right)\ldots\left(1 - \dfrac{r-1}{n}\right)(np)^r}{r!\left(1 - \dfrac{np}{n}\right)^r}\left(1 - \dfrac{np}{n}\right)^n$$

$\therefore \qquad P(r) =$ Probability of r successes in Poisson distribution

$$= \lim_{p\to 0, n\to\infty, np=m} p(x = r)$$

$$= \lim_{n\to\infty} \dfrac{\left(1 - \dfrac{1}{n}\right)\left(1 - \dfrac{2}{n}\right)\ldots\left(1 - \dfrac{r-1}{n}\right)}{r!} \dfrac{(np)^r}{\left(1 - \dfrac{np}{n}\right)^r} \times \left(1 - \dfrac{np}{n}\right)^n$$

$$= \lim_{n\to\infty} \dfrac{m^r}{\left(1 - \dfrac{m}{n}\right)^r r!}\left(1 - \dfrac{m}{n}\right)^n = \dfrac{m^r}{r!}\lim_{n\to\infty}\left(1 - \dfrac{m}{n}\right)^n$$

$$P(r) = \dfrac{m^r e^{-m}}{r!} \qquad\qquad \left[\because \lim_{n\to\infty}\left(1 - \dfrac{m}{n}\right)^n = e^{-m}\right]$$

which is Poisson distribution.

EXAMPLE 9. *Fit a Poisson distribution to the following data :*

(X)	0	1	2	3	4
(Y)	120	60	15	3	2

SOLUTION. Let $\qquad m = \bar{x} = \dfrac{\Sigma fx}{N} = \dfrac{0\times120 + 1\times60 + 2\times15 + 3\times3 + 4\times2}{200}$

$$= \dfrac{107}{200} = 0.5$$

Let us take the mean of the given distribution as the mean of the Poisson distribution we want to fit.

Let $\qquad k = e^{-0.5}$

$$\log k = (-0.5) \log e = (-0.5) (\log 2.7183)$$

$$= -0.5 \times 0.4343 = -0.21715$$

$$\log k = -\bar{1}.78285$$

$$k = antilog\ (\bar{1}.78285) \qquad\qquad\qquad [k = 0.6064]$$

Thus, $\qquad e^{-0.5} = 0.6064$

$$P(0) = e^{-0.5} = 0.6064$$

Using $\ P(r+1) = \dfrac{m}{r+1}.P(r)$, we get

$$P(1) = 0.5 \times P(0) = 0.5 \times 0.6064 = 0.3032$$

$$P(2) = \frac{0.5}{2} \times P(1) = \frac{0.5 \times 0.3032}{2} = 0.0758$$

$$P(3) = \frac{0.5}{3} \times P(2) = \frac{0.5 \times 0.0758}{3} = 0.01263$$

$$P(4) = \frac{0.5}{4} \times P(3) = \frac{0.5 \times 0.01263}{4} = 0.0015787$$

Now, $\qquad N = 200$

Let $\qquad F(r) = N \times P(r) = 200 \times P(r)$

Since the observed frequencies are non-negative integers, we convert the expected frequencies to nearest integers as shown below :

$$f(0) = 200 \times P(0) = 200 \times 0.6064 = 121.28 \approx 121$$

$$f(1) = 200 \times P(1) = 200 \times 0.3032 = 60.64 \approx 61$$

$$f(2) = 200 \times P(2) = 200 \times 0.0758 = 15.16 \approx 15$$

$$f(3) = 200 \times P(3) = 200 \times 0.01263 = 2.526 \approx 3$$

$$f(4) = 200 \times P(4) = 200 \times 0.0015787 = 0.31574 \approx 0.$$

EXAMPLE 10. *The probability that an injection manufactured by a company will be defective is $\dfrac{1}{10}$. If 12 such injections are manufactured, find the probability that.* **(RGPV–2005)**

    (a) Exactly two will be defective.   *(b) Atleast two will be defective.*

    (c) None will be defective.

SOLUTION.   It is given that the probability of defective injection is

$$p = \frac{1}{10}\ \text{and}\ n = 12$$

Now, $m = n \cdot p = \dfrac{1}{10} \times 12 = \dfrac{12}{10} = 1.2$

From Poission distribution, we know that

$$P(X = x) = \frac{e^{-m} \cdot m^x}{x!}$$

(a) $x = 2$

$$P(X = 2) = \frac{e^{-1.2} \cdot (1.2)^2}{2!} = \frac{0.3012 \times 1.44}{2} = 0.22$$

(b) $P(X = \text{at least two}) = P(2) + P(3) + \dots P(12)$

$$= 1 - [P(0) + P(1)] = 1 - \left[\frac{e^{-1.2}(1 \cdot 2)^0}{0!} + \frac{e^{-1.2}(1 \cdot 2)^1}{1!} \right]$$

$$= 1 - e^{-1.2} [1 + 1 \cdot 2] = 1 - 0.3012 \times 2.2 = 0.3373$$

(c) $P(X = 0) = \frac{e^{-1.2}(1.2)^0}{0!} = e^{-1.2} = 0.3012$

EXAMPLE 11. *In sampling a large number of capsules manufactured by a company, the mean number of defective in a sample of 20 is 2. Out of 1000 such samples, how many would be expected to contain at least 3 defective capsules?* **(RGPV–2004)**

SOLUTION. Here, it is given that

$$m = 2, N = 1000 \text{ and } n = 20$$

Then, we have

$$P(X = x) = \frac{e^{-m} m^x}{x!}, \text{ then}$$

For atleast 3 capsules are defective

$$P = P(3) + P(4) + \dots P(24) = 1 - [P(0) + P(1) + P(2)]$$

$$= 1 - \left[\frac{e^{-2} \cdot (2)^0}{0!} + \frac{e^{-2}(2)^1}{1!} + \frac{e^{-2}(2)^2}{2!} \right]$$

$$= 1 - e^{-2} [1 + 2 + 2] = 1 - 0.1353 \times 5 = 0.3233$$

Hence, the probability of defective capsules out of 1000 samples

$$= 1000 \times 0.3233 = 323.3$$

EXAMPLE 12. *Data were collected over a period of 10 years showing number of deaths from horse kicks in each of the 200 army corps. The distribution of deaths was as follows :*

No. of deaths	0	1	2	3	4	Total
Frequency	109	65	22	3	1	200

Fit a Poisson distribution to the data and calculate the theoretical frequencies.
(Given $e^{-0.61} = 0.5435$) **(UPTU–2003)**

SOLUTION. We, have

$$\text{Mean, } m = \frac{0 \times 109 + 1 \times 65 + 2 \times 22 + 3 \times 3 + 4 \times 1}{200}$$

$$= \frac{122}{200} = 0.61$$

It is given that $e^{-0.61} = 0.5435$

Now, Theoretical frequency of x deaths

$$= \frac{Ne^{-x}m^x}{x!} = \frac{200 \times 0.5435 \times (0.61)^x}{x!}$$

Computation Table

x	$P(X = x)$	$NP(X = x)$
0	$e^{-m} = 0.5435$	$200 \times 0.5435 = 109$
1	$me^{-m} = 0.3315$	$200 \times 0.3315 = 66$
2	$\dfrac{m^2 e^{-m}}{2!} = 0.1011$	$200 \times 0.1011 = 20$
3	$\dfrac{m^3 e^{-m}}{3!} = 0.0205$	$200 \times 0.0205 = 4$
4	$\dfrac{m^4 e^{-m}}{4!} = 0.0001$	$200 \times 0.0001 = 1$
	$= 0.9967$	

Exercise 23.5

1. If the probability of a bad reaction from a certain injection is 0.001, determine the chance that out of 2000 individuals, more than two will get a bad reaction.

   (VTU–2008, Kottayam–2005, rgpv–2002)

2. In a certain factory turning out razor blades, there is a small chance of 0.002 for any blade to be defective. The blades are supplied in packets of 10. Calculate the approximate number of packets containing no defective, one defective, two defective blades, respectively in a consignment of 10,000 packets.

   (Kurukshetra–2009S, Madras–2006, VTU–2004)

3. Fit a Poisson distribution to the following data

x	0	1	2	3	4
f	46	38	22	9	1

   (Kurukshetra–2009, Bhopal–2008, VTU–2003S)

4. If a random variable has a Poisson distribution such that $P(1) = P(2)$, then find
   (i) mean of distribution, and (ii) $P(4)$.

   (VTU–2003)

5. A car hire firm has two cars which it hires out day by day. The number of demands for a car on each day is distributed as a Poisson distribution with mean 1.5. Calculate the proportion of days
   (i) on which there is no demand
   (ii) on which demand is refused
   (Given, $e^{-1.5} = 0.2231$)

   (Bhopal–2008S; JNTU–2003)

6. Six coins are tossed 6400 times. Using Poisson distribution, obtain the approximate probability of getting 6 heads r times.

7. If X is a Poisson variate, such that $P(X = 1) = 2P(X = 2)$, find
   (i) Mean,   (ii) Variance,
   (iii) $P(X = 0)$.

8. The probability that a man aged 50 years will die within a year is 0.01125. What is the probability that out of 12 such men at least 11 will reach their fifty first birthday.

9. If X is Poisson variate such that
   $P(X = 2) = 9P(X = 4) + 90P(X = 6)$, find the mean of X.

10. Find the probability that at most 5 defective will be found in a lot of 200 if experience shows that 2% of such components are defective. Also find the probability of more than 5 defective components [given $e^{-4} = 0.018$].

11. A certain screw making machine produces an average of 2 defective screws out of 100 and pack them in boxes of 500. Find the probability that a box contain 15 defective screws. **(Kurukshetra–2006)**

12. An insurance company found that only 0.01% of the population is involved in a certain type of accident each year. If its 1000 policy holders were randomly selected from the population, what is the probability that not more than two of its clients are involved in such an accident next year. [given $e^{-0.1} = 0.9048$].

13. If the variance of the Poisson distribution is 2, find the probabilities for $r = 1, 2, 3, 4$ from the recurrence relation of the Poisson distribution. Also find $P(x \geq 4)$.

14. For Poisson distribution, Prove that $\mu_2 \gamma_1 \gamma_2 = 1$,

where symbols have their usual meanings.
(SVTU–2008)

15. Find the expectation of the function $\phi(x) = xe^{-x}$ in a Poisson distribution. **(VTU–2003)**

16. Fit a Poisson distribution to the following data given the number of yeast cells per square for 400 squares: **(SVTU–2007)**

No. of cells per squares	0	1	2	3	4	5
No. of squares	103	143	98	42	8	4
No. of cells per squares	6	7	8	9	10	
No. of squares	2	0	0	0	0	

ANSWERS

1. 0.32 **2.** 2 **3.** 44, 43, 27, 7, 1 **4.** (i) 2, (ii) 0.0902 **5.** (i) 0.2231, (ii) 0.1913

6. $P(X = r) = \dfrac{e^{-100}(100)^r}{r!}$, $r = 0, 1, 2, \ldots$ **7.** (i) 1, (ii) 1, (iii) $1/e$ **8.** 0.99166

9. Mean = 1 **10.** 0.216 **11.** $\dfrac{(10)^{15} e^{-10}}{15!} = 0.035$ **12.** 0.9998

13. 0.2706, 0.2706, 0.1804, 0.0902, $P(x \geq 4) = 0.1431$.

16. Theoretical frequencies are 109, 142, 92, 40, 13, 3, 1, 0, 0, 0, 0.

23.20 NORMAL DISTRIBUTION

Definition. *A random variable X is said to follow a normal distribution with parameter* μ *(mean) and* σ^2 *(variance) if its probability distribution* $f(x)$ *is given by*

$$f(x) = \frac{1}{\sigma\sqrt{2\pi}} e^{-(x-\mu)^2/2\sigma^2}, \quad -\infty < x < \infty$$

REMARKS

⟹ If a random variable X has a normal distribution with mean μ and variance σ^2, then X is said to be normal variate with mean μ and variance σ^2, and can be written as $X \sim N(\mu, \sigma^2)$.

⟹ It was discovered by Karl Pearson and De'Moivre.

⟹ It is a continuous distribution.

⟹ Any quantity whose variation depends on random causes is distributed according to the normal law.

⟹ The graph of $f(x)$ is a bell shaped curve, symmetrical about the mean μ. This curve is known as normal probability curve or normal curve.

⟹ Normal distribution is a limiting case of binomial distribution.

⟹ The equation of the normal curve is $y_x = y_0 e^{-x^2/2\sigma^2}$.

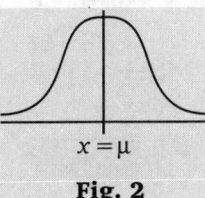

$x = \mu$

Fig. 2

23.21 STANDARD FORM OF THE NORMAL CURVE

Consider the equation of the normal curve $y_x = y_0 e^{-x^2/2\sigma^2}$, where origin is taken at mean and the value of y_0 is calculated in such a manner that the total frequency is unity.

Then

$$1 = y_0 \int_{-\infty}^{\infty} e^{-x^2/2\sigma^2} dx = 2y_0 \int_0^{\infty} e^{-x^2/2\sigma^2} dx$$

$$= 2y_0 \int_0^{\infty} e^{-t^2} \sigma\sqrt{2}\, dt \qquad \text{(Take } x = \sigma\sqrt{2}.t \Rightarrow dx = \sigma\sqrt{2}.dt)$$

$$= 2\sqrt{2}.\sigma y_0 \int_0^{\infty} e^{-t^2} dt = 2\sqrt{2}.\sigma y_0 \left(\frac{1}{2}\sqrt{\pi}\right) \qquad \left(\because \int_0^{\infty} e^{-t^2} dt = \frac{\sqrt{\pi}}{2}\right)$$

$$\Rightarrow \qquad y_0 = \frac{1}{\sigma\sqrt{2\pi}}$$

Hence, the standard form of the normal curve is $y = \dfrac{1}{\sigma\sqrt{2\pi}} e^{-x^2/2\sigma^2}$.

REMARKS

⟹ If the total frequency (or the total area under the curve is N, then the corresponding normal curve is given by $y = \dfrac{N}{\sigma\sqrt{2\pi}} e^{-x^2/2\sigma^2}$.

⟹ If the origin is taken at 0 success, then the equation of normal curve is given by $y = \dfrac{N}{\sigma\sqrt{2\pi}} e^{-(x-m)^2/2\sigma^2}$, where $m = np$ is the mean.

⟹ The point of inflexion of the normal curve is $x = \pm\sigma$.

23.22 PROPERTIES OF THE NORMAL DISTRIBUTION

Property (1). *The mean, median, mode of a normal distribution coincide at the origin.*

Proof. We have, Mean $= \int_{-\infty}^{\infty} y_0 e^{-x^2/2\sigma^2} . x\, dx$ $\qquad \left(\because y_0 = \dfrac{1}{\sigma\sqrt{2\pi}}\right)$

$$= y_0 \int_{-\infty}^{\infty} x e^{-x^2/2\sigma^2}.dx = 0 \qquad (\because \text{ it is an odd function of } x.)$$

Since $\qquad y = y_0 e^{-x^2/2\sigma^2}$

$$\Rightarrow \qquad \frac{dy}{dx} = \frac{-y_0}{\sigma^2} x e^{-x^2/2\sigma^2} \quad \text{and} \quad \frac{d^2y}{dx^2} = \frac{y_0}{\sigma^2}\left(1 - \frac{x^2}{\sigma^2}\right) e^{-x^2/2\sigma^2}$$

Therefore, $\qquad \dfrac{dy}{dx} = 0 \Rightarrow x = 0 \qquad$ also $\left(\dfrac{d^2y}{dx^2}\right)_{at\ x=0}$ = negative

$\Rightarrow y$ is maximum at $x = 0 \Rightarrow$ The mode is at $x = 0$.

Now, if x_1 be the median, then $\int_{x_1}^{\infty} \dfrac{1}{\sigma\sqrt{2\pi}} e^{-x^2/2\sigma^2}.dx = \dfrac{1}{2}$

$$\Rightarrow \qquad \frac{1}{2} = \frac{1}{\sigma\sqrt{2\pi}} \int_{x_1}^{\infty} e^{-x^2/2\sigma^2}.dx = \frac{1}{\sigma\sqrt{2\pi}} \int_{x_1/\sigma\sqrt{2}}^{\infty} e^{-t^2}.\sigma\sqrt{2}\, dt$$

(take $\sigma\sqrt{2} = t$)

$$= \frac{1}{\sqrt{\pi}} \int_{x_1/\sigma\sqrt{2}}^{\infty} e^{-t^2}.dt$$

or $\qquad \int_{x_1/\sigma\sqrt{2}}^{\infty} e^{-t^2}.dt = \dfrac{1}{2}\sqrt{\pi} = \int_{0}^{\infty} e^{-t^2} dt$

$\Rightarrow \qquad \dfrac{x_1}{\sigma\sqrt{2}} = 0 \Rightarrow x_1 = 0$

Hence, we have that mean, median and mode coincide at $x = 0$.

Property (2). *The curve is symmetrical about y-axis.*

Proof. Since the equation of the normal curve remains unchanged if x is replaced by $-x$, hence the curve is symmetrical about y- axis.

Property (3). *The point of inflexion of the normal curve are given by $x = \pm\sigma$.*

Property (4). *All moments of odd order about the mean vanish.*

23.22.1 MORE PROPERTIES OF NORMAL DISTRIBUTION.

(1) The probability density function is always non-negative.

(2) No part of the normal curve will lie below the x-axis, because the p.d.f. is always greater than or equal to zero.

(3) $\int_{-\infty}^{\infty} f(x)\,dx = 1$, which implies the total area bounded by normal probability curve and x-axis is 1.

(4) Mode and median of normal distribution is equal to m (mean).

(5) The x-axis is the asymptote of the normal curve, because it never tough the x-axis.

(6) The normal curve is concave near the mean value and convex near $x = \mu \pm 3\sigma$.

23.23 FITTING OF NORMAL DISTRIBUTION

To fit a normal distribution, first find its mean m and standard deviation σ.

Let us define $t = \dfrac{x-m}{\sigma}$, where x is the mid value of the class interval in case of grouped data. Then the normal curve fitted to the given data is

$$f(x) = \dfrac{1}{\sigma\sqrt{2\pi}}\, e^{-(x-\mu)^2/2\sigma^2}, \quad -\infty < x < \infty$$

REMARK

⟹ To find the expected frequencies, we may use any of the following method :
(i) Area method, (ii) Ordinate method.

23.24 MOMENT GENERATING FUNCTION OF NORMAL DISTRIBUTION

For normal distribution, the p.d.f. is given by $f(x) = \dfrac{1}{\sigma\sqrt{2\pi}}\, e^{-x^2/2\sigma^2}$

$\therefore \qquad M_0(t) = \dfrac{1}{\sigma\sqrt{2\pi}} \int_{-\infty}^{\infty} e^{tx}\, e^{-x^2/2\sigma^2}.dx = \dfrac{1}{\sigma\sqrt{2\pi}} \int_{-\infty}^{\infty}\left[\exp\left(tx - \dfrac{x^2}{2\sigma^2}\right)\right]dx$

$\qquad\qquad = \dfrac{1}{\sigma\sqrt{2\pi}} \int_{-\infty}^{\infty}\left[\exp\left\{-\dfrac{(x-t\sigma^2)^2}{2\sigma^2} + \dfrac{1}{2}t^2\sigma^2\right\}\right]dx$

$\qquad\qquad = \dfrac{1}{\sigma\sqrt{2\pi}}\, e^{t^2\sigma^2/2} \int_{-\infty}^{\infty}\left[\exp\left\{-\dfrac{(x-t\sigma^2)^2}{2\sigma^2}\right\}\right]dx$

$$= \frac{1}{\sigma\sqrt{2\pi}} e^{t^2\sigma^2/2} \int_{-\infty}^{\infty} e^{-z^2} \cdot \sigma\sqrt{2}\, dz \qquad \left(\text{Take } \frac{x - t\sigma^2}{\sigma\sqrt{2}} = z\right)$$

$$= \frac{1}{\sqrt{\pi}} e^{t^2\sigma^2/2} \int_{-\infty}^{\infty} e^{-z^2}\, dz \;=\; e^{\dfrac{t^2\sigma^2}{2}} \qquad \left(\because \int_{-\infty}^{\infty} e^{-z^2}\, dz = \sqrt{\pi}\right)$$

$$\Rightarrow \quad M_0(t) = e^{t^2\sigma^2/2} = 1 + \left(\frac{t^2\sigma^2}{2}\right) + \frac{1}{2!}\left(\frac{t^2\sigma^2}{2}\right)^2 + \ldots + \frac{1}{n!}\left(\frac{t^2\sigma^2}{2}\right)^n + \ldots$$

This involves only even powers of t and therefore the coefficient of t^{2n+1} in $M_0(t)$ is zero.

$$\Rightarrow \qquad \mu_{2n+1} = 0$$

Also, $\qquad \mu_{2n} = \text{Coefficient of } \dfrac{t^{2n}}{(2n)!} \text{ in the expression of } M_0(t)$

$$= \left(\frac{1}{2}\sigma^2\right)^n \frac{(2n)!}{n!} \;=\; \frac{1}{2^n}\sigma^{2n}\frac{2n(2n-1)(2n-2)\ldots 3.2.1}{n(n-1)(n-2)\ldots 2.1}$$

$$= \frac{1}{2^n}\sigma^{2n}\frac{[(2n-1)(2n-3)\ldots 3.1]\,[(2n)(2n-2)\ldots 4.2]}{n(n-1)(n-2)\ldots 2.1}$$

$$= \frac{1}{2^n}\sigma^{2n}\frac{[(2n-1)(2n-3)\ldots 3.1]\,[2^n.n(n-1)\ldots 2.1]}{n(n-1)(n-2)\ldots 2.1}$$

$$= 1.3.5\ldots(2n-1)\,\sigma^{2n}$$

In particular, $\mu_2 = 1.\sigma^2$ and $\mu_4 = 1.3\sigma^4 = 3\sigma^4$ and $\beta_2 = \dfrac{\mu_4}{\mu_2^2} = \dfrac{3\sigma^4}{(\sigma^2)^2} = 3$

Now, if probability function is given by $f(x) = \dfrac{1}{\sigma\sqrt{2\pi}} e^{\frac{1}{2}\left(\frac{x-m}{\sigma}\right)^2}$

Then, $\qquad M_0(t) = \int_{-\infty}^{\infty} e^{tx} \dfrac{1}{\sigma\sqrt{2\pi}} e^{\frac{1}{2}\left(\frac{x-m}{\sigma}\right)^2} \cdot dx = \dfrac{1}{\sigma\sqrt{2\pi}} \int_{-\infty}^{\infty} e^{tx}\, e^{-\frac{1}{2}\left(\frac{x-m}{\sigma}\right)^2} \cdot dx$

$$= \frac{1}{\sigma\sqrt{2\pi}} \int_{-\infty}^{\infty} e^{t(m+z\sigma)}\, e^{-z^2/2}\, \qquad \left(\text{Put } z = \frac{x-m}{\sigma} \Rightarrow x = m + z\sigma\right)$$

$$= \frac{e^{mt}}{\sqrt{2\pi}} \int_{-\infty}^{\infty} e^{\frac{1}{2}z^2 + t\sigma z} \cdot dz = \frac{e^{mt}}{\sqrt{2\pi}} \int_{-\infty}^{\infty} e^{\frac{1}{2}(z^2 - 2t\sigma z + t^2\sigma^2) + \frac{1}{2}t^2\sigma^2} \cdot dz$$

$$= \frac{e^{mt + \frac{1}{2}t^2\sigma^2}}{\sqrt{2\pi}} \int_{-\infty}^{\infty} e^{-\frac{1}{2}(z - t\sigma)^2} \cdot dz$$

$$= \frac{e^{mt + \frac{1}{2}t^2\sigma^2}}{\sqrt{2\pi}} \int_{-\infty}^{\infty} e^{-u^2} \cdot \sqrt{2}\, du \qquad \left(\text{Put } z - t\sigma = u\sqrt{2}\right)$$

$$= e^{mt + \frac{1}{2}t^2\sigma^2} \Rightarrow M_0(t) = e^{mt + \frac{t^2\sigma^2}{2}} \qquad \left(\because \int_{-\infty}^{\infty} e^{-u^2} = \sqrt{\pi} \right)$$

Also, the m.g.f. about mean m is given by

$$M_m(t) = e^{-mt} M_0(t) = e^{-mt} \left\{ e^{mt + \frac{1}{2}t^2\sigma^2} \right\} = e^{\frac{t^2\sigma^2}{2}}$$

23.25 CUMULANT GENERATING FUNCTION OF NORMAL DISTRIBUTION

Using above article, we have $M_0(t) = e^{mt + \frac{1}{2}t^2\sigma^2}$.

Now, $K_0(t) = \log M_0(t) = mt + \frac{1}{2}t^2\sigma^2$, which is the required cumulative function of normal distribution about origin.

Also, from $K_0(t) = mt + \frac{1}{2}t^2\sigma^2$, we get

$$K_1' = m, \ K_2' = \sigma^2, \ K_3' = 0, \ K_4' = 0, \ .... \text{ from coefficients of } \frac{t}{1!}, \frac{t^2}{2!}, \frac{t^3}{3!}, ...$$

The cumulative function about mean m is given in $K_m(t) = \log M_m(t)$

$$= \log [e^{t^2\sigma^2/2}] = \frac{t^2\sigma^2}{2}$$

which gives $K_1 = 0, \ K_2 = \sigma^2, K_3 = 0, \ K_4 = 0, ...$ from the coefficients of $\frac{t}{1!}, \frac{t^2}{2!}, \frac{t^3}{3!}, ...$

☛ RECAPITULATIONS
- The mean, median, mode of a normal distribution coincides at the origin.
- Normal curve is symmetrical about y-axis.
- The points of inflexion of the normal curve are given by $x = \pm\sigma$.
- All moments of odd order about the mean vanishes.

SOLVED EXAMPLES

EXAMPLE 1. For the normal distribution $N(\mu, \sigma^2)$ show that $\mu_{2r+2} = \sigma^2 \cdot \mu_{2r} + \sigma^3 \dfrac{d}{d\sigma} \mu_{2r}$, where μ_r is r^{th} moment about mean μ.

SOLUTION. Since, we know that

$$\mu_{2r} = \frac{1}{\sigma\sqrt{2\pi}} \int_{-\infty}^{\infty} (x - m)^{2r} \ e^{-(x-m)^2/2\sigma^2} .dx \qquad ...(1)$$

$$\Rightarrow \quad \frac{d}{d\sigma}(\mu_{2r}) = -\frac{1}{\sigma^2\sqrt{2\pi}} \int_{-\infty}^{\infty} (x - m)^{2r} \ e^{-(x-m)^2/2\sigma^2} .dx$$

$$+ \frac{1}{\sigma\sqrt{2\pi}} \int_{-\infty}^{\infty} (x - m)^{2r} \ e^{-(x-m)^2/2\sigma^2} .dx$$

$$\Rightarrow \quad \sigma^3 \frac{d}{d\sigma}(\mu_{2r}) = -\frac{\sigma^3}{\sigma^2\sqrt{2\pi}} \int_{-\infty}^{\infty} (x-m)^{2r} e^{-(x-m)^2/2\sigma^2} \, dx$$

$$+ \frac{1}{\sigma\sqrt{2\pi}} \int_{-\infty}^{\infty} (x-m)^{2r+2} \frac{2(x-m)^2}{2\sigma^2} e^{-(x-m)^2/2\sigma^2} \, dx$$

$$\Rightarrow \quad \sigma^3 \frac{d}{d\sigma}(\mu_{2r}) = -\sigma^2 \mu_{2r} + \mu_{2r+2} \qquad \text{(Using (1))}$$

$$\Rightarrow \quad \mu_{2r+2} = \sigma^2 \mu_{2r} + \sigma^3 \frac{d}{d\sigma} \mu_{2r}.$$

EXAMPLE 2. *In a normal distribution, 31% of the items are under 45 and 8% are over 64. Find the mean and standard deviation of the distribution.*

(VTU–2009, SVTU–2008, Kurukshetra–2007S)

SOLUTION. Let m and σ denote the mean and standard deviation, respectively. Since 31% of the items are under 45 means area to the left of ordinate $x = 45$.

When $x = 45$, let $t = t_1 \Rightarrow t_1 = \dfrac{45-m}{\sigma}$

$$\Rightarrow \quad \int_{-\infty}^{\infty} \phi(t) \, dt = 0.31$$

or $\int_{-\infty}^{0} \phi(t) \, dt - \int_{t_1}^{0} \phi(t) \, dt = 0.31$

Therefore

$$\int_{t_1}^{0} \phi(t) dt = \int_{-\infty}^{0} \phi(t) dt - 0.31$$

$$\Rightarrow \quad 0.5 - 0.31 = 0.19$$

Now, using table of normal curve
$$t_1 = -0.5.$$

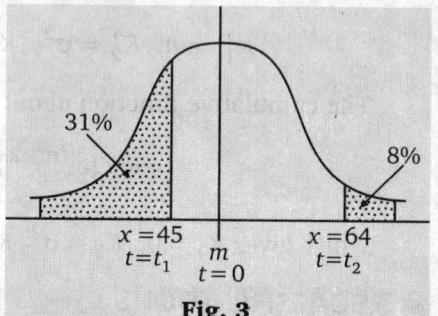

31%   8%

$x = 45$ $x = 64$
$t = t_1$ $t = t_2$
m
$t = 0$

Fig. 3

When $x = 64$, let $t = t_2 \Rightarrow t_2 = \dfrac{(64-m)}{\sigma}$

$$\therefore \quad \int_{t_2}^{\infty} \phi(t) \, dt = \int_{0}^{\infty} \phi(t) \, dt - 0.08$$

$$= 0.5 - 0.08 = 0.42$$

Again, using table, we get $t_2 = 1.4$.

Therefore, we conclude that

$$45 - m = -0.5\sigma \qquad \qquad \qquad \dots(1)$$

and $\qquad 64 - m = 1.4\,\sigma \qquad \qquad \qquad \dots(2)$

Now, solving (1) and (2) for m and σ,

we get $\qquad m = 50$ and $\sigma = 10$.

EXAMPLE 3. *Fit a normal curve to the following*

Length (in cm)	8.60	8.59	8.58	8.57	8.56	8.55	8.54	8.53	8.52
Frequency (f)	2	3	4	9	10	8	4	1	1

SOLUTION. Let the assumed mean be 8.56.

x_i	f_i	$d_i = (x_i - 8.56)/0.01$	$f_i d_i$	d_i^2	$f_i d_i^2$
8.60	2	4	8	16	32
8.59	3	3	9	9	27
8.58	4	2	8	4	16
8.57	9	1	9	1	9
8.56	10	0	0	0	0
8.55	8	−1	−8	1	8
8.54	4	−2	−8	4	16
8.53	1	−3	−3	9	9
8.52	1	−4	−4	16	16
	$\Sigma f_i = 42$		$\Sigma f_i d_i = 11$		$\Sigma f_i d_i^2 = 133$

Now mean,

$$\mu = 8.56 + \frac{11}{42} \times 0.01 = 8.56262$$

The normal curve is given by

$$y = \frac{N}{\sigma\sqrt{2\pi}} e^{-\left(\frac{1}{2}\right)\frac{(x-\mu)^2}{\sigma}}$$

Here, $N = 42$, $\mu = 8.563$

and

$$\sigma = i \times \sqrt{\left[\left\{\frac{\Sigma f_i d_i^2}{\Sigma f_i}\right\} - \left(\frac{\Sigma f_i d_i}{\Sigma f_i}\right)^2\right]}$$

$$= 0.01 \times \sqrt{\left[\left(\frac{133}{42}\right) - \left(\frac{11}{12}\right)^2\right]} = 0.0176$$

Hence, the equation of the normal curve fitted to the given data is

$$P(x) = \frac{N}{\sigma\sqrt{2\pi}} e^{-\frac{(x-\mu)^2}{2\sigma^2}}, \quad -\infty \leq x \leq \infty$$

EXAMPLE 4. *If the probability of committing an error of magnitude x is given by* $y = \frac{h}{\sqrt{\pi}} e^{-h^2 x^2}$.

Compute the probable error from the following data

$m_1 = 1.305$, $m_2 = 1.301$, $m_3 = 1.295$, $m_4 = 1.286$, $m_5 = 1.318$, $m_6 = 1.321$,

$m_7 = 1.283$, $m_8 = 1.289$, $m_9 = 1.300$, $m_{10} = 1.286$ **(KURUKSHETRA–2005)**

SOLUTION. Here, we observe that the data are normally distributed.

Now, we have,

$$\text{Mean} = \frac{1}{10} \Sigma m_i = \frac{12.984}{10} = 1.2984$$

Also, $\sigma^2 = \frac{1}{10} \Sigma (m_i - \text{mean})^2$

$$= \frac{1}{10}[(0.007)^2 + (0.003)^2 + (0.003)^2 + (0.012)^2 + (0.02)^2 + (0.023)^2$$
$$+ (0.015)^2 + (0.009)^2 + (0.002)^2 + (0.012)^2]$$
$$= 0.0001594$$

Hence, the probable error $= \frac{2}{3}\sigma = 0.0084$.

EXAMPLE 5. *If two normal universes have the same total frequency but standard deviation of one is k time that of the other, show that the maximum frequency of the first is $\frac{1}{k}$ that of the other.*

SOLUTION. Let the total number of frequencies be N and σ and $k\sigma$ be the standard deviation of the two universes. Then, equations of the normal curve are given by

$$y = y_0 e^{-x^2/2\sigma^2} \quad \text{and} \quad y = y_0' e^{-x^2/2k^2\sigma^2}$$

If the total frequencies for both are the same, then

$$\int_{-\infty}^{\infty} y_0 e^{-x^2/2\sigma^2}.dx = \int_{-\infty}^{\infty} y_0' e^{-x^2/2k^2\sigma^2}.dx$$

or $\int_0^\infty y_0 e^{-x^2/2\sigma^2}.dx = \int_0^\infty y_0' e^{-x^2/2k^2\sigma^2}.dx$ ...(1)

Putting $\frac{x}{\sqrt{2}.\sigma} = t \Rightarrow dx = \sqrt{2}\sigma\, dt$ in the LHS of (1), we get

$$\text{LHS} = y_0 \int_0^\infty e^{-t^2}\sqrt{2}\sigma\, dt = \sqrt{2}\sigma y_0 \frac{1}{2}\sqrt{\pi} \quad \text{...(2)}$$

Similarly for RHS, putting

$$\frac{x}{\sqrt{2}.k\sigma} = t \Rightarrow dx = \sqrt{2}k\sigma dt \text{ , we get}$$

$$\text{RHS} = \sqrt{2}k\sigma y_0' \frac{1}{2}\sqrt{\pi}$$

Using (2) and (3), we conclude that

$$y_0 = k y_0' \Rightarrow y_0' = \frac{1}{k}\cdot y_0$$

Hence, the maximum frequency of the first $\frac{1}{k}$ is that of the other.

EXAMPLE 6. *Show that for the normal distribution, the quartile deviation, the mean deviation from mean and standard deviation are approximately in the ratio 10 : 12 : 15.*

SOLUTION . For a normal distribution, we have

$$f(x) = \frac{1}{\sigma\sqrt{2\pi}} e^{-(x-\mu)^2/2\sigma^2}, \quad -\infty < x < \infty$$

Then, Quartile deviation, Q.D. $= \frac{2}{3}\sigma$

Mean deviation from mean, M.D. $= \frac{4}{5}\sigma$

where, σ = standard deviation = S.D.

Hence, Q.D. : M.D. : S.D. $= \frac{2}{3}\sigma : \frac{4}{5}\sigma : \sigma = 10\sigma : 12\sigma : 15\sigma = 10 : 12 : 15$

EXAMPLE 7. *For a certain normal distribution, the first moment about 10 is 40 and the fourth moment about 50 is 48. What is the arithmetic mean and variance of the normal distribution.*

SOLUTION . Let $f(x) = \frac{1}{\sigma\sqrt{2\pi}} e^{-(x-\mu)^2/2\sigma^2}, -\infty < x < \infty$

where, mean $= \mu$ and $\sigma =$ standard deviation.

Now, first moment about 10 is given by

$$\mu_1'(10) = E(X - 10) = E(X) - 10 = \mu - 10 = 40$$

$\Rightarrow \qquad \mu = 40 + 10 = 50 \qquad \qquad \qquad \qquad \qquad \qquad \qquad ...(1)$

Since mean = 50.

So, fourth moment about 50 = fourth moment about mean

$\Rightarrow \qquad \mu_4'(50) = \mu_4 = 48$

$\Rightarrow \qquad \mu_4 = 3\sigma^4$

$\Rightarrow \qquad \sigma^4 = \frac{48}{3} = 16 = 2^4$

$\Rightarrow \qquad \sigma = 2$

Hence, variance $= \sigma^2 = 4$.

EXAMPLE 8. *If skills are classified as A, B, C according to the length and breadth index as under : 75 between 75 and 80 and over 80, find approximately (assuming distribution is normal), the mean and standard deviation of a series in which A are 58%, B are 38% and C are 40% being given that if $f(t) = \frac{1}{2\sqrt{2\pi}} \int_0^t \exp(-t)^2 dt$.*

SOLUTION. Let m be the mean and σ be the standard deviation of the distribution. Since the total frequency is taken as unity frequency of skill A, whose length and breadth is under 75 is 0.58. The frequency of skill B whose index lies between 75 and 80 is 0.38 and frequency of skill C whose index is under 80 is 0.40.

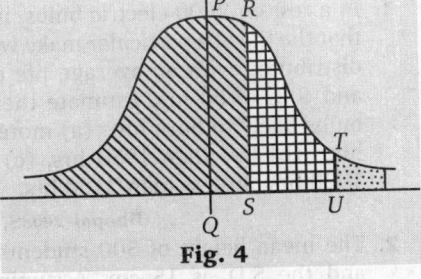

Fig. 4

Therefore, the total area to the left of ordinate RS is 0.58, area between RS and TU is 0.38 and the area to the right of TU is 0.04.

Area between origin and $x = 75 - m$,

i.e., area $\quad PQRS = 0.58 - 0.5 = 0.08$

and given $f(0.20) = 0.08$

Hence, $\dfrac{75 - m}{\sigma} = 0.20$ ...(1)

Again, area between origin and $(x = 80 - m)$.

i.e., Area $PQUT = 0.08 + 0.38 = 0.46$

and given $f(1.75) = 0.46$ ...(2)

Solving equations (1) and (2)

$$m = 74.3, \quad \sigma = 3.23.$$

EXAMPLE 9. *In a distribution exactly normal, 7% of the items are under 35 and 89% are under 63. What are the mean and standard deviation of the distribution?*

SOLUTION. Since 89% of the items are under 63, so 11% would be more than 63 from the standard normal table 43% (0.43) area is enclosed when $z = 1.4757$ and 39% (0.39) area is enclosed when $z = 1.2263$.

Let $z_1 = 1.4757,\quad z_2 = 1.2263,$

$$z_1 = \frac{\text{Mean} - 35}{\sigma}, \qquad z_2 = \frac{\text{Mean} - 63}{\sigma}$$

$$1.4757 = \frac{\text{Mean} - 35}{\sigma} \qquad \text{...(1)}$$

$$1.2263 = \frac{\text{Mean} - 63}{\sigma} \qquad \text{...(2)}$$

$$\text{Mean} - 35 = 1.47576\sigma \qquad \text{...(3)}$$

$$63 - \text{Mean} = 1.2263\,\sigma \qquad \text{...(4)}$$

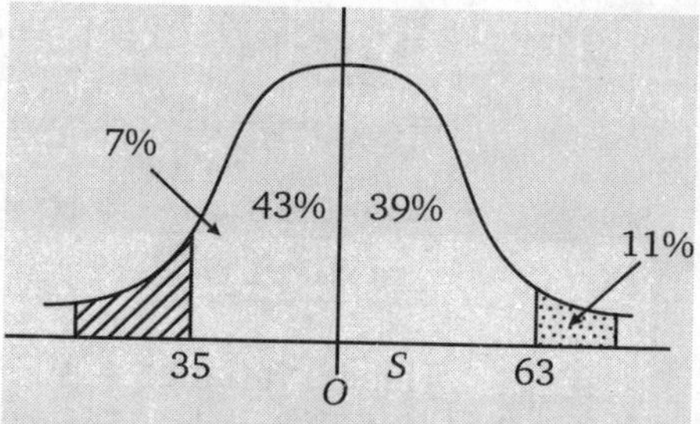

On adding $\sigma = 10.36$

Putting σ in equation (3)

$$\text{Mean} = 35 + 1.4757\,(10.36)$$

$$\text{Mean} = 50.288$$

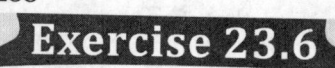

Fig. 5

Exercise 23.6

1. In a test on 2000 electric bulbs, it was found that the life of a particular make was normally distributed with an average life of 2040 hrs and S.D. of 60 hrs. Estimate the number of bulbs likely to burn for : (a) more than 2150 hrs, (b) less than 1950 hrs, (c) more than 1920 hrs but less than 2160 hrs.

   (Bhopal–2008S, UPTU–2008)

2. The mean height of 500 students is 151 cm and the S.D. is 15 cm. Assuming that the heights are normally distributed, find how many students height lie between 120 and 155 cm. **(Burdwan–2003)**

3. Show that the standard deviation for a normal distribution is approximately 25% more than the mean deviation.

4. In an examination taken by 500 candidates, the average and standard deviation of marks obtained are 40% and 10%. Find (a) How many will pass, if 50% is fixed as a minimum, (b) What should be the minimum if 350 candidates are to pass, (c) How many have scored marks above 60%.

5. Assuming that the diameter of 1000 brass plugs taken consecutively from a machine form a normal distribution with mean 0.7515 cm and standard deviation 0.0020 cm. How many of the plugs are likely to be rejected if the approved diameter is 0.752 ± 0.004 cm.

   (Bhopal–2002)

6. Find the equation of the best fitted normal curve to the following distribution

x	0	1	2	3	4	5
f	13	23	34	15	11	4

x	2	4	6	8	10
f	1	4	6	4	1

7. A random variable x has a standard normal distribution ϕ. Show that

$P(|x| < K) = 2\phi(K) - 1$.

8. Assume that the mean height of soldiers to be 68.22 inches with a variance of 10.8 inches square. How many soldiers in a regiment of 10,000 would you expect to be over 6 feet tall?

9. If X is a normal variate with mean 12 and standard deviation 4, find $P(0 \le X \le 12)$.

10. On a final examination in engineering, the mean was 72 and the standard deviation was 15. Determine the standard scores of students receiving grades (i) 60s, (ii) 93s, (iii) 72s.

11. A normal distribution with $\mu = 50$ and $\sigma = 10$ is given. Find the value of X that has (i) 13% of the area to its left, (ii) 14% of the area to its right.

12. The income distribution of workers in a certain factory was found to be normal with mean Rs. 500 and standard deviation Rs. 50. There were 228 persons getting above Rs. 600. How many workers were there in all?

13. The scores of candidates in a certain test are normally distributed with mean 500 and standard deviation 100. What percentage of candidates receives the scores between 350 and 550?

14. In an intelligence test administered to 1000 students, the average was 42 and S.D. 24. Find (i) the number of students exceeding a score 50, (ii) the number of children lying between 30 and 54.

15. Fit a normal curve to the following distribution. **(VTU–2001)**

16. The income of a group of 10,000 persons was found to be normally distributed with mean Rs. 750 p.m., and standard deviation of Rs.50. Show that of this group, about 95% had income exceeding Rs. 668 and only 5% had income exceeding Rs. 832. Also find the lowest income among the richest 100. **(UPTU–2004S)**

17. Mean height of 500 students is 160 cm. and S.D. is 5 cm.

   (i) What are the chances of heights above 175 cm. being normal. if height follows normal distribution?

   (ii) What percentage of boys will have height above 168 cm.?

   (iii) How many of the boys will have height between 168 and 175 cm? **(UPTU–2001)**

18. As a result of tests on 20,000 electric bulbs manufactured by a company, it was found that the life time of the bulbs was normally distributed with an average life of 2,040 hours and standard deviation of 60 hours. On the basis of the information, estimate the number of bulbs that are expected to last for

   (a) more than 2,150 hours and

   (b) less than 1,960 hours.

   Proportion of area under the normal curve :

Z	Area	Z	Area
1.23	0.3907	1.63	0.4484
1.33	0.4082	1.73	0.4582
1.43	0.4236	1.83	0.4664

(UPTU–2002)

ANSWERS

1. (a) 67,   (b) 184,   (c) 1904   2. 294   4. (i) 79, (ii) 35%, (iii) 11   5. 52

6. $y = \dfrac{100}{\sqrt{3.4\pi}} e^{-\frac{(x-2)^2}{3.4}}$ 8. 1251   9. 0.4987 10.   (i) – 0.8, (ii)   1.4,   (iii) 0

11. (i) 38.7,   (ii) 60.8   12. 1000   13. 62.47%   14. (i) 371,   (ii) 383

16. ₹ 866   17. 27   18. 8, 164

Objective Evaluation

☞ **MULTIPLE CHOICE QUESTIONS**

Choose the most appropriate one.

1. Probability is expressed as :
   (a) ratio   (b) proportion
   (c) percentage   (d) all the above

2. If A and B are two events, the probability of occurence of either A or B is given by :
   (a) $P(A) + P(B)$ (b) $P(A \cup B)$
   (c) $P(A \cap B)$ (d) $P(A)P(B)$

3. If A and B are two events, the probability of occurence of A and B simultaneously is given as :
   (a) $P(A) + P(B)$ (b) $P(A \cup B)$
   (c) $P(A \cap B)$ (d) $P(A)P(B)$

4. The probability of all possible outcomes of a random experiment is always equal to :
   (a) infinity   (b) zero
   (c) one   (d) none of the above

5. If $A \subset B$, the probability $P(A|B)$ is equal to :
   (a) zero   (b) one
   (c) $P(A)|P(B)$ (d) $P(B)|P(A)$

6. If $B \subset A$ the probability $P(A|B)$ is equal to :
   (a) zero   (b) one
   (c) $P(A)|P(B)$ (d) $P(B)|P(A)$

7. For any two events A and B, $P(A - B)$ is equal to :
   (a) $P(A) - P(B)$ (b) $P(B) - P(A)$
   (c) $P(B) - P(AB)$ (d) $P(A) - P(AB)$

8. If an event B has occured and it is known that $P(B) = 1$, the conditional probability $P(A|B)$ is equal to :
   (a) $P(A)$ (b) $P(B)$
   (c) one   (d) zero

9. If a bag contains 4 white balls and 3 black balls, two draws of 2 balls are successively made, the probability of getting 2 white balls at first draw and 2 black balls at second draw when the balls drawn at first draw were replaced is :
   (a) 3/7   (b) 1/7
   (c) 19/49   (d) 2/49

10. In tossing three coins at a time, the probability of getting at most one head is :
    (a) 3/8   (b) 7/8
    (c) 1/2   (d) 1/8

11. The mean and variance of a binomial distribution are 8 and 4 respectively. Then $P(x = 1)$ is equal to :
    (a) $\dfrac{1}{2^{12}}$ (b) $\dfrac{1}{2^4}$
    (c) $\dfrac{1}{2^6}$ (d) $\dfrac{1}{2^8}$

12. If for a binomial distribution $B(n, p)$, $n = 4$ and also $P(X = 2) = 3P(X = 3)$ the value of p is :
    (a) 9/11   (b) 1
    (c) 1/3   (d) none of the above

13. If for a binomial distribution $B(n, p)$, mean = 4, variance = 4/3 the probability $P(X \ge 5)$ is equal to :
    (a) $\left(\dfrac{2}{3}\right)^6$ (b) $\left(\dfrac{2}{3}\right)^5 \left(\dfrac{1}{3}\right)$
    (c) $\left(\dfrac{1}{3}\right)^6$ (d) $4\left(\dfrac{2}{3}\right)^6$

14. Am experiment succeeds twice as often as it fails. The chance that in the next six trials, there shall be at least four successes is :
    (a) $\dfrac{240}{729}$ (b) $\dfrac{489}{729}$
    (c) $\dfrac{496}{729}$ (d) none of the above

15. A manufacturer produces switches and experiences that 2 percent switches are defective. The probability that in a box of 50 switches, there are at most two defective is :
    (a) $2.5e^{-1}$ (b) e^{-1}
    (c) $2e^{-1}$ (d) none of the above

16. If $X \sim B\left(3, \dfrac{1}{2}\right)$ and $Y \sim B\left(5, \dfrac{1}{2}\right)$, the probability of $P(X + Y = 3)$ is :
    (a) 7/16   (b) 7/32
    (c) 11/16   (d) none of the above

17. If X and Y are two poission variates such $X \sim P(1)$ and $Y \sim P(2)$, the probability $P(X + Y < 3)$ is :
    (a) e^{-3} (b) $3e^{-3}$
    (c) $4e^{-3}$ (d) $8.5e^{-3}$

18. If X is a binomial variate with its mean $\mu = 4$ and third moment $\mu_3 = 4.8$, the value of Pearson's constant β_1 is :

(a) 2/3   (b) 5/6

(c) 0   (d) none of the above

19. A box contain 12 items out of which 4 are defective. A person selects 6 items from the box. The expected number of defective items out of his selected items is :

(a) 2   (b) 3

(c) 3/2   (d) none of the above

20. If X is a normal variate with mean 20 and variance 64 the probability that X lies between 12 and 32 is :

(a) 0.4332   (b) 0.1189

(c) 0.7475   (d) 0.5

☞ TRUE/FALSE

Write 'T' for True and 'F' for False statement.

1. If $B \subset A$, the relation between $P(A)$ and $P(B)$ is $P(A) \leq P(B)$. **(T/F)**

2. If two events A and B are disjoint then $P(A \cup B) = P(A) + P(B)$. **(T/F)**

3. For any two events A and B, $P(AB) = P(A)$ $P(B|A)$. **(T/F)**

4. In Bayes's Probability, we calculate inverse probability. **(T/F)**

5. For any two event A and B, the probability $P(A|B) + P(\bar{A}|B) = 1$. **(T/F)**

6. Suppose A and B are two events. Event B has occured and it is known that $P(B) < 1$ then

$$P(A|\bar{B}) = \frac{[P(A) - P(AB)]}{P(B)}.$$ **(T/F)**

7. If A_1 and A_2 are mutually exclusive events then $P(A_1 | A_1 \cup A_2) = \dfrac{P(A_1)}{P(A_1) + P(A_2)}$. **(T/F)**

8. In tossing a coin turning up of head and tail are exhaustive outcomes. **(T/F)**

9. If A is an arbitrary event, then $P(A|A) = 1$. **(T/F)**

10. If an event is not simple, it is a compound event. **(T/F)**

11. Two independent variables X and Y are said to be independent if $E(XY) = E(X) E(Y)$. **(T/F)**

12. If X and Y are two random variables such that their expectations exist and $P(x \leq y) = 1$ then $E(X) \geq E(Y)$. **(T/F)**

13. If X and Y two independent variables and their expected values are $\bar{X}$ and $\bar{Y}$ respectively, then $E(X - \bar{X})(Y - \bar{Y}) = 0$. **(T/F)**

14. If X is a random variable with its mean $\bar{X}$, the expression $E(X - \bar{X})^2$ represents second central moment. **(T/F)**

15. If X and Y are two random variables, then $E(XY)^2 \leq E(X^2) E(Y^2)$. **(T/F)**

16. For the normal variate within mean 20 and variance 64, the percentage of items between 24 and 44 is 19.27. **(T/F)**

17. If Z is a standard normal deviate, the proportion of items lying between $Z = -0.5$ and $Z = -3.0$ is 0.3072. **(T/F)**

18. If X is a normal variate representing the income in Rs per day with mean = 50 and S.D. = 10. If the no. of workers in a factory is 1200, then the number of workers having income more than Rs 62.00 per day is 138. **(T/F)**

19. Probability mass function for a binomial distribution with usual notations is $^nC_x p^n q^{n-x}$. **(T/F)**

20. If binomial random variable has mean = 4 and variance = 3 then its third central moment μ_3 is 3/2.

☞ FILL IN THE BLANKS

1. Intersection of two mutually exclusive events is a _____ event.

2. Mathematical probability is also known as _____ probability.

3. Two events A and B are equal if _____.

4. Multiplication theorem is applicable only if the events are _____.

5. If A and B are two events then $P(\bar{A} \cap B)$ is _____.

6. If A and B are independent then $P(A \cap B)$ is _____.

7. If the events A and B are independent then $P(B|A) = $ _____.

8. If $P(A \cap B) = P(A)P(B)$ then events A and B are _____.

9. If an event A is independent of the events B, $B \cup C$ and $B \cap C$ then $P(A \cap C) = $ _____.

10. For any two events A and B, the probability $P(A|B) + P(\bar{A}|B) = $ _____.

11. The mean of binomial distribution $B(n, p)$ is _____ and its variance is _____.

12. The mean of the binomial distribution is _____ than its variance.

13. If the mean and variance of binomial distribution are 2 and 1 respectively, then $P(x \leq 1)$ is equal to _____.

14. If on an average the rain falls on ten days in a month (30 days) the probability that the rain will fall on two days of the week is _____.

15. In a police control room there are on an average 3 calls per 10 minute interval. The probability of receiving 4 calls in a 10 minute interval is _____.

16. If the probability of a rocket hitting the target is 1/2 and the rockets are being launched one after the other, then the probability that the 10th launch being the 5th hit on the target is _____.

17. If a variable X follows poisson distribution with $P(x = 1) = P(x = 2)$, the probability $P(x = 1$ or $2)$ is _____.

18. If X is a poisson variate with $P(x = 1) = P(x = 2)$, the mean of the poisson variate equal to _____.

19. The normal curve is symmetrical about mean 50 and 5 percent values are greater than 70. The standard deviation of the distribution _____ [Given that $\phi(1.64) = 0.05$].

20. The standard deviation of a poisson variate is 2, the mean of the poisson variate is _____.

ANSWERS

⇨ MULTIPLE CHOICE QUESTIONS

1. (d)	**2.** (a)	**3.** (c)	**4.** (c)	**5.** (c)	**6.** (b)	**7.** (d)	**8.** (a)	**9.** (d)
10. (c)	**11.** (a)	**12.** (c)	**13.** (d)	**14.** (c)	**15.** (a)	**16.** (b)	**17.** (d)	**18.** (b)
19. (a)	**20.** (c)							

⇨ TRUE/ FALSE

1. F	**2.** T	**3.** T	**4.** T	**5.** T	**6.** F	**7.** T	**8.** T	**9.** T
10. T	**11.** T	**12.** F	**13.** T	**14.** F	**15.** T	**16.** F	**17.** T	**18.** T
19. F	**20.** T							

⇨ FILL IN THE BLANKS

1. null	**2.** a priori	**3.** $A \subset B$	**4.** independent	**5.** $P(B) - P(A \cap B)$
6. $P(A) \times P(B)$	**7.** $P(B)$	**8.** independent	**9.** $P(A)P(C)$	**10.** 1
11. np; npq	**12.** greater than	**13.** 5/16	**14.** 224/729	**15.** $\dfrac{27}{8} e^{-3}$
16. 63/512	**17.** $4e^{-2}$	**18.** 2	**19.** $\sigma = 12.2$	**20.** 4

❑❑❑❑❑

CHAPTER 24

Test of significance

24.1 INTRODUCTION

Generally the population parameters are unknown and estimate through the sample values. If the sample value are exactly the same as parameter then we accept it and if it is far from our parameter then we reject it. But the problem arises when the sample value neither exactly equal to the parametric value nor too far. In this situation one has to develop some procedure which enables us to decide whether to accept a hypothetical value or not on the basis of sample values. Such a procedure is known as testing of hypothesis.

Population : Any collection of individuals under study is said to be population (or universe). The individuals often called the members or the units of the population.

Sample : A part or small section selected from the population is called a sample and this process of selection is called sampling.

24.2 TYPES OF POPULATION

(i) **Hypothetical population :** The population of concrete objects is called an existent population while a hypothetical population may be defined as the collection of all possible ways in which a specified event can happen. The population of 1 2, 3, 4, 5, 6 obtained by rolling a die an infinite numbers of times is a hypothetical one.

(ii) **Real population :** A population of concrete individuals is called a existent or real population.

(iii) **Finite population :** A population containing a finite number of individuals is called a finite population.

(iv) **Infinite population :** A population containing an infinite number of individuals is called an infinite population.

24.3 STATISTICAL HYPOTHESIS

A statistical hypothesis is a statement or assertion about a parameter of a population or parameters of two or more populations is denoted by H.

Parametric hypothesis : A statistical hypothesis refers only to the values of unknown parameters of population is usually called a parametric hypothesis.

Simple and Composite Hypothesis : If statistical hypothesis completely specifies a distribution, it is known as simple hypothesis otherwise known as composite hypothesis for example, a normal population $N(\mu, \sigma^2)$ with σ^2 is known. The hypothesis

(i) $H : \mu = \mu_0$ (ii) $H : \mu = 25$

(iii) $H : \mu = 30$

are simple hypothesis and

(i) $H : \mu > \mu_0$ (ii) $H : \mu < \mu_0$

(iii) $H : \mu \neq \mu_0$

are composite hypothesis.

24.4 NULL AND ALTERNATIVE HYPOTHESIS

The hypothesis which is natural and non-committal attitude of the statistician or decision makes before the sample observation is known as null hypothesis.

According to Ronald, A. Fisher, "Any hypothesis tested for its possible rejection is called a null hypothesis." The null hypothesis is denoted by H_0 and the hypothesis which provides an alternative of the null hypothesis is known as alternative to the null hypothesis. It is denoted by H_1 or H_A.

For example : Two manufacturing company produces bulbs. Each company claims that the average life of its bulbs is larger to that of the other than there are three hypothesis.

(i)   First manufacturing company produces bulbs of larger average life than second.

(ii) First manufacturing company produces bulbs of lesser average life than second.

(iii) Both manufacturing company have same average life of the produce bulbs.

The first two statements appear to be biased since they reflect a preferential attitude to one or other of the two processes. Hence, the best course is to adopt the hypothesis of no difference as stated in (iii).

24.5 TESTS OF SIGNIFICANCE

Procedure which enables us to decide, on the basis of sample information whether to accept or reject the hypothesis or to determine whether observed sampling results differ significant from expected results are called tests of significance, rules of decision or test of hypothesis.

24.6 LEVEL OF SIGNIFICANCE

The probability level below which will reject the hypothesis is called the level of significance. The levels of significance usually employed in testing of hypothesis are 5% and 1 %.

24.7 CRITICAL REGION AND ACCEPTANCE REGION

A region (corresponding to a statistic) is called the sample space. The part of sample space which amounts to rejection of null hypothesis H_0, is called critical region or region of rejection.

If $X = (x_1, x_2, ..., x_n)$ is the random vector observed and W_e is the critical region of the sample space W, then

$$W_a = W - W_e$$

of the space is called the acceptance region.

WORKING PROCEDURE

Step 1. State the null hypothesis H_0 and alternative hypothesis H_1.

Step 2. Make some assumption such as the sample is random, the population is normal, etc.

Step 3. Find the most appropriate test statistic together with its sampling distribution.

Step 4. On the basis of sampling distribution make a decision to either accept or reject the null hypothesis H_0.

Step 5. Take a random sample and compute the test statistic. If the calculated value of the test statistic falls is the acceptance region, then accept the null hypothesis H_0. If it falls in the region of rejection, reject the null hypothesis and accept H_1.

24.8 TYPE–I AND TYPE–II ERRORS

There are two types of errors in testing of hypothesis : Type I and Type II errors. When a statistical hypothesis is tested, there are four possible results.

1. The hypothesis is true but our test rejects it.
2. The hypothesis is false but our test accepts it.
3. The hypothesis is true and our test accepts it.
4. The hypothesis is false and our test rejects it.

Obviously, the first two possibilities lead to errors. If we reject a hypothesis when it should be accepted (possibility no.1) we say that a type I error has been made. On the other hand if we accept a hypothesis when it should be rejected (possibility no. 2) we say that a type II error has been made. In either case wrong decision or error in judgement has occurred.

Decision	H_0 : True	H_0 : False
Accept H_0	Correct decision	Type II error
Reject H_0	Type I error	Correct decision

Fig. 1. Kinds of error

The probability of committing a type I error is designated as α and is called the level of significance. Therefore

$$\alpha = \Pr [\text{Type I error}] = \Pr \ [\text{Rejecting } H_0 \ / \ H_0 \text{ is true}]$$

must be complement of

$$((1 - \alpha) = \Pr [\text{Accepting } H_0 \ / \ H_0 \text{ is true}]$$

This probability $(1 - \alpha)$ corresponds to the concept of $100(1 - \alpha)\%$ confidence interval. Our efforts would obviously be to have a small probability of making a type I error, Hence, the objective is to construct the test to minimize α.

Similarly, the probability of committing a type II error is designated by β.

Then $$\beta = \Pr [\text{Type II error}] = \Pr \ [\text{Accepting } H_0 \ / \ H_0 \text{ is false}]$$

and $$(1 - \beta) = \Pr [\text{Rejecting } H_0 \ / \ H_0 \text{ is false}]$$

This probability $(1 - \beta)$ is known as the power of a statistical test.

The following table gives the possibility associated with each of the four cells shown in the previous table :

The decision is to	The null hypothesis is	
	True	False
Accept H_0	$(1 - \alpha)$ Confidence level	β
Reject H_0	α	$(1 - \beta)$ Confidence level
Sum	1.00	1.00

In order for any tests of hypothesis or rules of decisions to be good, they must be designed as to minimize errors of decision, However, this is not a simple matter, since for a given sample size, an attempt to decrease one type of errors is accompanied in general by an increase in other type of error. The probability of making type I error is fixed in advance by the choice of level of significance employed in the test. We can make the type I error as small as we please, by lowering the level of significance. But by doing so we increase the chance of accepting a false hypothesis i.e., of making a type II error. It follows it is impossible to minimize both errors simultaneously, In the long run errors of type I are perhaps more likely to prove serious research programmes in social science that are errors of type II.

🐦 **REMARKS**

➠ However, α and β are not independent of each other, nor are they independent of the sample size n. When n is fixed, if α is lowered, then β normally rises and vice versa.

➠ If n is increased, it is possible for both α and β decrease.

24.9 BEST CRITICAL REGION

In testing the hypothesis $H_0 : \theta = \theta_0$ against the alternative $H_1 : \theta = \theta_1$ the critical region is best if the types II error is minimum or the power is maximum when compared to every other possible critical region of size α.

A test defined by this critical region is called most powerful test.

24.10 ONE TAIL AND TWO TAIL TESTS

Two tailed test is that where the hypothesis about the population mean is rejected for value of $\overline{X}$ falling into either tail of the sampling distribution. When the hypothesis about population mean is rejected only for value of $\overline{X}$ falling into one of the tails of sampling distribution then it is known as one-tailed test.

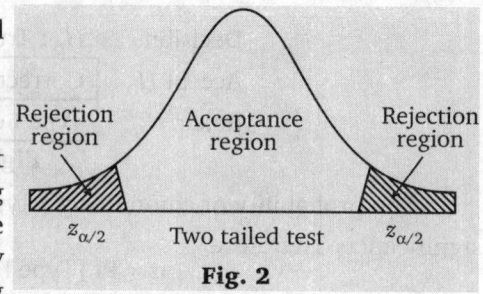

Fig. 2

If it is the right tail, then it is called right tailed test or one-sided alternative to the right and if it is one of the left tailed to the left and called left tailed test.

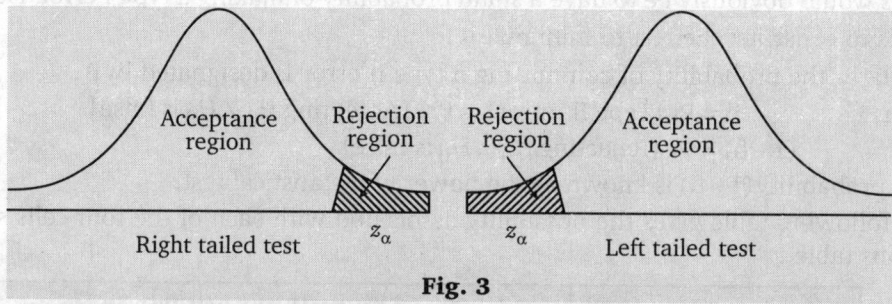

Fig. 3

SOLVED EXAMPLES

EXAMPLE 1. It is desired to test a hypothesis $H_0 : P = P_0 = 1/2$ against the alternative hypothesis $H_1 : P = P_1 = 3/4$ on the basis of tossing a coin, where P is the probability of getting a head in a single trial and agreeing to accept H_0 if a tail appears and to accept H_1 otherwise. Find the values of α and β.

SOLUTION. We have

$$\alpha = \text{probability of rejecting } H_0 \text{ when } H_0 \text{ is true}$$
$$= P(\text{reject } H_0 / H_0) = P(\text{head appears} / P = P_0)$$
$$= [P]_{P = P_0 = 1/2} = 1/2$$

Again,

$$\beta = \text{probability of accepting } H_0 \text{ when } H_1 \text{ is true}$$
$$= P(\text{accept } H_0 \,/\, H_1) = P(\text{tail appears} \,/\, P = P_1)$$
$$= [P]_{P = P_1 = 3/4} = 1 - \frac{3}{4} = \frac{1}{4}.$$

EXAMPLE 2. *Given a binomial distribution*

$$f(x, p) = \begin{cases} {}^n C_x \, p^x \, q^{n-x}, & x = 0, 1, 2, 3, 4 \\ 0 & , \quad \text{else where} \end{cases}$$

It is desired to test $H_0 : P = P_0 = 1/3$ against $H_1 : P = P_1 = 1/2$ by agreeing to accept H_0 if $x \le 2$ in four trials and to reject otherwise.
What are the probabilities of committing
(a) type I error,   (b) type II error.

SOLUTION. (a) $P(\text{type I error}) = P(\text{reject } H_0 \,/\, H_0 \text{ is true})$

$$= P(x \le 2 \,|\, P = 1/3) = \sum_{x=3}^{4} {}^4 C_x \left(\frac{1}{3}\right)^x \left(\frac{2}{3}\right)^{4-x}$$

$$= {}^4 C_3 \left(\frac{1}{3}\right)^3 \left(\frac{2}{3}\right)^{4-3} + {}^4 C_4 \left(\frac{1}{3}\right)^4$$

$$= 4 \times \frac{2}{3^4} + \frac{1}{3^4} = \frac{1}{3^2} = \frac{1}{9}$$

Hence, the hypothesis $H_0 : P = \dfrac{1}{3}$ is being tested at the level of significance.

(b) $\beta = P \text{ (type II error)} = P(\text{accept } H_0 \,|\, \text{when } H_1 \text{ is true})$
$$= P \,(x \le 2 \,|\, P = 1/2)$$

$$\Rightarrow \qquad P = \sum_{x=0}^{2} {}^4 C_x \left(\frac{1}{2}\right)^x \left(\frac{1}{2}\right)^{4-x}$$

$$= {}^4 C_0 \left(\frac{1}{2}\right)^4 + {}^4 C_1 \left(\frac{1}{2}\right)^1 \left(\frac{1}{2}\right)^3 + {}^4 C_2 \left(\frac{1}{2}\right)^2 \left(\frac{1}{2}\right)^2 = \frac{11}{2^4}.$$

EXAMPLE 3. *An urn contains 4 balls of which 1 or 2 are white and the rest are red. To test the hypothesis H_0 : one ball is white, balls are drawn one after another until a white ball appears. Suggest a good critical region for this test and find the value for α and β.*

SOLUTION. Let X be the number of red balls drawn until a white ball appears. Then the possible test may be to accept H_0 if $X = 3$ and reject otherwise.

$$\alpha = P \text{ (reject } H_0 \text{ when } H_0 \text{ is true)}$$
$$= P \,(X \ne 3 \,/\text{one ball is white})$$
$$= P \,(X = 0 \text{ or } 1 \text{ or } 2/\text{one ball is white})$$
$$= \frac{1}{4} + \frac{3}{4} \times \frac{1}{3} + \frac{3}{4} \times \frac{2}{3} \times \frac{1}{2} = \frac{3}{2}$$

$$\beta = P \text{ (accept } H_0, \text{ when } H_1 \text{ is true)}$$
$$= P(X = 3/\text{two balls are white}) = 0.$$

The second test may be to accept H_0 if $X = 2$ or 3 and reject otherwise.

$$\alpha = P(X = 0 \text{ or } 1/\text{one ball is white})$$
$$= \frac{1}{4} + \frac{3}{4} \times \frac{1}{3} = \frac{1}{2}.$$
$$\beta = P \text{ (accept } H_0 \text{ when } H_1 \text{ is true)}$$
$$= P[(X = 2 \text{ or } 3)/\text{two balls are white}]$$
$$= \frac{3}{4} \times \frac{1}{3} \times \frac{2}{3} + 0 = \frac{1}{6}.$$

24.11 TEST OF SIGNIFICANCE OF LARGE SAMPLES

Suppose a large number of samples is classified according to the frequencies of an attribute. It gives rise to a binomial distribution which tends to a normal distribution for large values of n, the number in the sample. It follows therefore a great majority of its members lie within a range $\pm 3\sigma$ on each side of the mean, i.e., of $\pm 3\sqrt{npq}$ on each side of the value np. If the number of successes in a large samples of size n differs from the expected value np by more

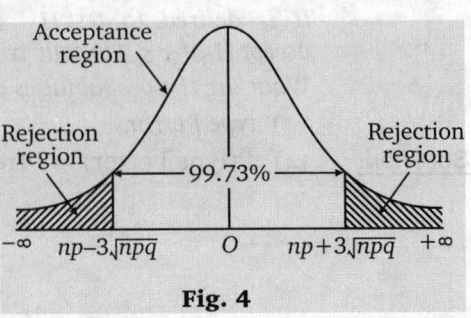

Fig. 4

then $3\sqrt{npq}$, we call the difference highly significant and the truth of the hypothesis is very improbable. Generally we accept the hypothesis as correct and then we calculate np and $\sqrt{npq}$ and apply the above test.

Thus, by test of significance, we mean a test of hypothesis where we decide to accept or reject a hypothesis on the basis of a random sample. If null hypothesis is rejected we say that the difference $x \sim np$ is significant etc.

24.12 STANDARD ERROR

The standard deviation of simple sampling is called standard error. In real sense, the term standard error has wider meaning than merely the standard deviation of simple sampling.

If the difference between the actual and observed frequencies is more than three times the standard error, the difference is said to be significant, and such a difference could not have arisen due to fluctuations of sampling, or the probability of such a difference arising due to change is very low. If the difference is less than three times the standard error, it could have arisen due to fluctuations of sampling. If the difference is less than twice the standard error, the probability of its arising due to chance is fairly high, and it can be ignored. If the difference is more than twice the standard error but less than thrice the standard error, then it could have arisen due to sampling fluctuation, but the low probability of its arising due to chance is very low.

24.12.1 STANDARD ERROR OF A PERCENTAGE

The formula for the estimated standard error of a percentage (S_p) is $S_p = \sqrt{\dfrac{pq}{n}}$

where, S_p = Estimated standard error of a percentage.

         p = Percentage of the universe with the characteristic being studied.

         $q = (100\% - p)$.

         n = Size of sample.

The standard formula for difference between two sample percentage is

$$S_{\text{difference}} = \sqrt{S_a^2 + S_b^2}$$

where, $\quad S_{\text{difference}}$ = Estimated standard error of the difference.

$\qquad\qquad S_a$ = Estimated standard error of a sample percentage usage rate.

$\qquad\qquad S_b$ = Estimated standard error of b sample percentage usage rate.

24.12.2 Standard Error of the Mean

When constructing a confidence interval estimate, researchers must also know how the standard error of the mean is related to the standard deviation of the universe. In a simple random sample, the relationships :

$$\sigma_{\bar{x}} = \frac{\sigma}{\sqrt{n}}$$

where, $\qquad\sigma_{\bar{x}}$ = Standard error of the mean;

$\qquad\qquad\sigma$ = Standard deviation of the universe.

and $\qquad\qquad n$ = Number of observation in the sample.

This formula applies if less than 5% of the universe is included in the sample.

If more than 5% of the universe is included in the sample, then the standard error may be computed as:

$$\sigma_{\bar{x}} = \frac{\sigma}{\sqrt{n}} \sqrt{\frac{N-n}{N}}$$

where, $\qquad N$ = Universe size; n = Sample size

Factor $\sqrt{\dfrac{N-n}{N}}$ recognizes that a sample from a finite universe partially exhausts that universe.

This relationship between the standard error of the mean ax and the standard deviation of the universe (σ) implies that when large sample size (n) are used the standard error of the mean will be a smaller value, and, therefore, the mean of a given sample is likely to be closer to the universe mean.

24.13 PROBABLE ERROR

Instead of standard error same authors have used a quantity called probable error, which is 0.67449 times the standard error.

24.14 TEST OF SIGNIFICANCE IN CASE OF ATTRIBUTES (LARGE SAMPLES)

The procedure for a test of significance in case of attributes (large samples) is as follows: .

(i) Assume null hypothesis H_0 and alternative hypothesis H_1 (though not necessary).

(ii) Define a test statistic

$$z = \frac{x - np}{\sqrt{npq}} \sim N(0, 1) \qquad \text{or} \qquad z = \frac{n}{\sqrt{pq/n}} \sim N(0, 1)$$

and calculate the value of z.

(iii) Decide α the level of significance

(a) for α = 5%, we reject H_0 if $|z| > 1.96$, two sided

(b) for α = 1 %, we reject H_0 if $|z| > 2.58$, two sided.

(c) for α = 0.27%, we reject H_0 if $|z| > 3$, two sided.

In testing a given hypothesis or a test of significance, the maximum probability with which we would be willing to risk an error is called the level of significance of the test. In practice a level of significance 0.05 or 0.01 is usually taken.

24.15 CONFIDENCE LIMITS OF UNKNOWN p

Let x = no. of success, n = sample size, P = probability of success or proportion is the population, $Q = 1 - P$.

$$p = \frac{x}{n} = \text{sample proportion.}$$

Then, $\quad z = \dfrac{p - P}{\sqrt{PQ/n}} \sim N(0,1)$

Now, $|z| \le 3 \Rightarrow \left| \dfrac{p - P}{\sqrt{PQ/n}} \right| \le 3 \Rightarrow P - 3\sqrt{\dfrac{PQ}{n}} \le p \le P + 3\sqrt{PQ/n}$

These are called confidence limits for P at $\alpha = 0.0027$.

However, the limits of np are

$$nP - 3\sqrt{nPQ} \le nP + 3\sqrt{nPQ} \quad \text{or} \quad x - 3\sqrt{nPQ} \le nP \le x + 3\sqrt{nPQ}$$

SOLVED EXAMPLES

EXAMPLE 1. *A certain cubical die was thrown 9000 times, and a 5 or a 6 was obtained 3240 times. On the assumption of random throwing do the data indicate an unbiased die?*

(VTU–2010)

SOLUTION. On the hypothesis of an unbiased die, the chance of throwing a 5 or 6 is

$$= \frac{2}{6} = \frac{1}{3} = p \text{ and hence } q = \frac{2}{3}.$$

∴ The expected number of successes

$$= np = 9000 \times \frac{1}{3} = 3000$$

The standard error of number of success

$$= \sqrt{npq} = \sqrt{9000 \times \frac{1}{3} \times \frac{2}{3}} = 44.72$$

The difference between observed frequency and expected frequency = 3240 − 3000 = 240 which is nearly 5.4 times the standard error and is therefore most unlikely to appear as a result of simple sampling. We therefore conclude that die is certainly biased one.

Alternative Method:

$$Z = \frac{x - np}{\sqrt{npq}} = \frac{3240 - 3000}{44.72} = \frac{240}{44.72} = 5.37$$

Since $|z| > 3$. Hence, we reject H_0 at 0.27% level of significance. That is the difference $x - np$ is highly significant.

EXAMPLE 2. *A sample of 900 days is taken from meteorological records of a certain district and 100 of them are found to be foggy. What are the probable limits to the percentage of foggy days in the district ?*

SOLUTION. Here $p = \dfrac{100}{900} = \dfrac{1}{9}$ so that $q = \dfrac{8}{9}$

Standard error

$$= \sqrt{npq} = \sqrt{900 \times \frac{1}{9} \times \frac{8}{9}} = \sqrt{88.2} = 9.4$$

$$3 \text{ S.E.} = 3 \times 9.4 = 28.2$$

$\therefore$ Limits are $x \pm 3\sqrt{npq}$, i.e., 100 ± 28.2, i.e., 128.2 and 71.8.

These are the limits out of 900 days. Hence, the limits in percentage are $\frac{128.2}{900} \times 100 = 14.2$ and $\frac{71.8}{900} \times 100 = 8$ nearly.

EXAMPLE 3. *In some dice throwing experiment, Mohan threw a die 49152 times, and of these 25145 yielded 4, 5 or 6. Is this consistent with the hypothesis that the die was unbiased?*

SOLUTION. The probability of throwing 4, 5 or 6 with one die

$$p = \frac{3}{6} = \frac{1}{2}, q = 1 - \frac{1}{2} = \frac{1}{2}; n = 49152, x = 25145$$

$\therefore$ Expected value of number of success

$$E(x) = np = 49152 \times \frac{1}{2} = 24576$$

Standard error $= \sqrt{npq} = \sqrt{49152 \times \frac{1}{2} \times \frac{1}{2}} = 110.9$

Test statistic, $Z = \frac{x - np}{\sqrt{npq}} = \frac{25145 - 24576}{110.9} = \frac{569}{110.9} = 5.1373$

Since $|z| > 3$, hence hypothesis $H_0 : P = \frac{1}{2}$ is rejected. The difference is not due to sampling fluctuations. Hence, the data is not consistent with the hypothesis that the die was unbiased.

EXAMPLE 4. *In a locality of 18,000 families a sample of 840 families was selected. Of these 840 families, 206 families were found to have a monthly income of Rs. 250 or less. It is desired to estimate how many out of the 18000 families have a monthly income of Rs. 250 or less. With in what limits would you place your estimate?*

SOLUTION. Here, p = proportion of families having monthly income of Rs. 250 or less

$$= \frac{206}{840} = 0.245$$

So that $q = 1 - p = 1 - 0.245 = 0.755$

S.E. of the proportion of families having monthly income of Rs. 250 or less

$$= \sqrt{\frac{pq}{n}} = \sqrt{\frac{0.245 \times 0.755}{840}} = 0.015 = 1.5\%$$

Hence, the required limits are 24.5 percent $\pm (3 \times 15)$ percent, i.e., 20 percent and 29 percent

$\therefore$ The limits for the number of families out of 18000 families are $\frac{20}{100} \times 18000$ and $\frac{29}{100} \times 18000$, i.e., 3600 and 5220 families.

24.16 TEST OF SIGNIFICANCE OF MEAN (LARGE SAMPLE OF VARIABLES)

Let X be a random variable with mean μ and standard deviation σ. Let $x_1, x_2, ..., x_n$ be a random sample of size n.

Let $\bar{x}$ = sample mean $= \dfrac{1}{n}\sum_{i=1}^{n} x_i$. Then, $Z = \dfrac{\bar{x}-\mu}{\sigma/\sqrt{n}} \sim N(0,1)$

If Z calculated greater than 3, we reject the null hypothesis $H_0 : \mu = \mu_0$, where $Z = \dfrac{\bar{x}-\mu_0}{\sigma/\sqrt{n}}$.

SOLVED EXAMPLES

EXAMPLE 1. *A sample of 1600 members found to have a mean 5.7. Could it be reasonably regarded as a simple sample from a large population whose mean is 4.5 and standard deviation is 2.8.*

SOLUTION. Here, $n = 1600, \bar{x} = 5.7, \mu = 4.5, \sigma = 2.8$.

The standard error (S.E.) of the mean for a simple sample of 1600 from such a

population is $\dfrac{\sigma}{\sqrt{n}} = \dfrac{2.8}{\sqrt{1600}} = \dfrac{2.8}{40} = 0.07$·

The deviation of the sample mean from that of the population mean = 5.7 – 4.5 = 1.2, which is 17.1 times the standard error (S.E.)

$$|\bar{x}-\mu| > 3 \text{ S.E.}$$

or $\qquad |Z| = \dfrac{|\bar{x}-\mu|}{\text{S.E.}} > 3$

Hence, the deviation is highly significant.

The given sample cannot be regarded as a simple sample from a large population with mean $\mu = 4.5$ and standard deviation 2.8.

EXAMPLE 2. *A sample of 400 male students have a mean height of 168 cm. Can it be reasonably regarded as a sample from a large population with mean height 167.8 cm and standard deviation 3.25 cm.*

SOLUTION. Here $n = 400, \bar{x} = 168$ cm, $\mu = 167.8, \sigma = 3.25$ cm

$$|Z| = \dfrac{\bar{x}-\mu}{\sigma/\sqrt{n}} = \dfrac{168-167.8}{3.25/\sqrt{400}} = \dfrac{0.2 \times 20}{3.25} = 1.23$$

Since $|Z| < 3$, the deviation of sample mean from population mean is not significant. Hence, it can reasonably be regarded as a simple sample from a large population with mean 167.8 cm and standard deviation 3.25 cm.

EXAMPLE 3. *A random sample of 400 months has an average length of 10 cm. Can this be regarded as a sample from a large population with mean of 10.2 cm and a standard deviation of 2.25 cm?*

SOLUTION. Here $N = 400, \bar{x} = 10$.

(1) Null Hypothesis (H_0) : The sample has been drawn from the normal population with mean (μ) = 10.2 cm and standard deviation (σ) = 2.25 cm *i.e.*, $\mu = 10.2$. Alternative hypothesis (H_1) $= \mu \neq 10.2$ (Two-tailed).

(2) Competition of test statistic: Under H_0 (since sample size is large).

$$Z = \frac{\bar{x} - \mu}{\sqrt{\sigma^2 / n}} = \frac{10 - 10.2}{\sqrt{(2.25)^2 / 400}} = \frac{-0.2 \times 20}{2.24} = -1.777$$

(3) Level of significance : $\alpha = 0.05$

(4) Critical value : At 0.05 level of significance, the critical value of $Z = \pm 1.96$ (from the table).

(5) Decision : Since the computed value of $(z) = 1.77$ is less than critical value of $Z = 1.96$, it falls in the acceptance region. Hence, the facts are consistent with the null hypothesis which is accepted with 95% confidence and it is concluded that the sample has been drawn from the normal population with mean of 10.2 cm and a standard deviation of 2.25 cm.

EXAMPLE 4. *A research worker wishes to estimate the mean of a population by using sufficiently large samples. The probability is 95 percent that the sample mean will not differ from the true mean by more than 25% of the standard deviation. How large a sample should be taken?*

SOLUTION. In usual notation

$\bar{x}$ = Sample mean, μ = true mean, σ = standard deviation of the population

$$Z = \frac{\bar{x} - \mu}{\sigma / \sqrt{n}} \sim N(0, 1).$$

According to the condition given, we have to find $\bar{x}$, so that

$$\frac{|\bar{x} - \mu|}{\sigma / \sqrt{n}} \leq 1.96$$

$$\Rightarrow \qquad |\bar{x} - \mu| \leq 1.96 \frac{\sigma}{\sqrt{n}} \qquad \qquad \text{...(1)}$$

$$\text{Again} |\bar{x} - \mu| = 25\% \text{ of } \sigma = \frac{25}{100} \sigma = \frac{\sigma}{4} \qquad \qquad \text{...(2)}$$

From (1) and (2), we have

$$1.95 \frac{\sigma}{\sqrt{n}} \leq \frac{\sigma}{4}$$

$$\sqrt{n} \geq 1.96 \times 4$$

$$n \geq (7.84)^2$$

$$n \geq 61.4656, \text{ i.e., } n = 62.$$

EXAMPLE 5. *The mean life time of sample of 100 aquaria fishes is computed to be 1,570 hours with a standard deviation of 120 hours. The shopkeeper claims that the average life of fishes is 1600 hours. Using the level of significance of 0.05, is the claim acceptable?*

SOLUTION. Here, $n = 100$, $\bar{x} = 1570$, and s = sample standard deviation $= 120$.

1. Null Hypothesis (H_0) : $\mu = 1600$, *i.e.*, average life of the fish is 1600 hours. Alternative Hypothesis (H_1) : $\mu_1 \neq 1600$ (two tailed test)

2. Test statistic: Under H_0, the test statistic is

$$Z = \frac{\bar{x} - \mu}{\sqrt{\sigma^2 / n}} = \frac{\bar{x} - \mu}{\sqrt{s^2 / n}}$$

$$Z = \frac{1570 - 1600}{\sqrt{(120)^2 / 100}} = -\frac{300 \times 10}{120} = -2.5$$

3. **Level of Significance :** $\alpha = 0.05$

4. **Critical Value:** At 0.05 level significance, the critical value of $Z = \pm 1.96$ (As per table).

5. **Inference :** Since the critical value of $Z = 2.5$ is greater than the critical value of $Z = 1.96$, it falls in the rejection region. Hence, the null hypothesis is rejected and it is concluded that the shopkeeper claim of average life of the fishes being 1600 hours is not acceptable.

24.17 TESTING THE SIGNIFICANCE OF THE DIFFERENCE BETWEEN THE MEANS OF TWO LARGE SAMPLES

If two random sample of n_1 and n_2 members respectively have been taken from the sample population of S.D. σ, we would like to know if the difference of their means $(\bar{x}_1 - \bar{x}_2)$, is due to fluctuations of sampling.

On the assumption that the samples are independent and have been taken from the sample population, the S.E. ε of the difference of their means is given by

$$\varepsilon^2 = \varepsilon_1^2 + \varepsilon_2^2$$

where ε_1 and ε_2 are the S.E. of the means of two samples given as

$$\varepsilon_1^2 = \sigma^2 / n_1, \ \varepsilon_2^2 = \sigma^2 / n_2$$

Hence,
$$\varepsilon^2 = \sigma^2 \left(\frac{1}{n_1} + \frac{1}{n_2} \right).$$

If the difference $\bar{x}_1 \sim \bar{x}_2$ exceeds 3ε, the difference cannot said to be due to functions of

sampling and the assumption $\mu_1 = \mu_2$ is unlikely to be correct.

Alternative method:

$$Z = \frac{(\bar{x}_1 - \bar{x}_2) - (\mu_1 - \mu_2)}{\sqrt{\dfrac{\sigma^2}{n_1} + \dfrac{\sigma^2}{n_2}}} \sim N(0, 1)$$

Hence $H_0 : \mu_1 = \mu_2$,
$$Z = \frac{(\bar{x} - \bar{x}_2)}{\sqrt{\dfrac{\sigma^2}{n_1} + \dfrac{\sigma^2}{n_2}}}$$

Alternative hypothesis may be $H_1 : \mu_1 > \mu_2, \ \mu_1 < \mu_2 \text{ or } \mu_1 \neq \mu_2$

With the corresponding critical regions

$$Z \geq 2.33, \ Z \leq -2.33 \quad |Z| \geq 2.58 \text{ for } \alpha = 0.01$$

$$Z \geq 1.645, \ Z \leq -1.645 \quad |Z| \geq 1.96 \text{ for } \alpha = 0.05$$

SOLVED EXAMPLES

EXAMPLE 1. *Number of fishes from one pond was taken 1000 and from second pond 2000. The mean weight of fishes of two ponds are 67.5 and 68.0 respectively. Test of equality of means of the two populations each with, S.D. 2.5 (No credit if the H_0 and H_1 are not stated) Assumptions should be clearly stated.*

SOLUTION. As we have been given

$$n_1 = 1000, n_2 = 2000, \bar{x}_1 = 67.5, \bar{x}_2 = 68,$$
$$\sigma_1 = \sigma_2 = \sigma = 2.5$$

1. **Null Hypothesis :** (H_0) $H_0 : \mu_1 = \mu_2$

    Alternative hypothesis : $H_0 : \mu_1 \neq \mu_2$

    [Two tailed test]

2. **Computation of test statistic :** Under H_0 the test statistics (for large samples is)

$$Z = \frac{\bar{x}_1 - \bar{x}_2}{\sqrt{\dfrac{\sigma_1^2}{n_1} + \dfrac{\sigma_2^2}{n_2}}} = \frac{\bar{x}_1 - \bar{x}_2}{\sqrt{\dfrac{\sigma^2}{n_1} + \dfrac{\sigma^2}{n_2}}} = \frac{67.5 - 68.0}{\sqrt{(2.5)^2 \left\{ \dfrac{1}{1000} + \dfrac{1}{2000} \right\}}}$$

$$= \frac{-0.5}{2.5\sqrt{\dfrac{3}{2000}}} = \frac{-0.5}{2.5 \times 0.039} = \frac{-0.5}{0.0975} = -5.13$$

3. **Level of significance :** $\alpha = 0.05$

4. **Critical value:** At 5% level of significance ($\alpha = 0.05$) the critical value of $Z = \pm 1.96$.

5. **Interpretation :** Since the computed value of $Z = 5.13$ is greater than critical value of $Z = 1.96$, it falls in the rejection region. Hence, the null hypothesis is rejected.

EXAMPLE 2. *Random samples of 500 and 400 have means 11.5 and 10.9 respectively. Can the samples be regarded as drawn from the population of the standard deviation 5.*

(Madras–2002)

SOLUTION. With usual notations, where $n_1 = 500, n_2 = 400, \bar{x}_1 = 11.5, \bar{x}_2 = 10.9, \sigma = 5$

$$\varepsilon^2 = \frac{\sigma^2}{n_1} + \frac{\sigma^2}{n_2} = \frac{25}{500} + \frac{25}{400} = \frac{225}{2000} = \frac{9}{80}$$

$$\varepsilon = 0.335$$

Difference of means $= 11.5 - 10.9 = 0.6$.

This difference is 1.79 times the S.E. Hence, it is not significant, *i.e.*, there is no reason to doubt the hypothesis.

EXAMPLE 3. *A potential buyer of light bulbs bought 50 bulbs of each of two brands. Upon testing these bulbs, he found that brand A had a mean life of 1282 hours with a S.D. of 80 hours, where as brand B had a mean life of 1208 hours, with a S.D. of 94 hours. Can the buyer be quite certain that two brands differ in quality ?*

SOLUTION. Here $n_1 = 50 = n_2, \sigma_2^2 = 94^2, \bar{x}_1 = 1282, \bar{x}_2 = 1208$.

Hence $\varepsilon = \sqrt{\dfrac{80^2}{50} + \dfrac{98^2}{50}} = 17.5$ hours.

This difference of means $\bar{x}_1 - \bar{x}_2 = 74 = 4.24\varepsilon$ is greater than 2 times of its S.E. Hence, the difference cannot be said due to the fluctuations of sampling. There is much difference in two brands. Hence, brand A is to be preferred.

EXAMPLE 4. *The mean yield of Soyabean from a district A was 210 kgs with standard deviation 10 kgs per acre from a sample of 100 plots. In another district 'B' the mean yield was 220 kgs, with standard deviation 12 kgs. from a sample of 150 plots. Assuming that the standard deviation of the yield in the entire state was 11 kgs, test whether there is any significance difference between the mean yield of Soya bean is the two districts.*

SOLUTION. We have

For District A: $n_1 = 100, \bar{x}_1 = 210, S_1 = 10, \sigma_1 = 11$
For District A: $n_2 = 150, \bar{x}_2 = 220, S_2 = 12, \sigma_2 = 11$

1. **Null hypothesis (H_0) :** $H_0 : \mu_1 = \mu_2$, *i.e.,* there is no significant difference between the mean yield of Soyabean in the two districts. That means the population means of Soyabean yield in the two districts are equal.

   Alternative Hypothesis: $H_1 : \mu_1 \neq \mu_2$ (Two-tailed test).

2. **Computation of test statistics:** Under H_0, the test statistic for large samples is

   $$Z = \frac{\bar{x}_1 - \bar{x}_2}{\sqrt{\dfrac{\sigma^2}{n_1} + \dfrac{\sigma^2}{n_2}}} \sim N(0, 1)$$

   Now, $Z = \dfrac{210 - 220}{\sqrt{11^2 \left(\dfrac{1}{100} + \dfrac{1}{150} \right)}} = \dfrac{-10}{11 \times \sqrt{\dfrac{3+2}{300}}} = -7.05$

3. **Level of significance :** $\alpha = 0.05$

4. **Critical value:** At 5% level of significance ($\alpha = 0.05$) the critical value of $Z = 1.96$.

5. **Inference:** Since the compound value of $Z = -7.05$ is less than the table value of $Z = -1.96$, it falls in the rejection region. Hence, the null hypothesis is rejected and it may be concluded that there is a significant difference between yield of Soyabean in the two districts.

EXAMPLE 5. *A random sample of 150 villages was taken from district A and the average population per village was found to be 440 and standard deviation 32. Another random sample of 250 villages from the same district gave an average population 480 per village with a standard deviation of 56. Is the difference between the average of two samples statistically significant? Give reasons.*

SOLUTION. Here, $n_1 = 150, \bar{x}_1 = 440, s_1 = 32$
                        $n_2 = 250, \bar{x}_2 = 480, s_2 = 56z$

Here, population standard deviations σ_1 and σ_2 are not known or we use s_1 in place of σ_1 and s_2 in place of σ_2

$$Z = \frac{(\bar{x}_1 - \bar{x}_2) - (\mu_1 - \mu_2)}{\sqrt{\dfrac{\sigma_1^2}{n_1} + \dfrac{\sigma_2^2}{n_2}}} = \frac{440 - 480}{\sqrt{\dfrac{32^2}{150} + \dfrac{56^2}{250}}}, \text{ where } H_0 : \mu_1 = \mu_2$$

$$= \frac{-40}{\sqrt{6.83 + 12.54}} = \frac{-40}{\sqrt{19.37}} = \frac{-40}{4.4} = -9.09$$

$|Z| = 9.09 > 3$.

Hence, the difference between sample means is highly significant. This difference cannot be due to fluctuations of sampling.

EXAMPLE 6. *In a survey of incomes of two classes of workers, two random samples gave the following results :*

Sample	Size	Mean Annual Income (In Rs.)	Standard Deviation (In Rs.)
I	100	582	24
II	100	546	28

Examine whether the difference between the means are significant.

[CCSU(B.Sc. Biotech)–2000]

SOLUTION. As we have been given

$n_1 = 100, n_2 = 100, \bar{x}_1 = 582, \bar{x}_2 = 546, \sigma_1^2 = (24)^2, \sigma_2^2 = (28)^2$

1. **Null hypothesis:** $(H_0) H_0 : \mu_1 = \mu_2$

   Alternative hypothesis: $H_0 : \mu_1 \neq \mu_2$ (Two tailed test)

2. **Computation of test statistic :** Under H_0 the test statistics.

$$|Z| = \frac{|\bar{x}_1 - \bar{x}_2|}{\sqrt{\dfrac{\sigma_1^2}{n_1} + \dfrac{\sigma_2^2}{n_2}}} = \frac{|582 - 546|}{\sqrt{\dfrac{24^2}{100} + \dfrac{28^2}{100}}}$$

$$= \frac{36}{\sqrt{\dfrac{1360}{100}}} = \frac{36}{\sqrt{13.6}} = \frac{36}{3.69} = 9.756$$

$$|Z| = 9.756$$

3. **Level of Significance:** $\alpha = 0.05$
4. **Critical Value:** At 5% level of significance the critical value of $Z = 1.96$.
5. **Interpretation:** Since the compound value of $Z = 9.756$ is greater than critical value of $Z = 1.96$, it falls in rejection region. Hence, the null hypothesis is rejected.

EXAMPLE 7. *Random samples drawn from two countries gave the following data relating to height of adult male:*

	Country A	Country B
Mean height (in inches)	67.42	67.25
Standard deviation	2.58	2.50
Number of samples	1000	1200

Is the difference between the standard deviation significant.   [CCSU(B.Sc. B.Tech)–2001]

SOLUTION.   As we have been given

$$n_1 = 1000, n_2 = 1200, \sigma_1 = 2.58, \sigma_2 = 2.50$$

1. **Null hypothesis:** $H_0 : \mu_1 = \mu_2$

   Alternative hypothesis: $H_0 : \mu_1 \neq \mu_2$

2. **Computation of test statistic:**

$$\frac{\sigma_1 - \sigma_2}{\sqrt{\dfrac{\sigma_1^2}{2n_1} + \dfrac{\sigma_2^2}{2n_2}}} = \frac{2.58 - 2.50}{\sqrt{\dfrac{(2.58)^2}{2 \times 1000} + \dfrac{(2.50)^2}{2 \times 1200}}}$$

$$= \frac{0.08}{\sqrt{\dfrac{6.66}{2000} + \dfrac{6.25}{2400}}} = 0.08 \times \sqrt{168.52}$$

$$= 0.08 \times 12.98 = 1.038$$

3. **Level of significance:** 0.01

4. **Critical value:** At 1% level of significance the critical value of $Z = 2.58$.

5. **Interpretation:** Since the compound value of $Z = 1.038$ is less than the critical value of $Z = 2.58$. Hence, null hypothesis is accepted and we conclude that difference of two standard deviation is not significant.

24.18 TEST OF SIGNIFICANCE OF DIFFERENCE BETWEEN THE SAMPLE PROPORTION

If two samples are drawn from different populations, we may be interested in finding out whether the difference between the proportion of success is significant or not. In this case our hypothesis will be that the difference between proportion of success in one sample (p_1) and the proportion of success in another sample, is due to fluctuations of random sampling. The standard error of the difference proportions is calculated as follows :

$$\text{S.E.} (p_1 - p_2) = \sqrt{pq \left(\frac{1}{n_1} + \frac{1}{n_2} \right)} \qquad \text{where,} \ p = \frac{n_1 p_1 + n_2 p_2}{n_1 + n_2} \ \text{ or } \ p = \frac{x_1 + x_2}{n_1 + n_2}$$

p is pooled estimate of the actual proportion in the population x_1 and x_2 stand for number of occurrences in the two samples of sizes n_1 and n_2 respectively.

Thus $Z = \dfrac{p_1 - p_2}{\text{S.E.}}$ is less than 1.96 S.E.

(5% level of significance), the difference is not significant and the null hypothesis is accepted and the difference is regarded as due to random sampling variation.

SOLVED EXAMPLE

EXAMPLE.   *In a simple random sample of 600 men taken from a big city, 400 are found to be smokers. In another simple random sample of 900 men taken from another city 450 are smokers. Do the data indicate that there is significant difference in the habit of smoking in the two cities?*   (JNTU–2003)

SOLUTION.   Let us take null hypothesis (H_0) that there is no significant difference in the habit of smoking in the two cities.

$$\text{S.E.} (p_1 - p_2) = \sqrt{pq \left(\frac{1}{n_1} + \frac{1}{n_2} \right)}$$

$$p_1 = \frac{400}{600} = 0.667, \quad p_2 = \frac{450}{900} = 0.5$$

$$p = \frac{x_1 + x_2}{n_1 + n_2}, \quad q = 1 - p$$

where, $\quad n_1 = 600, n_2 = 900, x_1 = 400, x_2 = 450$

$$p = \frac{400 + 450}{600 + 900} = \frac{850}{1500} = \frac{17}{30}$$

$$q = 1 - \frac{17}{30} = \frac{13}{30}$$

$$\text{S.E. } (p_1 - p_2) = \sqrt{\frac{17}{30} \times \frac{13}{30} \left(\frac{1}{600} + \frac{1}{900} \right)}$$

$$= \sqrt{\frac{17 \times 13}{90} \times \frac{1500}{600 \times 900}} = 0.026$$

Thus, $\quad Z = \dfrac{\text{Difference}}{\text{S.E. } (p_1 - p_2)} = \dfrac{0.667 - 0.50}{0.026} = 6.42 .$

24.19 TEST OF SIGNIFICANCE OF DIFFERENCE BETWEEN THE STANDARD DEVIATION (LARGE SAMPLES)

In case of two large random samples each drawn from a normally distributed population, the S.E. of the difference between the standard deviations

$$\text{S.E. } (\sigma_1 - \sigma_2) = \sqrt{\frac{\sigma_1^2}{2n_1} + \frac{\sigma_2^2}{2n_2}}$$

When population standard deviation are not given :

$$\text{S.E. } (s_1 - s_2) = \sqrt{\frac{s_1^2}{2n_1} + \frac{s_2^2}{2n_2}}$$

The problem is to test if $\sigma_1 = \sigma_2$, *i.e.*, if the sample S.D. differ significantly or not.

Under the null hypothesis (H_0) : that sample standard deviation do not differ significantly, the test statistic is given by

$$Z = \frac{\sigma_1 - \sigma_2}{\text{S.E. } (\sigma_1 - \sigma_2)} = \frac{s_1 - s_2}{\text{S.E. } (s_1 - s_2)} .$$

SOLVED EXAMPLE

EXAMPLE. *One town has poultry farm of State Govt. It is suspected that the efficiency in the poultry is not alike, so a test is carried out by ascertaining the variability of life of the chickens produced by each poultry. The results are as follows :*

	Poultry farm A	Poultry farm B
No. of chickens in samples	100 (n_1)	200 (n_2)
Average life	1100 $(\bar{x}_1)$ hrs	900 $(\bar{x}_2)$ hrs
Standard deviation	240 (s_1)	220 (s_2)

From the above data, determine whether the difference between the variability of life of chickens from each sample is significant (take 1% level of significance).

SOLUTION.   1. **Null hypothesis :** $H_0 : \sigma_1 = \sigma_2$

                Alternative hypothesis: $H_1 : \sigma_1 \neq \sigma_2$

       2. **Test statistic :** Under H_0, test statistic is

$$Z = \frac{s_1 - s_2}{\sqrt{\dfrac{\sigma_1^2}{2n_1} + \dfrac{\sigma_2^2}{2n_2}}} = \frac{s_1 - s_2}{\sqrt{\dfrac{s_1^2}{2n_1} + \dfrac{s_2^2}{2n_2}}}$$

$$(\sigma_1^2 = s_1^2 \text{ and } \sigma_2^2 = s_2^2 \text{ for large samples})$$

$$Z = \frac{240 - 220}{\sqrt{\dfrac{(240)^2}{200} + \dfrac{(220)^2}{400}}} = \frac{20}{\sqrt{288 + 121}} = 0.96$$

       3. **Level of significance :** $\alpha = 0.01$

       4. **Critical value :** To examine whether σ_1 is greater or lesser than σ_2 we require a two tailed test whether at 0.1 significant level is the extreme 0.005 of the distribution on each side of mean. Therefore critical value of $Z \neq 2.58$ (from table).

       5. **Inference :** Since computed value of $Z(0.9)$ fall in acceptance region, the null hypothesis must be accepted. The result is therefore, not significant, so that it may be concluded that the variability of life of each of the poultry farm is not appreciably different but, on an average, the chickens from one farm have longer life than that of the other farm.

24.20 STUDENTS 't' TEST

      Student's t-distribution was given by W.S. Gosset (1908) who wrote under the pen-name of student. The quantity t is defined as

$$t = \frac{\overline{x} - \mu}{S / \sqrt{n}}$$

where $n = $ the number of observations in the sample.

             $\overline{x} = \dfrac{1}{n} \sum_{i=1}^{n} x_i$ is the sample mean.

             $\mu = $ The mean of the population from which the sample has been drawn.

             $S = \sqrt{\dfrac{1}{n-1} \sum_{i=1}^{n} (x_i - \overline{x})^2}$ is the standard deviation of the sample.

24.20.1 APPLICATIONS OF THE t-DISTRIBUTION

The t-distribution has a wide number of applications in statistics, some are given below :

    (i) To test the significance difference between the sample mean $\overline{x}$ and the population mean μ.

   (ii) To test the significance difference between two sample mean.

24.20.2 ASSUMPTION FOR STUDENT'S t-TEST

The following assumptions are made in the t-test :

    (i) The population from which the sample is drawn in normal.

(ii) All observations in the sample are independent.

(iii) The standard deviation (σ) of the population is unknown.

(iv) The sample values are correctly measured and recorded.

(v) Test for testing the significance difference between same mean μ and the population mean.

Let $x_1, x_2, ..., x_n$ be a random sample of size n ($n < 30$) taken from normal population with mean μ and variance σ^2 (unknown).

Null hypothesis: There is no significance difference between the sample mean and population mean or the sample has been drawn from the population whose mean is μ_0.

$$H_0 : \mu = \mu_0$$

Alternative hypothesis : H_0 is not true.

Test statistic : Under null hypothesis

$$t_{cal} = \frac{|\bar{x} - \mu_0|}{S / \sqrt{n}}$$

where

$$\bar{x} = \frac{1}{n} \sum_{i=1}^{n} x_i , S^2 = \frac{1}{n-1} \sum_{i=1}^{n} (x_i - \bar{x})^2$$

t-Tabulated : Tabulated value are taken from the table at $(n-1)$ degrees of freedom at given level of significance generally 1 %, 5%, 10%.

Conclusion : If $t_{cal} \geq t_{tab}$ then the null hypothesis is accepted.

Degree of freedom : The degree of freedom can be defined as the number of independent variables which make up the statistic (*e.g.*, t, ψ^2). It is denoted by ν (read as Nu).

In general, the number of degrees of freedom is the total number of observations less the number of independent constraints imposed on the observations.

If we have to choose any four numbers whose sum is 70, we can exercise our independent choice for any three numbers only, the fourth being 70 minus the total of the three numbers selected.

Thus, through we were to choose any four numbers, our choice was reduced to three because of one condition imposed. There was only one restriction on our freedom and our degrees of freedom were $4 - 1 = 3$. If two restrictions are imposed, our freedom to choose will be further curtailed and degrees of freedom will be $4 - 2 = 2$ rejected at given level of significance, *i.e.*, there is a significance difference between sample mean and population mean. If $t_{cal} < t_{tab}$, then the null hypothesis may be accepted.

SOLVED EXAMPLES

EXAMPLE 1. *A manufacturer of dry cells claimed that the life of their cells is 24.0 hours. A sample of 10 cells bad mean life of 22.5 hours with a standard deviation of 3.0 hours. On the basis of available information, test whether the claim of the manufacturer is correct at 5% level of significance.*

SOLUTION. Here we are given

$\mu = 24.0$ hours, $x = 10, \bar{x} = 22.5$ hours, $S = 3.0$ hours.

Null hypothesis H_0: The claim of manufacturer is correct, *i.e.*, $\mu = 24.0$ hours.

Alternative hypothesis: $H_1 : \mu \neq 24.0$

Test statistic, under null hypothesis

$$t = \frac{|\bar{x} - \mu|}{S / \sqrt{n}} = \frac{|22.5 - 24.0|}{3.0 / \sqrt{10}} = 1.58.$$

t-tabulated at 5% level at significance with $(10 - 1) = 9$, degree of freedom $= 2.26$.

Conclusion: Since $t_{cal} < t_{tab}$, therefore null hypothesis may be accepted at 5% level of significance, i.e., the claim of manufacturer is correct.

EXAMPLE 2. *A machinist is making engine parts with axle diameter of 0.700 inch. A random sample of 20 parts shows a mean diameter of 0.742 inch with a standard deviation of 0.40 inch. Compute statistic you would use to test the whether the work is meeting the specification at 1% level of significance.* (VTU–2009)

SOLUTION. Here we are given

$$\mu = 0.700 \text{ inch}, n = 20$$

$$\bar{x} = 0.742 \text{ inch}, S = 0.40$$

Null Hypothesis : The product is confirming to specification, i.e., $\mu = 0.700$ inch
Alternative hypothesis H_1 : $\mu \neq 0.700$
Test statistic : Under H_0

$$t = \frac{|\bar{x} - \mu|}{S / \sqrt{n}} = \frac{|10.742 - 0.700|}{0.40 / \sqrt{20}} = 0.4696$$

t-tabulated at 1% level of significance with $(20 - 1)$, i.e., 19, d.f. $= 2.86$.
Conclusion: Since $t_{cal} < t_{tab}$, so H_0 may be accepted at 1% level of significance, i.e., the product is confirming to specification.

EXAMPLE 3. *Ten individuals are chosen at random from a population and their heights are found to be in inches 63, 63, 24, 65, 66, 69, 69, 70, 70, 71. Discuss the suggestion that the mean height in the inverse is 65 inches given that for 9 d.f. the value of student's that 5% level of significance is 2.262.*

SOLUTION. **Null hypothesis H_0:** The mean height of the universe is 65

$$H_0 : \mu = 65$$

Alternative hypothesis : $H_1 : \mu \neq 65$
Table for calculation of sample mean and S.D.

x	$d = x - 69$	d^2
63	−6	36
63	−6	36
64	−5	25
65	−4	16
66	−3	9
69	0	0
69	0	0
70	1	1
70	1	1
71	2	4
	$\Sigma d = -20$	$\Sigma d^2 = 128$

$$\bar{x} = A + \frac{\Sigma d}{n} = 69 - \frac{20}{10} = 67$$

$$S^2 = \frac{1}{n-1}\left[\Sigma d^2 - \frac{(\Sigma d)^2}{n}\right]$$

$$= \frac{1}{9}\left[128 - \frac{400}{10}\right] = \frac{1}{9} \times 88$$

$$S = \sqrt{88/9} = 3.13$$

Test statistic : Under H_0

$$t = \frac{|\bar{x} - \mu|}{S/\sqrt{n}} = \frac{|67 - 65|}{3.13} \times \sqrt{10} = 2.02.$$

t-tabulated at 1 % level of significance with $(10 - 1)$, *i.e.,* 9, d.f. = 3.25.

Conclusion: Since $t_{cal} < t_{tab}$, so H_0 may be accepted at 1% level of significance, *i.e.,* the mean height in the universe is 65 inches.

EXAMPLE 4. *Find the student's t for following variables values in a sample of eight.*
   – 4, – 2, – 2, 0, 2, 2, 3, 3.
taking mean of the universe to be zero.

SOLUTION.

S. No.	x	$x - \bar{x}$	$(x - \bar{x})^2$
1	–4	–4.25	18.0625
2	–2	–2.25	5.0625
3	–2	–2.25	5.0625
4	0	–0.25	0.0625
5	2	1.75	3.0625
6	2	1.75	3.0625
7	3	2.75	7.5625
8	3	2.75	7.5625
	$\Sigma x = 2$		49.5000

$$\bar{x} = \text{mean} = \frac{\Sigma x}{n} = \frac{2}{8} = 0.25$$

$$S = \sqrt{\frac{\Sigma(x - \bar{x})^2}{n-1}} = \sqrt{\frac{49.5}{7}}$$

$$= \sqrt{7.071428} = 2.659$$

H_0 **:** The mean of the universe $M = 0$, we get

Student's t

$$= \frac{(\bar{x} - M)}{S}\sqrt{n} = \frac{(0.25 - 0)}{2.659}\sqrt{8} = 0.27$$

EXAMPLE 5. *A drug given to each of the 12 persons resulted in the following changes in the blood pressure from normal −3, 2, 8, −1, 3, 0, 7, −2, 1, 5, 0, 4. Calculate by student's t-test whether changes is significance or not.*

SOLUTION.

$$\bar{x} = \frac{\Sigma x}{n} = \frac{-3+2+8-1+3+0+7+(-2)+1+5+0+4}{12}$$

$$= \frac{24}{12} = 2$$

Now

x	$x - \bar{x}$	$(x - \bar{x})^2$
−3	− 3 − 2 = −5	25
2	2 − 2 = 0	0
8	8 − 2 = 6	36
− 1	− 1 − 2 = −3	9
3	3 − 2 = 1	1
0	0 − 2 = −2	4
7	7 − 2 = 5	25
−2	− 2 − 2 = −4	16
1	1 − 2 = −1	1
5	5 − 2 = 3	9
0	0 − 2 = −2	4
4	4 − 2 = 2	4
$n = 12$		$\Sigma(x - \bar{x})^2 = 134$

$$\text{S.D.} = \sqrt{\frac{(x - \bar{x})^2}{n-1}} = \sqrt{\frac{134}{11}} = \sqrt{12.18} = 3.49$$

Now

$$'t' = \frac{\bar{x} \times \sqrt{n}}{\text{S.D.}} = \frac{2 \times \sqrt{12}}{3.49}$$

$$= \frac{2 \times 3.464}{3.49} = \frac{6.928}{3.49} = 1.98$$

Degree of freedom = $n - 1 = 12 - 1 = 11$.

The calculated value of 't' is less than the given value of 't' in table. Hence, the difference between the two means is not significant.

EXAMPLE 6. *Below are given the gain in weight (in kgs.) of pigs fed on two diets A and B.*

Gain in weight														
Diet A	25	32	30	34	24	14	32	24	30	31	35	25		
Diet B	44	34	22	10	47	31	40	30	32	35	18	21	29	35	22

Test of the two diets differ significantly as regards their effect on increase in weight.

SOLUTION. **Null hypothesis (H_0):** There is no significance difference between the mean increase in weight due to diets A and B, $\mu_x = \mu_y$

Altenative Hypothesis (H_1) $= \mu_x \neq \mu_y$

Diet A			Diet B		
x	$(x-\bar{x})$	$(x-\bar{x})^2$	y	$(y-\bar{y})$	$(y-\bar{y})^2$
25	−3	9	44	14	196
32	4	16	34	4	16
30	2	4	22	− 8	64
34	6	36	10	− 20	400
24	− 4	16	47	17	289
14	− 14	196	31	1	1
32	4	16	40	10	100
24	− 4	16	30	0	0
30	2	4	32	2	4
31	3	9	35	5	25
35	7	49	18	− 12	144
25	− 3	9	21	− 9	81
			35	5	25
			39	− 1	1
			22	− 8	64
336		380	450		1410

$$n_1 = 12, n_2 = 15, \bar{x} = \frac{\Sigma x_i}{n_1} = \frac{336}{12} = 28 \text{ and } \bar{y} = \frac{\Sigma y_i}{n_2} = \frac{450}{15} = 30$$

$$S^2 = \frac{1}{n_1 + n_2 - 2}[\Sigma(x_i - \bar{x})^2 + \Sigma(y_i - \bar{y})^2]$$

$$= \frac{1}{12 + 15 - 2}[380 + 1410] = 71.6$$

Test statistic

$$t = \frac{\bar{x} - \bar{y}}{\sqrt{S^2\left[\frac{1}{n_1} + \frac{1}{n_2}\right]}} = \frac{28 - 30}{\sqrt{71.6\left(\frac{1}{12} - \frac{1}{15}\right)}} = \frac{-2}{\sqrt{10.74}} = -0.609$$

Tabulated value of t for $(12 + 15 - 2) = 25$ degrees of freedom at 5% level of significance is 2.06.

Conclusion: Since $|t|_{cal}$ is less than t tabulated at 5% level of significance so null hypothesis may be accepted, *i.e.*, the two diets differs significantly as regards their effect on increase in weight.

EXAMPLE 7. *Two horses A and B were tested according to the time (in seconds) to run a particular track with the following results :*

Horse A	28	30	32	33	33	29	34
Horse B	29	30	30	24	27	29	

Test where you can discriminate between two houses. You can use the fact that 5 percent value of t for 11 degree of freedom is 2.20.

SOLUTION. Here $n_1 = 7, n_2 = 6$

Mean of A is $\bar{x} = \dfrac{\Sigma x}{n_1} = \dfrac{219}{7} = 31\dfrac{2}{7};$

Mean of B is $\bar{y} = \dfrac{\Sigma y}{n_2} = \dfrac{169}{6} = 28\dfrac{1}{6}.$

	A			B	
x	$(x-\bar{x})$	$(x-\bar{x})^2$	y	$(y-\bar{y})$	$(y-\bar{y})^2$
28	$-3\dfrac{2}{7}$	529/49	29	5/6	25/36
30	$-1\dfrac{2}{7}$	81/49	30	$1\dfrac{5}{6}$	121/36
32	5/7	25/49	30	$1\dfrac{5}{6}$	121/36
33	$1\dfrac{5}{7}$	144/49	24	$-4\dfrac{1}{6}$	625/36
33	$1\dfrac{5}{7}$	144/49	27	$-1\dfrac{1}{6}$	49/36
29	$-2\dfrac{2}{7}$	256/49	29	5/6	25/36
34	$2\dfrac{5}{7}$	361/49			
219	0	1450/49	169	0	966/36

Standard deviation of the combination is given by

$$s = \sqrt{\frac{\Sigma(x-\bar{x})^2 + \Sigma(y-\bar{y})^2}{n_1 + n_2 - 2}} = \sqrt{\frac{\dfrac{221}{7} + \dfrac{161}{6}}{7+6-2}}$$

$$= \sqrt{5.29} = 2.3$$

Under the null hypothesis $\mu_1 = \mu_2$

$$t = \frac{\bar{x} - \bar{y}}{x} \times \sqrt{\frac{n_1 n_2}{n_1 + n_2}} = \frac{31\frac{2}{7} - 28\frac{1}{6}}{2.3} \times \sqrt{\frac{7 \times 6}{7 + 6}}$$

$$= \frac{131}{24 \times 2.3} \times \sqrt{\frac{42}{13}} = \frac{131}{2.3\sqrt{42} \times 13} = 2.43$$

Number of degrees of freedom = 7 + 6 – 2 = 11

Five percent value of t for 11 degrees of freedom is 2.20 which is less than the calculated value of t. Hence the difference is significance, i.e., we can discriminate between two horses.

EXAMPLE 8. *Application of fertilizers were tested for the yield of rice grown in 10 plots. Another seed of 10 plots of similar size and condition were taken as control. Test the effect of fertilizer.*

Plot No	1	2	3	4	5	6	7	8	9	10
Fertilizer Applied	16	14	18	15	13	17	16	15	14	13
Fertilizer not Applied	10	12	11	9	13	13	12	14	13	11

SOLUTION. **Null hypothesis:** No significant effect of fertilizer on yield of rice grown.
Alternative hypothesis: Significant effect of fertilizer on yield of rice grown.

Plot No.	Fertilizer Applied x	$(x - \bar{x})$	$(x - \bar{x})^2$	Fertilizer not Applied y	$(y - \bar{y})$	$(y - \bar{y})^2$
1	16	0.90	0.81	10	– 1.8	3.24
2	14	– 1.1	1.21	12	0.2	0.04
3	18	2.9	8.41	11	– 0.8	0.64
4	15	– 0.1	0.01	9	– 2.8	7.84
5	13	– 2.1	4.41	13	1.2	1.44
6	17	1.9	3.61	13	1.2	1.44
7	16	0.9	0.81	12	0.2	0.04
8	15	– 0.1	0.01	14	2.2	4.84
9	14	– 1.1	1.21	13	1.2	1.44
10	13	– 2.1	4.41	11	0.8	0.64
	151		24.90	118		21.60

$$\bar{x} = \frac{151}{10} = 15.1 \quad \text{and} \quad \bar{y} = \frac{118}{10} = 11.8$$

$$\text{Standard error} = \sqrt{\frac{\Sigma(x - \bar{x})^2 - \Sigma(y - \bar{y})^2}{n_1 + n_2 - 2}} = \sqrt{\frac{24.9 + 21.6}{18}}$$

$$= \sqrt{\frac{46.5}{18}} = \sqrt{2.584} = 1.607$$

$$t = \frac{\bar{x} - \bar{y}}{s} \times \sqrt{\frac{n_1 \times n_2}{n_1 + n_2}} = \frac{15.1 - 11.8}{1.607} \times \sqrt{\frac{10 \times 10}{10 + 10}}$$

$$= \frac{3.3}{0.718} = 4.596.$$

Level of significance: 5% level of, *i.e.*, 0.05.

Critical value: Tabulated value at 0.05 for $10 + 10 - 2 = 18$ is 2.10.

Hence the calculated value of t is greater than the tabulated value.

So, the null hypothesis is rejected, *i.e.*, there is significant difference between the control and fertilizer used rice flow growth.

EXAMPLE 9. 13 *children (all male) were given mother's milk while the second group of 12 children (all male) of same age group were given dairy milk. After one year the gain in weight in kg was noted. The noted weight in both groups is given below.*

First Group	5	3	4	3	2	6	3	2	3	6	7	5	3
Second Group	1	2	3	4	2	1	3	4	3	2	2	3	

Find the two groups of children is significant or not.

SOLUTION. For I group $\bar{x} = \frac{\Sigma x}{n_1} = \frac{52}{13} = 4$;

For II group $\bar{y} = \frac{\Sigma y}{n_2} = \frac{30}{12} = 2.5$

	For Ist Group			For IInd Group	
x	$(x - \bar{x})$	$(x - \bar{x})^2$	y	$(y - \bar{y})$	$(y - \bar{y})^2$
5	1	1	1	− 1.5	2.25
3	− 1	1	3	0.5	0.25
4	0	0	2	− 0.5	0.25
3	− 1	1	4	1.5	2.25
2	− 2	4	2	− 0.5	0.25
6	2	4	1	− 1.5	2.25
3	− 1	1	3	0.5	0.25
2	− 2	4	4	1.5	2.25
3	− 1	1	3	0.5	0.25
6	2	4	2	− 0.5	0.25
7	3	9	2	− 0.5	0.25
5	1	1	3	0.5	0.25
3	− 1	1	–	–	–
50		32	30		11

The standard error of this difference

$$s = \frac{\Sigma(x-\bar{x})^2 + \Sigma(y-\bar{y})^2}{n_1 + n_2 - 2} = \sqrt{\frac{32+11}{13+12-2}}$$

$$= \sqrt{\frac{43}{23}} = \sqrt{1.869} = 1.367$$

We have $t = \dfrac{\bar{x}-\bar{y}}{s}\sqrt{\dfrac{n_1 \times n_2}{n_1 + n_2}} = \dfrac{4-2.5}{1.367}\sqrt{\dfrac{13\times 12}{13+12}} = \dfrac{1.5}{1.367\sqrt{0.16}} = 2.739$

Degrees of freedom $= n_1 + n_2 - 2 = 13 + 12 - 2 = 23$

The calculated value of t comes to 2.739. The table value of t at 23 on 0.1 level is 2.5. The calculated value is higher therefore, we can say that mother's milk has slight better influence on the growth of children in positive direction.

EXAMPLE 10. *The body weight (kg) of 8 adult males and of eight adult females are presented in respectively in the first and second columns of table. Find out whether or not the mean weight of males is significantly than that of females.*

Weight (in kg.)	
Males x	**Females y**
50	49
58	52
60	51
55	56
59	55
56	53
54	52
64	48
456	416

SOLUTION. **Null hypothesis:** Mean weight of males are not significantly higher than females.

Alternative hypothesis: Mean weight of males are significantly higher.

x	$(x-\bar{x})$	$(x-\bar{x})^2$	y	$(y-\bar{y})$	$(y-\bar{y})^2$
50	−7	49	49	−3	9
58	1	1	52	0	0
60	3	9	51	−1	1
55	−2	4	56	4	16
59	2	4	55	3	9
56	−1	1	53	1	1
54	3	9	52	0	0
64	7	49	48	4	16
456		126	416		52

$$\bar{x} = \frac{456}{8} = 57 \quad \text{and} \quad \bar{y} = \frac{416}{8} = 52$$

Standard error

$$s = \sqrt{\frac{\Sigma(x-\bar{x})^2 + \Sigma(y-\bar{y})^2}{n_1 + n_2 - 2}} = \sqrt{\frac{126+52}{8+8-2}} = \sqrt{\frac{178}{14}} = \sqrt{12.714} = 3.57$$

$$t = \frac{\bar{x} - \bar{y}}{s} \times \sqrt{\frac{n_1 n_2}{n_1 + n_2}} = \frac{57-52}{3.57} \times \sqrt{\frac{8 \times 8}{8+8}} = \frac{5}{1.785} = 2.80$$

Level of significance = 5%, i.e., 0.05

Critical value = Tabulated value at 0.05 for d.f. 16 − 2 = 14 is 2.14.

Hence, the calculated value of $|t| = 2.80$

$\qquad |t| \; 2.80 > t_{0.05} \; 2.14$

So, the null hypothesis rejected, i.e., mean weight of males are not significantly higher than females.

EXAMPLE 11. *In the rat feeding experiment, the following results were obtained.*

Diet	Gain in weight in gm.										
High Protein	13	14	10	11	12	16	10	8	11	12	9	12
Log Protein	7	11	10	8	10	13	9					

Investigate if there is any evidence of superiority of one diet over the other. The value of t for 17 degrees of freedom at 5% level of significance = 2.11.

Solution.

High Protein			Low Protein		
x	$x - \bar{x}$	$(x-\bar{x})^2$	y	$(y-\bar{y})$	$(y-\bar{y})^2$
13	1.5	2.25	7	$-2\frac{5}{7}$	361/49
14	2.5	6.25	11	$1\frac{2}{7}$	81/49
10	− 1.5	2.25	10	2/7	4/49
11	− 0.5	0.25	8	$-1\frac{5}{7}$	144/49
12	0.5	0.25	10	2/7	4/49
16	4.5	20.25	13	23/7	52/49
10	− 1.5	2.25	9	− 5/7	25/49
8	− 3.5	12.25			
11	− 0.5	0.25			
12	0.5	0.25			
9	− 2.5	6.25			
12	0.5	0.25			
	0	53.00	68	0	1148/49

Mean gain in weight on high protein $= \bar{x} = \dfrac{138}{12} = \dfrac{23}{2} = 11.5$

Mean gain in weight on low protein $\bar{y} = \dfrac{68}{7} = 9\dfrac{5}{7}$

The standard error of this differences

$$s = \sqrt{\dfrac{(x - \bar{x})^2 - (y - \bar{y})^2}{n_1 + n_2 - 2}}$$

$$= \sqrt{\dfrac{53 + \dfrac{1148}{49}}{12 + 7 - 2}} = \sqrt{\dfrac{2597 + 1148}{49 \times 17}}$$

$$= \sqrt{\dfrac{3745}{833}} = \sqrt{\dfrac{535}{119}}$$

substituting these value, we get

$$t = \dfrac{\bar{x} - \bar{y}}{s} \sqrt{\dfrac{n_1 n_2}{n_1 + n_2}} = \dfrac{\dfrac{23}{2} - \dfrac{64}{7}}{\sqrt{\dfrac{535}{119}}} \sqrt{\dfrac{12 \times 7}{12 + 7}}$$

$$= \dfrac{25}{14} \sqrt{\dfrac{84}{19} \times \dfrac{119}{535}} = \dfrac{25}{14} \sqrt{\dfrac{9996}{10165}}$$

$$= \dfrac{25}{14} \times \sqrt{0.9834} = \dfrac{25}{14} \times 0.99 = 1.77$$

The number of degree of freedom $= 12 + 7 - 2 = 17$

The value t for 17 degrees of freedom at 5% level of significance $= 2.11$.

The calculated value of t for 17 degrees of freedom is less than this. Hence, the difference is not significant.

i.e., there is no evidence of superiority of one diet over the other.

For this value of t and for degree of freedom $(10 + 10 - 2)$, i.e., 18 the value of P is 0.0072. This is less than 3.034. Hence there is significant difference between the yields on the two plates.

EXAMPLE 12. *10 fishes from tank I and 10 fishes tank II of a species of fish were produced and measured in cms.*

Tank I	20	24	20	28	22	20	24	32	24	26
Tank II	12	10	8	10	6	4	14	20	10	6

Calculate the mean difference in body length between the two ponds of fishes is significant or not.

SOLUTION. For tank I $\quad \bar{x} = \dfrac{\Sigma x}{n} = \dfrac{240}{10} = 24$

For tank II $\quad \bar{y} = \dfrac{\Sigma y}{n} = \dfrac{100}{10} = 10$

	Tank I			Tank II	
x	$(x-\bar{x})$	$(x-\bar{x})^2$	y	$(y-\bar{y})$	$(y-\bar{y})^2$
20	-4	16	12	2	4
24	0	0	10	0	0
20	-4	16	8	-2	4
28	4	16	10	0	0
22	-2	4	6	-4	16
20	-4	16	4	-6	36
24	0	0	14	4	16
32	8	64	20	10	100
24	0	0	10	0	0
26	2	4	6	-4	16
240		136	100		192

$$s = \sqrt{\frac{\Sigma(x-\bar{x})^2 + \Sigma(y-\bar{y})^2}{n_1 + n_2 - 2}} = \sqrt{\frac{136+192}{10+10-2}} = \sqrt{\frac{328}{18}} = 4.268$$

$$t = \frac{\bar{x}-\bar{y}}{s}\sqrt{\frac{n_1 n_2}{n_1 + n_2}} = \frac{24-10}{4.268}\sqrt{\frac{10 \times 10}{10+10}} = \frac{14}{4.268 \times 0.1447} = 7.337$$

The value of t at 18 degrees of freedom at 0.01 level of significance is 2.552. Thus the calculated value is much greater than the table value. Therefore, the difference between two means of two population is not significant.

EXAMPLE 13. *The nine items of a sample had the following values 45, 47, 50, 52, 48, 47, 49, 53, 51. Does the mean of the nine items differ significantly from the assumed population mean of 47.5 given that*

$$v = \begin{cases} p = 0.945 & \text{for } t = 1.8 \\ p = 0.953 & \text{for } t = 1.9 \end{cases}$$

SOLUTION. It is given that mean, $M = 47.5$

x	$x - \bar{x}$	$(x-\bar{x})^2$
45	-4	16
47	-2	4
50	1	1
52	3	9
48	-1	1
47	-2	4
49	0	0
53	4	16
51	2	4
$\Sigma x = 442$		$\Sigma(x-\bar{x})^2 = 55$

Now, $\quad s = \sqrt{\dfrac{\Sigma(x-\bar{x})^2}{n-1}}$

where $\quad n = 9, \bar{x} = \dfrac{442}{9} = 49$ (app.)

$\qquad\qquad = \sqrt{\dfrac{55}{8}} = 2.622$

$\qquad t = \dfrac{(\bar{x}-M)}{s}\sqrt{n} = \dfrac{(49-47.5)}{2.62}\sqrt{9} = 1.717$

Now, degree of freedom $= n-1 = 9-1 = 8$

and $\qquad p = 0.945$ for $t = 1.8, p = 0.953$ for $t = 1.9$.

Hence, $\qquad p = 0.945$ for $t = 1.717$.

Therefore, the chance of getting a value of t greater than observed is $1 - 0.95 = 0.5$.

Here, we observe that the probability of getting t greater in absolute value is $2 \times 0.05 = 0.10$ which is greater 0.05. So, the value of t is not significant. Hence, it may be random sample from a normal population of mean 47.5.

24.21 PAIRED t-TEST FOR DIFFERENCE OF MEANS

This test is used when the sample are not independent but the sample observaions are paired together.

If $(x_1, y_1), (x_2, y_2), \ldots (x_n, y_n)$ be a set of n paired observations. For example suppose we want to test the effect of diet A and B on the same group of pigs and let x_i and y_i ($i = 1, 2, \ldots n$) be the readings of the gain the weight (in kgs.) of pigs.

Let $d_i = x_i - y_i$ be the difference between x_i's and y_i's ($i = 1, 2, \ldots n$).

Null hypothesis (H_0): Therefore is no significant difference between the mean increase in weight due to diet A and B.

$$H_0 : \mu_x = \mu_y$$

Alternative hypothesis (H_1): $\mu_x \neq \mu_y$ (two tailed)

$$\mu_x > \mu_y \text{ (right tailed)}$$

$$\mu_x < \mu_y \text{ (left tailed)}$$

Test statistic: $\quad t = \dfrac{\bar{d}}{S/\sqrt{n}}$

where $\qquad \bar{d} = \dfrac{1}{n}\sum_{i=1}^{n} d_i, S^2 = \dfrac{1}{n-1}\sum_{i=1}^{n}(d_i - \bar{d})^2$

Test statistic t follows t-distribution with (1) degree of freedom.

Conclusion: If calculate $|t|$ is greater than or equal to tabulated value then null hypothesis (H_2) is rejected at $\alpha\%$ level of significance otherwise H_0 may be accepted.

SOLVED EXAMPLES

EXAMPLE 1. *Nine patients to whom a certain drink was administered registered the following increments in blood pressures.*

$$7, 3, -1, 4, -3, 3, -5, 6, -4$$

Do the data indicate that

(i) *drink was responsible for an increase in blood pressure,*

(ii) *blood pressure charges significantly after drinking?*

SOLUTION. **Null hypothesis (H_0):** There is no efficient of drink on the blood pressure.

Alternative hypothesis (H_1): For case I.

Drink was responsible for an increase in blood pressure (one tailed).

For Case I. There is significance effect on the blood pressure after drinking (two tailed).

$$\bar{d} = \frac{\Sigma di}{n} = \frac{1}{9}(7+3-1+4-3+3-5+6-4) = \frac{10}{9} = 1 \text{ (say)}$$

$$S^2 = \frac{1}{n-1}\Sigma(d_i - \bar{d})^2 = \frac{1}{n-1}\left[\Sigma di^2 - \frac{(\Sigma di)^2}{n}\right]$$

$$= \frac{1}{8}\left[181 - \frac{10 \times 10}{9}\right] = \frac{353}{18}$$

Test statistic: $t = \dfrac{\bar{d}}{S/\sqrt{n}} = \dfrac{10/9}{\sqrt{\dfrac{353}{18} \times \dfrac{1}{9}}} = \dfrac{10}{\sqrt{1765}} = 0.77$

Tabulated value t for $(9-2) = 7$ degrees of freedom at 5% level of significance for single tail is 1.86 and for two tailed is 2.306.

Conclusion: Since calculated t is less than tabulated value in both the cases so null hypothesis may be accepted at 5% level of significance, *i.e.*, the drinking makes no significant effect upon blood pressure.

EXAMPLE 2. *An I-Q test was administered to 5 persons before and after they are trained the result are given below :*

Candidates	I	II	III	IV	V
IQ before training	110	120	123	132	125
IQ after training	120	118	125	136	121

Test where there is any change in IQ after programme.

It is given that $t_{0.01} = 4.6$ for d.f.4.

SOLUTION. **Null hypothesis H_0:** No change in IQ after the training (assume).

Alternate hypothesis H_1 : change in IQ after the training (assume).

Candidates	IQ before training	IQ after training	Difference d	$\bar{d}^2$
I	110	120	– 10	100
II	120	118	+ 2	04
III	123	125	– 2	04
IV	132	136	– 4	16
V	125	121	4	16
			$\Sigma d = -10$	140

Mean difference $\bar{d} = \dfrac{-10}{5} = -2$

Standard error $= \sqrt{\dfrac{d^2 - \dfrac{(\Sigma d)^2}{n}}{n-1}} = \sqrt{\dfrac{140 - \dfrac{(-10) \times (-10)}{5}}{5-1}}$

$= \sqrt{\dfrac{140 - \dfrac{100}{5}}{4}} = \sqrt{\dfrac{120}{5}} = \sqrt{30}$

Standard error of difference $= \dfrac{S}{\sqrt{n}} = \dfrac{\sqrt{30}}{\sqrt{5}} = \sqrt{6}$

$$t = \dfrac{\bar{d}}{\text{S.E.}} = \dfrac{-2}{-2.45} = -0.816$$

$|t| = 0.816$

Level of significance $= 0.01$

Critical value of 't' at 0.05 for d.f. 4, i.e., 5 − 1 is 4.6.

Hence, the calculated value of $t = 0.816 < t_{0.01} = 4.6$.

So the null hypothesis is accepted, i.e., no significant difference after training programme.

24.22 F-TEST OF EQUALITY OF POPULATION VARIANCE

Let $x_1, x_2, ..., x_{n1}$ be a random sample of size n_1 and $y_1, y_2, ..., y_{n_2}$ be another independent random sample of size n_2 taken from two normal populations with mean μ_x and μ_y respectively and having variance σ_x^2 and σ_y^2.

Null Hypothesis (H_0) : The sample have been drawn from normal population with the same variance i.e.,

$$H_0 : \sigma_x^2 = \sigma_y^2 = \sigma^2.$$

Alternative Hypothesis (H_1) : $\sigma_x \neq \sigma_y$ or σ_x (two tailed)

or $\qquad\qquad \sigma_x > \sigma_y$ (right tailed) $\qquad\qquad$ or $\qquad\qquad \sigma_x > \sigma_y$ (left tailed)

Test statistics:

$$F = \dfrac{S_x^2}{S_y^2}$$

where $\qquad\qquad S_x^2 = \dfrac{1}{n_1 - 1} \sum_{i=1}^{n_1} (x_i - \bar{x})^2 \qquad$ and $\quad S_y^2 = \dfrac{1}{n_2 - 1} \sum_{j=1}^{n_2} (y_i - \bar{y})^2$

where $\bar{x}$ and $\bar{y}$ are mean samples defined as

$$\bar{x} = \dfrac{1}{n_1} \sum_{i=1}^{n_1} x_i, \quad \bar{y} = \dfrac{1}{n_2} \sum_{j=1}^{n_2} y_j.$$

F-statistics follows F distribution with $(n_1 - 1, n_2 - 1)$ degrees of freedom.

F-Tabulated: Tabulated value is taken from the table at $(n_1 - 1, n_2 - 1)$ degrees of freedom at given level of significance (generally 1%, 5% and 10%).

Conclusion: If $F_{cal} \geq F_{tab}$ then the null hypothesis is rejected at given level of significance, *i.e.*, there is a significance difference between population variances. Other hand if $F_{cal} \geq F_{tab}$ then the null hypothesis may be accepted.

Assumptions of F-test: The following assumptions are made in the F-test :

(i) The population from which the samples are drawn should be normal.

(ii) The samples should be drawn in random manner.

(iii) The observations should be independent.

(iv) The ratio S_x^2 / S_y^2 should be greater than 1, *i.e.*, $S_x^2 / S_y^2 > 1$

SOLVED EXAMPLES

EXAMPLE 1. *Two samples of size 9 and 8 give the sum of squares of deviations from their respective means equal to 160 inches square and 91 inches square respectively. Can they be regarded as drawn from the normal population with same variance?*

(Mumbai–2004, VTU–2002)

SOLUTION. Here, we are given

$$n_1 = 9, n_2 = 8, \Sigma(x_i - \overline{x})^2 = 160, \Sigma(y_i - \overline{y})^2 = 91$$

Null Hypothesis (H_0): The samples are drawn from the normal populations with same variance, *i.e.*, $H_0 : \sigma_x^2 = \sigma_y^2$

Alternative Hypothesis (H_1): H_0 is not true,

i.e., $\qquad\qquad H_1 : \sigma_x^2 \neq \sigma_y^2$

Test statistic:

$$F = \frac{S_x^2}{S_y^2}$$

$$S_x^2 = \frac{1}{n_1 - 1} \sum_{i=1}^{n_1} (x_i - \overline{x})^2 = \frac{1}{9 - 1} \times 160 = 20$$

$$S_y^2 = \frac{1}{n_2 - 1} \sum_{i=1}^{n_2} (y_i - \overline{y})^2 = \frac{91}{8 - 1} = 13$$

$$F = \frac{S_x^2}{S_y^2} = \frac{20}{13} = 1.54 .$$

F-tabulated at 5% level of significance with $(n_1 - 1, n_2 - 1)$, *i.e.*, $(8, 7)$ degrees of freedom = 3.73.

Conclusion: Since $F_{cal} < F_{tab}$, so null hypothesis may be accepted at 5% level of significance *i.e.*, the samples are drawn from normal populations with same variance.

EXAMPLE 2. *It is known that the mean diameters of eggs produced by two hens X and Y are practically the same but the standard deviation may differ. From 22 eggs produced by hen X, the standard deviation is 2.9 mm while for 16 eggs produced by hen Y, the standard deviation is 3.8 mm. Compute the statistic you would use to test whether the eggs of hen X have the same variability as those of hen X and test its significance.*

SOLUTION. We are given :

$$X : n_1 = 22, \ S_1 = 2.9 \, \text{mm}$$
$$Y : n_2 = 16, \ S_2 = 3.8 \, \text{mm}$$

Null hypothesis (H_0) : $\sigma_x^2 = \sigma_y^2$ is the products of both the hen X and Y have the same variability.

We have, $\ S_x^2 = \dfrac{n_1 S_1^2}{n_1 - 1} = \dfrac{22 \times (2.9)^2}{21} = 8.805$

$$S_y^2 = \frac{n_1 S_2^2}{n_1 - 1} = \frac{16 \times (3.8)^2}{21} = 15.393$$

which follows F-distribution with (15, 21), d.f. tabulated $f_{0.05}$ (15, 21) = 2.20 (App.).

Since calculated F is less the tabulated F, it is not significant at 5% level of significance and the hypothesis of equal variability may be accepted.

EXAMPLE 3. *Can the following two samples be regarded as coming from the same normal population?*

Sample	Size	Sample mean	Sum of squares of Deviation from the mean
A	10	12	120
B	12	15	314

SOLUTION. A normal population has two parameters, mean (μ) and variance (σ^2). To test if two independent samples have been drawn from the same normal population, We have the test, "the equality of population variance" and the "equality of population of mean."

Equality of mean will be tested by applying 't' test and equality of variance will be tested by applying F-test.

Null hypothesis (H_0) : The two samples have been drawn from the same normal population *i.e.*, $H_0 : \sigma_x^2 = \sigma_y^2$ (in F-test).

We are given that $n_1 = 10, n_2 = 12, \ \bar{x} = 12, \ \bar{y} = 12$

$$\Sigma (x - \bar{x})^2 = 120, \quad \Sigma (y - \bar{y})^2 = 314 ,$$

$$S_x^2 = \frac{\Sigma (x - \bar{x})^2}{n_1 - 1} = \frac{120}{10 - 1} = 13.33 ,$$

$$S_y^2 = \frac{\Sigma (y - \bar{y})^2}{n_2 - 1} = \frac{314}{12 - 1} = 28.55$$

Since $S_y^2 < S_x^2$ under $H_0 : \sigma_x^2 = \sigma_y^2$.

The test statistic is tabulated $F_{0.05}$ (11, 9) = 3.10 (app.).

Since calculated F is less then tabulated F it is not significant, hence null hypothesis of equality of population variance may be accepted.

EXAMPLE 4. *Two random samples are drawn from two normal populations*

Sample I	20	16	26	27	23	22	18	24	25	19		
Sample II	27	33	42	35	32	34	38	28	41	43	30	37

Obtain the estimates of the variances at the population and test whether the two populations have the same variance [given F and 11 and 9 d.f. at 5% level of significance = 311]

SOLUTION. **Null Hypothesis (H_0).** Two populations have the same variance.

$$H_0 : \sigma_x^2 = \sigma_y^2$$

Alternative Hypothesis (H_1) : H_0 is not true, *i.e.*, $\sigma_x^2 \neq \sigma_y^2$

Test Statistics: $F = S_x^2 / S_y^2$

Sample (x)	$(x - \bar{x})$	$(x - \bar{x})^2$	Sample (y)	$(y - \bar{y})$	$(y - \bar{y})^2$
20	−2	4	27	−8	64
16	−6	36	23	−2	4
26	4	16	42	7	49
27	5	25	35	0	0
23	1	1	32	−3	9
22	0	0	34	−1	1
18	−4	16	38	3	9
24	2	4	28	−7	49
25	3	9	41	6	36
19	−3	9	43	8	64
			30	−5	25
			37	2	4
Total 220		120	420		314

$$\bar{x} = \frac{\Sigma x_i}{n_1} = \frac{220}{10} = 22, \quad \bar{y} = \frac{1}{n_1} \Sigma y_i = \frac{1}{12} \times 420 = 35,$$

$$S_x^2 = \frac{1}{n_1 - 1} \Sigma(x_i - \bar{x})^2 = \frac{1}{9} \times 120 = 13.33,$$

$$S_y^2 = \frac{1}{n_2 - 1} \Sigma(y_i - \bar{y})^2 = \frac{1}{11} \times 314 = 28.55$$

So, $$F_{cal} = \frac{S_x^2}{S_y^2} = \frac{13.33}{28.55} < 1$$

So we take $$F_{cal} = \frac{S_x^2}{S_y^2} = \frac{28.55}{13.33} = 2.14$$

F_{tab} at $(n_2 - 1, n_1 - 1) = (11, 9)$ d.f. is 3.11.

Conclusion: Since $F_{cal} < F_{tab}$, so null hypothesis may be accepted at 5% level of significance *i.e.*, the samples are drawn from normal populations with same variance.

EXAMPLE 5. *Two random samples (length of fish) wave drawn from two species of fishes in cm. Both species of fishes represent the normal population.*

x	66	67	75	76	82	84	88	90	92		
y	64	66	74	78	82	85	87	92	93	95	97

Test whether the two populations of fishes have the same variance at 5% level for $V_1 = 10$ and $V_2 = 8$.

SOLUTION. **Null hypothesis $H_0 : \sigma_x^2 = \sigma_y^2$** (i.e., two population have the same variance).

x	$(x - \bar{x})$	$(x - \bar{x})^2$	y	$(y - \bar{y})$	$(y - \bar{y})^2$
66	−14	196	64	−19	361
67	−13	169	66	−17	289
75	−5	25	74	−9	81
76	−4	16	78	−5	25
82	2	4	82	−1	1
84	4	16	85	+2	4
88	8	64	87	4	16
90	10	100	92	9	81
92	12	144	93	10	100
			95	13	169
			97	14	196
720	0	734	913	0	1298

$$\bar{x} = \frac{720}{9} = 80 \text{ and } \bar{y} = \frac{913}{11} = 83,$$

$$S_x^2 = \frac{\Sigma(x - \bar{x})^2}{n_1 - 1} = \frac{734}{9 - 1} = 91.75,$$

$$S_y^2 = \frac{\Sigma(y - \bar{y})^2}{n_2 - 1} = \frac{1298}{11 - 1} = 129.8$$

Since $S_y^2 > S_x^2$ under $H_0 : \sigma_x^2 = \sigma_y^2$ the test statistic is $F = \dfrac{S_y^2}{S_x^2} = \dfrac{129}{91.75} = 1.415$

For $V_1 = 10$ and $V_2 = 8 = F_{0.05} = 3.36$ (Tabulated value).
Hence, the calculated value of F is less than the table value, the hypothesis is rejected.

Thus, it may be said the two population of fishes have the same variance.

Exercise 24.1

1. In one sample of 8 observations the sum of squares of the deviations of the sample values from the sample mean was 84.4 and in another sample of 10 observations, it was 102.6. Test whether the two samples have been drawn from two normal population with same variance 5% (For 7 and 9 d.f. at 5% level of significance = 3.29).

2. Two independents sample of 8 and 7 items respectively had the value of the variable:

Sample I :	9	11	13	11	15	9	12	14
Sample II:	10	12	10	14	9	9	10	

Do the two estimate of population variance differ significantly? Given that for (7, 6) d.f. the value of F at 5% level of significance is 4.20 nearly.

3. The boxes are choosen at random from a godwon and there weights are found to be in kgs. 15.75, 15.75, 16.0, 16.25, 16.5, 17.25, 17.5, 17.5, 17.75. Discuss the suggestion that mean weight in the universe is 16.25 kg. Given that for 9 degrees of freedom the value of student t at 5 percent level of significance is 2.262.

4. Ten soldiers tour a.m.f./range once in a week for two successive weeks. Their scores in the first week were 67, 24, 57, 55, 63, 54, 56, 68, 33, 43. Their scores in the second week (in the same order), were 70, 38, 58, 58, 56, 67, 68, 75, 42, 38. Is there any significant improvement?

5. In a test examination given in two groups of students, the marks obtained were as follows:

First group	18	20	36	50	49	36	34	49	41
Second group	29	28	26	35	30	44	46		

Examine the significance of difference between the arithmetic average of marks secured by students of the above two groups (using the value of 't' for 14 d.f. at 5% level of significance is 2.14.) **(CCSU(B.Sc. Biotech)–1993)**

ANSWERS

1. Yes   2. No   4. No   5. 2.14

24.23 ANALYSIS OF VARIANCE (ANOVA)

The analysis of variation is a powerful tool for tests of significance. If we want to test the significance difference between two sample means that we use t-test but if we have to test the significance difference among three or more sample means then we required a repetition of t-test. This is time consuming. This can be done by another technique known as "Analysis of variance".

The term "Analysis of variance" was developed by Prof. R. A. Fisher in 1920's.

Definition. According to Prof. R. A. Fisher, Analysis of Variance (ANOVA) is the "*separation of variation ascribable to one group of causes from the variation ascribable to other group.*"

So in this technique we split up the total variance into two parts :

(i) Variance between samples,   (ii)   Variance within samples

Analysis of variance based on F-statistic which can be defined as the ratio of variance between samples and variance with in sample is symbolically and this ratio always greater than unity.

$$F = \frac{\text{Variance between samples}}{\text{Variance within samples}}$$

If this ratio is less than unity then interchange these variance. So we take numerator value always greater than denominator value.

Assumptions: The following assumptions are made for validity of the F-test in ANOVA :

(i) The sample observations are independent.

(ii) The population from which the observation are taken is normal.

(iii) Various effects are additive in nature,

24.23.1 TECHNIQUE OF ANALYSIS OF VARIANCE FOR ONE WAY CLASSIFICATION

When the observations are classified according to anyone factor (for example, the application of one or more type of diets may be considered on several man).

Let us suppose that N observations $X_{ij}(i = 1, 2, ..., k; j = 1,2,...,n_i)$ are grouped in k classes of sizes $n_1, n_2, ..., n_k$ respectively such that

$$N = \sum_{i=1}^{k} n_i$$

						Total	Means
X_{11}	X_{12}	...	X_{1j}	...	X_{1n_1}	T_1	$\bar{X}_1$
X_{21}	X_{22}		X_{2j}		X_{2n_2}	T_2	$\bar{X}_2$
X_{i1}	X_{i2}		X_{ij}		X_{in_i}	T_i	$\bar{X}_i$
X_{k1}	X_{k2}		X_{kj}		X_{in_k}	T_k	$\bar{X}_k$

The total variation in the observation can be split in to the following two components :
(i) The variation between the classes or samples.
(ii) The variation within the classes or samples.
These are two methods for analysis of one way classification
(i) Direct method   (ii) Short-cut method
First of all we set up the null hypothesis as

$$H_0 : \mu_1 = \mu_2 = ... = \mu_k$$
$$H_1 : \mu_1 \neq \mu_2 \neq ... \neq \mu_k$$

where μ_i $(i = 1, 2, ..., k)$ are mean of the ith class in the population.

24.23.2 DIRECT METHOD

This method have following steps:

(a) Variance among the samples.

(i) Obtain the mean of each sample (column)

$$\bar{X}_1, \bar{X}_2, ..., \bar{X}_k \text{ where, } \bar{X}_1 = \frac{X_{i1} + X_{i2} ... X_{in_1}}{n_i}$$

(ii) Find the grand mean, *i.e.*, mean of samples means as

$$\bar{\bar{X}} = \frac{\bar{X}_1 + \bar{X}_2 + ... + \bar{X}_k}{n_1 + n_2 + ... + n_k} = \frac{\sum\limits_{i=1}^{k} \bar{X}_i}{\sum\limits_{i=1}^{k} n_i}$$

(iii) Find the deviation of sample means from the grand mean then square these deviations and multiply by the number of items or units in the samples,

i.e., $\quad n_i (\bar{X}_1 - \bar{\bar{X}})^2 + n_2(\bar{X}_2 - \bar{\bar{X}})^2 + ... + n_k (\bar{X}_k - \bar{\bar{X}})^2$

This is known as squariance or deviance.

(iv) This sum of square is divided by its degree of freedom which is $(k-1)$ where k denotes the number of samples. It is called mean sum of squares among samples.

(b) Variance within samples

(i) Find the mean of each sample
$$\bar{X}_1, \bar{X}_2, ..., \bar{X}_k.$$

(ii) Find the deviation of various items of the sample from the mean value of that sample these deviation and then add.

(iii) For all the samples repeat this procedure and obtain the total of these deviations.

$$\sum_{j=1}^{n_1} (X_{1j} - \bar{X}_1)^2 + \sum_{j=1}^{n_2} (X_{2j} - \bar{X}_2)^2 + ... + \sum_{j=1}^{n_k} (X_{kj} - \bar{X}_k)^2$$

So we obtain the sum of squares of variation with in sample it is also known sum of squares of error (SSE).

(iv) Now divide the sum of squares of error (SSE) by the degrees of freedom for error which would be

$$(n_1 - 1) + (n_2 - 1)...(n_k - 1) = n_1 + n_2 + ... + n_k - k$$

where N be the total number of items or units in all the samples. In this way we find the mean sum of squares of error (MSE).

(c) **Total sum of squares of variation (SST).** The total sum of squares of variation can be obtained to adding the sum of squares of deviation between the sample (SSC) and sum of squares of deviation within sampler (SSE).

So,   SST = SSC + SSE

(d) **Analysis of variance table.** Analysis of variance table have five columns as

Source of variation	Sum of squares (SS)	Degree of Freedom (df)	Mean Sum of Squares (MS)	F-ratio
Between samples	SSC	$k - 1$	$MSC = \dfrac{SSC}{k-1}$	$F = \dfrac{MSC}{MSE}$
Within samples (error)	SSE	$N - C$	$MSE = \dfrac{SSE}{N-k}$	
Total	SST	$N - 1$		

24.23.3 VARIANCE RATIO OF F

The F ratio can be obtained by dividing the mean sum of squares of between samples by mean sum of squares of within samples

$$F = \frac{\text{Variance between samples}}{\text{Variance within samples}} = \frac{MSC}{MSE}$$

24.23.4 STATISTIC IS DISTRIBUTED AS F DISTRIBUTION WITH (K − 1, M − K)

Definition. *If this ratio is less than unity then interchange the mean sum of squares, i.e.,*

$F = \dfrac{MSE}{MSC}$ *and degrees of freedom are adjusted, i.e., F-statistic is distributed as F-distribution with*

$(N - k, k - 1)$ *degrees of freedom.*

Tabulated value of F: From the F-table we find the tabulated value of F or $(k - 1, N - k)$ degrees of freedom at given level of significance. Generally we take 1%, 5%, 10%.

(e) **Interpretation of F.** If $F_{cal} \geq F_{tab}$ then the null hypothesis is rejected and the difference between the means is significant at given level of significance. Otherwise we may accept the null hypothesis.

SOLVED EXAMPLES

EXAMPLE 1.   *The following table gives the yields of 6 fields. In three fields the variety A of seeds and last three fields the variety B is used*

A	20	32	22
B	20	10	16

Prepare the analysis of variance table and discuss there is significance difference between variety of seeds A and B.

SOLUTION. **The null hypothesis:** There is no significance difference between the yields due to variety of A and B i.e.,

$$H_0 : \mu_A = \mu_B, \quad H_1 : \mu_A \neq \mu_B,$$

$$\bar{X}_1 = \frac{30+32+22}{3} = \frac{84}{3} = 28,$$

$$\bar{X}_2 = \frac{20+18+16}{3} = \frac{54}{3} = 18,$$

$$\bar{\bar{X}} = \frac{\bar{X}_1 + \bar{X}_2}{k} = \frac{28+18}{2} = 23$$

Sum of squares between the samples

$$= n_1(\bar{X}_1 + \bar{\bar{X}})^2 + n_2(\bar{X}_2 + \bar{\bar{X}})^2$$

$$= 3(28-23)^2 + 3(18-23)^2$$

$$= 2 \times 25 + 3 \times 25 = 150$$

Sum of squares within the samples

$$= \Sigma(X_{1f} - \bar{X}_1)^2 + \Sigma(X_{2f} - \bar{X}_2)^2$$

$$= (30-28)^2 + (32-28)^2 + (22-28)^3$$

$$+ (20-18)^2 + (18-18)^2 + (16-18)^2$$

$$= 4+16+36+4+0+4 = 64$$

Analysis of Variance Table

Source of variation	Sum of squares	Degree of Freedom (df)	Mean Sum of Squares (MS)	F-ratio
Between samples	150	$k-1 = 2-1 = 1$	$\frac{150}{1} = 150$	$F = \frac{150}{16}$
Within samples	64	$N-k = 5-1 = 4$	$\frac{64}{4} = 16$	$= 9.4$
Total		$N-1 = 6-1 = 5$		

F_{tab} at 5% level of significance

$$F_{2.12}(0.05)_{tab} = 3.89$$

$$F_{cal} = 8.51$$

Conclusion: Since $F_{cal} > F_{tab}$ so we reject H_0 at 5% level of significance and conclude the there is a significance difference between the average production of the three varieties A, B and C.

EXAMPLE 2. *To assess the significance of possible variation in body wt. of fishes of different ponds of a town a common survey was conducted for a number of fishes taken at random from the four ponds. The results are as follows :*

Ponds			
A	**B**	**C**	**D**
8	12	18	13
10	11	12	9
12	9	16	12
8	14	6	16
7	4	8	15

SOLUTION.

Pond A		Pond B		Pond C		Pond D	
X_1	X_1^2	X_2	X_2^2	X_3	X_3^2	X_4	X_4^2
8	64	12	144	18	324	13	169
10	100	11	121	12	144	9	81
12	144	9	81	16	256	12	144
8	64	14	196	6	36	16	256
7	49	4	16	8	64	15	225
ΣX_1 $= 45$	ΣX_1^2 $= 421$	ΣX_2 $= 50$	ΣX_2^2 $= 558$	ΣX_3 $= 60$	ΣX_3^2 $= 824$	ΣX_4 $= 65$	ΣX_4^2 $= 875$

The sum of all items of various samples

$$= \Sigma X_1 + \Sigma X_2 + \Sigma X_3 + \Sigma X_4 = 45 + 50 + 60 + 65 = 220$$

Correction factor

$$= \frac{T^2}{N} = \frac{(220)^2}{20} = \frac{18400}{20} = 2420$$

Total sum of square

$$= \Sigma X_1^2 + \Sigma X_2^2 + \Sigma X_3^2 + \Sigma X_4^2 - \frac{T^2}{N} = 421 + 558 + 824 + 875 - 2420$$

$$= 2678 - 2420 = 258$$

Sum of squares between the sample

$$= \frac{(\Sigma X_1)^2}{N} + \frac{(\Sigma X_2)^2}{N} + \frac{(\Sigma X_3)^2}{N} + \frac{(\Sigma X_4)^2}{N} - \frac{T^2}{N}$$

$$= \frac{(45)^2}{5} + \frac{(50)^2}{5} + \frac{(60)^2}{5} + \frac{(65)^2}{5} - 2420$$

$$= \frac{2025}{5} + \frac{2500}{5} + \frac{3600}{5} + \frac{4225}{5} - 2420$$

$$= \frac{12350}{5} - 2420 - 2470 - 2420 = 50$$

Sum of squares within samples :

= Total sum of squares – Sum of squares between sample

= 258 – 50 = 208

Analysis of Variance (ANOVA) table

Source of variance	Sum of square	Degree of freedom	Mean square
Between sample	50	$(C-1) = (4-1) = 3$	16.7
		$(N-C) = (20-4) = 16$	13.0

$$F = \frac{\text{Variance between samples}}{\text{Variance within samples}}$$

$$= \frac{16.7}{13.0} = 1.235$$

at df (3, 16) the table value of F at 5% level of significance = 3.24.
The calculated value of F is less than the table value hence the difference in mean values of sample is not significant, i.e., the samples could have come from the same universe.

EXAMPLE 3. *The following figures related to the production in kg of three varieties, A, B and C of vegetables in 12 plots :*

A	14	16	18	–	–
B	14	13	15	22	–
C	19	16	19	19	20

Is there any significant difference in the population of three varieties?

SOLUTION. **Null hypothesis (H_0):** There is no significant different in the production of three varieties $(H_0 : \sigma_1 = \sigma_2 = \sigma_3)$.

By subtracting 12 from each figure we prepare the following table and squaring each coded data. Let A, B and C are X_1, X_2 and X_3 respectively.

$A = X_1$	X_1^2	$B = X_2$	X_2^2	$C = X_3$	X_3^2
2	4	2	4	6	36
4	16	1	1	4	16
6	36	3	9	7	49
		10	100	7	49
				8	64
ΣX_1 = 12	ΣX_1^2 = 56	ΣX_2 = 16	ΣX_2^2 = 114	ΣX_3 = 32	ΣX_3^2 = 214

Sum of all items of various samples $= \Sigma X_1 + \Sigma X_2 + \Sigma X_3 = 12 + 16 + 32 = 60$

Correction factor $= \dfrac{T^2}{N} = \dfrac{(60)^2}{12} = 300$

Total sum of squares (SST) = 384 – 300 = 84

Sum of squares between the sample (SSC) would be

$$= \frac{(\Sigma X_1)^2}{N} + \frac{(\Sigma X_2)^2}{N} + \frac{(\Sigma X_3)^2}{N} - \frac{T^2}{N} = \frac{(12)^2}{3} + \frac{(16)^2}{4} + \frac{(32)^2}{5} - 300$$

$$= 48 + 64 + 204.8 - 300 = 16.8$$

Sum of squares with in the sample (SSE) would be

$$= \text{SST} - \text{SSC} = 84 - 16.8 = 67.20$$

Analysis of variance (ANOVA) table

Source of variance	Sum of square	Degree of freedom(df)	Mean square	F-ratio
Between sample	16.8	$(C-1)$ $=(3-1)=2$	8.4	$\dfrac{8.4}{7.467}$ $=1.125$
Within samples	67.2	$(N-C)$ $=(12-3)=9$	7.467	
Total	84	$N-1=6-1=5$	11	

For V_1, V_2 (d.f.)(2, 9) at 5% level of significance the table value of F = 19.4. The calculated value of F is less than the table value hence the difference in mean of sample is not significant, i.e., the sample could have come from the same universe. Thus the H_0 holds true.

EXAMPLE 4. *The following data give the yields of 12 plots of land in three samples under three varieties of fertilizers*

A	B	C
25	20	24
22	17	26
24	16	30
21	19	20

Is there any significant difference in the average yields of land under the three varieties of fertilizers?
Given that F at df(2, 9) at 5% level = 4.26.

SOLUTION. **Null hypothesis (H_0) :** There is no significant difference in the average yields under the three varieties.

Alternative hypothesis (H_1) : The difference in average yields is significant.

Sample A		Sample B		Sample C	
X_1	X_1^2	X_2	X_2^2	X_3	X_3^2
25	625	20	400	24	576
22	484	17	289	26	676
24	576	16	256	30	900
21	441	19	361	20	400
ΣX_1 $=92$	ΣX_1^2 $=2126$	ΣX_2 $=72$	ΣX_2^2 $=1306$	ΣX_3 $=100$	ΣX_3^2 $=2552$

Now,

$$T = \Sigma X_1 + \Sigma X_2 + \Sigma X_3 = 92 + 72 + 100 = 264$$

$$\text{Correcting factor} = \frac{T^2}{N} = \frac{(264)^2}{4+4+4} = \frac{264 \times 264}{12} = 5808$$

The total sum squares (SST)

$$= \Sigma X_1^2 + \Sigma X_2^2 + \Sigma X_3^2 - \frac{T^2}{N}$$

$$= 2126 + 1306 + 2552 - 2808$$

$$= 5984 - 5808 = 176$$

Sum of squares between the sample (SSB)

$$= \frac{(\Sigma X_1)^2}{N} + \frac{(\Sigma X_2)^2}{N} + \frac{(\Sigma X_3)^2}{N} - \frac{T^2}{N}$$

$$= \frac{(90)^2}{4} + \frac{(72)^2}{4} + \frac{(100)^2}{4} - 5808$$

$$= 2116 + 1296 + 2500 - 5808$$

$$= 5912 - 5808 = 104$$

Degree of freedom (df) $= k - 1 = 3 - 1 = 2$

Mean square between the sample (MSB)

$$= \frac{\text{Sum squares between the samples}}{V_1} = \frac{104}{2} = 52$$

Sum square within the samples (SSW)

$$= SST - SSE = 176 - 104 = 72$$

Degree of freedom (dfw)

$$= N - k = (n_1 + n_2 + n_3) - k = 12 - 3 = 9$$

Mean square within the samples (MSW) $= \dfrac{SSW}{V_2} = \dfrac{72}{9} = 8$

Analysis of variance (ANOVA) table

Source of variance	Sum of square (SS)	Degree of freedom (df)	Mean square (MS)	Test Statistic
Between sample	SSB = 104	dft = 2	$MSB = \dfrac{SSB}{V_1} = 52$	$F = \dfrac{MSB}{MSW}$ $= \dfrac{52}{8}$ $= 0.5$
Within samples	SSW = 72	dfw = 9	$MSW = \dfrac{SSW}{V_2} = 8$	

Total SST $= 176$, $N - 1 = 11$.

The calculated value of F at 0.05 at degrees of freedom $V_1 = 2$ and $V_2 = 9$ is 6.5. It is much greater than the given value of $F = 4.26$. Hence, we reject the null hypothesis (H_0) at 0.05 level and conclude the difference in average yields under the three varieties is significant.

24.23.4 TWO WAY CLASSIFICATION

If the observation are classified according to the circles, we get a two way classification for example if we may consider the yields of three varieties of wheat using three different type of fertilizers. In such experiment the observation are classified according to two criteria, the wheat variety and the type of fertilizer.

Let us suppose that N observations x_{ij} ($i = 1, 2, \dots k$ and $j = 1, 2, \dots n$) are grouped into k rows and columns. Then two way classification model can be represented as shown in the following table.

Two-way classifications

Rows	Columns						Total T_1^*	Mean $\bar{X}_1^*$
	1	2	...	j	...	n		
1	X_{11}	X_{12}		X_{1j}		X_{1n}	T_1^*	$\bar{X}_1^*$
2	X_{21}	X_{22}		X_{2j}		X_{2n}	T_2^*	$\bar{X}_2^*$
$\vdots$								
i	X_{i1}	X_{i2}		X_{ij}			T_i^*	T_i^*	$\bar{X}_i^*$
$\vdots$								
k	X_{k1}	X_{k2}		X_{kj}			T_k^*	T_k^*	$\bar{X}_k^*$
Total T_i^*	T_1^*	T_2^*				T_n^*	T^{**}	
Mean $\bar{X}_i^*$	$\bar{X}_1^*$	$\bar{X}_2^*$		$\bar{X}_j^*$		$\bar{X}_n^*$		$\bar{X}^{**}$

where T_i^* and T_j^* represent the total of ith row and j^{th} column respectively $(i = 1, 2, ..., k, \ j = 1, 2, ..., n)$.

$\bar{X}_i^*$ and $\bar{X}_j^*$ represent the mean of the ith row and j^{th} column respectively, T^{**} and $\bar{X}^{**}$ represent the total and mean of all the nk observations.

Null Hypothesis:

(a) $H_0^* : \mu_1^* = \mu_2^* ... = \mu_k^*$ row means are equal

$H_1' : \mu_{i*}^*$ are not equal

(b) $H'' : \mu_1^* = \mu_2^* ... = \mu_n^*$ column means are equal

$H_1'' : \mu_j^*$ are not equal.

The various steps of analysis of two way classification.

(i) **Correction Factor (C.F.) :** First of all we find the correction factor as $C.F. = \dfrac{T^{*2}}{k}$

(ii) **The Total Sum of Squares (TSS) :** This is obtained by subtracting the correction factor from the sum of squares of all the items

$$TSS = \sum_{i=1}^{k} \sum_{j=1}^{n} X_{ij}^2 - C.F.$$

where, $\sum_{i=1}^{k} \sum_{j=1}^{n} X_{ij}^2 - (X_{11}^2 + X_{12}^2 + ... + X_{1n}^2) + (-X - 2)^2 + ... + X_{2n}^2 + ... + (X_{k1}^2 + ... X_{kn}^2)$

(iii) **Sum of Squares between Columns (SSE) :** This is obtained by subtracting the correction factor from the sum of squares of all the items

$$SSC = \sum_{i=1}^{k} \dfrac{T_i^{*2}}{n} - C.F.$$

(iv) **Sum of Square between Rows (SSR):** As sum of squares between columns the SSR obtained by subtracting the correction factor from the sum of squares of the column total when divided by k

$$SSR = \sum_{j=1}^{n} \dfrac{T_j^{*2}}{n} - C.F.$$

(v) Error Sum of Square (SSE) : This is obtained by subtracting the SSC and SSR from SST as
$$SSE = TSS - SSC - SSR$$

(vi) Degree of Freedom.

D.F. associated with TSS $= (nk - 1)$

D.F. associated with SSR $= (k - 1)$

D.F. associated with SSC $= (n - 1)$

D.F. associated with SSE $= (k - 1)(n - 1)$

(vi) ANOVA Table for a Two-way Classification

Source of variation	Sum of squares (SS)	D.F.	Mean Squares (MS)	Calculated F-ratio
Row	SSR	$k - 1$	$\dfrac{SSR}{(k-1)} = V_1$	$F_1 = \dfrac{V_1}{V_3}$
Column	SSC	$n - 1$	$\dfrac{SSC}{(n-1)} = V_2$	$F_2 = \dfrac{V_2}{V_3}$
Error	ESE	$(k-1)(n-1)$	$\dfrac{ESS}{(k-1)(n-1)} = V_3$	
Total		$nk - 1$		

F_{tab} at 5% level of significance $= 7.71$.

Conclusion. Since $F_{cal} > F_{tab}$ at 5% level of significance so the null hypothesis is rejected, *i.e.*, the difference between yields due to A and B is significant.

24.23.5 SHORT CUT METHOD

The direct method of Analysis of variance is very tedious and time consuming. In general short cut method is used the various steps of this method are as follows:

(i) Total of sample items and sum of squares of items: First of all find the sum of each sample, *i.e.*,

$$\sum_{j=1}^{n_1} X_{1j}, \ \sum_{j=1}^{n_2} X_{2j}, \ ....., \ \sum_{j=1}^{n_k} X_{kj}$$

and sum of squares of items

$$\sum_{j=1}^{n_1} X_{1j}^2, \ \sum_{j=1}^{n_2} X_{2j}^2, \ ....., \ \sum_{j=1}^{n_k} X_{kj}^2 .$$

(ii) Correction Factor (C.F.): Calculate the correction factor as

$$C.F. = \frac{T^2}{N} \quad \text{where, } T^2 = (T_1^* + T_2^* + ... + T_k^*)^2, N = n_1 + n_2 \, ... n_k .$$

The Total Sum of Squares (TSS): TSS is found by subtracting the correction factor from the sum of squares of all the items of the samples

$$TSS = \sum_{i=1}^{k} \sum_{j=1}^{n_k} X_{ij}^2 - \frac{T^2}{N}$$

where, $\displaystyle\sum_{i=1}^{k} \sum_{j=1}^{n_k} X_{ij}^2 = \left[(X_{11}^2 + X_{12}^2 + ... + X_{1n_1}^2) + (X_{21}^2 + ... + X_{2n_2}^2) + ... + (X_{k1}^2 + ... X_{n_k}^2) \right]$

Sum of Squares Between Samples: This is obtained by subtracting the correction factor from the sum of squares of the total when divide by number of item in each sample.

$$\text{SSC} = \frac{(T_1^*)^2}{n_1} + \frac{(T_2^*)^2}{n_2} + \ldots + \frac{(T_k^*)^2}{n_k} - \frac{T^2}{N}$$

Sum of Squares within Samples: This is obtained by subtracting the sum of squares between samples from total sum of squares

$$\text{SSE} = \text{TSS} - \text{SSC}$$

ANOVA Table and Interpretation: The Analysis of variance table and interpretation are same as in direct method.

SOLVED EXAMPLES

EXAMPLE 1. *The following figures relate to the production in kg of three varieties of wheat A, B and C used on 15 plots.*

Wheat Variety	Yields (Kg)					
A	14	17	16	16		
B	15	11	13	15	13	14
C	18	16	18	19	15	

Test whether there is any significance difference in the production of three varieties.

SOLUTION. **Null Hypothesis (H_0).** There is no significance difference between the production of the three varieties.

i.e., $H_0 : \mu_1 = \mu_2 = \mu_3$

Wheat Variety	Yields (Kg)(X_{ij})						Total
A	14	17	16	16			$63 = \Sigma x_{1_i}$
B	15	11	13	15	13	14	$81 = \Sigma x_{2_j}$
C	18	16	18	19	15		$86 = \Sigma x3_j$
							$230 = \Sigma\Sigma x_{ij}$

$$\text{C.F.} = \frac{T^2}{N} = \frac{\Sigma(X_{1j} + X_{2j} + X_{nj})^2}{N} = \frac{(230)^2}{15} = 3526.67$$

$$\Sigma\Sigma x_{ij}^2 = (14^2 + 17^2 + \ldots + 16^2) + (15^2 + 11^2 + \ldots + 14^2) + (18^2 + 16^2 + \ldots 15^2)$$

$$= 3592$$

Total sum of squares (TSS)

$$\text{TSS} = \Sigma X_{ij}^2 - \text{C.F.} = 3592 - 3526.67 = 65.33$$

Sum of squares between samples

$$\text{SSC} = \left[\frac{(T_1^*)^2}{n_1} + \frac{(T_2^*)^2}{n_2} + \ldots + \frac{(T_k^*)^2}{n_k}\right] - \text{C.F.}$$

$$= \frac{(63)^2}{4} + \frac{(81)^2}{6} + \frac{(86)^2}{6} - 3526.67 = 3564.95 - 3526.67 = 38.28$$

Sum of squares within samples

$$\text{SSE} = \text{TSS} - \text{SSE} = 65 - 33 - 38.28 = 27.05.$$

ANOVA

Source of variance	Sum of square (SS)	d.f.	MSS	F-ratio (Calculated)
Between varitety	38.28	$k-1$ $= 3-1$ $= 2$	$\dfrac{38.28}{2}$ $= 19.14$	$F = \dfrac{19.14}{2.25}$ $= 3.51$
Within variety	27.05	$N-k$ $= 15-3$ $= 12$	$\dfrac{27.05}{12}$ $= 2.25$	
Total	66.33	$N-1$ $= 15-1$ $= 14$		

(viii) Interpretation of F: The decision about H_0' and H_0'' are taken as

If $F_{1\,cal} \geq F_{1\,tab}$, then H_0' is rejected otherwise H_0' may be accepted if $F_{2\,cal} \geq F_{2\,tab}$, then H_0'' is rejected otherwise H_0'' may be accepted.

EXAMPLE 2. *Set up ANOVA table for the following per hectare yield for three varieties of wheat each grown on four plots*

Plot of Land	Variety of Wheat		
	A	*B*	*C*
1	16	15	15
2	17	15	14
3	13	13	13
4	18	17	14

SOLUTION. Since the data have been classified in respect of two criteria, the plot of land and wheat variety. So we prepare two way ANOVA.

Null Hypothesis :

(i) $H_0' : \mu_1^* = \mu_2^* = ... = \mu_k^*$ row mean are equal

(ii) $H_0'' : \mu_1^* = \mu_2^* = ... = \mu_n^*$ column mean are equal

$H_1'' : \mu_1^* \neq \mu_2^* \neq ... \neq \mu_n^*$

Plot of Land	Variety of Wheat			Total(T_1)
	A	*B*	*C*	
1	16	15	15	$46 = T_1^*$
2	17	15	14	$46 = T_2^*$
3	13	13	13	$36 = T_3^*$
4	18	17	14	$49 = T_4^*$
Total (T_j)	$T_1^* = 66$	$T_2^* = 60$	$T_3^* = 56$	$T^{**} = 1890$

Correction factor (C.F.) $= \dfrac{T^{**2}}{nk} = \dfrac{(180)^2}{4 \times 3} = 2700$

$$\sum_{i=1}^{k} \sum_{j=1}^{n} X_{ij}^2 = (16^2 + \dots + 15^2) + (17^2 + \dots + 14^2) + (13^2 + \dots + 13^2)$$

$$+ (18^2 + \dots + 14^2) = 2732$$

$$\text{TSS} = \sum_{i=1}^{k} \sum_{j=1}^{n} X_{ij}^2 - \text{C.F.} = 2732 - 2700 = 32$$

$$\text{SSR} = \dfrac{\sum\limits_{i=1}^{k} T_i^{*2}}{k} -$$

$$\text{C.F.} = \dfrac{(46^2 + 46^2 + 39^2 + 49^2)}{5} - 2700 = 2718 - 2700 = 18$$

$$\text{SSC} = \dfrac{\sum\limits_{j=1}^{n} T_j^{*2}}{k} -$$

$$\text{C.F.} = \dfrac{(64)^2 + (60)^2 + (56)^2}{4} - 2700$$

$$= 2708 - 2700 = 8$$

$$\text{SSE} = \text{TSS} - \text{SSC} = 32 - 18 - 8 = 6.$$

Two-way ANOVA Table

Source of variation	SS	d.f.	MSS	Calculated F
Row	18	$4 - 1 = 3$	$18/3 = 6$	$F_1 = 6/1 = 6$
Column	8	$3 - 1 = 2$	$8/2 = 4$	$F_2 = 4/1 = 4$
Error	6	6	$6/6 = 1$	
Total	32	11		

F_{tab} at 5% level of significance

$$F_{3,\,6(0.05)} = 4.76$$

Conclusion: (i) $F_{1\,cal} > F_{(3,\,6)\,tab}$, so H_0' is rejected at 5% level of significance, *i.e.*, there is a significance difference between two means.

(ii) As $F_{2\,cal} < F_{(2,6)\,tab}$, so H_0'' is rejected at 5% level of significance, *i.e.*, column mean do not differ significantly.

EXAMPLE 3. *Prepare a two-way ANOVA on the data given below :*

Ponds	Treatment			
	A	B	C	D
I	38	40	41	39
II	35	42	49	36
III	40	38	42	42

Use coding method subtracting 40 from the given numbers.

SOLUTION. **Null hypothesis (H_0) :** There is no significant difference in the treatment and ponds. By using Analysis of variance technique, we get

By coding the data (subtracting 40 from each value).

Correction factor $= \dfrac{T^2}{N} = \dfrac{(12)^2}{12} = 12$

Sum of square between treatments

$$= \frac{(3)^2}{3} + \frac{(0)^2}{3} + \frac{(12)^3}{3} + \frac{(-3)^2}{3} - 12 = 3 + 0 + 48 + 3 - 12 = 42$$

Sum of square S between pond

$$= \frac{(-2)^2}{4} + \frac{(-12)^2}{4} + \frac{(2)^2}{4} - 12 = 1 + 36 + 1 - 12 = 26$$

Total sum of squares

$$= (-2)^2 + (5)^2 + (0)^2 + (0)^2 + (2)^3 + (-2)^2 + (1)^2 + (9)^2 + (2)^2$$

$$+ (-1)^2 + (-4)^2 + (2)^2 - \frac{T^2}{N}$$

$$= 4 + 25 + 0 + 0 + 4 + 4 + 1 + 8 + 81 + 4 + 1 + 16 + 4 - 12 = 132$$

Analysis of Variance (ANOVA) table

Source of variation	Sum of squares	Degree of Freedom	Mean Squares	F-ratio
Between Column	42	$(C-1)$ $= 4 - 1 = 3$	14	$\dfrac{14}{10.67}$ $= 1.312$
Between Row	26	$(r-1)$ $= 3 - 1 = 2$	13	$\dfrac{13}{10.67}$ $= 1.218$
Residuals	64	$(C-1)(r-1)$ $= (4-1)(3-1) = 6$	10.67	$\dfrac{13}{10.67}$ $= 1.218$
Total (SST)	132	$(N-1)(12-1) = 11$		

For (3, 6) d.f. $F_{0.05} = 4.76$ and For (2, 6) d.f. $F_{0.05} = 5.14$.
The calculated value of F is less than the table value at 5% level of significance. The hypothesis is accepted. Hence, there is no significant difference between treatment and ponds.

Exercise 24.2

1. The following table gives the yields on 15 sample plots under three varieties of seed A, B and C:

Varieties of School	Yields				
A	19	19	21	14	18
B	16	18	15	23	13
C	23	26	20	26	30

Test whether the average yield of land under different varieties of seed show significant differences. [**Hint.** One-way classification]

2. A tea company appoints four salesman and observers their sales in three seasons — summer, winter and monsoon. The figure (in lakhs) are given in the following table :

Seasons	Salesman				Totals
	A	B	C	D	
Summer	30	30	15	29	104
Winter	22	23	15	26	96
Monsoon	20	22	23	23	88
Total	72	75	63	78	288

Carry out on Analysis of variance.
[**Hint.** Two way ANOVA].

3. The following figures relate to the production in kg. of three varieties I, II, III of wheat sown in 12 plots :

Variety I	14	16	18	
Variety II	14	13	15	22
Variety III	16	16	19	20

Is there any significance difference in the population of three varieties? Given the tabulated value of F for $v_1 = 2$ and $v_2 = 9$ at 5% level of significance is 4.26.

4. Set up ANOVA table for the following per hectare yield (in kg) for three varieties of heat each grown on four plots

Plot of Land	Variety of Wheat		
	A_1	A_2	A_3
1	6	5	5
2	7	5	4
3	3	3	3
4	8	7	4

Also find F-ratio.

5. The following data represent the number of units of production per day turned out by 5 different workmen using different types of machines

Workman	Machines Type			
	A	B	C	D
1	44	38	47	36
2	46	40	52	43
3	34	36	44	32
4	33	38	46	33
5	38	42	49	38

(a) Test whether the mean productivity is the same for the four different machine type.

(b) Test whether 5 men differ with respect to productivity.

6. On an arithmetic reasoning of 31 ten-years old boys and 42 ten year old girls, the following score are made:

	Mean	S.D.	N
Boys	40.39	8.69	31
Girls	35.81	0.33	42

Perform an analysis of variance and show whether mean difference between boys and girls is significant. Given that table value of Fat 5% and at 1 % is 7.01.

7. A manufacturing company purchased three new machines of different makes and wishes to determine whether one of them is faster than the others in producing a certain out put. Five hourly production figures are observed at random from each machine and results are given below.

Observations	A_1	A_2	A_3
1	25	31	24
2	30	39	30
3	36	38	28
4	38	42	25
5	31	35	28

Use ANOVA and determine whether the machines are significantly different in their mean speed. (Given at 5% level $F_{2, 12} = 3.89$)

(UPTU–2002)

Objective Evaluation

∞ MULTIPLE CHOICE QUESTIONS

Choose the most appropriate one.

1. A hypothesis may be classified as :
   - (a) simple
   - (b) composite
   - (c) null
   - (d) all the above

2. The hypothesis under test is :
   - (a) simple hypothesis
   - (b) alternative hypothesis
   - (c) null hypothesis
   - (d) none of the above

3. Whether a test is one-sided or two sided depends on :
   - (a) alternative hypothesis
   - (b) composite hypothesis
   - (c) null hypothesis
   - (d) simple hypothesis

4. Wrong decision about H_0 leads to :
   - (a) one kind of error
   - (b) two kind of error
   - (c) three kind of error
   - (d) four kind of error

5. Power of a test is related to :
   - (a) type I error
   - (b) type II error
   - (c) type I and II errors both
   - (d) none of the above

6. If θ is the true parameter and β the type II error, the function $\beta(\theta)$ in known as :
   - (a) power function
   - (b) power of the test
   - (c) operating characteristic function
   - (d) none of the above

7. Level of significance is the probability of :
   - (a) type I error
   - (b) type II error
   - (c) not committing error
   - (d) any of the above

8. In terms of type II error β and θ, the true parameter, the function $1 - \beta(\theta)$ is called :
   - (a) power of the test
   - (b) power function
   - (c) OC function
   - (d) none of the above

9. Out of the two types of errors in testing, the more severe error is :
   - (a) type I error
   - (b) type II error
   - (c) both (a) and (b)
   - (d) none of the above

10. Area of the critical region depends on :
    - (a) size of type I error
    - (b) size of type II error
    - (c) value of the statistic
    - (d) number of observation

∞ TRUE/FALSE

Write 'T' for True and 'F' for False statement.

1. Type second error is more severe than type first error. **(T/F)**
2. Second kind of error is minimised for a prefixed level of first kind of error. **(T/F)**
3. Probability of first kind of error is called the size of the test. **(T/F)**
4. Probability of type I error is called level of significance. **(T/F)**
5. Rejecting H_0 when H_0 is true is type I error. **(T/F)**
6. Accepting H_0 when H_0 is false is a type I error. **(T/F)**
7. When a test is one-sided or two sided depends on alternative hypothesis. **(T/F)**
8. The test statistic remains same in case of one tailed test and two tailed test. **(T/F)**
9. If β is the probability of type II error, the power of the test is $1 - \beta$. **(T/F)**
10. With usual notation the function $\beta(\theta)$ is known as operating characteristic. **(T/F)**

∞ FILL IN THE BLANKS

1. There can be _____ procedure(s) to test the same hypothesis.
2. The theory of testing parametric hypothesis was first originated by _____ in _____.

3. Besides Neyman, the other pioneer worker in the field of Testing of hypothesis was _____ .

4. A hypothesis is an _____ about the parameter of a population.

5. The hypothesis which is under test for possible rejection is called _____ hypothesis.

6. A hypothesis contrary to null hypothesis is known as _____ hypothesis.

7. The idea of alternative hypothesis was populated by _____ .

8. The hypothesis $H_0 : \theta = \theta_0$ vs $H_1 : \theta = \theta_1$ is a _____ hypothesis against _____ alternative hypothesis.

9. The hypothesis $H_0 : \theta > \theta_0$ is a _____ hypothesis.

10. There can be only _____ types of errors in taking a decision about H_0.

ANSWERS

➣ MULTIPLE CHOICE QUESTIONS

1. (d) **2.** (c) **3.** (a) **4.** (b) **5.** (b) **6.** (c) **7.** (a) **8.** (b) **9.** (b)
10. (a)

➣ TRUE/ FALSE

1. T **2.** T **3.** T **4.** T **5.** T **6.** F **7.** T **8.** T **9.** T
10. T

➣ FILL IN THE BLANKS

1. more than one	**2.** J. Neyman; 1928	**3.** Karl Pearson	**4.** assertion	**5.** null
6. alternative	**7.** J. Neyman	**8.** simple; simple	**9.** composite	**10.** two

❑❑❑❑❑❑

Coefficient of Contingency and χ^2-Test

25.1 INTRODUCTION

Various test of significance, such as 'z' and 't' test are parametric test and applied to only quantitative data like length, weight, height, Hb%, consumption, number of seeds per pod etc. These tests were based on the assumption that the samples were drawn from the normally distributed populations. There are many situations in which it is not possible to make any dependable assumption about the distribution from which samples have been drawn. In biological experiments, we also get qualitative data like colour, health, intelligence, curve responses of drug, etc., in which observations are classified in a particular category, class or group. For these qualitative data a non-parametric test called chi-square test is commonly used. In many studies, especially genetic studies, it becomes necessary to test the significance of overall deviation between the observed data and expected frequencies.

Chi-square test was developed by Prof. R. A. Fisher in 1870. Karl Pearson improved Fisher's Chi-square test in its modern form in the year 1900. Chi-square is derived from the Greek Letter (Chi-χ) and pronounced as Ki and Ksy without s).

25.2 CHI-SQUARE (χ^2) TEST

A statistical test to determine if the "observed" numbers deviate from those "expected" or "theoretical" number under a particular hypothesis

$$\chi^2 = \sum_{i=1}^{n} \left[\frac{(O-E)^2}{E_i} \right]$$

where, O = observed frequency and E = expected frequency.

25.3 APPLICATIONS OF χ^2 DISTRIBUTION

χ^2-distribution is used:

(1) to test the significance of sample variances.
(2) to test the independence of attributes in a contingency table.
(3) to compare a number of frequency distribution.
(4) to test the goodness of fit.

WORKING PROCEDURE

Step 1. Calculate all the expected frequencies, *i.e.*, *E* for all value $i = 1, 2, 3, ... n$.

Step 2. Take the difference between each observed frequency '*O*' and the corresponding expected frequency '*E*' for each value of, *i.e.*, $(O - E)$

Step 3. Square the difference for each value of *i*, *i.e.*, $(O - E)^2$ for all value of $i = 1, 2, 3, ... n$.

Step 4. Divide each square difference by the corresponding expected frequency, *i.e.*, calculate $\dfrac{(O - E)^2}{E}$ for all values of $i = 1, 2, ... n$.

Step 5. Add all these quotients obtained in the 'step 4'.

$$\chi^2 = \sum_{i=1}^{n} \left[\frac{(O - E)^2}{E} \right]$$

It is the required value of χ^2.

25.3.1 CHARACTERISTICS OF CHI-SQUARE

1. The value of chi-square is always positive as each pair is squared up.

2. χ^2 (chi-square) will be zero if each pair is zero and it may assume any value extending to infinity, when the difference between the observed frequency and expected frequency in each pair are unequal. Thus chi-square lie between 0 and ∞.

3. Chi-square is a statistic but a parametric.

SOLVED EXAMPLE

EXAMPLE. *The standard deviation of a certain dimension of articles produced by a machine is 7.5 over a long period. A random sample of 25 articles gave a standard deviation of 10.0. Is it justifiable to conclude that the variability has increased? Use 5% level of significance.*

SOLUTION. $H_0 : \sigma^2 = 7.5$

$H_1 : \sigma^2 > 7.5$ (Variability has increased)

Test statistic : $\quad \chi^2 = \dfrac{\Sigma(x_i - \bar{x})^2}{\sigma^2} \sim \chi^2$ with $(n - 1)$ d.f.

Computation : $\quad \chi^2 = \dfrac{np^2}{\sigma^2}$, under H_0

$$= \frac{25 \times 10^2}{(7.5)^2} = 44 \qquad \left[\because s^2 = \frac{1}{n}\Sigma(x_i - \bar{x})^2 \right]$$

Critical region : Right hand side tail of a χ^2–distribution with *x* degrees of freedom. Here $x = 24$, $d = 0.05$.

$$\chi^2_{0.05}(24) = 36.415$$

(from table of χ^2 at different probability level)

Conclusion: Since the value of χ^2 calculated is greater than $\chi^2_{0.05}(24)$, H_0 is rejected. This means there is justification for believing that the variability has increased.

25.4 CHI-SQUARE (χ^2) TEST FOR GOODNESS OF FIT

When a coin is tossed 100 times, the theoretical considerations expect 150 heads and 50 tails. But in practice these results are rarely achieved. The magnitude of discrepancy (difference) between theory and experiment (observation) described through a Greek letter χ^2 read as chi-Square. If value of χ^2 is zero then the observed and expected frequencies completely coincide. The greater discrepancy between the observed and expected frequencies, the greater is the value of χ^2. This test is given by Karl Person in 1900.

If $Q_1, Q_2, ..., Q_n$ is a set of observed (experimental) frequencies and $E_1, E_2, ..., E_n$ is the corresponding set of expected (theoretical or hypothetical) frequencies, then null hypothesis (H_0). There is no significance difference between observed and expected (theoretical) frequencies.

Or there is no significance difference between theory and experiment.

Alternative Hypothesis : H_0 is not true.

Test statistic:

$$\chi^2 = \sum_{i=1}^{n} \frac{(O_i - E_i)^2}{E_i}$$

where χ^2 follows χ^2 distribution with $(n-1)$ degree of freedom.

Conclusion: If $\chi^2_{cal} \geq \chi^2_{tab}$ then the null hypothesis (H_0) is rejected at $\alpha\%$ level of significance otherwise H_1 may be accepted.

25.4.1 CONDITIONS FOR APPLYING χ^2 TEST

Following are the conditions which should be satisfied before χ^2 test can be applied:

 (i) The sample observation should be independent.
 (ii) The constraints on the cell frequencies if, any, should be linear, i.e., $\Sigma O_i = \Sigma E_i$.
 (iii) N, the total numbers of frequencies should be large. It is difficult to say what constitutes largeness but as an arbitrary figure, we may say that N should be at least 50.

 (iv) No expected frequency should be very small.

Here it is difficult to say that constitute smallness, but 5 should be regarded as the very minimum and 10 is better. In case where the expected frequencies fall below these limits, they are to be amalgamated in a single cell and degrees of freedom properly adjusted.

SOLVED EXAMPLES

EXAMPLE 1. Calculate χ^2 for the following data

Class	A	B	C
Observed frequency	37	44	19
Expected frequency	31	38	31

SOLUTION. We have

$$\chi^2 = \sum_{i=1}^{n} \frac{(O_i - E_i)^2}{E_\infty} = \frac{(37-31)^2}{31} + \frac{(44-38)^2}{38} + \frac{(19-31)^2}{31}$$

$$= \frac{36}{31} + \frac{36}{33} + \frac{144}{31} = 6.76$$

EXAMPLE 2. The following table gives the numbers of aircraft accidents that occurred during the various days of the week. Find whether the accidents are uniformly distributed over the week.

Day	Sun	Mon	Tue	Wed	Thu	Fri	Sat	Total
No. of accidents	14	16	8	12	11	9	14	84

(χ^2 for 4 d.f. at 5% level of significance = 9.41)　　　　　　　**(Hissar–2005)**

SOLUTION.　　**Null Hypothesis.$(H_0)'$:** The accidents are uniformly distributed over the week.

Alternative Hypothesis $(H_1)'$: H_0 is not true.

Expected frequencies of accidents on any day $= \dfrac{84}{7} = 12$.

Day	Sun	Mon	Tue	Wed	Thu	Fri	Sat	Total
Observed no. of accidents	14	16	8	12	11	9	14	84
Expected no. of accidents	12	12	12	12	12	12	12	84

Since frequencies in some classes are less than 10, we regroup the data

Day	Sun	Mon	Tue and Wed	Thu	Fri and Sat	Total
Observed no. of accidents	14	16	20	11	23	84
Expected no. of accidents	12	12	24	12	24	84

Test statistic :

$$\chi^2 = \sum_{i=1}^{n} \frac{(O_i - E_i)^2}{E_i}$$

$$= \frac{(14-12)^2}{12} + \frac{(16-12)^2}{12} + \frac{(20-24)^2}{24} + \frac{(11-12)^2}{12} + \frac{(23-24)^2}{24}$$

$$= \frac{1}{24}(8 + 32 + 16 + 2 + 1) = \frac{59}{24} = 2.46$$

Degrees of freedom = 5 – 1 = 4.

Tabulated value of χ^2 for 4 degrees of freedom at 5% level of significance = 9.41.

Conclusion: Since calculated value of χ^2 is less tabulated value at 5% level of significance so null hypothesis (H_0) may be accepted, i.e., the accidents are uniformly distributed over the week.

EXAMPLE 3.　　*In 120 throws of a single die, the following distribution of faces was obtained :*

Faces	1	2	3	4	5	6	Total
f_0	30	25	18	10	22	15	120

Do these results constitute of regulation of the equal probability null hypothesis?

SOLUTION. On the basis of principle of equal probabilities $p = 1/6$, the theoretical frequencies for each face is $NP = 120 \times \dfrac{1}{6} = 20$. Thus we have

Faces	1	2	3	4	5	6	Total
f_0	30	25	18	10	22	15	120
f_e	20	20	20	20	20	20	120

Since no theoretical frequency is less than 10. Hence,

$$\chi^2 = \sum \frac{(f_0 - f_e)^2}{f_e}$$

$$= \frac{(30-20)^2}{20} + \frac{(25-20)^2}{20} + \frac{(18-20)^2}{20} + \frac{(10-20)^2}{20} + \frac{(22-20)^2}{20} + \frac{(15-20)^2}{20}$$

$$= \frac{1}{20}(100 + 25 + 4 + 100 + 4 + 25) = \frac{258}{20} = 12.9$$

Degrees of freedom $v = 6 - 1 = 5$.

$$\chi^2_{0.05}(5) = 11.070 \qquad \text{(From table)}$$

Since $\chi^2_{cal} < \chi^2_{0.05}(5)$

Hypothesis of equal probabilities is rejected.

EXAMPLE 4. *Twelve dice were thrown 4096 times and a throw of 6 was respond as a success, the observed frequencies were as*

No. of Success	0	1	2	3	4	5	6	7 and over	Total
Frequencies	447	1145	1181	796	380	115	24	8	4096

Find the value of χ^2 on the hypothesis that the dice were unbiased and hence show that the data are consistent with the hypothesis so far as the χ^2 test is concerned.

SOLUTION. The probability of getting 6 in a throw of a single die, *i.e.*, probability of success $= 1/6$ and the probability of failure $= 1 - \dfrac{1}{6} = \dfrac{5}{6}$.

Hence, the theoretical frequencies of 0, 1, 2, 3, ... 12 successes with 12 dice 4096 times are respectively the successive terms of the binomial expansion

$$4096\left(\frac{5}{6} + \frac{1}{6}\right)^{12} = 4096\left[\left(\frac{5}{6}\right)^{12} + {}^{12}C_1\left(\frac{5}{6}\right)^{11}\left(\frac{1}{6}\right) + {}^{12}C_2\left(\frac{5}{6}\right)^{10}\left(\frac{1}{6}\right)^2 + \right.$$

$$\left. ... + {}^{12}C_{12}\left(\frac{1}{6}\right)^{12}\right]$$

$$= 459 + 1102 + 1212 + 809 + 365 + 116 + 27 + 6$$

Thus we have

f_0	447	1145	1181	796	380	115	24	8
f_e	459	1102	1212	809	365	116	27	6

$$\chi^2 = \sum \frac{(f_0 - f_e)^2}{f_e}$$

$$= \frac{(447-459)^2}{459} + \frac{(1145-1102)^2}{1102} + \frac{(1181-1212)^2}{1212} + \frac{(796-809)^2}{809}$$
$$+ \frac{(380-365)^2}{365} + \frac{(115-116)^2}{116} + \frac{(240-27)^2}{27} + \frac{(8-6)^2}{6}$$
$$= \frac{144}{459} + \frac{1849}{1102} + \frac{961}{1212} + \frac{169}{809} + \frac{225}{365} + \frac{1}{116} + \frac{9}{27} + \frac{4}{5}$$
$$= 0.314 + 1.670 + 0.793 + 0.209 + 0.616 + 0.009 + 0.333 + 0.667$$
$$= 4.619 \text{ (nearly)}$$

Degree of freedom $v - 1 = 8 - 1 = 7$

For 7 degrees of freedom at 5% level of significance

$$\chi^2_{0.05}(7) = 14.027 \qquad \text{(From table)}$$

Since the calculated value (4.619) of χ^2 is much less than $\chi^2_{0.05}(7)$, hence the data are consistent with the hypothesis, the dice were unbiased so far as the χ^2-test is connected.

EXAMPLE 5. *Five dice were thrown 192 times and the numbers of times, 4, 5 or 6 were as follows :*

No. of dice throwing 4, 5, 6	5	4	3	2	1	0
f	6	46	70	48	20	2

Calculate χ^2.

SOLUTION. Probability of throwing 4, 5 or 6 is $\frac{3}{6} = \frac{1}{2}$. Therefore, the theoretical frequencies getting 5, 4, 3, 2, 1, 0 success with 5 dice are respectively the successive terms of $192\left(\frac{1}{2} + \frac{1}{2}\right)^5$

which are as follows: 6, 30, 60, 60, 30, 6 respectively.

Since for the application of χ^2-test no frequency should be less than 5, hence on regrouping the table can be arranged as :

No. of dice throwing 4, 5, 6	5	4	3	2	1 or 0
f_0	6	46	70	48	22
f_e	6	30	60	60	36

$$\chi^2 = \frac{(6-6)^2}{6} + \frac{(46-30)^2}{30} + \frac{(70-60)^2}{60} + \frac{(48-60)^2}{60} + \frac{(22-36)^2}{36}$$
$$= 8.53 + 1.66 + 2.4 + 5.44$$

EXAMPLE 6. *200 digits were chosen at random from a set of tables. The frequencies of the digits were*

Digit	0	1	2	3	4	5	6	7	8	9
Frequency	18	19	23	21	16	25	22	20	21	25

Use the χ^2-test to assess the correctness of the hypothesis that the digits were distributed in equal numbers in the table from which these numbers were taken. The value of χ^2 for 8 degrees of freedom at 5% level is 16.919.

SOLUTION. **Null hypothesis (H_0):** Digits were distributed in equal number in the table, i.e., $p = \dfrac{1}{10}$.

$\therefore$ Expected frequency for each digit $= NP = \dfrac{1}{10} \times 200 = 20$

Computation of χ^2

Digits	Observed frequency, f_0	Expected frequencies	$f_0 - f_e$	$(f_0 - f_e)^2$	$\dfrac{(f_0 - f_e)^2}{f_e}$
0	18	20	−2	4	4/20
1	19	20	−1	1	1/20
2	23	20	3	9	9/20
3	21	20	1	1	1/20
4	16	20	−4	16	16/20
5	25	20	5	25	25/20
6	22	20	2	4	4/20
7	20	20	0	0	0/20
8	21	20	1	1	1/20
9	25	20	5	25	25/20
	200			86	$\dfrac{86}{20} = 4.30$

Degrees of freedom $\nu = n - h = 10 - 1 = 9$

$$\chi^2_{0.05} = 16.919 \text{ (given)}$$

Since $\chi^2_{cal} = (= 4.30)$ is less than $\chi^2_{tab} = (= 16.919)$, H_0 is accepted.

EXAMPLE 7. *Frequency distribution of the Earthworm length in cm. of 10000 sample is given below. Fit the normal distribution and test the goodness of fit.*

Length in cm.	60-65	65-70	70-75	75-80	80-85	85-90	90-95	95-100
Frequency	3	21	150	355	326	135	26	4

SOLUTION. Prepare a table of 7 columns and calculate mean and standard deviation.

Length (cm) Classes	Mid Point X	Frequency f	$f \cdot x$	$m - \bar{X} = x$	Σfx	fx^2
60-65	62.5	3	187.5	− 17.44	52.32	912.46
65-70	67.5	21	1417.5	− 12.44	261.24	3249.82
70-75	72.5	150	10875	− 7.44	142.5	1060.2
75-80	77.5	335	25962.5	− 2	670	1340.0
80-85	82.5	32	26895	+ 2.56	834.56	2136.47
85-90	87.5	135	11812.5	+ 7.56	1020.6	7715.73
90-95	92.5	26	2405	+ 12.56	326.56	4101.59
95-100	97.5	4	390	+ 17.55	70.2	1232.01
		1000	$\Sigma f \cdot x = 79945$		$\Sigma fx = 3377.98$	$\Sigma fx^2 = 28991.12$

$$\bar{X} = \frac{\Sigma f \cdot X}{\Sigma f} = \frac{79945}{1000} = 79.95$$

Standard deviation $(\sigma) = \sqrt{\dfrac{\Sigma f x^2}{\Sigma f}} = \sqrt{\dfrac{28991.2}{1000}} = \sqrt{28.9} = 5.38$

The formula to calculated normal deviate

$$Z = \frac{X - \bar{X}}{\sigma}$$

For expected frequency

Score	Upper Class Limit	Observed Frequency	$\dfrac{X - \bar{X}}{\sigma}$	Table Value of Z	Expected Relative Frequency	Expected Frequency
X	L_1	f	Z	Table Z	Exp. Rel. Freq.	Exp. Frequency
60-65	65	3	− 2.74	0.0037	0.0037	0.037 × 1000 = 3.70
65-70	70	21	− 1.83	0.0336	0.0299	29.90
70-75	75	150	− 0.91	0.1814	0.1478	147.8
75-80	80	335	− 0.05	0.5000	0.3189	318.6
80-85	85	326	+ 0.93	0.1762	0.3258	326.8
85-90	90	135	+ 1.84	0.0329	0.1433	143.3
90-95	95	26	+ 2.76	0.0029	0.300	30.00
95-100	100	4	+ 3.68	0.001	0.0028	2.80
		1000			0.0021	0.10

Calculation of χ^2 value

X	f_0	f_e	$f_0 - f_e$	$(f_0 - f_e)^2$	$X^2 = \dfrac{(f_0 - f_e)^2}{f_e}$
60-65	3	3.70	− 0.7	0.49	0.132
65-70	21	29.90	− 8.9	79.21	2.649
70-75	150	147.80	+ 2.2	4.84	0.033
75-80	335	318.60	+ 16.4	268.96	0.844
80-85	326	323.80	+ 2.20	4.84	0.015
85-90	135	143.30	− 8.30	68.89	0.481
90-95	26	30.00	− 4.0	16.00	0.533
95-100	4	2.80	+ 1.2	1.44	0.514
					$X^2 = 5.697$

Conclusion: $d \cdot f = 9 - 2 = 7$

The calculated value of χ^2 comes to 5.697, which is less than the table (14.07) at 7 d.f. for 5% level of significance. Therefore, the length of earthworm is not in the normal distribution.

EXAMPLE 8. *The following table gives the number of aircraft accidents that occurred during various days of the week. Test whether the accidents are uniformly distributed over the week.*

Days	Mon.	Tue.	Wed.	Thu.	Fri.	Sat.
No. of Accedents	14	18	12	11	15	14

Given $\chi_5^2(0.5) = 11.07$

[CCSU(B.Sc. Biotech)–1997]

SOLUTION. **Null hypothesis:** Digits were distributed in equal number in the table, *i.e.*,

$$p = \frac{1}{6}.$$

Expected frequency for each digit $= NP = \dfrac{1}{6} \times 84 = 14$

Computation of χ^2

S.No.	Observed Frequency (f_0)	Expected Frequency (f_e)	$f_0 - f_e$	$(f_0 - f_e)^2$	$\dfrac{(f_0 - f_e)^2}{f_e}$
1	14	14	0	0	0.0000
2	18	14	4	16	1.1429
3	12	14	– 2	4	0.2857
4	11	14	– 3	9	0.6429
5	15	14	1	1	0.0714
6	14	14	0	0	0.0000
Total	**84**	**84**			**2.1429**

Now,
$$\chi^2 = \Sigma \left[\frac{(f_0 - f_e)^2}{f_e} \right] = 2.1429$$

Degrees of freedom $\nu = n - h = 6 - 1 = 5$

$$\chi_{0.05}^2 = 11.07$$

Since $\chi_{cal}^2 = (2.1429)$ is less than $\chi_{tab}^2 = (11.07)$,

H_0 is accepted.

25.5 χ^2–TEST FOR INDEPENDENCE OF ATTRIBUTES

If N individuals or items are classified according two attributes A and B. There are m classification $A_1, A_2, ..., A_i, ..., A_m$ in A and n classifications $B_1, B_2, ..., B_j, ..., B_n$ in B. Then we have a two-way table known as contingency table such as :

B \ A	B_1	B_2	...	B_j	...	B_n	
A_1	O_{11}	O_{12}	...	O_{1j}	...	O_{1n}	R_1
A_2	O_{21}	O_{22}	...	O_{2j}	...	O_{2n}	R_2
⋮	⋮	⋮	⋮	⋮	⋮	⋮	⋮
A_i	O_{i1}	O_{i2}	...	O_{ij}	...	O_{in}	R_i
⋮	⋮	⋮	⋮	⋮	⋮	⋮	⋮
A_m	O_{m1}	O_{m2}	...	O_{mj}	...	O_{mn}	R_m
	C_1	C_2	...	C_j	...	C_n	N

where O_{ij} represent the number of individual or item (observed frequency) belonging to A_i and B_i, $(i = 1, 2, ..., m, \ \ j = 1, 2, ..., n)$

R_i : Total of ith row

C_j : Total of jth column

Here

$$\sum_{i=1}^{m} R_i = N = \sum_{j=1}^{n} C_j$$

Null hypothesis (H_0). The attributes are independent.
There is no association between the attributes under study.
Alternative hypothesis (H_1) : H_0 is not true.
Test statistic :

$$\chi^2_{(m-1) \times (n-1)} = \sum_{i=1}^{m} \sum_{j=1}^{n} \frac{(O_{ij} - E_{ij})^2}{E_{ij}}$$

where E_{ij} represent the expected frequency corresponding to O_{ij} and calculated as

$$E_{ij} = \frac{R_i \times C_j}{N} = \frac{\text{Total of row in which it occurs} \times \text{Total of column in which it occurs}}{\text{Total number of observation}}$$

Conclusion: If calculated value of χ^2 is greater than or equal to tabulated value of $\alpha\%$ level of significance then null hypothesis (H_0), is rejected otherwise null hypothesis (H_0) may be accepted.

SOLVED EXAMPLES

EXAMPLE 1. *Two sample polls of votes for two candidates A and B for a public office are taken, one from among the residents of the rural areas. The results are given in the adjoining table. Examine*

Area	Votes for		Total
	A	**B**	
Rural	620	380	1000
Urban	550	450	1000
Total	1170	830	2000

whether the nature of the area is related to voting performance in this elections.

SOLUTION. H_0 : The nature of the area is independent of the voting preference in the election.

H_1 : H_0 is not true.

Expected frequencies :

$$E_{(620)} = E_{11} = \frac{R_1 \times C_1}{N} = \frac{1000 \times 1170}{2000} = 585,$$

$$E_{(380)} = E_{12} = \frac{R_1 \times C_2}{N} = \frac{1000 \times 830}{2000} = 415$$

$$E_{(550)} = E_{21} = \frac{R_2 \times C_1}{N} = \frac{1000 \times 1170}{2000} = 585,$$

$$E_{(450)} = E_{22} = \frac{R_2 \times C_2}{N} = \frac{1000 \times 830}{2000} = 415$$

Test statistic :

$$\chi^2 = \sum_{i=1}^{2} \sum_{j=1}^{2} \frac{(O_{ij} - E_{ij})^2}{E_{ij}}$$

$$= \frac{(620 - 585)^2}{585} + \frac{(380 - 415)^2}{415} + \frac{(550 - 585)^2}{585} + \frac{(450 - 415)^2}{415}$$

$$= 10.089$$

Tabulated value for $(2-1) \times (2-1) = 1$ degrees of freedom is 3.841.

Conclusion: Since calculated χ^2 is greater than at 5% level of significance tabulate value so null hypothesis is at rejected at α% level of significance *i.e.,* nature of area is related to voting preference in the election.

EXAMPLE 2. *A random sample of 300 persons are classified as under*

	Male	Female	Total
Smokers	40	60	100
Non-smokers	110	90	200
Total	150	150	300

It there a relationship between sex and smoking?

SOLUTION. **Null hypothesis (H_0):** There is no relationship between sex and smoking or sex and smoking are independent.

Alternative hypothesis (H_1): H_0 is not true.

Expected frequencies :

$$E_{(H_0)} = E_{11} = \frac{R_1 \times C_1}{N} = \frac{100 \times 150}{300} = 50,$$

$$E_{(60)} = E_{12} = \frac{R_1 \times C_2}{N} = \frac{100 \times 150}{300} = 50$$

$$E_{(110)} = E_{21} = \frac{R_2 \times C_1}{N} = \frac{150 \times 200}{300} = 100,$$

$$E_{(90)} = E_{22} = \frac{R_2 \times C_2}{N} = \frac{200 \times 150}{300} = 100$$

Test statistic :

$$\chi^2 = \sum_{i=1}^{2} \sum_{j=1}^{2} \frac{(O_{ij} - E_{ij})^2}{E_{ij}}$$

$$= \frac{(40-50)^2}{50} + \frac{(60-50)^2}{50} + \frac{(110-100)^2}{100} + \frac{(90-100)^2}{100} = 2+2+1+1 = 6$$

Tabulated value of χ^2 for $(2-1) + (2-1) = 1$ degrees of freedom at 5% level of significance = 4.841.

Conclusion: Since χ_{cal}^2 calculate χ^2 is greater than tabulated value so null hypothesis is rejected at α% level of significance, i.e., sex and smoking are not independent.

EXAMPLE 3. *In experiments on pea-bruding, Mendel obtained the following frequencies of seeds*

Round and yellow	Wrinkled and yellow	Round and green	Wrinkled and green	Total
315	101	108	32	556

Theory predicts that the frequencies should be in proportions 9 : 3 : 3 : 1. Examine the correspondence between theory and experiment. Given that the value of ψ^3 for 3 degree of freedom at 5% level and significance is 7.815.

SOLUTION. **Null Hypothesis (H_0):** Theory fits well into the experiment.

Alternative Hypothesis (H_1) : H_0 is not true.

Sum of proportion $= 9 + 3 + 3 + 1 = 16$

So expected frequency of round and yellow (E_1)

$$= \frac{\text{proportion}}{\text{sum of proportion}} \times N = \frac{9}{16} \times 556 = 313 \text{ (approx.)}$$

Expected frequency at wrinkled and yellow (E_2)

$$= \frac{3}{16} \times 556 = 104 = 104 \text{ (approx.)}$$

Expected frequency of round and green seeds in (E_3)

$$= \frac{3}{16} \times 556 = 104 \text{ (approx.)}$$

Expected frequency of wrinkled and green seeds is

$$= \frac{1}{16} \times 556 = 35 \text{ (approx.)}$$

					Total
Observed Frequency	315	101	108	32	556
Expected Frequency	313	104	104	35	556

Test Statistic :

$$\chi^2 = \sum_{i=1}^{2} \frac{(O_j - E_1)^2}{E_i} = \frac{(315-313)^2}{313} + \frac{(101-104)^2}{104} + \frac{(108-104)^2}{104} + \frac{(32-35)^2}{35}$$

$$= 0.013 + 0.09 + 0.15 + 0.26 = 0.513$$

$\chi^2_{tab} = 7.815$ for 3 d.f. at 5% level of significance.

Conclusion : Since $\chi^2_{cal} = \chi^2_{tab}$ at 5% level of significance so hypothesis may be accepted. Hence, there is much correspondence between theory and experiment.

EXAMPLE 4. *The following table gives the classification of 200 fishes according to the sex and the helminth infection. Test whether infection is independent of the sex of the fish.*

Sex	Infected	Uninfected
Males	65	45
Females	35	55

SOLUTION. Prepare table of observed frequency in column and row wise.

Observed frequency

	Column 1	Column 2	Total
Row 1	65	45	110
Row 2	35	55	90
Total	100	100	200

Expected frequency $= \dfrac{RT \times CT}{N}$

E of cell (1) $= \dfrac{100 \times 110}{200} = 55$

E of cell (2) $= \dfrac{100 \times 110}{200} = 55$

E of cell (3) $= \dfrac{100 \times 90}{200} = 45$

E of cell (4) $= \dfrac{100 \times 90}{200} = 45$

Make a table of expected frequency

55	55	110
45	45	90
100	**100**	**200**

Applying the χ^2 test

Observed (O)	Expected (E)	$(O - E)^2$	$(O - E)^2/E$
65	55	100	$\dfrac{100}{55} = 1.81$
35	45	100	$\dfrac{100}{45} = 2.22$
45	55	100	$\dfrac{100}{55} = 1.81$
55	45	100	$\dfrac{\Sigma(O - E)^2}{E} = 8.06$

Degree of freedom = (No. of rows – 1)(No. of columns – 1) = (2 – 1)(2 – 1) = 1.

For degree of freedom = 1, χ^2 at 5% level = 3.84 (table value). The calculated value (8.06) is much greater than table value (3.84). Hence, the result of the experiment support the hypothesis, *i.e.*, infection is independent of sex.

EXAMPLE 5. *The following table shows the result of inoculation against cholera*

	Not Attacked	**Attacked**
Inoculated	431	5
Non-inoculated	291	9

Is there any significant association between inoculation and attack? Given that

$$x = 1 \, ; \quad \begin{cases} P = 0.047, & for \ \chi^2 = 3.2 \\ P = 0.069, & for \ \chi^2 = 3.3 \end{cases}$$

SOLUTION. Let us find theoretical frequencies and arrange in the following order :

	Not Attacked	**Attacked**	**Total**
Inoculated	431(427.7)	5(8.3)	436
Non-inoculated	291(294.3)	9(5.7)	300
Total	722	14	736

The theoretical frequencies have been shown in brackets.
The first theoretical frequency has been obtained

$$= \frac{436}{736} \times 722 = 427.7$$

$$\chi^2 = (3.3)^2 \left[\frac{1}{427.7} + \frac{1}{8.3} + \frac{1}{294.3} + \frac{1}{5.7} \right] = 3.28$$

Number of degree of freedom
$$= (2-1)(2-1) = 1$$

Since $\chi^2 = 3.2$ corresponds to $P = 0.074$ and $\chi^2 = 3.3$ corresponds to $P = 0.069$ hence by interpolation
$\chi^2 = 3.28$ corresponds to $P = 0.0706$ approximately.

Thus, if hypothesis is true, our data gives results which would be obtained about 7 times in hundred trails. This is infrequent but not very infrequent. We may be unjustified in rejecting the hypothesis but this leads us somewhat to believed that in osculation and attack are associated.

EXAMPLE 6. *In a cross between black male and gray female Drosophilae the offspring obtained were 25 black and 35 gray. Calculate the χ^2 and give your inference on the ratio of black and gray offspring. Expected number is calculated from the fact that gray body colour is dominant and the expected ratio of the nature is 1 : 1. [Total number of offspring over 60].*

SOLUTION. **Null hypothesis (H_0) :** It is presumed that the black and gray segregate in 3 : 1 ratio.

	Black	**Gray**
Observed number	25	35
Expected number	30	30

Deviation $(O - E)$	$25 - 30 = -5$	$35 - 30 = +5$
Deviation^2 $(O - E)^2$	$(25 - 30)^2 = 25$	$(35 - 30)^2 = 25$

Putting values in the formula

$$\chi^2 = \frac{\Sigma(f_0 - f_e)^2}{f_e} = \frac{25}{30} + \frac{25}{30} = \frac{5}{6} + \frac{5}{6} = 1.66$$

The table value of χ^2 at d.f. 1, $P = 0.05$ is 3.84. The obtained value 1.66 at same d.f. and P. This indicates that the observed ratio of black and gray Drosophilae is not in agreement with the theoretical ratio of 1 : 1.

EXAMPLE 7. *In a grassland, the earthworm population was sampled from ten randomly located plots of 1 m^2 area. The following table gives the number of earthworms obtained. Examine the distribution pattern of earthworms.*

Area	1	2	3	4	5	6	7	8	9	10
No. of earthworms/m^2	25	32	17	23	15	39	27	19	22	26

SOLUTION. **Null hypothesis (H_0) :** It is presumed that the earthworm population in ground is normally distributed.

$$\Sigma X = 245, \quad \overline{X} = \frac{\Sigma X}{N} = \frac{245}{10} = 245$$

The expected number of earthworms is determined taking into consideration that the population is equally distributed in all quadrates. Thus the expected number of earthworms in each quadrates is the mean number of earthworms. Following table gives the summary of the results :

Observed	25	32	17	23	15	39	27	19	22	26
Expected	24.5	24.5	24.5	24.5	24.5	24.5	24.5	24.5	24.5	24.5
Difference	0.5	7.5	−7.5	−1.5	−9.5	14.5	2.5	−5.5	−2.5	1.5

$$\chi^2 = \frac{(0.5)^2}{24.5} + \frac{(7.5)^2}{24.5} + \frac{(-7.5)^2}{24.5} + \frac{(-1.5)^2}{24.5} + \frac{(-9.5)^2}{24.5} + \frac{(14.5)^2}{24.5} + \frac{(2.5)^2}{24.5} + \frac{(-5.5)^2}{24.5}$$
$$+ \frac{(-2.5)^2}{24.5} + \frac{(1.5)^2}{24.5}$$

$$= 0.01 + 2.29 + 2.29 + 0.09 + 3.6 + 8.58 + 0.25 + 1.23 + 0.25 + 0.09$$
$$= 18.68$$

Here df $= 10 - 1 = 9$.

Significance: The table value of χ^2 at df 9 on $P = 0.05$ is 16.92 while obtained value in 18.68 which is much higher. This shows that the two series of frequencies, observed and expected are different, indicating that the population of earthworms is not distributed normally.

EXAMPLE 8. *Mendel self fertilized pea plants with round and yellow peas. In the next generation he recovered the following number of peas :*

315 round and yellow peas, 108 round and green peas, 101 wrinkled and yellow peas, 32 wrinkled and green peas.

What is your hypothesis about the genetic control of the phenotype? Do the data support this hypothesis?

SOLUTION. (i) Hypothesis about the genetic control, *i.e.*,
Null hypothesis = 9 : 3 : 3 : 1.

Observed (O)	Expected (E)	O – E	$(O-E)^2$	$(O-E)^2/2$
315	$\dfrac{9}{10} \times 556 = 312.75$	$315 - 312.75$ $= 2.25$	5.06	$\dfrac{5.06}{312.75} = 0.016$
108	$\dfrac{3}{16} \times 556 = 104.25$	$108 - 104.25$ $= 3.75$	14.06	$\dfrac{14.06}{104.25} = 0.135$
101	$\dfrac{3}{16} \times 556 = 104.25$	$101 - 104.2$ $= -3.25$	10.56	$\dfrac{10.56}{104.75} = 0.101$
32	$\dfrac{1}{16} \times 556 = 34.56$	$32 - 34.75$ $= -2.75$	7.56	$\dfrac{7.56}{34.75} = 0.218$
Total = 556				$\chi^2 = 0.470$

Degree of freedom = 3.

Critical Value: The control value of chi-square at 0.05 for 3 df is 7.82.

Decision: Since calculated value of $\chi 2$ at 3 d.f. is 0.470 it is less than the critical chi-square value, i.e., $0.470 < \chi^2_{0.05}, 3 = 7.82$ so the null hypothesis is accepted.

EXAMPLE 9.   *A survey of 320 families with 5 children shows the following distribution*

No. of boys and girls	5 Boys and 0 Girl	4 Boys and 1 Girl	3 Boys and 2 Girls	2 Boys and 3 Girls	1 Boy and 4 Girls	0 Boy and 5 Girls	Total
No. of families	18	56	110	88	40	8	320

Given that value of χ^2 for 5 df are 11.0 and 15.1 at 0.05 and 0.01 significance level respectively, test the hypothesis that male and female births are equally probable.

SOLUTION.   **Null Hypothesis (H_0):** Male and female births are equally probable.

Alternative Hypothesis (H_1) : H_0 is not true.

The probability of male births $(p) = 1/2$ and female births $(q) = 1/2$.

So, the expected frequency of r male births in a family of 5 out of 320 families

$$= N \,{}^5C_r \, p^r \, q^{5-r} \qquad \text{(From binomial distribution)}$$

$$= 320 \,{}^5C_r \, (1/2)^r \, (1/2) 5 . r$$

$$= 320 \,{}^5C_r \, (1/2)^5 = 10 \,{}^5C_r$$

So the expected frequency of 5 boys and 6 girls

$$E_1 = 10 \times {}^5C_5 = 10$$

The expected frequency of 4 boys and 1 girl

$$E_2 = 10 \times {}^5C_4 = 50$$

The expected frequency of 3 boys and 2 girls

$$E_3 = 10 \times {}^5C_3 = 100$$

The expected frequency of 2 boys and 3 girls, $E_4 = 10 \times {}^5C_2 = 100$

The expected frequency of 1 boy and 4 girls, $E_5 = 10 \times {}^5C_1 = 50$

The expected frequency of 0 boy and r girls, $E_{16} = 10 \times {}^5C_0 = 10$

So

							Total
Observed Frequency	18	56	110	88	40	8	320
Expected Frequency	10	50	100	100	50	10	320

Test Statistic :

$$\chi^2 = \sum_{i=1}^{2} \frac{(O_i - E_i)^2}{E_i}$$

$$= \frac{(18-10)^2}{10} + \frac{(56-50)^2}{50} + \frac{(110-100)^2}{100} + \frac{(88-100)^2}{100} + \frac{(40-50)^2}{50}$$

$$+ \frac{(8-10)^2}{10}$$

$$= 6.4 + 0.72 + 1 + 1.44 + 2 + 0.4 = 11.96.$$

Degrees of freedom

$$= (n-1) = (6-1) = 5 \text{ given } \chi^2_{tab} \text{ at 5\% level of significance} = 11.0.$$

Conclusion: (i) Since $\chi^2_{cal} = \chi^2_{tab}$ at 5% level of significance so null hypothesis is rejected at 5% level of significance, *i.e.*, boys and girls births are not equally probable.

Given ψ_{tab} at 1 % of significance = 15.1.

(ii) Since $\psi^2_{cal} < \psi^2_{tab}$ at 1 % level of significance so null hypothesis may be accepted at 1 % level of significance, *i.e.*, boys and girls births are equally probable.

Exercise 25.1

1. Find the value of χ^2 for the following data :

Class	A	B
Observed frequency	8	29
Expected frequency	7	24

2. Find the value of χ^2 for the following data :

Observed frequency	10	4	15	18	20
Expected frequency	10	7	10	15	25
Observed frequency	15	5	2	3	
Expected frequency	10	5	5	5	

3. In 120 throws of a single die the following distribution of faces was obtained:

Faces	1	2	3	4	5	6	Total
Frequency	30	25	18	10	22	15	120

Find χ^2 on the basis that the die was unbiased.

4. 200 digits were chosen at random from a set of tables. The frequencies of the digits were:

Digit	0	1	2	3	4
Frequency	18	19	23	21	16
Digit	5	6	7	8	9
Frequency	25	22	20	21	15

Use the chi-square test to assess the correctness of the hypothesis that the digits were distributed in the equal number in the table from which these were chosen (χ^2 for 9 d.f. at 5% level = 16.919).

5. Applying the χ^2 test of goodness of fit to the following data :

Observed frequency	:	1	5	20	28	42	22	15	5	2
Expected frequency	:	1	6	18	22	40	25	18	6	1

(χ^2 for 6 d.f. at 5% level = 12.59)

6. The theory predicts the proportion of beans in the four groups A, B, C and D should be 9 : 3 : 3 : 1. In an experiment among 1600 beans, the numbers in the four groups were 882, 213, 287 and 118. Does the experimental result support the theory (χ^2 for 3 d.f. at 5% level = 7.815).

7. In an experiment on immunization of cattle from tuberculosis, the following results were obtained:

	Affected	Unaffected
Inoculated	12	28
Non-inoculated	13	7

Examine the effect of vaccine in controlling the incidence of the disease.

8. The following data is collected on two characters :

	Cinegores	Non-cinegores
Inoculated	83	57
Non-inoculated	45	68

Based on this, can you conclude that there is no relation between the habit of cinema going and literacy ?

9. In a locality 100 persons are randomly selected and asked about their educational attainments. The results are as under:

	Education			
	Middle	High School	College	Total
Male	10	15	25	50
Female	25	10	15	50
Total	35	25	35	100

Does education depend on sex

($\chi^2_{0.045}(2)5.99, \psi^2_{0.01}(2) = 9.21$).

10. In a certain sample of 2,000 families : 1,400 families are consumers of tea, out of 1,800 Hindu families, 1,236 families consume tea. Use χ^2-test and state whether there is any significant between consumption of tea among Hindu and non-Hindu families.

11. 1072 college students were classified according to their intelligence and economic conditions. Test whether there is any associated between intelligence and economic conditions :

		Excellent	Good	Mediocre	Dull
Economic Conditions	Good	48	199	181	82
	Not Good	81	185	190	106

12. A chemical extraction plant processes sea-water to collect sodium chloride and magnesium. It is known that sea-water contains sodium chloride, magnesium and other elements in the ratio 62 : 4 : 34. A sample of 200 tonnes of sea-water has resulted in 130 tonnes of sodium chloride and 6 tonnes of magnesium. Are these data consist with the known composition of sea-water at 5% level of significance?

(Given that the critical value of $\chi^2 = 5.991$) for two degree of freedom) **(UPTU–2003)**

ANSWERS

1. $\chi^2 = 6.76$ **2.** $\chi^2 = 8.80$ **3.** $\chi^2 = 85.81$ **4.** $\chi^2 = 4.3$ accepted

5. $\chi^2 = 12.59$ fit is good **6.** $\chi^2 = 4.7266$ support theory **7.** $\chi^2 = 13.89$

12. 5.991, accepted

□□□□□□

GLOSSARY

Absolute Error : If x^A is the approximate value of exact number x^T, then the absolute error denoted by E_a is defined by

$$E_a = \Delta x = |x^T - x^A|$$

Analysis of Variance : Analysis of variance is the separation of variation ascribable to one group of causes from the ascribable to other group.

Antilogarithm : If $\log x = n$, then x is called the antilogarithm of n and is written as

$$x = \text{antilog } (n)$$

Approximate Numbers : These are numbers which are not exact.

Arithmetic Mean : Arithmetic mean is the amount obtained by dividing the sum of values of the items in a series by their number.

Bimodal Frequency Curve : A curve having two maximum point is known by bimodal frequency curve.

Binomial Distribution : The probability of r successes in a series of n trials is given by $^nC_r p^r q^{n-r}$, where r takes any integral value for 0 to n. The probabilities of 0, 1, 2, ..., r,..., n successes are given by

$$q^n, \ ^nC_1 pq^{n-1}, \ ^nC_2 p^2 q^{n-2}, \ ..., \ ^nC_r p^r q^{n-r}, ..., p^n.$$

Then the probability of the number of successes so obtained is called the binomial distribution.

Binomial Expression : An algebraic expression which contains only two terms is called a binomial expression.

Bivariate Data : If for each unit of observation, we record the values of two variables, then the observation forms a bivariate data.

Cardinal Number of a set : The number of distinct elements contained in a finite set A is called cardinal number of A and is denoted by $n(A)$.

Cartesian Product of Two Sets : The set of all ordered pairs of elements (a, b), $a \in A$, $b \in B$ is called the cartesian product of two sets A and B. It is denoted by $A \times B$, i.e., $A \times B = \{(a, b) : a \in A, b \in B\}$

Cartogram : These are known as the map-diagram. It is the geographical representation of the given distribution.

Class Size : The difference between any two consecutive class marks is known as class size.

Collinear Points : Three points A, B, C are said to be collinear if they lie on the same straight line.

Column Matrix or Column Vector : A matrix having only one column is called a column matrix or column vector.

Combined Mean : If $\bar{x}_1$ and $\bar{x}_2$ are the mean of two groups of sizes n_1 and n_2, then the mean $\bar{x}$ is the mean of two groups, given by

$$\bar{x} = \frac{n_1 \bar{x}_1 + n_2 \bar{x}_2}{n_1 + n_2}$$

Comparable Matrices : Two matrices A and B are said to be comparable if they are of same order.

Complimentary Events : If A denotes an event, then $\bar{A}$ denote an event which includes all the sample points not included in A.

Compound Event : When two or more events occur in relation with each other, they are known as compound events.

Congruence Modulo 'm' : Let m be an arbitrary but fixed integer. If $x - y$ is divisible by m, then two integers x and y are said to be congruence modulo m of one another.

i.e., $\qquad\qquad\qquad x \equiv y \pmod{m}$ is $x - y$ divisible by m.

Correlation : When the relationship is of quantitative nature, the appropriate statistical tool for discovering and measuring the relationship and expressing it in brief formula is known as correlation.

Cumulative Frequency : The cumulative frequency of a class interval is the sum of the frequencies upto and including that class.

Decile and Percentile : In an ascending or descending series, if we divide the complete series into ten equal parts, then each part of the series is known as decile and if we divide the whole series into 100 equal parts, then each part of the series is known as percentile or centile.

Definite Integral : If $f(x)$ is a continuous and non-negative function over a closed interval $[a, b]$, then $\int_a^b f(x)\,dx$ is called the definite integral of $f(x)$ between the limits a and b $(b > a)$.

Degree of a Differential Equation : The degree of a differential equation is the degree of the highest order differential coefficient appearing in it.

Degree of an Equation : The degree of an equation is the index of the highest power of variable quantity involved in the equation, when the equation has been expressed to the rational integral form (radical free form).

Derivative of A function : Derivative of a function $f(x)$ is the limiting value of $\dfrac{\delta y}{\delta x}$ as $\delta x \to 0$ provided the limit exist finitely and it is denoted by $\dfrac{dy}{dx}$.

Diagonal Matrix : A square matrix which has all its elements are zero except the diagonal elements, is said to be diagonal matrix.

Difference of Two Sets : If A and B are two sets, then the set of all elements which belong to A but do not belong to B is called the difference of sets A and B and is denoted by $A \sim B$. The set of all elements which belong to B but do not belong to A is called the difference of sets B and A and is denoted by $B \sim A$.

Therefore, $\qquad\qquad A \sim B = \{x : x \in A \text{ and } x \notin B\} = A \cap B'$

and $\qquad\qquad\qquad B \sim A = \{x : x \notin A \text{ and } x \in B\} = B \cap A'$

Differential Equations : An equation is said to be differential equation if it contains not only the dependent variable and independent variables but their differential coefficients also.

Disjoint Sets : When two sets have no common elements, they are called disjoint sets. Thus, if $A \cap B = \phi$, then A and B are disjoint.

Elementary Matrix : Elementary matrix is obtained on applying only one elementary operation on identity matrix.

Empty Set : A set containing no elements is called empty set and is denoted by the symbol ϕ.

Equal Sets : Two sets are said to be equal if they contain exactly the same elements.

Equality of Matrices : Two matrices A and B are said to be equal (*i.e.,* $A = B$) if they are of same order and their corresponding elements are equal.

Equally Likely Events : When there is no reason to except the happening of one event in preference to the other, then events are known as equally likely events.

Equivalence Relations : A relation R on a set E is said to be equivalence if it is

  (i) Reflexive,   (ii) Symmetric   and   (iii) Transitive

Equivalent Matrices : Two matrices A and B are said to be equivalent if one is obtained from other by applying finite no. of elementary operations. It is written as $A \sim B$.

Equivalent Sets : Two finite sets are said to be equivalent if they have the same cardinal number.

Exact Numbers : The numbers in which, there is no uncertainity and no approximation, is said to be exact numbers.

Exhaustive Events : The total number of possible outcomes in any trial is known as exhaustive events.

Favourable Events : The outcomes which make necessary the happening of an event in a trial are called favourable events.

Finite Population : A population containing a finite number of individuals is called a finite population.

Finite Sets : A set is said to be finite if it consists of only finite number of elements.

Functions : Let A and B be two sets, then the rule or corresponding, which associates each element of A to a unique element to B, is called a function from set A to set B.

Geometric Mean : Let $x_1, x_2, x_3, ..., x_n$ be the n variates of a variable x, then the geometric mean G of n variables is defined by

$$G = (x_1 \cdot x_2 \cdot x_3, ..., x_n)^{1/n}$$

Half-life Period : The half-life period T of C^{14} gives the value of disintegration constant k

from the equation $k = \dfrac{0.693}{T}$. The disintegration equation is $k = \dfrac{2.303}{t} \log \dfrac{a}{a-x}$.

Harmonic Mean : The harmonic mean of any series is given by the reciprocal of the arithmetic mean of the reciprocals of the variables.

Histogram : These graphs represent the time series. In these types of graphs, the data is plotted against the corresponding period.

Homogeneous Differential Equation : A differential equation of the form $\dfrac{dy}{dx} = \dfrac{f(x,y)}{g(x,y)}$

is said to be homogeneous if $f(x, y)$ and $g(x, y)$ are homogenous function of same degree.

Horizontal Matrix : A matrix of order $m \times n$ is called a horizontal matrix if $m < n$, *i.e.,* if number of rows is less than the number of columns.

Identity : An expression involving equality and a variable is called an identity, if it is satisfied by every value of the variable.

Independent Events : Two events are said to be independent if the occurence of one does not depend upon the occurence of the other.

Infinite Population : A population containing an infinite number of individuals is called an infinite population.

Infinite Sets : A set which is not finite, i.e., it contains infinite number of elements.

Integers : The numbers ... , –3, –2, –1, 0, 1, 2, 3, ... are called integers. We represent the set of integers by **Z**.

i.e., $\qquad\qquad$ **Z** = { ... –3, –2, –1, 0,1,2,3,.. }

Intersection of Two Sets : Let A and B be two sets. Then intersection of A and B, denoted by $A \cap B$ is the set of all those elements, which belongs to both A and B.

$$A \cap B = \{x : x \in A \text{ and } x \in B\}$$

Interval : A subset S of $\mathbf{R}$ is called an interval if $a, b \in S, x \in \mathbf{R}$ such that $a < x < b$ implies $x \in S$.

Into Function : A function $f : A \to B$ is called an into function, *i.e.,* if there is at least one element of set B which has no pre-image in the set A.

Inverse of a Matrix : Inverse of a square matrix A is a square matrix which gives identity matrix on multiplying with A.

Irrational Numbers : Any number which is not rational, is called an irrational number.

Left Hand Limit : A function $f(x)$ is said to approaches the limit l as x approaches a from the left, if corresponding to an arbitrary positive number ε there exists a positive number $\delta > 0$ such that

$$|f(x) - l| < \varepsilon \text{ whenever } |x - a| < \delta$$

Limit of A Sequence : Let $S_1, S_2, ...$ be the given sequence. Then a number A is said to be the limit of this sequence if the difference between S_n and A can be made as small as we please by making n large. Symbolically, it can be written as $\lim_{n \to \infty} |S_n| = A$.

Line Graph : Line graph is the graph in which we mark a point corresponding to the given data and then draw a line by joining them.

Linear Differential Equations : A differential equation of the form

$$\frac{dy}{dx} + Py = Q$$

where P and Q are function of x only or constant is called linear differential equation.

Linear Equation : An equation of the form $ax + b = 0$ is called a linear equation of x, where x is unknown variable (quantity) and a and b are any constants.

Locus : The curve described by a point which moves under given condition or conditions is called its locus.

Logarithms : Let there be a number $a > 0$ and $a \neq 1$. A number x is called the logarithm of another variable $y > 0$ to the base a if $a^x = y$.

$\therefore$ $\qquad\qquad$ $a^x = y \Leftrightarrow x = \log_a y.$

Lower Triangular Matrix : A square matrix $A = [a_{ij}]$, is called a lower triangular matrix if $a_{ij} = 0$ for all $i < j$. Therefore, a square matrix is said to be lower triangular matrix if all the elements above the main diagonal are equal to zero.

Magnitude of the Class Interval : Having fixed the number of classes, divide the range by it and nearest integer to this value gives the magnitude of the class interval.

Many-One Function : A function $f : A \to B$ is called many-one, if at least one element of co-domain B has two or more than two pre-images in domain A.

Many-One Into Function : A function $f : A \to B$ which is both many-one and into function is called a many-one into function, *i.e.,* two or more points in A are joined to some points in B and there are some point in B which are not joined to any point in A.

Many-One Onto Function : If function $f : A \rightarrow B$ is both many-one and onto function is called a many one onto function, *i.e.*, in B one point is joined to at least one point in A and two or more points in A are joined to some points in B.

Matrix : A set of mn numbers $a_{11}, a_{12},..., a_{mn}$ arranged in a rectangular array of m rows and n columns is called a matrix of order $m \times n$. Generally it is denoted by [ ] or ( ) or ‖ ‖.

Median : If we arrange the whole data in ascending or descending order, then the value of the middle variable is known as median.

Mode : The variable whose frequency is maximum, is known as the mode of the distribution.

Multimodal Frequency Curve : A curve having more than two maximum point is known by multimodal frequency curve.

Multivariate Data : If for each unit of observation, we record the values of more than one variable, then this observation forms a multivariate data.

Mutually Exclusive Events : The events are said to be mutually exclusive if the happening of any one of them precludes the happening of all others, *i.e.*, if no two or more of them can happen simultaneously in the same trial.

Natural Number : The numbers 1, 2, 3, ... are called natural numbers. We represent the set of natural numbers by **N**.

i.e., $\qquad\qquad N = \{1,2,3, ...\}$

Non-Singular Square Matrix : A square matrix is said to be non-singular if value of its determinant is non-zero, *i.e.*, if $|A| \neq 0$ then A is non-singular.

Normal Distribution : A random variable X is said to follow a normal distribution with parameter μ (mean) and σ^2 (variance) if its probability distribution $f(x)$ is given by

$$f(x) = \frac{1}{\sigma\sqrt{2\pi}} e^{-(x-\mu)^2/2\sigma^2}, -\infty < x < \infty$$

Null Function : A function $N(t)$ of t such that $\int_0^t N(t)dt = 0, \forall t > 0$ is called null function.

One-One Function : A function f from A to B, *i.e.*, $f : A \rightarrow B$ is said to be one-one (or injective) iff distinct elements of A have distinct images.

One-One Into Function : A function $f : A \rightarrow B$ is called a one-one into function, if it is both one-one and into, *i.e.*, the different points in A are joined to different points in B and there are some points in B which are not joined to any point in A.

One-One Onto Function : A function $f : A \rightarrow B$ is both one-one and onto, *i.e.*, the different points in A are joined to different points in B and no point in B is left vacant.

Onto Function : A function $f : A \rightarrow B$ is called an onto function, if there is no element of B which is not an image of some element of A.

Order of a Differential Equation : The order of a differential equation is the order of the derivative of the heighest order appearing in it.

Ordered Pair : An ordered pair is a pair of entries whose components occur in a specific order. It is written by listing the two components in the specific order, separating them by a comma and enclosing the pair in parentheses.

Ordinary Differential Equation : A differential equation is said to be ordinary differential equation if it contains only one independent variable.

Orthogonal Matrix : A matrix A is said to be orthogonal if $A'A = I$ where A' is the transpose of A.

Parametric Hypothesis : A statistical hypothesis refers only to the values of unknown parameters of population is usually called a parametric hypothesis.

Partial Differential Equation : A differential equation is said to be partial differential equation if it contains more than one independent variable and derivatives.

Percentage Error: The percentage error in x^A, which is the approximate value of x^T is

$$E_p = 100 \times E_r = 100 \times \left| \frac{x^T - x^A}{x^T} \right|$$

Pictogram : In this diagram, the data is represented by the picture.

Pie graph : In the pie graph, we use a circle divided into the required number of parts to represent the data.

Poisson Distribution : A variable x is said to have the Poisson distribution if it takes the values 0, 1, 2, ... to ∞ with probabilities $e^{-m}, \dfrac{m}{1!}e^{-m}, \dfrac{m^2}{2!}e^{-m}, \dfrac{m^3}{3!}e^{-m}, ...$ to ∞ respectively. The probability distribution function for Poisson distribution is given by

$$P(x,m) = P(X = x) = \begin{cases} \dfrac{e^{-m} \cdot m^x}{x!}, & x = 0,1,2 \\ 0 & , \text{ otherwise} \end{cases}$$

Polynomial Equation : Let $f(x)$ be a polynomial, then $f(x) = 0$ is called the polynomial equation.

Polynomial: A function f defined by

$$f(x) = a_0 + a_1 x + a_2 x^2 + ... + a_n x^n, x \in R$$

where $a_0, a_1, a_2, ..., a_n \in R$ is called a polynomial of a real variable with real coefficients.

Primary Data : The primary data are the data which are wholly collected by the investigator. Primary data contain all the new facts.

Quantiles : If we arrange all the variates into ascending or descending order and divide this series into five equal parts, then each part of this series is known as quantiles.

Quartiles : If we arrange all the variates into ascending or descending order and this series is divided into four equal parts, then each part of this series is known as a quartile.

Random Variable : The variate which can take certain values depending on chance is called a random variable. It is also known as 'chance variate' or 'stochastic variate'.

Range : The difference of maximum and minimum value of variables is known as range.

Rational Numbers : Any number of the form p/q, where $p, q \in Z, q \neq 0$ and p, q have no common factor (except ± 1) is called a rational number.

The set of rational numbers is denoted by Q.

$$\therefore \qquad Q = \left\{ \frac{p}{q} ; p, q \in Z, q \neq 0 \right\}$$

Real Number : A number which is either rational or irrational is called a real number. The set of real numbers is denoted by R.

Real population : A population of concrete individuals is called a existent or real population.

Reciprocal Equations : An equation which remains unchanged when x is changed to $\dfrac{1}{x}$

is called reciprocal equation.

Reflexive Relation : A relation is said to be reflexive if $(x, x) \in R \; \forall \; x \in A$, *i.e.*, $x \, R x \; \forall \; x \in A$.

Regression : Let us suppose that the scatter diagram indicates some relationship between the two variables x and y, the dots of the scatter diagram will be more or less concentrated round a curve. This curve is called the curve regression.

Relation : Let A and B be two sets. Then a relation R from A to B is a subset of $A \times B$.

Relative Error : The relative error is the absolute error divided by the true value of the given quantity. It is denoted by E_r and defined as

$$E_r = \left| \frac{x^T - x^A}{x^T} \right|$$

Right Hand Limit : A function f is said to approaches the limit l as x approaches a from the right, if corresponding to an arbitrary positive number ε then there exists a positive number $\delta > 0$ such that

$$|f(x) - l| < \varepsilon \text{ whenever } |x - a| < \delta$$

Roots of an Equation: Let $f(x) = 0$ be a quadratic equation. A real or complex numbers α is said to be a root or solution of a quadratic equation $f(x) = ax^2 + bx + c = 0$, if $f(\alpha) = a\alpha^2 + b\alpha + c = 0$, *i.e.*, α satisfies the given quadratic equation.

Row Matrix or Row Vector : A matrix having only one row is called a row matrix or row vector.

Sampling : A manner in which we can form a sample from the parent universe (population) is known as sampling.

Scalar Matrix : A square matrix is said to be scalar if all elements along the principal diagonal are equal and non-diagonal elements are zero.

Secondary Data : Secondary data are the data which are investigated by a investigator and after some time when a new investigator, investigate a new thing, he assumes all these data true.

Singleton set : Set containing only one element is a singleton set. The set $\{a\}$ is a singleton set.

Singular Square Matrix : A square matrix is said to be singular if value of determinant is zero, *i.e.*, if $|A| = 0$, then A is singular.

Skew-Symmetric Matrix : A square matrix $A = [a_{ij}]$ is said to be skew-symmetric if $a_{ij} = - a_{ji}$ for all possible values of i and j.

Solution Set : The set of all roots of an equation, is called the solution set of the given equation.

Square Matrix : A matrix in which number of rows is equal to number of columns is called square martix. In such type of matrix $m = n$.

Standard Variable : A variable with mean 0, variance unity and defined by $Z = \dfrac{X - M}{\sigma}$

is known as standard variable, in which M and σ are the mean and standard deviation of the variable X.

Statistical Series : The arrangement of data in a particular manner is called the statistical series.

Sub-Matrix : The matrix obtained by leave some row and column of the given matrix is called sub matrix.

Subset : The set A is said to be a subset of the set B if every element of A is also an element of B, i.e., $A \subseteq B$.

Symmetric Difference of Two Sets : If A and B are two sets, then the symmetric difference of two sets A and B is denoted by $A \triangle B$

$$A \triangle B = (A \sim B) \cup (B \sim A)$$

is given by $\qquad \triangle B = \{x : (x \in A \text{ and } x \notin B) \text{ or } (x \in B \text{ and } x \notin A)\}$

Symmetric Matrix : A square matrix $A = [a_{ij}]$ is said to be a symmetric matrix if $a_{ij} = a_{ji}$, for all possible values of i and j.

Symmetric Relation : A relation R on a set A is said to be symmetric if $(y, x) \in R$ whenever $(x, y) \in R \ \forall \ x, y \in R$, i.e., $x R y \Leftrightarrow y R x \ \forall \ x, y \in R$

The Inverse Laplace Transform : If the Laplace transform of a function $F(t)$ is $f(p)$, i.e., if $L\{F(t)\} = f(p)$. Then, $F(t)$ is known as inverse Laplace transform of $F(p)$.

The Laplace Transform : If $F(t)$ be a function of t defined for all values of t, then Laplace transform of $f(t)$, denoted by $L\{f(t)\}$ or $f(p)$ is defined by

$$L\{F(t)\} = f(p) = \int_0^\infty e^{-pt} F(t) dt$$

the two variables x and y, the dots of the scatter diagram will be more or less concentrated round a curve. This curve is called the curve regression.

Transitive Relation : A relation R on a set A is said to be transitive iff $(x, y) \in R$ and $(y, z) \in R \Rightarrow (x, z) \in R \ \forall \ x, y, z \in A$, i.e., $x R y, y R z \Rightarrow x R z$.

Transpose of A Matrix : Let A be the given matrix then the transpose of A denoted by A^T or A' is a matrix obtained by interchanging row and columns of A.

Unimodal Frequency Curve : A curve having only one maximum point is known by unimodal frequency curve.

Union of Two Sets : Let A and B be two sets. Then union of A and B, denoted by $A \cup B$ is the set of all those elements, which either belongs to A or B or both A and B.

$$A \cup B = \{x : x \in A \text{ and } x \in B\}$$

Unit or Identity Matrix : A square matrix $A = [aij]$ is a unit matrix if

$$a_{ij} = \begin{cases} 1 & \text{if} \quad i = j \\ 0 & \text{if} \quad i \neq j \end{cases}$$

Upper Triangular Matrix : A square matrix $A = [a_{ij}]_{m \times n}$ is called an upper triangular matrix if $a_{ij} = 0 \ \forall \ i > j$. Therefore, a square matrix is said to be upper triangular if all the elements below the main diagonal are equal to zero.

Vertical Matrix : A matrix of order $m \times n$ is called a vertical matrix if $m > n$, i.e., if number of rows is greater than the number of columns.

Zero or Null Matrix : A matrix in which every element is zero is called the null or zero matrix. It is denoted by O.

❏❏❏❏❏❏

APPENDIX

APPENDIX

TABLE 1. LOGARITHMS

	0	1	2	3	4	5	6	7	8	9	Mean Differences								
											1	2	3	4	5	6	7	8	9
10	0000	0043	0086	0128	0170	0212	0253	0294	0334	0374	4	8	12	17	21	25	29	33	37
11	0414	0453	0492	0531	0569	0607	0645	0682	0719	0755	4	8	11	15	19	23	26	30	34
12	0792	0828	0864	0899	0934	0969	1004	1038	1072	1106	3	7	10	14	17	21	24	28	31
13	1139	1173	1206	1239	1271	1303	1335	1367	1399	1430	3	6	10	13	16	19	23	26	29
14	1461	1492	1523	1553	1584	1614	1644	1673	1703	1732	3	6	9	12	15	18	21	24	27
15	1761	1790	1818	1847	1875	1903	1931	1959	1987	2014	3	6	8	11	14	17	20	22	25
16	2041	2068	2095	2122	2148	2175	2201	2227	2253	2279	3	5	8	11	13	16	18	21	24
17	2304	2330	2355	2380	2405	2430	2455	2480	2504	2529	2	5	7	10	12	15	17	20	22
18	2553	2577	2601	2625	2648	2672	2695	2718	2742	2765	2	5	7	9	12	14	16	19	21
19	2788	2810	2833	2856	2878	2900	2923	2945	2967	2989	2	4	7	9	11	13	16	18	20
20	3010	3032	3054	3075	3096	3118	3139	3160	3181	3201	2	4	6	8	11	13	15	17	19
21	3222	3243	3263	3284	3304	3324	3345	3365	3385	3404	2	4	6	8	10	12	14	16	18
22	3424	3444	3464	3483	3502	3522	3541	3560	3579	3598	2	4	6	8	10	12	14	15	17
23	3617	3636	3655	3674	3692	3711	3729	3747	3766	3784	2	4	6	7	9	11	13	15	17
24	3802	3820	3838	3856	3874	3892	3909	3927	3945	3962	2	4	5	7	9	11	12	14	16
25	3979	3997	4014	4031	4048	4065	4082	4099	4116	4133	2	3	5	7	9	10	12	14	15
26	4150	4166	4183	4200	4216	4232	4249	4265	4281	4298	2	3	5	7	8	10	11	13	15
27	4314	4330	4346	4362	4378	4393	4409	4425	4440	4456	2	3	5	6	8	9	11	13	14
28	4472	4487	4502	4518	4533	4548	4564	4579	4594	4609	2	3	5	6	8	9	11	12	14
29	4624	4639	4654	4669	4683	4698	4713	4728	4742	4757	1	3	4	6	7	9	10	12	13
30	4771	4786	4800	4814	4829	4843	4857	4871	4886	4900	1	3	4	6	7	9	10	11	13
31	4914	4928	4942	4955	4969	4983	4997	5011	5024	5038	1	3	4	6	7	8	10	11	12
32	5051	5065	5079	5092	5105	5119	5132	5145	5159	5172	1	3	4	5	7	8	9	11	12
33	5185	5198	5211	5224	5237	5250	5263	5276	5289	5302	1	3	4	5	6	8	9	10	12
34	5315	5328	5340	5353	5366	5378	5391	5403	5416	5428	1	3	4	5	6	8	9	10	11
35	5441	5453	5465	5478	5490	5502	5514	5527	5539	5551	1	2	4	5	6	7	9	10	11
36	5563	5575	5587	5599	5611	5623	5635	5647	5658	5670	1	2	4	5	6	7	8	10	11
37	5682	5694	5705	5717	5729	5740	5752	5763	5775	5786	1	2	3	5	6	7	8	9	10
38	5798	5809	5821	5832	5843	5855	5866	5877	5888	5899	1	2	3	5	6	7	8	9	10
39	5911	5922	5933	5944	5955	5966	5977	5988	5999	6010	1	2	3	4	5	7	8	9	10
40	6021	6031	6042	6053	6064	6075	6085	6096	6107	6117	1	2	3	4	5	6	8	9	10
41	6128	6138	6149	6160	6170	6180	6191	6201	6212	6222	1	2	3	4	5	6	7	8	9
42	6232	6243	6253	6263	6274	6284	6294	6304	6314	6325	1	2	3	4	5	6	7	8	9
43	6335	6345	6355	6365	6375	6385	6395	6405	6415	6425	1	2	3	4	5	6	7	8	9
44	6435	6444	6454	6464	6474	6484	6493	6503	6513	6522	1	2	3	4	5	6	7	8	9
45	6532	6542	6551	6561	6571	6580	6590	6599	6609	6618	1	2	3	4	5	6	7	8	9
46	6628	6637	6646	6656	6665	6675	6684	6693	6702	6712	1	2	3	4	5	6	7	7	8
47	6721	6730	6739	6749	6758	6767	6776	6785	6794	6803	1	2	3	4	5	5	6	7	8
48	6812	6821	6830	6839	6848	6857	6866	6875	6884	6893	1	2	3	4	4	5	6	7	8
49	6902	6911	6920	6928	6937	6946	6955	6964	6972	6981	1	2	3	4	4	5	6	7	8
50	6990	6998	7007	7016	7024	7033	7042	7050	7059	7067	1	2	3	3	4	5	6	7	8
51	7076	7084	7093	7101	7110	7118	7126	7135	7143	7152	1	2	3	3	4	5	6	7	8
52	7160	7168	7177	7185	7193	7202	7210	7218	7226	7235	1	2	2	3	4	5	6	7	7
53	7243	7251	7259	7267	7275	7284	7292	7300	7308	7316	1	2	2	3	4	5	6	6	7
54	7324	7332	7340	7348	7356	7364	7372	7380	7388	7396	1	2	2	3	4	5	6	6	7

TABLE 1. LOGARITHMS (Contd.)

	0	1	2	3	4	5	6	7	8	9	1	2	3	4	5	6	7	8	9
											Mean Differences								
55	7404	7412	7419	7427	7435	7443	7451	7459	7466	7474	1	2	2	3	4	5	5	6	7
56	7482	7490	7497	7505	7513	7520	7528	7536	7543	7551	1	2	2	3	4	5	5	6	7
57	7559	7566	7574	7582	7589	7597	7604	7612	7619	7627	1	2	2	3	4	5	5	6	7
58	7634	7642	7649	7657	7664	7672	7679	7686	7694	7701	1	1	2	3	4	4	5	6	7
59	7709	7716	7723	7731	7738	7745	7752	7760	7767	7774	1	1	2	3	4	4	5	6	7
60	7782	7789	7796	7803	7810	7818	7825	7832	7839	7846	1	1	2	3	4	4	5	6	6
61	7853	7860	7868	7875	7882	7889	7896	7903	7910	7917	1	1	2	3	4	4	5	6	6
62	7924	7931	7938	7945	7952	7959	7966	7973	7980	7987	1	1	2	3	3	4	5	6	6
63	7993	8000	8007	8014	8021	8028	8035	8041	8048	8055	1	1	2	3	3	4	5	5	6
64	8062	8069	8075	8082	8089	8096	8102	8109	8116	8122	1	1	2	3	3	4	5	5	6
65	8129	8136	8142	8149	8156	8162	8169	8176	8182	8189	1	1	2	3	3	4	5	5	6
66	8195	8202	8209	8215	8222	8228	8235	8241	8248	8254	1	1	2	3	3	4	5	5	6
67	8261	8267	8274	8280	8287	8293	8299	8306	8312	8319	1	1	2	3	3	4	5	5	6
68	8325	8331	8338	8344	8351	8357	8363	8370	8376	8382	1	1	2	3	3	4	4	5	6
69	8388	8395	8401	8407	8414	8420	8426	8432	8439	8445	1	1	2	2	3	4	4	5	6
70	8451	8457	8463	8470	8476	8482	8488	8494	8500	8506	1	1	2	2	3	4	4	5	6
71	8513	8519	8525	8531	8537	8543	8549	8555	8561	8567	1	1	2	2	3	4	4	5	5
72	8573	8579	8585	8591	8597	8603	8609	8615	8621	8627	1	1	2	2	3	4	4	5	5
73	8633	8639	8645	8651	8657	8663	8669	8675	8681	8686	1	1	2	2	3	4	4	5	5
74	8692	8698	8704	8710	8716	8722	8727	8733	8739	8745	1	1	2	2	3	4	4	5	5
75	8751	8756	8762	8768	8774	8779	8785	8791	8797	8802	1	1	2	2	3	3	4	5	5
76	8808	8814	8820	8825	8831	8837	8842	8848	8854	8859	1	1	2	2	3	3	4	5	5
77	8865	8871	8876	8882	8887	8893	8899	8904	8910	8915	1	1	2	2	3	3	4	4	5
78	8921	8927	8932	8938	8943	8949	8954	8960	8965	8971	1	1	2	2	3	3	4	4	5
79	8976	8982	8987	8993	8998	9004	9009	9015	9020	9025	1	1	2	2	3	3	4	4	5
80	9031	9036	9042	9047	9053	9058	9063	9069	9074	9079	1	1	2	2	3	3	4	4	5
81	9085	9090	9096	9101	9106	9112	9117	9122	9128	9133	1	1	2	2	3	3	4	4	5
82	9138	9143	9149	9154	9159	9165	9170	9175	9180	9186	1	1	2	2	3	3	4	4	5
83	9191	9196	9201	9206	9212	9217	9222	9227	9232	9238	1	1	2	2	3	3	4	4	5
84	9243	9248	9253	9258	9263	9269	9274	9279	9284	9289	1	1	2	2	3	3	4	4	5
85	9294	9299	9304	9309	9315	9320	9325	9330	9335	9340	1	1	2	2	3	3	4	4	5
86	9345	9350	9355	9360	9365	9370	9375	9380	9385	9390	1	1	2	2	3	3	4	4	5
87	9395	9400	9405	9410	9415	9420	9425	9430	9435	9440	0	1	1	2	2	3	3	4	4
88	9445	9450	9455	9460	9465	9469	9474	9479	9484	9489	0	1	1	2	2	3	3	4	4
89	9494	9499	9504	9509	9513	9518	9523	9528	9533	9538	0	1	1	2	2	3	3	4	4
90	9542	9547	9552	9557	9562	9566	9571	9576	9581	9586	0	1	1	2	2	3	3	4	4
91	9590	9595	9600	9605	9609	9614	9619	9624	9628	9633	0	1	1	2	2	3	3	4	4
92	9638	9643	9647	9652	9657	9661	9666	9671	9675	9680	0	1	1	2	2	3	3	4	4
93	9685	9689	9694	9699	9703	9708	9713	9717	9722	9727	0	1	1	2	2	3	3	4	4
94	9731	9736	9741	9745	9750	9754	9759	9763	9768	9773	0	1	1	2	2	3	3	4	4
95	9777	9782	9786	9791	9795	9800	9805	9809	9814	9818	0	1	1	2	2	3	3	4	4
96	9823	9827	9832	9836	9841	9845	9850	9854	9859	9863	0-	1	1	2	2	3	3	4	4
97	9868	9872	9877	9881	9886	9890	9894	9899	9903	9908	0	1	1	2	2	3	3	4	4
98	9912	9917	9921	9926	9930	9934	9939	9943	9948	9952	0	1	1	2	2	3	3	4	4
99	9956	9961	9965	9969	9974	9978	9983	9987	9991	9996	0	1	1	2	2	3	3	3	4

TABLE 2. ANTILOGARITHMS

	0	1	2	3	4	5	6	7	8	9	Mean Differences								
											1	2	3	4	5	6	7	8	9
.00	1000	1002	1005	1007	1009	1012	1014	1016	1019	1021	0	0	1	1	1	1	2	2	2
.01	1023	1026	1028	1030	1033	1035	1038	1040	1042	1045	0	0	1	1	1	1	2	2	2
.02	1047	1050	1052	1054	1057	1059	1062	1064	1067	1069	0	0	1	1	1	1	2	2	2
.03	1072	1074	1076	1079	1081	1084	1086	1089	1091	1094	0	0	1	1	1	1	2	2	2
.04	1096	1099	1102	1104	1107	1109	1112	1114	1117	1119	0	1	1	1	1	2	2	2	2
.05	1122	1125	1127	1130	1132	1135	1138	1140	1143	1146	0	1	1	1	1	2	2	2	2
.06	1148	1151	1153	1156	1159	1161	1164	1167	1169	1172	0	1	1	1	1	2	2	2	2
.07	1175	1178	1180	1183	1186	1189	1191	1194	1197	1199	0	1	1	1	1	2	2	2	2
.08	1202	1205	1208	1211	1213	1216	1219	1222	1225	1227	0	1	1	1	1	2	2	2	3
.09	1230	1233	1236	1239	1242	1245	1247	1250	1253	1256	0	1	1	1	1	2	2	2	3
.10	1259	1262	1265	1268	1271	1274	1276	1279	1282	1285	0	1	1	1	1	2	2	2	3
.11	1288	1291	1294	1297	1300	1303	1306	1309	1312	1315	0	1	1	1	2	2	2	2	3
.12	1318	1321	1324	1327	1330	1334	1337	1340	1343	1346	0	1	1	1	2	2	2	2	3
.13	1349	1352	1355	1358	1361	1365	1368	1371	1374	1377	0	1	1	1	2	2	2	3	3
.14	1380	1384	1387	1390	1393	1396	1400	1403	1406	1409	0	1	1	1	2	2	2	3	3
.15	1413	1416	1419	1422	1426	1429	1432	1435	1439	1442	0	1	1	1	2	2	2	3	3
.16	1445	1449	1452	1455	1459	1462	1466	1469	1472	1476	0	1	1	1	2	2	2	3	3
.17	1479	1483	1486	1489	1493	1496	1500	1503	1507	1510	0	1	1	1	2	2	2	3	3
.18	1514	1517	1521	1524	1528	1531	1535	1538	1542	1545	0	1	1	1	2	2	2	3	3
.19	1549	1552	1556	1560	1563	1567	1570	1574	1578	1581	0	1	1	1	2	2	3	3	3
.20	1585	1589	1592	1596	1600	1603	1607	1611	1614	1618	0	1	1	1	2	2	3	3	3
.21	1622	1626	1629	1633	1637	1641	1644	1648	1652	1656	0	1	1	2	2	2	3	3	3
.22	1660	1663	1667	1671	1675	1679	1683	1687	1690	1694	0	1	1	2	2	2	3	3	3
.23	1698	1702	1706	1710	1714	1718	1722	1726	1730	1734	0	1	1	2	2	2	3	3	4
.24	1738	1742	1746	1750	1754	1758	1762	1766	1770	1774	0	1	1	2	2	2	3	3	4
.25	1778	1782	1786	1791	1795	1799	1803	1807	1811	1816	0	1	1	2	2	2	3	3	4
.26	1820	1824	1828	1832	1837	1841	1845	1849	1854	1858	0	1	1	2	2	3	3	3	4
.27	1862	1866	1871	1875	1879	1884	1888	1892	1897	1901	0	1	1	2	2	3	3	3	4
.28	1905	1910	1914	1919	1923	1928	1932	1936	1941	1945	0	1	1	2	2	3	3	4	4
.29	1950	1954	1959	1963	1968	1972	1977	1982	1986	1991	0	1	1	2	2	3	3	4	4
.30	1995	2000	2004	2009	2014	2018	2023	2028	2032	2037	0	1	1	2	2	3	3	4	4
.31	2042	2046	2051	2056	2061	2065	2070	2075	2080	2084	0	1	1	2	2	3	3	4	4
.32	2089	2094	2099	2104	2109	2113	2118	2123	2128	2133	0	1	1	2	2	3	3	4	4
.33	2138	2143	2148	2153	2158	2163	2168	2173	2178	2183	0	1	1	2	2	3	3	4	4
.34	2188	2193	2198	2203	2208	2213	2218	2223	2228	2234	1	1	2	2	3	3	4	4	5
.35	2239	2244	2249	2254	2259	2265	2270	2275	2280	2286	1	1	2	2	3	3	4	4	5
.36	2291	2296	2301	2307	2312	2317	2323	2328	2333	2339	1	1	2	2	3	3	4	4	5
.37	2344	2350	2355	2360	2366	2371	2377	2382	2388	2393	1	1	2	2	3	3	4	4	5
.38	2399	2404	2410	2415	2421	2427	2432	2438	2443	2449	1	1	2	2	3	3	4	4	5
.39	2455	2460	2466	2472	2477	2483	2489	2495	2500	2506	1	1	2	2	3	3	4	5	5
.40	2512	2518	2523	2529	2535	2541	2547	2553	2559	2564	1	1	2	2	3	4	4	5	5
.41	2570	2576	2582	2588	2594	2600	2606	2612	2618	2624	1	1	2	2	3	4	4	5	5
.42	2630	2636	2642	2649	2655	2661	2667	2673	2679	2685	1	1	2	2	3	4	4	5	6
.43	2692	2698	2704	2710	2716	2723	2729	2735	2742	2748	1	1	2	3	3	4	4	5	6
.44	2754	2761	2767	2773	2780	2786	2793	2799	2805	2812	1	1	2	3	3	4	4	5	6
.45	2818	2825	2831	2838	2844	2851	2858	2864	2871	2877	1	1	2	3	3	4	5	5	6
.46	2884	2891	2897	2904	2911	2917	2924	2931	2938	2944	1	1	2	3	3	4	5	5	6
.47	2951	2958	2965	2972	2979	2985	2992	2999	3006	3013	1	1	2	3	3	4	5	5	6
.48	3020	3027	3034	3041	3048	3055	3062	3069	3076	3083	1	1	2	3	4	4	5	6	6
.49	3090	3097	3105	3112	3119	3126	3133	3141	3148	3155	1	1	2	3	4	4	5	6	6

TABLE 2. ANTILOGARITHMS (Contd.)

	0	1	2	3	4	5	6	7	8	9	Mean Differences 1	2	3	4	5	6	7	8	9
.50	3162	3170	3177	3184	3192	3199	3206	3214	3221	3228	1	1	2	3	4	4	5	6	7
.51	3236	3243	3251	3258	3266	3273	3281	3289	3296	3304	1	2	2	3	4	5	5	6	7
.52	3311	3319	3327	3334	3342	3350	3357	3365	3373	3381	1	2	2	3	4	5	5	6	7
.53	3388	3396	3404	3412	3420	3428	3436	3443	3451	3459	1	2	2	3	4	5	6	6	7
.54	3467	3475	3483	3491	3499	3508	3516	3524	3532	3540	1	2	2	3	4	5	6	6	7
.55	3548	3556	3565	3573	3581	3589	3597	3606	3614	3622	1	2	2	3	4	5	6	7	7
.56	3631	3639	3648	3656	3664	3673	3681	3690	3698	3707	1	2	3	3	4	5	6	7	8
.57	3715	3724	3733	3741	3750	3758	3767	3776	3784	3793	1	2	3	3	4	5	6	7	8
.58	3802	3811	3819	3828	3837	3846	3855	3864	3873	3882	1	2	3	4	4	5	6	7	8
.59	3890	3899	3908	3917	3926	3936	3945	3954	3963	3972	1	2	3	4	5	5	6	7	8
.60	3981	3990	3999	4009	4018	4027	4036	4046	4055	4064	1	2	3	4	5	6	6	7	8
.61	4074	4083	4093	4102	4111	4121	4130	4140	4150	4159	1	2	3	4	5	6	7	8	9
.62	4169	4178	4188	4198	4207	4217	4227	4236	4256	4256	1	2	3	4	5	6	7	8	9
.63	4266	4276	4285	4295	4305	4315	4325	4335	4345	4355	1	2	3	4	5	6	7	8	9
.64	4365	4375	4385	4395	4406	4416	4426	4436	4446	4457	1	2	3	4	5	6	7	8	9
.65	4467	4477	4487	4498	4508	4519	4529	4539	4550	4560	1	2	3	4	5	6	7	8	9
.66	4571	4581	4592	4603	4613	4624	4634	4645	4656	4667	1	2	3	4	5	6	7	9	10
.67	4677	4688	4699	4710	4721	4732	4742	4753	4764	4775	1	2	3	4	5	7	8	9	10
.68	4786	4797	4808	4819	4831	4842	4853	4864	4875	4887	1	2	3	4	6	7	8	9	10
.69	4898	4909	4920	4932	4943	4955	4966	4977	4989	5000	1	2	3	5	6	7	8	9	10
.70	5012	5023	5035	5047	5058	5070	5082	5093	5105	5117	1	2	4	5	6	7	8	9	11
.71	5129	5140	5152	5164	5176	5188	5200	5212	5224	5236	1	2	4	5	6	7	8	10	11
.72	5248	5260	5272	5284	5297	5309	5321	5333	5346	5358	1	2	4	5	6	7	9	10	11
.73	5370	5383	5395	5408	5420	5433	5445	5458	5470	5483	1	3	4	5	6	8	9	10	11
.74	5495	5508	5521	5534	5546	5559	5572	5585	5598	5610	1	3	4	5	6	8	9	10	12
.75	5623	5636	5649	5662	5675	5689	5702	5715	5728	5741	1	3	4	5	7	8	9	10	12
.76	5754	5768	5781	5794	5808	5821	5834	5848	5861	5875	1	3	4	5	7	8	9	11	12
.77	5888	5902	5916	5929	5943	5957	5970	5984	5998	6012	1	3	4	5	7	8	10	11	12
.78	6026	6039	6053	6067	6081	6095	6109	6124	6138	6152	1	3	4	6	7	8	10	11	13
.79	6166	6180	6194	6209	6223	6237	6252	6266	6281	6295	1	3	4	6	7	9	10	11	13
.80	6310	6324	6339	6353	6368	6383	6397	6412	6427	6442	1	3	4	6	7	9	10	12	13
.81	6457	6471	6486	6501	6516	6531	6546	6561	6577	6592	2	3	5	6	8	9	11	12	14
.82	6607	6622	6637	6653	6668	6683	6699	6714	6730	6745	2	3	5	6	8	9	11	12	14
.83	6761	6776	6792	6808	6823	6839	6855	6871	6887	6902	2	3	5	6	8	9	11	13	14
.84	6918	6934	6950	6966	6982	6998	7015	7031	7047	7063	2	3	5	6	8	10	11	13	15
.85	7079	7096	7112	7129	7145	7161	7178	7194	7211	7228	2	3	5	7	8	10	12	13	15
.86	7244	7261	7278	7295	7311	7328	7345	7362	7379	7396	2	3	5	7	8	10	12	13	15
.87	7413	7430	7447	7464	7482	7499	7516	7534	7551	7568	2	3	5	7	9	10	12	14	16
.88	7586	7603	7621	7638	7656	7674	7691	7709	7727	7745	2	4	5	7	9	11	12	14	16
.89	7762	7780	7798	7816	7834	7852	7870	7889	7907	7925	2	4	5	7	9	11	13	14	16
.90	7943	7962	7980	7998	8017	8035	8054	8072	8091	8110	2	4	6	7	9	11	13	15	17
.91	8128	8147	8166	8185	8204	8222	8241	8260	8279	8299	2	4	6	8	9	11	13	15	17
.92	8318	8337	8356	8375	8395	8414	8433	8453	8472	8492	2	4	6	8	10	12	14	15	17
.93	8511	8531	8551	8570	8590	8610	8630	8650	8670	8690	2	4	6	8	10	12	14	16	18
.94	8710	8730	8750	8770	8790	8810	8831	8851	8872	8892	2	4	6	8	10	12	14	16	18
.95	8913	8933	8954	8974	8995	9016	9036	9057	9078	9099	2	4	6	8	10	12	15	17	19
.96	9120	9141	9162	9183	9204	9226	9247	9268	9290	9311	2	4	6	8	11	13	15	17	19
.97	9333	9354	9376	9397	9419	9441	9462	9484	9506	9528	2	4	7	9	11	13	15	17	20
.98	9550	9572	9594	9616	9638	9661	9683	9705	9727	9750	2	4	7	9	11	13	16	18	20
.99	9772	9795	9817	9840	9863	9886	9908	9931	9954	9977	2	5	7	9	11	14	16	18	20

TABLE 3. POISSON DISTRIBUTION

x/λ	0.1	0.2	0.3	0.4	0.5	0.6	0.7	0.8	0.9	1.0
0	.9048	.8187	.7408	.6703	.6065	.5488	.4966	.4493	.4066	.3679
1	.0905	.1637	.2222	.2681	.3033	.3293	.3476	.3595	.3659	.3679
2	.0045	.0164	.0333	.0536	.0758	.0988	.1217	.1438	.1647	.1839
3	.0002	.0011	.0033	.0072	.0126	.0198	.0284	.0383	.0494	.0613
4	.0000	.0001	.0002	.0007	.0016	.0030	.0050	0.0077	.0111	.0153
5	.0000	.0000	.0000	.0001	.0002	.0004	.0007	.0012	.0020	.0031
6	.0000	.0000	.0000	.0000	.0000	.0000	.0001	.0002	.0003	.0005
7	.0000	.0000	.0000	.0000	.0000	.0000	.0000	.0000	.0000	.0001

x/λ	1.1	1.2	1.3	1.4	1.5	1.6	1.7	1.8	1.9	2.0
0	.3329	.3012	.2725	.2466	.2231	.2019	.1827	.1653	.1496	.1353
1	.3662	.3014	.3543	.3452	.3347	.3230	.3106	.2957	.2842	.2707
2	.2014	.2169	.2303	.2417	.2510	.2584	.2640	.2678	.2700	.2707
3	.0738	.0867	.0998	.1128	.1255	.1378	.1496	.1607	.1710	.1804
4	.0203	.0260	.0324	.0395	.0471	.0551	.0636	.0723	.0812	.0902
5	.0045	.0062	.0084	.0111	.0141	.0176	.0216	.0260	.0309	.0361
6	.0008	.0012	.0018	.0026	.0035	.0047	.0061	.0078	.0098	.0120
7	.0001	.0002	.0003	.0005	.0008	.0011	.0015	.0020	:0027	.0034
8	.0000	.0000	.0000	.0001	.0001	.0002	.0003	.0005	.0006	.0009
9	.0000	.0000	.0000	.0000	.0000	.0000	.0001	.1001	.0001	.0002

x/λ	2.1	2.2	2.3	2.4	2.5	2.6	2.7	2.8	2.9	3.0
0	.1225	.1108	.1003	.0907	.0821	.0743	.0672	.0608	.0550	.0498
1	.2572	.2438	.2306	.2177	.2052	.1931	.1815	.1703	.1596	.1494
2	.2700	.2681	.2652	.2613	.2505	.2510	.2450	.2384	.2314	.2240
3	.1890	.1966	.2033	.2090	.2138	.2176	.2205	.2225	.2237	.2240
4	.0992	.1082	.1169	.1254	.1336	.1414	.1488	.1557	.1622	.1680
5	.0417	.0476	.0538	.0602	.668	.0735	.0804	.0872	.0940	.1008
6	.0146	.0174	.0206	.0241	.0278	.0319	.0362	.0407	.0455	.0504
7	.0044	.0055	.0068	.0083	.0099	.0118	.0139	.0163	.0188	0.216
8	.0011	.0015	.0019	.0025	.0031	.0038	.0047	.0057	.0068	.0081
9	.0003	.0004	.0005	.0007	.0009	.0011	.0014	.0068	.0022	.0027
10	.0001	.0001	.0001	.0002	.0002	.0003	.0004	.0005	.0006	.0008
11	.0000	.0000	.0000	.0000	.0000	.0001	.0001	.0001	.0002	.0002
12	.0000	.0000	.0000	.0000	.0000	.0000	.0000	.0000	.0000	.0001

TABLE 3. POISSON DISTRIBUTION (CONTD.)

x/λ	3.1	3.2	3.3	3.4	3.5	3.6	3.7	3.8	3.9	4.0
0	.0450	.0408	.0369	.0334	.0302	.0273	.0247	.0224	.0224	.0183
1	.1397	.1304	.1217	.1135	.1057	.0984	.0915	.0850	.0789	.0733
2	.2165	.2087	.2008	.1929	.1850	.1771	.1692	.1615	.1539	.1465
3	.2237	.2226	.2209	.2186	.2158	.2125	.2087	.2046	.2001	.1954
4	.0734	.1781	.1823	.1858	.1888	.1912	.1931	.1944	.1951	.1954
5	.1075	.1140	.1203	.1264	.1322	.1377	.1429	.1477	.1522	.1563
6	.0555	.0608	.0662	.0716	.0771	.0826	.0881	.0936	.0989	.1042
7	.0246	.0278	.0312	.0348	.0385	.0425	.0466	.008	.0551	.1595
8	.0095	.0111	.0129	.0148	.0159	.0191	.0215	.0241	.0269	.0298
9	.0033	.0040	.0047	.0056	.0066	.0076	.0089	.0102	.0116	.0132
10	.0010	.0013	.0016	.0019	.0023	.0028	.0033	.0039	.0045	.0053
11	.0003	.0004	.0005	.0006	.0007	.0009	.0011	.0013	.0016	.0019
13	.0000	.0000	.0000	.0000	.0001	.0001	.0001	.0001	.0002	.0002
14	.0000	.0000	.0000	.0000	.0000	.0000	.0000	.0000	.0000	.0001

x/λ	4.1	4.2	4.3	4.4	4.5	4.6	4.7	4.8	4.9	5.0
0	.0166	.0150	.0136	.0123	.0111	.0101	.0091	.0082	.0074	.0067
1	.0679	.0630	.0583	.0540	.0500	.0462	.0427	.0395	.0365	.0337
2	.1393	.1323	.1254	.1188	.1125	.1063	.1005	.0948	.0894	.0842
3	.1904	.1852'	.1798	.1743	.1687	.1631	.1574	.1517	.1460	.1404
4	.1951	.1944	.1933	.1917	.1898	.1875	.1849	.1820	.1789	.1775
5	.1600	.1633	.1662	.1687	.1708	.1725	.1738	.1747	.1753	.1755
6	.1093	.1143	.1191	.1237	.1281	.1323	.1362	.1398	.1432	1462
7	.0640	.0686	.0732	.0778	.0824	.0869	.0914	.0959	.1002	.1044
8	.0328	.0360	.0393	.0428	.0463	.0500	.0537	.0537	.0614	.0653
9	.0150	.0168	.0188	.0209	.0232	.0255	.0280	.0307	.0334	.0363
10	.0061	.0071	.0081	.0092	.0104	.0118	.0132	.0147	.0164	.0181
11	.0023	.0027	.0032	.0037	.0043	.0049	.0056	.0064	.0073	.0082
12	.0008	.0009	.0011	.0014	.0016	.0019	.0022	.0026	.0030	.0034
13	.0002	.0003	.0004	.0005	.0006	.0007	.0008	.0009	.0011	.0013
14	.0001	.0001	.0001	.0001	.0002	.0002	.0003	.0003	.0004	.0005
15	.0000	.0000	.0000	.0000	.0001	.0001	.0001	.0001	.0001	.0002

TABLE 3. POISSON DISTRIBUTION (CONTD.)

x/λ	5.1	5.2	5.3	5.4	5.5	5.6	5.7	5.8	5.9	6.0
0	.0061	.0055	.0050	.0045	.0041	.0037	.0033	.0030	.0027	.0025
1	.0311	.0287	.0265	.0244	.0225	.0207	.0191	.0176	.0162	.0149
2	.0793	.0746	.0701	.0659	.0618	.0580	.0544	.0509	.0477	.0446
3	.1348	.1293	.1239	.1185	.1133	.1082	.1033	.0985	.0938	.0892
4	.1719	.1681	.1641	.1600	.1558	.1515	.1472	.1428	.1383	.1339
5	.1753	.1748	.1740	.1728	.1714	.1697	.1678	.1656	.1632	.1606
6	.1490	.1515	.1537	.1555	.1571	.1584	.1594	.1601	.1605	.1606
7	.1086	.1125	.1163	.1200	.1234	.1267	.1298	.1326	.1353	.1377
8	.0692	.0731	.0711	.0810	.0849	.0887	.0925	.0962	.0998	.1033
9	.0362	.0423	.0454	.0486	.0519	.0552	.0586	.0620	.0654	.0688
10	.0200	.0220	.0241	.0262	.0285	.0309	.0334	.0359	.0386	.0413
11	.0093	.0104	.0116	.0129	.0143	.0157	.0173	.0190	.0207	.0225
12	.0039	.0045	.0051	.0058	.0065	.0073	.0082	.0092	.0102	.0113
13	.0015	.0104	.0021	.0024	.0028	.0032	.0036	.0041	.0046	.0052
14	.0006	.0007	.0008	.0009	.0011	.0013	.0015	.0017	.0019	.0022
15	.0002	.0002	.0003	.0003	.0004	.0005	.0006	.0007	.0008	.0009
16	.0001	.0001	.0001	.0001	.0001	.0002	.0002	.0002	.0003	.0003
17	.0000	.0000	.0000	.0000	.0000	.0000	.0001	.0001	.0001	.0001

x/λ	6.1	6.2	6.3	6.4	6.5	6.6	6.7	6.8	6.9	7.0
0	.0022	.0020	.0018	.0017	.0015	.0014	.0012	.0011	.0010	.0000
1	.0137	.0126	.0116	.0106	.0098	.0090	.0082	.0076	.0070	.0064
2	.0417	.0390	.0364	.0340	.0318	.0296	.0276	.0258	.0240	.0223
3	.0848	.0806	.0765	.0726	.0688	.0652	.0617	.0584	.0552	.0521
4	.1294	.1249	.1203	.1162	.1118	.1076	.1034	.0992	.0952	.0912
5	.1579	.1549	.1519	.1487	.1454	.1420	.1385	.1349	.1314	.1277
6	.1605	.1601	.1595	.1586	.1575	.1562	.1546	.1529	.1511	.1490
7	.1399	.1418	.1435	.1450	.1462	.1472	.1480	.1486	.1489	.1490
8	.1066	.1099	.1130	.1160	.1188	.1215	.1240	.1263	.1284	.1304
9	.0723	.0757	.0791	.0825	.0858	.0891	.0923	.0954	.0985	.1014
10	.0441	.0469	.0498	.5285	.0558	.0588	.0618	.0679	.0679	.0710
11	.0245	.0265	.0285	.0307	.0330	.0353	.0377	.0401	.0426	.0452
12	.0124	.0137	.0150	.0164	.0179	.0194	.0210	.0227	:0245	.0264
13	.0058	.0065	.0073	.0081	.0089	.0098	.0108	.0119	.0130	.0142
14	.0025	.0029	.0033	.0037	.0041	.0046	.0052	.0058	,0064	.0071
15	.0010	.0012	.0014	.0016	.0018	.0020	.0023	.0026	.0029	.0033
16	.0004	.0005	.0005	.0006	.0007	.0008	.0010	.0011	.0013	.0014
17	.0001	.0002	.0002	.0002	.0003	.0003	.0004	.0004	.0005	.0006
18	.0000	.0001	.0001	.0001	.0001	.0001	.0001	.0002	.0002	.0002
19	.0000	.0000	.0000	.0000	.0000	.0000	.0000	.0001	.0001	.0001

TABLE 4. AREA UNDER THE STANDARD NORMAL
DISTRIBUTION FOR NEGATIVE VALUES OF Z

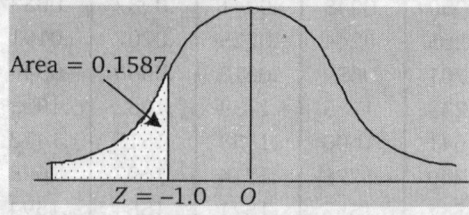

Area = 0.1587

$Z = -1.0$ O

Z to First	Second Decimal									
Decimal	.00	.01	.02	.03	.04	.05	.06	.07	.08	.09
−3.0	.0014	.0013	.0013	.0012	.0012	.0011	.0011	.0011	.0010	.0010
−2.9	.0019	.0018	.0018	.0017	.0016	.0016	.0015	.0015	.0014	.0014
−2.8	.0026	.0025	.0024	.0023	.0023	.0022	.0021	.0021	.0020	.0019
−2.7	.0035	.0034	.0033	.0032	.0031	.0030	.0029	.0028	.0027	.0026
−2.6	.0047	.0045	.0044	.0043	.0041	.0040	.0039	.0038	.0037	.0036
−2.5	.0062	.0060	.0059	.0057	.0055	.0054	.0052	.0051	.0049	.0048
−2.4	.0082	.0080	.0078	.0075	.0073	.0071	.0069	.0068	.0066	.0064
−2.3	.0107	.0104	.0102	.0099	.0096	.0094	.0091	.0089	.0087	.0084
−2.2	.0139	.0136	.0132	.0129	.0126	.0122	.0119	.0116	.0113	.0110
− 2.1	.0179	.0174	.0170	.0166	.0162	.0158	.0154	.0150	.0146	.0143
−2.0	.0228	.0222	.0217	.0212	.0207	.0202	.0197	.0192	.0188	.0183
− 1.9	.0287	.0281	.0274	.0268	.0262	.0256	.0250	.0244	.0238	.0233
− 1.8	.0359	.0352	.0344	.0336	.0329	.0322	.0314	.0307	.0300	.01.94
− 1.7	.0446	.0436	.0427	.0418	.0409	.0401	.0392	.0384	.0375	.0367
− 1.6	.0548	.0537	.0526	.0516	.0505	.0495	.0485	.0475	.0465	.0455
− 1.5	.0668	.0655	.0643	.0630	.0618	.0606	.0594	.0582	.0570	.0559
− 1.4	.0808	.7938	.0778	.0764	.0749	.0735	.0722	.0708	.0694	.0681
−1.3	.0968	.0951	.0934	.0918	.0901	.0855	.0869	.0853	.0838	.0823
− 1.2	.1151	.1131	.1112	.1093	.1075	.1056	.1038	.1020	.1003	.0985
−1.1	.1357	.1335	.1314	.1292	.1271	.1251	.1230	.1210	.1190	.1170
−1.0	.1587	.1562	.1529	.1515	.1492	.1469	.1446	.1423	.1401	.1379
−0.9	.1841	.1814	.1785	.1762	.1736	.1711	.1685	.1660	.1635	.1611
−0.8	.2119	.2090	.2061	.2033	.2005	.1977	.1949	.1922	.1894	.1867
−0.7	.2420	.2389	.2358	.2327	.2297	.2266	.2236	.2206	.2177	.2143
−0.6	.2743	.2709	.2676	.2643	.2611	.2578	.2546	.2514	.2483	.2451
−0.5	.3085	.3050	.3015	.2981	.2946	.2912	.2877	.2843	.2810	.2776
−0.4	.3446	.3409	.3372	.3336	.3300	.3264	.3228	.3192	.3156	.3121
−0.3	.3821	.3783	.3745	.3707	.3669	.3632	.3594	.3557	.3520	.3483
−0.2	.4207	.4168	.4129	.4090	.4052	4013	.3974	.3936	.3897	.3859
− 0.1	.4602	.4562	.4522	.4483	.4443	.4404	.4364	.4325	.4286	.4247
−0.0	.5000	.4960	.4920	.4880	.4840	.4801	.4761	.4721	.4681	.4641

TABLE 5. AREA UNDER THE STANDARD NORMAL DISTRIBUTION FROM EXTREME LEFT TO POSITIVE VALUES OF Z

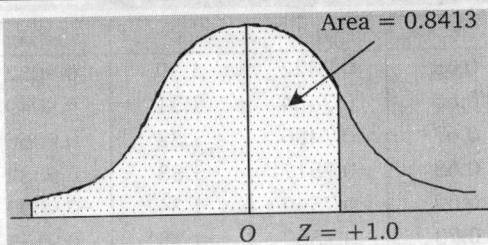

Area = 0.8413

O   Z = +1.0

Example. To find the area under the curve from extreme left $Z = -\infty$ to a point $Z = 1.0$ to the right of the mean, look up the value opposite 1.0 in the table; 0.8413 of the area under the curve lies between the extreme left ($Z = -\infty$) to a z value.

Z to First Decimal	Second Decimal									
	.00	.01	.02	.03	.04	.05	.06	.07	.08	.09
0.0	.5000	.5040	.5080	.5120	.5160	.5199	.5239	.5279	.5319	.5359
0.1	.5398	.5438	.5478	.5517	.5557	.5596	.5636	.5675	.5714	.5753
0.2	.5793	.5832	.5871	.5910	.5948	.5987	.6026	.6064	.6103	.6141
0.3	.6179	.6217	.6255	.6293	.6331	.6368	.6406	.6443	.6480	.6517
0.4	.6554	.6591	.6628	.6664	.6700	.6736	.6772	.6808	.6841	.6879
0.5	.6915	.6950	.6985	.9019	.7054	.7088	.7123	.7157	.7190	.7224
0.6	.7257	.7291	.7324	.7357	.7389	.7422	.7454	.7406	.7517	.7549
0.7	.7580	.7611	.7642	.7673	.7703	.7734	.7764	.7794	.7823	.7852
0.8	.7881	.7910	.7939	.7967	.7995	.8023	.8051	.8078	.8106	.8133
0.9	.8159	.8186	.8212	.8238	.8264	.8289	.8315	.8340	.8365	.8389
1.0	.8413	.8438	.8461	.8485	.8508	.8531	.8554	.8577	.8599	.8621
1.1	.8643	.8665	.8686	.8708	.8729	.8749	.8770	.8790	.8910	.8830
1.2	.8849	.8869	.8888	.8907	.8925	.8944	.8962	.8980	.8997	.9015
1.3	.9032	.9049	.9066	.9082	.9099	.9115	.9131	.9147	.9162	.9177
1.4	.9192	.9207	.9222	.9236	.9251	.9265	.9278	.9292	.9306	.9319
1.5	.9332	.9345	.9357	.9370	.9382	.9394	.9406	.9418	.9430	.9441
1.6	.9452	.9463	.9474	.9485	.9495	.9505	.9515	.9525	.9535	.9545
1.7	.9554	.9564	.9573	.9582	.9591	.9599	.9608	.9616	.9625	.9633
1.8	.9641	.9649	.9656	.9664	.9671	.9678	.9686	.9693	.9700	.9706
1.9	.9713	.9719	.9726	.9732	.9783	.9744	.9750	.9756	.9762	.9767
2.0	.9772	.9778	.9783	.9788	.9793	.9798	.9803	.9808	.9812	.9817
2.1	.9821	.9826	.9830	.9834	.9838	.9842	.9846	.9850	.9854	.9857
2.2	.9861	.9865	.9868	.9871	.9874	.9878	.9881	.9884	.9887	.9890
2.3	.9893	.9896	.9898	.9901	.9904	.9906	.9909	.9911	.9913	.9916
2.4	.9918	.9920	.9922	.9924	.9926	.9928	.9930	.9932	.9934	.9936
2.5	.9938	.9940	.9941	.9943	.9944	.9946	.9932	.9949	.9951	.9952
2.6	.9953	.9955	.9956	.9957	.9958	.9960	.9961	.9962	.9963	.9964
2.7	.9965	.9966	.9967	.9968	.9969	.9970	.9971	.9972	.9973	.9974
2.8	.9974	.9975	.9976	.9977	.9977	.9979	.9979	.9979	.9980	.9981
2.9	.9981	.9982	.9982	.9983	.9984	.9984	.9985	.9985	.9986	.9986
3.0	.9986	.9987	.9987	.9988	.9988	.9988	.9989	.9989	.9990	.9990

TABLE 6. PROPOSITIONS OF TOTAL AREA UNDER THE NORMAL CURVE FROM ∞ TO t,

WHERE $t = (x - m\mu)\sigma$

t	$\psi(t)$	t	$\psi(t)$	t	$\psi(t)$	t	$\psi(t)$
0.00	0.5000	0.65	0.7422	1.30	0.9032	1.95	0.9744
0.01	0.5040	0.66	0.7454	1.31	0.9049	1.96	0.9750
0.02	0.5080	0.67	0.7486	1.32	0.9066	1.97	0.9756
0.03	0.5120	0.68	0.7517	1.33	0.9082	1.98	0.9761
0.04	0.5160	0.69	0.7549	1.34	0.9099	1.99	0.9767
0.05	0.5199	0.70	0.7580	1.35	0.9115	2.00	0.9772
0.06	0.5239	0.71	0.7611	1.36	0.9131	2.02	0.9783
0.07	0.5279	0.72	0.7642	1.37	0.9147	2.04	0.9793
0.08	0.5319	0.73	0.7673	1.38	0.9162	2.06	0.9803
0.09	0.5359	0.74	0.7703	1.39	0.9177	2.08	0.9812
0.10	0.5398	0.75	0.7734	1.40	0.9192	2.10	0.9821
0.11	0.5438	0.76	0.7764	1.41	0.9207	2.12	0.9830
0.12	0.5478	0.77	0.7794	1.42	0.9222	2.14	0.9838
0.13	0.5517	0.78	0.7823	1.43	0.9236	2.16	0.9846
0.14	0.5557	0.79	0.7852	1.44	0.9251	2.18	0.9854
0.15	0.5596	0.80	0.7881	1.45	0.9265	2.20	0.9861
0.16	0.5636	0.81	0.7910	1.46	0.9279	2.22	0.9868
.0.17	0.5675	0.82	0.7939	1.47	0.9292	2.24	0.9875
0.18	0.5714	0.83	0.7967	1.48	0.9306	2.26	0.9881
0.19	0.5753	0.84	0.7995	1.49	0.9319	2.28	0.9887
0.20	0.5793	0.85	0.8023	1.50	0.9332	2.30	0.9893
0.21	0.5832	0.86	0.8051	1.51	0.9345	2.32	0.9898
0.22	0.5871	0.87	0.8078	1.52	0.9357	2.34	0.9904
0.23	0.5910	0.88	0.8106	1.53	0.9370	2.36	0.9909
0.24	0.5948	0.89	0.8133	1.54	0.9382	2.38	0.9913
0.25	0.5987	0.90	0.8159	1.55	0.9394	2.40	0.9918
0.26	0.6026	0.91	0.8186	1.56	0.9406	2.42	0.9922
0.27	0.6064	0.92	0.8212	1.57	0.9418	2.44	0.9927
0.28	0.6103	0.93	0.8238	1.58	0.9429	2.46	0.9931
0.29	0.6141	0.94	0.8264	1.59	0.9441	2.48	0.9934
0.30	0.6179	0.95	0.8289	1.60	0.9252	2.50	0.9938
0.31	0.6217	0.96	0.8315	1.61	0.9463	2.52	0.9941
0.32	0.6255	0.97	0.8340	1.62	0.9474	2.54	0.9945
0.33	0.6293	0.98	0.8365	1.63	0.9484	2.56	0.9948
0.34	0.6331	0.99	0.8389	1.64	0.9495	2.58	0.9951
0.35	0.6368	1.00	0.8413	1.65	0.9505	2.60	0.9953
0.36	0.6406	0.01	0.8438	1.66	0.9515	2.62	0.9956
0.37	0.6443	1.02	0.8461	1.67	0.9525	2.64	0.9959
0.38	0.6480	1.03	0.8485	1.68	0.9535	2.66	0.9961
0.39	0.6517	1.04	0.8508	1.69	0.9545	2.68	0.9963

Table 7. Propositions of Total Area Under the Normal Curve from ∞ to t, Where $t = (x - m\mu)\sigma$ (Contd.)

t	$\psi(t)$	t	$\psi(t)$	t	$\psi(t)$	t	$\psi(t)$
0.40	0.6554	1.05	0.8531	1.70	0.9554	2.70	0.9965
0.41	0.6591	1.06	0.8554	1.71	0.9564	2.72	0.9967
0.42	0.6628	1.07	0.8577	1.72	0.9573	2.74	0.9969
0.43	0.6664	1.08	0.8599	1.73	0.9582	2.76	0.9971
0.44	0.6700	0.09	0.8621	1.74	0.9591	2.78	0.9973
0.45	0.6736	1.10	0.8643	1.75	0.5999	2.80	0.9974
0.46	0.6772	1.11	0.8665	1.76	0.9608	2.82	0.9976
0.47	0.6808	1.12	0.8686	1.77	0.9616	2.84	0.9977
0.48	0.6844	1.13	0.8708	1.78	0.9625	2.86	0.9979
0.49	0.6879	1.14	0.8729	1.79	0.9633	2.88	0.9980
0.50	0.6915	1.15	0.8749	1.80	0.9641	2.90	0.9981
0.51	0.6950	1.16	0.8770	1.81	0.9649	2.92	0.9982
0.52	0.6985	1.17	0.8190	1.82	0.9656	2.94	0.9984
0.53	0.7019	1.18	0.8810	1.83	0.9664	2.96	0.9985
0.54	0.7054	1.19	0.8830	1.84	0.9671	2.98	0.99116
0.55	0.7088	1.20	0.8849	1.85	0.9678	3.00	0.99865
0.56	0.7123	1.21	0.8869	1.86	0.9686	3.20	0.99931
0.57	0.7190	1.22	0.8888	1.87	0.9693	3.40	0.99966
0.58	0.7190	1.23	0.8907	1.88	0.9699	3.60	0.999841
0.59	0.7224	1.24	0.8925	1.89	0.9706	3.80	0.999928
0.60	0.7257	1.25	0.8944	1.90	0.9713	4.00	0.999968
0.61	0.7291	1.26	0.8962	1.91	0.9719	4.50	0.999997
0.62	0.7324	1.27	0.8980	1.92	0.9726	5.00	0.999997
0.63	0.7357	1.28	0.8997	1.93	0.9732		
0.64	0.7389	1.29	0.9015	1.94	0.9738		

TABLE 8. VALUES OF χ^2 WITH VARIOUS VALUES OF p AND v

v \ P	0.99	0.95	0.50	0.10	0.05	0.01
1	0.0002	0.0039	0.455	2.706	3.841	6.635
2	0.0201	0.103	1.386	4.605	5.991	9.210
3	0.115	0.352	2.366	6.251	7.815	11.34
4	0.297	0.711	3.357	7.779	9.488	13.28
5	0.554	1.145	4.351	9.236	11.07	15.09
6	0.872	1.635	5.348	10.64	12.59	16.81
7	1.239	2.167	6.346	12.02	14.07	18.48
8	1.646	2.733	7.344	13.36	15.51	20.09
9	2.088	3.325	8.343	14.68	16.92	21.67
10	2.558	3.940	9.342	15.99	18.31	23.21
11	3.053	4.575	10.34	17.28	19.68	24.72
12	3.571	5.226	11.34	18.55	21.03	26.22
13	4.107	5.892	12.34	19.81	22.36	27.69
14	4.660	6.571	13.34	21.06	23.68	29.14
15	5.229	7.261	14.34	22.31	25.00	30.58
16	5.812	7.962	15.34	23.54	26.30	32.00
17	6.408	8.672	16.34	24.77	27.59	33.41
18	7.015	9.390	17.34	25.99	28.87	34.80
19	7.633	10.12	18.34	27.20	30.14	36.19
20	8.260	10.85	19.34	28.41	31.41	37.57
21	8.897	11.59	20.34	29.62	32.67	38.93
22	9.542	12.34	21.34	30.81	33.92	40.29
23	10.20	13.09	22.34	32.01	35.17	41.64
24	10.86	13.85	23.34	33.20	36.42	42.98
25	11.52	14.61	24.34	34.38	37.65	44.31
26	12.20	15.38	25.34	35.56	38.88	45.64
27	12.88	16.15	26.34	36.74	40.11	46.96
28	13.57	16.93	27.34	37.92	41.34	48.28
29	14.26	17.71	28.34	39.09	42.56	49.59
30	14.95	18.49	29.34	40.26	43.77	50.89

TABLE 9. VALUES OF F (F - DISTRIBUTION) FOR 5% AND 1% LEVEL, WHERE v_1 IS THE NUMBER OF DEGREE OF FREEDOM FOR GREATER ESTIMATE OF VARIANCE AND v_2 FOR THE SMALLER

v \\ P		1	2	3	4	5	6	8	12	24	∞
1	5%	161.4	199.5	215.7	224.6	230.2	234.0	238.9	243.9	249.0	254.0
	1%	4052	4999	5403	5625	5764	5849	5981	6016	6234	6366
2	5%	18.51	19.00	19.16	19.25	19.30	19.32	19.37	19.41	19.45	19.50
	1%	98.49	99.00	99.17	99.25	99.30	99.33	99.36	99.42	99.46	99.50
3	5%	10.13	9.55	9.28	9.12	9.01	8.94	8.84	8.74	8.64	8.53
	1%	34.12	30.82	29.46	28.71	28.24	27.91	27.49	27.05	26.60	26.12
4	5%	7.71	6.94	6.59	6.39	6.26	6.16	6.04	5.91	5.77	5.63
	1%	21.20	18.00	16.69	15.98	15.52	15.21	14.80	14.37	13.93	13.46
5	5%	6.61	5.79	5.41	5.19	5.05	4.95	4.82	4.68	4.53	4.36
	1%	16.26	13.27	12.06	11.39	10.97	10.67	10.27	9.89	9.47	9.02
6	5%	5.99	5.14	4.76	4.53	4.39	4.28	4.15	4.00	3.84	3.67
	1%	13.74	10.92	9.78	9.15	8.75	8.47	8.10	7.72	7.31	6.88
7	5%	5.59	4.74	4.35	4.12	3.97	3.87	3.73	3.57	3.41	3.23
	1%	12.25	9.55	8.45	7.85	7.46	7.19	6.84	6.47	6.07	5.65
8	5%	5.32	4.46	4.07	3.84	3.89	3.58	3.44	3.28	3.12	2.93
	1%	11.26	8.65	7.59	7.01	6.63	6.37	6.03	5.67	5.28	4.86
9	5%	5.12	4.26	3.86	3.63	3.48	3.37	3.23	3.07	2.90	2.71
	1%	10.56	8.02	6.99	6.42	6.06	5.80	5.47	5.11	4.73	4.31
10	5%	4.96	4.10	3.71	3.48	3.33	3.22	3.07	2.91	2.74	2.54
	1%	10.04	7.56	6.55	5.99	5.64	5.39	5.06	4.71	4.33	3.91
12	5%	4.75	3.88	3.49	3.26	3.11	3.00	2.85	2.69	2.50	2.30
	1%	9.33	6.93	5.95	5.41	5.06	4.82	4.50	4.16	3.78	3.36
14	5%	4.60	3.74	3.34	3.11	2.96	2.85	2.70	2.53	2.35	2.13
	1%	8.86	6.51	5.56	5.03	4.69	4.46	4.14	3.80	3.43	3.00
16	5%	4.49	3.63	3.24	3.01	2.85	2.74	2.59	2.42	2.24	2.01
	1%	8.53	6.23	5.29	4.77	4.44	4.20	3.89	3.55	3.18	2.75
18	5%	4.41	3.55	3.16	2.93	2.77	2.66	2.51	2.34	2.15	1.92
	1%	8.28	6.01	5.09	4.58	4.25	4.01	3.71	3.37	3.01	2.57
20	5%	4.35	3.49	3.10	2.87	2.71	2.60	2.45	2.28	2.08	1.84
	1%	8.30	5.85	4.94	4.43	4.10	3.87	3.56	3.23	2.86	2.42

Index